£100.90

Patty's Industrial Hygiene and Toxicology

THIRD REVISED EDITION
In Three Volumes

Volume I
GENERAL PRINCIPLES

Volume II
TOXICOLOGY

Volume III
THEORY AND RATIONALE OF INDUSTRIAL HYGIENE PRACTICE

Patty's Industrial Hygiene and Toxicology

THIRD REVISED EDITION

Volume I
GENERAL PRINCIPLES

GEORGE D. CLAYTON
FLORENCE E. CLAYTON
Editors

Contributors

M. C. Battigelli
A. R. Behnke
C. M. Berry
D. J. Birmingham
W. A. Burgess
G. D. Clayton
C. L. Crouch
B. D. Dinman
D. D. Douglas
J. F. Gamble
J. Grumer
P. D. Halley
A. M. Hyman
W. D. Kelley
M. Lippmann
W. E. McCormick
P. L. Michael
J. E. Mutchler
F. A. Patty
L. E. Renes
M. A. Shapiro
R. D. Soule
E. R. Tichauer
A. Turk
G. M. Wilkening
P. Wolkonsky
G. W. Wright

A WILEY-INTERSCIENCE PUBLICATION
JOHN WILEY & SONS, New York • Chichester • Brisbane • Toronto

Copyright © 1978 by John Wiley & Sons, Inc.

All rights reserved. Published simultaneously in Canada.

Reproduction or translation of any part of this work beyond that permitted by Sections 107 or 108 of the 1976 United States Copyright Act without the permission of the copyright owner is unlawful. Requests for permission or further information should be addressed to the Permissions Department, John Wiley & Sons, Inc.

Library of Congress Cataloging in Publication Data:

Patty, Frank Arthur, 1897–
Patty's Industrial hygiene and toxicology.

"A Wiley-Interscience publication."
Includes index.
CONTENTS: v. 1. General principles.—
1. Industrial hygiene. 2. Industrial toxicology.
I. Clayton, George D. II. Clayton, Florence E.
III. Battigelli, M. C. IV. Title. V. Title:
Industrial hygiene and toxicology.

RC967.P37 1977 613.6′2 77-17515
ISBN 0-471-16046-6

Printed in the United States of America

10 9 8 7 6 5

Contributors

MARIO C. BATTIGELLI, M.D., Professor of Medicine, Division of Health Affairs, School of Medicine, University of North Carolina, Chapel Hill, North Carolina

ALBERT R. BEHNKE, Jr., M.D., Clinical Professor of Preventive Medicine, University of California, San Francisco Medical Center, retired Captain M.C., U.S. Navy

CLYDE M. BERRY, Ph.D., Professor, College of Medicine, Institute of Agricultural Medicine, Department of Preventive Medicine and Environmental Health, University of Iowa, Oakdale, Iowa

DONALD J. BIRMINGHAM, M.D., Professor and Chairman, Department of Dermatology and Syphilology, Wayne State University, Detroit, Michigan

WILLIAM A. BURGESS, Associate Professor, School of Public Health, Harvard University, Boston, Massachusetts

GEORGE D. CLAYTON, Chairman of the Board, Clayton Environmental Consultants, Inc., Southfield, Michigan, and Fallbrook, California.

C. L. CROUCH, Director of Research, Illuminating Engineering Research Institute, New York, New York

BERTRAM D. DINMAN, M.D., Medical Director, Aluminum Company of America, Pittsburgh, Pennsylvania

DARREL D. DOUGLAS, Leader, Respirator Research and Development Section, Los Alamos Scientific Laboratory, University of California, Los Alamos, New Mexico

JOHN F. GAMBLE, Ph.D., ALFORD, Morgantown, West Virginia

JOSEPH GRUMER, Consulting Chemist, formerly Supervisory Research Chemist, Fire and Explosion Prevention, Pittsburgh Mining and Research Center, Bureau of Mines, U.S. Department of the Interior, Pittsburgh, Pennsylvania

PAUL D. HALLEY, Director, Environmental Health Services Division of Medical and Environmental Health Services Department, Standard Oil Company (Indiana), Chicago, Illinois

ANGELA M. HYMAN, Ph.D., The Rockefeller University, New York, New York

WILLIAM D. KELLEY, Deputy Director, Division of Biomedical and Behavioral Science, National Institute for Occupational Safety and Health, Department of Health, Education and Welfare, Cincinnati, Ohio

MORTON LIPPMANN, Ph.D., Associate Professor, Institute of Environmental Medicine, New York University Medical Center, New York, New York

WILLIAM E. McCORMICK, Managing Director, American Industrial Hygiene Association, Akron, Ohio

PAUL L. MICHAEL, Ph.D., Professor, Environmental Acoustics, Pennsylvania State University, University Park, Pennsylvania

JOHN E. MUTCHLER, formerly Director, Engineering Services, Clayton Environmental Consultants, Inc., Southfield, Michigan*

FRANK A. PATTY, formerly Director, Industrial Hygiene Services, General Motors Corporation, retired, Yuma, Arizona

LUCIAN E. RENES, formerly Director, Industrial Hygiene Services, Phillips Petroleum Corporation, Bartlesville, Oklahoma, Consultant, Fallbrook, California

MAURICE A. SHAPIRO, Ph.D., Professor, Environmental Health Engineering, University of Pittsburgh, Graduate School of Public Health, Pittsburgh, Pennsylvania

ROBERT D. SOULE, formerly Director, Industrial Hygiene Services, Clayton Environmental Consultants, Inc., Southfield, Michigan†

ERWIN R. TICHAUER, Sc.D., Professor, New York University Medical Center, Division of Biomechanics, New York, New York

AMOS TURK, Ph.D., Consulting Chemist, Danbury, Connecticut, and The City College, New York, New York

GEORGE M. WILKENING, Head, Environmental Health and Safety, Bell Laboratories, Murray Hill, New Jersey

PETER WOLKONSKY, M.D., Medical Director, Standard Oil Company (Indiana), Chicago, Illinois

GEORGE W. WRIGHT, M.D., Consultant, 460 South Marion Parkway, Denver, Colorado.

* Presently Director, Health and Safety, The Quaker Oats Company, Chicago, Illinois.

† Presently Associate Professor, Safety Sciences Department, Indiana University of Pennsylvania, Indiana, Pennsylvania.

Preface

Industrial hygiene is a profession that utilizes several basic disciplines, embracing the domains of the physicist, the chemist, the engineer, the toxicologist, and the physician. This book is illustrative of the expertise of these various scientific disciplines. During the late 1950s and 1960s attempts were made to change the name of the profession and the association that represents it. They did not succeed, however. Fortuitously, the impetus of the occupational safety and health legislation in 1970 gave further prestige to the name, and simultaneously intensified the demand for people well versed in this profession.

Volumes 1 and 2 of Patty's *Industrial Hygiene and Toxicology* have served as basic references books since they were first introduced in 1948 and subsequently revised in 1958. They have been a great contribution to the occupational health field.

The name "Patty" is held in high esteem in our profession. Among the pioneers of this field who were active before World War II, it was Frank Patty who recognized the importance of recording and documenting the work he and his cohorts were finding so challenging—in his case, at the U.S. Bureau of Mines and later as Director of Industrial Hygiene for General Motors Corporation. To assure that the wealth of information that hygienists in research, industry, and government were accumulating would be of benefit to future men of science, Frank Patty, with the assistance of his wife Ruth, a statistician and public health worker, began the work of recruiting other scientists to compile their experiences. This material was to constitute the two volumes. It is fitting that Frank Patty should write the first chapter of this third edition of Volume 1, "Retrospect and Prospect." He was there when much of what we are doing today had its beginnings; and now he has the advantage of detachment from managerial responsibility. He is in a position to give an overview of the past, and he has time to meditate on the future.

The revision of these volumes was undertaken not merely to update the books to incorporate developments of the past 15 years, but also to broaden the scope even more. In recent years certain aspects of agriculture, quality control, odors, and ergonomics have found their place in our family, and new hazards have emerged in other, familiar areas.

In 1948, in the preface to the first edition of Volume I, Frank Patty wrote:

> Industrial hygiene has been recognized and practiced from the time of Pliny down through the ages. It is the present concept of industrial hygiene that is relatively new . . . the concept of anticipating and recognizing potentially harmful situations and applying engineering control measures before SERIOUS injury results.

Today we would amend these words to state that the profession has made great strides during the past 25 years, and no longer is the professional looking only for the "serious injury." He is now concerned with *any* adverse reaction, however slight.

The primary concern of industrial hygiene is to protect the health and well-being of man at work. Our criterion is the World Health Organization definition of health, which goes beyond the simple reporting of time lost because of sickness. In earlier years, the aim was to protect the worker *only* in the workplace. This approach is now archaic; a person's health must be protected 24 hours a day, not just the 8 hours he is on the job. We recognize that the environment he is in and the stresses he encounters the *other* 16 hours influence his ability to handle stresses during his 8-hour work period.

Twenty-five years ago the working population was not as greatly concerned with healthful working conditions as it is today. Workers then wanted "hazardous pay" to compensate for hazardous occupations. In 1977 it is no longer a handful of scientists working alone for interested employers (in both industry and government), striving to improve the health conditions and safety of the workplace; the unions, the working man, and concerned citizens have joined the fight for a healthful environment.

Political maneuvering by government, industry, or organized labor has no place in industrial hygiene. The goal of all should be to maintain a healthful environment. The fundamental role of the federal government since the early 1900s has been to provide scientific information to permit formulation of sound legislation. The Williams-Steiger Occupational Safety and Health Act, of 1970, in addition to calling for research activities, gives enforcement powers to agencies of the federal government. We believe that these powers should be used judiciously and that the federal government should actively pursue local government involvement in policing industry, with the former providing research in all facets of the profession, as well as educational programs for local officials and concerned groups.

Standard setting and the resulting controversies about the appropriateness of the limits established point to an area of research that has been little explored to date, yet is of great concern—that is, determining the sensitivity of individuals. Such data could assist in the proper placement of people in the work environment, to assure their well-being, and at the same time to remove stifling restrictions from communities and industry.

Industrial hygiene has made great accomplishments during the past three-quarters century in some areas. It has been defined as "that science and art devoted to the recognition, evaluation and control of those environmental factors and stresses, arising in or from the workplace." To date we have accomplished much in the recognition and evaluation, but much remains to be done in the area of controls. Our expertise should be called on in designing equipment, processes, and building requirements. For example, in the machine tool industry, an industrial hygienist specializing in acoustics could make progress in designing unwanted sound out of machinery; or, in the design of chemical processes, the industrial hygienist could minimize the sources of air contaminants.

The greatest progress has been realized recently (1970–1977) and is primarily the result of increased public awareness of the undesirable effects of excess air contaminants and other stresses.

The individuals invited to participate in the revision of the Patty books are recognized authorities in their specialities. Each was aware of the impact the earlier volumes had made in our profession, and each was eager to contribute of his or her own expertise to help future industrial hygienists. These authors represent a variety of backgrounds: nine from universities, seven from consulting groups, seven from industry, one from government, and two from technical societies.

We hope our united effort to bring out the latest available data in our complex field will be beneficial in enabling readers to meet the future challenges of our profession and inspiring exploration in areas not yet studied.

George D. Clayton
Florence E. Clayton

Fallbrook, California,
November 1977

Contents

USEFUL EQUIVALENTS AND CONVERSION FACTORS

1 kilometer = 0.6214 mile
1 meter = 3.281 feet
1 centimeter = 0.3937 inch
1 micrometer = 1/25,4000 inch = 40 microinches = 10,000 Angstrom units
1 foot = 30.48 centimeters
1 inch = 25.40 millimeters
1 square kilometer = 0.3861 square mile (U.S.)
1 square foot = 0.0929 square meter
1 square inch = 6.452 square centimeters
1 square mile (U.S.) = 2,589,998 square meters = 640 acres
1 acre = 43,560 square feet = 4047 square meters
1 cubic meter = 35.315 cubic feet
1 cubic centimeter = 0.0610 cubic inch
1 cubic foot = 28.32 liters = 0.0283 cubic meter = 7.481 gallons (U.S.)
1 cubic inch = 16.39 cubic centimeters
1 U.S. gallon = 3.7853 liters = 231 cubic inches = 0.13368 cubic foot
1 liter = 0.9081 quart (dry), 1.057 quarts (U.S., liquid)
1 cubic foot of water = 62.43 pounds (4°C)
1 U.S. gallon of water = 8.345 pounds (4°C)
1 kilogram = 2.205 pounds
1 gram = 15.43 grains
1 pound = 453.59 grams
1 ounce (avoir.) = 28.35 grams
1 gram mole of a perfect gas $\approx$ 24.45 liters (at 25°C and 760 mm Hg barometric pressure)
1 atmosphere = 14.7 pounds per square inch
1 foot of water pressure = 0.4335 pound per square inch
1 inch of mercury pressure = 0.4912 pound per square inch
1 dyne per square centimeter = 0.0021 pound per square foot
1 gram-calorie = 0.00397 Btu
1 Btu = 778 foot-pounds
1 Btu per minute = 12.96 foot-pounds per second
1 hp = 0.707 Btu per second = 550 foot-pounds per second
1 centimeter per second = 1.97 feet per minute = 0.0224 mile per hour
1 footcandle = 1 lumen incident per square foot = 10.764 lumens incident per square meter
1 grain per cubic foot = 2.29 grams per cubic meter
1 milligram per cubic meter = 0.000437 grain per cubic foot

To convert degrees Celsius to degrees Fahrenheit: °C (9/5) + 32 = °F
To convert degrees Fahrenheit to degrees Celsius: (5/9) (°F − 32) = °C
For solutes in water: 1 mg/liter $\approx$ 1 ppm (by weight)
Atmospheric contamination: 1 mg/liter $\approx$ 1 oz/1000 cu ft (approx)
For gases or vapors in air at 25°C and 760 mm Hg pressure:
To convert mg/liter to ppm (by volume): mg/liter (24,450/mol. wt.) = ppm
To convert ppm to mg/liter: ppm (mol. wt./24,450) = mg/liter

CONVERSION TABLE FOR GASES AND VAPORS[a]

(Milligrams per liter to parts per million, and vice versa; 25°C and 760 mm Hg barometric pressure)

Molecular Weight	1 mg/liter ppm	1 ppm mg/liter	Molecular Weight	1 mg/liter ppm	1 ppm mg/liter	Molecular Weight	1 mg/liter ppm	1 ppm mg/liter
1	24,450	0.0000409	39	627	0.001595	77	318	0.00315
2	12,230	0.0000818	40	611	0.001636	78	313	0.00319
3	8,150	0.0001227	41	596	0.001677	79	309	0.00323
4	6,113	0.0001636	42	582	0.001718	80	306	0.00327
5	4,890	0.0002045	43	569	0.001759	81	302	0.00331
6	4,075	0.0002454	44	556	0.001800	82	298	0.00335
7	3,493	0.0002863	45	543	0.001840	83	295	0.00339
8	3,056	0.000327	46	532	0.001881	84	291	0.00344
9	2,717	0.000368	47	520	0.001922	85	288	0.00348
10	2,445	0.000409	48	509	0.001963	86	284	0.00352
11	2,223	0.000450	49	499	0.002004	87	281	0.00356
12	2,038	0.000491	50	489	0.002045	88	278	0.00360
13	1,881	0.000532	51	479	0.002086	89	275	0.00364
14	1,746	0.000573	52	470	0.002127	90	272	0.00368
15	1,630	0.000614	53	461	0.002168	91	269	0.00372
16	1,528	0.000654	54	453	0.002209	92	266	0.00376
17	1,438	0.000695	55	445	0.002250	93	263	0.00380
18	1,358	0.000736	56	437	0.002290	94	260	0.00384
19	1,287	0.000777	57	429	0.002331	95	257	0.00389
20	1,223	0.000818	58	422	0.002372	96	255	0.00393
21	1,164	0.000859	59	414	0.002413	97	252	0.00397
22	1,111	0.000900	60	408	0.002554	98	249.5	0.00401
23	1,063	0.000941	61	401	0.002495	99	247.0	0.00405
24	1,019	0.000982	62	394	0.00254	100	244.5	0.00409
25	978	0.001022	63	388	0.00258	101	242.1	0.00413
26	940	0.001063	64	382	0.00262	102	239.7	0.00417
27	906	0.001104	65	376	0.00266	103	237.4	0.00421
28	873	0.001145	66	370	0.00270	104	235.1	0.00425
29	843	0.001186	67	365	0.00274	105	232.9	0.00429
30	815	0.001227	68	360	0.00278	106	230.7	0.00434
31	789	0.001268	69	354	0.00282	107	228.5	0.00438
32	764	0.001309	70	349	0.00286	108	226.4	0.00442
33	741	0.001350	71	344	0.00290	109	224.3	0.00446
34	719	0.001391	72	340	0.00294	110	222.3	0.00450
35	699	0.001432	73	335	0.00299	111	220.3	0.00454
36	679	0.001472	74	330	0.00303	112	218.3	0.00458
37	661	0.001513	75	326	0.00307	113	216.4	0.00462
38	643	0.001554	76	322	0.00311	114	214.5	0.00466

CONVERSION TABLE FOR GASES AND VAPORS (Continued)

(*Milligrams per liter to parts per million, and vice versa; 25°C and 760 mm Hg barometric pressure*)

Molecular Weight	1 mg/liter ppm	1 ppm mg/liter	Molecular Weight	1 mg/liter ppm	1 ppm mg/liter	Molecular Weight	1 mg/liter ppm	1 ppm mg/liter
115	212.6	0.00470	153	159.8	0.00626	191	128.0	0.00781
116	210.8	0.00474	154	158.8	0.00630	192	127.3	0.00785
117	209.0	0.00479	155	157.7	0.00634	193	126.7	0.00789
118	207.2	0.00483	156	156.7	0.00638	194	126.0	0.00793
119	205.5	0.00487	157	155.7	0.00642	195	125.4	0.00798
120	203.8	0.00491	158	154.7	0.00646	196	124.7	0.00802
121	202.1	0.00495	159	153.7	0.00650	197	124.1	0.00806
122	200.4	0.00499	160	152.8	0.00654	198	123.5	0.00810
123	198.8	0.00503	161	151.9	0.00658	199	122.9	0.00814
124	197.2	0.00507	162	150.9	0.00663	200	122.3	0.00818
125	195.6	0.00511	163	150.0	0.00667	201	121.6	0.00822
126	194.0	0.00515	164	149.1	0.00671	202	121.0	0.00826
127	192.5	0.00519	165	148.2	0.00675	203	120.4	0.00830
128	191.0	0.00524	166	147.3	0.00679	204	119.9	0.00834
129	189.5	0.00528	167	146.4	0.00683	205	119.3	0.00838
130	188.1	0.00532	168	145.5	0.00687	206	118.7	0.00843
131	186.6	0.00536	169	144.7	0.00691	207	118.1	0.00847
132	185.2	0.00540	170	143.8	0.00695	208	117.5	0.00851
133	183.8	0.00544	171	143.0	0.00699	209	117.0	0.00855
134	182.5	0.00548	172	142.2	0.00703	210	116.4	0.00859
135	181.1	0.00552	173	141.3	0.00708	211	115.9	0.00863
136	179.8	0.00556	174	140.5	0.00712	212	115.3	0.00867
137	178.5	0.00560	175	139.7	0.00716	213	114.8	0.00871
138	177.2	0.00564	176	138.9	0.00720	214	114.3	0.00875
139	175.9	0.00569	177	138.1	0.00724	215	113.7	0.00879
140	174.6	0.00573	178	137.4	0.00728	216	113.2	0.00883
141	173.4	0.00577	179	136.6	0.00732	217	112.7	0.00888
142	172.2	0.00581	180	135.8	0.00736	218	112.2	0.00892
143	171.0	0.00585	181	135.1	0.00740	219	111.6	0.00896
144	169.8	0.00589	182	134.3	0.00744	220	111.1	0.00900
145	168.6	0.00593	183	133.6	0.00748	221	110.6	0.00904
146	167.5	0.00597	184	132.9	0.00753	222	110.1	0.00908
147	166.3	0.00601	185	132.2	0.00757	223	109.6	0.00912
148	165.2	0.00605	186	131.5	0.00761	224	109.2	0.00916
149	164.1	0.00609	187	130.7	0.00765	225	108.7	0.00920
150	163.0	0.00613	188	130.1	0.00769	226	108.2	0.00924
151	161.9	0.00618	189	129.4	0.00773	227	107.7	0.00928
152	160.9	0.00622	190	128.7	0.00777	228	107.2	0.00933

CONVERSION TABLE FOR GASES AND VAPORS (Continued)

(MIlligrams per liter to parts per million, and vice versa; 25°C and 760 mm Hg barometric pressure)

Molecular Weight	1 mg/liter ppm	1 ppm mg/liter	Molecular Weight	1 mg/liter ppm	1 ppm mg/liter	Molecular Weight	1 mg/liter ppm	1 ppm mg/liter
229	106.8	0.00937	253	96.6	0.01035	277	88.3	0.01133
230	106.3	0.00941	254	96.3	0.01039	278	87.9	0.01137
231	105.8	0.00945	255	95.9	0.01043	279	87.6	0.01141
232	105.4	0.00949	256	95.5	0.01047	280	87.3	0.01145
233	104.9	0.00953	257	95.1	0.01051	281	87.0	0.01149
234	104.5	0.00957	258	94.8	0.01055	282	86.7	0.01153
235	104.0	0.00961	259	94.4	0.01059	283	86.4	0.01157
236	103.6	0.00965	260	94.0	0.01063	284	86.1	0.01162
237	103.2	0.00969	261	93.7	0.01067	285	85.8	0.01166
238	102.7	0.00973	262	93.3	0.01072	286	85.5	0.01170
239	102.3	0.00978	263	93.0	0.01076	287	85.2	0.01174
240	101.9	0.00982	264	92.6	0.01080	288	84.9	0.01178
241	101.5	0.00986	265	92.3	0.01084	289	84.6	0.01182
242	101.0	0.00990	266	91.9	0.01088	290	84.3	0.01186
243	100.6	0.00994	267	91.6	0.01092	291	84.0	0.01190
244	100.2	0.00998	268	91.2	0.01096	292	83.7	0.01194
245	99.8	0.01002	269	90.9	0.01100	293	83.4	0.01198
246	99.4	0.01006	270	90.6	0.01104	294	83.2	0.01202
247	99.0	0.01010	271	90.2	0.01108	295	82.9	0.01207
248	98.6	0.01014	272	89.9	0.01112	296	82.6	0.01211
249	98.2	0.01018	273	89.6	0.01117	297	82.3	0.01215
250	97.8	0.01022	274	89.2	0.01121	298	82.0	0.01219
251	97.4	0.01027	275	88.9	0.01125	299	81.8	0.01223
252	97.0	0.01031	276	88.6	0.01129	300	81.5	0.01227

[a] A. C. Fieldner, S. H. Katz, and S. P. Kinney, "Gas Masks for Gases Met in Fighting Fires," U.S. Bureau of Mines, Technical Paper No. 248, 1921.

Patty's Industrial Hygiene and Toxicology

THIRD REVISED EDITION
In Three Volumes

Volume I
GENERAL PRINCIPLES

Volume II
TOXICOLOGY

Volume III
THEORY AND RATIONALE
OF INDUSTRIAL HYGIENE
PRACTICE

CHAPTER ONE

Industrial Hygiene: Retrospect and Prospect

FRANK A. PATTY

1 RETROSPECT

I have been involved in industrial hygiene for many years, beginning in 1928 as a physiological chemist at the U.S. Bureau of Mines in Pittsburgh. Our group of chemists, surgeons, and others was engaged in determining and recording the effects of dusts and the vapors of some organic compounds on dogs and guinea pigs. From this beginning, my industrial hygiene activities continued with an insurance company in New York, where I found that the information regarding chemical properties and effects of the materials studied was much needed and little understood. I also found that the engineering control of many exposures was possible, but not sufficiently implemented. The work was rewarding, and much satisfaction was derived from it.

During the early 1930s there were probably fewer than 50 people who were industrial hygienists working to protect the health of workers. The public was little interested in industrial hygiene, and some managers thought it was a source of trouble and a waste of money. A few physicians viewed it as an intolerable invasion of the doctor's domain and insisted that only medical doctors could express opinions regarding the effects of any material or any stress on the human body. Unions were more interested in getting hazardous pay than in controlling the environment, and the workers were kept in the dark, uninformed about the actual hazards. Gradually and gratifyingly, a light began to shine in the darkness. It has been a slow but positive evolution.

In 1939 the American Industrial Hygiene Association was formed to provide a means of assessing mutual problems, and through the association's efforts management began to realize that a *healthy* worker was a *productive* worker. Unions also came to understand that the health of their members should be of paramount concern to them. Government passed a series of workmen's compensation laws, culminating in the passage in 1970 of the Williams-Steiger Occupational Safety and Health Act. Thus today there are laws requiring industry to provide a healthful environment; unions assist in supplying information to their members concerning the hazards of the environment; and employees must be informed of the hazards of the products and materials with which they are working. We have witnessed the development of instrumentation from its crude beginnings to the present sophisticated, automated instrumentation.

Today the number of professionals dedicating their efforts to protecting the health of others is in the thousands. The American Industrial Hygiene Association, the leading technical association in this field, now has almost 3000 members. Many people feel that occupational diseases began with the advent of modern production of chemicals. To bring the proper perspective to our profession, let us review the history of man's working environment.

Historical accounts down through the ages are replete with allusions to the unattractive lot of the man who has earned his livelihood by manual labor. In ancient times a considerable portion of such work was done by slaves, a practice that has been extended in instances even into the twentieth century. Where slave labor has been employed, there has not been sufficient incentive or opportunity to study the adverse effects of occupational environments.

Early beliefs that unhealthful conditions are inherent in certain trades have now been disproved and are no longer expressed; when plant operators today are presented with clear-cut evidence of unhealthful working environments, they are almost universally ready to institute any practical control measures. One of the industries that rather recently joined the modern trend toward providing clean as well as healthful environments was the foundry industry, and even today some foundry operators still believe that considerable dirt, dust, and fumes are inherent in the industry. This is not an expression of conservatism among foundry management, but rather the result of the failure of the layout man, the industrial hygienist, and the ventilation engineer to provide information on the correct design of foundries and foundry equipment as well as on correct ventilation practices. No one wants to own or to work in an environment polluted with dust, metal fumes, smoke, or sulfur dioxide if it is easy, practical, and economically feasible to have a clean one. Neither does management want to spend huge sums for dust control only to find that the apparatus does not do what it was intended to do. Although there may still be isolated employers who regard labor as a commodity purchasable on the market and more easily replaced than preserved, the vast majority of workmen in the United States today are in a very enviable position with respect to the safeguards taken to protect them from occupational diseases and accidents.

2 HISTORICAL RÉSUMÉ

2.1 Medical and Industrial Hygiene Literature

Lead poisoning is the oldest recorded occupational disease. Hippocrates, who lived in the fourth century B.C. and is credited with lifting the practice of medicine from its basis of superstition and giving it a scientific foundation, has also been credited by some with being the first to record adverse effects on miners and metallurgists from exposure to lead.

Pliny the Elder, in the first century A.D., in his encyclopedia of natural science, refers to the use of bladders that "minium refiners" wore over their faces in an effort to avoid inhaling dust. "Minium" is the Latin word for cinnabar (red mercuric sulfide).

Medical history includes other brief references to occupational exposures, by such ancient authorities as Galen, Celsus, and others. However, it was not until the fifteenth century that further significant progress was recorded. In 1473 Ellenbog recognized that the vapors of some metals were dangerous, described the symptoms of industrial poisoning from lead and mercury, and suggested preventive measures. Agricola, a physician and mineralogist, in his *De re metallica* (1556) recognized "asthma" and ulceration of the lungs caused by the inhalation of certain kinds of dust. He stated that among the miners in the Carpathian Mountains some women married as many as seven husbands, each of whom succumbed to this disease.

Philippus Paracelsus (1493–1541), a Swiss chemist, physician, and professor of physics and surgery, with no academic degree, who was thoroughly disliked by the physicians of his time, was credited with being an independent thinker and investigator. He worked in Tyrol as a mining and smelting laborer for ten years, and then several years later returned to the mines to get material for his treatise on occupational diseases, in which he described various "miners' diseases," disturbances of the lungs, stomach, and intestines, that resulted from digging, smelting, and washing gold, silver, salt, alum, sulfur, lead, copper, zinc, iron, and mercury. He described chronic lung trouble of miners as "lung consumption, asthma, and dyspnea." He attributed the cause to vapors and emanations from the metals and advised that contact with them be avoided, as the condition was incurable.

Paracelsus pointed out fallacies of many medical theories then current, opposed the humoral theory of disease, and taught the use of specific remedies instead of indiscriminate bleeding and purging. Even though he introduced many new medicines, which are still in use today, he was regarded as a charlatan by physicians of his time—and still is in some circles. He died at the age of forty-eight. His untimely death has been variously attributed to "drunken debauchery," to his being thrown down a steep incline by assailants hired by his enemies, and to his contracting one of the occupational diseases he had described—perhaps the most plausible cause. His book was not published until 1567, more than a quarter-century after his death.

Ramazzini, in 1700, published the first book that could be considered a complete treatise on occupational diseases, *De morbis artificum diatriba.* A second and expanded edition appeared (in 1717) in which he discussed not only diseases resulting from exposure to dusts and metal fumes, but also those due to several chemicals. From personal observations, he accurately described scores of occupations with their attendant hazards and emphasized the necessity for the physician to inquire into the occupation of the patient. So logical were many of his observations that they remain substantiated today. In the majority of instances, however, the measures he advocated for the control of occupational diseases were therapeutic and curative rather than preventive, a methodology that persists to a certain extent today, although in the more progressive circles the control of industrial health is preventive, through hygiene, sanitation, and periodic health examinations.

K. B. Lehmann, in his experiments on the toxic effects of gases on animals, instituted a form of research about 1884 that has given us much of the information that serves as a guide for the control of industrial atmospheres today. A French pharmacist and chemist, Jean Baptiste Alphonse Chevallier (1793–1879), contributed liberally to the literature of industrial hygiene.

The first half of this century produced a number of publications reporting information on a wide range of subjects relating to our profession. These historical publications include:

Health of the Industrial Worker, E. L. Collis and Greenwood (1919).
Industrial Health, G. M. Kober and E. R. Hayhurst (1924).
Industrial Poisons in the United States, Alice Hamilton (1925).
Schädliche Gase, F. Flury and F. Zernik (1931).
Industrial Hygiene for Engineers and Managers, C. P. McCord and F. P. Allen (1931).
The Dermatergoses or Occupational Affections of the Skin, 4th ed., R. Prosser White (1934).
Occupation and Health, International Labor Office (1934).
Industrial Toxicology, W. N. McNally (1937).
Carbon Monoxide Asphyxia, C. K. Drinker (1938).
Toxicology and Hygiene of the Technical Solvents, K. B. Lehmann and F. Flury (1938).
Essentials of Industrial Health, C. O. Sappington (1943).
Industrial Medicine, F. J. Wampler (1943).
Manual of Industrial Hygiene, National Institutes of Health (1943).
Noxious Gases, Y. Henderson and H. W. Haggard (1943).
A Manual of Pharmacology, T. Sollman (1944).
Introduction to Industrial Medicine, T. Lyle Hazlett (1946).
The Industrial Environment and Its Control, J. M. DallaValle (1947).
Industrial Health Engineering, A. D. Brandt (1947).
Occupational Diseases of the Skin, 2nd ed., L. Schwartz, L. Tulipan, and S. M. Peck (1947).
Occupational Medicine and Industrial Hygiene, R. T. Johnstone (1948).
Analytical Chemistry of Industrial Poisons, Hazards, and Solvents, 2nd ed., M. B. Jacobs (1949).
Industrial Toxicology, L. T. Fairhall (1949).
Industrial Toxicology, 2nd ed., Alice Hamilton and H. L. Hardy (1949).
The Chemistry of Industrial Toxicology, H. B. Elkins (1950).
Handbook on Aerosols, Atomic Energy Commission (1950).
Handbook on Air Cleaning, S. K. Friedlander, L. Silverman, P. Drinker, and and M. W. First (1952).

Noise, University of Michigan School of Public Health (1952).
Toxicity of Industrial Organic Solvents, rev. ed., E. Browning (1953).
General Handbook for Radiation Monitoring, Robert F. Barker (1954).
Industrial Dust, 2nd ed., P. Drinker and T. F. Hatch (1954).
Plant and Process Ventilation, W. C. L. Hemeon (1955).
Rosenau, *Preventive Medicine and Public Health,* 8th ed., K. F. Maxcy (1956).

During the past quarter century there has been a virtual explosion of publications in the field of industrial hygiene and related areas of interest, with the federal government being a major producer of such publications. The reader is referred to the Government Printing Office, Washington, D.C., for additional information.

2.2 Labor Legislation

Labor legislation has played an important part in the progress and development of industrial hygiene, both in the United States and abroad. It is said (1) that the deplorable state of orphaned child workers in English cotton mills was a major factor in motivating reforms in working conditions in factories, and that the initial step was taken by Sir Robert Peel, a millowner, by acquainting the English Parliament with these conditions in 1802. Other reformers became interested. Statistics gathered showed that the average age of the working classes was 22 as compared with an average age of 44 among the wealthier classes, and that the death rate in the workingmen's districts was appreciably higher than the general rate. The Factory Act was passed in 1833, limiting the hours of work of children and providing for factory inspection in certain factories and industries. Several trades were brought under the control of the Factory Act in 1864, and in 1867 the act was broadened to include many industries and places employing more than 50 persons. This act prohibited the eating of meals in noxious plant atmospheres, provided for the guarding of machinery, and required mechanical ventilation for the control of injurious dusts. Medical inspection of factories was inaugurated in 1897, at which time the idea of compensation was adopted.

France and Germany also passed laws during the nineteenth century to regulate hours of labor and protect workmen in some of the more dangerous trades. Factory inspection was introduced in the United States by Massachusetts in 1877; soon thereafter Massachusetts, New Jersey, New York, Connecticut, Michigan, Missouri, and Minnesota passed laws requiring the removal of dusts and injurious gases by means of exhaust fans.

The United States became the last industrially important country to adopt compensation; in 1908 limited benefits were provided for U.S. Civil Service employees. New Jersey was the first state to pass such a law, in 1911. Other states followed rapidly in providing compensation for accidents. In 1911 California, Connecticut, Illinois, Michigan, New York, and Wisconsin enacted laws requiring the reporting of, and later compensation for, cases of occupational disease. In 1919 Wisconsin made occupational diseases compensable and by 1948 all states had laws requiring such compensation. Some of the states recognize only a few specific diseases, whereas others recognize any disease of occupational origin, that is, a disease arising out of or in the course of and peculiar to the occupation. The general coverage law, while it is said by some to be conducive to fraudulent claims that are expensive to refute, is obviously the only completely fair coverage for the workman, whom it presumably is designed to protect. As an example of the obviously unfair situation that can result under the limited coverage statutes, one state, which has been a leader in making occupational diseases compensable, at one time recognized diseases arising from exposure to "benzine (petroleum products) and its homologues" but did not recognize diseases arising from exposure to benzene (benzol), nor silicosis arising from silica-bearing dusts. Both these errors have since been corrected. Any fair-minded student of industrial hygiene appreciates the necessity of recognizing and controlling all exposures that adversely affect the health and well-being of employees, regardless of the nature of the exposure.

The first major public act that offered control of an occupational disease in the United States came in 1912 when a prohibitive federal tax levied on the use of yellow phosphorus (also called white phosphorus) in the manufacture of matches effectively stopped that practice.

What was perhaps the first significant major investigation (2) of occupational disease in this country was undertaken by the U.S. Public Health Service and the U.S. Bureau of Mines jointly to determine the cause and extent of pulmonary diseases among lead and zinc miners.

Within the last ten years Congress has passed three major pieces of legislation that involve the protection of the health of working men and women. These are as follows:

1. The Metal and Nonmetallic Mine Safety Act of 1966, defining health and safety standards for both metal and nonmetallic mines and providing for a mandatory reporting of all accidents, injuries, and occupational diseases of the miners.
2. The Federal Coal Mine Health and Safety Act of 1969, which attempts to attain the highest degree of health protection for the miner by "providing to the greatest extent possible, that the working conditions in each underground coal mine are sufficiently free of respirable dust concentrations in the mine atmosphere to permit each miner the opportunity to work underground during the period of his entire adult working life without incurring any disability from pneumoconiosis or any other occupation-related disease during or at the end of such period."
3. The Occupational Safety and Health Act of 1970, under which the federal government is authorized to develop and set mandatory occupational safety and health standards applicable to any business affecting interstate commerce. The responsibility for promulgating and enforcing the standards rests with the Department of Labor. The Department of Health, Education and Welfare is responsible for conducting research on which new standards can be based, and for implementing education and training programs to produce an adequate supply of manpower to implement and enforce the act. These responsibilities are carried out by the National Institute for Occupational Safety and Health (NIOSH). For more detailed information on legislation, see Chapter 2.

The most difficult function of the Department of Health, Education and Welfare (HEW) is supplying a list of all known toxic substances and the concentrations at which toxicity occurs. It is especially difficult, time-consuming, and costly to determine the lower level of toxicity or harmful effects of materials, processes, and stresses on the human body. As more health facts surface, it sometimes becomes necessary to lower the accepted standards of permissible quantities. This activity is being conducted presently (1977) by several official agencies such as the Environmental Protection Agency (EPA), the Occupational Safety and Health Administration (OSHA), the Center for Disease Control (CDC), the Federal Drug Administration (FDA), and the National Advisory Committee on Occupational Safety and Health (NACOSH). Unofficial agencies also involved include the American Industrial Hygiene Association, the American National Standards Institute (ANSI) Z-37 Committee, and the American Conference of Governmental Industrial Hygienists (ACGIH). The interests of the public are "represented" by Ralph Nader's Health Research Group and John Gardner's Common Cause.

Several of the drug and chemical manufacturers have their own toxicological laboratories with well-qualified staffs, who offer useful guidelines for exposure to their products. For instance, Dow Chemical Company adopted a guideline of 50 ppm for exposure to vinyl chloride in 1961, whereas the ACGIH had recommended 500 ppm. Dow's experience at the 50 ppm level for more than 10 years produced no history of cancer incidence among workers. After researchers elsewhere discovered a linkage of certain chemicals, (including vinyl chloride) with cancer, Dow lowered the limit to 10 ppm, and later to the limit of 1 ppm set by law, though questioning whether this is an unreasonable and unnecessarily low standard.

In this day and age it is difficult to understand how the manufacturers of the pesticide Kepone, a chlorinated hydrocarbon, could fail to notice the ill effects on their employees. It required a long delay, after many employees had become ill, and after various official agencies had made superficial inspections, before an employee's serious illness was correctly diagnosed by blood analysis showing a high level of Kepone. The state epidemiologist then took charge to stop the dangerous air pollution and hazards to exposed persons.

3 SIGNIFICANT EVENTS IN THE PROGRESS OF INDUSTRIAL HYGIENE IN THE UNITED STATES

3.1 Before 1942

The first national conference on industrial diseases was called at Chicago in 1910 by the American Association for Labor Legislation, and a commission consisting of representatives of medicine, engineering, and chemistry was assigned the task of investigating the magnitude of the problem and of proposing a method of attack in a warfare against industrial disease. About this time several groups began the study of occupational diseases. The U.S. Bureau of Mines was created in 1910. The U.S. Bureau of Labor, set up in 1885, became the federal Department of Labor in 1913. This department was charged with collecting "information upon the subject of labor, its relation to capital, the hours of labor, and the earnings of laboring men and women, and also upon the means of promoting their material, social, intellectual, and moral prosperity." The Department of Labor has been responsible for the collection and dissemination of much valuable information in the field of industrial hygiene.

The American Museum of Safety was created in New York in 1911 and later became known as Safety Institute of America, under which name it is still active along educational lines. The National Safety Council was organized in 1913. The American Public Health Association organized a section on Industrial Hygiene in 1914. The U.S. Public Health Service organized a Division of Industrial Hygiene and Sanitation in 1915. The American Association of Industrial Physicians and Surgeons was organized in 1916. The ill health and increased mortality accompanying the accelerated production of munitions and other war materials for World War I made many persons conscious of the necessity for technical guidance in the recognition and control of occupational diseases. The *Journal of Industrial Hygiene* was established in 1919 and was the leading publication in the field. It was acquired by the American Medical Association in 1949, underwent editorial policy changes, and is now published as the *AMA Archives of Industrial Health.* The *American Industrial Hygiene Association Quarterly,* published for the first time in 1946, has had wide acceptance and is now the leading source of information in the industrial hygiene field. In 1958 it was issued six times a year as the *American Industrial Hygiene Association Journal*; in 1975, it was changed in format and issued monthly as the *American Industrial Hygiene Association Journal.*

In 1918 the Harvard Medical School established a Department of Applied Physiology, which in 1922 became a part of the present Harvard School of Public Health as the Department of Physiology and the Department of Industrial Hygiene. This was the first time that instruction and research in industrial hygiene leading to advanced degrees had been offered anywhere in the world. This school cooperates with the graduate school of engineering: any of the courses offered in the School of Public Health may be elected by students working for a degree of master or doctor of science in engineering. The School of Public Health, which is open to graduates of schools of medicine, and graduates in arts and sciences with training in basic medical sciences or specialized training and experience in an important phase of public health work, offers the degree of master of public health and, to especially qualified persons, the degree of doctor of public health. The Harvard School of Public Health was the first place in the world where a qualified person could obtain scheduled, broad instruction in industrial hygiene regardless of whether his undergraduate training had been in medicine or in the sciences. Several universities, colleges, and technological schools now provide similar instruction to a growing number of students.

Many chemists and engineers, as well as physicians, have acquired specialized industrial hygiene knowledge and skills by association and in the "School of Hard Knocks." Official agencies, such as the Bureau of Mines, the Public Health Service, several state divisions of industrial hygiene, and a few city health department bureaus of industrial hygiene, have furnished valuable training stations for developing neophytes into full-fledged industrial hygienists. More recently, industry has been training chemists, engineers, and physicists for both specialized and general work in the field by giving them opportunities to work with trained industrial hygiene personnel.

The National Conference of Governmental Industrial Hygienists was organized in 1938 and has since furnished a valuable forum for the discussion of problems and experiences among this important group of industrial hygienists. The American Medical Association held its first Congress on Industrial Health in 1939. The American Industrial Hygiene Association was also organized in 1939, primarily as a specialized group to encourage and foster the exchange and dissemination of technical information in the basic sciences, such as chemistry, physics, mechanical engineering, and toxicology, as it applies to industrial hygiene. This organization held its first annual meeting in 1940. It became the nucleus around which industrial hygiene developed and gained recognition as a scientific profession. Its membership now includes both men and women from the several sciences and medicine who represent the many facets of this particular field of public health. Industrial hygiene has successfully withstood abortive efforts at absorption by both safety engineering and medicine. It borders on each of those fields, but differs in technique from both, and one of its primary functions is to get these two groups to work hand in hand. As long as men of the professional strength that has characterized the field since the early 1920s continue their interest, industrial hygiene will be likely to retain its identity as a separate and important part of public health. It has a specific job to do and the specialized training to do the job—qualifications that are becoming more and more widely recognized.

Until 1936 the industrial hygiene activities of the Public Health Service were relatively minor and were confined largely to medical and statistical research. From 1928 until 1932 toxicological research, such as the study of the effects of solvents, vapors, and gases on guinea pigs, rats, rabbits, dogs, and monkeys, as well as the development of the peritoneal injection method for evaluating proliferative action of dusts, was farmed out to the Bureau of Mines. Under the able guidance of R. R. Sayers and W. P. Yant, this research had become extensive and the Bureau of Mines was hailed by some in the chemical industries as an impartial, qualified research institute where industry could have relative toxicities determined in order that the manufacturer as well as the public might be intelligently guided regarding necessary precautions in the handling and use of chemical products. The studies were cooperative arrangements whereby industry absorbed costs connected with the study of their products. The amount of money involved began to be attractive, and protests against "subsidizing" the government and research caused the project to be discontinued. Later federal funds were made available to the Public Health Service through the Social Security Act, whereupon extensive research investigations and a program designed to establish active industrial hygiene units in the health departments of the various industrial states were inaugurated. By 1941 the facilities of the Industrial Hygiene Division, National Institutes of Health, of the Public Health Service had been greatly extended, and the personnel had increased from less than a score of persons in 1933 to well over 200. Many of these qualified persons were loaned to various states to help them establish their own divisions of industrial hygiene, assist in training local personnel, and "sell" the entire field of endeavor to the local officials and population.

Before 1936 there were only five state departments of health and three state departments of labor conducting industrial hygiene activities, and these activities were limited. Under the impact of World War II and generous federal grants of funds, all the industrial states established industrial hygiene units. In November 1946, there were 52 units, spread over 41 states. The enormous increase in harmful situations along with demands for manpower conservation accompanying World War II made investigational industrial hygiene surveys very much in demand and led to the hasty training of a number of men for this work. Many of these men had had only rudimentary training in the evaluation of exposures and little or no practical experience in control methods. Because of, or in spite of, the efforts of this group of novices and the relatively few seasoned men, however, industrial health was well promoted during World War II. Industrial health comparisons of munitions workers for 1917–1918 and 1941–1945, for instance, are striking. Under the stimulus of additional federal funds allocated to states and territories, the development of official industrial hygiene agencies reached an all-time high in 1949 and included all but two of the states (3). This peak was in turn followed by a five-year period of retrogression necessitated by economy drives and appropriation cuts, during which about half the trained personnel were absorbed by private enterprise and local health departments. In many instances they assumed positions of even greater responsibility in advancing industrial health objectives.

Other important organizations that have been active in the advancement of industrial hygiene include the American National Standards Association, the Industrial Health Foundation, Inc., the John B. Pierce Laboratory of Hygiene (New Haven, Connecticut), the Saranac Laboratories (Saranac Lake, New York), life insurance companies, as well as mutual and stock company compensation insurance carriers, and many of the larger private industries. Employee groups are also evidencing an active interest in industrial health. In 1945 a health institute was opened by the United Automobile Workers in Detroit to give diagnostic service to union members. Employees no longer regard X-ray examinations or industrial hygiene surveys with the suspicion that was common in earlier years. This change of attitude is due in part to improved practices, but mainly to educational successes: in short, the average employee now understands the purpose of such health control work and realizes its value to his well-being. Some states now require preemployment and periodic medical examinations for persons exposed to potentially harmful materials.

3.2 World War II and Industrial Hygiene Service

Industrial hygiene procedures have largely passed through the period of inquiry into the causes of ill health and are now devoted to anticipating and avoiding harmful situations before they have time to cause injury. Many different organizations played a part in this rapid development during the period of expanded production in 1941–1945.

3.2.1 Official Agencies

The army provided men to make surveys and recommend control measures in industrial installations owned and operated by the military. Similar service was rendered to government-owned, privately operated munitions plants by other government industrial hygiene groups cooperating with the army. The navy provided men for its shore establishments, and cooperated with the Maritime Commission in providing service for private shipyards. The Public Health Service supplied funds and loaned trained personnel to state and municipal health departments, and also conducted field surveys in industry, especially munitions works. The Department of Labor concentrated on educational activities. The Bureau of Mines greatly expanded its health program in relation to mining operations. During and following the period of reconversion to the manufacture of civilian goods, many of the trained men who had been active in government agencies were reabsorbed by industry.

3.2.2 Insurance and Industrial Groups

Life insurance and compensation carriers were rather active in the field well before 1941, and they continued their work as best they could despite the inroads into their personnel from all sides that occurred during the war production years. Later they expanded their activities and reemployed trained men.

During this period industry became much more active in the field; partly because of a newly awakened interest as a result of state and municipal activity and partly as a result of organized labor's demands for healthful working environments, but largely because of a desire on the part of management to make the working environment healthful and attractive. This came about not purely from altruistic motives but because management had begun to realize that the maintenance of a healthful environment pays dividends from the cost-saving viewpoint as well as in employee satisfaction and reduced turnover of the labor force. More and more industrialists came to the conclusion that industrial hygiene is a necessary adjunct to production, not something to be entrusted entirely to overworked government agencies.

From this period comes a striking example of the value of industrial hygiene expertise that, because of its fundamentality and at the same time dramatic departure from the daily routine of an industrial hygienist, may be worth describing without retouching. No tangible reward or token of appreciation was either anticipated or received and the only special remuneration was the feeling of satisfaction that goes with accomplishing any job, assumed or assigned; it was considered all in the day's work.

While busily engaged at my office one morning in the industrial hygienist's favorite occupation of pouring over survey reports, I received a telephone call: "This is Dr. ______, chemist over at ______ Dry Dock. Here's something I think you should know about and maybe you will want to come over and look around. We just sent a man to the hospital and we have two more who are laid up and in a serious condition. The ______ was docked here just two weeks ago for repairs and conversion and out of the 2000 men working on her, over 100 are affected with some sort of breaking out and itch; and all the men are threatening to quit if we don't find out what's wrong and correct it. The boat's been in the tropics for some time and the workmen fear some tropical disease is responsible for this outbreak."

Yes, I did want to look around and within an hour was aboard the ship and observed many of the afflicted men at work. There was considerable grumbling and an abundance of dirty looks. Several men wore bandages over vesicular patches, and on a few there was evidence of a generalized fine vesiculation. We went through the ship from stem to stern and from forecastle to bilge. It was like a beehive: men were cutting with torches, sawing out panels, knocking off plaster, shoveling out debris and filth, scrubbing, and removing the interior furnishings preparatory to refitting the boat completely. Admittedly the ship was dirty—in fact, in some areas it was filthy—but so what! Next we went to the first-aid room to see the attendant and find out what, if any, information could be obtained there. The place was crowded with patients, and while we waited to see the attendant first-aid man we could hear the grumblings of the waiting patients, who complained of "filthy working conditions" and said that if the health department were not called in to condemn the place, everyone should quit before they became ill—that the place was "infested with fleas" and that they didn't "want any tropical fevers." Finally the harassed attendant came to us, but he had little to offer except to comment that the cases and complaints were getting more numerous and that he certainly hoped we could do something.

We had seen the situation: a once luxuriant ship, somewhat filthy in spots, an explosive outbreak of dermatitis in nearly 10 percent of the 2000 men at work on the ship, no reported cases among the many thousands of workmen in adjacent areas of the same shipyard, and, unless something were done quickly, work on this desperately needed troop ship and possibly in the entire yard would stop.

The job looked interesting if not easy. We collected samples of everything we could get loose—plaster, upholstery, hair stuffing, sweepings, scrapings from panels and floors, and samples of the different woods and sawdusts, and returned to the laboratory. The samples were turned over to the chemists and microscopists to look for fumigants, insecticides, alkalies, and other common irritants, as well as any signs of insects or parasites. Two members of the staff were inveigled into joining me in making patch tests with some of the materials after saturating them with alcohol as a precautionary measure. While these tests were in progress a medical associate told me with an air of finality that our company had called in, as consultant, an authority on dermatology and that he planned to leave the matter entirely in the consultant's hands. When I took the story to my immediate superior he said it was very interesting, and sometime when he had more time I should "tell him all about it." I telephoned the consultant dermatologist and inquired if he had seen the cases and had any ideas of what might cause the difficulty. Yes, he had seen them, and, except for the fact that it didn't make sense, he would say that they resembled poison ivy cases. That agreed with what we had seen on the job, but we had not seen any poison ivy! After all the chemical and microscopic tests proved negative except for an insignificant amount of arsenic in the sweepings, there remained only the possibility that the patch tests might indicate an irritant. After 48 hours, and with all patch tests negative, the shipyard's chemist reported that there were several more cases, including an electrician working on the dock beside the boat, and that a walkout seemed imminent even though the ship was desperately needed for a troop transport.

The fact that an electrician who possibly had not been on the boat was affected gave an indication that the problem might be attacked from an epidemiological approach by personally interviewing some of the afflicted men. Discussion of personal matters with employees, especially shipyard workers, is ordinarily something to be avoided, but in this instance approval of such interviews was readily obtained from all parties concerned. Aided by the chemist and two safety engineers, we went out on the job and talked to some of the affected men, including the electrician, who had never been on the suspected ship, and a security policeman, who was seriously affected and had been aboard the ship only once for a few minutes. This introduced a strong element of doubt about the ship's being the contact source and diverted our attention to other possible sources of irritant material. It developed that nearly all cases were on the day shift. The electrician's case had been diagnosed "cable rash," but since the characteristics of his lesions were similar to those of the other workmen, we did not waste time examining the cables with which he had been working. Instead, we investigated all the possible exposure sources where the men spent their time when they were off the ship: where they ate their lunch, where they loafed, and the route they took to and from work. We found what appeared to be some damaged oil drums in a pile on a sand lot where several men had spent their lunch hour within the shipyard grounds, but about a block away from the dock to which the boat was tied. The drums were punctured or otherwise damaged and, for the most part, were empty, but three or four contained some dark oily liquid. Close examination of these drums revealed lettering still visible on one which read *Cashew Shell Liquid.*

The cause of the epidemic had been found.

At the shipyard, management quickly announced the facts over a public address system to the workmen so that their fears would be allayed. A strike was averted, and the ship was completed ahead of schedule. Further investigation revealed that cashew shell liquid had been spilled on the dock to some extent, in the street in considerable amounts, and was extensively scattered over the sand lot in question. Many of the men working at refinishing and repairing this ship had sat on the sand, or the drums themselves, during the noon hour while they ate lunch. Others had stretched out in the sun, bare-backed, on the contaminated sand. A check into the source of the drums revealed that a cargo of 1000 drums of the liquid had arrived and had been unloaded on the opposite side of the same dock about one month previously. This work had been done by an outside contractor who had dumped the damaged drums onto the unused lot.

At our suggestion, all suspicious-looking spots were covered with chlorinated lime, the dock was cleaned and scrubbed, the drums disposed of, and dirt filled in on the sand lot. The dermatitis, which by now was starting to appear in the homes of workmen, from contact with soiled clothing, quickly disappeared. My colleague the shipyard's chemist, however, in a sincere effort to convince himself and others of the potency of the cashew liquid, became impatient at negative results obtained in 24 hours with one tiny patch test and tried four more generous patches with samples from different drums. A few days later he joined those who had been hospitalized.

This is but one example of the thousands that have been reported by industrial hygienists over the years. It is illustrative, however, of the personal satisfaction that can be obtained by dedicated professionals. Private institutions, such as, for instance, the Industrial Health Foundation, Inc., encountered greatly expanded demands for their services during 1941–1945. Other organizations such as the National Sanitation Foundation later developed programs pointed toward air and stream pollution evaluation and control.

3.2.3 Educational Institutions

The colleges that had given any organized instruction in industrial hygiene before 1940 now provided short courses for greatly expanded classes in an effort to give a large number of interested persons sufficient knowledge to make them useful workers in the field during war-expanded production. During and following this period many medical schools set up short courses in industrial medicine and hygiene as orientation and refresher courses. A few institutions of advanced learning have provided special instruction in health and safety. The present plans include advanced training and education leading to degrees in health and safety engineering, and the opportunities for advanced instruction for chemists, physicists, engineers, and others who wish to prepare themselves for industrial hygiene engineering are improving. Today many universities offer degrees in industrial hygiene. In addition, there are a number of short, highly identifiable courses on specific subjects. For example, currently the American Industrial Hygiene Association (AIHA) has a grant from National Institute for Occupational Safety and Health (NIOSH) to provide one and two weeks courses on the fundamentals of industrial hygiene. These courses are given at strategic locations in the United States. For current information on educational opportunities, it is suggested the reader contact the Managing Director, AIHA, 66 South Miller Road, Akron, Ohio 44313.

3.2.4 Research Organizations

Much research into the toxicity of materials and into methods of analysis and control of harmful exposures was conducted during World War II by the Public Health Service, and some was done by industry. Several industries, seeing the need for such research, made plans to enter the field following the return to peacetime activities. The need for research into the cause and effects of many uncharted forms of air pollution is great. We can hardly expect such work to keep pace with industrial development, the manufacturing chemist, and the even faster moving physicist, but it can follow more closely than has been done in the past. Moreover, the individual homeowner needs to be made more aware of his contribution to a rapidly growing air pollution problem.

4 INDUSTRIAL HYGIENE IN FOREIGN COUNTRIES: 1920–1945

The best insight into what European countries had been doing in industrial hygiene is given in material regarding toxic limits and industrial health in wartime published by the International Labour Organization (ILO). According to reports, the rules and regulations in these countries concerning the use of poisonous materials in industry are based on the results of periodic medical examinations of the workers in plants handling such substances. At certain intervals prescribed by regulations, the worker must be examined by a physician. It is the task of the physician to *make an early diagnosis to recognize and evaluate the first signs of absorption of a poison* and if necessary to remove men from work temporarily. The ILO believed this to be a reliable method for the timely detection of endangered workers and also of the hazards in a specific plant. The shortcomings of such a plan of approach obviously are in evaluating the first sign of absorption—a trick physicians in the United States have not been able to turn to their complete satisfaction. The toxic materials that give rise to recognizable, specific, dependable clinical signs in advance of serious injury are disappointingly few.

The ILO report pointed out that periodic medical examinations were not compulsory in the United States. There is little room to doubt that regular and more frequent medical examinations are a desirable way of discovering dangerous exposures, especially in the absence of competent industrial hygiene engineering evaluation and control. When the two methods are properly coordinated, however, the most dependable safeguard against harmful exposures is provided.

4.1 England

In England the medical phases of industrial health have received more attention than the more technical and engineering phases of evaluation and control. Some support for this opinion is seen in the report of foundry experience (4). Only 150 sandblasters were engaged in steel foundries before the war, yet 31 sandblasters were given certificates covering death, total disablement, or suspension on account of silicosis in the 1932–1942 period. Congestion of operations, due to great demands on capacity, has been credited with increasing the number of silicosis cases by raising the concentration of dust-producing operations per unit of shop area. A committee appointed to study the situation recommended that (1) no silica be used in parting compounds; (2) no sand be used as an abrasive in "sandblasting"; and (3) no person under 18 work as a sandblaster, in repairing a sandblast plant, or within 20 feet of such a plant. Consideration was given to prohibiting the use of free silica in molding material as well as silica paint on cores. Certain of these regulations appear unnecessarily drastic and do not take into account the possibilities of engineering control.

E.R.A. Merewether, Senior Inspector of Factories in England, recognized this lack of engineering interest and skill when in 1942 he told an American Public Health Association audience that the time had arrived when "the doctor no longer knows everything about everything" and conceded American leadership in industrial hygiene engineering. Merewether urged the English chemists to instigate the formation of a society for "the holding of annual conferences and the publication and distribution of papers and literature on the subject of industrial health and safety" (5), much as is done by our own American Industrial Hygiene Association and to a limited extent by the American Public Health Association. The British Occupational Hygiene Society was established on April 27, 1953, and held its first conference in November of the same year.

4.2 U.S.S.R.

C.-E. A. Winslow, who visited Russia in 1917 and again in 1936, reported major advances in public health in that country. Emphasis had been placed on the health of industrial workers. Scientific studies of many occupational hazards and of means of promoting industrial health have been conducted by various research organizations, including the Institute of Industrial Diseases in Moscow, the Central Institute for Nutrition in Moscow, the Pavlov Institute, the Leningrad Institute of Safety, Hygiene and Technique, and the States' Scientific Institute of Labor Protection.

The Soviet state has given special attention to questions involving better labor conditions, looking toward the highest possible efficiency and production, as well as the mental and physiological well-being of the workers. Early permissible standards for air pollution in factories tended to be ultraconservative on the side of safety, and rather severe standards have been set for industry to meet. It has been suggested that various motives have contributed to the zeal of industrial hygiene investigators. Later, however, some standards were modified to conform with experience, and there have been indications that in general the requirements have been made somewhat less rigid (6). For instance, the maximum limit for carbon monoxide was 0.02 mg/liter, and is now 0.04 mg/liter (approximately 35 ppm), methyl alcohol was 0.03 mg/liter and is now 0.07 mg/liter (approximately 52 ppm), ethyl alcohol was 1.0 mg/liter and is now 1.5 mg/liter (approximately 800 ppm). Amyl and other alcohols, however, that were 1.0 mg/liter are now 0.5 mg/liter (approximately 280 ppm amyl alcohol and 330 ppm butyl alcohol). This trend, along with our own general trend downward, indicates that we may possibly reach common ground for some if not most of our thinking in regard to what constitutes a permissible and what a potentially harmful exposure. The disparity of standards for outdoor air may be somewhat greater (7). We have seen good information resulting from research into the control of industrial health in the U.S.S.R., and we hope to see more in the future. A free exchange of ideas among all countries regarding industrial hygiene can only benefit each participant.

4.3 Germany

In Germany in 1928 the task of collecting material and making practical observations in industrial hygiene was entrusted to the Medical Committee of the German Society for the Protection of Labor. A literature search was conducted as well as experimental investigations of the most important organic solvents. The German Health Office was established as a clearing house for information regarding chemical composition of solvents and the etiological factors of illness resulting from their use. The inspectors of factories were men with medical training but without engineering training, and as a result the control measures were largely of a medical nature. It has been stated by one of the German factory inspectors that it was easier and more routine to obtain biological specimens, including blood samples, from workmen than to obtain samples of air.

The literature on the results of pharmacological and toxicological investigations, on which much of our present practice is based, is replete with such names as Flury, Führer, Koelsch, Köster, Lehmann, Müller, Wirth, Zanger, and Zernik. Priceless instrumentation, records, and intellect have been reported lost as a result of World War II. Many rules and regulations for industry have been developed over the years and are presently applied in West Germany to the use or manufacture of certain chemicals, the conduct of specific operations, or workrooms below ground. These numerous regulations are pointed toward medical examinations and mechanical safeguards for the control of any exigency rather than toward the engineering control of the environment to conform with hygienic standards. The regulations are enforced by a host of factory inspectors.

4.4 Other Countries

Not only have many of the countries of Europe, Asia, and Latin America sent selected personnel to study industrial hygiene techniques in the United States and Canada—whose progress is somewhat similar to our own—but several of them have set up training centers at home. One of the most notable is that established at Lima, Peru, with the cooperation of the Institute of Inter-American Affairs (8). Japan has established an Industrial Hygiene Association. There is ample evidence of a widespread and growing interest in industrial hygiene and health engineering in nearly all countries.

5 INDUSTRIAL HYGIENE IN THE UNITED STATES

We have seen that the United States lagged behind other countries in providing control measures or compensation for occupational diseases, and that it was not until the late 1920s that industrial hygiene became more than an incipient dream in the minds of a relatively few individuals. A few farsighted pioneers in the field of public health saw the possibilities of making the control of industrial health a vocation and of enlisting the technical aid of professions other than medicine to do the job. Industrial hygiene in the United States has from the start been kept on a high professional plane and has therefore attracted people with a leaning toward technical research.

One of the outstanding reasons for the great success of industrial hygiene in controlling adverse environmental conditions and in preventing or controlling occupational disease is that pronounced rivalry has developed between the medical doctors and men and women of the sciences, as well as between the scientific professions involved, for instance, chemists and mechanical engineers. This rivalry, as has been pointed out before by one of the greatest industrial physicians of his time, Clarence D. Selby (9), has been wholesome and has resulted in advances in the field of industrial health that would have been improbable without it; competition and wholesome incentive have always inspired accomplishments. Advances in industrial process ventilation have been almost solely the achievement of the industrial hygienist: the heating and ventilating engineers, either from lack of interest or understanding, long remained aloof. As a result, many grotesque ventilating systems and ideas have appeared under the sponsorship of sheet metal men who did not even pretend to understand the physical laws applying to airflow and air movement but were attempting to satisfy a demand for protection of workmen as best they could.

6 PROSPECTIVE ROLES OF INDUSTRIAL HYGIENE

Industrial hygiene has laid aside its swaddling clothes and entered a vigorous stage of advancement. It is no longer seen by industry as the aimless effort of intellectuals to collect bottles filled with nothing, then preparing long and useless discourses that few would read or understand, and none would know how to implement. The safeguarding of industrial health is on a business basis of evaluation and control and is recognized as such by both labor and management. The purpose of the industrial hygienist is no longer merely to "lock the stable door after the horse has been stolen" but to anticipate and prevent harmful situations, or to control them before serious injury results.

The industrial hygienist has done much toward bringing medicine and safety engineering into close cooperation, partly because his interests overlap these two fields. For instance, the safety engineer looks to the industrial hygienist for assistance in problems involving the technical phases of the flammability or explosibility of solvents, gases, vapors, and dusts. Likewise, the technical differences between the classes of respiratory protection devices and the chemical absorbents used are somewhat confusing and require special study. Medical men with no engineering experience, on the other hand, will be more concerned with the toxicological aspects, and in their desire to prevent harmful exposures, physicians may consider only the promotion of health without due regard for practical economic measures or the problems of the plant production and maintenance engineers. It is the industrial hygienist's job to become acquainted with these engineers and to get the benefit of their reactions to any of his major recommendations so that his control measures will be more sound and acceptable. One of the most urgent needs is a common understanding between the heating and ventilating engineer and the industrial hygienist. The former thinks in terms of comfort and air conditioning and frequently is unable to grasp the problems involved in the control of gases, vapors, and dusts, which are designated in parts per million or million particles per cubic foot. The industrial hygienist must take the initiative in converting his control recommendations into terms of volume or rate of flow or other units more familiar to the heating and ventilating engineer. He should make it understood more generally that the control of all air contaminants, even excess heat, is something the heating and ventilating engineer needs to become acquainted with and reckon with—not a thing to be added as an afterthought or left to be planned by unskilled persons who may thereby upset the functioning of an otherwise carefully planned heating and ventilating or air-conditioning installation.

Then there are the architects, the plant layout experts, and building designers, as well as the designers of machine tools and equipment. The basic principles of industrial hygiene must be presented to these personnel so that industrial hygiene will start with the blueprints of factories and machines. All too frequently still industrial hygiene becomes a matter of telling management what their mistakes have been and how many tremendously expensive changes will have to be made, rather than getting these ideas into design where the most good will be accomplished easily and cheaply.

In one of the most modern aluminum foundries built during World War II, a facility frequently pointed out as a model of perfection, the mechanization was excellent, the sand-handling and dust-control equipment was good, but some fume, smoke, and heat sources were uncontrolled, and the general ventilation left much to be desired. No one had ever made the designer aware that dust- or fume-laden and heated air had to be moved out of any enclosure by removing all the contaminated or heated air. He had provided ample air supply and exhaust, but the inlet and exhaust ports were both in the relatively flat roof, and since they terminated at the ceiling, the air at the ceiling was given an ineffectual churning while fumes and smoke from pouring, cooling, and other incidental unhooded operations were cleared away only slowly. Obviously smoke and fumes should have been removed at their sources and the makeup air should have been distributed at or near the floor level, to ensure that contaminated and heated air would be displaced upward and exhausted through ports in or near the ceiling. Estimations of the cost of remedying the condition exceeded the cost of the original ventilating system, but strangely enough the same error is widespread today.

Until all schools of higher learning take up the teaching of some of the fundamentals of industrial hygiene in their engineering and other professional courses, the burden of this job will rest on the hygienist. This responsibility applies not only to plant design, but to machine design and plant layout as well. The opportunity to review plans should be sought in an effort to improve design so as to promote the correct application of all control measures, where such has not been done. Not only harmful exposures but also dirty occupations are on the way out. The worker has come to expect and demand a safe, healthful, and relatively clean and stress-free work place, and, having once worked in such surroundings, will not readily return to an excessively dangerous or dirty occupation. Widespread shifting of labor resulted in acquainting many workmen with the fact that control of the work environment is possible. Even the foundry has undergone a transmutation from a place with a dirty, dusty, smoke- and fume-filled atmosphere to one of comparative cleanliness. We have some of the cleanest work places in the world in the United States. We also have some that we do not care to talk about—they are the ones on which we should focus our attention until they have been cleaned up. American industry is aware of the benefits of more and better production derived from environmental control and health promotional activities and needs only to be guided in their application. If academic instruction in our engineering and other professional courses allied with this field adequately provides the principles and practice of industrial hygiene and safety, the trend will be toward the development of more competent and highly trained personnel to cover two fields now considered separately: industrial hygiene and safety engineering. For such instruction to be successful, it must provide a sufficient period of supervised practice in industry. Graduates of such courses would find common ground and interests with the industrial physician of the future, who will appreciate more and more the nature and value of engineering control practices that augment and simplify, in many cases are essential to, his own goal of preventive medicine and hygiene. Regardless of where or how it is taught—in medical schools, engineering schools, colleges, or trade schools—the engineering control of industrial health is sufficiently important to warrant its being placed on the same plane as civil, chemical, or other engineering courses. It should encompass a well-planned curriculum to include evaluation of the environment, toxicology, pharmacology, physiological effects, control methods including the principles and practice of ventilation, and the correction of all adverse environmental factors—not a course that can be picked up as an afterthought in the last semesters or as a three-month "refresher" course.

Industrial hygiene units now are found in industry, federal, state, and local governments, universities, insurance companies, and unions. Large industries sometimes have as many as 15 or 20 industrial hygiene employees, including several specialists in addition to a director and technicians. Small industries depend for their industrial hygiene expertise either on official governmental agencies or on private consulting firms. Those now in the fields of industrial hygiene or safety who have a broad understanding of their problems need have no concern about change. Those who do not have a broad viewpoint should perhaps lose no time in acquiring it or in finding a more dependable future.

With the momentum it now has, along with the rising awareness of large sections of the population, nothing can stop the movement for environmental improvement. But a little caution is perhaps in order to prevent the eager beavers from going too far too fast on costly changes that are sometimes relatively insignificant for controlling occupational hazards. The correct labeling of all chemicals is very desirable, and this includes the chemical name and necessary warnings for safe handling in addition to any trade name. There are now more than 4000 rules set down on 800 pages in the OSHA Code of Federal Regulations. This is a bit cumbersome for inspectors to cope with when visiting business establishments, especially in areas where new information is developing rapidly, as with carcinogenic action and newly emerging radiation-type exposures. Meantime we continue as we are. More schools are choosing their students from many professions and directing their instruction to the specialized field of industrial hygiene on the graduate-student level much as the Harvard School of Public Health has done, with the result that industrial hygiene will doubtless retain its identity and continue to work cooperatively with the medical and safety groups.

Considerable progress has been made during the past 50 years. Have all problems been solved? No. For one thing, the federal government is establishing time-weighted averages for various substances. Are these adequate to protect the health of *all* workers? No. There are persons who are highly susceptible to certain contaminants. These individuals are not being protected by current standards. At present there are no known ways of providing simple tests to protect individuals who are highly sensitive or highly allergic to specific contaminants. Much work must be done in this area.

Instruments are not now available to measure known exposures of people during a 168-hour weekly exposure. For example, noise is measured during an employee's 8-hour day. No measurements are made during his nonworking recreation periods, where noise levels at times may be much higher; in discotheques, for example, levels of 110 decibels are common. It is not *enough* to be concerned about the health of the worker only during the 40 hours of his work week, while many of his exposures actually cover 168 hours not 35 or 40. In the future, therefore, it is important for consideration to be given to the total environments—hours while working and at leisure, including the influence of air pollutants in the community.

Of major significance is the control of industrial hygiene hazards in the designing of machinery and building of new facilities that will reduce to the absolute minimum the exposure of employees. Too little work has been done in this area. Although this problem has been realized by industrial hygienists for many years, they should have been employed years ago in the design facilities to ensure that the problems are *designed out* of the equipment. Very few, if any, industrial hygienists have been employed by architectural firms to properly design and build factories. Too much emphasis is being placed on "Band-aid" measures today. There is a great need for more and more effort to be expended in the elimination of hazards and in providing adequate preventive protection for workers and their families. Very little emphasis has been placed on the families of the workers and their exposures.

In the future we shall see the ideals and the techniques of industrial hygiene extended and expanded along all present lines and into many more—perhaps some not now envisioned. The public is becoming more and more aware of the benefits our profession can give them throughout entire communities, not only in the working places, as important as those are. We must not—we shall not—fail them!

REFERENCES

1. L. B. Chenoweth, W. Machle, and H. Schneider, *Industrial Hygiene,* Crofts, New York, 1938.
2. A. J. Lanza and E. Higgins, U.S. Bureau of Mines Technical Paper No. 105, 1915.
3. V. M. Trasko, *Am. J. Pub. Health,* **45,** 39 (1955).
4. "Dust in Steel Foundries," *Engineering* (London), **158,** 152 (1944).
5. *Chem. Age* (London), **53,** 353, (1945).
6. A. Metsatunyan, *Gig. Sanit.* (*SSR*), **10,** 23 (1943).
7. V. A. Riazonov, "Sanitary Safeguarding of Atmospheric Air," Moscow 1954. Reviewed by F. S. Mallette, *Smog News,* (ASME) **2,** 54 (November 30, 1955).
8. J. J. Bloomfield and A. S. Landry, *Am. Ind. Hyg. Assoc. Quart.,* **16,** 65 (1955).
9. C. D. Selby, *Am. J. Pub. Health.* **30,** 1422 (1940).

Limerick Institute of Technology - LIBRARY
Institiúid Teicneolaíochta Luimnigh - LEABHARLANN
Class No. 614.7 PAT
Acc. No. 5293
Date:

CHAPTER TWO

Legislation and Legislative Trends

WILLIAM E. McCORMICK

More than 5000 years ago an unknown writer set down for the first time the relationship between an occupation and its potential for injury. Today we read this expression from the Bible: "He that diggeth a pit shall fall into it; and whoso breaketh an hedge, a serpent shall bite him. He that cleaveth wood shall be endangered thereby" (1). From that first recognition, slowly but inexorably, laws and regulations have been developed in society's efforts to improve the working environment. This chapter discusses current legislation concerned with occupational disease in the United States.

1 HISTORY OF OCCUPATIONAL DISEASE LEGISLATION

The beginnings of occupational disease legislation are unknown. Perhaps one of the earliest related to the use of a leather garment on the forearm of a bowman to protect him from injury from his bowstring. As industrialization of Europe proceeded, an appreciation for the need for better standards of the work place developed. Germany took the lead. We find a list published in 1895 by Lehmann, of the Munich Department of Hygiene, dealing with occupational health standards. Benjamin Franklin in America, as early as 1776, was writing about the "West Indian Gripes." This affliction was lead colic resulting from drinking Jamaican rum distilled through equipment made of lead. As a result, the Massachusetts legislature prohibited the use of lead-steel heads and worms for the making of rum. Table 2.1 presents the high points of occupational disease activities in the United States, beginning with the colonial period.

Table 2.1 High Points of Occupational Disease Activity in the United States

Date	Activity
1798	Establishment of the Marine Hospital Service, predecessor of the U.S. Public Health Service
1852	Massachusetts passes first safety law, dealing with steam engines
1877	Massachusetts passes a law requiring factory safeguards
1884	Bureau of Labor created in federal Department of the Interior
1898	Supreme Court decision, *Holden* vs. *Hardy*: protection of labor becomes public purpose
1908	Law enacted by Congress giving right to compensation to certain employees, for occupational injuries
1910	Bureau of Mines created as part of the Department of the Interior
1913	Department of Labor created
1917	Certain state workmen's compensation statutes held valid by Supreme Court
	First occupational disease legislation enacted by Hawaii
1918	Occupational disease legislation enacted by California
1919	Occupational disease legislation enacted by Connecticut and Wisconsin
1934	Department of Labor establishes Division of Labor Standards
1935	Passage of Social Security Act; great impetus for industrial hygiene funds made available to states
1936	Passage of Walsh-Healy Public Contracts Act: includes health and safety requirements
1938	American Conference of Governmental Industrial Hygienists (ACGIH) formed
1939	American Industrial Hygiene Association (AIHA) formed
1941	Passage of Federal Mine Inspection Act: Bureau of Mines authorized to inspect mines, recommend health and safety procedures
1941–1970	World War II and subsequent years; great expansion in industrial health activities in industry and government
1970	Passage of Public Law 91-596 (Occupational Safety and Health Act)

2 CURRENT OCCUPATIONAL DISEASE LEGISLATION

2.1 Walsh-Healy Act

In 1936 the Walsh-Healy Public Contracts Act was passed by Congress and signed by President Roosevelt. This legislation makes it mandatory for any organization supplying goods or services to the United States government to maintain a safe and healthful working environment. It took World War II to make the act really meaningful. Specific health standards as well as safety regulations were developed, and these, for the most part, became the startup standards under the Occupational Safety and Health Act of 1970.

2.2 Metal and Nonmetallic Mine Safety Act of 1966

Health and safety standards for metal and nonmetallic mines are delineated in the Mine Safety Act of 1966. A Review Board was created, consisting of five members who are appointed by the President, with the advice and consent of the Senate. Groundwork is laid for the creation of advisory committees to assist the Secretary of the Interior. These committees are expected to include an equal number of persons qualified by experience and affiliation to present the viewpoint of operators of such mines, and of persons similarly qualified to present the viewpoint of workers in the mines, as well as one or more representatives of mine inspection of safety agencies of the state. The act also provides for mandatory reporting of all accidents, injuries, and occupational diseases of the mines; and expanded programs are developed for the education of personnel in the recognition, avoidance, and prevention of accidents or unsafe or unhealthful working conditions. Finally, the act promotes sound and effective coordination between federal and state governments in mine inspection procedures.

2.3 Coal Mine Health and Safety Act of 1969

The Coal Mine Health and Safety Act of 1969 attempts to attain the highest degree of health protection for the miner. It delineates mandatory health standards and provides for the creation of an advisory committee to study mine problems. The federal government receives additional authority through this act to withdraw miners from any mine that is found to be in danger, and reentry into such mines is prohibited.

2.4 Occupational Safety and Health Act of 1970

In late December 1970, President Nixon signed Public Law 91-596, commonly known as the Occupational Safety and Health Act of 1970, the most far-reaching piece of legislation in the field of occupational health ever enacted in the United States. The basic philosophy of this legislation is stated in one of the early paragraphs of the act: "to provide for the general welfare to assure so far as possible every working man and woman in the nation safe and healthful working conditions and to preserve our human resources." The act applies to all businesses in the United States "affecting Interstate Commerce" and provides among other things for the Secretary of Labor to set mandatory occupational safety and health standards for such businesses. Under the act, two federal agencies, as well as all the states, play separate yet interlocking roles. Broadly defined, these are as follows:

1. *Occupational Safety and Health Administration (OSHA).* A part of the federal Department of Labor, this agency has as its basic responsibilities the promulgation and enforcement of health and safety standards.
2. *National Institute for Occupational Safety and Health (NIOSH).* A part of the U.S. Department of Health, Education and Welfare, this agency has as its principal role the carrying out of research for the development of new standards and the provision for training of personnel.
3. *The states' role.* Under Section 18 of the act, the right to assume the primary responsibility for enforcement of occupational safety and health regulations resides with an individual state, provided a suitable plan for such enforcement is approved by the

Secretary of Labor. Federal funding is also available to these states if the plan is approved and if it continues to meet the approval of the Department of Labor. In the event a state does not wish to accept such responsibility or if its plan is unacceptable, OSHA becomes the enforcement agency within that state. Under Section 7 of the act, OSHA is also authorized to contract with a state or an agent of that state to carry out the OSHA activities.

2.5 Standards

One of the basic provisions of the Occupational Safety and Health Act of 1970 authorizes the Secretary of Labor to establish standards for health and safety. Soon after the act had become law, the Secretary of Labor adopted by regulation the occupational health standards that existed under the Walsh-Healy Act. These were supplemented, where necessary, by adopting the 1968 Threshold Limit Values (TLVs) of the American Conference of Governmental Industrial Hygienists and the occupational health standards that existed as produced by the American National Standards Institute (2). New standards produced since these initial startup standards originate with NIOSH as criteria documents proposed to the Secretary of Labor. Appropriate input from all interested groups and through public hearings is considered before the adoption of a new standard by the secretary.

3 STATE ACTIVITY AND AUTHORITY

Historically, individual states have established their own occupational health laws and regulations. Consequently, over the years many variations existed between individual states in both type and quality of occupational health programs. With the advent of the Occupational Safety and Health Act of 1970, however, there has been much greater uniformity. As indicated in Section 2.4, the act presents three options to the individual states.

1. Under Section 18, a state agency assumes the responsibility for the act's enforcement under a plan approved by the Secretary of Labor. Federal funding is provided and OSHA monitors the program on a continuing basis.
2. OSHA contracts with an agency of the state as provided in section 7 to carry out OSHA's activities. The program is federally funded.
3. A state may refuse to be in either of the two categories just described, and hence becoming a "no plan" state. OSHA then enforces the provisions of the act with federal personnel. As of March 1976, there were 16 states in this category.

4 LEGISLATIVE OVERVIEW

In the last decade we have seen great progress in making the workplace safer. The Occupational Safety and Health Act of 1970, with all its compromises, does more than any other piece of legislation ever enacted in the United States to accomplish this objective. The basic substance of the act is sound; the problems arise in its administration.

Realistic—as distinguished from idealistic—standards for both health and safety must be devised and properly enforced. Most standards relating to *safety* are now fully developed. However those dealing with *health* are in their infancy. Many of the existing standards are based on opinion rather than fact. Hence new health standards—or revisions of present ones—must receive major emphasis. But the new standards must be realistic. They must take into consideration the economic impact (i.e., society's ability and willingness to pay for the benefits received) and practicality.

Proper health standards cannot be devised or enforced without definitions of harmful versus safe levels of exposure. This requires constant, ongoing research. This very important phase of occupational disease control must be increased if we are to keep pace with the newly developed materials of an expanding, technological society.

The enforcement of all occupational disease regulations requires professional judgment—not merely the reading of tables and numbers on a dial. Although many of the field measurements can and should be made by trained technicians, the ultimate responsibility for judging the safety of the working environment must reside with professionals—the industrial hygienist, the occupational health physician, and the safety professional. These personnel, and these alone in their respective fields of expertise, are qualified to make the final judgments relating to the existence or absence of health or safety hazards. Present legislation frequently does not recognize this necessity.

Based on past experience, it seems safe to predict that the next two decades will see new legislation enacted, not only in the United States, but in many countries throughout the world, to improve the health and safety aspects of man's working environment and to achieve the ultimate goal of every health and safety professional: to eliminate diseases and accidents from that environment.

REFERENCES

1. *Holy Bible* (King James Version): Ecclesiastes **10,** 8–9.
2. *Fed. Reg.,* **37,** 22139–22142, October 18, 1972.

CHAPTER THREE

Industrial Hygiene Records and Reports

PAUL D. HALLEY and
PETER WOLKONSKY, M.D.

1 INTRODUCTION

Industrial hygiene activities include surveys, measurements, evaluations, controls, and recommendations. Recordkeeping, as one important facet of such activities, serves as an essential tool for the industrial hygienist in managing, monitoring, and documenting his/her efforts in the evaluation and control of employee exposure. This, however, is not the only function of the industrial hygienist's records. Physicians may call on an industrial hygienist for employee exposure data if a causal relationship is suspected between exposure and illness. Corporate legal and employee relations staffs consider the industrial hygienist's records essential for health-related grievance and arbitration cases, and for compensation claims. Exposure records serve as the primary source for statistical considerations, epidemiological studies, and research. Properly documented exposure-illness records serve to determine safe or unsafe, healthy or unhealthy working conditions and limits of exposure. In addition, today's regulatory standards require that certain records be kept.

Recognizing that recordkeeping is an essential part of the industrial hygienist's job, it is natural to ask what constitutes "adequate" records. The novice industrial hygienist soon discovers that there is no one "adequate" recordkeeping system for all-purpose use. However the various sytems judged to be most effective all have certain characteristics in common. Records should be as detailed as possible for the data required (yet physically manageable), and they should be appropriately structured to relate to other available pertinent data (i.e., medical, personnel, air pollution, weather etc. records).

Records are only as good as the measurements they document. It is of primary importance not only to record data accurately, but also to describe the collection and analytical techniques employed for each sample result. This allows the user of the records to evaluate and consider more effectively the comparability of historical sample results.

Recordkeeping in itself obviously serves no useful purpose unless certain objectives have been defined and until the data recorded are translated into reports. The industrial hygienist must continually strive to present such data in a report format that is readily understandable, and in sufficient detail to permit the user to make adequate decisions.

The industrial hygiene report should reflect the special expertise of the industrial hygienist to interrelate all facets of the relationships of the worker and the worker's environment in evaluating the impact of the environment on the worker's health. It is not an assignment to be taken lightly. For example, inadequate assessment by the industrial hygienist could result either in the impairment of the health of the worker or in excessive costs for unnecessary control measures.

This chapter reviews the objectives of recordkeeping and reports and gives a few examples of workable recordkeeping systems; the role of the computer in recordkeeping systems is incorporated into the discussion.

2 INDUSTRIAL HYGIENE RECORDS

Any practical system for documenting industrial hygiene surveys and activities must be comprehensive, flexible, and simple.

A system of storage and retrieval should be developed that will accomplish the following functions:

1. Allow the user to obtain information in the form that is needed.
2. Cover all foreseeable areas of current and potential future interest.
3. Exclude extraneous data not required or not expected to be required in the future.
4. Minimize cost and inefficiency.

Because industrial hygiene studies produce large amounts of data of various types, and because consideration must be given to the later use of the data by others and for varying purposes, the data in the original industrial hygiene records cannot easily be presented in a uniform or constant format.

The permanent record from which the industrial hygiene report is developed should include notes logged in a field notebook or suitable survey form. The practice of jotting down random bits of information on pieces of scratch paper that may be lost cannot be too strongly condemned, and details not recorded on the spot are often forgotten. Reports written from memory, though they may be sufficiently accurate in certain instances for control purposes, are almost certainly inadequate as a legal record. For example, unless the original notes made on the job at the time of the survey can be produced, any record of atmospheric conditions in a plant offered as evidence in court may be open to question.

It is therefore just as important to record all data pertinent to the sample as it is to collect the sample. The experienced industrial hygienist will later compare his or her field observations with analytical data and judge whether the analytical data appear to be valid or whether a resurvey is needed. Records should be sufficiently detailed to permit another individual to locate the sampling area and duplicate the previous survey without the assistance of the person who originally recorded the data.

Last, the speed and accuracy of preparing a final report from survey records of an investigation depend largely on information recorded in the form of notes.

Photographs tell stories with a minimum of explanation needed. "Before and after" pictures can be particularly effective. Pictures serve as a permanent record. They may also provide information that one fails to record at the work site.

Evaluation of a potential health hazard may involve air sampling, laboratory analysis, calibration, evaluation of nonionizing physical stresses (noise, light, microwaves, heat), biological monitoring, and evaluation of ionizing radiation. Each of these may, in certain circumstances, require its own form or set of forms. When an industrial hygienist is called on to make many similar environmental surveys, however, it is often advisable to design an appropriate specific form to be filled out at the time of sampling.

In any event, the recording of data obtained in any survey activity requires the development of suitable forms and their orderly completion in a neat and efficient manner. Forms with spaces labeled for essential data help avoid the common failure to record needed information. The form itself may thus serve to determine whether all needed information has been obtained at the time of the survey. On the other hand, the use of a poorly designed or incomplete form will lead to incomplete reports or will require repetitive communication with field personnel or perhaps even a resurvey. Examples of some forms useful in industrial hygiene practice appear in Figures 3.1 to 3.28.

Recognition of potential health hazards involves an inventory of all materials and processes likely to create a health hazard by job or occupation or area. Typically an initial "walk-through" appraisal is made, for which forms such as those in Figures 3.1, 3.2, and 3.3 are designed.

Where chemicals and compounds are used, a health and safety material data sheet should be available for each, listing health and safety hazards and composition of mixtures. An example of an industrial hygiene toxicology and safety data sheet is presented in Figure 3.4.

Sample forms for air sampling field notes are given in Figures 3.5 and 3.6.

Records for the industrial hygiene laboratory should include adequate numbering and laboratory identification schemes for samples, as well as established calibration and quality control standards. The sample forms in Figures 3.7, 3.8, 3.9, and 3.10 are appropriate for this purpose.

Calibration data forms appear in Figures 3.11 and 3.12.

Nonionizing physical stress data forms include those for noise (Figures 3.13, 3.14,

Industrial Hygiene Presurvey Inspection

Client ____________________ Date ____________

Location ____________________ Investigator ____________

Department ____________________ CEC Job No. ____________

Supervisor ____________ Phone No. ________

Engineer ____________ Phone No. ________

____________ Phone No. ________

____________ Phone No. ________

Description of operations:

Process flow sheets ☐ Plot plans ☐

Subjective hazard survey:

Job Classification	Number of Workers				Environmental Stress	Severity			
	1	2	3	T		0	1	2	3

Figure 3.1

Industrial Hygiene Presurvey Inspection

Client ______________________ Department ______________________

Location ______________________

Inventory of chemicals

Substance	Occurrence				Exposure		Remarks
	R.M.	Int.	Byp.	Prt.	Skin	Inhal.	

Engineering controls:

Operation	Type of Control	Apparent Effectiveness	Corrective Action

Figure 3.1 (*Continued*)

Industrial Hygiene Presurvey Inspection

Client ______________________ Department ______________________

Location ______________________

Summary of personal protection program:

Protective Device	Selection and Fit	Education and Training	Evidence of Use	Comments
Respiratory				
Hearing				
Eyes				
Other				

*Summary of medical records of overexposure:

*Note. This does not refer to confidential medical information but rather to the types of medical information being recorded, such as special tests related to occupational exposures.

Per: ______________________

Overall subjective impression of conditions: ______________________

Figure 3.1 (*Continued*)

Industrial Hygiene Survey
Work Room Data

Name of plant: *J.S. Davidson*
Department: *Screw Machine*
Informant's name: *C. Olka*
Surveyed by: *C.S.*

Page *1* of *10*
Work room: *1*
Date: *3–10–77*

Occupation	Number of persons			Nature of Job	Raw Materials and Byproducts	Control Measures							Remarks and Potential Hazards
	Male	Fem.	Tot.			Local Exh.	Isolation	Wet Method	Gas Mask	Respirator	Press Hel.	Other	
Machine operators	20	–	20	Operate screw machines	Nonferrous metals Houghton #12 oils							X	
						Aprons							
Salvage men	3	–	3	Centrifuge and filter spent oil	Shavings – nonferrous Houghton cutting oil #12							X	Dermatitis Oil mist Noise
						Aprons Gloves							
													Dermatitis Oil mist Noise
Total:													

Form #

Figure 3.2 Based on Reference 3, page VI-5.

Exposures, Classified by Occupation and Type

Facility: ______________ Division: ______________ Dept.: ______________ Date: ______________

Department	Occupation	Dust — Fumes							Gases — Vapors — Mists									Physical Conditions						
		SiO_2 / FeO	Mn / Coke	Lime / Ore		Lead / Fluorides	Cr_2O_3 / NiO		HCN / NH_3	NO / H2S	CP / SO_2	BTX / HCl	Chromotos / CO	Pitch / Tar	H_2SO_4 / Alk.			IR / UV	Noise / Vib.	Heat				

Figure 3.3 Based on Reference 3, page XIV-3.

AMOCO

Industrial Hygiene Toxicology and Safety Data Sheet

Environmental Health Services
Medical and Health Services Department

Section I

Trade Name and Synonyms	
Manufacturer's Name	Emergency Telephone No.
Address (Number, Street, City, State and Zip Code)	
Chemical Name and/or Family	Formula

Section II Physiological Properties

Effects of Exposure
Acute:
Eyes ____
Skin ____
Respiratory System ____ Other ____

Chronic: ____
Sensitization Properties (Species): Skin: Yes ____ No ____ Unknown ____ Respiratory Yes ____ No ____ Unknown ____

Median Lethal Dose (LD_{50}, LC_{50}): (Species)	Irritation Index, Estimation of Irritation (Species)
Oral ____	Skin ____
Inhalation ____	Eyes ____
Dermal ____	Symptoms of Exposure
Other ____	

Section III Chemical and Physical Properties

Boiling Point (°F) Melting Point (°F)	Vapor Pressure (mm Hg at 20°C)	Vapor Density (Air = 1)	Specific Gravity (H_2O = 1)
pH	Appearance and Odor	Ignition Temp. °F	Flashpoint °F (Method)
Flammable Limits % Lower	Upper	Solubility in Water	

Products Evolved When Subjected to Heat or Combustion

Hazardous Polymerization: Occurs ____ Does not Occur ____

The Material Reacts Violently With:
Air ____ Water ____ Heat ____ Strong Oxidizers ____ Others (Specify) ____

List of All Toxic and Hazardous Components (%)		List of all Nuisance Components (%)	
____	____	____	____
____	____	____	____
____	____	____	____
____	____	____	____

This Product is Classified as: Hazardous ____ Nuisance (Only) ____

Figure 3.4

Section IV Control Procedures

A. Occupational

Protective Equipment (Type)

Eyes ____

Skin ____

Inhalation ____

Ventilation (Type Required) ____

Precautionary Label (may be attached)

Permissible Concentrations: Air Biological

Monitoring Procedures Air Biological

First Aid

Eyes ____

Skin ____

Ingestion ____

Inhalation ____

B. Environmental

Product is Biodegradeable ____

Not Biodegradeable ____

Product is a Pollutant if discharged into:

Water ____

Air ____

Waste Disposal Method

C. Recommended Fire Extinguishing Agents and Special Procedures

D. Unusual Fire and Explosive Hazards

E. Procedures in Case of Breakage or Leakage

F. Requirements for Transportation, Handling and Storage

G. NFPA Symbol

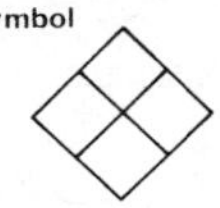

Section V U. S. Government and Other Regulatory Agency Controls

A. Marketing and use Regulated by (Specific Regulation)

FDA ____ USDA ____ EPA ____ Others (Specify) ____

B. State or Local Regulations Affecting the use of this Material (Restriction on Amount Released or Discharged into Air or Water, Etc.)

Section VI Comments

Additional Health and Safety Information Known About This Product

Information Supplied By: ____ Signature: ____

Title: ____ Date: ____

Figure 3.4 *(Continued)*

Air Sampling Data Sheet

Installation and location: ______________________ Unit, Dept., or area: ______________

Sample no.: ______________________ Sampling date: ______________

Evaluation of: ____

____

Process (source) sample: ______

Area sample: ______

Breathing zone: Handheld: ______ Personal sample: ______

Name: ______________________ Height: ______________

Job classification or title: ______________________

Sampling materials used: ______________________________

____

____

Pump no.: __________ Dial position (Sipin): ______ Detector tube lot no.: ________

Start sampling: Time ______ Rotameter or pump counter reading: ________

Stop sampling: Time ______ Rotameter or pump counter reading: ________

Instrument reading (s) ______________________________

Sampling location: ______________________________

____

____

Sampling conditions: Temp. of: ____ Humidity (%) ____ Air velocity: ____

Downwind: ____ Upwind: ____ Barometric pressure: ________

____

____

____

Industrial hygiene controls:

Ventilation: ______________________ Warning signs posted: ______________

Protective apparel: ______________________________

Respirator: ______________________________

____

Job/process description: ______________________________

____

____

Sampled by: ______________________ Date sample sent to laboratory: ________

Figure 3.5

Asbestos Sample Sheet

Plant: ____________________ Date received: ____________________

Location: ____________________ Results sent by telegram: ____________________

telephone: ____________________

Air Sampling								Analysis	
Date	Test No.	Dept.	Name and Social Security No.	Job Description	App. Resp.	Rate (ℓpm)	Time (min)	Date of Finding	Fibers per cc

Figure 3.6

HAZARD EVALUATIONS ONLY

SAMPLE DATA AND LABORATORY REPORT

Name of Plant__________ Address__________
Analyze for__________ Collected by__________ Date Collected__________
Send Report to__________

Laboratory Number*	Field Number	Type of Sample	Sample Volume	Description: Location, name, time, collecting medium and volume, etc. Use next two columns if convenient	Analysis*

Remarks (Interfering materials possibly present, purpose of samples, unusual circumstances, laboratory comments, etc. Use back for additional space.)__________

*To be filled out by laboratory. Please submit two copies of form with samples. One copy will be returned with results.

Date Received*__________ Date Reported*__________ Analyst*__________ __________
Chief, Laboratory Services

Figure 3.7

U.S. DEPARTMENT OF LABOR
Occupational Safety and Health Administration

SAMPLE IDENTIFICATION SHEET

1. SAMPLE NUMBER (REGIONAL)	2. DATE TAKEN		3. DATE SUBMITTED TO LAB
4. BY WHOM TAKEN			5. LABORATORY NUMBER
6. WHERE TAKEN	NAME OF ESTABLISHMENT		
	STREET ADDRESS		
	CITY	COUNTY	STATE
7. OPERATION			
8. SAMPLING TECHNIQUE	IMPINGER ____ BUBBLER ____ ESP ____ FILTER PAPER ____		
	MATERIAL ____ BAGS ____ SETTLED DUST ____ GRAB SAMPLE ____		
9. AIR VOLUME SAMPLED (IN LITERS AND CUBIC FEET)		BP (mm)	T (C)
10. POSSIBLE INTERFERENCES			
11. ANALYSIS REQUIRED			
12. DATE REPORTED FROM LAB	13. CHEMISTS		
14. REMARKS			

Form OSHA-23
Dec. 1971

Figure 3.8

Microscopic Fibrous Particulate Count Data

Date: ____________ Objective: Phaco 40X

Plant: ____________ Eyepiece: Periplan 10X

Process: ____________ Total magnification: 400X

Lab Number	Filter Number	Size of Field	Number of Fields Counted	Number of Fibers Counted	Number of Fibers Total	Fiber Concentration (fibers/ml)

Figure 3.9

Microscopic Sizing Data of Dust Particles

Date: ______________________
Plant: ______________________
Process: ______________________

Sample No.: ______________________
Objective: Phaco 40X
Eyepiece: Periplan 10X
Total magnification: 400X

Particle Size (μm)		Frequency of Particles, n	Size Distribution by Count		Size Distribution by Weight			
Size Range	Mean Size, d		Percentage of Total	Cumulative Percentage	d^3	nd^3	Percentage of Total	Cumulative Percentage
0.8–1.1	0.95				0.86			
1.1–1.6	1.35				2.5			
1.6–2.2	1.90				6.9			
2.2–3.1	2.65				18.6			
3.1–4.4	3.75				52.7			
4.4–6.2	5.3				149			
6.2–8.8	7.5				422			
8.8–12.5	10.6				1191			
12.5–17.7	15.1				3443			
17.7–25	21.3				9663			
Totals		______				______		

Figure 3.10

Calibration Data Sheet for Number

Collecting media: ______
Date: ______
GDC employee: ______
Method used: ______

Air temperature: ______
Barometric pressure: ______
Standard instrument: ______
Rotameter number: ______

Rotameter Reading	Elapsed Time	Manometer Start	Manometer Stop	Meter Reading Start	Meter Reading Stop	Volume	Flowrate

Collecting media: ______
Date: ______
GDC employee: ______
Method used: ______

Air temperature: ______
Barometric pressure: ______
Standard instrument: ______
Rotameter number: ______

Rotameter Reading	Elapsed Time	Manometer Start	Manometer Stop	Meter Reading Start	Meter Reading Stop	Volume	Flowrate

Flowrate

Rotameter Reading

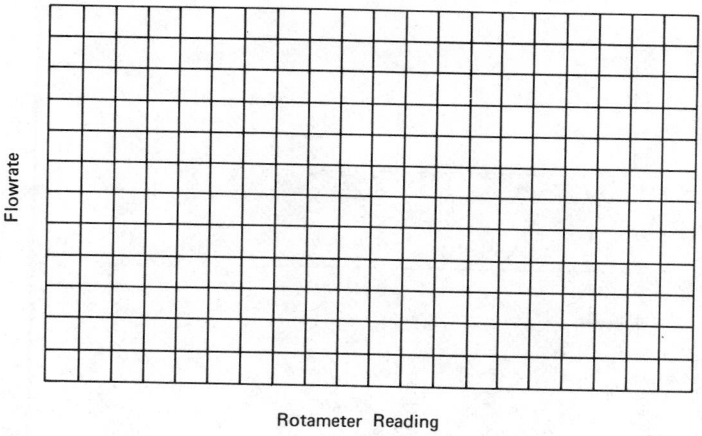

Figure 3.11

Sampler (Mfgr.)	Repairs		Calibration – Airflow											
	Date	Type	Date	Out	In	Date	Out	In	Date	Out	In	Date	Out	In
Model														
Serial or company number														
Date of purchase														
Remarks														

Figure 3.12 Typical log or schedule for equipment. After Reference 3, page X-2.

Sound Level Measurements

Company location: ______ Date: ______ Time: ______
(Plant, Refinery, Region, Division, etc.)
Wind direction: ______ Speed: ____ mph
Unit or plant area* ______ Temp.: ______ Humidity: ______
Operating conditions: ______ Sound level meter: Mfg. ______
Model: ______ Serial: ______
Accessories: ______
Calibrator model ______ Serial: ______
Calibration check: Start: ______ dB
End: ______ dB
Study by*: ______

Test No.	Location of Microphone*	Decibel Scale			Remarks*
		A	B	C	

*See directions.

Figure 3.13

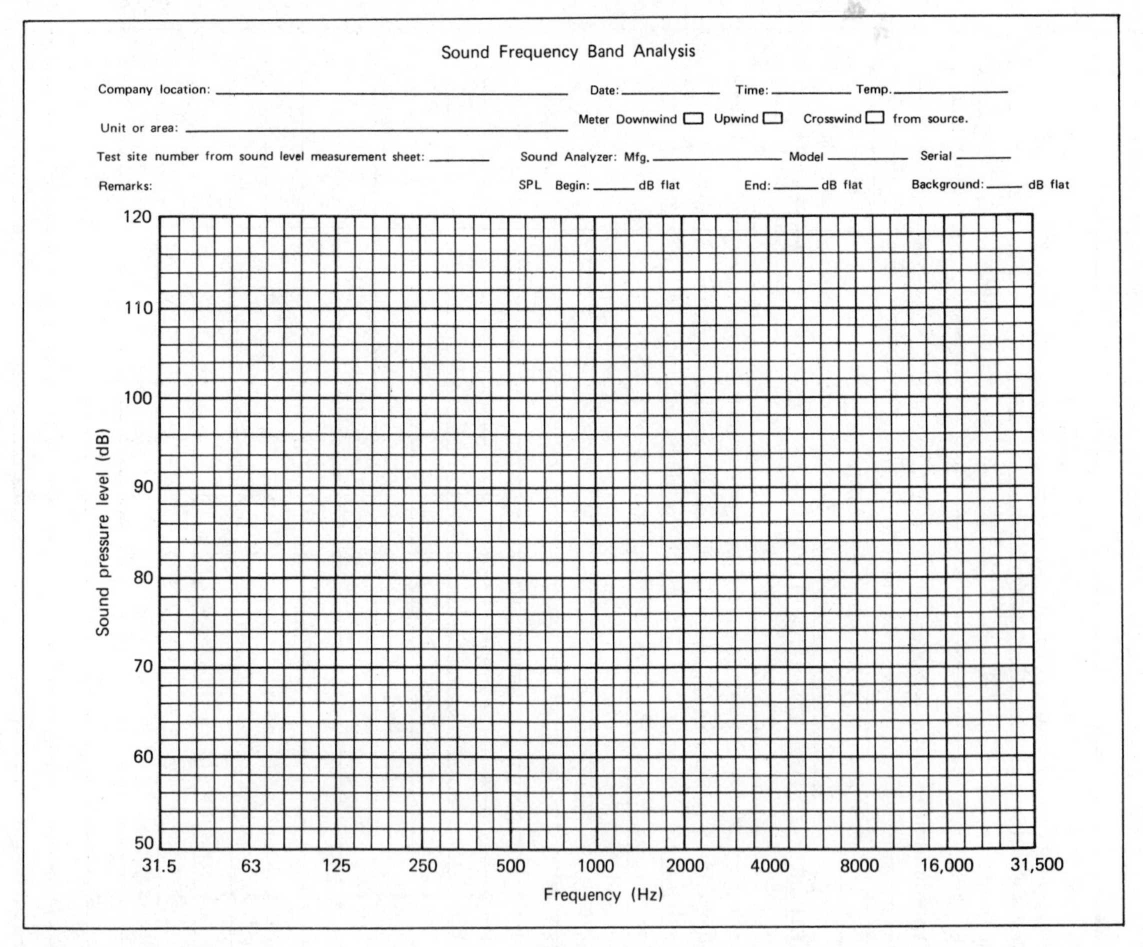
Sound Frequency Band Analysis

Company location: ______ Date: ______ Time: ______ Temp. ______

Unit or area: ______ Meter Downwind ☐ Upwind ☐ Crosswind ☐ from source.

Test site number from sound level measurement sheet: ______ Sound Analyzer: Mfg. ______ Model ______ Serial ______

Remarks: SPL Begin: ______ dB flat End: ______ dB flat Background: ______ dB flat

Figure 3.14

and 3.15), light (Figure 3.16), microwaves (Figures 3.17), and heat (Figures 3.18 and 3.19).

A biological monitoring form (Figure 3.20) is included, as well.

Evaluation of ionizing radiation exposures generates data that are recorded on forms like those in Figures 3.21 through 3.25.

Records on ventilation may be recorded on forms resembling those in Figures 3.26 and 3.27.

An example of a duty roster for making a historical record of employees entering areas of potential exposure to extremely toxic materials appears in Figure 3.28.

3 REPORT OF SURVEYS AND STUDIES

An effective industrial hygiene survey or study report conveys, accurately and efficiently, both the data and pertinent ideas developed by the investigation. These serve as a base for solving immediate problems and also document information for future reference. The report should present the facts, analyze and interpret the findings, and develop conclusions and recommendations. The writer must try to anticipate and answer questions, since feedback from the written communication may not be received. The report must be organized and written for the needs and understanding of a specific reader or group of readers. The content, approach, style of writing, and choice of words depend on the varied backgrounds of the intended readers.

Within the report writing process, these outlined steps are generally followed:

1. *Plan.* Outline the purpose of the study, define the problem, establish the scope of the report, and consider the manner in which information and materials will be presented.
2. *Organize.* Collect, tabulate, analyze, and interpret data, to permit developing conclusions and recommendations.
3. *Outline.* Present information in a form that will allow written discussion.
4. *Write.* Prepare a rough draft.
5. *Revise.* Rewrite and correct the first draft after an interval of time to allow a fresh viewpoint.

No standard format exists for all cases, but some technical writing books and company-style manuals do prescribe various report structures. In such cases, the final form selected by the writer should fit the needs of the organization, as well as those of the reader and the writer. One report format contains the following sections:

- A *summary* that concisely presents the work reported, including an abridgment of background information, conclusions, and recommendations.
- *Recommendations* that list all proposed changes, supported by brief comments on the reasons for the suggested courses of action.

Industrial Health
Dosimeter Sound Survey/Dust Survey

Field investigator:	Atmospheric conditions: Temperature ▶ Humidity ▶ Wind ▶		Date:	Page:
District/division:	Department	Superintendent	Operational status:	No. of employees in department

Dosimeter sound survey

Employee Name	Employee No.	Job Title	Dosimeter Serial No.	Calib.		Exposure Time			Over 115 dBa		Readout Serial No.	% Exposure	
				Bef.	Aft.	Time off	Time on	Net minutes	Yes	No		Reading	Pro-rated

Figure 3.15 After a form used by General Electric Company.

Illumination Survey

Plant: ______________________ Date: ______________

Department: ______________________ By: ______________

Test Number	Time	Location	Illumination (footcandles)		
			Measured		Recommended Average
			Range	Average	

Remarks:

GDC–7

Figure 3.16

Date: ____________________

TO: ____________________ ____________________

FROM: Occupational Health and Safety

SUBJECT: Microwave Oven Study, ____________________, _____ Div.

The microwave oven/s in the subject area/s was/were checked on __________ __________ as part of a semi-annual test. Results were:

Building:

Oven serial/ANL no.

Maximum leakage at door seal, mw/cm^2:

Maximum leakage at door screen, mw/cm^2:

Maximum leakage from cabinet, mw/cm^2:

Momentary maximum field indication at door latch when door was opened quickly while oven was in operation, mw/cm^2:

( ) All leakages were less than 10 mw/cm^2, which is the permitted maximum.

( ) Leakage exceeded the permitted level. We recommend servicing of the unit/s to correct this defect. Contact us for retesting after servicing.

Tests were conducted by ____________________, using a Narda 8200 microwave power meter.

Figure 3.17

Heat Stress Survey Summary

Area surveyed: ______________ Plant: ______________

Location	Date	Time	Air Temp. (° F)	Wet–Bulb Temp. (° F)	Black Globe Temp. (° F)	Air Velocity (fpm)	Radiant Heat Load (Btu/hr)	Convective Heat Load (Btu/hr)	Estimated Metabolic Rate (Btu/hr)	E_{req} (Btu/hr)	E_{max} (Btu/hr)	HSI	MST or SRT (hr)

GDC–10

Figure 3.18

Heat Stress Survey Summary

Plant: ______________________

Area surveyed: ______________________

Date	Time	Description	Air Temp. (°C)	Wet–Bulb Temp. (°C)	Black Globe Temp. (°C)	Air Velocity (fpm)	WBGT Index (°C)

Figure 3.19

Blood Lead Report

Date	Name (Type)	Clock	Check Correct Occupation*								Milligrams lead per 100 g blood
			Works with solder as:					Works on steel only as:			
			Torch Solderer	Filer	Booth Grinder	Repairman	Other (Specify)	Metal Finisher or Grinder	Assembler	Other (Specify)	

*Obtain accurate information on employee's occupation and classify correctly.

Figure 3.20

Form NRC-5
(2-75)
10 CFR 20

Approved by GAO
B-180225 (R004)
Expires – 4-30-77

U. S. NUCLEAR REGULATORY COMMISSION

CURRENT OCCUPATIONAL EXTERNAL RADIATION EXPOSURE

See Instructions on Back

IDENTIFICATION

1. NAME (PRINT – Last, first, and middle)	2. SOCIAL SECURITY NO.
3. DATE OF BIRTH (Month, day, year)	4. NAME OF LICENSEE

5. DOSE RECORDED FOR (Specify: Whole body; skin of whole body; or hands and forearms, feet and ankles.)	6. WHOLE BODY DOSE STATUS (rem)	7. METHOD OF MONITORING (e.g., Film Badge – FB; Pocket Chamber – PC; Calculations – Calc.) X OR GAMMA ______ BETA ______ NEUTRONS ______

8. PERIOD OF EXPOSURE (From – To)	DOSE FOR THE PERIOD (rem)				13. RUNNING TOTAL FOR CALENDAR QUARTER (rem)
	9. X OR GAMMA	10. BETA	11. NEUTRON	12. TOTAL	

LIFETIME ACCUMULATED DOSE

14. PREVIOUS TOTAL (rem)	15. TOTAL QUARTERLY DOSE date rem	16. TOTAL ACCUMULATED DOSE (rem)	17. PERM. ACC. DOSE 5(N-18) (rem)	18. UNUSED PART OF PERMISSIBLE ACCUMULATED DOSE (rem)

Figure 3.21

Form NRC-4
(2-75)
10 CFR 20

U. S. NUCLEAR REGULATORY COMMISSION

Approved by GAO
B-180225 (R0043)
Expires – 4-30-77

OCCUPATIONAL EXTERNAL RADIATION EXPOSURE HISTORY

See Instructions on the Back

IDENTIFICATION

1. NAME (PRINT – LAST, FIRST, AND MIDDLE)	2. SOCIAL SECURITY NO.
3. DATE OF BIRTH (MONTH, DAY, YEAR)	4. AGE IN FULL YEARS (N)

OCCUPATIONAL EXPOSURE – PREVIOUS HISTORY

5. PREVIOUS EMPLOYMENTS INVOLVING RADIATION EXPOSURE–LIST NAME AND ADDRESS OF EMPLOYER	6. DATES OF EMPLOYMENT (FROM–TO)	7. PERIODS OF EXPOSURE	8. WHOLE BODY (REM)	9. RECORD OR CALCULATED (INSERT ONE)

10. REMARKS

11. ACCUMULATED OCCUPATIONAL DOSE – TOTAL

13. CALCULATIONS – PERMISSIBLE DOSE
WHOLE BODY:

(A) PERMISSIBLE ACCUMULATED DOSE = 5(N–18) = ____REM

(B) TOTAL EXPOSURE TO DATE (FROM ITEM 11) = ____REM

(C) UNUSED PART OF PERMISSIBLE ACCUMULATED DOSE (A–B) = ____REM

12. CERTIFICATION: I CERTIFY THAT THE EXPOSURE HISTORY LISTED IN COLUMNS 5, 6, AND 7 IS CORRECT AND COMPLETE TO THE BEST OF MY KNOWLEDGE AND BELIEF.

EMPLOYEE'S SIGNATURE DATE

14. NAME OF LICENSEE

Figure 3.22

Radiation Survey Format

Date: ______________ Survey Conducted by: ______________

Geographical location: ______________

Background radiation (m rem/hr max deflection): ______________

Equipment: ______________

Cal. date: ______________

Location of Measurements	Scale Setting	Meter Reading (mR/hr)	Time Constant Mode	Conditions	Remarks

Figure 3.23

- A *discussion* that presents findings at length and makes and evaluates conclusions.
- An *appendix* that contains result data and material too detailed for inclusion in the discussion, providing supportive information and items of little interest to some readers.

A report longer than a few pages includes a *title page* and a *table of contents,* which can also serve as a report outline. As an aid to the reader, use headings and their subdivisions liberally. Include illustrations, tables, charts, diagrams, and photographs to reduce verbosity.

The industrial hygiene survey report should start with a summary statement telling why the survey was undertaken. Because most persons will want to see conclusions and recommendations as soon as possible, these should follow immediately. The report should tell what was done, what was seen, and what data were available to the industrial hygienist. Appropriate sections of the report should contain a brief description of the plant, department, or operation, and any significant changes that may have taken place since any previous survey; a discussion of control measures already implemented; potential health effects resulting from exposure to environmental stresses surveyed; sample, measurements, and test results, and an interpretation of these results; sampling, measuring, and analytical procedures; documentation of equipment calibration; and findings from other pertinent studies.

The written report should be aimed at the intended readers, directing their attention to the facts logically in the shortest possible time. Forceful writing accomplishes this by using plain words and proper grammar for clarity, and by omitting needless words and sentences. These characteristics will make the report more readable and will not distort the communication.

Books on writing style and on preparing technical reports are indispensable and should be part of a reference library. Writing skills improve through the use of such references, accompanied by practice.

Adequate follow-through is a prime factor in gaining acceptance of recommendations with a minimum of delay after a report has been issued. Follow-through methods include:

1. Offering assistance in the report transmittal letter.
2. Presenting the report in person and reviewing contents with plant management.
3. Providing assistance in carrying out recommendations.
4. Reviewing designs.
5. Conducting a follow-up survey.

4 LINKING INDUSTRIAL HYGIENE DATA TO HEALTH RECORDS

The health experience of workers in relation to exposure must be followed closely to achieve a complete occupational health program. Both the industrial hygienist and the

Bioassay Radioactivity Exposure Record

Name: ______________________ Payroll No.: __________ Supervisor: __________________ Division: ______________

Job Classification: ______________________ Date Form Completed: ______________ Information Obtained from: ______________

Form Completed by: ______________

Location of work: (Bldg.)				
Time spent in each:				

Check all applicable squares

Elements and/or Isotopes Used	Form								Quantity							Frequency of Exposure					
	Metal or Element	Compound	Massive Solid	Powdered Solid	Liquid	Gas	Sealed or Clad	Other	Gram	Milligram	Microgram	Curie	Millicurie	Microcurie	Other	Daily	Weekly	Monthly	Bi–Monthly	Semi–Yearly	Yearly
Normal Uranium																					
Enriched Uranium																					

Irradiated uranium																						
Depleted uranium																						
Plutonium																						
Irradiated plutonium																						
Tritium																						
Carbon–14																						
Iodine–131																						
Mixed fission products																						
Radium																						
Thorium																						
Strontium 89 or 90																						
Polonium																						
Protactinium																						
Technetium																						
Transplutonium																						
Other																						

If individual does not work directly with radioactive materials check here ☐

Remarks: ______________________________

Figure 3.24

To: ____________________ Date: ____________________

WIPE TEST REPORT

Division ____________________ Plant ____________________

Date of Wipe Test ____________________

Date of Counting ____________________

Wipe Number	Source & Source No.	Strength	Activity	Microcuries of Activity on Wipe

The microcurie figures listed are for comparison with the 0.005 microcurie figure, listed in A.E.C. regulations, which when exceeded, indicates objectionable leakage of a source. This figure is listed for use with gamma radiography sources in 10CFR 31, Radiation Safety Requirements for Radiographic Operations, and for use with alpha sources in Handbook 73 Protection Against Radiations from Sealed Gamma Sources.

A Nuclear Measurements Corporation Alpha-Beta-Gamma Proportional Counter, Model PC-4, was used to make the counts.

Figure 3.25

Plant________ Dept.________ Date________

Operation exhausted________ By________

Line sketch showing points of measurement

Date system installed________

Hood and transport velocity

Point	Duct		VP in. H_2O	SP in. H_2O	FPM (Fig. 6-16)	CFM $Q = VA$	Remarks
	D	Area (Fig. 6-18)					

Pitot traverse

Pitot readings - See Fig. 6-16 & Tables 9-1, 9-2, 9-3

Points	VP	Vel.	VP	Vel.	VP	Vel.
1						
2						
3						
4						
5						
6						
7						
8						
9						
10						
Total Vel.						
Average Vel.						
CFM						

Fan

Type________

Size________

Point	Dia.	SP	VP	TP	CFM
Inlet					
Outlet					

Fan SP________ (See Section 6)

Motor

Name________ Size________

HP____ E____ I____ W____

Collector

Type & size________

Point	Dia.	SP	ΔSP
Inlet			
Outlet			

Notes________

Industrial Ventilation — A Manual of Recommended Practice. American Conference of Governmental Industrial Hygienists

Figure 3.26 From *Industrial Ventilation—A Manual of Recommended Practice*, American Conference of Governmental Industrial Hygienists.

FIELD INSPECTION FORM

RESEARCH AND DEVELOPMENT DEPT. FUME HOODS

HOOD NO. ________________ ROOM NO. ________________ NE NW SE SW

TYPE: STAND. - SPEC. FACE VEL., DESIGN __________ FPM FAN NO. ______ , ______ CFM

INSPECTED BY: ____________________ DATE ____________________

Type of Fan Control:

__________ Controlled by switch located near hood.

__________ Two-speed fan controlled by switches located near hood.

__________ Continuously operating fan controlled from central control panel located in monitor.

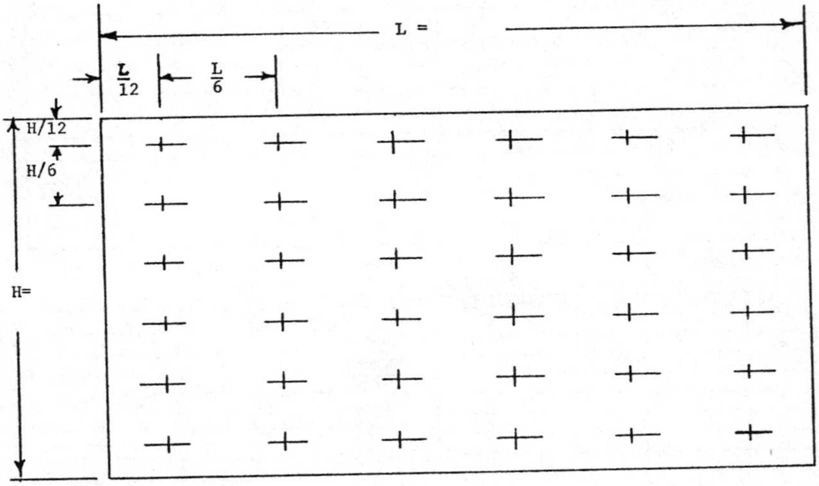

L =

L/12 L/6

H/12

H/6

H=

+	+	+	+	+	+
+	+	+	+	+	+
+	+	+	+	+	+
+	+	+	+	+	+
+	+	+	+	+	+
+	+	+	+	+	+

TOTALS

FACE VELOCITY, TEST ____________________ FPM

INSPECTOR'S COMMENTS:

SPECIAL CONDITIONS:

Figure 3.27

ROOF AREA REGISTRATION FORM

PERSONS ENTERING ROOF AREAS AND FAN LOFTS MUST REGISTER BELOW. DO NOT PASS BEYOND THIS POINT UNLESS ON OFFICIAL BUSINESS.

BLDG.________ DOOR NO.______

APPROX. TIME IN	DATE	APPROX. TIME OUT	NAME (Please Print)	DIV. OR DEPT.	PAYROLL NUMBER

(CONT. OTHER SIDE)

When spaces on both sides have been used, remove the new Form from in back of this sheet and slide into position in front.

IHS-47 (11-66)

Figure 3.28

physician are concerned with monitoring. One monitors the work environment, the other, the human body. The monitoring systems used are personal, environmental or area, biological and medical. Personal and environmental monitoring provide environmental exposure information necessary for the design of effective environmental controls and proper work practices. Biological and medical monitoring provides information on exposure only after absorption takes place. For adequate evaluation of the effect of the work environment, it may be necessary to use all four monitoring systems.

All "monitoring" personnel must keep in mind the value of reciprocated information. Employee exposure data may be useful in the diagnosis of occupational illness, or may indicate areas for medical surveillance. Medical findings may indicate areas for industrial hygiene study. Biological monitoring data may reveal exposure trends before illness symptoms.

Although of course medical surveillance should never be utilized as the primary means for monitoring employee exposure, it can be a supplementary tool in evaluating the effectiveness of a control program involving engineering or other control techniques, and/or personal protective control measures. Proper coordination of industrial hygiene and medical data can ultimately define worker exposure conditions that can be tolerated without any degradation of worker health.

Most physiological responses are nonspecific and the etiology of most pathological states is unknown. Accordingly, both medical and environmental quantification are needed to determine whether a causal relationship exists between the workplace and a worker's complaint or adverse health finding. For example, if a physician suspects that a worker's anemia is due to exposure to benzene, the results of personal and environmental monitoring will indicate whether sufficient benzene was present in the environment to produce this effect, given present knowledge of acceptable benzene levels. In comparable situations, personal monitoring can determine the need for biologic monitoring and periodic medical examinations. Noise level readings for a particular operation can be correlated with the results of audiometric examinations.

In cases where significant absorption of a chemical agent may occur through the skin, personal and environmental monitoring do not provide sufficient information. In such cases, medical and biological monitoring are required to fully evaluate the exposure. Accordingly, there must be a close and ongoing relationship between industrial hygiene and medicine to develop appropriate engineering and medical controls for the prevention of exposure to potentially hazardous substances. Figure 3.29 indicates the necessary flow of data involved in ultimate industrial hygiene and medical assessment of the worker.

5 OTHER RECORDS

Coordination of recordkeeping activities from many sources, including personnel records, medical data, environmental data, chemical audits, and so on, is essential if the linking of exposure data to individual employees is to be accomplished. Personnel records contain the cumulative summary of jobs held, with dates, department, and job

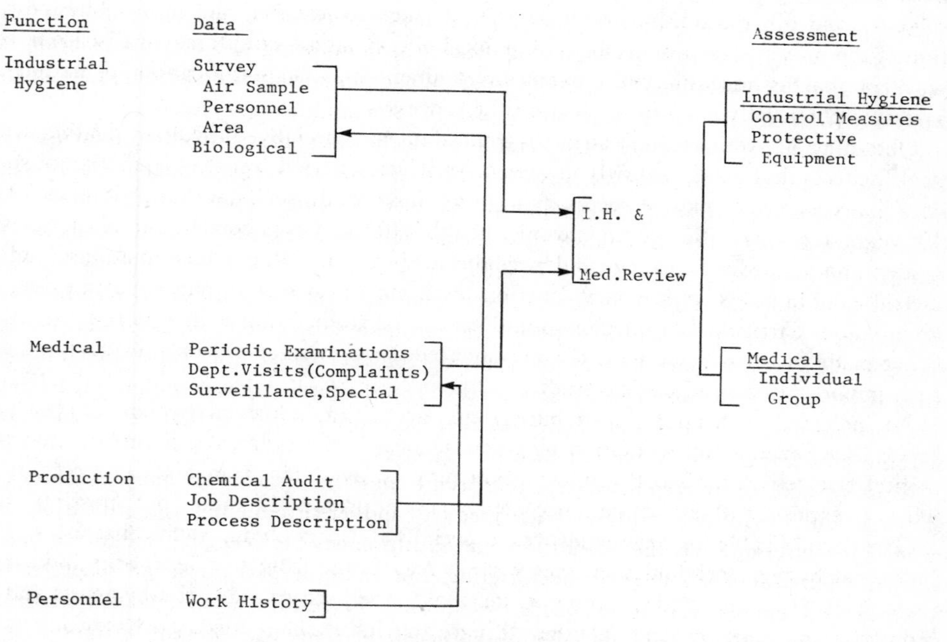

Figure 3.29

classifications. Periodic census lists will identify potentially exposed workers. Fundamental to the process is an audit identifying the chemical, biological, and physical agents in the work environment. It is essential that any process changes that could affect exposure conditions be reported promptly to the industrial hygienist for recording, and to the medical department for evaluation of the fitness of assigned employees to cope with the new work environment.

With the introduction of new processes, elimination of departments, modifications of jobs, and so on, business is dynamic; thus linking exposure measurement data to employees through the personnel record system can be extremely complicated without careful cross-referencing. Exposure records for a job may be assigned to all employees who held that job during a given time interval. In such cases it may be easier to retrieve the needed information if exposure data are cross-referenced with job and department designations, rather than to process and location in a department. Since employees frequently change jobs within a department, information regarding exposure levels for all jobs must be available. Since levels of exposure that are currently believed to be safe may later show a potential risk, all jobs within a department that offer a potential for such exposure should be evaluated.

Many factors in addition to work exposure can contribute to the cause of diseases. Among these are individual habits such as smoking or the use of alcohol, dietary preferences, exposure to air pollution or polluted water supplies, exposure to noise, solvents, and other conditions or materials, through avocations, and so on. The indication of the presence or absence of similar local effects, signs, symptoms, or disease in co-workers sharing essentially the same environment may be significant information in assessing potentially hazardous locations or processes.

Obtaining appropriate industrial hygiene data is especially difficult in dealing with health effects that develop slowly or after a long latent period, yet the acquisition of such data is necessary. It may be necessary to obtain and retain such data for as long as 30 to 40 years for correlation with chronic health effects. Cross-correlation of industrial hygiene data, medical history, and periodic medical surveillance examinations (audiometric examinations, pulmonary function tests, etc.) provides human exposure data of great value, particularly if large populations can be studied. More and more progress is being made in the development of computerized data retrieval methods for carrying out such important epidemiological studies.

As indicated in Section 1, an important part of an industrial hygiene survey and report is a review and evaluation of potential health effects from exposure to environmental stresses. Although the information may be available from historical records of worker exposure-illness experience, such an approach is often not possible nor necessarily desirable for new exposures, especially those involving chemical stresses. The industrial hygienist should then seek out data available from the files of the industrial toxicologist. In some instances one person may serve both as industrial hygienist and as industrial toxicologist. Among other things, the industrial toxicologist, working with research and development, manufacturing, industrial hygiene, and medical personnel, reviews proposed new products, usually beginning at the research stage and following through to marketing of the finished product. The toxicologist's files contain information and data on the new product, often varying from a preliminary evaluation based on his or her professional expertise through preliminary short-term and long-term chronic testing of the product in experimental animals. On occasion, animal testing is followed by carefully controlled testing in humans.

These data probably will never allow us to define the probable risk to humans with 100% certainty, but they do constitute an extremely valuable guide to the industrial hygienist in evaluating potential chemical stresses in the workplace.

Using the data from the toxicologist, the industrial hygienist evaluates the work environment in terms of levels of exposure found to be safe or unsafe in terms of experimental toxicological evaluations. Usual practice is to allow for some safety factor for human exposure as compared to animals. This should not be one fixed safety factor applicable to all chemical stresses but should, instead, be based on an evaluation of the type and degree of toxic effects.

The industrial hygienist, in addition to using such data for his own evaluations, can logically be expected to provide operating management with brief and concise information on toxicity, hazards, control, and first aid measures. These may take the form of short narrative-type reports or the more brief data sheets identified as material safety data sheets or industrial hygiene, toxicology, and safety data sheets (Figure 3.4).

After the industrial hygienist has evaluated the magnitude of the environmental stresses to which the worker is exposed, the data must be coordinated with medical records available on these same workers. Working with medical staffs and recommending special biological tests as indicated, the industrial hygienist can determine whether available yardsticks (e.g., permissible air concentrations of the contaminant) are adequate for evaluating a "safe" environment or whether they should be further refined.

Workmen's compensation or claims records provide information on employee exposures that have been allowed to develop to the stage of frank illness or disability. Sickness and disability records may provide clues to developing trouble. For example, a high incidence of absenteeism for respiratory problems in a particular department or plant as compared to other such locations may be an indication of excessive environmental stress from a particular chemical or condition. Such situations should be investigated thoroughly by the physician to determine whether a common medical problem exists, and then, if these suspicions are confirmed, by the industrial hygienist.

Personnel records can prove invaluable to the industrial hygienist, especially in investigating causative factors in demonstrated or suspected occupational illness. For example, if medical examination reveals conditions such as silicosis or hearing threshold shift, a review of the worker's job assignment history will tell whether the worker has ever been subjected to related exposures. A check of personnel records will assist the industrial hygienist in determining whether the worker's condition may have resulted from exposure at a previous job or if not, whether it may be attributable to the worker's avocation rather than occupation.

6 REQUIRED RECORDS

This chapter is primarily concerned with the occupational illness aspects of recordkeeping systems, not with injury records. However the fundamental similarities of regulatory and legal requirements for occupational illness and injury recordkeeping systems dictate that both be addressed here. Because of this, and because statistical consideration historically has been associated with the safety function only, the following commentary covers both the occupational illness and injury aspects of recordkeeping systems.

Governmental requirements to maintain occupational illness and injury records are becoming more prevalent in the industrialized nations of the world. The scope and detail of these requirements, however, vary considerably from country to country. Perhaps the most noteworthy and complete set of recordkeeping regulations has been promulgated by the U.S. Department of Labor under authority granted in the Williams-Steiger Occupational Safety and Health Act (OSHA), which became law in late 1970. Before this time, most recordkeeping systems both in the United States and around the world were maintained on a voluntary basis under guidelines set forth in American National Standards Institute (ANSI) Z-16.1, "Method of Recording and Measuring Work Injury Experience."

The purpose of ANSI Z-16.1, which was originally formulated in 1924, was to provide a uniform method of recording and weighting occupational injury experience. It contains a number of criteria to determine whether an injury is recordable, and, if so, its degree of severity. The standard recommends two basic disabling injury rates, frequency and severity, which can be used to compare safety performance in establishments or departments with comparable hazards, as well as to provide one measure of safety performance progress. These rates are defined as follows:

$$\text{disabling injury frequency rate} = \frac{\text{number of disabling injuries} \times 10^6}{\text{employee hours of exposure}}$$

$$\text{disabling injury severity rate} = \frac{\text{total days charged} \times 10^6}{\text{employee hours of exposure}}$$

where total days charged is obtained from a schedule in the standard.

Although ANSI Z-16.1 has served its purpose well, several factors in addition to a legal requirement imposed by the Williams-Steiger act prompted OSHA to establish its own recordkeeping system. Perhaps most salient of these is that Z-16.1 addresses itself almost solely to injuries and does not include criteria for occupational illness. An exception to this is that Z-16.1 does provide guides for recording and measuring industrial hearing loss and for considering skin irritations. OSHA, on the other hand, includes provisions for considering occupational skin diseases and disorders, pneumoconiosis and other respiratory conditions due to toxic agents, systemic poisoning, and disorders due to physical agents and repeated trauma. A second important factor is that ANSI Z-16.1 does not cover situations in which a particular injury does not result in lost time because the injured individual is reassigned to temporary, less exerting duties, or is transferred to another job.

Even though almost all employers in the United States are required to maintain the OSHA recordkeeping system, some also continue to maintain the ANSI system, at least temporarily. The primary reason for this is to facilitate comparison of safety performance from year to year, as well as to obtain a correlation between the two systems. An additional motivation is that the OSHA system does not provide for any type of severity rate, limiting itself to an incidence or frequency type of rate.

OSHA's regulations call for maintaining three basic records. These are a daily log of occupational illnesses and injuries, a supplemental record of each recordable occupational illness and injury, and an annual summary of occupational illnesses and injuries. The Bureau of Labor Statistics (BLS) of the Department of Labor, is responsible for developing and administering the recordkeeping program. The BLS often shares this responsibility with certain appropriate state agencies. None of the OSHA records considered here are required to be submitted to OSHA; however they must be maintained at the individual establishment for a period of 5 years. In addition, the annual summary of occupational illness and injury experience for a given year must be posted from February 1 to March 1 of the following year. All OSHA records can be inspected and reproduced by authorized government agents at any reasonable time. Failure to maintain OSHA records is grounds for citation and may be a criminal offense.

An OSHA recordable illness or injury is one that requires more medical attention than that normally regarded to be first aid. Cases are categorized as follows: (1) fatal, (2) lost workday with actual lost time, (3) lost workday with restricted activity, and (4) nonfatal. "Lost workday" is defined as a day on which an employee would have worked his normal duty but, because of the occupational illness or injury in question, either did not work at all or worked but as a result of the illness or injury did not perform his or her usual duties. The day on which the illness or injury is incurred or diagnosed is not counted as a lost workday. In cases where varying medical or other opinions cast doubt on the recordable nature of a case, it is the establishment manager's responsibility to decide the recordability of the case. The OSHA system provides for only one rate, the incidence rate, to be calculated as follows:

$$\text{incidence rate} = \frac{\text{total number recordable cases} \times 2 \times 10^5}{\text{employee hours of exposure}}$$

In considering either the OSHA incidence rate or the ANSI frequency and severity rates, several factors deserve comment, especially because these rates are often used as a basis for comparing occupational health and safety performance. First, there must be a clear understanding of the criteria for determining that a particular case is indeed recordable. If such an understanding does not exist, any comparison will probably be fallacious. Second, employee hours of exposure should be examined carefully to ascertain the percentage of the population that is actually at risk. For example, it would be improper to compare the incidence rates of two establishments if the work force in one entity is composed of 75 percent clerical and 25 percent operating personnel, while the second entity is staffed with 75 percent operating and 25 percent clerical. Should it be desirable to compare the performance of two such establishments, it would be well to adjust the employee hours of exposure to obtain a true picture of actual safety performance of those employees having comparable risks.

Finally, in comparing the OSHA incidence rate with the ANSI frequency rate, not only will the number of recordable cases differ, but there is a difference of a factor of 5 in the equation constant as well. It is generally not advisable to use rates as a sole indicator of industrial hygiene and safety performance. Statistical considerations should be only one of several factors used in developing and evaluating an industrial hygiene and safety program.

There are many more areas regarding the legal requirements of OSHA recordkeeping. Questions have arisen on many topics such as criteria for a recordable illness or injury, definition of employer and employee, type of employer covered, employee access to OSHA records, definition of work places or establishments, and government agency jurisdiction. Many of these have been resolved by the OSHA Review Commission, and some have been decided in the courts. There is precedence established in several areas but many more topics remain to be clarifed.

7 CONFIDENTIALITY OF RECORDS

It is now well recognized that for any employer, there exists between the employee and the company physician a patient-physician relationship dictating that all medical information be considered and retained in confidence by the physician. This confidentiality of medical records follows recognized standards of law and medical ethics. The role of a company-employed physician is further clarified by the provisions regarding proper conduct of an industrial physician including codes of ethics of the American Medical Association, the American Occupational Medical Association, and most state societies:

> The following described activities (of an occupational physician or clinic) are not to be considered as in violation of . . . the Principles of Professional Conduct . . . provided and only in the event that . . . accurate records are kept, the confidential character of such records including the result of health examinations is rigidly observed, the records remain in the exclusive custody and control of medical personnel, and no information from such records be given without the patient's request or consent, otherwise than as required by law. (Resolution of the Comitia Minora of the Medical Society of the County of New York, May 11, 1964.)

It is rightly held that management does not have a need to know of specific diagnoses, and so on, to fill a management function of assigning employees to specific tasks and operations. It follows, however, that since management is not privy to an employee's medical history and disabilities, the physician must have an adequate knowledge of the various work environments and their potential adverse effects on employees, and must advise management of any limitations in the placement of its employees. As an example, an employee with certain respiratory disabilities should not be assigned to operations presenting potential exposure to chemicals that are respiratory tract irritants. The industrial hygienist has a key role in defining to the physician the potential exposures in the work environments and the limitations for placement of employees.

Aside from the more readily defined personal medical history records of the employee, there are other records that do not fall into this category. The records of employee exposure as determined by monitoring of the work atmosphere are certainly open to review by nonmedical personnel. A growing trend is to make such data available not only to the physician and to management, but to the employee and the employee's union representative. New OSHA standards on health being promulgated will require that the employee or the employee's representative be so informed. Furthermore, in the United States these records must be made available on request to representatives of both OSHA and NIOSH (National Institute for Occupational Safety and Health).

Biological monitoring records of the employee, such as results of blood-lead analyses, are not quite as clear-cut. If, for example, the employee alleges illness due to exposure to lead and files for Workmen's Compensation award, or extended absence or medical treatment under any sickness and disability benefit plan, the employee forfeits any confidentiality of such medical and biological monitoring data. On the other hand, routine biological analytical data obtained either as an adjunct to environmental monitoring or as the sole monitoring device are not necessarily needed by management personnel to perform the management function. Such data are commonly made available to management, however, since they are solely related to the employee's work environment. Yet use of such data by management should be limited to instituting engineering or other control measures and to temporary assignment of the employee, if needed, to another area not presenting such an exposure.

When medical and environmental data are computerized, as discussed in the following section, it is important that the confidentiality of these restricted medical data be maintained. Only those privileged to see restricted medical data should be able to obtain this information from the computer. On the other hand, the industrial hygienist will have a continuing need for environmental and biological monitoring data and should have free access to it.

8 COMPUTERIZATION OF RECORDS

In the past, and even at present in many cases, numerous data on employee exposures were compiled in the form of narrative-style, written reports. These are used primarily to evaluate employee exposure as measured by some yardstick such as threshold limit values and to initiate control measures where indicated. Medical records, on the other hand, are largely kept on an individual employee basis. Under such circumstances epidemiological studies and other correlations between industrial hygiene and medical data were, and are at best, most difficult and time-consuming. One need only consider the problem of attempting to correlate hearing tests involving 14 measurements on each employee with noise level measurements, each being repeated as often as yearly, to recognize the complexity of manually correlating environmental and medical data.

With the use of the computer for industrial hygiene data, there lies ahead a much better prospect of industrial hygiene records becoming a dynamic force. There is now a much better opportunity and likelihood that adequate evaluation of the various occupational health parameters will be accomplished. This in the final analysis means better control of the work environment and early detection of any adverse health effects before irreparable damage has been done.

With an adequately designed computer program and necessary recording of data, the industrial hygienist can evaluate employee exposure to workplace contaminants to a degree heretofore impossible. Using benzene as just one example, data needed would include the employee's name, social security number, age, sex, company and plant location, specific workplace operation, use or nonuse of respiratory protection, air monitoring data (including dates), urinary phenol measurements, and clinical medical data and findings. In addition to personal monitoring data, area and source monitoring data would be included. From such data, a computer printout can be obtained of all employees potentially exposed to benzene, those having exposure above or below the allowed limit (or above or below other specific levels—depending on the standards set by the computer user), and correlation of exposure data with any adverse medical findings or with biological monitoring.

Finally, computerized records of employee exposures and biological monitoring data should be readily available to the examining physician at the time of the employee's periodic or special medical examination. This allows the physician to review the individual's medical findings in the light of environmental stresses, and it is an opportune time for the employee to be apprised of any exposures he may have received on the job.

In employing computer techniques, the industrial hygienist and physician must first establish goals and define accurately what will be needed in the way of computer reports. Without these two factors carefully thought out in advance, the data recorded may not be retrievable, or, if retrievable, may not be usable.

It is technically easy to enter data that will yield the usual statistical or other relatively simple information. What must be remembered is that the computer will not provide answers if the data input has not supplied the correct base for such answers, or if adequate formatting has not been provided to permit meaningful retrieval programming. The following examples illustrate this point.

1. For the purpose of cost and simplicity, the industrial hygienist could elect to record exposure data as below or above the allowed limit. This type of input would then require other, manually kept, records of any changes made in such limit. Furthermore, if medical examinations were to reveal adverse health effects for some employees having an exposure less than the allowed limit, it would not be possible to determine without going back to written records whether at a specific lower level, no effects are noted.

2. Emphasis today is on accumulation of individual or "personal" time-weighted average exposure sampling data, with less consideration for area and source monitoring data. It is, however, important that all three types of data be obtained and recorded. Then if adverse health effects are found in employees and "personal" time-weighted sampling data are found to be quite low, a review of short-term "personal" sample data and area and source environmental data may point to the short-term intermittent exposures as the causative factors. Unless the computer is properly programmed, however, such data may be unobtainable or irretrievable.

Typical examples of data collection forms appear in Figures 3.30 through 3.33. The individual location may add or delete various elements shown, but this approach generally can be used without significant alteration.

Typical examples of computer output reports are given in Figures 3.34 and 3.35.

8.1 Computer Software Specifications*

Development of computer software specifications to meet the objectives described earlier is a time-consuming and costly process, involving various professional skills. The necessary tasks and the required professional skills are set forth below. In addition, a summary, "milestone analysis," indicating system interrelationships, is shown in Figure 3.36. The task groups are as follows.

Task Group 1 Payroll, Personnel, Family History, and Physician's Examination

Skills required: clinical medicine, epidemiology, industrial hygiene, toxicology.
Requirements:

1. Determine data elements† to be obtained from the personnel/payroll system. Some of these data elements alternatively may be gathered on the personal and family history questionnaire.
2. Determine each question to be asked on the personal and family history questionnaire. Indicate whether that question is essential or desirable. Indicate all possible responses. Write out each question so that it stands alone.
3. Determine data elements to be recorded for the physician's examination. Indicate any coded responses. List code and/or responses.

Task Group 2 Laboratory Data and Measurements

Skills required: clinical medicine, epidemiology, industrial hygiene/toxicology.
Requirements:

1. Laboratory data. Determine data elements to be recorded for blood, urine, breath analyses, and any other laboratory determination.
2. Measurements and other elements. Determine data elements to be recorded for hearing, vision, X-ray, ECG, blood pressure, weight, height, and any other.
3. Immunizations. Determine data elements to be recorded for immunizations.

Task Group 3 Administrative and Reports

Skills required: medical director, administrative, industrial hygiene, safety.
Requirements:

1. Determine whether a frequency of periodic examination by type of employee is to be recommended. If so, specify. Specify minimum government requirements, if any, for

* A short glossary of terms is at the end of this section.
† Common data elements will allow inter- and intraindustry analyses.

INDUSTRIAL HYGIENE DATA

LOCATION ____________________ Page 1

FORM CODE - CASE NO.

EIS LOCATION CODE

0	0	0	0	0	0	0	0	0
1	1	1	1	1	1	1	1	1
2	2	2	2	2	2	2	2	2
3	3	3	3	3	3	3	3	3
4	4	4	4	4	4	4	4	4
5	5	5	5	5	5	5	5	5
6	6	6	6	6	6	6	6	6
7	7	7	7	7	7	7	7	7
8	8	8	8	8	8	8	8	8
9	9	9	9	9	9	9	9	9
		A	A	A	A	A	A	A
		B	B	B	B	B	B	B
		C	C	C	C	C	C	C
		D	D	D	D	D	D	D
		E	E	E	E	E	E	E
		F	F	F	F	F	F	F
		G	G	G	G	G	G	G
		H	H	H	H	H	H	H
		I	I	I	I	I	I	I
		J	J	J	J	J	J	J
		K	K	K	K	K	K	K
		L	L	L	L	L	L	L
		M	M	M	M	M	M	M

OCCUPATION/ACTIVITY

0	0	0	0	0	0
1	1	1	1	1	1
2	2	2	2	2	2
3	3	3	3	3	3
4	4	4	4	4	4
5	5	5	5	5	5
6	6	6	6	6	6
7	7	7	7	7	7
8	8	8	8	8	8
9	9	9	9	9	9
	A	A	A	A	A
	B	B	B	B	B
	C	C	C	C	C
	D	D	D	D	D
	E	E	E	E	E
	F	F	F	F	F
	G	G	G	G	G
	H	H	H	H	H
	I	I	I	I	I
	J	J	J	J	J
	K	K	K	K	K
	L	L	L	L	L
	M	M	M	M	M

SOCIAL SECURITY NUMBER

0	0	0	0	0	0	0	0	0
1	1	1	1	1	1	1	1	1
2	2	2	2	2	2	2	2	2
3	3	3	3	3	3	3	3	3
4	4	4	4	4	4	4	4	4
5	5	5	5	5	5	5	5	5
6	6	6	6	6	6	6	6	6
7	7	7	7	7	7	7	7	7
8	8	8	8	8	8	8	8	8
9	9	9	9	9	9	9	9	9

TYPE OF JOB

☐ Maintenance

☐ Production

PERSONAL PROTECTIVE EQUIPMENT

ADEQ.	INADEQ.	
☐	☐	Respiratory
☐	☐	Hearing
☐	☐	Eye and/or Face
☐	☐	Hands
☐	☐	Other

OCCUPATION/ACTIVITY AT TIME OF STUDY ________

SHIFT HOURS

☐ Shift 1 ☐ Shift 3

☐ Shift 2 ☐ Other

UNITS

AIR CONCENTRATION

☐ ppm ☐ mppcf

☐ ppb ☐ fibers/cc

☐ mg/M^3 ☐ Other

☐ μg/M^3

IONIZING RADIATION, FILMBADGE

☐ millirem

DURATION OF EXPOSURE

☐ 1 month ☐ 3 months

NOISE

☐ dBA

☐ % Exposure-Dosimeter

SAMPLE NUMBER			COMPONENT NUMBER	
0	0	0	0	0
1	1	1	1	1
2	2	2	2	2

AGENT I.D.

TEST CONDITIONS ☐ NORMAL ☐ ABNORMAL

0 0 0 0 0 0 - 0 0 - 0

1 1 1 1 1 1 - 1 1 - 1

2 2 2 2 2 2 - 2 2 - 2

EH LOCATION													
		N	N	N	N	N	N	N	N	N	N	N	N
		Ø	Ø	Ø	Ø	Ø	Ø	Ø	Ø	Ø	Ø	Ø	Ø
		P	P	P	P	P	P	P	P	P	P	P	P
0	0	Q	Q	Q	Q	Q	Q	Q	Q	Q	Q	Q	Q
1	1	R	R	R	R	R	R	R	R	R	R	R	R
2	2	S	S	S	S	S	S	S	S	S	S	S	S
3	3	T	T	T	T	T	T	T	T	T	T	T	T
4	4	U	U	U	U	U	U	U	U	U	U	U	U
5	5	V	V	V	V	V	V	V	V	V	V	V	V
6	6	W	W	W	W	W	W	W	W	W	W	W	W
7	7	X	X	X	X	X	X	X	X	X	X	X	X
8	8	Y	Y	Y	Y	Y	Y	Y	Y	Y	Y	Y	Y
9	9	Z	Z	Z	Z	Z	Z	Z	Z	Z	Z	Z	Z

3	3	3	3	3	3	3	3	3	3	3	-	3	3	-	3
4	4	4	4	4	4	4	4	4	4	4	-	4	4	-	4
5	5	5	5	5	5	5	5	5	5	5	-	5	5	-	5
6	6	6	6	6	6	6	6	6	6	6	-	6	6	-	6
7	7	7	7	7	7	7	7	7	7	7	-	7	7	-	7
8	8	8	8	8	8	8	8	8	8	8	-	8	8	-	8
9	9	9	9	9	9	9	9	9	9	9	-	9	9	-	9

LOCATION____________

☐ Effects are Additive
☐ Study Completed
☐ Delete This Study

EHS SURVEYOR		ANALYZER	DATE OF STUDY YEAR		MONTH		DAY	
0	0	0		0	0	0	0	0
1	1	1		1	1	1	1	1
2	2	2		2		2	2	2
3	3	3		3		3	3	3
4	4	4		4		4		4
5	5	5		5		5		5
6	6	6		6		6		6
7	7	7	7	7		7		7
8	8	8	8	8		8		8
9	9	9	9	9		9		9

SAMPLE DURATION			SAMPLING & ANALYTICAL TECHNIQUE		RESULTS ☐LESS THAN		☐MORE THAN		☐NONE DETECTED	
0	0	0	0	0	0	0	0	0	0	0
1	1	1	1	1	1	1	1	1	1	1
2	2	2	2	2	2	2	2	2	2	2
3	3	3	3	3	3	3	3	3	3	3
4	4	4	4	4	4	4	4	4	4	4
5	5	5	5	5	5	5	5	5	5	5
6	6	6	6	6	6	6	6	6	6	6
7	7	7	7	7	7	7	7	7	7	7
8	8	8	8	8	8	8	8	8	8	8
9	9	9	9	9	9	9	9	9	9	9

CONCENTRATION

Personal Sample
☐ Personal
☐ Wk Stn
☐ TWA
☐ Ceiling
☐ Peak
☐ Other

ASB Sample
☐ Area
☐ Source
☐ Bkgrnd
☐ Other

COMMENTS ____________________________

SIGNATURE -

Ⓒ AMOCO COMPUTER SERVICES COMPANY 1977

Figure 3.30

VISIT

Name____________________

FORM CODE

CASE NUMBER

*SOCIAL SECURITY NUMBER	*DATE OF INITIAL VISIT	*ONSET OF ILLNESS OR TIME OF INJURY	HOUR WORKED PRIOR
O / C	YEAR / MONTH / DAY	YEAR / MONTH / DAY / AM / PM / Estimated / HOURS / MINUTES	

***THIS RECORD STARTED AS A RESULT OF:**

- Personal visit to medical department
- Phone call
- Medical department visit at other location
- Other

***TYPE OF VISIT**

- Return to work
- Employee or Non-employee
- Illness or Injury or
- Other
- Occupational or Non-occupational
- Occurred on premises or Occurred off premises
- Recordable for OSHA
- Not seen by medical dept

ILLNESS OR COMPLAINT (one only)

- Occupational Skin disease or disorder
- Dust disease of the lung (Pneumoconiosis)
- Respiratory conditions due to toxic agents
- Poisoning (systemic effects of toxic materials)

DAYS LOST

CALENDAR DAYS LOST	WORKING DAYS LOST	HOSPITAL DAYS LOST	DAYS OF WORK RESTRICTION

LOCATION

SIDE
- Right
- Midline
- Left

UPPER BODY
- Head
- Face
- Mouth
- Teeth
- Ear
- Eye
- Nose
- Neck

ARM/SHOULDER
- Arm
- Elbow
- Wrist
- Finger
- Hand
- Forearm
- Shoulder

TORSO
- Chest
- Back

TORSO (continued)
- Abdomen
- Hip
- Pelvis
- Genitalia

LEG
- Thigh
- Knee
- Leg
- Ankle
- Foot
- Toe

OTHER
- Skin
- Other

*** SEVERITY**
- Major
- Minor

*** REQUIRED FIELD**

PRINCIPAL TYPE OF INJURY (one only)
- Abrasion
- Amputation
- Burn-chemical
- Burn-thermal
- Concussion
- Contusion
- Crushed
- Foreign body
- Fracture
- Infection
- Laceration
- Myositis
- Hernia
- Puncture
- Radiation injury
- Sprain/strain
- Dislocation
- Electrical injury
- Other

- Disorders due to physical agents (other than toxic materials)
- Disorder associated with repeated trauma
- All other occupational illnesses
- Dysmenorrhea
- Common cold
- Influenza
- Allergy
- Asthma
- Headache
- Dizziness
- Chest pain
- Indigestion
- Diarrhea and/or vomiting
- Urinary tract infection-Stones
- Pregnancy check
- Fever
- Fainting
- Emotional problems
- Blood pressure check
- All other non occupational illnesses

*** TREATMENT INITIAL VISIT**
- None
- Int. Med.
- Hydrotherapy
- Immobilization
- Laboratory
- C & D
- Suture
- Ice
- Irrigation
- Plaster
- F.B. removed
- Local heat
- Debridement
- Antibiotic
- Immunization
- Ext. Med.
- Diathermy
- Support
- I & D
- Counsel
- Other

DIAGNOSTIC PROCEDURE
- Back X-ray
- Chest X-ray
- Other X-ray
- ECG
- Limited phys. exam
- Other

*** INITIAL DISPOSITION**
- Limited duty
- Full duty
- Sent home
- Pers. Dr.
- Hospital
- Co. Specialist
- Other

*** SEEN BY INITIAL VISIT**
- Doctor
- RN
- Techn.
- First Aid
- Other

*** SUBSEQUENT TREATMENT**
- None
- Int. Med.
- Hydrotherapy
- Immobilization
- Laboratory
- C & D
- Suture
- Ice
- Irrigation
- Plaster
- F.B. removed
- Local heat
- Debridement
- Antibiotic
- Immunization
- Ext. Med.
- Diathermy
- Support
- I & D
- Counsel
- Other

*** DISABILITY**
- None
- Fatal
- Undetermined
- Temp partial
- Perm partial minor
- Perm partial major
- Temp total
- Perm total

*** FINAL DISPOSITION**
- Full duty
- Job change
- LTD Pension
- Other

By:
- MD
- RN

REVISITS

0	0
1	1
2	2
3	3
4	4
5	5
6	6
7	7
8	8
9	9

*** DOCTOR'S CODE**

0	0	0
1	1	1
2	2	2
3	3	3
4	4	4
5	5	5
6	6	6
7	7	7
8	8	8
9	9	9

Signature__________ Date__________

FORM 53-201 (8/76)

©AMOCO COMPUTER SERVICES COMPANY 1976

Figure 3.31

VISIT DIAGNOSIS OR SUPPLEMENTAL EXAMINATION DIAGNOSIS

Name ______________________

PAGE 2

FORM CODE

CASE NUMBER

DATE																		
YEAR		MONTH		DAY		DIAGNOSIS 1							PAGE		* DOCTOR'S CODE			
0	0	0	0	0	0	Y	0	0	0	0	0	*			0	0	0	
1	1	1	1	1	1	E	1	1	1	1	1	Exam	1	5	1	1	1	
2	2		2	2	2	R	2	2	2	2	2	or	2	6	2	2	2	
3	3		3	3	3	A	3	3	3	3	3		3	7	3	3	3	
4	4		4		4		4	4	4	4	4	Visit	4	8	4	4	4	
5	5		5		5		5	5	5	5	5				5	5	5	
6	6		6		6		6	6	6	6	6				6	6	6	
7	7		7		7		7	7	7	7	7	Completed exam/visit			7	7	7	
8	8		8		8		8	8	8	8	8				8	8	8	
9	9		9		9		9	9	9	9	9				9	9	9	

YEAR		MONTH		DAY		DIAGNOSIS 2					
0	0	0	0	0	0	Y	0	0	0	0	0
1	1	1	1	1	1	E	1	1	1	1	1
2	2		2	2	2	R	2	2	2	2	2
3	3		3	3	3	A	3	3	3	3	3
4	4		4		4		4	4	4	4	4
5	5		5		5		5	5	5	5	5
6	6		6		6		6	6	6	6	6
7	7		7		7		7	7	7	7	7
8	8		8		8		8	8	8	8	8
9	9		9		9		9	9	9	9	9

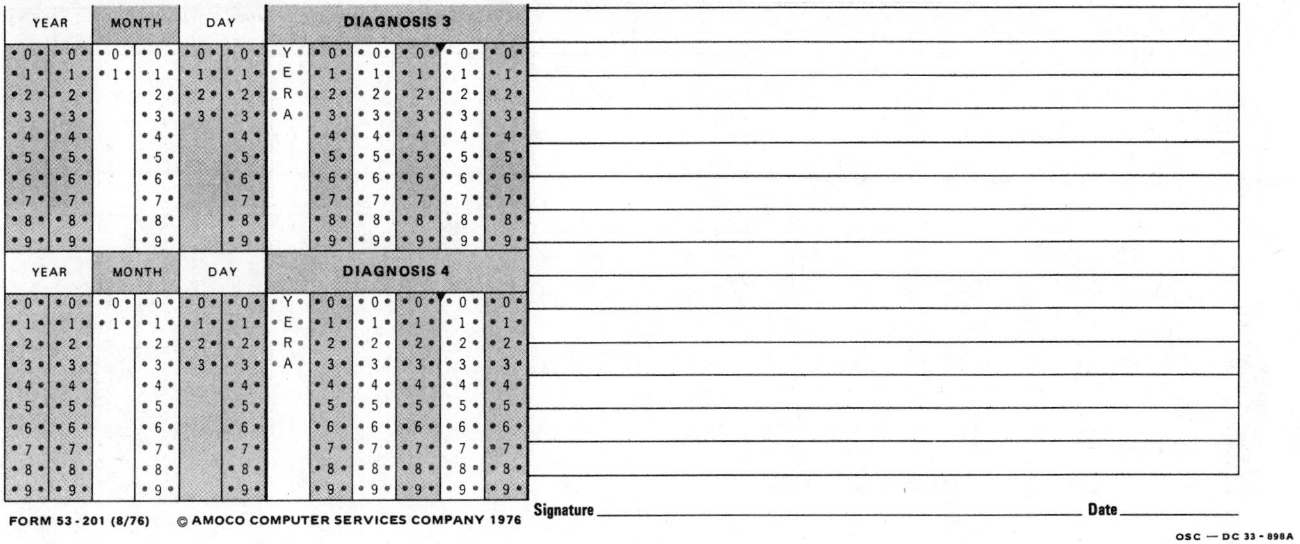

YEAR		MONTH		DAY		DIAGNOSIS 3					
0	0	0	0	0	0	Y	0	0	0	0	0
1	1	1	1	1	1	E	1	1	1	1	1
2	2		2	2	2	R	2	2	2	2	2
3	3		3	3	3	A	3	3	3	3	3
4	4		4		4		4	4	4	4	4
5	5		5		5		5	5	5	5	5
6	6		6		6		6	6	6	6	6
7	7		7		7		7	7	7	7	7
8	8		8		8		8	8	8	8	8
9	9		9		9		9	9	9	9	9

YEAR		MONTH		DAY		DIAGNOSIS 4					
0	0	0	0	0	0	Y	0	0	0	0	0
1	1	1	1	1	1	E	1	1	1	1	1
2	2		2	2	2	R	2	2	2	2	2
3	3		3	3	3	A	3	3	3	3	3
4	4		4		4		4	4	4	4	4
5	5		5		5		5	5	5	5	5
6	6		6		6		6	6	6	6	6
7	7		7		7		7	7	7	7	7
8	8		8		8		8	8	8	8	8
9	9		9		9		9	9	9	9	9

FORM 53-201 (8/76) © AMOCO COMPUTER SERVICES COMPANY 1976

Signature ______________________ Date __________

OSC — DC 33-898A

Figure 3.31

TIME IN AM PM
TIME OUT AM PM

PHONE
EXT

NAME

JOB TITLE

AGE SEX

SOCIAL SEC NO.

Not recorded in the computer

PATIENT'S COMPLAINT

Signature of patient

HOSPITAL

(name)

(address - no., street)

(address - city, state, zip)

PHYSICIAN (if referred)

(name)

(address - no., street)

(address - city, state, zip)

WHAT WAS THE EMPLOYEE DOING

HOW DID THE ACCIDENT OCCUR

NAME OBJECT OR SUBSTANCE WHICH INJURED EMPLOYEE

If Chemical code on page 2

SUPERVISOR'S NAME

PHONE EXT

COMPLETE IF NOT AVAILABLE FROM COMPUTER EMPLOYEE MAILING ADDRESS

(address - no., street)

(address - city, state, zip)

ADDRESS OF ACCIDENT OR EXPOSURE
(Always complete if accident not on premises)

(address - no., street)

(address - city, state, zip)

REVISITS

NOTES:

DATE	

Signature

Major injuries will include:
Those requiring the employee to leave the plant before the end of his scheduled work shift.
Those requiring a change of job, regardless of length of time involved.
Lacerations requiring any sutures, butterfly closure, or similar technique.
Eye injuries requiring an eye patch, and those where loss of sight is distinct possibility.
Burns more severe than first degree.
Injuries requiring a sling, cane or crutches, or otherwise interfering substantially with locomotion or use of limbs.
Hernia cases.
Injuries producing loss of consciousness or shock.
Fractures and amputations.
Others of comparable severity.
All injuries not considered "Major" by this definition should be coded as "Minor".

OSC—DC 33-897A

© AMOCO COMPUTER SERVICES COMPANY 1976

Figure 3.31 (*Continued*)

surveillance and for periodic examinations.

2. Determine data elements for administrative reports by report. Indicate how they are derived.
3. Determine data elements for OSHA reports. Indicate how they are derived.
4. Determine data elements for union report. Indicate how they are derived.

Task Group 4 Data Processing Environment

Skills required: systems.
Requirements:

1. Determine the data language for expression of company level data base (i.e., Mark IV, COBOL, FORTRAN, PL1, DL1).
2. Determine what assumption of input means will be used for company level system: keypunch/mail, teleprocess slow speed device–key entry nonconversational, optical mark reader, teleprocess high speed device–keypunch RJE.
3. Determine hardware-software configuration for company level systems including query language.
4. Determine the extent of security necessary for input, output reports, the data base.

Task Group 5 Diagnosis

Skills required: clinical medicine, epidemiology, industrial hygiene/toxicology.
Requirements:

1. Determine how many digits of ICD code will be used.
2. Determine any supplements to the ICD code that will be required. Name a specific code for each addition.
3. Determine a selected list of ICD code. This list will be for easy reference. Criteria for inclusion should be frequency of occurrence within employees and former employees or particular items of interest for occupational illness correlation.
4. Determine standards for recording death certificates.

Task Group 6 Visit, Accident, and Morbidity

Skills required: clinical medicine, safety, epidemiology, industrial hygiene/toxicology.
Requirements:

1. Determine data elements to be recorded for occupational illness and injury (medical).
2. Determine data elements to be recorded for nonoccupational illness and injury (morbidity).
3. Determine data elements to be recorded for accidents.

ACCIDENT REPORT

Name ______________________________

* SOCIAL SECURITY NUMBER OF INDIVIDUAL

0	0	0	0	0	0	0	0	0	0
C	1	1	1	1	1	1	1	1	1
	2	2	2	2	2	2	2	2	2
	3	3	3	3	3	3	3	3	3
	4	4	4	4	4	4	4	4	4
	5	5	5	5	5	5	5	5	5
	6	6	6	6	6	6	6	6	6
	7	7	7	7	7	7	7	7	7
	8	8	8	8	8	8	8	8	8
	9	9	9	9	9	9	9	9	9

* SUPERVISOR WAS:

- Permanent
- Temporary

* SUPERVISOR ACTION

Proper		Improper
▭	Attitude	▭
▭	Instruction	▭
▭	Assignment	▭
▭	Adherence to Regulations	▭
▭	Other	▭

* SOCIAL SECURITY NUMBER OF SUPERVISOR

0	0	0	0	0	0	0	0	0	0
C	1	1	1	1	1	1	1	1	1
	2	2	2	2	2	2	2	2	2
	3	3	3	3	3	3	3	3	3
	4	4	4	4	4	4	4	4	4
	5	5	5	5	5	5	5	5	5
	6	6	6	6	6	6	6	6	6
	7	7	7	7	7	7	7	7	7
	8	8	8	8	8	8	8	8	8
	9	9	9	9	9	9	9	9	9

* REQUIRED FIELD

FORM CODE

CASE NUMBER

E I S LOCATION CODE

0	0	0	0	0	0	0	0	0
1	1	1	1	1	1	1	1	1
2	2	2	2	2	2	2	2	2
3	3	3	3	3	3	3	3	3
4	4	4	4	4	4	4	4	4
5	5	5	5	5	5	5	5	5
6	6	6	6	6	6	6	6	6
7	7	7	7	7	7	7	7	7
8	8	8	8	8	8	8	8	8
9	9	9	9	9	9	9	9	9

NAME OF INDIVIDUAL: ______________________ PRINT ______ AGE

NAME OF SUPERVISOR: ______________________ PRINT

* DATE & TIME OF ACCIDENT

YEAR	MONTH		DAY			HOUR	
0	0	0	0	0	▭ AM	0	0
1	1	1	1	1		1	1
2		2	2	2			2
3		3	3	3	▭ PM		3
4		4		4			4
5		5		5			5

ACTIVITY

▭ Walking	▭ Climbing, Ascending
▭ Running	
▭ Standing	▭ Climbing, Descending
▭ Sitting	
▭ Pulling	▭ Diving
▭ Pushing	▭ Cutting
▭ Lifting	▭ Trimming
▭ Holding	▭ Bending

ACCIDENT WAS: *

- ▭ Actual
- ▭ Near-miss

INJURY WAS:

- ▭ Occupational
- ▭ Non-occupational
- ▭ Persons Injured?

EMPLOMENT STATUS

Driving
Loading
Unloading
Sampling
Other

Employee
Casual
Contractor Employee
Non-employee

*** ACCIDENT TYPE**

Slip
Fall
Struck By
Struck Against
Touched Against
Caught Under
Caught Between
Caught In
Falling Object
Flying Object
Wind Blown Object
Bite
Spray
Splash
Leak
Spill
Electric Contact
Chem/Biol. Ag. Contact
Chem/Biol. Ag. Inhalation
Radiation Exposure
Temp.Extremes
Vehicle,Traffic
Vehicle, Mat.Handling
Aircraft Crash
Train Crash
Fire
Explosion
Immersion
Other

Cause – Non-Traffic

Another Person
Horseplay
Communication
Lack of Alertness
Overexertion
Fatigue
Inexperience
Inadequate Training
Failure to Use Personal Protective Device
Safety Device Removed or Broken
Inadequate Light

FAILURE TO FOLLOW ESTABLISHED PROCEDURE
Lockout
Hot Work
Vessel Entry
Other

Housekeeping
Defective Tool
Wrong Tool
Improper Design
Equipment Failure
Maintenance
None of Above

WEATHER WAS:

Clear
Fog/Haze
Rain/Drizzle
Snow/Sleet
Hurricane
Tornado
High Wind

REPORT COMPLETE ? *

Yes
No

***MEDICAL LOCATION**

CLAIMS NUMBER

FORM 53-210 (01-77) PAGE 1
© AMOCO Computer Services Company 1977

DC 35-827

Figure 3.32

ACCIDENT REPORT — PAGE 2

Name ________________________________

	CHEMICAL HAZARD (CAS REGISTRY NUMBER)									OTHER HAZARD ▭ Physical ▭ Biological			EQUIPMENT INVOLVED			PAYMENT AMOUNT								PREPRINTED CASE NO. (From Page 1)				
[0]	[0]	[0]	[0]	[0]	[0]	[0]	[0]	[0]	[0]	[0]	[0]	[0]	[0]	[0]	[0]	[0]	[0]	[0]	[0]	[0]	[0]	[0]	[0]	[0]	[0]	[0]	[0]	[0]
[1]	[1]	[1]	[1]	[1]	[1]	[1]	[1]	[1]	[1]	[1]	[1]	[1]	[1]	[1]	[1]	[1]	[1]	[1]	[1]	[1]	[1]	[1]	[1]	[1]	[1]	[1]	[1]	[1]
[2]	[2]	[2]	[2]	[2]	[2]	[2]	[2]	[2]	[2]	[2]	[2]	[2]	[2]	[2]	[2]	[2]	[2]	[2]	[2]	[2]	[2]	[2]	[2]	[2]	[2]	[2]	[2]	[2]
[3]	[3]	[3]	[3]	[3]	[3]	[3]	[3]	[3]	[3]	[3]	[3]	[3]	[3]	[3]	[3]	[3]	[3]	[3]	[3]	[3]	[3]	[3]	[3]	[3]	[3]	[3]	[3]	[3]
[4]	[4]	[4]	[4]	[4]	[4]	[4]	[4]	[4]	[4]	[4]	[4]	[4]	[4]	[4]	[4]	[4]	[4]	[4]	[4]	[4]	[4]	[4]	[4]	[4]	[4]	[4]	[4]	[4]
[5]	[5]	[5]	[5]	[5]	[5]	[5]	[5]	[5]	[5]	[5]	[5]	[5]	[5]	[5]	[5]	[5]	[5]	[5]	[5]	[5]	[5]	[5]	[5]	[5]	[5]	[5]	[5]	[5]
[6]	[6]	[6]	[6]	[6]	[6]	[6]	[6]	[6]	[6]	[6]	[6]	[6]	[6]	[6]	[6]	[6]	[6]	[6]	[6]	[6]	[6]	[6]	[6]	[6]	[6]	[6]	[6]	[6]
[7]	[7]	[7]	[7]	[7]	[7]	[7]	[7]	[7]	[7]	[7]	[7]	[7]	[7]	[7]	[7]	[7]	[7]	[7]	[7]	[7]	[7]	[7]	[7]	[7]	[7]	[7]	[7]	[7]
[8]	[8]	[8]	[8]	[8]	[8]	[8]	[8]	[8]	[8]	[8]	[8]	[8]	[8]	[8]	[8]	[8]	[8]	[8]	[8]	[8]	[8]	[8]	[8]	[8]	[8]	[8]	[8]	[8]
[9]	[9]	[9]	[9]	[9]	[9]	[9]	[9]	[9]	[9]	[9]	[9]	[9]	[9]	[9]	[9]	[9]	[9]	[9]	[9]	[9]	[9]	[9]	[9]	[9]	[9]	[9]	[9]	[9]

▭ Cause – Traffic

- ▭ Another Vehicle
- ▭ Another Person
- ▭ Pedestrian
- ▭ Improper Turn
- ▭ Improper Parking
- ▭ Driver Inattention
- ▭ Driver Fatigue
- ▭ Driver Physical Condition
- ▭ Sickness, Alcohol, Drugs
- ▭ Railroad Train

TRAFFIC ACCIDENT TYPE:

- ▭ Head On
- ▭ Side Swipe
- ▭ Right Angle
- ▭ Front End
- ▭ Rear End
- ▭ Roll Over
- ▭ Jack Knife
- ▭ Backing
- ▭ Other

D.O.T. ACCIDENT?

- ▭ Yes
- ▭ No

City ____________________

____

State ____________________

Zip ____________________

NOTES: ____________________

FORM CODE

CASE NUMBER

- Improper Backing
- Following too Close
- Cutting or Crowding
- Failure to Obey Traffic Signal
- Failure to Obey Traffic Sign
- Failure to Grant Right of Way
- Excessive Speed
- Too Fast For Conditions
- Unbalanced Load
- Overload
- Obstructed Vision

- Slippery Road
- Gravel Road
- Bumpy Road
- Defective Brakes
- Defective Steering
- Defective Tires
- Defective Wheels
- Defective Lights
- Defective Wipers
- Other Defective
- None of Above

CONTENTS OF VEHICLE:

- Gasoline
- Distillates
- Other Petroleum
- L.P. Gas
- Anhydrous Ammonia Fertilizer
- Chemicals
- Packaged Goods
- Miscellaneous Cargo
- Passengers
- Other

*** REQUIRED FIELD**

REPORT COMPLETE? *

- Yes
- No

COMMENTS:

Signature ____________ Date ____________

FORM 53-210 (01-77) PAGE 2

© AMOCO Computer Services Company 1977

DC 35-828

Figure 3.32 *(Continued)*

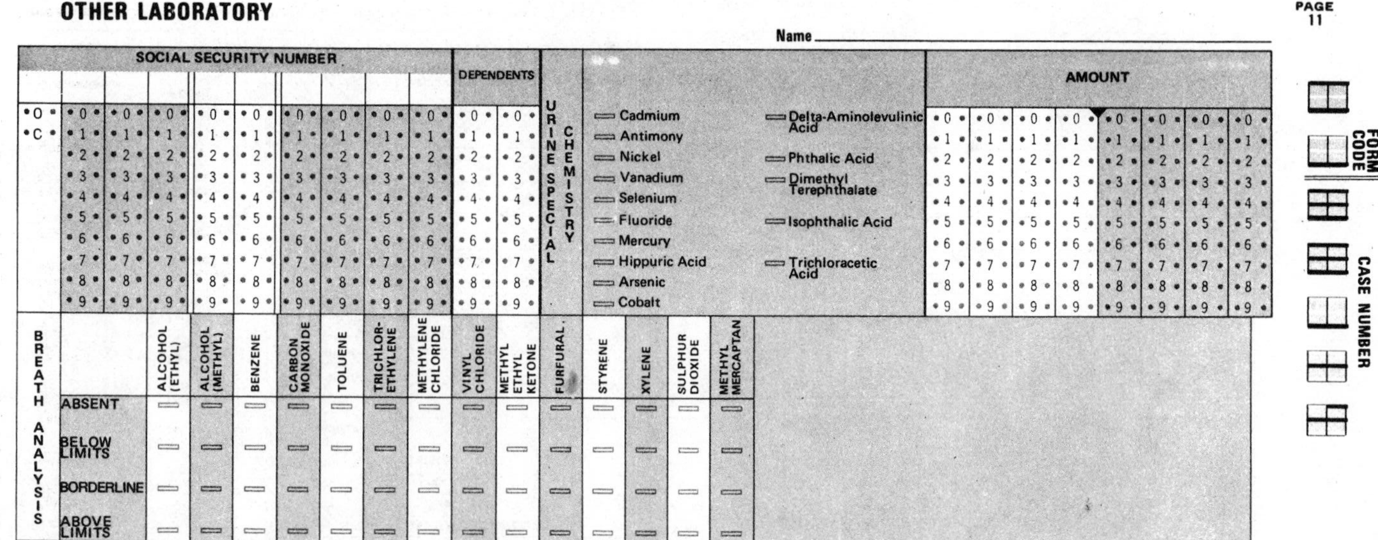

OTHER LABORATORY

PAGE 11

FORM CODE

CASE NUMBER

Name ____________________

SOCIAL SECURITY NUMBER

•O •C

•0 • to •9 • (9 columns)

DEPENDENTS

•0 • to •9 • (2 columns)

URINE SPECIAL CHEMISTRY

- Cadmium
- Antimony
- Nickel
- Vanadium
- Selenium
- Fluoride
- Mercury
- Hippuric Acid
- Arsenic
- Cobalt
- Delta-Aminolevulinic Acid
- Phthalic Acid
- Dimethyl Terephthalate
- Isophthalic Acid
- Trichloracetic Acid

AMOUNT

•0 • to •9 • (8 columns)

BREATH ANALYSIS

	ALCOHOL (ETHYL)	ALCOHOL (METHYL)	BENZENE	CARBON MONOXIDE	TOLUENE	TRICHLOR-ETHYLENE	METHYLENE CHLORIDE	VINYL CHLORIDE	METHYL ETHYL KETONE	FURFURAL	STYRENE	XYLENE	SULPHUR DIOXIDE	METHYL MERCAPTAN
ABSENT														
BELOW LIMITS														
BORDERLINE														
ABOVE LIMITS														

SPECIAL PROCEDURE

DATE PERFORMED						SPECIAL PROCEDURE	
YEAR		MONTH		DAY			
	0	0	0	0	0	0	0
	1	1	1	1	1	1	1
	2		2	2	2	2	2
	3		3	3	3	3	3
	4		4		4	4	4
	5		5		5	5	5
	6		6		6	6	6
7	7		7		7	7	7
8	8		8		8	8	8
9	9		9		9	9	9

QUESTIONS

1
2
3
4
5
6

AMOUNT 1							
0	0	0	0	0	0	0	0
1	1	1	1	1	1	1	1
2	2	2	2	2	2	2	2
3	3	3	3	3	3	3	3
4	4	4	4	4	4	4	4
5	5	5	5	5	5	5	5
6	6	6	6	6	6	6	6
7	7	7	7	7	7	7	7
8	8	8	8	8	8	8	8
9	9	9	9	9	9	9	9

AMOUNT 2							
0	0	0	0	0	0	0	0
1	1	1	1	1	1	1	1
2	2	2	2	2	2	2	2
3	3	3	3	3	3	3	3
4	4	4	4	4	4	4	4
5	5	5	5	5	5	5	5
6	6	6	6	6	6	6	6
7	7	7	7	7	7	7	7
8	8	8	8	8	8	8	8
9	9	9	9	9	9	9	9

AMOUNT 3							
0	0	0	0	0	0	0	0
1	1	1	1	1	1	1	1
2	2	2	2	2	2	2	2
3	3	3	3	3	3	3	3
4	4	4	4	4	4	4	4
5	5	5	5	5	5	5	5
6	6	6	6	6	6	6	6
7	7	7	7	7	7	7	7
8	8	8	8	8	8	8	8
9	9	9	9	9	9	9	9

FORM 53-209

© Copyright Amoco Comput[illegible] Company 197[illegible] DC 33-888

Figure 3.33

RESULTS OF SURVEY 1-20-77 thru 1-27-77

XXXXXXXXXX Company
XXXXXXXXXXXXXXXXX
#1 Unit

Soc. Sec. No.	Name of Employee	Occupation or Activity	Date	Samp No.	Samp Type	Duration Minutes	Hazard	LMN	Result	Units	Shf	Opr Cnd	Pers Pro Eq 1	2	3	4	5
XXXXXXXXX	Doe, J.	Operator	1-20-77	371	PN	120	Benzene		0.1	PPM	1	N					
		Operator		371	PN	120	Toluene		1.0	PPM	1	N					
		Operator		372	PN	120	Benzene		0.2	PPM	1	N					
		Operator		372	PN	120	Toluene		1.2	PPM	1	N					
		Operator		373	PN	120	Benzene	L	0.2	PPM	1	N					
		Operator		373	PN	120	Toluene		1.4	PPM	1	N					
					PT		Benzene		0.2	PPM	1	N					
					PT		Toluene		1.2	PPM	1	N					
		Drawing Process Samples	1-23-77	395	PN	20	Benzene		0.2	PPM	1	N	A				A
XXXXXXXXX	Doeks, A.	Operator	1-21-77	374	PN	120	Benzene		0.1	PPM	2	N					
		Operator		375	PN	120	Benzene		0.2	PPM	2	N					
		Drawing Process Samples		376	PN	20	Benzene		0.3	PPM	2	N	I				A
		Operator			PN	120	Benzene		0.2	PPM	2	N					
					PT		Benzene		0.2	PPM	2	N					
			1-20-77	415	A	0	Hydrogen Sulfide	L	1	PPM	1	N					
			1-20-77	416	A	0	Hydrogen Sulfide		2	PPM	1	N					
			1-20-77	417	S	0	Hydrogen Sulfide		3	PPM	1	N					
			1-24-77	407	A	480	Total Dust		2.0	mg/M^3	1	N					
			1-27-77	408	A	480	Total Dust		2.5	mg/M^3	1	N					

Sample Type Code Descriptions			LMN:	Personal Protective Equipment		Operating Conditions
A = Area	PN = Personal Normal	WN = Work Station Normal	L = Less Than	(1) Respiratory		N = Normal
S = Source	PC = Personal Ceiling	WC = Work Station Ceiling	M = More Than	(2) Hearing	A = Adequate	AB = Abnormal
B = Background	PP = Personal Peak	WP = Work Station Peak	N = None Detected	(3) Eye and/or Face	I = Inadequate	
Ø = Other, Non-Personal	PT = Personal TWA	WT = Work Station TWA	LN = Less Than and None Detected	(4) Hands	* = Not Answered	
	PØ = Personal Øther	WØ = Work Station Øther	* = Not Answered	(5) Øther		

Figure 3.34

RESULTS FOR NOISE STUDY TAKEN 01/01/77 THROUGH 02/28/77

XXXXXXXXXX Company
XXXXXXXXXXXXXXXXXX
#2 Process

Soc. Sec. No.	Name & Occup/Activity	Date	Samp No.	Samp Type	Samp Mins	Results Extrapolated to 8-hr. Read	DBA Results	Shf	Opr Cnd	Pers-Pr-Eq 1	2	3	4	5
XXXXXXXXX	Doe, John	01/15/77	001	% Exp-Dos	480	70		1	AB		A			
	Operator #1	01/16/77	002	% Exp-Dos	420	30		1	N		A			
		01/16/77	003				84	1	N		A			
		01/18/77	004	% Exp-Dos	480	50		1	N		A			
						50 Avg.								
XXXXXXXXX	Doekes, Jane	01/05/77	006	% Exp-Dos	420	30		3	N		A			
	Operator #2	01/06/77	007	% Exp-Dos	420	50		3	N		A			
		01/18/77	008	% Exp-Dos	450	45		3	N		A			
						42 Avg								
		01/18/77	009				82							
XXXXXXXXX	Doen, John T.	01/15/77	010	% Exp-Dos	480	30		3	N		A			
	Foreman	01/16/77	011	% Exp-Dos	480	40		3	N		A			
		01/18/77	012	% Exp-Dos	480	55		3	N		A			
						42 Avg								

Personal Protective Equipment

(1) Respiratory
(2) Hearing
(3) Eye and/or Face
(4) Hands
(5) Øther

A = Adequate
I = Inadequate
* = Not Answered

Operating Conditions

N = Normal
AB = Abnormal

Figure 3.35

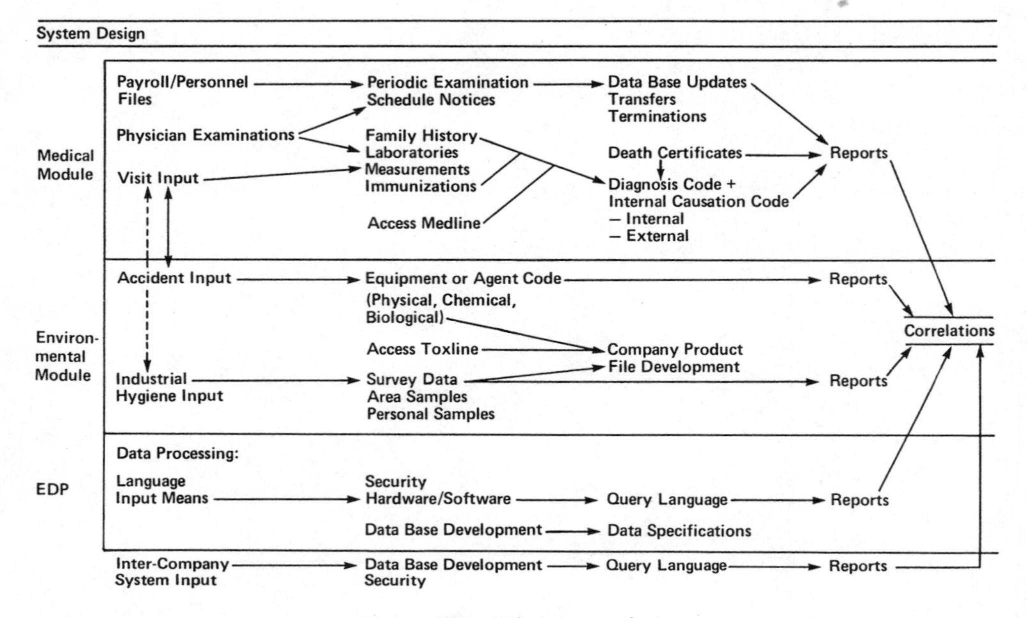

Figure 3.36 Milestone analysis.

Task Group 7 Chemical Substance Code

Skills required: epidemiology, industrial hygiene/toxicology, systems.
Requirements:

1. Determine what code set is to be used to identify chemical substances. The code will be used for (*a*) recording specific noticed exposures of employees, (*b*) recording workplace environmental data, (*c*) recording of toxicology information. Possibilities are *American Chemical Society Chemical Abstract Service Registry Number* (Chemical Substance Code) and *National Institute for Occupational Safety and Health Toxic Substances List.*
2. Review data available from Toxline.

   American Chemical Society
   1155 Sixteenth St., N.W.
   Washington, D.C. 20036

   National Institute for Occupational Safety and Health
   5600 Fishers Lane
   Rockville, Maryland 20852

Task Group 8 Biostatistics

Skills required: clinical medicine, industrial hygiene/toxicology, epidemiology.
Requirements:

1. Determine data elements to be recorded concerning the workplace and concerning individual jobs within the workplace; that is,
   a. The concentrations of gases and particulates in the air.
   b. Temperature—mean and extremes.
   c. Noise levels.
   d. Physical, biological exposure levels.
   e. Are chemicals physically handled, if so, which ones?
   f. Protective clothing or equipment worn or used.
   g. Industrial function performed at that location, and so on. The Chemical Abstract Service Registry would be used for recording data.

Task Group 9 Industrial Environment

Skills required: industrial hygiene/toxicology, epidemiology.
Requirements:

1. Determine the types of correlation of different data element groups for the kinds of studies expected to be done against the data base.
2. Give examples.

Task Group 10 Toxicology

Skills required: toxicology, epidemiology.
Requirements:

1. Provide as much information as possible to help determine whether the file of individual company products should be kept separately from the chemical substance file. The company products file might list components identified by the chemical substance code. What are the characteristics of company products? Would your management allow them to be put into a central file if the release of data were controlled by you? What is your position concerning a central chemical substances file?

Task Group 11 Synthesis, Basic Module, Company Level

Skills required: systems, clinical medicine, medical director, administrative, safety, epidemiology, industrial hygiene/toxicology.

Requirement:

1. Take the results for each task group (groups 1 through 8) and consider the requirements as a total system. Consider the volume of input data that would be required. Delete unnecessary items or move them from the essential/desired list.
2. Determine whether all the data elements on the output are available from the input and the data base; if not, add them.
3. Write up the results and determine data base structures.

Task Group 12 Synthesis, Industrial Environment, and Toxicology

Skills required: systems, industrial hygiene/toxicology, epidemiology, clinical medicine, medical director, administrative, safety.
Requirements:

1. Take the results for task groups 9 and 10 and develop the integrated data base with the basic module.
2. Define the necessary input.

Task Group 13 Synthesis, Basic Intercompany Module

Skills required: systems, clinical medicine, industrial hygiene/toxicology, epidemiology.
Requirements:

1. Determine specifically what data elements in the company level data base will have to be obscured to destroy individual identification.
2. Structure an intercompany data base.
3. Define the extraction program at the company level to feed this central data base.
4. Evaluate possible alternatives for processing central data.

Task Group 14 Forms Design, Basic Module

Skills required: systems, forms designer.
Requirements:

1. Design forms for basic module.

Task Group 15 Forms Design, Industrial Environment, and Toxicology

Skills required: systems, forms designer.
Requirements:

1. Design forms for industrial environment and toxicology.

Task Group 16 Write Systems Specification, Basic Module—Company Level

Skills required: systems.
Requirements:

1. Write system specification basic module.
2. Define input, data base, and so on.
3. Define linkages between data elements on reports, data base, and input.

Task Group 17 Write Systems Specifications, Basic Module—Intercompany

Skills required: systems.
Requirements:

1. Write systems specifications basic module—intercompany.
2. Define input, data base, and output.
3. Define linkages between data elements on reports, data base and input.

Task Group 18 Write Systems Specification, Industrial Environment, and Toxicology

Skills required: systems.
Requirements:

1. Write systems specification, industrial environment and toxicology module.
2. Define input, data base, and output.
3. Define linkages between data elements on reports, data base, and input.

Task Group 19 Complete System Specifications for Intercompany Module

Skills required: systems.
Requirements:

1. Write systems specification to complete intercompany module.
2. Define input, data base, and output.
3. Define linkages between data elements on reports, data base, and input.

Glossary of Computer Terms

BASIC MODULE: Main computer program.

COBOL: *CO*mmon *B*usiness *O*riented *L*anguage. A business data processing language.

COMPUTER SOFTWARE: A set of computer programs that operate the data processing system within the computer (which is hardware—physical equipment). These programs control the execution of application programs for a particular system, such as the medical history system.

CONVERSATIONAL: Pertaining to a system that carries on a dialogue with a terminal user, alternately accepting input, then responding to the input quickly enough for the user to maintain a train of thought.

DATA BASE: A collection of data fundamental to an enterprise.

DATA ELEMENTS: Specific information in the data base.

DLI: A particular data processing language used to construct the computer program.

FORMS DESIGN: Process of designing forms that will be used to enter data into a system.

FORMS DESIGNER: Person who designs the forms used to enter data into a system.

FORTRAN: *FOR*mula *TRAN*slating system. A data processing language primarily used to write computer programs by arithmetic formulas.

HARDWARE: The physical equipment used in a data processing system.

ICD CODE: *I*nternational *C*ode of *D*iseases Code. A standard code for a disease, obtained from the ICD Book.

IMS: *I*nformation *M*anagement *S*ystem. A general purpose software programs system that allows users to access a computer-maintained data base through remote terminals.

INPUT DATA: Data to be processed by a system.

INTERCOMPANY MODULE: Data shared among companies for statistical purposes.

KEY ENTRY: The process of entering information by way of the keyboard device attached to the terminal.

KEYPUNCH: A keyboard-actuated device that punches holes in a card to represent data.

KEYPUNCH RJE: The process of entering information or a request by way of the keyboard device attached to the terminal and teleprocessing the information via remote job entry (RJE), a computer program package, from a remote terminal.

MARK IV: A particular business data processing language.

MODULE: A computer program unit.

NONCONVERSATIONAL: Refers to key entry process in which the user enters data into the system and the system does not respond quickly to the data.

OPTICAL MARK READER: A device that reads handwritten or machine printed marks on special forms into a computing system.

OUTPUT REPORT: Report that is prepared by a computing system from the data processed.

PLI: A high level programming language, designed for use in a wide range of commercial and scientific computer applications.

QUERY LANGUAGE: A high level programming language designed for use by the user of a system to enable him/her to access information in the system for one-time reports.

RJE: See keypunch RJE.

SYSTEMS: Refers to a skill in which an activity is analyzed to determine precisely what must be accomplished and how to accomplish it.

SYSTEMS SPECIFICATION: Refers to the process of analyzing an activity to determine what must be accomplished and how to accomplish it.

TELEPROCESS SLOW SPEED DEVICE: Device that transmits data to and receives data from remote locations at slow speeds by way of telecommunication lines.

9 CONCLUSIONS

It should be clear that for industrial hygiene to fulfill its obligation and goal of the betterment of the health of workers, adequate records and reports are essential and must be employed to the fullest. The extent and sophistication of such records and reports, of course, varies with the individual situation.

The industrial hygiene report is the industrial hygienist's most forceful and enduring communication with employer and employee. Done properly, and in clear and concise language, understandable to those who are not industrial hygienists, it can effect the changes and improvements necessary to protect the health of the worker. If the report is done poorly, the effort that went into the original industrial hygiene survey record and evaluation may have been for naught.

Industrial hygiene records, though recognized as a necessary and valuable part of an industrial hygiene report, can also go far beyond the immediate need in the report of verifying environmental stresses that exist for the moment. Careful and full use of such records and data will allow the industrial hygienist to accomplish the following:

1. Evaluate the adequacy of current acceptable levels of exposure.
2. Establish acceptable levels of exposure for chemicals, and so on, for which none currently exist.
3. Show trends in exposures, controls, and so on, by time series.
4. Conduct retrospective studies if occupational illnesses occur, especially those of a more chronic nature.
5. Confirm or refute the validity of any compensation claims.
6. Compare data for like operations or work places, especially for the purpose of evaluating the effectiveness of differing approaches to control of the work environment.
7. In cooperation with medical staff, detect any early evidence of developing occupational health problems and effect controls at an early stage.

Finally, industrial hygiene records and reports must be retained long enough to ensure that they have fulfilled their intended purpose. Industrial hygiene engineering control recommendations, for example, may not be needed again after they have served their original purpose of effecting control of the work environment. Environmental measurements of work stresses, on the other hand, may be useful 20, 30, or even 40 years after serving their initial purpose. In some instances, retention of such records for a given period may be required by law. In the final analysis, the industrial hygienist, with his background of training and experience, can best determine the point at which such records and data are no longer of value in protecting and promoting the health of the worker.

REFERENCES

1. J. J. Bloomfield, "Industrial Health Records: The Industrial Hygiene Survey," *Am. J. Pub. Hlth,* **35,** 559 (1945).
2. P. DeFalco, "Data Acquisition Systems in Water Quality Control," U.S. Public Health Service Publication No. 999-AP-15.
3. *Environmental Health Monitoring Manual,* United States Steel Corp., Pittsburgh. Copyright 1973.
4. R. Gunning, *The Technique of Clear Writing,* McGraw-Hill, New York, 1952.
5. H. M. Quackenbas, "Creative Report Writing—Part I", *Chem. Eng.,* July 10, 1972, pp. 94–98.
6. R. R. Rathbone, and J. B. Stone, *A Writer's Guide for Engineers and Scientists,* Prentice-Hall, Englewood Cliffs, N.J., 1962.
7. W. Strunk, *The Elements of Style,* 2nd ed., revised by E. B. White, Macmillan, New York, 1966.
8. J. T. Siedlecki, "How Medicine and Industrial Hygiene Interact," *Int. J. Occu. Health Saf.,* September–October, 1974.
9. M. G. Ott, H. R. Hoyle, R. R. Langner, and H. C. Scharnweber, "Linking Industrial Hygiene and Health Records," *Am. Ind. Hyg. Assoc. J.,* **36,** 760 (1975).
10. H. J. Tichy, *Effective Writing for Engineers, Managers, Scientists,* Wiley, New York, 1966.

CHAPTER FOUR

The Industrial Hygiene Survey and Personnel

LUCIAN E. RENES

1 INTRODUCTION

Industrial hygiene surveys are scientific investigations conducted within the premises of an industrial establishment, primarily to determine the nature and extent of any conditions that may adversely affect the well-being of the persons employed. Usually the principal benefit of such a survey is the obtaining of information necessary to the development of engineering and medical control measures for avoiding or eliminating potentially harmful conditions.

Industrial hygiene preserves for employed adults their basic asset—health. It is the primary task of the industrial hygienist to maintain and safeguard the physical and mental capacities of workers at their highest peak and to prolong their productive lives. Although there is a fundamental moral obligation for an employer to preserve his workers' well-being, the enactment by Congress of the Occupational Safety and Health Act of 1970 (1) imposes a legal obligation as well.

Ramazzini (2) is the first outstanding industrial hygienist of whom we have a considerable record. His investigations about A.D. 1700 into the causes of occupational diseases were remarkably enlightening. Like many who came after him, however, he dwelled at some length on therapeutic measures for the cure of occupational diseases and gave little consideration to preventing them. Following this early auspicious beginning, it is remarkable that not until the third decade of the present century was there any widespread realization or acceptance of the next stage in the development of industrial hygiene; that is, the utilizing of engineering methods of control for the actual prevention of disease. Before the 1950s industrial hygiene was practiced largely on a complaint-investigation basis; when serious diseases developed, industrial hygiene investigations were undertaken to develop control measures for preventing a recurrence.

Modern industrial hygiene practice is to recognize harmful exposures, and to bring them under control before workmen experience injury or evidence any adverse signs or symptoms. This can be done by measuring exposures, evaluating their probable effects by existing toxicological and hygienic standards, and utilizing sensitive biological examinations of exposed persons to discover the entrance of harmful materials into the human system in advance of injury. The professional industrial hygienist recognizes his moral obligation to do his work well. This obligation precludes his accepting responsibilities beyond his capabilities; nor will he allow his findings to be influenced or colored by incompetent persons or selfish interests. The wise businessman recognizes the value of keeping the industrial hygiene service on such a plane.

2 TYPES OF SURVEY

In the current practice of industrial hygiene there are three types of survey: (1) the reconnaissance or observational survey; (2) the investigational or appraisement survey, which has as its chief purpose the evaluation and control of potentially harmful situations; and (3) the combined industrial hygiene and medical survey, which integrates the investigational survey with medical examinations of the workmen.

2.1 The Preliminary Industrial Hygiene Survey

The preliminary or observational industrial hygiene survey is usually the immediate forerunner of a survey employing technical instruments. It is made for the purpose of selecting locations in a plant where exposures or hazards are later to be evaluated by analytical studies, to determine whether additional control is necessary. During this survey pertinent data should be collected, such as the number of male and female workers employed at various operations or processes; safety supervision, first aid, and medical services; availability of accident and sickness records; the different types of operation conducted; raw materials, processing aids, products, and recognized by-products; measures employed for dust, fume, and vapor or gas control; and methods of solid, liquid, or gaseous waste disposal. If a plant layout showing the location of the operations or process equipment in the industrial establishment is immediately available or can be obtained in a short time, it will be invaluable in helping to orient the surveyor to the strangeness of an industrial complex. The preliminary survey will also determine the apparatus necessary for the investigational survey. Depending on the size and complexity of the operations, the time required for the preliminary survey may vary from one day in a small shop to a week or more in large industrial plants. In a small plant the preliminary survey may immediately precede the investigational survey, and both may be accomplished on the same day by the same individual.

The preliminary survey is usually made with no equipment other than the five senses—or possibly six. It should be made by a person who is familiar with the type of industry involved, especially the chemistry of its products and by-products, and is well grounded in the field of industrial hygiene. The surveyor may or may not be the person who is to make the final technical analysis, but he or she should be at least equally familiar with the problems involved in recognizing and evaluating exposures to potentially harmful materials. The survey is best accomplished by following the industrial process through the plant from raw materials to finished products. The industrial hygienist should be accompanied by the production superintendent, or some other qualified plant employee, to explain any process or steps in manufacture that are not evident to the surveyor. Among the plant personnel who are best suited to the role of guide for the investigator are the production superintendent, the chief process engineer, the plant chemist, the foreman of the department under investigation, and the safety director. The plant physician or nurse may be able to point out departments that cause trouble. It hardly needs to be stated that to be of much help to the hygienist, a guide must be prepared to supply or obtain information about any processes involved.

Since the surveyor ordinarily does not use equipment other than perhaps a smoke tube at this time, a most necessary asset is a well-developed and trained sense of smell. The person should be able to recognize and identify all the common gases and vapors that possess characteristic odors, tastes, or irritant effects, as well as to judge roughly whether the concentration of a substance may be exceeding permissible levels. He or she should be alert to any signs in the personal appearance of the workers that may indicate adverse occupational exposures. This statement does not suggest that nonmedical industrial hygienists should make an obvious examination of the exposed portion of a workman's body for occupational marks or signs, or ask an individual to display such marks. It should not even be evident that the surveyor is making such observations, but at the same time there is abundant reason for keeping alert to any indications that a workman may have been exposed to conditions sufficiently adverse to leave a mark or sign. In this way it may be possible to identify a serious exposure that might otherwise be overlooked because of the intermittent or obscure nature of its causation.

Frequently there are contributory or intermittent operations not in evidence at the time of the survey. These may include preliminary treatment of raw materials, disposal of by-products or waste products, "turn-around" or periodic maintenance and rebuilding, and warehouse operations. The surveyor may learn of these through experience in the industry, from a knowledge of similar operations, or by an extended discussion with a competent guide. One should always inquire into the location and method of conducting operations of such a nature, as well as into proposed new operations or pilot plant work, where these exist. It is most important for the surveyor not to hurry, or be hurried, through a plant, but to take sufficient time to permit the recognition of the more obscure situations.

The preliminary survey should determine the presence or absence of control measures and provide an opinion about (1) the probable need for, or effectiveness of, control; (2) the type of personnel, in terms of training, skill, or care; and (3) the attitude of management supervising staff, and workmen toward health control measures. At the end of the preliminary survey it may be desirable to give the management a report of findings and plans. Where an investigational survey is to be made without delay, it may not be necessary to discuss the situation with anyone other than the guide. In any case, having completed the preliminary survey, and having decided where all investigations are to be made, the surveyor now selects the necessary equipment and undertakes the technical analysis.

2.2 The Investigational Industrial Hygiene Survey

2.2.1 Surveying the Plant

The investigational survey, or the technical industrial hygiene analytical survey, involves the evaluation of all exposures to potentially harmful situations and the development of control measures. This is what is ordinarily referred to in the profession as an *industrial hygiene survey.* If no previous survey has been made of the plant, it is best to leave any equipment in a safe place and conduct a preliminary survey before undertaking the detailed investigation. It is sometimes unnecessary to evaluate exposures by exact measurements. When control is the only objective, and it is obvious to the investigator that contaminants are excessive, it is frequently possible to devise satisfactory controls without the benefit of quantitative measurements.

When there is any question about the necessity of control, however, or when facts for the record are desired because of medical, legal, or other needs, samples of the atmosphere must be collected and evaluated by chemical or instrumental methods. In surveys to determine the merits of a claim for compensation for injury allegedly arising from harmful exposures in industry, it is essential to obtain quantitative information about exposures, to ensure the establishment of facts, rather than the recording of opinions, even though it may be obvious to the experienced industrial hygienist that the exposures are either insignificant or excessive. In a system handling a potentially harmful chemical, even though the investigator is completely satisfied that control is ample, sampling may be desirable for its psychological effect; it eliminates personal factors and gives the uneasy employees assurance of a safe environment. Under the terms of the Occupational Safety and Health Act, records of periodic analyses of atmospheric contamination are required when the concentration of a chemical agent is borderline with respect to the threshold limit value for an 8-hour daily exposure. More frequent analyses are required when control measures are to be installed to reduce the air contamination to acceptable levels. The detailed regulatory requirements of the act with respect to chemical air contaminants are still being developed, and some of the procedures specified for air sampling, work practices, and other administrative matters are controversial and subject to dispute. However there is still room for the knowledgeable hygienist to exercise his judgment in evaluating exposures and creating a safe working environment.

The analytical survey can be accomplished with the cooperation of the production superintendent, the safety director, or the foremen of the particular departments in which investigations are to be made. If the results are to be representative, air sampling or other quantitative measurements should be made during normal, as well as the most unfavorable, operating conditions, and over a period of time long enough to yield the complete exposure picture. The foremen can assist in maintaining the normal conditions as well as in simulating abnormal conditions if such are required for a true evaluation. It is good policy to visit the physician or nurse on duty, if there is one, to establish cordial cooperation and to review any records of ill health. The plant physician may wish to accompany the hygienist during the survey of the plant, and this may prove advantageous to both; but it is an exceptional physician who can supply needed information about plant processes.

In the present social climate, it is a rare employee indeed who has no inkling of the possible occupational health hazards of a particular industry or a specific workplace. Both worker and management are constantly bombarded by articles on industrial health hazards that appear in the daily newspapers, union periodicals, trade journals, and the scientific literature. Some of the information is factual and straightforward, whereas some articles have an inflammatory tone. In addition to such educational efforts, some of the regulations promulgated by the Occupational Safety and Health Administration require that employee representatives be informed of the nature and purpose of an industrial hygiene survey and be granted the opportunity to observe air sampling, instrumental measurements, and other procedures in the conduct of the survey carried out by the employer, his agent or a regulatory officer. Furthermore, during the inspection or survey made by an OSHA compliance officer, the inspector has the legal right to interrogate employees in private concerning their knowledge of adverse working conditions and complaints of ill health due to alleged occupational exposure.

In view of these unsettling influences, how should a professional industrial hygienist carry out his responsibilities without creating any friction between the workers and their employer, or contributing to any further apprehension among the workers concerning the healthfulness of their occupation? It is advisable for management to explain the purpose of the survey to the employees and/or the union, thereby creating a feeling of partnership with the workers in their health protection efforts. At the same time, the element of surprise due to the presence of strangers or relatively unfamiliar persons making unfamiliar measurements will have been removed. A request can be made for management to assign a temporary guide to answer workers' questions at the different locations surveyed in the establishment. The normal exchange of greetings and courtesies between the fellow employees and the hygienist should be maintained. The plant escort can be useful in answering questions regarding the survey. Although all possible sources of friction between humans cannot be prevented, a friendly attitude will have a calming influence on many hostile situations.

Air samples may be taken at one or several locations to determine individual exposures, general room air conditions, and sources of contamination. Many variable factors should be considered at the time of sampling, in relation to normal, adverse, and optimal conditions. Single samples, though not to be scorned, should not be accepted as true indications of existing conditions because not only does the concentration of contaminants vary from minute to minute, day to day, and season to season, but also errors in sampling and analysis are certain to creep in occasionally, even with experienced, and most careful, investigators. The more samples taken, the more evident will be the reasons for any variance in results, trends, peak concentrations, and other factors. There are arguments on each side of the question of the short-interval sample versus the extended-time sample; but there can be no doubt that a large number of properly spaced short-interval samples tell a more comprehensive story than a single sample taken over a given 4-hour period. Here again, the trained observer, from the senses of smell, taste, or feeling, may be able to evaluate an exposure accurately enough to establish either the necessity for its control or its insignificance. In large measure, this depends on previous observations of known amounts and similar magnitudes of the material to be evaluated, or so to speak, a calibration of individual sensitivity, since there is some variation among persons in sense of smell (see Chapter 16, Section 3.14). Three is the minimum number of samples to be taken to evaluate any one operation in a series of successive steps of a process or job assigned to a worker. If the job consisted of four different tasks successively repeated, 12 air samples should be collected to express confidence in the results.

All shifts of a similar process should be observed, since a night shift may create an exposure at some operation that the day shift conducts safely, or vice versa. I recall a survey made many years ago in which an unusual night operation created a very high overexposure to toxic substances. The process involved the smelting of zinc sulfide ore in gas-fired retorts, and the molten zinc metal was collected in cylindrical clay condensing units. A certain number of condensers were removed and replaced each night, successively. This was a hot and dirty job that was rotated among all smelter employees. Tests for employee exposure to airborne heavy metal contaminants, such as lead, antimony, arsenic, and cadmium, were below acceptable levels during the day and "graveyard" shifts. However the night shift workers who performed this special operation for 4 to 5 hours had very high exposures to these agents, which offered an explanation for the subjective symptoms and clinical signs of illness among the employees.

The opportunity for employee exposure to airborne chemical contaminants is low in the chemical and petroleum processing industries, since the processing is carried out in closed systems. When such systems are shut down for periodic safety inspections, or cleaning and replacement of worn equipment, gross exposure to chemicals may occur unless the work is planned and safety precautions followed. During these periods, exposures can occur to the process chemicals, cleaning agents, and air contaminants arising from such construction activities as welding in confined spaces.

During the course of the technical survey it may be desirable to trace the air contaminants from their origin to the point or points of dispersion. Because of the possible synergistic effect of exposures to combined aerial contaminants or the enhancing influence of abnormal temperatures on chemical exposures, the industrial hygienist should also search for the unusual circumstances or combinations of factors that may give rise to adverse physiological responses.

Up to this point our discussion has centered on a survey designed to quantitate aerial contaminants and evaluate employee exposure to these agents. However associated matters, such as the types of emergency respiratory protective device issued for use against a specific air contaminant, also should be examined. The survey should determine that the device conforms with the NIOSH approval schedules, making sure that it will provide protection against the concentration of contaminant that may be encountered. No respiratory protective device is presently permitted to be used for continuous protection against air contaminants in lieu of engineering control. Nevertheless, certain devices—namely, mechanical filter-type respirators and chemical cartridge-type respirators—may be used temporarily over protracted periods of time while engineering controls are being installed.

2.2.2 Developing Control Procedures

If the industrial hygienist or his associates have knowledge and skills in the design of mechanical ventilation systems that may be suitable for controlling or reducing the dispersion of a chemical agent from a certain operation, advice and assistance in the form of elementary design specifications can be offered to the plant management. To be more specific, the hygienist should at least choose between general ventilation and exhaust at the source of contamination, the size, shape, and face velocity requirements of hoods, type of ventilating fan, the necessity of a dust collector, suitable types of dust collector, location of dust collector inside or outside, and the possibility of recirculation of the air. The judicious use of a smoke tube or velometer to locate and measure air currents and cross drafts may be used to demonstrate to the plant engineer the ineffective control of a present exhaust system.

At the end of this investigational survey, the hygienist is in possession of many accumulated facts that must be correctly applied; otherwise they become of academic interest only and serve little purpose in the promotion or control of health. The hygienist should develop methods or procedures for controlling overexposures and should have such ideas sufficiently crystallized to present to the plant management. Any unusual hazards found during the plant survey should be pointed out to and discussed with the physician or nurse, to ensure that by cooperation and knowledge of the complete picture, both the industrial hygienist and the plant medical department are aided in their work. It is important that the medical director, as health officer for the plant, have knowledge of all information developed as a result of the survey.

At the end of the survey, if not before, the plant manager, personnel director, the production superintendent, the safety engineer, and the maintenance engineer should be acquainted with all significant findings, along with recommendations or suggestions for control. Objections may be raised: some of them can be overcome, whereas others may indicate that alternatives to some of the original proposals are more practical. It is obviously desirable that any major control program be discussed in detail and its merits carefully explained. Since federal law requires preventing excessive exposure to potentially harmful agents, it is not a matter of "selling" the need for a control program but rather the "selling" of a system that is most effective and most economic for the specific situation. The professional industrial hygienist should not attempt to coerce management to undertake any corrective action nor allow any negative management attitudes to discourage reporting adverse findings or to color the conclusions and recommendations that are made.

2.2.3 Records and Reports

A written confirmation report of the survey, complete with findings and recommendations, should be delivered to plant management and other appropriate levels of a corporate establishment. The report should contain the results of all technical measurements in tabulated form. Interpretation of the findings in each table should be presented in an adjacent narrative, along with the surveyor's conclusions and any necessary recommendations for improving the working environment. A summary of the findings, conclusions, and recommendations should be located in the beginning of the report to draw the reader's attention to information of special importance.

The report may vary from a brief, unbound set of typewritten pages to a voluminous manuscript, in an attractive and expensive binding, depending on the needs of the surveyor and the desires of the plant management. The report is more likely to be read when brief, but it should always be sufficiently detailed to establish the necessity for the accompanying recommendations. Notes should have been made in a field notebook as a permanent record from which to develop the report. The industrial hygienist may wish to use forms prepared to accommodate personal needs on which to record each sample collected and all environmental and work information pertinent to the sample. Calculated air concentration of the chemical agent derived from the laboratory analysis of the sample can also be recorded on such a form. The reader is referred to Chapter 3 for more detailed information on industrial hygiene reports and record keeping.

A permanent record of observations made during the course of the survey should be entered into a notebook on the job for reference during preparation of the report; these notes will also serve as admissible evidence in any possible court action. The surveyor should not jot down random bits of information on loose pieces of paper that may be lost. Details that are not recorded on the spot are often forgotten. If management permits, snapshots of specific operations, taken with a camera that uses self-developing film, are extremely helpful for refreshing memory in report preparation. A decision to include photographs in a report should be made after careful consideration for their need, but only with prior approval of management.

It is important that the report of a survey reach the plant management while the situation is fresh in mind, never later than 30 days afer the survey and preferably within 2 weeks. The value of the report is enhanced if it is delivered in person by the surveyor, or this effect may be achieved by a personal call as soon as the management has read the report, in time to explain or support any findings or recommendations that may be questioned. Even though a recommendation was agreed on at the time of the survey, if it arrives at an inopportune time—or so late that it has been forgotten, or interest has waned—management may assume an antagonistic attitude, necessitating another personal contact. Reports or copies of reports should never be presented to anyone other than the person requesting the survey without the knowledge and permission of that person.

The written report should be delivered personally by the chief industrial hygienist, to ensure that the findings are discussed at the appropriate management level. A report sent by mail, or delivered by someone with little experience, detracts from the importance of the survey and can also lead to misinterpretation of the recommendations.

2.3 The Combined Industrial Hygiene and Medical Survey

The combined industrial hygiene and medical survey is a comprehensive study of the environment and of the health status of the workers. Its purpose is to determine any relationship between exposure to atmospheric contaminants and the workers' health. Depending on the size of the plant and the number of different kinds of agent to which the employees may be exposed, this type of survey may require a week to several months or even longer to complete. In addition to in-depth studies of the various environmental factors, this kind of investigation includes examination of individual workmen by physicians and medical technicians. It will include tests to observe the various organs of the body and to measure their physiological functions by microscopic and chemical examination of the blood and urine as well as X-ray and vital capacity tests of the pulmonary tracts. Other special examinations such as the electrocardiogram or en electroencephalogram may be required.

The interpretation and evaluation of medical information belong in the field of medicine and call for the services of physicians qualified by training in the fields just mentioned. Likewise, the environmental findings should be handled by persons technically trained in the scientific principles of industrial hygiene, preferably with long experience in this field. Since the purpose of the combined survey is to determine any possible relation of exposure to the environmental factors and any detectably adverse effects among the workers, another important professional, namely, the statistician or epidemiologist, completes the survey team.

In years past such studies evolved from the discovery of unexplained illnesses in certain occupations, but at present they are more likely to originate from a desire to learn whether the known presence of air contaminants with ill-defined or low toxic properties may exert little known or obscure adverse effects among occupationally exposed workers. Many of the maximum permissible concentrations of exposure for atmospheric contaminants, notably free-silica-bearing dusts, mercury vapor, lead dust and fumes, and carbon monoxide, were derived from such classical, comprehensive industrial hygiene studies (3) carried out in the 1920s and 1930s by the Division of Industrial Hygiene of the U.S. Public Health Service. The standards or "safe limits" established by these studies have been modified in recent years because of new evidence based on advances in analytical techniques and more sensitive clinical test procedures. For that reason, in all reports proposing safe working standards, the environmental and clinical methods of analysis should be explained and documented. At some later time, more accurate or sensitive analytical test methods may reveal new findings that show an accepted figure to be erroneous.

3 THE INDUSTRIAL HYGIENE UNIT OR ORGANIZATION

3.1 Recent Pattern of Growth

For a span of about 30 years, beginning in the middle 1930s, the centers of industrial hygiene activity were concentrated in the governmental units of about 25 states. The federal industrial hygiene activity was located in the Public Health Service in its Division of Industrial Hygiene, which provided leadership in research, grants in aid, and loans of personnel to the various states. The size of these industrial hygiene units in the states varied from one or two persons up a dozen or more. At the zenith of its field investigational activity, from about 1942 to about 1955, the Industrial Hygiene Division of the Public Health Service had a staff of several hundred technical persons. Obviously the smaller organizational units could not accomplish a great deal in the way of technical surveys, and much of their time was devoted to educational efforts and the investigation of complaints.

During the same period there was a slow but steady growth of industrial hygiene activity in private industry, primarily among the larger corporations. These units ranged from one to three or four persons and, in a few instances, as many as 10 hygienists. Most of the units were staffed by experienced industrial hygienists who had "won their spurs" in state or federal employment. Many of these companies made notable advances in researching areas of potential hazard to their employees and initiating corrective action to protect workers against adverse effects to their health. All industrial hygiene units in private industry have environmental evaluation and control as their primary objective, but their activities vary greatly depending on the views of management and the size of the establishment. Several types of industrial hygiene consulting organization were also created during this period. The Industrial Health Foundation, (formerly the Industrial Hygiene Foundation) is an organization having a general, educational, and investigational interest, whereas the industrial hygiene units of insurance companies were principally engaged in investigational surveys and individual control programs for their insurance clients.

The Occupational Safety and Health Act of 1970 has exerted great influence on the field of industrial hygiene and its centers of activity. At the present writing, governmental activity in industrial hygiene is primarily centered in the organizational structure of the Occupational Safety and Health Administration of the U.S. Department of Labor, and in the National Institute for Occupational Safety and Health of the U.S. Department of Health, Education and Welfare. Although quite a few of the state units are attempting to retain their industrial hygiene identity, the continued existence of most state units is problematical for many reasons, including lack of financing, lack of personnel, and lack of legislative interest. The federal industrial hygiene activities are divided into two broad objectives; inspection and compliance activities of OSHA, and the research and investigational activities of NIOSH. Each of these organizations employs about the same number of industrial hygienists or a total of about 150 technically trained personnel.

Unquestionably, the largest current employer of industrial hygienists is private industry, and its need for manpower will continue to grow to satisfy compliance with OSHA regulations on physical and chemical hazards of the work place. The number of industrial hygienists required for an individual plant or company will depend on a number of factors: the number of employees, the number of different chemical agents and physical factors providing opportunity for exposure, the number of such physical or chemical agents that may require periodic evaluation during the year, and the physical dimensions of the plant. Plants with 350 to 500 employees who may have opportunity for exposure to three or four agents or factors requiring semiannual evaluation may be able to utilize one professional industrial hygienist on a full-time basis. Larger plants may require a proportionate number of professional industrial hygiene personnel, and one ought to be experienced in the analysis of industrial hygiene air samples. Multiplant corporations may have an industrial hygiene staff of six or more professional persons providing services from a central location, or they may be located in selected company regions from which they can best serve their employers' needs.

Insurance carriers each employ at least one industrial hygienist and usually several, located in regional areas, to advise their clients on the hazards of various chemical agents and to provide minimum services. A client with 50 to 150 workers usually finds it necessary to employ a private consultant to obtain adequate evaluation of exposures in his establishment.

A number of competent, private consulting organizations devoted to industrial hygiene and air pollution activities have been in existence for 20 or more years. These organizations are generally staffed with professional industrial hygienists of long experience who carry out industrial hygiene evaluations and may supervise the work of younger, less experienced technical people. Consultants in this category provide complete in-house services and are prepared to undertake a job of any magnitude. Newer organizations, equally competent in industrial hygiene, have been started in more recent times and may not be as well known in their field. Since 1970, the number of industrial hygiene consultants has mushroomed. Many are retired professional industrial hygienists who operate privately and obtain analytical services from special consulting laboratories. Among this large group of capable consultants are a few organizations with little knowledge of industrial hygiene and who lack professional abilities and judgment. Employers seeking to utilize the services of consultants for industrial hygiene purposes should inquire into their qualifications just as they would ask about the qualifications of a product research organization or other types of management service.

3.2 Administrative Location in Organizational Structure

The administrative location of an industrial hygiene unit in the organizational structure of an establishment can exert a marked influence on its effectiveness and accomplishments. Private firms, entirely engaged in providing consulting services in industrial hygiene, air and water pollution control, and similar environmental matters, integrate such functions into a cohesive body, to utilize all available talents. In governmental units, industrial hygiene has attained equal status with other preventive health and safety functions that are necessary in today's society. The activity and effectiveness of such units depends on the needs of the public and the support of the legislative entity. At this writing, governmental activity in industrial hygiene is rapidly becoming centralized in the federal sphere, with laboratory and field research located in NIOSH, whereas authority for the framing of regulatory standards and enforcement activities resides in OSHA.

The industrial hygiene functions in industry show the widest variation in their location within the organizational structure, and thus is reflected in their functioning. Some units may report directly to the head of manufacturing and may or may not be combined with safety and fire protection. Frequently such a unit serves no function beyond monitoring employee exposures for in-company compliance purposes. The industrial hygiene function may be located in the central engineering department of a large establishment, and its obvious focus of attention is on measuring the effectiveness of control systems. In large corporations, especially those with multiplant activities, industrial hygiene generally is structured within the corporate headquarters. However branch units may be found in individual large plants or may be located to serve a number of smaller plants in regional areas of the corporation. Even the research departments of several large corporations have successfully accommodated the centralized industrial hygiene functions of their company. Among the corporate industrial hygiene units that have been operating for 15 years and longer, units tend to be located in the medical department of the establishment. Although there are honest differences of opinion regarding the most desirable administrative location of the industrial hygiene unit, the success of the large number of such units in medical departments appears to favor this location. It provides flexibility in functions, helps bridge normal departmental barriers, and provides access for communicating with top management. Functionally, this location permits the unit to engage in comprehensive medical–industrial hygiene studies, environmental surveys, training of plant personnel in monitoring techniques, research in sampling and measurement of exposures, and other activities. Traditionally, the physician who is trained and experienced in occupational medicine considers the industrial hygiene unit an important and equal partner in the integrated occupational health team.

4 QUALIFICATIONS AND TRAINING OF PERSONNEL

4.1 Director of Industrial Hygiene

The qualifications for the administrator of an industrial hygiene unit represent a personal opinion molded from observations and experience of numerous persons who have served in this capacity in the field of industrial hygiene during the last 25 years. The director should have at least a master's degree from an accredited college or university, and a basic education in science, preferably in physical chemistry or chemical engineering. His graduate training should include courses in physiology, toxicology, biochemistry, ventilation and air conditioning, and the practical application of statistical parameters to numbers obtained in scientific measurements. Certification as a diplomate by the American Board of Industrial Hygiene should be a minimum requirement for the occupant of administrator's position. A prerequisite to certification is 5 years of practical field experience followed by the passing of a comprehensive written examination. Although these are the basic requirements, the administrator should have at least 10 additional years of broad experience in industry, including involvement with the type of manufacturing over which he will maintain environmental supervision. In the process industries, the absence of such experience would be a serious obstacle to successful performance of the administrative function.

The director should be able to understand and discuss all phases of industrial hygiene and toxicology and should be interested in public health and industrial relations. He should be able to set up systems for keeping records of results achieved and for planning future activities. He should be an avid reader and should devise a systemic bibliography of references covering the field of his interests and activities. He should be able to locate the cause of cases of industrial ill health presented by a physician, to make comprehensive analyses of health conditions in industries, to draw adequate conclusions, and to prepare clear and informative reports for publication. Tact, good judgment, and courage of personal convictions are also important assets. He must do sufficient field work to keep the viewpoint of the man in the field, and it is necessary to maintain a wide personal acquaintance with the men and women active in promoting industrial health. The administrator should support and attend public meetings where current problems in the field are discussed, and he should develop the ability to address a group of peers as well as the public.

4.2 Associate Industrial Hygienists

The industrial hygienists reporting to the director should have received basic education in the scientific disciplines similar to that of the administrator. A degree in chemistry or chemical engineering is desirable, and a degree in mechanical engineering provides greater knowledge in the disciplines required for designing control measures. A graduate degree that includes training in illumination, ventilation and air conditioning, energy conservation, principles of air sampling, and the toxicology of the commonly used industrial chemicals, is essential. If the unit is so large that one of the industrial hygienists acts in the capacity of a supervisor, this person should be certified by the American Board of Industrial Hygiene or should possess sufficient experience to command respect of the junior members.

Depending on their individual experience, the associates must be able to recognize and evaluate a potentially harmful situation, to judge the necessity for carrying out a specific study, and of course they must have the ability to conduct a survey. The knowledge required for developing a control measure will vary directly with experience and the type of basic education. However one of the most important abilities required of each industrial hygienist is that of preparing a clear report of the survey findings, bearing a definitive conclusion. Since so much of the industrial hygienist's time is spent in studying plant situations, he must have the ability to communicate his ideas and thoughts to production personnel, safety engineers, maintenance supervisors, and plant physicians.

4.3 Other Unit Personnel

To operate effectively, an industrial hygiene unit may find it desirable to add such specialists to its staff as an industrial toxicologist, an industrial hygiene analytical chemist, a ventilation engineer, a statistician, a health physicist, and perhaps survey monitors. Each one will have a personal set of qualifications, and each will make special contributions to the output of the unit. Occasionally, a person may start out as a specialist and by close association broaden the scope of his activities and interests, to be able to properly claim the title of industrial hygienist.

Many young persons are now being graduated from recognized colleges with bachelor's degrees in environmental health, air and water conservation, ecology, and other general fields relating to the biosphere in which man lives. On examining numerous employment applications of persons with such degrees, one finds that their studies have emphasized the social aspects of environmental conservation with relatively brief exposure to the basic sciences. These young persons can be trained to carry out routine procedures such as exposure monitoring, sample collection, and other uncomplicated assignments as technicians or scientific aides. It could be a serious error for an employer to place the responsibility of a new industrial hygiene program in the hands of an inexperienced person on the basis of a degree with an attractive sounding title. Helpful advice and assistance in seeking competent industrial hygienists can be obtained from the office of the manager of the American Industrial Hygiene Association (8).

REFERENCES

1. Public Law 91-596, 91st Congress, S.2193, December 29, 1970.
2. B. Ramazzini, *De Morbis Artificum Diatriba,* 1700 and 1713.
3. *Health of Workers in a Portland Cement Plant,* Public Health Service Bulletin No. 176, U.S. Government Printing Office, Washington, D.C., 1926.
4. "Exposure to Siliceous Dust in the Granite Industry," Public Health Service Bulletin No. 187, U.S. Government Printing Office, Washington, D.C., 1929.
5. "Lead Poisoning in a Storage Battery Plant," Public Health Service Bulletin No. 205, U.S. Government Printing Office, Washington, D.C., 1933.
6. "Mercurialism and Its Control in the Felt-Hat Industry," Public Health Bulletin No. 263, U.S. Government Printing Office, Washington, D.C., 1941.
7. "Effect of Exposure to Known Concentrations of Carbon Monoxide," *J. Am. Med. Assoc.* **118,** 585–588 (1942).
8. American Industrial Hygiene Association, 66 S. Miller Road, Akron, Ohio 44313.

CHAPTER FIVE

Epidemiology

JOHN F. GAMBLE, Ph.D. AND
MARIO C. BATTIGELLI, M.D.

1 DEFINITION, AIMS, AND USES OF EPIDEMIOLOGY IN THE INDUSTRIAL SETTING

Epidemiology is the study of the configuration, distribution, and determinants of health and its disorders. In its applications to disorders from environmental determinants, epidemiology has spanned over its entire range of activities and methods. New entities have been described through its inquiries (*descriptive epidemiology*), and experiments have been conducted to evaluate the importance and/or effectiveness of a variety of factors (*interventional or experimental epidemiology*). No doubt the cause-effect relationship has been one of the most common objectives of epidemiological studies.

In defining etiological agents, a measure of exposure is needed to identify the risks a population is facing. With this information, appropriate health measurements are searched to permit the establishment of dose–response curves. It is on this base that the environment can be controlled and the worker's health protected. However achieving these goals has met traditionally with great difficulties. Measurements both of health outcome and exposure are intimately intermixed with innumerable confounding factors. Quite often jobs in industry involve exposure to a variety of chemicals, and fluctuation of intensity of exposure is always wide. Questions of changing technology, of internal migration, of limited or nonexistent records, of past health and environmental exposure, of inconsistent terminology of occupational position, and many more, multiply these difficulties. The classic question of Ramazzini, asking a patient about his work, finds neither simple nor satisfying answers. The whole problem of measuring health or health derangement remains to date a baffling, expensive, and universally unfinished task.

1.1 Descriptive Epidemiology

The studies relating observed effect to significant determinants are commonly grouped within the so-called *etiological epidemiology*. Recent legislative decisions have greatly increased the interest and the need for toxicological information on factors ranging from technological additives to pesticides and other contaminants of soil, air, and water. The formulation of threshold limit values, for instance, depends critically on epidemiological information. If the main role of occupational epidemiology is to establish meaningful dose–response relations, a wide field of applications remains fully open for its uses. In view of the multifactorial causes and nature of occupational diseases, and of the complexity of the industrial environment, occupational epidemiology is especially useful for these applications. Obviously the ultimate task of such a function is to allow the identification and implementation of a safe working environment.

Historically, the description of extent and distribution of health outcome in different populations was the earliest method employed. The most common descriptive variables relate to the following:

1. Time (year, season, day of week, etc.). The periodicity of the byssinosis "Monday" recurrence is a fitting example.
2. Place (county, state, urban-rural, department, technology, etc.). Lung cancer, associated with asbestos exposure, occurs more commonly in the areas with high cancer rates.
3. Person (age, sex, race, education, marital status, family history, smoking history, economic status, etc.). Chronic bronchitis, for instance, is more common in the wives of coal miners than in the wives of other workers.

Demography is obviously part of descriptive epidemiology, or vice versa. It is probably as old as recorded history, with enumerations of population reported in the earliest records, which date in the Chinese and Mesopotamic archaeology to 2500 B.C. In the United States, the first organized census was initiated in 1790, and it has been updated every 10 years since then.

1.2 Analytical Epidemiology

Indeed, epidemiological studies have received particular impulse and popular applications in recent years with the renewed interest in occupational toxicology. Their most probing procedures may be conveniently grouped under the term "*analytical epidemiology*," a branch making wide use of biostatistical methodologies.

1.3 Sample and Estimates

The conclusions derived from the study of a subset of a population (*sample*) may be extended and applied to the parent population (*universe*) within specified constraints (*confidence limits*). Central to the purview of analytical epidemiology is the question of inference. Any measurement, however carefully obtained and precisely calibrated, reflects dimension(s) of a fraction of the original population, limited in number, geography, or time. This is a sample. To extend these dimensions and measurements to the parent population, *inferential analysis* is needed to estimate the characteristics of the related universe.

1.4 Variables and Estimates

The inference process, in analytical epidemiology as in biostatistics, includes two fundamental steps, namely, the estimation of variables and the formulation of a hypothesis to be tested. The sample variables are accepted, if adequate, as the estimates of the corresponding parameters of the parent population. Measurements of any sample variable (point estimates) need to be defined in their quantitative credibility (confidence interval). This interval states the boundaries enclosing, within a range, probabilistically, the corresponding parent population parameter.

1.5 Hypothesis

The prime task of analytical epidemiology coincides with the test of a hypothesis. In its most elementary mode, the test contrasts two conditions or samples, distinguishable by one or more defined variables, and attempts to accept or reject the *null hypothesis*—namely, that the difference is due to chance (sampling), not to differing parent populations. A hypothesis can be ranked in order of merit by the following items: (1) accuracy of its prediction, (2) number of observations explained, (3) verification by alternative tests, and (4) logical connection with previously unrelated phenomena.

2 METHODS AND PROCEDURES OF EPIDEMIOLOGY

The most common situation studied by an epidemiological inquiry is a two-variable relationship. An *independent variable* (potential cause) is tested for its association with a characteristic or *dependent variable* (i.e., effect), or a group of independent variables are tested in their relation to one or more dependent characteristics.

2.1 Study Modes: Cohort, Cross Section, and Case Study

The selection of one or another or both types of variables in the definition of a group of individuals generates three distinct plans of study. The study mode that defines the independent variable and follows a group of individuals to examine the *incidence* (attack rate) of a condition(s) belongs to the *cohort type*. The definition of a disease (or condition) in a group of individuals contrasted to a similar group of nonaffected individuals (noncases, or controls), for the purpose of tracing back the significant "causative" factors, labels this plan a *case-control* mode. Finally, a definition of a group in terms of both disease (dependent) and characteristic (independent) variables qualifies a *prevalence* or *cross-sectional* study.

The terms "prospective" and "retrospective," used in the past to distinguish cohort from case-control studies, are not entirely satisfactory, since a cohort could be studied historically, therefore retrospectively. In the retrospective case-control study, one selects persons "with" (cases) and "without" (control) the health outcome of interest, and compares their exposure history. In the forward-looking prospective cohort study, all individuals without the disease but with known exposures are selected. The health experiences of the exposed and the nonexposed are compared. If records are adequate to establish a cohort in the past, a historical prospective study can reduce the waiting time for contemporary investigators. It is only in a prospective study that cause-effect relations can be definitely assessed.

A case-control study has the disadvantage that the population at risk from which the cases are drawn is not known. (It also suffers from the same problem as the prevalence study—it is not known whether the disease or the exposure came first.) The prospective study often requires a long time for the development of a number of cases sufficient to permit the assessment of the effect of exposure.

A prevalence study examines part or all of a population existing at some point in time. Though quick and inexpensive, it is merely the examination of a population of survivors. Comparisons are not possible concerning cause-effect relations because one does not know whether the disease or the exposure came first. Cross-sectional studies are usually descriptive morbidity studies. They provide information on the prevalence of the disease; and they can identify cases for a case-control study, and provide the cohort for a prospective study.

The essential comparison made in these studies can be represented in a 2 × 2 contingency table (Table 5.1), and the relative merits are summarized in Table 5.2. In a cohort study the population is selected on the basis of the independent variable and the incidence of the dependent variable is compared. In a case-control study the population is selected on the basis of the dependent variable and the presence of the exposure in cases and controls is compared.

In the cohort study individuals with and without exposure (independent variable) are entered in the study when free of the disease (dependent variable). At the conclusion of the observation period, the same population is classified as illustrated in Table 5.1, according to the presence (a and c) or absence (b and d) of the disease. The *incidence* of a phenomenon (disease) occurring (beginning) in the exposed population, during a given interval, is expressed by the ratio $a/(a \text{ and } b)$, and it may be conveniently compared to the incidence in the nonexposed $c/(c \text{ and } d)$.

Table 5.1. Contingency Table

Independent Variable (e.g., Exposure)	Dependent Variable (e.g., Health Outcome)	
	Present	Absent
Present	a	b
Absent	c	d

Table 5.2. Comparison of Three Types of Epidemiological Study

Methodological Attributes	Type of Study		
	Cohort	Case-Control	Cross-Sectional
Initial classification	Exposure-nonexposure	Disease-nondisease	Either one
Time sequence	Prospective	Retrospective	Present time
Sample composition	Nondiseased individuals	Cases and controls	Survivors
Comparison	Proportion of exposed with disease	Proportion of cases with exposure	Either one
Rates	Incidence	Fractional (%)	Prevalence
Risk index	Relative risk-attributable risk	Relative odds	Prevalence
Advantages	Lack of bias in exposure; yields incidence and risk rates	Inexpensive; small number of subjects; rapid results; suitable for rare diseases; no attrition	Quick results
Disadvantages	Large number of subjects required; long follow-up; attrition; change in time of criteria and methods; costly; inadequate for rare diseases	Incomplete information; biased recall; problem in selecting control and matching; yields only relative risk—cannot establish causation; population of survivors	Cannot establish causation (antecedent consequence); population of survivors; not good for rare diseases

In the cross-sectional study, *prevalence* of the dependent variable (phenomenon, disease, etc.) is computed in terms of number of cases existing ("active") at one time or through a given interval or duration.

It appears obvious that incidence and prevalence may readily differ depending on length of a disorder "activity" (e.g., duration) and length of observational interval.

2.2 Case-Control Studies

The case-control study follows a design similar to that indicated in Table 5.1 for cohorts. Here the frequencies of the independent variable in the two groups are compared. A higher frequency of cases in the exposed group suggests an association of exposure with the disease. The measure of this association is the relative *odds ratio* introduced by Cornfield (1). This is defined as the proportion of exposed individuals with the disease divided by the proportion of nonexposed individuals with the disease. By approximation in situation of disorders of uncommon occurrences (e.g., equal or less than 5 percent), this ratio is obtained by taking the quotient of the two cross products (ad/bc).

2.3 Selection of Control Group

It is obvious that in an epidemiological investigation, as well as in any endeavor requiring measurement, a reference or paragon unit is needed. The definition of an adequate reference against which the test population must be compared is a critical task. A reference may be the expected value for a given function, the analogous measurement (i.e., frequency of a disorder) observed in a parallel population, or a baseline value observed on the test population before the independent variable "acts." Methodology of sampling, of measurement, of comparative adequacy, of adjustment (matching, normalizing, etc.), are necessary procedures to be implemented as the case applies, to obtain appropriate and valid "controls."

2.4 Sampling: Size

For the sake of economy, it is seldom possible and never convenient to examine the entire population. Selecting a representative subset of a population is called sampling. A sample adequately represents a parent population whenever it satisfies numerically and qualitatively certain requirements. Size and representativeness are basic qualifications of an adequate sample. Size needs to be adequate for different reasons. The nature of a measurement, whether involving a continuous or a discrete variable, affects the numerical adequacy.

The repetition of any quantitative observation produces a certain variability or spread of measurements. This spread is particularly obvious in measuring a continuous variable, that is, a dimension characterized by infinite decimal. In this case a measurement error is, by definition, inevitable. The error may be minimized, but it cannot be eliminated entirely. The reproduction of a measurement within a narrow interval verifies the precision of it (*measurement adequacy*). On the other hand, in case of a discrete variable (i.e., count of living members of a population, etc.), the sample size matters for entirely different reasons. Here the size is adequate when it offers an even chance for the composition of the parent population to be reflected by the composition of the sample (*sample adequacy*). For instance, if the parent population is composed of several different groups, a sample of fewer units would be quite inadequate.

Enlarging of the sample size is the most direct approach to reduce variability, since the variance σ^2/N bears inverse proportionality to size N. The selection of an appropriate size is provided by a variety of approaches well described in the literature (2).

2.5 Sampling: Representativeness

Selection procedures in defining a subset of a population need to be carefully identified, in view of common and important errors associated with this technique.

We have just seen how size per se may directly limit representation. Additional and more complex considerations of adequacy must be followed in selecting appropriate samples. A common device to minimize artifactual distortion in selection of a sample has been traditionally provided by *randomization* (selection at random), *stratification* (adjustment for variables), and *blinding* (assignment to test by chance). "Stratified random" and "cluster random, multistage sample" are combination varieties of these procedures that have received successful applications.

3 ADJUSTMENT PROCEDURES FOR COMMON CONFOUNDING FACTORS

The analysis of significant groups or samples for comparative purposes calls for control of confounding factors. In fact, nonspecific etiological factors other than the variable(s) of interest may be associated with both the disease (or health outcome) and the specific independent variable. The careful selection of a sample certainly provides the most important means of control of spurious association in the comparison. The adjustment for age composition is one of the common corrections applied in the comparison of a variety of indices. These adjustments form the central methodologic issue in epidemiologic research. Age adjustment, smoking adjustment, and other processes of standardization are done not only to make the contrasted populations more homogeneous, therefore of more comparable quality, but also to control the independent effect of these nonspecific variables on health outcome. Uncontrolled, these variables confound our ability to seek alternative and possibly more valid explanations for differences noted in disease frequency between populations.

To assess the risk of an occupational exposure, it is best to compare a population at risk (PAR) with a control group of workers similar in every respect to the group under study except for exposure to the agent of interest. This premise is probably impossible to fulfill. For example, if the agent of interest is a given dust, which produces symptoms that result in a self-selection process biasing the sample, it is not easy to get a comparison (effect-free) group undergoing the same selection process. The best possible effort is needed to obtain the most realistic comparison group.

Recognizing that age is a powerful factor of mortality (and of morbidity), the correction for age of any index of health effect is mandatory in the interpretation of any comparative study. The standardized mortality (or morbidity) rate (SMR), for example, is a common example of the application of this correction.

The problem of adjustment for age is not always solved by an easy and straightforward process. For instance, in expressing a datum of pulmonary function as related to predicted value (e.g., percentage of expected), adjustment may engender the false security of an ideal computation. A discrepancy from the expected or predicted value, however, may be all too precipitously related to the exposure variable. One forgets, in this context, that the expected value is actually a single point of a range defined by the regression equation and is usually characterized by a substantial standard error of the mean. For instance if a male age 60, 6 feet high, has an expected vital capacity of 4 liters, the actual range is ±15 percent, and a range of values rather than a single value should be kept in mind to reach a reasonable interpretation of his expected value.

Furthermore, the expected value may not apply to the subject tested if the latter had additional etiologically important characteristics (e.g., cigarette smoking). The predicted value, in fact, was most probably derived from a population of nonsmokers or mixed smokers. Smokers' expected lung function values for example, are generally less than nonsmokers. Therefore to assess an effect of occupational exposure on lung function one must adjust for the effect of smoking by comparing the lung function of occupationally exposed smokers with nonexposed smokers, exposed nonsmokers with nonexposed nonsmokers, and so on.

Enterline has recently commented on common confounding factors that may limit a comparability assessment on matter of exposure (3). The limitations associated with an assessment of dose are probably the most disturbing factors in reaching a definite dose–response relationship. The levels of exposure to an agent are more often available in qualitative rather than quantitative terms. The appropriate selection of reference cases (*matching*) greatly assists in obtaining meaningful comparison between diverging control-test variables (4).

3.1 Age Adjustment

To make a meaningful comparison, two contrasted entities (groups, ratio, rates, etc.) must be congruous. One of the most common factors of discrepancy which has fouled quite a few observers, is certainly that of difference in age composition of two groups.

To reach a stage of realistic and useful contrast, populations to be compared must be made compatible, or as stated in the technical jargon, they must be age adjusted. Several methods are available. To illustrate the details of them the reader is referred to Reference 2. In brief, two alternative methods are available, the direct and the indirect.

3.1.1 Direct Method

In the direct method the age rates obtained from the study group are used to generate an "expected" or hypothetical number of events (death, illnesses) to be applied to the reference (control) population. This is realized by multiplying these age-specific rates of the study population with the number of individuals in the respective age group of the reference population. The hypothetical deaths are summed, and the total is divided by the total population of the standard. A standardized rate is therefore obtained. Table 5.3 reports an example of this procedure.

3.1.2 Indirect Method

The crude rates for each age group may be available from the standard (reference) population. In the indirect method, these are applied to the study population, classified by each age group. The expected age-specific mortality rate is obtained in this fashion. The ratio of observed to expected ratio for each age group is weighed therefrom by the crude (overall) rate (i.e., mortality) of the standard population, to obtain the *standard age-adjusted rate.* Again let it be stated that since the reference to "standard rates" is fraught with danger of limited comparability, the assumptions considered in drawing such an adjustment should be clearly stated.

A fixed relationship exists between the (direct) SMR and indirect ratio IR according to Chiazze (5):

$$\text{SMR} = IR \frac{1}{Ms} \tag{1}$$

where Ms is the crude mortality rate in the standard population.

Table 5.3. Direct Age Adjustment: Chronic Airways Disorder (Bronchitis)

Age Group	Observed Prevalence (%)	Number of Individuals in Reference Population	Calculated Events in PAR
30–39	5	100	5
40–49	10	200	20
50–59	15	100	15
60–69	20	200	40
		600	80

Age-adjusted rate = 80/600 = 13.3%

The SMR index, though providing an informative summary of the mortality force, cannot be interpreted without important qualifications. First of all, the SMR cannot be read as an indicator of population longevity. In fact, populations with different life expectancies may have identical SMRs, and vice versa. It is important to restate that the adjustment applied in formulating the standard mortality rate is merely one of age; many other confounding factors remain after age standardization is applied. Since standard mortality rates are statements of probability, the legitimacy of their application resides in the characteristics differentiating observed from standard populations. If the correspondence in such characteristics is limited or inappropriate, the assessment will be intrinsically biased, and the possibility that it may be worthless or misleading is proportionally increased.

An additional and important limitation of the SMR has been well illustrated by Gaffey (6), namely, that the magnitude of this index bears no direct relationship to the magnitude of the risk, and at older ages the SMR has limited value. Furthermore, the SMRs of two different populations are not directly comparable as a different standard population is used for adjustment.

Table 5.4 Rate, Ratios, and Proportions

$$\text{annual crude death rate} = \frac{\text{total number of deaths during year (Jan. 1–Dec. 31)}}{\text{total population as of July 1}} \cdot 1000$$

$$\text{specific death rates (annual)} = \frac{\text{total number of deaths in a specific group during year}}{\text{total population in the specific group as of July 1}} \cdot 1000$$

adjusted or standardized death rates = rate adjusted to reference population
(See Table 5.1., also applies to birthrate, attack rates, morbidity rates, etc.)

$$\text{cause-of-death ratio} = \frac{\text{number of deaths due to a specific disease during year}}{\text{total number of deaths due to all causes during year}} \cdot 100$$

$$\text{proportional mortality rate} = \frac{\text{number of deaths of persons 50 years of age and older}}{\text{total number of deaths}} \cdot 100$$

$$\text{attack or incidence rate} = \frac{\text{total number of new cases of a specific disease during year}}{\text{total population as of July 1}} \cdot 10^4 \text{ or } 10^5 \text{ or } 10^6$$

$$\text{(instantaneous) prevalence rate} = \frac{\text{total number of cases, new or old, existing as of an instant}}{\text{total population as of that instant}} \cdot 100{,}000$$

$$\text{case-load prevalence rate} = \frac{\text{total number of cases, new or old, occurring during an interval in time (e.g., day, week, month)}}{\text{average population during that time interval} \quad 1} \cdot 100{,}000$$

$$\text{case-fatality ratio} = \frac{\text{total number of deaths due to a disease}}{\text{total number of cases due to the disease}} \cdot 100 =$$

$$\frac{\text{number of deaths due to disease}}{\text{total population as of July 1}} \cdot \frac{\text{total population as of July 1}}{\text{number of new cases of disease}}$$

The mortality rate is just an example of a vast number of quotients, relating a significant numerator (i.e., death) to a denominator (i.e., population at risk to die). Table 5.4 reports a list of common rates used in epidemiological measurements. Note that in actuarial studies an *attrition rate* is often computed with the denominator corrected to a hypothetical midyear by subtracting from the population at risk at the beginning of the period half the deaths occurring in that period.

There is no doubt that mortality indices have the unequivocal attributes of a final event and provide powerful assessments of effects. Even today these measurements continue to dominate measures of health outcome in environmental health. However the value of this information varies widely in its accuracy in identifying cause and effect relations. Mortality rates cannot satisfy the demands for measurement of ill health or health effects sought by contemporary industrial society.

3.2 Life Tables

A life table is a statistical instrument expressing the experience (i.e., mortality) of a cohort, followed during a definite interval of time. Here the adjustment is brought to a maximum because rate and indices are generated by each cohort studied. This is probably the most sophisticated study of a cohort.

In brief, a life table consists of the following basic information: (1) number of subjects alive or at risk at any given interval O_x, (2) deaths during any given interval d_x, and (3) losses (other than deaths) during same intervals W_x. The computation provides the following indices: (4) probability of dying q_x; ratio of number of death d_x to the number of subjects at risk O_x at beginning of period x, corrected by half the losses other than death W_x, (5) death rate m_x, (6) chance of surviving $(1 - q_x = p_x)$, (7) cumulative chance of surviving, and (8) standard error of cumulative chance.

The relationship between the probability to die q_x and the death rate m_x (= D/P, where D is the number of deaths and P is the population at risk) is well portrayed by the following relationship:

from $m_x = \frac{D}{P}$ (per year), $D = Pm_x$*

Substituting in

$$q_x = \frac{D}{P + \frac{1}{2}D} \tag{2}$$

one obtains

$$q_x = \frac{m_x}{1 + \frac{1}{2}m_x} = \frac{2m_x}{2 + m_x}$$

Again it should be noted that the specific denominator in q_x (population plus half the deaths) is used to approximate a more accurate rate. The assumption accepted here is

* In homogeneous population (i.e. indentical sex, age, etc.).

that the deaths observed during a given period are evenly distributed, therefore occur in equal number over the two halves of that period (e.g., a year).

3.3 Ratio, Rates, and Proportions

Traditionally, proportions and ratios of proportions have formed the core of the information provided by epidemiological studies. Rates and ratios have been widely used, often interchangeably. Both allude to quotients resulting from two specified quantities. Elandt-Johnson has recently drawn a separation between these terms (7). Ratios relate to the quotient between separate and distinct quantities that are mutually exclusive (e.g., sex ratio, number of females versus number of males). Rates, on the contrary, reflect change of quantity per unit of another quantity on which it depends. Rate is erroneously confused with proportion, which is a quotient whose numerator is included in the denominator (e.g., number of females in the population).

3.4 Risks

The incidence of a health phenomenon or, as the early British epidemiologists used to say, the attack rate, carries per se a connotation of risk. Like many proportions, it summarizes the expected frequency or probability that a given phenomenon will occur. The indices of mortality, a direct offshoot of incidence in fact carry a message of probability, hence risk. Many other proportions and ratios have been proposed and used in epidemiology within similar purview. The risk ratio (RR) is the quotient of two proportions ("rates"), usually the frequencies of a given phenomenon in the test and control populations, respectively. When this ratio has been obtained from standardized quantities, a standardized risk ratio (SRR) is the result (8).

Since cases may occur in nonexposed populations, it is legitimate to ask what fraction of cases observed in the exposed population is attributable to the exposure itself. The attributable risk is defined as the difference in incidence between exposed and nonexposed divided by total incidence among exposed individuals (2). This central measure in epidemiologic research has also been called the etiologic fraction. Table 5.5 defines some of the most common risk indices in current use.

4 QUALITY OF MEASUREMENT

Measurement is the cornerstone of epidemiology, and the opportunity for error is vast. To discuss some of the major pitfalls it is convenient to partition instances of error into three categories: statistical, diagnostic, and interpretive error.

4.1 Statistical Error

The main purpose of statistical design and analysis is the calibration of error. The precision of a measurement indicates the repeatability of an observation and its related

Table 5.5. Risk Indices (9)

Relative risk, R (odds ratio, risk ratio, cross-product ratio): the ratio of the incidence rate in the population exposed to the risk factor to the incidence rate in the population not exposed.

Attributable risk, AR_e (simple difference): The rate of disease in individuals exposed to the risk factor that is attributed to the exposure.

Attributable risk percentage, $AR_e\%$ (etiologic fraction among exposed): the proportion of the disease rate in the exposed population which is attributable to exposure to the risk factor.

Relative difference, D_e: the relative excess risk associated with exposure to a risk factor among persons who could have been affected by the risk factor in question.

Population attributable risk, AR_p: the rate of disease in the entire population that can be attributed to exposure to the risk factor.

Population attributable risk percentage, $AR_p\%$ (etiologic fraction in population): the proportion of the disease rate in the total population which is attributable to exposure to the risk factor.

Prevented fraction: the proportion of the potential total load of a disease in a population which is prevented by a protective factor (e.g., intervention program) and/or factors associated with it.

spread. The coefficient of variation (standard deviation/mean) is a popular assessment of it. Accuracy—namely, proximity of a measurement to its true value—is controlled by statistical analysis only indirectly, since it requires the independent verification of a true value (God's opinion of early epidemiology).

Error due to selection bias (see Chapter 17 on sampling) may also be discussed under statistics, since the statistical design of a study critically depends on the parallel derivation of both test and control samples. Validity is also a statistical function in that it relates a measurement procedure to the congruity of the dimension to be assessed. Large areas of overlap obviously exists between statistical, diagnostic, and interpretive errors.

4.2 Diagnostic and Classification Error

Any laboratory procedure or measurement assessing a variable in man has a classification value, therefore a diagnostic connotation. Diagnosis and classification, in this context, are terms of interconnected meaning. Most powerfully, diagnostic criteria and classification schemes affect the validity, therefore the statistics and the interpretation, of epidemiological data. A good example in point is provided by the reading of abnormality in chest X-ray films.

Observer's error, specificity, and sensitivity of diagnostic information are important parameters to be used in gauging the validity of a diagnostic decision (10). Indeed, quite often errors stem from inconsistent criteria of classification and from the inadequacy of diagnostic instrumentation. A classic example of classification inconsistency is illustrated in Figure 5.1, describing the discrepancy in prevalence of chronic bronchitis when the syndrome is assessed by nonuniform criteria (11). The prevalence in any age group of a common disease (bronchitis) may appear to vary greatly, when the actual source of this difference is the divergence in diagnostic criteria. For example, the entity called "chronic bronchitis syndrome" is defined by the presence of productive cough, associated with

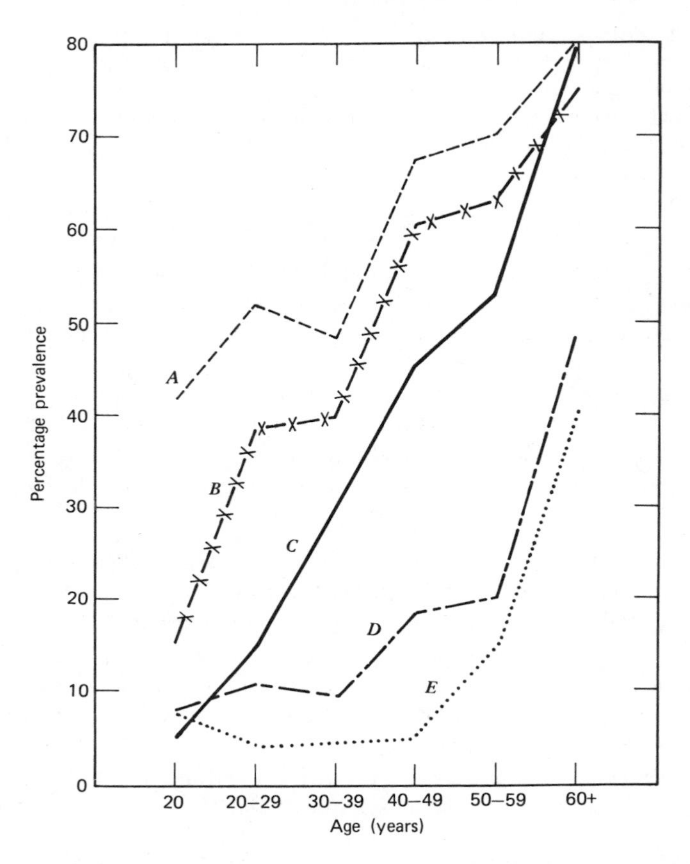

Figure 5.1 Effect of alternative diagnostic criteria on prevalence of chronic bronchitis. Modified from Ashford (11).

dyspnea on effort and episodes of pneumonia in the past. The entity called "simple chronic bronchitis" is defined by the presence of morning cough in winter months. The discrepancy possible in classifying the same patient by these criteria is quite obvious.

Errors associated with diagnostic tools and diagnostic procedural inadequacy are epitomized by X-ray assessment of minimal pulmonary tuberculosis (12). The validity of diagnostic tests has been extensively analyzed in the literature, with most noteworthy contributions by British epidemiologists. The requirements of a good test are described by Fletcher (13) in the following items: validity (correspondence between measurement and reality), simplicity, repeatability, and discrimination. "Discrimination" is the ratio of the mean difference of two groups (affected and nonaffected) to the common variance (scatter of the values), that is, the efficiency of a test in distinguishing between those with the characteristic and those without it.

Student's test has been used to evaluate the discriminative effectiveness (difference in mean values between affected and normal populations) of various spirometric measurements (13). This particular approach probably should receive wider attention in evaluating health measurements. The selection of cutoff limits, using continuous measurement of a variable spanning normal and abnormal ranges, requires considerations of *specificity* and *sensitivity.**

A final and not negligible diagnostic source of error is identified in the description of the natural history of an effect. In chronic diseases and in disease when the evolution of the subclinical process is prolonged in duration (i.e., neoplasia), the latency period is a common source of error. When the beginning of exposure is not known or reported, cause and effect relationships are often equivocal, and derived conclusion may appear contradictory. For instance, a carcinogenic dose–effect proportionality may be missed entirely (3).

4.3 Errors of Interpretation in Epidemiological Measurements

In his presidential address to the Royal Society of Medicine, Sir Austin Bradford Hill asked "How in the first place do we detect these relationships between sickness, injury, and conditions of work?" (14). His reply outlined nine rules of evidence, which are listed in Table 5.6.

The strength of *association* obviously is largely a probabilistic consideration, not often settled by tests of statistical significance. Questions of significance must be answered always with caution, in view of the tremendous variability of many biological events and their measurements.

Repeatability or *consistency* has been recognized to be fundamental in any measuring process, biological as well as statistical. If a factor of exposure has realistic etiological importance, its effects should be observable by alternative measurements by independent observers, and they should be found in unrelated populations. Ideally consistency should be observable in terms of the dose–effect relationship, so that the demonstration of a *biological gradient* strengthens the evidence of a cause and effect relationship.

Plausibility and coherence are part and parcel of any biological association whenever causal factors are etiologically related to a given event.

Verification by experiment, though most desirable and well sanctioned since Koch's postulates, is not always readily available. Therefore arguments by analogy provide additional indirect verification. The significance of specificity and temporality hardly needs to be emphasized.

4.3.1 Assessing the Environment (Dose)

The concept of dose, both in pharmacological and environmental analysis, rests on the assessment of intensity and duration. The latter is usually more easily measured,

* "Sensitivity" refers to the proportion correctly identified positive; "specificity" refers to the proportion correctly identified negative.

Table 5.6. Bradford Hill's Rules of Evidence of Cause and Effect Relationships

1. Strength of association.
2. Consistency (reproducibility in time and space).
3. Specificity (uniqueness in quality or quantity).
4. Temporality (congruous sequency in time).
5. Biological gradient (proportionality between exposure and outcome).
6. Plausibility (biological possibility).
7. Coherence (biological compatibility).
8. Experimental verification (biological modeling).
9. Analogy (biological extrapolation).

whereas the former often remains a matter of speculation. The scarcity of information in this regard has been noted in the recent literature (15).

To simplify the process of defining pertinent exposure, an approach used in some industries is to describe the industrial population by grouping jobs on the basis of operation or product. In this manner, a descriptive index of exposure can be defined and used, even though quantitative measurements of exposure may not be obtainable. For example, in an industry as chemically complex as the rubber industry, the process of tire manufacturing can be outlined (16). Each individual process can be inventoried and analyzed, permitting all jobs involved to be partitioned accordingly and categorized into specified occupational titles (OT). In this manner, the specific exposure connected to each OT can be practically assessed. Relevant toxic agents are enumerated and inventoried on this basis, and relevant derangements of health are appropriately scrutinized in relation to a given exposure as defined by the occupational title(s). The analysis of a complete work experience becomes manageable from a statistical point of view, since large numbers of jobs are reduced to smaller and more homogeneous categories. The ability to detect adverse health effects by this procedure is much facilitated and is made more sensitive. Gradients of hazard and/or risk may be identified, and the detection of health effects associated with specific technological operations may be correlated with the measurements of significant exposure. Thus controlled measures can be initiated even though a specific causative agent may not be known or quantitatively determined.

4.3.2 Assessing the Effect

The accuracy and precision of measurements of health (or ill health) are common limiting factors in any epidemiological study. It is regretable that traditional medicine has considered epidemiology a marginal subject, peripheral to the mainstream of medical practice. Epidemiology is indeed part and parcel of medical intelligence and of medical judgment. As vigorously commented by Feinstein, any act or decision in applied medicine is an exercise in probability, largely based on epidemiological consideration

(17). A methodological model of diagnosis is depicted in Figure 5.2. The importance of diagnostic criteria is illustrated in Figure 5.1. In general, measurements of a biological variable are subject to random as well as systematic (bias) errors. According to the type of measurement made, errors occur in grouping, in comparing, in sizing, and in labeling human subjects and their functions. The same errors incurred in any biological measurement may become important in epidemiology.

To identify, and ideally measure, any "effect," a standard reference of "normality" is needed. The complexity of defining normality stems most directly from the fact that the "normal" condition is not a unique state but rather a range of states (18). Indeed, the concept of normal is not definable in any absolute sense, and it should be approached through "limits" (or range) of a function rather than single values. The definition of *health,* meaning good health, has met head on with these difficulties, and a recent editorial attempting to reach a consensus definition of the term well portrays this. It has been said that "all life represents some measure of health" (19). Terris suggests a four-level definition of the health–disease continuum, with perceptive and functional input woven into descriptive elements (Figure 5.3) (20). Note the overlap involving maximal level of health status (4+) with incipient history of disease.

In interpreting biological measurements, the statistical concept of abnormality assumes that skewed distributions are meaningful. This has been severely criticized by Cochrane (21). Oversimplification may lead to the unrealistic assumption of a distinct and bimodal distribution of a quantitative diagnostic variable (Figure 5.4).

4.3.3 Questionnaire

One prime source of medical input in epidemiological studies is the data derived from questionnaires. Whether by interview or self-administration, however, questionnaires

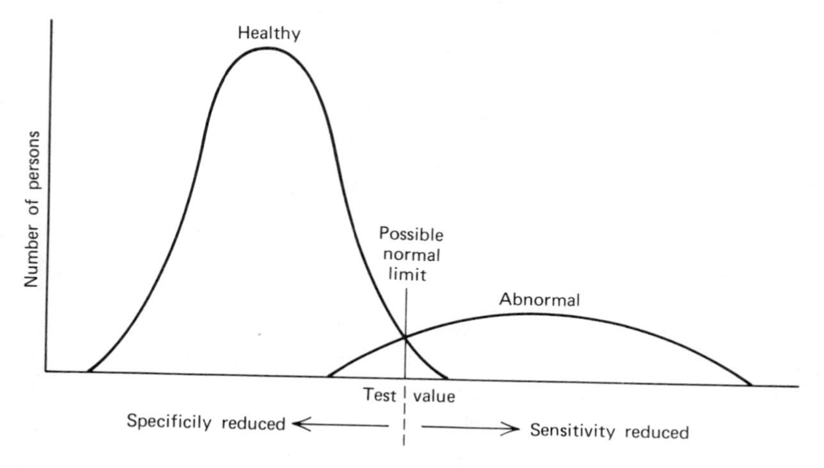

Figure 5.2 The discriminant value of a diagnostic test. Modified from Cochrane (21).

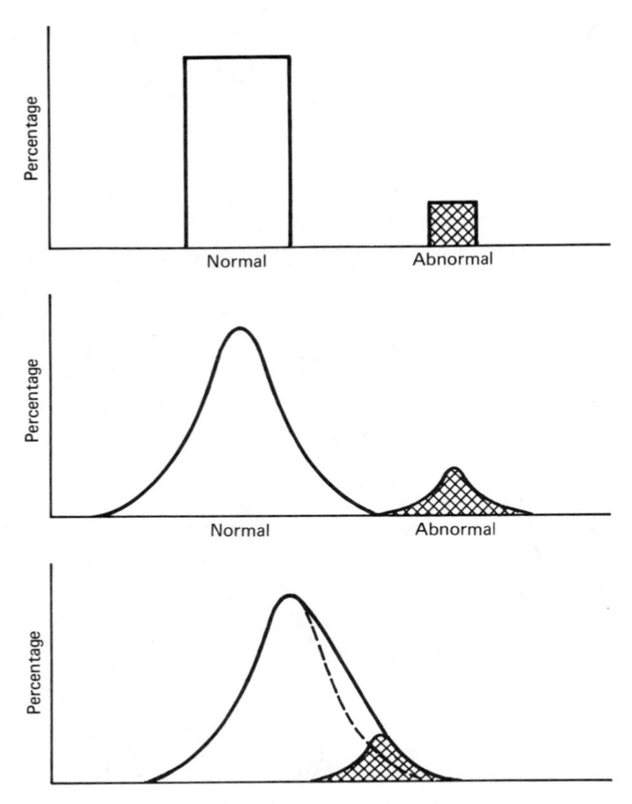

Figure 5.3 The health–disease continuum: subjective and objective aspects. From Terris (20).

provide a large source of error that must be guarded against. Admittedly, complaints and the patient's own perception of ill health are a fundamental source of information in patient care. These sources are heavily modulated by subjective interpretation. A good clinician usually adjusts the interpretation of these complaints to the patient equation, correcting for biases. A stoic patient may report a substantial weight loss as a minor problem, perhaps negligible in the patient's own evaluation. A more reactive patient may raise a fuss over a change in weight of a pound or two. Epidemiological surveys, utilizing medical history, are subject to similar biases, perhaps with more subtle opportunities for error.

Although there is little doubt that a well-standardized and tested questionnaire can be useful and valid instrument in the process of diagnosis, it may be the source of serious errors, as well. For instance, it has been established that the rate of reply from mailed questionnaires crucially depends on emotional factors and/or biological characteristics. Cigarette smokers tend to procrastinate in replying to a questionnaire inquiry about effects of smoking, whereas sick individuals behave in the opposite fashion (22). This differential may conceivably affect the results of an inquiry if the cutoff deadline for tabulating results happens to exclude an important fraction of the late respondents. Wording and sequencing of questions are important sources of discrepancy in estimates of health outcome assessed questionnaires (23). The assistance by a trained interviewer in the process of filling out a questionnaire may possibly correct this type of bias.

However the effect of the interviewer may also be to bias the formulation of a reply, as demonstrated recently by Choi and Comstock (24). A set of recommendations has been prepared by these authors, to compensate for this interviewer-mediated bias (Table 5.7).

The guidelines of Table 5.7 are not excessively heavy. The liability to error presented by a questionnaire is substantial. In a comparative study ascertaining the reliability of a questionnaire (self-administered), a persistent error occurred with several questions in a population repeatedly canvassed over several months (25). Inconsistency as high as 100 percent for questions related to symptoms have been encountered by administering questionnaires twice, independently, under supervision, to a group of older subjects (26).

Gandevia and Ritchie provide an excellent example of bias in reporting symptoms (27). They studied two groups of grain handlers doing identical tasks under the same working conditions. In one group there was good correlation between symptoms and

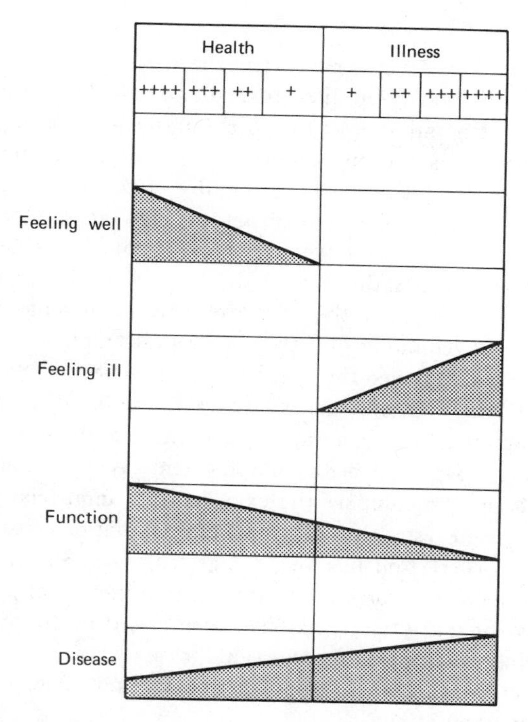

Figure 5.4 The health–disease gradient. Modified from Cochrane (21).

Table 5.7. Requirements for Interviewer Standardization (24)

1. Interviewers should be selected from the background and with the characteristics of the populations studied.
2. Interviewers should be highly trained.
3. Questions must be formulated in simple language.
4. Periodical comparison of interviewer objectivity should be obtained.
5. Rotation of interviewers should be planned and implemented.
6. Interviewers' objectivity should be checked repeatedly during the study.

exposure, but not in the other group. The difference was attributed to circumstances: the first group of easy-going, contented rural workers had no interest in union activities or industrial politics, and no concern for the hazard of the job. The second group of city workers, however, was frightened that the grain dust had caused two sudden deaths, and they were politically active and mistrustful of their employers. As a result, there was no correlation between symptoms and exposure of the two groups.

4.3.4 Confounding Factors

Bradford Hill's rules of evidence well review the many pitfalls inherent in epidemiological findings of associations. The literature abounds with examples illustrating these errors. Poor selection of population or unrepresentativeness of sample observations may lead to colossal errors of conclusion. The common attempt to compensate for intrinsic deficiencies with a large sample size is not generally successful (5). Since groups are most frequently compared in terms of exposure parameters, the common error reflected by this can be identified in sampling variation, inherent differences between the two groups, and differences in evaluating (processing) the respective groups. Although statistics can readily correct for error in the first case, only intelligent care can minimize the other sources of error. A common bias that tends to deemphasize cause and effect correlations has been recently labeled the "healthy worker effect" (28). An industrial population—that is, a population of workers actively employed—often has more favorable health indicators than the general population. The bias is introduced when an industrial population is screened before employment. Work demands contribute additionally to this selection. To compare the experience of industrial population to that of the general population (i.e., statistics derived from population census, etc.) may be inappropriate. A standard correction has been suggested for applications across the board, that is, the standard mortality rates when applied to a industrial population ought to be decreased by a constant tax. Objections have been raised to this standard correction on the basis of the variation on mortality effect due to age, race, work status, and period of observation; in other words, the so-called health worker effect may not apply evenly throughout a studied population.

5 CONCLUSIONS: STATISTICAL VERSUS LOGICAL SIGNIFICANCE

This brief review has stressed the problems encountered in drawing conclusions from epidemiological information. It has emphasized how the measurements obtained from a sample may be used to estimate, within certain limits of credibility, the actual parameters of the parent population (point and interval values). The importance of partitioning some of the sources of error has been brought out, with mention of a variety of techniques designed to minimize these errors (randomizing, blinding, stratification, adjustment, etc.).

Adequacy of measurement, of sample size and representation, of test validity, and of classification appropriateness, are no doubt critical for the significance of a conclusion. The limitation of dose assessment and the difficulties met in gauging health status or health responses have been also stressed, in the consideration of cause and effect problems. The rule of evidence of these relationships recapitulates the care and consideration needed in establishing the degree of confidence to be placed on any epidemiological conclusion.

In the final analysis, however, there is certainly no analytical tool or inferential methodology that can displace any interpretation that uses adequate doses of common sense. It is just as important to avoid the superficiality of dismissing a causative role because it appears to be weakly documented (error of omission), as it is to give in and accept an implausible factor, garnished by the frills of numerical seduction (error of commission). To quote Bradford Hill, "the more anxious we are to prove that a difference between groups is the result of some particular action we have taken or observed, the more exhaustive should be our search for an alternative and equally reasonable explanation of how that difference has arisen" (29).

REFERENCES

1. J. Cornfield, "A Method of Estimating Comparative Rates from Clinical Data: Applications to Cancer of the Lung, Breast, and Cervix," *J. Nat. Cancer Inst.,* **11,** 1269–1275, February 23, 1951.
2. B. MacMahon and T. F. Pugh, *Epidemiology: Principles and Methods,* Boston, Little, Brown, 1970.
3. P. E. Enterline, "Pitfalls in Epidemiological Research: An Examination of the Asbestos Literature," *J. Occup. Med.,* **18,** 150–156 (March 1976).
4. O. S. Miettinen, "Matching and Design Efficiency in Retrospective Studies," *Am. J. Epidemiol.,* **91,** 111–118 (1970).
5. L. Chiazze, Jr., "Problems of Study Design and Interpretation of Industrial Mortality Experience," *J. Occup. Med.,* **18,** 169–170 (March 1976).
6. W. R. Gaffey, "A Critique of the Standardized Mortality Ratio," *J. Occup. Med.,* **18,** 157–160 (March 1976).
7. R. C. Elandt-Johnson, "Reviews and Commentary, Definition of Rates: Some Remarks on their Use and Misuse," *Am. J. Epidemiol.,* **102,** 267–271 (October 1975).
8. O. S. Miettinen, "Standardization of Risk Ratios," *Am. J. Epidemiol.,* **96,** 383–388 (December 1972).

9. M. A. Lavenhar, "The Risk Factor Numbers Game," paper presented at the Statistics Section at the Annual Meeting of the American Public Health Association, Chicago, November 18, 1975.
10. L. H. Garland, "Studies on the Accuracy of Diagnostic Procedures," *Am. J. Roentgenol. Radiat. Ther. Nuclear Med.,* **82,** 25–38 (1958).
11. J. R. Ashford, "Comparison of Two Symptoms: Questionnaires in Comparability in International Epidemiology," R. M. Acheson, Ed., Milbank Memorial Fund, 1965, p. 96.
12. H. B. Zwerling, E. R. Miller, J. T. Harkness, and J. Jerushalmy, "Clinical Importance of Lesions Undetected in Mass Radiographic Survey of the Chest," *Am. Rev. Tuberc.,* **64,** 249–255 (1951).
13. C. M. Fletcher, "Difficulties in Definition and Observer Error in Medical Surveys," in: *Recent Studies in Epidemiology,* J. Pemberton, Ed., Blackwell, Oxford, 1958, p. 37.
14. A. Bradford Hill, "The Environment and Disease: Association or Causation," *Proc. Roy. Soc. Med.* **58,** 295–300 (1965).
15. C. H. Powell, and H. E. Christensen, "Development of Occupational Standards," *Arch. Environ. Health,* **30,** 171–173 (1975).
16. J. F. Gamble, R. Spirtas, and Peggy Easter, "Application of a Job Classification System in Occupational Epidemiology," *Am. J. Pub. Health,* **66,** 768–772 (1976).
17. A. R. Feinstein, "Clinical Epidemiology. II. The Identification Rates of Disease," *Ann. Intern. Med.,* **69,** 1037–1071 (1968).
18. E. A. Murphy, "The Normal," *Am. J. Epidemiol.,* **98,** 403–411 (1973).
19. F. Brokington, *World Health,* Churchill & Livingstone, Edinburgh, 1975, p. 5.
20. M. Terris, "Approaches to an Epidemiology of Health," *Am. J. Pub. Health,* **65,** 1037–1045 (1975).
21. A. L. Cochrane, "The History of the Measurement of Ill Health," *Int. J. Epidemiol.,* **1,** 89–92 (1972).
22. T. W. Oakes, G. D. Friedman, and C. C. Seltzer, "Mail Survey Response by Health Status of Smokers, Non-Smokers and Ex-Smokers," *Am. J. Epidemiol.,* **98,** 50–55 (1975).
23. W. W. Holland, J. R. Asford, J. R. T. Colley, Morgan D. Crooke, and N. J. Pearson, "A Comparison of Two Symptoms Questionnaires," *Br. J. Prev. Soc. Med.,* **20,** 76–96 (1966).
24. I. C. Choi and G. W. Comstock, "Interviewer Effect on Response to a Questionnaire Relating to Mood," *Am. J. Epidemiol.,* **101,** 84–92 (1975).
25. M. F. Collen, J. L. Cutler, A. B. Siegelaub, and R. L. Cella, "Reliability of a Self-Administered Questionnaire," *Arch. Intern. Med.,* **123,** 664–681 (1969).
26. J. S. Milne, K. Hope, and J. Williamson, "Variability in Replies to a Questionnaire on Symptoms of Physical Illness," *J. Chron. Dis.,* **22,** 805–810 (1970).
27. B. Gandevia and B. Ritchie, "Relevance of Respiratory Symptoms and Signs to Ventilatory Capacity After Exposure to Grain Dust and Phosphate Rock Dust," *Br. J. Ind. Med.,* **22,** 187–195 (1965).
28. A. J. McMichael, "Standardized Mortality Ratios and the 'Healthy Worker Effect': Scratching Beneath the Surface," *J. Occup. Med.,* **18,** 165–168 (1976).
29. A. Bradford Hill, *Principles of Medical Statistics,* Oxford University Press, New York, 1971.

CHAPTER SIX

The Mode of Entry and Action of Toxic Materials

BERTRAM D. DINMAN, M.D.

1 INTRODUCTION

The toxicity of a material and the hygienic standard or threshold limit applying to its use are of considerable interest, but they must not be confused with the hazards of using the material. Gases may replace all the air in the atmosphere, but the amount of vapor in the atmosphere at any time is limited by the vapor pressure of the liquid or solid from which it arises. However the amount of suspended particulate matter in the atmosphere (e.g., dust, mist, or fume) is not limited by vapor pressure. There is also evidence that particulates may adsorb gases and vapors, thereby altering an in some instances increasing their physiological action. This is logical because gas or vapors sorbed on particulates must follow the same physical principles governing the deposition of such particles in the pulmonary tree. Since particulates are more likely than gases to impinge on various portions of respiratory tract surfaces, their physiological consequences are more significant than they would be if the gas or vapor molecules did not impact and were exhales.

To plan the prevention of injury from toxic materials in industry, it is essential that we have a clear understanding of how these materials enter the body, how they act therein, and how they are eliminated. To understand better these processes, we should understand respiration and circulation and their roles in absorption and elimination. This, in turn, necessitates comprehension of the gas laws, with an ability to apply them to the solution of gases in liquids and, specifically, in the body fluids. We need also to know how different materials act on the body and to understand the different types and degrees of physiological response.

2 CLASSIFICATION OF CONTAMINANTS

The earth is surrounded by a gaseous atmosphere of rather fixed composition: 78.09 percent nitrogen, 20.95 percent oxygen, 0.93 percent argon, 0.03 percent carbon dioxide, insignificant amounts of neon, helium, and krypton, and traces of hydrogen, xenon, radioactive emanations, oxides of nitrogen, and ozone, with which may be mixed up to 5.0 percent water vapor. Any of these gases in greater proportions than usual, or any other substance present in the atmosphere, may be regarded as a contaminant, or as atmospheric pollution. The possibilities of contamination are legion, but we may classify them according to their physical state, their chemical composition, or their physiological action.

2.1 Physical Classifications

2.1.1 Gases and Vapors

Although strictly speaking a gas is defined as a substance above its critical temperature, and a vapor as the gaseous phase of a substance below its critical temperature, the term "gas" is usually applied to any material that is in the gaseous state at 25°C and 760 mm Hg pressure; "vapor" designates the gaseous phase of a substance that is ordinarily liquid or solid at 25°C and 760 mm Hg pressure. The usage distinction between gas and vapor is not sharp, however; for example, hydrogen cyanide, which boils at 26°C, is always referred to as a gas, but hydrogen chloride, which boils at −83.7°C, is sometimes referred to as an acid vapor.

2.1.2 Particulate Matter

There are at least seven forms of particulate matter.

1. *Aerosol.* A dispersion of solid or liquid particles of microscopic size in a gaseous medium—for instance, smoke, fog, and mist.
2. *Dust.* A term loosely applied to solid particles predominantly larger than colloidal and capable of temporary suspension in air or other gases. Derivation from larger masses through the application of physical force is usually implied.
3. *Fog.* A term loosely applied to visible aerosols in which the dispersed phase is liquid. Formation by condensation is implied.
4. *Fume.* Solid particles generated by condensation from the gaseous state, generally after volatilization from melted substances and often accompanied by a chemical reaction, such as oxidation. Popular usage sometimes includes any type of contaminant.

5. *Mist.* A term loosely applied to dispersion of liquid particles, many of which are large enough to be individually visible without visual aid.
6. *Smog.* A term derived from "smoke" and "fog" and applied to extensive atmospheric contamination by aerosols arising from a combination of natural and man-made sources.
7. *Smoke.* Small gas-borne particles resulting from incomplete combustion and consisting predominantly of carbon and other combustible materials.

2.2 Chemical Classifications

Chemical classifications are variously based on the chemical composition of the air contaminants and may vary widely depending on the aspect of the composition to be emphasized. The classification used in volume 2 is an example of chemical classification.

2.3 Physiological Classifications

As has been pointed out by Henderson and Haggard (1) the physiological classification of air contaminants is not entirely satisfactory because with many gases and vapors, the type of physiological action depends on concentration. For instance, a vapor at one concentration may exert its principal action as an anesthetic, whereas a lower concentration of the same vapor may, with no anesthetic effect, injure the nervous system, the hematopoietic system, or some visceral organ. Although it is frequently impossible to place a material in a single class correctly, a physiological classification may be suggested.

2.3.1 Irritants

Irritant materials are corrosive or vesicant in their action. They inflame moist or mucous surfaces. They have essentially the same effect on animals as on men, and the concentration factor is of far greater significance than the time (duration of exposure) factor. Some representative irritants are as follows:

1. Irritants affecting chiefly the upper respiratory tract: aldehydes (acetaldehyde, acrolein, formaldehyde, paraform), alkaline dusts and mists, ammonia, chromic acid, ethylene oxide, hydrogen chloride, hydrogen fluoride, sulfur dioxide, sulfur trioxide.
2. Irritants affecting both the upper respiratory tract and the lung tissues: bromine, chlorine, chlorine oxides, cyanogen bromide, cyanogen chloride, dimethyl sulfate, diethyl sulfate, fluorine, iodine, ozone, sulfur chlorides, phosphorus trichloride, phosphorus pentachloride, and toluene diisocyanate.
3. Irritants affecting primarily the terminal respiratory passages and air sacs: arsenic trichloride, nitrogen dioxide and nitrogen tetroxide, and phosgene. (To the extent that their action frequently terminates in asphyxial death, lung irritants are related to the chemical asphyxiants.)

2.3.2 Asphyxiants (Anoxia-Producing Agents)

Strictly speaking, "asphyxia" should be restricted to descriptions of agents that produce oxygen lack and increased carbon dioxide tension in the blood and tissues. Our concern with effects of chemicals on oxygen availability to the body requires that we refer to agents producing this effect as producing anoxia (i.e., lack of oxygen). Many of the chemical agents noted below produce effects at multiple loci (e.g., carbon monoxide affects hemoglobin as well as various tissue respiratory catalysts such as cytochrome P-450). However the classification to follow considers the major sites of action of such chemical agents.

This group can be subdivided into three classes, depending on how they cause anoxia within the body:

1. *Anoxic anoxia.* Lack of oxygen availability to the lungs and blood stems from simple displacement or dilution of atmospheric oxygen; this may occur even in the case of physiologically inert gases. This, in turn, results in reduction of the partial pressure of oxygen required to maintain oxygen saturation of the blood sufficient for normal cellular respiration. Agents producing anoxic anoxia include ethane, helium, hydrogen, nitrogen, and nitrous oxide.
2. *Anemic anoxia.* This implies a total or partial lack of availability of the blood pigment hemoglobin for oxygen carriage by the red blood cell. Whereas simple hemorrhage reduces the total loading of blood oxygen in proportion to red blood cell loss, arsene has an effect similar to hemorrhage by producing red blood cell breakdown, making it unavailable for oxygen carriage. In addition, numerous chemical agents can produce a similar effect by impairing or blocking the oxygen uptake and carrying capacity of hemoglobin in the lungs. Examples of this group are carbon monoxide, which combines with hemoglobin to form carboxyhemoglobin and aniline, dimethyl aniline, and toluidine, which forms methemaglobin.
3. *Histotoxic anoxia.* This condition results from the action of agents that impair or block the action of cellular catalysts necessary for tissue oxidative metabolism. Because the ability of hemoglobin to take up oxygen is not necessarily altered, and since the tissues cannot avail themselves of this oxygen, the capillary and venous oxygen saturation of the blood usually is higher than normal. Examples in this group are cyanogen, hydrogen cyanide, and nitrites. Hydrogen sulfide blocks cellular oxidation at the respiratory center controlling respiration and directly stops pulmonary air moving action. The ultimate effect of this agent is to secondarily cause anoxic anoxia.

2.3.3 Anesthetics and Narcotics

The anesthetics and narcotics group exerts its principal action as simple anesthesia without serious systemic effects, and the members have a depressant action on the central nervous system governed by their partial pressure on the blood supply to the brain. The following examples are arranged in the order of their decreasing anesthetic action compared with other actions: (*a*) acetylene hydrocarbons (acetylene, allylene, crotonylene), (*b*) olefin hydrocarbons (ethylene to heptylene), (*c*) ethyl ether and isopropyl ether, (*d*) paraffin hydrocarbons (propane to decane), (*e*) aliphatic ketones (acetone to octanone), (*f*) aliphatic alcohols (ethyl, propyl, butyl, and amyl), and (*g*) esters (not particularly anesthetic, but placed here for want of a better classification)—they hydrolyze in the body to organic acids and alcohols.

2.3.4 Systemic Poisons

The following substances are classified as systemic poisons:

1. Materials that cause organic injury to one or more of the visceral organs; the majority of the halogenated hydrocarbons.
2. Materials damaging the hematopoietic system: benzene, phenols, and, to some degree, naphthylene.
3. Nerve poisons: carbon disulfide, methyl alcohol, thiophene.
4. Toxic metals: lead, mercury, cadmium, antimony, manganese, beryllium, and so on.
5. Toxic nonmetal inorganic materials: compounds of arsenic, phosphorus, selenium, and sulfur; fluorides.

2.3.5 Sensitizers

Strictly speaking, materials that produce an antigen-antibody complex resulting in an allergic-type reaction in the body are referred to as sensitizers. Agents capable of producing this reaction have usually been complete or partial (i.e., haptenic) proteins. The end result of their interaction with the body may be the release of histamine, the formation of reagin antibody (IgE) or complement-fixing complexes of antigen and precipitating antibody (IgG, IgM). The reaction may be immediate (IgE mediated) or delayed for up to 12 hours (associated with IgG, IgM immunoglobulins).

It has become apparent recently that other pathways may be activated, eventually resulting in the same end results. Thus chemicals that are not protein in nature (e.g., the isocyanates) or are simple inorganics (e.g., SO_2) may precipitate reactions similar to those caused by antigen-antibody reactions. These reactions most clearly occur in the lungs and result in complex biochemical alterations of the tracheobronchial airway smooth muscle tone. It has become evident that the relative balance between cyclic adenosine monophosphate (cAMP) and cyclic guanosine monophosphate (cGMP) in these smooth muscle cells will determine their state of contractibility. Thus when smooth muscle cell cAMP concentration is less than cGMP, such cells will contract. The net result in the lungs is constriction of the breathing tubes and resistance to air movement. Multiple other agents present in such cells may also control muscle tone (e.g., the prostaglandins, kinins, histamine), although their interrelations are as yet unclear.

As previously indicated, many sensitizing chemicals are proteinaceous; in this category are castor oil, pollen, and other ill-defined agents causing farmer's lung, byssinosis, and bagassosis. The nonproteinaceous materials that may precipitate disorders similar to true allergic reactions are the isocyanates, sulfur dioxide, and so on.

2.3.6 Particulate Matter Other than Systemic Poisons

There are several categories of particulate matter not classified as systemic poisons that can produce toxic effects.

1. Fibrosis-producing dusts; silica, asbestos.
2. Inert dusts: carborundum, carbon, emery.
3. Proteolytic enzymes: certain laundry detergents containing enzymes.
4. Irritants: acids, alkalies, fluorides, chromates.
5. Bacteria and other microorganisms.

3 RESPIRATION

3.1 Mechanics of Ventilation

During inspiration, air is forced by atmospheric pressure into the nasal openings; it passes through the pharynx, larynx, trachea, bronchi, and bronchioles through the terminal bronchioles, the respiratory bronchioles, and the alveolar ducts into the air sacs or alveoli, filling the void created by involuntary muscular expansion of the thoracic cavity. Expiration may be either by muscular effort or by elastic contraction of lung tissue, but in quiet breathing contraction is thought to be entirely passive. The normal functions of respiration are to supply atmospheric oxygen through the alveolar walls to the blood for distribution to the tissues and to remove carbon dioxide resulting from oxidation within the cells. The oxygen and carbon dioxide exchange in the tissues is sometimes referred to as tissue or internal respiration, as distinguished from the aeration of the lungs or external respiration.

3.2 Lung Structure, Functional Lung Compartments and Lung Clearance

The tracheobronchial tree as described by Bouhuys (2) is supported by cartilage, the trachea being a series of open cartilaginous rings or crescents connected on the posterior side by bundles of muscle. The rings are joined to each other by a dense layer of connective tissue. The result is a rather rigid ribbed tube capable of some adjustment in diameter. This tube divides into the right and left main bronchi extending downward at approximately 20° for the right branch and 40° for the left. The bronchi are further divided and subdivided into smaller and smaller bronchial passages. There is a muscular network essentially transverse to these passages, and as the passages decrease in diameter, the cartilaginous rings become less complete, finally losing their crescent shape to become merely small plates of cartilage. As the cartilage decreases, there is an increase in the proportion of muscular tissue. Cartilaginous support finally disappears toward the outer end of the bronchioles, where the diameter is about 0.5 mm. These terminal bronchioles connect with respiratory bronchioles, which branch into alveolar ducts, terminating in the alveoli. These bronchioles range from 1.5 to 0.2 mm in length and are about 0.02 to 0.4 mm in diameter.

Normally the lungs fill about 80 percent of the total chest cavity's capacity (Figure 6.1). An average man can inhale about 3.5 liters of air by forced effort after an ordinary exhalation. Likewise, he can forcibly exhale about 1 liter after an ordinary exhalation. The sum of these, about 4.5 liters, is called the vital capacity. There is in addition about 1 to 1.5 liters of residual air that cannot be expelled by forced exhalation but remains in the lungs to aerate the blood during exhalation (residual volume). The amount of air that can be forcibly expelled from the lungs is referred to as the expiratory reserve volume; that amount which can be forcibly inhaled into the lung is known as the inspiratory reserve volume (Figure 6.1). Thus there is a total volume in excess of 5.5 liters, of which less than 10 percent is used in normal breathing. The approximately 500 ml of air inhaled and exhaled during normal respiration is called tidal air, and of this 500 ml, about 150 ml is required to fill the tracheobronchial tree or anatomical dead space—that portion of the respiratory tract consisting of thick-walled passages through which no interchange of gases between blood and air can occur. During an ordinary 500

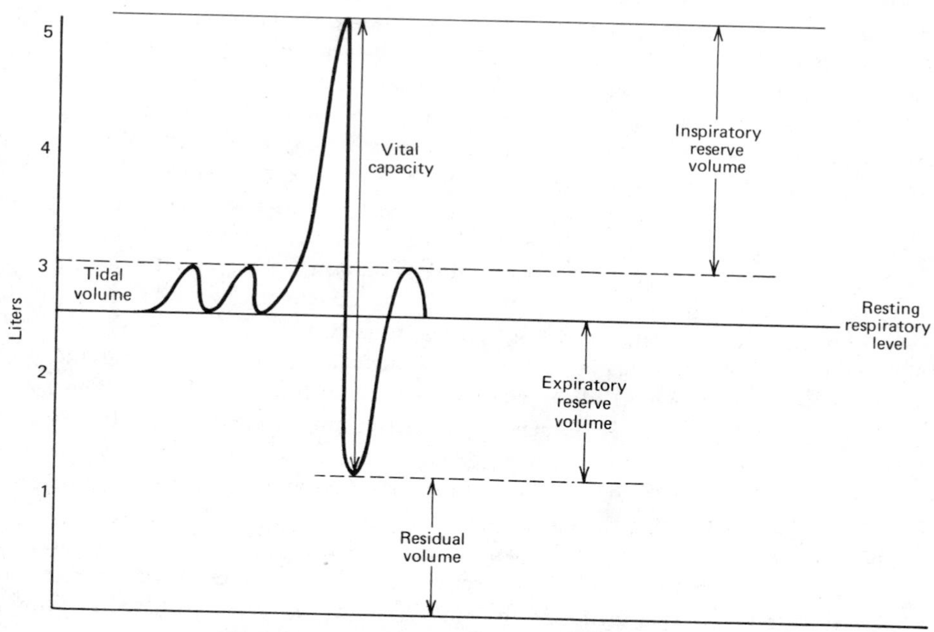

Figure 6.1 Functional pulmonary compartments.

ml inspiration, then, 150 ml of practically unaltered atmospheric air fills the dead space, while 350 ml enters the alveoli and respiratory exchange area of the lungs. Then during expiration, alveolar air containing 5 to 6 percent carbon dioxide and 12 to 14 percent oxygen, is forced from the alveoli through the dead space, leaving this dead space filled with alveolar air. The first part of the exhalation is essentially atmospheric air and the last part alveolar air, with the mixture growing richer in alveolar air throughout expiration. Undiluted alveolar air can be obtained consistently only from air forcibly exhaled at the end of an ordinary respiration. Failure to consider this fact can lead to serious errors in sampling expired air—for instance, in the determination of the radon or organic solvent content of the alveolar air. The maximum rate of physical activity that can be maintained by an individual is probably more dependent on his maximum breathing capacity (minute ventilation) than on lung volume. This minute ventilation is partially conditioned by frictional resistance along the air passages, due to impediments such as constrictions and abrupt changes in direction, or air passages merging at too great an angle.

The cells lining the airway passages are differentiated into two major types: a columnar-type ciliated epithelium, and secretory cells. The majority of the cells lining the tracheobronchial tree are columnar; on the surface facing the airway, they have hairlike projections. There are about 200 cilia per cell, beating to and fro, the rapid phase being in the direction of nose and throat, and the slow return beat in the opposite direction. The two different mucus-secreting cells are relatively fewer in number than the columnar ciliated airway lining cells; they produce a watery to thick secretion, which finds its way to a continuous liquid layer above the ciliated cells. The beating action of the cilia thus causes this mucus layer to move gradually up from the lung in escalator fashion. Thus foreign particulates or any other pieces of matter that are deposited in the air passages are eventually transported by way of this mucocilary escalator to be removed from the lung. The particles so transported eventually reach the throat and are swallowed in this normal mucus flow. The integrity of this protective mechanism obviously is vital to the lung's ability to rid itself of foreign matter. Any chemical that impairs the ability of the ciliated epithelium to beat well prolongs the time of residence of that agent or others in the lung. This can lead to enhancement of toxic effects in the lung. Thus although inhalation of a carcinogen (e.g., benzo-α-pyrene) produces little effect when inspired alone, when inhaled with sulfur dioxide, this simple lung irritant damages mucociliary clearance sufficiently to permit the carcinogen to remain in the lung long enough to exert its neoplasm-promoting effect. The rate of transport is susceptible to many influences (e.g., irritation, cold), thus is variable.

Since the ciliated cells extend only down to the respiratory bronchioles, other mechanisms need exist to clear the lung of inhaled foreign particles. In the healthy lung, protection is afforded by freely mobile cells that engulf deposited particles. These cells, the alveolar phagocytes, carry the foreign matter in some cases up to the mucus ladder for disposition as described previously. In other cases they migrate to lymph vessels and are carried to regional lymph nodes; in some cases they may reach the outer lining of the lung (i.e., the pleura).

Particles so engulfed by the phagocytes may be destroyed (e.g., bacteria); however in other cases (e.g., silica) this foreign matter may destroy the phagocyte. Evidently the breakdown products of these macrophages can play a major role in initiating adverse tissue reactions; this train of events is believed to play a major role in the pathogenesis of silicosis (see Chapter 7).

3.3 Regulation and Control of Respiration

The respiratory movements are regulated and controlled chemically as well as by the voluntary and involuntary nervous systems. Carbon dioxide acts as a stimulus to the respiratory center in the brain stem, and the CO_2 tension in the blood is by far the most significant factor of control. It is thought to act by raising the hydrogen ion concentration of the respiratory center. Lactic acid resulting from strenuous exercise is believed to have a similar effect. Oxygen—or, rather, lack of it—is likewise a stimulus, but under normal conditions does not come into play because the partial pressure of oxygen in the alveolar air must fall markedly before the oxygen tension of the arterial blood is significantly affected. This is easily understood by examining the oxyhemaglobin dissociation curve,, as in Figure 6.2. Men with long experience in testing atmospheres in the holds of ships say that they can recognize atmospheres deficient in oxygen by this stimulating effect; but reliance on this perception is not recommended even for the well initiated because the margin between irrespirable atmospheres and those causing respiratory distress in the form of hyperpnea and dyspnea is too narrow. The suddenness of the onset of weakness and unconsciousness is so very marked that if a person enters an oxygen-deficient atmosphere by means of a ladder, the probability of his being able to remount the ladder unassisted, after becoming aware of distress, is not favorable. The tension or partial pressure of oxygen in normal air is 20.95 percent of an atmosphere, which equals 159.2 mm Hg at sea level. Significant symptoms of distress do not occur until the tension falls below 16 percent; unconsciousness may occur at concentrations below 11 percent, and breathing soon stops if the oxygen falls below 6 percent.

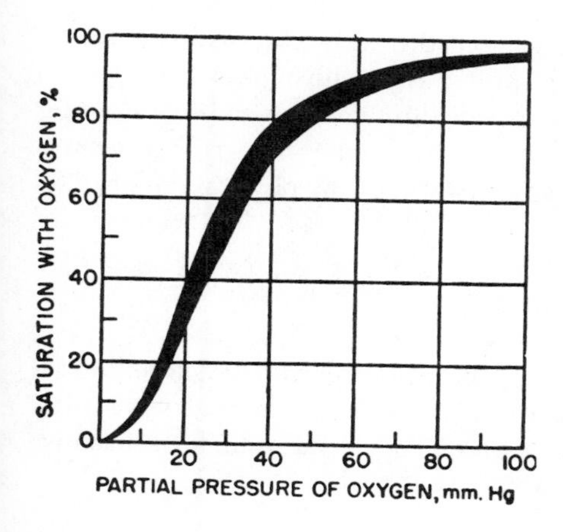

Figure 6.2 Dissociation curve of oxyhemoglobin for human blood expressed as the normal range of percentage of saturation with oxygen. (After Barcroft.)

The respiratory centers control and coordinate by way of the spinal cord the activities of the muscles between the ribs (i.e., intercostals) and the movement of the diaphragm and abdominal muscles. In addition to receiving signal feedback from these muscles, these centers also receive information from chemoreceptors (aortic, corotid body), which detect changes in oxygen and carbon dioxide tension in arterial blood. Finally, because voluntary activity (e.g., singing, speaking) arising from higher portions of the brain may also affect ventilation, the brain stem respiratory centers must process, store, and integrate this information with other signals to produce a coordinated ventilatory process.

3.4 Function of Hemoglobin

The transportation of oxygen from the lungs to the cellular tissues is accomplished largely by means of the hemoglobin in the red cells carried in the liquid plasma phase of the blood. Likewise, the carbon dioxide diffusing from the tissues is transported chiefly by the same cells. Each 100 ml of arterial blood contains about 15 g of hemoglobin, which combines with 19 or 20 ml of oxygen. In blood that is in equilibrium with air at normal atmospheric pressure, about 1 percent of the total oxygen is in solution in the plasma. Normal venous blood carries 55 to 60 volume percent carbon dioxide and deposits about 10 percent of this as it takes up its oxygen supply. The amount of carbon dioxide in simple solution is about 3 volume percent in the venous blood and 2.5 volume percent in the arterial blood. Since in making its circuit the blood flows through the capillaries of the lungs in approximately one second (3) and through active tissue in a similar time, it is evident that mere solution of oxygen and carbon dioxide cannot account for the exchange of these gases, even though the equilibration in the lungs is highly efficient. The speed of exchange is due to the light combination of these gases with hemoglobin in the oxyhemoglobin and carbamino reactions, respectively, as well as to the presence of catalysts in the red cells that play an important part in the oxygen and carbon dioxide exchange. Carbonic anhydrase greatly increases in either direction the reversible chemical reaction $H_2CO_3 \rightleftharpoons CO_2 + H_2O$, and oxidation and reduction in the tissues are greatly accelerated by oxygen-activating catalysts or by dehydrogenases.

It must be recognized that the transport of oxygen or other gases is strictly a passive, partial pressure gradient phenomenon. Transport by diffusion either from alveolar air sacs to blood or from tissue cells to blood involves passage through cell membranes. Since oxygen and gases other than carbon dioxide never move counter to partial pressure gradients, it may be said that such exchanges are governed simply by physical laws of diffusion and solubility.

3.5 Circulation as a Factor and Its Regulation

Since the oxygen in the lungs cannot take part in the metabolic process until it has entered the bloodstream and then been transported to the tissues, it is evident that the circulation rate is an important factor in respiration. The blood volume is usually slightly greater than the lung capacity. For a normal man it is approximately 6 to 7 liters.

The time required for a complete circuit of the blood varies with exercise, the part of the body supplied, and many other factors, but averages somewhat less than 1 minute. The normal heart stroke is about 60 to 70 ml, with a resting pulse of 68 to 72 per minute, giving a minute volume of 4 to 5 liters. Exercise can increase the heart stroke to 100 or 200 ml and the pulse rate to 170 or 180. Environmental temperatures above 30°C increase the minute volume 5 to 30 percent. Digestion of food increases the basal minute volume by 30 to 40 percent in about 3 hours. Emotional strain causes an increase of 15 to 25 percent. Changes in pH have an effect, as do both naturally occurring and foreign chemicals having a vasoconstrictor or vasodilator reaction. Also, just as the amount and content of the venous supply to the lungs play a major role in regulating the breathing required to keep the arterial blood constant, the circulation depends on breathing, and any retardation of breathing will be quickly followed by a change in circulation. Henderson (4, 5) strikingly demonstrated the control of circulation by breathing when he showed that excessive pulmonary ventilation could induce failure of the circulation. The part played by respiration in the action of atmospheric contaminants on the body is discussed later.

4 ABSORPTION, DISTRIBUTION, AND ELIMINATION

4.1 Modes of Expressing Concentrations

The extent of atmospheric contamination in the gaseous phase is frequently expressed as parts per million (ppm), denoting units of volume in 1,000,000 volumes, and is usually corrected to 25°C and 760 mm Hg pressure. This is a convenient mode of expression and, for routine examinations of workroom atmospheres, the actual conditions are frequently not sufficiently removed from this standard to require temperature–pressure corrections. Since comparisons in ppm are on a volume—hence molecular—basis, they have no direct relation to the weight–volume expression of milligrams per cubic meter generally used to express concentrations of particulate matter. The milligrams per cubic meter unit has the advantage of facilitating comparison between different chemical species and/or materials in different physical states. On the whole, the ppm unit has a longer tradition of use, expecially when referring to materials in a gaseous state; however, of late, the milligram per cubic meter unit has become more frequent in usage for both particulate and gaseous states. Variations in the expression of parts by volume are: percent by volume, parts per thousand, parts per ten thousand, parts per hundred thousand, parts per hundred million, and parts per billion. The excuse for employing so many units is the desire to use small whole numbers. However this multiplicity of units is confusing, and we might better confine ourselves to not more than three units: percent, parts per million, and parts per billion. Percent by volume is generally used when the concentrations are on the order of 0.1 percent or greater. Parts per million is a logical unit for concentrations ranging from 1000 down to 1 or even 0.1 ppm. For concentrations less than 1 ppm, parts per billion (ppb) appears to be the logical unit. One million ppb = 1000 ppm = 0.1 percent = a partial pressure of 0.76 mm Hg at sea level. One gram mole of a perfect gas or vapor has a volume of 24.45 liters at 25°C and a pressure of 760 mm Hg. Therefore under these conditions

$$1 \text{ ppm} = \frac{\text{molecular weight}}{24.45} \text{ mg/cubic meter}$$

and

$$1 \text{ mg/cubic meter} = \frac{24.45}{\text{molecular weight}} \text{ ppm}$$

Another method of expressing gas concentrations is the partial pressure method, using the unit of the pressure exerted by one millimeter of mercury. This is convertible directly to percent by volume of multiplying by 100 and dividing by the barometric pressure; or to ppm by multiplying by 1,000,000 and dividing by the barometric pressure:

$$\frac{\text{partial pressure of one constituent}}{\text{total barometric pressure}} \times 1{,}000{,}000 = \text{ppm of constituent}$$

When the contaminant is dispersed in the atmosphere in solid or liquid form, as a mist, dust, or fume, its concentration is usually expressed on a weight per volume basis; "particles per unit volume" are less generally utilized except to describe concentrations of fibrous material in air. Liquids and toxic solids are usually expressed as milligrams per cubic meter, or per 10 m^3 to approximate the maximum amount inhaled in one 8-hour day by a workman doing moderately heavy work. This standard was originally based on 20 liters/min × 60 minutes × 8 hours = 9600 liters = 9.6 m^3. Henderson and Haggard (1) estimate the average breathing rate to be 8 liters/min or about 4 m^3 per 8-hour day. Since not only the minute volume, but also retention, the physiological dead space, and other factors, must be considered in computing daily systemic intake, the validity of the figure 10 m^3, is open to question. Dusts that exert their effects largely within the lungs, rather than throughout the system as a whole, are usually counted microscopically and are expressed as million particles per cubic foot of air. Outdoor air contaminants and stack effluents are frequently expressed as grams, milligrams, or micrograms per cubic meter, ounces per thousand cubic feet, pounds per thousand pounds of air, and grains per cubic foot. There is generally a trend toward the use of metric units, because most data are collected in these units; but there is also justification for the use of English units (sic) because of existing data, especially specifications in such units for air-moving equipment. Conversion factors should always be provided. Expressions containing both English and metric units should not be used. The ppm unit (weight to volume) used in water analysis (milligrams per liter) must not be confused with the ppm (volume to volume) unit used in the analysis and expression of atmospheric constituents.

4.2 Volume-Pressure-Temperature Relations

Boyle's law, which states that a perfect gas without change of temperature varies inversely in volume with the pressure to which it is subjected, and Charles' law, which states that at constant pressure the volume of a gas varies directly with the absolute temperature, are usually combined in one operation to convert observed gas or vapor volumes to normal conditions (25°C and 760 mm Hg pressure) as follows:

$$\text{observed volume} \times \frac{298}{\text{observed absolute temperature (°C)}} \times \frac{\text{observed barometric pressure (mm Hg)}}{760} = \text{volume at normal conditions}$$

This formula may be used, of course, to convert from any observed temperature and pressure to any desired temperature and pressure by substituting the proper values for the normal condition values as given. These industrial hygiene "normal conditions" are not sufficiently different from the "standard air" used in air conditioning and ventilation work (70°F and 29.92 in. Hg pressure) to require correction factors in most instances. Likewise, these "normals" are close enough to the temperatures and pressures of most occupied spaces to permit their use without correction in much industrial hygiene work. However the "standard temperature and pressure" (STP) used by chemists (0°C and 760 mm Hg) is sufficiently different (9 percent) to introduce significant error unless temperature corrections for air volumes are made.

4.2.1 Partial Pressures

Each gas in a mixture exerts a pressure in direct proportion to its percent by volume. In a sample of dry air, for example, at sea level the percentage composition and partial pressures (pp) are as follows:

oxygen: 20.95% × 760 mm = 159.22 mm pp

nitrogen, carbon dioxide, and inert gases: 79.05% × 760 mm = 600.78 mm pp

If, without change in barometric pressure, this air should become humidified to the extent of 50 percent relative humidity at 25°C, the composition would be altered as follows. Water vapor, 11.88 mm partial pressure, would appear in the mixture, and since the total pressure does not change, the composition of the atmosphere would become:

water vapor = 11.88 mm pp

oxygen: 20.95% × (760 − 11.88) mm = 156.73 mm pp

nitrogen, etc.: 79.05% × (760 − 11.88) mm = 591.39 mm pp

Should a sample of this 50 percent humidified air be transported to a higher altitude and opened for analysis, the partial pressures would decrease, but the percentage composition would remain the same. At 750 mm barometric pressure, for instance, this humidified air would have the following partial pressure composition:

$$\text{water vapor} = \frac{750}{760} \times 11.88 \text{ mm} = 11.72 \text{ mm pp}$$

$$\text{oxygen: } 20.95\% \times (750 - 11.72) \text{ mm} = 154.67 \text{ mm pp}$$

$$\text{nitrogen, etc.: } 79.05\% \times (750 - 11.72) \text{ mm} = 583.61 \text{ mm pp}$$

Gases are ordinarily reported on a dry basis, but variations in volume due to temperature, pressure, and water vapor changes may be computed by the formula:

$$V_1 = \frac{V_2(P_2 - W_2)(273 + T_1)}{(P_1 - W_1)(273 + T_2)}$$

where V_2 is the observed volume of gas at the observed temperature T_2; V_1 is the calculated volume of gas at temperature T_1; temperatures T_1 and T_2 are in degrees celsius, W_2 and W_1 are mm Hg vapor tension of water at observed and calculated conditions; and P_2 and P_1 mm Hg are the observed and the calculated barometric pressures, respectively. Air and vapors inhaled into the lungs are rapidly saturated with water vapor at body temperature to approach a partial pressure of 47 mm Hg and, in computing the partial pressure of any vapor in alveolar air, this partial pressure of water must be considered.

4.3 Role of Solubility in Absorption of Gases and Vapors

4.3.1 Solubility Coefficient

Any gas confined in contact with a liquid dissolves at the surface of the liquid and enters the liquid until its partial pressure in the vapor phase above the liquid is in equilibrium with the gas dissolved in the liquid; that is, the rates of its dissolving and evaporating have become equal. The speed with which equilibrium is reached is a function of the solubility and the relationship of contact surface to volume of liquid—the greater the proportion of surface area, and the lower the solubility, the sooner equilibrium will be reached. At the same time the liquid vaporizes into the space above it until its partial pressure in the atmosphere is in equilibrium with its evaporation pressure in the liquid phase. Henry's law, *the concentration of a gas that will dissolve in a liquid is directly proportional to its concentration in the free space above the liquid,* may be expressed as $C_1/C_2 = K$, where C_1 and C_2 are the molar concentrations of the gas in the liquid and vapor phases, respectively, and K is the solubility coefficient. This coefficient is different for each vapor, each liquid, and each temperature.

Henderson and Haggard (1) have proposed that this law be expressed in a more convenient form when applied to the absorption of gases through the lungs into the blood. They suggest $C/C_1 = D$, where D is the coefficient of distribution, C is the concentration in the fluid phase (blood), and C_1 is the concentration in the vapor phase in alveolar air, both concentrations to be expressed on a weight per volume basis (milligrams per liter). This form of the law is particularly convenient for the purpose intended, and it is to be hoped that additional data will be collected to establish the coefficient of distribution of more gases and vapors between the atmosphere and the circulating blood. For a miscible or highly soluble material, the coefficient of distribution cannot be approximated by calculation, but for a less readily soluble vapor it can be approximated from its vapor pressure and solubility at body temperature 37°C. The coefficient of distribution is constant for the same vapor and liquid at any given temperature regardless of the concentration of vapor in the atmosphere.

The coefficient is known for the solubility of relatively few vapors and gases in blood at body temperature, this being a field that has been insufficiently explored in toxicity studies. It is to be hoped that in the future such studies will include the partial pressure of the vapor or gas under observation in alveolar air, and in the body fluids, at time intervals during accumulation and elimination of vapors, as well as at equilibrium. In computing a coefficient of distribution from room air vapor concentration data, corrections for changes in temperature and partial pressure of water vapor should be made if comparisons are desired with coefficients based on alveolar air, because alveolar air approaches saturation with water vapor at 37°C (47 mm Hg partial pressure). Some of the coefficients that have been established are given in Table 6.1.

Table 6.1. Distribution Coefficients

Solvent	Coefficient of distribution[a]	Reference
Methyl alcohol	1700	1
Ethyl alcohol	1300	1
Isoamyl alcohol	836	6
Primary *n*-amyl alcohol	804	6
Secondary isoamyl alcohol	550	6
Acetone	330	7
Methyl *n*-propyl ketone	167	6
Diethyl ketone	157	6
Methyl isopropyl ketone	101	6
Ethyl ether	15	1
Benzene	6.58[b]	8
Carbon disulfide	5	9

[a] The coefficient of distribution here is the ratio of milligrams of solvent per liter of blood to milligrams of solvent per liter of alveolar air.

[b] Based on room air rather than alveolar air.

4.3.2 Body Saturation

From the foregoing it is evident that any gas or vapor in the air we breathe tends to pass through the lungs into the bloodstream and to be distributed throughout the body. The respiratory tissue in the lungs, which has been estimated to have a surface area of about 55 m^2, acts as the exchange surface between blood and air. The blood in the capillaries is said to be separated from the air in the alveoli by two membranes of the utmost delicacy, perhaps only one cell thick; thus it becomes evident that equilibrium between the blood in the lungs and the alveolar air is reached rapidly. The accumulation of the foreign gas or vapor in the body depends on a number of factors, including, the concentration in the air, the solubility of the material in the blood and tissues, the length of exposure, the rate of breathing, the rate of circulation, and whether the material is reactive.

If a gas is very soluble in the blood, saturation of the body is slow (requiring days), is largely dependent on the ventilation of the lungs, and is only slightly influenced by changes in circulation; with a very slightly soluble gas such as nitrogen, however, saturation is very rapid (being nearly complete within a few minutes), is chiefly dependent on the rate of circulation, and is little influenced by the rate of breathing. With gases and vapors that are freely soluble in water and nonreactive or only slowly reactive in the body, such as acetone and methanol, absorption and distribution throughout the body are dependent on the water content of the tissues. Although the fat-soluble vapors may be transported chiefly by the aqueous content of the blood, they tend to concentrate in the fatty tissues. With any nonreactive vapors, although the blood and tissue concentrations, short of equilibrium, are proportional to the vapor concentration and functions of the time of exposure, the rate of saturation is, with constant circulation and breathing, a relation peculiar to each vapor and independent of the concentration of the vapor. In other words, the same time is required for the same vapor to reach a given percentage saturation of the body, regardless of the concentration of the vapor in the atmosphere breathed.

It is not practical to speak of the exact time required for complete saturation, since this is a variable and too indefinite. It is, however, practical to state a time of some percentage saturation such as 50, 60, 70, 80, or even 90 percent of saturation, but past this point the absorption curve for a gas or vapor of low reactivity becomes rather flat. With a relatively water-insoluble vapor such as benzene, blood saturation occurs so rapidly that even the venous blood may reach 70 to 80 percent of saturation within 30 minutes, yet relatively complete saturation may require as much as 2 or 3 days. This can be explained by the fact that the fatty tissue, which has the greater affinity for benzene, removes and stores the benzene carried by the blood, but this fatty tissue has in many instances a very meager blood supply and therefore requires a longer period to attain equilibrium.

Methanol is typical of vapors highly soluble in water. About 24 hours inhalation of this vapor is required before the blood is 70 percent saturated (about 50 times as long as for a similar percentage saturation with benzene); yet relatively complete saturation here requires little longer than with benzene, the difference being that the fatty tissues, with their smaller blood supply, are not a reservoir for methanol, and distribution throughout the body at equilibrium is directly proportional to the water content of each tissue. Both benzene and methanol are examples of very slightly reactive vapors.

Carbon disulfide, approximately 90 percent of which has been found to be metabolized, may be cited as an example of a moderately reactive, relatively water-insoluble vapor. McKee and associates (9) have reported on the blood saturation with carbon disulfide in inhalation exposures of dogs. From its physical and chemical properties, carbon disulfide would be expected to have an absorption curve resembling that for benzene except for the effect of a higher rate of metabolism, and McKee's data indicate that this is the case. If the points on his graph for the blood saturation of a dog breathing 50 ppm carbon disulfide are connected by an exponential curve, the data indicate a saturation of around 90 percent at the end of a 3½-hour exposure. The coefficient of distribution of CS_2 between blood and room air at this percentage saturation was 4.3 on a milligrams per liter basis (10). If this were corrected for alveolar air temperature and humidity, the value at equilibrium would be approximately 5. Desaturation was rapid and, for single exposures, more or less complete within 2 to 6 hours. This is similar to the desaturation curve for benzene. During accumulation or absorption, the arterial blood is saturated to a higher degree than venous blood, whereas at equilibrium or saturation the concentration is the same in each, and during elimination the concentration of benzene in the arterial blood is lower.

The difference in vapor concentration between arterial and venous blood is dependent upon the solubility or coefficient of distribution of the particular vapor. With a very soluble vapor the difference is negligible; but with a slightly soluble vapor, such as benzene, the difference is marked, and during accumulation the arterial blood, once the lungs are filled with vapor by the first few inhalations, is equilibrated with air only slightly less in concentration of vapor than the air inhaled. In other words, the blood passing through the lungs is not sufficient to absorb an appreciable percentage of the vapor from the alveolar air. This has an important bearing on brief exposures to high concentrations of slightly soluble vapors. The arterial blood supply to the brain is large: therefore brief exposures to high concentrations of anesthetic gases or vapors of low solubility produce rapid effects, even though the degree of saturation of the body is very low. The administration of ether is a good example of this point. Moderate concentrations may be given that will induce unconsciousness after a half-hour or more, and the effects will be somewhat prolonged, or concentrations may be given sufficiently high to induce unconsciousness quickly and have recovery occur shortly after removal from exposure, provided the exposure has not been prolonged. This also has a practical bearing in degreasing work, where a short exposure to high concentrations of trichloroethylene, or another similar solvent, may not materially raise general body concentration but may induce anesthesia. It is therefore important to know peak as well as average vapor concentrations.

The curves in Figure 6.3 illustrate the differences in rate of absorption by inhalation and in approach to equilibrium in the circulating blood for different classes of gases and

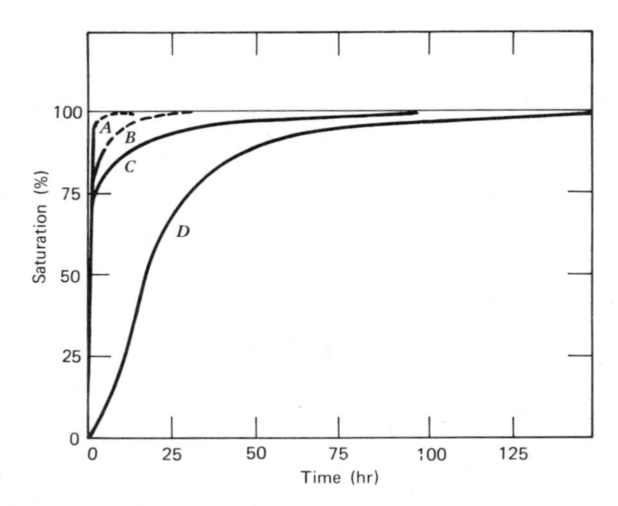

Figure 6.3 Curves illustrating the absorption of some representative gases and vapors into the blood of living dogs: *A*, nitrogen; *B*, carbon disulfide; *C*, benzine; *D*, methanol.

vapors. Nitrogen, nonreactive and slightly soluble, but with some specific affinity, approaches equilibrium rapidly. Carbon disulfide, moderately reactive and slightly soluble, approaches saturation rapidly, but data are lacking on the period of complete saturation. Benzene, very slightly reactive and slightly soluble, reaches 70 percent saturation rapidly but does not attain complete equilibrium even after a period of 3 days. Methanol, a slightly reactive, highly soluble vapor, gradually approaches saturation, requiring about 24 hours for 70 percent saturation and about 5 days for essentially complete saturation. Ethanol (not shown) attains saturation slightly more slowly than does methanol because it is metabolized fairly rapidly.

Figure 6.4 illustrates the relative rates of elimination for benzene after exposures for varying lengths of time. These curves indicate that the more severe the exposure to benzene, especially with regard to the time factor, the more slowly is it eliminated from fatty reservoirs.

Figure 6.5 plots the elimination of carbon disulfide after a brief single exposure to a low concentration, as well as typical elimination curves for ethanol and for methanol. Ethanol is more readily eliminated than methanol because rapid oxidation augments the elimination by expired air, while, as has been pointed out by Haggard and Greenberg (11), more than 70 percent of the methyl alcohol is eliminated in the expired air.

4.3.3 Effect of Intermittent Exposures

What is the effect of repeated 8-hour exposures, each followed by a 16-hour period in uncontaminated air? Dogs have been so exposed to 500 ppm methanol (12). Absorption ranged from 10 to 20 percent of saturation in the 8-hour period and was followed by 50

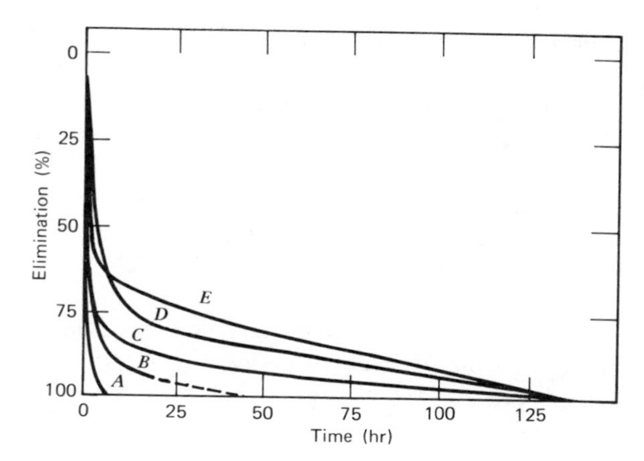

Figure 6.4 Curves illustrating the elimination of benzene from the blood of dogs after varying degrees of inhalation exposures: *A*, 800 ppm for 2 hours; *B*, 800 ppm for one 8-hour period; *C*, 800 ppm 4 hours per day for 65 days; *D*, 1300 ppm for one 37.5-hour exposure; *E*, 800 ppm 8 hours daily for 158 days.

to 100 percent desaturation. This resulted in erratic pictures in which some animals gradually accumulated methanol to the extent of 52 percent of saturation for the particular air concentration (52 percent of the amount that could accumulate from a continuous exposure). From this high degree of saturation they gradually excreted until they again reached a low level of less than 20 percent of saturation. Exercise or increased activity greatly speeded the change in either direction, but appeared to have no definite effect upon the direction.

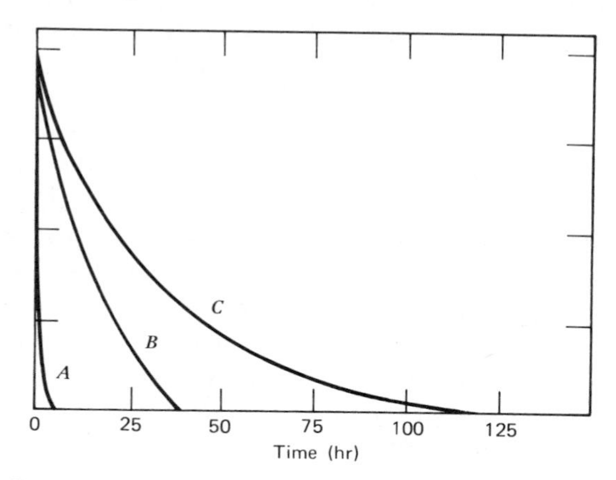

Figure 6.5 Curves illustrating the elimination of carbon disulfide, methyl alcohol, and ethyl alcohol from the blood of dogs after inhalation exposures: *A*, carbon disulfide, 50 ppm for 3.5 hours; *B*, ethanol, 4000 ppm for 5 days, continuous; *C*, methanol, 4000 ppm for 5 days, continuous.

In intermittent exposures to 800 ppm benzene, the blood of dogs exposed 2 hours per day reached an average of about 60 percent of saturation at the end of exposure, and the animals eliminated 82 percent of this during the ensuing 22 hours. With a 4-hour daily exposure, the blood reached about 85 percent saturation, and the animals eliminated an average of 85 percent of this within the following 20 hours. With a daily exposure of 8 hours, the blood reached an average saturation of 92 percent, while elimination in the ensuing 16 hours was only 76 percent of this. This frequently resulted in the animals apparently reaching complete saturation, from which high level they would again recede.

With a highly water-soluble vapor, methanol, we see that from daily 8-hour exposures the vapor may accumulate over a period of time to as much as 5 times the amount resulting from a single 8-hour exposure, but it never reaches more than about half the amount that could result from a continuous exposure, which is a function of the coefficient of distribution for that particular vapor. With a water-insoluble vapor such as benzene, however, there is relatively little accumulation, and the concentration of benzene in the blood is largely proportional to the concentration just having been inhaled at any time.

4.3.4 Relative Concentrations of Vapor in Blood, Tissues, and Expired Air

We have seen how the concentration of a vapor in the blood is controlled by the concentration of vapor in the inspired air, the length of exposure, and the solubility of the vapor in blood. Let us consider the concentrations of vapor in the tissues, in the body fluids other than blood, and in the expired air and how they are controlled. Henderson and Haggard (1) state that the concentration of any vapor in the urine is regulated by the concentration in the arterial blood passing through the kidney at the moment of secretion, differs from the concentration in the blood only by the ratio of solubilities of the vapor in the two fluids (approximately 1 : 1 for most vapors), and is a composite of the varying concentrations over the period during which it is secreted. This relation would be expected from the standpoint of physics, and appears to hold true for several vapors, but the concentration of benzene in the urine averaged 20 times that in the blood (8). This concentration has not been explained and certainly cannot be on the basis of the ratio of solubilities of benzene in blood and in urine, but it would seem to indicate that the kidneys have the power to extract benzene from the blood in circulation.

For the tissues in general, the relative amounts of a vapor they contain at equilibrium are directly proportional to the solubility of the vapor in the various tissues, but tissues with the greater blood supply and lower solvent capacity saturate, or reach equilibrium, before the tissues of higher solvent capacity and lower blood supply. Likewise the same tissues with the greater blood supply and lower solvent capacity desaturate more rapidly.

The concentration of vapor in the alveolar air during accumulation, equilibrium, or desaturation regulates the concentration of vapor in the arterial blood leaving the lungs at the moment, therefore can be used as an indication of arterial blood concentration. At complete saturation the vapor concentration in the alveolar air is the same as that in the room air, with due correction for temperature and humidity. With the more soluble vapors, alveolar air concentration at any point during accumulation may be used to estimate the average room air concentration, given the length of exposure, the saturation curve for the particular vapor, and its coefficient of distribution. An interesting variation, of much practical usefulness, is met in the case of radon: its concentration in the alveolar air may be used as an indication of radium storage or deposition, provided sufficient time has elapsed for relatively complete desaturation (8 hours or more after the end of an inhalation exposure).

The physical laws governing the rate of gas exchange between the blood and the respired atmosphere are the bases of methods (13) for measuring pulmonary functional capacity in terms of lung ventilation, gas transfer by diffusion across the alveolar wall, and pulmonary blood flow.

4.4 Dusts and Fumes

4.4.1 The Deposition of Particulate Matter

Dusts, fumes, and smokes represent solid particulate materials differing in quality, form, and particle size. The process of their accumulation, distribution, and elimination is much more complicated than the same process with gases. Although much information regarding lung deposition is available concerning particles larger than 1 μm, knowledge of those smaller than this dimension is still incomplete.

The physical forces governing deposition of particulates of 0.1 μm size and greater in the lung and its passages are inertial, gravitational, and diffusional. The last of these forces is insignificant in this size range. With the flow of air in and out of the tracheobronchial tree, inertial forces cause deposition, especially where flow lines change direction in the nasopharynx and at branches of the air-conducting tubes. Since the effectiveness of inertial deposition varies directly with air velocity, this force will come into play to a larger extent in the upper air passages than in the depths of the respiratory system. By contrast, gravity will play a larger role in particle deposition by settling where longer residence times obtain (i.e., in relatively still air). Accordingly, it is within the deepest portions of the lungs that gravitational settlement plays its greatest role (14).

In addition to these physical forces, the nature of the particles aside from size per se determines in part their deposition characteristics. Long, irregular particles such as asbestos fibers settle much less than would be expected from their total fiber mass. Other irregularly shaped particles (e.g., quartz, coal) are aerodynamically equivalent to spherical particles one-half to three-quarters of their measured diameters. Most models consider particles in terms of unit density spherical shapes (i.e., as aerosols), to reach a reasonable agreement between theoretical predictions and experimental observations. In addition to shape, the density of a particle determines its deposition characteristics.

Using several test substances varying in density from unity up to 11, investigators predictably found that the resulting deposition values varied directly as a function of their density.

The hygroscopic properties of the particles also affect pulmonary deposition characteristics. It has been clearly demonstrated by Dautrebande and Walkenhorst (15) that hygroscopic sodium chloride crystals suspended in air increase in size as a result of water absorption. Given the relatively high water vapor pressure of lung air, it is not surprising that such hygroscopic particles absorb water and increase in size in the order of 300 times their original volume. Accordingly, droplets of sulfuric acid mists are found to be deposited to a greater degree than is the case for different mineral particles of equal (dry) diameters.

It should be kept in mind that all deposition models are susceptible to errors due to the effects of pulsatile flow, intrapulmonary mixing of new and residual lung air, and nonuniform distribution of new air in the lungs. Nevertheless there is reasonably good agreement between directly measured values of deposition and those predicted by mathematical equations.

Although as previously indicated, all deposition models are susceptible to error, and our knowledge of deposition characteristics of submicron-sized particles is open to question, some useful descriptions of this phenomena are available. Figure 6.6 indicates the

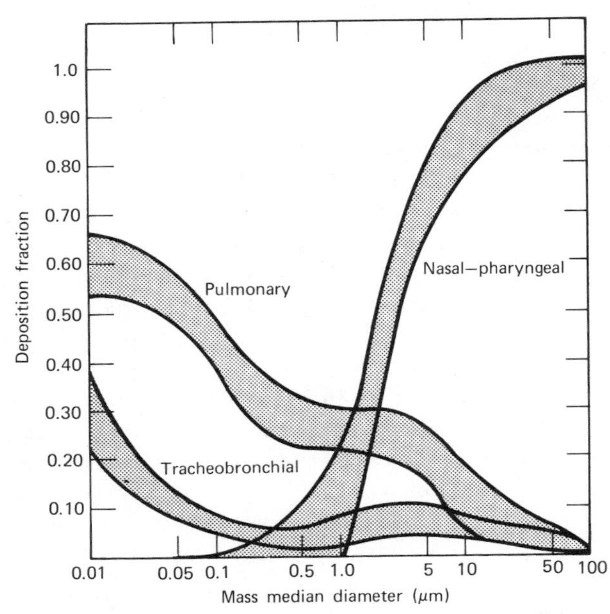

Figure 6.6 EAch shaded area (envelope) indicates the variability of deposition for a given mass median (aerodynamic) diameter (μm) in each compartment when the distribution parameter varies from 1.2 to 4.5 and the tidal volume is 1450 ml. From Reference 16, p. 181.

range of variability of deposition for a given aerodynamic diameter in each lung compartment (16). These curves and the envelopes about them are specific for a 1450 ml tidal volume (moderate to heavy activity) at the rate of 15 respirations per minute. As the respiratory tidal volume increases, the predicted gravimetric deposition shifts upward; conversely, at lower tidal volumes, predicted deposition decreases. Caution must be exercised in using these predictions, and the following factors militate against too-literal application of the data: variations in the mode of breathing (e.g., mouth, nose), the need for idealized models of lung anatomy and airflow patterns in making these estimates, and the occurrence of nonuniform ventilatory distribution. Furthermore, data are rapidly developing that may well alter such deposition models, especially as they apply to particles of less than 0.1 μm. Nevertheless, the use of these relationships to provide approximations of pulmonary deposition patterns is warranted insofar as their approximate nature is appreciated.

4.4.2 Action of Suspended Particulate Matter

The toxic dusts may dissolve and enter the circulation by absorption from the respiratory tract, or they may be absorbed after being swallowed. Their action is discussed under the material involved. Dusts producing "pneumoconiosis" may exert their effects after lodging in the alveoli. Other particulate matter may cause allergic reactions or, in the case of disease germs, infections. The possibility of the sorption of gases and vapors by particulate matter has been discussed. Also, it has been demonstrated that the normal locus of action of water-soluble gases may be transferred from the upper to the lower respiratory tract through sorption on particles, thereby increasing the physiological effect of such gases. Conversely, when gases less soluble in water, such as nitrogen oxides, phosgene, and ozone, are mixed with aerosols, their effects may be lessened because of sorption. For a complete discussion of the role of dusts and the evaluation of disability arising therefrom, see Chapter 7.

4.4.3 Fate of Deposited Particles

As previously indicated, particles deposited in the tracheobronchial segments of the lungs above the respiratory bronchioles are removed by the mucociliary escalator. Also, particles deposited below this level may be engulfed by phagocytic cells; these cells move either up to this escalator for transport or are the lymphatics to regional lymph nodes. In either of the first two cases, such particulates eventually reach the gastrointestinal tract, where systemic absorption by the bloodstream may occur.

However water-soluble particles that impact on the alveolar wall may go into solution here just as gas molecules that reach this air–blood carrier. For such freely diffusible material, the rate of solution and absorption has a consistent half-time of less than 10 minutes. Whether this action is due to simple solvation is open to question, but intuitively one would expect more absorbable materials to experience both lung and gastrointestinal absorption, whereas less absorbable materials would be subject to a relatively greater opportunity for gastrointestinal absorption by way of the mucociliary ladder and swallowing (16). Furthermore, for reactive materials, the possibility must be considered that bindings with protein and other ligands may occur. This in turn would lead to formation of reversible or irreversible complexes, which might enhance or hinder further transit from the reactive locus.

The mode of entrance into alveolar lining cells may be by pinocytosis or phagocytic responses of such cells. Some of these in turn may apparently become migratory phagocytes; others remain in place with the engulfed particles (16). Ultimately these particles may be translocated to the lung interstitium. In addition, there is some evidence that particulates and fibers may penetrate the alveolar wall, although the data require further confirmation. Whether this effect is due to forces exerted on the particulates by fluctuating respiratory pressure or to respiratory excursions (14) or microerosions (17) is at present not understood. In any event, it is clear that by whatever mechanism, inhaled particulate and gaseous chemicals ultimately can appear in the general circulation at variable times (10 minutes to 30 hours) after inhalation. This may result either from direct alveolar capillary uptake, or it may be a secondary consequence of lymphatic uptake and ultimate drainage into the bloodstream.

4.5 Mists

Mists may be inhaled to reach all parts of the lungs and from there may be absorbed. Concentrations of a slightly volatile material far above those possible from partial pressures of the gaseous phase at room temperature may be encountered. Therefore a solvent whose vapors may be considered harmless at ordinary temperatures can be very dangerous as a mist.

4.6 Other Means of Absorption

4.6.1. Ingestion

Toxic materials can be absorbed by means other than inhalation. Many other sources, such as contaminated food, tobacco, or beverages, can result in ingestion, as can putting fingers or other contaminated objects into the mouth, or licking the lips. Compared with inhalation, ingestion plays a minor role in the absorption of most toxic materials in industry. However progressive employers realize that occupational diseases can occur from the careless use of contaminated workrooms as a place in which to eat lunch—all too often without previously washing the hands.

4.6.2 Absorption Through the Skin

Many gaseous and liquid materials are absorbed to a limited extent through the intact skin by way of the air spaces in the hair follicle to the sebaceous glands and the gland cells. Through the sweat gland ducts, any substance may reach the secretory surface of the sweat gland cells after having passed the straight and the convoluted parts of the ducts.

Most electrolytes and water do not penetrate the skin in significant amounts. Alkaloids, phenols, oxalic and salicylic acids and esters, lead acetate, and lead oleate are absorbed in appreciable amounts. Salts of lead, tin, copper, arsenic, bismuth, antimony, and mercury are said to penetrate by combining with the fatty acid radical of the sebum. Ammoniated mercury, however, is said not to be absorbed through the skin as such. Nicotine, strychnine, and opium are absorbed readily, but their salts are not. Slight amounts of hydrogen sulfide and rapidly dangerous amounts of hydrogen cyanide may be absorbed from contaminated air. Nitrobenzene, dinitrobenzene, nitrotoluene, dinitrotoluene, trinitrotoluene (probably), aniline, dimethylaniline, triorthocresyl phosphate, and nitroglycerin are readily absorbed from skin contact with these materials or their solutions. Alcohols, aldehydes, and acetone are said to be readily absorbed, but from a practical viewpoint there seems to be little significant hazard from this source. Benzene, toluene, xylene, the chlorinated hydrocarbons and other fat solvents are absorbed to a degree, but the quantities of these materials accumulated in the body in this manner are probably not of a significant order, although tetrachloroethane and other unusually toxic solvents may be exceptions to this statement. Absorption through lesions of the epidermis is much more rapid than through the intact skin (see Chapter 8).

4.7 Application of Foregoing Principles to Industrial Exposures

All the foregoing data may be applied to chemical exposures encountered in industry to provide an approximation of uptake, transport, and excretion of such agents. A case in point as proposed by Dinman et al. (18) is the model for absorption of fluoride particulates and hydrogen fluoride (HF) associated with aluminum oxide reduction.

The data available under these circumstances indicated that Al_2O_3 particulates occurred in the following size distribution: greater than 5 μm, 15 percent; 1 to 5 μm, 25 percent; and less than 1 μm, the remaining 60 percent. Although HF gas was also present, theoretical considerations (i.e., the hygroscopicity of HF) and the observed data suggested that this was readily adsorbed (on Al_2O_3?) or absorbed (on water vapor?), thus this gas deposition behavior followed those of particulate dynamics. Examination of Figure 6.7 indicates the distribution of these particulates in the various portions of the tracheobronchial tree and alveoli: the approximate percentage of each particle size that could be expected to be deposited or exhaled from the lung is shown. The direct transfer of the adsorbed fluoride to the blood either directly from the alveolus or indirectly by way of the gastrointestinal tract is demonstrated. Oral intake by hand-to-mouth ingestion was grossly estimated; the percentage absorbed from gastrointestinal tract followed absorption data for finely divided fluoride particulate. That which was not absorbed in the gastrointestinal tract is represented as leaving the body with the feces. Distribution throughout the body could be expected to present the absorbed fluoride to the organ that most avidly takes up this ion; since the men tested were in a phase of relative equilibrium between blood and bone fluoride, however, only a small amount was deposited in

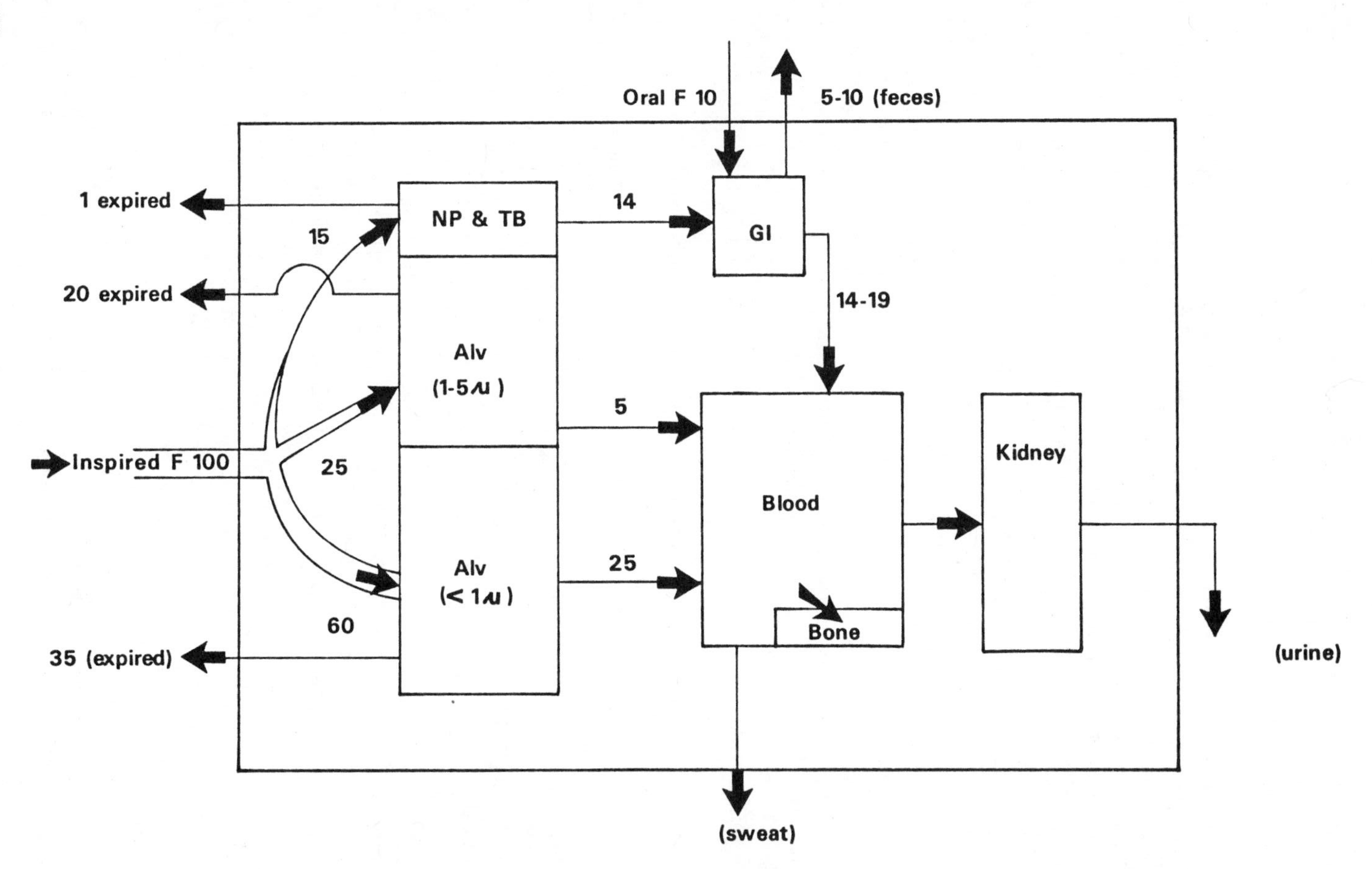

Figure 6.7 Model for absorption, transfer, and excretion of fluoride particulates and HF associated with aluminum oxide reduction.

the bony sink. Another excretory pathway from the blood permits fluoride loss from the skin; an approximation of the rate of excretion by this route was obtained from experimental data. Because of the relatively low uptake of fluoride by other tissue, most of the blood fluoride is presented to the kidneys for excretion. Readily available experimental data provided information on rural clearance rates for fluoride sufficient to predict urinary output of this ion under conditions of relative blood–bone equilibrium. In effect, this model predicts that for every 100 units of fluoride found in the air of this specific workplace (as particulate and gaseous fluoride), 40 units (or 40 percent) of this element will appear in the urine.

Given the approximate nature of this model and the assumptions of oral intake, gastrointestinal absorption, and sweat excretion of fluoride, it is of interest that this model could be confirmed by experimental data. The wearing of breathing zone personal samplers for 8-hour work shifts provided a satisfactory estimate of total fluoride absorbed through the respiratory tract. Collection of 24-hour urines during and after the 8-hour work shift provided a quantitative estimate of total urinary fluoride excretion. Despite the approximations and assumption inherent in this exercise, the predictive model closely paralleled the observed data.

Although this illustrates the practical use to which present knowledge of deposition, absorption, and excretory processes may be put, it should be emphasized that such exercises represent—at best—gross approximations of reality. Only further, more meticulous research will permit more precise quantitative depictions of the realities of these multiple, complex processes.

5 STANDARDS OF PHYSIOLOGICAL RESPONSE

To facilitate the comparison of data on the physiological response to the inhalation of atmospheric contaminants, many investigators have tabulated their data in degrees of response such as the following:

- Amount producing detectable or other degree of odor.
- Amount producing detectable or other degree of irritation.
- Maximum amount for repeated daily 8-hour exposures without injury.
- Maximum amount for a single 8-hour exposure without serious disturbance.
- Maximum amount for a single 1-hour exposure without serious disturbance.
- Amount dangerous to life within several (4 to 8) hours.
- Amount dangerous to life within ½ to 1 hour.
- Amount causing death within a few minutes.

Levels set for industrial practice have been variously labeled as threshold limits, maximum allowable concentration, toxic limit, and maximum permissible concentration. The most desired value, of course, is the maximum concentration to which an individual may be exposed throughout his or her working day, for an indefinite period of time, without suffering any ill effects. Unfortunately this value has been established satisfactorily for relatively few single substances. However the human body is fortified with defense, or detoxication, mechanisms able to cope with all toxic materials in amounts below a certain level, called the threshold of intoxication. As long as this threshold of daily intake is not exceeded, there are no harmful effects; but as soon as this level is exceeded, injurious effects result. The threshold of intoxication is not a definite and fixed point for any material, and it varies not only with individuals, but also with bodily conditions of the same individual. There are many degrees or shades of toxicity among the materials encountered in industry. For example, all the volatile solvents are toxic if absorbed into the system in excess, their toxicity being a matter of degree, and none merits the classification "nontoxic." How are we to set permissible or safe standards of air contamination for industry?

With very few exceptions, all our "safe" limits are at best estimates. At their best, they are educated guesses rather than actual safe limits. They are limits set after careful consideration of available toxicity data, largely obtained from recording the effects of exposure of animals, which further research or experience may or may not substantiate. It is foolhardy to assume that we know the maximum safe limit (except possibly for carbon monoxide and other asphyxiants, as well as some extensively studied heavy metals, e.g., lead, mercury), when there is a paucity of data to establish the point. Even with the asphyxiants, we perhaps are not positive of the effects of prolonged inhalation of concentrations too low to cause unconsciousness. Regarding the safe limit for lead, evidence points to a need for specifying the physical state and chemical combination of the lead. It seems logical that as a guide to the uninitiated, and a check on the alarmist—and possibly the unscrupulous—we should set suggested maximum concentrations for materials as soon as they attain industrial importance, if not before. We should call these maxima hygienic standards, recommended good practice, threshold limits, or suggested maximum concentration standards until we learn through experience or research the true maximum safe limit; then, if necessary, we can change our standards accordingly.

6 STANDARDS FOR PERMISSIBLE ATMOSPHERIC CONTAMINATION AND HOW THEY ARE SET; OCCUPATIONAL SAFETY AND HEALTH ACT OF 1970

Our present standards of practice regarding atmospheric contaminants are very helpful bench marks, as long as we regard them for what they are and do not try to set them up as absolute safe limits to be used as established facts comparable to, or in the same class with, flammable limits, flash points, or other properties that can be determined accurately. Rather, the idea should be fostered that the limits are flexible and are to be adjusted to conform with any new, properly weighted evidence of facts regarding physiological effects. It must also be remembered that just as everyone is subject to occasional ill health regardless of work environment, everyone is at one time more easily affected by foreign materials than at another. There are probably too many uncontrollable factors to allow us to set a positively safe limit, applying to all people at all times, and still have the limit appreciably different from that of pure air. No matter how carefully our standards are established, if they are maintained at a level compatible with industrial practice, there may be occasional ill health in the especially susceptible individual. We shall always need periodic medical examinations by thorough and competent physicians.

Over the years, the U.S. Public Health Service has been the outstanding contributor to our knowledge of maximum permissible limits for atmospheric contaminants, and some of the states have cooperated in the toxicological and chemical studies required to establish such limits. The Bureau of Mines, the Department of Labor, and a few universities and other institutions of higher learning, as well as private research organizations, have added to the store of information.

With the passage of the Occupational Safety and Health Act of 1970, the entire approach toward protection of the worker has changed. Whereas previously the American Conference of Governmental Industrial Hygienists (ACGIH) indicated that their standards were designed to protect *nearly all workers* repeatedly exposed to airborne substances on a daily, 40-hour week basis, the new federal statute requires protection of *all* workers so exposed. Previously the threshold limit values issued by the ACGIH were considered to be *guides* in the control of hazards and were not to be used (with some exceptions) on fine lines between safe and dangerous concentrations (19). Presently threshold limit values embedded in Occupational Safety and Health Administration regulations have the force of law. Thus given the legal approach, they have by legislative fiat been converted from guidelines for action to absolute limits defining finite safe and nonsafe values (i.e., given a limit of 50 mg/m^3, 49 mg/m^3 is "safe" and 51 mg/m^3 is not "safe"). Finally, the ACGIH indicated that these limits were "intended for use . . . and (interpretation) only by persons trained . . . in industrial hygiene practice." Presently they are the currency of the variously trained or, indeed, alien professions (e.g., lawyers and administrators).

Throughout these changes, an element of arbitrary usage has intruded itself. The legal demand for absolutes (e.g., protection, levels) begs the reality that no human activity has yet—or will ever—achieve the absolute. Presently the element of arbitrary setting of "action levels" (levels 50 percent or less than the TLV) above which surveillance and testing programs are mandated has been promulgated. This apparently ignores the fact that the original TLVs incorporated various margins of safety between lowest effect levels and the TLV guideline.

How these contradictions will be resolved cannot be perceived at this time. The Occupational Safety and Health Act certainly represents a significant step toward providing more widespread worker protection; its administration must ultimately recognize the imprecision inherent in all human knowledge and act within this constraint.

REFERENCES

1. Y. Henderson and H. W. Haggard, *Noxious Gases,* 2nd ed., Reinhold, New York, 1943.
2. Arend Bouhuys, *Breathing, Physiology, Environment and Lung Disease,* Grune & Stratton, New York, 1974.

3. C. N. Davies, Ed., *Inhaled Particles and Vapours,* Vol. II, Pergamon Press, Oxford, 1967, pp. 121–131.
4. Y. Henderson, *Am. J. Physiol.,* **24,** 310 (1910).
5. Y. Henderson and H. W. Haggard, *J. Biol. Chem.,* **33,** 355 (1918).
6. H. W. Haggard, D. P. Miller, and L. A. Greenberg, *J. Ind. Hyg. Toxicol.,* **27,** 1 (1945).
7. A. P. Briggs and P. A. Schaffer, *J. Biol. Chem.,* **48,** 413 (1921).
8. H. H. Schrenk, W. P. Yant, S. J. Pearce, F. A. Patty, and R. R. Sayers, *J. Ind. Hyg. Toxicol.,* **23,** 20 (1941).
9. R. W. McKee, C. Kiper, J. H. Fountain, A. M. Riskin, and P. Drinker, *J. Am. Med. Assoc.,* **122,** 217 (1943).
10. R. W. McKee, private communication.
11. H. W. Haggard, and L. A. Greenberg, *J. Pharmacol.,* **66,** 479 (1939).
12. R. R. Sayers, W. P. Yant, H. H. Schrenk, J. Chornyak, S. J. Pearce, F. A. Patty, and J. G. Linn, U.S. Bureau of Mines, Report of Investigation No. 3617, 1942.
13. J. H. Comroe, Jr., R. E. Forster II, A. B. Dubois, Jr., W. A. Briscoe, and E. Carlsen, *The Lung,* 2nd ed., Year Book Medical Publishers, Chicago, 1962.
14. T. F. Hatch and P. Gross, *Pulmonary Deposition and Retention of Inhaled Aerosols,* Academic Press, New York, 1964.
15. L. L. Dautrebande and W. W. Walkenhorst, in: *Inhaled Particles and Vapours,* Vol. II, C. N. Davies, Ed., Pergamon Press, Oxford, 1967, p. 116.
16. Task Group on Lung Dynamics, International Commission on Radiologic Protection, *Health Phys.,* **12,** 173 (1966).
17. H. Von Hayek, *Anat. Anz.* **93,** 149 (1942).
18. B. D. Dinman, et al., *J. Occup. Med.,* **18,** 1 (1976).

CHAPTER SEVEN

The Pulmonary Effects of Inhaled Inorganic Dust

GEORGE W. WRIGHT, M.D.

1 INTRODUCTION

Dust-laden air is the normal environment of man. There is continuous exposure to inhaled particles that vary from time to time in number and chemical and physical properties. It seems likely that the remarkable ability of the respiratory apparatus to cope with inhaled particles accounts in part for the survival and evolution of man, since we developed through our various stages in an environment that over long periods of time was in all probability even dustier than now as the result of volcanic activity and the scouring effect of wind flowing across the land. Fortunately the respiratory apparatus intercepts a large portion of inhaled particles as they move through the nose or mouth and along the course of air-conducting channels before they reach the more delicate distal parts of the lung. Moreover the lung has a large capacity to remove deposited dust by way of the mucociliary escalator and macrophage system. It also is important that the lung cells comprising the surfaces in contact with air normally have a rapid turnover or replacement rate, hence partially damaged surface cells are quickly replaced by new and normal cells. In addition, mechanisms are available for the repair of injured tissue in the lung. As is true of all other organ systems, however, the capacity for self-protection and repair of injury can be exceeded, and excessive dust deposition can cause adverse effects within the breathing apparatus.

Depending on the intrinsic chemical and physical nature of the inhaled particles and also the chemicals adsorbed onto their surfaces, the biological response may be noninjurious, slight or serious, or even fatal. Of equal importance governing the biological outcome are the quantity of particles deposited and the amount retained. The weight of evidence thus far overwhelmingly supports the conviction that all biological responses to nonliving agents are dose related.

Dust particles in contact with tissue cells evoke a response specific to each particular kind of cell. This response may be fleeting or temporary with no persistent or serious cell injury. On the other hand, persistent injury or death of the cell may occur, leading to secondary tissue alterations of varying degrees of gravity. Because of variations in the number of cells responding, the capacity or normality of cell replacement and repair and immunologic cell surveillance, and the hyperreactivity or sensitivity of the reacting cells, the overall tissue response can be expected to differ considerably in magnitude among individuals exposed in quite similar ways.

There are many kinds of respirable particles in the total environment of man. Some are essentially peculiar to the occupational environment, but others occur commonly in both the general and occupational environment. Thus biological effects of the occupational environment must always be related to the sum of the particles in the total environment, including those of nonoccupational origin. There is also the possibility that coexisting particles of different kinds and from various sources may act not only additively but also in a synergistic or even an inhibiting manner.

Those interested in industrial toxicology and hygiene or in occupational medicine quite properly are concerned with the biological reactions of humans exposed to dust comprised of various agents acting alone or in combination and in varying concentrations. One must be aware not only of the biological effects but also of the mechanisms underlying and controlling their occurrence. Such toxicological knowledge provides the *raison d'être* for industrial hygeine programs aimed at prevention of injury.

This chapter discusses principles relating the lung response to inorganic dust particles occurring in occupational environments. Specific agents are used to exemplify various kinds of reaction, but the biological effects of all the kinds of particle coexisting with other gases and fumes throughout all sorts of industrial environments cannot be covered. The reader is urged to consult more comprehensive texts for greater medical detail or to rectify omissions of the effect of specific agents. Among such texts, the recent one by W. Raymond Parkes, *Occupational Lung Disorders* (16) is recommended.

2 PERTINENT ANATOMY OF THE LUNGS

The primary purpose of the respiratory apparatus is to act as a gas exchange mechanism. A very thin layer of venous blood, pumped into the lungs by the right ventricle via the pulmonary artery system, is circulated through the pulmonary capillaries, which are arranged in a network over the surface of approximately 300 million air-containing alveoli. The total effective gas exchange surface of the alveoli has been estimated at ± 70 m^2. The tissue barrier separating the blood from the gas phase is the alveolocapillary wall, which averages 0.55 μm in thickness. During passage of venous blood over the surface of this barrier, the diffusion gradient favors the rapid movement of molecular oxygen from the air into the blood and of carbon dioxide out of the blood into the air phase. Thus the blood becomes arterialized in its transit over the alveolocapillary surface.

As can readily be appreciated, the partial pressure of these gases in the blood and gas phase would quickly become equal unless the air phase is continuously maintained at partial pressures of O_2 and CO_2 favoring the gradient for diffusion. External respiration brings ambient air low in CO_2 and high in O_2 content to the gas phase overlying the alveolocapillary membrane by mass movement and continuous replacement of air within the lungs. Mass movement of air does not go all the way to the surface of the alveolar wall, and there is an interface, probably within the alveolar sac, across which the final pathway for molecular delivery to the alveolar membrane occurs by diffusion.

After moving through the nose or mouth, inspired air enters the larynx and then the trachea, which divides into two main bronchi, each leading into a lung. Each main bronchus divides into 16 further generations of branches that serve solely the purpose of air conduction (Figure 7.1). There are seven subsequent generations of branching having alveoli in their walls. Counting limbs and twigs, there are approximately 116,300 solely air-conducting bronchi including the terminal bronchioles. Additional air-conducting branches that are also alveolus bearing, the so-called transitional and respiratory zone, bring the total number of branched limbs and twigs through which air moves to about 82 million. The main bronchi measure 1.2 cm in diameter. By the eleventh generation the lumen is reduced to 1 mm, and by the sixteenth and subsequent generations to 0.5 mm (1). These dimensions and numbers have major implications with respect to deposition and clearance of particles entering the lung while entrained in the respired air. The opportunity for impingement of air-entrained particles at the myriad of bifurcations before reaching any alveoli or gas exchange area is enormous. This means that particles of aerodynamic size favoring deposition by impingement, sedimentation, or interception will fall on and be removed by the mucociliary escalator, a structure that extends through the first 16 generations of branching. Hence the probability that a particle suspended in the air bolus of a single breath will actually reach and be deposited in a specific alveolus-bearing portion of the lung is small. Moreover, because of the extremely large number of branched units and their small individual volume, the probability that the moiety of air entering a single unit during a single breath will contain an entrained particle is very low by the time the bolus of air has passed beyond the seventh or eighth generation of branching and is small indeed by the time it has reached or passed the eighteenth generation. This type of distribution reduces the likelihood of overloading any single small conduit during prolonged exposure and helps maintain a high efficiency of dust removal.

Equal in importance to an understanding of the morphometric aspects of the respiratory apparatus is knowledge of the histologic or cellular structure of the lung. These components determine the manner in which dust particles entering the lung are disposed of and also what kind of biological reaction does or does not ensue. For details of the normal lung histology, the reader is referred to standard texts of histology and to the special texts on the lung (2, 3).

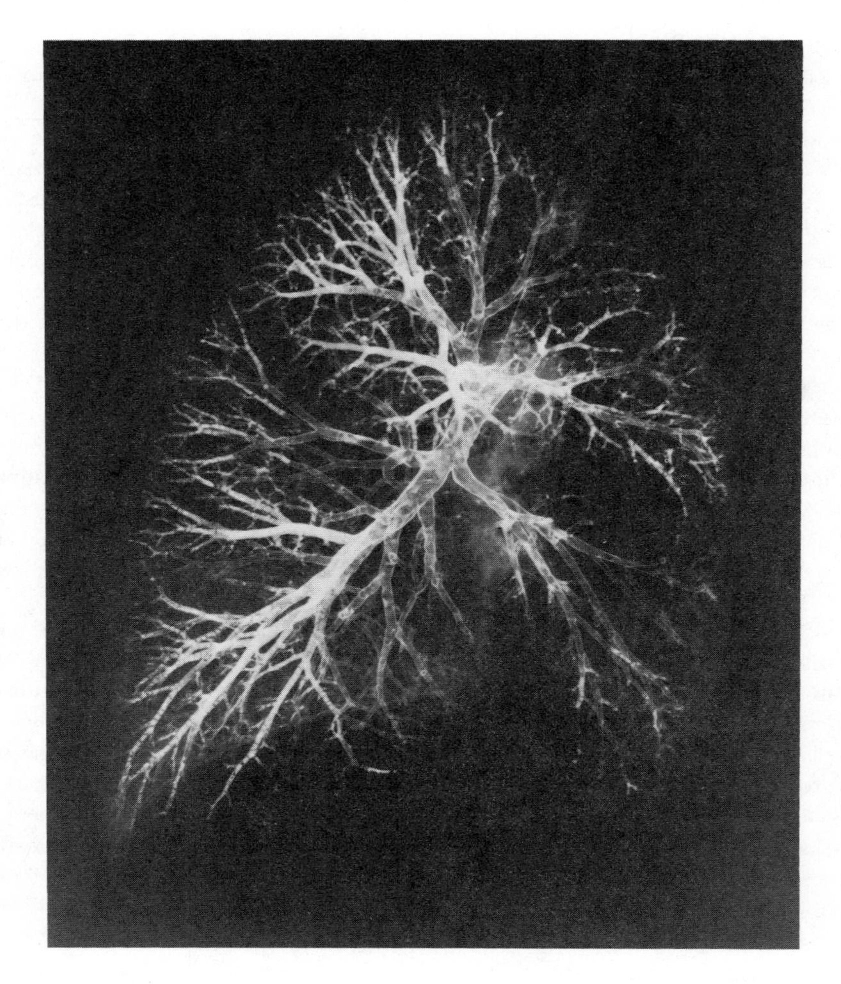

Figure 7.1 Radiograph of airway of distended human lung; right main bronchus with lobar and a few subsequent branchings (probably not beyond eighth generation) as outlined by contrast media.

The first few generations of the air-conducting system, the trachea and bronchi, have rather thick connective tissue walls containing cartilaginous plates that support the wall except in the posterior aspect where the encircling plates are incomplete. There is also a thin layer of smooth muscle located just beneath the mucous membrane lining these conduits. The muscle bundles are arranged in a spiral fashion encircling the conducting tube. As the air tubes further divide, the cartilaginous elements disappear and the walls become thinner, but the spiraling smooth muscle becomes proportionately larger. As the last few generations of conducting airways are reached, the wall becomes still thinner, and by the time the terminal bronchiole, the sixteenth generation, is reached it is very thin. In the seventeenth generation of branches an occasional small outpouching appears in the wall. This cup-shaped outpouching is the alveolus, the site of molecular gas exchange between air and blood. The airways having only a few alveoli protruding from the walls are termed resiratory bronchioles, but when the alveoli make up most of the wall they are called alveolar ducts. When the final division is reached and the walls are entirely formed by alveoli, the passage is termed an alveolar sac or atrium. Elastic fibers, coursing roughly parallel to the air tubes, exist from the trachea all the way to the alveolus, where they form a network within the alveolar wall.

Lining the surface of the air tubes is a membrane comprised of an epithelial layer supported on a base of interlacing reticulin, collagen, and elastic fibers embedded in a proteinaceous matrix, the ground substance, containing cells such as fibroblasts and macrophages. The epithelial surface of the trachea and bronchi as far out as the terminal bronchioles is composed of columnar and cuboidal ciliated cells; interspersed between these are mucus-secreting goblet cells, discharging onto the surface of the airways. In addition, there are clusters of large mucus-secreting glands scattered throughout the subepithelial region of the bronchi out to the ninth or tenth generation. These glands discharge their contents by way of ducts opening onto the surface of the airway lining. Their discharge plus that form the goblet cells form a carpet of mucus that is distributed rather evenly over the surface of the ciliated cells, which extend out to the terminal bronchiole or sixteenth generation of airways. This carpet of mucus is continuously moved by the action of the underlying cilia from the places of manufacture up through the trachea, where the mucus is then swallowed or expectorated. This system, known as the mucociliary escalator, is the mechanism whereby particles deposited on the airway surface are removed.

Beyond the terminal bronchiole the surface lining of the airways is no longer ciliated and becomes thinner as the cuboidal cells grow shorter and assume the shape of progressively flatter cells. The alveolus is the terminal unit, whether it arises from the wall of the respiratory bronchiole, from the alveolar duct, or from the lumen of the alveolar sac. The alveoli are clustered in a tightly packed manner, abutting upon themselves and the walls of conducting airways of their own and adjacent units. Each alveolus is lined by large, flat, very thin cells, the type I pneumocytes. These lie on a thin layer of reticulin plus elastic fibers that cover a network of capillaries. The capillaries in turn are separated from the reticulin and elastic fibers of the interstitium by large flat cells making up the endothelium that lines the capillaries. A scant amount of ground substance plus occasional connective tissue cells and macrophages lie in the interstices of the reticulin network. The alveoli are so closely packed that a single pulmonary capillary usually serves two or more alveoli abutting upon the pulmonary capillary network. The net pressure of the blood within the capillaries is below that in the interstitial space; thus liquid normally does not pass into the alveolar spaces.

Other cells of importance are found in the alveolar unit. The type II pneumocyte is found scattered at various places in the alveolocapillary wall as part of the surface lining cells. It apparently secretes surfactant, a material that affects the surface tension of the ultra-thin liquid layer located at the actual air–tissue interface of the alveolus. Alevolar macrophages (4), large free cells of the reticuloendothelial system measuring 10 to 50 μm in diameter, are located both in the interstitial tissue and within the alveolar spaces lying free on the alveolar surface. These are large mobile cells whose most spectacular function is phagocytosis. They also are involved in immunologic reactions. The number of these cells can be greatly augmented by the presence of dust particles within the alveoli. Particles not exceeding 10 μm in greatest dimension are readily ingested by the macrophage. The fate of such particles is discussed later.

The major branches of the pulmonary artery and veins follow closely the branching of the airways. In addition to these there are very thin walled vessels, the lymphatics, which begin within the interstitial spaces at the level of the respiratory bronchiole and alveolar ducts and ascend in a merging fashion with lymphatic channels from other regions of the lung along the course of the arteries and veins. Similar vessels underlie the pleura, the membrane that covers the lungs and lines the chest cage. These channels are for the purpose of removing excess interstitial fluid. Along their course are collections of lymphoid tissue, the smallest being arranged in cuffs around the respiratory bronchiole and adjacent arterioles. The lymph fluid flows toward the lung root, and larger aggregates of lymphoid tissue are found along the course of these lymphatics as the channels coalesce and become larger. The nodular clusters of lymphoid tissue in the root of each lung are quite large. The various regional aggregates of lymphoid cells serve as collecting points where particles penetrating to the interstitial fluid can be sieved or filtered out of the liquid phase and accumulated in the lymphoid tissue, thus minimizing introduction of such particles into the circulating blood by way of the main lymph vessels that spill their contents into the innominate veins.

Each lung is covered by a thin membrane except at its root, where the main bronchi, blood vessels, and lymph channels pass into or out of the lung. This membrane, the pleura, reflects off the lung root and passes onto the surface of the inner aspect of the thoracic cage, thus forming a closed sac bordered by the lung, the chest wall, and the mediastinum. It contains a small amount of constantly exchanged fluid. Covering the surface of the pleura are large flat cells of mesothelial origin lying on a thin layer of loose areolar connective tissue containing blood and lymph vessels, nerves, and a well-developed layer of elastic tissue along with its complement of macrophages and fibroblasts. Of particular importance for this discussion is the fact that alveoli abut directly onto the inner aspect of the elastic layer and supporting structures of the pleura. Hence there is the possibility that particles reaching these alveoli might pass directly on into the pleura. In this way particles in addition to those that move by way of the lymphatics might ultimately come into contact with cells of the pleura.

3 BEHAVIOR AND FATE OF PARTICLES THAT ENTER THE RESPIRATORY SYSTEM

Dust deposition is to a large degree dependent on the concentration and the physical nature of the specific dust particles in the air. A substantial body of information exists with respect to the respirability of particles of differing dimensions and shapes, their deposition in various compartments of the air-conducting and distal spaces of the lung, and the speed and effectiveness with which rapid and slow clearances occur (5–7).

If, by suitably gentle technique, one digests the lungs of older individuals who have worked in the dusty trades, a residue will be obtained that can be assumed to have come from exogenous sources by way of the airborne route over the years (8). These tiny particles have a most interesting size range. Approximately half will be smaller than 0.5 μm in diameter. Of those that are larger, almost all will be between 0.5 and 5.0 μm in diameter. Fewer than 0.2 percent of the total will be larger than 5.0 μm in diameter, and less than 0.002 percent will be larger than 10 μm in diameter. If a fiber is defined as a particle having parallel sides and an aspect ratio of 3 or more, fibers for the most part will be observed to be less than 50.0 μm long, although some may be as long as 200 μm. The diameters of these fibers will be less than 3 μm. In contrast, in samples of the ambient air to which the general public or those who work in the dusty trades are exposed, one finds particles of these dimensions, but in addition, many of much larger diameter and length. Why is the long-term retention of particles limited to the sizes just described, even though millions of particles of greater diameter or length become airborne, therefore have the potential for entry into the respiratory system? The explanation arises from our knowledge of the behavior of particles suspended in air (aerosols) and the anatomical structure of the lung as described in the preceding paragraphs.

Particles can vary markedly in shape and density, and both these factors play a role in the behavior of particles in air suspension. For our purposes we consider all particles as being spheres of unit density, with the understanding that there could be some variation between particles with respect to speed of settling, depending on their shape and density.

Several physical forces are conducive to the removal of particles from an air suspension and their deposition on surfaces of the respiratory system (5). Particles suspended in a moving airstream possess inertial forces tending to maintain the direction of motion of the particle. When the air column changes its direction, as at a branching point of the conducting system, or in the tortuous passages of the nose, the entrained particle tends to continue in its previous direction and to be cast on the surface. This effect is directly proportional to the size of the particle and the speed of the airstream, thus of the particle, and inversely proportional to the radius of the tube. Gravitational forces also remove particles from the airstream and precipitate them on the surface of the respiratory system. The terminal settling velocity of a particle is directly related to its density, the gravitational constant, and the square of the particle diameter. It is inversely related to air viscosity. Since the gravitational constant and air viscosity are the same at all times, the terminal settling velocity is predominantly related to the other two factors, plus the distance through which the particle must fall and the time permitted for the event to occur. Deposition of particles by diffusion is limited to those of a diameter smaller than 0.5 μm and predominantly to those smaller than 0.1 μm. The smaller the particle, the more rapid the diffusion movement that can occur. The electron microscope size particles are relatively uninfluenced by any deposition force other than that of diffusion, and the fact that such large numbers of particles of this size are found in the lung residue indicates that diffusion can play a major role in their deposition. It has been suggested that electrostatic and thermal forces may play a role in deposition of particles in the lungs, but this is still uncertain.

On the basis of known physical behavior of particles in air suspension and the anatomical arrangement of the conducting tubes, it can be predicted that particles larger than 10 μm in diameter would be removed completely in the passage of the airstream through the nose and upper airways and that the particles between 5 and 10 μm would be deposited primarily in the upper airways on the mucociliary escalator. Only the particles in the range of 1 to 2 μm would be likely to penetrate into the deeper portions of the lung, where deposition in the alveoli could occur by gravity and diffusion. On the basis of these calculations, particle deposition by sedimentation would be least for particles having a diameter of 0.5 μm; but deposition of particles smaller than this might be increased by diffusion, particularly in the most distal portions of the air system. Numerous actual experimental determinations have confirmed this general distribution for location of deposition. Particles larger than 10 μm in diameter are almost completely removed in the nose, and few, if any, reach the smaller conducting tubes of the lungs. Some smaller particles also are deposited in the nose, but the majority pass through and then are deposited, dependent primarily on their diameters, along the upper or lower airways. Particles greater than 3 μm in diameter have very little opportunity to penetrate beyond the mucociliary apparatus and to be deposited in the most distal alveolated portions of the conducting tubes. Since almost all the particles larger than 3.0 μm in diameter fall on the mucociliary escalator and are removed, there is a reasonable explanation for the very small number of particles of larger size that are found in the lung residue after a lifetime of exposure to aerosols of ambient air that undoubtedly contains particles of larger size.

A fiber represents a special case in terms of deposition. As is true of other particles, the settling velocity of a fiber is dependent primarily on its diameter. In a moving airstream, fibers tend to align their length parallel to the direction of air flow. The fibers that are straight and rigid therefore present an end-on aspect essentially that of their actual diameter. Fibers that are curved, curled, or bent in a U-shape have an end-on aspect equal to the width of the curl or curvature. Insofar as interception is concerned, there is thus a much greater chance for deposition of the nonstraight fibers, a factor of considerable importance in the narrow airways and along the boundaries of air flow close to the surface. It has been demonstrated that curly fibers penetrate to the deeper portions of the lung much less readily than do straight fibers of equivalent diameter (9). Length becomes important also to the degree that the fibers are distributed in a random way in the moving airstream. Thus a fiber 100 μm long oriented at right angles to the direction of flow will have a much greater chance of impacting on the surface than will fibers oriented parallel to the direction of flow. Although occasionally a fiber 200 μm long is observed in the lung dust residues, by far the majority are shorter than 50 μm.

The particles deposited beyond the terminal bronchiole and mucociliary escalator experience one of five fates. It has been suggested that the mucus sheet of the terminal bronchiole is contiguous with a more distal, slow-moving liquid layer manufactured by the Clara cell and the type II pneumocyte. Particles falling on this sheet could slowly be pulled up onto the mucociliary system, thus be removed. As a second fate, there is abundant evidence that particles deposited in the alveolus-bearing part of the lung can quickly be ingested by macrophages. The small size of the particles deposited in this area favors ingestion by the macrophage except for the long fibers thin enough to penetrate to this area but too long to be completely engulfed. The limiting length has not been proved, but fibers up to 10 μm long are readily ingested, and even longer fibers can become surrounded by the spreading membrane of the macrophage. By the mechanism of ingestion, particles of various shapes are taken into a cell whose function is to house and process foreign material. In addition, many of the mobile macrophages with their ingested particles move up onto the mucociliary escalator and appear in the sputum or are swallowed. Other macrophages filled with particles simply live out their life and die, discharging the particles, which are reingested by new macrophages. This process is repeated indefinitely. The life of the macrophage under ordinary circumstances is measured in terms of weeks and perhaps even a month or more. Its life can, of course, be shortened if the ingested particles are especially toxic, as is the case with free crystalline silica, which kills the macrophage in a period of hours or days. Materials such as coal, iron, asbestos, and glass do not appear to substantially alter the longevity of the macrophage.

A third fate of particles involves their movement, either naked or inside the macrophage, across the alveolar surface and into the interstitial substance. Once having moved there, some particles appear to remain in this space either free or within macrophages, and others enter the lymphatics and are sieved out in the regional lymphoid tissue, where they may remain indefinitely. A few particles pass on through by way of the lymph into the systemic blood, which then carries them to other organs of the body. The precise mechanism by which particles move through the alveolar surface may involve more than one method, and there is little agreement on this question. A fourth fate of particles is simply to remain free on the surface of the alveolus.

The fifth fate of particles is to be dissolved partially or completely or to disintegrate into smaller particles over a period of time. The small diameter of the respirable particles, hence the large ratio of surface area to weight, promotes the rate at which these reactions can occur. Some inorganic particles such as those comprised of $CaSO_4$ (gypsum) or $CaCO_3$ (limestone, marble, dolomite) are sufficiently soluble in body fluids to be more or less rapidly removed. Fibers of asbestos or glass, thin enough to be respirable, commonly are thought of as being impervious to the influence of body fluids and enzymes such as those contained within macrophages. Evidence is accumulating, however, to show that fibers of this nature fracture transversely into shorter fragments. Moreover, there now is reason to believe that both glass and asbestos fibers of diameters below 0.5 μm actually may dissolve within the lung tissue.

The self-cleaning mechanism of the lung appears to be efficient. Although there is some variation of this capacity between individuals, it has been reported that less than 1 percent of the total calculated amount of dust inhaled remains more or less permanently in the lung (5, 10). The paucity of studies bearing on this effect in humans warrants some caution in accepting the figure of 1 percent as applying under all conditions. Nevertheless, the mechanism for removing deposited particles is remarkably effective. It should be borne in mind that the process of lung cleansing adds a portion to the total gastrointestinal tract exposure to exogenous particulates, since a substantial part of the dust brought up on the mucociliary apparatus is swallowed.

4 THE RESPIRATORY FUNCTION OF THE LUNGS

A continuous supply of oxygen to and removal of carbon dioxide from the environment of tissue cells is an essential requirement for life. The demand varies from the resting state to the 10 times or more higher metabolic rates encountered during heavy exercise. The respiratory and cardiovascular systems are intimately associated in this accomplishment. The mass movement of air into and out of the lungs required to satisfy this wide range of demands by the tissue cells is accomplished by the rhythmic application of muscle force to the thoracic boundaries, causing the thorax to enlarge and diminish its volume in the familiar act of breathing. The lungs are attached to the thoracic boundaries by a thin layer of liquid that permits them to follow the excursions of the thoracic walls, thus change their volume in a spatially even manner. The lungs not only are stretchable but in addition are elastic and can recoil by virture of the abundant elastic tissue in the walls of the airways, alveoli, and blood vessels, and also because of the tension in the ultra-thin layer of liquid overlying the surface of the alveoli. This overlying liquid layer is elastic by reason of its simulation of a "bubble," with consequent surface tension properties. In fact, this force would collapse the lung if not counteracted by phospholipid particles, the "surfactant" manufactured by the type II pneumocyte and discharged into the thin liquid layer (11).

The rate of air exchange within the lungs reflects the depth and frequency of respiration, which in turn is controlled by the respiratory center. Under normal circumstances the ratio between alveolar ventilation and capillary blood flow is essentially the same throughout the lungs, and the arterialized blood is maintained at a remarkably constant oxygen and hydrogen ion concentration over a wide range of physical activity. There is, however, an inequality of distribution of air and blood flow to the individual units of the lung, the lower or more dependent units receiving a larger portion of the total air and blood flow than do the units higher on the vertical axis.

Since both the respiratory and cardiovascular systems are intimately involved in the movement of oxygen and carbon dioxide between the ambient air and the body cells, it is important to know which of these systems is the limiting factor. Without going into the details, the evidence supports the view that the cardiovascular system limits maximum ability for physical performance in normal persons. Because of this, in an overall sense the respiratory apparatus has a larger reserve capacity for function than does the cardiovascular system. Therefore a substantial impairment of pulmonary function can occur before there is a recognizable loss of capacity for physical effort, especially since the lungs are made up of literally millions of identical units and most types of pulmonary injury involve only some of the units while sparing the others.

5 REACTION OF THE LUNG TO INHALED PARTICLES OF DUST

Utilizing knowledge of the cellular components and organization of the lungs plus the manner in which cells respond to stimulation or injury, one might anticipate the various reactions this organ would display to the deposition of dust. As stated earlier, whether such reactions will develop depends on the nature and number of specific particles deposited and retained, as well as the influence of coexisting inhaled agents and the reactivity of the host or individual.

That dust particles can stimulate the smooth muscle distributed in a circular fashion within the thin-walled portions of the airways, thus narrowing the lumen and raising the resistance to airflow, has been demonstrated to occur as a reversible reaction following the inhalation of high concentrations of dust (12). This response in most persons requires high concentrations of dust and is not recognized to occur commonly in the workplace. In persons who for one or another reason are hyperreactive, however, lower concentrations of dust may evoke a recognizable response.

The deposition of dust on the mucociliary apparatus normally stimulates a flow of mucus. If the production of mucus is excessive, or if it is not removed adequately, it can accumulate in the airways, thus reducing the lumen of the conducting tubes and elevating the resistance to airflow. Furthermore, prolonged stimulation of the mucus-secreting glands and cells can lead to hypertrophy or enlargement of these structures. The enlarged subepithelial glands can encroach upon the lumen and cause persistent narrowing of the airways and elevation of resistance to airflow.

Particles of dust lodging on and beneath the surface of the alveoli stimulate the recruitment and accumulation of macrophages in this area. There can be enough of these cells to partially or completely fill some alveoli, but because of the cells' ability to move out onto the mucociliary escalator or into the lymphatics, this is unusual. Some particles are always found lying free on the surface of the alveolus, but whether these have never been phagocytized or whether they are waiting for reingestion cannot be determined. Presumably particles could stimulate any of the lung cells they come directly or indirectly into contact with, or they may act through the intermediary of cells such as the macrophage. Some particles lodged on or just beneath the alveolar surface appear to induce a proliferative reaction and become overgrown by the surface cells—at least they are found embedded in a mass of cells involving the surface and the immediate substructure of the alveoli. This reaction can evolve to include cells such as lymphocytes, macrophages, and polymorphonuclear leucocytes, as well as other components of the interstitial or connective tissue and the regional lymphoid collections of the lung. At times such reactions are transitory, but they may become chronic or persistent.

Since fibroblasts are present in the interstitium of the lung, these cells may be stimulated to form excessive amounts of reticulin or collagen. Excessive collagen formation is likely to accompany prolonged or chronic inflammation in most organs of the body. This is a part of the familiar formation of scar tissue, whether it be in the skin or in deeper portions of the body or in the lung. Pulmonary fibrosis is a common sequela of chronic pulmonary inflammation. The pathogenesis of pulmonary fibrosis caused by deposition of some kinds of dust is not entirely clear. It has been postulated in recent years that since the macrophage contains small vesicles holding potent enzymes, these substances may be released by certain ingested particles, but not all kinds, thus altering the cell milieu and causing premature death of the macrophage. The products released by the macrophage are then believed to stimulate the fibroblasts to lay down excessive amounts of collagen, thus promoting the development of tissue fibrosis (13). It has also been suggested that fibers not completely engulfed within the macrophage but protruding through the wall permit leakage of the macrophage content, and this material might also stimulate the fibroblasts to increased activity (14). These concepts are interesting because some dust particles are known to be relatively inert and not capable of causing an excess of fibrous tissue production. Such particles, though readily ingested by the macrophage, do not cause the premature death of the macrophage. In contrast, free crystalline silica, a known fibrogenic dust, has been demonstrated to exert this effect.

Malignant transformation can be the fate of any cell that can divide, and those of the lung are no exception. The cells lining the surface of all the airways and alveoli have a particularly fast turnover rate, hence are probably more vulnerable to carcinogenic alteration by a dust or other agents. In addition, particles lodged on the surface of the air-conducting tubes might promote an even higher turnover rate of the surface cells, thus making them unusually vulnerable to carcinogenic substances. Toxic agents adhering to particles could leach off onto the cells with a similar effect. A combination of these effects could lead to metaplasia or could play a role directly or indirectly in malignant transformation of the cells lining the conducting airways.

On the basis of the foregoing considerations, one could anticipate that dust deposited in the lungs might induce:

1. Little or no reaction of any kind.
2. Hyperproduction of mucus secretion.
3. Hypertrophy of mucus-secreting glands.
4. Macrophage recruitment and ingestion of particles.
5. Chronic proliferative or inflammatory reaction.
6. Reticulinosis.
7. Fibrosis.
8. Cell metaplasia or malignant transformation.

6 INFLUENCE OF ALTERED LUNG STRUCTURE ON PULMONARY FUNCTION

Narrowing of the airways, whether caused by muscle contraction, accumulation of mucus, or hypertrophy of the mucous glands, will increase the resistance to airflow through the involved tubes. It should be noted that resistance to airflow through a tube is a function not only of the length of the tube but of the fourth power of the radius (Poiseuille's law). For this reason, relatively small changes in the radius over a substantial length of the tube have a marked influence on the resistance to airflow. If the increased resistance to airflow occurs more or less uniformly throughout all the lung, the work of breathing is commensurately increased, and the volume of air moved per unit of effort is reduced. This not only decreases the maximum available capacity for mass movement of air but causes the expenditure of unusual respiratory effort at all volumes of air movement. Breathlessness or "shortness of breath" is closely related to the awareness of respiratory effort; hence the increased use of respiratory force to move each unit of air leads to the development of breathlessness at lower than normal levels of exercise and to a reduction of the maximum capacity for physical effort. If, as is usually the case, there is focal or regional variation of airway resistance, there will be a shunting of airflow away from the alveoli supplied by the airways experiencing the higher resistance. This leads to regional alveolar underventilation and produces a mismatch between ventilation and capillary blood perfusion in the lung region involved, causing the blood leaving the underventilated area to have a subnormal oxygen and an elevated carbon dioxide content. As a consequence, the arterial blood going to the tissues will be below normal in oxygen content. To some degree, coexisting overventilated alveoli can compensate by increased carbon dioxide removal; but if the regional hypoventilation is extensive, there can also be an accumulation of carbon dioxide in the arterial blood returning to the tissues.

The body attempts, with partial success, to correct this mismatch between ventilation and perfusion in the following manner. The low alveolar oxygen tension developing in the hypoventilated regions produces an effect on the alveolar blood capillaries causing partial closure of the capillaries, thus an increase in capillary blood flow resistance and a shunting of blood from the hypoventilated alveoli to those that are more normally ventilated. If the number of underventilated alveoli involved is relatively small, this is an effective mechanism; but if many alveoli are involved, the narrowing of the alveolar capillaries produces an elevated resistance to blood flow in the entire pulmonary circuit, with consequent pulmonary artery hypertension and the bad effects that flow from it. From these comments it can be seen that airway narrowing can lead to rather serious malfunction of the cardiorespiratory system and can be the cause of severe physical impairment, ultimately leading to death.

Chronic intrapulmonary inflammatory and fibrotic changes cause distortions of pulmonary architecture leading to functional derangements. In regions where the cellular infiltrate or fibrosis destroys functioning lung tissue *en bloc,* the effect is similar to that of simply removing this amount of lung. If the volume of tissue involved is relatively small, there is little or no measurable alteration of pulmonary function. If the volume of destroyed tissue is large, the remaining normal lung will be much smaller and will behave in an overall less compliant fashion than would a larger volume of lung. In addition, there will be a measurable loss of pulmonary vascular bed or diffusing surface. Moreover, because of the loss of arterioles and capillaries, more force will be needed to drive the normally required total amounts of blood flow through the pulmonary circulation. This in turn leads to pulmonary artery hypertension and the abnormalities consequent to that condition. If the fibrotic or inflammatory reaction is arranged as ribbons running diffusely throughout the lung, the entrapped normal lung units will be less compliant and the work required to change lung volume and produce a unit of mass movement of air will be increased. The work of breathing will be augmented, and the maximum ability to move air into and out of the lungs will be curtailed. Some portions of the entrapped normal lung between the ribbons of diseased tissue will be underventilated, and a mismatch between ventilation and perfusion will ensue.

It is of interest to observe that although an elevated resistance to airflow and a decrease in pulmonary compliance affect lung mechanics differently, the end results in terms of overall physiological impairment of the respiratory apparatus are quite similar. Both cause an increase in work of breathing, a loss of maximum ability to move air into and out of the lungs, a mismatch of the ventilation perfusion relationship, with consequent arterial hypoxia, and an elevation in the pulmonary artery pressure. One difference is that in airway obstruction there is carbon dioxide retention and an elevation of arterial pCO_2, whereas in the abnormalities associated with loss of compliance caused by chronic inflammation or fibrosis, carbon dioxide transfer is interfered with relatively slightly or not at all and an elevated pCO_2 usually does not develop.

7 CRITIQUE OF THE METHODS USED TO STUDY THE PULMONARY EFFECTS OF INHALED INORGANIC DUST

Knowledge about the way in which humans will respond to inhaled dust should be derived from the demonstrable effects of inhaled particles as observed in humans. Extrapolation from other modes of study such as animal experimentation or cell culture is speculation, even though this may be warranted or even necessary under some circumstances. Unfortunately the technical conditions imposed on observations of humans, especially when retrospective observations are made or new substances are being considered, at times limit the availability of the human study modality. Therefore experimental animals or isolated organ or cell systems are often utilized to explore the possibilities of inducing tissue reaction in the lung, the factors that control the reaction, and the biological mechanisms involved. The animal and cell modalities are used most appropriately for examining aspects of the overall question that cannot be approached by human study alone. This does not, however, assure the validity of extrapolating from animal studies to human experience.

The advantages provided by animal or cell experiments are several. The exposure can be controlled with respect to varying its concentration, its duration, and whether single agents or combinations are presented. The opportunity to follow the course of events by serial sampling of the tissue is available only with animal or cell experiments. Also it is possible to repeat experiments under controlled conditions in animals. The shorter life span of small animals is an advantage in that the process of carcinogenesis is telescoped into a shorter period of time, thus shortening the duration of the experiment. Although the methods used for examining tissues anatomically or biochemically are essentially the same whether animals or humans are being examined, there are some advantages to the use of animal tissue. Since dust particles are distributed in a nonhomogeneous way, the effects have a patchy distribution also. There is an advantage then to being able to examine whole lung slices and to do biochemical analysis, such as for hydroxyprolene on whole lungs of animals.

These manifest advantages explain why the animal or cell models are so widely used. There are, however, some very important and virtually insurmountable disadvantages associated with the use of these models if one intends to extrapolate to human experience. With respect to isolated cell systems, there are two important disadvantages. First, the systems are not under the normal influence of humoral and cellular controls provided by neighboring cells or emanating from other organs and circulating cells that exist in the intact animal or man. Second, long-term experiments cannot be done, since cell survival is rather brief in any event.

Animal experiments avoid these two difficulties but are limited in several other ways. There is always the influence of species difference with respect to carcinogenesis. Humans are but a single species of animal and do not invariably mimic the tissue reactions of other species. Also, experimental animals usually are kept under environmental conditions quite different from those of humans. For example, animals other than man do not smoke tobacco, a common biologically active coexisting agent in human respiratory exposure patterns. Although a 2-year-old rat may be equivalent to a 60-year-old human with respect to cell line aging, it will have had 58 fewer years of exposure to the coexisting agents usually being studied. Moreover, physical and emotional stress induced by caging conditions can influence the biological response of animals. How similar are the environmental conditions of man throughout his lifetime to those of the experimental animal in this respect?

The exposure concentrations used in animal experiments are usually at a level very much higher than those occurring in human experience. Understandably, this is done deliberately to guarantee an effect in animals if it is at all possible to produce one. Unfortunately, when a biological effect is produced, there is all too often little effort to determine or take into account the response at lower concentrations—the dose–response curve.

There is also a difficulty posed by the number of animals available for use in experimental protocols. Contrary to popular belief, an unlimited number of animals—or, in many experiments, the number actually needed for the kinds of extrapolation commonly made—is not realistically available for most animal experiments. This is especially true of the experiments seeking a no-effect level, as well as when the biological response of any type is meager. As a matter of fact, the number of individuals at risk and as controls is usually superior in human epidemiologic studies to the number available for most animal experiments. Few research laboratories can afford to buy or breed, feed, and house the number of animals needed to establish epidemiologic validity in their studies. Because of this fundamental defect, the conclusions reached by some animal studies are of questionable validity.

Observation of the effects of inhaled dust in humans has been the conventional way for establishing whether a biological reaction occurs. Clinical and autopsy or histological examination of the lungs of men working in the presence of the agent under study is the mainstay of this approach. A difficulty arises, however, because the cellular components of the lung will react in the same manner to a number of different kinds of stimulating agent. For example, the development of a biological abnormality in a person employed in a particular way might suggest the possibility of a relationship between an agent in the occupational environment and the biological disturbance observed. On the other hand, since the same biological abnormality of the lung can develop from various causes, the single isolated case may in fact be caused by something other than the agent present in the specific dust exposure. Therefore, although the clinical and histological approach is of great importance in determining whether an abnormality is present, other modalities to *prove its relationship to a specific agent must be used.* For this purpose the techniques of epidemiology are useful, and the reader is referred to Chapter 5 for discussion of this discipline.

To return to the advantages and disadvantages inherent in observation of human populations, it cannot be stressed too strongly that the greatest advantage of this approach is that no extrapolations are needed. Moreover, the exposure is that of a real-life circumstance with respect to the multiple other factors that may be playing a role. There is an advantage also in the greater possibility for observation of larger numbers of persons both at risk and in the control population. These advantages sometimes are offset by other problems that pose large difficulties. The first of these is the difficulty of making retrospective estimates of the total exposure of the individual to the specific agent in question, expressed in quantitative terms over long periods of time. At times the only available expression is that of duration of employment. Undocumented changes in the environment of the workplace over the many years of employment and the lack of records or measurements of the various agents present during the same period cannot be corrected for. Under such circumstances one can only compare the overall clinical or mortality experience of the observed group to that of some other group, usually the population at large. From such observations it is impossible to derive information usable for establishing the dose–effect relationship. Under such circumstances one also cannot be sure of just which of several coexisting agents is the cause of the biological reaction. Nor can one be sure that cofactors are not operating as the ultimate biological stimulus. To attribute the biological reactions observed to specific agents under such circumstances, though at times a temptation, is a practice of questionable validity.

A second disadvantage associated with human epidemiologic studies is that whereas in animals the lungs may be examined by microscope or chemical methods at any point in time relative to the exposure, these modalities are available in human epidemiologic studies only at death. Since only a portion of the deaths from any population are autopsied, there is a strong possibility of bias posed by autopsy selection, and epidemiologic data derived from autopsy studies must be accepted with some caution. In addition, autopsies rarely are done with the epidemiologic study in mind, and important observations may be lacking, whereas the opposite is true in animal experiments. To the extent that epidemiological studies are dependent on death certificates, they are subject to faults such as erroneous diagnosis, "styles" or "fashions" of diagnosis, incomplete diagnosis, and very commonly, partial unavailability of death certificates. These faults become particularly grievous when small numbers of a highly localized population are compared to the population of a state or of the entire country.

Because of the difficulties attending the use of mortality data, an alternate epidemiological approach is to relate the data derived by medical or physiological examinations during life to exposure up to that point in time. This approach does not take into account the results that might occur over a full life-span, and it is particularly limited with respect to the search for relationships between dust exposure and carcinogenesis. Nevertheless, with respect to ill health or impairment of function, a properly designed epidemiologic study during life can in a very useful way serve to compare the frequency of types of impairment or disease in the study group versus an internal control population ranked by the severity of exposure or to an external control population of persons not exposed to the agent in question. This approach has a high degree of validity.

It is certain that biological reactions too minor in extent or intensity to be discovered occur in lung tissue before evolving to an extent recognizable by our methods of searching for these changes during life. How should one view the importance of these subclinical alterations of tissue? The goal is to *prevent disease,* with its associated impairment of function leading to a reduced capacity for normal activity and in some instances to shortened life expectancy. Therefore it is important to distinguish between a small and limited area of tissue injury and what we mean by disease. The differentiation hinges on whether impairment of vital function will occur or has occurred. Cell death and orderly replacement by replication of cells characterizes the human lung from birth to death. Physical factors and external agents such as viruses and bacteria active in small areas of the lung commonly influence the orderly process of cell death and replacement during the state we call "good health." When one examines the body of an adult who has given no prior evidence of lung disease, some pulmonary fibrosis representing the end stage of injury is invariably found. Since these effects are minor in extent and occur in an organ possessing more reserves than can be used by normal persons, such minor scars do not in any way interfere with pulmonary function or life expectancy. There is, therefore, a quantitative as well as a qualitative aspect to body impairment and disease. In the light of these comments, it is imperative to realize that there is no sharp line of demarcation between being well or ill, normal or diseased, or injured or uninjured. We can speak in rather broad terms of the way in which gases and particles may or may not affect the lung, but insofar as impairment or disease is concerned, the quantitative aspects are probably more important under most circumstances than are the qualitative.

With the exception of sputum cytology, microscopic examination of the lungs of humans is not available for the purpose of early recognition of change in the individual. Epidemiologic studies during life are limited, therefore, to clinical, radiologic, pulmonary function, and sputum cytology modalities. Advantageous as these are, their inherent weaknesses must be recognized if the validity of conclusions derived from their use is to be evaluated.

The clinical history is subjective and difficult to confirm. Physical examination, chest radiography, and pulmonary function measurement, although more objective, exhibit a rather large variation of values among normal persons. Therefore, the recognition of "change" in an individual is difficult unless the observation can be made before as well as after the onset of exposure. This difficulty can be lessened by comparing the average of observations of one group to that of another group serving as a control. The various techniques for these examination modalities require skilled personnel and equipment not always available. In spite of these difficulties, virtually without exception these modalities, when properly used, will permit recognition of deviation from normal before there is interference with vital body functions, thus they are of great use for studying the tissue reaction to inhaled dust during life.

The chest radiogram exhibits shadows cast because of the greater absorption or deflection of X-ray photons by tissues of greater density. Highly cellular or fibrotic tissue is more dense than normal tissue, and when the difference is great enough or when the shadow is located where it should not exist, the normal shadows are altered. A difficulty arises because there is considerable variation between normal persons in the shadows cast by normal lung structure such as blood vessel and supporting lung tissue. Grossly abnormal lungs are readily recognized by alteration of the normal pattern of shadows, but minor tissue alterations make either no or very slight change in the pattern of densities. Therefore there is always a gray area of differing opinions between readers of the films from persons experiencing slight tissue alteration: are the shadows present in the radiogram those of normal or abnormal structure? For epidemiologic studies, these differences can be lessened by the use of multiple readers using the UICC system of notation and standard reference films (15).

Measurements of pulmonary function are aimed primarily at recognition and quantification of changes in pulmonary airway resistance, pulmonary compliance, the alveolar ventilation–perfusion relationship, and the total effective area of the alveolocapillary membrane. Using rather sophisticated methods, these aspects of pulmonary mechanics can be measured rather accurately. Recognition that an alteration of any of these functions has occurred is hampered by the wide range of values existing between normal persons. One of the less difficult, thus more widely used measurements, is the fast vital capacity (FVC), or volume of air expelled by prolonged, explosively applied maximum expiratory force beginning at the position of full inspiration. This not only indicates the stroke volume of the respiratory pump but also, because maximum expiratory effort is applied rapidly, it reflects the speed with which the air can be pushed through the distribution system, and this in turn reflects airway resistance. For this purpose, the volume expelled in the first second (FEV_1), expressed either as a percentage of the predicted volume or as a percentage of the total volume expelled, serves as the reflection of airway resistance. Other portions of the expiratory curve can be used for the same purpose.

With a few exceptions, such as the presence of large space occupying masses within the thorax or extreme obesity, pulmonary compliance is in part reflected by the total lung volume at maximum inspiration. Accurate measurement of total volume requires the use of a tracer gas or a plethysmograph, and it is therefore seldom used. An indirect expression of compliance can be obtained in the following manner. Both an elevated airway resistance and a lessened compliance will cause a reduction in FVC, thus when looked at alone will not differentiate between the two conditions. When airway resistance is the cause of the subnormal FVC, however, the FEV_1/FVC ratio is substantially lower than normal. In contrast, when the subnormal FVC is caused by a lessened compliance, this ratio is either normal or slightly reduced, or it may actually be increased.

The foregoing methods of measurement are useful in epidemiologic studies when a contrast between normal and moderately severe alteration of pulmonary function is expected. They will not suffice for recognition of minor changes in airway resistance or pulmonary compliance. This calls for more sophisticated measurement, utilizing body plethysmography.

In most circumstances the ventilation perfusion relationship is reflected by the arterial blood–gas measurements. Alteration in the size and character of the alveolocapillary membrane is best measured by the diffusion coefficient for carbon monoxide. Since these measurements require carefully controlled technical procedures that are not always available, they are rarely used in epidemiologic studies. Nevertheless, when one searches for biological reactions to dust in the lungs, these modalities should be used at some point in the overall study.

8 BIOLOGICAL EFFECTS OF SPECIFIC INHALED INORGANIC DUSTS

The variety of inorganic dusts to which humans are exposed either in pure or mixed form is large. A prototype of the effects of several will demonstrate the biological effects produced.

8.1 Nonfibrogenic or "Inert" Dust

Although some dusts are classified as biologically inert, the designation can be misleading. A more correct implication would be that such dusts do not cause pulmonary fibrosis, physical impairment, or disease. In that context, the classification is useful.

All durable inorganic dusts will evoke mobilization of macrophages. This response is a daily occurrence in all walks of life. Depending on the number of particles entering the alveoli over a period of time and the resistance of such particles to destruction by the macrophage, there will be an accumulation of these particles either free or inside macrophages within and around the alveoli. Particles classified as being inert (16) appear to be slightly if at all toxic to the macrophage in which they come to reside. The host macrophage dies after a normal or near-normal cell life-span, and the ingested particles are released but then undergo reingestion by another macrophage. Repetition of this cycle appears to have no limit.

After a long period of exposure to durable nonfibrogenic particles of respirable size, the particles will be found distributed within alveoli; within the interstitial tissue supporting the alveoli, airways, and blood vessels; and within the small and large collections of lymphoid tissues scattered throughout the lung. The particles at any point in time may be found free or within macrophages at any one of these locations. The particles attached to the alveolar surface at times appear to be overgrown by the type I alveolar pneumocyte, forming a plaque within the alveolar wall. In locations where the particles accumulate in small collections, within either the supporting connective or the lymphoid tissue, there is a slight proliferation of monocytic cells accompanied by a slight increase of interlacing reticulin fibers. Collagen fibers, thus genuine fibrosis, do not develop either in these small dust foci or within the lymphoid tissue more centrally located along the course of the lymphatics. The size of the focal collection of dust particles is directly related to the intensity and duration of the exposure.

Insofar as an effect on pulmonary function or longevity of life is concerned, it is of paramount importance that these particles evoke neither fibrosis or alteration of the basic lung structure. The tiny particles are present, but no destruction or replacement of normal tissue occurs. Since the fundamental mechanical quality of the lung is that of distensibility and elasticity based on the arrangement of tissue fibers having elastic qualities combined with an intact surface lining of the alveoli, one can understand why the mere physical presence of small foci of particles scattered sparsely in the interstices or on the surface of the alveolus would not affect the physical quality of the lung. It is only when the normal structure of the lung is destroyed or immobilized by replacement with nondistensible fibrous tissue that measurable impairment of pulmonary function ensues.

When the nonfibrogenic dust has a low atomic number, the foci of accumulated particles do not interrupt or deflect sufficient X-ray photons to cast a shadow, and the radiogram exhibits no unusual pattern of densities. In contrast, foci of nonfibrogenic particles having a higher atomic number do interrupt and deflect X-ray photons sufficiently to cast shadows, even though they do not evoke an inflammatory or fibrotic reaction within the tissues. They cause the radiogram to have a striking appearance when sufficient quantities of the particles have been accumulated. Among the inert dusts having a high atomic number are iron, tin, barium, and antimony.

Workers may be exposed to oxides of iron either as a dust or as a fume, as during welding, and the biological result is referred to as siderosis even though no fibrosis is induced (17). When the lungs are examined, they are rust-brown as a result of the focal depositions of iron, and microscopic examination reveals particles of iron oxide distributed in the manner common to all respirable particles. Characteristically, the particles are collected in discrete foci or macules associated with lymphoid tissue and around the smaller blood vessels. There is the usual minor accumulation of reticulin fibers and monocytes in the focal deposits of particles, but no true fibrotic reaction occurs. The dust foci of this material of higher atomic number cast a distinct, small, rounded nodular shadow on the film, tending to be no greater than 1 to 2 mm in diameter. When materials of an even higher atomic number such as tin or barium are inhaled, the shadows cast by the focal deposits may have a very sharp margin, and the large collections of particles in the lymph nodes at the lung root may cause these structures to be unusually radioopaque because of their metal content, but they are not enlarged as is so often the case in silicosis. There is no evidence of any impairment of cardiopulmonary function or any reduction of life expectancy in individuals with siderosis.

An excess of malignant transformation manifested as bronchogenic cancer has been reported among iron ore miners, but unusually high levels of radon have been reported in the same mines, and this is considered to be the more likely cause (18). The evidence to incriminate iron oxide or any of the inert dusts as a carcinogen is not convincing. As is the case with other of the inert dusts, exposure to iron oxide, especially in mining, is often combined with exposure to free crystalline silica, and siderosilicosis is the resulting entity. It should be borne in mind that silica is the etiologic or active agent in any impairment that ensues.

The possibility must be recognized that deposition of inert particles on the surface of the upper airways where the mucus-secreting apparatus is located might stimulate that mechanism to an inordinant degree. Such stimulation might lead to excessive production of mucus and to hypertrophy of the mucous glands or to impairment of the ciliary system. Thus although inert dusts do not produce tissue fibrosis, they could lead to excessive production of mucus and mucous gland hypertrophy and to increased airway resistance. The question of whether chronic bronchitis is an outcome of heavy exposure to dusts of various kinds over a long period of time is discussed later in this chapter.

8.2 Fibrogenic Dusts

8.2.1 Free Crystalline Silica

In contrast to the inert dusts, particles of free crystalline silica (FCS) have a strikingly different biological end effect (19). Silicon, the element, does not exist free in nature. It occurs most commonly as silica in the combined form of crystalline or amorphous silica (SiO_2) or as SiO_2 combined with one or another cation as a simple or complex crystalline or amorphous silicate.

FCS exists as quartz, tridymite, cristobalite, coesite, and stishovite. Of these, stishovite is essentially nonfibrogenic, whereas cristobalite and tridymite are more fibrogenic than quartz. In nature, quartz occurs as the most common of the three forms and is the one most frequently encountered during occupational exposures. It is converted to tridymite when heated to between 860 and 1470°C, and above 1470°C to it becomes cristobalite.

When the intensity and duration of exposure are too high to be defended against effectively, respirable sized particles of FCS accumulate in the alveolated regions of the lung. As is the case with inert dust particles, at any point in time the FCS particles will be distributed throughout the lung tissue in the manner described earlier for all respirable particles. Thus focal deposits of FCS, either free or within macrophages, will form in a discrete fashion throughout the lungs and lymphoid tissue. It has been suggested that FCS particles move more easily, therefore in larger numbers, through the alveolar wall to the interstitium and the lymph nodes than is the case with inert particles. This may be because particles of FCS are toxic to the macrophage. The life of the host macrophage is thus greatly shortened, their number diminished, and the defense against movement of particles into the interstitium and lymphoid tissue is thereby reduced.

The subsequent course of events within the foci of FCS particles is strikingly different from that characterizing similar foci of inert particles. At first, but with greater speed of development, fibroblasts increase in number and reticulin fibers are laid down in an interlacing fashion throughout the focal deposit. Instead of stopping at the point of reticulin development, which is the case with inert particles, the foci of FCS go on to develop masses of interlacing collagen fibers, and mature fibrosis characterized by the deposition of hyalin ensues.

With the passage of time the focus or nodule undergoing fibrosis enlarges, the active process going on at the periphery leaving behind a progressively larger central mass of mature fibrous tissue. The mature nodule has a characteristic whorled fibrous tissue appearance almost devoid of cells at its center. This mass is surrounded by a zone of reticulin and collagen fibers containing a few fibroblasts, and surrounding this layer is a peripheral zone of cells in which can be seen the process of laying down reticulin and collagen by fibroblasts, mixed in with macrophages and other mononucleated cells, in the process of forming fibrous tissue. At first the nodules are spaced rather far apart, the intervening lung or lymphoid tissue being free of abnormality. As the number of nodules increases, there is a tendency for them to coalesce and form a larger multicentric nodule, which may reach several millimeters in diameter. At all times large amounts of normal lung tissue persist between the individual or clustered nodules. In many individuals the reaction does not progress beyond this point, and the lung can be described as being the site of simple discrete nodular fibrosis. In other cases, however, there develops massive conglomeration of nodules, merging in a cement of dense fibrosis. These masses can become quite large, occupying in some cases as much as half of each lung. On the immediate periphery of the larger conglomerate masses and even some of the large single discrete nodules, the airspaces of the lung are larger than normal. This type of perifocal emphysema should not be confused with diffuse generalized emphysema of the lung, which is not a characteristic part of the anatomical alterations accompanying the interaction between FCS and lung tissue.

In most cases the fibrotic tissue development in the advanced nodular stage and especially in the conglomerate stage will extend to involve the pleura. As a result, the pleural space commonly is bridged by adhesions or is obliterated over large parts of the lung surface. The pleura itself may be unusually thickened, sometimes markedly so, over a large portion of the lung.

There is no precise location in the lung structure at which a silicotic nodule can be said to begin. A cluster of alveoli and adjacent subalveolar tissue, including small aggregations of lymphoid tissue, become involved virtually simultaneously. The air passages and pulmonary capillary bed caught up in this reaction are destroyed in the process of formation of the nodule. The lung tissue intervening between the nodules remains free of any discernable change. This is true even when large conglomerate masses also are present. Although there is undoubtedly some migration of the particles of FCS from their locus of deposition into aggregates or foci, the distribution of the nodules and the freedom from abnormality in the intervening tissues strongly suggests that the FCS particles taken into the lung are not dispersed in a completely even manner. One must assume that some regions of the lung receive far fewer particles than do others, and the lung cleansing mechanism keeps the total particles in such areas below a critical level.

In a similar manner, although the small regional lymph collections are usually destroyed in the fibrotic mass of the nodule, in the lymph nodes at the lung root the development of silicotic nodulation with sparing of intervening lymhpoid tissue is the rule. The involved lymph nodes become quite large and firm or hard as a result of the process. In some cases the nodules, whether located in lung or lymph node, may become calcified, producing a striking radiograph.

A number of explanations for the mechanism whereby FCS stimulates fibrogenesis have been put forth over the years. The most compelling, and the one having the greatest experimental support, is that developed by Heppleston (13). Leaving aside the question of just how the silica surface disrupts cell membranes, it can be stated that the ingested particles of silica rather quickly kill the host macrophage. In cell culture this happens wihin hours, but in the normal milieu of the body it probably takes considerably longer, perhaps several days. Heppleston deduces from his studies that "on present evidence it may be concluded that silica in some way reacts with cell constituents, probably lysosomal, to produce or activate a relatively soluble substance capable of stimulating collagen formation, possibly acting in low concentrations." Since the fibrogenic action of FCS is dependent on a continuous supply of macrophages in the foci where the particles have accumulated, a mechanism to promote this recruitment is needed. On the basis of his own work and that of others, Heppleston further suggests that the lipid material abundantly released from the dying macrophage killed by FCS acts as a stimulant for the accelerated production and delivery of macrophages to the foci containing FCS.

There is evidence that interaction between the membranes of the macrophage and the surface of the silica particles is the key to the death of the host macrophage. Alteration of the surface of the FCS particles by polyvinylpyridine-*N*-oxide and other agents, including dusts such as clay or iron, which commonly coexists with silica in occupational exposures considerably lessens or may even prevent the fibrogenic response.

On the basis of both human epidemiologic study and animal experiments, as is true of other toxicological reactions, the fibrogenic activity of FCS appears to be a dose-related phenomenon. One can envision the following train of events. A critical amount of FCS must accumulate and be ingested by a sufficient number of macrophages to supply the necessary amount of stimulation to fibroblasts before the process is pushed beyond the stage of reticulin formation into that of collagen formation. The development of hyalin, to reach the mature stage of the nodules, is more complex and appears in part to involve immunological processes. The rate at which the entire process goes on, once it has reached the critical concentration of FCS and macrophages, is related to the speed with which silica particles accumulate and the rate at which the macrophages die, plus a still ill-defined immune response. The rate of FCS accumulation in the various foci is dependent on the balance between deposition and removal from the lung. The entire process is in constant change, even after all external exposure ceases. Some particles are removed from contact with the macrophages within the focal deposit by being sequestered within the central mass of acellular fibrotic tissue and in addition, there is a continuous migration of particles away from the pulmonary foci to the lymph nodes and even to the mucociliary escalator.

According to the concept just expressed there is a level of silica accumulation below which the fibrogenic train of events is not set in motion. This is the tolerable dose. Moreover, once the critical level is reached, acceleration of the whole process will occur if the dose of the effective factor, probably the free bonds on the surface of the crystals, is rapidly augmented. This is consistent with the facts because if FCS accumulation occurs rapidly, the biological response, though fundamentally the same, has a distribution different from the one just described. In these circumstances, it appears that not all the particles have time to be transported to foci, and they accumulate along the way with a more diffuse macrophage mobilization and consequent fibrogenesis distributed in a more widespread, nonnodular manner. This produces a linear, mixed with a nodular, fibrotic reaction that has been called acute silicosis but is probably better termed accelerated silicosis, since it requires several years to develop. The accelerated process appears to be further aggravated if the particles are not only abundant but also especially small, less than 1 μm in diameter, thus providing an unusually large surface area per unit of weight.

A third variety of reaction to FCS has been reported under circumstances of extremely severe exposure, for example, in the process of sand blasting. This reaction shows combined interstitial fibrosis plus lipoproteinosis-like lesions in which the alveoli are filled with a pink-staining, relatively acellular material. This reaction may develop within less than a year, proceeding rapidly to a fatal outcome, and it merits the term "acute silicosis" (20).

As might be anticipated from knowledge of the effects of FCS on lung tissue, the evidences observable during life that these reactions have occurred will vary. At slight stages of development there are no observable evidences of tissue alteration except by direct examination of the tissue itself. The earliest sign during life is observed in the chest roentgenograms as an increase in the size and visibility of the shadows cast by the bronchovascular structures located in the outer reaches of the lungs. Presumably these changes are caused by accumulation of FCS and the slight cellular reaction produced. When the nodules within the lung develop sufficient size and density as a result of the mature cellular and fibrotic reaction, they cast abnormal shadows ranging from barely perceptible, multiple, small, rounded densities to larger and conglomerate densities plus enlargement of the lung root shadows. The number of densities varies from scant to myriads, and the nodules are predominantly located in the upper half of each lung. The very large conglomerate shadows are almost invariably in this region. Shadows cast by thickened pleura can also be seen.

Even at the stage of easily recognized discrete nodular silicosis, there usually are no accompanying symptoms and no detectable evidence of pulmonary or overall physical impairment unless the volume of lung involved is unusually large or other nonrelated causes of impairment coexist. In part, this reflects the fact that the functions of the lung are not the limiting factor for physical performance. Of greatest importance, however, is the fact that in pure simple nodular silicosis, the intervening tissue is normal. Physically the diseased areas have destroyed the lung in a focal or nodular fashion, and they comprise a relatively minor part of the total lung. The normal remaining portion of the lung tissue can readily accommodate the cyclic changes in thoracic volume during respiration; in addition, it can accept an increase in blood perfusion without an elevation of arterial pressure. In this fashion, cardiopulmonary function in the range that is needed during physical exercise is maintained. It must be said that if measurements of pulmonary function could be made in each individual before occupational exposure begins and after the nodulations appear, the person could serve as his own control, and alterations of pulmonary function might be recognized at an earlier stage than is possible when the first examination occurs after the nodules are already present.

Individuals showing overt radiographic evidence of nodular silicosis who also have a lessened capacity for physical effort because of unusual breathlessness on exertion are found in most instances to have a restrictive type of functional impairment. In others, however, there is an obstructive type or a combination of the two. The restrictive type of impairment can readily be accounted for by loss of lung tissue consequent to the extensive fibrosis, and it is most readily demonstrated in those who exhibit a combined massive conglomerate and linear type of disease. When severe, this kind of impairment can lead to pulmonary artery hypertension and cardiac failure. In some persons, though by no means all, impairment of oxygen transfer with associated arterial hypoxia develops.

Restrictive impairment is consistent with the specific pathology of silicosis. An obstructive type of impairment, though sometimes observed in individuals with simple discrete nodular silicosis, is difficult to reconcile with the specific pathology attributable to FCS per se. Either chronic bronchitis or obstructive emphysema or a combination of the two are the common causes of airway obstruction in the general population. In view of the great frequency with which an obstructive impairment occurs in men over age 50 among the general population, when an obstructive impairment is observed in persons who have silicosis, it would seem to be more consistent with the facts to attribute it to causes other than FCS per se. Cigarette smoking and the nonspecific effects of FCS or other coexisting dusts and fumes in the workplace play an etiologic role in the production of chronic bronchitis and ultimately may be shown to influence the development of obstructive emphysema. Cognizance of this probability directs our attention to the need for controlling the total dust, fume, and irritant gas exposure, not just that of FCS in the workplace.

It is rather remarkable in view of the multifocal fibrosis characteristic of silicosis that there is no convincing evidence of malignant transformation or carcinogenic effect. At the same time it must be said that a properly conducted epidemiologic study designed to explore the question of carcinogenesis in those exposed to FCS has not been made.

From what has been presented thus far, it could be concluded that silicosis in and of itself is not a severe crippling or life-shortening disease unless the fibrosis is very extensive. This, in fact, is the case. Why then has silicosis been an important cause of morbidity and death? The answer lies not in the discrete nodules characteristic of silicosis but in the very adverse influence of the tissue reaction to FCS on the tissue defenses against the tubercle bacillus and similar mycobacteria (21, 22).

In years past there was little chance that a person with coexisting silicosis and tuberculosis would recover. The infection slowly or rapidly but relentlessly progressed to a fatal outcome. Moreover, in the presence of quiescent or inactive tuberculosis, the development of silicosis often converted the dormant tuberculosis to an active, spreading disease. Since until recently almost all members of the adult population harbored one or more inactive tuberculosis foci, the development of silicosis fell on fertile ground and the likelihood of reactivating the tuberculosis followed by its relentless progression was high. This is no longer the case in the United States. With the advent of specific drugs to combat the tubercle bacillus and public health measures for tuberculosis control, the proportion of the adult population harboring latent tuberculosis is now very small. Moreover, specific antituberculosis drugs are quite effective, even in the presence of silicosis.

Since the reaction of lung tissue to FCS has a demonstrable dose–response relationship, reducing the severity of occupational exposure offers the possibility of abolishing the disease. Measures for environmental control in the workplace have been effective when instituted properly. New cases in metal mines where the levels of FCS have been kept for the most part below 5 million particles per cubic foot have been few (23). One large mine under my surveillance employs several hundred underground miners, and no new cases of silicosis have developed up to the present (1976) in men initially employed after 1940. Over this period a program of dust measurement shows the exposures to FCS have been kept below 5 million particles per cubic foot with few and temporary exceptions. In the same mine prior to 1938 and before installation of extensive dust control procedures, approximately 25 percent of the underground miners had radiographic evidence of silicosis. If the even more stringent current ACGIH-recommended threshold limit value for FCS is scrupulously adhered to silicosis as a disease should be abolished. Unfortunately and regrettably, exposure to excessive levels of FCS in some places of work still exists, and silicosis continues to be induced. This is not because the recommended TLV is too high but because it is not complied with.

8.2.2 Carbonaceous Dust

Exposure to respirable particles of carbon may occur in several occupations, the most common being those associated with the mining, processing, and handling of coal. Exposure also occurs in the mining and use of graphite and in the production of carbon black and carbon electrodes. In all these cases the biological effects are similar but not identical, and only the consequences of exposure as a coal worker are discussed. "Coal workers' pneumoconioses" (CWP) is the proper term to apply to the abnormalities caused by the dust inhaled in this class of occupation. "Black lung," which is a socioeconomic term, has no meaning in the scientific or medical context.

Respirable free crystalline silica in varying concentrations from slight to high commonly coexists with coal particles in the miner's environment. Some is in the coal seam, some is released by roof bolting where sandstone or hard shale is present, and in mines of the United States sand used on the haulageways is ground to respirable size. In the process of mining, the exposure commonly is to a mixed dust, though in subsequent handling of the commercial product the contamination by FCS is much less.

Pure coal or graphite inhaled by experimental animals produces dust foci characterized by reticulin fibers but little or no collagen, and in all respects it can be classed as an inert dust reaction (24). In general, coal particles come to reside in the same localities in the lung as already described under inert and FCS particles. In humans, the number of coal particles retained in the lung usually exceeds by far the number observed as a result of other inert or FCS exposures.

The "coal macule" is different in appearance from the FCS nodule previously described (25). It takes on a more stellate form and is localized more intimately around the respiratory bronchiole, the alveoli of which are filled with dust-laden macrophages. The normal architecture of the alveolus persists, even though it may be shrunken. In other words, there is no actual destruction of lung tissue. Alveoli become effaced by a combination of shrinkage and filling with macrophages. A sleeve of dust-laden tissue interlaced by reticulin fibers and a few collagen fibers forms a cuff, obliterating all structures except the lumen of the respiratory bronchiole. This bronchiole may become abnormally dilated, producing a form of focal or centrilobular emphysema. It is noteworthy that not every macule is associated with the development of focal emphysema. This raises the unanswered question of why focal emphysema develops at all. The mechanism is not clear.

The coal macule is a discrete focus, the intervening tissue remaining normal. The stellate densities may be 1 to 2 mm or more in diameter. Ingestion of coal particles by the macrophage does not prematurely kill the cell; thus the coal macule is fundamentally the same as that described earlier for inert materials. The larger area involved by the coal macule and obliteration of alveolar spaces appear to be the result of overwhelming the cleansing mechanism of the lung by very heavy exposure and retention of large amounts of coal. It has been shown that the number and size of the macules in the lung is closely related to the total exposure to coal dust (26). Although the coal macule may have a slight amount of mature collagen fiber interlaced throughout, characteristically this is a rather minor development, and reticulin fibers constitute the vast majority of the fibrillar tissue. Quite often, however, there is more than the usual scant amount of collagen laid down in an irregular fashion in the structure that ordinarily would be termed a coal macule. The components of the reaction are closer to that ascribed to FCS, but the organization is quite different in that the whorled arrangement is absent. Some have referred to these structures as nodules in contrast to macules. Obviously, when collagen is present in this amount the reaction can no longer be thought of as characteristic of an inert substance. Whether this nodular form is in fact a type of mixed silicotic reaction or is a response solely to coal particles has not been determined. Since FCS exists so commonly in the coal mine workplace, the former is a likely probability.

In some individuals, superimposed on the background of simple discrete coal macules, a marked coalescence of macules and proliferation of mature fibrous tissue causes large masses to develop. These are referred to as progressive massive fibrosis (PMF). Since the coal macule in contrast to the silicotic nodule has rather indeterminate borders, the macules lose their identity when they are caught up in this large mass and do not exhibit the agglomerated separate structures that characterize the appearance of conglomerate silicosis. The mechanism of PMF development is not clear. It does not appear to be closely related to the coexisting FCS, although some role for this substance in the development of PMF cannot be excluded. It seems likely that PMF is related in part to the extraordinarily high content of coal dust in the involved areas. On the whole, in pure silicosis the tissue effects are induced by a lung burden of 2 to 5 g of FCS. The lung burden of coal in simple CWP or PMF is 10 or more times greater than this. It is tempting to believe that uncomplicated or simple CWP is basically a moderate to severe example of an inert dust reaction and that PMF would develop in all types of inert dust exposure if it was heavy enough. This probably is too simplistic an approach. It has been suggested that immunological factors may play a role in PMF.

A quite different kind of massive lesion, Caplan's syndrome, may develop rapidly in some persons having simple CWP (27). A large proportion of these individuals have rheumatoid arthritis, a disease having a large immunological component. The intrapulmonary mass characteristic of this syndrome does not have the histological appearances of PMF but does have some of the characteristics associated with immunological reactions. This entity clearly raises the question of whether an immunological response plays some role in the development of PMF, even though PMF is definitely not the same lesion as that characterizing Caplan's syndrome.

In contrast to the adverse effect on tissue resistance to the tubercle bacillus caused by the silicotic reaction, the coal macule does not appear to influence the body defense against this organism. When tuberculosis and simple CWP coexist, the infection may play a role in causing PMF, but rapid acceleration of tuberculosis does not occur.

For the most part, the radiographic evidences of CWP including PMF are similar to those of silicosis and need not be repeated. The shadows cast by CWP are a mixture of round and irregular shapes measuring 1 to 5 mm in diameter. The attack rate and severity of this stage of the disease is related closely to the amount of dust in the lung. Many individuals never progress beyond this stage to that of PMF, which is characterized by large, irregular shadows, usually developing in the upper lung areas.

Measurement of pulmonary function in persons having CWP shows essentially the same range of alteration with respect to type and degree as discussed for simple and conglomerate silicosis. There is less evidence of a restrictive impairment in CWP except when severe PMF exists. In simple CWP, impairment of function is so minor that it does not interfere with physical effort or shorten the life-span (28, 29). In some individuals having severe PMF, however, there is marked restrictive impairment of pulmonary function, reduction in capacity for physical effort, and shortened life-span.

Several studies searching for evidence that CWP is related to an excess risk of bronchogenic cancer have failed to reveal such an association.

It should be emphasized that some occupations in coal mines (e.g., a motorman, a roof bolter, or a development man) are characterized by a predominance of FCS exposure. For this reason a coal miner can develop pure nodular silicosis. If he changes jobs during his total employment, his lungs may develop coexisting coal macules, coal nodules, and pure nodular silicosis. These combinations of abnormality led, particularly in the United States, to the term anthracosilicosis, before it was demonstrated that coal in the virtual absence of FCS could produce the coal macule and PMF. Since simple CWP is strongly related to total exposure or the dose of coal, and since PMF virtually always develops on a background of simple CWP, control of the dust exposure on the mine premises will prevent CWP.

8.3 Chronic Bronchitis and Emphysema

Examination of large numbers of persons who have been exposed to inert, fibrogenic, or mixed dusts invariably discloses a paradox. One is faced with the observation that among a group of persons demonstrating mild radiologic or histologic evidences of tissue alteration—for example, simple CWP or nodular silicosis—most have no physical complaint and demonstrate no impairment of function. Nevertheless, a few will complain of moderate to severe symptoms of cough and sputum production or breathlessness on exertion and will have moderate to severe measurable alteration of pulmonary function. Almost without exception such individuals have either chronic bronchitis or destructive emphysema or both, at a stage associated with increased airway resistance. We then ask whether these two diseases are caused or influenced in their natural course by the exposure to the specific dust.

Emphysema, defined as pathological enlargement of the air spaces distal to the terminal bronchioles, has been found to involve more than 5 percent of the lung in approximately 50 percent or more of all males over age 50 at autopsy. The individuals come from all walks of life, and there is no preponderance of men from the dusty trades. Possibly 10 to 20 percent of these individuals from the population at large have severe enough involvement to exhibit moderate to severe increase of airway resistance and to have experienced abnormal breathlessness on exertion (dyspnea).

Chronic bronchitis, defined as periodic or persistent bouts of coughing caused by the production of excessive bronchial mucus, is a common condition in males after the age of 30 and is one of the rather frequent causes of death. The prevalence among adult men may be as high as 15 to 20 percent. By far the most potent factor, though not the only one causing and aggravating this disease, is cigarette smoking. In view of the high prevalence of these two conditions in the adult male population at large, it is not surprising to meet the paradox just described. Efforts to disclose the role of occupational exposure to dust as a causative factor in the development of chronic bronchitis or emphysema have been made. There is surprisingly modest support for the concept that industrial dust or the irritating fumes commonly also present play a role in the development of aggravation of chronic bronchitis or emphysema. It is difficult to accept the belief that no role is played by these agents, particularly with respect to chronic bronchitis. What seems likely is that the powerful effect of cigarette smoking, the influence of social class with respect to place of residence, medical care, and so on, and general air pollution, overshadow any demonstrable strong influence of industrial dust, fumes, or gases (30). In studies in which these other factors are corrected for, there is some suggestion that occupational dust per se may contribute at least to chronic bronchitis, even though it is not the most potent factor.

9 ASBESTOS

"Asbestos" is a term customarily reserved for commercially mined and milled, naturally occurring crystalline fibers of the serpentine and amphibole family of minerals. Chrysotile, the fibrous form of serpentine, is by far the most abundantly produced. The amphiboles crocidolite, amosite, and anthophyllite are produced and used as such on a lower scale, tremolite is found primarily as a contaminant of talc, and actinolite is rarely encountered in industry.

In their natural state, the fibers are found massed together, forming a layer of cross or slip fiber within parent rock of the same chemical composition. In the process of separation from their parent rock or of "opening" during milling and manufacturing uses, the airborne fibers of asbestos vary enormously in length and diameter, but some ultimately become thin and short enough to be of respirable size. Since the aerodynamic behavior of fibers is determined by their diameter rather than length, fibers more than 3 μm thick do not penetrate into the alveolated regions of the lung. As discussed earlier, length and shape as well as diameter play a role in governing deposition. Although fibers as long as 200 μm may occasionally be found in the lung, the length of the majority is less than 50 μm.

As observed in humans or experimental animals, all the ultimate fates described earlier for particles entering the upper airways occur also for fibrous particles but with some important differences. Fiber deposited on the mucociliary apparatus is removed and expectorated or swallowed. Those less than 10 μm long can be ingested readily by the macrophage. Longer fibers, though thin enough to be respirable, are not readily ingested by single macrophages; yet occasional fibers 20 to 50 μm or longer may be seen with several macrophages attached and spread out over the fiber surface. At times two or more macrophages clump together, forming a giant cell totally surrounding one or more fibers. In contrast to respirable-sized nonfibrous particles, any one of which can be ingested by a macrophage, fibers longer than 20 μm are not readily ingested, and macrophage defense against this length of fiber is impaired or not available. In spite of this, asbestos fibers free on the surface of the alveoli are relatively scant. Almost always, fibers remaining in the lung either are within macrophages or appear to be embedded within the thickened wall of the alveolus or within the interstitial tissue, which then becomes the site of fibroblast proliferation and collagen fiber deposition. Fibers longer than 10 μm within the regional lymph nodes are unusual, and even shorter ones are not found in large numbers.

Some of the fibers within both the lungs and lymph tissue become coated with a proteinaceous iron-containing substance and are referred to as "ferruginous" or "asbestos" bodies. It is believed that the coating is laid down in some way by contact with the macrophage. This coating persists virtually indefinitely, but it is of interest that these bodies can undergo fragmentation. Curiously, even in persons who experience moderate to heavy exposure, many and perhaps most fibers are free of this coating even after long periods of residence within the lung. There is some evidence to indicate that the coated fibers are nonfibrogenic and that the coating process is a mechanism whereby fibers entering the lung are neutralized. It should also be pointed out that ferruginous bodies may be formed by any kind of durable fiber that enters the lung, including such things as glass and vegetable fibers of silicious composition.

It has been shown experimentally that asbestos fibers shorter than 5 to 10 μm whether coated or naked, behave essentially like inert particles (31, 32). This length of fiber is readily ingested by the macrophage, and in contrast to the effect of FCS, the host macrophage is not materially damaged by the ingested fiber. Its life cycle is not shortened. Experimental animals exposed solely to asbestos fibers shorter than 5 to 10 μm do not develop pulmonary fibrosis. If longer fibers in sufficient numbers are introduced into the lungs of animals, they produce a diffuse, nonnodular type of fibrosis of the mature variety. The fibrotic reaction spreads out along the supporting structures of the lung, obliterating the alveoli and associated vascular units. The diffuse spreading nature of this involvement stands in striking contrast to the discrete nodular fibrosis that is characteristic of silicosis. Although some areas of normal-appearing lung tissue persist between the areas of fibrosis, the volume of residual normal lung tissue is less, and the amount of lung tissue entrapped between strands of fibrosis is greater than is the case in silicosis (33).

The mechanism of fibrogenesis induced by asbestos is not clear. In animals, the early stages suggest an organizing alveolitis closely associated with the presence of long fibers. Eventually the fibrosis completely destroys the alveolar architecture and its intimately related vascular structures. Commonly the extensive mature fibrotic tissue contains long asbestos fibers, singly or in clusters, distributed throughout, but they are sparse and far apart. The fibrotic reaction seems to be excessive for the few fibers visible. Nevertheless, fibrogenesis must have something to do with the fibers and specifically with their shape because nonfibrous particles having identical chemical and crystalline structures will not produce fibrosis when introduced into experimental animals. The relative paucity of long fibers in mature fibrotic tissue may be the result of fragmentation and destruction of long fibers after they have induced the fibrotic reaction. The concept that asbestos fibers are biologically indestructible cannot be supported.

It has been suggested that long fibers induce alveolitis and subsequent fibrosis, while short fibers or particles of the same chemical nature do not, because the fibers are too long to be taken entirely into the macrophage. As a result, the macrophage leaks at the point of fiber entry, and the discharge of some of the macrophage contents leads to fibrogenesis (32). It is of interest that there is little if any fibrogenesis associated with the relatively few asbestos fibers that reach the lymph nodes. It would appear that the lymphoid tissue treats even the long fibers as inert material.

That immunologic processes may play a role in fibrogenesis consequent to inhalation of asbestos or FCS has been suspected not only because of the potential role of this mechanism for governing the actual tissue response but also because there appears to be some variation of individual reactivity to these dusts. Measurement of antinuclear antibodies, rheumatoid factor, and the W27 antigen of the HL-A system indicates higher values in those having asbestos-induced pulmonary fibrosis than among controls or those exposed but without fibrosis (34, 35). The importance of these studies, especially with regard to recognizing prior to exposure those who would be hyperreactors, remains to be determined.

In contrast to particles of FCS, asbestos fibers tend to remain on the surface of the alveoli; the cellular reaction to them develops inside the alveolus, and this alveolitis ultimately goes on to organization, with obliteration of the alveoli and the diffuse fibrotic reaction flowing from that. Perhaps it is this difference in particle transport that explains the diffuse rather than the nodular tissue reaction to asbestos. Also, in contrast to the long-term effects of free crystalline silica, the fibrosis caused by asbestos develops more extensively in the lower half of the lungs, and it is unusual for large conglomerate masses surrounded by normal tissue to develop.

In addition to the diffuse intrapulmonary tissue reaction to asbestos, in some exposed persons there is also a fibrotic reaction of two kinds involving the pleura. First, there may be slight to widespread fibrosis of the visceral pleura associated with the underlying pulmonary fibrosis, especially in severely involved cases. Obliteration of the pleural space may occur in these areas. Second, there may be small to large plaque formation involving virtually only the parietal pleura. The plaques are thickened tissue having quite sharp, raised margins, and they are composed of fibrous hyalinized tissue that at times becomes calcified. The parietal pleura over these plaques usually is not adherent to the visceral pleura, and the space is not obliterated. When visible in the radiogram, the pleural plaques are striking, especially when they are calcified. Plaque formation can develop without pulmonary fibrosis, and the mechanism, especially the reason for its predilection for the parietal pleura, is unknown.

The pathology as seen in experimental animals and in humans is essentially the same and explains reasonably well the impairment of function and health consequent to the reaction. The diffuse fibrous bands radiating throughout the lung, especially in the lower regions, reduces the distensibility of the normal lung tissue entrapped between the strands of fibrous tissue. This plus actual loss of lung substance by replacement with fibrous tissue causes a restrictive type of pulmonary impairment. As a result, the total lung volume, vital capacity, and residual volume are reduced without an increase of resistance to airflow in the proximal conducting system. If the restrictive impairment is severe, it causes an increase of the work of breathing and a lessened maximum capacity for ventilation. Regional hypoventilation of entrapped lung plus loss of total vascular surface may lead to impaired oxygen transfer and arterial hypoxia. The volume of air breathed in response to exercise may become abnormally high, and this plus the increased work of breathing leads to abnormal breathlessness and dyspnea. If the vascular bed is sufficiently reduced, the pulmonary artery pressure rises, especially during exercise, and right ventricle hypertrophy and heart failure may ensue.

With mild asbestosis, the radiogram is not as distinctly abnormal as is the case with silicosis because the tissue changes are diffuse and the well-demarcated densities so characteristic of silicosis are not present. The usual appearance of asbestosis in the radiogram is that of poorly defined, small, irregular densities that obscure the normally sharp margins of the bronchovascular structures, particularly in the lower parts of both lungs. Thickened pleura may cast a diffuse haze over both lungs and often obscures the intrapulmonary densities. When the disease is severe the radiographic abnormalities are readily recognized; but in the slight stages, the question of whether the radiogram is in fact abnormal is commonly disputed and may be answered differently by the same reader on different occasions.

Physical examination in the early and subsequent stages often reveals persistent dry crackling sounds (rales), audible over the lung bases posteriorly. Clubbed fingers with or without cyanosis may or may not be present.

In any given case all the abnormalities thus far described may exist, but commonly only one or two are present. Moreover, all the physical, radiographic, and altered pulmonary function measurements are characteristic of chronic interstitial lung disease, no matter what its etiology may be. Although idiopathic chronic interstitial lung disease is rather uncommon, it occurs frequently enough to pose a diagnostic problem, especially in persons exposed very slightly to asbestos fiber—that is, for the intermittent user of this material or the general public. The diagnosis of asbestosis therefore requires exclusion of other causes of observable abnormalities plus establishment of the occurrence of adequate exposure to respirable asbestos.

In contrast to the experience with exposure to inert particles, to free crystalline silica, and to coal dust, exposure to respirable asbestos fiber is associated with an excess risk of cancer (36, 37). The excess risk of bronchogenic cancer is virtually confined to those having a long history of cigarette smoking plus heavy exposure to asbestos. The greatest risk appears to be in those exposed in the insulation working trade, where the "total" exposure to multiple factors by reason of coexisting agents in their variable working conditions may play an important role. On balance, asbestos per se is a low order carcinogen for the bronchial cells, and its role is almost certainly that of a promoter, not an initiator.

There is evidence to suggest that an excess of bronchogenic cancer occurs only among those having diffuse pulmonary fibrosis or asbestosis (38, 39). Thus since fibers shorter than 5 to 10 μm do not cause diffuse pulmonary fibrosis, it seems likely they also do not play a role in the causation of an excess bronchogenic cancer risk in persons exposed to asbestos fiber in the workplace.

Some studies, but not all, suggest an increased risk of gastrointestinal cancer among workers exposed to respirable asbestos. The excess risk appears to be of a low order and confined to those heavily exposed (36). Curiously, there is inconsistency of excess risk for different parts of the gastrointestinal tract among the various studies.

An excess risk of developing either pleura or peritoneal mesothelioma has been observed among workers exposed to chrysotile, crocidolite, and amosite, but not to anthophylite or tremolite (39). The dose at which the excess risk occurs appears to be lower than for the other kinds of asbestos-related cancers. Some believe that exposure to crocidolite carries a greater risk than that for chrysotile or amosite, but this attitude is not uniformly accepted (40). As is true for fibrogenesis, the risk of carcinogenesis related to respirable asbestos shows a dose–response relationship including that for mesothelioma. Reconstruction of dose data for crocidolite exposure and associated mesothelioma risk has never been presented in a fashion to exclude dose as the factor responsible for what in some circumstances appears as a higher risk of mesothelioma among those exposed to crocidolite.

Mesothelial tumors can be produced in experimental animals by placing asbestos fibers directly into the pleural or peritoneal space (41–43). Using this technique, it has been shown that long, very thin fibers induce the greatest tumor response. If fibers of thicker or shorter dimensions are used, tumor induction decreases. When the fibers used to not exceed 5 μm in length, tumor production, if it does occur, is not statistically significant.

Since all the manifestations of impaired health and function related to respirable asbestos are dose related, it should be possible to continue the use of this material under adequate controls.

10 FIBROUS GLASS

Glass, an amorphous silicate containing various other ingredients, can be produced in fiber form from the molten material by drawing, as in textile glass, or by blowing or centrifugation, as in glass wool manufacture. In view of the biological effects of asbestos fibers, concern for possible effects of fibrous glass has often been voiced.

Fibrous glass began to be produced in large amounts only after 1935–1940. Since the fibers of drawn glass characteristically were thicker than 5 μm and the same was true of most of the early produced blown glass fibers, it was believed that these would be nonrespirable. In addition, mineral wool, a blown glass, has been in use since early in this century; and although those exposed to it have not been studied by modern epidemiological methods, their clinical experience did not lead observers to believe that there was an unusual health risk. Early animal experiments using glass fibers ground to respirable lengths were not entirely satisfactory because their large diameter interfered with carrying out adequate inhalation experiments. When these were supplemented by intratracheal injection, however, the results were those of an inert dust reaction, no mature fibrosis being produced. On these grounds, fibrous glass was considered to be an inert or nuisance dust. Radiographic, pulmonary function, and clinical plus autopsy studies performed since 1965 confirmed the absence of recognizable pulmonary alteration other than those of an inert nature in persons having experienced 25 to 30 years of exposure to fibrous glass in the workplace (44–46).

More recently, technological developments and requirements have led to the production for special purposes of thinner blown glass fibers having diameters as thin as 1 μm or less. In addition, electron microscope studies of the workplace where conventional glass fiber is being produced show that a small number of fibers of this dimension are present in the air where nominally thicker glass fibers are being manufactured. It is apparent that fibers thin enough to be respired have been in the workplace for some time and are now being manufactured for specific purposes.

In contrast to the earlier animal experiments showing only an inert dust reaction, contemporary animal experiments using repeated intratracheal instillations of dilute suspensions of glass fibers thinner than 2 μm and longer than 25 μm have revealed the development of fibrosis in the interstitial tissues around the respiratory bronchioles (32). This reaction, though qualitatively similar, is much less severe in comparison to that resulting from asbestos fiber, thus strongly suggesting that something in addition to fiber shape and dimension plays a role in the induction of fibrosis. The same experiments also indicate that glass fibers of identical composition and diameter but shorter than 10 μm produced only an inert dust reaction (32).

Experiments over the past several years have shown that introduction of long glass fibers thinner than 3 μm directly into the pleura and peritoneum of experimental animals has produced mesothelial tumors (47). Though such experiments are important in the study of mechanisms of carcinogenesis, it should be pointed out that such implantation of large numbers of fibers of this size directly into the pleura or peritoneum completely bypasses the defense mechanisms available to minimize the movement of fibers to the pleura by the usual inhalation route in the intact respiratory system. Furthermore, the number of fibers introduced in these experiments is doubtless several orders of magnitude higher than might conceivably result by way of the inhalation route. As was the case with asbestos, these experiments show that as the length of the fiber is reduced, the yield of tumors is lessened, and when all the fibers are shorter than 8 μm, no statistically significant excess of tumors is induced (47). One would anticipate that by the normal inhalation route, the long fibers would be less likely than short ones to reach the subpleural regions, and this would tend to further lessen the number of long thin fibers ultimately reaching the pleura by the inhalation route. Although these animal experiments suggest the possibility that glass fibers thinner than 2 μm and longer than 10 μm might cause pulmonary fibrosis or induce mesothelioma in man if sufficient numbers reached the alveolar areas of the lung and the pleura, they are so artificial in their manner of placement of the fibers within the animal that they have questionable relevance to human exposures. The more relevant inhalation experiments in animals remain to be done, and firm conclusions about the potential hazard posed by long, thin fibers of glass should be postponed until these investigations have been carried out.

Human experience with exposure to these unusually thin, long fibers is quite limited, but some data are available to place animal experiments such as those just referred to in perspective. As indicated earlier, fibers of these long, thin dimensions have been present in small numbers in the air of glass wool manufacturing plants for the past 30 to 35 years. In fact, for several years they were deliberately manufactured for the production of flotation materials during World War II and for a short time thereafter. Recent epidemiologic studies of workers in manufacture of fibrous glass where long, very thin fibers were present 30 or more years ago have shown neither an excess of cancer, including mesothelioma, nor of pulmonary fibrosis in workers exposed over this period of time (48).

11 PNEUMOCONIOSIS?

The reader must wonder why the term "pneumoconiosis" has been used only with regard to coal workers' pneumoconiosis (CWP). The author used it in that connection because in this particular circumstance it has a rather specific, well-entrenched meaning. As generally used, "pneumoconiosis" is defined as the circumstance of deposition of dust in lung tissue and the tissue reaction thereto. In this sense it has no specific implications, since all adults, no matter what their work experience, fall within this category. Moreover, some of the pneumoconioses have serious biological implications while others do not, and all too often there is a false inference that all pneumoconioses are harmful. The term is so ambiguous that its use should be discouraged, and more specific designations should be used for each of the circumstances.

REFERENCES

1. E. R. Weibel, *Morphometry of the Human Lung,* Academic Press–Springer-Verlag, New York, 1963.
2. H. von Hayek, *The Human Lung* (translated by V. E. Krahl), Hafner, New York, 1960.
3. H. Spencer, *Pathology of the Lung; Excluding Pulmonary Tuberculosis,* 3rd ed., Pergamon Press, Oxford, 1976.
4. B. Vernon-Roberts, *The Macrophage,* Cambridge University Press, Cambridge, 1972.
5. T. F. Hatch and P. Gross, *Pulmonary Deposition and Retention of Inhaled Aerosols,* Academic Press, New York, 1964.
6. W. H. Walton, Ed., *Inhaled Particles,* Vol. 3, Unwin Bros., Ltd., Gresham Press, Old Woking, Surrey, England, 1971.
7. M. Newhouse, J. Sanchis, and J. Bienenstock, *Lung Defense Mechanisms* (2 parts), *New Engl. J. Med.,* **295,** 990, 1045 (1976).
8. J. Cartwright and G. Nagelschmidt, "The Size and Shape of Dust from Human Lungs and Its Relation to Relative Sampling," in: *Inhaled Particles and Vapours,* C. N. Davies, Ed., Pergamon Press, New York, 1961.
9. V. Timbrell, "Inhalation and Biological Effects of Asbestos," in: *Assessment of Airborne Particles: Fundamentals, Applications and Implication to Inhalation Toxicity,* T. Mercer, P. Morrow, and W. Stäber, Eds., Thomas, Springfield, Ill., 1972, pp. 429–445.
10. C. N. Davies, "Deposition of Dust in the Lungs: A Physical Process," in: *Industrial Pulmonary Diseases,* King and Fletcher, Eds., Little Brown, Boston, 1960.
11. E. M. Scarpelli, *The Surfactant System of the Lung,* Lea & Febiger, Philadelphia, 1968.
12. J. Kaufman and G. W. Wright, "The Effect of Nasal and Nasopharyngeal Irritation on Airway Resistance in Man," *Am. Rev. Resp. Dis.,* **100,** 626 (1969).
13. A. G. Heppleston, "The Fibrogenic Action of Silica," *Br. Med. Bull.,* **25,** No. 3, 282 (1969).
14. J. Bruch, "Response of Cell Cultures to Asbestos Fibers," *Environ. Health Perspect.* **9,** 253 (1974).
15. ILO U/C 1971, *International Classification of Radiographs of the Pneumoconioses,* in: *Medical Radiography and Photography,* G. J. Jacobson and W. S. Lainhart, Eds., Vol. 28, No. 3, Eastman Kodak Co., Rochester, N.Y., 1972.
16. W. R. Parkes, *Occupational Lung Disorders,* Butterworths, London, 1974.
17. M. Kleinfeld et al., "Welders Siderosis," *Arch. Environ. Health,* **19,** 70 (1969).

18. J. T. Boyd et al., "Cancer of the Lung in Iron Ore (Haematite) Miners," *Br. J. Ind. Med.*, **27**, 97 (1970).
19. M. Ziskind, R. N. Jones, and H. Weill, "Silicosis," *Am. Rev. Resp. Dis.*, **113**, 643 (1976).
20. H. A. Buechner and A. Ansari, "Acute Silico-Proteinosis," *Dis. Chest*, **55**, 274 (1969).
21. L. U. Gardner, "Studies in Experimental Pneumoconiosis: Reactivation of Healing Primary Tubercles in Lung by Inhalation of Quartz, Granite and Carborundum," *Am. Rev. Tuberc.*, **20**, 833 (1929).
22. L. U. Gardner, "Etiology of Pneumoconiosis," *J. Am. Med. Assoc.*, **111**, 1925 (1938).
23. "Silicosis in the Metal Mining Industry, A Revolution," U.S. Public Health Service Publication No. 1076, Government Printing Office, Washington, D.C., 1963.
24. S. H. Zaidi, *Experimental Pneumoconiosis*, Johns Hopkins Press, Baltimore, 1969, pp. 187–198.
25. A. G. Heppleston, "The Pathological Anatomy of Simple Pneumoconiosis in Coal Miners," *J. Pathol. Bacteriol.*, **66**, 235 (1953).
26. S. Rae, "Pneumoconiosis and Coal Dust Exposure," *Br. Med. Bull.*, **27**, No. 1, 53 (1971).
27. A. Caplan, R. Payne, and J. Withey, "A Broader Concept of Caplan's Syndrome Related to Rheumatoid Factors," *Thorax*, **17**, 205 (1962).
28. W. K. C. Morgan et al., "Respiratory Impairment in Simple Coal Workers' Pneumoconiosis," *J. Occup. Med.*, **14**, 839 (1972).
29. C. E. Ortmeyer et al., "The Mortality of Appalachian Coal Miners," *Arch. Environ. Health*, **29**, 67 (1974).
30. I. T. T. Higgins, "The Epidemiology of Chronic Respiratory Disease," *Prevent. Med.*, **2**, 14 (1973).
31. P. Gross, "Is Short-Fibered Asbestos Dust a Biological Hazard?" *Arch. Environ. Health*, **29**, 115 (1974).
32. G. W. Wright and M. Kuschner, *Proceedings of the Fourth International Symposium on Inhaled Particles and Vapours*, British Industrial Hygiene Society, 1975, in press.
33. A. J. Vorwald, T. M. Durkan, and P. C. Pratt, "Experimental Studies of Asbestosis," *Arch. Ind. Hyg. Occup. Med.*, **3**, 1 (1951).
34. M. Turner-Warwick and W. R. Parkes, "Circulating Rheumatoid and Antinuclear Factors in Asbestos Workers," *Br. Med. J.*, **3**, 492 (1970).
35. J. A. Merchant et al., "The HL-A System in Asbestos Workers," *Br. Med. J.*, **1**, 189 (1975).
36. J. C. McDonald et al., "The Health of Chrysotile Asbestos Mine and Mill Workers of Quebec," *Arch. Environ. Health*, **28**, 61 (1974).
37. I. J. Selikoff, E. C. Hammond, and J. Churg, "Asbestos Exposure, Smoking and Neoplasia," *J. Am. Med. Assoc.*, **204**, 106 (1968).
38. H. Bohlig, G. Jacob, and H. Müller, *Die Asbestose der Lungen*, Georg Thieme, Stuttgart, 1960, p. 60.
39. J. C. Wagner, J. C. Gibson, G. Berry, and V. Timbrell, "Epidemiology of Asbestos Cancers," *Br. Med. Bull.*, **27**, 71 (1971).
40. I. Webster, "Asbestos and Malignancy," *S. Afr. Med. J.*, **47**, 165 (1973).
41. M. F. Stanton, "Fiber Carcinogenesis: Is Asbestos the Only Hazard?" *J. Nat. Cancer Inst.*, **52**, 633 (1974).
42. M. F. Stanton and C. Wrench, "Mechanisms of Mesothelioma Induction with Asbestos and Fibrous Glass," *J. Nat. Cancer Inst.*, **48**, 797 (1972).
43. J. C. Wagner and G. Berry, "Mesothelioma in Rats Following Inoculation with Asbestos," *Br. J. Cancer*, **23**, 567 (1969).
44. G. W. Wright, "Airborne Fibrous Glass Particles: Chest Roentgengrams of Persons with Prolonged Exposure," *Arch. Environ. Health*, **21**, 175 (1968).
45. R. T. de Treville and H. M. Utidgian, "Fibrous Glass Manufacturing and Health. Part I. Report of an

Epidemiological Study. Part II. Results of a Comprehensive Physiological Study," Transactions Bulletin, Industrial Health Foundation, Pittsburgh, 1970, pp. 98–111.

46. J. W. Hill et al., "Glass Fibers: Absence of Pulmonary Hazard in Production Workers," *Br. J. Ind. Med.,* **30,** 174 (1973).
47. M. F. Stanton et al., "The Carcinogenicity of Fibrous Glass: Pleural Response in the Rat in Relation to Fiber Dimension," *J. Nat. Cancer Inst.,* **58,** 1977.
48. D. L. Bayliss et al., "Mortality Patterns Among Fibrous Glass Workers," *Ann. N.Y. Acad. Sci.,* **271,** 324 (1976).

CHAPTER EIGHT

Occupational Dermatoses

DONALD J. BIRMINGHAM, M.D.

Skin diseases of occupational origin outnumber all other work-incurred illnesses. They are common to almost all industrial pursuits and range from mild transient conditions to complex impairments. Labeled usually industrial dermatitis, occupational contact dermatitis, or professional eczema, they can be identified causally as cement dermatitis, chrome ulcers, oil acne, rubber itch, tar warts, and so on. Because of the diverse skin changes caused by work, however, these diseases are properly titled "occupational dermatoses," a term inclusive of any cutaneous abnormality resulting directly from or aggravated by the work environment.

1 HISTORICAL

There can be little doubt that occupational dermatoses were experienced soon after man began to work. Having to fashion tools and weapons from stones, flint, and trees, he probably acquired blisters as well as calluses, fissures, and similar traumas. Dermatoses of various types caused by the elements and by chemical, plant, and biological agents probably occurred, but there were no records of such diseases until A.D. 100 when Celsus described ulcerations of the skin from corrosive materials. Centuries later Paracelsus and Agricola, physicians interested in diseases of miners, described cutaneous ulcerations caused by working with metallic salts (1). Thus for several centuries cutaneous ulceration was the major occupational dermatosis of record. Various metals were encountered in mining, smelting, tool and weapon making, creating objects of art, glass making, gold and silver coinage, and other occupations of that time. Additionally, there were stone cutters and masons, bakers, bath attendants, carpenters, cooks, domestics, farmers, and other tradesmen who doubtless experienced skin problems at work. Yet little documentation concerning these occupational diseases was available until 1700, when Bernardino Ramazzini produced his treatise on diseases of tradesmen (2). His first edition contained information about diseases among 42 different classes of workmen. Thirteen years later he published the second edition, which contained 12 more groups.

An outstanding contribution to occupational medicine occurred in 1775 when Sir Percival Pott published the first account of an occupationally induced cancer in which he described cancer of the scrotum among chimney sweeps who came into contact with soot (3).

The surge of industrial development in England, France, Italy, and Germany in the nineteenth century brought widespread interest in occupational skin diseases. Materials such as shoemaker's wax, sugars, spices, lime, tobacco, arsenic, coal tar, chrome, and sunlight were demonstrated as causal agents in occupational skin disease. This interest continued into the twentieth century on the European continent and in Great Britain and was highlighted in two excellent texts on occupational dermatology by L. Prosser White (1) in England and Oppenheim et al. (4) in Germany. Concern about these disorders in the United States began during World War I and has continued with the industrial expansion over the past 60 years. Numerous dermatologists, industrial physicians, practitioners, and allied scientists have added to our present knowledge in causalities, clinical manifestations, treatment, and the prevention of these troublesome diseases (5–13).

1.1 Incidence

During the period 1940–1970, records compiled by state health departments and state compensation boards showed that affections of the skin accounted for 60 to 80 percent of all reported occupational diseases (14). During the past 5 to 6 years there has been a decline in frequency figures in the United States. California statistics for 1972 indicate that skin diseases amounted to 39 percent of the reported occupational illnesses (15) (Table 8.1). Similarly, the U.S. Bureau of Labor Statistics for 1972 showed that of 210,500 occupational diseases cases incurred in several American industries, 41 percent were skin disorders. The BLS reported also that 25 percent of all workdays lost in 1972 were the result of occupational dermatoses (16). It has been advanced by the National Institute for Occupational Safety and Health that the incidence decline is the result of increasing automation, enclosure of industrial processes, and better educational efforts to prevent occupational diseases in general. Some though not all of these changes were stimulated by the passage of the Occupational Safety and Health Act in 1970, requiring employers to comply with regulations concerning the use of protective devices, including clothing and proper cleansing materials, for preventing occupational diseases. Federal legislation relating to hazardous substances also influences prevention because workmen must be made aware of the hazardous materials they encounter at work through proper labeling and instruction.

Despite the decline in the number of cases of occupational skin disease in the United States, the opposite trend has been experienced in Great Britain where interest and

Table 8.1. Reports of Occupational Diseases by Disease Group, California, 1972 (15)

Disease Group	Number	Percentage
Skin conditions	14,337	39.2
Eye conditions	10,881	29.7
Chemical burns	3,640	9.9
Respiratory conditions due to toxic materials	1,767	4.8
Digestive and other symptoms due to toxic materials	1,232	3.4
Ear conditions other than loss of hearing	1,066	2.9
Other and unspecified	961	2.6
Systemic effects of toxic materials	745	2.0
Heart and other circulatory conditions	638	1.7
Infective and parasitic diseases	447	1.2
Effects of environmental conditions	394	1.1
Respiratory conditions, mainly infectious	339	0.9
Loss of hearing	156	0.4
Pneumoconioses	9	—[a]
Total	36,612	100.0

[a] Less than 0.1 percent.

knowledge in occupational medicine is well established. The incidence in that country has risen (17). Furthermore, an increasing number of cases can be expected in the developing countries, where industrial hygiene and health practices rarely keep pace with the industrial growth (18).

1.2 Anatomy and Physiology of Natural Defense

Given the proper circumstances of exposure, any kind of work can cause an occupational dermatosis, yet of the 80 million or more employees in this country, the majority remain free of troublesome and disabling work-incurred skin disorders. A major reason for this seeming immunity can be attributed to the natural defense provided by the anatomic, physiologic, and chemical properties of skin. Through these inherent qualities, skin helps to control body heat, secrete and excrete body sweat, receive various sensations, manufacture pigment, and restock its own cell layers for the benefit of protection.

The anatomic structure of skin is composed of two main levels—epidermis and dermis. Epidermis is quite thin except where it covers the palms and soles, but in company with the elastic and collagenous connective tissue of the dermis, it acts as a resillient guardian of the blood and lymph vessels, nerves, appendages, and muscles. As such, it protects against blunt trauma. This flexibility also permits the return of stretched skin to its normal positional pattern (19).

Epidermis or outer skin has two layers: the keratin or stratum corneum composed of dead cells, and the living epidermal cells, which eventually become stratum corneum cells. Outer keratin cells, though methodically shed, are always being replaced through the replicating function of the skin. Stratum corneum, the staunchest part of the cutaneous defense, is a tightly packed, tough, fibrous protein barrier that resists the mass entrance and exit of water and electrolytes. It responds to physical stimuli of friction and pressure by thickening and, thereby, forming protective callus. Exposure to sunlight or sources of artificial ultraviolet light similarly cause it to thicken. The keratin barrier is somewhat resistant to the action of acids, but it offers much less protection against organic and inorganic alkaline materials. The latter substances soften the keratin cells and dissolve the cellular content; but keratin cell walls have a remarkable, if not unlimited, degree of resistance to chemical insult. Water content of the keratin layer is essential for the maintenance of the barrier effect. Therefore any physical or chemical force such as lowered temperature, humidity, solvents, and desiccating agents that cause water loss, necessarily weakens the defense action (20).

Living epidermis arises from the basal or germinative cells, which reside in palisade fashion within the basal layer or lowest level of epidermal cells. In other words, basal cells are the replacement force that supplies the epidermis with cellular continuity.

Located also in the basal layer are the pigment cells or melanocytes, which furnish a major defense against injury from sunlight. Melanin, the pigment manufactured by the melanocytes, is the end product of an enzyme reaction that takes place within the melanocytes. When the pigment granules have matured they are taken up by the epidermal cells, which move upward and eventually are shed by way of the outer keratin layer. Melanin protects against sunlight because it has the capacity to absorb photons of natural or artificial ultraviolet (21). Pigment formation is stimulated by exposure to sunlight and materials as coal tar, pitch, certain chlorinated hydrocarbons, and petroleum products. Conversely, pigment formation can be inhibited by the absorption of certain phenols and members of the quinone family (5, 6, 12, 22–27).

The dermal portion of skin is considerably thicker than the epidermis. It is composed of a connective tissue matrix made up of elastic and collagen fibers, and ground substance. Located within the dermis are sweat glands and their ducts, which channel sweat to the surface of the skin, as well as hair follicles, sebaceous or oil glands that excrete sebum or natural oil by way of the hair follicles, blood and lymph vessels, and nerves (19).

Heat control of the body is regulated by the excretion of sweat, the circulation of blood, and the nerve centers in the brain. Body temperatures and the circulating blood are physiologically stabilized at a constant temperature despite exposure to wide ranges of climatic variations. Sweat gland function and delivery of sweat, which is evaporated on the skin surface, aid greatly in the control of body heat. Simultaneously, heat loss is facilitated by dilation of the blood vessels within the vascular bed. In contrast, exposure to cold causes the blood vessels to contrast and conserve heat (19).

Nervous elements within the skin constitute a network of nerve endings and fibers that receive and conduct various stimuli, later recognized as heat, cold, pain, and other sense perceptions (19).

Secretory properties of the skin are contained within the sweat and sebaceous glands. Sweat is composed largely of water, but it contains numerous metabolic by-products. The function and delivery of sweat is essential to physiologic normality. However the delivery of excessive or insufficient sweat can be harmful not only to skin but to the total physiologic requirement of the body. Sebaceous glands reside within the dermis in proximity to hair follicles with which they connect. Their product, sebum, is extruded on the surface of the skin by way of the follicular duct and orifice. Overfunction of the sebaceous glands is associated with adolescent acne, but these glands are also target sites for occupational acne caused by oils, tars, and certain chlorinated hydrocarbons (19).

When sweat and sebum are delivered to the surface of the skin, they mix and form a waxy coating. The level of protection that the secretory layer provides is minimal, but it probably helps to maintain the required level of hydration within the keratin cells (19, 20).

By experience in the laboratory and in industry we know that skin provides a fairly strong defense against the entrance of most materials encountered in the work environment. Nonetheless, percutaneous absorption can occur when the continuity of the skin has been damaged by laceration or by denudation of the keratin layer. Skin can absorb materials transepidermally, transfollicularly, and through the sweat orifices and ducts. However the appendegeal routes are believed to provide entry only during the early stage of absorption. The major route is through skin itself (28–30). Numerous substances used in agriculture and industry have caused systemic intoxication by way of cutaneous penetration. Best known among these are the organophosphate and certain chlorinated hydrocarbon pesticides (31), the cyanides, aromatic and amino nitro compounds (32), mercury, tetraethyl lead, and a few other substances (33). Although systemic intoxication by percutaneous absorption occurs infrequently, it does not diminish the importance of skin as an entry route. It is a constant remainder that skin is not a perfect barrier.

2 INDIRECT CAUSALITIES

Despite the importance of recognizing the direct cause of an occupational dermatosis, various predisposing factors often influence the induction of these diseases. Such factors include race, type of skin, age, perspiration, season of the year, the presence of other skin diseases, and level of personal hygiene (1, 5–8, 11, 12).

2.1 Race

No one is racially immune to contracting an occupational dermatosis. It has been stated often that black skin is more resistant to the industrial environment, but this is true only in part. Because of its rich melanin content, black skin is unquestionably better equipped to resist the effects of sunlight. Consequently, malignant lesions caused by ultraviolet light rarely occur in blacks. Additionally, black skin is better protected against the combined effects of light and tar and its products. Nevertheless, blacks do develop photodermatoses from various photoreactive agents, though with less frequency than is seen in whites (34). Whether black employees actually are less susceptible to chemical dusts, solvents, and alkalies, among other substances, has not been proved (5).

Light-complexioned individuals, particularly redheads or blonds with blue eyes, have less resistance to actinic radiation (34). They seem also to be more susceptible to contact dermatitis from a multitude of chemicals, including solvents. Placing these individuals in outdoor work may lead to the early induction of precancerous lesions or frank cutaneous malignancies (5, 6, 8, 12).

2.2 Type of Skin

Caucasians differ in their susceptibility to industrial irritants. Individuals with swarthy, oily skin are more resistant to the action of soaps, solvents, and soluble cutting fluids. Hairy arms and legs are favored embedment sites for insoluble oils, greases, tars, and waxes (1, 5, 6, 8, 12, 35).

A worker with innately dry skin (ichthyosis, xerosis, etc.) is at greater risk when contacting such chemical dehydrators as alkalies, acids, detergents, and solvents (1, 5, 6, 11, 12).

2.3 Age

Young workers, particularly adolescents, often develop acute contact dermatitis. This occurs more because of disregard for the use of protective measures than from any inherent susceptibility associated with age. Older workers by experience are generally more careful at work. Conversely, they are at greater risk in outdoor work affording exposure to sunlight, wet environments, and temperature extremes (1, 5, 12).

2.4 Perspiration

Normal sweating is a physiologic necessity. Most often sweat is protective because it dilutes and washes away irritants. Excessive sweating can be injurious because it causes maceration of the skin in the axillae, groins, and other sites of skin apposition. Similarly, sweat can be detrimental when it solubilizes a chemical that then becomes an irritant, as when calcium oxide is changed to calcium hydroxide by sweat or moisture (1, 5). Sweat can also solubilize, at least in small amounts, certain metals such as cobalt, copper, and nickel, thus making more ions available for percutaneous absorption (10). A common problem associated with excessive perspiration is that of prickly heat. Most often this disorder undergoes normal resolution when the climatic or environmental situation improves and less sweat is being delivered. Prickly heat can be complicated by secondary infection or supervening contact dermatitis induced by various remedies applied to the skin (1, 5, 6, 8, 11, 12).

2.5 Season

As a general rule, occupational skin disease is more prevalent in the warmer months for two reasons. First, warm weather increases perspiration and the problems associated with it. Second, protective clothing (sleeves, gloves, caps, etc.) is more uncomfortable in warm weather, therefore is often discarded. This provides freer contact of the skin with irritant materials otherwise avoided by the use of protective devices. Cold weather also may add to the dermatitis frequency among workers who handle chemicals because many of them are less inclined to take showers or use washing facilities at the plant during the cold months (5, 7, 11, 12).

2.6 Presence of Other Skin Disease

Employees who have preexisting skin disease when hired become potential dermatitis cases. For example, adolescents with acne almost always find that the condition is made worse if they are placed in jobs affording contact with cutting oils, waxes, greases, and tars. The same individuals are at increased risk if they contact such chloracnegenic chemicals as the biphenyls, triphenyls, and chlornaphthalenes (5, 12).

If people with a history or presence of atopic dermatitis are hired and placed in work situations that afford ambient temperature changes, or exposure to greasy compounds or dusty chemicals, an aggravation of their atopic state generally results (5, 6, 8, 11, 12). This does not mean such people should be excluded from the work described, it simply means that more care must be exercised in job placement. Psoriasis, another chronic dermatosis, is often made worse by frictional and chemical contacts at work (5, 6, 8, 11, 12).

Hyperhidrosis of palms and soles provides an added hazard to workmen who handle chemicals that become irritants when solubilized. Furthermore, the use of impervious gloves or heavy footgear can add to the sweating and aggravate maceration of the palms or soles (5, 6, 8, 11, 12).

Sensitivity to sunlight as displayed by those with vitiligo, lupus erythematosus, polymorphous light eruption, and porphyria, is a deterrent to work affording exposure to natural or artificial ultraviolet light (12, 34).

2.7 Poor Hygienic Practices

The most efficient way of preventing an occupational dermatosis is to keep the skin relatively free of contact with irritants and allergenic materials. This is accomplished by wearing proper protective clothing and by cleansing the skin at regular intervals during the course of the workday. Most workers adhere to these basic precepts of prevention, but almost all plants have a few employees who not only resent wearing protective clothing but also do not take time out to wash. The same workmen tend to wear soiled work clothing over prolonged periods of time, which increases the chance of incurring an occupational dermatosis (5, 8, 11, 12).

3 DIRECT CAUSES OF OCCUPATIONAL SKIN DISEASE

The number of substances or conditions capable of inducing a cutaneous disorder in the workplace is unlimited. Each year the potential increases and new causes are reported, and most can be classified in one of the following categories: chemical, mechanical, physical, and biological (1, 5, 6, 8, 11–13).

3.1 Chemical Agents

Chemical agents always have been and will continue to be a leading cause of work-associated skin disease. They are divided into organic and inorganic substances, which act as primary skin irritants, cutaneous sensitizers, or photosensitizing agents.

3.1.1 Primary Irritants

A primary irritant is a substance that can cause a demonstrable effect on the skin at the site of contact if it is permitted to act in sufficient concentration and quantity for a sufficient length of time. In other words, a primary skin irritant can injure the skin of anyone. It must be understood, however, that primary irritants can differ in the speed and severity of their action. Many of them are vigorous in their attack and produce an effect within moments to a few hours, as would occur following contact with sulfuric or nitric acid, sodium hydroxide, ethylene oxide gas, or chloride of lime. Chemicals capable of such activity on skin are best termed "absolute or strong irritants." Conversely, some chemicals may take several days before a demonstrable effect occurs, and these are termed "marginal or mild irritant agents." This class includes such materials as soluble cutting fluids, solvents such as acetone, ketone, and alcohol, and soap and water (1, 5, 6, 8, 11–13, 35).

3.1.2 Clinical Effects

Although we have long recognized the clinical effects produced by primary irritants, there is little scientific knowledge concerning the nature of their action on skin. General behavior of many chemicals in the laboratory or in industrial processes is fairly well known, and at times the nature of their chemical action can be applied theoretically and, in some instances, actually, to chemical action on skin. For example, it is well accepted that organic and inorganic alkalies attack keratin (1, 5, 6, 11, 19, 20); that organic solvents dissolve surface lipids and remove lipid components from keratin cells (5, 6, 19); that salts of arsenic and chromium, among other heavy metals, precipitate protein and cause it to denature (1, 5, 6); that salicylic acid, oxalic acid, and urea, among other acid substances, can chemically and physically reduce keratin (1, 5); that arsenic, tar, methyl cholanthrene, and other known carcinogens stimulate skin to take on abnormal growth patterns (5). Yet little is known about the actual mechanism of primary irritation. Despite these shortcomings, it is known that most occupational dermatoses (80 percent) result from contact with primary irritants (5). These substances cause many of the occupational dermatoses because they are commonly encountered in agricultural, manufacturing, and service pursuits. Hundreds of substances within the chemical classes of acids, alkalies, gases, organic materials, metal salts, solvents, resins, and soaps (including synthetic detergents) can cause absolute or marginal irritation (Table 8.2).

Table 8.2 Typical Primary Irritants

Acids	
Inorganic	**Organic**
Arsenious	Acetic
Chromic	Acrylic
Hydrobromic	Carbolic
Hydrochloric	Chloracetic
Hydrofluoric	Cresylic
Nitric	Formic
Phosphoric	Lactic
Sulfuric	Oxalic
	Salicylic
Alkalies	
Inorganic	**Organic**
Ammonium	Butylamines
Carbonate	Ethylamines
Hydroxide	Ethanolamines
Calcium	Methylamines
Carbonate	Propylamines
Cyanamide	Triethanolamine
Hydroxide	
Oxide	
Potassium	
Carbonate	
Hydroxide	
Sodium	
Carbonate (soda ash)	
Hydroxide (caustic soda)	
Silicate	
Trisodium Phosphate	
Cement	
Soaps	
Metal Salts	
Antimony trioxide	
Arsenic trioxide	
Chromium and alkaline chromates	
Mercuric chloride	
Zinc chloride	

Table 8.2 (continued)

Solvents	
Alcohols	**Ketones**
Allyl	Acetone
Amyl	Methyl ethyl
Butyl	Methyl cyclohexanone
Ethyl	
Methyl	
Propyl	
Chlorinated	**Petroleum**
Carbon tetrachloride	Benzene
Chloroform	Ether
Dichlorethylene	Gasoline
Epichlorhydrin	Stoddard
Ethylene chlorhydrin	Varsol
Perchlorethylene	White spirit
Trichlorethylene	
Coal Tar	**Turpentine**
Benzol	Pure oil
Naphtha	Turpentine
Toluol	Turpineol
Xylol	Rosin spirit

Source. L. Schwartz, L. Tulipan, and D. J. Birmingham, *Occupational Disease of the Skin,* 3rd ed., Lea & Febiger, Philadelphia, 1957.

3.2 Cutaneous Sensitizers

Any chemical can act as a cutaneous sensitizer, but the frequency of contact dermatitis caused by allergic sensitization is much less than that resulting from primary irritants. This is fortunate because work-induced allergy is frequently attended by chronic recurrence, prolonged absences from work, and eventual job transfer because of the persistence of the allergic base. The phenomenon is a specifically acquired alteration in the capacity to react, brought about by an immunologic mechanism. The antigenic substance is usually a simple chemical that does not cause visible change in the first or perhaps after several contacts with the skin. However following a period of incubation (5 or more days while sensitization is taking place), subsequent contact with the same or a closely related substance induces a dermatitic reaction. (Some primary irritants can also act as allergens.) It is currently held that the antigenic material traverses the keratin or outer layer, thereby gaining access to epidermal cells and possibly other sites within the skin. At one or perhaps more sites, the antigenic material combines or conjugates with protein carrier to become a complete antigen. After the conjugate is formed, lymphocytes appear on the scene and recognize the antigen as foreign. How this takes place is not known, but the "informed" lymphocytes then journey to the regional nodes, where cellular proliferation occurs and more cells with the ability to recognize the antigen (i.e., sensitized cells) enter the circulation. When these circulating lymphocytes encounter the antigenic conjugate in the skin, they react and release mediators, which participate in an inflammatory reaction. The cutaneous expression of this allergenic reaction is an acute inflammatory one and may closely resemble a primary irritant reaction. Distinguishing between irritation and sensitization is highly dependent on the time span because the allergic reaction usually requires a longer induction period than is the case with primary irritation (36–40). As a rule, cutaneous sensitizers are not prone to affect large groups of workers. However this maxim does not hold when dealing with epoxy resins, phenol-formaldehyde plastics, poison ivy or poison oak, and cashew nut shell oil (5, 9, 13, 36, 41). Other industrial chemicals well known for their sensitizing property are potassium dichromate by itself or contained in cement, nickel sulfate, hexamethylene tetramine, mercaptobenzothiozole, and tetramethylthiuram disulfide (5, 6, 8–13) (Table 8.3).

3.3 Plants and Woods

A broad variety of plants and woods cause injury to the skin through primary irritation or allergic sensitization. However photosensitivity and mechanical effects also occur. Although the chemical identity of many of the plant toxins has not been established, it is well known that the irritant or allergenic agent can be present in the leaves, stems, flowers, bark, and other components of the plant (42). Employees at risk in contracting dermatitis from poisonous plants include agriculture and construction workers, electric and telephone linemen, florists, gardeners, lumberjacks, pipeline installers, and road builders, among others who work outdoors (5, 11–13, 43). Poison ivy and poison oak are the foremost offenders, particularly in California, where contact with poison oak causes some 4000 cases of occupational dermatitis each year (44). Poison oak, ivy, and sumac are members of the Anacardiaceae, which also includes such chemically related allergens as cashew nut shell oil, Indian marking nut oil, and mango (5, 42–44). The chemical principle or toxicant contained within the family Anacardiaceae is a phenolic (catechol), and sensitization to one member generally confers sensitivity or cross-reactivity to the others (45). Additional plants for which the dermatitis-producing tendency is well established include carrot, castor bean, celery, chrysanthemum, hyacinth bulbs, oleander, primrose, ragweed, tulip bulbs, and wild parsnip. Many other plants, including vegetables, have been reported to be causal in contact dermatitis (5, 10, 13, 42, 43, 46, 47).

Many woods have been catalogued as causal in occupational dermatoses. As compared to contact dermatitis from plants, however, the frequency is considerably lower. Carpenters, cabinet makers, furniture builders, lumberjacks, lumber yard workers, and model makers, among others, can acquire primary irritant, allergic contact dermatitis or

Table 8.3. Common Sensitizers

- Metals
  - Beryllium salts
  - Chromates (alkaline)
  - Mercurial salts
  - Nickel salts
- Plants
  - Anacardiaceae
    - *Anacardium occidentale* (cashew)
    - *Toxicodendron diversilobum* (poison oak, West)
    - *Toxicodendron quercifolium* (poison oak, East)
    - *Toxicodendron radicans* (poison ivy)
    - *Toxicodendron vernix* (poison sumac)
  - Compositae
    - *Ambrosia artemisiifolia* (ragweed)
    - *Chrysanthemum coccineum* (painted daisy)
    - *Chrysanthemum parthenium* (feverfew)
  - Euphorbiaceae
    - *Hippomane mancinella* (manchineel tree)
    - *Ricinus communis* (castor bean)
  - Liliaceae
    - *Allium cepa* (onion)
    - *Allium sativum* (garlic)
    - *Hyacinthus orientalis* (hyacinth)
    - *Tulipa spp.* (tulip)
  - Rutaceae
    - *Citrus aurantium* (sour orange)
    - *Citrus limon* (lemon)
- Resin systems
  - Acrylic
  - Epoxy
  - Melamine formaldehyde
  - Phenol formaldehyde
  - Urea formaldehyde
- Rubber chemicals
  - Accelerators
    - Diphenylguanidine
    - Hexamethylene tetramine
    - Mercaptobenzothiazole
    - Tetramethylthiuram disulfide
  - Antioxidants
    - *N*-Isopropyl-*N*-phenylparaphenylene diamine (IPPDA)
    - Phenyl-β-naphthylamine
    - Monobenzyl ether of hydroquinone

traumatic injury from the wood being handled. Chemical impregna, sawdust, or wood spicules may cause irritant effects. Allergic dermatitis is caused by the oleoresins, the natural oil or chemical additives. Best known for causing contact dermatitis are acacia, ash, beech, birch, cedar, maple, mahogany, pine, and spruce (5, 10, 13, 42, 43, 48). Other materials of concern are the chemical impregnates used for preservative purposes. These may include arsenicals, chlorphenols, creosote, and copper compounds (5, 43, 49).

3.4 Photosensitivity

Dermatitis resulting from photosensitivity is an untoward cutaneous reaction to light. Two types are recognized: phototoxic and photoallergic. Phototoxic reactions are the most common, the simplest example being a burn from natural or artificial light. Thousands of employees—construction workers, fishermen, gardeners, farmers, road builders, forestry workers, electric linemen, and many others—are exposed to natural light sources; artificial light exposures are experienced by electric furnace and foundry operators, glass blowers, photoengravers, steel workers, and welders, among others (50). In addition to the phototoxic reactions to natural and artificial light are the photobiologic effects produced by the interaction of light with chemicals, plants, or drugs (5, 10–13, 34, 43, 51).

In industry, coal tar distillation may afford exposure to anthracene, phenanthrene, and acridine, all known photoreactive chemicals. Related products as creosote, pitch, roof paint, road tar, and pipeline coatings are well-known causes of hyperpigmentation resulting from the interaction of coal tar agents and sunlight (5, 12, 34, 35, 52, 53) (Table 8.4).

The best-known photoreactive plants are members of the family Umbelliferae, which includes wild parsnip, cow parsnip, celery (infected with the pink rot fungus), wild carrot, fennel, and dill. The essential photoreactive materials contained within these plants are furocoumarins, notably the psoralens (54). Some examples appear in Table 8.5.

The problem of occupational photosensitivity is further complicated by a number of topically applied and ingested drugs that have the capacity of interacting with specific wavelengths of light to produce either a phototoxic or a photoallergic phenomenon. Among the agents known to produce this effect are certain halogenated compounds that

Table 8.4. Photosensitizers

Antibacterials	Coal Tar
Dibromsalicylanilide	Acridine
Hexachlorophene	Anthracene
Sulfonamides	Certain chlorinated hydrocarbons
Tribromsalicylanilide	Creosote
	Phenanthrene
	Pitch

Table 8.5. Photosensitizing Plants

Moraceae
Ficus carica (fig)
Rutaceae
Citrus aurantifolia (lime)
Dictamus albus (gas plant)
Ruta graveolens (rue)
Umbelliferae
Anthriscus sylvestris (cow parsley)
Apium graveolens (celery, pink rot)
Daucus carota var. *sativa* (carrot)
Heracleum spp. (cow parsnip)

have been incorporated in soaps, drugs related to sulfonamides, certain antibiotics, and tranquilizers of the phenothiazine group (34, 51, 53, 55, 56).

3.5 Mechanical

Trauma accompanying work may be subtle, moderate, or severe. Contact with small spicules of fiberglass, copra, hemp, and so on, can induce irritation and can stimulate severe itching and scratching (5, 11, 12). Skin can react to friction by forming a blister or a callus; to pressure by changing color or becoming thickened; to shearing or sharp external force by denudation or puncture wound (5, 11, 12, 57). Secondary infection may complicate blisters, calluses, or any minor wound that breaks the skin. Thousands of workmen use air-powered and electric tools that operate at variable frequencies. Exposures to vibration in the 40–300 Hz range can produce painful finger, a Raynaud-like disorder associated with spasm of the blood vessels in the tool-holding hand. Slower frequency tools, like jack hammers, may cause boney, muscular, and tendonous trauma (58, 59).

3.6 Physical

Work exposures to excess heat, cold, electricity, ultraviolet light (natural and artificial), and sources of radiation can lead to cutaneous and systemic effects. Thermal burns are common among welders, lead burners, metal cutters, roofers, molten metal workers, and glass blowers. Elevated temperatures and humidities induce miliaria (prickly heat), which is the result of waterlogging of the keratin layer and subsequent occlusion of the sweat ducts and their skin openings. Heat cramps, heat exhaustion, and heat stroke are more severe systemic effects of excess heat (60).

Cold injury to skin is most commonly noted as frostbite, which can result in permanent vessel injury. Frostbite usually attacks the fingers, toes, ears, and nose of outdoor workers (e.g., construction workers, farmers, fishermen, highway workers, linemen, military personnel, or those employed in frozen food and storage facilities) (60).

Electricity can cause severe cutaneous burns of local or widespread proportion (5).

Sunlight is a hazard of outdoor work to which thousands are exposed, but in the United States those who work in the southwest are at greatest risk because of the intensity of sunlight in that area. Artificial ultraviolet light is encountered in metal pouring in foundries and steel plants, glass blowing, photoengraving, plasma torch operations, and welding, among other situations (60).

Ionizing radiation has many applications in modern industry and technology. For example, the production and use of fissionable materials, radioisotopes, X-ray diffraction apparatus, X-ray, electron beam, and cobalt therapy in medicine, and industrial X-ray for the examination of metal flaws. Severe cutaneous and systemic injury can result from the accidental exposure to these sources (5, 12, 60).

3.7 Biological

Occupationally induced infection can happen in any occupation as a result of exposure to bacteria, viruses, fungi, or parasites. A simple laceration from the sharp edges of an envelope can become secondarily infected with staphylococci or streptococci. A thorn, a wood splinter, or a metal slug acting as a foreign body can pave the way for secondary infection of the skin. Most infectious dermatoses can be readily diagnosed and effectively treated with antibiotics or other appropriate measures. Nonetheless, certain occupations provide a greater risk for contracting bacterial infections. For example, one finds anthrax among sheepherders, hide processors, and wool handlers; erysipeloid disorders among butchers, fishermen, and fowl dressers; and folliculitis among machinists, garage workers, candy makers, sanitation and sewage employees, and those exposed to coal tar products (5, 9, 11–13, 61).

Fungi can cause purely cutaneous or systemic disease or both. Superficial infection of the skin from *Candida* occurs among bartenders, dishwashers, cannery workers, kitchen employees, laundry workers, nurses, or anyone exposed to wet work. Sporotrichosis is seen among garden and landscape workers, florists, and miners. Ringworm from animals can be transmitted to farmers, veterinarians, laboratory workers, or anyone who is frequently in contact with animals (5, 12, 61).

Certain parasitic mites inhabit cheese, grain, and other foodstuffs and will attack bakers, grain harvesters, grocers, and longshoremen. Similarly, mites that thrive on animals and fowl may parasitize humans, as seen in human infestation with dog scabies and bird mites. Another animal parasite that causes cutaneous disease among humans, particularly in southeastern United States, is animal hookworm. Construction workers, farmers, plumbers, and anyone whose work requires contact with sandy soil can contact the larvae deposited in the soil by infected dogs and cats (5, 12, 61).

Several occupational diseases are associated with virus infections—for example, Q fever, Newcastle disease, and ornithosis; however the occupational dermatoses best known as caused by virus infection are ecthyma contagiosum (orf) contracted from infected sheep, milker's nodules from infected cows, chicken pox from infected children, and certain cases of Herpes zoster following trauma (61, 62).

Rare, but real, dermatoses can result from animal bites—both domestic and wild (e.g., snake bites; stings of bees, hornets, and wasps; marine injury from shark bite and from various poisonous fish) (5, 12).

4 CLINICAL ASPECTS OF OCCUPATIONAL SKIN DISEASE

Unlimited numbers of animate and inanimate agents comprise the hazardous potential of the work environment. Effects they produce on the skin vary in clinical appearance and histopathologic pattern. Often the nature of the lesions and the sites of involvement provide a clue to the class of materials, but only in rare instances does clinical appearance indicate the precise causal agent. Except for a few strange and unusual clinical effects, most of the occupational dermatoses can be classified in one of the following clinical types. Several agents known to be causal for each clinical type are included.

4.1 Acute Eczematous Contact Dermatitis

Most of the occupational dermatoses fall into the acute eczematous contact dermatitis group. Heat, redness, swelling, vesiculation, and oozing are the clinical signs; and itch, burning, and general discomfort are the major symptoms experienced. The back of the hands, the inner wrists, and the forearms are the usual sites of involvement; but acute contact dermatitis can occur anywhere on the skin. When forehead, eyelids, face, ears, and neck are the sites most affected, dust and vapors usually are at fault. The prime cause of generalized contact dermatitis is the continued wearing of contaminated clothing.

Usually a contact dermatitis can be recognized as such, but whether the eruption is the result of contact with a primary irritant or a cutaneous sensitizer can be ascertained only through a detailed history, a working knowledge of agents within the environment, and the proper application of diagnostic tests. Profoundly severe blistering or marked tissue destruction generally indicates a strong irritant, but it is the history that reveals the precise agent (5–13).

Acute contact dermatitis is associated with hundreds of irritant and sensitizing chemicals, plants, and photoreactive materials, among which are some examples:

Acids, dilute	Herbicides	Resin systems
Alkalies, dilute	Insecticides	Rubber accelerators
Anhydrides	Liquid fuels	Rubber antioxidants
Detergents	Metal salts	Soluble Emulsions (Oil)
Germicides	Plants and woods	Solvents (5–13, 62–65)

4.2 Chronic Eczematous Dermatitis

Hands, wrists, and forearms are the characteristic sites affected by chronic eczematous dermatitis. Dry, thickened, scaling skin with cracking and fissuring of the digits and palms and chronic nail dystrophy best describe this chronic dermatitis. Acute weeping lesions may appear at any time. A large number of contact agents perpetuate the marked dehydration that accompanies this chronically recurrent dermatosis. Among these are the following:

Abrasive dusts (pumice, sand)
Alkalies
Cement
Cleansers (industrial)
Cutting fluids (soluble)
Chronic fungal infections
Oils
Resin systems
Solvents
Wet work (1, 4–13, 61–65)

4.3 Folliculitis, Acne, and Chloracne

Lesions involving the hair follicles favor such sites as the face, forearms, back of hands and fingers, buttocks, thighs, and lower abdomen. They can accompany any kind of work that entails heavy soilage. Comedones (blackheads) and follicular infections are common among garage mechanics, machine tool operators, tar workers, and tradesmen engaged in dusty occupations.

Acne caused by industrial sources usually occurs on the face, arms, upper back and chest; but when the exposure is severe, lesions occur on the abdomen, buttocks, and thighs. Machinists, mechanics, oil field refinery workers, road builders, and roofers exposed to tar are at risk (5–8, 11–13, 35, 62–67).

Chloracne is characterized by the appearance of folliculitis, blackheads, acne, cysts, and scar formation. The disease may be mild or severe, localized or widespread, but under any circumstance it is a disorder with a prolonged course. Severe exposure can produce toxic liver injury, including porphyria cutanea tarda. Lesions (blackheads and cysts) of chloracne first appear on the sides of the forehead and around the lateral aspects of the eyelids. Cystic lesions are frequently present behind the ears. As exposure continues, lesions can affect widespread areas, except for the palms and soles (5–7, 68–72).

Folliculitis	Acne	Chloracne
Asphalt	Asphalt	Chlornaphthalenes
Creosote	Creosote	Chlorinated diphenyls
Greases	Crude oil	Chlorinated triphenyls
Lubricants	Insoluble cutting oil	Hexachlorodibenzo-*p*-dioxin
Oils	Pitch	Tetrachloroazoxybenzene
Pitch	Tar	Tetrachlorodibenzodioxin
Tars		

4.4 Pigmentary Abnormalities

Color changes in skin stem from percutaneous absorption, from inhalation, or from a combination of both entry routes. The abnormality may represent a chemical discoloration from a dye that fixes to keratin, or an increase or decrease in epidermal pigment (melanin) (5–8).

Hypermelanosis results from stimulation of the melanocytes. It can follow an inflammatory dermatosis, exposure to sunlight, or an exposure of a combination of sunlight and one of several photoreactive chemicals or plants. Loss of pigment (leukoderma) results from direct injury to the epidermis and melanin cells by burns, chronic dermatitis, trauma, or by chemical interference with the tyrosine–tyrosinase reaction, which produces melanin. Antioxidant chemicals used in adhesives, cutting fluids, sanitizing agents, and rubber have caused chemical pigment loss or leukoderma. Inhalation or percutaneous absorption of certain toxicants such as aniline and other members of the aromatic nitro and amino compounds cause methemoglobinemia. Jaundice can result from hepatic injury by carbon tetrachloride or trinitrotoluene, among other agents (5–8, 11, 12, 22–25, 32, 52).

4.4.1 Pigmentation

	Melanin Abnormalities	
Discolorations	Plus	Minus
Arsenic	Chloracnegens	Antioxidants
Certain amines (org.)	Chronic dermatitis	Hydroquinone
Carbon	Coal tar products	Monobenzyl ether of hydroquinone
Dyes	Petroleum oils	
Mercury	Photoreactive chemicals	Tertiary amyl phenol
Picric acid	Photoreactive plants	Tertiary butyl catechol
Silver	Radiation (sunlight)	Tertiary butyl phenol
Tetryl	Radiation, ionizing	Burns
Trinitrotoluene		Chronic dermatitis
		Trauma

4.5 Neoplasms

Proliferative diseases of the skin can be benign or malignant. Such benign lesions as asbestos warts and petroleum and tar warts are well-known examples. Malignant growths include basal cell epithelioma and squamous cell carcinoma. Several chemical and physical agents are classified as industrial carcinogens, but only a few are frequent causes of skin cancer. It is true that more cancers occur on the skin than at any other site, but how many of these are occupational in origin is not known. The agents best known to cause skin cancer are listed below and of these, sunlight is the major cause. More working people are exposed to sunlight than any other carcinogen encountered in the following pursuits: agriculture, construction, fishing, forestry, gardening and landscaping, oil drilling, road building, and telephone and electric line installation.

4.6 Known Carcinogens

Actinic rays
Anthracene
Arsenic (inorganic)
Asphalt
Burns (thermal)
Coal tar
Coal tar hydrogenates
Creosote oil
Mineral oils
Oils (crude)
Pitch
Radium and roentgen rays
Shale oil
Soot
Tar
Ultraviolet light

In European countries shale oil and paraffin pressmen experienced a high frequency of carcinomatous lesions attacking the scrotum and lower extremities. Similar experiences with paraffin were encountered in the United States, but better hygienic practices have all but eliminated the difficulties. It is important to remember that cancers noted on workmen who contact carcinogenic agents are not necessarily occupational. A certain number of people develop skin cancer irrespective of their jobs; for instance, residence in Australia or in the southwestern United States is associated with a high frequency of skin cancer due to sunlight. Ascertaining whether a skin cancer is of occupational origin in individuals residing in these areas can be a difficult task (5, 12, 73–82).

4.7 Ulcerations

The earliest skin changes observed among metal miners and craftsmen were cutaneous ulcerations. In 1827 Cumin reported on skin ulcers produced by chromium. Presently the chrome ulcer or "hole" caused by chromic acid or concentrated alkaline chromate is the most familiar chemically produced ulceration of the skin. Workmen in chrome reduction plants and chrome plating operations experience a high frequency of nasal perforation and cutaneous ulceration following sufficient contact with the vapors. Other agents known to produce punched-out ulcers of the skin are arsenic trioxide, calcium arsenate, calcium nitrate, and slaked lime (1, 5, 78, 83–86). Nonchemical ulcerations may be associated with trauma or may be due to such nonoccupational sources as diabetes, pyogenic infections, vascular insufficiency, and sickle cell anemia (87).

The clinical types of ulceration just described constitute the most frequently observed forms of occupational skin disease. However a number of disorders affecting the skin, hair, and nails do not fit into these classifications. Some examples are nail discoloration and destruction by alkaline agents (5) and by fungal infections (5, 11–13, 61); hair loss caused by chloroprene dimers (88); facial flush resulting from the combination of certain chemicals such as butyral doxime (89) or trichloroethylene (90) following the ingestion of alcohol; granulomas of the skin caused by the embedment of beryllium salts or silica (91–94), foreign bodies such as animal bone and coral, or infections caused by atypical mycobacteria and fungi (95); Raynaud's phenomenon induced by high frequency power tools (96); and acroosteolysis accompanied by Raynaud's phenomenon among certain workmen exposed to polymerization products of vinyl chloride (97, 98). New and unusual clinical patterns caused by various agents in the work environment are continuing to appear. However with the discovery and application of reliable and practical tests that predict toxicologic potential of new chemicals, some of these disturbing clinical problems can be averted.

5 DIAGNOSIS

Anyone who works can contract a skin disorder, but not all skin disorders are occupational. Nonetheless, it is commonly assumed by employees that any skin disease incurred is caused by their work. Sometimes the assumption is correct, but often there is no true relation to the work situation. Arriving at the correct diagnosis is not generally difficult, but it is more than a routine exercise. The industrial physician has the distinct advantage of knowing the nature of the work environment and the specific agents encountered. The dermatologist may find the diagnosis elusive if he is unfamiliar with agents within the work environment. The practitioner with little or no interest in such problems, if he is also lacking in dermatologic skill, may find making a correct diagnosis most difficult. At any rate, the attending physician, specialist or otherwise, should satisfy certain fundamental criteria to arrive at a diagnosis.

5.1 Appearance of the Lesions

The dermatosis should fall into one of the accepted clinical types with respect to its morphologic appearance. Most occupational skin diseases are of the acute or chronic eczematous type; but other clinical forms such as follicular, acneform, pigmentary, neoplastic, and ulcerative can occur, and these deserve due consideration as occupational lesions.

5.2 Sites of Involvement

Usually the common contact sites are the hands, wrists, and forearms; but forehead, eyelids, ears, and neck also may be affected. At times the eruption can be generalized, assuming severe proportions. The dorsal surfaces of the hands, fingers, and webs are common attack sites; but palms also can be affected. Volar and extensor wrists and forearms are frequent areas of involvement. Widespread dermatitis generally implies heavy exposure to dust, because of inadequate protective clothing or poor hygienic habits.

5.3 History and Course of the Disease

Only through detailed questioning can the proper relationship between cause and effect be established. History should divulge a complete account of work experience. Information concerning the present job should reveal length of time in job, the substances handled, when the eruption began, the areas involved, how the eruption looked, what medication had been applied, cleansing habits at work and at home, and whether protective gear or barrier creams were used. Questioning should uncover whether other employees had similar difficulties. Information regarding substances contacted in hobbies, house and garden pursuits, and "moonlighting" should be elicited. The importance of obtaining a good history concerning drug intake is obvious. The behavior of the eruption on the weekends and during vacation or sick leave may be most helpful in assessing the occupational component.

5.4 Laboratory Tests

Laboratory tests should be used when indicated for detecting bacteria, fungi, and parasites. These may include direct microscopic examination and bacterial and fungal culture, skin biopsy of one or more lesions for histopathologic diagnosis, and patch tests to detect any occupational or nonoccupational allergens, including photosensitizers (5–13).

5.5 Patch Tests

The diagnostic patch test procedure has repeatedly proved its value when properly performed and interpreted. The test is based on the theory that if an acute or chronic eczematous dermatitis is caused by an allergic sensitization to a given substance, application of that material to an area of unaffected skin for 24 to 48 hours will cause an inflammatory reaction at the application site. When an employee is working with known primary irritants and his fellow employees are also affected with dermatitis, the cause is obvious and patch testing is unnecessary (99). Exception to this rule occurs when patients have been contacting irritants that can also act as potent sensitizers (epoxy and acrylic systems, resin hardeners of the aliphatic or aromatic amine group, formalin, chromates, nickel, etc.). If only one employee is affected among a group who handle the same materials, and the history and course of the disease suggest an allergic reaction, the test is indicated.

Materials capable of causing allergic eczematous contact dermatitis are limitless, and since allergic contact dermatitis represents about 20 percent of occupational skin disease, the test is performed with relative frequency (100).

The person performing the test must have a clear understanding of the difference between a primary irritant and a sensitizer. Testing with strong and even marginal irritants will produce a reaction, yet this does not mean that a patch test cannot be performed with dilute solutions of a known primary irritant (99). An abundance of published data on recommended test concentrations and proper vehicles is readily available, and the material is constantly being updated (5, 9–11, 13, 100, 101).

The test technique is relatively simple. Liquids, powders, or solids can be applied as open or closed tests as the situation requires. Suspected agents in nonirritant concentration in the appropriate vehicles are applied to an area of normal skin, usually the back. The North American Contact Dermatitis Group and the International Contact Dermatitis Group are advocating standardized test concentrations applied to a standardized patch (AL-test) which can be applied in vertical rows to the back, then covered with hypoallergenic tape. Some investigators and clinicians prefer to use patches that permit larger areas of contact with the allergen than occur with the AL-test. The patch is left in place for 24 to 48 hours before removal, unless severe itching and burning are experienced. When the patch is removed after a 24 to 48 hour application, a period of at least 30 minutes should elapse before the result is read. Subsequent readings should be made at 72 hours and at 96 hours, if necessary.

Interpreting the degree of reaction and its implication requires experience. The levels of reaction currently used are as follows: doubtful = ±; erythema = +; erythema, edema, papules = ++; erythema, edema, and vesicles = +++; bullous reaction = ++++. When a reaction is not present at the time of patch removal but appears at a later time (less than 5 days after initial application), the reaction is considered to be a delayed type. True allergic reactions tend to increase for 24 to 48 hours following test removal, whereas irritant reactions generally subside within the same period (99–105).

Interpreting the test results correctly is essential. A positive test can result from contact with an irritant or a sensitizer. If a positive test is the result of sensitization, it means that the patient was reactive to that substance at the time of the test. If the positive test correlates with the contact history, it can be considered to be good substantiating evidence. Conversely, false positive tests can take place if the patient was tested (*a*) during an active phase of the eruption, and he reacted in a nonspecific manner; (*b*) with a marginal irritant; (*c*) with a sensitizer to which previous sensitization had occurred, although it has no current relationship to the present occupation; or (*d*) with a cross-reacting chemical unrelated to the agents contacted at work.

A negative test, if correct, indicates the absence of a primary irritant or a sensitization reaction. However a negative test can be misleading for the following reasons: (*a*) the test substance may not have contained the actual allergenic agent, (*b*) the test substance may have been lacking in strength and quantity, (*c*) certain physical and mechanical factors (adherence and occlusion) may have been absent, or (*d*) hyposensitivity may have supervened (99–105).

6 TREATMENT

The immediate therapeutic management of an occupational dermatosis does not differ essentially from that required for a similar eruption of nonoccupational origin. In either case, treatment should be first directed to provide a reasonable level of comfort for the patient, fairly rapidly. The choice of treatment agents depends on the nature and severity of the dermatosis. In 70 to 80 percent of cases the patients have contracted acute or chronic eczematous dermatitis. Most of these can be managed as ambulatory cases; however there should be no hesitancy in hospitalizing an employee when the severity and extent of the eruption warrant this measure.

Acute eczematous dermatoses of contact origin generally respond well to wet dressings and to topical steroids, but systemic corticosteroid therapy should be used when necessary. Corticosteroids have provided remarkably lessened morbidity in the management of the acute and chronic eczematous dermatoses caused by work. Once the dermatosis is under good control, clinical management must be directed toward:

1. Ascertaining the cause.
2. Returning the patient to his job as soon as the skin condition warrants.
3. Instructing the patient thoroughly in the means required to minimize or prevent contact at work with the causative material.

In all contact dermatoses it is essential to establish the causal materials or situations contributing to the disease. Follicular or acneform cutaneous lesions, especially chloracne, are notoriously slow in responding to treatment. Pigmentary change, similarly, may defy the usual therapeutic agents and remain active for many months. New growths should be removed by an appropriate method required for the lesion. Ulcerations inevitably lead to scar formation; thus the importance of preventing further occurrences.

Almost all cases of occupational skin disease respond to appropriate therapy. Unless a strenuous effort is made to detect the cause and to minimize, if not entirely eliminate contact with the offending agent, however, prompt recurrence of the dermatosis can be expected (5–13, 106, 107).

6.1 Prolonged and Recurrent Dermatoses

It is generally accepted that an occupational dermatosis can be expected to disappear or to be under good control within a period of 4 to 8 weeks after beginning treatment. However when dealing with dermatoses caused by cement, chromium, nickel, mercury, and certain plastics, chronic eczematous dermatitis often ensues. Such dermatoses often are complicated by supervening infection or unwanted effects from various treatment agents. The ubiquitous nature of such materials as chromium and nickel makes it even more difficult to control the levels of contact with these substances. Additionally, there are numerous instances of patients with accepted or alleged occupational skin disease who undergo prolonged and recurrent attacks. When this happens, one or more of the following situations may be paramount:

1. Incorrect diagnosis.
2. Failure to establish the cause.

3. Failure to eliminate the cause even when correct cause was detected.
4. Improper treatment.
5. Poor hygienic habits at work.
6. Supervening secondary infection.
7. Cross-reactions with related chemicals.
8. Malingering.

Prolonged and recurrent dermatoses lend themselves to medicolegal action because the diagnosis is not clear-cut and medical opinions differ regarding why the dermatosis continues long after employment has ceased (108).

7 PREVENTION

Eliminating or, at least, minimizing skin contact with potential irritants and sensitizers is the key to preventing occupational skin disease. That prevention is possible is well documented by the 25 percent decrease in frequency of these disorders over the past 20 years. Notwithstanding the lower number of cases, occupational dermatoses still account for a great deal of lost time.

Preventive measures to control occupational diseases are most readily accomplished in large industrial plants. Conversely, many small plants have neither the money nor the necessary personnel to initiate and monitor effective preventive measures. Furthermore, most cases of occupational dermatoses occur in the small plants.

Time tested control methods known to prevent occupational diseases are classed as (*a*) environmental, (*b*) medical, and (*c*) personal (including protective or barrier creams) (5, 8, 11, 12, 109–112).

7.1 Environmental

Preventive engineering should begin when a new plant is on the drawing board. There is no better time to plan hygienically oriented processing systems, automation, and ventilation control of toxic gases, dust, vapors, and fumes. Elaborate control systems are expensive, but they cost less during construction of a new plant than when installed as an afterthought. Large industrial plants generally can afford these controls, whereas small plants find them too costly. Even so, small plants can select certain equipment, such as ventilated hoods, suction apparatus, fans, and similar hygienic devices, to control and remove noxious material from the workplace. In short, there should be no question about the importance of good engineering control in maintaining hygienic working conditions.

7.2 Medical

Strong medical programs also aid greatly in preventing occupational illness and injury. New applicants for work should be examined thoroughly for heart, lung, muscle, bone, eye, hearing, and other abnormalities, and a detailed examination of the skin is no less important. When the preplacement examination detects the presence or history of chronic eczema, psoriasis, hyperhidrosis, acne vulgaris, discoid lupus, chronic fungal disease, dry skin, or similar disorders, greater care must be exercised in job placement to avoid overt aggravation of preexisting cutaneous disease. New employees should be instructed thoroughly about the potential and general skin and health hazards associated with the intended job, as well as in the use of safety equipment, including clothing, and the means for maintaining cleanliness that are furnished (109–111).

Plant medical and nursing personnel should make periodic inspections of the operations to detect the presence of skin disease and the breaches in hygiene that predispose to occupational dermatitis. Employees should be encouraged to report promptly to the first aid or the medical departments all skin irritations, no matter how trivial.

Plant medical and industrial hygiene personnel should have constant surveillance over the introduction of new materials into industrial operations. Failure in this regard leads to the unwitting use of toxic agents, sometimes with serious results.

7.3 Personal

7.3.1 Protective Clothing

Protective clothing is used to shield the skin against injurious mechanical, physical, biological, and chemical agents. Well-designed, comfortable, attractive, and readily laundered clothing is an effective means of protecting the skin against irritant and sensitizing chemicals. The work apparel must be reasonably impervious to the chemical hazards and should be designed to protect the exposed parts. Good control over the use and cleanliness of protective clothing works best when management purchases, distributes, repairs, and launders the clothing. Worn and threadbare clothing should be discarded. Workmen who are required to purchase their own protective gear generally buy the cheaper equipment. Additionally, work clothing laundered at home can cause contamination of family wearing apparel with industrial chemicals, fiberglass, or other substances.

In many modern factories in which there are both closed processing and good industrial ventilation, workmen also wear protective clothing to avoid contamination during spills, breakages, or explosions. Proper protective clothing is especially needed by maintenance crews who enter tanks and vats containing harmful chemicals. Coveralls, aprons, caps, sleeves, boots, and gloves can be obtained in natural or synthetic fibers to protect against chemicals, heat, cold, and moisture. Protective clothing should be light weight and capable of acting as a shield. Cotton fabrics must be closely woven to prevent the entry of dusts. If protection is required against solvents, acids, or alkalies, impervious films of synthetic rubber or plastic are required. Neoprene, for example, is more resistant to oils, acids, and alkalies than is natural rubber. Vinyls and polyethylenes can be used, depending on their ability to shield out the chemical being handled (5, 8, 107, 109–111).

Gloves are an important item of protective gear because the hands are valuable instruments at work. Leather gloves, though expensive, provide good protection against forceful trauma (friction, abrasive dusts, etc.). Cotton gloves are satisfactory for light jobs, but their wear life is short. Neoprene- and vinyl-dipped cotton gloves provide satisfactory general protection against trauma, chemicals, solvents, and dusts. Unlined rubber and plastic gloves often cause maceration and sometimes contact dermatitis because of chemical leachout from the wearer's sweat in the glove. Part of the problem can be overcome by wearing either cotton-lined, impervious gloves or replaceable cotton liners that can be changed three or four times daily. Of no small importance in choosing gloves is the protection intended. Much time and money can be saved by reviewing the descriptive catalogs provided by the manufacturers of protective equipment, particularly gloves.

Protective sleeves can be impervious or nonimpervious, depending on the need. Protection against mild trauma and dusts is often furnished adequately by surgical stockinettes. In any event, sleeves should fasten or fit over the glove cuff to avoid entrance of chemical into the glove cuff. Furthermore, they are safer to wear if they tear readily when caught in machinery (5, 8, 107, 109–111).

Aprons are available in denim, leather, rubber, or synthetic films. An apron should be designed to cover the chest and to extend below the knees, while still providing freedom of movement.

Nowadays almost any refinement can be built into protective gear—for example, against fire, heat, cold, acids, alkalies, oils, and moisture. Despite such availability, protective clothing is most effective in promoting safety when it is kept clean and in good condition.

7.3.2 Industrial Cleansers

Of all the measures used to prevent occupational dermatoses, personal cleanliness is probably paramount. Ventilating systems and good maintenance are important in controlling the environment, but there is no substitute for washing the hands and keeping clean. Prompt and efficient removal of harmful industrial soils from the skin prevents prolonged action on skin. To do this, employees must have access to conveniently placed washing facilities with hot and cold running water, proper cleansing agents, and disposable towels. When a workman must walk a considerable distance to wash his hands, soap and water cleansing is often replaced by solvents or perhaps low grade waterless cleansers at the workbench.

Skin cleansers must be chosen for the work soil to be removed. Ordinary soap and washbasins may adequately fill the need, but in certain situations special soap, gloves, and lockers may not only be necessary but compulsory. There is no one all-purpose, industrial cleanser. Secretarial workers do not require strong abrasive cleansers, and machinists, garage attendants, and coal tar handlers find little use for a perfumed bar soap. In other words, hand cleansers should be selected to meet the need.

Several varieties of acceptable cleansers are available and include conventional soaps of liquid, cream, powdered, or bar type, synthetic detergents, and waterless cleansers. Conventional soaps are used daily by millions of people and are generally safe. Liquid varieties, mostly potassium salts of fatty acids, are satisfactory for light soil removal. They are popular and easily dispensed from metal or plastic containers. Powdered soaps are designed to remove heavy soil, and for this purpose may contain mineral or vegetable abrasive agents as pumice or wood flour. Bar soap, similarly, may contain an abrasive agent for soil removal; but bar soap has the disadvantage of being easily carried home. As a rule, workmen like to use powdered soap because the frictional factor provides a sense of thorough soil removal. Synthetic detergents (syndets) are generally more expensive than conventional industrial cleansers and are used only for selective purposes (5, 111).

Waterless cleaners are popular and have gained widespread usage in the workplace and at home. In plants or in field operations where water and washing facilities are not readily available, they are quite useful. Most of them are oil in water emulsions and vary in their content of petroleum solvents, petroleum oils, wetting agents, ammonia, amines, and synthetic detergents. Because of their capacity to remove tenacious soil rapidly, they are sometimes used too often, thus harming the skin. Such effects can range from extreme dryness and cracking to frank contact dermatitis of an irritant nature. Nevertheless, a waterless cleaner can serve as a fairly good substitute for cleansing with raw solvents such as naphtha and kerosene. Management should take an active interest in providing good cleansing facilities and cleansing agents. Procuring cheap cleansers can easily result in costly lost time from soap-induced dermatitis. If the following guidelines are used in choosing an industrial cleanser, little trouble need be expected:

1. The cleanser should have good cleansing properties.
2. It should not harmfully dehydrate the skin by normal usage.
3. It should not harmfully abrade the skin.
4. It should not be unpleasant to use.
5. It should not contain known sensitizers.
6. It should flow easily through dispensers.
7. It should not clog plumbing.
8. It should resist insect invasion (5, 109–111).

7.3.3 Barrier Agents

Protecting the skin with a barrier cream, lotion, or ointment is a common practice in industry. Easy application and removal, relative comfort on the skin, and the accompanying subjective sense of protection account for their population acceptance. Obviously a thin film of protective cream is less effective than good environmental control or an appropriate protective glove, sleeve, or other garment. Nonetheless, there are certain circumstances in which a barrier cream may be the preferred means of protecting the skin—for instance, when wearing gloves inhibits manual dexterity or when wearing gloves while working with machinery becomes too hazardous. Additionally, a worker may be able to do his job with less inconvenience with an application of cream than when wearing a plastic face shield.

There is no all-purpose protective cream or ointment; consequently, most manufacturers formulate different types for specific protection, including the following:

1. *vanishing cream* type, which contains soap, thereby facilitating easy removal of surface soils.
2. *water-repellent type,* which covers the skin with one or more water repellents such as beeswax, lanolin, petrolatum, or silicone, which provide some protection against mild acids, alkalies, and water-soluble chemicals.
3. *oil-repellent type* cream or ointment, which is usually water soluble but protects against oils and mildly against solvents: these creams may contain beeswax, lanolin, methyl cellulose, Borax, zinc oxide, preservative, and so on.
4. *solvent-repellent type* gel, which forms a dry coating on the skin, affording some protection against solvent, despite being water soluble: these repellents contain acacia or tragacanth as the gel barrier.

Creams or lotions made for such special purposes as protection against sunlight, coal tar, insects, and poisonous plants are available. When barrier creams are furnished to employees, instructions about their use must be explicit. The barrier must be applied to clean skin at the start of the work shift, removed at the break, and reapplied before returning to work. The same holds true for the lunch hour and the afternoon break. The barrier should be removed at the end of the shift.

A protective cream can be obtained from a dispenser or from an individual container issued to the workman and kept at the workbench or in the locker if it is located near the wash area. If dispensers are used, regular servicing is essential.

Suppliers of barrier creams should provide plant medical or safety personnel with test data that show the following:

1. The cream is nonirritant and nonsensitizing.
2. Actual protection is provided with use.
3. The product is easy to apply and to remove with washing.
4. The product will not separate or deteriorate in storage (5, 107, 109–112).

REFERENCES

1. R. P. White, *The Dermatergoses or Occupational Affections of the Skin,* 4th ed., H. K. Lewis & Company, London, 1934.
2. B. Ramazzini, *Diseases of Workers,* (translated from the Latin text, *De Morbis Artificum,* 1713, by W. C. Wright), Hafner, New York–London, 1964.
3. P. Pott, *Cancer Scroti,* Chirurgical Works, London, 1775, pp. 734; 1790 ed., pp. 257–261.

4. K. Ullman, M. Oppenheim, and J. Rille, *Die Schadidungen der Hant,* Parts 1, 2, and 3, Vol. II, Leopold Voss, Leipzig, 1926.
5. L. Schwartz, L. Tulipan, and D. J. Birmingham, *Occupational Diseases of the Skin,* 3rd ed., Lea & Febiger, Philadelphia, 1957.
6. R. R. Suskind, "Occupational Skin Problems. I. Mechanisms of Dermatologic Response. II. Methods of Evaluation for Cutaneous Hazards. III. Case Study and Diagnostic Appraisal," *J. Occup. Med.,* **1** (1959).
7. Advisory Committee on Occupational Dermatoses of the Council on Industrial Health, *Occupational Dermatoses* (a series of five reports), American Medical Association, Chicago, 1959.
8. M. H. Samitz, "Industrial Dermatoses," in: S. Moschella, D. Pillsbury, and H. Hurley Eds., *Dermatology,* Saunders, Philadelphia, 1975.
9. K. E. Malten and R. L. Zielhius, "Industrial Toxicology and Dermatology," in: the *Production and Processing of Plastics,* Elsevier, New York, 1964.
10. A. A. Fisher, *Contact Dermatitis,* 2nd ed., Lea & Febiger, Philadelphia, 1973.
11. R. M. Adams, *Occupational Contact Dermatitis,* Lippincott, Philadelphia, 1969.
12. G. A. Gellin, *Occupational Dermatoses,* Department of Environmental, Public and Occupational Health, American Medical Association, 1972.
13. N. Hjorth and S. Fregert, "Contact Dermatitis," in: *Textbook of Dermatology,* 2nd ed., Vol. 1, A. Rook, D. S. Wilkinson, and F. J. G. Ebling, Eds., 1972, Chapter 14.
14. V. M. Trasko, "Occupational Skin Disease Statistics," *Cutis,* **5,** 157–160 (1969).
15. *Occupational Disease in California, 1971–1972,* Bureau of Occupational Health and Environmental Epidemiology, California Department of Health, Berkeley, 1974.
16. *Occupational Injuries and Illnesses by Industry, 1972,* U.S. Department of Labor Bulletin 1830, Government Printing Office, Washington, D.C., 1974.
17. M. L. Newhouse, "Trends in Morbidity due to Industrial Dermatitis," *Proc. Roy. Soc. Med.,* **66,** 257–260 (1973).
18. A. Bell, "Focus on Health of Asian Workers," *Int. J. Occup. Health Saf.,* **43,** 36–37 (1974).
19. S. Rothman, *Physiology and Biochemistry of the Skin,* University of Chicago Press, Chicago, 1954.
20. A. M. Kligman, "The Biology of the Stratum Corneum," in: *The Epidermis,* W. Montagna and W. C. Lobitz, Jr., Eds., Academic Press, New York, 1964, pp. 387–433.
21. T. B. Fitzpatrick, "Biology of the Melanin Pigmentary System," in: *Dermatology in General Medicine,* T. B. Fitzpatrick et al., Eds., McGraw-Hill, New York, 1971.
22. L. Schwartz, "Occupational Pigmentary Changes in the Skin," *Arch. Dermatol. Syphilol.,* **56,** 592–600 (1947).
23. A. Rook, "Occupational Melanosis," *Br. J. Dermatol.,* **63,** 159–160 (1951).
24. G. Kahn, "Depigmentation Caused by Phenolic Detergent Germicides," *Arch. Dermatol.,* **102,** 177–187 (1970).
25. G. A. Gellin, P. A. Possick, and V. B. Perone, "Depigmentation from 4-Tertiary Butyl Catechol—An Experimental Study," *J. Invest. Dermatol.,* **55,** 190–197 (1970).
26. K. Malten, E. Seutter, and I. Hara, "Occupational Vitiligo due to *p*-Tertiary Butyl Phenol and Homologues," *Trans. St. John Hosp. Dermatol. Soc.,* **57,** 115–134 (1971).
27. C. D. Calnan, "Occupational Leukoderma from Alkyl Phenols," *Proc. Roy. Soc. Med.,* **66,** 258–260 (1973).
28. I. H. Blank and R. J. Scheuplein, "The Epidermal Barrier," in: *Progress in the Biological Sciences in Relation to Dermatology,* Vol. 2, A. Rook and R. A. Champion, Eds., Cambridge University Press, Cambridge, 1964.

29. F. Malkinson, "Permeability of the Stratum Corneum," in: *The Epidermis,* W. Montagna and W. C. Lobitz, Jr., Eds., Academic Press, New York, 1964.
30. R. B. Stoughton, "Some *In Vivo* and *In Vitro* Methods for Measuring Percutaneous Absorption," In: *Progress in the Biological Sciences in Relation to Dermatology,* Vol. 2, A. Rook and R. A Champion, Eds., Cambridge University Press, Cambridge, 1964.
31. W. J. Hayes, Jr., *Toxicology of Pesticides,* Williams & Wilkins, Baltimore, 1975.
32. D. O. Hamblin, "Aromatic Nitro and Amino Compounds in Industrial Hygiene and Toxicology," in: *Toxicology,* 2nd ed., Vol. 2, F. Patty, Ed., Wiley-Interscience P, New York, 1963.
33. D. J. Birmingham, "Cutaneous Absorption and Systemic Toxicity," in press.
34. J. Epstein, "Adverse Cutaneous Reactions to the Sun," in: *Year Book of Dermatology,* F. D. Malkinson and R. W. Pearson, Eds., Year Book Publishers, Chicago, 1971.
35. G. Hodgson, "Cutaneous Hazards of Lubricants," *Trans. Soc. Occup. Med.,* **19,** 9–15 (1969).
36. A. Rostenberg, Jr., "Primary Irritant Versus Allergic Eczematous Contact Dermatitis," in: *Dermatoses due to Environmental and Physical Factors,* R. B. Rees, Ed., Thomas, Springfield, Ill., 1962.
37. A. Rostenberg, Jr., "Concepts of Allergic Sensitization," *J. Ind. Med. Surg.,* **23,** 1 (1954).
38. R. L. Baer and V. H. Witten, Eds., "Allergic Eczematous Contact Dermatitis," in: *Year Book of Dermatology,* 1956–1957, Part I, and 1957–1958, Part II, Year Book Publishers, Chicago, Part I: pp. 7–38; Part II: pp. 7–46.
39. A. L. de Weck, "Contact Eczematous Dermatitis," in: *Dermatology in General Medicine,* T. B. Fitzpatrick et al., Eds., McGraw-Hill, New York, 1971.
40. J. L. Turk, "The Mechanism and Mediators of Cellular Hypersensitivity," in: *Clinical Aspects of Immunology,* 3rd ed., P. G. H. Gell, R. R. A. Coomes and P. J. Lachmann, Eds., Blackwell, Oxford, 1975, pp. 847–857.
41. L. Schwartz et al., "Skin Hazards in the Manufacture and Use of Cashew Nut Shell Liquid–Formaldehyde Resins," *Ind. Med.,* **14,** 500–506 (1945).
42. K. F. Lampke and R. Fagerstrom, *Plant Toxicity and Dermatitis* (*A Manual for Physicians*), Williams & Wilkins, Baltimore, 1968.
43. M. M. Key, "Plant and Wood Hazards," in: *Occupational Diseases—A Guide to Their Recognition,* W. M. Gafafer, Ed., U.S. Public Health Service Publication No. 1097, Government Printing Office, Washington, D.C., 1964, pp. 307–309.
44. G. A. Gellin, C. R. Wolf, and T. H. Milby, "Poison Ivy, Poison Oak and Poison Sumac—Common Causes of Occupational Dermatitis," *Arch. Environ., Health,* **22,** 280 (1971).
45. A. M. Kligman, "Poison Ivy (Rhus) Dermatitis," *Arch. Dermatol.,* **77,** 149 (1958).
46. E. Bluemink, J. C. Mitchell, and J. P. Nater, "Contact Dermatitis due to Chrysanthemums," *Arch. Dermatol.,* **108,** 220–222 (1973).
47. N. Hjorth and D. S. Wilkinson, "Contact Dermatitis. IV. Tulip Fingers, Hyacinth Itch and Lily Rash," *Br. J. Dermatol.,* **80,** 696–698 (1968).
48. L. F. Weber, "Dermatitis Venenata due to Native Woods," *Arch. Dermatol.,* **67,** 388 (1953).
49. *Wood Preservation Around the Home and Farm,* Forest Products Laboratory Publication No. 1117, Canadian Department of Forestry, Ottawa, 1966.
50. D. J. Birmingham, "Natural and Artificial Light Sources in Industry," paper presented at the American Industrial Hygiene Association Meeting, Pittsburgh, April 1966.
51. L. C. Harber, H. Harris, and R. L. Baer, "Photoallergic Contact Dermatitis," *Arch. Dermatol.,* **94,** 255 (1966).
52. K. D. Crow et al., "Photosensitivity due to Pitch," *Br. J. Dermatol.,* **73,** 220 (1961).
53. D. J. Birmingham, "Photosensitizing Drugs, Plants, and Chemicals," *Mich. Med.,* **67,** 39–43 (1968).

54. M. Pathek, F. Daniels, Jr., and T. B. Fitzpatrick, "The Presently Known Distribution of Furocoumarins (Psoralens) in Plants," *J. Invest. Dermatol.*, **39,** 225–239 (1962).
55. D. S. Wilkinson, "Photodermatitis due to Tetrachlorsalicylanilide," *Br. J. Dermatol.*, **73,** 213 (1961).
56. C. D. Calnan, R. Harman, and G. Wells, "Photodermatitis from Soap," *Br. Med. J.*, **2,** 1266 (1961).
57. D. J. Birmingham, "Abrasions, Contusions, Lacerations and Penetrating Wounds," in: *Traumatic Medicine and Surgery for the Attorney,* P. Cantor, Ed., Butterworths, Washington, D.C., 1962.
58. E. A. Drogicina and I. K. Raxumov, "Vibration," in: *Occupational Health and Safety,* Vol. 2 (L–Z), International Labour Office, Geneva, Switzerland, 1972, pp. 1463–1465.
59. N. Williams, "Biological Effects of Segmental Vibration," *J. Occup. Med.*, **17,** 37–39 (1975).
60. M. M. Key, T. H. Milby, D. A. Holaday, and A. Cohen, "Physical Hazards," in: *Occupational Diseases—A Guide to Their Recognition,* W. M. Gafafer, Ed., U.S. Public Health Service Publication No. 1097, Government Printing Office, Washington, D.C., 1964.
61. M. M. Key, "Biologic Hazards," in: *Occupational Diseases—A Guide to Their Recognition,* W. M. Gafafer, Ed., U.S. Public Health Service Publication No. 1097, Government Printing Office, Washington, D.C., 1964.
62. D. J. Birmingham and M. M. Key, "Occupational Dermatoses," in: *Occupational Diseases—A Guide to Their Recognition,* W. M. Gafafer, Ed., U.S. Public Health Service Publication No. 1097, Government Printing Office, Washington, D.C., 1964.
63. J. V. Klauder and M. K. Hardy, "Actual Causes of Certain Occupational Dermatoses: Further Study of 532 Cases with Special Reference to Dermatitis Caused by Certain Petroleum Solvents," *Occup. Med.*, **1,** 168–181 (1946).
64. J. V. Klauder and B. A. Gross, "Actual Causes of Certain Occupational Dermatoses: A Further Study with Special Reference to Effect of Alkali on the Skin, Effect of Soap on pH of Skin, Modern Cutaneous Detergents," *Arch. Dermatol. Syphilot.*, **63,** 1–23 (1951).
65. J. V. Klauder, "Actual Causes of Certain Occupational Dermatoses," *Arch. Dermatol.*, **85,** 441–454 (1962).
66. M. Kleinfeld, "Hazards of Exposure to Coal Tar, Asphalt, and Black Top in Roofing and Road-Repair Work," *J. Am. Med. Assoc.*, **224,** 1654 (1973).
67. M. M. Key, E. J. Ritter, and K. A. Arndt, "Cutting and Grinding Fluids and Their Effects on Skin," *Am. Ind. Hyg. Assoc. J.* **27,** 423–427 (1966).
68. J. Kimmig and K. H. Schultz, "Occupational Chloracne Caused by Aromatic Cyclic Ethers," *Dermatologica,* **115,** 540 (1957).
69. K. D. Crow, "Chloracne," *Br. J. Dermatol.*, **83,** 599 (1970).
70. J. Bleiberg, M. Wallen, R. Brodkin, and I. L. Applebaum, "Industrially Acquired Porphyria," *Arch. Dermatol.*, **89,** 793–797 (1964).
71. N. E. Jensen, I. B. Sneeden, and A. F. Walker, "Tetrachlorobenzodioxin and Chloracne," *Trans. St. Johns Hosp. Dermatol. Soc.*, **58,** 172 (1972).
72. J. S. Taylor, "Chloracne—A Continuing Problem," *Cutis,* **13,** No. 4, 585–591 (1974).
73. J. Belisario, *Cancer of the Skin,* Butterworths, London, 1959.
74. E. A. Emmett, "Occupational Skin Cancer—A Review," *J. Occup. Med.*, **17,** 44–49 (1975).
75. F. C. Combs, *Coal Tar and Cutaneous Carcinogenesis in Industry,* Thomas, Springfield, Ill., 1954.
76. I. Berenblum and R. Schoental, "Carcinogenic Constituents of Shale Oil," *Br. J. Ex. Pathol.*, **24,** 232–239 (1943).
77. R. E. Eckhart, *Industrial Carcinogens,* Grune & Stratton, New York, 1959).
78. W. D. Buchanan, *Toxicity of Arsenic Compounds,* Elsevier, Amsterdam, 1962.
79. M. A. Goldblatt, "Occupational Carcinogens," *Br. Med. Bull.*, **14,** 136–141 (1958).

80. E. Bingham, A. V. Horton, and R. Tye, "The Carcinogenic Potential of Certain Oils," *Arch. Environ. Health,* **10,** 449–451 (1965).
81. R. E. W. Fisher "Cancer of the Skin Caused by Tar," *Trans. Assoc. Ind. Med. Officers,* **15,** 122–130 (1965).
82. W. C. Hueper, "Occupational Cancer," in: *Occupational Safety and Health,* Vol. 1 (A–K), International Labour Office, Geneva, 1971.
83. W. Cumin, "Remarks on the Medicinal Properties of MADAR and on the Effects of Bichromate of Potassium on the Human Body," *Edinb. Med. Surg. J.,* **28,** 295–302 (1827).
84. S. R. Cohen, D. M. Davis, and R. S. Kozamkowski, "Clinical Manifestations of Chromatic Acid Toxicity: Nasal Lesions in Electroplate Workers," *Cutis,* **13,** 558–568 (1974).
85. D. J. Birmingham, M. M. Key, D. A. Holaday, and V. B. Perone, "An Outbreak of Arsenical Dermatoses in a Mining Community," *Arch. Dermatol.,* **91,** 457–464 (1965).
86. D. J. Birmingham, "Leg Ulcers Caused by Contact with Calcium Nitrate Solution," unpublished information.
87. M. D. Samitz and A. S. Dana, *Cutaneous Lesions of the Lower Extremities,* Lippincott, Philadelphia, 1971.
88. W. L. Ritter and A. S. Carter, "Hair Loss in Chloroprene Manufacture," *J. Ind. Hyg. Toxicol.,* **30,** 192 (1948).
89. W. Lewis and L. Schwartz, "An Occupational Agent (*N*-Butyraldoxime) Causing Reaction to Alcohol," *Med. Ann. D.C.,* **25,** 485–490, 1956.
90. R. D. Stewart, C. L Hake, and I. E. Peterson, "'Degreaser's Flush,' Dermal Response to Trichlorethylene and Ethanol," *Arch. Environ. Health,* **29,** 1–5 (1974).
91. R. S. Grier et al., "Skin Lesions in Persons Exposed to Beryllium Compounds," *J. Ind. Hyg. Toxicol.,* **30,** 228–237 (1948).
92. F. R. Dutra, "Experimental Beryllium Granulomas of the Skin," *AMA Arch. Ind. Hyg. Occup. Med.,* **3,** 81–89 (1951).
93. E. Epstein, "Silica Granuloma of the Skin," *Arch. Dermatol. Syphilol.,* **71,** 24–35 (1955).
94. W. B. Shelley and H. J. Hurley, "The Pathogenesis of Silica Granulomas in Man: A Non-Allergic Colloidal Phenomenon," *J. Invest. Dermatol.,* **34,** 107–122, 1960.
95. D. J. Birmingham and N. Olivier, "Occupational Granulomas," paper presented at the Pacific Dermatological Association Meeting, Honolulu, Hawaii, September 1963.
96. J. M. Agate and H. A. Druett, "A Study of Portable Vibrating Tools in Relation to the Clinical Effects Which They Produce," *Br. J. Ind. Med.,* **4,** 141 (1947).
97. R. H. Wilson, W. G. McCormick, C. F. Tatum, and J. L. Creech, "Occupational Acroosteolysis: Report of 31 Cases," *J. Am. Med. Assoc.,* **201,** 577–581 (1967).
98. D. K. Harris and W. G. F. Adams, "Acroosteolysis Occurring in Men Engaged in the Polymerization of Vinyl Chloride," *Br. Med. J.,* **3,** 712–714 (1967).
99. M. B. Sulzberger and F. Wise, "The Contact or Patch Test in Dermatology," *Arch. Dermatol. Syphilol.,* **23,** 519 (1931).
100. L. Schwartz and S. Peck, "The Patch Test in Contact Dermatitis," *U.S. Pub. Health Rep.,* **59,** 546–557 (1944).
101. R. L. Baer and H. Witten, "Allergic Eczematous Contact Dermatitis: A Review of Selected Aspects for the Practitioner," in: *Year Book of Dermatology and Syphilology* 1956–1957, R. L. Baer and V. H., Eds., Year Book Publishers, Chicago, 1957, pp. 7–38.
102. B. Magnusson, S. Fregert, N. Hjorth, et al., "Routine Patch Testing. V. Correlation of Reactions to the Site of Dermatitis and the History of the Patient," *Acta Dermatovener.* **41,** 556–563 (1969).
103. H. I. Maibach, "Patch Testing—An Objective Tool," *Cutis,* **13:** 4 (1974).

104. E. J. Rudner et al., "Epidemiology of Contact Dermatitis in N. America, 1972," *Arch. Dermatol.*, **108**, 537 (1973).

105. S. Fregert, *Manual of Contact Dermatitis,* Munksgaard, Copenhagen, 1974.

106. M. H. Samitz, "Occupational Dermatoses," in: *Current Therapy,* H. F. Conn, Ed., Saunders, Philadelphia, 1966.

107. D. J. Birmingham, "Occupational Dermatoses," Unit 19-12, in: *Clinical Dermatology,* Vol. 4, D. J. Demis, R. G. Crounse, R. L. Dobson, and J. McGuire, Eds., Harper & Row, New York, 1972, pp. 1–22.

108. G. E. Morris, "Why Doesn't the Worker's Skin Clear Up? An Analysis of Factors Complicating Industrial Dermatoses," *Arch. Ind. Hyg. Occup. Med.,* **10** (1954).

109. D. J. Birmingham, "Prevention of Occupational Skin Disease," *Cutis,* **5,** 153–156 (1969).

110. G. A. Gellin, "Prevention of Occupational Skin Disorders," *Cutis,* **13:** 4 (1974).

111. D. J. Birmingham, Ed., *The Prevention of Occupational Skin Diseases,* Soap & Detergent Association, New York, 1975.

112. W. J. Beach, "Skin Protective Preparations," *Cutis,* **13:** 4 (1974).

CHAPTER NINE

Physiological Effects of Abnormal Atmospheric Pressures

ALBERT R. BEHNKE, Jr., M.D.

1 INTRODUCTION

Man, in capsule atmospheres with variations in pressure of 0.25 to 60 atmospheres (ATA), can live for many weeks in the vacuum of space and in undersea habitats, where exploitation of resources on the ocean floor focuses on medical and physiologic aspects of new environmental frontiers. Incredible engineering achievements have placed men on the moon and in the deepest recesses of the ocean. Man in the relatively low pressure space capsule (~ 0.33 ATA) can adapt to a zero *g* environment, and on reentry into the earth's atmosphere he can readapt after a number of days to his normal 1 *g* environment (1). It is in diving and tunnel operations that pressure forces have taken their toll, chiefly during and after exit from the hyperbaric environment. In contrast to engineering achievement that provides underground rapid transit and sewage conduits in various parts of the world, is the deficit in physiologic knowledge of inert gas transport and elimination from body tissues. This chapter discusses the more difficult problems of the hyperbaric rather than the altitude environment.

1.1 Outline of Advances

1. Rapid compression to depths in excess of 400 ft gives rise to arthralgia, probably related to fluid shift in the joints attending initial osmotic dysequilibrium. Arthralgia is not reported if compression is slow.

2. Compression in helium atmospheres at depths in excess of 800 ft elicits tremors in man, unsteadiness, and somnolence, designated as high pressure nervous syndrome (HPNS). In animals, HPNS at high pressures (+60 ATA, 2000 ft equivalent diving depth) progresses from unsteadiness and tremors to generalized tonic-clonic convulsions. The etiology of this syndrome appears to be the effect of pressure per se on nervous tissue; it is not dependent on the helium atmosphere.

3. Ultrasonic surveillance of bubbles in the large veins and heart, notably detection of "silent" bubbles (i.e., not giving rise to symptoms) during the early stages of decompression.

4. Phenomenon of "counterdiffusion" supersaturation under isobaric conditions, characterized by bubble evolution in the subcutaneous fat layer of body tissue, when, for example, a switch from helium to air is made during decompression such that air is inhaled and helium envelopes the body.

5. Elucidation of blood–bubble interactions during decompression, notably those affecting blood platelets.

6. Relative safety of saturation-excursion diving in experimental wet-dry chambers and in the ocean.

7. The potential in utilizing oxygen for decompression of tunnel workers.

8. The potential in providing a pressurized habitat for tunnel workers with "no-decompression" excursions to work shift pressures analogous to deep-sea habitat diving.

9. Surveillance of and measures to maintain fitness of personnel working and living in the hyperbaric environment.

10. An adaptation (acclimatization) is observed as resistance to decompression sickness (DCS) during the course of repetitive decompressions of tunnel workers and divers from hyperbaric atmospheres.

1.2 Stresses and Problems in Current Operations

Man-in-the-sea saturation diving is routine at depths of 600 ft, and excursions from undersea habitats can be made 150 ft deeper with minimal decompression on return to habitat depth. An example of progress in experimental diving, which in France has attained an equivalent depth of 2000 ft, was the U.S. Navy dive to 1600 ft (49.5 ATA) conducted at Taylor Diving and Salvage Company, New Orleans. The compression rate (in contrast to earlier descent rates of 100 ft/min) was 5 ft/min from 14 to 400 ft, 40 ft/hr from 400 to 1000 ft, and 30 ft or less from 1000 to 1600 ft.

Seven days were spent at 1600 ft, where tasks involving moderate exercise were performed. Seventeen days were spent at 1000 ft or deeper. Decompression lasted 19 days, and the total duration of the dive was 32 days. The chief problems arose from confinement of vigorous, young adults in a helium–oxygen atmosphere conducive to rapid heat loss, from high pressure phenomena of tremor and unsteadiness, and from extended, tedious decompression, which did not free the divers entirely from symptoms after return to normal pressure.

In relatively shallow air dives, thousands of industrial divers employ either helmet or self-contained underwater breathing apparatus (SCUBA). The scuba gear is employed by hundreds of thousands of sport divers and underwater photographers, including a limited number of marine scientists. Revived interest in hyperbaric medicine results in the routine exposure of physicians and paramedical personnel in the hyperbaric environment.

In the sea, apart from the prime problem of decompression, are stresses of cold, darkness, spatial disorientation, impaired communication, at times marine hazards, and always the stringent requirements of gas supply and adequacy and reliability of equipment. An unresolved problem of the deep-sea diver working from oil rigs 100 miles offshore is the need to furnish definitive medical and surgical care when injury occurs under pressure. On the other hand, the scuba diver is generally confronted with lack of on-site recompression chambers for treatment of the myriad manifestations of decompression sickness. Always in imminent danger of drowning (breathing unnaturally through a mouthpiece with head surrounded by water), the scuba diver may compound aqueous hazards by physical impairments, emotional immaturity, and psychic instability. Not infrequently he is obsessed with the urge to establish depth records with diving gear which, for the novice, should be restricted to "free-ascent" depths routinely limited to 60 ft and occasionally not greater than 130 ft. On any particular day he may be in poor physical condition, or under the influence of prescribed or self-administered drugs and ethanol. Underwater and confronted with heavy exertion, his limited endurance and inadequate pulmonary ventilation predispose to hypercarbia. Underwater, the normal stimulus to respiration from carbon dioxide accumulation may be blunted. The unpredictable reactions of the inexperienced scuba diver facing an emergency may result in serious injury to himself and others.

In a dry hyperbaric environment, daily decompressions of the tunnel worker following relatively long work shifts in compressed air may lead to avascular necrosis of hip and shoulder joints and deficits in the central nervous system. Remedial measures to counter these difficulties are discussed in this chapter.

1.3 Pressure Equivalents

A standard atmosphere (ATA) is equal to pressure exerted by a column of mercury 760 mm (torr) high, and in slightly rounded values, to 33 ft of seawater (fsw), 10.06 m, 14.7 lb/in.^{2} gauge pressure (psig). Absolute pressure (unity) is 1 ATA added to the gauge pressure or equivalent sea depth. Pounds per sequare inch and feet or meters are used interchangeably for designation of pressure or depth.

feet	psi	meters
10	4.45	3.05
20	8.9	6.10
30	13.4	9.15
40	17.8	12.20
50	22.2	15.25
60	26.7	18.3
70	31.1	21.4
80	35.6	24.4
90	40.0	27.5
100	44.5	30.5

1 ft = 0.3048 m, 1 m = 3.281 ft, 1 psi = 2.245 ft.

2 HABITABILITY OF THE PRESSURIZED ATMOSPHERE

Three types of pressurized environment are (1) the relatively innocuous, dry hyperbaric medical facility, (2) the potentially hazardous environment of the tunnel worker, and (3) the wet environment of the diver, which in helium atmospheres presents difficulties in temperature control and maintenance of comfort.

2.1 The Dry Hyperbaric Facility

The ventilation engineer has provided a comfortable environment with respect to temperature and humidity control based on standard principles of heat transfer and air conditioning. The surgeon operates in chambers pressurized to 4 ATA; at higher pressures, nitrogen, because of its narcotic effect, is replaced by helium. In one hyperbaric facility, gas exchange at 31.2 ATA (1000 ft equiv) was normal in an atmosphere (99.1 percent He–0.9 percent O_2) 4.4 times more dense than air at 1 ATA. Psychometric conditions were apparently so satisfactory that carefully documented protocols made no mention of temperature (actually 88°F) or humidity. It can be stated that in the dry hyperbaric environment (in which helium is recirculated to conserve gas), pressures as high as 31 ATA are conducive to normal mentation and physical performance, provided temperature is maintained at a relatively high level in the narrow range of 85 to 90°F, with humidity at about 65 percent.

2.2 The Pressurized Tunnel

Compressed air workers subjected to pressures as high as 4 ATA (to prevent ingress of water in subsurface and subaqueous operations) may labor in a cold, damp atmosphere distal to the excavating shield at the face of the tunnel. High temperatures prevail in the area near the shield, which is currently automated, and a combination of radiant heat and high humidity subjects the workers to profuse sweating. Although enormous quantities of air are forced into the tunnel, there is no provision for mechanical exhaust. Body cooling may be inadequate, and welding fumes in combination with other contaminants tend to accumulate in pockets. Apart from danger of flooding or fire (at a pressure of 3 ATA, fire burns with a twofold intensity compared with combustion in free air), there may be an untoward or surreptitious hazard. Thus in certain soils oxygen is depleted such that when the pressure in the tunnel is reduced (in the absence of replenishment with outside air), influx of oxygen-deficient air from soil beds subjects workers to hypoxia.

During decompression in the medical lock, 3 hours may be spent in an atmosphere heavily laden with cigarette smoke. It is not comfort but rather a reasonable degree of safety that concerns the tunnel worker.

2.3 The Wet Environment of the Diver

As mentioned previously, cold at deep depths intensified by use of helium, with thermal conductivity 6 times that of air, is a deterrent to efficient performance and comfortable conditions at rest. At about 19 ATA (water temperature, 40°F) in a helium atmosphere, respiratory heat loss is equal to metabolic heat generated at any level of activity. Electrically heated underwear (employed in 1939 during the *U.S.S. Squalus* salvage operations) has been inadequate to supply the requisite heat to counteract the cold, dense helium atmosphere, notably in the personnel transfer capsule (PTC) from habitat to surface. The heat requirement amounting to 1000 to 3000 W (860 to 2580 cal/hr) led to the development of a tubing suit through which water at a temperature of about 104°F was circulated.

In the successful SEALAB habitat at an ocean depth of 200 ft, temperature was maintained between 85 and 89°F and relative humidity was about 76 percent. The elevated temperature was conducive to comfort, but the high humidity coupled with high diffusivity of helium caused electrical short-circuits. Although it was possible after a few days for the aquanauts to sleep, there were reports of paradoxical shivering of divers entering the relatively warm habitat after being in cold water. One explanation of the shivering is recirculation of "core" blood through the previously vascular-constricted cold tissues.

3 GENERAL COMPRESSION EFFECTS

3.1 Arthralgia

It is remarkable that pressure per se has so little discernible effect on mammals in the range of 0.10 ATA (equivalent to an altitude of 52,000 ft) to 16 ATA (500 fsw). However divers report arthralgic pain in connection with rapid descents to depths greater than 500 ft. In an earlier era, dives were made to depths of 500 ft within 5 minutes without reports of compression pain. Occasionally "crackling" of joints (probably related to cavitation) was manifest. All divers participating in a current U.S. Navy experimental program to simulated or actual depths of 600 ft or greater, have reported crepitation and stiffness in joints. Myalgia is common, affecting quadriceps muscles of the thigh, and both weight-bearing and non-weight-bearing joints are involved. Slow compression at a rate equivalent to 40 ft or less per hour prevents onset of pain. The etiology of this discontinuous response is obscure. A plausible hypothesis of Kylstra et al. (2) relates pain and cavitation to fluid shifts in response to osmotic pressure gradients between synovial lining and joint cartilage. Hills (3) has extended this consideration to clinical implications of gas-induced osmosis with reference to gaseous anesthesia, gouty arthritis, aseptic bone necrosis, and edematous separation of tissues in general.

3.2 High Pressure Nervous Syndrome (HPNS)

It has long been known that pressure per se, if sufficiently elevated (+100 ATA), can influence biological processes. Thus viscosity of protoplasm is increased, gels tend to become sols, and rates of biological reactions are altered. Yet in diving to 500 ft, it could be stated that if pressure were equalized in sinal and aural spaces, there were no gross physiologic effects attributed to hydrostatic pressure in the range of 2 to 200 psi. In 1965 Bennett (4), in a helium–oxygen chamber test, reported that at 800 ft (356 psig) there were marked decrements in motor and intellectual performance accompanied by dizziness, nausea, vomiting, and marked tremor of the hands, arms, and torso. Subsequently the extensive studies of Brauer et al. (5) revealed in addition to tremor, marked changes in the electroencephalogram (slow waves in the theta band and depression of alpha activity) and behavioral changes that included bouts of somnolence (microsleep). In monkeys and mice hyperexcitability progressed to grand mal seizures. It was the conclusion of Brauer (6) that "compression of primates to pressures of the order of 50 to 60 atm gives rise to a complex syndrome which we have called the high pressure neurological syndrome, and which cannot be attributed to anything other than the effects of hydrostatic pressure. . . ."

In experimental chamber dives to 2000 ft, if compression is slow, there is some adaptation tending to limit tremors, unsteadiness, and impaired performance. The addition of several atmospheres of nitrogen tends to suppress HPNS, but it appears that hydrostatic forces (to which man is more sensitive than lower primates or small mammals) pose a formidable barrier to diving at depths greater than 2000 ft.

3.3 Barotrauma

Aerootitis, aerosinusitis, and aeroodontolgia, generally regarded as minor impairments, are the most frequent adverse reactions associated with unequalized pressure. These ailments arise when tubal or sinal openings are partially or wholly occluded, or if an encapsulated gas pocket is present in dental pulp. At one time, 10 to 25 percent of submarine trainees in initial tests were unable to accommodate rapidly to application of chamber pressure because of subclinical infection of the upper part of the respiratory tract. Workers in pressurized tunnels frequently develop aerootitis, varying in degree from redness of the tympanic membrane to frank hemorrhage in the middle ear and occasional rupture of the drum. Resolution of the pathologic changes is usually uneventful and complete.

One would anticipate that the delicate mechanisms responsible for hearing and equilibrium would be disrupted by barotrauma. Yet rarely is there damage to such structures as the round window of the cochlea, attended by loss of hearing and equilibrative disturbances. Fortunately this serious complication, if recognized, can be relieved by surgical repair of the defect.

Various studies confirm the statement that barotraumatic hearing losses are moderate, almost always reversible, and rarely of medical consequence. Apart from rupture of the round window membrane, however, acoustic levels potentially damaging to the ear exist underwater. The inner ear (underwater) handles acoustic energy much as it deals with airborne or vibratory energy, but the conduction routes through the head are different in kind and/or degree (7). Airborne hearing exceeds underwater hearing acuity by 30 to 60 decibels (dB), being greater at high frequencies. The following case was reported by Edmonds et al. (8).

Case. The diver, who had been exposed to gunfire in the past, experienced considerable pain and difficulty in equalizing both middle ears during a dive to 10 meters. He continued to dive despite the pain, and it required him to perform forceful autoinflation. He noted tinnitus and also experienced pain and vertigo during ascent.

Otoscopic examination revealed the effects of barotrauma, and the diver progressively became more deaf, with episodes suggestive of vertigo. As both ears were involved, it was essential that exploratory surgery be performed. A fistula of the right window was observed with outpouring of perilymph fluid into the middle ear. The round window was packed, and subsequent audiograms over the following month revealed considerable improvement in hearing. A similar operative procedure was performed on the other ear five days later, with the same beneficial result.

Occasionally intense pain is felt in the middle ear on decompression or during diving ascent as the tympanic membrane is pushed outward by "overpressure" in the middle ear. "Alternobaric vertigo" is the term Lundgren (9) used to describe this condition, with its serious concomitants, particularly in relation to scuba divers.

"Ear block" or ear infection excluded 3.7 percent, or 49 out of 1324 applicants for tunnel work (10). In daily pressurized work shifts it was frequently necessary to use an oral decongestant with minimal side effects.

3.4 Effect of Hydrostatic Pressure Gradients

In recent years a great deal of information has been obtained from studies of simple head-out immersion in water. Some effects of the hydrostatic pressure gradient from the neck to the lower thorax are diuresis, natriuresis, and suppression of the renin-aldosterone mechanism (11). Increased stroke volume accompanies decrease in heart rate, and inert gas transport from tissues is facilitated (12, 13). These effects are mediated in part by an immersion-induced redistribution of blood volume to induce central hypervolemia (14). The increase in central blood volume is about 0.7 liter, and in cardiac output the increase is about 32 percent. The shift in blood is accompanied by lower systemic vascular resistance and an increase in peripheral circulation, measured by ^{133}X clearance from calf musculature.

Adverse effects of immersion to the neck are cardiac extrasystoles, and over long periods orthostatic hypotension supervenes; this condition is analogous to the circulatory deconditioning associated with weightlessness in space. Furthermore the hydrostatically induced redistribution of blood has been compared to the effects of negative pressure breathing with reduced expiratory lung volume. Dahlback and Lundgren (15) suggested that airway closure and air trapping induced by immersion in the upright position might add to the danger of lung rupture in "free ascent." This problem is discussed subsequently.

3.5 Breath-Hold Diving

Man's deficiency as a diving mammal is best appreciated when one considers that over millions of years, whales, dolphins, and porpoises have adapted so harmoniously to living in water, with ability to dive to 1000 ft and deeper, to remain at 1000 ft for as long as 30 minutes, to exist without the need for fresh water, to bear young in the ocean, and to make the wide sea their home in both temperate and icy waters. In 1934 Irving (16) suggested that during submersion, a reflex mechanism produces a redistribution of blood flow in the sea mammal, so that virtually no blood flows to tissues resistant to asphyxia (e.g., skin, muscle, gut), flowing instead to the heart and brain, which are more sensitive to lack of oxygen. The benefit of such a mechanism would be reduction to a minimum of oxygen utilization during submersion.

Man possesses an atavistic diving reflex such that apnea occurs in response to water touching the face. The responses associated with the diving reflex, namely, bradycardia, peripheral vasoconstriction, and maintained or elevated blood pressure, are qualitatively similar to those observed in diving mammals but quantitatively far less pronounced. Gooden (17) states, "It must surely be one of Nature's greatest ironies that man should spend the first nine months of his existence continuously surrounded by water, but the rest of his life with an inherent fear of submersion." Gooden further points out that in accidental submersion, as the victim's face is immersed, reflex apnea ensues immediately, and if some water enters the larynx, glottic spasm also results. The sensory stimulus of water touching the face, plus the reflex inhibition of the respiratory centers, combine to trigger the cardiovascular response.

Grave consequences arise in connection with breath-hold (skin) diving, notably in attempts to overstay or break records. Following surface hyperventilation, then breath-hold water, chemoreceptors are not responsive, and as oxygen rapidly diminishes, heart muscle becomes hypoxic. Even trained swimmers may develop cardiac irregularities during the course of a hyperventilation breath-hold dive. In descent to depths, the hypoxia becomes acute following ascent to the surface and reexpansion of the chest. Trained breath-hold divers have descended to depths in excess of 240 ft, but calculated total oxygen available in the lungs and blood of less than 3 liters is inadequate after 1.5 minutes to satisfy metabolic requirement.

By contrast, with oxygen in the lungs (and following hyperventilation) breath-hold may be prolonged for 10 minutes or longer, since metabolic carbon dioxide is not eliminated in the lungs but is dissolved and buffered in blood and body fluids. In patients at normal pressure, adequate oxygenation of blood has been maintained for periods of 15 to 55 minutes during the course of apnea but with the complication of respiratory and metabolic acidosis (18).

4 PHENOMENA ASSOCIATED WITH INCREASED GAS PRESSURES

4.1 Nitrogen Narcosis

"Air" nitrogen exerts a narcotic effect at higher pressures aptly characterized as "rapture of the deep," from a free translation of Cousteau's expression "*l'ivresse des grandes profondeurs*." The readily reversible impairment is featured by changes in mood (predominantly euphoria), muscular incoordination, slowing of mental activity, and fixation of ideas (19). The threshold for this type of narcosis is difficult to establish by reason of individual variability, but critical judgment may be impaired at pressures as low as 3 ATA in some individuals. A clarifying concept is that routine procedures can be performed faultlessly if effort is increased and individuals are aware of the impairment. Thus emotionally stable persons are able to perform *assigned and rehearsed* tasks (with deliberation) until consciousness is lost at depths usually greater than 300 ft. When compression was exceptionally rapid (within 30 seconds) to 600 ft, British Royal Navy divers have made "air" escapes from cruising submarines from depths successively graded to 600 ft.

The unstable person in a compressed air atmosphere is incapable of purposeful effort and gives way to emotional aberrations characteristic of some alcoholics. There are tests that demonstrate parallelism in behaviors induced by pressurized nitrogen and by alcohol.

When inhaled with 20 percent oxygen at 10 ATA, argon, a chemically nonreactive gas, induces an even greater degree of stupefaction than inhalation of air at the same pressure (20). By contrast, helium is singularly nonnarcotic, and convulsions observed in animals in high pressure helium atmospheres are attributable to the hydrostatic-induced high pressure nervous syndrome previously described.

Narcosis induced by nitrogen and other physiologically inert gases (e.g., xenon, krypton) has been studied intensively by Bennett (21). There is a "family of physical properties," of which solubility in lipid is prominent, that correlate with the phenomenon. An intriguing fact is reversal of narcosis at high pressures. Mice, anesthetized by nitrogen at 45 ATA, regain consciousness by further compression with helium (21). Basic mechanisms relate to the observation that the nerve cell membrane increases in volume in contact with anesthetic gases and decreases in volume when pressure is applied. Stern and Frisch (22) relate anesthetic potency of gases not only to lipid solubility but also to thermal expansivity and compressibility of the lipid phase, as well as environmental temperature and hydrostatic pressure.

In the all-important matter of diver performance, it can be concluded that stable individuals can accomplish planned tasks in compressed air at depths deeper than 200 ft if predive assignment is adhered to underwater. If a task cannot be accomplished within the allotted time, the dive must be terminated.

4.2 Oxygen Tolerance and Limiting Reactions

Pure oxygen is toxic if breathed continuously. At relatively low pressures (above 0.5 ATA and 400 mm Hg), oxygen induces respiratory irritation; at pressures of about 3 ATA and higher, nervous signs and symptoms predominate and culminate in convulsive seizures. A recent finding is that superoxide (O_2^-), a highly reactive free radical produced by one electron reduction of oxygen, is reduced to a less toxic form by an enzyme, *superoxide dismutase* (23). Harmful effects have not been identified as a result of extensive employment of hyperbaric oxygen to expedite transport of inert gases from tissues during decompression, nor in its employment as a prime modality in the therapy of decompression sickness.

Oxygen tolerance at and below 2.8 ATA (60 ft equiv) is approximately doubled if there are interspersed short (5 min) periods of air breathing. In tests on volunteers, Hendricks (24) found that symptoms of tracheal irritation and slight burning on inspiration occurred after 6 to 9 "oxygen hours" of exposure at 2 ATA (three 20-minute periods on oxygen interspersed with 5-minute intervals on air = 1 oxygen hour). Tests were terminated when a significant reduction in vital capacity of about 10 percent was incurred.

A severe limitation is that exercise (notably underwater) and elevated carbon dioxide levels render oxygen more toxic. Thus inhalation of pure oxygen by the working diver at depths deeper than 30 ft has elicited convulsive seizures; see Table 9.1. Apart from exercise as a limiting factor to inhalation of oxygen under dry and wet conditions, there is the disturbing variability in individual response. This is evident from the extensive tests of Donald (26) conducted by the British Royal Navy during World War II (Table 9.2).

4.3 Operational Applications of Oxygen Inhalation

Oxygen permits inert gas clearance of tissues at a pressure head sufficiently high (equivalent to 60 to 40 ft) to prevent excessive bubble evolution. In decompression during the course of helium–oxygen (heliox) dives, the last stop is at 40 ft and oxygen is inhaled for 100 minutes or longer. In diving incident to the salvage of the *U.S.S.*

Table 9.1. Oxygen Tolerance Tests[a] During of Exposure Relative to Depth in a Dry Chamber and Underwater; Oxygen Was Inhaled Through Open Circuit Apparatus (25)

Activity	Depth	Pressure (psig)	Number of Tests	Duration (min)		
				120	60	60
Dry Chamber						
Rest	60	26.7	20	20	—	—
Work	60	26.7	13	—	—	13
Rest	80	35.6	46	10	10	26
Rest	100	44.5	26	—	3	23
Underwater						
Rest	60	26.7	107	11	64	32
Rest	80	35.6	53	—	5	48
Rest	100	44.5	20	—	1	19

[a] Reactions terminating tests: nausea (40), twitching (21), anxiety (6), paresthesia (toes, fingers) 6, visual (6), convulsions (4).

Squalus, oxygen was used routinely in "surface" decompression at 60 to 40 ft levels. Surface decompression refers to the practice of bringing divers out of the water rapidly, usually after one or two stops, and completing decompression in a surface chamber. The interval for safe transfer from water to deck chamber is at least 5 minutes.

In diving operations on pure oxygen, the *U.S. Navy Diving Manual* (1973) stipulates the following conservative depth–time limits for *work* dives.

Normal Operations		Exceptional Operations	
Depth (ft)	Time (min)	Depth (ft)	Time (min)
10	240	30	45
15	150	35	25
20	110	40	10
25	75	—	—

These restricted values emanate from tests in which scuba-type gear (mouthpiece in place of a helmet) was employed, and conditions were conducive to buildup of carbon dioxide.

4.4 Carbon Dioxide Accumulation Relative to Work Underwater

Distinction is drawn between work underwater with conventional diving helmet and dress, and work with scuba equipment. In the well-ventilated diving helmet and suit,

Table 9.2. Variability in Oxygen Tolerance of 36 Men Breathing Through a Mouthpiece in a Dry Chamber at 3.13 ATA (90 ft equiv) (26)

Number of Exposures	Tolerance Time (min)	Signs and Symptoms at End Point (number of men affected)
12	6–17	Lip twitching (23), vertigo (5), nausea (4),
11	17–27	convulsions (3), numbness, confusion, vis-
5	30–35	ual and auditory hallucinations, stupefac-
8	50–96	tion (4)

there is no impairment to respiration other than increased density of the breathing mixture, and there are limited pressure gradients, which are not conducive to intrathroacic pooling of blood. In the tests of Lanphier (27) with subjects wearing scuba gear, the striking finding was inadequate response of pulmonary ventilation to exertion. Some of the divers became known as CO_2 retainers, and in such divers average alveolar pCO_2 rose from 42 mm Hg at rest to more than 48 mm Hg at work. The CO_2 level in the "high" man attained a rather incredible postexercise level of 57 mm Hg. These high levels of alveolar CO_2 were tolerated without distress. By contrast, in earlier submarine tests (1 ATA) in which the ambient recirculated air contained 5 percent CO_2 (associated with a rise in alveolar pCO_2 of 47 mm Hg), dyspnea was observed in resting men, and the average volume of breathing was 35 liters/min. Lanphier (27) states that individuals with characteristics of CO_2-retaining divers are found occasionally in the general population of athletic young men, but the proportion among divers at the Experimental Diving Unit (Washington Navy Yard) was exceptionally high. By all usual medical standards, these men were healthy and had a high tolerance level for exercise.

It is apparent that overexertion can give rise to serious difficulties in scuba diving. A preventive or rather ameliorative measure (without alteration of equipment) is to ensure that divers are physically fit as assessed by swimming tests, running performance, or maximal oxygen uptake. A laboratory measure of fitness is the respiratory quotient (RQ) under standard conditions of exercise. In trained persons the RQ rises from about 0.7 at rest to 1.10 or slightly higher during the course of 5-minute bouts of heavy exertion. In untrained persons, the RQ under similar conditions may attain a level of 1.4 or higher. The CO_2 response to exercise in the unfit is "like pouring acid on marble," as the accelerated rise in lactic acid reacts with blood and tissue buffers. The activity of the scuba diver and "CO_2-retaining" divers whose ventilation is inadequate should be restricted to no more than moderate work.

In the 1600-foot U.S. Navy dry-wet chamber dive at New Orleans, the divers using a modified closed-circuit underwater breathing apparatus were able to perform light work comfortably. With increase in work load the divers experienced respiratory distress, usually described as inability to breath in enough gas. However analysis of arterial blood gases at the time that the divers' work capability was exceeded revealed no significant retention of carbon dioxide. This result differs from previous tests at 1000 ft and less, where failure during work was accompanied by CO_2 retention.

A striking effect of elevated CO_2 tensions is dilation of cerebral blood vessels and increase of blood flow in the brain. Hence a given pressure of nitrogen in the lungs is rendered more narcotic if CO_2 tension is elevated; likewise, a given pulmonary oxygen tension is rendered more toxic. The augmentative role of CO_2 in relation to oxygen toxicity is illustrated by the classical data of Shaw et al. (28), summarized in Table 9.3.

Characteristic of recovery from elevated CO_2 atmospheres is rapid reversal of headache, and giddiness when a normal atmosphere is inhaled. Transient giddiness and at times the taste of ammonia were experienced by submarine personnel when outside air replaced recirculated air in which CO_2 had accumulated to a level of 3 percent over a period of 8 hours. Headaches may persist for about one-half hour, but this relatively short period of disagreeable condition is in contrast to the throbbing headache or "hangover" from hypoxia, which may persist for many hours.

5 DECOMPRESSION PROBLEMS RELATED TO PRESSURE

5.1 Generalized Pulmonary Overinflation

The occurrence of submarine disasters in relatively shallow waters initiated the development of an appliance to facilitate escape from submarines in 1929 by Momsen, Tibbals, and Hobson. The Momsen "lung" consists of a rebreathing bag containing a CO_2 absorbent canister (Figure 9.1). The appliance is charged with air or oxygen, and excess pressure in the lungs and appliance during ascent is relieved by escape of gas through a ventrally placed flutter valve. Training towers of 100 ft were constructed at New

Table 9.3. Alveolar CO_2 Level and O_2 Toxicity in Anesthetized Dogs (28)

Inhaled O_2 Tension (mm Hg)	Alveolar CO_2 Tension (mm Hg)	Duration of Exposure (min)	Time (min) of Onset of Symptoms: Fall in Blood Pressure	Time (min) of Onset of Symptoms: Seizures
2986	22[a]	164	—	—
3000	22[a]	180	—	—
3018	27[a]	159	152	—
2935	60[b]	124	20	57
2935	68[b]	133	10	64
190	85[b]	150	—	—

[a] Hyperventilation of anesthetized dog in Drinker respirator.
[b] CO_2 added to inhaled O_2.

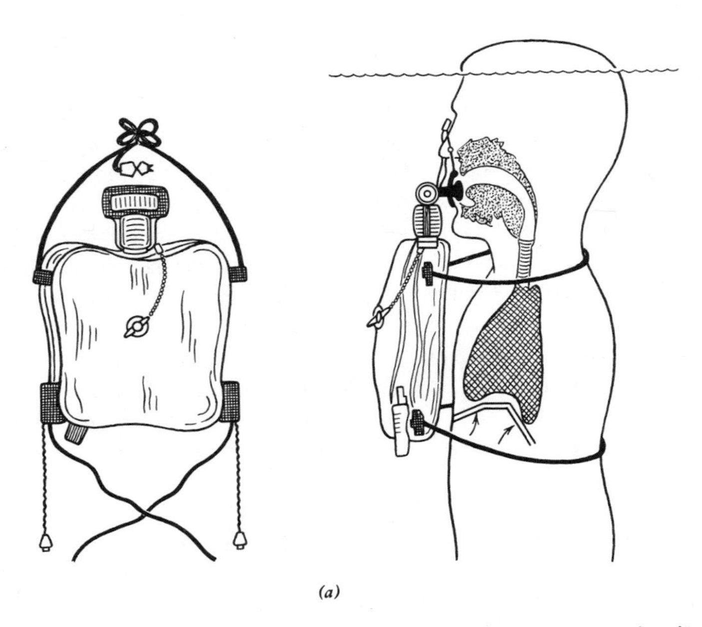

(a)

Figure 9.1 (*a*) Individual submarine escape appliance (Momsen "lung"). The distance from the mouthpiece of the relief (flutter) valve on the bottom of the bag determines the static expiratory pressure in the bag and lungs of the trainee. (*b*) The Momsen "lung" has been superseded by the hood and accessory life jacket or immersion suit. British Royal Navy divers have made escapes from depths to 600 ft with this type of equipment.

London, Connecticut, and at Pearl Harbor to indoctrinate trainees in a technique of escape from a sunken submarine. Thousands of supervised practice ascents have been made from depths of 18, 60, and 90 ft, but the training program has been marred occasionally by accidents. Virtually all have involved air embolism or, consequences attributable to overinflation of the lungs. Death of a trainee using the Momsen "lung" followed ascent from a diving bell submerged in cold water to a depth of only 15 ft.

To produce overdistension of the lungs and pulmonary damage during breath-hold and rapid ascent in water, it is necessary that the glottis remain spastically closed. As a subject in early tests both underwater and in a pressure chamber, a medical officer found that he could not hold his breath voluntarily to the point of injury. Substernal distress, actually a sensation of stretching, renders exhalation mandatory. Under conditions of panic and/or aspiration of water, however, spasm of laryngeal muscles closes the larynx, and during rapid ascent, excessive intrapulmonic pressure gives rise to interstitial edema, pneumothorax, and air embolism.

The mechanism of interstitial emphysema associated with increased intrapulmonic pressure involves creation of a pressure gradient between marginal alveoli and perivascular tissue. If the pressure difference is sufficient, gas escapes through ruptured alveoli and perivascular tissue into vascular sheaths leading to the hilar area of the

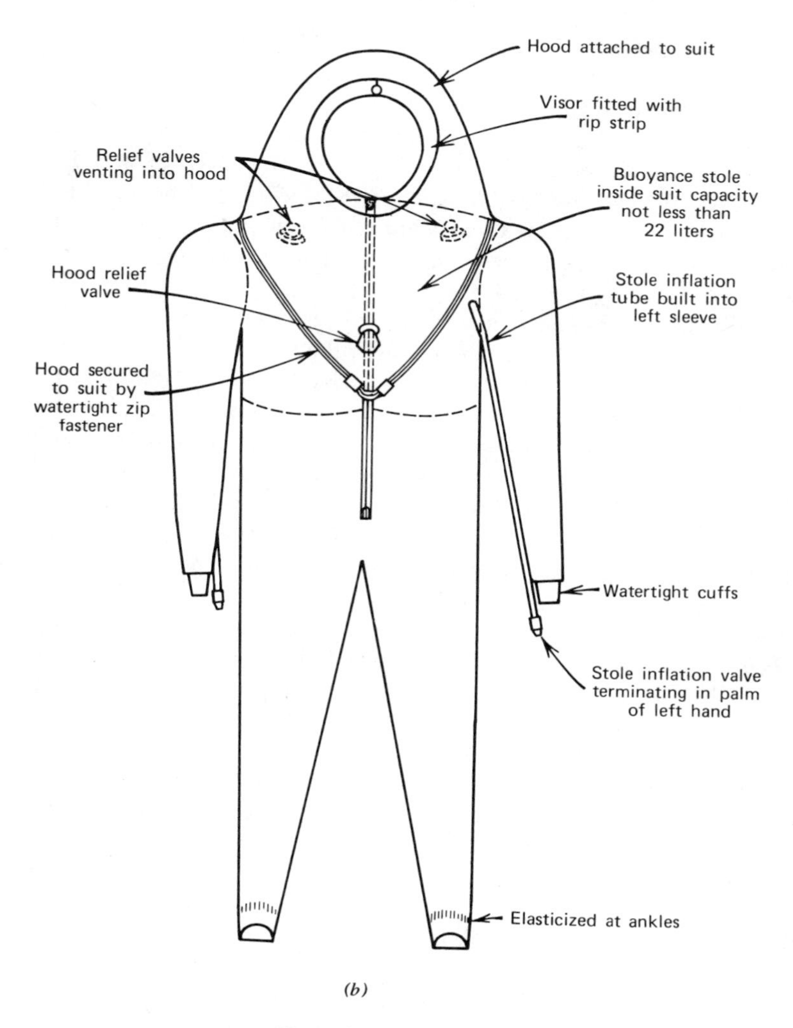

(b)

Figure 9.1 (Continued)

lungs. From the hilar area, gas dissects connective tissue planes to enter the pleural space and tissues of the neck. In one diver with a semiclosed fistula, increased intrapulmonic pressure incident to lifting heavy objects from the sea floor was sufficient periodically to force gas down the arm to the thenar area of the hand. Surgical correction removed this handicap.

In the presence of overdistended lungs, air can enter ruptured capillaries to embolize cerebral vessels. If lung cysts are present, gas may penetrate directly the pleural space to produce pneumothorax. Rupture of pulmonary capillaries is less likely to occur in the presence of pneumothorax, which serves as a "splint" for the collapsed lung. Treatment of air embolism is discussed in connection with decompression sickness. The importance of recompression cannot be overemphasized. As an example, the complication of air embolism in a submarine trainee became apparent some moments after he had reached the surface, leading up to the fatal accident from a depth of 15 ft. The trainee had been instructed to close the valve of his "lung" upon surfacing, and he did this. No abnormality was noted at the time. He then swam a few feet to the ladder, started to climb up, but fell backward into the water and died despite resuscitative measures, *which at the time did not include recompression*.

Although during breath-hold in the course of rapid ascent the lungs may be overdistended to the point of rupture of blood vessels, the entrance of air into the circulation apparently occurs when the individual breaks surface. With the initial breath, air presumably is aspirated into pulmonary capillaries when blood flow, previously obstructed by pressure, is resumed. Anesthetized dog experiments of Polak and Adams (29) support this assumption. Intrapulmonic pressures of 60 mm Hg maintained for 10 seconds caused systemic pressure to fall and pulmonary venous pressure to rise. Upon release of pressure, bubbles were not observed in a gas trap in the carotid artery. Following release of intrapulmonic pressures of 80 mm Hg and higher, however, bubbles appeared in the carotid gas trap.

Prevention of generalized pulmonary overdistension was effected in the dog by application of thoracoabdominal binders, despite a rise of intratracheal pressure to 180 mm Hg or higher. In man, voluntary abdominal splinting serves to prevent injury in routine tests to determine maximal expiratory effort (MEF). In young men, a representative mean value of MEF of 114 mm Hg (range, 60 to 350 mm Hg) is well above the uncompensated intrapulmonic pressures that injure the anesthetized dog. Momentary unconsciousness may occur occasionally in such tests, but the rapid recovery precludes anything more serious than a Valsalva response. Rupture of the round window of the chochlea may be a rare complication.

5.2 Localized Pulmonary Overinflation

In nonfatal cases of air embolism, large bullae have been observed on X-ray at the base of the lungs. In a fatal case of air embolism, a broncholith of tuberculous origin was found in a subsegmental bronchus (30). Data from Dahlback and Lundgren (15) are pertinent. During the course of head-out immersion in the vertical position, the expiratory reserve volume was reduced by an average of 1.55 liters (from 2.09 liters in the dry to 0.54 liter immersed). If the subjects (immersed to neck level) then exhaled to residual or near-residual volume (RV) and subsequently restricted their tidal volume to small fractions of vital capacity (VC), it was possible to induce terminal bronchiolar closure and trap air. The condition of air trapping persisted until the amplitude of tidal volume was 40 percent or more of vital capacity. During rapid ascent, regional overexpansion of the trapped air leads to embolization of blood vessels and emphysema of lung tissue.

5.3 Overdistension of Abdominal Viscera

In ascent from deep diving depths and, more frequently, in simulated altitude ascents, expansion of gas trapped in the stomach and segments of the large bowel constitutes an unquestionable impediment to further decompression. Once having been distended with gas, a viscus loses much of its motility. In overdistension of the stomach, for example, the cardiac and pyloric sphincters maintain spastic contractility to the point where overdistension pressure is sufficient to rupture the gastric wall.

6 DECOMPRESSION SICKNESS

There is an emergency aspect to debilities associated with decompression from high pressure atmospheres, or from normal pressure to substratospheric altitudes. The diver and the caisson worker, the aviator and the wind-tunnel investigator, and with the renascence of hyperbaric therapy, hospital personnel, are all subjected to potentially grave injury when pressure is reduced. The calculated risk is reflected in simulated altitude ascent (38,500 ft, 0.2 ATA), where more than 50 percent of the trainees without preoxygenation develop symptoms of decompression sickness. In deep-sea diving the incidence of symptoms may be as high as 5 percent, and in compressed air tunneling, about 1 to 4 percent.

6.1 Classification and Features

Usage has sanctioned the terms "decompression sickness" (DCS) following decompression from higher than normal atmospheres, and "dysbarism," for maladies associated with too rapid altitude ascent. The clinical connotation of sickness, although applicable to *complications* arising from intravascular bubbles, is a misnomer for the painful impairment (bends) mainly confined to the extremities and readily amenable to recompression. Type I DCS (bends) comprises 75 to 90 percent of all cases of DCS; Type II DCS comprises all serious manifestations arising from cardiopulmonary derangements and involvement of the nervous system. Delay in treatment of Type I cases may give rise to Type II symptomatology. Signs and symptoms, both acute and chronic, are outlined in Table 9.4.

The derangements are unique from an industrial point of view in that they are man-created through faulty decompression, and they are resolved by recompression. The multiple manifestations are recognized in a variety of clinical conditions as migraine, Meniere's syndrome, pulmonary embolism, circulatory shock, tabes dorsalis, and other neurologic pathology. The nascent bubble is capable not only of deranging rheology and morphology both of vessels and of formed elements of blood as platelets, but the bubble is able to assume various guises to mask prime pathogenicity. Thus it has been concluded that "missed diagnosis, together with lack of pressurized facilities, has been

Table 9.4. Signs and Symptoms of DCS

Acute DCS		
Type I (75 to 90 percent of cases)	Type II (10 to 25 percent of cases)	
	Cardiopulmonary	Nervous System
Extremities		
Pain (bends)	Substernal distress	Unconsciousness
Numbness	Paroxysmal coughing	Headache—migraine
Paresthesia	Dyspnea (chokes)	Visual (teichopsia)
Weakness	Asphyxia	Dizziness, ataxia, vertigo, deafness
Edema	Trachypnea (animals)	
Systemic	Circulatory obstruction	
Rash, mottled skin	Pulmonary tamponade	
Pruritus	Shock (pallor, dizziness, nausea)	Aphasia
Fatigue	Hemoconcentration	Slurred speech
Fever	Platelet sequestration	Spinal Cord
Sweating	Decrease clotting time	Motor and sensory
		Paralysis
Chronic DCS		
Avascular bone necrosis		
Paralysis following spinal cord lesions		
Focal lesions of the brain		

Source. Adapted from Reference 31, Table 1.

responsible for failure to apply recompression, and for serious ensuing complications in the absence of recompression" (31). Therapeutic error has extended over a period of 100 years beginning with the construction of the Eads Bridge across the Mississippi at St. Louis. In this project there were 8 deaths out of 78 serious cases, and the attending physician, Alphonse Jaminet, was paralyzed following decompression at the work site. Although recommended by Paul Bert, recompression had not yet been established as a therapeutic procedure. In recent years, both in England and in the United States, aviators in unsealed jet aircraft have had typical symptoms of DCS which have been ascribed to toxic effect of oxygen or to rapid acceleration.

6.2 Etiology and Pathogenesis

During the past 25 years papers on the etiology and pathogenesis at DCS have emphasized our lack of knowledge of basic mechanisms. Apparently the secular studies have not advanced our understanding in regard to specific mechanisms underlying, for example, the pain of bends. Nevertheless, the gross picture of the role of bubbles in producing the symptoms of DCS was suggested by the vacuum pump experiments of Robert Boyle in 1670. Convincing experimental data were compiled by Paul Bert (32) in about 1880. Thus far it has not been possible to demonstrate the specific manner in which intravascular bubbles produce symptoms, although ischemia, distention of the vascular walls to stimulate perivascular sensory fibers, and pressure from extravascular bubbles, have been postulated.

A most likely site giving rise to bends is bone, particularly the marrow of long bones, which have high absorption coefficients for nitrogen. Furthermore, the sluggish, sinusoidal circulation in marrow, and narrow, fixed lumina for exit of intravascular bubbles, serve to make bone a trap for bubbles disseminated from the general circulation or forming *in situ* in the marrow spaces. From the point of view of body economy, bone is the organ that renders man unsuited for long exposures in compressed air.

Many bubbles are present in the marrow of animals that have been rapidly decompressed. Their presence in man after decompression is inferred from intensified pain experienced during rapid recompression. This type of pain is postulated to arise from a "squeeze" of bone marrow when the decreased volume of the compressed bubbles is not compensated for by the circulation. Pressure within bone marrow can induce pain in man similar to bends. However typical bends pain occurs also after relatively short dives when bubbles are in the vascular bed of muscles and other soft tissues, as observed in too rapid decompression in helium atmospheres.

6.3 Macroscopic and Microscopic Observations

In anesthetized dogs rapidly decompressed from high pressure (65 psig after 1.75 hours in compressed air) small bubbles are observed, usually within 20 minutes, to circulate rapidly through subcutaneous arteries and veins. The bubbles move out of the relatively high pressure arterial tree and accumulate in the large veins, the right ventricle, and the pulmonary vascular bed. At this time many bubbles are present in mixed venous blood withdrawn from the right ventricle by insertion of a glass cannula by way of the external jugular vein. A characteristic triad of rapid breathing (tachypnea in the dog), bradycardia, and fall in arterial blood pressure follows *but does not precede* the appearance of bubbles in subcutaneous vessels. Despite respiratory rates greater than 100 per minute, the lungs are largely nonfunctional (33). There then supervenes a striking anoxemia, retention of CO_2, a progressively increasing arterial-venous oxygen difference, hemoconcentration, cell packing, and difficulty in withdrawing blood as clotting time is decreased. Elevation of pulmonic blood pressure is a consequence of pulmonary tamponade conducive to cor pulmonale and reversal of lesser and greater circulation pressures. The now elevated pulmonic pressure reembolizes the arterial tree, and in the absence of recompression, respiration fails (Table 9.5).

Recompression on oxygen (30 psig) relieves tachypnea and asphyxia, but a residual paralysis of the hind legs is a frequent complication (33).

Venous stasis in addition to arterial embolization contributes to injury of the spinal

Table 9.5 Type II DCS Cases: Cardiopulmonary Complications

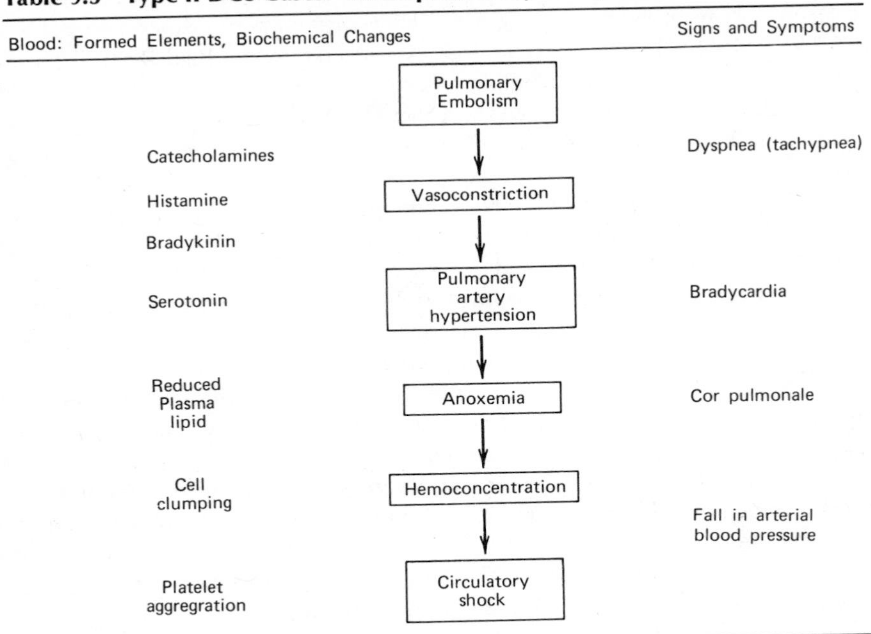

cord. Haymaker and Johnston (34) postulated that nitrogen bubbles from retroperitoneal fat depots might gain entrance to the spinovertebral plexus of veins, and by obstructive action, retard venous return. Under extreme conditions (presence of "chokes"), engorgement of the spinovertebral system may be aggravated through back pressure from bubble-laden pulmonary vessels. Thus embolic lesions in the cord appear to be dependent on retardation of blood flow on the venous side, as well as on the arteriolar-capillary bed. In recent experiments by Bove et al. (35), the nature and distribution of cord lesions appear to favor venous rather than arterial obstruction as the primary cause of spinal cord damage. Remarkably, however, the chokes (hence venous back pressure) is relatively common in altitude DCS; but spinal cord paraplegia rarely occurs.

6.4 Multiple Factors and Complications in DCS

Initially, bubbles attending decompression may be "silent" (31), that is, not associated with symptoms. As bubbles increase in number to obstruct peripheral circulation, Type I DCS (bends) supervenes, and this frequent malady is readily amenable to recompression. As bubbles increase in number and move into the large veins, the right side of the heart, and the pulmonary bed, a condition of true sickness (Type II) supervenes, manifest as shock and injury to the nervous system. Philp (36) has applied newer knowledge of the role of platelets in arterial thrombosis to the situation in Type II DCS. Essentially, platelets rich in serotonin and adenosine diphosphate (ADP) are aggregated by agents such as collagen and adrenaline, as well as serotonin and ADP. These powerful vasoconstrictors, in combination with smooth muscle acting bradykinins (37) are capable of inducing the complications of Type II DCS (Table 9.5).

Fat embolization may also complicate the primary and secondary effects of bubbles. Fat emboli have been found in the lungs of experimental animals and of caisson workers, and in cerebral vessels of two aviators (the anomaly of patent foramen ovale was present, which permitted transit of fat from the right to left circulatory shunt). Whitenack and Hausberger (38) found that a pressure of only 50 to 100 mm H_2O in medullary bone (tibia) of rabbits was sufficient to cause intravasation of fat: "Our results do not exclude the possibility that physical changes in the stability of blood lipids and formation of larger droplets contribute to the occurrence of pulmonary fat embolism."

7 SERIOUS (TYPE II) CLINICAL ENTITIES

7.1 Cardiopulmonary (Chokes, "Walking Shock")

Following incomplete decompression in early chamber tests from 4 ATA, there frequently supervened incipient chokes manifest on deep inspiration as substernal soreness, which progressed to bouts of paroxysmal coughing. Either recovery was spontaneous or the respiratory distress culminated in frank asphyxia, promptly relieved by recompression. Chokes were frequently experienced in altitude chamber tests during World War II. In one series of such tests, there were 754 exposures of 51 subjects to simulated altitude of 35,000 ft (0.25 ATA) for a projected period of 3 hours. The following adverse reactions were recorded: bends (519), chokes (155), skin lesions (67), and scotoma, including migraine headaches (30). Syncopal reactions (weakness, dizziness, pallor, sweating, nausea, fall in blood pressure, confusion, unconsciousness) occurred 91 times and affected 37 subjects.

Some of these cases and those reported later in pressurized operations are unique in medicine in that previously healthy men tolerate a condition of ambulatory "walking shock" over a period of many hours; for example, in the altitude chamber, the subject (a physician) was pale, dizzy, and nauseated; his blood pressure was 80/70. At ground level (i.e., following decompression from 0.25 to 1 ATA), subject was less dizzy and walked out of the chamber (B.P., 80/60). After 15 minutes the subject reported scintillating scotoma followed by headache, and EEG showed slow waves limited to the left occipital cortex. *Subject walked to his office.* Table 9.5 outlines events associated with pulmonary embolism following decompression from hyperbaric atmospheres (and from normal to hypobaric altitudes) and the complications involving formed elements of blood.

7.2 Central Nervous System Disturbances

Convulsive seizures, loss of consciousness, and signs of focal cerebral involvement characterize traumatic air embolism emanating from overdistension of the lungs. In DCS, by contrast, disorientation, dizziness ("the staggers"), aphasia, and visual symptoms (scintillating scotomata, teichopsia) are prominent.

There has been clinical evidence of brain involvement in approximately 10 percent of cases of divers and caisson workers stricken with DCS; yet curiously enough, in the century-long investigation of caisson workers and divers' sickness, very few changes have been noted in the brain (34). Because of its prolific blood supply and collateral circulation, the brain is less likely to be injured by disseminated bubbles. Damage may be indirect, however, as a result of pulmonary asphyxia. Chronic neurologic defects have been reported by Rózsahegyi (39) in Hungarian caisson workers. In his neurologic survey of 179 persons who had DCS and were examined after a period of 4 years or more, sequelae of lesions attributable to decompression (improper decompression) could be found in 130 cases, and in half of these cases the sequelae were serious. Rózsahegyi concluded that men who have had DCS affecting the central nervous system should not return to work in a pressurized environment. Divers who have had spinal cord lesions, however, have engaged subsequently both in experimental and deep-sea diving. The critical test of fitness to resume diving, in this author's experience, is ability to run.

Liske et al. (40) reported 13 cases of focal neurologic disorders that occurred in pilots either in aircraft or in altitude chambers. Although the majority made excellent recovery, there was residual focal deficit in two subjects. Interestingly, examinees who exhibited deficits were in the age range of 31 to 48 years. None had experienced focal lesions during the course of similar hypobaric exposures at a younger age.

Paralysis of spinal cord origin, spastic in type and involving white matter chiefly, is a serious and not infrequent complication of decompression sickness. Areas of hypersensitivity are present above and below the site of the lesion, and there is paresthesia or sensory paralysis below. There is frequently incontinence of urine and feces. Recovery of function in varying degree follows prolonged recompression. Prognosis for ultimate or partial recovery is uncertain, since improvement is usually observed over a period of 2 years following injury.

Acute auditory-vestibular dysfunction manifest as "apoplectiform" deafness, vertigo, nausea, and vomiting (Meniere-type syndrome) was reported in about 6 percent of DCS cases in earlier caisson workers at the turn of the century. Kennedy (41), in his review, states that many disorders in diving can be attributed to vestibular involvement. In U.S. Navy records (1960–1969) of 1050 cases of DCS, there were 7 cases of vertigo alone, 6 with hearing loss, and 3 with combined hearing loss and vertigo. In diving to exceptional depths, symptoms are present in the early stage of decompression and progress in severity as the diver approaches the surface. The etiology of the malady is not entirely clear, but embolic interference with blood supply to the labyrinth is confirmed by the prompt recovery of equilibrium and hearing attending immediate recompression. The inner ear derangements frequently coincide with a shift during decompression from helium to air, and the adverse effects of "counterdiffusion" may play a role. Idicula et al. (42), who recently elucidated this phenomenon, state:

> In recent simulated diving experiments, subjects have experienced intense itching, gross maculopapular skin lesions and a severe vestibular derangement with vertigo and nystagmus. These effects have been observed when a gas mixture containing nitrogen or neon was being breathed while a second inert gas, helium, was present in the surrounding environment.

Visual disorders are reported frequently following both pressure decompressions and altitude ascent. Symptoms described are blurring of vision, wavy lines, heat waves, glistening lights, and spots before the eyes (teichopsia). Remarkably, these symptoms are evanescent and are usually not mentioned by examinees unless specific inquiry is made in connection with Type I or Type II symptomatology. If such fleeting phenomena result from passage of "seed" bubbles through retinal vessels, this mechanism does not lead to the grave complication of clinical arterial embolism. In contrast to deafness or vestibular dysfunction, blindness has not been reported in connection with decompression sickness.

Migraine is implicated in the syndrome of blurred vision, scotoma, headache, and nausea, since vascular spasm of posterior cerebral vessels may be induced by gas embolization. In a series of 1361 chamber exposures to 30,000 to 38,000 ft, 17 of 155 subjects (medical students and doctors) reported scotomata 46 times. All the subjects who reported symptoms implicating migraine had other symptoms of DCS during or before the onset of visual disorders. Headache began about the time of disappearance of scotomata and was dull, pulsating, or aching. Subjects who previously had migraine reported that the headaches were similar but less intense. In contrast with clinical migraine, visual field defects may persist for several months.

7.3 Avascular Bone Necrosis

Bone infarction has become a major industrial compensation problem chiefly because X-ray visualization reveals islands of dense calcification in the shafts of long bones and perichondral "snowcaps" of the heads of the femur and humerus. In contrast to lesions of the spinal cord, and partly because the lesions are both painless and insidious (unless there is joint involvement), these effects are identified as sequelae of DCS mainly on the basis of occupational history. The roentgen manifestations in a full-blown case are almost pathognomonic because of predilection and occurrence in the lower femoral diaphysis, the upper tibial diaphysis, the humeral head and neck, and the femoral head and neck. The multiple lesions tend to be symmetrical bilaterally, and there is a characteristic geographic maplike appearance of the diaphyseal lesions, rendering one case almost identical to and a duplicate of others (43).

Only the heads of the humerus and femur are involved in joint disability following breakdown of cartilaginous surface. These joints derive nourishment from a few primary vessels, principally terminal arteries with arborization but poor collateral circulation

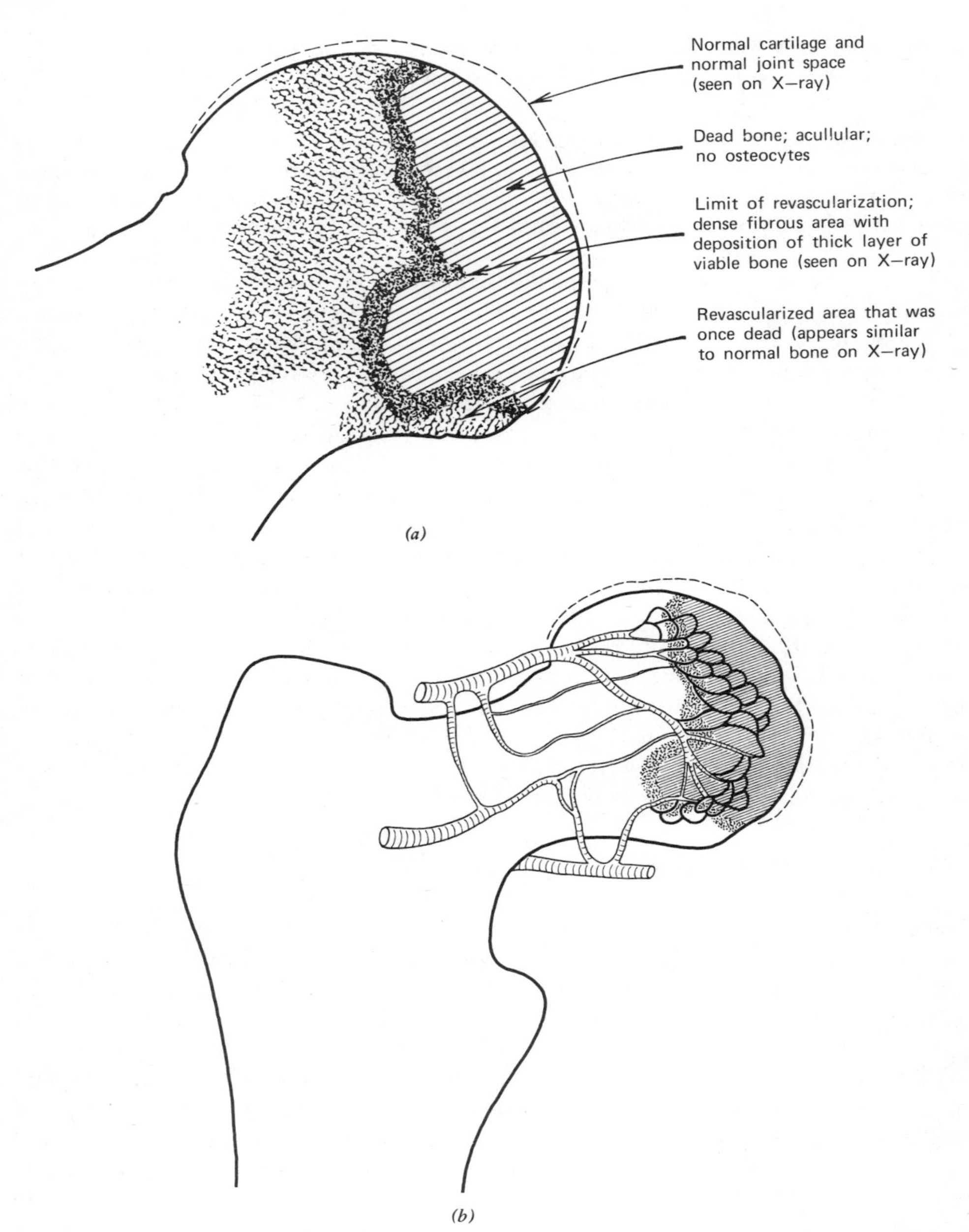

Figure 9.2 (a) Infarcted head of the femur. A dense fibrous band separates dead bone from proximal revascularized tissue; the cartilaginous surface is "folding." Courtesy of Drs. Griffiths and Walter, New Castle upon Tyne. (*b*) The superimposed blood supply shows arborization of nonanastomosing vasculature prone to embolization by gaseous and particulate emboli. Adapted from Prof. Jaffrès and Dr. Mérer, Brest, France.

(Figure 9.2). Remarkably, the knee joint, which is the common site of bends pain, is not disabled.

Data presented in Table 9.6 reveal high incidence of bone lesions in Japanese shellfish divers who disregard safe decompression practice, in contrast to divers in the British Royal Navy, who adhere to the stipulations of standard decomposition tables. Bone lesions appear to be the chronic complication of inadequate decompression. It should be noted, however, that half the homogeneous population of Japanese divers did not have disabling lesions. The unresolved problem is to explain this individual variation in susceptibility to decompression stress.

In recent years state codes in the United States have been revised to extend the time of decompression by a factor of 2 or more. It appears at this time that longer decompression, although not effective in reducing incidence of DCS (usually about 1 to 2 percent of man-decompressions) has eliminated or drastically reduced crippling disability. Work shifts in the United States are considerably shorter than those in England. Shortening of work shifts (to prevent saturation of bone marrow) may be more important than lengthening decompression time.

8 RECOMPRESSION THERAPY

At the turn of the century it was stated that recompression (RC) was the chief means of treating compressed air illness and in combination with oxygen, the only means.

Table 9.6. Incidence of Bone Lesions in Japanese Shellfish Divers Compared with British Royal Navy Divers; Bone Lesions in Relation to Age, Japanese Divers

Divers	Total Number of Divers	Divers with Lesions		Number of Divers with Lesions in Various Sites		
		Number	Percent	Humerus (head)	Femur (head)	Elsewhere
Japanese	301	152	50	92	58	81
Royal Navy	250	13	6	6	3	9

Divers	Age Group	Number of Men with Lesions	Number of Men Without Lesions	Percentage of Incidence
Japanese	16–19	3	16	15.8
	20–29	49	92	35
	30–39	62	29	67
	40–49	28	8	78
	50 and over	10	4	71

Source. Excerpt of tabular data from a compilation by Elliott and Harrison (45).

However oxygen RC therapy was not implemented in 1900. There were the usual barriers between plausible and established practice: (1) conclusive evidence about the value of oxygen was lacking, (2) man's tolerance for oxygen was not known, and (3) facilities were not available for safe administration of oxygen. By 1937 there were data to support a schema for projected oxygen therapy (Figure 9.3). It was an initial objective to limit pressure to 30 psig; however the frequent occurrence of paralysis in dogs decompressed in 15 seconds from 65 to 75 psig, then recompressed on oxygen to 30 psig, pointed to the need for higher pressures. Yet the requisite experiments to safeguard animals from paralysis still have not been carried out.

An innovation at the time, if there were residual symptoms following the outlined schedule, was prolonged (24 hours or more) recompression in air. The practice was referred to colloquially as the "overnight soak."

Minimal oxygen RC therapy is current practice [U.S. Navy treatment (58) Tables 5, 6, 5A, and 6A] and its value has been confirmed by Goodman and Workman (46). Cases of air embolism are treated by brief excursion to 165 ft (Tables 5A, 6A). The same oxygen format can be adapted for treatment of DCS that occurs under pressure, provided the partial pressure of oxygen at deeper depths does not exceed the value stipulated for inhalation of pure oxygen. A summary of authoritative experience with *minimal pressure oxygen therapy* is given in Reference 47.

Recompression on pure oxygen in cases of altitude dysbarism has been highly effective in resolving residual effects or serious complications. In the absence of "overpressure" oxygen therapy, convalescence has followed a stormy course, marked by fatalities.

8.1 Treatment of Compressed Air Workers

Pressurized tunnel work in connection with construction of the San Francisco Bay Area rapid transit (BART) project has been successfully completed (10). Recompression

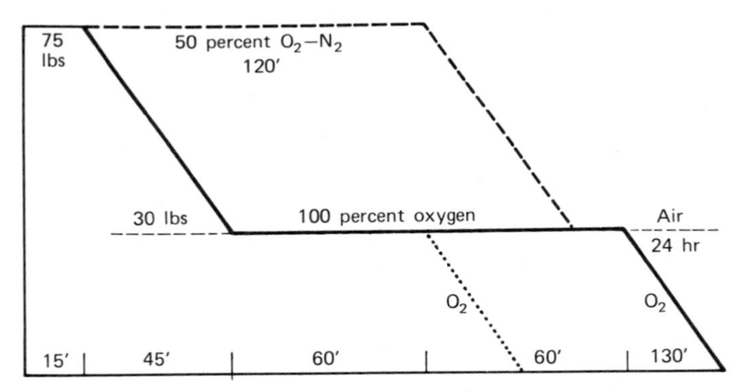

Figure 9.3 "Anlage" of oxygen recompression therapy. Type I cases were recompressed routinely to 44.5 psig (100 ft equiv). Type II cases were recompressed to 75 psig. Residual "altitude DCS" cases were recompressed to pressures no higher than 30 psig. All recurrences were recompressed in air (20 to 30 psig) for 24 hours or longer. From Behnke and Shaw (33); also Behnke (32).

Table 9.7. Type II Cases of DCS, BART Project

Tunnel Job (age of worker)	Pressure, Shift, and DC[a]	Elapsed Time[b] (hr)	Signs and Symptoms	Response to O_2 Recompression
Shifter (25)	*13.5* psig; 6 hr shift; 6 min DC	3.7	Blurred vision Headache, nausea Vomiting Skin rash Leg pain	Relief at 26 psig Slow resolution of symptoms *Therapy time*: 229 min
Electrician (36)	*13.0* psig; 0800–1200; 6 min DC 1330–1400; 6 min DC 1430–170; 5 min DC	3.2	Ataxia Slurred speech Extreme fatigue	Relief of all disability at 26 psig *Therapy time*: 155 min
Inspector (38)	*21.0* psig; 0305–0420; 16 min DC 0525–0825; 15 min DC	2.3	Scotomata Ringing in ears Nausea, vomiting	Relief at 26 psig *Therapy time*: 215 min
Miner (34)	*21.5*psig; 6 hr shift; 103 min DC	4.0	Dizziness, nausea, vomiting Pain in both knees (Bends, and viral infection (?))	After 175 min oxygen therapy, dizziness persists *Total therapy* time: 397 min
Miner (25)	*30.0* psig; 4 hr shift; 168 min DC	7.0	Scotomata Numbness lips, arm Pain in both knees	Relief at 26 psig *Therapy time*: 150 min
Miner (40)	*29.5*psig; 4 hr shift; 180 min DC	8.8	Blurred vision Headache Chest pain Pain in both legs Syncope	Relief in 90 min at 26 psig; second recompression *Total therapy time*: 415 min

[a] Decompression time.
[b] Elapsed time from onset of symptoms before reporting for treatment.

therapy following stipulations in U.S. Navy treatment, Tables 5 and 6, was highly satisfactory in treatment of 135 cases involving 85 workers, of whom 7 had Type II disability (Table 9.7). Irrespective of the time of onset of symptoms, there was an average elapsed time of more than 5 hours before workers reported for treatment (Table 9.8). With few exceptions, workers resumed their next regular work shift. Sometimes only a few hours elapsed between completion of therapy and return to the job. The data implicate 33 of 85 men as especially susceptible to bends; 10 of these workers had a second attack of bends within 48 hours following their next work shift. However resumption of the regular work shift by 75 of the 85 workers suggests that if "silent" bubbles persisted following treatment, such unstable condition did not render the men more susceptible to a second attack of decompression sickness.

In some patients, it was possible to shorten the time stipulated for oxygen recompression (Table 9.8). Although such abridgement appeared to be satisfactory, it is questionable therapy in view of possible painless bone infarction, regardless of whether a shorter schedule is advisable. An innovation based on earlier experience simplified oxygen therapy by maintaining essentially two pressure levels, one at 25 to 27 psig, the other at 15 psig. This modification (essentially an adaptation from Figure 9.3) facilitates therapy and maintains an average higher level of oxygen pressure more conducive to continued compression and elimination of bubbles.

8.2 Monitoring the Results of Oxygen RC Therapy

There are no criteria in our examinations (in the absence of an altitude decompression test) to separate men who are susceptible to DCS from those (about 75 percent) who are seemingly resistant, or at least do not report for treatment. In regard to age, weight, and blood pressure, there were no gross differences in the BART project between a susceptible and a nonsusceptible population (Table 9.9). The tabular data further tend to support the innocuousness of temporary, though painful impairment characteristic of Type I DCS. Recompression on oxygen did not appreciably alter blood pressures and pulse rates.

Table 9.8. Oxygen Recompression Therapy: BART Project (10)

Number of Cases	Onset of Symptoms Following DC (hr)	Elapsed time of onset of symptoms to therapy (hr)		Recompressions Time (min.)	
		Mean	Range	Mean	Range
29	<0[a]–0.5	5.8	1.3–11.2	144	90–275
46	0.6–2.0	4.7	0.5–13.0	150	90–255
38	2.1–5.0	5.6	0.8–13.7	140	60–365
23	5.1–12.0	5.3	0.3–10.0	165	95–415
2	>12	—	—	113	70–156

[a] In some cases, symptoms occurred during DC.

Table 9.9. Comparative Physiologic Data on 100 Cases of DC Involving 85 Men Before and Following Oxygen RC Therapy, and 100 Randomly Selected Shift Workers

	Age Groups							
	18–24	25–29	30–34	35–39	40–45	45+		
85 Men, (percentage distribution):	17	30	17	21	11	4		
100 Shift workers:	25	20	27	14	8	6		
Relative weight[a]:	70–79	80–89	90–94	95–100	101–105	106–110	111–119	>119
85 Men (percentage distribution):	2.4	5.9	24.7	31.7	16.5	10.6	8.2	0
100 Shift workers:	1	9	13	14	20	18	19	6
Blood pressure								
Systolic (mm Hg):	<100	100–110	111–120	121–130	131–135	136–140	141–146	>146
100 Cases DCS								
Pretherapy:	1	25	27	19	4	12	1	11
Posttherapy:	1	31	23	30	3	7	1	4
100 Shift workers:	1	29	26	28	1	8	4	3
Diastolic (mm Hg):	<60	60–69	70–79	80–89	90–96	96–100	>100	
100 Cases DCS								
Pretherapy:	7	15	29	35	5	4	5	
Posttherapy:	2	13	36	39	6	1	3	
100 Shift workers:	2	11	27	45	11	4	0	
Pulse rate (resting):	<60	60–69	70–79	80–89	90–96	96–100	>100	
100 Cases DCS								
Pretherapy:	8	19	47	12	9	4	0	
Posttherapy:	4	9	29	52	5	0	0	

[a] Relative weight = body weight/mean weight (for age and height) × 100.

The format of Table 9.9 merits further comment. It is feasible to monitor the physical condition of a large body of workers by selection of parameters that can be accurately assessed on a sufficiently large subsample (e.g., 100 men) to ensure validity of results. If time permits, additional parameters are highly informative, such as vital capacity, residual air volume, maximal expiratory force, response to submaximal exercise, hematocrit, and white and differential blood counts.

The format of Table 9.9 was also employed to assess any significant changes in three simple parameters—namely, blood pressure, pulse rate, and weight gain or loss in 100 pressurized tunnel workers on their initial examination and on reexamination a year later. There were no significant shifts in mean values of the basic parameters after a year of work in compressed air. However follow-up examinations were indicated for a number of workers who showed excessive weight loss or gain.

8.3 Recompression at High Pressures

Saturation diving confines men in pressurized habitats for weeks at a time at depths in excess of 400 ft. An extended period of time measured in days is required for safe decompression from habitat depth to the surface. It is this arduous technique that is employed in offshore drilling at deep depths. This type of diving, insofar as decompression is concerned, is relatively safe. In more than 1000 navy saturation dives, only Type I DCS was observed.

By contrast, the vulnerable dive is the one performed by the surface-supplied diver who works in the ocean for periods of 20 to 60 minutes in the depth range of 400 to 600 ft. In an earlier era of helium–oxygen diving, symptoms of DCS were occasionally reported as deep as 190 ft following ascent from bottom depth of 500 ft. At the present time, symptoms are reported at much greater depths. A case described by Elliott (48) illustrates a decompression–recompression problem in modern diving.

Case. A Royal Navy (U.K.) deep sea diver, age 25, was exposed 20 minutes in the sea at a depth of 600 ft by means of a submersible chamber. Decompression was carried out aboard ship in a dry chamber after transfer from the submersible chamber. At an equivalent depth of 250 ft, and four hours after initiation of decompression, the diver reported dizziness. He appeared sallow; he was nauseated, sweating, and deaf in the left ear with nystagmus to the left.

He was immediately recompressed to 300 ft (maximum depth of the ship's chamber), but there was no improvement. He was then transferred to the submersible chamber, which was lowered in the sea to a depth of 450 ft. At this depth, the patient had sudden and complete relief of all symptoms. Subsequent decompression was uneventful.

8.4 Importance of RC Therapy in "Late" Cases

There are cases of air embolism and DCS usually involving scuba divers who are far distant from recompression chambers. Paralysis of lower extremities is a frequent complication in survivors who are brought to the recompression chamber after a delay of many hours and even after several days. It is indisputable that the earlier recompression is applied, the better the outcome. Nevertheless, recompression, although delayed, remains a prime therapeutic modality in the effort to eliminate residual bubbles and to supply oxygen to hypoxic tissue. The following case, treated by Captain George Bond (49), is an example of recovery from spinal cord paralysis following inadvertent delay in administration of RC therapy.

Case. A dentist, age 54, surfaced in Martinique waters without decompression following a scuba dive to a depth of 230 ft, duration 18 minutes. He developed spasticity of the lower extremities, incapacitating ataxia, and paralysis of bowel and bladder function. A period of *60 hours elapsed* before he could be transported in an unpressurized aircraft from Martinique to the RC chamber at the Submarine Base, New London, Connecticut. With recompression (U.S. Navy Table 4) on air–oxygen over a period of 38 hours, there was rapid improvement. Ten days after discharge, the previously paralyzed scuba diver was working full time at the dental chair.

9 THERAPY ANCILLARY TO RECOMPRESSION

In the treatment of 135 BART cases (7 were Type II), recompression on oxygen alone was adequate. Ancillary therapy depends essentially on whether the patient is in an RC chamber or in an intensive care ward. With recompression on oxygen and recourse to higher pressures (if required), there should be reversal of circulatory and pulmonary malfunction. Intravenous fluid replacement (e.g., Dextran) is required only occasionally; the need for sedation may take precedence.

Without recompression, on the other hand, measures to counteract hypovolemia and the need to employ such drugs as isoproteronol to "unlock" a constricted microlung may be restorative. Lambertsen (47) has provided guidelines for pararecompression therapy with reference to pharmacologic agents.

In the administration of intravenous fluid, judicious consideration must be accorded the condition of the heart, since the right ventricle, filled with foamy, agglomerated blood, may become hypoxic, dilated, and disrupted in function. Measures to restore normal flow and composition of blood have been effective in animal experiments; see Cockett and Nakamura (50). In man, infusion of fluid on occasion has been restorative following ineffective recompression. The following illustrative case is from Barnard et al. (51).

Case. An engineer, age 26, developed pain in his legs after working his first full tunnel shift. Two recompressions *in air* were not effective. Upon transfer to a naval facility, the patient was cyanotic and in shock. A third RC *on air* led to some relief but blood pressure could not be recorded and urine output was scanty. *Three bottles of plasma* and one liter of normal saline were administered.

Patient was transferred to a hospital where examination revealed nonpitting swelling of the lower limbs and trunk. *Additional transfusion of three bottles of plasma* was followed by restoration of pulse and blood pressure, as well as reversal of peripheral vasoconstriction.

In the hyperbaric chamber, the stress on the heart in Type II cardiopulmonary DCS could be greatly reduced during the course of oxygen inhalation if the patient's frothy blood could be withdrawn by catheter and replaced by gas-free plasma, Dextran, or one of the newer electrolyte-nutrient perfusates. This procedure remains to be implemented.

A critical medical problem in the hyperbaric environment is treatment of injuries. Measures to carry out intensive care under pressurization have at times been heroic but generally futile. A solution to this difficult problem is transfer of patients in a portable chamber to a "surgical" compression chamber, where the usual competence of the surgeon can be utilized.

10 PREVENTION OF DECOMPRESSION SICKNESS

10.1 Safe Decompression Tables (Diving)

There are no systematic quantitative data with reference to the manner of gas transport during decompression. Reliance in formulation of tables rests on empirical observations derived from systematic tests, for the most part performed in dry-wet chambers. Calculation of decompression schedules has been based on the Haldanian model in which some degree of excess pressure is maintained between inert gas in tissues and ambient pressure. It has long been held by this author that what appears to be a "state of tolerated supersaturation" is in reality a tolerance of the body for gas separated from solution as bubbles. Up to a critical number, the bubbles are "silent" and are not productive of symptoms. Thus hundreds of thousands of relatively short ("no-decompression") dives have been remarkably safe. An adaptation of the formula of Hempleman (52) is convenient for calculation of "no-decompression" dives up to 100 minutes duration,

$$\text{time of stay (min)} = \left(\frac{500}{\text{depth}}\right)^2$$

From the formula, the calculated dive times for 250, 100, and 50 ft would be 4, 25, and 100 minutes, respectively.

In the surface-supplied dive of longer duration at a depth that requires decompression stops, no entirely satisfactory tables (i.e., for ascents not attended by DCS) are available. By means of the Doppler technique, a notable advance in diving technology by Mackay and Rubissow (53) and Spencer and Clarke (54), "silent" bubbles have frequently been detected in the precordial area during the first stage in decompression from bottom depth to first stop. Empirically, too rapid ascent to the first stop has long been held to be conducive to bubble evolution, but there was no proof that this was true before the development of a reliable monitoring technique. The following data are indicative of frequent occurrence of DCS under pressure, notably in dives from great depths.

	Onset of Symptoms (Cases)	
Diving Depth	Under Pressure	After Surfacing
to 300 ft	5	44
300 to 600 ft	30	35

Decompression tables for dives of 30 minutes duration have been formulated from tests on land in dry-wet chambers. When tested at sea, however, the schedules are attended by decompression sickness—a long-standing and disappointing experience for investigators in this field.

Excursion dives from habitats have been highly successful. At habitat depths of 150 to 300 ft, "no-decompression" excursions could be made to deeper depths of 100 ft (100 minutes stay), and to 150 ft for a stay of 60 minutes. Decompression to the surface from habitat saturation depth is relatively safe (i.e., not attended by Type II DCS), if conducted in accord with the following schedule (*U.S. Navy Diving Manual,* 1973).

Ascent Schedule		Daily Routine	
Depth	Rate (ft/hr)	6 Hour Period	Mode
1000–200	6	1600–2400	Ascend
200–100	5	2400–0600	Hold
100–50	4	0600–1400	Ascend
50–0	3	1400–1600	Hold

This prolonged schedule tends to conform to the isobaric principle of desaturation; that is, during the course of decompression from hyperbaric atmospheres following saturation, inert gas can be eliminated from tissues at a pressure isobaric with ambient pressure. This is possible because oxygen utilization by tissues permits an exchange with inert gas. The "oxygen window" approximates the pO_2 difference between arterial and venous blood (55).

10.2 Toward Safe Decompression Tables for Tunnel Workers

One of the paradoxes in decompression practice is that decrease in incidence of bends is not proportional to lengthening of decompression time. A work shift (one of two daily shifts) of 3 to 3.5 hours at 40 psig in 1910 (in the construction of the Pennsylvania Railroad tunnels) was accorded 48 minutes decompression. For the same work shift, British ("Blackpool") tables stipulate 170 minutes, and Washington State Tables (adopted by

California and other states), 183 minutes. The reported incidence of DCS is about the same for the several schedules.

Current tables, despite lengthened decompression time, do not prevent decompression sickness. Although total time (Washington State Tables) appears to be adequate, there is insufficient time accorded nitrogen clearance from rapidly desaturating tissues. Furthermore, a great deal of time is spent in decompression between 4 and 0 psig. Hempleman's "Blackpool" tables, for work shifts not in excess of 4 hours, tend to remove these objections. In these tables, however, the decompression time accorded shifts from 4 to 8 hours long appears to be inadequate.

There are two procedures for reducing or eliminating present difficulties. One is fabrication of a saturation habitat for tunnel work with "excursions" to shift pressures analogous to diving from sea habitats. The other measure is oxygen decompression. The value of oxygen inhalation has been established in decompression of divers and in recompression therapy. Preliminary tests give the following results.

		Decompression Time (min)	
Tunnel Pressure (psig)	Work Shift (hr)	Total	per Work hour
20	6	64	10
30	6	105	18
36	6	117	20
40	4	120	30

There are two objections to oxygen inhalation: *fire hazard,* which makes stringent monitoring mandatory, and the possibility of some long-term injury to the central nervous system, even though oxygen is inhaled at therapeutic levels (15 to 27 psig).

Too little attention has been accorded physiologic factors in decompression, particularly in regard to body movement and percutaneous diffusion of ambient gas into body tissues. Vigorous exercise produces a third gas, carbon dioxide to enlarge bubbles, and the associated stretching and muscular tension favor cavitation and bubble evolution. However walking between decompression locks, as occurred in early caisson work (Pennsylvania Railroad tunnels), may well have expedited clearance of nitrogen, helping to explain the relative safety of the restricted time accorded decompression. On the other hand, it is certain that prolonged decompression time (up to 3 hours daily) is not conducive to optimal clearance of tissue nitrogen if workers are relatively immobile. During the course of card games with workmen in the sitting position, the legs are "splinted" and circulation is retarded. The reality of this impairment is evident from the precaution taken by the orthopedic surgeon in operations on the hip, to rotate the leg externally to prevent postoperative restriction of blood flow to the head of the femur.

11 FITNESS FOR ABNORMAL PRESSURE ENVIRONMENTS

11.1 Individual Variation and Acclimatization (Habituation)

The puzzling difference in susceptibility to chronic bone lesions in a homogeneous population of Japanese shellfish divers has been mentioned. With reference to saturation diving, there may be a threefold difference in total (elapsed) decompression time from a depth of 100 ft. It is highly probable that we have not assessed the deterioration that begins with compression confinement and is enhanced by cold, hypercarbia, disturbed circadian rhythms, inadequate rest, isolation, monotony, and at times apprehension concerning diving hazards. *The diver we decompress is not the physically conditioned individual in a normal atmosphere.* Several factors routinely implicated in assessment of susceptibility to DCS are lack of acclimatization, age, weight, and body fat.

Walder (56) systematized data pertaining to the puzzling observation that men who expose themselves day after day to work in compressed air become less susceptible to attack of DCS as time goes on. Acclimatization appears to take about 14 days before maximum effect is obtained. Ceasing to work in compressed air leads to deacclimatization. "It therefore appears that for maximum protection against decompression sickness, it is best . . . to continue to work regularly each day rather than to work in compressed air sporadically" (56).

11.2 Role of Age, Weight, and Fat

Gray's analysis (57) of thousands of *altitude* decompressions in chambers during World War II revealed that "relative susceptibility" to dysbarism based on *mean* values, increased about 11 percent per year between ages 18 and 28 years. Similarly *mean* susceptibility was highly correlated with increase in weight per unit of height (linear density) over the same age range. Weight, however, did not refer specifically to fatness.

A wide variation in body fat of a group of divers would require formulation of decompression tables on an individual basis for *saturation* diving. If 10 percent (an arbitrary, "one unit") of body weight consists of lipid in adipose tissue, then a 70 kg man will have about 490 ml of nitrogen per ATA dissolved in fat. If blood flow to adipose tissue approximates 0.4 liter/min, half-time (one exponential time unit) is about 2 hours, and 6 time units is 12 hours for 98.5 percent clearance of nitrogen N_2 from tissues. As body fat increases, blood perfusion remains about the same; thus a diver on the borderline of obesity (3 units, 30 percent fat) would require some 6 hours for half-time nitrogen clearance and 36 hours for 98.5 percent desaturation. The slow elimination of nitrogen from excess quantities of fat dictates the need for prolonged decompression *following saturation exposures.* The time required for helium desaturation (since helium is less soluble in fat) is about 40 percent of the time required for nitrogen elimination.

The amount of fat (and evaluation of body build) can be quantified by anthropometric, radiologic techniques, and by underwater weighing to determine body volume. The percentage of fat is inversely proportional to body density.

11.3 Further Aspects of Selection and Fitness

Respiratory and cardiac parameters can be assessed reliably, and valid tests of an important parameter in diving, namely, standing steadiness, are available. An essential requirement in diving is adherence to a program of training that monitors daily fitness for undersea work, and informative tests can be carried out by paramedical personnel. Assessment is made *and recorded* of vision, hearing, speech, mental state (i.e., drowsy, awake, alert), pulse rate, blood pressure, neuromuscular coordination, and cardiovascular response to a platform step test. The time required for such tests is not more than 10 minutes.

In the initial dry-wet chamber tests at the Experimental Diving Unit to confirm the value of helium in diving, divers were conditioned by regular exercise, which consisted of a ball game in place of a noon meal. Exercise served to ameliorate appetite (mobilization of free fatty acids from adipose tissue and adrenergic stimulation), control weight, and most important, afford observation of neuromuscular coordination. Currently sophisticated techniques are available to provide computerized readout of gas exchange, pulse rate, and blood pressure during exercise of graded intensity. Pulmonary ventilation *underwater* must be adequate to prevent buildup of carbon dioxide. High respiratory quotients (1.2 to 1.45) in the unconditioned person following maximal test effort predispose divers to narcotic effects of nitrogen and toxic action of oxygen.

11.4 Group Isolation During Diving Operations

Group isolation shields divers from disruptive, extraneous infection, chiefly respiratory and gastrointestinal ailment. Isolation is conducive to a program of successive dives for the purpose of acclimatization. Especially important is the need to engender the dedication and group concentration essential to peak performance and safety. Mandatory in any safety program is "good housekeeping" and *orderly procedure*. Some measures that have served to eliminate accidents are embodied in the following experience.

During rescue efforts incident to the *U.S.S. Squalus* disaster (1939) and subsequently during the salvage work conducted over a period of months, there was not a single man-day loss of time due to injury. An effective safety measure was the dry run prior to a dive, which amounted to a rehearsal of each task to be performed underwater. Pressure was not applied to divers to accelerate normal or customary work effort, or to extend work time beyond prescribed limits. Finally, continual monitoring of diving fitness by brief but systematic tests served to detect chronic fatigue and deterioration. Relative isolation aboard ship accelerated adaptation to performance underwater and protected divers against intercurrent infection and erosion of job fitness.

REFERENCES

1. C. A. Berry, *Aviat. Space Environ. Med.,* **47,** 418 (1976).
2. J. A. Kylstra, I. S. Longmuir, and M. Grace, *Science,* **161,** 289 (1968).
3. B. A. Hills, *Arch. Intern. Med.,* **129,** 356 (1972).
4. P. B. Bennett in: *The Physiology and Medicine of Diving,* 2nd ed., P. B. Bennett and D. H. Elliott, Eds., Baillière Tindall, London, 1975, p. 248.
5. R. W. Brauer, R. O. Way, M. R. Jordan, and D. E. Parrish, in: *Underwater Physiology,* C. J. Lambertsen, Ed., Academic Press, New York, 1971, p. 498.
6. R. W. Brauer, in: *Barobiology and the Experimental Biology of the Deep Sea,* R. W. Brauer, Ed., University of North Carolina Press, Chapel Hill, 1972, pp. 4–5.
7. J. D. Harris, Naval Submarine Medical Research Laboratory Report 746, May 29, 1973, Box 600, New London Naval Submarine Base, Groton, Conn. 06340.
8. C. Edmonds, P. Freeman, R. Thomas, J. Tonkin, and F. A. Blackwood, *Otological Aspects of Diving,* Australasian Medical Publications, Glebe, New South Wales, 1973, p. 47.
9. C. E. G. Lundgren, *Br. Med. J.,* **1,** 511 (1965).
10. A. R. Behnke, *J. Occup. Med.,* **12,** 101 (1970).
11. M. Epstein, D. C. Duncan, and L. M. Fishman, *Clin. Sci.,* **43** (2), 275 (1972).
12. U. I. Balldin and C. E. G. Lundgren, *Aerosp. Med.,* **43,** 1101 (1972).
13. U. I. Balldin, in: *Proc. European Undersea Medical Society, Forsvarsmedicin,* C. M. Hesser and D. Linnarsson, Eds., **9,** 239 (1973).
14. M. Arborelius, Jr., U. I. Balldin, B. Lilja, and C. E. G. Lundgren, *Aerosp. Med.,* **43,** 592 (1972).
15. G. O. Dahlback and C. E. G. Lundgren, *Aerosp. Med.,* **43,** 768 (1972).
16. L. Irving, *Sci. Mon., N.Y.,* **38,** 422 (1934).
17. B. A. Gooden, *Med. J. Aust.* **2,** 583 (1972).
18. M. J. Frumin, *Anesthesiology,* **20,** 789 (1959).
19. A. R. Behnke, R. M. Thomson, and E. P. Motley, *Am. J. Physiol.,* **112,** 554 (1935).
20. A. R. Behnke and O. D. Yarbrough, *J. Physiol.,* **126,** 409 (1939).
21. P. B. Bennett, in: *The Physiology and Medicine of Diving,* 2nd ed., P. B. Bennett and D. H. Elliott, Eds., Baillière Tindall, London, 1975, p. 207.
22. S. A. Stern and H. L. Frisch, *J. Appl. Physiol.,* **34,** 366 (1973).
23. I. Fridovich, *Acct. Chem. Res.* **5,** 321 (1972).
24. P. L. Hendricks, Master's thesis, "The Effect of Intermittent Exposure on the Onset and Rate of Development of Pulmonary Oxygen Toxicity at 2 ATA in Man," University of Pennsylvania, 1975.
25. O. D. Yarbrough, W. Welham, E. S. Brinton, and A. R. Behnke, Experimental Diving Unit Report No. 1, Washington, D.C., 1947.
26. K. W. Donald, *Br. J. Med.,* **1,** 722 (1947).
27. E. H. Lanphier, in: *The Physiology and Medicine of Diving,* 2nd ed., P. B. Bennett and D. H. Elliott, Eds., Baillière Tindall, London, 1975, p. 102.
28. L. A. Shaw, A. R. Behnke, and A. C. Messer, *Am. J. Physiol.,* **108,** 652 (1934).
29. B. Polak and H. Adams, *U.S. Naval Med. Bull.,* **30,** 165 (1932).
30. A. E. Liewbow, J. E. Stark, J. Vogel, and K. E. Schaefer, *U.S. Armed Forces Med. J.,* **10,** 265 (1959).
31. A. R. Behnke, *Aerosp. Med.,* **42,** 255 (1971).

32. P. Bert, *Barometric Pressure, Researches in Experimental Physiology,* M. A. and F. A. Hitchcock, Trans., Columbus Ohio Book Co., Columbus, 1943.
33. A. R. Behnke and L. A. Shaw, *U.S. Naval Med. Bull.,* **35,** 61 (1937).
34. W. Haymaker and A. D. Johnston, *Mil. Med.,* **117,** 285 (1955).
35. A. A. Bove, J. M. Hallenbeck, and D. H. Elliott, *Undersea Biomed. Res.,* **1,** 207 (1974).
36. R. B. Philp, *Undersea Biomed. Res.* **1** (2), 117 (1974).
37. C. Chryssanthou, F. Teichner, G. Goldstein, and J. Kalberer, Jr., *Aeros. Med.,* **41,** 43 (1970).
38. S. H. Whitenack and F. X. Hausberger, *Am. J. Pathol.,* **65,** 335 (1971).
39. I. Ròzsahegyi, in: *Decompression of Compressed Air Workers in Civil Engineering,* R. I. McCallum, Ed., Oriel Press, Newcastle Upon Tyne, 1967, p. 127.
40. E. Liske, W. J. Crowley, Jr., and J. A. Lewis, *Aerosp. Med.,* **38,** 304 (1967).
41. R. S. Kennedy, *Undersea Biomed. Res.,* **1,** 73 (1974).
42. J. Idicula, D. J. Graves, J. A. Quinn, and C. J. Lambertsen, in: *Fifth Symposium on Underwater Physiology,* C. J. Lambertsen, Ed., Federation of American Societies of Experimental Biology, Bethesda, Md., 1976.
43. M. H. Poppel and W. T. Robinson, *Am. J. Roentgenol.,* **76,** 74 (1956).
44. R. I. McCallum, D. N. Walder, R. Barnes, M. E. Catto, J. K. Davidson, D. I. Fryer, F. C. Golding, and W. D. M. Paton, *J. Bone Joint Surg.,* **48B,** 207 (1966).
45. D. H. Elliott and J. A. B. Harrison, *J. Roy. Nav. Med. Serv.,* 1970.
46. M. W. Goodman and R. D. Workman, U.S. Navy Experimental Diving Unit Report, 5-65, 1965.
47. *Modern Aspects of Treatment of Decompression Sickness,* Symposium on Undersea—Aerospace Medicine, *Aerosp. Med.,* **39** (10), 1968.
48. D. H. Elliott, *J. Bone Joint Surg.,* **49B,** 588 (1967).
49. G. F. Bond, unpublished communication.
50. A. T. K. Cockett and R. M. Nakamura, *Am. Surg.,* **30,** 447 (1964).
51. E. E. P. Barnard, J. M. Hanson, M. A. Rowton-Lee, A. G. Morgan, O. Polak, and D. R. Tidy, *Br. Med. J.,* **2,** 154 (1966).
52. H. V. Hempleman, "Decompression Theory: British Practice," in: *The Physiology and Medicine of Diving,* 2nd ed., P. B. Bennett and D. H. Elliott, Eds., Baillière Tindall, London, 1975, p. 331.
53. R. S. Mackay and G. Rubissow, "Detection of Bubbles in Tissues and Blood," in: *Underwater Physiology,* C. J. Lambertson, Eds., Academic Press, New York, 1971, p. 151.
54. M. F. Spencer and H. F. Clarke, *Aerosp. Med.,* **43,** 762 (1972).
55. A. R. Behnke, "Early Quantitative Studies of Gas Dynamics in Decompression," in: *Physiology and Medicine of Diving,* 2nd ed., P. B. Bennett and D. H. Elliott, Eds., Baillière Tindall, London, 1975, p. 392.
56. D. N. Walder, *Ann. Roy. Coll. Surg. Engl.,* **38,** 288 (1966).
57. J. S. Gray, "Constitutional Factors Affecting Susceptibility to Decompression Sickness," in: *Decompression Sickness,* J. Fulton, Ed., Saunders, Philadelphia, 1951.
58. *U.S. Navy Diving Manual,* 1975.

CHAPTER TEN

Industrial Noise and Conservation of Hearing

PAUL L. MICHAEL, Ph.D.

1 INTRODUCTION

The harmful effects of noise on humans have been categorized as follows:

1. Exposure to high level noise over a significant period of time, which may cause both temporary and permanent damage to hearing (1–3).
2. Noise that interferes with speech communication and warning signals (4–12).
3. Noise that interferes with work performance (13–19).
4. Noise that disturbs relaxation and sleep (20).
5. Stress-causing noise that may contribute to heart disease, ulcers, and other stress-related problems (13, 21–26).

Concern about industrial noise problems has continued to grow during the past two decades as a result of increasing levels of noise exposure and greater understanding of the effects of noise, and because of new rules and regulations limiting noise exposures. Valuable information on noise problems has been gained through research; however the magnitude of this research remains relatively small compared with the amount of research on other environmental problems.

Increasing noise levels are to be expected in any society that consistently strives for faster and more economical ways to work, travel, or play; therefore noise exposure levels can be expected to increase rapidly unless concerted efforts are made on both the technical and political fronts. Manufacturers of noisy equipment must be motivated to produce reasonably quiet products, and adequate noise ordinances must be enacted and enforced. In situations where people must be exposed to noise levels that are too high for health and well-being, hearing conservation measures must be undertaken immediately, and efforts must be continued to lower these noise levels as much as feasible. The solution to hearing conservation or noise control problems depends, to a large extent, on a complete understanding of the problems by everyone involved; hence educational programs should be given a high priority, and these programs should be continued at a level that will remind exposed persons of the need for their cooperation.

2 PHYSICS OF SOUND

The sensation of sound is a result of oscillations in pressure, stress, particle displacement, and particle velocity, in any elastic medium that connects the sound source with the ear. In air, sound is usually described in terms of propagated changes in pressure that alternate above and below atmospheric pressure. These pressure changes are produced when vibrating objects (sound sources) cause regions of high and low pressure that propagate from the sources.

The characteristics of the sound received by an ear depend on the rate at which the sound source vibrates, the amplitude of the vibration, and the characteristics of the conducting medium. A sound may have a single rate or frequency of pressure alternations; most sounds, however, have many frequency components, and each frequency may have a different amplitude.

2.1 Terminology

ABSORPTION COEFFICIENT: The sound absorption coefficient of a given surface is the ratio of the sound energy absorbed by the surface to the sound energy incident upon that surface.

ACOUSTIC INTENSITY: see Sound Intensity.

ACOUSTIC POWER: see Sound Power.

ACOUSTIC PRESSURE: see Sound Pressure.

AMBIENT NOISE: Ambient noise may be defined as the overall composite of sounds in a given environment.

AMPLITUDE: The amplitude of sound may be described as the quantity of sound produced at a given location away from the source, or it may be designated in terms of the overall ability of the source to emit sound. The amount of sound at a location away from the source is generally described by the sound pressure or sound intensity, whereas the ability of the source to produce sound is described by the sound power of the source.

ANECHOIC ROOM: An anechoic room has essentially no boundaries to reflect sound energy generated therein. Thus any sound field generated within an anechoic room is referred to as "free field" (see definition).

AUDIOGRAM: An audiogram is a graphic recording of hearing levels referenced to a statistical normal sound pressure level as a function of frequency.

AUDIOMETER: An audiometer is an instrument for measuring hearing sensitivity.

CONTINUOUS SPECTRUM: A continuous spectrum of a sound has components continuously distributed over a known frequency range.

CRITICAL BAND: A critical band is a bandwidth within a continuous spectrum noise that has a sound power equal to that of a single-frequency tone centered in the critical band and just audible in the presence of the critical bandwidth of noise.

CYCLE: A cycle of a periodic function, such as a single-frequency sound, is the complete sequence of values that occur in a period.

CYCLE PER SECOND: see Frequency.

DECIBEL: The decibel (dB) is a convenient means for describing the logarithmic level of sound intensity, sound power, or sound pressure above arbitrarily chosen reference values (see text).

DIFFUSE SOUND FIELD: A diffuse sound field has sound pressure levels that are essentially the same throughout, and the directional incidence of energy flux is randomly distributed.

EFFECTIVE SOUND PRESSURE: The effective sound pressure at a given location is found by calculating the root-mean-square (rms) value of the instantaneous sound pressures measured over a period of time at that location.

FREE FIELD: A free field exists in a homogeneous isotropic medium free from boundaries. In a free field, sound radiated from a source can be measured accurately without influence from the test space. True free-field conditions are rarely found, except in expensive anechoic test chambers; however approximate free-field conditions exist in any homogeneous space where the distance from reflecting surfaces to the measuring location is much greater than the wavelengths of the sound being measured.

FREQUENCY: The frequency of sound describes the rate at which complete cycles of high and low pressure regions are produced by the sound source. The unit of frequency is the cycle per second (cps) or preferably, the hertz (Hz). The frequency range of the human ear is highly dependent on the individual and on the sound level, but an ear with normal hearing will have a frequency range of approximately 20 to 20,000 Hz at moderate sound levels. The frequency of a propagated sound wave heard by a listener will be the same as the frequency of the vibrating source if the distance between the source and the listener remains constant; however the frequency detected by a listener will increase or decrease as the distance from the source is decreasing or increasing (Doppler effect).

HERTZ: see Frequency.

INFRASONIC FREQUENCY: Sounds of an infrasonic frequency are below the audible frequency range.

INTENSITY: see Sound Intensity.

LEVEL: The level of any quantity, when described in decibels, is the logarithm of the ratio of that quantity to a reference value in the same units as the specified quantity.

LOUDNESS: The loudness of sound is an observer's impression of its amplitude, which includes the response characteristics of the ear.

NATURAL FREQUENCY: see Resonance.

NOISE: The terms "noise" and "sound" are often used interchangeably, but generally, sound is descriptive of useful communication or pleasant sounds, such as music, whereas noise is used to describe dissonance or unwanted sound.

NOISE REDUCTION COEFFICIENT: The noise reduction coefficient (NRC) is the arithmetical average of the sound absorption coefficients of a material at 250, 500, 1000, and 2000 Hz.

OCTAVE BAND: An octave band is a frequency bandwidth that has an upper band-edge frequency equal to twice its lower band-edge frequency.

PEAK LEVEL: The peak sound pressure level is the maximum instantaneous level that occurs over any specified time period.

PERIOD: The period T is the time (in seconds) required for one cycle of pressure change to take place; hence it is the reciprocal of the frequency.

PITCH: Pitch is a measure of auditory sensation that depends primarily on frequency but also on the pressure and wave form of the sound stimulus.

POWER: see Sound Power.

PURE TONE: A pure tone is a sound wave with only one sinusoidal change of level with time.

RANDOM INCIDENCE SOUND FIELD: see Diffuse Sound Field.

RANDOM NOISE: Random noise is made up of many frequency components whose instantaneous amplitudes occur randomly as a function of time.

RESONANCE: A system is in resonance when any change in the frequency of forced oscillation causes a decrease in the response of the system.

REVERBERATION: Reverberation occurs when sound persists after direct reception of the sound has stopped. The reverberation of a space is specified by the "reverberation time," which is the time required, after the source has stopped radiating sound, for the rms sound pressure to decrease 60 dB from its steady state level.

ROOT-MEAN-SQUARE SOUND PRESSURE: The root-mean-square (rms) value of a changing quantity, such as sound pressure, is the square root of the mean of the squares of the instantaneous values of the quantity.

SOUND: see Noise.

SOUND INTENSITY: The sound intensity I at a specific location is the average rate at which sound energy is transmitted through a unit area normal to the direction of sound propagation. The units used for sound intensity are joules per square meter per second [J/(m²)(sec)]. Sound intensity is also expressed in terms of a level (sound intensity level L_I) in decibels referenced to 10^{-12} W/m².

SOUND POWER: The sound power P of a source is the total sound energy radiated by the source per unit time. Sound power is normally expressed in terms of a level (sound power level L_p) in decibels referenced to 10^{-12} W.

SOUND PRESSURE: Sound pressure p normally refers to the rms value of the pressure changes above and below atmospheric pressure when used to measure steady state noise. Short-term or impulse-type noises are described by peak pressure values. The units used to describe sound pressures are newtons per square meter (N/m²), dynes per square centimeter (dyn/cm²), microbars (μbar), or pascals (Pa). Sound pressure is also described in terms of a level (sound pressure level, L_p) in decibels referenced to 2×10^{-5} N/m².

STANDING WAVES: Standing waves are periodic waves that have a fixed distribution in the propagation medium.

TRANSMISSION LOSS: Transmission loss TL of a sound barrier may be defined as 10 times the logarithm (to the base 10) of the ratio of the incident acoustic energy to the acoustic energy transmitted through the barrier.

ULTRASONIC: The frequency of ultrasonic sound is higher than that of audible sound.

VOLUME UNIT: The volume unit VU is used for expressing the magnitude of a complex wave form such as that of speech or music.

VELOCITY: The speed at which the regions of sound-producing pressure changes move away from the sound source is called the velocity of propagation. Sound velocity c varies directly with the square root of the density and inversely with the compressibility of the transmitting medium, as well as with other factors; in a given medium, however, the velocity of sound is usually considered constant under normal conditions. For example, the velocity of sound is approximately 1130 ft/sec in air, 4700 ft/sec in water, 13,000 ft/sec in wood, and 16,500 ft/sec in steel.

WAVELENGTH: The distance required to complete one pressure cycle is called one wavelength λ. The wavelength, a very useful tool in noise control, may be calculated from known values of frequency f and velocity c: $\lambda = c/f$.

WHITE NOISE: White noise has an essentially random spectrum with equal energy per unit frequency bandwidth over a specified frequency band.

2.2 Units for Noise Measurements

2.2.1 Sound Pressure and Sound Pressure Level

The range of sound pressures commonly encountered is very wide. Sound pressures well above the pain threshold (about 20 N/m²) are found in many work areas; also of interest, particularly to audiologists, are pressures down to the threshold of hearing (about

0.00002 N/m²). This range of more than 10^6 N/m² cannot be scaled linearly with a practical instrument, because to obtain the desired accuracy, such a scale would have to be many miles long. To cover this wide range of sound pressure with a reasonable number of scale divisions and to provide a means to obtain the required measurement accuracy at extreme pressure levels, the logarithmic decibel scale was selected. By definition, the decibel is a dimensionless unit: the logarithm of the ratio of a measured quantity to a reference quantity. The decibel is commonly used to describe levels of acoustic intensity, acoustic power, hearing thresholds, electric voltage, electric current, electric power, and so on, as well as sound pressure levels; thus it has no meaning unless a specific reference quantity is specified.

Most sound-measuring instruments are calibrated to provide a reading of rms sound pressures on a logarithmic scale in decibels. The reading taken from such an instrument is called the sound pressure level L_p. The term "level" is used because the measured pressure is at a particular level above a given pressure reference. For sound measurements in air, 0.00002 N/m² or 20 micropascals (μPa) commonly serves as the reference sound pressure. This reference is an arbitrary pressure chosen many years ago because it was thought to approximate the normal threshold of human hearing at 1000 Hz. Mathematically, L_p is written as follows:

$$L_p = 20 \log \frac{p}{p_0} \text{ dB} \tag{1}$$

where p is the measured rms sound pressure, p_0 is the reference sound pressure, and the logarithm (log) is to the base 10. Thus L_p should be written in terms of decibels referenced to a specified pressure level. For example, in air the notation for L_p is commonly abbreviated as "dB re 20 μPa."

Figure 10.1 shows the relationship between sound pressure (in micropascals) and sound pressure level (in dB re 20 μPa) and illustrates the advantage of using the dB scale rather than the wide range of direct pressure measurements. It is of interest to note that any pressure range over which the pressure is doubled is equivalent to 6 dB, whether at high or low levels. For example, a range of 20 to 40 μPa, which might be found in hearing measurements, and a range of 10 to 20 Pa, which might be found in hearing conservation programs, are both ranges of 6 dB.

The L_p referenced to 20 μPa may be written in any of the following six forms:

$$L_p = 20 \log \left(\frac{p}{0.00002}\right)$$

$$20\{\log p - \log 0.00002\}$$

$$20\{\log p - (\log 2 - \log 10^5)\}$$

$$20\{\log p - (0.3 - 5)\}$$

$$20(\log p + 4.7)$$

$$20 \log p + 94 \text{ re } 20\ \mu\text{Pa}$$

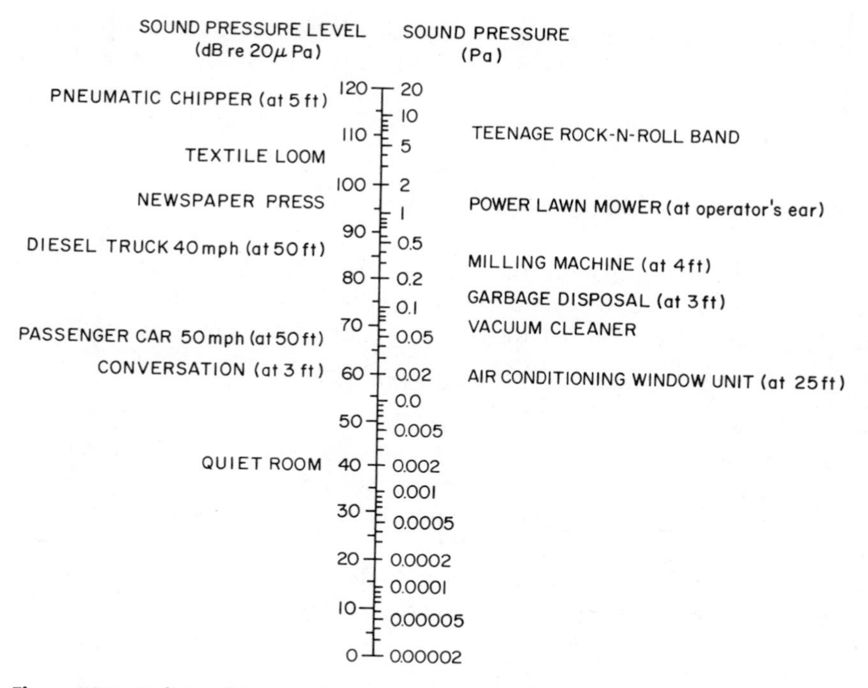

Figure 10.1 Relation between A-weighted sound pressure level and sound pressure.

2.2.2 Sound Intensity and Sound Intensity Level

Sound intensity at any specified location may be defined as the average acoustic energy per unit time passing through a unit area that is normal to the direction of propagation. For a spherical or free-progressive sound wave, the intensity may be expressed by

$$I = \frac{p^2}{\rho c} \tag{2}$$

where p is the rms sound pressure, ρ is the density of the medium, and c is the speed of sound in the medium. It is obvious from this definition that sound intensity describes, in part, characteristics of the sound in the medium, but does not directly describe the sound itself.

Sound intensity units, like sound pressure units, cover a wide range, and it is often desirable to use dB levels to compress the measuring scale. To be consistent with equations 1 and 2, intensity level is defined as

$$L_I = 10 \log \frac{I}{I_0} \text{ dB} \tag{3}$$

where I is the measured intensity at some given distance from the source and I_0 is a reference intensity. The reference intensity commonly used is 10^{-12} W/m^2. In air, this reference closely corresponds to the reference pressure 20 μPa used for sound pressure levels.

2.2.3 Sound Power and Sound Power Level

Sound power P is used to describe the sound source in terms of the amount of acoustic energy that is produced per unit time. Sound power may be related to the average sound intensity produced in free-field conditions at a distance r from a point source by

$$P = I_{\text{avg}}\, 4\pi r^2 \tag{4}$$

where I_{avg} is the average intensity at a distance r from a sound source whose acoustic power is P. The quantity $4\pi r^2$ is the area of a sphere surrounding the source over which the intensity is averaged. It is obvious from Equation 4 that the intensity will decrease with the square of the distance from the source; hence the well-known law of the inverse square.

Power units are often described in terms of decibel levels because of the wide range of powers covered in practical applications. Power level L_P is defined by

$$L_P = 10 \log \frac{P}{P_0} \tag{5}$$

where P is the power of the source and P_0 is the reference power. The arbitrarily chosen reference power commonly used is 10^{-12} W. Figure 10.2 shows the relation between sound power in watts and sound power level in dB re 10^{-12} W.

2.2.4 Relation Between Sound Power, Sound Intensity, and Sound Pressure

Many noise control problems require a practical knowledge of the relationship between pressure, intensity, and power. For example, consider the prediction of sound pressure levels that would be produced around a proposed machine location from the sound power level provided for the machine.

Example: The manufacturer of a pneumatic chipping hammer states that the hammer has an acoustic power output of 1.0 W. Predict the sound pressure level at a location 100 feet from the hammer.

From Equations 2 and 4, in a free field and for an omnidirectional source, we have

$$P = I_{\text{avg}}\, 4\pi r^2 = \frac{p_{\text{avg}}^2\, 4\pi r^2}{\rho c} \tag{6}$$

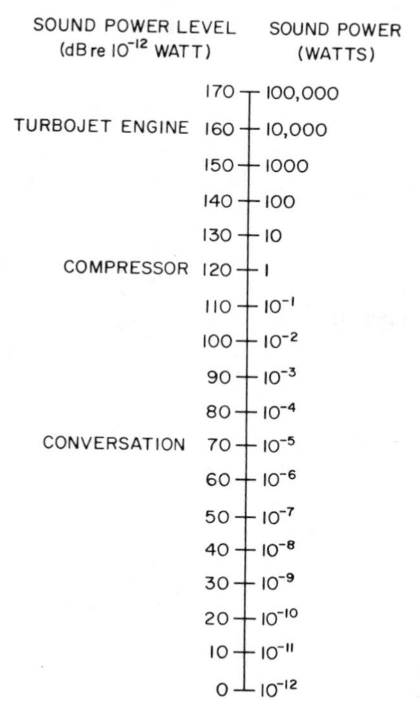

Figure 10.2 Relation between sound power level and sound power.

where

$$p_{\text{avg}} = \left(\frac{P\rho c}{4\pi r^2}\right)^{1/2} \tag{7}$$

If P is given in watts, r in feet, and p in pascals, then with standard conditions, Eq. 7 may be rewritten as

$$p_{\text{avg}} = \left(\frac{3.5P \times 10^2}{r^2}\right)^{1/2}$$

and, for this example,

$$p_{\text{avg}} = \left(\frac{3.5 \times 1.0 \times 10^2}{(100)^2}\right)^{1/2} = 0.187 \text{ Pa}$$

The sound pressure level may be determined from Equation 1:

$$L_p = 20 \log \frac{0.187}{0.00002} = 79.4 \text{ dB re } 20 \ \mu\text{Pa}$$

Noise levels in locations that are reverberant can be expected to be higher than that predicted because of the sound that is reflected back to the point of measurement.

2.2.5 Combining Sound Levels

It is often necessary to combine sound levels—for example, the combining of frequency band levels to obtain the overall or total sound pressure level. Another example is the estimation of total sound pressure level resulting from adding a machine of known noise spectrum to a noise environment of known characteristics. Simple addition of individual

Table 10.1. Table for Combining Decibel Levels of Noises with Random Frequency Characteristics

Numerical Difference Between Levels (dB)	Amount to be Added to the Higher Level (dB)
0.0–0.1	3.0
0.2–0.3	2.9
0.4–0.5	2.8
0.6–0.7	2.7
0.8–0.9	2.6
1.0–1.2	2.5
1.3–1.4	2.4
1.5–1.6	2.3
1.7–1.9	2.2
2.0–2.1	2.1
2.2–2.4	2.0
2.5–2.7	1.9
2.8–3.0	1.8
3.1–3.3	1.7
3.4–3.6	1.6
3.7–4.0	1.5
4.1–4.3	1.4
4.4–4.7	1.3
4.8–5.1	1.2
5.2–5.6	1.1
5.7–6.1	1.0
6.2–6.6	0.9
6.7–7.2	0.8
7.3–7.9	0.7
8.0–8.6	0.6
8.7–9.6	0.5
9.7–10.7	0.4
10.8–12.2	0.3
12.3–14.5	0.2
14.6–19.3	0.1
19.4–∞	0.0

sound pressure levels, which are logarithmic quantities, constitutes multiplication of pressure ratios; therefore the sound pressure corresponding to each sound pressure level must be determined and added with respect to existing phase relationships.

For the most part, industrial noise is broad band, with nearly random phase relationships. Sound pressure levels of random noises can be added by converting the levels first to pressure, then to intensity units that may be added arithmetically, and reconverting the resultant intensity to pressure and, finally, to sound pressure levels in decibels. Equations 1 and 2 can be used in free-field conditions for this purpose.

By using Table 10.1, the sound pressure levels from two separate random noise sources can be added very easily. The use of this table may be illustrated by the following example.

Example: The sound pressure level produced by a random noise source is measured in nine discrete octave bands. The measurements are 85, 88, 91, 94, 95, 100, 97, 90, and 88 dB, respectively. To add this series of decibel values, it is best to begin with the highest levels, since when the lower values are reached, they do not add significantly to the total and calculations may be discontinued. In our example, then, the levels 100 and 97 dB differ by 3 dB; therefore opposite the range 2.8–3.0 in the left-hand column of Table 10.1, read a value of 1.8 in the right-hand column and add this value to the higher of the two levels: 100 + 1.8 = 101.8 dB. The result is now added to the next highest level by repeating the process: 101.8 − 95 = 6.8; from the table read an amount to be added of 0.8. Thus 101.8 + 0.8 = 102.6 dB. This procedure is continued with each reading to arrive at an overall sound pressure level of 104 dB.

A simplified set of values for this procedure is given in Table 10.2. Since the error encountered in using this table seldom exceeds 1 dB, it is usable for most situations, and the limited number of values makes it relatively easy to commit to memory.

The overall sound pressure level calculated in the example above corresponds to the value that would be found by reading a sound level meter at this location with the frequency weighting set so that each frequency in the spectrum is weighted equally. Common names given to this frequency weighting are flat, linear, 20 kHz, and overall.

The corresponding A-weighted sound pressure level (dBA) found in many noise regulations may also be calculated from octave band values such as those in the foregoing example if the adjustments given in Table 10.3 are first applied. For example, the octave band (OB) levels with A-weighting corresponding to the example would be:

OB center frequency (Hz)	31.5	63	125	250	500	1000	2000	4000	8000
Sound pressure level (A-weighted) (dB)	45.8	61.9	77.8	85.4	91.7	100	98.2	91.0	86.9

These octave band levels with A-frequency weighting can be added by the procedure already described to obtain the resultant A-weighted level, which is about 103 dBA.

Most industrial noises have random frequency characteristics, and they may be combined as described in the previous paragraphs. In a few cases of noises with pitched or major pure tone components, however, these calculations will not hold, and phase relationships must be considered. In areas where pitched noises are present, standing waves often can be recognized by rapidly varying sound pressure levels over short distances. It is not practical to try to predict levels in areas where standing waves are present.

When the sound pressure levels of two pitched sources are added, it might be assumed that the resultant sound pressure level L_p (R) will be less as often as it is greater than the level of a single source; in almost all cases, however, the resultant L_p (R) is greater than either single source. The reason for this may be seen if two pure tone sources are added at several specified phase differences (Figure 10.3). At zero phase difference, the resultant of two like pure tone sources is 6 dB greater than either single level. At a phase difference of 90°, the resultant is 3 dB greater than either level. Between 90° and 0°, the resultant is somewhere between 3 and 6 dB greater than either level. At a phase difference of 120°, the resultant is equal to the individual levels; and between 120° and 90°, the resultant is between 0 and 3 dB greater than either level. At 180° there is complete cancellation of sound. Obviously, the resultant L_p (R) is greater than the individual levels for all phase differences from 0° to 120°, but less than individual levels for phase differences from 120° and 180°. Also, most pitched tones are not single tones but combinations thereof; thus almost all points in the noise fields have pressure levels that exceed the individual levels.

The most common frequency bandwidth used for industrial noise measurements is the octave band. A frequency band is said to be an octave in width when its upper band-edge frequency f_2 is twice the lower band-edge frequency f_1:

$$f_2 = 2f_1 \tag{8}$$

Octave bands are commonly used for measurements directly related to the effects of noise on the ear, and for some noise control work, because they provide the maximum amount of information in a reasonable number of measurements.

Table 10.2. Simplified Table for Combining Decibel Levels of Noise with Random Frequency Characteristics

Numerical Differences Between Levels	Amount To Be Added to the Higher Level
0–1	3
2–4	2
5–9	1
>10	0

Table 10.3. Sound Level Meter Random Incidence Relative Response Level as a Function of Frequency for Various Weightings

Frequency (Hz)	A-Weighting Relative Response (dB)	B-Weighting Relative Response (dB)	C-Weighting Relative Response (dB)
10	−70.4	−38.2	−14.3
12.5	−63.4	−33.2	−11.2
16	−56.7	−28.5	− 8.5
20	−50.5	−24.2	− 6.2
25	−44.7	−20.4	− 4.4
31.5	−39.4	−17.1	− 3.0
40	−34.6	−14.2	− 2.0
50	−30.2	−11.6	− 1.3
63	−26.2	− 9.3	− 0.8
80	−22.5	− 7.4	− 0.5
100	−19.1	− 5.6	− 0.3
125	−16.1	− 4.2	− 0.2
160	−13.4	− 3.0	− 0.1
200	−10.9	− 2.0	0
250	− 8.6	− 1.3	0
315	− 6.6	− 0.8	0
400	− 4.8	− 0.5	0
500	− 3.2	− 0.3	0
630	− 1.9	− 0.1	0
800	− 0.8	0	0
1,000	0	0	0
1,250	+ 0.6	0	0
1,600	+ 1.0	0	− 0.1
2,000	+ 1.2	−0.1	− 0.2
2,500	+ 1.3	−0.2	− 0.3
3,150	+ 1.2	−0.4	− 0.5
4,000	+ 1.0	−0.7	− 0.8
5,000	+ 0.5	− 1.2	− 1.3
6,300	− 0.1	− 1.9	− 2.0
8,000	− 1.1	− 2.9	− 3.0
10,000	− 2.5	− 4.3	− 4.4
12,500	− 4.3	− 6.1	− 6.2
16,000	− 6.6	− 8.4	− 8.5
20,000	− 9.3	−11.1	−11.2

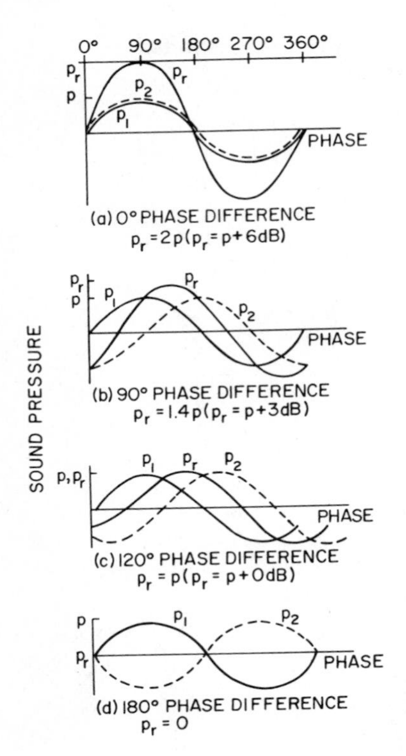

Figure 10.3 Combinations of two pure tone noises, (p_1 and p_2) phase differences.

When more specific characteristics of a noise source are required (e.g., for pinpointing a particular noise source in a background of other sources), it is necessary to use frequency bandwidths that are narrower than octave bands. Half-octave, third-octave, and narrower bands are used for these purposes. A half-octave bandwidth is defined as a band whose upper band-edge frequency f_2 is the square root of twice the lower band-edge frequency f_1:

$$f_2 = \sqrt{2}\, f_1 \tag{9}$$

A third-octave bandwidth is a band whose upper band-edge frequency f_2 is the cube root of twice the lower band-edge frequency f_1:

$$f_2 = \sqrt[3]{2}\, f_1. \tag{10}$$

The center frequency f_m of any of these bands is the square root of the product of the high and low band-edge frequencies (geometric mean):

$$f_m = \sqrt{f_2 f_1} \tag{11}$$

It should be noted that the upper and lower band-edge frequencies describing a frequency band do not imply abrupt cutoffs at these frequencies. These band-edge frequencies are conventionally used as the 3-dB-down points of gradually sloping curves that meet the American National Standard Specification for Octave, Half-Octave, and Third-Octave Band Filter Sets (S1.11-1966).

Noise measurement data (rms) taken with analyzers of a given bandwidth may be converted to another given bandwidth if the frequency range covered has a continuous spectrum with no prominent changes in level. The conversion may be made in terms of sound pressure levels by

$$L_p(A) = L_p(B) - 10 \log \frac{\Delta f(B)}{\Delta f(A)} \tag{12}$$

where $L_p(A)$ is the sound pressure level (dB) of the band having a width $\Delta f(A)$ Hz, and $L_p(B)$ is the sound pressure level (dB) of the band having a width $\Delta f(B)$ Hz. Sound pressure levels for different bandwidths of flat, continuous spectrum noises may also be converted to spectrum levels. The spectrum level describes a continuous spectrum, wide-band noise in terms of its energy equivalent in a band 1 Hz wide, assuming there are no prominent peaks. The spectrum level $L_p(S)$ may be determined by

$$L_p(S) = L_p(f) - 10 \log \Delta f \tag{13}$$

where $L_p(f)$ is the sound pressure level of the band having a width of Δf Hz.

It should be emphasized that accurate conversion of sound pressure levels from one bandwidth to another by the method just described can be accomplished only when the frequency bands have flat continuous spectra.

3 THE EAR

The normal human ear has a remarkable range of hearing that covers a frequency range from about 20 to 20,000 Hz at common loudness levels. The audibility and sensitivity characteristics of the ear over this wide frequency range are extremely complex because of large differences between individuals and differences in individual responses. An individual's response to noise may change as a result of physical or mental conditions, the sound level, and other factors.

The normal healthy human ear effectively transduces a remarkable range of sound levels. It is sensitive to very low sound pressures that produce a displacement of the ear-drum no greater than the diameter of a hydrogen molecule. At the other extreme, it can transduce sounds whose sound pressures are more than 10^6 times greater than the ear's lower threshold value; however exposure to high level sounds will eventually cause permanent damage to the ear. This is an insidious problem because a person may have experienced no pain before learning that severe hearing damage has taken place. The ability of the ear to withstand exposures to high sound pressures without suffering damage differs considerably from one individual to another. Thus it is impossible to set a particular exposure level as being safe, or unsafe, short of setting the exposure level on the basis of the most susceptible individual. Many everyday noise levels such as those from lawn mowers, subways, airplanes, or loud music may be harmful to some noise-sensitive individuals if they are exposed over an extended period of time.

The ear may be divided into three sections (Figure 10.4). Sound incident upon the ear travels through the ear canal to the eardrum, which separates the external and middle ear sections. The combined alternating sound pressures that are incident upon the eardrum cause the eardrum to vibrate with the same relative characteristics as the sound source(s). The mechanical vibration of the eardrum is then coupled through the three bones of the middle ear to the oval window of the inner ear. The vibration of the oval window is then coupled to the fluid contained in the inner ear. During its vibration, when the oval window moves inward, it pushes the fluid through the cochlea, which causes the round window to bulge outward. During the outward motion of the oval window, the round window bulges inward to compensate for the fluid flowing toward the oval window. The movement of the inner ear fluid is detected by thousands of transducers located within the cochlea. This complex information is then transmitted to the brain, through the eighth nerve, for further analysis and readout.

3.1 External Ear

The auricle, sometimes called the pinna (Figure 10.5), plays a significant part in the hearing process only at very high audible frequencies, where its size is large compared with that of a wavelength. It helps direct these high frequency sounds into the ear canal, and it assists the overall hearing system in determining the direction from which the sound comes. The ear canal is seldom as straight as is indicated in Figure 10.5, and the

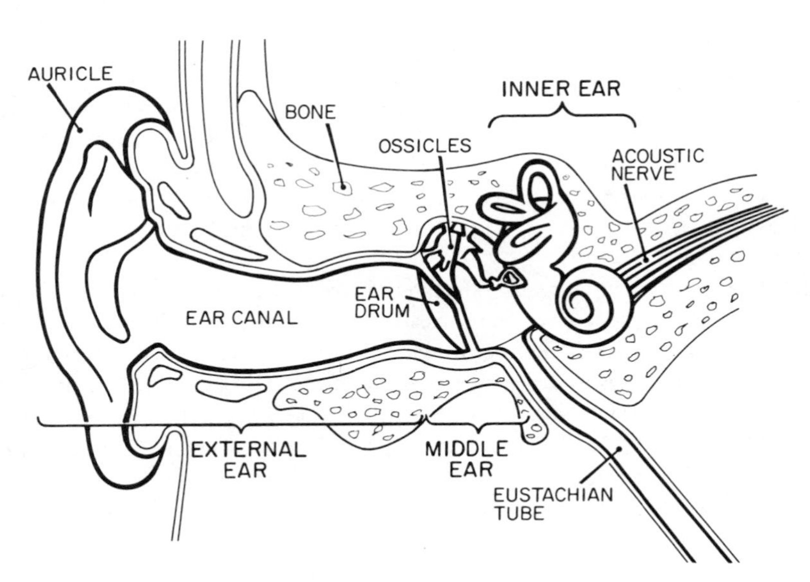

Figure 10.4 Cross section of the ear showing the external, middle, and inner ear configurations.

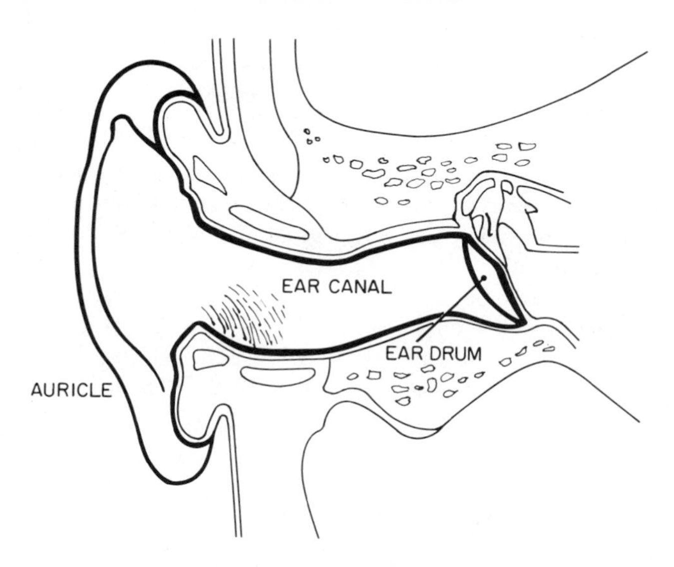

Figure 10.5 The external ear.

shape and size of ear canals differ significantly between individuals and even between ears of the same individual. The average length of the ear canal is about one inch, and when closed at one end by the eardrum, it functions in free field to produce a quarter-wavelength resonance at a frequency of about 3000 Hz. At a frequency of 3000 Hz, this resonance increases the response of the ear by about 10 dB.

The hairs at the outer end of the ear canal help to keep out dust and dirt; further into the canal are the wax-secreting glands. Normally, ear wax flows toward the entrance of the ear canal, carrying with it the dust and dirt that accumulate in the canal. The normal flow of wax may be interrupted by changes in body chemistry that can cause the wax to become hard and to build up within the ear. Too much cleaning or the prolonged use of ear plugs may cause increased production of wax, and when the wax builds up to the point of occluding the canal, a conductive loss of hearing will result. Any buildup of wax deep within the ear canal should be removed very carefully by a well-trained person, to prevent damage to the eardrum and middle ear structures.

During welding or grinding operations it is not uncommon for a spark to enter the ear canal and burn the canal or a large portion of the eardrum. Although very effective surgical procedures have been developed to repair or replace the eardrum, this painful and costly accident can be prevented by wearing ear protectors.

The surface of the external ear canal is extremely delicate and easily irritated. Cleaning or scratching with match sticks, nails, hairpins, and other objects, can break the skin and cause a very painful and persistent infection. Infections can cause swelling of the canal walls and, occasionally, a loss of hearing when the canal swells shut. An infected ear should be given prompt attention by a physician.

3.2 Eardrum

The eardrum is a very thin and delicate membrane that responds to the very low sound pressures at the lower threshold of normal hearing; yet it is seldom damaged by common high level noises. Although an eardrum may be damaged by a large displacement resulting from the force of an explosion or a rapid change in ambient pressure, the often repeated statement "the noise was so loud it almost burst my eardrums" is rarely true for common, steady state noise exposures. When an eardrum is ruptured, the attached middle-ear ossicle bones may be dislocated; therefore the eardrum should be carefully examined immediately after the injury occurs to determine whether realignment of the ossicle bones is necessary. In a high percentage of cases, surgical procedures are successful in realigning dislocated ossicles, so that little or no significant loss in hearing acuity results from this injury.

3.3 Middle Ear

The air-filled space between the eardrum and the inner ear is called the middle ear (Figure 10.6). The middle ear contains three small bones—the malleus (hammer), the incus (anvil), and the stapes (stirrups)—that mechanically connect the eardrum to the oval window of the inner ear. The eardrum has an area about 20 times that of the oval window, and the ossicles provide a mechanical advantage of about 3; therefore the

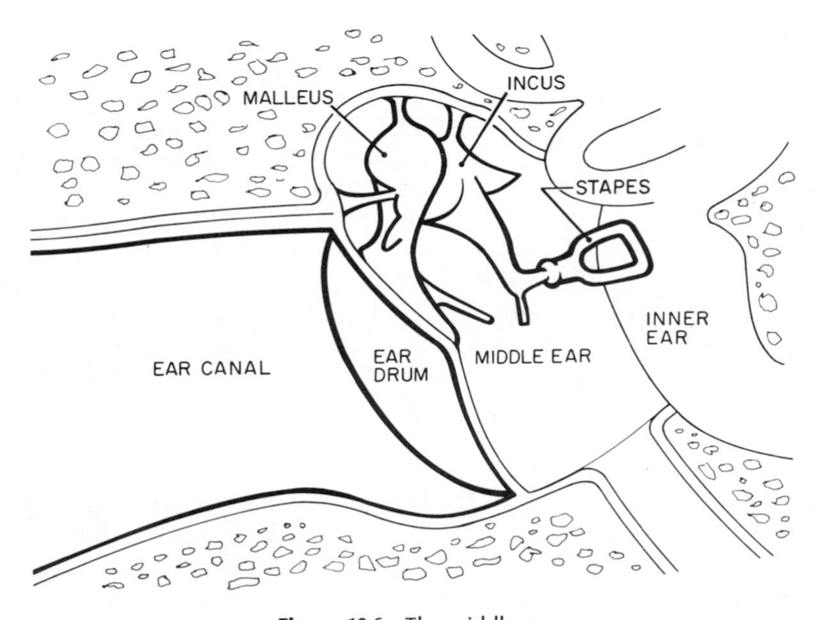

Figure 10.6 The middle ear.

overall mechanical advantage of the middle-ear structure is $3 \times 20 = 60$ at the natural resonant frequency of the system. This complex system also acts as an ear protector by mismatching impedances through the involuntary relaxation of coupling efficiency between the ossicle bones. The reaction time for the middle-ear system is approximately 10 msec.

The most common problem encountered in the middle ear is infection. This dark, damp, air-filled space is completely enclosed except for the small eustachian tube that connects the space to the back of the throat; thus it is very susceptible to infection, particularly in teenagers. If the eustachian tube is closed as a result of an infection or an allergy (Figure 10.6), the pressure inside the middle ear cannot be equalized with that of the surrounding atmosphere. In such an event, a significant change in atmospheric pressure, such as that encountered in an airplane or when driving in mountainous territory, may produce a loss of hearing sensitivity and extreme discomfort as a result of the displacement of the eardrum toward the low pressure side. Even a healthy ear may suffer a temporary loss of hearing sensitivity if the eustachian tube becomes blocked, but this loss of hearing can be restored simply by swallowing or chewing gum to momentarily open the eustachian tube.

Another middle-ear problem may result from an abnormal bone growth (otosclerosis) around the ossicle bones, restricting their normal movement. The cause of otosclerosis is not totally understood, but heredity is considered to be an important factor. The conductive type of hearing loss that results from otosclerosis is generally observed first at low frequencies, it then extends to higher frequencies, and eventually may result in a severe overall loss in hearing sensitivity. Hearing aids may often restore hearing sensitivity lost as a result of otosclerosis, but effective surgical procedures have been refined to such a point that they are now often recommended. An important side benefit of an effective hearing conservation program is the early detection of such hearing impairments as otosclerosis.

3.4 Inner Ear

The inner ear is completely surrounded by bone (Figure 10.7). One end of the space inside the bony shell of the inner ear is shaped like a snail shell; it contains the cochlea. The other end of the inner ear has the shape of three semicircular loops. The fluid-filled cochlea serves to detect and analyze incoming sound signals and to translate them into nerve impulses that are transmitted to the brain. The semicircular canals contain sensors for balance and orientation.

In operation, sound energy is coupled into the inner ear by the stapes, whose base is coupled into the oval window of the cochlea. Both the oval window and the round window located below it are covered by a thin, elastic membrane to contain the few drops of fluid within the cochlea. As the stapes forces the oval window in and out with the dynamic characteristics of the incident sound, the fluid of the cochlea is moved with the same characteristic motions. Thousands of hair cells located along the two and one-

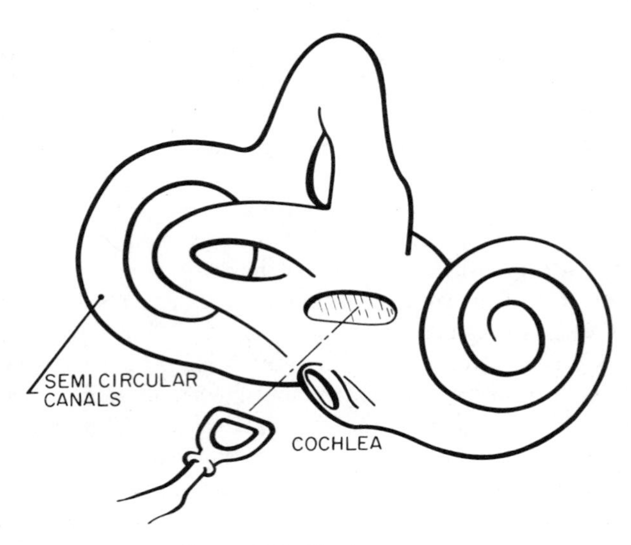

Figure 10.7 The inner ear.

half turns of the cochlea detect and analyze these motions and translate them into nerve impulses, which are transmitted to the brain for further analysis and interpretation.

The hair cells within the cochlea may be damaged by old age, disease, certain drugs, and exposure to high levels of noise. Unfortunately the characteristics of the hearing losses resulting from these various causes are often very similar, and it is impossible to determine the etiology of a particular case.

3.5 How Noise Damages Hearing

Noise-induced hearing loss may be temporary or permanent depending on the level and frequency characteristics of the noise, the duration of exposures, and the susceptibility of the individual. Usually temporary losses of hearing sensitivity diminish, and the original sensitivities are restored within about 16 hours (1–3). Permanent losses are irreversible and cannot be corrected by conventional surgical or therapeutic procedures.

Noise-induced damage within the inner ear generally occurs in hair cells located within the cochlea. Hearing acuity is generally affected first in the frequency range from 2000 to 6000 Hz, with most affected persons showing a loss or "dip" at 4000 Hz. If high level exposures are continued, the loss of hearing will further increase around 4000 Hz and spread to lower frequencies.

3.6 The Problem: At Work

Comprehensive data are not available for an accurate determination of the number of people who have some degree of noise-induced hearing impairment. However studies of relatively small groups indicate that workers in many industrial areas have sufficient noise exposures to cause significant hearing impairments (1–3).

The best estimates of the number of persons who have significant hearing impairment as a result of overexposure to noise are based on a comparison of the number of those with hearing impairments found in high noise work areas and members of the general population, who have relatively low noise exposures (25). These studies show that significant hearing impairments for industrial populations are 10 to 30 percent greater for all ages than for general populations that have relatively low level noise exposures. At age 55, for example, 22 percent of a group that has had low noise exposures may show significant hearing impairment, whereas in an industrial high noise exposure group, the figure is 46 percent. Significant hearing loss is defined in all cases to be greater than 25 dB hearing level (referenced to the American National Standards Institute (ANSI) Specifications (53.6, 1969), averaged at 500, 1000, and 2000 Hz.

Noise-induced hearing loss is a particularly difficult problem because a person usually does not know that he is having his hearing damaged, and as a rule the damage develops over a long period of time; thus the loss of hearing may not be apparent until a considerable amount of damage has been done. Even after a significant amount of damage, a person with noise-induced hearing loss is able to hear common, low frequency (vowel) sounds very well, but he misses the high frequencies (consonants) in speech. He can hear people talking at loudness levels that are nearly normal, but he may be unable to understand what they are saying. A noise-induced hearing loss becomes particularly noticeable when speech communication is attempted in noisy places, such as in a room where many people are talking or a radio is playing loudly, or in a car moving at high speed with the windows open.

3.7 The Problem: Away from Work

An additional concern to the supervisors of industrial health and safety programs is that many employees may be exposed to harmful noises while away from work. Many people are often exposed to potentially hazardous noises such as might be found from shooting guns, working with power tools, mowing lawns, riding airplanes, riding subways, driving in or observing automobile or motorcycle races, listening to loud music, or even riding in a car at high speed with open windows. Obviously a hearing conservation program cannot be effective unless each individual observes the rules limiting noise exposure both at work and away from work. Therefore all hearing conservation programs must emphasize the need for constant awareness of noise hazards, and this can be done only by a continuing program of education and enforcement.

3.8 Hearing Testing

The only way to monitor the overall effectiveness of a hearing conservation program is to check periodically the hearing acuity of all persons exposed to potentially hazardous noises. For industrial hearing conservation programs, hearing acuity is checked by the measurement of pure tone, air conduction hearing thresholds. Accuracy in the measurement of air conduction thresholds requires an accurately calibrated audiometer, a quiet test environment, and a well-trained audiometric technician or hearing conservationist. Requirements for audiometers and for test environments have been specified in standards published by the American National Standards Institute (27), and guidelines for training hearing conservationists have been established by the Council for Accreditation in Occupational Hearing Conservation (28). Thus these standards and guidelines should be used to establish procedures for accurate hearing threshold measurements.

3.8.1 Audiometers

The instrument used for measuring pure tone, air conduction hearing thresholds is called an audiometer. An audiometer may be designed for manual or self-recording (sometimes called automatic) operation. Both the manual and self-recording instruments must be operated by properly trained persons (28).

The ANSI Standard for Audiometers (27) provides all the information necessary to keep the instruments in proper operating condition. Calibration is often neglected because inaccuracies are seldom obvious, and there is a strong tendency to accept dial readings as being accurate, even if the instrument has not been checked for a long time. Audiometers may lose their specified accuracy very quickly if they are handled roughly; earphones are particularly susceptible to damage from rough handling. Since rough handling during shipment could cause unseen damage, a new instrument may be out of calibration when the customer receives it. In addition to the normal aging of components, heat may also cause audiometer inaccuracy; and a common cause of overheating is carrying an audiometer in a closed car on a warm day.

Dust or dirt inside the audiometer can cause switches to become noisy and to wear excessively, and sometimes poor electrical contact that affects instrument accuracy is produced. Dust covers should be used to protect the instrument when not in use, and the exterior of the instrument should be cleaned periodically to prevent dust and dirt from getting inside the case.

High humidity, salt air, and acid fumes may corrode electrical contacts in switches within an audiometer. The increased resistance that results from corroded contacts may cause electrical noise or affect the accuracy of the instrument.

If the operating characteristics of the audiometer change suddenly and considerably, it is an obvious indication that the instrument should be serviced. Many changes in the instrument occur slowly, however, and they may not be obvious; thus inaccurate measurements may result if the instrument is not calibrated periodically. An audiometer should be sent to a qualified laboratory for a complete calibration check whenever there is reason to suspect its accuracy; in any case, it should be checked at least once each year. Since much time can be wasted if hearing measurements are taken with an instrument that is out of calibration, the technician should check the audiometer daily using the following procedure:

1. All control knobs on the audiometer should be checked to be sure that they are tight on their shafts and not misaligned.

2. Earphone cords should be straightened so that there are no sharp bends or knots. Worn or cracked cord covers should be replaced. (Recalibration is not necessary when earphone cords are replaced.)

3. Earphone cushions should be replaced if they are not resilient or if cracks, bubbles, or crevices develop.

4. The audiometer calibration should be checked by measuring the hearing threshold at each test frequency of a normal-hearing person whose hearing levels are well known. If the hearing threshold of this test subject shows a change of 10 dB or more that cannot be traced to other causes, recalibration is indicated. Hearing thresholds may drop as much as 20 dB temporarily because of colds or other causes; therefore it is recommended that at least two normal-hearing persons be made available for this test. The hearing conservationist may serve as a test subject if he or she has normal hearing.

5. With the tone control set on 2000 Hz, the linearity of the hearing level control should be checked by listening to the earphone while slowly increasing the hearing level from threshold. Each 5 dB step should produce a noticeable increase in intensity without changes in tone quality or audible extraneous noise.

6. With the dials set at 2000 Hz and 60 dB, test the earphone cords electrically by listening while bending the cords along their length. Any scratching noise, intermittency, or change in test tone indicates a need for new cords.

7. With dials set at 2000 Hz and 60 dB, test the operation of the tone interrupter by listening to the earphones and operating the interrupter several times. No audible noises (such as clicks or scratches) or changes in tone quality should be heard when the interrupter switch is used.

8. With the hearing level control set at 60 dB and the test earphone jack disconnected from the audiometer, check by listening for extraneous noises from the case and earphone. While wearing the earphones, no noise should be heard when the tone control is switched to each test tone.

When an audiometer is sent to a laboratory for calibration, an understanding should be reached with the laboratory of the kind of calibration to be performed. Some laboratories calibrate audiometers by checking only a single hearing level for the various test tones, but this is not enough to assure accurate hearing threshold measurements. Hearing level accuracy should be checked for each test tone at each 5 dB interval throughout the operating range. For industrial applications, the range should cover from 10 to 70 dB on the ANSI 1969 hearing level scale. In addition, the several other specifications listed in the ANSI Standard for Audiometers (27) should be checked. These include checks on the attenuator linearity, interrupter operation, test tone accuracy and purity, and masking noises.

Many audiometers in use today have not been calibrated for several years, and a very high percentage of these instruments will not meet pertinent ANSI Standard Specifications. Furthermore, differences of several decibels can be found in threshold measurement data taken with two audiometers that meet specifications. Thus the need for dependable calibration services that will provide accurate adjustments and correction data is emphasized. The pertinent performance specifications may not be well understood by some of the laboratories offering calibration services; therefore a statement that the audiometer meets specifications should not be accepted without an explanation of the calibration procedure and a copy of the calibration data.

3.8.2 Test Rooms

The sound pressure level of the background noise in rooms used for measuring hearing thresholds must be limited to prevent masking effects that cause misleading, elevated threshold values. The maximum allowable sound pressure levels specified in an ANSI Standard (29) limit threshold elevation to no more than one decibel for one-ear listening. Limiting sound pressure levels are presented in the ANSI Standard in various bandwidths for ears covered with earphones mounted in MX-41/AR cushions and for ears uncovered. Normally the pertinent limiting sound pressure level data for industry is the octave-band sound pressure levels for ears covered with earphones mounted in MX-41/AR cushions (Table 10.4).

The maximum allowable sound pressure levels listed in Table 10.4 are very low and may be below the practical limit for room noise in many industrial locations. Thus it

Table 10.4. Maximum Allowable Sound Pressure Levels for One-Ear Listening for no more than a One-Decibel Threshold Elevation re the Reference Threshold Levels Given in ANSI S36-1969 (rounded to nearest 0.5 dB)

Test Tone Frequency (Hz)	Octave Band Levels (dB re 20 μPa)
125	34.5
250	23.0
500	21.5
750	22.5
1000	29.5
1500	29.0
2000	34.5
3000	39.0
4000	42.0
6000	41.0
8000	45.0

has been the accepted practice for industry to use the 10 dB hearing level as the lowest required hearing threshold measurement level (27). If it is determined that the low sound pressure levels (shown in Table 10.4) cannot be obtained in a practical way at the desired hearing test location to permit 0 dB hearing threshold levels, these background noise limits should be adjusted upward by 10 dB each, so that 10 dB hearing levels are the lowest levels measured.

In addition to the noise level requirements of Table 10.4, subjective tests should be made inside the closed test booth on location to determine that no noises (e.g., talking and heel clicking) are audible. Any noises audible in the test room will interfere with 0 dB hearing threshold measurements, and they must be eliminated. Short impulse-type noises, in particular, may be heard even though the measured sound pressure levels are below the limits in Table 10.4.

The noise reduction provided by a good prefabricated audiometric test booth should be adequate to permit limited range threshold measurements (at and above 10 dB hearing level) in most areas selected for hearing tests. However it is recommended that the burden of on-location performance be placed on the supplier, to be sure that the booth is erected properly. In any purchase agreement, there should be a guarantee that the booth will provide the required test environment when installed at the specified test site. An agreement of this kind will obviate any unexpected problems that might result from such factors as poor vibration isolation or faulty assembly.

In addition to the attenuation characteristics and the cost of prefabricated test rooms, other features should be considered. Size and appearance are important. Will opening and closing the door result in wearing of contact material, necessitating frequent replacement? Are the interior surfaces durable and easily cleaned? Is the door easily opened from the inside, so that subjects will not feel "locked in"? The observation window must be located to permit a subject's arm or hand response to be seen easily. If the booth is equipped with a ventilation system, the noise should be below the limits specified previously. Portability of the room is seldom an important factor, because test rooms are rarely moved.

A room with heavy masonry walls, floors, and ceilings will provide good noise isolation, provided care is taken to prevent leakage paths such as small cracks that might be found around windows, doors, electrical fixtures, and pipes. A 1.5 × 1.5 inch hole in a wall will transmit about the same amount of acoustical energy as 100 ft_2 of wall area that has a transmission loss of 40 dB. Wherever possible, all holes and cracks should be sealed permanently. Frequently used doors or windows should be sealed with flexible gaskets.

Leakage and reradiation of noise can occur through thin or light sections such as single-pane windows and doors. Additional noise reduction can be obtained by installing double doors or double-pane windows.

Structure-borne vibration can be transmitted through heavy walls and reradiated into the air of an enclosed space. If noise levels from structure-borne sources are high, "room within a room" construction may be required; this work should be undertaken only by experienced acousticians, however.

Although noise-absorbing materials within the room have little effect on the amount of noise leaking into the room from the outside, they will prevent noise levels from building up. For rooms of moderate size, adequate sound absorption is provided by a carpet on the floor and full drapes on two walls. If sound-absorbing treatment is used, it should be distributed on the ceiling and two adjoining walls for maximum effectiveness.

Two general principles are normally applied in the design of ventilation systems for hearing test rooms: one is to use long inlet and outlet ducts that are heavily lined with noise-absorbing material; the other is to use low air velocities that will minimize noise caused by air turbulence. The amount of noise reduction resulting from increased duct size, increased thickness of absorption materials, and reduced air velocity may be found in other literature.

3.8.3 Hearing Threshold Measurements

The purpose of hearing testing in industrial hearing conservation programs is to monitor the effectiveness of the various hearing conservation measures used. Since normally there is no need for diagnostic information that would require the use of sophisticated audiometric techniques, pure tone, air conduction hearing thresholds have been specified for industrial monitoring purposes (27).

Either manual or self-recording audiometers may be used to monitor hearing thresholds in industry. The recommended procedure for manual threshold testing is as follows:

1. Check the audiometer to be sure that the earphone cords and electrical power connections are properly made.
2. Before the earphones are placed on the subject, make the following adjustments on the audiometer:
    a. Place the power switch in the "on" position.
    b. Set the interrupter control switch to the "normally off" position.
    c. Set the "hearing level" control to 0 dB.
    d. Set the "frequency" or "tone" control to 1000 Hz.
    e. Set the earphone selector to the "right ear" position.
3. Seat the subject facing away from the audiometer so that he cannot see the instrument or the operator. Concise but complete instructions should be given regarding the responses expected during the test procedures. For example, the subject may be instructed to raise a finger or hand *immediately* when he hears a test tone, to hold this response until he can no longer hear the tone, then *immediately* lower his finger or hand. It is advisable to memorize a set of instructions to ensure that no aspect of the instructions will be omitted. During the instructions it is helpful to slowly raise and lower the hearing level control while the earphones are held about one foot from the subject's ear to illustrate the nature of a test tone. The subject should be advised that the pitch and level of the test tone will be changed during the test procedures.

4. Carefully place the earphones on the subject; they should be comfortably positioned directly over the entrance to the external ear canal. Be sure that the right earphone (red) is over the right ear and the left earphone (blue) is over the left ear. Earrings, hearing aids, glasses, and other obstacles must not be worn, and hair must be removed from between the ears and the earphones.
5. Reset the hearing level control to 0 dB.
6. Press and hold the interrupter switch (to present the tone to the subject) and slowly increase the test tone level (by turning the hearing level control) until the subject signals that the tone is heard.
7. Release the tone interrupter switch (tone off). The subject should signal that he no longer hears the tone.
8. Press the interrupter switch and present the tone again at the same level for about 3 seconds. The subject should signal that he hears the tone; if not, return to step 5.

In the following steps all adjustments of tone level must be made with the tone off. Also, all presentations of the tone shall be for a duration of about 3 seconds.

9. After the initial level of response has been established by two responses from the subject, lower the tone level about 10 dB to the nearest multiple of 5 on the hearing level scale (e.g., if the initial level is 32 dB, lower the hearing level control to 20 dB).
10. Present the test tone at the lower level found in step 9.
11. If the subject does not respond to the tone at this level, raise the level by 5 dB and present the tone again. If the subject responds to the tone at the level established in step 9, lower the tone level by 5 dB and present the tone again. Presenting the tone once at each of these levels is sufficient. Additional attempts to improve the threshold values should not be made.
12. Continue the bracketing procedure in 5 dB steps as described in step 11 until two positive and two negative responses are found between consecutive 5 dB test levels (e.g., two positive responses are found at a 15 dB hearing level and two negative responses are found at a 10 dB hearing level). The threshold level for the 1000 Hz test tone is taken as the level corresponding to the positive responses (i.e., 15 dB).
13. Record the threshold value at 1000 Hz on the audiogram record (Figure 10.8). Conventionally, a red "O" is used for recording thresholds for the right ear and a blue "X" for the left ear if a graphical presentation is to be used. If the values are to be recorded in tabular form, the right-ear thresholds are conventionally recorded in red and the left-ear values in blue.
14. Repeat steps 5 through 13 to determine and record thresholds for the right ear at test tones of 2000, 3000, 4000, 6000, 8000, 500, and 1000 Hz, in that order.
15. Set the earphone selector to the "left-ear" position and repeat steps 5 through 14.
16. Remove the earphones and reset the audiometer as indicated in step 2.

The recommended procedure for using the self-recording audiometer is as follows:

ENVIRONMENTAL ACOUSTICS LABORATORY

AUDIOGRAM RECORD

ENVIRONMENTAL ACOUSTICS LABORATORY
E.A.L.
THE PENNSYLVANIA STATE UNIVERSITY

Name ______________________ Title ______________________

Soc. Sec. No. ______________________ Birth date __________ Sex ______

Date	Tester	Right Ear							Left Ear							Comments
		1000	2000	3000	4000	6000	8000	500	1000	2000	3000	4000	6000	8000	500	

Job Assignment	Date	C/T	Job Assignment	Date	C/T

Figure 10.8 Serial-type audiogram form.

1. Check the audiometer to be sure that the earphone cords and the electrical power connections are properly made.
2. Before the earphones are placed on the subject, make the following adjustments on the audiometer:

   a. Place the power switch in the "on" position.
   b. Place the audiogram recording chart on the recording audiometer (Figure 10.9).
   c. Place the recording pen at the start position at the upper left-hand corner of the chart.

3. Seat the subject so that he can see neither the audiometer nor the operator. Concise but complete instructions should be given regarding the responses expected during the test procedures. For example, the subject should be instructed to push the control button when he hears a test tone and to hold it until he can no longer hear the tone, then *immediately* remove his finger from the button. It is advisable to memorize a set of instructions to ensure that no aspect of the instructions is left out. During the instructions it is helpful to slowly raise and lower the hearing level control while the earphones are held about one foot from the subject's ear to illustrate the nature of a test tone. The subject should be advised that the pitch and level of the test tone will be changed during the test procedures and that the test tone will automatically be transferred to his other ear during the test.

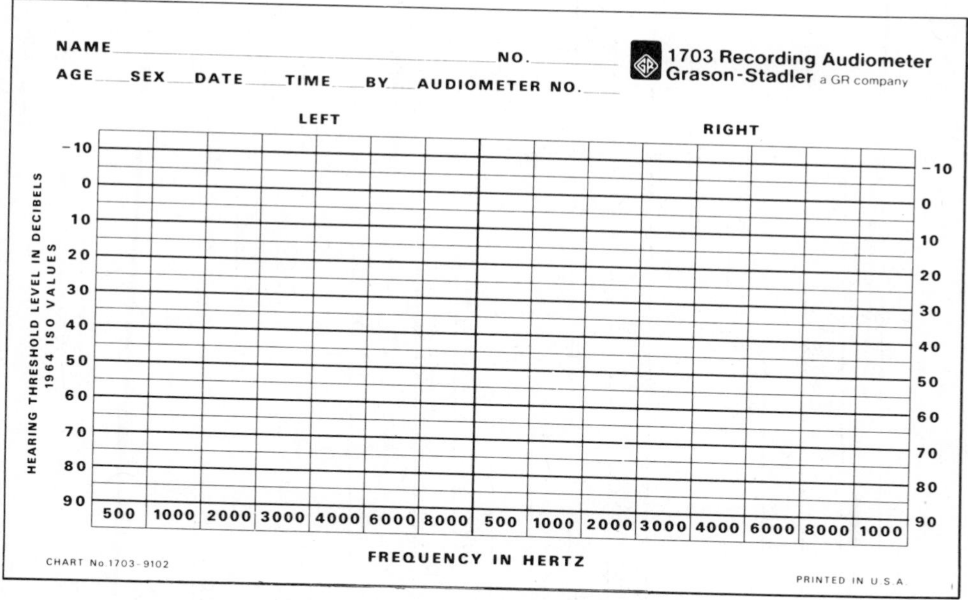

NAME ______ NO. ______
AGE ___ SEX ___ DATE ___ TIME ___ BY ___ AUDIOMETER NO. ___

1703 Recording Audiometer
Grason-Stadler a GR company

LEFT | RIGHT

HEARING THRESHOLD LEVEL IN DECIBELS
1964 ISO VALUES

-10, 0, 10, 20, 30, 40, 50, 60, 70, 80, 90

Left: 500 1000 2000 3000 4000 6000 8000
Right: 500 1000 2000 3000 4000 6000 8000 1000

FREQUENCY IN HERTZ

CHART No. 1703-9102

PRINTED IN U.S.A.

Figure 10.9 Recording-type audiogram form.

Name______ Date______ Recorded by______ Employee initial______

HISTORY	YES	NO	Comments
Have you had a previous hearing test?			
Have you ever had hearing trouble?			
Do you now have any trouble hearing?			
Have you ever worked in a noisy industry?			
Do you think you can hear better in your Right ear?			
or Left ear?			
Have you ever had noises in your ears?			
Have you ever had dizziness?			
Have you ever had a head injury?			
Has anyone in your family lost his hearing before age 50?			
Have you ever had measles, mumps, or scarlet fever?			
Do you have any allergies?			
Are you now taking or have you regularly taken drugs, antibiotics, or medication?			
Have you ever had an earache?			
Have your ears ever run? Right ear?			
Left ear?			
Have you been in the Military service? Describe			
Have you been exposed to any sort of gunfire? Describe			
Do you have a second job? Explain			
What hobbies do you have?			

Figure 10.10 Medical history form.

4. Carefully place the earphones on the subject; they should be comfortably positioned directly over the entrance to the external ear canal. Be sure that the right earphone (red) is over the right ear and the left earphone (blue) is over the left ear. Earrings, hearing aids, glasses, and other obstacles must not be worn, and hair must be removed from between the ears and the earphones.
5. Reset the recording pen to the start position.
6. Press the start button.
7. Observe the test continuously to be sure that the instructions are being followed.
8. When the test is completed, remove the earphones and recording chart, then reset the recording pen to the start position.

3.8.4 Records

The hearing conservationist's records must be complete and accurate if they are to have medicolegal significance. Records should be kept in ink, without erasures. If a mistake is made in recording, a line should be drawn through the erroneous entry, and the initials of the person making the recording should be placed above the line along with the date. In addition, the model, serial number, and calibration dates of the instruments used should be recorded. Records should be kept of periodic noise level measurements in the test space.

In addition to the threshold levels, an audiogram should have space provided for the recording of pertinent medical and noise exposure information. Some typical questions are listed on the form given in Figure 10.10. Additional materials may be found in References 30 to 40.

4 NOISE MEASUREMENT EQUIPMENT

A wide assortment of equipment is available for measuring sound that is propagated through the air, and this equipment must be properly selected if the objectives of the measurement are to be satisfied. The microphone must function reliably and accurately over the range of sound frequencies and levels that are to be measured. The operation of the microphone must not be impaired by the temperature and humidity conditions that may be encountered, and it must be readily adaptable for use with any instruments selected for the measurements. Of course, such equipment as sound level meters, amplifiers, attenuators, recorders, and various analyzers must also provide accurate and reliable measurements over the expected frequency and amplitude ranges.

Generally, complete measurement packages are available in the form of sound level meters and various analyzer types that are satisfactory for most purposes when the sound has reasonably steady and continuous characteristics. If the sound levels change significantly in a short time, however, the conventional sound level meters will not respond fast enough to afford accurate readings (41); therefore special instrumentations—such as impulse meters or oscilloscopes—are required. In all cases, accurate sound level measurements require a well-trained operator and accurate instruments.

4.1 The Sound Level Meter

The basic sound level meter consists of a microphone, an amplifier-attenuator circuit, and an indicating meter. The microphone transforms airborne acoustic pressure variations into electrical signals and feeds them to a carefully calibrated amplifier-attenuator circuit. The electrical signals are then directed through a logarithmic weighting network to an indicating meter, where the sound pressure is displayed in the form of levels above 20 μPa.

Most sound measuring instruments present the sound pressure levels in terms of its rms value, which is defined as the square root of the mean-squared displacements during one period. The rms value is useful for hearing conservation purposes because it is related to acoustic power and it correlates with human response. Also, the rms value of a random noise is directly proportional to the bandwidth; hence the rms value of any bandwidth is the logarithmic sum of the rms values of its component narrow bands. For example, octave band levels may be added logarithmically to find the overall level for the frequency range covered. The rms value of pure tones or sine waves is equal to 0.707 times the maximum value (Figure 10.11).

The rms values cannot be used to describe prominent peak pressures of noise that extend several decibels above a relatively constant background noise; maximum, or peak, values are used for this purpose. On the other hand, peak readings are of relatively little value for measuring sustained noises unless the wave form is known to be sinusoidal, because the relationship between the peak reading and the acoustic power changes with the complexity of the wave. As the wave form becomes more complex, the peak value can be as much as 25 dB above the measured rms value.

In addition to the rms and peak values, a rectified average value of the acoustic pressure is sometimes used for noise measurements. A rectified average value is an average taken over a period of time without regard to whether the instantaneous signal values are positive or negative. The rectified average value of a sine wave is equal to 0.636 times the peak value. For complex wave forms, the rectified average value may fall as much as 2 dB below the rms value. In some cases rectified average characteristics have been used in sound level meters by adjusting the output to read 1 dB above the rms level for sine wave signals, to ensure the average reading is always within 1 dB of the true rms value. Figure 10.11 compares rms, maximum, and rectified average values of a sinusoidal wave.

4.1.1 Meter Indication and Response Speed

The indicating meter of a sound level meter may have ballistic characteristics that are not constant over its entire dynamic range, or scale, and these will result in different readings, depending on the attenuator setting and the portion of the meter scale used.

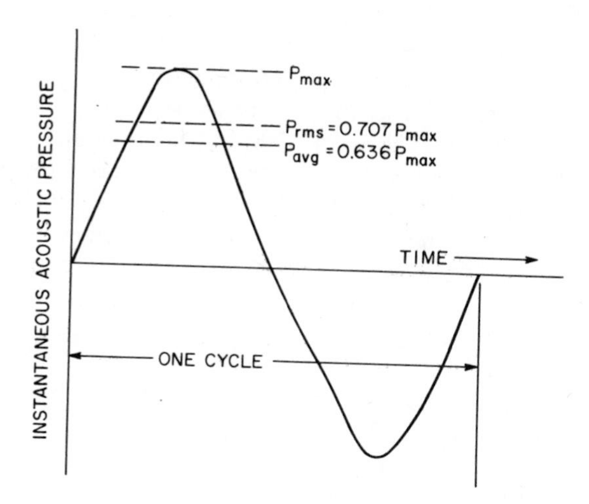

Figure 10.11 A comparison of maximum $p_{\max}$ or peak, root-mean-square p_{rms}, and rectified-average p_{avg} values of acoustic pressure.

When a difference in readings is noted, the reading using the higher part of the meter scale (the lowest attenuator setting) should be used, since the ballistics are generally more carefully controlled in this portion of the scale.

Most general-purpose (type II) sound level meters have fast and slow meter response characteristics that may be used for measuring sustained noise (42). The fast response enables the meter to reach within 4 dB of its calibrated reading for a 0.2-second pulse of 1000 Hz; thus it can be used to measure, with reasonable accuracy, noise levels that do not change substantially in periods less than 0.2 second. The slow response is intended to provide an averaging effect that will make widely fluctuating sound levels easier to read; however this setting will not provide accurate readings if the sound levels change significantly in less than 0.5 second.

4.1.2 Frequency-Weighting Networks

General-purpose sound level meters are normally equipped with three frequency-weighting networks (A, B, and C) that can be used to approximate the frequency distribution of noise over the audible spectrum (43). These three frequency weightings, given in Table 10.3, were chosen because (1) they approximate the response characteristics of the ear at various sound levels, and (2) they can be easily produced with a few common electronic components. A linear, flat, or overall response weights all frequencies equally.

The A-frequency weighting approximates the response characteristics of the ear for low level sound (i.e., below about 55 dB re 0.00002 N/m^2). The B-frequency weighting is intended to approximate the response of the ear for levels between 55 and 85 dB, and the C-frequency weighting corresponds to the response of the ear for levels above 85 dB.

In use, the frequency distribution of noise energy can be approximated by comparing the levels measured with each of the frequency weightings. For example, if the noise levels measured using the A and C networks are approximately equal, it can be reasoned that most of the noise energy is above 1000 Hz, because this is the only portion of the spectrum in which the weightings are similar. On the other hand, a large difference between these readings indicates that most of the energy will be found below 1000 Hz.

Many specific uses have been made of the individual weightings in addition to the frequency distribution of noise. In particular, the A-network has been given prominence in recent years as a means for estimating annoyance caused by noise and for estimating the risk of noise-induced hearing damage.

4.2 Microphones

The three basic types of microphone commonly used for noise measurements are piezoelectric, dynamic, and condenser. Each type has advantages and disadvantages that depend on the specific measurement situation; all three types can be made to meet the American National Standard Specification for Sound Meters (S1.4.1971) (44). The dynamic and piezoelectric microphones are normally less expensive than the condenser type and are provided as standard equipment with most low and medium-priced sound measuring instruments.

4.2.1 Frequency Response Characteristics

Piezoelectric microphones used with sound measuring equipment have reasonably flat response characteristics for low and middle audible frequencies; however the high frequency responses are not as flat, nor do they extend as far as dynamic or condenser types. Piezoelectric microphones of good quality are normally acceptable for a frequency range from about 15 to 8000 Hz. A few dynamic microphones have a somewhat more uniform frequency response than piezoelectric type above 1000 Hz, and their upper range may extend above 12,000 Hz. Condenser microphones of various sizes may be used with sound measuring instruments to cover a very wide frequency range from a few hertz to well over 100,000 Hz. Tolerance limits for sound level meters (including the microphone) are included in the ANSI Specification for Sound Level Meters (44).

4.2.2 Sound-Level Range

Most of the one-inch microphones supplied as standard equipment with sound level meters have a dynamic range of about 20 to 145 dB re 20 μPa, but the circuitry of the sound level meter may not permit measurements below levels of 40 dB. Special-purpose microphones designed for use at high noise levels usually have approximately the same dynamic range, but their lower overall sensitivity shifts the upper limit to higher levels than are possible with the standard microphones. For example, the ½, ¼, and ⅛ inch condenser microphones may have dynamic ranges of about 30 to 160 dB, 50 to 175 dB, and 60 to 185 dB, respectively.

4.2.3 Temperature and Humidity Effects

The Rochelle salt microphones that were supplied with older sound measuring equipment are easily damaged by extreme heat or humidity. These piezoelectric microphones can be permanently damaged if left in a closed car on a hot day; therefore extreme care should be taken in their use and storage. The barium titanate or lead-zirconate-titanate microphones that are furnished with most sound measuring equipment today are not damaged by exposure to normal ranges of temperature and humidity; however temporary erroneous readings may result from condensation when they are moved from very cold locations to warm, humid areas.

The response characteristics of the dynamic microphone are somewhat dependent on the ambient temperature, but over most of the audible frequency range the variation is less than about 1 dB for a 50°F change in temperature. The dynamic microphone is affected relatively little by humidity extremes, except for temporary erroneous readings that may result from condensation. Specific temperature-correction information for each microphone should be available from the manufacturer.

The condenser microphone is not permanently damaged by exposure to humidity extremes, but high humidity may cause temporary erroneous readings. The variation of condenser microphone sensitivity with temperature is approximately –0.04 dB/°F. Again, correction information for a specific microphone should be obtained from the manufacturer.

4.2.4 Microphone Directional Characteristics

Most noises encountered in industry are produced from many different noise sources and from their reflected energies; at any given position in these areas, therefore, noise is coming from many directions and often may be considered to be randomly incident upon any plane where a microphone diaphragm might be placed. For this reason, microphones are sometimes calibrated for randomly incident sound; depending on the design and purpose of the microphone, however, they may be calibrated for grazing incidence, perpendicular incidence, or for use in couplers (pressure calibration). Thus for accurate measurements, care must be taken to use the microphone in the manner specified by the manufacturer.

Microphones commonly used with sound measuring equipment are nearly omnidirectional for frequencies below 1000 Hz; however directional characteristics become important for frequencies above 1000 Hz (Figure 10.12). Therefore when measurements are to be made of high frequency noise produced by a directional noise source (i.e., where a high percentage of the noise energy is coming from one direction), the orientation of the microphone becomes very important—even though the microphone may be described as omnidirectional. A microphone calibrated with randomly incident sound should be pointed at an angle to the major noise source that is specified by the manufacturer. An angle of about 70° from the axis of the microphone is often used to produce characteristics similar to randomly incident waves, but the angle for each microphone should be supplied by the manufacturer. A free-field microphone is calibrated to

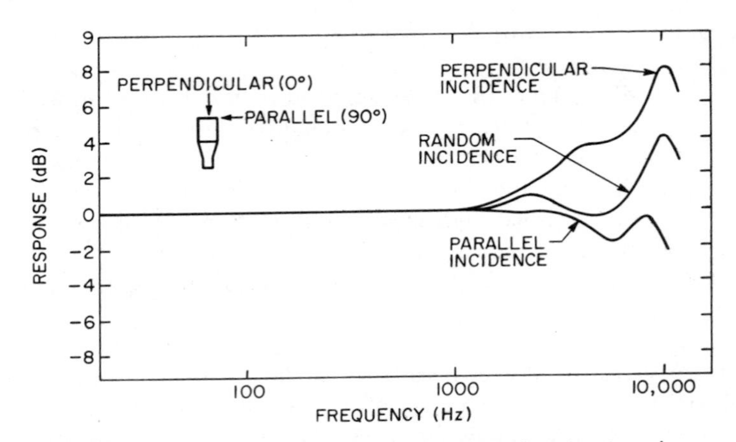

Figure 10.12 Directional characteristics of a piezoelectric microphone.

measure sounds perpendicularly incident to the microphone diaphragm; thus it should be pointed directly at the source to be measured. A pressure-type microphone is designed for use in a coupler such as those used for calibrating audiometers; however this microphone can be used to measure noise over most of the audible spectrum if the noise propagation is at grazing incidence to the diaphragm and the microphone calibration curve is used.

Directional characteristics of microphones may be used to advantage at times. For example, an improved signal-to-noise ratio may be obtained for sound pressure level measurements of a given source by using 0° incidence when high background levels are being produced by sources at other locations. Erroneous readings caused by reflected high frequency sound emitted by other sources but coming from the same direction may be checked with directional microphones by rotating the microphone about an axis coincident with the direction of incident sound. Reflected energy will be evidenced by a variation in level as the microphone is being rotated. The microphone orientation that corresponds to the lowest reading should be chosen, since the reflection error would be a minimum at this position.

Special-purpose microphones with sharp directional characteristics may be used to advantage in some locations. These microphones are particularly useful for locating specific high frequency noise sources in the presence of other noise sources.

4.2.5 Microphone Cables

Standard microphone cable with shielded and twisted wires should be used with a dynamic microphone to minimize electrical noise pickup. Usually no correction is needed when this cable is used between a dynamic microphone and its matching transformer unless the cable is longer than 100 ft.

Cable corrections may or may not be required for condenser microphones, depending on the preamplifier design and the overall calibration. Some condenser microphones are calibrated when mounted directly on the sound measuring equipment; others are calibrated with cables attached. Instructions with the microphones should provide this information.

A correction is normally required when a titanate-type piezoelectric microphone is used with a cable unless the microphone has a built-in circuit to lower its output impedance. When there is no built-in impedance-reducing circuit, a correction of about +7 dB must be added to the titanate microphone output when used with a 25-foot cable. The exact correction factor should be supplied with the microphone.

A correction factor that is a function of temperature must be applied when a Rochelle salt microphone is used with a cable. These correction factors are found in instruction manuals supplied by the instrument manufacturers.

4.3 Frequency Analyzers

In many instances the rough estimate of frequency-response characteristics that are provided by the weighting networks of the sound level meter does not give enough information. In such cases the output of the sound level meter can be fed into a suitable analyzer that will provide more specific frequency distribution characteristics of the sound pressure (45). The linear network of the sound level meter should be used when the output is to be fed to an analyzer. If the sound level meter does not have a linear network, the C-network may be used for analyses over the major portion of the audible spectrum (see Table 10.3).

4.3.1 Octave Band Analyzers

The octave band analyzer is the most common type of filter used for noise measurements related to hearing conservation. Octave bands are the widest of the common bandwidths used for analyses; thus they provide information of the spectral distribution of pressure with a minimum number of measurements.

An octave band is defined as any bandwidth having an upper band-edge frequency f_2 equal to twice the lower band-edge frequency f_1. The center frequency (geometric mean) of an octave band, or other bandwidths, is found from the square root of the product of the upper and lower band-edge frequencies. The specific band-edge frequencies for octave bands are arbitrarily chosen. The older instruments usually have a series of octave bands extending from 37.5 to 9600 Hz (37.5 to 75, 75 to 150, 150 to 300, . . . , 4800 to 9600). Newer octave band analyzers may be designed for octave bands centered at 31.5, 63, 125, 250, 500, . . . , 8000 Hz according to American Standard Preferred Frequencies for Acoustical Measurements (S1.6-1967) (46). Octave band-edge frequencies corresponding to the preferred center frequencies can be calculated using two equations with two unknowns. The first equation comes from the definition of an octave band: the upper band-edge frequency is equal to twice the lower band-edge frequency

($f_2 = 2f_1$). The second equation describes the center frequency f_c in terms of the band-edge frequencies ($f_c = \sqrt{f_1 f_2}$). For example, the band-edge frequencies corresponding to a center frequency of 1000 Hz can be calculated as follows:

$$f_c = \sqrt{f_1 f_2} \tag{14}$$

and

$$f_2 = 2f_1 \tag{15}$$

From Equation 14, $f_c = 1000 = \sqrt{f_1 f_2}$ and, substituting Equation 15 into Equation 14, $1000 = \sqrt{f_1 \times 2f_1} = f_1\sqrt{2}$. Therefore $f_1 = 1000/1.414 = 707$ Hz and, from Equation 15, $f_2 = 2 \times 707 = 1414$ Hz.

Most combinations of sound level meter and octave band analyzer have separate attenuators on each instrument. In these cases it is always important to first measure the overall noise, and leave the attenuator setting of the meter at this position for all analyzer measurements. This procedure prevents overloading of the sound level meter and resulting erroneous readings. If the overall level changes significantly during a series of measurements, the entire procedure must be repeated.

4.3.2 Half-Octave and Third-Octave Analyzers

When even more specific information about the pressure spectral distribution is desired than that provided by octave bands, narrower band analyzers must be used. The number of measurements necessary to cover the overall frequency range will be directly related to the bandwidth of the analyzer; thus a compromise must be reached between the resolution required and the time necessary for the measurements.

Half-octave and third-octave filters are the next steps in resolution above octave band analyzers. A half-octave is a bandwidth with an upper edge frequency equal to the $\sqrt{2}$ times its lower edge frequency. A third-octave has an upper edge frequency that is $\sqrt[3]{2}$ times its lower edge frequency.

4.3.3 Adjustable Bandwidth Broad Band Analyzers

Some analyzers are designed with independently adjustable upper and lower band-edge frequencies. This design permits a wide selection of bandwidths in octaves, multiples of octaves, or fractions of an octave. The smallest fraction of an octave usually available on these adjustable bandwidth analyzers is about one-tenth, and the largest extends up to the overall reading.

In addition to the obvious advantage of being able to choose the proper bandwidth for a particular job, these analyzers permit the selection of any octave band rather than a preselected series of octaves. For example, they can be adjusted to the older series of octaves (75–150, 150–300, etc.) or to bands with preferred center frequencies (125, 250, 500, etc.) as described in recent ANSI Standards (46). The disadvantage of these instruments is their relatively large size.

4.3.4 Narrow Band Analyzers

Analyzers with bandwidths narrower than tenth-octaves are normally referred to as narrow band analyzers. Narrow band analyzers are usually continuously adjustable, and they are classified either as constant percentage bandwidth or as constant bandwidth types.

The constant percentage, narrow band analyzer is similar to the broad band fractional octave analyzer in that its bandwidth varies with frequency. As its name indicates, the bandwidth of the constant percentage analyzer is a constant percentage of the center frequency to which it is tuned. Typically, a bandwidth of about $\frac{1}{30}$-octave might be selected with these analyzers.

The bandwidth of a constant bandwidth analyzer remains constant for all center frequencies over the spectrum. Provision may be made on some instruments to vary the bandwidth, but typically the bandwidth of a constant bandwidth analyzer remains constant at a few hertz.

The constant bandwidth analyzer normally provides a narrower bandwidth and better discrimination outside the passband than the constant percentage analyzer; therefore it is often the best choice when discrete frequency components are to be measured. Also, it usually covers the entire spectrum with a single dial sweep, thus facilitating coupling to recorders for automatic analysis. Most constant percentage analyzers require band switching to sweep the audible spectrum. On the other hand, caution must be exercised to avoid serious errors when constant bandwidth analyzers are used to analyze noises that have frequency modulation, or warbling, of components (47). Frequency-modulated noises are commonly produced by reciprocating-type noise sources in some machinery. Frequency-modulated noise is not a major problem if constant percentage analyzers are used.

4.4 Measurement of Impulse or Impact Noise

The inertia of the indicating meters of general-purpose sound level meters prevents accurate, direct measurements of single-impulse noises that change level significantly in less than 0.2 second. Typical noises with short time constants are those produced by drop hammers, explosives, and other noises with short, sharp, clanging characteristics. A low inertia device such as an oscilloscope must be used to measure these impulse-type noises if detailed information is required.

Impulse noise characteristics may be measured directly from a calibrated oscilloscope with a long persistance screen, or photographic accessories may be used to obtain permanent records. The oscilloscope is usually connected to the output of a sound level meter having a wide frequency response and calibrated with a known sound level of sinusoidal characteristics. The screen of the oscilloscope can be calibrated directly in decibels (rms) by comparing the oscilloscope deflection produced by a sinusoidal signal with the reading of the sound level meter. Several calibration points may be fixed on the oscilloscope screen by providing various signal levels into the sound level meter, or the scale may be determined from a single calibration level by using linear equivalents to decibels. For example, for a sine wave signal, half of a given deflection on the oscilloscope will be equivalent to a 6 dB drop in level, and 0.316 times the deflection will be the equivalent to a 10 dB drop in level. These equivalent values may be calculated from the equation

$$\text{drop in Level (dB)} = 20 \log_{10} \frac{d_1}{d_2}$$

where d_1 and d_2 are the small and large linear screen deflectors being compared. It should be noted that this calibration, using a sine wave, is for convenience, and that a constant factor of 3 dB must be added to the rms calibration to obtain the true instantaneous peak values for sine waves. The relationship of rms to peak values is more complex for nonsinusoidal waves (41–43).

While using an oscilloscope driven by a sound level meter, care should be taken to prevent errors resulting from overloading. If the oscilloscope deflections show a sharp clipping action at a given amplitude, the attenuator settings on one or both instruments may require adjustment upward. Also, a check should be made to determine whether the indicating meter of the sound level meter affects the wave form produced on the oscilloscope. This may be done by switching the meter out of the circuit, to a battery-check position, and observing the wave form. If the oscilloscope wave form is changed in any way by the indicating meter, it should be removed from the circuit each time a deflection is measured on the oscilloscope.

The oscilloscope is inconvenient to use in many field applications because it is relatively large and complex. Also, most oscilloscopes require ac power, and the supply voltage may vary in the field, causing changes in calibration. For these applications it is often convenient to use peak-reading impact noise analyzers, which may be connected to the output of a sound level meter. These battery-driven instruments do not provide as much information as the oscilloscope trace, but they are often adequate. The electrical energy produced by an impulse noise is stored for a short time by these instruments in capacitor-type circuits, permitting the acquisition of information on the maximum peak level, on the average level over a period of time, and on the duration of the impact noise. As with the oscilloscope, care must be taken not to overload the sound level meter driving these impact noise meters.

4.5 Magnetic Field and Vibration Effects

The response of sound level meters and analyzers may be affected by the strong alternating magnetic fields found around some electrical equipment. Dynamic microphones, coils, and transformers are particularly susceptible to hum pickup from these fields, but other types of microphone are also susceptible. Some of the newer dynamic microphones have hum-bucking circuits that minimize this pickup, but caution should be used in all cases. To test for hum pickup, disconnect the suspected component and check for a drop in level on the indicating meter. It is good practice to follow the equipment manufacturer's procedure for this check.

The magnetic fields produced by dynamic microphones may attract meter filings that will change the frequency-response characteristics; therefore dynamic microphones are not a good choice for measurements in metal-working shops.

Vibration of the microphone or measuring instrument may cause erroneous readings, and in some cases strong vibrations may permanently damage the equipment. It is always good practice to mechanically isolate sound measuring equipment from any vibrating surface. Holding the equipment in the hands or placing it on a foam rubber pad is satisfactory in most cases. To determine whether the equipment is being affected by vibration, observe the meter reading while the noise source is shut off, if this can be done without changing the vibration. If the meter reading drops by more than 10 dB, the effects of vibration are not significant. If the noise cannot be shut off without changing the vibration, the same result can be obtained by replacing the microphone cartridge with a dummy microphone. The equipment manufacturer can supply information on building a dummy microphone.

4.6 Tape Recording of Noise

It is sometimes convenient to record a noise to be analyzed at a later date. This is particularly helpful when lengthy narrow band analyses are to be made, or when very short, transient-type noises are to be analyzed. To avoid errors, however, extreme care must be taken in the calibration and use of the recorder. Also, direct sound pressure measurement and analysis should be made during the recording procedure so that the operator will know when additional measurements or data are necessary.

Many of the professional or broadcast-quality tape recorders are satisfactory for noise-recording applications; however the microphone must have the proper characteristics. Frequently the specifications given for a tape recorder do not include the microphone characteristics, and the microphone may be of very poor quality. When the tape recorder is not specifically built for measuring noise, it is usually good practice to connect the input of the recorder to the output of a properly calibrated sound level meter. As is the case when attaching any accessory equipment to sound level meters, it is important that the impedances be properly matched. The bridging input of a tape recorder is satisfactory for the output circuits of most sound level meters.

When a tape recorder is used to record noises that have no high prominent peaks, the recording level usually should be set so that the recorder meter (VU meter) reads between −6 and 0 dB. This setting assumes that a sinusoidal signal reading of +10 dB on the VU meter will correspond to about 2 or 3 percent distortion according to standard recording practice.

If the recorded noise has prominent peaks, it is good practice to make at least two additional recordings with the input attenuator set, giving recording levels between −6 and 0, −16 and −10, and −26 and −20 VU. If there is less than 10 dB between any two of these adjacent 10 dB steps, overloading has occurred at the higher recorded level, and the lower of the two steps should be used.

It is important to calibrate the combination of tape recorder and sound level meter at known level and tone control settings throughout the frequency range before the recordings are made. Before each series of measurements, a pressure level calibration should be made by noting the overall sound pressure level reading corresponding to the recording, along with the tape recorder dial settings. Also, it is good practice to note for each recording the type and serial numbers of the microphone and sound level meter, the location and orientation of the microphone, the description of the noise source and surroundings, and other pertinent information. It is often convenient to record these facts orally on the tape to prevent the information from being lost or confused with other tapes.

4.7 Graphic Level Recording

A graphic level recorder may be coupled to the output of a sound level meter or analyzer to provide a continuous written record of the output level. Recent graphic level recorders provide records in the conventional rms logarithmic form used by sound level meters; thus the data may be read directly in decibels. Some older recorders use rectified average response characteristics, and corrections must be made to convert these recordings to true rms values. As with sound level meters, these recorders are intended primarily for the recording of sustained noises without short or prominent impact-type peak levels. The equipment manufacturer or instruction manuals should be consulted to determine the limitations of each graphic level recorder.

4.8 Instrument Calibration

If valid data are to be obtained, it is essential that all equipment for the measurement and analysis of sound be in calibration. When equipment is purchased from the manufacturer, it should have been calibrated to the pertinent ANSI specifications (44, 45). However it is the responsibility of the equipment user to keep the instrument in calibration by periodic checks.

Most general-purpose sound measuring instruments have built-in calibration circuits that may be used for checking electrical gain. Most sound level meters have built-in or accessory acoustical calibrators that may be used to check the overall acoustical and electrical performance at one or more frequencies. These electrical and acoustical calibrations should be made according to the manufacturer's instructions at the beginning and at the end of each day's measurements. A battery check should also be made at these times. These calibration procedures cannot be considered to be of high absolute accuracy, nor will they allow the operator to detect changes in performance at frequencies other than that used for calibration; they will serve to warn of the most common instrument failures, however, thus avoiding many invalid measurements.

Periodically sound measuring instruments should be sent back to the manufacturer or to a competent acoustical laboratory for a complete overall calibration at several frequencies throughout the instrument range. These calibrations require technical competence and the use of expensive equipment. How frequently the more complete calibrations should be made depends on the purpose of the measurements and how roughly the instruments have been handled. In most cases it is good practice to have a complete calibration performed every 6 months, or at least once a year. A complete calibration should be made automatically if any unusual change (more than 2 dB) is seen in the daily calibration.

5 NOISE CONTROL PROCEDURES

Noise control procedures may take such varied forms as engineering, personal protection, and administrative approaches. As long as they are both practical and economically feasible, engineering procedures are by far the most desirable because they deal with predictable inanimate objects rather than relying on the uncertain cooperation of people. Unfortunately it is not always possible to obtain enough noise reduction with engineering procedures, and it may be necessary to employ a mixture of all these approaches to attain the desired exposure limits.

5.1 Engineering Procedures

Engineering noise control measures can be used most effectively at the design stage of potentially noisy equipment. Unfortunately there has not been a strong demand for quiet equipment, and available technology has not been used to full advantage in product design. The purchase orders for potentially noisy equipment should contain adequate specifications to supply an incentive for the design of quiet products.

The use of engineering control procedures on noisy equipment already in operation may be difficult, and, in many cases, ineffective. However, it is not economically feasible to replace many long-lived, noisy machines with quiet units; therefore it is imperative that the most efficient use be made of existing quieting techniques. These techniques include the use of noise-absorbing materials, enclosures, or barriers, mechanical isolation, damping, reduced driving force, driving system modifications, and muffling.

5.1.1 Absorption

Machines that contain cams, gears, reciprocating components, and metal stops are often located in large, acoustically reverbant areas that reflect and build up noise levels in the room. Frequently the noise levels in adjoining areas can be reduced significantly by using sound-absorbing materials. The type, amount, configuration, and placement of absorption materials depend on the specific application; however the choice of absorbing materials can be guided by the absorption coefficients listed in Table 10.5.

Table 10.5. Sound Absorption Coefficients[a] of Materials (48)

Materials	Frequency (Hz)					
	125	250	500	1000	2000	4000
Brick	0.01	0.01	0.01	0.01	0.02	0.02
Glazed	0.03	0.03	0.03	0.04	0.05	0.07
Unglazed	0.01	0.01	0.02	0.02	0.02	0.03
Unglazed, painted						
Carpet						
Heavy, on concrete	0.02	0.06	0.14	0.37	0.60	0.65
On 40 oz hairfelt or foam rubber (carpet has coarse backing)	0.08	0.24	0.57	0.69	0.71	0.73
With impermeable latex backing on 40 oz hairfelt or foam rubber	0.08	0.27	0.39	0.34	0.48	0.63
Concrete block						
Coarse	0.36	0.27	0.39	0.34	0.48	0.63
Painted	0.10	0.05	0.06	0.07	0.09	0.08
Poured	0.01	0.01	0.02	0.02	0.02	0.03
Fabrics						
Light velour: 10 oz/yd² hung straight, in contact with wall	0.03	0.04	0.11	0.17	0.24	0.35
Medium velour: 14 oz/yd² draped to half-area	0.07	0.31	0.49	0.75	0.70	0.60
Heavy velour: 18 oz/yd² draped to half-area	0.14	0.35	0.55	0.72	0.70	0.65
Floors						
Concrete or terrazzo	0.01	0.01	0.015	0.02	0.02	0.02
Linoleum, asphalt, rubber, or cork tile on concrete	0.02	0.03	0.03	0.03	0.03	0.02
Wood	0.15	0.11	0.10	0.07	0.06	0.07
Wood parquet in asphalt on concrete	0.04	0.04	0.07	0.06	0.07	—
Glass						
Large panes of heavy plate glass	0.18	0.06	0.04	0.03	0.02	0.02
Ordinary window glass	0.35	0.25	0.18	0.12	0.07	0.04
Glass fiber						
Mounted with impervious backing: 3 lb/ft³, 1 in. thick	0.14	0.55	0.67	0.97	0.90	0.85
Mounted with impervious backing: 3 lb/ft³, 2 in. thick	0.39	0.78	0.94	0.96	0.85	0.84
Mounted with impervious backing: 3 lb/ft³, 3 in. thick	0.43	0.91	0.99	0.98	0.95	0.93
Gypsum board: ½ in. nailed to 2 × 4's, 16 in. o.c.	0.29	0.10	0.05	0.04	0.07	0.09
Marble	0.01	0.01	0.01	0.01	0.02	0.02
Openings						
Stage, depending on furnishings			0.25–0.75			
Deep balcony, upholstered seats			0.50–1.00			
Grills, ventilating			0.15–0.50			
Grills, ventilating to outside			1.00			

Table 10.5. (Continued)

Materials	Frequency (Hz)					
	125	250	500	1000	2000	4000
Plaster						
Gypsum or lime, smooth finish on tile or brick	0.013	0.015	0.02	0.03	0.04	0.05
Gypsum or lime, rough finish on lath	0.14	0.10	0.06	0.05	0.04	0.03
With smooth finish	0.14	0.10	0.06	0.04	0.04	0.03
Plywood paneling: $\frac{3}{8}$ in. thick	0.28	0.22	0.17	0.09	0.10	0.11
Sand						
Dry, 4 in. thick	0.15	0.35	0.40	0.50	0.55	0.80
Dry, 12 in. thick	0.20	0.30	0.40	0.50	0.60	0.75
14 lb H_2O/ft^3, 4 in. thick	0.05	0.05	0.05	0.05	0.05	0.15
Water	0.01	0.01	0.01	0.01	0.02	0.02

[a] The absorption coefficient *a* of a surface that is exposed to a sound field is the ratio of the sound energy absorbed by the surface to the sound energy incident upon the surface. For instance, if 55 percent of the incident sound energy is absorbed when it strikes the surface of a material, the *a* value of that material would be 0.55. Since the *a* of a material varies according to many factors, such as frequency of the noise, density, type of mounting, and surface conditions, be sure to use the *a* for the exact conditions to be encountered and from performance data listings such as shown below. For a more comprehensive list of the absorption coefficients of acoustical materials, refer to the bulletin published yearly by the Acoustical Materials Association, 335 East 45th Street, New York, N.Y. 10017.

Example: The noise produced by 10 wire-cutting machines around the periphery of a 20 × 60 × 75 foot reverberant room was reduced as shown below by the installation of absorption material above the machines.

OB center frequency (Hz)	31.5	63	125	250	500	1000	2000	4000	8000
NR (dB)	—	—	—	2	5	5	10	12	10

Example: Several motor-generator sets were producing excessive noise levels in a large reverberant room. Noise levels at significant distances away from the generators were reduced as shown below by hanging 6 lb/ft^3 Fiberglas baffles in rows just above the level of lights on 3-foot centers. (These baffles can often be completely encased in polyethylene or Mylar film without significantly reducing their effectiveness.)

OB center frequency (Hz)	31.5	63	125	250	500	1000	2000	4000	8000
NR (dB)	—	4	7	9	10	7	8	8	3

5.1.2 Noise Barriers and Enclosures

The noise reduction that can be attained with barriers depends on the characteristics of the noise source, the configuration and materials used for the barrier, and the acoustical environment on each side of the barrier. The material used for noise barriers may be described generally in terms of its transmission loss TL (Table 10.6), but all other factors must be considered for specific problems.

The amounts of noise reduction achieved by various configurations of specific barrier or enclosure materials may vary significantly. Generally, a single-wall barrier with no openings between the source and the person exposed might result in a 2 to 5 dB reduction in the low frequencies and a 10 to 15 dB reduction in the high frequencies. If the noise source and the observer are close to the barrier, higher noise reduction values are possible. The effects of two- or three-sided barriers are difficult to predict on a general basis; however well-designed partial enclosures may provide a noise reduction of about 5 to 10 dB in the high frequencies. Complete enclosures of practical designs may provide a noise reduction in excess of 10 to 15 dB in the low frequencies and in excess of 30 dB in the high frequencies. Precautions must be taken with any barrier or enclosure to be sure that there are no unnecessary openings. Figure 10.13 shows the average TL of a single barrier as a function of barrier mass and percentage of open area.

Example: An operator positioned close to a punch press that uses jets of compressed air to blow foreign particles from the die was exposed to excessive noise

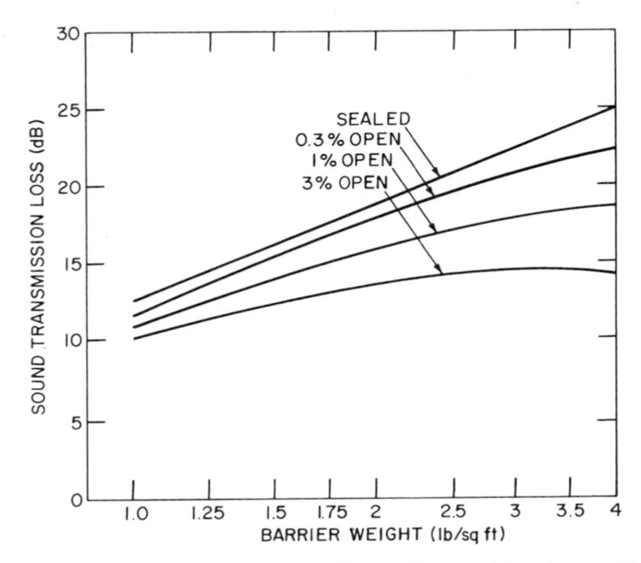

Figure 10.13 Average sound transmission loss of a single sound barrier as a function of barrier mass and percentage of open area.

Table 10.6. Sound Transmission Loss (dB) of General Building Materials and Structures (48)

The sound attenuation provided by a barrier to airborne diffuse sound energy may be described in terms of its sound transmission loss TL; TL is defined (in dB) as 10 times the logarithm to the base 10 of the ratio of the acoustic energy transmitted through a barrier to the acoustic energy incident upon its opposite side. It is a physical property of the barrier material, not of the construction techniques used.

Material or Structure	Frequency (Hz)								
	125	175	250	350	500	700	1000	2000	4000
Doors									
Heavy wooden door: special hardware; rubber gasket at top, sides, and bottom; 2.5 in. thick; 12.5 lb/ft^2	30	30	30	29	24	25	26	37	36
Steel-clad door: well-sealed at door casing and threshold	42	47	51	48	48	45	46	48	45
Flush: hollow core; well-sealed at door casing and threshold	14	21	27	24	25	25	26	29	31
Solid oak: with cracks as ordinarily hung; 1.75 in. thick	12		15		20		22	16	
Wooden door (30 × 84 in.), special soundproof construction: well sealed at door casing and threshold; 3 in. thick; 7 lb/ft^2	31	27	32	30	33	31	29	37	41
Glass									
0.125 in. thick; 1.5 lb/ft^2	27	29	30	31	33	34	34	34	42
0.25 in. thick; 3 lb/ft^2	27	29	31	32	33	34	34	34	42
0.5 in. thick; 6 lb/ft^2	17	20	22	23	24	27	29	34	24
1 in. thick; 12 lb/ft^2	27	31	32	33	35	36	32	37	44
Walls, homogeneous									
Steel sheet: fluted; 18 gauge stiffened at edges by 2 × 4 wood strips; joints sealed; 4.4 lb/ft^2	30	20	20	21	22	17	30	29	31
Asbestos board: corrugated, stiffened horizontally by 2 × 8 in. wood beam; joints sealed; 7.0 lb/ft^2	33	29	31	34	33	33	33	42	39

Table 10.6. (Continued)

Material or Structure	Frequency (Hz) 125	175	250	350	500	700	1000	2000	4000
Sheet steel									
30 gauge; 0.012 in. thick; 0.5 lb/ft²	3	6		11		16		21	26
16 gauge; 0.598 in. thick; 2.5 lb/ft²	13	18		23		28		33	38
10 gauge; 0.1345 in. thick; 5.625 lb/ft²	18	23		28		33		38	43
0.25 in. thick; 10 lb/ft²	23	28	38	33	41	38	46	43	48
0.375 in. thick; 15 /ft²	26	31	39	36	42	41	47	41	51
0.5 in. thick; 20 lb/ft²	28	33		38		43		48	53
Sheet aluminum									
16 gauge; 0.051 in. thick; 0.734 lb/ft²	5	8		13		18		23	28
10 gauge; 0.102 in. thick; 1.47 lb/ft²	8	14		19		24		29	34
Plywood									
0.25 in. thick; 0.73 lb/ft²		20		19		24		27	22
0.5 in. thick; 1.5 lb/ft²	8	14		19		24		29	34
0.75 in. thick; 2.25 lb/ft²	12	17		22		27		32	37
Sheet lead									
0.0625 in. thick; 3.9 lb/ft²			32		33		32	32	32
0.125 in. thick; 8.2 lb/ft²			31		27		37	44	33
Glass fiber board: 6 lb/ft³; 1 in. thick; 0.5 lb/ft²	5	5	5	5	5	4	4	4	3
Laminated glass fiber (FRP): 0.375 in. thick			26		31		38	37	38
Walls, nonhomogeneous									
Gypsum wallboard									
Two 0.5 in. sheets cemented together; joints wood battened; 1 in. thick; 4.5 lb/ft²	24	25	29	32	31	33	32	30	34
Four 0.5 in. sheets cemented together; fastened together with sheet metal screws; dovetail-type joints paper taped; 2 in. thick; 8/9 lb/ft²	28	35	32	37	34	36	40	38	49
0.25 in. plywood glued to both sides of 1 × 3 studs 16 in. o.c.; 3 in. thick; 2.5 lb/ft²	16	16	18	20	26	27	28	37	33

Same as above, but 0.5 in. gypsum wallboard nailed to each face; 4 in. thick; 6/6 lb/ft^2	26	34	33	40	39	44	46	50	50
0.25 in. dense fiberboard on both sides of 2 × 4 wood studs, 16 in. o.c.; fiberboard joints at studs; 0.80 in. thick; 3.8 lb/ft^2	16	19	22	32	28	33	38	50	52
Soft-type fiberboard (0.75 in.) on both sides of 2 × 4 wood studs, 16 in. o.c.; fiberboard joints at studs; 5 in. thick; 4.3 lb/ft^2	21	18	21	27	31	32	38	49	53
0.5 in. gypsum wallboard on both sides of 2 × 4 wood studs, 16 in. o.c.; 4.5 in. thick; 5.9 lb/ft^2	20	22	27	35	37	39	43	48	43
Two 0.375 in. gypsum wallboard sheets glued together and applied to each side of 2 × 4 wood studs, 16 in. o.c.; 5 in. thick; 8.2 lb/ft^2	27	24	31	35	40	42	46	53	48
2 in. glass fiber (3 lb/ft^3) + lead vinyl composite; 0.87 lb/ft^2			4		4		13	26	31
0.375 in. steel + 2.375 in. polyurethane foam (2 lb/ft^2) + 0.0625 in. steel			38		52		55	64	77
Same as above, but 2.5 in. glass fiber (3 lb/ft^3) instead of foam			37		51		56	65	76
0.25 in. steel + 1 in. polyurethane foam (2 lb/ft^2) + 0.055 in. lead vinyl composite; 1.0 lb/ft^2			38		45		57	56	67
Concrete block, dense aggregate									
4 in. hollow, no surface treatment	30	36	39	41	43	44	47	54	50
4 in. hollow, one coat cement base paint on face	30	36	39	41	43	44	47	54	49
6 in. hollow, no surface treatment	37	46	50	50	50	53	56	56	46
6 in. hollow, one coat resin-emulsion paint each face	37	50	54	52	53	55	57	56	46
8 in. hollow, no surface treatment	40	47	53	54	54	56	58	58	50
8 in. hollow, two coats resin-emulsion paint each face	38	50	54	54	55	58	60	38	49

levels. A ¼ inch thick safety glass provided good visibility and access to the work position and gave the following noise reduction at the operator's head position.

OB center frequency (Hz)	31.5	63	125	250	500	1000	2000	4000	8000
NR (dB)	—	—	1	2	3	9	14	20	22

Example: A sheet metal guard installed around a high speed, rubber-toothed drive belt achieved the following noise reduction.

OB center frequency (Hz)	31.5	63	125	250	500	1000	2000	4000	8000
NR (dB)	—	—	—	—	—	—	7	9	19

Example: An electric motor–gear drive assembly was enclosed with ⅛ inch thick steel with welded joints and lined with Fiberglas (No. 615) board 1 inch thick. Silencers for intake and exhaust ventilation air were constructed of 12-inch parallel plates 1 inch apart. The noise reduction achieved is as follows:

OB center frequency (Hz)	31.5	63	125	250	500	1000	2000	4000	8000
NR (dB)	—	5	6	12	14	25	35	24	23

Example: A complete enclosure for the production testing of large sirens was made of sheet steel lined with Fiberglas; the inner side of this material; in turn, was covered with an open-mesh protective surface. The noise reduction is as follows:

OB center frequency (Hz)	31.5	63	125	250	500	1000	2000	4000	8000
NR (dB)	—	—	—	15	13	27	33	38	43

Example: An operator of a pneumatic system that included compressors and ducts for conveying pellets spent a large portion of his time at a central location. An enclosure for the operator was made of wood framing with ⅝ inch thick gypsum board inside and out. The open spaces between the boards were sealed. Double-glazed windows were provided for observation of equipment on all sides. The noise reduction is as follows:

OB center frequency (Hz)	31.5	63	125	250	500	1000	2000	4000	8000
NR (dB)	8	6	8	15	15	10	14	18	19

5.1.3 Impact Noise, Noise Radiation, and Vibration Control Examples

Example: A high speed film-rewind machine (15 hp) produced excessive noise from the metal-to-metal impacts between gear teeth. Fiber gears were substituted, and the gears were flooded in oil. The noise reduction is as follows:

OB center frequency (Hz)	31.5	63	125	250	500	1000	2000	4000	8000
NR (dB)	—	10	6	5	5	8	20	16	14

Example: An 8 foot diameter hopper with an electric solenoid-type vibrator coupled solidly to a bottom bin was causing excessive noise. A live-bottom bin made by Vibra Screw was installed that required less vibratory power since only the cone is vibrated. Also, the new system had less radiation area and there were no metal-to-metal impacts. The noise reduction achieved is as follows:

OB center frequency (Hz)	31.5	63	125	250	500	1000	2000	4000	8000
NR (dB)	—	7	6	20	22	16	12	12	9

Example: Screw machine stock tubes constructed of steel usually make excessive noise because there is nearly continuous impact between the tube and the screw stock. New tubes, such as the Corlett Turner silent Stock tube, constructed as a sandwich with an absorbent material between the outer steel tube and an inner helically wound liner, provide significantly lower noise levels. At 4000 rpm with ½-inch hexagonal stock, the following noise reduction is achieved with the new tube:

OB center frequency (Hz)	31.5	63	125	250	500	1000	2000	4000	8000
NR (dB)	—	12	15	15	14	20	29	34	30

5.1.4 Acoustical Damping Examples

Example: A metal enclosure around a rubber compounding mill vibrated freely, thus amplifying the motor, gear, and roll noises of the mill. An application to the inner surface of the metal enclosure of vibration-damping material (¼-inch Aquaplas F 102A) reduced the noise as follows:

OB center frequency (Hz)	31.5	63	125	250	500	1000	2000	4000	8000
NR (dB)	10	9	9	13	9	7	8	10	11

Example: The guards and exhaust hoods of a 10-blade gang ripsaw was coated with 3M Underseal (EC-244). The following noise reduction was attained while the saw was idling:

OB center frequency (Hz)	31.5	63	125	250	500	1000	2000	4000	8000
NR (dB)	6	7	10	7	5	3	3	5	6

Example: The ⅜ inch thick steel casing and the 1 inch thick steel base of a 2000 hp extruder gear were causing unwanted noise. Accelerometer measurements showed the casing and the base to be vibrating at about the same level. The casing was damped with a ¼-inch felt (No. 11, Anchor Packing Co.) plus an outer covering of ¼-inch steel. The felt–steel sandwich was bolted together on 8-inch centers. Since the steel base had ribs 9 inches deep that made the felt–steel damping impractical, the base was damped by a thick cover of sand. The following noise reduction was attained:

OB center frequency (Hz)	31.5	63	125	250	500	1000	2000	4000	8000
NR (dB)	—	—	—	—	4	17	26	24	18

5.1.5 Reduced Driving Force

The noise produced by an eccentric or imbalanced rotating member will increase as the rotational speed is increased. One obvious noise control procedure is to dynamically balance all rotating parts. Also, these pieces should rotate concentrically. Proper maintenance of all bearing and other rotating contact surfaces is essential to keep equipment running quietly.

No machine should be operated at an unnecessarily high speed. In many instances a significant reduction in noise can be achieved by using a larger machine that can do the same job while operating at lower speeds.

The reduction of driving force in almost any form is an effective noise control procedure. In many instances, a reduction of driving force will provide the additional advantage of reduced radiation area.

Example: A blower exhaust system running at 705 rpm, 6-inch static pressure, and 13,800 cfm was badly out of balance, and bearings needed replacing. After new bearings were installed and the system balanced, the following noise reduction was found:

OB center frequency (Hz)	31.5	63	125	250	500	1000	2000	4000	8000
NR (dB)	—	3	3	11	12	11	10	8	10

Example: An oversized (36 inch) propeller-type fan mounted in the wall of a large reverberant room produced excessive noise when operated at 870 rpm. By reducing the fan speed from 870 to 690 rpm, it was possible to get the significant noise reduction shown below and still provide sufficient ventilation:

OB center frequency (Hz)	31.5	63	125	250	500	1000	2000	4000	8000
NR (dB)	—	3	7	8	12	9	8	6	4

Example: Small metal parts were dropped several inches into a metal chute, where they were moved by gravity onto another operation. The dropping distance and weight of the pieces could not be changed, but the chute surface was covered with a layer of paperboard 1/16 inch thick, and this layer was covered by 18-gauge steel. The noise reduction produced by this sandwich covering was as follows:

OB center frequency (Hz)	31.5	63	125	250	500	1000	2000	4000	8000
NR (dB)	4	4	4	2	7	9	12	14	16

Example: Steel balls tumbling against the steel shell of a ball mill were producing excessive noise. The steel shell was lined with resilient material (rubber) to achieve the following noise reduction:

OB center frequency (Hz)	31.5	63	125	250	500	1000	2000	4000	8000
NR (dB)	—	3	4	6	7	11	12	15	19

5.1.6 Muffler and Air Noise Generation Control Examples

Example: The noise produced by an air-driven impact gun can be easily reduced by piping (with rubber hose) to a remote location. An internal muffler can also be used. The following noise reductions were achieved with an air gun running free:

OB center frequency (Hz)	31.5	63	125	250	500	1000	2000	4000	8000
NR (dB), Muffler	—	—	2	2	4	15	9	6	7
NR (dB), Rubber hose	—	—	19	17	30	42	29	28	28

Example: The air intakes of reciprocating air compressors often create objectionable low frequency noise. An intake filter muffler, such as Burgess Manning model Delta P—SDF, can reduce the noise the 63 Hz octave band by as much as 23 dB.

Example: The discharge of a Gast Air Motor (models 4 AM and 6 AM) created excessive noise. A Burgess Manning Delta P—CA type muffler installed on the discharge outlet produced the following noise reduction:

OB center frequency (Hz)	31.5	63	125	250	500	1000	2000	4000	8000
NR (dB)	—	2	7	7	9	10	23	29	23

Example: Excessive blower noise was being produced by a pneumatic conveying system handling synthetic fiber fluff. An absorbing-type muffler was not desired because of the possibility of snagging and plugging. A resonant-type muffler supplied by Universal Silencer Corporation gave the following noise reduction:

OB center frequency (Hz)	31.5	63	125	250	500	1000	2000	4000	8000
NR (dB)	—	12	23	13	11	10	—	—	—

Example: The air intake of a gas turbine operating at 5800 rpm and 6200 hp created excessive noise. A parallel baffle muffler consisting of six plates, each 3½ inches wide, filled with Fiberglas and faced with 18-gauge perforated sheet steel, was attached to the intake; then the baffle was connected to a duct of unlined steel plate ¼ inch thick. The cross section of the duct was 7 × 8 feet. The following noise reduction was achieved:

OB center frequency (Hz)	31.5	63	125	250	500	1000	2000	4000	8000
NR (dB)	—	—	10	16	22	33	35	27	26

Example: The noise produced by a tube reamer was reduced by the following values by mounting a Wilson 8,500 muffler on the exhaust:

OB center frequency (Hz)	31.5	63	125	250	500	1000	2000	4000	8000
NR (dB)	2	2	3	10	23	26	28	16	18

5.1.7 Drive System Modification Examples

Example: A rubber-toothed belt used to drive a pump was replaced by a V-belt drive, yielding the following noise reduction:

OB center frequency (Hz)	31.5	63	125	250	500	1000	2000	4000	8000
NR (dB)	—	5	4	4	2	—	8	17	18

Example: An edger-planer for trimming foamed plastic created noise levels as high as 102 dB in the 250 Hz octave band. The noise was caused primarily by the cutter blades chopping the conveying air stream. The clearance between the cutter blades and the casing was increased from 3/32 to 1 inch, thereby lowering the air velocity and reducing the noise level to 84 dB in the 250 Hz band.

A single noise control procedure that is ineffective by itself may produce significant results when coupled with one or more other procedures. As an example, a typical noise source having a frequency spectrum in which all octave band pressure levels are essentially the same may have the following noise reduction values for the seven noise control procedures listed below.

	NROB Frequency (Hz)								
Procedure	31.5	63	125	250	500	1000	2000	4000	8000
1. Mounted on vibration isolators	11	7	3	—	—	—	—	—	—
2. Single-Wall barrier	—	—	3	5	6	6	6	6	7
3. Complete enclosure of absorbing material	—	—	—	4	5	5	6	7	7
4. Complete enclosure of solid material with no absorption inside	—	2	5	14	18	26	26	27	29
5. Complete enclosure of solid material with no absorption inside, mounted on vibration isolators	11	8	7	16	21	29	34	35	40
6. Complete enclosure of solid material with no absorption inside, mounted on vibration isolators	11	11	13	25	32	38	40	42	45
7. Complete procedure no. 6, mounted on vibration isolators and enclosed in solid material with absorption inside	20	17	22	44	50	57	57	59	64

Many of the examples of noise control in this chapter were taken from material prepared for the American Industrial Hygiene Association (AIHA) Noise Manual (48). A more extensive listing of examples and a more complete general discussion of engineering noise control can be found in the AIHA manual.

5.2 Personal Protection Devices

An effective personal hearing protector acts as a barrier between the noise and the inner ear, where noise-induced damage to hearing may occur. Hearing protecting devices usually take the form of earmuffs worn over the external ear, providing an acoustical seal against the head, or ear plugs that provide an acoustical seal at the entrance to the external ear canal (49).

The only certain means for evaluating the effectiveness of personal protectors is to measure periodically the hearing thresholds of the user. If no hearing losses are observed—other than those due to the aging process—the program may be considered to be successful. However a hearing monitoring program may take years to become meaningful because noise-induced hearing loss usually develops slowly; therefore protector noise reduction characteristics, or attenuation values, must be used as a short-term guide in the selection of the personal hearing protection equipment for particular noise exposure patterns.

5.2.1 Performance Characteristics and Limitations of Hearing Protectors

The effectiveness of a hearing protector depends on its design and on several physiological and anatomical characteristics of the wearer. Sound energy may reach the inner ears of persons wearing protectors by four different pathways: (1) by passing through bone and tissue around the protector; (2) by causing vibration of the protector, which in turn generates sound into the external ear canal; (3) by passing through leaks in the protector; and (4) by passing through leaks around the protector. These pathways are illustrated in Figure 10.14.

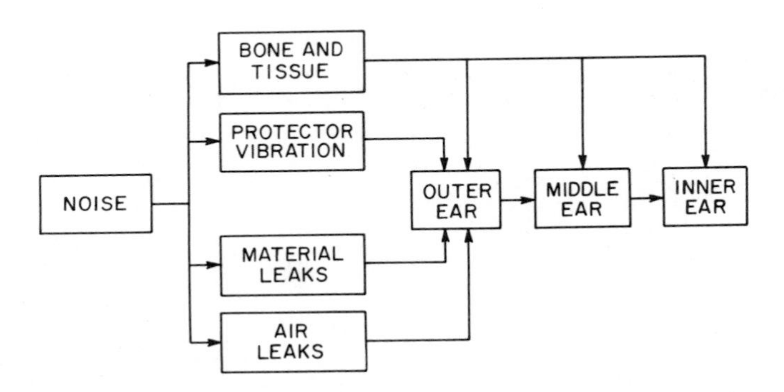

Figure 10.14 Noise pathways to the inner ear.

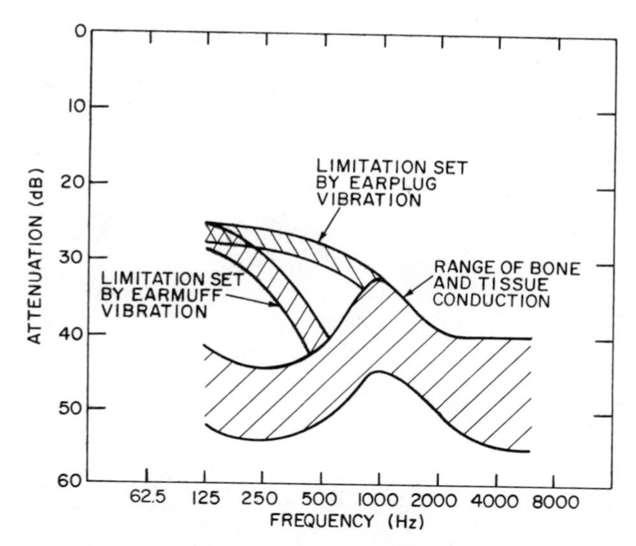

Figure 10.15 Practical protection limits for plugs and muffs.

Even if the protector has no acoustical leaks through or around it, some noise will reach the inner ear by one or both of the first two pathways if the levels are sufficiently high. The practical limits set by the bone- and tissue-conduction threshold and the vibration of the protector vary considerably with the design of the protector and with the wearer's physical makeup, but approximate limits for plugs and muffs are given in Figure 10.15 (50,51).

If hearing protectors are to provide noise reduction values approaching the practical limits appearing in Fig. 10.15, acoustical leaks through and around the protectors must be minimized. Figure 10.16 demonstrates that the mean attenuation values of a well-fitted, imperforate, plastic ear plug are considerably greater than those of a dry cotton plug, which is porous. Figure 10.17 illustrates the effects of leaks around a poorly fitted imperforate ear plug and of a leak through a small orifice in the center of a well-fitted ear plug.

5.2.2 Types of Hearing Protectors

Ear Plugs. Ear canals differ widely in size, shape, and position among individuals and even between the ears of the same individual; therefore ear plugs must be chosen which are adaptable to a wide variety of ear canal configurations.

Ear canals vary in cross-sectional diameter from about 3 to 14 mm, but most are between 5 and 11 mm. Most ear canals are elliptically shaped, but some are round, and many have only a small slitlike opening. Some ear canals are directed in a straight line

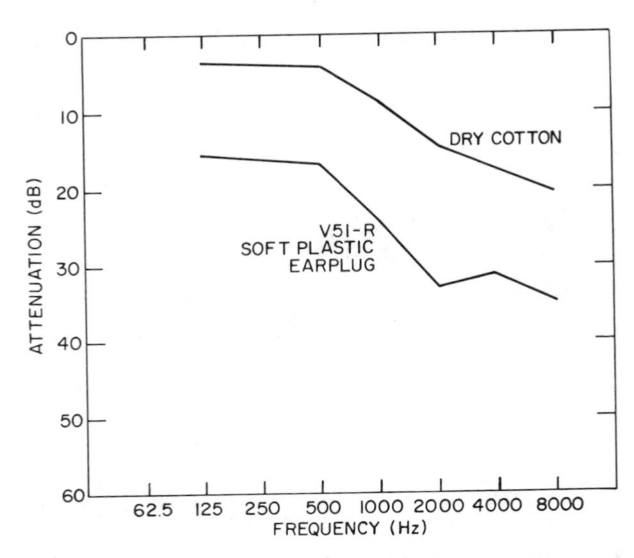

Figure 10.16 Mean attenuation characteristics of a well-fitted, imperforate ear plug and an ear plug made of dry cotton.

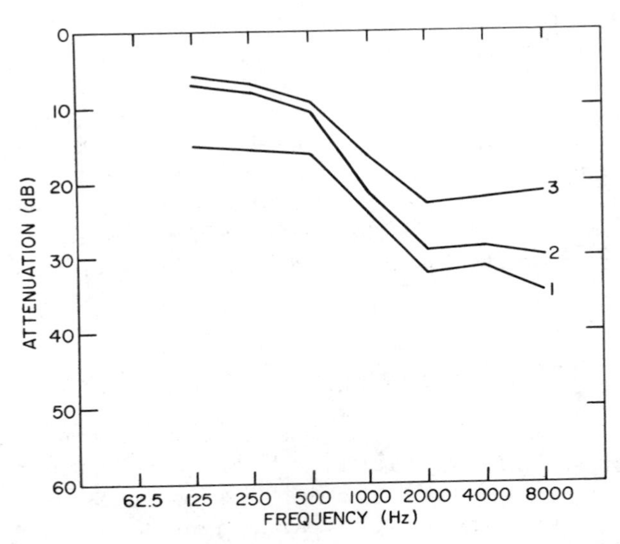

Figure 10.17 Mean attenuation characteristics of an ear plug: 1, well-fitted and imperforate; 2, well-fitted but with a small, uniform hole through its center; 3, poorly fitted and imperforate.

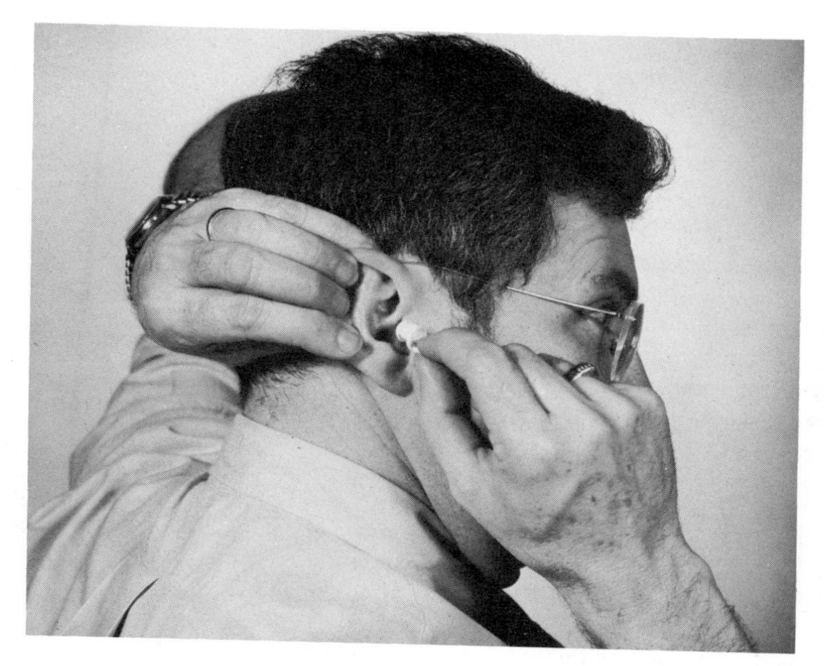

Figure 10.18 Recommended method for inserting ear plugs.

toward the center of the head, but most bend in various ways and are directed toward the front of the head.

In many cases, there is only a small space available to accommodate an ear plug, but almost all entrances to ear canals can be opened and straightened by pulling the external ear directly away from the head (Figure 10.18), making it possible to seat an ear plug securely. For comfort and plug retention, the canals must return to their approximate normal configuration once the protector is seated.

Sized Ear Plugs. Few single-sized molded ear plugs have proved to be very effective in attenuating noise when used for a large range of ear canal sizes and shapes; therefore most of the more widely accepted molded ear plugs come in two or more sizes. The best molded ear plugs are made of soft and flexible materials that will conform readily to the shape of the ear canal for a snug, airtight, comfortable fit. These ear plugs must be nontoxic, and they should have smooth surfaces that are easily cleaned with soap and water. Ear plugs should be made of materials that will retain their size and flexibility over long periods of time; however some ear waxes may cause changes in size and flexibility of protectors after extended periods of use.

The comfort and performance of ear plugs is usually very dependent on goodness of fit, and because ear canals differ widely in size and shape, it is generally advisable to stock more than one type of molded ear plug. Offering a choice of protectors also improves program acceptance.

The size distribution of ear canals in a large group of males in terms of hearing protector sizes will be approximately as follows: 5 percent extra small, 15 percent small, 30 percent medium, 30 percent large, 15 percent extra large, and 5 percent larger than those supplied by most ear plug manufacturers. This size range, showing an equal percentage of wearers for medium and large sizes, indicates that many persons are being fitted with ear plugs that are too small. If an individual is permitted to fit himself, in most cases he will pay more attention to comfort than to attenuation, and the size distribution will shift toward smaller sizes.

Some ear canals increase in size with the regular use of ear plugs, and ear plugs may shrink in size; therefore, if a given ear falls between two plug sizes, it is advisable to choose the larger of the two. It is good practice to check the fit of ear plugs periodically for these reasons.

A common (and often valid) complaint is that the case costs more than the ear plugs; however a good case keeps the ear plugs clean, in good condition, and readily available when needed. Also, the total cost of sized plugs and container is generally below the price of one set of replacement cushions for earmuffs, and an earplug container of good quality should outlast several pairs of plugs.

Malleable Ear Plugs. Malleable ear plugs are made of materials such as cotton, paper, plastics, wax, glass wool, and mixtures of these materials. Typically, a small cone or cylinder of the material is hand-formed and inserted into the ear with sufficient force to make the material conform to the shape of the canal and hold itself in position.

Malleable ear plugs should be formed and inserted with clean hands, because dirt or foreign objects placed in the ear may cause irritation or infection. Malleable plugs, then, should be carefully inserted at the beginning of a work shift and if removed, they should not be reinserted during the work period unless the hands are cleaned. For this reason, in a dirty work area malleable plugs (and to a somewhat lesser extent, all ear plugs) may be a poor choice of ear protector if the area is subjected to intermittent high level noises or if it may be desirable to remove and reinsert protective devices during the work period. However malleable plugs have the obvious advantage of fitting any ear canal, eliminating the need to keep a stock of various sizes, as for most molded ear plugs.

Earmuffs. Most muff-type protectors are similar in design. Sealing materials that contact the skin are nontoxic for the most part, and the fit, comfort, and general performance of comparable models do not vary widely.

If maximum protection is required, the protector earcups must be formed of a rigid, dense, imperforate material. Generally the size of the enclosed volume within the muff shell is directly related to the low frequency attentuation. The ear seals should have a small circumference, so that the acoustic seal takes place over the smallest possible irregularities in head contour, and leaks caused by jaw and neck movements are minimized. Obviously there must be a compromise in the selection of seal circumference to permit the accommodation of all sizes of ears.

The inside of each earcup should be partially filled with an open cell material to absorb high frequency resonant noises and to dampen any movement of the shell. The material placed inside the cup should not contact the external ear; otherwise discomfort to the wearer and soiling of the lining may result.

Earmuff cushions are generally made of a smooth, plastic envelope filled with a foam or fluid material. Since skin oil and perspiration have adverse effects on cushion materials, the soft and pliant cushions may tend to become stiff and to shrink after extended use. Fluid-filled cushions occasionally have the additional problem of leakage. For these reasons, most earmuffs are equipped with easily replaceable seals.

The acoustic seal materials used for earmuffs provide maximum protection when placed on relatively smooth surfaces; therefore less protection should be expected when muffs are worn over long hair, glasses, or other objects. Glasses with close-fitting, average-sized, plastic temples will cause about 5 to 10 dB reductions in attenuation in most cases, but this loss of protection can be reduced substantially if small, close-fitting wire temples are used. Acoustic seal covers that are sometimes provided to absorb perspiration also reduce the amount of attenuation by several decibels, because noise can leak through the porous material.

Since the loss of protection is directly proportional to the size of the obstruction under the seal, every effort should be made to minimize these obstructions. If long hair or other obstructions cannot be avoided, it must be realized that the muffs may not be able to provide the claimed attenuations, and it may be advisable to use ear plugs.

The force applied by the muff suspension is another factor directly related to the amount of noise attenuation provided. In choosing the suspension forces, a compromise must be made on the basis of performance versus comfort. Suspensions should never be deliberately sprung to reduce the applied force, if maximum protection is desired. The applied force should be measured periodically to be sure of obtaining the expected performance.

Concha-Seated Ear Protectors. Protectors that cannot be strictly classified as insert- or muff-types include individually molded ear pieces and others that provide an acoustic seal in the concha or at the entrance to the ear canal. Early models of individually molded ear pieces that are held in position by the external ear were not widely used because of their high cost and relatively poor performance. In recent years, new materials have made this type of protector competitive in price, and its performance is quite good *if* the devices are molded carefully according the manufacturers' instructions.

Another protector design in this class makes use of a narrow headband to press two soft plastic, conical caps against the entrance to the external ear canal. This protector, which is reasonably comfortable, provides average performance in some ears but gives below-average results in others.

5.2.3 Amount of Protection Provided in Practice

A comparison of the noise analysis of a particular noise exposure and the levels specified by the chosen hearing conservation criterion should be used to determine the amount of noise reduction required. When the hearing conservation criterion is expressed in octave bands, the amount of noise reduction required can be determined by subtracting the sound pressure levels (in decibels) specified by the criterion from the exposure levels (in decibels) measured in corresponding octave bands. When ear protectors are to be used to provide the necessary noise reduction, the attenuation that is provided within each of these bands must equal or exceed the noise reduction requirements (in decibels).

When hearing conservation criteria are based on the A-frequency weighting and exposure levels are available in octave bands, the following procedure should be used:

Step 1. Take sound level measurements in octave bands at the point of exposure.

Step 2. Subtract from the octave band levels (in decibels) obtained in step 1 the center-frequency adjustment values for the A-frequency weighting given in Table 10.3 (arithmetically).

Step 3. Subtract from the A-weighted octave bands calculated in step 2 the attenuation values provided by the protector for each corresponding octave band to obtain the A-weighted octave band levels that reach the ear while the ear protector is being worn (arithmetically).

Step 4. Calculate the equivalent A-weighted noise level reaching the ear while the ear protector is being worn by adding the octave band levels as shown in Table 10.1 or 10.2. An alternate method for adding decibels is as follows:

$$\text{equivalent dBA} = 10 \log_{10} \left(\text{antilog}_{11} \frac{L_{125}}{10}\right) + \text{antilog}_{10} \frac{L_{250}}{10} + \cdots + \text{antilog}_{10} \frac{L_{8000}}{10}$$

where L_{125} is the sound pressure level of the A-weighted octave band centered at 125 Hz; L_{250} the A-weighted octave band level at 250 Hz, and so on.

Example: The first and second columns in Table 10.7 contain octave band sound pressure level data measured in a textile mill weaving room. The third column shows the same octave band data, but with an A-frequency weighting (using Table 10.3). The mean and mean minus 2 standard deviations (SD) noise reduction values that can be expected from a good muff-type ear protector are listed under the Hearing Protector Attenuation column headings (by subtracting 2 standard deviations from the mean at each test frequency, the 98 percent confidence limit is determined). With the ear protector in place, the octave band levels reaching the ear canal are listed for mean and mean minus 2 SD attenuations, under the Resultant Exposure to Inner Ear columns. The octave band exposure levels given in the last two columns may be added (using Table 10.2) to determine the exposure level in dBA. In this example, the exposure level for those receiving the mean attenuation values from the protector would be 66 dBA; for those receiving minimum values of protection, the exposure level would be 75 dBA.

Table 10.7. Protection Provided by Hearing Protectors Worn in a Weaving Room

OB Center Frequency (Hz)	Weaving Room Spectra, L_p (dB)		Weaving Room Spectrum with, A-Weighting, L_p (dB)	Hearing Protector Attenuation (dB)		Resultant Exposure to Inner Ear (dB)	
				Mean	Mean minus 2 SD	With Mean Attenuation	With Mean minus 2 SD Attenuation
125	90	(less 16 =)	74	16	9	58	65
250	92	(less 9 =)	83	21	15	62	68
500	94	(less 3 =)	91	31	23	60	68
1000	95	(less 0 =)	95	42	30	53	65
2000	97	(less 1 =)	98	43	32	55	66
4000	95	(less 1 =)	96	45	35	51	61
8000	91	(less 1 =)	90	34	22	56	68
Overall	103 dB		102 dBA			66	75 dBA

Two factors must be carefully considered when selecting ear protectors to meet the noise reduction requirements:

1. The ear protector attenuation values determined according to the ANSI standards (52, 53) are for single frequencies or for one-third octave bands located at the center of each octave band. These values are not always an accurate representation of the noise reduction capability throughout the octave band. In addition, the noise exposure spectrum may not be flat throughout the band. Therefore the effectiveness of protectors can be only approximated, and a program for monitoring the hearing levels of all persons wearing ear protectors is strongly recommended wherever the margin of protection is small.
2. The protection provided by ear protectors can vary considerably between wearers and between protectors of the same model. Standard deviations of 2 to 5 dB are commonly found in subjective measurements of protector attenuation of any test frequency; therefore attenuation values may have a range of ±4 to ±10 dB for 95 percent confidence limits. Obviously standard deviations must be considered along with the mean attenuation values when selecting an ear protector for a particular application. Figures 10.19 and 10.20 plot mean attenuations and standard deviations for a good muff and a good ear plug, respectively.

Fortunately, a high percentage of work areas have noise levels lower than those listed in the foregoing example; thus the use of good personal protective devices can provide adequate protection for a large majority of noise exposures.

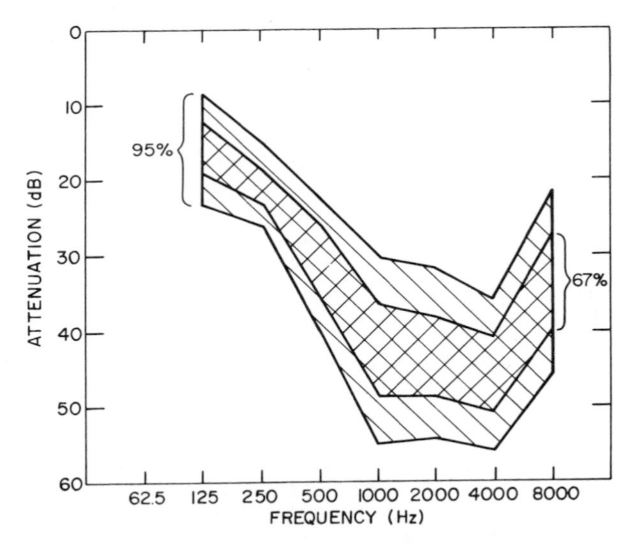

Figure 10.19 Mean attenuation characteristics of a good muff-type protector plotted with 1 and 2 standard deviations (shaded areas). Attenuation values determined according to the ANSI specifications using pure tone threshold shift techniques (ANSI Z24.22-1957).

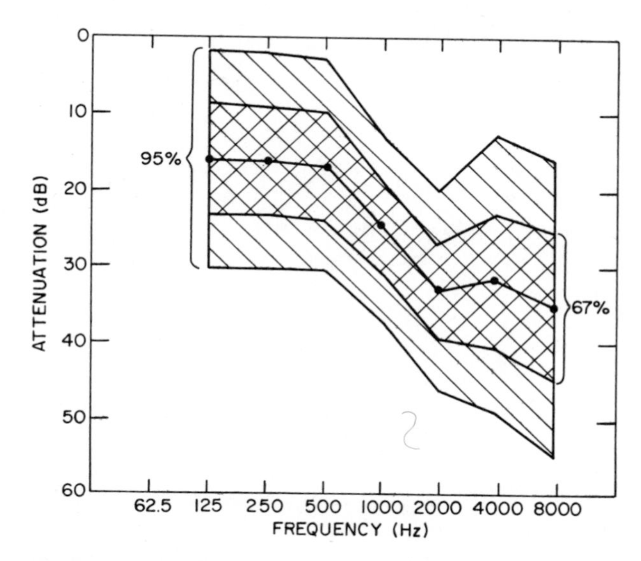

Figure 10.20 Mean attenuation characteristics plotted with 1 and 2 standard deviations (shaded areas) for a well-fitted imperforate ear plug. Attenuation values determined according to the ANSI specification using pure tone threshold shift techniques (ANSI Z24.22-1957).

If adequate protection cannot be provided by a single muff or ear plug protector, both can be worn together to provide additional protection (Figure 10.21). The combined attenuation of any two protectors cannot be predicted accurately because of complex coupling factors; at most frequencies, however, the resultant attenuation from two good protectors may be estimated to average about 6 dB greater than the higher of the two individual values.

In summary, the use of a good ear protector can provide adequate protection in most work environments where engineering control measures cannot be used. For the relatively few persons who may be exposed over long periods of time to noise levels in excess of 115 dBA, special care should be taken to obtain the best protectors, to ensure that they are used properly, and to monitor hearing thresholds regularly. For those exposed to higher levels over extended periods, it may be necessary to use a combination of insert- and muff-type protectors and/or limit the time of exposure. Ear protectors will provide adequate protection for only a small percentage of wearers when worn in noise levels greater than 125 dBA over long periods of time.

5.2.4 Noise Reduction and Communication

Workers must be able to communicate with each other and to hear warning signals in many high noise environments where the wearing of ear protectors can aggravate communication problems. The effect of noise on communication depends to a large extent on the spectrum of the noise, and it is most significant when the noise has high level

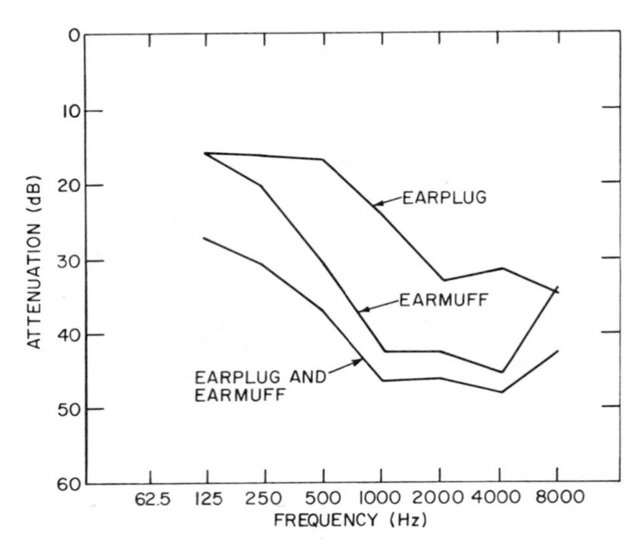

Figure 10.21 Mean attenuation characteristics of a muff and an ear plug worn separately and together.

components in the speech frequency range from about 400 to 3000 Hz. Speech interference studies (53, 54) show that conversational speech begins to be difficult when the speaker and listener are separated by about 2 ft in areas where noise levels are about 88 dBA.

5.2.5 Communication While Wearing Protectors

Ear protective devices obviously interfere with speech communication in quiet environments; however in noise levels above about 90 dB in octave bands (or about 97 dBA for flat spectra), ear plugs or muffs should not interfere and indeed may even improve speech intelligibility for normal-hearing ears, since speech-to-noise ratios are kept nearly constant and the protected ear does not distort because of overdriving caused by the high speech and noise levels (55, 56). Research studies on this subject have been restricted to normal-hearing ears, and it is not known how the wearing of ear protectors affects the intelligibility scores of persons with hearing impairments.

This concept of blocking the ear to improve hearing in a noisy environment can be difficult for a worker to accept, and he is likely to resist wearing protectors because of anticipated difficulties in communication. This opinion may even be enforced if he first tries the protector in a quiet environment. Often workers are attracted to ear protectors advertised as providing a "filter" that allows the low frequencies in the speech range to pass but, at the same time, blocks the noise. Some of these filter-type devices do enable better communication in quiet environments and may be applicable where there are intermittent exposures to moderately high noises. However in steady state (constant) noise levels greater than about 90 dB in octave bands, the conventional insert- or muff-type protectors enable equally good communication (56), and they furnish additional protection that may be needed in higher exposure levels.

5.2.6 Communication Through Radio Headsets

The wearing of protectors may help to improve the clarity of speech where electronic communication systems are used in high level noise environments (i.e., where insert-type protectors are worn under muff-type communication headsets or earphones). The improved perception of the speaker's own voice allows him to modulate his voice better and to avoid much of the distortion that often accompanies loud speech and shouting. Also, the listener can adjust his electronic gain or volume control to obtain the level of undistorted speech that will give him the best reception. An exception may be found in noise levels above about 130 dB overall, where speech perception is not always improved by the use of earplugs under communication headsets. In such cases it is often desirable to improve the signal-to-noise ratio at the microphone as well as at the receiver. One common method for increasing the signal-to-noise ratio at the microphone is to use noise cancellation principles. A noise-canceling microphone picks up ambient noise through two apertures, one on either side of its sensing element and a portion of the low frequency noise is canceled. The sensing element of the microphone is oriented so that the speech signal enters mainly through one aperture when the microphone is held close to the mouth; thus speech is not canceled.

An electronic noise-canceling technique has also been used with limited success to increase intelligibility in noisy environments (57, 58). In this technique, the noise picked up by the microphone has its phase changed 180° electronically, and the phase-adjusted signal is then fed into the headset.

In noise fields above about 120 dB re 20 μPa overall, noise may be attenuated at the microphone by noise shields that encase the sensing element. Efficient noise-attenuating shields tightly held around the mouth may be used with microphone systems to transmit intelligible communication in wide band noise levels exceeding 140 dB.

5.2.7 Guides for the Selection of Protector Types

Both insert- and muff-type protectors have distinct advantages and disadvantages. Some of the features of each type are listed below.

Insert-Type Protectors

Advantages

1. Small and easily carried.
2. Can be worn conveniently and effectively with no interference from glasses, head-gear, earrings, or hair.

3. Relatively comfortable to wear in hot environments.
4. Do not restrict head movement in close quarters.
5. Cost of sized ear plugs, except for some hand-formed and molded protectors, is significantly less than muffs.

Disadvantages

1. Sized and molded insert protectors require more time and effort for fitting than do muffs.
2. The amount of protection provided by a good earplug may be less and more variable between wearers than that provided by a good muff-type protector.
3. Dirt may be inserted into the ear canal if ear plugs are inserted with dirty hands.
4. It is difficult to monitor groups wearing ear plugs because they cannot be seen.
5. Ear plugs can be worn only in healthy ear canals, and even then, acceptance may take some time.

Muff-Type Protectors

Advantages

1. The noise attenuation provided by a good muff-type protector is generally greater and less variable between wearers than that of good ear plugs.
2. One size fits most heads.
3. It is easy to monitor groups wearing muffs because they are easily seen at a distance.
4. At the beginning of a hearing conservation program, muffs are usually accepted more readily than are ear plugs.
5. Muffs can be worn despite minor ear infections.
6. Muffs are not easily misplaced or lost.

Disadvantages

1. Uncomfortable in hot environments.
2. Not easily carried or stored.
3. Not convenient to wear without interference from glasses, headgear, earrings, or hair.
4. Usage or deliberate bending of suspension band may reduce protection to substantially less than expected.
5. May restrict head movement in close quarters.
6. More expensive than most insert-type protectors.

It is doubtful whether either the insert- or the muff-type protector alone can satisfy all needs. The obvious advantages of each should be used wherever possible to make a hearing conservation program more acceptable and more effective.

6 COMMUNITY NOISE

In communities, as well as in industry, the most serious effect of high level noise exposures is permanent hearing loss. This problem is discussed in detail in Section 3 of this chapter. Fortunately most community noise problems do not involve noise-induced hearing loss. Common problems do entail annoyance, communication, and the general well-being of humans.

The effects of noise on communication can be measured with a reasonable degree of accuracy (59–61); however other psychological and physiological responses to noise are extremely complicated and difficult to measure in a meaningful way. Attempts to correlate noise exposure levels and annoyance, or the general well-being of humans, are complicated by such factors as the attitude of the listener toward the noise source, the history of individual noise exposures, the activities of the listeners and stresses on the listeners during the noise exposures, the hearing sensitivities of the listeners, and other factors related to individual variability. For the most part, old rules and regulations were based on vaguely defined nuisance factors (62) instead of more precise measured sound pressure level limits because of these complex variables.

6.1 Nuisance-Type Regulations

Nuisance-type regulations often take the form "there shall be no unnecessary nor disturbing noise. . . ." An obvious weakness of this form of regulation is failure to specify the conditions of how, when, and to whom noise is unnecessary or disturbing. Thus innumerable arguments may result in the interpretation of these laws when an attempt is made to enforce them.

Recently enacted zoning codes have been established that specify maximum noise levels for various zones. These so-called *performance* zoning codes are more objective and easier to enforce than the *nuisance* laws, but even so, many problems must be considered individually on a nuisance basis for the widely varying conditions of noise exposure in individual situations.

6.2 Performance-Type Regulations

Groups under pressure to provide practical state-of-the-art recommendations, standards, regulations, or guidelines for exposures to noise levels below the level of physiological damage have found it necessary to consider community responses rather than individual annoyance reactions to noise. This decision has been made because the group statistics involved in describing "community responses" to noise avoid many of the variables that are extremely difficult to account for in individual annoyance reactions. Some performance-type documents with specific sound pressure level limits have provided reasonably good predictions of response to noise for particular exposure situations; however these rules or regulations have been based for the most part on specific exposures—such as those from aircraft or from ground transportation—and caution must be used in applying these results to other kinds of noise exposure.

Most of the new performance-type noise codes specify limits of sound pressure level that are based on (1) a selected frequency weighting or noise analysis procedure, (2) the pattern of exposure times for various noise levels, (3) the ambient noise levels that would be expected in that particular kind of community without the offending noise source(s), and (4) a land-use zoning of the area.

The document "Model Community Noise-Control Ordinance," which was published in 1975 by the Environmental Protection Agency (EPA) (63), will provide needed guidance in the development of performance-type noise control ordinances. This EPA document does not constitute a standard, specification, or regulation, but it does present the best available technical knowledge that may be used in a uniform and practical way by communities of various sizes to tailor ordinances to their specific conditions and goals.

6.3 Frequency Weighting or Analysis of Community Noise

Many frequency-weighting and analysis procedures have been proposed for use in estimating community reaction to noise. In recent years, single sound pressure level measurements, such as those with A-frequency weighting, and more complete analyses using octave, one-third octave, and narrower bands have been used in noise reaction measurements (see Section 4.1.2).

The single readings of sound pressure level have the obvious advantages of short measurement times and overall simplicity in data manipulation. The A-frequency weighting, which is provided on most sound level meters, has been given considerable prominence in hearing conservation criteria (64–68), for speech interference measurements (59–61), and in procedures for estimating community reaction to noise (67–71). The readily available A-frequency weighting is considered by many investigators to be one of the most practical means of measuring noise available at this time for the estimation of community response to general wide band noises.

6.3.1 Aircraft Noise

Exposures to aircraft noise are often unique for such factors as spectral and level characteristics and for listeners' attitudes toward the source. These factors, plus the considerable work that has resulted from the strong pressure for control guidelines, have brought forth several reasonably good procedures for measuring community reaction specifically in relation to aircraft noise exposures.

A D-frequency weighting of sound pressure level has been proposed for use in procedures to estimate community reaction to jet aircraft noise. The single reading with a D-frequency weighting has been used in several European countries, and it is being considered by the International Electrotechnical Commission (IEC) for use in their reports related to community noise (11).

Other, more sophisticated noise analysis techniques include the following:

1. Perceived noise level (PNL), measured in perceived noise decibels (PN dB).
2. Effective perceived noise levels (EPNL), measured in effective perceived noise decibels (EPN dB).
3. Noise exposure forecast (NEF), using EPN dB as the basic unit.

PNL is defined as an instantaneous measure of the noise produced at a given location, based on octave or third-octave sound pressure levels (72, 73). EPNL values are derived from PNL instantaneous levels with added adjustments for high level, pure tone content and for flyover duration (72, 73). NEF values are derived from EPNL levels with adjustments added for aircraft type, mix of aircraft, number of operations, runway utilization, flight path, operating procedures, and time of day (72, 73).

Although the PNL, EPNL, and NEF procedures have been shown to be more accurate than single frequency weightings for the prediction of community reaction to aircraft noise, their relatively high degree of complexity cannot be justified for many other uses. Differences between aircraft and nonaircraft noise exposure factors—such as noise characteristics, attitudes of listeners toward the source, and community composition—are significant in many situations. Thus the complex measuring and data manipulation procedures of the PNL, EPNL, and NEF methods are not covered in this chapter. Complete details on the use of these aircraft noise measures can be found in Department of Transportation Reports (74), Environmental Protection Agency documents (71–73), and other sources (62, 65, 75–80).

6.3.2 Nonaircraft Noise

The A-weighted sound levels are now the most commonly used and the most practical means for measuring most reactions to nonaircraft noise. Certainly instrumentation with A-weightings is more readily available, and there is less chance for errors in measurement and data manipulation using this relatively simple procedure than with other more complicated methods. The simplicity of measurement, coupled with the lack of evidence that there are better overall measuring means, supports the choice of the A-weighting of nonaircraft noises at this time.

6.4 Noise Measurement Equipment

Sound level meters used for community noise measurements should meet the American National Standard Specification for Sound Level Meters (ANSI S1.4-1971, Type 2) (81), to assure accurate and legally admissible data. Such accessory equipment as tape recorders or graphic level records, to be used with sound level meters, should also conform with the pertinent sound level meter specifications.

6.5 Measurement Locations and Procedures

Sound pressure level measurements or recordings must be made at the place and time of annoyance. Outdoor measurements should be made at positions about 4 to 5 feet above the ground and, as far as possible, away from any solid structure that may reflect sound. Complete descriptions of all measurement positions should be recorded, and particular attention must be given to conditions that might influence measurements (see Section 4). The measuring equipment should be operated in the manner specified by the manufacturer, and care should be taken to prevent the measurements from being influenced by such factors as wind blowing over the microphone and noise from extraneous sources or electromagnetic energy. Extremes in climatic conditions may cause significant differences in measured levels; therefore measurements should be made under normal conditions, and a range of noise level variations should be obtained for extreme conditions.

6.5.1 Measuring Steady and Fluctuating Noises

Several noise measurement procedures may be required in making community noise measurements. If the noise is continuous, has little variation in level, and is composed of low frequency sounds, almost any measuring procedure using a simple sound level meter will be satisfactory. If the levels vary widely in a complicated manner, however, it may be necessary to record the noise over a long period of time and to perform complicated statistical analyses on the data to obtain an accurate determination of average values. The complicated procedures for determining average values of widely varying noise levels are not justified in many cases, because community reactions vary so widely. One relatively simple procedure for community noise measurement, which was proposed as a standard by an American National Standards Institute writing group (82), is as follows:

1. Observe the A-frequency level reading on the sound level meter (slow meter damping) for 5 seconds and record (a) the best estimate of central tendency and (b) the range of the meter deflections, during that 5-second period (in decibels).
2. Repeat the observations of step 1 until the number of central tendency readings equals or exceeds the total range (in decibels) of all the readings.
3. Find the arithmetic average of all the central tendency readings in steps 1 and 2, and call this estimate the community noise level for this particular measuring time and location.

6.5.2 Multiple Noise Sources

On occasion, several noise sources contribute to the measured A-weighted levels, and it may be desirable to determine the contributions of each source for purposes of noise control or for assessing the responsibility for the noise. The simplest way to obtain the contribution of each source is to measure the weighted levels produced by each individual source when the other sources are not operating. If it is not possible to obtain measures of the individual source contributions directly, it may be necessary to analyze the overall noise levels in narrow frequency bands, generally one-third octave band or narrower, and correlate the characteristics of the individual noise sources.

Since analysis procedures are time-consuming and not always very meaningful, the individual sources should be measured directly if at all possible. These measurements may be taken at night, between work shifts, between process changes, or by special arrangements with supervisory personnel in charge of the individual operations.

6.5.3 Indoor Measurements

Indoor measurements should be made at locations at least 3 feet from walls or other reflecting surfaces. If regions of high and low levels are found close together (standing waves), three or four measurements about a foot apart should be averaged arithmetically.

6.5.4 Impulse Noise Measurements

Noise exposure criteria based on A-weighted sound pressure levels usually hold for steady state noises only. The inertia of the sensing elements of conventional sound level meters prevent accurate readings of levels that change significantly within a short time. To obtain accurate peak factor readings for single pressure pulses that change level by more than 5 dB in less than 0.2 second, a peak-reading instrument or an oscilloscope must be used.

When the noise impulses occur at a rate not less than 10 per second, a conventional sound level meter can be used because the noise peaks do not drop significantly before the arrival of the following peak. If the individual impulses are very short, however, and the variation in sound pressure levels between pulses exceeds 6 dB, the sound level meter will indicate a low reading.

Before effort is expended in impulse measurements, it is important to know precisely how the information is to be used. Only a few rules and regulations have provisions for impulsive noise, and most of these documents simply specify an adjustment that is to be made to the reading of the conventional sound level meter for "impulsive or hammering noises." For such cases, therefore, it is not possible to make sure of accurate measures of impulsive noise.

6.6 A Guide for Community Noise Criteria

Acceptable noise levels vary from one community to another depending on the kind of community and other variables. Ideally, noise codes are tailored to the character and requirements of the particular community.

A typical noise code will have a steady state noise limit of about 35 to 55 dBA for outdoor, daytime exposures in rural residential, hospital, or other quiet areas. The particular value of this base limit of sound pressure level should be selected with regard to the living habits, present and past, of the people in the area. For example, in a young neighborhood that contains children, more noise may be tolerated than in an older neighborhood composed mostly of retired people. When the base level is properly selected, there will be no observable reaction of the community to noises at or below this level.

The level selected for the base criterion may be completely impractical for other locations and conditions; therefore special adjustments must be made for each situation. It is impossible to provide the necessary adjustments for all situations in a document of reasonable size, but any situation can be covered, for various classes of exposures, if adjustments for the following factors are considered:

1. The kind of community; that is; rural, suburban, urban (with some workplaces, businesses, or main roads), city (with heavy business and traffic, and heavy-industry noise).
2. The duration and time pattern of exposures:
   a. Steady state constant.
   b. Nonuniform or intermittent steady state.
   c. Impulsive noise.
   d. Pure tone or whine characteristics.
3. Time of day (i.e., daytime, evening, or night).
4. Indoors or outdoors.
5. Climatic conditions:
   a. Wind.
   b. Precipitation.
   c. Temperature.
6. Special conditions for the district.

The estimated community response to noise, compared to criteria prepared for a noise code as just described, might be summarized as in Table 10.8.

If the community response to noise with respect to the selected criteria does not follow the pattern given in Table 10.8, the criteria should be adjusted until this pattern is

Table 10.8. Estimated Community Response to Noise as Compared to the Noise Code Criteria

Difference Between Measure Noise and Adjusted Noise Criterion (dBA)	Estimated Community Response
+0	No observed reaction
+5	Occasional complaints
+10	Widespread complaints
+15	Threats of community action
+20	Vigorous community action

reached. Additional reference materials on community noise criteria may be found in References 83 to 107.

7 NOISE EXPOSURE LIMITS AND THE OSHA

The development of effective and practical requirements and procedures for assuring the health and safety of industrial workers who are exposed to high level noise is very complex. In addition to the very complicated technical aspects related to the effects of exposure to high level noise, the procedures for measuring noise dosage, and the procedures for hearing measurement and impairment assessment, there is the very important factor of the economic impact of any selected set of criteria. Furthermore, there is the problem of the large differences between individuals in their susceptibility to noise-induced hearing damage. Because of these complex problems, agreement was not reached on rules and regulations to limit noise exposures until the Walsh-Healy Public Contract Act was revised, May 20, 1969 (108).

The noise portion of the Walsh-Healy Act was closely patterned after the Threshold Limit Value of Noise developed by the American Conference of Governmental Industrial Hygienists (ACGIH) in 1968 and published in 1969 (109). The federal Occupational Safety and Health Act (110) made use of essentially the same noise dose limits as the Walsh-Healy Act, but the new law provides for much wider coverage throughout industry.

7.1 OSHA Limits

The noise exposure limits set forth in the Occupational Safety and Health Act are designed for both continuous and impulsive noises. The continuous noise limit is set at 90 dB measured with an A-frequency weighting for exposures of 8 hours per day, with higher levels being permitted over less time at the rate of 5 dB for halving of exposure time (i.e., 95 dBA for 4 hours, 100 dBA for 2 hours, and 105 dBA for 1 hour; 110 dBA for 30 minutes, 115 dBA for 15 minutes). Exposures to continuous noise levels greater than 115 dBA are not allowed under any circumstances. The limit to impulsive noise exposures is 140 dB peak sound pressure level.

When the daily noise exposure is composed of two or more periods of exposure at different (nonimpulsive) levels, their combined effect is determined by adding the individual contributions as follows:

$$\frac{C_1}{T_1} + \frac{C_2}{T_2} + \frac{C_3}{T_3} + \cdots + \frac{C_N}{T_N}$$

The values C_1 to C_N indicate the times of exposure to specified levels of noise, and the corresponding values of T indicate the total time of exposure permitted at each of these levels. If the sum of the individual contributions exceeds 1.0, the mixed exposures are considered to exceed the overall limit value. For example, if a man were exposed to 90 dBA for 5 hours, 100 dBA for 1 hour, and 75 dBA for 3 hours during an 8-hour working day, the times of exposure are $C_1 = 5$ hour, $C_2 = 1$ hour; and $C_3 = 3$ hour, and the corresponding OSHA limits are $T_1 = 8$ hour, $T_2 = 2$ hour, and T_2 = infinity. Therefore, the combined exposure dose for this man would be $5/8 + 1/2 + 3/\infty = 1.125$, which exceeds the specified limit of 1.0.

The impulsive noise exposure limit of 140 dB peak sound pressure level of the 1972 OSHA Rules and Regulations does not specify a limit for the number of impulses that a person can be exposed to in an 8-hour working day, but it can be expected that a limit such as 100 impulses for 8 hours may be set in a modification of the OSHA noise criteria. Perhaps different peak level limits will be specified for a greater number of impulses, such as 135 dB for exposures to 100 to 1000 impulsive sounds; 130 dB for exposures to 1000 to 10,000 impulsive sounds; and 125 dB for exposures to more than 10,000 impulsive sounds in 8 hours.

The noise exposure limits specified in the OSHA regulations are not intended to provide complete protection for all persons. They are set forth as the most restrictive limits that are deemed feasible with due consideration given to other factors, such as economic impact. Wherever feasible, therefore, hearing conservation measures should be initiated at levels considerably below those specified by OSHA. The ideal action point for initiating hearing conservation measures would be about 70 dBA, since a few very susceptible individuals may be permanently injured by sustained exposures at higher levels. However the economic impact of limits set at this low sound pressure level would not be feasible in most situations. Also, many activities away from the workplace cause noise exposures greater than 70 dBA, making it generally impractical to use this exposure limit unless the normal life-style of the country is changed radically. Certainly, every effort should be made to institute hearing conservation measures for extended exposures above 80 dBA.

The initiation of noise exposure limits that are below those specified in the OSHA regulations has very meaningful benefits (other than to avoid an OSHA citation). Of course the most important benefit is that more noise-induced hearing loss may be prevented. In addition, the lower levels generally afford better working conditions, which should reduce annoyance and improve communication; thus safety conditions and the general well-being of workers should be improved. Economic advantages should include increased production with the lower noise levels and a reduction in claims (in future years) for noise-induced hearing loss. Also, since the OSHA limits for noise exposure may be lowered in the future, it is generally more economical to have noise levels as low as is feasible now rather than to attempt to institute control measures twice.

8 HEARING CONSERVATION PROGRAMS

8.1 General

An effective hearing conservation program may require background information presented in the foregoing sections of this chapter. Since particular situations may have conditions that call for special emphasis to be placed on a measurement methodology, a noise control procedure, or a psychological approach, care must be taken to design the program for the given conditions.

It should be kept in mind in all cases, that the objective of a hearing conservation program is to prevent noise-induced hearing loss. Simple compliance with local, state, or federal rules and regulations generally will not prevent all noise-induced hearing loss in susceptible individuals because the exposure limits selected for compliance purposes have necessarily been compromised, based on the economic impact of control measure costs. For the most part, the compromises between exposure level limits and economic impact have been made from an overall viewpoint that includes a majority of industrial situations; clearly then, it is economically feasible in many industrial situations to select lower exposure limits that will be more protective. The lowest and safest feasible limits are obviously desirable for the well-being of the workers. In addition, these low limits will be of economic value to the employer in the long run because they will minimize compensation claims that can be expected as long as any noise-induced hearing impairment continues. Furthermore, it is likely that the reduced noise levels will result in better overall working conditions that may contribute to safety and increased production.

8.2 Specific Requirements

Hearing conservation programs should be the joint responsibility of a company's industrial hygiene, preventive medicine, and safety programs. An effective hearing conservation program should provide for the identification of noise hazard areas, the reduction of the noise exposure to safe levels, and the measurement of hearing.

8.2.1 Identification of Noise Hazard Areas

Noise hazard areas can be identified by means of sound pressure measurements with A-frequency weighting (see Section 4) after the action exposure levels have been established. Action levels should be at least as low as those specified by the OSHA regulations, but also as low as is economically feasible for the particular set of conditions in each work area.

The noise hazard dose must be described in terms of both A-weighted sound pressure level and time of exposure. Noise doses (i.e., the sum of the C/T values) may be measured using a sound level meter and a clock, or they can be measured directly with dosimeters.

8.2.2 Reduction of Noise Exposure Levels

As soon as a noise hazard area has been determined, ear protective devices should be provided to reduce the exposures to safe levels (Section 5.2). Engineering means (Section 5.1) should then be employed to reduce the noise exposure levels to the limits established for the particular hearing conservation program. Continued monitoring of the wearing of ear protector devices must be maintained until engineering and control procedures have been used successfully to reduce the noise exposures to safe levels.

8.2.3 Measurement of Hearing

Hearing measurements (Section 3.8) made periodically are the only way to determine whether a personal ear protector program is preventing noise-induced hearing loss. Therefore hearing measurements should be continued, at least annually and more often for very high noise exposures, until the noise exposures are reduced to safe levels. Hearing measurements are also useful for the following purposes:

1. To observe hearing threshold changes that may occur throughout the period of employment.
2. To determine the ability to communicate.
3. To assist in proper job placement.
4. To assist in establishing the existence of hearing losses in new employees.
5. To diagnose.
6. To follow up.

8.3 Program Implementation

Much of the success of a hearing conservation program depends on the competence and motivation of the hearing conservationist. In addition to the overall coordination of the hearing conservation program, the hearing conservationist usually has the following specific duties:

1. To perform accurate hearing measurements.
2. To maintain accurate and reliable records.
3. To refer employees who have abnormal hearing thresholds for examination and diagnosis.
4. To educate employees and management about the need for hearing conservation.
5. To instruct employees in the use and care of personal hearing protectors.

Most of the duties listed can be handled best by one qualified individual. However it is often desirable to have a plant hearing conservation committee to assist in the establishment of general policies and in making special important decisions. Personnel who may contribute directly to a plant committee include the industrial hygienist, the industrial relations manager, the labor relations supervisor, the manufacturing engineer, the plant engineer, the plant manager, the plant physician, the plant nurse, and the safety engineer. In addition to those who are immediately concerned and responsible for the health and safety of the employee, top management and all levels of supervision must be well aware of the problem, and they must cooperate in every way if a hearing conservation program is to be successful.

REFERENCES

1. Occupational Safety and Health Standards (Williams-Steiger Occupational Safety and Health Act of 1970), U.S. Department of Labor, Federal Register, May 29, 1971.
2. "Occupational Exposure to Noise," U.S. Department of Health, Education and Welfare, National Institute for Occupational Safety and Health, 1972.
3. "Information on Levels of Environmental Noise Requisite to Protect Public Health and Welfare with an Adequate Margin of Safety," U.S. Environmental Protection Agency, March 1974.
4. J. L. Flanagan and H. Levitt, "Speech Interference from Community Noise," American Speech and Hearing Association Reports, Report 4, pp. 167–174, 1969.
5. M. B. Gardner, "Effect of Noise on Listening Levels in Conference Telephony," J. Acoust. Soc. Am., **36,** 2354–2362 (1964).
6. M. B. Gardner, "Effect of Noise, System Gain, and Assigned Task on Talking Levels in Loudspeaker Communications," J. Acoust. Soc. Am., **40,** 955–965 (1966).
7. A. S. House, C. E. Williams, M. H. L. Hecker, and K. D. Kryter, "Articulation Testing Methods: Consonantal Differentiation with a Closed Response Set," J. Acoust. Soc. Am., **37,** 158–166 (1965).
8. R. G. Klumpp and J. C. Webster, "Physical Measurements of Equally Speech-Interfering Navy Noises," J. Acoust. Soc. Am., **35,** 1328–1338 (1963).
9. K. D. Kryter, The Effects of Noise on Man, Academic Press, New York, 1970, Chaps II and III.
10. J. C. Webster, "Effects of Noise on Speech Intelligibility," American Speech and Hearing Association Reports, Report 4, pp. 49–73, 1969.
11. J. C. Webster, "Updating and Interpreting the Speech Interference Level (SIL)," J. Audio Eng., April 1970.
12. J. C. Webster, "Effects of Noise on the Hearing of Speech," Proceedings on the Second International Congress on Noise as a Public Health Problem, to appear in American Speech and Hearing Association Reports (1969).
13. D. E. Broadbent, "Effects of Noise on Behavior," in: Handbook of Noise Control, C. M. Harris, Ed., McGraw-Hill, New York, 1957, Chapter 10.
14. A. Carpenter, "Effects of Noise on Performance and Productivity," in: Control of Noise, National Physical Laboratories, Her Majesty's Stationery Office, London, 1962, pp. 297–310.
15. M. Loeb, "The Influence of Intense Noise on Performance of a Precise Fatiguing Task," Army Medical Research Laboratory Report No. 268, Fort Knox, Ky., 1957.
16. A. Carpenter, "How Does Noise Affect the Individual?" Impulse, No. 24, 1964.
17. A. Cohen, "Effects of Noise on Performance," Proceedings of the International Congress on Occupational Health, Vienna, Austria, A IV: pp. 157–160, 1966.
18. A. Cohen, "Vigilance Performance as Affected by Noise and Heat," report in preparation, 1967.
19. J. S. Felton and C. Spencer, "Morale of Workers Exposed to High Levels of Occupational Noise," Am. Ind. Hyg. Assoc. J., **22,** 136–147 (1961).
20. K. D. Kryter, "Psychological Reactions to Aircraft Noise," Science, **151** 1346–1355 (1966).
21. E. Grandjean, "Physiologische und Psychophysiologische Wirkingen des Larms," Menschen Umwelt, No. 4, 1960.
22. N. N. Shetalov, A. D. Sartausv, and K. V. Glotov, "On the State of the Cardiovascular System Under Conditions of Exposure to Continuous Noise," Labor Hyg. Occup. Dis., **6,** 10–14 (1962).
23. A. Bell, "Noise. An Occupational Hazard and Public Nuisance," Public Health Paper No. 30, World Health Organization, Geneva, Switzerland, 1966.
24. Report to the President and Congress on Noise. U.S. Environmental Protection Agency Document No. 92-63, pp. 1–46 to 1–48, February 1972.

25. "Noise as a Public Health Hazard," American Speech and Hearing Association Reports, Report No. 4, February 1969, pp. 105–109.
26. J. Sataloff and P. Michael, *Hearing Conservation,* Thomas, Springfield, Ill., 1973, pp. 181–186.
27. American National Standard Specifications for Audiometers, S3.6-1969, American National Standards Institute, New York.
28. Course Outline for Course Leading to Accreditation as an Occupational Hearing Conservationist, approved by the Council for Accreditation in Occupational Hearing Conservation, 1619 Chesnut Avenue, Haddon Heights, N.J. 08035.
29. American National Standard Criteria for Permissible Ambient Noise During Audiometric Testing, S3-X-197 (not yet published). American National Standards Institute, New York.
30. Intersociety Committee on Industrial Audiometric Technician Training, "Guide for Training of Industrial Audiometric Technicians," Am. Ind. Hyg. Assoc. J., **27,** 303 (1966).
31. P. L. Michael, "Standardization of Normal Hearing Thresholds," J. Occup. Med., **10,** 67 (1968).
32. E. L. Eagles and L. G. Doerfler, "Hearing in Children: Acoustic Environment and Audiometer Performance," J. Speech Hear. Res., **4,** 149 (1961).
33. Environ. Health Lett., **6** (17), 7 (1967).
34. J. J. Knight, "Normal Hearing Threshold Determined by Manual and Self-Recording Techniques," J. Acoust. Soc. Am. **39,** 1184 (1966).
35. E. C. Riley, J. H. Sterner, D. W. Fassett, and W. L. Sutton, "Ten Years Experience with Industrial Audiometry," Am. Ind. Hyg. Assoc. J., **22,** 151 (1961).
36. A. Glorig, Noise and Your Ear, Grune & Stratton, New York, 1958.
37. S. Rosen, D. Plester, A. El-Mofty, and H. V. Rosen, "High Frequency Audiometry in Presbycusis: A Comparative Study of the Maaban Tribe in the Sudan with Urban Populations," Arch. Otolaryngol., **79,** 1 (1964).
38. I. S. Whittle, and M. E. Delaney, "Equivalent Threshold Sound-Pressure Levels for the TDH39/MX41-AR Earphone," J. Acoust. Soc. Am., **38,** 1187 (1966).
39. Paul L. Michael and G. R. Bienvenue, "Calibration Data for a Circumaural Headset Designed for Hearing Testing," J. Acoust. Soc. Am., **60,** 944–950 (1976).
40. Guide for Industrial Audiometric Technicians, Employers Insurance of Wausau, Wausau, Wis.
41. W. E. Snow, "Significance of Reading of Acoustical Instrumentation," Noise Control, **5,** 40 (September 1959).
42. L. L. Beranek, Acoustic Measurements, Wiley, New York, 1949.
43. C. M. Harris, Ed., Handbook of Noise Control, McGraw-Hill, New York, 1957.
44. American National Standard Specification for Sound Level Meters, S1.4-1971, American National Standards Institute, New York.
45. American Standard Specifications for Octave, Half-Octave, and Third-Octave Filter Sets: S1.11-1966, American National Standards Institute, New York.
46. American National Standard for Preferred Frequencies and Band Numbers for Acoustical Measurements, S1.6-1967, American National Standards Institute, New York.
47. H. H. Scott, "The Degenerative Sound Analyzer," J. Acoust. Soc. Am., **11,** 225 (1939).
48. Industrial Noise Manual, 3rd ed., 1975; American Industrial Hygiene Association, 66 S. Miller Road, Akron, Ohio 44313.
49. H. E. von Gierke and D. R. Warren, "Protection of the Ear from Noise: Limiting Factors," Benox Report, University of Chicago, December 1, 1953, pp. 47–60.
50. C. W. Nixon and H. E. von Gierke, "Experiments on the Bone-Conduction Threshold in a Free Sound Field," J. Acoust. Soc. Am., **31,** 1121–1125 (1969).

51. Unpublished work by Paul L. Michael and David F. Bolka at Pennsylvania State University, University Park.
52. American National Standard Method for the Measurement of the Real-Ear Protection of Hearing Protectors and Physical Attenuation of Earmuffs, S3.19-1974, American National Standards Institute, New York.
53. John C. Webster, "Updating and Interpreting the Speech Interference Level (SIL)," J. Audio Eng. Soc., **18,** 114–118 (1970).
54. I. Pollack and J. M. Pickett, "Masking of Speech by Noise at High Sound Levels," J. Acoust. Soc. Am., **30,** 127–130 (1958).
55. K. D. Kryter, "Effects of Ear Protective Devices on the Intelligibility of Speech in Noise," J. Acoust. Soc. Am. **18,** 413–417 (1946).
56. Paul L. Michael, "Ear Protectors—Their Usefulness and Limitations," Arch. Environ. Health **10,** 612–618 (1965).
57. W. F. Meeker, "Active Ear Defender Systems: Component Considerations and Theory," Wright Air Development Center Technical Report No. WADC TR 57-368 (I), Wright-Patterson Air Force Base, Ohio, 1958.
58. W. F. Meeker, "Active Ear Defender Systems: Development of a Laboratory Model," Wright Air Development Center Technical Report No. WADC TR 57-368 (II), Wright-Patterson Air Force Base, Ohio, 1959.
59. American National Standard Methods for the Calculation of the Articulation Index, ANSI S3.5-1969, American National Standards Institute, New York.
60. K. D. Kryter, "Speech Communication in Noise," AFCRCOTR054-52, Air Force Cambridge Research Center, Air Research and Development Command, Bolling Air Force Base, Washington, D.C., 1955.
61. J. C. Webster, "SIL—Past, Present and Future," Sound Vib. Mag. August 1969.
62. "Compilation of State and Local Ordinances on Noise Control," Congressional Record, October 29, 1969, pp. E9031–E9112.
63. "Model Community Noise Control Ordinance," U.S. Environmental Protection Agency, EPA 550/9-76-003, September 1975.
64. Walsh-Healy Public Contracts Act, Fed. Reg., **34** (96), May 20, 1969.
65. "Threshold Limit Values of Physical Agents," adopted by American Conference of Governmental Industrial Hygienists, 1969, 1014 Broadway, Cincinnati, Ohio 45202.
66. Occupational Safety and Health Act of 1970, Public Law 91-596, 91st Congress, S.2193, December 29, 1970.
67. "Draft ISO Recommendation: Noise Assessment with Respect to Community Response," No. 1996, International Organization for Standardization, November 1969.
68. R. W. Young and A. Peterson, "On Estimating Noisiness of Aircraft Sounds," *J. Acoust. Soc. Am.,* **45,** 834–838 (1969).
69. "Procedure for Describing Aircraft Noise Around an Airport," ISO Recommendation R507, International Organization for Standardization, June 1970.
70. R. W. Young, "Measurement of Noise Level and Exposure," in: Transportation Noises, University of Washington Press, Seattle, 1970, p. 45.
71. Laws and Regulatory Schemes for Noise Abatement, U.S. Environmental Protection Agency, NTID 300.4, December 31, 1971.
72. Karl D. Kryter, "Perceived Noisiness (Annoyance)," in: The Effects of Noise on Man, Academic Press, New York, 1970, p. 269.
73. "Community Noise," U.S. Environmental Protection Agency NTID 300.3, December 31, 1971.
74. A Study of the Magnitude of Transportation Noise Generation and Potential Abatement, Vol. III,

Report OST-ONA-71-1, Department of Transportation, Office of Noise Abatement, Washington, D.C., November 1970.

75. W. J. Galloway and A. C. Pietrasanta, "Land Use Planning Relating to Aircraft Noise," Technical Report No. 821, Bolt Beranek and Newman, Inc., published by the Federal Aviation Administration, October 1964. Also published by the Department of Defense as AFM 86-5, TM 5-365, NAVDOCKS P-98, "Land Use Planning with Respect to Aircraft Noise."
76. W. J. Galloway and D. E. Bishop, "Noise Exposure Forecasts: Evolution, Evaluation, Extensions, and Land Use Interpretations," FAA-NO-70-0, August 1970.
77. Wyle Laboratories Research Staff, "Supporting Information for the Adopted Noise Regulations for California Airports," WCR 70-3(R) Final Report to the California Department of Aeronautics, January 1971.
78. "The Adopted Noise Regulations of California Airports," Title 4, Register 70, No. 48-11-28-70, Subchapter 6, Noise Standards.
79. Community Reaction to Airport Noise, Final Report, Vol. 1, Tracor Document No. T-70-AU-7454-U, September 1970.
80. J. B. Ollerhead, "An Evaluation of Methods for Scaling Aircraft Noise Perception," Wyle Laboratories Research Staff Report WR70-17, Contract NASI-9527, May 1971.
81. American National Standard Specification for Sound Level Meters, ANSI S1.4-1971, American National Standards Institute, New York.
82. "Method for Measurement of Community Noise," Draft No. 3, February 17, 1969, revised May 27, 1969, editorial changes added November 11, 1969. American National Standards Institute, New York.
83. Report to the President and Congress on Noise, U.S. Environmental Protection Agency, December 31, 1971.
84. Federal Aviation Regulations, Part 36. Noise Standards: Aircraft Type Certification, November 1969.
85. C. H. G. Mills and D. W. Robinson, "The Subjective Rating of Motor Vehicle Noise, Appendix IX, Noise, presented to Parliament by the Lord President of the Council and Minister for Science by Committee on the Problem of Noise, July 1963; Her Majesty's Stationery Office, London, reprinted 1966.
86. "Selection of a Unit for Specification of Motor Vehicle Noise," Appendix A, Urban Highway Noise: Measurement Simulation and Mixed Reactions, Bolt, Beranek and Newman Report No. 1505, April 1967.
87. W. A. Rosenblith, K. N. Stevens, and the Staff of Bolt Beranek and Newman, Handbook of Acoustic Noise Control, Vol. 2, Noise and Man, Wright Air Development Center Technical Report No. WADC TR-52-204, Wright-Patterson Air Force Base, Ohio, 1953.
88. K. N. Stevens, W. A. Rosenblith, and R. H. Bolt, "Community Noise and City Planning," in: Handbook of Noise Control, C. N. Harris, Ed., McGraw-Hill, New York, 1957, Chapter 35.
89. H. O. Parrack, "Community Reaction to Noise," in: Handbook of Noise Control, C. N. Harris, Ed., McGraw-Hill, New York, 1957, Chapter 36.
90. K. N. Stevens and A. C. Pietrasanta, and the Staff of Bolt, Beranek and Newman, "Procedures for Estimating Noise Exposure and Resulting Community Reactions from Air Base Operations," WADC TN-57-10, Wright-Patterson Air Force Base, Ohio; Wright Air Development Center, 1957.
91. G. L. Bonvallet, "Levels and Spectra of Traffic, Industrial and Residential Area Noise," J. Acoust. Soc. Am., **23,** 435–439 (July 1951).
92. K. N. Stevens, "A Survey of Background and Aircraft Noise in Communities Near Airports," NACA Technical Note No. 3379, December 1954.
93. Ray Donley, "Community Noise Regulation," Sound Vib. Mag., February 1969.
94. Myles Simpson and Dwight Biship, "Community Noise Measurements in Los Angeles, Detroit, and Boston," Bolt, Beranek and Newman Report No. 2078, June 1971.

95. "Noise in Towns," in: Noise, Chapter IV, pp. 22–31, presented to Parliament by the Lord President of the Council and Minister for Science by Committee on the Problem of Noise, July 1963; Her Majesty's Stationery Office, London, reprinted 1966.

96. N. Olson, "Statistical Study of Traffic Noise," APS-476, Division of Physics, National Research Council of Canada, Ottawa, 1970.

97. I. D. Griffiths and F. J. Langdon, J. Sound Vib., **8,** 16 (1968).

98. Wyle Laboratories Research Staff, "Noise from Transportation Systems, Recreation Vehicles and Devices Powered by Small Internal Combustion Engines," WR71-17, Office of Noise Abatement and Control, Environmental Protection Agency, Washington, D.C., November 1971.

99. "Social Survey in the Vicinity of London (Heathrow) Airport," Appendix XI, in: Noise, presented to Parliament by the Lord President of the Council and Minister for Science by Committee on the Problem of Noise, July 1963; Her Majesty's Stationery Office, London, reprinted 1966.

100. P. N. Borsky, "Community Reactions to Air Force Noise," WADC Technical Report No. 60-689, Parts 1 and 2, Wright-Patterson Air Force Base, Ohio, March 1961.

101. W. J. Galloway and H. E. von Gierke, "Individual and Community Reaction to Aircraft Noise: Present Status and Standardization Efforts," International Conference on the Reduction of Noise and Disturbance Caused by Civil Aircraft, London, November 1966.

102. D. W. Robinson, "The Concept of Noise Pollution Level," National Physical Laboratory, Aerodynamics Division, NPL Aero Report Ac 38, March 1969.

103. D. W. Robinson, "Towards a Unified System of Noise Assessment," J. Sound Vib., **14** (3), 279–298 (1971).

104. C. G. Bottom and D. M. Waters, "A Social Survey into Annoyance Caused by the Interaction of Aircraft Noise and Traffic Noise," Department of Transport Technology TT-7102, Loughborough University of Technology.

105. A Study of the Magnitude of Transportation Noise Generation and Potential Abatement, Vols. 1–7, Department of Transportation, Office of Noise Abatement, Washington, D.C., November 1970.

106. "Noise and Vibration Characteristics of High Speed Transit Vehicles," Technical Report No. OST-ONA-71-7, Department of Transportation, Office of Noise Abatement, Washington, D.C., June 1971.

107. A. Cohen, "Noise and Psychological State," in: Proceedings of the National Conference on Noise as a Public Health Hazard, American Speech and Hearing Association Report No. 4, February 1969, pp. 89–98.

108. "Safety and Health Standards for Federal Supply Contracts (Walsh-Healy Public Contracts Act)," U.S. Department of Labor, Fed. Reg., **34,** 7948 (1969).

109. American Conference of Governmental Industrial Hygienists," Threshold Limit Values of Physical Agents," adopted by ACGIH for 1970.

110. Occupational Safety and Health Standards (Williams-Steiger Occupational Safety and Health Act of 1970), U.S. Department of Labor, Fed. Reg., **36,** 10518 (1971).

CHAPTER ELEVEN

Nonionizing Radiation

GEORGE M. WILKENING

1 INTRODUCTION

1.1 Current Interest in the Nonionizing Radiations

The development and proliferation, over the past decade, of electronic devices that either intentionally or inadvertently emit nonionizing radiation, have brought about immense interest in the subject. The promise of an increase in the number and use of such devices has concerned many persons who believe that the radiation hazards have not been sufficiently studied. Those who express concern about an inadequate understanding of the biological effects of nonionizing radiations point out that many electronic devices have already found their way into common use (e.g., microwave ovens, radar for pleasure boats, scanning lasers in supermarket checkout counters, near-ultraviolet radiation in fluorescent lighting fixtures, and a variety of high intensity light sources). Other concerns include the many infrared, ultraviolet, microwave, and laser devices that might produce excessive occupational exposures. Because of the heightened public interest in electromagnetic radiation hazards, Congress enacted the Radiation Control for Health and Safety Act (1). The declared purpose of the act is to establish a national electronic product radiation control program, including the development and administration of performance standards to control the emission of electronic product radiation. The act covers both ionizing and nonionizing electromagnetic radiations emitted from any electronic product. This includes X-rays and gamma rays, and particulate, ultraviolet, visible, infrared, millimeter wave, microwave, radiofrequency, and interestingly enough, sonic, infrasonic, and ultrasonic radiation. The administration of the act rests with the Bureau of Radiological Health (BRH) of the U.S. Department of Health, Education and Welfare (HEW). Since the inception of the act, BRH has conducted or funded research into the biological effects of radiation, with special emphasis on low level effects. Standards have been developed and promulgated for TV sets (receivers), medical X-rays (amendments to existing standard), cathode ray tubes, microwave ovens, and lasers. Calibration, measurement, and product testing laboratories have been established to ensure proper evaluation of accessible radiation from electronic products, and a compliance program has been developed to obtain manufacturers' adherence to established standards. During the course of standards development efforts, BRH has been required to consult with the Technical Electronic Product Radiation Safety Standards Committee (TEPRSSC), an advisory body established under the act. Unlike some federal advisory committees, TEPRSSC has the authority to develop and recommend its own standards directly to the Secretary of Health, Education and Welfare.

Other federal agencies that are actively concerned with nonionizing radiation hazards include the Occupational Safety and Health Administration (OSHA), the Environmental Protection Agency (EPA); the American Conference of Governmental Industrial Hygienists (ACGIH), and the National Institute of Environmental Health Sciences (NIEHS). The National Council on Radiation Protection and Measurements (NCRP) has recently enlarged its scope of interest to include nonionizing radiation; its first effort is Scientific Committee 39 on Microwaves. In 1968–1969 the Electromagnetic Radiation Management Advisory Council (ERMAC) was established to advise the director of the Office of Telecommunications Policy (OTP) "on possible side effects and the adequacy of control of electromagnetic radiation. . . ." The main function of ERMAC to date seems to have been to review the individually funded bioeffects research programs of various governmental agencies and to advise the director of OTP on appropriateness and priority. Several very useful reports (2) have been issued.

1.2 Nature of Electromagnetic Radiation

One important aspect of electromagnetic energy is that it may be considered to be wavelike; having characteristics such as frequency, wavelength, and velocity of propagation; or it may be considered to consist of discrete quanta or energy packets (photons) having characteristics similar to those of particles (energy, momentum, etc.). This wave–particle duality is necessary to explain such physical phenomena as the photoelectric effect or diffraction. The photoelectric effect is most readily understood by considering the incident radiation to consist of a flux of discrete photons, each with a characteristic energy E. Diffraction of electromagnetic energy, on the other hand, can be approached by assuming the incident radiation to be wavelike. However what at first appears to be a contradiction is resolved by Planck's law, which states that the photon energy E is directly proportional to the frequency of radiation ν; that is, $E = h\nu$, where E is expressed in joules, h is Planck's constant equal to 6.625×10^{-34} joule and ν is the frequency, expressed in reciprocal seconds (sec^{-1}).

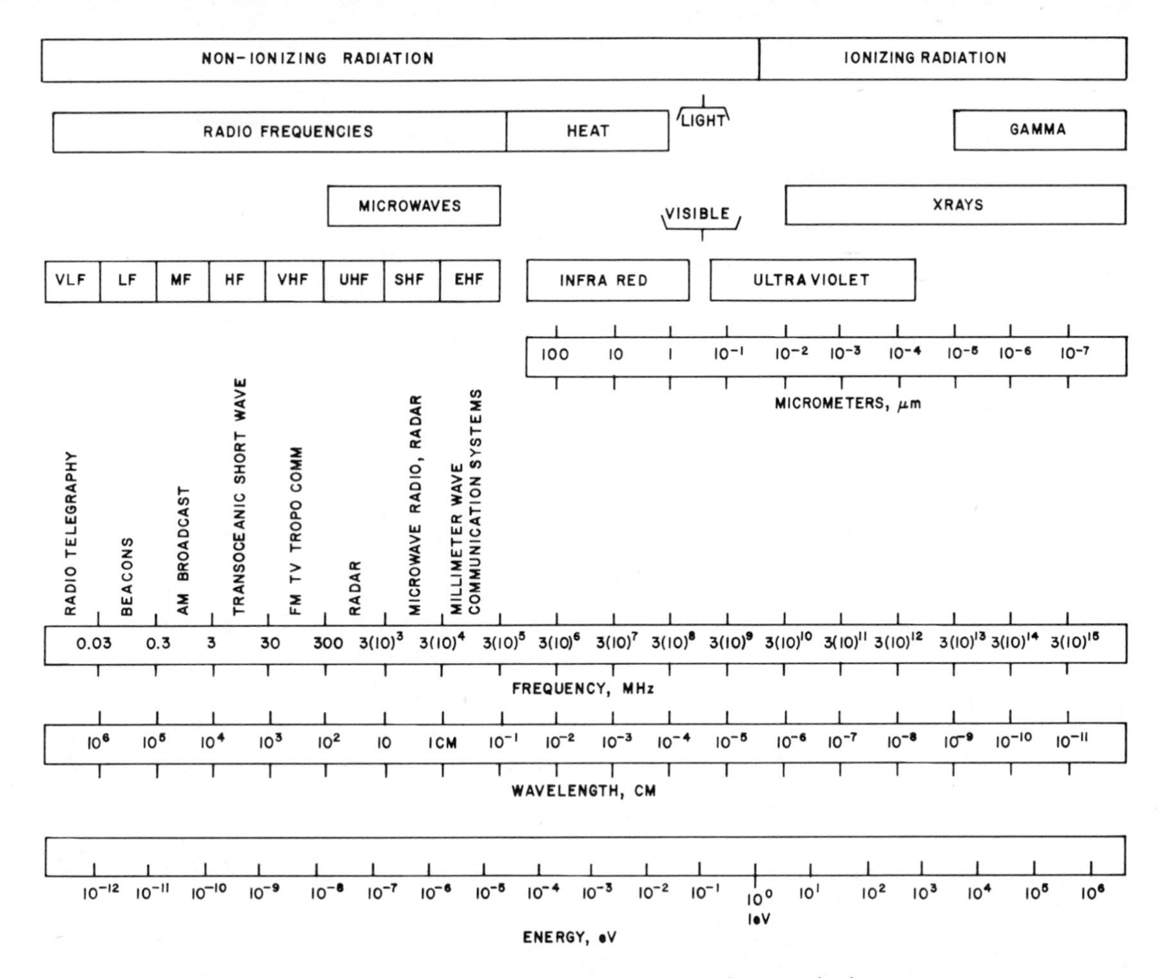

Figure 11.1 The electromagnetic spectrum. After Mumford (103).

In a more commonly used nomenclature, the photon energy is expressed in electron volts (eV), where one electron volt equals 1.602×10^{-19} joule.

As just indicated, the photon energies of electromagnetic radiations are proportional to the frequency of radiation, hence inversely proportional to wavelength. The electromagnetic spectrum extends over a broad range of wavelengths, from less than 10^{-12} cm to greater than 10^{10} cm. The shortest wavelengths, hence the highest energies ($>10^2$ eV), are associated with X- and gamma-rays, whereas the lower energies ($< 10^{-5}$ eV) are associated with radiofrequency and microwave radiation. Ultraviolet, visible, and infrared radiations occupy an intermediate position. For example, radiofrequency wavelengths may range from 3×10^{10} to 3×10^2 μm; infrared from 3×10^2 to about 0.7 μm; visible from approximately 0.7 to 0.4 μm, ultraviolet from approximately 0.4 to 0.1 μm, and gamma- and X-radiation below 0.1 μm. Figure 11.1 is a graphic representation of the entire electromagnetic spectrum.

The nuclear binding energy of protons may be of the order of 10^6 eV or greater; hence these energies are much greater than the binding energy of chemical bonds, which is of the order of 1 to 15 eV. The thermal energy associated with molecules at room temperature is even lower, that is, approximately 0.03 eV. Since the photon energy necessary to ionize atomic oxygen and hydrogen is of the order of 10 to 12 eV, it seems appropriate to adopt a value of approximately 10 eV as a lower limit at which ionization is produced in biological material. The electromagnetic radiations that do not produce ionization in biological systems may be presumed to have photon energies less than 10 to 12 eV and are therefore termed "nonionizing." Nonionizing radiation may be absorbed, causing changes in the vibrational and rotational energies of tissue molecules, thus leading to possible dissociation of the molecules; or, more often, the energy is dissipated in the form of fluorescence or heat (3, 4). Exposure to very high levels of nonionizing radiation may inflict considerable thermal injury, yet the damage will be due to nonionizing events.

This chapter is arbitrarily divided into sections on ultraviolet (UV), infrared (IR), visible, laser, radiofrequency (RF), microwave, and extremely low frequency (ELF) radiation. Laser radiation is included as a separate section because of the widespread special interest in lasers, even though lasers operate at wavelengths covered in the other sections. For this reason, the reader should consider the laser section to be an extension of material contained in the ultraviolet, visible, and infrared sections. In particular, the exposure limits contained in the laser section are generally applicable to UV, visible, and IR noncoherent sources.

2 ULTRAVIOLET RADIATION

2.1 Physical Characteristics

The ultraviolet wavelength range of interest extends from the vacuum UV (0.16 μm) to the near UV (0.4 μm) as follows:

UV Region	Wavelength Range (μm)	Photon Energy (eV)
Vacuum	0.16	7.7
Far	0.16–0.28	7.7–4.4
Middle	0.28–0.32	4.4–3.9
Near	0.32–0.4	3.9–3.1

The photon energy range for wavelengths between 0.1 and 0.4 μm lies between 12.4 and 3.1 eV.

A different but commonly used UV classification scheme, oriented toward biological effects, is as follows:

UV Region	Wavelength Range (μm)	Biological Implications
UV-A	0.4–0.32	Black light region Solar UV that reaches earth's surface UV-induced fluorescence
UV-B	0.32–0.28	Erythemal (sunburn) region
UV-C	<0.28	Germicidal region

In the wavelength region below 0.16 μm, UV radiation is completely absorbed by air and materials such as quartz; radiations can exist only in a vacuum. At somewhat longer wavelengths (0.16–0.2 μm) UV is poorly transmitted through air or quartz. At wavelengths in the range of 0.2 to 0.32 μm, UV is absorbed by ordinary window glass and by the epithelial layers of the skin and cornea; however transmission through air and water can occur. The ozone layer above the earth's surface absorbs solar radiation at wavelengths below 0.29 μm. At wavelengths in the range of 0.3 to 0.4 μm, UV is transmitted through air but is transmitted only partially through ordinary glass, quartz, and water.

2.2 Sources of UV Radiation

The sun represents the major source of UV energy at the earth's surface, even though the atmospheric ozone layer filters out all UV wavelengths below 0.29 μm. Significant man-made sources include high and low pressure mercury discharge lamps, plasma torches, and welding arcs. More than 85 percent of the radiation emitted by low pressure mercury vapor discharge lamps is at a wavelength of 0.2537 μm. At lamp pressures that are fractions of an atmosphere, the characteristic mercury lines predominate, whereas at high pressures (up to 100 atm), the lines broaden to produce a radiation continuum. In typical quartz lamps the amount of energy at wavelengths below 0.38 μm may be much greater than the radiated visible energy, depending on the mercury pressure. Other man-made sources include xenon discharge lamps, lasers, and certain fluorescent tubes that emit UV radiation at wavelengths above 0.315 μm, reportedly at an irradiance less than that measured outdoors on a sunny day (4).

The UV radiation in welding operations emits photons with sufficient energy (5 to 9.5 eV) to produce ozone and oxides of nitrogen from the dissociation of oxygen and nitrogen molecules (5).

2.3 Biological Interaction with UV Radiation

The penetration of UV radiation into human tissue is very limited. The deepest penetration may reach just below the epidermis if the wavelength is above 0.3 μm and the subject is essentially nonpigmented. The decrease in transmission through the skin as one proceeds from visible wavelengths to approximately 0.3 μm appears to be linear, indicating the lack of specific absorption bands in this wavelength range (6). Wavelengths between 0.28 and 0.32 μm penetrate appreciably into the corium or dermis; those between 0.32 and 0.38 μm are absorbed primarily in the epidermis, and those below 0.28 μm appear to be absorbed almost completely in the stratum corneum of the epidermis. The lens and tissues in the anterior segments of the eye may be exposed to UV if incident wavelengths are above 0.3 μm; the cornea, on the other hand, absorbs throughout the entire wavelength range of 0.1 to 0.4 μm, although much reduced between 0.32 and 0.4 μm. Since the cornea absorbs most of the energy up to 0.32 μm, radiation in this range generally produces corneal damage before lenticular effects are observed (7, 8).

The relative response of a biological system to irradiation at different wavelengths is called the "biological action spectrum" of that system. In some cases the action spectrum may be the same as the absorption spectrum; this is rarely true with complex biological systems, however. The maximum erythemal response at 0.29 μm in the biological action spectrum corresponds to absorption in tyrosine and other aromatic amino acids; the peak at 0.265 μm corresponds to absorption in pyrimidines and is thought to indicate changes in nucleic acids. The response at wavelengths longer than 0.3 μm shows no maxima or minima.

The advent of supersonic transport has raised speculation about detrimental effects to the atmospheric ozone layer caused by flight at exceptionally high altitudes. The same concern has been expressed about the interaction of halogenated chemicals (propellants) with the ozone layer. One estimate of damage is that a halving of the effective volume of the ozone layer would result in a two- to tenfold increase in the photobiological effects from UV (9).

2.4 Effects on the Skin

The immediate effects of UV radiation on the skin can be considered in terms of erythema, increased pigmentation (i.e., migration of melanin granules and the production of new melanin granules), darkening of pigment, and changes in cellular growth.

Depending on the total UV dose, the latent period for the production of erythema may range from 2 to several hours. Usually an immediate erythema appears, followed by a second, more intense erythema after a period of some 2 to 10 hours. The intensity of the erythema and the complicating sequelae (e.g., edema and blistering) is proportional to the UV radiation dose. Irradiation of the skin at wavelengths above 0.3 μm still produces an erythema, but the efficiency is low. Despite this relatively low efficiency, the amount of solar UV at wavelengths above 0.295 μm is sufficient to cause significant erythema in the population.

Photosensitizing agents may have biological action spectra in the UV range. The combined effect of skin contact with these agents and exposure to UV radiation may result in severe irritation and blistering. For example, it is common knowledge that workers who routinely expose themselves to coal tar products while working outdoors experience photosensitization of the skin.

The biological action spectrum for pigmentation changes is very similar to that of erythema. The first effect of UV irradiation appears to be a spreading of existing pigment granules, followed by the production of new pigment granules. Migration of melanin granules from the basal cells to the Malpighian cell layers of the epidermis may also cause a thickening of the horny layers of the skin. The process of increased pigmentation occurs immediately after irradiation. The mechanics of action is thought to be either oxidation of premelanin granules or free radical scavenger action of melanin. The period of increased pigmentation is usually followed by another period in which the skin is more vulnerable to an acute erythema induced at the same wavelength (10).

There appears to be no direct evidence that melanomas are produced by UV radiation; furthermore, melanomas are induced in experimental animals only with great difficulty. On the other hand, irradiation of persons suffering from xeroderma pigmentosum, an inherited disease caused by an insufficiency of DNA repair enzymes, will result in the early production of skin tumors (11).

Ultraviolet phototherapy has been used with infants to correct problems of icterus; however improper use has produced cases of erythema (12). There is widespread concern about the therapeutic use of UV radiation for Herpes, since this virus is already considered to be carcinogenic (13, 14).

In addition to erythema and increased pigmentation, slowing or cessation of cell mitoses in the basal and superficial layers of the epidermis often appears immediately after UV irradiation. An increase in mitoses is delayed for some 24 hours postirradiation (15), after which superfluous cellular material is sloughed off.

2.5 Effects on the Eye

The most common clinical sign of overexposure to UV at wavelengths shorter than 0.3 μm is photokeratitis, appearing some 2 to 24 hours after irradiation, along with acute hyperemia, photophobia, and blepharospasm if exposure is severe. The duration of such difficulties is usually of the order of 1 to 5 days. However there are usually no residual lesions.

2.6 Cancer

Although the premise that skin cancer may be induced by UV radiation is generally accepted in scientific circles, the evidence is largely indirect (i.e., epidemiologic and statistical). Firm quantitative data on a dose–response relationship do not exist. Also, the mechanism of UV-induced carcinogenesis is unknown. In spite of these uncertainties, however, certain epidemiologic studies have revealed a strong correlation between skin cancer and terrestrial solar UV-B levels found at given altitudes and ground elevations (16).

The types of skin cancer that are possibly related to UV radiation include basal cell carcinoma, spindle cell carcinoma, and melanoma, the latter generally considered to be a remote possibility. The induction of tumors is believed to occur as a result of irradiation at wavelengths below 0.3 μm (17), with the maximally effective wavelengths being in the 0.26 to 0.27 μm range, corresponding to specific interaction with nucleic acids.

The alleged high incidence of skin cancer in outdoor workers who come into contact with chemicals such as coal tar derivatives, benzpyrene, methylcholanthrene and other anthracene compounds raises questions about the role played by UV radiation in these cases.

There is no direct evidence that the UV radiation will produce tumors in the cornea or anterior chamber of the eye, however it is of some interest that melanoma of the eye is much more common in blue-eyed persons. Melanoma of the iris, for example, was found exclusively in blue-eyed persons in one study (18).

Abiotic effects from exposure to ultraviolet radiation occur in the spectral range of 0.24 to 0.31 μm, where most of the incident energy is absorbed by the corneal epithelium. Although the lens is capable of absorbing 99 percent of the energy below a wavelength of 0.35 μm, only a small portion of the radiation reaches the anterior lenticular surface. Photon energies of about 3.5 eV (0.36 μm) may excite the lens of the eye or cause the aqueous or vitreous humor to fluoresce, thus producing a diffuse haziness inside the eye that can interfere with visual acuity or produce eye fatigue. The phenomenon of fluorescence in the ocular media is not of concern from a bioeffects standpoint; the condition is strictly temporary and without detrimental effect.

An important conclusion about the effects of UV radiation on biological systems is that tissue damage appears to be dependent on the total energy absorbed (dose) rather than on the rate of energy absorption; hence the dose must be controlled to afford protection.

2.7 Threshold

Fairly good agreement exists on the amount of radiant exposure necessary to produce minimal photokeratitis. The threshold has been found to be 4 mJ/cm^2 for primates and humans when exposed to wavelengths between 0.220 and 0.310 μm. The minimal radiant exposure of 4 mJ/cm^2 was obtained at the most sensitive wavelength of 0.27 μm (19).

The energy necessary to elicit a barely perceptible reddening* of the skin is approximately 30 mJ/cm^2. Depending on skin type, the erythema thresholds may vary from 8 mJ/cm^2 for untanned skin to 50 mJ/cm^2 for tanned Caucasian skin. For white Caucasian skin with an average degree of tanning, the value for the minimal perceptible erythema is more on the order of 8 to 10 mJ/cm^2 when irradiation takes place at wavelengths between 0.24 and 0.29 μm. Any threshold number that is used should be identified with its wavelength, since some wavelengths are more effective than others in producing a given biological effect (action spectrum).

2.8 Exposure Criteria

The exposure criteria adopted by the Council on Physical Medicine of the American Medical Association based on erythemal thresholds for 0.2537 μm radiation are as follows: 0.5×10^{-6} W/cm^2 for exposure periods up to 7 hours (12.6 mJ/cm^2); 0.1×10^{-6} W/cm^2 for exposure periods up to and exceeding 24 hours (8.6 mJ/cm^2). Although these criteria are thought to be stringent, they have nevertheless been successfully used for many decades. Several important qualifications must be kept in mind when the AMA guidelines are used: (1) since the guidelines apply only to radiation at 0.2537 μm (germicidal), corrections must be made for the biological effectiveness at wavelengths other than germicidal, and (2) a tenfold safety factor has been applied to the germicidal photokeratitis threshold in deriving the recommended numerical values.

The ACGIH (20) recommended threshold limit values (TLV) for UV irradiation of unprotected skin and eyes for actinic wavelengths between 0.2 and 0.315 μm (200 and 315 nm) are given in Table 11.1. It should be recognized that these TLV have been derived on the basis of irradiating skin that has not received previous UV radiation. Note in this respect that the minimum TLV is somewhat below the photokeratitis threshold previously mentioned (i.e., 3 mJ/cm^2 rather than 4 mJ/cm^2). Conditioned or tanned individuals can tolerate skin exposure in excess of the TLV without erythemal effect. However such conditioning may not protect persons against possible skin cancer.

Exposure to wavelengths between 0.32 and 0.4 μm should not exceed the following criterion:

Wavelength, λ	TLV	Exposure Duration
0.32 to 0.4 μm	1 mW/cm^2	$>10^3$ sec
	1 J/cm^2	$<10^3$ sec

To determine the effective irradiance of a broadband source weighted against the peak of the spectral effectiveness curve (270 nm), the following weighting formula should be used:

* The minimal erythema dose (MED) is the smallest radiant exposure that produces a barely perceptible reddening of the skin that disappears after 24 hours. The MED is dependent on skin color, degree of pigmentation, age, and the irradiated body site.

$$E_{eff} = E_\lambda \, S_\lambda \, \Delta_\lambda$$

where E_{eff} = effective irradiance relative to a monochromatic source at 270 nm (0.270 μm)

E_λ = spectral irradiance [W/(cm^2)(nm)]

S_λ = relative spectral effectiveness (unitless)

Δ_λ = bandwidth (nm)

Table 11.2 shows relative spectral effectiveness as a function of wavelength.

Permissible exposure time in seconds for exposure to actinic ultraviolet radiation incident upon the unprotected skin or eye may be computed by dividing 0.003 J/cm^2 by E_{eff}, expressed in watts per square centimeter. The exposure time may also be determined using Table 11.3, which provides exposure times corresponding to effective irradiances in microwatts per square centimeter (W/cm^2 × 10^{-6}).

The TLV do not apply to lasers or to an exposure duration less than 0.1 second. Also no corrections have been made in these criteria for photosensitization to chemicals, drugs, and cosmetics (20).

2.9 Measurement of UV Radiation

One device that is particularly useful for measuring UV radiation is the thermopile. The coatings on the thermopile receiver elements are usually lampblack or goldblack, to

Table 11.1. Threshold Limit Values for Exposure to Ultraviolet Radiation (0.2–0.315 μm)

Wavelength (nm)	TLV (mJ/cm^{-2})
200	100
210	40
220	25
230	16
240	10
250	7.0
254	6.0
260	4.6
270	3.0
280	3.4
290	4.7
300	10
305	50
310	200
315	1000

Table 11.2. Relative Spectral Effectiveness by Wavelength

Wavelength (nm)	Relative Spectral Effectiveness, S_λ
200	0.03
210	0.075
220	0.12
230	0.19
240	0.30
250	0.43
254	0.5
260	0.65
270	1.0
280	0.88
290	0.64
300	0.30
305	0.06
310	0.015
315	0.003

Table 11.3. Permissible Ultraviolet Exposures

Duration of Exposure per Day	Effective Irradiance, $E_{eff}(\mu W/cm^2)$
8 hr	0.1
4 hr	0.2
2 hr	0.4
1 hr	0.8
30 min	1.7
15 min	3.3
10 min	5
5 min	10
1 min	50
30 sec	100
10 sec	300
1 sec	3,000
0.5 sec	6,000
0.1 sec	30,000

approximate the properties of a blackbody radiator. A special window material transparent to the UV is necessary to surround the thermopile element to minimize the detrimental effects of air currents. Commonly used materials for windows are quartz crystals, calcium and lithium fluoride, sodium chloride, and potassium bromide.

Other UV detection devices include (1) photodiodes, (e.g., silver–gallium arsenide, silver–zinc sulfide, and gold–zinc sulfide: peak sensitivity of these diodes is at wavelengths below 0.36 μm; the peak efficiency or responsivity is of the order of 50 to 70 percent); (2) thermocouples (e.g., Chromel–Alumel); (3) Golay cells; (4) superconducting bolometers; and (5) certain photomultiplier and vacuum photodiodes (21).

Care must be taken to use detection devices having the proper rise time characteristics (some devices respond much too slowly to permit the obtaining of meaningful measurement). Also, when measurements are being made special attention should be given to the possibility of UV absorption by many materials in the environment (e.g., ozone or mercury vapor), thus adversely affecting the readings. The possibility of photochemical reactions between ultraviolet radiation and a variety of chemicals also exists in the industrial environment.

Table 11.4 lists the sensitivity, impedance, and response time for certain junction detectors. It is possible to obtain a low radiation intensity calibration by exposing these and other detectors to a secondary standard furnished by the National Bureau of Standards.

Other devices that have been used to measure UV radiation include photovoltaic cells, photochemical detectors, photoelectric cells, and photoconductive cells. Photovoltaic and photoconductive cells are usually the same device operated in different modes (in series with an ammeter or across a load and voltmeter). It is common practice to employ selective filters, such as a monochrometer grating in front of the detecting device, to isolate the portion of the UV spectrum that is of interest to the investigator. Certain semiconductors such as copper or selenium oxide deposited on a selected metal develop a potential barrier between the layer and the metal. Light falling on the surface of the cell

Table 11.4. Sensitivity, Impedance, and Response Time of Junction Detectors

Type (circular)	Sensitivity [μV/(μW)cm^2)]	Impedance (Ω)	Response Time, 1/*e* (sec)
1 Junction: constantan-Mn	0.005	2	0.1
4 Junction: Cu-constantan	0.025	5	0.5
4 Junction: Bi-Ag	0.05	5	0.5
8 Junction: Bi-Ag	0.10	10	1.0
16 Junction: Bi-Ag	0.20	25	2.0
(linear)	(0.05)	(10)	(0.5)
12 Junction: Bi-Ag	0.05	10	0.5

causes the flow of electrons from the semiconductors to the metal. A sensitive meter placed in such a circuit will record a signal proportional to the intensity of radiation falling on the cell. Some photocells used for UV measurement take advantage of the property of certain metals to exhibit quantitative photoelectric responses to specific bands in the UV spectrum. Thus a photocell may be equipped with metal cathode surfaces that are sensitive to the UV wavelengths of interest.

One of the limitations of photocells is "solarization" or deterioration of the envelope, especially with long usage or following the measurement of high intensity radiation. Frequent calibration of the cell is necessary under these conditions. The readings obtained with these instruments are valid only when measuring monochromatic radiation or when the relationship between the response of the instrument and the spectral distribution of the source is known. A recent measurement technique for which the authors (22) claim some promise involves the use of beryllium oxide and lithium fluoride thermoluminescent crystals, which have been preirradiated with gamma-rays. The response of the detector, after several calibrated gamma predoses, appears to be a function of the UV exposure received from germicidal lamps.

For environmental measurements that are intended to correlate with a given biological effect, the ideal instrument would be designed in such a way that its spectral response closely approximated that of the biological action spectrum under consideration. Such an instrument is unavailable at this time. Since available photocells and filter combinations do not closely approximate the UV biological action spectrum, it is necessary to calibrate each photocell and meter at a number of wavelengths. Such calibrations are generally made at a great enough distance to ensure that the measuring device is in the "far field" of the source. In performing these calibrations, special care must be taken to control the temperature of the mercury lamps as secondary standards, since the spectral distribution of the radiation from the lamps is dependent on the pressure of the vaporized mercury.

2.10 Control of Exposure

Because ultraviolet radiations are so easily absorbed by a wide variety of materials, appropriate attenuation is accomplished in a straightforward manner. The information given in Section 2.8 (Exposure Criteria) should be used for the specification of shielding requirements. The data of Pitts (19) may be used for ultraviolet lasers because of the narrow band UV source employed in his experiments to determine the thresholds of injury to rabbit eyes. It is important to remember that photosensitization may be induced in certain persons at levels below the suggested exposure criteria.

Personal protection against radiation at wavelengths below 0.32 μm can be accomplished through the use of eyeglasses, goggles, plastic face shields, protective clothing, or sun screen creams or lotions. The wearing of tinted glasses or goggles is seldom recommended for protection against the UV intensities normally encountered in industry, however such devices may be used in a supplementary manner to reduce the high visible brightness that often accompanies UV emissions.

3 INFRARED RADIATION

Infrared (IR) wavelengths extend from 0.75 to approximately 10^3 μm. This range can be arbitrarily divided into three subareas known as the near infrared (0.75 to 3 μm), the middle infrared (3 to 30 μm), and the far infrared (30 to 10^3 μm).

Infrared radiation is emitted by a large variety of sources including the sun, heated metals, home electrical appliances, incandescent bulbs, furnaces, welding arcs, lasers, and plasma torches. The energy and wavelength characteristics emitted from IR sources are largely dependent on the source temperature. For example, at a temperature of 1000°C about 5 percent of the total energy is emitted at wavelengths below 1.5 μm, whereas at 1500 and 2000°C, the respective values are 20 and 40 percent (23).

Infrared radiation exchange between two heated bodies proceeds in accordance with the fourth power of the absolute tempeature of each body (Stefan-Boltzmann law). When persons are placed in the vicinity of large IR sources, the transfer of the heat load to the human body can represent a severe environmental stress. The problem of heat stress on the human body resulting from convective and radiative (as exemplified by large IR sources) heat loads is covered in detail in Chapter 20.

Infrared energy is not sufficiently energetic (< 1.5 eV) to cause the removal of electrons from orbital shells; therefore IR seldom enters directly into chemical reactions in biological systems and does not usually cause ionization. The usual mechanism for producing a biological effect is one involving absorption of energy with subsequent degradation to heat. However the possibility exists that IR energy at certain wavelengths and intensities may affect the vibrational or rotational properties of certain tissue molecules in such a way as to cause detrimental effects. Even in these cases the consensus is that the effects produced are largely thermal. The body organs at greatest risk from IR radiation are the eye and skin.

Most biological systems are considered to be opaque to wavelengths greater than 1.5 μm; transmission of IR through the ocular media at wavelengths from 1.3 to 1.5 μm is also poor. For skin, the wavelength region of high transmission is between 0.75 and 1.3 μm, with a maximum at 1.1 μm. At a wavelength of 1.1 μm, 20 percent of the energy incident upon the stratum corneum will reach a depth of 5 mm.

3.1 Eye

Thermal damage to the cornea usually results from the energy absorbed in the epithelium rather than in the deep stroma. The iris is especially susceptible to radiations below 1.3 μm. Since the iris can dissipate its heat load only to the surrounding ocular media, it is often regarded as a heat sink serving to mitigate the amount of radiation reaching the lens. It is thought that a radiant exposure of 4.2 J/cm^2 within a wavelength range of 0.8 to 1.1 μm will produce a minimally perceptible lesion on the iris. To obtain a radiant exposure of 4.2 J/cm^2 at the iris, it would be necessary to irradiate the cornea at a level of 10.8 J/cm^2 (24).

The transmission characteristics of the lens apparently vary with age and nuclear sclerosis. Selective IR absorption bands in the lens exist at wavelengths from 1.4 to 1.6 μm and from 1.8 to 2.0 μm (25).

Although the principal effect of infrared radiation on the eye and skin appears to be acute thermal injury, a controversy has raged throughout most of the twentieth century over the etiology of "glass blower's cataract," specifically, whether IR radiation is a cataractogenic agent. Although Dunn (26, 27) in 1950 was unable to uncover cases of cataract in employees who had been exposed to IR for more than 20 years, the evidence now seems to favor IR as the etiologic agent (28–32).

The formation of a cataract depends on initial heating of the anterior portions of the eye, especially the cornea and iris, followed by heat transfer from the iris to the lens epithelium. The elevation of temperature at the anterior portion of the lens is the primary etiologic factor in glass blower's cataract, according to Goldman (30, 31). In some recent studies of various glass and steel plants, a significant number of employees have shown lenticular changes after some 10 to 15 years of exposure to infrared sources (23). The reported range of irradiance to which employees were exposed was 0.08 to 0.4 W/cm^2, however no attempt was made to reconstruct exposure time–irradiance patterns for each employee. In practical cases where the eye is exposed to high temperature sources, there is often a significant near-infrared component, as well as some visible wavelength radiation. Special care is necessary to make distinctions between visible and IR emissions when industrial hygiene or epidemiological studies are made of infrared radiation and its possible relationship to cataract formation.

The retina is at risk from IR radiation only when near-infrared wavelengths are being generated by the source. The dissipation of energy within the retina is accomplished through conduction of heat to adjacent structures, notably the choroid. When the size of the retinal image and the rate of energy deposition are such that heat cannot be conducted away quickly enough to keep the tissue temperature below approximately 45°C, protein denaturation or destruction of tissue (microsteam generation and burn) may occur. More near-infrared energy than visible wavelength energy is required to produce a minimal lesion on the retina (threshold of damage as observed through an ophthalmoscope). For a discussion of maximum permissible exposure levels for infrared radiation, the reader is referred to Section 5 (Lasers).

3.2 Skin

Effects on the skin include vasodilatations of the capillary beds and increased pigmentation. The skin is normally able to dissipate a heat load imposed by IR radiation because of capillary bed dilatation, increased blood circulation, the production of sweat, and ambient air movement.

The perception of warmth by the skin is related to the rate at which the skin temperature is raised. For example, for skin temperatures of 32 to 37°C, the threshold of warmth perception is reached when the rate of skin temperature increase is of the order of 0.001 to 0.002°C per second. Such a perception threshold depends on the size of the skin area irradiated by IR as well as the density of thermoreceptors in that area. The thresholds for perceiving warmth decrease with increasing area of irradiated skin, but apparently the thresholds bear no relation to the absolute temperature of the skin. On the other hand, precooled skin can be rapidly heated without eliciting the sensation of warmth (3).

Unlike the thresholds of warmth sensation, the thresholds for pain appear to be directly dependent on skin temperature, despite wide individual variation in their values (17). Wertheimer and Ward (33) have recorded a range of skin temperature pain thresholds of 44.1 to 44.9°C, depending on body site. The experiments of Hardy et al. (34) indicate that the threshold of pain is reached when the skin temperature is raised to 45°C (44.5 ± 1.3°C). The same value also appears to be critical for producing a skin burn. Pain thresholds therefore appear to be related to skin temperature, whereas the extent of tissue damage is dependent on skin temperature and duration of exposure.

3.3 Measurement of Infrared Radiation

An almost unlimited variety of IR instrumentation is available to the industrial hygienist: spectrophotometers, radiation pyrometers, stationary or scanning radiometers, and other equipment. The various detectors usually fall into one of two categories:

1. *Thermal detectors,* where absorbed energy heats the detector, which in turn causes a change in the properties of the detector material. Response is to all IR wavelengths, but the response time is slow. Typical examples are thermopiles, thermocouples, liquid crystals, bolometers, and pyroelectric crystals.
2. *Quantum detectors,* where all incident photons are sufficiently energetic to free a bound electron. Response is to a limited range of IR wavelengths, but response time is fast. Examples of this type of detector are photodiodes, photomultiplier tubes, and semiconductor devices such as gallium arsenide diodes.

4 VISIBLE RADIATION

From the industrial hygiene standpoint the use of visible wavelength radiations generally involves two considerations: (1) proper lighting practices to achieve a pleasant visual environment and the efficient performance of tasks that require optimum visual acuity, and (2) the control of radiation to prevent damage to the human body, usually the eyes and skin. The first subject is covered in considerable detail in Chapter 13; the second is considered in this section.

The visible portion of the electromagnetic spectrum consists of a very narrow band of wavelengths between the ultraviolet and the infrared. The photon energies associated with the visible wavelength range of 0.38 to 0.75 μm lie between 3.1 and 1.65 eV.

These energy levels are rather low and innocuous in terms of producing direct biological effects on the skin surfaces of the human body, however radiation in the 0.38 to 0.75 μm range acts on a body organ that is uniquely and exquisitely designed to transduce the incident photons into intelligible vision by means of complex and incompletely understood photochemical reactions. This exquisite organ is the retina of the eye. The photoreceptors of the incident photons are the rods and cones. The differences in function of these two receptor systems are worthy of note. Cones respond to higher levels of ambient illumination than do the rods, hence their level of activity is normally greater during daylight hours. The cones also come into play for the discrimination of fine visual detail. Since the rods are in greater numerical abundance in the peripheral paramacular area of the retina, they provide a wide area peripheral view of the visual field. The rods and cones respond differently to different wavelengths: the spectral luminous efficiency for rods is at a wavelength of 0.510 μm, whereas the peak efficiency of the cones is at 0.555 μm (35). The retina is able to adapt to the wide range of light intensities, but the adaptation time varies with the luminous flux. Light adaptation occurs within a short time duration, usually of the order of seconds, whereas dark adaptation may require more than an hour.

Visible light is not hazardous to the eyes under ordinary circumstances, even when high intensity light sources are encountered. For example, the adaptation of the eye to high intensity sunlight by such mechanisms as restricted pupil size, light adaptation, partial closing of the eyelids (squinting), blinking, and shading of the eyes by the eyebrows and periorbital socket, are all usually sufficient to prevent excessive radiation from reaching the retina. On the other hand, certain conditions involving exposure to light may produce significant hazards for humans. For example, if light is pulsed or gated at a frequency near the alpha rhythm (brain function), certain light-sensitive individuals may experience an epileptiform seizure. The effect may be produced when such persons observe a pulsating light pattern such as flickering sunlight coming through a stand of trees as the light is observed from a moving vehicle, or when watching a pulsed light pattern from a television set. In some cases it is the pattern of light generation rather than its brightness that is critical to the production of seizures. Flash blindness is another condition that may result from exposure to high intensity light sources. Flash blindness is caused by the bleaching of visual pigments that produce an "afterimage" or temporary blind area (scotoma) in the field of vision. The greater the light intensity and time of exposure, the more persistent will be the afterimage. If the original stimulation by a high intensity light source is sufficient to completely obliterate focused perception, some protracted period of time may be required before normal visual function is restored. The sun may damage the retina when viewed without proper protection; eclipse blindness is a familiar example, where retinal damage is caused by exposing the eye to a combination of excessive visible and IR radiations.

On a lesser scale, ordinary glare caused by improper placement of light sources (fixtures) may produce severe visual discomfort and may represent a safety hazard because of interference with visual performance. A more serious condition is one in which a veiling glare may be temporarily superimposed on the retinal image, obscuring one's ability to discriminate details in the object being viewed. Such a condition exists on a more permanent basis in persons with cataracts.

High intensity light sources such as lasers, flashbulbs, spotlights, welding arcs, and carbon arcs can produce retinal burns under appropriate viewing conditions. Because of its coherent properties, a laser beam may be focused to an extremely small diameter. This results in the creation of highly intense beams of light, an obvious hazard to personnel.

Of the factors that have entered into the development of acceptable exposure levels for high intensity light sources, three seem to be of special importance.

1. The magnification factor of the eye as light passes through and is focused by the cornea and lens is of the order of 10^5. This means that the acceptable irradiance (W/cm^2) and the radiant exposure (J/cm^2) at the retina must have much greater numerical values than those present in front of the cornea where the industrial hygienist makes his field measurements. Permissible exposure levels are those which enter or impinge upon the eye at the cornea.
2. Acceptable exposure levels for wavelengths in the near infrared and infrared are somewhat less stringent than those for the visible wavelengths, whereas exposure levels for wavelengths in the blue and ultraviolet are more stringent, probably because of the photochemical mechanisms coming into play at the shorter wavelengths.
3. When exposed to a highly intense light source, most persons exhibit an "aversion response," that is, a tendency, to blink and turn the eyes and head. Presumably the elapsed time before the aversion response occurs gives some indication of the exposure time to be considered in establishing an exposure limit. In the case of the ANSI standard (36), the aversion time has been assigned a value of 0.25 second. The use of the aversion response concept in establishing safety guidelines for exposure to visible radiations has drawn criticism from some quarters because not all persons exhibit such a response. On the other hand, to establish safety standards strictly on the basis of experiments that irradiate the eyes of sedated animals whose eyelids are immobilized in an open position, to receive radiation pulses in excess of 0.25 second duration, means that practical conditions and normal physiologic response are ignored.

5 LASERS

Lasers are widely used for a variety of purposes such as alignment, welding, trimming, spectrophotometry, interferometry, flash photolysis, fiber optics communications systems, nuclear fusion experimentation, and surgical removal or repair procedures. A comprehensive review of medical and biological applications may be found in Wolbarsht (37).

The word "laser" is the acronym for "light amplification by stimulated emission of radiation." When an atom has been stimulated into an excited state, it may lose its excess energy through spontaneous decay, with the random emission of electromagnetic energy at a specific wavelength. When the atoms are raised to an excited state through an applied electromagnetic field, however, the radiated wave has the same phase, direction, and polarization as the stimulating (pumping) source. This process is known as stimulated emission. Since the population of atoms at a lower or ground state is normally greater than the number existing at a higher energy (excited) state, the energy of most radiations is usually absorbed by the atoms, thus tending to reduce or attenuate the field. However if the population of atoms can be reversed by stimulating a greater number to energy levels higher than those existing at ground state, the energy radiated by the stimulated emission will exceed the energy absorbed by the atoms; hence a phase coherent amplification of the incident radiation will occur. The net power difference (gain) will be proportional to the number of atoms that undergo a radiative decay to a lower energy level.

The energy stimulation and decay scheme just described is representative of a three-level (ground state, broad energy absorption band, and upper laser level) system, such as the ruby laser. Simply stated, atoms in a three-level system are pumped from the ground state to a broad energy absorption band, whereupon they rapidly decay by a radiationless transition into an upper laser level. Here they remain for a relatively long residence time (long compared with the time spent within the broad energy absorption band) before they return to the ground level, with an attendant emission of radiation at a specific frequency. The frequency of the radiation is equal to the difference in energy between the metastable upper laser level and the ground state divided by Planck's constant.

Many gas lasers and rare earth lasers (e.g., neodymium–yttrium-aluminum-garnet) operate in a four-level system, where a relatively small number of atoms (compared with a three-level system) are pumped from the ground state to achieve lasing.

Figure 11.2 depicts the energy diagram of a four-step laser system. Atoms are raised from ground state to a broad energy level (absorption band) through absorption of electromagnetic radiation, as in the ruby three-level system, or through electron impact as in the case of most gas lasers systems. Fast nonradiative relaxation occurs before the atoms are brought to level 2, the metastable level, where relaxation time is long and an accumulation of atoms occurs. Level 1, the lower laser level, has a fast relaxation time; in fact the dumping of atoms keeps the atom population low, to ensure population inversion. In passing from level 2 to level 1, radiation is emitted at the characteristic wavelength of the laser. The relative magnitude of the relaxation time at each level and the relative populations of atoms at each level are important parameters for achieving laser action.

Stimulated emission in gaseous systems was first reported in a helium–neon mixture (38). Since the early 1960s, lasing action has been reported at hundreds of wavelengths from the ultraviolet to the far infrared. Helium–neon lasers are typical of gas systems in which stable, single-frequency operation is important. The very popular helium–neon system can be operated in a pulsed or continuous wave mode at wavelengths of 0.6328, 1.15, or 3.39 μm, depending on resonator design. Typical power output for the helium–neon systems is of the order of 1 to 500 mW.

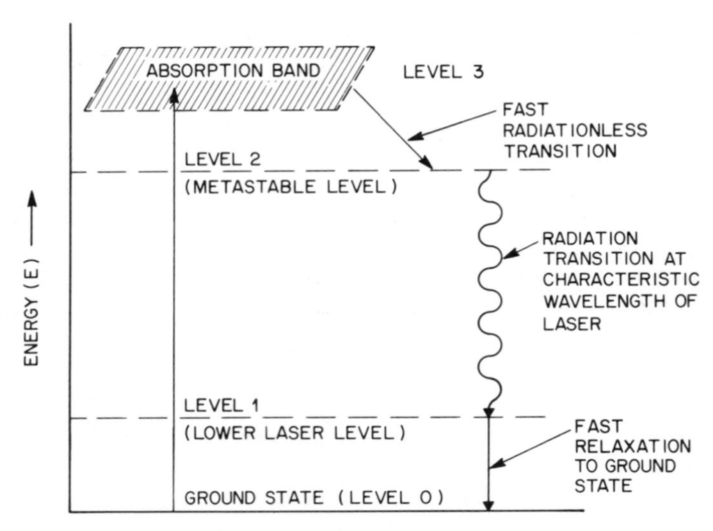

Figure 11.2 Four-level laser system.

The carbon dioxide (CO_2) laser belongs to a class known as molecular lasers. The energy for the system is derived from transitions between vibrational and rotational energy levels of CO_2 molecules. Such laser systems most frequently operate at wavelengths of 10.6 μm in either the continuous wave, pulsed, or *Q*-switched mode. A *Q*-switch is a device for enhancing the storage and quick discharge of energies to produce extremely high power pulses. The peak power output of a CO_2–N_2 system may range from several watts to greater than 10 kW. The CO_2 laser is attractive for terrestrial and extraterrestrial communication systems because of the low absorption window in the atmosphere between 8 and 14 μm. High powered DF–CO_2 (deuterium fluoride–carbon dioxide) transfer lasers are based on the combustion of carbon monoxide and oxygen, which subsequently thermally dissociates the fluoride before supersonic injection into a laser cavity. Both the DF-CO_2 laser and high powered CO lasers (design goal is to obtain more than 10 kJ output in subnanosecond pulses at an efficiency greater than 1 percent) are being used in laser fusion coupling experiments. The use of heavy ion beams, coupled with high energy accelerators of the storage ring type, seems to give a competitive edge over laser activation in experiments on inertial confinement fusion. However major activity continues with the development of high power transfer lasers for use in laser fusion research.

Of major significance from the personal hazard standpoint is the capacity of far infrared lasers to radiate enormous power at a wavelength that is invisible to the human eye.

The argon ion gas systems operate in the blue wavelengths, typically 0.4578, 0.488, and 0.5145 μm in either the continuous wave or pulsed mode. Power generation is usually less than 10 W.

Of the many ions in which laser action has been produced in solid state crystalline materials, perhaps neodymium (Nd^{3+}) in garnet, or glass and chromium (Cr^{3+}) in aluminum oxide (ruby) are most noteworthy (Table 11.5). Nd–glass lasers are already operational in the terawatt peak power range. Garnet or YAG (yttrium-aluminum-garnet) is an attractive host for the trivalent neodymium ion because the 1.06 μm laser transition line is sharper than in other host crystals. Frequently doubling to 0.53 μm, using lithium niobate crystals, may produce power approaching that available in the fundamental mode at 1.06 μm. Also, through the use of electrooptic materials such as potassium dihydrogen phosphate (KDP), barium–sodium niobate, or lithium tantalate, "tuning" or frequency scanning over wide ranges may be accomplished (39). Perhaps the most versatile device for obtaining a wide band of frequencies is the dye laser.

Semiconductor lasers are usually moderately low power devices (milliwatts to several watts) having relatively broad beam divergence, a factor that tends to reduce the potential ocular hazard. On the other hand, certain semiconductor lasers may be pumped with multikilovolt electron beams, thus introducing a potential ionizing radiation hazard (40). As a matter of fact, the X-rays from certain laser plasmas are already being considered for use in cancer detection.

Examples of certain ions that have exhibited lasing action appear in Table 11.5. The wavelength range of tunable infrared lasers is given in Table 11.6.

Through the use of carefully selected dyes, it is possible to tune through broad wavelength ranges. For example, if one is interested in producing lasing wavelengths in the vacuum ultraviolet (VUV), the beams from two dye lasers (one fixed at a double-quantum resonance wavelength, the other tuned over its tunable wavelength range) can be combined to obtain the VUV source. Through the use of four dye solutions, Sorokin et al. (41) were able to cover the range from 0.1578 to 0.1957 μm. Table 11.7 lists some additional lasers operating in the ultraviolet, and Table 11.8 gives the operating wavelengths and associated photon energies of certain lasers in common use.

Table 11.5. Certain Ions That Have Exhibited Lasing Action

Active Ion	Wavelength (μm)
Nd^{3+}	0.9–1.4
Ho^{3+}	2.05
Er^{3+}	1.61
Cr^{3+}	0.69
Tm^{3+}	1.92
U^{3+}	2.5
Pr^{3+}	1.05
Dy^{2+}	2.36
Sm^{2+}	0.70
Tm^{2+}	1.12

Table 11.6. Tunable IR Lasers

Laser Type	Wavelength (μm)
Diode	<1–34
Spin flip Raman	5–6.5
	9–14.6
Nonlinear device, parametric oscillators	<1–11
Difference frequency generator	2–6
Two-photon mixer	9–11
Four-photon mixer	2–25
Gas lasers	
Zeeman tuned	3–9
High pressure carbon dioxide	9–11

5.1 Biological Effects

Despite the dramatic benefits of technology, it is axiomatic to say that the improper use or design of any apparatus can produce undesirable effects. The laser is no exception. Nevertheless, although technical achievements often move in advance of any understanding of hazards, it has been encouraging to witness from the inception the serious attention given to the evaluation of laser radiation effects on biological systems (42).

The primary hazard from laser radiation is exposure of the eye and to a lesser extent, the skin. If the radiation levels are kept below those that damage the eye, there will be no harm to other body tissues. Therefore the material to follow emphasizes the effects of laser radiation on ocular tissue, the radiation exposure conditions required to produce detrimental effects (usually a threshold of injury), and some practical means of prevent-

Table 11.7. Recent Lasers at Wavelengths 360 nm and below (41)

Method	Medium	Wavelength (nm)	E/Pulse	Efficiency[a]
Nonlinear	$KB_5O_8 \cdot 4H_2O$	217.3–234.5	0.4 μJ	2×10^{-3} % (c)
	ADP (cooled)	208.0–212.4	0.5 mJ	5×10^{-3} % (c)
				50–60% (p)
	Rb–Xe	354.7	30 mJ	5–10% (p)
Molecular	H_2	109.8–161.3	5 kW/cm^2	
	Xe_2	172.0	0.76 J	1% (c)
	Xe_2[b]	172.0	30 μJ	
	$Ar–N_2$	357.7	45 kW/cm^2	2.3% (c)

[a] Abbreviations = c, conversion; p, photon.

[b] Continuous wave.

Table 11.8. Lasers in Common Use

Laser	Wavelength (μm)	Photon Energy (eV)
Ruby (Cr^{3+})	0.69	1.79
Nd–YAG	1.06	1.17
Nd–glass	1.06	1.17
GaAs	0.90	1.47
Dye	0.36–0.65	3.5–1.9
He–Ne	0.633–3.39	1.96 (λ:0.633)
He–Cd	0.325–0.4416	3.81 (λ:0.325)
CO_2	10.6	0.117
Ar	0.4519–0.5145	2.54 (λ:0.488)
CO	5.5	0.225
H_2	0.160	7.74
H_2O	7.0–220	0.0104 (λ:118.6)
HCN	773	0.0016

ing undue exposure. In describing the laser radiation that is incident upon tissue, it has become customary to use certain terms. For example, the output of pulsed lasers is described in terms of energy (joules), and that from continuous wave (CW) lasers in terms of power (watts). The corresponding beam density parameter of a pulsed laser is known as the "radiant exposure," expressed in joules per square centimeter (J/cm^2); that for a CW laser is known as "irradiance" and is expressed in watts per square centimeter (W/cm^2). Other parameters such as pulse duration, wavelength, spot size on the retina, pulse repetition rate (or pulse repetition frequency) are commonly used; still others are found in Table 11.A.1.

Effects of laser radiation on the eye can range from an annoying glare and mild bleaching of the photoreceptors to massive damage to the foveal region of the retina. The extent of injury depends on exposure conditions and the characteristics of the laser radiation incident upon the eye. In practical situations it is the short-term or intermittent type of exposure that is to be expected, rather than long-term, chronic exposure. Although the effects of long-term, chronic exposure have not been systematically studied, there is some indication that irreversible retinal effects may be produced when the eye is continuously bathed in a luminous flux, without interruption, for extended periods of time (43).

For the most part, work on the biological effects of laser radiation has been aimed at the elucidation of photobiological phenomena and the determination of tissue damage thresholds. "Damage" has usually been described in terms of changes that are grossly apparent or, in the case of the eye, observable with optical instruments such as microscopes and ophthalmoscopes.

More than 90 percent of the energy transmitted through the ocular media reaches the retina; and of the energy reaching the retina, most is absorbed in the neuroectodermal coat in the pigment epithelium. The cornea, lens, and ocular media are largely transparent in the visible region (0.380 to 0.750 μm). The greater part of visible radiation is absorbed in the melanin granules in the retinal pigment epithelium and choroid, which underlie the rods and cones. Figures 11.3 and 11.4 show the percentage transmission of various wavelengths of radiation through the ocular media and the percentage absorption in the retinal pigment epithelium and choroid, respectively. It is at the pigment epithelium that the greatest absorption of energy occurs; hence this thinly pigmented (10 μm) layer is the most susceptible to damage. Indeed, optical radiation damage may be produced in the pigment epithelium while the photoreceptor layer remains essentially unaffected. The graphs illustrate why the retina is the organ at risk with visible wavelength radiation, whereas the cornea and skin surfaces are at risk with IR and UV radiation.

Laser radiation damage may be attributed to thermal, thermoacoustic, or photochemical phenomena. For pulse durations of the order of 10^{-9} second and shorter, there is evidence that nonlinear mechanisms, Raman and Brillouin scattering, ultrasonic resonance, and acoustic shock waves may be brought into play (44). The wavelength shifts caused by scattering of electromagnetic radiation from acoustic modes (Brillouin) are small compared with scattering due to molecular vibrations and electron plasma oscillations (Raman). The precise mechanisms for producing damage under these short

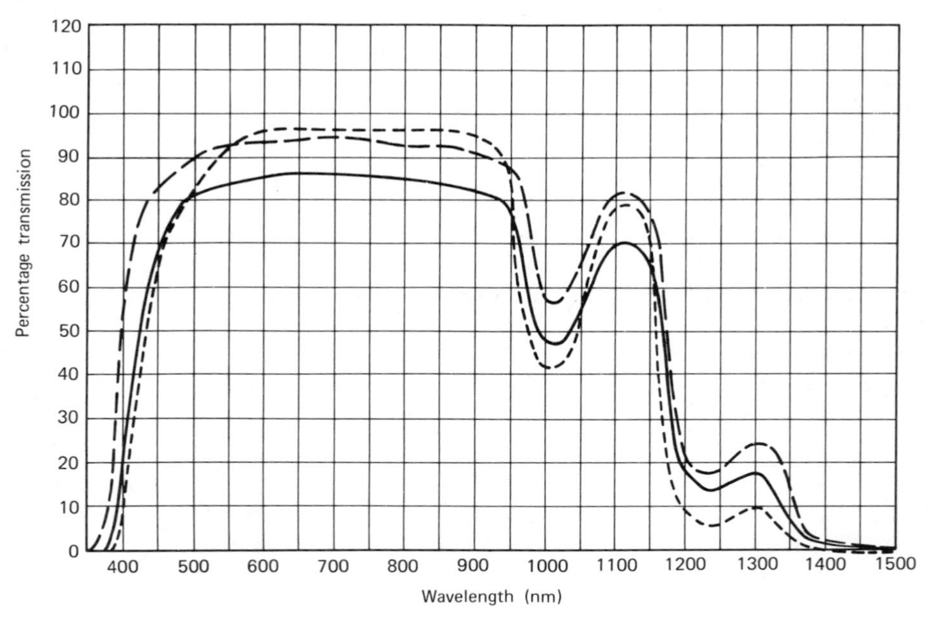

Figure 11.3 Percentage transmission for light of equal intensity through the ocular media of human (squares), rhesus monkey (solid curve), and rabbit (dashes). Curves are mean values.

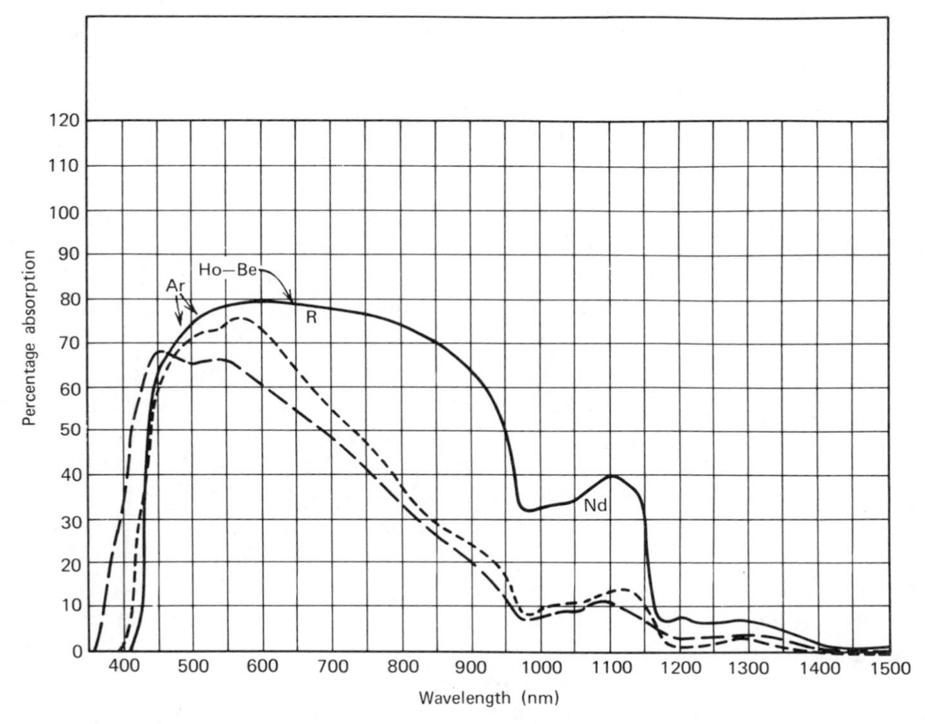

Figure 11.4 Percent absorption of light of equal intensity, incident on the cornea in the retinal pigment epithelium and choroid for Dutch chinchilla rabbits (dashes), rhesus monkey (solid curve), and man (dashes). Curves are mean values.

pulse conditions are not yet fully understood. The millisecond pulses from a solid state, visible wavelength laser may cause destruction of tissue by thermal means, whereas ocular and skin tissue may be damaged by photochemical mechanisms when exposed to UV and blue laser light. The consensus seems to be that several damage mechanisms exist. The mere stimulation of the photoreceptors caused by an irradiance that is only slightly higher than that presented to the eye under normal ambient conditions may be sufficient to increase cellular activity to the point of failure, especially if the irradiance is presented to the eye for an extended period of time (45).

When speaking of the thermal effects caused by laser radiation, one means the denaturation of protein through the absorption of energy and the consequent conversion to heat. Thermal injury is generally considered to be a rate process; in all probability, therefore, there is no single critical temperature at which injury takes place independent of exposure time. Also, since the molecules of the melanin granules in the pigment epithelium of the retina are relatively large, a broad spectral absorption would be expected to occur. Therefore the monochromaticity of laser radiation would not be expected to produce biological effects significantly different from those produced by radiation exposure to more conventional light sources. In fact, the coherence of the laser beam is not considered to be a significant factor in producing chorioretinal injury.

The biological response to UV laser radiation is expected to be similar to that produced by noncoherent UV sources. Photophobia, tearing, conjunctival discharge, surface exfoliation, and stromal haze accompanied by complaints of "sand in the eyes," are the expected consequences of excessive exposure. Damage to the corneal epithelium probably results from photochemical denaturation of proteins. Exposure to UV-C radiation (100 to 280 nm) and UV-B (290 to 315 nm) may produce photokeratitis. The latency period for photokeratitis varies from 30 minutes to as long as 24 hours, depending on the severity of the exposure. Chronic high level exposure to UV-A radiation (315 to 380 nm) may produce cataracts or, for that matter, retinal changes (8).

There are three wavelength regions in the infrared that have significance from a biological standpoint: IR-A (700 to 1.4 nm) IR-B (1.4 to 3 μm), and IR-C (3 μm to 1 mm). A transition zone occurs between the far end of the visible region where retinal effects are observed and the infrared region where corneal effects are produced. The biological effects in the IR-B region are primarily lenticular and corneal damage; in the IR-C region the damage is primarily corneal, since absorption by water (a primary constituent of the cornea) is very strong at these wavelengths. For wavelengths beyond the IR-C region, in the far infrared, absorption occurs in the peripheral layers of tissue. It is in this region as well as in the ultraviolet A and B regions that the threshold for damage to the cornea is comparable to that of the skin.

To translate these data into the potential hazards associated with specific laser systems, we can say that lasers operating in the visible wavelengths, such as ruby (0.6943 μm) or helium–neon (0.6328 μm), produce damage primarily in the pigment epithelium of the retina. Those that operate in the ultraviolet, such as helium–cadmium (0.3250 μm) and nitrogen (0.3371 μm), produce damage primarily in the cornea, although there is some recent evidence that effects may be produced in the lens and retina if the radiation levels are sufficiently high (8). Lasers that emit radiation in the near infrared, such as the semiconductor type, may produce injury to the lens, although the output power is usually too low to cause damage. Far infrared lasers such as the carbon dioxide (10.6 μm) have the potential for producing damage to the cornea. Unlike the visible wavelengths, there is no magnification of UV or far-infrared irradiation levels.

A retinal injury that occurs in the macula is more serious than one that occurs in the paramacula, since visual processes are more highly developed in the former. Injury to the paramacular region may have only a very minor effect on vision. Somewhat surprisingly, however, even when direct injury to the macula has occurred, near-normal visual acuity has apparently been restored in a matter of months.

Recent evidence (46) indicates that short wavelength light (0.4416 μm) produces retinal burns in primates by means of photochemical rather than thermal processes. Also, the thresholds of retinal burn were found to be considerably lower for irradiation at 0.4416 μm than for irradiation at 1.064 μm. For example, an irradiance of 24 W/cm^2 at a temperature of 23°C above the retinal ambient produced a threshold lesion in

1000 seconds using a wavelength of 1.06 μm (Nd–YAG laser). Using a wavelength of 0.4416 μm (He–Cd laser), however, required only 30 mW/cm^2, with a negligible temperature rise, to produce a lesion in a 1000-second exposure. The appearance of the lesion induced by the 0.4416 μm blue light was that of a light yellowish patch on the fundus, whereas the lesion induced by the 1.064 μm radiation had a typical center core "burn" characteristic. Table 11.9 supplies threshold lesion data for the wavelength range of 0.4416 to 1.064 μm (441.6 to 1064 nm).

Until recently the thresholds for producing retinal lesions as a result of irradiation at all visible wavelengths were considered to be of the same order of magnitude (i.e., 5 to 10 W/cm^2). However the data of Ham et al. (45) indicate a rather steep slope in the action spectrum at approximately 0.460 μm and below, particularly at exposure durations of 16 seconds or less. Figure 11.5 illustrates the effect of two factors, (decreasing wavelength and increasing exposure duration) on an increasing retinal sensitivity to laser radiation in the visible wavelengths. It is of considerable interest to note that although retinal sensitivity seems to increase as one shifts from the visible through the blue portions of the electromagnetic spectrum, there are fewer photons reaching the retina because of absorption in the lens and ocular media. Also, as expected, the site of absorption and damage changes from the retina to the lens.

Zuclich (8) has reported the UV induction of cataracts at energy doses that are lower

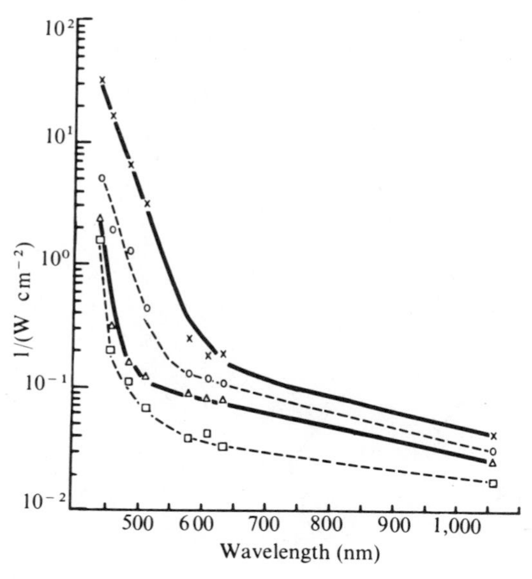

Figure 11.5 Action spectrum of retinal sensitivity to threshold damage in the rhesus monkey. Reciprocal of retinal irradiance 1/(W/cm^{-2}) is plotted semilogarithmically against wavelength for four different exposure durations. Beam diameter on the retina was 500 μm to the 1/e^2 points of the Gaussian distribution. Exposure durations (seconds): squares, 1; triangles, 16; circles, 100; crosses, 1000. From Ham et al. (46).

Table 11.9 Retinal Irradiance (W/cm^2) ± SD for Threshold Lesion as a Function of Wavelength and Exposure Duration[a] (46)

Laser	Wavelength (nm)	Exposure duration (s) 1	16	100	1,000	Transmittance (*T*)
Nd–YAG	1,064	56.1±5.3 55°	37.5±4.2 37°	32.5±3.1 32°	24.0±3.0 23°	0.76
He–Ne	632.8	29.9±1.4 49°	15.2±3.0 25°	8.4±2.8 14°	5.4±2.3 9°	0.93
Ar-pumped dye laser	610	22.0±1.9 36°	12.3±1.1 20°	8.1±1.1 13°	5.8±0.3 9.5°	0.92
	580	26.1±4.5 43°	11.5±2.4 19°	7.6±0.8 12.5°	4.0±1.7 6.6°	0.91
Ar	514.5	14.5±3.3 25°	10.3±2.3 18°	2.2±0.5 4°	0.32±0.1 1°	0.87
Ar	488	9.4±1.2 17°	6.1±0.8 11°	0.77±0.2 1°	0.15±.08 0.25°	0.83
Ar	457.8	5.1±0.8 10°	3.2±0.6 6°	0.52±0.1 1°	0.06±.02 0.1°	0.69
He–Cd	441.6	0.91±.03 2°	0.41±.02 1°	0.20±.02 0.4°	0.03±.01 0.05°	0.45

[a] Maximum retinal temperature (°C) above ambient estimated from mathematical model of Clarke; transmittance (*T*) through ocular media. Beam diameter on retina 500 m to $1/e^2$ points. Entire laser beam, TEM_{00} mode, enters eye of anesthetised animal with pupil dilated to 8 mm. A beam splitter and fundus camera are used to view and direct beam on the retina. Criterion for a threshold lesion is the appearance 24 hours after exposure of a funduscopically visible lesion.

than those required to cause corneal damage. The mechanism for inducing corneal damage or cataracts is believed to be photochemical, and that for the production of lenticular cataracts is thought to be thermal. The same investigator reports his finding that the cornea must be irradiated at a level tenfold its injury threshold to produce reversible clouding of the lens. On the other hand, immediate cataracts are produced from irradiation by nitrogen lasers operating at 0.3371 μm and 1 J/cm^2. This level of irradiation is reportedly an order of magnitude lower than the corneal injury threshold. Also, retinal lesions are produced using 0.3507 and 0.356 μm radiation reportedly at levels an order of magnitude below the corneal damage threshold. Therefore it would appear that the hazard to all eye structures is increased as one proceeds into the blue and ultraviolet wavelengths. Furthermore, the damage threshold appears to be dependent on the presence of oxygen in that a much higher incident energy is required to produce lesions when the cornea is flushed with nitrogen.

Exposure of rhesus monkeys to ultrashort single pulses of 25 to 35 picoseconds (psec) from an unfocused mode—locked Nd–YAG laser at a wavelength of 1.064 μm—resulted in a threshold injury to the retina at a mean energy of 13 $\pm$ 3 μJ (47). This compares with a threshold value of 68 $\pm$ 13 μJ determined by the same authors for a *Q*-switched Nd–YAG laser operating with pulse durations in the 9 to 30 nanosecond (nsec) range. Therefore it appears that shortening the pulse durations from the nanosecond to the picosecond range—or three orders of magnitude—results in a lowering of retinal injury threshold by a factor of approximately 5. The mechanism of injury appears to be the destruction of the retinal pigment epithelium through the combined effects of acoustic and shock wave pressure and the extremely high temperature rise in the melanin granules of the pigment epithelium.

Another study (48) using ultrashort pulse trains from Nd–glass gave much lower thresholds of injury than the Ham study: 1.7×10^{-8} J versus 1.3×10^{-5} J, using similar pulse durations. Since the threshold of 1.3×10^{-5} J (13 μJ) was obtained with single pulses rather than pulse trains, this study (48) will be repeated with single pulses. These results have complicated the setting of safety standards in that the present maximum permissible exposure (MPE) levels are largely based on pulse durations no shorter than nanoseconds.

Work has also been completed on the effect of ultrashort visible radiation pulses on rhesus monkeys (49). These results indicate that the melanin granules were much more severely damaged by the visible radiation than by 1.064 μm radiation, however the threshold of injury was higher for the visible wavelengths.

In future studies of the biological effects of laser radiation, greater attention should be given to techniques that detect functional as well as histologic changes in the eye structure. The development of such techniques may have an important bearing on the viability of the ophthalmoscope as a primary means for determining ocular change. Some investigators have already observed irreversible decrements in visual performance at exposure levels that are 10 percent below the threshold determined by observations through a ophthalmoscope. McNeer and Jones (50) found that at 50 percent of the ophthalmoscopically determined threshold, the ERG B wave amplitude was irreversibly reduced. Davis and Mautner (51) reported severe changes in the visually evoked cortical potential at 25 percent of the opththalmoscopically determined threshold. In addition to the problem of determining the true acute ocular injury threshold, there is a general deficiency of knowledge concerning the effect of long-term, chronic exposure to laser radiation.

5.2 Skin

The biological consequences of irradiating the skin with lasers are considered to be less than those caused by exposure of the eye, since skin damage is often repairable or reversible. On the other hand, exposure of the skin to high intensity radiation can cause depigmentation, severe burns, and possible damage to underlying organs. To keep the relative eye and skin hazard potential in perspective, one must not overlook possible photosensitization of the skin caused by use of drugs or cosmetic materials. In such cases the permissible exposure limits can be considerably below the currently recommended values.

Ultraviolet radiation in the wavelength region of 0.25 to 0.32 μm appears to be particularly injurious to the skin. Exposure to shorter (0.20 to 0.25 μm) and longer (0.32 to 0.40 μm) wavelengths is considered less harmful. Because of the structural inhomogeneities of the skin, any incident radiation will undergo multiple internal reflections; however most radiation is absorbed in the first 3.5 mm of tissue. The shorter wavelengths are absorbed in the outer dead layer of the epidermis (stratum corneum), whereas exposure to the longer wavelengths has a pigment-darkening effect. Chronic exposure to ultraviolet radiation accelerates skin aging and probably the development of skin cancer. Please refer to Section 2 on Ultraviolet Radiation.

5.3 Exposure Limits

5.3.1 Eye

The exposure limits developed by the American National Standards Institute (ANSI) (35) seem to have been generally accepted (with modification in certain cases) by the U.S. Department of Health, Education and Welfare and by international standards-setting organizations such as the World Health Organization (52) and the International Electrotechnical Commission (53). These limits are presented in Tables 11.10, 11.11, and 11.12. Table 11.10 illustrates the exposure limits for direct ocular exposure (intrabeam viewing) to a laser beam. Table 11.11 shows the exposure limits for viewing a diffuse reflection of a laser beam or an extended source laser, and Table 11.12 gives the exposure limits for skin.

In the ANSI standard, exposure limits are expressed in terms of the radiant exposure (J/cm^2) or irradiance (W/cm^2) required to produce a minimal lesion, usually a retinal lesion, which is visible through an ophthalmoscope 24 hours after exposure. All the ANSI criteria are based on acute exposure conditions, since very little information is

Table 11.10. Permissible Exposure Levels for Direct Ocular Exposures (intrabeam viewing) From a Laser Beam[a]

Wavelengths, λ	Exposure Duration, t (seconds)	Permissible Exposure Level[b]
Ultraviolet (μm)		
0.200–0.302	10^{-2}–3×10^{4}	3×10^{-3} J/cm^2
0.303	10^{-2}–3×10^{4}	4×10^{-3} J/cm^2
0.304	10^{-2}–3×10^{4}	6×10^{-3} J/cm^2
0.305	10^{-2}–3×10^{4}	1.0×10^{-2} J/cm^2
0.306	10^{-2}–3×10^{4}	1.6×10^{-2} J/cm^2
0.307	10^{-2}–3×10^{4}	2.5×10^{-2} J/cm^2
0.308	10^{-2}–3×10^{4}	4.0×10^{-2} J/cm^2
0.309	10^{-2}–3×10^{4}	6.3×10^{-2} J/cm^2
0.310	10^{-2}–3×10^{4}	1.0×10^{-1} J/cm^2
0.311	10^{-2}–3×10^{4}	1.6×10^{-1} J/cm^2
0.312	10^{-2}–3×10^{4}	2.5×10^{-1} J/cm^2
0.313	10^{-2}–3×10^{4}	4.0×10^{-1} J/cm^2
0.314	10^{-2}–3×10^{4}	6.3×10^{-1} J/cm^2
0.315–0.400	10^{-2}–10^{3}	1 J/cm^2
0.315–0.400	10^{3}–3×10^{4}	1×10^{-3} W/cm^2
Visible and near infrared (nm)		
400–700	10^{-9} to 1.8×10^{-5}	5×10^{-7} J/cm^2
400–700	1.8×10^{-5} to 10	$1.8\ (t/\sqrt[4]{t})$ mJ/cm^2
400–550	10 to 10^{4}	10 mJ/cm^2
550–700	10 to T_1	$1.8\ (t/\sqrt[4]{t})$ mJ/cm^2
550–700	T_1 to 10^{4}	$10C_B$ mJ/cm^{-2}
500–700	10^{4} to 3×10^{4}	C_B μW/cm^2
700–1059	10^{-9} to 1.8×10^{-5}	$5C_A \times 10^{-7}$ J/cm^2
700–1059	1.8×10^{-5} to 10^{3}	$1.8C_A\ (t/\sqrt[4]{t})$ mJ/cm^2
1060–1400	10^{-9} to 5×10^{-5}	5×10^{-6} J/cm^2
1060–1400	5×10^{-5} to 10^{3}	$9\ (t/\sqrt[4]{t})$ mJ/cm^2
700–1400	10^{3} to 3×10^{4}	$320C_A$ μW/cm^2
Far infrared		
1.4–10^{3} μm	10^{-9}–10^{-7}	10^{-2} J/cm^2
	10^{-7}–10	$0.56\sqrt[4]{t}$ J/cm^2
	>10	0.1 W/cm^2

[a] Note the following modification factors:

C_A = see Figure 11.6.

C_B = 1 for λ = 400 to 550 nm.

C_B = 10{exp[0.015(λ − 550 nm)]} for λ = 550 to 770 nm.

T_1 = 10 seconds for λ = 400 to 550 nm.

T_1 = 10 × 10{exp[0.02(λ − 550)]} for λ = 550 to 700 nm.

[b] The limiting aperture for all permissible exposure levels (PEL) for wavelengths greater than 0.1 mm and less than 1 mm is 1 cm. For all other skin PELs and for UV and IR-B, and IR-C ocular PELs the limiting aperture is 1 mm. For ocular PELs in the visible and near IR, the limiting aperture is 7 mm.

Table 11.11. Permissible Exposure Levels for Viewing a Diffuse Reflection of a Laser Beam or An Extended Source Laser[a]

Wavelengths, λ (nm)	Exposure Duration, t (seconds)	Permissible Exposure Level
Visible		
400–700	10^{-9} to 10	$10\sqrt[4]{t}$ J/(cm^2)(sr)
400–550	10 to 10^4	21 J/(cm^2)(sr)
550–700	10 to T_1	$3.83(t/\sqrt[4]{t})$ J/(cm^2)(sr)
550–700	T_1 to 10^4	21 C_B J/(cm^2)(sr)
400–700	10^4 to 3×10^4	$2.1\ C_B \times 10^{-3}$ W/(cm^2)(sr)
Near Infrared		
700–1400	10^{-9} to 10	$10C_A\sqrt[3]{t}$ J/(cm^2)(sr)
700–1400	10 to 10^3	$3.83C_A\ (t/\sqrt[4]{t})$ J/(cm^2)(sr)
700–1400	10^3 to 3×10^4	$0.64C_A$ W/(cm^2)(sr)

[a] Modification factors C_A, C_B, and T_1 are as given in note *a*, Table 11.10.

available on the long-term, chronic effects of laser radiation. For this reason caution must be exercised in applying such criteria to conditions when long-term viewing of optical sources is required. On the other hand, the use of the ANSI criteria should prove to be satisfactory for the overwhelming majority of practical situations in which accidental exposure to laser radiation occurs. Some questions have been posed regarding whether the so-called natural aversion response (the strong tendency to blink and look away from a dazzling or extremely bright light source) is truly protective under all circumstances. It is speculated that certain persons might be able to override the aversion

Table 11.12. Permissible Exposure Levels for Skin Exposure from a Laser Beam[a]

Spectral Region	Wavelengths, λ	Exposure Duration, t (seconds)	Permissible Exposure Level
UV	200–400 nm	10^{-3} to 3×10^4	Same as in Table 11.10
Visible and IR-A	400–1400 nm	10^{-9} to 10^{-7}	2×10^{-2} J/cm^2
	400–1400 nm	10^{-7} to 10	$1.1\sqrt[4]{t}$ J/cm^2
	400–1400 nm	10 to 3×10^4	0.2 W/cm^2
IR-B and IR-C	1.4 μm–1 mm	10^{-9} to 3×10^4	Same as in Table 11.10

[a] The limiting aperture for all skin PELs for wavelengths greater than 0.1 mm and less than 1 mm is 1 cm. For all other skin PELs and for UV and IR-B, and IR-C ocular permissible exposure levels the limiting aperture is 1 mm.

response if sufficiently well motivated; that is, some might be able to force themselves to stare at the extremely bright laser beam. The aversion response is generally thought to occur in a period of 0.1 to 0.25 second. In the ANSI Z136 standard, a value of 0.25 second has been used in calculating the permissible exposure limits for lasers operating at visible wavelengths.

The laser product standard proposed by the Bureau of Radiological Health establishes accessible radiation emission limits for different classes of lasers. These are given in Tables 11.13 through 11.16. A comparison of the BRH values with the ANSI figures reveals that the two are identical, with the exception of a less conservative ANSI limit for visible wavelength radiation. Other differences between these standards are discussed in the section on measurements.

The most recent revisions of the ANSI Z136.1 standard and the threshold limit values for lasers of the ACGIH call for a change in the exposure limits for visible and IR-A

Table 11.13. Wavelength-Dependent Correction Factors k_1 and k_2

Wavelengths, λ (nm)	k_1	k_2		
250–302.4	1.0	1.0		
>302.4–315	$10\left[\exp\left(\frac{\lambda - 302.4}{5}\right)\right]$	1.0		
>315–400	330.0	1.0		
>400–700	1.0	1.0		
>700–800	$\left[\exp\left(\frac{\lambda - 700}{515}\right)\right]$	If $t \leq 10{,}100/(\lambda - 699)$, then $k_2 = 1.0$	If $10{,}100/(\lambda - 699) < t \leq 10^4$, then $k_2 = [t(\lambda - 699)]/10{,}100$	If $t > 10^4$, then $k_2 = (\lambda - 699)/1.01$
>800–1060	$10\left[\exp\left(\frac{\lambda - 700}{515}\right)\right]$	If $t \leq 100$, then $k_2 = 1.0$	If $100 < t \leq 10^4$, then $k_2 = t/100$	If $t > 10^4$, then $k_2 = 100$
>1060–1400	5.0			
>1400–1535	1.0	1.0		
>1535–1545	$t \leq 10^{-7}$ $k_1 = 100.0$	1.0		
	$t > 10^{-7}$ $k_1 = 1.0$			
>1545–13,000	1.0	1.0		

Source. Fed. Reg., **40,** 148 (July 31, 1975).

Table 11.14. Limits for Accessible Radiation from Class 1 Lasers[a]

Wavelengths, λ (nm)	Emission Duration, t (seconds)	Class 1 Accessible Emission Limits		
>250 but ≤400	≤3.0×10^4	$2.4 \times 10^{-5} k_1k_2$ J*		
	>3.0×10^4	$8.0 \times 10^{-10} k_1k_2t$ J		
>400 but ≤1,400	>1.0×10^{-9} to 2.0×10^{-5}	$2.0 \times 10^{-7} k_1k_2$ J	Point Source	Choose Least Restrictive Value
	>2.0×10^{-5} to 1.0×10^1	$7.0 \times 10^{-4} k_1k_2t^{0.75}$ J		
	>1.0×10^1 to 1.0×10^4	$3.9 \times 10^{-3} k_1k_2$ J		
	>1.0×10^4	$3.9 \times 10^{-7} k_1k_2$ J		
	>1.0×10^{-9} to 1.0×10^1	$10k_1k_2\sqrt[3]{t}$ J/(cm²)(sr)	Extended Source	
	>1.0×10^1 to 1.0×10^4	$20k_1k_2$ J/(cm²)(sr)		
	>1.0×10^4	$2.0 \times 10^{-3} k_1k_2t$ J/(cm²)(sr)		
>1400 but ≤13,000	>1.0×10^{-9} to 1.0×10^{-7}	$7.9 \times 10^{-5} k_1k_2$ J		
	>1.0×10^{-7} to 1.0×10^1	$4.4 \times 10^{-3} k_1k_2t^{0.25}$ J		
	>1.0×10^1	$7.9 \times 10^{-4} k_1k_2t$ J		

Source. *Fed. Reg.*, **40,** 148 (July 31, 1975).

[a] Accessible radiation for wavelengths between 250 and 400 nm must not exceed accessible limits at wavelengths between 1400 and 13,000 nm with k_1 and k_2 equal to 1 and with comparable sampling intervals.

Table 11.15. Limits for Accessible Radiation from Class 2 Lasers

Wavelengths, λ (nm)	Emission Duration, t	Class 2 Accessible Emission Limits
>400 but ≤700	$>2.5 \times 10^{-1}$ second	$1.0 \times 10^{-3} k_1 k_2 t$ J

Source. *Fed. Reg.*, **40,** 148 (July 31, 1975).

laser radiation where the exposure duration is greater than 10 seconds. A change was also made in the permissible exposure level for skin when exposed to IR-A laser radiation. The pertinent amended exposure criteria are presented in Tables 11.17 to 11.19. Figure 11.6 provides the correction factors for wavelengths between 700 and 1400 nm. In Figure 11.6 the factor is known as corrections factor $A(C_A)$, but the values are identical to those used in the BRH and ANSI criteria.

If the revised ANSI and ACGIH figures are compared with the comparable BRH values, the revised permissible exposure levels are seen to be less restrictive in the 0.4 to 1.4 μm wavelengths than are the BRH values. However all other values are identical. Additional changes (more stringent permissible exposure levels) will doubtless be made by ANSI in the blue and UV portions of the spectrum at the time of the next revision of the Z136 standard.

5.4 Repetitive Pulses

There is still a degree of uncertainty about the proper correction factors to apply to multiple laser radiation. Until this question is fully resolved, however, it is recommended (36) that the following measures be adhered to:

Table 11.16. Limits for Accessible Radiation from Class 3 Lasers

Wavelengths, λ (nm)	Emission Duration, t (seconds)	Class 3 Accessible Emission Limits
>250 but ≤400	$\leq 2.5 \times 10^{-1}$	$3.8 \times 10^{-4} k_1 k_2$ J
	$>2.5 \times 10^{-1}$	$1.5 \times 10^{-3} k_1 k_2 t$ J
>400 but ≤1400	$>1.0 \times 10^{-9}$ to 2.5×10^{-1}	$10 k_1 k_2 \sqrt[3]{t}$ J/cm² to a maximum value of 10 J/cm²
	$>2.5 \times 10^{-1}$	$5.0 \times 10^{-1} t$ J
>1400 but ≤13,000	$>1.0 \times 10^{-9}$ to 1.0×10^{1}	10 J/cm²
	$>1.0 \times 10^{1}$	$5.0 \times 10^{-1} t$ J

Source. *Fed. Reg.*, **40,** 148 (July 31, 1975).

Table 11.17. TLVs for Direct Ocular Exposures (Intrabeam Viewing) from a Laser Beam[a]

Spectral Region	Wavelengths λ (nm)	Exposure Duration, t (seconds)	TLV
Visible	400–700	10^{-9} to 1.8×10^{-5}	5×10^{-7} J/cm^2
	400–700	1.8×10^{-5} to 10	$1.8(t/\sqrt[4]{t})$ mJ/cm^2
	400–549	10 to 10^4	10 mJ/cm^2
	550–700	10 to T_1	$1.8(t/\sqrt[4]{t})$ mJ/cm^2
	550–700	T_1 to 10^4	$10C_B$ mJ/cm^2
	400–700	10^4 to 3×10^4	C_B μW/cm^2
IR-A	700–1059	10^{-9} to 1.8×10^{-5}	$5C_A \times 10^{-7}$ J/cm^2
	700–1059	1.8×10^{-5} to 10^3	$1.8C_A(t/\sqrt[4]{t})$ mJ/cm^2
	1060–1400	10^{-9} to 10^{-4}	5×10^{-6} J/cm^2
	1060–1400	10^{-4} to 10^3	$9(t\sqrt[4]{t})$ mJ/cm^2
	700–1400	10^3 to 3×10^4	$320C_A$ μW/cm^2

[a] Modification factors as follows: C_A, see Figure 11.6; $C_B = 1$ for $\lambda = 400$ to 550 nm; $C_B = 10\{\exp[0.015(\lambda - 550)]\}$ for $\lambda = 550$ to 700 nm; $T_1 = 10$ seconds for $\lambda = 400$ to 550 nm; $T_1 = 10 \times 10\{\exp[0.02(\lambda - 550)]\}$ for $\lambda = 550$ to 700 nm.

1. Limit the exposure to a single pulse in a train to the exposure recommended for the comparable single pulse time limits.
2. Limit the average irradiance for a group of pulses to the permissible irradiance for a single pulse that has the same time duration as the entire group of pulses.
3. Whenever the instantaneous pulse repetition frequency (PRF) exceeds one per second for pulse widths less than 10^{-5} second, multiply the maximum permissible exposure (MPE) limit by the correction factor given in Fig. 11.7.

Table 11.18. TLVs for Viewing a Diffuse Reflection of a Laser Beam or an Extended Source Laser

Spectral Region	Wavelengths, λ (nm)	Exposure Duration, t (seconds)	TLV
Visible	400–700	10^{-9} to 10	$10\sqrt{t}$ J/(cm^2)(sr)
	400–549	10 to 10^4	21 J/(cm^2)(sr)
	550–700	10 to T_1	$3.83(t/\sqrt[4]{t})$ J/(cm^2)(sr)
	550–700	T_1 to 10^4	$21/C_B$ J/(cm^2)(sr)
	400–700	10^4 to 3×10^4	$2.1/C_B \times 10^{-3}$ W/(cm^2)(sr)
IR-A	700–1400	10^{-9} to 10	$10C_A\sqrt[3]{t}$ J/(cm^2)(sr)
	700–1400	10 to 10^3	$3.83C_A(t/\sqrt[4]{t})$ J/(cm^2)(sr)
	700–1400	10^3 to 3×10^4	$0.64C_A$ W/(cm^2)(sr)

[a] Modification factors C_A, C_B, and T_2 are as given in note a, Table 11.10.

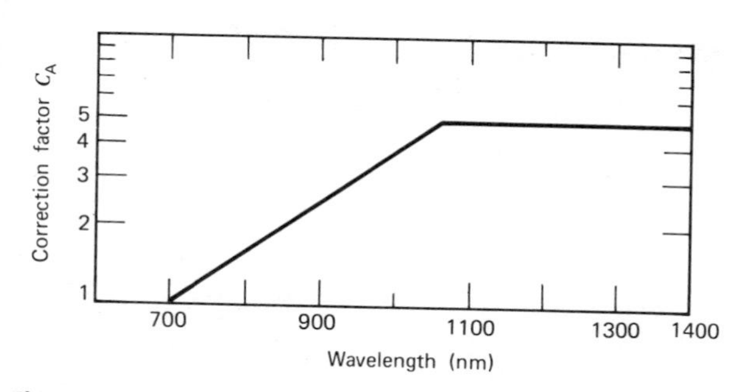

Figure 11.6 Correction factor C_A *for laser wavelengths (eye).*

5.5 Protection Guidelines

The organizations that have developed protection guidelines for lasers include the army, navy, and air force, ANSI, and the ACGIH. International standards are being prepared by the International Electrotechnical Commission (53) and the World Health Organization (52). In addition, a number of state governments have prepared codes or regulations on laser safety (54, 55).

A key approach to a laser control program is the proper classification of lasers in accordance with their potential hazards. The classification schemes that have been found to be generally acceptable are those developed by ANSI (35) and the Bureau of Radiological Health (56) in the United States. The provisions of the classification schemes for both ANSI and BRH are similar:

Class 1. Laser systems that are considered to be incapable of producing damaging radiation levels and are thereby exempt from control measures or other forms of surveillance. This is a no risk category.

Class 2. Visible wavelength laser systems that have a low hazard potential because of the expected aversion response. There is some possibility of injury if stared at for long periods of time. BRH requires the manufacturer to affix to the device a label cautioning the user not to stare at the beam. This is a low risk category.

Table 11.19. TLVs for Exposure of Skin to Laser

Spectral Region	Wavelengths, λ (nm)	Exposure Duration, t (seconds)	TLV[a]
Visible and IR-A	400–1400	10^{-9} to 10^{-7}	$2C_A \times 10^{-2}$ J/cm^2
	400–1400	10^{-7} to 10	$1.1C_A\sqrt[4]{t}$ J/cm^2
	400–1400	10 to 3×10^4	$0.2C_A$ W/cm^2

[a] $C_A = 1.0$ for $\lambda = 400$ to 700 nm; see Table 11.10 for greater wavelength values.

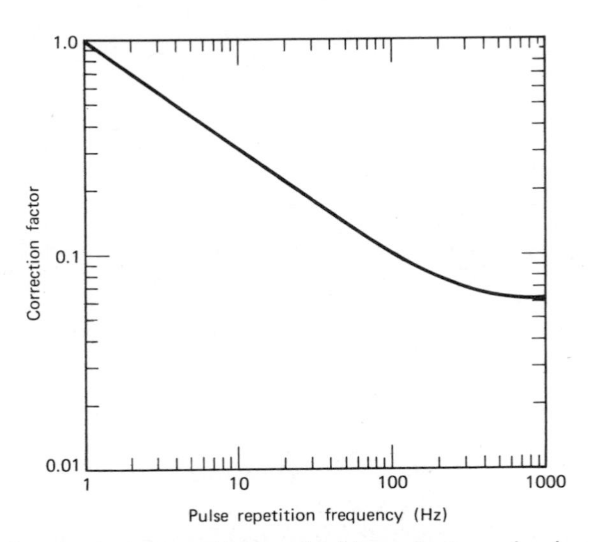

Figure 11.7 Correction factor for repetitively pulsed lasers having pulse durations less than 10^{-5} second. Maximum permissible exposure level for a single pulse of the pulse train is multiplied by the above correction factor. For pulse repetition frequencies greater than 1000 Hz the factor is 0.06.

Class 3. Laser systems in which intrabeam viewing of the direct beam or specular reflections of the beam may be hazardous. This is a moderate risk category. This class may be further subdivided into 3*a* and 3*b*; class 3*a* is considered to be hazardous only when the laser output beam is collected and focused by optical instruments. A label cautioning the user not to stare into the beam or view the direct beam with optical instruments will be required by BRH.

Class 4. Laser systems whose direct or diffusely reflected radiation may be hazardous, and where the beam may constitute a fire hazard. Class 4 systems (high power systems) require the use of controls that prevent exposure of the eye and skin to the direct or diffusely reflected beam, and the wording of the cautionary label should reflect that concern. This is a high risk category. When one considers the capacity of some laser systems for producing personal injury, it becomes absolutely mandatory to assign only well-informed and well-motivated persons to carry out the laser safety program.

5.6 Laser Devices Hazard Classification Definitions

Class 1: Nonrisk Laser Devices. Any laser, or system containing a laser, that cannot emit laser radiation levels in excess of specified exempt values of accessible radiant power or energy for a given exposure duration. The exemption from hazard controls for this class applies strictly to the emitted laser radiation hazard, not to other potential hazards. The term for exempt accessible radiant power is P_{exempt}; exempt accessible radiant energy is designated Q_{exempt}. The terms P_{exempt} and Q_{exempt} apply to the longest conceivable duration of exposure (exposure time) in a "worst case" situation or condition. Both P_{exempt} and Q_{exempt} are products of the intrabeam exposure limit (Table 11.10) and the area of the appropriate aperture. The appropriate aperture (opening) for visible wavelengths is $\pi d^2/4$ cm², where d is 0.7 cm, the diameter of the fully dilated pupil (worst case exposure conditions). For cases when the laser is not a point source (e.g., laser diode, arrays), P_{exempt} and Q_{exempt} are defined as the radiant power levels and radiant energy levels that do not exceed the extended source exposure limits (Table 11.11) when the source is viewed at the expected minimum viewing distance through a 0.7 cm exit aperture after the radiation has been collected by an 8 cm entrance aperture.

Class 2: Low Power Visible Laser Devices (Low Risk).

a. Visible (400 to 700 nm) CW laser devices. Devices that can emit accessible radiant power exceeding P_{exempt} for the classification duration (0.4 μW for the maximum expected exposure time greater than 0.25 second), but not exceeding 1 mW.

b. Visible (400 to 700 nm) scanning laser systems. These systems may be evaluated by specifying P_{exempt} at a point 10 cm from the point of emitted radiation. All repetitively pulsed laser devices and scanning systems are considered to be low risk if they do not exceed P_{exempt} for a 0.25-second exposure.

Class 3: Medium Power Laser Devices (Moderate Risk).

a. Infrared (1.4 μm to 1 mm) and ultraviolet (200 to 400 nm) laser devices. All such devices capable of emitting a radiant power in excess of the exempt quantity (P_{exempt}) for the classification duration, but incapable of emitting (1) an average radiant power in excess of 0.5 W for the maximum exposure time greater than 0.25 second, or (2) a radiant exposure of 10 J/cm² within an exposure duration of 0.25 second or less.

b. Visible (400 to 700 nm) CW or repetitively pulsed laser devices. All such devices that produce a radiant power in excess of the exempt power (P_{exempt}) for an 0.25-second exposure (1 mW for a CW laser) but cannot emit an average radiant power of 0.5 W for the maximum actual exposure time that exceeds 0.25 second.

c. Visible and near-infrared (400 to 1400 nm) pulsed laser devices. All such devices that can emit a radiant energy in excess of the exempt value (Q_{exempt}) but cannot emit a radiant exposure that exceeds either 10 J/cm² or that required to produce a hazardous diffuse reflection (Table 11.11).

d. Near-infrared (700 to 1400 nm) CW laser devices or repetitively pulsed laser devices. All such devices that can emit power in excess of the exempt quantity (P_{exempt}) for the classification exposure duration but cannot emit an average power of 0.5 W or greater for periods in excess of 0.25 second.

Class 4: High Power Laser Devices (High Risk).

a. UV (200 to 400 nm) and IR (1.4 μm to 1 mm) laser devices. All such devices that emit an average power in excess of 0.5 W for periods greater than 0.25 second or a radiant exposure of 10 J/cm^2 within an exposure duration of 0.25 second or less.

b. Visible (400 to 700 nm) and near-infrared (700 to 1400 nm) laser devices. All such devices that emit an average power of 0.5 W or more for periods greater than 0.25 second or a radiant exposure in excess of either 10 J/cm^2 or that required to exceed the hazardous diffuse reflection values (3.14 times the radiance values) given in Table 11.11.

Any laser device can revert to a lower risk classification by virtue of enclosure, but warning labels must indicate that a higher risk is encountered when access panels are removed.

An early consideration in the development of laser safety programs is the need for periodic eye and skin examinations for persons working with lasers. Until recently, the consensus seemed to be that eye examinations coupled with fundus photographs should be done semiannually. As experience has been gained with the eye examination protocol and with protective programs, the consensus has become that an initial eye examination upon entrance to work and a second examination upon termination of work should suffice for protective purposes, but only if a strong safety program has been instituted. This protocol should be supplemented with immediate eye examinations if emergency conditions should arise. The recommendation of ANSI is that workers receive eye examinations at approximately 3-year intervals. Another key consideration in protection programs is the preparation of written policies and practices outlining the responsibility of management, technical supervision, industrial hygiene, safety, and medical personnel. To the extent feasible, the policies and practices should be written and agreed upon before laser operations commence. Such policies and practices must emphasize the primary reliance to be placed on supervisory personnel for the safe conduct of laser operations and the preference for engineering controls rather than personal protective equipment in achieving goals. Only properly qualified persons should be placed in charge of laser operations.

5.7 Measurement of Laser Radiation

No single instrument can cover the required dynamic range, sensitivity, accuracy, precision, wavelength specificity, and stability of operation for proper measurement of the output from the wide variety of laser systems now in use. The wavelength range available in laser devices extends from the submillimeter to the vacuum ultraviolet; power ranges from microwatts to terawatts; and operational modes encompass both the CW and pulse durations from 10^{-1} to 10^{-13} second. The recent lesiglation promulgated by the Department of Health, Education and Welfare (56) requiring the manufacturer to indicate certain operating parameters and the classification of each laser device on a properly affixed label, tends to reduce the need for an elaborate measurement system. If such equipment is purchased, however, it is encouraging to note that high quality performance is readily obtainable. The output power of laser devices as specified by the manufacturer has rarely been found to differ from precision calibration data by more than a factor of 2 (4).

In choosing the appropriate detection and readout device for a specific measurement problem, one should strive to obtain (1) linear response over the wavelength range of interest, (2) uniform response over the area of the receiving aperture, (3) response time that is suitable for the measured signal (e.g., time constants that are much shorter than the rise and decay times of *Q*-switched or mode-locked pulses), and (4) detector materials whose saturation or damage thresholds are much higher than the irradiances (W/cm^2) or radiant exposures (J/cm^2) of incident laser beams.

High current vacuum photodiodes are useful for measuring the output of *Q*-switched systems and can operate with a linear response over a wide range of wavelengths (e.g., from the ultraviolet to the near-infrared). The response time is limited by the transit time of electrons from the photocathode to collector, but it is typically 0.5 nsec or less. Although the dark current is somewhat high (i.e, of the order of 5 nA at room temperature), the operation of photodiodes is very stable if they are used within the specified current limitations, and if the voltage of the power supply is well maintained. Since the sensitivity of the photocathode is usually nonuniform, a large fraction of the photocathode surface area must be well illuminated; otherwise measurement errors will be induced. The nonuniformity problem might be improved through the use of a planar photocell. Since the limiting incident power level of many vacuum photodiodes is of the order of one watt, it may become necessary to attenuate the beam. This can be accomplished by inserting optical interference filters, neutral density filters, or diffuse reflectors into the system, or by beam splitting.

Average power measurements of CW laser systems may be made with conventional thermopiles or photovoltaic cells. A typical thermopile can detect signals in the power range from 10 μW to about 100 mW. Since thermopiles are composed of many junctions, the response may be nonuniform unless the entire surface is fully illuminated by the laser beam. Bismuth–silver junctions are recommended for high sensitivity; copper–constantan junctions are better for rugged stability. Table 11.4 lists the sensitivity, impedance, and response time of typical junction detectors. Gold-black or platinum-black junctions provide the fastest response times. The use of the thermopile is not recommended for irradiance levels exceeding approximately 100 to 125 mW/cm^2. For the measurement of super high CW power (e.g., 1 to 5 kW), the use of a cone flow thermopile is recommended.

For high energy pulsed radiation, the ballistic thermopile should be considered. In this device the incident radiation travels down the inner surfaces of an active carbon cone, where absorption takes place. The active cone is balanced against an identical reference cone, to measure the difference in temperature and to minimize drift due to ambient temperature fluctuations. These devices employ a rather long response time characteristic (e.g., 1 to 1.5 minutes) to properly integrate the pulsed signals.

Photovoltaic cells consist of a junction of two materials that are specially selected to produce a contact potential. The two materials can be a semiconductor and a metal, or a type *n* and *p* semiconductor such as a silicon solar cell. The photovoltaic cells offer the following advantages: (1) the cell output is very stable with time, (2) external power supplies are not needed, and (3) the detector unit and associated circuitry is small enough to permit the construction of a portable device. Materials that are suitable as detector elements include indium arsenide and indium antimonide. In certain devices these detector elements are installed in glass Dewars flasks to cool them with liquid nitrogen. When this is done, the detectable wavelength band is expanded to include an upper limit of approximately 6 μm. Selenium cells are good for power measurements in the near ultraviolet; silicon solar cells are preferred for near-infrared measurements, and Schottky surface barrier silicon photodiodes are preferentially used for radiation in the blue wavelengths. Many calorimeters measure total energy, but they can also be used to measure total power if the time history of the radiation is known. Other calorimeters do not measure the temperature rise in the absorber; rather, they measure small changes in polarization, which in turn produce a potential difference across a detector. This "pyroelectric effect" occurs in ferroelectrics because such materials possess a permanent polarization that is temperature dependent. The pyroelectric detectors are especially suitable for the measurement of far-infrared radiation of the carbon dioxide and hydrogen cyanide lasers; however the input signals are usually chopped to prevent degradation of the sensing element. Table 11.20 summarizes certain characteristics of energy and power measuring devices.

For measurement throughout the infrared region, many quantum detectors are available. For example, germanium junction photodiodes are used for the near-infrared up to approximately 1.8 μm; PbS and InAs photodiodes may be used in the 1.8 to 3 μm range; PbTe, PbSe, and InSb are suitable for the 3 to 7 μm range, whereas germanium combined with a wide variety of metals (mercury, cobalt, zinc, copper, and gold) is preferred for the 7 to 40 μm range. Many of these detectors need liquid nitrogen cooling for proper performance.

Thermal detectors such as bolometers are still used where broad band spectral response is desired. A bolometer using a platinum ribbon sensor element gives a flat response over a wavelength range of 1 to 26 μm. The change in electrical resistance due to heating by the absorbed radiation is proportional to the absorbed radiant power in these devices. Since the time constants are usually of the order of 10^{-4} second, however, they are not recommended for the measurement of *Q*-switched pulses.

The type of readout device for the various detector systems varies with individual requirements. Microammeters and microvoltmeters are often used with CW systems; microvoltmeters or electrometers are frequently employed in conjunction with oscilloscopes to measure pulsed laser system parameters. All these devices may be coupled to recorders or panel displays. For purposes of calibration, tungsten ribbon filament lamps are available from the National Bureau of Standards as secondary standards of spectral radiance over the wavelength region of 0.2 to 2.6 μm. The calibration procedures using these devices permit comparisons within about 1 percent in the near ultraviolet and about a 0.5 percent in the visible. All radiometric standards are based on the Stefan-

Table 11.20. Devices for Energy and Power Measurements

Device	Range of Operation	Typical Response Time	Surface Damage[a]
Energy			
Cone calorimeters	10^{-2} to 2×10^{3} J	1–20 sec	10–20 mW/cm^2 in 50 nsec
Meter disk calorimeter	10^{-2} to 10 J	10 sec	50 mW/cm^2 in 50 nsec
Rat's nest calorimeter	10^{-3} to 10 J	10^{-4} sec	—
Wire calorimeter	10^{-3} to 0.5 J	10 sec	10 mW/cm^2 in 50 nsec
Liquid calorimeter	1–500 J	10–60 sec	—[b]
Torsion pendulum	0.5–500 J	60 sec	—
Integrating photocurrent	10^{-8} to 10^{-3} J	1 sec	1 mW/cm^2
Thermopile	10^{-6} to 1 J	10^{-1} sec	300 mW/cm^2
Copper sphere	5 10^{-4} to 10 J	180 sec	—
Power			
Phototube	10^{-8} to 10^{-3} W/cm^{-2}	3–10 nsec	1 mW/cm^2
Photodiode	10^{-4} to 6 W/cm^{-2}	0.3–4 nsec	10 W/cm^2 [c]
Nonlinear crystal	10^{3} to 10^{12} W/cm^{-2}	10^{-5} sec	10^{12} W/cm^2 [d]

Source. *Lasers in Industry*, S. S. Charschan, Ed., Van Nostrand Reinhold, New York, 1972, p. 521.
[a] Surface damage may occur at the indicated power density.
[b] Local boiling of the liquid should be avoided.
[c] Above this power, response of device is nonlinear.
[d] Breakdown of quartz at this power density.

Boltzmann and Planck laws of blackbody radiation. The spectral response of measurement devices should always be specified, since the ultimate use of the measurements is a correlation with the spectral response (action level) of the biological tissue receiving the radiation result.

The effective aperture (aperture stop) of any device that is used for measuring the irradiance (W/cm^2) or the radiant exposure (J/cm^2) in the visible wavelengths should closely approximate 0.7 cm (7 mm), the diameter of the fully dilated or dark-adapted pupil. Apertures other than 7 mm may be used for radiation at other wavelengths.

The specification of mutually inconsistent aperture diameters by two standards-setting organizations (ANSI and the BRH) has created some controversy. BRH and ANSI agree on the use of an 80 mm aperture for the purpose of collecting and focusing visible and near-infrared wavelength radiations down to a 7 mm aperture, to be able to estimate the hazard due to collecting optics. BRH and ANSI requirements are identical on the use of the 7 mm aperture. However ANSI specifies the use of a 1 mm aperture for measuring radiation in the far infrared and ultraviolet, and BRH specifies 7 mm.

The ANSI reasoning behind the recommended use of a 1 mm aperture is that a practical minimum aperture diameter should be employed to measure the radiant exposure or irradiance received by a relatively small area of tissue; yet the diameter should be large enough to impinge upon minimally acceptable sensing areas of transducing elements currently available in measuring instruments. Furthermore, there is no need to specify such a relatively large aperture size (7 mm), a value that is based on the magnification of visible light levels (not UV or far infrared) on their passage from the cornea to the retina. The absorption of UV and far infrared is, after all, a surface absorption phenomena. Finally, whereas a larger area of irradiation may be necessary for proper operation of thermal detectors (e.g., calorimeters), an impingement area of 1 mm (aperture) and even smaller is well within the acceptable operating requirements of state-of-the-art quantum detectors.

The BRH and ANSI measurement protocols for extended sources differ from each other and deserve further study. For example, the BRH document specifies an aperture of 7 mm to be used with an acceptance angle of 10^{-5} sr (approximately 3 mrad); the ANSI document specifies an aperture of 7 mm, but the angle of acceptance varies in relation to a quantity known as α_{min}. The term α_{min} in turn varies with exposure duration. The validity of having the angle of acceptance vary in accordance with exposure duration rather than source size (hence image size on the retina) is a matter to be investigated.

5.8 Precautions and Control Measures

The control measures prescribed in the ANSI Z136.1 standard on the safe use of lasers (1973, revised 1976), those required by the BRH, and those required in the draft of the proposed OSHA standard are presented in Tables 11.21 and 11.22. For further details on control measures, the reader is referred to source documents (36, 57–59). Some details on existing state regulations are given in Tables 11.23 and 11.24.

Probably the most significant hazard associated with lasers is electrical shock rather than exposure to optical radiation. Therefore all pertinent provisions of the National Electrical Code should be followed, including first aid training for all persons working with lasers, adequate grounding, proper discharge of energized circuits, well-designed interlocks and warning signals, and periodic educational programs.

6 MICROWAVE RADIATION

6.1 Physical Characteristics of Microwave Radiation

Microwave wavelengths vary from about 10 meters to 1 millimeter; the respective frequencies range from approximately 3 MHz to 300 GHz. Certain reference documents, however, define the microwave frequency range as beginning as low as 10 MHz and as high as 300 GHz. The region between about 30 kHz and the infrared is generally referred to as the radio frequency (RF) region. Reference may be made to Figure 11.1 to

Table 11.21. Laser Use Control Measures Required by ANSI and Proposed OSHA Standards[a]

	Class 1		Class 2		Class 3		Class 4	
Safe Use Requirement	ANSI	OSHA	ANSI	OSHA	ANSI	OSHA	ANSI	OSHA
Use properly marked eye protection when eye exposure possible	N/A	N/A		I	√	√	√	√
Do not use system if laser protective filters appear damaged		√		√		√		√
Only authorized operators								
Viewing optics must attenuate radiation to MPE or class 1 levels		√		√	√	√	√	√
No tracking of nontarget vehicles						√*		√
Eye exposure limited to ocular MPE	N/A	N/A		√	√	√	√	√
Skin exposure limited to skin MPE	N/A	N/A	N/A	N/A	√	√	√	√
Training of laser user					√		√	
Reduce laser output to class 1 levels when beam not in use				√		√		√
Use laser in "controlled area"					√		√	
Limit to spectators								
Visible or audible emission indicators						√*	√	√
Remove unessential specular surfaces from laser beam path						√*		√
Caution warning tags and labels			√	√	√†	√†		
Danger warning tags and labels					√*	√*	√	√
Restricted beam access and barriers					√	√*	√	√
Closed installation when indoors							√	√
Safety power shutoff switch								√
Posting of danger signs in laser area						√*		√
Remote firing and monitoring advisable							√	
Key switch master interlock							√	

[a] Key: checks, standard requirements; I, required only for intentional staring at beam; asterisk, Class 3b requirement only; dagger, requirement for class 3a only; MPE, maximum permissible exposure level.

Reprinted with permission from *1976 Laser Focus Buyers' Guide.*

Table 11.22. Laser Product Safety Requirements in ANSI and BRH Standards

Performance Requirement or Recommendation	Class 1		Class 2		Class 3		Class 4		Class 5
	ANSI	BRH	ANSI	BRH	ANSI	BRH	ANSI	BRH	ANSI 1973
Classification label or warning label		√	√	√	√	√	√	√	√
Protective housing and scanning safeguards for scanning lasers to limit accessible radiation to lowest achievable class required required for application		√		√		√		√	
Safety interlocks for protective housing to assure retention of hazard classification if cover(s) removed		√		√	—[a]	√	—[a]	√	√
Remote control connector to permit use of door interlocks in a closed installation or other optional ancillary safety interlocks						√	—[a]	√	
Key-actuated master control so that laser is inoperable when key is removed						√	√	√	
Laser-radiation emission indicator giving visible or audible warning:									
During emission (e.g. pilot light)				√		√	√	√	
During and prior to emission (e.g. pilot light with delayed warmup)						√		√	
Permanently attached attenuator to reduce output to class 1 levels other than main power (e.g.: shutter, rotatable polarizer, flip mirror, or cap on chain)				√		√		√	
Controls located to reduce chance of operator exposure				√		√		√	
Protective viewing optics so viewer exposed to only class 1 levels		√		√	√	√	√	√	√
Safety information to be furnished with laser system		√	√	√	√	√	√	√	√

[a] Such precautions are advisable but not required.

Table 11.23. States with some Form of Laser Safety Obligation

State	Agency	Title	Date
States with Current Regulations, Including Registration			
Alaska	Department of Environmental Conservation	Title 18, Arts. 7 and 8	10/71
Georgia	Department of Health	Ch. 270-5-27	9/ 1/71
Illinois	Department of Public Health	Registration Law	8/11/67
Massachusetts	Department of Public Health	Sect. 51, Ch. 111	10/ 7/70
New York	Department of Labor	Code Rule 50	8/ 1/72
Pennsylvania	Department of Environmental Resources	Ch. 203, Title 25, Part 1	11/ 1/71
Texas	Department of Health	Radiation Control Act, Parts, 50, 60, 70	7/ 2/74
States with Existing Regulations or Voluntary Regulations with No Registration Requirement			
Missouri	Department of Health	Existing Ionizing Regulation applies	
Montana[a]	Department of Health and Environmental Sciences	Reg: 92-003	
Virginia	Department of Health	Voluntary Program	
Washington	Department of Labor and Industry	Ch 296-62-WAC	

Table 11.23 (Continued)

State	Agency	Title	Date
	States with Enabling Legislation Passed		
Arizona[a]	—	HB-5	8/11/70
Arkansas[a]	Public Health	Act 460	
Florida[a]	Division of Health	Ch. 501-122	
Louisiana[a]	Division of Radiation Control	HB-1165	7/31/68
Mississippi[a]	Department of Health	HB-499	4/24/64
Oklahoma[a]	Health Department	HB-1405	4/14/69
	States with Existing OSHA–State Agreements[b]		
California	Labor Department		
Colorado	Labor Department		
Connecticut	Labor Department		
Minnesota	Labor Department		
North Carolina	Labor Department		

States Drafting or Awaiting Passage of Regulations[a]

Alabama	—	Pending
Iowa	—	Pending
Maine	Health Engineering	Pending
Michigan	—	Pending
Nebraska	—	Pending
New Hampshire	—	Pending
New Mexico	Environmental Improvement Agency	Pending
Oregon	Health Division	Pending
Wyoming	Division of Health	Pending

[a] New regulations now being drafted or pending passage.

[b] Not covered by other standards at state level.

[c] Enabling legislation not passed.

Reprinted with Permission of Laser Institute of America.

Table 11.24. Comparison of Existing State Laser Safety Standards[a]

State	Code	Registration Required	Exposure Criteria Basis	Controls Specified	Warning Signs Specified
Alaska	Laser Performance Standard Title 18, Arts. 7 and 8	Yes	Modified 1969 ACGIH TLVs	Yes	Yes
Georgia	Ch. 270-5-27 (9/1/71)	Yes	Registration law only		
Illinois	Laser Registration Law (8/1/67)	Yes	Registration law only		
Massachusetts	Sect. 51, Ch. 111 (10/7/70)	Yes	Recommendation only: modified 1969 ACGIH TLVs for visible, also 1940 AMA for UV also values for IR		No
Missouri	Based on existing ionizing regulation	No	No specific laser requirements		
Montana	Reg. 92-003	No	Uses 1969 ACGIH TLVs	As specified by ACGIH	
New York	Code Rule 50 (8/1/72)	Yes	Modified 1969 ACGIH TLVs	Yes[b]	Yes
Pennsylvania	Ch. 203, Title 25, Part 1 (11/1/71)		Registration law only		
Texas	Radiation Control Act, Parts 50, 60, 70 (6/2/74)	Yes	Based on ANSI/HEW	Yes	Yes
Virginia		Voluntary laser program			
Washington	Ch. 296-62 WAC	No	Uses 1969 ACGIH TLVs	As specified by ACGIH	

[a] Notes on table: (1) ACGIH, American Conference of Governmental Industrial Hygienists; (2) ANSI, American National Standards Institute, Standard Z-136.1, Safe Use of Lasers, 1973. (3) HEW, Department of Health, Education and Welfare, Food and Drug Administration proposed Laser Products Performance Standard, *Fed. Reg.* **38;** 236, Pt. III (December 10, 1973); (4) AMA, American Medical Association, 1948 Council on Physical Medicine [*J. Am. Assoc. Med.,* **137** 1600 (1948)].

[b] New York requires laser operator certification for all mobile lasers.

Reprinted with permission of Laser Institute of America.

Table 11.25. Letter Designation of Microwave Frequency Bands

Band	Frequency (MHz)
L	1,100–1,700
LS	1,700–2,600
S	2,600–3,950
C	3,950–5,850
XN	5,850–8,200
X	8,200–12,400
Ku	12,400–18,000
K	18,000–26,500
Ka	26,500–40,000

determine the position in the electromagnetic spectrum occupied by microwaves relative to other electromagnetic radiations. Certain bands of microwave frequencies have been arbitrarily assigned letter designations by industry (see, e.g., Table 11.25). However there is no universal agreement on these letter designations. Other discrete frequencies have been assigned by the Federal Communications Commission for industrial, scientific, and medical applications (labeled ISM bands) as shown in Table 11.26.

6.2 Sources of Microwave Radiation

Although microwave radiation has important applications in communications and navigational technology such as satellite communications systems, acquisition and tracking radar, air and traffic control radar, weather radar, and UHF TV transmitters, there is a growing number of commercial applications in other fields—microwave ovens, diathermy equipment, and industrial drying equipment. Some typical primary sources of microwave energy are klystrons, magnetrons, backward wave oscillators, and semicon-

Table 11.26. Industrial, Scientific, and Medical (ISM) Frequencies Assigned by the FCC

13.56 MHz ± 6.78 kHz
27.12 MHz ± 160 kHz
40,68 MHz ± 20 kHz
915 MHz ± 25 MHz
2450 MHz ± 50 MHz
5800 MHz ± 75 MHz
22,125 MHz ± 125 MHz

ductor transit time devices such as IMPATT diodes. Such sources may operate in a continuous mode (CW), as in the case of some communications systems, in an intermittent mode, as with microwave ovens, induction heating equipment, and diathermy equipment; or in a pulsed mode, as is the case with radar and digital communication systems. The various frequency designations and uses of microwave energy are given in Table 11.27. Of particular interest are the different designations used by the Eastern European countries, the USSR, and the Western countries (43).

Natural sources of RF and microwave energy can produce electric peak field strengths exceeding 100 V/m at ground level when cold fronts move through a given geographical area. Solar radiation intensities range from 10^{-18} to 10^{-17} W/(m^2)(H$_z$), but the integrated intensities at the earth's surface for the frequency range 0.2 to 10 GHz is approximately 10^{-8} mW/cm^2. This value is to be compared with an average value of 10^2 mW/cm^2 on the earth's surface attributable to the entire (UV, visible, IR, and microwave) solar spectrum.

6.3 Biological Effects of Microwave Radiation

When the human body is exposed to microwave radiation, the usual process of absorption, reflection, transmission, and scattering take place, depending on the wavelength and wave front characteristics. Microwave wavelengths less than 3 cm are absorbed mostly in the outer skin surface. Wavelengths between 3 and 10 cm penetrate more deeply (1 mm to 1 cm), and at wavelengths of 10 to 20 cm, penetration and absorption are sufficiently great that the potential for causing damage to internal body organs must be considered. The human body is thought to be essentially transparent to wavelengths greater than about 500 cm. At about 300 MHz (1 meter wavelength) the depth of penetration changes rapidly with frequency, declining to millimeter depths at frequencies of about 3000 MHz. For purposes of comparison, the skin penetration at 3000 MHz is approximately 16 mm, whereas that for 10,000 MHz it is approximately 4 mm. One of the difficulties in establishing the relationships between the characteristics of microwave fields and the bioeffects produced by those fields is our limited ability to measure and describe the incident fields as well as the special patterns of absorbed power within the tissue. Furthermore, when attempting to make environmental measurements one often encounters distortion of the field pattern when a measurement probe is inserted into that field. The reader is referred to the sections on Electromagnetic Fields and Absorption in Tissue and Measurement of Microwave Radiation (6.5 and 6.7, respectively) for a more detailed description of measurement techniques and special problems in quantifying the deposition of microwave energy in tissue.

A review of the evidence assembled by Western investigators concerning the biological effects of microwave radiation leads to the overall conclusion that almost all the reported effects are explainable on the basis of heat deposition in tissue. The photon energy of electromagnetic radiation at microwave frequencies is considered to be too low to cause the removal of electrons from molecules regardless of the number of quanta absorbed; hence ionization effects are excluded. One of the major sources of controversy is whether

Table 11.27. Radiofrequency and Microwave Band Designations

Band Designations				
United States	U.S.S.R.	Wavelengths	Frequencies	Typical Uses[a]
Radiofrequency bands				
Low frequency (LF)	Long (VCh)	10^4–10^3 m	30–300 kHz	Radio navigation, radio beacon
Medium frequency (MF)	Medium (HF)	10^3–10^2 m	0.3–3 MHz	Marine radiotelephone, Loran, Am Broadcast
High frequency (HF)	Short (UHF)	10^2–10 m	3–30 MHz	Amateur radio, worldwide broadcasting, medical diathermy, radio astronomy, citizen bands
Microwave bands				
Very high frequency (VHF)	Ultra-short (meter)	10–1 m	30–300 MHz	FM broadcast, television, air traffic control, radio navigation
Ultra high frequency (UHF)	Decimeter	1–0.1 m	0.3–3 GHz	Television, microwave point-to-point, microwave ovens, telemetry, tropo scatter and meteorological radar, mobile telephone
Super high frequency (SHF)	Centimeter (SHF)	10–1 cm	3–30 GHz	Satellite communication, airborne weather radar, altimeters, shipborne navigational radar, microwave point-to-point radio
Extra high frequency (EHF)	Millimeter	1–0.1 cm	30–300 GHz	Radio astronomy, cloud detection radar, space research, HCN (hydrogen cyanide) emission, millimeter wave communications

[a] Modified after Table 1 in Reference 65.

"nonthermal" effects can be induced in tissue as a result of microwave irradiation. The term "nonthermal" effect usually designates a biological change produced in the absence of any detectable rise in temperature in the test system. The phenomenon of "pearl chain" formation, an alignment of particles with the lines of force of an electromagnetic field, has been offered by some investigators as an example of a nonthermal effect. However Sher and Schwan (60) concluded that "the implications for pearl chain formation are that on no account can biological pearl chain occur for particles smaller than 3 μm in diameter without risking overheating of tissues. Particles smaller than about 30 μm would not form pearl chains; freely moveable particles of this size are not available in the body."

The current consensus is that pearl chain formation will not occur in the human body as a result of microwave irradiation. Frey (61) has reported what seems to be a direct auditory nerve response to microwaves. He reported "hearing" microwave pulses at an average power density as low as 100 μW/cm². The nature of the perception was described as a buzz or a ticking, depending on the pulse length and the pulse repetition rate. The subjects reported the sensation of sound within or behind the head. Selective shielding revealed that the most sensitive area was the region directly over the temporal lobe of the brain and that the greatest sensitivity was to the frequency range from 300 to 1200 MHz.

One explanation for such microwave audition is that the microwave energy does not exert direct action on the auditory nerve or brain centers; instead, the tissues of the head undergo rapid thermal expansion when the microwave energy is transformed into heat. Since the temperature rise is only of the order of 0.001°C, however, it is generally thought that the rate of energy deposition may be more important than a critical temperature reached by a body organ. More details on the mechanism of microwave audition may be found in a publication of A. W. Guy (62). To date there is no evidence that auditory sensations constitute a risk of injury.

Research performed in this country has generally concluded that microwave energy is capable of producing cataracts in exposed persons. This conclusion is based almost entirely on animal experimentation. Parenthetically, one experienced ophthalmologist has openly expressed his doubts about the possibility of producing cataracts in humans as a result of microwave exposure unless a massive, extremely high level of exposure occurs. Despite this, most knowledgeable persons concede the point that microwave radiation can induce cataracts.

One of the notable animal cataract studies just referred to was that of Carpenter and Van Ummersen, who investigated the effects of microwave radiation on the production of cataracts in rabbit eyes. Exposures to 2.45 GHz radiation were made at power densities ranging from 80 to 400 mW/cm² for different exposure durations. They found that repeated doses of 67 J/cm² spaced a day, a week, or 2 weeks apart produced lens opacities, even though the single threshold cataractogenic dose (at a power density 280 mW/cm²) was 84 J/cm². When the single expose dose was reduced to 50 J/cm², opacities were produced when the doses were administered 1 or 4 days apart, but when the interval between exposures was increased to 7 days, no opacification was noted even after five such weekly exposures. At a power density of 80 mW/cm² (dose of 29 J/cm²), the lowest exposure level used in the study, no effect developed; but when this dose was administered daily for 10 or 15 days, cataracts were produced. These experiments indicate a possible cumulative effect on the lens of the eye if exposures are repeated at a rate such that the lens does not have sufficient time to recover between exposures, even at levels below the single dose cataractogenic threshold.

The published allegation (63) that microwave radiation caused the death of a man has been challenged by medical authorities who claimed that the case in question was one of acute appendicitis that led to profound shock and death on the tenth postoperative day. The case was discounted as being due to microwave exposure in an Armed Forces Institute of Pathology Memorandum of July 25, 1957 (64).

A number of epidemiologic studies have been conducted on human populations that had at any time been exposed to microwave radiation. Usually these studies were designed to elicit information on the incidence of cataracts and general physiological effects. The studies of Zaret et al. (65, 66) have shown a small but statistically significant increase in the number of changes at the posterior polar surface of the lens; however cause and effect relationships in each individual case have not been subsequently elucidated. Barron and Baraff (67) compared the environmental and clinical data on 335 microwave workers to those of a controlled, nonexposed population. No significant differences were found in mortality, disease, sick leave, subjective complaints, or the results of clinical analyses. Czersky (68) has made a similar study of some 800 Polish workers who were exposed to microwaves during the course of their work. The incidence of functional disturbances could not be correlated with exposure level (0.2 to 6 mW/cm^2) or duration of exposure.

The investigations in the Soviet Union and Eastern European countries have largely stressed nonthermal microwave effects at the central nervous system and cellular levels: cataract formation in rabbits has been reported after a 60-minute exposure to 10 mW/cm^2 (74), a result that has not been produced in this country under the conditions described by the Soviets. It is the consensus of Soviet research that exposure to a microwave power density of the order of 10 mW/cm^2 for long periods of time constitutes a pathogenic factor (morphologic lesions in the nervous system, changes in reproductive function, and other borderline conditions). In response to American suggestions that some of the reported results may reflect adaptive or protective reactions to microwave radiation, the Soviets claim that their results are equivocal and cannot be regarded as harmless regulatory, adaptive, or compensatory reactions. Effects reported by the Soviets for exposure levels below 10 mW/cm^2 include lowered endurance, retarded weight gain, inhibition of conditioned reflexes, and neurosecretion and neurophysiologic disturbances. Even at intensities as low as 250 to 500 $\mu W/cm^2$, biological effects such as "change in the activity of the brain" and "immunobiologic resistance" are reported to occur. An interesting insight into Soviet research philosophy is contained in a recent statement (69):

> When speaking about criteria of significance of reactions, once you keep in mind that not all reactions are of importance but some, while lacking pathologic traits under certain conditions, should nevertheless be given attention when possible harmful consequences are considered; it is therefore

necessary to introduce the term "potentionally harmful reaction" as a main criterion of importance of a symptom, as opposed to either a threshold biologic (regulatory) reaction in general or a threshold pathologic reaction.

Assuming this philosophy is used for the development of standards of permissible exposure, it is relatively easy to understand why the present Soviet permissible exposure levels are so much more stringent than those in western countries.

Soviet *in vitro* studies of 10 cm irradiation of catalase and cholinesterase have been entirely negative, however *in vivo* investigations have shown changes in the activity of a whole range of enzymes including cholinesterase, at exposure levels significantly lower than 10 mW/cm^2 (i.e., down to a level of 1 mW/cm^2). According to the investigators, the difference between the *in vivo* and *in vitro* experiments "results not from a direct action on molecular structures but from changes in enzyme concentrations in the tissue which are apparently related to distrubances in the neurohormonal regulation of metabolic processes" (70, 71).

Another series of studies (72) on the effects of microwave radiation on cell membranes, particularly on the selective permeability of the membranes to potassium and sodium ions when subjected to a power density of 1 mW/cm^2, has revealed statistically significant changes in the rates of transport of potassium and sodium ions across all membranes. The transport rates were apparently unrelated to a rise in the temperature of the suspension or to the radiation frequency.

The effect of 2.45 GHz continuous wave radiation on rabbit erythrocytes and rat lymphocytes has been studied in two separate cell systems. Rabbit erythrocytes were exposed at a power density of 5 mW/cm^2 for 3 hours, thus duplicating the experimental conditions of a study by the Polish investigator Baranski, who reported an increase of some 1520 times in potassium efflux and a decrease in osmotic resistance. No changes were found in potassium efflux or osmotic resistance if the exposed and controlled cells were maintained at the same temperature. Although it was later learned that the cells in the Polish experiment had been washed and suspended in unbuffered saline, whereas buffered saline was used in the experiments at the U.S. National Institute of Environmental Health Sciences, (NIEHS), the repeated experiments conducted at NIEHS using unbuffered saline again showed no changes in potassium efflux or osmotic resistance as long as the exposed and controlled cells were maintained at the same temperature (73). The importance of this study is highlighted by the Soviets' hypothesis that the influence of microwaves on excitable cells such as erythrocytes is due to alterations in the permeability of the cell membranes and that these alterations in permeability are brought about by the transport of ions across the membranes, which in turn influences the membrane potential.

One of the experimental tools used by Soviet investigators to describe biological effects is the conditioned reflex. This fact accounts in part for the difference in threshold effects obtained by Western and Eastern European investigators. The Soviet investigator Livshits (74) comments on the dilemma (74):

In conditioned reflex investigations it has been discovered that irradiation induces various disturbances of the higher nervous activity. It must be mentioned that such an effect on radiation is observed only in experiments conducted according to the schemes used by I. P. Pavlov and his followers. Among the authors that have investigated the effects of radiation upon conditioned reflexes in different animals, there is no complete unanimity in the evaluation of the phenomena that they observed and the understanding of their mechanism. The opinion of certain foreign researchers that the change in the latent period and values of the conditioned reflex expresses only or primarily a change in the excitability of the unconditioned reflex centers, in general contradicts the considerable material obtained in the laboratories of I. P. Pavlov and his followers. In radiation pathology, in particular, this question is complex and will require a special investigation (74).

In commenting on Soviet research, Dodge has stated (75):

An often disappointing facet of the Soviet and East European literature on the subject of clinical manifestations of microwave exposure is the lack of pertinent data presented on the circumstances of irradiation, frequency, effective area or irradiation, orientation of the body with respect to the source, wave form (continuous or pulsed; modulation factors), exposure schedule and duration, natural shielding factors; and a whole plethora of important environmental factors (heat, humidity, light, etc.) are often omitted from clinical and hygienic reports. In addition, the physiological and psychological status of human subjects such as health, previous or concomitant medication, and mental status is also more often than not omitted. These variables, both individually and combined, affect the human response to microwave radiation (75, 3).

Whereas on the one hand Soviet researchers claim that low level ($\ll$ 10 mW/cm^2) nonthermal radiation produces behavioral effects in humans, McAfee indicates that it is thermal stimulation of the peripheral nervous system that produces the neurophysiological and behavioral changes (76, 77) and that the interaction between the peripheral nervous system and the central nervous system accounts for the heart rhythm and blood chemistry effects reported by the Soviets.

The debate over whether microwave radiation can produce cumulative effects continues unabated. As already indicated, the possibility of cumulative effects on the lens from repeated subthreshold cataractogenic exposures of rabbit eyes has been suggested by Carpenter and his associates (78). Others (79), however, point out that no cumulative rise in temperature can occur if the time between exposures exceeds the time required for the tissue to return to normal temperature. The reported cumulative effect may well be the accumulation of damage resulting from repeated exposures, each one capable of producing some degree of damage. If the repetitive exposures take place at time intervals shorter than those needed for the repair of damage, tissue damage is produced. Michaelson (3) claims that since this is the process that actually takes place, the suggestion of cumulative effects from microwave exposure is of questionable validity.

6.4 Skin Sensation

Questions are often raised about whether a person can "feel" microwaves or the heat that is produced by the incident radiation. Although it is not possible to calibrate one's sensation of heat, some quantitative information is available. Schwan et al. (80) found that if a person's forehead is exposed to a power density of 74 mW/cm^2 at 3000 MHz, the reaction time for experiencing a sensation of warmth varied between 15 and 73 seconds. The perception of warmth for a power density of 56 mW/cm^2 varied from 50 seconds to 3 minutes of exposure. Hendler's data (81, 82) show that when an area of the forehead of 37 cm^2 is exposed to microwaves, thermal sensation can be elicited within 1 second at power densities of 21 mW/cm^2 when using 10 GHz (10,000 MHz) radiation and 58.6 mW/cm^2 when using 3 GHz radiation. These data further indicate that if the entire face were exposed to 10 GHz radiation, the threshold for thermal sensation would 4 to 6 mW/cm^2 within 5 seconds, or approximately 10 mW/cm^2 within 0.5 second (3). The sensation of warmth or pain from microwave heating apparently differs little from those felt from infrared heating. In the experiments of Cook (83) an average skin temperature rise of 15°C is required to achieve pain when the initial skin temperature is approximately 32.5°C.

6.5 Electromagnetic Fields and Absorption in Tissue

It is most important for the industrial hygienist to remember that the amount of microwave energy deposited and absorbed in biological tissue depends on frequency and on body size, configuration, and homogeneity. Furthermore, it is not possible to classify any of the field quantities measured exterior to an exposed subject as being hazardous, nonhazardous, thermal, or nonthermal. First something must be known about the exposure conditions, frequency, subject size, and subject geometry, to permit determination of the specific absorption rate (SAR), formerly called the absorbed power density, in the subject. For subjects who are exposed to microwave fields, the equation for the time rate of change of temperature T (°C per second) per unit volume of subcutaneous tissue is

$$\frac{dT}{dt} = \frac{0.239 \times 10^{-3}}{c(W_a + W_m - W_c - W_b)}$$

where W_a is the specific absorption rate, W_m is the metabolic heating rate, W_c is the heat loss due to thermal conduction, W_b is the power dissipated by blood flow, all expressed in watts per kilogram, and c is the thermal conductivity, expressed in kilocalories per kilogram per degree celsius (84).

The specific absorption rate for tissue exposed to an electromagnetic field is

$$W_a = \frac{10^{-3}\,\sigma}{\rho\,E^2}$$

where σ is the electrical conductivity in mhos per meter, ρ is the tissue density in grams per cubic centimeter, and E is the rms value of the electric field in the tissue, in volts per meter. Before the tissue is exposed to electromagnetic fields, it is assumed that a steady state condition exists where $W_a = dT/dt = 0$, requiring $W_m = W_c + W_b$. Under normal conditions the metabolic rate W_m averages 1.3 W/kilogram of the whole body, 11 W/kg for brain tissue, and 33 W/kg for heart tissue (84).

According to the energy equation we would expect to see some change in tissue temperature due to applied electromagnetic fields if the specific absorption rate W_a were of the same order of magnitude as W_m or more. The safety standard of 10 mW/cm^2 of incident power is at least partially based on limiting the average value of W_a to the average resting value of W_m. Therefore specific absorption rates that are much greater than W_m would be expected to produce marked thermal effects, whereas specific absorption rates that are much less than W_m would not be expected to produce any significant thermal effects (84).

Peak values of specific absorption rate (50 W/kg $\leq W_a \leq$ 170 W/kg) have been used to provide vigorous local therapeutic heating of deep vasculated tissues in man treated with diathermy (85). Also, values of W_a that are equal to or much greater than 4 W/kg in the brain tissue of cats yielded measurable increases in brain temperature and decreases in the latency times of evoked potentials (86). Body temperature increases, accompanied by behavioral changes, have been observed in rats by Justesen (87) for values of W_a equal to or greater than 3.1 W/kg and by Hunt et al. (88) and Phillips (89) for W_a values equal to or greater than 6.3 W/kg averaged over the body of the animal (84). Cataracts have been induced in rabbits by acute exposures where the specific absorption rate in the eye was greater than 138 W/kg (90).

Figure 11.8 shows the variation of specific absorption rate with frequency in a spherical muscle model with the same mass as a 70 kg man. In the frequency range from 1 to 20 MHz the absorption varies as the square of the frequency and is due primarily to the magnetically induced electric fields. The maximum specific absorption rate induced by the incident H field is denoted by the curve marked with crosses, and that due to the incident E field is denoted by the curve with circles. In this range the maximum specific absorption rate is only 10^{-5} to 10^{-2} W/(kg)(mW)(cm^2) of incident power. In the frequency range of 100 to 1000 MHz internal reflections are significant for the man-sized sphere, and the average absorption attains a maximum of 2×10^{-2} W/(kg)(mW)(cm^2) of incident power at 200 MHz which remains relatively constant with frequency up to 10 GHz. The maximum specific absorption rate increases with frequency above 1000 MHz, approaching that produced by nonpenetrating radiation. The dashed lines illustrate roughly the frequency dependence of the total or average absorbed power and indicate how safety standards might be relaxed as a function of frequency if the absorption characteristics in man were the same as that for the sphere (84).

Figure 11.9 illustrates absorption patterns as a function of body size (radius of muscle sphere) for a sphere consisting of a central muscle core surrounded by concentric layers of subcutaneous fat and skin exposed to 2.45 GHz plane wave radiation at 1 mW/cm^2. Assuming that the spherical models have some relevance to the human body, Guy (84) makes the point that the peak specific absorption rate could be as high as 4.2 W/kg within the body or head of a small bird or animal but as low as 0.27 W/kg at the surface and 0.05 W/kg at a point 2.5 cm deep in the human body, when all are exposed to a 1 mW/cm^2, 918 MHz source. Therefore exposure to 10 mW/cm^2 could be of extreme significance, and 0.5 mW/cm^2 could be of mild thermal significance to the smaller animals in comparison with their metabolic rate. For the human model, on the other hand,

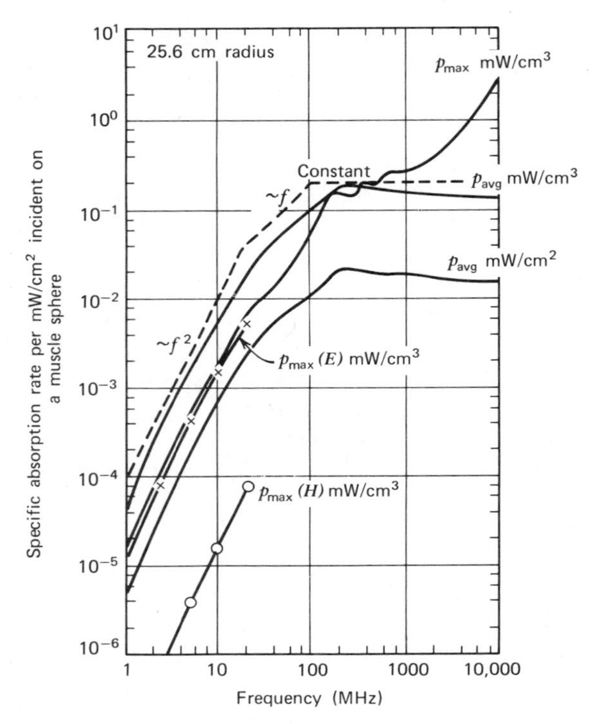

Figure 11.8 Specific absorption rate patterns versus frequency in spherical muscle model of 70 kg man exposed to plane wave 1 mW/cm² source. After Guy (84).

10 mw/cm² would appear to be minimally significant in terms of thermal insult; a power density of 0.5 mW/cm² would have negligible thermal significance.

6.6 Exposure Criteria

Schwan (91) in 1953 examined the thresholds for cataractogenesis and thermal damage to tissue. The minimum power density necessary for producing such changes was found to be approximately 100 mW/cm², to which he applied a safety factor of 10 to obtain a maximum permissible exposure level of 10 mW/cm². This number has been subsequently incorporated into many official standards. The American National Standards Institute C95.1 standard (92) requires a limiting power density of 10 mW/cm² for exposure periods of 0.1 hour or more; and a limiting energy density of 1 mWhr/cm² during any 0.1-hour period. The latter criterion permits intermittent exposure at levels above 10 mW/cm² on the basis that such intermittency does not produce a temperature rise greater than 1°C in human tissue.

More recently Schwan (93) has suggested that the permissible exposure levels be expressed in terms of current density rather than power density, especially when dealing with measurements in the near or reactive field, where the concept of power density loses its meaning. He proposes a permissible current density of approximately 3 mA/cm², since this value is comparable to a far field value of 10 mW/cm². At frequencies below 100 kHz this value should be somewhat lower, and for frequencies above 1 GHz it can be somewhat higher.

The recommendations of the ANSI C95.1 committee have been qualified as follows:

> Body temperature depends in part on sources of heat input such as electromagnetic radiation, physical labor and high ambient temperature and on heat dissipation capability as affected by clothing, humidity, etc. People who suffer from circulatory difficulties and certain other ailments are vulnerable. The power levels established by radiation guide numbers are related in a complicated way to power levels to which damage may occur. The guide numbers are appropriate for moderate environments. Under conditions of moderate to severe heat stress, the guide number given should be appropriately reduced. Under conditions of intense cold, higher guide numbers may also be appropriate after careful consideration is given to the individual situation. These formulated recommendations pertain to both whole body irradiation and partial body irradiation. Partial body irradiation must be included since it has been shown that some parts of the body, for example, the eyes, and testes, may be harmed if exposed to incident radiation levels significantly in excess of the recommended levels.

Mumford (94) has proposed a means for adjusting the 10 mW/cm² exposure criterion on the basis of heat stress. According to his recommendation the exposure limit would

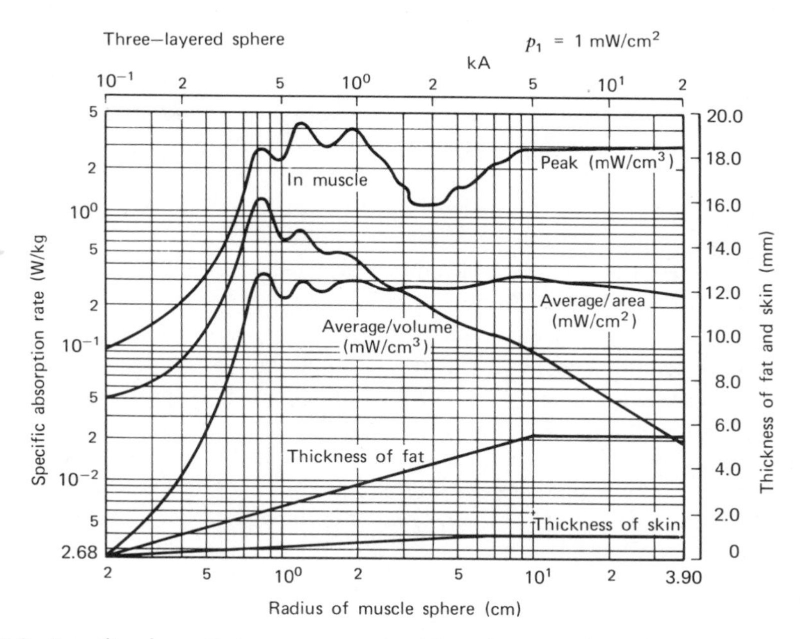

Figure 11.9 Specific absorption rate versus outside radius in spherical tissue layer model of animal exposed to 2.45 GHz 1 mW/cm² plane wave.

be 10 mW/cm^2 when the temperature–humidity index (THI) is equal to or less than 70. When the THI is between 70 and 79, the limit is (80 − THI) mW/cm^2. Thus when the THI is equal to or greater than 79, the exposure limit is 1 mW/cm^2.

The Soviet criterion (95) for exposure to microwave radiation states that the intensity of the radiation within the frequency range of 300 MHz to 300 GHz should not exceed a value of 10 μW/cm^2 during the working day. For exposures not longer than 2 hours per working day, the level should not exceed 100 μW/cm^2, and for exposures not longer than 15 to 20 minutes per working day, a power density of 1 mW/cm^2, provided the exposure level for the remainder of the working day does not exceed 10 μW/cm^2. Also, protective goggles need to be worn at the higher (1 mW/cm^2) level. From the public health standpoint, the Soviet standard states: "In the microwave range the intensity of radiation should not exceed 1 microwatt per square centimeter (1 μW/cm^2) for places where human occupancy occurs and for persons not occupationally exposed." According to Gordon (69), one of the main reasons for the relatively low permissible exposure levels is that clinical observations of microwave workers exposed for 6 to 8 years to power densities of the order of tenths of a milliwatt per square centimeter has in her opinion demonstrated functional disturbances, such as neural, vegetative, and asthenic syndromes.

The U.S.S.R. standard lists the following maximum permissible exposure limits for electric field strengths. For a frequency range of 100 kHz to 30 MHz, a value of 20 V/m is acceptable. For the range of 30 to 300 MHz, a value of 5 V/m is used. For magnetic field strengths in the frequency range of 100 kHz to 1.5 MHz, a value of 5 A/m is considered to be acceptable.

The microwave oven standard adopted by the U.S. Department of Health, Education and Welfare (96) to limit the permissible emission to 1 mW/cm^2 at a distance of 5 cm from any surface at the time of manufacture and to a value of 5 mW/cm^2 for the life of the product, should be considered a public health rather than an occupational standard. If one computes the actual power density to which the consumer public would be exposed while standing at a normal distance in front of microwave ovens conforming with the HEW standard, one may conclude that the exposure levels are generally of the order of 5 to 20 μW/cm^2 (64). These values approximate the most stringent (10 μW/cm^2) Soviet and Eastern European standards.

The Czechoslovakian standard as published and discussed by Marha et al. (97) calls for a value of 10 μW/cm^2 for long-term exposures to pulsed waves. For the "less risky" continuous wave devices, a value of 25 μW/cm^2 is acceptable. Under the Czechoslovakian standard the maximum permissible exposure (MPE) cannot exceed a guide number of 200 for occupational exposure to CW microwaves; for pulsed radiation exposures the guide number is 80. These guide numbers result in respective permissible power densities of 0.025 and 0.010 mW/cm^2 average power density over an 8-hour work day. The guide numbers for the general population assuming 24-hour exposures are 60 for continuous wave, and 24 for pulsed radiation. These guide numbers translate into values that are 10 times lower than the accepted occupational values, namely, 0.0025 and 0.001 mW/cm^2. "Continuous wave" is defined as emitted radiation with a ratio of on to off time greater than 0.1. The maximum peak power density for human exposure is limited to 1 kW/cm^2. This is a ceiling value that may not be exceeded.

The Polish standard on permissible microwave exposure as adopted by the Council of Ministers and the Ministers of Health and Social Welfare (98) is as follows: values of 10 and 100 μW/cm^2 are acceptable for exposure of the general population to a continuous (i.e., stationary fields) and intermittent (i.e., nonstationary fields), respectively. These values correspond to 0.1 W/m^2 and 1 W/m^2.

The permissible levels for occupational exposure are classified under the headings of stationary and nonstationary fields. For stationary fields four subcategories are defined. In the first or "safe zone," the mean power density cannot exceed 0.1 W/m^2, but human exposure time is unrestricted. In the second or intermediate zone, the minimum value is 0.1 W/m^2 and the upper limit is 2 W/m^2: exposure is allowed for a whole working day up to 10 hours. In the third or hazardous zone the minimum value is 2 W/m^2 and the upper limit is 100 W/m^2: exposure time for a 24-hour period is determined by the formula $t = 32/p^2$, where t is exposure time in hours and p is the mean power density in watts per square meter. In the fourth or dangerous zone, human exposure is forbidden for mean power densities in excess of 100 W/m^2 (10 mW/cm^2).

For intermittent exposure (i.e., exposure to nonstationary fields), four subheadings are likewise prescribed. In the first or safe zone, the mean power density does not exceed 1 W/m^2 (0.1 mW/cm^2). The second or intermediate zone has a minimum value of 1 W/m^2 and an upper limit of 10 W/m^2; exposure is allowed for a whole work day up to 10 hours. The third or hazardous zone has a minimum value of 10 W/m^2 and an upper limit of 100 W/m^2. Exposure time within a 24-hour period is determined by the formula $t = 800/p^2$, where t and p have the same units given previously. The fourth or dangerous zone forbids human exposure to a mean power density greater than 100 W/m^2 (10 mW/cm^2).

The original Bell System radiation protection guide (99) called for a three-step exposure criterion. First, power density levels in excess of 10 mW/cm^2 were considered to be potentially hazardous; hence personnel were not permitted to enter areas where major parts of the body might be exposed to such levels. Second, power density levels between 1 and 10 mW/cm^2 were permitted only for incidental, occasional, or casual exposure. And third, power density levels below 1 mW/cm^2 were considered to be acceptable for indefinite or prolonged exposure. The Bell System standard has since been changed to conform fairly closely to the ANSI criterion.

The U.S. Army standard (100) permits unlimited exposure to levels below 10 mW/cm^2, but exposure to power densities greater than 100 mW/cm^2 is considered dangerous. Within the range of 10 to 100 mW/cm^2, exposure is allowed for a limited time according to the formula $t = 6000/W^2$, where t is exposure time in minutes and W is the mean power density in milliwatts per square centimeter. Exposure to levels above 10 mW/cm^2 are usually avoided; however exposure is permissible if the foregoing guide is followed for levels between 10 and 100 mW/cm^2. The army standard calls attention to the fact that ionizing radiation hazards may exist in the vicinity of microwave equipment if such equipment is operating at accelerating voltages above 18 kV.

In applying any of the exposure guides, one should recall the differences in philosophies among Eastern European and Western countries, as described previously.

6.7 Measurement of Microwave Radiation

One of the key factors underlying the controversy over the reported biological effects of microwave radiation is the lack of standardization of (or in many cases even a means of comparing) the measurement techniques used to quantify the electromagnetic fields that impinge on experimental animals or humans. There seems to be little promise that such standardization will be realized in the near future.

An idealized conception of microwave propagation consists of a plane wave moving in an unbounded isotropic medium where the electric $|E|$ and magnetic field $|H|$ vectors are mutually perpendicular and both are perpendicular to the direction of wave propagation. Unfortunately the simple proportionality between the E and H fields in free space is valid only in the so-called far field (Fraunhofer region) of the radiating device. The far field region is sufficiently removed from the source that the angular field distribution is essentially independent of the distance to the source. The power density in the far field is inversely proportional to the square of the distance from the source and directly proportional to the product of $|E|$ and $|H|$. Therefore the measurement of either the $|E|$ and $|H|$ field vector in the far field is all that is needed to accurately determine the power density. Plane wave detection in the far field is well understood and easily obtained with equipment that has been calibrated for use in the frequency range of interest. Most hazard survey instruments have been calibrated in the far field to read in units of power density, usually milliwatts per square centimeter. A typical device for far field measurements is a suitable antenna coupled to a power meter.

To estimate the power density levels in the near field (Fresnel region) of large-aperture circular antennas, one can use the following simplified relationships (101):

$$W = \frac{16P}{\pi D^2} = \frac{4P}{A} \qquad \text{(near field)}$$

where P is the average power output, D is the diameter of the antenna, A is the effective area of the antenna, and W is power density. If this computation reveals a power density that is less than a specified limit (e.g., 10 mW/cm^2), no further calculation is necessary because the equation gives the maximum power density on the microwave beam axis. If the computed value exceeds the exposure criterion, one assumes that the calculated power density exists throughout the near field. The far field power densities are then computed from the Friis free space transmission formula:

$$W = \frac{GP}{4\pi r^2} = \frac{AP}{\lambda^2 r^2} \qquad \text{(far field)}$$

where λ is the wavelength, r is the distance from the antenna, G is the far field antenna gain, and W, P, and A are as in the near field equation.

The distance r from the circular antenna to the intersection of the near and far field is given by

$$r_1 = \frac{\pi D^2}{8\lambda} = \frac{A}{2\lambda}$$

These simplified equations do not account for reflections from ground structures or surfaces where the power density may be 4 times greater than the assumed free space value. If conditions are such that focusing effects can be produced, factors even greater than 4 can be achieved.

A number of uncertainties and errors accompany all measurements, but they are present particularly when the measurements are made in the near field. The region closest to the radiating source is called the reactive near field region where energy is stored. Between the reactive near field and the far field is the radiating near field (Fresnel region), where the phase and amplitude relationships between the E and H fields vary with distance from the source. The configuration of the source can also affect the pattern of radiation to a considerable extent. Multipath $|E|$ and $|H|$ vectors that exist in the reactive and radiative near fields may become so complex that the concept of power density is almost precluded. For this reason there is considerable merit in measuring $|E|^2$ or $|E|$ or $|H|^2$ or $|H|$. Since plane waves, $|E|^2$ and $|H|^2$ are simply related to W, there is no problem comparable to that in the near field. Other potential measurement errors may be brought about by the characteristics of the monitoring probe.

To make measurements in the near or reactive field, the measurement device should be able to separately measure $|E|$ and $|H|$ fields. The probe sensor itself must be sufficiently small compared with the wavelength or wavelengths in the field, to minimize perturbation of those fields. And the probe should be essentially nondirectional and capable of responding to all polarizations. The polarization characteristics of the device are particularly important in the near field because vertical linear, horizontal linear, elliptical, and circular polarizations may be present at any point in space regardless of source configuration. Circular polarization exists when two equal amplitude, linearly polarized waves (e.g., one in the x-direction and the other in the y-direction) are superimposed but 90° out of phase. An elliptically polarized wave consists of two unequal amplitude, orthogonal, linearly polarized waves that are 90° out of phase. The polarization of a wave largely determines its transmission, reflection, and scattering characteristics when impinging upon materials.

Many probes measure electric field strength in volts per meter by means of a thermocouple whose dc output is proportional to volts per square meter. This output is calibrated and translated into a power density reading, usually milliwatts per square centimeter, taking into account the space impedance Z_0 that relates the $|E|$ to the $|H|$ component of the propagated wave. This proportionality constant or impedance for the far field region has a value of 377 Ω. If the space impedance is not 377 Ω (the value at which the survey meter has been calibrated) at the point of measurement, a calibration error results.

Most microwave radiation protection guidelines have been formulated on the basis of measured average power. In the case of certain radar equipment, however, the ratio of peak to average power may be as high at 10^5. A reassessment of the average power concept is underway, and the ultimate result might mean a change in survey instrument design.

In general, all devices that are designed to measure microwave radiation for the purpose of assessing hazards to personnel should possess certain minimum characteristics: the antenna probe should be electrically small, to minimize perturbation of the field; the instrument impedance should be matched to the field, to prevent backscatter from the probe to the source; and the probe should behave like an isotropic receiver, be sensitive to all polarizations, possess a response time that is adequate for handling peak to average power, and exhibit a flat response over the frequency bands of interest.

Most microwave measuring devices are based on bolometry, calorimetry, voltage and resistance changes in detectors, rectification, and radiation pressure on a reflecting surface. The latter three methods are self-explanatory. Bolometry measurements are based on the absorption of power in a temperature-sensitive resistive element, the change in resistance being proportional to the absorbed power. For direct measurement of peak power, a bolometer is the element of choice because the typical time constant is of the order of 300 μsec. Bolometers have a positive temperature coefficient (i.e., the resistance increases with temperature); thermistors have a negative temperature coefficient. Thermistors are more rugged than bolometers and have a greater power-handling capacity. They are preferred for the measurement of average power of pulsed signals because of their log time constant (0.1 to 1 second).

Crystal diode detectors are useful for the measurement of microwave power, but care must be taken to avoid burnout of the diode in the vicinity of high powered equipment.

Through the use of three orthogonal dipoles, it is possible to design a probe that is both independent of polarization and isotropic in its response (102). The probe consists of resistive thin-film elements that are folded back in parallel on themselves with the output terminals at one end (Figure 11.10*a*).

The probe elements consist of thin films of overlapping antimony and bismuth deposited on a plastic substrate. Thermal energy is dissipated primarily at the narrow, high resistance strips, thereby producing an increase in temperature at these points (hot junctions). The low resistance, wider sections of the thermocouple act as cold junctions. The resultant dc voltage is directly proportional to the energy dissipated in the narrow, high resistance strips of the thermocouple. Variation in the sensitivity of the device due to changes in ambient temperature is reportedly less than 0.05 percent because of close spacing of the cold junctions (102). The placement of the three mutually orthogonal sensor elements within a polystyrene sphere, to obtain a 5 cm spacing between the sensor elements and the outer surface of the sphere, is illustrated in Figure 11.10*b*. A representative frequency response of the probe appears in Figure 11.10*c*. Note that the midfrequency bandwidth of interest (1 dB bandwidth) is 850 MHz to 15 GHz, thereby incorporating a flat response feature at the two microwave oven frequencies of 915 and 2450 MHz (2.45 GHz). This type of broad band radiation monitor may be used for the

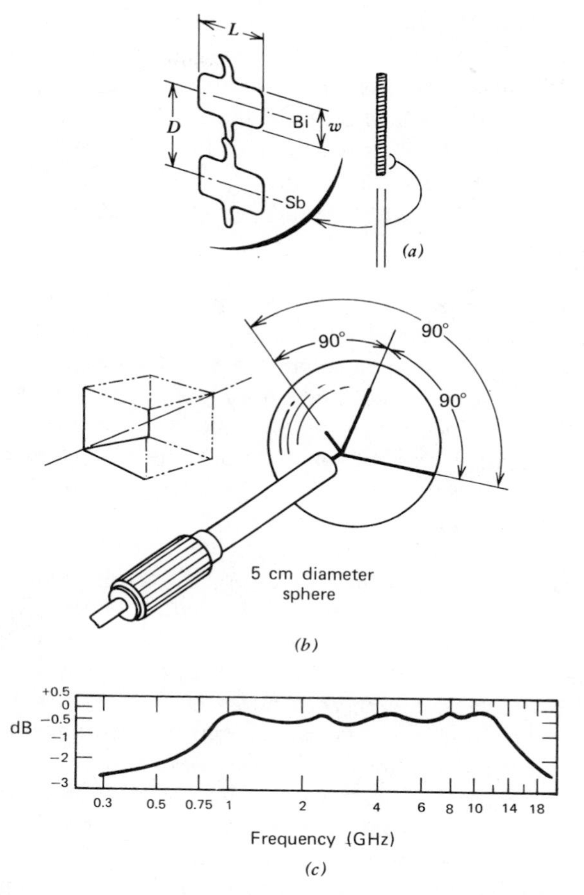

Figure 11.10 (*a*) Antenna probe element with distributed thermocouple film elements; tapered film leads reduce interaction between leads and elements. (*b*) Relationship of the three mutually orthogonal probe elements; probe elements are contained within a 5 cm sphere. (*c*) Typical frequency response characteristics. Reproduced with permission of the Narda Microwave Corporation.

measurement of radiation emitted from a wide variety of sources, but the original design was fashioned to provide a useful probe for measuring the leakage radiation from microwave ovens. The 5 cm electrically transparent Styrofoam spacing device was included to ensure that compliance with the microwave standard under Public Law 90-602 (Radiation Control for Health and Safety Act). By this standard, newly manufactured microwave ovens must not emit levels in excess of 1 mW/cm²; the permissible level throughout the useful life of the product is 5 mW/cm².

Certain broad band devices in the frequency range below 300 MHz have been designed to measure the $|H|$ field. The development of such instruments has been encouraged by some recent evidence (84) that the $|H|$ fields of low frequency radiation produce more significant biological effects than do $|E|$ fields. The sensing probes of the $|H|$ field monitors usually contain a series of mutually perpendicular loops. Low frequency radiation below 300 MHz is usually measured with loop or short whip antennas, tuned voltmeters, or field intensity meters.

6.8 Control Measures

Engineering control measures should be given consideration over any use of personal protective equipment. Engineering control measures may range from the restriction of azimuth and elevation settings on radar antennas to the complete enclosures found in microwave ovens. Since in most microwave installations, leaking of radiation adversely affects the proper performance of the system, high system performance and safety are mutually reinforcing goals. Personnel who work with microwave energy should have the benefit of orientation and training sessions on the potential hazards of microwaves and the proper means for controlling undue exposures. Safe work practices should be formally prepared and instituted. Emphasis must be placed on the safe design of equipment to include shielding, enclosures, interlocks to prevent accidental energizing of circuits, dummy load terminations, warning signals, and control of electrical shock potentials. Ionizing radiation (X-rays) may be produced by certain microwave generators because of the existence of accelerating voltages in excess of 16 to 18 kV.

Certain protective suits have been constructed of a nylon mesh on which has been deposited a layer of metallic silver. The mesh permits a fair degree of ventilation for the body. Since the silver deposits bestow good electrical conductivity properties to the mesh, special care must be taken to avoid direct contact with exposed electrical terminals. Before using the protective garment, one should have a thorough understanding of its attenuation characteristics as a function of radiation frequency, to be certain that the external field intensity does not exceed the safe level by an amount greater than the

Table 11.28. Attenuation Factors (dB) (Shielding) (104)

	Frequency (GHz)			
Material	1–3	3–5	5–7	7–10
60 × 60 mesh screening	20	25	22	20
32 × 32 mesh screening	18	22	22	18
16 × 16 window screen	18	20	20	22
0.25 in. Mesh (hardware cloth)	18	15	12	10
Window glass	2	2	3	3.5
0.75 in. Pine sheathing	2	2	2	3.5
8 in. Concrete block	20	22	26	30

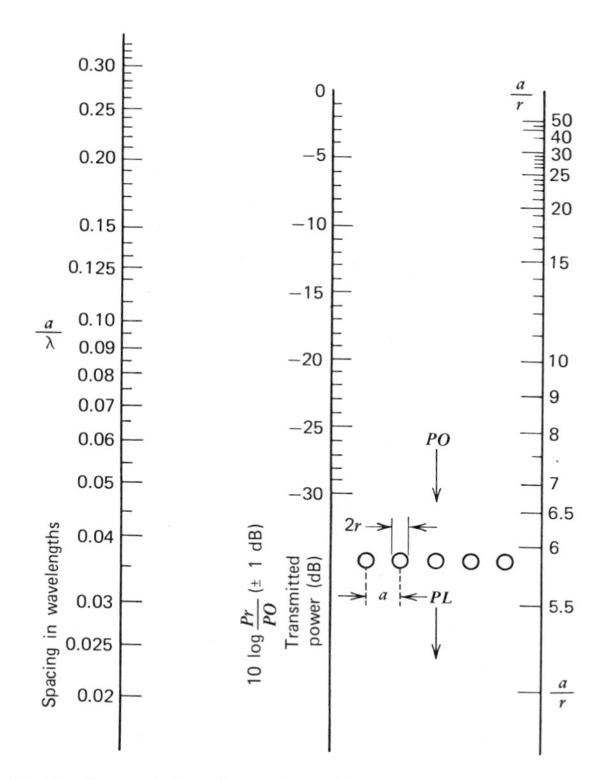

Figure 11.11 Transmission through a grid of wires of radius *r* and spacing *a*.

attenuation provided by the garment. In this connection, Figure 11.11 showing the transmission loss through a wire grid may prove useful (103). Similarly, the attenuation provided by various types of material as given in Table 11.28 may be of use in designing shields or enclosures (104).

Within the past few years it has been demonstrated that the function of certain cardiac pacemakers, particularly those of the demand type, may be seriously compromised, particularly by pulsed or intermittent microwave radiation. Furthermore, the radiation levels that cause interference with the pacemakers may be orders of magnitudes below the levels that cause detrimental biological effects. The most effective method of reducing the susceptibility of these devices to microwave interference seems to be improved shielding. Manufacturers of cardiac pacemakers have been engaged in major programs to minimize such interference.

The judicious use of appropriate signs and labels may prove useful in alerting people to the presence of microwave sources. Figure 11.12 illustrates the RF and microwave warning signs adopted by the American National Standards Institute C95 Committee (92).

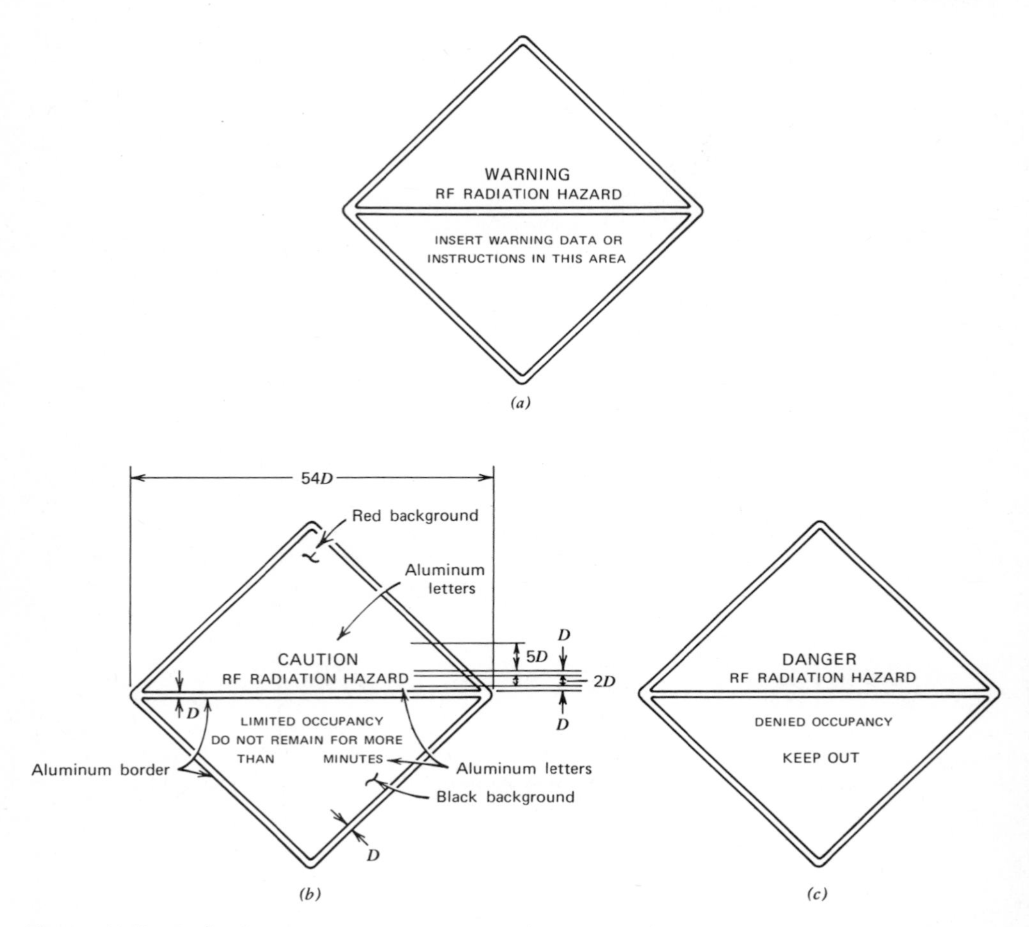

Figure 11.12 Radio frequency signs. (a) General warning. (*b*) Limited occupancy. (*c*) Denied occupancy. ANSI Instructions as follows: (1) Place handling and routing instructions on reverse side. (2) Use *D*-scaling unit. (3) Dimensions in *b* indicate lettering ratio of letter height to thickness of letter lines. (4) Symbol is square, triangle on right-angle isoscales. Derived from ANSI C95.2 Standard, American National Standards Institute, New York.

7 RADIOFREQUENCIES (RF) AND EXTREMELY LOW FREQUENCY (ELF) RADIATION

7.1 Radiofrequencies

Generally speaking, the potential biological hazard associated with radiofrequencies and extremely low frequencies is not as great as that posed by microwaves in the 800 to 3000 MHz frequency range. As a matter of fact, some thought (3) has been given to the advisability of raising the permissible exposure level to a value in excess of 10 mW/cm^2 for frequencies below 100 MHz. Since an electric field strength of approximately 200 V/m is equivalent to a power density of 10 mW/cm^2 for a plane wave (far field conditions), the suggestion for increasing the level would translate into some value above 200 V/m. Certain unofficial and unpublished estimates of a satisfactory electric field for continuous exposure to radiation frequencies below 30 MHz have ranged up to a maximum value of 1000 V/m.

The possibility of nonthermal effects of RF radiation has been suggested by experiments (105) at 27 MHz, where chromosomal and bacterial effects were demonstrated without a detectable rise in temperature of the experimental medium (water) and by the ability of inorganic particles to line up with an imposed electric field ("pearl chain" formation). Also considered by some to represent nonthermal effects is the observation that certain unicellular organisms line up and travel parallel to the lines of force created by RF radiation at 8.5 MHz (106). We still do not know whether these effects are truly nonthermal or whether they reflect a lack of analytical sophistication to detect thermal energy transfer on a micro scale.

Exposure of primates for one-hour periods to power densities ranging from 100 to 200 mW/cm^2 at frequencies of 10.5, 19.3, and 26.6 MHz failed to produce demonstrable effects on blood chemistry, metabolic activity, deep body temperature, or cardiac function, although thermal effects were evident when the power density was increased to 400 mW/cm^2 (107).

Irradiation of macromolecules at 13 MHz reportedly (108, 109) produced a constituent in gamma-globulin having an electrophoresis pattern and immunochemical properties different from those of nonirradiated gamma-globulin.

Because of the conflicting reports about observed biological effects of RF power density levels of 10 mW/cm^2 and below, it is difficult for the industrial hygienist to interpret the significance of such exposures. The reported effects of exposure to levels below 10 mW/cm^2 include changes in behavior, in central nervous system activity, in perception of electromagnetic energy, in metabolism, and in the hematopoietic system. Since most of these low level effects are reported from Eastern European countries and the Soviet Union, it is encouraging to note the recent United States–Soviet joint research program on the biological effects of RF and microwaves (73).

7.2 Extremely Low Frequencies

Most of the information on the biological effects of extremely low frequencies (ELF) is derived from studies of the effects of radiation produced by low power frequencies (50 to 60 Hz) and from the U.S. Navy's Project Sanguine (now called Seafarer). Novitskiy et al. (110) have reported a wide range of physiological effects resulting from exposure to ELF, including changes in heartbeat and respiration, anaphylactic shock, and even death. Field strengths that produced the effects varied from approximately 50 to 5000 V/cm, with most of the effects being reported between 200 and 2000 V/cm. On the other hand, Knickerbocker et al. (111) failed to show any effects on general health, behavior, or reproductive ability of mice exposed to a 60 Hz field of 4 kV/in. (1.5 kV/cm) for 1500 hours during the course of a 10.5-month study period. Soviet studies of occupations where employees have been exposed to ELF fields often report complaints such as listlessness, excitability, drowsiness, fatigue, headache, and the so-called neurasthenic syndrome. Ulrich and Ferin (112) have reviewed these studies and have concluded that the irregular shifts such as night work, dry air, and other factors were the probable causes of the signs and symptoms. Kouwenhoven et al. (113) have studied linemen in the United States who have worked on high voltage transmission lines for many years. Observations over a 3-year period of physical, mental, and emotional health parameters failed to reveal any detrimental effects.

Project Sanguine/Seafarer (114), was begun in the 1960s to investigate the design and operation of a large antenna system that would radiate in the ELF range as a means of communicating with nuclear submarines at depths of approximately 200 meters at any location in the world. Since the original proposal envisioned an antenna whose area was in the tens of thousands of square miles, the residents of the areas in which the antenna was likely to be installed became concerned about environmental and ecological effects. Later designs specified an overall area of approximately 3000 square miles. In 1969 the U.S. Navy initiated what is generally regarded as the most comprehensive research program ever attempted to determine whether biological or ecological effects could be expected from exposure to ELF radiation in the immediate vicinity of the antenna, and at distances from the antenna. The electric and magnetic field levels associated with the Seafarer system operating at a frequency of 75 Hz are approximately 0.07 V/m and 0.2 *g*, respectively, at the surface of the earth directly above a buried antenna.

Several studies have been conducted to determine whether exposure of fruit flies to ELF electric and magnetic fields would induce mutations on the X chromosome. A comprehensive fruit fly study performed by Mittler (115) in 1972, at field strengths of 10 V/m and 1.0 *g* at 45, 60, and 75 Hz, yielded the conclusions that there was no induction of sex-linked recessive lethal genes, no translocations between II and III chromosomes, no loss of the X or Y chromosomes, and no nondisjunctions or dominant lethal genes. The lethal genes reported in the initial study of Coate and Negherbon (116) were discounted. Exposure of *E. coli* cells to field strengths of up to 20 V/m and 2.0 *g* at 45 and 75 Hz has failed to produce any mutagenic effect on these bacteria (117). Onion roots exposed to 20 V/m and 2 *g* at 45 and 75 Hz did not yield any evidence of chromosomal damage (118).

Exposure of rats through the second generation at levels up to 20 V/m and 2 *g* at 45 and 75 Hz failed to show any significant differences in fertility between exposed and controlled populations (119). Factors measured during the study were mortality, body weight, and anatomical defects. A study (120) of blood pressure, body weight, and anatomical and physiological changes among canines was made under the same exposure conditions (i.e., 20 V/m, 2 *g*, at 45 and 75 Hz) with negative results except for a spurious increase in hypertension. The finding of hypertension was considered to be inconclusive.

Results were inconclusive on a study (121) of the thresholds of perception of ELF electric fields by marine animals at levels that were roughly 1000 times greater than levels proposed for Seafarer. A comprehensive study (122) of the behavior in pigeons and rats at field levels of up to 2 *g* and 100 V/m at 45, 60, and 75 Hz indicated no effect. Studies of reaction time in primates exposed to ELF fields showed the central nervous system to be relatively insensitive to a magnetic field of 10 *g* at 45 Hz (123, 124). Tests of immediate memory and general motor activity in primates exposed to 1.4 V/m and 10 *g* at 75 Hz had no influence on performance (125).

More recent studies have demonstrated the exquisite sensing apparatus of sharks to ELF at a frequency of 10 Hz (126). The detection threshold for ELF radiation at 75 Hz is being investigated, but it is expected to be higher than that at 10 Hz.

Since many biological functions exhibit a 24-hour periodicity, they are often referred to as circadian (*circa* = about; *dies* = day) rhythms. Although their mechanism remains obscure, circadian rhythms are endogenous to living organisms rather than the result of external influences. Also, there is some evidence that alteration of these internal timing systems may be associated with deterioration in psychological and physiological function.

7.3 Effects of Humans

When environmental time cues such as the normal day–night cycle have been eliminated, Wever (127) has demonstrated in humans a "free-running" rhythmic period of approximately 24 hours for activity, rest, and body temperature. When all known and recognized environmental cues have been eliminated from the test environment, the rhythm appears to be "free-running;" that is, it is not synchronized by environmental cues. Wever reports that under such conditions there is an increased tendency for the circadian rhythms to desynchronize from each other, producing what he terms "internal desynchronization." Such desynchronization apparently increases the normal circadian periods from approximately 25 hours to time periods that may be twice as great. The presence of natural or artificial fields such as ELF apparently strengthens the internal coupling of biological timing mechanisms, a presumably desirable effect. Therefore there is some implication in Wever's work that it is the absence of natural or artificial fields such as ELF, rather than their presence, that may produce undesirable physiological effects.

7.4 Effects on Plants and Animals

Halberg (128) undertook a series of investigations of ELF radiation effects on silk tree leaflets, flour beetles, and mice. Field conditions ranged from 45 to 75 Hz, 0.4 to 2 *g*, and 1 to 180 V/m, with the duration of exposure varying from a few days to several months. The effects of magnetic fields (60 Hz; 1 *g*) and electromagnetic fields (60 Hz, 0.4 *g*, 100 V/m; and 60 Hz, 1 *g*, 100 V/m) on the circadian rhythms of silk tree leaflets appeared to be negative or at best inconclusive. There was no effect of exposure to 60 to

70 Hz, 1 *g* fields on the circadian toxicity susceptibility of the flour beetle when exposed to an insecticide (Dichlorvos), nor did continuous exposure of mice to a 45 Hz, 136 V/m electric field for one week before and after ouabain injection increase susceptibility to the drug. No significant difference in mice mortality appeared to exist between control and experimental groups exposed to 75 Hz, 25 V/m electric fields. Body weight was found to be unchanged even when mice were continously exposed to 75 Hz, 1 *G* fields for several months. The effects of ELF fields ranging from 1 to 10 V/m and 0.5 to 2 *g*, singly and in combination, on the circadian temperature rhythm, food consumption, and survival rate, were negative. Even in cases of ELF fields reportedly increasing the synchronization of biorhythms, however, such results must likewise be considered to be suggestive rather than conclusive.

Although most investigations into the possible biological effects of ELF radiation have yielded negative data, a few findings bear careful follow up investigation. These include (1) the need to elucidate the reasons for an apparent increase in serum triglycerides in exposed navy personnel, (2) the need to follow up on the exquisite sensitivity of the shark (and possibly other marine life) to ELF radiation, and (3) the need to clarify the effect of ELF on bird migration patterns and avian biorhythms.

In 1976 the National Academy of Sciences appointed a committee on the Biosphere Effects of Extremely Low Frequency Radiation (129) to review and draw conclusions about existing knowledge and required research, with special relevance to the U.S. Navy Project Seafarer.

APPENDIX 11A

Table 11A.1. Useful Radiometric and Related Units

Term	Symbol	Description	Unit and Abbreviation
Radiant energy	O	Capacity of electromagnetic waves to perform work	Joule (J)
Radiant power	P	Time rate at which energy is emitted	Watt (W)
Irradiance or radiant flux density (dose rate in photobiology)	E	Radiant flux density	Water per square meter (W/m^2)
Radiant intensity	I	Radiant flux or power emitted per solid angle (steradian)	Watt per steradian (W/sr)
Radiant exposure (dose in photobiology)	H	Total energy incident on unit area in a given time interval	Joule per square meter (J/m^2)
Beam divergence	Φ	Unit of angular measure: 1 rad $\approx 57.3°$, 2π rad $= 360°$	Radian (rad)

Conversion Factors
Radiant Energy Units

To Find:	erg	joule	W-sec	μW-sec	g-cal
erg =	1	10^{-7}	10^{-7}	0.1	2.39×10^{-6}
jour =	10	1	1	10^6	0.230
W-sec =	10^7	1	1	10^4	0.239
μW-sec =	10	10^{-6}	10^{-6}	1	2.39×10^{-7}
g-cal =	4.19×10^7	4.19	4.19	4.19×10^6	1

Table 11A.1. (Continued)

Term	Symbol	Description	Unit and Abbreviation

Radiant Exposure (Dose) Units

To Find:	erg/cm^2	joule/cm^2	W-sec/cm^2	μW-sec/cm^2	g-cal/cm^2
erg/cm^2 =	1	10^{-7}	10^{-7}	0.1	2.39×10^{-6}
joule/cm^2 =	10^7	1	1	10^6	0.239
W-sec/cm^2 =	10^7	1	1	10^6	0.239
μW-sec/cm^2 =	10	10^{-4}	10^{-6}	1	2.39×10^{-7}
g-cal/cm^2 =	4.19×10^7	4.19	4.19	4.19×10^6	1

Irradiance (Dose Rate) Units

To Find:	erg/(cm^2)(sec)	joule/(cm^2)(sec)	W/cm^2	μW/cm^2	g-cal/(cm^2)(sec)
erg/(cm^2)(sec) =	1	10^{-7}	10^{-7}	0.1	2.39×10^{-8}
joule/(cm^2)(sec) =	10^7	1	1	10^6	0.239
W/cm^2 =	10^7	1	1	10^6	0.239
μW/cm^2 =	10	10^{-6}	10^{-6}	1	2.39×10^{-7}
g-cal/(cm^2)(sec) =	4.19×10^7	4.19	4.19	4.19×10^6	1

REFERENCES

1. Public Law 90-602, Radiation Control for Health and Safety Act of 1968, Washington, D.C.
2. "Report on Program for Control of Electromagnetic Pollution of the Environment: The Assessment of Biological Hazards of Non-ionizing Electromagnetic Radiation," issued March 1973, May 1974, April 1975, and June 1976. Office of Telecommunications Policy, Executive Office of the President, Washington, D.C.
3. S. M. Michaelson, "Human Exposure to Radiant Energy—Potential Hazards and Safety Standards," *Proc. IEEE,* **60,** April 1972.
4. G. M. Wilkening, in: *The Industrial Environment, Its Evaluation and Control,* Department of Health, Education and Welfare, NIOSH, Washington, D.C. 1973, chapter on "Non-Ionizing Radiations."
5. M. Kleinfeld, C. Giel, and I. R. Tabershaw, "Health Hazards Associated with Inert-Gas-Shielded Metal Arc Welding," *Am. Med. Assoc. Arch. Ind. Health,* **15,** 27–31 (1957).
6. K. G. Hansen, "On the Transmission Through Skin of Visible and Ultraviolet Radiation," *Acta Radiol.,* Suppl. 71 (1948).
7. S. Lerman, "Radiation Cataractogenesis," *N.Y. State J. Med.,* **62,** 3075–3085 (1962).
8. J. A. Zuclich and J. S. Connolly, "Ocular Damage Induced by Near Ultraviolet," Technology Inc., Life Sciences Division, 8531 New Braunfels Avenue, San Antonio, Tex. 78217.
9. T. M. Murphy, "Nucleic Acids: Interaction With Solar UV Radiation," in: *Current Topics in Radiation Research,* Vol. 10, 1975, p. 199.
10. I. Willis, A. Kligman, and J. Epstein, "Effects of Long Ultraviolet Rays on Human Skin: Photoprotective or Photoaugmentative?" *J. Invest. Dermatol.,* **59,** 419 (1973).
11. C. E. Keeler, *Albinism, Xeroderma Pigmentosum, and Skin Cancer,* National Cancer Institute, Monograph 10, 1963, p. 349.
12. R. L. Elder, "Phototherapy Warning," *J. Pediatr.,* **84,** 145 (1974).
13. L. E. Bockstahler, C. D. Lytle, and K. B. Hellman, "A Review of Photodynamic Therapy for Herpes Simplex: Benefits and Potential Risks," Publication No. (FDA) 75-8013, U.S. Public Health Service, Bureau of Radiological Health, Rockville, Md, 1975.
14. L. F. Mills, C. D. Lytle, F. A. Andersen, K. B. Hellman, and L. E. Bockstahler, "A Review of Biological Effects and Potential Risks Associated with Ultraviolet Radiation as Used in Dentistry," Publication No. (FDA) 76-8021, U.S. Public Health Service, Bureau of Radiological Health, Rockville, Md, 1976.
15. F. Daniels, *Ultraviolet Carcinogenesis in Man,* National Cancer Institute, Monograph 10, p. 407, 1963.
16. M. Faber, personal communication, 1975.
17. G. Funding, O. M. Henriques, and E. Rekling, *Ueber Lichtkanzer,* Verh 3., Int. Kongress für Lichtforsch, Wiesbaden, 1936, p. 166.
18. O. A. Jensen, "Malignant Melanomas of the Uvea in Denmark, 1943–1952," Thesis, Copenhagen, 1963.
19. D. G. Pitts, "The Ocular Ultraviolet Action Spectrum and Protective Criteria," *Health Phys. J.,* **25,** 559–566 (December 1973).
20. *Threshold Limit Values for Physical Agents,* published by American Conference of Governmental Industrial Hygienists, P.O. Box 1937, Cincinnati, Ohio 45201.
21. J. R. Richardson and R. D. Baertsch, "Zinc Sulfide Schottky Barrier Ultraviolet Detectors," in: *Solid State Electronics,* Vol. 12, Pergamon Press, New York, 1969, pp. 393–397.
22. P. Bassi, G. Busuoli, L. Lembo, and O. Rimonde, *G. Fis. Sanit. Prot. Contro Radiaz.,* **18:** 4, 137–142 (October–December 1974).
23. I. Matelsky, "Non-Ionizing Radiations" in: *Industrial Hygiene Highlights,* Vol. 1, L. V. Cralley and G. D. Clayton, Eds., Pittsburgh. Pa.

24. H. C. Weston, "Illumination and the Variation of Visual Performance with Age," *Proceedings of the International Commission on Illumination,* Vol. 2, Stockholm, 1951.
25. C. M. Edbrooke and C. Edwards, "Industrial Radiation Cataracts: The Hazards and the Protective Measures," *Arch. Occup. Hyg.,* **10,** 293–304 (1976).
26. K. L. Dunn, "Cataract from Infrared Rays. Glass Workers' Cataract—A Preliminary Study on Exposures," *Arch. Ind. Hyg. Occup. Med.,* **1,** 166–180 (1950).
27. K. L. Dunn, "A Preliminary Study on Glass Workers' Cataract Exposures," *Trans. Am. Acad. Ophthalmol. Otolaryngol.,* **54,** 597–605 (1950).
28. G. F. Keatinge, J. Pearson, J. P. Simons, and E. E. White, "Radiation Cataract in Industry," *Arch. Ind. Health,* **11,** 305–315 (1955).
29. D. G. Cogan, D. D. Donaldson, and A. B. Reese, "Clinical-Pathological Characteristics of Radiation Cataract," *Arch. Ophthalmol.,* **47,** 55–70 (1952).
30. H. Goldmann, H. Koenig, and F. Maeder, "The Permeability of the Eye Lens to Infrared," *Opthalmologica,* **120,** 193–205 (1950).
31. H. Goldmann, "The Genesis of the Cataract of the Glass Blower," *Ann. Ocul.,* **172,** 13–41 (1935); also in *Am. J. Ophthalmol.,* **18,** 590–591 (1935).
32. R. K. Langley, C. B. Mortimer, and C. McCullock, "The Experimental Production of Cataracts by Exposure to Heat and Light," *Am. Med. Assoc. Arch. Ophthalmol.,* **63,** 473–488 (1960).
33. M. Wertheimer and W. D. Ward, "Influence of Skin Temperature upon Pain Threshold as evoked by Thermal Radiation—A confirmation," *Science* **115,** 499–500 (1952).
34. J. D. Hardy, "Thermal Radiation, Pain and Injury," in: *Therapeutic Heat,* Vol. 2, S. Licht, Ed., New Haven, Conn., 1958, pp. 157–178.
35. *IES Lighting Handbook,* 4th ed., Waverly Press, Baltimore, 1966.
36. American National Standards Institute Standard for the Safe Use of Lasers, Z136.1, rev. 1976, ANSI, New York.
37. M. L. Wolbarsht, Ed., *Laser Applications in Medicine and Biology,* Vol. 1, Plenum Press, New York, 1971.
38. A. Javan, W. R. Bennett, and D. R. Herriott, "Population Inversion and Continuous Optical Laser Oscillation in a Gas Discharge Containing a He–Ne Mixture," *Phys. Rev. Lett.,* **6,** 106 (1961).
39. R. C. Miller and W. A. Nordlung, "Tunable Lithium Niobate Optical Oscillator with External Mirrors," *Appl. Phys. Lett.,* **10,** 53 (1967).
40. G. M. Wilkening, "The Potential Hazards of Laser Radiation," *Proceedings of Symposium on Ergonomics and Physical Environmental Factors,* Rome, September 1968, International Labor Office, Geneva, Switzerland, pp. 16–21.
41. Peter P. Sorokin, James J. Wynne, John A. Armstrong, and Rodney T. Hodgson, "Resonantly Enhanced, Nonlinear Generation of Tunable Coherent, Vacuum Ultraviolet (VUV) Light in Atomic Vapors," Third Conference on the Laser, *Ann. N.Y. Acad. Sci.,* **267,** 36 (1976).
42. G. M. Wilkening, "A Commentary on Laser Induced Biological Effects and Protective Measures," *Ann. N.Y. Acad. Sci.,* **68,** part 3, 621–626 (1970).
43. J. M. Osepchuk, R. A. Foerstren, and D. R. McConnell, "Computation of Personnel Exposure in Microwave Leakage Fields and Comparison with Personnel Exposure Standards," International Microwave Power Institute Symposium, Loughborough, England, September 10–13, 1973.
44. W. T. Ham, Jr., R. C. Williams, H. A. Mueller, D. Guerry, A. M. Clarke, and W. J. Geeraets, "Effects of Laser Radiation on the Mammalian Eye," *Trans. N.Y. Acad. Sci.,* **28:** 2, 517 (1965).
45. W. K. Noell, V. S. Walker, B. S. Kang, and S. Berman, "Retinal Damage by Light in Rats," *Invest. Ophthalmol.* **5,** 450 (1966).
46. W. T. Ham, Jr., H. A. Mueller, and D. H. Sliney, "Retinal Sensitivity to Damage from Short Wavelength Light," *Nature,* **260,** March 11, 1976.

47. A. I. Goldman, W. T. Ham, Jr., and H. A. Mueller, "Mechanisms of Retinal Damage Resulting from Exposure of Rhesus Monkeys to Ultrashort Laser Pulses," *Exp. Eye Res.*, **21**, 457–469 (1975).

48. J. Taboada and R. W. Ebbers, "Ocular Tissue Damage due to Ultrashort 1060 μm Light Pulses from a Mode-Locked Nd: Glass Laser," *Appl. Opt.*, **14**, 1759 (August 1975).

49. A. I. Goldman, report accepted for publication in *Experimental Eye Research*, 1976.

50. K. W. McNeer, M. Ghosh, W. J. Geeraets, and D. Guerry, "ERG After Light Coagulation," *Acta Ophthalmol.*, Suppl. 76, 94 (1963).

51. T. P. Davis and W. J. Mautner, "Helium–Neon Laser Effects on the Eye," Annual Report, Contract No. DADA 17-69-C 9013, U.S. Army Medical Research and Development Command, Washington, D.C., 1969.

52. L. Goldmann, W. Kapuscinska-Czerska, S. Michaelson, R. J. Rockwell, D. H. Sliney, B. Tengroth, and M. Wolbarsht, "Health Aspects of Optical Radiation with Particular Reference to Lasers," World Health Organization, 1976, in preparation.

53. Technical Committee on Laser Products, International Electrotechnical Commission (IEC), Geneva, Switzerland.

54. Radiation Control Act, parts 50, 60, 70, Texas State Health Department, Austin, adopted July 2, 1974.

55. Section 51, Chapter III, Massachusetts State Department of Public Health, Boston, adopted October 7, 1970.

56. U.S. Department of Health, Education and Welfare, Bureau of Radiological Health, Laser Products Performance Standard, *Fed. Reg.*, **40**, 148, 32252–32265 (July 31, 1975).

57. G. M. Wilkening, "Laser Hazard Control Procedures," in: *Electronic Product Radiation and the Health Physicist*, U.S. Department of Health, Education and Welfare Report BRH/DEP 70-26, October 1970, pp. 275–290.

58. American Conference of Governmental Industrial Hygienists, *Guide For Control of Laser Hazards*, ACGIH, P.O. Box 1937, Cincinnati, Ohio 45201, 1973.

59. Occupational Safety and Health Administration, Title 29, Code of Federal Regulations 1910, OSHA Standards 1972, under revision.

60. L. D. Sher and H. P. Schwan, "Mechanical Effects of AC Fields on Particles Dispersed in a Liquid, Biological Implications," Ph.D. dissertation (L. D. Sher), University of Pennsylvania, under Contract AF30(602) ONR Technical Report dissertation No. 37, 1963.

61. A. H. Frey, "Human Auditory System Response to Modulated Electromagnetic Energy," *J. Appl. Physiol.*, **17**, 689–692 (1962).

62. A. W. Guy et al., "Microwave Interaction with the Auditory Systems of Humans and Cats," IEEE G-MTT *International Symposium Digest of Technical Papers*, 1973, p. 321.

63. J. T. McLaughlin, "Tissue Destruction and Death From Microwave Radiation (Radar)," *Calif. Med.*, **86**, 336–339 (1957).

64. T. S. Ely, "Microwave Death," letter to the editor, *J. Am. Med. Assoc.*, **217**, 1394 (1971).

65. M. M. Zaret, S. Cleary, B. Pasternack, M. Eisenbud, and H. Schmidt, "A Study of Lenticular Imperfections in the Eyes of a Sample of Microwave Workers and a Control Population," New York University, Final Report No. RADC-TDR-63-125, 1963; ASTIA Doc. AD 413294.

66. M. M. Zaret and M. Eisenbud, "Preliminary Results of Studies of the Lenticular Effects of Microwaves Among Exposed Personnel," in: *Proceedings of the Fourth Annual Tri-Service Conference on Biological Effects of Microwave Radiating Equipment and Biological Effects of Microwave Radiations*, M. F. Peyton, Ed., Plenum Press, New York, 1961, pp. 293–308 (Technical Report No. RADC-TR-60-180).

67. C. I. Barron and A. A. Baroff, "Medical Considerations of Exposure to Microwaves (Radar)," *J. Am. Med. Assoc.*, **168**, 1194–1199 (1958).

68. P. Czerski, M. Siekierzynski, and J. Gidynski, "Health Surveillance of Personnel Occupationally Exposed to Microwaves," Parts I, II, III, *Aerosp. Med.*, 1137–1148, October 1974.

69. Z. V. Gordon, A. V. Roscin, and M. S. Byckov, "Main Directions and Results of Research Conducted in the U.S.S.R. on the Biologic Effects of Microwaves," *Proceedings of the International Symposium on Biological Effects and Health Hazards of Microwave Radiation,* Polish Medical Publishers, Warsaw, 1973.

70. S. V. Nikogosjan, in: *O biologiceskom deistvii EMP radiocastot.,* Vol. 3, Moskva, 1968, p. 97.

71. V. M. Stemler, in: *Gigiena truda i biologiceskoe deistvii elektromagnitnyk voln radiocastot.,* Moskva, 1968, p. 175.

72. V. M. Stemler, in: *Sbornik Bionika, 1973, Materialy 4 Vsisojuznoj konferenci po bionike,* Moskva, 1973, pp. 3, 87.

73. Fourth report on "Program for Control of Electromagnetic Pollution of the Environment: The Assessment of biological Hazards of Nonionizing Electromagnetic Radiation," Office of Telecommunications Policy, Executive Office of the President, Washington, D.C., June 1976, p. 21.

74. N. N. Livshits, "On the Causes of the Disagreements in Evaluating the Radiosensitivity of the Central Nervous System Among Researchers Using Conditioned Reflex and Maze Methods," *Radiobiology,* **7,** 238–261 (1967).

75. C. H. Dodge, "Clinical and Hygienic Aspects of Exposure to Electromagnetic Fields," in: *Biological Effects and Health Implications of Microwave Radiation, Symposium Proceedings,* S. F. Cleary, Ed., U.S. Department of Health, Education and Welfare, 1970, pp. 140–149.

76. R. D. McAfee, "Physiological Effects of Thermode and Microwave Stimulation of Peripheral Nerves," *Am. J. Physiol.,* **203,** 374–378 (1962).

77. R. D. McAfee, "The Neural and Hormonal Response to Microwave Stimulation of Peripheral Nerves," in: *Biological Effects and Health Implications of Microwave Radiation, Symposium Proceedings,* S. F. Cleary, Ed., U.S. Department of Health, Education and Welfare, 1970, 150–153.

78. R. L. Carpenter and C. A. Van Ummersen, "The Action of Microwave Radiation on the Eye," *J. Microwave Power,* **3,** 3–19 (1968).

79. S. M. Michaelson, "Biomedical Aspects of Microwave Exposure," *Am. Ind. Hyg. Assoc. J.,* **32,** 338–345 (1971).

80. H. P. Schwan, A. Anne, and L. Sher, "Heating of Living Tissues," U.S. Naval Air Engineering Center, Philadelphia, Report No. NAEC-ACEL-534, 1966.

81. E. Hendler, "Cutaneous Receptor Response to Microwave Irradiation," in: *Thermal Problems in Aerospace Medicine,* J. D. Hardy, Ed., Surrey, England, 1968, pp. 149–161.

82. E. Hendler, J. D. Hardy, and D. Murgatroyd, "Skin Heating and Temperature Sensation Produced by Infrared and Microwave Irradiation," in: *Temperature Measurement and Control in Science and Industry,* Part 3, *Biology and Medicine,* J. D. Hardy, Ed., Reinhold, New York, 1963, p. 221–230.

83. H. F. Cook, "The Pain Threshold for Microwave and Infrared Radiations," *J. Physiol.,* **118,** 1–11 (1952).

84. A. W. Guy, "Quantitation of Induced Electromagnetic Field Patterns in Tissue and Associated Biologic Effects," *Proceedings of the International Symposium on Biological Effects and Health Hazards of Microwave Radiation,* Polish Medical Publishers, Warsaw, 1973.

85. A. W. Guy, J. F. Lehmann, and J. S. Stonebridge, "Therapeutic Applications of Electromagnetic Power," *IEEE Proc. Ind., Sci., Med. Appl. Microwaves,* special issue, January 1974.

86. C. C. Johnson and A. W. Guy, *Proc. IEEE,* **60,** 692 (1972).

87. D. R. Justesen and N. W. King, "Behavioral Effects of Low Level Microwave Irradiation in the Closed Space Situation," in: *Biological Effects and Health Implications of Microwave Radiation,* S. F. Cleary, Ed. U.S. Department of Health, Education and Welfare, Report No. BRH/DBE 70-2 (PB 193 8 58), Rockville, Md., 1970, p. 154.

88. E. I. Hunt, "General Activity of Rats Immediately Following Exposure to 2450 MHz Microwaves," in *Digest, 1972 IMPI Symposium,* Ottawa, Canada, May 1972.

89. R. D. Phillips, N. W. King, and E. L. Hunt, "Thermoregulatory Cardiovascular and Metabolic Response of Rats to Single or Repeated Exposures to 2450 MHz Microwaves," in: *1973 Symposium on Microwave Power,* Microwave Power Institute, Ottawa, Canada, September 1973.

90. P. Kramar, A. F. Emery, A. W. Guy, and J. C. Lim, "Theoretical and Experimental Studies of Microwave Induced Cataracts in Rabbits," in: *Applications in the 70's,* 1973 IEEE G-MTT International Microwave Symposium, University of Colorado, Boulder, 1973, p. 265.

91. H. P. Schwan and K. Li, *Proc. IRE,* **41,** 1735 (1953).

92. C95.1 Committee of the American National Standards Institute, "Safety Level of Electromagnetic Radiation with Respect to Personnel," ANSI, New York.

93. H. P. Schwan, in: *Biological Effects and Health Implications of Microwave Radiation,* S. F. Cleary, Ed., U.S. Department of Health, Education and Welfare, Report No. BRH/DBE 70-2 (PB 193 8 58), Rockville, Md., 1970.

94. W. W. Mumford, "Heat Stress due to RF Radiation," *Proc. IEEE,* **57,** 171–178 (1969).

95. "Sanitarnyje normy i pravila pri raboti s istocuikami elektromagnitnyck polei vysokih, ultravipokih i sverhvysokih castot," Ministerstvo Zdravohranenija, SSSR 30, 03 NS 848-70, 1970.

96. "Performance Standard for Microwave Ovens," Title 42, Part 78, Suppl. C Sec. 78.212, Code of Federal Regulation; *Fed. Reg.,* **35** (194), 15642 (October 6, 1970).

97. "Jednotna metodika stanoveni intensity pole a ozareni elektromagneticRymi vlnami v pasma vysokych frekvenci a velmi vysokych frekvenci k hygienickym ucelum." Vynos hlavniho hygienika CSRcj. HE 344.5-3.2.70/Priloha c. 3k Informacnim zpravam z oboru hygieny prace a nemoci z povolani, Praha, kveten, 1970.

98. "Rozpovzadzenie Rady Ministrow z dnia 25.02.1972 w. spraure bezpieczenstwa i higieny pracy przy stosowaniu urzadzen wytroarzajacych pola elektromagnetyczne w zakresie mikrofalowym," Dziennik Ustaw PRL No. 21, II, Poz. 153, 1972.

99. M. M. Weiss and W. W. Mumford, "Microwave Radiation Hazards," *J. Health Phys.,* **5,** 160 (1961).

100. W. A. Palmisano and A. Peczenik, "Some Considerations of Microwave Hazards Exposure Criteria," *Mil. Med.,* **131,** 611 (1966).

101. C95.3 Committee of the American National Standards Institute, "Techniques and Instrumentation for the Measurement of Potentially Hazardous Electromagnetic Radiation at Microwave Frequencies," ANSI, New York, 1973.

102. Narda Microwave Corporation, Catalog No. 20, Plainview, N.Y., 1976, p. 124.

103. W. W. Mumford, "Some Technical Aspects of Microwave Radiation Hazards," *Proc. IRE,* **49,** 427 (1961).

104. W. A. Palmisano, U.S. Army Environmental Hygiene Agency, presented at the American Industrial Hygiene Conference, Akron, Ohio, 1967.

105. J. H. Heller and A. A. Teixeira-Pinto, "A New Physical Method of Creating Chromosomal Aberrations," *Nature* **183,** 905–906 (1959).

106. A. A. Teixeira-Pinto, L. L. Nejelski, J. L. Cuttler, and J. H. Heller, "The Behavior of Unicellular Organisms in an Electromagnetic Field," *Exp. Cell Res.,* **20,** 548–564 (1960).

107. J. N. Bollinger, "Detection and Evaluation of radiofrequency Electromagnetic Radiation-Induced Biological Damage in *Macaca mulatta,*" Southwest Research Institute, San Antonio, Tex., Final Report under Contract No. F41609-70-C-0025, SWR1 05-2808-01, February 1971.

108. S. A. Bach, "Biological Sensitivity to Radiofrequency and Microwave Energy," *Fed. Proc.,* **24,** Suppl. 14, 22–26 (1965).

109. S. A. Bach, A. J. Luzzio, and A. S. Brownell, "Effects of Radiofrequency Energy on Human Gamma Globulin," in Biological Effects of Microwave Radiation, M. F. Peyton, Ed., Plenum Press, New York, 1961, pp. 117–133.

110. Y. I. Novitskiy, Z. V. Gordon, A. S. Presman, and Y. A. Kholodov, "Radio Frequencies and Microwaves, Magnetic and Electric Fields," Nassau Technical Translation No. Nassau TTF-14.021, 1971.

111. G. G. Knickerbocker, W. B. Kouwenhoven, and H. C. Barnes, "Exposure of Mice to a Strong AC Electric Field—An Experimental Study," *IEEE Trans. Power Appar. Syst.,* **PAS-96,** 498 (1967).

112. L. Ulrich and G. Ferin, "The Effect of Working on Power Transmitting Stations upon Certain Functions on the Organisms (in Czechoslovakian), *Pracovni Lek.* (Prague), **11,** 500 (1959).

113. W. B. Kouwenhoven, O. R. Langworthy, M. L. Singleweld, and G. G. Knickerbocker, "Medical Evaluation of Man Working in AC Electric Fields," *IEEE Trans. Power Appar. Syst.,* **PAS-86,** 506 (1967).

114. *Project Seafarer,* Department of the Navy, Washington, D.C.

115. S. Mittler, "Low Frequency Electromagnetic Radiation and Genetic Aberrations," Northern Illinois University, DeKalb, Final Report, September 15, 1972 (AD 749959).

116. *Project Sanguine Biological Effects Test Program Pilot Studies,* Hazleton Laboratories, Final Report Contract No. N00039-69-C-1572, November 1970 (AD 717408).

117. R. A. Pledger and W. B. Coate, "Bacteria Mutagenesis Study," in: Reference 116, Chapter F.

118. W. B. Coate and S. S. Ho, "Plant Cytogenetic Study," in: Reference 116, Chapter G.

119. W. B. Coate and F. E. Reno, "Rat Fertility Studies," in: Reference 116, Chapter C.

120. W. R. Teeters and W. B. Coate, "Canine Physiological Study," in: Reference 122, Chapter D.

121. "Effects of Low Frequency Electrical Current on Various Marine Animals," Naval Air Systems Command Air Task Report No. 0410801, Work Unit 0100, June 1972 (AD 749335).

122. M. J. Marr et al., "The Effect of Low Energy, Extremely Low Frequency (ELF) Electromagnetic Radiation on Operant Behavior in the Pigeon and the Rat," Georgia Institute of Technology, Atlanta, February 28, 1973 (AD 759415).

123. J. D. Grissett et al., "Central Nervous System Effects as Measured by Reaction Time in Squirrel Monkeys Exposed for Short Periods to Extremely Low Frequency Magnetic Fields," Naval Aerospace Medical Research Laboratory Report No. NAMRL-1137, August 1971 (AD 731994).

124. J. D. Grissett, "Exposure of Squirrel Monkeys for Long Periods to Extremely Low Frequency Magnetic Fields: Central Nervous System Effects as Measured by Reaction Time," Naval Aerospace Medical Research Laboratory Report No. NAMRL-1146, October 1971 (AD 735456).

125. J. deLorge, "Operant Behavior of Rhesus Monkeys in the Pressure of Extremely Low Frequency–Low Intensity Magnetic and Electric Fields," Experiment 1, November 1972 (AD 754058).

126. A. J. Kalmijn, Woods Hole Marine Biological Laboratory, Woods Hole, Mass., personal Communication; also A. J. Kalmijn, "The Electric Sense of Sharks and Rays," *J. Exp. Biol.,* **55,** 371–383 (1971), and A. J. Kalmijn, "Electro-Orientation in Sharks and Rays: Theory and Experimental Evidence," Scripps Institution of Oceanography, University of California, San Diego, Report Contract No. N00014-69-A-0200-6030, Office of Naval Research, November 1973.

127. Rütger Wever, "Einfluss schwacher elektromagnetische Felder auf die circadiane Periodik des Menschen," *Naturwissenschaften,* **55,** 1, 29–32 (1968).

128. F. Halberg, "Circadian Rhythms in Plants, Insects and Mammals Exposed to ELF Magnetic and/or Electric Fields and Currents," University of Minnesota, Final Report to the Office of Naval Research, August 1975.

129. Committee on Biosphere Effects of Extremely Low Frequency (ELF) Radiation, National Academy of Sciences–National Research Council, Washington, D.C.

CHAPTER TWELVE

Ionizing Radiation

GEORGE M. WILKENING

1 INTRODUCTION

The mystique of ionizing radiation is undoubtedly derived from the fact that dangerous exposures may be experienced without adequate warning by the human sensory systems. This tragic lesson was learned by the early workers with radioisotopes and X-rays near the turn of the twentieth century.

Because of the widespread concern about potential detrimental genetic as well as somatic effects of ionizing radiation, there has been a steady flow of laws and regulations governing the manufacture, transport, use, and disposal of radioactive material and radiation-producing devices. The principal governmental agency responsible for the development of information on the biological effects of ionizing radiation and regulations for control, was the Atomic Energy Commission (AEC). However in 1975 the research functions of the AEC were transferred to the Energy Research and Development Administration (ERDA), and the regulatory functions were assumed by the Nuclear Regulatory Commission (NRC). The Environmental Protection Agency (EPA) has some responsibility for the regulation of nuclear power plant discharges, and the Food and Drug Administration (FDA) of the Department of Health, Education and Welfare is active in the development of standards on radiation limits for medical devices and electronic products. The Occupational Safety and Health Administration (OSHA) of the Department of Labor can inspect radiation facilities within the industrial workplace under the provisions of the Occupational Safety and Health Act of 1970.

The general standards for protection against ionizing radiation are contained in Title 10, Code of Federal Regulations, Part 20 (10 CFR 20). The material contained in 10 CFR 20 is the most generally useful information, on a day to day basis, to the practitioner in radiation protection. Regulations covering by-product nuclear materials are found in 10 CFR 30 through 36; those for source materials (e.g., uranium and thorium) are found in 10 CFR 40; those for special nuclear material (e.g., plutonium, ^{233}U or ^{235}U) are found in 10 CFR 70. The reader is urged to obtain copies of the appropriate Codes of Federal Regulations from the Government Printing Office in Washington, D.C.

In making practical distinctions between radiations that might produce ionization in tissue and those that will not, a value of 10 eV is often used (see Chapter 11). That is, radiant energies above 10 eV would be expected to produce ionization (10 eV = 1.6×10^{-12} erg). This should be kept in mind when reviewing the section on the interaction of radiation with matter (Section 2).

A condensed glossary of terms not otherwise defined or described in this chapter is included here:

BREMSSTRAHLUNG: The electromagnetic radiation associated with the deceleration and acceleration of charged particles.

CURIE (Ci): The special unit of activity. One curie equals 3.7×10^{10} disintegrations per second (dps).

DAUGHTER: A nuclide, stable or radioactive, formed by radioactive decay.

DOSE EQUIVALENT: The quantity for radiation protection purposes that expresses on a common scale for all radiations, the irradiation incurred by exposed persons. It is defined as the product of the absorbed dose and certain modifying factors.

FLUENCE (Φ): The number of particles that enter a sphere (or plane) of specified cross-sectional area:

$$\Phi = \frac{dN}{da}\left(\frac{\text{particles}}{\text{cm}^2}\right)$$

where dN is the number of particles and da is the cross-sectional area.

GENETICALLY SIGNIFICANT DOSE (GSD): The dose that, if received by every member of the population, would be expected to produce the same total genetic injury to the population as do the actual doses received by the various individuals.

IONIZING RADIATION: For radiation protection purposes, that radiation which may produce in tissue ionization directly attributable to ionizing charged particles (alpha particles, electrons, protons, etc.) and indirectly attributable to ionizing uncharged particles (photons, neutrons). Also included in the definition is any interaction with molecules that may initiate nuclear transformations.

ISOTOPES: Nuclides having the same atomic number but different mass numbers.

NUCLIDE: A species of atom characterized by its mass number, atomic number, and energy state of the nucleus, provided the mean life in that state is long enough to be observable.

PARTICLE FLUX DENSITY (φ): The number of particles per unit time that enter a sphere (or plane) of specified cross-sectional area.

$$\varphi = \frac{d\Phi}{dt}\left(\frac{\text{particles}}{\text{cm}^2\ \text{sec}}\right)$$

QUALITY FACTOR: A factor dependent on linear energy transfer by which absorbed doses are to be multiplied to obtain the dose equivalent.

RAD: The special unit of absorbed dose. One rad equals 100 ergs per gram.

REM: The unit of dose equivalent. For purposes of this chapter, the following is considered to be equivalent to a dose of one rem: (*a*) an exposure of one roentgen of X- or gamma-radiation. (*b*) a dose of one rad of X-, gamma-, or beta-radiation. (*c*) a dose of 0.1 rad due to neutrons or high energy protons. (*d*) a dose of 0.05 rad due to particles heavier than protons and with sufficient energy to reach the lens of the eye.

ROENTGEN (R): The special unit of exposure. One roentgen equals 2.58×10^{-4} coulomb (of ions of either sign) per kilogram (C/kg) of air.

2 INTERACTION OF RADIATION WITH MATTER

When photons interact with matter, they may be scattered, reflected, or absorbed. These processes reduce or attenuate the number of photons in a beam in an exponential fashion as expressed by the equation:

$$I = I_0\, e^{-\mu x}$$

where I is the fluence at a certain depth x, I_0 is the fluence rate incident at the surface, and μ is the attenuation coefficient. The attenuation coefficient may be expressed in terms of thickness as a linear attenuation coefficient (cm^{-1}) or as a mass attenuation coefficient (cm^2/g), the latter obtained by dividing the linear coefficient μ by the density ρ of the absorbing material. Hence the equation may read

$$I = I_0\, e^{-(\mu/\rho)x}$$

where μ/ρ is the mass attenuation coefficient. Mass attenuation coefficients for some common materials such as aluminum, iron, lead, water, and concrete are given in Table 12.1. Because of the presence of X-rays and secondary scattered photons, it has been found that in irradiated tissue there exist intensities of photons higher than one could predict solely on the basis of the foregoing equation, particularly at great depth. This increase in exposure rate, called the "buildup factor," should be calculated if accurate estimates or radiation dose are desired. More detailed information on buildup factor can be found elsewhere (2).

Table 12.1. Mass Attenuation Coefficients (1)

Photon Energy (MeV)	Mass Attenuation Coefficients (cm^2/g)				
	Aluminum	Iron	Lead	Water	Concrete
0.01	26.3	173	133	5.18	26.9
0.02	3.41	25.5	85.7	0.775	3.59
0.05	0.369	1.94	7.81	0.227	0.392
0.1	0.171	0.370	5.40	0.171	0.179
0.5	0.0844	0.0840	0.161	0.0968	0.087
1.0	0.0613	0.0599	0.0708	0.0707	0.0637
5.0	0.0284	0.0314	0.0424	0.0303	0.0290
10.0	0.0231	0.0298	0.0484	0.0222	0.0231

2.1 Absorption of Radiation

The absorption of photons occurs primarily through the photoelectric effect, the Compton effect, and pair production.

2.1.1 The Photoelectric Effect

At some energy threshold, incident photons are absorbed by matter through the transfer of energy to orbital electrons, causing these to be ejected, thus producing ionization in the medium. This is the so-called photoelectric effect. The ejected electrons possess kinetic energies equal to the difference between the incident photon energy and the binding energy of the shell from which the electron has been ejected. Subsequent to this transfer of energy, X-rays are emitted as the shell vacancies are corrected. The portion of photons interacting by the photoelectric process increases with increasing atomic number and decreasing photon energy. The photoelectric effect predominates when the photon energy is less than 0.5 MeV.

2.1.2 The Compton Effect

As the photon energy is increased above 0.5 MeV, the dominant photon interaction is incoherent scattering from electrons, causing the electrons to recoil. A fraction of the incident photon energy is imparted to the electron; the remaining energy is retained by the scattered photon. The energy of the scattered photon is expressed as follows:

$$h\nu' = \frac{m_0c^2}{m_0c^2/h\nu + 1 - \cos\theta}$$

whereas the energy of the recoil electron is

$$\frac{(h\nu)^2 (1 - \cos\theta)}{m_0c^2 + h\nu(1 - \cos\theta)}$$

where θ is the scattering angle for the photon. In all cases, h is Planck's constant, ν is the photon frequency, and c is the speed of light. The essential effect is that energy is imparted to a free or loosely bound electron, and at the same time there is a loss in frequency and energy in the scattered photon. As the energy of the incident photon decreases, the fraction of its energy imparted to the electron also decreases. The angular relationships between recoil electrons, scattered photons, and incident photons is largely dependent on the original incident photon energy. As the incident photon energies decrease to values approximating 0.5 MeV, the angles at which the electrons are scattered become greater.

2.1.3 Pair Production

Pair production occurs by the interaction of photons with the electric field surrounding a charged particle. When the incident photon energy is above 1.02 MeV, or twice the rest mass energy of an electron, the photon may be annihilated in the high electric field of the nucleus and converted into an electron–positron pair in accordance with the following relationship:

$$h\nu = 2m_0c^2 + E_{e^-} + E_{e^+}$$

where $h\nu$ is the incident photon energy and m_0c^2 is the rest mass energy of the electron or positron, E_{e^-} is the kinetic energy of the electron, and E_{e^+} is the kinetic energy of the positron. The positron produced in this reaction will rapidly interact with electrons, causing electron–positron annihilation, which in turn produces two gamma photons, each with an energy equal to 0.51 MeV. The proportion of incident photons that interact with the nucleus by pair production increases with increasing atomic number.

3 PARTICULATE MATTER

A very large percentage of available radionuclides emit beta radiation. Beta radiation, in turn, is frequently accompanied by X- or gamma radiation. The decay of the nuclide by way of beta particle emission is characterized by a continuous spectrum of beta energies, with the maximum energy being unique to the nuclide. The range of beta particle energies is quite wide, varying from a few kiloelectron volts to slightly over 4 MeV. Beta particles have a wide range of penetration depending on their initial energy and the density of the material through which they traverse. Figure 12.1 shows the range of beta particles in all materials as a function of their energy. By dividing the range of a beta particle by the density of the material being traversed, the thickness of material that will completely stop that particular beta radiation can be obtained. As beta particles traverse matter, they are decelerated; the lost energy is emitted as photons or electromagnetic radiation, sometimes called bremsstrahlung. Bremsstrahlung increases with increasing energy of the beta particles and atomic number of the absorber. The beta particles or electrons emit photons having a range of energy from zero up to the maximum energy of the beta particles being stopped. Only modest thicknesses of commonly available materials are sufficient to completely stop beta radiation.

Alpha particles are physically identical to helium nuclei in that they contain two neutrons and two protons. They are emitted spontaneously during the radioactive decay of certain radionuclides, primarily those of high molecular weight. Because of their comparatively large size and double positive charge, alpha particles do not penetrate matter readily. Even the most energetic alpha radiation is completely stopped by the skin. Once alpha particles are inhaled, ingested or otherwise absorbed into the body, however, they produce a high degree of ionization in tissue. Figure 12.2 illustrates the range of alpha particles in air as a function of energy. Although not revealed by Figure 12.2, the energy range of alpha particles is predominantly between 4 and 8 MeV.

Although not an ionized particle, the neutron readily interacts with matter to generate secondary sources of ionization, usually by inducing radioactivity in or ejecting particles from materials in its path. Neutron energies are often classified into three groups: slow, intermediate, and fast. A reasonable classification for the range of energies

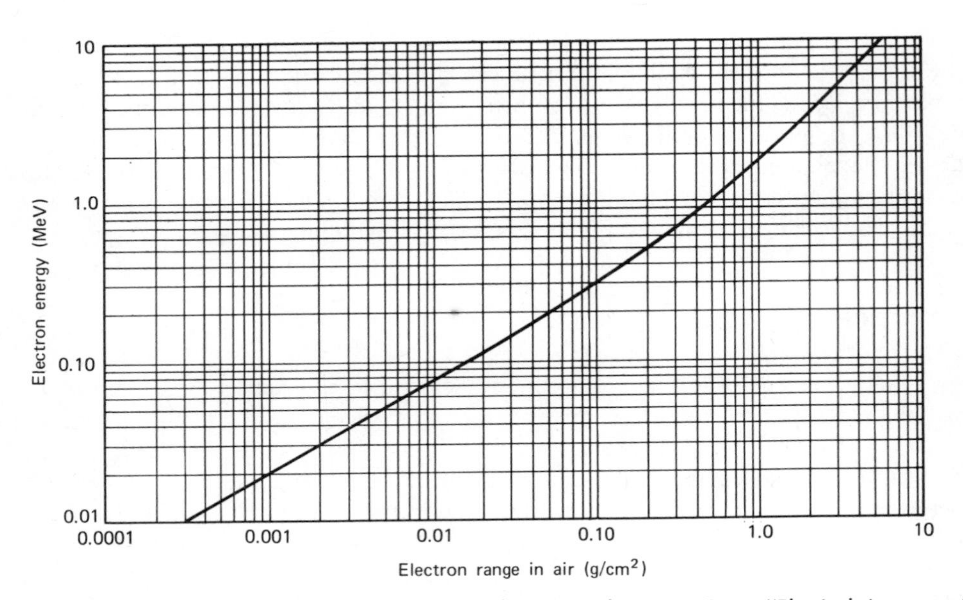

Figure 12.1 Range of beta particles in air as a function of energy. From "Physical Aspects of Irradiations," NBS Handbook No. 85, 1964.

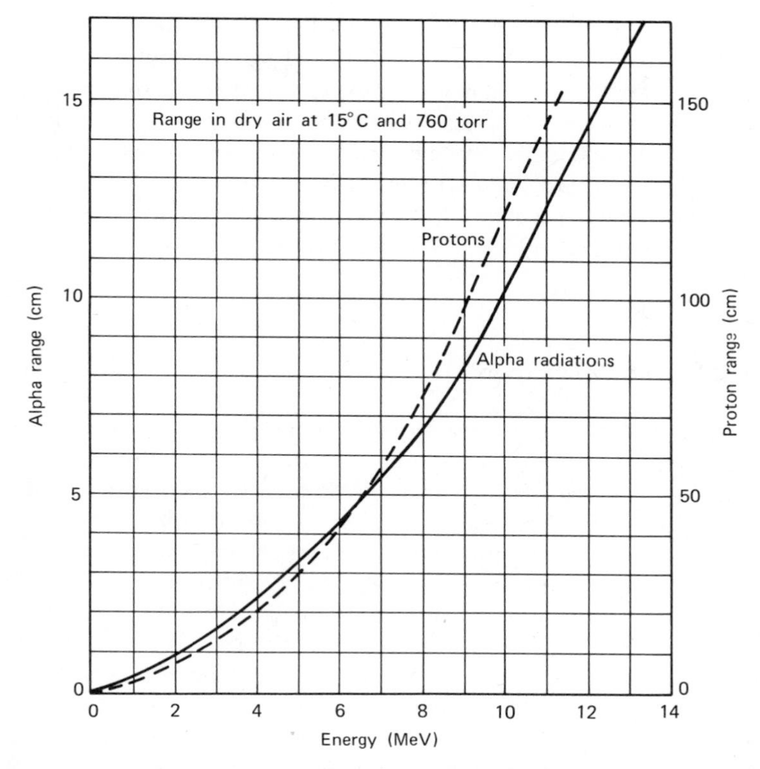

Figure 12.2 Range of alpha particles and protons as a function of energy. From E. F. Gloyna and J. O. Ledbetter, *Principles of Radiological Health,* Dekker, New York, 1969, p. 61.

for each group would be: less than 1 eV, 1 eV to 0.1 MeV, and greater than 0.1 MeV, respectively. Thermal neutrons are so named because they are in thermal equilibrium with the medium in which they are found. The mean value of thermal neutron energy is 0.025 eV at 20°C. Ordinarily neutrons are not emitted by radionuclides, except for a few that undergo spontaneous fission. The more common neutron sources depend on the interaction of alpha or gamma radiation with the nuclei of certain target materials. These sources emit neutrons with an energy spectrum characteristic of the radionuclide and target material. The attenuation of the neutron usually occurs as a result of elastic and inelastic scattering, capture, and induced nuclear reactions. Moderation or slowing down of the neutrons by elastic collisions progressively changes the energy spectrum. The probability of such interactions taking place is specified in terms of cross sections, expressed in units of barns (1 barn = 10^{-24} cm^2). Since the various neutron interactions result in the production of secondary radiations, particularly gamma rays, the person who is evaluating exposures in a neutron facility must not lose sight of the significance of secondary radiations.

At energies beginning at 5 to 13 MeV, except for ^{2}H and ^{9}Be, photons have enough energy to eject a neutron from the nucleus and produce a radioactive nucleus. The liberated neutron, in turn, interacts with surrounding matter to produce secondary ionization. Lower energy electrons can also produce ionization, especially in gases. Protons and heavier charged ions ionize the medium through which they pass; high energy protons interact with nuclei to induce radioactivity in target material and expel secondary particles which become sources of secondary ionization. The secondary ionization then continues until the energies of the secondaries are reduced below the ionization threshold.

4 X- AND GAMMA-RADIATION

X-Rays originate in the extranuclear part of the atom, whereas gamma rays are emitted from the nucleus during nuclear transitions or particle annihilation. X-Rays are commonly produced in an evacuated tube by accelerating electrons from a heated element to a metal target with voltages in excess of 16 kV. The electrons interact with orbital electrons of atoms in the target, causing energy level changes that result in the emission of characteristic X-rays, or the electrons may interact with the nucleus of the atom to produce electromagnetic radiation having a continuous spectrum (bremsstrahlung). The production of bremsstrahlung increases at higher electron accelerating voltages. The energy spectrum or quality of the X-Ray beam may be expressed either in terms of an "effective energy" or in terms of its half-value layer. The accelerating voltage may be constant or continuous or pulsed, thus producing different photon energy distributions. X-Ray tubes and their housings are arranged to provide shielding in all directions except for a window, where the useful beam is emitted. The solid angle and shape of the useful beam is determined by the size of the window and by collimating devices such as diaphragms and cones. Some low energy X-rays are absorbed in the target, others are removed from the useful beam by the material in the tube window, and still others are removed by filters that preferentially absorb the less penetrating radiation. This, in effect, hardens X-ray beams used for medical diagnostic purposes, preserving the more useful penetrating radiation.

Gamma rays are emitted by the nucleus of certain radionuclides during their decay scheme. Each of the nuclides emits one or more gamma rays having a specified energy. Gamma rays may also be produced by neutron interactions with nuclei. Typical gamma ray energies vary from 8×10^{3} to 10^{7} eV; the corresponding frequencies are 2×10^{18} to 2.5×10^{21} Hz.

5 SOURCES AND USES OF IONIZING RADIATION

In gaining some broad insight into the overall pattern of radiation insult to man, let us begin with an appreciation of the radiation received from the natural background. Most of this background is derived from cosmic radiations that vary in intensity as a function of altitude and latitude, and from terrestrial gamma radiations that vary in intensity with the amount of natural radioactive material present in the earth. Cosmic radiation, for example, increases by a factor of 3 in going from sea level to 10,000 ft and by 10 to 20 percent in going from 0° to 50° geomagnetic latitude (3). Internal exposures arise from body deposition of naturally produced radionuclides that have been inhaled or ingested; the major contributors are radon and its daughter products and ^{40}K, respectively. Other naturally occurring radiation sources include uranium, actinium, and thorium, and their decay products, ^{14}C, and tritium, the latter through neutron activation (neutrons supplied by cosmic radiation).

Nuclear reactors are prime examples of man-made radiation sources. In the operation of reactors, materials such as uranium, thorium, and plutonium break up into two or more fragments when bombarded with neutrons, thereby emitting more than one neutron, and the neutrons further sustain chain reactions. High intensities of neutron, beta, and gamma radiation exist within the reactor compartment. Fission weapons could be considered as separate and distinct sources of radiation, however they are actually special cases of uncontrolled fast burst reactors. Fission weapons and all other fission processes convert nuclear fuel into elements of higher atomic number in a series of exothermic reactions. Nonfissionable materials will generally absorb neutrons and become radioactive isotopes of the original material, thereby producing additional sources of radiation. Other well-known sources of ionizing radiation include the uranium industry (i.e., mines, mills, fabricating plants, and fuel reprocessing plants). With the increased attention to energy sufficiency and the need to develop multiple sources of power, it is likely that the level of activity in uranium mining and fabrication will be stepped up.

Another prominent man-made source of radiation is the high energy, charged particle accelerator. Such accelerators can present a variety of potential radiation hazards depending on the particle being accelerated, the interaction with the target, and the mode of acceleration. In addition to the hazards of an electron beam, for example, X-rays may be produced. Also, acceleration of heavy particles such as protons and dueterons, tritium, or heavy positively charged particles may produce secondary electrons that are accelerated in a direction opposite to that of the main beam toward the high voltage region of the device where X-rays can be produced. It has been noted that positive ion accelerators can produce X-rays in the vicinity of the target if the target is a good insulator (4). Normally, heavy ions produce negligible X-radiation as they are absorbed in the target. When the target is an insulator, however, the ions build up a charge on the target material; the resultant field accelerates electrons in its vicinity to produce significant localized radiation levels. Some accelerators produce neutrons upon interaction of the beam with the target material, causing activation of the target and adjacent components. Activation can also be produced directly by higher energy particles. In these instances, the operators of these devices encounter not only the machine-made transient radiation hazards but also material with long-term radioactivity.

Devices such as medical and dental X-ray tubes, X-ray diffractometers, and X-ray radiographic and fluorescence equipment are well-known examples of ionizing radiation sources. X-ray generators usually accelerate electrons onto a target material in a vacuum. As the electron is slowed in the target, it radiates energy in the form of X-rays. The maximum kinetic energy attained by the electrons before impinging on the target is equal to $\frac{1}{2}m_0v^2 = Ve = h\nu_{max} = hc/\lambda_{min}$, where m_0 is the electron mass, v is the maximum electron velocity, V is the anode to cathode potential difference, e is the charge on the electron, ν_{max} is the maximum X-ray frequency, λ_{min} is the minimum X-ray wavelength, h is Planck's constant, and c is the speed of light.

Any device that employs accelerating voltages above approximately 20 kV may generate X-rays of sufficient energy to penetrate the device envelope and present a source of potentially hazardous ionizing radiation. Magnetrons, klystrons, thyratrons, cathode ray tubes, and electron guns are examples. Many devices (e.g., alphatron gauges, electron tubes, fire alarms) have some radioactive materials sealed within their envelopes to improve their ionization characteristics. Gas and aerosol detectors such as those used in fire (smoke) detection and alarm systems employ the use of an alpha emitter to ionize the gas between two electrodes and cause the steady flow of a weak current. When smoke or thermal degradation products enter the air gap between the electrodes, the resulting change in current is relayed to a monitoring and alarm system. Most of the systems marketed for use in commercial buildings use ^{241}Am as the alpha source, in amounts ranging up to approximately 100 μCi, with an average of 50 μCi in each unit.

High voltage rectifiers, electron microscopes, radioluminous devices, and static eliminators represent additional sources of ionizing radiation. Radioactive static eliminators may use ceramic microspheres which contain ^{210}Po an alpha emitter. The size of the spherical particles is large enough to preclude inhalation and deposition in the alveolar spaces if the microspheres happen to get loose from the encapsulation. Reportedly (5) some loss of the microspheres actually occurs during use, but the biological implications appear to be negligible.

Both the manufacture and use of gas mantles may involve exposure to beryllium oxide and the daughters of radioactive thorium (6).

The following is a list of typical examples of the practical uses of ionizing radiation.

1. In radiography: X-rays and gamma rays; beta emitters for thin film and low density materials; neutrons.
2. In gauging levels in tanks, thickness of sheet materials and films, and moisture content.
3. In tracing the flow of materials in pipes.
4. In studying wear rates on metal parts.
5. As tracers in chemical and biological research.
6. In sterilization of food and medical supplies.
7. As cross-linking agents to improve the properties of plastics.
8. As energy sources for navigational beacons.
9. In implanted cardiac pacemakers.

10. In nuclear power reactors.
11. In chemical analyses (e.g., neutron activation).
12. In research.

6 BIOLOGICAL ASPECTS OF IONIZING RADIATION

The effects of ionizing radiation on exposed individuals can be divided into somatic and genetic, even though such effects cannot always be distinguished from each other, especially when irradiation of body tissues and germ plasm occurs simultaneously.

6.1 Somatic Effects

6.1.1 Chronic Exposure

The estimation of biological risk at relatively low radiation dose levels is made by extrapolation from data obtained at higher dose levels. Certain assumptions are then made about dose–response relationships, the probable mechanisms involved, and the susceptibility of the population at risk. Because of the uncertainties about the dose–response relationships, extrapolation is usually made of the linear dose–incidence function at high exposure levels down to the origin at zero dose, on the assumption that the incidence at zero dose is also a point on the extrapolated line (7). The most important chronic effect of radiation on human populations is carcinogenesis, including leukemogenesis; yet cancers induced by radiation are indistinguishable from those occurring naturally. Hence the existence of cancer can be inferred only in terms of an excess over what is regarded as the natural incidence. The natural incidence of cancer varies over several orders of magnitude depending on the type and the site of the neoplasm, age, sex, and other factors. The period of time that elapses before the appearance of a clinically detectable neoplasm is characteristically long—years, or even decades. Such a long induction time complicates the prospective follow-up observations of an irradiated population. It also complicates the retrospective evaluation of patients for possible history of relevant radiation exposure. Many of the existing human and animal data on radiation-induced tumors come from populations exposed to internally deposited radionuclides where the dose–incidence relation is obscured by nonuniform distribution of dose to body tissues. In certain instances, data have been derived from studies of therapeutically irradiated patients in whom the effects of radiation may have been complicated by effects of the underlying disease and medication (8).

Information derived from the Hiroshima and Nagasaki bombings leads to the conclusion that about one case of leukemia per million exposed persons per year per rem can be expected. The evidence derived from the bombings of Japanese cities suggests that susceptibility to the induction of leukemia is several times higher in persons irradiated *in utero,* during childhood, or late in adult life than it is in individuals of intermediate ages (9).

Various types of radiation, such as alpha, beta, gamma, X- or neutron radiation, produce different responses in tissue even if the individual doses are the same. This difference in response depends not only on the type of radiation but also on the way in which the energy is distributed within the tissue. This difference in the way energy is distributed or is transferred is known as the linear energy transfer (LET). The "relative biological effectiveness" (RBE), a term that is receiving less attention and usage, denotes the ratio between the doses of high LET and of low LET radiations that produce equivalent effects. The RBE value is reflected indirectly in the term quality factor *Q*, described below. Usually the RBE reference value of 1.0 is assigned to the biological effectiveness of 250 kV X-rays or ^{60}Co gamma rays. Examples of high LET radiations are alpha particles, protons, and fast neutrons; examples of low LET radiations are electrons and x- and gamma rays. Table 12.2 lists the average LET values for various nuclear particles.

The biological risk per rad of low LET radiations decreases to a greater extent with a decrease in dose equivalent and dose rate than does the risk with high LET radiations. Data obtained from sources other than those from Hiroshima and Nagasaki, (e.g., uranium miners, radium-treated patients, or radium dial painters) indicate that for cancer

Table 12.2. Average LET Values[a]

Particle	Mass (amu)	Average LET (keV/μm)	Tissue Penetration (μm)
Electron	0.00055	12.3	0.01
		2.3	1
		1.42	180
		1.25	5,000
Proton	1	90	3
		16	80
		8	350
		4	1,400
Deuteron	2	6	700
		1	190,000
Alpha	1	260	1
		95	365
		5	20,000

[a] The average amount of energy lost per unit of particle spur-track length: low LET, radiation characteristic of electrons, X-rays, and gamma rays; high LET, radiation characteristic of protons or fast neutrons. Average LET is specified to even out the effect of a particle that is slowing down near the end of its path and to allow for the fact that secondary particles from photon or fast neutron beams are not all of the same energy.

Table 12.3. Relation Between Linear Energy Transfer L_∞ and Quality Factor Q (11)

L_∞ (keV/μm)	Q
3.5 or less	1
7	2
23	5
53	10
175	20

induction, alpha particles that are delivered at relatively low dose rates may be some 10 times more effective per rad average tissue dose than X- or gamma rays delivered at high dose rates.

For purposes of radiation protection, it is important to develop a method for unifying all the complicating factors that determine a given biological effect in tissue. An indication of such an effect on an organ or tissue is given by the term "dose equivalent," which is basically an expression of absorbed dose that has been weighted or modified by certain factors. The dose equivalent H is equal to the product of D, Q, and N at a given point in tissue, where D is the absorbed dose, Q is the quality factor, and N is the product of any other modifying factors (10). A value of 1 is usually given to N for external sources. The value Q accounts for the influence of radiation quality on biological effect. It is usually thought to express the microscopic distribution of absorbed radiation energy, but it is also a function of the collision-stopping power of water. The collision-stopping power of water is synonymous with linear energy transfer, or L_∞. To determine H, one needs to know the spectrum of absorbed dose in L_∞ for all values of L_∞. When this spectrum is known for the specific mass of tissue of interest, an average value of Q ($\bar{Q}$) can be calculated for that mass of tissue.

$$\bar{Q} = \frac{1}{D}\int_0^\infty D_{L_\infty}\, dL_\infty$$

If one does not know the spectrum of absorbed dose, certain estimates are acceptable; for example, when the neutron energy spectrum is unknown, a $\bar{Q}$ value of 10 may be used to convert the measured absorbed dose into dose equivalent. The relationship between LET (L_∞) and Q is shown in Table 12.3.

When the dose equivalent has been determined for a particular exposure and one wishes to compare the result with primary protection limits for either internal or external radiation, it is important to remember that the protection limits differ from different organs and tissues in the body and for various combinations of these organs, that the organs and tissues are at different depths within the body, and that the dose equivalent DQN is a function of position in the body (12). The numerical values of the absorbed dose and the dose equivalent are equal if the linear energy transfer of the charged particles that deliver the dose is less than 3.5 keV/μm. This is almost always the case for electrons. It is also largely true for X- and gamma radiations with energies of up to 10 MeV. Above this value, nuclear reactions may contribute particles of high collision-stopping power.

A review of the sources of information on which the present radiation protection guidelines are based elicits the striking fact that the basic data are taken from cases of victims of the atomic bombs at Hiroshima and Nagasaki, the underground miners exposed to radon gas and radioactive decay products, populations such as the radium dial painters, with high body burdens of alpha-emitting radionuclides, and persons with ankylosing spondylitis who had received long-term treatment with X-rays. A point of interest is that most of the tumors observed in uranium miners have been located in the bronchial epithelium. The data from these populations indicate that the excess mortality from all forms of cancer correspond to roughly 50 to 165 deaths per million persons per rem during the first 25 to 27 years after irradiation (13). One of the most controversial subjects is the extrapolation of available mortality data from the high exposure level groups just mentioned to an estimate of probable mortality resulting from a continuing low level irradiation (0.10 to 0.20 rem) of the general population. However estimates of this type are bound to be made, despite complexities or lack of appropriate data. One such estimate is that the most likely number of deaths per 0.10 rem per year in the United States population would be expected to be of the order of 1350 to 3300 (13).

6.1.2 Acute Exposure

From considerations of low level radiation exposure levels and their long-term chronic effects, we now turn to acute somatic effects and radiation exposures that produce injury. Tables 12.4 and 12.5 present information on acute effects in humans: Table 12.4 lists the dose–effect relationships in man for acute whole body irradiation, and Table 12.5 presents symptoms and probable mortality rates as a function of acute whole body absorbed dose.

The dose entries in Table 12.4 should be taken as representative compromises only of a surprisingly variable range of values that would be offered by well-qualified observers asked to complete the right-hand column. This comes about in part because whole body irradiation is not a uniquely definable entity. Midline absorbed doses are used. The data are a mixed derivative of experience from radiation therapy (often associated with "free air" exposure dosimetry) and a few nuclear industry accident cases (often with more up-to-date dosimetry). Also, the interpretation of such qualitative terms such as "readily detectable" is a function of the conservatism of the reporter.

In connection with Table 12.5, it should be noted that little information is available on long-term protracted exposures of man. Animal experiments such as the continuous gamma irradiation of beagles at Argonne National Laboratories show that with daily exposures in the 10 to 40 R range, anemia is the major cause of death, and with daily exposures of 50 R or greater infection is the major cause of death. With protracted

Table 12.4. Representative Dose–Effect Relationships in Man for Whole Body Irradiation

Nature of Effect	Representative Absorbed Dose of Whole Body X- or Gamma Radiation (rads)
Minimal dose detectable by chromosome analysis or other specialized analysis but not by hemogram	5–25
Minimal acute dose readily detectable in a specific individual (e.g., one who presents himself as a possible exposure case)	50–75
Minimal acute dose likely to produce vomiting in about 10 percent of people so exposed.	75–125
Acute dose likely to produce transient disability and clear hematological changes in a majority of people so exposed	150–200
Median lethal dose for single short exposure	300

Reprinted with permission of the National Council on Radiation Protection and Measurements from NCRP Report No. 39, Washington, D.C., 1971.

exposure below 10 R per day, myeloproliferative diseases, potentially leukemic, are causes of death.

Spermatogonia are among the most radiosensitive cells in the body. A dose of 50 rems delivered in a single brief exposure may result in cessation of sperm formation (14). Because acute whole body irradiation has not been observed to cause permanent sterility, however, the sterilizing dose for man as well as for other male mammals is thought to exceed the lethal dose if applied to the whole body in a single brief exposure (15). In contrast to the testes, the ovary possesses its entire supply of germ cells early in life and lacks the ability to replace them as they are subsequently lost. Therefore ionizing radiation may cause a lasting reduction in the reproductive potential of the affected ovary. Data obtained from Japanese survivors of the atomic bomb and from investigations of women exposed to fallout in the Marshall Islands suggest that a minimum of 300 to 400 rems must be given in a single exposure to cause permanent sterility and that a larger dose, of the order of 1000 to 2000 rems, is required for sterilization if administered to young women in fractionated exposures over a period of 10 to 14 days (16).

The erythemal dose to skin is generally believed to be of the order of at least several hundred rads of X-rays; and the erythema appears in a matter of hours postirradiation. This threshold varies depending on the energy of the radiation, the dose rate, size of the area of skin irradiated, and the region of the body exposed. In order of increasing severity, the skin would be expected to react in the following sequence: erythema, dry desquamation, vesiculation, sloughing of the skin layers, and chronic ulceration.

Approximately 200 R of X- or gamma irradiation in a single dose is required to

Table 12.5. Somatic Injury Chart[a]

Exposure—Acute Whole Body (R)	Type of Injury	Medical Care Required	Able to Work	Probable Acute Mortality Rate During Emergency
10–50	Asymptomatic	No	Yes	0
50–200	Acute radiation sickness; see level I	No	No	Less than 5 percent
200–450	Acute radiation sickness; see level II	Yes	No	Less than 50 percent
>450 (450–900)	Acute radiation sickness; see level III	Yes	No	Greater than 50 percent
>600 (600–1000)	Acute radiation sickness; see level IV	Yes	No	100 percent
>1000–3000	Acute radiation sickness; see level V	Yes	No	100 percent

Level I symptoms	Less than half this group vomit within 24 hours of onset of exposure. Minor subsequent symptoms. Fewer than 5 percent require medical care. All others can perform their customary tasks.
Level II symptoms	More than half this group vomit soon after onset of exposure and are ill for a few days. This is followed by a period of 1 to 3 weeks with few symptoms. After the latent period, epilation is seen in more than half, followed by a moderately severe illness due to loss of white blood cells and infection. Most of the people in this group require medical care, more than half survive. However, essentially, all are unable to work.
Level III symptoms	A more serious version of the sickness described in level II: the initial period of gastric distress is more severe and prolonged; the main episode of illness is characterized by extensive hemorrhages and complicating infections. People in this group need medical care and hospitalization. Fewer than half survive, even with excellent medical care.
Level IV symptoms	An accelerated version of the sickness in level III. All in this group begin to vomit soon after the onset of exposure, and this continues for several days or until death. Damage to the gastrointestinal tract predominates, manifested by uncontrollable diarrhea which becomes bloody. Changes in the blood count occur early. Death occurs before the end of the second week and usually before the appearance of hemorrhages or epilation. Clinical problems resulting from the low exposure rate are related to the failure of the bone marrow. All in this group need care and few, if any, survive.
Level V symptoms	An extremely severe injury in which hypotensive shock secondary to vascular damage predominates. Symptoms and signs or rapidly progressing shock appear almost as soon as the dose has been received. Death occurs within a few days.

Reprinted with permission of National Council on Radiation Protection and Measurements; adapted from NCRP Report No. 42, Washington, D.C., 1974.

[a] The results reported above are for whole body irradiation; the effect is less for partial body

produce opacities in the human lens. The value of 200 R is probably close to the maximum single dose that produces opacities that do not interfere with vision. For multiple doses of low LET radiation, approximately 400 to 550 R is required to produce detectable opacities, depending on the fractionation of exposure time (10).

6.2 Genetic Effects

The two most notable genetic effects of ionizing radiation are the production of mutant genes and aberrations in chromosomes. Both profoundly affect the future of the human race. On the other hand, it is well to remember that mutations and chromosome aberrations also occur spontaneously. Therefore a major question for all persons, but particularly those in the radiation protection profession, is the extent to which man-made ionizing radiation should be permitted to add to the load of defective genes already present in the population. Should humans be exposed to man-made radiation that is above the background level of radiation? If so, to what extent?

Because of these profound considerations, the National Academy of Sciences Committee on the Biological Effects of Atomic Radiation (BEAR Committee) introduced the concept of regulating the dose to the population (17). Specifically, the committee recommended that man-made radiation be kept at such a level that the average individual exposure is less than 10 R before the mean age of reproduction (i.e., 30 years). In 1956 the genetically significant medical dose—that is, the prereproductive gonad exposure—was estimated to be about half of the recommended 10 R limit. However the Federal Radiation Council (18) did not include medical radiation in its calculations and recommended a 5 R average as the 30-year limit for the population. Using these values provides a limiting dose rate to the population of 0.17 R per year, or, 170 mR per year, the value now in effect. There is still no limitation on population exposures resulting from medical practice. In reaching its recommended level of exposure for the population, the Federal Radiation Council made certain assumptions including the following:

1. That mutations at any given dose rate increase in direct linear proportion to the genetically significant dose.
2. That mutations are irreparable and are almost always harmful.
3. That the mutations caused by man-made radiation are similar to those which occur naturally.
4. That there is no known threshold dose below which an effect will not occur.

The differences between genetic and somatic effects should be noted with care. There is unequivocal evidence that high radiation doses produce cancer and leukemia. That such effects are produced by low radiation doses is less clear. On the other hand, there is

---

exposures. Delayed effects may sometimes occur after the early effects have been ameliorated, the extent depending on the dose received. In addition to the possible loss of hair, one possible effect that is of general concern is sterility.

no direct evidence of any human genetic effects even at high radiation doses. The main basis for claiming genetic effects is the rather convincing studies of radiation effects in animals. As a result of these studies, there is general agreement that humans must be affected in the same way. The principal human concern about radiation-induced somatic effects is the possible induction of malignant diseases, whereas the concern about radiation-induced genetic effects is that all conceivable bizarre anatomical and physiological alterations may be passed on to future generations.

7 GENERAL RADIATION PROTECTION CONSIDERATIONS FOR INTERNAL AND EXTERNAL SOURCES

Some appreciation of the genetically significant radiation resulting from natural as well as man-made sources is prerequisite to a full appreciation of radiation protection requirements. The dose equivalent of the highly penetrating and uniformly distributed cosmic radiation has an average value of 28 mrem per year. The population weighted absorbed dose rate in air in the United States from external terrestrial radionuclides is estimated to be 40 mrad per year; however the absorbed doses must be corrected by a housing factor of 0.8 and a body screening factor of 0.8 to obtain a dose equivalent rate of 26 mrem per year. This dose is largely due to gamma and X-rays. External terrestrial radiation dose is largely determined by the concentrations of ^{40}K and the uranium and thorium series in the soil. To illustrate the variability in external terrestrial whole body dose equivalent rates in terms of geographical area, it is worth noting that the values for the Atlantic and Gulf Coastal plains, the noncoastal plains excluding Denver, and the Colorado plateau, are 15, 30, and 55 mrem per year, respectively. Adding the average dose equivalent rates from internal and cosmic radiation to the gonads (approximately 50 mrem per year), the total respective dose equivalents for each of the three geographical areas mentioned are found to be 65, 80, and 105 mrem per year. The dose equivalent rate for persons living at about the altitude of Denver, Colorado, would be increased by approximately 20 mrem per year from cosmic radiation. Therefore the highest whole body dose received in the United States from natural radiation would approximate 125 mrem per year (19).

If we consider the additional whole body radiation received by the United States population from man-made sources, we find that medical diagnostic radiology accounts for at least 90 percent of the total. One estimate of the average dose rate to the population attributable to medical and dental exposures is 73 mrem per year (20). By comparison, the average dose rate attributable to occupational exposures is 0.8 mrem per year, and that due to nuclear power is well below 1 mrem per year. This means that the average total whole body annual dose equivalent received by each person in the country is approximately 200 mrem, and of that, at least one-third is due to medical and dental X-rays. Although the medical and dental component of radiation received by the human population is not figured in the calculation of radiation protection guides (RPG), major educational efforts have been conducted for the benefit of physicians and dentists to reduce the patient dose to the minimum, consistent with diagnostic and therapeutic requirements.

The methods used for evaluating exposures to external radiation sources and those involving sources that have been deposited internally in the body are quite different. Significant variables in both cases are the characteristic properties of the radiation source, the absorbed energy, the quality factor, the duration of exposure, and the identity of critical organs. The assumption is usually made that any radiant energy from external sources is uniformly distributed over or throughout the part of the body that is irradiated. In the case of inhaled or ingested radioactive material, the usual result is a rather complex pattern of distribution throughout the body in accordance with biochemical rather than nuclear or radioactive properties. Questions of residence time in the body, concentration, and proximity of the radioactive material to organs and tissues become major considerations. Absorption through the skin may be significant for some compounds, such as tritium. If airborne radioactive particulate matter is inhaled, one would expect its deposition in the respiratory tract to follow well-established patterns, which are dependent primarily on the particle size distribution of the inhaled particulate and the construction of the respiratory system. Particles with diameters of less than a few micrometers are more important physiologically because they penetrate deeply into the lung to reach the alveolar spaces. Particles larger than this are preferentially deposited in the bronchioles, bronchi, and upper respiratory tract, including the nasal passages. Soluble particles that reach the alveolar spaces are usually circulated throughout the body by the bloodstream; however their subsequent deposition, retention, and clearance from the body are dependent on the chemical and toxicological properties of the element. The importance of the chemical properties of the inhaled or ingested material is illustrated by the fact that the biological half-life for both stable and radioactive isotopes is the same.

When exact knowledge of the internal distribution of the radioisotope is lacking, the International Commission on Radiological Protection (ICRP) has recommended the assumption that 25 percent of the inhaled aerosol or particulate matter has been exhaled, 50 percent has been deposited in the upper respiratory passages, and 25 percent has been deposited in the lungs. The ICRP recommends the further assumption that all the material in the upper respiratory passages has been swallowed, and that half of the amount assumed to be in the lungs has also been swallowed. In other words, 37.5 percent of the original is assumed to be in the gastrointestinal tract (21).

The dose equivalent received from radioactive materials that have been embedded in the skin surface is based on the assumptions that the irradiation is confined to a local body area and that biological transport through the body by way of the bloodstream is very slow. External alpha sources are of little significance in terms of exposure because of their limited range. Beta particles are usually localized at the skin surface or within the outer skin layers. The depth of penetration depends on the energy of the beta particle.

Although the dose equivalent received from external radioactive sources can be estimated by means of field measurements, the significance of exposure to radioactive materials that have been inhaled or ingested must be evaluated by other means. In the latter case, one must refer to the maximum permissible concentrations (MPC) of the appropriate radionuclide in air and water and compare the actual inhalation or ingestion conditions to these limiting concentrations. Sometimes special consideration of residence time in the body and metabolic factors is required. Values for the maximum permissible concentrations of radionuclides are shown in Appendix 12B. The use of the data in Appendix 12B is explained in the section on standards and regulations.

Many situations feature a combination exposure to external radiation and internal emitters. Since the various types of radiation affect organ systems in different ways, it is essential that the dose equivalent to the most sensitive organs be given primary consideration. In the case of whole body irradiation, the blood-forming organs such as the bone marrow, the lens of the eye, and the gonads are more susceptible to detrimental effects. Hence these organs are considered to be "critical organs" for external exposures. For internal sources, where the radiation distribution is nonuniform, the critical or limiting organs are generally considered to be the lung, the gastrointestinal tract, bone, muscle, fatty tissue, thyroid, kidney, spleen, pancreas, and prostate (22). The radioactivity level in the critical organ is the limiting factor, but the MPD for the critical organ indicates the total activity that should be permitted in the entire body. This value is designated as the maximum permissible body burden (MPBB).

7.1 Plutonium

The use of plutonium serves to illustrate some of the problems associated with internal emitters in the work environment. More is known about this alpha-emitting element than about others, and plutonium appears to be destined to play a role of continuing significance in satisfying the future energy requirements of the United States. The accidental release of plutonium from nuclear power reactor sites appears to be an extremely unlikely possibility; if such a situation ever develops, however, the plutonium would be released in the dioxide form.

If plutonium particles become lodged in the body tissue, bone sarcomas or liver tumors might eventually be produced. The maximum permissible body burden of plutonium has been deduced from the level established for ^{226}Ra; namely, 0.1 μg. A value of 5 μg was first established for ^{239}Pu, then successively reduced to 0.04 μCi of ^{239}Pu. Estimations of the amounts of plutonium in living human organisms are complicated by the fact that the X- and gamma emissions during decay are of very low energy, hence are almost completely absorbed in tissue. Photon measuring devices placed at or very near the surface of the body are severely handicapped in their ability to detect radiation. Urinary and fecal assay methods therefore have been useful in estimating body burdens, but continuous collection of all excrement over a period of time is necessary to detect very low levels.

In an analysis of occupational exposed groups, as well as those of the general population, Nelson (23) has found that concentrations of plutonium in all groups were highest in the tracheobronchial lymph nodes when compared with other organs and that liver concentrations in the occupationally exposed groups were higher than those found in the skeleton. There is evidence that the particle size of inhaled aerosol plays a role in determining the level of deposition not only in the respiratory systems but also in the systemic distribution. For example, larger particles tend to be trapped in the tracheobronchial lymph nodes, whereas smaller particles, which as a rule are more readily soluble, find their way to the bloodstream and to other organs such as the liver. Once the plutonium is in the bloodstream, the chemical and physical properties of the material are of paramount importance. Very small ($< 0.01\ \mu m$) particles tend to concentrate on bone surfaces; those with diameters in the 0.01 to 1 μm range tend to be taken up by the liver. Of the total amount of plutonium that is present in the bloodstream, 45 percent is deposited in the skeleton and 45 percent in the liver; 10 percent is deposited in other tissues or is excreted (24). The rate of clearance of plutonium from the human body is notoriously slow when compared with ^{226}Ra. For example, the plasma clearance is 0.1 to 0.2 liter per day, whereas approximately 100 liters of plasma per day is cleared on ^{226}Ra (25).

Detailed clinical studies (26, 27) have been made of a group of men who worked with plutonium during World War II. The calculated body burdens of this group ranged from 0.005 to 0.42 μCi, but apparently nothing remarkable has been found. The assumption is usually made that the critical body organ for plutonium is the skeleton, but there is still considerable uncertainty about this. Furthermore, although lung cancer has been produced experimentally in animals, no such cases have been demonstrated in humans, nor can any firm conclusions be drawn about long-term cumulative effects. Of those who have been exposed to plutonium, the major mode of entry has been inhalation. Although there has been no case of human injury due to plutonium exposure, the latent period for the induction of physiological changes may be of the order of 50 years or more, therefore no one can say with complete certainty that injuries or detrimental physiological effects will not be realized. A relatively small number of persons have lived for a considerable time with plutonium body burdens which are well above maximum permissible levels yet no detrimental effects have been observed.

A commonly employed technique for estimating the body burden of plutonium is the measurement of radioactivity in the urine (28). This can be done by X-ray spectrometry, proportional counting, or the counting of neutron-induced tracks in plastic. A rough assessment of the absorbed dose in tissue may be made by measuring the low energy X- and gamma rays from plutonium by means of instruments placed on the surface of the body. Also, alpha ray spectrometry of urine samples has found wide application as a plutonium bioassay technique.

Although it is generally conceded that the urine assay is the method of choice in assessing plutonium body burden, it is also true that the correlation between urine assays taken on living human subjects and analyses of tissue from the same subjects after death has not been particularly reassuring. Table 12.6, for example, illustrates one recent set of urine assays that predicted much higher body burdens than were actually found when quantitative analyses were made of tissue obtained at autopsy. The same study (29) showed a general correlation between the amount of plutonium in the body,

Table 12.6. Plutonium Body Burden Calculations (29)

Case Number	From Urine Assay (nCi)	From Tissue Analysis (nCi)	
		Systemic	Total
1-039	24.6	18.3	21.0
2-030	10.0	4.2	8.8
2-100	2.8	0.3	0.3
2-126	2.4	0.0	0.5
2-130	3.2	1.0	1.2
3-016	3.4	0.7	0.9
5-150	2.7	0.0	0.01
7-066	30	0.6	0.8
7-084	0.9	0.1	0.4

exclusive of the respiratory tract, and the amount of plutonium estimated by urine assay, especially at higher levels of body plutonium. The urine assay does not correlate well with plutonium burden in the respiratory tract.

Parenthetically, the situation is considerably better in the case of tritium. Where environmental levels of tritium are low, urine analysis is sufficient for personal monitoring; changes at exposure levels well below regulatory limits will be detected. Proportional counters are capable of measuring ^{239}Pu in amounts as low as approximately 4 nCi to values of the order of 40 nCi, depending on counting gas, chamber characteristics, and counting time.

7.2 Cardiac Pacemakers

Within the working population, some will always require the use of medical prostheses such as cardiac pacemakers. Certain of the pacemakers will be powered by batteries, but others may be powered by a radioactive source such as plutonium. According to one source (30), the effects of 10,000 plutonium-powered pacemakers would be as follows:

1. Exposure to family members living with the pacemaker wearer would not be expected to produce a dose equivalent greater than 7.5 mrem per year and a dose to the entire United States population of 128 man-rems per year. These values are to be compared with the average natural background dose equivalent of 102 mrem per year to individuals and the total natural background dose to the United States population of approximately 20×10^6 man-rems per year.
2. Doses to critical organs and the whole body received by pacemaker patients are well below the occupational exposure limit of 5 rems per year.
3. The surgical implantation and removal of pacemakers results in a very small radiation dose to medical personnel. (Surface dose rate from the pacemaker is on the order of 5 to 15 mrem/hr.)

Assuming that the plutonium-powered pacemakers perform safely and efficiently, the advantages, such as long-term, trouble-free maintenance, avoidance of surgical replacement operations, and the like, far outweigh any disadvantages.

The standards for permissible levels of plutonium or permissible body burden, assuming that bone is the critical organ, are not based on calculation of the maximum permissible dose rate to bone; rather they are based on a comparison of the relative toxicity of plutonium and radium in animals and on correlations with human radium exposure data. There is some concern among professionals in radiation protection that the quantification of dose and the criteria on which maximum permissible intake levels are based, are all too heavily founded on the radium experience. Even the newer exposure criteria seem to be mere modifications of the radium-based information and data rather than the result of tests made on the specific radionuclide in question.

8 RADIATION MEASUREMENT

From the radiation protection standpoint, the primary purpose of radiation measurement is to make quantitative assessments of potential environmental exposure conditions, to permit the installation of proper controls, particularly if the exposure conditions appear to exceed established standards. Such procedures clearly indicate that the primary role of instrumentation is to prevent the possible detriment to the health and safety of individuals working with radiation-emitting sources. Unfortunately preventive measures are not always completely effective; hence measurements may be required to assess environmental conditions or the probable absorbed dose after individuals have been irradiated.

Many techniques are employed to measure energy transfer, absorption, or ionization phenomena caused by incident radiation. Some of the instruments and devices used to measure such phenomena include ionization chambers, proportional counters, scintillation counters, Geiger-Müller detectors, luminescent detectors, photographic emulsions, chemical reaction detectors, and fissionable material and induced-radiation detectors. The choice of detector depends on the radioactive source and its emitted radiation, its temporal and spatial characteristics, and the quality of radiation. Since radiation-measuring instruments are often designed to measure a specific type of radiation under very specific conditions, a broad characterization of parameters that are important for proper measurements is in order.

8.1 General Considerations

When the radiation source consists of a mixture of unknown radionuclides, it is possible to differentiate the particles by means of their individual energy spectra. If one of the interfering isotopes is an isotope of the same element being measured, chemical methods will not distinguish one from the other. However all the different elements in the source can be distinguished from one another by chemical means. Gamma rays are often distinguished by spectrometry, using scintillation or semiconductor detectors, and the ability of gamma rays to ionize gaseous molecules within ionization chambers is often employed to measure exposure rate. Beta particles have continuous energy spectra that permit only limited discrimination among different radionuclides; for certain beta decay schemes, however, coincidence techniques may provide additional selectivity. Alpha particles are distinguished by gas ionization according to their energies (alpha spectrum), and by scintillation and semiconductor detection techniques. To avoid contributions to background from low LET (L_∞) radiation, one should avoid using detector elements such as anthracene, whose response to absorbed energy shows a pronounced inverse dependence on LET. Alpha particles are emitted by many naturally occurring radioisotopes (most common construction materials show alpha particle emission at freshly cut surfaces), but man-made isotopes are essentially limited to the transuranic elements, and of these, only the plutonium, americium, californium, and curium are potentially found in the work environment.

For a recent review of measurement techniques used for purposes of radiation protection, the reader is referred to Special Publication No. 456, "Measurements for the Safe Use of Radiation," issued by the National Bureau of Standards in November 1976.

8.2 Detection and Measurement

8.2.1 Thermoluminescent Detectors

Until recently, the only devices that had been employed for measuring external radiation exposure were film badges and pocket ionization chambers. The use of thermoluminescent dosimeters (TLD) for personnel monitoring is now a widespread practice. The TLD consists of a small crystalline detector, usually lithium fluoride, lithium borate, calcium fluoride, or calcium sulfate. Traces of other metal ions serve as activators. When the crystalline structure is exposed to radiation, energy is quantitatively absorbed in traps. When the material is subsequently heated, the stored energy is quantitatively released in the form of light, thus permitting an estimate of radiation exposure. Lithium fluoride is most frequently used because its response is relatively independent of the type and energy of incident radiation, and it has a useful range of about 0.003 to 10,000 rems (31). The general response and a readout cycle for a typical TLD chip are given in Figure 12.3. The sensitivity and accuracy of the TLD (Table 12.7) are adequate for personnel monitoring, and the reproducibility is excellent. A reproducible rate of heating must be used in measuring the light output from the TLD. By making simultaneous measurements using fluorides of ^{6}Li which is very sensitive to neutrons, and ^{7}Li which is less sensitive, it is possible to evaluate the neutron contribution to the overall dosage. A major advantage of the TLD over film is its energy-independent response. On the other hand, the energy-dependent response of film can be used

Table 12.7. Typical Measurement Precision and Accuracy of TLD Badge

Measurement Precision						
Exposure (mR):	5	10	50	300	3000	30,000
Standard deviation:	20%	10%	4%	4%	4%	4%

Accuracy—Personnel Badge

Type Radiation	100 mrem Dose Reported as Whole Body	Skin
Gamma, ^{226}Ra	94	114
Gamma, ^{137}Cs	100	101
Gamma, $^{60}C0$	90	110
X-ray, 175 keV	104	111
X-ray, 145 keV	102	113
X-ray, 100 keV	107	113
X-ray, 82 keV	111	119
X-ray, 58 keV	114	123
X-ray, 43 keV	129	143
X-ray, 23 keV	80	130
X-ray, 16 keV	21	116
Beta, ^{90}Sr, ^{90}Y	30	70
Beta, uranium slab	20	60

Courtesy Eberline Instrument Corporation

to obtain a rough estimate of the energy spectrum of the radiation. In most routine applications, however, the measurement of the energy spectrum is unnecessary. One word of caution is that the lithium fluoride disks exhibit energy-dependent directional sensitivity, which may be further complicated by badge-filter orientation when the disks are placed in personal monitoring badges (32).

Both TLDs and radiophotoluminescent (RPL) detectors are replacing film badges in many applications. A comparison study between film and thermoluminescent dosimetry for routine personnel monitoring revealed that film is less reliable than the thermoluminescent dosimetry for monitoring radiation exposures above 10 R (33).

Glass dosimeters using RPL have been developed. One system is thermally stimulated exoelectron emission (TSEE). The system is similar to TLD except that instead of measuring light that is emitted when the material is heated, electrons are measured with a Geiger-Müller tube. The emission of electrons may be stimulated by light or heat; also, ultraviolet radiation may be emitted following optical stimulation (34).

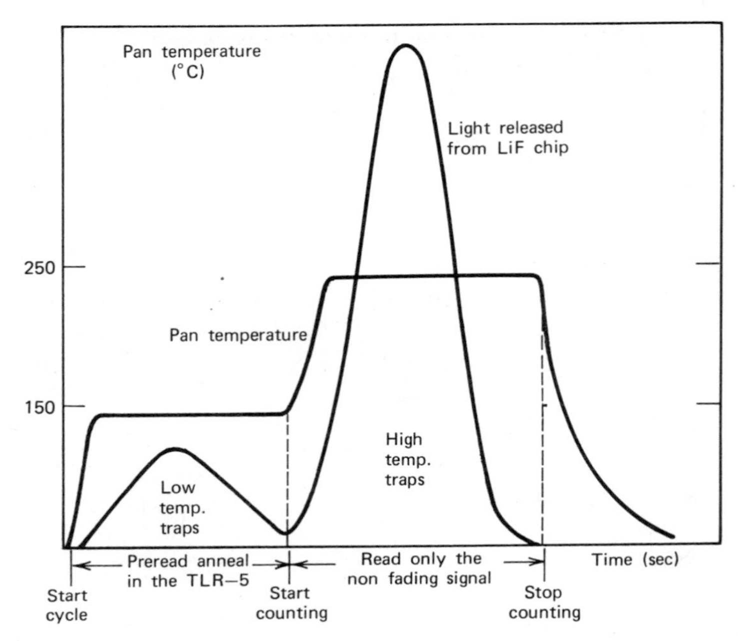

Figure 12.3 Typical "glow curves" for TLD reader (Eberline TLR-5). Data on typical measurement precision and accuracy of TLD badge appear Table 12.7. Courtesy of Eberline Instrument Corporation.

8.2.2 Neutron Dosimetry

Personal neutron dosimetry has traditionally been accomplished by counting proton recoil tracks in exposed special film. However a recent development has been the use of dielectric materials such as plastic or glass for detecting and measuring neutron flux. The plastic or glass material is covered with a layer of foil of fissionable material such as uranium. Neutrons cause fission action in the foil, and the highly energetic fission fragments cause damage in the dielectric material. When the material is chemically etched, the damage becomes apparent as visible tracks, or etch pits, which can be counted under the microscope. The proper measurement of neturons is highly complex, especially if the energy dependent spectral shape is not known.

One method for measuring neutron absorbed dose over the range of 50 to 500 rads has been developed (35) using fission fragment track etching. It was discovered that by using combinations of ^{232}Th or ^{237}Np as fissile materials and organic and inorganic track detectors, both automatic spark counting and visual track counting methods can be developed to cover the dose range mentioned. Similarly, neutron fluence distributions from ^{252}Cf sources have been measured in a phantom, using the track etching method. Thin polycarbonate sheets are first subjected to fragments from neutron-induced fission in foils of ^{235}U, ^{238}U, ^{232}Th, then counted under the microscope after etching (36). Fast neutron fission track dosimeters have appreciable directional dependence. In one study (37) the recorded ratio of sensitivities for 14 MeV neutrons incident normal and parallel to the plane of the detector was approximately 1.6. For fission neutrons and thorium radiators, the value was 1.4.

8.2.3 Pocket Dosimeters

The pocket dosimeter (ion chamber) is normally used in conjunction with the film badge for measuring X- and gamma radiation. The walls of the dosimeter, as well as the internal components and ionizable gas, are usually air equivalent or tissue equivalent. The indirect reading gamma dosimeter is similar to the direct reading dosimeter except that it must be read out on a charger-reader. The direct reading instrument merely needs to be held up to the light to make a reading. The reading is directly proportional to the discharge of current from a precharged value, and the discharge is proportional to the integrated radiation dose. The range of most pocket dosimeters is from 0 to 200 mR; some are available in the 0 to 200 R range. The energy dependence is usually less than 10 percent from 50 keV to 1.33 MeV (^{60}Co). The minimum measurable energy is approximately 10 to 15 keV. Although the models of dosimeters have vastly improved over the past two decades, they are still subject to accidental discharge.

Films are photographic emulsions that may be used to detect various types of radiation. One type of film blackens if exposed to beta, gamma, or x-radiation, whereas another type is used to record tracks that are produced by charged particles such as protons from fast or thermal neutron interactions. The blackening of the film by beta, gamma, and X-rays is not proportional to air or tissue absorbed dose at different energies. Therefore various types of moderating metal and plastic or loaded absorbing material are usually placed adjacent to film to minimize the nonproportionality when the film is used in a radiation monitoring badge. Shielding material placed over films may also permit the discrimination between gamma rays and X-rays and beta radiation.

The minimum energy measured by films is of the order of 20 keV for X-rays and 200 keV for beta radiation. Some of the main shortcomings of film for radiation protection purposes are as follows:

1. False readings may be produced if film is exposed to heat, pressure, or chemicals.
2. Variations in film quality occur from batch to batch.
3. Energy dependence for low energy X-rays is strong.

Chemical reaction detectors are systems in which radiation produces a chemical change in a material in such a manner that a chemical analysis or direct reading indicator will measure the amount of change. Such detectors are not used extensively because of their low sensitivity.

Induced radiation detectors are materials in which radiation interacts to form radionuclides whose radiation is subsequently measured. They are particularly useful for detecting or measuring neutron radiation.

8.2.4 Geiger-Müller Survey Meters

The detector tube in the Geiger-Müller (GM) survey meter is usually encased in a protective outer metal shield. Some tubes have openings at the extremities; others have openings along their length. These openings are usually covered with a thin sheet of mica or Mylar whose density does not exceed 2 mg/cm². By making provision for shielding these openings with metal, discrimination between beta and gamma radiation is possible. The output of the GM counters is not proportional to the exposure or absorbed dose rate unless the device is carefully calibrated for the specific radiation being measured. Significant errors can occur if the meter readings have not been specifically calibrated. GM tubes are usually halogen quenched to prevent undue avalanching of signals. The efficiency of beta measurement ranges from approximately 30 to 45 percent in GM tibes equipped with beta discrimination shields; gamma sensitivity with the shields closed is approximately 5000 counts per minute per milliroentgen. The minimum energy reliably measured is approximately 20 keV for X-rays and 150 keV for beta radiation. The main advantages of GM equipment for surveying purposes are reasonably high sensitivity and rapid response. Some of the disadvantages include saturation at high counting rates, strong energy dependence, and possible interference by ultraviolet and microwave radiation.

Typical full-scale readings of GM equipment range from 0.2 to 20 mR/hr; response time is from 2 to 10 seconds.

8.2.5 Ionization Chambers

Ionization chambers have been specifically designed to measure dose or dose rate from beta, gamma, and X-radiation. The principle of operation of the chamber is that ions are produced when radiation impinges upon a preselected gas contained within the chamber. The ions move in a field supplied by a continuously applied voltage to produce small measurable currents whenever ionization events occur because of incident radiation.

The voltage required to produce a saturation current in any chamber is proportional to the rate of ionization. At saturation, the ionization current is related to the product of the number of ion pairs per unit time and the electronic charge. Since the chamber is essentially an integrating device for a large number of ionization events, the time constant of the electrical current readout device is long, to make possible the averaging out of wide fluctuations. Another variation in detection schemes is to measure the rate of voltage change on a capacitor whose charge depends on the rate of ionization within the chamber. The number of ion pairs formed per centimeter of path length is proportional to the density of gas in the chamber.

Most gas-filled ion chambers operate in the proportional or gas multiplication region, which yields a nominal ±10 percent photon energy response from about 50 keV to 1.3 MeV. The minimum accurately measurable energy is usually of the order of 20 keV (for X-rays). Advantages include low energy dependence and the capability of measuring a wide range of air doses (i.e., 3 mR/hr to more than 500 R/hr). Disadvantages include slow response and relatively low sensitivity. In selecting an ion chamber for use in radiation surveys, certain desirable features should be kept in mind: drift-free response, solid state circuitry to eliminate warmup time, lightweight construction, compactness, capability of installing an audible alarm that is triggered at a level selected by the user, capability of accurately measuring X-ray dose rate independently of the spectral energy distribution, remote operation, and interchangeable ion chambers. For example, it would be desirable to be able to interchange a rate meter that measures 0 to 1000 mR/hr using a high sensitivity chamber with a low sensitivity chamber that measures in the 0 to 1000 R/hr range.

Ion chambers must be compensated for use in measuring tritium concentrations in high gamma fields because as little as 1 mR/hr gamma dose rate may produce a reading that is 10 times the MPC for tritium in air (38, 39).

8.2.6 Scintillation Detectors

Scintillation detectors operate on the basis of producing light from the interaction of ionizing radiation (X, beta, and gamma) with constituents in a scintillation crystal. Scintillation counters should have good pulse height resolution; for example, a 3 × 3 inch unit for ^{137}Cs should deliver at least a 7.5 percent resolution. Detector performance is closely related to crystal purity. Minimum measurable energy from X-rays is of the order of 20 keV; full-scale range of commercially available detectors is typically from 0.02 to 20 mR/hr. Special attention must be paid to obtaining a good light shield around the photomultiplier tube, as well as to the need for using antimagnetic mu metal in the housing to minimize changes in gain due to magnetic fields. A typical scintillation crystal consists of sodium iodide doped with tellurium ions. Light pulses from the phosphor are detected by means of a photomultiplier tube whose output voltage is measured on a voltmeter. The magnitude of the light signal is proportional to the energy absorbed in the scintillation crystal. The operating voltage of the detector is variable, depending on the type and intensity of radiation. In the case of alpha scintillation probes, the window thicknesses are usually of the order of 0.5 mg/cm^2 of aluminized Mylar. The maximum efficiency using this particular window thickness is of the order of 30 percent. If a fine wire mesh screen is placed over the Mylar for protection, however, the efficiency is reduced to approximately 20 percent. In general, scintillation counters are known for rapid response and high sensitivity, but they are also relatively fragile and expensive.

8.2.7 Proportional Counters

Proportional counters usually consist of a gas cylinder containing a central wire to which a potential is applied. The potential may be adjusted to render the output voltage proportional to the energy released by the radiation-induced ionization events in the chamber. Because of this feature, the type of radiation to be measured can be selected.

For example, the proportionality between alpha and beta particles can be determined because of the considerably greater ionization produced by the former. Proportional counters may be used for general surveys or for performing sophisticated spectrometric analyses, but in the majority of cases they are employed in the measurement of neutron and alpha radiation. In the case of gas flow proportional probes to measure alpha radiation, the window thickness if usually of the order of 0.5 to 0.9 mg/cm^2 of aluminized Mylar. The efficiency of these devices, expressed as a percentage of radiation received from a 2π surface, is of the order of 35 to 50 percent. A well-known tritium probe consists of a windowless gas flow proportional chamber for measuring the very low energy beta particles emitted by tritium. This device may be modified to measure alpha or high energy beta particles, but it can also be used for the assessment of alpha and beta swipe samples.

8.3 Calibration

Standard sources, such as those available through the National Bureau of Standards (NBS), should be used for purposes of instrument calibration. In the absence of an NBS source, a secondary standard that is directly traceable or relatable to NBS material may be used. In many cases, sources can be sent to the NBS for calibration. Another approach to instrument calibration is the development of a standard for use within a specified instrument facility, irrespective of whether the standard is relatable to an NBS primary standard.

When obtaining a calibration source for photon measuring instruments, the effective energy (usually in kilo- or megaelectron volts), half-life, and the specific exposure rate of that source must be ascertained. In general, the source should be much smaller than the intended detector, and the distance between the source and detector should be such that the detector measures essentially the radiation from a point source. It is common practice to make the source-to-detector distance at least 7 times that of the largest dimension of either the source or the detector and to keep the source and detector far removed from scattering surfaces. The calibration source should have an energy spectral distribution identical to, or at least similar to, that of the radiation to be measured. It should also have a half-life long enough to permit a reasonable number of calibrations to be performed before replacement. The response of the instrument to be calibrated to various radiation energies will establish its energy dependence characteristics. If X-rays are used as a calibration source, the excitation voltage, the spectral energy characteristics, and patterns of radiated energy must be known.

The instruments most often used for the standardization of photon fields include cavity ion chambers, free air ion chambers, and calorimeters. Detectors must be operated within specified pressure, temperature, and humidity limitations. Whenever possible, a determination should be made of the extent to which radiations other than those of primary interest might interfere with the sensitivity and accuracy of the device. Specific examples of interferences include radiofrequency and microwave radiations generated by a wide variety of sources. Some of these sources may escape the attention of the unsuspecting user because of their common everyday presence (e.g., electrical appliances, coils, electrical discharges, solar radiation, and microwave ovens).

When NBS standards cannot be used and for reasons of convenience it is preferred to employ a secondary standard, every effort should be made to keep the accuracy of the secondary source calibration to within ± 2 percent of the primary or NBS standard.

The following factors are important in the design of any calibration facility:

1. Background radiation levels shall be low and quantifiable during the actual calibration procedure.
2. Neutron and photon calibrations should use free space geometry.
3. Neutron sources shall be described in terms of effective energy and flux density.
4. Photon-emitting sources shall be described in terms of exposure rates at specified distances from the source.
5. Exposure of persons conducting the calibration should be kept as low as practicable but at least within permissible limits.
6. The calibration radiation field should closely approximate that found in the test condition, if possible.
7. Proper records shall be maintained on all calibrations.

When specifying the activity of a primary calibration source, especially when the source consists of an equilibrium mixture, the total activity of the radionuclides in equilibrium must be determined. For example, in the case of the two beta emitters ^{90}Sr and its daughter ^{90}Y, the two in equilibrium have double the activity of freshly separated ^{90}Sr. Within a period of five half-lives, the ^{90}Y activity is 95 percent of the ^{90}Sr activity.

8.3.1 Beta Calibration

Calibrations are often made by placing the detector window of the instrument in close proximity to a properly calibrated beta source such as ^{90}Sr. If an attempt is made to use a free space geometry calibration, one must be aware of the significant absorption of beta particles in air, that is, their limited range in air. Typical beta calibration sources include ^{85}Kr, ^{204}Tl, ^{90}Sr–^{90}Y, ^{14}C, and ^{42}K.

8.3.2 Alpha Calibration

Alpha calibration sources can be purchased commercially or from the NBS. Typical alpha sources include ^{241}Am, ^{210}Po, ^{252}Cf, ^{148}Gd, ^{230}Th, and ^{239}Pu. Since the half-life of ^{210}Po is 138.4 days, the usefulness of this isotope is limited.

8.3.3 Neutron Calibration

Three types of radionuclide neutron sources exist. First there are (α, n) emitters such as ^{210}Po, ^{238}Pu, ^{239}Pu, or ^{241}Am, which emit alpha particles in close contact with low Z elements, such as lithium, beryllium, bismuth, and fluorine. The energy involved with the alpha-emitting types ranges from 0 to tens of megaelectron volts, neutron yield is of the order of 10^7 to 10^8 neutrons per second (n/sec), and the sources are dangerous to handle unless sealed. Wipe tests are definitely required before handling. Second we have (γ, n) sources consisting of a high energy gamma emitter such as ^{124}Sb or ^{226}Ra placed in close proximity to low Z elements such as beryllium or deuterium. Monoenergetic neutrons are produced along with intense photon fields. Although ^{124}Sb has a relatively short half-life (160 days) and yields low neutron energies (30 keV), ^{226}Ra sources produce neturon yields in the 10^6 to 10^7 n/sec range. Third, there are spontaneous fission neutrons, such as those produced by ^{252}Cf; these are produced in yields of 10^8 to 10^9 n/sec, but the associated photon emission is of low intensity. Since ^{252}Cf can be fabricated in relatively small size, it acts as a point source; also NBS is in a position to calibrate ^{252}Cf sources, thereby rendering them derived standards.

Particle accelerators also produce intense neutron fields by accelerating charged particles such as deuterons, protons, or tritons onto low Z materials such as deuterium, lithium, and tritium. High neutron fluxes (10^{12} n/sec) are obtainable, but these are highly variable depending on the accelerator and target characteristics. Therefore the output of the accelerator must be monitored constantly if it is to be used for calibration purposes.

Neutron fields from reactors can vary from an unmoderated fission spectrum to heavily filtered slow neutrons from a thermal column. The flux densities [10^{10} n/(cm^2)(sec)] are high enough for calibration purposes, but they are highly variable and require constant monitoring.

Alpha-, beta-, and photon-emitting calibration sources may be obtained commercially and from such agencies as the National Bureau of Standards, Washington, D.C. (Gaithersburgh, Md.) and the International Atomic Energy Agency, Vienna, Austria.

Calibrating sources should have high activity per unit mass, high chemical and radiochemical purity, and sufficient half-life to suit the purposes of a good instrumentation program. Radionuclides that are suitable as calibration sources include the following.

1. *Alpha sources:* ^{148}Gd, ^{210}Po, ^{230}Th, ^{238}Pu, ^{239}Pu, ^{241}Am, ^{244}Cm, and ^{252}Cf.
2. *Beta sources:* ^{3}H, ^{14}C, ^{32}P, ^{35}S, ^{45}Ca, ^{85}Kr, ^{90}Sr–^{90}Y, ^{111}Ag, ^{185}W, ^{204}Tl, ^{210}Bi, and ^{238}U.
3. *Photon sources:* ^{24}Na, ^{51}Cr, ^{57}Co, ^{60}Co, ^{137}Cs, ^{241}Am, and ^{226}Ra.

In general the following points should be made with regard to the measurement of ionizing radiation:

1. Several methods of surveying and monitoring should be used rather than one. For example, one should not rely exclusively on film badges; a combination of film badges and direct reading dosimeters is preferred. A dangerous situation might be missed while waiting for the film badge readout if no other method is employed. Survey-type instru-

ments should be supplemented with fixed station monitoring equipment whenever possible.

2. The function of all equipment should be checked periodically with a calibrating source.

3. The energy response of the instruments should be continually checked.

4. Special care should be taken in the choice of instruments where pulsed sources are to be monitored. Instruments should have a time constant that permits accurate measurement of the pulse duration. For more detailed techniques and procedures, the reader is referred to other sources (40, 41).

9 EXPOSURE EVALUATION

The contribution of occupational exposures to the total ionizing radiation dose to the general population is far less than 1 percent (42). Except for cases involving the natural radioactive materials radon and radium, industrial operations have not caused any ill effects in humans from internal deposition of radioactive materials. This is a tribute to the excellent preventive programs established in industry. Also, the health of people employed in industry is generally better than that of comparable age groups in the population, and the employees are better trained than the average citizen in the safe use of radioactive sources. Since the general population is comprised of individuals from the fetus to the extremely old, including the sick or disabled, who may collectively be exposed for 24 hours a day, 365 days a year, it stands to reason that the permissible levels of radiation exposure for the general population must be set lower than those for radiation workers. The permissible exposure levels for the two groups (i.e., general population vs. radiation workers) usually differ by a factor of 10 or greater.

A complete tabulation of maximum permissible concentrations of radionuclides in air and water for occupational exposures can be found in NCRP report No. 22 (43) and in Appendix 12B (Part 20, 25 FR 10914) (44). As has been the case in past NCRP recommendations, radiation exposures of individuals resulting from necessary medical and dental procedures have not been included in the dose limiting recommendations, and such medical and dental procedures are presumed to have no effect on the radiation status of the individual.

Three basic assumptions have been made in the derivation of all radiation protection guides: (1) that the effects of all types of radiation, dose rates, and exposure durations are not precisely known; (2) that there is no radiation level below which biological damage will not occur; and (3) that there is a linear relationship between biological effect and radiation dose.

In estimating the significance of external body exposures to airborne radioactive materials, the concentration in the breathing zone averaged over a period of 40 hours a week is compared with the tabulated maximum permissible concentrations (MPC) for the radionuclides in question, as given in Appendix 12B (44). In deriving the MPC values, it has been assumed that the air or water has been inhaled or ingested over a working lifetime of 50 years. Many other assumptions regarding organ retention rates, breathing rates, clearance times, and metabolic mechanisms have entered into these calculations.

One method for estimating the retention of radionuclides that have been ingested or inhaled is to use an instrument that measures the gamma radiation being emitted from the whole body or a critical organ. The dosage to the thyroid, for example, can be estimated reasonably accurately by means of a crystal detector placed next to the thyroid. For practical purposes, the use of such instrumental techniques is limited to gamma or high energy beta emitters. Alpha radiation, which is the most serious type of radiation once inside the body, cannot be detected on the external surfaces of the body. Since in some cases beta emitters (e.g., plutonium) radiate small amounts of gamma and X-rays, devices can detect the presence of a particular radionuclide. A reasonable estimation may be made of absorbed dose equivalent in the case of internal contamination through the use of a whole body counter. This device consists of a battery of very sensitive detectors that have been well shielded to minimize the effect of background radiation. Under these conditions, a spectometric analysis may be made of radionuclides that are present in the body. In the case of external irradiation of the body tissues, contamination may be deposited on the skin or in the clothing. For such situations, suitable instruments equipped with probes that can monitor body surfaces and clothing may be used. The clothing is sometimes removed and the body washed, to help differentiate internal from external radiation.

10 STANDARDS AND REGULATIONS

Many of the basic standards on radiation protection have been formulated by the National Council on Radiation Protection and Measurements and the International Commission on Radiological Protection. The principal set of legally binding standards, however, consists of those issued and enforced by the Nuclear Regulatory Commission. Other regulatory agencies that have responsibilities for radiation protection include the Environmental Protection Agency, the Department of Health, Education and Welfare, the Occupational Safety and Health Administration, and state governments.

The permissible dose for individuals working in radiation (restricted) areas as specified by the Nuclear Regulatory Commission (44) in Part 20 of the Code of Federal Regulations is as follows:

1. Whole body; head and trunk; active blood-forming organs; lens of the eye; or gonads: 1.25 rems per calendar quarter.
2. Hands and forearms; feet and ankles: 18.75 rems per calendar quarter.
3. Skin of whole body: 7.5 rems per calendar quarter.

A higher dose to the whole body is permitted within a given calendar quarter provided a value of 3 rems is not exceeded. Also the dose to the whole body, when added to the accumulated occupational dose to the whole body, shall not exceed $5(N - 18)$ rems, where N is the person's age in years at the last birthday. Figure 12.4 plots of the radiation dose accumulated at the rate of 5 rems per year, beginning at age 18. All cumulative radiation doses must be below the diagonal line to meet the $5(N - 18)$ rem criterion.

The rem is a measure of the dose of any ionizing radiation to body tissue in terms of its biological effect relative to a dose of one roentgen of X-rays. The NRC considers the following doses to be equivalent to a dose of one rem:

1. A dose of 1 R due to X- or gamma radiation.
2. A dose of 1 rad due to X-, gamma, or beta radiation.
3. A dose of 0.1 rad due to neutrons or high energy protons.
4. A dose of 0.05 rad due to particles heavier than protons and bearing sufficient energy to reach the lens of the eye. One rem is assumed to be equivalent to 14×10^6 n/cm^2 incident upon the body. If data are available on the distribution of neutron energy, it is possible to estimate the neutron fluence that is equivalent to one rem. Table 12.8 provides the data necessary to make this conversion.

The NRC forbids any licensee from permitting individuals to be exposed to an average concentration of airborne radioactive materials in excess of the limits given in Appendix 12B, Table I. No allowance is made for the use of protective clothing or equipment or the particle size of the aerosol unless specifically authorized by the NRC. The limits in Appendix 12B are based on exposure to the specified concentrations for 40

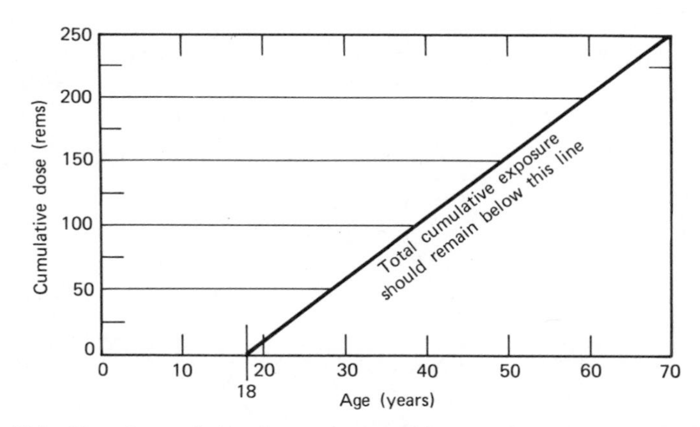

Figure 12.4 Plot of cumulative dose at permissible rate of 5 rems per year after age 18.

hours in any period of 7 consecutive days. For exposure periods greater or less than the 40 hours, the limits can be decreased or increased proportionately.

The exposure of minors in restricted areas shall not exceed 125 mrem per calendar quarter for whole body, head and trunk, active blood-forming organs, lens of the eye, or gonads; or 1875 mrem for hands and forearms, feet and ankles; or 750 mrem for the skin of the whole body. Also, minors shall not be exposed to airborne radioactive material in excess of the limits cited in Appendix 12B, Table II. Personal monitoring of minors is required if they are likely to receive radiation doses that are one-half of the aforementioned values.

Permissible levels of radiation received by any individual in unrestricted areas, as specified by the NRC, are 0.5-rem dose to a whole body in a calendar year, 2 mrem in any one hour if the individual is continuously present in the area, and 100 mrem in any 7 consecutive days, if the individual is continuously present in the area. The recent ruling by the NRC requires that special protection be afforded the fetus by keeping the exposure of pregnant females to a maximum of 0.5 rem during the period of gestation (45).

The dose-limiting recommendations of the NCRP are given in Table 12.9. The similarity of permissible doses between NCRP and those adopted by NRC is readily apparent. Table 12.10 presents some practical quality factors to be used in calculating dose equivalents. The data in Table 12.10 may be examined in terms of related data in Table 12.3.

Table 12.8. Neutron Flux Dose Equivalents

Neutron Energy (MeV)	Number of Neutrons per Square Centimeter Equivalent to a Dose of 1 rem n/cm^2	Average Flux Density to Deliver 100 mrem in 40 hours n/cm^2 sec
Thermal	970×10^6	670
0.0001	720×10^6	500
0.005	820×10^6	570
0.02	400×10^6	280
0.1	120×10^6	80
0.5	43×10^6	30
1.0	26×10^6	18
2.5	29×10^6	20
5.0	26×10^6	18
7.5	24×10^6	17
10	24×10^6	17
10–30	14×10^6	10

Table 12.9. NCRP Dose-Limiting Recommendations

Maximum Permissible Dose Equivalent for Occupational Exposure	
Combined whole body occupational exposure	
Prospective annual limit	5 rems in any one year
Retrospective annual limit	10–15 rems in any one year
Long-term accumulation to age N years	$(N - 18) \times 5$ rems
Skin	15 rems in any one year
Hands	75 rems in any one year (25 per quarter)
Forearms	30 rems in any one year (10 per quarter)
Other organs, tissues and organ systems	15 rems in any one year (5 per quarter)
Fertile women (with respect to fetus)	0.5 rem in gestation period
Dose Limits for the Public, or Occasionally Exposed Individuals	
Individual or occasional	0.5 rem in any one year
Students	0.1 rem in any one year
Population dose limits	
Genetic	0.17 rem average per year
Somatic	0.17 rem average per year
Emergency dose limits—life saving	
Individual (older than 45 years if possible)	100 rems
Hands and forearms	200 rems, additional (300 rems total)
Emergency dose limits—less urgent	
Individual	25 rems
Hands and forearms	100 rems, total
Family of radioactive patients	
Individual (under age 45)	0.5 rem in any one year
Individual (over age 45)	5 rems in any one year

Reprinted with permission of National Council on Radiation Protection and Measurements, from NCRP Report No. 39 Washington, D.C., 1971.

Table 12.10. Practical Quality Factors QQ

Radiation Type	Rounded QQ
X-Rays, gamma rays, electrons or positrons; energy >0.03 MeV	1
Electrons or positrons; energy <0.03 MeV	1
Neutrons; energy <10 keV	3
Neutrons; energy >10 keV	10
Protons	10
Alpha particles	20
Fission fragments, recoil nuclei	20

Reprinted with permission of the National Council on Radiation Protection and Measurements, NCRP Report No. 39, Washington, D.C., 1971.

11 ELEMENTS OF A RADIATION PROTECTION PROGRAM

The elements of any given radiation protection program vary according to the size, number, and type of radiation sources and their specific application(s), the facility in which the radioactive sources are housed, training level of personnel, complexity of the operations, and many other factors; therefore an attempt is made to describe principles that are common to radiation protection practice in general. The following elements are covered.

1. Administration.
2. Orientation and training.
3. Control measures.
4. Surveys and monitoring.
5. Emergency procedures.
6. Medical surveillance.
7. Records
8. Notification of incidents

11.1 Administration

The most important requirement for an effective radiation protection program is an informed executive officer or manager who has supervisory responsibility for the basic goal of human protection. This is essential to proper implementation of the program. The elements of the radiation protection program must be prescribed on a case by case basis, with the advice and counsel of professionally qualified individuals.

11.2 Orientation and Training

Of all the ingredients in a radiation protection program, the one that is most indispensable is informing the employee. Once the employee knows about radiation hazards, their potential biological effects, and necessary control, he is in a position to offer his cooperation to meet common objectives. Under these circumstances, the employee is sufficiently well motivated to use his ingenuity to keep exposure as low as practicable, even when environmental controls are less than ideal. The specifics of the training program should include nature of radioactivity, background radiation, interaction of radiation with matter, biological effects (acute and chronic, somatic and genetic, internal and external), exposure criteria (permissible doses), standards and regulations, monitoring procedures, control measures (time, distance, shielding, engineering design, cautionary procedures, protective equipment, warning signs and labels, waste disposal practices), emergency procedures, and medical surveillance. Most important is a declaration of strong support of the program by management.

The person or persons conducting the orientation and training should be qualified professionals in radiation protection. If the radiation facility warrants a full-time professional, such a person should be designated as the radiation safety officer (RSO) and given as one responsibility the training functions of the program. Part-time personnel or consultants in radiation protection should be professionally qualified.

11.3 Control Measures

Primary emphasis and reliance should be placed on engineering control measures whenever possible. For example, if radiation levels can be controlled at the source, many administrative or procedural practices may be eliminated.

11.3.1 Ventilation and Facility Layout

The layout and design of a facility should be such that there is minimal risk of contamination. For example, high radiation areas should be separated from lower level operations wherever possible. Ventilation systems can be designed to place the high radiation area hoods at the end of the line, nearest the blower; then "once-through" (no recirculation air) ventilation systems can entrain radioactive contaminants and pass them through high efficiency filters and/or scrubbers before discharge to the external environment. Ducts on the discharge (positive pressure) side of the blower should not be housed within buildings, since a leak in the discharge duct could release contaminants within the building.

For a more detailed treatment of ventilation system design, the reader is referred to Chapter 19 discussing ventilation, the *Industrial Ventilation Manual* published by ACGIH (46), and other references (47, 48).

11.3.2 Shielding

The thickness of a specified substance that when introduced into the path of a given beam of radiation, reduces the value of a specified radiation quantity by one-half is referred to as the "half-value layer" (HVL). The HVL is often used to characterize the effectiveness of shielding materials. Table 12.11 presents HVL and TVL (tenth-value layer) data on lead, concrete, and steel for attenuation of X-rays produced at various peak voltages. Also included are some HVL and TVL values for ^{137}Cs, ^{60}Co, and ^{226}Ra. Similar data for the gamma radiation from a number of radionuclides are presented in Table 12.12. The gamma ray constants in the right-hand column may be used to determine the dose rates at various distances from the source. In using HVL data, it should be remembered that a shield thickness of 2 HVL reduces the exposure rate by a factor of 4; a thickness of 3 HVL by a factor of 8, and so on.

The shielding designs should take into account such factors as human occupancy and use, but the overriding consideration in the design should be to reduce all possible exposures to the lowest practicable value, and in every case, to well below the maximum permissible dose.

Table 12.11. Half-Value (HVL) and Tenth-Value (TVL) Layers[a]

Peak Voltage (kV)	Attenuating Material					
	Lead (mm)		Concrete (in.)		Steel (in.)	
	HVL	TVL	HVL	TVL	HVL	TVL
50	0.05	0.16	0.17	0.06		
70	0.15	0.5	0.33	1.1		
100	0.24	0.8	0.6	2.0		
125	0.27	0.9	0.8	2.6		
150	0.29	0.95	0.88	2.9		
200	0.48	1.6	1.0	3.3		
250	0.9	3.0	1.1	3.7		
300	1.4	4.6	1.23	4.1		
400	2.2	7.3	1.3	4.3		
500	3.6	11.9	1.4	4.6		
1,000	7.9	26	1.75	5.8		
2,000	12.7	42	2.5	8.3		
3,000	14.7	48.5	2.9	9.5		
4,000	16.5	54.8	3.6	12.0	1.08	3.6
6,000	17.0	56.6	4.1	13.7	1.2	4.0
10,000	16.5	55.0	4.6	15.3		
^{137}Cs:	6.5	21.6	1.9	6.2	0.64	2.1
^{60}Co:	12	40	2.45	8.1	0.82	2.7
^{226}Ra:	16.6	55	2.7	9.2	0.88	2.9

Reprinted with permission of the National Council on Radiation Protection and Measurements, NCRP Report No. 34, Washington, D.C., 1970.

[a] Approximate values obtained at high attenuation for the indicated peak voltage values under broad beam conditions; with low attenuation these values will be significantly less.

The choice of shielding materials can be crucial to the proper attenuation of radiation. In the case of photon emissions, such as X-rays, the selection of materials usually becomes a choice between concrete or lead, or combinations of both, or combinations with other materials (e.g., lead-lined lath and wallboards, lead-lined concrete blocks, lead glass, lead–steel combinations, loaded concrete, loaded concrete–lead). The attenuation of photons in ordinary concrete is dependent more on the uniformity of material density than on a special chemical composition; in the case of loaded concretes, however, the addition of magnetite, steel, lead, ferrophosphorus, and the like, produces a considerably improved attenuation, particularly when these additives are uniformly distributed in the mix.

Variations in the chemical compositions of concrete can be critically important in the case of neutron shielding. Such variations in composition have a more pronounced effect

Table 12.12. Selected Gamma Ray Sources

Radio-isotope	Atomic Number	Half-Life	Gamma Energy (MeV)	Half-Value Layer[a]			Tenth-Value Layer[a]			Specific Gamma Ray Constant[b] [(R/cm^2)/mCi-h]
				Concrete (in.)	Steel (in.)	Lead (cm)	Concrete (in.)	Steel (in.)	Lead (cm)	
^{137}Cs	55	27 years	0.66	1.9	0.64	0.65	6.2	2.1	2.1	3.2
^{60}Co	27	5.24 years	1.17, 1.33	2.6	0.82	1.20	8.2	2.7	4.0	13.0
^{198}Au	79	2.7 days	0.41	1.6	—	0.33	5.3	—	1.1	2.32
^{192}Ir	77	74 days	0.13–1.06	1.7	0.50	0.60	5.8	1.7	2.0	5.0[c]
^{226}Ra	88	1622 years	0.047–2.4	2.7	0.88	1.66	9.2	2.9	5.5	8.25[d]

Reprinted with permission of National Council on Radiation Protection and Measurements, NCRP Report No. 34, Washington, D.C., 1970.

[a] Approximate values obtained with large attenuation.

[b] These values assume that gamma absorption in the source is negligible. Value in roentgens per millicurie-hour at 1 cm can be converted to roentgens per curie-hour at 1 meter by multiplying the number in this column by 0.10.

[c] This value is uncertain.

[d] This value assumes that the source is sealed within a 0.5 mm thick platinum capsule, with units of roentgens per milligram-hour at 1 cm.

on the gamma radiation produced secondarily to neutron interaction than on the transmission of neutrons through the shield. Also, when dealing with photoneutron sources, gamma shielding may actually prove to be more important from a protection standpoint than neutron shielding.

In any case, before any shielding material is installed, special attention must be given to such details as overlapping joints, eliminating voids or nonhomogeneities in the material, need for structural support for non-load-bearing material such as lead, need to ensure proper attenuation through notoriously leaky areas in the shield (e.g., glass windows, joints, seams, pipes, conduits, service boxes, and doors), need to be certain that the correct shielding materials are being used for the type of radiation in question, and the need for continuous maintenance of the shielding structure, to prevent deterioration.

For a more detailed treatment of shielding design, the reader is referred to References 49 to 52.

11.3.3 Protective Equipment

Every effort should be made to control the potential radiation exposure by engineering means, rather than through the use of personal protective equipment. Protective equipment does have a place in the program however. If respirators must be used, for example (usually for nonroutine, intermittent operations), it is important to ensure that such respirators meet the approval of the National Institute for Occupational Safety and Health. Special training is necessary for the proper use of protective equipment (e.g., the proper fit and maintenance, removal procedures that avoid recontamination, laundering, and disposal). Therefore such training responsibilities should be placed in the hands of a qualified radiation safety officer.

11.3.4 Radioactive Wastes

Radioactive wastes have half-lives ranging from minutes to thousands of years, complicating the procedure of delay and decay in achieving reduced radioactivity levels. Storage of radioactive solidified waste can be accomplished in excavated salt formations underground where heat can be generated and dissipated without producing seismic activity (53). For gaseous waste, the usual procedure is to delay (store) the material to permit decay. Filters have been used to collect radioactive particles that have been formed when a gaseous parent nuclide decays to a particulate radioactive daughter or becomes attached to other particles. Low temperature adsorption may be useful in providing a delay–decay mechanism for short-lived noble gases (54). Special attention must be paid to leaks around filters, particularly with materials such as ^{234}Pu; more than one filter bank may be necessary. A major objective of plant design and operation is to process and recycle all waste streams in a manner serving to minimize both volume and level of activity. Continuous monitoring of the effluent is necessary to be certain that permissible levels are not exceeded. The NRC does not permit releases of effluents to unrestricted areas exceeding the limits specified in Appendix 12B, Table II. Applications for higher limits must be made directly to the commission.

For the discharge of radioactive wastes into sanitary sewage systems, the material must be readily soluble or dispersible in water, and the average daily concentration must not exceed the larger of either (*a*) the limits specified in Appendix 12B, Table I, column 2, or (*b*) 10 times the quantity of such material specified in Appendix 12C.

The average monthly concentration of radioactive wastes cannot exceed the limits specified in Appendix 12B, Table I, column 2, nor can the gross quantity of radioactive material released into the sewerage system in any one year exceed one curie.

Burial of radioactive materials in soil is permitted (1) if the total activity at the time of burial does not exceed 1000 times the amount specified in Appendix 12C, (2) if the burial is at a minimum depth of 4 feet; (3) if the successive burials are separated by distances of at least 6 feet, and (4) if not more than 12 burials are made in any year.

Incineration may not be used as a means for disposing of radioactive materials unless the case is specifically approved by theNRC.

11.4 Surveys and Monitoring

11.4.1 Surveys

Periodic surveys should be conducted to evaluate any potential hazards associated with the production, use, release, disposal, or presence of radioactive materials and radiation-producing devices. Then all radioactive sources and devices producing ionizing radiation should be evaluated to determine the possible levels of radiation associated with their use. Areas should then be designated in accordance with the following NRC definitions.

Area	Definition (NRC)
"Radiation area"	Where dose in excess of 5 mrem in any hour is possible; or dose in excess of 100 mrem in any 5 consecutive days is possible.
"High radiation area"	Where a dose in excess of 100 mrem in any hour is possible

Radiation areas should be posted with the standard radiation symbol and the words "CAUTION, Radiation Area" or "CAUTION, High Radiation Area," as appropriate. High radiation areas require the establishment of special cautionary operating procedures, including the use of interlocks and alarms.

Wherever licensed radioactive material (exclusive of natural uranium and thorium) is used or stored in an amount exceeding 10 times the quantity specified in Appendix 12C, the area must be posted with a sign bearing the radiation caution symbol and the words "CAUTION, Radioactive Material(s)." The use or storage of natural uranium or thorium in an amount exceeding 100 times the quantity specified in Appendix 12C requires the identical posting format. The outside surfaces of containers of radioactive materials must be similarly labeled.

The periodic survey of work environments should include as assessment of gross contamination levels. For this purpose, contamination working counts (c) should be established. One example of such limits is the following.

Contamination	Limit per 100 cm^2 Area
Removable	
Alpha emitters other than uranium and thorium	100 c/m
Uranium and thorium	1000 c/m
Beta emitters other than ^{90}Sf and ^{129}I	1000 c/m
^{90}Sr and ^{129}I	250 c/m
Nonremovable	
Alpha emitters other than uranium and thorium	1 c/m per cm^2
Uranium and thorium	10 c/m per cm^2
Beta emitters other than ^{90}Sr and ^{129}I	0.2 mR/hr
^{90}Sr and ^{129}I	0.05 mR/hr

11.4.2 Monitoring

Various forms of monitoring and surveillance are required to ensure the continuing effectiveness of the radiation protection program. Some general principles and practices for a monitoring program follow.

1. Monitoring instruments should have sufficient sensitivity, precision, accuracy, response time, and dynamic range to accommodate the type of radiation being generated and its operational characteristics.
2. Radiation protection personnel must be alert to changes in the radiation environment, preferably before they have a chance to occur. This means being made aware of the introduction of new radiation sources or changes in the operating modes of existing radiation-producing devices. One cannot rely solely on the readout from monitoring devices to signal the advent of a changed environmental condition.
3. All monitors should be designed for a fail-safe response. This means that when instrument components fail, they fail in a mode that alerts people to an unsatisfactory condition (e.g., a visual and audible signal or alarm is used to signify a component failure).
4. The monitoring level on the instruments should be set high enough to avoid spurious signals but low enough to ensure the safety of personnel. Shielding or adjustments to the operating characteristics of monitoring instruments should enable the user of the instruments to eliminate interferences from external sources (e.g., radiofrequency, microwave, and ultraviolet radiation).

5. Periodic calibration and maintenance of all monitoring instruments are absolutely essential to the program.

6. Periodic judgments by a qualified radiation protection specialist will be needed on such matters as the following: (*a*) length of time personnel should wear film badges, personal dosimeters, nuclear accident dosimeters (NAD), thermoluminescent dosimeters, and so on; (*b*) frequency of environmental surveys; and (*c*) frequency of monitoring of internally deposited radionuclides; specific analytical measures to be taken (e.g., bioassay).

7. Periodic determination of whether the areas, sampling points, or conditions being monitored are the appropriate ones to monitor, or whether program changes are necessary.

11.5 Emergency Procedures

A stepwise emergency procedure should be established and posted in each radiation area. Since conditions change over time, the emergency procedure must be reviewed and updated periodically. The essential elements of an emergency procedure include simple, direct, stepwise instructions on the course of action to be taken; for example,

1. "In the event of fire or a radioactive spill, leave the room immediately. Close the door behind you."
2. "Call extension ____ on nearest phone and report conditions."
3. "Call Radiation Safety Officer on extension ____."

Prior arrangements should be established with local police and fire departments, hospitals, in-house and outside emergency squads, and the medical department. Evacuation routes and assembly points should have been designated. Periodic mock drills should help ensure the continuing effectiveness of emergency procedures.

11.6 Medical Surveillance

A preplacement examination for radiation workers is to be recommended, with emphasis on medical history, complete blood analysis, previous radiation exposure history, and eye examination, particularly if the employee has had previous exposure to neutrons or plans to work with neutrons in the future. The maintenance of complete medical records on each radiation worker is essential.

11.7 Records

The Nuclear Regulatory Commission requires each licensee to keep records on the radiation exposure of all individuals who are required to be monitored and on the disposal of radioactive wastes.

Records on the results of surveys should be kept for 2 years, with the exception of the following records, which must be kept until the NRC authorizes disposition:

1. Records of surveys used to determine compliance with exposure limits.
2. Records of surveys to determine external dose, assuming that personnel monitoring data are unavailable.
3. Records of surveys used to evaluate the release of radioactive effluents to the environment.

In general, records should be kept on the following:

1. Radiation surveys.
2. Airborne and waterborne effluents.
3. Exposure of all individuals, including visitors.
4. Bioassay data.
5. Surface contamination and removal.
6. Reviews of new designs and processes.

Appropriate records must be kept to control the use of radiation sources and to facilitate the investigation of accidents and incidents. Since the disease that is closely related to exposure to radiation may make its appearance one or two decades after significant exposures, it is generally a legal requirement to keep records well past the date of employment termination.

11.8 Notification of Incidents

Immediate notification of the NRC Regional Office is required in the event of the occurrence of any of the following situations.

1. Any individual receives a whole body radiation exposure of 25 rems or more, an exposure to the skin of the whole body of 150 rems or more, or an exposure to the feet, ankles, hands, or forearms of 375 rems.
2. The release of radioactive materials in concentrations that when averaged over a 24-hour period, would exceed 500 times the limits specified in Appendix 12B, Table II.
3. A loss of one work day or more has occurred in the affected facility.
4. Damage to property in excess of $1000 has occurred.

The NRC has special requirements for reporting overexposure of individuals and for monitoring reports. Also, procedures must be established to assure shipment or transportation of all radioactive materials in accordance with the rules and regulations of the U.S. Department of Transportation, the Coast Guard, the Federal Aviation Agency, the Nuclear Regulatory Commission, and the International Atomic Energy Agency.

REFERENCES

1. *Radiological Health Handbook,* Public Health Service, rev. ed. Washington, D.C., January 1970.
2. E. C. Barnes, "Ionizing Radiation," in *The Industrial Environment—Its Evaluation and Control,* U.S. Department of Health Education and Welfare, 1973.
3. "The Effects on Populations of Exposure to Low Levels of Ionizing Radiation," National Academy of Sciences–National Research Council, Washington, D.C., November 1972, p. 12.
4. M. M. Weiss, "X-Radiation from Positive Ion Beams Incident on Insulator Targets," *Health Phys.,* **15:** 4, 372 (October 1968).
5. M. K. Robertson and M. W. Randle, "Hazards from the Industrial Use of Radioactive Static Eliminators," *Health Phys.,* **26,** 245 (1974).
6. K. Griggs, "Toxic Metal Fumes from Mantle Type Camp Lanterns," *Science* **181,** 842 (1973).
7. K. Griggs, in: "The Effects on Populations of Exposure to Low Levels of Ionizing Radiation" (3), p. 86.
8. "Plutonium—Health Implications for Man," *Proceedings of the Second Los Alamos Life Sciences Symposium,* Los Alamos, N.M., May 22–24, 1974; *Health Phys.,* **29** (4) (October 1975).
9. See "The Effects on Populations of Exposure to Low Levels of Ionizing Radiation" (3), p. 87.
10. W. D. Norwood, *Health Protection of Radiation Workers,* Thomas, Springfield, Ill., 1975, p. 72.
11. "Radiation Quantities and Units," Report No. 19, International Commission on Radiological Units and Measurements, Washington, D.C., 1971.
12. "Conceptual Basis for the Determination of Dose Equivalent," Report No. 25, International Commission on Radiation. Units and Measurements, Washington, D.C., 1976.
13. See "The Effects on Populations of Exposure to Low Levels of Ionizing Radiations" (3), p. 89.
14. W. H. Langham, Ed., "Radiobiological Factors in Manned Space Flight," National Academy of Sciences–National Research Council Publication No. 1487, National Academy of Sciences, Washington, D.C., 1967.
15. United Nations Scientific Committee on the Effects of Atomic Radiation, Report General Assembly, Official Records, 17th Session, Supplement No. 16 (A5216) United Nations, New York, 1962.
16. T. G. Baker, "Radiosensitivity of Mammalian Oocytes with Particular Reference to the Human Female," *Am. J. Obstet. Gynecol.,* **110,** 746–761 (1971).
17. "Biological Effects of Atomic Radiation," National Academy of Sciences–National Research Council, National Academy of Sciences, Washington, D.C., 1956.
18. Federal Radiation Council, "Background Material for the Development of Radiation protection Standards," staff reports of the Federal Radiation Council, 1960, 1962.
19. "Natural Background Radiation in the United States," Recommendations of the National Council on Radiation Protection and Measurements, Washington, D.C., issued November 15, 1975.
20. See "The Effects on Populations of Exposure to Low Level Ionizing Radiation" (3), p. 50.
21. M. Eisenbud, *Environmental Radioactivity,* McGraw-Hill, New York, 1963, p. 36.
22. W. H. Langham, J. N. P. Lawrence, J. McClelland, and L. H. Hempelmann, "The Los Alamos Scientific Laboratory's Experience with Plutonium in Man," *Health Phys.,* **8,** 753 (1962).
23. I. C. Nelson, K. R. Heid, P. A. Fugua, and T. D. Mahony, "Plutonium in Autopsy Tissue Samples," *Health Phys.,* **22,** 925 (1972).
24. "The Metabolism of Compounds of Plutonium and Other Actinides," International Commission on Radiological Protection, Pergamon Press, Oxford, England, 1972.
25. W. H. Langham, S. H. Bassett, P. S. Harris, and R. E. Carter "Distribution and Excretion of Plutonium Administered to Man," Los Alamos Report No. 1151, Los Alamos Scientific Laboratory, Los Alamos, N.M., 1950.

26. L. H. Hempelmann, W. H. Langham, C. R. Richmond, and G. L. Voelz, "Manhattan Project Plutonium Workers: A Twenty-Seven Year Follow-Up Study of Selected Cases," *Health Phys.*, **25**, 461 (1973).
27. L. H. Hempelmann, C. R. Richmond, and G. L. Voelz, "A Twenty-Seven Year Study of Selected Los Alamos Plutonium Workers," Report No. LA-5148-MS, Los Alamos Scientific Laboratory, Los Alamos, N.M., 1973.
28. S. A. Beach, G. W. Dolphin, K. P. Dolphin, K. P. Duncan, and H. J. Dunster, "A Basis for Routine Urine Sampling of Workers Exposed to Plutonium-239," *Health Phys.*, **12**, 1671 (1966).
29. H. F. Schulte, "Plutonium: Assessment of the Occupational Environment," *Health Phys.*, **29**, 613–618 (1975).
30. R. L. Shoup, T. W. Robinson, and F. R. O'Donnell, "Generic Environmental Statement on the Routine Use of Plutonium Powered Cardiac Pacemakers," Annual progress report, period ending June 30, 1976, No. ORNL-5171, Health Physics Division, Oak Ridge National Laboratory. Oak Ridge, Tenn., 1976, p. 41.
31. Data obtained from Eberline Instrument Corporation, P.O. Box 2108, Santa Fe, N.M.
32. E. H. Dolecek and R. A. Wynveen, "Evaluation of Employing Lithium Fluoride–Teflon Dosimeters in a Personnel Monitoring Program," paper presented at 20th Annual Meeting of the Health Physics Society; *Health Phys.*, **29** (6), 906 (December 1975).
33. W. L. Beck, R. J. Cloutier, and E. E. Watson, "Personnel Monitoring with Film and Thermoluminescent Dosimeters for High Exposures," *Health Phys.*, **25**, 425 (1973).
34. "Radiation Protection Instrumentation and Its Application," Report No. 20, International Commission on Radiological Protection, Washington, D.C., 1971.
35. K. Becker and J. S. Jun, "Transfer Dosimeters for Fast Neutron Sources," *Health Phys.*, **29**: 6, 915 (December 1975).
36. R. A. Oswald, L. H. Lanzl, and M. Rozenfeld, "Use of Fission Track Detectors in a Tumor-Mouse Phantom Irradiated with ^{252}Cf Neutrons," *Health Phys.* **29**: 6, 916 (December 1975).
37. W. G. Cross and H. Ing, "Directional Dependence of Fast Neutron Fission Track Dosimeters," *Health Phys.* **29**: 6, 907 (December 1975).
38. W. R. Busch, "Assessing and Controlling the Hazard from Tritiated Water," AECL-4150, 1972.
39. R. V. Osborne and G. Cowper, "The Detection of Tritium in Air with Ionization Chambers," AECL-2604, 1966.
40. "Measurement of Low Level Radioactivity, "Report No. 22, International Commission on Radiological Units and Measurements, Washington, D.C., 1972.
41. American National Standards Institute "Radiation Protection Instrumentation Test and Calibration," Standard N 323, (N13/N42), ANSI, New York, 1976.
42. See "The Effects on Populations of Exposure to Low Level Ionizing Radiation" (3), p. 18.
43. Maximum Permissible Body Burdens and Maximum Permissible Concentrations of Radionuclides in Air and in water for Occupational Exposure, 1959 (includes Addendum 1 issued in August 1963).
44. "Standards for Protection Against Radiation," Part 20, *Fed. Reg.* **25**, 10914 (November 17, 1960). Nomenclature changes appear *Fed. Reg.* **40**, 8783 (March 3, 1975).
45. *Instruction Concerning Prenatal Radiation Exposure,* USNRC Regulatory Guide 8.13, Nuclear Regulatory Commission, Washington, D.C., March 1975.
46. Committee on Industrial Ventilation, American Conference of Governmental Industrial Hygienists, *Industrial Ventilation,* 13th ed., ACGH, Box 453, Lansing, Mich., 1974.
47. A. D. Brandt, *Industrial Health Engineering,* Wiley, New York, 1947.
48. W. C. L. Hemeon, *Plant and process Ventilation,* Industrial Press, New York, 1954.
49. T. D. Jones and F. F. Haywood, "Transmission of Photons Through Common Shielding Media," *Health Phys.*, **28**, 630 (1975).

50. E. D. Trout, J. P. Kelley, and G. L. Herbert, "X-Ray Attenuation in Steel—50 to 300 kVp," *Health Phys.*, **29,** 163 (1975).
51. *Concrete Radiation Shields for Nuclear Power Plants,* USNRC Regulatory Guide 1.69, Nuclear Regulatory Commission, Washington, D.C., 1975.
52. O. Bozyap and L. R. Day, "Attenuation of 14 MeV Neutrons in Shields of Concrete and Paraffin Wax," *Health Phys.*, **28,** 101 (1975).
53. R. L. Bradshaw, F. M. Empson, W. C. McClain, and B. L. Houser, "Results of a Demonstration and Other Studies on the Disposal of High Level Solidified Radioactive Wastes in a Salt Mine," *Health Phys.*, **18,** 63 (1970).
54. "Nuclear Power and the Environment," International Atomic Energy Agency, P.O. Box 590, A-1011, Vienna, 1972, p. 22.

APPENDIX 12A REPORTS OF NCRP AND ICRU

NCRP Reports

The NCRP Publications office distributes the NCRP Reports. Information on prices and how to order may be obtained by directing an inquiry to:

NCRP Publications
P.O. Box 30175
Washington, D.C. 20014

The extant NCRP Reports are listed below.

NCRP Report No.	Title
8	Control and Removal of Radioactive Contamination in Laboratories (1951)
9	Recommendations for Waste Disposal of Phosphorus-32 and Iodine-131 for Medical Users (1951)
10	Radiological Monitoring Methods and Instruments (1952)
12	Recommendations for the Disposal of Carbon-14 Wastes (1953)
14	Protection Against Betatron–Synchrotron Radiations up to 100 Million Electron Volts (1954)
16	Radioactive Waste Disposal in the Ocean (1954)
22	Maximum Permissible Body Burdens and Maximum Permissible Concentrations of Radionuclides in Air and in Water for Occupational Exposure (1959) [includes Addendum 1 issued in August 1963]
23	Measurement of Neutron Flux and Spectra for Physical and Biological Applications (1960)
25	Measurement of Absorbed Dose of Neutrons and of Mixtures of Neutrons and Gamma Rays (1961)
27	Stopping Powers for Use with Cavity Chambers (1961)
28	A Manual of Radioactivity Procedures (1961)
30	Safe Handling of Radioactive Materials (1964)

NCRP Report No.	Title
31	Shielding for High-Energy Electron Accelerator Installations (1964)
32	Radiation Protection in Educational Institutions (1966)
33	Medical X-Ray and Gamma-Ray Protection for Energies up to 10 MeV—Equipment Design and Use (1968)
35	Dental X-Ray Protection (1970)
36	Radiation Protection in Veterinary Medicine (1970)
37	Precautions in the Management of Patients Who Have Received Therapeutic Amounts of Radionuclides (1970)
38	Protection Against Neutron Radiation (1971)
39	Basic Radiation Protection Criteria (1971)
40	Protection Against Radiation from Brachytherapy Sources (1972)
41	Specification of Gamma-Ray Brachytherapy Sources (1974)
42	Radiological Factors Affecting Decision-Making in a Nuclear Attack (1974)
43	Review of the Current State of Radiation Protection Philosophy (1975)
44	Krypton-85 in the Atmosphere—Accumulation, Biological Significance, and Control Technology (1975)
45	Natural Background Radiation in the United States (1975)
46	Alpha-Emitting Particles in Lungs (1975)
47	Tritium Measurement Techniques (1976)
48	Radiation Protection for Medical and Allied Health Personnel (1976)
49	Structural Shielding Design and Evaluation for Medical Use of X-Rays and Gamma Rays of Energies up to 10 MeV (1976). (Full sized reproductions of the figures giving barrier requirements are available as an adjunct to the report.)

The following NCRP reports are now superseded and/or out of print.

NCRP Report No.	Title
1	X-Ray Protection (1931) [superseded by NCRP Report No. 3].
2	Radium Protection (1934) [superseded by NCRP Report No. 4].
3	X-Ray protection (1936) [superseded by NCRP Report No. 6].
4	Radium Protection (1938) [superseded by NCRP Report No. 13].
5	Safe Handling of Radioactive Luminous Compounds (1941) [out of print].
6	Medical X-Ray Protection up to Two Million Volts (1949) [superseded by NCRP Report No. 18].
7	Safe Handling of Radioactive Isotopes (1949) [superseded by NCRP Report No. 30].
11	Maximum Permissible Amounts of Radioisotopes in the Human Body and Maximum Permissible Concentrations in Air and Water (1953) [superseded by NCRP Report No. 22].
13	Protection Against Radiations from Radium, Cobalt-60, and Cesium-137 (1954) [superseded by NCRP Report No. 24].

NCRP Report No.	Title
15	Safe Handling of Cadavers Containing Radioactive Isotopes (1953) [superseded by NCRP Report No. 21].
17	Permissible Dose from External Sources of Ionizing Radiation (1954) including Maximum Permissible Exposure to Man, Addendum to National Bureau of Standards Handbook 59 (1958) [superseded by NCRP Report No. 39].
18	X-Ray Protection (1955) [superseded by NCRP Report No. 26].
19	Regulation of Radiation Exposure by Legislative Means (1955) [out of print].
20	Protection Against Neutron Radiation up to 80 Million Electron Volts (1957) [superseded by NCRP Report No. 37].
21	Safe Handling of Bodies Containing Radioactive isotopes (1958) [superseded by NCRP Report No. 37].
24	Protection Against Radiation from Sealed Gamma Sources (1960) [superseded by NCRP Reports Nos. 33, 34, and 40].
26	Medical X-Ray Protection up to Three Million Volts (1961) [superseded by NCRP Reports Nos. 33–36].
29	Exposure to Radiation in an Emergency (1962) [superseded by NCRP Report No. 42].
34	Medical X-Ray and Gamma-Ray Protection for Energies up to 10 MeV—Structural Shielding Design and Evaluation (1970) [superseded by NCRP Report No. 49].

The following statements of the NCRP were published outside the NCRP Report series:

"Blood Counts, Statement of the National Committee on Radiation Protection, *Radiology,* **63,** 428 (1954).

"Statements on Maximum Permissible Dose from Television Receivers and Maximum Permissible Dose to the Skin of the Whole Body," *Am. J. Roentgenol., Radium Ther. Nuclear Med.,* **84,** 152 (1960), *Radiology,* **75,** 122 (1960).

"X-Ray Protection Standards for Home Television Receivers," Interim Statement of the National Council on Radiation Protection and Measurements (National Council on Radiation Protection and Measurements, Washington, D.C., 1968).

"Specification of Units for Natural Uranium and Natural Thorium" (National Council on Radiation Protection and Measurements, Washington, D.C., 1973).

The following Scientific Committees currently are actively engaged in formulating recommendations.

Committee	Recommendation
SC-1	Basic Radiation Protection Criteria
SC-7	Monitoring Methods and Instruments
SC-11	Incineration of Radioactive Waste

Committee	Recommendation
SC-18	Standards and Measurements of Radioactivity for Radiological Use
SC-22	Radiation Shielding for Particle Accelerators
SC-23	Radiation Hazards Resulting from the Release of Radionuclides into the Environment
SC-24	Radionuclides and Labeled Organic Compounds Incorporated in Genetic Material
SC-25	Radiation Protection in the Use of Small Neutron Generators
SC-26	High Energy X-Ray Dosimetry
SC-28	Radiation Exposure from Consumer Products
SC-30	Physical and Biological Properties of Radionuclides
SC-31	Selected Occupational Exposure Problems Arising from Internal Emitters
SC-32	Administered Radioactivity
SC-33	Dose Calculations
SC-34	Maximum Permissible Concentrations for Occupational and Nonoccupational Exposures
SC-35	Environmental Radiation Measurements
SC-37	Procedures for the Management of Contaminated Persons
SC-38	Waste Disposal
SC-39	Microwaves
SC-40	Biological Aspects of Radiation Protection Criteria
SC-41	Radiation Resulting from Nuclear Power Generation
SC-42	Industrial Applications of X-Rays and Sealed Sources
SC-44	Radiation Associated with Medical Examinations
SC-45	Radiation Received by Radiation Employees
SC-46	Operational Radiation Safety
SC-47	Instrumentation for the Determination of Dose Equivalent
SC-48	Apportionment of Radiation Exposure
SC-50	Surface Contamination
SC-51	Radiation Protection in Pediatric Radiology and Nuclear Medicine Applied to Children
SC-52	Conceptual Basis of Calculations of Dose Distributions
SC-53	Biological Effects and Exposure Criteria for Radiofrequency Electromagnetic Radiation
SC-54	Bioassay for Assessment of Control of Intake of Radionuclides

ICRU Reports

ICRU Reports are distributed by the ICRU Publications' office. Information on prices and how to order may be obtained from:

ICRU Publications
P.O. Box 30165
Washington, D.C. 20014

The extant ICRU Reports are listed below.

ICRU Report No.	Title
10b	Physical Aspects of Irradiation (1964)
10c	Radioactivity (1963)
10d	Clinical Dosimetry (1963)
10e	Radiobiological Dosimetry (1963)
10f	Methods of Evaluating Radiological Equipment and Materials (1963)
12	Certification of Standardized Radioactive Sources (1968)
13	Neutron Fluence, Neutron Spectra, and Kerma (1969)
14	Radiation Dosimetry: X-Rays and Gamma-Rays with Maximum Photon Energies Between 0.6 and 50 MeV (1969)
15	Cameras for Image Intensifier Fluorography (1969)
16	Linear Energy Transfer (1970)
17	Radiation Dosimetry: X-Rays Generated at Potentials of 5 to 150 kV (1970)
18	Specification of High Activity Gamma-Ray Sources (1970)
19	Radiation Quantities and Units (1971)
19S	Dose Equivalent [supplement to ICRU Report No. 19] (1973)
20	Radiation Protection Instrumentation and Its Application (1971)
21	Radiation Dosimetry: Electrons with Initial Energies Between 1 and 50 MeV (1972)
22	Measurement of Low-Level Radioactivity (1972)
23	Measurement of Absorbed Dose in a Phantom Irradiated by a Single Beam of X- or Gamma Rays (1973)
24	Determination of Absorbed Dose in a Patient Irradiated by Beams of X- or Gamma Rays in Radiotherapy Procedures (1976)
25	Conceptual Basis for the Determination of Dose Equivalent (1976)

The following ICRU Reports have been superseded by the publications indicated and are now out of print:

ICRU Report No.	Title
1	"Discussion on International Units and Standards for X-ray Work," *Br. J. Radiol.*, **23,** 64 (1927).
2	"International X-Ray Unit of Intensity," *Br. J. Radiol.* (new series), **1,** 363 (1928).
3	"Report of the Committee on Standardization of X-Ray Measurements," *Radiology,* **22,** 289 (1934).
4	"Recommendations of the International Committee for Radiological Units," *Radiology,* **23,** 580 (1934).
5	"Recommendations of the International Committee for Radiological Units," *Radiology,* **29,** 634 (1937).

ICRU Report No.	Title
6	"Report of the International Commission on Radiological Protection and International Commission on Radiological Units," National Bureau of Standards Handbook 47 (Government Printing Office, Washington, D.C., 1951).
7	"Recommendations of the International Commission for Radiological Units," *Radiology,* **62,** 106 (1954).
8	"Report of the International Commission on Radiological Units and Measurements" 1956, National Bureau of Standards Handbook 62 (Government Printing Office, Washington, D.C., 1957).
9	"Report of the International Commission on Radiological Units and Measurements" 1959, National Bureau of Standards Handbook 78 (Government Printing Office, Washington, D.C., 1961).
10a	"Radiation Quantities and Units," National Bureau of Standards Handbook 84 (Government Printing Office, Washington, D.C., 1962).
11	"Radiation Quantities and Units" (International Commission on Radiation Units and Measurements, Washington, D.C., 1968).

APPENDIX 12B PERMISSIBLE CONCENTRATIONS IN AIR AND WATER ABOVE NATURAL BACKGROUND

With respect to the concentrations table, note the following.

In any case where there is a mixture in air or water of more than one radionuclide, the limiting values for purposes of this appendix should be determined as follows:

1. If the identity and concentration of each radionuclide in the mixture are known, the limiting values should be derived as follows. Determine, for each radionuclide in the mixture, the ratio between the quantity present in the mixture and the limit otherwise established in the table for the specific radionuclide when not in a mixture. The sum of such ratios for all the radionuclides in the mixture may not exceed "1" (i.e., "unity").

Example If radionuclides A, B, and C are present in concentrations C_A, C_B, and C_C, and if the applicable MPCs, are MPC_A, and MPC_B, and MPC_C, respectively, then the concentrations shall be limited so that the following relationship exists:

$$\frac{C_A}{MPC_A} + \frac{C_B}{MPC_B} + \frac{C_C}{MPC_C} \leqq 1$$

2. If either the identity or the concentration of any radionuclide in the mixture is not known, the limiting values for purposes of Appendix B shall be:

a. For purposes of Table I, Col. 1—6×10^{-13}.
b. For purposes of Table I, Col. 2—4×10^{-7}.
c. For purposes of Table II, Col. 1—2×10^{-14}.
d. For purposes of Table II, Col. 2—3×10^{-8}.

3. If any of the conditions specified below are met, the corresponding values specified below may be used in lieu of those specified in paragraph 2.

a. If the identity of each radionuclide in the mixture is known but the concentration of one or more of the radionuclides in the mixture is not known, the concentration limit for the mixture is the limit specified in the table for the radionuclide in the mixture having the lowest concentration limit; or

b. If the identity of each radionuclide in the mixture is not known, but it is known that certain radionuclides specified in the table are not present in the mixture, the concentration limit for the mixture is the lowest concentration limit specified in the table for any radionuclide that is not known to be absent from the mixture; or

c. If any of the following conditions hold:

Element (atomic number) and Isotope	Table I		Table II	
	Column 1 Air (μCi/ml)	Column 2 Water (μCi/ml)	Column 1 Air (μCi/ml)	Column 2 Water (μCi/ml)
If it is known that ^{90}Sr, ^{125}I, ^{126}I, ^{129}I, ^{131}I, (^{133}I, Table II only), ^{210}Pb, ^{210}Po, ^{211}At, ^{223}Ra, ^{224}Ra, ^{226}Ra, ^{227}Ac, ^{228}Ra, ^{230}Th, ^{231}Pa, ^{232}Th, thnat, ^{248}Cin, ^{254}C, and ^{256}Fm are not present	—	9×10^{-5}	—	3×10^{-6}
If it is known that ^{90}Sr, ^{125}I, ^{126}I, ^{129}I, (^{131}I, ^{133}I, Table II only), ^{210}Pb, ^{210}Po, ^{223}Ra, ^{226}Ra, ^{228}Ra, ^{231}Pa, Th-nat, ^{248}Cm, ^{254}C, and ^{256}Fm are not present	—	6×10^{-5}	—	2×10^{-6}
If it is known that ^{90}Sr, ^{129}I, (^{125}I, ^{126}I, ^{131}I, Table II only), ^{210}Pb, ^{226}Ra, ^{248}Cm, and ^{253}C are not present	—	2×10^{-5}	—	6×10^{-7}
If it is known that (^{129}I, Table II only), ^{226}Ra, and ^{228}Ra are not present	—	3×10^{-6}	—	1×10^{-7}
If it is known that alpha emitters and ^{90}Sr, ^{129}I, ^{210}Pb, ^{227}Ac, ^{228}Ra, ^{230}Pa, ^{241}Pu, and ^{249}Bk are not present	3×10^{-9}	—	1×10^{-10}	—
If it is known that alpha emitters and ^{210}Pb, ^{227}Ac, ^{228}Ra, and ^{241}Pu are not present	3×10^{-10}	—	1×10^{-11}	—
If it is known that alpha emitters and ^{227}Ac are not present	3×10^{-11}	—	1×10^{-12}	—
If it is known that ^{227}Ac, ^{230}Th, ^{231}Pa, ^{238}Pu, ^{239}Pu, ^{240}Pu, ^{242}Pu, ^{244}Pu, ^{248}Cm, ^{249}Cf and ^{251}Cf are not present	3×10^{-12}	—	1×10^{-13}	—

4. If a mixture of radionuclides consists of uranium and its daughters in ore dust prior to chemical separation of the uranium from the ore, the values specified below may be used for uranium and its daughters through ^{226}Ra, instead of those from paragraphs 1, 2, or 3.

a. For purposes of Table I, Col. 1—1×10^{10} μCi/ml gross alpha activity; or 5×10^{-11} μCi/ml natural uranium; or 75 micrograms per cubic meter of air natural uranium.
b. For purposes of Table II, Col. 1—3×10^{-12} μCi/ml gross alpha activity; or 2×10^{-12} μCi/ml natural uranium; or 3 micrograms per cubic meter of air natural uranium.

5. For purposes of this note, a radionuclide may be considered as not present in a mixture if (*a*) the ratio of the concentration of that radionuclide in the mixture (C_A) to the concentration limit for that radionuclide specified in Table II (MPC_A) does not exceed 1/10 (i.e., $C_A/MPC_A \leqq 1/10$) and (*b*) the sum of such ratios for all the radionuclides considered as not present in the mixture does not exceed ¼, that is,

$$\frac{C_A}{MPC_A} + \frac{C_B}{MPC_B} + \cdots \leqq 1/4$$

Appendix 12B. Permissible Concentrations in Air and Water Above Natural Background

			Table I Occupational		Table II Non-Occupational	
Element atomic number	Isotope[a]		Column 1 Air (μCi/ml)	Column 2 Water (μCi/ml)	Column 1 Air (μCi/ml)	Column 2 Water (μCi/ml)
Actinium (89)	^{227}Ac	S	2×10^{-12}	6×10^{-5}	8×10^{-14}	2×10^{-6}
		I	3×10^{-11}	9×10^{-3}	9×10^{-13}	3×10^{-4}
	^{228}Ac	S	8×10^{-8}	3×10^{-3}	3×10^{-9}	9×10^{-5}
		I	2×10^{-8}	3×10^{-3}	6×10^{-10}	9×10^{-5}
Americium (95)	^{241}Am	S	6×10^{-12}	1×10^{-4}	2×10^{-13}	4×10^{-6}
		I[b]	1×10^{-10}	8×10^{-4}	4×10^{-12}	3×10^{-5}
	^{242m}Am	S	6×10^{-12}	1×10^{-4}	2×10^{-13}	4×10^{-6}
		I	3×10^{-10}	3×10^{-3}	9×10^{-12}	9×10^{-5}
	^{242}Am	S	4×10^{-8}	4×10^{-3}	1×10^{-9}	1×10^{-4}
		I	5×10^{-8}	4×10^{-3}	2×10^{-9}	1×10^{-4}
	^{243}Am	S	6×10^{-12}	1×10^{-4}	2×10^{-13}	4×10^{-6}
		I	1×10^{-10}	8×10^{-4}	4×10^{-12}	3×10^{-5}
	^{244}Am	S	4×10^{-6}	1×10^{-1}	1×10^{-7}	5×10^{-3}
		I	2×10^{-5}	1×10^{-1}	8×10^{-7}	5×10^{-3}

Appendix 12B. (Continued)

			Table I Occupational		Table II Non-Occupational	
Element atomic number	Isotope[a]		Column 1 Air (μCi/ml)	Column 2 Water (μCi/ml)	Column 1 Air (μCi/ml)	Column 2 Water (μCi/ml)
Antimony (51)	^{122}Sb	S	2×10^{-7}	8×10^{-4}	6×10^{-9}	3×10^{-5}
		I	1×10^{-7}	8×10^{-4}	5×10^{-9}	3×10^{-5}
	^{124}Sb	S	2×10^{-7}	7×10^{-4}	5×10^{-9}	2×10^{-5}
		I	2×10^{-8}	7×10^{-4}	7×10^{-10}	2×10^{-5}
	^{125}Sb	S	5×10^{-7}	3×10^{-3}	2×10^{-8}	1×10^{-4}
		I	3×10^{-8}	3×10^{-3}	9×10^{-10}	1×10^{-4}
Argon (18)	^{37}A	Sub	6×10^{-3}	—	1×10^{-4}	—
	^{41}A	Sub	2×10^{-6}	—	4×10^{-8}	—
Arsenic (33)	^{73}As	S	2×10^{-6}	1×10^{-2}	7×10^{-8}	5×10^{-4}
		I	4×10^{-7}	1×10^{-2}	1×10^{-8}	5×10^{-4}
	^{74}As	S	3×10^{-7}	2×10^{-3}	1×10^{-8}	5×10^{-5}
		I	1×10^{-7}	2×10^{-3}	4×10^{-9}	5×10^{-5}
	^{76}As	S	1×10^{-7}	6×10^{-4}	4×10^{-9}	2×10^{-5}
		I	1×10^{-7}	6×10^{-4}	3×10^{-9}	2×10^{-5}
	^{77}As	S	5×10^{-7}	2×10^{-3}	2×10^{-8}	8×10^{-5}
		I	4×10^{-7}	2×10^{-3}	1×10^{-8}	8×10^{-5}
Astatine (85)	^{211}At	S	7×10^{-9}	5×10^{-5}	2×10^{-10}	2×10^{-6}
		I	3×10^{-8}	2×10^{-3}	1×10^{-9}	7×10^{-5}
Barium (56)	^{131}Ba	S	1×10^{-6}	5×10^{-3}	4×10^{-8}	2×10^{-4}
		I	4×10^{-7}	5×10^{-3}	1×10^{-8}	2×10^{-4}
	^{140}Ba	S	1×10^{-7}	8×10^{-4}	4×10^{-9}	3×10^{-5}
		I	4×10^{-8}	7×10^{-4}	1×10^{-9}	2×10^{-5}
Berkelium (97)	^{249}Bk	S	9×10^{-10}	2×10^{-2}	3×10^{-11}	6×10^{-4}
		I	1×10^{-7}	2×10^{-2}	4×10^{-9}	6×10^{-4}
	^{250}Bk	S	1×10^{-7}	6×10^{-3}	5×10^{-9}	2×10^{-4}
		I	1×10^{-6}	6×10^{-3}	4×10^{-8}	2×10^{-4}
Beryllium (4)	^{7}Be	S	6×10^{-6}	5×10^{-2}	2×10^{-7}	2×10^{-3}
		I	1×10^{-6}	1×10^{-3}	4×10^{-8}	2×10^{-3}
Bismuth (83)	^{206}Bi	S	2×10^{-7}	1×10^{-3}	6×10^{-9}	4×10^{-5}
		I	1×10^{-7}	2×10^{-3}	5×10^{-9}	4×10^{-5}
	^{207}Bi	S	2×10^{-7}	2×10^{-3}	6×10^{-9}	6×10^{-5}
		I	1×10^{-8}	1×10^{-3}	5×10^{-10}	6×10^{-5}
	^{210}Bi	S	$6 \times; 10^{-9}$	1×10^{-3}	2×10^{-10}	4×10^{-5}
		I	6×10^{-9}	1×10^{-2}	2×10^{-10}	4×10^{-5}
	^{212}Bi	S	1×10^{-7}	1×10^{-2}	3×10^{-9}	4×10^{-4}
		I	2×10^{-7}		7×10^{-9}	4×10^{-4}
Bromine (35)	^{82}Br	S	1×10^{-6}	8×10^{-3}	4×10^{-8}	3×10^{-4}
		I	2×10^{-7}	1×10^{-3}	6×10^{-9}	4×10^{-5}
Cadmium (48)	^{109}Cd	S	5×10^{-8}	5×10^{-3}	2×10^{-9}	2×10^{-4}
		I	7×10^{-8}	5×10^{-3}	3×10^{-9}	2×10^{-4}

Appendix 12B. (Continued)

Element atomic number	Isotope[a]		Table I Occupational Column 1 Air (μCi/ml)	Table I Occupational Column 2 Water (μCi/ml)	Table II Non-Occupational Column 1 Air (μCi/ml)	Table II Non-Occupational Column 2 Water (μCi/ml)
	^{115m}Cd	S	4×10^{-8}	7×10^{-4}	1×10^{-9}	3×10^{-5}
		I	4×10^{-8}	7×10^{-4}	1×10^{-9}	3×10^{-5}
	^{115}Cd	S	2×10^{-7}	1×10^{-3}	8×10^{-9}	3×10^{-5}
		I	2×10^{-7}	1×10^{-3}	6×10^{-9}	4×10^{-5}
Calcium (20)	^{45}Ca	S	3×10^{-8}	3×10^{-4}	1×10^{-9}	9×10^{-6}
		I	1×10^{-7}	5×10^{-3}	4×10^{-9}	2×10^{-4}
	^{47}Ca	S	2×10^{-7}	1×10^{-3}	6×10^{-9}	5×10^{-5}
		I	2×10^{-7}	1×10^{-3}	6×10^{-9}	3×10^{-5}
Californium (98)	^{249}Cf	S	2×10^{-12}	1×10^{-4}	5×10^{-14}	4×10^{-6}
		I	1×10^{-10}	7×10^{-4}	3×10^{-12}	2×10^{-5}
	^{250}Cf	S	5×10^{-12}	4×10^{-4}	2×10^{-13}	1×10^{-5}
		I	1×10^{-10}	7×10^{-4}	3×10^{-12}	3×10^{-5}
	^{251}Cf	S	2×10^{-12}	1×10^{-4}	6×10^{-14}	4×10^{-6}
		I	1×10^{-10}	8×10^{-4}	3×10^{-12}	3×10^{-5}
	^{252}Cf	S[b]	6×10^{-12}	2×10^{-4}	2×10^{-13}	7×10^{-6}
		I[b]	3×10^{-11}	2×10^{-4}	1×10^{-12}	7×10^{-6}
	^{253}Cf	S	8×10^{-10}	4×10^{-3}	3×10^{-11}	1×10^{-4}
		I	8×10^{-10}	4×10^{-3}	3×10^{-11}	1×10^{-4}
	^{254}Cf	S	5×10^{-12}	4×10^{-6}	2×10^{-13}	1×10^{-7}
		I	5×10^{-12}	4×10^{-6}	2×10^{-13}	1×10^{-7}
Carbon (6)	^{14}C (CO_2)	S	4×10^{-6}	2×10^{-2}	1×10^{-7}	8×10^{-4}
		Sub	5×10^{-5}	—	1×10^{-6}	—
Cerium (58)	^{141}Ce	S	4×10^{-7}	3×10^{-3}	2×10^{-8}	9×10^{-5}
		I	2×10^{-7}	3×10^{-3}	5×10^{-9}	9×10^{-5}
	^{143}Ce	S	3×10^{-7}	1×10^{-3}	9×10^{-9}	4×10^{-5}
		I	2×10^{-7}	1×10^{-3}	7×10^{-9}	4×10^{-5}
	^{144}Ce	S	1×10^{-8}	3×10^{-4}	3×10^{-10}	1×10^{-5}
		I	6×10^{-9}	3×10^{-4}	2×10^{-10}	1×10^{-5}
Cesium (55)	^{131}Cs	S	1×10^{-5}	7×10^{-2}	4×10^{-7}	2×10^{-3}
		I	3×10^{-6}	3×10^{-2}	1×10^{-7}	9×10^{-4}
	^{134m}Cs	S	4×10^{-5}	2×10^{-1}	1×10^{-6}	6×10^{-3}
		I	6×10^{-6}	3×10^{-2}	2×10^{-7}	1×10^{-3}
	^{134}Cs	S	4×10^{-8}	3×10^{-4}	1×10^{-9}	9×10^{-6}
		I	1×10^{-8}	1×10^{-3}	4×10^{-10}	4×10^{-5}
	^{135}Cs	S	5×10^{-7}	3×10^{-3}	2×10^{-8}	1×10^{-4}
		I	9×10^{-8}	7×10^{-3}	3×10^{-9}	2×10^{-4}
	^{136}Cs	S	4×10^{-7}	2×10^{-3}	1×10^{-8}	9×10^{-5}
		I	2×10^{-7}	2×10^{-3}	6×10^{-9}	6×10^{-5}
	^{137}Cs	S	6×10^{-8}	4×10^{-4}	2×10^{-9}	2×10^{-5}
		I	1×10^{-8}	1×10^{-3}	5×10^{-10}	4×10^{-5}

Appendix 12B. (Continued)

Element atomic number	Isotope[a]		Table I Occupational		Table II Non-Occupational	
			Column 1 Air (μCi/ml)	Column 2 Water (μCi/ml)	Column 1 Air (μCi/ml)	Column 2 Water (μCi/ml)
Chlorine (17)	^{36}Cl	S	4×10^{-7}	2×10^{-3}	1×10^{-8}	8×10^{-5}
		I	2×10^{-8}	2×10^{-3}	8×10^{-10}	6×10^{-5}
	^{39}Cl	S	3×10^{-6}	1×10^{-2}	9×10^{-8}	4×10^{-4}
		I	2×10^{-6}	1×10^{-2}	7×10^{-8}	4×10^{-4}
Chromium (24)	^{51}Cr	S	1×10^{-5}	5×10^{-2}	4×10^{-7}	2×10^{-3}
		I	2×10^{-6}	5×10^{-2}	8×10^{-8}	2×10^{-3}
Cobalt (27)	^{57}Co	S	3×10^{-6}	2×10^{-2}	1×10^{-7}	5×10^{-4}
		I	2×10^{-7}	1×10^{-2}	6×10^{-9}	4×10^{-4}
	^{58m}Co	S	2×10^{-5}	8×10^{-2}	6×10^{-7}	3×10^{-3}
		I	9×10^{-6}	6×10^{-2}	3×10^{-7}	2×10^{-3}
	^{58}Co	S	8×10^{-7}	4×10^{-3}	3×10^{-8}	1×10^{-4}
		I	5×10^{-8}	3×10^{-3}	2×10^{-9}	9×10^{-5}
	^{60}Co	S	3×10^{-7}	1×10^{-3}	1×10^{-8}	5×10^{-5}
		I	9×10^{-9}	1×10^{-3}	3×10^{-10}	3×10^{-5}
Copper (29)	^{64}Cu	S	2×10^{-6}	1×10^{-2}	7×10^{-8}	3×10^{-4}
		I	1×10^{-6}	6×10^{-3}	4×10^{-8}	2×10^{-4}
Curium (96)	^{242}Cm	S	1×10^{-10}	7×10^{-4}	4×10^{-12}	2×10^{-5}
		I[b]	2×10^{-10}	7×10^{-4}	6×10^{-12}	2×10^{-5}
	^{243}Cm	S	6×10^{-12}	1×10^{-4}	2×10^{-13}	5×10^{-6}
		I	1×10^{-10}	7×10^{-4}	3×10^{-12}	2×10^{-5}
	^{244}Cm	S	9×10^{-12}	2×10^{-4}	3×10^{-13}	7×10^{-6}
		I	1×10^{-10}	8×10^{-4}	3×10^{-12}	3×10^{-5}
	^{245}Cm	S	5×10^{-12}	1×10^{-4}	2×10^{-13}	4×10^{-6}
		I	1×10^{-10}	8×10^{-4}	4×10^{-12}	3×10^{-5}
	^{246}Cm	S	5×10^{-12}	1×10^{-4}	2×10^{-13}	4×10^{-6}
		I	1×10^{-10}	8×10^{-4}	4×10^{-12}	3×10^{-5}
	^{247}Cm	S	5×10^{-12}	1×10^{-4}	2×10^{-13}	4×10^{-6}
		I	1×10^{-10}	6×10^{-4}	4×10^{-12}	2×10^{-5}
	^{248}Cm	S	6×10^{-13}	1×10^{-5}	2×10^{-14}	4×10^{-7}
		I	1×10^{-11}	4×10^{-5}	4×10^{-13}	1×10^{-6}
	^{249}Cm	S	1×10^{-5}	6×10^{-2}	4×10^{-7}	2×10^{-3}
		I	1×10^{-5}	6×10^{-2}	4×10^{-7}	2×10^{-3}
Dysprosium (66)	^{165}Dy	S	3×10^{-6}	1×10^{-2}	9×10^{-8}	4×10^{-4}
		I	2×10^{-6}	1×10^{-2}	7×10^{-8}	4×10^{-4}
	^{166}Dy	S	2×10^{-7}	1×10^{-3}	8×10^{-9}	4×10^{-5}
		I	2×10^{-7}	1×10^{-3}	7×10^{-9}	4×10^{-5}
Einsteinium (99)	^{253}Es	S	8×10^{-10}	7×10^{-4}	3×10^{-11}	2×10^{-5}
		I	6×10^{-10}	7×10^{-4}	2×10^{-11}	2×10^{-5}
	^{254m}Es	S	5×10^{-9}	5×10^{-4}	2×10^{-10}	2×10^{-5}
		I	6×10^{-9}	5×10^{-4}	2×10^{-10}	2×10^{-5}

Appendix 12B. (Continued)

Element atomic number	Isotope[a]		Table I Occupational Column 1 Air (μCi/ml)	Table I Occupational Column 2 Water (μCi/ml)	Table II Non-Occupational Column 1 Air (μCi/ml)	Table II Non-Occupational Column 2 Water (μCi/ml)
	^{254}Es	S	2×10^{-11}	4×10^{-4}	6×10^{-13}	1×10^{-5}
		I	1×10^{-10}	4×10^{-4}	4×10^{-12}	1×10^{-5}
	^{255}Es	S	5×10^{-10}	8×10^{-4}	2×10^{-11}	3×10^{-5}
		I	4×10^{-10}	8×10^{-4}	1×10^{-11}	3×10^{-5}
Erbium (68)	^{169}Er	S	6×10^{-7}	3×10^{-3}	2×10^{-8}	9×10^{-5}
		I	4×10^{-7}	3×10^{-3}	1×10^{-8}	9×10^{-5}
	^{171}Er	S	7×10^{-7}	3×10^{-3}	2×10^{-8}	1×10^{-4}
		I	6×10^{-7}	3×10^{-3}	2×10^{-8}	1×10^{-4}
Europium (63)	^{152}Eu,	S	4×10^{-7}	2×10^{-3}	1×10^{-8}	6×10^{-5}
	t_2 = 13 years	I	3×10^{-7}	2×10^{-3}	1×10^{-8}	6×10^{-5}
		S	1×10^{-8}	2×10^{-3}	4×10^{-10}	8×10^{-5}
		I	2×10^{-8}	2×10^{-3}	6×10^{-10}	8×10^{-5}
	^{154}Eu	S	4×10^{-9}	6×10^{-4}	1×10^{-10}	2×10^{-5}
		I	7×10^{-9}	6×10^{-4}	2×10^{-10}	2×10^{-5}
	^{155}Eu	S	9×10^{-8}	6×10^{-3}	3×10^{-9}	2×10^{-4}
		I	7×10^{-8}	6×10^{-3}	3×10^{-9}	2×10^{-4}
Fermium (100)	^{254}Fm	S	6×10^{-8}	4×10^{-3}	2×10^{-9}	1×10^{-4}
		I	7×10^{-8}	4×10^{-3}	2×10^{-9}	1×10^{-4}
	^{255}Fm	S	2×10^{-8}	1×10^{-3}	6×10^{-10}	3×10^{-5}
		I	1×10^{-8}	1×10^{-3}	4×10^{-10}	3×10^{-5}
	^{256}Fm	S	3×10^{-9}	3×10^{-5}	1×10^{-10}	9×10^{-7}
		I	2×10^{-9}	3×10^{-5}	6×10^{-11}	9×10^{-7}
Fluorine (9)	^{18}F	S	5×10^{-6}	2×10^{-2}	2×10^{-7}	8×10^{-4}
		I	3×10^{-6}	2×10^{-2}	9×10^{-8}	5×10^{-4}
Gadolinium (64)	^{153}Gd	S	2×10^{-7}	6×10^{-3}	8×10^{-9}	2×10^{-4}
		I	9×10^{-8}	6×10^{-3}	3×10^{-9}	2×10^{-4}
	^{159}Gd	S	5×10^{-7}	2×10^{-3}	2×10^{-8}	8×10^{-4}
		I	4×10^{-7}	2×10^{-3}	1×10^{-8}	8×10^{-4}
Galium (31)	^{72}Ga	S	2×10^{-7}	1×10^{-3}	8×10^{-9}	4×10^{-5}
		I	2×10^{-7}	1×10^{-3}	6×10^{-9}	4×10^{-5}
Germanium (32)	^{71}Ge	S	1×10^{-5}	5×10^{-2}	4×10^{-7}	2×10^{-3}
		I	6×10^{-6}	5×10^{-2}	2×10^{-7}	2×10^{-3}
Gold (79)	^{196}Au	S	1×10^{-6}	5×10^{-3}	4×10^{-8}	2×10^{-4}
		I	6×10^{-7}	4×10^{-3}	2×10^{-8}	1×10^{-4}
	^{198}Au	S	3×10^{-7}	2×10^{-3}	1×10^{-8}	5×10^{-5}
		I	2×10^{-7}	1×10^{-3}	8×10^{-9}	5×10^{-5}
	^{199}Au	S	1×10^{-6}	5×10^{-3}	4×10^{-8}	2×10^{-4}
		I	8×10^{-7}	4×10^{-3}	3×10^{-8}	2×10^{-4}
Hafnium (72)	^{181}Hf	S	4×10^{-8}	2×10^{-3}	1×10^{-9}	7×10^{-5}
		I	7×10^{-8}	2×10^{-3}	3×10^{-9}	7×10^{-5}

Appendix 12B. (Continued)

Element atomic number	Isotope[a]		Table I Occupational Column 1 Air (μCi/ml)	Table I Occupational Column 2 Water (μCi/ml)	Table II Non-Occupational Column 1 Air (μCi/ml)	Table II Non-Occupational Column 2 Water (μCi/ml)
Holmium (67)	^{166}Ho	S	2×10^{-7}	9×10^{-4}	7×10^{-9}	3×10^{-5}
		I	2×10^{-7}	9×10^{-4}	6×10^{-9}	3×10^{-5}
Hydrogen (1)	^{3}H	S	5×10^{-6}	1×10^{-1}	2×10^{-7}	3×10^{-3}
		I	5×10^{-6}	1×10^{-1}	2×10^{-7}	3×10^{-3}
		Sub	2×10^{-3}	—	4×10^{-5}	—
Indium (49)	^{113m}In	S	8×10^{-6}	4×10^{-2}	3×10^{-7}	1×10^{-3}
		I	7×10^{-6}	4×10^{-2}	2×10^{-7}	1×10^{-3}
	^{114m}In	S	1×10^{-7}	5×10^{-4}	4×10^{-9}	2×10^{-5}
		I	2×10^{-8}	5×10^{-4}	7×10^{-10}	2×10^{-5}
	^{115m}In	S	2×10^{-6}	1×10^{-2}	8×10^{-8}	4×10^{-4}
		I	2×10^{-6}	1×10^{-2}	6×10^{-8}	4×10^{-4}
	^{115}In	S	2×10^{-7}	3×10^{-3}	9×10^{-9}	9×10^{-5}
		I	3×10^{-8}	3×10^{-3}	1×10^{-9}	9×10^{-5}
Iodine (53)	^{125}I	S	5×10^{-9}	4×10^{-5}	8×10^{-11}	2×10^{-7}
		I	2×10^{-7}	6×10^{-3}	6×10^{-9}	2×10^{-4}
	^{126}I	S	8×10^{-9}	5×10^{-5}	9×10^{-11}	3×10^{-7}
		I	3×10^{-7}	3×10^{-3}	1×10^{-8}	9×10^{-5}
	^{129}I	S	2×10^{-9}	1×10^{-5}	2×10^{-11}	6×10^{-8}
		I	7×10^{-8}	6×10^{-3}	2×10^{-9}	2×10^{-4}
	^{131}I	S	9×10^{-9}	6×10^{-5}	1×10^{-10}	3×10^{-7}
		I	3×10^{-7}	2×10^{-3}	1×10^{-8}	6×10^{-5}
	^{132}I	S	2×10^{-7}	2×10^{-3}	3×10^{-9}	8×10^{-6}
		I	9×10^{-7}	5×10^{-3}	3×10^{-8}	2×10^{-4}
	^{133}I	S	3×10^{-8}	2×10^{-4}	4×10^{-10}	1×10^{-6}
		I	2×10^{-7}	1×10^{-3}	7×10^{-9}	4×10^{-5}
	^{134}I	S	5×10^{-7}	4×10^{-3}	6×10^{-9}	2×10^{-5}
Iodine (53)	^{134}I	I	3×10^{-6}	2×10^{-2}	1×10^{-7}	6×10^{-4}
	^{135}I	S	1×10^{-7}	7×10^{-4}	1×10^{-9}	4×10^{-6}
		I	4×10^{-7}	2×10^{-3}	1×10^{-8}	7×10^{-5}
Iridium (77)	^{190}Ir	S	1×10^{-6}	6×10^{-3}	4×10^{-8}	2×10^{-4}
		I	4×10^{-7}	5×10^{-3}	1×10^{-8}	2×10^{-4}
	^{192}Ir	S	1×10^{-7}	1×10^{-3}	4×10^{-9}	4×10^{-5}
		I	3×10^{-8}	1×10^{-3}	9×10^{-10}	4×10^{-5}
	^{194}Ir	S	2×10^{-7}	1×10^{-3}	8×10^{-9}	3×10^{-5}
		I	2×10^{-7}	9×10^{-4}	5×10^{-9}	3×10^{-5}
Iron (26)	^{55}Fe	S	9×10^{-7}	2×10^{-2}	3×10^{-8}	8×10^{-4}
		I	1×10^{-6}	7×10^{-2}	3×10^{-8}	2×10^{-3}
	^{59}Fe	S	1×10^{-7}	2×10^{-3}	5×10^{-9}	6×10^{-5}
		I	5×10^{-8}	2×10^{-3}	2×10^{-9}	5×10^{-5}

Appendix 12B. (Continued)

Element atomic number	Isotope[a]		Table I Occupational Column 1 Air (μCi/ml)	Table I Occupational Column 2 Water (μCi/ml)	Table II Non-Occupational Column 1 Air (μCi/ml)	Table II Non-Occupational Column 2 Water (μCi/ml)
Krypton (36)	^{85m}Kr	Sub	6×10^{-6}	—	1×10^{-7}	—
	^{85}Kr	Sub	1×10^{-5}	—	3×10^{-7}	—
	^{87}Kr	Sub	1×10^{-6}	—	2×10^{-8}	—
	^{88}Kr	Sub	1×10^{-6}	—	2×10^{-8}	—
Lanthanum (57)	^{140}La	S	2×10^{-7}	7×10^{-4}	5×10^{-9}	2×10^{-5}
		I	1×10^{-7}	7×10^{-4}	5×10^{-9}	2×10^{-5}
Lead (82)	^{203}Pb	S	3×10^{-6}	1×10^{-2}	9×10^{-8}	4×10^{-4}
		I	2×10^{-6}	1×10^{-2}	6×10^{-8}	4×10^{-4}
	^{210}Pb	S	$\times 10^{-10}$	4×10^{-6}	4×10^{-12}	1×10^{-7}
		I	2×10^{-10}	5×10^{-3}	8×10^{-12}	2×10^{-4}
	^{212}Pb	S	2×10^{-8}	6×10^{-4}	6×10^{-10}	2×10^{-5}
		I	2×10^{-8}	5×10^{-4}	7×10^{-10}	2×10^{-5}
Lutetium (71)	^{177}Lu	S	6×10^{-7}	3×10^{-3}	2×10^{-8}	1×10^{-4}
		I	5×10^{-7}	3×10^{-3}	2×10^{-8}	1×10^{-4}
Manganese (25)	^{52}Mn	S	2×10^{-7}	1×10^{-3}	7×10^{-9}	3×10^{-5}
		I	1×10^{-7}	9×10^{-4}	5×10^{-9}	3×10^{-5}
	^{54}Mn	S[b]	4×10^{-7}	4×10^{-3}	1×10^{-8}	1×10^{-4}
		I	4×10^{-8}	3×10^{-3}	1×10^{-9}	1×10^{-4}
	^{55}Mn	S	8×10^{-7}	4×10^{-3}	3×10^{-8}	1×10^{-4}
		I	5×10^{-7}	3×10^{-3}	2×10^{-8}	1×10^{-4}
Mercury (80)	^{197m}Hg	S	7×10^{-7}	6×10^{-3}	3×10^{-8}	2×10^{-4}
		I	8×10^{-7}	5×10^{-3}	3×10^{-8}	2×10^{-4}
	^{197}Hg	S	1×10^{-6}	9×10^{-3}	4×10^{-8}	3×10^{-4}
		I	3×10^{-6}	1×10^{-2}	9×10^{-8}	5×10^{-4}
	^{203}Hg	S	7×10^{-8}	5×10^{-4}	2×10^{-9}	2×10^{-5}
		I	1×10^{-7}	3×10^{-3}	4×10^{-9}	1×10^{-4}
Molybdenum (42)	^{99}Mo	S	7×10^{-7}	5×10^{-3}	3×10^{-8}	2×10^{-4}
		I	2×10^{-7}	1×10^{-3}	7×10^{-9}	4×10^{-5}
Neodymium (60)	^{144}Nd	S	8×10^{-11}	2×10^{-3}	3×10^{-12}	7×10^{-5}
		I	3×10^{-10}	2×10^{-3}	1×10^{-11}	8×10^{-5}
	^{147}Nd	S	4×10^{-7}	2×10^{-3}	1×10^{-8}	6×10^{-5}
		I	2×10^{-7}	2×10^{-3}	8×10^{-9}	6×10^{-5}
	^{149}Nd	S	2×10^{-6}	8×10^{-3}	6×10^{-8}	3×10^{-4}
		I	1×10^{-6}	8×10^{-3}	5×10^{-8}	3×10^{-4}
Neptunium (93)	^{237}Np	S	4×10^{-12}	9×10^{-5}	$\times 10^{-13}$	3×10^{-6}
		I	1×10^{-10}	9×10^{-4}	4×10^{-12}	3×10^{-5}
	^{239}Np	S	8×10^{-7}	4×10^{-3}	3×10^{-8}	1×10^{-4}
		I	7×10^{-7}	4×10^{-3}	2×10^{-8}	1×10^{-4}
Nickel (28)	^{59}Ni	S	5×10^{-7}	6×10^{-3}	2×10^{-8}	2×10^{-4}
		I	8×10^{-7}	6×10^{-2}	3×10^{-8}	2×10^{-3}

Appendix 12B. (Continued)

Element atomic number	Isotope[a]		Table I Occupational Column 1 Air (μCi/ml)	Table I Occupational Column 2 Water (μCi/ml)	Table II Non-Occupational Column 1 Air (μCi/ml)	Table II Non-Occupational Column 2 Water (μCi/ml)
	^{63}Ni	S	6×10^{-8}	8×10^{-4}	2×10^{-9}	3×10^{-5}
		I	3×10^{-7}	2×10^{-2}	1×10^{-8}	7×10^{-4}
	^{65}Ni	S	9×10^{-7}	4×10^{-3}	3×10^{-8}	1×10^{-4}
		I	5×10^{-7}	3×10^{-3}	2×10^{-8}	1×10^{-4}
Niobium (columbium) (41)	^{93m}Nb	S	1×10^{-7}	1×10^{-2}	4×10^{-9}	4×10^{-4}
		I	2×10^{-7}	1×10^{-2}	5×10^{-9}	4×10^{-4}
	^{95}Nb	S	5×10^{-7}	3×10^{-3}	2×10^{-8}	1×10^{-4}
		I	1×10^{-7}	3×10^{-3}	3×10^{-9}	1×10^{-4}
	^{97}Nb	S	6×10^{-6}	3×10^{-2}	2×10^{-7}	9×10^{-4}
		I	5×10^{-6}	3×10^{-2}	2×10^{-7}	9×10^{-4}
Osmium (76)	^{185}Os	S	5×10^{-7}	2×10^{-3}	2×10^{-8}	7×10^{-5}
		I	5×10^{-8}	2×10^{-3}	2×10^{-9}	7×10^{-5}
	^{191m}Os	S	2×10^{-5}	7×10^{-2}	6×10^{-7}	3×10^{-3}
		I	9×10^{-6}	7×10^{-2}	3×10^{-7}	2×10^{-3}
	^{191}Os	S	1×10^{-6}	5×10^{-3}	4×10^{-8}	2×10^{-4}
		I	4×10^{-7}	5×10^{-3}	1×10^{-8}	2×10^{-4}
	^{193}Os	S	4×10^{-7}	2×10^{-3}	1×10^{-8}	6×10^{-5}
		I	3×10^{-7}	2×10^{-3}	9×10^{-9}	5×10^{-5}
Palladium (46)	^{103}Pd	S	1×10^{-6}	1×10^{-2}	5×10^{-8}	3×10^{-4}
		I	7×10^{-7}	8×10^{-3}	3×10^{-8}	3×10^{-4}
	^{109}Pd	S	6×10^{-7}	3×10^{-3}	2×10^{-8}	9×10^{-5}
		I	4×10^{-7}	2×10^{-3}	1×10^{-8}	7×10^{-5}
Phosphorus (15)	^{32}P	S	7×10^{-8}	5×10^{-4}	2×10^{-9}	2×10^{-5}
		I	8×10^{-8}	7×10^{-4}	3×10^{-9}	2×10^{-5}
Platinum (78)	^{191}Pt	S	8×10^{-7}	4×10^{-3}	3×10^{-8}	1×10^{-4}
		I	6×10^{-7}	3×10^{-3}	2×10^{-8}	1×10^{-4}
	^{193m}Pt	S	7×10^{-6}	3×10^{-2}	2×10^{-7}	1×10^{-3}
		I	5×10^{-6}	3×10^{-2}	2×10^{-7}	1×10^{-3}
	^{193}Pt	S[b]	1×10^{-6}	3×10^{-2}	4×10^{-8}	9×10^{-4}
		I	3×10^{-7}	5×10^{-2}	1×10^{-8}	2×10^{-3}
	^{197m}Pt	S	6×10^{-6}	3×10^{-2}	2×10^{-7}	1×10^{-3}
		I	5×10^{-6}	3×10^{-2}	2×10^{-7}	9×10^{-4}
	^{197}Pt	S	8×10^{-7}	4×10^{-3}	3×10^{-8}	1×10^{-4}
		I	6×10^{-7}	3×10^{-3}	2×10^{-8}	1×10^{-4}
Plutonium (94)	^{238}Pu	S	2×10^{-12}	1×10^{-4}	7×10^{-14}	5×10^{-6}
		I	3×10^{-11}	8×10^{-4}	1×10^{-12}	3×10^{-5}
	^{239}Pu	S	2×10^{-12}	1×10^{-4}	6×10^{-14}	5×10^{-6}
		I	4×10^{-11}	8×10^{-4}	1×10^{-12}	3×10^{-5}
	^{240}Pu	S	2×10^{-12}	1×10^{-4}	6×10^{-14}	5×10^{-6}
		I	4×10^{-11}	8×10^{-4}	1×10^{-12}	3×10^{-5}

Appendix 12B. (Continued)

Element atomic number	Isotope[a]		Table I Occupational Column 1 Air (μCi/ml)	Table I Occupational Column 2 Water (μCi/ml)	Table II Non-Occupational Column 1 Air (μCi/ml)	Table II Non-Occupational Column 2 Water (μCi/ml)
	^{241}Pu	S	9×10^{-11}	7×10^{-3}	3×10^{-12}	2×10^{-4}
		I	4×10^{-8}	4×10^{-2}	1×10^{-9}	1×10^{-3}
	^{242}Pu	S	2×10^{-12}	1×10^{-4}	6×10^{-14}	5×10^{-6}
		I	4×10^{-11}	9×10^{-4}	1×10^{-12}	3×10^{-5}
	^{243}Pu	S	2×10^{-6}	1×10^{-2}	6×10^{-8}	3×10^{-4}
		I	2×10^{-6}	1×10^{-2}	8×10^{-8}	3×10^{-4}
	^{244}Pu	S	2×10^{-12}	1×10^{-4}	6×10^{-14}	4×10^{-6}
		I	3×10^{-11}	3×10^{-4}	1×10^{-12}	1×10^{-5}
Polonium (84)	^{210}Po	S	5×10^{-10}	2×10^{-5}	2×10^{-11}	7×10^{-7}
		I	2×10^{-10}	8×10^{-4}	7×10^{-12}	3×10^{-5}
Potassium (19)	^{42}K	S	2×10^{-6}	9×10^{-3}	7×10^{-8}	3×10^{-4}
		I	1×10^{-7}	6×10^{-4}	4×10^{-9}	2×10^{-5}
Praseodymium (59)	^{142}Pr	S	2×10^{-7}	9×10^{-4}	7×10^{-9}	3×10^{-5}
		I	2×10^{-7}	9×10^{-4}	5×10^{-9}	3×10^{-5}
	^{143}Pr	S	3×10^{-7}	1×10^{-3}	1×10^{-8}	5×10^{-5}
		I	2×10^{-7}	1×10^{-3}	6×10^{-9}	5×10^{-5}
Promethium (61)	^{147}Pm	S	6×10^{-8}	6×10^{-3}	2×10^{-9}	2×10^{-4}
		I	1×10^{-7}	6×10^{-3}	3×10^{-9}	2×10^{-4}
	^{149}Pm	S	3×10^{-7}	1×10^{-3}	1×10^{-8}	4×10^{-5}
		I	2×10^{-7}	1×10^{-3}	8×10^{-9}	4×10^{-5}
Protoactinium (91)	^{230}Pa	S	2×10^{-9}	7×10^{-3}	6×10^{-11}	2×10^{-4}
		I	8×10^{-10}	7×10^{-3}	3×10^{-11}	2×10^{-4}
	^{231}Pa	S	1×10^{-12}	3×10^{-5}	4×10^{-14}	9×10^{-7}
		I	1×10^{-10}	8×10^{-4}	4×10^{-12}	2×10^{-5}
	^{233}Pa	S	6×10^{-7}	4×10^{-3}	2×10^{-8}	1×10^{-4}
		I	2×10^{-7}	3×10^{-3}	6×10^{-9}	1×10^{-4}
Radium (88)	^{223}Ra	S	2×10^{-9}	2×10^{-5}	6×10^{-11}	7×10^{-7}
		I	2×10^{-10}	1×10^{-4}	8×10^{-12}	4×10^{-6}
	^{224}Ra	S	5×10^{-9}	7×10^{-5}	2×10^{-10}	2×10^{-6}
		I	7×10^{-10}	2×10^{-4}	2×10^{-11}	5×10^{-6}
	^{226}Ra	S	3×10^{-11}	4×10^{-7}	3×10^{-12}	3×10^{-8}
		I	5×10^{-11}	9×10^{-4}	2×10^{-12}	3×10^{-5}
	^{228}Ra	S	7×10^{-11}	8×10^{-7}	2×10^{-12}	3×10^{-8}
		I	4×10^{-11}	7×10^{-4}	1×10^{-12}	3×10^{-5}
Radon (86)	^{220}Rn	S	3×10^{-7}	—	1×10^{-8}	—
		S	3×10^{-8}	—	3×10^{-9}	—
Rhenium (75)	^{183}Re	S	3×10^{-6}	2×10^{-2}	9×10^{-8}	6×10^{-4}
		I	2×10^{-7}	8×10^{-3}	5×10^{-9}	3×10^{-4}
	^{186}Re	S	6×10^{-7}	3×10^{-3}	2×10^{-8}	9×10^{-5}
		I	2×10^{-7}	1×10^{-3}	8×10^{-9}	5×10^{-5}

Appendix 12B. (Continued)

Element atomic number	Isotope[a]		Table I Occupational Column 1 Air (μCi/ml)	Table I Occupational Column 2 Water (μCi/ml)	Table II Non-Occupational Column 1 Air (μCi/ml)	Table II Non-Occupational Column 2 Water (μCi/ml)
	^{187}Re	S	9×10^{-6}	7×10^{-2}	3×10^{-7}	3×10^{-3}
		I	5×10^{-7}	4×10^{-2}	2×10^{-8}	2×10^{-3}
	^{188}Re	S	4×10^{-7}	2×10^{-3}	1×10^{-8}	6×10^{-5}
		I	2×10^{-7}	9×10^{-4}	6×10^{-9}	3×10^{-5}
Rhodium (45)	^{103m}Rh	S	8×10^{-5}	4×10^{-1}	3×10^{-6}	1×10^{-2}
		I	6×10^{-5}	3×10^{-1}	2×10^{-6}	1×10^{-2}
	^{105}Rh	S	8×10^{-7}	4×10^{-3}	3×10^{-8}	1×10^{-4}
		I	5×10^{-7}	3×10^{-3}	2×10^{-8}	1×10^{-4}
Rubidium (37)	^{86}Rb	S	3×10^{-7}	2×10^{-3}	1×10^{-8}	7×10^{-5}
		I	7×10^{-8}	7×10^{-4}	2×10^{-9}	2×10^{-5}
	^{87}Rb	S	5×10^{-7}	3×10^{-3}	2×10^{-8}	1×10^{-4}
		I	7×10^{-8}	5×10^{-3}	2×10^{-9}	1×10^{-4}
Ruthenium (44)	^{97}Ru	S	2×10^{-6}	1×10^{-2}	8×10^{-8}	4×10^{-4}
		I	2×10^{-6}	1×10^{-2}	6×10^{-8}	3×10^{-4}
	^{103}Ru	S	5×10^{-7}	2×10^{-3}	2×10^{-8}	8×10^{-5}
		I	8×10^{-8}	2×10^{-3}	3×10^{-9}	8×10^{-5}
	^{105}Ru	S	7×10^{-7}	3×10^{-3}	2×10^{-8}	1×10^{-4}
		I	5×10^{-7}	3×10^{-3}	2×10^{-8}	1×10^{-4}
	^{106}Ru	S	8×10^{-8}	4×10^{-4}	3×10^{-9}	1×10^{-5}
		I	6×10^{-9}	3×10^{-4}	2×10^{-10}	1×10^{-5}
Samarium (62)	^{147}Sm	S	7×10^{-11}	2×10^{-3}	2×10^{-12}	6×10^{-5}
		I	3×10^{-10}	2×10^{-3}	9×10^{-12}	7×10^{-5}
	^{151}Sm	S	6×10^{-8}	1×10^{-2}	2×10^{-9}	4×10^{-4}
		I	1×10^{-7}	1×10^{-2}	5×10^{-9}	4×10^{-4}
	^{153}Sm	S	5×10^{-7}	2×10^{-3}	2×10^{-8}	8×10^{-5}
		I	4×10^{-7}	2×10^{-3}	1×10^{-8}	8×10^{-5}
Scandium (21)	^{46}Sc	S	2×10^{-7}	1×10^{-3}	8×10^{-9}	4×10^{-5}
		I	2×10^{-8}	1×10^{-3}	8×10^{-10}	4×10^{-5}
	^{47}Sc	S	6×10^{-7}	3×10^{-3}	2×10^{-8}	9×10^{-5}
		I	5×10^{-7}	3×10^{-3}	2×10^{-8}	9×10^{-5}
	^{48}Sc	S	2×10^{-7}	8×10^{-4}	6×10^{-9}	3×10^{-5}
		I	1×10^{-7}	8×10^{-4}	5×10^{-9}	3×10^{-5}
Selenium (34)	^{75}Se	S	1×10^{-6}	9×10^{-3}	4×10^{-8}	3×10^{-4}
		I	1×10^{-7}	8×10^{-3}	4×10^{-9}	3×10^{-4}
Silicon (14)	^{31}Si	S	6×10^{-6}	3×10^{-2}	2×10^{-7}	9×10^{-4}
		I	1×10^{-6}	6×10^{-3}	3×10^{-8}	2×10^{-4}
Silver (47)	^{105}Ag	S	6×10^{-7}	3×10^{-3}	2×10^{-8}	1×10^{-4}
		I	8×10^{-8}	3×10^{-3}	3×10^{-9}	1×10^{-4}
	^{110m}Ag	S	2×10^{-7}	9×10^{-4}	7×10^{-9}	3×10^{-5}
		I	1×10^{-8}	9×10^{-4}	3×10^{-10}	3×10^{-5}

Appendix 12B. (Continued)

Element atomic number	Isotope[a]		Table I Occupational Column 1 Air (μCi/ml)	Table I Occupational Column 2 Water (μCi/ml)	Table II Non-Occupational Column 1 Air (μCi/ml)	Table II Non-Occupational Column 2 Water (μCi/ml)
	^{111}Ag	S	3×10^{-7}	1×10^{-3}	1×10^{-8}	4×10^{-5}
		I	2×10^{-7}	1×10^{-3}	8×10^{-9}	4×10^{-5}
Sodium (11)	^{22}Na	S	2×10^{-7}	1×10^{-3}	6×10^{-9}	4×10^{-5}
		I	9×10^{-9}	9×10^{-4}	3×10^{-10}	3×10^{-5}
	^{24}Na	S	1×10^{-6}	6×10^{-3}	4×10^{-8}	2×10^{-4}
		I	1×10^{-7}	8×10^{-4}	5×10^{-9}	3×10^{-5}
Strontium (38)	^{85m}Sr	S	4×10^{-5}	2×10^{-1}	1×10^{-6}	7×10^{-3}
		I	3×10^{-5}	2×10^{-1}	1×10^{-6}	7×10^{-3}
	^{85}Sr	S	2×10^{-7}	3×10^{-3}	8×10^{-9}	1×10^{-4}
		I	1×10^{-7}	5×10^{-3}	4×10^{-9}	2×10^{-4}
	^{89}Sr	S	3×10^{-8}	3×10^{-4}	3×10^{-10}	3×10^{-6}
		I	4×10^{-8}	8×10^{-4}	1×10^{-9}	3×10^{-5}
	^{90}Sr	S	1×10^{-9}	1×10^{-5}	3×10^{-11}	3×10^{-7}
		I	5×10^{-9}	1×10^{-3}	2×10^{-10}	4×10^{-5}
	^{91}Sr	S	4×10^{-7}	2×10^{-3}	2×10^{-8}	7×10^{-5}
		I	3×10^{-7}	1×10^{-3}	9×10^{-9}	5×10^{-5}
	^{92}Sr	S	4×10^{-7}	2×10^{-3}	2×10^{-8}	7×10^{-5}
		I	3×10^{-7}	2×10^{-3}	1×10^{-8}	6×10^{-5}
Sulfur (16)	^{35}S	S	3×10^{-7}	2×10^{-3}	9×10^{-9}	6×10^{-5}
		I	3×10^{-7}	8×10^{-3}	9×10^{-9}	3×10^{-4}
Tantalum (73)	^{182}Ta	S	4×10^{-8}	1×10^{-3}	1×10^{-9}	4×10^{-5}
		I	2×10^{-8}	1×10^{-3}	7×10^{-10}	4×10^{-5}
Technetium (43)	^{96m}Tc	S	8×10^{-5}	4×10^{-1}	3×10^{-6}	1×10^{-2}
		I	3×10^{-5}	3×10^{-1}	1×10^{-6}	1×10^{-2}
	^{96}Tc	S	6×10^{-7}	3×10^{-3}	2×10^{-8}	1×10^{-4}
		I	2×10^{-7}	1×10^{-3}	8×10^{-9}	5×10^{-5}
	^{97m}Tc	S	2×10^{-6}	1×10^{-2}	8×10^{-8}	4×10^{-4}
		I	2×10^{-7}	5×10^{-3}	5×10^{-9}	2×10^{-4}
	^{97}Tc	S	1×10^{-5}	5×10^{-2}	4×10^{-7}	2×10^{-3}
		I	3×10^{-7}	2×10^{-2}	1×10^{-8}	8×10^{-4}
	^{99m}Tc	S	4×10^{-5}	2×10^{-1}	1×10^{-6}	6×10^{-3}
		I	1×10^{-5}	8×10^{-2}	5×10^{-7}	3×10^{-3}
	^{99}Tc	S	2×10^{-6}	1×10^{-2}	7×10^{-8}	3×10^{-4}
		I	6×10^{-8}	5×10^{-3}	2×10^{-9}	2×10^{-4}
Tellurium (52)	^{125m}Te	S	4×10^{-7}	5×10^{-3}	1×10^{-8}	2×10^{-4}
		I	1×10^{-7}	3×10^{-3}	4×10^{-9}	1×10^{-4}
	^{127m}Te	S	1×10^{-7}	2×10^{-3}	5×10^{-9}	6×10^{-5}
		I	4×10^{-8}	2×10^{-3}	1×10^{-9}	5×10^{-5}
	^{127}Te	S	2×10^{-6}	8×10^{-3}	6×10^{-8}	3×10^{-4}
		I	9×10^{-7}	5×10^{-3}	3×10^{-8}	2×10^{-4}

Appendix 12B. (Continued)

Element atomic number	Isotope[a]		Table I Occupational Column 1 Air (μCi/ml)	Table I Occupational Column 2 Water (μCi/ml)	Table II Non-Occupational Column 1 Air (μCi/ml)	Table II Non-Occupational Column 2 Water (μCi/ml)
	^{129m}Te	S	8×10^{-8}	1×10^{-3}	3×10^{-9}	3×10^{-5}
		I	3×10^{-8}	6×10^{-4}	1×10^{-9}	2×10^{-5}
	^{129}Te	S	5×10^{-6}	2×10^{-2}	2×10^{-7}	8×10^{-4}
		I	4×10^{-6}	2×10^{-2}	1×10^{-7}	8×10^{-4}
	^{131m}Te	S	4×10^{-7}	2×10^{-3}	1×10^{-8}	6×10^{-5}
		I	2×10^{-7}	1×10^{-3}	6×10^{-9}	4×10^{-5}
	^{132}Te	S	2×10^{-7}	9×10^{-4}	7×10^{-9}	3×10^{-5}
		I	1×10^{-7}	6×10^{-4}	4×10^{-9}	2×10^{-5}
Terbium (65)	^{160}Tb	S	1×10^{-7}	1×10^{-3}	3×10^{-9}	4×10^{-5}
		I	3×10^{-8}	1×10^{-3}	1×10^{-9}	4×10^{-5}
Thallium (81)	^{200}Tl	S	3×10^{-6}	1×10^{-2}	9×10^{-8}	4×10^{-4}
		I	1×10^{-6}	7×10^{-3}	4×10^{-8}	2×10^{-4}
	^{201}Tl	S	2×10^{-6}	9×10^{-3}	7×10^{-8}	3×10^{-4}
		I	9×10^{-7}	5×10^{-3}	3×10^{-8}	2×10^{-4}
	^{202}Tl	S	8×10^{-7}	4×10^{-3}	3×10^{-8}	1×10^{-4}
		I	2×10^{-7}	2×10^{-3}	8×10^{-9}	7×10^{-5}
	^{204}Tl	S	6×10^{-7}	3×10^{-3}	2×10^{-8}	1×10^{-4}
		I	3×10^{-8}	2×10^{-3}	9×10^{-10}	6×10^{-5}
Thorium (90)	^{227}Th[b]	S	3×10^{-10}	5×10^{-4}	1×10^{-11}	2×10^{-5}
		I	2×10^{-10}	5×10^{-4}	6×10^{-12}	2×10^{-5}
	^{228}Th	S	9×10^{-12}	2×10^{-4}	3×10^{-13}	7×10^{-6}
		I	6×10^{-12}	4×10^{-4}	2×10^{-13}	1×10^{-5}
	^{230}Th	S	2×10^{-12}	5×10^{-5}	8×10^{-14}	2×10^{-6}
		I	1×10^{-11}	9×10^{-4}	3×10^{-13}	3×10^{-5}
	^{231}Th[b]	S	1×10^{-6}	7×10^{-3}	5×10^{-8}	2×10^{-4}
		I	1×10^{-6}	7×10^{-3}	4×10^{-8}	2×10^{-4}
	^{232}Th	S	3×10^{-11}	5×10^{-5}	1×10^{-12}	2×10^{-6}
		I	3×10^{-11}	1×10^{-3}	1×10^{-12}	4×10^{-5}
	Th natural[d]	S	6×10^{-11}	6×10^{-5}	2×10^{-12}	2×10^{-6}
			6×10^{-11}	6×10^{-4}	2×10^{-12}	2×10^{-5}
	^{234}Th	S	6×10^{-8}	5×10^{-4}	2×10^{-9}	2×10^{-5}
		I	3×10^{-8}	5×10^{-4}	1×10^{-9}	2×10^{-5}
Thulium (69)	^{170}Tm	S	4×10^{-8}	1×10^{-3}	1×10^{-9}	5×10^{-5}
		I	3×10^{-8}	1×10^{-3}	1×10^{-9}	5×10^{-5}
	^{171}Tm	S	1×10^{-7}	1×10^{-2}	4×10^{-9}	5×10^{-4}
		I	2×10^{-7}	1×10^{-2}	8×10^{-9}	5×10^{-4}
Tin (50)	^{113}Sn	S	4×10^{-7}	2×10^{-3}	1×10^{-8}	9×10^{-5}
		I	5×10^{-8}	2×10^{-3}	2×10^{-9}	8×10^{-5}
	^{125}Sn	S	1×10^{-7}	5×10^{-4}	4×10^{-9}	2×10^{-5}
		I	8×10^{-8}	5×10^{-4}	3×10^{-9}	2×10^{-5}

Appendix 12B. (Continued)

Element atomic number	Isotope[a]		Table I Occupational Column 1 Air (μCi/ml)	Table I Occupational Column 2 Water (μCi/ml)	Table II Non-Occupational Column 1 Air (μCi/ml)	Table II Non-Occupational Column 2 Water (μCi/ml)
Tungsten (Wolfram) (74)	^{181}W	S	2×10^{-6}	1×10^{-2}	8×10^{-8}	4×10^{-4}
		I	1×10^{-7}	1×10^{-2}	4×10^{-9}	3×10^{-4}
	^{185}W	S	8×10^{-7}	4×10^{-3}	3×10^{-8}	1×10^{-4}
		I	1×10^{-7}	3×10^{-3}	4×10^{-9}	1×10^{-4}
	^{187}W	S	4×10^{-7}	2×10^{-3}	2×10^{-8}	7×10^{-5}
		I	3×10^{-7}	2×10^{-3}	1×10^{-8}	6×10^{-5}
Uranium (92)	^{230}U	S	3×10^{-10}	1×10^{-4}	1×10^{-11}	5×10^{-6}
		I	1×10^{-10}	1×10^{-4}	4×10^{-12}	5×10^{-6}
	^{232}U	S	1×10^{-10}	8×10^{-4}	3×10^{-12}	3×10^{-5}
		I	3×10^{-11}	8×10^{-4}	9×10^{-13}	3×10^{-5}
	^{233}U	S	5×10^{-10}	9×10^{-4}	2×10^{-11}	3×10^{-5}
		I	1×10^{-10}	9×10^{-4}	4×10^{-12}	3×10^{-5}
	^{234}U[d]	S[e]	6×10^{-10}	9×10^{-4}	2×10^{-11}	3×10^{-5}
		I	1×10^{-10}	9×10^{-4}	4×10^{-12}	3×10^{-5}
	^{235}U[d]	S[e]	5×10^{-10}	8×10^{-4}	2×10^{-11}	3×10^{-5}
		I	1×10^{-10}	8×10^{-4}	4×10^{-12}	3×10^{-5}
	^{236}U	S	6×10^{-10}	1×10^{-3}	2×10^{-11}	3×10^{-5}
		I	1×10^{-10}	1×10^{-3}	4×10^{-12}	3×10^{-5}
	^{238}U[d]	S[e]	7×10^{-11}	1×10^{-3}	3×10^{-12}	4×10^{-5}
		I	1×10^{-10}	1×10^{-3}	5×10^{-12}	4×10^{-5}
	^{240}U	S	2×10^{-7}	1×10^{-3}	8×10^{-9}	3×10^{-5}
		I	2×10^{-7}	1×10^{-3}	6×10^{-9}	3×10^{-5}
	U natural[d]	S[e]	1×10^{-10}	1×10^{-3}	5×10^{-12}	3×10^{-5}
		I	1×10^{-10}	1×10^{-3}	5×10^{-12}	3×10^{-5}
Vanadium (23)	^{48}V	S	2×10^{-7}	9×10^{-4}	6×10^{-9}	3×10^{-5}
		I	6×10^{-8}	8×10^{-4}	2×10^{-9}	3×10^{-5}
Xenon (54)	^{131m}Xe	Sub	2×10^{-5}	—	4×10^{-7}	—
	^{133}Xe	Sub	1×10^{-5}	—	3×10^{-7}	—
	^{133m}Xe	Sub	1×10^{-5}	—	3×10^{-7}	—
	^{135}Xe	Sub	4×10^{-6}	—	1×10^{-7}	—
Ytterbium (70)	^{175}Yb	S	7×10^{-7}	3×10^{-3}	2×10^{-8}	1×10^{-4}
		I	6×10^{-7}	3×10^{-3}	2×10^{-8}	1×10^{-4}
Yttrium (39)	^{90}Y	S	1×10^{-7}	6×10^{-4}	4×10^{-9}	2×10^{-5}
		I	1×10^{-7}	6×10^{-4}	3×10^{-9}	2×10^{-5}
	^{91m}Y	S	2×10^{-5}	1×10^{-1}	8×10^{-7}	3×10^{-3}
		I	2×10^{-5}	1×10^{-1}	6×10^{-7}	3×10^{-3}
	^{91}Y	S	4×10^{-8}	8×10^{-4}	1×10^{-9}	3×10^{-5}
		I	3×10^{-8}	8×10^{-4}	1×10^{-9}	3×10^{-5}

Appendix 12B. (Continued)

Element atomic number	Isotope[a]		Table I Occupational		Table II Non-Occupational	
			Column 1 Air (μCi/ml)	Column 2 Water (μCi/ml)	Column 1 Air (μCi/ml)	Column 2 Water (μCi/ml)
	^{93}Y	S	2×10^{-7}	8×10^{-4}	6×10^{-9}	3×10^{-5}
		I	1×10^{-7}	8×10^{-4}	5×10^{-9}	3×10^{-5}
	^{92}Y	S	4×10^{-7}	2×10^{-3}	1×10^{-8}	6×10^{-5}
		I	3×10^{-7}	2×10^{-3}	1×10^{-8}	6×10^{-5}
Zinc (30)	^{65}Zn	S	1×10^{-7}	3×10^{-3}	4×10^{-9}	1×10^{-4}
		I	6×10^{-8}	5×10^{-3}	2×10^{-9}	2×10^{-4}
	^{69m}Zn	S	4×10^{-7}	2×10^{-3}	1×10^{-8}	7×10^{-5}
		I	3×10^{-7}	2×10^{-3}	1×10^{-8}	6×10^{-5}
	^{69}Zn	S	7×10^{-6}	5×10^{-2}	2×10^{-7}	2×10^{-3}
		I	9×10^{-6}	5×10^{-2}	3×10^{-7}	2×10^{-3}
Zirconium (40)	^{93}Zr	S	1×10^{-7}	2×10^{-2}	4×10^{-9}	8×10^{-4}
		I	3×10^{-7}	2×10^{-2}	1×10^{-8}	8×10^{-4}
	^{95}Zr	S	1×10^{-7}	2×10^{-3}	4×10^{-9}	6×10^{-5}
		I	3×10^{-8}	2×10^{-3}	1×10^{-9}	6×10^{-5}
	^{97}Zr	S	1×10^{-7}	5×10^{-4}	4×10^{-9}	2×10^{-5}
		I	9×10^{-8}	5×10^{-4}	3×10^{-9}	2×10^{-5}
Any single radionuclide not listed above with decay mode other than alpha emission or spontaneous fission and with radioactive half-life less than 2 hours		Sub	1×10^{-6}	—	3×10^{-8}	—
Any single radionuclide not listed above with decay mode other than alpha emission or spontaneous fission and with radioactive half-life greater than 2 hours			3×10^{-9}	9×10^{-5}	1×10^{-10}	3×10^{-6}

Appendix 12B. (Continued)

Element atomic number	Isotope[a]	Table I Occupational, Column 1 Air (μCi/ml)	Table I Occupational, Column 2 Water (μCi/ml)	Table II Non-Occupational, Column 1 Air (μCi/ml)	Table II Non-Occupational, Column 2 Water (μCi/ml)
Any single radionuclide not listed above, which decays by alpha emission or spontaneous fission		6×10^{-13}	4×10^{-7}	2×10^{-14}	3×10^{-8}

[a] S = soluble; I = insoluble; "Sub" means that values are given for submersion in a semispherical infinite cloud of airborne material. As amended 39 FR 29314.
[b] Amended 37 FR 23319.
[c] These radon concentrations are appropriate for protection from ^{222}Rn combined with its short-lived daughters. Alternatively, the value in Table I may be replaced by one-third "working level." (A "working level" is defined as any combination of short-lived ^{222}Rn daughters, ^{218}Po, ^{214}Pb, ^{214}Bi, and ^{214}Po, in one liter of air, without regard to the degree of equilibrium, that will result in the ultimate emission of 1.3×10^3 MeV of alpha particle energy.) The Table II value may be replaced by one-thirtieth of a "working level." The limit on ^{222}Rn concentrations in restricted areas may be based on an annual average.
[d] Amended 39 FR 23990.
[e] For soluble mixtures of ^{238}U, ^{234}U, and ^{235}U in air, chemical toxicity may be the limiting factor. If the percent by weight (enrichment) of ^{235}U is less than 5, the concentration value for a 40-hour work week, Table I, is 0.2 milligram uranium per cubic meter of air average. For any enrichment, the product of the average concentration and time of exposure during a 40-hour work week shall not exceed 8×10^{-3} SA μCi-hr/ml, where SA is the specific activity of the uranium inhaled. The concentration value for Table II is 0.007 milligram uranium per cubic meter of air. The specific activity for natural uranium is 6.77×10^{-7} curie per gram U. The specific activity for other mixtures of ^{238}U, ^{235}U, and ^{234}U, if not known, shall be:

$SA = 3.6 \times 10^{-7}$ curies per gram U U-depleted
$SA = (0.4 + 0.38E + 0.0034E^2)\ 10^{-6}$ $E \geq 0.72$

where E is the percentage by weight of ^{235}U expressed as percent. Amended 39 FR 25463.

Appendix 12C. Labeling, Posting and Disposal Quantities of Radioactive Materials[e]

Material	Quantity (μCi)	Material	Quantity (μCi)
Americium-241	0.01	Arsenic-74	10
Antimony-122	100	Arsenic-76	10
Antimony-124	10	Arsenic-77	100
Antimony-125	10	Barium-131	10
Arsenic-73	100	Barium-133[a]	10

Appendix 12C. (Continued)

Material	Quantity (μCi)	Material	Quantity (μCi)
Barium-140	10	Indium-115m	100
Bismuth-210	1	Indium-115	10
Bromine-82	10	Iodine-125	1
Cadmium-109	10	Iodine-126	1
Cadmium-115m	10	Iodine-129	0.1
Cadmium-115	100	Iodine-131	1
Calcium-45	10	Iodine-132	10
Calcium-47	10	Iodine-133	1
Carbon-14	100	Iodine-134	10
Cerium-141	100	Iodine-135	10
Cerium-143	100	Iridium-192	10
Cerium-144	1	Iridium-194	100
Cesium-131	1000	Iron-55	100
Cesium-134m	100	Iron-59	10
Cesium-134	1	Krypton-85	100
Cesium-135	10	Krypton-87	10
Cesium-136	10	Lanthanum-140	10
Cesium-137	10	Lutetium-177	100
Chlorine-36	10	Manganese-52	10
Chlorine-38	10	Manganese-54	10
Chromium-51	1000	Manganese-56	10
Cobalt-58m	10	Mercury-197m	100
Cobalt-58	10	Mercury-197	100
Cobalt-60	1	Mercury-203	10
Copper-64	100	Molybdenum-99	100
Dysprosium-165	10	Neodymium-147	100
Dysprosium-166	100	Neodymium-149	100
Erbium-169	100	Nickel-59	100
Erbium-171	100	Nickel-63	10
Europium-152, t_2 = 9.2 hr	100	Nickel-65	100
Europium-152, t_2 = 13 years	1	Niobium-93m	10
Europium-154	1	Niobium-95	10
Europium-155	10	Niobium-97	10
Fluorine-18	1000	Osmium-185	10
Gadolinium-153	10	Osmium-191m	100
Gadolinium-159	100	Osmium-191	100
Gallium-72	10	Osmium-193	100
Germanium-71	100	Palladium-103	100
Gold-198	100	Palladium-109	100
Gold-199	100	Phosphorus-32	10
Hafnium-181	10	Platinum-191	100
Holmium-166	100	Platinum-193m	100
Hydrogen-3	1000	Platinum-193	100
Indium-113m	100	Platinum-197m	100
Indium-114m	10	Platinum-197	100

Appendix 12C. (Continued)

Material	Quantity (μCi)	Material	Quantity (μCi)
Plutonium-239	0.01	Technetium-99	10
Polonium-210	0.1	Tellurium-125m	10
Potassium-42	10	Tellurium-127m	10
Praseodymium-142	100	Tellurium-127	100
Praseodymium-143	100	Tellurium-129m	10
Promethium-147	10	Tellurium-129	100
Promethium-149	10	Tellurium-131m	10
Radium-226	0.01	Tellurium-132	10
Rhenium-186	100	Terbium-160	10
Rhenium-188	100	Thallium-200	100
Rhodium-103m	100	Thallium-201	100
Rhodium-105	100	Thallium-202	100
Rubidium-86	10	Thallium-204	10
Rubidium-87	10	Thorium (natural)[b,c]	100
Ruthenium-97	100	Thulium-170	10
Ruthenium-103	10	Thulium-171	10
Ruthenium-105	10	Tin-113	10
Ruthenium-106	1	Tin-125	10
Samarium-151	10	Tungsten-181	10
Samarium-153	100	Tungsten-185	10
Scandium-46	10	Tungsten-187	100
Scandium-47	100	Uranium (natural)[b,d]	100
Scandium-48	10	Uranium-233	0.01
Selenium-75	10	Uranium-234, Uranium-235	0.01
Silicon-31	100	Vanadium-48	10
Silver-105	10	Xenon-131m	1000
Silver-110m	1	Xenon-133	100
Silver-111	100	Xenon-135	100
Sodium-24	10	Ytterbium-175	100
Strontium-85	10	Yttrium-90	10
Strontium-89	1	Yttrium-91	10
Strontium-90	0.1	Yttrium-92	100
Strontium-91	10	Yttrium-93	100
Strontium-92	10	Zinc-65	10
Sulphur-35	100	Zinc-69m	100
Tantalum-182	10	Zinc-69	1000
Technetium-96	10	Zirconium-93	10
Technetium-97m	100	Zirconium-95	10
Technetium-97	100	Zirconium-97	10
Technetium-99m	100		

[a] Amended 36 FR 16898.

[b] Amended 39 FR 23990.

[c] Based on alpha disintegration rate of ^{232}Th, ^{230}Th and their daughter products.

[d] Based on alpha disintegration rate of ^{238}U, ^{234}U, and ^{235}U.

CHAPTER THIRTEEN

Lighting for Seeing

C. L. CROUCH

1 INTRODUCTION

Most of life's activity is guided by visual cues that are being processed by the thousands and millions during the waking hours. Most of these are unconsciously handled, and many result in involuntary activity such as transition in bodily movements or transition from place to place. These unconscious cues must be seen for proper unconscious judgment and safe involuntary action to occur. A much smaller number of visual cues involve conscious processing in carrying out productive work. These are concerned with critical details that must be handled successfully for production and economic returns. There are a multitude of details in industrial processes that are critical for good products; there are a tremendous number of symbols to be recognized and individually processed in the office. The illuminating engineer calls each of these critical details the visual task. Each detail involves not only its own configuration but the background against which it is seen. The contrast of the detail with its background determines its ability to be seen.

The total ability to see the details depends on (1) the characteristics of the visual task, and (2) the characteristics of the observer. The characteristics of the task depend on the following:

1. The configuration of the detail.
2. The narrowest size of the configurated detail or gap in the detail.
3. The contrast of the detail with its background.
4. The time of exposure of the detail to the pause of the eye or the glance of the observer.
5. The luminance (or brightness) of the background to which the eyes become adapted.
6. The specularity of the detail and its background.

The characteristics of the observer depend on another set of factors:

1. The sensitivity of the individual visual system to
   a. Size.
   b. Contrast.
   c. Time.
2. Transient adaptation.
3. Glare.
   a. Disability.
   b. Discomfort.
4. Age.
5. Motivational factors.
6. Physiological factors.

2 THE LUMINOUS ENVIRONMENT

Man has been provided with a luminous environment—the vault of the sky, the greensward, the forest, the sand—all are sending light to the eyes of the observer. Even on the darkest night, light is radiating from all elements of the field of view. Since absolute black with no reflection does not exist, in providing for productive activity, how should the designer proportion the luminous radiation for optimum seeing of critical detail, and ideally provide for motivational values for energetic pursuit of the tasks?

Research is being carried out to find the answers in relation to the characteristics of the task and the characteristics of the observers, to ultimately yield the optimum visual environment (1). Currently much is known to provide an appropriate environment for quick accurate seeing.

3 LIGHTING TERMS AND UNITS

There are many terms and units used in illuminating engineering. However it is only necessary to have a clear understanding of certain basic ones.

CANDLEPOWER: A luminous intensity expressed in candelas. An ordinary candle about an inch in diameter produces about one candela in a horizontal direction.

CONTRAST (LUMINANCE): The relationship between the luminances of an object and its immediate background. It is equal to $(L_1 - L_2)/L_1$ or $(L_2 - L_1)/L_1$, where L_1 and L_2 are the luminances of the background and object, respectively. The form of the equation must be specified.

FOOTCANDLE (FC): The unit of illumination when the foot is the unit of length. It is the illumination on a surface one square foot in area, on which is uniformly distributed a flux of one lumen (lm). One footcandle equals one lumen per square foot.

FOOTLAMBERT (FL): The unit of photometric brightness and is equal to the uniform brightness of a perfectly diffusing surface emitting or reflecting one lumen per square foot. The brightness in footlamberts is equal to the illumination on a surface multiplied by its reflectance.

GLARE: The sensation produced by luminances within the visual field that are sufficiently greater than the luminances to which the eyes are adapted to cause annoyance, discomfort, or loss in visual performance and visibility. "Disability glare" results in reduced visual performance; "discomfort glare" produces discomfort but does not necessarily interfere with visual performance or visibility.

ILLUMINATION: The density of light flux incident upon a surface.

LUMEN: The unit of luminous flux emitted through a unit solid angle (steradian, sr) from a uniform point source of one candela. It is used to represent the total light output of light sources.

LUMINANCE or PHOTOMETRIC BRIGHTNESS: A measure of the light emitted by a luminous body or reflected from a surface. It may be expressed in terms of the lumens being emitted or reflected from a perfectly diffusing source. Used alone, the term "brightness" usually refers to the sensation that results from viewing surfaces from which light comes to the eye; it often is used interchangeably with luminance, since the context usually indicates whether the sensory or photometric meaning is intended.

LUMINANCE RATIO: The ratio between the luminances (photometric brightnesses) of any two areas in the visual field.

REFLECTANCE: The ratio of light reflected by an object to the incident light.

SPECULAR REFLECTION: That which is predominantly reflected as from a mirror.

VEILING REFLECTIONS: The mirrorlike reflections, superimposed on diffuse reflections from an object, that partially or totally obscure the details to be seen by reducing the contrast. This effect sometimes is called "reflected glare."

VISIBILITY: The quality or state of being perceivable by the eye. In many outdoor applications, visibility is defined in terms of the distance at which an object can be just perceived by the eye. In indoor applications it usually is defined in terms of the contrast or size of a standard test object, observed under standardized view conditions, having the same threshold as the given object.

VISUAL ANGLE: The angle subtended by an object or detail at the point of observation. It usually is measured in minutes of arc.

VISUAL FIELD: The locus of objects or points in space that can be perceived when the head and eyes are kept fixed. The field may be monocular or binocular.

VISUAL PERFORMANCE: The quantitative assessment of the performance of a task, taking into consideration speed and accuracy.

VISUAL TASK: Conventionally designates the details and objects that must be seen for the performance of a given activity, including the immediate background of the details or objects.

4 VISUAL PERFORMANCE

Visual performance is the measurement of the observer response to the characteristics of the task. It may be the response to the visibility of the details of the task or the overall response to carrying out a given sequence of acts involving the visibility of the details.

4.1 General Light Response to Different Parameters

It is helpful to know the general response of the visual system to each of the elemental characteristics of a task, its size, contrast, and time of seeing, and the luminance of the background of the detail. For more than a century, eye specialists have measured the response of human visual capability by having a person read the smallest line of letters on an "eye chart" bearing a graduated series of lines of different sized letters. A series of test objects that are symbolic of the details found in practical workplace tasks has evolved into the configurations shown in Figure 13.1. Each series gives a different curve of response, but the curves for each are largely parallel, showing the same general nature of the response (3). The ability to see decreasing size with changing illumination for black Landolt rings on white background (high contrast is represented in Figure 13.2. The ability to see decreasing contrasts with changing illumination for a disk with visual angle of 4′ subtense and 0.1-second exposure is shown in Figure 13.3. The ability to see in decreasing time with changing illumination can be understood from Figure 13.4. A further illustration of how these elemental characteristics influence the seeing ability with changing illumination is given by the eye charts in Figure 13.5, illustrating the effects of size, contrast, and reflectance of background. Of course reflectance of background makes a marked difference in seeing the detail because the sensitivity of the observer is dependent on the adaptation luminance of the background (7), as further illustrated in Figure 13.6.

4.2 Effect of Surroundings

The ability to see size and contrast is influenced not only by the luminance of the background surrounding the detail but by the differing luminance and size of immediate surroundings around the background of the task, as illustrated in Figures 13.7, 13.8, and

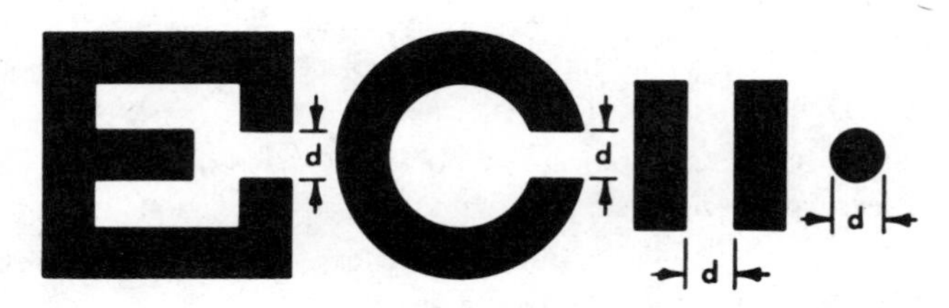

Figure 13.1 Commonly used test objects for determining size discrimination and visual acuity: Snellen E, Landolt ring, parallel bars, and disk (2).

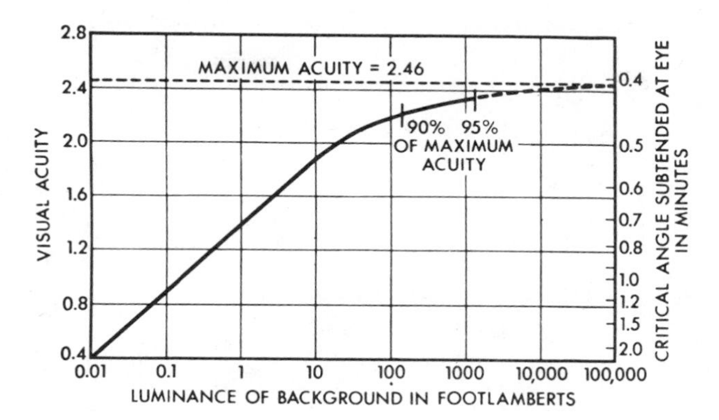

Figure 13.2 The variation is visual acuity and visual size with background luminance for a black object (Landolt ring) on a white background (4).

13.9. These effects appear to come from static viewing of the details involved, but it has been reasoned by some that the eyes could be stealing glances to the surroundings, and the effect of sensitivity losses because of the differing surround luminance could be due to transient adaptation, which is discussed later.

4.3 Simulating Practical Tasks

Generally the productivity of practical tasks in commerce and industry is related to the speed and accuracy of carrying out those tasks. Thus laboratory tasks simulating practical tasks have been designed to measure the speed of response to the exposure of test objects

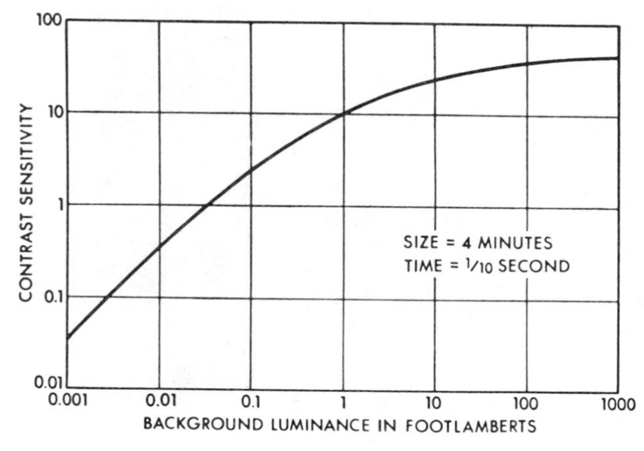

Figure 13.3 Variation of contrast sensitivity with background luminance (5).

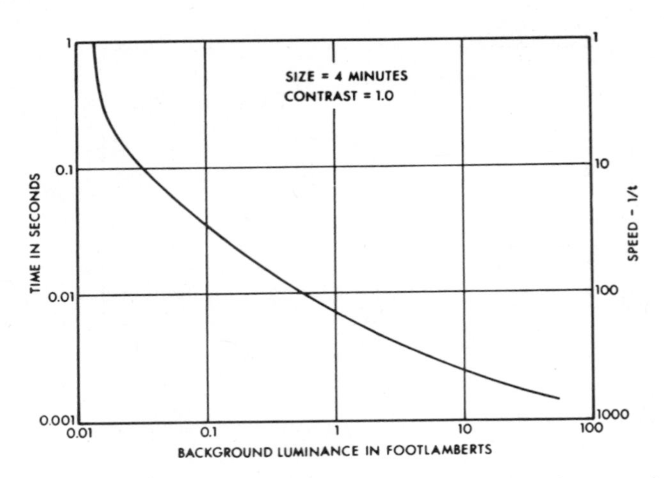

Figure 13.4 The effect of background luminance on the time required to see a high contrast test object; right-hand scale represents speed of vision (5).

that would approximate the size and contrast of the range of details found in the field. Such a series, under the auspices of the Committee on Industrial Lighting of the National Research Council (1920s), appears in Figures 13.10, 13.11, and 13.12. Unfortunately the study did not cover values of illumination beyond 100 footcandles.

A further practical simulation of a task vitally used in industry was the reading of a steel vernier rule to the nearest $\frac{1}{1000}$ inch (12). In carrying out the simulation, it was found that the speed of making the complete reading of the tenth-inch numbers, the

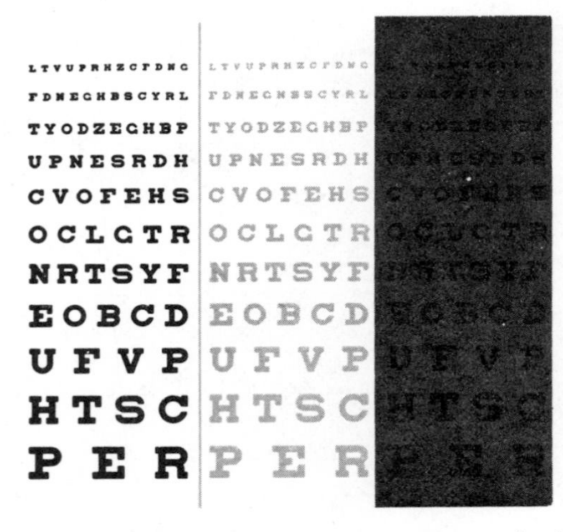

Figure 13.5 Series of eye charts showing effects of size, contrast, and reflectance. Black on white required 1 fc; gray on white, 120 fc; black on dark gray 525 fc (6).

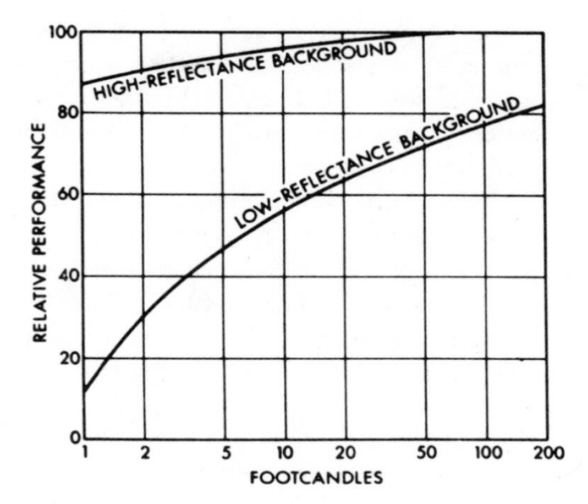

Figure 13.6 The effect of level of illumination on relative visual performance when the task involves high contrast (upper curve) and low contrast (lower curve) (7).

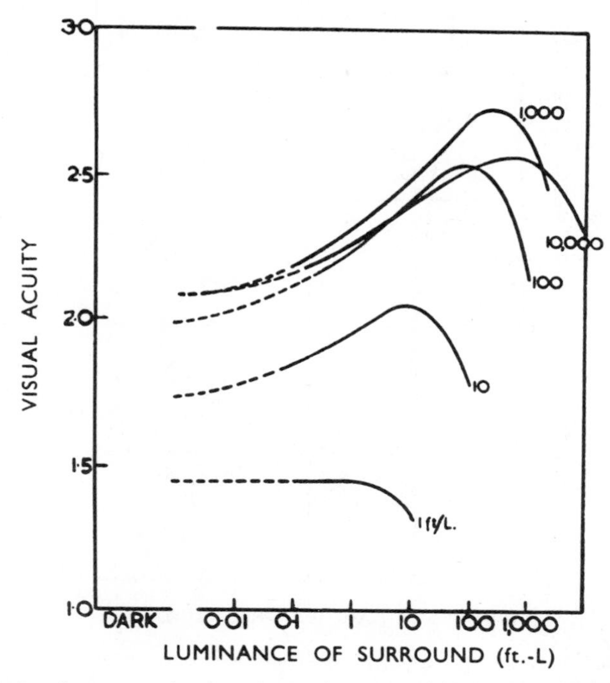

Figure 13.7 Relation between visual acuity and surround luminance for various central field luminances (120° surround to central field) (8).

quarter divisions within the tenth, and the exact position of the vernier divided into 25 parts varied with the luminance of the highlight on the background of the rule against which the black divisions were seen in bold relief (Figure 13.13). The speed of reading the vernier with luminance is shown in Figure 13.14. Again the study did not carry values of luminance beyond 200 footlamberts.

In the 1970s further work was provided on simulated industrial tasks. The threading of needles (Figure 13.15) represented fine detail work in industry; the reading of date digits on Lincoln pennies (Figure 13.16) represented metal working details; and the finding of symbols on a circuit board (Figure 13.17) illustrating intricate electronic assembly. Some of the results are given in Figures 13.18 through 13.20. When one considers labor costs, space ownership or rental, lighting costs, and electricity rates, it is

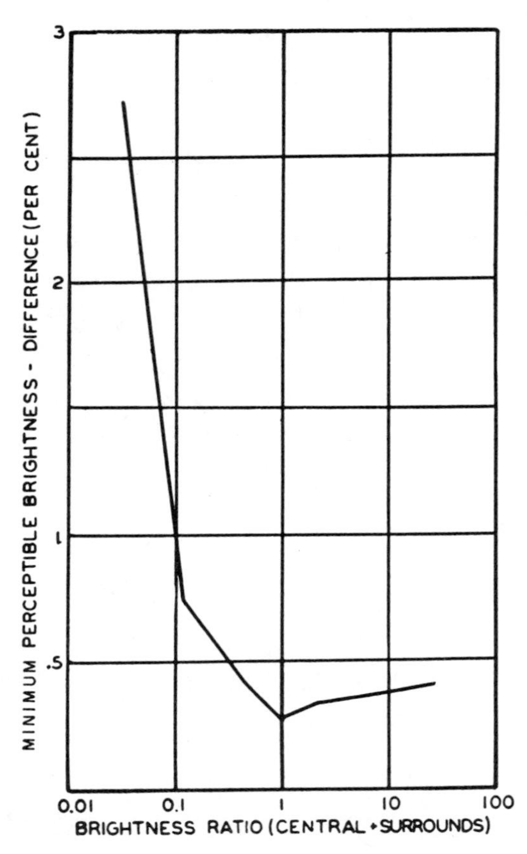

Figure 13.8 Relation between minimum perceptible brightness difference and the ratio of the brightness of the task to the brightness of the surrounding field (9).

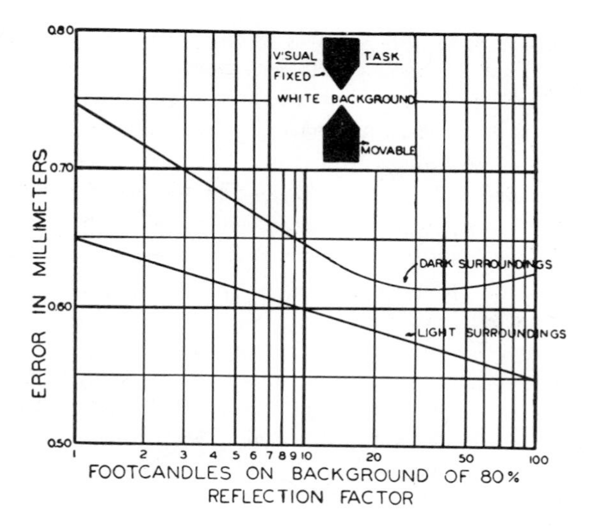

Figure 13.9 The effect of surroundings and illumination on the precision of performing a mechanical task guided by vision (10).

found in many cases that monetary savings in productivity and decrease in errors is positive in relation to lighting levels (14) (Figure 13.21).

All the productivity graphs shown are in positive relation to the visibility of the detail. In other words, the better the visual cues are seen, the faster the speed of processing and the fewer the errors, but at a diminishing rate until a maximum is reached. Figure 13.22 illustrates this principle, where visibility level (VL) represents the distance in terms of contrast (log contrast) above the laboratory threshold contrast. It is the degree of visibility above "just barely seeing" (50 percent of the times one looks at the task). The higher the VL value, the higher the object is above its threshold and, therefore, the more visible it is. Since the eyes become more sensitive to contrast as the luminance of the background of the detail is greater, the lower the threshold becomes and the greater the distance of the object's contrast above threshold; therefore its VL is higher (Figure 13.23).

There are field tasks that involve much mechanical activity or manipulation along with the visual piloting of the work. This activity may occupy so much of the time of doing the overall processing that the productivity curves (time or speed) reveal little or no change with illumination level or increasing VL. The designer is then faced with a dilemma. Either the emphasis will be placed on providing good visual efficiency, regardless of the output curves, or the visual efficiency will be depreciated on the basis of the output curve, letting the economics force a lower level of illumination. This would result in a lower VL than would be used if the visual component were the major work item. Unfortunately, the tests described are of short time duration, and data are not available on possible deterioration of performances over a long period, such as several hours or a day. Lowered visual efficiency induced by money-saving measures might result in lowered output over a period of time. This might be particularly true for older workers whose visual capability is much less than that of college students, who were the subjects of the test.

Also, there is the issue of the workers' morale. The same visual details may be processed at different locations in a plant or office with differing components of visual versus mechanical. Shall the designer furnish much more illumination to the workers where the visual is a large proportion and much less to those having to do more mechanical work on the same visual details?

The same concepts apply to office lighting. In the middle 1970s tests similar to those described earlier for simulated industrial tasks were conducted. These involved proofreading (14) of mimeographed material having good, fair, and poor quality copy (Figure 13.24), reading handwritten bank checks of varying quality (14) and comparing

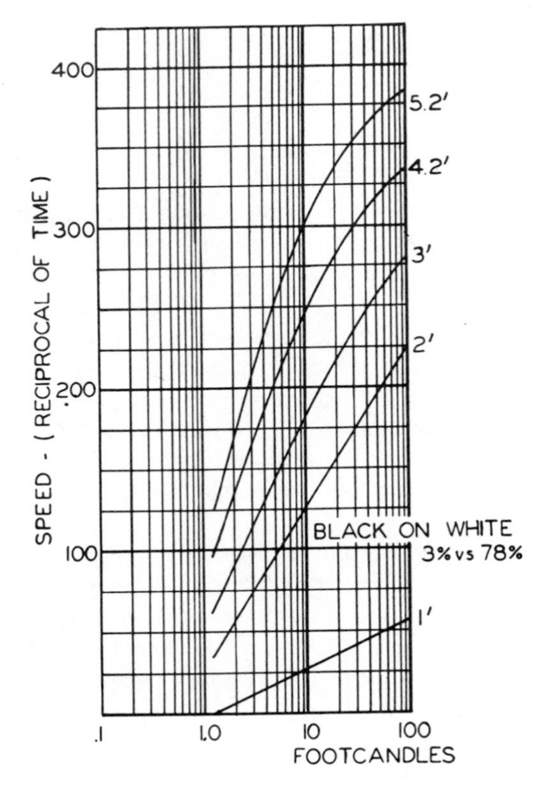

Figure 13.10 Speed of vision versus illumination for the task of identifying the position of a Landolt ring. Data from Ferree and Rand (11), under the auspices of the National Research Council.

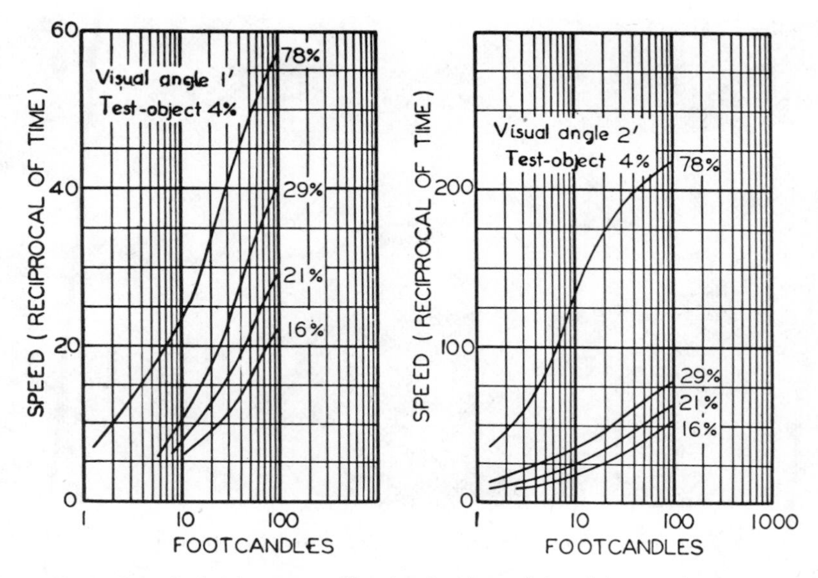

Figure 13.11 Speed of vision versus illumination for task involving varying contrasts (11).

the dollar amounts against an adding machine tape listing that included systematic errors (Figure 13.25), reading and typing (14) from printed copy having 12-point and 6-point characters printed in black on white and light gray on white (Figure 13.26), and reading the Davis Reading Test on black and white and poor reproduction (Figure 13.27) for comprehension and speed of comprehension. Some of the results are presented in Figures 13.28 through 30.

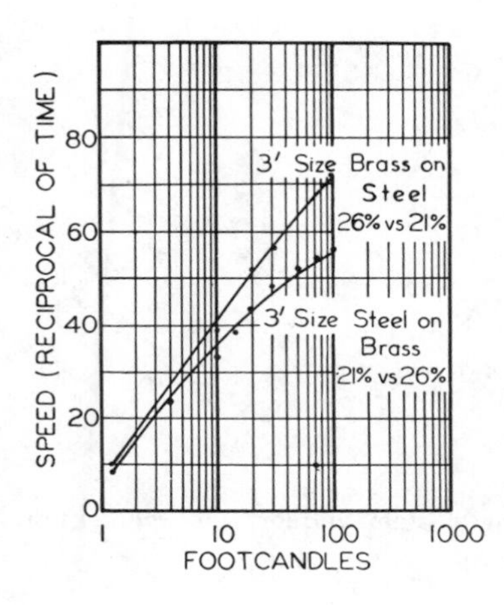

Figure 13.12 Discrimination of a brass test object on a steel background and a steel test object on a brass background (12).

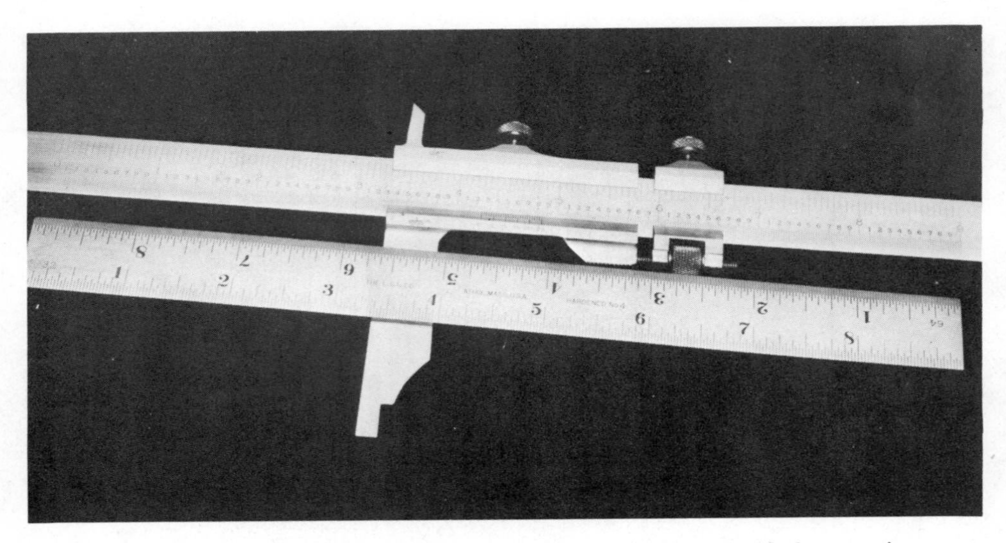

Figure 13.13 A steel vernier rule used in the practical simulation of a task down to the nearest 1/1000 inch.

The comprehension and speed of comprehension of the Davis Reading Test did not change with changes of illumination from approximately 1 to 454 footcandles, even for the poorer copy. Analysis of the results indicates that study and thinking occupied so much time that the gain in visual efficiency was masked. Here, again, the designer is faced with a dilemma. Since the overall performance remained the same under one footcandle as for 454, should the designer use one footcandle (used in street lighting); put in some amenity lighting, such as 10 to 30 footcandles, to make the task look lighted; or

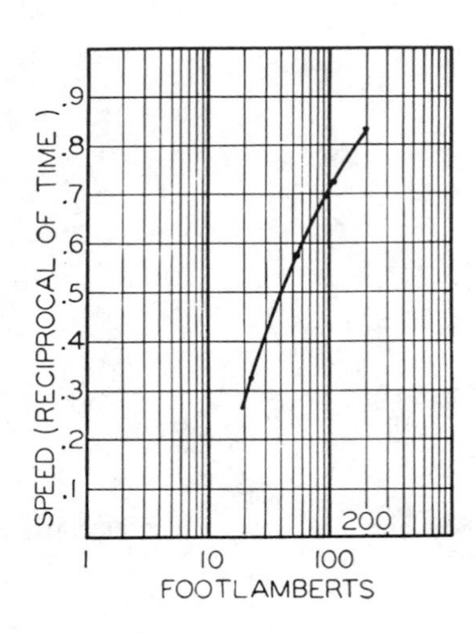

Figure 13.14 Reading a steel vernier rule: perception 1/1000 inch out of alignment.

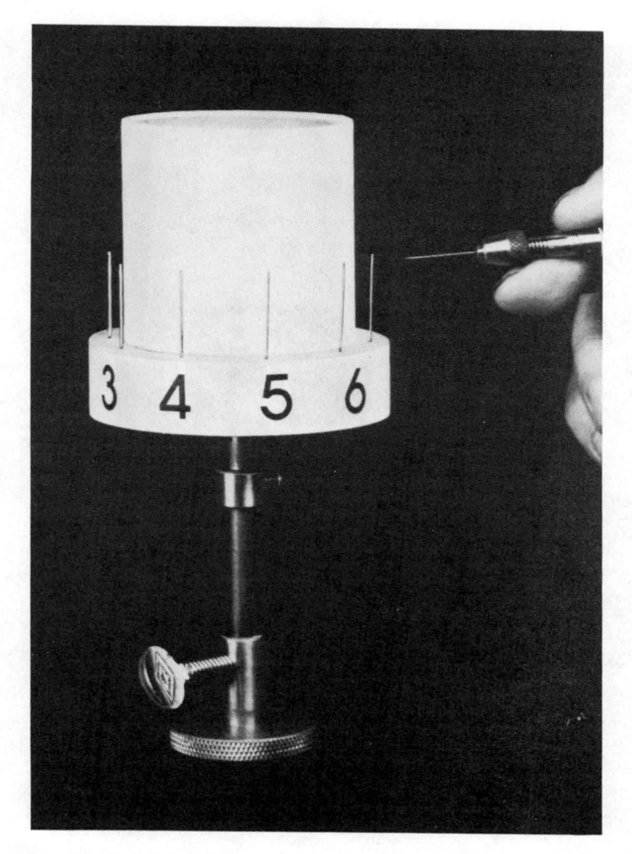

Figure 13.15 Simulated industrial task from 1974 research: the threading of needles represents fine detail work (13).

put in what visual efficiency would call for, 70 to 100 footcandles? Furthermore, should the designer place 1 to 10 footcandles on the desk with the Davis Reading type of work but put 70 to 100 footcandles on the desk with the "reading-typing" task that involves approximately the same visual details as the Davis Reading type of task? Compare Figures 13.26 and 13.27.

In the 1930s Weston faced the same dilemma in England. He was director of research in occupational optics for the Medical Research Council. For many years he had studied the role of illumination in various types of industry, but he came to the conclusion that one could not isolate the gain in visual performance from the many contributing factors to industrial plant productivity. Under the aegis of his advisory committee, therefore, he set up a series of laboratory tests that would simulate industrial tasks, eliminating the many extraneous factors and determining the true function of "visual efficiency" in relation to illumination (15). He recognized that mechanical or "action" time, which varies

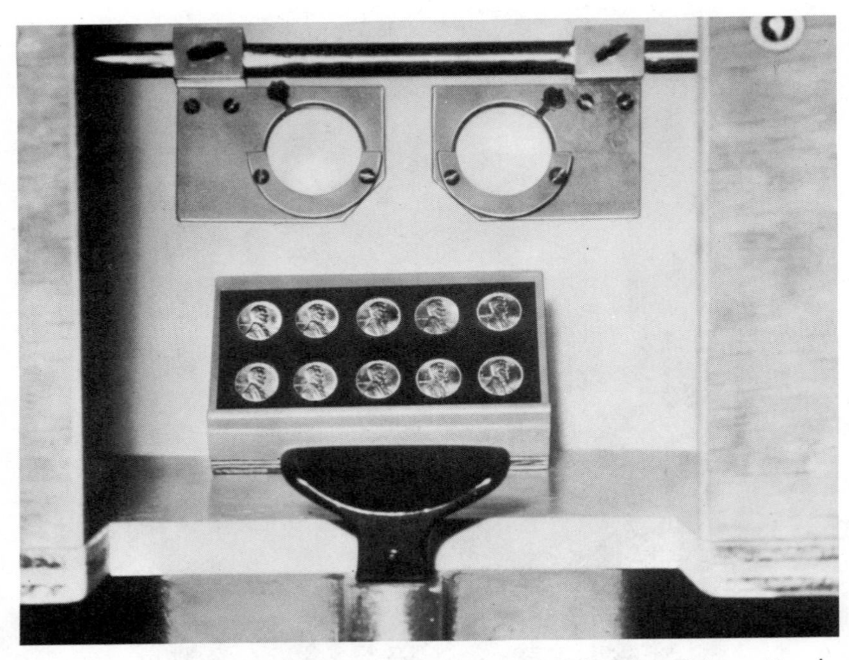

Figure 13.16 Simulated industrial task from 1974 research: reading date digits on Lincoln pennies represents metal working details (13).

very greatly in field tasks, diluted or masked the true visual efficiency; therefore he measured and subtracted this component to arrive at the net visual time and its relationship to illumination. This would be the true visual performance as related to illumination.

When the Illuminating Engineering Research Institute (IERI) started its research program on required illumination for good visual performance in 1950, it followed the Weston concept.

4.4 Current Illumination Determination Method

Sizes of details that are critical for seeing in commerce and industry have been found to be in the range of 1 to 10 minutes of arc subtended at the eye at the usual viewing distance. This is confirmed by the distance on the contrast scale between 1 and 10′ curves as compared to the distance between the 1 and 60′ curves in Figure 13.31. It has been found over the last 50 years (Ferree and Rand; Luckiesh, Cobb, and Weston) that 3 to 4′ size (*Readers' Digest* type or No. 2 pencil handwriting) represented the weighted average of the critical seeing tasks in working situations. The average eye pause in reading is 0.2 second. The curve of contrast versus luminance for a 4′ circular

Figure 13.17 Simulated industrial task from 1974 research: a circuit board represents an intricate electronic assembly (13).

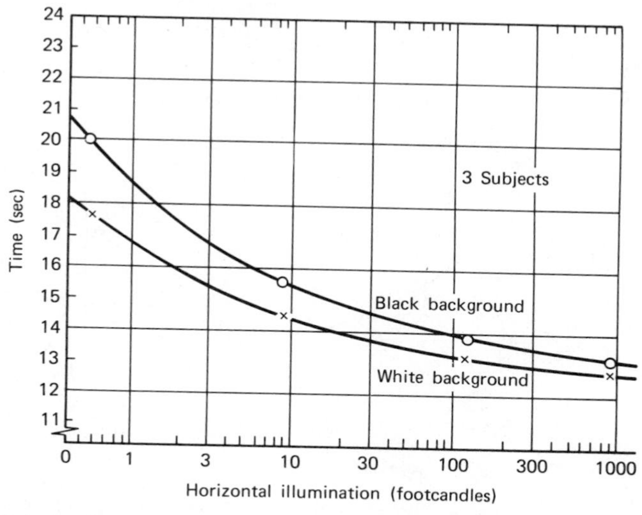

Figure 13.18 Relation between performance, time, and illumination to the needle probe task.

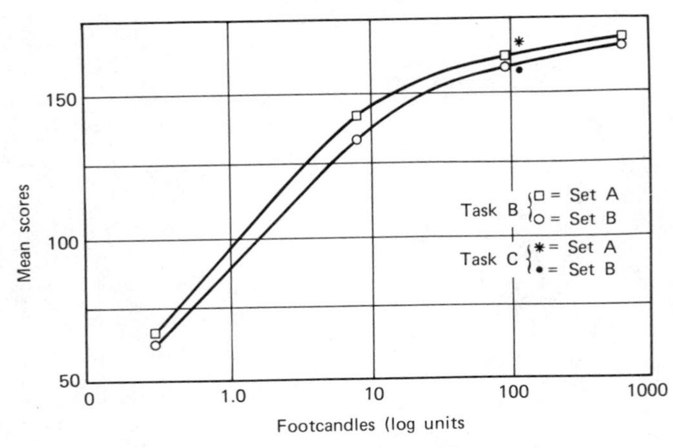

Figure 13.19 Results of response to the reading of date digits on Lincoln pennies.

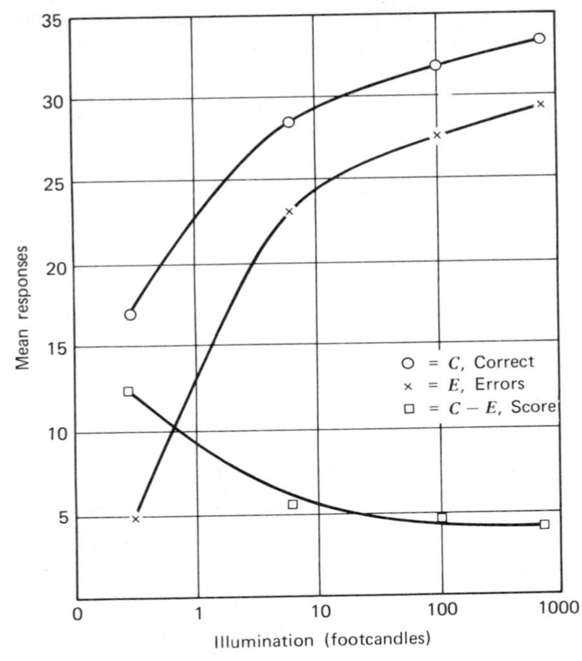

Figure 13.20 Results of response to the reading of a circuit board (finding of symbols vs. illumination).

disk exposed for 0.2 second formed the basis of the function for determination of illumination required for quite accurate seeing (5). The shape of the function was determined by laboratory measurements with static viewing at threshold representing 50 percent probability of correct seeing. The working curve was determined by experiments on a Field Task Simulator in which 50 4-inch disks on a large wheel were inspected for randomly located, 4′ round spot defects (5), and the resulting curve was found parallel with the threshold curve in the range of practical use. This determined the field factor between the 50 percent threshold curve, static viewing, and the scanning viewing involved on the wheel. After considerable study over several years, this factor has been determined to be 8. (The Field Task Simulator curve position did not change, but the laboratory static viewing did change.) These relationships (16) are shown in Figure 13.32.

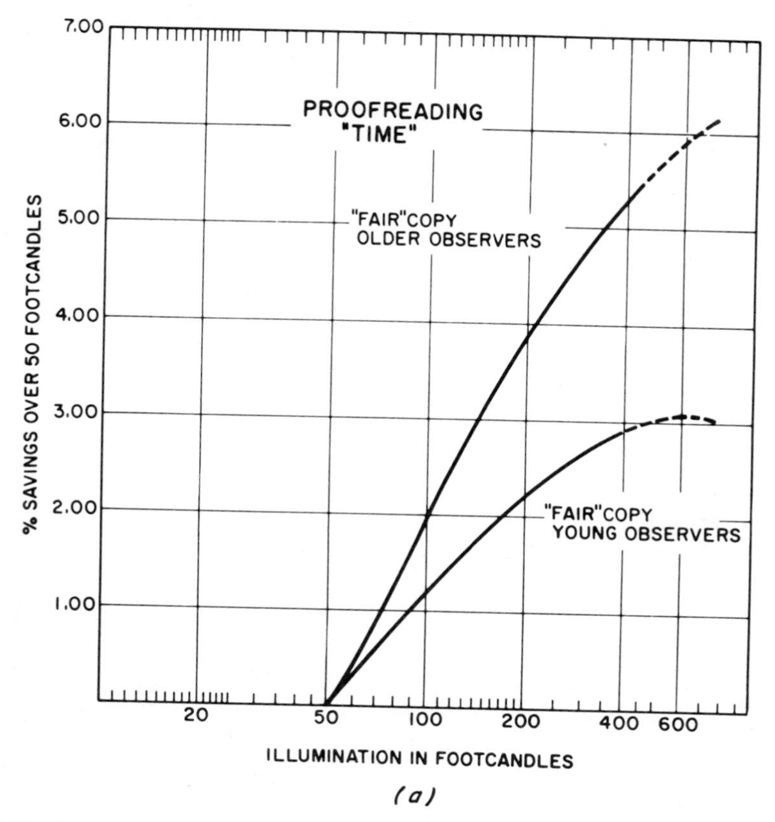

Figure 13.21 Percentage money saving from productivity response to lighting levels. (*a*) Proofreading time for older observers and young observers using "fair" copy. (*b*) Check reading time for young observers with fair visibility and good visibility.

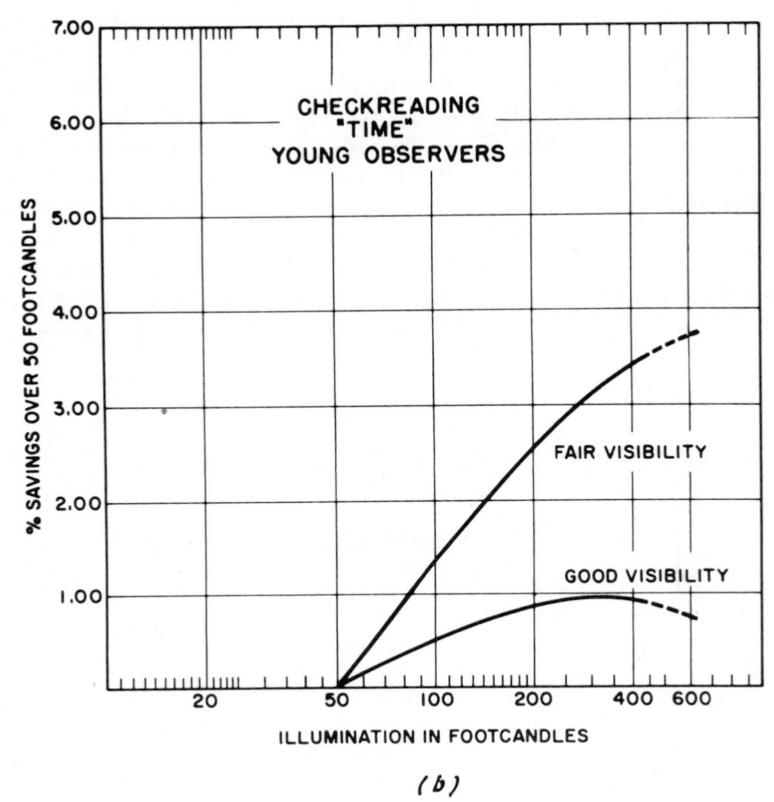

Figure 13.21 (Continued)

4.5 Field Application

A number of investigators (17) have found that the visibility of unknown details in the field can be assessed by using a contrast threshold meter. Such a meter puts a variable veiling luminance over the detail of regard until its contrast has been reduced to threshold. The amount of veil to reduce it to threshold denotes its degree of visibility (above threshold), provided the luminance to which the eyes are adapted is maintained constant. This is done by coupling an absorbing filter for the external view with the mechanism for increasing the veil, so that viewed luminance through the instrument stays constant. Furthermore, since the eyes in scanning pause to take a picture at 200 msec, it is important to determine the threshold of visibility on the basis of exposure of the field for 0.2 second. Once threshold has been reached for the field detail, then at the same luminance of the field object, 4′ luminous disk is reduced to threshold. Since all objects at threshold are equal in visibility, there is now an equivalency between the 4′ disk with its resulting contrast and the field detail. Such instruments are now available, as illustrated in Figure 13.33. The result of these measurements is to obtain an equivalent contrast of the standard 4′ test object designated $\widetilde{C}$. When this is obtained, the value of luminance required can be obtained by going to the working curve of contrast (log) versus luminance (log) (Figure 13.34), and on the contrast scale on the ordinate, locating the value of *C*, going horizontally to the curves, and dropping down. Divide this luminance by the reflectance of the background of the field task and find the footcandles to be used. (This footcandle illumination may need to be reconsidered because of unfavorable optical conditions, such as veiling reflections, discussed later.)

4.6 Limitations

Our discussion has represented the state of the art according to the best research knowledge available, yet the situation never remains constant because new knowledge is being obtained by continuing research. New improved techniques are being developed. Until these can be finished and validated, our method is subject to the knowledge of its

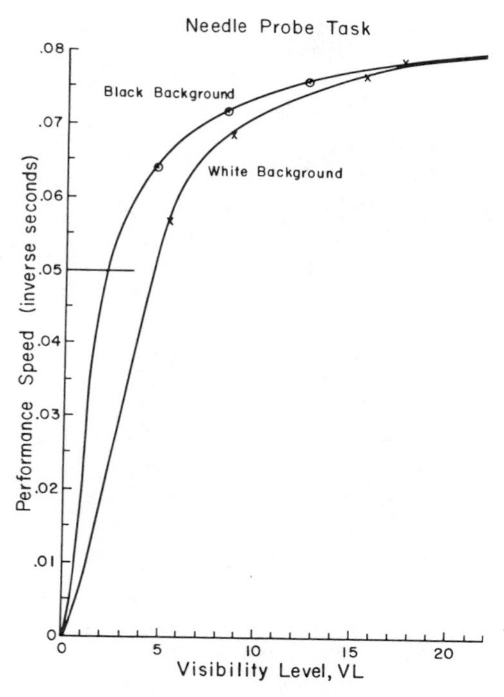

Figure 13.22 Relation between performance speed and visibility level (13).

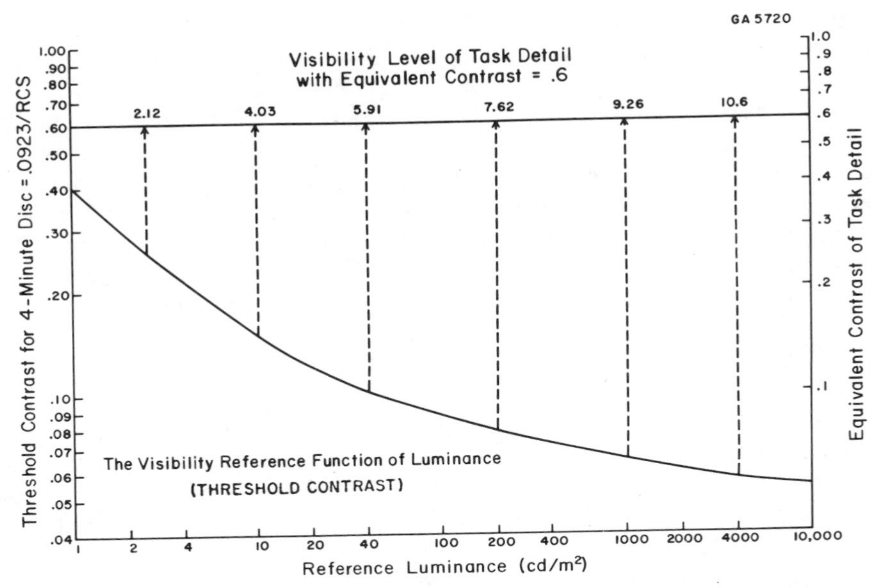

Figure 13.23 The distance of given object contrast above threshold contrast as background luminance increases.

limitation. The research data on which it is based apply only to young, normal-sighted adults, 20 to 30 years old (college students). That is, the data apply only to the average of the 20 to 30 year population (5, 18). This means that 50 percent will have a visual performance represented by the curves in Figures 13.32 and 13.34 or better, but 50 percent will have increasingly less performance. This is represented by Figure 13.35 and an interpretation is as follows:

Out of every 100 people (20 to 30 years) and every 100 times a test object was presented:

40 people missed 2 times the object was presented
25 people missed 5 times the object was presented
16 people missed 10 times the object was presented
11 people missed 15 times the object was presented
9 people missed 20 times the object was presented

Some tasks have such poor inherent visibility that no amount of ordinary lighting will give a VL of 8. In this case the contrast might be improved by special direction or color of light, to enhance the contrast. If there are indentations or raised characteristics to the detail, light striking these configurations obliquely will cause highlight and shadows that will increase the contrast. Magnification is another method of increasing the visibility.

4.7 Industrial Tasks

Since there are almost innumerable industrial processes and products, it would be difficult to describe all the critical details from start to finished products. A number of details in various industries are described in lighting literature (19). Although we have described the details, the evaluation of the illumination required by the recent research-based method just described has not been comprehensively done. In the meantime, representative values of illumination for various industries are given in the current edition of the *IES Lighting Handbook* (20).

4.7.1 Types of Task

The surface characteristics vary from matte to highly glossy finishes and from two-dimensional to three-dimensional shapes. Matte finishes do not present a lighting problem, although if there are three-dimensional objects to be seen, some unidirectional light beamed together with diffused overall lighting will bring out highlights and shadows

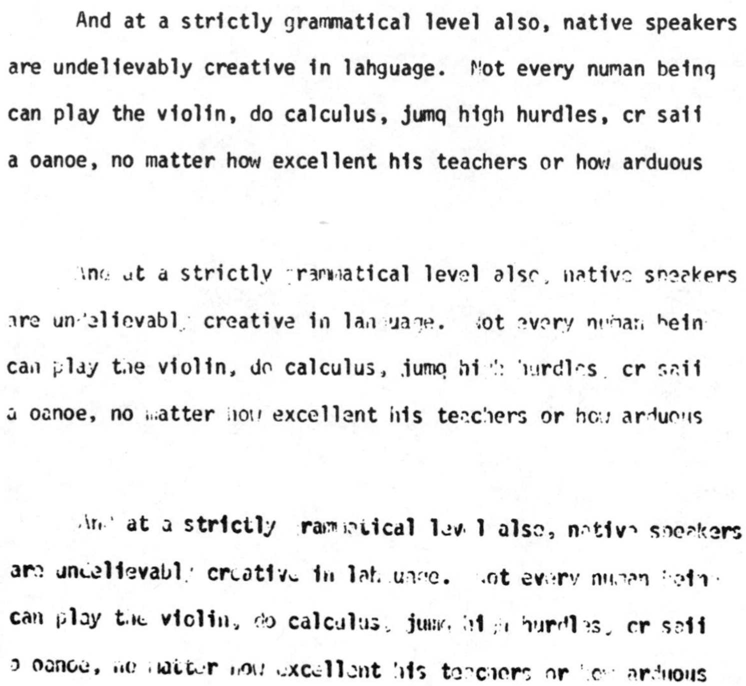

And at a strictly grammatical level also, native speakers
are undelievably creative in lahguage. Not every numan beinq
can play the violin, do calculus, jumq high hurdles, cr sail
a oanoe, no matter how excellent his teachers or how arduous

And at a strictly grammatical level also, native speakers
are undelievably creative in lan uage. Not every numan bein
can play the violin, do calculus, jumq hi h hurdles, cr sail
a oanoe, no matter how excellent his teachers or how arduous

And at a strictly rammatical lev l also, native speakers
are undelievabl creative in lah uage. Not every numan bein
can play the violin, do calculus, jumq hi h hurdles, cr sail
a oanoe, no matter how excellent his teachers or how arduous

Figure 13.24 Simulated office task showing good, fair, and poor quality copy.

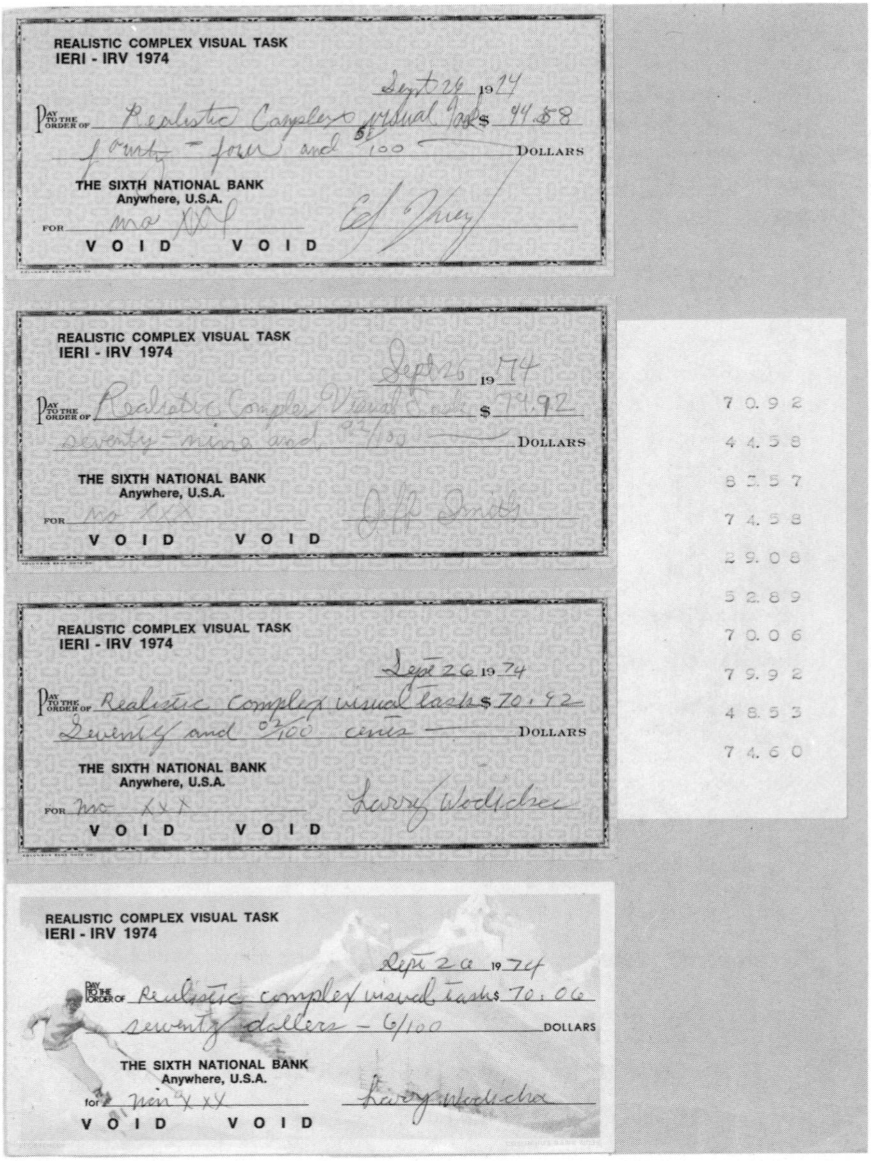

REALISTIC COMPLEX VISUAL TASK
IERI - IRV 1974
Sept 26 1974
PAY TO THE ORDER OF Realistic Complex visual Task $ 44.58
forty - four and 58/100 DOLLARS
THE SIXTH NATIONAL BANK
Anywhere, U.S.A.
FOR no XXX
VOID VOID

REALISTIC COMPLEX VISUAL TASK
IERI - IRV 1974
Sept 26 19 74
PAY TO THE ORDER OF Realistic Complex Visual Task $ 79.92
seventy-nine and 92/100 DOLLARS
THE SIXTH NATIONAL BANK
Anywhere, U.S.A.
FOR no XXX
Jeff Smith
VOID VOID

REALISTIC COMPLEX VISUAL TASK
IERI - IRV 1974
Sept 26 19 74
PAY TO THE ORDER OF Realistic complex visual task $ 70.92
Seventy and 02/100 cents DOLLARS
THE SIXTH NATIONAL BANK
Anywhere, U.S.A.
FOR no XXX
Larry Wodicka
VOID VOID

REALISTIC COMPLEX VISUAL TASK
IERI - IRV 1974
Sept 2a 19 74
PAY TO THE ORDER OF Realistic complex visual task $ 70.06
seventy dollars - 6/100 DOLLARS
THE SIXTH NATIONAL BANK
Anywhere, U.S.A.
for no XXX
Larry Wodicka
VOID VOID

70.92
44.58
85.57
74.58
29.08
52.89
70.06
79.92
48.53
74.60

Figure 13.25 Simulated office task: variation in visibility of handwritten checks being compared with adding machine tape.

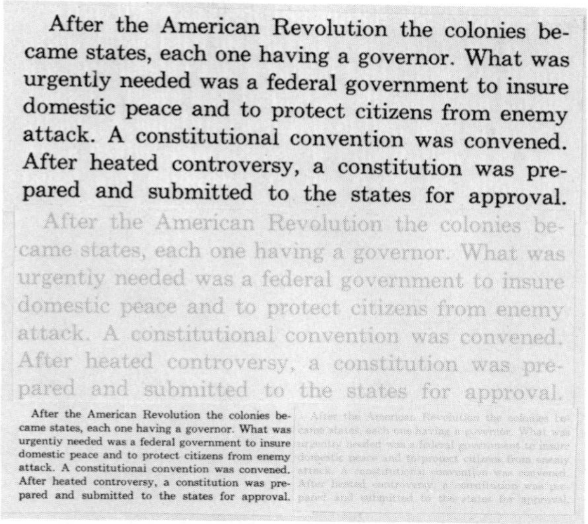

After the American Revolution the colonies became states, each one having a governor. What was urgently needed was a federal government to insure domestic peace and to protect citizens from enemy attack. A constitutional convention was convened. After heated controversy, a constitution was prepared and submitted to the states for approval.

After the American Revolution the colonies became states, each one having a governor. What was urgently needed was a federal government to insure domestic peace and to protect citizens from enemy attack. A constitutional convention was convened. After heated controversy, a constitution was prepared and submitted to the states for approval.

After the American Revolution the colonies became states, each one having a governor. What was urgently needed was a federal government to insure domestic peace and to protect citizens from enemy attack. A constitutional convention was convened. After heated controversy, a constitution was prepared and submitted to the states for approval.

Figure 13.26 Simulated office task showing variation in type size with contrast variability.

that identify the three-dimensional form. Some matte surfaces have slight wrinkles or ridges in them that do not show up with general overhead lighting but do appear when a light beam is directed very obliquely across the surface (Figure 13.36). Some objects have a surface gloss that overlays the grain or textured base material. The glossy reflections of light room surfaces near the line of sight interfere with the view of the texture beneath. In this case, light directed to the surface obliquely eliminates the reflections from the line of sight and penetrates the surface to illuminate the texture beneath. The metal working industry largely involves glossy surfaces that act like mirrors. In this case, the use of large, low brightness overhead luminaires serves to reflect low brightness highlights in the surface against which are seen any markings as interruptions of the highlight (Figure 13.37).

For the lighting solution of many industrial seeing tasks, one should refer to the American National Standard Practice for Industrial Lighting A 11 (21).

5 VEILING REFLECTIONS

In offices and similar locations where paper work is done, it has long been recommended that matte paper be used to avoid glossy or mirrorlike reflections that overlay the details to be seen and interfere with visibility. Research since the latter 1950s (22) has discovered that matte paper work continues to exhibit microscopic mirrorlike reflections. Overhead lighting with a heavy downward component of light rays (Figure 13.38) causes serious loss of visibility because of veiling reflections in the tiny facets of the paper fibers and the inks or pencil marks used for the printed or handwritten characters

DAVIS READING TEST SAMPLES

Scientists insist that, from their point of view, many musical instruments are faulty. For example, the violin is made of wood, is very fragile, and is affected by temperature. An unnecessary complication is that the player has not only to determine the pitch of a given note, but at the same time must control the quality. It is not adapted for mass production and, even though expensively made by hand, may turn out defective.

The same objections hold for the cello, and more so for the double bass, which doesn't really belong to the violin family, being the sole survivor of the otherwise obsolete family of viols. The viola is even worse. It is mechanically absurd, having a set of strings that is too long for its body, with the result that it emits a rather hoarse, hollow tone.

The piano is the worst of all. It, too, is made largely of wood. But worse than that, it not only gets out of tune easily, but even when it is in tune, it's out of tune. That is because it is tuned to the so-called tempered scale. This is a technical matter, which I wish to postpone for a while. Let us first consider some more criticisms of music, musicians, and musical instruments.

46. Which one of the following lists of musical instruments puts them in *ascending* order of merit, according to the scientific point of view?
A Violin, cello, viola, double bass
B Violin, double bass, viola, piano
C Piano, viola, double bass, cello
D Piano, double bass, viola, violin
E Cello, double bass, viola, piano

Scientists insist that, from their point of view, many musical instruments are faulty. For example, the violin is made of wood, is very fragile, and is affected by temperature. An unnecessary complication is that the player has not only to determine the pitch of a given note, but at the same time must control the quality. It is not adapted for mass production and, even though expensively made by hand, may turn out defective.

The same objections hold for the cello, and more so for the double bass, which doesn't really belong to the violin family, being the sole survivor of the otherwise obsolete family of viols. The viola is even worse. It is mechanically absurd, having a set of strings that is too long for its body, with the result that it emits a rather hoarse, hollow tone.

The piano is the worst of all. It, too, is made largely of wood. But worse than that, it not only gets out of tune easily, but even when it is in tune, it's out of tune. That is because it is tuned to the so-called tempered scale. This is a technical matter, which I wish to postpone for a while. Let us first consider some more criticisms of music, musicians, and musical instruments.

46. Which one of the following lists of musical instruments puts them in *ascending* order of merit, according to the scientific point of view?
A Violin, cello, viola, double bass
B Violin, double bass, viola, piano
C Piano, viola, double bass, cello
D Piano, double bass, viola, violin
E Cello, double bass, viola, piano

DAVIS READING TEST material, showing black-and-white and degraded copy.

Figure 13.27 The Davis Reading Test. *Right*: with poor reproduction; *left*: in original form.

reducing the contrast. Such microscopic reflections cause loss of visibility in accordance with Figure 13.39. If one were to work in the shade of a tree (Figure 13.40 *a*, 13.40*b*), these glossy reflections would be eliminated and light would come to the work from the sky beyond the tree. Reflections of this light would go away from the eyes of the worker, and the details would be seen ideally. A good compromise has been to reduce the downward components and have the light come from the ceiling and side walls as though they represented the hemisphere of the sky without sun (Figure 13.40*c*). Therefore the current recommendations for office tasks (20) are 70 to 200 footcandles "equivalent sphere illumination" (ESI) (spherical vault of the sky). If only black on white material is being used, the recommended level is 30 footcandles ESI. However a recent survey of the range of office tasks (23) indicated that a large majority were of poor contrast. Therefore 70 to 200 footcandles ESI, depending on the contrast and size of the details, should be used.

A veiling reflection study (24) indicates that commonly used luminaire systems have only 15 to 50 percent effectiveness in revealing the details of the task. Therefore the lighting system should provide a minimum of downward light immediately over the

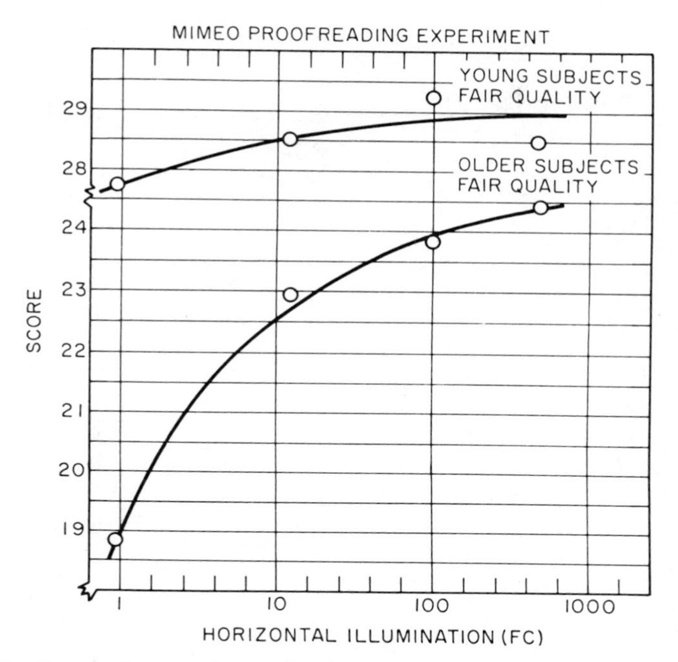

Figure 13.28 Results of mineograph proofreading experiment of Figure 13.24. Score is a combination of speed and penalty for errors.

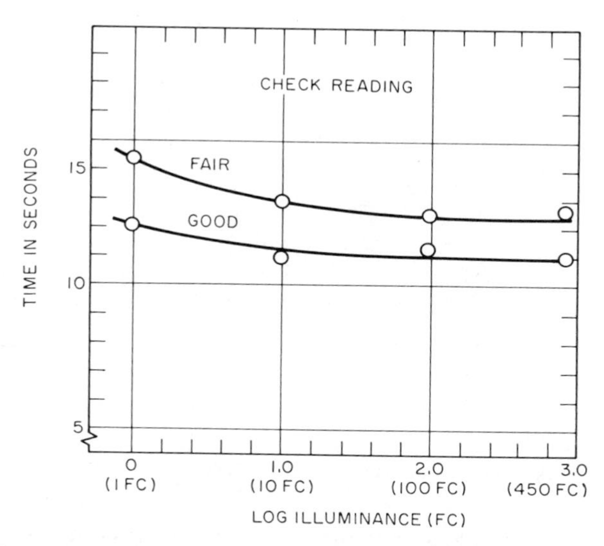

Figure 13.29 Results of check reading experiment of Figure 13.25.

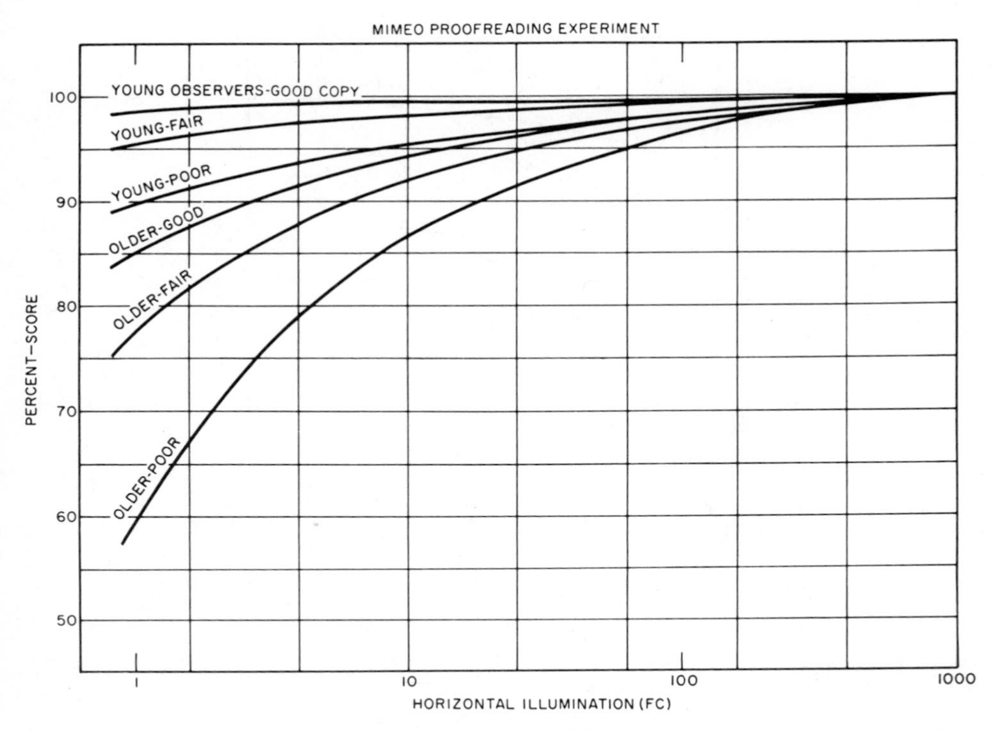

Figure 13.30 Relative results of "mimeograph" proofreading experiment from Figure 13.24 showing effect of quality of mineographed material and age of observers.

desks, in comparison to the light coming from wider angles (25, 26), which does not cause glossy reflection toward the eye (Figure 13.41). Study of the angular reflection of a typical task (pencil handwriting) has resulted in a determination of the effectiveness of light coming toward the task from various angles (27). This effectiveness is represented in Figure 13.42, where the length of the radial lines indicates the relative effectiveness of light approaching the task in revealing the details of that task. This average effectiveness holds for light approaching from every direction of the compass (i.e., 360° in azimuth including body shadow). It will be noted that the maximum occurs at 75° with the vertical, indicating that the light from walls and windows is very important in rendering more effective visibility of the task. (The reflectance of the walls should not be greater than 60 percent to avoid glare.)

6 EFFECT OF SURROUNDINGS: TRANSIENT ADAPTATION

The eyes are constantly scanning both the details of the task at hand and the immediate surroundings of the rest of the room. Time-lapse photography of office workers shows

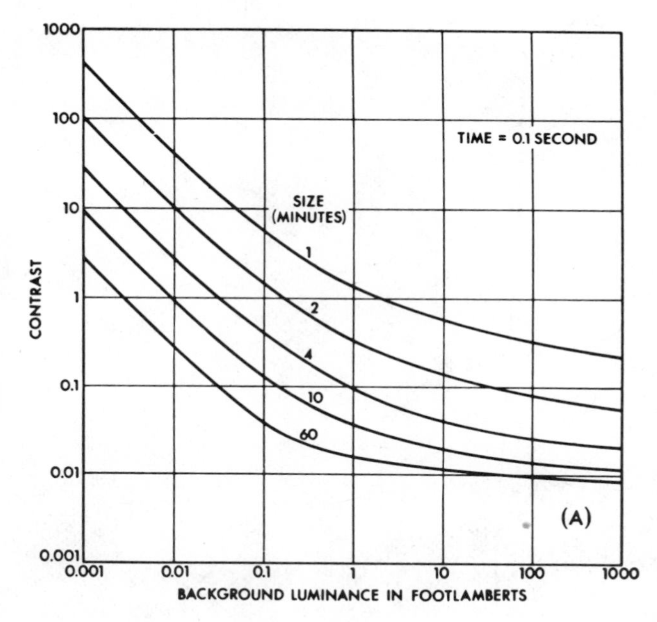

Figure 13.31 Relation between threshold contrast and background luminance for objects of various sizes (5).

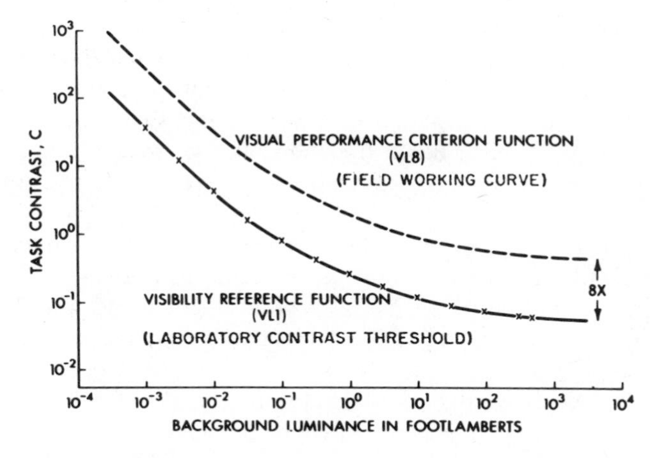

Figure 13.32 Plot of the Visibility Reference Function (solid curve) representing task contrast required at different levels of task background luminance to achieve threshold visibility for a 4′ luminous disk exposed for 0.2 second.The dashed curve represents the visual performance criterion function (values of the solid curve multiplied by 8).

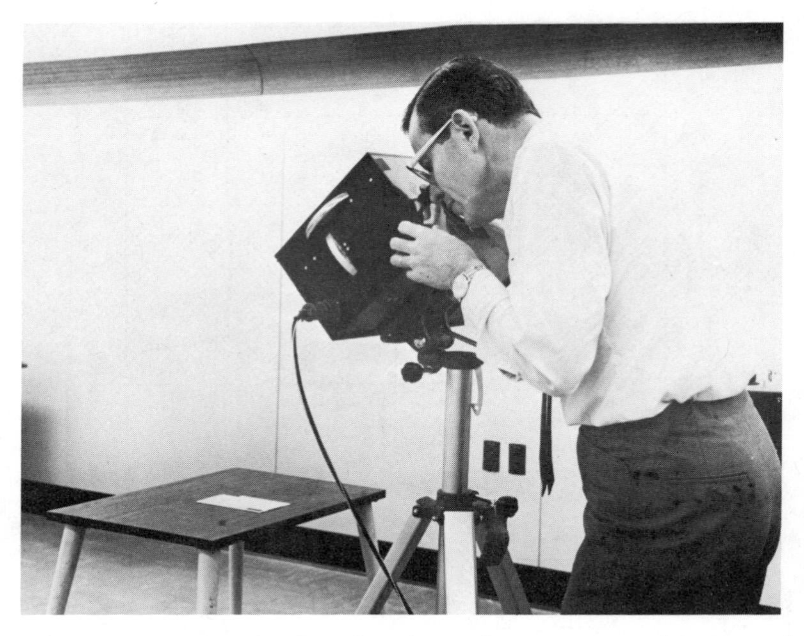

Figure 13.33 The Visual Task Evaluator (VTE) is designed to use 0.2 second exposures that permit more realistic field measurements.

that when persons look up from their desk tasks, the peak of the frequency of their angles of viewings is at 90°, or a horizontal line of sight (28). Thus they look around the room and then back at their desk tasks with their immediate surroundings. This involves looking at various surfaces with their different luminances (brightnesses) as in Figure 13.43. These sudden changes from one luminance to another causes a serious loss of visual sensitivity or visibility unless the luminances are kept within a suitable range of each other (29). The additional contrasts necessary to compensate for losses of sensitivity are plotted in Figure 13.44.

The time-lapse studies show that the relation between the luminance of the background of the details of the task to its immediate and more remote surroundings affects the ability to see the task after looking away from it (Figure 13.45). Furthermore, they demonstrate that the smaller the difference between the brightness of the visual surroundings and that of the visual task, the higher the efficiency of seeing. However equal brightness in all areas of the room would result in a marked feeling of monotony. Investigation has shown that the immediate surroundings should be less bright than the task, but not less than one-third as bright. The luminances of other areas in the field of view should not be less than one-third that of the task and should never be greater than 5 times that of the task. These ratios apply to large areas of the surroundings and can be achieved by the use of ordinary lighting equipment and light-colored room surfaces. (Small areas with accent color do not interfere with the transient adaptation.) Every

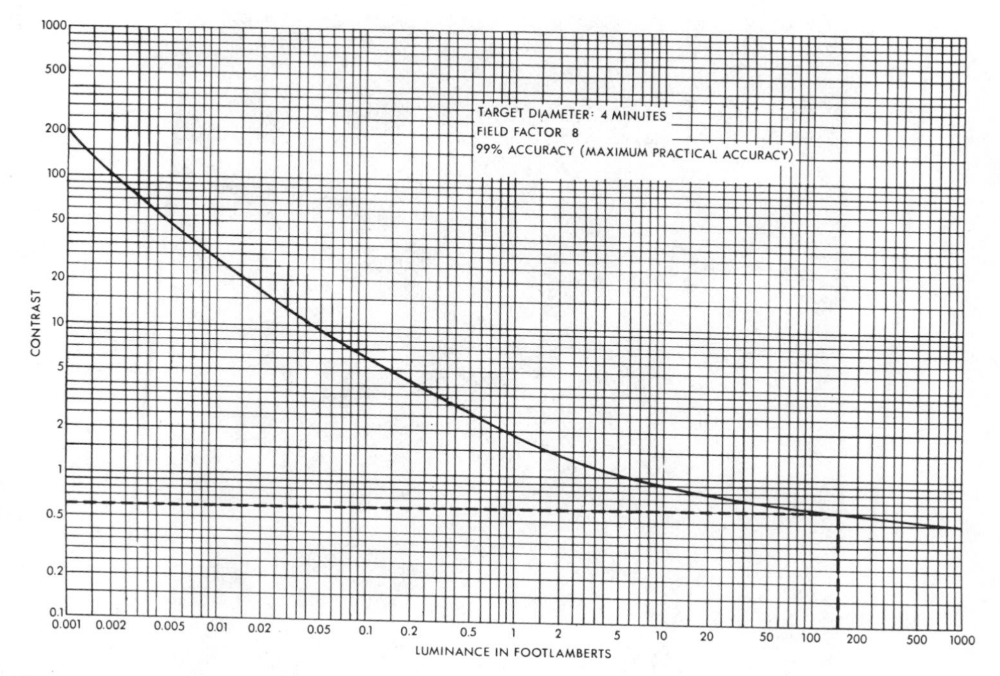

Figure 13.34 Relationship for the standard 4′ target (0.2 second exposure time, 99 percent accuracy, and a field factor 8.00) to which practical tasks ae equated.

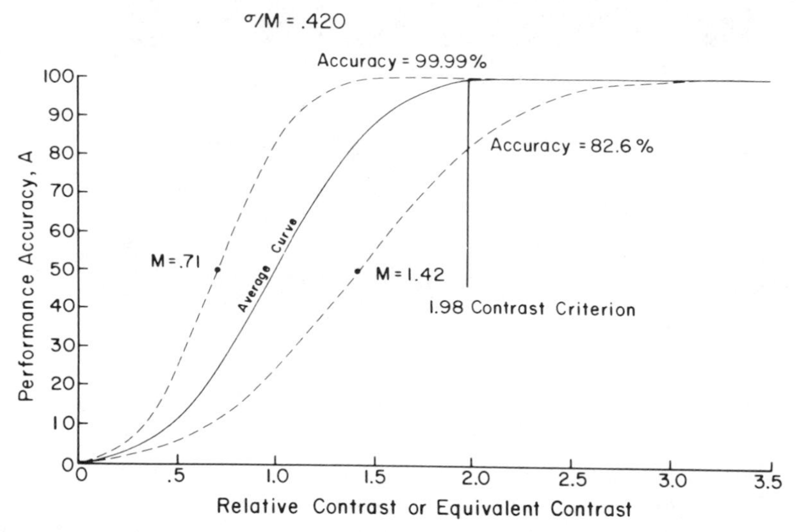

Figure 13.35 Variations in individual accuracy due to variations in *M*.) dashed curves represent limits of individual variability (i.e., limits of young population).

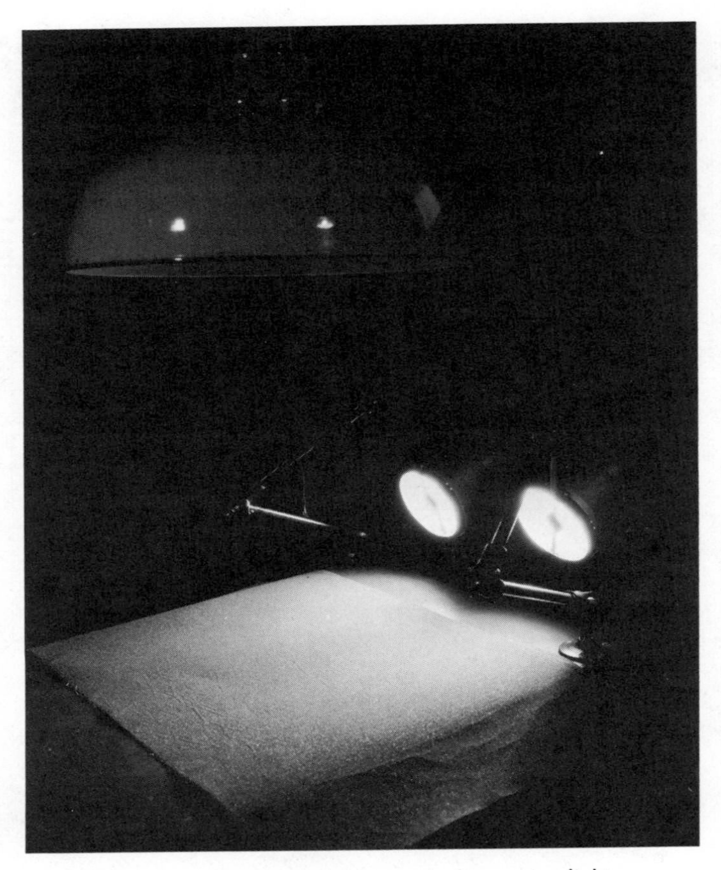

Figure 13.36 Ridges show up under grazing light.

object reflects some portion of the light it receives, and the percentage reflected is known as *reflectance*. The reflectances of floor and furniture should be in the range of 30 to 50 percent, the walls 40 to 60 percent, and the ceiling 70 to 90 percent (Figure 13.46).

7 ELIMINATION OF DISCOMFORT GLARE FROM OVERLY BRIGHT LIGHTING SYSTEMS

Recently it has been found that the feeling of discomfort is generated in the sphincter muscle controlling the opening of the iris of the eye (30). When one looks up from his or her work and encounters an overly bright lighting system, the sphincter muscle closes down the iris for protection. If one continues to look near the horizontal, the sphincter muscle first closes down too much, then the system desires more light. It opens up too much, and the luminaire overbrightness causes it to close down again. This alternation continues and the phenomenon is called "hippus."

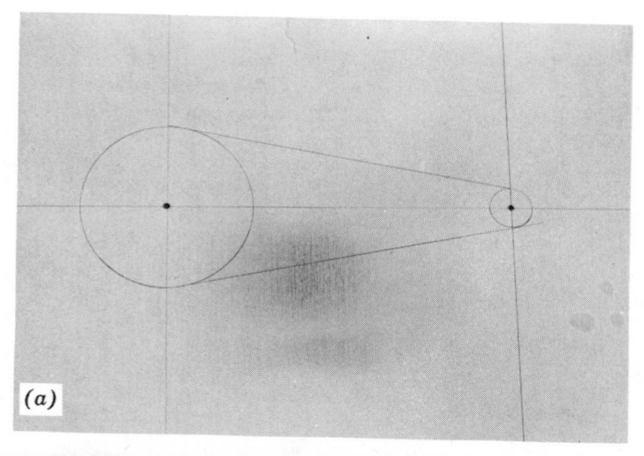

Figure 13.37 (*a*) Appearance of scribed lines on steel surface under large area, low brightness lighting equipment. (*b*) Highlight covering metal surfaces reveal contours and details. (c) Grid system of fluorescent produces broad highlights on metal surfaces.

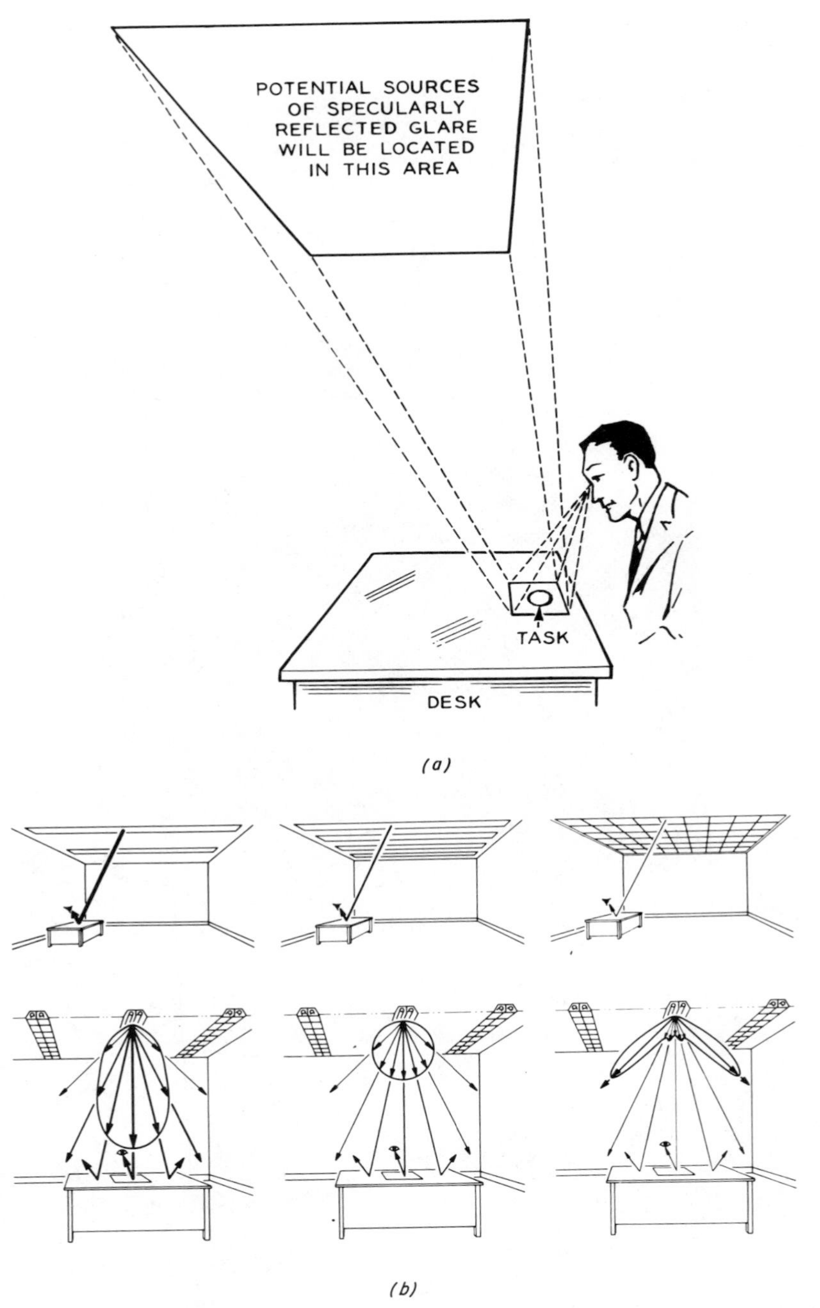

Figure 13.38 (a) Downward light from luminaires in this zone will cause veiling reflections and serious loss of visibility of office tasks. (*b*) Different luminaire downward components. The less the downward component in the reflection zone of the desk task, the greater the visibility of the task.

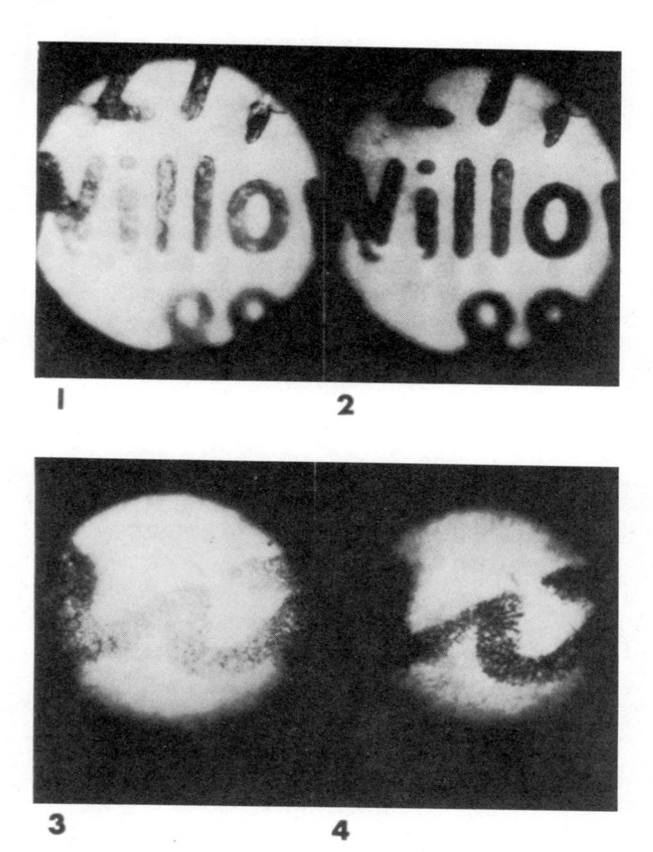

Figure 13.39 Veiling reflections reduce task contrast. (1) Magnified sample of type under a troffer causing veiling reflections. (2) Same sample with veiling reflection shielded out. (3) and (4) Similar situation with letters "a r t" (parts only shown) handwritten in pencil.

An age effect on the limiting luminance at the borderline of discomfort glare has been determined (31). It appears that on the average, the limiting luminance varies inversely as the age of the observer. This may require greater protection for older workers than the current recommendation.

Research on the sensitivity of people to peripheral brightness has developed a formula for discomfort glare that takes into account the luminances of the lighting units, the position of the lighting units in relation to the line of sight, the additive effect of a number of luminaires in the field of view, and the range of sensitivity of the population (32). The result of this research has brought about a standard known as visual comfort probability (VCP) for offices and classrooms. A VCP of 70 means that the visual comfort probability will be such that 70 percent of the occupants at the end or side of the room will be satisfied when the luminances of the luminaires are limited to a given

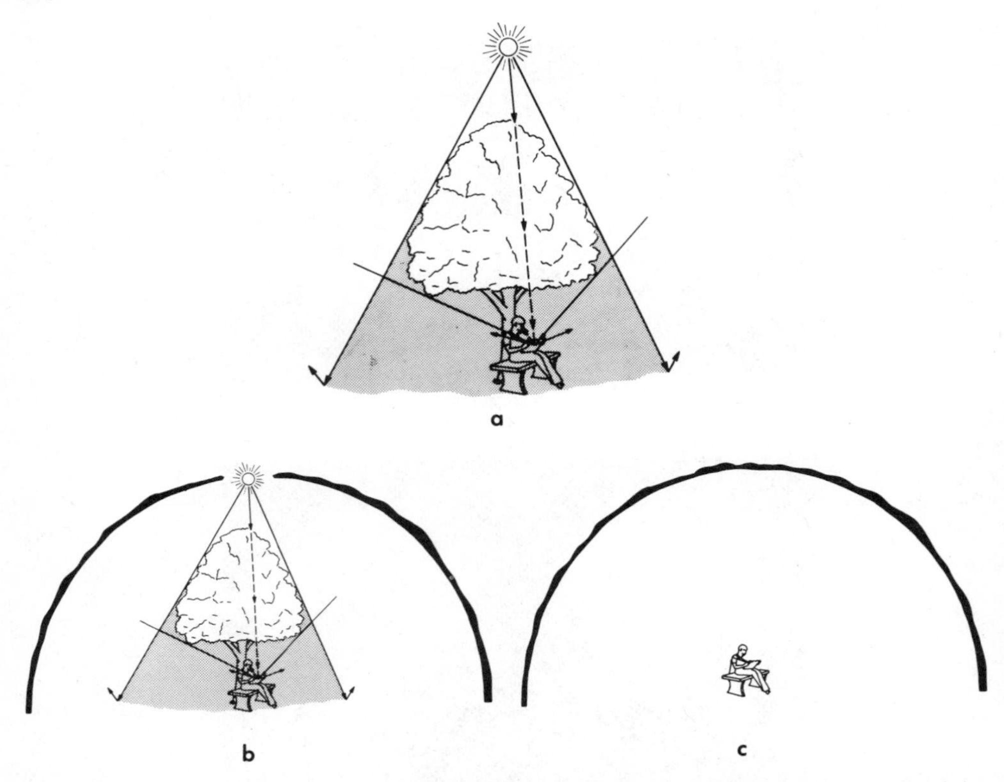

Figure 13.40 (a) Tree shields out veiling light. (*b*) Lights to reading task comes from vault of sky outside veiling zone. (c) Vault of sky with no sun produces good results with low component from the veiling zone overcome by majority of light outside the veiling reflection zone.

value. From a layman's practical viewpoint, this means that when the lighting system is suddenly exposed to view 70 percent of the observers (Figure 13.47) will obtain no shock of overbrightness.

8 DISABILITY GLARE: LOSS OF VISIBILITY

Various researchers (33) have found that a high brightness source away from the line of sight had the same effect in reducing visibility as a uniform veil of brightness overlying the details to be seen on the line of sight. This equivalent veiling luminance is expressed as $L_v = 10\pi[E_1\theta_1^{-2} + E_2\theta_2^{-2} + \cdots + E_n\theta_n^{-2}]$

where L_v = equivalent veiling luminance (footlamberts)
E = illumination on a plane through the center of the pupil perpendicular to the line of sight contributed by each glare source (footcandles)
θ = angular displacement between the line of sight and each glare source

If L_v is put in the formula for the resulting contrast of the object to be seen, we arrive at the resulting contrast:

$$C' = \frac{L_b - L_0}{L_b + L_v}$$

where C' = contrast under glare effect
L_b = luminance of background of detail
L_0 = luminance of the detail

The original contrast without glare is expressed as follows:

$$C = \frac{L_b - L_0}{L_b}$$

The effect of surroundings around the immediate background of the details was presented in Figures 13.7, through 13.9. When the surroundings became brighter than the background of the task detail, the visibility decreased rather rapidly with this increased luminance. This is a disability glare effect. Even with uniform and equal surroundings field, there is a 7 percent loss, but this is better than a dark surrounding field. As an illustration of the disability glare effect even with comfortable surroundings, consider that one cannot see inside a mountain tunnel in bright daylight. As one approaches the mountain with its face illuminated in the sun, the tunnel looks like a dark hole with no details visible. If one could look through a tube at the tunnel opening,

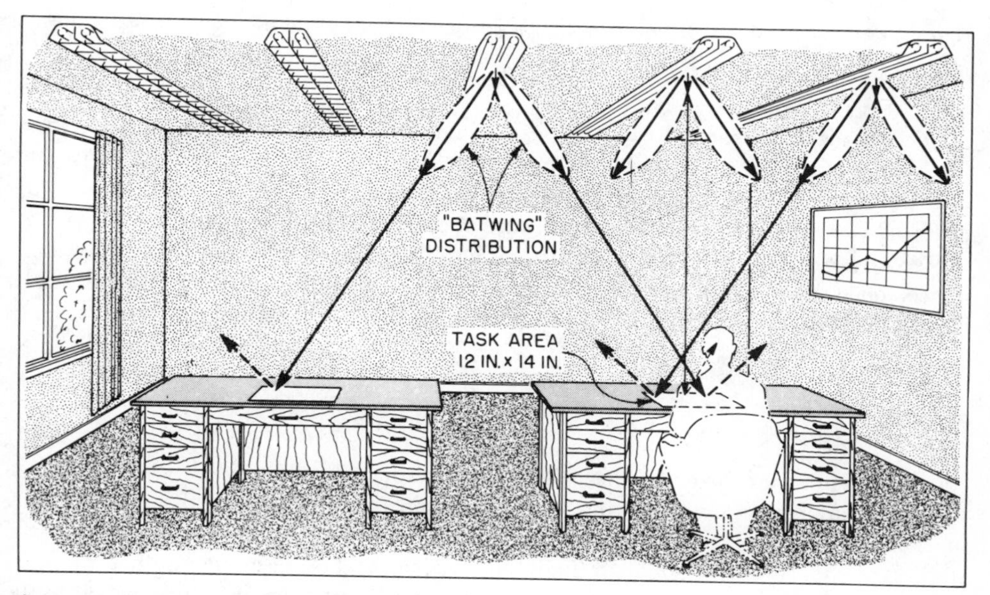

Figure 13.41 Batwing candlepower distribution of light from the luminaire minimizes light in veiling zone, distributes it from other luminaires outside the zone, and reflects it from the walls.

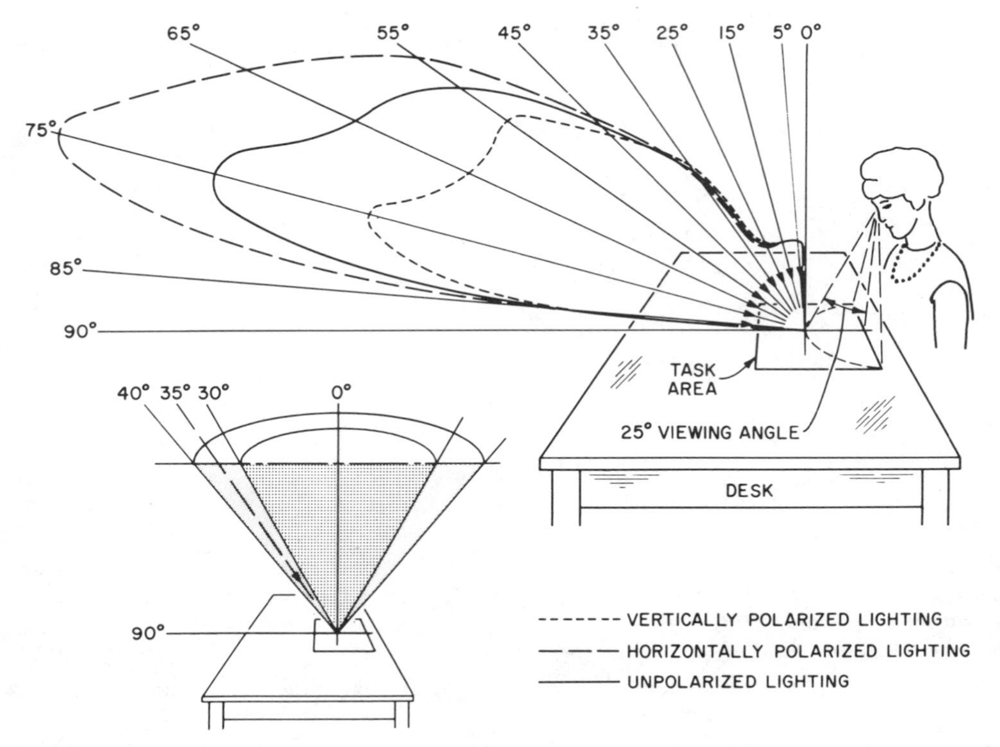

Figure 13.42 Average effectiveness of light approaching the task in different angular cones, of 360° azimuth.

the details inside the tunnel could be seen. If one could look out at the sunlit sky from the bottom of a well, one could see the stars still shining. This form of apparently comfortable lighting can be a real danger in dark industrial processes and out on the roadway where vital cues are missed because of overly bright surroundings.

Recently it has been discovered that age has a significant effect on the veiling luminance, since there was a threefold difference between younger and older observers (34). This probably will result in a factor of age being put in the L_v formula.

9 AGE

Age is a significant factor in the ability to carry out visually guided activity. This is extremely important in safety and accurate productivity. The age span is increasing markedly, as Figure 13.48 indicates. There has been a recent bringing together of studies of the effect of aging on light and vision. It has been found that the eyes deteriorate considerably in their ability to adjust the pupil opening in proportion to the light available. Figure 13.49 shows that this eye gate becomes smaller and smaller, even in the dark. The mean difference in pupil size between ages 20 to 29 and 80 to 89 was 2.6 mm in the dark, and 1.7 mm at a level of one footlambert. The reduction in pupillary area with increasing age is equivalent to approximately a 0.3 log unit reduction in luminance. This decrease in the amount of light reaching the retina for image-forming purposes is one of the underlying causes of loss in visual sensitivity with age (36). Critical detail in commerce and industry involves small size, and in many cases, poor contrast. Figure 13.50 demonstrates that the ability to see fine detail is seriously reduced with increasing age. Another closely related factor is the speed of ability to see small objects, to be able to maintain safety and accurate production. Weston (38) found that this performance speed was radically reduced with age (Figure 13.51).

A most vital factor in maintaining safety is the ability to see movement of objects representing a potential hazard out of the corner of the eye. Wolf (39) has found that

Figure 13.43 Marked differences in luminance in the field of view cause serious momentary losses of visibility, thereby scanning sensitivity.

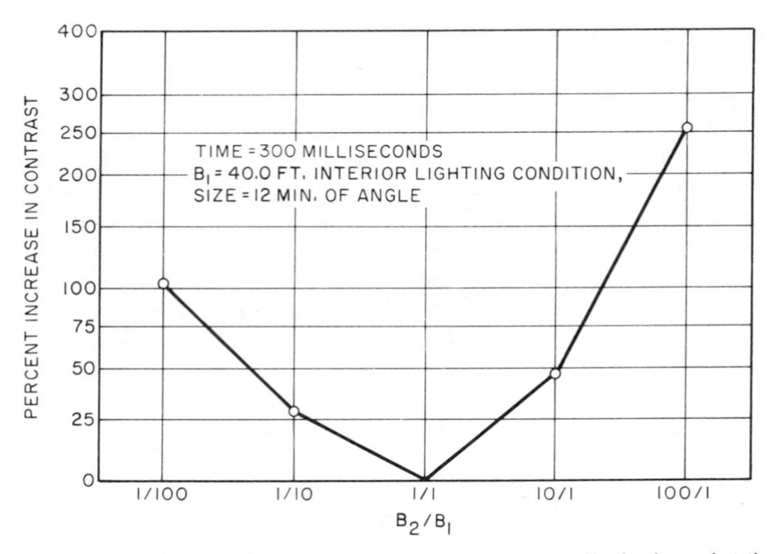

Figure 13.44 Percentage increase in contrast necessary to compensate for loss of visibility due to change from 40 footlamberts to 1/10 or 1/100 or to 10 or 100 times that luminance.

the ability to see flicker (movement) 40 to 80° out from the line of sight is reduced as much as 60 percent (Figure 13.52).

A young person sees very keenly in the presence of very little light, but this sensitivity to low brightness is reduced to approximately one-half at 80 to 85 years of age as shown by McFarland et al. (Figure 13.53). Furthermore, the same investigators found (Figure 13.54) that it takes 30 minutes for the 85-year-old to adapt to lower outdoor night

Figure 13.45 Serious loss of scanning sensitivity occurs when one encounters dark surrounding (or too bright surroundings).

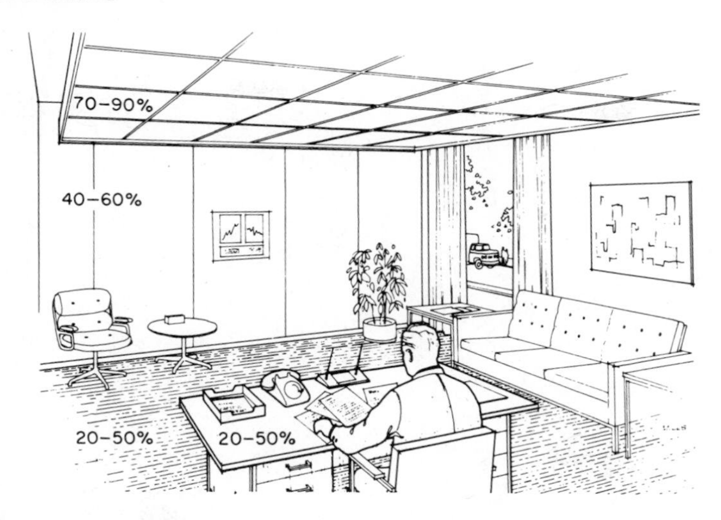

Figure 13.46 Reflectances of room surfaces to favor good scanning sensitivity.

Figure 13.47 Practical test for uncomfortable lighting units. (*a*) Stand or sit at end of room, looking straight ahead. Shade eyes so that no units are seen and hold for a few moments for adaptation. (*b*) Suddenly remove hand; if there is a feeling of overbrightness, the lighting units are too bright for comfortable seeing.

b

Figure 13.47 (Continued)

brightnesses. The shape of the curve shows that the older person would be partially blind for awhile after having gone outdoors at night.

Most startling of all is the loss of ability to see in the presence of glare sources. Unshielded or poorly shielded light sources with high candlepower directed toward the eyes cause severe losses of visibility, especially to older people (Figure 13.55). Yet inadequately shielded light sources are found in many industrial plants. This glare condition is found especially in night driving and in some sport arenas.

The increased need for more illumination for the older worker has been shown (Figures 13.28 and 13.30). Recent studies (42) have indicated that the working curve of contrast versus luminance (Figure 13.34) is displaced upward for every 10-year interval of age. The displacement is slight for the years 30 to 40, but it is increasingly larger with every succeeding decade of age (Figure 13.56). These displacements are in terms of static viewing (looking fixedly at detail to be seen), but studies are underway for determining the displacements for dynamic viewing such as the scanning of detail being car-

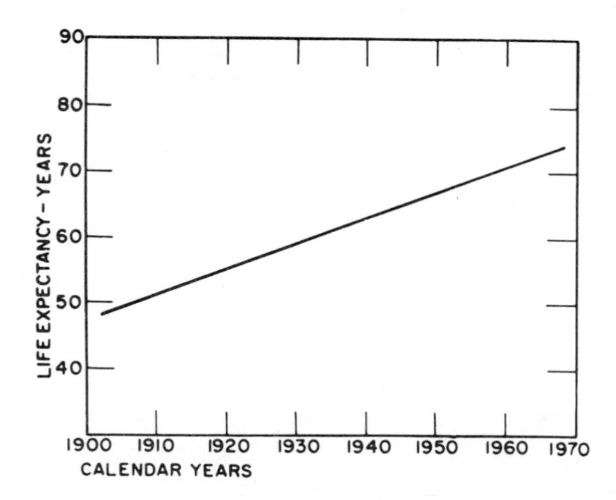

Figure 13.48 Life expectancy since 1900 (35).

ried out in commerce and industry. If the static viewing is taken as a possible criterion, the additional illumination needed would be as shown in Figure 13.57.

10 MOTIVATIONAL FACTORS

Architects, designers, and artists have been using light as an expression of grace, beauty, and art since the days of ancient Greece. First because of the lack of suitable man-made

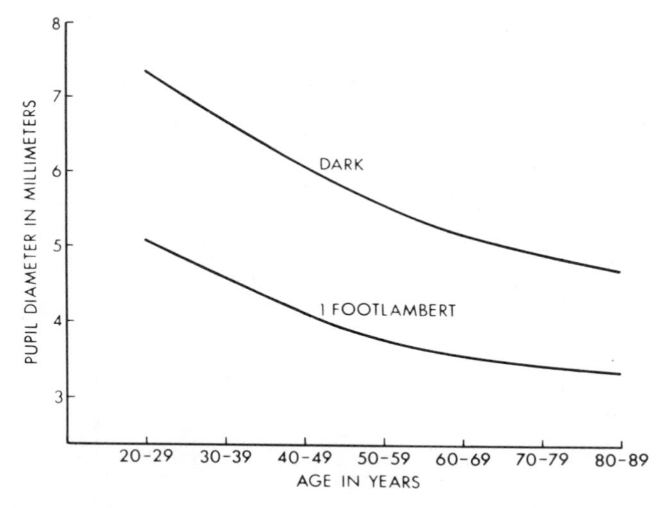

Figure 13.49 The change in pupil diameter with age as measured in the dark and at one foot-lambert for 222 observers (36).

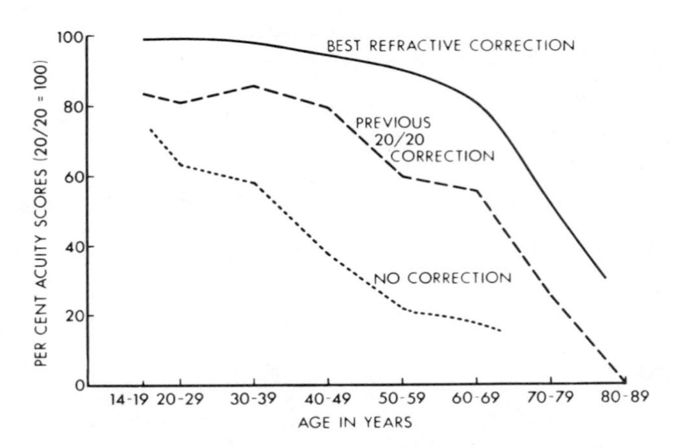

Figure 13.50 The reduction in visual acuity with age (37).

light sources, daylight, and particularly the sun, was used to create highlights and shadows, and to focus attention. Now a whole artistic palette of electric light sources is available to give highlights, shadow, diffusion, and color in innumerable patterns depending on the skill of the innovative designer. People are impressed with and artistically motivated by the beauty of the natural scenes about them, many scenes calling for activity, relaxation, or meditation. In the same way, the moods and behavior of people can be affected by the patterns of electric lighting combined with color and decor of reflecting surfaces. This concept opens up a whole vista of possibilities that seem to defy systematization short of the individual ingenuity of the designer. Some, however, have been inspired to make the attempt.

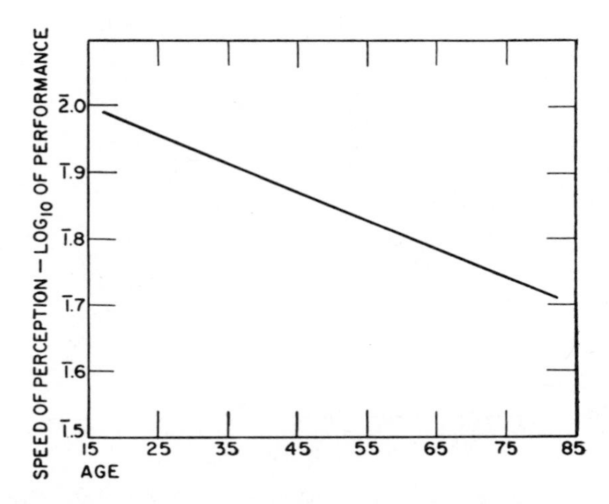

Figure 13.51 Speed of perception is reduced as the age of the eye increases.

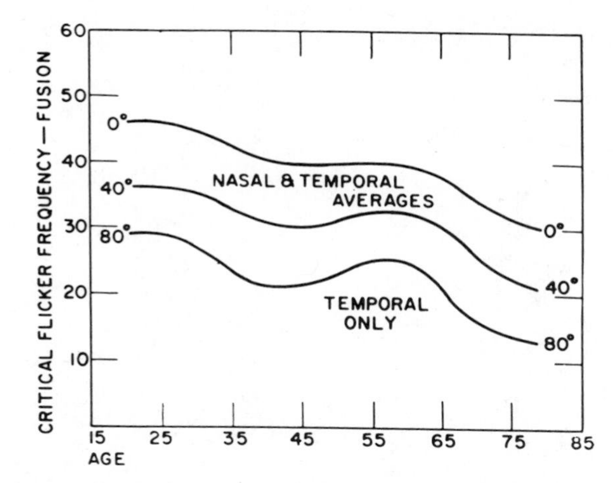

Figure 13.52 The ability to perceive movement at various degrees off center of view exhibits an undulating decline as age increases.

10.1 Color Appearance and Color Preferences

The Inter-Society Color Council has been a great aid in isolating and attacking color problems. Its leaders have been of great help to the Illuminating Engineering Research Institute on planning and executing researches in the application of color to the lighting field. As a result, Helson and Judd have made studies of change of color appearance from under average daylight to various electric lighting sources (43). From this study

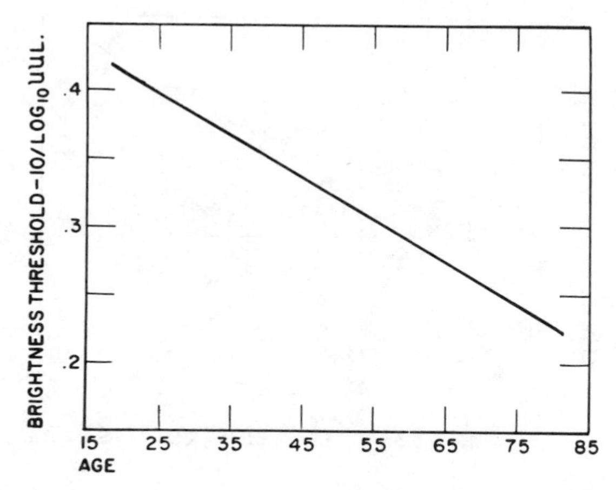

Figure 13.53 In moving from a bright to a dark area, the ability to see is sharply reduced with age (40).

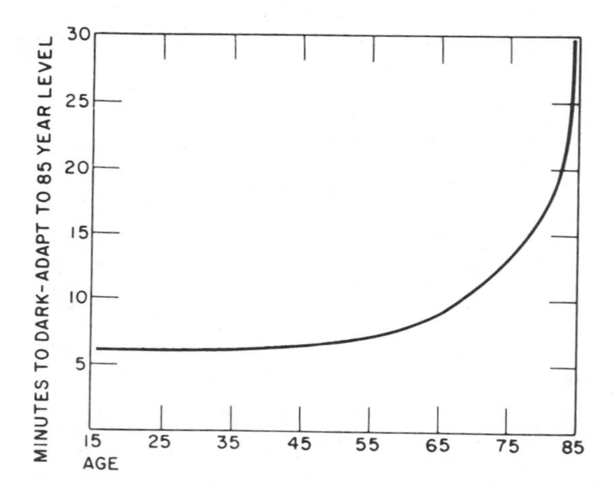

Figure 13.54 Eyes of "young normals" need time to adjust to brightness changes. Eyes aged 20 through 65 years gradually need more; eyes older than 65 years need a sharp increase (40).

and previous knowledge, an international system of indices for rating color rendering of various light sources was developed (44).

Helson and Lansford studied the preferences of a sample of population for various color combinations, using a psychological scaling on a rating of 1 to 10 of viewing samples of color space (125 Munsell color cards) against large Munsell-rated backgrounds (25 widely different chromatic samples). The researchers were able to make 156,000 judgments under five different current color emission light sources (average

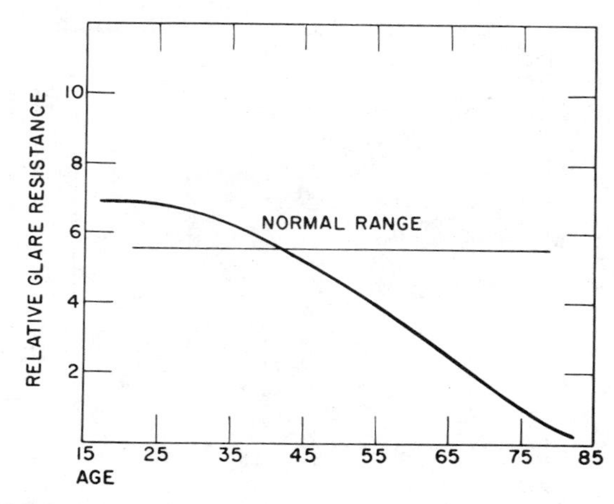

Figure 13.55 The older eye is much less able to resist glare that "washes out" the visual image (41).

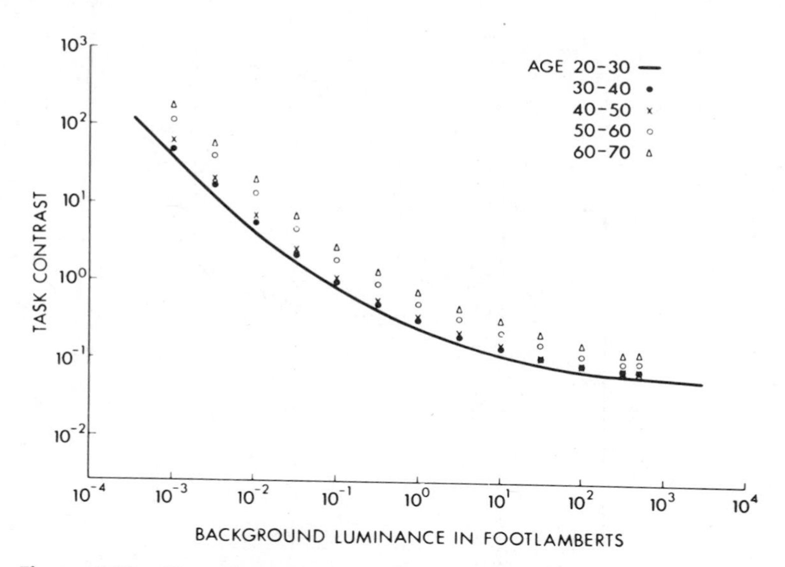

Figure 13.56 Effect of age on required task contrast for task performance.

daylighting to incandescent to three fluorescent lights having different color temperatures). Ratings above 5 indicate increasingly preferred color combinations, and those below 5 increasing lack of preference (45).

As a result of his studies and those of others, Judd developed a "flattery index" (46) for light sources in complimenting the appearance of human complexions and interior color schemes. This concept has been further developed by Jerome (47).

10.2 Helpful Color Tools

A series of standardized nomenclature, color measurement techniques, color specification systems, color temperature specification, colorimetry of light sources, color rendering, and color application has been developed to promote the effective use of color data. These tools are described in the current edition of the *IES Lighting Handbook* (48). An American Standard System of Specifying Color (D1535) designates the Munsell system as a standard of specifying color in terms of hue, value, and chroma. Collections of carefully standardized color chips in matte or glossy surface are available.* A very handy tool for the field is the "Munsell Value Scales for Judging Reflectance," which contain a good sampling of color chips from the color space (various hues, value, and chroma), permitting rough judgments of the color specification of unknown color in the field and its reflectance.

* Munsell Division of the Kollmorgen Corporation, P.O. Box 950, Newburgh, N.Y. 12550.

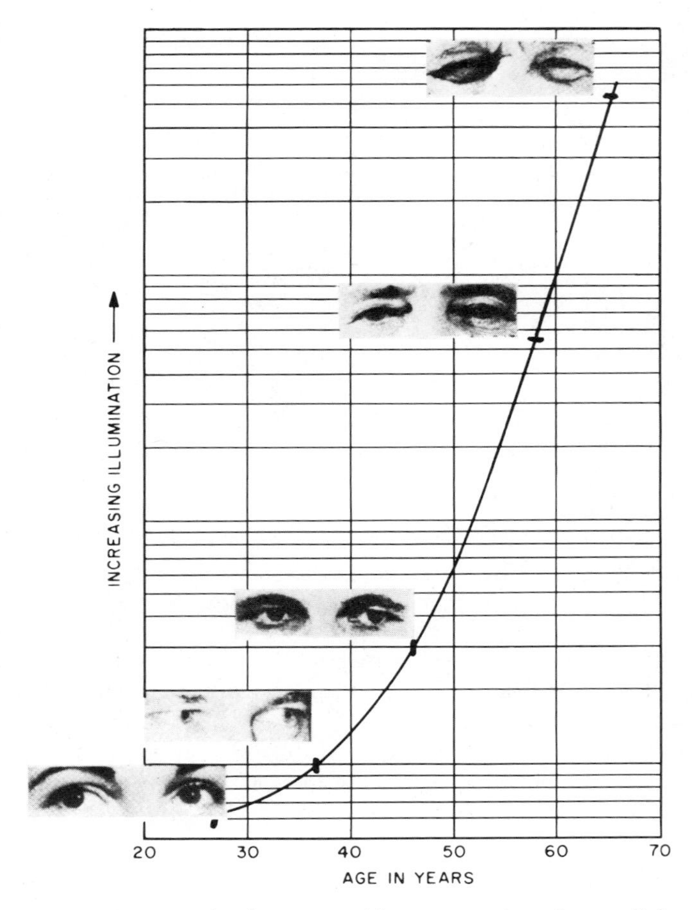

Figure 13.57 For fixed viewing, older eyes need much more light.

A handy guide for the application of color in the lighting field, produced by the Color Committee of the Illuminating Engineering Society, is entitled "Color and the Use of Color by the Illuminating Engineer" (49).

10.3 Subjective Feeling of Interiors

Every interior "has a feeling about it"—an impression, an impact that each occupant consciously or unconsciously senses. Naturally the senses record this impression, and the visual system shares a major responsibility. Architects and designers are constantly striving to create a stimulating and pleasurable experience, hoping for a reaction that will result in a behavioral pattern suitable to the function of the room. Psychological scaling has long been a method of rating people's feelings about objects or interiors, and a series of reactions is ranked on a numerical scale such as 1 to 10 or 1 to 100. This scaling procedure has now been developed into two approaches: semantic differential scaling (SDS) and multidimensional scaling (MDS). In semantic differential scaling a series of adjectives, both positive and negative, are used to describe the subjective characteristics of an interior. Such an approach employs a rating sheet with a series of positive adjectives on the left side and, the opposing negative adjectives on the right side, with a scale of 6 to 10 divisions in between. The observer, after becoming acclimated to the interior, is to mark the scale to indicate his reaction to each pair of opposite adjectives. For instance, is the interior "neutral" or "positive" or "negative"? If positive, how positive? Very positive or slightly positive? Or negative? How negative? By marking the scales in answer to these questions, the observer has recorded his impression of the feeling of the room. These answers are then put in a computer program, and the important characteristics are brought out. These now appear to be impressions of (1) visual clarity, (2) spaciousness, (3) complexity (or liveliness), (4) personal prominence or anonymity (public vs. private), (5) relaxation or tension, and (6) pleasantness.

Multidimensional scaling involves trying to determine the psychological dimensions that underly the judgment of differences or similarities between the impressions of different lighted interiors. In this approach one takes a given lighted room or lighted arrangement as representing a base figure such as 10 or 100. Then all other rooms or other lighting arrangements in a given room are rated as bearing a higher or lower value, and a computer program is set up to bring out the two, three, or four psychological dimensions that governed the judment of the higher or lower values.

Out of such studies (50) have come the following three dimensions:

1. The overhead–peripheral mode emphasizing vertical surfaces, unlike conventional overhead lighting, which emphasizes horizontal surfaces.
2. The uniform–nonuniform mode (sometimes specular–nonspecular) affecting the articulation or modeling of forms or objects in the room.
3. The bright–dim mode, affecting the perceived intensity of light on the horizontal activity plane.

The color tone of light tends to warp recognition of these dimensions. Light color might also be a fourth multidimensional scaling element if properly manipulated during the experiment. Glare appears to be another modifier of subjective ratings of light settings, but the evidence is ambiguous, and investigation continues.

Both SD and MD scaling provide reasonable consistency between spaces and between groups of subjects of similar backgrounds. Putting both methods together provides insight into the ways in which light settings affect judgments.

11 PHYSIOLOGICAL FACTORS

"Physiological" is defined as "characteristic of or appropriate to an organism's healthy or normal functioning." Therefore fatigue and health effects on organic responses not part of but related to the visual process, come under this classification.

11.1 Fatigue and Eyestrain

Early literature (1900–1920) indicated studies of fatigue (51) that dealt with low levels (less than 5 footcandles) of illumination. Between 1913 and 1924 Ferree and Rand (52) made studies of the degree of fatigue from different distributions of light (indirect versus direct), different spectral emission light sources, and different colored paper backgrounds. During the same early period there was much discussion of eyestrain, which appeared to be related to the low levels and poor quality of electric lighting. Weston (53), conducting a series of studies in industrial plants of different types for the British Industrial Fatigue Research Board, and later for the Industrial Health Research Board, became convinced that fatigue and eyestrain were due to poor lighting, and upon review of Sir Duke Elder, famous British ophthalmologist, described strain from visual factors, strain from oculomotor factors, and "posterial" eyestrain. Lancaster (54) describes eyestrain from faulty illumination as ocular discomfort and associated symptoms. Ocular discomfort shows up as "sandiness," tired aching feelings of the eyeball, orbit, or head. Associated symptoms appear in terms of headache and feelings of fatigue. If eyestrain is bad enough, it may have indirect symptoms of vertigo and digestive and psychic reactions. However both Weston and Lancaster were talking about levels comparatively lower than those being used in visual performance studies conducted by Blackwell and Smith. As a result there appears to be little or no recent discussion of eyestrain. On the contrary, some ophthalmolgists are saying that there is no such thing as eyestrain, which was medically called "asthenopia" by MacKenzie in 1843 and described in various ophthalmological textbooks throughout the years.

One of the forms of eyestrain mentioned by Weston is "eyestrain from oculomotor factors." Recently Hebbard (55) had a similar concept that there might be more oculomotor adjustment activity under lower levels of illumination than under higher levels. He discovered that the amplitude of the microsaccades was significantly greater for one footlambert luminance of the background than for 306 footlamberts (30 percent). Furthermore he found under sudden imposition of glare conditions (changing the surroundings from 0 to 306 footlamberts) that the frequency of the microsaccades increased significantly (200 percent). This would indicate that the oculomotor adjustment system has to work much harder under adverse lighting conditions; therefore, over a working period of one-half to one day's time, there would be fatigue, or under prolonged conditions, eyestrain such as was described in the earlier literature.

11.2 Ease of Seeing

There has always been a dramatic contrast between the levels of illumination of daylight and electric lighting levels. Weston (56) pointed out that lace makers took their tasks involving fine detail and poor contrast out of doors. Campbell et al. (57) described the early weavers' houses in Coventry fitted with tall windows for daylighting. They stated, "The eye is accustomed to the natural illumination provided by sunshine and outdoor daylight, and it is natural for us to seek the best possible daylight for the performance of work which is difficult to see." Sir Duke Elder (58) wrote:

In the care of the eyes the most important general principle is that they should have sufficient light whereby to work. It is to be remembered that the eyes of man were evolved to function in daylight; and although there is a popular superstition that artificial light is bad for them, the only fault lies in the fact that artificial light is rarely provided liberally or adequately distributed.

With the dramatic difference, it was natural for a feeling to arise that there was less fatigue (59) or a greater ease of seeing under daylighting and, by inference, under higher illumination levels. Luckiesh and Moss (60), by using a key for recording the completion of each page of reading (but really a pressure-sensitive key), were able to record the pressure exerted on the key during 308 half-hour sessions of 14 observers. The reading sessions were conducted under 1, 10, and 100 footcandles. The results showed a steady decrease of muscular tension as the results of the levels of illumination were examined (Figure 13.58). Glare was also introduced and made the nervous muscular tension greater. The same authors, in another series of experiments, used the blink rate as an index of mental tension (Figure 13.59) and found apparent consistent trends related to illumination levels, glare, severe ocular scanning, and simulated refractive errors (60). Others could not reproduce the same effects. Much of the work depends on the careful design of the lighting conditions, such as veiling reflections, subtle differences in discomfort glare, and unrealized transient adaptation effects (61). Some of the nonconfirming researchers were not aware of these effects, but knowledge of them has been developed in recent years (16).

11.3 Biological Effects

Steinmetz (62) stated in 1916 that illuminating engineering embraced not only light for seeing but a knowledge of the effect of radiation from light sources involving the physiologist. Gage (63) in 1930 reviewed the work of physiologists who had discovered in the 1920s that the overcoming of rickets, the strengthening of bones, the reduction of colds, and the treatment of tuberculosis all are affected by sunshine and simulated sources of sunshine such as mercury lamps. During the 1930s the lighting industry developed sunlamps S_1, S_2, RS, and later fluorescent sunlamps, which together with light control luminaires produced dual-purpose lighting—light for seeing and ultraviolet radiation in the zone found in sunlight from 280 nm to the visible region. Complaints regarding the use of the emerging fluorescent lamps for general lighting because of suspicions of harmful ultraviolet radiation (64) dampened the enthusiasm for dual-purpose lighting. Furthermore, some dermatologists became concerned with the carcinogenic effects of sunlight and urged that all sunlight be excluded. Recently medical discoveries and reemphasis on overcoming respiratory diseases have served to renew interest in the use of sunlight and electric simulated daylight (65, 66). Current light sources such as white or blue fluorescent lamps are being successfully used to overcome jaundice in the newborn, a condition that occurs in 15 to 20 percent of all premature infants. The portion of the spectrum responsible for photodegradation of bilirubin was found to be in the blue region (Figure 13.60). A leader in the United States in implementing the discovery of Cremer et al. (67) is Dr. Jerold F. Lucey (68), who first reported his work in 1968.

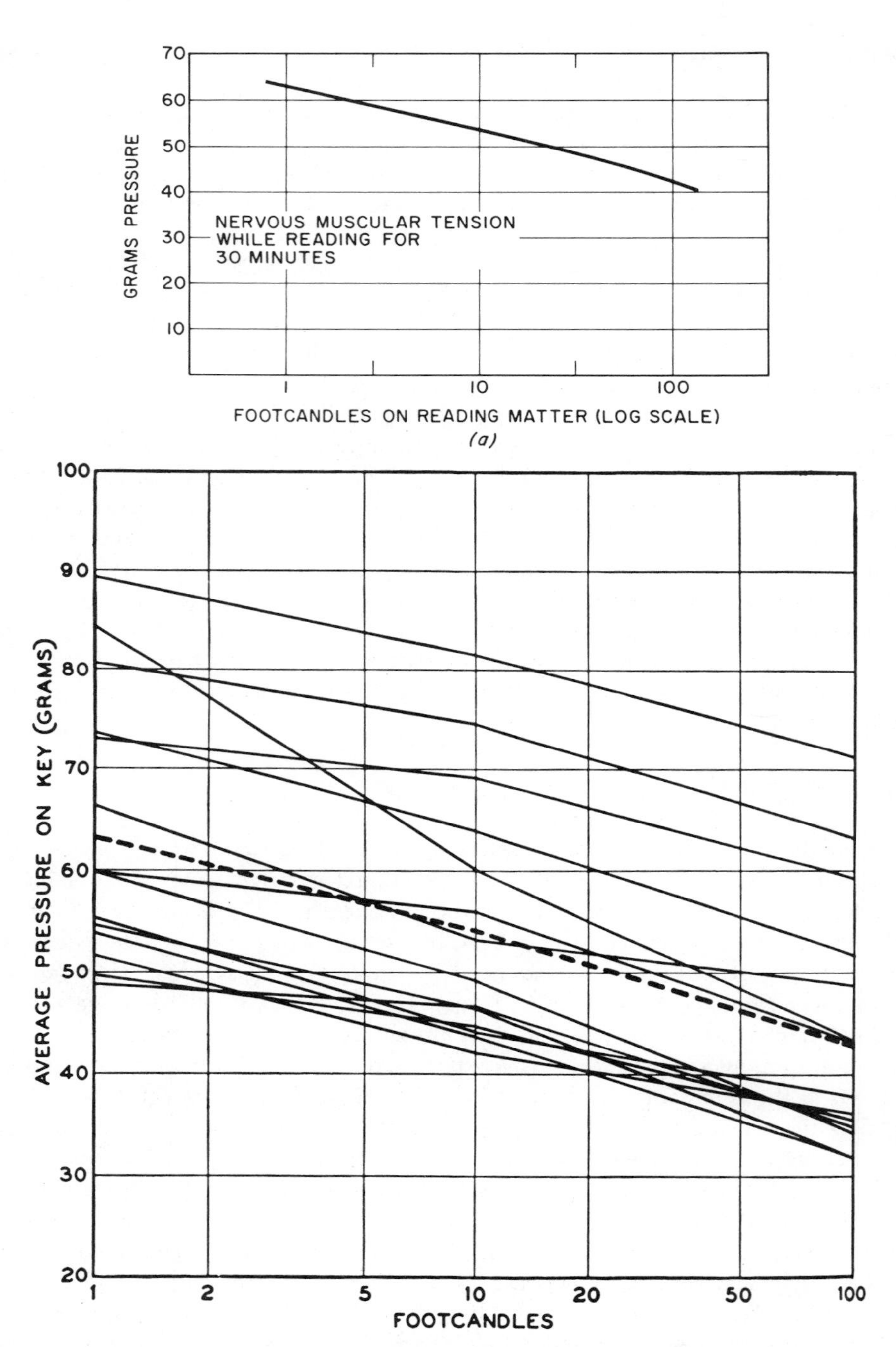

Figure 13.58 (*a*) Mean value of nervous muscular tension under three levels of illumination. (*b*) Relations between indicated nervous muscular tension and level of illumination for each of the 14 subjects; dashed line represents the geometrical mean of the results of the 14 subjects. The probable errors of the average values of nervous muscular tension are less than 1 percent.

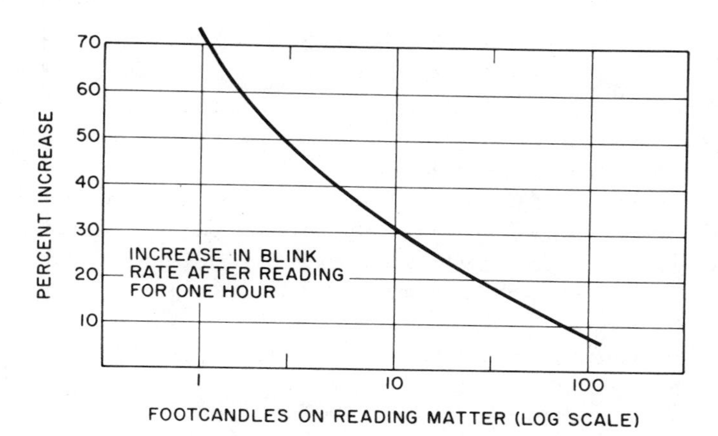

Figure 13.59 Increase in blink rate with lower levels of illumination.

Another study on calcium absorption to avoid osteoporosis was carried out by Neer et al. (69). It was found that the small ultraviolet component of currently available fluorescent lamps, when used in a level of 500 footcandles, stimulated significantly more calcium absorption. This effect appears to be related to the role of vitamin D_3 in curing rickets (70). When one considers that as people, especially women, grow older, they go out less and less for daylight exposure, and behind glass they receive no ultraviolet, it is little wonder that they absorb less calcium from their food and have weakness of bones (osteomalacia). Mechanical shock may break their bones, and the recovery is very slow or indefinite. Wurtman (65) describes a study by Jean Aaron in England whose autopsies in winter reveal more cases of osteomalacia than those in the summer.

Wurtman (65) further describes the evidence that light, its duration, and color affects the hormones, the bodily rhythms, and the sexual activity of animals and man. Recently it has been found that light has other curative values for disease. Psoriasis is an affliction characterized by the "eruptions of large rough red areas that are very itchy and disfigur-

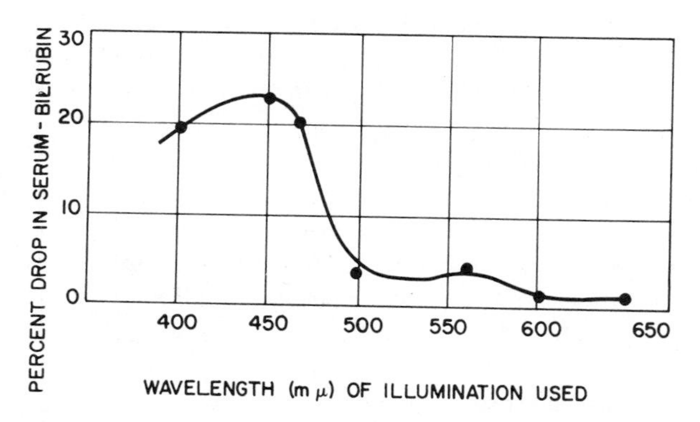

Figure 13.60 Wavelength effectiveness in reducing bilirubin in serum.

ing" (66). J. A. Parrish et al. (71) have developed a successful treatment consisting of ingesting a photoactive drug called Psoralen, and after a 2-hour time delay, irradiating the area of the psoriasis with ultraviolet near the visible spectrum, 320 to 400 nm (UV-A). The photoactive chemical is found in natural vegetable and fruits such as carrots, parsnips, celery, parsley, limes, lemons, and figs.

A very recent avenue of approach originated by Diamond (72) for treatment of cancer has used the property of the chemical hematoporphyrin, having been given in a dose, to stay in the malignant cells while clearing out of the normal cells. When hematoporphryin is excited with a beam of red light, it generates an excited state of the oxygen molecule (oxygen singlet), which destroys the cell in which it is located.

From all the studies above it would appear that light can very seriously affect the health and general welfare of man. Much more knowledge is needed in the establishment, through design, of the role of lighting for good health as well as good vision. All the evidence seems to indicate that ultraviolet radiation in the range of sunlight and daylight is beneficial in limited amounts but dangerous in unlimited use.

12 APPLICATION INFORMATION

12.1 Electric Light Sources*

The designer has a vast repertoire of available "bulbs and tubes" in various shapes, sizes, wattages, and colors. They vary all the way from a single wavelength of light (low pressure sodium) with its yellow light, to the blue-green of the mercury with its individual four-wavelength emissions, to the white multiwavelength emissions of metal halide and the orange high pressure sodium. They vary in efficacy from 8 lumens of light per watt of power (small incandescent) to 180 lm/w (low pressure sodium). There are two main categories of lamp: incandescent and gaseous discharge.

12.2 Incandescent Lamps

Hot tungsten wires that are high enough in temperature to glow with a pale yellowish white light are the basis of incandescent lights, which are available for general service in pear-shaped bulbs from 10 to 1500 W with an efficacy of 8 lm/W for the 10 W size to 24 lm/W for the 1000 W size. They are also available in parabolic and elliptical reflector bulbs. They are available in low voltages, medium voltages, and high voltages for special applications such as flashlights, bicycles, trains, locomotives, aircraft, and automotive vehicles. The life of the general service lamps varies from 750 to 1000 hours.

A longer life, concentrated incandescent lamp, called the tungsten halogen lamp, has been developed. In this design the filament is put in a small quartz tube with a negative-valence element such as iodine, bromine, chlorine, and fluorine. As this lamp burns, the

* Reference 2, Section 8, contains an excellent exposition.

tungsten gradually evaporates off the filament and normally would collect on the bulb wall as blackening; instead, it combines with the halogen that is in the quartz tube to form a gas. As this gas approaches the filament, it breaks down and deposits tungsten on the filament. These lamps maintain a good output over their life and last 1000 to 2000 hours depending on particular size and design. They are also available in reflector-type bulbs.

Architects and many designers like to use incandescent lamps because of the complimentary effects on human complexions and on warm colors. They are an extension of candlelight and the light of the fireplace. The emitted light emphasizes the yellows and reds but deemphasizes the blues and greens.

12.3 Fluorescent Lamps

Fluorescent lamps are essentially a mercury gaseous discharge in a long tube, coated inside with phosphor powder. The phosphor powder reacts to the invisible mercury emission spectral line of 253.7 nm and converts the invisible energy into visible light. By using various chemical phosphors usually in a mixture, colors of visible light are emitted, and the emission becomes a continuous spectral output. Aside from particular spectral color emphasis such as blue, green, gold, pink, and red, the phosphor mix most commonly used is such that the emission tends to follow the CIE (International Commission on Illumination) spectral luminous efficiency curve of the eye for photopic (or day) vision. By an appropriate shift in the mix, the emission emphasis moves toward the warm side (warm white), in having proportionately more yellow and red light, or cool white, with a greater proportion of blue and green. For good portrayal of human complexion and all colors, one should use the deluxe cool white or deluxe warm white, where a proportionately greater red emission is present.

Fluorescent lamps require a ballast to limit the current and in many instances to transform the supply voltage to a suitable level. Lamp performance is influenced by the character of the ballast and luminaire, line voltage, ambient temperature, burning hours per start, and air movement.

The efficacy of fluorescent lamps (73) varies from approximately 30 to 80 lm/W for lamps of standard cool white color, depending on bulb size and shape. Lamps range in size from 6 to 96 inches (0.5 to 244 cm) long. This does not include power losses in the ballast, typically in the order of 20 percent. Most fluorescent lamps are of conventional tubular design, but there are some special types such as circular (Circline), reflectorized, jacketed, and panel shaped.

12.4 High Intensity Discharge Lamps

High intensity discharge (HID) lamps (73) are also electric discharge sources. The basic difference between HID and fluorescent lamps is that the operating pressure of the arc is much higher in HID lamps. HID lamps include the groups commonly known as mercury, metal halide, and high pressure sodium lamps. Spectral characteristics are different because the higher pressure arc emits a large portion of its visible light. HID lamps produce full light output only when full operating pressure has been reached, which generally requires several minutes. Most HID lamps have both an inner and an outer bulb. The inner is made of quartz or polycrystalline aluminum, and the outer bulb is generally made of thermal shock-resistant glass. Light output is practically unaffected by surrounding temperatures. Like fluorescent lamps, HID lamps also require devices that limit the current.

Mercury lamps for general lighting are available with either "clear" or phosphor-coated bulbs of various sizes and shapes from 40 to 1000 W. There is also a 3000 W clear mercury lamp. Typical efficacies range from 30 to 63 lm/W, not including 6 to 20 percent power loss in the ballast. "Clear" mercury lamps produce light that is rich in yellow and green tones and almost entirely lacking in red. Various types of phosphor-coated lamp are available to provide improved color, and they are more popular than clear lamps for industrial lighting. A number of special mercury lamps are available, including semireflector, reflectorized, and self-ballasted.

Metal halide lamps are very similar in construction to the mercury lamp, the major difference being that the arc tube contains various metal halides in addition to mercury. They are available with either clear or phosphor-coated bulbs of various sizes and shapes from 175 to 1500 W. Typical efficacies range from 70 to 100 lm/W not including power loss in the ballast. The metal halide additives improve the efficacy and color compared to a clear mercury lamp. Further color improvement is achieved with the addition of phosphor coatings to the lamp.

In the high pressure sodium lamp, light is produced by electricity passing through sodium vapor. The lamps are available in wattages from 250 to 1000 W. Typical efficacies range from 100 to 130 lm/W, not including power loss in the ballast. This lamp produces a golden white light.

In low pressure sodium discharge lamps, the arc is carried through vaporized sodium. The starting gas is neon with small additions of argon, xenon, or helium. To maintain the maximum efficacy of the conversion of the electrical input to the arc discharge into light, the vapor pressure must be at a given level and constant, and the proper operating temperature must be maintained. The arc tube must be enclosed in a vacuum flask or in an outer bulb at high vacuum, to ensure that the temperature is maintained. The light produced by low pressure sodium arc is almost monochromatic, consisting of a double line in the yellow region of the spectrum at 589 and 589.6 nm.

Two types of arc tube construction are used in modern low pressure sodium lamps—the hairpin or "U" tube and the linear. The efficacy is 180 lm/W without ballast.

13 DAYLIGHTING

Over the centuries man has attempted to bring daylight to his indoor visual activities. Central courts open to the sky allowed the light to penetrate into the covered areas. In northern latitudes the churches and guild halls had high windows to admit daylight as far as possible into the interior, permitting work to be carried on in spite of inclement weather. All this was necessary because of the weakness of man-made sources. Since

World War II there has been decreasing dependence on daylighting because of the availability of far more efficient electric light sources (fluorescent lights and the developing use of high intensity discharge sources). The architects and designers have found that they can reduce costs of buildings by lowering the ceilings (reducing the height of the windows). The architects have continued to use large areas of glass, but they have learned that the workers near the windows have been overwhelmed by the high levels of daylight. Accordingly low transmission glass (neutral tinted with 15 to 50 percent transmission) has come into use. Furthermore, to reduce the amount of dirt accumulated, the windows were sealed and the interiors air conditioned. It was found that the daylighting, and particularly the sunload on the air conditioning, was appreciable. This has resulted in reflecting-type window glazing, where a large part of the heat is reflected back outside and low transmission of light is maintained.

Our discussion of transient adaptation pointed out that encountering higher luminances than the task luminance greatly reduces the immediate scanning sensitivity of the visual system (Figure 13.44). The sky luminances as seen through clear glass windows may range between 10 and 100 times the task luminance; therefore the scanning sensitivity is severely reduced. This and a significant degree of discomfort glare probably accounted for the trend toward low transmission glass. In the older industrial plants with large expanses of glass in the walls or the roof, it is common to see the panes covered with paint to reduce the outdoor luminances, with the aim of achieving better balance with interior luminances.

Nevertheless window expanses with appropriate limited luminance do provide a significant illumination to the peripheral areas of a building. For peripheral offices the light that approaches desk tasks at wide angles with the vertical does a most effective job of overcoming veiling reflections and producing high visibility. Illumination from windows is not significant at a distance beyond twice the height of the windows above the floor.

Reflected ground light can make an important contribution by being directed onto a white ceiling as soft indirect illumination. A scheme of having a low transmission vision strip opposite the eye level with clear glass above to admit ground light but with the view of the clear glass occluded by a horizontal shield or louvers is helpful (74).

Another system is to use silvered louvers that catch the sunlight (in high percentage sunlight areas) and direct the light out across the ceiling for indirect lighting (75).

Many structural forms such as unilateral, bilateral roof monitor, clerestory, sawtooth, and skylight have been studied as to the effective admission of daylight (76). A Recommended Practice of Daylighting (77), available from the Illuminating Engineering Society, has guidelines for designing for daylighting. Limitations of outdoor luminances with respect to transient adaptation and discomfort glare should be borne in mind.

14 ENERGY CONSERVATION

Since 1944 the Illuminating Engineering Research Institute, a trust affiliate of the Illuminating Engineering Society, has been dedicated to ascertaining the most efficient operation of the visual system for the most efficient use of human resources. The pattern of these university-determined results has been reflected in the previous pages of this chapter. The amount of light on the details of the visual task, the reduction or elimination of veiling reflections for optimum visibility, the proper brightness balance for optimum scanning sensitivity (transient adaptation), the avoidance of discomfort and disability glare, the motivation of good subjective feeling, and the provision of good biological health—all contribute to optimum ability to perform efficiently.

Having provided the optimum use of human resources, one can begin to plan the strategy of the use of lighting energy (78). The research data do not indicate the need of illumination for a given visual performance over all the habitable space, but only on the task area, as long as the surrounding luminances are in balance with task luminance to avoid serious transient adaptation losses, disability glare, or discomfort glare. Therefore there can be nonuniform lighting with less light in the nontask areas (79, 80). The IES has developed specific recommendations for nonuniform lighting, more efficacious light sources (fluorescent and HID), better and more efficient design of luminaires, combining daylight with electric lighting, providing controls for flexible use of lighting, using lighter finishes on room surfaces, and providing for better maintenance (80). To aid the architects and designers, there is the "IES Recommended Lighting Power Budget Determination Procedure," (EMS-1), supplemented with "An Interim Report Relating the Lighting Design Procedure to Effective Energy Utilization," (EM2), and "Example of the Use of the 'IES Recommended Lighting Power Budget Determination Procedure'" (EM3).

REFERENCES

1. "What is IERI?" Illuminating Engineering Research Institute, New York, 1974; "Illuminating Engineering Research Institute—Its Organization and Program," New York, 1970.
2. "Light and Vision," in: *IES Lighting Handbook,* 5th ed., Illuminating Engineering Society, New York, 1972, pp. 3–8.
3. S. K. Guth and J. F. McNelis, *Illum. Eng.,* **63,** 32 (1968); **64,** 99 (1969).
4. R. J. Lythgoe, "Measurement of Visual Acuity," Special Report No. 173, Medical Research Council, His Majesty's Stationery Office, London, 1932.
5. H. R. Blackwell, *Illum. Eng.,* **54,** 317 (1959).
6. M. Luckiesh, *Light Vision and Seeing,* 1st ed., Van Nostrand, New York, 1944, pp. 158–165.
7. S. K. Guth, *J. Am. Ind. Hyg.,* **23,** 359 (1962).
8. W. R. Stevens and C. A. P. Foxell, *Light and Lighting,* London, **48,** 419 (1955).
9. P. W. Cobb, *J. Exp. Psychol.,* **1,** 540 (1916).
10. P. W. Cobb and F. K. Moss, *J. Franklin Inst.,* **199,** 507 (1925).
11. C. E. Ferree and G. Rand, *Trans. IES,* **22,** 79 (1927).
12. C. E. Ferree and G. Rand, *Trans. IES,* **23,** 507 (1928).
13. S. W. Smith, in: 1974 Annual Report, Illuminating Engineering Research Institute, New York, pp. 10–22; *J. IES,* **5,** 235 (1976).
14. "Office Lighting Studies," Illuminating Engineering Society, New York 1977; 1975 Annual Report of IERI.

15. H. C. Weston, "The Relation Between Illumination and Industrial Efficiency: (1) The Effect of Size of Work," Joint Report of the Industrial Health Research Board (Medical Research Council) and the Illumination Research Committee (Department of Scientific and Industrial Research), Her Majesty's Stationery Office, London, 1935; "The Relation Between Illumination and Visual Efficiency—The Effect of Brightness Contrast," Joint Report of the Industrial Health Research Board (Medical Research Council) and the Illumination Research Committee (Department of Scientific and Industrial Research), Her Majesty's Stationery Office, London, 1944.
16. "Light and Vision," in: *IES Lighting Handbook,* 5th ed., Illuminating Engineering Society, New York, 1972, pp. 3–15.
17. L. A. Jones, *Phil. Mag.,* **39,** 96 (1920); C. Dunbar, *Trans. IES,* London, **5,** 33 (1940); C. L. Cottrell, *Illum. Eng.,* **46,** 95 (1951); A. E. Simmons and D. M. Finch, *Illum. Eng.,* **48,** 517 (1953); A. A. Eastman, *Illum. Eng.,* **63,** 37 (1968); H. R. Blackwell, *Illum. Eng.,* **65,** 267 (1970).
18. H. R. Blackwell, *J. IES,* **6** (1976).
19. American National Standard Practice for Industrial Lighting, (A 11), Illuminating Engineering Society, New York, 1973.
20. "Levels of Illumination," in: *IES Lighting Handbook,* 5th ed., Illuminating Engineering Society, New York, 1972, bibliographical references 5, 7, 11, 16–27, 34, 36, 37.
21. American National Standard Practice for Industrial Lighting, Illuminating Engineering Society, New York, 1973, pp. 27–33.
22. J. F. Chorlton and H. F. Davidson, *Illum. Eng.,* **54,** 482 (1959); D. M. Finch, *Illum. Eng.,* **54,** 474 (1959); H. R. Blackwell, *Illum. Eng.,* **58,** 161 (1963); W. Allphin, *Illum. Eng.,* **58,** 244 (1963); C. L. Crouch and J. E. Kaufman, *Illum. Eng.,* **58,** 277 (1963).
23. R. L. Henderson, J. F. McNelis, and H. G. Williams, *J. IES,* **4,** 150 (1975).
24. F. K. Sampson, "Contrast Rendition in School Lighting," Technical Report No. 4, Educational Facilities Laboratories, New York, 1970.
25. I. Lewin and J. W. Griffith, *Illum. Eng.,* **65,** 594 (1970); N. S. Florence and S. B. Glickman, *Illum. Eng.,* **66,** 149 (1971).
26. I. Goodbar, L. J. Buttolph, E. J. Breneman, and C. L. Crouch, *Lighting Design and Application,* Vol. 4, Illuminating Engineering Society, 45 (1974).
27. I. Goodbar, in: 1973 Annual Report, Illuminating Engineering Research Institute, New York, pp. 22–29.
28. L. J. Buttolph, in: 1973 Annual Report, Illuminating Engineering Research Institute, New York, p. 34.
29. R. M. Boynton, E. J. Rinalducci, and C. Sternheim, *Illum. Eng.,* **64,** 217 (1969).
30. G. A. Fry and V. M. King, *J. IES,* **4,** 307 (1975).
31. C. A. Bennett, "Discomfort Glare: Demographic Variables," Special Report No. 118, Kansas Engineering Experiment Station, Kansas State University, Manhattan, 1976.
32. "Outline of a Standard Procedure for Computing Visual Comfort Ratings for Interior Lighting," *J. IES,* **2,** 328 (1973); S. K. Guth, *Illum. Eng.,* **58,** 351 (1963); "An Alternate Simplified Method for Determining Acceptability of a Luminaire from the VCP Standpoint for Use in Large Rooms," *J. IES,* **1,** 256 (1972).
33. L. L. Holladay, *J. Opt. Soc. Am.,* **12,** 279 (1926); W. S. Stiles, *Proc. CIE,* 1928, p. 220; G. A. Fry, *Illum. Eng.,* **50,** 31 (1955).
34. A. J. Fisher and A. W. Christie, *Vision Res.,* **5,** 565 (1965).
35. "Public Lighting Needs," *Illum. Eng.,* **61,** 585 (1966); through *The World Almanac, New York World-Telegram,* New York, 1964.
36. J. E. Birren, R. C. Casperson, and J. Botwinick, "Age Changes in Pupil Size," *J. Gerontol.,* **5,** 216 (1950).

37. L. B. Zerbe and H. W. Hofstetter, "Prevalence of 20/20 with Best Previous and No Lens Correction," *J. Am. Optom. Assoc.*, **29,** 772 (1957–1958).
38. H. C. Weston, "The Effects of Age and Illumination upon Visual Performance with Close Sights," *Br. J. Ophthalmol.*, **32,** 645 (1948).
39. E. Wolf, "Effects of Age on Peripheral Vision," Highway Research Board Bulletin No. 336, 1962, p. 26.
40. R. A. McFarland, R. G. Domey, A. B. Warren, and D. C. Ward, "Dark Adaptation as a Function of Age and Tinted Windshield Glass," Highway Research Board Bulletin No. 255, 1960, p. 51.
41. E. D. Fletcher, "An Investigation of Glare Resistance and Its Relationship to Age," *Motorists' Vision,* **5,** 1 (1952).
42. O. M. Blackwell and H. R. Blackwell, "Visual Performance Data for 146 Normal Observers of Various Ages," *J. IES,* **1** (1971).
43. H. Helson, D. B. Judd, and M. H. Warren, "Object-Color Changes from Daylight to Incandescent Filament Illumination," *Illum. Eng.*, **47,** (1952).
44. D. Nickerson and C. W. Jerome, "Color Rendering of Light Sources: CIE Method of Specification and Its Application," *Illum. Eng.*, **50,** 262 (1965).
45. H. Helson and T. Lansford, "The Role of Spectral Energy of Source and Background Color in the Pleasantness of Object Colors," *Appl. Opt.*, **9,** 1513 (1970).
46. D. B. Judd, "A Flattery Index for Artificial Illuminants," *Illum. Eng.*, **52,** 593 (1967).
47. C. W. Jerome, "The Flattery Index," *J. IES,* **2,** 351 (1973).
48. "Color," in: *IES Lighting Handbook,* 5th ed., Illuminating Engineering Society, New York, 1972.
49. "Color and the Use of Color by the Illuminating Engineer," *Illum. Eng.*, **57,** 764 (1962).
50. "The Psychological Potential of Illumination," Environmental Design Research Association Conference, Pennsylvania State University, University Park, 1975.
51. L. T. Troland, "An Analysis of the Literature Concerning the Dependency of Visual Functions upon Illumination Intensity," *Trans. IES,* **26,** 107 (1931).
52. C. E. Ferree and G. Rand, "Tests for the Efficiency of the Eye Under Different Systems of Illumination and a Preliminary Study of the Causes of Discomfort," *Trans. IES,* **8,** 40 (1913); J. R. Cravath, "Some Experiments with the Ferree Test for Eye Fatigue," *Trans. IES,* **9,** 1033 (1914); C. E. Ferree and G. Rand, "The Efficiency of the Eye Under Different Conditions of Lighting: The Effect of Varying the Distribution Factors and Intensity," *Trans. IES,* **10,** 407 (1915); "Some Experiments on the Eye with Different Illuminants. Part I," *Trans. IES,* **13,** 50 (1918); "Some Experiments on the Eye With Different Illuminants. Part II," *Trans. IES,* **14,** 107 (1919); "The Effect of Variations in Intensity of Illumination on Functions of Importance to the Working Eye," *Trans. IES,* **15,** 769 (1920); "The Effect of Variation of Visual Angle, Intensity and Composition of Light on Important Ocular Functions," *Trans. IES,* **17,** 69 (1922); "Further Studies on the Effect of Composition of Light on Important Ocular Functions," *Trans. IES,* **19,** 424 (1924).
53. H. C. Weston, *Sight Light and Efficiency,* H. K. Lewis, London, 1949.
54. W. B. Lancaster, "Light and Lighting," in: *The Eye and Its Diseases,* 2nd ed., C. Berens, Ed., Saunders, Philadelphia, 1950, p. 81.
55. F. W. Hebbard, "Micro Eye Movements: Effects of Target Illumination and Contrast," *Illum. Eng.*, **64,** 199 (1969).
56. H. C. Weston, *Sight Light and Efficiency,* H. K. Lewis, London, 1949.
57. D. A. Campbell, W. J. B. Riddell, and A. S. MacNalty, *Eyes in Industry,* Longmans, Green, New York, 1951.
58. D. A. Campbell, W. J. B. Riddell, and A. S. MacNalty, *Eyes in Industry,* Longmans, Green, New York, 1951, p. 2.

59. M. Poser, "Eye Fatigue in Industry," *Trans. IES,* **16,** 431 (1921).
60. M. Luckiesh and F. K. Moss, *The Science of Seeing,* Van Nostrand, New York, 1937.
61. C. L. Crouch, "Discussion," *Illum. Eng.,* **47,** 344 (1952).
62. C. P. Steinmetz, "The Scope of Illuminating Engineering," *Trans. IES,* **11,** 625 (1916).
63. H. P. Gage, "Hygienic Effects of Ultraviolet Radiation in Daylight," *Trans. IES,* **25,** 377, (1930).
64. C. Berens and C. L. Crouch, "Is Fluorescent Lighting Injurious to the Eyes?" *Am. Ophthalmol.,* **45,** 47 (1958).
65. R. J. Wurtman, "The Effects of Light on the Human Body," *Sci. Am.,* **233,** 68 (1975).
66. T. P. Vogl, "Photomedicine," *Opt. News,* Spring 1976, p. 6.
67. R. J. Cremer, P. W. Perryman, and D. H. Richards, "Influence of Light on the Hyperbilirubinemia of Infants," *Lancet,* May 1958, p. 1094.
68. J. R. Lucey, "The Effects of Light on the Newly Born Infant," *J. Perinat. Med.,* **1,** 1 (1973).
69. R. M. Neer et al., "Stimulation by Artificial Lighting of Calcium Absorption in Elderly Human Subjects," *Nature,* **229** (1971).
70. W. F. Loomis, "Rickets," *Sci. Am.,* **223,** 77 (1970).
71. J. A. Parrish, T. B. Fitzpatrick, L. Tanenbaum, and M. A. Pathak, *New Engl. J. Med.,* **291,** 1207 (1974).
72. I. Diamond, A. F. McDonagh, C. B. Wilson, S. G. Granelli, S. Nielson, and R. Jaenicke, *Lancet,* **15,** 1175 (1972); S. G. Granelli, I. Diamond, A. F. McDonagh, C. B. Wilson, and S. L. Nielsen, *Cancer Res.,* **35,** 2567 (1975); T. J. Dougherty, G. B. Grindey, R. Fiel, K. R. Weishaupt, and D. G. Boyle, *J. Nat. Cancer Inst.,* **55,** 115 (1975); J. F. Kelly, M. E. Snell, and M. C. Berenbaum, *Br. J. Cancer,* **31,** 237 (1975).
73. "Light Sources," in: *IES Lighting Handbook,* 5th ed., Illuminating Engineering Society, New York, 1972, Section 8.
74. "Daylighting," in: *IES Lighting Handbook,* 5th ed., Illuminating Engineering Society, New York, 1972.
75. Dean Rosenfeld, "Efficient Use of Energy in Buildings, A Report on the 1975 Berkeley Summer Study," LBL 4411, Lawrence Berkeley Laboratory, University of California, Berkeley, 1976.
76. Bibliography, "Recommended Practice of Daylighting," *Illum. Eng.,* **57,** 517 (1962).
77. "Recommended Practice of Daylighting," *Illum. Eng.,* **57,** 517 (1962).
78. "Effective Seeing in an Era of Energy Conservation," Illuminating Engineering Research Institute, New York, 1973.
79. "Effective Seeing and Conservation Too," Illuminating Engineering Research Institute, New York, 1974.
80. J. E. Kaufman, "Optimizing The Use of Energy for Lighting" *Lighting Design and Application,* Vol. 4, Illuminating Engineering Society, 8 (1973).
81. "IES Recommended Lighting Power Budget Determination Procedure," (EMS-1); "An Interim Report Relating the Lighting Design Procedure to Effective Energy Utilization" (EMS-2); "Example of the Use of the IES Recommended Power Budget Procedure" (EMS-3), Illuminating Engineering Society, New York, 1975–1976.

APPENDIX 13A

The information in the table of currently recommended levels of interior illumination was developed by the Illuminating Engineering Society as one criterion to be used in lighting design. For the proper application of these levels, see the fifth edition of the *IES Lighting Handbook*.

For convenience of use, Table 13A.1 sometimes lists locations rather than tasks, but the recommended footcandle values have been arrived at for specific visual tasks. The tasks selected for this purpose have been the more difficult ones that commonly occur in the various areas.

To assure these values at all times, higher initial levels should be provided, as required by the maintenance conditions. Where tasks are located near the perimeter of a room, special consideration should be given to the arrangement of the luminaires in order to provide the recommended level of illumination on the task.

The illumination levels given in the table are intended to be minimum on the task irrespective of the plane in which it is located. In some instances, denoted by a superscript *c*, the values are for equivalent sphere illumination E_s. The commonly used lumen method of illumination calculation, which gives results only for a horizontal work plane, cannot be used to calculate or predetermine E_s values. The ratio of vertical to horizontal illumination will generally vary from 1 : 3 for luminaires having narrow distribution to 1 : 2 for luminaires of wide distribution. Where the levels thus achieved are inadequate, special luminaire arrangements should be used or supplemental lighting equipment employed.

Supplementary luminaires may be used in combination with general lighting to achieve these levels. The general lighting should be not less than 20 footcandles and should contribute at least one-tenth the total illumination level.

Table 13A.1. Levels of Interior Illumination Currently Recommended by the IES

Area	Footcandles on Tasks[a]	Dekalux[b] on Tasks[a]
Aircraft manufacturing		
Stock parts		
Production	100	110
Inspection	200	220
Parts manufacturing		
Drilling, riveting, screw fastening	70	75
Spray booths	100	110
Sheet aluminum layout and template work, shaping, and smoothing of small parts for fuselage, wing sections, cowling, etc.	100[j]	110[j]
Welding		
General illumination	50	54
Precision manual arc welding	1000[d]	1080[d]
Subassembly		
Landing gear, fuselage, wing sections, cowling, and other large units	100	110

Table 13A.1. (Continued)

Area	Footcandles on Tasks[a]	Dekalux[b] on Tasks[a]
Final assembly		
Placing of motors, propellers, wing sections, landing gear	100	110
Inspection of assembled ship and its equipment	100	110
Machine tool repairs	100	110
Aircraft hangars		
Repair service only	100	110
Assembly		
Rough easy seeing	30	32
Rough difficult seeing	50	54
Medium	100	110
Fine	500[d]	540[d]
Extra fine	1000[d]	1080[d]
Auditoriums		
Assembly only	15	16
Exhibitions	30	32
Social activities	5	5.4
Automobile showrooms (see Stores)		
Automobile manufacturing		
Frame assembly	50	54
Chassis assembly line	100	110
Final assembly, inspection line	200	220
Body manufacturing		
Parts	70	75
Assembly	100	110
Finishing and inspecting	200	220
Bakeries		
Mixing room	50	54
Face of shelves (vertical illumination)	30	32
Inside of mixing bowl (vertical mixers)	50	54
Fermentation room	30	32
Make-up room		
Bread	30	32
Sweet yeast-raised products	50	54
Proofing room	30	32
Oven room	30	32
Fillings and other ingredients	50	54
Decorating and icing		
Mechanical	50	54
Hand	100	110
Scales and thermometers	50	54
Wrapping room	30	32

Table 13A.1. (Continued)

Area	Footcandles on Tasks[a]	Dekalux[b] on Tasks[a]
Book binding		
Folding, assembling, pasting, etc.	70	75
Cutting, punching, stitching	70	75
Embossing and inspection	200	220
Breweries		
Brew house	30	32
Boiling and keg washing	30	32
Filling (bottles, cans, kegs)	50	54
Candy making		
Box department	50	54
Chocolate department		
Husking, winnowing, fat extraction, crushing and refining, feeding	50	54
Bean cleaning, sorting, dipping, packing, wrapping	50	54
Milling	100	110
Cream making		
Mixing, cooking, molding	50	54
Gum drops and jellied forms	50	54
Hand decorating	100	110
Hard candy		
Mixing, cooking, molding	50	54
Die cutting and sorting	100	110
Kiss making and wrapping	100	110
Canning and preserving		
Initial grading raw material samples	50	54
Tomatoes	100	110
Color grading (cutting rooms)	200[d]	220[d]
Preparation		
Preliminary sorting		
Apricots and peaches	50	54
Tomatoes	100	110
Olives	150	160
Cutting and pitting	100	110
Final sorting	100	110
Canning		
Continuous-belt canning	100	110
Sink canning	100	110
Hand packing	50	54
Olives	100	110
Examination of canned samples	200[e]	220[e]
Container handling		
Inspection	200[d]	220[d]
Can unscramblers	70	75
Labeling and cartoning	30	32

Table 13A.1. (Continued)

Area	Footcandles on Tasks[a]	Dekalux[b] on Tasks[a]
Central station		
Air-conditioning equipment, air preheater and fan floor, ash sluicing	10	11
Auxiliaries, battery rooms, boiler feed pumps, tanks, compressors, gauge area	20	22
Boiler platforms	10	11
Burner platforms	20	22
Cable room, circulator, or pump bay	10	11
Chemical laboratory	50	54
Coal conveyor, crusher, feeder, scale areas, pulverizer, fan area, transfer tower	10	11
Condensers, deaerator floor, evaporator floor, heater floors	10	11
Control rooms (see Control rooms)		
Hydrogen and carbon dioxide manifold area	20	22
Precipitators	10	11
Screen house	20	22
Soot or slag blower platform	10	11
Steam headers and throttles	10	11
Switchgear, power	20	22
Telephone equipment room	20	22
Tunnels or galleries, piping	10	11
Turbine bay sub-basement	20	22
Turbine room	30	32
Visitor's gallery	20	22
Water treating area	20	22
Chemical works		
Hand furnaces, boiling tanks, stationary driers, stationary and gravity crystallizers	30	32
Mechanical furnaces, generators and stills, mechanical driers, evaporators, filtration, mechanical crystallizers, bleaching	30	32
Tanks for cooking, extractors, percolators, nitrators, electrolytic cells	30	32
Clay products and cements		
Grinding, filter presses, kiln rooms	30	32
Molding, pressing, cleaning, trimming	30	32
Enameling	100	110
Color and glazing—rough work	100	110
Color and glazing—fine work	300[d]	320[d]
Cleaning and pressing industry		
Checking and sorting	50	54
Dry and wet cleaning and steaming	50	54
Inspection and spotting	500[d]	540[d]

Table 13A.1. (Continued)

Area	Footcandles on Tasks[a]	Dekalux[b] on Tasks[a]
Pressing	150	160
Repair and alteration	200[d]	220[d]
Cloth products		
Cloth inspection	2000[d]	2150[d]
Cutting	300[d]	320[d]
Sewing	500[d]	540[d]
Pressing	300[d]	320[d]
Clothing manufacture (men's)		
Receiving, opening, storing, shipping	30	32
Examining (perching)	2000[d]	2150[d]
Sponging, decating, winding, measuring	30	32
Piling up and marking	100	110
Cutting	300[d]	320[d]
Pattern making, preparation of trimming, piping, canvas, and shoulder pads	50	54
Fitting, bundling, shading, stitching	30	32
Shops	100	110
Inspection	500[d]	540[d]
Pressing	300[d]	320[d]
Sewing	500[d]	540[d]
Coal tipples and cleaning plants		
Breaking, screening, and cleaning	10	11
Picking	300[d]	320[d]
Control rooms and dispatch rooms		
Control rooms		
Vertical face of switchboards		
Simplex or section of duplex facing operator:		
Type A—large centralized control room 66 in above floor	50	54
Type B—ordinary control room 66 in above floor	30	32
Section of duplex facing away from operator	30	32
Bench boards (horizontal level)	50	54
Area inside duplex switchboards	10	11
Rear of all switchboard panels (vertical)	10	11
Emergency lighting, all areas	3	3.2
Dispatch boards		
Horizontal plane (desk level)	50	54
Vertical face of board (48 in above floor, facing operator):		
System load dispatch room	50	54
Secondary dispatch room	30	32

Table 13A.1. (Continued)

Area	Footcandles on Tasks[a]	Dekalux[b] on Tasks[a]
Cotton gin industry		
Overhead equipment—separators, driers, grid cleaners, stick machines, conveyors, feeders, and catwalks	30	32
Gin stand	50	54
Control console	50	54
Lint cleaner	50	54
Bale press	30	32
Dairy farms (see Farms)		
Dairy products		
Fluid milk industry		
Boiler room	30	32
Bottle storage	30	32
Bottle sorting	50	54
Bottle washers	—[e]	—[e]
Can washers	30	32
Cooling equipment	30	32
Filling: inspection	100	110
Gauges (on face)	50	54
Laboratories	100	110
Meter panels (on face)	50	54
Pasteurizers	30	32
Separators	30	32
Storage refrigerator	30	32
Tanks, vats		
Light interiors	20	22
Dark interiors	100	110
Thermometer (on face)	50	54
Weighing room	30	32
Scales	70	75
Dispatch boards (see Control rooms)		
Drafting rooms (see Offices)		
Electrical equipment manufacturing		
Impregnating	50	54
Insulating: coil winding	100	110
Testing	100	110
Electrical Generating Station (see Central Station)		
Elevators, freight and passenger	20	22
Engraving (wax)	200[d]	220[d]
Explosives		
Hand furnaces, boiling tanks, stationary driers, stationary and gravity crystallizers	30	32

Table 13A.1. (Continued)

Area	Footcandles on Tasks[a]	Dekalux[b] on Tasks[a]
Mechanical furnace, generators and stills, mechanical driers, evaporators, filtration, mechanical crystallizers	30	32
Tanks for cooking, extractors, percolators, nitrators	30	32
Farms—dairy		
Milking operation area (milking parlor and stall barn)		
General	20	22
Cow's udder	50	54
Milk handling equipment and storage area (milk house or milk room)		
General	20	22
Washing area	100	110
Bulk tank interior	100	110
Loading platform	20	22
Feeding area (stall barn feed alley, pens, loose housing feed area)	20	22
Feed storage area—forage		
Haymow	3	3.2
Hay inspection area	20	22
Ladders and stairs	20	22
Silo	3	3.2
Silo room	20	22
Feed storage area—grain and concentrate		
Grain bin	3	3.2
Concentrate storage area	10	11
Feed processing area	10	11
Livestock housing area (community, maternity, individual calf pens, and loose housing holding and resting areas)	7	7.5
Machine storage area (garage and machine shed)	5	5.4
Farm shop area		
Active storage area	10	11
General shop area (machinery repair, rough sawing)	30	32
Rough bench and machine work (painting, fine storage, ordinary sheet metal work, welding, medium benchwork)	50	54
Medium bench and machine work (fine woodworking, drill press, metal lathe, grinder)	100	110
Miscellaneous areas		
Farm office	70[c]	75[c]
Restrooms	30	32
Pumphouse	20	22

Table 13A.1. (Continued)

Area	Footcandles on Tasks[a]	Dekalux[b] on Tasks[a]
Farms—poultry (see Poultry industry)		
Flour mills		
Rolling, sifting, purifying	50	54
Packing	30	32
Product control	100	110
Cleaning, screens, man lifts, aisleways and walkways, bin checking	30	32
Forge shops	50	54
Food service facilities		
Dining areas		
Cashier	50	54
Intimate type		
Light environment	10	11
Subdued environment	3	3.2
For cleaning	20	22
Leisure type		
Light environment	30	32
Subdued environment	15	16
Quick service type		
Bright surroundings[f]	100	110
Normal surroundings[f]	50	54
Food displays—twice the general levels but not under	50	54
Kitchen, commercial		
Inspection, checking, preparation, and pricing	70	75
Entrance foyer	30	32
Marquee		
Dark surroundings	30	32
Bright surroundings	50	54
Foundries		
Annealing (furnaces)	30	32
Cleaning	30	32
Core making		
Fine	100	110
Medium	50	54
Grinding and chipping	100	110
Inspection		
Fine	500[d]	540[d]
Medium	100	110
Molding		
Medium	100	110
Large	50	54

Table 13A.1. (Continued)

Area	Footcandles on Tasks[a]	Dekalux[b] on Tasks[a]
Pouring	50	54
Sorting	50	54
Cupola	20	22
Shakeout	30	32
Garages—automobile and truck		
Service garages		
Repairs	100	110
Active traffic areas	20	22
Parking garages		
Entrance	50	54
Traffic lanes	10	11
Storage	5	5.4
Gasoline station (see Service station)		
Glass works		
Mix and furnace rooms, pressing and lehr, glassblowing machines	30	32
Grinding, cutting glass to size, silvering	50	54
Fine grinding, beveling, polishing	100	110
Inspection, etching and decorating	200[e]	220[e]
Glove manufacturing		
Pressing	300[d]	320[d]
Knitting	100	110
Sorting	100	110
Cutting	300[d]	320[d]
Sewing and inspection	500[d]	540[d]
Hangars (see Aircraft hangars)		
Hat manufacturing		
Dyeing, stiffening, braiding, cleaning, refining	100	110
Forming, sizing, pouncing, flanging, finishing, ironing	200	220
Sewing	500[d]	540[d]
Hospitals		
Anesthetizing and preparation room	30	32
Autopsy and morgue		
Autopsy room	100	110
Autopsy table	1000	1080
Museum	50	54
Morgue, general	20	22
Central sterile supply		
General, work room	30	32
Work tables	50	54
Glove room	50	54

Table 13A.1. (Continued)

Area	Footcandles on Tasks[a]	Dekalux[b] on Tasks[a]
Syringe room	150	160
Needle sharpening	150	160
Storage areas	30	32
Issuing sterile supplies	50	54
Corridor		
General in nursing areas—daytime	20	22
General in nursing areas—night (rest period)	3	3.2
Operating, delivery, recovery, and laboratory suites and service areas	30	32
Cystoscopic room		
General	100	110
Cystoscopic table	2500	2690
Dental suite		
Operatory, general	70	75
Instrument cabinet	150	160
Dental entrance to oral cavity	1000	1080
Prosthetic laboratory bench	100	110
Recovery room, general	5	5.4
Recovery room, local for observation	70	75
(EEG) encephalographic suite		
Office (see Offices)		
Work room, general	30	32
Work room, desk or table	100	110
Examining room	30	32
Preparation rooms, general	30	32
Preparation rooms, local	50	54
Storage, records, charts	30	32
Electromyographic suite		
Same as EEG but provisions for reducing level in preparation area to 1		
Emergency operating room		
General	100	110
Local	2000	2150
EKG, BMR, and specimen room		
General	30	32
Specimen table	50	54
EKG machine	50	54
Examination and treatment room		
General	50	54
Examining table	100	110
Exits, at floor	5	5.4

Table 13A.1. (Continued)

Area	Footcandles on Tasks[a]	Dekalux[b] on Tasks[a]
Eye, ear, nose, and throat suite		
Darkroom (variable)	0–10	0–11
Eye examination and treatment	50	54
Ear, nose, throat room	50	54
Flower room	10	11
Formula room		
Bottle washing	30	32
Preparation and filling	50	54
Fracture room		
General	50	54
Fracture table	200	220
Splint closet	50	54
Plaster sink	50	54
Intensive care nursing areas		
General	30	32
Local	100	110
Laboratories		
General	50	54
Close work areas	100	110
Linens (see Laundries)		
Sorting soiled linen	30	32
Central (clean) linen room	30	32
Sewing room, general	30	32
Sewing room, work area	100	110
Linen closet	10	11
Lobby (or entrance foyer)		
During day	50	54
During night	20	22
Locker rooms	20	22
Medical records room	100[c]	110[c]
Nurses' station		
General—day	70[c]	75[c]
General—night	30	32
Desk for records and charting	70[c]	75[c]
Table for doctor's making or viewing reports	70[c]	75[c]
Medicine counter	100[c]	110[c]
Nurses' gown room		
General	30	32
Mirror for grooming	50	54
Nurseries, infant		
General	30	32
Examining, local at bassinet	100	110

Table 13A.1. (Continued)

Area	Footcandles on Tasks[a]	Dekalux[b] on Tasks[a]
Examining and treatment table	100	110
Nurses station and work space (see Nurses' station)		
Obstetrical suite		
Labor room, general	20	22
Labor room, local	100	110
Scrub-up area	30	32
Delivery room, general	100	110
Substerilizing room	30	32
Delivery table	2500	2690
Clean-up room	30	32
Recovery room, general	30	32
Recovery room, local	100	110
Patients' rooms (private and wards)		
General	20	22
Reading	30	32
Observation (by nurse)	2	2.2
Night light, maximum at floor (variable)	0.5	0.5
Examining light	100	110
Toilets	30	32
Pediatric nursing unit		
General, crib room	20	22
General, bedroom	10	11
Reading	30	32
Playroom	30	32
Treatment room, general	50	54
Treatment room, local	100	110
Pharmacy		
Compounding and dispensing	100	110
Manufacturing	50	54
Parenteral solution room	50	54
Active storage	30	32
Alcohol vault	10	11
Radioisotope facilities		
Radiochemical laboratory, general	30	32
Uptake or scanning room	20	22
Examining table	50	54
Retiring room		
General	10	11
Local for reading	30	32
Solarium		
General	20	22
Local for reading	30	32

Table 13A.1. (Continued)

Area	Footcandles on Tasks[a]	Dekalux[b] on Tasks[a]
Stairways	20	22
Surgical suite		
Instrument and sterile supply room	30	32
Clean-up room, instrument	100	110
Scrub-up area (variable)	200	220
Operating room, general (variable)	200	220
Operating table	2500	2690
Recovery room, general	30	32
Recovery room, local	100	110
Anesthesia storage	20	22
Substerilizing room	30	32
Therapy, physical		
General	20	22
Exercise room	30	32
Treatment cubicles, local	30	32
Whirlpool	20	22
Lip reading	150	160
Office (see Offices)		
Therapy, occupational		
Work area, general	30	32
Work tables or benches, ordinary	50	54
Work tables or benches, fine work	100	110
Toilets	30	32
Utility room		
General	20	22
Work counter	50	54
Waiting rooms, or areas		
General	20	22
Local for reading	30	32
X-Ray suite		
Radiographic, general	10	11
Fluoroscopic, general (variable)	0–50	0–54
Deep and superficial therapy	10	11
Control room	10	11
Film viewing room	30	32
Darkroom	10	11
Light room	30	32
Filing room, developed films	30	32
Storage, undeveloped films	10	11
Dressing rooms	10	11
Ice making—engine and compressor room	20	22

Table 13A.1. (Continued)

Area	Footcandles on Tasks[a]	Dekalux[b] on Tasks[a]
Inspection		
Ordinary	50	54
Difficult	100	110
Highly difficult	200	220
Very difficult	500[d]	540[d]
Most difficult	1000[d]	1080[d]
Iron and steel manufacturing		
Open hearth		
Stock yard	10	11
Charging floor	20	22
Pouring slide		
Slag pits	20	22
Control platforms	30	32
Mold yard	5	5.4
Hot top	30	32
Hot top storage	10	11
Checker cellar	10	11
Buggy and door repair	30	32
Stripping yard	20	22
Scrap stockyard	10	11
Mixer building	30	32
Calcining building	10	11
Skull cracker	10	11
Rolling mills		
Blooming, slabbing, hot strip, hot sheet	30	32
Cold strip, plate	30	32
Pipe, rod, tube, wire drawing	50	54
Merchant and sheared plate	30	32
Tin plate mills		
Tinning and galvanizing	50	54
Cold strip rolling	50	54
Motor room, machine room	30	54
Inspection		
Black plate, bloom and billet chipping	100	110
Tin plate and other bright surfaces	200[g]	220[g]
Jewelry and watch manufacturing	500[d]	540[d]
Kitchens (see Food service facilities or Residences)		
Laundries		
Washing	30	32
Flat work ironing, weighing, listing, marking	50	54
Machine and press finishing, sorting	70	75
Fine hand ironing	100	110

Table 13A.1. (Continued)

Area	Footcandles on Tasks[a]	Dekalux[b] on Tasks[a]
Leather manufacturing		
Cleaning, tanning and stretching, vats	30	32
Cutting, fleshing and stuffing	50	54
Finishing and scarfing	100	110
Leather working		
Pressing, winding, glazing	200	220
Grading, matching, cutting, scarfing, sewing	300[d]	320[d]
Library		
Reading areas		
Reading printed material	30[c]	32[c]
Study and note taking	70[c]	75[c]
Conference areas	30[c]	32[c]
Seminar rooms	70[c]	75[c]
Book stacks (30 in. above floor)		
Active stacks	30[h]	32[h]
Inactive stacks	5[h]	5.4[h]
Book repair and binding	70	75
Cataloging	70[c]	75[c]
Card files	100[c]	110[c]
Carrels, individual study areas	70[c]	75[c]
Circulation desks	70[c]	75[c]
Rare book rooms—archives		
Storage areas	30	32
Reading areas	100[c]	110[c]
Map, picture, and print rooms		
Storage areas	30	32
Use areas	100[c]	110[c]
Audiovisual areas		
Preparation rooms	70	75
Viewing rooms (variable)	70	75
Television receiving room (shield viewing screen)	70	75
Audio listening areas		
General	30	32
For note taking	70[c]	75[c]
Record inspection table	100[d]	110[d]
Microform areas		
Files	70[c]	75[c]
Viewing areas	30	32
Locker rooms	20	22
Machine shops		
Rough bench and machine work	50	54

Table 13A.1. (Continued)

Area	Footcandles on Tasks[a]	Dekalux[b] on Tasks[a]
Medium bench and machine work, ordinary automatic machines, rough grinding, medium buffing, and polishing	100	110
Fine bench and machine work, fine automatic machines, medium grinding, fine buffing, and polishing	500[d]	540[d]
Extrafine bench and machine work, grinding, fine work	1000[d]	1080[d]
Materials handling		
Wrapping, packing, labeling	50	54
Picking stock, classifying	30	32
Loading, trucking	20	22
Inside truck bodies and freight cars	10	11
Meat packing		
Slaughtering	30	32
Cleaning, cutting, cooking, grinding, canning, packing	100	110
Municipal buildings—fire and police		
Police		
Identification records	150[c]	160[c]
Jail cells and interrogation rooms	30	32
Fire hall		
Dormitory	20	22
Recreation room	30	32
Wagon room	30	32
Offices		
Drafting rooms		
Detailed drafting and designing, cartography	200[c]	220[c]
Rough layout drafting	150[c]	160[c]
Accounting offices		
Auditing, tabulating, bookkeeping, business machine operation, computer operation	150[c]	160[c]
General offices		
Reading poor reproductions, business machine operation, computer operation	150[c]	160[c]
Reading handwriting in hard pencil or on poor paper, reading fair reproductions, active filing, mail sorting	100[c]	110[c]
Reading handwriting in ink or medium pencil on good quality paper, intermittent filing	70[c]	75[c]
Private offices		
Reading poor reproductions, business machine operation	150[c]	160[c]
Reading handwriting in hard pencil or on poor paper, reading fair reproductions	100[c]	110[c]

Table 13A.1. (Continued)

Area	Footcandles on Tasks[a]	Dekalux[b] on Tasks[a]
Reading handwriting in ink or medium pencil on good quality paper	70[c]	75[c]
Reading high contrast or well-printed materials	30[c]	32[c]
Conferring and interviewing	30	32
Conference rooms		
Critical seeing tasks	100[c]	110[c]
Conferring	30	33
Note-taking during projection (variable)	30[c]	33[c]
Corridors	20[c]	22[i]
Packing and boxing (see Materials handling)		
Paint manufacturing		
General	30	32
Comparing mix with standard	200[d]	220[d]
Paint shops		
Dipping, simple spraying, firing	50	54
Rubbing, ordinary hand painting and finishing art, stencil and special spraying	50	54
Fine hand painting and finishing	100	110
Extrafine hand painting and finishing	300[d]	320[d]
Paper-box manufacturing		
General manufacturing area	50	54
Paper manufacturing		
Beaters, grinding, calendering	30	32
Finishing, cutting, trimming, papermaking machines	50	54
Hand counting, wet end of paper machine	70	75
Paper machine reel, paper inspection, and laboratories	100	110
Rewinder	150	160
Plating	30	32
Polishing and burnishing	100	110
Power plants (see Central station)		
Post Offices		
Lobby, on tables	30	32
Sorting, mailing, etc.	100	110
Poultry industry (see also Farm—dairy)		
Brooding, production, and laying houses		
Feeding, inspection, cleaning	20	22
Charts and records	30	32
Thermometers, thermostats, time clocks	50	54
Hatcheries		
General area and loading platform	20	22
Inside incubators	30	32

Table 13A.1. (Continued)

Area	Footcandles on Tasks[a]	Dekalux[b] on Tasks[a]
Dubbing station	150	160
Sexing	1000	1080
Egg handling, packing, and shipping		
General cleanliness	50	54
Egg quality inspection	50	54
Loading platform, egg storage area, etc.	20	22
Egg processing		
General lighting	70	75
Fowl processing plant		
General (excluding killing and unloading area)	70	75
Government inspection station and grading stations	100	110
Unloading and killing area	20	22
Feed storage		
Grain, feed rations	10	11
Processing	10	11
Charts and records	30	32
Machine storage area (garage and machine shed)	5	5.4
Printing industries		
Type foundries		
Matrix making, dressing type	100	110
Font assembly—sorting	50	54
Casting	100	110
Printing plants		
Color inspection and appraisal	200[d]	220[d]
Machine composition	100	110
Composing room	100	110
Presses	70	75
Imposing stones	150	160
Proofreading	150	160
Electrotyping		
Molding, routing, finishing, leveling molds, trimming	100	110
Blocking, tinning	50	54
Electroplating, washing, backing	50	54
Photoengraving		
Etching, staging, blocking	50	54
Routing, finishing, proofing	100	110
Tint laying, masking	100	110
Professional offices (see Hospitals)		
Receiving and shipping (see Materials handling)		
Restaurants (see Food serfice facilities)		
Rubber goods—mechanical		
Stock preparation		

Table 13A.1. (Continued)

Area	Footcandles on Tasks[a]	Dekalux[b] on Tasks[a]
Plasticating, milling, Banbury	30	32
Calendering	50	54
Fabric preparation, stock cutting, hose looms	50	54
Extruded products	50	54
Molded products and curing	50	54
Inspection	200[d]	220[d]
Rubber tire manufacturing		
Banbury	30	32
Tread stock		
General	50	54
Booking and inspection, extruder, check weighing, width measuring	100[j]	110[j]
Calendering		
General	30	32
Letoff and windup	50	54
Stock cutting		
General	30	32
Cutters and splicers	100[j]	110[j]
Bead building	50	54
Tire building		
General	50	54
At machines	150[d]	160[d]
In-process stock	30	32
Curing		
General	30	32
At molds	70[d]	75[d]
Inspection		
General	100	110
At tires	300[j]	320[j]
Storage	20	22
Sawmills		
Grading redwood lumber	300[d]	320[d]
Service space (see also Storage rooms)		
Stairways, corridors	20	22
Elevators, freight and passenger	20	22
Toilets and wash rooms	30	32
Service stations		
Service bays	30	32
Sales room	50	54
Shelving and displays	100	110
Rest rooms	15	16
Storage	5	5.4

Table 13A.1. (Continued)

Area	Footcandles on Tasks[a]	Dekalux[b] on Tasks[a]
Sheet metal works		
Miscellaneous machines, ordinary bench work	50	54
Presses, shears, stamps, spinning, medium bench work	50	54
Punches	50	54
Tin plate inspection, galvanized	200[j]	220[j]
Scribing	200[j]	220[j]
Shoe manufacturing—leather		
Cutting and stitching		
Cutting tables	300[d]	320[d]
Marking, buttonholing, skiving, sorting, vamping, counting	300[d]	320[d]
Stitching, dark materials	300[d]	320[d]
Making and finishing, nailers, sole layers, welt beaters and scarfers, trimmers, welters, lasters, edge setters, sluggers, randers, wheelers, treers, cleaning, spraying, buffing, polishing, embossing	200	220
Shoe manufacturing—rubber		
Washing, coating, mill run compounding	30	32
Varnishing, vulcanizing, calendering, upper and sole cutting	50	54
Sole rolling, lining, making and finishing processes	100	110
Soap manufacturing		
Kettle houses, cutting, soap chip and powder	30	32
Stamping, wrapping and packing, filling and packing soap powder	50	54
Stairways (see Service space)		
Steel (see Iron and steel)		
Stone crushing and screening		
Belt conveyor tubes, main line shafting spaces, chute rooms, inside of bins	10	11
Primary breaker room, auxiliary breakers under bins	10	11
Screens	20	22
Storage battery manufacturing		
Molding of grids	50	54
Storage rooms or warehouses		
Inactive	5	5.4
Active		
Rough bulky	10	11
Medium	20	22
Fine	50	54
Structural steel fabrication	50	54

Table 13A.1. (Continued)

Area	Footcandles on Tasks[a]	Dekalux[b] on Tasks[a]
Sugar refining		
Grading	50	54
Color inspection	200[d]	220[d]
Testing		
General	50	54
Extrafine instruments, scales, etc	200[d]	220[d]
Textile mills—cotton		
Opening, mixing, picking	30	32
Carding and drawing	50	54
Slubbing, roving, spinning, spooling	50	54
Beaming and splashing on comb		
Gray goods	50	54
Denims	150	160
Inspection		
Gray goods (hand turning)	100	110
Denims (rapidly moving)	500[d]	540[d]
Automatic tying-in	150	160
Weaving	100	110
Drawing-in by hand	200	220
Textile mills—silk and synthetics		
Manufacturing		
Soaking, fugitive tinting, and conditioning or setting of twist	30	32
Winding, twisting, rewinding and coning, quilling, slashing		
Light thread	50	54
Dark thread	200	220
Warping (silk or cotton system)		
On creel, on running ends, on reel, on beam, on warp at beaming	100	110
Drawing-in on heddles and reed	200	220
Weaving	100	110
Textile mills—woolen and worsted		
Opening, blending, picking	30	32
Grading	100[d]	110[d]
Carding, combing, recombing and gilling	50	54
Drawing		
White	50	54
Colored	100	110

Table 13A.1. (Continued)

Area	Footcandles on Tasks[a]	Dekalux[b] on Tasks[a]
Spinning (frame)		
White	50	54
Colored	100	110
Spinning (mule)		
White	50	54
Colored	100	110
Twisting		
White	50	54
Winding		
White	30	32
Colored	50	54
Warping		
White	100	110
White (at reed)	100	110
Colored	100	110
Colored (at reed)	300[d]	320[d]
Weaving		
White	100	110
Colored	200	220
Gray-goods room		
Burling	150	160
Sewing	300[d]	320[d]
Folding	70	75
Wet finishing, fulling, scouring, crabbing, drying	50	54
Dyeing	100[d]	110[d]
Dry finishing, napping, conditioning, pressing	70	75
Dry finishing, shearing	100	110
Inspecting (perching)	2000[d]	2150[d]
Folding	70	75
Tobacco products		
Drying, stripping, general	30	32
Grading and sorting	200[d]	220[d]
Toilets and wash rooms	30	32
Upholstering—automobile, coach, furniture	100	110
Warehouse (see Storage rooms)		
Welding		
General illumination	50	54
Precision manual arc welding	1000[d]	1080[d]

Table 13A.1. (Continued)

Area	Footcandles on Tasks[a]	Dekalux[b] on Tasks[a]
Woodworking		
Rough sawing and bench work	30	32
Sizing, planing, rough sanding, medium quality machine and bench work, gluing, veneering, cooperage	50	54
Fine bench and machine work, fine sanding and finishing	100	110

[a] Minimum on the task at any time for young adults with normal vision and better than 20/30 corrected vision. Other general notes given elsewhere in this appendix.

[b] The dekalux is an SI unit equal to 0.929 footcandles; 1 dekalux = 10 lux.

[c] Equivalent sphere E_s illumination; see appendix text.

[d] Obtained with a combination of general lighting plus specialized supplementary lighting. Care should be taken to keep within the recommended luminance ratios. These seeing tasks generally involve the discrimination of fine detail for long periods of time and under conditions of poor contrast. The design and installation of the combination system must not only provide a sufficient amount of light, but also the proper direction of light, diffusion, color, and eye protection. As far as possible it should eliminate direct and reflected glare as well as objectionable shadows.

[e] Special lighting such that (1) the luminous area shall be large enough to cover the surface that is being inspected, and (2) the luminance shall be within the limits necessary to obtain comfortable contrast conditions. This involves the use of sources of large area and relatively low luminance in which the source luminance is the principal factor rather than the footcandles produced at a given point.

[f] Including street and nearby establishments.

[g] The specular surface of the material may necessitate special consideration in selection and placement of lighting equipment, or orientation of the work.

[h] Vertical.

[i] Or not less than one-fifth the level in adjacent areas.

[j] Localized general lighting.

CHAPTER FOURTEEN

Air Pollution

GEORGE D. CLAYTON

1 INTRODUCTION

The first edition of Patty's Volume I in 1948 did not contain a chapter on air pollution, indicating the contemporary lack of emphasis and community interest in the subject. However in October 1948, the Donora, Pennsylvania, episode (see below) occurred, kindling public interest. Thus the 1958 edition of Volume I included my contribution on the subject. The 18 years that have elapsed since then are interesting to contemplate. There has been a greater concentration of activity in science, medicine, and legislation relating to this subject during this period than during the entire preceding era. However the questions raised then are still valid and remain unanswered.

History cannot be changed; obviously we must continue to record it. Those who first became aware of "air pollution" in the past decade often tend to think that air pollution and its effects are of late twentieth century origin. On the contrary, air pollution is as old as the history of fuel, and the magnitude of the problem has increased along with the increase in the industrialization and growth of urban populations. An early prophetic clue of things to come was given in 361 B.C., when Theophrastus noted that "fossil substance called 'coals' burn for a long time, but the smell is troublesome and disagreeable." By 65 B.C., the poet Horace was lamenting that the shrines of Rome were blackened by smoke. In 1273 in the reign of Edward I, the first smoke abatement law was passed in England in response to the people's fears that the air pollutants were detrimental to health. During this period it was believed that food cooked over burning coals would cause illness and even death. In 1306 the people became so concerned that a Royal Proclamation was signed, prohibiting the burning of coal in London. Because an owner of an industry was caught disobeying this Royal Proclamation, he was tried, found guilty, and beheaded. This is the first recorded penalty given as a result of violating an air pollution code.

The effect of air pollution on health has been of concern to people since its inchoation. As mentioned previously, the first air pollution law was passed in an attempt to curb air pollutants because of their detrimental health effects. During the first half of this century four major occurrences have justified this concern. The four principal recorded air pollution disasters occurred in the Meuse Valley, Belgium, in 1930; in Donora, Pennsylvania, in 1948; in Poza Rica, Mexico, in 1950; and in London in 1952. Since 1952 there has been statistical evidence of increased deaths associated with abnormally high air pollution levels in various communities, such as in New York City (1, 2), and most recently in the autumn of 1975 in Pittsburgh, where there were 14 excess deaths (3) recorded during the period of inversion. The four major disasters are discussed below.

Meuse Valley. The first well-known air pollution disaster occurred between Seraning and Hug, in the Meuse Valley, Belgium. During the period December 1–5, 1930, a large number of persons were taken ill, and more than 60 died. Older persons with previously known disease of the heart or lungs accounted for the majority of the fatalities. The signs and symptoms, primarily those caused by a respiratory irritant, included chest pain, cough, shortness of breath, and irritation of the mucous membranes and of the eyes. The area of the Meuse Valley where the episode occurred is approximately 15 miles long, and 1½ miles wide and is surrounded by hills 330 feet high. Within this area at the time of the disaster, there were the following types of industry:

1. Four very large steel plants, each having coking installations, blast furnaces, rolling mills, welding furnaces, boilers, and other operations compatible with large steel plants.
2. Three large metallurgical works.
3. Four electric power generating plants.
4. Six glass works, ceramic plants, or brick works, which were equipped with coal or producer-gas heating furnaces.
5. Three groups of lime kilns.
6. Three zinc plants, each equipped with reduction furnaces, retort drying ovens, and other operations compatible with zinc plants.
7. One coking plant (other than those associated with steel plants).
8. One sulfuric acid plant with roasting ovens.
9. One concentrated fertilizer plant, with concentration and drying ovens.

In addition to these industries discharging pollutants into the atmoshphere, most of the homes and communities burned coal, thus adding considerably to the pollution load. Other sources of pollution at the time of the episode were railroads and automobiles.

The investigators considered the sources of pollution and drew up the following list of substances that could have been, and probably were, present in the atmosphere at the time of the disaster:

Hydrogen
Sulfur dioxide
Sulfur trioxide (sulfuric acid)
Hydrochloric acid
Hydrofluoric acid and its salts, ammonium fluoride and zinc fluoride
Carbon monoxide
Carbon dioxide
Hydrosulfuric acid
Nitric oxide
Nitrogen dioxide (NO_2)
Nitrous acid (HNO_2)
Nitric acid (HNO_3)
Ammonia and ammonium salts (thiocyanates, sulfides)
Ammonium sulfide
Saturated and unsaturated hydrocarbons, including natural gas and gasoline vapors
Odor of organic products (not identified) from phosphate plants
Drops of tar, phenol, naphthalene, and so forth
Soot
Cement dust
Lime dust
Metal dust
Zinc oxide
Lead
Arsenous acid anhydride
Arsine
Methyl alcohol
Ethyl alcohol
Formaldehyde
Zinc chloride
Silica from Bessemer steel works

The meteorological investigators found the conditions for Monday, December 1, through Friday, December 5, 1930, to have been anticyclonic, characterized by high atmospheric pressures and mild winds of a general easterly direction. During this period the fog became increasingly worse. There was considerable cooling during the night, often as much as 10°C. The wind was a very mild east wind, not exceeding 5 miles per hour and often dropping during the days of December 2, 3, and 4 to 1.6 miles per hour.

After the investigators had checked all the available data on the medical, meteorological, and environmental factors, efforts were made to determine the causative agent. Many of the contaminants could be eliminated immediately because their known toxicological action was not compatible with the medical findings. Other contaminants were eliminated because it was decided that the concentrations could not have been sufficient to cause the damage found. From all the findings the investigators concluded the contaminants most likely to have caused the symptoms were sulfur compounds; however they added that the synergistic effect of the other contaminants may have played an important part.

Donora *(5).* The second well-known air pollution disaster occurred in Donora, Pennsylvania, during the period of Wednesday, October 27, to Sunday, October 31, 1948. Within this period 20 people died and 6000 became ill.

At the time the investigation was being conducted, there were the following industries in Donora:

1. One large steel plant having blast furnaces, looping rod mills, wire drawing department, and wire finishing department.

2. One zinc plant equipped with reduction furnaces, retort drying ovens, and other operations compatible with zinc plants.
3. One sulfuric acid plant.

Other heavy industries in the nearby area included two steel companies and one by-product plant in Monessen, a steel and by-product plant in Clairton, a glass company in Charleroi, and a power company and a railroad in Elrama.

A study of the plant processes and analyses of the raw materials used in those industries led the investigators to analyze the atmosphere for the following constituents:

Total particulate matter	Acid gases	Carbon monoxide
Lead	Hydrogen	Oxygen
Cadmium	Arsine	Oxide of nitrogen
Zinc	Arsenic	Stibine
Iron	Sulfur dioxide	Manganese
Chloride	Total sulfur	Iron carbonyl
Fluoride	Carbon dioxide	

The investigators made the following conclusions:

> It seems reasonable to state that while no *single* substance was responsible for the October 1948 episode, the syndrome could have been produced by a combination or summation of the action of two or more of the contaminants. Sulfur dioxide and its oxidation products, together with particulate matter, are considered significant contaminants. However, the significance of the other irritants as important adjuvants to the biological effects, cannot be finally estimated on the basis of present knowledge.
>
> Information available on the toxicological effects of mixed irritant gases is meager, and data on possible enhanced action due to adsorption of gases on particulate matter is limited. Further, available toxicological information pertains mainly to adults in relatively good health. Hence, the lack of fundamental data on the physiologic effects of a mixture of gases and particulate matter, over a period of time, is a severe handicap in evaluating the effects of atmospheric pollutants on persons of all ages and in various stages of health.

Although sufficient data were not available to determine the exact causative agent, sufficient environmental data were available to make ten recommendations which, if fulfilled, would prevent a recurrence of the disaster.

Poza Rica (6). The third much-publicized incident occurred in Poza Rica, Mexico, on November 24, 1950, setting a record of 22 persons dead and 320 people hospitalized.

A review of the history of the incident shows that the sulfur-removal unit (Girbotol unit) began operation on November 21, 1950. A few days after the unit was in operation, trouble developed when the amine solution overflowed, partly plugging the gas lines to the pilot lights of the flares. However the flare appeared normal under the reduced rate of flow.

On November 24, 1950, at approximately 2:00 A.M., efforts were made to increase the rate of flow of gas to the plant's rated capacity of 60 billion cubic feet per day. The desired rate of flow was reached at approximately 2:30 A.M. At approximately 4:00 A.M. difficulty was encountered, and the gas flow began surging through the unit, with the probable result that unburned hydrogen sulfide escaped into the atmosphere.

The available meteorological data indicate a pronounced low altitude temperature inversion, a high concentration of haze and nuclei, and a very slight wind movement prevailing.

The epidemiological study indicated that the time of onset of the incident was about 4:50 A.M., and the acute phase was ended about 5:10 A.M. The victims were affected in a geographical area in direct proximity to the effluent stack of the Girbotol unit. During this short period the acute exposure to the atmospheric pollution caused the death of 22 people and the hospitalization of 320.

The onset, the symptoms and signs, as well as the pathological findings, are consistent with hydrogen sulfide poisoning, and there were no findings that conflicted with this diagnosis. Therefore it is to be concluded that the hydrogen sulfide that caused this morbidity and mortality came from the effluent stack of the Girbotol unit.

***London** (7).* During the period December 5–8, 1952, the fourth and worst air pollution disaster occurred. When these deaths were being recorded, the London area experienced periods of smog, culminating in one of much intensity. The onset was accompanied by meteorological factors of low temperature inversion and almost complete absence of wind or air movement. No radio or balloon-sonde ascents were made within the fog belt. Ascents at Cardington, 50 miles north of London, where visibility did not fall below 1000 yards during the period, show a continuous inversion from ground level up to heights between 500 and 1000 feet. Reports from aircraft landing at London Airport after the fog had cleared temporarily place the height of the inversion at 200 to 300 feet, with a haze layer extending to 2000 feet.

This is the first air pollution disaster for which air sampling was conducted before, during, and after the episode. The data collected showed that the particulate matter (obtained by sampling with filter paper and reporting the degree of color in milligrams of total weight per cubic meter of air) ranged from 0.30* to 0.84 mg per cubic meter of air, with an average of 0.50 for the 2-week period preceding the disaster. On the days when deaths were occurring, the values increased to a maximum of 4.46 mg per cubic meter of air, a ninefold increase over the previous average.

Sulfur dioxide ranged from 0.09 to 0.33 ppm, with an average of 0.15, for the 2-week period preceding the disaster. During the period when deaths were occurring, the SO_2 values increased to a maximum of 1.33 ppm, which was also a ninefold increase over the average of the preceding days.

During the weeks ending December 13 and December 20, 1952, at least 3000 more deaths occurred than would be expected. Although the increase was present in every age

* To compare this figure with United States measurements, use a factor of 1.8 for approximate relationships.

group, the greatest increase was in the age group of 45 years and over. More than 80 percent of these deaths occurred among individuals with known heart disease and respiratory disease. As in other disasters reported, the symptoms produced were associated with irritation of the respiratory tract.

The British Ministry of Health concluded: "The fog was, in fact, a precipitating agent operating on a susceptible group of patients whose life expectation, judging from their pre-existing diseases, must even in the absence of fog, have been short."

A study of the four major disasters reveals the following similarities:

1. People died, and many more became ill.
2. Each area was subject to an unusually long period of inversion.
3. Each area was subject to natural fog.
4. There was no new or unusual activity in the community before the disaster. Each area (except Poza Rica) had been essentially the same for many years with no unusual effect on the people.
5. The symptoms experienced were similar to those caused by a respiratory irritant.
6. Older people, and especially those with previous heart and respiratory disease (except in Poza Rica), were more severely affected than the younger age groups and normal health groups.
7. With the exception of Poza Rica, the agent or agents that caused the deaths are unknown.

Since 1958 there has been a plethora of activity in the field of air pollution, especially in the United States. The federal government has been extremely active since 1969 in passing legislation to control mobile and stationary sources of pollution, and in establishing standards for emission as well as for ambient air for six common pollutants (total particulates, carbon monoxide, sulfur dioxide, photochemical oxidants, nitrogen dioxide, and hydrocarbons). The Environmental Protection Agency (EPA) was established, having authority in all areas of pollution: air, water, and solid waste. All states have passed air pollution laws, as have most of the major cities. California has continued to be the most active state in its efforts to control air pollution.

Extensive studies since 1958 have been conducted on health, vegetation, and economic damage. Control procedures have been refined for both stationary and mobile sources. Numerous studies have been devoted to finding a viable alternative to the internal combustion engine. These efforts to date have not been successful. The auto industry has spent millions of dollars seeking alternate engines and reducing emissions from the reciprocating engine. Citizens' groups have been formed in recent years to actively pursue the goal of a cleaner environment. Many technical societies now devote a good portion of their committee activities and annual programs to reporting on the technical aspects of evaluating and controlling air pollution; for example, the Air Pollution Control Association is interested exclusively in this endeavor and has evidenced a phenomenal growth since 1958.

This tremendous activity has produced gratifying results. From the youngest school child in the nation to the president of the largest corporation, everyone is very much aware of the undesirable effects of air pollution. Billions of dollars have been spent on the evaluation and control of air pollution, just in the past decade. In the early 1970s people believed that air pollution should be controlled at any cost. This was explicit in the language of the legislation enacted in 1969. Legal requirements were established, and the courts subsequently ruled that the financial inability of a company to pay for controls was *not* a legitimate excuse for not installing controls. In 1976 the Supreme Court of the United States ruled that federal approval of state clean air programs may not be overturned on the grounds that the plans cost too much or require new types of equipment. In a unanimous opinion, the court stated that Congress intended to have tough standards that would force the development of new equipment and methods to alleviate air pollution. The 1970 amendments to the Clean Air Act required the states to design clean air plans and submit them for approval by the EPA. The ruling upheld the U.S. Court of Appeals in Chicago, which had refused to consider a case filed by an electric utility in a metropolitan area of Missouri. However it has become apparent that cost considerations must be weighed along with the benefits of the controls, if the economy of the United States is to continue to grow and provide the needed, new jobs for our growing population. The 1975–1976 Los Angeles County Grand Jury, expressing concern over the "economic consequences of overzealous use of air-pollution restrictions on industry," recommended relaxing the rules to encourage new plants to locate there. The chairman of the grand jury committee supplied the following reasoning to support the recommendation:

> There has been a tendency to prohibit any new industry from locating in Los Angeles County if that industry would be a source of air pollutants, no matter how relatively minute the quantities. The Committee believes there should be a balance between the economic and environmental concerns.

Thus we have a conflict between the idealism of the goals of the legislation passed by Congress, enthusiastically supported by environmentalists, and the practicalities of the economics of sustaining a growing population along with control of air pollutants.

The position of EPA is that before any *new* emissions can be added to an air basin, an equal or greater amount of existing emissions must be eliminated. "We cannot justify permitting a net increase in emissions in an area which is already violating the standards," stated Paul DeFalco, Jr., an EPA regional administrator (8).

The policies of the EPA are in keeping with the mandate from Congress. Leon Billings, a senior professional staff member of the subcommittee on Environmental Pollution of the Senate Public Works Committee, made a revealing statement in June 1976.

> To understand the evolution of the controversies it is useful and necessary to review implementation of the 1970 act. In 1970 the Congress set forth the basic goal of protection of public health from the adverse effects of air pollution, through achievement of fixed emission reductions by date certain. The goal was absolute. The timetables and tools for its achievement were not to be tempered by economic considerations. The purpose was to eliminate as quickly as humanly pos-

sible unhealthy and thus unreasonable levels of air pollution. The Clean Air Act of 1970 was in that respect a significant departure from previous pollution policy.

The year 1975 was the time by which all major air pollution sources would eliminate their contribution to the problem. The auto industry was to have achieved clean air standards, and both stationary sources and urban areas requiring additional transportation controls were required to complete their portion of the work toward the goal of healthful air by that year. The clean areas of the country were required to protect their air resource from degradation—a policy first articulated in the 1967 Air Quality Act and required by federal guidelines until 1971 when the requirement was dropped at the insistence of the White House.

In 1958 I (10) raised the question of whether control of air pollution could become stringent enough to ruin an established business, regulate the manner of doing business and the use of property, prohibit it altogether, or require large expenditures for equipment. As legislation has evolved during the ensuing years, all this has come to pass. It is extremely difficult if not impossible for certain types of industrial facility to be built. New power generating facilities, or planned expansions of existing facilities, offer prime examples; extensive impact statements are required, and these are both time-consuming and costly. Not infrequently opposition to new facilities can delay approval so long that they are no longer economically feasible, as exemplified by the proposed power generating plant in southern Utah, which was successfully blocked (1976) after many years of litigation and millions of dollars of expenditure in an effort to obtain the necessary permits. The New Source Review Program of EPA admittedly is a tough program, which in some cases has to say "no" to new industry because of the firm federal mandate.

The desirability of establishing *national* standards is now being questioned. The concept of local options above those established at the federal level is being weighed. At the present time EPA has no authority to permit an industry or a community to continue to pollute the atmosphere in excess of the established standards, irrespective of the economic damage that might result to a community if the industry were forced to cease operations. Thus it would appear advantageous to permit *local options* to allow some freedom to local communities in establishing their environmental requirements.

It is also becoming increasingly evident that engineering controls are often inadequate to meet the stringent air quality standards. To meet these goals, social changes must be made, changes that could markedly affect each person living in the United States. EPA has already proposed to curtail the sale of gasoline in the Los Angeles Basin, thereby materially affecting the livelihood of thousands of commuters. Similar programs have been endorsed by the EPA for some of the other more populated cities of the United States. The drastic action now being considered probably was not anticipated by the legislators when Congress passed the environmental laws in 1969 and 1970. Although EPA has been receiving the brunt of criticism for such actions as those mentioned previously, the agency is only carrying out its Congressional mandate. Through its diligence, however, and the passage of environmental legislation, the freedom of individual communities to be masters of their destinities has been seriously eroded.

2 LEGISLATION

2.1 Federal Level

The federal environmental legislation apparently reflects three major Congressional viewpoints: (1) that there is an atmospheric concentration *below* which there would be no adverse effects on people or vegetation; (2) that atmospheric pollution was primarily created by man's activity; and (3) that the cost of controls could be borne by the country's economy without adversely affecting its viability.

Today it is clear that these three tenets have a shaky, if not a false foundation, and each is discussed further in this chapter.

During the earlier years of abatement effort, the federal government had a very limited involvement. The Public Health Service had conducted investigations relating to the Donora incident under the authority of the Public Health Service Act. The federal program in air pollution, under the aegis of the Department of Health, Education and Welfare, was established by Public Law 84-159 in July 1955. This legislation assigned primary responsibility for the control of air pollution to the states and to local governments, and gave the main objectives of the federal program as providing leadership and assistance to control programs throughout the country.

Prior to the passage of this legislation, the automobile had emerged as a major source of pollution. Two subsequent Congressional actions, the Schenck Act of July 1960 and the Air Pollution Control Act of 1962, called for the Surgeon General to study the exhaust emissions from motor vehicles. With the enactment of the Clean Air Act of 1963, federal policy underwent a significant evolution. Federal responsibility to state and local programs was reinforced. The act further singled out for special attention two of the major unsolved air pollution problems: motor vehicle exhaust, and sulfur oxide pollution from the burning of fossil fuels.

The Clean Air Act of 1963 was amended in 1965 and again in 1966. These amendments resulted in expansion of state and local control programs through federal grant support. Federal abatement activities were also strengthened. Motor vehicle exhaust emission standards applicable to the 1968 model year were promulgated (under the 1965 amendments).

The Air Quality Act of 1967 is distinguished from preceding legislation by its focus on regional activities. This act places emphasis on regional air pollution control programs implemented by state and local authorities, with the Department of Health, Education and Welfare assuming a leadership role. The act is interpreted as a clear invitation for government, industry, and private organizations to join together in constructive action to abate pollution.

The National Environmental Policy Act of 1969 was amended in January 1970 and on August 9, 1975. The purposes of this act are as follows:

To declare a national policy which will encourage productive and enjoyable harmony between man and his environment; to promote efforts which will prevent or eliminate damage to the environment and biosphere and stimulate the health and welfare of man; to enrich the understanding of the ecological systems and natural resources important to the Nation; and to establish a Council on Environmental Quality.

Executive Order 11514, dated March 5, 1970, spelled out the responsibilities of the federal agencies and of the Council on Environmental Quality.

On July 9, 1970, the President transmitted to Congress Reorganization Plan No. 3 of 1970, establishing the Environmental Protection Agency. Heading the new agency would be an administrator, appointed by the President with the advice and consent of the Senate. The responsibilities assigned to the new agency were the following:

1. All functions vested by law in the Secretary of the Interior and the Department of the Interior that are administered through the Federal Water Quality Administration, all functions that were transferred to the Secretary of the Interior by Reorganization Plan No. 2 of 1966 (80 Sta. 1608), and all the functions vested in the Secretary of the Interior or the Department of the Interior by the Federal Water Pollution Control Act or by provisions of law amendatory or supplementary thereof.
2. The functions vested in the Secretary of the Interior by the act of August 1, 1958 (72 Stat. 479, 16 U.S.C. 742d-1; an act relating to studies on the effects of insecticides, herbicides, fungicides, and pesticides on the fish and wildlife resources of the United States), and the functions vested by law in the Secretary of the Interior and the Department of the Interior that are administered by the Gulf Breeze Biological Laboratory of the Bureau of Commercial Fisheries at Gulf Breeze, Florida.
3. The functions vested by law in the Secretary of Health, Education and Welfare or in the Department of Health, Education and Welfare that are administered through the Environmental Health Service, including the functions exercised by the following components thereof: (i) the National Air Pollution Control Administration, and (ii) the Environmental Control Administration: (*a*) Bureau of Solid Waste Management, (*b*) Bureau of Water Hygiene, and (*c*) Bureau of Radiological Health. Activities dealing with industrial health and related matters are excluded.
4. The functions vested in the Secretary of Health, Education and Welfare of establishing tolerances for pesticide chemicals under the Federal Food, Drug and Cosmetic Act, as amended (21 U.S.C. 346, 346a, 348), together with authority, in connection with the functions transferred

(*a*) to monitor compliance with the tolerances and the effectiveness of surveillance and enforcement.

(*b*) to provide technical assistance to the states and conduct research under the federal Food, Drug, and Cosmetic Act, as amended, and the Public Health Service Act, as amended.

In addition, many of the functions relating to the environment formerly vested in the Council on Environmental Quality and in the Atomic Energy Commission, as well as certain functions of the Secretary of Agriculture and the Department of Agriculture, were transferred to the EPA.

Periodically the laws relating to the environment are reviewed by Congress. In a series of hearings held during 1975 and 1976 the areas of greatest controversy were the timetable for automotive emissions and the nondegradation clause of the 1970 act. Congress adjourned without any action.

2.2 State and Local Programs

In this limited presentation no attempt is made to discuss or classify the air pollution codes of the states and cities. Suffice it to say that each state and all major cities have laws on the books relating to the environment, and the reader can consult the specific governmental unit of interest.

3 AMBIENT AIR QUALITY STANDARDS

Chapter 19 discusses emission standards, along with air pollution control methodology. We are concerned here with ambient air quality standards, which are the limits promulgated to protect the health of the public and to provide a quality of atmosphere conducive to our well-being. In the United States the approach has been to legislate such standards, establishing time periods for compliance to be achieved. In contrast, the World Health Organization and the Commission of European Communities have established goals to which they aspire over an indefinite period of time.

3.1 United States

The Clean Air Act amendments of 1970 set mid-1975 as the deadline for achieving public health (primary) standards for ambient air quality. The quality of air in the United States is better at this writing, than it was in 1970, but the deadline has not been fully met, and air pollution levels at many locations still range above the primary standards.

Earlier in this chapter reference was made to the national ambient air quality standards designated for six major pollutants: total suspended particulates, sulfur dioxide, carbon monoxide, photochemical oxidants, nitrogen dioxide and hydrocarbons—the "criteria" pollutants. Table 14.1 presents pertinent data for these pollutants. As of mid-1975, the standard levels had been fully achieved in only 91 of the nation's 247 Air Quality Control Regions. At monitoring sites in nearly two-thirds of the regions, the pollution levels designated by the standards were still being exceeded for one or more pollutants. The standards are exceeded for suspended particulates in 118 of the 247 regions, for sulfur dioxide in 34 regions; for carbon monoxide in 69 regions, for oxidants in 79 regions, and for nitrogen dioxide in 16 regions, according to EPA estimates.

Table 14.1. Six Ambient Air Quality Standards (11)

Pollutant[a]	Characteristics	Principal Sources	Principal Effects	Controls	National Ambient Standards[b] $\mu g/m^3$	
Total suspended particulates (TSP)	Any solid or liquid particles dispersed in the atmosphere, such as dust, pollen, ash, soot, metals, and various chemicals; the particles are often classified according to size as settleable particles: larger than 50 μm; aerosols: smaller than 50 μm; and fine particulates: smaller than 3 μm	Natural events such as forest fires, wind erosion, volcanic eruptions; stationary combustion, especially of solid fuels; construction activities; industrial processes; atmospheric chemical reactions	*Health:* directly toxic effects or aggravation of the effects of gaseous pollutants; aggravation of asthma or other respiratory or cardiorespiratory symptoms; increased cough and chest discomfort; increased mortality *Other:* soiling and deterioration of building materials and other surfaces, impairment of visibility; cloud formation; interference with plant photosynthesis	Cleaning of flue gases with inertial separators, fabric filters, scrubbers, or electrostatic precipitators; alternative means for solid waste reduction; improved control procedures for construction and industrial processes	Primary Secondary Alert	Annual=75 24-hour=260 Annual=60 24-hour=150 24-hour=375
Sulfur dioxide (SO_2)	A colorless gas with a pungent odor; SO_2 can oxidize to form sulfur trioxide (SO_3), which forms sulfuric acid with water	Combustion of sulfur-containing fossil fuels, smelting or sulfur-bearing metal ores, industrial processes, natural events such as volcanic eruptions	*Health:* aggravation of respiratory diseases, including asthma, chronic bronchitis, and emphysema; reduced lung function; irritation of eyes and respiratory tract; increased mortality *Other:* corrosion of metals; deterioration of electrical contacts, paper, textiles, leather, finishes and coatings, and building stone; formation of acid rain; leaf injury and reduced growth in plants	Use of low sulfur fuels; removal of sulfur from fuels before use; scrubbing of flue gases with lime or catalytic conversion	Primary Alert	Annual=80 24-hour=365 24-hour=800
Carbon monoxide (CO)	A colorless, odorless gas with a strong chemical affinity for hemoglobin in blood	Incomplete combustion of fuels and other carbon-containing substances, such as in motor vehicle exhausts; natural events such as forest fires or decomposition of organic matter	*Health:* reduced tolerance for exercise, impairment of mental function, impairment of fetal development, aggravation of cardiovascular diseases *Other:* unknown	Automobile engine modifications (proper tuning, exhaust gas recirculation, redesign of combustion chamber); control of automobile exhaust gases (catalytic or thermal devices); improved design, operation, and maintenance of stationary furnaces (use of finely dispersed fuels, proper mixing with air, high combustion temperature)	Primary Alert	8-hour=10,000 1-hour=40,000 8-hour=17,000

Photochemical oxidants (O_x)	Colorless, gaseous compounds that can comprise photochemical smog [e.g., ozone (O_3), peroxyacetyl nitrate (PAN), aldehydes, and other compounds]	Atmospheric reactions of chemical precursors under the influence of sunlight	*Health:* aggravation of respiratory and cardiovascular illnesses, irritation of eyes and respiratory tract, impairment of cardiopulmonary function *Other:* deterioration of rubber, textiles, and paints; impairment of visibility; leaf injury, reduced growth, and premature fruit and leav drop in plants	Reduced emissions of nitrogen oxides, hydrocarbons, possibly sulfur oxides	Primary Alert	1-hour=160 1-hour=200
Nitrogen dioxide (NO_2)	A brownish-red gas with a pungent odor, often formed from oxidation of nitric oxide (NO)	Motor vehicle exhausts, high temperature stationary combustion, atmospheric reactions	*Health:* aggravation of respiratory and cardiovascular illnesses and chronic nephritis *Other:* fading of paints and dyes, impairment of visibility, reduced growth and premature leaf drop in plants	Catalytic control of automobile exhaust gases, modification of automobile engines to reduce combustion temperature, scrubbing flue gases with caustic substances or urea	Primary Alert	Annual=100 24-hour=282 1-hour=1130
Hydrocarbons (HC)	Organic compounds in gaseous or particulate form (e.g., methane, ethylene, acetylene)	Incomplete combustion of fuels and other carbon-containing substances, such as in motor vehicle exhausts; processing, distribution, and use of petroleum compounds such as gasoline and organic solvents; natural events such as forest fires and plant metabolism; atmospheric reactions	*Health:* suspected contribution to cancer *Other:* major precursors in the formation of photochemical oxidants through atmospheric reactions	Automobile engine modifications (proper tuning, crankcase ventilation, exhaust gas recirculation, redesign of combustion chamber); control of automobile exhaust gases (catalytic or thermal devices); improved design, operation, and maintenance of stationary furnaces (use of finely dispersed fuels, proper mixing with air, high combustion temperature); improved control procedures in processing and handling petroleum compounds	Primary	3-hour=160

Source. Based on information compiled by Enviro Control, Inc.

[a] Pollutants for which national ambient air quality standards have been established.

[b] Primary standards are intended to protect against adverse effects on human health. Secondary standards are intended to protect against adverse effects on materials, vegetation, and other environmental values.

The highest levels of particulates and sulfur oxides, the major pollutants from stationary sources (smokestacks) occur mainly in the northeastern and north central states. California continues to have the most severe automotive pollution problems. Occurrences of poor air quality are not restricted to any region or state, however, and in such widely separated large cities as Los Angeles, Chicago, and Philadelphia, primary standards for all the criteria pollutants are still exceeded at times. There are also problems in many small cities, and even in some rural areas.

Efforts to reduce air pollution have encountered a number of obstacles. Scientific knowledge and available technological and legal tools are sometimes insufficient, and attempts to reduce the nation's economic and energy difficulties are not always compatible with environmental goals. Some facilities have been granted extensions or variances that permit the burning of coal or oil with a high sulfur content. The full effects of such variances on ambient air quality are still uncertain. However between 1970 and 1976 the nation's efforts to improve air quality have accomplished a great deal.

The main factors governing ambient air pollution levels are as follows:

- Types and quantities of pollutants produced by human activities and natural sources.
- Controls used to reduce emissions of pollutants.
- Geographical location of pollution sources and of emissions over a given period of time.
- Dispersal and movement of pollutants, which depend on meteorological conditions and topography.
- Various chemical reactions of pollutants in the atmosphere often resulting in the formation of different types of pollutants.
- Processes by which pollutants are removed from the atmosphere, such as gravity and precipitation.

The first two factors determine the emissions that occur; the other four determine how the emissions affect air quality.

3.2 World Health Organization

The World Health Organization Technical Report Series No. 506 (12) states that the degree of health protection to be selected above the minimum acceptable level is a matter for political decision. That is, the appropriate authorities must decide on the level of health protection desirable for their society. Increments of health protection above the minimum acceptable level are generally purchased at ever-increasing increments in cost of control. Furthermore, the costs of the control program are directly related to the deadline by which it is to be operational; for example, it is more expensive to achieve the desired goals in 3 years than in 10 years. The zone in which increased health protection (benefit) is obtained at increasing control costs (cross-hatched area in Figure 14.1)

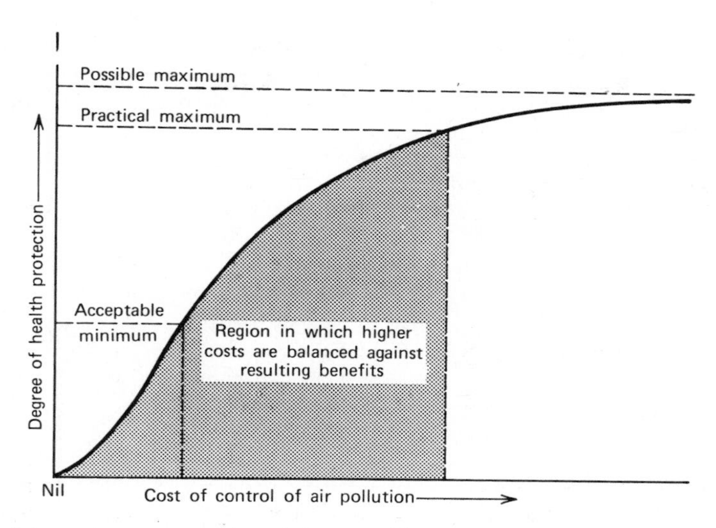

Figure 14.1 Schematic representation of degree of health protection as a function of cost of air pollution control.

is also the region of social decision making. Of course the level of protection desired must reflect awareness of the existing air pollution effects, but other considerations are also important, including general social, cultural, and economic factors, as well as the magnitude of other health problems.

Another issue faces environmental administrators responsible for making recommendations for the control of sulfur oxides and suspended particulates: the ratio between these pollutants varies from country to country, and there is no information documenting equivalent effects for the various concentrations of the two pollutants.

In the WHO report, effects are described as related to pollutant concentrations measured over short periods and over longer periods, indicating that different exposure times may be associated with different effects. This raises a problem for the air pollution control agency, which must be sure that the air quality standards adopted will protect from the effects of *both* short-term and long-term exposures. To solve this problem one must know, for example, the relation between the annual mean 24-hour value and the daily 24-hour values. If the effects against which protection is sought are known to be produced by exposure for 24 hours or less, any control measure stipulating an annual mean 24-hour value must take note of the variations expected, and must state the number of days per year on which the specified concentrations may be reached. Taking into consideration all the evidence available to it, the WHO committee agreed that in the light of present knowledge, the recommendations summarized in Table 14.2 could be offered as long-term goals intended to prevent undesirable effects from the air pollutants under discussion.

Table 14.2. Long-Term Goals Recommended by WHO

Pollutant and Measurement Method		Limiting Level ($\mu g/m^3$)
Sulfur oxides:[a] British Standard Procedure[b]	Annual mean	60
	98 percent of observations[c] below	200
Suspended particulates: [a] British Standard Procedure[b]	Annual mean	40[b]
	98 percent of observations[c] below	120
Carbon monoxide: nondispersive infrared[b]	8-hour average	10
	1-hour maximum	40
Photochemical: oxidant as measured by neutral buffered KI method expressed as ozone	8-hour average	60
	1-hour maximum	120

[a] Values for sulfur oxides and suspended particulates apply only in conjunction with one another.
[b] Methods are not those necessarily recommended but indicate those on which these units have been based. Where other methods are used an appropriate adjustment may be necessary. For example, to compare *annual* mean only to the United States procedure, multiply by 1.8 for appropriate comparison.
[c] The permissible 2 percent of observations over this limit may not fall on consecutive days.

3.3 Commission of the European Communities

In January 1976 the Health Protection Directorate, Commission of the European Communities (CEC) (13) prepared air pollution criteria and levels of action for the same pollutants studied by the World Health Organization. Essentially, their findings and philosophy were very similar to that of WHO. Because of the importance of sulfur dioxide and particulate matter as air pollutants, the data developed on these pollutants by CEC are presented in Table 14.3.

The criteria used to establish the relationships between given exposures and observable effects on man for sulfur dioxide and suspended particulate matter were as follows:

1. When sulfur dioxide and suspended particulate matter (determined as "black smoke"*) exceed simultaneously a mean value of 500 $\mu g/m^3$ for several days, excess mortality and increase in the number of hospitalizations among aged persons having in particular severe cardiovascular symptoms are observed.
2. When sulfur dioxide and suspended particulate matter exceed simultaneously concentrations of 250 $\mu g/m^3$ for several days, a subjective exacerbation of symptoms is

* The American high volume sampling method and the British method differ; although both express results in micrograms per cubic meter, they are not directly comparable.

Table 14.3. Health Protection Standards for Sulfur Dioxide and Suspended Particulates in Urban Atmospheres

Sulfur Dioxide

Reference Period	Maximum Concentrations	Associated Concentrations of Suspended Particulates
Year	Median of daily means, 80 $\mu g/m^3$	Annual median of daily means >40 $\mu g/m^3$
Year	Median of daily means, 120 $\mu g/m^3$	Annual median of daily means <40 $\mu g/m^3$
Winter (October–March)	Median of daily means, 130 $\mu g/m^3$	Winter median of daily means >60 $\mu g/m^3$
Winter (October–March)	Median of daily means, 180 $\mu g/m^3$	Winter median of daily means <60 $\mu g/m^3$
24 hours	Arithmetic mean, 250 $\mu g/m^3$	Arithmetic mean of concentration over 24 hours >100 $\mu g/m^3$
24 hours	Arithmetic mean, 350 $\mu g/m^3$	Arithmetic mean of concentration over 24 hours <100 $\mu g/m^3$

Suspended Particulates[a]

Reference Period	Maximuma Concentrations
Year	Median of daily means,[a] 80 $\mu g/m^3$
Winter (October–March)	Median of daily means, 130 $\mu g/m^3$
24 hours	Arithmetic mean, 250 $\mu g/m^3$

[a] To compare annual mean only to the United States procedure, multiply by 1.8 for appropriate comparison.

observed in patients having chronic bronchitis. This exacerbation is much less pronounced when only sulfur dioxide exceeds these levels.

3. For levels slightly lower than 250 $\mu g/m^3$ (daily concentrations) for sulfur dioxide and suspended particulate matter, there are indications that sensitive persons exhibit temporary changes in their pulmonary respiratory functions.

4. When sulfur dioxide and suspended particulate matter exceed simultaneously 100 $\mu g/m^3$ as long-term averages,* respiratory symptoms in the form of increased infection

* This amount is used in the British Standard Procedure.

of the lower respiratory tract and decrease in the maximum expiratory flow rates are observed in children.

3.4. Overview of Ambient Air Quality Standards

Air pollution is undesirable, and no reputable scientist to date has indicated that air contaminants are beneficial. The goal of reducing air contaminants to an absolute minimum should be encouraged, compatible with other national goals, such as employment, financial stability, and other health-related programs. Ambient air quality standards were promulgated to protect public health (primary standards) and to provide a more enjoyable environment (secondary standards). Since standards are nationwide in their impact, their enforcement will have a major influence on the future development of this country.

The primary standards are reviewed periodically, and each of these reviews has indicated concurrence with the present standards (14). However a number of health scientists believe that the primary standards are unnecessarily restrictive, thus creating an economic burden on and deterrent to national growth. They point out that all the epidemiological studies conducted thus far relate to people exposed to a multitude of contaminants as well as many other environmental factors, and they question the validity of relating injury and/or death to any individual contaminant.

During the 1970s, the EPA conducted an extensive investigation relating to community health entitled "Community Health and Environmental Surveillance System" (CHESS). The validity of the data acquired in these studies has been questioned by scientists throughout the world, however. Publicity concerning the CHESS studies led to a Congressional investigation of the various charges surrounding the CHESS studies, through the staffs of the Subcommittee on Special Studies, Investigations, and Oversight and the Subcommittee on the Environment and the Atmosphere. Their report states that the CHESS program has historical value as a first attempt at a broad-based definitive study relating air quality to health effects in a precise, quantitative manner, and that the program makes a contribution to the general field of air pollution epidemiological studies. However procedural problems were not solved in advance of the studies, resulting in too many inconsistencies in the data, and numerous technical problems led to large data uncertainties or errors associated with the aerometrics. Therefore the results of this program could not provide quantitative support for policy decisions. The 25 $\mu g/m^3$ lower sensitivity limit of the method used for most sulfur dioxide measurements coupled with the overall large error band on all measurements (possibly exceeding 100 percent) and the apparent bias of sulfur dioxide concentration data on the low side, make most of the numbers presented in the CHESS monograph unusable. In particular, the data are too imprecise and inaccurate in the range of the sulfur dioxide standard (80 $\mu g/m^3$) to allow the determination of any health effects threshold. Specifically, the staffs made the following criticisms:

1. The data and analysis in the 1974 CHESS Monograph do not provide a reliable, quantitative basis for relating SO_2 or sulfate levels in the air to adverse health effects.

2. The CHESS program did appear to demonstrate adverse health effects in association with local air pollution of undefined level and pollutant mix. Studies of chronic respiratory disease prevalence rates provided the most consistent results. These results are not translatable to communities other than those measured (an objective of the research) because the error uncertainties in the aerometric data made it impossible to arrive at reliable quantitative relationships.

The Congressional staffs performed a thorough job of evaluating the CHESS program. If one carefully studies their findings and relates their critique to many other epidemiological studies conducted in the United States and elsewhere, it becomes obvious that the conclusions could apply equally to certain other studies.

The impact on the United States of the requirement of meeting the six national air quality standards will be monumental. Billions of dollars will be required; growth and patterns of behavior of entire cities and states could be affected. In fact, the economical strength of this country could be influenced by compliance with the air pollution standards.

Because the standards are based primarily on health effects, I made an extensive evaluation of the health effects reported in the literature. Tables 14.4 through 14.9 present in abstracted form the data from the current literature on these six contaminants; pertinent comments on each follow.

3.4.1 Nitrogen Dioxide

Oxides of nitrogen are produced by the combustion of fossil fuels, and also by chemical and nitration industries. Most studies on the effects of oxides of nitrogen have focused on nitrogen dioxide (NO_2), since other oxides of nitrogen react in air to produce NO_2. However the affinity of nitric oxide (NO) has not been well defined, and may be found to be as important as NO_2.

The number of community studies on human health effects of NO_2 are limited. A review of Table 14.4 brings out several effects of NO_2. The odor threshold for this gas is approximately 1 to 3 ppm, and it is capable of producing irritation to the eyes and nose at concentrations ranging from 10 to 15 ppm (15, 44). One study reported an increase in airway resistance in humans at concentrations of 1.6 to 2.0 ppm (28). However the only major epidemiological studies involving oxides of nitrogen were carried out in Chattanooga, Tennessee, where a TNT factory was responsible for high concentrations of NO_2 in different areas of town. A survey of school children showed a significant decrease in the pulmonary function of youngsters exposed to high levels of NO_2 (0.06 to 0.10 ppm) when compared to individuals in a control area. Although SO_2 and "total particulates" were also present, these pollutants did not seem to account for the health effects observed (26). The same study showed that the incidence of respiratory illness in families was higher in the 0.06–0.1 ppm NO_2 area (25). There was an increased bronchitis rate in this area for children who had been exposed to 0.1 to 0.15 ppm of NO_2 for 2 to 3 years (29).

Animal studies have been carried out at different levels of NO_2 exposure, and the health effects generally produced by this contaminant were increased mortality from

Table 14.4. Critical Literature Review of NO_2, 1961–1973

Reference	Publication Date	Country	Type of Study: Epidem.	Type of Study: Lab.	Exposure	Contaminants Measured
15	1961	U.S.		×	Controlled exposure	NO_2
16	1962	U.S.		×	Controlled exposure	NO_2
17	1963			×	Controlled exposure	NO_2 and an aerosol of *K. pneumonia*
18	1964	U.S.		×	Controlled exposure	NO_2 and carbon particles
19	1965	U.S.		×	Controlled exposure	NO_2
20	1965			×	Controlled exposure	NO_2
21	1965			×	Controlled exposure	NO_2
22	1968 1969			×	Controlled exposure	NO_2
23	1968			×	Controlled exposure	NO_2
24	1969			×	Controlled exposure	NO_2

Concentrations[a]	Effects Studied	Effects Observed (Concentration)
1–50 ppm	NO_2 exposures on small number of volunteers	1–3 ppm Odor threshold 13 ppm Eye and nasal irritation 25 ppm For 5 min pulmonary discomfort 50 ppm For 1 min increased nasal irritation and pulmonary discomfort
0–104 ppm	Threshold concentrations for rats at various exposure times	Threshold concentrations based on borderline changes in lung to body weight ratios: 5 min, 104 ppm; 15 min, 65 ppm; 60 min, 28 ppm
0–3.5 ppm NO_2	Susceptibility of mice to *K. pneumonia* in presence of NO_2	2.5 ppm, no increased susceptibility; 3.5 ppm for 2 hr, doubled the mortality rate
100–1000 ppm NO_2	Effect of NO_2 on lungs of mice and carbon as a carrier of NO_2	250 ppm or greater of NO_2 alone, respiratory distress during and following exposure
5 ppm	Effect of continuous exposure of NO_2 for 90 days on 50 rats and 100 mice	5 ppm for 90 days: rats, 18% mortality; mice, 13% mortality
5–25 ppm	Intermittent exposures for 18 months on laboratory animals	5–25 ppm: no increased mortality, no changes in body weight, hemoglobin values, or biochemical indices
5–15 ppm	Effect of NO_2 on the lung tissues of guinea pigs	Daily exposures of 5–15 ppm resulted in increasing titers of lung tissue serum
0.8–25 ppm	Effect of NO_2 on respiratory tract of rats	0.8–2 ppm, rats grew and survived normally; 4.0 ppm for 16 weeks, terminal bronchiolar epithelium was broadened and hypertrophied; 10–25 ppm, developed large air-filled heavy lungs and died of respiratory failure
26 ppm	Effect on pulmonary function in lungs of dogs	26 ppm for 6 months, marked histopathologic changes in the lungs; however 225 days after cessation of NO_2 exposure, no significant pulmonary function alterations
0–40 ppm	Effect on germ-free mouse lungs	40 ppm (6–8 weeks), proliferation of epithelial cells of bronchi to form

Table 14.4. (Continued)

Reference	Publication Date	Country	Type of Study		Exposure	Contaminants Measured
			Epidem.	Lab.		
25	1970	U.S.	×		Total exposure	NO_2; suspended nitrates
26	1970	U.S.	×		Total exposure	NO_2; suspended nitrates
44	1971	Japan	×		Total exposure	SO_2, CO, NO_2,O_3, TSP
45	1972	U.S.		×	Controlled exposure	NO_2, SO_2, CO, PbClBr
27	1972	U.K.	×		Total exposure	NO_2, NO
28	1973			×	Controlled exposure	NO_2
29	1973	U.S.	×		Total exposure	NO_2
30	1973		×		Total exposure	NO_2, CO, HC
31	1973			×	Controlled exposure	NO_2

[a] Primary national standard (United States): ~0.05 ppm per year.

Concentrations[a]	Effects Studied	Effects Observed (Concentration)
		multicellular membrane projecting into lumen; also increased epithelial cells in alveolar spaces
NO_2, 0.062–0.109 ppm; SN, 3.8–7.2 μg/m^3	Acute respiratory illness in families	0.062–0.109 ppm (24 weeks), 18.8% excess in acute respiratory illness in this range
NO_2, 0.062–0.109 ppm; SN, 3.8–7.2 μg/m^3	Pulmonary function in elementary school children	0.062–0.109 ppm (24 weeks), decreased pulmonary function in children exposed to high levels of NO_2
Varying levels of all contaminants	Effect of NO_2 exposure on humans is reviewed	1 ppm, perceptibility threshold; 10–15 ppm, irritation of eyes, nose, and upper airway 25 ppm, safety threshold of short-term exposure
Varying levels of all contaminants	Effects on cynomolgus monkeys	6.78 ppm (NO_2 only), increased osmotic fragility of erythrocytes
88–167 ppm (NO_2 + NO)	Chronic lung disease in coal miners.	88–167 ppm, 84% of the men had emphysema based on residual volume of more than 150%
1–5 ppm (1880–9400 μg/m^3)	Pulmonary function of healthy and diseased adults	*Healthy adults*: 5 ppm (15 min), decrease in diffusion capacity *Diseased adults*: 4–5 ppm (60 min), decrease in arterial partial pressure of O_2 *Healthy and diseased adults*: 1–1.6 ppm (30 breaths); no effect; 1.6–2.0 ppm (30 breaths), increased airway resistance
0.10–0.15 ppm (188–282 μg/m^3)	Bronchitis rates of children exposed for 2–3 years	0.10–0.15 ppm, significantly higher bronchitis rates than children in low exposure areas
NO_2, 0.005–0.09 ppm; CO, 1–44 ppm; HC, 4.4–15.8 ppm	Chronic bronchitis in policemen	Increased chronic bronchitis, though not significantly, among policemen exposed for longer durations
2.9 ppm	Pulmonary effects on laboratory animals	2.9 ppm, significant pulmonary abnormalities

Table 14.5. Critical Literature Review of Photochemical Oxidants, 1960–1974

Reference	Publication Date	Country	Type of Study		Exposure	Contaminants Measured
			Epidem.	Lab.		
48	1960			×	Controlled exposure	O_3
49	1961	U.S.	×		Total exposure	Total oxidant levels
50	1963			×	Controlled exposure	O_3
51	1964			×	Controlled exposure	O_3
52	1965			×	Controlled exposure	O_3
53	1967			×	Controlled exposure	Synthetic photochemical smog
54	1968			×	Controlled exposure	O_3
55	1969	Japan		×	Controlled exposure	O_3 (SO_2 and NO_2 were measured separately)
56	1971			×	Controlled exposure	O_3
57	1971			×	Controlled exposure	O_3
44	1971	Japan	×		Total exposure	O_3, SO_2, NO_2, CO, TSP, HC
58	1972			×	Controlled exposure	O_3
59	1972			×	Controlled exposure	O_3
60	1973	Japan		×	Controlled exposure	O_3

Concentrations[a]	Effects Studied	Effects Observed (Concentration)
0.02–0.05 ppm	Odor detection in 10 volunteers	0.02–0.05 ppm, odor detection in 9 out of 10 subjects
25 ppm and over	Effect on asthmatics	0.25 ppm and over, significantly greater number of attacks in asthmatic patients
0.2–0.5 ppm	Effect on visual parameters	0.2–0.5 ppm, significant changes in visual parameters such as visual acuity
0.6–0.8 ppm	Exposure of 11 subjects	0.6–0.8 ppm, impaired diffusion capacity (DL_{CO})
1–3 ppm	Effect on 25 subjects	1–3 ppm, more profound effects and changes in pulmonary function than caused by smoking
Oxidant material, 0.50–0.75 ppm	Effect on the ultrastructure of mice	0.50–0.75 ppm, irreversible lesions of alveolar tissue
0.3–4.0 ppm	Effect on the pulmonary cells of rabbits	0.3–4.0 ppm, decreases in the percentage of alveolar macrophages and their ability to engulf streptococci
0.5 ppm	Effect on guinea pigs	0.5 ppm (2 hr), increase in air current resistance, increase in frequency of respiration, and decrease in tidal volume
1 ppm	Effect on mouse lungs	1 ppm, decrease in bacterial pulmonary deposition (67%), decrease in bactericidal activity
0.5 ppm	Effect on pulmonary alveoli in aging mice	0.5 ppm, lowered rates of DNA synthesis in alveolar cells
O_3, 0.1–1.0 ppm	Effect of O_3 exposure to humans reviewed	0.02 ppm, perceptibility threshold; 0.2–0.3 ppm, irritation of respiratory system; 0.1–1.0 ppm for 1 hr, resistance in the respiratory tract; 0.2–0.5 ppm for 3–6 hr, decrease in human sight
0.75 ppm	Effects on healthy volunteers, lightly exercising	0.75 ppm, substernal soreness, cough, some pharynaitis and dyspnea coupled with decreased pulmonary function
0.4 ppm	Effect of 10 month exposure on rabbits	0.4 ppm, emphysematous and vascular type lesions in the lung; small pulmonary arteries were thicker
1.0 ppm	Effect on erythrocytes of mice	1.0 ppm, increased resistance to erythrocyte hemolysis

Table 14.5. (Continued)

Reference	Publication Date	Country	Type of Study: Epidem.	Type of Study: Lab.	Exposure	Contaminants Measured
61	1973	Japan		×	Controlled exposure	O_3
62	1973			×	Controlled exposure	O_3 Simultaneous SO_2 exposure
63	1973			×	Controlled exposure	O_3
64	1974	U.S.	×		Total exposure	Photochemical oxidants
65	1974			×	Controlled exposure	O_3

[a] Primary national standard (United States): 0.08 ppm, 1 hr.

aerosol infection, histopathologic and cellular changes in the lungs, and respiratory distress. These effects were found at concentrations of NO_2 ranging from 2.0 to 40 ppm.

Since community studies of the effects of NO_2 on human health are limited, it is extremely difficult to make absolute conclusions about precise limitations. A World Health Organization expert committee states: "While biological activity in animals and plants at low concentrations has been demonstrated, the Committee believes that there is insufficient information upon which to base specific air quality guides at this time" (12).

3.4.2 Photochemical Oxidants

Photochemical oxidants result from the chemical reactions of oxides of nitrogen, reactive hydrocarbons, carbon monoxide, and water vapor in the presence of sunlight and in relatively still conditions. The result of these chemical reactions are colorless, gaseous compounds: ozone, peroxyacetyl nitrates (PAN), aldehydes, and other oxidizing products.

Ozone is probably the most important constituent of photochemical smog and it is a frequent measure of the overall severity of this condition. Therefore most of the studies

Concentrations[a]	Effects Studied	Effects Observed (Concentration)
1.0 ppm	Effect on the lungs of rabbits	1.0 ppm for 3 hr, 1–3 days after exposure, the animals had reduced vital capacity; 7 day after, the vital capacity was reduced only slightly
O_3, 0.37 ppm SO_2, 1000 $\mu g/m^3$	Effect on healthy subjects, lightly exercising	0.37 ppm, market decrease in pulmonary function among healthy subjects
0.37–0.75 ppm	Effect on healthy volunteers, lightly exercising	0.37–0.75 ppm, decrease in pulmonary function
—	Effect on student nurses, recorded in diaries	100 $\mu g/m^3$ (0.05 ppm), headache 300 $\mu g/m^3$ (0.15 ppm), eye discomfort 510 $\mu g/m^3$ (0.26 ppm), cough 580 $\mu g/m^3$ (0.29 ppm), chest discomfort
0.37 or 0.50 ppm	Adult male volunteers with exercise	0.37 or 0.50 ppm (2 hr), measurable physiological and biochemical changes; subjects felt physically ill; sensitive subjects, 0.37 (2 hr), respiratory symptoms

on the health effects of photochemical oxidants have been conducted using ozone. Table 14.5 reviews the health effects of photochemical oxidants. The symptoms observed are aggravation of respiratory illnesses, irritation of the eyes and respiratory tract, and impairment of cardiopulmonary function.

The perceptibility threshold and odor detection for ozone generally has been accepted at 0.02 to 0.05 ppm (44, 48). At this level most people regard the odor of ozone as irritating and unpleasant. In a study by Hammer et al. (64), student nurses in the Los Angeles area were asked to record symptoms using the diary system, to permit the calculation of dose–response thresholds for photochemical oxidants. The following thresholds were found: headache, 0.05 ppm; eye discomfort, 0.15 ppm; cough, 0.26 ppm; and chest discomfort, 0.29 ppm. The threshold for eye discomfort was similar to other findings, and the thresholds for cough and chest discomfort are in the range cited by other studies.

The acute effects of photochemical oxidants on mortality have been studied in the Los Angeles area, and there appears to be no relation between increased mortality and oxidant levels ranging from 0.50 to 0.90 ppm (128). A study by Schoettlin and Landau (49), however, revealed that there were a significantly greater number of attacks in

Table 14.6. Critical Literature Review of Carbon Monoxide, 1963–1974

Reference	Publication Date	Country	Type of Study: Epidem.	Type of Study: Lab.	Exposure	Contaminants Measured
32	1963			×	Controlled exposure	CO
33	1963			×	Controlled exposure	CO
137	1966	U.S.		×	Controlled exposure	CO
34	1967			×	Controlled exposure	CO
35	1969	U.S.		×	Controlled exposure	CO
36	1970	U.K.		×	Controlled exposure	CO
37	1970			×	Controlled exposure	CO
131	1970			×	Controlled exposure	CO
132	1970			×	Controlled exposure	CO
133	1970			×	Controlled exposure	CO
46	1970	U.S.		×	Controlled exposure	CO
47	1971	Germany		×	Controlled exposure	CO

Concentrations[a]	Effects Studied	Effects Observed (Concentration)
100 ppm	Effect of CO on healthy adults' nervous systems	100 ppm, maximum of 20.4% carboxyhemoglobin with no changes in any spinal or cranial nerve reflexes
		5% Carboxyhemoglobin, definite impairment in cognitive discrimination and psychomotor abilities
100 ppm	Effect of CO on hemoglobin	100 ppm, 16% inactivation of hemoglobin to carboxyhemoglobin
	Effect of CO on healthy nonsmokers	4% Carboxyhemoglobin saturation, increase in oxygen debt with exercise
50 ppm	Effect on the impairment of sound perception on adults	50 ppm for 90 min, impairs subjects' ability to discriminate between different durations of sound
10 ppm	Effect on the behavior and performance of humans	10 ppm, 2% carboxyhemoglobin level and degraded behavioral performance
500 ppm	Effect on the auditory flutter fusion threshold and critical flicker fusion threshold	500 ppm, no evidence of any depressant effect of CO; however auditory flutter fusion threshold increased a little
60 ppm	Effect on rat liver function	60 ppm, decreased ability of rat liver to metabolize 3-OH-benzo-α-purene
—	Effects of levels of carboxyhemoglobin on performance and perception tests	15–20% saturation, headache and impairment of manual coordination
—	Effects on vision and EEG activity	>20% carboxyhemoglobin level, modified visual evoked responses 33% carboxyhemoglobin level, EEG activity affected
—	Effect of carboxyhemoglobin level on patients with coronary disease	5–10% carboxyhemoglobin level, coronary artery blood flow accelerated; >6% saturation, significant myocardial changes
50–100 ppm	Effect on dogs	50–100 ppm (chronic exposures) may produce functional disorders and morphologic changes in the heart and brain
100 ppm	Effect on healthy volunteers	100 ppm for 2½ hr, significant decrease in visual perception, manual dexterity, and ability to learn and perform certain intellectual tasks

Table 14.6. (Continued)

Reference	Publication Date	Country	Type of Study		Exposure	Contaminants Measured
			Epidem.	Lab.		
38	1971			×	Controlled exposure	CO
39	1971			×	Controlled exposure	CO
40	1972			×	Controlled exposure	CO
41	1973	U.S.	×		Total exposure	Co, Pb, NO_x, TSP
42	1973	U.S.		×	Controlled exposure	CO
43	1973	U.S.		×	Controlled exposure	CO
134 135	1973			×	Controlled exposure	CO
136	1974			×		CO

[a] Primary national standard (United States): 9.0 ppm, 8 hr; 35.0 ppm, 1 hr.

Concentrations[a]	Effects Studied	Effects Observed (Concentration)
10,000 ppm	Survival of newly born chicks to CO	10,000 ppm, 50% of newly hatched chicks withstood exposure for 32 minutes; survival time to CO decreased with postnatal age; 1 day old, 10 min; 8–21 days old, 4 min
51, 96, 200 ppm	Effects of exposure on rats, guinea pigs, monkeys, and dogs	51, 96, and 200 ppm, no toxic effects were seen; only physiological change was increased in hemoglobin and hematocrit values for all species
90 ppm	Effect on pregnant rabbits	90 ppm (90 days), birthweights decreased 12%; exposure during first pregnancy resulted in carboxyhemoglobin 9–10%; neonatal mortality increased from 4.5 to 10%; mortality during following 21 days increased from 13 to 25%
CO, 63 ppm; Pb, 30.9 $\mu g/m^3$; NO_x, 1.38 ppm; TSP, 200 $\mu g/m^3$ (all values 30-day average)	Effect of pollution on bridge and tunnel workers.	63 ppm, high percentage had symptoms suggestive of chronic bronchitis; airway resistance was elevated in 33% of the workers; almost all bridge and tunnel workers had an increase in closing volume, suggesting small airway disease
<2–500 ppm	Effect on time perception of healthy adult volunteers	<2, 50, 100, 200, and 500 ppm, carboxyhemoglobin saturation as great as 20% but has no detrimental effect on man's time sense
100 ppm	Effect of CO on monkeys with induced myocardial infarctions	100 ppm (24 weeks), significant and persistent characteristic elevations in the hematocrit, hemoglobin and red blood counts after 3 weeks
—	Effect of CO on patients with atherosclerotic heart disease	2.5–3.0% carboxyhemoglobin level, patients experience chest pain earlier during exercise than at 1.0%
—	Effect of CO on men with peripheral atherosclerotic disease	3.0%, men with atherosclerotic disease develop leg pain after less walking

Table 14.7. Critical Literature Review of Hydrocarbons, 1957–1976

Reference	Publication Date	Country	Type of Study: Epidem.	Type of Study: Lab.	Exposure	Contaminants Measured
66	1957			×	Controlled exposure	Propionaldehyde, butyraldehyde, isobutyraldehyde
67	1960					Formaldehyde
68	1963–1964					
69	1960			×	Controlled exposure	Aldehydes
70	1960					
71	1960			×	Controlled exposure	Formaldehyde
72	1961		×		Total exposure	Formaldehyde
73	1961		×		Total exposure	Aldehydes
74	1960			×	Controlled exposure	Formaldehyde
75	1962			×	Controlled exposure	Acrolein
76	1963			×	Controlled exposure	Acrolein
77	1964			×	Controlled exposure	Formaldehyde
78	1965			×	Controlled exposure	Hexane Oxides of nitrogen; formaldehyde
79	1966			×	Controlled exposure	Acrolein
80	1966			×	Controlled exposure	Acrolein

Concentrations[a]	Effects Studied	Effects Observed (Concentration)
200 ppm, 134 ppm	Effects of hydrocarbons on humans	134 ppm (30 min) of propionaldehyde, mildly irritating to exposed mucosal surfaces; 200 ppm (30 min) butyraldehyde and isobutyraldehyde, almost nonirritating
0.1–1.0 ppm	Odor threshold	0.06–0.5 ppm, odor threshold
0.57–2.6 ppm	Effect on the eyes of rabbits	0.57 ppm, no apparent effect; 1.9–2.6 ppm (4 hr), enzyme alterations in eye tissues
1–3 ppm	Pulmonary effects on guinea pigs	1–3 ppm (1 hr), increased airflow resistance and respiratory work
—	Eye irritation of people working in paper processing	0.9–1.6 ppm, irritation of the eyes
0.035–0.35 ppm	Eye irritation in man exposed to smog in Los Angeles	0.035–0.35 ppm, direct relation to eye irritation
19 ppm	Toxicology of formaldehyde	19 ppm (10 hr), edema and hemorrhage in lungs and hyperemia of liver
0.25 ppm	Eye irritation of humans	0.25 ppm (5 min), irritation of the eyes
0.06 ppm	Physiologic effects on guinea pigs	0.06 ppm, total pulmonary resistance; tidal volume increased and respiratory rate decreased
3.5 ppm	Biochemical effects on rats	3.5 ppm (18 hr), increased alkaline phosphatase activity in liver
0.3–2.6 ppm, 0.1–1.1 ppm, 0.15–0.3 ppm	Eye irritation in a smog chamber	High correlation of eye irritation with formaldehyde; 0.15 ppm, concentration of formaldehyde at which eye irritation first occurred
200 ppm	Lung damage in rats	200 ppm (10 min, once a week for 10 weeks), residual lung damage
0.57 ppm	Effect of chronic exposure to rats	0.57 ppm, loss of weight, decrease in whole blood cholinesterase activity, decrease in urinary coproporphyrin excretion and change in conditioned reflex activities

Table 14.7. (Continued)

Reference	Publication Date	Country	Type of Study		Exposure	Contaminants Measured
			Epidem.	Lab.		
81	1967			×	Controlled exposure	Ethylene, propylene, isobutane, gas mixture and auto exhaust
82	1968		×		Acute exposure	Toluene (aromatic HCs)
83	1970	U.S.				Aliphatic HC
						Alicyclic HC
84	1972	U.S.	—	—	—	Particulate polycyclic organic matter
85	1973		×		Industrial exposure (total)	Formaldehyde
86	1974		×		Industrial exposure (total)	Benzene
87	1976	U.S.	—	—	—	Volatile HC

[a] Primary national standard (United States): .024 ppm, 3 hr.

Concentrations[a]	Effects Studied	Effects Observed (Concentration)
0.1–4.0 ppm	Eye irritation	0.1–4.0 ppm, high correlation of eye irritation with formaldehyde content
50–200 ppm	Effect of chronic exposure on man	50–100 ppm, no apparent effects; 200 ppm (8 hr), paresthesia, fatigue, confusion
	Criteria document	5000 ppm (4 min) heptane and octane, incoordination and vertigo; 10,000 ppm octane and 15,000 ppm heptane, produces narcosis in 30–60 min; 5500 ppm of ethylene, little or no effect; 350,000 ppm (5 min) acetylene, unconsciousness
	Criteria document	18,000 ppm (5 min) cyclohexane, slight muscle tremors in mice and rabbits; 3330 ppm (6 hr/day, 60 days) cyclohexane, no effect in rabbits; 1240 ppm cyclohexane, no effect in monkeys
—	Biologic effects of polycyclic organic matter	Hypothesis that a reduction of 1 μg of benzo[α]pyrene per 1000 m^3 of air will decrease the lung cancer death rate by 5%.
2–10 ppm	Effects on 10 new female employees	2–10 ppm, rapid loss of consciousness
150–210 ppm	Effect of chronic benzene exposure on Hodgkin's disease	150–210 ppm (mean of 11 years of exposure), six cases of Hodgkin's disease
—	Biologic effects of HC	Discussion on effects of all HC that are considered pollutants

Table 14.8. Critical Literature Review of SO_2, 1964–1975

Reference	Publication Date	Country	Type of Study: Epidem.	Type of Study: Lab.	Exposure	Contaminants Measured
88	1964	U.K.	×		Total exposure	SO_2 TSP
89	1966	U.K.	×		Total exposure	SO_2 TSP
126	1966	U.S.		×	Controlled exposure	
90	1967	U.K.	×		Total exposure	SO_2
91	1967	U.K.	×		Total exposure	SO_2 Smoke
92	1967	Holland	×			
93	1968	U.K.	×		Total exposure	SO_2 Smoke
94	1968	U.S.				
95	1969	U.S.	×		Total exposure	SO_2
96	1970	U.S.	×		Total exposure	SO_2 TSP
97	1970	U.S.		×	Controlled exposure	SO_2

Concentrations[a]	Effects Studied	Effects Observed (Concentration)
75–115 μg/m^3 (0.026–0.040 ppm); 80–160 μg/m^3 (annual mean)	Effects on lung cancer and bronchitis in men and women in two communities	115 μg/m^3 (0.040 ppm), increase in lung cancer mortality in men, and increase in bronchitis mortality in men and women when compared to SO_2 level of 75 μg/m^3 (0.026 ppm)
130–148 μg/m^3 (0.05–0.06 ppm) 91–138 μg/m^3 (annual mean)	Effect on respiratory illness in children	130–148 μg/m^3 (0.05–0.06 ppm), lower respiratory tract disease was increased
		Review by Hazelton Laboratories on the role and nature of SO_2, SO_3, H_2SO_4, and fly ash in air pollution
400–500 μg/m^3 (0.15–0.19 ppm) (24 hr average)	Effects of SO_2, O_2, and soot on man	400–500 μg/m^3 (0.15–0.19 ppm), increased mortality and disease
123–275 μg/m^3 97–301 μg/m^3 (24-hr average)	Patterns of respiratory illness in school children	200 μg/m^3, presence of both upper and lower respiratory infections significantly increased
	Review document on acceptable level of SO_2	500 μg/m^3 (0.19 ppm), excess mortality; 300–400 μg/m^3, increased respiratory illness as well as absenteeism
0.16 ppm–0.21 ppm (458–600 μg/m^3) 300–400 μg/m^3	Respiratory illness observations of 1000 men studied	0.16 ppm (458 μg/m^3), illness attack rates increased (weekly average); 0.21 ppm (600 μg/m^3), decrease in ventilatory lung function (daily)
—	Review article on the acute effects of SO_2	Author concludes that any effect air pollution might have on health, "does not appear to involve SO_2 in its mechanism"
—	Effects on patients with bronchial pulmonary disease	0.25–0.30 ppm (710–858 μg/m^3), illness rates increased significantly
119–500 μg/m^3 >260 μg/m^3 on about 15% of the days (24 hr average)	Effect on patients with chronic bronchitis	300–500 μg/m^3 (0.11–0.19 ppm), increased hospital admissions with respiratory illness; 119–249 μg/m^3 (0.05–0.09 ppm), substantial increase in illness of patients 55 years or older with more severe bronchitis
0.13, 1.01, 5.72 ppm	Long-term exposure of guinea pigs	0.13, 1.01, and 5.72 ppm (12 months), pulmonary function measurements indicated that no detrimental changes

Table 14.8. (Continued)

Reference	Publication Date	Country	Type of Study		Exposure	Contaminants Measured
			Epidem.	Lab.		
98	1970	U.S.		×	Controlled exposure	SO_2
102	1970	U.K.	×		Total exposure	SO_2 Smoke
99	1971	NATO/ CCMS	—	—	—	—
100	1971	U.S.		×	Controlled exposure	H_2SO_4 mist and particulate sulfates
101	1972	U.S.		×	Controlled exposure	SO_2
103	1974	U.S.	×		Total exposure	SO_2 TSP Suspended nitrates Suspended sulfates NO_2
104	1974	U.S.	—	—	—	—
105	1974	U.S.				

Concentrations[a]	Effects Studied	Effects Observed (Concentration)
		could be attributed to SO_2; 5.72 ppm increase in size of hepatocytes accompanied by cytoplasmic vacuolation in liver
650 ppm	Effect on Syrian hamsters with emphysema	650 ppm, mild bronchitic lesion with relatively minor changes in mechanical properties of lung
550 μg/m^3 250 μg/m^3 (24 hr average)	Effects on panels of bronchitis patients	500 μg/m^3, increased morbidity
—	Air Quality Criteria Document for SO_2	—
—		Literature review of H_2SO_4 mist and particulate sulfate toxicology studies
3 ppm (9000 μg/m^3)	Effect of SO_2 on humans	3 ppm (120 hr), increased small airway resistance and significant but minimal decrease in dynamic compliance of lung
40 μg/m^3 60 μg/m^3 2 μg/m^3 6 μg/m^3 30 μg/m^3	Effects on well adults and cardiopulmonary patients	40 μg/m^3, best judgment estimate of threshold for effect (24 hr average)
—	—	Toxicologic and epidemiologic review of the health effects of SO_x
	Health consequences of SO_x (CHESS studies)	*Best Judgment effects for long-term exposure:* 95 μg/m^3 (100 μg/m^3 TSP, 15 μg/m^3 sulfates), increased prevalence of chronic bronchitis, 95 μg/m^3 (102 μg/m^3 TSP, 15 μg/m^3 sulfates), increased acute lower respiratory disease in children; 106 μg/m^3 (151 μg/m^3 TSP, 15 μg/m^3 sulfates), increased frequency of acute respiratory disease in families; 200 μg/m^3 (100 μg/m^3 TSP, 13 μg/m^3 sulfates), decreased lung function of children *Best Judgment effects for short-term exposure:* >365 μg/m^3 (80–100 μg/m^3

Table 14.8. (Continued)

Reference	Publication Date	Country	Type of Study: Epidem.	Type of Study: Lab.	Exposure	Contaminants Measured
106	1974	Denmark U.S.		×	Controlled exposure	SO_2
107	1975	U.S.	×		Total exposure	SO_2 TSP Suspended sulfates

[a] Primary national standard (United States): 0.03 ppm, 1 year; 0.14 ppm, 24 hr.

asthmatic patients when oxidant levels were above 0.25 ppm. Various other acute effects have been reported within this range (i.e., 0.2 to 1.0 ppm). At ozone levels between 0.2 and 0.5 ppm, significant changes in such parameters as visual acuity and peripheral vision have been noted (44, 50). Young and Shaw (51) found pulmonary changes in the 0.2 to 1.0 ppm range. Their report cites significant differences in pulmonary alveolar diffusing capacity in 12 subjects following inhalation of ozone at 0.6 to 0.8 ppm for 2 hours.

Three important studies were conducted during 1969–1974 on the effect of ozone during short-term exposure. In two of the experiments, healthy volunteers were exposed to 0.37 or 0.75 ppm ozone for 2 hours. Both studies reported decreased pulmonary function, cough, dyspnea, and pharyngitis among the volunteers (58, 63). Bates (62) investigated the synergistic effects of ozone and sulfur dioxide and found a marked decrease in pulmonary function when healthy volunteers were exposed to 0.37 ppm ozone and 1000 $\mu g/m^3$ of sulfur dioxide while performing light exercise. It also was noted that healthy subjects doing light exercise and exposed to 0.37 or 0.5 ppm of ozone suffered measurable physiological and biochemical changes, and felt physically ill (65).

The effects of oxidants on athletic performance have also been recorded. In one study the times of long-distance runners were compared to previous competition times, to determine the effect of the oxidant levels on the runners. It was found that if the oxidant level was above 0.1 ppm during the hour before the event, there was an increase in running times, (i.e., a decrease in athletic performance). However no such association was found between the oxidant levels 2 to 3 hours before the race, or during the event (129). In another study, Smith (130) exposed 32 male college students to 0.3 ppm PAN while

Concentrations[a]	Effects Studied	Effects Observed (Concentration)
		TSP, 8–10 $\mu g/m^3$ sulfates), aggravation of pulmonary symptoms in elderly; 180–250 $\mu g/m^3$ (70 $\mu g/m^3$ TSP, 8–10 $\mu g/m^3$ sulfates), aggravation of asthma
1, 5, 25 ppm	Effect of 6 hr exposure on nasal mucus flow rate and airway resistance	5, 25 ppm, significant decrease in nasal mucus flow rate; 1, 5, 25 ppm, increased nasal airflow resistance
38–425 $\mu g/m^3$ 60–185 $\mu g/m^3$ 9–20 $\mu g/m^3$ (annual mean)	Acute respiratory effects on children	38–425 $\mu g/m^3$, excess acute lower respiratory disease morbidity in children

at rest and during a 5-minute bicycle exercise. Oxygen uptake was increased over the control during the period of bicycling, but not while at rest.

A few animal studies are included in Table 14.5. Coffin et al. (54) reported that exposure of rabbits to 0.3 to 4.0 ppm ozone enhances the susceptibility of the animals to infective aerosols. In other laboratory studies cited, a range of ozone between 0.4 and 0.75 ppm appears to be responsible for lesions in the lungs, lower DNA synthesis in alveolar cells, decreased tidal volume, increased frequency of respiration, and decreased pulmonary function (53, 55, 57, 59).

3.4.3 Carbon Monoxide

Carbon monoxide (CO) is a colorless, odorless gas produced by incomplete combustion. The principal source is motor vehicle exhaust, however industrial plants burning carbonaceous fuels contribute significant amounts, as well.

The predominant characteristic of CO is its great affinity for hemoglobin—about 240 times that of oxygen. Therefore CO impairs the transport of oxygen at the tissue level and interferes with the release of oxygen from the hemoglobin molecule. There is a background level of carboxyhemoglobin in the blood of about 0.4 percent, however smoking a pack of cigarettes a day will increase this level to about 5 percent (138). It should be recognized that CO exposure will not necessarily raise the level of carboxyhemoglobin in the blood because of the equilibrium that is finally established between the blood and air. For example, 25 ppm CO will result in 4 percent saturation regardless of the initial concentration in the blood (12). A person with less than 4 percent carboxyhe-

Table 14.9. Critical Literature Review of Total Particulates, 1964–1975

Reference	Publication Date	Country	Type of Study: Epidem.	Type of Study: Lab.	Exposure	Contaminants Measured
108	1964	U.K.	×		Total exposure	Smoke SO_2
109	1964	U.K.	×		Total exposure	Smoke SO_2
88	1964	U.K.	×		Total exposure	Smoke SO_2
110	1965	U.K.	×		Total exposure	Smoke SO_2
111	1965	U.K.	×		Total exposure	Smoke SO_2
112	1966	U.K.	×		Total exposure	Smoke SO_2
113	1967	U.S.	×		Total exposure	Smoke SO_2
91	1967	U.K.	×		Total exposure	Smoke SO_2
114 115 116	1967 1968 1969	U.S.	×		Total exposure	TSP
117	1969	U.S.	—	—	—	—
118	1970	U.K.	×		Total exposure	Smoke SO_2
119	1971	NATO/ CCMS	—	—	—	—

Concentrations[a]	Effects Studied	Effects Observed (Concentration)
>200 μg/m³ >200 μg/m³ (24 hr time)	Effects on mortality from lung cancer and bronchitis	>200 μg/m³, excess bronchitis mortality
1000 μg/m³ 715 μg/m³	Effects on mortality and morbidity	1000 μg/m³, increased mortality from all causes
80–160 μg/m³ 74–115 μg/m³ (annual mean)	Effects on lung cancer and bronchitis mortality	160 μg/m³ (115 μg/m³ SO_2), positive association between incidence of bronchitis and lung cancer and the level of TSP and SO_2
100–400 μg/m³ 150–400 μg/m³ (24 hr average)	Effects on absences due to bronchitis influenza, arthritis, and rheumatism	100–200 μg/m³ (150–250 μg/m³ SO_2), lowest bronchitis inception rates 400 μg/m³ (400 μg/m³ SO_2), highest bronchitis inception rates
120–200 μg/m³ 30–300 μg/m³(annual mean)	Prevalence of chronic respiratory disease symptoms in outdoor telephone workmen	Increased smoke concentration from 120–200 μg/m³, with an increase in SO_2, will increase the risk to older workers of poorer lung function and chronic respiratory disease
>130 μg/m³ >130 μg/m³ (annual mean)	Effect on respiratory infection of children	>130 μg/m³, increased frequency and severity of lower respiratory diseases in school children
2.0–8.2 cohs 0.10–1.00 ppm	Mortality and morbidity during an episode of high pollution	2.0–8.2 cohs, immediate rise in daily deaths due to all causes and rise in exacerbation of bronchitis and asthma
97–301 μg/m³ 123–275 μg/m³ (24 hr average)	Patterns of respiratory illness in school children (5–6 years old)	100 μg/m³ (>120 μg/m³ SO_2), increased association of respiratory infections
80–>135 μg/m³ (2 yr geometric mean)	Effect on total mortality and respiratory mortality in men	>135 μg/m³, death rate twice as high as at <80 μg/m³ >135 μg/m³, death rate due to stomach cancer almost twice as high as at <80 μg/m³
—	Air Quality Criteria based on health effects (API)	—
250 μg/m³ 500 μg/m³ (24 hr average)	Effect on exacerbation of bronchitis	250 μg/m³, minimum pollution level leading to a significant worsening of bronchitis patients
—	Air Quality Criteria Document for particulates	

Table 14.9. (Continued)

Reference	Publication Date	Country	Type of Study		Exposure	Contaminants Measured
			Epidem.	Lab.		
120	1973	U.S.	×		Total exposure	TSP
121	1973	U.S.	×		Total exposure	TSP Suspended sulfates SO_2
122	1973	U.S.	×		Total exposure	TSP SO_2
123	1973	U.S.	×		Total exposure	TSP SO_2 Suspended sulfates
124	1974	U.S.	×		Total exposure	TSP SO_2
125	1974	U.S.	×		Total exposure	TSP SO_2 Suspended nitrates Suspended sulfates
127	1974	U.S.	×		Total exposure	TSP SO_2 Suspended sulfates
103	1974	U.S.	×		Total exposure	TSP SO_2 Suspended sulfates NO_2
107	1975	U.S.	×		Total exposure	TSP SO_2 Suspended sulfates

Concentrations[a]	Effects Studied	Effects Observed (Concentration)
100–269 μg/m^3 (24 hr average)	Family surveys of symptoms during acute pollution episodes	100–269 μg/m^3, increased cough and chest discomfort; restricted activity
96–133 μg/m^3 8.9–10.1 μg/m^3 (annual mean) 39–57 μg/m^3	Effect on ventilatory function of school children	96–133 μg/m^3, decreased pulmonary function in school children
100–150 μg/m^3 Steady decrease until <80 μg/m^3 (annual mean)	Effect on chronic respiratory disease of military inductees	100–150 μg/m^3, increased chronic respiratory disease symptom prevalence in adults
66–151 μg/m^3 63–275 μg/m^3 2.4–20.3 μg/m^3 (annual mean)	Effect on acute respiratory disease in adults and children	>100 μg/m^3, increased frequency and severity of acute lower respiratory disease in school children
103–168 μg/m^3 <25 μg/m^3 (annual mean)	Survey of acute lower respiratory disease of children	103–139 μg/m^3, authors' best judgment regarding level that produces lower respiratory disease in children
81–168 μg/m^3 11–<25 μg/m^3 2–3 μg/m^3 10–13 μg/m^3	Survey of respiratory disease symptoms of adults	<150 μg/m^3, no strong relation to chronic respiratory disease symptoms
34–185 μg/m^3 22–425 μg/m^3 9–18 μg/m^3 (annual mean)	Survey of lower respiratory disease symptoms in school children	60–185 μg/m^3 (38–425 μg/m^3 SO_2, 9–20 μg/m^3 suspended sulfates), excess acute lower respiratory disease morbidity in children
<60–120 μg/m^3 <40–100 μg/m^3 <6–12 μg/m^3 <30–75 μg/m^3	Survey of a panel of well subjects and cardiopulmonary patients	70–120 μg/m^3 (significant SO_2), cardiopulmonary effects
34–185 μg/m^3 22–425 μg/m^3 9–18 μg/m^3 (annual mean)	Acute lower respiratory disease symptoms in children	85–110 μg/m^3 (175–250 μg/m^3 SO_2, 13–14 μg/m^3 suspended sulfates), "best judgment" estimate associated with excess childhood respiratory morbidity

moglobin will absorb the CO, whereas a person with a level greater than 4 percent will excrete it until he reaches an equilibrium at 4 percent. Table 14.6 indicates that the general health effects of CO are reduced tolerance for exercise, impairment of psychomotor function, impairment of fetal development, and aggravation of cardiovascular disease.

Carbon monoxide levels of 10 ppm produce carboxyhemoglobin saturation of about 2 percent; this level has been shown to decrease behavioral performance (35). At a concentration of 100 ppm, CO produces about 16 percent inactivation of hemoglobin to carboxyhemoglobin (33). This level has been reported not to have any effect on cranial or spinal nerve reflexes (Schulte, 32), yet carboxyhemoglobin levels of 5 percent and below were associated with the control of choice discrimination and psychomotor abilities, and with the degree of impairment related to increasing carboxyhemoglobin levels. The effects of varying levels of carboxyhemoglobin on performance and perception tests have also been evaluated. Values of 15 to 20 percent saturation were associated with headache and impairment of manual coordination (131). Bender et al. (47) exposed healthy volunteers to 100 ppm CO for 2.5 hours, then performed certain psychological tests. Significant decreases in visual perception, manual dexterity, and the ability to learn and perform certain intellectual tests were found. Studies by Hosko (132) have also shown effects on visual perception at levels above 20 percent saturation, although he found that EEG activity was not affected until 33 percent saturation. The effect of 500 ppm CO on auditory flutter fusion and critical flicker fusion thresholds was investigated by Guest et al. (36), but the only reported effect was a slight increase in the auditory flutter fusion threshold. In another study, however, CO at 50 ppm for 90 minutes was found to impair the ability of subjects to discriminate between different durations of sound (34). The aggravation of cardiovascular disease can also be observed when carboxyhemoglobin levels become too high. Ayers et al. (133) studied the effects of carboxyhemoglobin levels on patients with coronary artery disease. When carboxyhemoglobin was raised to 5 to 10 percent saturation, coronary artery blood flow was accelerated, and at levels above 6 percent, myocardial changes were seen. Other studies have demonstrated that patients with atherosclerotic heart diseases experience chest pain earlier during carboxyhemoglobin levels of 2 to 3 percent when compared to 1 percent carboxyhemoglobin levels (134, 135). Patients with the same disease have also developed leg pain after less walking at 3 percent carboxyhemoglobin saturation (136). This indicates that people most sensitive to carbon monoxide exposure are those with cardiovascular or respiratory conditions.

Preziosi (46) chronically exposed dogs to 50 to 100 ppm CO and found that these levels may produce functional disorders and morphological changes in the heart and brain. Other work revealed that CO levels of 60 ppm decreased the ability of rat livers to metabolize 3-OH-benzo-αpyrene (37). Pregnant rabbits exposed to 90 ppm CO during half the gestation period experienced carboxyhemoglobin levels of 9 to 10 percent. This resulted in decreased birth weights and increased neonatal and infant mortality (40). Several studies have also shown that CO exposures ranging from 51 to 200 ppm have increased hematocrit, hemoglobin, and red blood count values in rats, guinea pigs, dogs, and monkeys with induced myocardial infarctions (39, 43).

It has been generally accepted that carboxyhemoglobin levels of 4 percent and above are undesirable for humans. However many people in our society already have carboxyhemoglobin levels exceeding 4 percent because of smoking. Although it may be readily inferred that it is the impaired individual with cardiovascular disease who is most affected by high CO levels, it is difficult to decide which segment of the population the standard should protect. As the WHO report states, "One is confronted by a similar dilemma when defining the fraction of the population that must receive absolute protection at all costs, for it is obvious that in any urban community, there will be some patients in extremis to whom any stress will prove ultimately intolerable" (12).

3.4.4 Hydrocarbons

Hydrocarbons are organic compounds in gaseous or particulate form, such as methane, formaldehyde, benzene, and acrolein. They are produced by the incomplete combustion of fuels and other carbon-containing substances; typical sources are motor vehicle exhaust and coal-burning heating plants.

Hydrocarbons promote the formation of photochemical smog, although there is little evidence relating to the direct health effects of hydrocarbons. Therefore the standards have been based almost entirely on the role of hydrocarbons as precursors of other compounds of photochemical smog.

Aliphatic hydrocarbons are basically inert and have virtually no demonstrable effect on health except at extremely high concentrations. The Criteria Document for Hydrocarbons (83) reports that exposure to 5000 ppm heptane and octane for 4 minutes can cause incoordination and vertigo, and 10,000 ppm octane or 15,000 ppm heptane will produce narcosis in 30 to 60 minutes. Ethylene at 5500 ppm for several hours, however, has been reported to have little or no effect.

Alicyclic hydrocarbons have also very little effect on health, although at high concentrations they can act as depressants or anesthetics (87). The Criteria Document states that cyclohexane vapor at 18,000 ppm for 5 minutes can produce slight muscle tremors in mice and rabbits, and the same concentration for 25 to 30 minutes will produce muscular incoordination and paralysis. Although chronic exposures of smaller concentrations of cyclohexane (3300 ppm for 6 hours per day in rabbits and 1240 ppm for 6 hours per day in monkeys) have no noticeable effects.

Aromatic hydrocarbons are much more irritating than aliphatic or alicyclic hydrocarbons. The review of vapor phase organic pollutants by the National Academy of Sciences (87) states that chronic exposures to some aromatic hydrocarbons have been associated with leukopenia and anemia, and that benzene, toluene, or xylene at concentrations above 100 ppm may result in fatigue, weakness, confusion, skin paresthesias, and mucous membrane irritation; at concentrations above 2000 ppm prostration and unconsciousness could result. Askoy et al. (86) report six cases of

Hodgkin's disease among workers chronically exposed to 150 to 210 ppm benzene for 1 to 28 years. The validity of this study is questionable, however, since there was no mention of how many other workers were exposed, or to what other pollutants they were exposed. In other studies it was reported that toluene at 50 to 100 ppm produces no effect; but fatigue, confusion, and paresthesia have resulted after acute exposure to 220 ppm for 8 hours (82).

Aldehydes are formed through photochemical reactions in the atmosphere and have very irritating properties. Of all the aldehydes present in the atmosphere, formaldehyde and acrolein are the two most important. Formaldehyde is recognized as being highly irritating to the mucous membranes of the eyes, nose, and throat. The odor threshold for formaldehyde has been set at 0.06 to 0.5 ppm (67, 68). Tuesday et al. (78) investigated the effect of formaldehyde on eye irritation, using a smog chamber. It was found that eye irritation occurred at levels as low as 0.15 ppm. Morrill (72) puts the level of irritant action of formaldehyde at 0.9 to 1.6 ppm, and it has been reported that acrolein irritates the eyes at exposures to concentrations of 0.25 ppm for 5 minutes (75). In addition, Ahmad and Whitson (85) reported rapid loss of consciousness in 10 new workers who had been exposed to formaldehyde concentrations ranging from 2 to 10 ppm.

Various experiments have been performed using animals to observe the effects of formaldehyde and acrolein (Table 14.7). The effect of aldehydes on the eyes have been observed, and in one study rabbits exposed for 30 days to 0.57 ppm acrolein experienced no effect, but at 1.9 to 2.6 ppm for 4 hours, enzyme alterations in eye tissues occurred (69, 70). However Gusev et al. (80) have shown that concentrations of 0.57 ppm acrolein produce decreased weights, changes in conditioned reflexes, decreased urinary coproporphyrin excretion, and decreased cholinesterase activity in rats. At 0.06 ppm acrolein was seen to increase total pulmonary resistance and tidal volume and decrease the respiratory rate of guinea pigs (76). Catilina et al. (79) reported residual lung damage in rats after exposure to 200 ppm acrolein for 10 minutes once a week for 10 weeks. Formaldehyde has also been shown to affect the respiratory system, and increased airflow resistance and respiratory work in guinea pigs have been attributed (71) to concentrations of 1.0 to 3.0 ppm.

3.4.5 Sulfur Dioxide

Sulfur dioxide is a colorless gas with a pungent odor; it can be oxidized to form sulfur trioxide, sulfuric acid, and later, sulfate. The principal sources are the combustion of sulfur-containing fossil fuels, the smelting of sulfur-bearing metal ores, and industrial processes. Literature on SO_2 (Table 14.8) shows the main health effects to be aggravation of respiratory diseases (including asthma, chronic bronchitis, and emphysema), reduced lung function, irritation of the eyes and respiratory tract, and increased mortality. In epidemiological studies it is difficult to consider the health effects of SO_2 in isolation, since SO_2 and particulate matter tend to occur together in the same kinds of polluted atmosphere. The WHO states: "It follows, therefore, that unless the effect sought is highly specific, the use of epidemiologic techniques will seldom result in the attribution, with any degree of certainty, of an observed effect to a specific pollutant" (12).

The effect of SO_2 on animals and humans has been the subject of many toxicological experiments. Although studies conducted in controlled atmospheres do not simulate urban air pollution accurately, they do shed light on the physiological changes caused by known levels and durations of SO_2. This gas is a respiratory irritant that is absorbed in the nose and upper respiratory tract, but very high concentrations of it are needed to produce changes in the lungs. Exposures of Syrian hamsters to SO_2 levels of 650 ppm produced mild bronchitic lesions with relatively minor changes in the mechanical properties of the lungs (98). However in one human study, after 120 hours of exposure to 3 ppm SO_2, increased small airway resistance and significant but minimal decrease in dynamic compliance of the lung were observed (101). In a study using guinea pigs, long-term exposure (12 months) of levels up to 5.72 ppm produced no detrimental changes on pulmonary function measurements, and the only effects observed were increased size of hepatocytes and cytoplasmic vacuolation in the liver (97). Amdur (100) and Hazelton Laboratories (126) have performed many toxicological studies, and many of their results can be found in the review articles cited in Table 14.8. Animal data on the health effects of SO_2 can be very useful in helping to determine the mechanisms of action, but as one review on the health effects of sulfur oxides states, "The physiological responses found under controlled conditions give useful insights into mechanisms of action, but cannot be translated directly into adverse effects of exposures to contaminated urban air" (104).

Epidemiological techniques have also been used to observe the health effects of SO_2. These techniques are useful in considering the effects that might produce chronic respiratory disease. Some of the epidemiologic studies conducted are reviewed below; one should recognize the many difficulties encountered when trying to establish dose–response effects, such as standardizing measurement techniques, allowing for smoking, sex, age, and socioeconomic factors, and considering meteorological variables.

Wicken and Buck (88) examined the mortality rates in two English communities with different kinds of pollution. They found an increase in lung cancer mortality in men and increases in bronchitis mortality in men and women who were exposed to long-term SO_2 levels of 0.04 ppm. Another study revealed a significant correlation between SO_2 pollution levels and deaths or disease with a 24-hour mean SO_2 level of 0.15 to 0.19 ppm, when there was a high soot content (90). Fletcher and his colleagues (93) observed 1000 men aged 30 to 59 for 5 years; they discovered a significant relation between respiratory illness and SO_2 and smoke levels. Illness attack rates were increased when the weekly SO_2 levels exceeded 0.16 ppm with an accompanying smoke level of 400 $\mu g/m^3$. At daily SO_2 concentrations of 0.21 ppm and smoke concentrations above 300 $\mu g/m^3$, decreases in ventilatory lung function were observed. Major air pollution episodes were examined by Dutch scientists, and their review concluded that excess mortality resulted when 24-hour mean levels of SO_2 exceeded 500 $\mu g/m^3$ (0.19 ppm) for a few days. In addition, hospital admissions and absenteeism increased when SO_2 24-hour mean levels were 300 to 400 $\mu g/m^3$ (0.11 to 0.15 ppm) for 3 to 4 consecutive days (92).

Since SO_2 is known to aggravate respiratory symptoms, many of the epidemiologic studies have been carried out using bronchitic or asthmatic patients. Carnow et al. (95) performed a study of bronchitic patients in Chicago to determine what levels made their illness worse and found the critical level at which bronchitis illness rates increased to lie between 0.25 and 0.30 ppm. It should be noted, however, that although particulates were not measured, they were probably at high concentrations of about 148 $\mu g/m^3$ as determined by other studies. Carnow et al. (96) later reported that a group of patients with chronic bronchitis had significant increases in the age range of 55 years and older at daily levels of SO_2 between 0.05 and 0.09 ppm and in patients with more severe bronchitis. This relationship was not observed among patients under 55 years old, or with less severe bronchitis. Lawther et al. (102) studied a panel of bronchitic patients, who recorded their daily conditions of health in diaries. It was observed that the minimum pollution level leading to a significant response was about 500 $\mu g/m^3$ (0.19 ppm) of SO_2 with about 250 $\mu g/m^3$ of smoke.

A number of studies have been conducted using children, since they do not smoke cigarettes and might be more sensitive to pollution levels. Douglas and Waller (89), investigating the pollution effect on respiratory disease in London children, learned that lower respiratory disease was increased in areas with annual SO_2 levels of 0.05 to 0.06 ppm or higher, with accompanying levels of particulates. In another study, Lunn et al. (91) examined school children aged 5 years and 10 to 11 years, in four areas of Schefffield. Both upper and lower respiratory infections were found to be increased at annual SO_2 concentrations of 200 $\mu g/m^3$ (0.076 ppm) and smoke concentrations of 200 $\mu g/m^3$.

Recent epidemiological studies of EPA have been carried out under the CHESS program to develop dose–response information relating short-term and long-term exposures to adverse health effects. *Health Consequences of Sulfur Oxides: A Report from CHESS, 1970–1971* (105) reviews studies associated with sulfur dioxide exposure in New York City, the Salt Lake Basin area, five Rocky Mountain areas in Idaho and Montana, Chicago, and Cincinnati. The best judgment effects for long-term exposure were estimated at 95 $\mu g/m^3$ (0.036 ppm) for increased prevalence of chronic bronchitis and increased acute lower respiratory disease in children, 106 $\mu g/m^3$ (0.04 ppm) for increased acute respiratory disease in children and families, and 200 $\mu g/m^3$ (0.076 ppm) for decreased lung function of children. The best judgment effects for short-term exposure were cited at greater than 365 $\mu g/m^3$ (0.14 ppm) for aggravation of cardiopulmonary symptoms in the elderly, and 180 to 250 $\mu g/m^3$ (0.07 to 0.095 ppm) for aggravation of asthma. All these health effects, however, were accompanied with TSP levels greater than 70 $\mu g/m^3$ and sulfate levels greater than 8 to 10 $\mu g/m^3$. The validity of the CHESS studies has been questioned, as previously mentioned. The methods used, the techniques of sampling, and the internal inconsistency of the findings have been criticized. It should be noted that some of the recent evidence suggests that the epidemiological data may be more closely related to the presence and concentration of sulfates than to SO_2, from which sulfates are most commonly derived.

3.4.6 Particulates (TSP)

Particulate matter is any solid or liquid particle dispersed in the atmosphere, such as soot, dust, ash, and pollen. Table 14.9 gives the main health effects of TSP as the aggravation of asthma or other respiratory disease, increased cough and chest discomfort, and increased mortality. Therefore one recognizes the importance of deposition, retention, and disposition of particulates; for a detailed discussion of these factors the reader is referred to Chapter 7.

Since particulate matter and sulfur oxides tend to occur in the same kinds of polluted atmosphere, it is difficult to differentiate between the two pollutants in epidemiologic studies. Such factors as smoking, age, socioeconomic conditions, and sex are not always accounted for, and although particle size is important for the effectiveness of deposition, it is hardly ever considered in field studies. Therefore the results of many epidemiologic studies are of questionable value in establishing standards.

In several such studies, the effects particulates have on mortality were examined. Buck and Brown (108) studied 214 areas of the United Kingdom* to relate mortality ratios for 5 years to daily smoke and SO_2 concentrations greater than 200 $\mu g/m^3$. However mortality from lung cancer was not found to be positively associated with these smoke and SO_2 levels. Martin (109) reported an increase in mortality from all causes when smoke levels rose above 1000 $\mu g/m^3$ with SO_2 levels above 715 $\mu g/m^3$, but since his pollutant measurements were made in only one part of central London, a variety of concentrations must be considered to have contributed to the effects observed. In 1964 Wicken and Buck (88) conducted a study of bronchitis and lung cancer mortality in six areas of northeast England. They compared deaths from bronchitis and lung cancer to nonrespiratory disease deaths for a period of 10 years, adjusting for age, sex, social class, and smoking. However Eston was the only one of six areas in which SO_2 and TSP values were available. Eston was divided into two sections of pollution, and a positive association appeared between the incidence of bronchitis and lung cancer in the high pollution area of 160 $\mu g/m^3$ smoke and 115 $\mu g/m^3$ SO_2 when compared to smoke levels of 80 $\mu g/m^3$ and SO_2 levels of 74 $\mu g/m^3$.

Other epidemiologic studies have attempted to establish morbidity effects of particulates. In one conducted by the (British) Ministry of Pensions and National Insurance (110), a representative population was observed for sickness from bronchitis, influenza, arthritis, and rheumatism. There was a significant correlation between bronchitis incapacity and particulate and SO_2 levels. There was also more arthritis and rheumatism in areas with high smoke pollution. The lowest bronchitis inception rate observed in this study was set at smoke levels of 100 to 200 $\mu g/m^3$, with SO_2 levels of 150 to 250 $\mu g/m^3$. Holland et al. (111) studied outdoor telephone workers in London, rural England, and the east and west coasts of the United States to observe effects of SO_2 and particulates on health. They concluded that an increase in smoke concentrations from 120 to 200 $\mu g/m^3$ with an equivalent increase in SO_2 will increase the risk to older workers of deteriorated pulmonary function and chronic respiratory disease.

Since particulate pollution has been known to cause or aggravate respiratory diseases, many epidemiologic studies have been conducted using bronchitic or asthmatic patients. For example, Lawther et al. (118) evaluated conditions of a group of bronchitic patients

* The American high volume sampling method and the British method differ; although both express results in micrograms per cubic meter, they are not directly comparable.

through the diary system. Worsening of patients' conditions was related to daily smoke and SO_2 levels. With daily levels of 250 $\mu g/m^3$ of smoke and 500 $\mu g/m^3$ of SO_2, a significant worsening of bronchitic symptoms occurred. Children are also considered to be more susceptible to particulate pollution, and their health effects have been studied. In the earlier work by Douglas and Waller (112), illness data on upper and lower respiratory symptoms were collected from the mothers and school doctors. Upper respiratory tract infections were found not to be related to the amount of pollution, but lower respiratory infections were related. Increased frequency and severity of lower respiratory diseases were observed when smoke levels were about 130 $\mu g/m^3$. However in another study by Lunn et al. (91) levels of smoke of 100 $\mu g/m^3$ were associated with increased respiratory infections.

Subsequent evidence on the health effects of particulates has been obtained from the EPA and the CHESS programs. In many of the investigations children were used as the study populations. Shy et al. (121) examined the effects of particulates on the ventilatory function of school children. At TSP levels ranging from 96 to 133 $\mu g/m^3$ with accompanying SO_2 levels of 39 to 57 $\mu g/m^3$, and sulfate levels of 8.9 to 10.1 $\mu g/m^3$, they found a decrease in the pulmonary function of their subjects. At TSP levels of 103 to 109 $\mu g/m^3$ with SO_2 less than 25 $\mu g/m^3$, Hammer et al. (124) estimated increased production of lower respiratory disease in children. In another study of children, the best judgment estimate associated with excess childhood respiratory morbidity was levels of TSP of 85 to 110 $\mu g/m^3$, SO_2 of 175 to 250 $\mu g/m^3$, and sulfates of 13 to 14 $\mu g/m^3$ (107).

It is evident that there is no unanimity of results in the various studies. In one case TSP levels less than 150 $\mu g/m^3$ were found to have no strong relation to chronic respiratory disease symptoms (125), whereas in another TSP levels of 100 to 150 $\mu g/m^3$ were shown to increase chronic respiratory disease symptoms (122).

4 SUMMARY

The foregoing pages have pointed out that air pollution is an undesirable facet in our lives; on the other hand, we recognize that man has evolved through many centuries in contaminated atmospheres. In recent years, millions of dollars have been spent and hundreds of man-years have been devoted to determining the levels that people and vegetation can tolerate without adverse effects. Much of this work was summarized in developing the six national ambient air standards discussed at length in this chapter.

There is a dose–response relationship between air contaminants and human health. In industrial hygiene, the standards established are to protect normal, healthy working people, and as a result they are less stringent than in community air pollution, where one must consider the very young, the sick, those with allergies, and the aged. The dose–response value of any one contaminant, or a combination of contaminants, may have to reach essentially *zero* to afford protection to the entire population. Literature is replete with case histories of the influence of pollens on sensitive individuals, similarly, unmeasurable trace quantities of some substances, such as isocyanates, can cause undesirable reactions among the very sensitive. The obvious answer to reducing all pollutants to protect the most sensitive is impractical, since in addition to man's normal activities, nature provides a considerable source of air contaminants.

The ideal arrangement would be for scientists to develop more comprehensive data that would be provided to the political bodies (cities, states, or the federal government). Then the appropriate authorities would be in a position to give due consideration to all the facets of this complex problem—health, economy, social well-being, plant damage, and related factors. For highly sensitive persons who would not be protected by compliance with the standards, alternate methods should be prescribed.

REFERENCES

1. L. Greenburg et al., "Report of an Air Pollution Incident in New York City, November 1953," *Pub. Health Rep.*, **77**, 7 (1962).
2. L. Greenburg et al., "Air Pollution and Morbidity in New York City," *J. Am. Med. Assoc.*, **182**, 161 (1962).
3. J. H. Stebbings, D. G. Fogleman, K. E. McClain, and M. C. Townsend, "Effect of the Pittsburgh Air Pollution Episode upon Pulmonary Function in Schoolchildren," *J. Air Pollut. Control Assoc.*, **26**, 6 (1976).
4. *Meuse Valley Air Pollution,* Royal Academy of Medicine of Belgium, December 19, 1931 pp. 683–732.
5. H. H. Schrenk, H. Heimann, G. D. Clayton, and W. N. Gafafer, "Air Pollution in Donora, Pennsylvania," Public Health Bulletin No. 306, Government Printing Office, Washington, D.C.
6. L. C. McCabe and G. D. Clayton, "Air Pollution by Hydrogen Sulfide in Poza Rica, Mexico," *Arch. Ind. Hyg. Occup. Med.*, **6,** 199–213 (1952).
7. Ministry of Health, "Mortality and Morbidity During the London Fog of December 1952." Reports on Public Health and Related Subjects, Her Majesty's Stationery Office, London, 1954.
8. P. DeFalco, Jr., "Regional Goals," *J. Air Pollut. Control. Assoc.*, **26,** 9 (1976).
9. L. Billings, "Legislative Prospects," *J. Air Pollut. Control Assoc.*, **26,** 9 (1976).
10. G. D. Clayton, "Air Pollution," in: *Industrial Hygiene and Toxicology,* 2nd rev. ed., Vol. I, F. A. Patty, Ed., Wiley-Interscience, New York, 1958, pp. 435–437.
11. Sixth annual report of the Council on Environmental Quality, Government Printing Office, Washington, D.C., No. 040-000-00337-1, December 1975.
12. "Air Quality Criteria and Guides for Urban Air Pollutants," Report of a WHO Expert Committee, World Health Organization Technical Report Service, No. 506, Geneva, 1972.
13. "Sulphur Dioxide and Suspended Particulate Matter in Urban Environments," Health Protection Directorate, Commission des Communautés Européennes, Brussels, Belgium, January 1976.
14. "The Environmental Protection Agency's Research Program with Primary Emphasis on the Community Health and Environmental Surveillance System (CHESS): An Investigative Report," prepared for the Committee on Science and Technology, U.S. House of Representatives, 94th Congress, November 1976. Government Printing Office, Washington, D.C.
15. F. H. Meyers and C. H. Hine, "Some Experiences of NO_2 in Animals and Man," paper presented at the Fifth Air Pollution Medical Research Conference, Los Angeles, 1961.
16. T. R. Carson, M. S. Rosenholtz, F. T. Wilinski, and M. H. Weeks, *Am. Ind. Hyg. Assoc. J.*, **23,** 457 (1962).

17. M. R. Purvis and R. Ehrlich, *J. Infect. Dis.,* **113,** 72 (1963).
18. H. G. Boren, *Arch. Environ. Health,* **8,** 119 (1964).
19. K. C. Back, *Proc. 1st Ann. Conf. Atmospheric Contamination Confined Spaces,* Wright-Patterson Air Force Base, Ohio, 1965.
20. W. D. Wagner, B. R. Duncan, P. G. Wright, and H. E. Stokinger, *Arch. Environ. Health,* **10,** 455 (1965).
21. O. J. Balchum, R. D. Buckley, R. Sherwin, and M. Gardner, *Arch. Environ. Health,* **10,** 274 (1965).
22. G. Freeman et al., *Arch. Environ. Health,* **17,** 181 (1968); **18,** 609 (1969).
23. J. H. Riddick, Jr., K. I. Campbell, and D. L. Coffin, *Am. J. Clin. Pathol.,* **49,** 239 (1968).
24. R. D. Buckley and C. G. Loosli, *Arch. Environ. Health,* **18,** 588 (1969).
25. C. M. Shy et al., *J. Air Pollut. Control Assoc.,* **20,** 582 (1970).
26. C. M. Shy et al., *J. Air Pollut. Control Assoc.,* **20,** 539 (1970).
27. M. C. S. Kennedy, *Ann. Occup. Hyg.,* **15,** 285 (1972).
28. G. D. Von Nieding, H. Kreckler, R. Tuchs, H. M. Wagner, and K. Koppenhagen, *Int. Arch. Arbeitsmed.,* **31,** 61 (1973).
29. C. M. Shy, L. Niemeyer, L. Truppi, and T. English, "Re-evaluation of the Chattanooga School Children Studies and the Health Criteria for NO_2 Exposure," In-House Technical Report, National Environmental Research Center, Research Triangle Park, N.C., March 1973.
30. F. E. Speizer and B. G. Ferris, Jr., *Arch. Environ. Health,* **26,** 313, 325 (1973).
31. E. C. Arner and R. A. Rhoades, *Arch. Environ. Health,* **26,** 156 (1973).
32. J. H. Schulte, *Arch. Environ. Health,* **7,** 524 (1963).
33. J. R. Goldsmith, J. Terzaghi, and J. D. Hackney, *Arch. Environ. Health,* **7,** 647 (1963).
34. R. R. Beard and G. A. Wertheim, *Am. J. Pub. Health,* **57,** 2012 (1967).
35. National Academy of Sciences and National Academy of Engineering, "Effects of Chronic Exposure to Low Levels of Carbon monoxide on Human Health, Behavior and Performance," NAS–NAE, Washington, D.C., 1969.
36. A. D. L. Guest, C. Duncan, and P. J. Lawther, *Ergonomics,* **13,** 587 (1970).
37. D. Rondia, *C. R. Acad. Sci.* (*D*), **271,** 617 (1970).
38. J. J. McGrath and J. Jaeger, *Respir. Physiol.,* **12,** 46 (1971).
39. R. A. Jones, J. A. Strickland, J. A. Stunkard, and J. Siegel, *Toxicol. Appl. Pharmacol.,* **19,** 46 (1971).
40. Astrup et al., *Lancet,* **2,** 1220 (1972).
41. S. M. Ayres, R. Evans, D. Licht, J. Griesbach, F. Reimold, E. F. Ferrand, and A. Criscitiello, *Arch. Environ. Health,* **27,** 168 (1973).
42. R. D. Stewart, P. E. Newton, M. J. Hosko, and J. E. Peterson, *Arch. Environ. Health,* **27,** 155 (1973).
43. D. A. Debias, C. M. Banerjee, N. C. Birkhead, W. V. Harrer, and L. A. Kazal, *Arch. Environ. Health,* **27,** 161 (1973).
44. H. Hattori, *J. Sulphuric Acid Assoc.,* Tokyo, **24,** 13 (1971).
45. W. M. Busey, "Summary Report: Study of Synergistic Effects of Certain Airborne Systems in Cynomolgus," Hazelton Laboratories, Inc., Vienna, Va., Coordinating Research Project CAPM-6-68-5, June 1972.
46. T. J. Preziosi, "An Experimental Investigation in Animals of the Functional and Morphological Effects of Single and Repeated Exposures to High and Low Concentrations of CO," Preprint, New York Academy of Sciences, New York, 1970.
47. W. Bender, M. Goethert, G. Malorny, and P. Sebbesse, *Arch. Toxicol.* (*Berlin*), **27,** 142 (1971).

48. D. Henschler, A. Stier, H. Beck, and W. Neumann, *Arch. Gewerbepathol. Gerwerbehyg.*, **17**, 547 (1960).
49. C. Schoettlin and E. Landau, *U.S. Pub. Health Rept.*, **76**, 545 (1961).
50. J. M. Lagerwerff, *Aerosp. Med.*, **34**, 479 (1963).
51. W. A. Young and D. B. Shaw, *J. Appl. Physiol.*, **19**, 765 (1964).
52. W. Y. Hallett, *Arch. Environ. Health,* **10**, 295 (1965).
53. R. F. Bils and J. C. Romanovsky, *Arch. Environ. Health,* **14**, 844 (1967).
54. D. L. Coffin et al., *Arch. Environ Health,* **16**, 633 (1968).
55. Eiji Yokoyama, *Jap. J. Ind. Health,* **11**, 563 (1969).
56. E. Goldstein, W. Tyler, P. Hoeprich, and C. Eagle, *Arch. Intern. Med.*, **128**, 1099 (1971).
57. M. Evans, W. Mayr, T. Bils, and C. Loosli, *Arch. Environ. Health,* **22**, 450 (1971).
58. D. V. Bates, G. M. Bell, C. D. Burnham, M. Hazucha, J. Mantha, L. D. Pengelly, and F. Silverman, *J. Appl. Physiol.*, **32**, 176 (1972).
59. A. Pan, J. Beland, and Z. Jegier, *Arch. Environ. Health,* **24**, 229 (1972).
60. I. Mizoguchi, M. Osawa, Y. Sato, K. Makino, and H. Yagyu, *J. Japan. Soc. Air Pollut.*, **8**, 414 (1973).
61. E. Yokoyama, "The Effects of Low Concentration of Ozone on the Lung Pressure of Rabbits and Rats," Preprint, Japan Society of Industrial Hygiene 1973, pp. 134–135.
62. D. Bates, "Hydrocarbons and Oxidants," Clinical Studies, paper presented at Conference on Health Effects of Air Pollutants, National Academy of Sciences, Washington, D.C., October 4, 1973.
63. M. Hazucha, F. Silverman, C. Parent, S. Fields, and D. V. Bates, *Arch. Environ. Health,* **27**, 183 (1973).
64. D. I. Hammer, V. Hasselblad, B. Portnoy, and P. F. Wehrle, *Arch. Environ. Health,* **28**, 255 (1974).
65. J. D. Hackney, "Physiological Effects of Air Pollutants in Humans Subjected to Secondary Stress," Final Report January 1, 1973–June 30, 1974, State of California Air Resources Board Contract No. ARB 2-372.
66. V. M. Sim and R. E. Pattle, *J. Am. Med. Assoc.*, **165**, 1908 (1957).
67. V. P. Melekhina, "Maximum Permissible Concentration of Formaldehyde in Atmospheric Air," in: *Literature on Air Pollution and Related Occupational Diseases,* Vol. 3, B. S. Levine, Trans., Public Health Service, Washington, D.C., No. TT 60 21475, pp. 135–140.
68. V. P. Melekhina, "Hygienic Evaluation of Formaldehyde as an Atmospheric Air Pollutant," in: *Literature on Air Pollution and Related Occupational Diseases,* Vol. 9, B. S. Levine, Trans., (U.S.S.R.) Public Health Service, Washington, D.C., 1963–1964, pp. 9–18.
69. C. H. Hine, M. J. Hogan, W. K. McEwen, F. H. Meyers, S. R. Mettier, and H. K. Boyer, *J. Air Pollut. Control Assoc.*, **10**, 17 (1960).
70. S. R. Mettier, H. K. Boyer, C. H. Hine, and W. K. McEwen, *AMA Arch. Ind. Health,* **21**, 1 (1960).
71. M. O. Amdur, *Int. J. Air Pollut.*, **3**, 20 (1960).
72. E. E. Morrill, Jr., *Air Cond., Heat., Vent.*, **58**, 94 (1961).
73. N. A. Renzetti and R. J. Bryan, *J. Air Pollut. Control Assoc.*, **11**, 421 (1961).
74. H. Salem and H. Cullumbine, *Toxicol. Appl. Pharmacol.*, **2**, 183 (1960).
75. C. W. Smith, Ed., *Acrolein,* Wiley, New York, 1962.
76. S. D. Murphy, D. A. Klingshirn, and C. E. Ulrich, *J. Pharmacol. Exp. Ther.*, **141**, 79 (1963).
77. S. D. Murphy, H. V. Davis, and V. L. Zaratzian, *Toxicol. Appl. Pharmacol.*, **6**, 520 (1964).
78. C. S. Tuesday, B. A. D'Alleva, J. M. Huess, and G. J. Nebel, "The General Motors Smog Chamber," Research Publication No. GMR-490, General Motors Corp., Warren, Mich., 1965.

79. P. Catilina, L. Thieholt, and J. Champeix, *Arch. Mal. Prof.,* **27,** 857 (1966).
80. M. Gusev, A. I. Svechnikova, I. S. Dronov, M. D. Grebenskova, and A. I. Golovina, *Hyg. Sanit.* **31,** 8 (1966).
81. J. C. Romanovsky, R. M. Ingels, and R. J. Gordon, *J. Air Pollut. Control Assoc.,* **17,** 454 (1967).
82. J. M. Peters, R. L. H. Murphy, L. D. Pagnotto, and W. F. Van Ganse, *Arch. Environ. Health,* **16,** 642 (1968).
83. Department of Health, Education and Welfare, Public Health Service, Environmental Health Service, National Air Pollution Control Administration. *Air Quality Criteria For Hydrocarbons.* NAPCA Publication No. AP-64, Government Printing Office, Washington, D.C., 1970.
84. National Academy of Sciences, *Particulate Polycyclic Organic Matter,* in the series *Biologic Effects of Atmospheric Pollutants,* NAS, Washington, D.C., 1972.
85. Ahmad and T. C. Whitson, *Ind. Med. Surg.*; September 1973.
86. M. Absoy, S. Erdem, K. Dincol, T. Hepyüksel, and G. Dincol, *Blut Band,* **8,** 293 (1974).
87. National Academy of Sciences, *Vapor-Phase Organic Pollutants: Volatile Hydrocarbons and Oxidation Products,* in the series *Medical and Biologic Effects of Environmental Pollutants,* NAS, Washington, D.C., 1976.
88. A. J. Wicken, and S. F. Buck, "Report on a Study of Environmental Factors Associated with Lung Cancer and Bronchitis Mortality in Areas of North East England," Research Paper No. 8, Tobacco Research Council, London, 1964.
89. J. W. B. Douglas and R. E. Waller, *Br. J. Prev. Soc. Med.,* **20,** 1 (1966).
90. P. E. Joosting, *Ingenieur,* **79:**50, A739 (1967).
91. J. E. Lunn, J. Knowelden, and A. J. Handyside, *Br. J. Prev. Soc. Med.,* **21,** 7 (1967).
92. L. J. Brasser, P. E. Joosting, and D. van Zuilen, "Sulfur Dioxide—To What Level Is it Acceptable?" Research Institute for Public Health Engineering, Report No. G-300, Delft, Netherlands, July 1967.
93. C. M. Fletcher, C. M. Tinker, I. D. Hill, and F. E. Speizer, "A Five-Year Prospective Field Study of Early Obstructive Airway Disease," Current Research in Chronic Respiratory Disease," *Proceedings of the Eleventh Aspen Conference,* Department of Health, Education and Welfare, Public Health Service, Washington, D.C., 1968.
94. M. C. Battigelli, *J. Occup. Med.,* **10:**9, 500 (1968).
95. B. W. Carnow et al., *Arch. Environ. Health,* **16,** 768 (1969).
96; B. W. Carnow, R. M. Senior, R. Karsh, S. Wesler, and L. V. Avioli, *J. Am. Med. Assoc.,* **214:**5, 894 (1970).
97. Y. Alarie, C. E. Ulrich, W. M. Busey, H. E. Swann, and H. N. MacFarland, *Arch. Environ. Health,* **21,** 769 (1970).
98. I. P. Goldring, L. Greenburg, S. S. Park, and I. M. Ratner, *Arch. Environ. Health,* **21,** 32 (1970).
99. Committee on the Challenges of Modern Society NATO *Air Quality Criteria For Sulfur Oxides,* No. 7, November 1971.
100. M. O. Amdur, *Arch. Environ. Health,* **23,** 459 (1971).
101. F. W. Weir and P. A. Bromberg, "Further Investigation of the Effects of Sulfur Dioxide on Human Subjects," American Petroleum Institute, Project No. CAW C S-15, Washington, D.C., June 1972.
102. P. J. Lawther, R. E. Waller, and M. Henderson, *Thorax,* **25,** 525 (1970).
103. J. H. Stebbings and C. G. Hayes, "Frequency and Severity of cardiopulmonary Symptoms in Adult Panels: 1971–1972 New York Studies," National Environmental Research Center, Environmental Protection Agency, Research Triangle Park, N.C., August 1974.
104. D. P. Rall, *Environ. Health Perspect.,* **8,** 97 (1974).

105. *Health Consequences of Sulfur Oxides: A Report from CHESS, 1970–1971.* U.S. Environmental Protection Agency, Publication No. EPA-650/1-74-004 Research Triangle Park, N.C., 1974.
106. I. Anderson, G. R. Lundquist, P. L. Jensen, and D. F. Proctor, *Arch. Environ. Health,* **28,** 31 (1974).
107. D. I. Hammer, F. J. Miller, A. G. Stead, and C. G. Hayes, "Acute Lower Respiratory Disease in Children in Relation to Sulfur Dioxide and Particulate Air Pollution: Retrospective Survey in New York City, 1972," Human Studies Laboratory, National Environmental Research Center, EPA, Research Triangle Park, N.C., November 1975.
108. S. F. Buck and D. A. Brown, "Mortality from Lung Cancer and Bronchitis in Relation to Smoke and Sulfur Dioxide Concentration, Population Density, and Social Index," Research Paper No. 7, Tobacco Research Council, London, 1964.
109. A. E. Martin, *Proc. Roy. Soc. Med.,* **57,** 969 (1964).
110. "Report on an Enquiry into the Incidence of Incapacity for Work. Part II. Incidence of Incapacity for Work in Different Areas and Occupations," Ministry of Pensions and National Insurance, Her Majesty's Stationery Office, London, 1965.
111. W. W. Holland, D. D. Reid, R. Seltser, and R. W. Stone, *Arch. Environ. Health,* **10,** 338 (1965).
112. J. W. Douglas and R. E. Waller, *Br. J. Prevent. Soc. Med.,* **20,** 1 (1966).
113. M. Glasser, L. Greenburg, and F. Field, *Arch. Environ. Health,* **15,** 684 (1967).
114. W. Winkelstein, *Arch. Environ. Health,* **14,** 162 (1967).
115. W. Winkelstein, *Arch. Environ. Health,* **16,** 401 (1968).
116. W. Winkelstein, *Arch. Environ. Health,* **18,** 544 (1969).
117. M. C. Battigelli, Air Quality Monograph No. 69-2, American Petroleum Institute, New York, 1969.
118. P. J. Lawther, R. E. Waller, and M. Henderson, *Thorax,* **25,** 525 (1970).
119. Committee on the Challenges of Modern Society/NATO, "Air Quality Criteria for Particulate Matter," No. 8, November 1971.
120. C. J. Nelson, C. M. Shy, T. English, C. R. Sharp, R. Andleman, L. Truppi, and J. Van Bruggen, *J. Air Pollut. Control Assoc.,* **23:**2, 81 (1973).
121. C. M. Shy, V. Hasselblad, R. M. Burton, C. J. Nelson, and A. A. Cohen, *Arch. Environ. Health,* **27,** 124 (1973).
122. R. S. Chapman, C. M. Shy, J. F. Finklea, D. E. House, H. E. Goldberg, and C. G. Hayes, *Arch. Environ. Health,* **27,** 138 (1973).
123. J. G. French, G. Lowrimore, W. C. Nelson, J. F. Finklea, T. English, and M. Hertz, *Arch. Environ. Health,* **27,** 129 (1973).
124. D. I. Hammer, F. J. Miller, D. E. House, K. E. McClain, E. Tompkins, and C. G. Hayes, "Frequency of Acute Lower Respiratory Disease in Children; Retrospective Survey of Two S.E. Communities, 1968–1971, In-house technical report, U.S. Environmental Protection Agency, May 8, 1974.
125. W. Galke and D. House, "Prevalence of Chronic Respiratory Disease Symptoms in Adults: 1971–1972 Survey of Two S.E. United States Communities," In-house technical report, EPA, October 18, 1974.
126. W. O. Negherbon, "Sulfur Dioxide, Sulfur Trioxide, Sulfuric Acid and Fly Ash: Their Nature and Their Role in Air Pollution," Edison Electric Institute, EEI Publication No. 66-900, New York, 1966.
127. D. I. Hammer, F. J. Miller, A. G. Stead, and C. G. Hayes, "Air Pollution and Childhood Lower Respiratory Disease: Exposure to Sulfur Oxides and Particulate Matter in New York, 1972," paper presented at the American Medical Association Air Pollution Medical Research Conference, San Francisco, December 1974.
128. U.S. Department of Health, Education and Welfare, National Air Pollution Control Administration, Publication No. AP-63, 1970.

129. W. S. Wayne, P. F. Carroll, and R. E. Carroll, *J. Am. Med. Assoc.,* **199,** 901 (1967).

130. L. Smith, *Am. J. Publ. Health,* **55,** 1460 (1965).

131. R. D. Stewart, J. E. Peterson, E. D. Baretta, R. T. Bachand, M. J. Hosko, and A. Herrman, *Arch. Environ. Health,* **21,** 154 (1970).

132. M. J. Hosko, *Arch. Environ. Health,* **21,** 174 (1970).

133. S. M. Ayres, S. Giannelli, Jr., and H. Mueller, *Ann. N.Y. Acad. Sci.,* **174,** 268 (1970).

134. Aronow and Isbell, *Ann. Intern. Med.,* **79,** 392 (1973).

135. Anderson, Andelman, Strauch, Fortuin, and Knelson, *Ann. Intern. Med.,* **79,** 46 (1973).

136. Aronow, Stemmer, and Isbell, *Circulation,* **49,** 415 (1974).

137. R. B. Chevalier, R. A. Krumholz, and J. C. Ross, *J. Am. Med. Assoc.,* **198,** 1061 (1966).

138. R. L. Masters, "Air Pollution—Human Health Effects," in: *Introduction to the Scientific Study of Atmospheric Pollution,* McCormac, Ed., D. Reidel, Dordrecht, Holland, 1971, pp. 97–130.

CHAPTER FIFTEEN

Agricultural Hazards

CLYDE M. BERRY, PH.D.

1 THE AGRICULTURAL INDUSTRY

Until recent years agriculture has not been regarded as an industry. It varies widely in practice throughout the world. Some primitive tribes are still almost totally hunters and gatherers. Some have advanced to the slash-and-burn stage, moving on when the soil has been depleted of nutrients or when insects and parasites reduce the effectiveness of their agricultural efforts. The use of draft animals such as oxen, donkeys, and horses, which represents further progress in agricultural practices, continues in many areas.

Modern agriculture as performed in developed countries is observed to have many of the characteristics of industry:

- It is heavily mechanized and requires a heavy input of energy.
- It becomes more capital intensive as manual labor usage continues to diminish.
- Economic factors, such as cost of unit grown and regional and world markets heavily influence what is grown and how it is grown.

A distinction is made here between agriculture and agribusiness. Agricultural pursuits are the activities associated with the production of food and fiber. Agribusiness includes such activities as farm machinery manufacture, production of fertilizer and pesticides, food processing, textile-related activities, supplying of petroleum products, and banking and finance. Agriculture and agribusiness interface at many points and sometimes overlap.

Medical services are usually limited in agricultural areas, and prevention thus becomes an imperative. Safety, or the prevention of trauma, has probably received the greatest amount of professional attention. Industrial hygiene, in the traditional sense, has lagged far behind. Sanitation has not received a great deal of attention, and medical services have been largely confined to the control of infectious diseases by way of maternal and child health programs.

Industrial hygiene as practiced in a manufacturing facility must be adapted to the substantial differences that exist in agriculture. Some of the major differences are as follows:

- The workplace and the residence are the same in many instances.
- The work force is usually excluded from restrictions relating to age, sex, and other legal barriers.
- There is little or no distinction between labor and management.
- Activities of industrial hygiene concern are usually intermittent and are usually further affected by such parameters as geography, terrain, seasons, climate, and weather. Consequently a time-weighted average (TWA) is impossible to determine for any activity.
- Individuals performing an activity may not receive or understand instructions on proper work procedures.
- Maintenance of equipment may be poor; repair frequently involves improvisation under field conditions with concomitant violation of accepted good practice procedures.
- Facilities for the maintenance of good personal hygiene do not exist at the workplace.
- Appropriate personal protective equipment may not be available, and little help exists on its selection and use.

Identifiable trends will exacerbate concerns that have industrial hygiene aspects. Some of these are given below:

- An increase in the size and speed of machines will lead to more problems associated with noise, vibration, fatigue, human factors such as the location of controls, and others.
- A growing dependence on the use of chemicals; the quantity used will increase and many will be of increased toxicity. The environmental requirements for less persistent pesticides will mean more frequent application, with an associated increased potential for exposure.
- Fewer workers, increased specialization, and more material produced will result in more frequent exposures to higher concentrations and for longer periods of time than have been encountered in the past.
- More and more small farm operators will be employed in industry, and farm operations will be carried out after a day in the factory and on weekends. There may be exposure to potentiating factors.

- A growing involvement with consumer product safety concerns will lead the farmer to demand more assistance from the manufacturer or supplier of items he uses.
- A significant departure from heavy dependence on the extension service of land grant colleges and universities for technical information.

2 CHEMICAL HAZARDS

Construction, operation, or maintenance activities in an agricultural setting may confront the industrial hygienist with many exposures that would normally be associated with industry. Solvents, for example, are used in degreasing and painting. Welding and brazing are done, and these operations, which involve various exposures, may be carried out on surfaces covered with lead-based paint, on galvanized metal, or on equipment contaminated with pesticides. Rarely is mechanical exhaust ventilation provided.

Certain operations may be encountered that would occur only once or twice in a decade, such as the use of epoxies to reline a silo, or the foaming in place of urethanes for insulation, with attendant exposures to 2,4-toluene diisocyanate.

Internal combustion engines are operated for extended periods of time inside poorly ventilated buildings with problems arising from the carbon monoxide in the exhaust. Combustion products can create difficulty in the use of salamanders and petroleum-fueled space heaters, neither of which are commonly vented to the outside.

On nearly every farm chemically treated seeds are handled. There is ample opportunity for contact with these materials at planting time, that is, for dermal and respiratory exposures. The exposure is greater if the farmer does the seed treating himself.

A growing problem on farms is caused by anhydrous ammonia, its transportation, storage, and use. The handling of other fertilizer materials, some chemical and some natural, is less hazardous, but industrial hygiene concerns are present nonetheless.

There are thousands of pesticides to which the farmer might be exposed. Some are used in large quantities, alone or in combination with another pesticide or an adjuvant. Federal regulations do not adequately protect the farmer because registration required by government amounts almost entirely to a showing that the product will be effective in the use for which it is sold.

The degree of toxicity of pesticides varies tremendously. Sulfur as a fungicide has a very low toxicity, but the rodenticide sodium fluoroacetate is so toxic that there is no known antidote. Classification of the pesticides is difficult because some of them are used for dual purposes. Pentachlorophenol, for example, might be used as a wood preservative, for termite control, or as a defoliant.

Many chlorinated insecticides that were used in the past in large quantities have been banned by the Environmental Protection Agency because of their persistence in the environment and their alleged carcinogenicity. The trend is toward a greater use of organic phosphates and carbamates, and some of these compounds are very toxic. Their degradation after application can be affected by temperature and humidity. The same organic phosphate insecticide used on citrus in California has a different health potential when it is used in the more humid, higher rainfall areas of Florida. Standards and good practice guides that would be relevant in the manufacture and formulation of pesticides are completely inappropriate when dealing with application or postapplication exposure. The risk faced by the pilot of a crop-dusting plane is not comparable to that of the mixer and loader, and the flagger who helps guide the pilot over the field and the scout who goes into the field after the treatment to determine its effectiveness also represent two potential problems. All the foregoing situations are different from that of the worker who goes into the field later to cultivate or harvest. Reentry times seem to offer the only functional means of protecting the field worker.

Soil sterilants and growth regulators have not found extensive use in modern agriculture, but they are occasionally encountered. Fumigants, however, are rather commonly used where grain in storage must be protected against weevil and other insects. Very toxic exposures can result from the unprotected use of such materials as methyl bromide, carbon disulfide, and aluminum phosphide.

Data are currently inadequate to permit the appraisal of the toxic potential associated with some of the more commonly used herbicides such as 2,4-D and 2,4,5-T. Arsenicals are being phased out under regulatory pressure, and these materials are now less likely to be encountered. Perhaps one of the reasons for this official stance against arsenicals has been the frequency of animal deaths due to foraging on treated areas.

The contact the farmer may have with chemical additives in feed must be considered to offer a largely unknown potential. These substances can be hormones, such as diethylstilbestrol (DES), or arsenic compounds used as coccidiostats in chicken feed. Many feeds contain antibiotics, and the opportunity for sensitivities to develop is definitely present. Veterinary pharmaceuticals are another source of exposure. There is a growing trend toward the "recycling" of manure, an operation involving the drying and sterilization of manure and mixing it with fresh feed. Obviously there will be a concentration of additives, and the drying and sterilization operation will involve exposures of unknown significance.

Not to be ignored is the production in farming operations of toxic materials that had not been inherent in the materials initially handled. A good example of this is the exposure to "silo gas," presumed to be largely oxides of nitrogen. Ammonia and hydrogen sulfide are produced in certain confinement feeding operations at levels that exceed threshold limit values. Anoxia problems may result from the increasing use of hermetically sealed forage containers.

Protective measures that are common in industry are usually not available to the agricultural sector, or they might be inappropriate or impractical.

Personal respiratory equipment, if it were available, would not be fitted or maintained under professional supervision. Impervious clothing would be completely inapplicable to the case of someone working in the outdoors in high ambient temperatures, in the direct rays of the sun, and without any breeze because of surrounding lush crops.

3 BIOLOGICAL HAZARDS

Because agricultural pursuits involve close relationships to a wide spectrum of living organisms, they can result in disabilities that are either unique to agriculture to have a higher prevalence among farm workers. Probably the most attention has been given to maladies that are transmissible from animal to man, (i.e., the zoonoses).

Worldwide there are several hundred zoonoses, and fortunately most of them are rare. Most physicians are not alert to the possibility of zoonotic diseases, and the ailments often are not diagnosed. Frequently appearing as a nonfatal fever of unknown origin, they receive only symptomatic treatment. Efforts of control in the animal population will be reflected in reduced incidence in the human population. Brucellosis provides an example. Abortion and loss of fertility in a herd produces support for a control program that serves to reduce the number of human cases. Trichinosis has been almost eliminated in the human population with the requirement that garbage be cooked before it is fed to swine as a measure to control *Ecthyma* in the swine population.

Rabies is an almost universally fatal disease among warm-blooded animals. Consequently programs initiated and maintained primarily for economic reasons assure that very few cases occur each year in humans. A protracted, painful clinical regimen is required for prevention of rabies after an infected animal has bitten a person.

Psittacosis control is best effected among the parrot–parakeet pet population through the use of antibiotic additives in the bird food.

Leptospires are shed through the urine of infected animals and may contaminate water where swimming occurs, thus resulting in human cases of leptospirosis. Efforts to control this condition are usually directed at the elimination of the disease in animals, which automatically removes the possibility of the disease appearing in the human population.

Wild or migrating animals present a more difficult problem. Histoplasmosis may be contracted through the inhalation of aerosols associated with bird droppings. Migrating blackbirds and starlings present such a nuisance and health problem, and the damage they inflict on to crops is so great, that programs are initiated in some areas to reduce the bird population at their roosting location.

Plague and Rocky Mountain spotted fever are examples of zoonotic diseases associated with parasites. Ticks that have fed on diseased wild animals may move from the animal to the human as a result of a unique event in which man may become a temporary host.

In an agricultural activity there may be contact with plants, such as poison ivy and poison sumac, that can result in a disabling dermatitis. In certain cases the sensitivity to such plants may be so exquisite that causal and recreational activities must be restricted. Celery harvesters may develop photosensitivity. The fuzz from peaches can produce a dermatitis, particularly in the axilla or other areas of the body where perspiration and friction are combined.

Anthropods are not responsible for a large number of illnesses, but scorpion stings occur in certain arid portions of the world with some regularity, and black widow spiders have a wide range of habitat. Tarantulas are a problem unique to tropical areas except when they are imported into temperate regions by accident (e.g., in a stalk of bananas).

Insect bites and stings occur with a degree of frequency that makes them worthy of mention. Some of the stinging insects are domestic, such as the bee. Some are encountered in the wild state—wasps, hornets, and yellow jackets. The fire ant can be a considerable hazard to young domestic animals, and its mounds can be destructive to harvesting machinery, but these insects represent a small hazard to humans. Flies and mosquitoes, of course, are annoying for their biting, and some are infectious disease vectors for malaria and encephalitis.

Reptiles are not normally considered a substantial hazard in usual agricultural pursuits. Snakes, alligators, turtles, and other such reptiles might present problems where specialized activities were involved such as rearing them for sale.

4 PHYSICAL HAZARDS

Trauma is a highly significant occupational hazard for the agricultural worker. Much of the problem arises out of the use (and misuse) and maintenance of highly sophisticated machinery that is employed in the preparation for planting, in the tending of crops during the growing season, and in harvesting. Problems can also be associated with grinding and baling and other activities characteristic of feed preparation.

Agricultural machinery is designed to push, pull, shake, grind, pound, squeeze, separate, cut, or otherwise perform a function that is related to an agricultural pursuit. The energy input into these machines, usually substantial, is getting larger as the acres to be tilled by a single individual or a family increases and larger machines are used. The power source is usually an electric motor or an internal combustion engine, either gasoline or diesel.

Power transfer from the initial energy source is usually rotary. The equipment is then modified as needed to carry out the function desired through additional direct or belt-driven rotary devices, chains, sprockets, cams, hydraulic pistons, or other such approaches. Rotary devices, if exposed in a location where clothing can be caught in them, will twist the clothing in a manner that will break an extremity. If the clothing tears loose, emasculation may result. Gears, sprockets, and rollers are particularly dangerous because of their grip–squeeze effect, which results in serious mangling. Hydraulic systems can develop leaks and inject fluid directly into the tissues without any indication on the skin surface that the event has occurred. The resulting injury presents a difficult medical problem.

Transportation of large equipment from field to field in many cases involves use of a public highway. Vehicular movement is slow, and rear vision is often impeded by trailed equipment. Automobiles using the highway may strike the rear of such machinery, with substantial injury potential for the driver of the auto and the operator of the farm machinery. Recent legislation in most states requires the display on the rear of the slow-moving vehicle of a triangular sign that specifically designates such equipment.

High center of gravity and difficult terrain, as well as field obstacles, have caused frequent overturns that pin the operator under upset machinery. Tractors have been especially vulnerable to this kind of accident, and more and more manufacturers are fitting this prime mover with antiroll bars that are capable of withstanding an overturn, thus protect the driver.

Often repairs and adjustments must be made in the field, and the urgency of the situation and lack of appropriate tools and training may lead to an accident sequence that would not normally occur in an industrial setting.

Prolonged exposure to sunlight can be associated with premature aging of the skin and a higher incidence of skin cancer than is found in other occupations.

Strenuous physical effort must often be made in summer in direct sunlight when temperatures are high; there may be little air movement, and relative humidities are approaching saturation. Under such conditions heat stress may result in heat stroke, heat prostration, or dermatitis. The latter can be of the prickly heat type, or it can be manifest in skin folds or axilla, where sweating may be profuse and associated with friction.

Agricultural operations are usually at their lowest ebb in winter and at a minimum during times of severe cold. Certain operations, such as caring for livestock, may require exposure to cold of sufficient duration and intensity that frostbite occurs. Clothing appropriate for severe cold is now more readily available, and there is less likelihood of untoward conditions developing from working in frigid temperatures.

Vibration appears to be a growing problem. Chain saws and other vibrating tools such as impact wrenches and power-operated chisels are becoming more common, and extensive use results in Raynaud's phenomenon. The use of unsprung equipment such as tractors and combines during the autumn season when the ground is frozen seems to be accompanied by an increase of medically observed prostatitis. Whether this is associated with visceral resonance in the 5 to 15 Hz range is not certain.

Ergonomics is undoubtedly an important factor in the operation of agricultural machinery. Engineering design seems to have ignored many of the human factor aspects that would be considered important in industry, such as difficulty in climbing into, over, or around. Knobs, levers, and dials seem to have been positioned for design convenience without recognizing the limitations of the person who will be operating the machine. Seating has been improved in recent years, but the presence of cabs sometimes makes it impossible to stand, and spending many hours of operation in the seated position results in considerable fatigue. Skeletal dysfunction seems to be more prevalent among farmers and to appear at a younger age than in industry. (See Chapter 22.)

Fires and explosions are a hazard associated with the use of considerable amounts of flammable liquids and gases such as gasoline and compressed natural gas. Grain dusts are explosive, and accidents can occur in the drying of grains or the transfer of grain from storage into the market pipeline. Ignition can occur from a number of sources but is usually associated with an electric spark or an overheated bearing or exhaust manifold.

5 AIR POLLUTION

The smoke from the deliberate burning of fields after harvest and before plowing does not pose the problem it once did because of the recognized need to introduce organic matter into the soil to increase its humus content. However it is still used as a first step in the harvesting of sugar cane. Burning associated with cleaning fence rows and disposing of slashings harvested from woodlots contributes to air pollution.

Particulates may be introduced into the atmosphere as fugitive dust. This is most pronounced and on the largest scale in areas where rainfall is light, the ground is bare, and strong seasonal winds blow. Overgrazing, overcropping, or a lack of snow cover may exacerbate the situation.

Odors are a significant and growing problem related to confinement feeding operations, primarily of swine and cattle. Thousands of animals in a single, relatively small area, even though in a semiarid location, produce odors that are regarded as serious nuisances by governmental air pollution control personnel, and some residents living downwind from such operations indicate that they have been made physically ill.

Aerial spraying and dusting with pesticides constitutes a problem only when there is substantial drift. Usually these activities are performed only when wind velocity is low, but on occasions unfortunate results have occurred at a substantial distance from the location of the pesticide application. Usually the damage was identified with herbicides, and unusually sensitive plants such as grapes were affected; but accidents with insecticides have also brought about cases of human illness. The reader is referred to Chapter 23, Section 3.14.

6 WATER POLLUTION

Erosion of the land in the course of agricultural pursuits is responsible for enormous amounts of silting along streams and into lakes, reservoirs, and ponds. In impoundments this silting results in accelerated eutrophification with attendant loss of function of the body of water for recreation, flood control, as a source of water for irrigation, or as a supply for towns or cities.

Fertilizer runoff and leaching of chemicals from the soil into streams or impoundments are significant sources of water pollution. Phosphates from fertilizers stimulate undesirable algal blooms. In the case of pesticides, particularly the slowly biodegradable insecticides, there may be so much pollution that with biological magnification, whole bodies of water or segments of waterways must be closed to fishing, oyster harvesting, or other such activities.

Animal wastes represent a particularly difficult problem. Manure is usually placed on the land in relatively large amounts for relatively small areas, and with heavy rains the biological oxygen demand of the receiving waters can be so high that there is complete or nearly complete reduction in dissolved oxygen. This results in fish kills, and it places an enormous burden on municipal water purification plants. Bacterial counts may be very high. Break-point chlorination to kill bacteria produces tastes and odors in drinking water that make it quite unpalatable. Spreading on the land is usually done when the soil is sufficiently firm (either dry or frozen) to withstand the weight of the manure-conveying vehicle, and this maximizes the likelihood of substantial runoff with a heavy rain, thawing, or melting of snow cover.

Very large confinement operations are beginning to come under governmental environmental controls, and active research is underway to find better ways to handle large amounts of manure in the most nonpolluting way. The smaller sources of animal waste will continue to exist with consequent serious water pollution.

7 SOLID WASTES

Spreading on top of the land was just described as a disposal technique for manure that results in substantial water pollution. A less troublesome method of surface disposal is the knifing into the soil of liquid manure. Agronomists have yet to agree on the best way to effect the desired result, but one method is to have a tractor-drawn blade of the scarifier type open a trench in the ground continuously; the manure is discharged into the trench and is immediately covered. The fly, odor, and runoff problems are reduced substantially by such a technique. Much depends on the type of soil, its moisture content, scheduled subsequent soil preparation for planting, and other factors.

Composting is a method of solid waste disposal that is being studied, but to date it has not lived up to the claims of earlier proponents. After bacterial action and attendant heating are complete, the final material is relatively odor free and essentially sterile. The material needs to be aerated at periodic intervals, and most farmers have neither the time, the equipment, nor the inclination to become involved with the technique.

Lagooning is feasible in most instances, but it entails two main problems. If the lagoon is overloaded and becomes anaerobic, it becomes a source of air pollution. It is also a heavy user of land area that most farmers would rather employ for a purpose more closely related to production.

A most interesting aspect of animal waste disposal is that of recycling through the animal. When dried and sterilized, urine and feces, particularly feces, can be fed back to animals and can have a food value equivalent in certain instances to an equal weight of virgin feed. Most of the concern expressed to date lies in the heavy energy consumption necessary for drying and sterilizing, and in the need to build up antibiotics and minerals that were constituents in carefully formulated fresh feed.

8 FUTURE IMPLICATIONS

No predictions can be made, but the following discernible trends have a bearing on agricultural hazards.

8.1 Chemicals

Economics and world demand for food will make imperative the continued use of substantial amounts of chemicals. Future use will be more carefully targeted, however, and multiple approaches will be employed. There will be more emphasis on biodegradability. Less toxic chemicals will be sought, more information will be supplied to the user, and there will be more monitoring. Personal protective equipment approaches will not be directly transferred from industry. Medical services, particularly emergency services relating to occasional acute exposures, will be improved but not significantly. Disposal problems will remain, and regional incinerators may become the means of disposing of relatively large amounts of unusable material.

8.2 Biological Hazards

Biological hazards will remain but will respond to improved controls from such efforts as herd eradication procedures, immunology, and early diagnosis and treatment. Improved reporting, better statistics, and laboratory research will shape the direction of preventive measures that are technically and economically feasible.

8.3 Physical Hazards

Physical hazards will continue to be the most serious and obvious problem to those engaged in agricultural pursuits. Engineering improvements will be made in design as the full impact of product liability and consumer protection efforts are realized. Field adjustments and maintenance activities will be simplified. The smaller operator will move toward the use of specialists who will do custom work for a fee. This will ease the financial pressure on the individual as equipment tends to become bigger, costlier, and more specialized.

Human factors engineering principles that are in common use in industry will be extended to agriculture. Cabs will be climate controlled for winter or summer and will be made crushproof. Power transfer will utilize hydraulics to a greater degree, and mechanical transfer of power will be reduced. Pumps and lines now operate at higher pressures so that motors can be smaller, and this will permit simplified design.

Noise and vibration will continue to cause problems, but current high levels of noise will be reduced either through improved machine design or isolation of the operator in a sound-conditioned enclosure.

8.4 Air Pollution

Air pollution will be reduced as this area comes under government regulation. The legal concept of riparian rights will yield to societal demands for land stewardship. Some of the changes will have economic overtones, such as greater use of minimum tillage (with greater dependence on chemicals), terracing, grass waterways, and spring instead of fall plowing. Nuisance laws and their enforcement, along with zoning requirements, will result in a more judicious location for operations that have the effect of polluting the air.

Aerial seeding, fertilizing (particularly foliar), and application of chemicals will increase. Improved equipment design, more carefully trained pilots; and stiffer licencing procedures will reduce the problems that in the past have been associated with drift.

8.5 Water Pollution

Water pollution will be dramatically reduced through greater emphasis on grass waterways, terracing, minimum tillage, and impoundments such as farm ponds. Use of liquid manure will be emphasized, and it will be applied to the land during the growing season in a modified irrigation approach.

8.6 Solid Wastes

Solid wastes will yield to the recycling concept. Dead animals will go to rendering plants to be made into tankage instead of being buried or burned. Vegetable solid wastes will be partially processed for animal food as is currently being done with corn cobs (with the addition of urea). Manure will be processed and blended into fresh feed. Some of this activity will be due to governmental regulations, some the result of improved technology, and some will be for economic return.

8.7 Miscellaneous

Several miscellaneous aspects of future agricultural operations may have relevance to the hazards now encountered. Governmental regulations will exist to an equal or greater degree, but they will be more applicable and more functional than is the current approach. Standards, where they can be devised, will lean more toward standards of performance than of specification. For example, it will be found impractical to check exposures of migrant workers to pesticides, either through exposure or medical monitoring. Instead, emphasis will be placed on such approaches as reentry standards, and these will be specific for the chemical applied, the crop to which it is applied, the weather, the climate, and the specific task that is to be performed. Currently (1977) research is being conducted in this area by the Institute of Food and Agricultural Sciences at the University of Florida.

Educational efforts will be expanded considerably, probably patterned after the land grant extension approach, which has had great success for more than a century.

From the preceding discussion it is evident that the problems facing the agricultural industry are similar in many respects to the industrial hygiene problems in other types of industry, with one big difference: in the agricultural industry there is no production line—the exposures change with great frequency. In any one day the agricultural worker may be caring for animals, welding, spraying, plowing, and harvesting, and all these activities have a different stress on the individual. On the other hand, a production line worker having the same job for 8 hours will be exposed to the same stresses during that period. Controlling the stresses on a production line obviously is much simpler.

The only logical way to protect the health of the farm worker is through an extensive educational program, utilizing the federal, state, and county agencies, the farm bureau, the unions, and other organizations, to keep the farm owner informed and to ensure that he is able to educate those who work for and with him on his farm.

CHAPTER SIXTEEN

Odor Measurement and Control

AMOS TURK, PH.D., and
ANGELA M. HYMAN, Ph.D.

1 INTRODUCTION

The scope and applications of odor measurement and control are frequently confused by various uses of the word *odor*. In the past, "odor" has been used to mean either (*a*) the perception of smell, referring to the sensation, or (*b*) that which is smelled, referring to the stimulus. To eliminate confusion, *odor* should be used only for the former meaning, and *odorant* should be defined as any odorous substance. *Odor intensity,* then, is the magnitude of the olfactory sensation produced on exposure to an odorant. *Odor control* is a term that can be used to describe any process that makes olfactory experiences more acceptable to people. The perceptual route to this objective is usually, but not always, the reduction of odor intensity. An alternative route is the change of odor quality in some way that is considered to be an improvement.

When the reduction of odor intensity is accomplished by removal of odorant from the atmosphere, the process is equivalent to gas and vapor abatement, or to air cleaning, but with some special considerations. These include (*a*) problems related to the need to attain very low concentrations, often approaching threshold levels, (*b*) uncertainties with regard to the reliability of sensory or chemical analyses (see Section 4), and (*c*) difficulties associated with diffuse or sporadic sources.

When an odorous atmosphere is improved by the addition of another (usually pleasant) odorant under conditions in which chemical reactions are not involved, the process is called *odor modification,* referring to modification of the odor perception, not of the odorant. These and other methods of odor control are considered in some detail in Section 6.

2 OLFACTORY PERCEPTION

2.1 Anatomy of the Olfactory System

The olfactory system is unique among the sensory systems in that the cells responsible for the receipt and transduction of odorous stimuli also transmit the encoded information to the brain. In man, olfactory receptor cells can number in the hundreds of millions. The olfactory receptor cells are interspersed among supporting and basal cells in a yellow pigmented epithelium called the *olfactory mucosa.* The term *mucosa* refers to a mucus-secreting membrane; a layer of mucus 10 to 40 μm thick coats the olfactory mucosa, which lies in the upper part of the nasal cavity above the path of the main air currents that enter the nose with normal inspiration. To reach the olfactory mucosa, therefore, odorous molecules must either diffuse up to the olfactory receptor cells or be drawn up by sniffing.

Olfactory receptor cells are bipolar neurons; that is, they are nerve cells with two main processes. The *apical* or *dendritic* process of each cell reaches the surface of the mucosa, and several hairlike projections called *cilia* extend into the mucus. The cilia are typically 0.3 μm in diameter and 50 to 150 μm long. The basal process of each olfactory receptor cell forms an unmyelinated axon about 0.2 μm in diameter. This diameter is small, and propagation of nerve impulses along the length of these axons consequently is slow. The axons bundle together as they pass out of the olfactory mucosa and form the *olfactory nerve* or *cranial nerve I.* The human olfactory nerve is only a few millimeters long. The olfactory nerve courses centrally, enters the cranium through a series of perforations called the *cribiform plate,* and terminates in the surface layers of a region of cortex called the *olfactory bulb.* Thus olfactory bulb is the first synaptic relay center in the olfactory pathway. Within the olfactory bulb there is a complex pattern of connections and interconnections among various types of neuron. The synaptic organization of the bulb is thought to be related to the processing of olfactory information. The principal relay neurons of the bulb send their axons out from the bulb through the *olfactory tract* to project to several other parts of the brain. Olfactory information is believed to participate in complex patterns of behavior such as feeding and reproduction, and in the control of emotional responses such as fear, pleasure, and excitement. The olfactory bulb is merely two or three relays from nearly every other brain area. Information from other brain areas is likewise relayed to the olfactory bulb; thus there exists the potential for modulation of activity in the olfactory system.

Cellular replication has recently been observed in basal cells located within the olfactory mucosa of the adults of some nonhuman species. Some investigators believe that the olfactory receptor cell population turns over and suggest that basal cells divide to form precursors of new olfactory receptor cells.

2.2 Sensory Characteristics of Odors

An odor can be characterized by its absolute threshold, its intensity, its quality, and its affective tone or pleasantness–unpleasantness dimension. Determinations of thresholds of odor perception and measurements of odor intensity are considered in Sections 2.4 and 2.5, respectively. Discussion of measurements of odor quality and odor acceptability are reserved for Section 4.4. It is important to recognize throughout that any program of odor measurement should be related to realistic analytical objectives. For example, if the odor from a landfill operation is to be controlled by masking agents, it is irrelevant to appraise odor intensities or threshold levels; a more appropriate measurement would describe the changed odor quality or character and would assay its acceptability to the

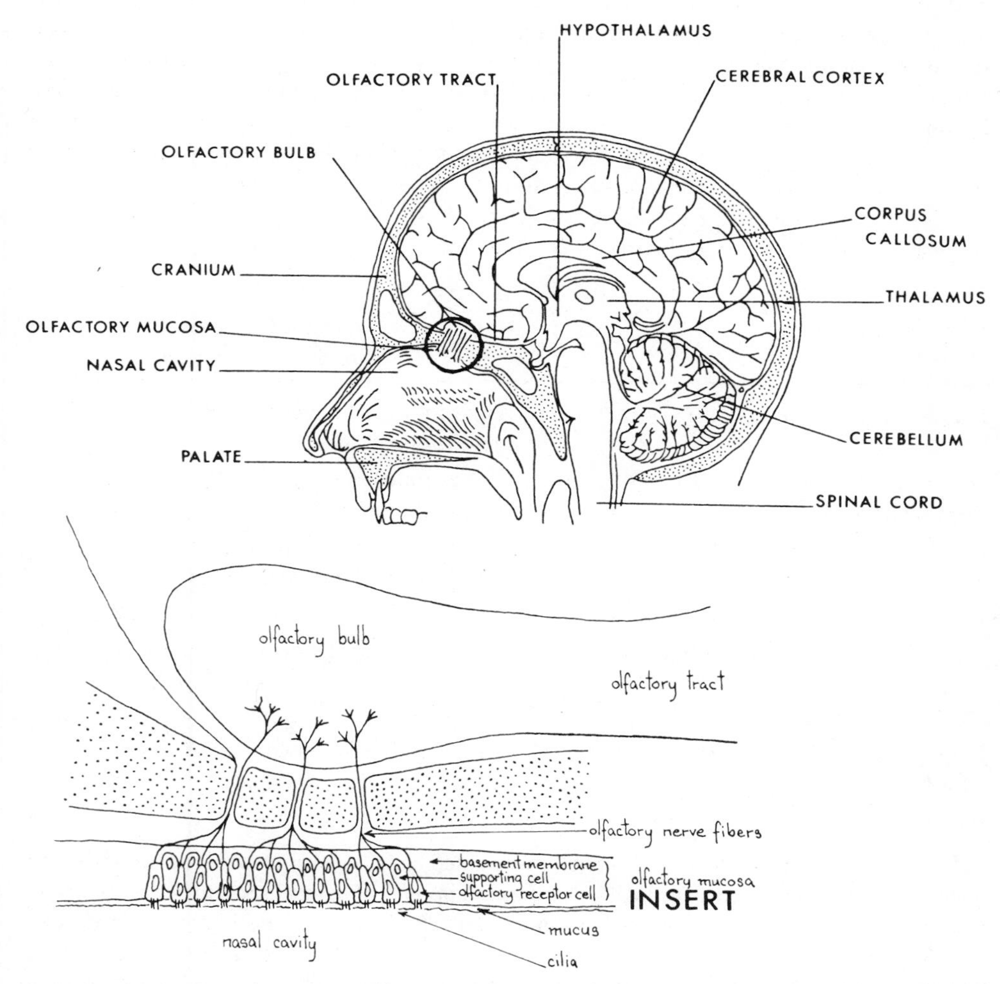

Figure 16.1 Schematic of human olfactory system in sagittal view showing location of olfactory receptor cells.

people in the community. On the other hand, a nonselective method of odor reduction, like ventilation or activated carbon adsorption, may well be monitored by measurements of intensity or by threshold dilution techniques.

2.3 Psychophysics and the Measurement of Odor Perception

G. T. Fechner is generally considered to be the founder of psychophysics, which he defined as the exact theory of the functionally dependent relations of body and soul, or more generally, of the physical and the psychological worlds (1). Fechner developed the concept of estimating or measuring the magnitude of sensations by assigning numerical values, reasoning that the relative magnitudes of sensations would be mathematically related to the magnitudes of the corresponding physical stimuli. In the years since Fechner's work, psychophysicists have endeavored to determine the laws relating physical stimuli to the resulting conscious sensations of the human observer. These efforts have concerned the responses of subjects to a number of basic questions about physical stimuli. Referring specifically to odors, these questions and the types of sensory evaluation needed to answer them can be tabulated.

Questions About an Odor Stimulus	Relevant Category of Sensory Evaluation
Is an odor present?	Determination of sensitivity to the detection of odor
Is a particular odor (e.g., that of phenol) present?	Determination of sensitivity to the recognition of odor
Is this odor different from that?	Determination of sensitivity to the discrimination between odors
How strong (intense) is this odor?	The scaling of odor intensity
What does this odor smell like, or what is the quality of this odor?	The scaling of the degree of similarity or dissimilarity between a given odor and each of a set of different odors
How pleasant (or unpleasant) is this odor?	The scaling of the hedonic, or like–dislike, response to an odor

Each except the last of the problems named has a counterpart in quantitative neurophysiology. Indeed, applications of psychophysical techniques in some fields of research have aided in the investigation of underlying neural mechanisms. However, electrophysiological signals, even if it were convenient to monitor them routinely from human subjects, are not direct expressions of subjective experience. The magnitudes of sensations are therefore measured along psychophysical scales, and for olfactory sensations the reference points on such scales consist of the perceptions experienced on exposure to standard odorants.

2.4 Decision Processes and Thresholds of Odor Perception

The word *threshold* means a boundary value, a point on a continuum that separates values that produce a physiological or psychological effect from those that do not. The *upper threshold* is the odorant concentration above which further increases do not produce increases in perceived intensity. The *detection* or *absolute threshold* is the minimum odorant concentration that can be distinguished from an environment free of that odor. Correspondingly, the *recognition threshold* is the minimum concentration at which an odorant can be individually identified. The recognition threshold of a particular odorant is never lower than its absolute threshold.

Thresholds are not firmly fixed values. Sensitivity fluctuates irregularly, and a certain odorant concentration will elicit a response at one time but not at another. Classical psychophysicists recognized the instability inherent in sensitivity, and they defined the absolute threshold statistically as that odorant concentration which is detected as often as not over a series of presentations. The probability of detection of such a stimulus is 50 percent. This definition of threshold has prevailed in the literature of air pollution and industrial hygiene. The theory of signal detection, originally engineered for telephone and radio communication systems by Shannon and Weaver (2) and later translated into a more general theory by Swets and others (3), contributed significantly to the development of modern psychophysics by specifying the dependence of threshold determinations on certain experimental variables. Detection theory has revealed the criteria employed by subjects in making perceptual judgments.

To illustrate how odor thresholds differ from other sensory evaluations of odor, refer to the questions about odor tabulated in Section 2.3 and note that the first three questions, which deal with thresholds, can be answered only by "yes" or "no." Since either answer can be right or wrong, four possibilities emerge when a subject is asked one of these questions. These possibilities are set out in a "response matrix" as follows:

	Response	
Odorant	Yes	No
Present	Hit	Miss
Absent	False alarm	Correct rejection

In such an experimental situation, the subject must decide whether in an observation period of set duration there is only *noise* (i.e., background interference either introduced by the experimenter or inherent in the sensory process), or whether a *signal* (i.e., the designated odor stimulus) is present as well. Two kinds of error can be made by the subject in such a situation: a *miss* (i.e., the failure to detect an odor stimulus when one is present) and a *false alarm* (i.e., the report of perception of an odor stimulus when one is absent). Detection theory emphasizes the relation between the occurrence of these two types of error. According to detection theory (4), a positive response by a subject is favored (*a*) by positive expectations that are enhanced by a high rate of stimulus presentation or suggested by the experimenter's instructions, and (*b*) by reluctance to miss the presence of an odorant in accordance with rewards obtained for hits and/or punishments incurred for misses. Conversely, a negative response is favored (*a*) by an expectation, perhaps resulting from experience, that no odor stimulus is present, and (*b*) by reluctance to score a false alarm, perhaps resulting from rewards obtained for correct rejections and/or punishments incurred for false alarms. These personal biases can be so strong that it is possible, in laboratory situations, to manipulate a subject's responses by manipulation of the variables influencing his decision process. Thus the experimental paradigm can drive the value determined for the "odor threshold" to a particular target. For example, a subject can avoid false alarms by not saying "yes"; this behavior generates a high value for the threshold. Conversely, a subject will not miss if he does not say "no"; the consequence of such a decision is the generation of a low value for the threshold. Young and Adams (5), Steinmetz et al. (6), and Johansson et al. (7) describe such effects of the manipulation of experimental variables on odorant threshold determinations.

It is against this background of psychological biases that determinations of the limits of sensory perception must be evaluated. The absolute thresholds of different odorants determined in the same investigation can vary over 6 orders of magnitude, for example, 0.00021 ppm for trimethylamine and 214 ppm for methylene chloride (8). Such wide ranges may reflect real differences among the odors of different substances, yet since some reported threshold concentrations are very low, we have the implication that analytical and calibration errors are large and that the accuracy of the threshold values is suspect. Furthermore, the signal detection model of threshold determinations suggests that the variance can reflect decision as well as sensory processes. This suggestion could account for the wide range of scatter found in values of the absolute threshold generated for the same odorant in different investigations, for example, 3.2×10^{-4} to 10 ppm for pyridine (9). Another reasonable explanation for such scatter is the possibility that different samples of nominally identical odorants may really be very dissimilar, owing to different contents of odorous impurities. For example, phosphine, for which values of the detection threshold reported in the literature range from 0.2 to 3 ppm, has been shown to be odorless when pure (10), the reported odors being due to impurities in the form of organic phosphine derivatives. Similarly, it has been demonstrated (11) that ultrapurification engenders radical changes in the odor of nominally "pure" samples. Interindividual differences in sensitivity are yet another possible cause of large differences among values of the absolute threshold generated for the same odorants in different investigations. Interindividual variation can be high, but typically this factor is ignored in data analyses for the determination of odorant thresholds. There is also the problem of diversity in experimental procedures, for example, the mode of odorant presentation and sampling, which could introduce adaptation effects (Section 2.7), odor masking (Section 6.7), and errors in values reported for odorant concentration (12).

2.5 Laws Relating Odor Intensity to Odorant Concentration

A major objective of psychophysics is to measure the dynamic properties of sensory systems. Odorant concentration is the one property of the olfactory stimulus that can be varied somewhat systematically, although its effective control at the level of the olfactory receptor cell or even the olfactory mucosa has yet to be demonstrated. In the past, odorant concentration has been estimated or measured at some distance from the olfactory receptor cells. Psychophysicists have strived to characterize the functional dependence of odor intensity on such values of stimulus magnitude.

Psychophysical scaling attempts to determine the mathematical form of the relation between odor intensity and odorant concentration. *Direct scaling* refers to methods in which direct assessments of psychological quantities are made on an equal-interval scale (i.e., a graduated series with a constant unit) or on a ratio scale (i.e., a graduated series with a constant unit and a true zero). Direct scaling procedures include *estimation methods* in which stimuli are manipulated by the experimenter and judged by subjects, and *production methods* in which stimuli are manipulated by the subject to achieve a defined relation. In *category estimation* a given segment of a sensation continuum (usually predefined by stimuli supplied as anchors by the experimenter) is partitioned by subjects into a predetermined number of perceptually equal intervals, and subsequently presented stimuli are distributed by subjects among these categories. In *ratio estimation* the apparent ratio corresponding to a pair of stimulus magnitudes is numerically estimated. *Magnitude estimation* is a method in which various magnitudes of a stimulus are individually presented, and subjects assign a number to each signal in proportion to perceived intensity. For a complete discussion of psychophysical scaling methods, refer to Engen (13). There are numerous important variations in these procedures, and psychophysical scales generated with different experimental methods often do not correspond (14).

Stevens (15) proposed the psychophysical relation in which perceived intensity is a power function of stimulus magnitude, specifically the relation $R = cS^n$, where R is the perceived intensity, S is the stimulus magnitude, and c and n are constants referring to the intercept and the slope, respectively, when R and S are plotted on log-log coordinates. This yields a linear relation between the logarithm of the magnitude of the stimulus and the logarithm of the magnitude of the sensation. This relation has been verified for a wide range of sensory continua (14, 16) and has been proposed to be the fundamental psychophysical law.

Not all olfactory psychophysical scaling data conform to the power function proposed by Stevens. Magnitude estimations of odor intensities reported by Engen (17) for diacetone alcohol, *n*-heptane, and phenylethyl alcohol diluted in benzyl benzoate, and by Cain (18) for air dilutions of *n*-butyl acetate and 1-propanol, depart from the psychophysical law at low odorant concentrations. One reasonable explanation for these departures is that control and manipulation of concentration become increasingly difficult at greater dilutions. Another consideration for work with benzyl benzoate is that the diluent itself is odorous, and contributions of the diluent to odor intensity are greatest at the lowest concentrations of test odorant. Yet another possibility is that dilution does not produce a universal, unidimensional psychophysical scale for all odorants. Changes in odorant concentration can produce changes in odor quality as well as in odor intensity, and these transitions may not be identical for all subjects for all or even any odorants. In the experiments cited, shifts in odor quality might not have been ignored by subjects in their assessments of odor intensity. Indeed, it is questionable whether human subjects in any experimental situation can reliably distinguish odor intensity from other attributes of the stimulus. There is also the problem of psychological impurity, which is discussed in Section 2.6.

For most odorants, however, perceived intensity is reported to be a power function of stimulus magnitude. Hyman (12) has tabulated values of the exponent of the psychophysical function for odor intensity for 33 odorants generated in different investigations. The values are typically less than 1.00, with some exceptions for some subjects. They range from 0.03 for 1-decanol (19) to 0.82 for 1-hexanol (20). Some of the results of the scaling of perceived intensity for the same and even for different odorants are quite similar, whereas considerable disagreement exists among the results of many other investigations of the intensity of different and even of the same odorants. Hyman (12) has discussed the factors capable of influencing the outcome of olfactory psychophysical scaling in the hope of elucidating possible sources of discrepancies. In brief, the factors concerning stimulus magnitude are (*a*) the use in different investigations of different physical scales that are not linearly related; (*b*) the use in some investigations of excessively strong stimuli or intertrial intervals so short that adaptation effects are introduced (see Section 2.7); (*c*) the use in some investigations of stimuli in concentrations so low that analytical and calibration errors are large; (*d*) the adsorption of vapors on the walls of gas-sampling vessels, which introduces errors in reported odorant concentrations; (*e*) the evaporation of volatiles from mixtures, which alters their composition during a given study; and (*f*) deviations from ideal behavior as given by Raoult's law, which is used to predict the vapor pressure of a dissolved substance on the basis of proportionality to solute concentration.

Deviations from Raoult's law readily explain differences among values of the exponent of the psychophysical function for odor intensity generated for the same as well as for different substances under different methods of dilution (21). Furthermore, because the extent of these deviations depends on the solvent and on concentration, the possible error in values of stimulus magnitude used in the determination of odor intensity function parameters also applies for odorants diluted in different liquids and for odorants diluted in the same liquid but at different ranges of concentration. There are also problems of psychological impurity (Section 2.6) and factors concerning chemical impurities and contaminants of stimuli that can introduce adaptation effects (Section 2.7) and odor interactions (Section 6.7) into the test situation.

With the exception of a few investigations (22–24) in which reported values of the exponent of the psychophysical function for odor intensity are the mean of values of individual subjects, published values of the exponent seem to be based on odor intensity assessments collected from a group of subjects, then pooled to form one regression plot from which the parameters of the psychophysical function are found. The assumption implicit in such a pooling procedure is that subjects are interchangeable psychophysical transducers, or that they represent a normal distribution of such transducers and that the distribution is the same for all odorants. But variation across individuals can be high (22, 25), and individual subjects can be consistent in the values of the exponent they generate for a particular odorant over repeated test sessions (25). It therefore seems improper to regard subjects as interchangeable for the purpose of pooling data in the indiscriminate manner traditionally employed for the calculation of odor intensity function parameters. Indeed, interindividual variation can be as large as differences between values of the exponent of the psychophysical function for different odorants derived from odor intensity assessments pooled from groups of subjects. Group measures may be obscuring important attributes of the olfactory system which these investigations are endeavoring to elucidate. Not only can the value of the exponent of the psychophysical function for odor intensity for any particular odorant vary greatly between subjects, but large intraindividual differences can exist between values of the exponent for different odorants (22, 25). Thus it seems appropriate to direct investigations to individual behavior over a variety of stimulus and response modes within the olfactory sensory continuum while concurrently striving to minimize variation in experimental procedures.

2.6 "Pure" and "Impure" Odor Perceptions

The experience of smell can be taken either to mean any perception that results from nasal inspiration or to refer only to sensations perceived by way of the olfactory receptor cells. An odorant that is sensed only by the olfactory receptor cells (e.g., air containing 0.01 ppm vanillin) is said to be *psychologically pure*. An odorant that stimulates both olfactory receptor cells and other sensitive cells (the so-called common chemical sense) is said to be *psychologically impure*; an example is air containing 50 ppm propionic acid. The common chemical sense includes sensations of heat, cold, pain, irritation, and dimensions of pungency described by our language inadequately. The trigeminal nerve, or cranial nerve III, is the primary mediator of common chemical sensations. The receptors responsible for the transduction of these perceptions remain incompletely characterized. The concept of psychological purity implies nothing about the chemical composition of the stimulus.

Differences between psychologically pure and impure odors are neglected in many odor measurements, especially in industrial hygiene and community air pollution applications, where irritants and stenches are grouped together as objectionable atmospheric contaminants that ought to be removed. It is not always operationally feasible to make distinctions among odors based on degrees of psychological purity. In establishing sensory measurement scales, however, it is important to recognize that "irritating odor" is not necessarily an extension in magnitude of "strong odor" but refers to a different type of sensation.

The transitions in perceived quality that can accompany changes in the perceived intensity of an odorant may be purely olfactory in origin, or they may be due to a change from purely olfactory stimulation at low odorant concentrations to olfactory plus trigeminal stimulations at high odorant concentrations. Evidence has accumulated in support of the notion that the trigeminal nerve contributes to the overall intensity of odorants. For human subjects with unilateral destruction of the trigeminal nerve, the magnitudes of perceived intensities for certain odorants are consistently lower through the deficient nostril, with the most dramatic differences at high odorant concentrations (26). Doty (27) reported the detection of odorous vapors by what he considered to be anosmic human observers; such detection presumably occurs by way of the trigeminal system. The olfactory and trigeminal systems can have different psychophysical functions for a particular odorant (28), and psychophysical scaling data reported for odor intensity could actually be some combination of the two. There is usually no effort to eliminate the trigeminal component from assessments of odor intensity. It is therefore plausible that excitation of trigeminal receptors is responsible, at least in part, for differences among values of the exponent of the psychophysical function for odor intensity and for departures from that relation reported for some odorants in some investigations.

2.7 Olfactory Adaptation

Olfactory adaptation is the decrement in sensitivity to an odorant following exposure to what is termed an "adapting" odorous stimulus. The rate and degree of loss in sensitivity and subsequent recovery depend on the adapting stimulus and on its concentration (29). Self-adaptation to an odorant affects the perceived intensity of the same odorant, whereas cross-adaptation affects the perceived intensity of other odorants. Olfactory adaptation is common, yet its physiological mechanism has not yet been elucidated.

Olfactory adaptation has been studied by measuring the time required for the odor of a continuously presented stimulus to disappear, and by determining absolute threshold values before and after exposure to an odorant for a given period. An increase in odorant concentration usually increases the time required for disappearance of an odor, and for a particular experimental design, absolute threshold values increase with an increase in the magnitude of the adapting odorous stimulus.

Self-adaptation affects the magnitude of the olfactory sensation as a function of stimulus magnitude. Values of the exponent of the psychophysical function for odor intensity are significantly greater after self-adaptation (20, 30). High concentrations of self-adapting stimuli generate steeper psychophysical functions for odor intensity than do self-adapting stimuli of low concentrations (30). The use of weak stimuli closely spaced in magnitude or long periods of exposure to an adapting odorous stimulus can permit resolution of the influence of duration of an adapting stimulation on the psychophysical function for odor intensity (31). However, without consideration of these factors, increasing the duration of self-adaptation produces only minor effects on the odor intensity function (32).

Cross-adaptation has approximately the same effect as self-adaptation on the form of the psychophysical function for odor intensity. It is important to recognize, however, that the effects of cross-adaptation within pairs of odorants can be asymmetric; for example, concentrations of 1-pentanol and 1-propanol closely matched for odor intensity have unequal cross-adapting effectiveness (30).

Olfactory adaptation may be responsible for some of the differences among values of the exponent of the psychophysical function for odor intensity reported for the same and even for different odorants. Adaptation effects can be introduced in an experiment by (*a*) the use of excessively strong stimuli, (*b*) short intertrial intervals, (*c*) the presence of odors apart from those generated by stimuli in intertrial intervals, and (*d*) the presentation of a standard reference stimulus immediately before test concentrations (32).

In the context of industrial hygiene and air pollution situations, two important questions arise. Does prolonged exposure to odors produce a permanent loss of ability to smell? And to what extent does temporary adaptation interfere with the sensory measurements of odor intensity carried out in the workspace or in the outside community?

With regard to the first question, there is simply no evidence on which a reliable answer can be based. In view of the many difficulties involved in quantifying sensory odor measurements under controlled test conditions, it would seem hopelessly unreliable to attempt any studies that depended on retrospective estimates of exposures to odors.

To answer the second question, we must consider patterns of recovery from adaptation and the opportunities for such recovery during typical odor survey programs. Köster (33) studied recovery times from adaptation to the odors of benzene and various alkylbenzenes, as well as to isopropyl alcohol, dioxane, cyclopentanone, and β-ionone. Typically, 60 to 70 percent recovery of sensory response to the odors was realized in about 2 minutes. The implication of these findings, which is in accord with practical experience, is that if a subject is downwind from a point source such as a stack and is exposed to the odor intermittently depending on changes in wind direction, he can make perfectly good sensory evaluations when the plume comes his way again after a lapse of 5 minutes or more. However a person in an odorous enclosure, who is subjected to unrelieved exposure to odor, will indeed become adapted to a point that can invalidate any sensory judgment.

2.8 Theories of Olfactory Mechanism

Theories of olfaction attempt (*a*) to delineate the mechanism by which odorant stimuli elicit responses in olfactory receptor cells, and (*b*) to correlate the sensory characteristics of odors with physicochemical properties of the odorants. The objective is to understand the physiological basis of olfactory sensory discriminations.

The perception of smell begins with the impingement of molecules upon the olfactory receptor. The molecular property responsible for the initiation of excitation in olfactory receptor cells has yet to be identified. Cross-adaptation has been applied as a basis for the classifying of odorants with the aim of isolating the physical correlate of olfactory reception, the idea being that odorants with a common characteristic for olfactory stimulation will cross-adapt. Almost every physicochemical property has been implicated (34). Theories based exclusively on molecular size and shape have been found to be inadequate. There have been attempts to combine several physicochemical properties into complex functions in the hope that a certain combination will correlate with the available data on odor quality and thresholds. Intermolecular interactions, electron affinities, and even bulk properties such as vapor pressure and molar volume have been considered. Such attempts to formulate an appropriate composite function are more realistic than attempts to discern a unitary determinant of olfactory reception, but the correct balance of molecular properties that will predict all odors has not yet been produced. Indeed, even these attempts may be futile in that the parameters of a composite function may be correlated with molecular properties not considered in the function.

Much attention has been given to intramolecular vibrations as the physical correlate of odors. Wright (35, 36) has proposed that the energy state of olfactory receptor molecules becomes altered when an odorant molecule of appropriate vibrational frequencies is encountered, and the change in state of the receptor molecule induces excitation in the olfactory receptor cell. This theory is rather vague, but as with other theories of olfaction, vibrational measures do successfully predict some odors. However the pertinent spectral features have not been identified, and many odorants that produce different odors have similar vibrational properties. Furthermore, optical enantiomers, which have identical spectra, can have different odors (37).

Although direct evidence is lacking, it is thought that the cilia that project from each olfactory receptor cell contain the sites at which chemical stimuli are received. It is generally believed that odorant molecules are adsorbed at specific receptor sites, where they interact with proteins or other macromolecular constituents of the cilia surface, causing conformational changes that somehow initiate the transduction of molecular energy into electrical energy. Davies (38, 39) has proposed that odorant molecules cross the interface between the surrounding medium and the nerve cell membrane, creating temporary defects in the nerve cell membrane through which ions can pass; the result is a collapse of the nerve membrane potential, which provides the drive for the development of action potentials. Many other theories on the generation of neural signals by odorant stimuli have been proposed, but each is supported by circumstantial evidence only.

Not only is the sequence of events leading to excitation of olfactory receptor cells not understood, but the encoding of sensory information in olfactory nerve signals remains a mystery. On the microscopic level, there are no discernible differences in olfactory receptor cells to account for sensory discrimination. However electrophysiological studies have revealed that individual olfactory receptor cells differ in their responses to odor stimuli with regard to the extent of their excitation and the odorant concentrations at which they become excited. Such a system could provide for sensory discrimination by the pattern of activity of many receptors. However the code has yet to be deciphered.

In sum, the functional parameters of the olfactory stimulus remain obscure, and knowledge of the physiological basis of olfactory stimulation and sensory discrimination is scanty. Even investigations of the relation between odor intensity and odorant concentration have not provided insight into the operating characteristics of the olfactory sensory system.

3 CHARACTERISTICS OF ODOROUS SUBSTANCES

3.1 Odor Sources

A vented storage tank being filled with liquid ethyl acrylate from a delivery truck discharges to the atmosphere a volume of air saturated with the acrylate vapor at the prevailing temperature of the liquid. A system of this type is a simple example of a confined odor source: the location, molecular aggregation, composition, concentration, and volumetric discharge of the source can be quantitatively specified. Such definite characterizations facilitate the establishment of relationships between the odor source and the odor measurements made in the workspace or the community. In general, an odor source may be said to be confined when its rate of discharge to the atmosphere can be measured and when the atmospheric discharge is amenable to representative sampling and to physical or chemical processing for purposes of odor abatement. For meteorological diffusion calculations, the location at which the odor is discharged into the atmosphere is assumed to be a point in space.

The characterization of a confined odor source should include (*a*) the volumetric rate of gas discharge (gas volume/time), (*b*) the temperature at the point of discharge, (*c*) the moisture content, (*d*) the location and elevation, and the area and shape of the stack or vent from which the discharge is emitted, and (*e*) a description of the state of aggregation (gas, mist, etc., including particle size distribution) of the discharge. Such information is useful when meteorological factors relating to odor reduction by dispersion and dilution are being evaluated. The characterization of an odor source in terms of chemical composition will depend on the relationships to be established between sensory and chemical measurements. If there are no technical or legal uncertainties regarding the source of a given community odor, a detailed chemical analysis is not needed; instead, a comparative method of appraising the effect of control procedures will suffice. On the other hand, chemical characterization may be helpful in tracing a community odor to one of several alleged sources, in relating variations in odor from a given source to changes in process conditions, or in appraising the effectiveness of chemical procedures that are designed for odor abatement.

The direct sensory characterization of a confined odor source is frequently impossible because extremes of temperature or of concentration of noxious components make the source intolerable for human exposure. Methods suitable for the dilution and cooling for sensory evaluation of odor sources are described in Section 4.2.

A drainage ditch discharges odorous vapors to the atmosphere along its length. The composition of the contents of the ditch changes from place to place, and the rate of discharge of odorants to the atmosphere is affected by wind and terrain. No single air sample is likely to be representative of the odor source at the time the sample is taken. Such a configuration is an unconfined odor source. Other examples are garbage dumps, settling lagoons, and chemical storage areas. An unconfined source may be represented by an imaginary emission point for the purpose of meteorological diffusion calculations. Such assignment may be made on the basis of supposing that if all the odor from the unconfined area were being discharged from the "emission point," the dispersion pattern would just include the unconfined source. Figure 16.2 schematically illustrates this procedure.

3.2 Odorous Gases and Vapors

In general, gases and vapors are odorous. The relatively few odorless or practically odorless exceptions include oxygen, nitrogen, hydrogen, steam, hydrogen peroxide, carbon monoxide, carbon dioxide, methane, and the noble gases. No precise relationship has been established between the olfactory quality of a substance and its physicochemical properties, although attempts have been made for many years. This topic is treated in more detail in Section 4.1. Most odorants encountered in the workspace or in the outside air are complex mixtures, however, and it cannot be assumed that the odor of the mixture is that of its major component. Thus a "phenolic" odor from the curing of a phenolic resin is not the same as the odor of pure phenol, and the pungent odor of burning fat is not the same as that of acrolein, though it is often so characterized. Even more dramatic are instances in which the ultrapurification of a supposedly "pure" substance produces a considerable change in odor (11).

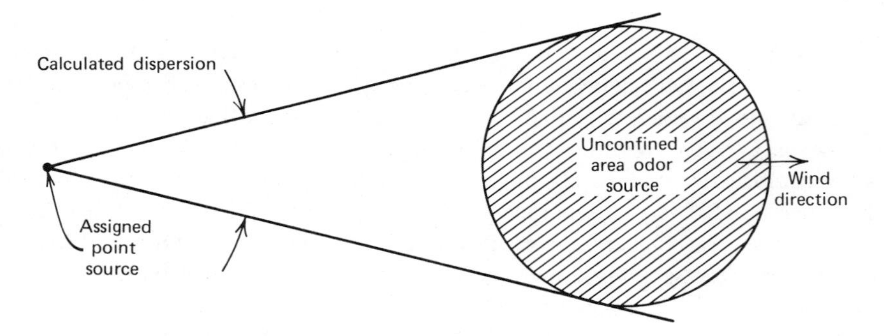

Figure 16.2 Assignment of emission point to an undefined area odor source.

3.3 Odors Associated with Airborne Particulate Matter

It is often assumed that all odorants are gases or vapors. However there is evidence that some particles can stimulate the sense of smell because the particles themselves are volatile or because they are desorbing a volatile odorant (40). There is also speculation that some particulate matter is capable of stimulating the sense of smell. Regardless of the mechanism by which particles may stimulate the olfactory sense, they definitely appear to be involved.

The idea that odors are associated with particles is supported by observations that filtration of particles from an odorous airstream can reduce the odor level. It has been shown (41) that the removal of particulate matter from diesel exhaust by thermal precipitation effects a marked reduction in odor intensity. The precipitation method was selected because it provides minimal contact between the collected particles and the gaseous components of the diesel exhaust stream. Thus, effects that could be produced by a filter bed, such as removal of odorous gases by adsorption in the filter cake, are eliminated. Therefore, the observed odor reduction must have resulted directly from the removal of particulate matter. Other more or less causal observations on the role of particulate matter in community odor nuisance problems appear occasionally in the literature (42). Several hypothetical mechanisms have been developed for the association of odor with particles.

3.3.1 Volatile Particles

Liquid or even solid aerosols may be sufficiently volatile that their vaporization on entering the nasal cavity produces enough gaseous material to be detected by smell. Such aerosols may be relatively pure substances such as particles of camphor, or they may be mixtures that release volatile components. The retention of the odorous properties of volatile aerosols, of course, depends on the prevailing temperature and on the length of time they are dispersed in air. In a cold atmosphere, the relatively greater temperature rise accompanying inhalation accelerates the production of gaseous odorant.

3.3.2 Desorption of Odorous Matter by Particles

The interaction between gas molecules and the surface of airborne particles has been discussed by Goetz (43). The theoretical considerations were directed to the question of transfer of toxicants by particles, but they are also applicable to odors. Even if a given aerosol is intrinsically odorless, it could act as an odor intensifier (*a*) if the sorptive capacity of the aerosol particles for the odorant were smaller than the affinity of the odorant for the nasal receptor, and (*b*) if at the same time, the sorptive capacity of the aerosol particles were large enough to produce an accumulation of odorant on the particle surface. Such aerosol particles would concentrate odorous molecules on their surfaces, but the odorous matter would be transferred to olfactory receptors when the aerosol entered the nasal cavity. The odorous matter would then be present at the receptor sites in concentrations higher than in the absence of the aerosol. The resulting effect would be synergistic. If an odorant is more strongly adsorbed by the aerosol particles than by the olfactory receptors, transfer of the odorant to the receptors would be impeded and the particles would actually attenuate the odor.

3.3.3 Odorous Particles

No study has rigorously defined the upper limit of particle size for airborne odorous matter. Particles up to about 8 or 10×10^{-4} μm in diameter are considered to be molecules that can exist in equilibrium with a solid or liquid phase from which they escape by vaporization. The vapor pressure decreases as the molecular weight increases, and particles above about 10^{-3} μm do not generally exist in significant concentrations in equilibrium with a bulk phase; hence we do not consider them to be "vapors." Nonetheless it is possible that odorant properties do not disappear when particle sizes exceed those of vapor molecules. Our knowledge about particles in size range of 1 to 5×10^{-3} μm (up to about the size of small viruses) is relatively meager, and we do not know whether they can be odorous, nor what effect an electrical charge might have on their odorous properties. Larger particles may also be intrinsically odorous, although their more significant role may be to contribute to odor by absorbing and desorbing odorous gases and vapors.

4 ODOR MEASUREMENT

4.1 Relation Between Sensory and Physicochemical Measurements

No one can predict the odor of a compound from its molecular structure, nor is it possible to specify the molecular structure of a substance that will yield a predicted odor. In the context of odorous air pollutants, this means that attempts to predict odor nuisance from a knowledge of sources and processes can serve as a guide to signal possible problems, but not as a substitute for direct sensory evaluation. Recent statistical attempts at multidimensional scaling of such attributes (44) lead to conclusions that explain the degree of similarity between different odors in terms of physicochemical variables. The information generated by such methods, however, seems to be of the type that is already common knowledge—for example, that compounds differing greatly in molecular weight are likely to have very different odors, or that sulfides do not smell like aldehydes. Since most organic chemists believe they can identify the functional group (alcohol, amine, ester, etc.) in a compound by smell, it is interesting to determine the degree to which such attempts are successful. Brower and Schafer (45) conducted such a study and found that for most representative compounds the functional group was correctly identified in 45 percent of the cases. The performance was poor for alcohols, ethers, and halides, and excellent for amines, sulfur compounds, esters, phenols, and carboxylic acids. When the subjects missed the functional group, they used the labels alcohol, ester, and ketone twice as often as average. The label "sulfur compound" was misapplied in only 1 percent of all cases. Bulky hydrocarbon groups near the functional group can weaken or obliterate the odor quality, but aliphatic amines and sulfur compounds are very resistant to steric hindrance. On the other hand, they are greatly weakened by electron-withdrawing groups. Aliphatic compounds bearing a multiplicity of methyl groups have the odor of camphor or menthol.

Of course, most odor problems are produced by complex mixtures of odorants, which severely complicates the task of predicting sensory effects. The literature from 1958 through March 1969 was surveyed for instances of instrumental–sensory correlation studies under the sponsorship of ASTM Committee E-18 on Sensory Evaluation of Materials and Products (46). Of the several thousand articles reviewed in 65 major technical journals, including journals devoted to air pollution, only 45 were judged to have any direct value toward progress in establishing standards for instrumental–sensory correlations. Of these, some 40 were concerned with foods, and the remainder with pure chemicals and other topics. None dealt with odorous air pollutants per se.

There have been significant advances, however, in the analyses of various organic substances that are known to be odorous. For example, the short-term analysis of malodorous sulfur-containing gases can be performed in the ambient air (47, 48). For most other odorous air pollutants, however, direct analytical methods that are demonstrably related to odor are not available.

4.2 Sampling for Odor Measurements

In the case of nuisance odors in the workspace or in the community, sensory judgments are usually made from evidence gained on direct exposure of human subjects to the odorous atmosphere in question. Where odor intensities are low enough to be tolerable and where they vary from time to time in accordance with atmospheric turbulence, changes in indoor air currents or outdoor wind direction, and changes in the odor source, it is much better for the observers to expose themselves directly to the odorous atmosphere rather than to a previously collected sample. Furthermore, when odor levels are variable, there are often enough intervals of low odor intensity to provide adequate recovery from odor fatigue. Under such circumstances, it would require considerable effort to collect samples that were representative of all the experiences a roving observer could accumulate in a single session of an hour or so.

In some instances, however, it is advantageous to collect a sample of odorant before presenting it for sensory measurements. Such instances occur (*a*) when the odorant must be diluted, concentrated, warmed, cooled, or otherwise modified before people can be exposed to it; (*b*) when it is necessary to have a uniform sample large enough to be presented to a number of judges; or (*c*) when the samples must be transported to another location at which sensory evaluations are made.

4.2.1 Grab Sampling

A grab sample places a volume of odorant at barometric pressure into a container from which it can subsequently be presented to judges for evaluation. The general principles of sampling for gases and vapors are described in Chapter 17. The subject has also been reviewed by Weurman (49). However the objective of preserving the integrity of a sample for odor measurement may present special problems because mass concentrations of odorants are often very low. As a result, adsorption or absorption by the container walls may alter the odor properties of the sample. It is therefore important to use containers that are known to be inert to the odorant and are large enough to minimize wall effects. A number of container materials have been used, including glass, stainless steel, and various inert plastics (50, 51).

When the grab sample is to be evaluated by human subjects, it must be expelled from the container into the space near the judge's nose. This expulsion may be effected by displacement of the sample with another fluid (usually water) or by collapsing the container. It must be recognized that when a person smells a jet of air from a small orifice, the aspiration of ambient air around the jet creates some additional dilution before the odorant reaches the subject. Springer (52) has been a conical funnel that is positioned in front of the subject's nose to prevent this effect.

4.2.2 Sampling with Dilution

Dilution procedures for sampling odor sources such as oven exhausts can reduce concentrations and temperatures to levels suitable for human exposure without permitting condensation of the odorant material. The dilution ratio must be specified; for this purpose it is convenient to define a dilution ratio Z such that $Z = C/C_t$, where C is the concentration of odorant at the source and C_t is the target concentration to be reached by dilution (53), and both concentrations are expressed in the same units.

The material of choice for sampling of hot odor sources with dilution is stainless steel. A suitable device is a 1000 in.^{3} (16-liter) stainless steel tank, fitted with valves, a vacuum gauge, and a pressure gauge. The tank is first evacuated, then filled through a metal probe to a desired pressure P_1, which must be low enough to prevent condensation when the gas warms to room temperature. This dilution is $Z_1 = 1$ atm$/P_1$ (atm). The tank is then pressurized with odor-free air to a new pressure P_2, which is above atmospheric pressure. The diluted, pressurized sample may then be sniffed by one or more judges for odor evaluation. The dilution that occurs when pressurized gas is released to the atmosphere is $Z_2 = P_2(\text{atm})/1$ atm. The overall dilution ratio is

$$Z = Z_1 \times Z_2$$

$$= \frac{1 \text{ atm}}{P_1 \text{ (atm)}} \times \frac{P_2 \text{ (atm)}}{1 \text{ atm}} = \frac{P_2}{P_1}$$

For example, if the evacuated tank is first filled to 0.1 atm, then pressurized to 2 atm (gauge), $P_1 = 0.1$ atm, and $P_2 = (2 + 1)$ atm (absolute), and $Z = 3/0.1 = 30$.

4.2.3 Sampling with Concentration

A dilute odorant may have to be concentrated before it can be adequately characterized by chemical analysis. Such circumstances are likely to arise when a community malodor is to be traced to one or more possible sources. The ratio of concentrations between the odor source and the odorant outdoors where it constitutes a nuisance may be in the range of 10^3 to 10^5. Under such circumstances the comparison of the odorant with the alleged source is greatly facilitated if the concentration is increased to approximate that of the source.

The most widely used methods of concentration are freezeout trapping and adsorptive sampling. Chapter 17 elaborates on both these methods.

When the concentration ratio must be high, the accumulation of water in a cold trap is a serious drawback, and adsorptive sampling is usually preferred. Activated carbon has been used, but it is so retentive to organic vapors that recovery of the sample often requires special methods (54, 55). More recently, various porous polymers (56) have been found to be both effective and convenient, especially when the concentrated sample is to be injected into a gas chromatograph.

4.3 Sensory Evaluation

4.3.1 General Conditions

Sensory testing requires concentration and high motivation on the part of human judges. Therefore outside interferences, such as noise, extraneous odors, and any other environmental distractions or discomforts, must be kept to a minimum (57). Odor control is generally achieved by use of air conditioning combined with activated carbon adsorbers. The testing room should be under slight positive pressure to prevent infiltration of odors. All materials and equipment inside the room should be either odor free or have a low odor level. Transite partitions have proved to be very effective as wall and ceiling material. If highly odorous products are to be examined or high humidities are anticipated, these partitions may be sprayed with an odorless, strippable, soft-colored coating that can be replaced if it becomes contaminated. Low odor asphalt tile has proved effective as floor material.

The panel moderator must exert every precaution to avoid bias with respect to any of the tests. Ideally, the moderator, like the judges, should be unaware of the identity of the samples, so that the test is "double blind."

4.3.2 Selection and Training of Judges

The general criteria for selection of judges, as outlined in an ASTM manual (57), involve the individuals' natural sensitivity to odors, their motivation, and their ability to work in a test situation. Wittes and Turk (58) have pointed out that three variables characterize the efficacy of any screening procedure: (*a*) the cost, as determined by the number of sensory tests, (*b*) the proportion of potentially suitable candidates rejected by the screen, and (*c*) the proportion of potentially unsuitable candidates accepted by the screen. These variables are functionally dependent; that is, specification of any two determines the value of the third. Therefore a screening procedure can be designed to favor any two of these variables, but not all three.

It is generally best, especially with novices, to start with familiar substances such as food flavors. The transition to malodors can come later. Screening tests are based on the ability of a judge to identify the odd sample in a group of three odors of which two are the same ("triangle" test), or to order samples in a sequence of increasing intensity, or to identify the components of a mixture. Details are given by Wittes and Turk (58).

After a group of judges has been selected, they must be trained. The screening tests should be repeated, and this time any errors should be discussed and analyzed. The judges are then introduced to the types of sensory tests they will carry out. If possible, the original measurements should be made on known standards.

4.4 Odor Surveys and Inventories

Because the ASTM "odor unit" is expressed in cubic feet and is thus an additive value, it is tempting to use it to quantify odorous emissions, as suggested by Hemeon (59). And since the ASTM "odor emission rate" is the number of odor units discharging per minute from a stack or vent, this value, too, is additive, and the total emission from a given industrial area can be taken to be the sum of all the odor emission rates (60). For a nonconfined source such as a lagoon or drainage ditch, some estimate of emission rate can be made on the basis of the odor concentration of the air in equilibrium with the source, and an assumed velocity of transfer of odor to the atmosphere by wind action. Then the total odor emission rate is the emission rate per square foot of surface times the total area in square feet. Of course, such procedures suffer from all the limitations of threshold measurements: they ignore considerations of odor quality or objectionability, as well as differences in the odor intensity exponent among different odorants; the threshold level itself depends strongly on the response criterion; and problems in preserving the integrity of odorant samples become severe when the dilutions approach threshold levels, as evidenced by the large differences obtained by different dilution devices.

An approach to odor surveying that distinguishes among sources of different qualities was used as early as the 1950s in Louisville, Kentucky (61). In that study a kit of 14 reference odorants was chosen to "represent the principal odors expected to occur" in the area being surveyed. Some of the reference standards were actual samples of presumed odorant sources, such as a creosote mixture used as a wood preservative in an operation that was known to generate an odor nuisance. In the actual survey, the observers reported daily the nature (by identification with the reference samples), the intensity (category scale), and the location of the odors they smelled. In later studies odor reference standards have been used in conjunction with improved methods of intensity scaling to yield quality–intensity inventories of odor sources of various kinds (60, 62).

Assessments of odor intensity and quality, however, do not serve as reliable predictors of human affective responses to unpleasant odors. There have been two approaches to direct surveying of such reactions. In one approach, developed by Karl Springer (52) and his co-workers for diesel exhaust odors, the odor source to be surveyed is set up in a transportable facility, under controlled conditions of dilution and presentation to judges. Responses are elicited from a large number of randomly selected, untrained people. For the diesel odor, the exhaust from an engine was diluted and piped into compartments in a trailer for sensory evaluations. The source–trailer combination was then used in several sites in various cities in the United States, to constitute a national survey. The questionnaire submitted to each judge was designed around the cartoon scale of Figure 16.3. Note that the facial expressions, the bodily actions, and the descriptive adjectives are all mutually reinforcing.

The second approach to measuring human affective responses to odors is a survey of annoyance reactions. Jönsson (63) has reviewed human reactions studied in different odor surveys and concludes that responses to questions about annoyance are more reliable than such other indices as willingness to sign a petition or to take direct action to modify the environment (e.g., by using household deodorants or installing air conditioning). A detailed survey procedure is described in "A Study of the Social and Economic Impact of Odors. Phase II" also known as the Copley Report (64). An important feature of this procedure is the use of an odor-free area as a control to account for the fact that in some instances annoyance to odor is expressed when no odor source exists. Even in an odorous area, complainants may call attention to the odor problem, but their opinions are

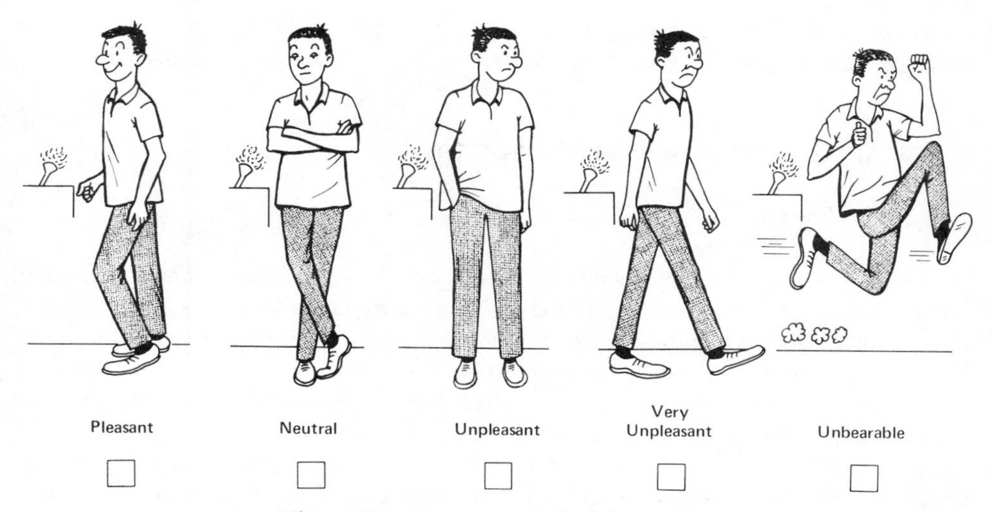

Figure 16.3 Cartoon scale for odor testing.

not generally typical of the majority of the community. Consumer and social research studies have found repeatedly that the likes and dislikes of persons who volunteer their opinions are different from the likes and dislikes of their neighbors who must be solicited for their opinions. The purpose of the survey, then, is to compare the attitudes of people residing in a community thought to be an odor problem area, with attitudes of similar people residing in an odor-free area. Attitudes of both groups are determined by conducting interviews by telephone with residents of both areas. The survey proceeds in the following sequence:

1. The first task is to define the possible odor problem area.
2. Next, a matching odor-free area is located.
3. Utilizing a street address (reverse order) telephone directory, a list of telephone numbers in each area is made, and a sample of these telephone numbers is selected at random.
4. Utilizing the questionnaire provided, telephone interviews are conducted with the adult occupant of the house for each telephone number included in the sample.
5. The total number of responses to key questions asked in both areas is tabulated and compared for problem identification.
6. If an odor problem is found, an odor problem index number is calculated.

The questionnaire itself explores the respondent's length of residence in the community, attitude to various categories of complaints about environmental problems, personal experiences of odor, degree of annoyance by odors, and opinion with regard to the origin of the odors; place of employment is also ascertained for all members of the household. The final calculated odor index expresses the degree to which it can be confidently stated that there is a statistically significant difference between the test area and the control area; such a difference is said to constitute an "odor problem identification."

5 SOCIAL AND ECONOMIC EFFECTS OF ODORS

5.1 Social Effects

For convenience in definition, social effects can be said to differ from economic effects in that the former cannot be measured directly in monetary terms (65). The social effects of odors include (*a*) interference with the everyday activities of the exposed individuals, (*b*) feelings of annoyance caused by offensive smell, (*c*) physical symptoms of physiological changes, (*d*) actual complaints to an authority, and (*e*) various forms of direct individual action to modify the environment other than through complaints.

All these effects are very difficult to quantify; for example, the average homeowner cannot say with assurance how many times an unpleasant odor has prevented him from using his yard to entertain friends. As described in the preceding section, the principal tool used to assess social effects has been the attitude survey, and the results yield only a statistical measure of the confidence that an odor problem is correctly identified, not a rating of the intensity of human reactions.

5.2 Economic Effects

Assuming that unpleasant odors annoy people, can we say that they also cause economic loss? If so, how can this loss be determined? Again, the most direct approach is to survey the affected area by asking a representative sample of persons in the community how much they would be willing to pay to get rid of the odors. Though theoretically valid, this approach runs into the practical difficulty of separating what people say they would pay from what they actually will pay if asked to comply with their own responses.

A more fruitful approach lies in determining what people actually have paid to obtain an odor-free environment. This requires recognition that the economic impact of odor pollution is most likely to manifest itself in reduced property values, reduced productivity of industrial companies, and reduced sales in commercial areas.

Economic theory states that if odors are bothersome, people should be willing to pay more to live in an odor-free area. Thus two similar properties in similar neighborhoods should sell for different prices if one area is affected by odors and the other is not. Thus we are assuming that all economic losses due to the presence of odors are capitalized negatively into property values and that buyers need only know that they prefer some properties to others and be willing to pay more for them.

Odors may affect industrial property values in somewhat the same way residential areas are affected. However another form of loss to commercial and industrial establishments is possible, namely, odors may reduce the productivity of employees because of induced illness or distraction from work assignments. Productivity losses are likely to be particularly noticeable if such odors are intermittent as well as strong. Such a situation would be offensive, while also tending to inhibit persons from becoming adapted to their work enviornment.

Commercial areas may suffer economic losses from odors in the form of a general loss of customers and reduced sales per customer.

The one serious attempt to measure such economic effects was carried out in Los Angeles in 1969 as part of the Copley Report (64). Unfortunately, the attempt was not successful, either because the methods were not sensitive enough to isolate economic effects caused by odors, or because transitory odors are not capitalized into property values.

6 ODOR CONTROL METHODS

6.1 Process Change and Product Modification

The modification of malodorous chemical processes merits first consideration because of the possibility that slight changes may yield significant results. Certainly a review of existing facilities and practices to increase process efficiency, separation efficiency, and collection efficiency—in other words, to reduce waste—can only be helpful. The same may be said of upgrading the quality of valves, pumps, drainlines, and other potential sources of leaks. Changes of process conditions, however, yield less easily predictable results. For example, an increase in process temperature may promote more complete oxidation to odorless products, or it may bring about more volatilization and cracking, to release more odorants. When such changes are contemplated, their possible effects must be carefully assessed.

It is also attractive to substitute chemicals with low odors for more highly odorous ones. Frequently such substitution can be made for solvent mixtures when different types are interchangeable in function, if not in cost. Differences may be less easy to accommodate with regard to change of product, but product substitution should be considered before making a major investment in odor abatement equipment.

6.2 Dispersal and Dilution Techniques

6.2.1 Ventilation Systems in Enclosed Spaces

A time-honored approach to controlling odors in enclosed spaces has been to exhaust the malodorous air to the outdoors. There are several fundamental limitations to such a remedy. First, the exhausted air may transfer its nuisance to the outdoors. Such instances are particularly troublesome in congested areas where people may be exposed to the exhaust, or where one building's exhaust becomes another's intake. Under some conditions such atmospheric short-circuiting may even occur between the vents of the same structure. Second, the exhausted air must be replaced, and if the makeup air requires heating or cooling, large consumptions of energy may be required. Moreover, the odorant concentration is not reduced to zero; rather, it approaches an equilibrium level in which generation and removal rates are equal, and $C_\infty = G/Q$, where C_∞ is the concentration of odorant at equilibrium (mg/m^3), G is the rate of generation of odor (mg/min), and Q is the ventilation rate (m^3/min). This approach occurs at an exponentially *decreasing* rate (66, 67). In addition, depending on the mixing characteristics of the space, there may be areas in which people are exposed to greater than average odorant concentrations. Finally, the perceived odor intensity does not decrease linearly with the decrease of odorant concentration (see Section 2), but more slowly, because the intensity exponents are typically less than unity.

6.2.2 Validity of Outdoor Dispersion Models

Some of the problems of ventilation described in the preceding section have their counterparts in outdoor dispersal methods. Thus exhausted ventilation air must be replaced, often with heating or cooling; and the perceived odor reduction does not match the physical dilution. In addition, the results of efforts to predict by conventional diffusion models the extent of odor travel from a source have not been very successful. Early efforts (68) revealed extreme discrepancies in which odors were experienced at far greater distances than were predicted. More recent approaches by Högström (69) have established a model to predict the frequency of occurrence, as a function of distance from the source, of instantaneous concentrations equal to or above an absolute odor threshold level. Högström reports fairly good agreements between observed and predicted frequencies up to several kilometers from the source, but at distances between 5 and 20 km the observed frequencies are larger than those calculated by a factor of 2 or 3. The reasons for these discrepancies are not well understood, but they may involve (*a*) failure of the dispersion model to account for peak concentrations of short duration, (*b*) irregularities in threshold responses, or (*c*) participation of particulate matter.

6.3 Adsorption Systems

6.3.1 General Principles

Any gas or vapor will adhere to some degree to any solid surface. This phenomenon is called *adsorption*. When adsorbed matter condenses in the submicroscopic pores of an adsorbent, the phenomenon is called *capillary condensation*. Adsorption is useful in odor control because it is a means of concentrating gaseous odorants, thus facilitating their disposal, their recovery, or their conversion to innocuous or valuable products. When an odorous airstream is passed through a fresh adsorbent bed, almost all the odorant molecules that reach the surface are adsorbed, and desorption is very slow. Furthermore, if the bed consists of closely packed granules, the distance the molecules must travel to reach some point on the surface is small, and the transfer rate is therefore high. In practice, the half-life of airborne molecules streaming through a packed adsorbent bed is of the order of 0.01 second, which means that a 95 percent removal can occur in about 4 half-lives, or around 0.04 second (70). This means that the very high efficiencies required to deodorize a highly odorous airstream may be achieved with a bed of moderate depth at reasonable airflow rates.

Disposal of adsorbed odorants may be effected in any of the following ways. (*a*) The adsorbent with its adsorbate may be discarded. Since even the saturated adsorbent is relatively nonvolatile, this step seldom entails difficult problems. (*b*) The adsorbate may be desorbed and recovered, if it is valuable, or discarded; the adsorbent is recovered in either case. (*c*) The adsorbate may be chemically converted to a more easily disposable product, preferably with preservation and recovery of the adsorbent.

6.3.2 Adsorbents

Adsorbents are most significantly characterized by their chemical natures, by the extent of their surfaces, and by the volume and diameter of their pores. The most important chemical differences among adsorbents are those of electrical polarity.

Activated carbon, consisting largely of neutral atoms of a single element, presents a surface with a relatively homogeneous distribution of electrical charge. As a result, carbon exhibits less preference for highly polar molecules such as water than for most organic substances; it is therefore suitable for the overall decontamination of an airstream that contains odorous organic matter.

Table 16.1 gives ranges of surface areas and pore volumes for several different adsorbents. Among these, activated carbon is generally highest in surface area and pore volume, and these are the properties that primarily determine overall adsorptive capacity.

The pore size distributions of activated carbons are important determinants of their adsorptive properties. Pores less than about 25 Å in diameter are generally designated as micropores; larger ones are called macropores. The distinction is important because most molecules of concern in air pollution range in diameter from about 4 to about 8.5 or 9 Å. If the pores are not much larger than twice the molecular diameter, opposite-wall effects play an important role in the adsorption process by facilitating capillary condensation. Maximum adsorption capacity is determined by the liquid packing that can occur in such small pores.

6.3.3 Equipment and Systems

When odorant concentrations are low, thin-bed adsorbers often provide a useful service life while offering the advantage of low resistance to airflow. Flat, cylindrical, or pleated bed shapes are retained by perforated metal sheets (Figures 16.4 to 16.6). Commercially available cylindrical canisters are designed for about 25 cubic feet of air per minute (cfm); the larger pleated cells handle 750 to 1000 cfm, and cells comprising aggregates of flat bed components handle 2000 cfm.

Thick-bed adsorbers are used when large adsorbing capacity is needed and when on-site regeneration is used. Bed depths are in the range of 1 to 6 feet (0.3 to 1.8 meters). Design airflow capacities are up to 40,000 cfm (67,960 m^3/hr). The ratio of weight of carbon to design airflow capacity is typically about 0.5 lb/cfm [0.27 kg/(m^3)(hr)]. Typical thick-bed adsorbers, such as are used in solvent recovery systems are shown in Figure 16.7. Other systems include fluidized, rotating, and falling bed adsorbers.

The service life of the adsorbent is limited by its capacity and by the contaminating load. Provisions must therefore be made for determining when the adsorbent is saturated

Table 16.1. Surface Areas and Pore Sizes of Adsorbents

	Activated Carbon	Activated Alumina	Silica Gel	Molecular Sieve
Surface area (m^2/g)	1100–1600	210–360	750	—
Surface area (m^2/cm^3)	300–560	210–320	520	—
Pore volume (cm^3/g)	0.80–1.20	0.29–0.37	0.40	0.27—0.38
Pore volume (cm^3/cm^3)	0.40–0.42	0.29–0.33	0.28	0.22—0.30
Mean pore diameter (Å)	15–20[a]	18–20	22	3—9

[a] Refers to micropore volume (<25 Å diameter); macropores (>25 Å) not included.

Figure 16.4 Aggregated flat cell thin-bed adsorber. The small test element located on the upstream side of the cell contains carbon that is to be analyzed after some period of service for degree of saturation, thereby to predict the remaining capacity of the cell. Photograph courtesy of Connor, Inc., Danbury, Conn.

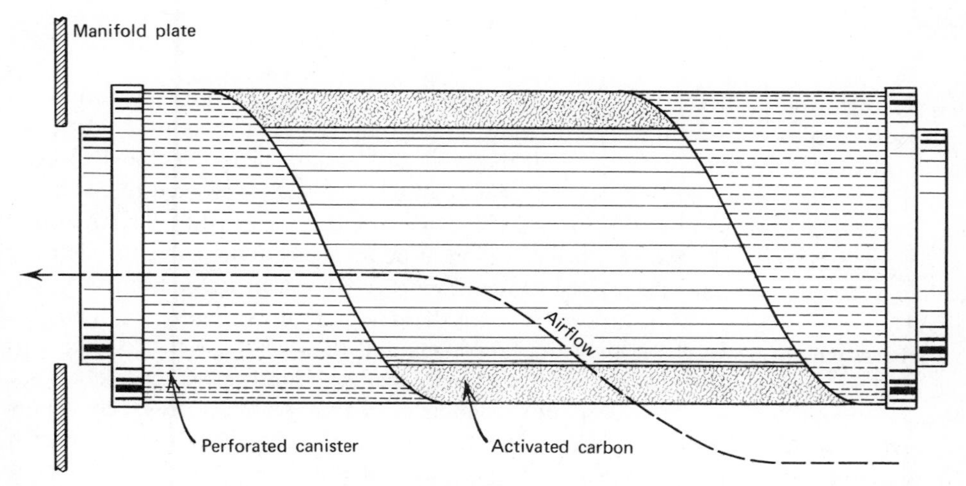

Figure 16.5 Cylindrical thin-bed canister adsorber. Schematic courtesy of Connor, Inc., Danbury, Conn.

Figure 16.6 Pleated cell, thin-bed adsorber. Photograph courtesy of Barneby-Cheney Co., Columbus, Ohio.

and for renewing it. The weight of an adsorbent is not a valid measure of its saturation because its moisture content, which depends on the relative humidity of the gas streaming through it, is likely to be variable. If mechanically feasible, a representative element or portion of the adsorbent bed may be removed and chemically analyzed to determine the degree of saturation of the entire bed (71). In many cases a schedule for renewal of adsorbent is determined by actual deterioration of performance (odor breakthrough), or it may be based on a time schedule calculated from previous performance history.

When the adsorbent is saturated it must be replaced or regenerated. Thin-bed adsorbers, which are used for light odorant loads and thus are expected to have long service lives, are normally replaced when they are exhausted. For thick-bed adsorbers and heavy contaminant loads, it is generally economical to regenerate the adsorbent by on-site stripping with superheated steam. The adsorbate is thereby also removed and may be recovered it if is valuable. When the adsorbate is not worth recovering, either because its intrinsic value is low or because the recovery procedure is too difficult or expensive, it may nonetheless pay to regenerate the adsorbent at the site. The desorbed matter is then disposed of or destroyed. In such applications the preferred regenerating gas is hot air, at either atmospheric or reduced pressure. The desorbate may then be removed from the effluent steam by incineration or scrubbing. In effect, the adsorber serves as a vapor concentrating medium. For example, benzene at a concentration of 150 ppm in air can be effectively stripped by a carbon bed and returned to a regenerating airstream at concentrations up to about 3 percent, or 30,000 ppm (70). This ratio represents a 200-fold magnification, which greatly reduces the cost of any subsequent treatment.

The oxidation of the adsorbate by air may also occur on the adsorbent surface, preferably in the presence of a catalyst. It has been shown (72, 73) that various oxide and noble metal catalysts are effective for such applications, that hydrocarbons and oxygenates can be completely oxidized before the carbon bed itself starts to oxidize, and that repeated cycles of adsorption and catalytic oxidation can be carried out without impairing the function of the carbon.

In general, activated carbon adsorption is the method of choice for deodorizing at ambient temperature an odorous airstream whose vapor concentrations are low (ppm range or below). At higher temperatures and concentrations, other methods as described

Figure 16.7 Thick-bed adsorbers used in a solvent recovery system. Photograph courtesy of Union Carbide Corp., New York.

in the following sections, become progressively more attractive, and the choice of activated carbon usually must be justified by some additional benefit such as recovery of a valuable solvent. When a less efficient but cheaper method can serve to remove the bulk of contaminant organic matter from an airstream, an activated carbon adsorbent may be used as a final stage to advance the cleanup to a condition of complete deodorization.

For low emission rates of malodorous air streams, earth filters are sometimes suitable and have been used in several European sewage plants (74).

6.4 Oxidation by Air

The complete oxidation of most odorants in air results in deodorization. Some final products are odorless (H_2O, CO_2), but others (SO_2, SO_3, NO, NO_2) have higher odor thresholds than their precursors. When malodorous waste gases containing halogens are oxidized, however, the products may include the free halogens, the halogen acids, or other toxic halogen compounds such as phosgene. All such substances must be removed by scrubbing before the gas stream is discharged to the atmosphere.

In addition, partial oxidation of hydrocarbons and oxygenates often yields intermediate products that are more highly odorous than their precursors. Unsaturated aldehydes, unsaturated ketones, and unsaturated carboxylic acids, all having highly pungent odors, are frequently encountered.

The system to be used depends on the reactivity of the contaminants with oxygen, their heat content, and the concentration of oxygen in the gas stream. Many malodorous vapors, especially those formed in decomposition reactions, are relatively easy to oxidize. These include such odorants as rendering plant emissions, cooking vapors, and coffee roasting effluents. On the other hand, many hydrocarbons, such as toluene, represent products of considerable chemical evolution and are much more resistant to oxidation.

The heat content of the oxidizable vapors determines the temperature rise of the gas stream during oxidation. It has been found (75) that this rise, ΔT, can be expressed in terms of the lower explosive limit (LEL) of the vapor. Such a relationship is convenient because combustible gas meters are scaled directly in "percentage of the LEL." The expression is

$$\Delta T\ (°\mathrm{F}) = 29 \times (\text{percentage of the LEL})$$

$$\Delta T\ (°\mathrm{C}) = 16 \times (\text{percentage of the LEL})$$

The choice of mode of operation depends on the various factors outlined previously. In flame incineration air and fuel are used to sustain a flame (Figure 16.8). This is in effect an enclosed flare, and it operates at temperatures of 2500°F (1371°C) or higher. The fuel cost is so high that the method is practical only when the combustible gas concentration exceeds 50 percent of the LEL, thus contributing at least about 1500°F (816°C) temperature rise. Such high vapor loadings often suggest that activated carbon solvent recovery may be the better choice.

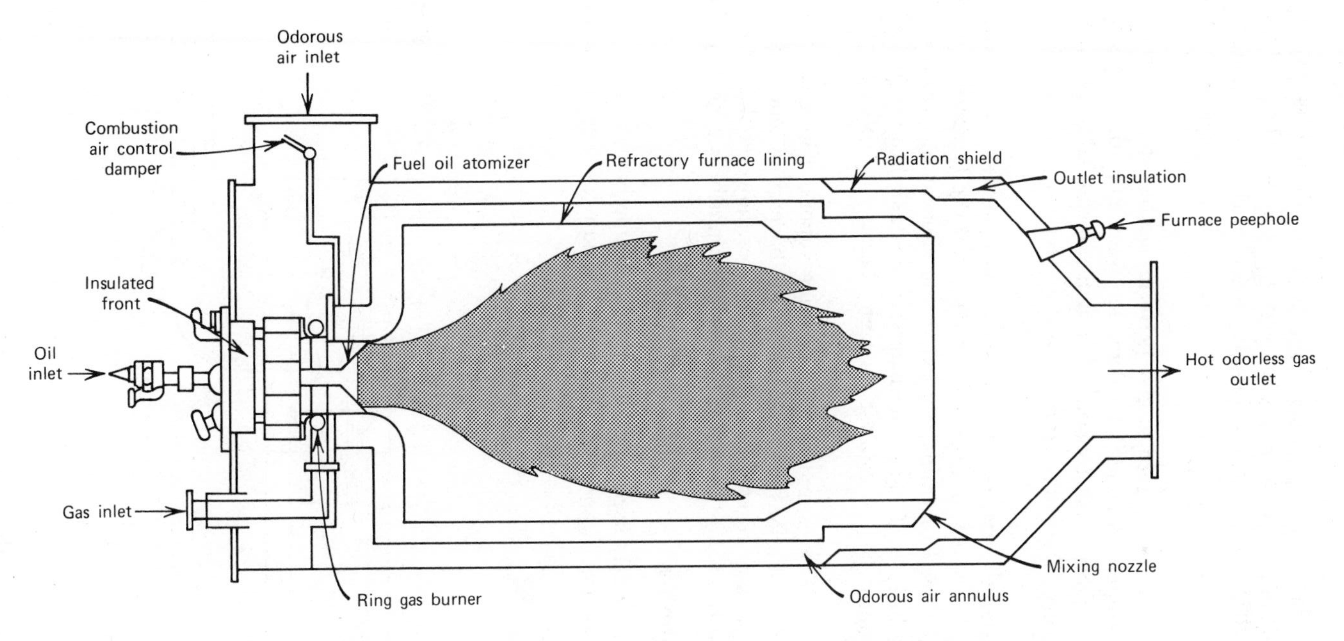

Figure 16.8 Direct-fired air heater. Schematic courtesy of Peabody Engineering Co.

In thermal incineration the operating temperature is about 1200 to 1400°F (649 to 760°C); the half-life of reactive odorants is about 0.1 second, and that of the more stable hydrocarbons is slightly longer. Consequently, a detention time of about 0.5 second is sufficient for most odor control objectives. Figure 16.9 schematically represents a thermal incinerator.

Catalytic incineration is designed to give performance like that of a thermal incinerator but at a lower temperature and faster detention time. The advantages result from the action of a solid catalyst that consists of a noble metal alloy or in some cases a metallic oxide mixture. Operating temperatures are typically in the 600 to 900°F (316 to 482°C) range.

The loss of catalyst activity, which determines catalyst life, hence equipment maintenance costs, is related to three major factors: (*a*) the presence of catalyst poisons (such as metallic or organometallic vapors) in the odorous air, (*b*) the obstruction of catalyst surface by deposit of inorganic materials such as silicates from silicone resins, and (*c*) the mechanical loss of catalyst through abrasion by solid particles in the air stream.

For air free from particles and metal-containing vapors, a long catalyst life may be realized, and some installations are reported (76) to have given more than 23,000 hours of service without catalyst regeneration. In other cases, however, loss of activity may be quite rapid. A pilot run before installing full-scale equipment is generally advisable.

6.5 Liquid Scrubbing

Liquid scrubbing is widely used for odor control. The mechanisms for its action include (*a*) solution of the odorous vapors into the scrubber liquid, (*b*) condensation of odorous

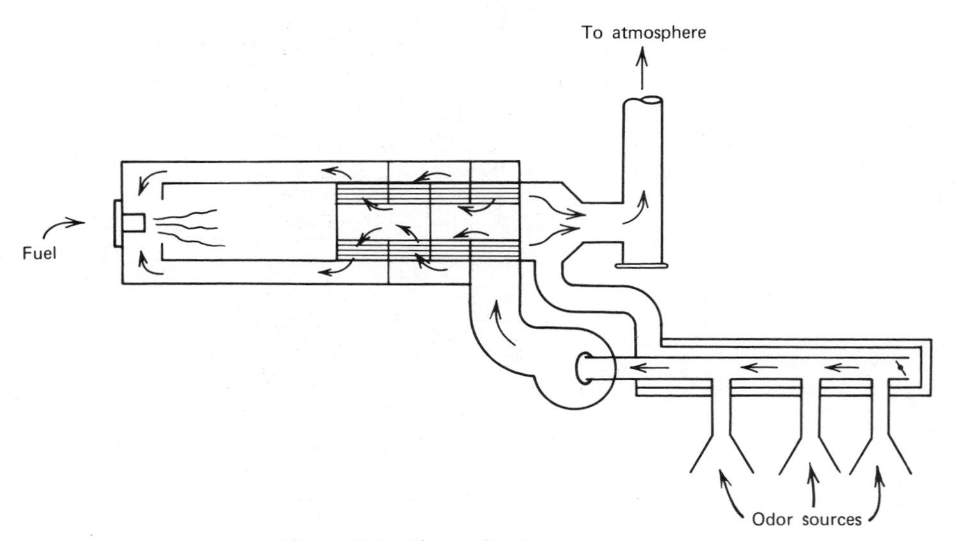

Figure 16.9 Thermal incinerator system.

vapors by the cooling action of the liquid, (*c*) chemical reaction of the odorants with the scrubber liquid to yield an innocuous product, and in some cases (*d*) adsorption of odorant onto particles suspended in the scrubber liquid.

The physical actions of solution, condensation, and absorption generally approach equilibrium conditions that still involve a significant partial pressure of the odorant vapors. Therefore such actions are only partially effective in deodorizing gas streams. For example, the water scrubbing of a gas stream containing ammonia and other nitrogenous odorants will remove much of the ammonia but only a small portion of some of the organic nitrogen compounds, which may be extremely odorous.

Chemical conversions in scrubbers are generally oxidations or acid–base neutralizations. Since the latter category involves very rapid proton exchange, the important determinant of effective action is the equilibrium condition. Thus soluble acidic odorants like hydrogen sulfide or phenol are effectively scrubbed by basic solutions, and basic odorants like ammonia or soluble amines can be neutralized by acids.

Reagents for chemical oxidation include potassium permanganate, sodium hypochlorite, chlorine dioxide, and hydrogen peroxide. In general, such oxidations are much slower than flame reactions and require considerably more residence time for effective odor control. Furthermore, the absolute rate of vapor removal is virtually independent of the chemical nature of the odorants in the case of physical adsorption, moderately dependent in the case of incineration, and extremely dependent in the case of ambient temperature oxidation. These differences are significant because a ratio of, say, 5 : 1 in the rates of removal of two vapors has only marginal significance when dealing with hundredths of seconds in a granular bed; it has more significance in the ranges of tenths of seconds in a flame, and it can well be of overriding importance in the much slower, ambient temperature reactive systems.

The various oxidants cited earlier differ in their reaction pathways for different odorants, and it is meaningless to specify which one is "best." Potassium permanganate (77) is generally used under mildly alkaline conditions (pH $\approx$ 10), where its reduction product is the insoluble manganese dioxide (MnO_2), which poses a waste disposal problem. However the MnO_2 slurry acts as an oxidation catalyst (78), and thus makes it possible for air to participate in the oxidation of some very reactive odorants, such as mercaptans and some amines. Sodium hypochlorite has been found to react more rapidly with rendering plant emissions and has been recommended for such applications (79).

Figure 16.10 represents a typical scrubber installation for control of rendering plant odors. The malodorous gases enter the first stage of the system, where they pass through water in a venturi scrubber. This step removes particulate matter and cools and saturates the gas stream. The gases are then passed through a packed bed, where they contact a countercurrent stream of scrubbing liquid containing a permanganate or hypochlorite oxidant. Malodorous gases are absorbed and oxidized. The scrubbed gas stream leaves the packed bed, flows through a mist elimination section, and is exhausted to the atmosphere. The depleted scrubbing liquid is collected and recycled to the scrubber. A portion of the depleted scrubbing solution is continuously removed from the recycle stream and replaced with makeup water and chemicals. The bleed stream is

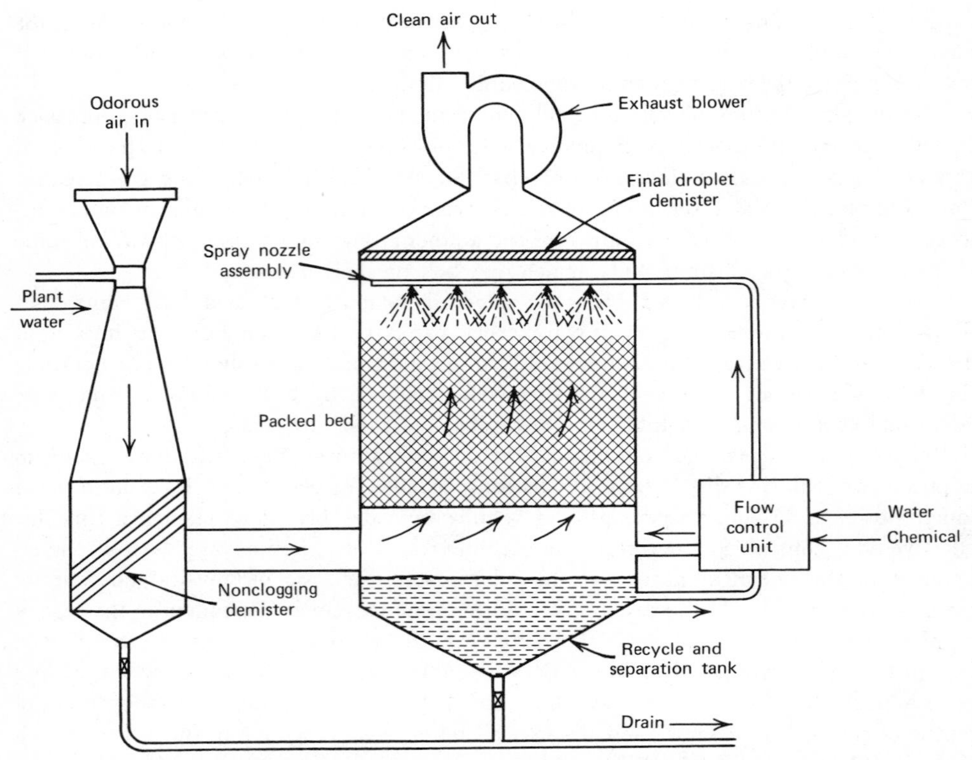

Figure 16.10 Two-stage chemical scrubbing system. Schematic courtesy Environmental Research Corp., St. Paul, Minn.

combined with the wastewater from the venturi scrubber and sent to a sewage treatment facility.

6.6 Ozonation

Ozone is a reactive ambient temperature oxidizing agent that has been used for gas phase conversion of malodorants to less offensive products. The toxicity of ozone renders it unfit for use in occupied spaces, however. Ozone-producing devices have been offered for indoor use, but they generate such low concentrations that their effect in controlling malodorants is nil and the injury they cause is probably too little to be evident.

For controlling odorants before they are discharged to the outdoors, ozone is introduced into the odorous airstream in concentrations of 10 to 30 ppm, and a reaction time of 5 seconds or more is provided during passage of the stream through a stack or special detection chamber. Ozonation is chemically selective and is not equivalent to thermal incineration or catalytic oxidation, which converts malodorants to their ultimate oxidation products. Ozone reacts with mercaptans to produce sulfones and sulfoxides, with amines to produce amine oxides, and with unsaturated hydrocarbons to produce aldehydes, ketones, and carboxylic acids. The reactions with mercaptans and amines are often so rapid that the available detention times are sufficient to provide considerable conversion, and any degree of oxidation of these malodorous compounds is a great improvement. Consequently ozonation has been reported (80) to be effective in odor control from sources such as sewage treatment plants. For unsaturated hydrocarbons or oxygenates, however, odor control performance is less reliable. For example, ozonation of a mixture of styrene and vinyl toluene from resin operations yields a mixture of aromatic aldehydes with a distinct cherrylike odor. Attempts to deodorize acrylic esters by ozonation during the detention time available in typical industrial stacks have been unsuccessful.

6.7 Odor Masking and Counteraction

When a mixture of odorants is smelled, the odor qualities of the components may be perceived separately or may blend into one quality such that the individual components cannot be recognized. The odor mixture may be perceived as stronger than, equal to, or less than the sum of the odor intensities of the components. Likewise, the odor intensity of any single component of such a mixture may be stronger than, equal to, or less than the odor intensity of that component smelled alone. When referring to the mixture as a whole, these effects are designated hyperaddition, complete addition, and hypoaddition, respectively. When referring to any individual component, the effects are called synergism, independence, and antagonism, respectively.

Interaction effects on odor intensity have been studied for some two-component mixtures (81). For example, the perceived intensity of vapor phase mixtures of various concentrations of pyridine and a second component such as linalyl acetate, linalool, or lavandin oil, is less than the sum of the perceived intensities of the two components smelled alone. The addition of the second component to a relatively weak stimulus of pyridine causes an increase in overall odor intensity, but the addition of the same amount of the second component to a relatively intense stimulus of pyridine causes a reduction in overall odor intensity. Mixtures of 1-propanol and *n*-amylbutyrate have been reported to interact similarly (82). These data suggest the existence of complex interactions in the perceived intensity of odorous mixtures.

When mixtures of odor components are perceived as a single blend, a vector summation model of odor interaction has been suggested as a means of predicting the odor intensity of mixtures of malodorants such as dimethyl disulfide, dimethyl monosulfide, hydrogen sulfide, methyl mercaptan, and pyridine (83–85). For components equal in perceived intensity when smelled alone, a direct proportionality has been reported (83, 84) between odor intensity of the mixtures and the arithmetic sum of the odor intensities of the components.

The interpretation of the application of these phenomena to practical odor control objectives presents difficulties, and the common industrial terminology does not make matters easier. *Counteraction* has been used to connote reduction of intensity, although it is not always clear whether this refers to the blend or to the malodorant alone. *Cancellation* means reduction to zero intensity, a phenomenon that has never been convincingly documented. *Masking* refers to a change in odor quality that makes the malodorant unrecognizable; the connotation of concealment has made the term unpopular. In spite of this variety of terminology, the odor control practices to which the words refer are operationally indistinguishable. The materials used are selected from industrially available high intensity odorants, often from by-product sources. They may be applied in undiluted form or as an aqueous emulsion. They may be incorporated into the process or product that constitutes the malodorous source, sprayed into a stack or over a stack exit, or vaporized over a large outdoor area.

The general method has the important practical advantages of low initial equipment costs, negligible space requirements, and greater freedom from the necessity of confining the atmosphere into a closed space for treatment. It is not applicable when irritation or toxicity accompany odor.

Clearly it is very difficult to estimate the effectiveness of this category of odor control methods. Not the least of the problems is that of choosing a criterion for evaluation. Furthermore, industrial or commercial installations are not designed to be controlled experiments. Instead, they are generally combined with other beneficial actions, such as improvements in sanitation and general housekeeping, to maximize the opportunities for odor reduction. As a result, information concerning the performance of such systems consists entirely of descriptions of actual operations and other anecdotal reports.

6.8 Epilogue: Current Status of Odor Technology

Odor control has long been considered to be more or less equivalent to the reduction of gaseous emissions that happen to be odorous. Furthermore, if the criterion of odor control is the reduction of odorant concentration to the threshold level, the efficiency required of a control device can readily be calculated. It is true that the ratio of high source concentrations to some very low odor threshold levels led to rather unprecedented requirements for the performance of gas-cleaning devices, but the usual remedy was to count on atmospheric dispersal to help solve the problem. Then, in the schematic diagram of Figure 16.11, we assume an odorant concentration C_s at a source that is treated by an abatement device (dashed lines) of efficiency E_a to discharge the abated concentration C_a to the atmosphere. Atmospheric dispersal of efficiency E_d further reduces the odorant concentration to the target or threshold value C_t before it reaches ground level. Then,

$$E_a = \frac{C_s - C_a}{C_s}$$

$$E_d = \frac{C_a - C_t}{C_a}$$

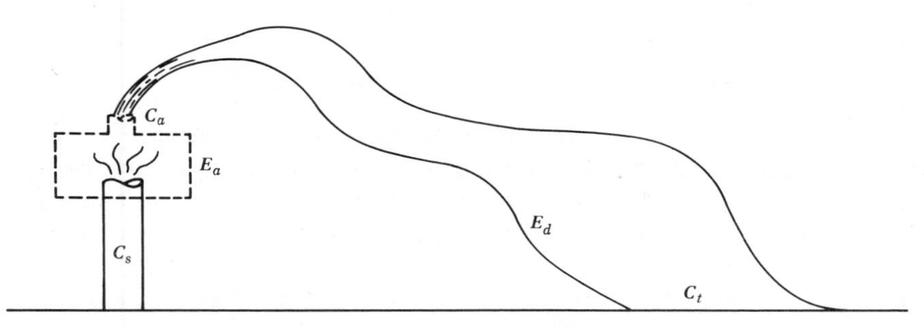

Figure 16.11 Overall efficiency of control of gaseous emission is $E = E_a + E_d - E_aE_d$, where E_a is the efficiency of a vapor control device and E_d is the efficiency of vapor dilution by atmospheric dispersal.

and the overall efficiency is

$$E = \frac{C_s - C_t}{C_s}$$
$$= E_a + E_d - E_aE_d$$

Now, if we assume that the required overall efficiency can be measured by human judges exposed to source samples of progressively greater dilution, and that the atmospheric dilutions can be predicted by dispersion models, we can readily calculate the required design efficiency of the abatement equipment to be applied to the source.

We have pointed out the many deficiencies in this sequence, such as the inadequacies of dispersion models, the inconstancy of the odor threshold, the variability of the exponent of odor intensity functions and the influence of odor quality on human reactions to odor. The technical and anecdotal literature offers many instances (68) in which this strategy fails to predict human responses to unpleasant odors, either in terms of the distance at which odors can be detected or the affective reactions or overt social initiatives they elicit.

The recognition that odor control effectiveness is most truly manifested by its reduction of human annoyance has also, unfortunately, been of little help. Although such recognition does allow for the consideration of odor masking and counteraction methods that other strategies do not accommodate, it fails to offer a basis for establishing design criteria for odor abatement systems. Furthermore, the measurement of economic effects of odors, such as the depression of property values or the reduction in work efficiency, has not been successfully accomplished in any precise manner.

On the other hand, methods of controlling gaseous emissions are being improved continually by means such as the design of more efficient scrubbers, the development of new catalysts, and advances in the effectiveness of systems for heat recovery and the regeneration of adsorbent beds. In addition, the continued refinement of instrumented monitoring systems serves all these abatement methods. Thus although there has been no spectacular breakthrough such as the discovery of a universal ambient temperature oxidation catalyst, the options available for odor control are better than ever.

The overall result has been that odor control strategies are determined in large measure by the available technology, which is often very good. Attempts to predict how the control of gaseous emissions reduces odors are described by two terms currently in vogue. The first is *dose–response relationships,* which purport to relate odorous emissions at the source to odor problems in the community or the workspace. However we have seen that such relationships have not been quantitatively established. Instead, they are generally approximated in the give-and-take fashion that is typical of political and social processes in an open society—by persuasion, by negotiation, and sometimes by legal adjudication. The second popular term is *cost–benefit analysis,* which purports to predict how much odor control each dollar will buy. It should be obvious that human "benefit" from odor control is no less difficult to quantify than human "response." However it is much easier to compare costs of alternative abatement methods that yield the same reductions of odorant concentrations, and such comparisons are valid and important. The main findings of the latter approaches have been that as the costs of energy continue to increase, the search for more sophisticated approaches to odor control, starting with a full review of the role of the process itself, becomes ever more urgent.

REFERENCES

1. G. T. Fechner, *Elemente der Psychophysik,* Breitkopf and Härterl, Leipzig, 1860. English translation of Vol. 1 by H. E. Adler (D. H. Howes and E. G. Boring, Eds.), Holt, Rinehart and Winston, New York, 1966.
2. C. E. Shannon and W. Weaver, *The Mathematical Theory of Communication,* University of Illinois Press, Urbana, 1949.
3. J. A. Swets, W. P. Tanner, Jr., and T. G. Birdsall, *Psychol. Rev.,* **68,** 301 (1961).
4. T. Engen, "Psychophysics. 1. Discrimination and Detection," in: *Woodworth and Schlosberg's Experimental Psychology,* 3rd ed., J. W. Kling and L. A. Riggs, Eds., Holt, Rinehart and Winston, New York, 1971, pp. 11–46.
5. F. A. Young and D. F. Adams, *Proceedings of the 74th Annual Convention of the American Psychological Association,* New York, 1966, 75.
6. G. Steinmetz, G. T. Pryor, and H. Stone, *Percept. Psychophys.,* **6,** 142 (1969).
7. B. Johansson, B. Drake, B. Berggren, and K. Vallentin, *Lebensm.-Wiss. Technol.,* **6,** 115 (1973).
8. G. Leonardos, D. Kendall, and N. Barnard, *J. Air Pollut. Control Assoc.,* **19,** 91 (1969).
9. E. M. Adams, "Physiological Effects," in: *Air Pollution Abatement Manual,* Manufacturing Chemists Association, Washington, D.C., 1951, Chapter 5.
10. E. Fluck, *J. Air Pollut. Control Assoc.,* **26,** 795 (1976).
11. A. Turk and J. Turk, "The Purity of Odorant Substances," in: *Methods in Olfactory Research,* D. G. Moulton, A. Turk, and J. W. Johnston, Eds., Academic Press, New York, 1975, pp. 63–73.
12. A. M. Hyman, *Sensory Processes,* **1,** 273 (1977).

13. T. Engen, "Psychophysics. II. Direct Scaling Methods," in: *Woodworth and Schlosberg's Experimental Psychology,* 3d ed., J. W. Kling and L. A. Riggs, Eds., Holt, Rinehart and Winston, New York, 1971, pp. 47–86.
14. S. S. Stevens, in: *Psychophysics: Introduction to Its Perceptual, Neural and Social Prospects,* G. Stevens, Ed., Wiley, New York, 1975.
15. S. S. Stevens, *Psychol. Rev.,* **64,** 153 (1957).
16. G. Ekman and L. Sjöberg, *Ann. Rev. Psychol.,* **16,** 451 (1965).
17. T. Engen, "Report from the Psychological Laboratory," University of Stockholm, 1961, p. 106.
18. W. S. Cain, *ASHRAE Trans.,* **80,** 53 (1974).
19. K. E. Henion, *Psychon. Sci.,* **22,** 213 (1971).
20. H. Stone, G. T. Pryor, and G. Steinmetz, *Percept. Psychophys.,* **12,** 501 (1972).
21. H. G. Haring, "Vapor Pressures and Raoult's Law Deviations in Relation to Odor Enhancement and Suppression," in: *Human Responses to Environmental Odors,* A. Turk, J. W. Johnston, Jr., and D. G. Moulton, Eds., Academic Press, New York, 1974, pp. 199–226.
22. B. Berglund, U. Berglund, G. Ekman, and T. Engen, *Percept. Psychophys.,* **9,** 379 (1971).
23. M. J. Mitchell and R. A. M. Gregson, *Percept. Mot. Skills,* **26,** 720, (1968).
24. M. J. Mitchell and R. A. M. Gregson, *Quart. J. Exp. Psychol.,* **22,** 301 (1970).
25. M. J. Mitchell and R. A. M. Gregson, *J. Exp. Psychol.,* **89,** 314 (1971).
26. W. S. Cain, *Ann. NY Acad. Sci.* **237,** 28 (1974).
27. R. L. Doty, *Physiol. Behav.,* **14,** 855 (1975).
28. W. S. Cain, *Sens. Processes* **1,** 57 (1976).
29. G. Steinmetz, G. T. Pryor, and H. Stone, *Percept. Psychophys.,* **8,** 327 (1970).
30. W. S. Cain, *Percept. Psychophys.,* **7,** 271 (1970).
31. G. T. Pryor, G. Steinmetz, and H. Stone, *Percept. Psychophys.,* **8,** 331 (1970).
32. W. S. Cain and T. Engen, "Olfactory Adaptation and the Scaling of Odor Intensity," in: *Olfaction and Taste,* Vol. 3, C. Pfaffmann, Ed., Rockefeller University Press, New York, 1969, pp. 127–141.
33. E. P. Köstor, *Adaptation and Cross-Adaptation in Olfaction,* Bronder-Offset, Rotterdam, 1971.
34. A. Dravnieks, "Theories of Olfaction," in: *Chemistry and Physiology of Flavors,* H. W. Schultz, E. A. Day, and L. M. Libbey, Eds., Avi Publishing Co., Westport, Conn., 1967, pp. 95–118.
35. R. H. Wright, *J. Appl. Chem.,* **4,** 611 (1954).
36. R. H. Wright, *Ann. NY Acad. Sci.,* **116,** 552 (1964).
37. E. E. Langenau, "Correlation of Objective-Subjective Methods as Applied to the Perfumery and Cosmetics Industries," in: *Correlation of Subjective-Objective Methods in The Study of Odors and Taste,* American Society for Testing and Materials Special Publication No. 440, ASTM, Philadelphia, 1968, pp. 71–86.
38. J. T. Davies, *Symp. Soc. Exp. Biol.,* **16,** 170 (1962).
39. J. T. Davies, *J. Theor. Biol.,* **8,** 1 (1965).
40. W. R. Roderick, *J. Chem. Educ.,* **43,** 510 (1966).
41. A. T. Rossano and R. R. Ott, "The Relationship Between Odor and Particulate Matter in Diesel Exhaust," paper presented at the Pacific Northwest Section Meeting of the Air Pollution Control Association, Portland, Ore., November 5–6, 1964.
42. W. A. Quebedeaux, *Air Repair,* **4,** 141 (1954).
43. A. Goetz, *Int. J. Air Water Pollut.,* **4,** 168 (1961).
44. S. S. Schiffman, *Science,* **185,** 112 (1974).

45. K. R. Brower and R. Schafer, *J. Chem. Educ.,* **52,** 538 (1975).
46. American Society for Testing and Materials, *Reviews of Correlations of Objective-Subjective Methods in the Study of Odors and Taste,* Special Technical Publication No. 451, ASTM, Philadelphia, 1969.
47. R. A. Schmall, *Atmospheric Quality Protection Literature Review—1972,* Technical Bulletin No. 65, National Council of the Paper Industry for Air and Stream Improvement, New York, 1972.
48. R. A. Rasmussen, *Am. Lab.* **4:**12, 55 (1972).
49. C. Weurman, "Sampling in Airborne Odorant Analysis," in: *Human Responses to Environmental Odors,* A. Turk, J. W. Johnston, and D. G. Moulton, Eds., Academic Press, New York, 1974, pp. 263–328.
50. C. A. Clemons and A. P. Altshuller, *J. Air Pollut. Control Assoc.,* **14,** 407 (1964).
51. W. D. Connor and J. S. Nader, *Am. Ind. Hyg. Assoc. J.,* **25,** 291 (1964).
52. K. Springer, "Combustion Odors," in: *Human Responses to Environmental Odors,* A. Turk, J. W. Johnston, and D. G. Moulton, Eds., Academic Press, New York, 1974, pp. 227–262.
53. A. Turk, *Atmosph. Environ.,* **7,** 967 (1973).
54. A. Turk, J. I. Morrow, and B. E. Kaplan, *Anal. Chem.,* **34,** 561 (1962).
55. A. Turk, J. I. Morrow, S. H. Stoldt, and W. Baecht, *J. Air Pollut. Control Assoc.,* **16,** 383 (1966).
56. A. Dravniecks and B. K. Krotoszynski, *J. Gas Chromatogr.,* **6,** 144 (1968).
57. American Society for Testing and Materials, *Manual on Sensory Testing Methods,* Special Publication No. 434, ASTM, Philadelphia, 1968.
58. J. Wittes and A. Turk, "The Selection of Judges for Odor Discrimination Panels," in: *Correlation of Subjective–Objective Methods in the Study of Odors and Taste,* American Society for Testing and Materials Special Publication No. 440, ASTM Philadelphia, 1968, pp. 49–70.
59. W. C. L. Hemeon, *J. Air Pollut. Control Assoc.,* **18,** 166 (1968).
60. A. Turk, *Pollut. Eng.,* **4,** 22 (1972).
61. *The Air Over Louisville,* Technical Report of the Public Health Service, U.S. Department of Health, Education and Welfare, 1956–1957.
62. A. Turk, J. T. Wittes, L. R. Reckner, and R. E. Squires, *Sensory Evaluation of Diesel Exhaust Odors,* National Air Pollution Control Administration, Publication No. AP-60, 1970.
63. E. Jonsson, "Annoyance Reactions to Environmental Odors," in: *Human Responses to Environmental Odors,* A. Turk, J. W. Johnston, D. G. Moulton, Eds., Academic Press, New York, 1974, pp. 330–333.
64. Copley International Corp., "A Study of the Social and Economic Impact of Odors, Phase I, 1970; Phase II, 1971, Phase III, 1973," U.S. Environmental Protection Agency, Report of Contract No. 68-02-0095.
65. R. D. Flesh and A. Turk, "Social and Economic Effects of Odors," in: *Industrial Odor Technology Assessment,* P. N. Cheremisinoff and R. A. Young, Eds., Ann Arbor Science Publishers, Ann Arbor, Mich., 1975, pp. 57–74.
66. A. Turk, *J. ASHRAE.,* October 1963.
67. A. Turk, "Concentrations of Odorous Vapors in Test Chambers," in: *Basic Principles of Sensory Evaluation,* American Society for Testing and Materials Special Publication No. 433, ASTM Philadelphia, 1968, pp. 79–83.
68. H. C. Wohlers, *Int. J. Air Water Pollut.,* **7,** 71 (1963).
69. U. Högstrom, "Transport and Dispersal of Odors," in: *Human Responses to Environmental Odors,* A. Turk, J. W. Johnston and D. G. Moulton, Eds., Academic Press, New York, 1974, pp. 164–198.
70. A. Turk, "Adsorption," in: *Air Pollution,* Vol. 4, 3rd ed., A. C. Stern, Ed., Academic Press, New York, 1977.
71. A. Turk, H. Mark, and S. Mehlman, *Mater. Res. Stand.,* **9,** 24 (1969).

72. J. Nwanko and A. Turk, *Ann. NY Acad. Sci.,* **237,** 397 (1974).

73. J. Nwanko and A. Turk, *Environ. Sci. Technol.,* **9,** 846 (1975).

74. H. L. Bohn, *J. Air Pollut. Control Assoc.,* **25,** 953 (1975).

75. R. J. Ruff, "Catalytic Method of Measuring Hydrocarbon Concentrations in Industrial Exhaust Fumes," in: American Society for Testing and Materials Special Publication No. 164, ASTM, Philadelphia, 1954, p. 13.

76. R. J. Ruff, *Am. Ind. Hyg. Assoc. Quart.,* **14,** 183 (1953).

77. H. S. Posselt and A. H. Reidies, *Ind. Eng. Chem. Prod. Res. Develop.,* **4,** 48 (1965).

78. D. F. S., Natusch and J. R. Sewell, "A New Solid Filter for Industrial Odors," in: *Proceedings of the Second International Clean Air Congress,* Academic Press, New York, 1971, p. 948.

79. T. R. Osag and G. B. Crane, *Control of Odors from Inedibles—Rendering Plants,* U.S. Environmental Protection Agency Publication No. 450/1-74-006, 1974.

80. C. Nebel, W. J. Lehr, H. J. O'Neill, and T. C. Manley, *Plant Eng.,* March 21, 1974.

81. W. S. Cain and M. Drexler, *Ann. NY Acad. Sci.,* **237,** 427 (1974).

82. W. S. Cain, *Chem. Sens. Flav.,* **1,** 339 (1975).

83. B. Berglund, U. Berglund, and T. Lindvall, *Acta Psychol.,* **35,** 255 (1971).

84. B. Berglund, U. Berglund, T. Lindvall, and L. T. Svensson, *J. Exp. Psychol.,* **100,** 29 (1973).

85. B. Berglund, *Ann. NY Acad. Sci.,* **237,** 35 (1974).

CHAPTER SEVENTEEN

Industrial Hygiene Sampling and Analysis

ROBERT D. SOULE

1 INTRODUCTION

Industrial hygiene sampling is done for a variety of reasons: to identify and quantitate specific contaminants present in the environment, to determine exposures of workers in response to complaints, to determine compliance status with respect to various occupational health standards, or to evaluate the effectiveness of engineering controls installed to minimize workers' exposures. The purpose of the sampling will dictate to some extent the sampling strategy which should be used. Sampling can be conducted in a grid pattern throughout a plant to document the environmental characteristics of the workplace, or personal breathing zone samples can be obtained to document actual exposure conditions. The substances being evaluated will determine the type of sampling devices to be used, and the analytical requirements will specify time and perhaps flowrate of sampling. The occupational health standard will indicate whether continuous or grab sampling is required. In short, consideration must be given to a number of questions pertaining to the fundamental purpose of the sampling.

Many analytical methods available to the industrial hygienist have been so standardized and simplified that they require relatively little thought or experience. On the other hand, many seemingly simple tests call for a basic understanding of solubility and gas laws, partial pressures, and chemical reactions. In many instances questions arising from such considerations can be answered only by qualified specialists. The ultimate methods of analysis to be used will depend on the problem at hand rather than mere application of a "standard method." The trend in recent years has been toward development of methods that give relatively prompt results with a high degree of accuracy. The latter aspect has been given increased importance because of the greater legal significance given to the occupational health standards promulgated under authority of the Occupational Safety and Health Act of 1970. The National Institute for Occupational Safety and Health (NIOSH) and the Occupational Safety and Health Administration (OSHA) are developing specific methods for sampling and analyzing atmospheric contaminants in the workplace. These procedures typically require an accuracy of ±25 percent with 95 percent confidence at the occupational health standard limit.

This chapter discusses the development of proper strategy for sampling in the workplace, the statistical bases for industrial hygiene sampling, the sampling techniques for gases and vapors and particulates, the techniques available for analyzing atmospheric samples and biological monitoring as it relates to industrial hygiene sampling. Relatively complete descriptions of detailed methods of analyses for specific contaminants have been published elsewhere (1–3).

2 GENERAL CONSIDERATIONS

The magnitude of chemical and physical stresses can be evaluated in various ways. One form of evaluation is qualitative, using one or more of the human senses but without taking any actual measurements. Often this kind of inspection and evaluation of a work situation is very beneficial. Another form of evaluation is quantitative, involving collection and analysis of samples that represent actual exposures of workers. Generally, this type of evaluation is most desirable, and necessary in many cases particularly when the purpose of the sampling is to determine compliance with occupational health standards or to form the basis for designing engineering controls.

2.1 Preliminary Subjective Survey

An experienced professional industrial hygienist often can evaluate, quite accurately and in some detail, the magnitude of chemical and physical stresses associated with an operation without benefit of any instrumentation. In fact, the professional uses this qualitative evaluation every time he makes a survey, whether it is intended to be the total effort of his work or a preliminary inspection prior to actual sampling and analysis of potential stress. The qualitative evaluation of an operation can be applied by anyone familiar with an operation, from the worker to the professional investigator, to ascertain some of the potential problems associated with work activities.

The first step in evaluating the occupational environment is to become as familiar as possible with particular operations. The person evaluating the operation should be aware of the types of industrial process and the chemical, raw materials, by-products, and contaminants encountered. He should also know what protective measures are provided, how engineering controls are being used, and how many workers are exposed to contaminants generated by specific job activities.

The number of chemical and physical agents capable of producing occupational illnesses is increasing steadily. New products that require the use of new raw materials or new combinations of familiar substances are continually being introduced. This is particularly true in the chemical industries, where new chemicals and products and the operations for their processing are being developed. It is becoming increasingly important that the responsible industrial hygienist establish and maintain a list of the chemical and physical agents encountered in his particular area of jurisdiction. The list should include the composition of the products and by-products and as many of the associated contaminants and "undesirables" as possible. Usually this means that the industrial hygienist must obtain complete information on the composition of various commercial products. In most instances the desired information can be obtained from descriptive material provided by the suppliers. At the very least, the labels on the containers of the material should be read carefully. Typically, however, labels do not give complete information, and further investigation of the composition of the materials is necessary.

After this inventory is obtained, it is necessary to determine the toxicity of the chemical substances. Information of this type can be found in several excellent reference texts on toxicology and industrial hygiene (4–6).

During a qualitative walk-through evaluation, many potentially hazardous operations can be detected visually. Operations that produce large amounts of dusts and fumes can be spotted, although they are not necessarily hazardous: airborne dust particles that *cannot* be seen by the unaided eye normally are most hazardous, since they are more likely to be inhaled into the lungs. Often concentrations of respirable dust must reach extremely high levels before they are visible. Thus the absence of a visible cloud of dust is not a guarantee that a "dust-free" atmosphere exists. However in most operations the activities that generate dust can be spotted visually and are likely to warrant implementation of additional controls.

In addition to the sense of sight, the sense of smell can be used to detect the presence of many vapors and gases. Trained observers are able to estimate rather accurately the concentration of various gases and solvent vapors present in the workroom air. For many substances the odor threshold concentration (i.e., the lowest concentration that can be detected by smell) is greater than the generally permissible safe exposure level. In these cases, there is an indication of excessive levels if the substances can be detected by its odor. However many substances, such as hydrogen sulfide, can cause olfactory fatigue (i.e., anesthesis of the olfactory nerve endings) to the extent that even dangerously high concentrations cannot be detected by odor. A detailed presentation of evaluation and control of odors is contained in Chapter 16.

Although it is usually possible to determine the presence or absence of potentially hazardous physical agents at the time of the qualitative evaluation, the potential hazard cannot be evaluated without the aid of special instruments. In most cases the sources of physical agents such as radiant heat, abnormal temperatures and humidities, excessive noise, improper or inadequate illumination, ultraviolet radiation, microwaves, and various other forms of radiation, can be noted.

An important aspect of the qualitative evaluation is an inspection of the types of control measure in use in a particular operation. In general, the control measures include such features as shielding from radiant or ultraviolet energy, local exhaust and general ventilation provisions, respiratory protection devices, and other personal protective measures. General indices of the relative effectiveness of these controls are the presence or absence of accumulated dust on floors, ledges, and other work surfaces, the condition of ductwork for the ventilation systems (i.e., whether there are holes or badly damaged sections of ductwork, whether the fans for ventilation systems appear to provide adequate control of contaminants generated by the processes), and the manner in which personal protective measures are accepted and used by the workers.

2.2 Strategy for Representative Quantitative Surveys

Although the information obtained during a qualitative evaluation or walk-through inspection of an industrial activity is important and always useful, only by measurement can the hygienist document the actual level of chemical or physical agent associated with a given operation. Of course the strategy used for any given air sampling program depends to a great extent on the purpose of the study. The specific objectives of any sampling program may include one or more of the following:

1. To provide a basis on which unsatisfactory or unsafe conditions can be detected and the sources identified.
2. To assist in designing controls.
3. To provide a chronicle of changes in operational conditions.
4. To provide a basis for correlating disease or injury with exposure to specific stresses.
5. To verify and assess the suppression of contaminants by methods designed to do so.
6. To document compliance with health and safety regulations.

These objectives can be condensed into two major categories:

1. Sampling for industrial health engineering surveillance, testing, or control.
2. Sampling for health research or epidemiological purposes.

A sampling program for engineering purposes should be designed to yield the specific information desired. For example, one might need only single samples before and after a change in ventilation to determine whether the change has had the desired effect. On the other hand, industrial hygiene primarily is directed at predicting the health effects of an exposure by comparing sampling results with hygienic guides, determining compliance with health codes or regulations, or defining as precisely as possible environmental factors for comparison with observed medical effects.

Regardless of the objective or objectives of the sampling program, the investigating industrial hygienist must answer the following questions to be able to implement the correct strategy.

1. *Where* should samples be obtained?
2. *Whose* work area should be sampled?
3. For *how long* should the samples be taken?
4. *How many* samples are needed?
5. Over what period of work activity should the samples be taken?
6. *How* should the samples be obtained?

In answering these questions, the importance of adequate field notes must be emphasized. While the sample is being obtained, notes should be made of the time, duration, location, operations underway, and all other factors pertinent to the sample, and the exposure or condition it is intended to define. Printed forms with labeled spaces for essential data help to avoid the common failure to record needed information. Industrial hygiene recordkeeping is discussed in detail in Chapter 3.

2.2.1 Where Should Samples Be Obtained?

There are three general locations in which air samples can be collected: at a specific operation, in the general workroom air, and in a worker's breathing zone. The choice of sampling location is dictated by the type of information desired, and combination of the three types of sampling may be necessary. Most frequently, the sampling is intended to determine the level of exposure of a worker or group of workers to a given contaminant throughout a work day. To obtain this kind of information it is necessary to collect samples at the worker's breathing zone as well as in the areas adjacent to his particular activities. When the purpose of the survey is to determine sources of contamination or to evaluate engineering controls, a strategic network of area sampling would be more appropriate.

2.2.2 Whose Work Area Should Be Sampled?

Samples should be collected in the vicinity of workers directly exposed to contaminants generated by their own activities. In addition, samples should be taken from the breathing zone of workers in nearby work areas, not directly involved in the activities that generate the contaminant, and from those of workers remote from the exposure who have either complained or have reason to suspect that the contaminants have been drawn into their work areas.

2.2.3 For How Long Should The Samples Be Taken?

In most cases minimum sampling time is determined by the time necessary to obtain an amount of the material sufficient for accurate analysis. The duration of the sampling period therefore is based on the following considerations: the sensitivity of the analytical procedure, the acceptable concentration of the particular contaminant in air, and the anticipated concentration of the contaminant in the air being sampled.

The sampling period should represent some identifiable period of time of the worker's exposure, usually a minimum of one complete cycle of his activity. This is particularly important in studying nonroutine or batch-type activities, which are characteristic of many industrial operations. Exceptions include operations that are highly automated and enclosed operations where the processing is done automatically and the operator's exposure is relatively uniform throughout the workday. In many cases it is desirable to sample the worker's breathing zone for the duration of the full shift. This is particularly important if sampling is being done to determine compliance status relative to occupational health standards.

Evaluation of workers' daily time-weighted average exposures is best accomplished, when analytical methods will permit, by allowing the worker to work his full shift with a personal breathing zone sampler attached to his body. The concept of full-shift integrated personal sampling is much preferred to that of short-term or general area sampling if the results are to be compared to standards based on time-weighted average concentrations. When methods that permit full-shift integrated sampling are not applicable, time-weighted average exposures can be calculated from alternative short-term or general area sampling methods.

The first step in calculating the daily, time-weighted exposure of a worker or a group of workers is to study the job descriptions obtained for the persons under consideration, and to ascertain how much time during the day they spend at various tasks. Such information usually is available from the plant personnel office or foreman on the job. In many situations the investigator must make time studies himself to obtain the correct information. Information obtained from plant personnel should be checked by the investigator because in many situations job activities as observed by the investigator do not fit official job descriptions. From this information and the results of the environmental survey, a daily, 8-hour time-weighted average exposure can be calculated, assuming that a sufficient number of samples have been collected, or measurements obtained with direct reading instruments under various plant operating conditions to represent accurately the "exposure profiles."

Where sampling for the purpose of comparing results with airborne contaminants whose toxicological properties warrant short-term and ceiling limit values, it is necessary to use short-term or grab sampling techniques to define peak concentrations and estimate peak excursion durations. For purposes of further comparison, the time-weighted average, 8-hour exposures can be calculated using the values obtained by short-term sampling.

2.2.4 How Many Samples Are Needed?

The number of samples collected depends to a great extent on the purpose of sampling. For example, two samples may be sufficient to estimate the relative efficiency of control methods, one sample being taken while the control method is in operation and the other while it is off. On the other hand, several dozen samples may be necessary to define accurately the average daily exposure of a worker who performs a variety of tasks. For any given task, the number of samples depends to some extent on the concentrations encountered. If the concentration is quite high, a single sample may be sufficient to warrant further action. If the concentration is somewhat near the acceptable level, a minimum of three to five samples usually is desirable for each operation being studied.

There are no set rules regarding the duration of sampling or the number of samples to be collected. These decisions usually can be reached quickly and reliably only after much experience in conducting such studies. For a discussion of the statistical basis of industrial hygiene sampling for compliance purposes, refer to Section 3.

2.2.5 Over What Work Period Should the Samples Be Taken?

The answer to this question depends on the type of information desired and the particular operations under study. If, for example, the operation continued for more than one shift, it is usually desirable to collect air samples during each shift. The airborne concentrations of toxic chemicals or exposure to physical agents may be quite different for each shift. It usually is desirable to obtain samples during both summer and winter months, particularly in plants located in areas where large temperature variations occur during different seasons of the year. In such areas there generally is more natural ventilation provided to dilute the airborne contaminants during the summer months than during the winter.

2.2.6 How Should the Samples Be Obtained?

In general, the choice of a particular sampling instrument or instruments depends on a number of factors.

1. The portability and ease of use.
2. The efficiency of the instrument or method.
3. The reliability of the equipment under various conditions of field use.
4. The type of analysis or information desired.
5. The availability of the instrument.
6. The "personal choice" of the industrial hygienist based on past experience and other factors.

No single, universal air sampling instrument is available today, and it is doubtful that such an instrument will ever be developed. In fact, the present trend in the profession is toward a greater number of specialized instruments.

The sampling instruments used in the field of industrial hygiene generally can be classified by type as follows: (*a*) direct reading; (*b*) those which remove the contaminant from a measured quantity of air, and (*c*) those which collect a fixed volume of air. The three methods are listed in order of their general application and preference in use today.

The industrial hygienist must consider a proposed sampling program in relation to his own familiarity with the sampling and analytical method. As a rule, a method should never be relied on unless and until the individual has personally evaluated it under controlled conditions, such as by the following.

1. Sampling a synthetic atmosphere from a proportioning apparatus or gas-tight impervious chamber of sufficient size to permit making and sampling mixtures without introducing significant errors.
2. Introducing measured amounts of contaminants into a device attached to the sampling arrangement in a manner that utilizes the entire amount deposited.
3. Comparing performance of the method with a device of proved performance by sampling from a common manifold over the same period of time.

In other words, the sampling program must be preceded by appropriate calibration of all equipment to be used. This subject is discussed in detail in Chapter 25.

Regardless of the sampling instrumentation selected for use in conducting industrial hygiene surveys, it is critical and imperative that the *actual* performance characteristics be known. This requires that various types of calibration be done periodically or at any time that the performance of the device is questioned.

3 STATISTICAL BASIS FOR INDUSTRIAL HYGIENE SAMPLING

Since the implementation of the Occupational Safety and Health Act, statistical tests for noncompliance are being applied more consistently when environmental data are used to make a decision concerning a worker's exposure to a contaminant (7). The decision options concerning compliance and noncompliance are given in Table 17.1. In statistical

Table 17.1 Decisions Options for Compliance Versus Noncompliance

TRUE STATE / ACTION	COMPLIANCE WITH STANDARD	NONCOMPLIANCE WITH STANDARD
DECLARE COMPLIANCE	No Error	Type II Error
DECLARE NONCOMPLIANCE	Type I Error	No Error

terms, these options correspond to the null hypothesis that the worker is in compliance with the industrial hygiene standard. Our detailed discussion of procedures is presented from the compliance officer's viewpoint. With his approach, samples are collected to see whether it is possible to reject the hypothesis that compliance exists with appropriate certainty. The type I error is the probability of declaring noncompliance given that the true state is compliance. This probability is a measure of any uncertainty that noncompliance does exist, and it should be kept small to ensure that noncompliance decisions are correct.

The employer's responsibility, on the other hand, is the protection of the employees. The employers goal is to keep the type II error, which is the probability of declaring compliance when the true state is noncompliance, as small as possible. In statistical decision terms, therefore, the employer wants to assume for a given worker that he is in a state of noncompliance (null hypothesis). Then data are collected to show that the hypothesis can be rejected, with a goal of keeping as small as possible the probability of wrongly rejecting the null hypothesis.

Three questions are of primary importance when considering a statistical basis for industrial hygiene sampling. Over what time span should each sample be taken? How many samples should be obtained? At what times during the workday should the samples be obtained?

3.1 Sample Duration

For some types of sampling unit, such as the common colorimetric detector tubes, the sampling period is predetermined. In other cases, such as sampling for asbestos, the sampling period is defined fairly closely by the requirements of the analytical procedure. However in most cases the industrial hygienist has a choice over a wide range of sampling times from a few seconds to a complete work shift. The industrial hygienist's first and intuitive reaction might be to maximize the sampling period, expecting that this would increase significantly the reliability of the data—that is, that it would ultimately provide a "better answer." This is not true for short-term or "grab" samples, however. In such cases the primary consideration in arriving at sampling duration should be the requirements of the analytical method. Each analytical procedure requires some mimimum amount of material for reliable analysis. This should be known in advance by the industrial hygienist, and the sampling period should be selected accordingly. Any increase in the duration of the sampling past this minimum time required to collect an adequate amount of material is both unnecessary and unproductive.

When attempting to make a decision on a possible noncompliance situation, as the statistical discussion in this chapter is directed, it is better to take shorter samples because this allows more samples to be taken in a given time period. It is much more important to collect several samples of short duration than to collect one medium-length (partial-shift) sample covering the same total sampling period. This is true because the random sampling and analytical errors can be averaged out, along with the longer term environmental fluctuations during the sampling period, by taking a mean of random independent short-term samples. If sampling can be conducted over essentially 100 percent of the work period, either by a single sample or several consecutive samples, a better estimate of the true average exposure of the employee can be obtained. This is because only the sampling and analytical errors must be contended with, and these are typically much smaller than the environmental fluctuations that affect short-term samples.

Thus there is a marked advantage in using a single full-period sample (or several consecutive samples over the full work period) when attempting to demonstrate noncompliance with an occupational health standard. It is much more difficult to demonstrate noncompliance using the mean of several grab samples because the additional variability due to environmental fluctuations lowers the lower confidence limit (LCL) of the sampling result. This is demonstrated in Figure 17.1, which presents the effect of

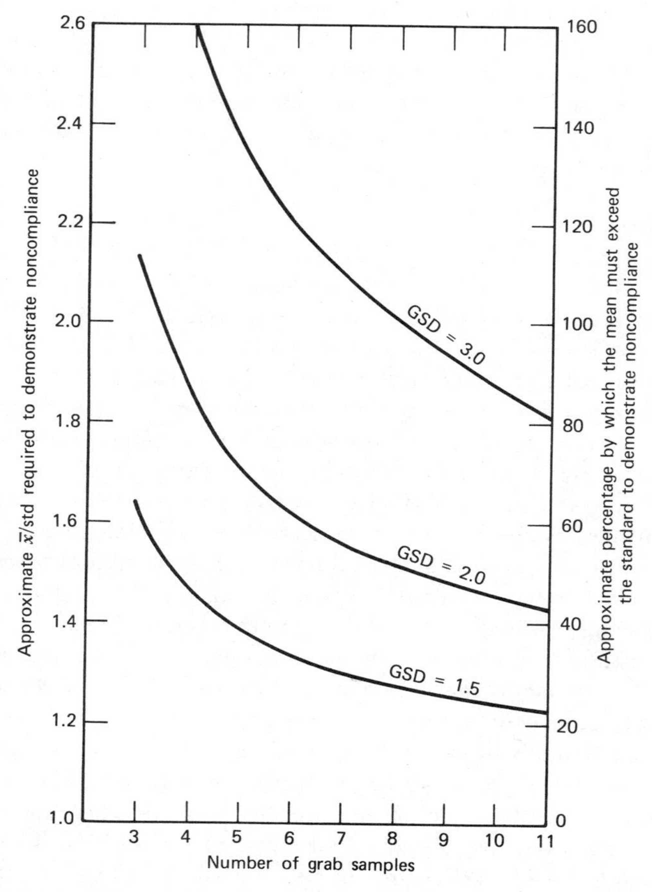

Figure 17.1 Effect of small grab sample sizes on requirements for demonstration of noncompliance; the three GSDs reflect the amount of variability in the environment.

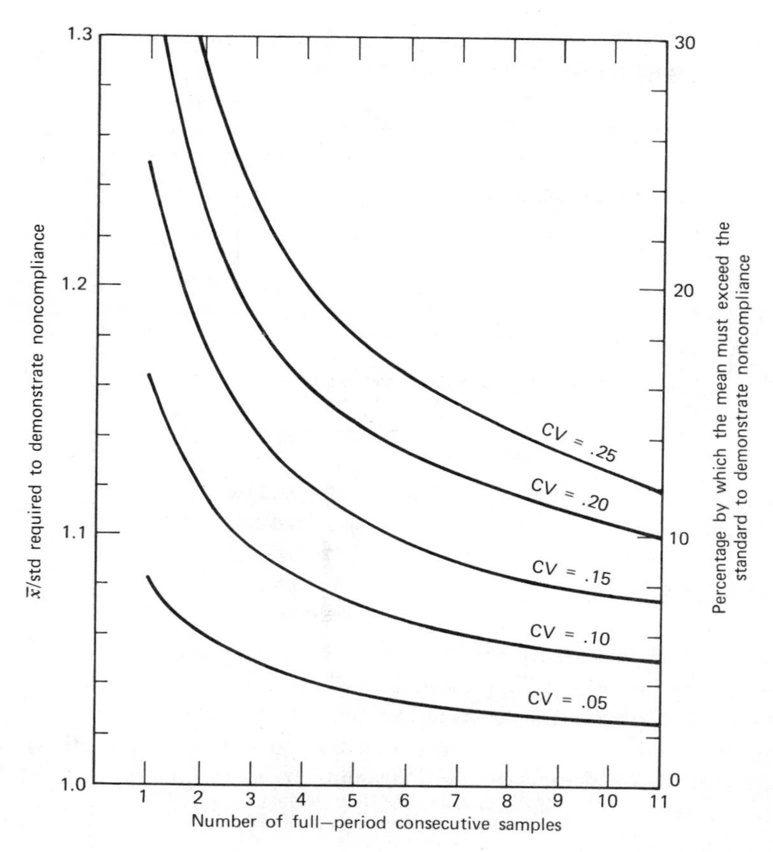

Figure 17.2 Effect of full-period consecutive sample size on requirements for demonstration of noncompliance.

number of grab samples on statistical requirements for demonstrating noncompliance, using three levels of data variability, as expressed by the geometric standard deviation (GSD). In much industrial hygiene data the geometric standard deviations range between 1.5 and 2.5. Within this range, and for sample sizes of 3 to 10, it generally is necessary to obtain a measured mean exposure of 150 to 250 percent of the occupational health standard to demonstrate noncompliance.

If a full-period sample, (or a series of several consecutive samples) is used in attempting to demonstrate noncompliance the degree to which the observed mean must exceed the occupational health standard is much lower. Figure 17.2 illustrates the effect of sample size for consecutive samples covering a full period on requirements for demonstrating noncompliance. The curves represent different sampling and analytical coefficients of variation (CVs) ranging from 0.05 to 0.25. Table 17.2 summarizes typical coefficients of variation. As Figure 17.2 indicates, for sample sizes of 1 to 4, it is necessary only that the

Table 17.2 Coefficients of Variation (CV) for Some Sampling-Analytical Procedures

SAMPLING/ANALYTICAL METHODS	C.V.
Asbestos (sampling and counting)	0.22
Charcoal Tubes (sampling and analytical)	0.10
Colorimetric Detector Tubes	0.14
Gross Dust (sampling and weighing)	0.05
Respirable Dust (sampling and weighing)	0.09

average concentration exceed the occupational health standard by 5 to 14 percent to demonstrate noncompliance.

3.2 Number of Samples

The second question of primary importance to the industrial hygienist is, How many samples should be obtained? This question is vital because it relates directly to the confidence that can be placed in the resulting estimate of the airborne concentration of the contaminant in question and subsequently the employee exposure. The effects of sample size on requirements for demonstrating noncompliance were illustrated in Figures 17.1 and 17.2. As these figures demonstrate, the curves for geometric standard deviations of 1.5 to 2.5 change relatively slowly after sample sizes of 7 or 8 (Figure 17.1), and a similar leveling off occurs in the curves relating the value of the "measured mean-to-standard" ratio necessary to demonstrate noncompliance versus sample size.

For full-period consecutive samples, Figure 17.2 shows that an appropriate number of samples is between 4 and 7. Practical considerations include costs of sampling and analyses, and the impossibility of running some long-duration sampling methods for longer than about 4 hours per sample. Thus most full-period consecutive sampling strategies result in at least two samples when an 8-hour, time-weighted average standard is being applied. As the curves demonstrate for a sampling and analytical technique with a coefficient of variation of 10 percent, the degree to which the estimate of the mean concentration must exceed the standard drops from 12 to 6 percent with an increase from 2 to 7 samples, respectively. The relatively small decrease in percentage, with sample size exceeding 7, normally cannot be justified when compared to the time and effort required to obtain the additional samples. Thus on a cost–benefit basis, it can be concluded that two consecutive samples covering a full work period (e.g., two consecutive 4-hour measurements) is the best number of samples to be obtained when comparison is made to an 8-hour, time-weighted average standard.

For grab samples Figure 17.1 shows that the estimate of the mean exposure concentration must exceed the standard by unreasonably large amounts to demonstrate noncompliance when less than 4 grab samples are obtained. As was discussed earlier, there is a point beyond which little is gained in attempting to reduce errors in the mean by taking more than 7 grab samples. Since the level of variability in the mean of grab samples usually is much higher than for the same number of full-period samples, however, it might be necessary to take more than 7 grab samples to attain the same level of precision afforded by even 4 or fewer full-period samples. Thus on the basis of the statistical criterion that can lead to economies in sampling by permitting reduction in the sampling effort with a calculable degree of confidence, it can be concluded that the optimum number of grab samples to be taken over the time period appropriate to the standard is between 4 and 7.

Figure 17.3 represents the procedure to be used when consecutive samples are obtained in a series over only a portion of the total period for which the standard applies; that is, the full work period is not included in the consecutive sampling period. The effect of sample size and total time covered by all samples on the requirement for demonstrating noncompliance is shown, using a typical sampling and analytical coefficient of variation of 10 percent.

Obviously the taking of partial-period consecutive samples is a compromise between the preferred full-period sampling and the less desirable approach with grab samples. Comparison of the curves in Figures 17.2 and 17.3 reveals that a geometric standard deviation of 2.5 (Figure 17.2) would be comparable to a curve of approximately 5½ hours in Figure 17.3 for the same ratios of estimates of the mean to the standard necessary to judge noncompliance. Thus, in general, if it is not possible to sample for at least 70 percent of the time period appropriate to the standard (e.g., 5½ hours for an 8-hour standard) it is better to use the grab sampling strategy.

3.3 Sampling Periods for Grab Sampling

The last of the three statistical questions to be answered concerns the periods of exposure during which grab sampling should be conducted. The accuracy of the probability level for the test depends on assumptions of the log-normality and independence of sample results that are averaged. These assumptions are not highly restrictive if precautions are taken to avoid any bias when selecting the sampling times over the period for which the standard is defined. Thus it usually is preferable to choose the sampling period in a statistically random manner. For a standard that is defined as a time-weighted average concentration over a period significantly longer than the sampling interval, an unbiased estimate of the true mean can be assured by taking samples at random. On the other hand, it is valid to sample at regular intervals if the contaminant level varies randomly about a constant mean and any fluctuations are of short duration relative to the length of the sampling interval. If means and their confidence limits are to be calculated from samples taken at regularly spaced intervals, however, biased results could occur if the industrial operation being monitored were

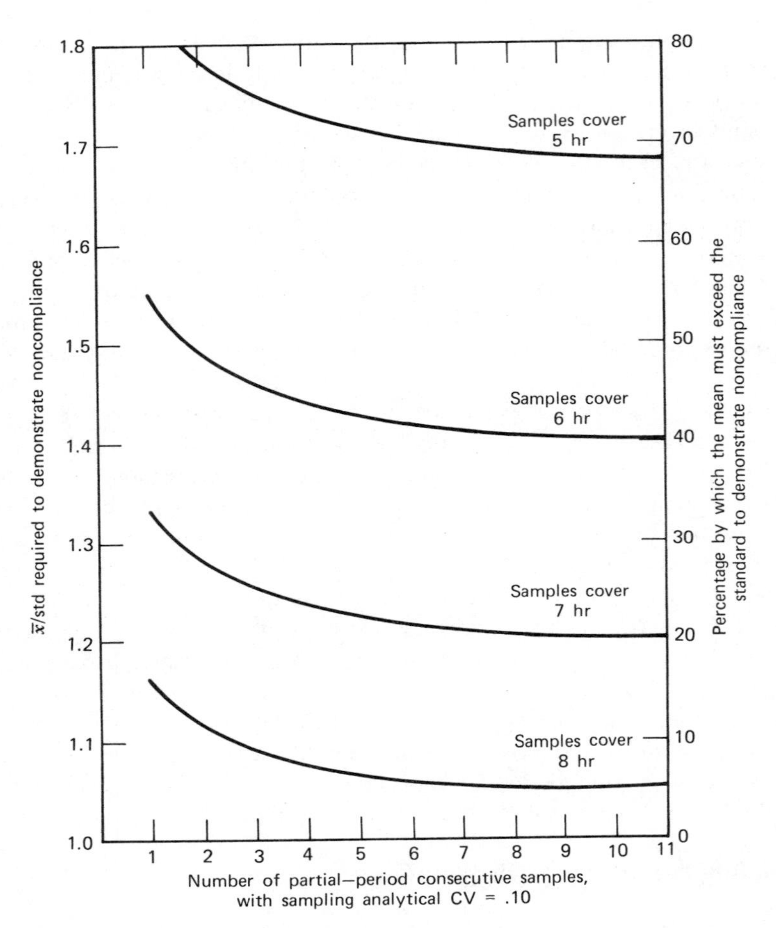

Figure 17.3 Effect of partial-period consecutive sample size on requirements for demonstration of noncompliance.

cyclic and in phase with the sampling periods. Results from random sampling would be valid nevertheless, even when cycles or trends occurred during the period of the standard. In this context, the word "random" refers to the manner of selecting the sample, and any particular sample could be the outcome of a random sampling procedure. A practicable way of defining random sampling is "a strategy by which any portion of the work shift has the same chance of being sampled as any other."

Strictly speaking, sampling results are valid only for the portion of the work period during which measurements were obtained. However if it is not possible to sample during the entire workday or the entire length of a particular operation, professional judgment may permit inferences to be made about concentrations during other unsampled portions of the day. Reliable knowledge concerning the operation obviously is required to make these types of extrapolations.

3.4 Statistical Procedures to be Used for Industrial Hygiene Sampling

The following procedures should be used by the industrial hygienist when comparing sampling results with the applicable occupational health standard. As is the case with other material in this section, the statistics have been oriented toward determining whether noncompliance with the time-weighted average, ceiling, or excursion standard exists.

3.4.1 Single Sample for Full Period

The following procedure can be used to determine noncompliance with either a time-weighted average or a ceiling standard. It is used when only one sample is being tested. For a time-weighted average standard, the sample must have been taken for the entire period for which the standard is defined (usually 8 hours). The variability of the sampling, expressed either as a standard deviation or as a coefficient of variation, and the analytical methods used to collect and analyze the sample, must be well known from previous measurements. The statistical test given is the "one-sided comparison-of-means" test using the normal distribution at the 95 percent confidence level.

Only if the lower confidence limit of the sample exceeds the standard is there 95 percent confidence that the true average concentration exceeds the standard, thus that a condition of noncompliance exists. The lower confidence limit can be expressed as

$$\mathrm{LCL} = x - 1.645\sigma \tag{1}$$

where x = measurement being tested
1.645 = critical standard normal deviate for 95 percent confidence
σ = standard deviation of sampling-analytical method

If the coefficient of variation (CV) is known, the LCL can be computed from

$$\mathrm{LCL} = x - [(1.645)\,(\mathrm{CV})\,(\mathrm{standard})] \tag{2}$$

The nomogram in Figure 17.4 can be used to aid this calculation. Some coefficients of variation are available from Table 17.2.

3.4.2 Consecutive Samples for Full Period

The procedure involving consecutive samples for a full period should be used to determine noncompliance with either a time-weighted average or a ceiling standard. That is, several consecutive samples are taken for the entire time period for which the

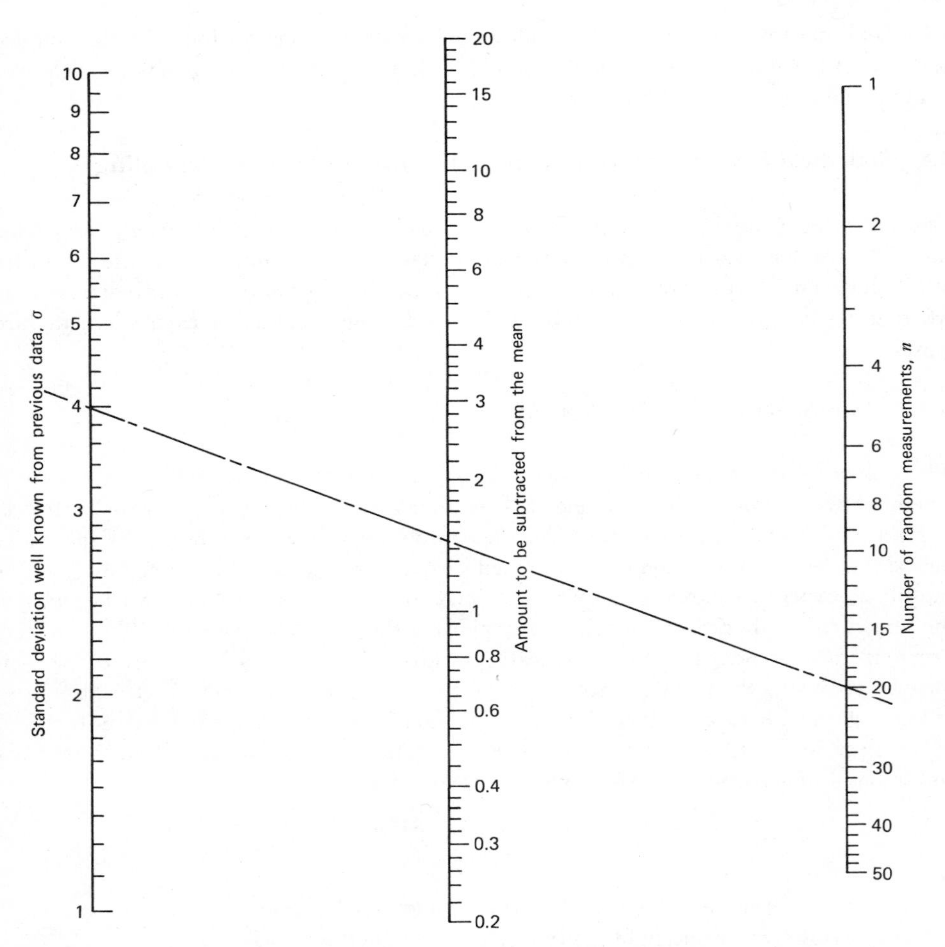

Figure 17.4 Nomogram for full-period or partial-period procedures using a well-known standard deviation; both σ and the amount can be multiplied by the same power of 10 (7).

standard is defined. If the samples do not cover the entire time period of the standard, refer to the procedure described in Section 3.4.3. The variability (standard deviation or coefficient of variation) of the sampling and analytical methods used to collect and analyze the samples must be well known from previous measurements. The statistical test given is the "one-sided comparison-of-means" test using the normal distribution at the 95 percent confidence level.

Only if the lower confidence limit of the mean of the consecutive samples exceeds the standard, is there 95 percent confidence that the true average concentration exceeds the standard and that a condition of noncompliance exists.

$$\text{LCL} = \bar{x} - 1.645\ \sigma_{\bar{x}} \tag{3}$$

where $\bar{x}$ = time-weighted average of n samples

$$= \frac{1}{T}(T_1x_1 + T_2x_2 + \cdots + T_nx_n)$$

T_i = duration of ith sample

x_i = measurement of concentration in ith sample

$T = T_1 + T_2 + \cdots + T_n$ = total of durations for the n consecutive samples

$$\sigma_{\bar{x}} = \frac{\sigma}{T}(T_1^2 + T_2^2 + \cdots + T_n^2)^{1/2}$$

$$= \frac{\sigma}{(n)^{1/2}} \quad \text{if} \quad T_1 = T_2 \cdots = T_n$$

If the coefficient of variation is known, σ can be computed from

$$\sigma = (\mathrm{CV})\ (\mathrm{standard}) \tag{4}$$

Again, Figure 17.4 can be used to aid this calculation for the case of n equal-duration samples. Some coefficients of variation are given in Table 17.2.

3.4.3 Consecutive Samples for Partial Period

One sample or a series of consecutive samples, collected over less than the period for which a standard is defined, is referred to as a "partial-period" sample(s). Since the concentration during the period not covered by the sample could not be less than zero, the full-period standard can be multiplied by a factor to obtain a conservative "partial-period standard." The lower confidence limit then can be calculated as in the previous section and compared to the "partial-period standard."

$$\text{factor} = \frac{(\text{time period of the standard})}{(\text{actual time of the sample(s)})}$$

Thus for an 8-hour, time-weighted average standard, typical factors would be as follows:

Total Time of Sample(s) (hr)	Factor
8.00	1.00
7.75	1.032
7.50	1.067
7.25	1.103
7.00	1.143
6.75	1.185
6.50	1.231
6.25	1.280
6.00	1.333

3.4.4 Analysis of Grab Sample Data

As stated previously, if full-period samples of industrial contaminant concentrations are available, the best method of modeling the uncertainties of the result is with a normal distribution. When only a set of grab samples is available, however, the log-normal distribution best describes the uncertainties of the process. Recently a procedure for analyzing grab samples by estimating the average concentration of a contaminant and making a decision on the level of the contaminant was developed by Systems Control, Inc., under a NIOSH-funded contract (8, 9). A detailed presentation of the theory and procedure would be beyond the scope of this book, but several advantages offered by the technique can be listed.

1. It is contaminant independent. Thus it becomes possible to use only a single decision chart for any contaminant.
2. It is capable of both the noncompliance and the "no-action decision." Each of these decisions is subject to a predetermined probability of type I or type II error. A type I error is said to occur if the noncompliance decision is wrongly asserted. A type II error is said to occur if the no-action decision is wrongly asserted.
3. The estimation and decision procedures are implemented by a simple, straightforward nomographic method. For estimation, the procedure yields the best estimate of the actual average contaminant level.

A minor disadvantage of the procedure is that the lower confidence limit is not directly computed. However the simplicity of the calculations required and the plotting of a single point on the decision chart far outweigh any advantages that direct calculation of a lower confidence limit would yield.

4 SAMPLING FOR GASES AND VAPORS

For purposes of definition a substance is considered to be a gas if at 70°F and atmospheric pressure, the normal physical state is gaseous. A vapor is the gaseous state of a substance in equilibrium with the liquid or solid state of the substance at the given environmental conditions. This equilibrium results from the vapor pressure of the substance causing volatilization or sublimation into the atmosphere. The sampling techniques discussed in this section are applicable to a substance in "gaseous" form, regardless of whether it is technically a gas or a vapor.

4.1 General Requirements

As discussed in the following section on sampling for particulate, particulate substances can be readily scrubbed or filtered from sampled airstreams because of the larger relative physical dimension of the contaminant and the operation and interaction of agglomerative, gravitational, and inertial effects. On the other hand, gases and vapors form true solutions in the atmosphere, thus requiring either sampling of the total atmosphere using a gas collector or the use of a more vigorous scrubbing mechanism to separate the gas or vapor from the surrounding air. Sampling reagents can be chosen to react chemically with the contaminants in the airstream, thus enhancing the collection efficiency of the sampling procedure. In the development of an integrated sampling scheme it is necessary to consider the following basic requirements:

1. The method must have an acceptably high efficiency of collecting the contaminant of interest.
2. A rate of airflow that can provide a sufficient sample for the required analytical procedure, maintain the acceptable collection efficiency, and be accomplished in a reasonable time period must be available.
3. The collected gas or vapor must be kept in the chemical form in which it exists in the atmosphere under conditions that maintain the stability of the sample before analysis.
4. The sample must be submitted for analysis in a suitable form and medium.
5. A very minimal amount of analytical procedure in the field must be associated with the overall method.
6. To the extent practicable, the use of corrosive (e.g., acidic or alkaline) or relatively toxic (e.g., benzene) sampling media should be avoided.

Of these general requirements, perhaps the most important is the first, that is, knowing the collection efficiency of the sampling system chosen or anticipated. This efficiency information can be obtained either from published documented data (10) or as a result of independent evaluations as an essential part of planning the industrial hygiene survey. In making such evaluations, known concentrations of the gases must be prepared by means of either dynamic or static test systems. Once the known atmosphere is generated, the sampling device should be used as anticipated or intended and the efficiency defined in terms of such variables as characteristic of the sampler, rate of airflow, stability of the sample during collection, and apparent losses by adsorption on walls on the sampling device.

The collection of sufficient sample for the subsequent analytical procedure is a matter that must be clearly understood not only by the laboratory analyst but by the field investigator as well. Understanding can best be promoted by discussions between the two professionals before the industrial hygiene field survey. The field man must discuss as fully as possible with the chemist the nature of the process involved in the survey so that together they may select the best combination of sampling and analytical methods to satisfy the sensitivity requirements of the analytical method, minimize effects of potential interferences, and complete the sampling within a time frame consistent with processing conditions or potential exposure. Chapter 4 reviews the general considerations in evaluating the industrial environment.

4.2 Collection Techniques

There are two basic methods for collecting samples of gaseous contaminants: instantaneous or "grab" sampling, and integrated or long-term sampling. The first involves the use of a gas-collecting device, such as an evacuated flask or bottle, to obtain a fixed volume of a contaminant-in-air mixture at known temperature and pressure. This is called "grab" sampling because the contaminant is collected almost instantaneously, that is, within a few seconds or minutes at most; thus the sample is representative of atmospheric conditions at the sampling site at a given point in time. This method commonly is used when atmospheric analyses are limited to such gross contaminants as mine gases, sewer gases, carbon dioxide, or carbon monoxide, or when the concentrations of contaminants likely to be found are sufficiently high to permit analysis of a relatively small sample. However with the increased sensitivity of modern instrumental techniques such as gas chromatography and infrared spectrometry, instantaneous sampling of relatively low concentrations of atmospheric contaminants is becoming feasible.

The second method for collection of gaseous samples involves passage of a known volume of air through an absorbing or adsorbing medium to remove the contaminants of interest from the sample airstream. This technique makes it possible to sample the atmosphere over an extended period of time, thus "integrating" the sample. The contaminant that is removed from the airstream becomes concentrated in or on the collection medium; therefore it is important to establish a sampling period long enough to permit collection of a quantity of contaminant sufficient for subsequent analysis.

4.2.1 Instantaneous or Grab Sampling

Various devices can be used to obtain instantaneous or grab samples. These include vacuum flasks or bottles, gas- or liquid-displacement collectors, metallic collectors, glass bottles, syringes, and plastic bags. The temperature and pressure at which the samples are collected must be known, to permit reporting of the analyzed components in terms of standard conditions (normally 25°C and 760 mm Hg) for industrial hygiene purposes.

Grab samples are usually collected when analysis is to be performed on gross amounts of gases in air (e.g., methane, carbon monoxide, oxygen, and carbon dioxide). The samplers should not be used for collecting reactive gases such as hydrogen sulfide, oxides of nitrogen, and sulfur dioxide unless the analyses can be made directly in the field. Such gases may react with dust particles, moisture, wax sealing compound, or glass, altering the composition of the sample.

The introduction of highly sensitive and sophisticated instrumentation has extended the applications of grab sampling to low levels of contaminants. In areas where the atmosphere remains constant, the grab sample may be representative of the average as well as the momentary concentration of the components; thus it may truly represent an integrated equivalent. Where the atmospheric composition varies, numerous samples must be taken to determine the average concentration of a specific component. The chief advantage of grab sampling methods is that collection efficiency is essentially 100 percent, assuming no losses due to leakage or chemical reaction preceding analysis. The more common types of grab sampling devices are discussed in detail below.

Evacuated Containers. Evacuated flasks are heavy-walled glass containers, usually of 250 or 300 ml capacity, but frequently holding as much as 500 or 1000 ml, from which 99.97 percent or more of the air has been removed by a heavy-duty vacuum pump. The internal pressure after the evacuation is practically zero. The neck is sealed by heating and drawing during the final stages of evacuation. These units are simple to use because no metering devices or pressure measurements are required. The pressure of the sample is taken as the barometric pressure reading at the site. After the sample has been collected by breaking the heat-sealed end, the flask is resealed with a ball of wax and transported to the laboratory for analysis. A typical evacuated container is depicted in Figure 17.5.

A variation of this procedure with evacuated flasks is to add a liquid absorbent to the flask before it is evacuated and sealed to preserve the sample in a desirable form following collection.

Partially evacuated containers or vacuum bottles are prepared with a suction pump just before sampling is performed, although frequently they are evacuated in the laboratory the day before a field visit. No attempt is made to bring the internal pressure to zero, but temperature readings and pressure measurements with a manometer are recorded after the evacuation, and again after the sample has been collected. Examples of this type of collector are heavy-walled glass bottles and metal or heavy plastic containers with tubing connectors closed with screw clamps or stopcocks.

Displacement Collectors. Gas or liquid displacement collectors include 250 to 300 ml glass bulbs fitted with end tubes that can be closed with greased stopcocks or with rubber tubing and screw clamps. They are used widely in collecting samples containing oxygen, carbon dioxide, carbon monoxide, nitrogen, hydrogen, or other combustible gases for analysis by an Orsat or similar analyzer. Another device operating on the liquid displacement principle is the aspirator bottle, which has exit openings at the bottom through which the liquid is drained during sampling. Figure 17.6 shows a typical displacement-type collector.

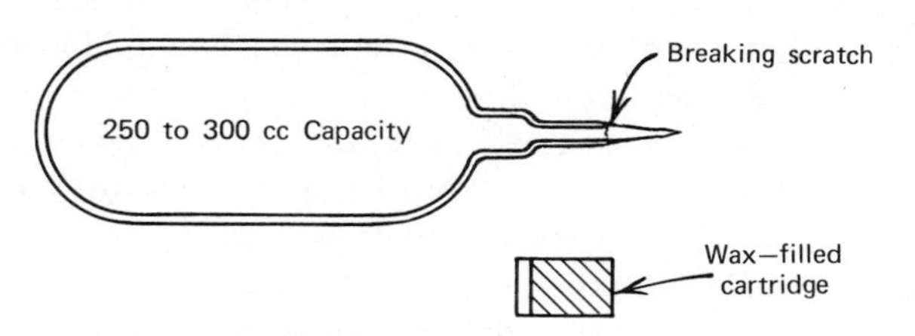

Figure 17.5 Evacuated flask.

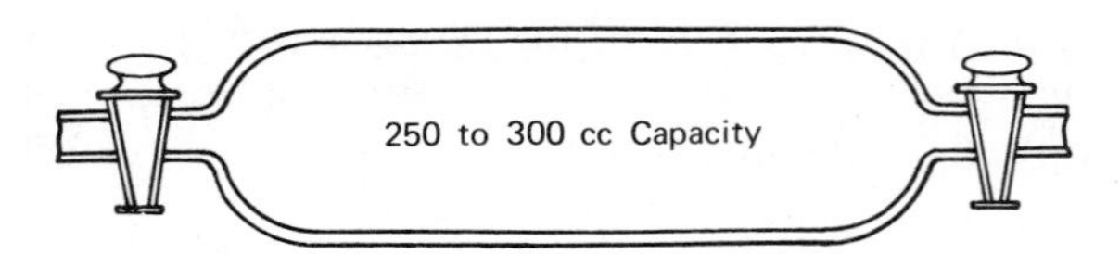

Figure 17.6 Gas or liquid displacement collector.

In applying the gas displacement technique, the samplers are purged conveniently with a bulb aspirator, hand pump, small vacuum pump, or other suitable source of suction. Usually satisfactory purging can be achieved by drawing a minimum of 10 air changes of the test atmosphere through the gas collector.

The devices used for gas displacement collectors can also be filled by liquid displacement, the most frequently employed liquid being water. In sampling, liquid in the container is drained or poured out slowly in the test area and replaced by air to be sampled. Of course, this method is limited to gases that are insoluble in and nonreactive with the displaced liquid. The solubility problem can be minimized by using mercury or water conditioned with the gas to be collected. Mercury, however, must be used with caution, since it may create an exposure problem if handled carelessly.

Flexible Plastic Bags. Flexible plastic bags can be used to collect air and breath samples containing organic and inorganic vapors and gases in concentrations ranging from parts per billion to more than 10 percent by volume in air. They also are convenient for preparing known concentrations of gases and vapors for equipment calibration. The bags are available commercially in a variety of sizes, up to 9 cubic feet, and may also be made in the laboratory.

Bags can be manufactured from various plastic materials, most of which can be purchased in rolls or sheets cut to the desired size. Some materials, such as Mylar, may be sealed with a hot iron using a Mylar tape around the edges. Others, such as Teflon, require high temperature and controlled pressure in sealing. Certain plastics, including Mylar and Scotchpak, may be laminated with aluminum to seal the pores and reduce the permeability of the inner walls to sample gases and the outer walls to moisture. Sampling ports may consist of a sampling tube molded into the fabricated bag and provided with a closing device or a clamp-on air valve.

Plastic bags have the advantages of being light, nonbreakable, and inexpensive, and they permit the entire sample to be withdrawn without the difficulty associated with dilution by replacement air, as is the case with rigid containers. However they must be used with caution because generalization of recovery characteristics of a given plastic cannot be extended to a broad range of gases and vapors. Important factors to be considered in using these collectors are absorption and diffusion characteristics of the plastic material, concentration of the gas or vapor, and reactive characteristics of the gas or vapor with moisture and with other constituents in the sample. Information on the storage properties of gases and vapors in plastic containers has been published (11, 12).

The bags must be leak tested and preconditioned for 24 hours before they are used for sampling. Preconditioning consists of flushing the bag 3 to 6 times with the test gas, the number of times depending on the nature of the bag material and the gas. In some cases it is recommended that the final refill remain in the bag overnight before the bag is used for sampling. Such preconditioning usually is helpful in minimizing the rate of decay of a collected gas. At the sampling site the air to be sampled is allowed to stand in the bag for several minutes, if possible, before removal and subsequent refilling of the bag with a sample. Once collected, the interval between sampling and analysis should be as short as practicable.

4.2.2 Integrated Sampling

Integrated sampling of the workroom atmosphere is necessary when the composition of the air is not uniform, when the sensitivity requirements of the method of analysis necessitate sampling over an extended period, or when compliance or noncompliance with an 8-hour, time-weighted average standard must be established. Thus the professional observations and judgment of the industrial hygienist are called on in devising the strategy for the procurement of representative samples to meet the requirements of an environmental survey of the workplace. This deliberation was discussed earlier in this chapter.

Sampling Pumps. Integrated air sampling requires a relatively constant source of suction as an air-moving device. A vacuum line, if available, may be satisfactory. However the most practical source for prolonged periods of sampling is a pump powered by electricity. These pumps come in various sizes and types and must be chosen for the sampling devices with which they will be used.

If electricity is not available or if flammable vapors present a fire hazard, aspirator bulbs, hand pumps, portable units operated by compressed gas, or battery-operated pumps are suitable for sampling at rates up to 3 l/min. The latter have become the workhorses of the industrial hygiene profession, particularly in judging compliance with health standards. For higher sampling requirements, ejectors using compressed air or a water aspirator may be employed.

When compressed air or batteries are to be used as the driving force for a pump, the length of the sampling period is important in relation to the supply of compressed air or the life of the rechargeable battery. These units must not be allowed to run unattended, and periodic checks on the airflow must be made.

The common practice in the field is to sample for a measured period of time at a constant, known rate of airflow. Direct measurements are made with rate meters such as rotameters and orifice or capillary flowmeters. These units are small and convenient to use, but at very low rates of flow their accuracy decreases. The sampling period must be timed carefully with a stopwatch.

Many pumps have inlet vacuum gauges or outlet pressure gauges attached. These gauges, when properly calibrated with a wet or dry gas meter, can be used to determine the flowrate through the pump. The gauge may be calibrated in terms of cubic feet per minute or liters per minute. If the sample absorber does not have enough resistance to produce a pressure drop, a simple procedure is to introduce a capillary tube or other resistance into the train behind the sampling unit.

Samplers are always used in assembly with an air-moving device (source of suction) and an air metering unit. Frequently, however, the sampling train consists of filter, probe, absorber (or adsorber), flowmeter, flow regulator, and air mover. The filter serves to remove any particulate matter that may interfere in the analysis. It should be ascertained that the filter does not also remove the gaseous contaminant of interest. The probe or sampling line is extended beyond the sampler to reach a desired location. It also must be checked to determine that it does not collect a portion of the sample. The meter that follows the sampler indicates the flowrate of air passing through the system. The flow regulator controls the airflow. Finally at the end of the train the air mover provides the driving force.

Absorbers. Four basic types of absorber are employed for collecting gases and vapors: simple gas washing bottles, spiral and helical absorbers, fritted bubblers, and glass-bead columns. The absorbers provide sufficient contact between the sampled air and the liquid surface to ensure complete absorption of the gaseous contaminant. In general, the lower the sampling rate, the more complete is the absorption (Figure 17.7).

Simple gas washing bottles include Drechsel types, standard Greenburg-Smith devices, and midget impingers. The air is bubbled through the liquid absorber without special effort to secure intimate mixing of air and liquid, and the length of travel of the gas through the collecting medium is equivalent to the height of the absorbing liquid. These scrubbers are suitable for gases and vapors that are readily soluble in the absorbing liquid or react with it. One or two units may be enough for efficient collection, but in some cases several may be required to attain the efficiency of a single fritted glass bubbler. Advantages of these devices are simplicity in construction, ease of rinsing, and the small volume of liquid required.

Spiral and helical absorbers provide longer contact between the sampled air and the absorbing solution. The sample is forced to travel a spiral or helical path through the liquid 5 to 10 times longer than that in the simpler units. In fritted glass bubblers air passes through porous glass plates and enters the liquid in the form of small bubbles. The size of the air bubbles depends on the liquid and on the diameter of the orifices from which the bubbles emerge. Frits are classified as fine, coarse, or extra coarse, depending on the number of openings per unit area. The extra-coarse frit is used when a more rapid flow is desired. The heavier froth generated by some liquids increases the time of contact of gas and liquid. These devices are more efficient collectors than the simple gas washing bottles and can be used for the majority of gases and vapors that are soluble in the reagent or react rapidly with it. Flowrates between 0.5 and 1.0 l/min are used commonly. These absorbers are relatively sturdy, but the fritted glass is difficult

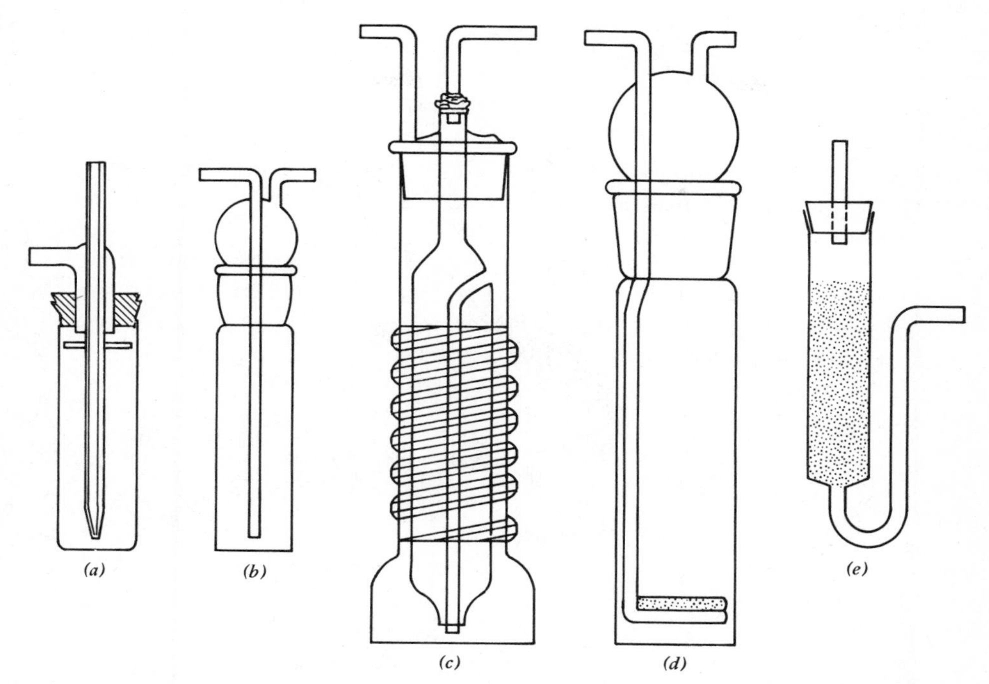

Figure 17.7 Absorbers. (a) and (*b*) Simple gas-washing absorbers. (c) Helical absorber. (*d*) Fitted bubbler. (e) Glass-bead column. From "Tentative Methods of Sampling Atmospheres for Analyses of Gases and Vapors," American Society for Testing and Materials, Philadelphia, July 24, 1956.

when used for contaminants that form a precipitate with the reagent, in which cases a simple gas washing bottle would be preferable.

Packed glass-bead columns are used for special situations calling for a concentrated solution. Glass pearl beads, wetted with the absorbing solution, provide a large surface area for the collection of a sample and are especially useful when a viscous liquid is required. The rate of sampling is low, usually 0.25 to 0.5 liter of air per minute.

Adsorption Media. When it is desired to collect insoluble or nonreactive vapors, an adsorption technique frequently is the method of choice. Activated charcoal and silica gel are common adsorbents (Figures 17.8 and 17.9). Solid adsorbents require less manipulative care than do liquid absorbents; they can provide high collection efficiencies, and with improved adsorption tube design and a better definition of desorption requirements, they are becoming increasingly popular in industrial hygiene surveys.

Activated charcoal is an excellent adsorbent for most vapors boiling above 0°C; it is moderately effective for low boiling gaseous substances (between −100 and 0°C), particularly if the carbon bed is refrigerated, but a poor collector of gases having boiling points below −150°C. Its retentivity for sorbed vapor is several times that of silica gel.

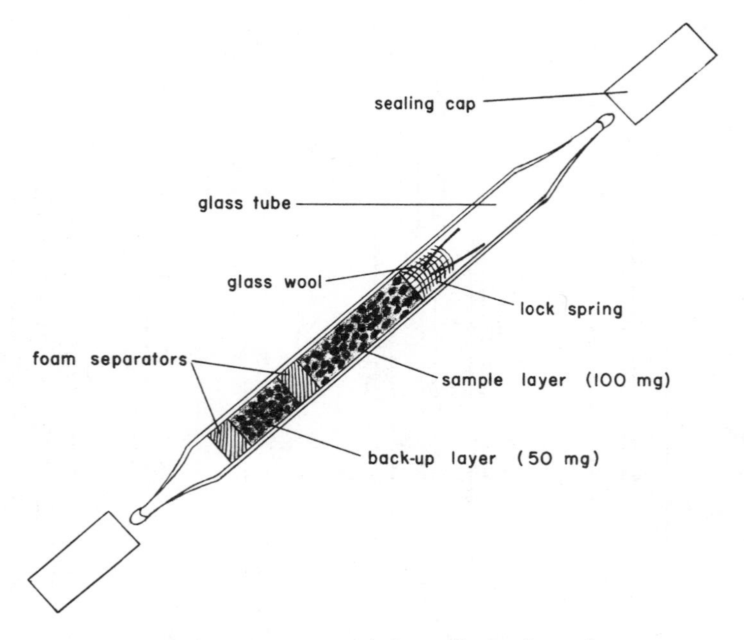

Figure 17.8 Activated charcoal sampling tube.

Because of its nonpolar characteristics, organic gases and vapors are adsorbed in preference to moisture, and sampling can be performed for long periods of time.

Silica gel has been used widely as an adsorbent for gaseous contaminants in air samples. Because of its polar character it tends to attract polar or readily polarizable substances preferentially. The general order of decreasing polarizability or attraction is: water, alcohols, aldehydes, ketones, esters, aromatic compounds, olefins, and paraffins. Organic solvents are relatively nonpolar in comparison with water, which is strongly adsorbed onto silica gel; such compounds will be displaced by water in the entering airstream. Consequently the volume of air sampled under humid conditions may need to be restricted. Despite this limitation, silica gel is a very useful adsorbent.

Condensation. In condensation methods, vapors or gases are separated from sampled air by passing the air through a coil immersed in a cooling medium, dry ice and acetone, liquid air, or liquid nitrogen. The device is not considered to be a portable field technique ordinarily. It may be necessary to use this method when the gas or vapor may be altered by collecting in liquid or when it is difficult to collect by other techniques. A feature of this method is that the contaminating material is obtained in concentrated form. The partial pressure of the vapor can be measured when the system is brought back to room temperature.

Collection Efficiency. The collection efficiency (the ratio of the amount of contaminant retained by the absorbing or adsorbing medium to that entering it) need not be 100 percent as long as it is known, constant, and reproducible. The minimum acceptable collection performance in a sampling system is usually 90 percent, but higher efficiency is certainly desirable. When the efficiency falls below the acceptable minimum, sampling may be carried out at a lower rate, or in the case of liquid absorbers, at a reduced temperature by immersing the absorber in a cold bath to reduce the volatility of both the solute and solvent.

Frequently the relative efficiency of a single absorber can be estimated by placing another in series with it. Any leakage is carried over into the second collector. The absence of any carryover is not in itself an absolute indication of the efficiency of the test absorber, since the contaminant may not be stopped effectively by either absorber. Analysis of the various sections of silica gel or activated charcoal tubes used in sampling a contaminant is a useful check on the collection efficiency of the first section of the tube.

Another valuable technique is the operation of the test absorber in parallel or in series with a different type of collector having a known high collection efficiency (an absolute collector if one is available) for the contaminant of interest. By running the test absorber at different rates of flow, the maximum permissible rate of flow for the device can be ascertained.

4.3 Direct Reading Techniques

Various direct reading techniques that can be used to evaluate airborne concentrations of gases and vapors are available to the industrial hygienist. These include instruments

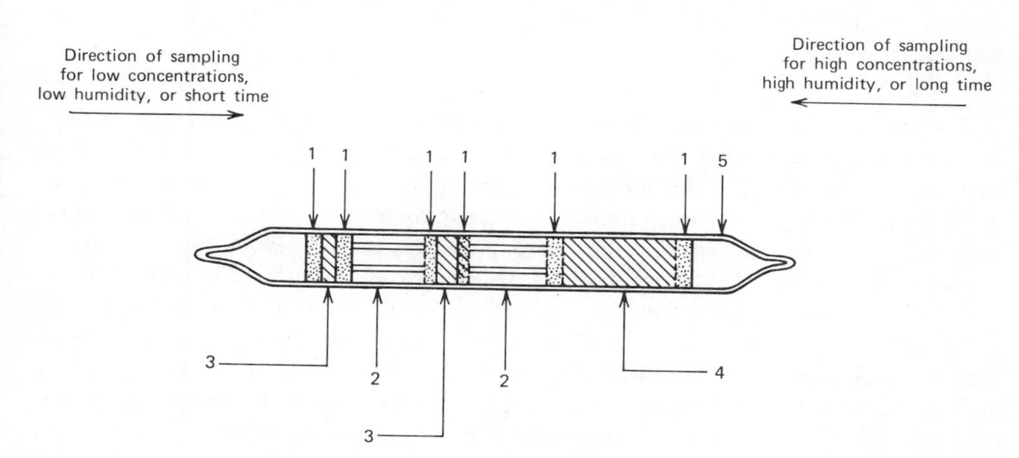

Figure 17.9 Silica gel sampling tube for aromatic amines: 1, 100 mesh stainless steel screen plugs; 2, 12 mm glass tube separator; 3, 150 mg silica gel section, 45/60 mesh; 4, 700 mg silica gel section, 45/60 mesh; 5, 8 mm i.d. glass tube.

capable of direct response to airborne contaminants, various reagent kits that can be used for certain substances, colorimetric indicator (detector) tubes, and, recently, "passive dosimeters."

4.3.1 Instruments

A direct reading instrument, for purposes of this discussion, is described as "an integrated system capable of sampling a volume of air, making a quantitative analysis and displaying the result." Direct reading instruments can be portable devices or fixed-site monitors. Generally these devices are characterized by disadvantages that limit their application for measuring the low concentrations of significance to the industrial hygienist.

Direct reading instruments are used commonly for on-site evaluations for a variety of reasons, depending primarily on the understood purpose of the survey. Direct reading instruments are useful in the following applications.

1. To find *sources* of emission of hazardous substances "on the spot."
2. To determine the performance characteristics of specific operations or control devices, usually by comparing results of "before and after" surveys.
3. As a *qualitative* industrial hygiene monitoring instrument to ascertain whether specific air quality standards are being complied with.
4. As continuous monitoring devices, by establishment of a network of sensors at fixed locations throughout a plant. Readout from such a system can be used to activate either an alarm or an auxiliary control system in the event of process upsets, or to obtain permanent recorded documentation of concentrations of contaminants in the workroom atmosphere.

The advantages of having direct reading instruments available for industrial hygiene surveys are obvious. Such on-site evaluations of atmospheric concentrations of hazardous substances make possible the immediate assessment of unacceptable conditions and enable the industrial hygienist to initiate immediate corrective action in accordance with his judgment of the seriousness of the situation without causing further risk of injury to the workers. It cannot be overemphasized that great caution must be employed in the use of direct reading instruments and, more important, in the interpretation of their results. Most of these instruments are nonspecific, and before recommending any action the industrial hygienist often must verify his on-site findings by supplemental sampling and laboratory analyses to characterize adequately the chemical composition of the contaminants in a workroom area and to develop the supporting quantitative data with more specific methods of greater accuracy. Such precautions become mandatory if the industrial hygienist or other professional investigator conducting the sampling has not had extensive experience with the process in question or when a change in the process or substitution of chemical substances may have occurred.

The calibration of *any* direct reading instrument is an absolute necessity if the data are to have any meaning. Considering this to be axiomatic, it must also be recognized that the frequency of calibration is dependent on the type of instrument. Certain classes of instruments, because of their design and complexity, require more frequent calibration than others. It is also recognized that peculiar "quirks" in an individual instrument produce greater variations in its response and general performance, thus requiring a greater amount of attention and more frequent calibration than other instruments of the same design. Direct personal experience with a given instrument serves as the best guide in this matter.

Another unknown factor that can be evaluated only by experience is the environmental variability of sampling sites. For example, when locating a particular fixed-station monitor at a specific site, consideration must be given to the presence of interfering chemical substances, the corrosive nature of contaminants, vibration, voltage fluctuations, and other disturbing influences that may affect the response of the instrument.

Finally, the required accuracy of the measurements must be determined initially. If an accuracy of ± 3 percent is needed, more frequent calibration must be made than if ± 25 percent accuracy is adequate in the solution of a particular problem.

As indicated earlier, direct reading instruments for atmospheric contaminants are classified as devices that provide an immediate indication of the concentration of contaminants by a dial reading, a stripchart recording, or a tape printout. When properly calibrated and used with full cognizance of their performance characteristics and limitations, these services can be extremely helpful to industrial hygienists who are engaged in on-site evaluations of potentially hazardous conditions. Many types of instrument depend on certain physical or chemical principles for their operation. They are discussed briefly later in this chapter and are described in detail in other publications (13, 14).

In general, the advantages of direct reading instruments include the following:

1. Rapid estimations of the concentration of a contaminant, permitting on-site evaluations and implementation of appropriate measures.
2. Provision of permanent 24-hour records of contaminant concentrations when used as continuous monitors.
3. Easy incorporation of alarm systems onto continuous monitoring instruments to warn workers of buildup of hazardous conditions.
4. Reduction of the number of manual tests needed to accomplish an equivalent amount of sampling.
5. Similarly, reduction of the number of laboratory analyses.
6. Provision of evidence of monitoring environmental conditions for presentation in litigation proceedings.
7. Reduced cost per sample of obtaining data.

The disadvantages of direct reading instruments usually include at least one of the following:

1. High initial cost of instrumentation and, if used as a continuous monitor, installation of sensing network.
2. Need for frequent calibration.
3. General lack of adequate calibration facilities.
4. Lack of portability.
5. Lack of specificity, the most critical negative factor.

The remainder of this presentation is devoted to descriptions of several types of direct reading instrument that can be used if necessary precautions are taken and the limitations of the devices are understood. In the following discussions, when the term "0–?" range is used, the investigator should recognize that "0" is a relative term and relates only to the sensitivity of the instrument.

Colorimetry. Colorimetry is the measurement of the relative power of a beam of radiant energy in the visible, ultraviolet, or infrared region of the electromagnetic spectrum, which has been attenuated as a result of passing a suspension of solid or liquid particulates in air or other gaseous medium, or a photographic image of a spectral line or an X-ray diffraction pattern on a photographic film or plate. Such photometers contain a lamp or other generating source of energy, an optical filter arrangement to limit the bandwidth of the incident beam of radiation, and an optical system to collimate the filtered beam. The beam then is passed through the sample system contained in a cuvette or gas cell to a photocell, bolometer, thermocouple, or pressure-sensor type of detector, where the signal is amplified and fed to a readout meter or to a stripchart recorder. The more sophisticated technique, termed spectrophotometry, employs prisms made of glass (visible region), quartz (ultraviolet), and sodium chloride or potassium bromide (infrared), or of diffraction gratings, instead of optical filters, to provide essentially monochromatic radiation as the source of energy.

Although spectrophotometers are used mostly in laboratories for highly specific and precise analytical determinations, field-type colorimetric analyzers have been designed to function primarily as fixed-station monitors for active gases such as oxides of nitrogen, sulfur dioxide, "total oxidant," ammonia, aldehydes, chlorine, hydrogen fluoride, and hydrogen sulfide. These instruments require frequent calibration with zero and span gases at the sampling site to assure generation of reliable data. However built-in automated calibration systems, which standardize regularly zero and span controls against pure air and calibrated optical filter, are now available for several gases including nitrogen oxides, sulfur dioxide, and aldehydes.

Heat of Combustion. With these instruments a combustible gas or vapor mixture is passed over a filament heated above the ignition temperature of the substance of analytical interest. If the filament is part of a Wheatstone bridge circuit, the resulting heat of combustion changes the resistance of the filament, and the measurement of the imbalance is related to the concentration of the gas or vapor in the sample mixture. The method is basically nonspecific, but it may be made more selective by choosing appropriate filament temperatures for individual gases or vapors or by using an oxidation catalyst for a desired reaction, such as Hopcalite for carbon monoxide.

Combustible gas indicators must be calibrated for their response to the anticipated individual test gases and vapors. These instruments are definitely portable and they are valuable survey meters in the industrial hygienist's collection of field instruments. Readings are in terms of 0 to 1000 ppm or 0 to 100 percent lower explosive limit (LEL). However industrial atmospheres rarely contain *one* gaseous contaminant, and these indicators will respond to all combustible gases present. Hence supplementary sampling and analytical techniques should be used for a complete definition of environmental conditions.

Electrical Conductivity. Electrical conductivity instruments function by drawing a gas–air mixture through an aqueous solution. Gases that form electrolytes, such as vinyl chloride, produce a change in the electroconductivity as a summation of the effects of all ions thus produced. Hence the method is nonspecific. If the concentrations of all other ionizable gases are either constant or insignificant, the resulting changes in conductivity may be related to the gaseous substance of interest. Temperature control is extremely critical in conductance measurements; if thermostated units are not used, electrical compensation must enter into the measurements to allow for the 2 percent/C° conductivity temperature coefficient average for many gases.

The electrical conductivity method has found its greatest application in the continuous monitoring of sulfur dioxide in ambient atmospheres. However a lightweight portable analyzer that uses a peroxide absorber to convert SO_2 to H_2SO_4 is now available; this battery-operated instrument can provide an integrated reading of the SO_2 concentration over the 0 to 1 ppm range within one minute. A larger portable model which may be operated from a 12 V automobile battery is also available for the high concentration ranges of SO_2 encountered in field sampling.

Thermal Conductivity. The specific heat of conductance of a gas or vapor is a measure of its concentration in a carrier gas such as air, argon, helium, hydrogen, or nitrogen. However thermal conductivity measurements are nonspecific, and the method has had only limited application as a primary detector. It has found its greatest usefulness in estimating the concentrations of separately eluted components from a gas chromatographic column. This method operates by virtue of the loss of heat from a hot filament to a single component of a flowing gas stream, the loss being registered as a decrease in electrical resistance measured by a Wheatstone bridge circuit. This method is applied mainly to binary gas mixtures, and uses are based on the electrical unbalance produced in the bridge circuit by the difference in the filament resistances of the sample and reference gases passed through the separate cavities in the thermal conductivity cell.

Flame Ionization. The hydrogen flame ionization detector (FID) typically consists of a stainless steel burner in which hydrogen is mixed with the sample gas stream in the base of the unit. Combustion air or oxygen is fed axially and diffused around the jet through which the hydrogen–gas mixture flows to the cathode tip, where ignition occurs. A loop of platinum, which set about 6 mm above the tip of the burner, serves as the collector electrode. The current carried across the electrode gap is proportional to the number of ions generated during the burning of the sample. The detector responds to essentially all organic compounds, but its response is greatest with hydrocarbons and diminishes with increasing substitution of other elements, notably oxygen, sulfur, and chlorine. Its low noise level of 10^{-12} A provides a high sensitivity of detection and it is capable of the wide linear dynamic range of 10^7. Its usefulness is enhanced by its insensitivity to water, the permanent gases, and most inorganic compounds, thus simplifying the analysis of aqueous solutions and atmospheric samples. It serves to great advantage in both laboratory and field models of gas chromatographs (an application discussed in detail subsequently), as well as in hydrocarbon analyzers that are set up as fixed-station monitors of ambient atmospheres in the laboratory or field.

Hydrocarbon analyzers, operating with an FID detector, are literally carbon counters; their response to a given quantity of a typical C_6 hydrocarbon is 6 times to that of methane, at a fixed flowrate of the sample stream. Thus the instrumental characteristics such as sensitivity are usually given as methane equivalent. In addition to hydrocarbons, these analyzers respond to alcohols, aldehydes, amines, and other compounds, including vinyl chloride, which will produce an ionized carbon atom in the hydrogen flame.

Gas Chromatography. Gas chromatography, a physical process for separating components of complex mixtures, is now being used as a portable technique for in-plant studies. A gas chromatograph has the following components:

- A carrier gas supply complete with a pressure regulator and flowmeter.
- An injection system for the introduction of a gas or vaporizable sample into a port at the front end of the separation column.
- A stainless steel, copper, or glass separation column containing a stationary phase consisting of an inert material such as diatomaceous earth, used alone as in gas–solid chromatography (GSC) or as a support for a thin layer of a liquid substrate, such as silicone oils, in gas–liquid chromatography (GLC).
- A heater and oven assembly to control the temperature of the column(s), injection port, and detector unit.
- A detector.
- A recorder for the chromatograms produced during the separations.

The separations are based on the varied affinities of the sample components for the packing materials of a particular column, the rate of carrier gas flow, and the operating temperature of the column. Improved separations are made possible by the use of temperature programming. The sample components, as a consequence of their varied affinities for a given column, are eluted sequentially; thus they evoke separate responses by the detection system, from which the signal is amplified to produce a peak on the strip chart recorder. The height and area of the peak are proportional to the concentration of the eluted sample component. Calibrations can be made using known mixtures of the pure substance in a gas–air mixture prepared in a flexible plastic bag or other suitable container. The time of retention on the column and supporting analytical techniques (e.g., infrared spectrophotometry) can be used in qualitative analysis of the individual peaks of a chromatogram. The method is capable of providing extremely clean-cut separations and is one of the most useful techniques in the field of organic analysis. It is sensitive to fractional part per billion concentrations of many organic substances. The most commonly used detectors include flame ionization, thermal conductivity, and electron capture.

Rugged, battery-operated, portable gas chromatographs have been refined sufficiently to be practical for many field study applications. These instruments may be obtained with a choice of thermistor, thermal conductivity, flame ionization, and electron capture detectors. Complete with gas sampling valve, rechargeable batteries, appropriate columns, and self-contained supplies of gases, these chromatographs have much to offer the industrial hygienist engaged in on-site analyses of trace quantities of organic compounds and the permanent gases. The gas lecture bottles provide 8 to 20 hours of operation, dependent on the flowrates, and they must be recharged using high pressure gas regulators. The retention times of the compounds of analytical interest must be determined in the laboratory for a given type of column, as is true for the laboratory type chromatographs.

Other Principles of Operation. Various additional types of direct reading instrument are available, although their range of applicability is more limited. The principles of operation include chemiluminescence, coulometry, polarography, potentiometry, and radioactivity. Table 17.3 summarizes the types of direct reading instrument available, along with typical operating characteristics and examples of gases and vapors for which the instruments have been used successfully.

4.3.2 Reagent Kits

Direct reading colorimetric techniques are available which utilize the chemical properties of an atmospheric contaminant for the reaction of the substance with a color-producing reagent. Detector kit reagents may be in either a liquid or a solid phase or supplied in the form of chemically treated papers. The liquid and solid reagents are generally supported in sampling devices through which a measured amount of contaminated air is drawn. On the other hand, chemically treated papers are usually exposed to the atmosphere, and the reaction time for a color change to occur is noted.

Liquid reagents come in sealed ampoules or in tubes for field use. Such preparations are provided in concentrated or solid form for easy dilution or dissolution at the sampling site. Representative of this type of reagent are the *o*-tolidine and the Griess-Ilsovay kits for chlorine and nitrogen dioxide, respectively. Although the glassware

Table 17.3. Direct Reading Physical Instruments

Principle of Operation	Applications and Remarks	Code[a]	Range	Repeatability (precision)	Sensitivity	Response Time
Aerosol photometry	Measures, records, and controls particulates continuously in areas requiring sensitive detection of aerosol levels; detection of 0.05 to 40 μm diameter particles. Computer interface equipment is available.	A and B	10^{-3} to 10^{2} μg/liter	Not given	10^{-3} μg/liter (for 0.3 μm DOP)	Not given
Chemiluminescence	Measurement of NO in ambient air selectively and NO_2 after conversion to NO by hot catalyst. Specific measurement of O_3. No atmospheric interferences.	B	0–10,000 ppm	±0.5–±3%	Varies: 0.1 ppb to 0.1 ppm	ca 0.7 sec, NO mode and 1 sec, NO_x mode; longer period when switching ranges
Colorimetry	Measurement and separate recording of NO_2-NO_x, SO_2, total oxidants, H_2S, HF, NH_3, Cl_2 and aldehydes in ambient air.	A and B	ppb and ppm	±1–±5%	0.01 ppm (NO_2, SO_2)	30 sec to 90% of full scale
Combustion	Detects and analyzes combustible gases in terms of percent LEL on graduated scale. Available with alarm set at $\frac{1}{3}$ LEL.	A	ppm to 100%	—	ppm	<30 sec
Conductivity, electrical	Records SO_2 concentrations in ambient air. Some operate off a 12-V car battery. Operate unattended for periods up to 30 days.	A and B	0–2 ppm	<±1–±10%	0.01 ppm	1–15 sec (lag)

Coulometry	Continuous monitoring of NO, NO_2, O_3 and SO_2 in ambient air. Provided with stripchart recorders. Some require attention only once a month.	A and B	Selective: 0–1.0 ppm overall, or to 100 ppm (optional)	±4% of full scale	varies: 4–100 ppb dependent on instrument range setting	<10 min to 90% of full scale
Flame ionization (with gas chromatograph)	Continuous determination and recording of methane, total hydrocarbons, and carbon monoxide in air. Catalytic conversion of CO to CH_4. Operates up to 3 days unattended.	B	Selective: 0–1 ppm; 0–100 ppm	±1% of full scale	Not given	5 min (cycle time)
	Separate model for continuous monitoring of SO_2, H_2S, and total sulfur in air. Unattended operation up to 3 days.	B	0–20 ppm	±4% of full scale	0.005 ppm (H_2S); 0.01 ppm (SO_2)	5 min (cycle time)
Flame ionization (hydrocarbon analyzer)	Continuous monitoring of total hydrocarbons in ambient air; potentiometric or optional current outputs compatible with any recorder. Electronic stability from 32 to 110°F.	B	0–1 ppm as CH_4; ×1, ×10, ×100, ×1000 with continuous span adjustment	±1% of full scale	1 ppm to 2% full scale as CH_4; 4 ppm to 10% as mixed fuel	<0.5 sec to 90% of full scale
Gas chromatograph, portable	On-site determination of fixed gases, solvent vapors, nitro and halogenated compounds, and light hydrocarbons. Instruments available with choice of flame ionization, electron capture, or thermal conductivity detectors and appropriate columns for desired analyses. Rechargeable batteries.	A	Depends on detector	Not given	<1 ppb (SF_6) with electron capture detector; <1 ppm (HCs)	—

Table 17.3. (Continued)

Principle of Operation	Applications and Remarks	Code[a]	Range	Repeatability (precision)	Sensitivity	Response Time
Infrared analyzer (photometry)	Continuous determination of a given component in a gaseous or liquid stream by measuring amount of infrared energy absorbed by component of interest using pressure sensor technique. Wide variety of applications include CO, CO_2, Freons, hydrocarbons, nitrous oxide, NH_3, SO_2, and water vapor.	B	From ppm to 100% depending on application	±1% of full scale	0.5% of full scale	0.5 sec to 90% of full scale
Photometry, ultraviolet (tuned to 253.7 mμ)	Direct readout of mercury vapor; calibration filter is built into the meter. Other gases or vapors that interfere include acetone, aniline, benzene, ozone, and others that absorb radiation at 253.7 mμ.	A	0.005–0.1 and 0.03–1 mg/m^3	±10% of meter reading or ± minimum scale division, whichever is larger	0.005 mg/m^3	Not given
Photometry, visible (narrow-centered 394 mμ band pass)	Continuous monitoring of SO_2, SO_3, H_2S, mercaptans, and total sulfur compounds in ambient air. Operates more than 3 days unattended.	B	1–3000 ppm (with airflow dilution)	±2%	0.01–10 ppm	<30 sec to 90% of full scale
Particle counting (near forward scattering)	Reads and prints directly particle concentrations at 1 of 3 preset time intervals of 100, 1000 or 10,000 seconds, corresponding	B	Preset (by selector switch); particle size ranges:	±0.05% (probability of coincidence)	—	Not given

	to 0.01, 0.1, and 1 cubic foot of sampled air.		0.3, 0.5, 1.0, 2.0, 3.0, 5.0, and 10.0 μm; counts up to 10^7 particles per ft^3 (35×10^3/liter)			
Polarography	Monitor gaseous oxygen in flue gases, auto exhausts, hazardous environments, and in food storage atmospheres and dissolved oxygen in wastewater samples. Battery operated, portable, sample temperature 32 to 110°F, up to 95% relative humidity. Potentiometric recorder output. Maximum distance between sensor and amplifier is 1000 feet.	A	0–5 and 0–25%	±1% of reading at constant sample temperature	Not given	20 sec to 90% of full scale
Radioactivity	Continuous monitoring of ambient gamma and X-radiation by measurement of ion chamber currents, averaging or integrating over a constant recycling time interval, sample temperature limits 32 to 120°F; 0 to 95% relative humidity (weatherproof detector); up to 1000 feet remote sensing capability. Recorder and computer outputs. Complete with alert, scram, and failure alarm systems. All solid state circuitry.	B	0.1–10^7 mR/hr	±10% (decade accuracy)	—	<1 sec

Table 17.3. (Continued)

Principle of Operation	Applications and Remarks	Code[a]	Range	Repeatability (precision)	Sensitivity	Response Time
Radioactivity	Continuous monitoring of beta- or gamma-emitting radioactive materials within gaseous or liquid effluents; either a thin-wall Geiger-Müller tube or a gamma scintillation crystal detector is selected depending on the isotope of interest; gaseous effluent flow, 4 cfm; effluent sample temperature limits 32 to 120°F using scintillation detector and 65 to 165°F using G-M detector. Complete with high radiation, alert and failure alarms.	B	10–10^6 counts/min	±2% full scale (rate meter accuracy)	<10^{-7} μCi of ^{131}I per cc of air and 10^{-7} μCi of ^{137}Cs per cc of water using a scintillation detector	0.2 sec at 10^6 counts/min (ratemeter)
Radioactivity	Continuous monitoring of radioactive airborne particulates collected on a filter tape transport system; rate of airflow, 10 scfm; scintillation and G-M detectors, optional but a beta-sensitive plastic scintillator is provided to reduce shielding requirements and offer greater sensitivity. Air sample temperature limits 32 to 120°F; weight 550 pounds. Complete with high and low flow alarm and a filter failure alarm.	B	10 to 10^6 counts/min	±2% of full-scale (rate-meter accuracy)	10^{-12} μCi of ^{137}Cs per cc of air using a scintillation detector	0.2 sec at 10^6 counts/min (ratemeter)

[a] Codes: A, portable instruments; B, fixed monitor or "transportable" instruments.

needed for these applications may be somewhat inconvenient to transport to the field, methods based on the use of liquid reagents are more accurate than those which use solid reactants. This is due to the inherently greater reproducibility and accuracy of color measurements made in a liquid system.

Papers impregnated with chemical reagents have found wide applications for many years for the detection of toxic substances in air. Examples include the use of mercuric bromide papers for the detection of arsine, lead acetate for hydrogen sulfide, and a freshly impregnated mixture of *o*-tolidine and cupric acetate for hydrogen cyanide. When a specific paper is exposed to an atmosphere containing the contaminant in question, the observed time of reaction provides an indication of the concentration of that substance. For example, a 5-second response time by the *o*-tolidine–cupric acetate paper is indicative of a concentration of 10 ppm HCN in the tested atmosphere.

Similarly, sensitive detector crayons have been devised for the preparation of a reagent smear on a test paper for which response to a specific toxic substance in a suspect atmosphere may then be timed to obtain an estimation of the atmospheric concentration of a contaminant. Crayons for phosgene, hydrogen cyanide, cyanogen chloride, and lewisite (ethyl dichloroarsine) have been developed for this purpose (15).

4.3.3 Colorimetric Indicator Tubes

Colorimetric indicating tubes containing solid reagent chemicals provide compact direct reading devices that are convenient to use for the detection and semiquantitative estimation of gases and vapors in atmospheric environments. Presently there are tubes for nearly 200 atmospheric contaminants on the market; five major companies manufacture and/or distribute these devices.

Whereas it is true that the operating procedures for colorimetric indicator tubes are simple, rapid, and convenient, there are distinct limitations and potential errors inherent in this method of assessing atmospheric concentrations of toxic gases and vapors. Therefore dangerously misleading results may be obtained with these devices unless they are used under the supervision of an adequately trained industrial hygienist who (1) enforces rigidly the periodic (as required) calibration of individual batches of each specific type of tube for its response to known concentrations of the contaminant, as well as the refrigerated storage of all tubes, to minimize their rate of deterioration; (2) informs his staff of the physical and chemical nature and extent of interferences to which a given type of tube is subject and limits the tube's usage accordingly; and (3) stipulates how and when other independent sampling and analytical procedures will be employed to derive needed quantitative data. A manual describing recommended practice for colorimetric indication tubes (16) discusses in detail the principles of operation, applications, and limitations of these devices. A brief summary is given below.

Colorimetric indicating tubes are filled with a solid granular material, such as silica gel or aluminum oxide, which has been impregnated with an appropriate chemical reagent. The ends of the glass tubes are sealed during manufacture. When a tube is to be used, its end tips are broken off, the tube is placed in the manufacturer's holder, and the recommended volume of air is drawn through the tube by means of the air-moving device provided by the manufacturer. This device may be one of several types, such as a positive displacement pump, a simple squeeze bulb, or a small electrically operated pump with an attached flowmeter. Each air-moving device must be calibrated frequently, for example, after sampling 100 tubes as an arbitrary rule, or more often if there are reasons to suspect changes due to effects of corrosive action from contaminants in the tested atmospheres. An acceptable pump should be correct to within ± 5 percent by volume; with use, its flow characteristics may change. It should also be checked for leakage and plugging of the inlet after every 10 samples.

In most cases a fixed volume of air is drawn through the detector tube, although in some systems varied amounts of air may be sampled. The operator compares either an absolute length of stain produced in the column of the indicator gel or a ratio of the length of stain to the total gel length against a calibration chart, to obtain an indication of the atmospheric concentration of the contaminant that reacted with the reagent. To make estimates using another type of tube, a progressive change in color intensity is compared with a chart of color tints. In a third type of detector, the volume of sampled air required to produce an immediate color change is noted; it is intended that this air volume be inversely proportional to the concentration of the atmospheric contaminant. Basic mathematical analyses of the relationships among the variables affecting the length of stain (i.e., the concentration of test gas, volume of air sample, sampling flowrate, grain size of gel, tube diameter, and other variables) have been published (16, 17). These sources should be consulted for a full appreciation of the complex interrelationships among the factors affecting the kinetics of indicator tube reactions. It is sufficient to point out here that the length of stain is proportional to the logarithm of the product of gas concentration and air sample volume as follows:

$$\frac{L}{H} = \ln(CV) + \ln\frac{K}{H} \tag{5}$$

where L = length of stain (cm)
C = gas concentration (ppm)
V = air sample volume (cc)
K = a constant for a given type of indicator tube and test gas
H = a mass transfer proportionality factor (cm) known as the height of a mass transfer unit

If this mathematical model is correct for a given indicator tube, a linear plot of L versus the logarithm of the CV product, for a fixed constant flowrate, will yield a straight line with slope equal to H. The significance of this equation is the implication that it is important to control the flowrate, which may produce a greater effect on the length of stain than does the concentration of the test gas. In the optimal design of an indicator tube, therefore, it is desirable that the reaction rate be rapid enough to permit the establishment of equilibrium between the indicating gel and the test gas, thus producing a stoichiometric relationship between the volume of stained indicating gel and the quantity of the absorbed test gas. Such equilibrium conditions may be assumed to exist when stain lengths are directly proportional to the volume of sampled air and are not affected by the sampling flowrate. With this situation a log-log plot of stain length versus concentrations for a fixed sample volume may be prepared in the calibration of a given batch of tubes.

From the preceding discussion of the complexity of the heterogeneous phase kinetics of indicator tube reactions, the quality control problems associated with their manufacture and storage, and the difficulties posed by interfering substances, it is obvious that frequent, periodic calibration of these devices should be made by the user. Dynamic dilution systems for the reliable preparation of low concentrations of a test gas or vapor are recommended for this purpose. The Department of Health, Education and Welfare has issued regulations for the certification of gas detector tube units (18). The performance requirements for these units were developed by the NIOSH Division of Laboratories and Criteria Development, with the cooperation and assistance of members of the Joint Direct Reading Gas Detecting Systems Committee of the AIHA and the ACGIH. Table 17.4 lists the detector tubes that have been certified to date (March, 1977) by this program, and the manufacturer or supplier.

4.3.4 Passive Personal Monitors

In recent years significant progress has been made in technology for monitoring of gases and vapors using binary diffusion. The monitors utilize Brownian motion to control the sampling process into a collection media. This technology is particularly well suited for personal monitoring devices, resulting in lightweight, low cost monitors that require no power source.

The monitors rely on a concentration gradient across a static or placid layer of air to induce a mass transfer. The following equation, based on Fick's law, gives the steady state relationship for the rate of mass transfer.

$$W = D\left(\frac{A}{L}\right)(C_1 - C_0) \tag{6}$$

where W = mass transfer rate
D = diffusion coefficient
A = frontal area of static layer
L = length or depth of static layer
C_1 = ambient concentration
C_0 = concentration at collection surface

It can be seen that by choosing an effective collection surface, such that C_0 is essentially zero, the mass transfer or collection rate is proportional to the ambient vapor concentration C_1. It may also be noted that the units of $D(A/L)$ are volume per unit time, the same as for the volumetric flow in a pump monitoring system. The rate of

Table 17.4 NIOSH-Certified Detector Tubes[a]

SUBSTANCE	MANUFACTURER			
	Dräger Model 31	Gastec Model 400	Kitagawa Model 400	MSA Universal
Acetone	-	-	-	460423
Ammonia	CH20501	3M	105Sc, 105c	460103
Benzene	67-26641	121	—	—
Carbon Dioxide	CH 23501	2L	126Sa, 126a	85976
Carbon Monoxide	CH25601, CH20601	1La	106S, 100	91229
Carbon Tetrachloride	—	134	—	—
Chlorine	—	8La	—	460225
Hydrogen Cyanide	CH25701	—	112Sb	—
Hydrogen Sulfide	67-19001	4LL	120b	460058
Nitric Oxide	CH 31001	10	—	—
Nitrogen Dioxide	CH 30001	9L	—	93099
Sulfur Dioxide	CH31701	5La	103Sd, 103d	92623
Toluene	CH23001	122	—	—
Trichloroethylene	CH24401	132H	—	—

[a] Names and addresses of manufacturers and distributors of detector tubes:

Bendix Corporation
1400 Taylor Avenue
Baltimore, Md. 21204
(Gastec distributor)

Matheson Gas Products
1275 Valley Brook Avenue
Lyndhurst, N.J. 07071
(Kitagawa Distributor)

Mine Safety Appliances Company
400 Penn Center Boulevard
Pittsburgh, Pa. 15235

National Mine Service Company
3000 Koppers Building
Pittsburgh, Pa. 15219
(Dräger distributor)

sampling of the contaminant is then the product of the $D(A/L)$ term and the average ambient concentration.

Figure 17.10 illustrates the construction of a monitor for mercury vapor in which the collection surface is a gold layer (19). A similar configuration (Figure 17.11) has been developed for nitrogen dioxide (20). In the case of the mercury monitor, mercury vapor in the atmosphere passes through the microporous barrier film and continues through the static air column according to Fick's law. The average time required to reach the collection surface is

$$T = \frac{L^2}{2D} \text{ seconds} \tag{7}$$

For example, the average time for a mercury atom to progress to the collection surface when the depth L is 0.65 cm can be shown, using $D = 0.12$ cm²/sec, to be 1.75 seconds. Thus the sampling occurs rapidly when thin static air columns are used.

The accuracy and precision of the sampling process are functions of the measured exposure time, velocity effects, and temperature effects. The accuracy and precision of the reported concentration are functions of the calibration standards, collection media, and analytical method used. Of these factors, the potential velocity and temperature effects distinguish this type of monitoring device from the conventional dynamic or flow monitor. All other factors are common to both methods.

Regarding velocity effects, the thickness of the attached boundary layer on the outer surface of the barrier film is a function of the velocity of air movement over the face of the monitor. Sampling that is independent of this air velocity can be achieved when the L term is large compared to the average boundary layer thickness. For temperatures between 50 and 88°F, the temperature factor is constrained to ± 1.8 percent. For use at higher or lower temperatures, the temperature data may be rerecorded to allow corrections to be made.

This type of personal monitor is attached to the worker in his breathing zone. The total exposure time is noted, and analysis results give the amount of vapor collected. These data provide an average mass collection rate, which can then be used to calculate the time-weighted average concentration. The physical parameters of the sampler design are chosen according to desired exposure time and the substance to be monitored.

Corroborative testing of both the mercury and nitrogen dioxide monitors and comparison of results to reference methods have shown excellent agreement. With the

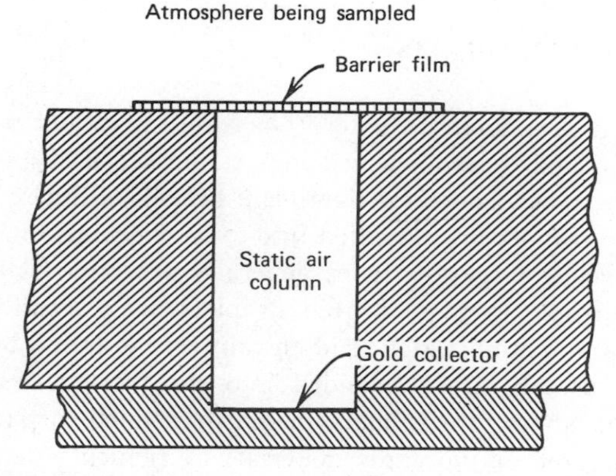

Figure 17.10 Cross-sectional view of mercury detector.

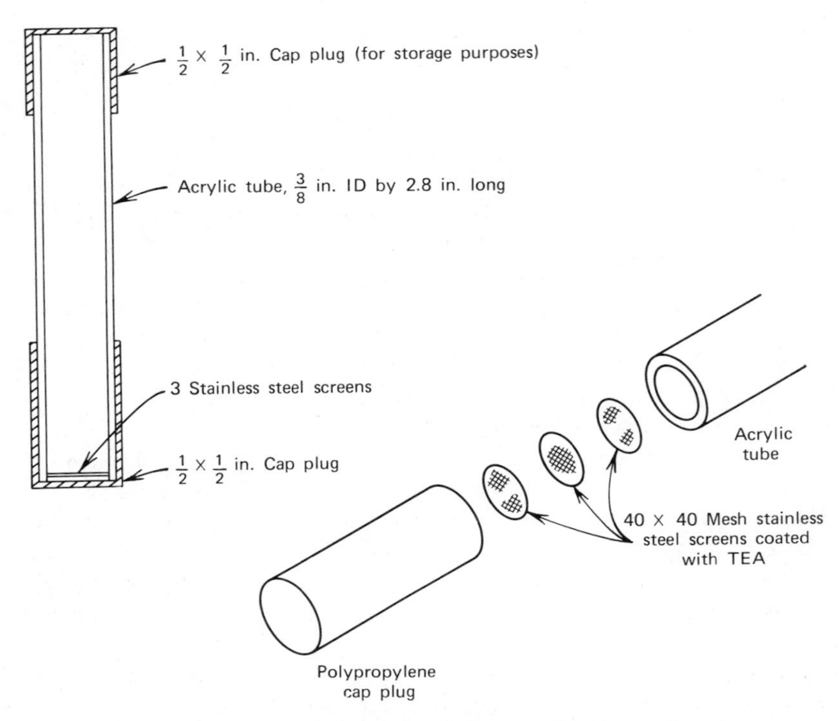

Figure 17.11 Exploded view of nitrogen dioxide monitor.

increasing emphasis on development of specific methods for monitoring workers' exposures to contaminants, it is likely that the passive monitor concept will become the vital basis for a new generation of industrial hygiene monitoring equipment.

5 SAMPLING FOR PARTICULATES

In classifying airborne particulates, the term "aerosol" normally refers to any system of liquid droplets or solid particles dispersed in a stable aerial suspension. This requires that the particles remain suspended for significant periods of time.

Liquid particulates usually are classified into two subgroups, mists and fogs, depending on particle size. The larger particles generally are referred to as mists, whereas small particle sizes result in fogs. Liquid droplets are normally produced by such processes as condensation, atomization, and entrainment of liquid by gases.

Solid particulates usually are subdivided into three categories: dusts, fumes, and smoke, the distinction among them being primarily related to particle size. Dusts are formed from solid organic or inorganic materials by reducing their size through some mechanical process such as crushing, drilling, or grinding. Dusts vary in size from the visible to the submicroscopic, but their composition is the same as the material from which they were formed. Fumes are formed by such processes as combustion, sublimation, and condensation. The term is generally applied to the oxides of zinc, magnesium, iron, lead, and other metals, although solid organic materials such as waxes and some polymers may form fumes by the same methods. These particles are very small, ranging in size from 1 μm to as small as 0.001 μm in diameter. Smoke is a term generally used to refer to airborne particulate resulting from the combustion of organic materials (e.g., wood, coal, or tobacco). The resulting smoke particles are all usually less than 0.5 μm in diameter.

Thus the nature of the airborne particulate dictates to a great extent the manner in which sampling of the environment is to be accomplished. Sampling is performed by drawing a measured volume of air through a filter, impingement device, electrostatic or thermal precipitator, cyclone, or other instrument for collecting particulates. The concentration of particulate matter in air is denoted by the weight or the number of particles collected per unit volume of sampled air. The weight of collected material is determined by direct weighing or by appropriate chemical analysis. The number of particles collected is determined by counting the particles in a known portion or aliquot of the sample.

5.1 General Requirements

The general requirements discussed earlier for gases and vapors apply, for the most part, to particulate contaminants as well. There are several aspects of sampling that apply only for particulates, however, because of the wide range of particle sizes of airborne particulates confronting the industrial hygienist in most industrial settings.

In general, the sampling train for particulates consists of the following components: air inlet orifice, particle separator (and preselector, if used to classify the total particulate), flowmeter, flowrate control valve, and air mover or pump. Of these, by far the most important is the particle separator. Both the efficiency of the device and its reliability must be high. The pressure drop across the separator should be low, to keep the size of the required pump to a minimum. The separator may consist of a single element, such as a filter or impinger, or there may be two or more elements in series, to classify the particulate into different size ranges. The need for proper selection, care, and calibration of the other components of the sampling train, particularly the flowmeter and flow control mechanism, is discussed elsewhere in this book.

It is important that the sampling method not alter chemical or physical characteristics of the particles collected. For example, if the material is soluble, it cannot be collected in a medium capable of dissolving it. If the particles have a tendency to agglomerate, and it is important to be able to distinguish individual particles, deep-section collection on a filter should not be used.

An additional consideration, which is of much greater concern with particulate than with gases and vapors, is the variation of concentration in space. Many cases have been reported of significant differences in concentrations being documented with sampling units placed equidistant from a source of contaminant generation. Similarly, with personal breathing zone sampling, it is not uncommon for simultaneous samples obtained on both shoulders of a worker to indicate substantially different concentrations. The importance of these observations lies in the understanding that particulate sampling results by themselves are indicative of conditions within a short distance of the sampling unit and should be augmented with additional information, such as studies of airflow patterns within the workroom.

5.2 Collection Techniques

Although the concept of grab sampling for particulates is not as valuable as for gases or vapors, there are methods for collecting instantaneously a sample of airborne particles. One such device is the konimeter, which draws a small, measured volume of air into the instrument and literally blasts it at high velocity against a glass plate on which the particles are deposited. The particles then can be examined microscopically and counted or defined in other terms. Since there are millions of particles per cubic foot of most industrial workrooms, a very small volume of air is needed for this technique. Another means of obtaining an instantaneous sample for particulates is the settling chamber. With this device, a chamber is opened in the atmosphere being tested and is closed rapidly to trap the sample. The particles in the air then are allowed to settle by gravitational forces and are collected on a glass plate for subsequent microscopic analysis. Of course the most meaningful sampling for particulate is done over extended periods of time with various collection techniques, depending on the material being collected and the availability of sampling equipment. The most common collection techniques are filtration, impingement, impaction, elutriation, electrostatic precipitation, and thermal precipitation.

5.2.1 Filtration

Filtration is the most common method of collecting airborne particulate. The fibrous type of filter matrices consists of irregular meshes of fibers, usually about 20 μm or less in diameter. Air passing through the filter changes direction around the fibers and the particles impinge against the filter, where they are retained. The largest particles (30 μm and greater) deposit to some extent by sieving action; the smaller particles (submicrometer sizes) also deposit through their Brownian motion, which carried them into the filter material. Efficiency of collection generally increases with airstream velocity, density, and particle size for particles greater than 0.5 μm in diameter. Deposition by diffusion dominates for the smallest particle sizes and decreases as the diameter of the particle increases. Thus there is a size at which the combined efficiency by impingement and diffusion is a minimum; this is usually 0.1 μm diameter. Since the total weight of particles less than 0.1 μm diameter is usually less than 2 percent of the total collected dust, deposition by diffusion can practically be ignored. Of course there are exceptions, such as samples of freshly formed metal fumes, for which the diffusion deposition is significant.

Filters are available in a wide variety of matrices including cellulose, glass, asbestos, ceramic, carbon, and polystyrene fibers (see Table 17.5). Filters made of these fibrous materials consist of thickly matted fine fibers and are small in mass per unit face area, making them useful for gravimeteric determinations. Of these, cellulose fiber filters are the least expensive, are available in a wide range of sizes, have high tensile strengths, and are relatively low in ash content. Their greatest disadvantage is their hygroscopicity, which can present problems during weighing procedures. The filters made of synthetic fibers, particularly glass, are becoming more common, partly because stable tare weights can be determined easily.

Membrane filters, microporous plastic films made by precipitation of a resin under controlled conditions, are used to collect samples that are to be examined microscopically, although they can also serve for gravimetric sampling and for specific determinations using instrumentation. Thus the cellulose ester membrane filters are the most commonly used filters for sampling for such substances as asbestos (analyzed by fiber count) and metal fumes (analyzed by atomic absorption techniques).

5.2.2 Impaction and Impingement

Impactors use a sudden change in direction in airflow and the momentum of the dust particles to cause the particles to impact against a flat surface. Usually impactors are

Table 17.5 Common Applications of Filters

FILTER MATRIX	COMMON APPLICATIONS
Cellulose Ester	asbestos counting, particle sizing, metallic fumes, acid mists
Fibrous Glass	total particulate, oil mists, coal tar pitch volatiles
Paper	total particulate, metals, pesticides
Polycarbonate	total particulate, crystalline silica
Polyvinyl chloride	total particulate, crystalline silica, oil mists
Silver	total particulate, coal tar pitch volatiles, crystalline silica
Teflon	special applications (high temp.)

constructed in several stages, to separate dust by size. The particles adhere to the plate, which may be dry or coated with an adhesive, and the material on each plate is weighed or analyzed at the conclusion of sampling. It is imperative that the impactors be calibrated for the particular material of interest, since the manufacturer's calibration typically is based on a uniformly sized particle that may not give accurate results for the substance of interest.

Impingers also utilize inertial properties of particles to collect samples. Although the interest in, and application of, impingers is waning, they still play an important role in industrial hygiene sampling. The impinger consists of a glass nozzle or jet submerged in a liquid, usually water. Air is drawn through the nozzle at high velocity, and the particles impinge on a flat plate, lose their velocity, are wetted by the liquid, and become trapped. Gases that are soluble in the liquid are also collected. Usually the contents of the impinger samples are analyzed microscopically, gravimetrically, or in a few cases, by specific methods.

The principles of collection for impaction and impingement are quite similar. The primary distinction between them is that with impaction, the particles are directed against a dry or coated surface, whereas a liquid collecting medium is used in impingement.

5.2.3 Elutriation

Elutriators have been essential elements in the sampling trains used to characterize dust levels in many mineral dust surveys, usually as preselectors at the front of the sampling train. Elutriators can have either a horizontal or a vertical orientation.

Elutriators are quite similar to inertial separators in the theoretical basis of operation. The primary difference is that elutriators operate at normal gravitational conditions, whereas the inertial collectors induce very high momentum forces to achieve collection of particles. Horizontal elutriators make use of the fact that as air moves across a horizontal channel with laminar flow, particles of greatest mass tend to cross the streamlines and settle because of gravity. The smaller particles remain airborne by resistance forces of the air for longer times and distances, depending on their size and mass. It is imperative that elutriators be used and operated exactly as described by the manufacturer to avoid disturbances at the inlet, to ensure laminar flow along the elutriator, and avoid risk of redispersion of settled particles.

Vertical elutriators utilize the same dependence on gravitational forces to separate the dust into fractions, except that with the vertical elutriator the natural force works in a direction opposite to the induced airflow instead of normal to it. The vertical elutriator is recognized as a practical device to sample for cotton dust when it is desired to avoid collecting the lint or fly (21).

5.2.4 Electrostatic Precipitation

Electrostatic precipitators have been used for many decades for industrial air analysis in workrooms. These systems have the advantage of negligible flow resistance, no clogging, and precipitation of the dust onto a metal cylinder or foil liner whose weight is unaffected by humidity. In most units the "wire and tube" system is used. A stiff wire, supported at one end, is aligned along the center of the tube and serves as the charged electrode. The tip of the wire is sharpened to a point, and a high voltage (10 to 25 kV dc) is applied to the electrode. The corona discharge from the tip charges particles suspended in the air that is drawn through the tube. The electrical gradient between the wire and the wall of the cylinder (or foil) causes the charged particles to migrate to the inside surface of the tube.

The migration velocity of the charged particles greater than one micrometer in diameter increases in proportion to particle diameter. On the other hand, migration velocity is approximately independent of particle diameter for particles smaller than about one micrometer. Therefore very high separation efficiencies are attainable with electrostatic precipitators, and they have become particularly well suited for particles of submicrometer size, such as those in metal fumes.

5.2.5 Thermal Precipitation

A particle in a thermal gradient in air is directed away from a high temperature source by the differential bombardment from gas molecules around it. This action is taken advantage of in the design of thermal precipitation units. Air is drawn past a hot wire or plate, and the dust collects on a cold glass or metal surface opposite the hot element. Since a high thermal gradient is needed, the gap between the wire or plate and the deposition surface is kept very small, typically less than 2 mm. The migration velocity induced by the thermal gradient is small and is very nearly independent of particle diameter. Because of the severe limitations on maximum flowrate possible with high deposition efficiency, however, these units have been used only for collecting sufficient particulate matter for examination under a microscope. An additional limitation is the inappropriateness of these devices for sampling mists or other liquid particulates, unless their boiling points are high enough to ensure that the liquid is not volatilized by the operating temperatures of the instrument.

5.2.6 Centrifugal Collection

Within recent years there has been increasing interest in and use of preselectors in conjunction with industrial hygiene sampling in the documentation of concentrations of "respirable" dust. This is most commonly done with centrifugal separators, such as the cyclone. Air enters the cyclone tangentially through an opening in the side of a cylindrical or inverted cone-shaped unit. The larger particles are thrown against the side of the cyclone and fall into the base of the assembly. The particles are drawn toward the center of the unit, where they swirl upward along the axis of a tube extending down from the top. The air in the cyclone rotates several times before leaving, and consequently the dust deposits as it would in a horizontal elutriator having an area several times that of the cyclone's outer surface. Thus the volume of a cyclone is much smaller than a horizontal elutriator, or other inertial collector with the same flowrate and efficiency.

Cyclones used to sample for respirable dust, such as in the course of determining compliance with the respirable mass standard for silica-containing dusts, should have performance characteristics meeting the criteria of the ACGIH (Figure 17.12). The orientation of the cyclone is not as critical as for the elutriators, and small 10 mm diameter cyclones have become commonplace for personal breathing zone sampling.

5.3 Direct Reading Techniques

Although there has been substantial development of methods for continuous monitoring of particles in air, the applications have been more in the field of air pollution than industrial hygiene. In general, these instruments have limitations on their sensitivity from two primary sources: the random property fluctuations of the accompanying gas molecules, and the noise level of the electronic circuitry, which converts the physical change to a measurable signal. Instruments that read out particle "sizes" are often calibrated using well-characterized aerosols not necessarily representative of the particles sampled; thus the accuracy of these instruments may be highly variable. To be sure, the response of the instruments is nonspecific; that is, the devices respond to a *property* of the substance (size, shape, or mass), not to the material itself. However there are needs of the industrial hygienist that can be met by direct reading instruments for particulates, and an increasing number of such applications can be expected.

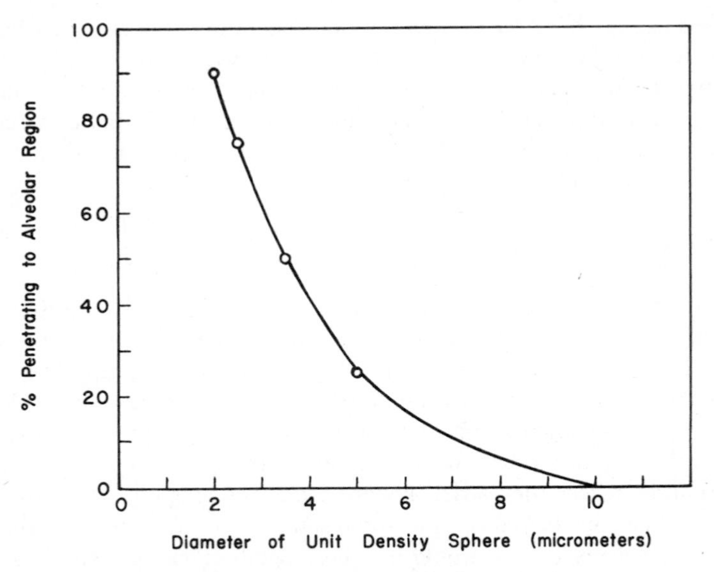

Figure 17.12 ACGIH respirable dust criteria.

5.3.1 Aerosol Photometry (Light Scattering)

Instruments using light scattering techniques are based on generation of an electrical pulse by a photocell that detects the light scattered by a particle. A pulse height analyzer estimates the effective diameter. The number of pulses is related to the number of particles counted per unit flowrate of the sampled medium. Instruments that give a size analysis based on the measurement of total particle concentration in a large illuminated volume are used in monitoring particulate concentrations in experimental rooms or exposure chambers.

Aerosol photometry can usually provide only an approximate analysis of particulate, classified according to particle size in industrial surveys. This is because calibrating the instrument with each type of particulate to be measured in not practicable. The great variations in shape, size, agglomerative effects, and refractive indices of the various components in a given dust suspension make such calibrations virtually impossible. Whereas aerosol photometry can indicate the particulate concentration in the different size ranges of interest, it usually is necessary to perform size distribution analyses by microsieving or microscopic procedures.

5.3.2 Respirable Mass Monitors

A relatively recent addition to the industrial hygienist's array of sampling equipment is the direct reading sampler for respirable mass. Two different types of instrument are presently available, and others are in various stages of development.

The first respirable mass monitor to be introduced to the industrial hygiene field was based on the attenuation of beta radiation resulting from collection of a sample of dust between comparative readings of a beta-radiation source. The instrument uses a two-stage collection system. The first stage consists of a cyclone precollector for the retention of the nonrespirable fraction of the dust. The precollector retains the larger particles and allows those of respirable size to pass to the second stage, which consists of a circular nozzle impactor and beta absorption assembly with a polyester impaction disk. The dust collected by impaction of the thin plastic film increasingly absorbs the beta radiation reaching a Geiger-Müller detector from a ^{14}C source. The penetration of low energy beta radiation depends almost exclusively on the mass per unit area of the absorber and the maximum beta energy of the impinging electrons; it is independent of the chemical composition or the physical characteristics of the collected, absorbing matter. This unit, and modifications thereof, incorporate a digital readout of the respirable mass concentration.

A second type of unit is based on the change in resonant frequency of a piezoelectric quartz crystal accompanying deposition of particulate on the face of the crystal. This unit uses an impactor inlet to separate the dust into respirable and nonrespirable fractions. Modifications have utilized electrostatic precipitation for particle collection as well as multistage impactors. The mass concentration frequency relationship is linear, and the instrument has sufficient sensitivity to measure ambient particulate mass concentration in a few seconds sampling at 1 to 3 liters/min.

These monitoring devices, like all direct reading instrumentation, are not personal monitoring devices. Also, since the readings are instantaneous, they cannot be used directly to determine compliance with 8-hour, time-weighted average standards. Thus such units cannot be used as compliance instruments. However the usefulness of direct reading equipment for analyzing environmental conditions "on the spot" has been great, and continued development and refinement is likely.

6 ANALYSIS OF SAMPLES

The continuing advances and improvements in analytical capabilities have made it possible to measure minute quantities of specific compounds, ions, or elements. The industrial hygienist can process a very small sample of air and accurately determine the presence of suspected contaminants. As mentioned earlier in this chapter, the field industrial hygienist should work very closely with the industrial hygiene chemist, to become familiar with the limitations of the analytical equipment of interest, thus being able to plan his sampling strategy with maximum efficiency.

It is beyond the scope of this chapter to outline specific procedures for industrial contaminants. Instead, we give brief descriptions of the various analytical methods and techniques that have been applied to industrial hygiene samples, with the expectation that the reader will consult more detailed sources for the better understanding of analytical requirements for particular substances of interest (1–3).

6.1 Gravimetric Techniques

Perhaps the most frequently employed analytical method for industrial hygiene samples is gravimetric analysis of filters or other collection media to determine the weight gain. This requires careful handling and processing of the media before collecting the sample, as well as the conditioning of the media after collection of the sample in the exact manner as used to obtain tare weights. In so doing, any necessary correction for the "blank" can be incorporated into the analysis.

Another type of gravimetric technique involves the formation of a precipitate by combining the sample solution with a precipitating agent, and subsequent weighing of the solid precipitate.

6.2 Titrametric Methods

Acid–base and oxidation–reduction volumetric procedures are outstanding examples of simple but useful analytical methods still widely employed in the analysis of industrial hygiene samples. Hydrogen chloride gas and sulfuric acid mist can be collected in an impinger containing a standard sodium hydroxide solution and quantitated by back titration with a standard acid. Ammonia and caustic particulate matter can be collected in acid solution with similar apparatus, and the airborne concentration is determined by titration with a standard base. Oxidation–reduction titrations, principally iodimetric, are useful for measuring sulfur dioxide, hydrogen sulfide, and ozone. Improved volumetric methods utilize electrodes to indicate acid–base null points, and amperometric methods are available for oxidation–reduction titrations. These electrical techniques increase analytical precision and speed up the analyses but do not affect the sensitivity appreciably.

6.3 Optical Methods

As mentioned in the discussion of particulate sampling, much of the sampling for dust requires analysis by microscopic techniques. The use of light microscopy for "dust counting" is decreasing, being replaced by more specific, and more reproducible, mass sampling techniques. However, it is frequently necessary to determine the particle size distribution of airborne particulate, and optical methods offer an effective way of doing this.

In addition to the more classic counting applications of microscopy, the present sampling and analysis method for asbestos is based on actual fiber counting at 400 to 450× magnification using phase contact illumination. As with all optical methods, the analytical results (i.e., actual counts) are somewhat analyst dependent, since much of the technique requires subjective analysis by the microscopist.

6.4 Colorimetric Procedures

Changes of color intensity or tone have been the bases of many useful industrial hygiene analytical methods. For example, the use of Saltzman's reagent in a fritted glass bubbler to determine the airborne concentrations of nitrogen dioxide has been a classic application of such methods. Under controlled conditions of sampling, the concentration of NO_2 in the air is inversely proportional to the time required to produce the color change. Titrations employing acid-base and iodimetric reactions with color indicators are conducted in similar fashion.

Usually such titrations of air samples lack the accuracy and precision obtainable with careful laboratory procedures, but they are adequate for most field studies and have the great advantage of giving a direct and immediate indication of the environmental concentrations. Relatively sensitive and specific analyses of many contaminants can be made, using the spectrophotometers available as both laboratory and field instruments.

In any case, colorimetric methods involve analytical reactions to produce a color in proportion to the quantity of the contaminant of interest in the sample. For example, in the determination of metals, the dithizone extraction method is able to determine selectively the various metallic elements depending on the pH of the solution.

6.5 Spectrophotometric Methods

In addition to the colorimetric procedures, which take advantage of spectrophotometry operating in the visible range, infrared and ultraviolet spectrophotometers have considerable application in the industrial hygiene area. The interaction of electromagnetic radiation with matter is the basis for such analytical techniques. Principles of operation extend from the infrared radiation spectra, to the ultraviolet and, in fact, to the X-ray region. The latter can be used to provide information on elementary composition (fluorescence) and crystal structure (diffraction).

In most cases the sample, whether gas, liquid, or solid, is exposed to radiation of known characteristics and the specific wavelengths (fluorescence), and the fractions transmitted or scattered are determined and quantitated. Color production, turbidity, and fluorescence are examples of properties determined by electromagnetic radiations that are widely used for quantitating industrial hygiene air samples.

6.6 Spectrographic Techniques

Since the smallest trace of materials can be detected by the spectrograph, spectrographic procedures may be employed for small amounts of metallic ions and elements when other procedures cannot be utilized. The chief limitations are the high cost and the need for a highly trained technician, having access to a rather complete spectrographic laboratory, to do quantitative work. Generally, the degree of sensitivity afforded by these units is not required by the industrial hygienist, although it frequently is desirable to obtain a complete elemental analysis of a sample of unknown composition as a starting point for an elaborate analytical program.

In applying emission spectroscopy, a solid sample is vaporized in a carbon arc, causing the formation of characteristic radiation, which is dispersed by a grating or a prism, and the resulting spectrum is photographed. Each metallic or metallike element can be identified from the spectra that are formed. Elemental analyses of body tissues, dust, ash, and air samples can be qualitatively analyzed by this technique.

With mass spectroscopy, gases, liquids, or solids are ionized by passage through an electron beam. The ions thus formed are projected through the analyzer by means of an electromagnetic or electrostatic field, or simply by the time necessary for the ions to travel from the gun to the collector. Each compound has a characteristic ionization pattern that can be used to identify the substance. This analytical tool, in conjunction with gas chromatography, has become a powerful technique for separating and identifying a wide range of trace contaminants in industrial hygiene and ambient air samples.

6.7 Chromatographic Techniques

The development of chromatographic methods of analysis has given the industrial hygienist an extremely versatile means of quantitating low concentrations of airborne contaminants, particularly organic compounds. Gas chromatography utilizes the selective absorption and elution provided by appropriately chosen packings for the columns to separate mixtures of substances in an air sample or in a desorption solution. The various compounds in the air sample have different affinities for the material in the column, thus "slowing" some of the constituents more than others, with the result that as the individual compounds reach the detector associated with the chromatograph, they can be quantitated by running standards of known concentration of the various substances along with the unknowns. This separation is achieved without any change of the entities; thus as mentioned earlier, the chromatograph can serve as an analytical technique in its own right by attaching an appropriate detector. Thermal conductivity, flame ionization, and electron capture detectors are commonly used for this purpose. The chromatograph can also serve to "purify" a sample by separating the constituents and selecting a narrow portion of the eluted sample. This portion then can be subjected to other more sophisticated types of analysis, such as mass spectroscopy.

6.8 Atomic Absorption Spectrophotometry

In absorption flame photometry (atomic absorption), monochromatic radiation from a discharge lamp containing the vapor of a specific element, such as cadmium, passes through a flame into which the sample is aspirated. The absorption of the monochromatic radiation is measured by a double-beam method, and the concentration is determined.

This technique permits rapid determination of almost all metallic elements. Solutions of the metals are aspirated into the high temperature flame, where they are reduced to free atoms. The absorption generally obeys Beer's law in the parts per million range, where quantitative determinations can be made. The characteristic absorption gives this technique high selectivity. Most interferences can be overcome by proper pretreatment of the samples. Atomic absorption methods have found substantial use in industrial hygiene in the determination of both major and trace metals in industrial hygiene samples, as well as in blood, urine, and other body fluids and tissues.

6.9 Other Techniques

Literally dozens of additional analytical methods are available to the industrial hygienist and analytical chemist for application to specific qualitative and quantitative needs. For detailed information on particular methods or procedures, the reader is advised to consult the references listed at the end of this chapter and, more important, to keep abreast of current developments in the industrial hygiene analytical field by subscribing to journals or routinely reviewing the wealth of new information constantly coming forth. Of particular interest in this respect are the criteria documents for specific contaminants being published by NIOSH, and the sampling and analytical methods that are being validated as part of the Standards Completion Program undertaken recently by OSHA and NIOSH.

Table 17.6 Sampling-Analytical Methods

Substance	Sampling Medium					Analytical Method				
	charcoal / silica gel	direct-reading unit	filter	solvent / reagent	std. acid / base	atomic absorption	gas chromatography	gravimetric/colorim.	ion-selective electrode	X-ray diffraction
Acid mists					X				X	
Alcohols	X						X			
Ammonia					X				X	
Asbestos			X							X
Carbon monoxide		X								
Coal tar pitch volatiles			X					X		
Hydrocarbon	X						X			
Hydrogen sulfide				X				X		
Metals			X			X				
Oil mists			X					X		
Organic vapors	X						X			
Pesticides			X	X			X			
Phenol					X		X			
Silica			X							X
Solvents	X						X			

As a general guide to the industrial hygienist, a simple summary of sampling and analytical techniques appropriate for a variety of contaminants commonly encountered in industry is presented in Table 17.6.

7 BIOLOGICAL MONITORING

In the final analysis, the degree of success associated with attempts to provide a safe work environment must be determined by assessing the amount of the contaminant of concern that has been actually absorbed by the workers. Regardless of the degree of sophistication applied to environmental and personnel monitoring, the extent to which workers have absorbed the contaminant must be determined by some clinical measurement on the individual. In the industrial hygiene profession this can be derived from direct quantitative analysis of body fluids, tissue, or expired air for the presence of the substance or metabolite. An indirect determination of the effect of the substance on the body can be made by measurements on the functioning of the target organ or tissue. With the possible exception of carcinogenic substances, even the most hazardous materials have some "no effect" level below which exposure can be tolerated by most workers for a working lifetime without incurring any significant physiological injury.

An ideal approach to establishment of no-effect or tolerance levels would require a classical chemical engineering materials balance applied to man and his environment. Simply stated, the amount of any substance entering the body must equal the products and by-products leaving the system, plus any accumulation within the body. A material balance such as this was established for lead by and through the efforts of Dr. Robert Kehoe and his co-workers (22). The study involved quite elaborate test facilities for analyzing food, beverages, air intake, and urinary, fecal, and expired breath outputs, as well as a closely controlled environment in which volunteer subjects were willing to put in 40-hour weeks under conditions that closely simulated actual work environments. The studies indicated that lead did not undergo metabolism to other forms within the body. This made the determination of the overall material balance relatively straightforward. Unfortunately, such is not the case with most occupational contaminants of interest today, particularly the organic materials. Material balances are much more difficult to establish for compounds that undergo metabolic change to other chemical structures within, and during passage through, the human body.

The following general presentation of potential applications for biological monitoring should be a useful supplement to an industrial hygiene monitoring program.

7.1 Urine Analysis

Perhaps the most common biological fluid analyzed in attempts to determine the extent to which individual workers have been exposed to contaminants is urine. The following procedure for collection and analysis of urine samples for phenol as an indication of exposure to benzene is illustrative of procedure, since it encompasses many of the considerations that must be made in processing such samples (23). Deviations or alternatives available for urine samples for other contaminants are indicated in parentheses.

For urinary phenol samples, "spot" urine specimens of about 10 ml are collected as close to the end of the working day as possible. (For other contaminants, it is preferable to collect the specimen at the beginning of the work shift or perhaps composite a 24-hour sampling.) If any worker's urine phenol level exceeds 75 mg/liter, procedures should be instituted immediately to determine the cause of the elevated levels and to reduce the exposure of the worker to benzene. Weekly specimens are collected as described until three consecutive determinations indicate that the urinary phenol levels are below 75 mg/liter.

To collect the sample, the workers should be instructed to thoroughly wash their hands with soap and water and to provide a sample from a single voiding into a clean, dry specimen container having a tight closure and at least 120 ml capacity. The containers may be of glass, wax-coated paper, or other disposable materials if desired. After collection of the specimens, a milliliter of a 10 percent copper sulfate solution should be added to each sample as a preservative. (Other contaminants will require the addition of specific preservatives to maintain the stability of the sample.) The samples should be stored immediately under refrigeration, preferable below 4°C. Such refrigerated specimens will remain stable for approximately 90 days. (The stability of specimens for other metabolites or contaminants will vary; this factor should be ascertained before establishing a biological monitoring program.) If shipment of samples is necessary, the most rapid method available should be employed, using acceptable packing procedures specified by the carrier. Each specimen must include proper identification, including the worker's name, the date, and the time of collection.

The urine samples for phenol are analyzed by treating the specimen with perchloric acid at 95°C to hydrolize the phenol conjugates, phenyl sulfate, and phenyl glucuronide formed as detoxification products following absorption of benzene. The total phenol is extracted from the urine with diisopropyl ether, and the phenol concentration is determined by gas chromatographic analysis of the ether extract.

Although the specific requirements of methods for other compounds may differ significantly from those just outlined, it should be emphasized that the specific concentrations indicating excessive buildup of contaminants or metabolites within the body is not known for most compounds with any certainty. The biological threshold limit values, that is, concentrations indicative of excessive exposure for some compounds, have been determined (22). Table 17.7 lists representative organic and metallic compounds for which biological threshold limit values in urine have been determined. The substances in this table are analyzed directly in the urine; that is, metabolic products are not evaluated, as was the case for exposure to benzene and consequent analysis for phenol in the urine. The first concentration presented in the table is intended to serve as a warning that exposures to the indicated compounds are approaching the limits for acceptable absorption of the substance into the body; the second concentration is a level at which medical intervention and/or treatment of the individual is advised (22).

7.2 Blood Analysis

Collection of samples of workers' blood and subsequent analysis for specific indications of exposure, either for the contaminant in question directly or for key metabolites, is another useful biological monitoring technique. The following procedure for collecting and analyzing blood samples for lead is presented as an illustration of the technique for this method (24).

A 10 ml sample of whole blood is collected using a vacutainer and a sterilized, stainless steel needle. (It is important that the vacutainers have been determined to be free of the contaminant of interest, since any leaching of the material from the vacutainers into the blood samples would distort the analytical result. The possibility of such distortion has been of particular concern in the analysis of blood samples for lead.) In the laboratory, the sample is transfered to a tared, leadfree beaker; no aliquoting of the sample is

Table 17.7. Biological Threshold Limit Values for Urine

SUBSTANCE	BLV, mg/l	
	Warning	Intervention
Aniline	10	20
Arsenic	0.3	0.6
Cadmium	0.05	0.1
Lead	0.10	0.15
Mercury	0.05	0.1
Nitrobenzene	25	50

permissible, since most of the lead is present in the clotted portion. The weight of the blood sample is determined to the nearest 0.01 gram, weighing rapidly to minimize evaporation of the sample. A system for ashing the sample is employed to permit the analyst to handle the sample easily, since the blood clot will break up readily and smoothly. The sample then is evaporated just to dryness and, after cooling, additional acid is added and ashing continued, until the residue fails to darken upon additional heating. The residue then is put into solution and analyzed directly by atomic absorption spectrophotometry or another suitable method.

Another useful application of blood analysis as an index of exposure to contaminants is the determination of carboxyhemoglobin in the blood as an indication of exposure to carbon monoxide. Extensive studies of the concentration of carbon monoxide in the air and consequent level of carboxyhemoglobin in the blood have been made (25). Figure 17.13 presents a family of curves giving the length of time of exposure to various concentrations of carbon monoxide in air necessary to achieve 5 percent carboxyhemoglobin levels in the blood as a function of the relative activity of the workers (i.e. sedentary, light work, or heavy work). Table 17.8 summarizes typical biological threshold limits for compounds, or their metabolic indicators, in blood.

7.3 Expired Breath Analysis

A biological monitoring technique that is increasing in application is analysis of expired air for contaminants for which equilibrium between the body and respired air can be used as an indication of the concentration of the contaminant in the workroom air to which the individual has been exposed. For the most part, this type of analysis has been limited to the chlorinated hydrocarbon solvents such as methylene chloride, carbon

Table 17.8 Blood Biological Limit Values (BLV)

SUBSTANCE	BLV
Carboxyhemoglobin, as CO index	~10 %
Ethanol	0.08 %
Lead	0.08 mg/100 g blood
Methemoglobin, as nitro-amino index	~7 %
Methyl bromide	10 mg/100 g blood
Organophosphates	15 % inhibition of cholinesterase activity

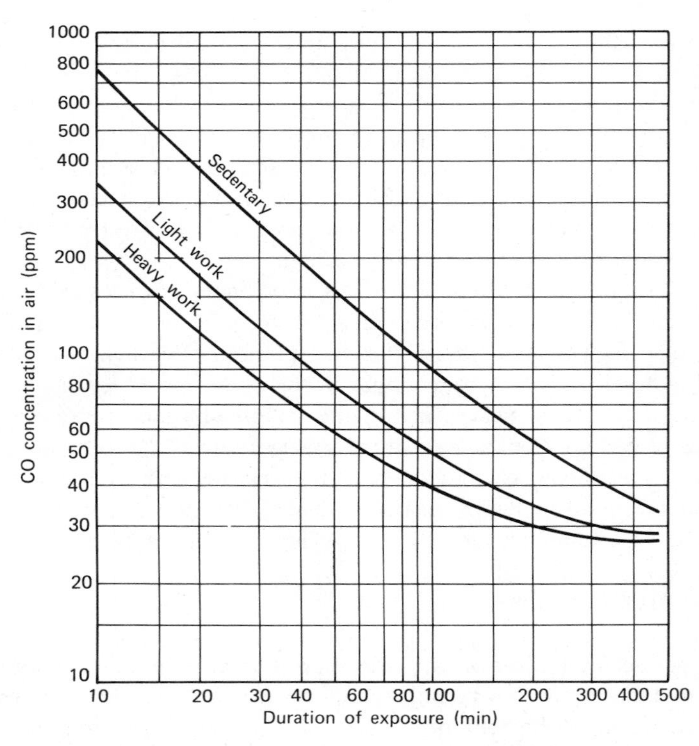

Figure 17.13 Length of time to achieve 5 percent carboxyhemoglobin at various concentrations of carbon monoxide (25).

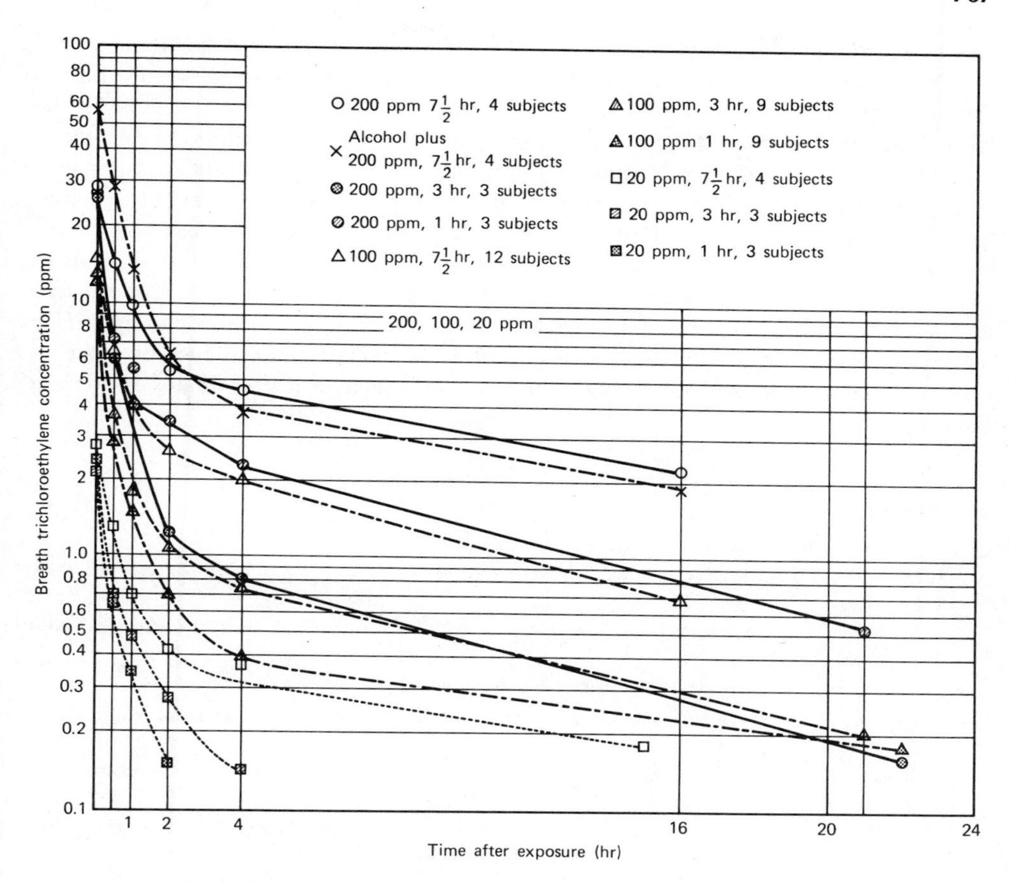

Figure 17.14 Trichloroethylene postexposure breath decay curves.

tetrachloride, vinyl chloride, 1,1,1-trichloroethane, trichloroethylene, tetrachloroethylene, and some of the Freons (22).

In collecting the sample, the worker is instructed to take several deep breaths and direct the expired air into a flexible bag or to pass it through a glass tube after flushing the contents several times with expired air. In either case, the container is tightly sealed and the contents subsequently analyzed directly for the substance, usually using gas chromatographic techniques. The analytical result is compared to "breath decay curves" (Figure 17.14) to determine the concentration of the contaminant to which the worker had been exposed recently (25).

7.4 Future Considerations

It is likely that with the increasing interest in providing safe working conditions for persons exposed to a tremendously wide range of contaminants, development of cause and effect relationships between the level of exposure to a particular contaminant and the amount of the substance remaining in the body will become a more integral part of the total occupational health monitoring procedure. Accordingly, greater emphasis will be placed on the medical aspects of the total monitoring effort in attempts to document, as an end result, that workers have not been exposed to excessive levels of contaminants. Where the results of biological monitoring indicate excessive exposures, and the environmental monitoring aspects indicate that concentrations were within acceptable limits, analysis of individuals' work practices may be required, or at least specific analysis of an individual's work activity, in attempts to ascertain the cause of the elevated readings in a given individual. As such, the biological monitoring program is a viable and extremely useful supplement to the ongoing environmental and medical surveillance program in the industrial settings and for the contaminants for which such coordinated efforts are possible.

REFERENCES

1. M. B. Jacobs, *The Analytical Toxicology of Industrial Inorganic Poisons,* Wiley Interscience, New York, 1967.
2. F. A. Patty, Ed. *Industrial Hygiene and Toxicology,* Vol. 2, 2nd rev. ed., Wiley, Interscience, New York, 1962.
3. H. B. Elkins, *The Chemistry of Industrial Toxicology,* 2nd ed., Wiley, New York 1959.
4. A. Hamilton, and H. Hardy, *Industrial Toxicology,* 3rd ed., Publishing Sciences Group, Acton, Mass., 1974.
5. National Institute for Occupational Safety and Health, *The Industrial Environment—Its Evaluation and Control,* U.S. Department of Health, Education and Welfare, Cincinnati, Ohio, 1973.
6. American Conference of Governmental Industrial Hygienists, *Documentation of the Threshold Limit Values,* 3rd ed., ACGIH, Cincinnati, Ohio, 1971.
7. N. A. Leidel and K. A. Busch, *Statistical Methods for the Determination of Noncompliance with Occupational Health Standards,* U.S. Department of Health, Education and Welfare, Publication No. 75-159, Cincinnati, Ohio, 1975.
8. Y. Bar-Shalom, A. Segall, D. Budenaers, and R. B. Shainker, *Statistical Theory for Sampling of Time-Varying Industrial Atmospheric Contaminant Levels,* Systems Control, Inc., Report to NIOSH Contract HSM-99-73-78, Palo Alto, Calif., 1974.
9. D. Budenaers, Y. Bar-Shalom, A. Segall, and R. B. Shainker, *Handbook for Decisions on Industrial Atmospheric Contaminant Exposure Levels,* U.S. Department of Health, Education and Welfare, Cincinnati, Ohio, 1975.
10. L. D. Pagnotto and R. G. Keenan, "Sampling and Analysis of Gases and Vapors," in: *The Industrial Environment—Its Evaluation and Control,* U.S. Department of Health, Education and Welfare, Cincinnati, Ohio, 1973.
11. F. J. Schuette, "Plastic Bags for Collection of Gas Samples," *Atmos. Environ.* **1,** 515 (1967).
12. G. O. Nelson, *Controlled Test Atmospheres, Principles and Techniques,* Ann Arbor Sciences Publishers, Ann Arbor, Mich. 1971.
13. American Conference of Governmental Industrial Hygienists, *Air Sampling Instruments for Evaluation of Atmospheric Contaminants,* 4th ed., ACGIH, Cincinnati, Ohio, 1972.

14. C. D. Yaffe, D. H. Byers, and A. D. Hosey, *Encyclopedia of Instrumentation for Industrial Hygiene,* University of Michigan Press, Ann Arbor, 1956.

15. B. Witten and A. Prostak, "Sensitive Detector Crayons for Phosgene, Hydrogen Cyanide, Cyanogen Chloride, and Lewisite," *Anal. Chem.,* **29,** 885–887 (1957).

16. American Industrial Hygiene Association, *Direct Reading Colorimetric Indicator Tubes Manual,* AIHA, Akron, Ohio, 1976.

17. B. E., Saltzman, Direct Reading Colorimetric Indicators, Section S, in: *Air Sampling Instruments for Evaluation of Atmospheric Contaminants,* 4th ed., American Conference of Governmental Industrial Hygienists, Cincinnati, Ohio 1972.

18. *Fed. Reg.,* **38,** 11458 (May 8, 1973), incorporated into *Code of Federal Regulations* as Title 42, CFR Part 84.

19. D. L. Braun, "Personal Monitoring of Mercury Vapor Exposure," paper presented at the International Conference on Environmental Sensing and Assessment, Las Vegas, 1975. (to be published.)

20. G. H. Schnakenberg, "A Passive Personal Sampler for Nitrogen Dioxide," U.S. Department of the Interior, Bureau of Mines, Technical Progress Report, Pittsburgh, 1976.

21. National Institute for Occupational Safety and Health, *Criteria for a Recommended Standard . . . Cotton Dust,* U.S. Department of Health, Education and Welfare, Cincinnati, Ohio, 1973.

22. A. Linch, *Biological Monitoring for Industrial Chemical Exposure Control,* CRC Press, Cleveland, 1974.

23. National Institute for Occupational Safety and Health, *Criteria for a Recommended Standard . . . Benzene,* U.S. Department of Health, Education and Welfare, Cincinnati, Ohio, 1974.

24. National Institute for Occupational Safety and Health, *Criteria for a Recommended Standard . . . Inorganic Lead,* U.S. Department of Health, Education and Welfare, Cincinnati, Ohio, 1972.

25. National Institute for Occupational Safety and Health, *Criteria for a Recommended Standard . . . Carbon Monoxide,* U.S. Department of Health, Education and Welfare, Cincinnati, Ohio, 1972.

26. R. D. Stewart et al., *Biological Standards for the Industrial Worker by Breath Analysis: Trichloroethylene,* U.S. Department of Health, Education and Welfare, Cincinnati, Ohio, 1974.

CHAPTER EIGHTEEN

Industrial Hygiene Engineering Controls

ROBERT D. SOULE

1 INTRODUCTION

Industrial hygiene is the science of preventing occupational disease through proper control of the work environment. Although diseases suspected of being related to the work environment were recognized more than 2500 years ago (e.g., lead poisoning of workers engaged in mining operations), the systematic application of industrial hygiene engineering controls is a recent technical development. Certainly the profession of industrial hygiene is less than 50 years old, coming into prominence only since the 1930s.

Evidence of the importance of engineering control of the work environment among the various alternative solutions to industrial hygiene problems is found in every current industrial hygiene text: all list the possible solutions in priority fashion as engineering controls, administrative controls, and as a last resort, use of personal protective equipment. In fact, recently proposed rule making by the Occupational Safety and Health Administration makes it clear that engineering control of specific substances must be attempted even when complete control is not feasible by engineering means alone. In other words, the use of personal protective equipment alone cannot be justified simply by analyzing a problem and determining that complete engineering control cannot be provided. Instead, the language of new federal regulations clearly mandates that engineering controls be provided *to the extent feasible* and, in the event these are not sufficient to achieve acceptable limits of exposure, use of personal protection or other corrective measures may be considered.

Thus it is clear that the importance of applying engineering controls, as a component of the total responsibility of the industrial hygienist, will continue to increase tremendously in the foreseeable future. Therefore all professionals in the field of industrial hygiene must be cognizant of the principles of engineering controls, and the engineering professional must be familiar with the details of implementing the control concepts. Accordingly, this chapter presents material useful to the industrial hygiene engineer and nonengineer alike in applying their resources to the industrial hygiene problems occurring in their respective settings.

2 PRINCIPLES OF CONTROL

The industrial hygiene engineering control principles are deceptively few: substitution, isolation, and ventilation, both general and localized. In a technological sense, an appropriate combination of these principles can be brought to bear on any industrial hygiene control problem to achieve satisfactory quality of the work environment. It may not be, and usually is not, necessary or appropriate to apply all these principles to any specific potential hazard. A thorough analysis of the control problem must be made, to ensure that a proper choice from among these methods will produce the proper control in a manner that is most compatible with the technical process, is acceptable to the workers in terms of day-to-day operation, and can be accomplished with optimal balance of installation and operating expenses.

It must be kept in mind that with advancing technology and the tendency for acceptable limits of exposure to become increasingly narrow, the industrial hygiene engineer's function must include an ongoing analysis of installed control measures. Aside from problems associated with operation and maintenance of these provisions, the concept of the control provisions itself might become outdated by changes in process or regulations. The engineer then must be able to develop effective control methods, and he must have the capability to continue to evaluate the effectiveness of these methods regularly.

2.1 Substitution

Substitution, although frequently one of the most simple engineering principles to apply, often is overlooked as an appropriate solution to an industrial hygiene problem. There is a tendency to analyze a particular problem from the standpoint of correcting rather than eliminating it. For example, the first inclination in considering a vapor exposure problem in a degreasing operation is to provide ventilation of the operation rather than consider substituting a solvent having a much lower degree of hazard associated with its use. However substitution of less hazardous substances, changing from one type of process equipment to another, or in some cases even changing the process itself, may provide an effective control of a hazard at minimal expense.

Examples of substituting materials of lower toxicity are many, and some of them are classics in the history of industrial hygiene. The case most commonly referred to is the substitution of red for white phosphorus in the manufacture of matches. Although this substitution was done primarily because of a tax law, not in direct response to the toxicity problem, the result was a markedly reduced potential hazard. A series of substitutions can be cited in the case of degreasing solvents from petroleum naphtha to carbon tetrachloride to chlorinated hydrocarbons to fluorinated hydrocarbons. It should be emphasized that each of these substitutions alleviated one problem but resulted in a new potential hazard. This underscores a basic problem associated with using substitution as a control method in that one hazard can be replaced by another inadvertently.

Often it is not possible to substitute the process materials, although different equipment can be used or substituted, or at least modifications can be made. To be sure, the substitution or modification of process equipment is usually the result of an obvious potential problem. This concept is often used in conjunction with safety equipment: substituting safety glass for regular glass in some enclosures, replacing unguarded equipment with properly guarded machines, replacing safety gloves or aprons with garments made of materials more impervious to the chemicals being handled. Since substitution of equipment frequently is done as an immediate response to an obvious problem, it may not be recognized as an engineering control, even though the end result is every bit as effective.

Substituting one process for another may not be considered except in major modifications of a process. In general, a change in any process from a batch to a continuous type of operation carries with it an inherent reduction in potential hazard. This is true primarily because the frequency and duration of potential contact of workers with the process materials is reduced when the overall process approach becomes one of continuous operation. The substitution of processes can be applied on a fundamental basis. For example, substitution of airless spray for conventional spray equipment can reduce the exposure of a painter to solvent vapors. Substitution of a paint dipping operation for the paint spray operation can reduce the potential hazard even further. In any of these cases the automation of the process can further reduce the potential hazard.

2.2 Isolation

Application of the principle of isolation is frequently envisioned as consisting of installation of a physical barrier between a hazardous operation and the workers. Fundamentally, however, this isolation can be provided without a physical barrier by appropriate use of distance and in some situations time.

Perhaps the most common example of isolation as a control measure is associated with storage and use of flammable solvents. The large tank farms with dikes around the tanks, underground storage of some solvents, the detached solvent sheds, and fireproof solvent storage rooms within buildings, all are commonplace in American industry. Although the primary reason for the isolation of solvents is the risk of fire and explosion, the principle is no less valid as an industrial hygiene measure.

Frequently the application of the principle of isolation maximizes the benefits of additional engineering concepts such as local exhaust ventilation. For example, the charging of mixers is the most significant operation in many processes that use formulated ingredients. When one of the ingredients in the formulation is of relatively high toxicity, it is worthwhile to isolate the mixing operation, that is, install a mixing room, thereby confining the airborne contaminants potentially generated by the operation to a small area rather than having them influence the larger portion of the plant. In ensuring isolation thus, the application of ventilation principles to control of this contaminant at the source (i.e., the mixer) is much more effective.

2.3 Ventilation

Ventilation can be defined as a method for providing control of an environment by strategic use of airflow. The flow of air may be used to provide either heating or cooling of a workspace, to remove a contaminant near its source of release into the environment, to dilute the concentration of a contaminant to acceptable levels, or to replace air exhausted from an enclosure. Ventilation is by far the most important engineering control principle available to the industrial hygienist. Applied either as general or local control, this principle has industrial significance in at least three applications: the control of heat and humidity primarily for comfort reasons, the prevention of fire and explosions, and most important to the industrial hygienist, the maintenance of concentrations of airborne contaminants at acceptable levels in the workplace.

Detailed discussions of the principles and application of general and local exhaust ventilation appear in subsequent sections of this chapter. Application of these principles to industrial problems, whether general or localized, requires a basic understanding of the fundamentals of airflow. The scientific laws that define the complete motion of air or any fluid are complex and except in the relatively simple case of laminar flow, we know relatively little about them. Nevertheless there are fundamental relationships that must be understood and conscientiously applied by the industrial hygiene engineer. These are described briefly below; more elaborate discussions of these principles are given in several of the general reference texts identified at the end of this chapter.

2.3.1 Conservation of Mass

Perhaps the most basic principle of airflow is the continuity equation or principle of conservation of mass. This relation states that the mass rate of flow remains constant along the path taken by a fluid. For any two points in the fluid stream therefore, we can write

$$A_1 v_1 \delta_1 = A_2 v_2 \delta_2$$

where A = cross-sectional area (ft^2 or m^2)
v = velocity (ft/min or m/sec)
δ = specific weight (lb/ft^3 or g/m^3)

In most applications in industrial ventilation, δ is relatively constant because the absolute pressure within a ventilation system usually varies over a very narrow range and the air remains relatively incompressible. Therefore,

$$A_1 v_1 = A_2 v_2 \qquad \text{and} \qquad Q_1 = Q_2$$

where $Q = Av$, the volumetric rate of airflow (ft^3/min or m^3/sec).

2.3.2 Conservation of Energy

The basic energy equation of a frictionless, incompressible fluid for steady flow along a single streamline is given by Bernoulli's theorem:

$$H + \frac{P}{\delta} + \frac{v^2}{2g} = C$$

where H = the evaluation above any arbitrary datum plane (ft or m)
P = absolute pressure (lb/ft^2 or g/m^2)
δ = specific weight (lb/ft^3 or g/m^3)
v = velocity (ft/sec or m/sec)
g = gravitational acceleration (ft/sec^2 or m/sec^2)
C = a constant, different for each steamline

Each of the three variable terms in Bernoulli's equation has the unit "foot-pounds per pound" (gram-meters per gram) of fluid, or feet of fluid, frequently referred to as elevation head, pressure head, and velocity head, respectively.

The elevation term H usually is omitted when Bernoulli's equation is applied to industrial exhaust systems, since only relatively small changes in elevation are involved. Since all streamlines originate from the atmosphere, a reservoir of nearly constant energy, the constant C, is the same for all streamlines; and the restriction of the equation to a single streamline can be removed. Furthermore, since the pressure changes in nearly all exhaust systems are at most only a few percent of the absolute pressure, the assumption of incompressibility may be made with negligible error.

2.3.3 Velocity Pressure

Air in motion exerts a pressure called velocity pressure. Velocity pressure maintains air velocity and is analogous to kinetic energy. It exists only when air is in motion, it acts in the direction of air flow, and it is always positive in sign. In Bernoulli's equation, the term $v^2/2g$ represents the velocity head. The relationship between the velocity of air and velocity pressure is

$$v = \sqrt{2gh}$$

where v = velocity (ft/sec or m/sec)
g = gravitational acceleration (ft/sec^2 or m/sec^2)
h = head of air (ft or m)

2.3.4 Static Pressure

Static pressure produces initial air velocity, overcomes the resistance in a system caused by friction of the air against the duct walls, and overcomes turbulence and shock caused by a change in direction or velocity of air movement. Static pressure is analogous to potential energy, and it exists even where there is no air motion. It acts equally in all directions and tends either to collapse the walls of the duct upstream from the fan or to expand the walls of the duct on the downstream side. Static pressure is usually negative in sign upstream from a fan and positive downstream. It is measured as the difference between duct pressure and atmospheric pressure. The most common unit of static pressure is "inches of water" (in. H_2O).

2.3.5 Total Pressure

The driving force for the airflow is a pressure difference that is required to start and maintain flow. This pressure is called total pressure (TP) and has two components, velocity pressure (VP) and static pressure (SP); the three are interrelated as follows:

$$SP + VP = TP$$

Figure 18.1 shows the relationship among static, velocity, and total pressures at different points in a duct system.

If gas flowing through a duct system undergoes an increase in velocity, a part of the available static pressure is used to create the additional velocity pressure necessary to accelerate the flowing gas. Conversely, if the velocity is reduced at some point in a duct system, a portion of the kinetic energy or velocity pressure at that point is converted into potential energy or static pressure. Static pressure and velocity pressure are, therefore, mutually convertible. However this conversion is always accompanied by a net loss of total pressure due to turbulence and shock; that is, the conversion is never 100 percent efficient.

2.3.6 Energy Losses

Air in motion encounters resistance along any surface confining the flowing volume. Consequently, some of the energy of the air is given up in overcoming this friction and is

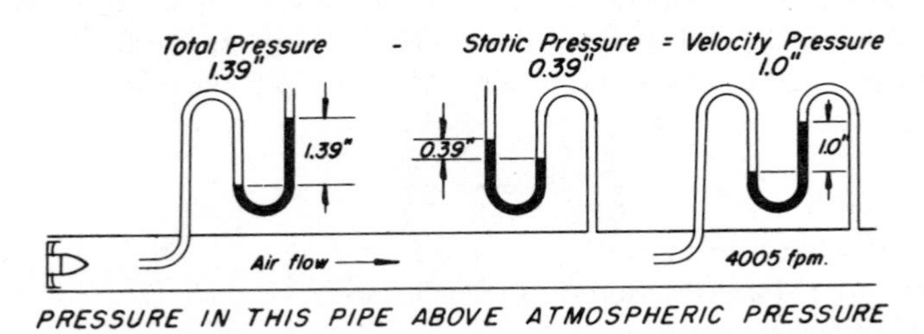

Figure 18.1 Relation between velocity pressure, static pressure, and total pressure.

transformed into heat. The rougher the surface confining the flow or the higher the flow rate, the higher the frictional losses will be.

Frictional loss in a duct varies directly as the length, inversely as the diameter, and directly as the square of the velocity of air flowing through the duct. This loss can be calculated from charts (1) using the Fanning friction factor, which is an empirical function of Reynolds number, duct material, and type of construction.

Another type of energy loss encountered in airflow results from turbulence caused by a change in direction or velocity within a duct. The pressure drop in a duct system due to dynamic losses increases with the number of elbows or angles and the number of velocity changes within the system. The resulting pressure drop from these energy losses is expressed in units of "equivalent length." For example, an elbow of 10-inch diameter and 20-inch centerline radius is said to have an equivalent length of 14, meaning that the loss through the elbow will be the same as the loss through 14 feet of straight pipe with the same diameter operating under the same conditions (2).

Another method of defining the losses due to turbulence and friction is to express the losses in terms of velocity pressure. For example, a loss of $0.28\,VP$ in a transition or elbow means that the incremental pressure drop is equal to 0.28 of the velocity pressure of the airstream at that point (3).

Acceleration and hood entrance loss is a drop in pressure caused by turbulence when air is accelerated from rest to enter a duct or opening. Turbulence losses of this type vary with the type of opening (4).

This entry loss plus the acceleration energy required to move the air at a given velocity (one VP) comprise the hood static pressure SP_h, which is expressed algebraically as

$$SP_h = h_e + VP$$

The SP_h can be measured directly at a short distance downstream from the hood entrance. The calculation of SP_h is the first step in the design or evaluation of a local exhaust system, discussed later.

The coefficient of entry C_e is a measure of how efficiently a hood entry is able to convert static pressure to velocity pressure. The coefficient of entry is the ratio of rate of flow by the hood static pressure to the theoretical flow if the hood static pressure were completely converted to velocity pressure.

The result of the friction and dynamic losses to air flowing through ductwork is a pressure drop in the system. Bernoulli's theorem can be restated in a simplified expression of conservation of energy as follows:

$$SP_1 + VP_1 = SP_2 + VP_2 + \text{energy losses}$$

Static pressure plus velocity pressure at a point upstream in the direction of airflow equals the static pressure plus velocity pressure at a point downstream in the direction of airflow plus friction and dynamic losses.

3 GENERAL VENTILATION

The term "general ventilation" normally is applied to the practice of supplying and exhausting large volumes of air throughout a workspace. It is used typically in industry to achieve comfortable work conditions (temperature and humidity control) or to dilute the concentrations of airborne contaminants to acceptable limits throughout the workspace. Properly used, general ventilation can be effective in removing large volumes of heated air or relatively low concentrations of low toxicity contaminants from several decentralized sources.

General ventilation can be provided by either natural or mechanical means; often the best overall result is obtained with a combination of mechanical and natural air supply and exhaust.

3.1 Natural Ventilation

Natural ventilation may be provided either by gravitational forces (being motivated primarily by thermal forces of convection) or by anemotive forces (created by differences in wind pressure). These two natural forces operate together in most cases, resulting in the natural displacement and infiltration of air through windows, doors, walls, floors, and other openings in an industrial building. Unfortunately, the wind currents and thermal convection profiles on which natural ventilation is dependent are erratic and frequently unpredictable. Thus it is perhaps a misnomer to refer to natural ventilation as a "control" method, since to employ this technique requires dependence on, rather than control of, natural forces.

On the other hand, there are applications for general ventilation. The pressure exerted on the upwind side and concurrent suction exerted on the downwind side of a building as a result of wind movement can be predicted for flat terrain fairly reliably. Thus the wind forces exerted on an isolated building in a relatively flat area permit the prediction of natural ventilation forces. In the more common complex industrial buildings, however, the effects of the presence of one building on the others normally cannot be calculated; thus use of wind pressure models in the development of general ventilation systems is not feasible.

The amount of natural ventilation provided to a building is dependent both on the wind pressure profiles and on the thermal effects occurring within the structure. The warmer inside air tends to rise and leave the building through any available openings in the upper structure; cooler air tends to infiltrate the building by the reverse process in the lower structure. These thermal effects, which are much more predictable than the external wind forces; can be useful in design of a general ventilation scheme (5).

Because of the lack of control inherent in systems of natural ventilation, the remainder of this chapter, dealing with the concepts of ventilation or quality control of the work environment, is presented in terms of mechanically controlled ventilation systems. To be sure, many of the principles associated with mechanical ventilation systems apply equally well to natural ventilation. However the remainder of this chapter focuses on concepts and design principles for which it is assumed there is, indeed, control.

3.2 Mechanically Controlled Ventilation

A modern industrial complex characterized by a low profile building structure (large floor space, low height) as well as the multistory buildings of masonry and glass construction, present ventilation problems that were not found among the older industrial plants. For the most part the industrial facilities constructed 30 or more years ago incorporated by design many features that permitted—in fact, expected—natural ventilation of the workspace. By contrast, the modern buildings generally defy the exertion of natural ventilation forces, and mechanical ventilation must be relied on almost completely. Mechanical ventilation exhausts contaminated air by mechanical means (exhaust fans), with the concomitant use of appropriate air supply to replace the exhausted air. The best method of achieving this in modern closed buildings is to supply air through a system of ductwork, distributing the air into the work areas in a manner that will provide optimum benefit to the worker for both comfort and control of contaminants.

3.3 Applications of General Ventilation

As referred to earlier, general ventilation is applied to achieve one of two end results: an environment that is comfortable to the worker, or one that is free of harmful concentrations of airborne contaminants.

3.3.1 Comfort Ventilation

Ventilation for comfort (principally heat relief) includes certain aspects of what is commonly considered to be air conditioning engineering, that is, the treating of air to control simultaneously temperature, humidity, cleanliness, and distribution to meet the requirements of the conditioned space. In the typical residential or office building these requirements are primarily associated with comfort for the occupants. In many industrial situations, however, comfort conditions are impractical if not impossible to maintain, and the chief function of ventilation for comfort control is to prevent acute discomfort and the accompanying adverse physiological effects.

General exhaust ventilation also may be used to remove heat and humidity if a source of cooler air is available. If it is possible to enclose the heat source, as in the case of ovens or furnaces, a gravity or forced-air stack may be all that is necessary to prevent excessive heat from entering the workroom.

Air Conditioning. Complete air conditioning improves the control of temperature, humidity, radiation, air movements, or drafts, as well as the cleanliness and purity of the air. Under ordinary circumstances of comfort and health the body temperature is kept approximately constant at its normal level of 98.6°F, through a balance between the internal production of heat (metabolism) and thermal equilibrium with the environment. The exchange of heat between man and the environment is discussed in detail in Chapter 20. According to the various equations that have been developed to express the thermal exchange between the human body and the environment, four external physical factors are of primary importance: air temperature, radiant temperature, air velocity, and the moisture content of the air. These four environmental factors are independently variable and may be combined in many different ways to produce the same thermal effect on man. Thus an increasing velocity can offset a rise in air temperature, extreme radiant heat can be compensated somewhat by lowering the air temperature, and a considerable elevation of air temperature can be tolerated if the relative humidity is low. Combinations of thermal conditions that induce the same sensations are called thermoequivalent conditions. The effective temperature charts of the American Society of Heating, Refrigeration and Air-Conditioning Engineers (ASHRAE) give the combinations of equivalent conditions for clothed sedentary individuals and for workers stripped to the waist. These charts are presented for illustrative purposes in Figures 18.2 to 18.4. Although many indices of heat stress and subjective response to this stress have been developed over the years, the effective temperature charts have been the primary tool of the air conditioning engineer. These scales were derived by experiments with test subjects exposed to a variety of thermal conditions by repeating trials with combinations of temperature, humidity, and air motion. Despite its limited subjective basis, subsequent laboratory and field experience with the effective temperature scale has indicated that it is useful in defining thermal comfort conditions and, to a great extent, it correlates with objective measurements of physiological response to heat.

Any other features of ventilation for purposes of comfort are similar to those associated with control of airborne contaminants and are discussed later in this chapter. For aspects of engineering control unique to the thermal problem, see Chapter 20.

3.3.2 Dilution Ventilation

The primary function of dilution ventilation is to maintain the concentration of airborne contaminants at acceptable levels either in terms of potential fire and explosion or from occupational health considerations. If enough clean air is mixed with contaminated air, the concentration of the contaminant can be diluted to any reasonable level. Of course, dilution ventilation for purposes of fire and explosion control should not be employed in areas occupied by workers, since the concentrations of concern from an occupational health (i.e., employee exposure) standpoint invariably are orders of magnitude below those of concern for explosive limits. Exposures to atmospheres controlled to concentrations below the lower explosive limit or even a fraction thereof could cause narcosis, severe illness, or even death. Therefore it is extremely important not to confuse dilution ventilation requirements for health hazard control with those for fire and explosion prevention.

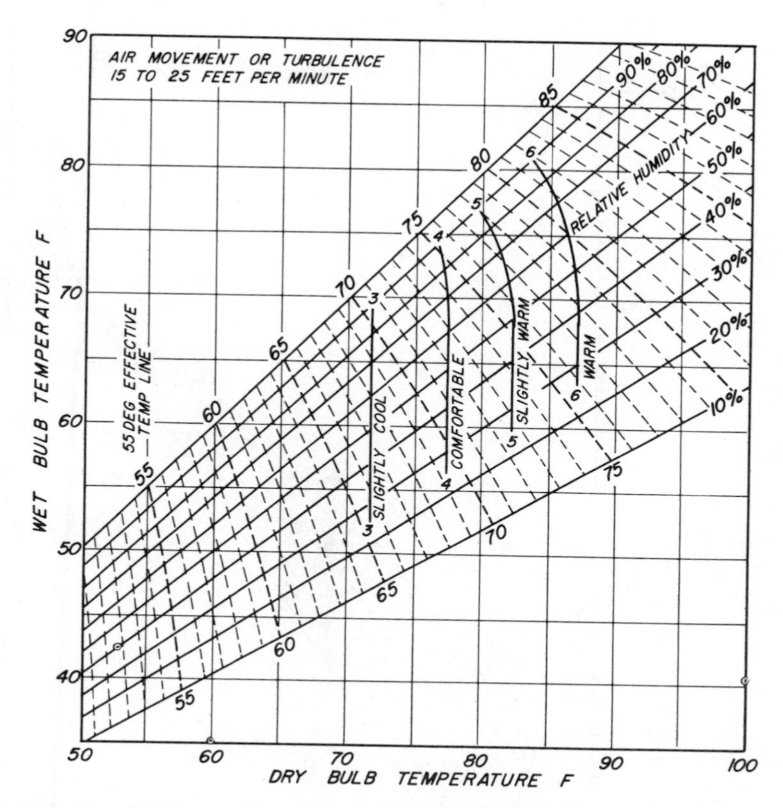

Figure 18.2 ASHRAE comfort chart for normally clothed people at rest in still air. Effective temperature (broken) lines indicate sensation of warmth immediately after entering conditioned space. Solid lines (3 to 6) indicate sensations experienced after 3-hour occupancy.

When considering whether dilution ventilation is appropriate for health hazard control, it should be remembered that dilution ventilation has four limiting factors:

1. The quantity of contaminant generated must not be excessive, or the air volume necessary for dilution would be impractical.
2. Workers must be far enough away from contaminant evolution, or the contaminant must be of sufficiently low concentrations, that the workers will not be exposed above acceptable limits.
3. The toxicity of the contaminant must be relatively low.
4. The evolution or generation of the contaminant must be reasonably uniform and consistent.

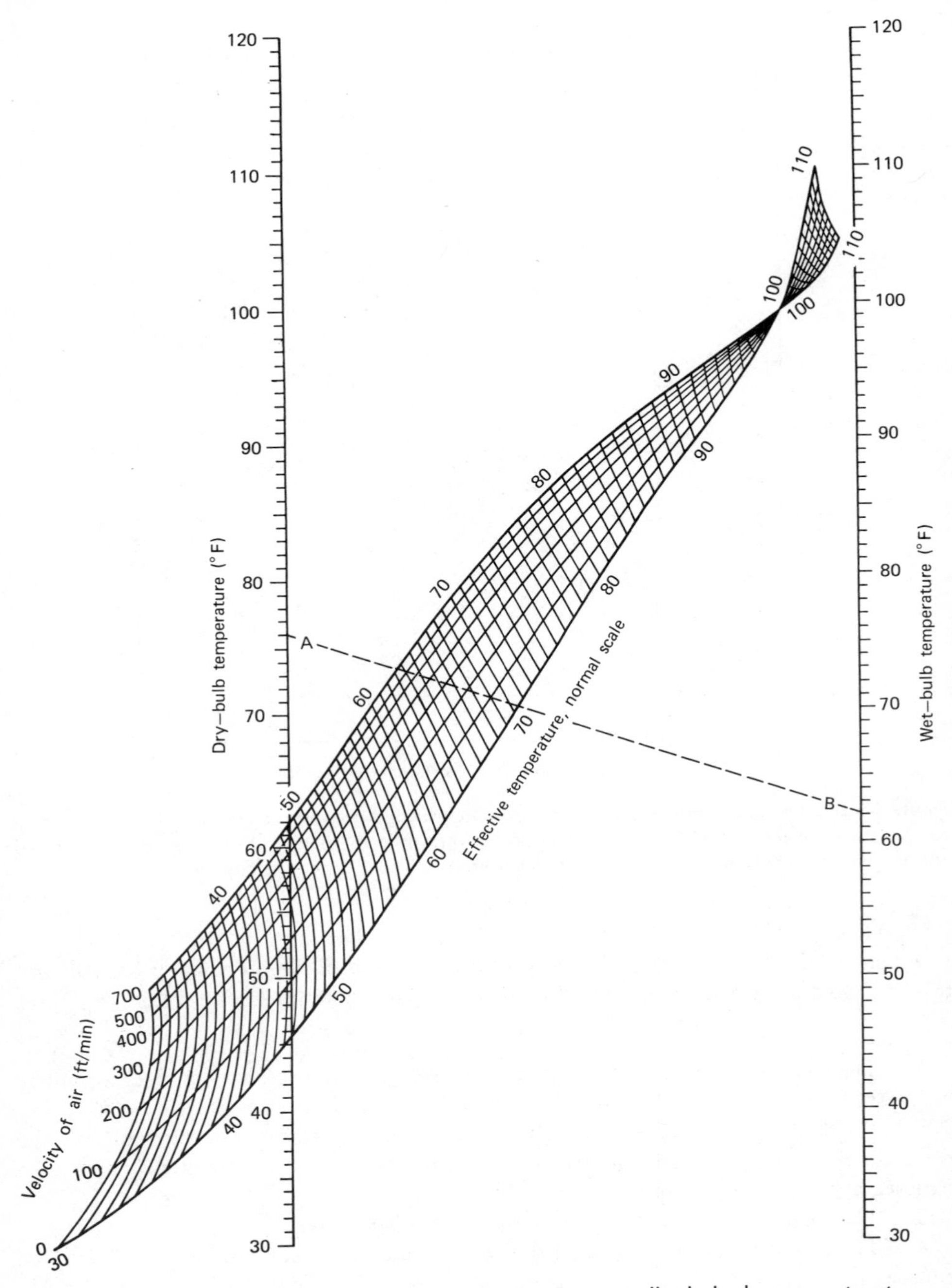

Figure 18.3 ASHRAE Effective Temperature Scale for normally clothed persons at rest.

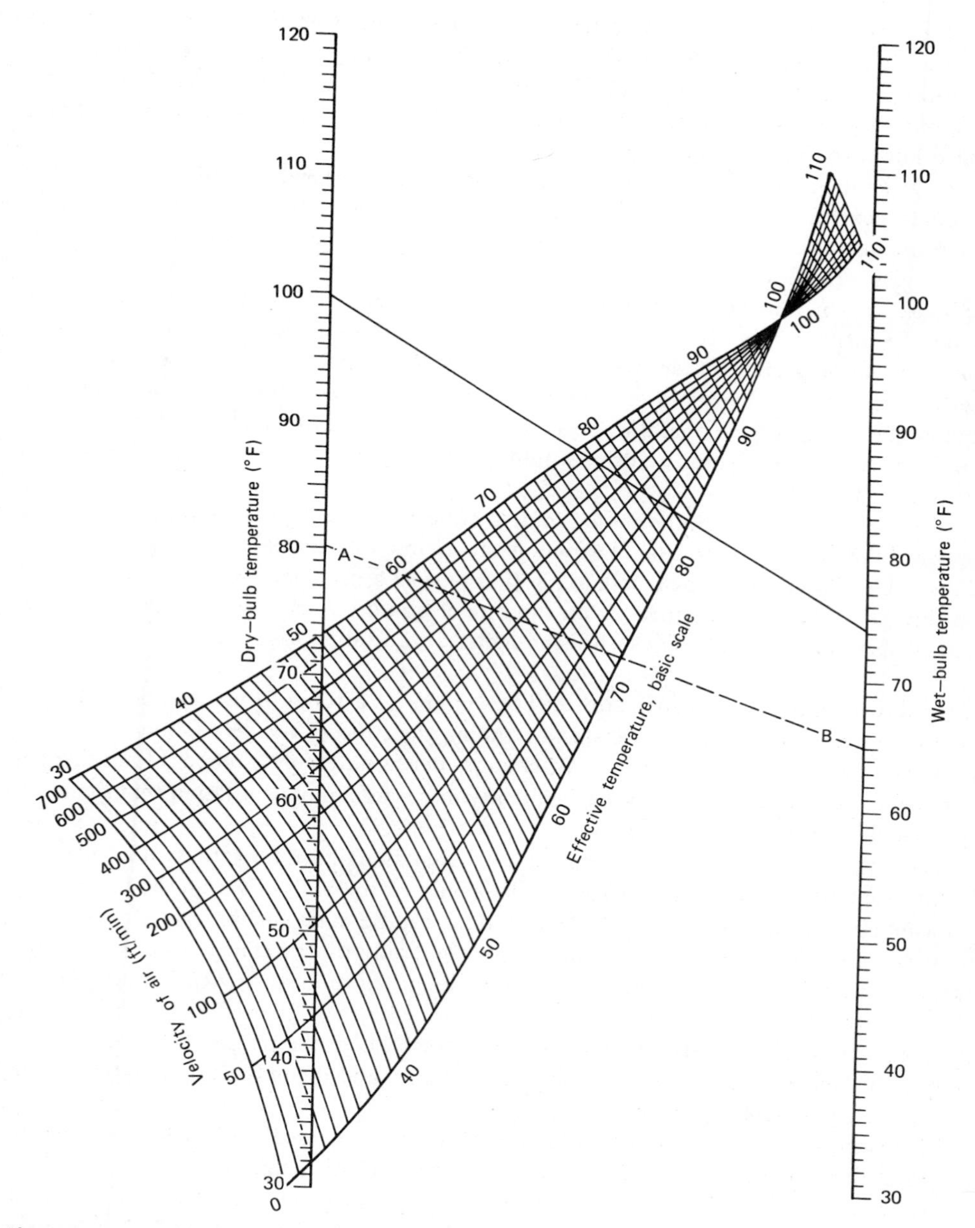

Figure 18.4 ASHRAE Effective Temperature Scale for men stripped to the waist; at rest or doing light physical work in rooms heated by convection methods.

A review of these factors indicates clearly that dilution ventilation is not normally appropriate for control of fumes and dust, since the high toxicity often encountered requires excessively large quantities of dilution air. Moreover the velocity and rate of evolution usually are very high, resulting in locally high concentrations, thus rendering the dilution ventilation concept inappropriate.

In general, dilution ventilation is not as satisfactory as local exhaust ventilation for primary control of health hazards. Occasionally, however, dilution ventilation must be used because the operation of the process prohibits local exhaust. Dilution ventilation sometimes provides an adequate amount of control more economically than a local exhaust system. However this condition is an exception rather than the rule, and it should be kept in mind that the economical considerations of the long-term use of dilution ventilation systems often overshadow the initial cost of the system because such a system invariably exhausts large volumes of heated air from a building. This workload can easily result in huge operating costs because of the need for conditioned makeup air, and the general ventilation scheme would be much more expensive over an extended period of time.

In practice, dilution ventilation for control of health hazards is used to best advantage in controlling the concentration of vapors from organic solvents of relatively low toxicity. To successfully apply the principle of dilution to such a problem, data must be available on the rate of vapor generation or on the rate of liquid evaporation. Usually such data can be obtained from the plant records on material consumption.

Calculating Dilution Ventilation. The volume of a room to be ventilated and the ventilation rate are frequently related by taking the ratio of the ventilation rate to the room volume to yield a "number of air changes per minute" or "number of air changes per hour." These terms, used quite frequently in discussion of ventilation requirements, are usually employed incorrectly.

Ventilation requirements based on room volume alone have no validity. Calculations of the required rate of air change can be made only on the basis of a material balance for the contaminant of interest. Similar calculations can be made for the rate of concentration increase or decrease; however they require not only the air change rate but also the rate of generation of contaminant. In the design of industrial ventilation, "x air changes" has valid application only very rarely. The term is useful when applied to meeting rooms, offices, schools, and similar spaces, where the purpose of ventilation is simply the control of odor, temperature, or humidity, and the only contamination of the air is a result of activity of people.

Dilution ventilation requirements should always be expressed in cubic feet per minute or some other absolute unit of airflow, not in "air changes per minute."

The concentration of a gas or vapor at any time can be expressed by a differential material balance which, when integrated, provides a rational basis for relating ventilation to the generation and removal rates of a contaminant. Let C be concentration of gas or vapor at time t, and take the following additional quantities:

G = rate of generation of contaminant
Q = rate of ventilation
K = design distribution constant, allowing for incomplete mixing
$Q' = Q/K$ = effective rate of ventilation, corrected for incomplete mixing
V = volume of room or enclosure

Starting with a fundamental material balance, and assuming no contaminant in the air supply, we write

$$\begin{bmatrix} \text{rate} \\ \text{of} \\ \text{accumulation} \end{bmatrix} = \begin{bmatrix} \text{rate} \\ \text{of} \\ \text{generation} \end{bmatrix} - \begin{bmatrix} \text{rate} \\ \text{of} \\ \text{removal} \end{bmatrix}$$

$$V\,dC = G\,dt - Q'C\,dt$$

This basic relationship can be applied in various ways.

Concentration Buildup. Rearranging the differential material balance, the buildup of contaminant can be expressed as follows

$$\int_{C_1}^{C_2} \frac{dC}{G - Q'C} = \frac{1}{V} \int_{t_1}^{t_2} dt$$

$$\ln\left(\frac{G - Q'C_2}{G - Q'C_1}\right) = -\frac{Q'}{V}(t_2 - t_1)$$

if $C_1 = 0$ at t_1,

$$\ln\left(\frac{G - Q'C}{G}\right) = -\frac{Q'}{V}\,t$$

or

$$\frac{G - Q'C}{G} = e^{-Q't/v}$$

Rate of Purging. When a volume of air is contaminated but further contamination or generation has ceased, the rate of decrease of concentration over a period of time is as follows:

$$V\,dC = -Q'C\,dt$$

$$\int_{C_1}^{C_1} \frac{dC}{C} = \frac{-Q'}{V} \int_{t_1}^{t_2} dt$$

$$\ln\frac{C_2}{C_1} = \frac{-Q'}{V}(t_2 - t_1)$$

Maintaining Acceptable Concentrations at Steady State. At steady state, $dC = 0$, we have

$$G\,dt = Q'C\,dt$$

$$\int_{t_1}^{t_2} G\,dt = \int_{t_1}^{t_2} Q'C\,dt$$

at a constant concentration C, and uniform generation rate G,

$$G\,(t_2 - t_1) = Q'C\,(t_2 - t_1)$$

$$Q' = \frac{G}{C}$$

$$Q = \frac{KG}{C}$$

Therefore the rate of flow of uncontaminated dilution air required to reduce the atmospheric concentration of a hazardous material to an acceptable level can be easily calculated, if the generation rate can be determined. Usually the acceptable concentration is considered to be the threshold limit value (TLV) or acceptable 8-hour time-weighted average (TWA) concentration. For liquid solvents the steady state dilution ventilation requirement can be conveniently expressed as

$$Q = \frac{(6.71)\,(10^6)\,(SG)\,(ER)\,(K)}{(MW)\,(TLV)}$$

where Q = actual ventilation rate (ft^3/min)
SG = specific gravity of volatile liquid
ER = evaporation rate of liquid (pints/hr)
MW = molecular weight of liquid
K = design safety factor for incomplete mixing
TLV = threshold limit value (ppm)

Specifying Dilution Ventilation. The foregoing discussion introduced the concept of a "design safety factor" K for calculating dilution ventilation requirements. The K factor is based on several considerations:

1. The efficiency of mixing and distribution of makeup air introduced into the room or space being ventilated.
2. The toxicity of the solvent. Although TLV and toxicity are not synonymous, the following guidelines have been suggested for choosing the appropriate K value:

Slightly toxic	TLV $>$ 500 ppm
Moderately toxic	TLV 100–500 ppm
Highly toxic	TLV $<$ 100 ppm

3. A judgment of any other circumstances the industrial hygienist determines to be of importance, based on experience and the individual problem. Included in these criteria are such considerations as (*a*) seasonal changes in the amount of natural ventilation; (*b*) reduction in operation efficiency of mechanical air-moving devices; (*c*) duration of the process, operational cycle, and normal location of workers relative to sources of contamination; (*d*) location and number of points of generation of the contaminant in the workroom or area; and (*e*) other circumstances that may affect the concentration of hazardous material in the breathing zone of the workers.

The K value selected usually varies from 3 to 10 depending on the foregoing considerations.

Table 18.1 lists the air volumes required to dilute the vapors of common organic solvents to the TLV level based on the liquid volume of solvent evaporated per unit time. These values must be multiplied by a K factor to allow for variations in uniformity of air distribution, and other considerations. Hemeon (6) includes a table of recommended dilution rates for 53 organic solvents. The "Ventilation Design Concentrations" in this table are not based on threshold limit values alone; they also incorporate the effects of odor. All the concentrations in this table are lower than the threshold limits, but the substances that are especially toxic or have a very disagreeable odor have the greatest safety factors.

It must be emphasized that threshold limit values are subject to revision, and the dilution values estimated from such tables become obsolete if the TLV's are lowered; therefore such a table should be used with caution, and with reference to the latest TLV list (7).

3.3.3 General Ventilation Airflow Patterns

One of the important disadvantages of general ventilation is that it permits the occupied space to become in effect a large settling chamber for the contaminants, even though the concentration may be within acceptable limits as far as potential health hazard is concerned. In some cases the settling or separation of contaminants from the air may represent condensation of materials that were vaporized by high temperature processes and accumulated on surfaces after condensation. The undesirability of the settling of contamination onto surfaces within the plant has been dramatically demonstrated in plants handling highly combustible dusts, where although the dust concentrations at any one time were not sufficient to constitute a hazard, accumulation over many months resulted in disastrous explosions (8). It is therefore necessary to maintain an effective ongoing good housekeeping program in conjunction with any broad-based engineering control such as general ventilation.

The design of general ventilation systems for a plant is not complete without consideration of the routes by which the air will enter and leave the workspace. The routes are of critical concern not only to the occupants of the building but potentially to

Table 18.1. Dilution Air Volumes for Vapors[a]

Liquid	(Air Volume (ft^3, STP) Required for Dilution to TLV[b]) Per Pint Evaporation	Per Pound Evaporation
Acetone (1000)	5,500	6,650
n-Amyl acetate (100)	27,200	29,800
Isoamyl alcohol (100)	37,200	43,900
Benzol (25)	Not recommended	
n-Butyl acetate (150)	20,400	22,200
Butyl cellosolve (50)	61,600	65,600
Carbon disulfide (20)	Not recommended	
Carbon tetrachloride (10)	Not recommended	
Cellosolve (2-ethoxyethanol) (100)	41,600	43,000
Cellosolve acetate (2-ethoxyethyl-acetate) (100)	29,700	29,300
Chloroform (25)	Not recommended	
1–2 Dichloroethane (50) (ethylene dichloride)	Not recommended	
1–2 Dichloroethylene (200)	26,900	20,000
Ethyl acetate (400)	10,300	11,000
Ethyl alcohol (1000)	6,900	8,400
Ethyl ether (400)	9,630	13,100
Gasoline	Requires special consideration	
Methyl acetate (200)	25,000	26,100
Methyl alcohol (200)	49,100	60,500
Methyl cellosolve (25)	Not recommended	
Methyl cellosolve acetate (25)	Not recommended	
Methyl ethyl ketone (200)	22,500	26,900
Methyl isobutyl ketone (100)	32,300	38,700
Methyl propyl ketone (200)	19,000	22,400
Naphtha (petroleum)	Requires special consideration	
Nitrobenzene (1)	Not recommended	
n-Propyl acetate (200)	17,500	18,900
Isopropyl alcohol (400)	13,200	16,100
Isopropyl ether (250)	11,400	15,100
Stoddard solvent (100)	30,000	40,000
1,1,2,2-Tetrachloroethane (5)	Not recommended	
Tetrachloroethylene (100)	39,600	23,400
Toluol (Toluene) (100)	9,500	10,500
Trichloroethylene (100)	45,000	29,400
Xylol (xylene) (100)	33,000	36,400

[a] American Conference of Governmental Industrial Hygienists, Committee on Ventilation. *Industrial Ventilation—A Manual of Recommended Practice,* 13th ed. ACGIH, Lansing, MI, 1974, p. 2-2 (modified).

[b] The tabulated dilution air quantities must be multiplied by the selected *K* value. TLVs for 1977.

the neighbors and the community at large. In general, the relative locations of air inlets and outlets should be considered in the implementation of a general ventilation system.

Air Inlets. Sufficient combined or total inlet area should be planned to accommodate the required volume of makeup air during the heating or cooling season, whichever needs the larger volume. This will prevent excessive inlet velocities that consume power, create drafts on workers, stir up dust, and interfere with local exhaust hoods. The chosen inlet velocity is likely to represent an engineering compromise between low rates, which require large inlet areas, and high rates, which facilitate rapid dilution of contaminants.

If widespread hazardous operations are controlled by dilution or general ventilation, inlets should be well distributed around the building to provide uniform circulation.

If contaminants are controlled by localized dilution, or by forceful diffusion into the general room-air reservoir, inlets may be purposely located to give nonuniform air supply distribution.

Inlets should be located to take full advantage of any thermal or convection effects within the building. For this purpose the designer must avoid the location of inlets (and outlets) near the "neutral zone" of inside–outside pressure differentials, which is approximately midway between the floor and roof.

Inlets should be located remote from stacks or ventilators discharging contaminants from the same or neighboring structures.

Air Outlets. Outlets should be located as far as possible from air inlets to prevent "short-circuiting." This advice holds for both natural and mechanical ventilation systems.

As in the case of air inlets, if widely scattered operations are controlled by general dilution, outlets should be uniformly distributed around the buidling. Similarly, if contaminants are controlled by localized dilution, air outlets may be purposely located to short-circuit a corner of a large room, with the intent of creating a high rate of air change there without involving the atmosphere of the entire space. This method of localized space ventilation requires a higher rate of exhaust air than supply air within the area to be controlled, to prevent the spread of contaminants throughout the room.

Outlets should be placed to take full advantage of thermal effects. They should be protected against the direct force of prevailing winds, which reduce the capacity of any exhaust fans or destroy the anticipated airflow route planned on the basis of thermal effects inside the building.

In general, dilution air should enter the workspace at approximate breathing zone height, passing through the workers' breathing zone, then through the zone of contamination; afterward, if it is not exhausted, it should enter a space of higher relative contamination. A useful concept in the interest of air-handling economy is "progressive ventilation." The air removed from an industrial plant frequently may be directed in a way that will provide both local and general ventilation. In fact, the air may be routed through several areas in succession, always in the direction of increasing air contamination, as long as the exhaust from one area is acceptable as the supply for the next area. Progressive ventilation saves energy in the form of airborne heat and horsepower to move the air.

3.3.4 General Ventilation Versus Local Exhaust Ventilation

The art of industrial process ventilation has been somewhat evolutionary from the natural general ventilation achieved by correct design of industrial buildings for hot, humid, and dusty operations to the highly effective local exhaust hoods in current use. Local ventilation prevents the spread of air contaminants throughout the building atmosphere with surprisingly small quantities of airflow in comparison with the volumes of air required by general ventilating systems. In spite of this characteristic air conservation of local ventilation, there are highly persuasive reasons for the selection by many industrial managements of general ventilation: (1) simplicity and economy of natural general ventilation, (2) relatively low first cost of mechanical general ventilation, (3) absence of interference with manufacturing operations, (4) flexibility in plants with constantly changing layouts, (5) existence of contaminating processes throughout the entire plant, (6) desire for large volumes of air circulation in hot weather, and (7) discovery that local exhaust systems do not eliminate the necessity of supplying large volumes of heated air in the wintertime to replace that escaping by exfiltration from loosely constructed buildings. This volume of air change may be more than ample to control process air contaminants by the method of dilution, if means are provided to disperse contaminants from the immediate vicinity of the workers.

4 LOCAL EXHAUST VENTILATION

A local exhaust ventilation system is one in which the contaminant is controlled by capturing it at or near the place where it is generated and removing it from the workspace. Local ventilation relies more heavily on mechanical methods of controlling airflow than does general ventilation. A local exhaust system usually includes all the following components: a hood or enclosure, ductwork, an air-cleaning device (where necessary for air pollution abatement purposes), and an air-moving device (usually an exhaust fan) to draw the contaminated air through the exhaust system and discharge it to the outside air. In general, local exhaust systems consist of more individual components than do general exhaust systems; since they also offer more operational parameters that must be controlled within acceptable ranges, they require more maintenance and involve higher operating expenses, as well.

When the primary purpose of the ventilation is to provide control of airborne contaminants, local exhaust systems generally are much superior to general ventilation. The advantages of the local exhaust ventilation system over general exhaust for any particular application will include many of the following.

1. If the system is designed properly, the capture and control of a contaminant can be virtually complete. Consequently, the exposure of workers to contaminants at the sources exhausted can be prevented. With general ventilation the contaminant is diluted when the exposure occurs, and at any given workplace this dilution may be highly variable, therefore inadequate at certain times.

2. The volumetric rate of required exhaust is less with local ventilation; as a result, the volume of makeup air required is less. Local ventilation offers savings in both capital investment and heating costs.

3. The contaminant is contained in a smaller exhausted volume of air. Therefore if air pollution control is needed, it is less expensive because the cost of air pollution control is approximately proportional to the volume of air handled.

4. Local exhaust systems can be designed to capture large settlable particles or at least confine them within the hood, thus greatly reduce the effort needed for good housekeeping.

5. Auxiliary equipment in the workroom is better protected from such deleterious effects of the contaminant as corrosion and abrasion.

6. Local exhaust systems usually require a fan of fairly high pressure characteristics to overcome pressure losses in the system. Therefore the performance of the fan system is not likely to be affected adversely by such influences as wind direction or velocity or inadequate makeup air. This is in contrast to general ventilation, which can be greatly affected by seasonal factors.

4.1 Components of a Local Exhaust System

The four components of a simple local exhaust system are illustrated in Figure 18.5: (*a*) a hood, (*b*) ductwork, (*c*) an air-cleaning device (cleaner), and (*d*) an air-moving device

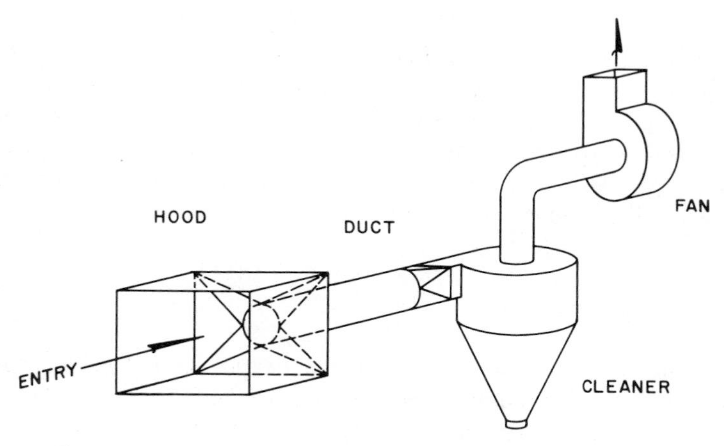

Figure 18.5 Interrelated components of a local exhaust system.

(fan). Typically the system is a network of branch ducts connected to several hoods or enclosures, main ducts, air cleaner for separating the contaminants from the airstream, exhaust fan, and discharge stack to the outside atmosphere.

4.1.1 Hoods

A hood is a structure designed to partially enclose a contaminant-producing operation and to guide airflow in an efficient manner to capture the contaminant. The hood is connected to the ventilation system with ductwork that removes the contaminant from the hood. The design and location of the hood is one of the most critical aspects in the successful operation of a local exhaust system.

4.1.2 Ductwork

The ductwork in an exhaust system provides a path for flow of the contaminated air exhausted from the hood to the point of discharge. The following points are important in design of the ductwork.

1. In the case of dust, the duct velocity must be high enough to prevent the dust from settling out and plugging the ductwork.
2. In the absence of dust, the duct velocity should strike an economic balance between ductwork cost and fan, motor, and power costs.
3. The location and construction of the ductwork must furnish sufficient protection against external damage, corrosion, and erosion to maximize the useful life of the local exhaust system.

4.1.3 Air Cleaner

Most exhaust systems installed for contaminant control need an air cleaner. Occasionally the collected material has some economic reuse value, but this is seldom the case. To collect and dispose of the contaminant is usually inconvenient and certainly represents an added expense. Yet the growing concern with air pollution control, and the need to comply with legal restrictions on discharges from sources of atmospheric emissions, place new importance on the air-cleaning device within a local exhaust system.

4.1.4 Air-Moving Device

Fans, usually of the centrifugal type, are the heart of the local exhaust systems. Wherever practicable a fan should be placed downstream from the air cleaner so that it will handle uncontaminated air. In such an arrangement, the fan wheel can be the backward curved blade type, which has a relatively high efficiency and lower power cost. For equivalent air handling, the forward curved blade impellers run at somewhat lower speeds, and this may be important when noise is a factor. If chips and other particulate matter have to pass through the impeller, the straight blade or paddlewheel-type fan is best because it is least likely to clog.

Fans and motors should be mounted on substantial platforms or bases and isolated by antivibration mounts. At the fan inlet and outlet the main duct should attach through a vibration isolator (i.e., a sleeve or band of very flexible material, such as rubber or fabric).

When the system has several branch connections, consideration should be given to using a belt drive instead of a direct-connected motor. Then if increased airflow is required at a future date, the need can be accommodated, to some degree, by adjusting the fan speed.

The importance and interaction of the various components of the exhaust system will become evident in the discussion on design of a ventilation system later in this chapter. Because of the importance of the role played by the exhaust hood in the overall success of the ventilation system, and since it is the component of the system that typically permits most innovation in concept, specific discussion of the application of the various types of hood follows.

4.2 Applications of Local Exhaust Hoods

The local exhaust hood is the point at which air first enters the exhaust ventilation system. As such, the term can be used to apply to any opening in the exhaust system regardless of its shape or physical disposition. An open-ended section of ductwork, a canopy-type hood situated above a hot process, and a conventional laboratory booth-type hood all could be called hoods in the context of this discussion.

The hood captures the contaminant generated by a particular process or operation and causes it to be carried through the ductwork to a convenient discharge point. The quantity of air required to capture and convey the air contaminants depends on the size and shape of the hood, its position relative to the point of generation of the contaminant, and the nature and quantity of the air contaminant itself. It should be emphasized that there is not necessarily a "standard hood" that is correct for all applications of a particular operation, since the methods of processing are unique to each operating plant. On the other hand, standard concepts for exhaust hooding have been developed and have been recommended for specific types of operations (9).

Hoods can be classified conveniently into four categories: enclosures, booth-type hoods, receiving hoods, and exterior hoods.

4.2.1 Enclosure Hoods

Enclosure hoods normally surround the point of emission or generation of contaminant as completely as practicable. In essence, they surround the contaminant source to such a degree that all contaminant dispersal action takes place within the confines of the hood. Because of this, enclosure hoods generally require the lowest rate of exhaust ventilation, therefore are economical and quite efficient. They should be used whenever possible, and they deserve particular consideration when a moderately or highly toxic contaminant is involved.

4.2.2 Booth-Type Hoods

Booths, such as the common spray-painting enclosure, are a special case of enclosure-type hoods. These are typified further by the common laboratory hood in which one face of an otherwise complete enclosure is open for ready access. Air contamination takes place within the enclosure, and air is exhausted from it in such a way, and at such a rate, that an average velocity is induced across the face of the opening sufficient to overcome the tendency of the contaminant to escape from within the hood. The three walls of the booth greatly reduce the exhaust air requirements, although not to the extent of a complete enclosure.

4.2.3 Receiving Hoods

The term "receiving" refers to a hood in which a stream of contaminated air from a process is exhausted by a hood located near the source of generation of the contaminants specifically for purposes of control. Two examples of this type of hood are canopies situated above hot processes and hoods attached to grinders, positioned to take advantage of centrifugal and gravitational forces to maximize control of the dust generated by the process. Canopy hoods, frequently located above hot processes, are similar to exterior hoods in that the contaminated air originates beyond the physical boundaries of the hood. The fundamental difference between receiving and exterior hoods is that in the former the hood takes advantage of the natural movement of the released contaminant, whereas in the latter, air is induced to move toward the exterior hood.

In practice, receiving hoods are positioned to be in the pathway of the contaminant as normally released by the operation. If hood space is limited by the process, baffles or shields may be placed across the line of throw of the particles to remove their kinetic energy. Then the particles may be captured and carried into the hood by lower air velocities. Additional examples of receiving hoods include those associated with many hand tool operations (surface grinders, metal polishers, stone cutters, sanding machines, etc.).

4.2.4 Exterior Hoods

Exterior hoods must capture air contaminants generated from a point outside the hood itself, sometimes at quite a distance. Exterior hoods therefore differ from enclosure or receiving hoods in that their sphere of influence must extend beyond their own dimensions in capturing contaminants without the aid of natural forces such as natural drafts, buoyancy, and inertia. In other words, directional air currents must be established adjacent to the suction opening of the hood to provide adequate capture. Thus they are quite sensitive to external sources of air disturbance and may be rendered completely ineffective by even slight lateral movement of air. They also require the most air to control a given process and are the most difficult to design of the various hoods. Examples of exterior hoods include the exhaust slots on the edges of the tanks or surrounding a workbench such as a welding station, exhaust grilles in the floor or workbench below a contaminated process, and the common propeller-type exhaust fans frequently mounted in walls adjacent to a source of contamination.

4.3 Fundamental Concepts in Local Exhaust Ventilation

4.3.1 Capture and Control Velocities

All local exhaust hoods perform their function in one of two ways: capture or control. One approach is to create air movement that draws the contaminant into the hood. When the air velocity that accomplishes this objective is created at a point outside a nonenclosing hood, it is called "capture velocity." Some exhaust hoods essentially enclose the contaminant source and create an air movement that prevents the contaminant from escaping from the enclosure. The air velocity created at the openings of such hoods is called the "control velocity."

The determination of the two quantities, control velocity and capture velocity, is the basis for the successful design of any exhaust hood. The air velocity, which must be developed by the exhaust hood at the point or in the area of desired control, is based on the magnitude and direction of the air motion to be overcome and is not subject to direct and exact evaluation (Table 18.2). Many empirical ventilation standards, especially concerning dusty equipment like screens and conveyor belt transfers, are based on parameters such as "cubic feet per minute per foot of belt width." These so-called exhaust rate standards are usually based on successful experience, are easily applied, and usually give satisfactory results if not extrapolated too far. In addition, they minimize the effort and uncertainty involved in calculating the fan action of falling material, thermal forces within hoods, and external air currents. However, such standards have three major pitfalls.

1. They are not of a fundamental nature; that is, they do not follow directly from basic "laws."
2. They presuppose a certain minimum quality of hood or enclosure design, although it may not be possible or practical to achieve the same quality of hood design in a new installation.
3. They are valid only for circumstances similar to those which led to their development and use. It should be clear that the nature of the process generating the contaminant will be an important determinant of the required capture velocity.

Table 18.2. Range of Recommended Capture Velocities[a]

Condition of Dispersion of Contaminant	Examples	Capture Velocity (ft/min)
Released with practically no velocity into quiet air	Evaporation from tanks; degreasing, etc.	50–100
Released at low velocity into moderately still air	Spray booths; intermittent container filling; low speed conveyor transfers; welding; plating; pickling	100–200
Active generation into zone of rapid air motion	Spray painting in shallow booths; barrel filling; conveyor loading; crushers	200–500
Released at high initial velocity into zone of very rapid air motion	Grinding; abrasive blasting, tumbling	500–2000

Lower End of Range	Upper End of Range
1. Room-air currents minimal or favorable to capture.	1. Disturbing room-air currents.
2. Contaminants of low toxicity or of nuisance value only.	2. Contaminants of high toxicity.
3. Intermittent, low production.	3. High production, heavy use.
4. Large hood—large airmass in motion.	4. Small hood—local control only.

[a] American Conference of Governmental Industrial Hygienists, Committee on Ventilation, *Industrial Ventilation—A Manual of Recommended Practice,* 13th ed. ACGIH, Lansing, MI, 1974, p. 4-5 (modified).

4.3.2 Airflow Characteristics of Blowing and Exhausting

The flow characteristics at a suction opening are much different from the flow pattern at a supply or discharge opening. Air blown from an opening maintains its directional effect in a fashion similar to water squirting from a hose and, in fact, is so pronounced that it is often called "throw." However if the flow of air through the same opening is changed so that the opening operates as an exhaust or intake with the same volumetric rate of air flow, the flow becomes almost completely nondirectional and its range of influence is greatly reduced. As a first approximation, when air is blown from a small opening, the velocity 30 diameters in front of the plane of the opening is about 10 percent of the velocity at the discharge. The same reduction in velocity is achieved at a much smaller distance in the case of exhausted openings, such that the velocity equals 10 percent of the face velocity at a distance of only one diameter from the exhaust opening.

Figure 18.6 illustrates this point. Therefore local exhaust hoods must not be applied for any operation that cannot be conducted in the immediate vicinity of the hood.

4.3.3 Airflow into Openings

Airflow into round openings was studied extensively by Dalla Valle (10). His theory of airflow into openings is based on a point source of suction that draws air from all directions. The velocity at any point in front (distance X) of such a source is equivalent to the quantity of air Q flowing to the source divided by the effective area of the sphere of the same radius. Conversely,

$$Q = VA$$
$$A = 4\pi X^2$$

Thus

$$Q = V(12.57X^2)$$

where Q = airflow, (ft^3/min or m^3/sec)
V = velocity at point X, (ft/min or m/sec)
X = centerline distance (ft or m)
A = pipe area (ft^2 or m^2)
π = 3.1416, dimensionless constant

Postulating that a point source is approximated by the end of an open pipe, Brandt (11) and Dalla Valle determined the actual velocity contours for a circular opening (Figure 18.7). These contours, or lines or constant velocity, are best described by the following equation:

$$Q = V(10\,X^2 + A)$$

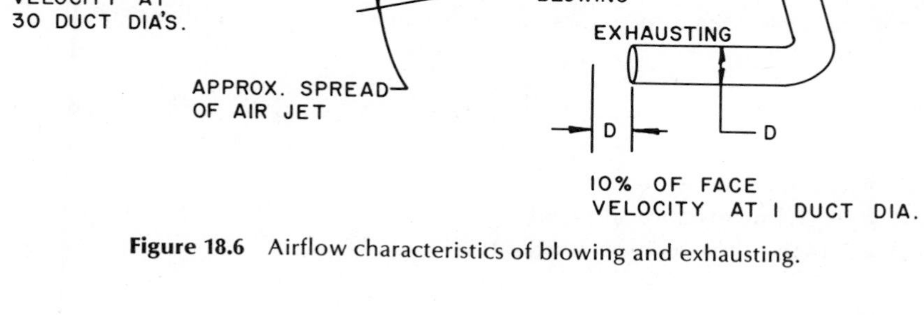

Figure 18.6 Airflow characteristics of blowing and exhausting.

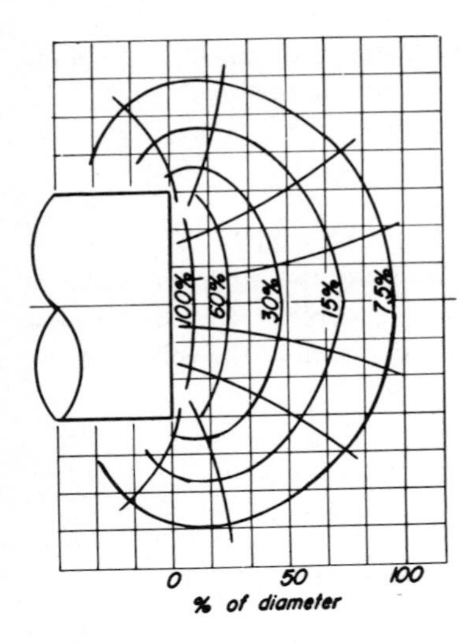

Figure 18.7 Velocity contours for unflanged circular opening.

4.3.4 Effects of Flanging

Flanges surrounding a hood opening force air to flow mostly from the zone directly in front of the hood. Thus the addition of a flange to an open duct or pipe improves the efficiency of the duct as a hood for a distance of about one diameter, as shown by the following equation:

$$Q = 0.75V\,(10X^2 + A)$$

For a flanged opening on a table or bench, we write

$$Q = 0.5V(10\,X^2 + A)$$

4.3.5 Slots

Caution must be used in applying the generalized continuity equation when the width to length ratio (aspect ratio) of an exhaust opening approaches 0.1, since the opening behaves more like a slot. Using a line of reasoning similar to that of Dalla Valle, Silverman (12) considered the slot to be a line source of suction. Disregarding the end, the area of influence then approaches a cylinder and the velocity is given by

$$V = \frac{Q}{2\pi XL}$$

where L = length of slot, (ft or m)
X = centerline distance (ft or m)
π = 3.1416, dimensionless constant

Correcting for empirical versus theoretical considerations, the design equation that best applies for freely suspended slots is

$$V = \frac{Q}{3.7XL}$$

Since flanging the slots will give the same benefits as flanging an open pipe, only 75 percent of the air is required to produce the same velocity at a given point. Therefore, for a flanged slot, we have

$$V = \frac{Q}{2.8XL}$$

4.3.6 Air Distribution in Hoods

To provide efficient capture with a minimum expenditure of energy, the airflow across the face of a hood should be uniform throughout its cross section. For slots and lateral exhaust applications, this can be done by incorporating external baffles. Another method of design is to provide a velocity of 2000 to 2500 ft/min into the slot with a low velocity plenum or large area chamber behind it. For large, shallow hoods, such as paint spray booths, laboratory hoods, and draft shakeout hoods, the same principle may be used. In these cases unequal flow may occur, with resulting higher velocities near the takeoffs. Baffles provided for the hood improve the air distribution and reduce pressure drop in the hood, giving the plenum effect. When the face velocity over the whole hood is relatively high or the hood or booth is quite deep, baffles may not be required.

4.3.7 Entrance Losses in Hoods

The negative static pressure that is exhibited in the ductwork a short distance downstream from the hood is called the hood static pressure, SP_h. This term represents the energy needed to accelerate the air from ambient velocity (often near zero) to the duct velocity and also to overcome the frictional losses resulting from turbulence of the air upon entering the hood and ductwork. Therefore,

$$SP_h = VP + h_e$$

where VP = velocity pressure in the duct
h_e = hood entry loss

The hood entry loss h_e is expressed as a function of the velocity pressure VP. For hoods of most types $h_e = F_h VP$, where F_h is the hood entry loss factor. For plain hoods, where the hood entry loss is a single expression $F_h VP$, the new VP referred to is the duct velocity pressure. The hood static pressure can be expressed as

$$SP_h = VP_{\text{duct}} + h_e$$

or

$$SP_h = VP_{\text{duct}} + F_h VP_{\text{duct}} = (1 + F_h)VP_{\text{duct}}$$

However for slot, plenum, or compound hoods there are two entry losses; one through the slot and the other into the duct. Thus

$$SP_h = h_{e_{\text{slot}}} + VP_{\text{duct}} + h_{e_{\text{duct}}} = (F_{\text{slot}})(VP_{\text{slot}}) + (VP_{\text{duct}}) + (F_{\text{duct}})(VP_{\text{duct}})$$

The velocity pressure resulting from acceleration through the slot is not lost as long as the slot velocity is less than the duct velocity, and usually this is the case.

Another constant used to define the performance of a hood is "coefficient of entry" C_e. This is defined as the ratio of the actual airflow to the flow that would exist if all the static pressure were present as velocity pressure. Thus we write

$$C_e = \frac{Q_{\text{actual}}}{Q_{VP} = SP_h} = \frac{KA\,(VP)^{1/2}}{KA\,(SP_h)^{1/2}} = \left(\frac{VP}{SP_h}\right)^{1/2}$$

This quantity is constant for a given shape of hood and is very useful for determining the flow into a hood by a single hood static pressure reading. The coefficient of entry C_e is related to the hood entry loss factor F_h by the following equation, only where the hood entry loss is a single expression:

$$C_e = \left(\frac{1}{1 + F_h}\right)^{1/2}$$

Listings of the entry loss coefficients C_e and the entry losses h_e in terms of velocity pressure VP are available (13). Most of the more complicated hoods have coefficients that can be obtained by combining some of the simpler shapes illustrated.

4.3.8 Static Suction

One method of specifying the air volume for a hood is to give the hood static pressure SP_h and duct size. For example, the hood static pressure at a typical grinding wheel hood is 2 inches of water. This reflects a conveying velocity of 4500 ft/min and entrance coefficient C_e of .78. For other types of machinery where the type of exhaust hood is relatively standard, a specification of the static suction and the duct size can be found in various reference sources (14). Specification of the static suction without duct size is, of course, meaningless because decreased size increases velocity pressure and static suction, while actually decreasing the total flow and the degree of control. Therefore static suction measurements for standard hoods or for systems in which the airflow has been measured previously are quite useful to estimate, in a comparative way, the quantity of air flowing through the hood.

4.3.9 Duct Velocity for Dusts and Fumes

The air velocity for transporting dusts and fumes through ductwork must be high enough to prevent the particles from settling and plugging the ducts. This minimum velocity, called "transport velocity," is typically 3500 to 4000 linear feet per minute. At these velocities, frictional loss from air moving along the surface of the ducts becomes significant; therefore all fittings, such as elbows and branches, must be wide-swept and gradual, having smooth interior surfaces. The cross-sectional area of the main duct generally equals the sum of the areas of cross sections for all branches upstream, plus a safety factor of approximately 20 percent. When the main duct is enlarged to accommodate an additional branch, the connection should be tapered, not abrupt.

Local exhaust systems for gases and vapors may have lower duct velocities (1500 to 2500 ft/min) because there is little to settle and plug the ducts. Lower velocities reduce markedly the frictional and pressure losses against which the fan must operate, thereby realizing a saving in power cost for the same airflow.

4.4 Principles of Local Exhaust Ventilation

When applying local exhaust ventilation to a specific problem, control of the contaminant is more effective if the following basic principles are followed.

4.4.1 Enclose the Source

A process to be exhausted by local ventilation should be enclosed as much as possible. This generally provides better control per unit volume of air exhausted. Nevertheless, the requirement of adequate access to the process must always be considered. An enclosed process may be costly in terms of operating efficiency or capital expenditure, but the savings gained by exhausting smaller air volumes may make the enclosure worthwhile.

4.4.2 Capture the Contaminant with Adequate Velocities

Air velocity through all hood openings must be high enough not only to contain the contaminant but to remove the contaminant from the hood. The importance of optimum capture and control velocity was discussed in the preceding sections.

4.4.3 Keep the Contaminant Out of Workers' Breathing Zone

Exhaust hoods that do not completely enclose the process should be located as near to the point of contaminant generation as possible and should provide airflow in a direction away from the worker toward the contaminant source.

This item, which is closely related to the characteristics of blowing and exhausting from openings in ductwork, was considered in more detail in the preceding sections.

4.4.4 Provide Adequate Air Supply

All the air that is exhausted from a building or enclosure must be replaced, to keep the building from operating under negative pressure. This applies to local exhaust systems as well as general exhaust systems. Additionally, the incoming air must be tempered by a makeup air system before being distributed inside the process area. Without sufficient makeup air, exhaust ventilation systems cannot work as efficiently as intended. Makeup air requirements are discussed later in this chapter.

4.4.5 Discharge the Exhausted Air Away from Air Inlets

The beneficial effect of a well-designed local exhaust system can be offset by undesired recirculation of contaminated air back into the work area. Such recirculation can occur if the exhausted air is not discharged away from supply air inlets. The location of the exhaust stack, its height, and the type of stack weather cap all can have a significant effect on the likelihood of contaminated air reentering through nearby windows and supply air intakes.

5 DESIGN OF VENTILATION SYSTEMS

With the increasing interest in and concern for environmental conditions in the American workplace, the design of ventilation systems has advanced from the "art" it was in the recent past to a true science of engineering design. To be sure, many of the empirical guidelines or rules of thumb still are applied appropriately through continuing experimentation and satisfactory experience. These rules and guidelines have developed into special forms and charts published for the direct purpose of facilitating the design of ventilation systems.

In the industrial workplace there are two basic types of ventilation system: general and local exhaust. These concepts have been described earlier in this chapter. It is noted here that the industrial hygienist's primary concern in the design of ventilation systems is to provide a safe and healthful environment for the worker. Although he usually deals mainly with local exhaust ventilation considerations, general ventilation is an integral part of his concern as well.

5.1 General Ventilation

As described earlier, "general ventilation" refers to the general flushing of a work environment with a supply of fresh air, to maintain desirable temperature and humidity conditions or to dilute a contaminant or contaminants to acceptable levels.

5.1.1 Office Buildings

The general ventilation considerations for production areas usually are focused on maintaining the contaminant level at acceptable limits, whereas in the nonproduction areas such as office buildings the principal concern is worker comfort. Accordingly, it is necessary as a preliminary phase of general ventilation design for office space to consider the conditions that may affect worker comfort: air temperature, humidity, the radiant heat sources, concentrations of tobacco smoke or other potentially offensive odors, the possibility of body odors, and general airflow patterns desirable in the office space.

Various attempts have been made to develop indices that describe a "comfort zone," that is, a definition of a combination of parameters (usually temperature, humidity, and air velocity) that result in a "comfortable" environment. The comfort zone defined in the effective temperature charts (Figure 18.2) is one of the most frequently used indices for this purpose. These comfort conditions can be met by proper selection of air conditioning equipment for office spaces. Details of the design of air conditioning systems are presented in several of the references listed at the end of this chapter. The appropriate air exchange rates are dictated to some extent by the concentrations of various odors within a room in addition to the environmental factors (temperature and humidity). The concentrations of odors obviously are affected by the air supply, the space provided per occupant, and the odor-absorbing capacity of the air conditioning process, as well as the temperature and humidity of the air.

In summary, the design of general office space ventilation requires that an analysis be made of the worker comfort zone parameters (primarily temperature and humidity), the tolerable level of odor, the equivalent space allowed per person in the room, and the operating characteristics of the air conditioning unit to be installed. Table 18.3 summarizes the requirements for ventilation under a variety of conditions typical of office-type space.

Table 18.3. Recommended Quantities of Outdoor Air Supplied to Occupied Space[a]

Outdoor Air (ft^3/min per person)	Type of Space or Occupancy
5–10	High-ceiling space such as: bank, auditorium, church, department store, theater; room with no smoking
10–15	Apartment, barber shop, beauty parlor, hotel room; room with light smoking
15–20	Cafeteria, drug store with lunch counter, general office, hospital room, public dining room, restaurant; room with moderate smoking
20–30	Broker's board room, private office, tavern; room with heavy smoking
30–60	Conference room, night club; crowded room with heavy smoking

[a] Air-cleaning or odor-removing devices not used. (Space not less than 150 ft^3 per person or floor area not less than 15 ft^2 per person.) State and city codes should be consulted to make certain that minimum outdoor air requirements are followed.

5.1.2 Industrial Process Buildings

The general ventilation systems typically employed in process operations can be natural or mechanically controlled. As discussed earlier in this chapter, the two forces available for use in natural ventilation are wind pressure and thermal or convective currents produced within the building. Although many of the older production facilities were designed with features such as the sawtooth or monitor-type roofs to provide maximum ventilation, these applications are becoming less common, particularly in light of the relatively straightforward application of mechanically controlled ventilation systems. In some industrial buildings, such as warehouses and pump rooms, natural ventilation is sufficient to provide adequate movement of air, particularly since the number of people normally employed in such situations is limited. Despite the economic incentive associated with natural ventilation, however, it is somewhat restricted in its application, whereas ventilation by mechanically controlled devices is virtually without limit, particularly when used in conjunction with ductwork to distribute the air strategically throughout the space. As noted earlier, a primary reason for applying general ventilation within a process area is to secure the dilution of the contaminant level to acceptable limits. Although this is accompanied by benefits, usually in terms of worker comfort, the chief concern is generally the elimination of a potential health hazard. In that respect, dilution ventilation has application chiefly for vapors or gases, and only under unusual circumstances is it appropriate to consider general ventilation as a control means when the contaminant of concern is a particulate material. In these instances application of local exhaust ventilation would be more appropriate. Table 18.1 summarizes the air volumes necessary to dilute the concentrations of specifically identified vapors to acceptable limits. These values, used in conjunction with the considerations presented earlier in this chapter concerning the airflow patterns created by the general ventilation system, form the basis for the design of dilution ventilation systems.

5.2 Design of Local Exhaust Ventilation

The industrial hygiene engineer usually looks to local exhaust ventilation as the primary control measure where enclosure or other form of isolation or substitution is not appropriate or has been used to the extent practicable. With proper application of local exhaust ventilation, the potential workroom contaminant is controlled at or near the source of generation or release of the substance. It should be kept in mind that local exhaust systems are not necessarily designed to be 100 percent effective in capturing and controlling contaminants released by the process. It is because of this residual or background level of contamination throughout the workroom that the general ventilation provisions are included as a complementary aspect of the total ventilation within a plant.

Many approaches to the overall design of the local exhaust ventilation system are available to the engineer. However they all consist of a series of similar steps. The following section presents the sequential procedure to be used in the design of local exhaust systems. For illustrative purposes, an example of design of a local exhaust system is carried through the complete procedure:

1. Prepare a sketch or layout of the workroom and operations within it. This will facilitate preparation of a sketch and/or layout of the contemplated exhaust system, using single-line drawings for ductwork. Changes in elevation should not be shown at this point. Although it is helpful if the layout is approximately to scale, subsequent steps in the procedure, such as calculating proper duct sizes, can be accomplished from even a very simple drawing. For purposes of illustration, we design a local exhaust system for a manual grinding station and a welding operation. The concept for the ventilation system to provide control of contaminants at these operations is illustrated in Figure 18.8.

2. A rough design or sketch of the desired hood for each operation, indicating direction and elevation of the outlet for the duct connections, should be prepared. The type of hood should be selected with an eye to performing the job with the least amount of air. This requires some ingenuity in designing enclosures that will not interfere with production or are equipped with movable sections. It should be emphasized that there may not be a "standard hood" available for a particular application.

For convenience, the more or less standard design of local exhaust hoods for portable hand grinding and bench welding operations are used. Figures 18.9 and 18.10, respectively, depict the configuration and design parameters for these hood arrangements (15).

3. Estimate the required control or capture velocity for each operation and type of hood or enclosure considered. This part of the process involves some judgment, particularly with hoods of nonstandard design. For our illustrative problem, the information presented in Figures 18.9 and 18.10 is used in making preliminary estimates of appropriate velocities and volumetric flowrates. Accordingly, we use an airflow rate of 200 ft^3/min per square foot (cfm/ft^2) of bench area for the grinding table and a slot velocity of 1000 (linear) ft/min for the welding bench.

4. Estimate or compute the "sphere of influence" over which the selected capture

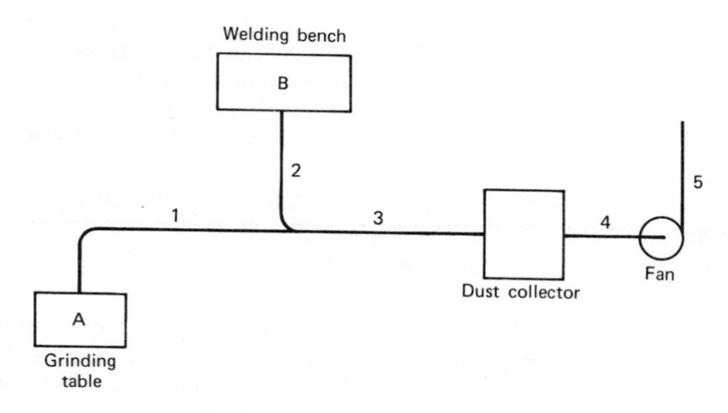

Figure 18.8 Local exhaust system.

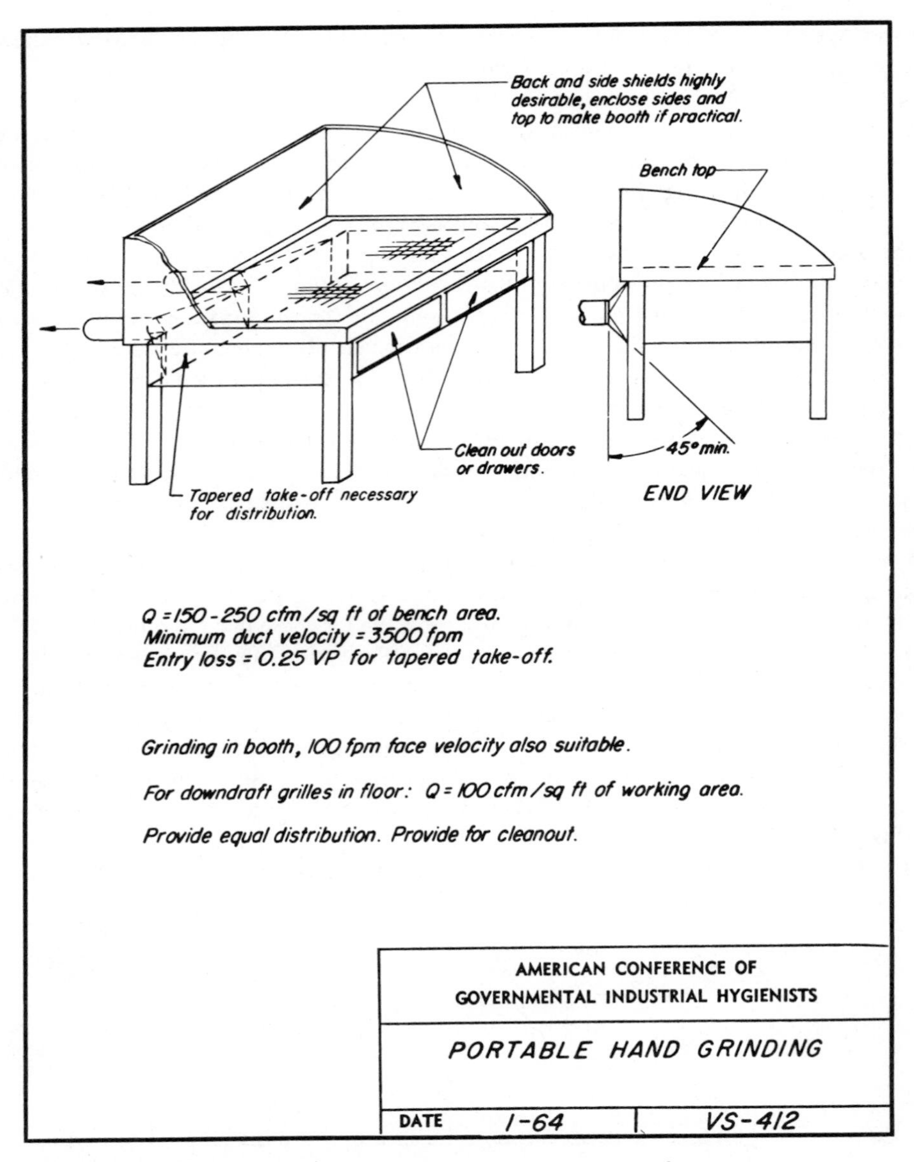

Figure 18.9 ACGIH-recommended hood for portable hand grinding station.

velocities must function. It will be advantageous to consider the hood arrangements in three dimensions. In the illustrated problem, assume that work at both the grinding table and the welding bench will be conducted on a flat surface 2 × 4 feet and that most of the work will be conducted within 2 inches of the table surface.

5. Compute the total volumetric airflow needed for each hood and indicate this on

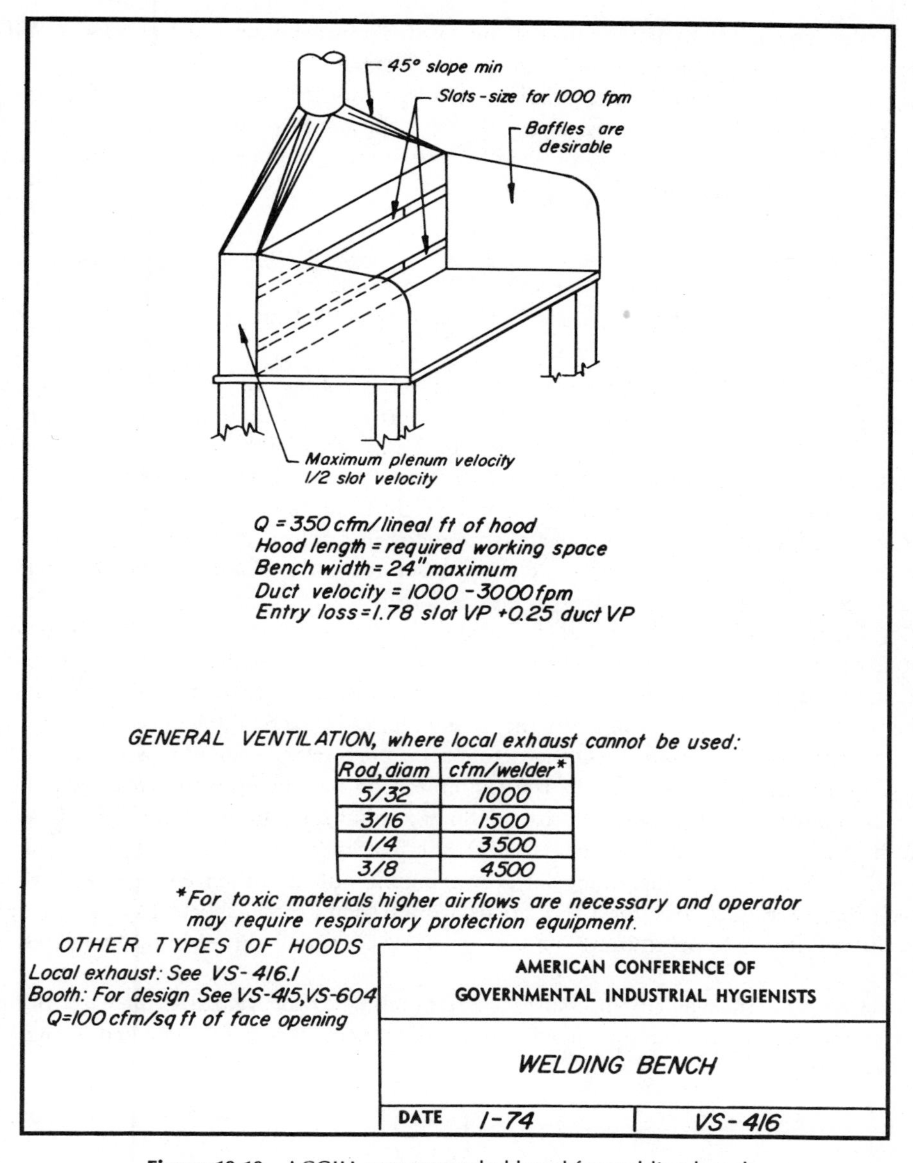

Figure 18.10 ACGIH-recommended hood for welding bench.

the working sketch of the system or in an appropriate worksheet. From the information supplied in Figures 18.9 and 18.10, the required volumetric airflows for the two hoods can be calculated. For this example, a volumetric flowrate of 200 ft^3/min per square foot of work surface and a bench area of 8 $ft.^2$ have been assumed. Thus a volumetric flowrate of 1600 ft^3/min will be required for hood A, the grinding table. For the welding bench, the required volumetric rate will be 350 ft^3/min per linear foot of hood × 4 feet (length of hood) or a total of 1400 ft^3/min.

6. Summarize the volumetric airflow requirements for each section of the branch and main ductwork on the work sheet, or enter the air flows directly on the working sketch of the system. For this simple illustration, the summation of volumetric rate requirements is straightforward.

Duct section 1 = 1600 ft^3/min
Duct section 2 = 1400 ft^3/min
Duct sections 3, 4, and 5 = 3000 ft^3/min

7. Select the minimum effective transport velocities for the contaminant-laden air being moved through the ductwork. For transporting particulate matter, data such as those in Table 18.4 can be used. Velocities in the range of 1000 to 2000 linear feet per minute commonly are used for gases and vapors, although in these cases duct velocities can be as low as is consistent with space and weight limitations.

The recommended transport velocities presented in Table 18.4 indicate values of 4000 and 3500 linear feet per minute for ductwork servicing the grinding table and welding bench hoods, respectively.

8. Figure the nearest practical duct diameter necessary to carry the required volumetric rate of air at the proper velocity. It is often necessary at this juncture to consider the standard sizes of ducts available from the fabricator, to avoid the unnecessary expense of special fabrication. If the available duct sizes differ greatly from the theoretically required diameters, the actual transport velocity should be computed for use in subsequent design steps. In the example problem, dividing the volumetric flowrates by the desired transport velocities (and assuming that standard ductwork is available in full-inch diameter increments only), the nearest duct diameter to the calculated necessary diameters is 8 inches for both the grinding table and the welding bench. Using 8-inch diameter ductwork for these hoods would produce transport velocities of 4580 and 4010 linear feet per minute for the grinding hood and the welding hood, respectively.

9. Compute the velocity pressures corresponding to air velocities in each branch duct connected to an exhaust opening. Velocity pressures are calculated using the following formula:

$$VP = \left(\frac{V}{4005}\right)^2$$

10. Estimate the entry loss for each hood in percentage of velocity pressure in the branch duct. In the example problem, these relationships are given in Figures 18.9 and 18.10.

11. From the information developed in items 9 and 10, compute the entrance losses for all hoods. Using the relationships presented in Figures 18.9 and 18.10, the entrance

Table 18.4. Typical Recommended Transport Velocities

Material, Operation, or Industry	Minimum Transport Velocity (ft/min)	Material, Operation, or Industry	Minimum Transport Velocity (ft/min)
Abrasive blasting	3500–4000	Lead dust	4000
Aluminum dust, coarse	4000	With small chips	5000
Asbestos carding	3000	Leather dust	3500
Bakelite molding powder dust	2500	Limestone dust	3500
Barrel filling or dumping	3500–4000	Lint	2000
Belt conveyors	3500	Magnesium dust, coarse	4000
Bins and hoppers	3500	Metal turnings	4000–5000
Brass turnings	4000	Packaging, weighing, etc.	3000
Bucket elevators	3500	Downdraft grille	3500
Buffing and polishing		Pharmaceutical coating pans	3000
Dry	3000–3500	Plastics dust (buffing)	3800
Sticky	3500–4500	Plating	2000
Cast iron boring dust	4000	Rubber dust	
Ceramics, general		Fine	2500
Glaze spraying	2500	Coarse	4000
Brushing	3500	Screens	
Fettling	3500	Cylindrical	3500
Dry pan mixing	3500	Flat deck	3500
Dry press	3500	Silica dust	3500–4500
Sagger filling	3500	Soap dust	3000
Clay dust	3500	Soapstone dust	3500
Coal (powdered) dust	4000	Soldering and tinning	2500
Cocoa dust	3000	Spray painting	2000
Cork (ground) dust	2500	Starch dust	3000
Cotton dust	3000	Stone cutting and finishing	3500
Crushers	3000 or higher	Tobacco dust	3500
Flour dust	2500	Woodworking	
Foundry, general	3500	Wood flour, light dry sawdust	
Sand mixer	3500–4000	and shavings	2500
Shakeout	3500–4000	Heavy shavings, damp sawdust	3500
Swing grinding booth exhaust	3000	Heavy wood chips, waste,	4000
Tumbling mills	4000–5000	green shavings	
Grain dust	2500–3000	Hog waste	3000
Grinding, general	3500–4500	Wool	3000
Portable hand grinding	3500	Zinc oxide fume	2000
Jute			
Dust	2500–3000		
Lint	3000		
Dust shaker waste	3200		
Pickerstock	3000		

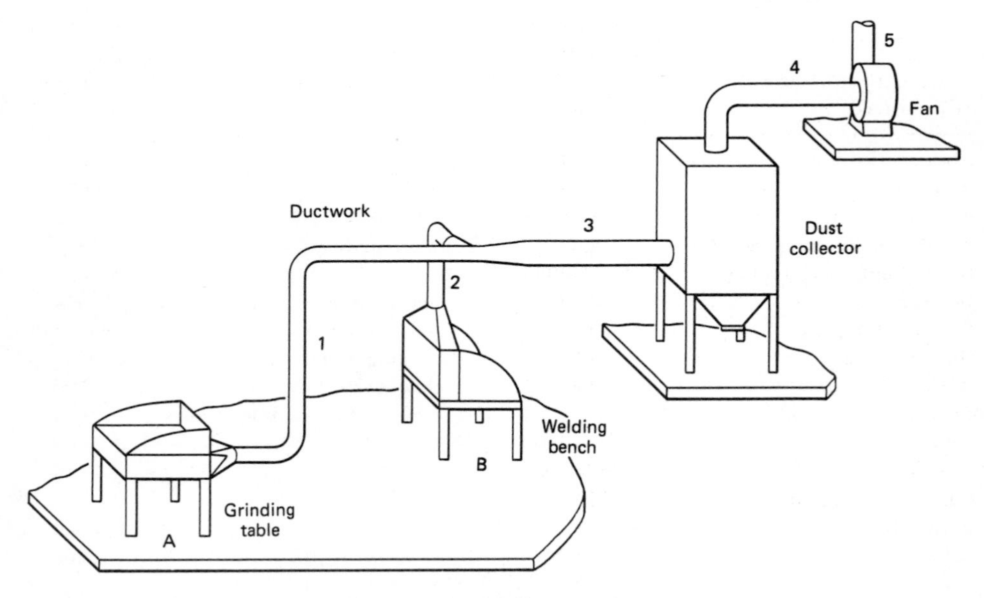

Figure 18.11 Ventilation system.

losses for the grinding table hood and the welding bench hood can be calculated to be 0.327 and 0.357 in. H_2O, respectively.

12. Compute the losses in elbows, the branch-to-main connections, or other transition points in the duct system. This can be done in terms of either equivalent straight-duct lengths or velocity pressures, whichever form corresponds with the data being used. The equivalent straight-duct method has the advantage that such transition losses can be added directly to the straight-duct sections for use with friction charts. The velocity-head method may be preferred by those who are especially familiar with the fundamentals of fluid mechanics. For the example problem, refer to Figure 18.11 for the relative arrangement of equipment in the exhaust system. In computing friction and transition losses assume that branches 1 and 2 contain two 90° elbows, that the centerline radius of the elbow turns is twice the duct diameter, that branch 2 merges with duct section 3 on a 45° angle of entry, and that duct section 4 contains one 90° elbow also with a centerline radius of 2 diameters. On the basis of these assumptions, and using information such as that presented in Table 18.5, it can be shown that each of the elbows in duct sections 1 and 2 is equivalent to 10 feet of straight duct, that the elbow in duct section 4 is equivalent to 17 feet of straight duct, and that the transition point from duct section 2 to 3 is equivalent to 11 feet of straight duct.

13. Obtain the duct friction losses from an appropriate chart, table, or formula. A friction chart like the one illustrated in Figure 18.12 can be used at this point. For the various sections of ductwork associated with the example problem, the following friction losses can be derived by referring to Figure 18.12 or similar charts:

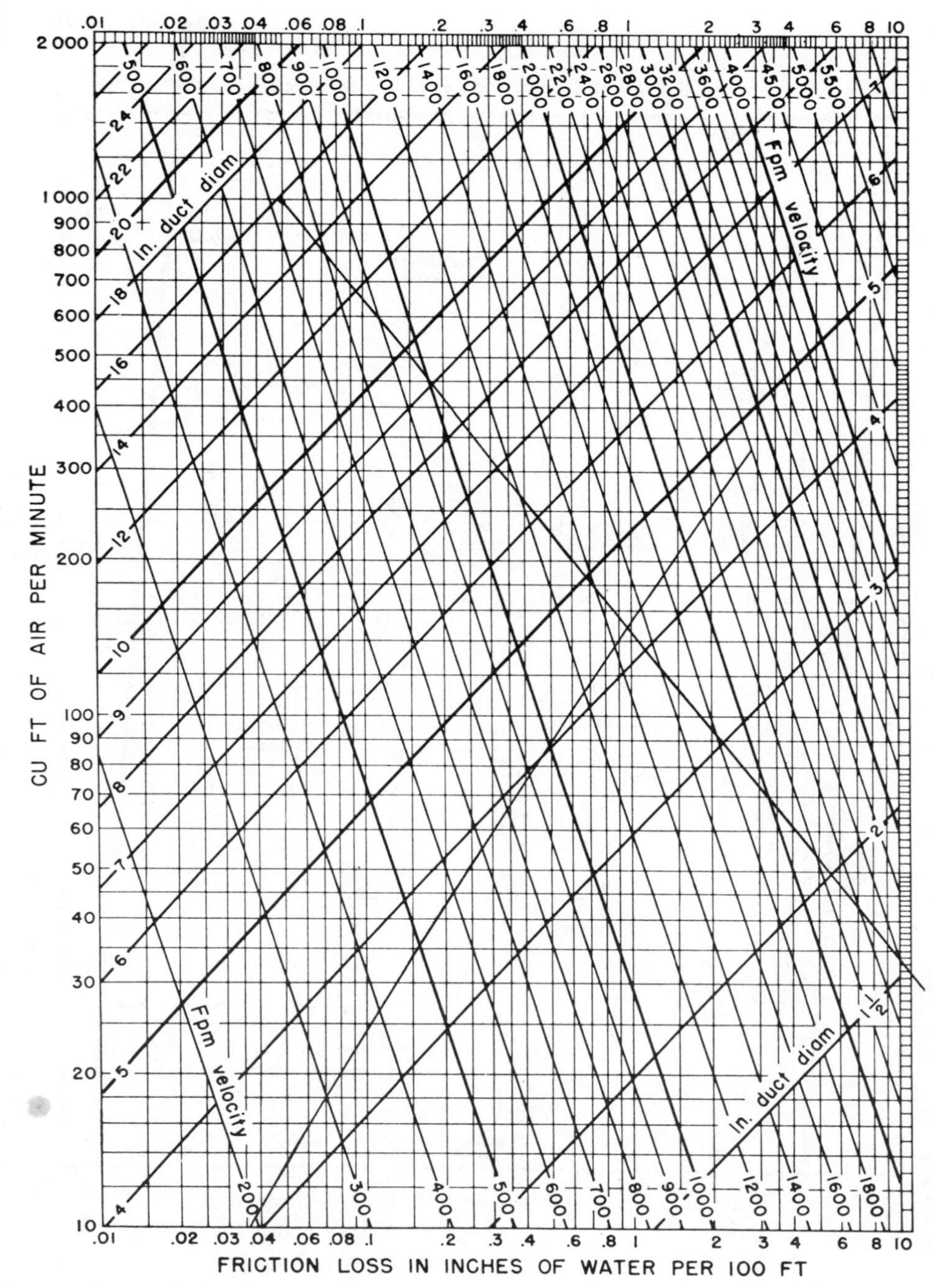

Figure 18.12 Friction of air in straight ducts. Reprinted from Am. Conf. of Govt. Ind. Hygienists, Committee on Ventilation, *Industrial Ventilation—A Manual of Recommended Practice,* 13th ed., Lansing, Md., 1974 pg. 6–31.

Table 18.5 Equivalent Resistance of Elbows and Duct Connections

EQUIVALENT RESISTANCE IN FEET OF STRAIGHT PIPE

Dia. of Pipe	90° Elbow * Centerline Radius			Angle of Entry		H, No of Diameters		
	1.5 D	2.0 D	2.5 D	30°	45°	1. D	.75 D	.5 D
3"	5	3	3	2	3	2	2	9
4"	6	4	4	3	5	2	3	12
5"	9	6	5	4	6	2	4	16
6"	12	7	6	5	7	3	5	20
7"	13	9	7	6	9	3	6	23
8"	15	10	8	7	11	4	7	26
10"	20	14	11	9	14	5	9	36
12"	25	17	14	11	17	6	11	44
14"	30	21	17	13	21	7	13	53
16"	36	24	20	16	25	9	15	62
18"	41	28	23	18	28	10	18	71
20"	46	32	26	20	32	11	20	80
24"	57	40	32			13	24	92
30"	74	51	41			17	31	126
36"	93	64	52			22	39	159
40"	105	72	59					
48"	130	89	73					

* For 60° elbows —— x.67
* For 45° elbows —— x.5

AMERICAN CONFERENCE OF GOVERNMENTAL INDUSTRIAL HYGIENISTS

DUCT DESIGN DATA

DATE 1–74 Fig. 6-11

Duct section 1	4.0 in. H_2O/100 ft of duct
Duct section 2	2.9 in. H_2O/100 ft of duct
Duct sections 3, 4, 5	1.0 in. H_2O/100 ft of duct

14. Summarize the losses for each branch to determine where to increase or decrease duct sizes to ensure that a balance of pressure drop will be obtained. The total pressure

drop from any points in the system to each upstream opening or exhaust point should be the same if balance is to be achieved. In actual operation, the airflow quantities through all open branches automatically adjust themselves to fulfill this requirement of balance. If the flow through any branch is too high, further closing of the damper increases the friction or energy loss through that branch, thus making it possible for a smaller volume of airflow to create the necessary pressure drop for a balanced condition.

In summing the entry losses for the hoods and the friction losses in the ductwork for duct sections 1 and 2, it can be shown that these two sections are balanced by virtue of the total pressure drop at the common point (where duct section 2 merges with the main duct) being the same, namely 1.60 in. H_2O.

15. After the system has been balanced on paper to a reasonable degree by adjusting round duct sizes, any necessary conversion to rectangular duct sizes can be made using a chart such as that in Figure 18.13. In the example problem, this step is not applicable.

16. Determine the adjusted total pressure drop of the branch and main duct system by adding the pressure losses, beginning with any one exhaust opening and proceeding toward the fan or dust collector. If the branch duct system has been reasonably balanced (in accordance with item 14), it will be a simple matter to select the starting point for totaling the pressure losses. However if no balancing has been attempted, the losses must be estimated for the single run of duct that is believed to have the greatest total resistance. Experienced designers, after a few moments of inspection, usually can determine the longest equivalent run in terms of friction loss—that is, the course that offers the greatest resistance to airflow, not necessarily the longest one in terms of physical length.

In summing the total pressure drop of the branch and main duct system, it can be calculated that the total pressure drop up to the point of entry into the dust collector is 1.91 in. H_2O.

17. Add the resistance of accessory equipment such as air cleaners and weather caps. For the most part, this information should be obtained directly from the manufacturer, since equipment varies greatly in resistance to airflow. For this example, assume that the pressure drop across the dust collector is 2.0 in. H_2O.

18. On the basis of the total airflow in cubic feet per minute and the total system resistances or static pressure, select a fan of proper type, size, and speed to give the highest operating efficiency, or lowest horsepower, consistent with other considerations. In selecting a fan, it is necessary to be familiar with the fan laws and system curves, that is, how the static pressure and volumetric airflow are related. A fan curve or system curve shows graphically the possible combinations of volumetric flow and static pressure for a given application. Because the fan and system each can operate only at a single point on its own curve, the combination can operate only where these curves intersect (Figure 18.14). Thus a fan must be chosen on the basis of its characteristics and the requirements of the system in which it will be applied. Each fan is characterized by the volume of gas flow, the pressure at which the flow is produced, the speed of rotation, the power required, and the efficiency. These quantities are measured by manufacturers with standard testing methods, and the results are plotted to furnish the characteristic

fan curves of most manufacturers. Tables such as Table 18.6 are available to assist the design engineer in selecting a fan.

The fan static pressure is calculated from the following equation:

$$\begin{aligned} \text{fan } SP &= SP \text{ (fan inlet)} + SP \text{ (fan outlet)} - VP \text{ (fan inlet)} \\ &= 4.20 + 0.10 - 0.91 \\ &= 3.39 \text{ in. } H_2O \end{aligned}$$

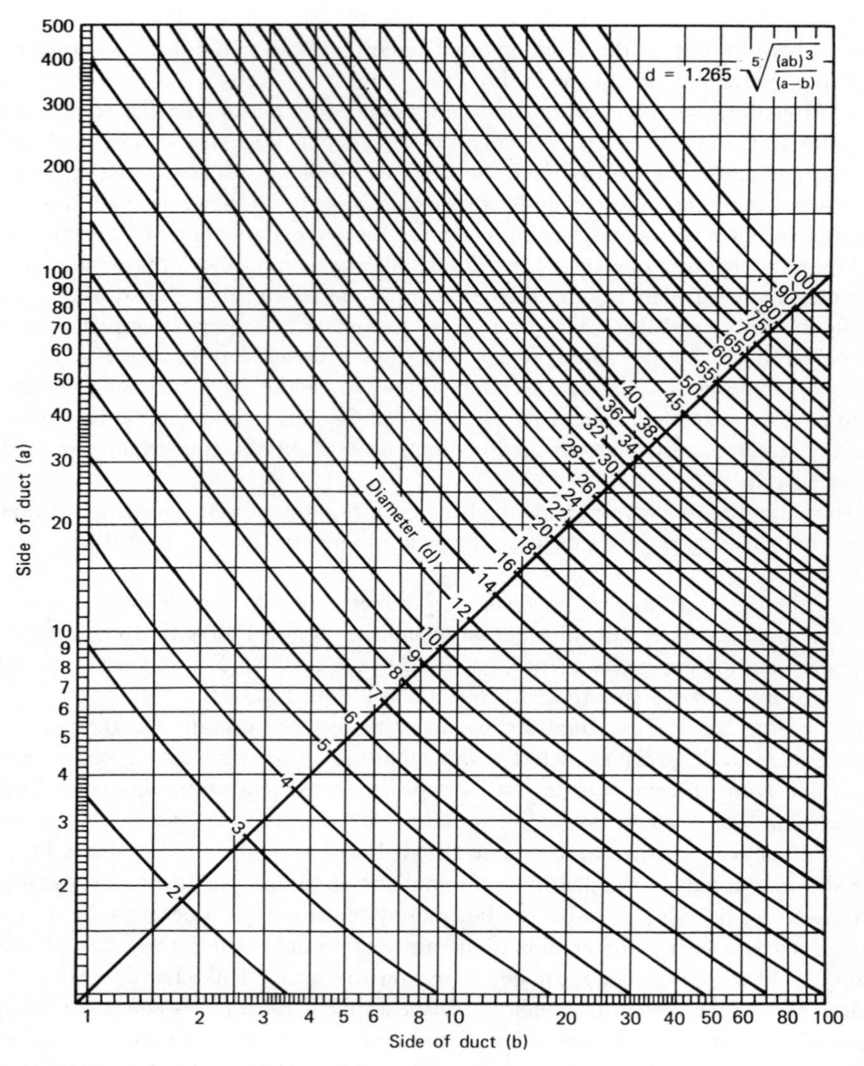

Figure 18.13 Chart for converting round to rectangular duct sizes having equivalent friction losses. Reprinted from Am. Conf. of Govt. Ind. Hygienists, Committee on Ventilation, *Industrial Ventilation—A Manual of Recommended Practice,* 13th ed., Lansing, Md., 1974 pg. 6–42.

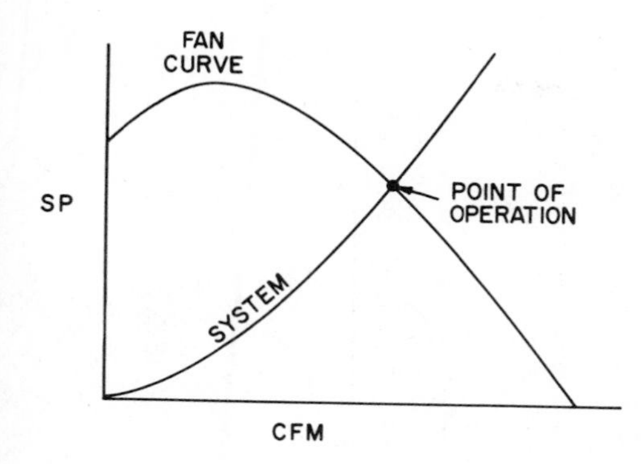

Figure 18.14 Fan and system curves.

Thus the fan and motor selected for this system should be capable of handling a static pressure of 3.5 to 4.0 in. H_2O.

5.3 Ventilation Regulations

It should be noted here that some applications of ventilation, both general and local exhaust, are covered specifically by occupational safety and health regulations. The federal standards as established by the Occupational Safety and Health Administration (OSHA) now include specific ventilation requirements for the following applications.

Industrial Application	Standard Reference
Abrasive blasting	29 CFR 1910.94
Dip tanks containing combustible or flammable liquids	29 CFR 1910.108
Flammable or combustible liquids in storage rooms and enclosures	29 CFR 1910.106
Gaseous hydrogen	29 CFR 1910.252
Grinding, polishing, and buffing	29 CFR 1910.94
Open surface tanks	29 CFR 1910.94
Oxygen	29 CFR 1910.252
Spray-finishing operations	29 CFR 1910.107
Welding, cutting, and brazing	29 CFR 1910.252

For the most part, these OSHA standards are based on the consensus-type standards developed by organizations such as the American National Standards Institute (ANSI) and the National Fire Protection Association (NFPA). In all likelihood the number and degree of specificity of ventilation standards will increase in the future in the form of regulations promulgated by OSHA and other regulatory agencies.

Table 18.6. Typical Multirating Table for Fans

Volume (ft^3/min)	Outlet Velocity, (ft/min)	Velocity Pressure (in. water)	Static Pressure, SP (in H_2O)																	
			1 in.		2 in.		3 in.		4 in.		5 in.		6 in.		7 in.		8 in.		9 in.	
			rpm	bhp	rpm	bhp	rpm	bhp	rpm	bhp	rpm	bhp	rpm	bhp	rpm	bhp	rpm	bhp	rpm	bhp
2,520	1000	0.063	437	0.63	595	1.27	728	2.00	837	2.66										
3,120	1200	0.090	459	0.85	610	1.55	735	2.30	842	3.10										
3,530	1400	0.122	483	1.05	626	1.87	746	2.72	847	3.57	943	4.60								
4,030	1600	0.160	513	1.33	642	2.18	759	3.17	858	4.12	950	5.21	1030	6.29						
4,530	1800	0.202	532	1.61	666	2.56	774	3.63	876	4.63	964	5.82	1040	6.92	1125	8.18				
5,040	2000	0.250	572	2.00	688	2.97	797	4.12	890	5.30	976	6.50	1052	7.75	1134	8.96	1208	10.15	1270	11.67
5,540	2200	0.302	603	2.36	712	3.43	816	4.66	910	5.93	999	7.38	1068	8.60	1145	9.93	1210	11.18	1279	12.82
6,040	2400	0.360	637	2.79	746	3.99	840	5.33	926	6.73	1017	8.17	1088	9.50	1160	10.88	1230	12.25	1288	13.92
6,550	2600	0.422	670	3.27	762	4.62	866	6.05	954	7.83	1032	9.08	1095	10.50	1171	11.98	1245	13.50	1298	15.10
7,060	2800	0.489	708	3.81	795	5.32	892	6.72	963	8.78	1050	9.97	1125	11.60	1188	13.06	1257	14.70	1310	16.48
7,560	3000	0.560	746	4.42	833	6.05	920	7.70	993	9.32	1068	11.00	1142	12.75	1210	14.28	1277	15.98	1328	17.80
8,060	3200	0.638			866	6.96	943	8.71	1020	10.40	1097	12.10	1168	14.02	1228	15.50	1292	17.36	1340	19.15
8,560	3400	0.721			900	7.93	964	9.80	1053	11.48	1120	13.30	1188	15.35	1248	16.93	1310	19.00	1360	20.90
9,070	3600	0.808					1010	11.00	1078	12.70	1148	14.65	1213	16.70	1270	18.42	1335	20.75	1380	22.60
9,570	3800	0.900					1038	12.25	1108	14.15	1170	14.90	1240	18.80	1292	19.46	1355	22.35	1405	24.40
10,080	4000	0.998					1162	13.60	1138	15.40	1200	17.35	1270	19.70	1320	21.70	1380	23.15	1430	26.40
10,580	4200	1.100							1168	16.90	1230	19.05	1283	21.50	1348	23.50	1405	26.10	1450	28.45
11,100	4400	1.210							1198	18.58	1258	20.55	1322	22.50	1373	25.40	1430	27.95	1478	30.60
11,600	4600	1.310							1232	20.30	1290	22.50	1355	23.80	1405	27.40	1450	30.15	1500	32.90
12,100	4800	1.450							1270	21.00	1321	24.40	1383	25.65	1432	29.60	1482	32.40	1528	35.20
12,600	5000	1.570							1301	24.20	1355	26.40	1410	28.80	1462	31.80	1513	34.60	1555	37.80
15,120	6000	2.230													1622	45.90	1670	49.00	1702	51.50

As noted earlier, many of the occupational health standards currently proposed by OSHA are couched in language that makes it clear that engineering control of the specific contaminants covered by the standard is the primary responsibility of the employer. In fact, engineering controls (chiefly ventilation in the case of airborne contaminants) must be applied to the extent feasible even if they are not sufficient by themselves to reduce the airborne concentration of the contaminant to the specified acceptable level. Thus in addition to the occupational health standards dealing with specific aspects of ventilation, there are indirect references to the requirement for installation of ventilation systems to maintain the work environment within stated acceptable limits. Accordingly, the need for application of sound engineering in the control of workplace contaminants undoubtedly will continue to increase.

6 MAKEUP AIR REQUIREMENTS

An adequate supply of outside air is necessary to ensure proper operation of ventilation systems. Properly designed exhaust systems by themselves will remove toxic contaminants; however they should not be relied on to draw air into the building to replace that which is exhausted, since this will result in a negative pressure within the building, leading to undesirable effects. Mechanically controlled systems for supplying the air to the workspace are preferred and in most modern industrial facilities are necessary, to achieve the overall desired ventilation. Most outside air supply systems currently installed in industrial plants have been placed in operation only to replace the exhausted air, with little or no regard to the overall positive effects that can be achieved by the air supply system in the total environmental control system. There is a definite trend toward consideration of the environmental control potential of air supply systems, that is, ability to affect beneficially the air quality, temperature, and humidity of the workplace. To be sure, control of total air volume, velocity, and temperature is important for a satisfactory work environment, and a properly implemented air supply system can provide both replacement air and some degree of environmental control.

6.1 The Need for Adequate Makeup Air

Air will enter a workspace in an amount equal to the volume actually exhausted, regardless of whether mechanical provisions are made for replacement of the air. Of course in some cases, particularly when the total exhaust is small, the replacement air enters the building with no adverse effects, even in absence of a mechanical makeup air system. When the exhaust volumes are relatively large in comparison to the size of free inlet area of the building, however, undesirable effects will be experienced. Older industrial buildings with large void spaces in the enclosing walls may provide significant openings, and air leakage into the building can be quite pronounced. Modern windowless plants of masonry construction, however, can be essentially airtight, with the result that the building is starved for air if there is a significant volume of exhaust ventilation. In a building that is relatively open, an influx of outside air is particularly undesirable in northern climates, where the incoming air cools the perimeter of the building, workers exposed to the incoming air are subjected to drafts, temperature gradients are produced within the building, and the internal heating system usually is overtaxed. Although the incoming air eventually may be tempered to acceptable conditions by mixing with interior air, it is an ineffective heat transfer process, usually resulting in wasted energy.

Reasons for considering an adequate supply of makeup air in conjunction with a total ventilation system include the following:

1. As indicated previously, lack of adequate makeup air creates a negative pressure within the building; this increases the static pressure that must be overcome by the exhaust fans if they are to function properly. This effect can result in a reduction of the total volume of air exhausted from the enclosure, and it is particularly serious with low pressure fans such as the common wall fans and roof exhausters of the propeller type.

2. Cross-drafts created by the infiltration of uncontrolled makeup air interfere with proper operation of exhaust hoods and also may disperse contaminated air from one area in the building to another; moreover, they can seriously interfere with proper operation of process equipment. The relatively high velocity cross-drafts through windows and doors created by the infiltration of air can even dislodge settled material on beams and other horizontal surfaces, resulting in resuspension of solid contaminants in the workroom.

3. If the negative pressure created within the building is significant enough, there may be actual backdraft on flues and other natural draft stacks, posing a potential health hazard from release of combustion products into the workroom. Such backdrafting can occur in natural draft stacks at very low negative pressures. Secondary problems associated with the backdraft phenomenon include difficulty in maintaining pilot lights in burners, erratic operation of temperature controls, and potential corrosion damage in stacks and heat exchangers due to condensation of water vapor from the flue gases.

4. The cold drafts resulting from infiltration of uncontrolled makeup air can produce worker discomfort, leading to reduced work efficiency attributable to lower ambient temperatures and the drafts impinging on the workers.

5. If the amount of infiltrated makeup air is high enough, the differential pressures created may make doors difficult to open or close and, in some extreme cases, can pose a potential safety hazard when doors are moved by the force of the pressure gradient itself in an uncontrolled manner.

6. Without adequate makeup air, the conditions of cooler ambient temperatures near the building perimeter often lead to a decision to attempt to correct the problem by installing more heating equipment in those areas. The heaters warm the air only after it has entered the building, usually too late to be of benefit to the workers in the perimeter. Then the overheated air moves toward the building interior, making those areas often uncomfortably warm. Often attempts then are made to alleviate this problem by

installation of more exhaust fans to remove the excess heat, but this further aggravates the problem of temperature gradients within the building. The net result is the unnecessary wasting of heat without solving the problem. The fuel consumption associated with attempts to achieve comfortable temperatures within the building by this approach is much higher than it would be with a properly installed system for providing makeup air.

6.2 Principles for Supplying Makeup Air

The volume of makeup air provided to a workspace generally should equal the total volume of air removed from the building by exhaust ventilation systems, process venting, and combustion processes. The determination of the actual amount of air removed from the building usually requires no more than a simple inventory of air exhaust locations, with measurements of the amount of air being exhausted and compilation of the exhaust volumes. It frequently is desirable to incorporate any reasonable projections for additional exhaust requirements in the near future, certainly if process changes or additions or modifications to the plant are being contemplated. Often it is desirable to install an air supply system of a capacity slightly larger than actually needed, to allow for future needs. In most cases the drive mechanism on the air supply system can be modified to supply only the desired quantity of air.

The locations of the makeup air units, with the consequent distribution of air and creation of airflow patterns within the building, often do not receive sufficient forethought before installation, and in many instances this aspect of the system is neglected altogether. However successful operation of the total ventilation system is dependent on proper distribution of makeup air. The following principles for supplying makeup air should be incorporated into the design of an air supply system.

1. The inlet for the fresh air should be located away from any sources of contamination, such as exhaust stacks or furnace exhausts. It is advisable to filter the fresh air, to protect equipment and to ensure maximum efficiency of heat exchange in the tempering system.
2. A mechanically operated air mover (i.e., a fan) must be incorporated into the air supply system to prevent the development of negative pressure within the room or building.
3. The locations of the makeup air entries into the workspace should be assigned in a manner that allows maximum utilization of the air. Properly located makeup air supply inlets can be used to provide general dilution ventilation first, after which the air provides makeup for exhaust systems operated within the building. Obviously this concept does not apply for specialized types of makeup air such as those installed for spot-cooling purposes, where the air is introduced at temperatures significantly below room temperature, and at specific work stations. In general, the air distribution pattern created by the combination of makeup air supply and process exhaust must be engineered carefully to provide effective coverage within the building without creating

excessive drafts, which could interefere with process operations or compromise the comfort of workers.

4. To the extent practicable, makeup air should be introduced into the plant at a height of approximately 8 to 10 feet. In this way the air is used first by the workers, maximizing the results of general or dilution ventilation possibilities. Such an arrangement for distribution of the incoming air also permits closer control of the ambient workplace temperature.

5. It usually is necessary to either heat or cool the makeup air to approximately the desired inside temperature. Accordingly, the temperature range for incoming air typically varies between 65 and 80°F, depending on the geographic location of the building and the climatological and meteorological conditions characteristic of the location and time of year.

6.3 Recirculation of Air from Exhaust Systems

A typical application of local exhaust ventilation includes discharge of the contaminated air to the atmosphere, with or without benefit of an air-cleaning device. In recent years industries have begun actively investigating exhaust systems that will clean the air of contaminants so effectively that recirculation of air directly to the workplace is possible. This approach is desirable because air-cleaning or pollution control devices have been employed to substantial degrees to meet air quality or emission standards. Additionally, the heating and/or cooling of makeup air needed to replace exhausted air is an expensive item, particularly since costs of energy are increasing and in some cases energy to operate new installations is not even available. Unfortunately, there are no established criteria for design and operation of such systems.

The acceptability of recirculating air systems obviously depends to a great extent on the health hazard associated with the contaminant being exhausted. It has been a fairly consistent policy of official occupational health agencies not to condone the recirculation of exhausted air if the contaminant may have an adverse effect on the health of the workers, mainly because even though the air cleaner used is efficient enough to clean the air sufficiently for health protection, incorrect operation or poor maintenance of the system would result in return of the contaminated air to the workers.

6.3.1 Circumstances Under Which Recirculation May Be Permitted

Recirculation of exhausted air is sometimes a feasible method of supplementing ventilation within the workplace. In some situations the economic and energy conservation factors may be important enough to warrant the additional capital and operating expenditures of assuring the safe provision of recirculated air. In general, recirculation of exhausted air may be permitted under the following circumstances:

1. The ventilation system must be furnished with an air cleaner system efficient enough to provide an exit concentration (i.e., concentration of contaminant in the air recirculated to the workroom) not exceeding an allowable value, which may be calculated for equilibrium conditions using the following equation:

$$C_R = \frac{1}{2}(TLV - C_0) \times \frac{Q_T}{Q_R} \times \frac{1}{K}$$

where C_R = concentration of contaminant in exit air from the collector before mixing
Q_T = total ventilation flow through the affected workspace (ft^3/min)
Q_R = recirculated air flow (ft/min)
K = a mixing factor, usually varying between 3 and 10 (3 = good mixing conditions)
TLV = threshold limit value of contaminant
C_0 = concentration of contaminant in workers' breathing zone with local exhaust discharged outside

This air-cleaning system is referred to as the primary system.

2. A secondary air-cleaning system of efficiency equal to or greater than the primary system should be installed in series with the primary cleaner. As an alternative to this approach, a reliable monitoring device may be installed to furnish a representative sample of the recirculated air. This monitoring system must be fail-safe with respect to failure of the power supply, environmental contamination, or the obvious results of poor maintenance.

3. A warning signal should be provided to indicate the need for attention to the secondary air-cleaning system, or the air monitor should indicate when the concentrations have exceeded predetermined allowable limits.

4. There should be a mechanism for immediate bypass of the recirculated air to the outdoors, or complete shutdown of the contaminant-generating process, to become operative under conditions that activate the system's warning device.

Although the application of recirculated air systems in industry has been quite limited, the potential for and anticipated future applications of this approach are quite extensive. A recent study (16) indicates that 356 of the 514 compounds currently comprising the ACGIH list of TLVs are potentially recirculatable. The compounds excluded from consideration were primarily those identified as having carcinogenic properties and those for which a ceiling exposure limit had been established. This preliminary analysis of design and operating criteria for recirculation systems emphasized that air monitoring equipment should be on-line, automatic, and specific for the individual contaminant. Although a preliminary effort, this study does indicate that use of recirculated air in industrial workplaces will become much more commonplace than it is today.

6.3.2 Design Considerations for Recirculated Air Systems

With the full expectation that interest in and development of refined recirculated air systems will continue, the following factors and considerations should be incorporated into analysis of the appropriateness of recirculated air systems in any particular setting.

1. It usually is necessary to provide general ventilation air in addition to that recirculated, to ensure continual dilution of the contaminants in the recirculated airstream. When it is proposed or possible that all the supplied ventilation air be that recirculated from exhaust systems, all possible contaminants in the airstream must be evaluated, not just the most significant. Of particular concern would be minor contaminants in the airstream that might pass through the primary and secondary air-cleaning devices. For example, a fabric filter–high efficiency particulate filter combination, although providing essentially complete collection of particulate contaminants, could allow the concentration of gases and vapors to build up during the course of operation from relatively insignificant to potentially hazardous levels.
2. Wherever practicable, the recirculating air system should be designed to permit bypass to the outdoors when weather conditions permit. For example, if the system is intended to conserve heat during winter months, it can discharge outdoors in warmer weather if windows and doors are designed to permit sufficient makeup air when opened. Of course continuous operation of the bypass mode would not be desirable where the workspace is conditioned or where mechanically supplied makeup air is required at all times.
3. Air recirculated through wet collectors may pose a problem in that the humid air from such equipment usually causes uncomfortably high humidity within the workplace and possible condensation problems. Excessive humidity may be prevented through the use of auxiliary ventilation equipment.
4. Design data and testing programs preceding installation of recirculating air systems should consider all operational time periods of the system, since it can be reasonably expected that the concentration of contaminants exiting typical collectors will vary over time.
5. As with design and layout of any air supply system, the ductwork from the recirculating air system should provide adequate mixing with other air supplies, to avoid uncomfortable drafts on workers or air currents that could adversely affect the performance of local exhaust hoods.
6. In general, even when a monitoring device is available, the installation of a secondary air cleaner system is preferable because it usually lends a greater degree of reliability to the system and requires a less sophisticated maintenance program.
7. Although the primary concern for the quality of the recirculated air is on acceptable exposure limits, odors or nuisance values for contaminants should be considered as well. In fact, in some locations, adequately cleaned recirculated air, provided by a system with appropriate safeguards, can be of better quality than the makeup air supply entering the building from outside.

8. Routine testing, maintenance procedures, and records should be developed for the recirculating air system. In addition, the workroom air should be periodically tested.

The foregoing considerations for recirculating air systems do not appear to impose any insurmountable obstacles to continuing development of proper criteria for design and operation of such systems. With the probable continuing concern for energy conservation and air quality (both within and outside the industrial plant), it is likely that the next generation of industrial hygienists will deal routinely with recirculating air systems.

REFERENCES

1. R. H. Perry, Ed., *Chemical Engineers' Handbook,* 4th ed., McGraw-Hill, New York, 1963.
2. American Conference of Governmental Industrial Hygienists, Committee on Industrial Ventilation, *Industrial Ventilation—A Manual of Recommended Practice,* 13th ed., ACGIH, Lansing, Mich., 1974, p. 6–27.
3. American Conference of Governmental Industrial Hygienists, Committee on Industrial Ventilation, *Industrial Ventilation—A Manual of Recommended Practice,* 13th ed., ACGIH, Lansing, Mich., 1974, pp. 6–28, 6–29.
4. American Conference of Governmental Industrial Hygienists, Committee on Industrial Ventilation, *Industrial Ventilation—A Manual of Recommended Practice,* 13th ed., ACGIH, Lansing, Mich., 1974, p. 4–12.
5. American Society of Heating, Refrigeration and Air-Conditioning Engineers, *ASHRAE Guide and Data Book—Fundamentals and Equipment,* ASHRAE, New York, 1963.
6. W. C. L. Hemeon, *Plant and Process Ventilation,* 2nd ed., Industrial Press, New York, 1963.
7. American Conference of Governmental Industrial Hygienists, *Threshold Limit Values for Chemical Substances and Physical Agents in the Workroom Environment,* ACGIH, Cincinnati, Ohio, 1975.
8. Cited by W. N. Witheridge, in: "Ventilation," *Industrial Hygiene and Toxicology,* Vol. 1, 2nd ed., Wiley-Interscience, New York, 1958.
9. American Conference of Governmental Industrial Hygienists, Committee on Industrial Ventilation, *Industrial Ventilation—A Manual of Recommended Practice,* 13th ed., ACGIH, Lansing, Mich., 1974, Chapter 5.
10. J. M. Dalla Valle, *Exhaust Hoods,* Industrial Press, New York, 1944.
11. A. D. Brandt, *Industrial Health Engineering,* Wiley, New York, 1947.
12. L. Silverman, "Velocity Characteristics of Narrow Exhaust Slots," *Ind. Hyg. Toxicol. J.,* **24,** 267 (1942).
13. American National Standards Institute, "Fundamentals Governing the Design and Operation of Local Exhaust Systems," ANSI Z9.2 Committee, New York, 1971.
14. J. L. Alden, *Design of Industrial Exhaust Systems,* Industrial Press, New York, 1949.
15. American Conference of Governmental Industrial Hygienists, Committee on Industrial Ventilation, *Industrial Ventilation—A Manual of Recommended Practice,* 13th ed., ACGIH, Lansing, Mich., 1974, pp. 5–48, 5–52.
16. National Institute for Occupational Safety and Health, *Recirculation of Exhaust Air,* NIOSH Report No. 76-186, Cincinnati, Ohio, 1976.

CHAPTER NINETEEN

Air Pollution Controls

JOHN E. MUTCHLER

This chapter focuses on the requirements for air pollution controls, principles of air cleaning, selection of suitable control methods, and economic and energy resources needed to install and operate emission control systems of various types.

1 RELATION OF ATMOSPHERIC EMISSIONS TO WORKPLACE AIR QUALITY

Although the close relation between air pollution controls and industrial hygiene should be obvious to a practicing industrial hygienist, it is worthwhile to review the basis for this important concept. In this day of specialization, some tend to think of air pollution engineering as a field of endeavor somewhat removed from industrial hygiene. This unfortunate compartmentalization, found all too often in governmental and industrial organizations, can detract significantly from a full understanding of air pollution and industrial hygiene, fostering less than a total approach to problem solving in either air cleaning or occupational health. Only when an industrial hygienist applies his skills to achieve the optimum *overall* control strategy will due regard be given to controlling occupational exposures and maintaining suitable ambient air quality.

The most basic conceptual entity common to air pollution control and industrial hygiene is the "source-process." The source-process represents a common denominator between workroom air quality and atmospheric emissions. With this concept as a basis, the interrelating effect of the source-process falls into two general categories: operations with and without direct atmospheric exhaust.

1.1 Operations Exhausted Directly to Atmosphere

When an industrial operation (process) is exhausted directly to the atmosphere, as with a hooded metallurgical furnace, workroom air quality is affected directly by the design and performance of the exhaust system. An improperly designed hood or a hood evacuated with a less than sufficient volumetric rate of air will contaminate the occupational environment and affect workers in the vicinity of the furnace. This is a simple but powerfully symbolic representation of one form of the close relation between atmospheric emissions and occupational exposure.

Another example of the close relation between atmospheric emissions and workplace air quality is the use of coke-side sheds adjacent to coke-oven batteries. These shed structures, when properly designed and evacuated, can provide effective capture of virtually all particulate and gaseous emissions emanating from the coke side of coke-oven batteries: door leaks, pushing emissions, and emissions from the coke car in transit to the quenching station. On the other hand, the same structures present a semiconfined space for containing coke-side emissions, and under conditions of inadequate exhaust, they can restrict the dilution and dispersion of emissions from the coke ovens, thereby jeopardizing the quality of the work environment for persons within the shedded structure. In this instance, the shed serves as the first of four components of the model "local" exhaust system: hood, ductwork, air cleaner, and air-moving device (1).

1.2 Operations Exhausted Indirectly to Atmosphere

In some situations the first step in the eventual and ultimate outdoor emission of materials generated from a source-process first is the dispersion of the contaminant throughout an enclosed workplace, followed by significant release to the atmosphere through the general ventilation system, natural or mechanical. This delayed, decentralized mechanism for atmospheric release has been a reality in industry for decades, but it draws increasing attention as our regulatory standards impinge on the generation and emission of toxic materials, including those deemed "hazardous air pollutants" by the administrator of the Environmental Protection Agency (EPA) (2). Materials such as beryllium, asbestos, and vinyl chloride are officially labeled "air pollutants" without regard to the location or mechanism of atmospheric release—whether it be through mechanically exhausted local systems or through workplace dispersion and subsequent release through general ventilation, including natural exfiltration through doors, windows, and other openings in a building structure.

1.3 Recirculation of Exhaust Air

The recirculation of exhaust air also provides a clear representation of the close relation between atmospheric emissions and workplace air quality. Industrial hygienists have been concerned for some time about the undesirable recirculation of exhaust air, notably due to the entrapment of atmospheric emissions in the cavity wake of a structure as a result of exhaust gas discharged at relatively low elevations. When such entrapment occurs, it can often affect the workplace air quality as contaminated air flows as "air supply" back into the workroom by virtue of natural or mechanical ventilation.

The deliberate recirculation of exhaust air not only provides a direct link between atmospheric emissions and workplace air quality, but deserves and draws an increasing amount of attention and interest as the costs of energy rise sharply.

There has been a long-standing tradition in industrial hygiene engineering to disallow recirculation of any exhaust air, even after cleaning, if the air contains toxic materials. Nevertheless, this maxim is being modified and qualified steadily as a result of the energy crisis and the increasingly significant cost of heating and cooling makeup air. In some cases recirculation is the only really feasible method of providing suitable general ventilation. In such cases, economic and energy conservation considerations may be important enough to warrant the additional capital and operating expenses required to safely provide the desired air recirculation. It is now generally accepted (3) that recirculation may be permitted under the following circumstances:

1. A primary air-cleaning system must provide an exit concentration (workroom air supply) not more than the allowable value C_R.
2. A secondary air-cleaning system of equal or greater efficiency than the primary system must be installed in series with the primary system; or a reliable air-monitoring device must be used to provide a representative sample of the exhaust (recirculated) air from the primary system. This monitoring device must be fail-safe with respect to interruption of power supply, environmental contamination, or typical results of poor maintenance.
3. A warning signal must be provided to indicate the need for attention to the secondary air-cleaning system or above-limits concentration detected by the monitor.
4. Provision must be made for the immediate bypass of recirculated air to the outdoors or complete shutdown of the contaminant-generating process if conditions occur that activate the warning device.

The permissible concentration of contaminant in recirculated air under steady state conditions may be calculated by the following equation (3):

$$C_R = \frac{1}{2}(TLV - C_0)\frac{Q_T}{Q_R}\frac{1}{K} \tag{1}$$

where C_R = concentration of contaminant in (recirculated) exit air from the collector system before mixing

TLV = threshold limit value of the contaminant (4)

C_0 = concentration of contaminant in the worker's breathing zone without recirculation of local exhaust

Q_T = total ventilation flowrate through affected space

Q_R = recirculated air flowrate

K = "effectiveness of mixing" factor, usually between 3 ("good mixing") and 10

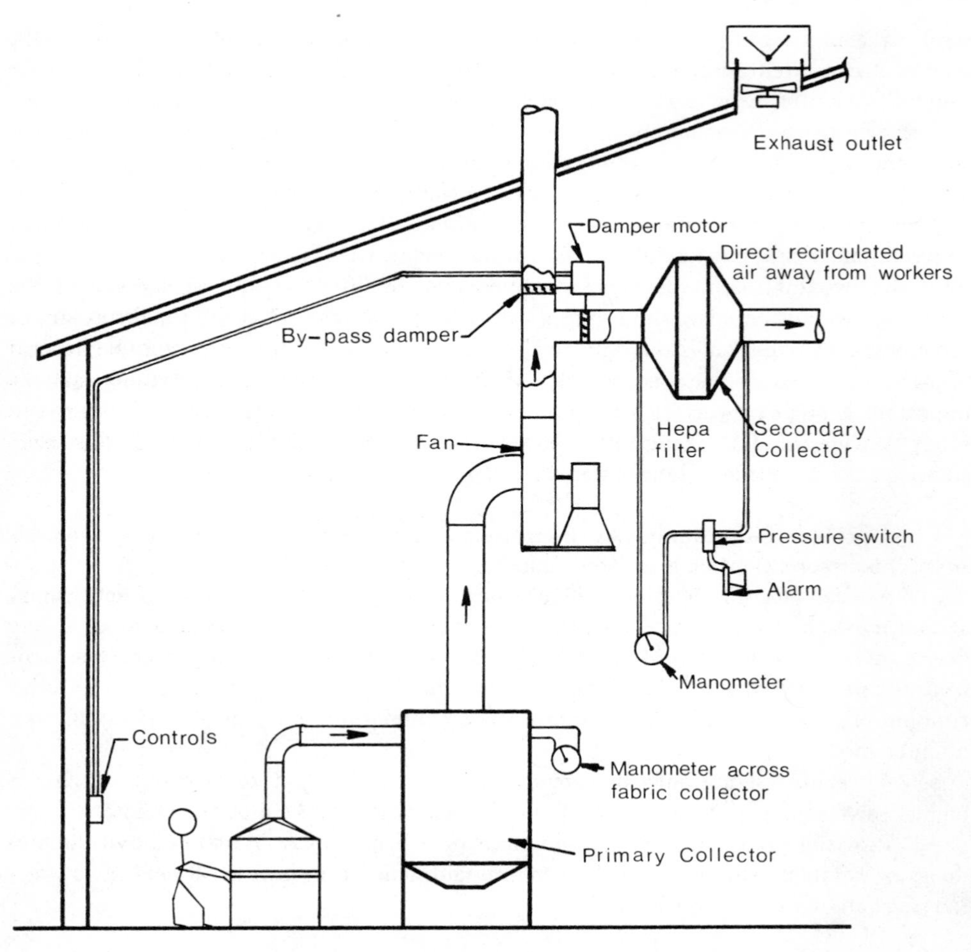

Figure 19.1 Secondary air-cleaning system with monitoring device (3).

Figure 19.1 gives a simple example of a suitable secondary air-cleaning system with a reliable monitoring device.

2 EMISSION CONTROL REQUIREMENTS

Requirements for emission controls generally find their basis in two broad types of overlapping expectation—regulatory standards and engineering performance specifications. Most air pollution regulations in the United States relate to or, at a minimum, conform with, federal regulations based on the Clean Air Act as amended in 1970 (2). Engineering specifications for air pollution controls have developed into increasingly precise stipulations that often reference the regulatory constraints.

2.1 Federal Clean Air Act

The Clean Air Act of 1970 provided the EPA with broad powers to adopt and enforce air pollution emission regulations. The agency subsequently promulgated National Primary and Secondary Air Quality Standards that set maximum ambient concentrations for oxidants, carbon monoxide, nitrogen dioxide, sulfur dioxide, nonmethane hydrocarbons, and particulate matter. Under a revised national strategy for air pollution control set forth by the Clean Air Act, each state was required to develop an EPA-approved "Implementation Plan" directed at source control, ensuring that the ambient air standards would be met according to a stipulated timetable. The plans had to meet with EPA approval. "Emergency Episode Plans" were also required, and a lawsuit by environmentalists resulted in the requirement to draw up "Significant Deterioration Regulations" (5).

Emergency Episode Plans apply in areas of the country where the ambient air quality does not meet the Secondary Air Quality Standards promulgated by the EPA. Such plans include phased shutdowns during periods when air contaminants reach high levels.

Significant Deterioration Regulations apply to areas where ambient air quality is superior to the Secondary Standards. They are set forth in increasingly restrictive levels that limit industrial growth. Accordingly, permission for the construction or expansion of industrial plants is contingent on increasing the ambient sulfur dioxide and particulate concentrations by no more than the levels given in Table 19.1.

The most direct and explicit requirements for emission controls set in motion by the Clean Air Act are those for a growing list of major sources and for highly toxic substances.

Table 19.1. Significant Deterioration Limitations ($\mu g/m^3$) (5)

	Area Designation	
Pollutant	Class I	Class II
Particulate matter		
Annual geometric mean	5	10
24-hour maximum	10	30
Sulfur dioxide		
Annual arithmetic mean	2	15
24-hour maximum	5	100
3-hour maximum	25	700

"New Source Performance Standards" are specific source-emission limitations set by the EPA and applicable to new or modified stationary sources in several major industrial categories. This provision of the law was intended to ensure that new stationary sources be designed, built, equipped, and maintained to reduce emissions to a minimum regardless of whether the sources are located in a clean area or in an area that requires strict controls.

"Hazardous Air Pollutants" are substances for which no ambient air standards have been set, but because they are defined by the EPA as presenting imminent health hazards, they are subject to emission standards under Section 112 of the amended Clean Air Act (2). The list of such hazardous substances now includes asbestos, beryllium, mercury, and vinyl chloride.

2.2 Typical Regulatory Requirements

Within today's multiplicity of regulatory constraints for abating air pollution from stationary sources, three basic requirements emerge to affect the operator or owner of emission sources. These are quantitative emission standards, subjective prohibitions, and installation and operating permits (air use approval).

2.2.1 Quantitative Emission Standards

Local, state, and federal regulations provide a complex spectrum of emission standards, ranging from general specifications of engineering standards to very precise emission limitations in specific concentration or mass-effluent units. Increasingly, these emission limitations are related to specific sampling and analytical methods. Typical of this class of regulations are those referenced to process-weight, effluent grain loading, and efflux weight limitations.

2.2.2 Subjective Prohibitions

Plume Visibility. Regulations covering plume appearance are usually expressed in terms of an allowable equivalent opacity or Ringelmann number, a subjective measure of visual obscuration. The original use of plume visibility as a regulatory tool covered the evaluation of the density of black smoke. Today most visual plumes are neither black nor soot containing; their visibility is due to fly ash, dust, fumes, and condensed water vapor, and they often are white or light colored.

Although plume visibility is widely used by enforcement agencies as a convenient index for monitoring emission sources, certain inconsistencies in the equivalent opacity concept are obvious to the objective mind. These include the dependence of plume opacity on particle size, shape, and refractivity, the typical lack of correlation between opacity and the amount of particulate matter in a plume, and the dependence of the opacity on background and relative position of the observer with respect to the sun.

To decrease the subjective element in plume appearance reading, the EPA and other agencies have developed "smoke schools" where observers are certified on their ability to read "calibrated" dark and light plumes accurately and consistently.

Nuisance and Trespass. Typical of the nuisance and trespass class of restriction are the portions of air pollution regulations prohibiting the emission of an air contaminant that causes or "will cause" detriment to the safety, health, welfare, or comfort of any person, or that causes or "will cause" damage to property or business. Such sections of regulations are typically known as "nuisance clauses," and they supplement the quantitative or objective restrictions, as well as providing for administrative and legal relief to receptors when emissions jeopardize aesthetic values, comfort, or well-being.

2.2.3 Installation and Operating Permits

It is very common today, especially at the state and local agency levels, for air pollution regulations to require at least one and usually two permits before a newly operating emission source or emission control system is certified to be in compliance with regulatory requirements. An "installation permit" is a sanction granted by the regulatory agency before construction begins and after consideration by the agency of the emission sources and the nature and engineering characteristics of the control system. It is at this stage that predictive dispersion modeling, even if rudimentary, can help assure maintenance of acceptable air quality downwind from the proposed installation.

Predictive Air Quality Modeling. The well-known technique of air quality dispersion modeling has become a widespread requirement for predicting the impact of new or modified stationary emission sources and emission control systems in terms of expected downwind concentrations of the contaminant. Preconstruction dispersion modeling examines the proposed emission or control system configuration and any alternatives being considered, to estimate the maximum downwind concentrations as they relate to the national ambient air quality standards and, where appropriate, to the EPA Significant Deterioration Regulations for sulfur dioxide and suspended particulate matter. Such modeling is mandated by the EPA for the various state agencies as they administer the EPA-approved Implementation Plans designed to achieve the ambient concentrations specified in the national air quality standards and to restrict significant increases in the atmospheric burden in regions of superior air quality.

Certification. After installation of the source or control system is complete and the system begins to operate, a second permit must be obtained in many jurisdictions, attesting that the source operates in a manner satisfactory to the control agency. Often an operating permit is granted after inspection and observation of the operating facility by a representative from the air pollution control agency. Nevertheless, in many instances, compliance source testing is required to document the level of effluent with reference to the applicable emission limitations.

2.3 Engineering Performance Specifications

Today's requirements for an emission control system usually include engineering performance specifications. It is technically sound, in general, that the performance specifications of air or gas cleaning equipment for a given application include any or all of the following factors:

1. Range of air-cleaning capacity in terms of volumetric flowrate.
2. The exit concentration of material in the system exhaust gases.
3. Mass emission rate of discharged material.
4. Efficiency of control system based on mass flowrate.
5. Acceptance test methods.

The value of specifying these parameters among such traditional specifications as materials of construction, fabrication, and inspection is obvious, yet these factors are important enough to warrant review.

2.3.1 Range of Air Cleaning Capacity

An emission control system must be designed to operate at sufficient capacity to exhaust all the gases that come from the contaminant-generating source. The application of even the latest technology to a given source is rather meaningless if the control unit is undersized and unable to capture and exhaust all the contaminants. Examples of inadequate system design are commonplace in industry, especially in primary and secondary metallurgical industries.

2.3.2 Exit Concentration of Material

If the object of an air-cleaning system is the recovery of suspended materials, it is important to know the percentage or amount of material collected or retained in the equipment. Usually, however, even though the recovery of material is an important consideration, the percentage of material collected is only incidental. If the object is to produce air of better quality, either for supply to occupied spaces or for discharge outdoors, performance should be specified in terms of the concentration of contaminant in the outgoing airstream.

2.3.3 Emission Rate of Discharged Material

The emission rate of material, mass per unit time, is defined by the concentration of material in the exit gas stream and the volumetric flowrate of the gas stream. Specification of only one or the other of these two factors is insufficient to characterize the expected performance of an air pollution control system. In many applications in industry poor maintenance or other factors result in an exit flowrate from a control system that is greater than the inlet flowrate; specification of the concentration alone could be misleading because of a dilution effect caused by the increased outlet flowrate.

Furthermore, the real impact of air pollution must be measured in terms of the emission rate to the atmosphere. Only this parameter provides a meaningful index of mass contribution to the atmospheric burden.

Finally, it is important to specify the emission rate of an air pollution control system either directly or indirectly (by flowrate and by concentration) because emission regulations are invariably expressed in terms of mass emission rate or emission factors that depend on both the mass emission rate and the process rate. When combined, mass emission rate and process rate yield such dimensional emission terms as pounds per ton of product produced, or pounds per ton of feed material.

2.3.4 Collector Efficiency

The term "efficiency" in air pollution control does not refer to the proportion of energy used effectively in the cleaning process. Rather, "efficiency" is used almost exclusively to indicate the proportion of material removed from an airstream, without regard to the amount of power required. Energy consumption, an increasingly important consideration in air pollution control, is treated separately in Section 3.

If a material were completely removed from an airstream by an air-cleaning device, the efficiency would be rated as 100 percent. The general definition of "efficiency" with respect to the degree of separation effected by an air pollution control system is the "emission rate entering the collector minus the emission rate leaving the collector divided by the emission rate entering the collector." For example, if the mass flowrate of material in a contaminated airstream entering a collector is 100 kg/hr and the emission rate is 5 kg/hr the "efficiency" of the control device is

$$100\left[\frac{100-5}{100}\right], \qquad \text{or 95 percent}$$

2.3.5 Acceptance Test Methods

Measurement of the control efficiency of a collector is accomplished with source sampling equipment operating as a miniature-scale separating and collection system with very high inherent effectiveness. The measured emissions, however, are very much dependent on the method of sampling. Therefore the performance of an air or gas cleaning system must be specified with reference to the sampling method to be used to determine the nature, amount, and mass flux of contaminants in the inlet and outlet of the system. Accordingly, the specific method of testing should be named in a statement of expected performance. If the specified method is not a standard method, such as an ASTM or EPA method, the sampling device or method should be described in enough detail to permit duplication.

3 ECONOMIC IMPACT OF AIR POLLUTION CONTROLS

Air pollution represents a problem for which the goals of public policy are "noneconomic," yet have important economic aspects. Federal, state, and local governments have major programs aimed at the prevention and abatement of atmospheric pollution. Nevertheless, many economic issues related to these programs are inadequately understood or documented. For example, would it be cheaper and more effective to disperse activities that pollute the atmosphere, to burn higher quality fuel, or to remove pollutants from process materials before effluents are discharged to the atmosphere? Current answers to these and similar questions are generally inadequate.

A macroeconomic analysis of air pollution control issues, although a very interesting and relevant topic, is not the subject of this section. Rather, the objective is to present some microeconomic considerations in air pollution control in terms of the cost–benefit relationships, the resources necessary to apply certain types of control, and the minimum requirements of an economic feasibility analysis needed to specify the most "economic" control methods for a given stationary source of air pollution from among suitable technical alternatives.

Many air pollution control problems can be solved by more than one alternative measure; but to identify the optimum method for controlling emissions, each technically adequate solution must be evaluated carefully before implementation. Sometimes basic measures such as the substitution of fuels or raw materials and the modification or replacement of processes provide the most cost-effective solution, and these avenues should never be overlooked (see Section 4).

3.1 Cost-Effectiveness Relationships

A cost-effectiveness "variable" measures all costs associated with a given project as a function of achievable reduction in pollutant emissions. For example, when estimating the total cost of an emission control system, we must consider raw materials and fuel used in the control process, needed alterations in process equipment, control hardware, auxiliary equipment, disposal or reuse of collected materials, and similar factors.

Figure 19.2 plots a typical cost-effectiveness relationship in stationary source emission control. The cost of control is represented on the vertical axis, and the quantity of material discharged is represented on the horizontal axis. Point P indicates the uncontrolled state at which there are no control costs. As control efficiency improves, the quantity of emissions is reduced and the cost of control increases. In most cases the marginal cost of control is smaller at lower levels of efficiency (higher emissions) near point P of the curve. This cost-effectiveness curve demonstrates an important reality: as the degree of control increases, greater increments of costs are usually required for corresponding increments in emission abatement.

Obviously cost-effectiveness information is needed for sound decision making in emission control. Although several technically feasible measures may be available for controlling an emission source, in most cases the least-cost solution for a source can be

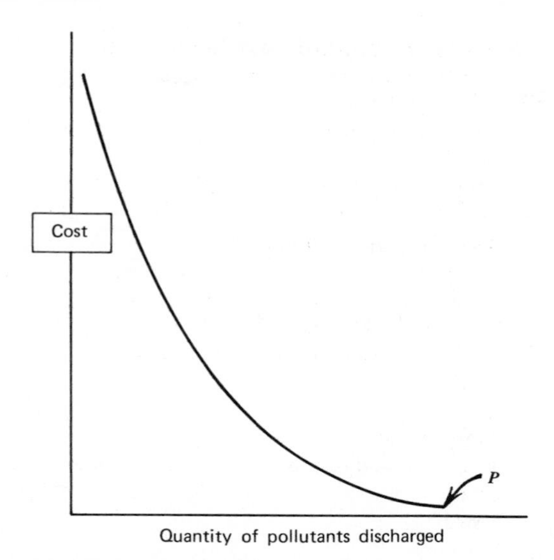

Figure 19.2 Cost-effectiveness relationship in stationary source emission control (6).

calculated at various levels of control. After evaluating each alternative, and after considering future process expansions and the likelihood of more rigid control restrictions, sufficient information should be available on which to base an intelligent control decision.

3.2 Control Cost Elements

The cost of installing and operating an air pollution control system depends on many direct and indirect cost factors. An accurate analysis of control costs for a specific emission source must include an evaluation of all relevant factors. The definable control costs are those directly associated with the installation and operation of control systems. These cost elements have been compiled into a logical organization for accounting purposes and are presented in Table 19.2. The control cost elements that receive attention in this discussion are capital investment, maintenance and operating costs (including energy consumption), and conservation of material and energy in the design and operation of air pollution control systems.

3.3 Capital Investment

The *installed cost* quoted by a manufacturer of air pollution control equipment should result from an analysis of the specific emission source. This cost usually includes three of the eight capital investment items listed in Table 19.2: control hardware costs, auxiliary equipment costs, and costs for field installation. Basic control hardware usually includes built-in instrumentation and pumps.

Table 19.2. Control Cost Elements (6)

Capital investment
Engineering studies
Land
Control hardware
Auxiliary equipment
Operating supply inventory
Installation
Startup
Structure modification
Maintenance and operation
Utilities
Labor
Supplies and materials
Treatment and disposal of collected material
Capital charges
Taxes
Insurance
Interest

Purchase costs are the amounts charged by a manufacturer for equipment constructed of standard materials. Purchase costs depend to a large degree on the size and collection efficiency of the control device. In addition, equipment fabricated with special materials for extremely high temperatures or corrosive applications generally costs much more than equipment constructed from standard materials.

3.4 Maintenance and Operating Costs

The costs of operating and maintaining air pollution controls depend not only on the inherent characteristics of the different control methods, but also on a wide range of quality and suitability of control equipment, the user's understanding of its operation, and his commitment to maintaining it in reliable operation.

Maintenance costs represent the expenditures required to sustain the operation of a control device at its designed efficiency, using a scheduled maintenance program and necessary replacement of any defective component parts. Simple control devices of low efficiency have low maintenance costs, and complex, highly efficient control systems generally have high maintenance costs.

Annual operating costs refer to the yearly expense of operating an emission control system at its designed collection efficiency. These costs depend directly on the following factors:

1. Volumetric rate of gas cleaned.
2. Pressure drop across the control system.

3. Annual operating time.
4. Consumption rate and cost of electricity.
5. Mechanical efficiency of the air-moving device and, where applicable, two additional items:
6. Scrubbing liquor consumption and replacement costs.
7. Consumption rate and cost of fuel.

The energy requirement for air pollution control equipment is an important and increasingly significant component of operating costs. The increased costs of fuel point to a growing emphasis on the role of energy conservation in industry generally, and in pollution control equipment specifically. Traditionally, the energy requirements for emission control equipment have been included in an overall consideration and expression of operating costs, but the current emphasis on energy costs leads logically to the isolation of this cost element for at least two purposes: (1) to relate operating costs to energy costs, and (2) to locate potential areas for energy conservation.

Because air pollution control equipment ranges from natural-draft stacks to combinations of collectors (cyclones, filters, electrostatic precipitators, scrubbers, and stacks with induced or forced-draft fan units), the energy consumption requirements of the various types of pollution control equipment must be examined separately. Table 19.3 gives the most commonly used air pollution control equipment and the associated power requirements. Power requirements are listed in kilowatts, since most pollution control equipment operates on electricity. Pressure drop is expressed in inches of water. The heating requirements for afterburners and regeneration of adsorption beds are given in Btu per hour. The typical requirements in Table 19.3 can be added together directly to obtain an estimate of total energy, pressure drop, and heating requirements for a complex chain of combined equipment. The total pressure drop can then be used to estimate the total fan power requirement.

As an example, consider a new 120-MW coal-burning power plant (7). The boiler exhaust emission controls could include an electrostatic precipitator for removing fly ash particulates, an alkali scrubber system for removing sulfur dioxide, and a fan and a tall stack to help achieve acceptable air quality by dispersion. The equipment power requirements for this system could then be estimated as follows:

Item	Electrical Energy (kW/1000 scfm)
Electrostatic precipitator	0.3
Limestone scrubber	12.1
Tall stack	0
Fan	0.1
Total exhaust system power	12.5

Table 19.3. Power Requirements for Air Pollution Equipment (7)

Item	Pressure Drop (in. H_2O)	Power[a] Required (kW per 1000 scfm), Typical Range	Remarks
Exhaust stacks	0	0	—[b]
Filters and separators			
Baghouses, cloth filters	1–30	0.1–1.0	Power function of rapping rate, particle size, cloth and dust-layer pressure drop
Impingement and gravitational separators	0.1–1.5	None	
Cyclones Single	0.1–2	None	ΔP = 1 to 20 inlet velocity heads
High efficiency	2–10	None	
Dry centrifugal	2–4	None	Combined separator–fan effect
Packed bed	1–10	None	
Electrostatic precipitators			
High voltage	0.1–1	0.2–0.6	
Low voltage	0.1–1	0.01–0.04	
Scrubbers		1–12	Power requirements for fan and pump
Centrifugal, mechanical	2–8	None	
Venturi High pressure drop	10–60	None	3 gal/min per 1000 cfm typical liquor rate
Low pressure drop	0.5–100	None	3 gal/min per 1000 cfm typical liquor rate

Spray tower	1–2	None	
Impingement and entrainment	4–20	None	
Packed-bed absorption	0.5 per foot of thickness	None	
Plate absorption plus liquor circulation pump[c]	1–3 per plate 1–60 ft	None (see manufacturer's pump curve)	1–20 gal/min liquor, some sprays to 600 psi
Cyclone	0.1–10	None	
Adsorption beds	1–30	100–200 Btu/lb adsorbent to heat to 600°F without heat exchange	Regeneration rate dependent on adsorbent and pollutant concentration; fan power may also be required for hot gas regeneration
Acoustic fume afterburners	0.1–2 (0.05–1 per in. of bed)	2 to 14 × 10^5 Btu/hr	1500°F, with 80% heat exchanger (2 × 10^5 Btu/hr); 1500°F, no regeneration, heat exchanger (14 × 10^5 Btu/hr)
Catalytic oxidizers	0.5–2 (0.05–1 per in. of bed)	0.1 to 1.0 × 10^6 Btu/hr	300–900°F, 80% heat exchanger to 900°F, without heat exchanger depends on inlet temperature

[a] "None" indicates that no direct power is required; the only power requirement is indirect from fans (due to pressure drop in equipment).
[b] The rise of hot gas in the stack creates a pressure head of about 0.5 in. H_2O, but this is counteracted by friction losses in the stack plus the desired velocity head leaving the stack for dispersions.
[c] Add scrubber pump power to scrubber power requirements.

For a 120-MW power plant with a stack flowrate of 300,000 standard cubic feet per minute (scfm) the electrical power required would be $300 \times 12.5 = 3750$ kW.

3.5 Material Conservation

Historically, except for precious metal applications, most air pollution control systems have collected material of little economic worth. Nevertheless, as resource recovery becomes more feasible because of current and expected shortages of many material resources, the economic worth of material collected by emission controls will gain in significance in the economic analysis of air pollution controls.

The alternatives for handling collected particulate remain as follows:

1. Recycle material to the process.
2. Sell material as collected.
3. Convert material to salable products.
4. Discard material in the most economical (and acceptable) manner.

In some process operations collected material is sufficiently valuable to warrant its return to the process, and the value of the recovered material can partially or wholly pay for the collection equipment. In many applications, however, the cost of the high efficiency control systems necessary to achieve desired ambient air quality would be greater than the revenue returned for recovery of the material collected.

The cement industry provides an example of collected material being routinely returned to the process. Not only does recovered dust, in such situations, have value as a raw material, its recovery also reduces disposal costs, as well as costs related to the preparation of raw materials used in the process.

Although material collected by air pollution control equipment may be unsuitable for return to a process within that plant, it may be suitable for another manufacturing activity. Hence it may be treated and sold to another firm that can use the material. Untreated, pulverized fly ash, for example, cannot be reused in a furnace, but it can be sold as a raw material to a cement manufacturer. It can also be used as a soil conditioner, as an asphalt filler, or as landfill material. When treated, pulverized fly ash can yield an even more valuable product. Some utilities, for example, sinter pulverized fly ash to produce a lightweight aggregate that can be used to manufacture bricks and lightweight building blocks.

3.6 Energy Recovery

Although there has been much complaining and comment about the high cost and scarcity of fuel, the fact remains that most exhaust stacks in industry are very hot, suggesting that much energy is being wasted.

The most economical approach to energy conservation and environmental control is a combined approach whenever possible. Real savings can be gained by combining heat recovery with air pollution control. Whereas in the past low fuel costs made it uneconomical to install heat recovery equipment in many industrial applications, such as on stacks of direct-fired furnaces, current shortages and the higher cost of fuel turn the spotlight on fuel economy now and in the foreseeable future.

A major advantage to the combined approach to energy and air pollution controls is the drop in temperature, caused by the recovery of sensible heat energy from an airstream, that can lead to a dramatic reduction in the volumetric exhaust rate and the requisite cost and size of air pollution control equipment. Thus it is possible to save not only on capital expenses but also on operating costs, as well as in the consumption of scarce and increasingly expensive fuel. Recovered heat can be used to reduce fuel requirements by preheating primary combustion air, preheating air going to another process, or preheating other fuel; or it can be used in heating buildings.

3.7 Economic Analysis Models

We present cost estimates for the installation and operation of five general types of control equipment:

1. Gravitational and inertial collectors.
2. Fabric filters.
3. Wet scrubbers.
4. Electrostatic precipitators.
5. Afterburners.

3.7.1 Gravitational and Inertial Collectors

In general, the only significant cost for operating mechanical collectors is the electric power cost, which varies with the unit size and the pressure drop. Since pressure drop in gravitational collectors is low, operational costs associated with these units are considered to be insignificant. Maintenance cost includes the costs of servicing the fan motor, replacing any lining worn by abrasion, and, for multiclone collectors, flushing the clogged small diameter tubes.

The theoretical annual cost G of operation and maintenance for centrifugal collectors can be expressed as follows (6):

$$G = S\left(\frac{0.7457PHK}{6356E} + M\right) \tag{2}$$

where S = design capacity of the collector (actual cubic feet per minute: acfm)
P = pressure drop (in. H_2O)
E = fan efficiency, assumed to be 60 percent (expressed as 0.60)
0.7457 = a constant (1 hp = 0.7457 kW)
H = annual operating time (assumed 8760 hours)
K = power cost (dollars per kilowatt-hour)
M = maintenance cost (dollars per acfm)

For computational purposes, the cost formula can be simplified as follows (6):

$$G = S\,(195.5 \times 10^{-6}\, PHK + M)$$

3.7.2 Fabric Filters

Operating costs for fabric filters include power costs for operating the fan and the bag cleaning device. These costs vary directly with the size of the equipment and the pressure drop. Maintenance costs include costs for servicing the fan and shaking mechanism, emptying the hoppers, and replacing the worn bags.

The theoretical annual cost G for operation and maintenance of fabric filters is as follows (6):

$$G = S\left(\frac{0.7457\ PHK}{6356E} + M\right) \tag{3}$$

units are defined as in (2).

For computational purposes, the cost formula can be simplified as follows (6):

$$G = S\,(195.5 \times 10^{-6}\, PHK + M)$$

3.7.3 Wet Collectors

The operating costs for a wet collector include power and scrubbing liquor costs. Power costs vary with equipment size, liquor circulation rate, and pressure drop. Liquor consumption varies with equipment size and stack gas temperature. Maintenance includes servicing the fan or compressor motor, servicing the pump, replacing worn linings, cleaning piping, and any necessary chemical treatment of the liquor in the circulation system.

The theoretical annual cost G of operation and maintenance for wet collectors can be expressed as follows (6):

$$G = S\left[0.7457HK\left(\frac{P}{6356E} + \frac{Q_g}{1722F} + \frac{Q_h}{3960F}\right) + WHL + M\right] \tag{4}$$

where S = design capacity of the wet collector (acfm)
0.7457 = a constant (1 hp = 0.7457 kW)
H = annual operating time (assumed 8760 hours)
K = power costs (dollars per kilowatt-hour)
P = pressure drop across fan (in. H_2O)
Q = liquor circulation (gallons per acfm)
g = liquor pressure at the collector (psig)
h = physical height liquor is pumped in circulation system (ft)
W = makeup liquor consumption (gallons per acfm)

L = liquor cost (dollars per gallon)
M = maintenance cost (dollars per acfm)
E = fan efficiency, assumed to be 60 percent (expressed as 0.60)
F = pump efficiency, assumed to be 50 percent (expressed as 0.50)

This equation can be simplified according to Semrau's total "contacting power" concept (8). Semrau shows that efficiency is proportional to the total energy input to meet fan and nozzle power requirements. The scrubbing (contact) power factors in Table 19.4 were calculated from typical performance data listed in manufacturers' brochures. These factors are in general agreement with data reported by Semrau. Using Semrau's concept, the equation for operating cost can be simplified as follows (6):

$$G = S\left[0.7457HK\left(Z + \frac{Q_h}{1980}\right) + WHL + M\right] \tag{5}$$

where Z = contact power; that is, total power input required for collection efficiency (hp per acfm; Table 19.4). It is a combination of:

1. Fan horsepower per acfm

$$\left(= \frac{P}{6356E}\right) \tag{6}$$

2. Pump horsepower per acfm

$$\left(= \frac{Q_g}{1722F} \text{ the power to atomize water through a nozzle}\right) \tag{7}$$

The pump horsepower $Q_h/1980$ required to provide pressure head is not included in the contact power requirements.

3.7.4 Electrostatic Precipitators

The only operating cost considered in running electrostatic precipitators is the power cost for ionizing the gas and operating the fan. Depending on the length and design of

Table 19.4. Contact Power Requirements for Wet Scrubbers (6)

	Scrubber Efficiency		
Parameter	Low	Medium	High
"Scrubbing" (contact) power (horsepower per acfm)	0.0013	0.0035	0.015

the upstream and downstream ductwork, the fan energy requirements may be relatively small. Since the pressure drop across the equipment is usually less than 0.5 in. H_2O, the cost of operating the fan is assumed to be negligible if minimal ductwork is used. The power cost varies with the efficiency and the size of the equipment.

Maintenance usually requires the services of an engineer or highly trained operator, in addition to regular maintenance personnel. Maintenance includes servicing fans and replacing damaged wires and rectifiers.

The theoretical annual cost G for operation and maintenance of electrostatic precipitators is as follows (6):

$$G = S\,(JHK + M) \tag{8}$$

where S = design capacity of the electrostatic precipitator (acfm)
J = power requirements (kilowatts per acfm)
H = annual operating time (assumed 8760 hours)
K = power cost (dollars per kilowatt-hour)
M = maintenance cost (dollars per acfm)

3.7.5 Afterburners

The major operating cost item for afterburners is fuel. Fuel requirements are a direct function of the gas volume, the enthalpy of the gas, and the difference between inlet and outlet gas temperatures. For most applications, the inlet gas temperature at the source ranges from 50 to 400°F. Outlet temperatures may vary from 1200 to 1500°F for direct flame afterburners and from 730 to 1200°F for catalytic afterburners. The use of heat exchangers may bring about a 50 percent reduction in the temperature difference.

The equation for calculating the operation and maintenance costs G is as follows (6):

$$G = S\left(\frac{0.7457\,PHK}{6356E} + HF + M\right) \tag{9}$$

where S = design capacity of the afterburner (acfm)
P = pressure drop (in. H_2O)
E = fan efficiency, assumed to be 60 percent (expressed as 0.60)
0.7457 = a constant (1 hp = 0.7457 kW)
H = annual operating time (assumed 8760 hours)
K = power cost (dollars per kilowatt-hour)
F = fuel cost (dollars per acfm per hour)
M = maintenance cost (dollars per acfm)

For computational purposes, the cost formula is simplified as follows (6):

$$G = S\,(195.5 \times 10^{-6}\,PHK + HF + M)$$

4 PROCESS AND SYSTEM CONTROL

Process and system control implies a careful review of a production unit operation within the context of air pollution control to examine whether the manufacturing process is optimum, considering the emission rate or necessary control and treatment of the process effluents.

A fundamental notion in air pollution control asserts: "The problem can be solved best if it is solved at the source of emissions." This simple rule, if applied consistently, could greatly reduce the number and complexity of emission problems in industry. Methods for total or partial "control at the source" include elimination of emissions, minimization of emissions, and concentration of contaminants before discharge.

4.1 Elimination of Emissions

Too often, "emission control" denotes the addition of some device or system to do the controlling. Nevertheless, substitution of materials, processes, or equipment may be the least expensive as well as the most positive method for abating an atmospheric emission.

4.1.1 Fuel Substitution

An excellent example of the potentially beneficial effect of material substitution is conversion from a "dirty" fuel to a "clean" fuel, such as a switch from coal to gas in utility boilers. In the late 1960s and early 1970s many state and local pollution control agencies enacted regulations to curb sulfur dioxide emissions from coal- and oil-burning power plants. The effect of such regulations was a trend toward the use of low sulfur oil and coal and, where possible, to the use of the cleanest fossil fuel, natural gas. Because of the shortages and higher costs of natural gas and low sulfur coal and oil, however, the achievement and maintenance of air quality goals cannot be accomplished solely by switching to naturally available low sulfur fuels. It is apparent that coal with medium to high sulfur content will have to be burned in increasing quantities by power plants to meet growth requirements and to safeguard an adequate supply of clean fuels for residential, commercial, and industrial use.

4.1.2 Process Changes

Process changes can be as effective as material substitution in eliminating air polluting emissions. It has often been possible and even profitable in the chemical and petroleum industries to control the loss of volatile organic materials to the atmosphere by condensation, leading to the "reuse" of otherwise fugitive vapors. The compressors, absorbers, and condenser units on petroleum product process vessels provide a good example of such process modification.

In the case of burning fossil fuels, process change can also be construed to encompass the addition of plants auxiliary to the main process that generate substitute fuels. As coal gasification processes assume a larger role in the industrial complex, the low sulfur "synthetic" fuels enable the use of high sulfur fuels through a process change that may be feasibly interwoven with main process streams. One familiar example is the recycling of distilled volatiles in coke production. When high sulfur coals are distilled destructively, they produce gas streams that when treated, are useful energy sources, producing low sulfur oxides emissions on burning. Recycling a portion of this fuel to the coke ovens supplies energy for further fuel gas production from the next charge of coal in the ovens while also providing product coke.

The addition of a step to reduce atmospheric emissions has been commonplace in brass foundry practice when indirect fired furnaces are employed. A fluxing material applied to the surface of the molten brass serves as an evaporation barrier and reduces the evolution of brass fumes; this additional step was developed strictly as an air pollution control measure.

4.1.3 Equipment Substitution

An example of the benefit to be derived from equipment substitution is the trend in the polyvinyl chloride (PVC) industry to use large polymerization reactors, which not only boost productivity but help to stem fugitive emissions of vinyl chloride monomer. A large PVC reaction vessel has only half the possible fugitive emission leak points of two smaller units of the same total capacity (9). As late as 1972 the PVC industry relied on reactors of capacity smaller than 7500 gallons for 85 percent of its output. Now 20,000-gallon vessels and even some massive 55,000-gallon reactors are employed in the PVC industry to control vinyl chloride emissions (9).

4.2 Reduction of Emissions

One way to reduce air pollution control costs is to minimize, if not eliminate, release of the contaminant. Such action is cost effective both for fugitive emissions and process emissions.

4.2.1 Fugitive Emissions

Leaking conveyor systems can be modified to eliminate spilling and dusting. Airtight enclosures can be built around conveyors, or cover housings can be equipped with flanges and soft gaskets. Shaft bearings can be redesigned and relocated inside the conveyor housing to prevent dust leakage around the shaft.

Tanks and bins should have airtight covers and joints sealed with cemented strips of plastic or rubber sheeting. Each bin or tank normally should have only one vent where temperature difference can create a natural draft from the vent; other openings should be kept closed.

Where possible, interconnecting a series of vents from several tanks should be considered. While one tank is being filled, another can be emptied, thus reducing the need for exhausting air to the atmosphere. It may also be possible to have one small dust collector serve a number of units.

4.2.2 Process Emissions

When contaminants are a by-product of a production operation, generation can often be minimized by changes in operating conditions. Lower combustion temperatures will reduce nitrogen oxides formation. Levels of fluorine-containing compounds, released when some ores are heated, frequently can be decreased if waper vapor is eliminated from the atmosphere.

Reducing excess air in coal- or oil-fired boiler systems can result in a substantial reduction in the conversion of sulfur dioxide to sulfur trioxide, thus reducing sulfuric acid emissions.

4.3 Concentrating Pollutants at the Source

One major phase of process and system control involves centralizing and decreasing the number of emission points and reducing the effluent volume of the exhaust gas to be treated.

4.3.1 Reducing the Number of Emission Points

Many industrial operations, such as machining in metalwork industries, are characterized by the replication of identical or similar worker-operated unit operations. Examples include grinding, sawing, and cutting. Control of worker exposures and effective, economical emission control can often be achieved by clustering the operations together and exhausting them locally and simultaneously to a common collection system, rather than exhausting each operation to an individual collector. The concept of centralized collection and abatement offers the advantage of economy of scale for collection systems; this tends to decrease control costs per unit weight of material collected. This advantage is due in part to the relatively high concentration of pollutants at a combined source compared to the more dilute concentrations, which would be experienced from collecting and exhausting materials from each similar operation independently.

4.3.2 Volume Reduction by Cooling

Many dust collectors such as bag filters and precipitators are sized more on a volumetric flow basis than on a mass flow basis. If the effluent is a hot, dusty gas, cooling can reduce the volume appreciably. Cooling by the addition of cold air is the poorest method from the standpoint of reducing cost. Radiation panels, fin surfaces, waste heat boilers, forced convection, heat interchange, and direct spray cooling with water all are possible means of effecting volume reduction.

4.3.3 Optimal Local Exhaust Ventilation

Well-designed and properly installed local exhaust systems are an important aspect of control measures to minimize exhaust flowrate and maximize concentrations in exhaust gases to be cleared before discharge to the atmosphere. By careful design, local exhaust hoods can be applied to assure complete capture of all contaminants. Close-fitting hoods not only ensure better control of materials that may escape into the work environment, they also provide minimal exhaust ventilation of processes that produce contaminants. Accordingly, local exhaust is preferable to general ventilation with respect to emission abatement.

In control systems abating, for example, solvent vapors in air using carbon bed adsorption, both capital and operating costs can be reduced by increasing solvent concentration by optimal use of local exhaust. Very dilute solvent concentrations require massive beds of activated carbon relative to the amount of solvent recovered per pound of carbon, because of the low "driving force" that determines the rate of adsorption for a given solvent. The initial capital investment cost for such a system could be reduced measurably by reducing, if possible, the air volume used to collect and carry the solvent vapors produced by the process. Smaller air volumes translate to smaller equipment and increased "driving force," which also decreases required equipment size. Clearly any decrease in air volume also reduces the amount of electric power consumed by air movers; the yearly savings in operating costs can be surprisingly substantial.

5 TECHNICAL CRITERIA FOR SELECTING AIR-CLEANING METHODS

The cost effectiveness of an air pollution control system is closely related to its degree of control, its capacity, and the type of control system. Often more than one type of control could be used to solve a specific emission problem. Nevertheless, there is generally only one type that is optimal from both technical and economic standpoints.

Some of the most basic technical factors affecting selection of equipment are as follows:

1. System volumetric flowrate.
2. Concentration (loading) of contaminant.
3. Size distribution of particulate contaminants.
4. Degree of cleaning required.
5. Conditions of air or gas stream with reference to temperature, moisture content, and chemical composition.
6. Characteristics of the contaminant, such as corrosiveness, solubility, reactivity, adhesion or packing tendencies, specific gravity, surface, and shape.

This section focuses on these and other technical criteria that are helpful in selecting emission controls.

5.1 Performance Objectives

The prime factor in the selection of control equipment is the maximum amount or rate of contaminant to be discharged to the atmosphere. Knowledge of this amount, together with knowledge of the amount of contaminant entering a proposed collection system, defines the required collection efficiency; this begins the technical selection of air-cleaning methods.

If the material to be collected is particulate, it is essential to understand that collectors have different efficiencies for different sized particles. Therefore the particle size distribution of the emission must be known before the collector efficiency required for each particle size range can be determined. Collector efficiency also varies with gas flowrate and with properties of the carrier gas, which may fluctuate with flowrate or time. Such variations need to be considered carefully in determining collector efficiency.

When the material to be collected is a gas or vapor, it is necessary to know to what extent the material is soluble in the scrubbing liquid or retainable on the adsorbing surfaces. The determinations must be made for the concentrations expected in the collector inlet and outlet streams and for the conditions of temperature, pressure, and flowrate expected.

5.2 Contaminant Properties

Contaminant properties, as distinguished from carrier gas properties, comprise both chemical and physical characteristics of the material to be removed from an exhaust airstream.

5.2.1 Loading

Contaminant loading from many processes varies over a wide range for an operating cycle. Variations of an order of magnitude in concentration are not uncommon. Well-known examples of such variation include the basic oxygen furnace in steel making and soot blowing in a steam boiler. Contaminant loading may also vary with the carrier gas flowrate. Particularly in the case of gases, concentration is all-important in predicting removal efficiency and specifying system design parameters.

5.2.2 Composition

The composition of a contaminant affects both physical and chemical properties. Chemical properties, in turn, affect physical properties. For example, if collected material is to be used in a process or shipped in a dry state, a dry collector is indicated.

If the collected material has a very high intrinsic value, a very efficient collector is called for.

Since chemical and physical properties vary with composition, a collector must be able to cope with both expected and unexpected composition changes. For example, in the secondary aluminum industry a collector must be able to deal with the evolution of aluminum chloride during chlorine "demagging." The aluminum chloride levels vary widely throughout the cycle; peak levels last for only a few minutes, but continue to develop in decreasing amounts throughout a full cycle of 16 hours or more.

Further examples include adsorption, in which solubility may be important to the ease with which the adsorbent may be regenerated, and scrubbing to remove particulate matter, in which wettability of the dust by the scrubbing liquor aids the collector mechanisms and the basic separating forces that determine scrubber performance.

5.2.3 Combustibility

Generally, it is not desirable to use a collection system that permits accumulation of "pockets" of contaminant when the contaminant collected is explosive or combustible. Systems handling such materials must be protected against accumulation of static charges. Electrostatic precipitators are not suitable for such contaminants because of their tendency to spark. Wet collection by scrubbing or absorption methods is especially appropriate. Some dusts, however, such as magnesium, are pyrophoric in the presence of small amounts of water. In combustion (with or without a catalyst), explosibility must always be considered.

5.2.4 Reactivity

Certain obvious precautions must be taken in the selection of equipment for the collection of reactive contaminants. In filtration, selection of the filtering media can present a special problem. In adsorption, since certain applications require that the adsorbed contaminant react with the adsorbent, the degree of reactivity is important. Where scrubbers are used to remove corrosive gases or particulate, the potential corrosiveness must be balanced against the potential savings when using corrosion-resistant construction. The decision is thus whether to use more expensive materials or to incur higher maintenance costs.

5.2.5 Electrical and Sonic Properties

The electrical properties of the contaminant can influence the performance of several types of collector. Electrical properties are considered to be a factor influencing the buildup of solids in inertial collectors. In electrostatic precipitators, such electrical properties of the contaminant as the resistivity are of paramount importance in determining collection efficiency and precipitator size, and they influence the ease with which collected particulate is removed by periodic cleaning of the collection surfaces. In fabric filtration, electrostatic phenomena may have direct and observable effects on the process of cake formation and the subsequent ease of cake removal. In spray towers, where liquid droplets are formed and contact between these droplets and contaminant particles is required for particle collection, the electrical charge on both particles and droplets is an important parameter in determining collection efficiency. The process is most efficient when the charges on the droplet attract rather than repel those on the particle. Sonic properties are significant where sonic agglomeration is employed.

5.2.6 Toxicity

The degree of contaminant toxicity influences collector efficiency requirements and may necessitate the use of equipment that will provide ultrahigh efficiency. Toxicity also affects the means for removal of collected contaminant from the collector and the means of servicing and maintaining the collector. However toxicity of the contaminant does not influence the removal mechanisms of any collection technique.

5.2.7 Particle Size, Shape, and Density

Size, shape, and density are three properties of particulate matter that determine the magnitude of forces resisting movement of a particle through a gas stream. These are the major factors determining the effectiveness of removal in inertial collectors, gravity collectors, venturi scrubbers, and electrostatic precipitators. In these collectors resisting forces are balanced against some removal force (e.g., centrifugal force in cyclones) that is applied in the control device, and the magnitude of the net force tending to remove the particle determines the effectiveness of the equipment.

Size, shape, and density of a particle can be related to terminal settling velocity, which is a useful parameter in the selection of equipment for particulate control. Settling velocity is derived from Stokes' law, which equates the velocity at which a particle will fall at constant speed (because of a balance of the frictional drag force and the downward force of gravity) to the properties of the particle and the viscosity of the gas stream through which it is settling.

Terminal settling velocities can be determined by any of a number of standard techniques and used in the quantitative evaluation of the difficulties to be anticipated in designing particulate removal equipment.

Since particle size is related to the ease with which individual particles are removed from a gas stream, it is apparent that size distribution largely determines the overall efficiency of a particular piece of control equipment. Generally, the smaller the particle size to be removed, the greater the expenditure required for power or equipment or both. To increase the efficiencies obtainable with scrubbers, it is necessary to expend additional power either to produce high gas stream velocities, as in the venturi scrubber, or to produce finely divided spray water. Cyclones call for the use of a larger number of small units for higher efficiency in a given situation. Both the power cost and equipment cost are increased. Achieving higher efficiencies for electrostatic precipitators necessitates the use of a number of units or fields in series because there is an approximately inverse logarithmic relationship between outlet concentration and the size of collection equipment. A precipitator giving 90 percent efficiency must be approximately doubled in size to give 99 percent efficiency and approximately tripled to give 99.9 percent efficiency.

5.2.8 Hygroscopicity

Although not specifically related to any removal mechanism, hygroscopicity may be a measure of how readily particulate will cake or tend to accumulate in equipment if moisture is present. If such accumulation occurs on a fabric filter, it may completely blind it and prevent gas flow.

5.2.9 Agglomerating Characteristics

Collectors are sometimes used in series, with the first collector acting as an agglomerator and the second collecting the particles agglomerated in the first one. In carbon black collection, for example, extremely fine particles are first agglomerated; then they can be collected practicably.

5.2.10 Flow Properties

Flow properties of the material are mainly related to the ease with which the collected dust may be discharged from the collector. Extreme stickiness may eliminate the possibility of using equipment such as fabric filters. Hopper size and shape depend in part on the packing characteristics or bulk density of the collected material and its flow properties. Hygroscopic materials, or collected dust tending to cake, exhibit flow properties that change with time as the dust remains in the hopper. Hopper heaters may be required.

5.3 Carrier Gas Properties

Carrier gas properties are important insofar as they affect the selection of control equipment, especially with reference to composition, reactivity, conditions of temperature, pressure, moisture, solubility, condensability, combustibility, and toxicity.

5.3.1 Composition

Gas composition affects physical and chemical properties, which are important to the extent that there may be chemical reactions between the contaminants and the collector, either in its structure or in its contents. One common reaction between gaseous components and equipment is the corrosion of metallic parts of collectors when gases contain sulfur oxides and water vapor.

Composition, concentration, and chemical reaction properties of the inlet stream determine the collection efficiency in packed-tower scrubbers removing gaseous or vapor phase contaminants.

5.3.2 Temperature

The temperature of the carrier gas principally influences the volume of the carrier gas and the materials of construction of the collector. The volume of the carrier gas influences the size and cost of the collector and the concentration of the contaminant per unit volume; concentration in turn is the driving force for removal. In addition, viscosity, density, and other gas properties are temperature dependent. Temperature also affects the vapor–liquid equilibria in gaseous contaminant scrubbing such that scrubbing efficiency for partially soluble gases decreases with increasing temperature.

Adsorption processes are generally exothermic and are impracticable at higher temperature, the adsorbability being inversely proportional to the temperature (when the reaction is primarily physical and is not influenced by an accompanying chemical reaction). Similarly, in absorption (where gas solubility depends on the temperature of the solvent), temperature effects may have significance if the concentration of the soluble material is such that appreciable temperature rises result. When combustion is used as a means for contaminant removal, the gas temperature affects the heat balance, which is the vital factor in the economics of the process. In electrostatic precipitation, both dust resistivity and the dielectric strength of the gas are temperature dependent.

Wet processes cannot be used at a temperature that would cause the liquid to freeze, boil, or evaporate too rapidly. Filter media can be used only in the temperature range within which they are stable. The filter structure must retain structural integrity at the operating temperature.

Finally, low temperature gases flowing from a stack downstream of control equipment disperse in the atmosphere less effectively than do high temperature gases. Consequently, benefits derived from partial cleaning accompanied by cooling may be offset if the cooler exhaust gas cannot be well dispersed. This is a factor of importance in wet cleaning processes for hot gases, where the advantage gained by cleaning is sometimes offset by downwash from the stack near the plant because the exhaust gas is cooled. In the case of wet collection devices, the effluent gases may present a visually objectionable steam plume or even "rain out" in the stack vicinity. Raising the discharge gas temperature may eliminate or reduce these problems.

5.3.3 Pressure

In general, carrier gas pressure much higher or lower than atmospheric pressure requires that the control equipment be designed as a pressure vessel. Some types of equipment are much more amenable to being designed into pressure vessels than others. For example, catalytic converters are incorporated in pressure processes for the production of nitric acid and provide an economical process for the reduction of nitrogen oxides to nitrogen before release to the atmosphere.

Pressure of the carrier gas is not of prime importance in particulate collection except for its influence on gas density, viscosity, and electrical properties. It may, however, have importance in certain special situations—for example, when the choice is between high efficiency scrubbers and other devices for collection of particulate. The available source pressure can be used to overcome the high pressure drop across the scrubber, reducing the high power requirement that often limits the utilization of scrubbers. In adsorption, high pressure favors removal and may be required in some situations.

5.3.4 Viscosity

Viscosity is of importance to collection techniques in two respects. First, it is important to the removal mechanisms in many situations (inertial collection, gravity collection, and electrostatic precipitation). Particulate removal technique involves migration of the particles through the gas stream under the influence of some removal force. Ease of migration decreases with increasing viscosity of the gas stream. Second, viscosity influences the pressure drop across the collector, thereby becoming a major parameter in power requirement computation.

5.3.5 Density

Density of the carrier gas, for the most part, has no significant effect in most real gas cleaning processes, although the difference between particle density and gas density appears as a factor in the theoretical analysis of all gravitational and centrifugal collection devices. Particle density is so much greater than gas density that the usual changes in gas density have negligible effects. Carrier gas density does influence fan power requirements, therefore is important to fan selection and operating cost. Furthermore, special precautions must be taken in "cold startup" of a fan designed to operate at high temperature, to ensure that the motor capacity is not exceeded.

5.3.6 Humidity

Humidity of the carrier gas stream may affect the selection and performance of control equipment in any of several basically different ways. High humidity may contribute to accumulations of solids and lead to the caking and blocking of inertial collectors as well as the caking of filter media. It can also result in cold spot condensation and aggravation of corrosion problems. In addition, the water vapor may act on the basic mechanism of removal in electrostatic precipitation and greatly influence resistivity. In catalytic combustion it may be an important consideration in the heat balance that must be maintained. In adsorption it may tend to limit the capacity of the bed of water preferentially or concurrently adsorbed with the contaminant. Even in filtration it may influence agglomeration and produce subtle effects.

The above-mentioned considerations are the main limitations on the utilization of evaporation cooling in spite of its obvious power advantage. When humidity is a serious problem for one of the foregoing reasons, scrubbers or adsorption towers may be particularly appropriate devices. Humidity also affects the appearance of the stack exhaust gas discharged from wet collector devices. Because steam plumes are visually objectionable, sometimes it is necessary to heat exhaust gases to raise their dew point before discharge. This can add considerable operating expense to the total system.

5.3.7 Combustibility

The handling of a carrier gas that is flammable or explosive requires certain precautions. The most important is making sure that the carrier gas is either above the upper explosive limit or below the lower explosive limit for any air admixture that may exist or occur. The use of water scrubbing or adsorption may be an effective means of minimizing the hazards in some instances. Electrostatic precipitators are often impractical, since they tend to spark and may ignite the gas.

5.3.8 Reactivity

A reactive carrier gas presents special problems. In filtration, for example, the presence of gaseous fluorides may eliminate the possibility of high temperature filtration using glass fiber fabrics. In adsorption, carrier gas must not react preferentially with the adsorbents. For example, silica gel is not appropriate for adsorption of contaminants when water vapor is a component of the carrier gas stream. Also, the magnitude of this problem may be greater when a high temperature process is involved. On the one hand, devices that use water may be eliminated from consideration if the carrier gas reacts with water. On the other hand, scrubbers may be especially appropriate in that they tend to be relatively small and require small amounts of construction material, permitting the use of corrosion-resistant components, with lower relative increase in cost.

5.3.9 Toxicity

When the carrier gas is toxic or is an irritant, special precautions are needed in the construction of the collector, the ductwork, and the means of discharge to the atmosphere. The entire system, including the stack, should be under negative pressure, and the stack must be of tight construction. Since the collector is often under "suction," special means such as "airlocks" must be provided for removing the contaminant from the hoppers if collection is by a dry technique. Special precautions may also be required for service and maintenance operations on the equipment.

5.3.10 Electrical and Sonic Properties

Electrical properties are important to electrostatic precipitation because the rate or ease of ionization will influence removal mechanisms.

Generally speaking, intensity of Brownian motion and gas viscosity increase with gas temperature. These factors are important gas stream characteristics that relate to the "sonic properties" of the stream. Increases in either property tend to increase the effectiveness with which sonic energy can be used to produce particle agglomeration.

5.4 Flow Characteristics of Carrier Gas

5.4.1 Volumetric Flowrate

The rate of evolution from the process, the temperature of the effluent, and the degree and the means by which it is cooled, if cooling is used, fix the rate at which carrier gases must be treated, therefore the size of removal equipment and the rate at which gas passes through it. For economic reasons it is desirable to minimize the size of the equipment. Optimizing the size and velocity relationship involves consideration of two effects: (1) reduction in size results in increased power requirements for handling a given amount of gas because of increased pressure loss within the control device, and (2) velocity exerts an effect on the removal mechanisms. For example, higher velocities favor removal in inertial equipment up to the point of turbulence, but beyond this, increased velocity results in decreased efficiency. In gravity settling chambers, flow velocity determines the smallest size that will be removed. In venturi scrubbers efficiency is proportional to velocity through the system. In absorption, velocity affects film resistance to mass transfer. In filtration, the resistance of the medium often varies with velocity because of changes in dust cake permeability with flow. In adsorption, velocity across the bed should not exceed the maximum that permits effective removal. Optimum velocities have not generally been established with certainty for any of the control processes because they are highly influenced by the properties of the contaminant and carrier gas as well as by the design of the equipment.

5.4.2 Variations in Flowrate

Rate variations result in velocity changes, thereby influencing equipment efficiency and pressure drop. Various control techniques differ in their abilities to adjust to flow changes. When rate variations are inescapable, it is necessary to (1) design for extreme conditions, (2) employ devices that will correct for flow changes, or (3) use a collector that is inherently positive in its operation. Filtration is most easily adapted to extreme rate variations because it presents a positive barrier for particulate removal. This process is subject to pressure drop variations, however, and generally the air-moving equipment will not deliver at a constant rate when pressure drop increases. In most other control techniques, variations in flow produce a change in the effectiveness of removal.

One means of coping with rate variation is the use of two collectors in series, one that improves performance with increasing flow (e.g., multicyclone) and one whose performance decreases with increasing flow (e.g., electrostatic precipitator). Some venturi scrubbers are equipped with automatically controlled, variable size throats. Changes in gas flowrate are automatically sensed, and the throat's cross-sectional area is changed correspondingly to maintain a relatively constant pressure drop and efficiency over a relatively wide range of conditions.

6 CLASSIFICATION OF AIR POLLUTION CONTROLS

In a broad sense, air pollution controls for stationary source emissions include process and system control, air cleaning methods, and the use of tall stacks. Process and system control was covered in Section 4. This section focuses on the other types of air pollution control and offers an overview of the role of each method. More detailed information on air-cleaning methods follows in Sections 7 through 12.

6.1 Tall Stacks

In the context of air pollution control, a tall stack has the simple function of discharging exhaust gases at an elevation high enough to put the maximum concentration of exhaust gas material experienced at ground elevation within acceptable levels. Increasingly, however, it is unwise or illegal to rely only on tall stacks for solving air pollution problems. Nevertheless, there are several instances in which atmospheric dispersion and reliance on natural processes is sufficient to safeguard against excessive ground level concentrations. One example is that of sulfur dioxide (SO_2) discharged from power plants that burn fossil fuels. Because coal desulfurization and flue gas scrubbing are not economically feasible at this time for large boilers, the use of tall stacks, with heights determined by dispersion modeling, is an acceptable interim method of achieving compliance with the federal ambient air quality standards.

Ground level concentrations depend on the strength of the emission source, the physical and chemical nature of materials discharged, atmospheric conditions, stack and exhaust gas parameters, topography, and the aerodynamic characteristics of the physical surroundings. The state of the art in atmospheric dispersion modeling continues to advance, and air quality modeling is embodied in many state and local air pollution regulations and is sanctioned as well by the Environmental Protection Agency (see Section 2.2).

The computational details of atmospheric dispersion modeling are outside the scope of this discussion. Several excellent publications (10–12) illustrate the rationale and methods for estimating ground level concentrations downwind from stationary sources.

6.2 Gravitational and Inertial Separation

Gravitational and inertial separation comprise the simpler forms of particulate collectors, but several different types and configuations are included. Each type is built to incorporate a hopper into which the collected material will eventually settle. The performance of this type of collector depends either on velocity reduction to permit settling or on the application of centrifugal force that increases the effective mass of particles. The common types of mechanical collectors include settling chambers, cyclones, and multiple cyclones.

6.2.1 Settling Chambers

A settling chamber may be nothing more than a long, straight, bottomless duct over a hopper, or it may consist of a group of horizontal passages formed by shelves in a chamber. A baffle chamber consists simply of a short chamber with horizontal entry and exit, using single or multiple vertical baffles. The inertia of the entrained particles causes them to strike the plates, enabling the dust to fall into the collection hopper. Settling chambers require a large space and are effective only on particles with diameters greater than 50 to 100 μm. However their resistance to airflow, as measured by the pressure drop, is low. Collectors of this class are frequently used as precleaners for coarse particles, preceding other, more efficient types of collector.

6.2.2 Cyclones

In a cyclone, a vortex created within the collector propels particles to locations from which they may be removed from the collector; the devices can be operated either wet or dry. Cyclones may either deposit the collected particulate matter in a hopper or concentrate it into a stream of carrier gas that flows to another separator, usually of a different, more efficient type, for ultimate collection. As long as the interior of a cyclone remains clean, pressure drop does not increase with time. Up to a certain limit, both collection efficiency and pressure drop (usually less than 2 in. H_2O, gauge) increase with flowrate through a cyclone. This type of collector is best applied in the removal of coarse dusts.

Cyclones are frequently used in parallel but seldom in series. When they are used in series it is to accomplish a special objective, such as to provide a backup in case the dust discharge of the primary cyclone fails to function. Small diameter cyclones are more effective than larger ones; centrifugal force for a given tangential velocity varies inversely with the radius. Accordingly, multiple cyclones, banks of small cyclones arranged in parallel, are used commonly to maximize particulate control efficiency within the inertial collector concept. Whereas single cyclones can be several feet in diameter, multiple cyclones are often less than 12 inches in diameter; pressure drop across this configuration is typically 3 to 8 in. H_2O, gauge.

6.2.3 Other Devices

Other types of inertial separator include impingement collectors, consisting of a series of nozzles, orifices, or slots each followed by a baffle or plate surface, and dynamic collectors, consisting of a power-driven centrifugal fan with the provisions and housing for skimming off a layer of gas in which dust has been concentrated.

6.3 Filtration

Filters are devices used for removal of particulate matter from gas streams by retention of particles in and around a porous structure through which the gas flows. The porous structure is most commonly a woven or felted fabric, but pierced, woven, or sintered metal can be used, as well as beds of a large variety of substances (fibers, metal turnings, coke, slag roll, sand, etc.).

Unless operated wet to keep the interstices clean, filters in general improve in retention efficiency as the interstices in the porous structure begin to be filled by collected material. These collected particles form a pore structure of their own, supported by the filter, and because of the increased surface area, they have the ability to intercept and retain other particles. This increase in retention efficiency is accompanied by an increase in pressure drop through the filter. To prevent decrease in gas flow, therefore, the filter must be cleaned continuously or periodically, or else replaced after a certain length of time.

For controlling air pollution, the fabric filter collector, commonly known as a "baghouse," together with its fan or air mover, can be likened to a giant vacuum cleaner. Baghouses utilize fabrics, woven or felted, from natural or synthetic fibers. In its most usual configuration, a fabric filter consists of a series of cylindrical bags or tubes, vertically mounted, and dust is deposited on either the inner or outer surfaces of the fabric. Fabric filters are generally useful where:

1. Very high particulate control efficiencies are desired.
2. Temperatures are below 550°F.
3. Gas temperatures are above the dew point.
4. Valuable material is to be collected dry.

6.4 Electrostatic Precipitation

Electrostatic precipitators are devices in which one or more high intensity electrical fields cause particles to acquire an electrical charge and migrate to a collecting surface. The collecting surface may be either dry or wet. Since the collecting force is applied only to the particles and not to the gas, the pressure drop of the gas is only that due to flow through a duct having the configuration of the collector. Hence pressure drop is very low and does not tend to increase with time. In general, collection efficiency increases with the length of passage through an electrostatic precipitator. Therefore replicate precipitator sections are employed in series to obtain higher collection efficiency.

Electrostatic precipitators may incorporate pipes or flat plates as grounded electrodes onto which electrically charged dust particles are deposited for subsequent removal. In any electrostatic precipitator, three functional units exist by virtue of three specific operations tending to occur simultaneously:

1. Charging of particulate.
2. Collection of particulate.
3. Removal and transport of collected particulate.

Electrostatic precipitators find their greatest use in situations where:

1. Gas volumes are relatively large.
2. High efficiencies are required on fine particulate.
3. Valuable materials are to be collected dry.
4. Relatively high temperature gas streams must be cleaned.

6.5 Liquid Scrubbing

The prime means of collection in wet scrubbers is a liquid introduced into the collector for contact with the contaminant aerosol. Scrubbers are used both to remove gases and vapor phase contaminants and to separate particulate from the carrier gas. The scrubber liquid may be wet, or it may dissolve or react chemically with the contaminant collected.

Methods of effecting contact between scrubbing liquid and carrier gas include the following:

1. Spraying the liquid into open chambers or chambers containing various forms of baffles, grilles, or packing.
2. Flowing the liquid into these structures over weirs.
3. Bubbling the gas through tanks or troughs of liquids.
4. Using gas flow to create droplets from liquid introduced at a location of high gas velocity.

Scrubber liquid frequently can be recirculated after the collected contaminant is partially or completely removed. In other cases, all or part of the liquid must be discarded. In general, as long as the interior elements of the scrubber remain clean, the pressure drop does not increase with time. Collection efficiency tends to increase with increasing gas flowrate, provided the liquid feed keeps pace with gas flow and carryout or entrainment of liquid with the effluent gases is prevented effectively.

Scrubbers are generally either of the low energy or high energy type. Examples of low energy scrubbers are spray chambers, centrifugal units, impingement scrubbers, most packed beds, and the submerged nozzle type of scrubber. High energy scrubbers, especially the venturi scrubber, find wide application in situations calling for the removal of fine particulate matter at high efficiency levels. In the most common design, a venturi scrubber consists of a venturi-shaped air passage with radially directed water jets located just before or at the high velocity throat. The water is broken into fine droplets by the action of the high velocity of the gas stream. The particulate matter is deposited on the water droplet by impaction, diffusion, and condensation. Coarse droplets of water, together with the entrained dust, are readily separated by a comparatively simple demister section on the venturi discharge.

Wet scrubbers can handle high temperatures and are often used simultaneously as particulate collectors and heat transfer devices. Wet scrubbers generally are applicable and should be considered in a feasibility analysis if any one of the following situation descriptions applies.

1. The gaseous vapor phase contaminant is soluble in water (particulate need not be soluble).
2. The exhaust gas stream contains both gaseous and particulate contaminants.
3. Exhaust gases are combustible.
4. Cooling is desirable with an increase in moisture content satisfactory.

6.6 Gas–Solid Adsorption

Adsorbers are devices in which contaminant gases or vapors are retained on the surface of a porous medium through which the carrier gas flows. The medium most commonly used for adsorption is activated carbon. The design of an adsorber parallels that of a filtration device for particulate matter in that the gas flows through a porous bed. In the case of an adsorber, however, the adsorption bed should be preceded by a filter, to protect it from plugging due to particulate matter.

In true adsorption, there is no irreversible chemical reaction between the adsorbent and the adsorbed gas or vapor. Therefore the adsorbed gas or vapor can be driven off the adsorbent by heat, vacuum, steam, or other means. In some adsorbers the adsorbent is regenerated in this manner for reuse. In other applications the spent adsorbent is discarded and replaced with fresh adsorbent. Pressure drop through an adsorber that does not handle gas contaminated by particulate matter does not increase with time but does increase with gas flowrate. The relation between adsorption efficiency and gas flowrate depends entirely on adsorber design and on the characteristics of the material being collected.

Carbon adsorption is generally carried out in large horizontal fixed beds with depths of 3 to 25 feet. Such units can handle from 2000 to about 50,000 cfm and are often equipped with blowers, condensers, separators, and controls. Typically, an installation includes two carbon beds; one is onstream while the other is being regenerated.

Although molecular sieves have reached the commercial stage in several new applications, activated carbon remains the most important dry adsorbent in gaseous emission control. Molecular sieves, like activated carbon, usually require fixed-bed units with sequence valves for switching the beds from adsorption to regeneration. The key element in these systems is molecular sieve-adsorbent blends: synthetic crystalline metallic–alumina silicates that are highly porous adsorbents for some liquids and gases.

6.7 Combustion

Many processes produce gas streams that bear organic materials having little recovery value or containing toxic or odorous materials that can be oxidized to less harmful combustion products. In such cases thermal oxidation may be the optimum control route, especially if the gas streams are combustible. There are three methods of combustion in common use today: thermal oxidation, direct flame incineration, and catalytic oxidation.

Thermal oxidizers or afterburners can be used when the contaminant is combustible. The contaminated airstream is introduced to an open flame or heating device, then it goes to a residence chamber where combustibles are oxidized to carbon dioxide and water vapor. Most combustible contaminants can be oxidized at temperatures between 100 and 1500°F. The residence chamber must provide sufficient dwell time and turbulence to allow complete oxidation.

Direct combustors differ from thermal oxidizers by introducing the contaminated gases and auxiliary air into the burner itself as fuel. Auxiliary fuel, usually natural gas or oil, is generally necessary for ignition and may or may not be required to sustain burning.

Catalytic oxidizers may be used when the contaminant is combustible. The contaminated gas stream is preheated, then passed through a catalyst bed that promotes oxidation of the combustibles to carbon dioxide and water vapor. Metals of the platinum family are commonly used catalysts, and they promote oxidation at temperatures between 700 and 900°F.

To use either thermal or catalytic oxidation, the combustible contaminant concentration must be below the lower explosive limit. Equipment specifically designed for control of gaseous or vapor contaminants should be applied with caution when the airstream also contains solid particles. Solid particulate can plug catalysts and, if noncombustible, it will not be converted in thermal oxidizers and direct combustors.

Thermal and direct combustors usually have lower capital cost requirements but higher operating costs because an auxiliary fuel is often required for burning. This cost is offset when heat recovery is employed. The catalytic approach has high capital cost because of the relatively expensive catalyst, but it needs less fuel. Either method will provide a clean, odorless effluent if the exit gas temperature is sufficiently high.

6.8 Combination Systems

Within the wide spectrum of stationary emission sources, there are several applications for coupling complementary control devices into an effective combination system. Selection of the best combination and sequence of control methods depends to some degree on past experience, but excellent results can be achieved by careful analysis of the contaminated air or gas to be treated.

A good example of a combination system is the inertial collector–electrostatic precipitator system, applicable when a wide range of particle sizes occurs in the gas, or where dust loadings are high. Examples are blast furnaces and coal-fired boilers.

Another innovative combination uses the carbon adsorber and incinerator to minimize operating costs. In this control system, organic vapors are adsorbed for several hours on activated carbon; during this time, the incinerator is turned off. When the carbon bed is regenerated, the desorbent vapors, including the steam used to heat the bed, can be sent directly to the incinerator for destruction. Not only is auxiliary fuel saved by this intermittent operation, but the desorbent may be self-sustaining during incineration. Furthermore, incinerator size is reduced dramatically compared with direct-fired equipment. Capital costs may sometimes be offsetting, however.

Two principal arrangements of multiple-unit air-cleaning systems are recognized: parallel combination and serial combination. Parallel combination is simply a means of providing a wide range of flexibility in adjusting the capacity of a system to the airflow rate. It permits small, highly effective elements, such as centrifugal tubes (multiple cyclones), to be assembled in parallel arrangement for handling equal shares of the total gas flow. Serial combination presents the problem of determining the best sequence, if different types of cleaning units are to be used, or the best combination of preparation and separation units to meet the range of contaminants that must be removed from the airstream.

6.8.1 Two-Stage Cleaning Systems

Two-stage systems comprise a preparation stage and a separation stage. Examples of this arrangement include the following:

1. Two-stage electrostatic precipitators, with particle charging preceding precipitation.
2. Venturi scrubbers, with simultaneous agitation and liquid injection preceding cyclone separation.
3. Sonic agglomeration systems, with simultaneous agglomeration and gravitational precipitation followed by inertial or centrifugal separation.

6.8.2 Sequence of Separating Stages

The selectivity of particulate separators, or their fractionating tendencies, makes it advisable to combine stages of cleaning that are complementary. The sequence of treatment is generally, but not invariably, from the types of separator least able to remove submicrometer particles to those at the final stages most likely to succeed at this removal.

The agglomerating action of certain types of separator results in large aggregates of flocs blowing intermittently from dry collecting surfaces into the airstream. When this condition occurs, because of excessive air velocities or poor retentivity at the contact surfaces, it is good practice to interpose coarse particle separators to reduce the load on subsequent stages of air cleaning.

Some of the variation in airborne matter is gradually suppressed or damped as the air progresses through a multistage system of cleaning. The process fluctuations may not be eliminated, but they frequently are modified so greatly in the initial stages of air treatment that the final, and most critical, stages are protected against disturbing variations in the character of materials traveling with the air.

6.8.3 Concentration and Subdivision of Airstreams

As the air passes through a cleaning system, it may be handled as a single stream from inlet to outlet. There are numerous installations, however, in which the airstream is subdivided one or more times, to specialize the task of cleaning and to make each stage more effective. A common example is conveyance of the small volume, high concentration effluent from the apex of the periphery of a cyclone to a secondary separator generating greater centrifugal force.

7 GRAVITATIONAL AND INERTIAL SEPARATION

The simplest type of particulate collector is commonly known as a "mechanical collector." Mechanical collectors utilize either gravity or inertia to remove relatively large particles from suspension in a moving gas stream. In these collectors the gas stream is made to flow in a path that either enhances the gravitational separation or changes direction such that particles cannot easily follow because of their inertia. Most mechanical collectors operate in a dry condition, although water is sometimes used in conjunction with a mechanical collector to aid in continuous cleaning of the control device.

7.1 Range of Performance

Gravitational and inertial separation devices range in capture efficiency from less than 50 to about 90 percent, depending on particle size and type of collector. Generally, as the velocity through the collector increases, control efficiency increases, except for gravitational settling chambers where the opposite effect occurs.

Pressure drop across an inertial collector of this type is one indicator of relative efficiency. Although a direct relationship between pressure drop and efficiency cannot always be calculated easily, there is usually a significant correlation between the two; control efficiency usually increases with pressure drop. One important aspect of this phenomenon is that pressure drop increases require greater amounts of energy to operate the collection system.

Table 19.5 compares characteristics of various kinds of gravitational and inertial separators.

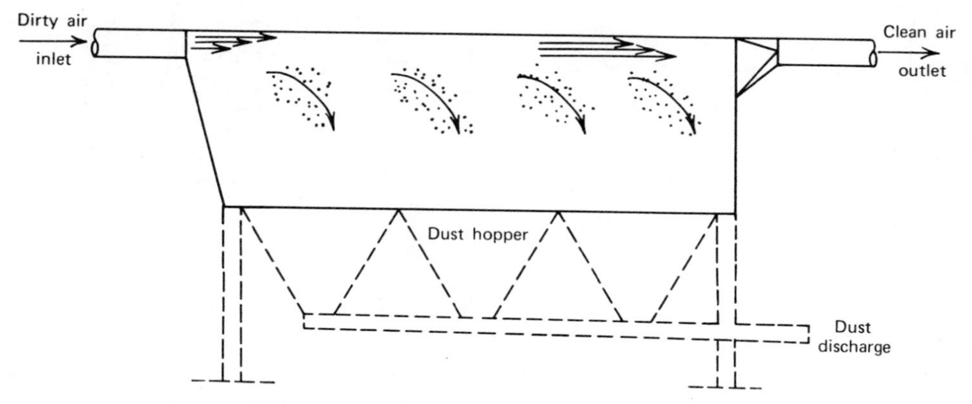

Figure 19.3 Gravity dust-settling chamber (13).

7.2 Settling Chambers

Settling chambers represent the oldest form of air pollution control device. Because of their simplicity of design and relatively low operating costs, they are most commonly used today as precleaners for more efficient particulate control methods that consume higher amounts of energy.

Settling chambers use the force of gravity to separate dust and mist from the gas stream by slowing down the gas stream so that particles will settle out into a hopper (Figure 19.3) or onto shelves from which they can be removed (Figure 19.4).

From a practical standpoint, settling chambers are not employed for the removal of particles less than 50 μm in diameter because of the excessive size of the equipment required for collection below this size limit. To prevent reentrainment of the separated

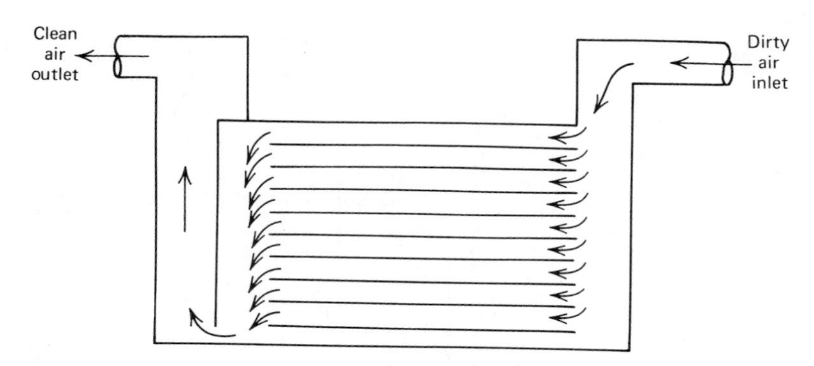

Figure 19.4 Multiple-tray dust collector (6).

Table 19.5. Characteristics of Mechanical Collection Equipment (13)

Collector Type	Space Require-ments	Volume Range	Efficiency by Weight	Pressure Loss[a] (in. H_2O)	Temperature Limitations	Power[b] (hp per 1000 cfm gas)	Application Areas
Settling chamber	Large	Space available only limitation	Good above 50 μm	0.2–0.5	700–1000°F, limited only by materials of construction	0.04–0.12	Precollector for fly ash metallurgical dust, can be used for any large size dust particles above 50 μm
Conventional cyclone	Large	Normal range up to 50,000 cfm	Approximately 50% on 20 μm	1–3	700–1000°F, limited only by materials of construction	0.24–0.73	Woodworking, paper, buffing fibers, etc.; well suited for dry dust particles in 20 μm and above range
High efficiency cyclone	Medium	Normal range up to 12,000 cfm	Approximately 80% on 10 μm	3–5	700–1000°F, limited only by materials of construction	0.73–1.2	Woodworking, material conveying, product recovery, etc.; well suited for dry dust particles in 10 μm and above range

Multitube cyclone	Small	Normal range up to 100,000 cfm	90% on 7.5 μm	4.5	700–1000°F	1.1	Precollector for electrostatic precipitator on fly ash, product recovery, etc.; well suited for dry dust particles in 5 μm and above range
Dynamic precipitator	Small	17,000 cfm	80% on 15 μm	No loss (true fan)	700°F	Power consumption will depend on selection point, mechanical efficiency in usual selection range from 40–50%	Woodworking, nonproduction buffing, metal working, etc.; well suited for dry dust particles in 10 μm and above range
Impingement separator	Small	Space available only limitation	90% on 10 μm	1–5	700°F	0.24–1.2	Certain types used for collecting coarse particles in boiler fly ash and cement clinker cooler; recent designs used for cleaning atmospheric air to diesel engines and gas turbines

[a] Pressure drop is based on standard conditions.

[b] Power consumption figured from: horsepower = cfm × Total Pressure/6356 × ME; mechanical efficiency (ME) assumed to be 65 percent.

particles, the velocity of the gas stream entering the settling chamber should not exceed 600 fpm.

The efficiency of a settling chamber can be calculated by the following equation (14):

$$E = \frac{100U_t\, WL}{Q} \tag{10}$$

where E = efficiency, weight percent of particles of settling velocity U_t (dimensionless)
U_t = terminal settling velocity of dust (ft/sec)
L = chamber length (ft)
W = chamber width (ft)
Q = gas flowrate (ft^3/sec)

This efficiency model assumes good distribution at the inlet and outlet of the chamber. Combining this equation with Stokes' law, the minimum particle size that can be completely separated from the gas stream can be calculated by the following equation (15):

$$d_p = \left[\frac{18\mu HV}{gL(\rho_p - \rho_g)}\right]^{1/2} \tag{11}$$

where d_p = minimum size particle collected completely (ft)
μ = gas viscosity (lb/ft-sec)
H = chamber height (ft)
V = gas velocity (ft/sec)
g = gravitational constant (32.2 ft/sec^2)
ρ_p = particle density (lb/ft^3)
ρ_g = gas density (lb/ft^3)

If there are horizontal plates or trays in the chamber, the effective settling distance is reduced, and the efficiency and minimum particle size that can be completely separated are given by Equeations 12 and 13, respectively (15):

$$E = \frac{NU_tWL}{Q} \tag{12}$$

$$d_p = \left[\frac{18\mu Q}{gNWL\,(\rho_p - \rho_g)}\right]^{1/2} \tag{13}$$

where N is the number of plates and other units are as in Equations 10 and 11.

The chief advantages of settling chambers are their low operating cost, relatively low initial cost, simple construction, low maintenance costs, and low pressure drop in operation.

Their main disadvantages are large space requirements and low collection efficiency for smaller particles. When using shelves to increase efficiency, this characteristic also causes a problem with cleaning the settled particulate off the shelves; this is often done by rinsing the plates with water.

7.3 Inertial Separators

Inertial separators represent one of the simplest pollution control devices used today. They take up less room than settling chambers and are more efficient. Inertial separators are relatively inexpensive to acquire and operate; but they are not very efficient, and like settling chambers, they are used basically for precleaning of gas streams. Inertial separators include all dry-type collectors that utilize the relatively great inertia of the particles to effect particulate–gas separation. Two basic types of inertia separation equipment are simple impaction separators, which employ incremental changes of the carrier gas stream direction to exert the greater inertial effects of the particles, and cyclonic separators, which produce continuous centrifugal force as a means of exerting the greater inertial effects. The cyclones are discussed in a subsequent section.

Impingement or impaction separation occurs when the gas stream suddenly changes its direction because of the presence of an obstructing body. The impingement efficiency, defined as the fraction of particles in the gas volume swept by the obstructing body which will impinge on that body, is given as follows:

$$E = \frac{U_t V}{D_B} \tag{14}$$

where E = separation efficiency (dimensionless)
U_t = terminal settling velocity of particle (ft/sec)
V = gas stream velocity (ft/sec)
D_B = equivalent diameter of obstructing body (ft)

Impingement separators vary with the configuration of the obstructing bodies and include baffle type, orifice impaction type, high velocity gas reversal type, and louver type.

The baffled chamber (Figure 19.5) is the simplest of the impingement separators. It uses one or two plates as impingement sites to stop larger particles and cause them to fall into a dust-collecting bin. This equipment can remove particles larger than 20 μm in diameter with pressure drops varying from 0.5 to 1.5 in. H_2O.

The orifice impaction collector (Figure 19.6) may consist of many nozzles, slots, or orifices followed by a plate or baffle for an impingement surface from which the particles can fall into the dust bin. This device can remove particles greater than 2 μm in diameter. It is more efficient than the simple baffle type but more expensive. Normal velocities through the orifices are about 50 to 100 ft/sec and the pressure drop is approximately 2.5 orifice velocity heads.

The louvered impingement separator uses many impingement surfaces on an angle, to rebound the particles into a secondary airstream and allow the cleaner air to pass through the louver. The particulate is thus concentrated into a secondary airstream whose flowrate is 5 to 10 percent that of the primary airstream. This concentrated airstream is usually passed through another more efficient cleaner for discharge into the atmosphere. The efficiency of this type of collector is basically a function of the louver

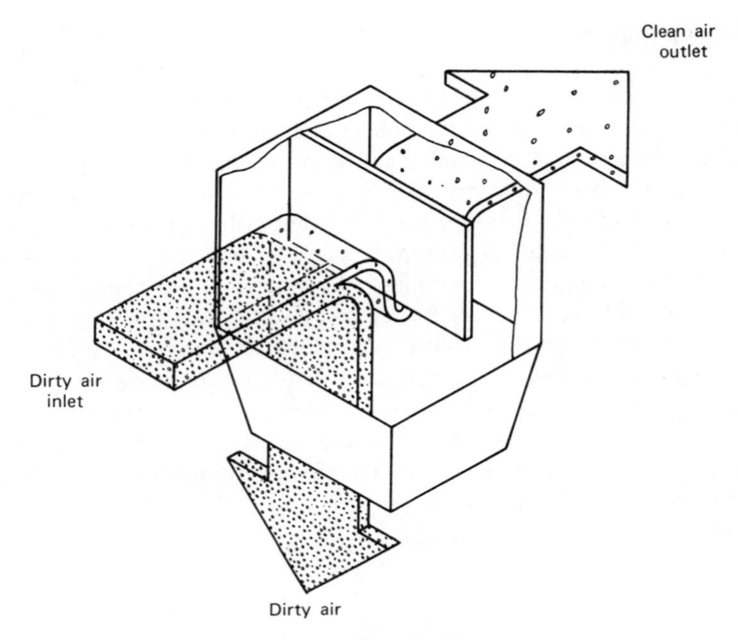

Figure 19.5 Baffled chamber.

spacing, closer spacings producing higher efficiencies. The two principal arrangements of louvers are the flat louver impingement separator (Figure 19.7) and the conical louver impingement separator (Figure 19.8).

7.4 Dynamic Separators

Dynamic separators, sometimes termed rotary centrifugal separators, are the newest gravitational and inertial devices used for collecting particulate. Dynamic precipitators use centrifugal force to separate particulate from an airstream. The precipitator concentrates the dust around the impellers, and it is dropped into a hopper. These devices are about 80 percent efficient on particles larger than 15 μm in diameter.

The major advantage of the dynamic separator is its small size, which is most helpful when a facility needs many independently operated precipitators but has only limited space. Since this separator is a true fan, there is no pressure drop across the device. Its major limitation is its tendency toward plugging and imbalance as well as wearing of the blades.

A dynamic precipitator (see Figure 19.9) separates particles by first drawing the gas stream into it; then as the particulate impacts the impeller, centrifugal force throws the heavier particles to the periphery of the housing. The lighter particles are impacted on the blades and glide along the blade surface to the outside edge where they are also thrown to the periphery of the housing. The particles are then discharged out of the annular slot to the dust collection bin. The cleaned air is discharged into a scroll-shaped discharger.

7.5 Cyclones

Cyclones have long been regarded as one of the simplest and most economical mechanical collectors. They can be used as precleaners or as final collectors. The primary elements of a cyclone are a gas inlet that produces the vortex, an axial outlet for cleaned gas, and a dust discharge opening. Several different types of cyclones (Figures 19.10 to 19.13) are as follows:

1. Tangential inlet with axial dust discharge (most common).
2. Tangential inlet with peripheral dust discharge.
3. Axial inlet through swirl vanes with axial dust discharge.
4. Axial inlet through swirl vanes with peripheral dust discharge.

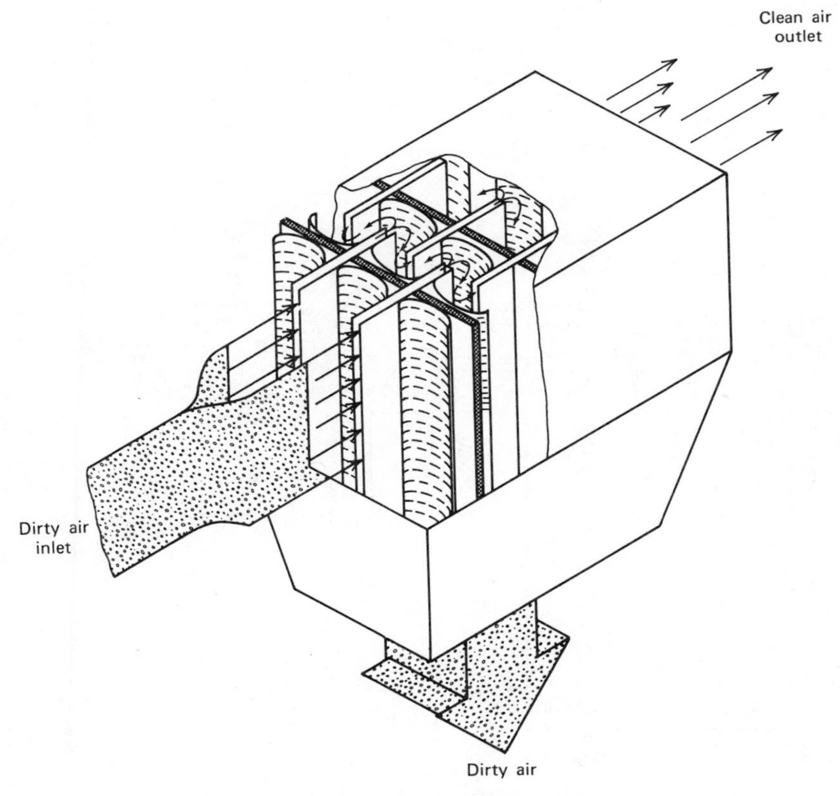

Figure 19.6 Orifice impaction collector.

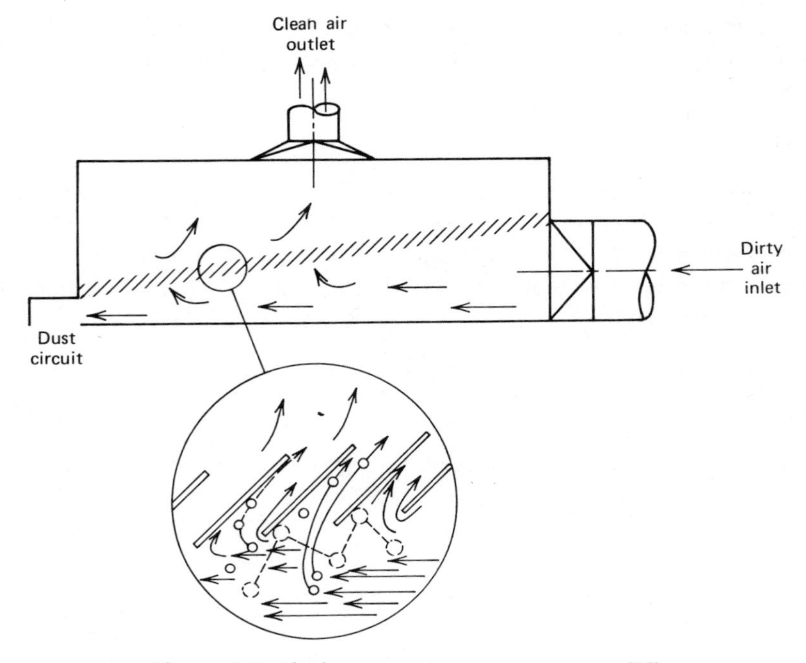

Figure 19.7 Flat louver impingement separator (16).

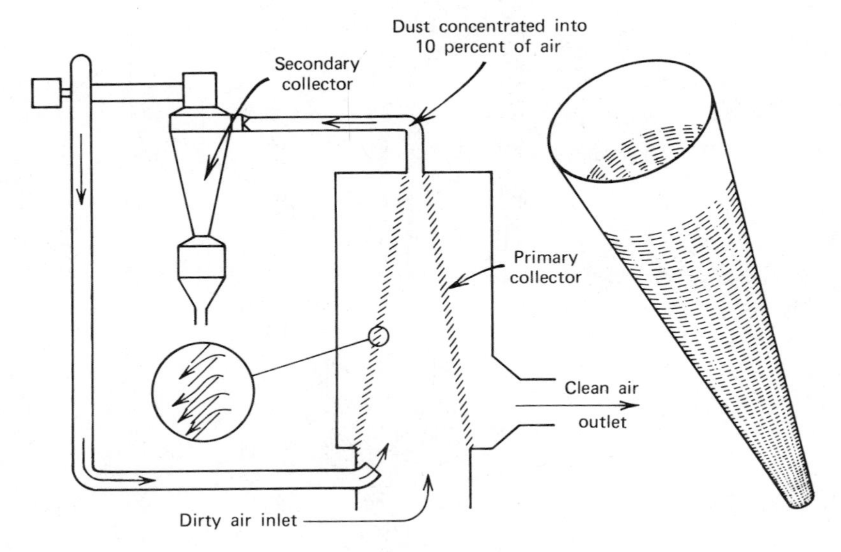

Figure 19.8 Conical louver impingement separator (16).

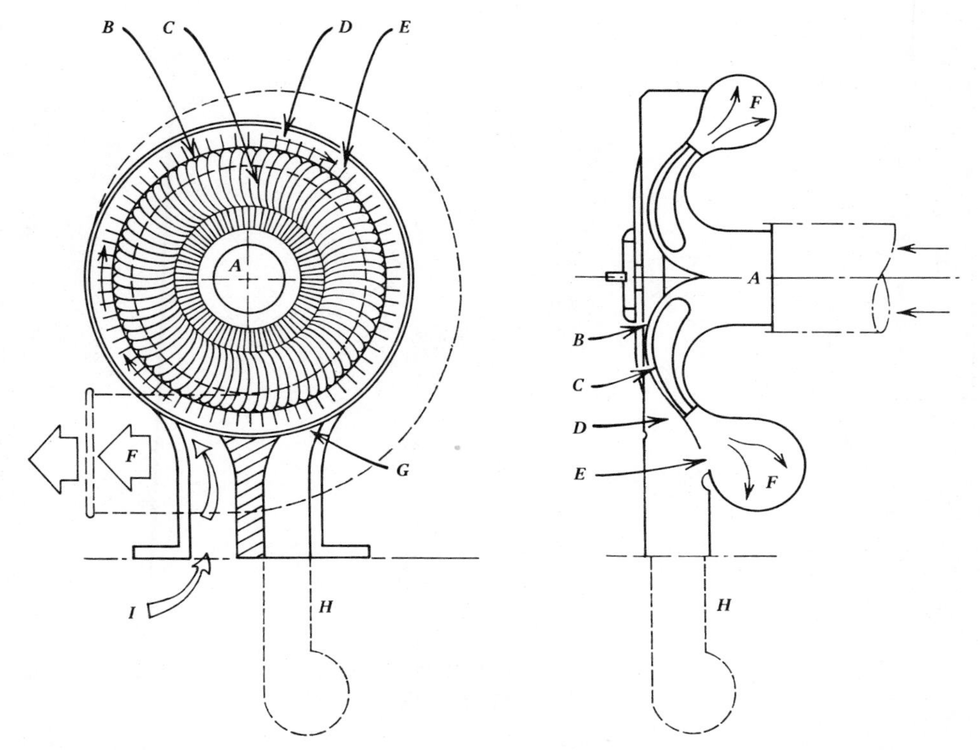

Figure 19.9 Dynamic separator: *A*, center inlet; *B*, interception disk; *C*, impeller blades; *D*, opening for dust escape; *E*, chamber; *F*, scroll-shaped discharge chamber; *G*, point of deflection into hopper; *H*, hopper; *I*, secondary air return (16).

Centrifugal force is the precipitating force for particulate and droplets. The air to be cleaned usually enters at the top of a cyclone, either tangentially or axially. Incoming air is made to swirl into a vortex by a tangential entry, by tangential flow, or by curved vanes in an axial inlet. As the air turns, it flows down the cyclone; near the bottom, it reverses direction and flows up the center of the cyclone through the axial exhaust port. Because air is always spinning in the same direction, suspended particles are being forced to the outside wall, where they slide down to the discharge chute (Figure 19.14). Standard cyclone design dimensions are given in Table 19.6.

Efficiency of a cyclone is defined as the fractional weight of particles collected. The major parameter in the prediction of collection efficiency is particle size. The particle that can be removed from the inlet gas stream at an efficiency of 50 percent in a cyclone is defined as the "particle cut size" d_{pc} and is represented in the following relationship (18):

$$d_{pc} = \left[\frac{9\mu W}{2NV(\rho_p - \rho_g)\,\pi} \right]^{1/2} \tag{15}$$

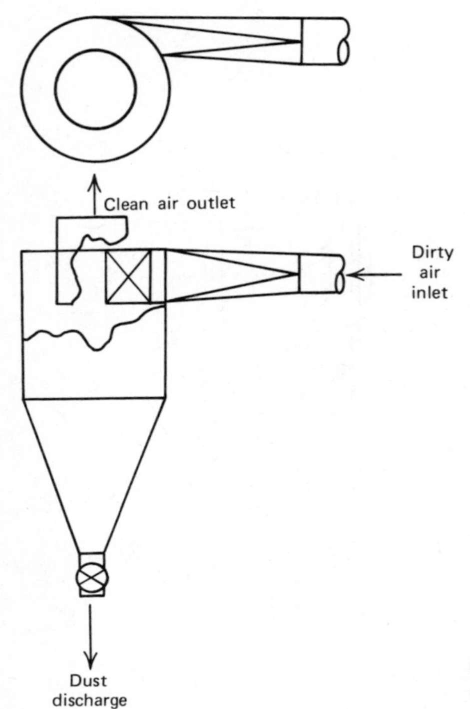

Figure 19.10 Cyclone with tangential inlet and axial discharge (15).

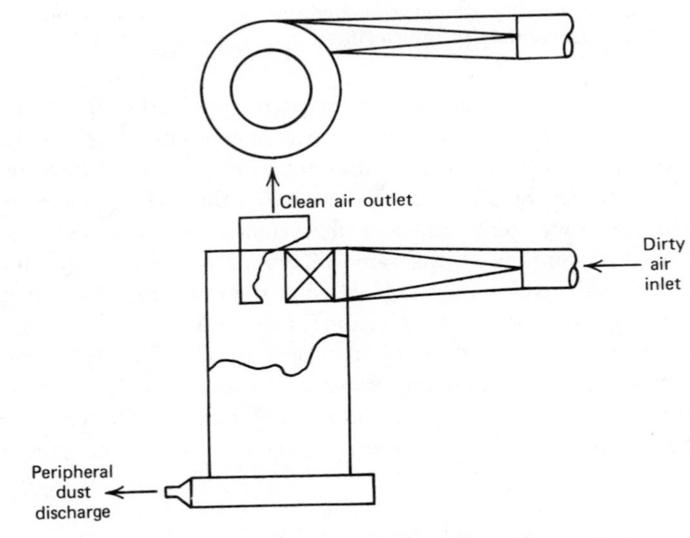

Figure 19.11 Cyclone with tangential inlet and peripheral discharge (15).

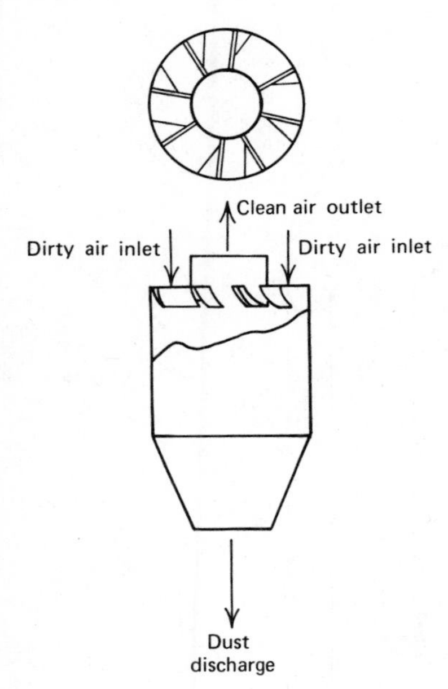

Figure 19.12 Cyclone with axial inlet and axial discharge (15).

where d_{pc} = diameter cut size particle collected at 50 percent efficiency (ft)
μ = gas viscosity (lb mass/sec-ft = centipoise $\times 0.672 \times 10^{-3}$)
W = cyclone inlet width (ft)
N = effective number of turns within cyclone
V = inlet gas velocity (ft/sec)
ρ_p = true particle density (lb/ft^3)
ρ_g = gas density (lb/ft^3)
π = constant, 3.1416

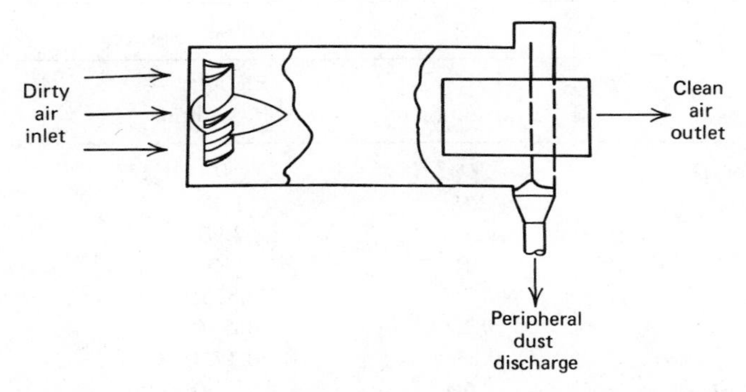

Figure 19.13 Cyclone with axial inlet and peripheral discharge (15).

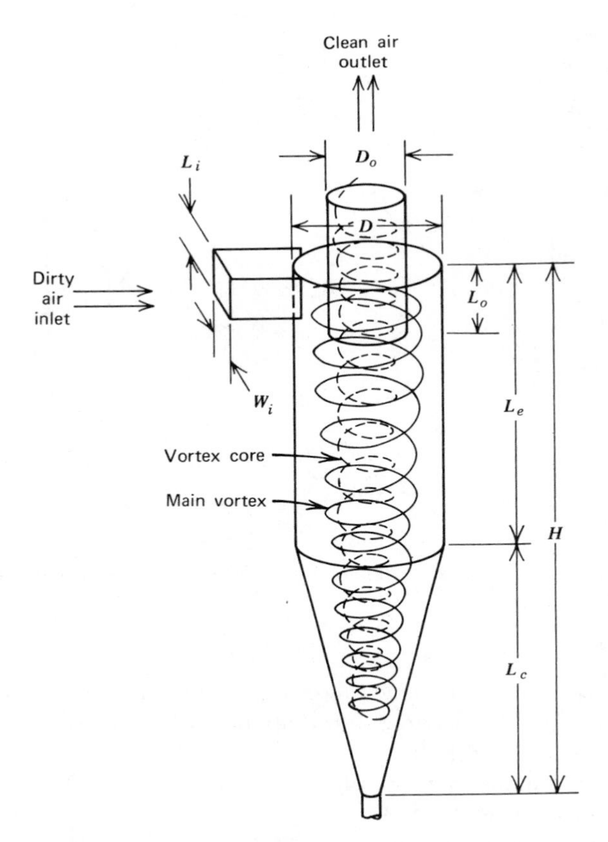

Figure 19.14 Typical cyclone; for abbreviations and standard dimensions, see Table 19.6 (17).

Table 19.6. Standard Basic Cyclone Design Dimensions

	Cyclone Type		
Parameter	Conventional	High Throughput	High Efficiency
Cyclone diameter, D	D	D	D
Cyclone length, L_e	$2D$	$1.5D$	$1.5D$
Cone length, L_c	$2D$	$2.5D$	$2.5D$
Total height, H	$4D$	$4D$	$4D$
Outlet length, L_o	$0.675D$	$0.875D$	$0.5D$
Inlet height, L_i	$0.5D$	$0.75D$	$0.5D$
Inlet width, W_i	$0.25D$	$0.375D$	$0.2D$
Outlet diameter, D_o	$0.5D$	$0.75D$	$0.5D$

This equation, together with Figure 19.15, permits the accurate prediction of the collection efficiency of a cyclone when the particle size is known. The design factor having the greatest effect on collection efficiency is cyclone diameter. For a given pressure drop, the smaller the diameter of the unit, the higher the collection efficiency obtained. The efficiency of a cyclone also increases with increased gas inlet velocity, cyclone body length, and ratio of body diameter to gas outlet diameter. Conversely, efficiency will decrease with increased gas temperature, gas outlet diameter, and inlet area.

The pressure drop across cyclones is another factor affecting collection efficiency. It varies between approximately 2 and 8 in. H_2O, gauge; pressure drop increases with the square of the inlet velocity. Collection efficiency also increases with the square of the velocity, but not as rapidly as does the pressure drop.

The inlet velocity can be increased only to the point (70 ft/sec) at which the turbulence is not causing reentrainment of the particles being separated. After this point, the efficiency decreases with increased inlet velocity. Efficiency also increases as the dust loading increases, as long as the dust does not plug the cyclone or cause it to be eroded severely. In addition, the size of the particles affects the efficiency: the larger the particle the higher the collection efficiency (Table 19.7).

One of the greatest advantages of cyclones is that they can be made of many materials, thus are able to handle almost any type of contaminant. Other advantages of cyclones include their ability to handle high dust loads, relatively small size, relatively low initial and maintenance costs, and, excluding mechanical centrifugals, temperature and process limitations imposed only by materials and construction. Because cyclones have larger pressure drops than settling chambers or inertial separators, their cost of operation is higher, but still lower than that of most other cleaners. A disadvantage of cyclones is that efficiency drops off with decreased dust loading. Plugging can also occur, but if the particle characteristics are taken into consideration, this problem can be avoided. Low collection efficiencies for particles below 10 μm in diameter is another disadvantage.

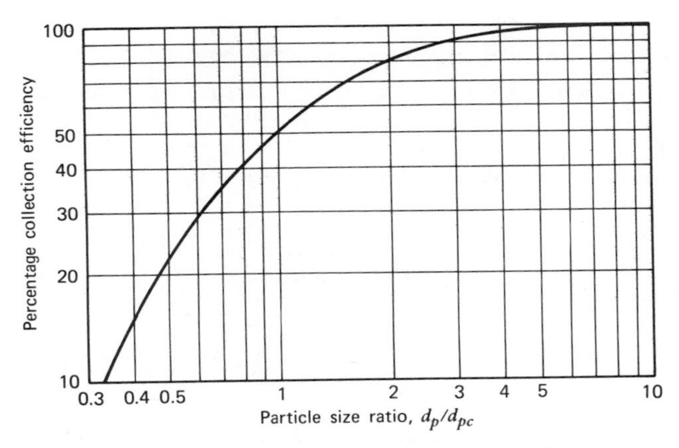

Figure 19.15 Cyclone collection efficiency as a function of particle size ratio (18).

7.6 Wet Cyclones

A wet cyclone is nothing more than a dry cyclone with an inlet for water or some other fluid to be impacted with the incoming particles, rinsing the particles from the cyclone. It costs more to operate than a dry cyclone because of the increased cost for the fluid introduced into the cyclone and for requisite cleaning of the discharge fluid to remove suspended or dissolved particulate.

The forces acting to remove the particles are the same as for dry cyclones. The smaller particles are made larger by impaction with fluid droplets, to increase the efficiency for smaller particles. Since the cyclone walls bear a liquid film, the chance for particulate reentrainment is reduced; thus efficiencies are greater.

Wet cyclones offer the ability to take the liquid from any of several points other than the axial part of the vortex cone, which means that the vortex itself is not disturbed. A wet cyclone usually has fewer erosion problems than a dry cyclone, but if corrosive dust or gas is present, water may activate a serious corrosion problem. Disposal or cleaning of the contaminated water or liquid is another problem to be solved in wet operations.

Wet cyclones entail some design problems that are not present with dry cyclones. First, the liquid tends to creep along the walls of the cyclone to the outlet, where droplets are sheared off into the outlet stream. This can be corrected by installing a skirt on the exit (Figure 19.16) so that the droplets go into the vortex rather than the exit gas stream. If the velocity is too high in the cyclone, very small droplets cannot be recollected. The recommended inlet velocity is not to exceed 150 ft/sec at atmospheric pressure and for injected water; this estimate varies with different gases and liquids but is a general rule.

7.7 Multiple Cyclones

Multiple cyclones are simply many cyclones put in parallel or series. Cyclones in series are not used very often. Cyclones in parallel constitute the most common configuration for multiple cyclones. Space requirements are moderate for multiple cyclones, but these devices are more expensive than regular single cyclones to purchase and design. Opera-

Table 19.7. Efficiency Range of Cyclones (15)

Particle Size Range (μm)	Efficiency Range (wt. % collected)	
	Conventional	"High Efficiency"
<5	<50	50–80
5–20	50–80	80–95
15–40	80–95	95–99
>40	95–99	95–99

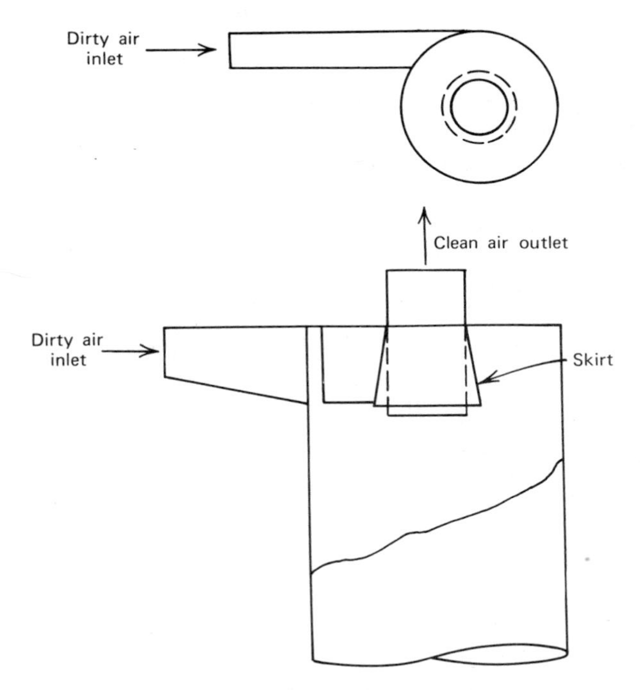

Figure 19.16 Gas outlet skirt for wet cyclone (15).

tional costs for multiple cyclones are about the same as those for single cyclones. The devices are usually used as a final control but also serve as precleaners in some applications.

Multiple cyclones make use of the same forces to separate the particles from the airstream as the single cyclone. The multiple cyclones usually consist of smaller cyclones that have more "separation force" for a given overall energy input. When many small (3- to 12-inch diameter) cyclones are used in parallel, they are more efficient than one larger cyclone of the same capacity. Their pressure drops are usually between 3 and 6 in. H_2O.

Multiple cyclones offer the most significant advantage of ability to handle large airflows with heavy dust loadings and still exhibit good efficiency. When properly designed inlet ductwork provides even flow to all the individual cyclones, the multiple cyclone performs at peak design efficiency. This also ensures that there will not be any backflow through some of the cyclones, causing particles to flow from the dust bin to the exit plenum. Many multiple cyclones have several gas inlet and exit plenums and several dust exit valves; this allows any part of the multiple cyclone to be shut off from another part in the event that a cyclone plugs or needs repair.

8 FILTRATION

Filtration represents the oldest, and inherently the most reliable, of the many methods by which particulate matter (dusts, mists, and fumes) can be removed from gases. Filters are especially desirable for extracting particulate matter from gases produced by industrial operations because they generally offer very high collection efficiencies with only moderate power consumption. Initial investment costs and maintenance expenditures can range from relatively low to comparatively high, depending on the size and density of the particulate matter being collected, as well as the quantity and temperature of the dusty gas to be cleaned.

Filters are most readily classified according to filtering media. For particulate emissions control, the different media can be broadly categorized as (1) woven fibrous mats and aggregate beds, (2) paper filters, and (3) fabric filters.

8.1 Range of Performance

Filtration is highly effective for small particulate, even down to less than 0.05 μm in diameter, if the flow through the filter is kept low enough. Filtration can be used on almost any process emitting small particles of dust. Limiting factors, however, are the characteristics of the gas stream. If the unfiltered stream is too hot, it may be necessary to precool the gas stream. If the gas stream is too heavily loaded, a precleaner may be necessary to rid the stream of larger particles. The cost of auxiliary equipment required for such pretreatment may equal or even exceed that of the filter itself. There are also several causes for filter failure, including blinding, caking, burning, abrasion, chemical attack, and aging.

Designs of filters and filter enclosures are many. Filters can be configured as mats, panels, tubes, or envelopes. When tubular fabric filters are enclosed, the structure and its contents are often called a "baghouse."

8.2 Fibrous Mats and Aggregate Beds

Fibrous mats and aggregate-bed filters are characterized by high porosity; both are comprised largely (97 to 99 percent) of void spaces. Such void space is usually much larger than the particles being collected, thus the mechanisms of sieving and straining of particles are of no significance in these filters. The predominant forces functioning in the cleaning of a gas stream using large void filters include impaction, impingement, and surface attraction. Impaction and impingement are effective when the particles are made to change direction quickly and impact or impinge on the surface of the filter medium. Surface-attractive forces are mostly electrostatic, and before they contribute significantly to total collection, the dust particles must come within several particle diameters.

Efficiencies can be extremely high with very low dust loadings. The U.S. Atomic Energy Commission of Hanford, Washington, has reported collection efficiencies of 99.7 percent for submicrometer particles using sand in the filtering medium (19). Glass fiber beds have been installed instead of sand, and efficiencies of 99.99 percent have been reported for dust loadings of 0.00002 to 0.0004 grain per cubic foot (20).

Advantages of aggregate-bed filters include a longer life without frequent cleaning, high dust storage capacity with a modest increase in airflow resistance, and application to high temperature emissions. One disadvantage is difficulty in cleaning; many fibrous mats are simply discarded. Large space requirements also pose a problem when these filters are used.

These filters range from very inexpensive mats that are changed and discarded to very expensive beds that are almost never replaced. Deep beds and fibrous mats are relatively inexpensive to operate because of low pressure losses (between 0.1 and 1 in. H_2O, gauge) in most operations.

Designs for deep-bed filters are numerous. Almost any material can be chosen to make up the bed. Sand has been used most often, but glass fiber beds have been used recently. Sulfuric and phosphoric acid plants use what are known as "coke boxes" for collection of acids. A lead- or ceramic-lined box is filled with several feet of $\frac{1}{40}$- to $\frac{1}{2}$-inch diameter coke to collect the mechanically produced mist at an efficiency of 80 to 90 percent. Condensed mist is generally too small to be collected by the coke box.

The mats are made of many materials. Glass fibers, stainless steel, brass, and aluminum are all used in mats for different types of airstream. Fibrous mats need to be cleaned when their pressure loss becomes too high or the flowrate becomes too low. Cleaning can be accomplished by removing a portion of the filter bed continually and replacing it with new or cleaned portions. Some cleaning is accomplished by reversing the airflow and vibrating the bed or using shock waves. Some mats are self-cleaning and some are continuous cleaning.

Designs for self-cleaning beds are numerous. For example, an automatic viscous filter (Figure 19.17) uses an oil film to ensure that the particles are not reentrained in the airstream. The airstream must pass through the filter twice. The plates on the conveyor belt open and close at the top and bottom of the filter. At the bottom, the plates are plunged into an oil reservoir and cleaned by agitation. Upon emerging from the oil bath, they are cleaned further and reoiled.

Another type of self-cleaning mat is the water spray or wet filter (Figure 19.18). The mats are sprayed continuously while the airstream flows through the filter. The particles are dislodged by the water and flow with it to the sump.

8.3 Paper Filters

Paper filters are of relatively recent design for air pollution control and find service chiefly where ultrahigh efficiencies are needed. These filters have come into wide use where very clean air is essential, as in "white rooms" of hospitals, data processing centers, the space industry, food processing plants, and atomic energy installations.

Paper filters can be made of minerals, asbestos, or glass microfibers, with or without binders that add strength, formability, or water resistance. Glass microfilters are the most popular because of their fire resistance and availability. The frames that contain a

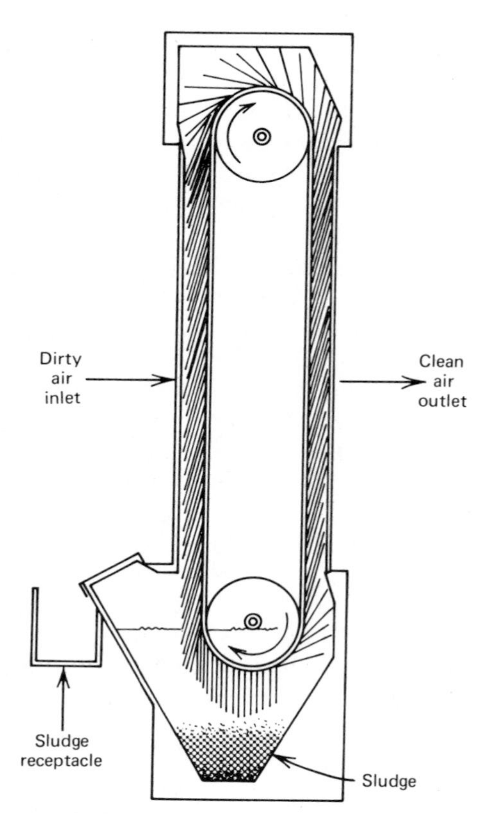

Figure 19.17 Self-cleaning automatic viscous filter (20).

set of filters, known as "packs" or "plugs," can be made of steel, aluminum, hardboard, or plywood, depending on the application. These frames are normally fitted with gaskets on both sides to ensure a good seal; the seals are made of appropriate materials for the application (e.g., neoprene, mineral wool, asbestos, or rubber).

Paper filters may be flat, cylindrical, or any shape that is practical for the application and space available. Fluting of the paper medium increases the surface area of a filter compared to that of a flat filter; this can save appreciably in equipment size used to house the filtering elements.

Paper filter systems require moderate initial capital investment, and operating costs are relatively low. Variations in design, however, can make capital investment cost very high, depending on the application.

Paper filters use all mechanisms of particle capture and retention. The most significant is diffusion. Paper filters are often used as final filters to remove very small particles or dusts in very low concentrations. Any larger particles still present in the gas stream may be captured, but the life of a filter is shortened measurably if too many large particles are captured on a continuous paper filter. Many times high efficiency, inexpensive precleaners are used before the paper filter to make the paper medium last longer. Since the filtration is mostly a surface action phenomenon, the dust storage capacity is a function of the surface area.

The efficiency of paper filter media is very high, 99.97 percent by weight for the best commercial cellulose–asbestos paper filters. The dioctyl phthalate (DOP) method for calculating the efficiency of a filter is employed for paper filters. The test method is the U.S. Army Chemical Corps DOP Smoke Penetration and Air Resistance Test No. MIL-STD-282, Method 102.9.1 (21).

To ensure good collection efficiencies throughout their lifetime, paper filters must have larger pressure losses when new compared with other filter types. This means that the ratio of pressure loss in a new filter to a spent filter is larger. Flow velocity through the paper is usually around 5 ft/min.

Paper filters are advantageous in that they have a long life expectancy, usually 1 to 2 years, and are relatively inexpensive to replace. Paper filters cannot be cleaned and reused. Other disadvantages include the necessity to provide low flowrates and low dust loadings to the paper filtering element.

8.4 Fabric Filters

One of the most positive methods for removing solid particulate contaminants from gas streams is filtration through fabric media. A fabric filter is capable of providing a high

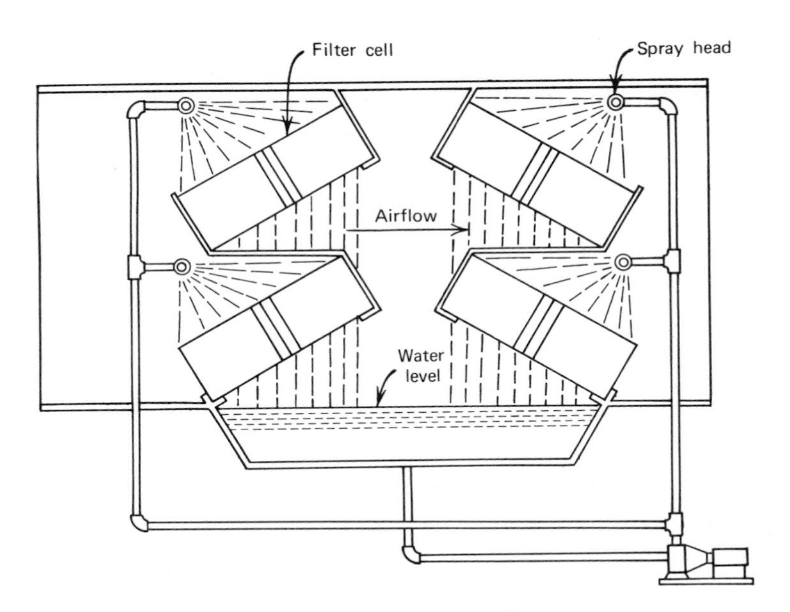

Figure 19.18 Self-cleaning wet filter (20).

collection efficiency for particles as small as 0.5 μm and will remove a substantial quantity of particles as small as 0.01 μm.

Fabric filters are usually tubular (bags) or flat and made of woven or felted synthetic fabric. The dirty gas stream passes through the fabric, and the particles are collected on the upstream side by the filtration of the fabric. The dust retained on the fabric is periodically shaken off and falls into a collecting hopper.

The structure in which the bags hang is known as a baghouse. The bags have an average life of 18 to 36 months. The number of bags may vary from one to several thousand. The baghouse may have one compartment or many, making it possible to clean one while others are still in service.

Removal of the particles from the gas stream is not a simple filtration or sieving process, since the pores of the fabric employed in fabric filters are normally many times the size of the particles collected, sometimes 100 times larger or more. The collection of particles takes place through interception and impaction of the particles on the fabric filters, and Brownian diffusion, electrostatic attraction, and gravitational settling within the pores. Once a mat or a cake of dust is accumulated, further collection is accomplished by the mechanism of mat or cake sieving, as well as by the foregoing mechanisms. Periodically the accumulated dust is removed, but some residual dust remains and serves as an aid to further filtering.

Direct interception occurs whenever the gas streamline, along which a particle approaches a filter element, passes within a distance one-half the particle diameter from the filter element. Inertial impaction occurs when a particle unable to follow the streamline curving around an obstacle comes closer to the filter element than it would have come if it had approached along the streamline. Small particles, usually less than 0.2 μm, do not follow the streamline because collision with gas molecules occurs, resulting in a random Brownian motion that increases the chance of contact between the particles and the filter element. Electrostatic attraction results from electrostatic forces drawing particles and filter elements together whenever either or both possess a static charge.

The major particulate removal mechanisms as they apply to a single fiber in a fabric filter are shown in the following tabulation:

Primary Collection Mechanism	Diameter of Particle (μm)
Direct interception	>1
Impingement	>1
Diffusion	<0.01–0.2
Electrostatic attraction	<0.01
Gravity	>1

Fabric filter systems (i.e., baghouses) are characterized and identified according to the method used to remove collected dust from the filters. This is accomplished in a variety of types, including shaker (Figure 19.19), reverse flow (Figure 19.20), reverse jet (Figure 19.21), and reverse pulse (Figure 19.22).

The fundamental criterion in applying any baghouse to any application is the air to cloth ratio, defined as

$$A/C = \frac{Q}{A} \tag{16}$$

where A/C = air to cloth ratio (ft/min)
Q = volumetric gas flowrate (acfm)
A = net cloth area (ft^2)

Air to cloth ratio is equal to the superficial face velocity of the air as it passes the cloth. Shaker and reverse airflow baghouses usually operate at an air to cloth ratio of from 1 to 3, whereas a reverse pulse baghouse operates at about 3 to 6 times this range.

A second important factor in baghouse application is the type of filter material.

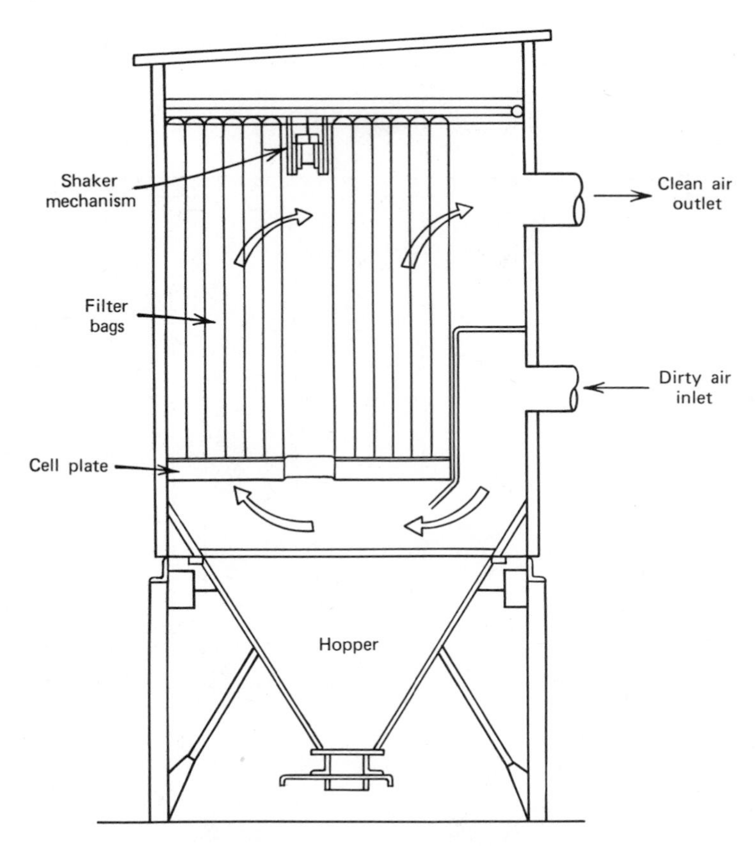

Figure 19.19 Shaker-type baghouse (13).

Woven cloth is used in shaker and reverse airflow baghouses; felted cloth is selected for reverse pulse baghouses. One of the characteristics of filter fabrics is the permeability, which is expressed as the air volume, in actual cubic feet per minute, passing through a square foot of clean cloth with a pressure differential of 0.5 in. H_2O, gauge. The overall range of permeability is from 10 to 110 acfm per square foot, but it usually ranges from 10 to 30 acfm per square foot for all types of baghouses.

Another important operating characteristic is the filter drag, given as

$$S = \frac{\Delta PA}{Q} \tag{17}$$

where S = filter drag (in. H_2O/per fpm)
ΔP = pressure drop across filter (usually 2 to 6 in. H_2O)
A = net cloth area (ft^2)
Q = volumetric gas flowrate (acfm)

Figure 19.23 shows the effect of filter drag on outlet concentration (22).

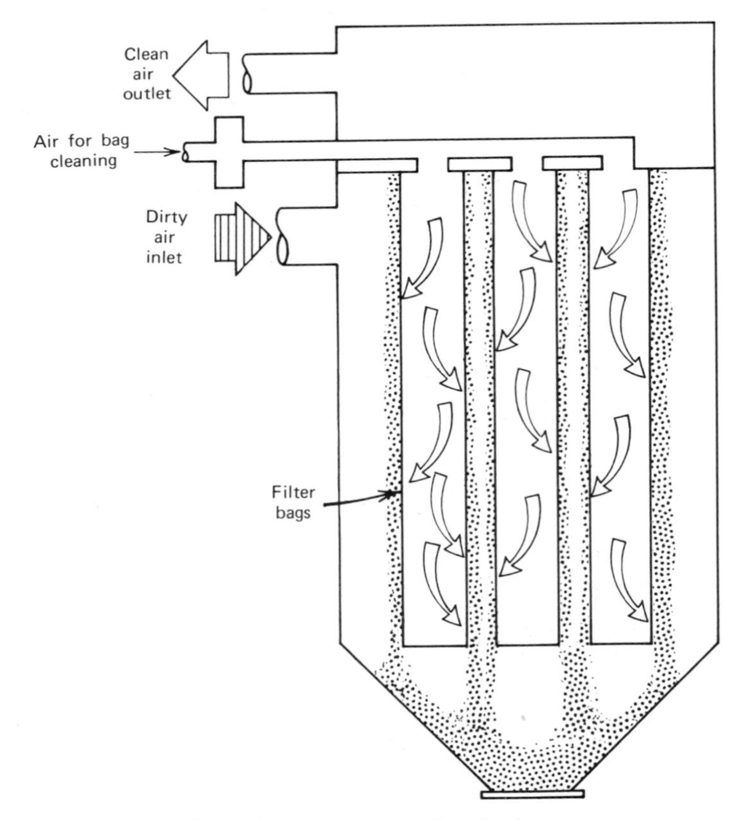

Figure 19.20 Reverse airflow baghouse.

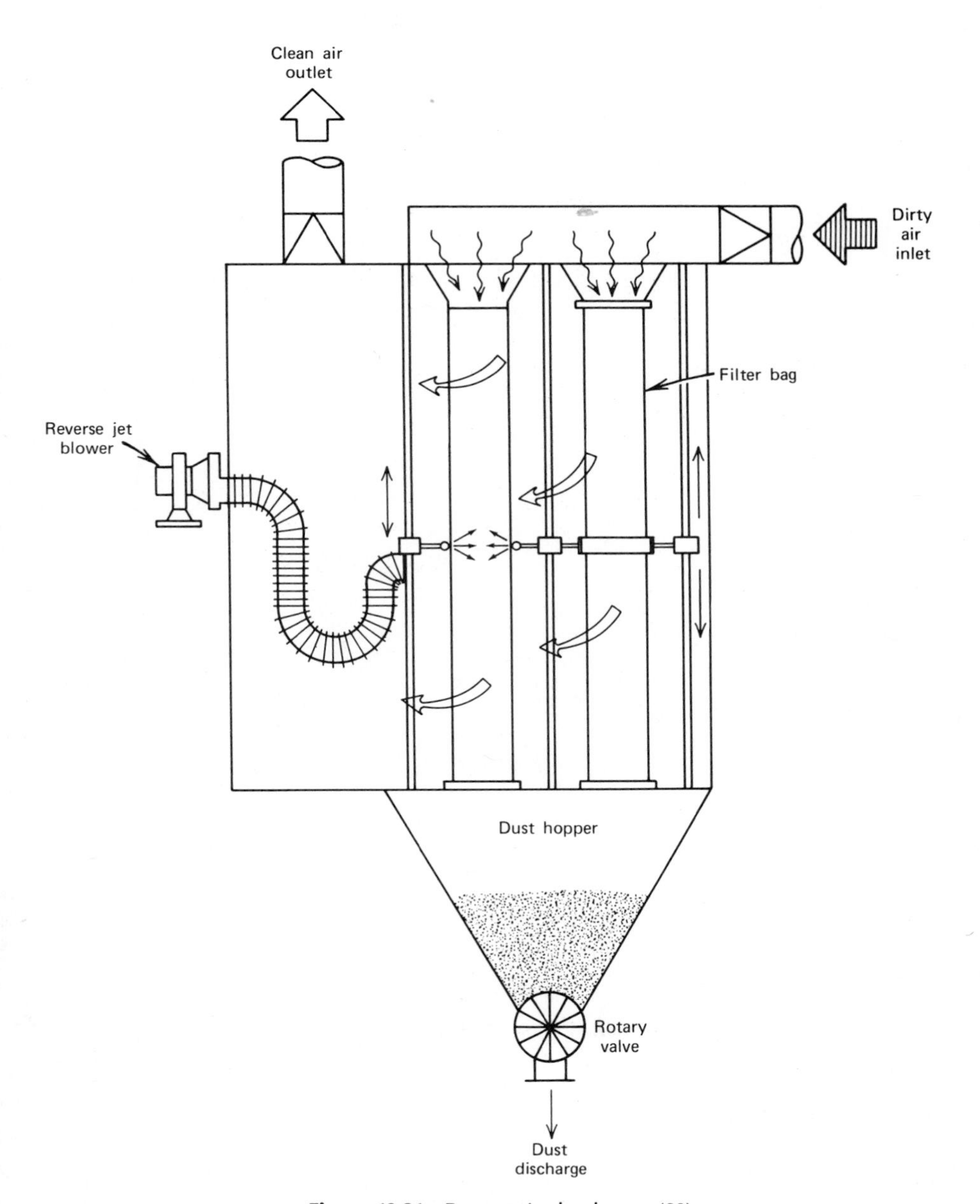

Figure 19.21 Reverse jet baghouse (20).

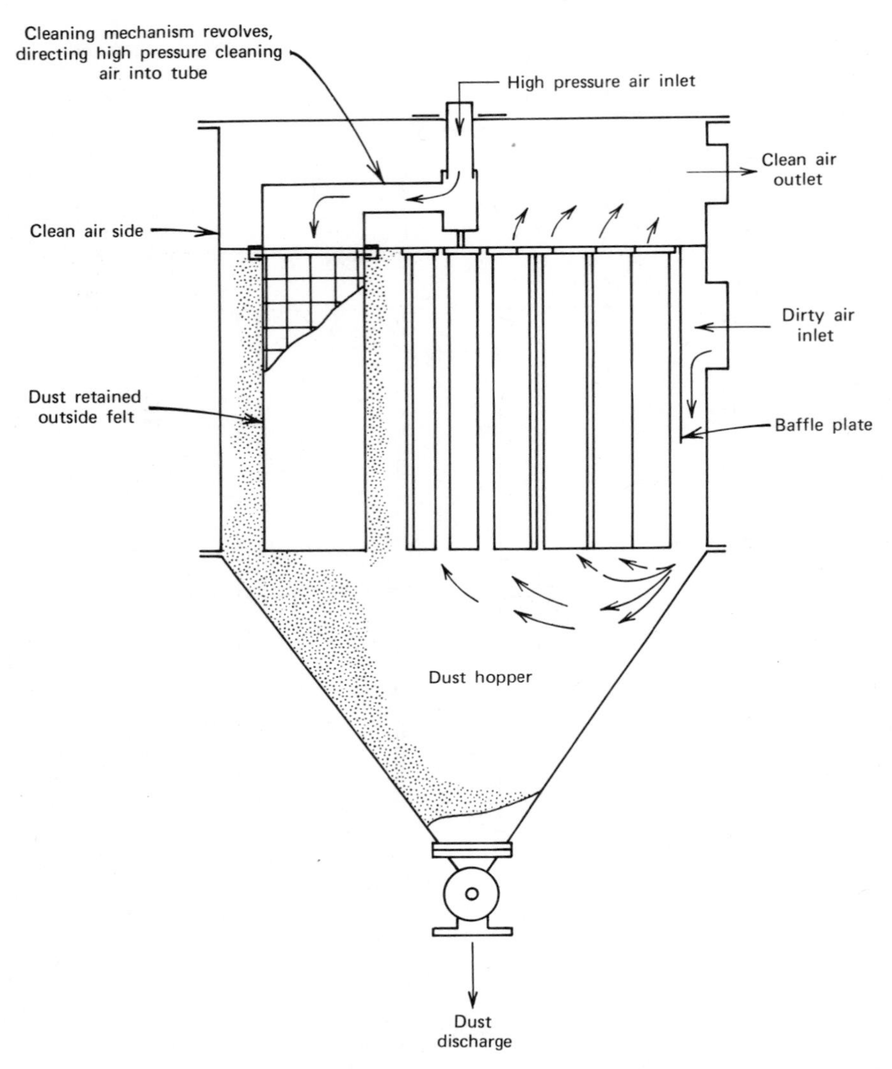

Figure 19.22 Reverse pulse baghouse (13).

9 LIQUID SCRUBBING

"Liquid scrubbing" denotes a process whereby soluble gases or particulate contaminants are removed from a carrier gas stream by contacting the contaminated gas stream with a suitable liquid to decrease the concentration of the contaminant. Scrubber geometry, contacting media, and the scrubbing liquor are design variables that have been the subject of years of investigation to optimize scrubber performance in a variety of applications.

Although the liquid used in liquid scrubbing is generally recirculated, the need to discharge some portion of the scrubbing liquor can create water pollution problems complex enough to render liquid scrubbing infeasible. Nonetheless, the application of liquid scrubbers to air pollution abatement strategies finds optimal use when soluble gaseous contaminants or fine particulate contaminants must be removed to high efficiency levels, when the gases involved are combustible, when cooling is desired, and when increased moisture content can be tolerated. When one or more or these prerequisites is met, liquid scrubbing may very well provide the only applicable air pollution abatement strategy within the contexts of economic and technical feasibility.

9.1 Range of Performance

The type of scrubber used in a particular application depends mostly on the characteristics of the contaminants being scrubbed and the degree of control required. Almost any device in which good contact is promoted between the scrubbing liquor and a

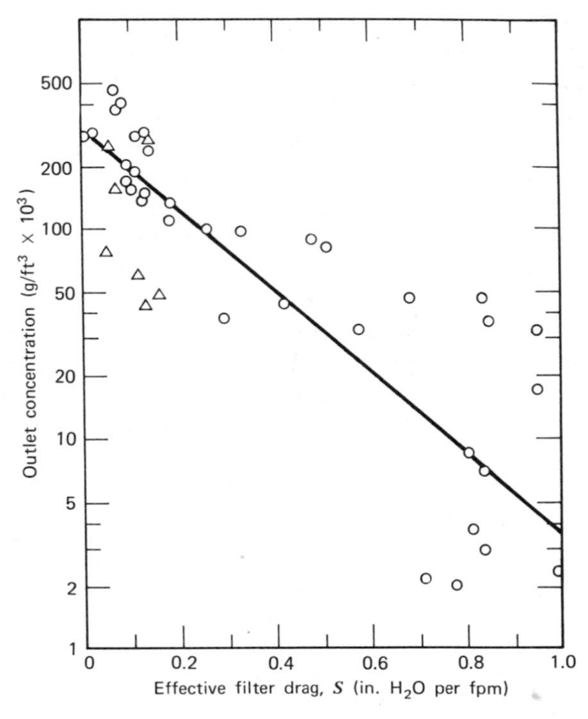

Figure 19.23 Baghouse outlet concentration as a function of filter drag (22).

contaminated gas stream will absorb to some degree both gaseous and particulate contaminants. The question of what type of scrubber to specify in an individual case hinges on the efficiency desired, the properties of the contaminant, and the merging of these two factors within the context of minimum operating costs and energy consumption. For example, although a high energy venturi scrubber using water as the scrubbing liquor will efficiently absorb some water-soluble gases, the energy requirement of the venturi scrubber can be an order of magnitude greater than that of a packed tower to promote the same degree of mass transfer.

The design parameters and established correlations useful in specifying and sizing liquid scrubbers can be subdivided into two general classes: (1) mass transfer dynamics for gases dissolving and/or reacting with the scrubbing liquor, and (2) momentum transfer dynamics for particulate matter colliding with and being entrained in the liquid scrubbing medium. In the first class, gas–liquid equilibrium data and mass transfer coefficients must either be established by experience with the actual species of chemical constituents being encountered, or existing data for chemically similar scrubbing system applications may be extrapolated, in the hope that useful design information can be extracted and the new system design parameters thereby established. In the second class, semitheoretical considerations provide useful design criteria in specifying scrubbers that utilize inertial impaction or Brownian motion to extract particulate matter. Even in this case, however, empirical design data supply the soundest base on which to specify the type and size of liquid scrubber needed to ensure efficient collection of particulate matter at minimum feasible capital and operating cost.

Packed towers using water as a scrubbing medium have found wide applications in the chemical industry to absorb water-soluble gases and vapors, both for pollution abatement and for product recovery. In the power industry, scrubbers are frequently used in the absorption of sulfur dioxide, and they function efficiently as long as the partial pressure of the sulfur oxides in the gas stream is greater than the vapor pressure of the sulfur oxides existing at the surface of the scrubbing liquor. Such a concentration difference provides a driving force to promote mass transfer of sulfur oxides from the gas into the liquid stream. Because the solubility of most water-soluble gases is limited by equilibrium relationships, the amount of scrubbing liquor specified per unit volume of contaminated gas is determined using mass balance calculations and equilibrium relationships, and giving due consideration to any chemicals added to the scrubbing liquor to increase absorption efficiency.

A wide variety of scrubbers, operating in pressure drop ranges of 0.5 to more than 80 in. H_2O have been used successfully to absorb particulate matter ranging in size from tens of micrometers down to material of submicrometer size. In any particulate scrubber, the design must provide as large a constant area as possible for scrubbing liquor and particulate matter to interface. In a venturi scrubber, for example, the submicrometer dust impacts on the surfaces of dispersed liquid droplets and penetrates into and becomes part of the liquid droplets, which are subsequently agglomerated, coalesced, and removed from the gas stream.

Although contact area is a prerequisite to particulate absorption in liquid scrubbing, the nature of the dust and its physical properties in relation to the scrubbing liquor determine whether the dust can be wetted by the scrubbing liquid or whether it can even penetrate the surface of the scrubbing liquor. For example, in the scrubbing of iron oxide dust using water, the relatively high surface tension of water does not permit iron oxide to be "wetted" well enough to assure permanent retention in the scrubbing water. Addition of "surfactants" to the scrubbing water decreases the surface tension, thereby permitting the particles to penetrate the "skin" of the droplets at the gas–liquid interface.

9.2 Spray Chambers

Spray-type scrubbers are useful in collecting particulate matter and gases in liquid droplets dispersed in the chamber using spray nozzles to atomize the liquid. The geometry of the spray chamber can vary from a simple straight cylinder to configurations designed to provide maximum contact area and contact time between droplets and gases passing through the atomized sprayer.

In the simplest type of spray chamber, a vertical tower is the container in which droplets and gas meet. In vertical spray chambers, the droplet velocity eventually becomes the terminal settling velocity for a given size droplet. The droplet velocity relative to the particulate collision velocity determines efficiency. Characteristically, spray chambers exhibit low pressure drop, utilize high relative rates of scrubbing liquid per unit volume of gas, and are generally very low in cost because of their simple design. Generally, spray chambers exhibit high collection efficiency for particles that are larger than 10 μm in diameter.

Usually the gases are introduced to the lower section of a spray tower, and the sprays, positioned at the top of the tower, inject scrubbing liquid at some velocity determined by the nozzle orifice (Figure 19.24). The residence time and the average relative velocity difference between the upward traveling gas and the downward traveling droplets determines the amount of contaminant removal that can be expected for a given tower size and ratio of scrubbing liquid to gas flowrate.

Some spray towers introduce the gases at the base of the tower in tangential fashion, thereby imparting a spiraling motion to the gases. Not only is longer contact time obtained in this manner, but spray droplets are driven to the walls of the spray tower, where they impinge and drain down the walls as a liquid film.

In spray chambers, collection efficiency of particulate matter increases as droplet size decreases. Therefore the use of high pressure atomizing nozzles to produce fine droplets will enhance collection efficiency near the nozzle. These small droplets quickly decelerate, however, because of aerodynamic drag. Therefore the relative velocity difference between the spray droplets and the particulate traveling with the gas falls off so rapidly that the collection efficiency is almost zero in the rest of the tower a short distance from the nozzle orifices. An optimum droplet size is 500 to 1000 μm. These

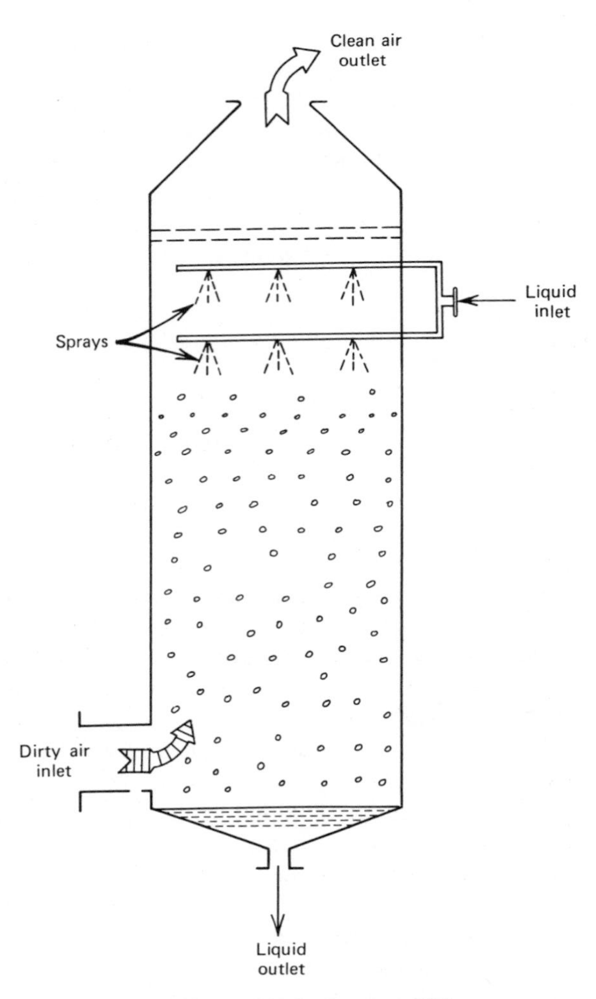

Figure 19.24 Spray tower (23).

droplets in spray chambers scrub relatively efficiently over long enough path lengths throughout the spray tower to provide an optimum collection efficiency.

Because of the relatively simple design, spray towers also serve a useful function as precoolers or quench chambers for relatively quickly dropping the temperature of hot gases as they are emitted from a wide variety of processes. Very often these hot gases contain a relatively high solids content, since the spray chamber is most likely to be the first piece of gas conditioning equipment the stream encounters before final discharge to the atmosphere. The spray chamber can handle recycled scrubbing liquid with relatively high solids concentrations suspended in the scrubbing liquid because of its simple design and because the clearance in the spray nozzle orifice is relatively large. Therefore the spray chamber is ideally suited both as an initial solid particulate remover and as a gas conditioner useful in preparing the gas for subsequent downstream processing equipment, which is probably more complex in design and more efficient in particulate extraction.

Specification of pertinent design parameters for a particular application depends to a large extent on the type of contaminant being scrubbed in the spray chamber. A rule of thumb, however, would indicate that in simple spray towers from 5 to 20 gallons of scrubbing liquor per 1000 cubic feet of contaminated gas provides optimum liquid gas contact.

In applications featuring the scrubbing of gases, equilibrium considerations come into play and computation of the number of mass transfer units is necessary. These calculations are theoretical and contain proportionality constants that depend on the physical and chemical properties of the contaminants being scrubbed; therefore empirical data are required to provide reasonably accurate specifications. The calculations are similar to those made in sizing packed-tower scrubbers.

9.3 Packed Towers

Scrubbers containing packing material such as Pall rings, Berl or Interlox saddles, or Raschig rings are termed packed-bed scrubbers or packed towers. The packing provides high surface area of liquid films ideal for mass transfer. In general, packed-bed scrubbers can be classified as countercurrent, cocurrent, and cross flow. In the countercurrent packed-bed scrubber (Figure 19.25) scrubbing liquid, injected at the top of the packing, trickles down through the packing material as gases pass up. Countercurrent scrubbers generally cannot be operated at liquid or gas flowrates as high as those used with cocurrent packed-bed scrubbers, where the gas and liquid phases flow in the same direction. Cocurrent scrubbers, in general, provide lower mass transfer per cubic foot of packing than countercurrent units.

Packed-bed scrubbers are normally used to remove gaseous contaminants from airstreams. In general, when high loading of particulate is a possibility in such applications, the packing media may be fouled by deposited particulate matter, resulting in eventual blockage of the unit. Because gaseous contaminant removal is limited by equilibrium considerations, the efficiency of the unit, as well as the absolute concentration of contaminant in the gas stream leaving the packed-bed scrubber, is affected by whether a cocurrent or countercurrent scrubber is specified.

The cross-flow scrubber is the most capable of the three types of packed-bed scrubber in coping with deposited solids on the tower packing. In these units, scrubbing liquid trickles through the packing from the top of the packing bed, while the dirty gas moves horizontally through the bed. This arrangement provides good washing of the media and simultaneously is a very stable configuration, difficult to flood.

Packed-bed scrubbers have been used historically in chemical operations in a wide variety of applications. By adding various chemical substances to the scrubbing liquor,

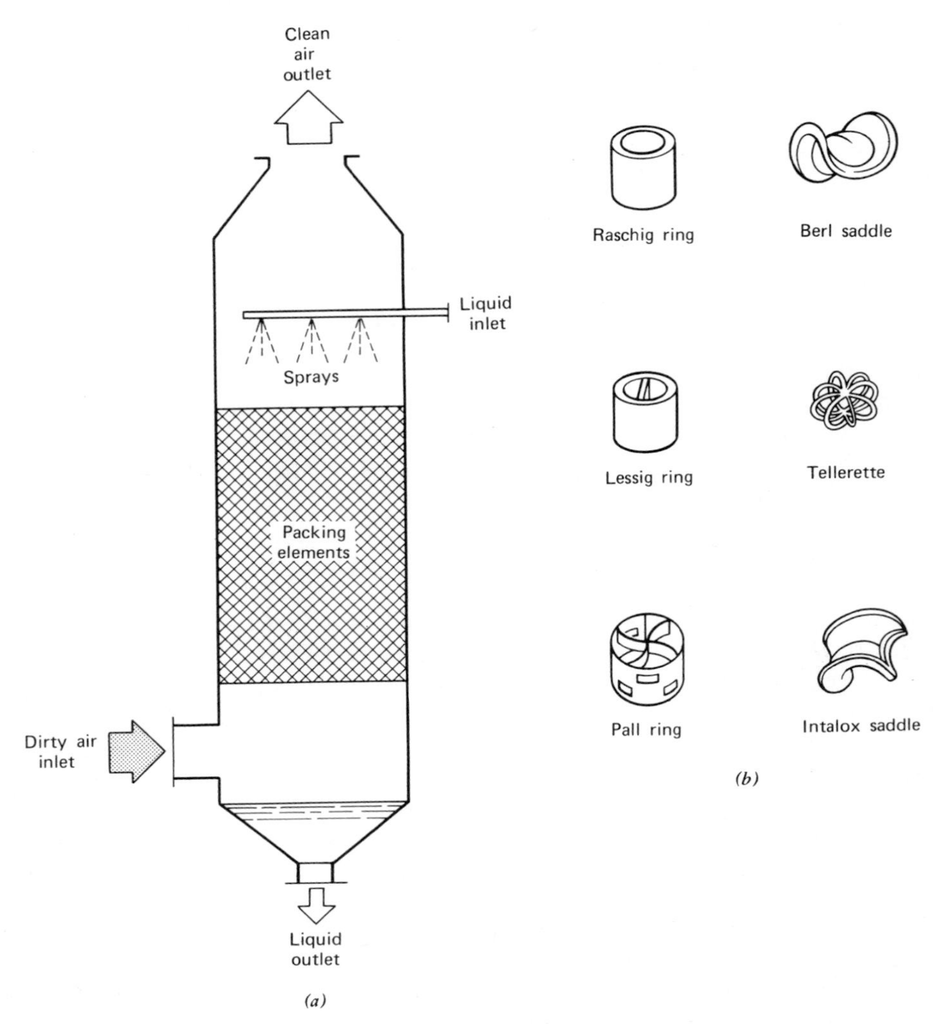

Figure 19.25 (*a*) Countercurrent packed-bed scrubber and (*b*) packings (23).

pollutant gases can be removed to very low outlet concentrations if sufficient fan power is provided to overcome the bed pressure drop and the absolute volume of packing and tower height are sufficiently large. Typically, empirical design data are necessary to determine mass transfer coefficients useful in predicting efficiencies for particular combinations of contaminant and scrubbing liquid. Pressure drops through packed-bed towers can be quite low. Although the pressure drop depends on the flowrate through the tower and the liquid–gas ratio, as well as the type of packing, typically 0.5 in. H_2O per foot of packing is the pressure loss in packed-bed towers.

Packed beds offer the added advantage of being intrinsic mist eliminators. Because flowrate is characteristically low, entrainment of liquid into the gas leaving the tower is much less than in scrubbers of other types where the liquid phase is dispersed in fine droplets. Furthermore, since tower height is somewhat greater than packing height, scrubbing efficiency of the packed tower can be increased by simply adding more packing, at minimal cost to the overall system. Thus packed-bed scrubbers are relatively flexible to process changes that may alter the concentration or nature of gaseous contaminants in their inlet streams.

Section 9.6 discusses the pertinent design parameters and applicable equations necessary to size a packed-bed scrubber for a variety of conditions. In general, pertinent parameters include the concentration of the contaminant in the gas stream admitted to the scrubber, the required efficiency of the unit, the ratio of scrubbing liquor flowrate to inlet gas flowrate, an estimate of the mass transfer area per cubic foot of packing in the packed bed, and equilibrium data specifying the concentration of contaminant in the gas stream that is in equilibrium with a given concentration of contaminant in the scrubbing liquor.

9.4 Orifice Scrubbers

Orifice scrubbers use the carrier gas velocity to promote dispersal of the scrubbing liquid and turbulent contact of the contaminated gas stream with the dispersed liquid (Figure 19.26). Both the efficiency and the pressure drop depend on the gas flowrate through the scrubber and the ratio of the scrubbing liquor injected per unit volume of gas. Orifice scrubbers make use of internal geometric designs that attempt to supply scrubbing liquid uniformly across the cross section of gas flowing through the unit. The kinetic energy of the gas supplies the energy needed to disperse the liquid phase. In the turbulent contacting zone, usually a short distance downstream of the orifice, the intensely turbulent motions provide violent contact between particulate matter and the larger scrubbing liquid droplets, which possess high inertia.

Downstream of the contact zone, mist eliminators agglomerate and coalesce the droplets of scrubbing liquor containing the absorbed contaminant. Depending on the energy supplied to the unit, scrubbing efficiencies can range from 85 percent to more than 99.5 percent, depending on the nature and physical properties of the contaminant being removed. The optimum use of input energy to effect a given degree of contaminant removal is the subject of many innovative designs applied to a wide variety of industrial process emissions.

There are numerous geometries and internal designs for orifice scrubbers. This general classification includes all such scrubbers that depend on contacting by passing the gas and liquid phases through some type of opening. In the more sophisticated designs, the opening may be varied to accommodate changes in requisite removal efficiency or gas flowrate. The basic principle of an orifice scrubber having a fixed orifice size dictates that scrubbing efficiency will fall off markedly if gas velocity through the orifice is decreased. By providing pressure drop sensors to the variable pressure drop

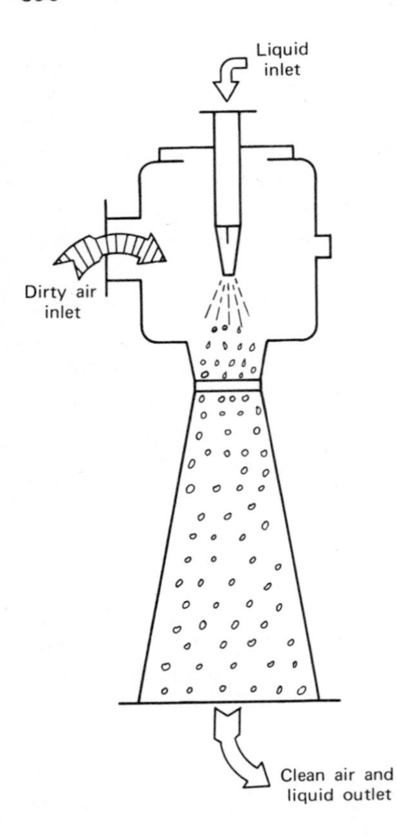

Figure 19.26 Ejector-type orifice scrubber (23).

orifice type scrubber, automatic operation can compensate for wide variations in gas flowrate, thus ensuring relatively constant collection efficiency.

Another special type of orifice scrubber is the venturi scrubber (Figure 19.27). Relatively high velocities between the gas and liquid droplets promote high collection efficiency in this unit. Such high velocities require a large energy input, raising operating costs to high levels, especially in applications processing large gas volumes. Venturi scrubbers typically use gas velocities of 200 to 400 ft/sec. At the venturi throat, the intensely turbulent action and shearing forces produce extremely fine water droplets that collect particulate matter very efficiently. At the instant of formation, these droplets move relatively slowly, and the fast-moving dust particles collide with nearly 100 percent efficiency with any droplets they may encounter. In the diverging section of venturi scrubbers, the droplets are slowed in preparation for removal in downstream mist eliminators.

Ejector venturis use spray nozzles and high pressure liquid to collect both particulate and gaseous contaminants and also to move the gas through the unit. These scrubbers have no fan to provide requisite gas flow. Thus fan power costs are eliminated, but relatively high energy consumption is still necessary because of the relatively high scrubber liquid pumping costs. Recent innovations have featured sonic nozzles in the ejector venturi to give high collection efficiency while conserving somewhat on high liquid pumping costs.

The collection efficiency of the venturi scrubber increases with increased energy input. Usually a venturi scrubber operating in a pressure drop range of 30 to 40 in. H_2O is capable of an almost total collection of particles ranging from 0.2 to 1.0 μm. In one application a venturi scrubber operating at 80 to 100 in. H_2O pressure drop scrubbed zinc oxide particulate matter (average diameter approximately 0.3 to 0.5 μm) with a scrubbing efficiency in excess of 99 percent (24). Efficiencies of this magnitude are normally obtained for this size particulate matter only using fabric filtration and pressure drops of nearly 10 in. H_2O. In the application cited, however, the hot gas stream precluded the use of fabric filtration without somewhat sophisticated and expensive precooling equipment, which would require a total pollution abatement system whose annual capital costs would be nearly equal to the operating cost of the high energy venturi system.

9.5 Mist Eliminators

Depending on the type of scrubber used, the number and size of water droplets entrained in the gas stream emerging from the scrubber may present objectionable stack gas discharge. If adequate mist elimination is not practiced, "raining" of contaminated scrubbing liquid droplets near the discharge stack can mean noncompliance with both mass emission and opacity regulations.

Normally the droplets emerging from a scrubber with inadequate mist elimination contain dissolved gases and/or captured particulate matter. Depending on the size of these droplets, fallout may occur a considerable distance beyond the point of discharge. In some cases, depending on ambient conditions and on the degree of saturation of the plume emerging from the scrubber, the droplets either grow because of further condensation, or evaporate, thus liberating the captured contaminant.

Generally droplets formed from tearing of liquid sheets of water (e.g., in orifice-type scrubbers) are 200 μm in diameter and larger. In extremely high energy applications, droplets as small as 40 μm may be produced. In low energy systems using coarse spray

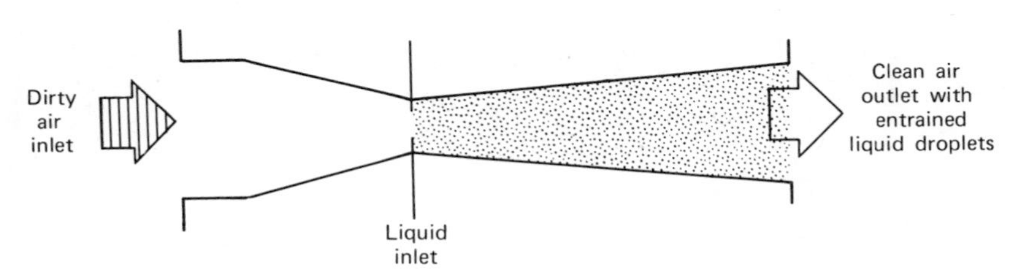

Figure 19.27 Venturi scrubber (23).

nozzles, the droplets may be as large as 1000 to 1500 μm. Clearly the type of device that must be used to ensure adequate mist elimination is partly a function of the size of the droplets.

Mist eliminators are devices that are designed to use physical separation processes, which depend on gravitational and inertial forces to "knock out" the droplets from the emerging gas stream. One such device appears in Figure 19.28. Devices used include packed beds, cyclone collectors, gravitational settling chambers, and chambers containing baffles that make a tortuous "zigzag" path for the gases to follow, while more inert water droplets impinge on the solid surfaces. No matter what type of device is used, once the scrubbing liquid droplets are captured, if the mist eliminator is operated at proper velocities, the droplets will coalesce and drain down to some quiescent zone in the mist eliminator for subsequent treatment.

One recent study attempts to provide correlations useful in specifying mist eliminators (25). The model relates the mist eliminator efficiency E to the zigzag design in a baffle-type mist eliminator as follows:

$$E = 1 - \exp\left[-\frac{u_t}{u_g}\frac{nw\theta}{b \tan\theta}\right] \tag{18}$$

where b = distance between baffles normal to gas flow (ft)
u_t = droplet terminal velocity (ft/sec)
u_g = superficial gas velocity (ft/sec)
n = number of rolls of baffles
w = width of baffle (ft)
θ = angle of baffle from flow direction (radians)

The droplet terminal velocity is that defined by normal Stokesian drag correlations, which inherently contain droplet density and size.

In this study, collection efficiency in baffle-type mist eliminators ranged from 95 to 100 percent droplet elimination for air velocities between 200 and 600 cm/sec. Beyond

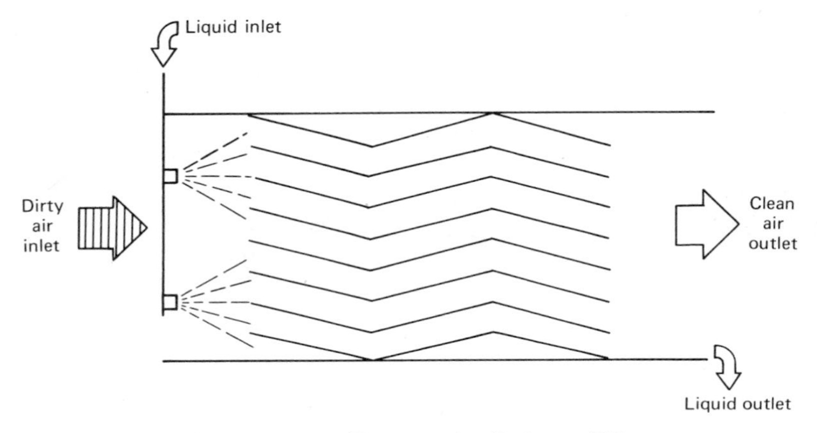

Figure 19.28 Baffle-type mist eliminator (23).

600 cm/sec the efficiency fell off very rapidly, dropping to less than 80 percent at approximately 700 cm/sec.

One common problem with some mist eliminator designs is reentrainment. If liquid load is too great for the mist eliminator, the onset of reentrainment begins at lower than normal gas velocities. That is, for a given gas velocity, as the amount of liquid removed from the mist eliminator increases, the efficiency remains relatively constant up until the capacity of the mist eliminator is exceeded; then efficiency drops markedly.

Cyclone-type mist eliminators function basically on the same principles delineated in Sections 7.5 and 7.6. Some departure from theory occurs, depending on reentrainment or shearing effects at the cyclone walls, causing the liquid film to be sheared and reconverted to atomized droplets if cyclone operation is improper.

The fiber-type mist eliminator is a relatively new device. The fibrous bed provides a tortuous path for the carrier gas laden with entrained droplets. In one type of application using fibrous mist eliminators, an annular cylinder is used within a concentric shell. The entering wet gases pass readily through the annulus, and the liquid drains down the inner walls of the annular cylinder. Gases pass up through the center and exit through the top of the mist eliminator. The water is collected at the bottom of the outer shell and is drained off.

Specification of any mist eliminator must involve consideration of the pressure drop that the mist eliminator will add to the total control system. Frequently manufacturers' specifications are the best source of such empirical data. When the emerging gas stream is saturated and still hot, liquid entrainment may be a problem when the gases are discharged to a relatively cool ambient environment. In such cases, it may be necessary to heat the exhaust gases to minimize production of steam plumes or condensation before the gases have a chance to mix and disperse in the atmosphere.

9.6 Design Parameters

The previous sections discussed some design parameters for the particular types of scrubber delineated. Because of the largely empirical correlations, frequently rules of thumb are necessary to specify scrubbers either for unique applications or when the state of the art and data correlation for the particular unit are not yet sufficiently refined to permit accurate prediction of collection efficiencies.

There do exist, however, some concepts fundamental to nearly any type of wet scrubber designed to remove the general gaseous and/or particulate contaminant from the carrier gas stream. In the case of gaseous contaminant removal, the concepts of "transfer units" and "theoretically equivalent height of packing" form the basis of these correlations. In the case of particulate removal, inertial impaction parameters form the basis of predicting the size of droplets required to collect a dust of known size distribution.

9.6.1 Particulate Scrubbing

Particulate contaminant removal collects particulate matter by liquid droplet impaction and by mechanisms that depend on inertial forces. Statistical probability and impaction correlations are the basis of mathematical modeling in spray collectors. Assumptions are made about interactions between large populations of droplets and the aerosol particles. The impaction parameter is the basis of nearly all collection mechanisms that involve inertial impaction of particles on droplets and account for departure of solid particles from gas streamlines in passing around or colliding with liquid droplets. In general, according to impaction correlations, the higher the impaction parameter, the higher the collection efficiency. For particles whose diameters are considerably greater than the molecular mean free path of the carrier gas, the inertial impaction parameter K is defined as follows (23):

$$K = \frac{v_r \rho_p D_p^2}{9 \mu_g D_d} \tag{19}$$

where D_d = droplet diameter
D_p = particle diameter
v_r = relative velocity difference between particles and droplets
μ_g = gas viscosity
ρ_p = particle density

All units should be dimensionally consistent so that K is dimensionless.

The functional relationship between the impaction efficiency E_i for each particle size category of a given particle size distribution and the impaction parameter depends on the collector geometry. For spherical droplets, the function $E_i = f(K)$ is completely specified and empirical. A semilogarithmic plot of collection efficiency versus the logarithm of the impaction parameter yields an S-shaped curve. The curve is asymptotic to zero efficiency for K less than 0.4 and asymptotic to unit efficiency for K greater than 40.

Having the capture efficiency E_i for particles of a given size from this correlation, the overall capture efficiency is a function of the path length through which particles move relative to the cocurrently moving droplets. This determines the integrated effect of the impaction efficiency over the contact length to give the capture efficiency. For a given particle size D_p an estimated efficiency is calculated from inertial impaction parameter correlations using the following expressions (23):

$$E_{\text{cap}_{\text{cocurrent}}} = 1 - \exp\left(-\frac{2 Q_w \rho_d D_d f E_i}{55 A \mu_g}\right) \tag{20}$$

$$E_{\text{cap}_{\text{cross current}}} = 1 - \exp\left(-\frac{3}{2}\frac{Q_w h E_i}{Q_g D_d}\right) \tag{21}$$

$$E_{\text{cap}_{\text{countercurrent}}} = 1 - \exp\left(-\frac{3 Q_w h E_i u_t}{2 Q_g D_d v_r}\right) \tag{22}$$

where Q_w = volumetric water flowrate
ρ_d = density of scrubbing liquid
D_d = average droplet diameter
f = empirical atomization factor (0.4, normally)
E_i = impaction efficiency (as a function of K)
A = cross-sectional area of scrubbing zone
μ_g = gas viscosity
h = distance of cross-current scrubbing action
Q_g = volumetric gas flowrate
D_p = particle diameter
u_t = terminal velocity of the particle
v_r = relative velocity difference between particles and droplets

All units should be dimensionally consistent so that the exponential expressions are dimensionless.

From these relations, the capture efficiency of each particle size range can be determined. To calculate scrubber efficiency for all particles, the capture efficiency in each particle size range should be mathematically "weighted" by the particulate mass in that size range as obtained from particle size distribution data.

It is often necessary to estimate droplet diameter to use any of the correlations above. In gas atomization of liquids in high velocity jets, a liquid film that is considered to be nearly at rest is contacted by a high velocity gas stream. The film distends and atomizes, with a resultant droplet size distribution whose average droplet diameter, the Sauter-mean diameter, is a function of liquid surface tension and viscosity, liquid and gas density, and gas velocity.

For the normal velocity regime present in atomizing scrubbers, the following correlation can be used to estimate the average droplet diameter (23):

$$D_d = \frac{283{,}000 + 793(Q_w/Q_g)^{1.992}}{v_g^{1.602}} \tag{23}$$

where Q_w = liquid flowrate (gallons/min)
Q_g = gas flowrate (10^3 acfm)
v_g = gas velocity (ft/sec)
D_d = Sauter-mean droplet diameter produced by atomization mechanisms (μm)

The droplet size distribution for pressure nozzles (as distinguished from gas atomizing nozzles) is a function of surface tension, liquid viscosity, orifice diameter, and pressure drop across the atomizing nozzle. Of course, smaller orifice diameters produce smaller droplets. Manufacturers' specifications are useful in determining droplet sizes in the use of atomizing nozzles.

9.6.2 Gaseous Contaminant Scrubbing

In gaseous contaminant scrubbing, the design equations are not based on particle dynamics. In its most general form, the design equation relating the height of the packed rm a given gaseous separation is (23):

$$lz = \frac{Q_g}{(K_g a)\, S\, (1 - y)_{LM}} \int_{y_1}^{y_2} \frac{(1 - y)_{LM}\, dy}{(1 - y)\,(y^* - y)} \tag{24}$$

g height

Q_g = incoming flowrate of nonabsorbable carrier gas

$K_g a$ = mass transfer coefficient

S = tower cross-sectional area

y = vapor phase concentration (molar) of gaseous contaminant being absorbed at bottom of tower (subscript 1) and at top of tower (subscript 2)

$(1 - y)_{LM}$ = log-mean average over inlet and outlet of $1 - y$, the molar concentration of nonabsorbable gas

y^* = equilibrium vapor phase concentration of gaseous contaminant at any point in the tower

Application of this equation to various systems in which equimolar counterdiffusion does and does not occur is the subject of the development of further simplifications to this equation. In general, a required height of a packed column is a function of the height of a "transfer unit" multiplied by "the number of transfer units." If it is assumed that the gaseous phase is rate controlling in the absorption process, the equation simplifies to (23):

$$VP = \frac{Q_g\,(y_1 - y_2)}{(K_g a)} \quad \frac{1}{(y^* - y)_{LM}} \tag{25}$$

where

$$(y^* - y)_{LM} = \frac{(y_1^* - y_1) - (y_2^* - y_2)}{\ln\,[(y_1^* - y_1)/(y_2^* - y_2)]}$$

and it has been assumed that dilute solutions exist in both the gaseous and liquid phases. In terms of partial pressures, the equations can be written (23):

$$VP = \frac{N}{(K_g a)\,(p^* - p)_{LM}} \tag{26}$$

where N = moles of contaminant absorbed in tower per unit time

p = partial pressure of contaminant at any point in the tower

p^* = equilibrium partial pressure of contaminant at any point in the tower

$$(p^* - p)_{LM} = \frac{(p_1^* - p) - (p_2^* - p_2)}{\ln\,[(p_1^* - p_1)/(p_2^* - p_2)]}$$

and $K_g a$ has units consistent with its empirical value such that the equation is dimensionally consistent.

To simplify the design equation to this relationship, it has been assumed that dilute solutions exist and that the operating and equilibrium lines are approximately linear. This is a legitimate assumption for the scrubbing conditions encountered in this study, where the equilibrium and operating relationships are plotted on a mole–ratio basis (23).

Again, assuming dilute solutions,´ the operating line relating the inlet and outlet gaseous and liquid concentrations is given by

$$Q_1y_1 + L_2x_2 = Q_2y_2 + L_1x_1 \tag{27}$$

where L, Q = molar flowrates of liquid and gas
x, y = molar concentrations of contaminant in liquid and gaseous phases, respectively

If recycle is used, and if an amount B of the effluent scrubbing liquor is discarded (blowdown) and an equivalent amount of fresh makeup water is added (before tower reentry but after blowdown), the inlet and outlet concentrations are related as follows:

$$x_2 = \left(1 - \frac{B}{L_1}\right)x_1 \tag{28}$$

where x_1 = molar concentration of contaminant in the blowdown
x_2 = molar concentration of contaminant in the recycled scrubbing liquor

These relationships indicate that for a given scrubbing efficiency, inlet gas flowrate, and entrained acid content, the volume of packing required is completely determined if the scrubbing liquor flowrate is known and the mass transfer coefficient K_ga is defined. The equilibrium partial pressure at both inlet and outlet gaseous streams is determined by the Henry's law relationship, which states that the equilibrium partial pressure is a linear function of the molar concentration of that component in the liquid phase. The slope of the relationship is known as the Henry's law constant:

$$p^* = Hx \tag{29}$$

where p^* = equilibrium partial pressure of the contaminant in the gas phase at the point of interest
H = Henry's law constant for that contaminant
x = molar concentration of the contaminant in the liquid phase at the point of interest

If it is assumed that the Henry's law constant is extremely small, the equilibrium partial pressure can be neglected. This is tantamount to neglecting the "back pressure" (equilibrium partial pressure) at both the scrubber inlet and outlet relative to the actual existing concentration of contaminant in the gaseous phase. When this is assumed, the equation for $(p^* - p)_{LM}$ becomes

$$p_{LM} = (p^* - p)_{LM} \cong \frac{p_1 - p_2}{\ln (p_1/p_2)} \tag{30}$$

The assumption that the equilibrium partial pressure is approximately zero is valid for constituents that are relatively nonvolatile when in solution or for scrubbing liquors that react chemically with the absorbed gas. In general, for dilute solutions, acids that ionize readily, such as sulfuric and nitric, are retained within the scrubbing liquor.

10 ELECTROSTATIC PRECIPITATION

Electrostatic precipitation is a process by which particulate matter is separated from a carrier gas stream using electrostatic forces and deposited on solid surfaces for subsequent removal. Electrostatic precipitators find wide application in many segments of pollution abatement systems. One of the largest users of electrostatic precipitation is the electric power industry. The need to remove particulate matter from large volumes of combustion-exhaust gases produced by coal-fired boilers calls for a collector that can extract particulate matter efficiently with small consumption of energy. This all-important advantage renders electrostatic precipitation an extremely attractive control alternative from the viewpoint of low operating cost. Unfortunately the relatively large initial capital investment cost needed to construct and install these massive and complex steel structures may render the application somewhat less attractive from the point of view of annual cost.

10.1 Range of Performance

The high voltage electrostatic precipitator has been applied at more installations emitting large gas volumes than any other class of high efficiency particulate collecting device (6). Although the device can efficiently extract even submicrometer material, commonly the precipitator treats dusts whose particle size distributions indicate a maximum diameter of approximately 20 μm. In those cases, more coarse dusts are also included in the contaminated gas stream and the precipitator is often preceded by mechanical collectors.

Precipitators applied to the electric power industry have been designed to handle gas volumes ranging from 50,000 to 2×10^6 cfm (6). Both "cold-side" and "hot-side" electrostatic precipitators have been installed which efficiently remove particulate matter from gas streams ranging from ambient temperatures to 1650°F, respectively (26). Depending on the location of the precipitator in the control system (i.e., whether gas flows through the precipitator by way of an induced or forced-draft fan), the unit may function under negative or positive pressures ranging from several inches of water, gauge, to 150 psig (6). Characteristically, the pressure drop through electrostatic precipitators, mostly open chambers with nonrestrictive gas passages, is on the order of 0.5 to 2 in. H_2O, gauge.

Typically, one stage of an electrostatic precipitator provides from 70 to 90 percent collection efficiency. By providing sequential stages, the overall precipitator efficiency can be increased to very high values. For example, in typical installations abating fly ash of moderate resistivity, one electrostatic precipitator stage is capable of producing nominally 80 percent collection efficiency. The remaining 20 percent of the entering particulate matter can be removed to an efficiency of 80 percent, approximately, in a second stage. The removal then, of 80 percent plus 16 percent in the second stage (80 percent of the dust pentrating the first stage) provides an overall collection efficiency of 96 percent for the two-stage precipitator. Additional stages can be installed. Using this estimate of collection efficiency, a six-stage precipitator would provide a precipitator collection efficiency in excess of 99 percent.

10.2 Mechanisms of Precipitation

The electrostatic precipitator relies on three basic mechanisms to extract particulate matter from entering carrier gases. First, the entrained particulate matter must be electrically charged in the initial stages of the unit. Second, the charged particulate matter is propelled by a voltage gradient (induced between high voltage electrodes and the grounded collecting surfaces), causing the particulate matter to collect on the grounded surface. Third, the particulate matter must be removed from the collection surfaces to some external container such as a hopper or collecting bin.

The high voltage potential across the discharge and collecting electrodes causes a corona discharge to be established in the region of the discharge electrode, and this forms a powerful ionizing field. When the particles in the gas stream pass through the field, they become charged and begin to migrate toward the collecting surfaces under the influence of the potential gradient existing between the electrodes.

Once the dust is deposited on the grounded surface, easy removal from the collecting surface depends on many factors. Dust resistivity is one prime factor dictating the collection efficiency and degree of reentrainment that can be expected under a given set of design conditions. If dust resistivity is too low, particles once deposited on the grounded surface may lose their charge and reenter the gas stream from the electrode collecting surface. If resistivity is too high, particles once deposited can cause "corona quenching" or "back corona." In this condition, the voltage gradient between dust deposited on the plates and the high voltage electrode begins to diminish because of the large voltage drop in the dust layer accumulated on the plate. This decreases collection efficiency, since the voltage gradient is all-important in determining the rate of migration of particulate matter toward the collecting surface.

In many electrostatic precipitators the dust is "rapped" from the collection surface by imparting a sharp blow to some area of the surface. The dust then cascades in a "sheeting action" down the collecting surface into a hopper. The extent to which this dust is removed in large pieces determines the degree of reentrainment upon rapping. Plate design, rapper design, and use of modular sections of precipitators are among the widely varying approaches employed by precipitator manufacturers.

10.3 Plate-Type Precipitators

Two basic types of electrostatic precipitator have been used: plate type and pipe type. In plate-type devices the collecting surface consists of a series of parallel, vertical grounded steel plates between which the gas flows (Figure 19.29). Alterations to the geometry of the plates are implemented by various manufacturers in an attempt to provide better collection and more efficient removal of collected dust upon rapping. Some plate-type precipitators are equipped with shielding-type structures installed vertically at incremental distances along the plate surface. These structures represent an attempt to minimize reentrainment as the dust sheet falls behind the structure. The design of these protuberances affects the geometry and intensity of the electrostatic field set up between the high voltage electrodes and the plates. Thus the spacing, size, and shape have been the subject of years of developmental work and research sponsored by precipitator manufacturers.

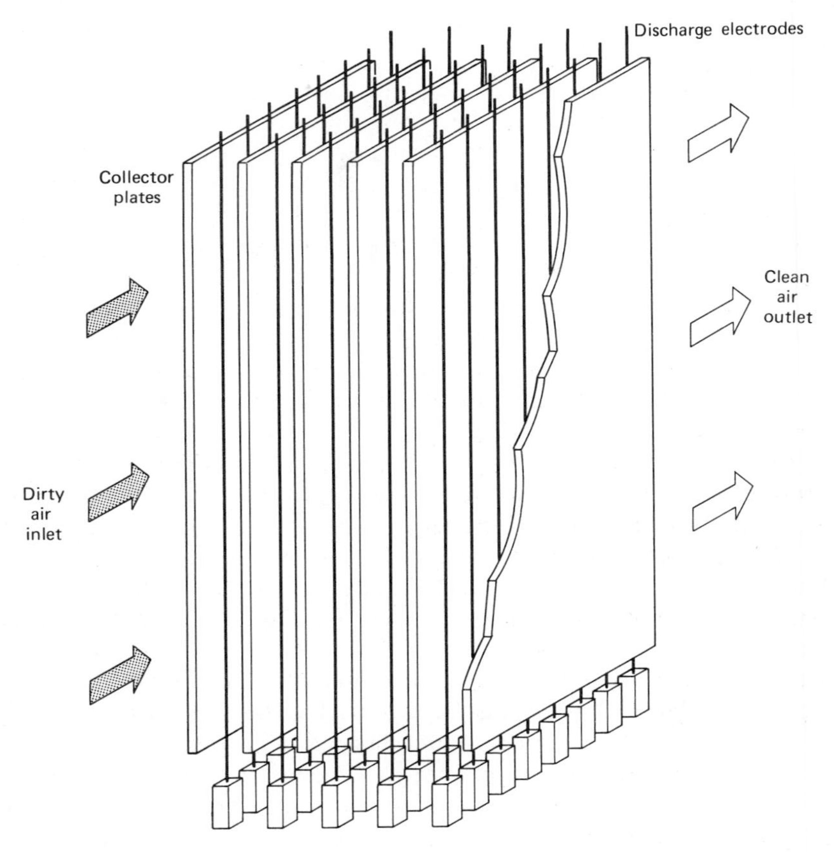

Figure 19.29 Plate-type electrostatic precipitator.

Most commonly, the plate-type precipitator finds application in the collection of solid particulate matter or dust. Plate-type precipitators consume approximately 200 W per 1000 cfm. This extremely low power requirement is affected also, however, by the resistivity of the dust and the geometry of the precipitator.

Sometimes the dust is removed from plate-type precipitators by washing. Strategically placed nozzles outside the electrostatic field are timed for actuation at various intervals, supplying irrigation to flush away dust such as sticky or oily particulate matter that is especially difficult to remove.

One disadvantage of plate-type precipitators with smooth plates (i.e., not equipped with a certain type of protuberance) is that they permit the gradual creepage of deposited particulate toward the precipitator outlet. Collected dust may be gradually moved by aerodynamic drag from the traveling gas stream, or particles may "jump" along the plate, depending on dust resistivity. In general, capital investment costs of the plate-type precipitator may be lower because of the simplicity of plate design over pipe-type precipitators. The exact design, however, dictates actual assembly and construction costs.

10.4 Pipe-Type Precipitators

Pipe-type precipitators generally feature high voltage electrodes positioned at the centerline of a grounded pipe (Figure 19.30). The gas flows through a bank of these pipes parallel to the high voltage wire. Dust deposited on the inner walls of the grounded pipe is removed in much the same manner as in the plate-type precipitator, that is, by rapping.

Some studies indicate that the pipe-type precipitators should have superior electrical characteristics (16). The pipe-type precipitator is prone, however, to creepage of collected dust toward the clean gas exit. In general, the design parameters for the specification of pipe-type precipitators may vary in the actual numerical values used to specify the number of pipes. The theoretical models on which sizing correlations are based, however, still include consideration for dust resistivity, total plate collection area, and gas velocity through the pipe.

10.5 Design Parameters

Precipitator size is normally based on a predetermined estimation of the resistivity of the dust, knowledge of the volumetric flowrate of gas to be processed by the precipitator, and an estimate of the dust loading. Depending on dust resistivity, the cross-sectional area of the precipitator may be sized for the known gas flowrate to provide a line velocity through the precipitator parallel to the collection surface of 2 to 6 ft/sec. The number of collecting plates (i.e., plates that form gas passages or channels through which the dirty gas flows parallel to the plates) is fixed by this calculated cross-sectional area and by the plate spacing. In conventional precipitators the plate spacing is predetermined by the maximum voltage to which the high voltage electrode, located in

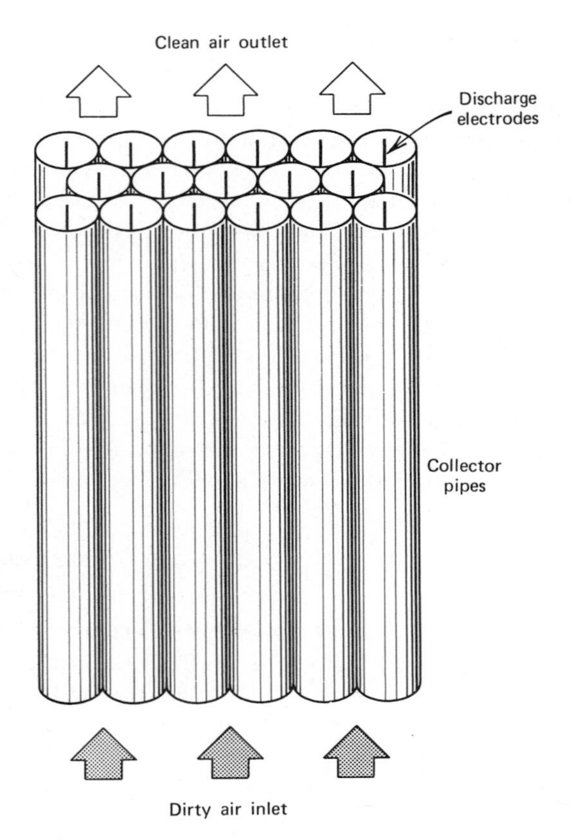

Figure 19.30 Pipe-type electrostatic precipitator.

the center of the channels, is subjected. Typically, 45,000 V applied to these centerline high voltage electrodes, located typically 4½ inches from the grounded electrode surface, produces 10 kV/in. voltage gradient. This rule-of-thumb design parameter then forms the basis for the determination for the total number of gas passages in the precipitator and all necessary hardware needed to support and power the high voltage electrodes.

Manufacturers claim higher efficiency for a given area of collecting surface depending on the design of the collecting electrodes, the design of the high voltage electrodes, the wave form of the high voltage imparted to the electrodes, and the mechanism of particulate removal from the collecting surfaces. The collection efficiency of an electrostatic precipitator, however, is estimated using the Deutsch equation:

$$E = 100\left[1 - \exp\left(\frac{-AW}{Q}\right)\right] \tag{31}$$

where E = collection efficiency (percent)
A = area of collecting electrodes (ft^2)
Q = gas flowrate (acfm)
W = particle migration velocity (ft/min)

Generally, the migration velocity is an empirically determined sizing factor for a given application, grain loading, and dust resistivity. This experience factor is often a closely guarded number among precipitator designers and manufacturers.

The design parameter of dust resistivity is a prime component in determining the migration velocity W. Dust resistivity measurings are frequently made on site. Resistivities vary from 10^9 to 10^{15} ohm-centimeters and are a function of the nature of the dust and the temperature. Some precipitator designers analyze collected dust in the laboratory for dust resistivity, yet dust resistivity measurements directly from the gas stream are indispensable in providng accurate design information. For example, it is not uncommon to find 1 to 3 orders of magnitude difference in measured dust resistivity when comparing field measurements to laboratory measurements of collected dust.

Rules of thumb indicate that the migration velocity W varies from 2 to 14 cm/sec, depending on the aforementioned design parameters. Typical values for coal-fired boilers range from 6 to 10 cm/sec. Clearly, from the foregoing equation, the required collecting plate area for a given gas volume and collection efficiency is inversely proportional to the migration velocity. Therefore, choosing a conservative migration velocity of 6 as opposed to 10 cm/sec requires approximately 1.7 times the plate area.

11 GAS–SOLID ADSORPTION

Adsorption denotes a diffusion process whereby certain gases are retained selectively on the surface, or in the pores of interstices, of specially prepared solids. The principle has been used successfully for many years in the separation of mixtures and in product improvement through removal of odorous contaminants or through decolorizing.

In many applications the primary purpose of adsorption is to render an operation economical by recovery of valuable materials for reuse or resale. A good illustration of this is the long-standing practice in several industries of recovering organic solvent vapors by adsorption. In such applications, the adsorbed materials may be "desorbed" by a variety of methods and concentrated and purified for reuse. The adsorbing solid is regenerated and can be reused for many such cycles.

11.1 Range of Performance

In atmospheric pollution control as practiced today, the principle of adsorption is employed primarily to prevent odorous and offensive organic vapors from escaping into populated areas. A typical example is the removal of vapors generated by printing plants in large cities. Occasionally, however, adsorption is used to remove trace concentrations of highly toxic gaseous materials, such as radioactive iodine vapors.

Fixed-bed adsorbers are used when trace concentrations are encountered, whereas for heavier concentrations, regenerative adsorbers are needed. Product recovery from regenerative adsorbers (e.g., solvent recovery in the dry-cleaning field) may pay for the emission control system and its operation, or at least offset part of the annual cost.

Recently carbon bed adsorption techniques have been applied to the goal of lessening the impact of the energy crisis on processes that rely on incineration techniques to abate combustible process emissions. In earlier applications of incinerators, airstreams laden with vapors containing organic material have been directly incinerated; some of the incinerators use heat recovery to minimize the amount of auxiliary fuel required to support combustion and provide efficient abatement. In practice, either the organic vapor concentration in the airstream's incinerator was low due to process dynamics, or the concentration of organic vapors was necessarily limited to comply with safety code requirements (e.g., maintaining the concentration of combustible vapors at 25 percent of the lower explosive limit).

In recent applications, carbon bed adsorbers have been used to adsorb organic vapors over prolonged periods (e.g., 2 to 8 hours). Upon saturation, the bed can be desorbed using saturated or superheated steam. The effluent from the bed, consisting mostly of water vapor and concentrated organic compounds, can be sent to an intermittently fired incinerator, which is designed to operate only during bed desorption. In most cases the concentration of organic material has been raised sufficiently so that by mixing auxiliary air at the point of combustion, the combustion process sustains itself with little or no auxiliary fuel. Furthermore, since the desorbed stream consists of steam and organic compounds, the danger of explosion is eliminated in most cases (because the amount of air in the desorbed stream is very small, the concentration of organics is therefore above the upper explosive limit).

The nuisance aspect of industrial odorous emissions is becoming a more significant factor in the network of developing emission regulations. Characteristically, odors can be produced by organic compounds at very low concentrations and emission rates, which means that the odorous emission is not limited by organic compound emission regulations. Such low emissions are ideally suited to adsorption processes, especially those utilizing activated carbon. Furthermore, because the absolute amount of odorous substances is usually so low, the adsorbent can be used for extended periods and discarded with no significant impact on system operating cost. Thus adsorption systems have found wide use in rendering plants and industrial processes emitting high molecular weight hydrocarbons.

In general, the range of performance of gas–solid adsorption systems is related to many factors controlling the effectiveness of physical adsorption in dynamic systems. These factors include surface area of the adsorbent, the affinity of the adsorbent for the adsorbate, the density and vapor pressure of the adsorbate, the concentration of the adsorbate, the system temperature and pressure, the dwell time, the bed packing geometry, the thermodynamics of adsorption of solvent mixtures, and the mechanism of removal of heat of adsorption (13). All these factors are interrelated, requiring careful analysis and specification before installing or even considering a carbon adsorption system as a viable pollution abatement system in a particular application.

Carbon adsorption has been used in many areas of the food processing industry: the handling and blending of spices, canning, cooking, frying, and fermentation processes, for example. In the manufacture and use of chemicals, process discharges have been abated by carbon adsorption to minimize objectionable emissions in the manufacture of odorous substances such as pesticides, glue, fertilizers, and pharmaceutical products, in paint and varnish production, and in the release of odorous vapors emitted to the atmosphere from tanks during filling operations (27).

Aside from activated carbon, gas–solid adsorbers use other materials for reversible or irreversible removal of gaseous constituents. Among these are alumina, bauxite, and silica gel (28). Activated carbon, however, constitutes one of the most popular types of solid adsorbent, mostly because of its low cost and its ability to be regenerated.

11.2 Physical Adsorption

Physical adsorption denotes a type of adsorption process wherein multiple layers of adsorbed molecules accumulate on the surface of the adsorbing solid. In this process, the number of layers that can be made to accumulate on the activated carbon surface is related to the concentration of the adsorbate in the gas stream (29).

Physical adsorption processes generally entail reversible processes. In the case of adsorption of organic solvents, for example, on the surface and interstices of activated carbon, the rate at which adsorbate molecules are "driven" from the relatively homogeneous gas phase into the pores of the activated carbon depends to some extent on the concentration of the gas. At the surface of the adsorbing solid, molecules are attracted and deposited, providing a concentration gradient useful in causing migration of solvent molecules into the activated carbon pores. From the standpoint of residence time and the ease of saturation of activated carbon, therefore, solvent concentration in the gas phase partly determines the amount of activated carbon required and the optimum placement or geometry of the carbon relative to the flowing gas stream. The more molecular layers that can be accumulated on the activated carbon interstitial area, the less carbon is required per pound of solvent to be adsorbed. Conversely, for a given residence time of contaminated gas in an activated carbon bed, the higher the concentration, the higher will be the extraction efficiency. These considerations apply in physical adsorption processes, whether the adsorbing solid is activated carbon, a siliceous adsorbent, or a metallic oxide adsorbent.

In the case of activated carbon adsorption, the micropore structure of the carbon surface provides an extremely large surface area for a given weight of carbon; typically, several hundred thousand square feet of available surface is distributed throughout the interstices of one pound of activated carbon (27). Generally the adsorptive capacity of a particular activated carbon is measured by its ability to retain a given chemical entity at various concentrations of that chemical species. These empirical data are normally obtained by laboratory investigations performed by the suppliers of activated carbon. The data are useful for specifying the equilibrium concentrations theoretically obtainable at a given set of conditions, assuming that sufficient residence time is available in the bed (i.e., that sufficient carbon is present for a specified gas flowrate and velocity). From adsorption "isotherms" (supplied by activated carbon manufacturers), the minimum amount of carbon required can be determined for a given adsorption cycle time. The amount of solvent adsorbed on the carbon at this "saturation point" can be theoretically achieved, but only at the expense of low collection efficiency near the end of the adsorption cycle. Adsorption isotherm data do not necessarily specify adsorption efficiency, since efficiency may vary with time since last regeneration or composition of mixtures being adsorbed, as well as with bed geometry and solvent-laden gas flowrate through the bed.

As a rule, the particular geometrical design and arrangement of activated carbon in an activated carbon bed adsorber is best specified by well-informed engineers who specialize in adsorption equipment. Even then, pilot plant work may be required before a full-scale carbon bed adsorber can be specified or designed. Generally, bed geometries can be arranged in layers, beds, cylindrical canisters, or in a variety of other shapes to provide optimum contact time for a given concentration of material and a given gas flowrate.

11.3 Polar Adsorption

Polar adsorption denotes a gas–solid adsorption process whereby molecules of a given adsorbate are attracted to and deposited on the surface of an adsorbing solid by virtue of the polarity of the adsorbate. Some siliceous adsorbents depend on the polarity of gas molecules for selective removal from a gas stream. For example, synthetic zeolites are adsorbents that can be produced with specific uniform pore diameters, giving them the ability to segregate molecules in the liquid or gas stage on the basis of the shape of the adsorbent molecule (27). Adsorbents of this type will not even adsorb organic molecules of the same size as their pores from a moist airstream, since the water molecules are differentially adsorbed because of the chemical structure of the adsorbing solid (27).

11.4 Chemical Adsorption

Chemical adsorption takes place when a chemical reaction occurs between the adsorbed molecule and the solid adsorbent, resulting in the formation of a chemical compound (29). Chemical adsorption may be contrasted with physical adsorption, where the forces holding the adsorbate to the solid surface are weak. In chemical adsorption, usually only one molecular layer is formed, and the chemical reaction is irreversible except in cases when the energy of reaction can be applied to the solid adsorbent to reverse the chemical reaction.

Many of the same factors useful in sizing and specifying physical adsorption equipment apply in chemical adsorption. As in applications featuring the removal of low concentration, high molecular weight, odorous substances, the contaminated adsorption bed is usually discarded or shipped for external regeneration after its efficiency has degraded. The cost of operating chemical adsorption equipment may be prohibitive in applications involving either high flowrate or a concentration of adsorbate in the incoming gas stream that causes the bed to deteriorate quickly. Alternatively, chemical adsorption systems can be applied to streams where the contaminant is present in high concentrations but is emitted only intermittently.

11.5 Design Parameters

One of the first steps in specifying the quantity of activated carbon required for a given cycle time in an activated carbon adsorber is to establish what chemical species and concentrations are present in the incoming gas stream. Adsorption isotherms are available from activated carbon suppliers either for individual chemical constituents being adsorbed, or for chemical species that are similar enough in chemical structure or physical properties to enable reasonable approximations regarding adsorbability and retention. Such empirical data, established by the suppliers, apply for the single chemical species being adsorbed on a given type of activated carbon. When mixtures of organic species are being processed, it can be expected that the higher molecular weight substances are preferentially adsorbed, and lower molecular weight species are adsorbed initially to a lesser efficiency than would be predicted if the lower molecular weight species were adsorbed alone. As the adsorption cycle proceeds, the effluent can be expected to be rich in low molecular weight compounds from the mixture, and high molecular weight compounds will be adsorbed at the highest efficiency.

Having calculated the minimum bed size based on a given cycle time and equilibrium retention relation, the bed geometry and actual amount of carbon must be determined by consideration of the retention time in the adsorber for a given adsorption efficiency, sufficient capacity to ensure an economical service life, low resistance to airflow to minimize overall system pressure drop, uniformity of airflow distribution over the bed to ensure full utilization, pretreatment of the air to remove particulate matter that would gradually impair and poison the bed, and provision for some manual or automatic mode for bed regeneration (27).

The geometry and depth of the adsorbing bed depend primarily on the superficial gas velocity through the bed, the concentration of contaminants in the incoming airstream, and the cycle time. As adsorption continues, the portion of the activated carbon bed first encountered by the incoming gas becomes rich and eventually fully saturated in adsorbate. Deeper areas of the bed are progressively less and less saturated with adsorbate until some small but finite level of penetration of adsorbate can be observed at the bed exit. As the adsorption cycle continues, this "adsorption wave" proceeds through the bed until the concentration of the exit gas reaches a finite value in excess of allowable emissions. At this point "breakthrough" is said to have occurred, and it is necessary to regenerate the bed. By arranging the given necessary amount of activated carbon in an optimum geometry, breakthrough can be postponed to the point at which minimum utilities are required for bed desorption and the maximum practical level of adsorbate is present in the desorption stream.

It is not possible to set down rules of design encompassing all applications of carbon bed adsorption to pollution abatement processes. Because of the wide variation in possible solvent mixtures and concentrations, the bed geometry is usually specified from previous experience or extrapolated from performance data obtained in a similar type of installation.

12 COMBUSTION

Combustion, sometimes termed "incineration," oxidizes combustible matter entrained in exhaust gases, thus converting undesirable organic matter to less objectionable products, principally carbon dioxide and water.

In many instances combustors are the most practical means for bringing equipment into compliance with air pollution control regulations. These devices have been employed to reduce or eliminate smoke, odors, and particulate matter in cases when such emissions might exceed the limits set by law.

Equipment in operation today falls essentially into two classifications: direct flame combustion and catalytic combustion. Direct flame incineration has been more widely used and has, in general, been more successfully applied. However properly designed catalytic incineration units that perform satisfactorily have been constructed.

Catalytic burners differ from direct flame burners in that they allow organic vapors to be oxidized at temperatures considerably below their autoignition point and without direct flame contact upon passage through certain catalysts.

With the increased cost of gas and oil, efficient heat recovery systems must be employed with either catalytic or direct flame burners to make them economically sound. The arrangement of these devices is limited only by the type of operation and by other plant facilities. Effluent gases from these units often are used in dryoff ovens, water heaters, space heaters, and many other types of heat exchanger, to make use of their heat content.

12.1 Range of Performance

Prior to the energy crisis, any organic emissions not economically recoverable using the control techniques outlined in previous sections were often subjected to direct incineration. With the advent of higher energy costs and the unavailability of auxiliary fuel to support the combustion, the range of performance of combustion equipment has become the subject of innovative designs and recent investigations.

The myriad of combustible organic compounds emitted to the atmosphere from our numerous manufacturing processes calls for the application of carefully designed combustion equipment. The design must be based on the particular contaminants being combusted, the concentration of these contaminants in the carrier gas stream, and the flowrate of the gas stream, which affects the physical size of the combustion equipment.

The specifically designed and engineered systems used to control the process emissions are as diverse as the industrial operations emitting combustible compounds. Typical examples of processes requiring combustion as a means for the control of hydrocarbon emissions are industrial drying processes, baking of paints, application of enamels and printing ink, application of coatings and impregnates to paper, manufacturing of fabric and plastic, and manufacturing of paints, varnishes, and organic chemicals, as well as synthetic fibers and natural rubber (30). In all these processes, flares, furnaces, and catalytic combustors have found suitable application, and the materials emitted may be oxidized to odorless, colorless, and innocuous carbon dioxide and water vapor (30).

The incineration of gases and vapors has found wide application in the control of both gaseous and liquid wastes. In the area of gaseous emissions, use of incineration to minimize odorous pollutants is often the only available control alternative. Because most odorous compounds are characteristically low in concentration, the choice of carbon bed adsorption, for example, may prove to be economically or even technologically infeasible because of low control efficiencies for certain compounds.

The opacity of plumes emitted from various processes has historically been reduced by passing the otherwise highly visible emission through a combustion apparatus. Examples include application to coffee roasters, smoke houses, and some enamel baking ovens. In addition, afterburners are sometimes used to heat otherwise wet and highly visible stack gases that would emit a harmless, but visually objectionable, steam plume.

In some applications the organic gases and vapors being emitted from a process are classified as "reactive hydrocarbons." These compounds, when discharged to the atmosphere, are known to contribute to the formation of smog and otherwise irritating compounds. Afterburners can be used to convert most of these reactive hydrocarbons into carbon dioxide and water vapor.

Combustion equipment has been used extensively in refineries and chemical plants as a method of disposal for unusable waste and as a method of reducing explosion hazards. In these applications, ultimate destruction using a combustion-based process must be carefully designed to avoid creating a hazard to nearby, highly flammable storage tanks or to the process itself.

12.2 Flares

A flare is a method for the efficient oxidation of combustible gases when these are present in a stream that is within or above the limits of flammability (30). Flares usually find their widest application in petrochemical plants, especially in the disposal of waste gases that are often mixed with other inert gases such as nitrogen or carbon dioxide. Additionally, many of the chemical processing plants that produce, use, or otherwise discharge highly dangerous or toxic gases are often equipped with flares designed to be activated under emergency conditions, in the event that such toxic gases should require immediate discharge.

Generally flares are designed after thorough analysis of the gas stream being incinerated. Flare heights are established by taking into account the heat and light emitted from the flare, and they are supposed to ensure sufficient mixing to prevent unburned organic compounds that may penetrate the flare from presenting a hazard or a nuisance to the surrounding area.

Often when flares are applied to installations where the concentration of combustible gases is not sufficient to ensure or maintain the persistence of the flare, auxiliary fuel must be added to provide for efficient combustion and sustained burning.

In some applications, where the materials being combusted are difficult to burn, auxiliary air must be introduced at the point of combustion at the top of the flare, using steam jets or other satisfactory means to ensure good fuel–air mixing.

One design parameter in the sizing and specification of flares is the ratio of hydrogen to carbon in the materials being combusted. As an example, in low molecular weight, aliphatic hydrocarbons (e.g., methane) the high ratio of hydrogen to carbon guarantees burning with very little, if any, soot or smoke. In contrast, double-bonded or triple-bonded hydrocarbon molecules (e.g., acetylene) burn with a very sooty flame because the ratio of hydrogen to carbon is low. Normally the discharge from a flare should be limited by design to carbon dioxide and water vapor and should certainly exclude the emissions of visually objectionable products of combustion such as carbonaceous soot.

When condensable mists are present in the gases to be flared, inertial separators should be provided at the base of the flare to permit automatic separation and automatic drainage of these liquefiable compounds, which may otherwise flow back through the flare, down its walls, causing flashback and/or an explosion.

12.3 Direct Flame Combustion

Direct combustors represent a broad classification of combustion equipment in which auxiliary fuel is added and heat recovery may or may not be practiced to ensure efficient combustion and minimization of operating costs. The designs of direct combustors are as varied as the manufacturer's experience. Generally, direct combustors rely on three factors to provide efficient conversion of the organic compounds to carbon dioxide and water.

1. Sufficient time in the incinerator to allow complete conversion to carbon dioxide and water vapor, all other conditions being specified.
2. Sufficient fuel to ensure combustion temperatures high enough to permit complete combustion within the allowable residence time in the incinerator.
3. Sufficient turbulence in the incinerator to provide for good mixing and contacting of the combustible materials with the active flame front.

Direct combustors equipped with heat-exchange units are often used to preheat the incoming gases to a temperature high enough to ensure more complete and efficient combustion and to save on total energy consumption needed to preheat the incoming gases to the combustion temperature. This heat exchange can be derived in a variety of ways, but it essentially causes the removal from the incinerator discharge of waste heat that would otherwise pass into the atmosphere, and recycles this heat to the incoming gases by way of some physical medium. Heat exchange efficiencies range from 20 percent to more than 90 percent depending on the design of the heat exchanger, the pressure drop through the device, and the materials of construction; these variables in design also affect, of course, the initial capital investment cost and the operating costs.

Direct combustors are also useful in the removal of combustible particulate matter, particularly that which is odorous. When such particulate matter is of submicrometer size, this method of control is often the only available or most technologically and economically feasible technique.

Incinerators are also quite widely used because they occupy relatively little space, have simple construction, and generate low maintenance requirements (6).

Multiple-chamber direct combustors provide a useful means for increasing combustion efficiency and overall conversion to carbon dioxide and water for a given quantity of fuel input. The multiple chambers provide additional residence time, therefore additional combustion efficiency, as it may be required to convert difficult-to-burn organic compounds to carbon dioxide and water.

12.4 Catalytic Combustion

In catalytic combustors reactions occur that convert organic compounds into carbon dioxide and water on the surface of a "catalyst," usually composed of platinum or palladium, with the end result that auxiliary fuel costs are minimized.

Catalytic combustion has come into wide use in installations where concentrations of hydrocarbons are relatively low and large amounts of auxiliary fuel would be required to sustain combustion. The systems do not find wide applicability in processes involving an inlet gas stream that can contain materials capable of "poisoning" the catalyst, thus rendering it ineffective. Additionally, systems containing high amounts of particulate matter often cannot be incinerated using catalytic systems, or else they must be equipped with high efficiency precleaners to prevent the particulate matter from coating the surface of the catalyst, making it ineffective.

Basically a catalytic system includes a preheat burner, exhaust fan, and catalyst elements, as well as control and safety equipment (31). The preheat burner usually raises the incoming gas stream to a temperature of 700 to 900°F. As the heated gases pass through the catalyst bed, the heat of reaction raises the temperature further to levels comparable to those found in direct combustors. The obvious advantage of minimal fuel input renders catalytic incineration an attractive alternative where applicable.

Typically, for a 10,000-scfm system, capital investment costs for catalytic incinerators may be approximately the same as direct flame incinerators; however annual fuel costs may be reduced to about 2 percent of flame-type incinerators depending on hydrocarbon concentration and incoming gas temperature (31).

Catalytic incinerators may also be equipped with heat recovery devices, which may be used to recapture otherwise wasted heat and inject it into the incoming gas stream, thereby minimizing the already low preheat fuel requirements.

Even when well-controlled with precleaners, catalytic incinerators require periodic washing and maintenance to remove particulate matter that accumulates after long operating periods. Additionally, catalysts often require periodic reactivation since materials such as phosphorus, silica, and lead, even when present in trace amounts, shorten the active life of the catalysts (31).

12.5 Design Parameters

Depending on the type of incinerator selected, the sizing and specification of the unit relies primarily on thermodynamic calculations of auxiliary fuel requirements and system residence times. In general, the calculations are based on the simple concept that all incoming materials must be preheated to the combustion temperature, generally in excess of 1200°F and usually 1500°F, so that conversion to carbon dioxide and water vapor will occur if sufficient residence time is provided for the heated constituents to interact with the provided oxygen.

When concentrations of hydrocarbons in the gas stream are low, the specific heat of the gas stream may be regarded as that of air. In general, the rate of heat input Q required to raise the temperature of the incoming gas from inlet conditions to 1200°F is expressed simply by

$$Q = SCFM\ \rho_g\ C_p(1200 - T) \tag{32}$$

where Q = required rate of heat input (Btu/min)
$SCFM$ = gas flowrate at standard conditions, (cfm)
ρ_g = gas density at standard conditions (lb/ft^3)
C_p = average specific heat for air over temperature range (Btu/(lb)(°F))
T = incoming gas temperature (°F)

To provide the requisite heat input, the total amount of auxiliary fuel may be determined by considering the heating value of the fuel being used as well as the heating value of any organic compounds present in the incoming stream. Assuming that the incoming stream has an average concentration C_a of hydrocarbons (lb/ft^3), and that this stream of hydrocarbons provides Q_{hc}, the quantity of heat per pound of hydrocarbons, the available heat from the organic content of the stream Q_{oc} is calculated by the following equation:

$$Q_{oc} = SCFM\ C_a Q_{hc} \tag{33}$$

Now, subtracting this available energy from the preheating requirements, the total amount of heat required may be computed if the heating value of the available fuel Q_f is known:

$$Q_{\text{req}} = Q - Q_{oc} = AF \times Q_f \tag{34}$$

where AF = auxiliary fuel rate required (lb/min)
Q_f = heat content of auxiliary fuel (Btu/lb)

If heat recovery equipment is available, the quantity of available heat may be subtracted directly from the heat computed in Equation 34 and the fuel requirements thereby proportionally reduced. In practice, even though the heating requirements are small or even negative according to these simplistic energy balances, it is likely that auxiliary fuel will still be required because no account has been made in this analysis for heat losses in the total system.

Thermodynamic calculations, therefore, are useful in arriving at fuel requirements; however the specification of total residence time or degree of turbulence involves an experience factor available from the supplier of the combustion equipment. As a rule, though, a residence time of 0.5 to 1 second in the primary incinerator with at least 0.2 second in its secondary chambers for direct flame combustion methods is considered to be adequate reaction time to convert oxidizable substances to carbon dioxide and water vapor, presuming sufficient or excess oxygen. The residence time required for efficient combustion in catalytic incinerators is reduced by at least one-fifth to one-tenth of the time required for direct flame combustion (32). This residence time must be computed knowing the cross-sectional area of the incinerator and the volume of gases in *actual* cubic feet per minute as determined from the combustion temperature.

To provide adequate turbulence, the internal design of the incinerator often features baffles and/or relatively tortuous turns. This allows good mixing while minimizing system pressure drop. The gas velocity required for good mixing is considered to be about 2100 ft/min (32).

13 SELECTING AND APPLYING CONTROL METHODS

The previous sections of this chapter have presented and summarized the regulatory, economic, and technical considerations involved in the selection of the optimal method of emission control for a specific stationary source. The application of these basic factors in an orderly sequence of decisions may require not only a carefully conceived and applied methodology, but a perspective broader than this chapter presents.

The selection of the "best" control method for a specific emission source can be a very simple and obvious choice, or it can entail a complex decision-making procedure, especially if more than one type of control seems to be appropriate. Strategic selection is especially important when the expected cost of control—usually related to the magnitude and strength of the emission—will be large relative to the financial resources of the owner or operator of the source.

Because of technical uncertainties and ever-present alternatives, it would be very desirable to have an inclusive, systematic approach to the selection of a control system for a given application. Such a strategy should combine the type of information and methods presented in this chapter with an even more basic understanding of certain very crucial steps in a whole sequence of decisions necessary to ensure success of any emission cleanup effort. The ultimate performance of an emission control system rests very heavily on the avoidance of oversights during one of the planning steps, as well as the proper use and understanding of the basic fundamentals for selecting optimum control methods. Sound and accurate emissions data are essential, and skillful interpretation of all available data can spell the difference between a control system that functions efficiently and one that performs only marginally.

The overall strategy that should be incorporated when addressing the need to control a stationary emission entails (33): defining the emission limit, identifying all related emission sources, investigating process modifications, defining the technical aspects of control problems, and selecting the optimum control system.

13.1 Defining the Emission Limit

The emission limit that applies for any given air pollution control problem is the most basic information and, indeed, a building block with which to develop and specify the optimum control system. It is important to understand that emissions are not always covered by an established legal or regulatory limitation. At this writing, the EPA has established emission standards for hazardous air pollutants and also has specified maximum permissible emission levels of the "criteria pollutants" (particulate matter, carbon monoxide, sulfur dioxide, nitrogen oxides, and hydrocarbons) for only about two dozen categories of new sources. Although these and related standards will continue to be promulgated, it is highly unlikely that federal standards will cover many of the industrial situations for which emission "controls" are needed or being planned. State and local standards also may not apply specifically to a given plant emission. Nevertheless, virtually all emissions require control under widely applicable and generally phrased "nuisance," "opacity," or "odor" regulations. Others logically demand control if they constitute a substance capable of being a potential toxic hazard.

In these cases, the responsible design engineer will have to determine an emission level without simple reference to a regulatory action. Sources of emissions that are likely in violation of nuisance, opacity, or odor limitations call for expert analysis, since all three areas could necessitate the application of the most recent technology to assess the magnitude of the problem and to specify a viable abatement strategy. In some cases the engineer should seek the advice and counsel of a toxicologist or industrial hygienist. Such professional specialists can sometimes give a reasonable exposure level for the general population that reflects the stability of the material and the possible chronic effects from environmental exposure. A note of caution should be injected here with respect to threshold limit values published by the American Conference of Governmental Industrial Hygienists (4). These values were derived and are intended for the control of worker exposures; they cannot be used reliably to determine relative toxicities, to evaluate air pollution problems, or to assess the effects of continuous exposures. The relation between industrial hygiene and air pollution is indeed a close one, but it does not include an indiscriminate cross-referencing of TLVs with respect to ambient levels or emission limits for stationary source controls.

13.2 Identifying All Related Emission Sources

Many times when emission controls are specified, it is found that all emission sources have not been included in the "process envelope" of the control system because of insufficient care in the planning stage.

Normally, it is not enough to attach a well-designed control device to the main exhaust vent emitting the pollutant. With the very low emission levels required for the more toxic materials, emissions from sources other than the main process vent can, in total, overshadow even those from the main exhaust. It is good practice, particularly when dealing with toxic pollutants, to study the entire process and identify all points of emission and all possible solutions to the control problem. Some frequently overlooked emission points that have been found to contribute heavily to process emissions are as follows (33):

1. Accidental releases:
   a. Spills.
   b. Relief valve operation.
2. Uncollected emissions:
   a. Tank breathing.
   b. Packing gland or rotary seal leakage.
   c. Vacuum pump discharges.
   d. Sampling station emissions.
   e. Flange leaks.
   f. Manufacturing area ventilation systems.
3. Reemission of collected materials:
   a. Vaporization from water wastes in ditches or canals.
   b. Vaporization from aeration basins.
   c. Reentrainment or vaporization from landfills.
   d. Losses during transfer operations.

13.3 Process Modification

Our discussion of various air-cleaning methods began, as it always must, with a suggestion that process and system control represent a sound first step in making decisions in the selection and specification of emission controls. If the problem can be solved at its source, this approach offers several advantages, including a typical economic advantage, over the installation of some added or incremental type of control system.

Process modification is usually a most economical way to reduce emissions because little or no capital is needed to purchase control equipment. In addition, there can be improvements in operating efficiency that will reduce material or energy losses, and the cost of a terminal control system, if one is ever required, can also be cut because it has to handle less material. It is important to retain the option of process modification from the very beginning to avoid costly repetition of an engineering analysis once the add-on controls have been specified and the cost has been estimated to be exorbitantly expensive. Some of the techniques (Section 4) that can be helpful include the following (33):

1. Substitution of a less toxic or less volatile solvent for the one being used.
2. Replacement of a raw material with a purer grade, to reduce the amount of inerts vented from the process or the formation of undesirable impurities and by-products.
3. Changing the process operating conditions to reduce the amount of undesirable by-products formed.
4. Recycling process streams to recover waste products, conserve materials, or diminish the formation of an undesirable by-product by the law of mass action.
5. Enclosing certain process steps to lessen contact of volatile materials with air.

13.4 Defining the Technical Aspects of the Control Problem

An important final step before selecting a control method is to define the properties of the exhaust gas stream; the basic data needed were reviewed in Section 5. Such data can come from several sources, however, and the information obtained must be viewed with objectivity if not skepticism. Laboratory data, for example, are never quite the same as results reported under actual plant conditions. Empirical prediction of emission factors or rates cannot replace actual source testing of stack gases. Particle size measurement is more reliable than literature data citing "typical" size distributions from various processes. In addition, the method of sampling where source testing is employed is a most important consideration that can yield differences both in reported results and the interpretation of those results.

Some common pitfalls to avoid in the technical definition of a control problem include failure to recognize the presence of another phase, failure to recognize the presence of fine particulates, and failure to recognize variations in the characteristics of the emission (33).

If particulates, especially fines, are present when only a gaseous type of effluent is expected—or vice versa—serious control efficiency problems can follow, since most devices designed for removing gaseous pollutants are far less efficient for particulates, and most particulate control devices are inefficient at removing gases. Only a few devices do both jobs well. Some packed-tower scrubbers or carbon adsorption units can be rendered inefficient by particulate matter, and baghouse filters can be corroded by trace amounts of gases.

The problem is particularly complex when the particulate is an aerosol resulting from the condensation of a relatively nonvolatile material. In this case, a sample at the only available sample point may give one answer, whereas a sample further upstream might show more gaseous material and one downstream might show more particulate. Often sulfur trioxide (SO_3) presents this type of difficulty (33). Dry SO_3 is gaseous and can be absorbed in 98 percent sulfuric acid (H_2SO_4) or 80 percent isopropanol. When SO_3 contacts water, however, it hydrates to H_2SO_4 and condenses to a submicrometer aerosol that is not collected well by either reagent. In this form, it should be collected with a dry filter. The EPA Sampling Method 8 (34) for this substance employs both an absorber and a filter. In the plant, however, dual control devices are usually prohibitively expensive.

The failure to recognize the presence of fine particles in an emission can lead to great difficulty. Until recently, this was a minor problem, but now fine particles in the atmosphere are being cited as a serious environmental hazard.

The fineness of particulate matter is particularly important in the treatment and control of toxic pollutants for two reasons (33). First, fine particles stay airborne longer than coarse ones, thus increasing the chance of exposing the surrounding population to the pollutant. Second, the low emission levels permissible for toxic materials generally require the removal of nearly all the fine, difficult-to-remove material from the emissions.

The presence of fine particles can sometimes be detected by the hazy appearance of a plume, except when coarse particles or steam are also present. Unless one has experience with a similar emission, the particle size distribution of the emission should be determined with a cascade impactor or an optical device. Care should be taken to extend the test into sufficiently fine sizes, to ensure that all particles that may have to be removed by the control device will be accounted for.

Failure to recognize variations in the characteristics of the emission is the third problem that can lead to serious design errors. Virtually all emissions vary as the temperatures of both the air and the process change. Emissions from continuous processes vary from day to day, and some, particularly from batch processes, vary from hour to hour. The ranges of such variations should be determined to evaluate the effect on a potential control device. If the control device is designed for the worst possible case, it will usually handle less severe cases. Nevertheless, problems peculiar to the control device should be recognized, such as the sensitivity of the venturi scrubber to flowrate, or of the electrostatic precipitator to dust resistivity (33).

13.5 Selection of Controls with a Technical–Economic Decision Model

Clearly the basic strategy for the selection of the optimum control method must be an integrated approach that includes, at a minimum, identification of suitable technical alternative control methods, and selection of the alternative that is most attractive economically from among the suitable alternatives available.

It is not too much to expect the development of a computer program utilizing all relevant technical and cost-related information needed to select not only the most technically appropriate control methods, but also the most attractive economic choice. The need to use an overall systems analysis approach to the problem is obvious; indeed it is relatively simple to identify and interrelate the various major types of information needed to achieve a logical stepwise set of decisions that will lead to the choice of the "best"

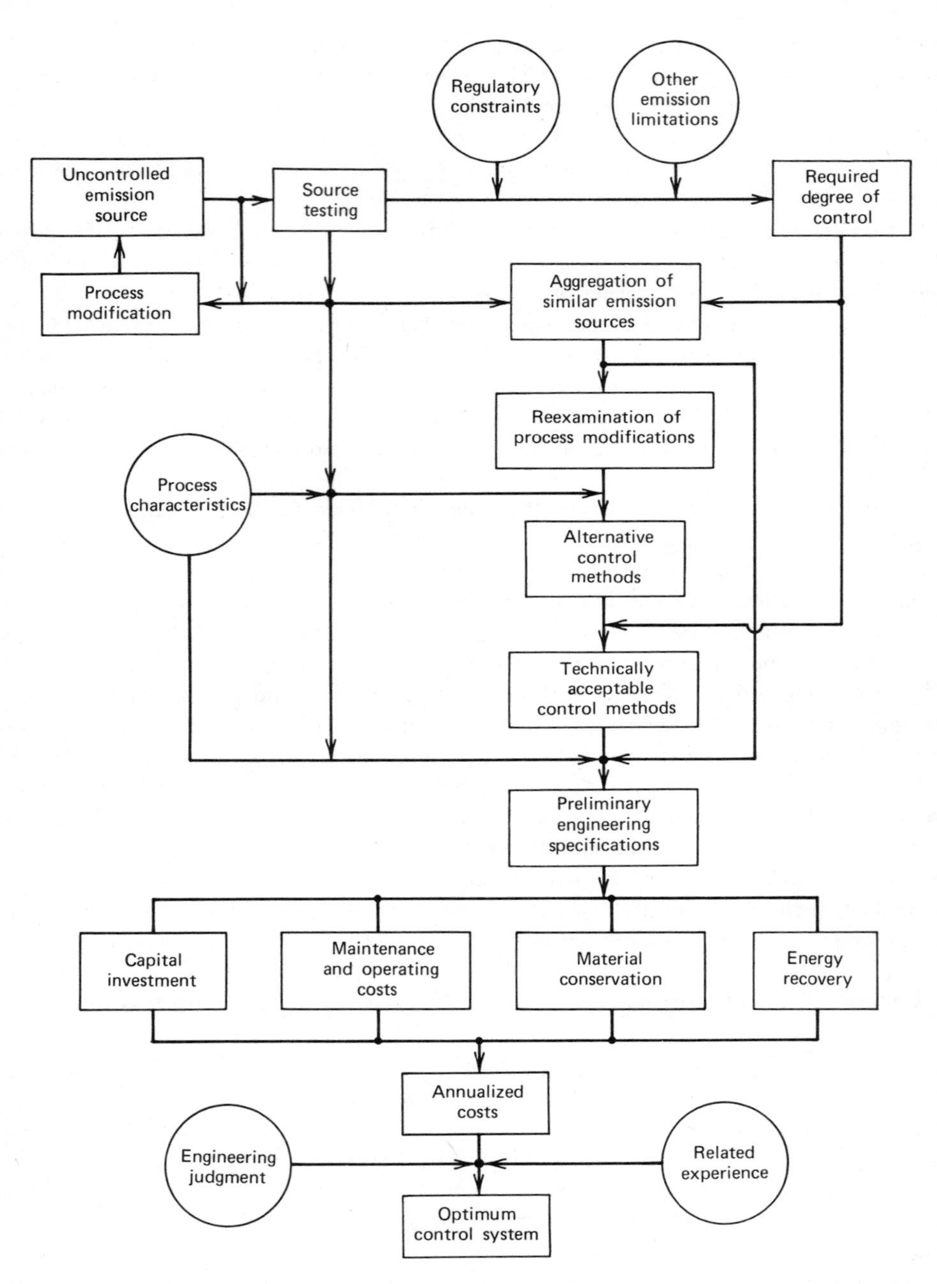

Figure 19.31 Technical–economic decision model for selecting emission controls; circled items denote required inputs of information and data.

method of control for a given emission problem. One such approach is outlined schematically in Figure 19.31.

REFERENCES

1. J. E. Mutchler, "Principles of Ventilation," in: *The Industrial Environment—Its Evaluation and Control,* U.S. Department of Health, Education, and Welfare, NIOSH, Cincinnati, Ohio, 1973.
2. Clean Air Act, as amended through June 24, 1974, 42 U.S.C. 1857 et seq., *Environment Reporter,* S-210, 71:1101, 1974.
3. American Conference of Governmental Industrial Hygienists, *Industrial Ventilation—A Manual of Recommended Practice,* ACGIH, Lansing, Mich., 1976.
4. American Conference of Governmental Industrial Hygienists, *Threshold Limit Values for Chemical Substances and Physical Agents in the Workroom Environment with Intended Changes for 1975,* ACGIH, Cincinnati, Ohio, 1975.
5. Environmental Protection Agency Regulations on Approval and Promulgation of Implementation Plans, as amended through December 22, 1976, 40 CFR 52, *Environment Reporter,* S-333, 125:0201, 1977.
6. *Control Techniques for Particulate Air Pollutants,* Department of Health, Education and Welfare, National Air Pollution Control Administration, Washington D.C., 1969.
7. F. I. Honea, *Chem. Eng. Deskbook,* **81,** 55–60 (1974).
8. K. T. Semrau, *J. Air Pollut. Control Assoc.,* **13,** 587–594 (1963).
9. "PVC Makers Move to Mop Up Monomer Emissions," *Chem. Eng.,* **82,** 25–27 (1975).
10. D. B. Turner, *Workbook of Atmospheric Dispersion Estimates,* U.S. Environmental Protection Agency, Office of Air Programs, Publication No. AP-26 (revised), 1970.
11. G. A. Briggs, "Diffusion Estimation for Small Emissions," Air Resources Atmospheric Turbulence and Diffusion Laboratory, National Oceanographic and Atmospheric Administration, ATDL Contribution File No. 79, 1973.
12. U.S. Nuclear Regulatory Commission, "Methods for Estimating Atmospheric Transport and Diffusion of Gaseous Effluents in Routine Releases from Light Water Reactors," NRC, Regulatory Guide No. 1.DD, 1975.
13. American Industrial Hygiene Association, *Air Pollution Manual,* Part II, AIHA, Detroit, 1968.
14. A. H. Rose, D. G. Stephan, and R. L. Stenburg, "Control by Process Changes or Equipment," in: *Air Pollution,* Columbia University Press, New York, 1961.
15. K. J. Caplan, "Source Control by Centrifugal Force and Gravity," in: *Air Pollution,* Vol. 3, 2nd ed., A. C. Stern, Ed., Academic Press, New York, 1968.
16. C. J. Stairmand, *J. Inst. Fuel,* **29,** 58–81 (1956).
17. H. E. Hesketh, *Understanding and Controlling Air Pollution,* 2nd ed., Ann Arbor Publishers, Ann Arbor, Mich., 1974.
18. C. E. Lapple, *Chem. Eng.,* **58,** 145–151 (1951).
19. U.S. Atomic Energy Commission, Hanford, Wash., 1948.
20. K. Iinoya and C. Orr, Jr., "Source Control by Filtration," in: *Air Pollution,* Vol. 3, 2nd ed., A. C. Stern, Ed., Academic Press, New York, 1968.
21. B. Goyer, R. Gruen, and V. K. Lamer, *J. Phys. Chem.,* **58,** 137 (1954).
22. D. C. Drehmel, "Relationship between Fabric Structure and Filtration Performance in Dust Filtration," U.S. Environmental Protection Agency, Control Systems Laboratory, Research Triangle Park, N.C., 1973.

23. S. Calvert, J. Goldshmid, D. Leith, and D. Mehta, *Wet Scrubber System Study,* Vol. 1, U.S. Environmental Protection Agency, Control Systems Division, Office of Air Programs, Research Triangle Park, N.C., 1972.
24. T. A. Loch, Clayton Environmental Consultants, Inc., Southfield, Mich., private communication, 1976.
25. S. Calvert, I. L. Jashnani, and S. Yung, *J. Air Pollut. Control Assoc.,* **24,** 971–975 (1974).
26. A. B. Walker, *Pollut. Eng.,* **2,** 20–22 (1970).
27. A. Turk, "Source Control by Gas-Solid Adsorption," in: *Air Pollution,* Vol. 3, 2nd ed., A. C. Stern, Ed., Academic Press, New York, 1968.
28. J. A. Danielson, Ed., *Air Pollution Engineering Manual,* 2nd ed., U.S. Environmental Protection Agency, Office of Air and Water Programs, Research Triangle Park, N.C., 1973.
29. P. N. Cheremisinoff and A. C. Moressi, *Pollut. Eng.,* **6,** 66–68 (1974).
30. H. J. Paulus, "Nuisance Abatement by Combustion," in: *Air Pollution,* Vol. 3, 2nd ed., A. C. Stern, Ed., Academic Press, New York, 1968.
31. G. L. Brewer, *Chem. Eng.,* **75,** 160–165 (1968).
32. L. Thomaides, *Pollut. Eng.,* **3,** 32–33, 1971.
33. W. L. O'Connell, *Chem. Eng. Deskbook,* **83,** 97–106 (1976).
34. Environmental Protection Agency Regulations on Standards of Performance for New Stationary Sources, as amended through November 29, 1976, 40 CFR 60, *Environment Reporter,* S-331, 121:0401, 1977.

CHAPTER TWENTY

Heat Stress: Its Effects, Measurement, and Control

JOHN E. MUTCHLER

1 INTRODUCTION

This chapter covers the health effects, environmental aspects, engineering control, and management of heat stress in industry. Heat represents one of the classical physical stresses in the occupational environment. Although overexposure to heat has been a problem for centuries, only in modern times have the physiological and environmental characteristics of heat exposure been related to develop a good understanding of the limits of exposure and suitable control methods for work in hot environments.

1.1 Significance of Heat Stress in Industry

As industry has developed, from the early days of the Industrial Revolution to our present highly technological society, on-the-job potential for injury and illness from acute exposure to heat has increased far beyond that known earlier to home-centered craftsmen. Among the more dangerous original industrial vocations were those involving molten materials such as glass and metal. In these first "hot industries," in addition to the ever-present danger of burns, explosions, and spills of molten material, illness and death were well-known and accepted potential results of very hard physical work in excessively hot environments.

Historically, except for slave laborers in conquered ancient lands, heat stress likely manifested itself first as a serious "occupational" hazard among armies operating in warm climates. Interestingly, our present-day standards of good practice and steadily emerging regulatory requirements for control of heat stress in the workplace are based in large measure on experience in the armed forces, including studies at military training centers in the United States (1–4).

The mining industry has also provided much information on heat disorders and tolerance limits, not only because of traditionally substantial mortality and morbidity from overexposure, but also because of concern for reduced productivity that accompanies excessive thermal exposure (5–9).

In many ways, the workers in hot jobs traditionally have been a highly select population. Those who cannot cope with the prevailing hot conditions seek less demanding employment. As a result of this selection process, the majority of workers in hot jobs perform at a high level and are highly adaptable to work in heat (10).

In addition to ever-increasing levels of mechanization, there are widespread, often undocumented, work practices in industry that are unofficial, yet aimed at relieving workers of heat strain on excessively hot days (10). Such practices include the following:

1. Performing only the unavoidable operations and postponing other less important jobs.
2. Reassigning workers in auxiliary jobs to help those who work in the hottest areas.
3. Substituting younger and more fit workers to relieve the older and less fit.

In the "hot-dry" industries, such as steel mills, forge shops, and glass manufacturing plants, the thermal load on the worker is increased by the sensible heat that escapes from process equipment and operations into the surrounding workspace. The major factor is radiant heat from the surfaces of tanks, hot metals, and the like; the heat load contributed by the surrounding hot air is secondary. In this class of hot industry, little or no moisture is added to the air, except by changing weather conditions; thus there is little or no decrement in the evaporative cooling capacity of the work environment. Nevertheless, the requirements for sweating and evaporation are considerable because of the added radiant and convective loads.

In the "warm-moist" industries, such as laundries, papermaking, and mines in which large amounts of water are used for dust control, water vapor is added to the air from wet processes or as escaping steam. Although relatively little sensible heat is transferred from the process to the work environment, high relative humidities greatly reduce the evaporative cooling capacity. Consequently, the worker may be unable to dispose adequately of his metabolic heat as well as the small additional heat gained from the environment.

Today we would expect deaths and heat-related illnesses from involuntary overwork to be rare in the United States. It is true that one of the characteristic features of industrial exposure to heat today is the relatively rare occurrence of frank illness. This is largely due to the body's capability to maintain a balance with the thermal environment, including voluntary reduction in the metabolic generation of heat.

Universal as heat may be, heat stress in modern industry tends to be either a potentially serious factor, especially in the so-called hot industries, or a problem that only rarely requires industrial hygiene expertise. Although definitively accurate data are not available on the frequency of heat casualties (11), and although heat stress is widely recognized as a serious potential hazard, most industrial hygienists give heat exposure relatively low priority in their day-to-day consideration of worker health hazards.

As our technology advances, we are also discovering that the interaction of heat and other chemical and physical stresses imposes a concern that would not be anticipated for exposure to heat alone. In part, these combined or interactive effects are attributable to certain peculiar characteristics of heat on physiological mechanisms. Such effects seem to be contributing to a more general and growing concern among environmental investigators that the combined action of two or more physical and/or chemical stresses can lead to adverse effects that are clearly unacceptable, both in view of today's standards of comfort and in the context of burgeoning legislative prohibitions against health impairment. Studies of such interactions not only indicate the complexity of multiple stresses, they also raise fundamental questions about the validity of exposure standards for physical stresses or toxic substances based on single-stress investigations.

1.2 Development of Heat Stress Standards

1.2.1 Current Status—OSHA

The only federal standard in the United States for occupational exposure to heat stress is that implicit in the "general duty clause" of the Occupational Safety and Health Act of 1970 (12). This legislation stipulates that each employer has a general duty to furnish each of his employees "employment free from recognized hazards causing or likely to cause physical harm." The Occupational Safety and Health Administration (OSHA) has received recommendations from both the National Institute for Occupational Safety and Health (NIOSH) and a statutory Standards Advisory Committee on Heat Stress, but a standard covering exposure to heat has not yet been proposed.

1.2.2 Historical Development

The first American standard for work in hot environments dates back to 1947, when the Committee on Atmospheric Comfort published its report "Thermal Standards in Industry" (13). The criteria and permissible exposure limits in this standard were intended only as a guide. The committee recommended that each industry develop its own standard because of the complexity of industrial work and the variation in individual tolerance to heat among workers. As general criteria, the committee cited the comfort and health of individuals, their work output, and their physiological and psychological reaction to work. It utilized an index known as "effective temperature"

(ET) as the basis of its guidelines, but the guidelines were limited to exposure limits for only two levels of work. The higher level was 432 kcal/hr, and the lower was given only in the qualitative terminology, "light sedentary activities."

In 1971 the American Conference of Governmental Industrial Hygienists (ACGIH) published a notice of intent to establish threshold limit values (TLVs) for heat stress (14). These proposed TLVs were based on the assumption that acclimatized, fully clothed workers whose deep body temperatures are maintained at 38°C or less are not subjected to excessive heat stress (15). The ACGIH adopted the TLVs in 1974.

Because it is not practical to monitor every worker's deep body temperature, industrial hygienists must rely on the measurement of certain environmental factors and hope that these correlate with physiological response to heat. Of the several environmental indices of heat stress currently in common use, the wet bulb–globe temperature (WBGT) was, in the opinion of ACGIH, the most suitable index for assessing heat stress for exposed workers.

The proposed TLVs for heat stress are listed in Table 20.1. The TLVs emphasize the significance of work load level and the percentage of time spent working. The specific WBGT values in Table 20.1 were derived from data in the literature correlating deep body temperatures with work load and environmental conditions. In formulating the proposed TLVs, the Physical Agents Committee of ACGIH pointed out the following facts (16):

1. The permissible exposure limits in the TLVs are based on the assumption that the WBGT value of the resting place is the same or very close to that of the working place. If the resting place is air conditioned and its climate is kept at or below 24°C

Table 20.1 Permissible Heat Exposure TLVs, Given in °C (°F) WBGT (16)

Work—Rest Regimen	Work Load		
	Light	Moderate	Heavy
Continuous Work	30.0 (86)	26.7 (80)	25.0 (77)
75% Work—25% Rest, Each hour	30.6 (87)	28.0 (82)	25.9 (79)
50% Work—50% Rest, Each hour	31.4 (89)	29.4 (85)	27.9 (82)
25% Work—75% Rest, Each hour	32.2 (90)	31.1 (88)	30.0 (86)

(75°F) WBGT, the allowable resting time may be reduced by 25 percent. Also, higher exposure limits are permitted if additional resting time is allowed.

2. When a job is self-paced, workers often spontaneously limit their hourly work load to 30 to 50 percent of maximum physical performance capacity. This is done by either setting an appropriate work speed or by interspersing unscheduled breaks. Thus the worker's daily average metabolic rate seldom exceeds 330 kcal/hr (1300 Btu/hr). However, within an 8-hour work shift there may be periods when the worker's hourly average metabolic rate is higher.

3. The TLVs are valid for light summer clothing worn customarily by workers when employed under hot environmental conditions. If special clothing is required and it impedes sweat evaporation or has higher insulation value, the worker's heat tolerance is reduced and the TLVs are not applicable.

In July 1972 NIOSH published its *Criteria for a Recommended Standard . . . Occupational Exposure to Hot Environments* (10). Again WBGT was specified as the index to be used in determining requirements for "work practices" designed to reduce the risk of harmful effects from interactions between excessive heat and toxic chemical and physical agents.

The Criteria Document differs from the ACGIH TLVs in several respects. By assuming continuous heavy work for a period of one hour, a threshold WBGT was specified at 79°F (26°C). The NIOSH document recommends that when exposure exceeds this level, any one or a combination of work practices be initiated to ensure that body core temperature does not exceed 100.4°F (38.0°C).

In addition, the Criteria Document outlines other requirements for management when exposures exceed 79°F (26°C) WBGT. These include preemployment and periodic medical examinations, WBGT profiles for each workplace, and detailed recordkeeping.

The Criteria Document received widespread negative response from the hot industries, especially those in the southern states, where outdoor WBGT values exceed 79°F nearly every day during the warm seasons (17).

In 1973 the Secretary of Labor appointed a Standards Advisory Committee on Heat Stress in an attempt to resolve some of the discrepancies in and objections to the Criteria Document relative to industrial experience. This panel, chaired first by Professor Belding and subsequently by Professor Ramsey, issued its "Recommended Standard for Work in Hot Environments," reported to the Secretary of Labor in January 1974 (18).

The Standards Advisory Committee modified the recommendations set forth in the NIOSH Criteria Document by differentiating between threshold WBGT values on the basis of both work load and air speed. Table 20.2 summarizes the threshold WBGT values recommended by the Standards Advisory Committee.

In addition, the stipulated work practices were revised and some alternative practices were included. Although the work practices proposed by the committee generally follow those in the Criteria Document, they are more detailed and differentiate further between mandatory work practices, special work practices for special conditions, and work

Table 20.2 Threshold WBGT Values Recommended by the Standards Advisory Committee (18)

Workload	Threshold WBGT Values Degrees F (Degrees C)	
	Low Air Velocity ≤300 fpm (≤1.5 m/sec)	High Air Velocity >300 fpm >1.5 m/sec
Light (Level 2) ≤200 kcal/hr (≤795 BTU/hr)	86 (30)	90 (32)
Moderate (Level 3) 201–300 kcal/hr (800–1190 BTU/hr)	82 (28)	87 (31)
Heavy (Level 4) >300 kcal/hr (>1190 BTU/hr)	79 (26)	84 (29)

practices for extreme heat exposure. Other sections of the recommended standard deal with required medical examinations, training of employees, monitoring of the environment, and recordkeeping.

Perhaps the most important aspect of the Standards Advisory Committee's work was the status of the final draft report: it was not a consensus recommendation. The committee voted 10 to 5 that the draft be the basis of a proposed (OSHA) rule to be published by the Secretary of Labor (17). There was substantial support for the idea that a heat stress standard was not appropriate at that time, and in fact, there has been no subsequent proposed rule making by OSHA for occupational exposure to heat.

However it seems likely that any heat stress standard promulgated by OSHA will be a "work practices standard." This means that when the value of an environmental index, such as WBGT, reaches a threshold level based on various levels of work load and possibly other environmental criteria, certain actions and procedures will be required to minimize the effect of exposure. These work practices will undoubtedly include the following:

1. Modifying the bodily heat production (metabolic rate) for the task being performed.
2. Limiting the number and duration of exposures.

3. Reducing heat exchange with the thermal environment, first by feasible engineering controls, and then, if necessary and appropriate, by protective clothing.

2 PHYSIOLOGY OF HEAT STRESS

Our current understanding of the physiological effects of working under hot conditions has developed and emerged from a large number of significant laboratory and field studies. The modern era of such investigations began in the 1920s with the work of the American Society of Heating and Ventilating Engineers. Since then extremely important contributions have been made by such investigators as Bedford (5, 6), Yaglou (1, 19, 20), Leithead (21, 22), Lind (21, 23–25), Minard (2–4, 26, 27), Hatch (28–30), Haines (29), Hertig (31, 32), Belding (26, 30, 32), and Wyndham (8, 33, 34), among others.

Regardless of the thermal environment, man attempts through a set of involuntary compensating mechanisms, to maintain steady heat content and body temperature. In that context, physiological factors influence body temperature regulation, thermal exchange with the environment, manifestations of heat strain, heat disorders, and tolerance factors.

2.1 Body Temperature Regulation

Man is a homeotherm; he must maintain an internal body temperature within a narrow range and near 37.0°C if he is to remain healthy and efficient (35). If the "core temperature" of the body falls below 35.0°C, hypothermia results, and death is likely at core temperatures of 27.0°C and below. Hyperthermia results when core temperature exceeds 40.6°C with the absence of sweating; death will occur when the core temperature exceeds about 42°C.

The inherent thermal homeostasis of man allows him to regulate internal body temperature within suitable limits by the involuntary control of blood flowrate from sites of metabolic heat production (muscles and deep tissues) to the cooler body surface (skin), where heat is dissipated through the mechanisms of radiation, convection, and evaporation. Thus man reacts dynamically to changes in his thermal environment, always striving to maintain a core temperature within the critical range.

2.1.1 Metabolic Heat Production

The energy required to sustain life and all bodily functions is released in the body by the exothermic oxidation, under enzymatic control, of food-derived fuel: carbohydrates, fats, and proteins (36). Although such oxidation occurs at rates and temperatures much different from those of the equivalent combustion in the external environment, the thermodynamics of the oxidation are the same in both instances. Thus the same quantities of food substrates yield essentially the same thermal energy when oxidized at low temperature inside the body or at high temperature outside the body. This equivalency provides a basis for "indirect calorimetry" whereby metabolic heat production can be measured indirectly by the rate of oxygen uptake of the body (36). One liter of oxygen is approximately equivalent to a metabolic heat output of 5 kcal. An average man at rest consumes about 0.3 liter of oxygen per minute, equivalent to a metabolic heat production of 1.5 kcal/min or about 90 kcal/hr (36).

Individual variation in metabolic heat production is relatively small for resting, but wide differences exist for maximal physical effort, dependent largely on body size, muscular development, physical fitness, and age (36). Table 20.3 compares oxygen uptake rates and metabolic heat equivalents for several typical work activities.

2.1.2 Hypothalamic Regulation of Temperature

Man's temperature-regulating center lies in a region at the base of the brain known as the hypothalamus. The anterior portion of the hypothalamus includes the "heat loss" center, which responds to increases in its own temperature as well as to incoming nerve impulses from warm receptors in the skin (36). It activates heat loss through increased blood flow to the skin and sweating. Such physiological reactivity to elevated body

Table 20.3 Oxygen Uptake, Body Heat Production, and Relative Energy Cost of Work in 70 kg Men

Activity	Oxygen Uptake (l/min.)	Body Heat Production (M) (kcal/hr)	Maximum Oxygen Uptake (l/min) Low 2.5	Medium 3.0	High 3.5
			Uptake Required, % of Maximum		
Rest (seated)	0.3	90	12	10	8.5
Light Machine Work	0.66	200	26	22	19
Walking (3.5 mph on level)	1.0	300	40	33	28
Forging	1.3	390	52	43	37
Shoveling (depends on rate, load and lift)	1.5–2.0	450–600	60–80	50–66	43–58
Slag Removal	2.3	700	92	77	66

temperature leads to interaction with the thermal environment to offset the increase, thus maintaining that temperature within an acceptable range.

2.2 Worker-Environment Thermal Exchange

Man's need to maintain a nearly uniform core temperature, even under elevated ambient temperature, causes the exchange of thermal energy between a worker and his environment. This heat loss occurs at a rate proportional to the rate of metabolic heat production and the degree of discomfort from elevated ambient conditions.

Heat exchange between man and his environment occurs by three primary routes: convection, radiation, and evaporation.

2.2.1 Convection

Convection C is a function of both the temperature gradient between the skin and the ambient air, and the movement of air past the skin surface. As revised by Hatch (37, 38), the expression becomes

$$C = 1.0\ V^{0.6}\ (TA - TS)$$

where C = convection (kcal/hr)
V = air speed (m/min)
TA = air temperature (°C)
TS = skin surface temperature (°C)

When TA is higher than TS, C will be positive; thus heat gained by the worker will be positive, and heat lost will be negative. The empirically obtained exponent, 0.6, applies to forced convection over vertical cylinders, the geometric configuration that best corresponds to man in the working environment (39). The coefficient 1.0 includes the surface area for an "average" man (1.8 m^2). For an individual whose surface area differs from average, the value of C may be adjusted by multiplying by the ratio of actual area to 1.8 (11).

2.2.2 Radiation

Radiation R is a function primarily of the gradient between the mean radiant temperature of the solid surroundings and skin temperature. Actually, radiative heat exchange is a function of the fourth power of absolute temperature, but a first-order approximation is sufficiently accurate for estimating R in the applications outlined here (29).

$$R = 11.3\ (MRT - TS)$$

where R = radiation (kcal/hr)
MRT = mean radiant temperature of the solid surroundings (°C)
TS = skin surface temperature (°C)

2.2.3 Evaporation

Evaporation E is a function of air speed and the difference between the vapor pressure of perspiration on the skin (vapor pressure of water at skin temperature) and the partial pressure of water in the air. In hot-moist environments, evaporative heat loss may be limited by the capacity of the ambient air to accept additional moisture, in which case we have

$$E_{max} = 2\ V^{0.6}\ (VPS - PPA)$$

where E_{max} = maximum evaporative capacity (kcal/hr)
V = air speed (m/min)
VPS = vapor pressure of water on the skin (mm Hg)
PPA = partial pressure of water in air (mm Hg)

In hot-dry environments, E may be limited by the amount of perspiration that can be produced by the worker. The maximum sweat production that can be maintained by the average man throughout an 8-hour shift is slightly above one liter per hour, equivalent to an evaporative heat loss of about 600 kcal/hr (2400 Btu/hr) (11).

From the foregoing considerations it is evident that four environmental factors define thermal exchanges: TA, MRT, PPA, and V. The equations for C, R, and E_{max} in units of degrees fahrenheit, feet per minute, and Btu per hour are, respectively:

$$C = 1.08\ V^{0.6}(TA - TS)$$

$$R = 25(MRT - TS)$$

$$E_{max} = 4.0V^{0.6}(VPS - PPA)$$

Even under favorable ambient conditions, 25 percent of metabolic heat M at rest is transferred from the skin surface to the cooler air by convection C, 50 percent by radiative transfer to cooler surfaces of the surroundings R, and the remaining 25 percent by warming inspired air and by evaporation of 20 to 30 g/hr of moisture lost through the nonsweating skin (36). Because respiratory heat loss is relatively insignificant in temperature regulation, it is customary for investigators in this field to consider only heat transfer through the skin.

2.2.4 Steady State Thermal Exchange

The foregoing discussion can be summarized with a heat balance equation that defines steady state with respect to the exchange of thermal energy between a worker and his environment:

$$E = M \pm C \pm R$$

It is informative to examine this expression of thermal balance in the context of both "hot-dry" and "warm-moist" environments. In a hot-dry environment, if the solid objects surrounding the worker are hotter than skin temperature, the radiant heat gain may exceed the capacity of the sweating mechanism to provide sufficient cooling, and body temperature will rise. In a warm-moist environment, heat load from radiation may not be great, but the humid environment severely limits the sweating-convection mechanism. Again, when the maximum evaporative capacity is insufficient to permit dissipation of the body's heat load, body temperature must rise.

Examination of the steady state heat transfer model also shows the primary importance of sweat evaporation E in regulating heat loss to maintain body temperature. Such evaporation is enhanced by air movement and the removal of clothing. In addition, as indicated in the definition of E, the efficiency of evaporation depends on the ambient partial pressure of water vapor, which may be the limiting factor in elevated environmental temperatures.

2.3 Indices of Heat Strain

Heat strain is a reactive physiological manifestation of environmental heat stress. The burden of balancing bodily heat gain and heat loss—a balance needed to prevent elevation in body temperature—falls on the sweating mechanism and the circulatory system. Several indices of heat strain have been identified and studied in the ongoing evaluation of the relationships between heat stress and strain with respect to the thermal environment. The most common indices of heat strain include sweat rate, sweat evaporation rate (skin temperature), heart rate, and core temperature.

2.3.1 Sweat Rate

Sweating occurs over a wide range of thermal exposure in amounts just sufficient to achieve enough evaporative cooling E to offset the total heat load, represented by $M + R + C$. The rate of secretion of individual sweat glands as well as the number of active glands determine the sweat rate. Under maximum thermal insult, some 2.5 million sweat glands secrete at peak rates of more than 3 kg/hr for up to an hour; highly acclimatized men can maintain rates of 1 to 1.5 kg/hr for several hours (36). When sweat evaporates freely from the skin, that is, when the sweat rate equals the rate of evaporation, evaporative cooling E is regulated under steady state conditions of work and heat exposure to balance the heat load ($M + R + C$). The central drive for sweating is determined by metabolic work rate M; however the actual sweat output is modulated by skin temperature to meet the evaporative requirements up to the limits for sweating capacity.

2.3.2 Sweat Evaporation Rate (Skin Temperature)

The thermodynamics of water vaporization suggest that one gram of sweat can eliminate 0.58 kcal of body heat upon evaporation. The efficiency of cooling by sweat depends on the rate of evaporation E, which is determined by the gradient between vapor pressure of the wetted skin *VPS* and the partial pressure of water vapor in the ambient air *PPA,* multiplied by a root function of effective air velocity at the skin surface $V^{0.6}$ and the fraction of body surface that is wetted.

When the evaporation of sweat is restricted, skin temperature *TS* rises above that observed under less humid conditions with the same air temperature. The heat loss center (hypothalamus) responds by recruiting more sweat glands, thereby increasing the fraction of wetted body surface. If such a response achieves the degree of cooling needed to balance the heat load ($M + R + C$), the body core temperature remains essentially unchanged (36).

At higher levels of *PPA* or lower air velocities, the fraction of the body surface that is wetted increases until the body is completely wetted. Any further increase in sweat production does not contribute to cooling because the liquid perspiration drips off the body and is wasted as a coolant. Under higher levels of PPA with further restrictions on evaporation, body heat is not dissipated, thus raising the temperature of the skin *TS* as well as the deep body temperature. The response is a greater central drive for sweating, with *VPS* and evaporative rate *E* increasing as well. As a result, a new thermal balance may be reestablished, but at a higher body temperature and at a cost of increased thermoregulatory strain. A modest rise is physiologically acceptable, but any substantial increase in body temperature is accompanied by serious strain (36).

Therefore sweat rate in the zone of free evaporation changes in proportion to heat load. In the zone of restricted evaporation, when the wetted body surface area approaches total body area, the sweat rate exceeds the evaporative capacity and is proportional to the increase in the core temperature and the skin temperature.

Sweat rate is thus an index of heat stress over the entire range of compensation. Furthermore, it is also an index of heat strain in the zone of time-limited compensation, where its rise parallels deep body temperature and heart rate. Sweat rate serves as a time-weighted average of heat stress and is measured by the difference in body weight over a given time period, corrected for weight gained by water and food intake and weight lost in urine and feces (36).

2.3.3 Heart Rate (Circulatory Strain)

To facilitate a thermal balance without rise in body core temperature, a physiological response occurs to increase the transport of metabolic heat to the skin. This requires augmented blood flow through the dermal vessels, which under extreme conditions may be sufficient to tax the capacity for cardiac output.

Heart rate is responsive both to the increased cardiac output required by working muscles and the added circulatory strain imposed by heat exposure; therefore it is a useful index of total heat load. In fact, heart or pulse rate is one of the most reliable indices of heat strain. The heart rate of a subject reflects the combined demands of environmental heat, work level, elevation of body temperature, and individual cardiovascular fitness. As Figure 20.1 indicates, heart rate increases disproportionately with heat load (40). This pattern reflects the body's increasingly futile effort to avoid a rise in body core temperature as heat load is increased. As the external heat load rises, skin temperature

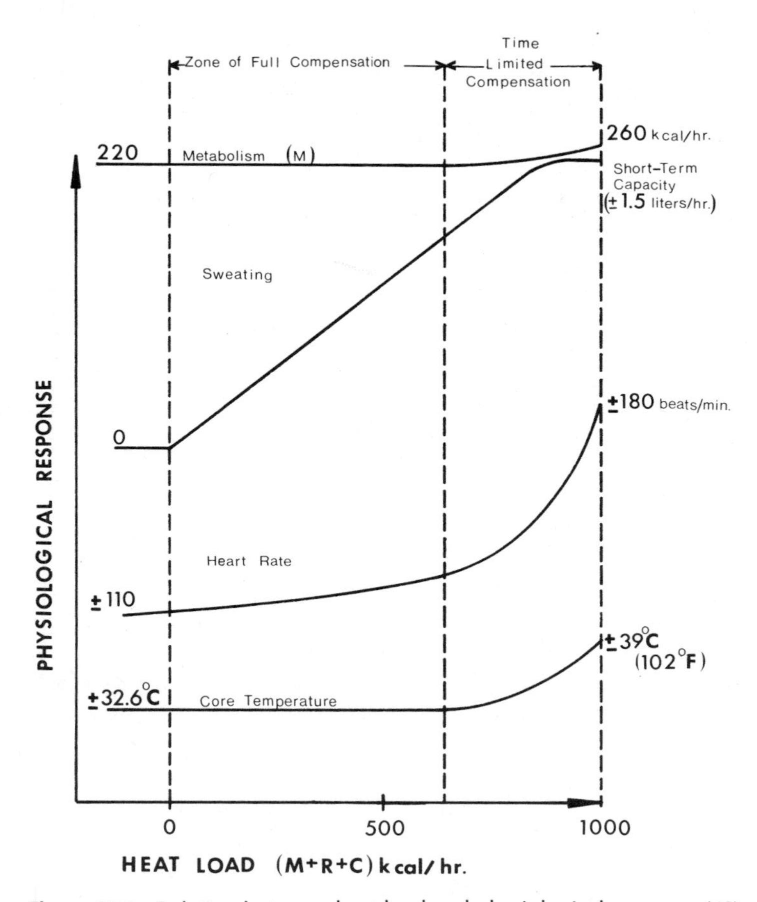

Figure 20.1 Relation between heat load and physiological response (40).

TS also rises, decreasing the thermal gradient between body core and skin temperature and thereby demanding an increased volume of dermal blood flow.

Because heart rate reflects the combined influence of environmental heat, work level, elevation of body temperature, and individual cardiovascular fitness, a given heart rate has a relatively consistent physiological meaning among workers. For example, a rate of 180 to 200 beats per minute represents the maximum capacity for most adults and is sustainable for only a few minutes. In the same manner, a given heart rate signifies approximately the same level of strain regardless of the degree of an individual's fitness, but a person with high cardiovascular fitness will achieve a given heart rate at a higher level of thermal stress than an unfit person.

The detrimental effect of heat stress on work performance is indicated by an increase in cardiac count both at work and in recovery. Brouha (41) observed that progressive deterioration of pulse rate does not occur when the pulse rate at ½ to 1 minute after a work period ends does not exceed 110 beats per minute and when the rate falls at least 10 beats per minute within the next 2 minutes. Brouha proposed a simple guide to ensure that men performing intermittent work in heat will remain in thermal balance for a full work shift without cumulative effects of strain. Using his guidelines:

> Pulse rate is counted for the last 30 seconds of the first three minutes after rest begins. If the first recovery pulse (i.e., from 30 to 60 seconds) is maintained at 110 beats per minute or below and deceleration between the first and third minute is at least 10 beats per minute, no increasing strain occurs as the work day progresses.

An integrated or time-weighted heart rate level observed during an entire work shift reflects sustained elevations in peak rates as well as recovery and resting rates, thus can serve as a guide in assessing circulatory strain. Electronic devices for integrating total heart rate are now generally available and provide a valuable tool for refining investigations of the relation between heat stress and heat strain.

2.3.4 Body Core Temperature

The temperature control center in the brain (hypothalamus) reacts to adjust internal temperature to a level determined by an individual's metabolic rate M. As total heat load $(M + R + C)$ increases, sweat rate and blood flowrate increase proportionally, with body core temperature maintained at a uniform level determined only by metabolic rate. Lind (23) termed this level of thermal equilibrium the "prescriptive zone," to indicate the range of thermal environments in which men can work without strain on homeostatic control of body core temperature.

The data of Robinson (42) and others (43–46) show that the upper limit of the prescriptive zone in highly acclimatized men working at 300 kcal/hr is 31 to 32°C effective temperature. Lind (47) recommends an ET of 27.5°C as a realistic limit for nonacclimatized men of varying physical fitness working at this level. The wide difference in these two limits symbolizes the perplexing nature of the problem of establishing rational standards for industrial heat stress (36).

In the prescriptive zone the mechanisms of circulatory control and sweating are so effective that an acceptable core temperature is maintained over a wide range of environmental temperatures. When the body core temperature exceeds a critical level, injury to the temperature regulatory center may result, causing reduced sweating and a dangerous further temperature rise. Thus an increase in body core temperature is a very significant index of heat strain and one that has received much attention in efforts to define acceptable levels of heat stress.

2.4 Heat Disorders and Tolerance Factors

Environmental heat and the inability to remove metabolic heat lead to well-known reactions in man, including increased cardiovascular activity, sweating, and increased body core temperature. Viewed from a wider perspective, three overlapping responses to heat occur in man: psychological, psychophysiological, and physiological.

In addition to the pathological effects of heat, subpathological effects may modify performance or behavior or potentiate response to other stresses imposed simultaneously. There is no demonstrable effect on health at low levels of thermal stress. As heat load increases, there appears to be a higher order of psychophysiological disturbance, an increase in the frequency of errors, a higher frequency of accidents, and a reduction in efficiency in the performance of skilled physical tasks. At even higher levels of exposure there is an increasingly well-defined disturbance of physiological well-being, with strain on the heart and circulatory system, and overloading of the mechanisms of salt and water balance in the body (10). Nevertheless, one of the most striking features of heat stress and tolerance to such stress is the very high degree of individual variability in susceptibility to heat. This section examines in more detail the various heat disorders and focuses on the several factors that influence heat tolerance.

2.4.1 Psychophysiological Effects

Under conditions of heat stress, increased demands are made for blood flow to the periphery of the body, diverting some of the cardiac output and rendering it unavailable to active muscles. Accordingly, as upper limits of tolerance are reached, work output must be reduced (48). Wyndham et al. (33) demonstrated that real limits of endurance exist. Other investigators have reported heat-limited work situations suggesting upper limits for unrestricted work (21, 26, 49).

Other effects of heat stress at levels lower than limiting levels are lack of efficiency in performing heavy tasks (50) and interference with skilled manipulations or psychological tasks (48). Peppler has reviewed the effects of heat on skilled tasks and mental tasks; qualitatively, there is no doubt that heat interferes with these types of activity (51, 52). Furthermore, it is common experience that heat exposure accelerates the onset of fatigue. One study showed that the production of delicately assembled items (based on incentive pay) included higher scrap rates when the subjects were working in heat, and although production of good pieces was maintained, lower efficiency resulted (53). In another study, Peppler (54) showed that relatively slight increases in environmental temperature affect classroom learning adversely. Increased bodily temperature and discomfort also increase irritation, anger, and other emotional states that may induce workers to commit rash acts or otherwise exhibit emotional reactions (21, 55).

2.4.2 Interactive Effects

Much of our recent insight into the combined effects of heat and other stresses, such as noise and vibration, comes from the space program and related aerospace medical research. Crew members in aircraft and space vehicles are exposed to combinations of environmental stresses that traditionally have been studied singly in laboratory research.

Thus various combinations of stresses that occur in flight have been studied in recent years (56, 57).

The combination of heat and carbon monoxide (CO) has also been shown to have a deleterious effect greater than that due to either stress alone (31). The subjects reported persistent headaches, anorexia, irritability, depression, and general malaise; the effects were more pronounced in women than in men. These symptoms were markedly more severe after exposure to heat and CO than after exposure to either alone. It is interesting to note that these physiological disturbances were more severe in the hours *following* the exposures.

Renshaw (58) investigated the effects of noise and heat on performance on a five-choice serial reaction task. The effect of heat was found to be statistically significant. Performance was poorer at 90°F effective temperature than at 72°F effective temperature when the noise level was held constant.

2.4.3 Heat-Induced Illnesses

Three major clinical disorders can result from excessive heat stress to susceptible workers:

1. Heatstroke from failure of the thermoregulatory center.
2. Heat exhaustion from depletion of body water and/or salt.
3. Heat cramps from salt loss and dilution of tissue fluid.

These disorders occur when the normal response of increased skin blood flow and sweat production are not adequate to meet the needs for body heat loss or when the thermoregulatory mechanisms fail to function properly.

Heatstroke. Heatstroke occurs when the mean temperature of the body is such that the continued functioning of some vital tissue is endangered. This condition represents a marked failure of the thermoregulatory system to maintain a proper balance. Some of the factors that bring about heatstroke may be classified as follows (59):

1. Reduced heat loss because of lack of sweat glands, inhibition of sweating, inadequate peripheral circulation, high environmental temperature, and high humidity with restricted convection.
2. Increased heat reception because of radiant energy absorption or environmental temperatures above skin temperature.
3. Increased heat production because of muscular exercise, overactivity of the thyroadrenal apparatus, or elevated body temperature.
4. Interference with the heat regulating center from infection.

Heat Exhaustion. Heat exhaustion is a state of collapse brought about by an insufficient blood supply to the cerebral cortex. The failure is not so much that of heat regulation but an inability to meet the price of heat regulation. The critical event is low arterial blood pressure caused partly from inadequate output of blood by the heart and partly from widespread vasodilation. The chief factors leading to heat exhaustion can be classified as follows (59):

1. Increased vascular dilation and decreased capacity of circulation to meet the demands for heat loss to the environment, exercise, and digestive activities.
2. Decreased blood volume due to dehydration, gravitational edema, adrenalin insufficiency, or lack of salt.
3. Reduced cardiac efficiency because of emotion, malnutrition, lack of physical training, infection, intoxication, or cardiac failure.

Heat Cramps. Heat cramps are spasms in the voluntary muscles following a reduction in the concentration of sodium chloride in blood below a certain critical level. Chloride loss is due to high sweating rates and lack of acclimatization. Depletion of chloride reserves is facilitated by low dietary intake of salt and by adrenal-cortical insufficiency.

The abdominal as well as the skeletal musculature may be affected, but the exact site may not necessarily be related to the preceding exercise. Heat cramps can be prevented by taking salt whenever heavy work is to be executed in hot-dry environments, especially by unacclimatized persons.

Table 20.4, based on nomenclature prepared jointly by committees representing the United Kingdom and the United States, shows the clinical symptoms of disorders resulting from failure to adapt to heat stress, along with methods of treatment and prevention (21). Several minor disorders not discussed in detail in this section are also included.

2.4.4 Acclimatization

Acclimatization refers to a set of physiological adjustments that occur when an individual accustomed to working in a temperate environment undertakes work in a hot environment. These progressive adjustments occur over periods of increasing duration and reduce the strain experienced on the initial exposure to heat. This enhanced tolerance allows a person to work effectively under conditions that might have been unendurable before acclimatization. Acclimatization is acquired by working in the heat; much of the adjustment takes place during a week of prolonged or intermittent periods of physical activity, although the biggest improvement occurs during the first two days.

Any worker, however healthy, well conditioned, and well motivated, when exposed for the first time to occupational heat stress, will develop signs of significant strain with abnormally high body temperature, pounding heart, and other signs of heat intolerance. On each succeeding day of heat exposure, however, his ability to work at the same level of heat stress improves as signs of discomfort and strain diminish. Thus after a period of 1 to 2 weeks, a worker can perform without difficulty.

Table 20.4. Classification, Medical Aspects, and Prevention of Heat Illness (adapted from Reference 21)

	Heatstroke and Heat Hyperpyrexia
Clinical features	*Heatstroke:* (1) hot dry skin: red, mottled, or cyanotic; (2) high and rising core temperature, 40.5°C and over; (3) brain disorders: mental confusion, loss of consciousness, convulsions, or coma, as core temperature continues to rise. Fatal if treatment delayed. *Heat hyperpyrexia:* milder form; core temperature lower; less severe brain disorders; some sweating.
Underlying physiological disturbance	*Heatstroke:* failure of the central drive for sweating (cause unknown) leading to loss of evaporative cooling and an uncontrolled accelerating rise in core temperature. *Heat hyperpyrexia:* partial rather than complete failure of sweating.
Predisposing factors	(1) Sustained exertion in heat by unacclimatized workers, (2) obesity and lack of physical fitness, (3) recent alcohol intake, (4) dehydration, (5) individual susceptibility, (6) chronic cardiovascular disease in the elderly.
Treatment	*Heatstroke:* immediate and rapid cooling by immersion in chilled water with massage, or by wrapping in wet sheet with vigorous fanning with cool dry air. Avoid overcooling. Treat shock if present. *Heat hyperpyrexia:* less drastic cooling required if sweating still present and core temperature <40.5°C.
Prevention	Medical screening of workers. Selection based on health and physical fitness. Acclimatization for 8 to 14 days by graded work and heat exposure. Monitoring workers during sustained work in severe heat.
	Heat Syncope
Clinical features	Fainting while standing erect and immobile in heat.
Underlying physiological disturbance	Pooling of blood in dilated vessels of skin and lower parts of body.
Predisposing factors	Lack of acclimatization.
Treatment	Remove to cooler area. Recovery prompt and complete.
Prevention	Acclimatization. Intermittent activity to assist venous return to heart.
	Heat Exhaustion
Clinical features	(1) Fatigue, nausea, headache, giddiness; (2) skin clammy and moist, complexion pale, muddy, or with hectic flush; (3) may faint on standing, with rapid thready pulse and low blood pressure; (4) oral temperature normal or low but

Table 20.4. (Continued)

	rectal temperature usually elevated (37.5 to 38.5°C). Water-restriction type: urine volume small, highly concentrated. Salt-restriction type: urine less concentrated, chlorides less than 3 g/liter.
Underlying physiological disturbance	(1) Dehydration from deficiency of water and/or salt intake, (2) depletion of circulating blood volume, (3) circulatory strain from competing demands for blood flow to skin and to active muscles.
Predisposing factors	(1) Sustained exertion in heat, (2) lack of acclimatization, (3) failure to replace water and/or salt lost in sweat.
Treatment	Remove to cooler environment. Administer salted fluids by mouth or give intravenous infusions of normal saline (0.9 percent) if patient is unconscious or vomiting. Keep at rest until urine volume and salt content indicate that salt and water balances have been restored.
Prevention	Acclimatize workers using a breaking-in schedule for 1 or 2 weeks. Supplement dietary salt only during acclimatization. Ample drinking water to be available at all times and to be taken frequently during work day.
Heat Cramps	
Clinical features	Painful spasms of muscles used during work (arms, legs, or abdominal). Onset during or after work hours.
Underlying physiological disturbance	Loss of body salt in sweat. Water intake dilutes electrolytes. Water enters muscles, causing spasm.
Predisposing factors	(1) Heavy sweating during hot work, (2) drinking large volumes of water without replacing salt loss.
Treatment	Salted liquids by mouth, or more prompt relief by intravenous infusion.
Prevention	Adequate salt intake with meals. In unacclimatized men, provide salted (0.1 percent) drinking water.
Heat Rash	
Clinical features	Profuse tiny raised red vesicles (blisterlike) on affected areas. Pricking sensations during heat exposure.
Underlying physiological disturbance	Plugging of sweat gland ducts, with retention of sweat and inflammatory reaction.
Predisposing factors	Unrelieved exposure to humid heat with skin continuously wet with unevaporated sweat.
Treatment	Mild drying lotions. Skin cleanliness to prevent infection.
Prevention	Cooled sleeping quarters to allow skin to dry between heat exposures.

Table 20.4. (Continued)

	Heat Fatigue—Transient
Clinical features	Impaired performance of skilled sensorimotor, mental, or vigilance tasks, in heat.
Underlying physiological disturbance	Discomfort and physiological strain.
Predisposing factors	Performance decrement greater in unacclimatized and unskilled men.
Treatment	Not indicated unless accompanied by other heat illness.
Prevention	Acclimatization and training for work in the heat.

The factor that induces acclimatization appears to be sustained elevation of body core temperature and skin temperature above the levels that would exist if the same work were accomplished in a cool environment. Such levels must be sustained for an hour or more per day for 1 or 2 weeks.

It is well established that acclimatization to dry heat increases tolerance to wet heat, and vice versa. The reason is not well understood, however, because the increased sweat output, which may nearly double, is largely wasted. Heat conduction through the skin is enhanced, though, which suggests a change in the distribution of blood to the skin (60).

Workers employed in hot industrial activities acquire levels of acclimatization commensurate with their average exposure. Unusual demands for work or sudden spells of hot weather may overload their thermoregulatory capacity, leading to signs of strain. Heat acclimatization seems to require periodic reinforcement such as occurs daily during the work week. Thus workers may show some loss of acclimatization after returning from an absence as brief as 2 days. After absences of 2 weeks or longer, the detriment in acclimatization is substantial, and several days at work is required before heat tolerance is fully restored. Acclimatization persists for as long as a week after the last heat exposure.

Physiological adjustments during acclimatization include changes in sweat composition. Sweat is a dilute solution of electrolytes, principally sodium chloride. In unacclimatized subjects, sodium chloride concentration in sweat (3 to 5 g/kg) is about half the concentration in blood plasma (36). In acclimatized subjects sweat is not only more abundant but more dilute. The salt concentration is reduced to 1 to 2 g/kg, reflecting an adaptive change in hormonal balance through secretion of aldosterone, which acts to conserve body salt by the kidneys and the sweat glands (36).

2.4.5 Surface Area to Weight Ratio

Another heat tolerance factor is obesity, measured by the ratio of body surface area to weight. Obese or stocky individuals possess a relatively low surface area to weight ratio.

Thus an inherent handicap exists for performing sustained work in heat because heat loss is a function of surface area and heat production is a function of weight. If lacking in acclimatization, therefore, physically unfit and obese workers are at a greater risk of succumbing to heat strain.

2.4.6 Age and Disease

Older, yet healthy workers perform well on hot jobs if allowed to proceed at a worker-regulated pace. Under demands for sustained work output in heat, however, an older worker is at a distinct disadvantage compared with younger men. First, the maximum oxygen uptake for maximum aerobic work capacity declines 20 to 30 percent between ages 30 and 65, leaving an older worker with less cardiocirculatory reserve capacity. Second, under levels of heat stress above the prescriptive zone, an older worker compensates for heat loads less effectively than do younger men, as indicated by higher core temperature and peripheral blood flow for the same work output (25). This occurs because there is a delay in the onset of sweating and a lower sweat rate capacity in older men, thus resulting in greater heat storage during work and longer time required for heat recovery.

For similar reasons, degenerative diseases of the heart and circulatory system intensify the aging effect on heat tolerance by limiting the circulatory capacity to transport heat from the body core to the surface. Elderly men and women with chronic diseases of aging account for much of the excess mortality reported in large northern cities during sustained heat waves (61). As evidence that workers with long experience in hot industries seem to be less at risk of death from cardiovascular and other diseases than workers of similar age without a work history of exposure, a recent epidemiological study of nearly 13,000 open hearth steel workers showed that the mortality rate for arteriosclerotic heart disease and respiratory disease (as well as overall mortality rate) was significantly less than that for the entire population of nearly 59,000 steel workers (62). Nevertheless, in any such study, it is possible that a self-selection process is masked that would eliminate those of low physical fitness and heat tolerance from jobs on the open hearth furnace operation.

2.4.7 Water Balance

Sustained and effective work performance in heat requires, among other things, a replacement of body water and salt lost through sweating. A fully acclimatized worker weighing 70 kg secretes a maximum of 6 to 8 kg of sweat per 8-hour shift. If this water is not replaced by drinking, continued sweating will draw on water reserves from both tissues and body cells, leading to dehydration, with the symptoms of shriveled skin, dry mouth and tongue, and sunken eyes (36).

Water loss from sweating of one kilogram (1.4 percent of body weight) can be tolerated without serious effects. Water deficits of 1.5 kg or more during work in the heat reduce the volume of circulating blood, resulting in signs and symptoms of increasing heat strain, including elevated heart rate and body temperature, thirst, and severe heat discomfort. At water deficits of 2 to 4 kg (3 to 6 percent of body weight), work performance is impaired, and continued work under such conditions leads to heat exhaustion. To tolerate heat in the sense of avoiding excessive depletion of body water, therefore, sweating workers should drink at intervals of 30 minutes or less (36).

Here again, acclimatization is an important consideration, for well-acclimatized men tend to achieve a better water balance than unacclimatized men, even when sweating at high rates.

2.4.8 Salt Balance

Sodium chloride losses in sweat must be replaced by ingestion, normally at mealtime. A typical American diet (10 to 15 grams per day of salt) would suffice to replace the needs of an acclimatized worker producing 6 to 8 kg of sweat during a single shift, since a kilogram of sweat contains 1 to 2 g of salt (36). For the period of acclimatization, however, workers with no previous heat exposure require additional salt. Although maximal sweating rates in unacclimatized men are lower, salt concentrations are higher than after acclimatization. Thus an unacclimatized man may lose 18 to 30 g of salt. Salt supplements in the form of tablets (preferably impregnated, to avoid gastric irritation) may be ingested if ample water is available. A better practice is to use salted water (one tablespoon per gallon) or to increase salt intake at meals (36).

Depletion of body salt may occur in unacclimatized men exposed to heat who replace water losses without adequate salt intake. This too leads to progressive dehydration because thermoregulatory controls in the human body are geared to maintain a balance between electrolyte concentration in tissue fluids and body cells. Deficient salt intake with continued intake of water dilutes the tissue fluid, which in turn suppresses the antidiuretic hormone (ADH) of the pituitary gland. The kidney then fails to reabsorb water and excretes dilute urine containing little salt (36).

Under these conditions, homeostasis maintains the electrolyte concentration of body fluids but at the cost of depleting body water with ensuing dehydration. Under continued heat stress, the symptoms of heat exhaustion (elevated heart rate and body temperature and severe discomfort) develop similarly to those resulting from water restriction, but signs of circulatory insufficiency are more severe and there is notably little thirst. An excellent diagnostic tool for salt deficiency is the presence of a very low level of chloride (less than 3 g/liter) in the urine (36).

On a short-term basis, sweating men drinking large volumes of unsalted water may develop heat cramps, extremely painful spasms of the muscles used while working, such as arms, legs, and abdomen. The dilution of tissue fluid around the working muscle results in transfer of water into muscle fibers, causing the spasms. Treatments of the clinical symptoms of water and salt depletion are similar, namely, the replacement of water and/or salt by oral ingestion of salted liquids in mild cases or the intravenous infusion of saline in more serious cases. Any excess of salt or water over actual needs is readily controlled by kidney excretion.

2.4.9 Alcohol Consumption

Many investigators have reported an excessive alcohol intake by patients hours or a few days before the onset of heatstroke. Other investigators have noted a striking reduction in heat tolerance on the day after an alcoholic "binge." It is well known that alcohol suppresses the antidiuretic hormone, leading to loss of body water in urine; hence dehydration may be a primary factor (36).

2.4.10 Physical Fitness

Physical conditioning alone does not yield heat acclimatization. Physical training without heat exposure, however, does improve heat tolerance as indicated by somewhat lower heart rates and core temperatures when men are exposed to heat after conditioning as compared with before. Sweat rates do not increase, and skin temperature remains high. Physical conditioning enhances heat tolerance by increasing the functional capacity of the cardiocirculatory system. This occurs by two important changes. First, an increase occurs in the number of capillary blood vessels relative to muscle mass, providing a larger interface between blood and muscle for the exchange of oxygen and waste products. Second, the increased tone of small veins from tissue other than muscle reduces their capacity during exercise, thus increases pressure on large central veins returning blood to the heart. Therefore cardiac output can increase during work with less need to accelerate the heart. Thus a physically conditioned man, by virtue of having an increased maximum ventilatory capacity, has a wider margin of safety in coping with the added circulatory strain of working under heat stress. Although the extent to which men might gain in heat tolerance by acclimatization is not well defined in an absolute sense, those with a high level of physical fitness have a distinct advantage (36).

3 MEASUREMENT OF THE THERMAL ENVIRONMENT

Assessment of heat stress always includes some theoretical or empirical combination of environmental variables to describe the severity of the thermal environment and often the capacity of that environment to facilitate heat exchange with the workers. As discussed in Section 2, both the degree of discomfort and the level of stress caused by environmental heat depend on air temperature, relative humidity, velocity of air movement, and temperature of the surrounding surfaces with which the body exchanges heat by radiation. Investigators in this field have devised indices of heat stress involving as simple a parameter as dry-bulb temperature, as well as algebraic and multiplicative combinations of environmental variables, some with correction factors and modifications. Rather than trace the development of a large number of these indices, this section provides descriptions and applications of the most widely used and technically meritorious of the environmental heat stress parameters.

3.1 Indices of Heat Stress

Several indices of thermal stress have survived or emerged as those most commonly encountered in the evaluation and control of industrial heat stress. Included are effective temperature (ET), defined by Houghton and Yaglou (63); equivalent effective temperature corrected for radiation (ETCR), which is a modification of ET to make allowance for radiant heat (65); the predicted 4-hour sweat rate (P4SR) (66); the heat stress index (HSI) (30); and the wet bulb–globe temperature (WBGT) index (1).

Investigators who have tried to develop optimal heat stress indices by combining environmental measurements have approached the problem in three general ways. The first approach is based on the thermometric scale, the second on the rate of sweating, and the third on calculations of heat load and of the evaporative capacity of the environment.

3.1.1 Effective Temperature

Effective temperature (ET) has the longest history and probably has been the most widely applied index of thermal comfort. Effective temperature emerged in the search for design criteria for thermal comfort in occupied space and was introduced in 1923 by Houghton and Yaglou (63). Effective temperature is an empirical sensory index combining dry- and wet-bulb temperatures and air speed to yield a thermal sensation equivalent to that at a given temperature of still, saturated air. To develop the effective temperature scale, subjects were exposed to one combination and then another of the various parameters of air temperature, air motion, and humidity. On the basis of a large number of trials, nomograms were developed which characterized equivalent environments, expressed in terms of the temperature of a still, saturated environment.

Wyndham, in studying gold miners, concluded that below 26°C ET the risk of heatstroke was negligible and that the risk of fatal heatstroke began at 28°C ET (9). Up to 32°C ET there was a slow, steady rise in the risk of both fatal and nonfatal cases of heatstroke, but above 33°C ET the risk increased sharply. Similar values were observed by Brief and Confer in a comparison of heat stress indices (64).

3.1.2 Equivalent Effective Temperature Corrected for Radiation

Bedford showed that the ET scale could be corrected to account for radiation from hot surroundings by using the black globe temperature (see Section 3.2.4) in place of the dry-bulb temperature (65). Present use of the equivalent effective temperature corrected for radiation (ETCR) scale incorporates such modifications into a single chart. Figure 20.2 shows the normal effective temperature scale, which is used for people wearing lightweight summer clothing similar to workers' uniforms.

In spite of widespread use, the ET and ETCR have serious limitations as indices of heat stress (21).

1. ET was developed on the basis of transient thermal sensations. This emphasis tended to neglect the importance of the sorption and desorption of moisture from the subject's clothing.
2. The ET scale was developed using clothed subjects in the dress of that day.
3. The subjects were sedentary. Later modifications were made to include the effective metabolic rate. Nevertheless, the scale was designed primarily for environments reasonably near the comfort zone, and extrapolation to thermally stressful environments is tenuous.

3.1.3 Predicted Four-Hour Sweat Rate

The predicted 4-hour sweat rate scale (P4SR) is based on a long series of observations of sweat rate under various combinations of environmental conditions (66). Such a choice of sweat rate as a criterion of stress was later supported by Hatch (38). As P4SR rises above 4.5 liters, it has been stated that an increasing number of fit, well-acclimatized men will be unable to tolerate 4 hours of exposure (45). Lower P4SR values have never been very clearly related to physiological strain. It has been suggested that the safe limit for exposure of unacclimatized men may be 2.5 to 3 liters.

P4SR requires the measurement of ventilated wet-bulb and globe temperatures, air velocity, and the metabolic rate of the workers. The index works fairly well for temperatures hot enough to induce moderate sweating, but because of the marked difference in acclimatization of industrial workers relative to the subjects examined by the original investigators, its validity suffers under other conditions.

3.1.4 Heat Stress Index

The heat stress index (HSI), developed by Belding and Hatch at the University of Pittsburgh in the mid-1950s (30), combines the environmental heat-exchange components radiation R and convection C with metabolic heat M in an expression of stress in terms of the required sweat evaporation E_{req}. Stated algebraically

$$E_{req} = M \pm R \pm C$$

The resulting physiological strain is determined by the ratio of the stress (E_{req}) to the maximum evaporative capacity of the environment (E_{max}). Thus HSI is defined as follows

$$\text{HSI} = 100\left(\frac{E_{req}}{E_{max}}\right)$$

When HSI exceeds 100, body heating occurs, and when HSI is less than 100, body cooling occurs. Values of E_{req} and E_{max} may be computed by means of the equations below or by a more convenient method following a nomogram developed by McKarns and Brief (67). This nomogram (Figure 20.3) is based on further revisions in the method which pro-

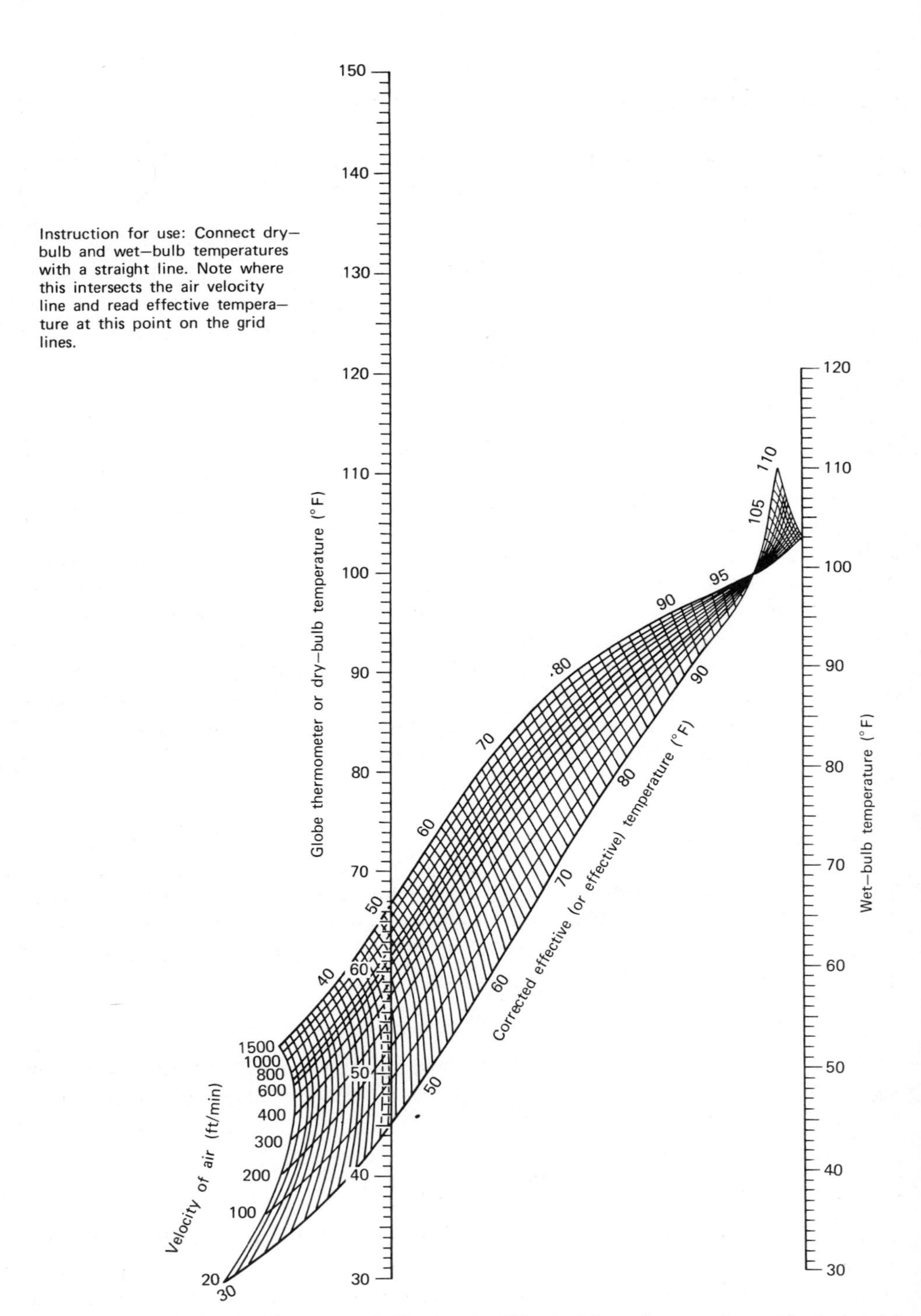

Figure 20.2 Normal scale of Corrected Effective (or Effective) Temperature. From *The Industrial Environment—Its Evaluation and Control,* 2nd ed., C. H. Powell and A. D. Hosey, Eds., Public Health Service Publication No. 614, Government Printing Office, Washington, D.C., 1965.

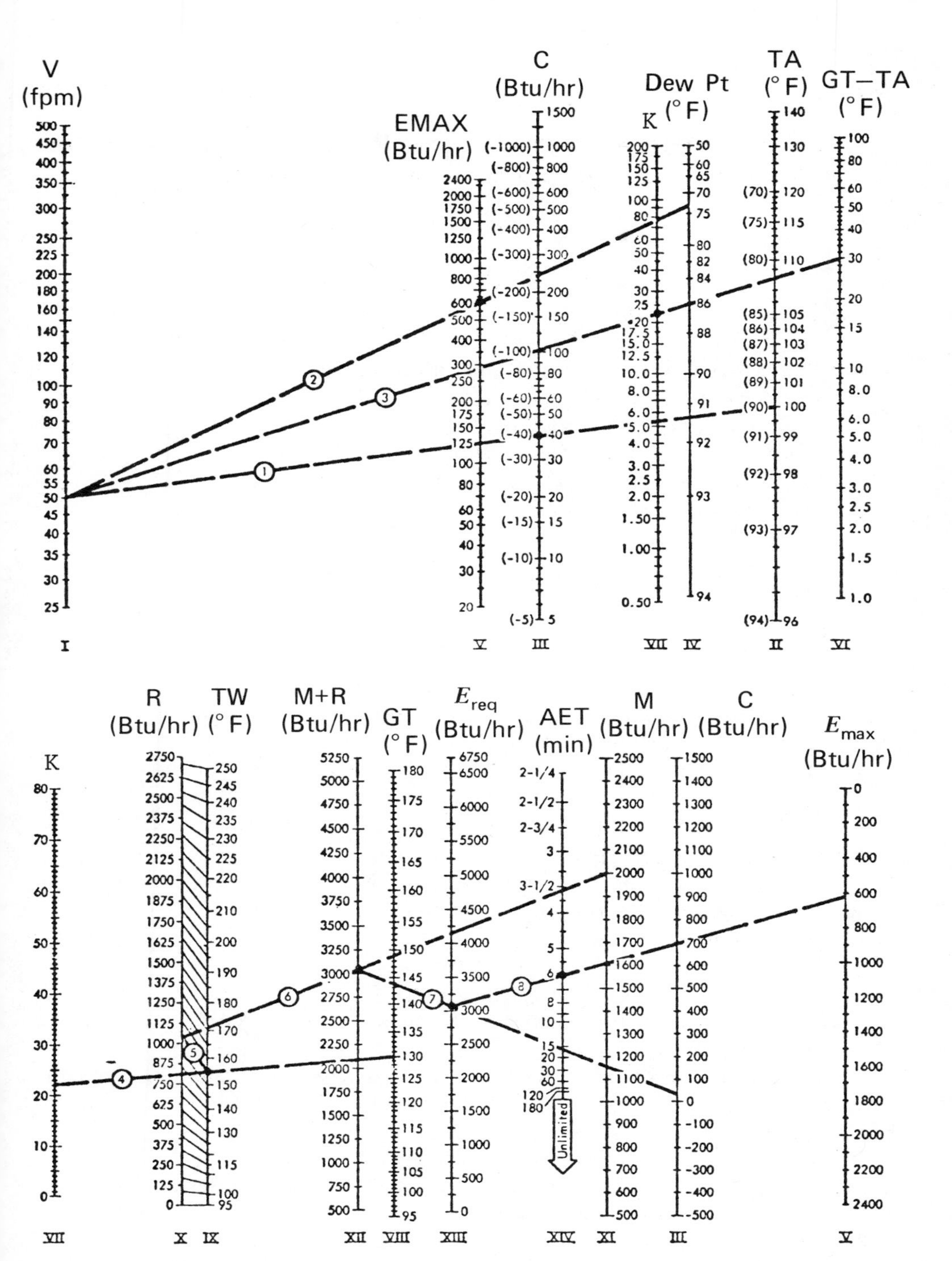

Figure 20.3 Nomogram for the determination of Heat Stress Index (HSI) and Allowable Exposure Time (AET). Reprinted from *Heating/Piping/Air Conditioning*, © 1976 by Reinhold Publishing Co., Inc., Penton/IPC. All rights reserved.

vided a 30 percent reduction in R, C, and E_{max} for the average man wearing light clothing. The following equations were used in the nomogram development:

$$R = 17.5\ (MRT - 95)$$

$$C = 0.756\ V^{0.6}(TA - 95)$$

$$E_{max} = 2.8 V^{0.6}\ (42 - PPA)$$

where PPA = partial pressure of water vapor in air (mm Hg)
MRT = mean radiant temperature (°F)
R = radiative heat exchange (Btu/hr)
C = convective heat exchange (Btu/hr)
E_{max} = maximum evaporative heat loss (Btu/hr)
TA = air temperature (°F)
V = air velocity (ft/min)

To use the nomogram, follow these steps:

1. Determine the convective heat load by extending a line from the air velocity scale (column I) to the air temperature scale (column II). The convective load is indicated at the intersection of this line with column III. Note whether the heat load is positive or negative as indicated by an air temperature above or below 95°F, respectively.
2. Obtain the maximum available evaporative cooling E_{max}, from column V by extending a line from the air velocity scale (column I) to the dew point temperature scale (column IV). The dew point temperature is obtained from a psychrometric chart at the intersection of the lines for wet-bulb and dry-bulb air temperatures.
3. Determine K in column VII by extending a line from the air velocity scale to the temperature difference scale (column VI). Transfer the value of K to column VII in the second set of alignment charts.
4. Extend the line from the K scale to the globe temperature scale (column VIII) and read the radiant wall temperature in column IX.
5. Project this value to the radiation scale (column X) by extending a line parallel to the given slanting lines.
6. Estimate an appropriate metabolic rate and locate this value of M on the metabolism scale (column XI). Connect columns X and XI and determine the sum of metabolism and radiation in column XII.
7. Transfer the convective load value determined in step 1 to column III noting whether it is positive or negative. Connect this point with the sum in column XII and read the required rate of evaporation in column XIII.
8. Locate the available evaporative cooling in column V. Extend a line from this point to column XIII. The approximate allowable continuous exposure time (AET) is indicated at the intersection of this line with column XIV.

Table 20.5. Evaluation of Values in Belding and Hatch HSI (11)

Index of Heat Stress (HSI)	Physiological and Hygienic Implications of 8-hour Exposures to Various Heat Stresses
−20 −10	*Mild cold strain*. This condition frequently exists in areas where men recover from exposure to heat.
0	*No thermal strain.*
+10 20 30	*Mild to moderate heat strain.* For a job that involves higher intellectual functions, dexterity, or alertness, subtle to substantial decrements in performance may be expected. In performance of heavy physical work, little decrement is expected unless ability of individuals to perform such work under no thermal stress is marginal.
40 50 60	*Severe heat strain,* involving a threat to health unless men are physically fit. Break-in period required for men not previously acclimatized. Some decrement in performance of physical work is to be expected. Medical selection of personnel desirable because these conditions are unsuitable for those with cardiovascular or respiratory impairment or with chronic dermatitis. These working conditions are also unsuitable for activities requiring sustained mental effort.
70 80 90	*Very severe heat strain.* Only a small percentage of the population may be expected to qualify for this work. Personnel should be selected (*a*) by medical examination and (*b*) by trial on the job (after acclimatization). Special measures are needed to assure adequate water and salt intake. Amelioration of working conditions by any feasible means is highly desirable and may be expected to decrease the health hazard, while increasing efficiency on the job. Slight "indisposition" that in most jobs would be insufficient to affect performance may render workers unfit for this exposure.
100	*Maximum strain* tolerated daily by fit, acclimatized young men.

Table 20.5 shows the physiological and hygienic implications of 8-hour exposures to various levels of the HSI.

The heat stress index finds application in the engineering analysis of occupational heat exposure and also as a predictor of the AET (67). For an average man, AET is given by the equation

$$\mathrm{AET} = \frac{250 \times 60}{E_{\mathrm{req}} - E_{\mathrm{max}}}$$

HSI and its components offer an excellent starting point for specifying corrective measures when heat exposure is excessive. The relative values of the convective exchange C, radiant exchange R, and evaporative capacity E_{max} provide not only a rigorous way to estimate heat stress but also a basis for a rational approach to corrective engineering measures. This subject is developed further in Section 5.

3.1.5 Wet Bulb–Globe Temperature

The WBGT index was intended originally as a simple expression of heat stress for use in military training where men were exercising outdoors in conditions of high solar radiation (1). It proved very successful in monitoring heat stress and minimizing heat casualties in the United States and has been widely adopted. The WBGT index provides a convenient method to assess quickly, with a minimum of operator skills, conditions that pose threats of thermal strain. Because of its simplicity and close correlation with ETCR, it was adopted in 1971 as the principal index for the tentative TLV for heat stress by the American Conference of Governmental Industrial Hygienists (14).

Fundamentally, the WBGT index is an algebraic approximation of the effective temperature concept. As such, it has all the built-in limitations of the effective temperature, but it also has the advantage that wind velocity does not have to be measured directly to calculate the intensity of WBGT. For outdoor use with solar load, the index is derived from the formula

$$\text{WBGT} = 0.7NWB + 0.2GT + 0.1TA$$

where NWB = natural wet-bulb temperature
GT = globe temperature
TA = dry-bulb (air) temperature

For indoor use, the weighted expression becomes

$$\text{WBGT} = 0.7NWB + 0.3GT$$

Although direct measurement of air movement is not required for the computation of WBGT, allowances are made for this factor by the use of the naturally convected wet-bulb sensor.

The National Institute for Occupational Safety and Health also proposed that WBGT be used as a measure of severity of occupational exposures to heat stress (10). The main criteria used by NIOSH for the selection of a suitable index were (1) the measurements and calculations must be simple, and (2) index values must be predictive of physiological strains of heat exposure.

With respect to the first criterion, there is little dispute that the WBGT index is convenient and simple to use; but WBGT clearly is not a perfect predictor of physiological strain, as required by item 2. Ramanathan and Belding (68), among others, have shown clearly that environmental combinations yielding the same WBGT levels result in different physiological strains in individuals working at a moderate level. This suggests that WBGT has limitations as a heat stress index, especially at high levels of severity.

3.2 Sensing Instruments

The four environmental factors that determine the rate of heat transfer between man and his surroundings are air temperature, humidity, air velocity, and thermal radiation. This section discusses the instruments used and techniques involved in measuring these four variables. We include only the instruments that are widely available and easily adapted to field use.

3.2.1 Thermometry

Air temperature can be measured by a variety of instruments, but each has advantages under specific circumstances.

Liquid-in-Glass Thermometers. The most widely used instrument for measuring air temperature is a glass thermometer in which mercury or alcohol is the expanding liquid. This type of instrument is relatively inexpensive and is available in various temperature ranges and with varying degrees of accuracy. The response time of a thermometer depends primarily on the bulb size and the air velocity at the bulb. Transient temperature readings should not be recorded until the thermometer reaches steady state.

Bimetallic Thermometers. A bimetallic thermometer consists of thin, bonded strips of two different metals, each having a unique coefficient of expansion. Unequal expansion occurs when there is a change in temperature, causing the elements to bend and producing a displacement of the free end. This displacement is transmitted through a suitable linkage to a needle indicator that moves across a scale to indicate the temperature. The bimetallic thermometer element is widely used in the common dial thermometer and in many inexpensive, self-contained temperature recorders.

Thermocouples. A thermocouple is a junction of wires of two dissimilar metals. Such thermojunctions release an electromotive force (emf) that varies with the temperature of the junction. In a circuit containing two thermocouples, the emf in the unit depends on the temperature difference between the two junctions. If one of the junctions is held at a constant temperature, the temperature of the other junction can be determined by the measured emf on the circuit. The emf in a thermocouple circuit may be measured accurately with a potentiometer or a high resistance millivolt meter. Thermocouples of copper and constantan are commonly used for environmental temperatures ranging to 700°F.

Although use of thermocouples entails the initial cost of a potentiometer, together with its bulk, thermocouples are advantageous because they can give remote readings. In such applications, simultaneous readings may be taken at one place for several worksites. In addition, the equilibration time required with varying temperatures is much less than that of mercury-in-glass thermometers.

Thermistors. A thermistor (thermal resistor) is a semiconductor that exhibits a substantial change in resistance with a small change in temperature. Because the resistance of a thermistor is on the order of thousands of ohms, incremental resistance added by lead wires up to about 25 meters is immaterial. Therefore, like thermocouples, thermistors offer the advantage of remote monitoring.

Other advantages of thermistors include their simplicity and the availability of an output signal for recording. Disadvantages are the initial cost of a suitable potentiometer, as well as the cost of thermistor probes.

Difficulties with Air Temperature Measurements. Thermal radiation can cause serious errors in the measurement of air temperature. When the surrounding surfaces are warmer or cooler than the air in a space, radiation effects will cause a sensor used to measure air temperature to become warmer or cooler than the air. Because of their smaller size, thermocouples and thermistors are less affected by radiation than mercury-in-glass thermometers. The temperature distortion due to radiation can be reduced by shielding, by increasing the velocity of air movement over the sensor, or by a combination of the two. Heavy aluminum foil positioned loosely around the sensor provides a simple, yet effective, shield. It is important, however, that a shield not restrict the free flow of air around the sensor element.

3.2.2 Air Velocity Measurements

Any instrument used to measure air movement in the ambient work environment must be capable of measuring very low air velocities and must be nondirectional. These two requirements can be fulfilled only by some type of thermal anemometer. A thermal anemometer has an electrically heated, temperature-sensing element that is cooled by air passing over it. Because a thermal anemometer responds to mass flow over the sensor, a correction is required when the gas density differs greatly from standard air density.

Heated Thermocouple Anemometer. A heated thermocouple anemometer consists of heated and unheated thermocouples connected in series and exposed to the airstream. The resultant emf (voltage) is a function of temperature difference between the heated sensor and the air. The response of the instrument is calibrated to indicate air velocity.

Hot-Wire Anemometer. The sensor in a hot-wire anemometer is a resistance thermometer element of fine wire that is heated. Its temperature and electrical resistance vary with the velocity of air passing over it. The air velocity is calibrated to and determined by the measurement of resistance.

Heated Bulb Thermometer Anemometer. The heated bulb thermometer anemometer includes two matched mercury-in-glass thermometers. The temperature of one thermometer is raised by passing a known current through a fine resistance wire wound around the bulb. The two thermometers are exposed to the air simultaneously and their temperatures are read. The air velocity is calibrated to and determined by the temperature difference.

3.2.3 Humidity Measurement

Humidity refers to water vapor in the atmosphere. The level of humidity can be expressed quantitatively in several different ways, each expression being applicable to certain processes or problems. The terms "specific humidity" or "dew-point temperature" are useful to an air conditioning engineer. In heat stress calculations, "vapor pressure" and "partial pressure" are convenient terms. Only "dew-point temperature" and "relative humidity" can be measured directly by ordinary instrumentation. The other expressions of humidity must be derived from other humidity-related measurements.

Psychrometers. All thermodynamic properties of mixture of air and water vapor can be determined readily from knowledge of the dry- and wet-bulb temperatures. These temperatures are measured with a psychrometer, an instrument consisting of two sensors, one of which is enclosed in a wetted wick or sock. The sensor in the wetted wick is cooled by evaporation, depressing its temperature below that of the dry-bulb sensor. The air velocity over the wetted sensor must be at least 900 ft/min to reach the true wet-bulb temperature (69). The psychrometric wet-bulb temperature is sometimes referred to as the "vented wet-bulb."

Natural wet-bulb temperature (NWB) is used in the determination of the wet bulb–globe temperature (WBGT) index. The NWB is obtained by exposing a thermometer with a wetted wick to *natural* air movement without regard to the minimum air velocity. Therefore NWB is always less than or, in the limit, equal to psychrometric wet-bulb temperature.

A sling psychrometer uses a pair of liquid-filled thermometers, one being covered by a wetted cotton sock. The operator holds the instrument by a handle and whirls the pair of thermometers to effect the required air velocity over the wetted bulb. After repeatedly whirling the sling psychrometer until no further depression of wet-bulb temperature is noted, the final readings of the two thermometers define dry-bulb and (psychrometric) wet-bulb temperatures.

Aspirated psychrometers include two stationary temperature-sensing elements, one dry and one covered with a wetted wick. Airflow over them is produced by a small fan or a squeeze bulb. The sensors of an aspirated psychrometer are usually shielded to preclude radiation-related errors in the indicated temperatures.

Hygrometers. Hygrometers are classified as one of two types: "organic" and "dew point." An organic hygrometer is a direct-reading device for relative humidity. "Relative humidity" refers to the percentage saturation of water vapor in air at a given air temperature. The sensing element of a hygrometer is made of an organic material that undergoes dimensional changes as a function of the relative humidity. This type of sensor is widely used in self-contained humidity recorders and in control instruments.

Dew-point hygrometers provide a direct measure of dew-point temperature. This is accomplished by noting the temperature of a highly polished surface at the time condensation starts to form on it. The condensing plate can be cooled thermoelectrically or by other means, such as the evaporation of a refrigerant at the back of the plate. The onset of condensation can be determined by visual inspection or by use of a photoelectric cell and light source. The temperature at the surface is usually measured by thermocouples affixed to it. The dew-point hygrometer is not a widely used field instrument for measuring humidity in industrial plants. However it is useful in the laboratory and finds specific application in estimating the moisture content of air or other gases at elevated temperatures.

3.2.4 Radiant Heat Exchange

The radiant heat exchange between a worker and the surrounding surfaces of his workspace can be computed, at least theoretically, if the absolute temperatures and emissivities of all surfaces are known. In practical applications, however, such an analytical solution is extremely tedious. Therefore an empirical approximation of radiant heat exchange has been developed and is used widely.

Globe Thermometer. The black globe thermometer is a popular method for assessing thermal radiation in the environment (70, 71). A temperature sensor is located at the center of a thin-walled copper sphere. The outer surface of the globe is painted flat black. When the globe thermometer is used in a space in which surface temperatures are higher than the air temperature, the globe temperature rises above the air temperature because of the radiant heat adsorbed by its black surface. As this occurs, the globe begins to lose heat by convection. Its temperature varies until the rate of heat gained by radiation equals the rate of heat lost by convection. Usually a minimum of 15 minutes is required for the globe temperature to reach steady state.

Mean radiant temperature (MRT) represents the uniform blackbody temperature of an imaginary enclosure, equivalent to the actual environment, with which man exchanges radiative heat. The MRT can be calculated from the globe thermometer temperature, air velocity, and air temperature using the following equation:

$$MRT = [(GT + 460)^4 + 1.03 \times 10^8 V^{0.5} (GT - TA)]^{0.25} - 460$$

where MRT = mean radiant temperature (°F)
GT = black globe temperature (°F)
V = air velocity (ft/min)
TA = dry-bulb (air) temperature (°F)

MRT is a very localized index and refers only to the specific point at which the measurements (GT, V, TA) were taken.

3.3 Integrating Instruments

Many attempts have been made to develop a single-reading heat stress instrument for assessing and integrating the environmental factors of air temperature, air speed, humidity, and radiant temperature. Among the instruments that have been proposed are the globe thermometer (72), the heated globe thermometer (73), the wet Kata thermometer (74), the eupathescope (75), the thermointegrator (76), and the wet globe thermometer (77). Although most of these instruments have been largely abandoned, the search continues for a suitable single-reading instrument.

3.3.1 Requirements and Limitations

In an excellent review of the design requirements and limitations of a single-reading heat stress meter, Professor Hatch showed that despite the mathematical equivalence of man and instrument, the capacity of heat exchange by convection and evaporation is inherently greater for a physical instrument of simple geometric shape and small size than for a much larger man (78).

Single-reading integrating instruments have the appeal of convenience that surely will continue, but their use in the context of today's state of the art in heat stress evaluation offers convenience at the cost of knowing the levels of the four stress components of the thermal environment. Often there is a dual purpose in assessing heat stress: (1) to determine the level of stress, and (2) to determine in what manner and to what extent control procedures can be employed to reduce the stress to an acceptable level. There is a parallel in industrial hygiene with respect to exposure levels estimated on the basis of integrated samples, such as full-period personal samples, and exposure estimates constructed from discrete exposure intervals based on the temporal and spatial variability of contaminant levels in the work environment. Although a single, time-weighted exposure level obtained from an integrated sample is convenient, the convenience is enjoyed at the expense of the diagnostic information needed to assess the causes for exposure and the basis for a rational approach to controls.

3.3.2 Wet Globe Thermometer

Botsford has developed a wet globe instrument consisting of a small copper sphere fitted with a black cotton wick and a water reservoir (77). This device not only integrates the effects of air temperature, mean radiant temperature, and air motion, it includes the effect of fractional wetness. The objective of the "Botsball" is to provide a model that simulates man's exchange of heat with the environment by convection, radiation, and evaporation. Although the relative values of the three components of heat transfer from the wet globe and the human body may differ considerably, several investigators have reported good correlation between wet globe temperature and other well-known heat stress indices (64, 77, 79).

3.3.3 WBGT Index

The TLVs for heat stress as well as the recommendations to OSHA from both NIOSH and the Standards Advisory Committee specify the use of the WBGT index. Based on the equations for calculating WBGT (Section 3.1.5), one must know the natural wet-bulb temperature (*NWB*), the globe temperature (*GT*), and, in the case of a solar load, the dry-bulb temperature (*TA*). The instruments and techniques for measuring these parameters have been discussed in Section 3.2. A suggested arrangement for manual measurement of WBGT appears in Figure 20.4.

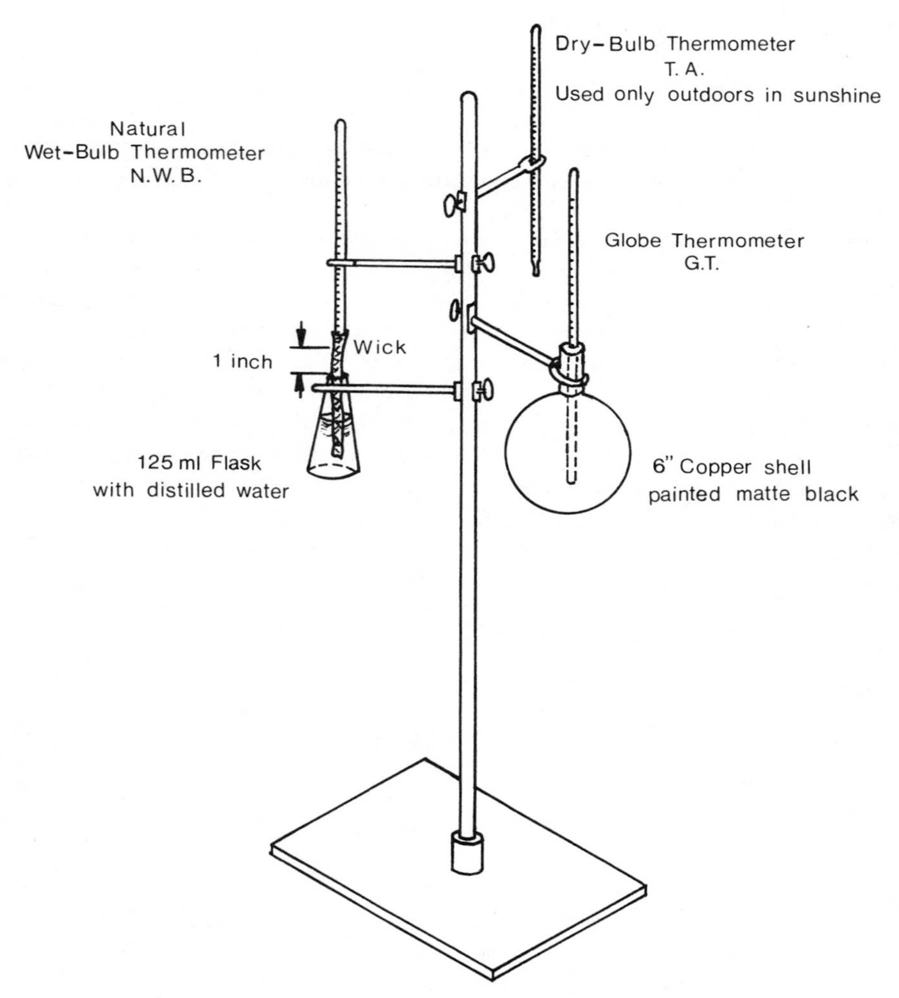

Figure 20.4 Arrangement for manual measurement of WBGT.

Integrated WBGT instruments have been developed to sense and indicate dry-bulb, natural wet-bulb, and globe temperatures, as well as to integrate the values to yield WBGT. The literature describes several instruments that are designed to give a direct readout of WBGT (80–84).

3.4 The Effect of Weather

Even in the hot industries the severity of heat stress is determined largely by weather conditions. The prevailing meteorological conditions, the onset of hot weather, and heat wave episodes can have a very significant effect on the heat stress experienced by workers.

Attempts have been made by several industries to establish qualitative or semiquantitative relationships between outdoor weather conditions and the degree of environmental heat stress imposed on workers. Most of these studies have been limited to the consideration of specific locations and, until recently, have resulted mainly in subjective correlations.

An early study (85) offers a limited functional relationship between outdoor and indoor climates developed on the basis of afternoon measurements of climatic variables taken on 4 days during the summer of 1966. Curves of the temperature differences between indoors and outdoors were developed for individual work areas in a shop, forming a family of hyperbolas. A "temperature parameter," independent of outside climatic conditions, was developed based on the indoor and outdoor dry-bulb temperatures. This parameter is determined as the mean value of temperature differential of two pairs of measurements and is dependent on the heating value of the work area and the rate of ventilation in the shop. Limitations of the model include insensitivity to rapid changes in outside temperature and a narrow applicable time span (1 to 4 P.M.) during which deterministic measurements are made.

Other investigators (86, 87) used a thermal circuit concept to estimate indoor air temperature and relative humidity as these are affected by outdoor air temperature and solar radiation. Since no indoor heat source was considered, however, inside climate was estimated directly from outdoor meteorological conditions, considering building structure and materials to be the only deterministic factors.

Recently NIOSH commissioned a major study to develop predictive models that can be used to calculate WBGT from estimates of a minimum number of conventional meteorological variables (88). The models were based on data acquired during 27 week-long surveys conducted over a 2-year period in 15 representative hot industries in three climatic regions of the United States.

Using the recommended methodology and three sets of relatively simple equations, WBGT can be calculated directly from independent variables that quantify outdoor weather conditions. The first set of models is predictive in nature and is designed to estimate *future* levels of WBGT from weather forecasts. The second series can be used to estimate *current* levels of WBGT at a worksite from current values of meteorological variables reported routinely by a nearby airport station of the National Weather

Service. The third set also estimates *current* levels of WBGT, but uses outdoor meteorological variables measured *locally* in the immediate vicinity of the workplace of interest.

For each of the three sets of models, one of seven equations is most appropriate for any given worksite, depending on the range of differences between inside and outside dry-bulb temperature. The appropriate model is then used to estimate quantitatively the magnitude of WBGT from either direct knowledge of, or forecasts on, weather conditions (88). This modeling technique should simplify the determination of WBGT, leading to increased heat stress monitoring in industry.

4 ASSESSMENT OF HEAT STRESS AND STRAIN

The specification of workplace standards or guidelines for limiting heat stress logically requires an understanding of the relation between environmental heat stress and physiological strain. Although several studies show generally good correlation between some index of environmental heat stress and one or more indices of physiological strain, no single stress index and no single strain index have proved sufficient to characterize adequately the stress–strain relationship over the entire range of conditions expected in industry.

4.1 Correlation of Stress and Strain

As research continues on the environmental and physiological aspects of heat stress, it is desired to define the criteria of physiological limitations in the context of the industrial environment, and to translate acceptable physiological limits to corresponding levels of environmental heat stress. Many investigators share the credit for remarkable advances in our level of understanding.

The approaches that have been used or could be used to correlate heat stress and physiological strain fall into three types: theoretical or semitheoretical calculations, empirical evaluations under controlled conditions, and epidemiological-type studies conducted on workers exposed in occupational settings.

4.1.1 Calculated Limits of Heat Exposure

Gagge et al. (89) defined an upper limit of tolerable heat exposure at which the body heat balance is just maintained, at a given work rate, without a significant rise in skin temperature. Starting with the fundamental heat balance, Haines and Hatch (29) derived a thermal balance line for skin temperature on the basis of the ratio of E_{req}, the rate of evaporative cooling required to maintain thermal balance, to E_{max}, the maximum rate of evaporative cooling possible in a given environmental situation, using either the "operative temperature" (90) or the Vernon globe temperature (91).

Belding and Hatch (30) combined the two limiting requirements to define the heat stress index, the percentage ratio of E_{req} to E_{max}. They related several levels of HSI to expected physiological or psychological effects. Unfortunately, the HSI did not provide the complete solution to the problem of measuring and predicting heat stress and strain (40). Experience showed that the effects did not always occur as predicted. Modifications were therefore attempted to increase the accuracy of HSI and to use the heat balance concept to derive better correlative indices of heat stress and strain.

4.1.2 Empirical Studies in Controlled Conditions

Although several new or modified heat stress indices have appeared in the literature, they usually have been tested in experimental programs designed to evaluate their relation to only a few physiological responses. As Peterson (92) pointed out, many physiological parameters have been related to limited measures of heat stress but not necessarily to any of the heat stress indices.

In a comprehensive project illustrative of the empirical approach identified earlier, Peterson designed and executed a study to relate several heat stress indices to several measures of physiological response. His objective was to identify an optimum indicator of strain as well as the physiological responses that are sensitive to change as a function of heat stress. Although WBGT was not included among the heat stress indices considered, the results demonstrated that at least three indices of strain were necessary to describe the probable response of man to his thermal environment. Sweat rate was the biothermal strain best correlated with all stress indices except the equivalent operative ambient temperature. This suggests that sweat rate is a suitable parameter in terms of total strain. However neither the actual nor predicted sweat rate correlated well with ear, rectal, or skin temperatures, suggesting circumstantially that the total strain concept probably has little utility (92).

Another recent study of three fit subjects under controlled environmental conditions evaluated the relations among environmental parameters, physiological strain parameters, and heat stress indices (68); WBGT was included as a heat stress index at levels of 68, 85, and 89°F WBGT. The authors concluded that although a given level of WBGT has meaning dependent on environmental conditions, higher levels of the WBGT scale do not always signify greater strain, especially when the environment is relatively dry. Furthermore, the WBGT values failed to parallel the increased physiological strain of varying humidity, air speed, or radiant heat. A more serious inconsistency was that in dry environments a higher WBGT level did not produce greater strain than a lower WBGT level. Therefore WBGT would seem to have limited value as a predictor of physiological strain at the higher heat stress levels that may be encountered in industry (68).

The same study showed that the heat stress index differentiates between environmental conditions in the correct order of physiological strain, although the resulting strain seems to be overemphasized. However the wet globe temperature did not differentiate between two widely different levels of physiological strain in the participants (68).

4.1.3 Exposed Workers in Occupational Settings

In the late 1930s researchers began to focus on the relations between stress and strain for workers in hot industries (5, 13, 93–96). Nevertheless, few epidemiological studies have been reported that relate the length and intensity of heat exposure to the long-term health experience of workers. Health experience statistics, although appearing in numerous studies, tend to relate qualitative differences in some dependent variable representing morbidity or mortality to unique levels of some single categorical or independent environmental variable (6, 7, 21, 97–100).

Our present understanding of the effects of heat on health and safety suggests that standards of exposure should be based on the acute effects; yet it is clear that the physiological consequences of exposure to heat are not directly proportional to the intensity of heat stress over the entire tolerable range. Physiological strain increases exponentially in the upper range of heat stress; thus a small incremental increase at high levels of heat stress can result in a large increase in strain. Furthermore, the presence of many factors mitigates the relation between exposure and effects. All environmental variables, as well as age, physical fitness, acclimatization, motivation, physical capability, and emotional stability, act to mitigate this relation. In addition, "job factors" such as the complexity of a task, physical load, perceptual motor load, and the level of skill required, are of great importance. Any exposure limits stipulated for heat stress must reflect recognition that the specific environment–worker–job situation defines a total stress that must be maintained within acceptable levels by adequate control of one or a combination of factors.

4.2 Rationale for a Heat Stress Standard

Over the years there have been various expressions of permissible exposure limits for working in hot environments. These recommendations or standards have been expressed in terms of several indices of heat stress, but most of them account in some manner for the four crucial climatic factors: air temperature, humidity, radiant heat, and air velocity. Several indices consider the work load as well. A recent World Health Organization panel of experts attested to the shortcomings in virtually all the indices, as well as the proposed upper limits set forth for each of these parameters (15).

Unfortunately, most of the recommended standards for thermal exposure have been derived from laboratory experience rather than from studies on industrial workers. This is a shortcoming because the experimental subjects are not necessarily representative of an employed workforce in age and fitness. In addition, the severity of physiological strain is known to correlate poorly with various levels of each of the heat stress indices. In a study previously referenced, the use of at least three indices simultaneously was recommended to obtain an adequate evaluation of the heat strain of an exposed worker (92).

The WHO panel of experts recommended that a deep body temperature of 38°C be considered the limit of permissible exposure for work in heat (15). This is consistent with the observation that body core temperature in excess of 38°C increases the likelihood that a heat disorder or illness will occur (23, 101).

4.2.1 Criteria for a Standard Index

The NIOSH Criteria Document sets forth five very reasonable criteria for any heat stress index to be used in industry (10):

1. The applicability should be proved in industrial use.
2. All important factors should be included.
3. The measurements and calculations should be simple.
4. The environmental factors should have a valid weighting in relation to the total physiological strain.
5. The index should be applicable and feasible for specifying regulatory limits.

Four indices satsify the first criterion: ET, WBGT, HSI, and P4SR. Because an estimate of the work load is included in HSI and P4SR, they have an advantage over WBGT and ET with respect to the second criterion. However the measurements and calculations are much simpler for ET and WBGT than for HSI and P4SR, and WBGT is somewhat simpler than ET. With respect to the fourth criterion, all four indices have shortcomings. Considering the applicability for use in regulatory limits (criterion 5), HSI has the advantage of being used to calculate an allowable exposure time and minimum recovery time for a given heat stress condition. However, WBGT is the most convenient index with which to monitor levels of exposure because it can be read directly from an integrating instrument or computed easily from sets of two or three environmental measurements.

Even though there is no obvious significant advantage to the use of WBGT, it appears to be the preferred environmental index, with HSI offering several distinct supplementary advantages with respect to the engineering analysis of a given heat stress exposure.

4.2.2 Prescriptive and Environment-Driven Zones

The "prescriptive zone" (PZ) refers to the range of environmental conditions in which deep body temperature is determined by work intensity only. The prescriptive zone has been defined empirically in terms of effective temperature by a series of experiments (rectal vs. effective temperature) at three levels of metabolic rate M: 180, 300, and 420 kcal/hr (23). Such studies show clearly that steady state rectal temperature is not a function of effective temperature until critical levels of effective temperature (dependent somewhat on M) are reached, at which point rectal temperature increases sharply. The range of environmental heat stresses within which the body core temperature rises sharply with elevations in climatic conditions is called the "environment-driven zone" (EDZ). The value of an environmental heat stress index at the interface of the PZ and the EDZ is called the "upper limit of prescriptive zone" (ULPZ). The value of the ULPZ varies for different individuals. It is higher for workers who are acclimatized to heat, but it is reduced when more clothing is worn (10).

To be certain that 95 percent of a heat-acclimatized population wearing worker uniforms will not have rectal temperatures in excess of 38°C, the level of environmental heat stress at which a 5-percentile man will reach his ULPZ must be established (10). This value must then be corrected for the level of acclimatization and clothing. Lind and Liddell found the ULPZ to be 80.5°F ET for a group of 128 men of average physical fitness (102).

Permissible exposure limits for heat stress cannot be based on 8-hour average values because exposures in excess of 1 hour may cause the worker to accumulate enough heat in his body to induce an acute heat disorder. Thus in continuous heat exposure, hourly averages are necessary. If the exposure is intermittent, however, the accumulation of heat will be slowed, allowing the use of a 2-hour average exposure (10).

Another study by Lind justifies the use of time-weighted average hourly work load values for intermittent work (24). The results indicate that from the point of view of the ULPZ, a certain hourly amount of work can be performed either at a lower rate continuously or at a higher rate interrupted with rest periods. In another study, the ULPZ was found to be the same for men of different ages; thus no correction for age is required (103). However when older men are exposed to a strenuous heat load, increased caution is advisable because their physiological capacities are lowered and susceptibility to disease is increased.

4.2.3 Work Practices Versus Environmental Standard

Lind's work on the prescriptive zone represents a sound basis for the development of an environmental heat stress standard. Not only does it combine both the climatic and work load conditions that are imposed on the worker, but it can be monitored with a convenient index such as WBGT. Nevertheless, a number of practical shortcomings and unresolved questions remain (10).

It is important that the upper limit of prescriptive zone concept be validated with data from a representative industrial workforce. More data are required on the age and sex distribution of the workforce. Also to be clarified is the effect of the natural selection process that normally occurs in an industrial situation when the worker himself determines his ability to endure high levels of heat stress. This consideration may have resulted in past heat stress standards that were unrealistic for an industrial population. The lack of data regarding intermittent exposures to heat is another major unresolved question of the effect of heat stress on the workforce. In addition, differences in sweat loss under a wide variety of industrial conditions still have not been thoroughly studied. The very different work loads and the intermittency of work loads that are normal in industrial operations may also have a major effect on heat stress (10).

The answers to these questions require additional research to validate the techniques presently proposed for the evaluation of heat stress conditions. In the interim, our current level of understanding justifies a "work practices standard" rather than an environmental standard for heat stress (10). Clearly the environmental measurements we have are sufficient to estimate threshold limits of heat stress that can be utilized to initiate work practices to protect the industrial worker adequately against heat.

4.3 Options for Control

Before initiating control measures it is crucial to identify the components of heat stress to which workers are exposed currently or expected to be exposed in new operations. Only then, by examining the alternatives, can the best means of control be selected.

Heat stress control for an individual worker depends on several elements of behavior control and environmental control. The most important elements are as follows:

1. Bodily heat production.
2. Number and duration of exposures.
3. Heat exchange components as affected by environmental factors (*MRT*, E_{max}, *V*, *TA*, and *PPA*).
4. Thermal conditions of the rest area.
5. Clothing.

The third element is covered in greater detail in Section 5, Engineering Control of Heat Stress. It is important, however, to be aware of the full range of options for control in a specific situation. Table 20.6 provides a concise summary of actions that may be considered alternatives in the control of heat stress (104).

Metabolic heat *M* can comprise a large fraction of the total heat load. However the amount by which this factor may be reduced by control is often quite limited. This is because an average-sized worker who is pushing buttons will produce heat at a rate of 100 kcal/hr, whereas one who is manually transferring heavy materials at a steady pace will seldom have a metabolic rate in excess of 250 to 300 kcal/hr (105). Obviously control measures such as partial mechanization can reduce the metabolic rate only for steady work by 100 to 200 kcal/hr. However mechanization can also make it possible to isolate the worker from the heat source, perhaps in an air conditioned booth.

Tasks such as shoveling, which result in metabolic heat production at rates up to 500 or 600 kcal/hr, require that rest be taken one-half to two-thirds of the time simply because of the physical demands of the labor. Thus the hourly contribution of *M* to heat load will seldom exceed 300 kcal/hr. It is obvious that mechanization of such work can increase worker productivity by making possible a decrease in the amount of time needed for rest (10).

Table 20.6 Alternative Measures for Controlling Heat Stress (104)

ITEM	POSSIBLE ACTIONS
1 Heat Components	
M	Reduce by • Mechanization of some or all tasks • Sharing workload with others (particularly during peak heat periods) • Increasing rest time
R	Reduce by • Minimizing line-of-sight to source • Insulating furnace walls • Using reflective screens • Wearing reflective aprons (particularly valuable when workers face source) • Covering exposed parts of body
C	If air temperature is above 95°F, reduce C by • Lowering air temperature • Lowering air velocity • Wearing clothing If air temperature is below 95°F, reduce C by • Lowering air temperature • Increasing air velocity • Removing clothing
E_{max}	Increase by • Increasing air velocity • Decreasing humidity
2 Work Schedule	Duration • Shorten duration of each exposure • Use more frequent rest periods Recovery • Use nearby air conditioned space for rest area • Adjust V in rest area for effective cooling Other • Allow worker to self-limit exposure on basis of signs and symptoms of heat strain • Provide cool, potable water containing 0.1% salt
3 Clothing	• For extreme conditions, use cooled (by vortex tube or other means) clothing • Wear type of clothing to obtain $E_{max} > E_{req}$ with minimum sweating

The second element that may be modified is the number and duration of exposures. When the task in a hot environment involves work that is a regularly scheduled part of the job, the combined experience of workers and management will have resulted in an arrangement that makes the work tolerable most of the time for most of the workers. For example, a task that involves the manual transfer of hot materials may involve two workers who alternate at intervals from 5 minutes up to an hour. Under such conditions, overall strain for an individual will be less if the cycles are short (106). When a standardized quota of hot work exists for each man, it is sometimes accomplished at the beginning of the shift. This arrangement may be preferred by workers in cooler weather; however there is evidence that the strain may become excessive on hot days. The total strain will be less, evidenced by fewer heart beats, if the work is spread out (10).

The stress of hot jobs is also dependent on variations in the weather. A hot spell or an unusual rise in humidity may create overly stressful conditions for a few hours or days during the summer. Nonessential tasks should be postponed during such emergency periods in accordance with a prearranged plan (Section 3.4). Also, the assignment of an extra helper can reduce the heat exposure of individual members of a working team (10).

Many of the critical exposures to heat faced by employees in industry are incurred irregularly, as in furnace repair or emergencies, when levels of heat stress and physical effort are high and largely unpredictable, and values for the components of the stress cannot readily be assessed. Usually such exposures force a progressive rise in body temperature. Ideally, physiological measurements such as body temperature and heart rate should be monitored and used as criteria for limiting such exposures on an ad hoc basis. Practically, however, tolerance limits may be based on the experience of the worker as well as of his supervisor. Fortunately, perception of fatigue, faintness, or breathlessness may usually be relied on to bring individual exposures to a safe ending (10).

A third potential modification is the thermal conditions of the rest area. Brouha states, "It is undeniable that the possibility of rest in cool surroundings reduces considerably the total cost of work in the heat" (106). No definitive information exists on the optimum thermal conditions for such areas, but there are data to support setting the temperature near 25°C. This feels chilly upon first entry from the heat, but adaptation is rapid (10). The placement of these areas is also important. The farther they are from the workplace, the more likely that they will be used infrequently or that individual work periods will be lengthened in favor of prolonged rest periods (10).

A final possible modification is clothing. Heat stress often can be altered substantially by being selective with clothing. In the heat, as in the cold, the thermal function of clothing is to reduce heat transfer between the individual and his environment (107). Clothing may alter heat transfer by radiation, by convection, and by evaporation of sweat.

Conventional work clothing interferes substantially with heat loss by radiation and convection. A man doing moderately hard physical work (300 kcal/hr) and wearing only shorts has a comfort temperature of about 20°C. In work clothing, his comfort temperature might well be 13°C. If the environmental temperature were 20°C, the cost of wearing clothing, in terms of heat stress, would be equivalent to an added sweat rate of at least one-half pint per hour (10).

Laboratory studies clearly indicate that ordinary work clothing will reduce radiant heat transfer by 30 to 40 percent (32). Theory yields a similar reduction for transfer by convection. Other recent studies demonstrate that conventional work clothing will reduce the potential for evaporating sweat by about 40 percent (10).

Long winter-weight underwear has been adopted by many workers who move in and out of very hot environments. The extra layer provides a substantial buffer against extremes of heat. In humid summer weather the practice has less justification, unless there is ready access to air conditioned recovery areas, because the underwear interferes with evaporation of sweat from the skin. More sweat must be produced to maintain heat balance, and little or no more can be evaporated (10).

Special clothing may take various forms. For example, infrared reflecting face shields may be necessary when radiant heat is high. When hot materials are handled, it is good practice to provide several pairs of wide-gauntlet, oversized insulated gloves, which can be put on easily without using both hands (10).

For very hot exposures, as in relining furnaces, thick insulative clothing that acts as a heat "sponge" is appropriate. This sponge may be more effective if made of high density materials (asbestos in the recent past) because of the higher heat capacity, but insulation with minimum weight is best imparted by a layer of trapped, still air. For relatively long intervals of exposure, high density and highest feasible thickness should be sought. The protective value of such clothing is enhanced by aluminizing its surface and sometimes by interlining with foil between insulative layers (10).

When shielding against radiant heat loads cannot be accomplished by fixed barriers (Section 5.1), aluminized clothing or clothing components may often be used to advantage. Layers of coated fabrics used near radiant sources above 300°C can result in reflection of 90 percent of incident energy (108). The efficiency of protection is somewhat lower with decreasing radiant heat. When the coverage with reflective clothing is only partial, there is much more opportunity for evaporation of sweat (10).

Several techniques have been incorporated in thermally conditioned clothing for maintaining comfort in extreme heat. Some systems supply cool air from a mechanical refrigerator to points under a jacket or coveralls. When air from a remote source is used, there are two problems. One is the gain of heat through the walls of the supply tubing. This problem has been solved in some cases by using porous tubing that will leak an appropriate amount of supply air to keep the walls suitably cool. The other problem is distribution of the air through the suit. With a simple, single orifice it is difficult to cool a sufficient area of skin, and the area cooled may be too cold. Provision of several orifices, though better, will create bulk and restrict mobility (10).

The vortex tube has been used successfully as a source of cool air in some situations (109). The device is carried on the belt, and air is introduced tangentially at high velocity and forced into a vortex resulting in two separable streams of air, one cold, which is distributed under the suit, the other hot, which is discarded. Compressed air requirements to operate the vortex system are large.

Self-contained sources of conditioned air that can be backpacked have also been developed. One uses a liquid refrigerant, sealed into a finned container. After being cooled in a deep freeze, the container is placed in the pack. A small battery-driven fan circulates air across the fins and into the suit. A single charging of this device may extend tolerance while relining furnace walls from several minutes to 30 or 60 minutes (10).

The nuisance factor must be considered for all cooling devices. Workers will not go to the trouble of using them unless they recognize more than a marginal advantage. On the other hand, such devices have sometimes changed hot tasks that required long rest periods and multiple workers into single-worker, continuous-duty operations (10).

5 ENGINEERING CONTROL OF HEAT STRESS

The control of heat in industry to ensure that worker exposure falls within acceptable limits requires the application of feasible engineering procedures. Cost-effective engineering control of heat stress calls for an understanding of the physiological response of man in a hot environment. The underlying worker–environment thermal balance (Section 2) suggests that when a worker is exposed to elevated temperatures, his rate of heat loss will decrease. The role of engineering controls is to help sustain a rate of bodily heat loss equal to the imposition of heat from the environment. In a general sense this can be accomplished by increasing the velocity of air across the body, a technique that is useful within certain limits of temperature and humidity; above such limits it is necessary to reduce the surrounding temperature. In practice, it is often necessary to combine these two approaches to achieve an acceptable thermal work environment.

The important role of engineering methods in regulating heat stress is underscored in the recommendations of the Department of Labor Standards Advisory Committee on Heat Stress (18). In its report, the committee sets forth "special work practices" needed to bring the 2-hour average heat exposure level within the limits of its recommended threshold WBGT values. The first such practice is that "the employer should adopt engineering controls which are appropriate for reducing and controlling the level of heat exposure" (18).

An engineering approach to the reduction of heat stress generally parallels the same control strategy applied for other environmental stresses such as airborne gases, vapors, and dusts. Among the alternative control methods are substitution, control at the source, local controls, and general ventilation. Another approach to an analysis of engineering control of heat stress is to focus, in turn, on the important environmental components of the thermal balance model: convection, evaporative capacity, and radiation. This section combines both viewpoints and presents control methods that not only reflect a traditional approach to engineering control alternatives, but add needed emphasis to the aspects of heat control that characterize this specific physical stress.

The first and most fundamental approach to engineering control of heat is to examine options for eliminating heat at its point of generation. Occasionally, one feasible alternative is to change the operation or substitute a process component of lower temperature for one of higher temperature. One well-known example is the use of induction heating rather than direct-fired furnaces for certain forging operations.

5.1 Control at the Source

As with most environmental stresses, heat is controlled most effectively if it is regulated at the source, to prevent "contamination" of the space occupied by workers. In general, the options for control of heat at the source are isolation, reduction in emissivity, insulation, radiation shielding, and local exhaust ventilation.

5.1.1 Isolation

The most practical method for limiting heat exposure from hot processing operations that are difficult to control or for operations that are extremely hot is to isolate the heat source. Such operations might be partitioned and separated from the rest of the facility, located in a separate building, or relocated outdoors with minimal shelter. A typical example is an industrial boiler, invariably segregated from the other operations in the same facility.

5.1.2 Reduced Emissivity of Hot Surfaces

Often the rate at which heat is radiated from the surface of a hot source can be lowered if the emissivity of the source is reduced through surface treatment to one of lower emissivity. When an oven, boiler, or other hot surface is covered with aluminum paint, the reduced emissivity of its surface offers two advantages. Not only is less heat radiated to workers nearby, but heat is conserved inside the unit, representing an increasingly substantial saving in energy costs (108).

The emissivity of a hot source can also be lowered by sheathing or by covering the source with sheet aluminum. Other metals, such as galvanized steel, have relatively low emissivities but are more expensive to use than aluminum. The emissivity of galvanized steel increases faster with aging than does that of aluminum sheet (108).

Because of reduced emissivity, structural steel members will radiate less heat if they are painted with aluminum paint. Even though aluminum-painted surfaces have a higher emissivity than aluminum sheet, they do not radiate as much heat as oil-based painted steel at a given temperature. Table 20.7 shows the emissivity and reflectivity (Section 5.1.4) of various common materials (108).

5.1.3 Insulation

"Insulation" is not mutually exclusive of "isolation" in the context of engineering control of heat stress, for in a general sense insulation prevents the escape of sensible and

Table 20.7 Relative Efficiencies of Common Shielding Materials (108)

Surface	Reflectivity of Radiant Heat Incident Upon Surface (%)	Emissivity of Radiant Heat from Surface (%)
Aluminum, bright	95	5
Polished aluminum	92	8
Polished tin	92	8
Zinc, bright	90	10
Aluminum, oxidized	84	16
Varnished aluminum	80	20
Varnished tin	80	20
Zinc, oxidized	73	27
Aluminum paint, new, clean	65	35
Iron, sheet, smooth	45	55
Aluminum paint, dull, dirty	40	60
Iron, sheet, oxidized	35	65
Steel and Iron	10-----20	80-----90
Brick	4-----20	80-----96
Lacquer, black	10	90
Wood	4-----8	92-----96
Asbestos board	4-----8	92-----96
Oxide paints, all colors	6	94
Lacquer, flat black	3	97

radiant heat into the work environment. A well-known example of insulation that also has implications with respect to energy conservation is that of pipe-covering insulation on steam lines. By reducing the escape of heat into the environment, such insulation clearly helps conserve energy and fuel resources.

In addition to reducing radiative exchange, insulation reduces the convective heat transfer from hot equipment to the work environment by minimizing local convective currents that form when air that comes in contact with very hot surfaces is heated. Insulation of hot surfaces in the workplace is a measure consistent with the concern about energy shortages and the rapidly accelerating prices for fuel and energy.

5.1.4 Radiation Shielding

Shielding against radiant heat represents an extremely important control measure. The characteristics of radiant heat are quite different from those of high air temperatures, and the difference must be understood if engineering controls are to be effective. Radiant heat passes through air without heating the air; it heats only the objects in its path that are capable of absorbing it. Shielding radiant heat sources means putting a barrier between the worker and the source to protect the worker from being a targeted receptor of the radiant energy. Radiation shielding can be categorized into reflecting, absorbing, transparent, and flexible shields.

Reflective Shields. Reflective shields are constructed from sheets of aluminum, stainless steel, or other bright-surface metallic materials. Aluminum offers the advantage of 85 to 95 percent reflectivity. It is used also as shielding in the form of foil with insulative backing, and in aluminized paint, with reduced effectiveness. Reflectivity of other shielding materials is shown in Table 20.7.

Successful use of aluminum as shielding requires an understanding of certain principles (108):

1. There must be an aluminum-to-air surface; the shield cannot be embedded in other materials.
2. The shield should not be painted or enameled.
3. The shield should be kept free of oil, grease, or dirt, to maximize reflectivity.
4. When used to enclose a hot source, the shield should be separated from the source by several inches.
5. Corrugated sheeting should be arranged so that the corrugations run vertically rather than horizontally, to help maintain a surface free of foreign matter.

Absorptive Shielding. Absorption shielding absorbs infrared radiation readily. This type of shielding, preferably flat black, is constructed typically of two or three sheets separated by air spaces. Heat can then be removed by causing water to flow between two metal plates in the shield, transferring heat from the shield by conduction (108). The surface(s) of absorptive shielding exposed to work areas should be constructed of aluminum or aluminized to reduce emissivity.

Transparent Shielding. Transparent shielding consists of two general types: special glass and metallic mesh. Special glass reduces transmission of infrared radiation because it is either "heat absorptive" or "infrared reflecting." Infrared reflecting glass is used commonly in the windows of control rooms amid excessive heat sources. Metallic mesh shielding involves the use of chains and wire mesh to provide partial reflectance and to help reduce the amount of radiant heat reaching an operator (108). Such use is warranted where manual operations preclude use of a solid barrier.

Flexible Shielding. Flexible shielding utilizes fabrics treated with aluminum. When worn as aprons or other items of clothing, they protect against radiant heat by reflecting up to 90 percent. Reflective garments are useful for protection against very localized and directional radiant sources (Section 4.3).

5.1.5 Local Exhaust Ventilation

Canopy hoods with natural draft or mechanical exhaust ventilation are used commonly over furnaces and similar hot equipment. Although the benefits of such an application are obvious, especially since heated air has a natural tendency to rise, it must be remembered that local ventilation removes only convective heat. Radiant energy losses, whose magnitude often overrides convective losses, are not controlled by local exhaust hooding. Radiation shielding must be used as well to control what is likely to be the larger fraction of the total heat escaping from the hot process (110).

5.2 Localized Cooling at Work Stations

When practical considerations limit the feasibility of heat control at the source or throughout the general work area, relief can be provided at localized areas. In such instances, cool air is introduced in sufficient quantity to surround the worker with an "independent" atmospheric environment. This local relief, often termed "spot cooling," can serve two functions, depending on the relative magnitudes of the radiant and convective components in the total heat load.

If the overall load is primarily convective in the form of hot air surrounding the worker, the only function of the local relief system is to displace the hot air immediately around the individual with cooler air having a higher velocity. If such air is available at a suitable temperature and is introduced without mixing with the hot plant air, no further cooling of the worker is necessary.

When there is a significant radiation load, however, the local relief system must provide some actual cooling to offset the radiant energy that penetrates the mass of air surrounding the worker. Thus the temperature of the supplied air must be low enough to make the convection component C negative in the heat exchange model to offset R, the radiation component.

Air movement within the local relief zone is desirable to maintain adequate evaporative cooling. However the use of high air velocities (increased evaporative cooling) to offset radiation may be counterproductive if the convective load is increased (where air temperature exceeds skin temperature), causing a greater sweating demand. This defines a rationale for not using simple man-cooling fans, which nevertheless are still found throughout industry.

A major problem in the design of localized cooling systems is the introduction of supplied air in such a way that minimum mixing takes place with the surrounding hot air. High velocity jets encourage mixing, yet are necessary to "throw" the air into the work zone from an offside supply duct.

In situations of extreme heat, workers should be stationed inside an insulated, locally cooled observation booth or relief room to which they can return after brief periods of high exposure. The practical usefulness of such a protected room will vary from industry to industry. Air conditioned crane cabs represent one application of this concept now in increasingly common use.

5.3 General Ventilation

A common method for heat removal in the hot industries is general ventilation, making use of wall openings for the entrance of cool outside air and roof openings, commonly of the gravity type, for the discharge of heated air.

Although a higher fraction of heat loss from hot sources occurs by radiation, such radiant heat is absorbed by walls and other solid structures. These heated surfaces become secondary sources of convective heat and also act as secondary reradiators. Eventually, most of the heat must be removed from the enclosed space by ventilation; the balance is lost through walls. Therefore general ventilation is an essential part of an overall heat control strategy. Unfortunately it cannot offset direct radiant heat exposures and, even in its secondary role, often fails to function as needed, mainly because of inadequate supply of makeup air. Insufficient area of openings and poor location of inlets often result in improper distribution or too little air within the building (110).

5.3.1 Fresh Air Supply

Although the use of uncooled outside air is sometimes effective in controlling heat, the more practical approach for a new facility is to design a general ventilation system with the capability of cooling outside air before distributing it throughout the plant.

Evaporative Cooling. Evaporative cooling is a well-established method for lowering the dry-bulb (air) temperature. An evaporative cooler provides intimate contact between the incoming air and water by using sprays or wetted filters, and upon such contact, the air is cooled adiabatically. This means that no appreciable amount of heat is transferred to or from the air. By converting sensible heat in the airstream to latent heat, dry-bulb temperature is caused to drop as the relative humidity increases. Even though additional humidity is added to the incoming air, the effect is not necessarily negative because the new dry-bulb temperature may be sufficiently below skin temperature to allow the worker to be cooled by convection.

Chilled Coil System. In a chilled coil system, air passes over coils containing a medium whose temperature is sufficiently below that of the air to result in satisfactory cooling. These systems are usually one of the following types (111).

Water-Cooled Coil. In this system, which is the simplest, water is circulated directly through the coil while air from the conditioned space is passed over the coil to remove heat and moisture. As a general rule, water must enter the coil at not more than 11°C, and it is wasted after leaving the coil unless it is used for process work.

Cooled Water System. When water is not available from a supply of sufficient quantity or at a satisfactory temperature, it can be cooled artificially and recirculated. This system is often used when the coolant must be piped long distances and a variety of loads must be overcome. Usually the heat is removed from the water by mechanical refrigeration, but in some intermittent operations it is removed by the melting of ice.

Refrigerated Coils. As the name implies, the coils are cooled by the direct vaporization and expansion of a liquid refrigerant. This method is widely used in small units as well as large central systems.

5.3.2 Distribution of Makeup Air

Convective flow around hot bodies is naturally upward. For an ideal system of general ventilation, combined with radiation shielding, the inlet air should enter near floor level, be directed toward the workers, and flow toward the hot equipment. In this way the coolest air available is received by the workers before its temperature is increased by mixing with warm building air or circulation over hot processes. This air then flows toward the hot equipment and, as its temperature increases, rises and escapes through vent openings in the roof. The combination of the motion and the cool temperature of the air yields the maximum comfort that can be obtained for the worker without some method of artificially cooling the air. Contrary to general practice, provisions for proper distribution of the air supply should receive the same careful consideration that is given to the selection of exhaust equipment (110).

5.3.3 Removal of Heated Air

Increasingly, buildings are constructed to provide for the natural removal of heated air. For example, glass manufacturing facilities usually have large gravity roof ventilators for exhaust purposes. Unfortunately many hot industries are characterized by large, flat building configurations in which case exhaust fans or gravity ventilators without a forced air supply may not provide satisfactory ventilation patterns in the building's workspace.

The basic strategy when general ventilation is used to remove heated air is to locate the exhaust openings, either natural draft or mechanically operated, above the sources of heat and as close as practical to them (108).

5.4 Moisture Control

The importance of moisture control is underscored in the warm-moist industries where high temperatures and high relative humidities prevail. In some industries relative humidity is deliberately maintained at a high level to maintain product quality or to prevent static electricity. Examples are textile mills, munitions plants, coal mines, and flour mills. Nevertheless, whenever feasible, dehumidification clearly helps to offset heat stress because it reduces the partial pressure of water vapor in air, which in turn increases the evaporative capacity E_{max} (Section 2.2).

In a general sense, moisture control includes both prevention of increased humidity and the use of dehumidification procedures. Such unsophisticated yet effective controls as enclosing hot water tanks, covering drains carrying hot water, and repairing leaky joints and valves in steam piping, constitute direct measures that serve to alleviate heat stress in warm-moist industries.

Dehumidification, aimed at reducing workplace humidity rather than preventing its increase, can be accomplished by refrigeration, absorption, or adsorption. In the context of occupational heat exposures, refrigeration is the most widely used technique to condition the air of relief areas, operating booths, or other local or regional portions of an industrial facility. Although it is clearly impractical to air condition some operations, such as a hot rolling mill in a steel plant, the use of air conditioning is increasing as a heat control method, especially in combination with a work–rest regimen for a given job function. The concept of mechanical refrigeration to reduce the temperature (and humidity) of supply air is mentioned explicitly in the list of engineering controls of the Standards Advisory Committee on Heat Stress (18).

5.5 Typical Examples

The first step in the engineering control of heat stress is to identify the magnitude of the various components of stress to which the workers are exposed in existing or expected conditions. Only by knowing the relative magnitude and the factors contributing to the heat stress components, can rational methods be selected to help alleviate the problem with a cost-effective program.

5.5.1 Increasing Ventilation to Offset High Humidity

Table 20.8 gives environmental data typical for a hypothetical laundry, where a small fan on the wall moves air at a relatively low velocity. Figures for both semiclothed and clothed workers indicate the added stress due to wearing clothing in high humidity environments. Here WBGT is at 32.5°C, indicating that some control measures are

Table 20.8. Use of Ventilation to Offset High Humidity (104)

Data	Before		After Increased Ventilation: Semiclothed
	Semiclothed	Fully Clothed	
GT (°C)	35.0	35.0	35.0
TA (°C)	35.0	35.0	35.0
VWB (°C)	30.3	30.3	30.3
NWB (°C)	31.4	31.4	31.1
PPA (mm Hg)	30	30	30
V (m/min)	15	15	45
WBGT (°C)	32.5	32.5	32.3
MRT (°C)	35.0	35.0	35.0
R (kcal/hr)	0	0	0
C (kcal/hr)	0	0	0
M (kcal/hr)	200	200	200
E_{req} (kcal/hr)	200	200	200
E_{max} (kcal/hr)	130	80	250
Sweat rate (liter/hr)	0.33[a]	0.33[a]	0.33

[a] Dripping.

necessary. Note that R and C are both zero because of the simplified thermal conditions assumed, namely, air and skin temperature are both 35°C.

In this case, increasing air speed from 15 to 45 m/min produces an E_{max} greater than E_{req}, a necessary condition to avoid bodily storage of heat and its consequent strain. In the "before" situation, the workers are sweating at a near-maximum rate but cannot achieve thermal balance because a deficit of 70 kcal/hr exists between E_{req} and E_{max}. With clothing, the deficit increases to 120 kcal/hr. At 120 kcal/hr, the limit of bodily tolerance would be reached within one hour.

Tripling the ventilation rate approximately doubles the evaporative cooling, and under such conditions E_{max} exceeds E_{req}. It should be noted that even under these conditions WBGT is at 32.3°C, a level that would be considered unacceptable based on the NIOSH Criteria Document (10) and the Report of the Standards Advisory Committee (18). Nevertheless, an acclimatized workman could continue in this environment, and the 0.33 liter/hr sweat rate would not create undue strain. The increased ventilation has created a thermal condition that allows the sweat to be evaporated efficiently enough that the skin is no longer dripping wet. At higher air velocities, additional cooling would occur, but there may be an upper limit of air speed that could disrupt laundry operations.

A more effective and permanent control technique would be to reduce the moisture content of the room air. This should be done as close as possible to the source of water vapor when the source is within the room. If the source of humidity were outside the workroom, mechanical air conditioning would probably be required to reduce the temperature and moisture level of the room air (104).

5.5.2 Use of Evaporative Cooling in a Hot-Dry Environment

Table 20.9 presents data for a hot-dry environment where the presence of high air speed becomes a liability and the wearing of clothing is advantageous. Since the air speed is already at 110 m/min, modest increases in air velocity would not be expected to make major improvements in evaporative capacity. If the workers were seminude rather than clothed, the increased thermal load from both R and C created by a greater exposure of skin would account for an increase in sweat rate from 0.68 to 0.91 liter/hr. The added benefit of increased evaporative capacity (E_{max}) from clothing removal does not improve the situation. It follows that when MRT and TA are above 35°C and VWB is low, the wearing of full clothing can reduce heat stress. The type and weight of clothing can be optimized for different hot work situations.

With low humidity (PPA = 10 mm Hg), a better solution to the problem would be the installation of an evaporative cooler. Assume that in the situation described, the inside temperature is usually 5°C hotter than the outside due to process heat and building insulation. If the outside temperature does not exceed 40°C and PPA is about 10 mm Hg, outside air could be drawn through a water spray and the temperature reduced to that of the prevailing outdoor VWB (23°C). Then PPA would be increased to 18.5 mm Hg and the temperature of the workspace could be reduced to 29°C. The "after" calculation of the heat components shows a negative value for the convective heat component and an E_{req} of only 190 kcal/hr. Since E_{max} exceeds E_{req}, the sweat rate of 0.32 liter/hr could be easily accomplished without bodily strain (104).

5.5.3 Use of Radiant Heat Shielding

When the main component of heat stress is from a radiant source, such as a furnace wall, the first control method should be shielding of the radiant energy. Increased air velocity does little to improve such situations. Table 20.10 represents a "before and after" situation taken from the work of Leihnard, et al. (112). Here the task is skimming dross from molten bars of aluminum. Manipulation of a ladle at a fixed station, with moderate use of shoulder and arm muscles, requires a metabolic rate of about 200 kcal/hr. Air is directed at the worker from an overhead duct at a velocity of 240 m/min. The humidity is quite high, and WBGT is 46.9°C, an extremely stressful situation. Calculation of the heat components confirms this; the worker would have to take numerous work breaks to maintain thermal equilibrium.

Despite clothing and a face shield, the workers were able to perform this task only a few minutes at a time, and heat exhaustion was not uncommon (112). Control of the heat exposure problem was achieved by interposing a finished aluminum sheet between the heat source and the worker. Infrared reflecting glass at face level enabled the workers to see the work. Spaces were left in the aluminum sheet to permit workers to operate the ladle. After shielding was installed, *GT* and *TA* were reduced to 43.3°C

Table 20.9. Use of Evaporative Cooling in a Hot-Dry Environment (104)

Data	Before: Fully Clothed	Before: Semiclothed	After Evaporative Cooling Installed: Fully Clothed
GT (°C)	46.1	46.1	35.0
TA (°C)	42.8	42.8	29.4
VWB (°C)	22.2	22.2	23.3
NWB (°C)	24.2	24.2	24.4
PPA (mm Hg)	10	10	18.5
V (m/min)	110	110	110
WBGT (°C)	30.8	30.8	27.6
MRT (°C)	54.4	54.4	41.7
R (kcal/hr)	130	220	50
C (kcal/hr)	80	130	−60
M (kcal/hr)	200	200	200
E_{req} (kcal/hr)	410	550	190
E_{max} (kcal/hr)	650	1090	480
Sweat rate (liter/hr)	0.68	0.91	0.32

Table 20.10. Use of Radiant Heat Shielding (104)

Data	Before: Clothed	After Shielding Installed: Clothed
GT (°C)	71.7	43.3
TA (°C)	47.8	43.3
VWB (°C)	30.6	29.7
NWB (°C)	36.3	29.8
PPA (mm Hg)	24.5	24.5
V (m/min)	240	240
WBGT (°C)	46.9	33.8
MRT (°C)	159	43.3
R (kcal/hr)	850	60
C (kcal/hr)	210	140
M (kcal/hr)	200	200
E_{req} (kcal/hr)	1260	400
E_{max} (kcal/hr)	630	630
Sweat rate (liters/hr)	2.1[a]	0.65

[a] Dripping.

and WBGT was 33.8°C. However, E_{max} was now greater than E_{req}, and it was possible for the worker to evaporate approximately 0.65 liter of sweat per hour, thereby maintaining thermal balance. The previous sweat rate was an impossible-to-sustain level of 2.1 liters/hr. Although there may be large errors in the estimate of *R* at extremely high globe temperatures, in the case cited the maximum relief that could be expected from shielding was actually achieved (112).

It should be noted that a polished metallic surface will not maintain its shielding effectiveness (reflectivity) if it is allowed to become dirty. This is true even for fabrics coated with very fine metallic particles. A thin layer of grease or oil can change the emissivity of a polished surface from 0.1 to 0.9. To provide shielding that does not interfere with performance of the work task, a curtain of metal chains can be installed; the chains reduce radiation but can be parted as required. Another approach is the use of a mechanically activated door, which is opened only during ejection or manipulation of a product. Also, remote-operated tongs can be used. Taking advantage of the fact that radiant heating from an open portal is limited to the line of sight, partial barriers can also be used effectively (104).

6 MANAGEMENT OF EMPLOYEE HEAT EXPOSURE

As discussed in Section 4.3, control of heat stress in industry requires a multifaceted approach because it involves environmental parameters, human factors, and job factors.

Work practices are now the accepted mechanism of control, and work practices will likely be the basis of any new legal standards. Clearly the definition and implementation of such practices requires the enlightened participation of industrial management to safeguard against excessive exposure to heat.

Accordingly, the control of heat stress in industry should include each of the following elements:

1. Medical supervision.
2. Employee training and education.
3. Acclimatization.
4. Work–rest regimen.
5. Salt and water provisions.
6. Environmental monitoring.
7. Forecast of episodes.

6.1 Medical Supervision

Medical supervision in the management of employee heat exposure includes both the process of selection and the periodic examination of workers. The work practices specified by the NIOSH criteria (10) and the Standards Advisory Committee (18), include a preplacement medical evaluation to determine a worker's fitness for work in heat, with special emphasis on the cardiovascular system, and a review of his medical history with reference to cardiovascular and heat disorders. Periodic medical examinations are recommended for all employees working in conditions of extreme heat exposure, with an annual examination of all heat-exposed workers over the age of 45.

6.1.1 Medical Examination

A preplacement examination for a worker applying for a hot job serves to determine his mental, physical, and emotional qualifications to perform the job assignment with reasonable efficiency and without risk to his own health and safety or to that of fellow employees (113).

The NIOSH Criteria Document suggests that the examining physician should seek to discern possible evidence of intolerance to heat, either occupational or off the job. A history of successful adaptation to heat exposure on previous jobs is perhaps the best basis for predicting the effectiveness of a worker's future performance under heat stress, assuming that levels of work load and heat exposure are equivalent and that no significant alteration has occurred in health status since the time of the previous employment (10).

The NIOSH document also suggests that the medical examination be conducted during the summer. After the age of 45, physical and laboratory examinations should be designed to detect the onset of chronic impairments of the cardiocirculatory and cardiorespiratory systems and to detect metabolic, skin, and renal disease. For all employees on hot jobs undergoing periodic examinations, any history of acute illness or injury, either occupational or nonoccupational, during the interval between examinations should be evaluated carefully. Repeated accidental injuries on the job or frequent absence due to illness should alert the physician to possible heat intolerance or the presence of an aggravating stress such as carbon monoxide in combination with the exposure to heat. Nutritional status should be noted, and advice to correct overweight should be offered (10).

6.1.2 First Aid

The NIOSH Criteria Document recommends that during working hours a person be available who is trained in first aid and in the recognition of the signs and symptoms of any heat disorder (10). The Standards Advisory Committee recommended among its work practices for extreme heat exposure that workers be under the observation of a trained supervisor or fellow worker who can note any early signs of heat effects (18).

Supervisors and selected personnel should be trained to recognize the signs and symptoms of heat disorders and to administer first aid. The most serious emergency is heatstroke, signaled by dry, hot, red or mottled skin, mental confusion, delirium, convulsions or coma, and a high and rising rectal temperature, usually 41°C, but occasionally as low as 40°C (10).

6.2 Employee Education and Training

Exposure to excessively hot environmental conditions can lead to primary heat illnesses, to unsafe acts, or to increased susceptibility to toxic chemicals and physical substances. Through the application of basic health and safety procedures, the individual may reduce the likelihood of ill effects from a hot work environment. Each employee who may be exposed to heat, as well as each supervisor, should receive a safety training program to be made aware of the following points, as a minimum (10, 18):

1. Information concerning heat acclimatization (Section 6.3).
2. Information concerning water intake for replacement purposes (Section 6.5).
3. Information concerning salt replacement (Section 6.5).
4. Instruction on the recognition of the symptoms of heat disorders and illnesses.
5. Instruction on the possible combined effects of heat and one or more of the following: (*a*) alcoholic beverages, (*b*) prescription or nonprescription drugs, (*c*) toxic agents, (*d*) other physical agents.
6. Information concerning the use of appropriate clothing.

6.3 Acclimatization

Acclimatization refers to the adaptive process that results in a decrement in the physiological response produced by the application of a constant environmental stress. Upon initial exposure to a hot environment, there is a noticeable impairment in the ability to work, with physiological strain. However if the exposure is repeated on several successive days, there is a gradual improvement in the ability to work and a corresponding decrease in strain. After 4 to 7 days of the acclimatization process, subjective discomfort almost disappears, body temperature and heart rate are lower, sweat is more profuse and dilute, and there is a substantial improvement in the ability to perform work.

NIOSH recommends that unacclimatized workers be acclimatized over a period of 6 days, beginning with 50 percent of the anticipated total work load and time exposure on the first day. This should be followed by daily 10 percent increments, building up to the total exposure on the sixth day (10). Acclimatized workers who return to work after 9 or more consecutive days of absence should undergo a 4-day acclimatization period. This schedule should again begin with 50 percent of the anticipated total exposure on the first day, followed by daily 20 percent increments (10).

Finally, acclimatized workers who return to work following 4 consecutive days of illness should have medical permission to return to the job and should undergo a 4-day reacclimatization in the same manner as employees who return from leaves of 9 or more days (10).

6.4 Work–Rest Regimen

The Standards Advisory Committee (18) recommended that the duration of work periods, the frequency and length of rest pauses, and the pace and tempo of work should be adjusted to avoid heat strain. The appropriateness and length of rest periods obviously depends on the severity and duration of heat stress. Rest periods should be taken before excessive fatigue develops and should be long enough to reduce the pulse rate to below 100 beats per minute (108).

The NIOSH Criteria Document recommends that a work and rest regimen be implemented to reduce the peaks of physiological strain and to improve recovery during rest periods (10).

Brouha (41) showed that the heart rates of men who rest in an air conditioned room after working in hot environments not only drop much faster but also fall to a lower level than is recorded in hot rest areas. Air conditioned rest areas and low humidity encourage recovery markedly.

6.5 Salt and Water Provisions

It is important that water intake during a work period approximate the amount of sweat produced. Work in hot environments could result in a sweat production of 1 to 3 gallons per day. If this loss is not replaced, dehydration will result. Unfortunately thirst is not an adequate drive to stimulate replenishment. Nevertheless, an ample supply of cool water should be readily available, and workers should be encouraged to drink every 15 to 20 minutes.

The NIOSH Criteria Document recommends that the employer provide a minimum of 8 quarts of cool, potable water with 0.1 percent sodium chloride or 8 quarts of cool, potable water and salt tablets for each worker on each shift (10).

Workers exposed to heat should be encouraged to salt their food abundantly, particularly during hot spells. Large amounts of salt may be lost in the sweat, particularly by the individual not acclimatized to heat; this loss must be replaced daily to prevent illness (heat cramps) due to salt deficiency.

6.6 Environmental Monitoring

Knowledge of the levels of stress is critical in the management of heat stress in industry. In view of our current understanding with respect to environmental indices and recommended standards, management in a hot industry should have, as a minimum, a WBGT profile for the hot jobs and the hot areas throughout a facility. In addition, an engineering assessment of heat stress by measurement of the values of the various environmental components should be made using the concepts and measurements required for the heat stress index.

Furthermore, the use of work practices, regardless of whether sanctioned officially by a federal standard, requires some "action level" based on a measurement or estimate of an environmental heat stress index such as WBGT. The NIOSH Criteria Document (10) suggests that a WBGT profile be established for each workplace for winter and summer seasons to serve as a guide for deciding when work practices should be initiated. After such initial profiles have been established, monitoring should be conducted during the warmest part of each succeeding year.

The Standards Advisory Committee specified in its recommendation (18) that a worker's thermal exposure be based on the hottest 2-hour period of the work shift in which regular work is performed. Accordingly, maximum WBGT values during the hottest period would be used to characterize the temperature level at that workplace. If no single work station accounts for more than half of this 2-hour period, the WBGT value that is compared with the recommended threshold levels (Table 20.2) must be determined as a time-weighted average as described below:

$$WBGT = \frac{1}{120} \sum_{i=1}^{h} (WBGT)_i \times T_i$$

where $WBGT$ = 2-hour, time-weighted wet bulb–globe temperature
$(WBGT)_i$ = $WBGT$ during interval i
i = a discrete exposure interval
T_i = duration of a discrete exposure interval (minutes)

In the case of the Standards Advisory Committee procedure, it is also necessary to estimate the employee's work load during the hottest 2 hours of a work shift to classify it as light, moderate, or heavy. As with WBGT, this work load level can be time weighted (18).

In addition, air speed must be estimated or measured to implement the recommendation of the Standards Advisory Committee because the threshold levels of WBGT are based on a division of air velocity at 300 ft/min (18). It should be noted that air velocity is not required to assess WBGT, which depends on natural wet-bulb temperature and black globe temperature for indoor exposures.

6.7 Forecasting Episodes

In many hot industries, heat stress tends to be a seasonal phenomenon. For other facilities, it may be a significant problem only during the hottest episodes of the summer season. Sudden heat waves early in the summer can create acute heat strain situations (10, 18).

Accordingly, alert management should be interested in any such episodes that are forecast by the National Weather Service or other forecasting agencies. Depending on the severity of the measured or expected levels of heat stress, additional work practices such as the distribution of total work load, scheduling of hottest jobs for the coolest part of the work shift, or other administrative measures may be appropriate. Although many of the hot industries have subjective correlations to relate weather with the intensity of heat stress in the occupational environment, there now exists a systematic modeling technique to relate WBGT at any worksite to external weather conditions, current or predicted (88). However such modeling is more convenient, and certainly more reliable, if WBGT profiles have been developed at the work stations where significant heat stress can be anticipated.

REFERENCES

1. C. P. Yaglou and D. Minard, *Arch. Ind. Health,* **16,** 302–316 (1957).
2. D. Minard, Research Report No. 4, Contract No. MR 005.01-0001.01, Naval Medical Research Institute, Bethesda, Md., 1961.
3. D. Minard and R. L. O'Brien, Research Report No. 7, Contract No. MR 005.01-0001.01, Naval Medical Research Institute, Bethesda, Md., 1964.
4. D. Minard, *Mil. Med.,* **126,** 261–272 (1961).
5. T. Bedford and C. G. Warner, *J. Ind. Hyg. Toxicol.,* **13,** 252 (1931).
6. H. M. Vernon, T. Bedford, and C. G. Warner, Industrial Fatigue Research Board Report No. 62, Her Majesty's Stationery Office, London, 1931.
7. C. P. Yaglou, *J. Ind. Hyg. Toxicol.,* **19,** 12–43 (1937).
8. C. H. Wyndham, *Am. Ind. Hyg. Assoc. J.,* **35,** 113–136 (1974).
9. C. H. Wyndham, *Ergonomics,* **5,** 434–444 (1962).
10. *Criteria for a Recommended Standard . . . Occupational Exposure to Hot Environments,* U.S. Department of Health, Education and Welfare, NIOSH, HSM-72-10269, 1972.
11. B. A. Hertig, "Thermal Standards and Measurement Techniques," in: *The Industrial Environment—Its Evaluation and Control,* U.S. Department of Health, Education and Welfare, NIOSH, 1973.

12. Occupational Safety and Health Act PL91-596, 91st Congress, S.2193, U.S. Department of Labor, Washington, D.C., 1970.
13. C. P. Yaglou, "Thermal Standards in Industry," in: *Report of the Committee on Atmospheric Comfort,* Year Book Publishers, Chicago, 1947.
14. American Conference of Governmental Industrial Hygienists, *Threshold Limit Values for Chemical Substances and Physical Agents in the Workroom Environment with Intended Changes for 1971,* ACGIH, Cincinnati, Ohio, 1971.
15. World Health Organization, *Health Factors Involved in Working Under Conditions of Heat Stress,* WHO, Geneva, 1969.
16. American Conference of Governmental Industrial Hygienists, *Threshold Limit Values for Chemical Substances and Physical Agents in the Workroom Environment with Intended Changes for 1975,* ACGIH, Cincinnati, Ohio, 1975.
17. B. A. Hertig, "Work in Hot Environments: Threshold Limit Values and Proposed Standards," in: *Industrial Environmental Health—The Worker and the Community,* Academic Press, New York, 1975.
18. "Recommendations for a Standard for Work in Hot Environments," Draft No. 5, Standards Advisory Committee on Heat Stress, Department of Labor, Washington, D.C., 1974.
19. C. P. Yaglou, *J. Ind. Hyg.,* **9,** 297 (1927).
20. C. P. Yaglou, *Am. J. Pub. Health,* **40,** 131 (1950).
21. C. S. Leithead and A. R. Lind, *Heat Stress and Heat Disorders,* F. A. Davis, Philadelphia, 1964.
22. C. S. Leithead, *Bull. WHO,* **38,** 649–657 (1968).
23. A. R. Lind, *J. Appl. Physiol.,* **18,** 51–56 (1963).
24. A. R. Lind, *J. Appl. Physiol.,* **18,** 57–60 (1963).
25. A. R. Lind, *J. Appl. Physiol.,* **28,** 50 (1970).
26. D. Minard, H. S. Belding, and J. R. Kingston, *J. Am. Med. Assoc.,* **165,** 1813–1818 (1967).
27. D. Minard, Naval Medical Research Institute, Research Report No. 6, Bethesda, Md., 1964.
28. T. F. Hatch, *Heat. Pip. Air Cond.,* **23,** 140 (1951).
29. G. F. Haines, Jr. and T. F. Hatch, *Heat. Vent.,* 1952.
30. H. S. Belding and T. F. Hatch, *Heat. Pip. Air Cond.,* **27,** 129–135 (1955).
31. B. A. Hertig, D. W. Badger, P. J. Schmitz, and L. D. Siler, *Am. Ind. Hyg. Assoc. J.,* **32,** 4 (1971).
32. H. S. Belding, B. A. Hertig, and M. L. Reidesel, *Am. Ind. Hyg. Assoc. J.,* **21,** 25 (1960).
33. C. H. Wyndham, N. B. Strydom, H. M. Cooke, and J. S. Maritz, "Studies on the Effects of Heat on Performance of Work," Applied Physiology Laboratory Reports, Transvaal and Orange Free State Chamber of Mines, South Africa, 1959.
34. C. H. Wyndham, *Ergonomics,* **5,** 115 (1962).
35. J. D. Walters, *Ann. Occup. Hyg.,* **17,** 255–264 (1975).
36. D. Minard, "Physiology of Heat Stress," in: *The Industrial Environment—Its Evaluation and Control,* U.S. Department of Health, Education and Welfare, NIOSH, 1973.
37. W. Machle and T. F. Hatch, *Physiol. Rev.,* **27,** 200–227 (1947).
38. T. F. Hatch, "Assessment of Heat Stress," in: *Temperature: Its Measurement and Control in Science and Industry,* Vol. 3, Part 3, J. D. Hardy, Ed., Reinhold, New York, 1963, p. 307.
39. R. W. Powell, *Trans. Inst. Chem. Eng.* (London), **18,** 36 (1940). Quoted by N. Nelson, L. W. Eichna, S. M. Horvath, W. B. Shelley, and T. F. Hatch, *Am. J. Physiol.,* **151,** 626 (1947).
40. H. S. Belding, "Work in Hot Environments," in: *Industrial Hygiene Highlights,* Vol. 1, Industrial Hygiene Foundation of America, Pittsburgh, 1968.
41. L. A. Brouha, *Physiology in Industry—Evaluation of Industrial Stresses by the Physiological Reactions of the Worker,* 2nd ed., Pergamon Press, Elmsford, N.Y., 1967.

42. S. Robinson, "Physiological Adjustments to Heat," in: *Physiology of Heat Regulation and the Science of Clothing,* Saunders, Philadelphia, 1949.
43. L. W. Eichna, W. F. Ashe, W. B. Bean, and W. B. Shelley, *J. Ind. Hyg. Toxicol.,* **27,** 59 (1945).
44. L. W. Eichna, C. R. Park, N. Nelson, S. Horvath, and E. D. Palmes, *Am. J. Physiol.,* **5,** 299 (1952).
45. C. H. Wyndham, W. M. Bouwer, M. G. Devine, H. E. Paterson, and D. K. C. MacDonald, *J. Appl. Physiol.,* **5,** 299 (1952).
46. E. Kamon and H. S. Belding, *J. Appl. Physiol.,* **31,** 472 (1971).
47. A. R. Lind, *J. Appl. Physiol.,* **28,** 57 (1970).
48. H. S. Belding, "Resistance to Heat in Man and Other Homeothermic Animals," in: *Thermobiology,* A. H. Rose, Ed., Academic Press, London, 1967.
49. N. B. Strydom, C. H. Wyndham, H. M. Cooke, J. S. Maritz, G. A. G. Bredell, J. F. Morrison, J. Peter, and C. G. Williams, *Fed. Proc.,* **22,** 893 (1963).
50. C. W. Suggs and W. E. Splinter, *J. Appl. Physiol.,* **16,** 413 (1961).
51. R. D. Pepler, "Performance and Well-Being in Heat," in: *Temperature—Its Measurement and Control in Science and Industry,* Vol. 3, Part 3, J. D. Hardy, Ed., Reinhold, New York, 1963.
52. R. D. Pepler, "Psychological Effects of Heat," in: *Heat Stress and Heat Disorders,* F. A. Davis, Philadelphia, 1964.
53. B. C. Duggar, M. S. thesis, University of Pittsburgh, "Trials of an Assembly Job for Evaluating the Effects of Heat on Human Performance," 1956.
54. R. D. Pepler, "Variations in Students' Test Performance and in Classroom Temperatures in Climate Controlled and Non-Climate Controlled Schools," *ASHRAE Transactions,* Part II, 1971, pp. 35–42.
55. J. B. Moses, *Ind. Med. Surg.,* **22,** 20 (1960).
56. W. F. Grether, C. S. Harris, G. C. Mohr, C. W. Nixon, M. Ohlbaum, H. C. Sommer, V. H. Thaler, and J. H. Veghte, *Aerosp. Med.,* **42,** 1092–1097 (1971).
57. R. D. Dean, C. L. McGlothlen, and J. L. Monroe, *Effects of Combined Heat and Noise on Human Performance, Physiology, and Subjective Estimates of Comfort and Performance,* Boeing Company, Seattle, 1964.
58. F. Renshaw, *Am. Ind. Hyg. Assoc. J.,* Abstr. Suppl., **38,** 1971.
59. A. F. Henschel, "Heat Stress," in: *The Industrial Environment—Its Evaluation and Control,* 2nd ed., C. H. Powell and A. D. Hosey, Eds., Public Health Service Publication No. 614, Government Printing Office, Washington, D.C., 1965.
60. H. S. Belding and T. F. Hatch, *Fed. Proc.,* **22,** 881 (1963).
61. A. Henschel, L. L. Burton, L. Margolies, and J. E. Smith, *Am. J. Pub. Health,* **59,** 2232 (1969).
62. J. Gustin, M. S. thesis, University of Pittsburgh, "Disease Specific Mortality Patterns for Open Hearth Workers," 1971.
63. F. C. Houghton and C. P. Yaglou, *J. Am. Soc. Heat. Vent. Eng.,* **29,** 165–176 (1923).
64. R. S. Brief and R. G. Confer, *Am. Ind. Hyg. Assoc. J.,* **32,** 11–16 (1971).
65. T. Bedford, Medical Research Council War Memo No. 17, London, 1946.
66. B. McArdle, W. Dunham, H. E. Holling, W. S. S. Ladell, J. W. Scott, M. L. Thomson, and J. S. Weiner, Medical Research Council R. N. P. Report 47, London, 1947.
67. J. S. McKarns and R. S. Brief, *Heat. Pip. Air Cond.,* **38,** 113 (1966).
68. N. L. Ramanathan and H. S. Belding, *Am. Ind. Hyg. Assoc. J.,* **34,** 375–383 (1973).
69. World Meteorological Organization, *Guide to Meteorological Instrument and Observing Practices,* WMO, Geneva, 1971.
70. T. Bedford, *Basic Principles of Ventilation and Heating,* K. J. Lewis & Co., London, 1948.

71. T. Bedford and C. G. Warner, *J. Hyg.,* **34,** 458 (1934).
72. H. M. Vernon, *J. Physiol.,* **70,** 15 (1970).
73. C. P. Yaglou, *J. Ind. Hyg.,* September 1935.
74. L. Hill, Medical Research Council Special Report No. 32, Part I, London, 1919.
75. A. F. Dufton, *Physiol. Mag.,* **9,** 858 (1930).
76. C. E. A. Winslow and L. Greenburg, *Am. Soc. Heat. Vent. Eng. Trans.,* **41,** 149 (1935).
77. J. H. Botsford, *Am. Ind. Hyg. Assoc. J.,* **32,** 1 (1971).
78. T. F. Hatch, *Am. Ind. Hyg. Assoc. J.,* **34,** 66–72 (1973).
79. J. E. Mutchler and J. L. Vecchio, "Empirical Relationships Among Heat Stress Indices in 14 Hot Industries," *Am. Ind. Hyg. Assoc. J.,* **38,** 253–263 (1977).
80. J. D. Walters, *Br. J. Ind. Med.,* **25,** 235–240 (1968).
81. N. B. G. Taylor, L. A. Kuehn, and M. R. Howat, *Can. J. Physiol. Pharmacol.,* **47:** 3 (1969).
82. L. A. Huehn, DRET Report No. 692, Defense Research Establishment, Toronto, Defense Research Board, Canada.
83. "'Light' Miniature Wet Bulb Globe Thermometer Index Meter," Catalog of Light Laboratories, 10 Ship Street Gardens, Brighton 1, Sussex, England.
84. L. A. Kuehn, R. A. Stubbs, and L. E. MacHattie, *J. Phys. E: Sci. Instrum.,* **4,** 1971.
85. V. Basus, "A Temperature Parameter Useful in Evaluation of Summer Microclimatic Conditions in Spacious Manufacturing Shops and Smaller Individual Working Places," STS Translation, from *Zdrav. Tech. Vzduchotech.* (Prague), **10:** 1, 29–36 (1967).
86. A. G. Loudon, *J. Inst. Heat. Vent. Eng.,* London, **37,** 280–292 (1970).
87. G. V. Parmelee, *Bull. Am. Meteorol. Soc.,* **36,** 256–264 (1955).
88. J. E. Mutchler, D. D. Malzahn, J. L. Vecchio, and R. D. Soule, *Am. Ind. Hyg. Assoc. J.,* **37,** 151–164 (1976).
89. A. P. Gagge, L. P. Herrington, and C.-E. A. Winslow, *Am. J. Hyg.,* **26,** 84 (1937).
90. C.-E. A. Winslow and L. P. Herrington, *Temperature and Human Life,* Princeton University Press, Princeton, N.J., 1949.
91. H. M. Vernon, *J. Ind. Hyg. Toxicol.,* **19,** 498 (1937).
92. J. E. Peterson, *Am. Ind. Hyg. Assoc. J.,* **31,** 305–317 (1970).
93. H. C. Bazett and J. B. S. Haldane, *J. Physiol.,* **4,** 252 (1921).
94. D. B. Dill, *Life, Heat, and Altitude,* Harvard University Press, Cambridge, Mass., 1938.
95. C. K. Drinker, *J. Ind. Hyg. Toxicol.,* **18,** 471 (1936).
96. J. H. Talbott, *Medicine* (Baltimore), **14,** 323 (1935).
97. H. M. Vernon, T. Bedford, and C. G. Warner, Industrial Fatigue Research Board Report No. 51, Her Majesty's Stationery Office, London, 1928.
98. H. M. Vernon and E. A. Rusher, Industrial Fatigue Research Board Report No. 5, Her Majesty's Stationery Office, London, 1920.
99. R. H. Britten and L. R. Thompson, U.S. Public Health Service Bulletin No. 163, 1926.
100. C. R. Bell, in: *Hot Environments and Performance in the Effects of Abnormal Physical Conditions at Work,* E. M. Davis, P. R. Davis, and F. H. Tyrer, Eds., Livingston, London, 1967.
101. F. H. Fuller and L. Brouha, *ASHRAE J.,* **39,** 1966.
102. A. R. Lind and F. D. K. Liddell, "The Influence of Individual Variation on the Prescriptive Zone of Climates," National Coal Board Memo, London, 1963.
103. A. R. Lind, P. W. Humphreys, K. Collins, and K. Foster, "The Influence of Aging on the Prescriptive Zone of Climates," National Coal Board Memo, London, 1963.

104. R. S. Brief and R. G. Confer, *Med. Bull.,* **33,** 229–253 (1973).

105. *Am. Ind. Hyg. Assoc. J.,* **32,** 560 (1971).

106. L. Brouha, *Physiology in Industry,* 2nd ed., Pergamon Press, New York, 1967.

107. *Physiology of Heat Regulation and the Science of Clothing,* L. G. Newburgh, Ed., Hafner, New York, 1968.

108. American Industrial Hygiene Association, *Heating and Cooling for Man in Industry,* 2nd ed., AIHA, Akron, Ohio, 1975.

109. B. A. Hertig, "Control of the Thermal Environment," in: *Ergonomics and Physical Environmental Factors,* International Labor Office, Geneva, Occupational Safety and Health Series No. 21, 1970.

110. T. F. Hatch, "Heat Control in the Hot Industries," in: *Industrial Hygiene and Toxicology,* Vol. 1, F. A. Patty, Ed., Wiley-Interscience, New York, 1958.

111. K. E. Robinson, "Air Conditioning," in: *Industrial Hygiene and Toxicology,* Vol. 1, F. A. Patty, Ed., Wiley-Interscience, New York, 1958.

112. W. F. Leihnard, R. S. McClintock, and J. F. Hughes, "Appraisal of Heat Exposures in an Aluminum Plant," Paper No. 8-3, presented at the 13th International Congress on Occupational Health, New York, 1960.

113. *Scope, Objectives, and Functions of Occupational Health Programs,* American Medical Association Council on Occupational Health, Chicago, revised December 1971.

CHAPTER TWENTY-ONE

Respiratory Protective Devices

DARREL D. DOUGLAS

1 INTRODUCTION

In the field of health and safety, a respirator is a device that protects the wearer from toxic materials in the atmosphere. It must cover the nose and mouth and either filter out the toxic material or provide air of acceptable quality to the wearer. The respirator can vary in size from one that shields only the nose and mouth to one that covers the entire body.

The use of respirators has been known in history as early as Roman times (1). Scattered mention of them occurs in reports of industrial processes during the Middle Ages. Until the nineteenth century, all devices appear to be air-purifying devices intended to prevent the inhalation of aerosols. They were varied in design, ranging from animal bladders or rags wrapped around the nose and mouth to full face masks made of glass with air inlets covered by particulate filters. During the 1800s masks were produced that combined aerosol filters and vapor sorbents. These advances were made primarily for fire fighters (2).

When chemical warfare agents were introduced during World War I, attention was focused on the need for adequate masks to protect people from toxic gases and aerosols. In the United States this work was successfully carried out for the army under the direction of the Bureau of Mines. After the war, misuse of surplus army gas masks by civilians highlighted the need for respiratory protection standards. Again the Bureau of Mines was given the responsibility of setting these standards. Out of this effort grew the federal respirator approval system that is still in effect. Many of the respirator developments made during the decade following 1920 remain in use today (3).

2 RESPIRATOR CLASSIFICATION

In general, there are two types of respirators—air purifying and air supplying, (see Figure 21.1). Each type may have tight-fitting facepieces and loose-fitting hoods that cover the head and may cover the body. An important aspect of respirator operation and classification is the pressure within the facepiece. The facepiece pressure may be above or below the outside air pressure. If facepiece pressure is lower than the outside air pressure, it is classified as negative; if above, it is positive. The concept of negative and positive pressure operation is extremely important when considering potential contaminant leakage into the respirator. A brief description of facepieces and hoods is given below; and, the detailed description of respirator function is given in later sections.

2.1 Tight-Fitting Respirators

The tight-fitting facepiece usually is molded rubber or plastic that is intended to adhere tightly to the skin of the wearer. Customarily, it is available in three categories: quarter mask, half mask, and full face mask. The quarter mask covers the mouth and nose; the half mask covers the mouth, nose, and chin; and the full face mask covers the entire face from chin to hairline and from ear to ear (see Figure 21.2). In use, a variety of air-purifying or supplied air units may be attached to the facepiece (Figure 21.3).

In recent years a new type of respirator has been introduced called a single-use or disposable respirator. In contrast to the facepieces just above, it contains no replaceable parts (Figure 21.4). When the filter or sorbant is no longer usable, the entire respirator is discarded and a new one obtained. Maintenance problems are reduced with such devices. These respirators may be categorized as quarter, half, or full face masks, depending on their design. If properly selected for an intended application, there is no difference in their function from respirators designated as reusable.

One device that does not fit into the foregoing categories is the mouthpiece respirator. A mouthpiece is held in the wearer's mouth, and a clamp is placed over his nostrils. The lips are pursed tightly around the mouthpiece, and all air comes through the filtering device attached to the mouthpiece. This device is designed to be carried with the person and used for emergency escape (Figure 21.5).

2.1.1 Respirator Suspension

An important part of the facepiece is its suspension, or attachment to the head. Full face masks have a head harness attached to the facepiece at five or six points. The large sealing surface of the full face mask and the distribution of the head band attachment assists in maintaining a stable facepiece with little slippage. Half and quarter masks may be secured by one strap attached to each side of the facepiece, known as two-point suspension, or two straps may be used, attached to the facepiece at two points on each side of the facepieces; this is known as four-point suspension (Figure 21.2). The respirators in Figure 21.2*a* to 21.2*d* use two-point suspension and those in Figure 21.2*e* and 21.2*f* use

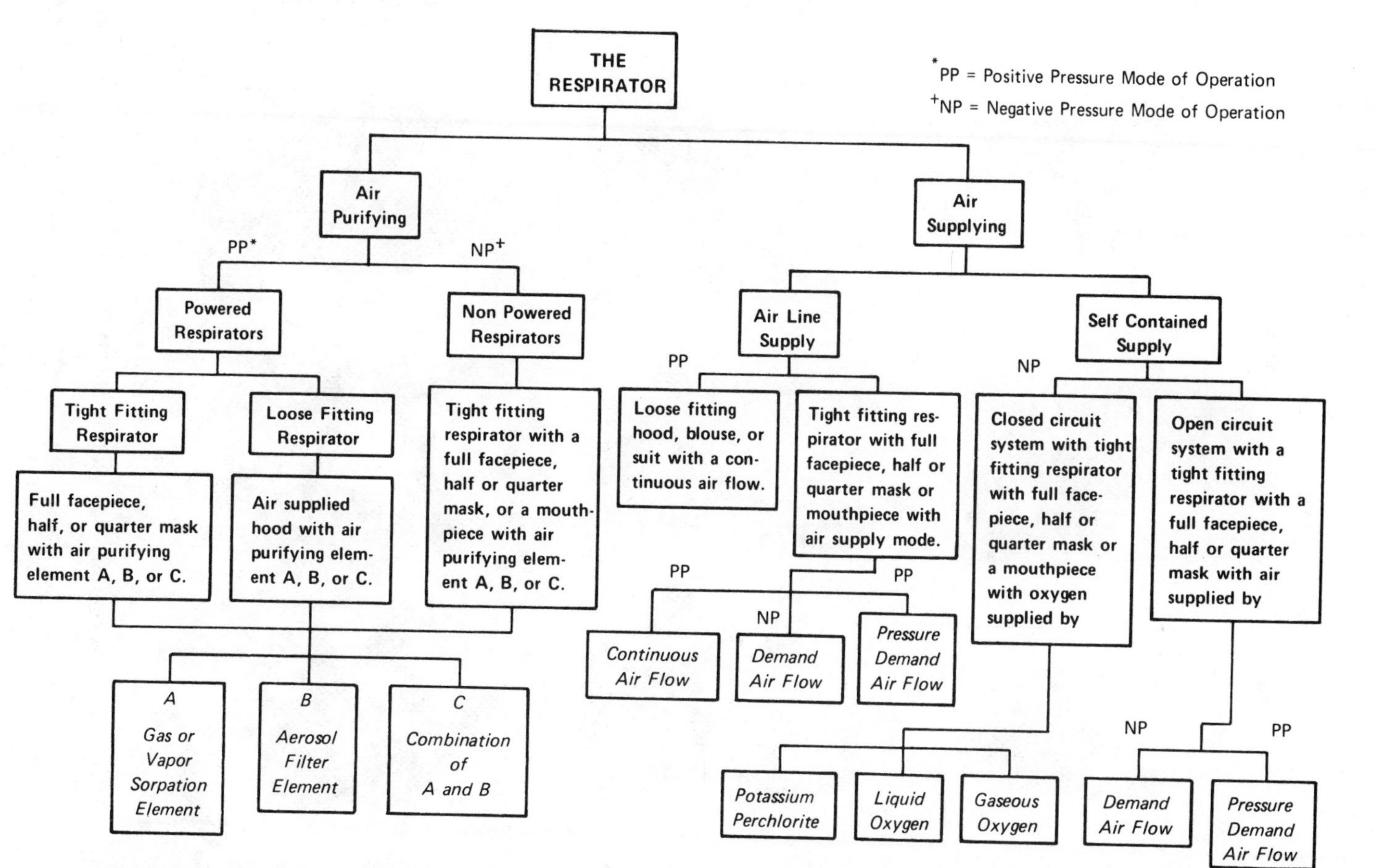

Figure 21.1 Classification of respirators: PP, positive pressure mode of operation; NP, negative pressure mode of operation.

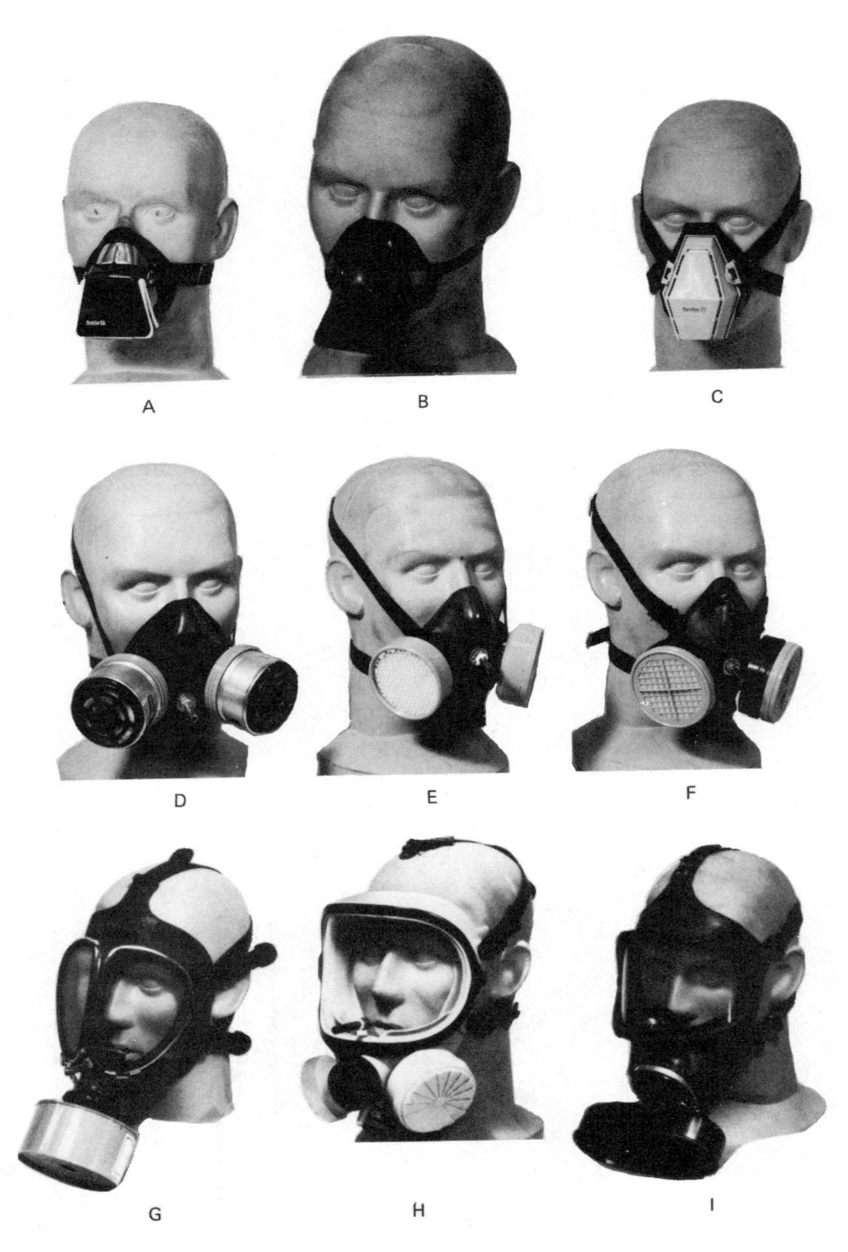

Figure 21.2 (*a*)–(*c*) Quarter masks. (*d*)–(*f*) Half masks. (*g*)–(*i*) Full face masks.

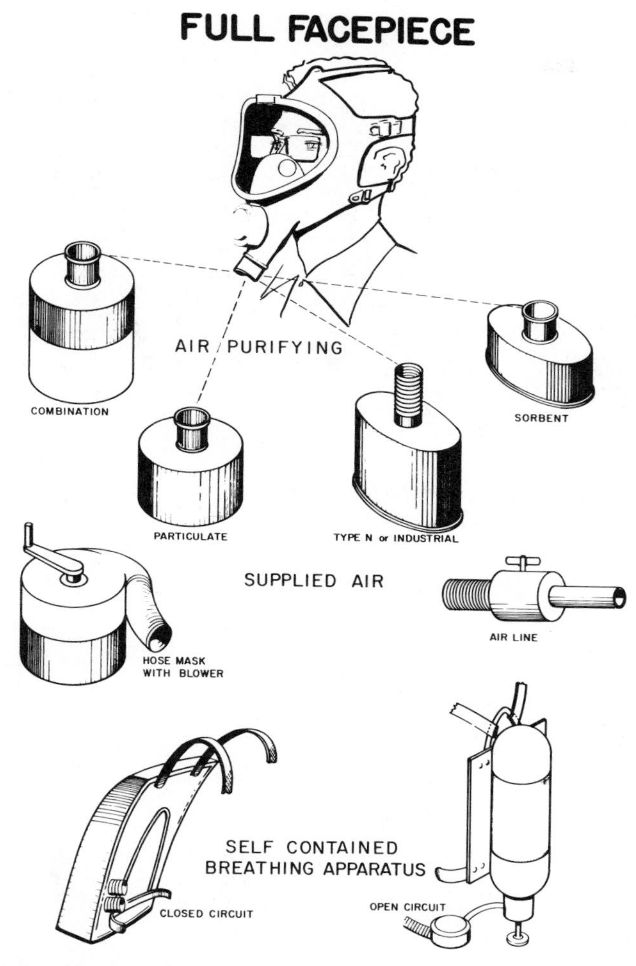

Figure 21.3 A facepiece may be used with varied air-purifying and air-supplied equipment.

four-point suspension. The latter provides greater stability, thus assisting in providing a better facepiece seal. In general, the full face mask is considered useful in concentrations of toxic materials higher than can be overcome by the quarter or half masks.

2.1.2 Loose-Fitting Respirators

The best-known loose-fitting respirator is the sandblaster's hood, which covers the head, neck, and upper torso (Figure 21.6). The wearer is provided breathing air by a hose leading into the hood. Since the hood is not tight fitting, it is important that sufficient air be provided to ensure that there is always an outward flow of air. This prevents

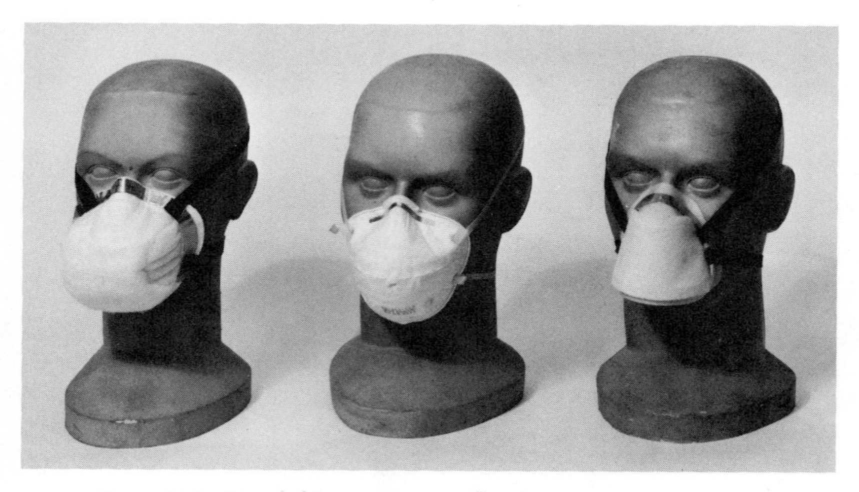

Figure 21.4 Disposable respirators, no replaceable parts required.

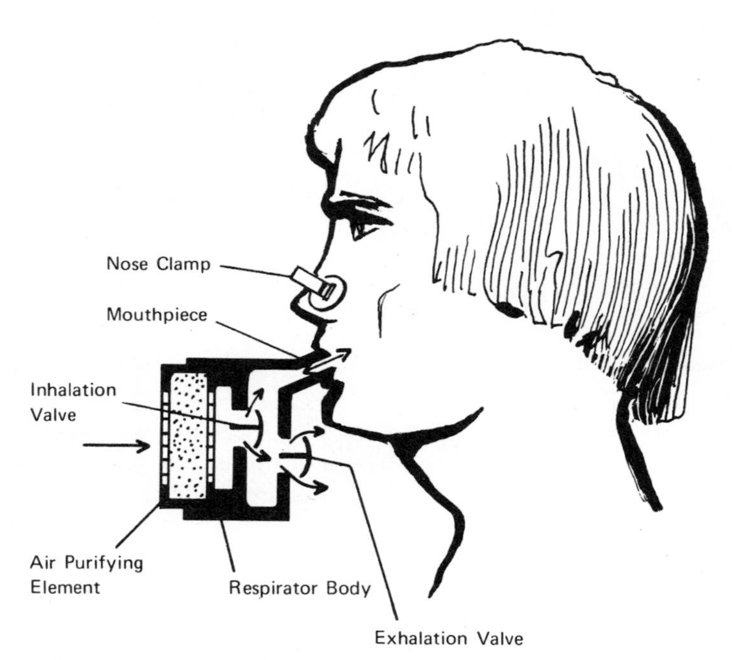

Figure 21.5 Mouthpiece respirator, used for escape under emergency conditions.

contaminants from entering the interior of the hood. Variations of this device are found in the air-supplied blouse and the air-supplied suit. Air-supplied suits can be obtained in one- and two-piece styles (Figure 21.7). In both cases air is supplied to the top and bottom areas of the suit. Exhalations ports may be provided in the one-piece suit to vent the exhaust air.

It was pointed out in connection with tight-fitting facepieces that the degree of security tends to increase with the size of the facepiece. This is not necessarily true with loose-fitting respirators. The outside atmosphere may enter by aspiration when the user makes movements that increase the volume of the hood or suit. Sufficient air is not always supplied to prevent aspiration from occurring.

Figure 21.6 Air-supplied hood: front (*a*) and side (*b*). A protective screen is placed over the window for sand blasting.

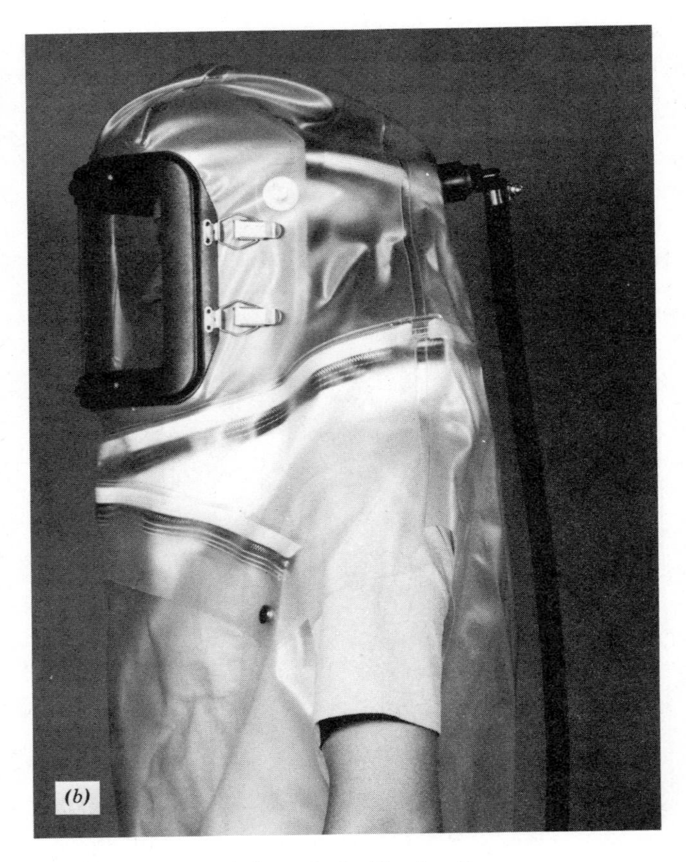

Figure 21.6 (Continued)

3 NEGATIVE PRESSURE RESPIRATORS

A negative pressure respirator must have a tight-fitting facepiece. Air-purifying respirators are the most common negative pressure devices, but some air-supplied respirators also operate in the negative pressure mode.

3.1 Air-Purifying Respirators

The facepieces used for negative pressure air-purifying devices have been described in preceding sections. They can be quarter, half, full face, or mouthpiece respirators.

3.1.1 Respirator Facepiece Valves

The respirator inhalation and exhalation valves were not mentioned in the foregoing facepiece description. With the exception of some disposable respirators, all facepieces use inhalation and exhalation valves.

Inhalation Valves. Inhalation valves are normally not considered essential for respirators (Figure 21.8). By preventing exhaled air from coming in contact with sorbants and filters, they protect sorbants and aerosol filters from possible detrimental effects. Exhaled air tends to increase breathing resistance of the filters and to cause them to be changed more frequently. In the case of electrostatic filtering media, it is essential

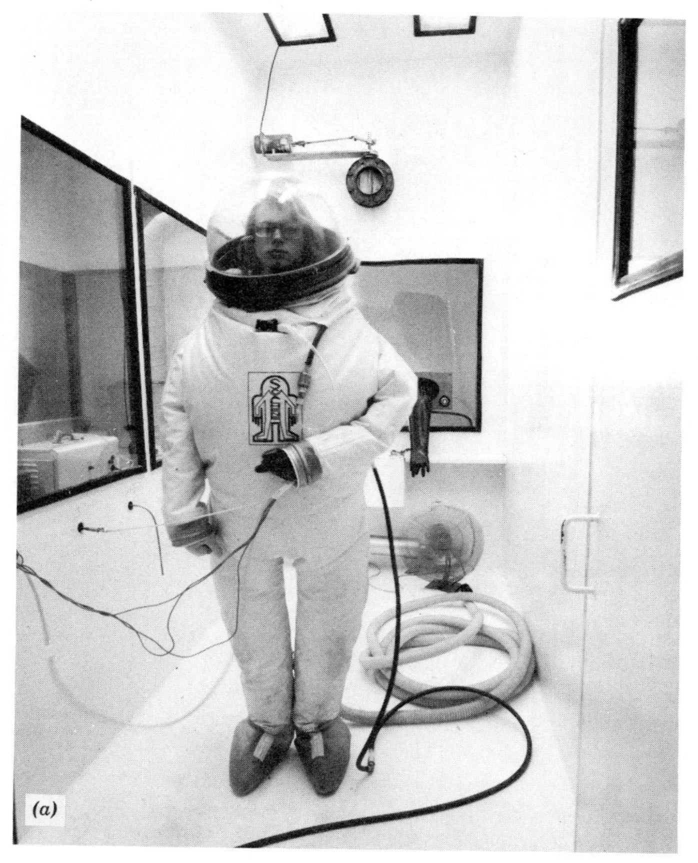

Figure 21.7 (*a*) One-piece reusable air-supplied unit. (*b*) Two-piece disposable or reusable air-supplied unit.

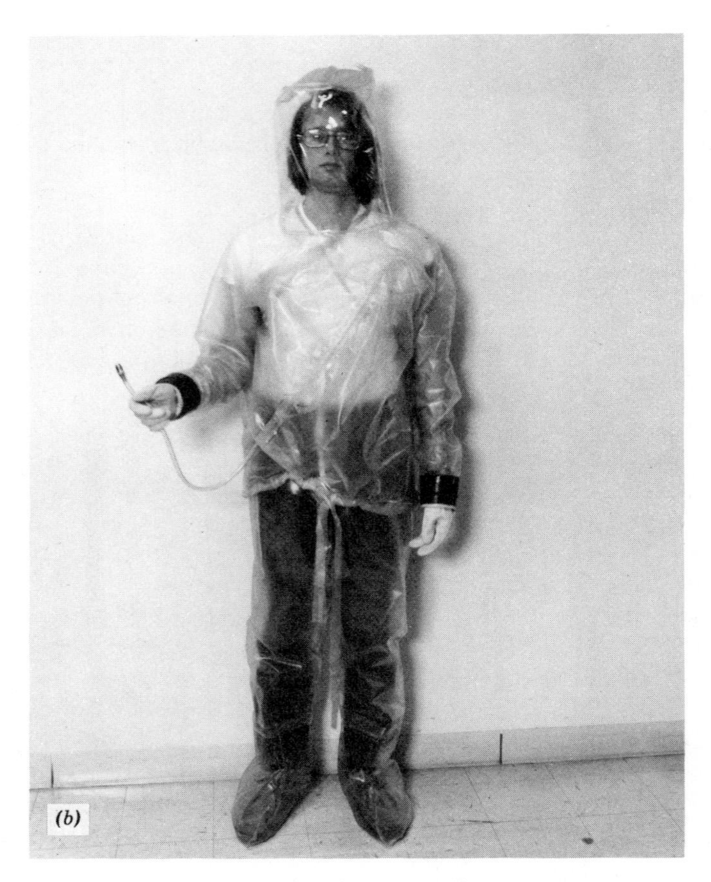

Figure 21.7 (Continued)

that the moisture-laden air be prevented from contacting the filter, since moisture destroys the efficiency of this type of filter (see Section "Dust and Mist Filters").

Exhalation Valves. An exhalation valve provides an exit for exhaled air (Figure 21.9). With the exception of some disposable respirators, NIOSH regulations require the provision of exhalation valves. This important part of a mask, which is often neglected, should be frequently checked to ensure that it is working. A malfunctioning exhalation valve can render useless both positive and negative pressure respirators. If an obstruction such as a hair or a dirt particle comes between the valve and the valve seat, a serious leak can occur. A properly fitting respirator coupled with an efficient filter is of no avail when the toxic material may enter through a defective valve.

Exhalation valves must also have a valve cover. This serves the dual purpose of protecting the valve from physical damage and retaining a small reservoir of exhaled air. A

Figure 21.8 Inhalation valves.

Figure 21.9 Exhalation valve and cover.

finite time elapses between the beginning of inhalation and the closure of the exhalation valve. During this period, the clean, recently exhaled air prevents toxic material from entering the valve. Two types of exhalation valve are illustrated in Figure 21.10.

3.1.2 Removal of Gases, Vapors, and Aerosols

With the exception of single-use respirators, the air-purifying elements used with respirators can be removed from the facepiece and replaced. Air-purifying units can be divided into two major classes: aerosol removing, and vapor and gas removing. In the former case the removal process depends primarily on the size of the aerosol, regardless of the aerosol composition. In the latter case the vapor or gas is sorbed onto a chemical, which may be selective in the material sorbed. The sorbant must be carefully selected to ensure compatibility with the material for which protection must be provided.

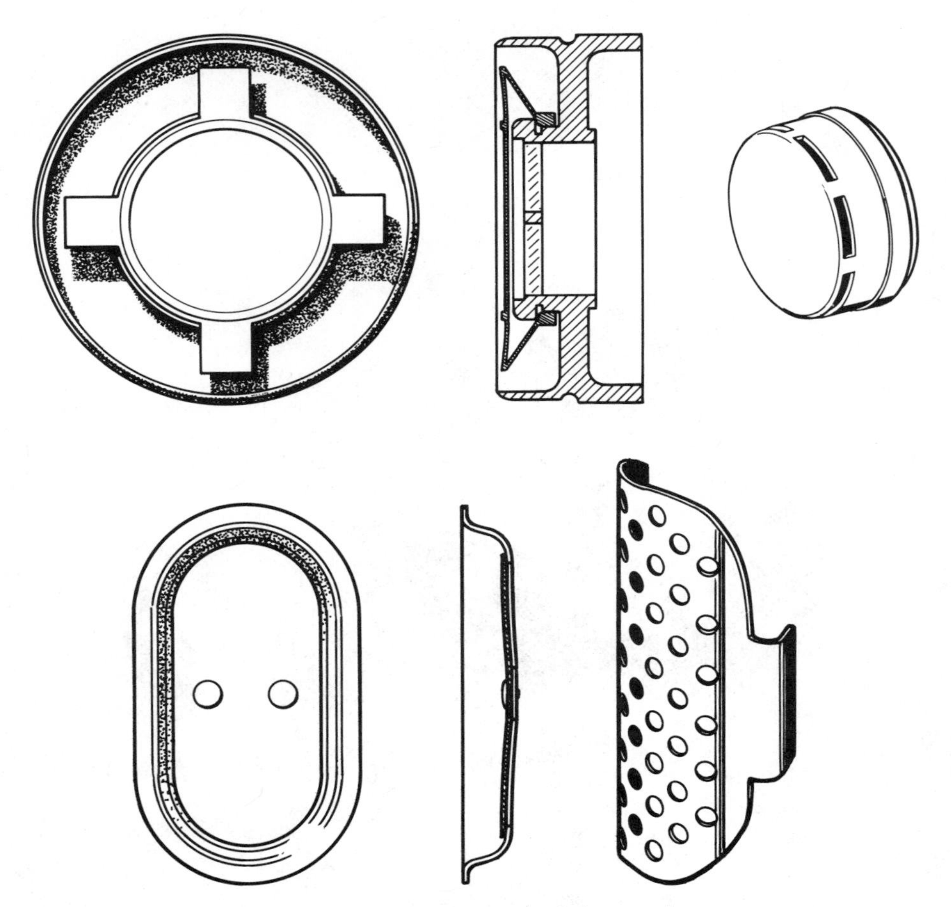

Figure 21.10 Two types of exhalation valve.

Aerosol Capture. The removal of particulates from the breathing air is accomplished using a variety of filtration mechanisms. All these mechanisms require that the air pass through a fibrous media. The efficiency of aerosol removal by the various mechanisms of impaction, interception, and diffusion is related to the size of the aerosol and the filter fiber medium used. The efficiency of all mechanisms is improved as the filter fiber diameter is decreased. The resistance to air flow usually increases as the fiber diameter decreases, which may be a limiting factor because the respiratory system provides the force moving the air through the filter. Breathing resistance in air-purifying respirators usually does not exceed 85 mm of water pressure. The efficiency of filters tends to improve with use as the particulates lodge on the filter surface. This also increases breathing resistance.

In addition to the mechanical methods of entrapment such as impaction, additives to the filter medium can increase its efficiency. The most common additive is a resin with a high dielectric constant. After the material is added to the filter medium, a high electrostatic charge is imparted to the resin at the factory. Properly stored, this charge will remain unchanged for a period of years. The advantage of the electrostatic filter is that the charge on the filter assists in aerosol collection, which reduces the importance of particle size and filter fiber diameter. The fiber diameter of the electrostatic filter can be larger without decreasing filter efficiency, and this results in lower resistance and less filter area.

Presently filters are produced in four classes: dust, dust and mist, fume, and high efficiency.

Dust and Mist Filters. Figure 21.11 presents some dust and mist filters. The large filters are a fiberglass mat measuring 1/2 to 1 inch in thickness. It has low resistance and good efficiency to dust particles above 1 μm. The most common dust filter in use is the electrostatic felt filter. Wool or synthetic fiber is felted into a thin mat less than 1/4 inch thick and impregnated with a high dielectric constant resin.

A problem with electrostatic felt filters is that exposure to high heat or high humidity causes the charge to dissipate, thus reducing filter efficiency (Figure 21.12 and Table 21.1). It is important that these filters be stored to protect them from high heat and humidity.

A mist is a liquid aerosol in the same particle size range as dust particles. For filtration purposes they can be considered to be solid particles. With means other than electrostatic filters, they are captured with the same efficiency as solid particles. Mist aerosols tend to dissipate the charge of electrostatic filters, thus reducing the efficiency (Figure 21.13). It is important that such filters not be used in the presence of mists.

Fume Filters. The fume filter must be 99 percent efficient when tested with lead fumes (5). These devices can be used for protection from metal fumes such as those generated by welding. The high efficiency filter is now selected for many applications that formerly used fume filters. In use, the fume filter resistance does not increase as rapidly as the high efficiency filter. Examples of fume filters appear in Figure 21.14.

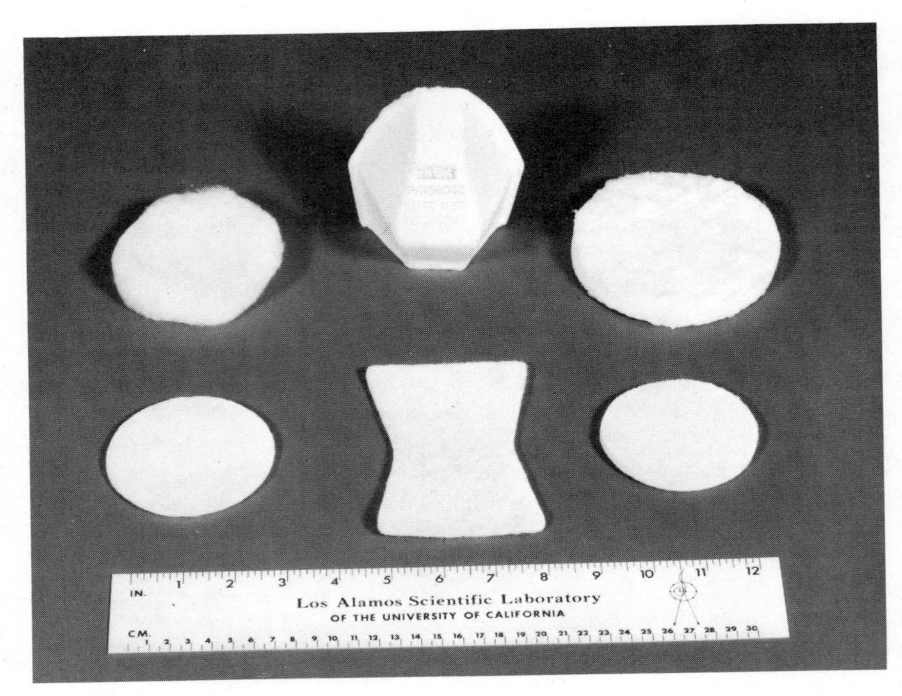

Figure 21.11 Dust and mist filters. (*Back row*) Fiberglass filters. (*Front row*) Electrostatic filters.

High Efficiency Filters. These filters are thin sheets of filter material with small fiber diameter and high resistance to flow per unit area. The high efficiency filter must be 99.97 percent efficient against dioctyl phthalate aerosol, with an MMAD of 0.3 μm (6). The high efficiency filter was originally designed for use in atmospheres containing radioactive particulates; however its efficiency has made it popular for use with all highly toxic particulates.

Table 21.1 Effects of Exposure to Elevated Temperatures on Aerosol Penetration[a,b] (4)

	Incident Aerosol Mass Passing Filter (%) After 4 Hours Exposure to Indicated Temperature				
Test flowrate (liters/min)	22°C (72°F)	50°C (122°F)	88°C (190°F)	116°C (240°F)	143°C (290°F)
16	5	5	30	31	37
32	10	12	45	48	51
38	12	13	49	51	51
77	22	24	56	58	61

[a] Filter Material: Resin impregnated, needles felt, half wool, half acrylic.
[b] Aerosol: Polydisperse NaCl, MMAD = 0.6μ, $\sigma g = 2$, concentration = 15 mg/m^3.

As mentioned previously, filter use presents problems associated with breathing resistance. It is necessary to decrease the airflow per unit area until the resistance is within an acceptable range for the respiratory system. The low resistance characteristics of the electrostatic felt filters make it possible to have filters with a small filtering surface, which results in a smaller respirator. The high efficiency filter must have a larger surface area than the dust filter. To provide this large surface area, yet keep the filter container size small enough to be acceptable to the consumer is the challenge. This has led to some ingenious high efficiency filter holders that fold the filters in complex patterns to expose a large surface area for use (Figure 21.15).

Gas and Vapor Sorption. Gases and vapors are captured by passing them through a particulate bed of chemicals where they are adsorbed or absorbed. When absorptive

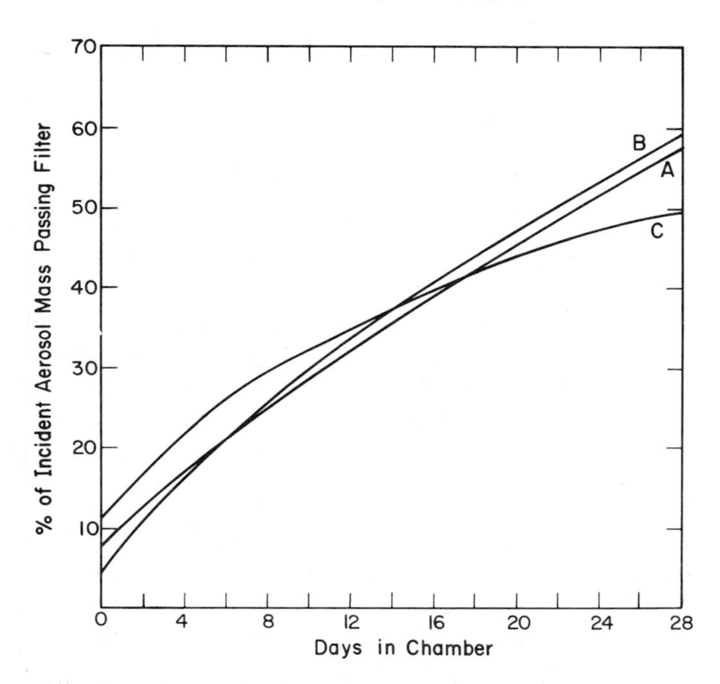

Figure 21.12 Effects on electrostatic filters of storage at 32°C and 90 percent relative humidity.

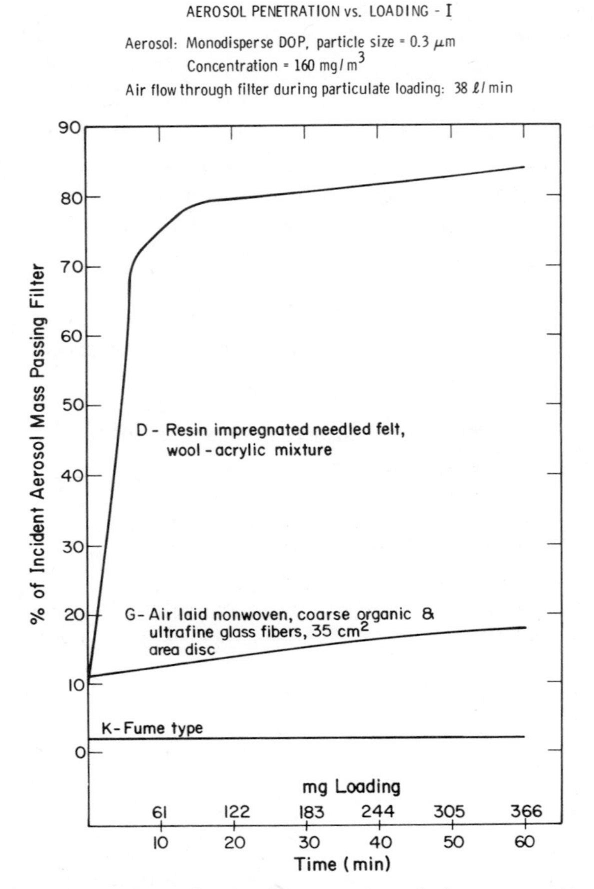

Figure 21.13 Liquid aerosol penetration of electrostatic filters.

materials are used, gases passing through the particulate bed are absorbed on the surface and then penetrate throughout the particle. Absorption materials are not widely used. The majority of gases and vapors are trapped by adsorption. Adsorptive materials are porous and have extremely large surface area per unit weight—up to 1500 m^2/g. As the gas passes through the particulate bed, the molecules are adsorbed on the surface and remain there.

The adsorbent most widely used is activated carbon or charcoal. Alone it is an excellent adsorbent of many organic vapors. Impregnating it with specific materials increases its retention efficiency for some gases and vapors. For example, impregnated with iodine it will capture mercury vapor; impregnated with certain metal oxides, it will capture acid gases; and impregnated with certain metallic salts it will capture ammonia. Other adsorptive agents widely used are silica gel and molecular sieves (Figure 21.16).

Effectiveness of Gas and Vapor Canisters and Cartridges. Gas and vapor canisters are reviewed by the NIOSH Testing and Certification Branch (TCB) to determine their effectiveness. This review system is discussed in Section 7. The official approval system does not involve a check of every gas and vapor that may be faced in industry. In fact, only a few such gases and vapors, are tested (Table 21.2). Nelson has done extensive work in determining service life for organic vapor cartridges (7). He exposed commercial organic vapor cartridges to a variety of organic vapors and measured the time to breakthrough. The results of his work are given in Table 21.3. Note that of the 107 materials tested, 18 do not meet the minimum life of 50 minutes used in organic vapor cartridge testing. Eleven of the 18 failures occur in the chlorinated hydrocarbon group, indicating that chemical sorption respirators should be used with caution.

Carbon Monoxide. Carbon monoxide canisters do not function by adsorption or absorption. The canister is filled with a mixture of manganese and copper oxides, widely known under the trade name of Hopcalite. This mixture catalytically converts carbon monoxide to carbon dioxide. The reaction does not occur in the presence of moisture. To prevent the entry of moisture, carbon monoxide canisters have layers of drying agents on both sides of the catalytic agent. The indicator window on carbon monoxide

Figure 21.14 Fume filters, showing sandwich construction with different filter media.

Figure 21.15 High efficiency filter holders; note the large surface area of these filters when unfolded.

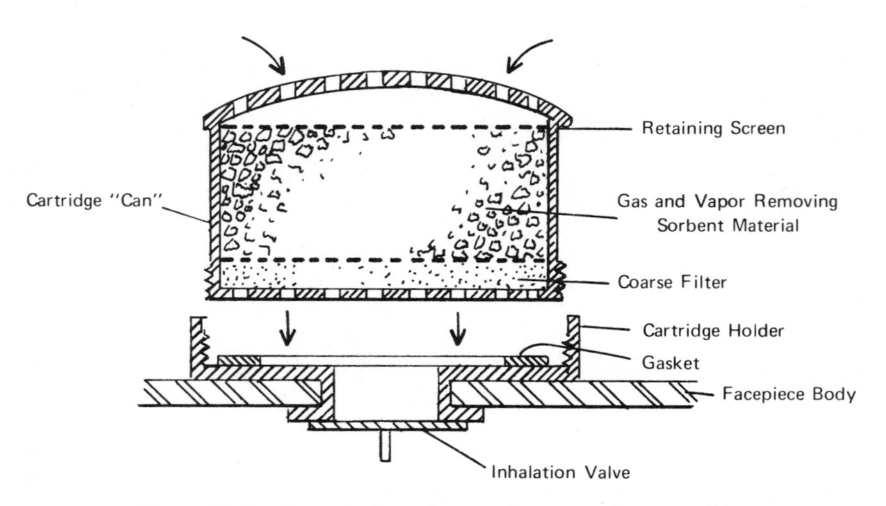

Figure 21.16 Cross section of gas and vapor sorbent cartridge.

and universal canisters indicates the condition of the drying agent. When the drying agent is saturated with moisture, the useful life of the canister is considered to be at an end.

Universal Canisters. Universal canisters (Figure 21.17) were at one time considered to be satisfactory for use against all hazards. Since they are tested by only five gases, however, as shown in Table 21.2, they are hardly universal. They do offer protection from some acid gases, some basic gases, some organic vapors, carbon monoxide, and highly toxic particulates. This protection from a variety of hazards is at the expense of the length of canister life. Table 21.2 reveals that the minimum service life for all components except carbon monoxide is less for a universal canister than for a canister made to protect against a single constituent. Universal canisters should be used with caution and replaced after each use. They should not be used if only one specific contaminant is expected.

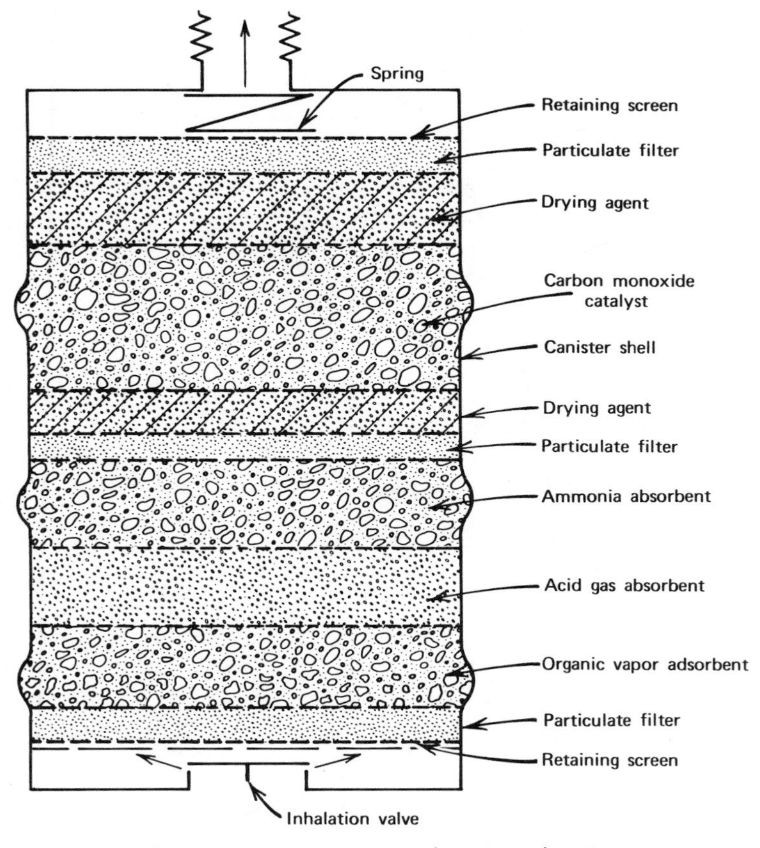

Figure 21.17 Cross section of a universal canister.

Table 21.2 Test Agents for Gas Masks and Chemical Cartridge Respirators as Listed in 30 CFR 11

30 CFR 11 Subpart	Gas Test Agent	Concentration Tested[a] (ppm)	Minimum Service Life (min)
Subpart I gas masks			
Chest canisters	Acid gases		
	Sulfur dioxide	20,000	12
	Chlorine	20,000	12
	Organic vapors		
	Carbon tetrachloride	20,000	12
	Ammonia	30,000	12
	Carbon monoxide	20,000	60
Type N chest canisters	Sulfur dioxide	20,000	6
	Chlorine	20,000	6
	Carbon tetrachloride	20,000	6
	Ammonia	30,000	6
	Carbon monoxide	20,000	60
Chin-style canisters and escape mask canisters	Acid gases		
	Sulfur dioxide	5,000	12
	Chlorine	5,000	12
	Organic vapors		
	Carbon tetrachloride	5,000	12
	Ammonia	5,000	12
	Carbon monoxide[b]	10,000	60
Subpart L, chemical cartridges	Ammonia	1,000	50
	Chlorine	500	35
	Hydrogen chloride	500	50
	Methyl amine	1,000	25
	Organic vapors		
	Carbon tetrachloride	1,000	50
	Sulfur dioxide	500	30
Subpart M, pesticide canisters and cartridges	Test agent in all cases is carbon tetrachloride		
Chest canisters		20,000	12
Chin-style canisters		5,000	9
Chemical cartridge		1,000	50
Powered air-purifying respirator		1,000	50
Subpart N, special use respirators			
Vinyl chloride			
Cartridges	Vinyl chloride	10	120
Canisters	Vinyl chloride	25	360
Powered air-purifying respirators	Vinyl chloride	25	360

Pesticide Cartridges and Canisters. At one time the U.S. Department of Agriculture (USDA) performed tests on respirator chemical cartridges to determine their efficiency against specified pesticides (8). This program is no longer in existence. As noted in Table 21.2, pesticide cartridges and canisters are tested only with carbon tetrachloride vapors. Pesticide cartridges, however, must furnish aerosol protection in addition to the requirement to protect against organic vapor.

Combination Cartridges and Canisters. Experience with the universal canister and the pesticide cartridges and canisters demonstrated that it is possible to combine two or more filters or sorbents to provide multiple protection. The most common of the multiple-sorbent and filters units is the one issued for paint spraying; this device (Figure 21.18) combines an aerosol filter with an organic vapor cartridge. Because of the size and weight of multiple-filter and sorbent cartridges, face seal leakages may occur with half mask facepieces if fitting is not carefully done (Figure 21.19).

3.2 Air-Supplied Respirators

Air-supplied respirators do not have air-purifying filters and cartridges; instead they depend on air delivered from an external source to the breathing zone.

3.2.1 Hose Masks, Powered and Nonpowered

The oldest air-supplied unit is the hose mask. It consists of a large diameter hose, anchored in uncontaminated air and connected to the wearer's facepiece. The diameter of the hose is large, thereby lowering the requirement for negative pressure in the facepiece to draw in fresh air.

Another version of the hose mask has a powered air mover connected to the fresh air source to assist in supplying air. The maximum flow of approved hose masks is 150 liters/min (9). This flow will not maintain a positive pressure within the facepiece during periods of exertion by the wearer; thus the powered hose mask is also considered to be a negative pressure device.

3.2.2 Air Line Respirators

The air line respirator differs from the hose mask in that the air is delivered to the facepiece under pressure. It can be obtained in demand (Figure 21.20), pressure demand,

---

[a] The gas and vapor concentrations listed pertain to bench tests under laboratory conditions. Maximum use concentrations will be governed by the efficiency of the respirator face fit in addition to the cartridge and canister efficiency. Extrapolation of maximum use times at other concentrations should not be done without additional information. See original tables, 30 CFR 11, and Table 21.3.

[b] Carbon monoxide canister available only for use with an escape mask.

Table 21.3 Effect of Solvent Vapor on Respiratory Cartridge Efficiency[a] (7)

Solvent	Time to Reach 1% Breakthrough, 10 ppm (min)	Solvent	Time to Reach 1% Breakthrough, 10 ppm (min)
Aromatics[b]		1,4-Dichlorobutane	108
Benzene	73	*o*-Dichlorobenzene	109
Toluene	94	Trichlorides	
Ethyl benzene	84	Chloroform	33
m-Xylene	99	Methyl chloroform	40
Cumene	81	Trichloroethylene	55
Mesitylene	86	1,1,2-Trichloroethane	72
Alcohols[b]		1,2,3-Trichloropropane	111
Methanol	0.2	Tetra- and pentachlorides[b]	
Ethanol	28	Carbon tetrachloride	77
Isopropanol	54	Perchloroethylene	107
Allyl alcohol	66	1,1,2,2,-Tetrachloroethane	104
n-Propanol	70	Pentachloroethane	93
sec-Butanol	96	Acetates[b]	
Butanol	115	Methyl acetate	33
2-Methoxyethanol	116	Vinyl acetate	55
Isoamyl alcohol	97	Ethyl acetate	67
4-Methyl-2-pentanol	75	Isopropyl acetate	65
2-Ethoxyethanol	77	Isopropenyl acetate	83
Amyl alcohol	102	Propyl acetate	79
2-Ethyl-1-butanol	76.5	Allyl acetate	76
Monochlorides[b]		*sec*-Butyl acetate	83
Methyl chloride	0.05	Butyl acetate	77
Vinyl chloride	3.8	Isopentyl acetate	71
Ethyl chloride	5.6	2-Methoxyethyl acetate	93
Allyl chloride	31.	1,3-Dimethylbutyl acetate	61
1-chloropropane	25.	Amyl acetate	73
1-chlorobutane	72	2-Ethoxyethyl acetate	80
Chlorocyclopentane	78	Hexyl acetate	67
Chlorobenzene	107	Ketones[c]	
1-chlorohexane	77	Acetone	37
o-chlorotoluene	102	2-Butanone	82
1-chloroheptane	82	2-Pentanone	104
3-(chloromethyl heptane)	63	3-Pentanone	94
Dichlorides[b]		4-Methyl-2-Pentanone	96
Dichloromethane	10	Mesityl oxide	122
trans-1,2-Dichloroethylene	33	Cyclopentanone	141
1,1-Dichloroethane	23	3-Heptanone	91
cis-1,2-Dichloroethylene	30	2-Heptanone	101
1,2-Dichloroethane	54	Cyclohexanone	126
1,2-Dichloropropane	65	5-Methyl-3-heptanone	86

Table 21.3 (Continued)

Solvent	Time to Reach 1% Breakthrough, 10 ppm (min)	Solvent	Time to Reach 1% Breakthrough, 10 ppm (min)
3-Methylcyclohexanone	101	Isopropyl amine	66
Diisobutyl ketone	71	Propyl amine	90
4-Methylcyclohexanone	111	Diethyl amine	88
Alkanes[c]		Butyl amine	110
Pentane	61	Triethyl amine	81
Hexane	52	Dipropyl amine	93
Methylcyclopentane	62	Diisopropyl amine	77
Cyclohexane	69	Cyclohexyl amine	112
Cyclohexene	86	Dibutyl amine	76
2,2,4-Trimethylpentane	68	Miscellaneous materials[c]	
Heptane	78	Acrylonitrile	49
Methylcyclohexane	69	Pyridine	119
5-Ethylidene-2-norbornene	87	1-Nitropropane	143
Nonane	76	Methyl iodide	12
Decane	71	Dibromomethane	82
Amines[c]		1,2-Dibromoethane	141
Methyl amine	12	Acetic anhydride	124
Ethyl amine	40	Bromobenzene	142

[a] These cartridge pairs were tested at 1000 ppm, 50 percent relative humidity, 22°C, and 53.3 liters/min (equivalent to a moderately heavy work rate). Cartridges were preconditioned at room temperature and 50 percent relative humidity for at least 24 hours before testing.
[b] Mine Safety Appliances Co. cartridges.
[c] American Optical Corp. cartridges.

and continuous flow modes of operation. Only the demand mode is considered in this section.

Demand Mode of Operation. The demand air line respirator has an airflow regulation valve between the facepiece and the air supply. This valve (Figure 21.21) is an integral part of the demand air line respirator. The negative pressure created during inhalation activates the diaphragm, which opens the compressed air supply through the lever system. The negative pressure required to activate the air supply diaphragm may be as much as 50 mm of H_2O. This is comparable to the inhalation negative pressure encountered when using an air-purifying respirator.

A demand air line respirator may have a quarter, half, or full facepiece. In each case, because of the negative pressure in the facepiece, the degree of facepiece leakage will be comparable to that of air-purifying respirator with the same facepiece. The NIOSH

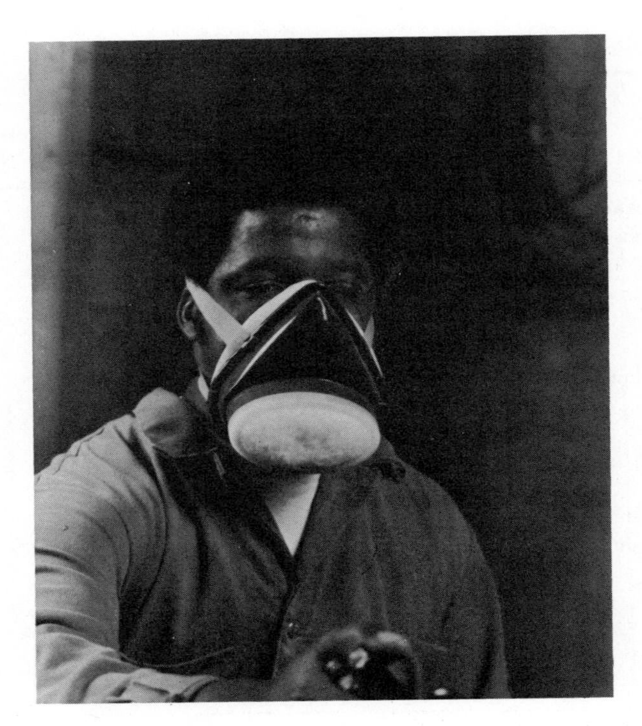

Figure 21.18 Paint spray respirator; behind the aerosol filter is a layer of activated charcoal.

approval regulations discussed in Section 7 require that a minimum of 115 liter/min be delivered to the facepiece on demand.

3.2.3 Self-Contained Breathing Apparatus (SCBA)

The SCBA type of respirator carries an air supply of compressed air, compressed or liquid oxygen, or oxygen-generating chemicals. No external connection to an air supply is necessary.

Open Circuit SCBA. The most widely used SCBA is a demand open circuit device, consisting of a compressed air tank, an air line and regulator, and a facepiece. The air supply may last from 5 to 30 minutes, depending of the size of the air tank. The customary SCBA for use in hazardous atmospheres is one with an air supply rated at 30 minutes; however SCBA units for escape only may have air supplies of 5 to 15 minutes. A 30-minute SCBA is illustrated in Figure 21.22. The normal operation of the demand SCBA is the same as the demand air line respirator just described. In case of malfunction of the regulator, a bypass circuit allows a continuous flow of air to the facepiece (Figure 21.21). Air supply ratings given with this equipment are nominal and should not be regarded as absolute. High work rates combined with the stress of emergency conditions will reduce the actual use time of the air tank. The problems of facepiece leakage are the same as those encountered with air-purifying respirators.

Closed Circuit SCBA. Instead of exhausting the exhaled air, as is done with the open circuit devices, the exhaled air from a closed circuit SCBA is reused after removal of the carbon dioxide. Closed circuit devices are available with up to a 4-hour service life. This contrasts with the open circuit device which now has a maximum service life of 30 minutes. Closed circuit equipment obtains this long life by utilizing an oxygen supply that is in a liquid, gaseous, or solid form.

Figure 21.23*a* diagrams an oxygen supply system utilizing gaseous oxygen. Exhaled air is recirculated through a chemical to remove the carbon dioxide, then returned to the facepiece. The volume of air is decreased by the carbon dioxide removed. When the air volume is not sufficient to meet the wearer's demands, a negative pressure is created in the breathing bag, and oxygen is added.

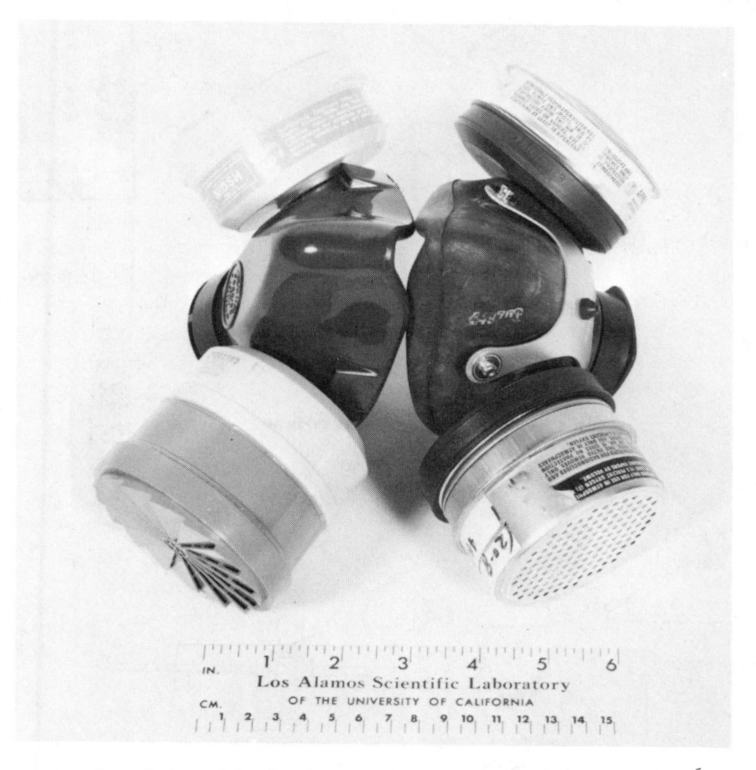

Figure 21.19 Combined cartridges for vapor and aerosol protection, bottom. Single cartridges, top.

Oxygen may also be stored as a chemical compound. The chemical is stored in canisters slightly larger than an air-purifying canister. It is advantageous to use such a device when storage and procurement of gaseous oxygen is a problem. The chemical, usually a type of peroxide, reacts in the presence of moisture and carbon dioxide to produce pure oxygen. Since moisture and carbon dioxide are both components of human breath, a usable breathing cycle is formed by recirculating exhaled breath through the oxygen-rich chemical. The carbon dioxide is removed as it becomes part of the reaction to release the oxygen. The air then enters a breathing bag for mixing before being

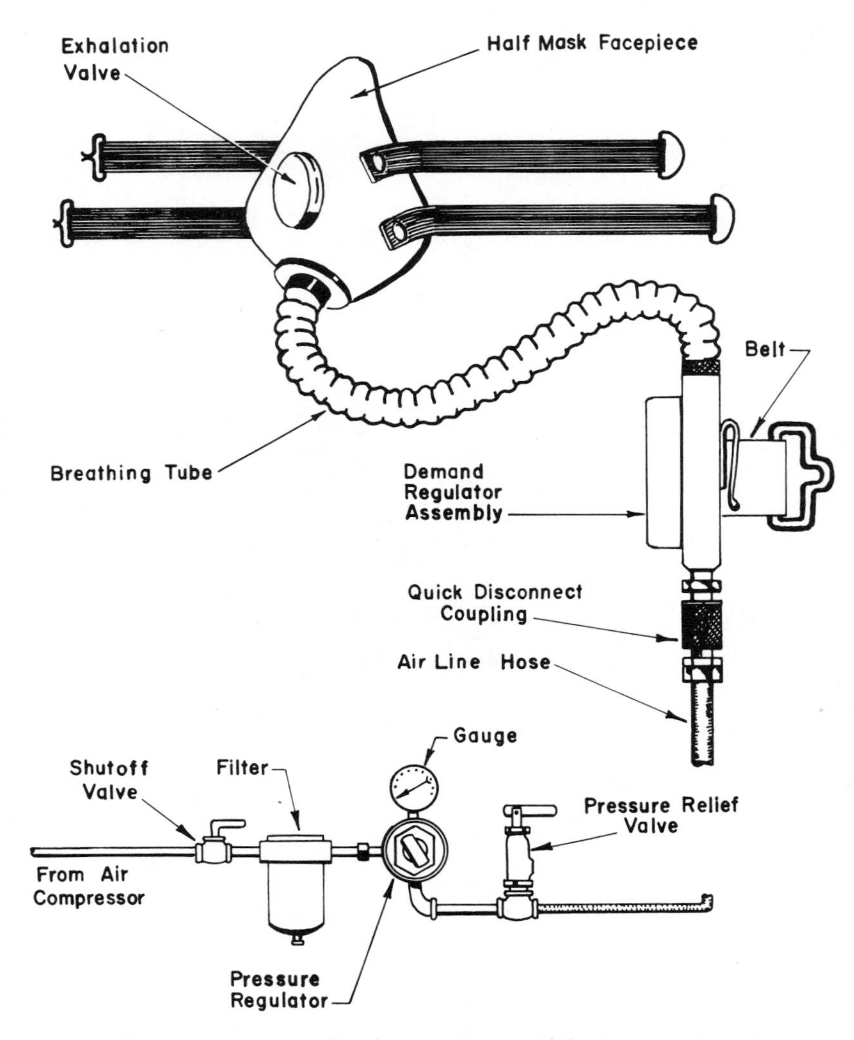

Figure 21.20 Demand air line respirator with basic components.

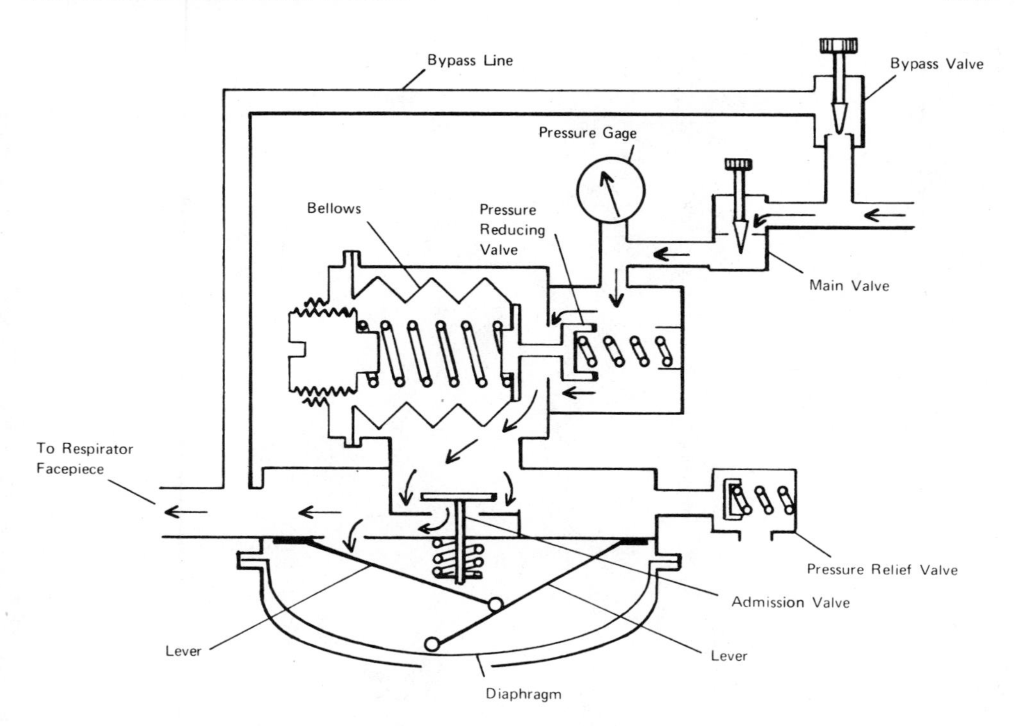

Figure 21.21 SCBA demand air regulation valve. The air line demand valve does not have a bypass line. The low pressure demand valve does not have the pressure-reducing valve.

drawn into the facepiece. A disadvantage of this unit is that once the reaction has been started in the canister, it cannot be terminated until the chemical reaction is compete. Canisters are available for 1- and 2-hour periods of duration (Figure 21.23*b*).

Liquid oxygen supplies are little used in the United States because the logistics of providing refills is complex.

4 POSITIVE PRESSURE RESPIRATORS

The problem of obtaining a facepiece seal with negative pressure devices, mentioned frequently in the preceding sections, has led to the development of the concept of positive pressure within the facepiece, which greatly reduces leakage into the respirator.

Positive pressure devices can be obtained with tight-fitting facepieces, identical to those of negative pressure respirators with the exception of the exhalation valve, which must be modified for pressure demand use. Positive pressure can be maintained by a continuous airflow, such as that used for air-supplied hoods and suits. In addition to the air-supplied respirators mentioned earlier, positive pressure air-purifying respirators are available.

Figure 21.22 Thirty-minute SCBA open circuit device.

4.1 Powered Air-Purifying Respirators

Essentially the powered device are air-purifying respirators with an electrically operated blower inserted in the breathing tube to provide the energy necessary to force air through the air-purifying unit and into the facepiece. It provides excess air, to ensure that any leakage will be outward from the facepiece. This type of respirator can be operated with a stationary powerpack and filter unit, feeding air to the user through an air line, or as a portable unit with the powerpack and filters mounted on the wearer's body (Figure 21.24). These units are advantageous for use in areas where it is difficult to provide air for air-supplied respirators. For example, heavy equipment operators exposed to toxic dusts or vapors may need clean air for extended periods. It should be remembered that these respirators are air-purifying units only and must not be used

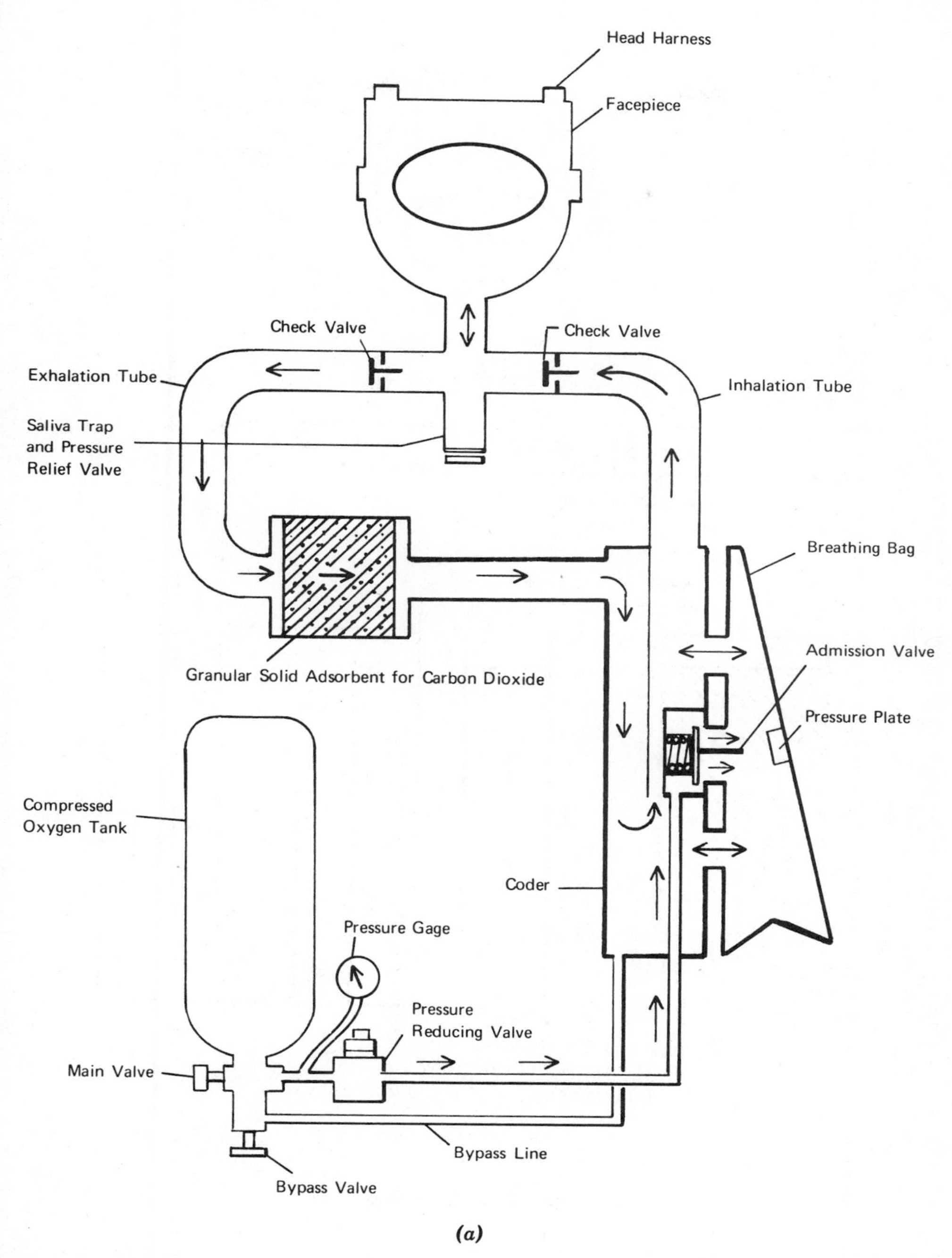

(a)

Figure 21.23 SCBA closed circuit devices. (*a*) Using gaseous oxygen. (*b*) Using a type of peroxide as an oxygen source.

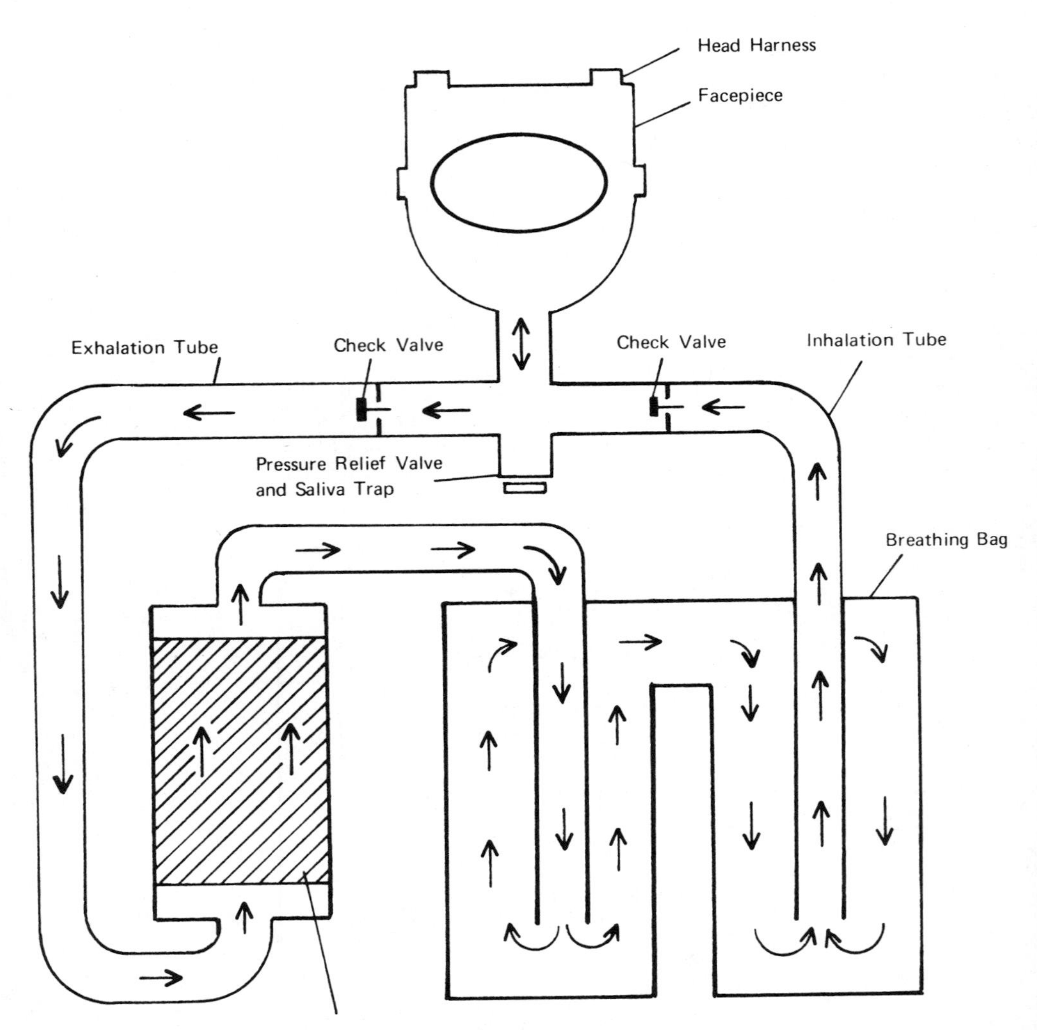

(b)

Figure 21.23 (Continued)

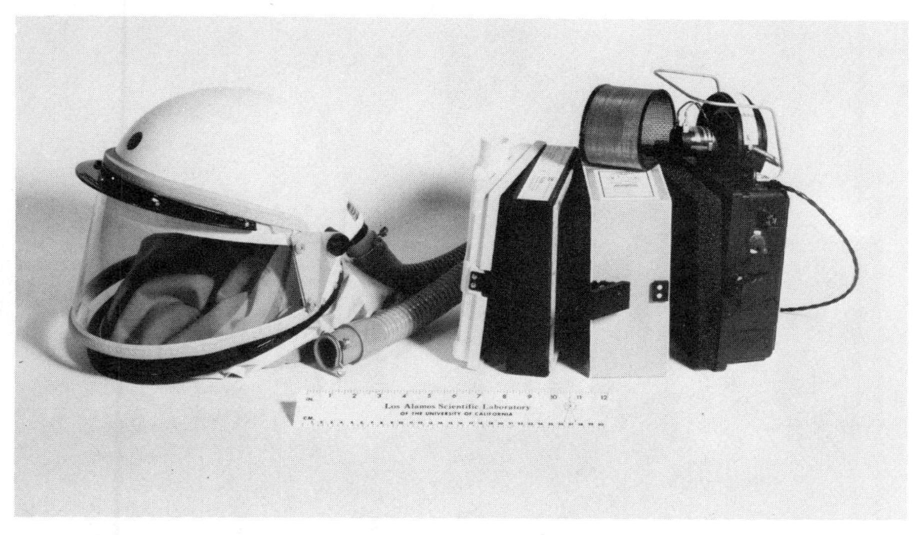

Figure 21.24 Powered air-purifying respirator; note hood, filter, and blower.

when oxygen deficiency is a factor or in the presence of atmospheric contaminants for which the filters or sorbents are not designed.

4.2 Pressure Demand Self-Contained Breathing Apparatus

The pressure demand SCBA is identical to the demand SCBA already described, except that the exhalation valve on the facepiece is different, and there is a slight change in the air regular unit. The air regulation valve is designed to provide air to the facepiece when the ambient pressure in the facepiece drops to 59 mm H_2O above atmospheric. If a normal exhalation valve were used, a continual supply of air would be dumped out the exhalation valve (Figure 21.25). The exhalation valve, combined with the air regulator valve, gives an atmosphere inside the facepiece that is continually above atmospheric pressure. Normally when inhalation occurs, the positive pressure inside the facepiece is reduced, but it does not reach a negative condition.

It is possible under high work rates for the user to demand a higher instantaneous airflow than the SCBA can deliver. Under this condition, negative pressure may occur in the facepiece for a short period of time.

Under normal conditions exhalation is a passive response, involving little muscular effort. Exhalation while wearing the pressure demand SCBA requires more muscular effort than with the demand SCBA, since the wearer must overcome a resistance of 51 mm H_2O before air can be exhaled. Initial use of pressure demand equipment concerns some users because of the exhalation effort required. With training, normal individuals rapidly become accustomed to the pressure demand masks.

Figure 21.25 Normal exhalation valve (top) and pressure demand exhalation valve.

A problem of greater concern to many is the possibility of losing the air supply through face seal leaks. Leaks will occur with a poor face seal; however the condition is noticeable when air escapes through the leak. This assists the wearer in adjusting the mask to obtain a better seal. When the facepiece is properly adjusted, there should be no difference in the length of service provided by either pressure demand or demand. This device is considered the safest to wear in an atmosphere immediately dangerous to life.

4.3 Pressure Demand Air Line Respirators

Basically the pressure demand air line respirator is the same as the SCBA pressure demand unit just described. The air supply regulator is similar, and together with the pressure demand exhalation valve, it performs the same functions as the pressure demand SCBA (Figure 21.25).

4.4 Continuous Flow Air Line Respirators

There are tight-fitting and loose-fitting continuous flow air line respirators. The tight-fitting respirator is simply a facepiece with a continual flow of air through the mask. A minimum of 115 liters of air per minute must be supplied.

The loose-fitting air line respirator may consist of an air-supplied hood, blouse, or full suit. The materials of construction for such hoods and suits vary widely, ranging from light guage plastic placed over a lightweight frame, to garments with built-in safety hats and coverings designed to withstand the rigors of sandblasting. These units offer protection for the head and shoulders in addition to providing fresh breathing air. They also penalize the wearer with additional weight, awkward movement, and in most cases, restricted vision. The minimum flow of air for approved hoods is 170 liters/min. As discussed later, it is important that this minimum air supply be maintained if proper protection is to be obtained.

Air-supplied suits are widely used in the nuclear industry. Lightweight, disposable suits are favored because of the difficulty of removing radioactive contaminants from reusable suits. There are no standards or official regulations governing the design, sale, and use of air line suits.

Much is unknown about the design and use of air line suits, but leakage of atmospheric contaminants into air line suits can be excessive. The traditional figure of 170 liters/min for air-supplied hoods may not give sufficient airflow to prevent inward leakage for most air line suits; see below. Air line suits should be used with caution in atmospheres of high toxicity.

5 RESPIRATOR EFFICIENCY

The efficiency of any respiratory protective system can be measured by the amount of outside contaminant that enters the wearer's respiratory system.

5.1 Respirator Fitting Tests

It is important to determine that the respirator fits satisfactorily before entry into a contaminated atmosphere. Fit can be evaluated by the use of a qualitative or quantitative fitting test. The qualitative fitting tests, though not as accurate as the quantitative test, can provide a satisfactory estimate of respirator integrity if performed with care.

5.1.1 Qualitative Fitting Tests

Positive and Negative Pressure Tests. The negative pressure test is performed on tight-fitting facepieces only, by covering the air inlet lightly and inhaling slightly. If a leak exists, the air can be felt as it enters. The positive pressure test is performed by blocking the exhalation valve and exhaling lightly. Again, air leakage can be felt if a leak is evident. If such leaks are found, the respirator is adjusted and retested.

Neither the positive or negative pressure test is considered to be a satisfactory initial fitting test. These tests are useful when donning a respirator; however the position of the respirator on the face may be affected by touching the facepiece to block the air inlets and exits.

Vapor Fit Test. Air-purifying respirators must be equipped with organic vapor cartridges for this test. Isoamyl acetate (IAA), which is the commonly used test agent, has an odor of bananas. To use, the respirator is donned and a carrier saturated with the liquid is passed close to the sealing surface of the respirator. If the odor cannot be detected by the wearer, it is assumed that the fit is satisfactory. The wearer should perform a series of exercises during the test.

Certain disadvantages accompany the use of this test. Since some individuals cannot smell isoamyl acetate, subjects must be tested with a low concentration of isoamyl acetate before the test is begun. Care must be taken to prevent exposure to high concentrations of isoamyl acetate before the test, since the nose is easily fatigued when smelling it. Such olfactory fatigue may produce spurious results when the respirator is donned. Subjects must have nasal passages open during the test, to ensure use of the sense of smell. Since the vapor is pleasant smelling, test subjects may not signal immediately upon detecting the vapor. The cooperation and complete understanding of the test subject is vital, and the importance of this factor cannot be overemphasized.

Aerosol Fit Test. Stannic chloride and titanium tetrachloride sublime upon exposure to moist air to produce smoke composed of intensely irritating fine particles (approximately 1 μm or less in diameter). These materials are available as smoke tubes that are sold to check ventilation systems. Applied with caution, they can be used to conduct an excellent qualitative fit test.

The respirator must be equipped with high efficiency filters before starting the test. The subject adjusts the respirator and closes his eyes, to begin the test. Smoke is blown toward the respirator from a distance of 2 feet. If no leakage is detected, the smoke tube is moved toward the subject. One must be alert for any sign that the smoke has penetrated the mask (Figure 21.26). The respirator wearer should perform a series of exercises during the test. If no smoke is detected, the fit is considered to be satisfactory.

A disadvantage of this test in the intensely irritating effect of the smoke, and the test must be administered with care to prevent a high concentration inside the mask. The irritation is also an advantage, however, in that the subject reacts immediately on smelling the smoke, permitting the operator to determine immediately whether there is a leak.

Coal Dust Test. This test has been in use for many years and depends on spraying the respirator wearer with find coal dust. When the respirator is removed, any leakage is indicated by dark lines on the facial area that was covered by the device (Figure 21.27). A disadvantage of this test is the reluctance of personnel to be sprayed with coal dust.

Talc Test. Talc can be sprayed on the wearers of respirators in the same manner described for the coal dust test, and this material has the advantage of being better accepted than coal dust. Its disadvantage is that it is harder to see a streak of talc on the skin. Oblique lighting is suggested to assist in detecting talc on the face.

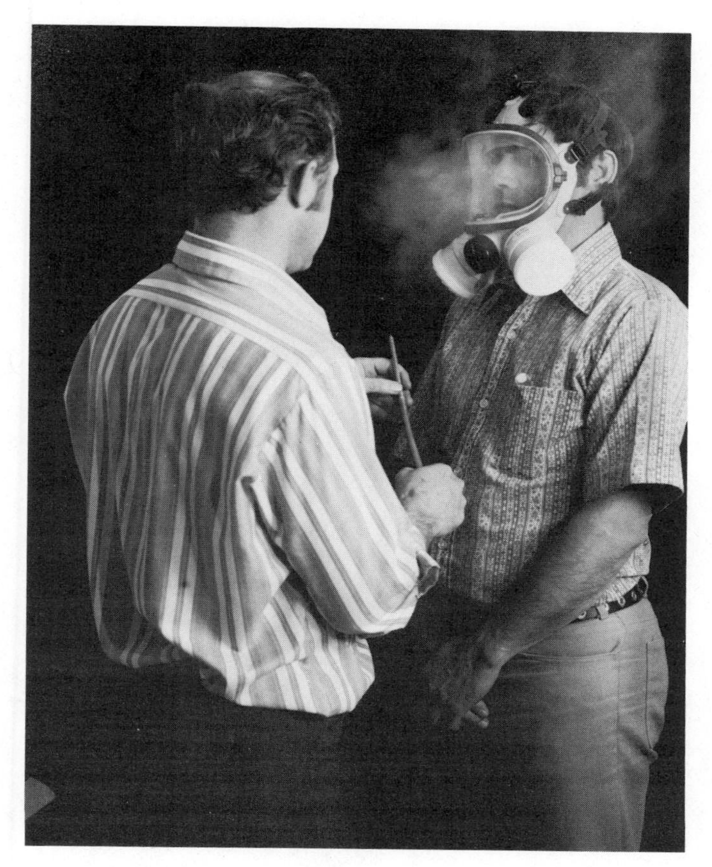

Figure 21.26 The irritant fume qualitative fitting test.

5.1.2 Quantitative Fitting Tests

The qualitative test determines whether leaks exist, whereas the quantitative test determines the degree of leakage. The latter is a more expensive type of test to run. All the tests described below have the following procedure (Figure 21.28) in common.

1. A test atmosphere is generated inside a chamber, and a means of constantly monitoring the concentration is provided.
2. The subject enters the chamber wearing a respirator modified to permit the removal of an air sample.
3. During the test an air sample is constantly being withdrawn from the facepiece and analyzed. It is customary to express the concentration inside the facepiece as a percentage of the test atmosphere inside the chamber.

Figure 21.27 The coal dust qualitative fitting test; note leakage at corner of mouth.

4. The subject performs a series of exercises, which may change depending on the respirator being worn.

Quantitative Fit Testing Agents. Quantitative fitting tests can be conducted with a gas or an aerosol. Aerosols have the advantage of a detector system with such rapid response that when sampling inside a mask, the contaminant penetration can be determined with each inhalation. The gas or vapor detection system has a slower responding detector that gives an average value of the contaminant inside the facepiece. Advocates of the aerosol system contend that there is little difference in results when using a gas or an aerosol, since the aerosol size is small compared to the size of the openings at the face to facepiece seal.

The two aerosols commonly used for quantitative fit tests are dioctylphthalate (DOP) and sodium chloride. The detection system for DOP is a forward light scattering photometer; the sodium chloride system uses a flame to vaporize the salt and a phototube sensitive to the sodium line. The readout from the photometer and phototube must be fed into a stripchart recorder for future data analysis.

DOP. DOP is used as an aerosol to inspect high efficiency filters before use. These filters must collect 99.97 percent of all particles 0.3 μm in diameter and greater. When heated, DOP can be controlled to produce an aerosol of single size. Quantitative man testing has been done using thermogenerated DOP, also known as monodispersed DOP. Although this substance is an excellent test aerosol, thermal generation leads to decomposition of the DOP into a variety of products, and some may be toxic to man. This has led to the abandonment of thermally generated monodispersed DOP and the adoption of air-generated DOP, also known as polydispersed DOP. Polydispersed DOP does not have the toxic properties of monodispersed DOP, nor does it produce the single sized

Figure 21.28 A quantitative respirator fitting test using a test hood with a sodium chloride aerosol.

aerosol. The size of aerosol can be controlled to produce a given MMAD. The lower limit of detection is 0.005 percent of the challenge concentration when using the foward light scattering photometer.

Sodium Chloride. Sodium chloride is the newer of the aerosols in use for quantitative man testing. It has been employed in England for many years, but has been used in the United States only since 1971. At that time, research into the use of NaCl was initiated at the Los Alamos Scientific Laboratory (LASL). The sodium chloride generator and readout systems developed by LASL are now commercially produced. The size of the solid aerosol produced can be controlled and is reproducible. An advantage of sodium chloride over DOP is the degree of toxicity. The former is classed as a nuisance particulate with a TLV of 10 mg/m^3, but DOP is an oil mist with a TLV of 5 mg/m^3. Both the aerosols are usually used with an MMAD of 0.6 μm. This requires the use of high efficiency filters when testing air-purifying respirators.

Dichlorodifluoromethane. The gas that has been used as a quantitative fit test agent it dichlorodifluoromethane (Freon 12). Two disadvantages of this system are lack of sensitivity of the detector and lack of an adsorption agent for use with air-purifying respirators. When air-purifying respirators are used, they must be adapted to allow fresh air to be drawn in through the air inlets. This requires the attachment of tubing to the respirator, which can cause problems during tests.

5.2 Problems Affecting Respiratory Efficiency

The problems affecting respirator efficiency vary depending on the type of resirator in use.

5.2.1 Tight-Fitting Respirators

Difficulties are encountered with respect to the ability of the wearer to obtain an efficient seal with the facepiece of a tight-fitting respirator and improper use practice also create problems, as described below.

Anthropometric Considerations for Respirators. Despite the obvious fact, that the unit consisting of the human head and face comes in a variety of shapes and sizes, people persist in using one size respirator to fit a variety of facial sizes. The average person seldom knows his facial size and shape and is unaware of the important facial dimensions to be considered when choosing a respirator.

Figure 21.29 shows a variety of facial measurements that have been used in relation to respirator design. The principal dimensions that have been studied in relation to respirator face fit are face length (Menton–Nasal root depression length), face width (bizygomatic breadth), and lip length. The dimensions of face length and face width are used for full face masks, and face length and lip length are used for half and quarter

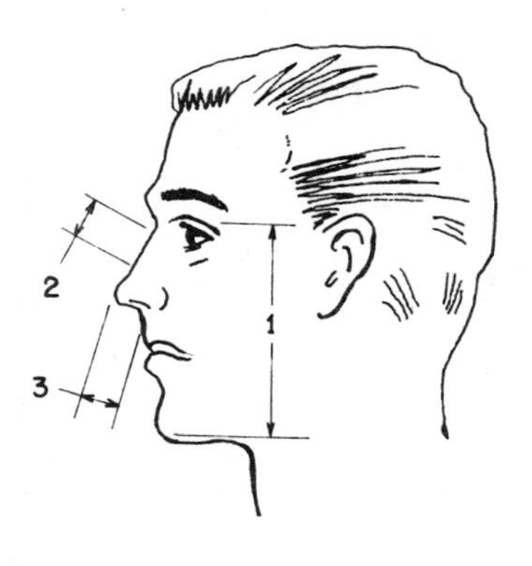

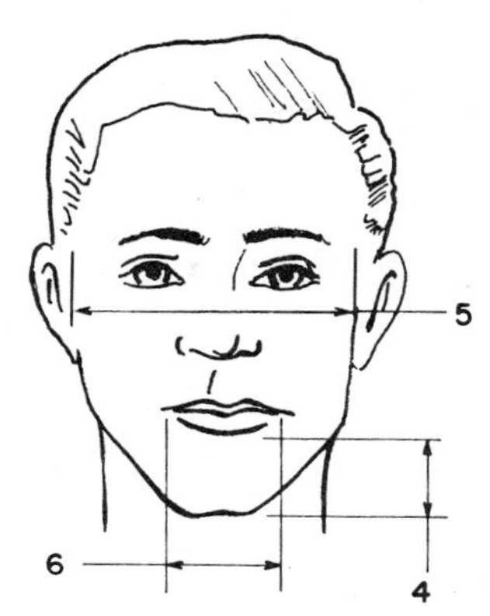

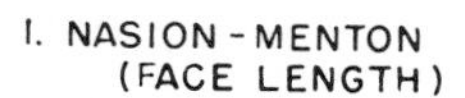

Figure 21.29 Anthropometric landmarks for facial measurements.

masks. These are the dimensions that can be easily measured, and they are understood readily by the respirator user. Disadvantages of using only these facial dimensions to relate respirator efficiency to facial size are as follows:

1. They may not provide all the information necessary to predict how a respirator will fit an individual.
2. They are two-dimensional only.
3. They do not make allowances for abnormalities, such as hollow temple areas and unusual nose size and shape.

To determine how well existing respirators fit the working population, data on facial measurements of the population must be obtained. The principal source for such data has been anthropometric studies carried out by the U.S. Air Force (10). The Air Force surveys, accomplished in 1967 and 1968, are among the most complete studies performed on American adults.

Utilizing the study just cited, representative sample groups have been established to represent the working population. Figure 21.30 presents the population distribution as related to face length and face width. A similar presentation can be made to show the relation of face length and lip length (10). Utilizing these data, Hack et al. (11) constructed panels of 25 people to represent the general population (Figure 21.31). At the Los Alamos Scientific Laboratory these panels have been utilized to determine how

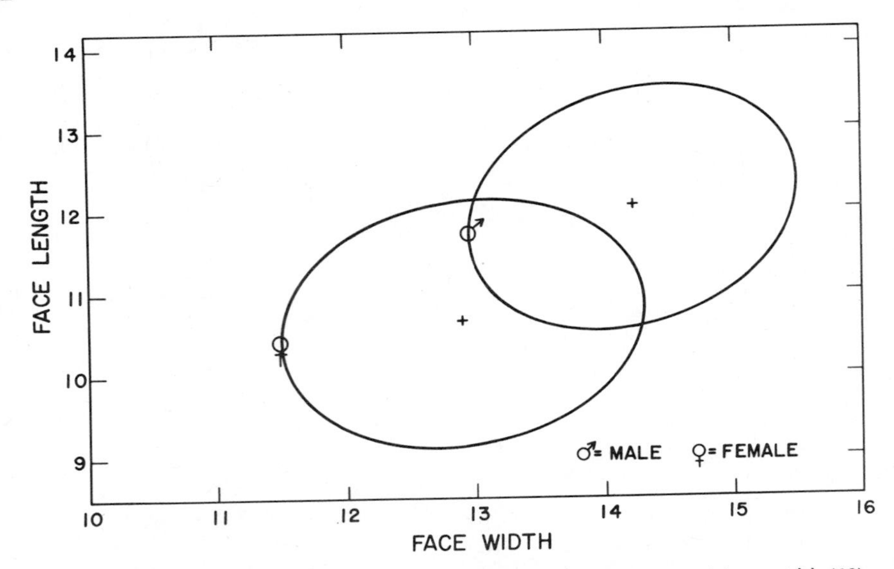

Figure 21.30 Adult population distribution based on face length and face width (10).

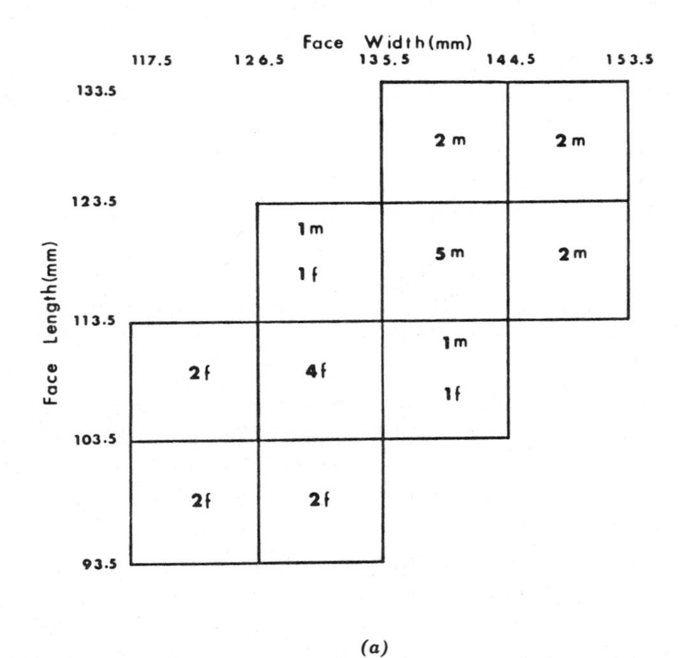

(a)

Figure 21.31 Sample 25-person test panels, male and female, for respirator testing. (a) For testing full face masks. (*b*) For testing half masks.

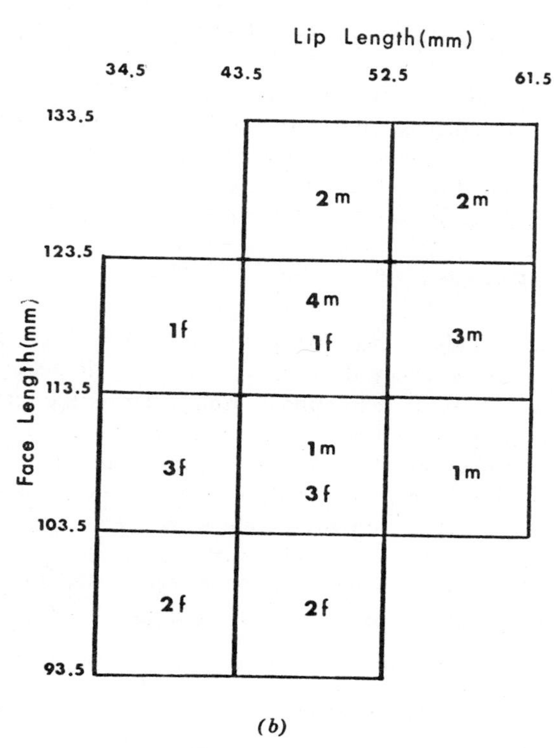

(b)

Figure 21.31 (Continued)

well the respirators fit the general population. Results of such tests are discussed in Section 5.3.

Improper Use Practices. The problems listed below, which seriously affect respirator efficiency can be prevented by instituting a program of respirator training and maintenance.

The device most significantly affected is the nonpowered air-purifying respirator. Since this respirator creates a negative pressure in the facepiece, any opening in the respirator is a potential source of leakage. The principal openings are the exhalation valve, the filter or cartridge holder, and the filter or cartridge proper.

1. An improperly seated exhalation valve is a source of enormous leakage. Davis has shown that when properly installed, exhalation valves have only minor leakage (4). They should be checked frequently for damage or impediments to proper closure.
2. Improperly installed filters and cartridges, and damaged filter–cartridge receptacles, may be sources of leakage. They are eliminated by training the wearer to inspect

the holders for damage and ensuring that he knows how to install filters and cartridges properly.

3. Improper installation of flexible filters has been found to pose a problem with dust masks. It is easy to install such filters carelessly, with the result that leaks occur.

4. The installation of the wrong filter or cartridge unit for the hazard at hand can cause serious leakage. This is not a fault of the respirator, but reflects inadequate training of the wearer and improper assessment of the hazard.

5.2.2 Loose-Fitting Respirators

Air-supplied hoods are widely used for protection when high concentrations of contaminant are encountered. They depend on a flow of air sufficient to provide a constant outward flow to prevent contaminants from entering the interior of the hood. The problem of leakage into loose-fitting respirators does not occur if the user is standing still and sufficient air is being supplied to satisfy his breathing requirements. Problems arise when he moves. Any movement that tends to move the hood from the head increases the volume, and unless the airflow is high, outside air may enter the hood. In the case of air-supplied suits, movements such as bending may decrease the suit volume, however when the wearer straightens up, the volume is again increased and aspiration of outside air will occur. The results of an efficiency study of air-supplied hoods conducted at LASL are discussed in Section 5.3.3.

5.3 Protection Factors For Respirators

5.3.1 Assignment of Protection Factors to Respirators

The protection factor (PF) represents the efficiency of a respirator. Major work in the determination of respirator efficiency has been carried out by Edwin C. Hyatt. Much of his investigative work has been summarized in the publication "Respirator Protection Factors" (12). The formula for calculation of the PF, along with a summary of Hyatt's PF recommendations, are reproduced in Tables 21.4 and 21.5. Table 21.5 provides information to guide the user in determining the suitable respirator for various levels of exposure. It also gives a view of the relative efficiency of respirators.

The use of PFs presupposes that a qualitative fitting program is carried out for personnel assigned to wear respirators. There is then reasonable assurance that the respirator fit is adequate to utilize the PF recommended.

If a quantitative fit test program is in operation, Hyatt's recommendations for allowable PFs (Table 21.6) can be used. Note that the respirator wearer cannot use this table if the average penetration of the quantitative test exercises exceeds the allowable penetration listed.

At the present time, Hyatt's recommendations are the best data published. There are however, some limitations associated with the use of his data.

Table 21.4 Protection Factor Formula (12)

$$\text{protection factor (PF)} = \frac{\text{ambient air concentration}}{\text{concentration inside facepiece or enclosure}}$$

Protection Factor	Respirator Efficiency (%)	Facepiece Penetration (%)	Selection Guide for Maximum Use Concentration (× TLV)
5	80	20	5
10	90	10	10
20	95	5	20
50	98	2	50
100	99	1	100
200	99.5	0.5	200
500	99.8	0.2	500
1,000	99.9	0.1	1,000
2,000	99.95	0.05	2,000
5,000	99.98	0.02	5,000
10,000	99.99	0.01	10,000

1. Although these data were gathered through the use of anthropometric panels, the panels were restricted to men. If women form part of the workforce in question, care must be taken to ensure that they receive good fitting respirators. It is possible that PFs for women are lower than those for men.
2. Hyatt had to extrapolate facepiece leakage data in certain cases where no data existed at that time. He states that PFs based on this interim method should only be used until adequate quantitative man-test data are available for the device in question.
3. All respirators are judged as a class. This means that all respirators in a class bear the PF that can be met by the poorest performing member in the class.

Despite these problems, Hyatt's recommendations form the best overall evaluation of respirator efficiency available.

5.3.2 Respirator Fit Test Data

Data on the fit of respirators on men and women were recently published by LASL (4). Table 21.7 gives information on half mask air-purifying respirators tested on an anthropometeric panel using the design proposed by Hack (11). Only one of the six respirators listed can provide a PF of 10 for 95 percent of the total panel. If males only are

Table 21.5 Respiratory Protection Factors[a] (12)

Type Respirator	Facepiece[b] Pressure	Protection Factor
I. Air purifying		
A. Particulate[c] removing		
Single-use,[d] dust[e]	–	5
Quarter mask, dust[f]	–	5
Half mask, dust[f]	–	10
Half or quarter mask, high efficiency[h]	–	10
Half or quarter mask, fume[g]	–	10
Full facepiece, high efficiency	–	50
Powered, high efficiency, all enclosures	+	1,000
Powered, dust or fume, all enclosures	+	—[i]
B. Gas and vapor removing[j]		10
Half mask	–	50
Full facepiece	–	
II. Atmosphere supplying		
A. Supplied air		
Demand, half mask	–	10
Demand, full facepiece	–	50
Hose mask without blower, full facepiece	–	50
Pressure demand, half mask[k]	+	1,000
Pressure demand, full facepiece[l]	+	2,000
Hose mask with blower, full facepiece	–	50
Continuous flow, half mask[k]	+	1,000
Continuous flow, full facepiece[l]	+	2,000
Continuous flow, hood, helmet, or suit[m]	+	2,000
B. Self contained breathing apparatus (SCBA)		
Open circuit, demand, full facepiece	–	50[n]
Open circuit, pressure demand full facepiece	+	10,000
Closed circuit, oxygen tank-type, full facepiece	–	50
III. Combination respirator		
A. Any combination of air-purifying and atmosphere-supplying respirator	Use minimum protection factor listed above for type of mode of operation.	
B. Any combination of supplied-air respirator and an SCBA		

Exception: Combination supplied-air respirators, in pressure demand or other positive pressure mode, with an auxiliary self-contained air supply, and a full facepiece, should use the PF for pressure demand SCBA.

[a] The overall protection afforded by a given respirator design (and mode of operation) may be defined in terms of its protection factor (PF). The PF is a measure of the degree of protection afforded by a respirator, defined as the ratio of the concentration of contaminant in the ambient atmosphere to that inside the enclosure (usually inside the facepiece) under conditions of use. Respirators should be selected so that the concentration inhaled by the wearer will not exceed the

appropriate limit. The recommended respirator PFs are selection and use guides, and should only be used when the employer has established a minimal acceptable respirator program as defined in Section 3 of the ANSI Z88.2-1969 Standard.

[b] In addition to facepieces, this includes any type of enclosure or covering of the wearer's breathing zone, such as supplied-air hoods, helmets, or suits.

[c] Includes dusts, mists, and fumes only. Does not apply when gases or vapors are absorbed on particulates and may be volatilized or for particulates volatile at room temperature. Example: Coke oven emissions.

[d] Any single-use dust respirator (with or without valve) not specifically tested against a specified contaminant.

[e] Single-use dust respirators have been tested against asbestos and cotton dust and could be assigned a PF of 10 for these particulates.

[f] Dust filter refers to a dust respirator approved by the silica dust test and includes media of all types, that is, both nondegradable mechanical-type media and degradable resinimpregnated wool felt or combination wool–synthetic felt media.

[g] Fume filter refers to a fume respirator approved by the lead fume test. All types of media are included.

[h] High efficiency filter refers to a high efficiency particulate respirator. The filter must be at least 99.97 percent efficient against 0.3 μm DOP to be approved.

[i] To be assigned, based on dust or fume filter efficiency for specific contaminant.

[j] For gases and vapors, a PF should be assigned only when published test data indicate that the cartridge or canister has adequate sorbent efficienty and service life for a specific gas or vapor. In addition, the PF should not be applied in gas or vapor concentrations that are (1) immediately dangerous to life, (2) above the lower explosive limit, and (3) cause eye irritation when using a half mask.

[k] A positive pressure supplied-air respirator equipped with a half-mask facepiece may not be as stable on the face as a full facepiece. Therefore, the PF recommended is half that for a similar device equipped with a full facepiece.

[l] A positive pressure supplied-air respirator equipped with a full facepiece provides eye protection but is not approved for use in atmospheres immediately dangerous to life. It is recognized that the facepiece leakage, when a positive pressure is maintained, should be the same as an SCBA operated in the positive pressure mode. However, to emphasize that it basically is not for emergency use, the PF is limited to 2000.

[m] The design of the supplied-air hood, suit, or helmet (with a minimum of 170 liters/min of air) may determine its overall efficiency and protection. For example, when working with the arms over the head, some hoods draw the contaminant into the hood breathing zone. This may be overcome by wearing a short hood under a coat or overalls. Other limitations specified by the approval agency must be considered before using in certain types of atmospheres.

[n] The SCBA operated in the positive pressure mode has been tested on a selected 31-man panel and the facepiece leakage recorded as <0.01 percent penetration. Therefore, a PF of 10,000+ is recommended. At this time, the lower limit of detection 0.01 percent does not warrant listing a higher number. A positive pressure SCBA for an unknown concentration is recommended. This is consistent with the 10,000+ that is listed. It is essential to have an emergency device for use in unknown concentrations. A combination supplied-air respirator in pressure demand or other positive pressure mode, with auxiliary self-contained air supply, is also recommended for use in unknown concentrations of contaminants immediately dangerous to life. Other limitations, such as skin absorption of HCN or tritium, must be considered.

Table 21.6 Maximum Respirator Protection Factors for Individual Wearers Based on Overall Performance Criteria Measured by Quantitative Man Tests (12)

Type of Respirator	Criteria Allowable max av Penetration (%)	Maximum Permissible PF
I. Air purifying		
A. Particulates		
Type, filter facepiece		
Single-use, dust, and mist	10	10
Dust, mist, or fume, half or quarter mask	5	20
High efficiency, half or quarter mask	2	50
High efficiency, full facepiece	0.2	500
B. Gases and vapors		
Half mask facepiece	2	50
Full facepiece	0.2	500
II. Atmosphere supplying		
A. Supplied-air, mode, facepiece		
Demand, half mask	2	50
Demand, full facepiece	0.2	500
Hose mask without blower, full facepiece	0.2	500
Atmosphere supplying		
B. SCBA, mode, facepiece		
Open circuit, demand, full facepiece	0.2	500
Closed circuit, full facepiece	0.2	500

considered, the results appear to be better. Four of the masks provide a PF of 10 for more than 95 percent of the males tested. For females, only one mask could provide a PF of 10 for 95 percent of the female test subjects. Not surprisingly, the facepieces seem to be reasonably well designed for males. Females have difficulty fitting facepieces that are large, whereas males can readily obtain a good fit from a small facepiece. It appears easier to stretch a small facepiece over a large face than to shrink a large facepiece to a small face.

5.3.3 Air-Supplied Hoods

A study of air-supplied hoods recently completed by LASL provides information not available to Hyatt when he made his recommendations (13). The study of 16 approved supplied-air hood configurations indicates that although a PF of 2000 is achieved at the high flow rate of 400 liters/min, it is not achieved by all hoods at the low flat rate of 170 liters/min. At 170 liters/min, two hoods did not achieve a PF of 2000, one did not achieve a PF of 1000, and all achieved a PF of 500.

The hoods were tested using an exercise regimen intended to duplicate the normal movements of personnel wearing them. The PF for the hoods were influenced by the airflow and by the movements of the hood on the head during exercise. As the hood rises from the head, a negative pressure is created which allows the contaminated atmosphere to enter. Although increased airflow decreases the inward leakage, it is possible to reduce the leakage by (1) ensuring that the hood is firmly attached to the head and no motion independent of the head occurs, and (2) tying the inner collar as snugly as possible. The latter requirement is of primary importance. It should be pointed out that attaching the hood tightly to the head will not eliminate all negative pressure problems. Many of the hoods have aprons that extend down the front, back, and sides. Movement of the arms in these hoods will cause a movement of the hood, and sometimes creating a negative pressure.

Problems were also encountered in achieving the minimum required flow of 170 liters/min following the manufacturer's instructions for proper operation. The customary way of setting flows is through the use of a pressure gauge on the plant air line. It is suggested that a flow measuring device be installed to ensure that the actual flow desired is obtained. The hoses, valves, and fitting for hoods are neglected frequently, and scant attention is paid to whether they are in good working order, or whether the equipment is designed to be used with the hood. The importance of paying attention to such factors cannot be overstressed. An air line that is longer than design length, or has a

Table 21.7 PFs for Half Mask Air-Purifying Respirators (4)

Respirator	Protection Factor	Panel Members Achieving PF, Entire Panel Represented (%)	Panel Members Achieving PF, Male Portion of Panel Only (%)	Panel Members Achieving PF, Female Portion of Panel only (%)
1	10	74	88	65
2	10	94	96	92
	20	90	92	89
	50	83	87	84
5	10	89	96	83
	20	80	91	74
	50	74	83	74
6	10	86	96	78
	20	77	83	65
	50	71	83	65
7	10	43	61	22
8	10	97	100	96
	20	91	91	90
	50	80	83	87

smaller diameter or both, can result in a drastic reduction in the air supply with the same setting of the pressure gauge.

5.3.4 Air-Supplied Suits

Little research has been done of the efficiency of air line suits. Suits considered to be disposable are commonly supplied in two pieces. The air distribution system is rudimentary, consisting mainly of a halo air manifold above the head with a single tube down into the pants. The following recommendations should assist in preventing penetration when using two-piece suits:

1. Use a minimum airflow greater than 170 liters/min.
2. The blouse should be worn outside the pants: (*a*) a 4-inch overlap of blouse and pants is advisable; however the blouse should not extend to the crotch of the pants; (*b*) the blouse should be tightly tied at the bottom.
3. The air volume in the suit should be kept to a minimum. For small persons, the excess suit material should be confined by folding and taping.

These things should assist in reducing the aspiration of contaminant into the suit when the wearer moves.

In addition to the problems of contaminant leakage, the heat load that builds up inside a suit must be considered. Air-supplied suits are impervious costumes that protect the wearer from skin contact as well as protecting the respiratory system. The distribution of air to the extremities helps to reduce the heat build up; however in hot climates a vortex tube should be used to reduce the heat load inside the suit. This device, which requires a modest increase in the compressed air supply, is able to cool the breathing air well below the ambient temperature.

6 PHYSIOLOGICAL CONSIDERATIONS IN THE USE OF RESPIRATORS

Any respirator produces undesirable effects on the wearer. It is uncomfortable, and it may reduce his field of vision, may require him to carry extra weight, may place an additional burden on his respiration system to obtain breathing air, may cause a feeling of claustrophobia, and may result in a general feeling of anxiety, since he is entering a toxic atmosphere. The two areas of greatest interest as far as the physiological effects are concerned have been the respiratory system and the cardiovascular system.

Studies have been carried out to determine the physiological effect of external breathing resistance on both the cardiovascular and respiratory systems. The results indicate that the normal breathing resistances found in approved respirators do not represent a serious stress upon the body. Work rates of 266 kg-m/min for filter masks and 450 kg-m/min for gas masks can be carried out without exceeding the maximum breathing resistances recommended in the NIOSH approval standards (14). Work rates in excess of these could exceed the breathing resistances recommended; however the main effects to be noted would be those of discomfort, not physiological distress.

The research efforts cited used as subjects persons in good health. Questions are raised concerning persons who have impaired respiratory and/or cardiovascular systems. At this time there are no guidelines available, but such individuals should be medically examined before being assigned to use respirators. The examining physician should be given information about the equipment to be used. He should know whether it produces additional inspiratory and expiratory stress, whether it represents an additional weight, as self-contained equipment does, and whether it may cause an increase in the metabolic heat load, as an air-supplied suit does.

7 RESPIRATOR CERTIFICATION

7.1 Approved Respirators

The term "Approved Respirator" has become a part of the language of industrial hygiene. For a period of 50 years "Approved Respirator" designated approval by the Bureau of Mines (BM). This agency has had an interest in the use of respirators in mines since the early 1900s. As mentioned earlier, this experience was used during World War I to develop gas masks for the army. After the war, interest remained in the development of adequate respiratory protection of civilian use. Finally the bureau was assigned the task of setting standards for respiratory protection.

In 1920 the bureau began to publish "Approval Schedules" for specific types of respirator. Manufacturers could voluntarily submit their devices for testing by the bureau, and, if the equipment was satisfactory, it could be sold as BM approved. Since this system proved to be of value to the consumer, industrial health regulatory agencies began to make it mandatory that only BM-approved respiratory equipment be used.

While the federal Occupational Health and Safety Administration was being set up, most of the respiratory approval testing was assumed by NIOSH. The present approval program is carried out in the NIOSH TCB in Morgantown, West Virginia. The BM, under the Mining Enforcement Safety Act (MESA), jointly approves respiratory equipment.

7.2 NIOSH Approval Requirements

The specific regulations concerning respirator approval are to be found in the Code of Federal Regulations (30 CFR 11). These regulations have been gradually changed through the years and have had the effect of upgrading respirator performance. They should be studied carefully because the methods of approval have limitations that affect the use of an approved respirator.

All respirators must pass a fit test to determine that the face to facepiece seal is effective, and in the case of hoods and helmets, to ensure that they are not likely to develop a negative pressure during use. The fit test procedures are given in Table 21.8. Note that

Table 21.8 Respirator Fit Test Requirements for Contaminant Leakage into the facepiece

30 CFR 11 Subpart	Type of Facepiece	Test Atmosphere	Number of People Required for Test	Total Time of Test (min)	Exercises Required
H, Self-contained breathing apparatus (11.85-19)	Half mask, full facepiece, and mouthpiece	1000 ppm isoamyl acetate (IAA)	6	2	None required
I, Gas masks	Half mask	100 ppm IAA	Not specified	8	Four 2-minute exercises specified
	Full face mask	1000 ppm IAA	Not specified	8	Four 2-minute exercises specified
J, Supplied-air respirators[a]	Half masks, full face masks, hoods, and helmets	1000 ppm IAA	Not specified	10	Two 5-minute exercises specified
	Respirators approved for sandblasting	—[b]	Not specified	30	Sandblasting exercises specified

K, Dust, fume, and mist respirators	Half masks, full face masks, hoods, powered and nonpowered	For dusts and mists, no fit test required			
TLV above 0.05 mg/m^3		Fumes: 100 ppm IAA	Not specified	2	Not specified
	Single-use respirators for dust and mist		No fit test required		
TLV below 0.05 mg/m^3	Half mask	100 ppm IAA	Not specified	5	2- and 3-minute exercises are specified
	Full face and hoods, powered and nonpowered, and mouthpieces	1000 ppm IAA	Not specified	5	Same as half mask, above
L, Chemical cartridge respirators	Half masks	100 ppm IAA	Not specified	8	Four 2-minute exercises specified
	Full face, hood, mouthpieces, powered and nonpowered	1000 ppm IAA	Not specified	8	Same as half mask, above
M, Pesticide respirators	Same as Subpart L, Chemical cartridge respirators				

[a] Tests must be made as follows: (1) 4 cfm for tight-fitting facepieces, 6 cfm for hoods and helmets, and (2) 15 cfm for both types.

[b] Silica dust is generated under conditions that duplicate sandblasting. Silica dust concentration is not specified, but dust generation procedures and subject activities are specified so the procedure can be duplicated. Maximum allowable SiO_2 dust within the hood = 0.52 mg/m^3 (TLV for 100 percent SiO_2 = 0.29 mg/m^3).

except for sandblasting hoods, the test atmosphere is isoamyl acetate. The means of detection used is the wearer's sence of smell. Isoamyl acetate fit tests require careful use if the results are to be meaningful. Subjects should be highly motivated and trained before using this equipment.

Significant points of interest for each approval category are listed below.

7.2.1 Subpart H: Self-Contained Breathing Apparatus

Approvals are given for demand and pressure demand modes of operation, and tests are conducted to determine that the oxygen and carbon dioxide concentrations in the air supply are within the prescribed limits. The length of time the air supply lasts is tested under a moderate work load of 622 kg-m/min. SCBA can be approved for periods of 3 to 240 minutes, with 30 minutes being the most common. Units giving less than 45 minutes service cannot be used for underground work.

7.2.2 Subpart I: Gas Masks

Gas masks are defined as full facepieces attached to a canister that can be mounted on the front, back, or chin. An exception is the escape mask which may be a half mask or a mouthpiece regulator. Table 21.2 lists the canister test agents. Note that the canisters are tested against only five materials.

The test agent for organic vapor canisters is carbon tetrachloride. As Nelson's work in Table 21.3 shows, the collection efficiencies of different organic vapors vary greatly in length of service. Caution should be exercised in using approved organic vapor respirators for organic vapors with unknown sorbent properties.

The universal canister is tested with all five of the materials specified. Note, however, that the minimum service life for each component is only 6 minutes, with the exception of carbon monoxide, which must have a minimum service life of 60 minutes. Dust, fume, and/or mist filters may be attached to the gas or vapor canisters; however they must pass the tests listed under Subpart K-Dust, Fume, and Mist Respirators.

7.2.3 Subpart J: Supplied-Air Respirators

Supplied-air respirators are divided into three classes. Classes A and B refer to the hose mask discussed in Section 3.2.1. These are not widely used. Class C refers to half masks, full face masks, and hoods or helmets supplied by a hose with air under pressure. Half masks and full face masks may be obtained in demand, pressure demand, or continuous flow modes. A minimum air flow of 115 liters/min for tight-fitting facepieces and 170 liters/min for hood and helmets are required. A subclassification, CE, refers to devices used in sandblasting. The outer covering of CE devices is designed to withstand the rigors of sandblasting.

Hoses for class C devices can be obtained in 15, 25, or 50 foot lengths up to a maximum of 300 feet. Added hose length requires more pressure to deliver the minimum airflow.

7.2.4 Subpart K: Dust, Fume, and Mist Respirators

Approvals are granted for all types of facepiece under this section. In addition to half and full facepieces, hoods and helmets can be used with powered air-purifying units.

The filters attached to the facepieces may be for protection against dust, and/or fume and/or mist. The approval depends on the aerosol that is used for the test. Table 21.9 summarizes the test aerosols and tells how they are used. Note that the table is divided into two sections, by the TLV of 0.05 mg/m^3. Materials with a TLV below 0.05 mg/m^3 are considered to be highly toxic. Only high efficiency filters that pass the DOP test can be used for such aerosols.

Single-use respirators are approved under this subsection for use against dust producing pneumoconiosis and fibrosis or other dusts of low toxicity. Note that a fit test is not required. The filters must pass the standard silica dust and silica mist test.

7.2.5 Subpart L: Chemical Cartridge Respirators

Chemical cartridge respirators are basically similar to the gas masks in subpart I, but the canister for vapor and gas adsorption is smaller. Chemical cartridge respirators are available as half masks, as full face masks, and as hoods or helmets. Although gases and vapors are the main concern, these respirators may also be approved for use in lacquer and enamel mists. An organic vapor filter is combined with aerosol filter to provide protection from paint spray droplets and the vapor from the paint vehicle.

The gases and vapors for which chemical cartridge respirators are tested are not the same as the materials used to test gas masks (see Table 21.2). Chemical cartridge respirators are not available for carbon monoxide, but they are available for hydrogen chloride and methyl amine. Combination cartridges for protection from vapors and aerosols may also be approved.

7.2.6 Subpart M: Pesticide Respirators

Essentially, a pesticide respirator is composed of a facepiece, an organic vapor canister, and an aerosol filter. The organic vapor used for testing is carbon tetrachloride, as in subparts I and L. As in subpart L, facepieces can be half masks, full face pieces, or helmets and hoods. Although the organic vapor portion of the protective canister remains the same regardless of the facepiece, the aerosol filter must change. As Table 21.9 reveals, pesticide cartridges must pass the silica dust and lead fume tests; chin style canisters, too, must pass these tests. Gas masks with chest- or back-mounted canisters (a category that includes powered air-purifying devices) must pass the DOP aerosol test. During application, many pesticides are in an aerosol form or are impregnated on an aerosol. The particulate filters should provide adequate protection against such aerosols.

Table 21.9 Respirator Filter Test Requirements, 30 CFR 11

	TLV above 0.05 mg/m^3			TLV below 0.05 mg/m^3
	Dust	Mist	Fume	Dust, Mist, and Fume
Aerosol test conditions	Silica dust with a geometric mean diameter of 0.4–0.6 μm; concentration: 50–60 mg/m^3	Aqueous solution of silica that passes 270 mesh; concentration: 20–25 mg/m^3 as silica	Molten lead with oxygen gas torch impinging on surface; concentration: 15–20 mg/m^3 as lead	100 $\mu g/m^3$ of DOP, 0.3 μm diameter
Respirator filters, 30 CFR 11 subpart K				
Respirator with reusable filters	Test for 90 min at 32 liters/min; dust through filter must not exceed 0.5 mg/m^3	Test for 312 min at 32 liters/min; mist through filter must not exceed 0, 25 mg/m^3	Test for 312 min at 32 liters/min; lead through filter must not exceed 0.15 mg/m^3	DOP through filter must not exceed 0.03% of DOP concentration
Powered respirator, tight fitting	Test for 240 min at 115 liters/min; dust through filter must not exceed 0.52 mg/m^3	Test for 240 min at 115 liters/min; mist through filter as silica must not exceed 0.25 mg/m^3	Test for 240 min at 115 liters/min; lead through filter must not exceed 0.15 mg/m^3	DOP through filter must not exceed 0.03% of DOP concentration

Powered respirator, loose fitting	Test for 240 min at 170 liters/min; dust through filter must not exceed 0.52 mg/m^3	Test for 240 min at 170 liters/min; mist through filter as silica must not exceed 0.25 mg/m^3	Test for 240 min at 170 liters/min; lead through filter must not exceed 0.15 mg/m^3	DOP through filter must not exceed 0.03% of DOP concentration
Single-use respirators	Test for 90 min at 40 liters/min cyclic flow; dust through filter as silica must not exceed 0.5 mg/m^3	No test scheduled	No test scheduled	No test scheduled
Respirator filters, 30 CFR 11 subpart M (pesticide respirators)				
Cartridge filters Must pass fume test	All tests to be conducted as required for Subpart K	No test scheduled	Test for 90 min at 32 liters/min; lead through filters must not exceed 0.15 mg/m^3	No test scheduled
Canister filters for chest or back mount Must pass dop test	All tests to be conducted as required for Subpart K	No test scheduled	Test for 90 min at 32 liters/min; lead through filter must not exceed 0.15 mg/m^3	DOP through filter must not exceed 0.03% DOP concentration
Powered respirators, tight and loose fitting Must pass dop test	All tests to be conducted as required for Subpart K	No test scheduled	All tests to be conducted as required for subpart K	No test scheduled

The usefulness of the organic vapor cartridge against all pesticides is questionable, and its use for protection against pesticides should be considered with caution.

7.2.7 Subpart N: Special-Purpose Respirators

Only vinyl chloride respirators are classified as special-purpose devices at this time. Two features of interest are (1) the requirement for an end of service life indicator, and (2) the fact that the length of service life for the canister can be specified by the manufacturer. TCB will then test the canister to determine if the service life is met.

8 SELECTION OF THE PROPER RESPIRATOR

Respirator selection all too often consists of obtaining a device of modest cost that is light in weight, low in breathing resistance, and above all else, "something the men will wear." Such a method may result in serious harm to the users, since a false sense of security may be bolstered by the assignment of respirators that are not adequate for the task at hand.

8.1 Preselection Information

Though not considered an "engineering control," respirators should be engineered for the environment in which they are to be used. It is essential that certain information be obtained before a respirator is chosen for protection. The information required to choose a respirator is no more and no less than that required for any other type of industrial hygiene control.

1. Is the contaminant a dust, mist, fume, vapor, or gas?
2. What is the concentration of the air contaminant?
3. What is the TLV?
4. Is the atmosphere oxygen deficient?
5. Is the material readily detectable below the TLV, and does it irritate the skin, nose, or eyes?
6. Does the concentration found approach that which is considered to be immediately dangerous to life?
7. If the air contaminant is a gas or vapor, can it be absorbed by an available gas or vapor canister?
8. Is the material readily absorbed through the skin?

8.2 Respirator Selection

Respirator selection can begin when the information just outlined has been assembled. The initial type of device can be chosen by following the steps in Figure 21.32. This

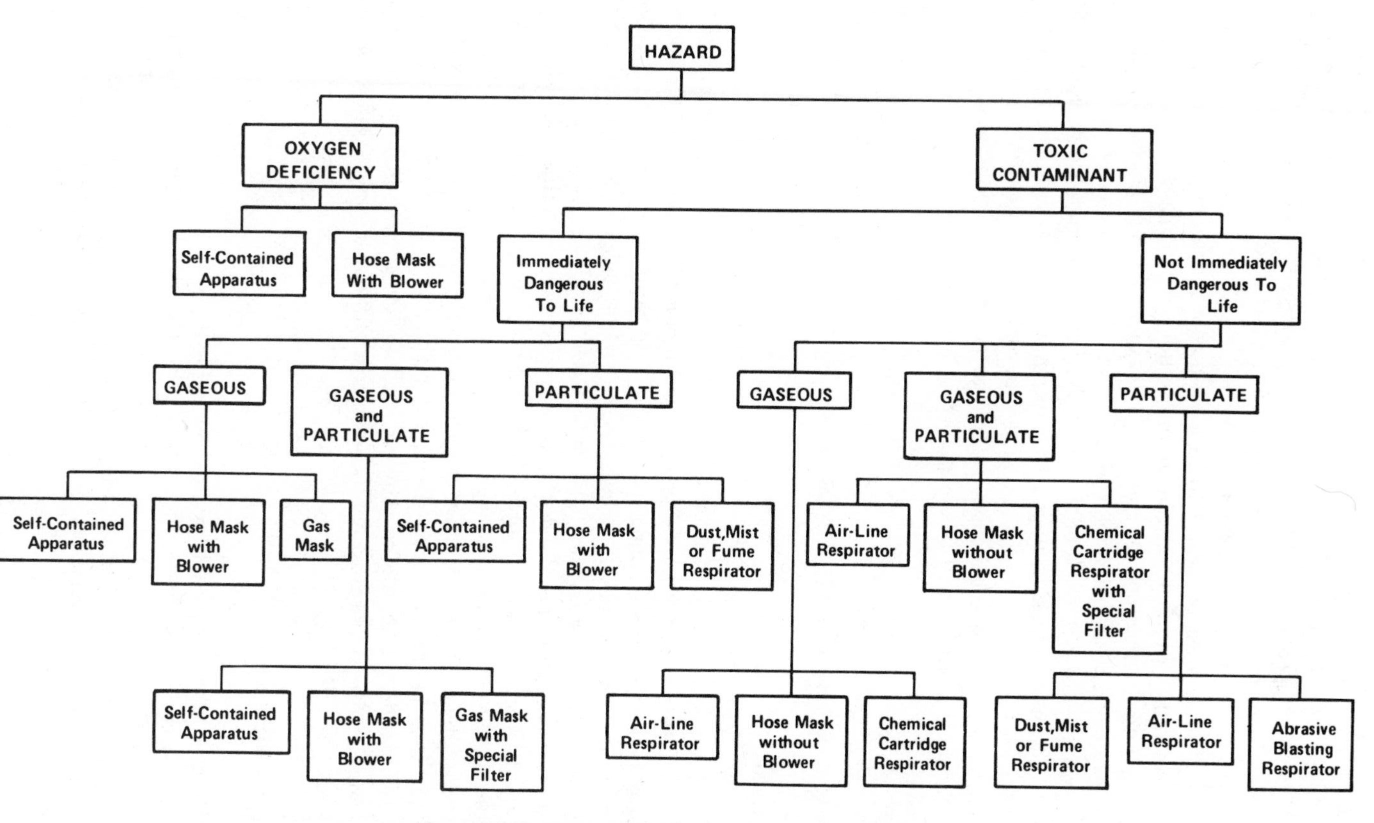

Figure 21.32 Respirator selection according to hazard.

narrows the types of respirators that can be used, but we must determine which devices provide the necessary protection factor. This can be done by using Table 21.5. Finally, it is necessary to determine whether eye irritation is a factor. If so, only full face respirators, which provide eye protection, can be used.

The selection guide must be used with care. Although the use of air-purifying devices is permissible in IDLH (immediately dangerous to life or health) atmospheres under this system, such use must be tempered by such considerations as length of canister life at the concentrations involved, and lack of warning properties at high concentrations. The reader is urged to consult ANSI Z88.2, "Practices for Respiratory Protection" (15) for more detailed information on this subject. In addition, regulations promulgated by regulatory agencies forbid the use of respirators in certain conditions. After making a respirator selection that is satisfactory to the customer, the governing regulations must be reviewed to ensure that the equipment will be acceptable to the appropriate regulatory agency for the intended use.

9 TRAINING IN THE USE OF RESPIRATORS

9.1 General

The proper respirator will be of no value if the wearer is not fitted and trained to its use. As was pointed out in Section 5.2, respirators do not automatically fit the wearer. It is essential that the wearer's training start with a respirator fit test. The preferred test is the irritant fume test, described earlier; successful passage of the qualitative fit test allows the use of the protection factors listed in Table 21.5.

Respirator users may be divided into three classes: routine, occasional, and emergency. All wearers must be introduced to the respirator they will use, receiving information on how the respirator works, the periodic maintenance to be done by the wearer, and problems that can occur from neglect and misuse of the respirator. For example, the user of the air-supplied respirator should be aware of the air pressure necessary to provide the minimum required airflow. He must also be aware that the diameter and length of hose cannot be changed without seriously affecting his air supply.

9.2 The Routine User

The routine user, after his initial training, should be checked periodically to ensure that he is using and caring for the respirator properly. He may be required to perform all cleaning and maintenance in small plants; or, in plants having many respirator users, his responsibility may end when the respirator is turned in at the end of the shift. He should be given periodic refresher instruction in the proper use and care of the respirator.

9.3 The Occasional User

The occasional user will require periodic training after the initial training, since he does not use the respirator routinely. The longer time between uses, the more important it is to have regular interim training sessions.

9.4 The Emergency User

The emergency user needs more training than the routine and occasional users. "Emergency" connotes hazardous exposures to toxic material during escape, during repair work, or during rescue efforts. Serious consequences will result for the user if the measures required by the emergency situation are not executed without mishap. Only persons whose services are necessary in the emergency should be trained in the use of emergency equipment. Such personnel are those who can repair or inactivate malfunctioning equipment, or can effect rescue of personnel in areas with hazardous atmospheric conditions. No one who does not fit into these categories should be allowed to wear emergency respiratory equipment. After the initial training for emergency personnel, it is imperative that there be retraining sessions. These periods should cover not only a checkout of the equipment, but wearing the equipment and performing exercises designed to simulate the actions necessary in an emergency. If such elaborate measures are felt to be beyond the efforts management can support, emergency equipment should not be purchased and employees should be instructed to rely in emergencies on the agencies equipped to carry out such missions—the county and municipal fire departments.

10 AIR SUPPLY FOR SELF-CONTAINED BREATHING APPARATUS AND AIR LINE RESPIRATORS

The advantage of the air-supplied respirator is the use of air separate from the contaminated environment. It is necessary to see that this air is of acceptable breathing quality, and remains so. Regulations require that breathing air meet the minimum standards in Table 21.10.

10.1 Breathing Air Sources

Breathing air can be supplied by the following sources:

1. A hose terminating in a fresh air area.
2. A low pressure compressor with the air intake in a fresh air area.
3. By air that is compressed into a receiver at high pressures (above 2000 psi).

Table 21.10 Standards for Breathing Air (16)

	Breathing Air Grade	
Limiting Characteristics	D	E
Percent O_2 (v/v)	19.5–23.5	19.5–23.5
Hydrocarbons, condensed (mg/m^3) as NTP	5	5
Carbon monoxide (ppm)	20	10
Carbon dioxide (ppm)	1000	500

10.2 Breathing Air Contamination

Breathing air may become contaminated in the following ways:

1. The air intake is located in an area with contaminated air.
2. The oil-lubricated air compressor emits oil mist and hydrocarbon vapors, which enter the air hose.
3. The compressor overheats, or the wrong grade of lubricating oil is used, resulting in partial combustion of the oil. Such partial combustion generates carbon monoxide.

10.3 Prevention of Contamination

The solution to some of the problems listed should be self-evident; however people still persist in placing air intakes in parking lots and near loading docks, or locating the compressor intake for a portable gasoline-powered air compressor next to the engine exhaust. It is suggested that this common-sense matter be given some thought.

There are filters available that will remove the oil mist and hydrocarbon vapor from the compressor airstream, and these must be installed on all oil-lubricated compressors. The removal of carbon monoxide from the compressor airstream is more difficult. There are filters available for this purpose, however they are more adaptable to high pressure compressors than to the low pressure compressors commonly found in industrial plants. If a carbon monoxide filter is not installed, the compressor must be equipped with a high temperature alarm that will shut down the compressor and notify personnel that the compressor has overheated. It is urged that a carbon monoxide monitor be installed to continuously read carbon monoxide concentration in the breathing air. It is a minimum requirement that the air be tested frequently when oil-lubricated compressors are used. Tests should monitor for oil mist, organic vapors, and carbon monoxide to be sure contaminants are not present and that filters are still functioning satisfactorily.

A solution to the problem of carbon monoxide production is to use an oilless compressor. For low pressure systems, the diaphragm pump will supply breathing air for a small number of air-supplied masks. For higher pressure systems, there are compressors available with graphite or Teflon piston rings, as well as water-lubricated compressors.

11 CLEANING, MAINTENANCE, AND STORAGE OF RESPIRATORS

To ensure that the respirator is serviceable beyond the first day it is used, a maintenance program must be put into effect. Respirators must be cleaned daily, and if used by different individuals, sanitized between uses. They must be inspected as they are cleaned, and all defective parts must be replaced before reuse. Finally, they must be stored with the goal of keeping them in prime condition until the next use.

There is no perfect way for these tasks to be accomplished. If the respirators used are few in number, it is preferable to give the user time and facilities to perform his own servicing. If there are many respirators in use, it is necessary and economical to set up a central respirator servicing section. This ensures that all the respirators receive the same service and reduces the surveillance tasks for management. It is important to realize that a definite time is necessary to carry out this service, and it must be made available. The cost of this respirator service is real. If no attention is paid to this phase of respiratory protection, respirators do not get cleaned, repaired, or properly stored. They reside on nails in the shop or paint booth or just outside the sandblast room. Under these circumstances the usefulness of filters, cartridges, or valves is questionable. Unless adequate time can be devoted to servicing, respiratory protection equipment should not be used.

12 SPECIAL PROBLEMS WITH RESPIRATORS

12.1 Respirators and Hair

Hair styles vary geographically and chronologically. During the present period, long, bushy hair, sideburns, mustaches, and beards are in vogue. Such adornment may be highly prized, however it is usually incompatible with the use of a respirator. It is imperative that clean, smooth skin be in contact with the respirator sealing surface. Even a mild growth of whiskers may interfere with this seat. This is illustrated in Figure 21.33. Even small beards and mustaches that fit entirely within the respirator facepiece may cause an exhalation valve to fail if a hair becomes lodged in it. Hair styles in which the hair is teased into a ball several inches thick create problems in maintaining proper tension of respirator head straps.

12.2 Cold Temperatures

In cold temperatures problems of facepiece flexibility, fogged vision ports, and frozen valves must be overcome.

12.2.1 Facepiece Flexibility

Work in cold environments must be planned ahead of time. This makes it possible to choose respirator facepieces that can function in the cold. The manufacturer must be consulted about flexibility problems.

12.2.2 Vision Port Fogging

It is important that the warm exhaled breath be kept from impinging on the vision port (Figure 21.34). The respirator on the right has a nose cup to keep the cold air away from the vision port. Nose cups can be obtained for all full face respirators. In loose-fitting hoods it may be necessary to provide a facepiece to remove the warmed exhaled breath to the outside air instead of releasing it inside the hood.

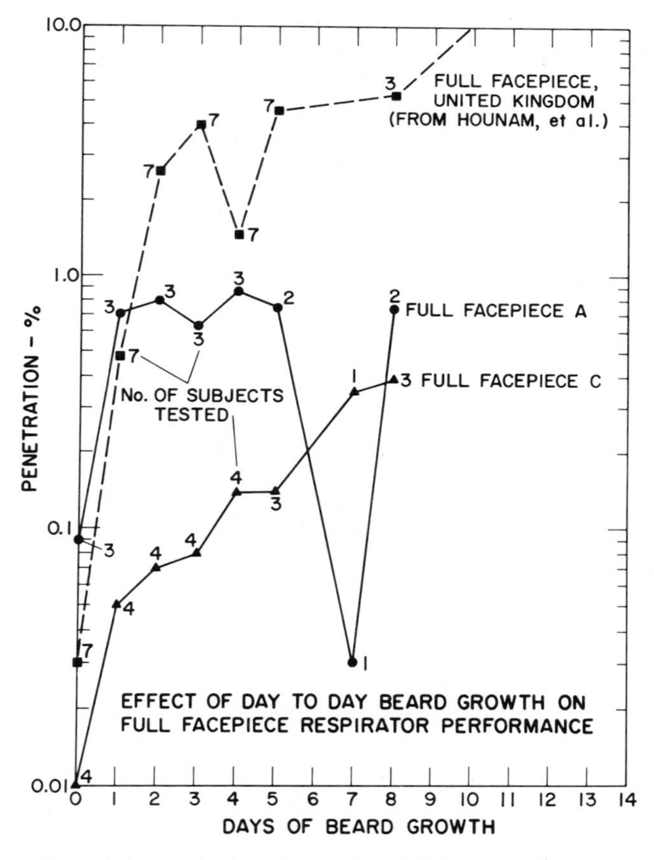

Figure 21.33 Effect of day-to-day beard growth on full facepiece respirator performance.

Figure 21.34 Respirator accessories. Both respirators are equipped with communication devices that penetrate the facepiece. The facepiece on the right has a nosecup to prevent warm breath from coming in contact with the visionport.

12.3 Corrective Lenses

Spectacles. Temple bars on spectacles interfere with the seal of full facepiece respirators and are not acceptable. Full facepieces are required to provide interior appurtenances for mounting spectacles. It is necessary to obtain spectacles designed for this purpose, since it is not practicable to adapt conventional frames.

Half and quarter masks may interfere with proper seating of spectacles. If spectacles are necessary, the respirator fit is not complete unless the spectacles are found to be compatible with the facepiece.

Contact Lenses. Contact lenses are not to be worn with full facepiece respirators. Donning a full facepiece tends to pull at the outer corner of the eyes, and this is the motion used to remove contact lenses. A lense dislodged while a worker is wearing a full facepiece necessitates a trip to an area where the mask can be removed with safety, and such conditions may not always exist.

12.4 Communications

Communication through a facepiece is difficult, and several devices are available to assist the wearer in communicating with others (Figure 21.34).

12.4.1 Facepiece-Mounted Systems

A hole can be made in the facepiece to allow a microphone that connects to an outside amplifier to be mounted inside. These systems are usually heavy and may affect the face seal. A hole in the facepiece may void the TCB approval.

12.4.2 Throat Microphones

Throat microphones do not damage the facepiece, and they may be attached to a variety of amplifiers. Clarity of speech can be a problem when throat microphones are used.

12.4.3 Ear Microphone–Speakers

An earpiece that functions as a microphone and a speaker is now available. The speech is transmitted to the microphone through the eustachian tube. Speech clarity is satisfactory. The device is lightweight and may be mated with any type of speech transmission system.

ACKNOWLEDGEMENTS

Figures 21.2 through 21.17 and 21.19 through 21.34 were provided by the Los Alamos Scientific Laboratory, Los Alamos, New Mexico. Figure 21.18 was provided by the Three M Company, St. Paul, Minnesota. The assistance of these organizations is appreciated.

REFERENCES

1. C. Plinius Secundus, *Historias Naturalis Lib.*, 33, Sec. 11.
2. C. N. Davies, "Fibrous Filters for Dust and Smoke," *Proceedings of the Ninth International Congress on Industrial Medicine,* London, September 12–17, 1948, p. 164.
3. W. P. Yant, "Bureau of Mines Approved Devices for Respiratory Protection," *J. Ind. Hyg.*, **15:** 6 473–480 (1934).
4. D. D. Douglas, W. Revoir, J. A. Pritchard, A. L. Hack, L. A. Geoffrion, T. O. Davis, P. L. Lowry, C. P. Richards, L. D. Wheat, J. M. Bustos, and P. R. Hesch, "Respirators Studies for the National Institute for Occupational Safety and Health, July 1, 1974 through June 30, 1975," Los Alamos Scientific Laboratory Report No. LA-6386-PR, August 1976.

5. Code of Federal Regulations, 30/CRF/11.140-6.
6. Code of Federal Regulations, 30/CRF/11.140-11.
7. G. O. Nelson and C. A. Harder, "Respirator Cartridge Efficiency Studies," *Am. Ind. Hyg. Assoc. J.*, **35:** 7 491–510 (1974).
8. R. A. Fulton, F. F. Smith, and R. L. Busbey, "Respiratory Devices for Protection Against Certain Pesticides," ARS-33-76, U.S. Department of Agriculture, 1962.
9. Code of Federal Regulations, 30/CFR/11.124-2.
10. U.S. Air Force Anthropometric Survey, 1967 and 1968, Anthropology Branch, Aerospace Medical Research Laboratory, Wright-Patterson Air Force Base, Ohio.
11. A. L. Hack, E. C. Hyatt, B. Held, T. Moore, C. Richards, and J. McConville "Selection of Respirator Test Panels Representative of U.S. Adult Facial Sizes," Los Alamos Scientific Laboratory Report No. LA-5488, March 1974.
12. E. C. Hyatt, "Respirator Protection Factors," Los Alamos Scientific Laboratory Report No. LA-6084-MS, January, 1976.
13. D. D. Douglas and P. E. Hesch, "Air Supplied Hood Report," Los Alamos Scientific Laboratory Report No. LA-NUREG-6612-MS, December, 1976.
14. American Industrial Hygiene Association and American Conference of Governmental Industrial Hygienists, *Respiratory Protective Devices Manual,"* AIHA–ACGHI, 1963, pp. 16–17.
15. American National Standards Institute, "Practices for Respiratory Protection" Z88.2-1969, ANSI, New York, 1969.
16. American National Standards Institute, "Commodity Specification for Air," Z86.1-1973, ANSI, New York, 1973.
17. E. C. Hyatt, J. A. Pritchard, C. P. Richards, and L. A. Geoffrion, "Effect of Facial Hair on Respirator Performance," *Am. Ind. Hyg. J.*, **34,** 4, 135–142.

CHAPTER TWENTY-TWO

Ergonomics

ERWIN R. TICHAUER, ScD

1 HISTORICAL BACKGROUND

The systematic study of the ill effects on man of poorly designed work situations is by no means of recent origin. Ramazzini (1), around 1700, elaborated on the disastrous effects of work stress and its disabling consequences for those engaged in physical labor. He wrote: "Manifold is the harvest of diseases reaped by craftsmen. . . . As the . . . cause I assign certain violent and irregular motions and unnatural postures . . . by which . . . the natural structure of the living machine is so impaired that serious diseases gradually develop." At that time, however, and for centuries to come, there existed no technique for the study of the anatomy of function of the living body. Also laborers were considered to be expendable, and occupational disease was ordinarily rewarded by dismissal.

Thus with few noteworthy exceptions, such as Thackrah (2), industrialists, physicians, and even the precursors of today's social scientists, remained insensitive to the exposure of the workforce to ergogenic disease produced by mechanical noxae until World War I, when labor became a scarce resource essential to the very survival of the warring nations. This stimulated physiologists (3), as well as psychologists, to embark on an intensive study of the effects of working conditions on human performance and well-being.

Experimental methods for these disciplines were already well established, and their application to problems of man at work was quite easy. One of the man-oriented life sciences lagged sadly behind, however: anatomy, which had remained a cadaver-based, geographic discipline. Thus physiological responses and behavioral reactions to the demand of work situations were investigated for several decades before the exploration of the structural and mechanical basis of musculoskeletal performance.

Change was pioneered by Adrian (4) who recorded electromyograms during movement, relating them to kinesiological events. The Great Depression slowed down progress in these avenues of scientific inquiry, but during World War II biological and behavioral scientists again had the opportunity to lead the quest for an improved utilization of human resources.

During the postwar period work physiology, industrial psychology, and other specialties consolidated into a broad discipline dedicated to the study of man at work: ergonomics. The Ergonomics Research Society was organized in 1949 to serve the need of those professionals in a variety of disciplines concerned with the effects of work on man.

The founders of the new society argued about the most suitable name, suggesting (5) "Society for Human Ecology" and "Society for the Study of Human Environment." Finally, "ergonomics" was adopted as a term neutral with respect to the relative importance of the behavioral sciences, physiology, and anatomy. An anatomical methodology for the study of work did not yet exist. Thus the thrust of ergonomics had to stem initially from the other two disciplines. However a speedy development of experimental biomechanics and novel instrumentation available added "live body anatomy" to the spectrum of research resources available to ergonomists.

Gradually, as interest grew in the structural basis of human performance, the electrogoniometer was perfected (6). This made it possible to use simultaneously goniometry and Adrian's myography when investigating the relationships between muscular activity and ensuing movements. Thus information about the functional anatomy of the living body could be applied by ergonomists to the problems germane to both occupational health and industrial productivity. One of the numerous pioneers was Lundervold (7), who related myoelectric signals obtained from typists to posture and hand usage (this was perhaps the first comprehensive biomechanical analysis of a common work situation). Thus electrophysiological kinesiology was developed.

Utilizing procedures of electrophysiological kinesiology, Tichauer (8) added the biomechanical profile to the techniques available for the study of interaction between worker and industrial environment. Meanwhile, behavioral scientists such as Lukiesh and Moss (9) pioneered research into the effects of light and illumination on human performance. Work physiologists like Belding and Hatch (10) described and explained the effects of climate on working efficiency and studied noise and its effects on workers, not only with respect to deafness, but also in relation to many other physiological parameters (11). The study of performance decrements due to adverse working conditions became the object of an entire new school of students of human fatigue (12).

Thus today, knowledge gathered from many tributary sciences has been blended into a unified discipline dealing with the effects of work on man.

1.1 Ergonomic Stress Vectors

The basic philosophy of ergonomics (Figure 22.1) considers man to be an organism subject to two different sets of laws: the laws of Newtonian mechanics, and the biological laws of life. It is part of this philosophy to postulate that man in work situations is sur-

Figure 22.1 A worker surrounded by external physiological and mechanical environment, which must be matched to his internal physiological and biomechanical environments, symbolizes the concept of modern ergonomics (biomechanics). From Reference 79.

rounded by the external physical working environment, and inside the human body the "internal biomechanical environment," an array of levers and springs also known as the musculoskeletal system, responds to the demands of the task. The stress vectors commonly acting upon man in the industrial environment have been set out in flow diagram form (Figure 22.2 and 22.3).

1.2 Definition of "Ergonomist"

An ergonomist is a professional trained in the health, behavioral, and technological sciences and competent to apply them within the industrial environment for the purpose of reducing stress vectors sufficiently to prevent the ensuing work strain from rising to pathological levels or producing such undesirable by-products as fatigue, careless workmanship, and high labor turnover. Ergonomics as a discipline aims to help the individual members of the workforce to produce at levels economically acceptable to the employer while enjoying, at the same time, a high standard of physiological and emotional well-being.

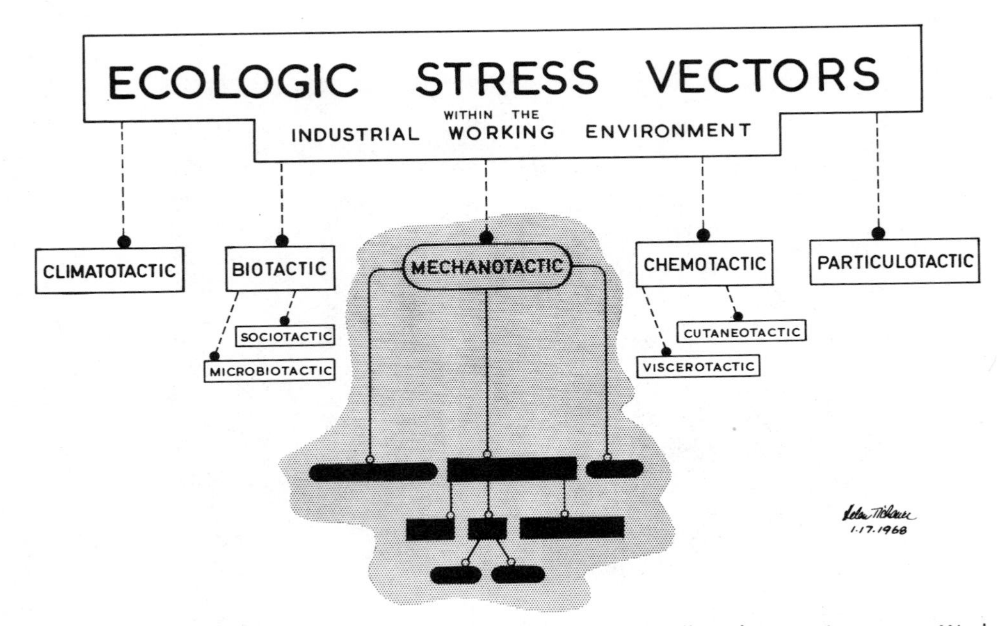

Figure 22.2 The scheme of ecologic stress vectors common to all working environments. Work stress is derived from contact with climate, contact with living organisms such as fellow man or microbe, contact with chemical elements or compounds, contact with hostile particles such as silica, or asbestos and finally, contact with mechanical devices. From Reference 13.

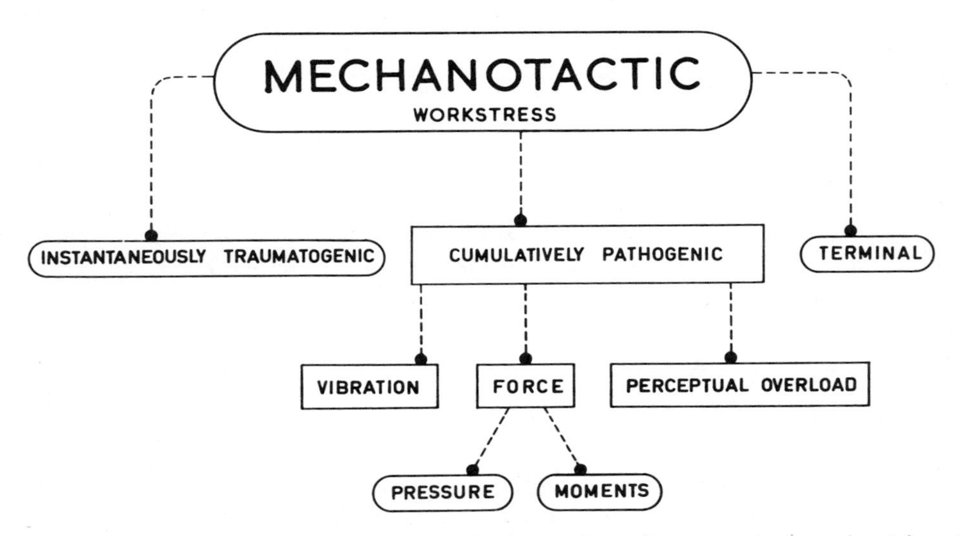

Figure 22.3 The mechanotactic stress vectors leading to hazard exposure in the industrial environment include instantaneous traumatogenesis (e.g., an arm is torn off); terminal (i.e., death occurs immediately) and, most frequently, cumulative pathogenesis. The latter term describes the gradual development of disability or disease through repeated exposure to mechanical stress vectors over extended periods of time. From Reference 13.

2 THE ANATOMY OF FUNCTION

Anatomy is concerned with the description and classification of biological structures. Systematic anatomy describes the physical arrangement of the various physiological systems (e.g., anatomy of the cardiovascular system); topographic anatomy describes the arrangement of the various organs, muscular, bony, and neural features with respect to each other (e.g., anatomy of the abdominal cavity); and functional anatomy focuses on the structural basis of biological functions (e.g., the description of the heart valves and ancillary operating structures; the description of the anatomy of joints). As distinct and different from the aforementioned categories, the anatomy of function is concerned with the analysis of the operating characteristics of anatomical structures and systems when these interact with physical features of the environment, as is the case in the performance of an industrial task. Whenever the "motions and reactions inventory" demanded by the external environment is not compatible with the one available from the internal biomechanical environment, discomfort, trauma, and inefficiency may arise (3).

The anatomy of function is the structural basis of human performance, thus it provides much of the rationale by which the output measurements derived from work physiology and engineering psychology can be explained.

2.1 Anatomical Lever Systems

The neuromuscular system is, in effect, an array of bony levers connected by joints and actuated by muscles that are stimulated by nerves. Muscles act like lineal springs. The velocity of muscular contraction varies inversely as the tension within the muscles. With very few exceptions, lever classifications and taxonomy in both anatomy and applied mechanics are identical. Each class of anatomical levers is specifically suited to perform certain types of movement and postural adjustment efficiently, with undue risk of accidents or injury but may be less suited to perform others. Therefore a good working knowledge of location, function, and limitation of anatomical levers involved in specific occupational maneuvers is a prerequisite for the ergonomic analysis and evaluation of most man–task systems.

First-class levers have force and load located on either side of the fulcrum acting in the same direction but opposed to any force supporting the fulcrum (Figure 22.4). This is exemplified by the arrangement of musculoskeletal structures involved in head movement when looking up and down. Then the atlantooccipital joint acts as a fulcrum of a first-class lever. The muscles of the neck provide the force necessary to extend the head. This is counteracted by gravity acting on the center of mass of the head, which is located on the other side of the joint, hence constitutes the opposing flexing weight.

First-class levers are found often where fine positional adjustments take place. When standing or holding a bulky load, static head movement in the mid-sagittal plane produces the fine adjustment of the position of the center of mass of the whole body, necessary to maintain upright posture. Individuals suffering from impaired head movement (e.g., arthritis of the neck), should not be exposed to tasks in which inability to

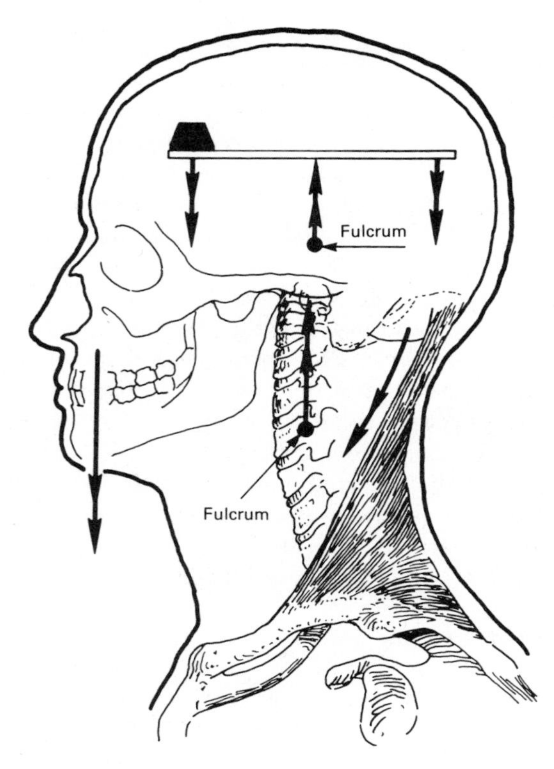

Figure 22.4 The action of the muscles of the neck against the weight of the head is an example of a first-class lever formed by anatomical structures. The atlantooccipital joint acts as a fulcrum. Adapted from Reference 48.

maintain postural equilibrium constitutes a potential hazard. Likewise, workplaces where unrestricted head movement is difficult should be provided with chairs or other means of postural stabilization. Special attention should be paid to any feature within the working environment that may cause head fixation (e.g., glaring lights). Sometimes even an unexpected acoustic stimulus, such as a friendly "hello" directed at an individual carrying a heavy and bulky load, may cause inadvertent sideways movement of the head, which can interfere with postural integrity and result in a fall.

Second-class levers have the fulcrum located at one end and the force acting at the other end, but in the same direction as the supporting part of the fulcrum. The weight acts on any point between fulcrum and force in a direction opposed to both of them. Second-class levers are optimally associated with ballistic movements requiring some force and resulting in modifications of stance, posture, or limb configuration. The muscles inserted into the heel by way of the Achilles tendon (i.e., force) and the weight of the body transmitted through the ankle joint and the weight of the big toe (i.e., fulcrum) are a good example of a second-class lever system used in locomotion (Figure

22.5). The movements of this type of lever are never very precise, therefore foot pedals should have adequately large surfaces and their movement should be terminated by a positive stop rather than by relying on voluntary muscular control. Another example of a second-class lever is provided by the structural arrangement of the shoulder joint. Here the head of the humerus acts as a fulcrum, the anterior and posterior heads of the deltoid provide the force; the "weight" is provided by the inertia of the mass of the arm. Hence it follows that a shoulder swing moves the hand to a rather indeterminate location; thus such a movement should be terminated by a positive stop or, alternatively, when an object has to be dropped at the end of a motion, the receptacle should be sufficiently large, or have a flared inlet, to overcome the kinesiological deficiency.

Third-class levers have the fulcrum at one end, and the weight acts on the other end in the same direction as the supporting force of the fulcrum. The "force" acts on any point between weight and fulcrum, but in a direction opposed to them both. Tasks that require the application of strong but voluntarily graded force are best performed by this type of anatomical lever system. Holding a load with forearm and hand when the bra-

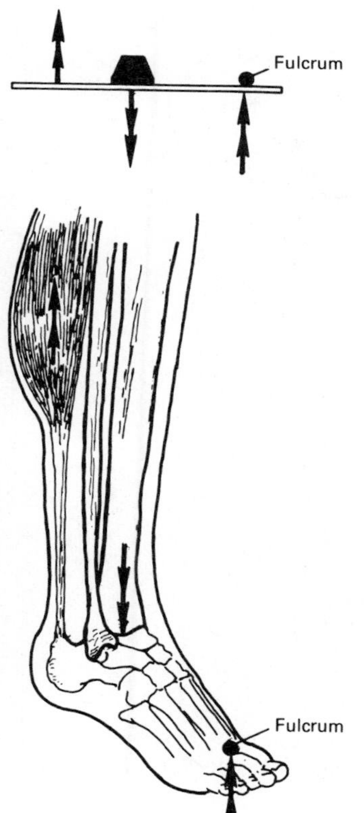

Figure 22.5 The ankle joint, as an example of an anatomical second-class lever system. The fulcrum is located at the base of the big toe. Adapted from Reference 48.

chialis muscle acts on the ulna, with the elbow joint constituting a pivot, is a typical example (Figure 22.6).

Torsional levers are a specialized case of third-class lever. Here the axis of rotation of a limb or long bone constitutes a fulcrum. The force generating muscle of the system is inserted into a bony prominence and produces rotation of the limb whenever the muscle contracts. The "weight" is represented by the inertia of the limb plus any external torque opposing rotation. An example is the supination of the flexed forearm (Figure 22.7). Here the fulcrum is the longitudinal axis of the radius; the force is exerted by the biceps muscle inserted into the bicipital tuberosity of the radius. The opposing load is made up by the inertia of the forearm and hand, plus the resistance of, for example, a screw being driven home. Tasks to be performed for strength and precision and at variable rates of speed are best assigned to torsional lever systems.

For identification and classification of other lever systems involved in specific occupational maneuvers, standard text books on kinesiology (14–16) should be consulted.

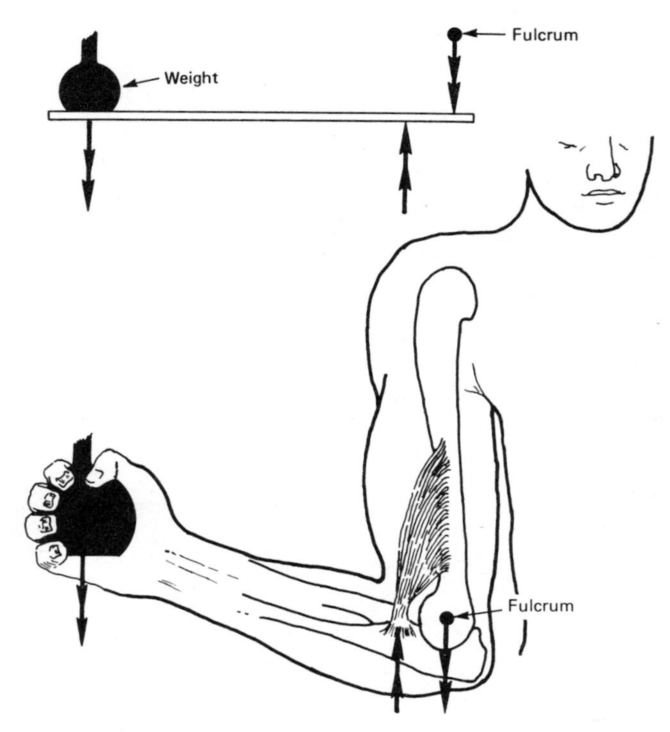

Figure 22.6 An anatomical third-class lever is formed between ulna and humerus. The brachialis muscle provides the activating force, the fulcrum is formed by the trochlea of the humerus. Adapted from Reference 48.

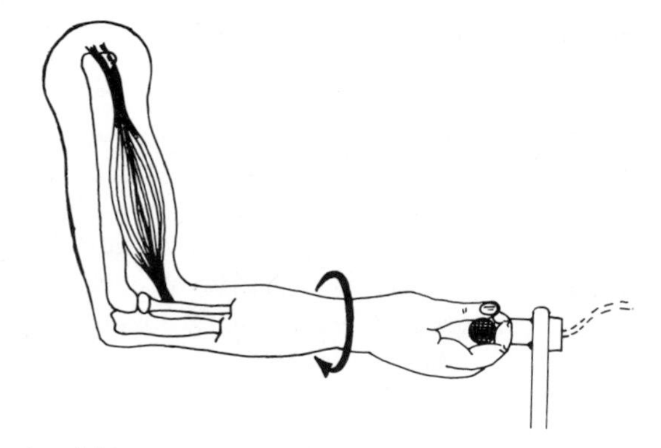

Figure 22.7 The torsional lever system involving radiohumeral joint, biceps, and a resistance against supination of the forearm can be employed with advantage in such operations as closing of valves.

2.2 Occupational Kinesiology

Occupational kinesiology is the discipline concerned with the basic study of human movement and its limitations in work situations. Unfortunately, with the exception of brief monographs (17), all texts and reference books in the field of kinesiology relate either to athletics or rehabilitation medicine.

Kinesiology describes the laws and quantitative relationships essential for the understanding of the mechanisms involved in human performance, either of individuals or of groups of individuals interacting with one another (i.e., a working population). Its basic tributaries are anatomy, physiology, and Newtonian mechanics. It describes and explains the behavior of the whole body, its segments, or individual anatomical structures in response to intrinsic or extrinsic forces.

The student of kinesiology should be thoroughly familiar with the nomenclature and organization of mechanics. Here statics is concerned with the generation and maintenance of equilibrium of bodies and particles. In the context of kinesiology, "bodies" are generally synonymous with anatomical structures and "particles" become anatomical reference points. Likewise, the biodynamic aspects of kinesiology are explained through kinematics, which is concerned with the geometry and patterns of movements, but not with causative forces producing motion. Kinetics, on the other hand, deals with the relation between vectors and forces producing motion and also with the output from body segments in terms of force, work, and power, including the resulting changes in temporal and spatial coordinates of anatomical reference points.

Before the kinematics basic to a specific kinesiological maneuver can be explored, the kinematic element involved must be identified. A kinematic element consists of bones, fibrous and ligamentous structures pertaining to a single joint inasmuch as they affect the geometry of motion. Because kinematics is not concerned with forces, muscles do not normally form part of a kinematic element. As an example, consider the kinematic element of forearm flexion, which consists of the humerus, the ulna, the joint capsule of the humeroulnar joint, and associated ligaments.

Kinematic elements can have several degrees of freedom of motion; the higher this number is, the greater the variety of movements that can be produced. However accurate movements produced by elements possessing a high degree of freedom require a proportionally higher level of skill: for example, it is easier to position the hand accurately by means of humeroulnar flexion than by a shoulder swing. Likewise, the higher the degree of freedom, the greater the influence of musculoskeletal configuration on the effectiveness of movement. As an example, the following are mentioned:

1. Humeroulnar joint: 1 degree of freedom (i.e., flexion); effectiveness quite independent of general musculoskeletal configuration.
2. Whole elbow joint: 2 degrees of freedom (i.e., flexion, pro/supination); effectiveness somewhat dependent on musculoskeletal configuration.
3. Hip joint: 3 degrees of freedom (i.e., flexion, ad/abduction, circumduction); the effectiveness of this chain actually, in a number of situations, such as walking, is quite dependent on musculoskeletal configuration.

Often in workplace layout, kinematic elements are considered in the initial planning of the geometry of the work situation; but when activity tolerance and other work and effort relationships are important, kinetic elements must be considered. These include constituents of kinematic elements, but in addition they incorporate muscular structures as well as the stimulating nerves and nutrient blood vessels because these affect the immediacy, strength, and endurance aspects of a specific kinesiological maneuver (Figure 22.8).

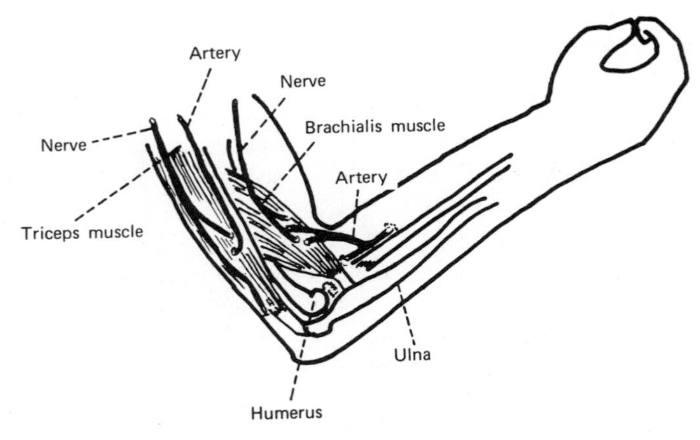

Figure 22.8 The kinetic element of forearm flexion and extension. From Reference 48.

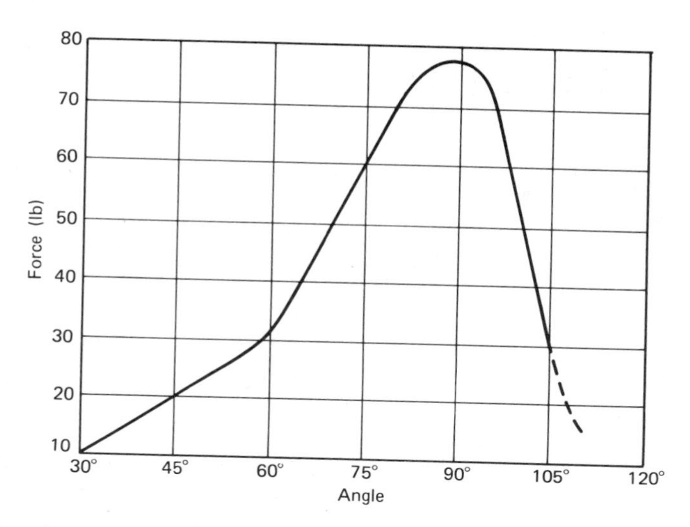

Figure 22.9 Length–tension diagram produced by flexion of the forearm in pronation. "Angle" refers to included angle between the longitudinal axes of forearm and upper arm. The highest parts of the curve indicate the configurations where the biomechanical lever system is most effective.

2.3 Application of Kinesiology to Workplace Layout

Kinesiological concepts can be applied with advantage to the design of work situations when it is essential to minimize physical stress and fatigue. When analyzing motion patterns incidental to the performance of industrial tasks, it is desirable to start with the preparation of a length–tension diagram (Figure 22.9). This is produced by making the protagonist muscle of the kinetic element performing the motion contract isometrically against measured resistance at different points of the motion's pathway. We plot on the x-axis the included angle between the major bony elements involved. On the y-axis the maximal force exerted during contraction in each position is recorded. Generally, the only range of joint movement to be utilized in workplace layout is that which coincides with the highest portion of this curve. The length–tension diagram will show that most kinetic chains can be utilized effectively only throughout a very narrow angle of joint movement, and it will identify the limits of this range.

The force–velocity diagram is another useful graphic representation of the effectiveness of joint movement (Figure 22.10). The rate of change in joint configuration is plotted against maximum forces developed by the muscle, forming roughly a negative exponential curve. The zero velocity value corresponds to the maximum of force. At maximum velocity the force exerted by a muscle approaches zero as a limit. Hence high velocity and high muscular forces are mutually exclusive. The plotting of a force–velocity curve is a complex undertaking not always feasible under field conditions or

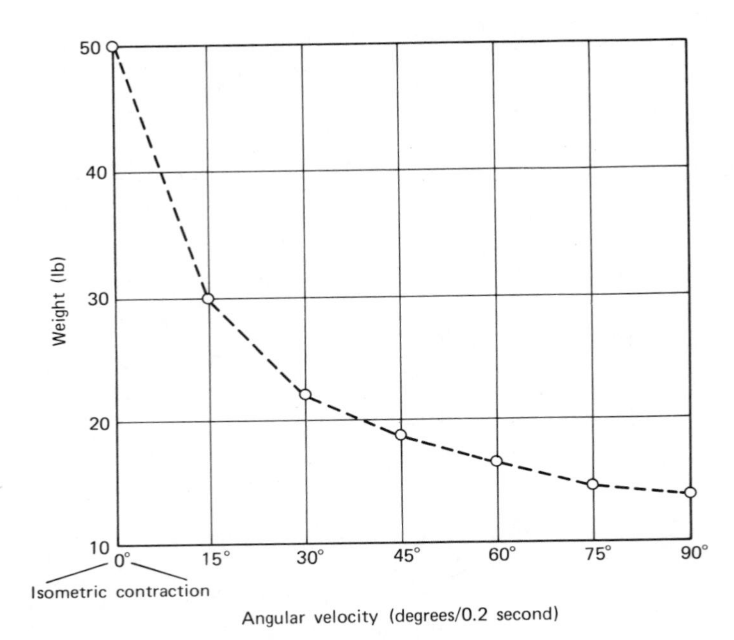

Figure 22.10 Force-velocity curve of elbow flexion; forearm in supination. Excursion of the limb through the narrow range between 75 and 110° of included angle between forearm and upper arm. The diagram shows that high strength and high velocity of movement are mutually exclusive conditions. Furthermore, the highest strength is developed under conditions of zero velocity (i.e., isometricity).

even in the laboratory. However an awareness of its general configuration often serves to protect the practicing ergonomist against the selection of ineffective musculoskeletal configurations in workplace layout.

2.4 Optimal Placement of Equipment Controls

In the operation of many equipment controls and other industrially used devices there is no noticeable displacement of anatomical reference points while muscles contract. Under such conditions, isometricity of movement may be assumed. This implies that forces exerted by protagonist and antagonist muscles acting on a limb are in equilibrium with each other, even when exerting maximal force. Occasionally—for example in the raising of a leg—the force of gravity also acts on a joint; in this case, the sum of all three forces, protagonist, antagonist, and gravity, must be zero.

Kinesiological analysis of anatomical lever systems is of special usefulness whenever it becomes necessary to optimize the position of apparently innocuous but nevertheless potentially traumatogenic equipment controls.

The analysis and computation of some of the forces generated within the kinetic element of forearm flexion during the operation of a pushbutton are described. This may serve as example of the step by step procedure to be followed under similar circumstances.

Step 1. Those anatomical structures absolutely essential to the performance of the task are identified and all others are eliminated from consideration. These are (Figure 22.8) the humerus, ulna, and brachialis muscle.
Step 2. Those forces that, if excessive, may lead to anatomical failure are identified. In this example these are tension in the brachialis muscle and thrust on the elbow joint.
Step 3. A force diagram (such as is used in mechanics) is drawn true to scale (Figure 22.11).

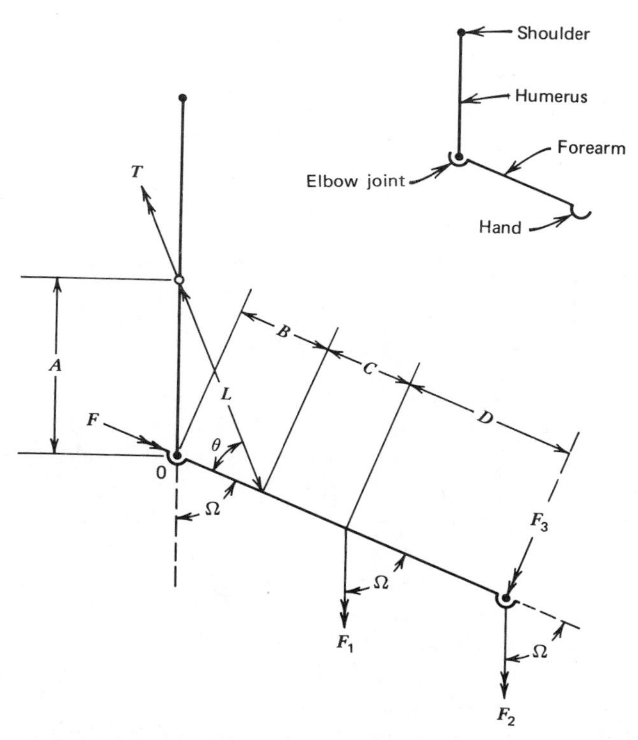

Figure 22.11 Vector diagram used in the computation of forces acting on elbow joint and tension generated in brachialis muscle when operating a push button. *A*, distance from the average origin of the brachialis to the center of rotation of the humero-ulnar joint; *B*, distance from the center of rotation to the average insertion of the brachialis on the ulna; *C*, distance from the insertion to the center of gravity of the forearm; *D*, distance from the center of gravity to the application of the load (either F_2 or F_3); F_1, weight of the forearm and hand; F_2, a weight in the hand; F_3, a force normal to the hand at all angles; *T*, tension in the brachialis; *F*, compressive force exerted on the elbow joint; *L*, distance between origin and insertion of brachialis; Ω, angle of flexion of the forearm; θ, angle of insertion of the brachialis.

Step 4. The mechanical assumptions essential to the solution of the problem are made. As the condition of isometricity exists, the sum of torques acting clockwise on the elbow joint must be equal to the sum of those acting counterclockwise. (The symbols used in the following computations are defined in the legend to Figure 22.11).

The following clockwise torques act on the elbow joint:

$$(B + C)F_1 \sin \Omega \tag{1}$$

$$(B + C + D)F_2 \sin \Omega \tag{2}$$

$$(B + C + D)F_3 \tag{3}$$

The counterclockwise torque is

$$(BT \sin \theta) \tag{4}$$

where T is the unknown tension.

Equating clockwise and counterclockwise torques and regrouping terms, we have

$$T = \frac{[(F_1 + F_2)(B + C) + DF_2] \sin \Omega + (B + C + F)F_3}{B \sin \theta} \tag{5}$$

The force thrusting the ulna against the humerus is expressed as

$$F = \frac{T\,(B + A \cos \Omega)}{(A^2 + B^2 + 2AB \cos \Omega)^{1/2}} - (F_1 + F_2) \cos \Omega \tag{6}$$

If the weight of forearm and hand is assumed to be approximately 10 lb and to act in a vertical direction on the center of mass of the limb, and when a push button control, such as is used on a crane—also estimated at 10 lb—acts normally on the palm of the hand, then depending on the included angle between forearm and upper arm, the tension in the brachialis muscle will vary from 2150 lb to as low as 170 lb. The thrust exerted on the elbow joint will vary from a high of 2140 lb to a low positive value of 5 lb to a strong negative (joint separation) force of 700 lb. It can be seen (Figure 22.12) that unless the push button is located so that the included angle between upper arm and the forearm is between 80° and 120°, the combined effects of tension in the muscle and thrust acting on the joint surfaces can create conditions conducive to joint injury.

If the safe range of joint movement has been exceeded, tensions in the muscles will increase rapidly. Likewise, under the same circumstances, dangerous thrust forces develop within the joint. Alternatively, poor workplace layout may also produce "separation" forces of substantial magnitude. These may not injure the surfaces of the synovial linings but can create conditions conducive to severe luxations. Quantitative biomechanical and kinesiological analysis of man–task systems is essential to the protection of the workforce from the deleterious influences of mechanical noxae.

Standard references (18) permit the rapid quantitative assessment of forces and stresses generated in anatomical lever systems by work. Once the analysis has been

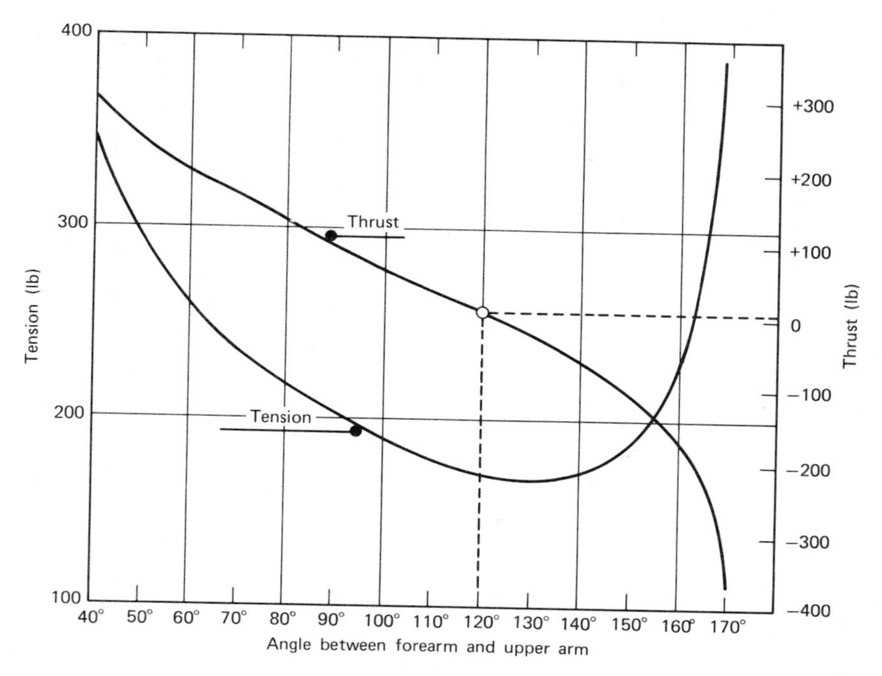

Figure 22.12 Tension in the brachialis muscle and thrust on the elbow joint generated by a push button requiring 10 pounds of pressure, applied normal to the longitudinal axis of the forearm, for its operation. The combined weight of forearm and hand is assumed to be 10 pounds. (X° = 180° minus Angle.)

made, standard reference tables (19) should be consulted to ascertain whether muscles are stressed to excess or are too weak to accomplish the task as planned.

The term "muscular strength" is defined as the maximum tension per unit area that can be developed within a muscle. A maximal effort resulting from the strongest of motivations will yield a tension of approximately 142 psi. However many authors (20–22) agree that under normal working conditions, heavy work would generate approximately 50 to 60 psi tension, while light to medium work would produce tension values of the order of 20 to 30 psi. All these values presume maximal shortening of the muscle and are applicable only to continued "ordinary" nonathletic work situations (Table 22.1).

Occasionally the force available from muscles involved in a kinetic chain is adequate to perform a given task but is unavailable because of a condition known as "muscular insufficiency." A state of active insufficiency exists whenever a muscle passing over two or more joints is shortened to such a degree that no further increase of tension is possible and the full range of joint movement cannot be completed. For example, when a person is seated too low, if the knee joint is hyperflexed it becomes impossible to plantar-flex the foot and operate a pedal. On the other hand, passive insufficiency exists when, in a

Table 22.1 Maximal Work Capacities of Flexors of Elbow (19–22)

	Forearm in Supination		Forearm in Pronation	
Muscle	Cross Section (in.^{2})	Maximal Work Capacity (lb-ft)	Cross Section (in.^{2})	Maximal Work Capacity (lb-ft)
Brachialis	1.0	28	1.0	28
Biceps	1.1	35	0.8	25
Pronator teres	0.5	9.11	0.65	12
Whole flexor forearm as lumped muscle mass (e.g., grip strength)	3.2	90		

particular limb configuration, the antagonists passing over one joint are extended to such a degree that it is impossible for the protagonist to contract further. A good example is the inability to close a fist and hold a rod while the wrist is hyperflexed (Figure 22.13).

3 PHYSIOLOGICAL MEASUREMENTS

Physiology is the discipline that deals with the qualitative and quantitative aspects of physical and chemical processes intrinsic to the function of the living body. As far as the study of man is concerned, the term "physiology" relates by general agreement to the function of the healthy body. The mechanisms of body functions specific to disease form the field of pathophysiology. Both these disciplines are extremely wide and cover almost all aspects of life or death.

Work physiology is a much narrower field. It is restricted to the effects of work and exercise on physiological function. In industrial practice, work physiologists tend to limit themselves to the study of two narrow aspects of this discipline: the assessments of man's capacity to perform physical work, and the study, as well as description, of the effects of fatigue. Most quantitative results obtained from work physiological evaluation are "output measurements." They constitute physical behavior resulting from the combined effects of a variety of inputs. Oxygen metabolism is not purely a function of severity of exercise or of the lean muscle mass involved; it is also determined, to some degree, by the obesity of the subject, postprandial status, and several emotional factors. The result, however, will be one single number: oxygen uptake expressed in milliliters per minute. Solely on the basis of this result, it will not be possible to make any meaningful statement about the magnitude or quality of the contributing vectors. Therefore most work physiology procedures measure effect but do not permit one to establish cause, except in rare cases of near-perfect technique and experimental design.

The following procedures are commonly employed in industry:

1. Metabolic and quasi-metabolic measurements.
2. Electromyography.
3. The measurement of cardiac performance.
4. The measurement of body temperature and heat loss from the body.

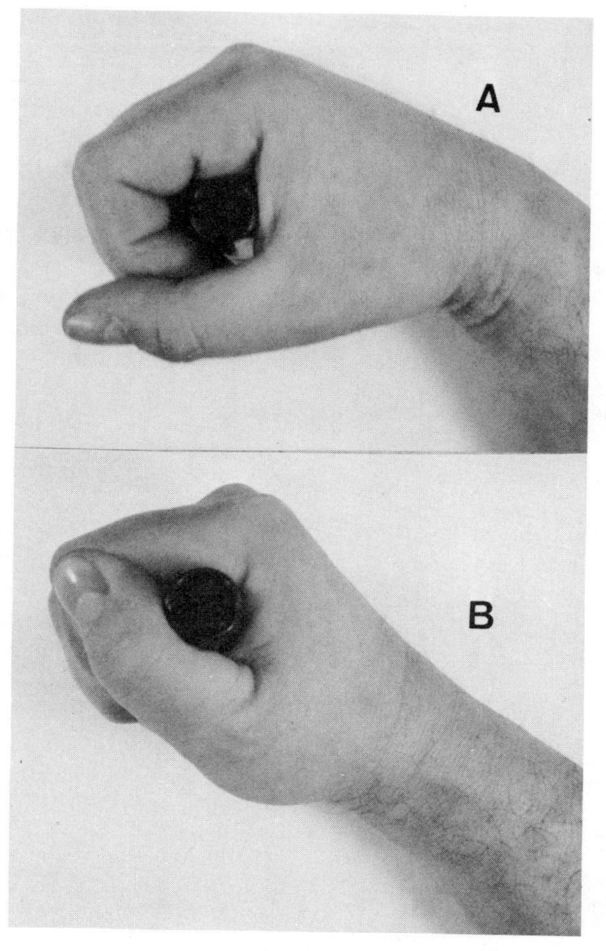

Figure 22.13 The flexed wrist (*a*) cannot grasp a rod firmly. (*b*) The straight wrist can grip and hold firmly. From Reference 48.

Procedures 3 and 4, however, fall within the purview of the trained specialist and are therefore not discussed here.

3.1 Metabolic and Quasi-Metabolic Measurements

It is common practice in some branches of industry to determine physiological energy expended in the performance of certain tasks by direct measurement or indirect estimation of oxygen consumption. During respiration, oxygen passes through the walls of the pulmonary alveoli and surrounding capillaries into the blood vessels. It is then taken up by the red blood corpuscles, which are pumped by the heart to body tissues, such as muscles. Oxygen is unloaded in the muscles to take part in physiological combustion processes. The bioenergetics may be represented as follows (23):

At rest

chemical energy in nutrients contained in body tissues + oxygen → heat (7)

At work

chemical energy in nutrients contained in body tissues + oxygen → heat (approximately 78%); → energy available for work (approximately 22%) (8)

1. One liter of oxygen consumed releases approximately 5 kcal.
2. 5 kcal ≈ 21,000 J.
3. 4600 J (22 percent of 21,000 J) is available for the performance of work.

Thus energy available per liter of oxygen consumed per minute is approximately 0.1 hp. However ambient fresh air contains approximately 20 percent oxygen, of which, roughly a quarter, or 0.05 liter, becomes available for metabolic activities.

Therefore, as a bench mark figure, it may be assumed that 20 liters of air per minute passing through the lungs over and above normal rest levels of pulmonary ventilation corresponds roughly to 0.1 hp expended in the pursuit of physiological work. This constitutes, however, an extremely crude and not always accurate approximation. The true level of net energy expenditure per liter of air passing through the lungs depends on a wide variety of physiological variables. Yet this has not deterred some industrial enterprises from basing computations of physiological effort, often under conditions of heavy work, on pulmonary ventilation (24). Most commonly, the measurement is performed by a knapsack gasometer, and various models are commercially available (Figure 22.14). To obtain readings that are of any use at all, it is necessary to measure total airflow through the lungs from the onset of the task to be investigated until the time after termination of work when pulmonary ventilation has returned to resting level. To obtain the net airflow ascribable to the demands of the task, an air volume corresponding to pulmonary ventilation at rest is then subtracted from the total.

Figure 22.14 Metabolic measurement with the knapsack gasometer. Adapted from Reference 80.

More accurate is a procedure developed by Weir (25). In its application, both the knapsack gasometer and an oxygen analyzer are employed. The volume of air passing per minute through the lungs is measured and reduced to the volume corresponding to standard temperature, pressure, and dryness. To compensate for the inaccuracies of the equipment used, this value *must* be further corrected by multiplication with a "calibration constant" specific for the actual individual instrument used. The result is obtained as follows:

$$\text{kcal/min} = \frac{1.0548 - 0.0504V}{1 + 0.082d} \times \text{liters vent/min} \qquad (9)$$

where V = percentage (vol) oxygen in expired air
d = decimal fraction of total dietary kilocalories from protein

Under most circumstances it is possible to ascertain the protein content of a workman's diet with reasonable accuracy. There is general consensus that this value is relatively constant for each individual.

Methods of metabolic measurement more accurate than those just outlined are complex and sophisticated and should be attempted only by experienced work physiologists. Therefore they are not discussed here. Practitioners active in enterprises where metabolic measurements are taken routinely should familiarize themselves with the basic theory and appropriate techniques through reliable references (26).

3.2 Electromyographic Work Measurement

The aforementioned changes in pulmonary or metabolic activity when measured and quantified by suitable instrumentation can be indicative of the level of effort demanded for the performance of a specific task. Changes in metabolism and pulmonary ventilation represent both dynamic and isometric work performed by muscles; therefore the accuracy of measurement is greatest whenever the musculature of the entire body, or at least large muscle masses such as those of the thighs or the back, must be applied to the successful completion of a job. However in light work, where only mild muscular activity takes place or only small muscle groups or single muscles are utilized, the percentage change in metabolic activity is proportional to the relationship

$$\frac{m}{M} \times 100 = \Delta u$$

where M = total lean muscle mass of body
m = mass of lean muscle applied to task (10)
Δu = percentage change of metabolic activity due to work

In many instances this percentage may be too small to permit a meaningful statement to be made about the level of effort expended, and under such circumstances experimental and computational error may render the result meaningless.

Whenever light work or effort expended by small kinetic chains is to be determined, electromyography can serve as a useful estimator of effort and fatigue.

The term "electromyography," as well as the purpose of the procedure, mean different things to ergonomists, anatomists, physiologists, and physicians. Electrodes, signal conditioning, and display instrumentation vary widely between professions.

In ergonomics, myoelectricity may be assumed to be the "by-product" of muscular contraction which makes it possible to estimate strength and sequencing of muscular activity through techniques noninvasive to the human body. The myogram is an analogue recording of this bioelectric activity.

Each individual muscle fiber maintains in its resting state a negative potential within its membrane wall. This is termed the "resting potential." Excitation produces a transient reversal of the resting potential, causing a characteristic "depolarization" pattern to appear. A discussion of natural and quantitative events related to changes in membrane potential is beyond the scope of this chapter, and specialized literature should be consulted (27).

Modern electrophysiological thinking assumes (28) that muscle fibers probably never contract as individuals. Instead, small groups contract at the same moment. It has been established that all the fibers in each contracting group are stimulated by the terminal branches of one single nerve fiber, the axon of a motor cell whose body is located in the grey matter of the spinal cord. The nerve cell body proper plus the axon, its branches, and the muscle fibers supplied by them, has been named a "motor unit" (Figure 22.15). Since an impulse from the nerve cell causes all muscle fibers connected to the axon to contract almost simultaneously, the action potential resulting from such contraction constitutes the elemental event basic to all electromyographic work. This signal can be picked up by a variety of electrodes and amplified and recorded.

By insertion of very fine needle or wire electrodes into the muscle, it is possible to display action potentials generated by a single motor unit. These individual action potentials, or a sequence of them, are mainly used for electrodiagnostic purposes in medicine. The transducers of choice for electromyographic work measurement are surface electrodes, either permanent or disposable disks of silver coated with silver chloride. More recently, conductive adhesive tape has become commercially available and is often preferred because of economy, flexibility, and ease of application (Figure 22.16).

Needle electrodes are best avoided in ergonomics. They are invasive, produce pain, and call for great skill in application, moreover, their use entails always a certain risk of infection, particularly under the circumstances prevailing on the shop floor or in the industrial laboratory. Even if the aforementioned difficulties could be overcome, however, the signal generated by needle electrodes is only of limited use to the ergonomist unless he specializes in the more basic aspect of work physiology or electrophysiological kinesiology. The study of single action potentials does not generally permit us to arrive at conclusions relating to the total effort expended by a muscle, and

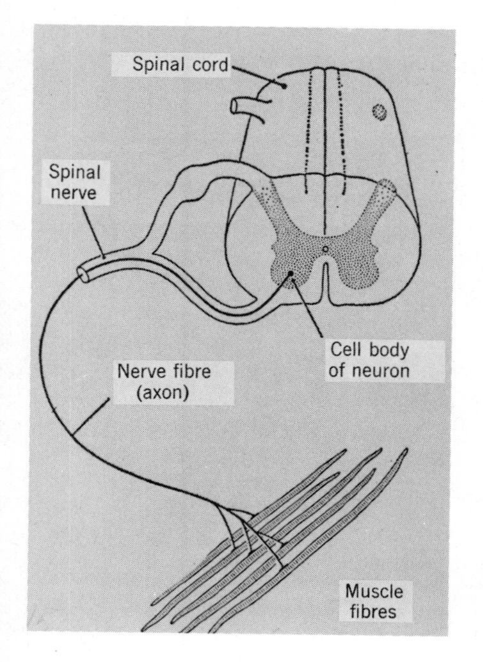

Figure 22.15 Scheme of motor unit. From Reference 28, by permission. © Williams and Wilkins, Baltimore, Md.

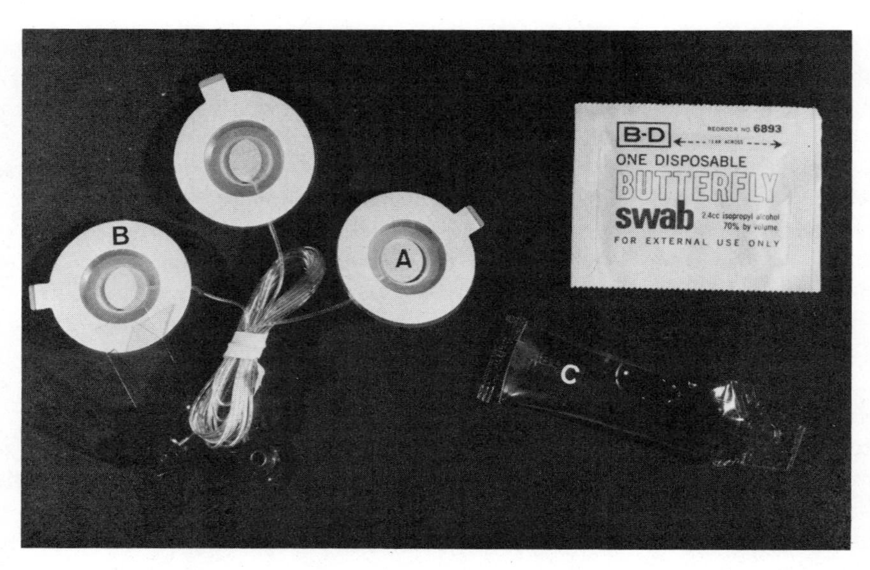

Figure 22.16 A disposable electrode kit, which is inexpensive, noninvasive, and avoids danger of cross-infection between subjects: *A*, electrode; *B*, adhesive collar; *C*, conductive jelly. The elimination of substantial amounts of time spent on the cleaning of permanent electrodes makes use of this kit very economical.

this is why adhesive surface electrodes are preferred in ergonomics. The principal argument against the use of surface electrodes is that unlike needles, they do not permit the study of single action potentials; rather, the signal gathered by them is merely a representation of the level of contractile activity within a relatively large volume of muscle considered to be a "lumped muscle mass" (29). This, however, is the specific advantage of surface myography in the study of the activity of whole muscles as opposed to single motor units.

Thus the surface electromyogram may be considered to be representative of the sum of electrical activity generated simultaneously by a large number of motor units. To understand in more detail the theoretical basis of the procedure, the reader should consult specialized reference works (30).

3.3 Electromyographic Technique

In usual practice the myoelectric signal, which is in the microvolt or millivolt range, is gathered by suitably placed electrodes and amplified by a factor of approximately 1000 prior to display by oscilloscope or oscillographic pen recorder.

Myographic apparatus embodying high gain operational amplifiers is generally inexpensive and requires only two electrodes for the production of a myogram. These characteristics make its use more economical, especially where a large number of readings are to be taken. However the operational amplifier, at the levels of magnification involved in myography, produces a noisy signal, often difficult to interpret. Furthermore, a two-electrode system does not permit the investigator to produce repeatable results easily.

Where simplicity of operation and noise-free signals under field conditions are desired, apparatus embodying differential amplifiers is definitely the equipment of choice. A differential amplifier uses three electrodes: one reference, and two active ones. However it augments only the difference between the two active electrodes based on the potential difference between each of them and the reference. Since any interference from external causes will produce identical changes in the potential of all three electrodes, the display signal will not change. This "common mode" rejection makes it imperative to use differential amplifiers in such settings, where electrical interference from fluorescent tubes, radio transmitters, and other equipment is abundant.

In most instances it is desirable (31) to record the myogram by means of one of the numerous commercially available oscillographic recorders, equipped with modular amplifiers and couplers provided by the manufacturer (Figure 22.17).

Most standard recording equipment has been specifically designed for short-term

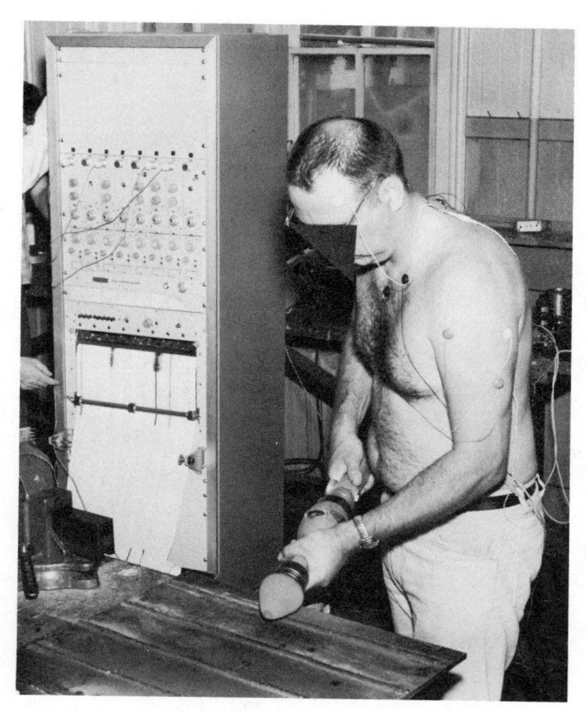

Figure 22.17 The recording of surface electromyograms by means of a commercially available oscillographic recorder for the purpose of physiological work measurement.

myography. Whenever data gathering exceeds one hour of operational time, large machines can become uneconomical, and it is recommended that modern miniaturized magnetic tape recorders be employed. These can be easily attached to the belt of a subject, and they record myograms continuously for up to 24 hours on miniature magnetic tape cassettes. In ergonomics the myogram serves three main purposes:

1. To determine the level of effort expended by a specific muscle mass.
2. To determine the nature of sequencing of protagonist and antagonist muscles involved in specific kinesiological maneuvers.
3. To identify or predict localized muscular fatigue.

Correct electroding technique is a prerequisite to successful myography. It starts with muscle testing to determine the location and surface relationships of the individual muscle to be investigated (Figure 22.18).

By way of example, we discuss the relationship between the biceps and the brachialis muscles. The brachialis is a short and stout muscle that originates from the lower third of the humerus and inserts into the ulna, just distal to the coronoid process (Figure 22.6). It is a powerful and precise flexor of the forearm (see Section 22.2.1). More superficially situated, and covering the brachialis, is the biceps. This muscle originates from the scapula and inserts medially into the proximal end of the shaft of the radius (Figure 22.19).

The biceps is a powerful supinator of the forearm but a comparatively weak flexor. Quite often, especially in materials handling, it is desirable to ascertain the relative magnitude of involvement of either the biceps or the brachialis in the maneuver under study.

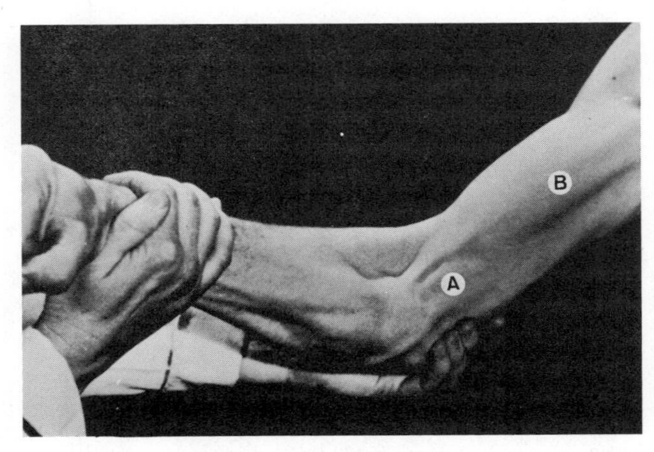

Figure 22.18 Correct procedure for muscle testing produces the contours and relationships of the biceps *B* as well as the brachialis muscle *A*. Adapted from Reference 32, by permission. © Williams and Wilkins, Baltimore, Md.

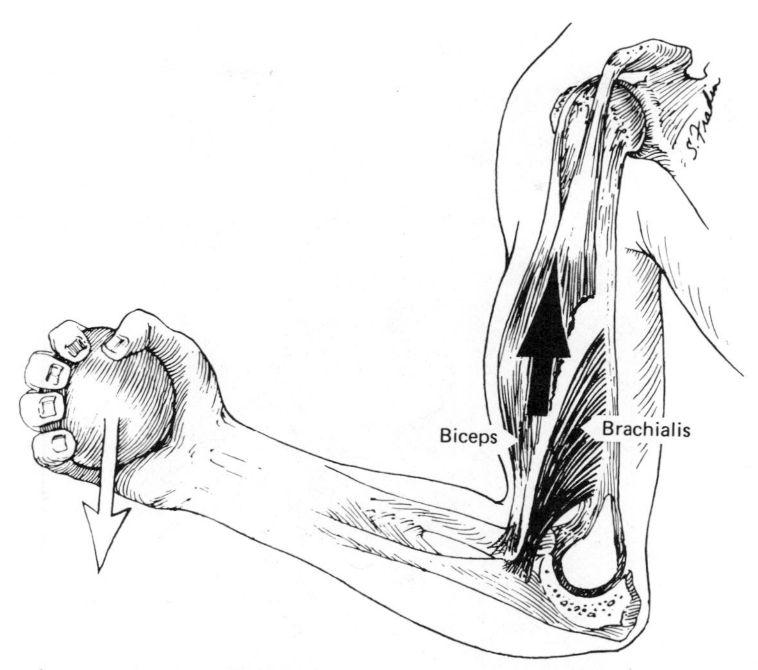

Figure 22.19 The topographic relationships between two muscles: the biceps and the brachialis. The biceps is superficial to the latter, inserts into the radius, is a powerful supinator but a less effective flexor of the forearm. The brachialis is a short but powerful muscle connecting humerus and ulna; because of the character of the humeroulnar joint as a hinge, this muscle is a very powerful flexor of the forearm. Separate electroding of the two muscles is desirable in electrophysiological work measurement. Adapted from Reference 81.

These two muscles are located so close together that it is not possible to obtain separate surface myograms for each one unless special precautions are taken. First the brachialis is palpated. This is done by asking the subject to flex the forearm to form an angle of 90° with the upper arm. Then the forearm is pronated strongly against resistance, while simultaneously the experimenter causes the brachialis to flex the forearm against powerful opposition (Figure 22.20). The outlines of the muscles are then palpated and the electrodes applied so that a maximum of muscle mass is triangulated by them. The reference electrode is placed conveniently over the triceps tendon. The subject is asked to supinate the flexed forearm strongly against external resistance, and the biceps is palpated and triangulated with an additional set of electrodes (31).

It is essential to provide a separate reference electrode for each muscle investigated. To verify the electrode placement, the subject is first asked to flex the forearm isometrically while the limb is being actively and strongly pronated. This yields a strong brachialis signal but little activity in the biceps (Figure 22.21). Subsequently the limb is supinated against resistance while being flexed strongly against an opposing force. This

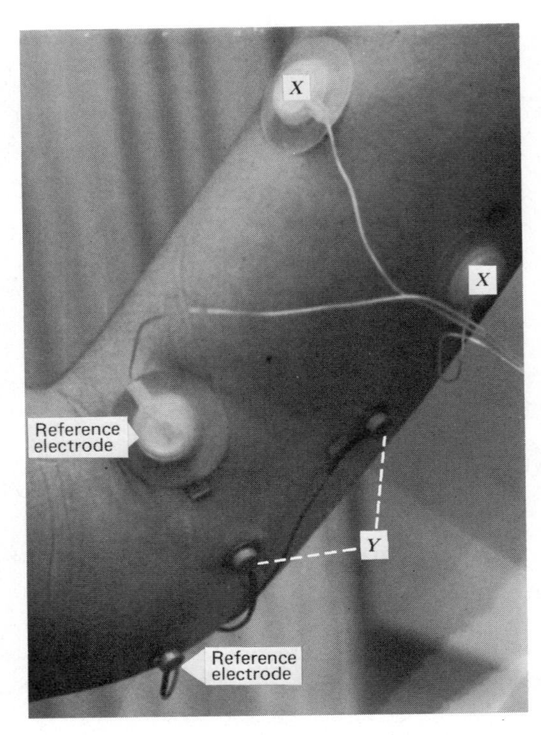

Figure 22.20 Differential electroding of biceps and brachialis: *X*, active electrodes of biceps; *Y*, active electrodes of brachialis.

produces a strong biceps myogram, concurrent with light to moderate myoelectric activity in the brachialis (Figure 22.21). Standard reference works (32) should be consulted if it is necessary to acquire proficiency in electroding the muscles contracted during the performance of common industrial elements of work.

3.4 Interpretation of Myograms

Once it has been established through muscle testing that the electrodes indeed produce a specific signal, representative of the level of activity in the single muscle under investigation, the real task is performed and the myogram interpreted.

An indispensable prerequisite to the correct interpretation of the signal is an understanding of the circuitry that produces the display and of the operating characteristics of the recording apparatus. Due to the nature of the procedure, myograms produced by surface electrodes are records of the summed signals from a number of action potentials generated simultaneously, near-simultaneously, or consecutively, within the muscle mass triangulated by the electrodes. Since the interval of time elapsing between the peaking of individual action potentials may be as little as a few microseconds, readout devices and pen recorders may be "overdriven." The signal is then not representative of the action potentials but is conditioned and distorted as a function of the quality of the recording device (Figure 22.22). Even a change in the viscosity of the recording ink may produce a drastic change in the pattern of a tracing from the same amplifier reproduced by the same recorder. Likewise, pen inertia and the quality of maintenance the instrument has received may cause a badly distorted signal, obfuscating the physiological status of the muscles studied.

To obtain a satisfactory resolution of the direct surface myogram, it is necessary in most instances to run paper recorders at speeds of 12.5 cm/sec. This procedure is uneconomical, and because of friction between pen and paper it gives a completely distorted signal. Therefore in ergonomics a conditioned type of myogram is employed, called the "integrated myogram." However no true integration has taken place. Integrators are circuits that produce a signal that essentially records the sum total of the action potentials counted over a sampling interval of time. This type of myogram is therefore representative of the total number of muscle fibers contracting at any instant. Because of the physiological "all or none" law, it is also representative of the effort expended at any instant. Therefore the area under the integrated myogram is proportional to the total effort made during the time interval under consideration. The generation of this signal requires a much slower recording speed, as low as one centimeter per second, and therefore it can be reliably reproduced by almost any recording device available (Figure 22.23).

A myogram must not only be read, it must be "interpreted." This entails recourse to

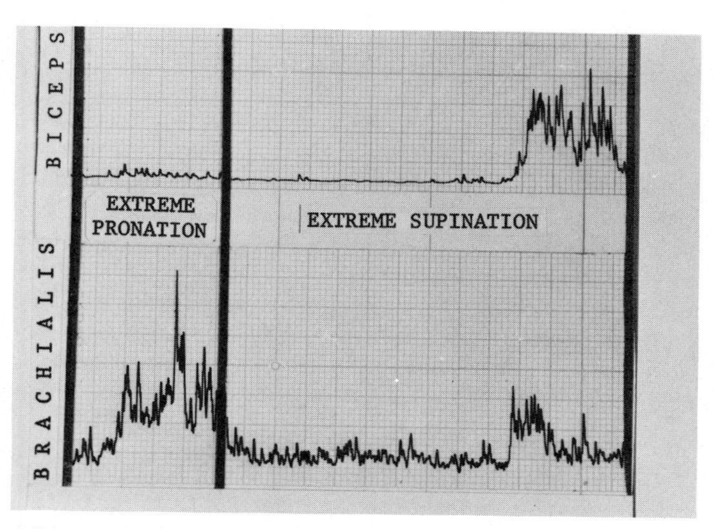

Figure 22.21 Differential myograms from both biceps and brachialis obtained while hand held 20 pounds of weight; the included angle between forearm and upper arm was approximately 100°.

a judgmental process that takes into consideration magnitude of pen excursion, general pattern of the tracing, and some features hard to define, such as fuzziness of recording.

The ability to interpret a myogram is an acquired skill, which can be developed within a short time, provided the ergonomist restricts himself, at least in the beginning, to work with tracings obtained from the same make of recorder. This has a twofold advantage. In the first instance, it is easy to develop an appreciation about the "soundness" of the tracing. Second, the integrated surface myogram, the only type of myoelectric readout considered in this chapter, constitutes a signal conditioned and transformed into shapes and patterns, which may vary for different designs of recording equipment.

Recording speed should be kept constant to facilitate recognition of patterns and to keep slopes and tracings uniform. Many practitioners prefer to work at a recording speed of one centimeter per second. Whenever an exchange of myographic information with other workers is planned, it is advisable for all to adopt the same paper speed. Recorders designed to produce integrated myograms take into consideration both the fre-

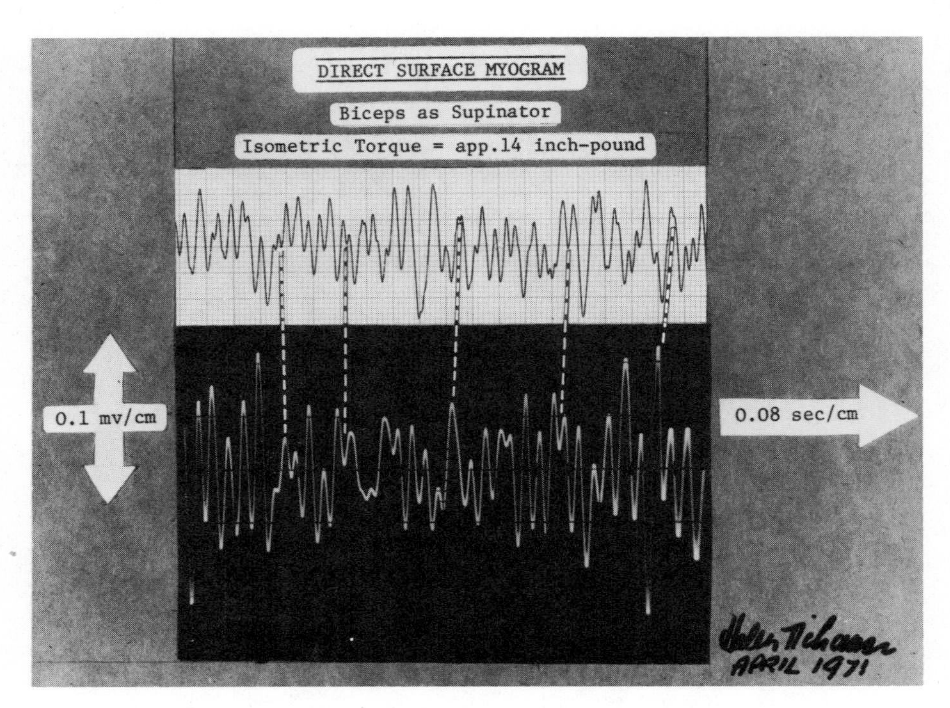

Figure 22.22 A simultaneous recording of the biceps muscle firing pattern displayed on a chart recorder (*upper half*) and an oscilloscope (*lower half*) at exactly the same sensitivity and speed. Only five points of similarity are evident because the signal speed of the myogram exceeds the rise time and slew rate of commercial chart recorders. From Reference 82, by permission. © American Institute of Industrial Engineers, Inc., Norcross, Ga.

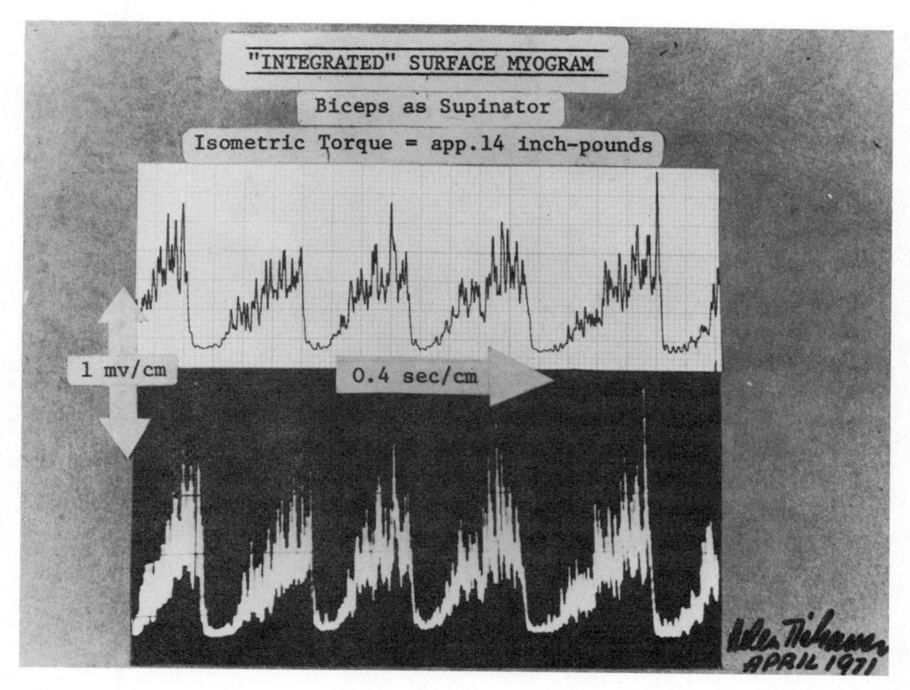

Figure 22.23 The same biceps contraction pattern shown in Figure 22.22: chart recorder (*upper half*), oscilloscope (*lower half*). However the signals here have been conditioned by summing all action potentials over a time constant so that the trace now represents the analogue of the firing rate, which is indicative of the total activity of the muscle mass at any instant during the sampling period. The signals are fully compatible with the frequency response of the chart recorder. The "integrated" myogram produces repeatable and very reliable measurements of muscular activity levels. From Reference 82, by permission. © American Institute of Industrial Engineers, Inc., Norcross, Ga.

quency of peaking of the raw signal and the amplitude of the action potential, but circuits developed by different manufacturers weigh each of these features differently; thus the appearance of myograms from the same muscles, tasks, and individuals, taken at the same occasion, differs considerably according to the make of recorder employed. It is therefore essential not to vary equipment from study to study.

The shape and quality of myograms may also be substantially affected by the following factors:

1. Loose electrodes.
2. Dry electrodes.
3. Loose or broken wiring.
4. Electrical interference from light fixtures or other machinery being used nearby.

It is essential to label myograms with the sensitivity settings and speed of the recording device, otherwise tracings obtained at different occasions cannot be compared. Likewise, the baseline should be clearly indicated, or amplitude of pen excursion cannot be quantified. When performing an isometric task (Figure 22.21), it is relatively easy to ascertain the degree to which each of the muscles investigated participates in the performance of the task, and how changes in musculoskeletal configuration produce a different distribution of work stress acting on the individual members of a kinetic chain. It is also quite simple, when the precondition of isometricity exists, as in static holding (Figure 22.24), to determine when a critical work stress level has been reached and when a relatively light increase in the severity of the task will produce an undesirably violent myoelectric response.

The area under the integrated myogram has the dimensions of force (volts), multiplied by time (paper speed), which is identical to the dimensions of "linear impulse" in physics and also to the "tension–time" concept used by work physiologists to quantify isometric work (33).

The interpretation of myograms of dynamic tasks, however, is a far more complex matter. The shape of the tracing, its slopes and troughs and amplitudes, are affected by a multitude of factors. These include force and velocity of contraction, tension within the muscle, and whether the contraction is eccentric or concentric.

Thus it is useless in most dynamic situations to even attempt the numeric quantification of the signal. The qualitative discussion, however, can yield information of considerable usefulness. This is the case, for instance, in the analysis of sequencing of different muscles during the performance of a kinesiological maneuver (Figure 22.25).

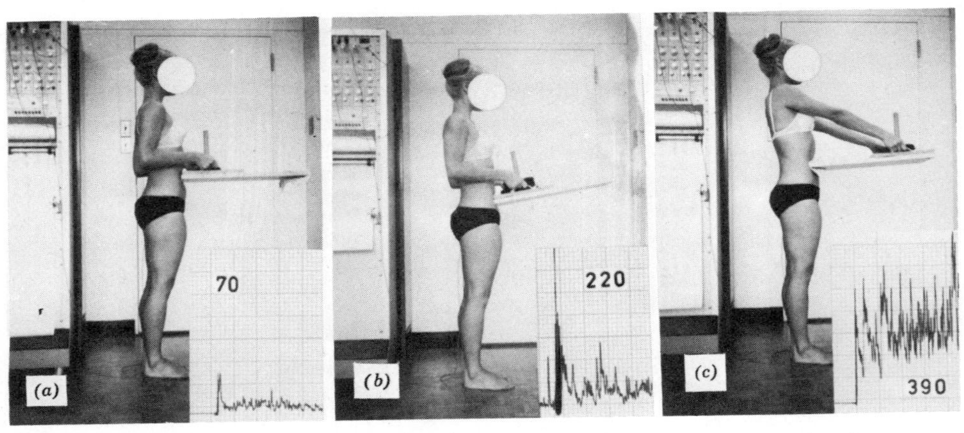

Figure 22.24 An isometric holding task. Numbers below the myograms represent incremental inch-pounds of torque, applied to the lumbosacral joint, which have elicited the electrophysiological signal. It can be seen that once a certain "critical" level of stress has been exceeded, the electrophysiological signal increases disproportionately as compared with the increment of stress. Under such conditions the subject is at risk. Adapted from Reference 31.

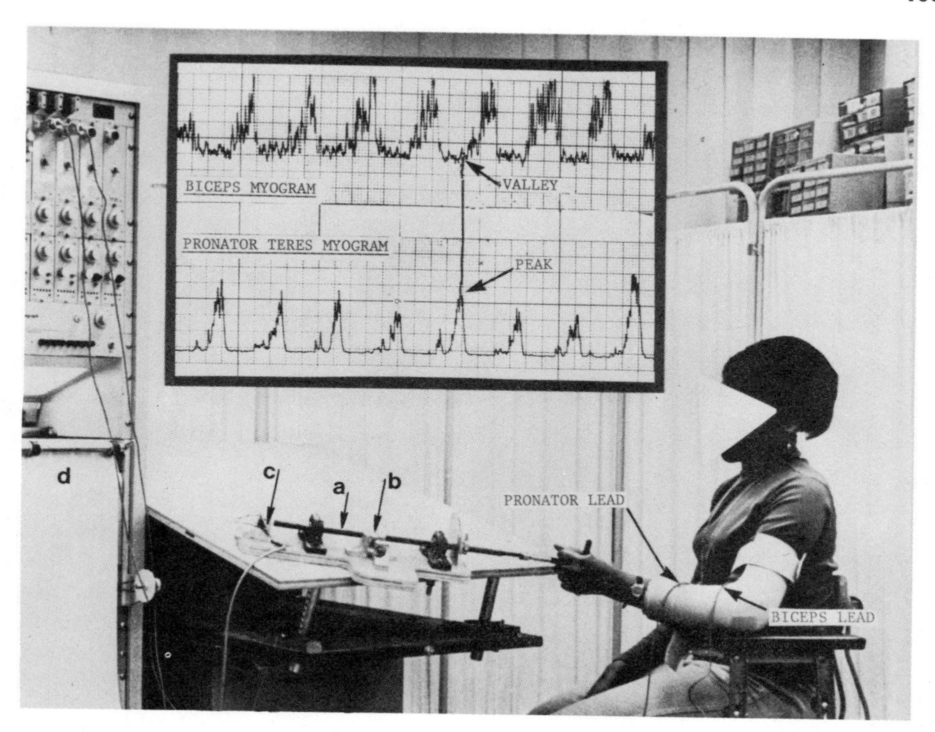

Figure 22.25 Kinesiometer and subject wired for two surface electromyograms; insert shows how antagonist myograms peak in proper sequence with one another. The kinesiometer consists of *A*, rotatable shaft; *B*, friction brake; *C*, potentiometer; *D*, recorder. Adapted from Reference 83.

In wrist rotation the biceps acts as protagonist during supination, while the pronator teres is the antagonist that reverses the movement. Therefore the integrated myograms of both muscles show peak and valley phasing in a nonfatigued, efficiently working individual. When the protagonist fires, the antagonist should be relaxed, and vice versa. In a state of fatigue, however, this clear phasing of muscular activity becomes blurred. A fatigued muscle has lost the ability to relax quickly; therefore the weaker of the two muscles, pronator teres, may fire simultaneously with the biceps, slowing down movement and bringing about undue exertion by the antagonist, increasing further the level of fatigue.

Whenever it is desired to ascertain whether a given musculoskeletal configuration is conducive to an undue expenditure of effort resulting in fatigue or potential trauma, the integrated myogram is the method of investigation of choice. A wire-cutting operation may be deemed to be a quasi-isometric work situation (Figure 22.26). Very often, when a tool such as a side-cutter is employed while the wrist is in ulnar deviation, the tendons of the flexor muscles bunch against each other inside the carpal tunnel. This produces

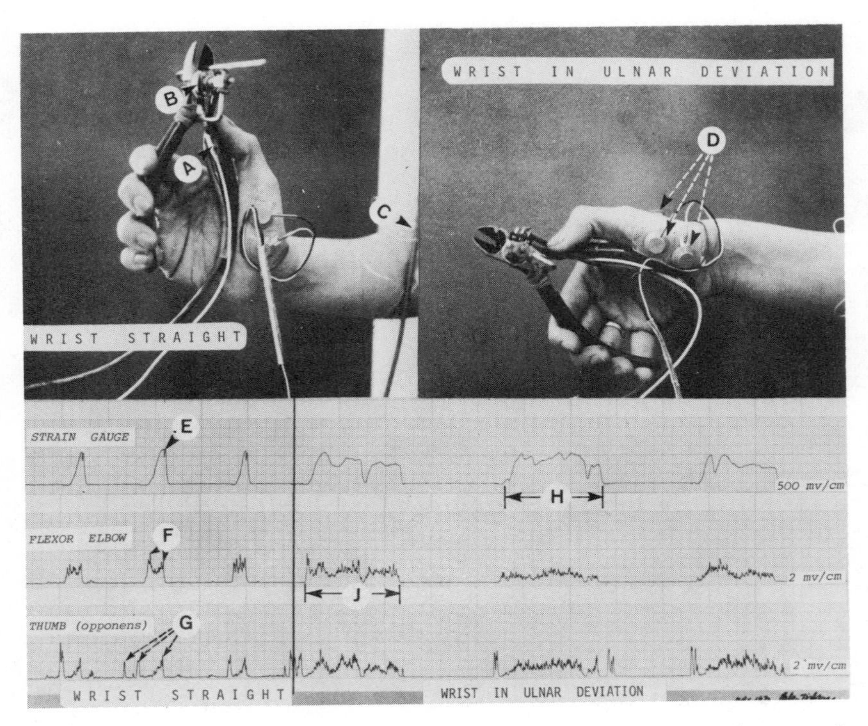

Figure 22.26 The profile of wire cutting (84). For explanation, see text. *A*, strain gauge; *B*, potentiometer; *C*, electrodes for common flexor myograms; *D*, electrodes for thenar myograms; *E*, "notched" strain signal; *F*, dicrotic flexor myogram; *G*, multicrotic thumb myogram; *H*, increased duration of strain signal; *J*, increased "tension time" of muscle. From Reference 84.

friction and necessitates a disproportionately large effort by the muscles of the flexors of the fingers and the thumb for effective performance. This excessive effort becomes immediately apparent on inspection of the myogram.

It is impossible to discuss in detail all the potential uses of electromyographic kinesiology, and specialized papers should be consulted (34–36). However the proper and imaginative use of electromyography constitutes one of the most elegant and useful techniques of ergonomic work measurement.

4 WORK TOLERANCE

Within the context of ergonomics, any action on the living body by any vector intrinsic to the industrial environment is termed "work stress." It is irrelevant whether these vectors are forces and produce movement, whether they merely cause sensory perception or whether, like heat, they increase metabolic activity. All physiological responses to work stress are identified as "work strain." Frequently work stress and the resulting strain occur in anatomical structures quite distant from each other, or the two effects may even be observed in separate physiological systems. High environmental temperatures are referred to as "heat stress," and the resulting increase in sweating rate is then correctly termed "heat strain." Likewise, in heavy physical work, the forces exerted on the musculoskeletal system are correctly identified as "work stress" and the resulting increase of metabolic activity is an example of "work strain."

The performance of any task, no matter how light, imposes some work stress, and consequently produces work strain in terms of physiological responses. Neither stress nor strain per se is undesirable unless it becomes excessive and diminishes work tolerance.

Work tolerance is defined as a state in which the individual worker performs at economically acceptable rates, while enjoying high levels of emotional and physiological well-being (13). It is common in industrial engineering practice to employ incentive schemes as inducements to increase production rates, thus reducing labor costs. Job enrichment as well as other procedures applied by behavioral and managerial scientists have their place in increasing job satisfaction and in enhancing the social well-being of the workforce. However no management technique available has been found to be successful in overcoming the results of physical discomfort and occupational disease resulting from a poorly designed work situation, ill-matched to the physical operating characteristics of man.

In a room illuminated by a defective spectrum, everyone is color blind. On a job where the motions and reactions inventory demanded by a task is not available from the musculoskeletal system of the worker, everybody is disabled (37), the physically impaired more so. The institution and maintenance of work tolerance has a high priority in the practice of industrial hygiene and ergonomics.

4.1 The Prerequisites of Biomechanical Work Tolerance

The 15 most important prerequisites of biomechanical work tolerance (38) have been arranged in the form of a table (Table 22.2) and can be employed as a checklist in industrial surveys. Proper use of the table can prevent workplace design from imposing physical demands that cannot be met by a wide range of individual workers. The use of

Table 22.2 Prerequisites of Biomechanical Work Tolerance

Postural		Engineering		Kinesiological	
P1	Keep elbows down.	**E1**	Avoid compression ischemia.	**K1**	Keep forward reaches short.
P2	Minimize moments on spine.	**E2**	Avoid critical vibrations.	**K2**	Avoid muscular insufficiency.
P3	Consider sex differences.	**E3**	Individualize chair design.	**K3**	Avoid straight-line motions.
P4	Optimize skeletal configuration.	**E4**	Avoid stress concentration.	**K4**	Consider working gloves.
P5	Avoid head movement.	**E5**	Keep wrist straight.	**K5**	Avoid antagonist fatigue.

this checklist may also help to avoid the generation of anatomical failure points, which may develop over a number of months, or years, as a result of cumulative work stress. Not all "prerequisites" are applicable to all work situations. However a correctly designed working environment will not violate many of them because this will, beyond doubt, lead to low productivity, poor morale, feelings of ill health and, sometimes, real occupational disease (13).

The "prerequisites" have been arranged in three sets of five statements. The first is concerned with postural integrity; the second relates to the proper engineering of the man–equipment interface; and the third set may be used to ensure that the motions demanded from the workforce are kinesiologically effective.

4.2 The Postural Correlates of Work Tolerance

P1 Keep the elbows down.

Abduction of the unsupported arm for long intervals may produce fatigue, severe emotional reactions, and also decrements in production rates. The need to keep the unsupported elbow elevated is often the result of poor workplace layout. For example, if the chair height of seated workers is poorly policed, a seat positioned only 3 inches too low with respect to the work bench will produce an angle of abduction of the upper arm of approximately 45° (Figure 22.27). When this is the case, wrist movement in the hori-

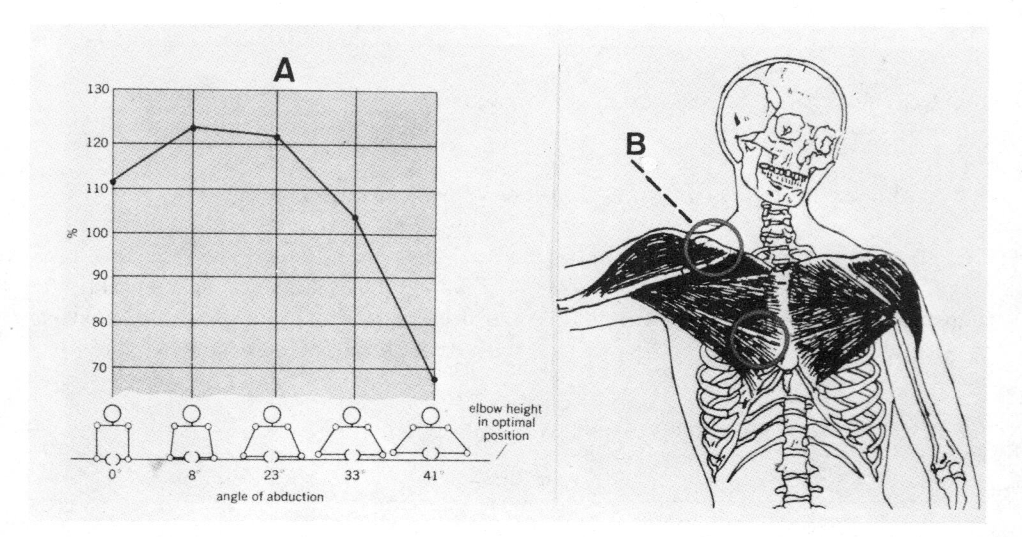

Figure 22.27 (*a*) Chair height determines the angle between upper arm and torso, also the moment of inertia of the moving limb. For example, a chair 3 inches too low may raise this moment to a level at which increased effort causes performance to drop to 70 percent of "standard." (*b*) Also, continued tension in the muscles stabilizing the arm in the raised posture may cause great discomfort at the shoulder over the breast bone. From Reference 37, by permission. © American Institute of Industrial Engineers, Inc., Norcross, Ga.

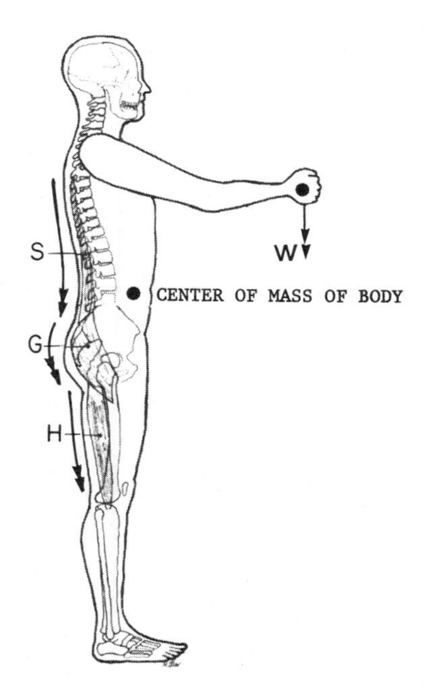

Figure 22.28 Even when no object is handled, very often a bending moment acts on the vertebral column because of the location of the center of mass of the body. The erector muscles of the trunk counteract this: *S*, sacrospinalis; *G*, glutei; *H*, hamstrings; *W*, weight. From Reference 51.

zontal plane, normally performed by rotation of the humerus, could require a physically demanding shoulder swing. The resulting fatigue over several hours may reduce the efficiency rating by as much as 50 percent. Also, when the seat is too low—especially in assembly operations—the left arm is frequently used as a vise, while the right hand is employed to manipulate objects. This may result in the left arm being held in abduction for several hours. After the elapse of an hour or two, particularly under incentive conditions, some vague sense of discomfort may be felt in the general region of the origin of the left pectoralis major and deltoid muscles, which stabilize the abducted arm. Elderly and overweight workers especially, or individuals with a history of cardiac disease, may develop an unjustified fear of an impending heart attack, and they themsevles, as well as all those around them, may suffer from the ensuing undesirable emotional difficulties (13).

P2 Minimize moments acting upon the vertebral column.

Lifting stress is not solely the result of the weight of any object handled. Its magnitude must be expressed in terms of a "biomechanical lifting equivalent" in the form of a "moment."

The location of the center of mass of the body proper causes a bending moment to be exerted on the axial skeleton even when no object is handled (Figure 22.28). The

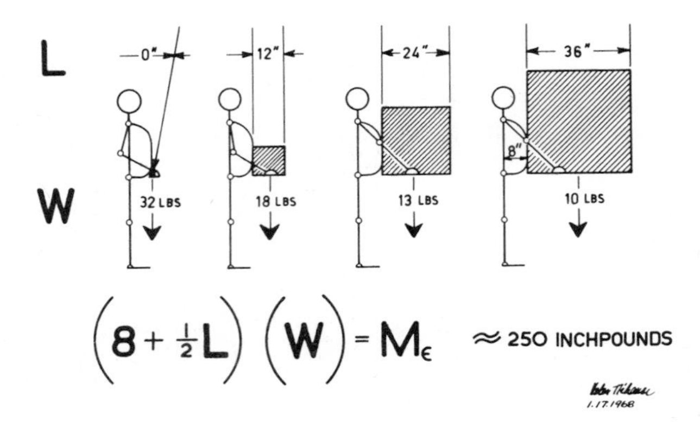

Figure 22.29 The "moment concept" applied to the derivation of biomechanical lifting equivalents. All loads represented produce approximately equal bending moments on the sacrolumbar joint, approximately 250 lb-in. Moments exerted by body segments are neglected (13). In the equation, 8 is the approximate distance (inches) from the joints of the lumbar spine to the front of the abdomen, a constant for each individual; L is the length (inches) of one side of the object; W is the weight (pounds) of the object; Me is the biomechanical lifting equivalent, approximately 250 lb-in. in this example. From Reference 13.

muscles erecting the trunk counteract the moment and thus help to maintain upright posture. Thus even simple variations in posture and trunk–limb configuration may modify and substantially increase or decrease, according to the circumstances, the forces exerted on the lumbar spine according to the contribution of the individual body segments to the total moment sum (39). The ergonomic effects of these posture-generated forces are discussed in Section 5. The present checklist is merely concerned with the additional moments imposed on the back by an external load. Very often a light, but bulky object (Figure 22.29) imposes a heavier lifting stress than a heavy load of great density. It should be remembered that the only way to reduce the lifting stress exerted by an object resides in devising a handling method that will bring the center of mass of the article as close to the lumbar spine as possible.

P3 Consider sex differences.

If employment opportunities for both sexes are to be equal, work environments must be engineered in a manner that takes cognizance of, and compensates for, any sex-dependent differences in anatomy that may affect work tolerance. In the context of lifting tasks, it must be appreciated that male hip sockets are located directly below the bodies of the lumbar vertebrae; in the female they are situated more forward (Figure 22.30). A line through the center of the sockets of both hip joints in a woman is located several inches in front of a vertical line passing through the center of gravity of the female body. This activates a force couple. Therefore any object handled by a woman exerts a moment on the back approximately 15 percent larger than if it were handled by a male of identical size or strength.

P4 Optimize skeletal configuration.

Through faulty workplace design, musculoskeletal configuration, especially angular relationships of long bones and muscles, may impose great stress on joints and produce physical impairment (Figure 22.31). Variations of a few inches horizontally in the distance between the chair and the workspace may make the difference between a productive working population and one that must perform under medical restriction because of great mechanical stress imposed on joint surfaces.

P5 Avoid the need for excessive head movement during visual scanning of the workplace.

It is not possible to estimate correctly, and/or easily, the true sizes or the relative distance of objects except under conditions of binocular vision, which can take place without head movement only within a visual cone of 60° of included angle. The axis of

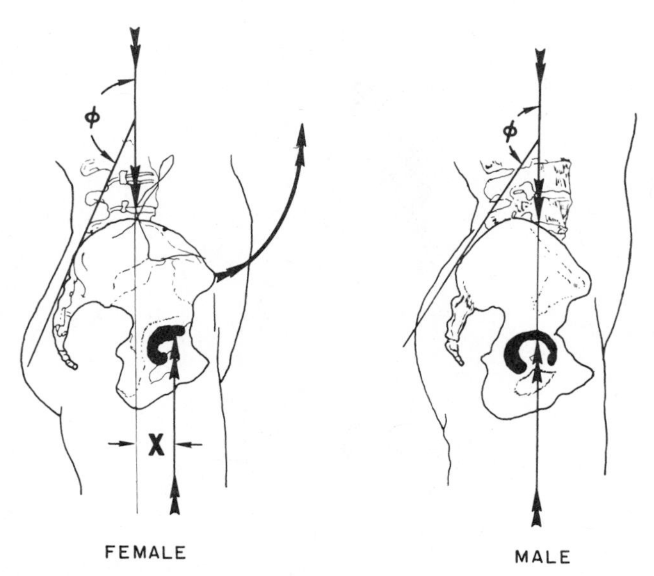

Figure 22.30 Hip sockets in the male are located directly under the bodies of the lumbar vertebrae in the same plane as the center of mass of the body. In the female, the sockets are located further forward, represented by the distance *X*. This produces a force couple so that the lifting stress in the back muscles in women, for the same object, can be as much as 15 percent higher than in males. From Reference 85, by permission, American Industrial Hygiene Association, Akron, Ohio.

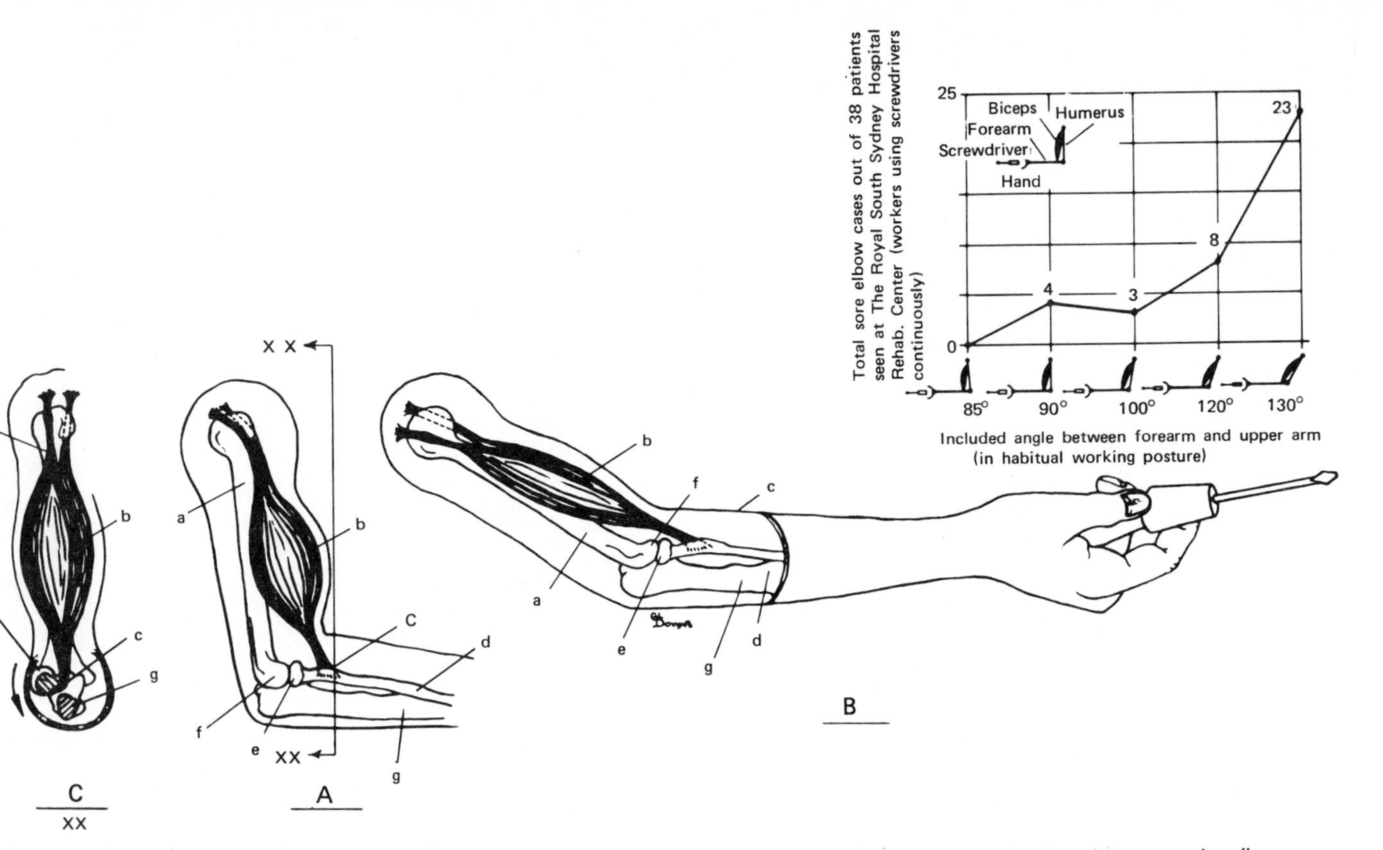

Figure 22.31 The mechanical advantage of the biceps depends on the angle of flexion of the forearm. This muscle is not only a flexor, but also, because of the mode of attachment, the most powerful outward rotator of the limb. The worker who sits too far away from his workplace has to overexert himself when using a screwdriver because the biceps operates at mechanical disadvantage. Sore muscles and excessive friction between the bony structures of the elbow joint are the results (80). (*A*) The forearm flexed at 90 degrees: *a*, humerus; *b*, biceps; *c*, attachment of biceps; *d*, radius; *e*, head of radius; *f*, capitulum of humerus; *g*, ulna. (*B*) The angle of the forearm extended when the efficiency of the biceps as an outward rotator is reduced. Here the muscles will pull the radius strongly against the humerus, causing friction and heat in the joint. (*C*) Cross section X–X through *A*, showing why the biceps is an outward rotator of the forearm. From Reference 80.

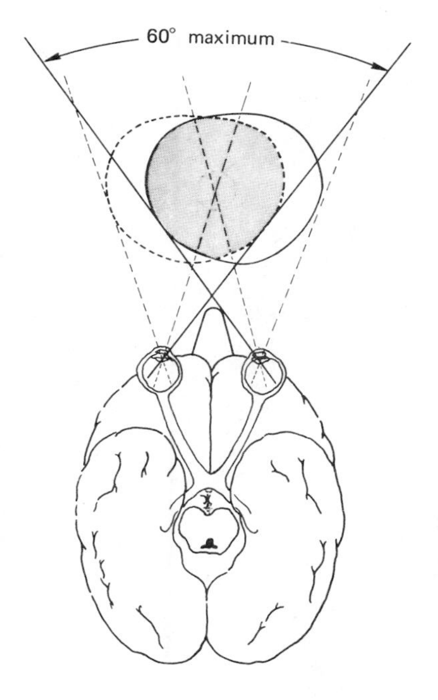

Figure 22.32 Eye travel and binocular vision. Whenever an object is located outside a binocular field of vision (shaded area) depth perception is impossible; head movement will automatically ensue to correct the deficiency. Heavy dotted lines indicate convergence at 16 inches. From Reference 48.

this cone originates from the root of the nose and is located in the mid-sagittal plane of the head (Figure 22.32). Head movement at the workplace is often invoked as a "protective reaction" (17) whenever it is necessary to reestablish binocular sight if the visual target is located outside of the cone. Simultaneous eye and head movements take much time, and this may produce a hazard whenever fast-moving equipment—such as motor vehicles, airplanes, or conveyors—are operated. Binocular vision without head movement can be instituted either by dimensioning the workplace appropriately or by changing the position of the operator or the adjustment of the working chair.

4.3 The Engineering of the Man–Equipment Interface

E1 Avoid compression ischemia.

The term "ischemia" describes a situation in which blood flow to the tissues is obstructed. It is essential that the designer as well as the evaluator of tools and equipment be familiar with the location of blood vessels vulnerable to compression. Improperly designed or misused, a piece of equipment may have the effect of a tourniquet. Of special importance is a knowledge of the location of blood vessels and other pressure-sensitive anatomical structures in the hand. A poorly designed or improperly held hand tool may squeeze the ulnar artery (Figure 22.33). This may lead to numbness and tingling of the fingers. The afflicted worker will put down his tools and may make

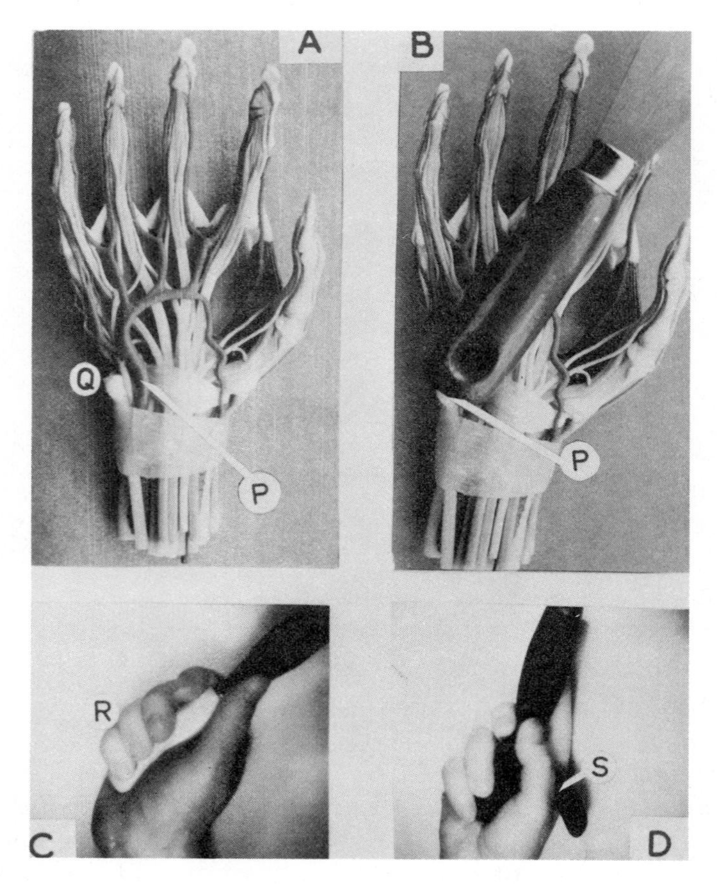

Figure 22.33 Ergonomic considerations in hand tool design. (*a*) The relations of bones, blood vessels, and nerves in the dissected hand. (*b*) A paint scraper is often held so that it presses on a major blood vessel *P* and directs a pressure vector against the hook of the hamate bone *Q*. (*c*) In the live hand, this results in a reduction of blood flow to, among others, the ring and little fingers, which shows as light areas on infrared film *R*. (*d*) A modification of the handle of the paint scraper causes it to rest on the robust tissues between thumb and index finger *S*, thus preventing pressures on the critical areas of the hand. From Reference 40, by permission. American Industrial Hygiene Association, Akron, Ohio.

use of any reasonable excuse to absent himself temporarily from the workplace as the only means of relief open to him. Apart from the resulting drop in productivity, the health of the working population under such circumstances is in serious jeopardy inasmuch as cases of thrombosis of the ulnar artery and other instances of permanent damage have been reported. Unless the engineering and medical departments are alerted to a possible mismatch between hand and tool, the complaint of numb and tingling fingers could be erroneously attributed to one of the numerous other causes of these symptoms, and the sufferer improperly diagnosed or treated (41). Generally speaking, handles of implements should be designed to make it impossible for the tools to dig into the palm of the hand or to exert pressure on danger zones.

E2 Avoid vibrations in critical frequency bands.

Vibrations transmitted at the man–equipment interface can easily lead to somatic resonance reactions. "White finger syndrome" or intermittent blanching and numbness of the fingers, sometimes accompanied by lesions of the skin, has been identified for many years as an occupational disease associated with the operation of pneumatic hammers and other vibrating tools (42, 43). It is cited as only one example of numerous diseases and injuries caused by exposure to vibrations. When exposed to critical ranges of vibration, various viscera, muscle masses, and bones may react in an undesirable manner. This can simulate a wide range of diseases that are commonly associated with musculoskeletal discomfort, including back pain, respiratory difficulties, and visual disturbances. The critical ranges of vibration that may produce such undesirable side effects are fortunately very narrow, and it is often easy, through recourse to such normal engineering procedures as construction of vibration-absorbing tool handles, to avoid exposure to noxious frequencies. An "epidemic" of otherwise inexplicable afflictions of the musculoskeletal system should always alert the ergonomist to consult a reliable reference work on industrial vibration (44).

E3 Individualize chair design.

The design of any seating device should match the need of individual work situations. The anthropometric and biomechanical basis of chair design and adjustment are discussed in section 7. When evaluating work situations on the shop floor, it should be remembered that universally useful "ergonomic work chairs" do not exist. Likewise, diagrams that show "standard dimensions" of the components of a work chair, or of seating height, are highly suspect unless they list a wide range of tolerances for each dimension. By way of example, it is mentioned that the pilot seat in an aircraft should support the trunk while the pilot is sitting still. On the other hand, the chair on an assembly line should give adequate lumbar support but at the same time permit the body to perform all necessary productive movements (45) (Figure 22.55). Working chairs should have an adjustable-height backrest that swivels about a horizontal axis, to be able to adapt to the demands of the contours of the back. It should be small enough not to interfere with the free movement of the elbows during work. The seat should be adjustable in height, and it should be complemented by an adjustable footrest to relieve pressure exerted by the edge of the seat on the back of the thigh.

E4 Avoid stress concentration on vulnerable bones and joints.

Sometimes features in tool and equipment design look deceptively advantageous but are in reality most dangerous to the integrity of the skeletal system. Finger-grooved tools are an example (Figure 22.34). They fit perfectly one hand—the hand of the designer. If gripped firmly by a hand too large or too small, the metal ridges of the handle may exert undue pressure on the delicate structures of the interphalangeal joints (46). This will make it painful to grip the tool firmly. Sometimes worse, the working population may show signs of discomfort and absenteeism and will be exposed to medical restriction of performance levels. Finally, under the worst of circumstances, permanent and disabling bone and joint disease may result. Sometimes simple shielding devices are quite effective to protect anatomical structures from stress concentration. The tailor's thimble is a good example of this—it is, perhaps, the oldest device in history protecting tissues against stress concentration.

E5 Keep the wrist straight while rotating forearm and hand.

Four wrist configurations, particularly when they approach the extremes of their range, are conducive to fatigue, discomfort, and sometimes disease. These are: (*a*) ulnar deviation, (*b*) radial deviation, (*c*) dorsiflexion, and (*d*) palmar flexion.

Especially unhygienic situations are those in which these positions alternate fairly rapidly during the work cycle, or occur in combination with each other. Unfortunately tools and workplaces are often so designed that the aforementioned movements are demanded as part of the normal work cycle. This affects both health and efficiency. The principal flexor and extensor muscles of the fingers originate in the elbow region, or

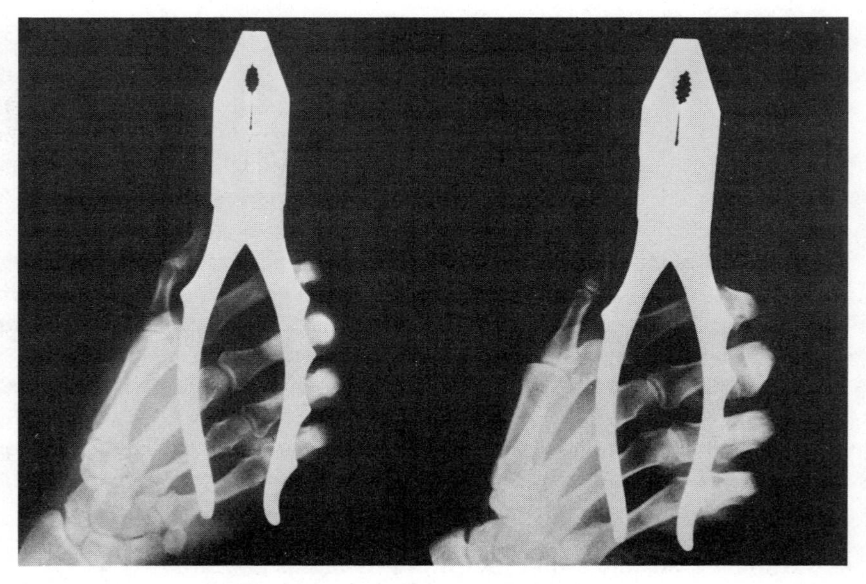

Figure 22.34 Form-fitting grips on hand tools may cause severe pressures on the finger joints of a person with a larger hand than the hand size for which the tool was designed. From Reference 46, by permission. © Journal of Occupational Medicine, Chicago, Ill.

from the forearm, and are connected to the phalanges by way of long tendons. The extensor tendons are held in place by the confining transverse ligament on the dorsum of the wrist, and the flexor tendons on the palmar side of the hand pass through the narrow carpal tunnel, which contains also the median nerve.

Failure to maintain the wrist straight causes these tendons to bend, to become subject to mechanical stress, and to traumatize such ancillary structures as the tendon sheaths and some ligaments. Ulnar deviation, combined with supination, favors the development of tenosynovitis. Radial deflexion, particularly if combined with pronation, increases pressure between the head of the radius and the capitulum of the humerus. This is conducive to epicondylitis or epicondylar bursitis. These conditions are frequently observed in those who operate hand tools. If hand tools like screwdrivers or pliers are pronated and supinated against resistance for only a few minutes during the day, no harm results. However continuous production jobs may constitute a hazard to the working population. Generally speaking, it is safer to bend the implement than to bend the wrist. Furthermore, both ulnar and radial deviations of the wrist, when combined with simultaneous pronation or supination, reduce the range of rotation of the forearm by more than 50 percent (46). Under these circumstances, the afflicted individuals are obliged to go through double the number of motions to perform a given task, such as looping a wire around a peg (i.e., they have to work twice as hard for half the output). This may dramatically increase personnel turnover and lead to massive dropouts of new workers during training (13) (Figure 22.35).

4.4 The Development of Effective Kinesiology

K1 Keep forward reaches short.

Numerous industrial engineering texts describe and define "normal" and "extended" reach areas, specifying that the motion elements "reach" and "transport" may safely be included in repetitive tasks for continuous work provided they do not exceed 25 inches (47). This assumption is fallacious. The protagonist muscles of forward flexion of the upper arm operate at biomechanical disadvantage. One of their principal antagonists is the large and powerful latissimus dorsi (Figure 22.36). Whenever an extended and fast reach movement in the sagittal plane away from the body is performed, a strong stretch reflex is produced in the latissimus dorsi. This, in turn, subjects the vertebral column to compressive and banding stresses. Frequent and rapid forward reach movements, especially when associated with the disposal of objects, require that the length of this motion be kept to somewhat less than 16 inches. The operative word here is "frequent." Extended reaches per se, especially when performed slowly and relatively rarely during the working day, are quite harmless.

K2 Avoid muscular insufficiency.

Sometimes a specific muscle is unable to produce an expected full range of joint movement. Such a situation is defined by the term "muscular insufficiency" (15). It occurs

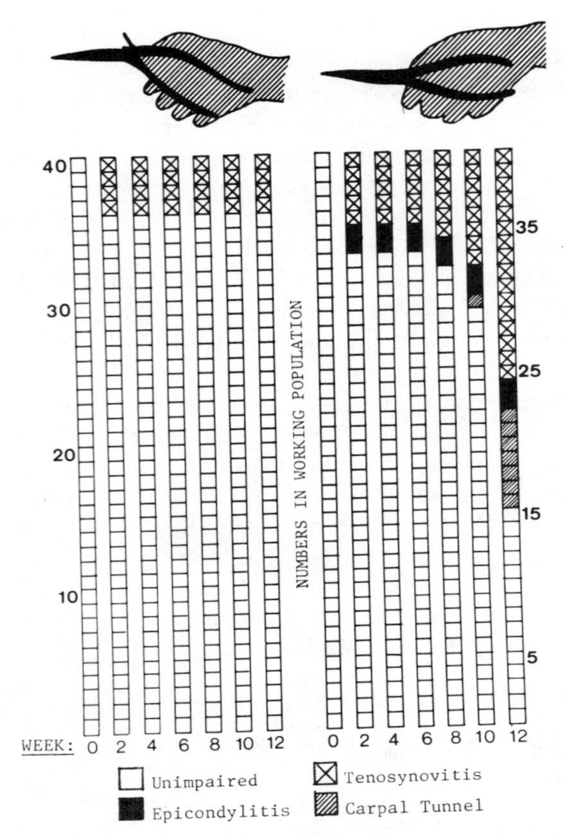

Figure 22.35 Comparison of two groups of trainees in electronics assembly shows that it is better to bend pliers than to bend the wrist (37). With bent pliers and wrists straight, workers become stabilized during second week of training. With wrist in ulnar deviation, a gradual increase in disease is observed. Note the sharp increase of losses in tenth and twelfth weeks of training: only 15 of 40 workers in the sample remained unimpaired. From Reference 37, by permission. © American Institute of Industrial Engineers, Inc., Norcross, Ga.

when the protagonist is contracted to such a degree that it cannot further shorten, or an antagonist has become hyperextended and impedes further joint movement. By way of example, when the wrist is fully flexed, the hand cannot grasp a rod firmly because the extensors of the fingers are overextended, and the flexors are overcontracted (Figure 22.37).

K3 Movements along a straight line should be avoided.

All joints involved in productive movements at the workplace are hinges. Therefore the pathway of an anatomical reference point is always curved when it results from the

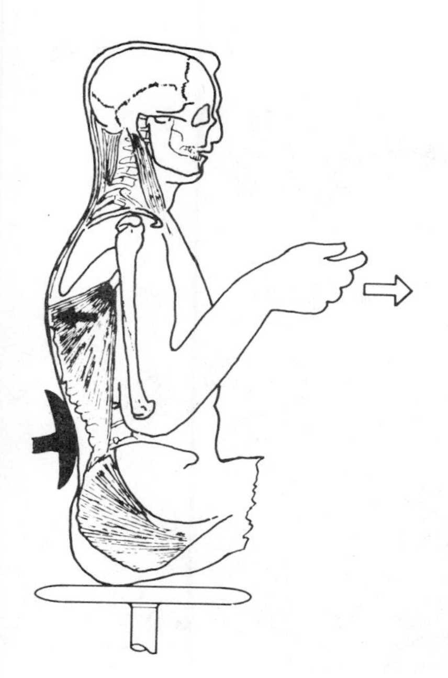

Figure 22.36 The disposal of objects in a direction away from the body and in the mid-sagittal plane is conducive to discomfort and early fatigue because of the strong antagonist activity in the latissimus dorsi muscle. From Reference 48.

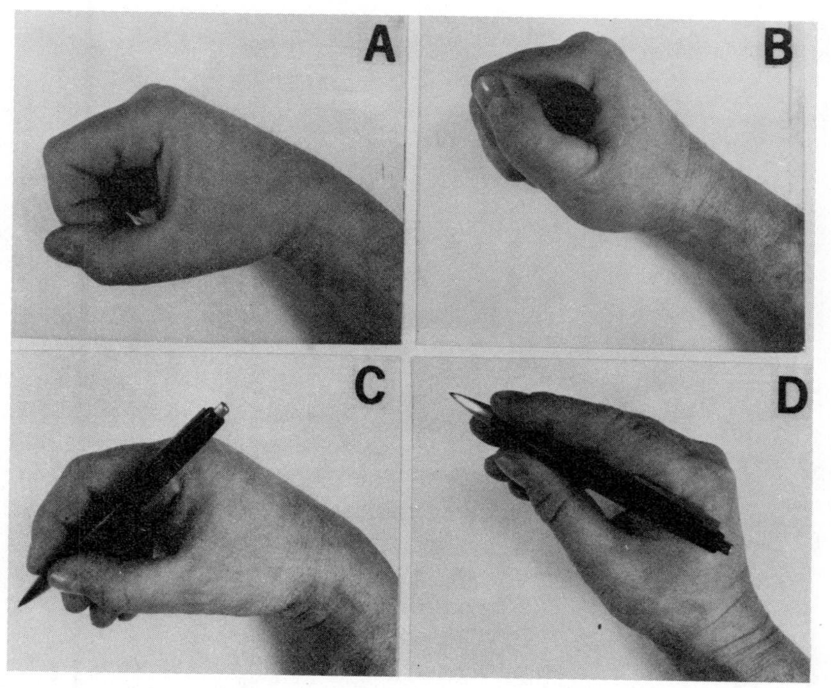

Figure 22.37 Flexed wrist (*A*) cannot grasp a rod firmly; the straight wrist (*B*) can grip and hold it firmly. Conversely, the flexed wrist (*C*) is well positioned for fine manipulation, but when extended (*D*), freedom of finger movement is severely limited. From Reference 48.

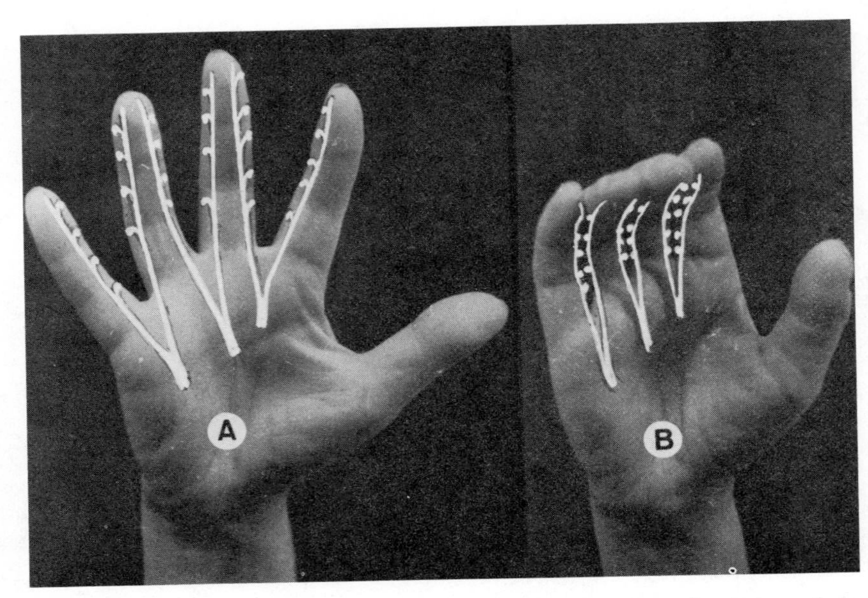

Figure 22.38 Nerve endings that provide feedback information about the degree of closure of the hand are located between the fingers (*A*). When the hand executes a gripping motion (*B*), the fingers abut and the nerve endings press against each other. Work gloves that are too thick may produce pressure against the interdigital surfaces too early and provide misleading information about the firmness of grip when holding a heavy or slippery object. From Reference 37, by permission. © American Institue of Industrial Engineers, Inc., Norcross, Ga.

simple movement of a single joint. Such curved movements can be produced by a contraction of one single muscle. On the other hand, motion along a straight line requires higher skills not always available from a subclinical impaired working population (e.g., the aged). Straight line movements generally require longer learning, produce early fatigue, and are less precise than simple ballistic motions (48).

K4 Working gloves should be correctly designed.

Often occupational hazards and operational inefficiencies are erroneously ascribed to improper tool or usage when, instead, the working glove is implicated. The end organs of the sensory nerves of the hand are, among others, also distributed along the interdigital surfaces of the fingers (49) (Figure 22.38). Since it is impossible to close the fist firmly while the fingers exert pressure against each other, it is difficult to maintain grip strength under such circumstances. If a working glove is too thick between the fingers, high pressure on the interdigital surfaces may be generated before the hand is firmly closed about a tool handle or equipment control. This results in an insecure grasp. Lack of awareness of this lowered capability may cause heavy objects to slip out of the hands of workers, resulting in accidents. It may also lead to inadequate control over cutting tools or dials operated in a cold climate.

K5 Antagonist fatigue should be considered in task design.

In the performance of many simple movements, the muscles that reestablish the initial condition after a specific maneuver has been performed are often weaker than those which bring about the primary movement. For example, when inserting screws by means of a ratchet screwdriver, the biceps is the outward rotator of the forearm, and this strong muscle will not easily fatigue even when operating against strong resistance. Even when operating against a resistance too small to be measurable, however, the opposing inward rotator of the forearm, the pronator teres, fatigues easily because of its small size (19). Physiological work stress should never exceed the capacity of the smallest muscle involved in a kinetic chain. Reliable tables relating name of muscle, function, cross section, and working capacity should be consulted by the practicing ergonomist (15).

The 15 prerequisites of biomechanical work tolerance just given are no substitute for a comprehensive knowledge of the theory and practice of ergonomics. They are merely a convenient expedient for the rapid and gross evaluation of work situations during discussions on the shop floor. Properly employed in discussions with first-line supervisors and engineering personnel engaged in the planning and design of work situations, these prerequisites can help to reduce, without access to complex esoteric and costly laboratory facilities, the incidence of occupational accident and disease right where it occurs: on the shop floor.

5 MANUAL MATERIALS HANDLING AND LIFTING

Almost one-third of all disabling injuries at work—temporary or permanent—are related to manual handling of objects (48). Many of these incidents are avoidable and are the consequence of inadequate or simplistic biomechanical task analysis. The relative severity of materials handling operations, and differences in lifting methods, can be evaluated only when all elements of a lifting task are considered together as an integral set (Table 22.3). All these elements have different dimensional characteristics, but

Table 22.3 The Elements of a Lifting Task (48)

Static Moments	Gravitational Components	Inertial Forces
Sagittal	Isometric	Acceleration
Lateral	Dynamic	Aggregation
Torsional	Negative	Segregation
	Frequency of task	

nevertheless have one basic property in common: any major change in magnitude of any element of a lifting task produces a change in the level of metabolic activity.

No matter what the dimensions of mechanical stress imposed on the human body during materials handling, the physiological response will always result in changes of energy demand and release, customarily expressed in kilocalories. Thus physiological response to lifting and materials handling stress has always the dimensions of work. Hence when the task is heavy enough, the measurement of metabolic activity provides a convenient experimental method for the objective comparison of the relative severity of materials handling chores.

Current consensus (26) assumes on the basis of an 8-hour working day that the limit for heavy continuous work has been reached when the oxygen uptake over and above resting levels approaches 8 kcal/min. The upper limit for medium-heavy continuous work, seems to be 6 kcal/min, and an increment of 2 kcal/min appears to be the dividing line between light and medium-heavy work. However in many instances the application of metabolic measurement is not feasible. The complex procedure may be too difficult to perform on the shop floor. Furthermore, the assessment of work stress through the analysis of respiratory gas exchange is, by definition, an ex post facto procedure. The job exists already, and the energy demands are merely computed to decide whether corrective action is indicated. It is, of course, much better to analyze a task objectively while both job and workplace layout are still in the design stage. Recourse must then be taken to "elemental analysis."

5.1 Elemental Analysis of Lifting Tasks

Any activity producing a moment that acts—no matter in which direction—on the vertebral column, must be classified as a "lifting task." A "moment" is defined as the magnitude of a force multiplied by the distance from the points of its application. In most instances ergonomic analysis of lifting tasks is concerned with moments acting on the lumbar spine or, when a specific task involves head fixation, with moments acting on the cervical spine.

There are three static moments to be considered (Table 22.3). In many instances they are easily computed by direct measurement or with the aid of drawings or photographs. Sometimes, however, it is more convenient to estimate them by speculative analysis or visual inspection of the work situation. They are conveniently expressed in pound-inches (lb-in.), or kilogram-centimeters (kg-cm). This value is obtained by multiplying the force acting on an anatomical structure with the distance from the point of maximal stress concentration.

The heaviest article normally handled by man at work is his own body, or its subsegments. Only rarely do workers handle objects weighing 150 pounds and, in most instances, the mass of an object moved is quite insignificant when compared with the weight of the body segment involved in the operation. For example, the majority of hand tools or mechanical components in industry weigh considerably less than 0.5 pound, but an arm, taken as an isolated body segment, weighs 11 pounds (50).

The *sagittal lifting moment* is the one most frequently encountered and easiest to compute. It is most conveniently derived by graphical methods. First the weights of the body segments involved in a specific task are obtained from reliable tables (18). Then a "stick figure" of proper anthropometric dimensions (Figure 22.39) is drawn, and the location of the center of mass for each body segment is marked, as well as the center of mass for the load handled. Finally, the sum of all moments acting on the selected ana-

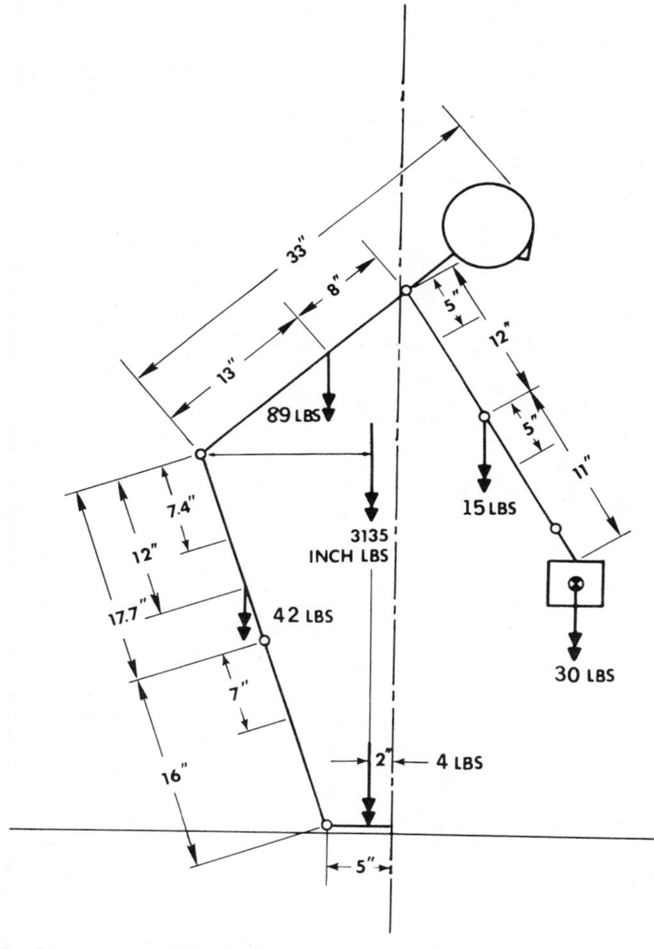

Figure 22.39 Graphic computation of the location of center of mass of whole body and body segments as well as of the sagittal moment acting on the lumbosacral joint can be conveniently accomplished through the use of stick figures. This example shows that in improper working posture, a load weighing only 30 pounds, combined with the mass of the various body segments involved in a lifting task, may produce a torque exceeding 300 in.-lb, which is the lifting equivalent of a very severe task. Illustration courtesy of Dr. C. H. Saran; from Reference 48.

tomical reference structure (in this case, the lumbosacral joint) is computed and becomes the sagittal biomechanical lifting equivalent of the specific task under consideration. Whenever the vector representing the moment sum of all gravitational moments is directed at a point on the floor located in front of or behind the soles of the feet, a prima facie hazard exists because of inherent postural instability and the ensuing likelihood of falls.

The estimation of sagittal lifting equivalents is of great practical usefulness in the comparison of work methods or in the assessment of the relative magnitude of lifting stress due to sex differences (Figure 22.40).

Males and females of approximately the same height and weight may be subject to different stresses when handling the same object. This is because of sex-dependent differences of the relative proportions of body segments. Moments acting on the lumbosacral joint during lifting depend largely on work surface height; therefore females

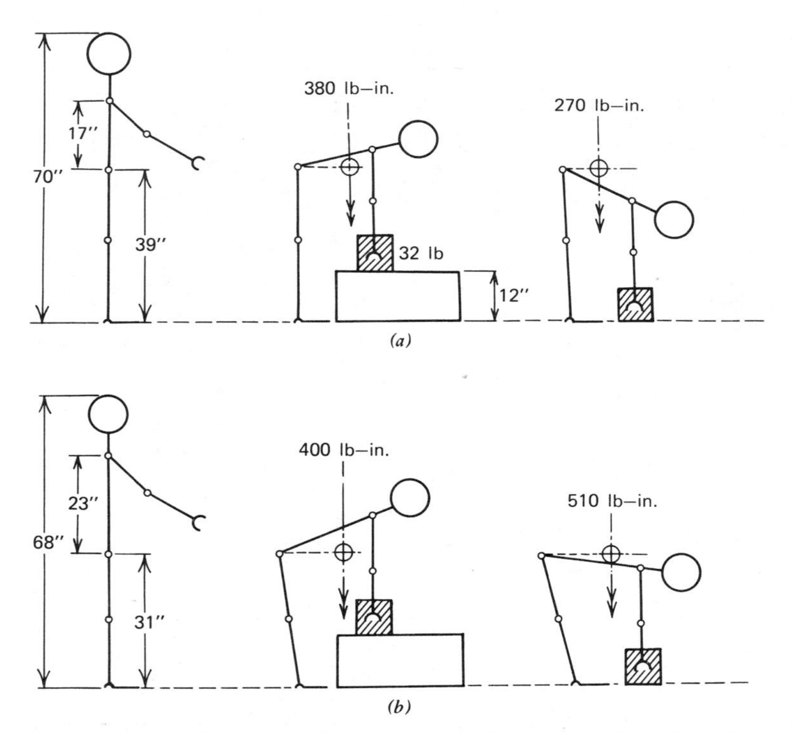

Figure 22.40 (*a*) Males and (*b*) females of approximately the same height and weight may be subject to quite different stresses when handling the same object. This is due to sex-dependent differences of the relative proportions of body segments. Because moments acting on the lumbosacral joint when lifting depend largely on work surface height, females are at a disadvantage in certain postures during load acquisition, whereas in others sex differences are minimal. Adapted from Reference 51.

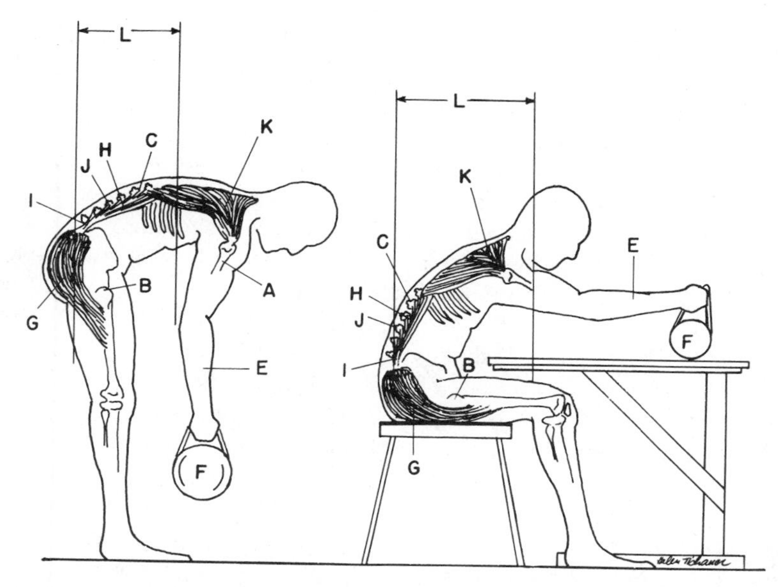

Figure 22.41 Awareness of the "hidden" lifting task should exist. Because of increase of dimension *L*, hence higher torques exerted on the lumbar spine, a seated job, instead of being "light work" may be the physiological equivalent of a severe lifting task. In seated work, the rule is: "get the job close to the worker." *A*, humerus; *B*, socket of hip joint; *C*, vertebral column; *D*, shoulder joint; *E*, arm; *F*, load; *G*, muscles of the buttocks (gluteus maximus); *H*, muscles of the back (sacrospinalis); *I*, lumbosacral joint; *J*, spinous process of a vertebra; *K*, trapezius muscle; *L*, distance from the center of mass of combined body-load aggregate to the joints of the lumbar spine. From Reference 58, by permission. © American Institute of Industrial Engineers, Inc., Norcross, Ga.

are often at a disadvantage when picking up objects from the floor, but a pallet 12 to 14 inches high reduces sex differences in lifting stress during load acquisition (51).

Sometimes it is necessary to decide whether a task would be better performed sitting, as opposed to standing (Figure 22.41). Then a sketch, true to scale, or a photograph, becomes a convenient aid to decision making. If the pictorial representation of the work situation is supplemented by the estimated weight of the body segments involved and the location of the respective centers of gravity, the relative lifting stress for both postures can be easily estimated. Since the procedure does not compare lifting stresses between different individuals but merely establishes how the same person is affected by changes in posture, it may be assumed that the body segments involved (i.e., the torso above the lumbosacral joint, plus neck, head, upper limb, and the object manipulated) are identical in both the seated and standing postures.

According to standard data (52) the body mass in the case of a 110-pound female would be 45 pounds. To this must be added the weight of the object handled—20 pounds in the example under consideration. For this example, the distance from the lumbosacral joint to the center of mass of body segments and load combined may be estimated to be 1½ feet when standing. This exerts a torque of approximately 98 foot-pounds (ft-lb). However if the individual is seated, this value increases to approximately 2½ feet because of the forward leaning posture of the trunk and the outstretched arm. Therefore the torque now exerted on the lumbar spine amounts to 146 ft-lb, or an increase of nearly 50 percent compared with the standing position. This explains why, in so many instances, emloyees complain—and rightly so—about much increased work stress when chairs are introduced unnecessarily into a work situation.

Analyzing lifting tasks routinely in terms of moments tends to develop in supervisors not only a "clinical eye" for the magnitude of a task but also a healthy and critical attitude toward "cookbook" rules of lifting. The principle of "knees bent—back straight—head up" is well enough known. However in many work situations concessions must be made to the influence of body measurements. In Figure 22.42 the male, long legged, short torsoed, does not benefit at all from the application of the standard lifting rule. A female, however, having a differently proportioned body, can get under the load and close to it.

Thus the sagittal lifting moment acting on the lumbosacral joint becomes much less, and work stress is approximately halved. Under such circumstances, provided the height of the work bench cannot be changed, the standard lifting rule may be applied to the female, whereas in the case of the male no benefit will be derived. Working according to

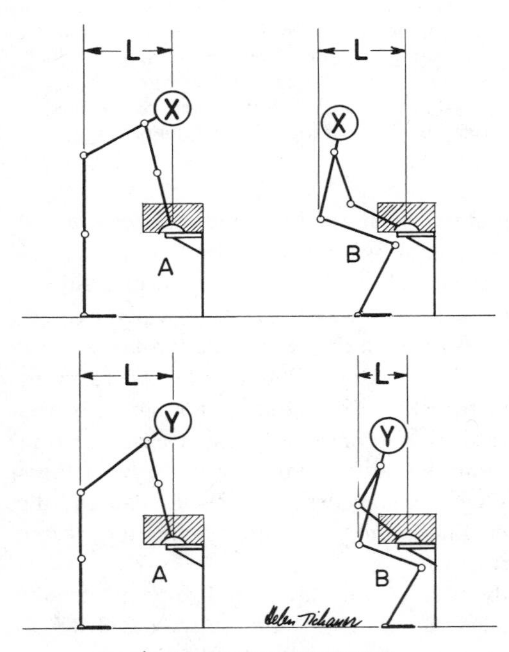

Figure 22.42 Postural corrections in training for lifting should be aimed at reducing torques acting on the spine: *X*, an anthropometric male, does not benefit materially from the "approved" lifting posture because *L*, the distance from the center of mass of load to the fourth lumbar vertebra, does not shorten materially; *Y*, an anthropometric female, does benefit from the "bent knees, straight back" rule because she can get under the load. When matching worker and task, the measurements of the individual worker as well as the dimensions of the workplace should be considered. From Reference 58, by permission. © American Institute of Industrial Engineers, Inc., Norcross, Ga.

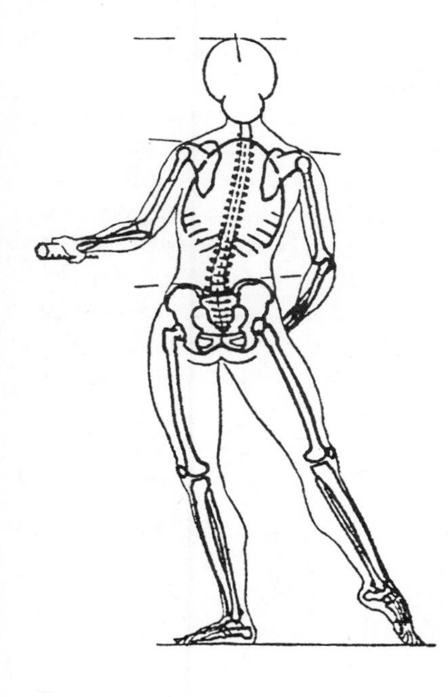

Figure 22.43 Sidestepping induces heavy lateral bending moments acting on the spine (86).

the "approved" lifting posture could lull the male worker into a sense of false security. Therefore it is always advisable to temper the categorical instruction of "knees bent—back straight—head up" with the additional explanation "provided it helps to get the load closer to your body."

It has already been described (Figure 22.29) how a light but bulky object often imposes a lifting stress much greater than the one exerted by a heavier article of greater density. A metal ingot held close to the body exerts a lesser sagittal bending moment on the spine than a box of equal weight containing small miniaturized components, such as transistors packed for shipping in a Styrofoam container. This age of miniaturization and containerization has added a serious and sinister overtone to the age-old joke: "Which is heavier, a pound of lead or a pound of feathers?" The feathers, of course; they are so much bulkier.

In all instances, however, the two other moments in addition to the sagittal one, must also be considered. *Lateral bending moments* are all-important whenever a job calls for "sidestepping" (Figure 22.43). This often occurs when, for example, the serving of food at lunch counters is involved, or when components have to be transferred to a tray from a jig mounted on a machine.

Generally speaking, bench work involves unnecessary sidestepping. Lateral bending moments can be of considerable magnitude, and in special cases (e.g., when an individual suffers from a mild nerve root entrapment syndrome), they may impose considerable hazard and suffering. A workplace designed for sidestepping is a workplace designed for trouble.

Consideration of *torsional moments* acting on the vertebral column becomes necessary when materials are transferred from one service or work bench to another (Figure 22.44). The "L"-shaped work surfaces appeal to both architect and industrial engineer because they combine aesthetic considerations with opportunities for performance efficiency. Whenever a lifting task requiring rotation of the torso about the vertical axis of the body is to be performed by a standing person, a serious hazard is not presented because of the interaction between ankle, knee, and hip joints. However when such a task is performed seated, the pelvis is securely anchored and the entire torsional moment must be absorbed by the lumbar and thoracic spine. This condition is conducive to great mechanical stress on the vertebral column; it may aggravate preexisting back pain caused by pathology in the lumbar or lumbosacral region, and it can cause distress, sometimes of a respiratory nature, to scoliotic or kyphotic individuals.

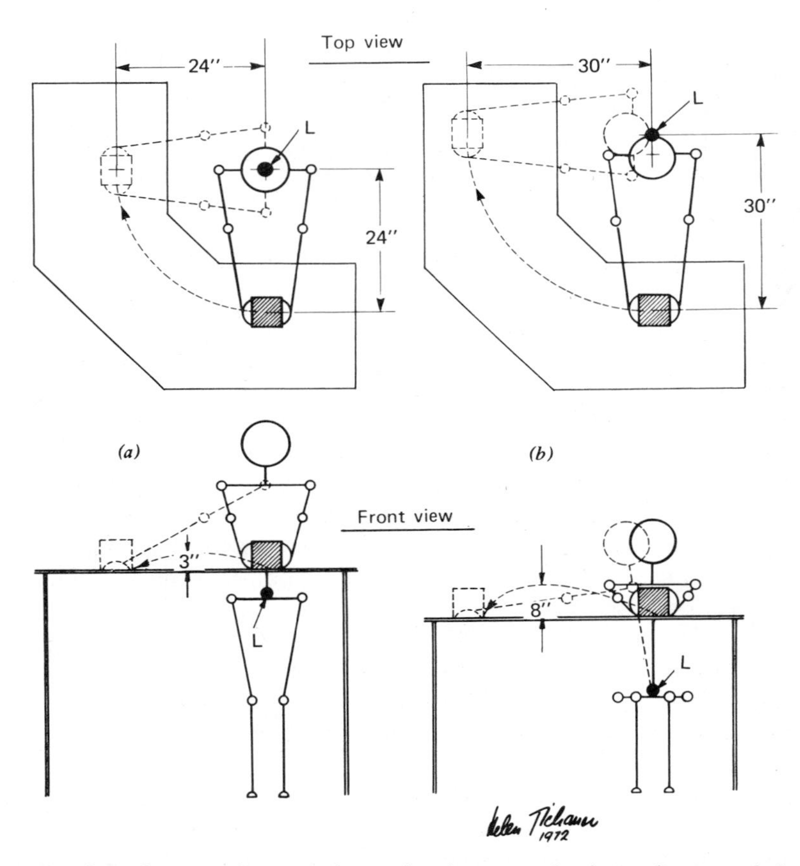

Figure 22.44 The schematic drawing shows that because of relative fixation of the pelvis, a seated lifting task may produce very heavy torsional bending moments acting on the lumbar spine; *L* is the lumbosacral joint. From Reference 48.

It is easy to avoid excessive torsional moments in seated work situations by providing well-designed swivel chairs. It is possible to determine numerically the severity of moment action on the human body during a lifting task by the use of computerized models (53). These have been developed to a fairly high degree of perfection, and their use constitutes the procedure of choice in mass production operations under conditions of rigidly controlled work methods. One of the principal advantages of the computerized lifting models resides in their capability to permit accurate estimation of moments exerted by the mass of various body segments on a wide variety of anatomical reference points. Sometimes, however, this accuracy and convenience must be sacrificed when the need occurs to make rapid ad hoc decisions under field conditions or when the small size of the workforce makes recourse to a large computational facility uneconomical.

It is for this reason that ergonomists should aim to develop the knack of "guesstimating" the magnitude of all three moments by looking at the worker, at motion pictures of an operation, or even at a drawing of the workplace layout. Then these guessed moments should be added, not algebraically, but vectorially.

By way of bench mark, it may be assumed that when the vector sum of all three moments is 350 in.-lb or less, the work is light and can be performed with ease by untrained individuals, male as well as female, irrespective of body build. Moments above this level, but below 750 in.-lb, classify a task as "medium-heavy," requiring good body structure as well as some training. Tasks above this moment but below 1200 in.-lb may be considered to be heavy, requiring selective recruitment of labor, careful training, and attention to rest pauses. Whenever the vector sum of moments exceeds those stated before, the work is very heavy, cannot always be performed on a continuing basis for the entire working day, and requires great care in recruitment and training.

It is my personal experience that the ability to guess the magnitude of moments exerted by the load on the body is easily acquired. The same does not hold true, however, for moments exerted by segments of the living body. Therefore the validity of the aforementioned bench mark values is somewhat limited because they do not take into consideration individual differences in lifting stress caused by body segments, but refer to moment increases caused by external loads only.

The *gravitational components* are elements of a lifting task closely related to the concept of "work" in the sense of physics. In mechanics, work is defined as the product of force multiplied by the distance through which it acts. Thus lifting 10 pounds against gravity to a height of 5 feet will constitute 50 ft-lb of work. Likewise, pushing an object horizontally for a distance of 5 feet when 10 pounds of pushing force is required will also result in 50 ft-lb of work. This definition, however, is not always applicable to a situation involving the human body at work. For example, if an individual pushes with all his force against a wall and moves neither his body nor the wall, he has not accomplished any work according to the rigid definitions of physics. Nevertheless, during the entire time, muscles have been under tension, metabolic activities increased, and the added energy demands of the living organism manifested themselves in the expenditure of additional calories which, in physics, are assigned the dimensions of work.

The effort expended on an activity that requires the application of force for a period of time without concurrent displacement of an object is called "isometric activity." Sometimes the term "tension time" is also applied to this kind of activity. "Isometric work" is assigned the physical dimensions of "impulse," which equals force multiplied by time. This makes mathematical processing somewhat difficult inasmuch as gravitational components of a lifting task have the dimensions of "work," which equals force multiplied by distance. Mathematical transformations to overcome this difficulty have been developed (54) and are useful whenever recourse to computerized models must be taken. For all practical purposes, it is often adequate to estimate *isometric work* by taking the weight of the object handled plus the estimated weight of the body segment involved in the task, and to multiply these by the time the muscles are under tension.

Dynamic work is defined as the product of the weight of an object handled, multiplied by the vertical distance through which it is lifted upward against gravity. It has the dimensions of work as defined in physics and can be generally computed with ease.

Negative work is performed whenever an object is lowered at velocities and accelerations of less than gravity so that work against the "*g*" vector is performed.

To avoid complex computations, it is practical to assume, under industrial working conditions (55), that one-third of the work that would have been expended when lifting the same object over the same distance in an upward direction is approximately equal to the negative work performed.

Finally, in the evaluation of a lifting task, the *inertial forces* must be considered. When an object handled is in motion, acceleration is generally insignificant as far as work stress is concerned. However the forces involved in *aggregation* and *segregation* of man and load may impose severe stresses on the human body.

To maintain the unstable equilibrium of upright posture, it is necessary that the center of mass of the body be located over a line connecting the sesamoid bones of the big toes. Whenever a load is lifted, object and human body become one single aggregate, and as soon as the load has left the ground, the body, through changes in postural configuration, must place the center of mass of this body–load aggregate over the area of support. This requires displacement of the center of mass of the body proper which, during normal acquisition of the load from the ground, takes place over a time interval of roughly 0.75 second. During this brief time interval "stress spikes" will be observable if myograms of the muscles of the back are taken (56). Acquisition stress can best be minimized by having the point of pickup of the load as close to the worker's body as possible.

A stress far more severe is experienced on segregation of the load. Since release of an object is normally fairly rapid, segregation may take place over as small an interval of time as 40 msec, or $\frac{1}{20}$ of the time involved in acquisition. This requires that postural adjustments be far more rapid during release than during pickup, which, in turn, produces great stress on the musculoskeletal structures involved. Electromyographic studies (31) have shown that stress experienced during release may be a multiple of the physical work stress generated during the rest of the work cycle. This has been confirmed by other investigators (56). Segregation stress can best be reduced by having the point of release of the load as high above floor level as possible. It is also essential that workers be made aware of the postural adjustments occurring during load release, including pelvic rotation, and be trained to perform these slowly.

Finally, *frequency of lift* is often one of the elements deserving considerable attention. The number of times a lifting task is performed during the day is equivalent to "productivity," and therefore it is determined by economic needs. Where bulky loads, such as television tubes, are handled, it may be quite impossible to control the severity of a lifting operation by attention to the frequency of lift, since this will not be amenable to modification. Yet often a number of relatively small units, several at a time, have to be handled, and the frequency of the operation then can be controlled by optimizing or by changing the amount to be handled at one time.

Gilbreth, at the beginning of the century (57), stated ". . . lifting 90 lb of brick on a packet (sic) to the wall will fatigue a bricklayer less than handling the same number of bricks one or two at a time. . . ." This remark was, of course, based on purely empirical and subjective observations because work physiological instrumentation available during the lifetime of Gilbreth simply did not permit the objective substantiation of such hypotheses. The bricklaying task was repeated by a group of volunteers during the 1950s (57).

In the Gilbrethian example, "lightness" of the task when handling 90 pounds of bricks (18 bricks) one at a time is illusory. Each time a 5-pound brick is handled, the worker must move, bend, erect, rotate, and so on, approximately 100 pounds of body mass. Thus the task load imposed by the handling of the body itself becomes much more severe than the work stress caused by the material being handled. Approximately 1800 pounds of lifting body mass must be maneuvered to shift 90 pounds of brick.

When 90 pounds of brick was handled at one time, according to the Gilbrethian intuitive prescription, the ratio between body mass moved and inanimate material handled was roughly 1 : 1. As was to be expected, the physiological response to the task, under the circumstances described, was quite moderate and was essentially a function of the rate of productive work, not of unnecessary and physiologically expensive body movements. However when the task load was increased to 120 pounds (i.e., 24 bricks) handled at the same time, the metabolic cost of the job rose out of proportion to the increment in the dynamic component of the task (Table 22.4). This apparent paradox was rationally explained at the time of the experiment as a result of oxygen debt. The significance of the element "frequency of lift," as shown previously, can easily cause the handling of numerous small loads to become the equivalent of a very stressful task. It is therefore essential to be aware of the concept of "optimal load" per lift.

5.2 Queueing Situations

In materials handling situations, loads are frequently passed from one worker to the other, or they arrive at the lifting station by way of conveyor belts and are then handled manually. Much useless materials handling takes place when conveyor belts, device pallets, trolleys, and other devices, or areas for temporary holding or storage of products, are too small. Bottlenecks occur, and emergency measures may have to be

Table 22.4 Results of a Replication of the Gilbrethian Lifting Experiment (57)

Frank Bunker Gilbreth, the father of motion study, was intuitively right when he stated that "... to lift 90 pounds of brick at a time is most advantageous physiologically as well as economically ..." (57).

	I	II	III
Bricks per lift	1	18	24
Weight per lift (lb)	5	90	120
Work per hour (kcal)	520	285	450
Bricks per hour	250	600	300

taken to clear the congested areas. The dimensions of all temporary storage areas and devices should be computed on the basis of queueing theory, assuming that the arrival at the handling station follows the Poisson distribution, and service times are exponentially distributed. Frequently the following formula can be used to advantage (58):

$$N = \frac{\log P}{\log R} \tag{11}$$

where N = required capacity of the area

P = greatest acceptable probability that the area will become temporarily overloaded; this principle is normally determined on the basis of a subjective management decision

R = mean arrival rate of units per time divided by the mean processing rate; these values are normally available from industrial engineers

It should not be forgotten that in addition to readily discernable or "overt" lifting tasks, workplace design may embody hidden or "covert" situations. A covert lifting task exists whenever a moment is exerted on the axial skeleton without an extraneous object being handled. This may be the case when a typist bends her head over the typewriter, or an arm has to be held out for extended periods of time because of the poor location of storage bins, or working shoes have heels that are too high, so that they increase the concavity of the lordotic curve of the lumbar spine (31). To assess the true severity of such a situation, textbooks of classical biomechanics (18) or computerized mathematical models should be consulted (39).

In conclusion, it is reemphasized that the elements of a lifting task are heterogeneous in their physical dimensions and their physiological effects; therefore, although all contribute to the level of work stress, the individual effects cannot be "summed." The reality of most materials handling situations will demand that each element be considered separately and that those which are amenable to control be reduced in magnitude and severity as far as possible.

Unfortunately, at the present state of the art, there are only partial and no total solutions available to eliminate hazards from manual lifting and materials handling tasks.

6 HAND TOOLS

Toolmaking is probably as old as the human race itself. Almost all the basic tools used today, as well as the "basic machines," were invented in the dawn of the prehistoric ages, evolving gradually and parallel with the development of technology until modern times.

However the current technology explosion proceeds too rapidly to permit the gradual development of tools to suit the new industrial processes. Now "instant" creation of new and specialized implements often becomes an acute economic necessity, if an industry wishes to accommodate itself to the rapid changes of workforce and technology. Hand tools are more often than not standardized to be "fairly acceptable" to the broadest possible spectrum of populations and activities. Only rarely are they designed to fit perfectly the specialized needs of a particular manufacturing pursuit or the anthropometric attributes of the workforce in a given plant.

Thus considering their widespread use, the number of varieties of hand tools is quite limited, and each species of tool is normally employed by vast numbers of users. Therefore hand-tool-generated work stress, trauma, and ergogenic disease may at times reach epidemic proportions, disabling numerous individuals, seriously impairing the productive capacity of a manufacturing plant, and having very detrimental effects on labor–management relations. All this is quite apart from the medical costs and the human suffering involved. Therefore the occupational health specialist should make the proper selection, evaluation, and usage of hand tools one of his major concerns.

6.1 Basic Considerations in Tool Evaluation

Though the shapes of tools are varied and the functions of the different classes of implement diverse, there are nevertheless many principles of biomechanics and ergonomics that are applicable to the prevention and solution of problems created by hand tools, no matter how different their fields of application.

Many of the principles enunciated in this section extend in application beyond the narrow field of hand tools proper. They are equally useful in the analysis of equipment controls, since the latter are merely the "tools" that permit the operation of machinery. The initial purpose of primeval tools, such as stone axes and scrapers, was to transmit forces generated within the human body onto inanimate materials, food, or live animals of prey. As the spectrum of artisanal and industrial pursuits widened, the basic purpose of tools became more varied, and today these implements are designed to extend, reinforce, and make more precise range, strength, and effectiveness of limb movement engaged in the performance of a given task.

In this context, the word "extend" does not solely imply a magnification of limb function. Often tools such as tweezers and screwdrivers also make possible far smaller and finer movements than the unarmed hands would be capable of performing. An even better example is the micromanipulator of the "master–slave" type, which serves as attenuator rather than amplifier of human force in motion. A third example is a suction tool that makes it possible to transport small, fragile, and soft workpieces without injury either to their dimensions or their surface finish.

Selection and evaluation of all hand tools, whether manual or power operated, should be based on all the following: technical, anatomical, kinesiological, anthropometric, physiological, and hygienic considerations (59).

In the course of technological development, tool designers were conditioned to focus their attention on a real or imagined need to maximize the tool force output obtainable from a minimal muscular force input from the hand. This, however, should not be overdone.

The operation of a tool should always require sufficient force to provide adequate sensory feedback to the musculoskeletal system in general, and the tactile surfaces of the hand in particular. This is frequently a process of optimization. For example, if a fine screw thread is tapped by hand and the handle of the threading tool is too large, the force acting on the tool becomes excessive, resulting in stripped threads, broken taps, or bruised knuckles. If, on the other hand, the ratio of force output to force input is too small, an unduly large number of work elements must be repeated, and this makes the job fatiguing. An example would be the pounding of a large nail with a very small hammer.

A tool should provide a precise and optimal stress concentration at a specific location on the workpiece. Thus, up to a certain limit, an ax should be as finely honed as possible to fell a tree with a minimum number of strokes, but the edge should not be so keen that it requires frequent resharpening or is fragile. Preferably the tool should be shaped so that it will be automatically guided into a position of optimal advantage where it will do its job best without bruising either hand or workpiece. The Phillips screwdriver, as compared with the ordinary, flat, blade tool, illustrates the latter point.

Hand tool usage causes a variety of stress vectors to act on the man–equipment interface. These may be mechanical, thermal, circulatory, or vibratory, and they are often propagated to quite distant points within the body, far from the actual locus of application of force. The cause of severe pain in the neck muscles or numbness and tingling in the fingers of the left hand may quite conceivably be due to the transmission of somatic resonance vibrations triggered by a vibrating hand tool held in an unergonomic configuration in the right hand. Numerous other examples could be cited. Whenever a single specific anatomical region becomes the locus of repeated manifestations of signs and symptoms of trauma, no matter how far away from the tool operating hand, the work situation should be carefully analyzed for the possible implication of hand tool design or usage as a traumatogenic vector.

Contact surfaces between the tool and the hand should be kept large enough to avoid concentration of high compressive stresses (Figure 22.45). Pressure and impact acting on

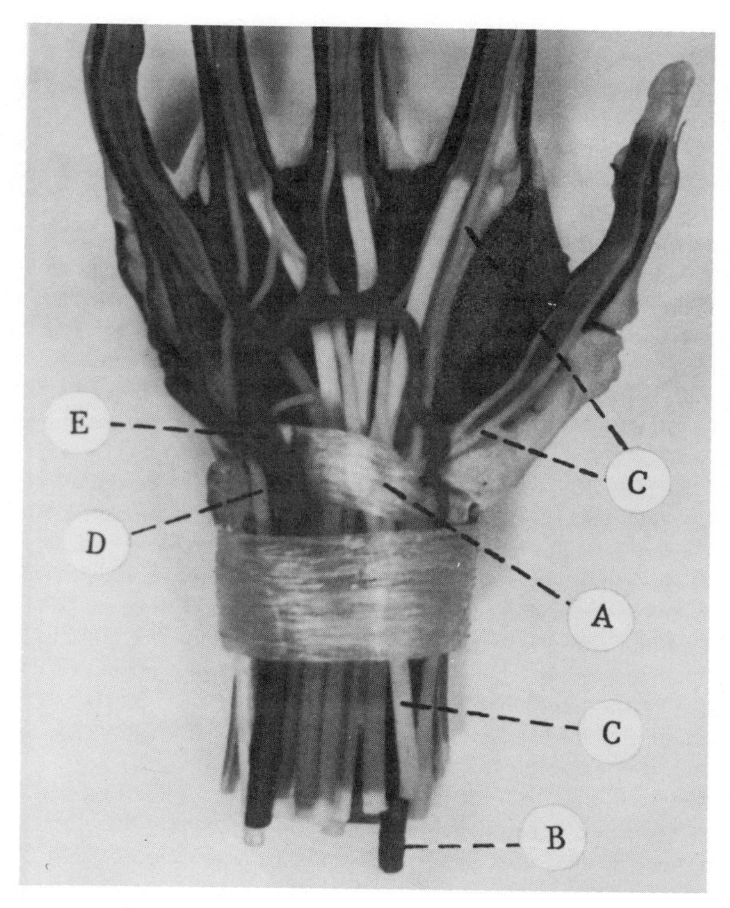

Figure 22.45 Through the carpal tunnel *A* pass many vulnerable anatomical structures: blood vessels *B* and the median nerve C. Outside the tunnel, but vulnerable to pressure, are the ulnar nerve *D* and the ulnar artery *E*. From Reference 48.

the hand may be transmitted either directly or by rheological propagation on vulnerable structures. A poorly designed or improperly held scraping tool may squeeze the ulnar artery and sometimes the ulnar nerve between the timber of the handle and the bones of the wrist. This may deprive the ring and little fingers of proper blood supply (Figure 22.33), which may cause numbness and tingling of the fingers. Under such circumstances, the afflicted worker will devise an excuse to leave the workplace temporarily, since this is the only means of relief open to him. Apart from the resulting drop in productivity, the health of the working population is in serious jeopardy. Literature (60) suggests that compressive stresses applied against the medical side of the hook of the hamate (Figure 22.50) can traumatize the ulnar artery and may result in thrombosis or other irreversible injury. The ulnar artery and nerve may also be injured indirectly by stress propagation whenever the palm of the hand is used as a hammer or repeatedly pushes a tool against strong resistance. The resulting lesion, be it vascular or nervous, is known as "hypothenar hammer syndrome" (61).

6.2 The Anatomy of Function of Forearm and Hand

Further chances of injury exist when the motions inventory demanded by specific features in the design of the tool is not readily available from arm and hand. The kinesiology of the upper extremity is basically determined by the structure and arrangement of the skeleton of arm and hand (Figure 22.46). There are two bones in the forearm, the ulna and the radius. The ulna is stout at its joint surface of contact with the humerus and forms a hinge bearing there. It is, however, slender at the distal end, where it articulates well with the radius but only poorly with the carpus formed by the bones of the root of the hand. The radius, on the other hand, is slender at its point of contact with the humerus and stout at its distal end. At the proximal end it forms a thrust bearing with the humerus and a journal bearing with the ulna. The distal end forms a joint with the carpal bones, the primary articulation between forearm and hand. The unique configuration of the mating surfaces of this joint permits movement in only two planes, each one at an angle of approximately 90° to each other. The first of these maneuvers is palmar flexion, or when performed in the opposite direction, dorsiflexion (Figure 22.47). The second set of wrist movements possible consists of either ulnar or radial deviation of the hand (Figure 22.48). The wrist joint does not allow rotation of the carpus about the longitudinal axis of the forearm. Swiveling the wrist without forearm rotation is not possible (Figure 22.49).

This geometry of joint movement causes the axis of longitudinal rotation of the forearm–hand aggregate to run roughly from the lateral side of the elbow joint through a point at the base of the ulnar side of the middle finger (Figure 22.46). The palmar aspect of the carpal bones forms a concave surface roofed by the transverse carpal ligament. The resulting channel is known as the "carpal tunnel." Through this conduit pass the tendons of the flexor muscles of the fingers, which originate from the medial side of the elbow. Some blood vessels and nerves also pass through this tunnel. A similar, albeit much shallower, passage for the extensor tendons and some nerves is formed on the dorsal surface by another ligament and the dorsal aspects of parts of the carpus, the radius, and the ulna (Figure 22.50). Each of the two passages is further divided into several longitudinal compartments, and this produces considerable friction and lateral pressures between tendons, tendon sheaths, adjacent nerves, and vascular structures. If the fist is forcefully opened, closed, or rotated, these forces, as well as the friction between anatomical structures, can become unduly high whenever the carpal tunnel and its homologue compartment on the dorsum of the hand are not properly aligned with the longitudinal axis of the forearm. Such alignment exists only if the wrist is kept perfectly straight, so that the metacarpal bone of the ring finger is reasonably parallel with the distal end of the ulna. A potentially pathogenic situation may exist whenever manipulative maneuvers,

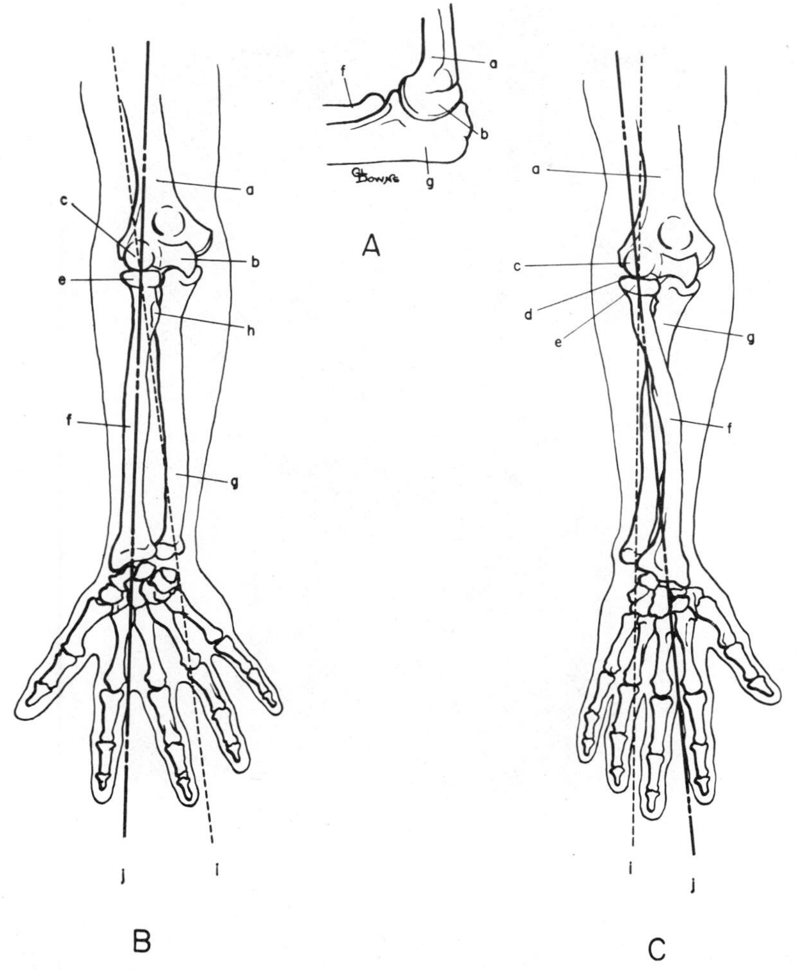

Figure 22.46 The construction of the skeleton of the forearm. (*A*) The hinge joint between ulna and humerus from the medial side. (*B*) The right forearm outwardly rotated. (C) The right forearm medially rotated: a, humerus; *b*, trochlea of humerus; c, capitulum of humerus; *d*, thrust bearing formed by capitulum and head of radius; e, head of radius; *f*, radius; *g*, ulna; *h*, attachment of biceps; *i*, axis of rotation of forearm; *j*, optimal axis for thrust transmission. From Reference 48.

especially forceful ones that require ulnar deviation, radial deviation, palmar- or dorsiflexion, either singly or in combination with each other, are performed. The misalignments of tendons at the wrist under such circumstances and their bunching up against each other multiply drastically the already high interstructural forces and frictions produced by the muscles operating the hand; early fatigue may result, among other undesirable manifestations (Figure 22.51). The long interval of time (several weeks, months,

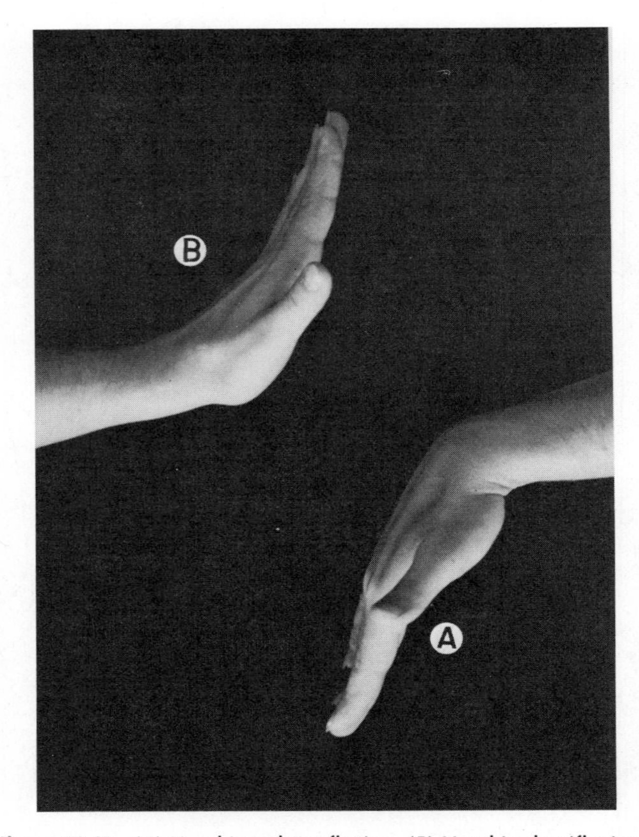

Figure 22.47 (*A*) Hand in palmar flexion. (*B*) Hand in dorsiflexion.

even years) of repetitive multiple work stress elapsing before occupational disease of forearm or hand becomes clinically evident, has militated against the definitive establishment of direct cause and effect relationships between the forearm–wrist configuration and specific pathology. Nevertheless, certain motion elements are almost inevitably associated with fairly narrow spectra of occupational disease of the hand (2).

6.3 Elemental Analysis of Hand Movements

The design of tool and handle determines, limits, and defines the motion elements that are necessary for the purpose of the productive process. This intimate relationship between man and hand tools affects occupational health and safety most directly.

It is almost impossible to describe innocuous as well as potentially pathogenic manipulative maneuvers, either verbally or by way of two-dimensional pictorial representation. As a somewhat simplistic yet practical alternative, recourse to the analysis of manipulative maneuvers in terms of their pertinence to "clothes wringing" (Figure 22.52) is recommended. "Clothes wringing" has been associated for more than a century with tenosynovitis and other undesirable conditions of the hand (2). Here we assume that the wringing is done by a clockwise movement of the right fist and counter-clockwise action of the left.

Whenever a detailed elemental analysis of hand motions is performed, it should be remembered that stress injurious to the hands is produced by four basic conditions:

1. Excessive use against resistance.
2. Hand use while in a potentially pathogenic configuration.
3. Repetitive maneuvers and cumulative work stress rather than "single episode" overexertion.
4. Use of the hand in an unaccustomed manner, as in training, for example (Figure 22.35).

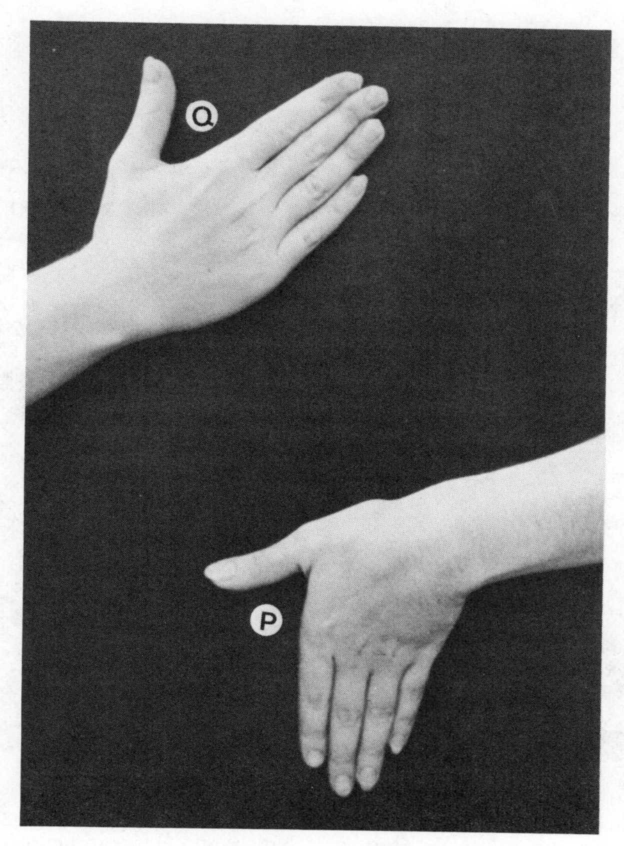

Figure 22.48 *P,* ulnar deviation; *Q,* radial deviation.

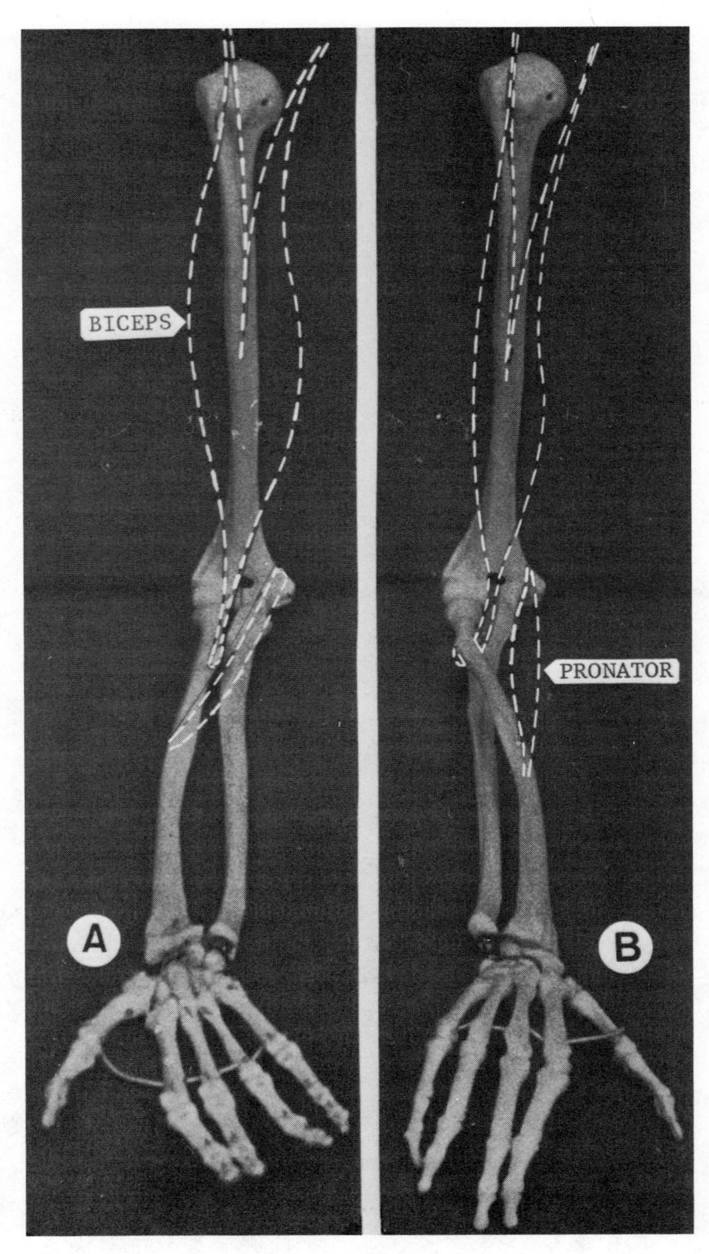

Figure 22.49 The bones of the wrist articulate with only one of the two long bones of the forearm, the radius, with which they form a firm aggregate. Therefore swiveling the wrist without rotating the forearm is impossible. (*A*) Forearm in supination. (*B*) forearm in pronation.

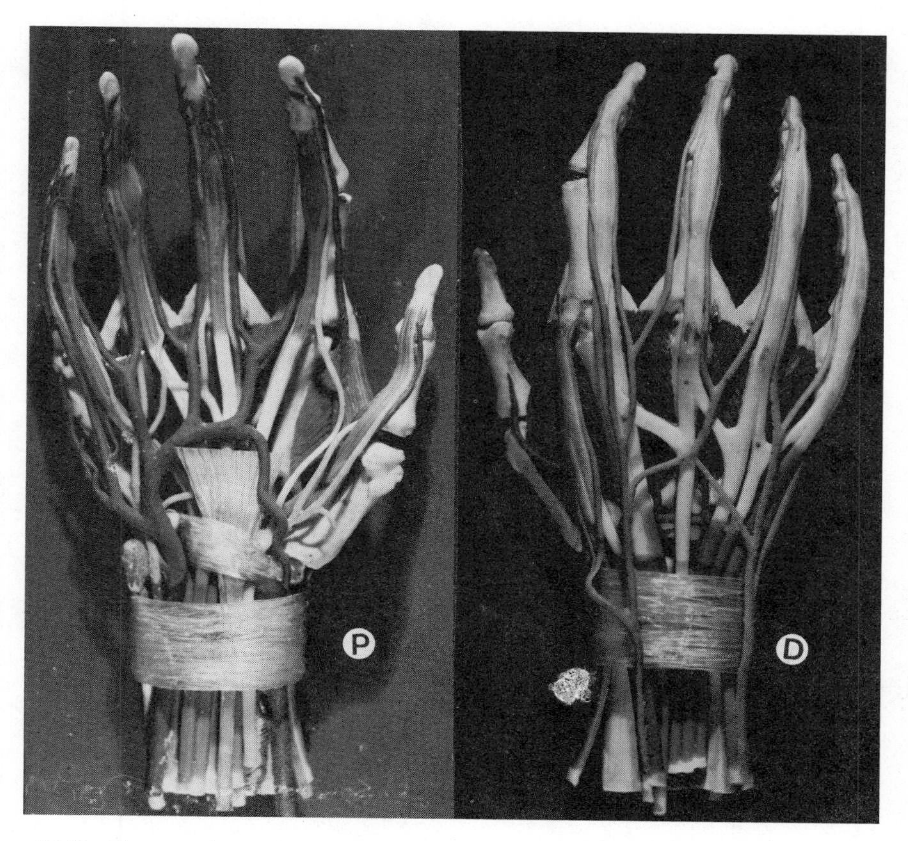

Figure 22.50 The complex arrangements of tendons, blood vessels, and nerves underneath the ligaments of the wrist: *P*, palmar aspect; *D*, dorsal aspect. Whenever the wrist is deviated, these bunch up against each other. The ensuing friction may lead to trauma and disease.

Figure 22.52 shows that in "clothes wringing" the right hand is engaged simultaneously in supination, ulnar deviation, and palmar flexion. This motion pattern is frequently associated with occurrences of tenosynovitis of the extensor tendons of the wrist or the abductors of the thumb, the latter affliction also known as De Quervain's disease. It is an inflammation of the synovial lining of the tendon sheaths or associated structures. It becomes frequently manifest under conditions of unaccustomed hand usage in new employees (Figure 22.35) and is therefore occasionally considered quite without justification to be "a training disease." However the basic cause of much exposure to excessive work stress resides not in the learning process but in the misuse of the hand, which has been forced by poor workplace or tool design into unergonomic configuration.

The similarity between the movement of the right hand motion in "clothes wringing" and the use of pliers when looping wires around pegs is quite evident. Since three factors

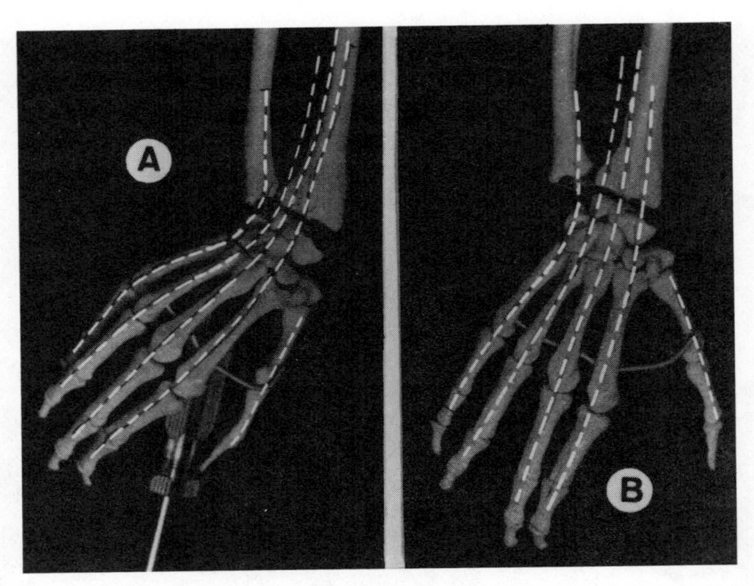

Figure 22.51 The need to align a tool with the axis of the forearm often forces the hand to deflect toward the ulna (*A*). The tendons operating the fingers get "kinked" and bunch up. This causes friction between these delicate anatomical structures which, in turn, produces discomfort and, occasionally, diseases of the wrist. When the wrist can be maintained in a straight configuration (*B*) by good tool design, the tendons are well separated, run straight, and can operate efficiently. From Reference 37, by permission. © American Institute of Industrial Engineers, Inc., Norcross, Ga.

Figure 22.52 Any manipulative motion element involving the wrist that may be considered to be part of a "clothes wringing" operation puts the worker at risk.

are implicated in the generation of the undesirable work stress, elimination of one of them may successfully reduce the stress to a nonpathogenic level. In this case ulnar deviation was eliminated by bending the tool handle (Figure 22.35).

In some other work situations involving the same wrist configuration, it may be more desirable, or easier, to reduce excessive palmar flexion by small modifications in workplace layout. A simple change of the distance between worker and work bench, location of the work chair, work surface height, or the degree of tilt of the workspace, may constitute adequate remedial action (40). The experienced practitioner will apply the same principles to the improvement of other pathogenic work situations, such as the manual insertion of screws, the manipulation of rotating switches, or the operation of electrical or air-powered nut setters suspended over the workplace.

In Figure 22.52 the left hand is engaged in pronation, radial deviation, and dorsiflexion of the wrist. This configuration is conducive to pressures between the head of the radius and the mating joint surface of the humerus (Figure 22.46). This posture should be strongly discouraged through proper tool design because it may produce a high incidence of the group of diseases known in the vernacular as "tennis elbow." The condition may occur in such tasks as overhead use of wire brushes in maintenance work. This is a good example of the basic principle that work strain may affecct a site distant from the location of work stress. The wrist is being stressed, but it is the elbow that gets sore (62). Strong and repeated dorsiflexion of the wrist, especially in combination with some other hand or forearm movement, is conducive to carpal tunnel syndrome—a disease that may also be provoked by direct trauma to the region of the hand over the carpal tunnel, compression of the median nerve in the tunnel through tenosynovitis and swelling of the flexor tendons, as well as several other causes that are not occupationally related (61). Implicated in unergonomic imposed patterns conducive to carpal tunnel syndrome are tossing motions, the operation of valves located overhead or on vertical walls, and a number of poor handle designs of power tools. This list, however, is by no means exhaustive.

Finally, the wrists of both hands can be severely stressed when a two-handled tool is designed so that the longitudinal axes of both handles coincide. Under such circumstances, the included angle between the axes of the handles should approximate 120°.

6.4 Trigger-Operated Tools

Occasionally the condition of "trigger finger" is encountered. The afflicted person typically can flex the finger but cannot extend it actively. It must be righted passively by external force. When snapped back in such a way, an audible click may be heard. This has been attributed to the generation of a groove in the flexor tendon which snaps into a constriction produced by a fibrous tunnel guiding the tendon along the palmar side of the finger. Small ganglia arising in the tendon sheath have also been implicated (63). This affliction has been observed under conditions of overusage of the index finger as an equipment control, such as the trigger of a tool embodying a pistol grip. The association between overusage of the index finger and the lesion seems to occur most frequently if the tool handle is so large that the distal phalanx of the finger has to be flexed while the middle phalanx must be kept straight. This can easily be the case when females, with relatively small hands, operate tools designed for males. A tool handle a shade too large may put a female working population to distress, but the tool designer who errs on the side of smallness will produce an implement that can be operated with equal effectiveness by both sexes (48). As a rule, frequent use of the index finger should be avoided, and thumb-operated controls should be put into tools and implements wherever possible, because for all practical considerations, the thumb is the only finger that is flexed, abducted, and, in addition to muscles crossing the wrist, opposed by strong, short muscles located entirely within the palm of the hand. It can therefore actuate push buttons and triggers repeatedly, strongly, without fatigue, and without exposure to undue hazard.

6.5 Miscellaneous Considerations

Numerous handheld tools, especially of the power-driven type, are advertised as "light." The lightest tool is not always the best tool. Optimal tool weight is dependent on a number of considerations. If a power tool houses vibrating components, it should be heavy enough to possess adequate inertia. Otherwise it may transmit vibrations of pathogenic frequency onto the body of the operator (see Section 4.3) (42, 43).

Heavy tools should be designed so that the center of mass of the implement is located as close as possible to the body of the person holding it. When this type of facilitation is inadequate, recourse should be taken to suspension mechanisms and counterweights. Strength of handgrip is the most important single factor limiting the weight of a tool that must be held without external assistance from supporting mechanisms. Under such circumstances, a reliable and specialized reference work should be consulted (64).

The design of the tool–hand interface should be based on carefully selected anthropometric considerations. There are a number of specialized anthropometric reference works available (65–67). However not all the hand dimensions identified in literature are necessarily representative of a working population of specialized age, sex, or ethnic origin. Furthermore, the geometry of manipulative movements imposed on the hand by the specifics of tool design or usage may force the user population to employ a geometry of hand movement based on special anthropometric parameters quite different from those available from the majority of reference works. Likewise, the use of working gloves changes the dimensions of the hand. Gloves vary in design and thickness of material used. This does affect grip strength and sensory feedback from the tool (see Section 4.4).

It is recommended that producers as well as consumers of handheld tools conduct their own anthropometric testing based on the hand and other body dimensions of the specialized working population under study, as well as of the tools and gloves to be used. Of course basic reference charts can be employed with advantage as the point of departure for anthropometric hand studies designed to suit specialized needs (68).

Gloves may affect the working hand in a number of additional ways. Mild trauma due to pressure may be aggravated through contacts with irritants entrapped unintentionally in working gloves. In the case of abrasives, such as metal chips or mineral particles, the foreign substances may be worked into the skin, producing in some cases benign tumors, such as talcum granuloma. Certain fat-soluble chemicals, such as many common solvents and detergents, may be soaked up in the material of the glove and transferred onto the skin, causing maceration, dermatoses, or other tissue damage. Sometimes it may be advisable to wear thin cotton gloves under these working gloves to absorb perspiration and improve hygiene.

Any study of work stress, occupational disease, or safety involving hand tools is incomplete unless the potential effects of working gloves are fully considered, inasmuch as they could modify fit between tool and hand, change the distribution of pressure, or become detrimental to the integrity of the skin.

7 CHAIRS AND SITTING POSTURE

Many jobs require performance in the seated posture, and chairs are among the most important devices used in industry. They determine postural configuration at the workplace as well as basic motion patterns. A well-constructed chair may add as much as 40 productive minutes to the working day of each productive individual (48). Furthermore, poorly designed seating and inadequate policing of seating posture constitute frequent and definitive occupational hazards. Properly designed working chairs are a prerequisite to the maintenance of occupational health and safety of many working populations, including individuals with preexisting disabilities of the back.

A number of reference works (69, 70) provide dimensions for working chairs and benches considered by many as optimal. However such optimality can never be general; it applies only to the restrictive parameters of a population defined by age, sex, ethnic origin, cultural background, and specific working conditions. Sliderulelike devices (71) are available that permit the adaptation of anthropometric data to a wide variety of body dimensions, and in some countries the anthropometric basis for industrial seating has been subject to national standardization (72, 73). However all dimensional aids to chair design must be personalized and revalidated for each individual application on the basis of certain aspects of functional and surface anatomy.

7.1 Anatomical, Anthropometric, and Biomechanical Considerations

It was realized almost a century ago, that ". . . our chairs almost without exception are constructed more for the eye than for the back . . ." (74). This statement was made in an article appearing in a journal of preventive medicine and stating the urgency of providing better lumbar support to seated workers. Unfortunately, in many instances, the same plea can be made today with respect to some of the most modern office and factory furniture. Somewhat later Strasser (75) quantified the forces exerted by the backrest and the seat on the lumbar region and the buttocks. He also showed how these changed drastically as a function of the slope of these features. The first comprehensive study of the biomechanics of seating was not conducted until 1948 (45). This existing body of knowledge should be put to good use in the design of working chairs.

To facilitate a description of seated work situations, a nomenclature of planes of reference suitable for the definition of relationships between postural configuration and the position of equipment controls, as agreed by convention, is employed throughout this section (Figure 22.53). The "axis of support" of the seated torso is a line in a coronal plane passing through the projection of the lowest point of the ischial tuberosities on the

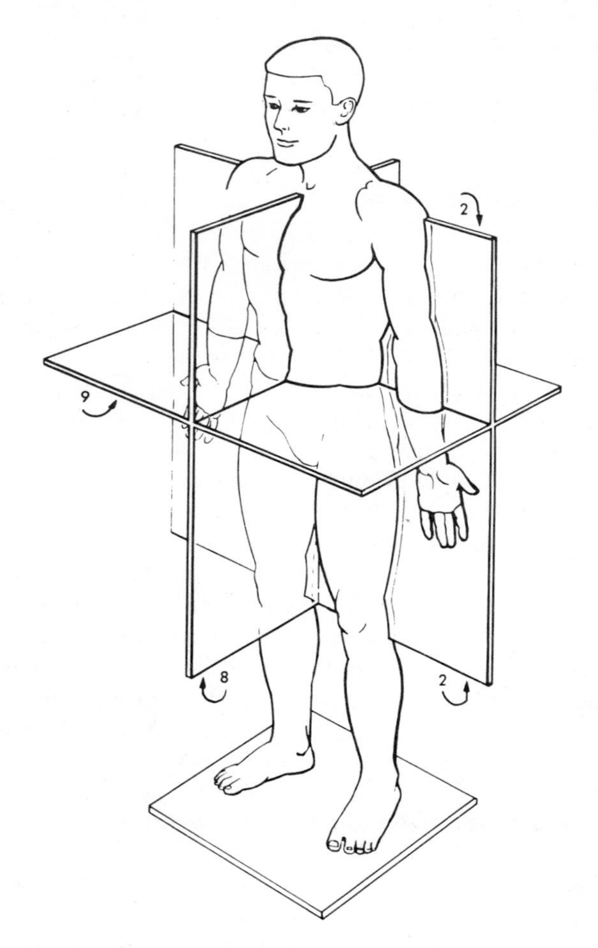

2 = CORONAL PLANE

8 = MID-SAGITTAL PLANE

9 = TRANSVERSE PLANE

Figure 22.53 The basic planes of reference for biomechanical and anatomical description. From Reference 76, by permission. © W. B. Saunders Company, Philadelphia, Pa.

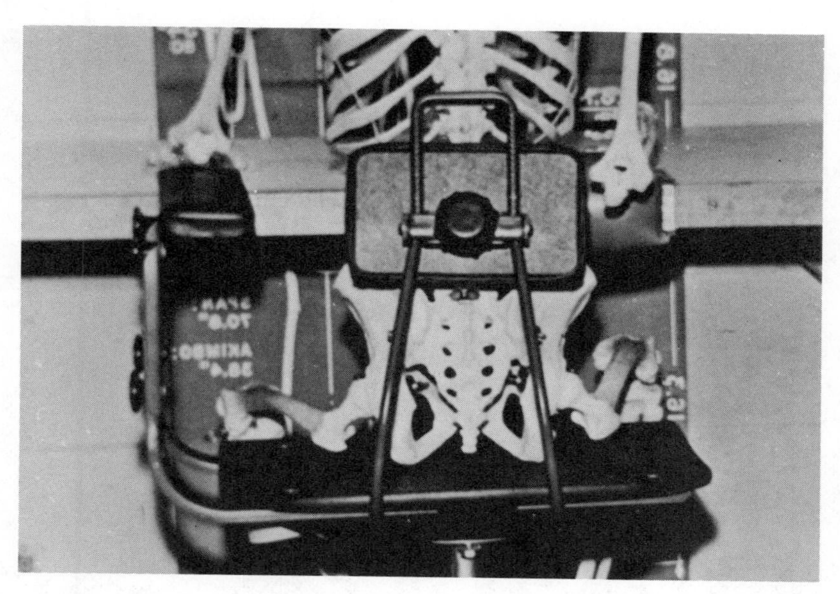

Figure 22.54 The structure of the pelvis and the location of the ischial tuberosities demand that the coronal sections of a chair be not contoured.

surface of the seat. This is, in essence, a "two-point support" (Figure 22.54). As a result, the compressive stresses exerted on the areas of the buttocks underlying the tuberosities is quite high and has been estimated as 85 to 100 psi. This, of course, varies with body weight and posture. Stress can be nearly double when a person is sitting cross-legged. These high pressures make it necessary to vary sitting posture and position on a seat frequently to provide the necessary stress relief for the body tissues. Therefore the seat of the chair should be approximately 25 percent wider than the total breadth of the buttocks. To further facilitate change of position, all coronal sections of the seating surface of the chair should be straight lines. A coronally contoured seating surface tends to restrict postural freedom at the workplace and, especially when poorly matched to the curvatures of the buttocks, may cause severe discomfort. Improper contouring may also interact with sanitary napkins and other devices worn by women during the menstrual period, with the ensuing further reduction in physical well-being during an already trying time.

The height of the seat has been the subject of much argument, and literature abounds with numerical data, most stating the relevant dimension as the vertical distance of the highest point of the seat from the floor. This is ergonomically wrong. The back of the thighs is ill-equipped to withstand pressure (Figure 22.55)—compression applied there may deform the limb severely, irritate important nerves and blood vessels, and interfere with the circulation in the lower extremity. Many people had the experience that a chair too high causes the leg "to go to sleep." This is an especially undesirable condition

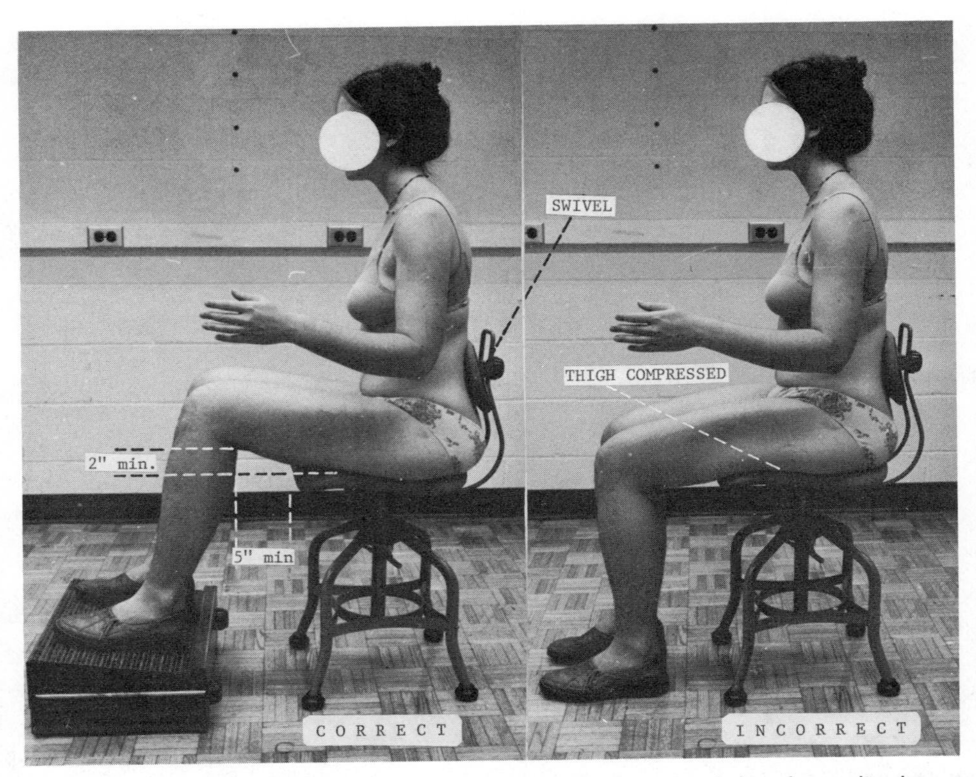

Figure 22.55 The highest point of the seat should be at least 2 inches below the popliteal crease of the worker. If necessary, this must be accomplished by a footrest. The backrest should swivel about the horizontal axis to align with the lumbar curve.

when the circulation is already impaired by preexisting disease, such as diabetes or varicose veins.

The maximum elevation of the seat above the surface supporting the feet—be this the floor or a footrest—should be 2 inches less than the crease at the back of the hollow of the knee, known as the popliteal crease. For this reason, the seat height of the chair should be adjustable in a limited number of discrete steps (Figure 22.55). The frontal end of the seating surface should terminate in a "scroll" edge, which does not cut into the back of the thigh. For further protection, there should be a distance of at least 5 inches between the scroll and the popliteal crease. The seating surface should have a backward slant of approximately 8° in a sagittal direction. This encourages the use of the backrest and prevents forward sliding of the buttocks. Since the thighs are tapered, any inadvertent forward sliding may produce compression of the limb between the lower edge of the work bench and the top of the chair. The seating surface should be slightly padded and covered with a porous, rough, fabric that "breathes" and facilitates adequate conduction of heat away from the contact area between buttocks and chair.

The backrest of a chair employed in manufacturing operations should provide lumbar support. It should be small enough not to exert pressure against the bony structures of the pelvis or the rib cage. At least the top of the backrest should be below all but the "false" ribs. Many work situations require continuous and rhythmic movement of the torso in the sagittal plane, and a backrest that is too high will produce bruises on the backs of a considerable number of the working population.

The best designed backrests are so small that they do not interfere with elbow movement and can swivel freely about a horizontal axis located in the coronal plane (Figure 22.55). Thus they fit well into the hollow of the lumbar region and provide the needed support for the lower spine without detrimental interference with soft tissues. When much materials handling and twisting of the torso in the seated position takes place a backrest that is overly wide will make frequent and repeated contact with the breasts of female workers (Figure 22.56). This can be most uncomfortable, especially during the premenstrual period, when the breasts of many women are quite tender.

Whenever backrests produce either bruising or even only slight discomfort, workers protect themselves. The painful effects of excessive interaction between chair and human body are reduced under such circumstances by ad hoc protective devices, such as pillows brought from home and strapped to the backrest. Work sampling studies show that

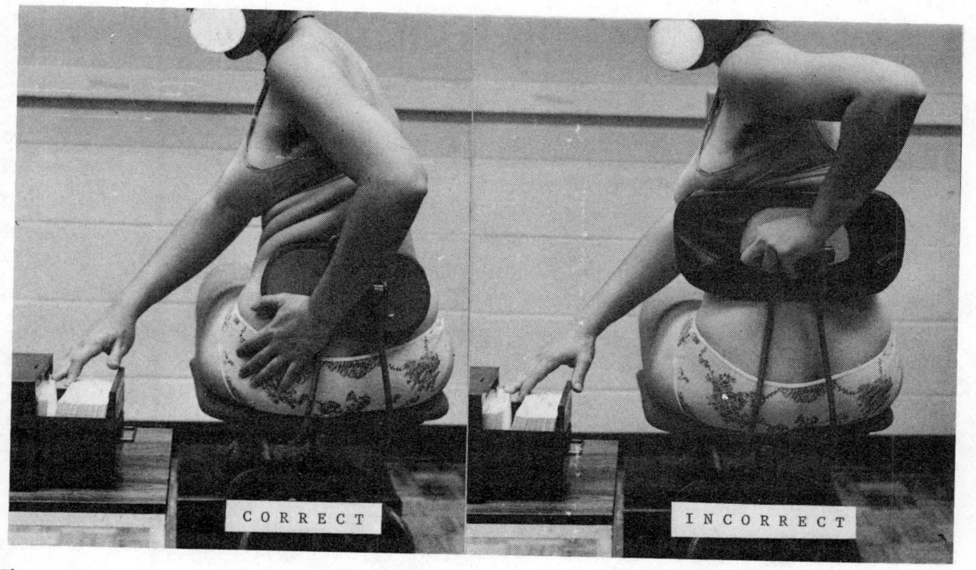

Figure 22.56 A poorly located backrest that is overly wide may severely traumatize the breasts of female workers during torsional movements. (77)

several hours per worker per week may be wasted by the effort needed to keep such improvised cushioning devices in place. Properly designed backrests are much cheaper than improvisation, both in the long run and in the short run.

Some working chairs are equipped with castors. These make possible limited mobility of the worker and permit materials handling without abandoning the seated posture. In many situations castors reduce unnecessary torsional moments acting on the lumbar spine. Also, in some cases of circulatory disturbance in the lower extremity, a chair that can be pushed around by leg movement, while the person maintains seated posture, can activate the "muscle pump," consequently improving circulation. The disadvantage of any chair mounted on rollers resides in the risk that the chair may roll accidentally away or be inadvertently removed while the user gets up for a brief time. If the same person afterward tries to sit down without being aware of the changed situation, a dangerous fall can result. Whenever possible under such circumstances, therefore, a cheap restraining device, such as a nylon rope, or chain or, sometimes, a more rigid linkage between chair and work bench should be considered.

7.2 Adjustment of Chairs on the Shop Floor

A frequent but sometimes dangerous response to chair-generated discomfort is temporary absenteeism from the workplace. When the level of personal tolerance has been exceeded, many individuals simply "take a walk." It is found occasionally that individuals involved in accidents are at the place of injury without authorization. Since no accident is possible unless victim and injury-producing agent meet at the same spot and at the same time, temporary absenteeism from the workplace may result in unnecessary exposure to potentially hazardous situations (40).

To guard against this, especially in new plants or when existing working chairs are being replaced by different models, the workforce should be polled and a subjective assessment of chair comfort established through attitudinal measurement techniques specifically designed for the evaluation of seating accommodations (78). Although subjective comfort evaluations decrease the longer a chair has been in use, it should also be remembered that a comfortable chair is not necessarily the "best chair," regardless of whether high levels of productivity or optimal physiological compatibility with the anatomy of the worker are the guiding considerations. Therefore it is essential that the necessary initial attitudinal measurement be either preceded or immediately followed up by rational adjustment of the seating accommodations as related to the work bench on the basis of biomechanical and ergonomic considerations.

Supervisors as well as workers should receive formal instruction in the proper adjustment of working chairs. First, the height between seating surface and top of the work bench is adjusted so that an angle of abduction of approximately 10° between the upper arms and the torso can be maintained during activity (Section 4.2). In specialized work situations, adjustment for an angle of abduction other than optimal may be necessary. It

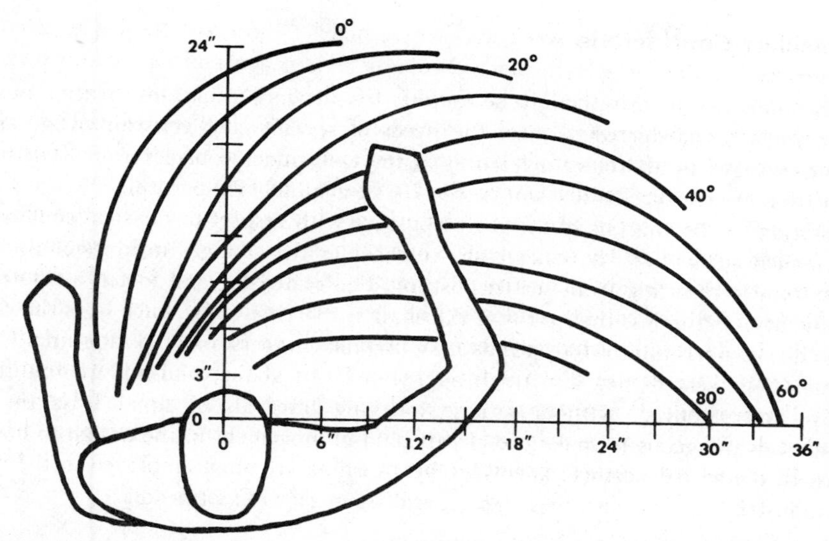

Figure 22.57 The natural motions pathway of the wrist changes with the angle of abduction. From Reference 48.

should be taken into consideration that the natural motions pathway of the wrist changes with the angle of abduction (Figure 22.57), and the implements of work—components, tools, jigs, and fixtures—should be arranged accordingly on the work bench, otherwise imprecise movements, poor eye–hand coordination, and early fatigue will result. Second, correct seat height with respect to the floor as well as the popliteal crease is established by means of an adjustable footrest. Footrails, because they cannot be easily adjusted, are less desirable. Occasionally drawers located underneath work benches may interfere with proper seat height adjustment, and these should be removed if necessary. Third, the height and position of the backrest should be arranged so that the minimal distance of 5 inches between the front edge of the seat and the popliteal crease is maintained. The adjustment of the backrest should enable the worker to have it low enough to permit freedom of trunk movement as demanded by the work situation, but not so low that it interferes with the bony structures of the pelvis.

Sometimes it is desirable to provide the opportunity to perform a job from either the seated or standing posture, or to alternate both postures in the course of the working day, according to the personal preference of the worker. Then the basic angle of abduction of the upper arm should be attained by a suitable height of the work bench. To maintain this for both postures, the chair and the footrest should be of the appropriate dimensions (Figure 22.58).

Finally, the distance of the backrest of the chair from the near edge of the work bench should be standardized to optimize skeletal configuration (Sections 4.2 and 4.4), to facilitate the development of the most effective kinesiology.

7.3 Ancillary Considerations

No single chair design can possibly be optimal for all work situations. Seating analysis should always be conducted to meet the needs of specific, not generalized situations, giving due weight to all relevant features of the task under consideration. It is highly desirable that standard reference works (68–70) be available for perusal.

In addition to the analysis of the seated posture with respect to physical comfort and biomechanical correctness, it is necessary to consider the changes in kinesiology of the lower extremity resulting from seated posture. The seated leg and foot can rotate only with difficulty, and unless the height of the chair is extremely low, such as is the case in motor vehicles, operation of foot pedals may become cumbersome and fatiguing. On the other hand, the seated "leg and thigh aggregate" can abduct and adduct voluntarily, precisely, and strongly, without fatigue for long intervals of time. It is therefore frequently advantageous to make use of this kind of movement in the design of machine controls. In the seated posture, knee switches, such as are often employed in the operation of industrial sewing machines, are generally superior to foot pedals.

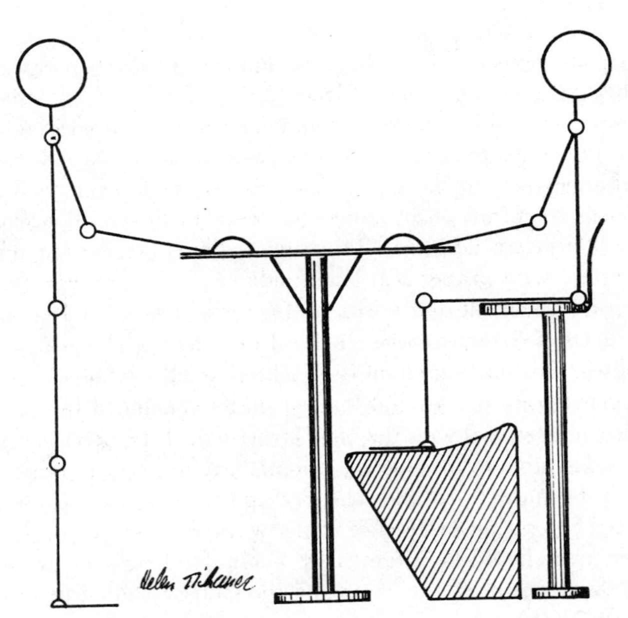

Figure 22.58 When work is possible in either seated or standing position, work bench and seating design should permit change of posture without change of musculoskeletal configuration. From Reference 48.

8 ERGONOMIC EVALUATION OF WORK SITUATIONS

Historically, all industrial and technological development in the United States has been triggered by the need to overcome and solve problems in the design, production, distribution, and use of manufactured articles. Traditionally, industry has subscribed to the "improvement approach" as the principal avenue toward economic efficiency and viability of American enterprise.

Before the development of industrial and human engineering, it was common practice to conceive products hastily, determine an adequate manufacturing process intuitively, and gradually remove deficiencies in product design or manufacturing methods during actual production. Such improvement extended over a period of several months or even years. This approach was to some extent acceptable in past decades, and the model T Ford remained in production for approximately 30 years. In today's fast-changing marketplace, this kind of policy often leads to economic disaster. By the time a product has been improved, it may have become redundant. Unfortunately the "improvement" approach to product design and manufacture is still maintained in many industries under the guise of the term "continuous cost reduction." This often conveniently excuses failure to design the product and process correctly before production.

In some industries "cost reduction" is expected to be practiced routinely by supervisory and engineering personnel during the first months or years of production of a new article. This tempts those who are responsible for "efficiency" to design new products and work methods initially imperfect, so that easy opportunities for later cost reduction are not lost.

Unfortunately, a similar attitude often prevails when the occupational health and safety of the working population is at stake. Too often industry waits for evidence of work-induced occupational disability before commissioning an occupational health specialist to identify the causes, which are then removed by a process of not always satisfactory gradualism.

Only too often the practitioner of ergonomics is challenged to justify the removal of pathogenic vectors from the working environment by a prediction of potential savings in medical costs, or an increase of productivity likely to accrue from his activities. It is therefore important that those who are engaged in the practice of ergonomics be ready to prove that the maintenance of occupational health and high levels of productivity are inseparable. Three main areas of ergonomic evaluation are of prime interest to the practitioner in industry: (*a*) historical evaluation, (*b*) analytical evaluation, and (*c*) projective evaluation.

8.1 Historical Evaluation

An active interest of management in ergonomics is often initially triggered by noticeable breakdowns in occupational health occurring in the performance of a well-defined, essential, and easily identifiable productive operation. This then results in increased manufacturing expense, high medical cost, potential retribution by a regulatory agency, and other undesirable side effects. Under such circumstances, a request for historical evaluation of past activities and events may be made. Such study is best conducted keeping in mind the "four big C's" of occupational health investigation:

1. Cause.
2. Consequence.
3. Cost.
4. Cure.

Frequently consequence and cost are known, and the cause and the cure remain to be discovered. Here theoretical analysis is the most expedient tool of research. Experimental methods are generally an unnecessary expense and quite superfluous in developing a critique of past events. Furthermore, a theoretical analysis offers a degree of confidentiality not available in experimentation with man.

It must be reemphasized that this book concentrates only on those narrower aspects of ergonomics (79) that are related to its subspecialty: biomechanics (80). To treat all the disciplines tributary to ergonomics and their applications within the confines of the space available would have led to a superficial discussion or sometimes mere mention of numerous important aspects of the industrial environment, its evaluation, and control.

It is important that no evaluation of work situations be undertaken unless the investigator is familiar with the ergonomic aspects of climate control, noise, vibration, illumination, circadian rhythms, and related topics. All these should be included in an environmental workup, which should always precede the general study of physical interaction between man and the implements of the workplace. Figure 22.2 should provide helpful guidance in the systematic conduct of such a preliminary workup. The next step would be to ascertain whether the prerequisites of biomechanical work tolerance (Section 4.1) were to some extent disregarded in the design of the product manufactured or the work method. A step-by-step checkout of each situation with Table 22.2 in hand is the best approach; afterward it should be possible to suggest some potential causes for the observed anatomical, physiological, or behavioral failure. Sometimes it may then be possible to suggest remedial action immediately. However the use of tables and "cookbooks" is no substitute for speculative analysis based on sound professional knowledge.

Superficial study of occupational accident or disease vectors based on mere guidelines may very often obfuscate cause and effect relationships. What may appear to be the cause of an accident could be, in fact, an effect produced by a less obvious mechanism. Discussion of an actual case may illustrate the need to probe energetically for the primary cause of occupational injury.

In a chemical factory the number of workers hit by forklift trucks while crossing aisles increased suddenly and dramatically without any apparent cause. Whenever such a "lost time" accident report was filed, either pedestrian error or driver error was listed as the cause. Subsequently, common human factors engineering approaches likely to eliminate the problem were explored. The trucks were made more visible by painting them in conspicuous colors, and illumination in the aisles was improved. At some of the places of greatest accident frequency, automatic warning horns were installed which signaled whenever a vehicle approached. When none of these measures proved to be successful, investigators began to try to discover why so many individuals were walking around the factory instead of remaining seated safely at their workplaces. Accident frequency was proportional to the number of pedestrians present in the aisles at any given time. Brief periods of absenteeism from the workplace suddenly increased dramatically, leading in turn to an increase in pedestrian traffic density.

The time of this change coincided with the introduction of a new tool (see Figure 22.33). An electrical brush used to clean trays was replaced by a much less expensive but equally effective paint scraper that produced insults to the ulnar artery (Section 4.3). This reduced blood supply to the ring and little fingers. The resulting numbness and tingling caused the individuals afflicted to lay down their tools occasionally and seek relief by exercising their hands. To avoid ensuing arguments with supervisors, workers were tempted to make use of every opportunity of brief absences from the job. Trips to the washroom, the toolroom, and so on, became much more frequent, and this was the true cause of increased exposure of the factory population to the risk of traffic accidents. Thus the first of the four big C's was identified. The cure: the handle of the paint scraper was redesigned. The result: the workers spent more time per day in productive activity; thus the output and economy of the operation increased, while at the same time, because of diminished risk exposure, the accident rate returned to normal.

Whenever the frequency of incidence of occupational ill health or accident increases after a manufacturing process has been in safe operation for some time, the following question should be asked: *what change in equipment, product design, tools used, working population employed, or work method applied has taken place immediately before the breakdown of occupational health?*

8.2 Analytical Evaluation

The procedures of analytical evaluation are called for whenever an existing manufacturing operation is generally satisfactory but has to be improved to make it more competitive, to reduce training time, or to eliminate operator discomfort and ill health.

Theoretical analysis (Section 8.1) is the initial step in all analytical evaluation. However this should often be followed by some additional procedures. The simplest and perhaps most effective aid in this kind of study is cinematography and subsequent frame-by-frame analysis. This permits a detailed evaluation of the workers' reaction to each event at the workplace and to each contact with tool, machine, or manufactured article. Slow-motion viewing of the operation not only reveals biomechanical or ergonomic defects but is instrumental in discovering reflex reactions to mild localized repetitive trauma that cannot be detected by the naked eye because of the brief duration of many such events. Stage magicians know that the hand is quicker than the eye. The current trend toward the use of video tape for work analysis should be discouraged because the tapes are inadequate for the detection of fine details of expression (81) or blanching of the skin, or frame-by-frame analysis. Furthermore, color video taping is exorbitantly expensive, whereas color film, especially in the Super-8 size, is economical and tells much more than a black and white picture. Finally, manufacture of a video tape from movie film is very inexpensive, whereas the converse (i.e., the manufacture of a movie film from video tape) is a very costly operation. Furthermore, being magnetic, video tape requires more careful storage, is sensitive to magnetic fields, and often is erased accidentally.

It is important for motion picture analysis of a work situation to allow the viewing of the workplace in at least two different planes, if necessary, with the aid of mirrors. When motion picture analysis alone is not adequate for process evaluation, recourse to other experimental technologies must be taken. Some investigative procedures, such as metabolic measurement, electromyography, and electromyographic kinesiology, have already been described (Section 3). This work can often be complemented by the production of biomechanical profiles, which are of special usefulness when the potential pathogenic effects of individual and brief motion elements of work are under discussion (37); therefore their main field of application resides in projective evaluation.

8.3 Projective Evaluation

Whenever possible, a job should be ergonomically evaluated while it is still in the planning phase. This makes it possible to "design out" of a task features, equipment, and maneuvers that are potentially traumatogenic. Projective evaluation should include reliable predictions with respect to the work tolerance of a specific population, estimated duration of training, and counseling procedures useful in overcoming difficulties in the training process.

All projective evaluation should include, when necessary, biomechanical profiles (82, 83), both as a means to establish reliable effort input–production work output relationships and to predict potential anatomical failure points.

The profile is a polygraphic recording produced during the performance of a standard element of work that includes displacement, velocity, and acceleration of at least one anatomical reference point. Such measurement is performed by kinesiometers—apparatus that permits the performance of a specific motion element against a known resistance. *Displacement* is indicative of range and pattern of motion; *velocity* serves as an index of speed as well as strength (slow joint movement is often associated with muscular weakness) (37).

Finally, *acceleration* reflects control over precision and quality of motion (84). Abnormal acceleration and deceleration signatures are invariably associated with imprecise and unsafe movements due to the inability to terminate a motion at the correct place and time. These biomechanical parameters are recorded simultaneously with integrated myograms of selected muscles involved in the performance of the task under study. The usefulness of the biomechanical profile becomes evident when a tossing motion is investigated. Many individuals have difficulty with wrist extension, but this does not interfere with their competence to perform assembly tasks in an entirely satisfactory manner. Once the task is finished, however, the individual may not be able to dispose of the article by tossing it into a bin (Figure 22.59). Inability to toss is more frequent than is commonly assumed. If caused by common industrial disorders, such as tenosynovitis, it will persist throughout the working day. Alternatively, when caused by fatigue, it may become evident only after several hours of work have elapsed.

To establish a cause and effect relationship between the muscular effort involved in wrist extension and the pattern and quality of the tossing motion produced, a biomechanical profile is constructed. The kinesiometer employed for this purpose appears in Figure 22.60. The subjects are tested before the start of the working day, then retested after several hours of normal productive work. The manner in which the task affects the performing individual can be established by comparison of the "prework" and "postwork" biomechanical profiles (Figure 22.61).

In the case of a tendency toward or a history of tenosynovitis, characteristic changes in the profile can be observed with ease. Displacement shows no change. However, the

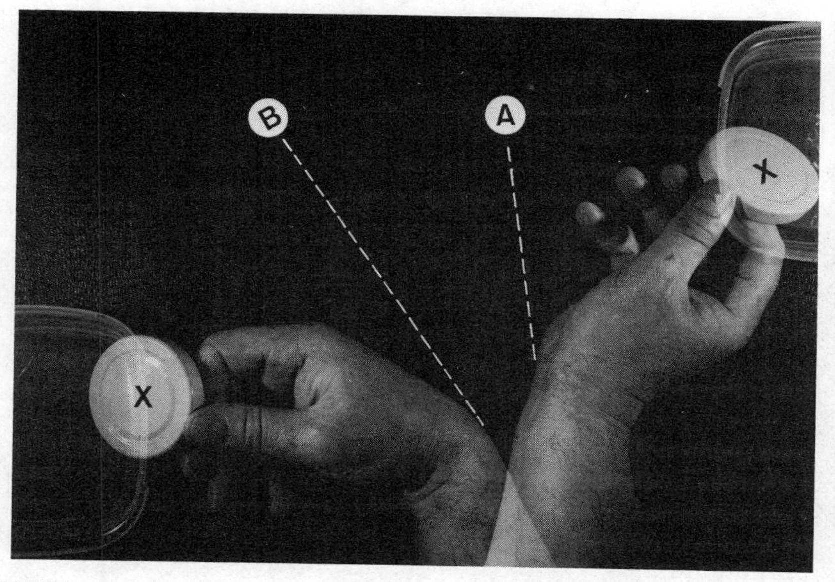

Figure 22.59 Hyperextension of the wrist is not a recognized element in motion study, but it is often essential for disposal of items at workplaces *X*. Hand and wrist disease are frequent occurrences, preventing the successful execution of this biomechanical motion, especially in women and aged workers. A simple change of location of the disposal bin may make the difference between occupational disability *A* or ability *B*. From Reference 37, by permission. © American Institute of Industrial Engineers, Inc., Norcross, Ga.

myograms are slightly stronger, indicative of a greater effort necessary to produce movement. The velocity curve displays notches at peak speed. This demonstrates that multiple, repetitive efforts must be made during each movement to achieve peak performance. The biomechanical profile cannot be used to make a definitive diagnosis such as tenosynovitis. The results obtained permit only the statement that some anatomicomechanical obstruction interferes with the tossing motion. This could be caused by a large number of conditions—orthopedic, arthritic, or others. Medical diagnosis, of course, is under the jurisdiction of a physician.

Fatigue changes the profile in quite a different fashion. No meaningful change whatsoever can be observed in the type of myogram used here. However the displacement tracing shows that it is not possible to maintain wrist extension for the necessary interval of time, and the acceleration signature displays evidence of muscular rigidity and fine tremors (Figure 22.61). This establishes that the task is fatiguing. Either the rhythm of the work cycle should be changed or the length of work periods and rest pauses should be significantly modified. A less expensive, less complex, and more feasible way to deal with both classes of disability may be a simple change in the position of the disposal bin, which would simply eliminate the necessity for wrist extension at the end of the work cycle (Figure 22.59). This demonstrates that the biomechanical profile, as distinct from other methods of performance measurement, not only indicates that quality and/or magnitude of performance are defective, but also makes it possible to

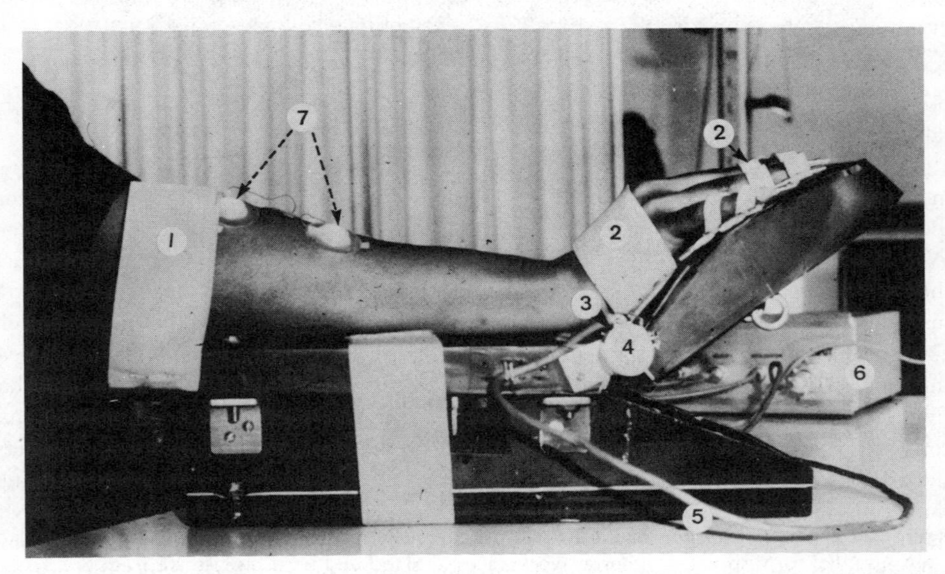

Figure 22.60 Wrist extension kinesiometer: 1, arm stabilizer; 2, wrist and finger stabilizers; 3, hinge under radiocarpal joint producing 10 lb-in.; 4, electrogoniometer; 5, cable to computation module; 6, analogue computation module generating biomechanical profile; 7, electrodes for extensor myogram. From Reference 84.

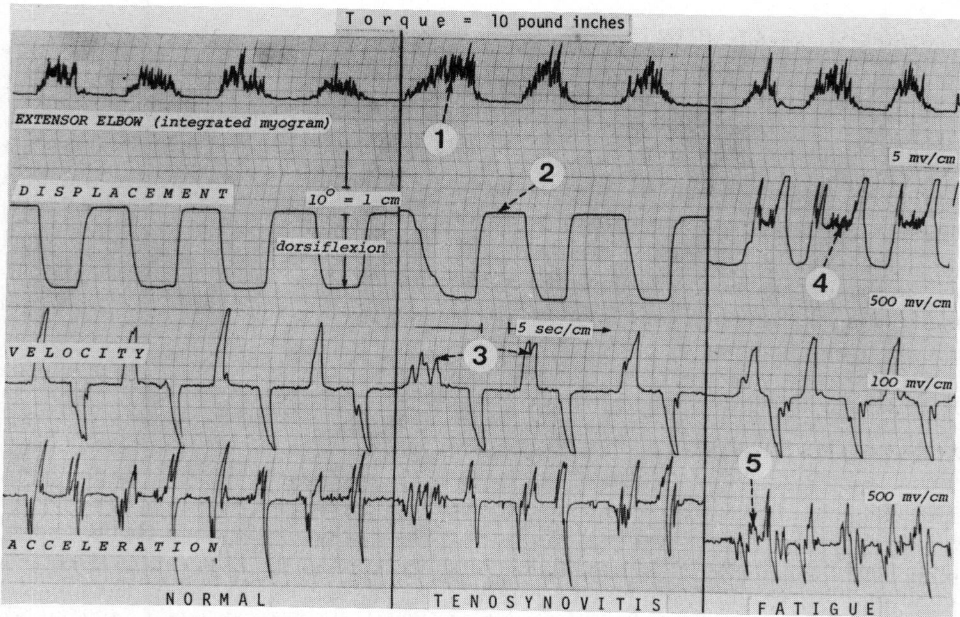

Figure 22.61 Biomechanical profile produced by the wrist extension kinesiometer in Figure 22.60: 1, increased extensor myogram; 2, normal displacement signature; 3, notched velocity peaks; 4, inability to maintain wrist extension; 5 signs of rigidity in acceleration signature. From Reference 84.

pinpoint the physical cause of the deficiency so that successful remedial action may be taken.

The biomechanical profile, one of the newest procedures in ergonomic work measurement, bridges the gap that has existed since the beginning of scientific management, industrial psychology, and work physiology, plaguing most practitioners in industry. Workers as early as the Gilbreths had fully established a scientific rationale on which the disciplines and biomechanics as practiced today are based. Nevertheless, they lacked instrumentation adequate to conduct experimental investigations into the physical effort expended by individual muscles in the performance of a specific task. The analytical thinking of these pioneers was simply 50 years ahead of the technology available at that time. The second industrial revolution has introduced productive processes that constrain the worker to a relatively rigid posture, which forces him to maintain repetitive motion patterns throughout a long working day. This has produced or aggravated numerous known, as well as previously unknown, elements and complaints. However the new technologies (85, 86), as a by-product, have produced instrumentation now available for effective ergonomic work measurement. These make possible the prevention of occupational disability, increasing levels of physiological and emotional well-being of the working population, as well as the productive capacity and the competitive posture of American enterprise.

9 PURPOSE OF CHAPTER

This does not purport to be a comprehensive treatise on the practice of the broad field of ergonomics. The discipline draws on so many aspects of the biological, behavioral, medical, and technological sciences that a complete coverage of the field within the confines of a chapter would have been impossible. Instead, the subject matter presented relates to the aspects of ergonomics pertinent to the biomechanics of work situations, very often requiring immediate decisions in the field. The analysis of problems created by the general industrial environment—including those stemming from noise, light, climate, and fatigue—has been omitted. Separate descriptive chapters are contained elsewhere in this volume.

ACKNOWLEDGMENTS

This chapter is based on three main references: 37, 38, and 48; it was developed with the partial support of the National Institute for Occupational Safety and Health and, to some extent, by the Social Rehabilitation Administration of the Department of Health, Education and Welfare. Additional information has been derived from my lecture notes distributed to my students at the University of New South Wales (Australia), Texas Tech University, and New York University.

Helen Tichauer contributed most of the graphic work and illustrations. Dr. H. Gage comtributed to the section on "Handtools." Special thanks are due to Audrey Lane for the painstaking editorial efforts taken with this manuscript.

REFERENCES

1. B. Ramazzini, *Essai sur les Maladies de Artisans* (translated from the Latin text *De Morbis Artificum* by M. de Fourcroy), Chapters 1 and 52, 1777.
2. D. Hunter, *The Diseases of Occupations,* 4th ed., Little, Brown, Boston, 1969, p. 120.
3. J. Amar, *Organization Physiologique du Travail,* H. Dunod, Paris, 1917.
4. E. D. Adrian, "Interpretation of the Electromyogram", *Lancet,* **2,** 1229–1233; 1283–1286 (1925).
5. O. G. Edholm and K. F. H. Murrell, *The Ergonomics Research Society, A History,* 1949–1970, Wykeham Press, Winchester, 1973.
6. P. V. Karpovich and G. P. Karpovich, "Electrogoniometer: A New Device for the Study of Joints in Action," *Fed. Proc.,* **18,** 311 (1959).
7. A. J. S. Lundervold, *Electromyographic Investigations of Position and Manner of Working in Typewriting,* A. W. Broggers Boktrykkeri A/S, Oslo, 1951.
8. E. R. Tichauer, "Electromyographic Kinesiology in the Analysis of Work Situations and Hand Tools", *Proceedings of the First International Conference of Electromyographic Kinesiology, Electromyography,* Supplement 1 to Vol. 8, 1968, pp. 197–212.
9. M. Lukiesh and F. K. Moss, "The New Science of Seeing," in: *Interpreting the Science of Seeing into Lighting Practice,* Vol. 1, General Electric Company, Cleveland, 1932, pp. 1927–1932.

10. H. S. Belding and T. F. Hatch, "Index for Evaluating Heat Stress in Terms of Resulting Physiological Stress," *Heat. Pip. Air Cond.*, **27,** 129 (1955).
11. B. L. Welch and A. S. Welch, Ed., *Physiological Effects of Noise,* Plenum Press, New York, 1970.
12. W. F. Floyd and A. T. Welford, Eds., *Symposium on Fatigue,* H. K. Lewis & Co., London, 1953.
13. E. R. Tichauer, "Potential of Biomechanics for Solving Specific Hazard Problems," in: *Proceedings 1968 Professional Conference,* American Society of Safety Engineers, Park Ridge, Ill., 1968, pp. 149–187.
14. P. J. Rasch and R. K. Burke, *Kinesiology and Applied Anatomy,* 3rd ed., Lea & Febiger, Philadelphia, 1967.
15. S. Brunnstrom, *Clinical Kinesiology,* 3rd ed., revised by R. Dickinson, F. A. Davis, Philadelphia, 1972.
16. D. L. Kelley, *Kinesiology: Fundamentals of Motion Description,* Prentice-Hall, Englewood Cliffs, N.J., 1971.
17. E. R. Tichauer, *Occupational Biomechanics (The Anatomical Basis of Workplace Design),* Rehabilitation Monograph No. 51, Institute of Rehabilitation Medicine, New York University Medical Center, New York, 1975.
18. M. Williams and H. R. Lisner, *Biomechanics of Human Motion,* Saunders, Philadelphia, 1962.
19. R. Fick, *Anatomie und Mechanik der Gelenke,* Vol. 3, Spezielle Gelenk und Muskelmechanik Fisher, Jena, 1911, pp. 318–389.
20. N. Recklinghausen, *Gliedermechanik und Lähmungsprothesen,* J. Springer, Berlin, 1920.
21. H. A. Haxton, "Absolute Muscle Force in Ankle Flexors of Man," *J. Physiol.,* **103,** 267–273 (1944).
22. R. W. Ramsey and S. F. Street, "Isometric Length-Tension Diagram of Isolated Skeletal Muscle Fibers in Frog," *J. Cell. Comp. Physiol.,* **15,** 11–34 (1940).
23. B. A. Houssay, *Human Physiology,* translated by J. T. Lewis and O. T. Lewis, McGraw-Hill, New York, 1955, p. 385.
24. L. Brouha, *Physiology in Industry,* 2nd ed., Pergamon Press, Oxford, 1960.
25. J. B. de V. Weir, "New Methods for Calculating Metabolic Rate with Special Reference to Protein Metabolism," *J. Physiol.,* **109,** 1–9 (1949).
26. C. F. Consolazio, R. E. Johnson, and L. J. Pecora, *Physiological Measurements of Metabolic Functions in Man,* McGraw-Hill, New York, 1963.
27. F. H. Norris, Jr., *The EMG: A Guide and Atlas for Practical Electromyography;* Grune & Stratton, New York, 1963.
28. J. V. Basmajian, *Muscles Alive,* 3rd ed., Williams & Wilkins, Baltimore, Md., 1974, p. 7.
29. H. H. Ju, "A Statistical Multi-Variable Approach to the Measurement of Performance Effectiveness of a Lumped System of Human Muscle," Master's thesis, New York University, New York, 1970.
30. J. F. Davis, *Manual of Surface Electromyography,* Wright Air Development Center Technical Report No. 59-184, Wright-Patterson Air Force Base, Ohio, 1959.
31. E. R. Tichauer, *Biomechanics of Lifting,* Report No. RD-3130-MPO-69, prepared for Social and Rehabilitation Service, U.S. Department of Health, Education and Welfare, Washington, D.C., 1970.
32. H. O. Kendall et al., *Muscles: Testing and Function,* 2nd ed., Williams & Wilkins, Baltimore, Md., 1971.
33. A. V. Hill and J. V. Howarth, "The Reversal of Chemical Reactions of Contracting Muscle During an Applied Stretch," *Proc. Roy. Soc.,* S.B. 151:169 (1959).
34. B. Jonsson, "Electromyographic Kinesiology, Aims and Fields of Use," in: *New Developments in Electromyographic and Clinical Neurophysiology,* Vol. 1, J. E. Desmedt, Ed., Karger, Basel, 1973, pp. 498–501.
35. B. Jonsson and M. Bagberg, "The Effect of Different Working Heights on the Deltoid Muscle," *Scand. J. Rehab. Med.,* Suppl. 3, 26–32 (1974).

36. E. Asmussen et al., *Quantitative Evaluation of the Activity of the Back Muscles in Lifting,* Communication No. 21, Danish National Association for Infantile Paralysis, Hellerup, 1965.

37. E. R. Tichauer, "Biomechanics Sustains Occupational Safety and Health," *Ind. Eng.,* February 1976.

38. E. R. Tichauer, "Occupational Biomechanics and the Development of Work Tolerance," in: *Biomechanics V-A,* P. V. Komi, Ed., University Park Press, Baltimore, Md., 1976, pp. 493–505.

39. D. B. Chaffin and W. H. Baker, "A Biomechanical Model for Analysis of Symmetric Sagittal Plane Lifting," *AIIE Trans.,* **2:** 1, 16–27 (1970).

40. E. R. Tichauer, "Ergonomics: The State of the Art," *Am. Ind. Hyg. Assoc. J.,* **28,** 105–116 (1967).

41. J. Hasan, "Biomedical Aspects of Low Frequency Vibration," *Work-Environment-Health,* **6:** 1, 19–45 (1970).

42. G. Loriga, in *Occupation and Health, Encyclopedia of Hygiene, Pathology and Social Welfare,* ILO, Geneva, 1934.

43. A. Hamilton, J. P. Leake, et al., Bureau of Labor Statistics Bulletin No. 236, Department of Labor, Washington, D.C., 1918.

44. D. E. Wasserman and D. W. Badger, *Vibration and the Worker's Health and Safety,* Technical Report No. 77, National Institute for Occupational Safety and Health, Government Printing Office, Washington, D.C., 1973.

45. B. Akerblom, *Standing and Sitting Posture,* A.-B. Nordiska Bokhandelns, Stockholm, 1948.

46. E. R. Tichauer, "Some Aspects of Stress on Forearm and Hand in Industry," *J. Occup. Med.,* **8:** 2, 63–71 (1966).

47. B. W. Niebel, *Motion and Time Study,* 4th ed., Irwin, Homewood, Ill., 1967, p. 169.

48. E. R. Tichauer, in: *The Industrial Environment—Its Evaluation and Control,* National Institute for Occupational Safety and Health, Department of Health, Education and Welfare, Washington, D.C., 1973, pp. 138–139.

49. M. Arnold, *Reconstructive Anatomy,* Saunders, Philadelphia, 1968, p. 391.

50. W. T. Dempster, "The Anthropometry of Body Action," in: *Dynamic Anthropometry,* R. W. Miner, Ed., *Ann. NY Acad. Sci.,* **63:** 4, 559–585 (1955).

51. E. R. Tichauer et al., *The Biomechanics of Lifting and Materials Handling,* Report No. HSM 99-72-13, submitted to the National Institute of Occupational Safety and Health, New York, 1974.

52. L. E. Abt, "Anthropometric Data in the Design of Anthropometric Test Dummies," in: *Dynamic Anthropometry,* R. W. Miner, Ed., *Ann. NY Acad. Sci.,* **63:** 4, 433–636 (1955).

53. J. B. Martin and D. B. Chaffin, "Biomechanical Computerized Simulation of Human Strength in Sagittal-Plane Activities," *AIIE Trans.,* 4(1): 19–28 (1972).

54. I. Starr, "Units for the Expression of Both Static and Dynamic Work in Similar Terms, and Their Application to Weight-Lifting Experiments," *J. Appl. Physiol.,* **4:** 21 (1951).

55. P. V. Karpovich, *Physiology of Muscular Activity,* Saunders, Philadelphia, 1959.

56. I. J. Schorr, *Changes in Myoelectric Activity of the Erector Spinae, Gluteus Maximus and Hamstring Muscles During Pick-Up and Release of Loads for Various Workplace Geometrics,* Master's thesis, New York University, New York, 1974.

57. F. B. Gilbreth, *Motion Study,* Van Nostrand, New York, 1911.

58. E. R. Tichauer, "Industrial Engineering in the Rehabilitation of the Handicapped," *J. Ind. Eng.,* **19:** 2, 96–104 (1968).

59. R. Drillis, D. Schneck, and H. Gage, "The Theory of Striking Tools," *Hum. Factors,* **5:** 5 (October 1963).

60. J. M. Little and A. F. Grant, "Hypothenar Hammer Syndrome," *Med. J. Aust.,* **1,** 49–53 (1972).

61. D. Briggs, "Trauma," in: *Occupational Medicine,* C. Zenz, Ed., Year Book Medical Publishers, Chicago, 1975, pp. 254 ff.
62. E. Grandjean, *Fitting the Task to the Man,* Taylor and Francis, London, 1969.
63. H. Bailey et al., *A Short Practice of Surgery,* H. K. Lewis & Co., London, 1956.
64. F. Fitzhugh, *Gripstrength Performance in Dynamic Tasks,* Technical Report, University of Michigan, Ann Arbor, 1973.
65. C. E. Clauser et al., *Anthropometry of Air Force Women,* AMRL-TR-70-5, Aerospace Medical Research Laboratory, Wright-Patterson Air Force Base, Ohio, 1972.
66. J. W. Garrett, "The Adult Human Hand: Some Anthropometric and Biomechanical Considerations," *Hum. Factors,* **13:** 2 (1971).
67. J. W. Garrett et al., *A Collation of Anthropometry,* AMRL-TR-68-1, 2, 2 vols., Wright Air Development Center, Wright-Patterson Air Force Base, Ohio, 1971.
68. H. Dreyfuss, *The Measure of Man,* 2nd ed., Whitney Library of Design, New York, 1967.
69. K. H. E. Kroemer, *Seating in Plant and Office,* AMRL-TR-71-52, Aerospace Medical Research Laboratory, Wright-Patterson Air Force Base, Ohio, 1971.
70. E. Grandjean, *Fitting the Task to the Man—An Ergonomic Approach,* Taylor and Francis, London, 1967.
71. N. Diffrient et al., *Humanscale 1/2/3,* Henry Dreyfuss Associates, MIT Press, Cambridge, Mass., 1974.
72. *Specification for Office Desks, Tables and Seating,* B. S. No. 3893, British Standards Institution, 1965.
73. *Anthropometric Recommendations for Dimensions of Non-Adjustable Office Chairs, Desks and Tables,* B. S. No. 3079, British Standards Institution, 1959.
74. F. Staffel, "Zur Hygiene des Sitzens," *Z. Allg. Gesundheitspflege,* **3,** 403–421 (1884).
75. H. Strasser, *Lehrbuch der Muskel- und Gesundheitspflege,* Vol. 2, J. Springer, Berlin, 1913.
76. S. W. Jacob and C. A. Francone, *Structure and Function in Man,* Saunders, Philadelphia, 1970, p. 8.
77. S. Slesin, "Biomechanics," *Ind. Design,* **18:** 3, 36–41 (1971).
78. B. Shackel, K. D. Chidsey, and Pat Shipley, "The Assessment of Chair Comfort," in: *Sitting Posture,* E. Grandjean, Ed., Taylor and Francis, London, pp. 155–192.
79. *The Origin of Ergonomics,* Ergonomics Research Society, Echo Press, Loughborough, England, 1964.
80. E. R. Tichauer, in: *Biomechanics Monograph,* E. F. Byars, R. Contini, and V. L. Roberts, Eds., American Society of Mechanical Engineers, New York, 1967, p. 155.
81. R. J. Nagoe and V. H. Sears, *Dental Prosthetics,* Mosby, St. Louis, 1958.
82. E. R. Tichauer, H. Gage, and L. B. Harrison, "The Use of Biomechanical Profiles in Objective Work Measurement," *J. Ind. Eng.,* **4,** 20–27 (1972).
83. E. R. Tichauer et al., "Clinical Application of the Biomechanical Profile of Pronation and Supination," *Bull. NY Acad. Med.,* 2nd ser., **50:** 4, 480–495 (1974).
84. E. R. Tichauer, in: *Rehabilitation After Central Nervous System Trauma,* H. Bostrom, T. Larsson, and M. Ljungstedt, Eds., Nordiska Bokhandelns Förlag, Stockholm, 1974.
85. E. R. Tichauer, M. Miller, and I. M. Nathan, "Lordosimetry: A New Technique for the Measurement of Postural Response to Materials Handling," *Am. Ind. Hyg. Assoc. J.,* **34,** 1–12 (1973).
86. C. Sparger, *Anatomy and Ballet,* A. & C. Black, London, 1960.

CHAPTER TWENTY-THREE

Potential Exposures in Industry—Their Recognition and Control

WILLIAM A. BURGESS

1 INTRODUCTION

This chapter describes the major health hazards associated with industrial materials and processes. When additional information is contained elsewhere in the book, pertinent citations are included. The first section of the chapter considers the occupational health hazards of common industrial unit processes; the second discusses the health hazards of major industries. Only selected industries could be covered, and the choice was based on the size of the work population, the potential hazards, and to a lesser degree, the interests of the author.

Toxicologic data are not included, nor are the threshold limit values cited, since this information is presented elsewhere in the book. The specific contaminant and its physical state are noted for each operation, but methods of air sampling and analysis are not included because this information is available in Chapter 17. I have chosen not to use the term "nuisance" to describe the potential health hazard of nontoxic industrial contaminants. Fire and explosion hazards are not normally covered unless this hazard would not normally be considered to result from a specific operation (see Chapter 27). Repetitive statements on the flammability of solvents, for example, would do little to enhance the value of the chapter.

A limited amount of detailed industrial process information is provided in the chapter where this information is especially important to an understanding of the origin of the contaminant, its physical state, and relative concentration. Other sources of information on industrial processes are available and are necessary reference materials for an industrial hygienist (1–5).

The type of control for specific operations is cited; detailed information on the application of various controls is available elsewhere in the book (chapter 18). The industrial ventilation manual published by the American Conference of Governmental Industrial Hygienists (6) is referenced extensively in this chapter.

2 INDUSTRIAL UNIT OPERATIONS

2.1 Abrasive Blasting

A wide range of abrasive blasting techniques are used in foundries, electroplating shops, welding, and other metal fabrication facilities to clean and prepare metal surfaces for finishing. These processes are designed to deliver an abrasive to the metal surface by air or water. A variety of nonmetallic and metallic abrasives are in common use. The nonmetallic materials may be composed of organic grit such as corn husks and pecan shells; nonorganic materials are sand, glass beads, silicon carbide, and aluminum oxide, among others. For heavy-duty cleaning either steel grit or shot is commonly used. Because of the large quantity of dust generated by abrasive blasting, the operation is usually conducted in exhausted enclosures. A variety of enclosures have been developed to serve specific industry needs, including abrasive blasting rooms, cabinets, barrels or mills, rotary tables, and tunnels (6, 7). These enclosures are designed by the manufacturer to provide effective exhaust. In many cases air-cleaning equipment is an integral part of the enclosure.

The abrasive blasting systems are designed to reuse the large particulate, and the respirable fines are collected by a suitable air cleaner. The reuse of the abrasive may result in the concentration of metal dust, and if the metal is toxic, this may present a greater hazard than the abrasive itself.

The ventilation requirements for abrasive blasting enclosures have evolved over several decades, and effective design criteria are now available. Systems to be used for organic abrasives require special design features because of the fire and explosion hazard. Operators should not be permitted to work inside such enclosures. The nature of the operation necessitates rugged enclosures and easily maintained curtains and seals to eliminate outward dust leakage. The minimum exhaust volumes are based on seals and curtains in new condition.

When the operator must work inside the enclosure, a continuous flow air line respirator supplied with respirable air must be worn. This equipment must be designed and approved for this service, and periodic maintenance is essential (8). Serious health hazards from an inadequate abrasive blasting respirator program have resulted in silicosis in one industry (9).

The use of sand containing high concentrations of free silica continues to present the major hazard to operators. A number of European nations have forbidden the use of silica sand for this application, but this effective measure has not been adopted in the United States. The other organic and inorganic abrasives normally present low hazard potential, however the base metal and its surface coating may represent a significant health hazard. In foundry operations the fused sand on the casting represents a silica hazard.

In evaluating the exposure of the worker, one must identify the base metal being cleaned, the nature of the surface contamination, and the abrasive in use; it is also necessary to know whether the operation is wet or dry, and details of the mechanical design of the enclosure and the controls in place. Air sampling and ventilation measurements are needed to evaluate the potential hazards from the abrasive, metal, and coating dust.

Periodic evaluation of the exhaust ventilation system is required. The minimum ventilation is usually 20 air changes per minute for enclosures and a face velocity of 200 fpm for curtained openings. The loading of parts should be monitored to make sure the "designed-in" airflow path is not obstructed. The condition of the metal enclosure and the various curtains and seals must be reviewed frequently.

2.2 Acid and Alkali Treatment

After the removal of major soils, oils, and grease, metal parts are often treated with acid or alkaline baths to condition the surface for electroplating or other finishes. The potential hazards in this series of operations include exposure to acid mist released because of air agitation and from misting due to the evolution of hydrogen and oxygen in electrolytic operations.

2.2.1 Acid Pickling

Acids such as sulfuric, hydrochloric, nitric, chromic, and nitric-hydrofluoric are used for pickling operations to remove oxide coatings and scale from various metals. The splash hazard and the corrosive action of the acids on skin, clothing, and machinery are well recognized. Acid mist and various acid gases are released (Table 23.1), depending on the bath temperature, the acid in use and its concentration, the metal being treated and its surface area, the current density (if the bath is electrolytic), and whether the bath contains an inhibitor. Inhibitors may work as a protective film, slowing down the action on the metal more than the oxide; as a producer of foam, thereby inhibiting the escape of mist; or as an additive that lowers the surface tension of the bath.

The principal control is local exhaust ventilation, and the exhaust rate can be determined from design criteria (6, 10).

2.2.2 Acid Bright Dips

Acid bright dips are usually mixtures of nitric and sulfuric acids employed to remove surface tarnish from aluminum, cadmium, magnesium, copper, copper alloys, monel, and silver. Nitrogen oxide gases are commonly emitted from the bath as indicated in

Table 23.1 Contaminants Released by Pickling Operations (6)

Type	Component of Bath That May Be Released to Atmosphere	Physical and Chemical Nature of Major Atmospheric Contaminant
Aluminum	Nitric acid	Nitrogen oxide gases
Aluminum	Chromic, sulfuric acids	Acid mists
Aluminum	Sodium hydroxide	Alkaline mist
Cast iron	Hydrofluoric-nitric acids	Hydrogen fluoride–nitrogen oxide gases
Copper	Sulfuric acid	Acid mist, steam
Copper	None	None
Duraluminum	Sodium fluoride, sulfuric acid	Hydrogen fluoride gas, acid mist
Inconel	Nitric, hydrofluoric acids	Nitrogen oxide, HF gases, steam
Inconel	Sulfuric acid	Sulfuric acid mist, steam
Iron and steel	Hydrochloric acid	Hydrogen chloride gas
Iron and steel	Sulfuric acid	Sulfuric acid mist, steam
Magnesium	Chromic-sulfuric, nitric acids	Nitrogen oxide gases, acid mist, steam
Monel and nickel	Hydrochloric acid	Hydrogen chloride gas, steam
Monel and nickel	Sulfuric acid	Sulfuric acid mist, steam
Nickel silver	Sulfuric acid	Acid mist, steam
Silver	Sodium cyanide	Cyanide mist, steam
Stainless steel	Nitric, hydrofluoric acids	Nitrogen oxide, hydrogen fluoride gases
Stainless steel	Hydrochloric acid	Hydrogen chloride gas
Stainless steel	Sulfuric acid	Sulfuric acid mist, steam
Stainless steel		
Immunization	Nitric acid	Nitrogen oxide gases
Passivation	Nitric acid	Nitrogen oxide gases

Table 23.2. Bright-dip baths require more efficient hooding and exhausting than do most other acid treatment operations. The parts removed from the tank must also be effectively exhausted until they are rinsed.

2.2.3 Molten Caustic Descaling Baths

Molten caustic is used for cleaning and descaling cast iron and alloy steel. The advantages claimed for this type of cleaning are that it prepares the parts for soldering and brazing and provides a good bond when the part is coated with lead, zinc, and vitreous enamel. These baths are also used to clean sand from castings.

Metal oxides of chromium, nickel, and iron are removed by reduction, and the base metal is not attacked. Oils are burned off, and graphite, carbon, and sand are removed.

The oxides and possibly other debris collect as sludge on the bottom of the bath. The other contaminants combine with the molten caustic, float on the surface of the bath, or are volatized as vapor, smoke, or fume.

These baths require well-defined operating procedures because the molten caustic can be hazardous to health. Local exhaust is necessary, and usually the tank is designed so that a complete enclosure protects the operator from violent splashing as the part is immersed in the bath.

2.3 Degreasing

The removal of surface grime, oil, and grease from metal is commonly done either by vapor phase degreasing or by soak cleaning. The soak cleaners can be straight solvents, emulsion cleaners, or solvent–emulsion cleaners. The significant problems associated with these processes are described below.

2.3.1 Cold Degreasing

Cold degreasing is practiced to a certain extent for the cleaning of various objects in small shops. The solvents used may vary from a high flash petroleum distillate to a mixture that includes aliphatic and aromatic chlorinated hydrocarbons, ketones, cellosolves, creosote, and cresylic acid. No generalities regarding control can be made other than to note that no readily volatile or fast-drying solvent should be used in large open

Table 23.2 Contaminants Released by Acid Dip Operations (6)

Type	Component of Bath that May be Released to Atmosphere	Physical and Chemical Nature of Major Atmospheric Contaminant
Aluminum bright dip	Phosphoric, nitric acids	Nitrogen oxide gases
Aluminum bright dip	Nitric, sulfuric acids	Nitrogen oxide gases, acid mist
Cadmium bright dip	None	None
Copper bright dip	Nitric, sulfuric acids	Nitrogen oxide gases, acid mist
Copper semibright dip	Sulfuric acid	Acid mist
Copper alloys bright dip	Nitric, sulfuric acids	Nitrogen oxide gases, acid mist
Copper matte dip	Nitric, sulfuric acids	Nitrogen oxide gases, acid mist
Magnesium dip	Chromic acid	Acid mist, steam
Magnesium dip	Nitric, sulfuric acids	Nitrogen oxide gases, acid mist
Monel dip	Nitric, sulfuric acids	Nitrogen oxide gases, acid mist
Nickel and nickel alloys dip	Nitric, sulfuric acids	Nitrogen oxide gases, acid mist
Silver dip	Nitric acid	Nitrogen oxide gases
Silver dip	Sulfuric acid	Sulfuric acid mist
Zinc and zinc alloys dip	Chromic, hydrochloric acids	Hydrogen chloride gas (if HCl attacks Zn)

containers without effective mechanical exhaust ventilation if workmen are to be exposed to its vapors for more than a few minutes a day. Skin contact with these materials should be avoided, and a face shield should be used to protect the eyes and face. Covered soak tanks with an adjacent mechanically ventilated work table offer satisfactory vapor control.

Under certain conditions covered soak tanks may be used successfully without ventilation if the tank has a water solution layer over the surface to retard the escape of solvent vapors. Low volatility materials such as mineral spirits or kerosene usually present minimal inhalation exposure, but the widely employed air blowoff operation may produce an objectionable irritant mist unless controlled.

When solvents are kept in a safety can or other suitable covered container and applied in small amounts by brushing and wiping, the inhalation hazard far exceeds any fire hazard; and where this kind of operation is found necessary, it is better to use petroleum distillates, ketones, and esters rather than chlorinated hydrocarbons. If chlorinated hydrocarbons of established moderate toxicity are found necessary to the operation, exhaust should be provided unless the solvent is applied very sparingly.

Spraying with high flash petroleum distillates such as Stoddard Solvent, mineral spirits, or kerosene is a widely used method of cleaning oils and grease from metals. Solvents with a flash point below 100°F should not be used for this purpose. The operation should always be provided with suitable mechanical exhaust ventilation. The hood may be of conventional spray-booth type and may or may not be fitted with a fire door and automatic extinguishers. The fire hazard attendant to spraying a high flash petroleum solvent is no more than that attendant to spraying many lacquers and paints.

2.3.2 Emulsion Cleaners

Emulsion cleaners containing petroleum and coal tar solvents are commonly used in power washers and soak tanks. When used in soak tanks at room temperature, no ventilation is required. When the cleaner is sprayed or used hot, the operation should be confined and ventilated.

Emulsion cleaners containing cresylic acid, phenols, or chlorinated halogenated hydrocarbons should be provided with ventilation. Skin contact should be carefully avoided.

2.3.3 Alkali Cleaning

Alkaline cleaners are used in soak tanks, dip tanks, and power washers. Alkaline baths may contain caustic soda, soda ash, trisodium phosphate, sodium pyrophosphate, sodium hexametaphosphate, and other soaps, wetting agents, and emulsifiers. Sometimes solvents such as *o*-dichlorobenzene, butyl cellosolve, pine oil, or petroleum distillates are added.

In general these baths are operated hot and may be electrolytic. They do require ventilation control, and skin and eye protection must be worn.

2.3.4 Vapor Phase Degreasing

A vapor phase degreaser is a tank containing a quantity of solvent that is heated to its boiling point by either electricity or steam. The solvent vapor rises and fills the tank to an elevation determined by the location of a cold water condenser. The vapor condenses at this point and returns to the liquid sump. The tank extends above the condenser to minimize air currents inside the tank.

The work is lowered into the vapor above the boiling solvent. The solvent condensing on the cool metal surface dissolves and washes away oil and grease film. This action is continued until the metal reaches the temperature of the solvent vapor, whereupon it becomes dry. The length of time required depends on the size, shape, surface area, temperature, and specific heat of the parts to be cleaned. If the material is not allowed to remain long enough to reach the temperature of the vapor, considerable solvent will be dragged out with the dripping parts.

Variations of the vapor phase degreaser are available for special purposes. In one type both a liquid immersion tank and a vapor phase tank are available. This equipment is useful when articles to be degreased have intricate shapes and are heavily contaminated with dirt and grease. Immersion of the part in boiling solvent prolongs solvent action and affords mechanical scrubbing by the boiling liquid.

If the insoluble matter cannot be removed by immersion in boiling solvent, spray degreasing may be necessary. In this case the work is first lowered into the vapor to remove oil and grease, then sprayed with warm solvent from a lance, and finally given a vapor rinse to remove all traces of oil and grease. To minimize solvent loss, it is essential that spraying be done below the vapor level of the degreaser in a manner that does not disturb the vapor level, and baffles or screens must be placed to prevent the rebound or ricochet of droplets of solvent into the area above the vapor level.

All these cleaning methods are applicable to either hand-operated or conveyorized machines. When the work is done in sufficient quantity to justify the cost, conveyor equipment is more satisfactory, since solvent loss is easier to control.

A number of solvents are used with vapor phase degreasers including trichloroethylene, perchloroethylene, methyl chloroform, ethylene dichloride, and certain Freons. The degreaser must be designed and applied in relation to the specific solvent in use. Commercial degreasing solvents sold under trade names usually contain a small amount of stabilizer. The purpose of the stabilizer, frequently an organic amine, is to neutralize any free acid that might result from (1) oxidation of the degreasing liquid in the presence of air, (2) hydrolysis in the presence of water, or (3) pyrolysis under influence of high temperatures.

One must remember that trichloroethylene, the most common degreasing solvent, is flammable at the elevated temperatures present in degreasers. When heated above 110°F the solvent has a narrow flammable range around 20 percent by volume. This range increases with temperature, and above 135°F the flammable range is from about 15 to 40 percent by volume. The ignition temperature is 770°F. These conditions do not ordinarily occur in plant atmospheres but may occur within a degreaser. Trichloroethylene vapors will not explode violently under any circumstances, but they may burn slowly to form dense smoke and gases such as chlorine, hydrogen chloride, and phosgene. Although perchloroethylene vapor will not ignite or burn, oils or greases accumulated in the degreaser will; therefore sources of ignition, especially overheating with gas or electric heaters, should be avoided during distillation for sludge removal. Also, welding on or in a degreaser when it contains solvent should be avoided.

Safety Features. Vapor phase degreasers should have adequate condensers in the form of water jackets, pipe coils, or both, extending around the tank. The condenser prevents the escape of the concentrated vapors into the room. The vertical distance between the lowest point at which vapors can escape from the degreaser machine and the highest normal vapor level is called the freeboard. The freeboard should be at least 15 inches and not less than half the width of the machine. The portion of the condenser above the vapor line should be maintained above room temperature and below 110°F. The effluent water should be regulated to this same range, and a temperature indicator or control is desirable.

All machines should have thermostatic controls located a few inches above the normal vapor level, to shut off the source of heat if the vapor rises above the condensing surface.

Local exhaust ventilation should be installed on all vapor phase degreasers. To ensure effective use of local exhaust, the units should be installed away from drafts from open windows, spray booths, space heaters, supply air grilles, and fans. When degreasers are installed in pits, mechanical exhaust ventilation should be provided at the lowest part of the pit. Open flames, electric heating elements, and welding operations should be divorced from the degreaser locations, since the solvent will be degraded by both direct flame and ultraviolet radiation, thereby producing toxic air contaminants. This problem is discussed in greater detail under Welding, Section 2.13.

Operation. An extensive list of installation and operating precautions in the use of vapor phase degreasers has been proposed (11, 12). The following, which are the most important, should be posted at all degreasers:

1. Condenser water should be turned on before the heat.
2. Water temperature must be automatically maintained between 100 and 110°F.
3. Work should be placed in and removed from the vapor as slowly as possible (no faster than 20 fpm) to minimize vapor loss.
4. The part must be kept in the vapor until it reaches vapor temperature and is dry.
5. Parts should be loaded to minimize pullout. For example, cup-shaped parts should be inverted.
6. Overloading should be avoided because it will cause displacement of vapor into the workroom.
7. The work should be sprayed below the vapor level.
8. Proper heat input must be available to ensure vapor recovery when large loads are placed in the degreaser.

9. A thermostat should be installed in the boiling solvent to prevent overheating of the solvent.

10. A thermostat vapor level control must be installed on the degreaser and set for the particular solvent in use.

11. The degreaser tank should be covered when not in use.

12. Hot solvent should not be removed from the degreaser, nor should garments be cleaned in the degreaser.

Sludge and metal chips should be removed as often as necessary to prevent their accumulation. The solvent should be distilled off until the heating surface or element is nearly but not quite exposed, or until the solvent vapors fail to rise to the collecting trough. After cooling, the oil and solvent should be drained off and the sludge removed. A fire hazard may exist during the cleaning of machines heated by gas or electricity because the flash point of the residual oil may be reached, and because trichloroethylene itself is flammable at elevated temperatures. After sludge and solvent removal the degreaser must be mechanically ventilated before any maintenance work involving flames or welding is undertaken. A person should not be permitted to enter a degreaser or place his head within one until all compartments have been blown free of vapors. Anyone entering a degreaser should wear a respirator suitable for conditions immediately hazardous to life, as well as a lifeline held by an attendant. In such circumstances, anesthetic concentrations of vapor may be encountered and oxygen concentration may be insufficient. Such an atmosphere may cause unconsciousness with little or no warning. A number of deaths have occurred because of failure to observe the foregoing precautions (13).

2.4 Electroplating

Metals, plastics, and rubber are plated to provide decoration, corrosion resistance, mechanical wearing properties, or electrical properties. Before parts are electroplated, rigorous surface cleaning and treatment is usually necessary. These preparatory procedures are discussed in Sections 2.2 and 2.3.

The basic electroplating tank consists of a cathode and an anode immersed in an electrolyte and powered by a low voltage dc supply. In most cases the part is the cathode, and the anode is a bar or slab of metal to be plated. The electrolyte contains ions of the metal to be deposited and other additives such as sulfuric acid to improve the electrical conductivity of the bath or the plating quality. In operation the metallic ions from the bath electrolyte are deposited on the cathode or pai to be plated. An equivalent amount of the metal is dissolved from the anode. In chromium plating an insoluble anode is used and the source of the metal to be plated is the electrolyte.

In addition to the actual plating process, considerable energy is consumed in the electrolytic decomposition of water in the bath, with the release of hydrogen at the cathode and oxygen at the anode. These gases rise to the top of the bath and break through the surface, forming a mist of the electrolyte, which becomes airborne. The rate of gassing varies with the current density of the particular bath. In addition to this source of contaminant, the bath electrolyte can be released to the workroom by air agitation and by bath operation at high temperature with resulting bath evaporation. The air contaminants released by various electroplating are listed in Table 23.3.

The misting at a given plating operation caused by such work practices can be reduced by lowering the bath temperature, reducing the current density, and covering the plating tank. In addition, misting can be reduced by the use of additives to the bath which will trap the released mist. Plastic balls or chips are also used to provide a trapping layer for the mist released from the bath. These procedures certainly are helpful; however they should not be relied on to replace local exhaust ventilation at the tank. A ventilation design procedure for electroplating baths and associated open surface tank operations has evolved (6, 10, 14, 15). This procedure permits one to determine a minimum control velocity based on the hazard potential of the bath and the rate of contaminant generation. The exhaust volume is calculated using this control velocity and tank measurements and geometry.

Following are some general rules to be observed in the design of exhaust ventilation systems:

1. Copper plating in an acid solution is normally conducted at low current densities and does not use air agitation. In such cases local exhaust ventilation is not required; but when high current densities and air agitation are used, significant air contamination occurs and the tank must be exhausted.
2. When a cyanide bath is used for strike and bright copper plate, both cyanide and alkali mist are released, and ventilation is required.
3. Neither sulfate nor chloride plating baths for nickel require exhaust ventilation, since gassing is negligible.
4. As in the case of acid copper plate, neither zinc nor cadmium plating solutions require exhaust ventilation unless high current densities and bath temperature exist.

Although certain baths do not call for local exhaust ventilation by the standard calculation method, it is wise to provide hoods and exhaust capacity on all new installations, since bath operating conditions change and usually water vapor and heat must be controlled.

In addition to the proper design and installation of good local exhaust ventilation, one must provide adequate makeup air, backflow dampers on any combustion devices to prevent carbon monoxide contamination of the workplace, and suitable air cleaning on chromic acid and alkali mist releases.

All cyanide materials should be stored under lock in a posted area that is divorced from acid storage. The drums should be raised from the floor to ensure that they will not be wet with water. Waste liquid treatment usually involves the mixing and reacting of acid and alkali wastes, the storage of cyanide waste in a holding tank with subsequent oxidation of the cyanide, and the reaction of chromium wastes with sodium bisulfite.

Adequate protective clothing in the form of eye, hand, skin, and foot gear are necessary because of the splash potential and the likelihood of contact with various solutions. The dermatitis risk is significant and can range from nickel itch to primary irritation from acids and alkalies.

2.5 Forging

The plastic deformation of metal by cold and hot forging is accomplished by means of a variety of forge presses (16). The principal occupational health hazards associated with forging include heat stress, noise exposure, and air contamination.

The heat load to the worker is due to a combination of high metabolic rate, radiation load from the hot furnaces, metals, and dies, and the convective load due to the high air temperature. Controls should include reduction of the work load by improved materials handling and plant layout, shielding of furnaces to reduce the radiation load, and spot cooling or cold air showers.

The noise hazard is due to impact noise and to a lesser degree aerodynamic noise from the steam and air cylinders and various air solenoids. The air release noise can be handled by mufflers; however the primary noise can be reduced only by die design. This forge shop noise problem, which has caused serious hearing loss in forge shop populations, is normally controlled by ear protection and a complete hearing conservation program.

The forge furnaces produce a carbon monoxide hazard when fired by gas or oil and provided with a flame curtain. A variety of oil-based lubricants are applied to the die surfaces by brush or spray. These substances may be mineral oil or heavy residual oils. Graphite and molybdenum disulfide are commonly added to the oil to increase die lubricity at high temperatures. When the hot part is forged, the oil burns off, resulting in a heavy particulate cloud containing oil mist, sooty particulate, trace metals such as vanadium, and polynuclear aromatic hydrocarbons (17, 18). Sulfur dioxide may also be formed from the combustion of residual oils. In the most recent forge shop study, the concentrations of respirable mass particulates (3.0 to 33.3 mg/m^3) and benzo-α-pyrene (1.6 to 2.9 $\mu g/m^3$) demonstrated that controls were required (18). In this study the concentrations of sulfur dioxide, aldehydes, and nitrogen dioxide were not excessive.

Dilution exhaust ventilation is a frequently used control for carbon monoxide and particulate control. Little attention has been given local exhaust ventilation, although this was under consideration in the cited study, as was the substitution of a water-based die lubricant. Additional air sampling is clearly required in forge shops to document the need for more critical controls.

The periodic viewing of the hot parts and the furnace interior presents a possible infrared radiation exposure that should be evaluated. Heat cataracts may develop depending on radiation intensity, time exposed, and age of the worker (19). Eye protection constitutes a suitable control for this hazard.

In general, the exposure to the metal fume is not significant; however the forging of such toxic, high vapor pressure metals as beryllium warrants close evaluation.

Table 23.3 Contaminants Released by Electroplating Operations (6)

Process	Metal	Component of Bath That May Be Released to Atmosphere	Physical and Chemical Nature of Major Atmospheric Contaminant
Electrodeless plating	Copper	Formaldehyde	Formaldehyde gas
	Nickel	Ammonium hydroxide	Ammonia gas
Electroplating, alkaline	Platinum	Ammonium phosphate, ammonia gas	Ammonia gas
	Tin	Sodium stannate	Tin salt mist, steam
	Zinc	None	None
Electroplating, fluoborate	Cadmium	Fluoborate salts	Fluoborate mist, steam
	Copper	Copper fluoborate	Fluoborate mist, steam
	Indium	Fluoborate salts	Fluoborate mist, steam
	Lead	Lead fluoborate–fluoboric acid	Fluoborate mist, hydrogen fluoride gas
	Lead–tin alloy	Lead fluoborate–fluoboric acid	Fluoborate mist
	Nickel	Nickel fluoborate	Fluoborate mist
	Tin	Stannous fluoborate, fluoboric acid	Fluoborate mist
	Zinc	Fluoborate salts	Fluoborate mist, steam
Electroplating, cyanide	Brass, bronze	Cyanide salts, ammonium hydroxide	Cyanide mist, ammonia gas
	Bright zinc	Cyanide salts, sodium hydroxide	Cyanide, alkaline mists
	Cadmium	None	None
	Copper	None	None
	Copper	Cyanide salts, sodium hydroxide	Cyanide, alkaline mists, steam

	Gold	Cyanide salts	Cyanide mist, steam
	Indium	Cyanide salts, sodium hydroxide	Cyanide, alkaline mists
	Silver	None	None
	Tin–zinc alloy	Cyanide salts, potassium hydroxide	Cyanide, alkaline mists, steam
	White alloy	Cyanide salts, sodium stannate	Cyanide, alkaline mists
	Zinc	Cyanide salts, sodium hydroxide	Cyanide, alkaline mists
Electroplating, acid	Chromium	Chromic acid	Chromic acid mist
	Copper	Copper sulfate, sulfuric acid	Sulfuric acid mist
	Gold	Cyanide salts	Cyanide mist
	Indium	None	None
	Indium	Sulfamic acid, sulfamate salts	Sulfamate mist
	Iron	Chloride salts, hydrochloric acid	Hydrochloric acid mist, steam
	Iron	None	None
	Nickel	Ammonium fluoride, hydrofluoric acid	Hydrofluoric acid mist
	Nickel and black nickel	None	None
	Nickel	Nickel sulfate	Nickel sulfate mist
	Nickel	Nickel sulfamate	Sulfamate mist
	Palladium	None	None
	Rhodium	None	None
	Tin	Tin halide	Halide mist
	Tin	None	None
	Zinc	Zinc chloride	Zinc chloride mist
	Zinc	None	None

2.6 Foundry Operations

Founding involves the production of a mold with a cavity of specific geometry into which is placed a refractory core. The mold is usually conveyed to a pouring station, where molten metal is poured into the cavity. After cooling, the mold is taken to a shakeout facility where the molding refractory material is removed to release the casting. The casting is cleaned, and extraneous cast metal is removed from it.

The potential health hazard in foundry operations are exposure to various air contaminants and physical conditions including noise, heat, and vibration; a survey of 281 foundries in Michigan in 1968 revealed that more than 7 percent of the employees were exposed to serious health hazards (20). It is reasonable to presume that conditions are at least that serious in other states.

This section discusses the occupational health hazards in molding, coremaking, melting, and pouring. Limited attention is given the cleaning operations, covered in Section 2.1. The discussion is based on a ferrous foundry operation. The casting of steel is similar in most respects to iron, but there probably is more fused sand on the casting, and some of the silica may be converted to tridymite or crystobalite. Nonferrous founding normally includes aluminum, magnesium, and copper-based alloys such as brass and bronze. The hazards specific to these operations are usually encountered in melting and pouring, as noted below.

2.6.1 Molding

The majority of ferrous castings are produced by green sand molds. The molding sand is usually a mixture of silica sand, bentonite clay, sea coal or finely ground coal, and various organic binders such as dextrine, wood flour, or oat hulls. Since the molding sand is reused, it also contains metallic impurities from previous pourings. This mixture is usually mixed in a sand muller.

The presence of quartz in the sand presents the major health hazard to foundry workers. As Table 23.4 indicates, the hazard from silica exposure varies with the physical state of the sand, its chemical composition, and the method of handling. Since the sand is initially dry, the mixing or mulling operation is dusty, and local exhaust ventilation is required. After this point the moisture content is high, and little dusting occurs during the molding operation. After casting the sand becomes friable and dusty; serious dust exposures may occur during shakeout and sand conditioning for reuse (21).

Molding involves placement of a pattern in the flask with the dusting of the pattern with a facing sandmold release material, which may be silica flour. Then the green backing sand is introduced manually, by gravity feed from hoppers, or in the case of large floor molds by sand slinger. In most cases these operations do not represent a silica hazard because the sand is moist. Both halves of the mold are prepared in this fashion and are ready for the setting of cores, and necessary metal paths are cut in place.

Paints used on cores and molds are commonly graphite suspended in resin or an alcohol. During the pour the alcohol burns forming a carbon coating, which provides a smooth metal parting surface.

Table 23.4 Factors Influencing Hazards from Foundry Sand (21)

Factors	Effect on Health
Physical states of sand	
New sand, screened and dustfree	Limited health hazard
New wet sand	Limited health hazard
New sand, fine and dry	Potential hazard
Used sand	Potential hazard
Chemical composition of sand	
Olivine (mixture of silicates and oxides)	Limited health hazard
Clays (mixture of silicates) and inert binders (dextrose)	Limited health hazard
Sand contaminated with toxic metallic particles	Potential hazard
Pure quartz (silica, SiO_2)	Potential hazard
Method of Handling	
Unloaded by vacuum system	Potential hazard
Compressed air in closed pipe system	Potential hazard
Belt conveyor	Potential hazard
Clamshell	Potential hazard
Mechanical shovels	Potential hazard
Work done outdoors	Potential hazard
Manual indoor work (sweeping, sand cutting)	Potential hazard

2.6.2 Core Production

The production of cores has changed considerably during the last decade and has been the one foundry area reflecting greatest technological advances (22, 23). The core must be mechanically strong to permit handling and to maintain its geometry when the molten metal flows into the mold, yet friable after pouring to allow its removal from inside the casting. Historically, cores have been made of sand with a binder, usually of linseed or vegetable oils, or other organic binders. This type of core requires curing in an oven, with the accompanying release of acrolein and a persistent odor.

New coremaking procedures involve a series of binder systems (Table 23.5). Certain binders require oven heating, others require gassing to cure the system, and others are no-bake systems. These systems may or may not release air contaminants during curing, but they most certainly do this during the casting operation. The use of urea and phenol formaldehyde, and isocyanate systems has added a new dimension to foundry core room exposures, as suggested by the variety of air contaminants released to the workplace (Table 23.5).

A major change in binders for core production has been the introduction of the sodium silicate–carbon dioxide system. This system involves the mixing of sand and sodium silicate, introduction of the mix to the core box, and the passing of carbon dioxide through the core to form silica gel and sodium carbonate, which "sets" the core in rigid form. This system presents minimal health hazard to the worker.

Table 23.5 Decomposition Products from Various Core Systems (23)

Core Type	Decomposition Products
Oven-baked binders	
Oleo resinous (core oils)	Acidic compounds Aldehydes Ammonia Carbon dioxide Carbon monoxide Hydrocarbons
Urea-formaldehyde resins (usually blended with oleoresins or phenol formaldehyde)	Amines Ammonia Carbon monoxide Low molecular weight polyureas
Phenol-formaldehyde resins	Aldehydes Ammonia Benzoic acid Carbon monoxide Cresols Hydrogen cyanide Phenol Polyhydric phenols Toluene, benzene
Heated corebox	
Urea-formaldehyde resins modified with furfuryl alcohol (furan resins) Furan resins modified with phenol to produce urea-phenol-formaldehyde resins (UPF resins)	Aldehydes Amines Ammonia Benzene Benzoic acid Carbon monoxide Cresols Hydrogen cyanide Phenol Polyhydric phenols
Gassed core: isocyanate cold box Phenolic resin Polyisocyanate Catalyst Triethylamine Dimethyl ethylamine	Amine catalysts Ammonia Carbon monoxide Hydrogen cyanide Methane Phenol

Several hot box core systems based on resin systems have been introduced. A sand binder and a catalyst are injected into a heated metal core box, and the heat initiates curing. The core can be removed intact. As Table 23.5 reveals, the most common binders used in this system are furfuryl alcohol, urea-formaldehyde, and phenol-formaldehyde. Air contaminants are released when the core is ejected from the core box, and ventilation may be required.

After the core is completed, it is coated with a spray frequently containing graphite, which prevents penetration of the metal and provides a better metal surface. The vehicle is usually an alcohol.

2.6.3 Molding

A full mold process has been introduced which eliminates the conventional pattern and in its place uses a polystyrene pattern. The pattern is positioned with a runner and resin system made of polystyrene, and as the molten metal flows, it vaporizes the polystyrene. The volatile gases escape through the sand. Although this system eliminates the air contamination from cores, it does present an exposure to styrene, benzene, carbon dioxide, and carbon monoxide.

Several other molding procedures are available, and these, too, are accompanied by specific hazards. A popular molding procedure is shell molding used for small castings. The mold is a light shell formed from a sand–resin mixture fused over a heated matchplate having the geometry of the part to be cast. Both urea- and phenol- formaldehyde resins are used in this system, and hexamethylene tetramine is also added. Under heat, the resins degrade to form the products listed in Table 23.5 for the heated core box system.

The investment or lost wax technique is a method used for precision casting. A wax form is made in the desired shape by injecting molten wax into a die. The solid wax form is then placed in a flask, and an investment consisting of an alcohol slurry of alumina, sand, and gypsum is poured into the flask. The investment is cured and the wax form is melted out, leaving the cavity.

2.6.4 Melting and Pouring

A variety of furnaces are used for melting depending on the type of metal and the size of the heat. As noted in Table 23.6, the principal hazard from metal fume exists in nonferrous foundries. The furnace charging and pouring operations present physical hazards from molten metal splashes, heat stress, and infrared radiation.

2.6.5 Shakeout

The cooled casting is removed from the mold at the shakeout position. At this point the sand is dry and friable and represents a major dust hazard. The dust exposure at

Table 23.6 Dust and Fume Exposures from Metal Melting and Pouring (21)

Metal	Dust or Fume	Occurrence
Iron and steel	Iron oxide	Common
	Lead, Leaded steel	Common
	Tellurium	Rare
	Silica	Common
	Carbon monoxide	Common
	Acrolein	Rare
Bronze and brass	Copper	Common
	Zinc	Common
	Lead	Common
	Manganese	Rare
	Phosphine	Rare
	Silica	Common
	Carbon monoxide	Common
Aluminum	Aluminum	Common
Magnesium	Magnesium	Common
	Fluorides	Common
	Sulfur dioxide	Common
Zinc	Zinc	Common
Cadmium	Cadmium	Common
Lead alloys	Lead	Common
	Antimony	Common
	Tin	Common
Beryllium	Beryllium	Common
Beryllium–copper	Beryllium	Common
Uranium	Uranium	Rare

shakeout may include not only silica but the residual metal particulate in the reconditioned sand, and the decomposition products from the core resin system. If sea coal is used in the system, there will be serious exposure to carbon monoxide during cooling and at shakeout.

After shakeout the casting is processed in the finishing room. Normal operations include removal of sprues by cutoff wheels or burning, removal of cores by hydraulic blasting or vibration, abrasive cleaning, and heat treating. There are potentials for exposure to abrasive dust, silica dust, and metal fume. Physical hazards include noise, heat, and vibration from pneumatic grinding and chipping tools.

2.7 Grinding, Polishing, and Buffing

Grinding operations involve significant removal of metal; polishing is an intermediate step to remove grinding marks and burrs; and buffing, the final operation in the series, is used to develop a high polish or sheen. These operations can be carried out on a number of materials, but this discussion is limited to metals. In all operations an abrasive is directed to the surface of the workpiece by a variety of wheels, disks, belts, and cloth. The type of particulate generated by these operations depends on the surface coating of the workpiece, the base metal, the abrasive, and the bonding material.

In grinding, the abrasive is normally silicon carbide, aluminum oxide, or industrial diamonds. In polishing operations fine grit silicon carbide and aluminum oxide may be used, in addition to emery, garnet, flint, chalk, pumice, tripoli, and iron oxide. In buffing operations, a low toxicity polishing rouge or emery is applied to cloth wheels.

The bonding of the abrasive is accomplished with an inorganic bonding agent such as the vitrified silicates or, more commonly, by a phenolic or rubber bond.

The hazards vary with the material released to the air. The fused sand removed in a foundry finishing room may constitute a major hazard, as may lead paint on a bridge structure being removed by grinding. The hazard is minor if the base metal is iron or steel, but if a beryllium part is being worked, the hazard is great. The hazard from the bonding agent is minor. In the case of an organic resin system, the bond will degrade and a characteristic phenolic smell will be noted.

The hazard from bursting wheels operated at high speed and the fire hazard from handling certain metals such as magnesium are not covered here, but these problems affect the design of hoods and the design of wet dust collection systems.

The principal control on these operations is ventilation (6, 24, 25). The hood designs are based on a tight hood enclosure with minimum wheel–hood clearance to provide dust control at a minimum exhaust volume. An adjustable tongue on the hood also reduces the air induced by wheel rotation at high speed. On buffing operations the hood is usually designed with a settling chamber to minimize duct plugging with wheel debris. Fixed location operations are handled by conventional systems, whereas portable tools can be efficiently exhausted by low air volume–high velocity systems. Specific ventilation controls have been proposed which handle a variety of operations.

As in other cases, the use of water on grinding operations does not eliminate the need for ventilation control. Water droplets can evaporate, leaving a respirable airborne particle.

2.8 Heat Treating

The heat treating process heats and cools metal under controlled conditions of temperature and atmosphere, to impart certain physical properties to the metal. In annealing operations the metal is heated to a given temperature and slowly cooled or quenched to reduce hardness. Hardness is imparted to the metal by quick cooling or quenching. Controlled atmospheres are also used in a variety of ways to impart surface characteristics to the metal. The principal furnaces and bath operations and the main occupational health hazards are given in Table 23.7.

Hardening is commonly done in controlled-atmosphere furnaces and salt and metal baths. The controlled atmosphere may consist of an inert gas containing either carbon

Table 23.7 Air Contaminants Released During Heat Treating

Process	Cycle	Air Contaminants
Annealing	Initial heating, slow cooling	Carbon monoxide
Patenting	Heat in furnace, then directly to lead bath (400–870°C)	Lead fume
Tempering	Heat, then direct quench in salt or lead bath	Caustic fume or mist, lead fume
Case hardening		
Flame hardening	Surface heating, direct application of gas torch	Carbon monoxide
Induction	Eddy current in metal causes temperature rise in furnace	None
Carburizing	Heat part in	
	(1) NaCN and $NaCO_3$ bath;	Sodium carbonate fume
	(2) CO (20%), H (50%) furnace atmosphere;	Carbon monoxide
	(3) methane and propane furnace atmosphere	Carbon monoxide, methane, propane
Nitriding	At temperature add C and N to metal surface in 30% NaCN bath	None
Carbon nitriding	Gaseous ammonia in furnace atmosphere	Ammonia

dioxide or carbon monoxide. Special atmospheres for nitriding employ ammonia. When carbon monoxide is used, care must be exercised to prevent its escape into the workroom. The inert gas producers should be checked for leakage when first used and after repairs, to make certain that all piping is tight. Good practice dictates the use of exhaust hoods above the furnaces and flame curtains at the openings. If an excess gas-producing capacity exists, as when only a portion of furnace capacity is used, the excess gas must be vented safely to the atmosphere outside the plant. Furnaces depending on gravity stack ventilation should be monitored for carbon monoxide leakage during the warmup period (26).

Ventilation should be employed on nitriding furnaces when ammonia gas is used. Tanks and lines containing liquid ammonia must be protected from mechanical damage. The release of liquid ammonia to the workroom could result in an explosive atmosphere.

It is common practice to provide exhaust ventilation for cyanide baths; but this is not because of the production of hydrogen cyanide. The fume consists essentially of sodium carbonate, which is somewhat irritating. Cyanides should be stored under lock and key, away from acid carboys.

Nitrate baths are exhausted as a precaution against irritant gases. Care should be taken to prevent organic matter from coming in contact with the hot salt because the mixture might be explosive. Neutral salt baths are also frequently exhausted, principally to control the heat during hot weather and to remove vapors corrosive to metal parts.

Lead baths are frequently held at temperatures between 1000 and 1500°F, therefore requiring exhaust ventilation. Oil quench tanks are often ventilated to remove the irritating smoke that is evolved during their use. Some tanks have cooling coils to control the temperature of the oil and to reduce both smoke production and the fire hazard.

Where sprinkler systems are used, canopies should be erected above all oil, salt, and metal baths to prevent water from cascading into them. Any workman who happened to be adjacent to a hot bath when water struck it would be in grave danger.

2.9 Industrial Radiography

The most valued tool in nondestructive testing of metal fabrications such as forgings, welded assemblies, and castings is industrial radiography. The technique utilizes both machine and radionuclide sources of radiation. Although conventional X-ray tubes have been the principal source for machine units, Van de Graaff and linear accelerators are now becoming commonplace. The common radionuclides, ^{60}Co, ^{192}Ir, and ^{226}Rad, are used in conjunction with remote projectors. Inspection at plants may be conducted directly at the assembly line by fluoroscopy or at a central inspection facility equipped with high energy X-ray, betatron machines and radionuclide sources.

In many cases a part or weldment must be inspected on site—possibly on board ship, at a power station under construction, or in a petroleum refinery high pressure piping system. High energy radioisotope sources are now available for field radiography. Portable low energy X-ray units have also been developed for this purpose. The field applications of these devices require well-trained personnel who conform to all necessary regulations to minimize exposure.

The precautions in handling radioisotope sources are described in detail in Chapter 12. These techniques require approved installation, detailed operating instructions, periodic inspection, and medical surveillance of personnel. All radiation sources must be tested for leaks.

2.10 Metal Machining

In most machining operations the cutting tool is usually designed to remove metal in an efficient way by cutting fine chips, it does not generate large quantities of respirable particulate. However respirable dust is generated on certain operations, such as grinding. Machining operations are commonly performed using a coolant that floods the point of contact between the cutting tool and the part.

Hazardous dust concentrations may occur during machining of toxic surface coatings on metal, such as cadmium, or toxic base metals that are friable, such as beryllium. The necessary controls for machining beryllium are presented in Section 3.9.

Occasionally a gaseous exposure is detected from a metal exposure. In one state it was found that ductile iron containing a copper or nickel alloy of magnesium released significant concentrations of phosphide because of the reaction of magnesium and phosphorus. The phosphide is hydrolized to phosphorus by moisture in the air, by the coolant, or in the respiratory tract.

In many cases the machining operations on such metals as magnesium and titanium may generate explosive concentrations of dust, and the operations must be conducted with suitable ventilation control and air cleaning.

The exhaust hoods commonly used to capture toxic dusts include conventional exterior hoods, high velocity–low volume capture hoods, and enclosures. A number of special hood designs of the enclosure type have been utilized for beryllium fabrication shops.

The principal occupational disease for machinists is dermatitis from grinding and cutting oils (27). A variety of oils are used, including mineral-based oils with special additives such as sulfur for heavy-duty applications, soluble oil–water emulsions, and aqueous-based synthetic coolants. The bulk of coolants presently in use are water based. The chief dermatitis problem associated with these coolants, as discussed in Chapter 8, is an eczematous contact dermatitis, although oil acne and chloracne are occasionally encountered. The means of controlling contact dermatitis from coolants and cutting oil is rigorous maintenance of the coolant system, with frequent changing and cleaning of the coolant system and pump, good housekeeping, and personal cleanliness, combined with personal protective clothing such as sleevelets and aprons.

During certain machining operations the cutting or grinding oil becomes an aerosol and presents an oil mist exposure that must be controlled by local exhaust ventilation in conjunction with air cleaning (6, 28, 29).

2.11 Metallizing

A number of metals can be sprayed as molten droplets on the surface of the part to be metallized. This is commonly done to provide corrosion protection, to build up a worn or corroded section on a part, to improve wear resistance, to reduce production costs, or to give a decorative surface. The metals commonly metallized are copper, zinc, lead, cadmium, aluminum, stainless steel, and brass or bronze alloys. The metal, usually in the form of a wire or powder, is fed into the flame and propelled to the part. In one common procedure the fuel is oxygen–acetylene or oxygen–propane with a compressed air atomizing stream. If powdered metals such as nickel-chromium-boron-silicon, tungsten carbide mixtures, and ceramics are used, the powder is fed from a hopper to an oxygen–acetylene gun. In a third common process the metal is melted in a plasma gun, thereby achieving temperatures up to 30,000°F and permitting the user to select from a wide range of metals and metallic oxides.

The principal hazard from these operations is the exposure to the metal being sprayed. Since the processes have deposition efficiencies from 60 percent to more than 90 percent, concentrations of metal may be high at the operator's station. In plasma gun operations oxides of nitrogen and ozone concentrations may be significant and should be monitored; as noted in the discussion of welding in Section 2.13, moreover, both noise and ultraviolet radiation are potential hazards.

The main control is local exhaust ventilation, with a hood face velocity of 125 fpm for nontoxic metals and 200 fpm for toxic metals (6). In addition, the operator should wear an airline respirator with continuous or pressure demand flow when spraying toxic metals. Care should also be taken in cleaning out spray booths used for toxic metals. The operators should be placed on a medical control program, and the ventilation and respiratory equipment should be checked frequently.

In vacuum metallizing the metal is vaporized in a vacuum chamber. Although this would appear to be a safe operation, one study reveals hazardous exposures to cadmium during loading and unloading the chamber (30).

2.12 Painting

Paints, enamels, varnishes, and lacquers may be applied by brushing, dipping, roller, electrocoating, and spraying. Spraying is the most common method and presents the greatest hazard (31). The three common spraying techniques are air atomization, airless spray, and electrostatic spray. The amount of overspray and resulting operator exposure varies considerably among these three methods (Table 23.8).

Paint consists of vehicles or binders, pigments, solvents or thinners, and additives such as driers and fungicidal agents as listed in Table 23.9. The binders may be natural materials such as shellac and linseed oil or one of a number of synthetic resins. A large number of metallic oxides are used for pigments, including several toxic compounds (Table 23.9). The use of lead compounds for pigments has largely disappeared, but lead is still used in yellow and green pigments. A wide range of solvents with varying toxicity are encountered, and a number of special purpose paints contain additives.

Special purpose paints have unique formulations. Emulsion paints prepared for general-purpose applications contain a suspension of synthetic latex or polyvinyl acetate in water and may contain pentachlorophenol as a fungicide. Automobile paints may be based on alkyd-melamine-formaldehyde, acrylic, and nitrocelluose complexes. Marine paints may be based on epoxy resin systems, and for many years they contained organic mercury compounds as a fungicide.

The operator is exposed to the solvent or thinner and the paint spray mist. The exposure to solvents occurs during paint mixing, spray painting, air drying, and oven baking. Each of these operations requires ventilation control, usually in the form of a spray booth, room, or tunnel provided with some type of paint spray mist arrestor (6, 26, 32). The degree of control necessary on spray painting varies with the type of operation—that is, whether it is air atomization, airless, or electrostatic spray painting (26). The latter two techniques call for somewhat lower exhaust volumes.

A common exposure to organic vapors occurs when the spray operator places the freshly sprayed part on a rack directly behind him. The air movement to the spray booth sweeps over the drying parts and past the breathing zone of the operator. Drying

Table 23.8 Overspray During Painting (26)

Method of Spraying	Percentage Overspray	
	Flat Surface	Table Leg
Air atomization	50	85
Airless	20–25	90
Electrostatic disk	5	5–10
Airless	20	30
Air atomization	25	35

stations and baking ovens must also be exhausted. The choice of exhaust control on tumbling and roll applications depends on the surface area of the parts and the nature of the solvent.

Dermatitis due to primary irritation and defatting from solvents or thinners, as well as sensitization from epoxy or urethane systems, is not uncommon. Skin contact must be minimized, rigorous personal cleanliness encouraged, and protective equipment supplied to the operator.

2.13 Welding

Welding has been defined as a process for joining metals in which coalescence is produced by heating the metals to a suitable temperature. The welding process, according to the American Welding Society, can be classed as pressure, nonpressure, or brazing. There are also a number of allied processes—such as oxygen cutting and metallizing—that have important industrial hygiene implications. The nonpressure techniques warrant major attention because they involve the fusing or melting together of parts. In brazing, the joining is accomplished by use of a filler metal, at a temperature below the melting point of the base metal (33, 34) but above that of the filler metal.

We now consider the most common of the welding techniques, assuming that all techniques are used on mild steel. The nature of the base metal certainly is important in evaluating the metal fume exposure, and occasionally it has impact in other areas. When such impact is important, it is discussed. It is also convenient to discuss the potential health hazards in terms of metal fume exposure, gas and vapor exposure, electromagnetic radiation, and other sources of hazard specific to the procedure. The information presented in the text is summarized in Table 23.10.

2.13.1 Arc Welding

Shielded Metal Arc Welding. This most common of all arc welding processes is commonly called "stick welding." In this process the part to be welded and the wire electrode represent two electrodes that are powered by a dc or ac power supply. When using dc, if the work is positive the process is termed straight polarity; if negative it is termed reverse polarity. The voltages on these operations are commonly low; however current density as calculated by current flow divided by the area of the electrode is an important consideration in assessing the potential hazard. When the electrode is brought close to the part, an arc is established and is maintained by the welder as he progresses along the weldment.

The function of the electrode used to be to establish the arc and deposit a metal bead identical to the base metal. Electrodes are now of specific alloy composition and are

Table 23.9 Materials Used in Compounding Paints

Compounds	Components
Binders	
Acrylics	
Alkyd-melamine-formaldehyde	
Alkyds	
Epoxies	
Linseed oil	
Nitrocellulose	
Polyesters	
Polyurethanes	
Shellac	
Vinyl	
Pigments	
Black	Carbon black
Blue	Ferric ferrocyanide, phthalacyanide blue
Colorless	Calcium carbonate, magnesium silicate, barium sulfate
Green	Chromium oxide
Red	Iron oxide, cadmium selenide
White	Titanium dioxide, zinc oxide, white lead, lithopone
Yellow	Cadmium sulfide, lead chromate
Solvents	
Alcohols	
Esters	
Glycol esters	
Hydrocarbons, aliphatic and aromatic	
Ketones	
Water	
Additives	
Driers	Lead octate, cobalt naphthenate, and manganese isodecanoate
Fillers, reinforcers	Silica, asbestos, talc, clay, calcium carbonate
Fungicides	Mercury aryl, pentachlorophenol
Plasticizers	Tricresyl orthophosphate

Table 23.10 Health Hazards from Welding[a]

	Process					
Hazard	Shielded Arc	Low Hydrogen	Gas Metal Arc, Gas Tungsten Arc Welding	Submerged Arc Welding	Plasma Arc Welding	Gas Welding
Metal fumes	Low	Low	High	Low	Very high	High
Gases	NO_2 in enclosures; negligible CO_2, CO, O_3	Coating contains 9% F; fumes contain 10–12% F, 50% is water soluble, HF	O_3 > 2000 Å; A > He; Al > Fe; $CO_2 \rightarrow CO$; NO_2 > 5 ppm (N_2–H_e)	F, HF	High O_3; high NO_2	NO_2, CO
Radiation	UV, flash burns	UV, flash burns	UV, 10–20 × conventional arc; A > He; Al > Fe; erythemal effect; clothing damage; $C_2HCl_3 \rightarrow$ HCl, Cl, $COCl_2$	Nil	Rich UV; ultra-soft X-ray	—
Other hazards	—	—	Thoria 1–2% in tungsten electrode	—	Noise > 110 dbA	Flux: Cl, F, Br compounds

[a] Hazard is assigned based on mild steel welding.

coated with a granular layer designed to provide a protective slag over the molten metal, provide a gas shield over the arc, and deposit a small alloying content in the weld. The coating may include titanium dioxide, calcium carbonate, cellulose, clay, talc, or asbestos. The composition of the coating and the wire alloy may be obtained from the National Electric Manufacturer's Association–American Welding Society (NEMA-AWS) identification number stamped directly on all rods or electrodes. The hazards from the conventional or shielded arc welding technique include the metal fume that is characteristic of the base metal being welded, and to a lesser degree the electrode alloy.

By far the greatest exposure to metal fume is from iron oxide generated during welding or oxygen cutting of mild steel or steel-based alloys. Iron oxide does cause a benign pneumoconiosis because of the deposition of the inert particulates. There is no fibrous tissue proliferation, but the possibility of pulmonary function disability is in dispute.

This high energy arc has the ability to fix nitrogen as nitrogen oxides, and significant concentrations have been measured in enclosed work locations; however this is not normally a problem in an open shop. Carbon monoxide and carbon dioxide may originate from the cellulose electrode covering, but the concentrations are negligible. Ozone is also noted in the workplace, but it is not a problem unless the current densities are extremely high.

The welding arc does generate a broad spectrum electromagnetic radiation exposure, including infrared (8000 to 13,000 Å), visible light (3500 to 8000 Å), and ultraviolet (2900 to 3000 Å). As discussed in Chapter 11, the wavelengths in the range of 2800 to 3100 Å will cause severe eye irritation, which the welder calls "flash burn," "arc eye," or "sand in the eye." The hazard from infrared radiation in welding operations has not received critical attention.

Low Hydrogen. A special modification of the conventional shielded arc welding procedure is the low hydrogen–rod technique. This procedure, designed for use on stainless steel and other alloys, utilizes fluorine compounds such as calcium fluoride in the coating to take certain nonferrous oxides into the slag. The coating may contain up to 9 percent fluorides, and the fumes may have 10 to 22 percent fluorides, with approximately half that amount present as soluble fluorides. Hydrogen fluoride is present in low concentration. Exposure to fumes from low hydrogen welding may prompt complaints of nose and throat irritation and chronic nosebleeds. There has been no evidence of systemic fluorosis from this exposure. Monitoring can be accomplished by both air sampling and urinary fluoride measurements.

Gas Metal Arc Welding. To improve the shielding of the molten metal and therefore the quality of the weldment, several techniques have been introduced in which the arc and molten metal are blanketed by an inert gas from a compressed gas supply.

A common method is gas metal arc welding (GMA), also called metal inert gas welding (MIG), which uses a welding torch with a center consumable wire that maintains the arc as it melts into the weldment. Around this electrode at the torch is an annular passage for the flow of an inert gas such as helium, argon, carbon dioxide, or a blend of these gases. The process may use a high current density, resulting in high metal fume concentrations. It does present a significant UV exposure depending on the current density, the shielding gas, and the parent metal (35). It is not uncommon to find UV spectral intensities 10 to 20 times greater than those associated with conventional stick welding. The effects may be skin and eye burns and disintegration of wool and cotton clothing. Ozone concentrations in excess of 0.05 ppm are easily reached outside the welding helmet. Nitrogen dioxide in the parts per million range is also noted, especially when a nitrogen–helium inert gas is in use. As one would expect, if the shielding gas is carbon dioxide, high concentrations of carbon monoxide are observed.

An additional vapor and gas hazard has been noted when the vapor of trichloroethylene is present in the welding area. A number of studies of the specific products resulting from the degradation of the vapor by UV radiation indicate that phosgene, hydrogen chloride, and chlorine are generated (36). In another study dichloroacetyl chloride was defined as the critical degradation product (37).

Gas Tungsten Arc Welding (GTA). In this process, the arc is established between a nonconsumable tungsten electrode in the welding torch and the workpiece. The inert gas flows around the electrode and blankets the arc and the molten metal in the same manner as the MIG process. A manually fed filler rod may or may not be used. The hazards are similar to those for the MIG process. An early study showed that the exposure from thorium oxide present in the tungsten electrode did not constitute a hazard to the welder, although electrode dressing operations on a grinder should be controlled by local exhaust.

Submerged Arc Welding. The arc is established between the electrode and the workpiece. Shielding is obtained by covering the arc with a granular material that contains soluble fluoride compounds. This process eliminates the UV exposure and reduces the metal fume emission, but it does release hydrogen fluorides. The hazards are similar to those described for low hydrogen welding.

Plasma Arc. In the plasma arc process the welding head is designed to provide flow of a gas such as argon through an orifice under a high voltage gradient, resulting in a highly ionized gas stream. A complex interaction of mechanical and electromagnetic forces results in temperatures as high as 60,000°F. This technique can be used for welding, cutting, and metallizing. The hazards of the process include high noise exposure (110 to 120 dBA), the fixing of ozone and nitrogen dioxide, and high metal fume exposures.

There is a series of other welding operations including atomic hydrogen, electroslag, electron beam, and flux-cored arc welding; the hazards are similar to those already discussed.

2.13.2 Gas welding

In gas welding a number of oxygen–fuel systems are used to generate a torch flame for welding, cutting, and metallizing. The metal fume hazard can be significant during oxygen cutting. When used on nonferrous metal, the filler rods and fluxes release significant metal fume, gases, and vapors. In brazing, a variation of gas welding, a significant cadmium exposure could occur.

Fluxes used on magnesium and aluminum to remove oxides and assist in a proper bond are generally halide salts of lithium, potassium, sodium, and magnesium.

A principal hazard exists in the use of gas-oxygen welding in enclosed spaces, and deaths have been documented from exposures to nitrogen dioxide. A lethal concentration (200 ppm for 1 hour) can be produced in a 20 m^3 space in 15 minutes.

The common metal fume exposures given in Table 23.11 are due to various base metals, alloys, and electrode coatings.

2.13.3 Related Processes

In the rough cutting of slab stock to prepare piece parts for a welding fabrication, it is common to use oxyacetylene torches for iron and steel and, more recently, plasma arc torches for aluminum and some other metals. The hazards involved in these operations are similar to those described for the basic welding techniques from which they derive. This is also the case for oxygen wash and arc-gouging techniques for the removal of defective metal from castings or assemblies. In these procedures very high concentrations of metal fume may be encountered. It has also been noted that the copper electrode coating is present as fume in high concentrations.

Welding controls vary with the process and the metal. If mild steel is worked, dilution ventilation may be adequate; in many cases local exhaust ventilation is required for control of gases. When toxic metals are welded or cut, local exhaust ventilation is mandatory. Standards are available for the design of such controls (6). An interesting new control technique is a low volume–high velocity capture system now available and undergoing extensive evaluation.

Personal protection equipment, an important control for the welder, includes helmet

Table 23.11 Hazards from Common Metal Coatings and Alloys

Coatings and paints
Zinc, galvanized steel
Cadmium plating
Mercury fungicidal paints
Lead paint
Alloying metals
Lead
Nickel
Chromium
Manganese
Beryllium

with proper-density lens, protective clothing usually including chrome leather gloves and jacket, respirators, and the use of portable flash shields to reduce the hazard of eye injury to workers in the area of the work.

Operating procedures on reworked material are extremely important. In shipbreaking and bridge demolition, surface coatings such as lead and zinc chromate should be removed before cutting is begun. If vessels are being repair welded, it is important that they be checked and the work area purged to ensure that there is no flammable liquid or gas residue (see Section 3.29).

3 PRODUCTION FACILITIES

3.1 Abrasives

The use of natural abrasives such as quartz in the form of sandstone, flint, tripoli, pumice, and diatomite, and aluminum oxide as emery and corundum in the preparation of grinding wheels and coated abrasive products has been reduced dramatically during the past two decades. The main hazard in the use of these natural materials is exposure to the quartz dust released during the crushing, sizing, and manufacture of the product. The principal control of such dust is ventilation, although in many cases water suppression may be useful.

The major abrasives now in use are aluminum oxide (Al_2O_3) and silicon carbide (SiC). The substitution of these synthetic abrasives has resulted in a major reduction in the health hazards within industry, and the use of these materials should be considered in place of quartz-type abrasives. Other synthetics, including boron carbide (B_4C) and industrial diamonds, represent a small fraction of the total abrasive market (38).

The raw materials for the manufacture of silicon carbide include silica sand, coke, salt, and sawdust, which are fused together in an electric arc furnace, then crushed, washed, and classified by size. A silica dust exposure may occur during the charging of the furnace. A potential carbon monoxide exposure may occur during charging and furnace operation. Crushing and size classification generate dust, which must be controlled by ventilation with emphasis on good hooding at material transfer points. The literature presents divergent opinions on the toxicity of silicon carbide, but dust control should be established in the manufacture of grinding wheels and coated abrasive products from this material.

Aluminum oxide (artificial corundum) is made by fusing bauxite in an electric furnace, crushing the fused mass, and sizing the particles for final product application. Lung changes have been noted in furnace operators where the bauxite is fused with coke, iron, and silica sand (Chapter 7). The principal air contaminants in that operation are aluminum oxide, silica, and mullite ($3Al_2O_3 \cdot 2SiO_2$). A sintered product is also made by firing a submicrometer bauxite powder.

The major exposure in the end product application of these materials is to the dust of the synthetic abrasives. The chief dust control is local exhaust ventilation, and design standards are available (6).

Synthetic abrasives are commonly employed in the manufacture of grinding wheels, coated paper, and textiles using various adhesives and bonding materials. During product manufacture there may be exposure to clays and fillers, such as feldspars in the manufacture of vitrified bonded wheels, and shellac, rubber, and various resins systems such as phenol in the production of solid bonded wheels. Strict ventilation control should be maintained during crushing, grinding, sifting, and screening, and the edging, facing, and shaving of grinding wheels to minimize dust exposure. A dermatitis problem may exist depending on the resin systems and adhesives in use.

3.2 Acids

3.2.1 Hydrochloric Acid

Hydrochloric acid is made by absorbing hydrogen chloride gas into water in absorption towers or scrubbers. The hydrogen chloride may be prepared as a by-product of chlorination of organic compounds, by the reaction of sodium bisulfite with sodium chloride in a retort, the burning of hydrogen in chlorine, or the reaction of sodium chloride with sulfuric acid in a furnace. The principal hazard in all these enclosed processes is the leakage of the gas from the system or tail gas from the scrubber. Present absorption tower design is adequate to minimize exposure of operating personnel and the neighborhood. The safety and handling procedures for hydrochloric acid have been published and include operating practice, protective clothing, eye protection, and respiratory protection (39).

3.2.2 Nitric Acid

The principal method for manufacturing nitric acid is the high pressure ammonia–oxidation process in which air and ammonia are passed over a heated platinum catalyst to yield nitric oxide (NO). In this process the concentration of ammonia must be kept below the lower limit of flammability (15.5 percent in the air, 14.8 percent in oxygen), to avoid an explosion. The nitric oxide oxidizes to nitrogen dioxide (NO_2) and is absorbed by water to yield nitric acid (HNO_3). The hazards are mainly associated with the potential exposure to nitrogen oxide and nitrogen dioxide. Exposure to nitrogen dioxide may occur if there are leaks in system piping and during on-stream sampling. The safety hazards in the handling of this acid are well documented and are in general those assigned to strong oxidizers.

Accidental exposure to ammonia may also occur in the process because of escape of gas from storage tanks, gauge glasses, valves, and process lines. In addition to the health hazard, a potential fire and explosion hazard may exist.

Nitric acid recovery plants may represent a significant exposure to nitrogen dioxide because of liquid or gas leaks. The emissions from exhaust stacks or nitrogen dioxide scrubbers may also result in hazardous concentrations at ground level.

The serious hazards from nitrogen dioxide and ammonia necessitate adequate emergency escape procedures and entry respiratory protection at these plants. The usual requirements for personal protective equipment, showers, and well-defined work practices apply to nitric acid plants (41).

3.2.3 Sulfuric Acid

The contact process for producing sulfuric acid has now replaced the lead chamber process, in which pyrites are burned. In the contact process sulfur dioxide (SO_2), along with an excess of air, is passed through vanadium or platinum catalysts. The SO_2 is oxidized to SO_3 with the evolution of heat. The sulfur trioxide is absorbed, and it is important that it not be allowed to come in contact with water vapor, since a persistent fog of finely divided droplets of sulfuric acid is formed. Sulfuric acid of 98 to 99 percent purity may permit sulfur trioxide to escape. Electrostatic precipitators have been used successfully to prevent the escape of sulfur trioxide fog.

Significant sulfur dioxide and sulfur trioxide air concentrations at sulfuric acid plants are attributable to fugitive leaks and off-gas emissions; therefore both emergency escape routes and entry respiratory protective equipment must be available.

Effective controls may be achieved by maintenance of the closed processing system and the use of materials handling to minimize employee contact with the operation. Retaining walls should be available in case of rupture of holding tanks and drums. Acid should not be stored near reducing agents because of the fire and explosion hazard. Impervious protective clothing, eye protection, and emergency showers should be available at the workplace (42–43).

The emission of ammonium vanadate or vanadium pentoxide from the contact process may present a hazard to workers.

3.3 Aluminum

The principal aluminum ores are gibbsite ($Al_2O_3 \cdot 3H_2O$) and boehmite ($Al_2O_3 \cdot H_2O$). The ore bodies normally include hydrated alumina, iron oxides, and silica (44). Bauxite is mined in open pits and the dust hazard is considered to be minor, since bauxite does not present a pneumoconiosis problem. However the free silica content of the ore should be identified, and if significant, this exposure must be evaluated.

After the bauxite is washed to remove clay and other waste, it is refined by dissolving the bauxite in a hot caustic solution, precipitating the purified aluminum hydroxide, and calcining the hydrate to form alumina. The handling of the ore and alumina creates a potential dust problem, which is readily controlled by engineering methods. The alkaline solution poses problems that may range from simple dermatoses to serious chemical burns. Rigorous housekeeping procedures must be established, personal protective equipment worn, and personal cleanliness encouraged.

The metal is produced electrolytically from a solution of alumina in molten cryolite. In the Hall-Heroult reduction process the alumina is dissolved in a cryolite bath (Na_3AlF_6) in a lead-lined pot. An array of these pots electrically in series is called a potline; potlines are located in potrooms. The prebaked aluminum reduction cell uses a formed carbon electrode as the anode, and the lead-lined pot is the cathode. In the Soderberg reduction cell, the uncured carbon anode paste is fed from bins to the cells. In the cell the paste is "baked out" during the process and continuously forms the consumed anode. The electrolysis results in the aluminum sinking to the bottom of the pot. Carbon monoxide and carbon dioxide are liberated at the carbon anode. The molten metal is periodically tapped to pigs. In addition to CO and CO_2, the cell generates fluorides, alumina, and aluminum dust. A heat stress problem may exist in potrooms.

The manufacture of the electrode material at the production plant is another source of workroom contamination. The electrodes are made from petroleum or coal coke, and soft petroleum or coal tar pitch. In the prebaked system the anodes are made by grinding the coke, mixing it with pitch, casting it in a form, and baking it until it has mechanical strength. During this operation the workers are exposed to dust with a significant concentration of benzene-soluble coal tar pitch volatiles (CTPV). In the Soderberg anode pot system, the anode is baked by convective heat from the bath. In this procedure the volatiles are released to the potroom as the electrode "cures."

A series of studies of worker exposure to fluorides in aluminum production shows that the concentration of gaseous and particulate fluorides are lowest in the prebake process, (45, 46). It has been proposed that both the prebake and Soderberg processes are amenable to control by local exhaust ventilation.

A recent study of four aluminum reduction facilities in the United States has shown significant CTPV concentrations in potrooms with the highest level in the Soderberg process (46). The time weighted average exposure for fluorides was below 2.5 mg/m^3, and carbon monoxide levels were not usually significant. Noise and heat stress are also potential problems in potrooms and warrant study.

The principal control for air contaminants in potrooms is effective local exhaust ventilation on the pots.

In general, the techniques used in subsequent aluminum fabrication do not cause unique occupational health problems. For example, the casting of aluminum alloys presents no problems more unusual than those that would be encountered in other nonferrous foundry operations. In welding on aluminum, however, the spectra generated by inert gas shielded arc welding procedures on aluminum are unique and result in a rich UV spectrum and high ozone concentrations.

There are few data to support the claim that aluminum dust generated during machining operations causes either pneumoconiosis or systemic toxic effects. However aluminum dust in air may present an explosion hazard, and of course this material has been used in the manufacture of pyrotechnics.

3.4 Ammonia

Ammonia is obtained from by-product coke ovens and from the Haber-Bosch process, which involves the synthesis of NH_3 from nitrogen and hydrogen using a catalyst under high temperature and pressure. Ammonia has wide application because of its available nitrogen in the production of fertilizers, nitric acid, explosives, and plastics, and as a chemical intermediate.

The principal hazard in its manufacture is represented by the accidental release of major amounts of the very irritating gas, because of failure of piping or a valve due to poor maintenance. Emergency escape and entry respiratory protection should be provided at ammonia plants. Fire and explosion hazards are a significant problem, and gas–air mixtures of 16 to 25 percent can cause violent explosions. Skin and eye protection must be worn where exposure may occur (47–49).

3.5 Art Work

The introduction of new materials and processes has provided the artist with new modes of expression, and unfortunately, new health hazards (50, 51). The painter has a number of new synthetic paint media, including acrylics, vinyl acrylics, vinyl chloride, ethyl silicate, vinyl acetate, and pyroxylin. The vehicles and solvents now in use present the normal hazards of primary irritation, defatting dermatitis, and systemic toxicity due to inhalation. The advent of airbrush techniques and protective sprays for the completed art present high air concentrations of contaminants for short periods. Use of silk screen techniques can also result in significant solvent evaporation.

Plastic resin systems including urethanes, epoxies, and polyesters are also in wide use. The resin-curing agents have the potential for causing contact dermatitis and respiratory sensitization.

The control measures required for these materials include good housekeeping, personal cleanliness, and proper exhaust ventilation. Siedlecki (50) recommends the use of disposable covers for work surfaces that may become contaminated, disposable gloves if skin contact is to be avoided, and the use of mild soap and water to remove resins and curing agents from the skin. Proper ventilation control is the most difficult to achieve. In the studio of a firm employing several artists, it is feasible, but for the one-artist studio, respiratory protection may be a necessary substitute for ventilation.

The sculptor has also adopted new materials and processes used in industry. The dust exposure from working conventional materials such as marble and granite has been extended by high energy power tools, which can remove large quantities of material in a short time but generate very high dust concentrations in the process. In addition, new quarrying techniques such as flame cutting present serious hazards from fume, dust, and noise.

Structural resin systems now found in the sculptor's studio include cement, polyester resins, acrylic, and epoxy resins. All the important casting systems discussed in Section 2.6 are now used by the sculptor. The introduction of welded sculpture brought with it a new set of problems including UV radiation, exposure to metal fume, and exposure to gases such as ozone and nitrogen dioxide as described in Section 2.13.

The inventory of hazards to the artist now parallels that noted in the various industrial processes of painting, founding, welding, and ceramics. An inventory of materials and processes must be made, as it is in industry. The recommendation for controls is difficult in the case of a single artist, however, who has an intermittent exposure and does not have the resources necessary to install controls.

3.6 Asbestos Products

The widespread application of asbestos to building materials, pipe insulation, fireproofing, textiles, and friction products makes it difficult to summarize completely the potential health hazards from this material. It is hoped that the deserved attention given this substance during the 1960s by occupational health researchers will supplement this brief coverage. The types of asbestos used commercially (Table 23.12) have differing biological activity; however chrysotile is the most widely used type, representing more than 90 percent of the total use.

The ore deposits contain from 3 to 30 percent asbestos. The ore is crushed, concentrated by magnetic separation, and milled, and the fibers are commonly recovered by air elutriation. The mill operations can be controlled by local exhaust ventilation. Burlap bags have been replaced by plastic bags for packaging of asbestos, to eliminate the hazard from contaminated burlap.

The textile operations involving asbestos include all conventional textile processes—carding, dry weaving, spinning, twisting, winding, and warping, which are all extremely dusty. The debagging of asbestos and its mixing, pressing, and curing in the manufacture of asbestos cement board and pipe are also dusty operations. The manufacture of friction products such as brake pads require sawing, filing, and grinding of the pads and can result in serious exposure to operators.

The control of asbestos dust can be accomplished at the primary manufacturing facility by proper choice of manufacturing method coupled with effective local exhaust ventilation, air cleaning, and work practices (52).

The field use of these asbestos products is another matter: installation of brake linings at a small garage facility is usually uncontrolled and can result in short-term exposure to high levels of asbestos. The field installation of asbestos insulation at stationary power plants and on shipboard is also difficult to control (53).

The control of asbestos during field installation of asbestos insulation taxes the inge-

Table 23.12 Types of Asbestos

Chrysotile
Amphiboles
Amosite
Anthophyllite
Crocidolite
Tremolite
Actinolite

nuity of the industrial hygienist. In many situations control can be achieved by using another material instead, as in the case of fiberglass and ceramic insulation substitutes for asbestos. The introduction of new techniques for doing a job can also be effective—for example, the use of a preweighed asbestos-cement blend packaged in a plastic bag. Water is added at the job, and the material is hand kneaded. This technique eliminates the open mixing of asbestos cement, which was one of the very dusty jobs in the industry. The introduction of portable low volume–high velocity exhaust systems has made possible the control of asbestos dust from cutting asbestos pipe cover and pads in field operations. Housekeeping and rigorous personal hygiene, including the use of clothing change and showering are also necessary to control asbestos. Vacuum cleaning should be used in place of dry sweeping, and other trades should be prohibited from the workspace during active insulation work.

A greater problem exists in the removal of installed asbestos material. Frequently bonded asbestos fiber is released to the air when vinyl-asbestos tile is sanded to provide a bonding surface before recovering. The application of sprayed asbestos for fireproofing structural steel in the 1960s has resulted in the covering of ceilings and walls with a loosely bonded asbestos blanket that can release asbestos. If this material is applied to the air supply plenum surfaces in a building, it may provide a low level asbestos exposure to all occupants. Removal of the material or effective bonding is difficult and expensive, but necessary. The removal or "rip-out" of degraded insulation from high pressure steam lines can also result in significant short-term exposures. The U.S. Navy has found water injection to be helpful in controlling this problem.

3.7 Asphalt

The term "asphalt" applies to a naturally occurring deposit or a product recovered from crude petroleum by a refinery process. The material can be separated into three groups: asphaltenes, resins, and oils. The resins form a cover over the particulate asphaltenes, and the paraffinic, naphthenic, or naphtha-aromatic oils are the suspending oils. The great bulk of asphalt is now obtained from petroleum, and the refinery process permits production of materials ranging from viscous liquids to solids.

Asphalt is principally used in paints, coatings on roads, aircraft runways, and roofing products, and as a saturant for insulation board. The principal hazards from asphalt, in addition to skin burns during hot application, are dermatitis, photosensitization, and melanoses. The carcinogenicity of asphalt has been demonstrated in animals in both inhalation and cutaneous tests.

The most significant exposure to asphalt occurs as emissions from air blowing of the residual oil fraction at 300 to 400°F. The asphalt products manufactured by this technique are commonly used in roofing products. The emissions from the process include various nitrogen and sulfur compounds and hydrocarbons. Control of off-gases is usually achieved by scrubbing and incineration (26).

A major exposure to asphalt-based contaminants occurs at asphalt batching plants, where dried and sized aggregate is mixed with asphalt and transferred to trucks for transportation to the road paving site. The hazard from the crushing, drying, and sizing of the aggregate is similar to that from any crushing operation and depends on the quartz content of the rock. A mineral filler such as finely ground rock, Portland cement, or limestone may also be added to the paving material. Here again the principal hazard is the dust from aggregate preparation, and local exhaust ventilation at equipment and transfer points is necessary, as well as adequate air cleaning (6).

A second major application of asphalt is the production of asphalt roofing. In the process, felted paper is impregnated with asphalt, coated with granules of mineral dust, and cut to form conventional roof shingles, roll covering, or side wall shingles. The felt is a paper with wood or cloth fiber added for strength; in special cases asbestos is added as a fire retardant. The impregnant is usually an airblown asphalt that is applied by spray and dip. The production of roof covering is similar. The mineral granules are applied to the felt after a hot coating of asphalt. The granules are pressed in place, and the roll is cooled for subsequent cutting. In these operations the asphalt releases dense white clouds of hydrocarbon oils, which must be controlled by local exhaust ventilation at the asphalt storage tanks, saturator, drying drums, and wet looper. Dust from the sand, mica, and talc used in the process also may require ventilation control at the dryer locations, although at this point the granules are large, and little dusting is noted during simple gravity feed from a hopper to the felt. The saturators, drying-in drums, and wet loopers are usually ventilated by enclosure-type hoods. Air cleaning in the form of incinerators and scrubbers is required.

3.8 Batteries

3.8.1 Lead–Acid Batteries

The lead–acid battery industry presents a significant risk of lead poisoning because a basic component of the product is made from pure lead or an antimony alloy of lead. The battery consists of a series of plates with the positive plates made of active lead peroxide and the negative plates of metallic lead. Lead is processed in a ball mill to form a lead–lead oxide particulate. The sized particulate is mixed with sulfuric acid and carbon black to form a paste. The paste is pressed into the lead grid, which is the structural member. The plates are assembled in an electric circuit and either are "formed" immediately in weak sulfuric acid or are "formed" after complete assembly. The plates are cleaned to remove excess paste, and the grids are assembled by melting them together. The assembly is then placed in a battery case, which is usually precast from hard rubber or plastics such as polystyrene, polypropylene, polyvinyl chloride, or acrylonitrile-butadiene-styrene. Coal or petroleum pitch, asbestos, or other minerals serve as fillers. Finally the top of the battery is sealed with asphalt or epoxy resin.

The lead smelting and reclamation operations associated with battery manufacture present a critical problem and require rigid ventilation control and good housekeeping. This is also true of the production of the oxide in ball mills. The oxide mixing facility must be equipped with a dust control system, and dispensing operations should be designed to minimize exposure to lead.

From the time the lead oxide paste is applied to the plates, the control of the exposure becomes more difficult. The paste is applied wet either by hand or machine. The application itself presents little difficulty, but as soon as the paste becomes dry it is easily dispersed into the air. This may occur when the plates are subsequently cleaned, transferred, racked, stacked, trimmed, and split. If the wet paste is allowed to fall on the floor or contaminate equipment, sooner or later it dries and adds to the airborne dust. Good housekeeping is absolutely essential in this industry.

Work tables with downdraft exhaust have proved to be very effective in controlling dust dispersion, as well as floor contamination. The plates should be handled manually as little as possible. The plates of ready-charged batteries are handled and shipped dry, ready for service as soon as the electrolyte is added, and these offer more serious dust-control problems than plates that are handled and shipped wet.

Controlling lead exposure in storage battery manufacture is not a piecemeal problem, but one that requires a well-rounded health maintenance program, including (1) competent engineering control, (2) preemployment examinations with biological monitoring, (3) periodic surveys of atmospheric contamination, (4) periodic blood-lead determinations of persons in potentially hazardous workplaces, (5) education of employees about personal hygiene, good health practices, and safe working procedures, and (6) transfer of overexposed workers to a position of minimum exposure (54, 55).

Floors must be surfaced and covered for easy cleaning. Vacuum cleaning must be used; protective clothing and suitable eating and sanitary facilities divorced from the production area are necessary.

The battery charging room facility is subject to contamination by hydrogen and acid mist, and since battery grids often contain from 5 to 10 percent antimony, production of stibine is at least theoretically possible. Dilution ventilation is the conventional control technique in the charging area.

3.8.2 Nickel-Cadmium Batteries

The basic material for the negative electrode of the nickel-cadmium battery is formed electrolytically in a ferrous sulfate bath using a cell with aluminum cathodes and iron and cadmium anodes. The positive battery electrode of the battery is made from nickel sulfate, cobalt sulfate, and sodium hydroxide. The plates are assembled in a battery shell with a potassium hydroxide electrolyte.

The principal cadmium dust exposures occur when the active negative plate material is scraped from the aluminum cathode, dried, and milled. From this point on in the manufacturing process the cadmium plate material is dry. Cadmium dust and fume exposure may occur during the manufacture of the plate strip, as well as during the connection of the plates, the welding of bus strips, and the finishing of the plates.

The batteries are usually assembled in steel or plastic cases. Eye hazards can occur during filling with electrolyte.

3.8.3 Dry Cell Batteries

The common dry cell battery consists of a carbon rod as the positive electrode and a zinc container that is the negative electrode. A paste of manganese dioxide, powdered graphite, ammonium chloride, and zinc chloride used as a depolarizer is packed around the center carbon rod. The conducting medium is then placed between the depolarizer and the zinc cathode. This mixture is a gelatinous paste made of starch, zinc chloride, and ammonium chloride, and zinc chloride is added. The filled cell is sealed with hot pitch or sealing wax.

The principal hazard in dry cell manufacture is attributed to manganese oxide dust generated from the initial mixing and later handling of the depolarizer mix. However there can also be exposure to ammonium and zinc chloride during weighing, mixing, and dispensing. Once the electrolyte is added, the dust exposure is minimized.

The controls required during preparation of the depolarizer include ventilation, wearing of personal protective clothing, and rigorous housekeeping based on vacuum cleaning of the area.

3.9 Beryllium

Beryllium is extensively used in industry because of its unique properties of high tensile strength, workability, and electrical and thermal conductivity. The major industrial uses of beryllium and the principal exposures are given in Table 23.13. The exposures occur in the processing plant, among the intermediate producers, and among the product manufacturers. It has been stated that there are 8000 plants handling beryllium in the United States, and 30,000 workers with potential exposure. Controls on beryllium producers were established by the U.S. Atomic Energy Commission in 1949. Although the toxicity of this metal has been well described, new cases of disease continue to occur.

The major sources of beryllium emissions have been identified as beryllium metal extraction plants, ceramic production plants, foundries, machining facilities, and propellant manufacturing plants (56). The extraction plants produce beryllium metal, beryllium oxide powders, and copper, nickel, and aluminum alloys of beryllium. The varied exposure to dust, fumes, and mists of beryllium in extraction plants must be controlled by well-maintained local exhaust ventilation systems and efficient air-cleaning devices (57). The Beryllium Case Registry has recorded cases of berylliosis in smelting and extraction that have been attributed to initial exposure after 1949, when control measures were adopted.

The production of beryllium ceramic has the greatest potential for exposure due to the loss from such operations as milling, screening, drying, sintering, and machining. These operations must normally be conducted in glove boxes or total enclosures to ensure adequate controls.

Foundries take master alloys and recast them into special purpose beryllium copper alloys of varying beryllium content. As in any other foundry job, exposure in these operations occurs as a result of fumes evolved during the melting and pouring of the

Table 23.13 Air Contaminants Released During Beryllium Processing (56)

Process	Contaminant
Extraction plants	
Ore crushing, milling, mulling	Beryl ore dust
Briquetting, crushing, and milling	Briquette dust
Sintering	Beryl dust, sinter dust
Beryllium hydroxide production	$Be(OH)_2$ slurry, H_2SO_4 fume
Beryllium metal production	$(NH_4)_2BeF_4$, $PbCrO_4$, CaF_2, HF, $Be(OH)_2$, BeF_2, NH_4F, Mg, Be, MgF_2, BeO, acid fume
Beryllium oxide production	BeO fume and dust
Beryllium copper alloy production	Be, Cu dust, BeO dust
Beryllium alloy machine shops	
Beryllium machining operations	Be dust
Beryllium copper foundries	
Melting ingots in primary crucibles	Be fume
Preheating transfer crucibles	Be fume
Drossing and dross handling	Be fume
Pouring molds	Be fume
Finishing operations	Be dust
Beryllium ceramics	
Spray drying of oxide	BeO dust
Kilns	BeO dust, binders
Machining	BeO dust, binders

alloy, and dust from drossing, shakeout, and metal finishing operations. In general, the conventional ventilation and air-cleaning techniques normally installed in foundries for metal fume are not adequate to control the exposure to beryllium.

Breslin has suggested a format that is useful in determining the controls required on beryllium operations (58):

1. The beryllium operations should be segregated from all other plant work. Access should be limited, and the area must be adequately posted.
2. Process techniques should be used that will minimize dust or fume release, and the manual transfer of material should be discouraged.
3. Almost all processes must be ventilated, even though the process is a wet one. In the case of beryllium alloys this may not be necessary, however air samples should be taken on all operations to ensure that no hazard is presented. The hood and system design must be generally superior to that normally seen in industry. When possible, enclosing hoods should be used because they give superior performance. Small bench operations can be controlled with table top enclosure or glove boxes. Conventional machining operations have been controlled by enclosures or by low volume–high velocity

exhaust systems. The advantages and disadvantages of these systems are discussed by Breslin.

4. The importance of work practices in control of beryllium exposure can be noted by monitoring different employees engaged in the same task. The care that one person may take in following specific task instructions can make a significant difference in his exposure.

5. Spills must be cleaned up immediately by vacuum systems. A high efficiency filter must be used on the vacuum system discharge. Housekeeping must be encouraged by proper design of the facility, including continuous floor covering, coved corners and joints, and regular surfaces.

6. Personal protection normally includes frequent issue of personal protective clothing, which is to be worn only in the plant. Locker rooms should have "clean" and "dirty" sections, to effectively isolate the production area and reduce clothing contamination.

7. Depending on the concentration, half or full facepiece air-purifying respirators, air line respirators, or self-contained breathing apparatus are required.

8. All controls must be adequately monitored by air sampling, and medical management of exposed workers is essential.

3.10 Chlorine

Chlorine gas is manufactured by the electrolytic decomposition of sodium chloride brine solution in a cell. Two methods are presently in use. In the diaphragm cell process, which produces approximately three-quarters of the United States supply, an asbestos diaphragm separates the anode and the cathode. The chlorine is formed at the anode, and sodium hydroxide and hydrogen are produced at the cathode.

The mercury cell process uses an electrolytic cell and a decomposer. The mercury and salt solutions flow through the cell. The mercury acts as the cathode, and the anode is a carbon electrode. The chlorine is released at the anode, and a mercury–sodium amalgam is formed at the flowing cathode. The amalgam is decomposed in a companion electrolytic cell, with the release of hydrogen and sodium hydroxide and the conversion of the amalgam to metallic mercury.

The major health hazard in the mercury cell process is due to the mercury present in the hydrogen stream released from the decomposer and the exhaust from the end boxes on the electrolytic cell. Fugitive leaks of chlorine occur in both processes. The principal control techniques are the removal of the mercury from these two streams by condensers, scrubbing systems, and treated, activated charcoal beds. A rigorous housekeeping program has been described to reduce fugitive leaks from the mercury cell. The problem can be eliminated by converting to diaphragm cell production, which does not present a mercury hazard (59).

In addition to the mercury exposure, there are the hazards due to the release of hydrogen and chlorine gas, as well as the risk to skin and eyes from handling the caustic.

3.11 Cement and Concrete

Cement is made by initially blending finely ground limestone with clay, shale, or slate, then calcining this mixture in a rotary kiln at approximately 1400°C. The cooled clinker thus formed is ground and mixed with certain additives, including gypsum, to form the cement, which is combined with sand and gravel to form concrete.

The hazard in the operation of cement plants is principally a dust hazard and varies with the quartz content of the clay, shale, or slate used to form the clinker. The rock prepared for calcining may contain significant quantities of quartz; however it is usually less than 1 percent in the finished product. Special products such as acid-resistant cements may use high concentrations of siliceous rock, and some are also blended with asbestos to form refractory and insulation cements. Occasionally cement may contain diatomaceous earth, which when calcined forms cristobalite, a biologically active mineral.

The first step in evaluating the exposure of operators of a cement plant is to determine the mineralogical composition of the raw materials, and if silica is present, air sampling must be carried out to demonstrate the respirable concentration of silica. The various dusty equipment, such as the crusher, grinder, and sizing screens, and various transfer points, must be properly exhausted and provided with effective air cleaning (6). The emissions from the kiln are usually handled by electrostatic filters and fabric collectors (26).

The operation of dryers and the calcining kiln results in serious heat stress, and radiant heat is a principal source of the load. Radiation shielding of the operators' work locations can assist in reducing this stress. The operation of crushers, grinders, and the rotary kiln also present a noise hazard that may necessitate isolation booths for operators. A carbon monoxide exposure may also occur from the firing of the kiln.

The alkalinity of the cement presents a dermatitis hazard, and a cement eczema has been reported; it is considered to be due to the hexavalent chromium in the finished product. Cement has been labeled an inert dust, but that practice is being reviewed.

3.12 Cotton

The principal occupational health hazard in cotton processing is the exposure to cotton dust released from operations involving the handling of the raw cotton up to the carding operations (60). Byssinosis, the lung disease resulting from this exposure is mentioned in Chapter 7. The major dust-producing equipment is now exhausted, and air is returned through an air cleaner to the workspace. High air recirculation rates are common in this industry because of the need for controlled temperature and humidity in the workroom. A study of airborne cotton dust in a large textile plant revealed that present ventilation practice results in the return of most of the respirable dust (61). Recirculation must be curtailed, or improved air cleaning is necessary to control this problem (62). In addition to local exhaust ventilation, a series of sensible work practices and good housekeeping rules must be followed.

Processing of cotton may involve the use of sodium hydroxide and sulfuric acid for scouring and mercerizing, sodium chlorite and hypochlorite for bleaching of cotton, and the use of a variety of synthetic resins.

3.13 Fertilizers

3.13.1 Natural Fertilizer

The processing of natural fertilizers from excreta of horses, cows, and poultry involves exposure of workers to biological hazards that may cause a variety of infections such as brucellosis, bovine tuberculosis, tularemia, psitticosis, and Q fever. In addition, handling the decomposed material in poorly ventilated areas may involve exposure to hydrogen sulfide, ammonia, and carbon dioxide. The principal controls are the mechanical handling of the product and the provision of adequate dilution ventilation in the workplace. Drying and bagging of the fertilizer may require local exhaust ventilation.

3.13.2 Mineral Fertilizers

Mineral fertilizers are based on the main soil nutrients, nitrogen, phosphorus, and potassium, and are obtained from the treatment of natural mineral product or produced synthetically.

Normal superphosphate is produced by reacting phosphate rock containing up to 4 percent fluorides with sulfuric acid to produce a material containing approximately 20 percent phosphoric anhydride (P_2O_5). Silicon tetrafluoride and sulfur dioxide are released during this operation, and ventilation controls and air cleaning must be implemented. The material is then cured, ground and bagged. Gaseous fluorides are released during curing. Fluoride concentrations of 1.5 to 3.1 mg/m^3 in one study occurred in the storage building where curing takes place (63, 64).

Triple superphosphate containing 40 to 49 percent P_2O_5 is prepared by the reaction between phosphoric acid and phosphate rock. This continuous operation involves drying the rock, grinding, treating the rock with concentrated phosphoric acid, and subsequent mixing and curing in large buildings. Silicon tetrafluoride, hydrogen fluoride, ammonia, ammonium chloride, and particulate are the principal air contaminants from these operations.

Since soluble and acidic fluorides can cause skin, eye, and respiratory irritation, adequate personal protective equipment must be made available, and showers and eye wash stations must be located in the work area.

Ammonium phosphate is manufactured by combining ammonia and phosphoric acid in a reactor vessel. The slightly acidic material is pumped to an ammoniator where additional ammonia is added, causing agglomeration. The agglomerate, diammonium phosphate, is dried, screened, and cooled for packaging. The principal air contaminants are fluorides and ammonia.

Ammonium nitrate is manufactured by the neutralization of nitric acid with gaseous ammonia. The product is concentrated to 95 percent NH_4NO_3, and a solid product called prill is produced by spraying the concentrated product in a tall tower and permitting it to solidify into small pellets during its fall. The product is then dried. The principal exposures to the worker are to ammonia and nitrogen oxides.

3.14 Food

The health hazards in the food industry are attributable to contact with possible biological hazards leading to infections from exposure to animal products or their waste; exposure to various chemicals used in food preparation and sanitation, which may cause respiratory or skin sensitization; and the hazards due to physical operating facilities, which may vary from high temperature processes including cooking, sterilization, and pasturization, to low temperature facilities such as cold meat cutting rooms or freezing lockers.

The biological hazard must be eliminated by close quality control on raw materials such as animal products, and by medical control of workers. The variety of chemicals used in preparation is expanding rapidly, and an inventory of exposures must be prepared.

A respiratory disease entity known as meat wrappers asthma has been identified and is presumed by the authors to be caused by airborne thermal degradation products from heat cutting, sealing, and labeling plastic meat wrapping materials (65). The air contaminants from the hot wire cutting of the plastic meat wrapping film can be eliminated by mechanical cutting or reduced by using a low temperature wire.

3.15 Garages

Automotive garage work involves a range of metal working operations including sheet metal fabrication, welding, painting, metal cleaning, and abrasive cleaning. These operations are discussed separately in Section 2.

The major hazard in service garages comes from the carbon monoxide emitted by internal combustion engines during the running of vehicles in no-load and dynamometer servicing operations. Control may be achieved by opening large service doors to achieve natural ventilation control, although the effectiveness of this measure varies, and during cold weather it is impractical. The second control is the use of a flexible exhaust hose discharging outdoors. Leakage around the connection between the tail pipe and the hose and back pressure due to the resistance of the hose makes the value of this control questionable. The best ventilation practice is a mechanical exhaust system with the exhaust volume varying from 100 to 400 cfm, depending on horsepower (6, 66).

Dilution ventilation is also needed to eliminate exhaust air contaminants due to vehicle movement in the garage. The ventilation standards for this problem are not as well developed; however it has been recommended that a minimum of 400 cfm per stall be exhausted. In our own experience it is rather commonplace to find service garage personnel exhibiting elevated carboxyhemoglobin levels because of significant carbon monoxide exposure.

An additional problem is commonly noted in repair garages that do body work. Lead or plastic fillers are used to fill dents or cracks. When solder is applied by torch to automobile bodies, there may be significant exposure to lead. Significant dust concentrations are easily generated when the solder is removed by a disk grinder. Rasping or scraping does not produce a hazardous amount of respirable dust, but it does present a housekeeping problem. Extensive body work with lead should be done in a properly designed, exhausted enclosure. If exposure is occasional and ventilation is not available, a suitable respirator should be worn.

The use of gasoline in open containers such as vats, pans, or buckets is hazardous and should be avoided. Proper degreasing techniques as described in Section 2.3 should be adopted. Sandblast cleaners for spark plugs, if operated for prolonged periods, should be checked for possible silica dust exposures.

With the gradual elimination of tetraethyl lead in gasoline, there is an increase in the aromatic content of this fuel. Initial studies have suggested this does not represent a hazard to garage personnel, but it should be considered in reviewing future garage operations.

3.16 Glass

Nearly every element of the periodic table has been utilized in modern glass technology, and the hazards associated with each have been experienced. The major ingredient of all glass, however, is still silica sand. The common glasses such as soda-lime-silica glass, lead-potash-silica glass, and borosilicate glass contain, in addition to silica sand, the following major constituents: limestone, soda ash, salt cake, lead oxide, boric acid, and crushed glass. Minor constituents include arsenic, antimony, fluoride salts, salts of chromium, cobalt, cadmium, selenium and nickel, fluorides, sodium silica fluoride, and rare earths.

Since the major component of each batch of glass is sand, this material would seem to present a potentially serious silicosis hazard. In most cases, however, washed sand is used, with a substantial portion of the fine particles removed. It is common to find that airborne dust from the mixed batch contains only from 1 to 5 percent free silica. Although silicosis is rare in modern glass plants, the methods of handling certain types of sand can still present a dust hazard. The manual unloading of dry sand from boxcars may produce dangerous quantities of fine silica dust. When wet washed sand is obtained in hopper cars and unloaded by gravity or pneumatically in a totally enclosed system with exhaust ventilation, there is minimal exposure. In modern plants, preblended batch materials are obtained in hopper cars, thereby eliminating dust exposure from in-plant handling and mixing.

In the manufacture of optical glass and certain decorative glasses, lead is an important source of employee exposure. Handling of this material, usually in the form of lead oxide, requires personal hygiene procedures and the use of local exhaust ventilation.

The other major constituents of glass do not normally present a health hazard, although dermatitis may occur.

The minor constituents have caused health effects, with arsenic the principal offender. Perforation of the nasal septum or severe skin effects due to exposure to arsenic and to highly alkaline constituents of the batch were common occurrences in the past. Modern methods of handling and ventilation control have eliminated most of this trouble.

The various types of glass manufactured in the modern glass industry are made by two processes: the pot process, and the more modern and now common tank method. The pot process now serves mainly for the manufacture of high quality glass, such as optical and mirror glass, and for small quantities of specialty glass. The pots vary in size up to 2 tons capacity, and in the past the manufacture of pots has been responsible for the greater portion of the silicosis in the glass industry from refractory dust.

Pot melting of glass necessarily introduces the hazard of hand shoveling and filling of the pots. Optical and specialty glasses frequently contain heavy metals such as lead, barium, and manganese. Close attention must be given to handling these toxic materials during the hand-filling process. With the introduction of the more efficient tank melting system, the hazards of the pot melting process are rapidly disappearing.

The tank process permits enclosed continuous feeding of batch ingredients, thereby reducing dust exposure at the batch end of the tank. The blocks and bricks used in the construction of the furnaces and tanks contain free silica. Silica brick contains tridymite as its principal constituent. When evaluating the dust exposure to such dusts, care should be taken to determine tridymite and cristobalite as well as quartz. In previous years furnace blocks and parts were cut to fit at the installation site. Now large, well-ventilated, mechanized shops are used to prefabricate refractory furnace parts, which are shipped to the furnace site for installation. Only occasional cutting should be done at the construction site under present-day methods.

Glass objects may be formed by blowing, pressing, casting, rolling, and drawing, and by a float method. After forming, all glass objects must undergo a process of annealing to reduce internal stresses in the formed object. This is accomplished in most cases by introducing the objects into long continuous annealing chambers called lehrs. Because of their size and the quantity of heat generated, the furnaces introduce a major heat problem.

After the glass has been formed and annealed, it is frequently finished by labeling, smoothing of rough edges, and the like. Glass grinding is done by wet processes, and abrasives such as silicon carbide are used. Polishing is accomplished by the use of revolving felt pads with rouge (iron oxide) as the agent. Glass dust itself is not toxic, since silica is in the combined or silicate form.

Abrasive blasting is sometimes done in enclosed exhausted cabinets with nonsiliceous abrasive materials. Application of decorative enamels by spraying or silk-screen processes introduces the possibility of exposure to solvent vapors, which must be controlled by exhaust ventilation.

In summation, the hazards in glass manufacture are principally associated with the handling of bulk materials in specialty or small-run pot production. Tank processes do

Table 23.14 Glass Additives

Process	Materials
Color	Salts of
	Chromium
	Cobalt
	Cadmium
	Manganese
	Nickel
	Selenium
Remove bubbles	Salts of
	Arsenic
	Antimony
Accelerate melting	Fluorine
	Calcium fluoride
	Sodium silicafluoride
Improve optical properties	Rare earth metals
	Thorium

not, in general, present hazardous dust concentrations. The major dust hazards are the quartz sand and special additives listed in Table 23.14.

The concentration of furnaces and the handling of the molten glass present a heat stress hazard in the industry, primarily because of radiation. This industry was one of the first to demonstrate the importance of reflective shielding to reduce heat stress.

The hazard from infrared radiation due to the viewing of incandescent glass warrants the wearing of eye protection on selected jobs.

3.17 Iron and Steel

Steel manufacturing, which involves almost all basic unit operations, presents one of the most diversified sets of occupational health problems of any industry (67). The modern plant or so-called integrated steel plant includes all operations from the initial handling of coal and ore to the loading of the finished product. The iron ore is reduced to provide pig iron in blast furnaces using coke as a fuel and reducing agent, and limestone as a flux. The pig iron is processed through open hearth furnaces, converters, electric furnaces, or basic oxygen furnaces to remove impurities and reduce the carbon content. Alloying metals are added when the "heat" is poured. The steel is poured in molds and the ingots are held at temperature in ovens called soaking pits before final processing in the mill. Table 23.15 lists the major air contaminants encountered in the industry.

The blast furnaces and converters may present a significant exposure to carbon monoxide, especially during maintenance operations. Concentrations should be monitored before maintenance operations commence, and self-contained breathing apparatus must be available for escape purposes.

Table 23.15 Major Air Contaminants in Iron and Steel Industry

Operation	Exposure
Dust	
Mining	Ore and coal dust
Ore sintering and pelletizing	Iron oxide
Coke ovens	Coke oven emissions
Refractory handling	Silica dust
Foundries	Silica sand
Metal fume	
Furnaces	Iron oxide
Scarfing operations	Iron oxide
Scrap preparation	Lead fume
Galvanizing	Flux fume, zinc
Leaded and ferromanganese steels	Lead and manganese fume
Gases and vapors	
Blast furnace	Fluorides, CO
Coking operation	CO, SO_2, H_2S
Welding	Ozone, oxides of nitrogen
Maintenance and cleaning motors	Solvent vapors
Mists	
Pickling	Sulfuric acid mist
Plating	Various
Spray painting	Lead paint spray mist

Major dust hazards may occur in sintering operations, and these require control by local exhaust ventilation. Metal fume exposures occur during oxygen injection at blast furnaces and in electric arc furnaces. These emissions call for ventilation control and air cleaning (6, 26).

The major silica hazard occurs during the installation of refractory brick materials containing high concentrations of quartz in the lining of furnaces and ovens.

A principal hazard in the industry is heat stress in coke oven operations, basic steel making, and in final mill operations. A major heat load is due to radiation from furnaces and the molten metal. The industry has utilized the following controls (see also Chapter 20.):

1. Protective shields for radiation.
2. Air conditioning of control stations, pulpits, and crane cabs.
3. Spot cooling of work sites.
4. Personal protective clothing such as aluminized garments equipped with vortex man coolers.

The noise hazards in this industry are pervasive and necessitate a variety of controls, including modification of equipment, mufflers for air exhaust and intakes, isolation and enclosures, and personal protective devices.

Lung cancer from coke oven operations has been associated with oven emissions, although the exact causative agent in that emission is not known. The coke oven worker is exposed to coal dust, particulate coke oven emissions with a significant benzene-soluble component, carbon monoxide, hydrogen sulfide, and sulfur dioxide. Heat stress is also a significant physical hazard from coke ovens.

By-product coke plants produce many valuable chemicals that must be carefully handled. Carbon monoxide, ammonia, benzene, carbon disulfide, and so on, are potentially harmful. Since most operations are in totally enclosed systems, the principal difficulty is unexpected leakage, resulting in high concentrations for brief periods. Plant maintenance, intelligent supervision, and thorough training are required to ensure safety with respect to chemicals.

A serious lead exposure occurs during the production of leaded steel, but control by local exhaust ventilation is possible. Special steels may contain nickel, bismuth, chromium, ferromanganese, tungsten, and molybdenum. Fluorides may be encountered in connection with certain iron ores.

The production of tin plate involves exposure to acids, and terneplate also involves lead. Zinc baths for galvanized steel have been found to be contaminated with lead in amounts sufficient to constitute a hazard. Iron and steel plants also may have plating, spray painting, and welding operations that could present hazards.

3.18 Leather

Anthrax is a much-cited biological hazard in the leather industry. This was a serious problem in the past, especially from skins imported from anthrax-infested areas, but regulations and improved methods of treating imported hides have done much to bring the exposure under control. Anthrax is usually acquired by contamination of wounds or abrasions by *Bacillus anthracis,* but the organism may be inhaled or ingested. Workers handling raw hides or skins should be included in a medical control program.

Dried animal hides received for processing have been treated with insecticides and possibly other materials such as salt to stabilize the hide and prevent decay. At the tanning plant, the hide is first treated with a disinfectant and a surfactant in a soak tank (soaking). Excess flesh is removed from the inside of the skin mechanically (fleshing), and the hides are then "limed" in milk of lime to loosen the epidermis and remove soluble protein and fat (unhairing). Various chemicals including sodium sulfide, sodium hydrogen sulfide, and dimethylamine may be added during "liming" to enhance the unhairing. Deliming is carried out in a solution of sulfurous acid or ammonium salts. Enzymes are added to remove undesirable constituents from the skin (bating). The hides are then pickled in solutions of sulfuric and hydrochloric acids to prepare the skin for tanning and 2-naphthol and *p*-nitrophenyl to prevent molding of the skin. These procedures are normally carried out in series in a large wooden tumbling barrel (68). A potential hazardous exposure to hydrogen sulfide seems likely from the addition of acid to sulfide solutions residual on the hides due to poor washing. Several deaths from hydrogen sulfide in leather plants have been described in European literature.

The tanning process is also conducted in a tumbling barrel using chromic acid, alkalies such as trisodium phosphate and borax, oxalic acid, formaldehyde, and natural and synthetic vegetable tanning minerals. In addition to exposure to these chemicals, the putrefaction of animal material may cause exposure to carbon monoxide and hydrogen sulfide.

The tanned hides then go through a series of finishing operations: pressing, splitting, shaving, sanding, buffing, finishing, waxing, oiling, and so on. Sanding and buffing result in an exposure to leather dust, and local exhaust ventilation must be provided for such operations. Certain machines may also result in a serious noise hazard.

To prevent the generation of carbon monoxide, the tanning solutions should be changed frequently. Hydrogen sulfide may be present as a product of putrefaction or as a result of the treatment with acid of hides containing sulfides. Good housekeeping must be stressed, and workers should be urged to minimize skin contact with hides and chemicals. Protective boots, aprons, and gloves should be worn, and adequate shower facilities must be available. All operators should be included in a periodic medical surveillance program. In the finishing plant a variety of surface coatings with associated volatile vehicles are encountered. The spray application of the finish and its drying must be performed where there is good local exhaust ventilation.

3.19 Lime

Calcium oxide or quicklime (CaO) is produced by calcining limestone ($CaCO_3$) in a vertical or rotary kiln. Water is added to quicklime in hydrators to form hydrated or slaked lime [$Ca(OH)_2$]. This material suspended in water is known as milk of lime. Dust exposures occur during all handling and transfer operations.

Protection of the eyes and the skin presents the major environmental control problem in the production of calcium oxide or quicklime. The material is of small particle size and is very irritating to mucous membranes and the moist skin. It combines with water with the evolution of heat to form calcium hydroxide, which is nearly as caustic as potassium hydroxide.

Air-slaked lime, which is almost 100 percent calcium carbonate, has mild causticity and usually attacks mucous membranes. If warm, it may cause dermatitis after prolonged exposure. Quicklime rarely affects the lungs, since it is so irritating to the upper respiratory tract that exposure for the necessary period does not occur. It rapidly produces coughing and sneezing, which limit further exposure.

Workmen at lime kilns may be exposed to dangerous concentrations of carbon monoxide, carbon dioxide, hydrogen sulfide, and arsine. The principal controls are local exhaust ventilation at lime hydration operations, conveying transfer points, and pulverizing and bagging stations. Eye and hand protection may also be required. As in all calcining operations, heat stress is a potential problem and must be handled by conventional controls.

3.20 Meat

The chief health problems for the meat processor are zoonotic disease contracted from diseased animals (e.g., anthrax, brucellosis, glanders, erysipelas, leptospirosis, tularemia, and Q fever) (69).

Because of processing demands, the work environment in these plants is usually characterized by high humidity and extreme temperature differences between work areas. Since sanitation is a production requirement, exposure to putrefaction gases such as carbon dioxide and hydrogen sulfide is not a problem.

Contact dermatitis is a common result of contact with such primary irritants as brine and various preservatives. Skin abrasions and cuts on the hand may furnish another mode of entry for pathological organisms.

3.21 Milling and Baking

Milling operations include grinding of various cereals and vegetables to produce fine flour. The major hazard is fire and explosion during grinding, conveying, collection, drying, and storage of the organic material (70). A series of respiratory problems including bronchial asthma, buccopharyngeal disorders, flour allergy, and chronic rhinitis are attributed to such dust exposures. Exposure to parasites in ground beans and certain cereals causes pruritis and papular skin lesions; some workers are allergic to certain flour molds. One must also consider insecticide residue in flour handling. Bleaching processes employ a variety of materials including dibenzoyl peroxide, and flour is enriched by various additives, including nicotinic acid and its amide.

In addition to an organic dust exposure, the baker experiences hot working conditions with highly variable temperatures in the workplace. In the preparation and use of powdered sugar icing, which is frequently mixed and applied by hand, sugar dermatitis occasionally results. Lard oil, when applied to pans by means of swabs, sometimes gives rise to skin irritation.

3.22 Paint

As discussed in Section 2.12, paint is a pigment suspended in a vehicle that may be oil, a varnish, or a natural or synthetic resin. Solvents are included in the system to adjust the viscosity of the paint for the specific application technique. The pigments provide opaqueness and the desired color. Numerous materials may be added to impart special properties of the paint. The common materials used in the preparation of paints are listed in Table 23.9.

The preparation of the paint is usually conducted in a multistory structure, and the materials flow by gravity from the initial blending stages on the top floor to the bulk dispensing and packaging on the ground floor. The pigments are first ground with the resin vehicle in a mill. This dispersed vehicle–pigment mix is dumped into a mixer with the solvent and other additives, mixed thoroughly, and finally filtered and dispensed into containers.

The principal hazard comes from exposure to dust during the bulk handling and dispensing of dry pigment in the initial milling operation. Later addition of vehicle or thinner, most of which are flammable, may result in a limited exposure to these organic vapors, depending on the integrity of the system. Primary control is by local exhaust ventilation at debagging stations, transfer locations, kettle loading, and dispensing locations. The use of silica and asbestos reinforcement and minerals must be carefully controlled during debagging. Good housekeeping and personal cleanliness are also necessary to minimize exposure and prevent dermatitis.

3.23 Petroleum

3.23.1 Exploration and Transportation

The occupational health problems encountered in exploration and extraction are associated with the severe environmental conditions under which drilling is conducted and the nature of the oil (71). Chemicals such as formaldehyde and hydrogen chloride may be pumped into the well during drilling. If the crude oil is a sour crude, hydrogen sulfide may be evolved. Other possible crude oil contaminants are vanadium and arsenic.

In the transport and storage of the oil, hydrogen sulfide may be encountered. The principal exposure occurs during the cleaning of shipboard and tank farm storage tanks. Tank cleaning procedures involving direct entry by workers may also constitute a hazard because of oxygen deficiency resulting from previous inerting procedures, rusting, and oxidation of organic coatings. Carbon monoxide may also be present in the inerting gas. In addition, depending on the characteristics of the product previously stored in the tanks, one may encounter hydrogen sulfide, metal carbonyls, arsenic, and tetraethyl lead. Detailed procedures on tank entry must be prepared for ventilation purging, environmental testing, and respiratory protection.

3.23.2 General Refinery Problems

Most petroleum products have low flash points; thus the principal hazards at refineries are those of fire and explosion. All operations are designed to prevent such catastrophes, but serious refinery fires nevertheless occur.

The principal exposures in refinery operations occur during shutdown, maintenance, and startup procedures. So-called plant turnarounds must be carefully scheduled and step-by-step procedures developed to ensure that operations are conducted safely. The exposures vary depending on the unit being maintained. Difficult control problems include abrasive blasting and welding in enclosed spaces. Asbestos insulation is still in use in refineries, and rip-out and repair of this material are short-term but potentially hazardous operations.

Because of the many pump and blowers, burners, fans, flares, and steam and air releases, a significant noise exposure may exist in certain parts of the refinery.

3.23.3 Specific Refinery Processes

One must review the basic processes to understand the sources of potential health hazards in refineries (72, 73). A modern refinery is a maze of furnaces, heat exchangers, pumps, tanks, fractionating columns, pipes, pipe fittings, and valves. However the refinery can be divided into a number of unit operations to allow the hygienist to make a reasonable inventory of health hazards.

The initial refinery process, crude distillation, involves separation of the oil into various fractions or "cuts," which have specific boiling temperature ranges. These fractions lack a definite chemical formula, being defined solely by their boiling ranges.

The oil is classified as sweet or sour depending on its hydrogen sulfide content. Sweet crudes contain less than 50 ppm H_2S, and sour crudes contain more than 50 ppm. The main hazard of the initial distillation is exposure to H_2S. Hydrogen sulfide can be recovered as elemental sulfur or burned to SO_2, depending on the amounts present and the economics of sulfur production.

A single distillation does not produce the desired quantities or quality of each product. Therefore less desirable products must be transformed to more desirable products by splitting, uniting, or rearranging the original molecular structures. This is done in a series of refining processes. An integrated refinery merely carries on these processes simultaneously.

The coverage of individual processing begins with a discussion of the highest boiling fractions and ends with the lowest boiling fractions.

Reduced Crude Fraction. The highest boiling fraction of crude oil is transferred to a vacuum distillation unit where it is further fractionated. Lubricants, waxes, asphalts, and heavy fuel oils are the end products. The principal hazard is skin contact, with resulting dermatoses. Asphalt, in particular, can be a photosensitizer. A high incidence of scrotal cancer was previously related to the wax preparation process, which has been changed from chilling and mechanical filtration to solvent extraction with toluene.

Heavy Oil Fraction. The heavy oil fraction is transferred to a catalytic cracker where larger molecules are converted to smaller molecules. This is one of the key processes at the refinery and permits doubling of gasoline yields.

The process is operated at about 500°C in the presence of a catalyst. There are many commercial catalysts of proprietary compositions, such as alumina-silicates. The catalysts used in this operation are usually nonhazardous.

The cracked product is next "sweetened" by converting mercaptans to disulfides by treatment with lead oxide, saturated caustic, cuprous chloride, or sodium hypochlorite. The hazards depend on the process being used.

Middle Distillate. The middle distillates and raw kerosene are treated in the hydrotreating plant. The organic sulfur compounds are converted to H_2S and are removed by burning to SO_2. In addition, hydrocarbons and diolefins are converted to more stable compounds.

The operation is carried out using high pressure hydrogen in the presence of cobalt and molybdenum catalysts and may lead to the formation of toxic metal carbonyl compounds.

Heavy Naphtha. The re-former produces high octane gasoline, blending components through a rearrangement of molecular structures to highly branched compounds. Thus the quality of the product is improved during this operation, not the quantity. The procedure is carried out at 480°C under high pressure in the presence of a catalyst.

The hydrogen by-product from this process is used in hydrotreating plants. The "wet gas" that is given off is processed further. "Wet gas" is a vapor containing high proportions of hydrocarbons that are recoverable as liquids.

Straight Run Naphtha. The lowest boiling fraction is first processed in the gas plant where liquid hydrocarbons found in the wet gas are separated from fuel gases, such as propane and butane. Some of the hydrocarbons are run straight through to the gasoline blending plant, but others must be carried through the alkylation process.

In alkylation, isobutane and butylenes or other light olefins are combined in the presence of a hydrogen fluoride catalyst. Exposure to hydrogen fluoride presents the main hazard of this process.

The final processing involves blending of the various materials to obtain a specific product. There may be exposure to tetraethyl lead where that material is still used and to hydrocarbons from leaks in storage tanks and pump and valve leakage.

Table 23.16 is a résumé of the major air contaminants encountered in refinery operations.

3.24 Pottery

The manufacture of tableware, bathroom fixtures, and industrial ceramics involves the preparation of raw materials and shaping the product by throwing, casting, and mechanical spreading in a mold, or dry pressing. The part is subsequently dried, shaped, and biscuit fired. A glaze is then applied by dipping or spraying, and the part is fired again. In the case of bathroom fixtures the glaze is applied and the part fired only once. Any decoration is applied after firing.

The principal hazard in this industry is pneumoconiosis due to exposure to silica dust. The principal exposures to dust occur during the crushing, screening, and preparation of the silica, feldspar, kaolin, and other materials, the secondary shaping of the part, and the spraying of the slip and the glazes. A major portion of the airborne dust is attributable to resuspension from equipment, floors, and clothing. The control measures for silica exposure include reducing the percentage of quartz in the raw

Table 23.16 Refinery Air Contaminants

Principal Air Contaminants	Sources
Hydrocarbon vapors	Transfer and loading operations
	Storage tanks
	Flares
	Cracking unit regeneration
	Boilers
	Pumps, valves
	Cooling towers
	Treating operations
Sulfur dioxide	Boilers
	Cracking unit regeneration
	Flares
	Treating operations
Carbon monoxide	Cracking unit regeneration
	Flares
	Boilers
Nitrogen dioxide	Flares
	Boilers
Hydrogen sulfide	Sour crudes
	Liquid waste
	Pumps
	Hydrocracker
	Hydrogenation
Particulates	Cracking unit regeneration

materials, substituting nonsiliceous material, maintaining all materials in a wet state, and providing local exhaust ventilation, good housekeeping, and personal protective clothing.

The preparation and application of glazes may present a potential hazard if glazes based on lead and other heavy metals are used. In the United Kingdom, the hazard from lead has been reduced by specifying that the glazes may not contain more than 5 percent soluble lead. This material specification was backed by an aggressive program of good housekeeping, clothing changes; locker and shower facilities were provided, smoking and eating at the workplace were prohibited, and local exhaust ventilation was installed at the application locations.

3.25 Plastics

There are more than a hundred polymers in production and many hundreds of copolymer systems, however the polymers listed in Table 23.17 represent the bulk of commonly used materials.

Since the plastic production often uses petroleum intermediates as raw material, as Table 23.17 indicates, the production facilities may be located at or near a refinery. The production operations are similar to refinery activities in that the processes are normally conducted in a closed system. In many cases the process is continuous, although batch operations are still common. The large automated plants require relatively few operating personnel.

The plant personnel must tour the plant to check on pump operation, make sight glass readings, and take intermediate and final product samples. The major exposures occurring during such inspections are usually due to fugitive leaks from piping, valves, drains, and pump seals. Maintenance personnel may encounter acute exposure situations.

Many of the polymer production operations are quite straightforward, neither employing hazardous materials nor resulting in the formation of such materials. The polyolefin plastics are in this "safe" category. In other resin production systems, such as phenol-formaldehyde, the raw materials have known toxicity; however the facilities are usually totally enclosed and do not present a source of hazardous exposure. In the case of polyvinyl chloride the principal exposure to the carcinogenic monomer occurs in the polymer production facility.

Table 23.17 Chemicals Used in Plastics Production

	Principal Chemicals Used in Production of	
Polymer/Resin	Monomer	Polymer or Copolymer
Acrylic resins	Hydrocyanic acid	Various catalysts
	Acetone	
	Methyl or isopropyl alcohol	
Alkyd resins	Glycerol	Phenol or urea formaldehyde
	Phthalic anhydride	Aliphatic hydrocarbons
	Maleic anhydride	Styrene
	Linseed, tung, castor oil	Phenols
		Formaldehyde
	Litharge	
	Sodium hydroxide	
	Various solvents	
Epoxy resins	Allyl chloride	Aliphatic amines
	Epichorohydrin	Organic acids and anhydrides
	Diglycidyl ether	Hydroxy compounds
		Polyamides
Phenolic resins (or substituted phenol or aldehyde)	Ammonia or lime	Fillers such as asbestos, various silicate minerals, and cellulose
	Hexamethylenetetramine	
	Hydrazine or various amines	
	Phenol	
	Formaldehyde	
	Methyl alcohol	

Table 23.17 Continued

Polymer/Resin	Principal Chemicals Used in Production of	
	Monomer	Polymer or Copolymer
Amino resins	Ammonia Urea Formaldehyde Methyl alcohol	Same as for phenolic resins
Polyurethane	Ethylene oxide Ethylene glycols Propylene oxide	Toluene diisocyanate (TDI) Hexamethylene diisocyanate (HDI) Diphenylmethane diisocyanate (MDI) Polymethylene polyphenylisocyanate (PAPI) 4,4-methylene-bis-2-chloraniline (MOCA) Organic tin compounds Hydrocarbon blowing agents Polyglycols
Polyamides		Hexamethylene diamine Adipic acid Caprolactum Diphenyl ether–diphenyl mixture
Polyesters	Propylene glycol Maleic anhydride Styrene Hydroquinone or *t*-butyl catechol	Organic peroxides Reinforcement such as fiber glass Dimethylaniline
Polyolefins	Ethylene Propylene Isobutylene	Polyethylene Polypropylene Polyisobutylene Catalysts Heavy metals Boron trifluoride Ammonium chloride Aluminum alkyls
Polyvinyl chloride	Hydrogen chloride Acetylene or ethylene dichloride Chlorine Vinyl chloride	Vinyl chloride Benzoyl peroxide Dioctyl phthalate
Polystyrene	Benzene Styrene	Styrene

There may be a potential problem caused by residual monomer in the polymer, as was true of vinyl chloride. Up to the 1970s this exposure was significant, but stripping techniques have been introduced to minimize the presence of the unreacted monomer. The fabricator should be informed of the concentration of the residual monomer in cases of monomer having known toxicity, to allow fabrication plants handling the polymer to provide adequate controls.

A principal problem with certain resins such as the epoxies is skin and respiratory sensitization. In the epoxy system the vapor pressure of the resin is very low; thus the resin itself usually does not present an airborne hazard. The curing agent, however, frequently causes dermatitis and sensitization. In the case of polyurethane systems, the isocyanates used as catalysts produce an asthmalike condition.

The production of many low toxicity resins involves the addition of various plasticizers, antioxidants, and stabilizers that are physiologically active and may require control (74). The thermal stability of the parent polymer and the additives are of importance in considering the application of the product (75). Fire retardant components are one of the more recent additives whose toxicity must be considered.

As in the case of refinery processing, a review of the health hazards of plastics production must be based on a flow diagram of the process and data on the raw materials. The initial, intermediate, and final product releases to the atmosphere must be taken into account. Maintenance operations, especially on pumps, reaction vessels, and valves, may offer acute exposures, and specific instructions must be proposed to protect personnel.

3.26 Pulp and Paper

Pulp is produced both by mechanical and chemical processes. The chemical methods produce more than 80 percent of the pulp used today. In the kraft or sulfate process described in Table 23.18, chipped wood is digested with steam in tanks using a solution of sodium sulfide and sodium hydroxide (white liquor). Relief gases are periodically vented from the digester to relieve the pressure buildup. When the digestion is complete,

Table 23.18 Kraft (Sulfate) Pulp Process

Process Step	Formula
1. White liquor	$NaS_2 + NaOH$
2. Black liquor	$NaS_2 + NaOH$ + dissolved lignin (15% solids) + Na_2SO_4
3. Products of combustion of recovery furnace	$NaS_2 + NaCO_3$
4. Green liquor smelt tank	$NaS_2 + NaCO_3$ in H_2O
5. Causticizer	$NaS_2 + Na_2CO_3 + Ca(OH)_2$ (quicklime) $\rightarrow NaS_2 + NaOH + CaCO_3 \downarrow$
6. Lime kiln	$CaCO_3 \xrightarrow{\Delta} CaO + CO_2$
7. Slaked	$CaO + H2O \rightarrow Ca(OH)_2$ (for causticizer)

the load is dumped to the blow tank and the gases vent from the pulp and digestion liquid. The spent cooking liquid (black liquor) is drained off, and the pulp is washed, screened, and bleached. The chemicals are recovered from the spent liquor by concentrating it in multiple-effect evaporators. Salt cake is added, and the mixture is sprayed into the recovery furnace; here water is removed, the remaining liquor is burned, and the chemicals are recovered. The chemicals are dissolved in water in the smelt tank, and quicklime is added to convert the sodium carbonate to sodium hydroxide. The calcium carbonate thus formed is converted to calcium oxide in the lime kiln. This product is slaked with water to produce calcium hydroxide for the causticizer (76).

The principal exposure to operators occurs when the bottom of the digester is opened and the contents are dumped. The released gases include hydrogen sulfide, methyl mercaptan, dimethyl sulfide, dimethyl disulfide, and sulfur dioxide. The effluent from the recovery furnaces includes organic mercaptans and sulfides, hydrogen sulfide, and sulfur dioxide.

The most significant effort at air pollution control has consisted of the oxidation of the black liquor before the multiple-effect evaporation. In this process the sulfur compounds are oxidized to produce less volatile materials. Incineration of sulfur off-gases has also been tried, by collecting the gases in a gas holder and burning them in the furnace. Other air pollution control techniques are in use that also reduce exposure of workers.

The soda pulp process is similar to the sulfate technique except that sodium carbonate is used for chemical makeup in the furnace. The digestion is carried out with a sodium hydroxide cooking liquor.

In the sulfite process the digester liquor is an aqueous solution of sulfurous acid mixed with lime or other base to form bisulfites. The sulfur dioxide is obtained either as a compressed gas or from the burning of sulfur or the roasting of pyrite ores. The relief gas in this process contains high concentrations of sulfur dioxide, which must be recovered for economical operation. This is accomplished by separators and coolers.

Bleaching of the pulp is usually accomplished with chlorine, followed by extraction with sodium hydroxide, then calcium or sodium hypochlorite, and finally a chlorine dioxide treatment. Chloride hydrate may form when gaseous chlorine enters the vat and is carried to the surface, where it releases chlorine into the atmosphere. As a rule, however, the exposure to chlorine is not difficult to control by local exhaust ventilation.

Paper is coated by coating machines of various types, and the materials used include clay, mica, talc, casein, soda ash, dyes, plastics, gums, varnishes, linseed oil, organic solvents, and plastics. The principal exposures arising from these operations involve (1) acrolein and other aldehydes resulting from the atmospheric oxidation of linseed oil, and (2) solvent vapors from the coating and subsequent drying of the paper. When the coating and drying are done in air conditioned rooms, the environmental control problems become difficult.

Lime exposure may be excessive in both the sulfate and soda processes during handling of the lime. The digestion pit and the first washing cycle should be exhausted to eliminate sulfur dioxide, hydrogen sulfide, and mercaptans. A serious sulfur dioxide exposure may occur throughout the entire sulfate process. This gas can be controlled by ventilation at the sulfur burners, thereby achieving negative pressure on the acid towers, ventilation on the digesters, and remote operation of blowdown valves.

3.27 Rayon

In the rayon-making process, pads or sheets of cellulose primarily prepared from wood pulp are steeped in sodium hydroxide to form a "soda" cellulose. The sheet material is shredded, aged, and mixed with carbon disulfide in an xanthating churn. The cellulose xanthate, in solution in alkali, is a brown syrupy liquid known as viscose. After filtration, aging, and deaeration, the viscose is forced through small holes in a nozzle or "spinerette" submerged in a sulfuric acid bath. The stream emitted from this spinning operation contacts the bath, and the cellulose is regenerated to form a continuous fiber. Tension on the fiber is established using two rollers called "godets," which operate at slightly different speeds. This process orients the yarn fibers in position parallel to the yarn axis. The yarn is usually chopped into short elements at this point for additional processing.

The principal hazard in this industry is the exposure to carbon disulfide in the xanthation, spinning, and "godet" and cutting house operations. The xanthation process usually can be controlled, since it is an enclosed operation. The open spinning baths release carbon disulfide and hydrogen sulfide, and operations should be monitored and ventilation control applied. The stretching and processing of the fiber at the "godets" is a principal exposure point because carbon disulfide is released from the fiber. Moreover, if the continuous fiber breaks, the fiber or "tow" must be pulled manually, resulting in a serious carbon disulfide exposure; thus this operation must be controlled by local exhaust ventilation.

The complete range of conventional controls must be employed in the rayon industry to minimize exposure to carbon disulfide. Work practices must be carefully defined; housekeeping and handling procedures must be encouraged to reduce the deposits of waste viscose, bath solution, and tow on the floor; and it must be recognized that effective ventilation controls on spinning and cutter house operations are essential. Medical control by periodic examination of workers and biological monitoring is necessary.

3.28 Rubber

The first man-made rigid polymer, natural rubber, has been joined by at least two dozen synthetic rubber polymers used in a variety of industrial applications (77, 78). The commonly used rubber polymers and copolymers are given in Table 23.19. Many applications require a blending of natural and synthetic rubbers, to take advantage of the unique characteristics of each. This is the case in tire manufacture, which represents the largest single rubber application. Natural rubber is converted at its source to dry rubber or to a latex concentrate. In either case a preservative is added, such as

Table 23.19 Types of Rubbers in Order of Appearance (78)

Natural rubber
Polysulfide polymers
Polychloroprene
Nitrile rubber
Styrene-butadiene rubber
Butyl rubber
Polybutadiene
Silicone rubber
Acrylic rubber
Chlorosulfonated polyethylene
Polyurethanes
cis-Polybutadiene
cis-Polyisoprene
Fluorine-containing elastomers
Epichlorohydrin elastomers
Ethylene-propylenediene elastomers

ammonia, formaldehyde, or sodium sulfite. The material is shipped in bales or in barrels. The synthetic polymers are received in solid bale form.

A variety of rubber processing chemicals are presently in use, to permit fabrication into finished product and to ensure specific properties for the product. Usually organic materials, these chemicals are added in relatively small quantities to the rubber stock formula. The chemicals and production facilities employed in processing many of these rubber materials are similar and can be discussed in general fashion.

3.28.1 Materials

Vulcanizing Agents. Still the most important vulcanizing agent, sulfur is used either as elemental sulfur or in one of many organic forms. The common vulcanizing materials appear in Table 23.20.

Table 23.20 Rubber Vulcanizing Agents (78)

Tetramethylthiuram disulfide
Tetrathiuram disulfide
Dipentamethylene thiuram tetrasulfide
4,4^1-Dithiodimorpholine
Selenium diethyldithiocarbamate
Aliphatic polysulfide polymer
Alkylphenol disulfides

Table 23.21 Commercial Accelerators (78)

Aldehyde–amine reaction products
Arylguanidines
Dithiocarbonates
Thiuram sulfides
Thiazoles
Sulfenamides
Xanthates
Thioureas

Accelerators. Since the early 1900s chemicals have been added to rubber systems to hasten vulcanization. Initially inorganic lead compounds were tried, then aniline, and finally a series of various organic compounds. The principal accelerators in current use are listed in Table 23.21. At present, the thiazole accelerators such as benzothiazole sulfenamides are most common.

Activators. It is common to use zinc oxide and fatty acid in conjunction with accelerators to achieve a given property. Other activators such as litharge, magnesium oxide, amines, and amine soaps are also employed.

Antioxidants. Stabilizers are designed to protect the polymer during extended storage before its end use in manufacturing, whereas antioxidants are designed to protect the finished product. The important antioxidants are arylamines and phenols, as given in Table 23.22.

Table 23.22 Commercial Antioxidants (78)

Arylamines
Aldehyde-amines
Aldehyde-imines
Ketone-amines
p-Phenylenediamines
Diarylamines
Alkylated dairylamines
Ketone-diarylamines
Phenols
Substituted phenols
Alkylated bisphenols
Substituted hydroquinones
Thiobisphenols

Antiozonants. The principal antiozonants are symmetrical *p*-phenylenediamine, unsymmetrical *p*-phenylenediamines, dihydroquinolines, and dithiocarbonate metal salts. Paraffin and microcrystalline waxes are also used in rubber components.

Plasticizers. The viscosity, therefore the workability, of the polymers can be improved by adding organic lubricants or physical softeners to the rubber. These additives may include coal tar, petroleum, ester plasticizers, liquid rubbers, fats and oils, and synthetic resins.

Pigments. Pigments in rubber occasionally have as their principal contribution the addition of color, but usually they are applied as reinforcing pigments, fillers, or extenders. The common pigments are carbon black, zinc oxide, clay, and silicates.

Processes. The processing techniques are similar throughout the rubber industry. It is common to have a defined processing area where weighing and, initially, processing are conducted. A normal tread tire stock recipe (Table 23.23) gives an idea of the quantities of materials involved. In some cases the bulk natural or synthetic polymer must be worked in a breakdown or mastication mill to make the stock more flexible and easy to work. The various rubber chemicals specified in the formula are then weighed out for a specific batch size. This operation involves opening bagged and barrelled material and placing it in hoppers for weighing. Since most of the compounds are solid and granular, this usually is a dusty operation, and weigh stations should be exhausted. Where possible, small quantities of chemicals should be prepackaged in plastic bags that can be placed directly in the batch, thereby eliminating a dust exposure in emptying bag contents.

The individual components are either mixed in an open two-roll mill or an internal mixer such as a Banbury. In either case dust is released during initial mixing. Standard

Table 23.23 Natural Rubber Tire Tread Compounds (78)

Material	Parts
Natural rubber	100
Sulfur	3
Accelerator	1
Zinc oxide	5
Stearic acid	1
Antioxidant	1
Softener	5
Pigment	50
Total	166

local exhaust designs are available for both types of equipment, and such apparatus should be installed (6). Of the two processes, better control can be established on the Banbury, since a more effective enclosure can be fabricated.

The rubber batch is processed from the Banbury to a drop mill or from the mixing mill to a sheeting mill. During this transfer the hot stock may release volatile fractions of oils, high vapor pressure organic components, or possibly degradation products of the worked material. If biocidal additives are in the formula, they may be released. At the drop mill the stock is blended, sheeted, cooled, and cut for racking.

At this point the stock is usually dusted or dipped in a talc slurry or other material to reduce sticking. If talc or soapstone is used, it should be analyzed to ensure that it is a nonasbestos form and has negligible free silica. The wet slurry process is obviously the most effective way to add the talc, since the level of dustiness is much lower than dry dusting.

This batch can include the vulcanizing agent or, if the rubber is to be stored, a master batch can be produced which includes all components except a part or all of the curing agent. The basic rubber stock is then ready for a variety of processes. Specific shapes, such as tire treads, may be formed by extrusion. Dimensional rubber sheeting may be produced using a multiroll calender, which smears the rubber into a fabric to make reinforced rubber sheets such as tire ply. If materials are "layed up" for sheet fabrication, as in the case of tarpaulins or clothing, exposure to rubber cement may occur. The type of solvent depends on the application. Exhausted work benches are usually required for these operations (6). At one time the use of benzene was common and constituted a significant hazard, but white gasoline-type solvents are now commonly substituted for aromatic hydrocarbons. In recent years the aromatic content of white gasoline has been increased, and this factor should be monitored by air sampling.

The assembled products usually require curing with heat. These procedures vary and may present specific hazards. Tires are cured in molds, with the aluminum molds heating the outside and water or steam-filled bladders heating the inside of the tire. Large conveyor belts are cured in flat, steam-heated presses. Water hose is cured by extruding a lead sheath over the hose and directing steam inside to cure the part. Continuous bath and drum curing is also conducted. Certain problems are common to all curing operations: (1) release of curing fume, whose composition and concentration depend on the rubber stock and the curing temperature, (2) heat stress due to the steam release and convective load, and (3) noise exposure from air and steam release.

3.29 Shipbuilding and Repair

The modern shipyard is a complex facility incorporating machining, welding, founding, painting, electroplating, abrasive blasting, and electronic repair operations. The control of occupational health problems is complicated because the activities are not confined to the shops but are often carried out on shipboard.

Before a vessel enters a drydock for construction work, repairs, or alterations of any kind, all tanks, compartments, or lines that have contained flammable liquids should be cleaned and freed of flammable vapor to comply with the code of the National Fire Protection Association. The atmosphere in all unventilated areas or compartments should be checked for harmful or flammable gases and for oxygen deficiency by an industrial hygienist or qualified marine chemist. Each compartment should be examined before workmen are permitted to enter.

Tankers that have carried gasoline or volatile crude oils require periodic checks even after a "gas-free" status has been established. Rust on the bulkheads or decks of compartments may continue to dissipate flammable vapor. The pumping of ballast also may introduce flammable vapor from some inaccessible part of the pipelines or storage tanks. When work is conducted in tanks used for the transport of gasoline, the lead exposure involved in welding or cutting operations on rust-coated surfaces must be evaluated.

Repairs on pipelines should be undertaken only after all "hot work" on the hull and in tanks and compartments has been completed. Heat must never be applied to any closed line or section of line; all lines must be opened by cold operation for examination; and if any welding or torch cutting is to be done on a line that may contain flammable materials, air should be blown through it before and during the operation.

The most widespread exposures in shipbuilding and repair are those connected with welding and flame cutting. There are many opportunities for welding in confined spaces on a ship, areas into which a person must crawl and where there is no ventilation except what is supplied mechanically. Under such circumstances the air concentration of nitrogen dioxide produced by a gas torch can reach fatal levels in a matter of minutes in the absence of ventilation. Properly distributed mechanical ventilation is the control method of choice, complemented by supplied-air respirators.

In cutting, burning, or welding operations that involve a potential toxic metal fume exposure, as in cutting galvanized parts or lead paint surfaces during shipbreaking, ventilation control must be provided.

Spray painting of the tanks and compartments requires not only control of inhalation exposures to thinners, oils, and pigments, but also prevention of fire and explosion. Control can be accomplished by the application of mechanical air supply and exhaust. Where necessary, this can be supplemented by air-line or air-purifying respirators, depending on the nature and amount of air contamination. Painting of the hull poses problems not only of respiratory protection but of visibility, because uncontrolled paint mist quickly covers goggles or face shields. Where natural or artificially induced air movement cannot be used to advantage to prevent the inhalation of paint mists, spray nozzles mounted on long pipes have been used; but brush painting has been found to be the most satisfactory answer to the control problem in many instances.

The compartments of floating drydocks should be tested periodically for flammable materials, and vents should be protected by flashback arresters.

To be able to evaluate the hazardous nature of cargoes and to establish the proper precautions to be exercised in their handling requires a rather broad understanding of industrial toxicology. Oxygen deficiency, resulting from fermentation, dry ice refrigeration, or displacement of air by gases other than carbon dioxide, has probably caused more fatalities on cargo ships than any atmospheric contamination except explosions of flammable vapors. Skin irritation from cargoes is also common.

3.30 Smelting and Refining

The processing of lead includes all common smelting and refining techniques. The crushed and ground ore, which may contain 6 to 10 percent lead, is concentrated using differential flotation. The concentrate is mixed with limestone, silica sand, and iron ore, and is pelletized. The pelletized product is sintered, and in the process the sulfur content is reduced by oxidization to sulfur dioxide. The sulfur dioxide gas is recovered by an acid plant. Sinter and coke are charged to a blast furnace for smelting, and the lead is reduced to metallic lead. The molten lead and slag are run off, and the two materials are separated by gravity. The slag, which is rich in zinc that was present in the ore, is removed to the zinc plant. The molten lead bullion is cooled, the slag, matte, and speiss are removed for metal recovery, and the bullion, purified by the addition of sulfur that forms a copper sulfide matte, is sent to a plant for recovery of the copper. The lead bullion is cast in blocks for additional refining. In the electrolytic refining process, precious metals and impurities are removed.

The second step is the refining of the lead produced by the smelting operation. There are several processes available for refining the bullion. In one process the bullion is charged to a reverberatory furnace in an oxidizing atmosphere, and arsenic, antimony, and tin are removed. Zinc and silver are recovered by other techniques. In another, lead is refined electrolytically or by using molten sodium hydroxide and sodium nitrate as a substitute for furnace softening. Silver and gold are recovered from the zinc slag.

Copper is extracted from its ore by smelting and is then refined electrolytically. Blister copper slabs become the anodes and pure copper slabs the cathodes, in an electrolytic bath of copper sulfate and sulfuric acid. Acid mist is generated by the release of hydrogen at the cathode. A slime containing selenium, tellurium, gold, and silver is deposited at the bottom of the tank. The zinc oxide from the furnace operation is refined electrolytically.

Industrial hygiene hazards are for the most part common to similar operations, irrespective of the metal being processed. In the handling of ore, crushing and grinding create a dust hazard that must be controlled by suitable local exhaust ventilation and the use of water for dust suppression where possible. These operations can also constitute severe noise hazards. On the other hand; the concentration by flotation process creates little dust exposure, and few workers are involved.

Sulfuric acid is used in leaching the copper from the ore, and since this is a wet process, there is no dust exposure. A sulfuric acid mist may exist however, requiring control by local exhaust ventilation. The acid dissolves out the CuO. This process is not used with a sulfide ore, since the acid leaching of such ores would generate hydrogen sulfide.

The roasting of ores at high temperature to release the sulfur as SO_2 results in an exposure in the workplace. The hazards in roasting operations are exposure to lead, copper, and zinc fumes, carbon monoxide, sulfur dioxide, and heat stress. Arsenic is also released in this operation. The roasting operations can produce the same exposure as smelting.

The major exposures in this industry are dust (selenium, tellurium, cadmium, arsenic) metal fumes (lead, zinc, copper, cadmium), arsenical compounds as gases (AsH_3) and particulates, heat stress, noise, and sulfur dioxide. The principal controls are ventilation, water for wetting down, use of wet processes, and personal protective equipment.

3.31 Quarrying

Quarrying refers to the open pit removal of mineral products. The well-known operations involve the bulk removal of such common materials as limestone, clay, and gravel, and dimensional stone products such as granite and marble. In the first case the material can be removed by conventional earthmoving equipment without major occupational health hazards. The principal hazards occur during quarrying operations that require preparatory drilling and blasting and in some cases preliminary crushing at the quarry site (79). It is not uncommon to find dry drilling proceeding without dust control in the belief that since this is an outside job, dust concentrations are negligible. At a minimum, all drilling should be conducted wet. The blasting operations are commonly done with ammonium nitrate–fuel oils (an-fo) and do not result in a significant exposure in open pits.

In the early quarrying of dimensional stone, the material was removed by drilling and blasting operations. This was quite wasteful of stock, and a variety of quarrying techniques have been developed, some of which present occupational health hazards. The degree of hazard in quarrying dimensional stone also depends on the characteristics of the parent rock and, most important, on its free silica content. In the early 1900s core drilling was used to block out an island of granite for subsequent removal. This procedure has been replaced by flame cutting using a fuel oil–air burner that cuts through the stone by sloughing it off. This process results in a serious noise hazard and exposure to a particulate cloud consisting of a crystalline respirable dust, a submicrometer rock fume, and a fused micrometer sized aerosol (80). After this cut has been made, the island of stone is cut into slabs or blocks using a convoluted wire saw that carries an abrasive such as silicon carbide to the cut. This operation does not present a health hazard, although erecting the cutting towers can be dangerous. In-quarry cutting provides a semifinished product that can be removed from the quarry using lifting eyes inserted into full holes drilled in the individual slabs or blocks. If this drilling is done dry, a significant dust exposure may occur.

The dimensional stone is usually processed at a mill located close to the quarry. The production techniques vary, but a consideration of architectural products covers most of the common hazards. In small job shop operations the quarry stone is cut to shape by a circular or pit saw using steel shot as an abrasive. Dimensional cuts can be made by diamond saws. Small wire saws are also used for dimensional cutting. Until recently, granite road curbing was cut by first drilling with gang drills, then splitting the stone with wedges. Now high production of such items as curbstone is done by guillotine splitters. All these operations may be dusty, although the saw cutting, wire saw, and gang drilling can usually be controlled by wet methods. Guillotine cutting is done dry, and the work area must be equipped with local exhaust ventilation.

The finishing operation in the mills includes surfacing with pneumatic tools, small fuel oil–oxygen burners, and polishing tables using various wet abrasives. Sculpting is done by burners, pneumatic tools, and abrasive blasting. The hazard varies with the quartz content of the parent rock. Most of these finishing operations require excellent local ventilation control, and effective guidelines have been proposed. Flame surfacing has not been adequately studied and warrants monitoring. Polishing does not require ventilation control.

Bulk and dimensional stone quarries have a crushing plant for preparation of the final product in the case of a bulk quarry or the recovery of scrap in a dimensional stone quarry. The crushers, transfer points, and screening operations all require ventilation control. The dustiest operation is usually bagging. Modern bagging equipment now available offers integral hooding.

The present ventilation standards for granular material transfer points are deficient; however a proposed induced flow calculation method provides improved control (81).

In evaluating exposures in quarries, it is extremely important that personal monitoring be conducted; experience has shown little relationship between fixed location and personal monitoring. In siliceous dust such as that encountered in granite and slate quarries, the silica exposure should be measured by respirable mass determinations on personal air samples. It is not uncommon for the parent rock to contain 30 percent free silica and the respirable sample only 10 percent free silica.

3.32 Underground Mining

A great variety of minerals are mined, both underground and in surface or open cast mines. This discussion is limited to underground mining operations. The health hazards include dust exposures to the mineral being recovered and its associated rock, natural and man-made gases and vapors, and a gamut of physical hazards. The subject is too great to be covered in depth; however a variety of common occupational problems arise in underground mining operations.

A principal problem is usually associated with the mineral being extracted or the materials associated with the ore body. In the case of asbestos and mercury, the recovered material may present the major hazard, whereas the high concentrations of quartz in hard rock mining for lead and zinc ore present the major dust hazard in those mines. One should know the complete geological characteristic of the ore body, not merely the main mineral constituents, to be able to evaluate this dust exposure properly.

Coal mining presents special dust hazards when high energy mining methods are employed. The conventional mining procedures of cutting, drilling, mining, and loading the coal have been generally replaced by more energy intensive high production systems. A common coal mining method in the United States is the use of a continuous miner, which accomplishes all the foregoing operations with a single piece of equipment. Long wall mining is popular in the United Kingdom, and to a lesser degree in the United States, because of differences in coal seam geometry. These mining methods generate considerable dust and warrant special dust control techniques. In the open hearings on the Coal Mine Safety and Health Act of 1969 it was claimed that dust control technology in coal mines would not be adequate to achieve the proposed standard at 5 mg/m^3. However 6 years later, tremendous strides in dust control in coal mines have been made, and most working faces meet the early standard.

A number of exposures to mine gases and vapors are listed in Table 23.24. The natural gaseous emission in coal mines and in certain hard rock mines are methane, carbon dioxide, and nitrogen. Because of its wide flammability range (5 to 15 percent), methane presents a major hazard. Firedamp explosions in coal mines can initiate subsequent violent dust explosions, accompanied by the release of carbon monoxide.

In addition to natural sources, man's underground activity generates other toxic gases and vapors. Straight dynamite (100 percent nitroglycerin) is not employed underground, but a blend of dynamite and ammonium nitrate or other explosives compounded to minimize the release of toxic gases is in common use. Recent practice is to use ammonium nitrate prills saturated with fuel oil and fired with dynamite caps. The principal gaseous contaminants continue to be carbon monoxide and nitrogen dioxide.

The widespread use of diesel engines underground contributes exhaust gases containing aldehydes, nitrogen dioxide, and carbon monoxide. Scrubbers are frequently used on such equipment to remove aldehydes. Urethane foam systems for sealing brattice and leaks have been introduced underground, and this practice results in exposure to diphenyl methane diisocyanate (MDI).

Oxygen deficiency, especially in old workings when a reducing ore exists or organic material is decomposing, can be a major hazard. Significant radon daughter concentrations occur not only in uranium mines but in other underground works, and excellent ventilation controls are required in such cases. In mineral mines high radon concentrations are encountered in areas that have not been worked recently. The physical hazards in mining include temperature and humidity extremes, poor lighting, and noise and vibration. The rock temperatures underground increase at a rate of 1°C for each 100 meters of depth. The high rock temperature in deep mines and the extensive use of

Table 23.24 Common Names of Mine Gases

Type of Gas	Common Name
Methane	Firedamp
Carbon monoxide	Whitedamp
Hydrogen sulfide	Stinkdamp
Oxygen deficiency	Blackdamp
Gases from explosives	Afterdamp

water for dust suppression may cause a serious heat stress problem. It is not uncommon to place air conditioning plants underground to cool the air delivered to working faces. Personal cooling systems have also been developed.

The widespread use of percussion drills and other compressed air tools presents a serious noise hazard to the underground miner. This problem can be fully controlled only by equipment redesign, prompted by rigorous purchasing specifications. Serious loss of hearing has been noted in mining populations. A study of vibration disease in this population has not been attempted to date.

Certain necessary controls of occupational health hazards have been cited. The foremost hazard, dust exposure to the toxic mineral dust or silica exposure from the host rock, calls for special attention. Wet methods are a principal control, and water is used with percussion drills to infuse working faces at the cutting head of continuous miners and to wet down loose rock. Wetting agents are used to improve the effectiveness of the wet procedures. To minimize dust explosions from settled coal dust, an inert dust such as limestone is employed to dust all surfaces. Equipment redesign, especially of cutting tools, can be helpful in reducing dustiness. The effects of dust from blasting is minimized by conducting this operation when the men are not in the area. Wetting down with water before and after blasting is also important.

The principal control is, of course, ventilation. Occasionally local exhaust with air cleaning and direct return of cleaned air can be accomplished underground at specific dusty locations, such as crushers and conveyor transfer points. For the most part, however, dilution ventilation is necessary for both methane and dust control.

3.33 Woodworking

The major domestic woods do not present a health hazard; however several local woods and a large number of tropical woods (Table 23.25) used in the production of wood products are known to be dermatologically active and allergens. The common complaints are skin irritation, skin allergy, conjunctivitis, asthma, and irritation of the upper respiratory tract.

Table 23.25 List of Common Toxic Woods

Boxwood
Cashew
Mahogany
Red cedar
Rosewood
Satinwood
Teak
Yew

The principal exposures occur either in the harvesting of the wood products or in the mill operations. In the latter case control can usually be achieved by the application of local exhaust ventilation techniques. Occasionally personal protective clothing may be necessary. The exposures in certain mill operations include a spectrum of glues and adhesives for assembly and paint, lacquers, and enamel systems for finishing.

REFERENCES

1. J. A. Kent, Ed., *Riegel's Handbook of Industrial Chemistry,* 7th ed., Van Nostrand Reinhold, New York, 1974.
2. D. M. Considine, Ed., *Chemical and Process Technology Encyclopedia,* McGraw-Hill, New York, 1974.
3. *Encyclopedia of Polymer Science and Technology,* Vol. 12, Wiley-Interscience, New York, 1970.
4. R. H. Perry, Ed., *Chemical Engineers Handbook,* 5th ed., McGraw-Hill, New York, 1973.
5. T. Baumeuter, Ed., *Standard Handbook for Mechanical Engineers,* 7th ed., McGraw-Hill, New York, 1967.
6. Committee on Industrial Ventilation, American Conference of Governmental Industrial Hygienists, *Industrial Ventilation: A Manual of Recommended Practice,* 14th ed., ACGIH, Lansing, Mich., 1976.
7. J. L. Goodier, E. Boudreau, G. Coletta, and R. Lucas, "Industrial Health and Safety Criteria for Abrasive Cleaning Operations," U.S. Department of Health, Education and Welfare Publication No. (NIOSH) 75-122, Cincinnati, Ohio, 1974.
8. A. Blair, *Am. Ind. Hyg. Assoc. J.,* **34,** 61 (1973).
9. M. Ziskind, H. Weill, A. E. Anderson, B. Samini, A. Neilson, and C. Waggenspack, "Silicosis in Shipyard Sandblasters," paper presented at the International Shipyard Health Conference, University of Southern California Medical School, Los Angeles, December 13–15, 1973.
10. American National Standards Institute, "American National Standard Practices for Ventilation and Operation of Open-Surface Tanks," Z9.1, ANSI, New York, 1971.
11. "Hand and Automatic Degreaser Operations," *Mich. Occup. Health,* **3,** 4 (Summer 1958).
12. "Are you Alert to the Hazards of Solvent Degreaser?" *Mich. Occup. Health,* **15,** 1 (Fall 1969).
13. M. W. First, *ASSE J.,* **14,** 11 (1969).
14. "Ventilation for Electroplating Plants," *Mich. Occup. Health,* **12,** 4 (Summer 1967).
15. L. J. Flanigan, S. G. Talbert, D. E. Semeones, and B. C. Kim, "Development and Design Criteria for Exhaust Systems for Open-Surface Tanks," NIOSH Research Report Contract No. HSM 099-71-61, Battelle-Columbus Laboratories, Columbus, Ohio, October 1974.
16. American Society for Metals, *Metals Handbook,* Vol. 5, ASM, New York, 1970.
17. J. T. Siedlecki, *J. Am. Med. Assoc.,* **215,** 1676 (1971).
18. A. H. Goldsmith, K. W. Vorpahl, K. A. French, P. T. Jordan, and N. B. Jurinski, *Am. Ind. Hyg. Assoc. J.,* **37,** 214 (1976).
19. I. Matelsky, in: *Industrial Hygiene Highlights,* Vol. 1, Industrial Hygiene Foundation of America, Pittsburgh, 1968.
20. "Health Risks in the Foundry," *Mich. Occup. Health,* **13,** 4 (Winter 1968).
21. American Foundrymen's Society, *AFS Foundry Environmental Control,* Vol. 1, AFS, Des Plaines, Ill., 1972.
22. R. F. Boddey, *Ann. Occup. Hyg.,* **10,** 231, (1967).

23. C. E. Bates and L. D. Scheel, "A Survey of Processing Emissions and Occupational Health in the Ferrous Foundry Industry," Southern Research Institute, undated.
24. American National Standards Institute, "Ventilation Control of Grinding, Polishing, and Buffing Operations," 243.1-1966, ANSI, New York, 1966.
25. E. K. Bastress et al., "Ventilation Control for Grinding, Buffing and Polishing Operations," Report No. 0213, IKOR Inc., Burlington, Mass., June 1973.
26. J. Danielson, Ed., *Air Pollution Engineering Manual,* 2nd ed., Publication no. AP-40, Government Printing Office, Washington, D.C., 1973.
27. M. M. Keys, E. J. Ritter, and K. A. Arndt, *Am. Ind. Hyg. Assoc. J.,* **27,** 423 (1966).
28. N. V. Hendricks, G. H. Collings, A. E. Dooley, J. T. Garrett, and J. B. Rather, *Arch. Environ. Health,* **4,** 139 (1962).
29. T. S. Ely, S. F. Pedley, F. T. Hearne, and W. T. Stille, *J. Occup. Health,* **12,** 7 (1970).
30. T. S. Virgil, *Am. Ind. Hyg. Assoc. J.,* **32,** 203 (1971).
31. R. Piper, *Br. J. Ind. Med.,* **22,** 247 (1965).
32. American National Standards Institute, "American Standard Safety Code for the Design, Construction, and Ventilation of Spray Finishing Operations," A9.3-1964, ANSI, New York, reaffirmed 1971.
33. American Welding Society, "The Welding Environment: A Research Report on Fumes and Gases Generated During Welding Operations," AWS, Miami, Fla., 1973.
34. *Hobart Pocket Welding Guide,* Hobart Brothers Co., P.O. Box EW-434, Troy, Ohio 45373, 1973.
35. L. Silverman and H. Gilbert, *Welding J.,* **33,** 218 (1954).
36. P. J. R. Challen, D. E. Hickish, and J. Bedford, *Br. J. Ind. Med.,* **15,** 276 (1958).
37. J. A. Dahlberg and L. M. Myrin, *Ann. Occup. Hyg.,* **14,** 269 (1971).
38. L. Coes, Jr., *Abrasives,* Springer-Verlag, Bonn, 1971.
39. Manufacturing Chemists' Association, "Hydrochloric Acid, Aqueous, and Hydrogen Chloride, Anhydrous," Chemical Safety Data Sheet No. 5D-39, MCA, Washington, D.C., 1951.
40. "Nitric acid—HNO_3," Hygiene Guide Series, *Am. Ind. Hyg. Assoc. J.,* **25,** 426 (1964).
41. Manufacturing Chemists' Association, "Properties and Essential Information for Safe Handling and Use of Nitric Acid," Chemical Safety Data Sheet No. SD-5, MCA, Washington, D.C., 1961.
42. O. T. Fasullo, *Sulfuric Acid, Use and Handling,* McGraw-Hill, New York, 1965.
43. "Criteria for a Recommended Standard—Occupational Exposure to Sulfuric Acid," U.S. Department of Health, Education and Welfare Publication No. (NIOSH) 74-128, 1974.
44. K. R. Van Horn, Ed., *Aluminum,* Vols. 1–3, American Society for Metals, Metals Park, Ohio, 1967.
45. N. L. Kaltreider, M. J. Elder, L. V. Cralley, and M. O. Colwell, *J. Occup. Med.,* **14,** 531 (1972).
46. P. J. Shuler and P. J. Bierbaum, "Environmental Surveys of Aluminum Production Plants," U.S. Department of Health, Education and Welfare Publication No. (NIOSH) 74-101, 1974.
47. Manufacturing Chemists' Association, "Anhydrous Ammonia," Chemical Safety Data Sheet No. SD-8, MCA, Washington, D.C., 1960.
48. American National Standard Institute, "Standard for the Handling and Storage of Anhydrous Ammonia," K61.1-1972, ANSI, New York, 1972.
49. "Criteria for a Recommended Standard—Occupational Exposure to Ammonia," U.S. Department of Health, Education and Welfare Publication No. (NIOSH) 74-136, 1974.
50. J. T. Siedlecki, *J. Am. Med. Assoc.,* **204,** 1176 (1968).
51. M. McCann, *Health Hazards Manual for Artists,* Foundation for the Community of Artists, New York, 1975.
52. "Criteria for a Recommended Standard—Occupational Exposure to Asbestos," U.S. Department of Health, Education and Welfare, Report HSM 72-10267 NIOSH, 1972.

53. J. L. Balzer and W. C. Cooper, *Am. Ind. Hyg. Assoc. J.,* **29,** 222 (1968).
54. S. Tola, S. Heinberg, J. Nikkanen, and S. Valkonen, *Work-Environment-Health,* **8,** 81 (1971).
55. "Criteria for a Recommended Standard—Occupational Exposure to Inorganic Lead," U.S. Department of Health, Education and Welfare Publication No. HSM 7-11010, 1973.
56. "Control Techniques for Beryllium Air Pollutants," U.S. Environmental Protection Agency Publication No. AP-116, EPA, Research Triangle, N.C., February 1973.
57. "Criteria for a Recommended Standard—Occupational Exposure to Beryllium," U.S. Department of Health, Education and Welfare Report HSM 72-10268, NIOSH, 1972.
58. A. Breslin, in: *Beryllium—Its Industrial Hygiene Aspects,* H. E. Stokinger, Ed., Academic Press, London and New York, 1966.
59. "Control Techniques for Mercury Emissions from Extraction and Chlor-Alkali Plants," U.S. Environmental Protection Agency Publication No. AP-118, 1973.
60. "Criteria for a Recommended Standard—Occupational Exposure to Cotton Dust," U.S. Department of Health, Education and Welfare Publication No. NIOSH 75-118, 1975.
61. Y. Y. Hammad and M. Corn, *Am. Ind. Hyg. Assoc. J.,* **32,** 662 (1971).
62. "Cotton Dust Controls in Yarn Manufacture," U.S. Department of Health, Education and Welfare Publication No. NIOSH 74-114, 1974.
63. W. A. Rye, "Fluorides and Phosphates—Clinical Observations of Employees in Phosphate Operations," *Proceedings of the 13th International Congress on Occupational Health,* July 25–29, 1960, pp. 361–364.
64. O. M. Derryberry, M. D. Bartholemew, and R. B. L. Fleming, *Arch. Environ. Health,* **6,** 503 (1963).
65. W. N. Sokal, et al., *J. Am. Med. Assoc.,* **226,** 639 (1973).
66. "Garage Ventilation," *Mich. Occup. Health,* **8,** 2 (Winter 1962–1963).
67. American Iron and Steel Institute, *Steel Mill Ventilation,* AISI, Washington, D.C., 1965.
68. *Leather Facts,* New England Tanners Club, P.O. Box 371, Peabody, Mass., 1973.
69. "Occupational Diseases Acquired from Animals," University of Michigan Education Service, School of Public Health, Ann Arbor, 1964.
70. K. N. Palmer, *Dust Explosions and Fires,* Chapman and Hall, New York, 1973.
71. American Petroleum Institute, "Safe Practices in Drilling Operations," API Recommended Practice 2010, 3rd ed., API, New York, 1967.
72. *International Petroleum Encyclopedia,* Petroleum Publishing Co., Tulsa, 1967.
73. American Petroleum Institute, *API Manual on Refinery Hazards,* in press.
74. K. E. Malten and R. L. Zielhuis, *Industrial Toxicology and Dermatology in the Production and Processing of Plastics,* Elsevier, Amsterdam, 1964.
75. R. A. De Gresero, *Ann. Occup. Hyg.,* **17,** 123 (1974).
76. R. G. MacDonald and J. N. Franklin, Eds., *Pulp and Paper Manufacture,* 2nd ed., Vol. 3, McGraw-Hill, New York, 1969.
77. *The Vanderbilt Rubber Handbook,* R. T. Vanderbilt, New York, 1968.
78. *Encyclopedia of Polymer Science and Technology, Plastics, Resins, Rubbers, Fibers,* Wiley-Interscience Vol: 12, New York, 1970.
79. "International Report on the Prevention and Suppression of Dust in Mining, Tunneling, and Quarrying (1970)," Occupational Safety and Health Series No. 24, International Labour Office, Geneva, 1970.
80. W. A. Burgess and P. Reist, *Am. Ind. Hyg. Assoc. J.,* **30,** 107 (March–April 1969).
81. D. Anderson, *Ind. Med. and Surg.,* **133,** 68 (February 1964).

CHAPTER TWENTY-FOUR

Quality Control

WILLIAM D. KELLEY

1 INTRODUCTION

The philosophy of quality control in industrial hygiene must reflect the professional goal of the industrial hygiene program, health protection through prevention of occupational diseases. Quality control's contribution to the real value of the industrial hygiene program is measured in terms of its contribution to the performance, cost, or schedule of the program. Quality control is then an extension of the role of management and assures effective management of the industrial hygiene program. Incorporation of the quality control plan into the management of an industrial program will trigger timely, consistent actions, and will provide evidence of the industrial hygiene program's effectiveness not only internally but externally. The outside parties may include facility management, the workers being protected, evaluation and accreditation groups, and regulatory agencies.

Quality control in industrial hygiene is analogous to quality control for plant processes. The industrial hygiene program itself is the embodiment of the principles and practices of quality control. The identification, evaluation, and control of health hazards, the triad of industrial hygiene, is the quality control approach or philosophy. A plant process produces certain occupational exposures to the workers from the noise, heat, dust, gases, and vapors, and so on, being produced by the process. These exposures can be considered as defects in a perfect production process. The perfect process would be so engineered that the loss of energy, the formation of by-products or contaminants, and the wear or breakdown of equipment would not occur. We must, however, deal with reality. The industrial hygiene engineer contributes to the design of optimum, if not perfect, plant processes. The monitoring program validates the effectiveness of the process design and control techniques. It detects the impending or recent failure of these control measures. The cause of the "out-of-control" situation is determined, and corrective action is taken to put the plant operation back into control—that is, no unacceptable occupational exposures. Monitoring is resumed. The level of monitoring is proportional to the magnitude and frequency of problems or "out-of-control" situations. According to Hagan, a good ground rule to follow might be: "Let the level of detail be equal to the value of preventing problems" (1).

This chapter provides guidance needed by the formulators of the industrial hygiene program for incorporating efficient quality control practices. These practices will assure timely and consistent actions to achieve the program's goal. The material is presented as an overall plan or system. The order of presentation follows the sequence that the industrial hygiene program would use to consider and install the various elements of the system. Initial consideration must be given to the scope and application of quality control in the industrial hygiene program. The program management section includes such elements as objectives and policies, organization, planning, operating procedures, chain of custody, records, corrective actions, training, and costs. The section on equipment, standards, and facilities covers the areas of sampling equipment, direct reading instruments, analytical instruments, calibration, preventive maintenance, reagents and reference standards, control of purchases, and facilities. The laboratory analysis control section is analogous to process quality control and deals with sample identification and control, intralaboratory control, and interlaboratory testing. The area of sample handling, storage, and delivery is considered in light of the logistical needs of the industrial hygiene program. The section on statistical quality control presents aspects of control charts, data validation, data analysis techniques, and the use of sampling plans. The final section discusses the third element of management responsibility, that is, evaluation in the context of a quality control program audit.

2 SCOPE OF THE QUALITY CONTROL PROGRAM

2.1 Program Elements

The elements of an industrial hygiene quality control program include (1) statement of objective, (2) policy statements, (3) organization, (4) quality planning, (5) standard operating procedures, (6) chain-of-custody procedures, (7) recordkeeping, (8) corrective acction, (9) quality training, (10) quality costs, (11) document control, (12) calibration, (13) preventive maintenance, (14) reagent and reference standards, (15) procurement control, (16) sample identification and control, (17) laboratory analysis and control, (18) intra- and interlaboratory testing programs, (19) sample handling, storage, and delivery, (20) statistical quality control, (21) data validation, and (22) system audits.

Most of these program elements are already operational in a more or less adequate and documented system. Evaluation of the degree of conformity of the program's operations to program needs is the key to fulfilling the program's responsibility.

2.2 Application of Program Elements

The application of program elements to the industrial hygiene program must be accomplished in consonance with the program's operational needs and resources. A systematic approach to program evaluation and development is an investment whose payoff is the reduction of embarrassments (serious errors in reported results and recommendations). In a more positive tone, the quality control aspects of the industrial hygiene program contributes value in achieving the performance, cost, and schedule commitments of the program. A principal aspect of performance relates to the professional quality rather than quantity of results and decisions of the industrial hygiene program.

3 QUALITY CONTROL PROGRAM MANAGEMENT

Several areas or elements of a quality control program are part of the overall management plan. These elements include objectives, policies, organization, planning, operating procedures, chain of custody, recordkeeping, corrective actions, training, and costs.

3.1 Objective

The objective of the quality control function of the industrial hygiene program is to "assure the medical and or scientific reliability of data and subsequent decisions of the program." The subobjectives include the following: (1) the use of rugged method meeting the program's needs, (2) routine determination of the level of performance of the program, (3) making item 2 compatible with item 1, (4) monitoring routine performance to assure long-term adequacy, and (5) validation of performance by comparison with peer groups. The National Institute for Occupational Safety and Health (NIOSH) has published a quantified objective statement for its Industrial Hygiene Analytical Laboratory (2).

3.2 Policies

Quality policies provide the framework of procedures that the industrial hygiene program uses to accomplish the foregoing objective(s). The policies are based on good quality control practices and compliance with applicable regulations, and they will assure the implementation of the quality control program elements just outlined. They may also specify the manner of implementation and frequency of implementing certain procedures such as preventive maintenance, calibration and checking of field and sampling equipment, laboratory internal quality control procedures, and participation in interlaboratory quality control or evaluation programs.

3.3 Organization

The quality control function of the industrial hygiene program must be carried out by both the line supervision and by nonline monitoring of the total effort. This situation is no different from that of the quality control function for the plant or facility or for the total industrial hygiene program itself. The quality control staff function will be a "collateral-duty" or part-time responsibility in most programs. The quality control coordinator should be a designated individual, having defined responsibilities (3). These responsibilities require the authority and organizational freedom to identify and evaluate quality problems and to initiate, recommend, or provide solutions. The job description of the quality control coordinator will include such areas as developing and carrying out quality control programs, monitoring quality control activities of the program, and advising management with respect to the quality aspects of industrial hygiene work.

3.4 Planning

The special controls, methods, equipment, and skills necessary for upcoming work should be identified and provided for in timely fashion. The planning allows for necessary research or development work to provide analytical methods or instrumentation or correlation among these. Compatibility of program's needs with both field and laboratory capabilities is assured.

3.5 Operating Procedures

The operating procedures necessary to the industrial hygiene program include survey protocols, sampling procedures (4) (sampling data sheets), analytical procedures (5–7) (may include sampling), instrumentation calibration and maintenance procedures, and quality control procedures (3) to ensure the validity of the industrial hygiene program's data.

Once such operating procedures are established, they must be communicated to the appropriate staff members. For the smallest of programs, a single volume for each type of adopted operating procedure may suffice. Generally, several sets of operating procedures need to be maintained. The industrial hygiene program manager must institute some means of assuring himself that all personnel are using the current adopted procedures. Each procedure or section should be uniquely identified. The document control system provides for updating by having the program manager issue memos describing adopted changes. Any member of the staff can and should initiate such updates, but they must be adopted and promulgated by the manager to assure consistency and awareness of the entire staff of the changes. Formalized systems (3, 8) can be used. Each section or procedure should contain the following elements: Section No. XX, Revision 0 (for initial issue), date of issue, and page Y or Z. A single page can then be revised, if appropriate, later. The manager will need to know who has received operating procedures, so that revisions also can be distributed to the staff.

3.6 Chain of Custody

Consideration of chain-of-custody procedures benefits the industrial hygiene program in several ways. If the samples or measurements taken may be part of a court case, the need for chain-of-custody procedures is self-evident. Since such procedures guarantee the integrity of the evidence collected, data developed for standards setting or for regular program work will benefit by such assurances. The manager should consciously assess the "benefits versus risk" of not following a chain-of-custody requirement. The laws or regulations of the state applicable to chain-of-custody should be complied with. A discussion with legal personnel would provide the basic requirements of that state.

1. A general discussion (9) of the chain-of-custody procedure follows. A sample is in your custody if (*a*) it is in your actual physical possession, or (*b*) it is in your view, after being in your physical possession, or (*c*) it was in your possession and you locked it up in a manner that would prevent any tampering.

Chain-of-custody record tags should be prepared before the worksite work and should contain as much information as possible to minimize clerical work by field personnel. The source of each sample should also be written on the container itself, if feasible, prior to the field work. Field log sheets, if used, should also be completed to the extent practicable before arriving at the site.

If more than one person is involved in the survey, all participants should receive a copy of any study plan and should become acquainted with its contents before the survey. A presurvey briefing and a postsurvey debriefing should be held. The debriefing should determine adherence to chain-of-custody procedures.

2. Specific points applicable to the industrial hygiene survey include the following:

a. To the extent achievable, as few people as possible should handle the sample.
b. Standard sampling techniques should be used.
c. The chain-of-custody record should be attached to the sample container at the time the complete sample is collected, and it should contain the following information: sample number, time taken, date taken, source of sample (includes type of sample and name of person or area sampled), preservative, if applicable (e.g., for biological samples), analyses required, name of person taking sample, and names of witnesses. The record should be signed, timed, and dated by the person doing the sampling. The records must be legibly filled out in ballpoint (waterproof ink).
d. Blank samples should also be taken. They will be analyzed by the laboratory to exclude the possibility of sample contamination.
e. A preprinted, bound field data record should be maintained to record field measurements and other pertinent information necessary to refresh the sampler's

memory in the event he later testifies about his work during the evidence-gathering activity. A separate set of field records should be maintained for each survey and stored in a safe place where they can be protected and accounted for at all times. The entries should be signed by the field sampler. The preparation and conservation of the field records during the survey will be the responsibility of the lead industrial hygienist. Once the survey is complete, field records will be retained by the head industrial hygienist, or his designated representative, as a part of the permanent record.

f. The field sampler is responsible for the care and custody of the samples collected until properly dispatched to the receiving laboratory or turned over to an assigned custodian. He must assure that each sample container is in his physical possession or in his view at all times, or is locked in such a place and in such a manner that no one can tamper with it.

g. If colored slides or photographs are taken to substantiate any conclusions of the survey, written documentation on the back of the photo should include the signature of the photographer, time, date, and site location. Photographs of this nature, which may be used as evidence, should also be handled in accordance with chain-of-custody procedures to prevent alteration.

3. Some considerations for transfer of custody and shipment are as follows:

a. When turning over the possession of samples, the transferee must sign, date, and time the chain-of-custody record. If a third person takes custody, he must fill in a second "receipt of sample" record. An additional custody record must be completed by persons who thereafter take "custody." Therefore the number of custodians in the chain should be as small as possible. Additional custody records should be numbered consecutively.

b. The field custodian or sampler has the responsibility of properly packaging and dispatching samples to the laboratory for analysis. The sample identification and sample accountability portions of the chain-of-custody record must be completed, dated, and signed.

c. Samples must be properly packed in shipment containers to avoid breakage. The shipping containers are sealed for shipment to the laboratory.

d. All packages must be accompanied by the "sample identification" and "sample accountability record." A copy of each is mailed directly to the laboratory.

e. If sent by mail, the package is registered with return receipt requested. If sent by common carrier, a bill of lading should be obtained. Receipts from post offices and bills of lading are sent to and retained by the laboratory custodians as part of the permanent chain-of-custody documentation.

f. If samples are delivered to the laboratory when appropriate personnel are not there to receive them, the samples must be locked in a designated area within the laboratory, to prevent tampering. The same person must then return to the laboratory, unlock the samples, and transfer custody to the appropriate custodian.

4. Laboratory operations must be so organized that the principles and practices of

chain-of-custody procedures are observed. Clearly these procedures require the application of the best principles of management and scientific investigations.

a. The laboratory must designate one employee as a "sample custodian," and another as alternate. In addition, the laboratory must set aside a "sample storage security area." This should be a clean, dry, isolated area that can be securely locked from the outside.

b. All samples should be handled by the minimum possible number of persons.

c. All incoming samples are to be received only by the custodian, who will indicate receipt by signing the sample transmittal sheets accompanying the samples and retaining the sheets as permanent records.

d. Immediately upon receipt, the custodian places the sample in the sample storage area, which is locked at all times except when samples are removed or replaced by the custodian. To the maximum extent possible, only the custodian should be permitted in the sample storage area.

e. The custodian must ensure that heat-sensitive or light-sensitive samples, or other sample materials having unusual physical characteristics or requiring special handling, are properly stored and maintained.

f. Only the custodian can distribute samples to personnel who are to perform tests. The custodian must enter into a permanent logbook the laboratory sample number, time and date, and the name of the recipient.

g. The analyst will record in his laboratory notebook or worksheet the name of the person from whom the sample was received, whether it was sealed, identifying information describing the sample (by origin and sample identification number), the procedures performed, and the results of the testing. The notes should be signed and dated by the person performing the tests and retained as a permanent record in the laboratory. In the event that the person who performed the tests is not available as a witness at time of trial, the laboratory may be able to introduce the notes in evidence under the Federal Business Records Act.

h. To the extent possible, standard methods of laboratory analyses are to be used. If laboratory personnel deviate from standard procedures, they should be prepared to justify their decision during cross-examination.

i. Laboratory personnel are responsible for the care and custody of the sample once it is handed over to them and should be prepared to testify that the sample was in their possession and view, or securely locked up, at all times from the moment it was received from the custodian until the tests were run.

j. Once the sample testing is completed, the unused portion of the sample, together with all identifying tags and laboratory records, should be returned to the custodian, who will make the appropriate entries in his log. The returned, tagged sample will be retained in the sample room until it is required for trial. Stripcharts and other documentation of work are also turned over to the custodian.

k. Samples, tags, and laboratory records of tests may be destroyed only on the order of the laboratory director, who first confers with his legal adviser to make certain that the information is no longer required or that the samples have deteriorated. Several industrial hygiene analytical laboratories currently have a 5-year record retention time.

3.7 Recordkeeping

Records are considered to be a principal form of objective evidence of the operation and effectiveness of the quality control system. Any laboratory needs a system of records to establish and maintain control over the samples received for analysis. An ultimate system would certainly provide confirmation that chain-of-custody requirements have been met. The previous section indicates the nature of such requisite records. Any laboratory also needs administrative records. Since these are not unique to industrial hygiene or to a laboratory operation, however, they are not discussed further.

Quality control records are an integral part of an effective and economical quality control program. The records provide the assurance that calibrations, "blind sampling," recycled sampling, and similar procedures have been carried out in accordance with the laboratory's plan. The records of the day-to-day system and instrumental quality control checks document routine control and the initiation of necessary corrective actions. Control charts for instrument performance and check samples perform an efficient and economical recordkeeping function. As each instrumental check is made, for instance, the result should be posted immediately on the control chart near or on the instrument itself. The analyst has access to the record without difficulty and can make an instantaneous decision about the action to be taken. If the chart shows a pattern of expected variability, no action is necessary. If a real trend is developing or an out-of-control point is plotted, corrective action is initiated. This achievement of real-time decision making is hard to match, even with the more sophisticated computer systems.

The test of the quality control records system is analogous to a certified public accountant's audit of a company's financial statement and its bookkeeping system. A knowledgeable quality control person should be able to audit the records system and come to the same conclusion given by the laboratory personnel about the "state of quality control." An affidavit analogous to that appearing with the annual financial report attesting to the validity of the laboratory's claim about its real state of quality control represents the ultimate compliment to the records system.

3.8 Corrective Actions

Problems are bound to occur with any analytical system. For example, components of the system wear out. When a problem or defect begins to affect the performance of the analytical system, corrective action should be initiated. For major problems, management must initiate the corrective action plan and assure that all affected individuals cooperate in a well-planned program. In many instances, the analyst can detect erratic behavior or changed response level, either based on experience or from out-of-control points on a control chart. The analyst should check the operational parameters likely to cause such "nonroutine" response. Contamination, which will cause the blank or baseline to shift, is not likely to be evident in the higher level standards. The small change in concentration or instrumental response value will be lost in the normal variability at the higher levels. Calibration-type problems will cause a constant percentage shift in concentration or instrumental response that will be evident in higher level standards but lost in the normal variability of the blank or baseline determination. Seeking out the case of a problem and correcting it requires analytical or technical expertise rather than statistical knowledge. One of the advantages of the analyst or instrument operator being the principal in the bench level quality control program is that he or she on many occasions can initiate corrective actions before problems become major.

When there are significant (out-of-control) differences between the observed and expected results on quality control test samples, notification should be sent to the analyst. The analyst should investigate, determine the cause of difference, if possible, take necessary corrective action, and report results to the laboratory management. If the required corrective action is beyond the responsibility and authority of the analyst (e.g., in the case of an incompatibility between the sampling and analytical procedures) the appropriate level of laboratory management must decide what type of corrective action is necessary and ensure its implementation.

As the time increases between the "out-of-control" incident and its investigation, the probability of the analyst finding the cause rapidly decreases. In some cases, the original sample may have been discarded already, or the reagents changed, or the instrument reprogrammed for another analysis. In these cases, the quality control system rather than the analyst must be faulted for what will usually be a futile effort to determine an appropriate corrective action.

3.9 Training

The personnel of the industrial hygiene program must have sufficient training and/or experience to develop the necessary attitudes, knowledge, and skills to proficiently perform their job functions. In most cases, on-the-job or "buddy-system" training is the principal mechanism used to develop new capabilities in the staff. More formal programs should be used to supplement the baseline training as appropriate. There are a number of training resources available to the industrial hygiene program (10, 11). The effectiveness of the training should be evaluated by the program management. The person trained should be able to perform the new function completely and accurately. Performance measures should relate to the degree of proficiency of the staff already performing the function. Values for precision and recovery for an analytical procedure are relatively easy to develop and provide an objective measure of performance. Other functions are more difficult to quantify. In some cases the performance test is the skill with which the staff person can perform the given function—for example, organizing and conducting a plant survey independently.

The quality control function of the industrial hygiene program can perform the function of auditing the effectiveness of the training program. It is self-evident that personnel charged with the responsibility for the quality control function must also acquire the appropriate training to develop and implement the quality control system.

3.10 Costs

The industrial hygiene program allocates financial and personnel resources among the various elements of the quality control system. These elements can be grouped into four categories. Prevention costs involve elements associated with planning, implementing, and maintaining the quality control program itself—for example, objectives, policies, organization, planning, operating procedures, chain-of-custody procedures, recordkeeping, training, equipment calibration, preventive maintenance, reference standards, control of purchases, and facilities.

Appraisal costs are those entailed in efforts to evaluate and maintain the quality performance levels of the industrial hygiene program. The elements included in appraisal consist of sample identification and control, intralaboratory quality control testing, interlaboratory proficiency testing programs (including accreditation program costs), control charts, data analysis, and data validation. The reports generated on the effectiveness of the quality control system are also included in appraisal costs.

Internal failure costs are the costs to the industrial hygiene program attributable to defective materials, reagents, instruments, sampling, and/or analytical procedures, which cause data to be discarded and repeated or lost. Corrective actions to determine and correct these problems are included in internal failure costs. The costs of having to resurvey a plant, including the travel and personnel time costs, would also be included if the survey had been done by the industrial hygiene program. If the resurvey is to be done by an outside group, the cost would belong to the next category as an external failure cost.

Also included in the external failure costs category are those associated with investigating complaints from defective data that have already been reported. The repetition of procedures or effort spent in validating or revalidating procedures because of reported defective data also constitutes an external failure cost. The costs to other programs or departments may be considerable when defective data are reported. Should the defective data be discovered during an enforcement proceeding, the indirect costs will greatly outweigh the direct costs to plant or agency management. The loss of credibility to the industrial hygiene program cannot be calculated in terms of dollars but may include the very existence of the industrial hygiene program.

The value of categorizing costs is to allow and encourage a programmed and budgetary apportionment of these costs. As the costs can be concentrated in the prevention category, the total costs will decrease. The cost effectiveness of the quality control system and of the industrial hygiene program itself will also improve. In his book *Total Quality Control,* A. V. Feigenbaum provides a comprehensive framework for quality control costs allocations.

4 EQUIPMENT, STANDARDS, AND FACILITIES

4.1 Sampling Equipment

The variability or error associated with sampling equipment is considerable. In a collaborative test on the sampling and analysis of solvents using personal sampling pumps, charcoal adsorption tubes, carbon disulfide elution, and gas chromatographic analysis, the sampling equipment error was of the same order of magnitude as the analytical errors; the relative standard deviation ranged from 5 to 14 percent (12). This error or variability existed when all participating laboratories calibrated their sampling pumps with a single calibration system. The report indicates that the procedure was used to reduce the calibration variability apparent with each laboratory having calibrated its own pumps at its own facility before meeting at the site for the collaborative test. The variability was measured by sampling from a homogeneous atmosphere. No spatial or time variability was involved in sampling from the controlled concentration level test chamber. No real plant sampling situation can possibly approach the homogeneous conditions in a test chamber.

The appropriate frequency of calibration depends on the handling and use (abuse) the sampling pump has undergone. Specifically, pumps should be recalibrated after suspected abuse (such as dropping), when received from the manufacturer, and when repaired. Experience will determine an appropriate frequency of recalibration. A control chart approach can efficiently develop the information necessary for such decisions. At least until such experience is gained, pumps should be calibrated before being used in the field and at intervals during the field work if numerous samples are taken. The calibration should be checked upon return from the field.

The accuracy of the calibration depends on the calibration system itself. The choice of calibration system depends on the location at which the calibration is made and the facilities available to the industrial hygiene program (13). In the laboratory, a 1- or 2-liter burette or wet test meter is generally used. Other standard calibrating systems such as a spirometer, Marriott's bottle, or dry-gas meter could also be used.

The procedure for calibration with the soap bubble flowmeter follows (14). The calibration setup for personal sampling pumps with a cellulose membrane filter follows standard procedures. Since the flowrate indicated by the flowmeter of the pump is dependent on the pressure drop across the sampling device, the pump flowmeter must be calibrated while operating with a membrane filter and appropriate backup pad in the line.

1. While the pump is running, the voltage of the pump battery is measured with a voltmeter to assure that the battery is charged adequately for calibration.
2. The cellulose membrane filter with backup pad is placed in the filter cassette.
3. The calibration setup is assembled so that the air flows from room atmosphere through the soap bubble meter to the cassette, through the cassette to the sampling

pump, and to room atmosphere. A U-tube water manometer monitors the pressure drop between the cassette and the pump.

4. The pump is turned on, and the inside of the soap bubble meter is moistened by immersing the burette in the soap solution and drawing bubbles up the tube until they are able to travel the entire length of the burette without bursting.
5. The pump is adjusted to provide a flowrate of 2.0 liters/min.
6. The water manometer is checked to ensure that the pressure drop across the sampling train does not exceed 13 in. H_2O at 2 liters/min.
7. A soap bubble is started up the burette, and the time it takes the bubble to travel a minimum of 1.0 liter is measured with a stopwatch.
8. The procedure in step 7 is repeated at least 3 times, the results are averaged, and the flowrate is calculated by dividing the volume between the preselected marks by the time required for the soap bubble to travel the distance.
9. Data for the calibration should include the volume of air measured, elapsed time, pressure drop, air temperature, atmospheric pressure, serial number of the pump, date, and name of the person performing the calibration.

The specific requirements for the calibration procedure, the frequency of its use, and maintenance and operational notes must be permanently recorded since these records form an integral part of the quality control system and are invaluable in any court proceeding. Requirements for the calibration system itself are discussed later.

4.2 Direct Reading Instruments

The calibration of direct reading instruments involves not only flow but concentration level response calibration. For guidance on the general technical problems, consult Lippmann (13) Saltzman (15), and the air sampling instruments manual (16) of the American Conference of Governmental Industrial Hygienists (ACGIH). In addition, a number of evaluation reports have been developed; Johnson on NO_2 meters (17), McCammon on portable, direct reading combustible gas meters (18), and Parker and Strong on CO meters (19). These reports provide detailed performance evaluations for those classes of instrument. The calibration of direct reading instruments for physical agents is handled in a manner equivalent to that for chemical agents. Guidance for specific instruments should be available from the manufacturer or the literature. In some cases a calibration capability, such as for ultraviolet radiation, must be developed (20).

4.3 Analytical Instruments

The calibration system records and approach are equivalent for sampling equipment, direct reading instruments, and analytical instruments. The following discussion on the calibration system has been adapted from the NIOSH quality control manual TR-78 (3).

All equipment to be calibrated should have a permanently affixed record number (serial or property number). A calibration control card or sheet for each instrument should show identification number, description (including manufacturer, model, and serial number), location of use or storage, calibration procedure used, calibration interval, date of last calibration, signature of the person performing the calibration, due date for next calibration, and values obtained during the calibration. Calibration reports and compensation or correction figures should be filed with the calibration control card.

All equipment should be calibrated in a room in which the laboratory has provided controls for environmental conditions to the degree necessary to assure measurements of the specified and required accuracy. The calibration area should be reasonably free of dust, vibration, and radiofrequency interferences and should not be located near equipment that produces acoustical noise or vibration. Isolation of pressure, mass, and acceleration equipment from vibrations is particularly essential; isolation mounts, seismic masses, and other means of protection should be provided.

The laboratory calibration area should have adequate temperature and humidity controls. A temperature from 68 to 73°F and a relative humidity of 35 to 55 percent normally provide a suitable environment. A filtered air supply is a necessity in the calibration area. Dust particles are more than just a nuisance; they can be abrasive, conductive, and damaging to instruments. A measure of dust filtration can be provided in the air conditioning system by the washing action of sprays and atomizers, but this may need to be supplemented by electrostatic and/or mechanical filters of the activated charcoal, oil-coated or ribbon types.

Recommended requirements for electrical power within the laboratory should include voltage regulation of at least 10 percent (preferably 5 percent), low values of harmonic distortion, minimum line transients as caused by interaction of other users on the main line to the laboratory (separate input power if possible), and a suitable grounding system established to assure equal potentials to ground throughout the laboratory (or, isolation transformers may be used to operate individual pieces of equipment). Adequate lighting (suggested values, 80 to 100 foot candles) is necessary for work bench areas. The lighting may be provided by overhead incandescent or fluorescent lights. Fluorescent lights should be shielded properly to reduce electrical noise.

All instruments should be calibrated and checked by qualified personnel. An outside calibration service group may schedule and perform these checks, but this does not relieve the laboratory of the responsibility for controlling, monitoring, and identifying calibration intervals and having the checks made on time.

All measurements or calibrations performed by or for the laboratory for the calibration program should be traced, directly or indirectly, through an unbroken chain of properly conducted calibrations (supported by reports or data sheets) to some ultimate or national reference standard maintained by a national organization such as National Bureau of Standards; an example of proper calibration procedure for standard quartz cuvettes appears in the literature (21). The ultimate reference standard can also be an independent, reproducible standard (i.e., a standard that depends on accepted values of natural physical constants). A typical example is the cesium beam type of microwave frequency standard.

There should be an up-to-date report for each reference standard (except independent reproducible standards) used in the calibration system, and for any subordinate standards or measuring and test equipment if their accuracy requires supporting data. If calibration work is contracted out to a commercial laboratory or facility, copies of reports issued by the subcontractor should be kept available.

All reports should be kept in the calibration system file and should contain the following information: (1) identification or serial number of standard to which the report pertains, (2) conditions under which the calibration was performed (temperature, relative humidity, etc.), (3) accuracy of standard (expressed in percentage or other suitable terms), (4) deviation or corrections, (5) report number or designation; in addition (6) reports for the highest level standards of sources other than NBS or a government laboratory should bear a statement that comparison has been made with national standards at periodic intervals using proper procedures and qualified personnel, and (7) corrections that must be applied if standard conditions of temperature, gravity, air buoyancy, and so on, are not met, or if they differ from those at place of calibration, must be noted.

The description of the calibration program indicates that an extensive effort may be required. If the data to be generated by the industrial hygiene program are sufficiently important that related decisions may be challenged in court or elsewhere, the complete calibration program is cost effective. For other situations, the program manager must decide on an appropriate allocation of his resources among this and the other elements of the quality control system.

4.4 Preventive Maintenance

The preventive maintenance program complements the calibration program. An effective preventive maintenance program increases the reliability of measurement systems and decreases downtime. An inadequate preventive maintenance program will be responsible for increasing unscheduled downtime, a probable increase in total maintenance costs, and possibly a lack of trust in the validity of the data being generated.

A schedule of preventive maintenance should be keyed to the calibration program schedule. The frequency of the preventive maintenance program should be based on the equipment manufacturer's recommendations and cumulative experience. All records from the maintenance program should be kept as part of the records system and used to assure the validity of data generated by the industrial hygiene program.

4.5 Reagent and Reference Standards

A number of reagent standards suitable for laboratory programs are available from commercial sources. Standard reference materials (SRMs) are available from the National

Bureau of Standards. Out of approximately 900 SRMs available, about 40 have been prepared for environmental use. Another half-dozen have been specifically prepared for industrial hygiene use, and several more are to be released.

Calibration gases may be obtained from commercial suppliers. A cross-check program comparing the old and new tanks or tanks from several suppliers will serve to validate the calibration gases for the program. Such checking may be necessary because the responsibility for valid data lies with the industrial hygiene program itself.

The use of all available commercial analytical standards still leaves most of the industrial hygiene program's analytical needs without standards. The only recourse is the internal preparation of needed working standards for calibration and quality control purposes. The development of a diverse range of working standards for many elements in wide concentration ranges has been covered by Hill (22), who uses the working standards in a ratio of 1 : 10 where the unknowns cover a narrow concentration range and in even greater ratio for more diverse samples. Hill states: "Having calibrated our laboratory, the standards are still used daily on every set run as a datum and an early warning system to detect any problems that may develop. Any set exhibiting poor results on the included standards is immediately rerun, and the problem is identified and eliminated." Some literature (3) uses the term "control sample" in much the same sense as Hill used the term "working standard." As Hill concludes: "Millions of dollars may be spent upon the results of a large suite of chemical analyses. There is no mystery to obtaining the good analytical results needed to make the right decisions the first time; one must get or make the needed standards and use them continuously."

4.6 Control of Purchases

Common laboratory reagents and chemicals are generally bought to a specified grade or quality such as "ACS Reagent" grade or "Spectroquality." Experience will indicate differences between such materials from various suppliers. The main assurance of the quality of such materials is the routine use of reagent blanks in the analytical procedures. The shelf life should be determined and stocks rotated on a first in–first used basis.

The purchase of equipment may offer the program a greater opportunity to set performance requisites in the purchase specifications. Acceptance testing against the performance specifications protects the program.

In many cases the industrial hygiene program is obliged or desires to purchase analytical measurements from another laboratory. Since the data reported by the program are the program's responsibility regardless of where the analyses were made, the program must obtain valid data from outside sources. It must assure that such data are being generated by a laboratory whose quality control system is at least as effective as its own. A routine technique is to submit split samples, spiked samples, and blanks as blind samples.

4.7 Facilities

The role of facilities in assuring the quality of an industrial hygiene program's results is flexible. The best possible building and physical facility does not guarantee "good work"; on the other hand, "good work" may be produced under poor conditions. Good facilities certainly simplify and minimize an additional set of variables that the program does not want to have affecting the quality of the data produced. Downtime and time-consuming corrective action investigations should be minimized with adequate facilities. A number of books and manuals on the design of laboratory facilities can be consulted. It is important to minimize occupational safety and health problems in the design and/or development of laboratory space for the industrial hygiene program.

5 LABORATORY ANALYSIS CONTROL

The quality control system to provide for precise and accurate industrial hygiene measurements has been discussed in earlier sections for the program management and for its physical elements. Sections 5 and 6 cover the operational elements.

5.1 Sample Identification and Control

All samples taken or received for analysis must be logged in and identified with respect to type and source of the sample. The sample identity must be carefully and clearly maintained throughout the sampling and analysis–reporting cycle of events. The earlier discussions on chain-of-custody and record keeping (Sections 3.6 and 3.7) apply.

5.2 Intralaboratory Control

The principal operational element of the industrial hygiene program's quality control system is its intralaboratory control program.

5.2.1 Method Adaption and Validation

The adaption and validation of reference or other methods to the specific needs of the program comprise one of the most important elements. These specific needs may require validation because of matrix, interference, sensitivity, or other operational aspects unique to the sampling and analysis problems to be expected. Numerous texts (23–25), articles (26, 27), and other publications (12, 28–33) discuss the approaches and techniques of the collaborative test evaluation of sampling and/or analytical methods. The individual laboratory must assure itself that the methods it uses are "rugged" enough for the conditions of use entailed by its problems. Youden (23) extensively discusses "ruggedness testing" and its value to the analytical laboratory. Linch (34), Linning et al. (35), and Wilson (36–38) describe the types of error that may afflict a method, as well as programs to detect, quantitate, and eliminate or correct for these determinate errors.

When the method is ready for intralaboratory validation, any number of protocols may be used. One such protocol (6) divides the overall program into two phases. The analytical variability and bias, if any, are evaluated by spiking samples for laboratory analyses. Six spiked samples at each of three levels (0.5, 1, and 2 times the permissible limit for routine sampling techniques) are analyzed by the laboratory. If the precision and bias are acceptable, the second phase is initiated. In the second phase, contaminant atmospheres are generated, sampled using the prescribed collecting techniques, and analyzed. Experimentally, it has been found desirable to determine sampling pump variability separately. Laboratory air-moving devices with low variability can then be used in the sampling of the contaminant. Independent verification of the contaminant concentration in the atmosphere being sampled provides a measure of the collection efficiency of the sampling device. To maintain symmetry, again six samples at each of the three levels specified previously are taken and analyzed by the analytical method.

By reasonably straightforward mathematical and statistical operations, the variability due to the individual components of analysis, of sample collection, and of pump error can be determined. These operations also can be combined to provide an overall determination of variability when a homogeneous atmosphere is being sampled. It is recognized that neither currently available statistical theory nor experience provides a validated quantitative estimate of the variability due to "representative" sampling. The bias of the sampling and analytical method can also be determined under the test conditions by comparing the values determined in the second phase with the concentrations of the contaminant "known" to be in the chamber at the time of sampling.

5.2.2 Instrumental Quality Control

The industrial hygiene laboratory many times finds itself using a basic analytical technique for a wide variety of materials. Atomic absorption spectrophotometry, X-ray analysis, and gas chromatography are common examples. In considering the quality control aspects of such procedures, emphasis should be directed to the common points, rather than the differences.

As an example, a common sampling and analytical technique for many organic materials involves charcoal tube adsorption, desorption with carbon disulfide (CS_2), and gas chromatography instrumental analysis (39). The sampling and analytical method has been validated by collaborative testing for seven solvents (12). The method has been evaluated for application to numerous materials. There are still materials that have not been checked out. Each new challenge is not unique, with the result that prior developmental and operational information does not apply. By checking the operation of the parts of the procedure and by using control samples of a similar "evaluated" material, the laboratory can with great confidence analyze for the new "unknown." The desorption efficiency must be specifically determined for the laboratory's operating condition for all materials. The gas chromatograph must be calibrated for the "unknown" for quantitative analysis. However these operations can be done routinely rather than on a research project basis (40). The use of control charts for instrumental quality control is most appropriate. The monitoring of a critical performance indicator assures the analyst that the instrument is operating reliably. Control charts are developed in a later section.

5.2.3 Routine Samples

One of the most generally applicable and easily applied statistical quality control techniques is the Shewhart control chart. The Shewhart control chart technique can be applied to almost any area of somewhat repetitive measurements, including monitoring critical instrumental performance characteristics, analytical blanks, instrument calibration standards, and total analysis control samples, as well as monitoring the desorption efficiency determination for consistent technique, and ultimately, plant environmental control monitoring (41). For routine analytical procedures, control charts should be set up for the "control samples" or "working standards" discussed by Hill (22). The control chart for the "control sample" (3) will monitor the recovery of the procedure, thereby detecting errors that cause a shift in the process average (42). These sources of errors are due to calibration-type problems or changes in concentration of reagents (43).

A control chart for blank determinations will aid in detecting contamination-type problems. Reagent or sample contamination or certain instrumental operational problems will tend to cause a small constant absolute value shift in the determination. Such a shift would be difficult to detect at concentration levels up in the working range, but it is more easily found using blanks because the absolute value would be lost in the usual variability at higher concentrations. Excellent summaries of general precautions and techniques applicable to such problems are presented by the Intersociety Committee on Air Sampling and Analysis (44) and by Linch (45, 46).

If the "control sample" used in a determination is not a realistic simulation of regular environmental samples, it will be of value in determining the precision for recycled routine samples. The use of a "mercuric chloride in distilled water" (3) control sample because of instability problems associated with a "pooled mercury in urine" control sample is such a case. Such recycled samples should be more affected by interferences and matrix effects, therefore will yield a more realistic value, within laboratory precision or repeatability (47) on routine samples. Special considerations and techniques are applicable when automated procedures are used for high production routine work (48–50).

5.2.4 Single Samples or Infrequently Employed Procedures

The single-time or infrequently employed procedures present the greatest problems to the laboratory analyst. The analyst's unfamiliarity with the procedure to be used and probable unfamiliarity with the sample matrix and interferences presented to the procedure make him deservedly wary. Since almost certainly no SRMs or reasonable substitutes will be readily available, the analyst must rely on his generally good technique (as validated with other procedures) and a limited assessment of recovery (51) and interferences on the subject sample. The basic recovery procedure (34) is summarized below.

The recovery procedure applies the analytical method to (1) a reagent blank, (2) a series of standards (covering the expected range of concentration of the sample), (3) the sample itself (at least in duplicate), and (4) the recovery samples. The recovery samples are prepared by adding known quantities of the substance sought to separate portions of the sample itself. Each portion of the sample should be equal to the size of sample taken for the run. The substance sought should be added in sufficient quantity to overcome the limits of error of the analytical method, but without causing the total in the sample to exceed the range of the standards used.

The results are corrected by subtracting the reagent blank from each of the other determined values. The resulting standards are then graphed by plotting readout (y-axis) versus concentration (x-axis). From this graph the amount of contaminant (or substance) in the sample alone is determined. This determined value is then subtracted from each of the recovery samples consisting of sample plus known added substance. The resulting amount of substance divided by the known amount originally added, and multiplied by 100, gives the percentage recovery.

The basic recovery procedure may be applied to a colorimetric, flame photometric, fluorimetric, titrimetric, gravimetric, and other analytical techniques.

An example of the recovery procedure as applied to a lead in blood analysis appears in Table 24.1.

In other methods, the internal standard procedure may be effectively used (34, 52). The internal standard technique serves primarily for emission spectograph and polarographic procedures. This procedure enables the analyst to compensate for electronic and mechanical fluctuations within the instrument.

The internal standard method involves the addition to the sample of known amounts of a substance that will respond to the instrument in a similar manner to the contaminant being analyzed. The ratio of the measurement of the internal standard to the measurement of the contaminant is the value used to determine concentration of contaminant present in the sample. Changes in instrumental conditions during analyses should affect the internal standard and the contaminant in the same manner and degree, and the determination will compensate for such changes. The internal standard should be of similar chemical reactivity to the contaminant and approximately the same concentration anticipated for the contaminant. It should be a substance as pure as possible. Other constraints may be more critical for specific analytical techniques.

Since the occasional sample presents the greatest uncertainty and potential problems to the industrial hygiene analytical laboratory, major efforts of senior staff are required to ensure the quality of such work. The difficulties, the time involved, and the infrequency of such samples also present a special challenge to the person responsible for the reporting of such efforts. Elwell and Lawton (53) have reported on a "relative value structure" to aid in such reporting.

Table 24.1 Lead in Blood Analysis (34)

Basis. A 10.0-gram sample of blood from the blood bank pool, ashed and lead determined by double extraction, mixed color, dithizone procedure.

Lead Added (pg)	Optical Density	Lead Found (pg)		Recovery (%)
		Total	Recovered	
None—blank	0.0969	—	—	—
5: Calibration point	0.2596	—	—	—
None	0.1427	1.6	—	—
None	0.1337	1.3	—	—
None	0.1397	1.4	—	—
None	0.1397	1.4	—	—
Average	0.1389	1.4	—	—
2.0	0.1805	2.9	1.5	75
4.0	0.2636	5.4	4.0	100
6.0	0.3372	7.8	6.4	107
8.0	0.3925	9.4	8.0	100
10.0	0.4437	11.4	10.0	100
30.0 total	—	36.9	29.9	96

Mean error = 36.9 − (30.0 + 5 × 1.4) = 0.1 μg for entire set
= 2.9 − (2.0 + 1.4) = 0.5 μg for 2 μg spike
Relative error = (0.1 × 100)/37.0 = 0.27% for entire set
= (0.5 × 100)/3.4 = 14.7% for 2 μg spike

5.3 Interlaboratory Testing

It has been recognized that external proficiency analytical testing on a continuing basis is essential to assure quality (54). The American Industrial Hygiene Association (AIHA) laboratory accreditation program (55) includes satisfactory continued performance in a proficiency analytical testing program as an integral requirement for accreditation. Such a program offers the laboratory director confirmation of the effectiveness of the internal quality control program. Ratliff (56) has reported the development of the National Institute for Occupational Safety and Health Proficiency Analytical Testing (PAT) program used by NIOSH and OSHA, and subsequently AIHA.

The industrial hygiene program will need to supplement such a program by splitting or exchanging samples with one or several other laboratories interested in the same problem. These informal programs should be documented and appropriate credit taken for the value of the data so generated.

6 SAMPLE HANDLING, STORAGE, AND DELIVERY

The logistical aspects of sample handling, storage, and delivery impose a significant constraint on some industrial hygiene programs. When the industrial hygienist must rely on common transportation systems, whether plane, truck, mail, or parcel delivery, his options are limited. The use of bubblers and absorbing solutions become a "method of last resort." The risk of loss of samples, or of instability of the collected samples over a several-day transport time, and the attendant cost of doing a resurvey, make other alternatives attractive despite any inherent limitations or difficulties. The industrial hygienist who can personally, or by courier, transport his samples (collected in bubblers) back to the laboratory in a few hours may be able to choose more specific, more accurate, or more economical sampling and analytical methods.

The industrial hygiene program manager should ensure that operating procedures, chain-of-custody procedures, and records are adequate for the specific needs of the program. Sampling and shipping procedures are available for many industrial hygiene samples (4). Perishable samples, even milk (57), can also be safely and routinely shipped. The shipping procedures must consider Interstate Commerce Commission rules and other applicable shipping regulations.

7 STATISTICAL QUALITY CONTROL

Statistical quality control involves the application of statistical techniques to the appropriate areas of the industrial hygiene quality control program. For routine samples or for instrumental quality control programs, the Shewhart control chart is the simplest and most appropriate technique (42, 58, 59).

7.1 Control Charts

The industrial hygiene program should institute a control chart for an instrument or a procedure (control sample, or blank or recycled samples) when it is more efficient to consider the procedure to be semiroutine than to use the single or occasional sample procedures outlined earlier and advocated by Linch (34, 45, 46).

7.1.1 Purposes

The control chart is both a diagnostic and a reliability tool. Since the control chart differentiates between the usual pattern of random variation (from indeterminate errors) and mistakes or biases (due to determinate errors), it can be used to trouble-shoot a procedure or to make it more rugged. In the process, the weak points of the procedure, whether due to interferences, instability of sample, reagents or equipment, or required operator judgments, are identified (or determined) and eliminated. The resulting procedure has become "ruggedized" by this use of the control chart. The second use of the control chart is to establish the normal operational precision and stability of the sampling and/or analytical method or measurement system. As experience is gained, a reliable, valid value of routine precision will be developed. This value has built into it the "real world" sources of variability that the industrial hygiene program routinely faces. The evaluation of the measuring process stability (accuracy or bias) permits the program to make validated decisions with respect to whether plant processes have actually changed environmental or biological concentration levels or whether an apparent change is due to the inability of the industrial hygiene program to make reliable measurements. Such knowledge may obviously affect management decisions on hazard control equipment investments. The third use of control charts is as a monitoring instrument on the procedure or measurement system itself. The control chart provides immediate objective evidence of reliable operation or detects impending (trends) or actual problems (out-of-control points) with the procedure or measurement system. Corrective action can then be initiated by the industrial hygiene program, making the corrective action costs (Section 3.10) internal and not external (including loss of credibility).

7.1.2 Parameters to be Controlled

The calibration of a direct reading instrument such as a combustible gas meter can be checked by the use of commercial calibration kits. A calibration check on each day of use, before and after the day's survey, guarantees valid data throughout the day and backup if the measurements are challenged. A number of radiation survey instruments have "check sources" on the instrument. The surveyor performs the check and can even plot the result directly on a small control chart taped to the side of the instrument. These calibration checks are not a substitute for full individual contaminant response calibrations, but they do provide an objective means of determining their appropriate frequency and offer assurance of reliable measurements between such calibrations.

Laboratory instruments, whether used to calibrate sampling equipment or to measure analytical response, such as spectrophotometers, gas chromatographs, atomic absorption instruments, or radiation spectrometers, can be monitored for continued calibration and reliable performance. If an instrument's response can be completely calibrated for a given procedure in a few minutes, it may be more efficient to recalibrate each time a procedure is run. If, however, the instrument can be calibrated only infrequently by some reference or transfer standard, if calibration is quite expensive, involving weeks of effort, such as for a gamma radiation spectrometer, if economies can be achieved by checking a couple of points on a calibration curve, or if a critical response parameter can be monitored, a control chart is appropriate. Not only may economies in operational costs be realized, but greater reliability and comparability of data over longer periods of time will be achieved.

Total analysis quality control uses the control chart for routine analytical procedures to monitor the procedure's performance on "control samples," blanks, and "recycle samples" where used.

7.1.3 Type of Control Chart

The Shewhart control chart measures both measurement process variability and calibration stability. The Shewhart or $\bar{X}$ and R chart uses the range R to estimate process variability. The average of these individual measurements provides the $\bar{X}$ for monitoring the stability of the procedure or indicating how well it is maintaining its calibration. There are other types of charts that use the standard deviation to measure variability or a moving average, or individual values, to measure process stability (60). These are special application tools that may be more appropriate after the procedure or instrument is known to be in a "state of control" through use of the $\bar{X}$ and R chart. The cumulative sum or cu-sum chart is more complex to institute and is not as rugged for situations not already in a "state of control." In most cases the basic Shewhart or $\bar{X}$ and R chart can be easily and effectively used.

7.1.4 Trial Control Charts

The industrial hygiene program often has some historical data on which to develop "trial control limits." These data are invaluable in getting a start on an individual problem. As experience is gained, the control limits are recalculated, and decisions based on these control limits are better. The advantage of initially using historical data is the gain of several weeks to several or many months in developing effective control limits.

7.1.5 Calculation of Control Limits

The calculation of control limits is a straightforward exercise. Grant (42), Linch (34), and others (3, 58, 60–62) have shown the mechanics of such calculations. Another presentation (8) develops the following format:

1. Calculate R for each set of measurements.
2. Calculate $\bar{R}$ from the sum of R divided by the number of sets (or subgroups) k.
3. Calculate the upper control limit (approximately $3s$) on the range by $UCL_R = D_4\bar{R}$. There will be no lower control limit where there are six or fewer values in each set (subgroup) k. The value of D_4 for duplicate analyses from Table 24.2 is $D_4 = 3.27$.
4. Calculate the upper warning limit (approximately $2s$) on the range by $UWL_R = D_5\bar{R}$. The value of D_5 for duplicate analyses from Table 24.2 is 2.51.
5. Construct the control lines and plot the consecutive analyses on graph paper. Circle or highlight any values outside the control limits. Corrective action to investigate and eliminate the causes will be taken after each out-of-control point as it happens. Less

Table 24.2 Control Chart Lines Factors (58)

Factor	Two Measurements per Set	Three Measurements per Set
D_4	3.27	2.57
D_5	2.51	2.05
A_2	1.88	1.02

vigorous follow-up should happen for each value outside the warning limit but within the control limit.

6. Calculate $\overline{X}$ for each set of measurements.
7. Calculate $\overline{\overline{X}}$ or the mean of the $\overline{X}$'s for the set of measurements.
8. Calculate the upper and lower control limits by $UCL_{\overline{X}} = \overline{\overline{X}} + A_2\overline{R}$ and $LCL_{\overline{X}} = \overline{\overline{X}} - A_2\overline{R}$. The value of A_2 for duplicate analyses from Table 24.2 is 1.88.
9. Calculate the upper and lower warning limits by $UWL_{\overline{X}} = \overline{\overline{X}} + (2/3)\ A_2\overline{R}$ and $LWL_{\overline{X}} = \overline{\overline{X}} - (2/3)\ A_2\overline{R}$.
10. Construct control lines and plot the consecutive analyses on graph paper. See step 5.

The development of trial control limits from the laboratory's own experience is most appropriate and realistic. However other sources of information can be used. Collaborative test data can be used for developing these trial limits until the laboratory's own experience is developed. The data of Table 24.3 is taken from a NIOSH report.

The calculation of trial control limit values from the data of Table 24.3 is set up in Table 24.4.

The calculation of control chart lines using the information from Table 24.4 and factor values from Table 24.2 are given in Table 24.5.

7.1.6 Interpretation of Control Limits

Standardized control chart calculation and graphing formats are available from many sources (3, 8, 42). The trial control limits were calculated for one concentration range. Other parts of the working range may require different control limits. The NIOSH collaborative test data (12) indicate that at least for all levels except the level near the detection limit of the method, a single percentage value could be used to express the variability. In such a case it would be easy to convert all ranges into a percentage value and calculate for all values a "$\%R$." The accuracy or bias plot then becomes useful over the working range by plotting the percentage deviation of the average of each set of control tubes from its known or nominal value. The control tubes would ideally be generated using known concentrations of airborne contaminant but could also be made

Table 24.3 Benzene Found in Duplicate Analyses (12)

Laboratory	Benzene (mg/tube)
1.	1.50, 1.70
2.	1.53, 1.73
3.	1.56, 1.57
4.	1.67, 1.88
5.	1.40, 1.98
6.	1.33, 1.42
7.	1.45, 1.62
8.	1.49, 1.69
9.	1.42, 1.61
10.	1.12, 1.27
11.	1.36, 1.53
12.	1.43, 1.57
13.	1.59, 1.59
14.	1.38, 1.55
15.	1.17, 1.44

Table 24.4 Data for Calculation of Control Chart Lines

Laboratory	Measurements		Average, $\bar{X}$	Range, R
	1st	2nd		
1	1.50	1.70	1.60	0.20
2	1.53	1.73	1.63	0.20
3	1.56	1.57	1.57	0.01
4	1.67	1.88	1.78	0.21
5	1.40	1.98	1.69	0.58
6	1.33	1.42	1.38	0.09
7	1.45	1.62	1.54	0.17
8	1.49	1.69	1.59	0.20
9	1.42	1.61	1.52	0.19
10	1.12	1.27	1.20	0.15
11	1.36	1.53	1.45	0.17
12	1.43	1.57	1.50	0.14
13	1.59	1.59	1.59	0.00
14	1.38	1.55	1.47	0.17
15	1.17	1.44	1.31	0.27
Σ			22.82	2.75
$\bar{\bar{X}}$			1.52	
$\bar{R}$				0.18

Table 24.5 Calculation of Control Chart Lines

Formula	Example
1. $\bar{R} = \Sigma R \div k$	1. $\bar{R} = 2.75 \div 15 = 0.18$
2. $UCL_R = D_4\bar{R}$	2. $UCL_R = 3.27 \times 0.18 = 0.59$
3. $UWL_R = D_5\bar{R}$	3. $UWL_R = 2.51 \times 0.18 = 0.45$
4. $\bar{\bar{X}} = \Sigma\bar{X} \div k$	4. $\bar{\bar{X}} = 22.82 \div 15 = 1.52$
5. $UCL_{\bar{X}} = \bar{\bar{X}} + A_2\bar{R}$	5. $UCL_{\bar{X}} = 1.52 + 1.88 \times 0.18 = 1.86$
6. $LCL_{\bar{X}} = \bar{\bar{X}} - A_2\bar{R}$	6. $LCL_{\bar{X}} = 1.52 - 1.88 \times 0.18 = 1.18$
7. $UWL_{\bar{X}} = \bar{\bar{X}} + \frac{2}{3}A_2\bar{R}$	7. $UWL_{\bar{X}}1.52 + \frac{2}{3} \times 0.34 = 1.75$
8. $LWL_{\bar{X}} = \bar{\bar{X}} - \frac{2}{3}A_2\bar{R}$	8. $LWL_{\bar{X}} = 1.52 - \frac{2}{3} \times 0.34 = 1.29$

by spiking a known quantity of the liquid contaminant onto the charcoal in the sampling tube. The latter technique is essentially that used for the determination of desorption efficiency. Analogous techniques for producing control samples can be developed for other measurement systems.

As the industrial hygiene program gains experience with a measurement system, the current data are used to calculate new control limits to replace the trial control limits. Periodically, the validity of the control limits should be checked and recalculated if appropriate.

7.2 Data Analysis Techniques

Numerous data analysis techniques are useful in the industrial hygiene program. Standard textbooks and handbooks such as the NBS handbook by Natrella (63) and quality control manuals (3) should be consulted for approaches and example problems. Consultation with a statistician can be of great help in selecting and applying the most appropriate approach to a specific data analysis problem.

7.3 Data Validation

Data validation is accomplished by a critical review of a set of data based on previously determined criteria. For large amounts of data from automatic measurement systems, a computer program may be used (8). For more usual industrial hygiene program sets of data, the analyst or surveyor and supervisor should determine whether the set of data conforms to the program's requirements for precision and accuracy. At least a spot check to assure the accuracy of calculations and conversions should be made. The measurement or analytical report (may be a report form) should be checked for transcription errors, omissions, and mistakes. Any outlier values or values widely different from the expected, based on a history of measurements from the area of the survey, should be specifically checked out to assure the industrial hygiene program that a real change in plant operations rather than a measurement error has taken place. This determination of consistency with past (expected) values may not be possible. An efficient way of automatically alerting the industrial hygiene program to such a situation is to develop a control chart based on operational experience. Trends and changes in plant operations may be more readily detected and interpreted using such an approach. Linch (34) discusses such a case, a lead in urine analysis program.

7.4 Sampling Plans

Sampling plans for acceptance sampling of such items as detector tubes can be an efficient tool for the industrial hygiene program. Their use in developing a sampling program for environmental or biological monitoring is discussed elsewhere in this book.

8 QUALITY CONTROL PROGRAM AUDIT

8.1 Internal Evaluation

The industrial hygiene program should periodically seek assurance that the elements of the quality control system are living up to the program's expectations for the "return on invested resources." Other program elements not directly related to quality control, such as compliance with statutory rules and regulations, and other controlling programmatic standards such as company or agency policies, should also be audited.

Auditing of operating procedures can be accomplished having a supervisor or an operator/analyst other than the person conducting the routine measurements or analyses perform the procedures. A second set of calibration equipment and reference standards can be acquired for checking sampling equipment calibration procedures and operational equipment. A "transfer standard" such as NBS uses for some basic measurements could be employed by several programs in a cooperative program.

8.2 Self-Appraisal Quality Control Check List

Ratliff (56) has adapted standard manufacturing quality program audits and vendor qualification procedures to the needs of the analytical laboratory. The following checklist has been modified to more broadly cover the needs of the overall industrial hygiene program laboratory functions. Ratliff used a standard of 3.8 average score as acceptable. An average score of 2.5 to 3.7 indicated a need for improvement. It was felt that an average score of less than 2.5 indicated a risky situation, requiring immediate correction.

INDUSTRIAL HYGIENE PROGRAM SELF-APPRAISAL QUALITY CONTROL CHECKLIST

Score

1. The acceptance criteria for the level of quality of the industrial hygiene laboratory's routine performance are:

a. Clearly defined in writing for all key or critical characteristics. 5
b. Defined in writing for some characteristics, and some are dependent on experience, memory, and/or verbal communication. 3
c. Only defined by experience and verbal communication. 1 ______

2. Acceptance criteria for the level of quality of the industrial hygiene program's routine performance are determined by:

a. Monitoring program performance in a structured program of inter- and intralaboratory evaluations. 5
b. Program determination of what is technically feasible. 3
c. Program determination of what can be done using currently available equipment, techniques, and manpower. 1 ______

3. The quality control coordinator has the authority to:

a. Affect the quality of measurements by inserting controls to assure that the methods meet the user's needs for precision, accuracy, sensitivity, and specificity. 5
b. Reject suspected results and stop any method that produces high levels of discrepancies. 3
c. Submit suspected results to laboratory management for a decision on disposition. 1 ______

4. Accountability for quality is:

a. Clearly defined for all program elements and their chiefs where their actions have an impact on quality. 5
b. Vested with the quality control coordinator, who must use whatever means possible to achieve quality goals. 3
c. Not defined. 1

5. "Quality" in the industrial hygiene program long-range planning:

    a. Is considered an important factor with regard to changing user demands, new applications, legal considerations, and technical advances in control and methods. 5
    b. Is considered part of the technology or service. 3
    c. Is not considered a factor for planning purposes. 1 ______

6. Calibration, measuring, gauging, and analytical instruments are:

    a. Maintained operative, accurate, and precise by regular checks and calibrations against stable standards which are traceable to the National Bureau of Standards. 5
    b. Periodically checked against a zero point or other reference and examined for evidence of physical damage, wear, or inadequate maintenance. 3
    c. Checked only when they stop working or when excessive defects are experienced that can be traced to inadequate instrumentation. 1 ______

7. Reagents and chemicals (critical items) and sampling system components such as detector tubes are:

    a. Procured from suppliers who must submit samples for test and approval before initial shipment. 5
    b. Procured from suppliers who certify that they can meet all applicable specifications. 3
    c. Procured from suppliers on the basis of price and delivery only. 1 ______

8. Reagents, chemicals, and sampling systems (or components) are:

    a. Checked 100 percent against specification and quantity, and for certification where required, and accepted only if they conform to all specifications. 5
    b. Spot-checked for proper quantity and for shipping damage. 3
    c. Released to program personnel by the receiving clerk without being checked as described in a or b. 1 ______

9. Discrepant purchased systems and materials are:

    a. Submitted to a review by quality control and chief chemist for disposition. 5

b. Submitted to the operational program elements for determination on acceptability. 3
c. Used because of scheduling requirements. 1 ______

10. Inventories are maintained on:

a. First-in, first-out basis. 5
b. Random selection in stock room. 3
c. Last-in, first-out basis. 1 ______

11. Inventories are:

a. Identified with respect to type, age, and acceptance status. 5
b. Identified with respect to material only. 3
c. Not identified in writing. 1 ______

12. Reagents and chemicals and sampling system components (e.g., detector tubes) that have limited shelf life are:

a. Identified with respect to shelf life expiration date and systematically issued from stock only if they are still within that date. 5
b. Issued on a first-in, first-out basis, expecting that there is enough safety factor that the expiration date is rarely exceeded. 3
c. Issued at random from stock. 1 ______

13. The operating conditions of the methods are:

a. Clearly defined in writing in the method for each significant variable. 5
b. Controlled by supervision based on general guidelines. 3
c. Left up to the field personnel or bench chemist/analyst. 1 ______

14. Operational procedures are checked:

a. During the measurements for conformity to operating conditions and to specifications. 5
b. After the measurements or analyses to determine acceptability of the results. 3
c. Not at all. 1 ______

15. Revisions to technical operational procedures and sampling/analytical methods are:

a. Clearly spelled out in written form and distributed to all parties affected on a controlled basis, which assures that the change will be implemented and permanent. 5
b. Communicated through memoranda to key people who are responsible for effecting the change through whatever method they choose. 3
c. Communicated verbally to operating personnel, who depend on experience to maintain continuity of the change. 1 ______

16. Changes to methods and other operational procedures are:

a. Analyzed to make sure that any harmful side effects are known and controlled before revision implementation. 5
b. Installed on a trial or gradual basis, monitoring the product to see whether the revision has a net beneficial effect. 3
c. Installed immediately with action for correcting side effects taken if they are evident in the final results. 1 ______

17. Revisions to operational procedures and sampling/analytical methods are:

a. Recorded with respect to date, serial number, and so on, when the revision becomes effective. 5
b. Recorded with respect to the date the revision was made on written specifications. 3
c. Not recorded with any degree of precision. 1 ______

18. The capability of the measurement or sampling/analytical method to produce within specification limit is:

a. Known through method capability analysis ($\overline{X}$–R charts) to be able to produce consistently acceptable results. 5
b. Assumed to be able to produce a reasonably acceptable result 3
c. Unknown. 1 ______

19. Measurement or sampling/analytical method determination discrepancies are:

a. Analyzed immediately to seek out the causes and apply corrective action. 5

b. Checked out when time permits. 3
c. Not detectable with present controls and procedures. 1 ______

20. Decisions on acceptability of questionable results are made by:

a. A review group consisting of the chief chemist, the chief field industrial hygienist, quality control personnel, and others who can render expert judgment. 5
b. An informal assessment by quality control. 3
c. The bench chemist/analyst or field personnel. 1 ______

21. Final acceptance of the results is made:

a. By replicating statistically adequate samples. 5
b. By routine acceptance because of lack of complaints. 3
c. On faith. 1 ______

22. Follow-up action is:

a. Taken to identify assignable causes for "suspect" determinations. 5
b. Taken to make sure that method errors have been corrected. 3
c. Not considered necessary. 1 ______

23. Data reports on quality are distributed to:

a. All levels of management. 5
b. One level of management only. 3
c. Quality control only. 1 ______

24. Quality reports contain:

a. Information on trends, required action, and danger spots. 5
b. Information on suspected analyses and their causes. 3
c. Number of analyses per month. 1 ______

25. Quality control analysis is performed to:

a. Seek out the optimum levels of operation. 5
b. Provide for highlighting trouble spots. 3
c. Fix the blame for substandard results. 1 ______

26. When key personnel changes occur:

a. Specialized knowledge and skills are retained in the form of documented methods and descriptions. 5
b. Replacement people can acquire the knowledge of their predecessors from co-workers, supervisors, and detailed study of specifications and memoranda. 3
c. Knowledge is lost and must be regained through long experience or trial and error. 1 ______

27. The people who have an impact on quality, bench chemists, industrial hygiene field personnel, supervisors, and so on, are:

a. Trained in the reasons for and the benefits of standards of quality and the methods by which high quality can be achieved. 5
b. Told about quality only when their work falls below acceptable levels. 3
c. Reprimanded when quality deficiencies are directly traceable to an individual's work. 1 ______

28. Training of new employees is accomplished by:

a. A programmed system of training where elements of training, including quality standards, are incorporated in a training checklist; the employee's work is immediately rechecked by supervisors for errors or defects, and the information is fed back instantaneously for corrective action. 5
b. On-the-job training by the supervisor, who gives an overview of quality standards; details of quality standards are learned as normal results are fed back to the employee. 3
c. On-the-job learning, with training on the rudiments of the job by senior co-workers. 1 ______

29. Auditing of the quality control program is:

a. Performed on a random but regular basis, to verify that all quality procedures are being implemented and are effective. 5
b. Performed whenever a suspicion arises that there are areas of ineffective performance. 3
c. Never performed. 1 ______

30. If the costs of quality are known, the major portion of the expenditure is in:

a. Prevention. 5
b. Appraisal. 3
c. External or internal failure (duplicate determinations to correct errors; reruns). 1 ______

31. Corrective action to reduce failure costs is:

a. An ongoing program with specific objectives, measurements, and target dates. 5
b. Taken when deficient results threaten schedules. 3
c. Taken after considerable losses have occurred in the laboratory or in the field. 1 ______

32. The management, through the NIOSH Proficiency Analytical Testing program or other external evaluation mechanisms, regards the laboratory's performance quality as:

a. Significantly better than peer laboratories. 5
b. About the same as peer laboratories. 3
c. Significantly worse than peer laboratories. 1 ______

33. Support for laboratory quality goals and results is indicated by:

a. A clear statement of quality objectives by the top executive, with continuing visible evidence of sincerity to all levels of the organization. 5
b. Periodic meetings among the section heads of service, field operations, research and development, and quality assurance on quality objectives and progress toward their achievement. 3
c. A "one-shot" statement of the desire for product quality by the top executive, after which the quality control staff is on its own. 1 ______

34. The quality control system is:

a. Formalized and documented by a set of procedures clearly describing the activities necessary and sufficient to achieve desired quality objectives from initial design through to final delivery to the user. 5

b. Contained in operational methods and procedures or is implicit in those procedures; experience with the materials, product, and equipment is needed for continuity of control. 3

c. Undefined in any procedures and left to the current managers or supervisors to determine as the situation dictates. 1 ______

Summary

Strong points:

Weak points:

Improvement goals:

8.3 External Evaluations

External evaluations of environmental programs such as that for water laboratories (64) have been in existence for many years. Clinical laboratories have been involved in such programs at least since the passage of the Clinical Laboratories Improvement Act of 1967. The American Industrial Hygiene Association program has initiated (55) and validated (65) an accreditation program for industrial hygiene analytical laboratories. The Occupational Health and Safety Program Accreditation Commission (66) (administrative aspects are being handled by the AIHA) was set up by sponsoring occupational health and safety professional societies and has developed and validated (67) program standards and audit criteria for evaluating the overall quality of the occupational health and safety function of an employer (whether a private company, a nonprofit institution, or a governmental agency). Such external evaluation programs will assure industrial hygiene program management and facility management that the safety and health function compares favorably with peer groups, or will objectively identify areas of needed improvement to achieve that status. These programs are continuing, thus ensuring that operational capabilities are maintained over a long period of time.

9 SUMMARY

This chapter has covered the philosophy and scope of quality control in the industrial hygiene program. The elements of such a program have been grouped and developed for the areas of program management; equipment, standards, and facilities; laboratory analysis control; sample handling, storage, and delivery; statistical quality control techniques; and program audits. A high score on the Industrial Hygiene Program Self-Appraisal Quality Control Checklist indicates that an industrial hygiene quality control program is effective. External evaluation, however, provides corroborative evidence that may carry more weight with the users of the industrial hygiene program's efforts.

REFERENCES

1. John T. Hagan, *A Management Role for Quality Control.* American Management Association, New York, 1968, p. 18.
2. National Institute for Occupational Safety and Health, NIOSH Specification, "Industrial Hygiene Laboratory Quality Control Program Requirements," NIOSH, Cincinnati, Ohio, 1976.
3. National Institute for Occupational Safety and Health, *Industrial Hygiene Service Laboratory Quality Control Manual, TR No. 78,* NIOSH, Cincinnati, Ohio, 1974.
4. National Institute for Occupational Safety and Health, *NIOSH Manual of Sampling Data Sheets,* NIOSH, Cincinnati, Ohio, 1974.
5. J. V. Crable and D. G. Taylor, *NIOSH Manual of Analytical Methods.* U.S. Department of Health, Education and Welfare Publication No. (NIOSH) 75-121, Cincinnati, Ohio, 1975.
6. Stanford Research Institute, "Laboratory Validation of Air Sampling Methods Used to Determine Environmental Concentrations in Work Places," NIOSH Contract No. CDC-99-7445, NIOSH, Cincinnati, Ohio, 1976.
7. American Society for Testing and Materials, *1975 Annual Book of ASTM Standards,* Part 26, *Gaseous Fuels; Coal and Coke; Atmospheric Analysis.* ASTM, Philadelphia, 1975.
8. Quality Assurance and Environmental Monitoring Laboratory, *Quality Assurance Handbook for Air Pollution Measurement Systems,* Vol. 1, *Principles,* U.S. Environmental Protection Agency, Research Triangle Park, N.C., 1975.
9. R. L. Crim, Chairman, Water Monitoring Task Force, Ed. *Model State Water Monitoring Program,* Environmental Protection Agency, No. EPA-44019-74-002, Washington D.C., 1975.
10. Division of Training and Manpower Development, NIOSH, *Announcement of Courses,* U.S. Department of Health, Education and Welfare Publication No. (NIOSH) 75-170, Cincinnati, Ohio, 1975.
11. Division of Training and Manpower Development, NIOSH, *Training Resources Manual,* NIOSH, Cincinnati, Ohio, 1975.
12. L. R. Rechner and J. Sachdev, "Collaborative Testing of Activated Charcoal Sampling Tubes for Seven Organic Solvents," NIOSH Contract No. HSM 99-72-98, U.S. Department of Health, Education and Welfare Publication No. (NIOSH) 75-184, Cincinnati, Ohio, 1975.
13. M. Lippmann, "Instruments and Techniques Used in Calibrating Sampling Equipment," in: *The Industrial Environment: Its Evaluation and Control,* Stock No. 1701-00396, Government Printing Office, Washington, D.C., 1973, Chapter 11.
14. National Institute for Occupational Safety and Health, "Criteria for a Recommended Standard . . . Occupational Exposure to Sodium Hydroxide," U.S. Department of Health, Education and Welfare Publication No. (NIOSH) 76-105, Rockville, Md., 1975.
15. B. E. Saltzman, "Preparation of Known Concentrations of Air Contaminants," in: *The Industrial Environment: Its Evaluation and Control,* Stock No. 1701-00396. Government Printing Office. Washington, D.C., 1973, Chapter 12.
16. American Conference of Governmental Industrial Hygienists, *Air Sampling Instruments,* 4th ed., ACGIH, Cincinnati, Ohio, 1972.

17. B. A. Johnson, "Evaluation of Portable, Direct Reading NO_2 Meters," U.S. Department of Health, Education and Welfare, Publication No. (NIOSH) 74-108, Cincinnati, Ohio, 1974.
18. C. S. McCammon, "Evaluation of Portable, Direct-Reading Combustible Gas Meters," U.S. Department of Health, Education and Welfare Publication No. (NIOSH) 74-107, Cincinnati, Ohio, 1974.
19. C. D. Parker, and R. B. Strong, "Evaluation of Portable Direct-Reading Carbon Monoxide Meters," NIOSH Contract No. HSM-99-73-1 (T.O. No. 1), U.S. Department of Health, Education and Welfare Publication No. (NIOSH) 75-106, Cincinnati, Ohio, 1974.
20. R. P. Madden, "Ultraviolet Transfer Standard Detectors and Evaluation and Calibration of NIOSH UV Hazard Meter," Interagency Agreement No. NIOSH-IA-73-20, U.S. Department of Health, Education and Welfare Publication No. (NIOSH) 75-131, 1975.
21. R. Mavrodineanu and J. W. Lazor, "Standard Reference Materials: Standard Quartz Cuvettes for High-Accuracy Spectrophotometry," *Clin. Chem.* **19:** 9, 1053–1057 (1973).
22. W. E. Hill, Jr., "Analytical Standards for Quality Controlled Analysis of Ore and Pilot Plant Products," *Am. Lab.* **8:** 2, 65–67 (1976).
23. W. J. Youden and E. H. Steiner, *Statistical Manual of the Association of Official Analytical Chemists,* AOAC, Washington, D.C., 1975.
24. American Society for Testing and Materials, *ASTM Manual for Conducting an Inter-Laboratory Study of a Test Method,* ASTM Special Technical Publication No. 335. Available from University Microfilms, Ann Arbor, Mich., 1963.
25. H. H. Ku, Ed., "Precision Measurement and Calibration," NBS Special Publication No. 300, Vol. 1, Government Printing Office, Washington, D.C., 1969.
26. American Society for Testing and Materials, Committee D-19 on Water, "Practice for Determination of Precision of Methods of Committee D-19 on Water—D-2777," in: *1975 Annual Book of ASTM Standards,* Part 31, *Water,* ASTM, Philadelphia, 1975.
27. American Society for Testing and Materials, "Practice for Developing Precision Data on ASTM Methods for Analysis and Testing of Industrial Chemicals—E180," in: *1974 Annual Book of ASTM Standards,* Part 30, *General Methods,* ASTM, Philadelphia, 1974.
28. J. F. Foster and G. H. Beatty, "Interlaboratory Cooperative Study of the Precision and Accuracy of the Measurement of Sulfur Dioxide Content in the Atmosphere Using ASTM Method D 2914," ASTM, Publication No. DS 55-S1, Philadelphia, 1974.
29. H. F. Hamil and D. E. Camann, "Collaborative Study of Method for the Determination of Nitrogen Oxide Emissions from Stationary Sources. U.S. Environmental Protection Agency Contract No. 68-02-0623, EPA, Research Triangle Park, N.C., 1973.
30. H. F. Hamil, "Laboratory and Field Evaluations of EPA Methods 2, 6 and 7," U.S. Environmental Protection Agency Contract No. 68-02-0626, EPA, Research Triangle Park, N.C., 1973.
31. H. F. Hamil and D. E. Camann, "Collaborative Study of Method for the Determination of Sulfur Dioxide Emissions from Stationary Sources," U.S. Environmental Protection Agency, Contract No. 68-02-0623, EPA, Research Triangle Park, N.C., 1973.
32. J. A. Winter and H. A. Clements, "Interlaboratory Study of the Cold Vapor Technique for Total Mercury in Water," in: *Water Quality Parameters,* American Society for Testing and Materials, Special Technical Publication No. 573, ASTM, Philadelphia, 1975, pp. 566–580.
33. J. Mandel and T. W. Lashof, "Interpretation and Generalization of Youdens' Two-sample Diagram," *J. Qual. Technol.,* **6:** 1, 22–36 (1974).
34. A. L. Linch, "Quality Control for Sampling and Laboratory Analysis," in: *The Industrial Environment: Its Evaluation and Control,* Stock No. 1701–00396, Government Printing Office, Washington, D.C., 1973, Chapter 22.
35. F. J. Linning, J. Mandel, and J. M. Peterson, "A Plan for Studying the Accuracy and Precision of an Analytical Procedure," *Anal. Chem.* **26:** 7, 1102–1110 (1954).

36. A. L. Wilson, "The Performance-Characteristics of Analytical Methods, I," *Talanta,* **17,** 21–29 (1970).
37. A. L. Wilson, "The Performance-Characteristics of Analytical Methods, II," *Talanta,* **17,** 31–44 (1970).
38. A. L. Wilson, "The Performance Characteristics of Analytical Methods, III," *Talanta,* **20,** 725–732 (1973).
39. National Institute for Occupational Society and Health, "Organic Solvents in Air," P&CAM No. 127, in: *NIOSH Manual of Analytical Methods,* U.S. Department of Health, Education and Welfare Publication No. (NIOSH) 75-121, 1975.
40. G. M. Bobba and L. F. Donaghey, "A Microcomputer System for Analysis and Control of Multiple Gas Chromatography," *Am. Lab.* **8:** 2, 27–34 (1976).
41. S. A. Roach, "Sampling Air for Particulates," in: *The Industrial Environment: Its Evaluation and Control,* Stock No. 1701-00396, Government Printing Office, Washington, D.C., 1973, Chapter 13.
42. E. L. Grant, *Statistical Quality Control,* 3rd ed., McGraw-Hill, New York, 1964.
43. B. A. Punghorst, "Methods for Detection and Elimination of Determinate Errors," in: *Statistical Method—Evaluation and Quality Control for the Laboratory,* Training Course Manual in Computational Analysis, Department of Health, Education and Welfare, Public Health Service, Washington, D.C., 1968.
44. Intersociety Committee on Air Sampling and Analysis, *Methods of Air Sampling and Analysis,* American Public Health Association, Washington D.C., 1972.
45. A. L. Linch, *Evaluation of Ambient Air Quality by Personnel Monitoring,* CRC Press, Cleveland, 1974.
46. A. L. Linch, *Biological Monitoring for Industrial Chemical Exposure Control,* CRC Press, Cleveland, 1974.
47. J. Mandel, "Repeatability and Reproducibility," *Mater. Res. Stand.,* **11:**8, 8–16 (1971).
48. D. A. B. Lindberg and H. J. Van Peenen, "The Meaning of Quality Control with Multiple Chemical Analyses," paper presented at Technicon Symposium, in: *Automation in Analytical Chemistry,* New York, 1965.
49. M. H. Shamos, in: *Proceedings of the 1974 Joint Measurement Conference,* Instrument Society of America, Pittsburgh, 1974.
50. R. B. Conn, "Effects of Automation on Clinical Laboratory Operations," in: *Proceedings of the 1974 Joint Measurement Conference,* Instrument Society of America, Pittsburgh, 1974.
51. American Industrial Hygiene Association, Analytical Committee, *Quality Control for the Industrial Hygiene Laboratory,* AIHA, Akron, Ohio, 1975.
52. Intersociety Committee on Air Sampling and Analysis, *Methods of Air Sampling and Analysis,* American Public Health Association, Washington, D.C., 1972, p. 70.
53. G. R. Elwell and H. E. Lawton, "A Relative Value Structure Helps Laboratory Management Fight the Numbers Racket," *Health Lab. Sci.,* **10:** 3, 203–208 (1973).
54. Clinical Laboratories Improvement Act of 1967—Notice of Effective Date, 42 CFR Part 74, *Fed. Reg.* **33:** 253, December 31, 1968; as amended.
55. L. J. Cralley et al., "Guidelines for Accreditation of Industrial Hygiene Analytical Laboratories," *Am. Ind. Hygiene Assoc. J.,* **31,** 335 (1970).
56. T. A. Ratliff, Jr., "Laboratory Quality Program Requirements," paper presented at 30th Annual Technical Conference Transactions, American Society for Quality Control, Milwaukee, 1976.
57. C. B. Donnelly, et al., "Containers, Refrigerants and Insulation for Split Milk Samples," *J. Milk Food Technol.,* **21:** 5, 131–137 (1958).
58. American Society for Testing and Materials, *ASTM Manual on Quality Control of Materials,* Special Technical Publication 15-C, ASTM, Philadelphia 1951.

59. C. C. Craig, "The $\bar{X}$- and R-Chart and Its Competitors," *J. Qual. Technol.,* **1:** 2 (1969).
60. W. D. Kelley, Ed., *Statistical Method—Evaluation and Quality Control for the Laboratory,* Training Course Manual in Computational Analysis, Department of Health, Education and Welfare, Public Health Service, Washington, D.C., 1968.
61. Analytical Quality Control Laboratory, *Handbook for Analytical Quality Control in Water and Wastewater Laboratories,* U.S. Environmental Protection Agency, Cincinnati, Ohio, 1972.
62. F. S. Hillier, "$\bar{X}$- and R-Chart Control Limits Based on a Small Number of Subgroups," *J. Qual. Technol.,* **1:** 1, 17–26 (1969).
63. M. G. Natrella, *Experimental Statistics,* NBS Handbook No. 91, Government Printing Office, Washington D.C., 1963.
64. "Evaluation of Water Laboratories—Recommended by the U.S. Public Health Service," Public Health Service Publication No. 999-EE-1, Washington, D.C., 1966.
65. American Industrial Hygiene Association, "Development of a Laboratory Accreditation Program for Occupational Health Laboratories," NIOSH Contract No. HSM99-72-58, NIOSH, Cincinnati, Ohio, 1975.
66. Occupational Health Programs Accreditation Commission, *Transactions of the 37th Annual Meeting of the American Conference of Governmental Industrial Hygienists,* ACGIH, Cincinnati, Ohio, 1975, pp. 105–108.
67. Occupational Health Institute, "Develop and Validate Criteria for Performance Standards of Occupational Health Programs," NIOSH Contract No. HSM-99-72-109, NIOSH, Washington, D.C., 1975.

CHAPTER TWENTY-FIVE

Calibration

MORTON LIPPMANN, Ph.D.

1 INTRODUCTION

The proper interpretation of any environmental measurement is dependent on an appreciation of its accuracy, precision, and whether it is representative of the condition or exposure of interest. This chapter is concerned primarily with the accuracy and precision of industrial hygiene measurements. Although considerations of the location, duration, and frequency of measurements may be equally or even more important in the evaluation of potential hazards, they require knowledge of the process variables, the kinds of hazard and/or toxic effect that may result from exposures, and their temporal variations. Such considerations, which require the exercise of professional judgment, are beyond the scope of this chapter.

The accuracy of a given measurement is dependent on a variety of different factors including the sensitivity of the analytical method, its specificity for the agent or energy being measured, the interferences introduced by cocontaminants or other radiant energies, and the changes in response resulting from variations in ambient conditions or instrument power levels. In some cases the influence of these variables can be defined by laboratory calibration, providing a basis for correcting a field sample or instrument reading response. In cases such as the effect of variable line voltage on an instrument's response, they can be avoided by modifications in the circuitry, or the addition of a constant voltage transformer. When the effects of the interferences cannot be controlled or well defined, it may still be desirable to make field measurements, especially in range-finding and exploratory surveys. The interpretation of any such measurements is greatly aided by an appreciation of the extent of the uncertainties. Laboratory calibrations can provide the basis for such an appreciation.

2 TYPES OF CALIBRATIONS

Occupational health problems can arise from exposure to airborne contaminants, heat stress, excessive noise, vibration, ionizing radiations, and nonionizing electromagnetic radiations. Each of these types of exposure involves a different set of measurement variables and calibration considerations, and they are considered separately. Other types of measurements requiring calibration are associated with the evaluation of ventilation systems used to control exposures to airborne contaminants.

2.1 Air Sampling Instruments

Air samples are collected to determine the concentrations of one or more airborne contaminants. To define a concentration, the quantity of the contaminant of interest per unit volume of air must be ascertained. In some cases the contaminant is not extracted from the air (i.e., it may simply alter the response of a defined physical system). An example is the mercury vapor detector, where mercury atoms absorb the characteristic ultraviolet radiation from a mercury lamp, reducing the intensity incident on a photocell. In this case the response is proportional to the mercury concentration, not to the mass flowrate through the sensing zone; hence concentration is measured directly.

In most cases, however, the contaminant is either recovered from the sampled air for subsequent analysis or is altered by its passage through a sensor within the sampling train, and the sampling flowrate must be known to be able to ultimately determine airborne concentrations. When the contaminant is collected for subsequent analysis, the collection efficiency must also be known, and ideally it should be constant. The measurements of sample mass, of collection efficiency, and of sample volume are usually done independently. Each measurement has its own associated errors, and each contributes to the overall uncertainty in the reported concentration.

The sample volume measurement error often is greater than that of the sample mass measurement. The usual reason is that the volume measurement is made in the field with devices designed more for portability and light weight than for precision and accuracy. Flowrate measurement errors can further affect the determination if the collection efficiency is dependent on the flowrate.

Each element of the sampling system should be calibrated accurately before initial field use. Protocols should also be established for periodic recalibration, since the performance of many transducers and meters changes with the accumulation of dirt, as well as with corrosion, leaks, and misalignment due to vibration or shocks in handling, and so on. The frequency of such recalibration checks should be high initially, until experience indicates that it can be reduced safely.

2.1.1 Flow and/or Volume

If the contaminant of interest is removed quantitatively by a sample collector at all flowrates, the sampled volume may be the only airflow parameter that need be recorded. On the other hand, when the detector response is dependent on both the flowrate and sample mass, as in many length-of-stain detector tubes, both quantities must be determined and controlled. Finally, in many direct reading instruments, the response is dependent on flowrate but not on integrated volume.

In most sampling situations the flowrates are, or are assumed to be, constant. When this is so, and the sampling interval is known, it is possible to convert flowrates to integrated volumes, and vice versa. Therefore flowrate meters, which are usually smaller, more portable, and less expensive than integrated volume meters, are generally used on sampling equipment even when the sample volume is the parameter of primary interest. Normally little additional error is introduced in converting a constant flowrate into an integrated volume, since the measurement and recording of elapsed time generally can be performed with good accuracy and precision.

Flowmeters can be divided into three groups on the basis of the type of measurement made: integrated volume meters, flowrate meters, and velocity meters. The principles of operation and features of specific instrument types in each group are discussed in succeeding pages. The response of volume meters, such as the spirometer and wet test meter, and flowrate meters, such as the rotameter and the orifice meter, are determined by the entire sampler flow. In this respect they differ from velocity meters such as the thermoanemometer and the Pitot tube, which measure the velocity at a particular point of the flow cross section. Since the flow profile is rarely uniform across the channel, the measured velocity invariably differs from the average velocity. Furthermore, since the shape of the flow profile usually changes with changes in flowrate, the ratio of point-to-average velocity also changes. Thus when a point velocity is used as an index of flowrate, there is an additional potential source of error, which should be evaluated in laboratory calibrations that simulate the conditions of use. Despite their disadvantages, velocity sensors are sometimes the best indicators available–for example, in some electrostatic precipitators, where the flow resistance of other types of meters cannot be tolerated. Velocity sensors are also used in measurements of ventilation airflow and to measure one of the components in the determination of heat stress.

2.1.2 Calibration of Collection Efficiency

A sample collector need not be 100 percent efficient to be useful, provided its efficiency is known and consistent and is taken into account in the calculation of concentration. In practice, acceptance of a low but known collection efficiency is reasonable procedure for most types of gas and vapor sampling, but it is seldom if ever appropriate for aerosol sampling. All the molecules of a given chemical contaminant in the vapor phase are essentially the same size, and if the temperature, flowrate, and other critical parameters are kept constant, the molecules will all have the same probability of capture. Aerosols, on the other hand, are rarely monodisperse. Since most particle capture mechanisms are size dependent, the collection characteristics of a given sampler are likely to vary with particle size. Furthermore, the efficiency will tend to change with time because of loading; for example; a filter's efficiency increases as dust collects on it, and electrostatic precipitator efficiency may drop if a resistive layer accumulates on the collecting electrode. Thus aerosol samplers should not be used unless their collection is essentially complete for all particle sizes of interest.

2.1.3 Recovery from Sampling Substrate

The collection efficiency of a sampler can be defined by the fraction removed from the air passing through it. However the material collected cannot always be completely recovered from the sampling substrate for analysis. In addition, the material sometimes is degraded or otherwise lost between the time of collection in the field and recovery in the laboratory. Deterioration of the sample is particularly severe for chemically reactive materials. Sample losses may also be due to high vapor pressures in the sampled material, exposure to elevated temperatures, or reactions between the sample and substrate or between different components in the sample.

Laboratory calibrations using blank and spiked samples should be performed whenever possible to determine the conditions under which such losses are likely to affect the determinations desired. When it is expected that the losses would be excessive, the sampling equipment or procedures should be modified as much as feasible to minimize the losses and the need for calibration corrections.

2.1.4 Sensor Response

When calibrating direct reading instruments, the objective is to determine the relation between the scale readings and the actual concentration of contaminant present. In such tests the basic response for the contaminant of interest is obtained by operating the instrument in known concentrations of the pure material over an appropriate range of concentrations. In many cases it is also necessary to determine the effect of such environmental cofactors as temperature, pressure, and humidity on the instrument response. Also, many sensors are nonspecific, and atmospheric cocontaminants may either elevate or depress the signal produced by the contaminant of interest. If reliable data on the effect of such interferences are not available, they should be obtained in calibration tests. Procedures for establishing known concentrations for such calibration tests are discussed in detail in Sections 4.6 and 4.7.

2.2 Ventilation System Measurements

2.1.1 Air Velocity Measurements

Most ventilation performance measurements are made with anemometers (i.e., instruments that measure air velocities), and with the exception of the Pitot tube, all require periodic calibration. Instruments based on mechanical or electrical sensors are sensitive to mechanical shocks and/or may be affected by dust accumulations and corrosion. Calibration requirements are indicated in Table 25.1.

Anemometers are usually calibrated in a well-defined flow field that is relatively large in comparison to the size of meter being calibrated. Such flow fields can be produced in wind tunnels, which are discussed in Section 4.5.

2.2.2 Pressure Measurements

Although the standard Pitot tube and a water-filled U-tube manometer may not require calibration, Pitot tubes and other flowmeters may be used with pressure gauges that do. Many direct reading gauges can give false readings because of the effects of mechanical shocks and/or leakage in connecting tubes or internal diaphragms. For pressures of approximately 1 in. H_2O or greater, a liquid-filled laboratory manometer should be an adequate reference standard. For lower pressures, it may be necessary to use a reference whose calibration is traceable to a National Bureau of Standards (NBS) standard. Calibration requirements for pressure gauges are given in Table 25.2.

2.3 Heat Stress Measurements

The four environmental variables used in heat stress evaluations are air temperature, humidity, radiant temperature, and air velocity. The measurement of air velocity is discussed in detail elsewhere in this chapter. Liquid-in-glass thermometers used to measure dry-bulb, wet-bulb, or globe thermometer temperatures should have calibrations traceable to certified NBS, but should not need recalibration. Other temperature sensors will require periodic recalibration.

2.4 Electromagnetic Radiation Measurements

The electromagnetic spectrum is a continuum of frequencies whose effects on human health are discussed in Chapters 11, 12, and 13. Most of these effects are frequency dependent, and some are attributable to narrow bands of frequency. Thus many of the instruments used to measure frequency and/or intensity are designed to operate over specific frequency regions. Other instruments, known as bolometers, measure the total incident radiant flux over a wide range of frequencies.

The literature on calibration techniques for the measurement of electromagnetic radiation is too extensive to adequately summarize in the space available. Instead, the reader is directed to Chapters 11 and 12 and to the reference works cited in those chapters.

2.5 Noise-Measuring Instruments

The accuracy of sound-measuring equipment may be checked by using an acoustical calibrator (Figure 25.1) consisting of a small, stable sound source that fits over a microphone and generates a predetermined sound level within a fraction of a decibel. The acoustical calibration provides a check of the performance of the entire instrument,

Table 25.1 Testing of Ventilation Systems: Characteristics of Air Meters

Instruments	Range (fpm)	Hole Size (for ducts) (in.)	Tempera-ture Range[a]	Dust, Fume Difficulty	Calibration Requirements	Ruggedness	General Usefulness and Comments
Pitot tubes with inclined manometer	600 and up	$^3/_8$	Wide	Some	None	Good	
Standard	600 and up	$^3/_8$	Wide	Some	None	Good	Good except at low velocities
Small size	600 and up	$^3/_{16}$	Wide	Yes	Once	Good	Good except at low velocities
Double	500 and up	$^3/_4$	Wide	Small	Once	Good	Special
Swinging vane anemometers							
Alnor Velometer	25–10,000	$^1/_2$–1	300°F	Some	Frequent	Fair	Good
Rotating vane anemometers							
Conventional	30–10,000	Not for duct use	Narrow	Yes	Frequent	Poor	Special; limited use
Electronic (airflow development)	25–200 25–500 25–2000 25–5000	Not for duct use	Narrow	Yes	Frequent	Poor	Special; can record; direct reading
Digital (airflow development)	200–5000	Not for duct use	Narrow	Yes	Frequent	Poor	Special
Bridle vane anemometers							
Florite Air Velocity Meter	50–2500	Not for duct use	Narrow	Yes	Frequent	Fair	Special

Heated-wire anemometers							
Anemotherm model 60	10–8000	3/8	Medium	Some	Frequent	Poor	Good
Anemotherm Gas Flow Meter	10–5000	1	300°F	No	Frequent	Good	Not portable; for permanent station airflow
Flowtronic Air Meter 55A	0–1000 1000–2000 2000–4000	1/2	Medium	Yes	Frequent	Poor	Good
Heated Thermocouple Anemometers							
Alnor Thermoanemometer model 8500	10–2000 2 scales	5/16	Narrow	Yes	Frequent	Poor	Good
Hasting Precision Air Meter B-22	10–500 500–10,000	5/16	Narrow	Yes	Frequent	Poor	Good
Flow Corporation Series 800	10–4000	1/2	Narrow	Yes		Poor	Good
Alnor Air Velocity Transducer System (AVT)	20–500 50–1000 100–2000	5/16	Narrow	Yes	Frequent	Poor	Special; for permanent station use
Variable area Meter							
Airmeter F. W. Dwyer Co.	200–1200 1000–4000	Not for duct use	Narrow	Yes	Occasional (needs cleaning)	Good	Satisfactory for estimates of flow

Source. Industrial Ventilation, courtesy of the American Conference of Governmental Industrial Hygienists.
[a] Narrow, 20–150°F; medium, 20–300°F; wide, 0–800°F.

Table 25.2 Characteristics of Pressure Measuring Instruments: Static Pressure, Velocity Pressure, and Differential Pressure

Instrument	Range (in. H_2O)	Manufacturer's Stated Precision (in. H_2O)	Comments
Liquid manometers			
Verticle U-tube	No limit	0.1	Portable; needs no calibration.
Inclined 10:1 slope	Usually up to 10	0.005	Portable; needs no calibration; must be leveled.
Hook gauge	0–24	0.001	Not a field instrument; tedious, difficult to read, for calibration only
Micromanometer (Meriam model 34FB2TM)	0–10 0–20	0.001	Heavy; need to locate on vibration-free surface; not difficult to read; uses magnifier
Micromanometer (Vernon Hill type C)	0.001–1.2	0.0004	Small; portable; uses magnifier; need experience to read to manufacturer's precision; calibration needed
Micromanometer (Electric Microtonic, F. W. Dwyer, Mfg.)	0–2	0.0003	Portable; needs vibration-free mount; no magnifier; slow to use; no eyestrain; no calibration needed
Diaphragm and mechanical			
Diaphragm-Magnehelic Gauge	0–0.5 0–1 0–4	0.01 0.02 0.10	Calibration recommended; no leveling, no mounting needed. Direct reading
Swinging vane anemometer Alnor Velometer	0–0.5 0–20	5% scale	Calibration recommended; no leveling, no mounting needed; use manufacturer's exact recommendation for size of SP hole
Pressure transducers and electronic instruments			
Pressure transducers	0.05–6	0.3%	Must be calibrated; remote reading responds to rapid change in pressure

Source. Industrial Ventilation, courtesy of the American Conference of Governmental Industrial Hygienists.

Figure 25.1 Sound level calibrator.

including microphone and electronics. Some sound level meters have internal means for calibration of electronic components only. Sound level calibrators should be used only with the microphones for which they are intended. Manufacturers' instructions should be followed regarding the use of calibrators and indications of malfunction of instruments.

3 CALIBRATION STANDARDS

Calibration procedures generally involve a comparison of instrument response to a standardized atmosphere or to the response of a reference instrument. Hence the calibration can be no better than the standards used. Reliability and proper use of standards are critical to accurate calibrations. Reference materials and instruments available from, or calibrated by the National Bureau of Standards should be used whenever possible. Information on calibration aids available from NBS is summarized in Table 25.3.

Test atmospheres generated for purposes of calibrating collection efficiency or instrument response should be checked for concentration using reference instruments or sampling and analytical procedures whose reliability and accuracy are well documented. The best procedures to use are those which have been referee or panel tested, that is, methods that have demonstrably yielded comparable results on blind samples analyzed by different laboratories. Such procedures are published by several organizations, listed in Table 25.4. Those published by the individual organizations have been supplemented in recent years by those approved by the Intersociety Committee on Methods for Air Sampling and Analysis, a cooperative group formed in March 1963 and composed of representatives of the Air Pollution Control Association (APCA), the American Conference of Governmental Industrial Hygienists (ACGIH), the American Industrial Hygiene Association (AIHA), the American Public Health Association (APHA), the American Society for Testing and Materials (ASTM), the American Society of Mechanical Engineers (ASME), and the Association of Official Analytical Chemists

Table 25.3 National Bureau of Standards Standard Reference Materials[a] for Calibration of Air Analysis Instruments and Procedures

SRM No.	Type	Description
1625	Permeation tube	Sulfur dioxide, 10 cm, 2.8 μg/min @ 25°C
1626	Permeation tube	Sulfur dioxide, 5 cm, 1.4 μg/min @ 25°C
1627	Permeation tube	Sulfur dioxide, 2 cm, 0.56 μg/min @ 25°C
1629	Permeation tube	Nitrogen dioxide, 0.5–1.5 μg/min @ 25°C
1604a	Gas cylinder	Oxygen in N_2, 1.5 ppm
1607	Gas cylinder	Oxygen in N_2, 212 ppm
1609	Gas cylinder	Oxygen in N_2, 20.95 mol %
1665	Gas cylinder	Propane in air, 2.8 ppm
1666	Gas cylinder	Propane in air, 9.5 ppm
1667	Gas cylinder	Propane in air, 48 ppm
1668	Gas cylinder	Propane in air, 95 ppm
1669	Gas cylinder	Propane in air, 475 ppm
1673	Gas cylinder	Carbon dioxide in N_2, 0.95%
1674	Gas cylinder	Carbon dioxide in N_2, 7.2%
1675	Gas cylinder	Carbon dioxide in N_2, 14.2%
1677	Gas cylinder	Carbon monoxide in N_2, 9.74 ppm
1678	Gas cylinder	Carbon monoxide in N_2, 47.1 ppm
1679	Gas cylinder	Carbon monoxide in N_2, 94.7 ppm
1680	Gas cylinder	Carbon monoxide in N_2, 484 ppm
1681	Gas cylinder	Carbon monoxide in N_2, 947 ppm
1683	Gas cylinder	Nitric oxide in N_2, 50 ppm
1684	Gas cylinder	Nitric oxide in N_2, 100 ppm
1685	Gas cylinder	Nitric oxide in N_2, 250 ppm
1686	Gas cylinder	Nitric oxide in N_2, 500 ppm
1687	Gas cylinder	Nitric oxide in N_2, 1000 ppm
1571	Solid material	Orchard leaves: certified for content of lead, mercury, and 17 others
1577	Solid material	Bovine liver: certified for content of lead, mercury, and 10 others
1579	Solid material	Powdered lead-based paint, lead: 11.187 wt. %
1631	Solid material	Coal: certified for sulfur content
1632	Solid material	Coal: certified for content of lead, mercury, and 120 others
1633	Solid material	Coal fly ash: certified for content of lead, mercury, and 10 others
1621	Residual fuel oil	Sulfur content, 1.05 wt. %
1622	Residual fuel oil	Sulfur content, 2.14 wt. %
1623	Residual fuel oil	Sulfur content, 0.268 wt. %
1624	Distillate fuel oil	Sulfur content, 0.211 wt. %
1634	Residual fuel oil	Trace elements
1641	Water	Mercury, 1.49 μg/ml
1642	Water	Mercury, 1.18 μg/ml
2671	Freeze-dried urine	Fluorine (two human urine samples; one low and one elevated in fluoride ion content)

Table 25.3 (Continued)

SRM No.	Type	Description
2672	Freeze-dried urine	Mercury (two human urine samples; one low and one elevated in mercury)
2675	Membrane filter	Beryllium salts with contents representative of industrial hygiene air samples
2676	Membrane filter	Quartz, three filters with content representative of industrial hygiene air samples

[a] Available from: Office of Standard Reference Materials
Room B 311 Chemistry Building
National Bureau of Standards
Washington, D.C. 20234

(AOAC). "Tentative" methods endorsed by the Intersociety Committee have been published at random intervals since April 1969 in *Health Laboratory Science,* a publication of APHA. These "tentative" methods become "standard" methods only after satisfactory completion of a cooperative test program. Published "tentative" and "standard" methods for air sampling and analysis are summarized in Table 25.5.

Table 25.4 Organizations Publishing Recommended or Standard Methods and/or Test Procedures Applicable to Air Sampling Instrument Calibration

Abbreviation	Full Name	Mailing Address
APCA	Air Pollution Control Association	440 Fifth Avenue Pittsburgh, Pa. 15213
ACGIH	American Conference of Governmental Industrial Hygienists	P. O. Box 1937 Cincinnati, Ohio 45201
AIHA	American Industrial Hygiene Association	475 Wolf Ledges Akron, Ohio 44311
ANSI	American National Standards Institute, Inc.	1430 Broadway New York, N.Y. 10018
ASTM	American Society for Testing and Materials, D-22 Committee on Sampling and Analysis of Atmospheres	1016 Race Street Philadelphia, Pa. 19103
EPA	Environmental Protection Agency, Office of Air Programs	5600 Fischer's Lane Rockville, Md. 20852
ISC	Intersociety Committee on Methods for Air Sampling and Analysis	250 West 57th St. New York, N.Y. 10019

Table 25.5 Summary of Recommended and Standard Methods Relating to Air Sampling and Instrument Calibration

Organization	Number of Methods	Types of Method	Panel Tested
ACGIH	19	Analytic methods for air contaminants	Yes
AIHA	93	Analytical guides	No
ISC	4	Recommended methods of air sampling and analysis	Yes
ISC	121	Tentative methods of air sampling and analysis	No[a]
ASTM	37	Sampling and analysis of atmospheres	No[b]
ASTM	7	Recommended practices for sampling procedures, nomenclature, etc.	NA[c]
APCA	3	Recommended standard methods for continuing air monitoring for fine particulate matter	NA[c]
ANSI	1	Sampling airborne radioactive materials	NA[c]
EPA	9	Reference and equivalent methods for air contaminants	No

[a] All tentative methods will be panel tested before advancing to reference methods.
[b] Methods are undergoing panel validation under ASTM Project Threshold.
[c] Not applicable.

4 INSTRUMENTS AND TECHNIQUES

4.1 Cumulative Air Volume

Many air sampling instruments utilize an integrating volume meter for measurement of sampled volume. Most of them measure displaced volumes that can be determined from linear measurements and geometric formulas. Such measurements usually can be made with a high degree of precision.

4.1.1 Water Displacement

Figure 25.2 is a schematic drawing of a Mariotti bottle. When the valve at the bottom of the bottle is opened, water drains out of the bottle by gravity, and air is drawn by way of a sample collector into the bottle to replace it. The volume of air drawn in is equal to the change in water level multiplied by the cross section at the water surface.

4.1.2 Spirometer or Gasometer

The spirometer (Figure 25.3) is a cylindrical bell with its open end under a liquid seal. The weight of the bell is counterbalanced so that the resistance to movement as air moves in or out of the bell is negligible. The device differs from the Mariotti bottle in that it measures displaced air instead of displaced liquid. The volume change is calculated in a similar manner (i.e., change in height times cross section). Spirometers are available in a wide variety of sizes and are frequently used as primary volume standard (1).

4.1.3 "Frictionless" Piston Meters

Cylindrical air displacement meters with nearly frictionless pistons are frequently used for primary flow calibrations. The simplest version is the soap bubble meter illustrated in Figure 25.4. It utilizes a volumetric laboratory burette whose interior surfaces are wetted with a detergent solution. If a soap-film bubble is placed at the left side, and suction is applied at the right, the bubble will be drawn from left to right. The volume displacement per unit time (i.e., flowrate) can be determined by measuring the time required for the bubble to pass between two scale markings which enclose a known volume.

Soap-film flowmeters and mercury-sealed piston flowmeters are available commercially from several sources (1). In the mercury-sealed piston, most of the cylindrical cross section is blocked off by a plate that is perpendicular to the axis of the cylinder. The plate is separated from the cylinder wall by an O-ring of liquid mercury, which

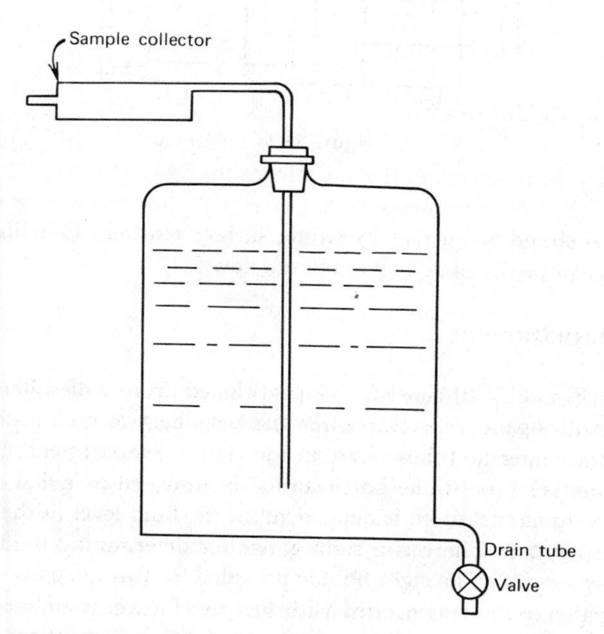

Figure 25.2 Mariotti bottle. From *The Industrial Environment—Its Evaluation and Control,* 2nd ed., C. H. Powell and A. D. Hosey, Eds., Public Health Service Publication No. 614, Government Printing Office, Washington, D.C., 1965.

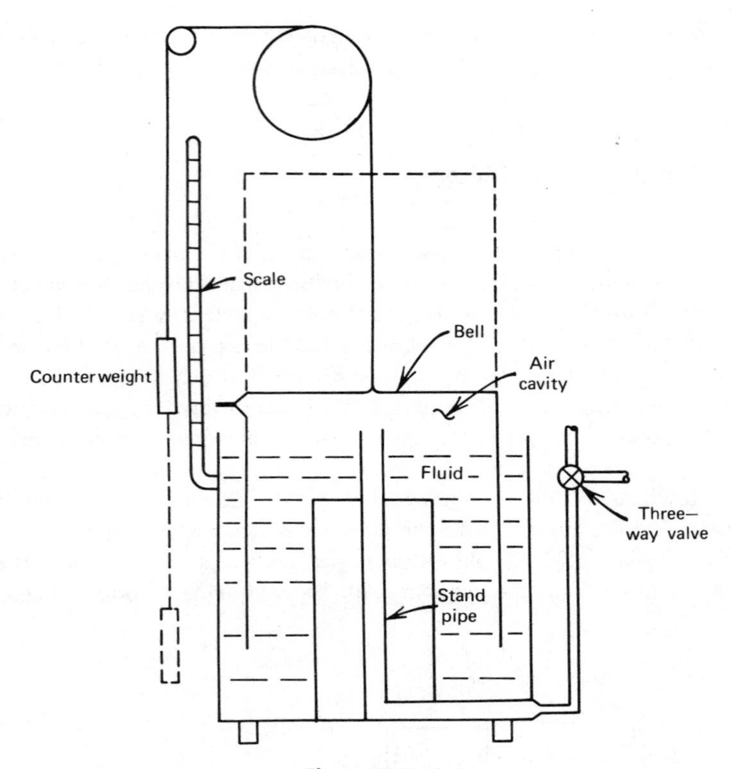

Figure 25.3 Spirometer.

retains its toroidal shape because of its strong surface tension. This floating seal has a negligible friction loss as the plate moves up and down.

4.1.4 Wet Test Meter

A wet test meter (Figure 25.5) consists of a partitioned drum half-submerged in a liquid (usually water), with openings at the center and periphery of each radial chamber. Air or gas enters at the center and flows into an individual compartment, causing the drum to rise, thereby producing rotation. This rotation is indicated by a dial on the face of the instrument. The volume measured is dependent on the fluid level in the meter, since the liquid is displaced by air. There is a sight gauge for determining fluid height, and the meter is leveled by screws and a sight bubble provided for this purpose.

Several potential errors are associated with the use of a wet test meter. The drum and moving parts are subject to corrosion and damage from misuse, there is friction in the bearings and the mechanical counter, inertia must be overcome at low flows (< 1 rpm), whereas at high flows (> 3 rpm), the liquid might surge and break the water seal at the

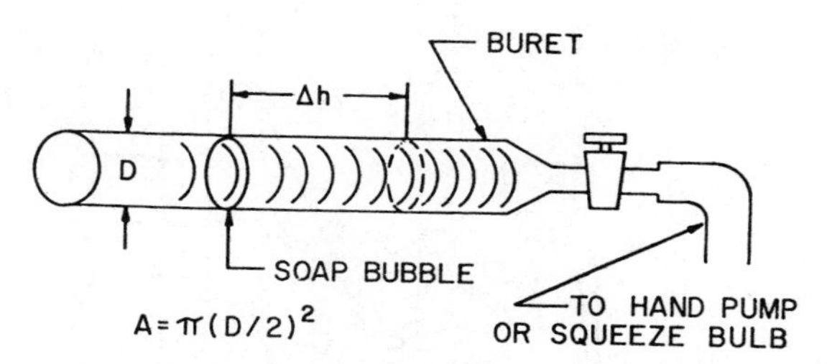

Figure 25.4 Bubble meter. From *The Industrial Environment—Its Evaluation and Control,* 2nd ed., C. H. Powell and A. D. Hosey, Eds., Public Health Service Publication No. 614, Government Printing Office, Washington, D.C., 1965.

inlet or outlet. In spite of these factors, the accuracy of the meter usually is within 1 percent when used as directed by the manufacturer.

4.1.5 Dry Gas Meter

The dry gas meter shown in Figure 25.6 is very similar to the domestic gas meter. It consists of two bags interconnected by mechanical valves and a cycle-counting device. The air or gas fills one bag while the other bag empties itself; when the cycle is completed, the valves are switched, and the second bag fills while the first one empties. Any such device has the disadvantages of mechanical drag, pressure drop, and leakage; however the advantage of being able to use the meter under rather high pressures and

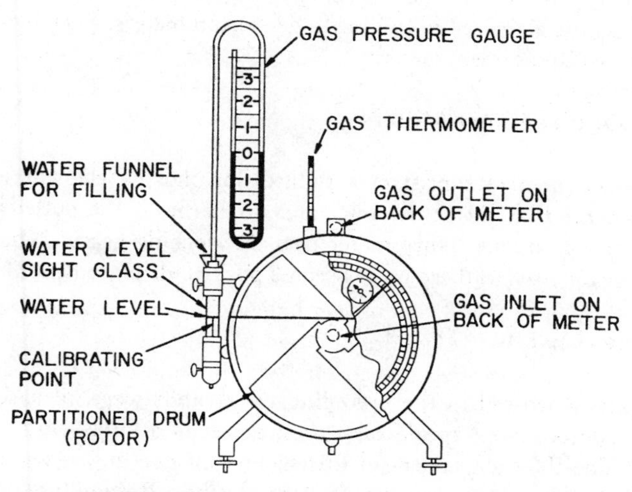

Figure 25.5 Wet test meter. From *The Industrial Environment—Its Evaluation and Control,* 2nd ed., C. H. Powell and A. D. Hosey, Eds., Public Health Service Publication No. 614, Government Printing Office, Washington, D.C., 1965.

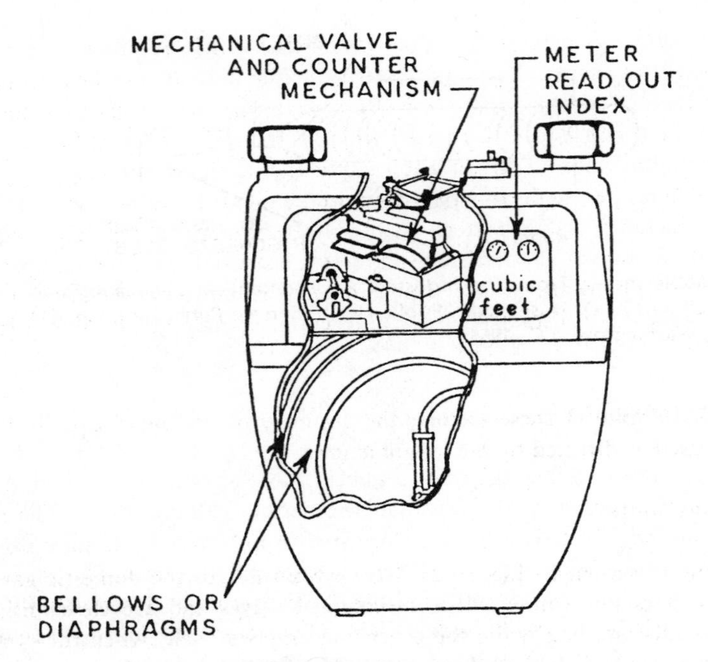

Figure 25.6 Dry gas meter.

volumes often outweighs the disadvantages created by the errors, which can be determined for a specific set of conditions. The alternate filling of two chambers as the basis for volume measurement is also used in twin-cylinder piston meters. Such meters can also be classified as positive displacement meters.

4.1.6 Positive Displacement Meters

Positive displacement meters consist of a tight-fitting moving element with individual volume compartments that fill at the inlet and discharge at the outlet parts. Another multicompartment continuous rotary meter uses interlocking gears. When the rotors of such meters are motor driven, these units become positive displacement air movers.

4.2 Volumetric Flowrate

The volume meters discussed in the preceding paragraphs were all based on the principle of conservation of mass—specifically, the transfer of a fluid volume from one location to another. The flowrate meters in this section all operate on the principle of the conservation of energy; more specifically, they utilize Bernoulli's theorem for the exchange of potential energy for kinetic energy and/or frictional heat. Each consists of a flow restriction within a closed conduit. The restriction causes an increase in the fluid velocity, therefore an increase in kinetic energy, which requires a corresponding decrease in potential energy (i.e., static pressure). The flowrate can be calculated from a knowledge of the pressure drop, the flow cross section at the constriction, the density of the fluid, and the coefficient of discharge, which is the ratio of actual flow to theoretical flow and makes allowance for stream contraction and frictional effects.

Flowmeters that operate on this principle can be divided into two groups. The larger group, which includes orifice meters, venturi meters, and flow nozzles, have a fixed restriction and are known as variable-head meters because the differential pressure head varies with flow. The other group, which includes rotameters, are known as variable area meters, because a constant pressure differential is maintained by varying the flow cross section.

4.2.1 Variable Area Meters (Rotameters)

A rotameter consists of a "float" that is free to move up and down within a vertical tapered tube that is larger at the top than the bottom (Figure 25.7). The fluid flows upward, causing the float to rise until the pressure drop across the annular area between the float and the tube wall is just sufficient to support the float. The tapered tube is usually made of glass or clear plastic, and the flowrate scale is etched directly on it. The height of the float indicates the flowrate. Floats of various configurations are used. They are conventionally read at the highest point of maximum diameter, unless otherwise indicated.

Most rotameters have a range of 10:1 between their maximum and minimum flows. The range of a given tube can be extended by using heavier or lighter floats. Tubes are made in sizes from about ⅛ to 6 inches in diameter, covering ranges from a few cubic

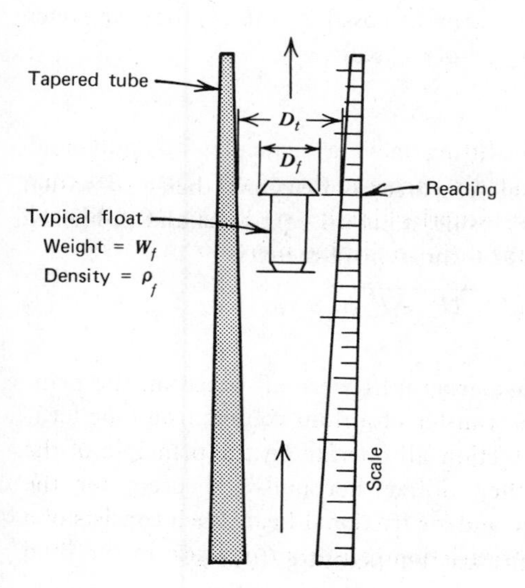

Figure 25.7 Schematic of rotameter.

centimeters per minute to more than 1000 cfm. Some of the shaped floats achieve stability by having slots that make them rotate, but these are less commonly employed than previously. The term "rotameter" was first used to describe such meters with spinning floats but now serves generally for tapered metering tubes of all types.

Rotameters are the most commonly used flowmeters on commercial air samplers, especially on portable samplers. For such sampler flowmeters, the most common material of construction is acrylic plastic, although glass tubes may also be used. Because of space limitations, the scale lengths are generally no more than 4 inches and most commonly nearer to 2 inches. Unless they are individually calibrated, the accuracy is unlikely to be better than ± 25 percent. When individually calibrated, ± 5 percent accuracy may be achieved. It should be noted, however, that with the large taper of the bore, the relatively large size of the float, and the relatively few scale markers on these rotameters, the precision of the readings may be a major limiting factor.

Calibrations of rotameters are performed at an appropriate reference pressure, usually atmospheric. Since good practice dictates that the flowmeter be located downstream of the sample collector or sensor, however, the flow is actually measured at a reduced pressure, which may also be a variable pressure if the flow resistance changes with loading. If this resistance is constant, it should be known; if variable, it should be monitored, to permit adjustment of the flowrate as needed, and the making of appropriate pressure corrections for the flowmeter readings.

Craig (2) has shown that the change in calibration with air density cannot be made by simple computation, especially for the small diameter rotameter tubes and floats commonly used on air sampling equipment. Figure 25.8 gives experimental calibration data at various suction pressures for a specific glass rotameter. Clearly it is not practical to generate such a family of empiric calibration curves for each rotameter. Craig recommends (1) that a pressure gauge be used at the inlet to the rotameter, (2) that the flowrates used be those which give as low a pressure drop as possible, and (3) that the meter size be selected to give readings near the upper end of the scale.

4.2.2 Variable Head Meters

When orifice and venturi meters are made to standardized dimensions, their calibration can be predicted with $\sim \pm 10$ percent accuracy using standard equations and published empirical coefficients. The general equation (3) for this type of meter is

$$W = q_1 \rho_1 = KYA_2 \sqrt{2g_c(P_1 - P_2)\rho_1} \tag{1}$$

where $K = C/\sqrt{1 - \beta^4}$

C = coefficient of discharge (dimensionless)

A_2 = cross-sectional area of throat (ft^2)

g_c = 32.17 ft/sec^2

P_1 = upstream static pressure (lb/ft^2)

P_2 = downstream static pressure (lb/ft^2)

q_1 = volumetric flow at upstream pressure and temperature (ft^3/sec)

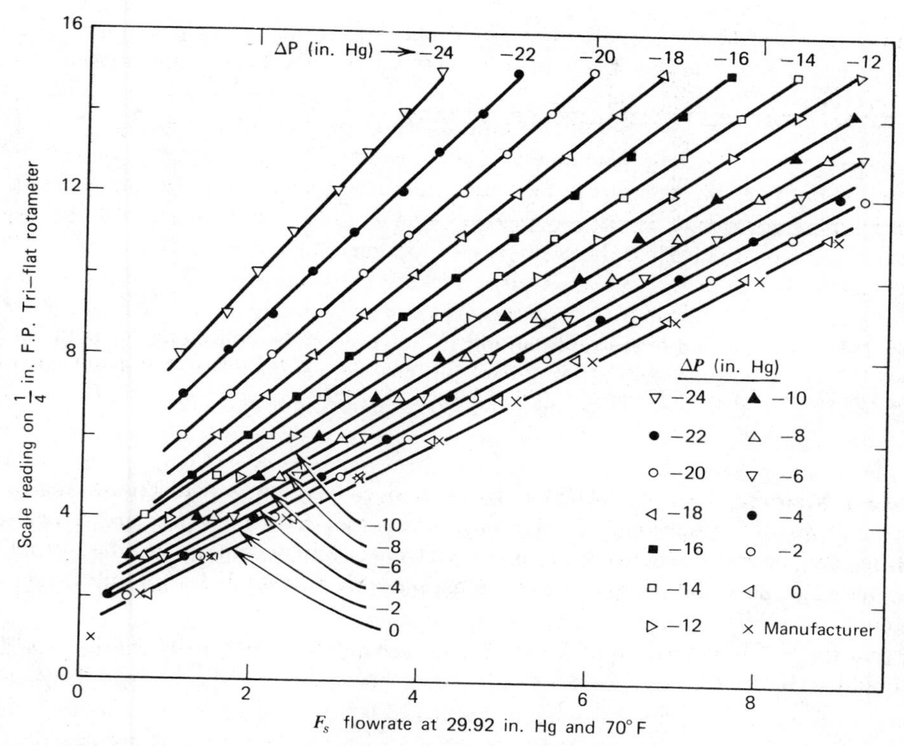

Figure 25.8 Rotameter reading versus airflow rate under standard conditions for various pressure gauge readings at rotameter.

W = weight rate of flow (lb/sec)
Y = expansion factor (see Figure 25.10)
β = ratio of throat diameter to pipe diameter (dimensionless)
ρ_1 = density at upstream pressure and temperature (lb/ft³)

Orifice Meters. The simplest form of variable head meter is the square-edged or sharp-edged orifice illustrated in Figure 25.9. It is also the most widely used because of its ease of installation and low cost. If it is made with properly mounted pressure taps, its calibration can be determined from Equation 1 and Figures 25.10 and 25.11. However even a nonstandard orifice meter can serve as a secondary standard, provided it is carefully calibrated against a reliable reference instrument.

Although the square-edged orifice can provide accurate flow measurements at low cost, it is inefficient with respect to energy loss. The permanent pressure loss for an orifice meter with radius taps can be approximated by $(1 - \beta^2)$ and often exceeds 80 percent.

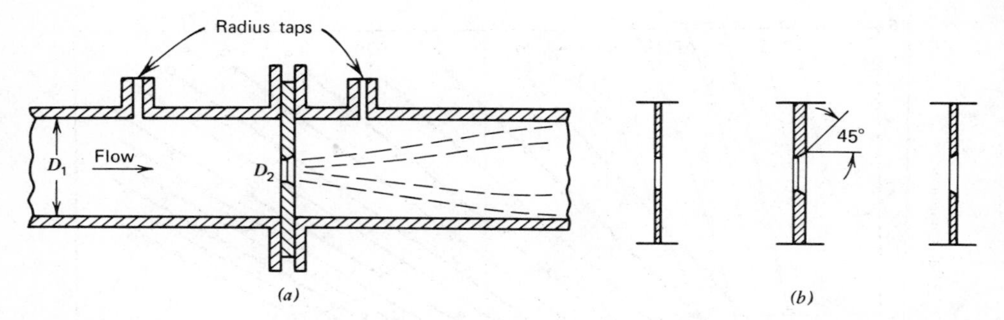

Figure 25.9 Square-edged or sharp-edged orifices. The plate at the orifice opening must not be thicker than 1/30 of pipe diameter, 1/8 of the orifice diameter, or 1/4 of the distance from the pipe wall to the edge of the opening. (*a*) Pipeline orifice. (*b*) Types of plate.

Venturi Meters. Venturi meters have optimal converging and diverging angles of about 25° and 7°, respectively, which means that they have high pressure recoveries; that is, the potential energy that is converted to kinetic energy at the throat is reconverted to potential energy at the discharge, with an overall loss of only about 10 percent.

For air at 70°F and 1 atm and for $\frac{1}{4} < \beta < \frac{1}{2}$, a standard venturi would have a calibration described by

$$Q = 21.2\ \beta^2 D^2 \sqrt{h} \tag{2}$$

where Q = flow (cfm)
β = ratio of throat to duct diameter (dimensionless)
D = duct diameter (in.)
h = differential pressure (in. H_2O)

Other Variable Head Meters. The characteristics of various other types of variable head flowmeters (e.g., flow nozzles, Dall tubes, centrifugal flow elements) are described in standard engineering references (3, 4). In most respects they have similar properties to the orifice meter, the venturi meter, or both.

One type of variable head meter that differs significantly from all the foregoing is the laminar flowmeter. These devices are seldom discussed in engineering handbooks because they are used only for very low flowrates. Since the flow is laminar, the pressure drop is directly proportional to the flowrate. In orifice meters, venturi meters and related devices, the flow is turbulent and flowrate varies with the square root of the pressure differential.

Laminar flow restrictors used in commercial flowmeters consist of egg-crate or tube bundle arrays of parallel channels. Alternatively, a laminar flowmeter can be constructed in the laboratory using a tube packed with beads or fibers as the resistance

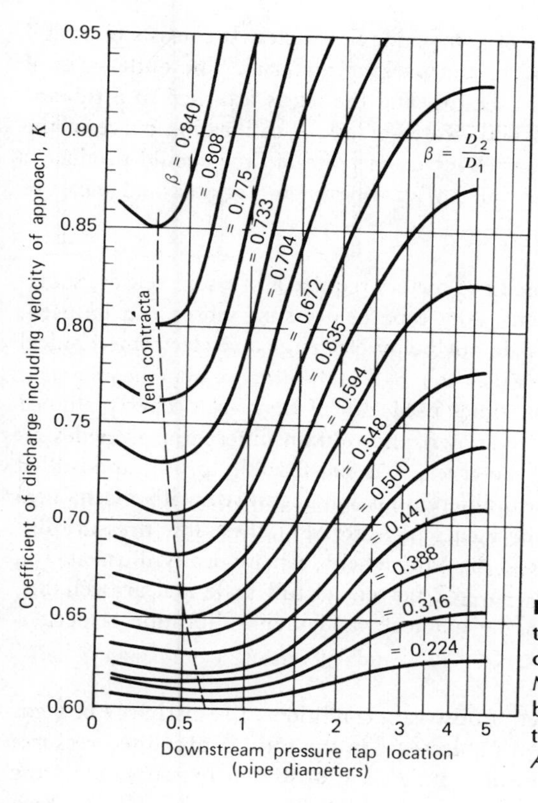

Figure 25.10 Downstream pressure tap location in pipe diameters. Coefficient of discharge for square-edged circular orifices for $N_{\mathrm{Re}_2} > 30{,}000$ with the upstream tap location between one and two pipe diameters from the orifice position. From Spitzglass, *Trans ASME,* **44,** 919 (1922).

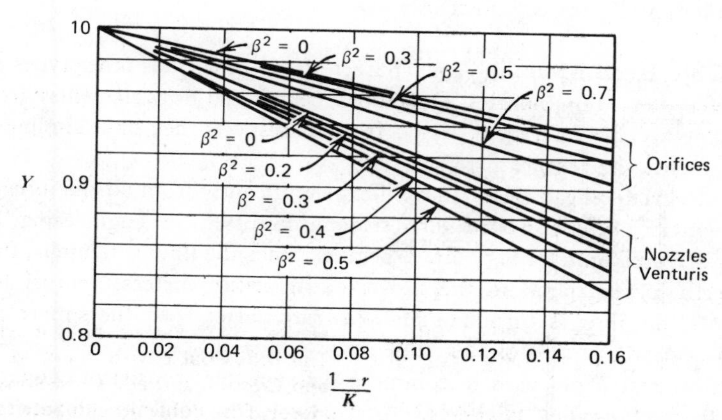

Figure 25.11 Values of expansion factor Y for orifices, nozzles, and venturis.

element. Figure 25.12 illustrates this kind of homemade flowmeter. It consists of a "T" connection, pipette or glass tubing, cylinder, and packing material. The outlet arm of the "T" is packed with material such as asbestos, and the leg is attached to a tube or pipette projecting into the cylinder filled with water or oil. A calibration curve of the depth of the tube outlet below the water level versus the rate of flow should produce a linear curve. Saltzman (5) has used tubes filled with asbestos to regulate and measure flowrates as low as 0.01 cm^3/min.

Pressure Transducers. All the variable head meters require a pressure sensor, sometimes referred to as the secondary element. Any type of pressure sensor can be used, although high cost and fragility usually rule out many electrical and electromechanical transducers.

Liquid-filled manometer tubes are sometimes used, and if they are properly aligned and the density of the liquid is accurately known, the column differential provides an unequivocal measurement. In most cases however, it is not feasible to use liquid-filled manometers in the field, and the pressure differentials are measured with mechanical gauges with scale ranges in centimeters or inches of water. For these low pressure differentials the most commonly used gauge is the Magnehelic, schematically illustrated in Figure 25.13. These gauges are accurate to $\pm$ 2 percent of full scale and are reliable, provided they and their connecting hoses do not leak, and their calibration is periodically rechecked.

Critical Flow Orifice. For a given set of upstream conditions, the discharge of a gas from a restricted opening will increase with a decrease in the ratio of absolute pressures P_2/P_1, where P_2 is the downstream pressure and P_1 the upstream pressure, until the velocity through the opening reaches the velocity of sound. The value of P_2/P_1 at which the maximum velocity is just attained is known as the critical pressure ratio. The

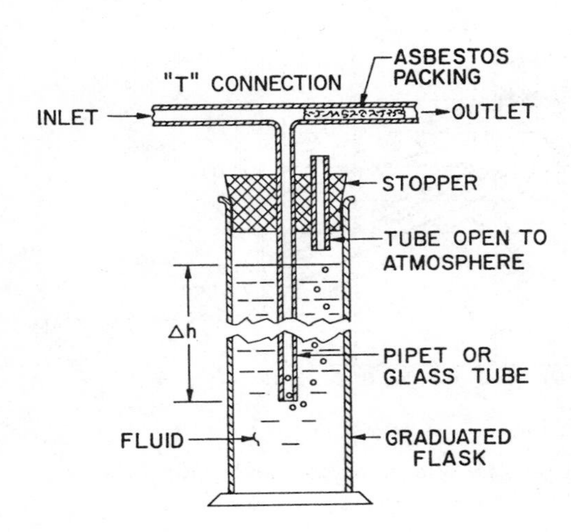

Figure 25.12 Packed plug flowmeter. From *The Industrial Environment—Its Evaluation and Control,* 2nd ed., C. H. Powell and A. D. Hosey, Eds., Public Health Service Publication No. 614, Government Printing Office, Washington, D.C., 1965.

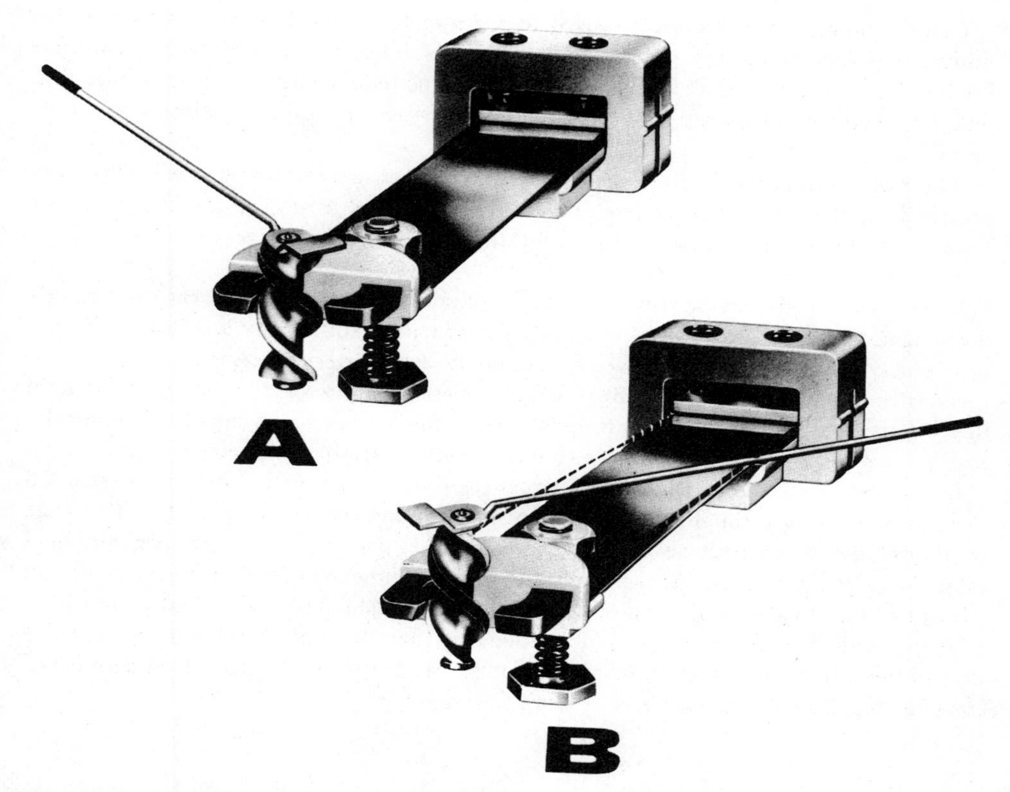

Figure 25.13 Workings of the magnetic linkage. Photograph courtesy of Dwyer Instruments, Inc., Michigan City, Ind., Bulletin No. A-20.

pressure in the throat will not fall below the pressure at the critical point, even if a much lower downstream pressure exists. When the pressure ratio is below the critical value, therefore, the rate of flow through the restricted opening is dependent only on the upstream pressure.

It can be shown (3) that for air flowing through rounded orifices, nozzles, and venturis, when $P_2 < 0.53\ P_1$, and $S_1/S_2 > 25$, the mass flowrate w is determined by

$$w = 0.533\ \frac{C_v S_2 P_1}{T_1}\ \text{lb/sec} \tag{3}$$

where C_v = coefficient of discharge (normally ~ 1)
S_1 = duct or pipe cross section (in.^{2})
S_2 = orifice area (in.^{2})
P_1 = upstream absolute pressure (psi)
T_1 = upstream temperature (°R)

Critical flow orifices are widely used in industrial hygiene instruments such as the midget impinger pump and squeeze bulb indicators. They can also be used to calibrate flowmeters by using a series of critical orifices downstream of the flowmeter under test. The flowmeter readings can be plotted against the critical flows to yield a calibration curve.

The major limitation in their use is that the orifices are extremely small when they are used for flows of 1 cfm or less. They become clogged or eroded in time, therefore require frequent examination and/or calibration against other reference meters.

Bypass Flow Indicators. In most high volume samplers, the flowrate is strongly dependent on the flow resistance, and flowmeters with a sufficiently low flow resistance are usually too bulky or expensive. A commonly used metering element for such samplers is the bypass rotameter, which actually meters only a small fraction of the total flow—a fraction, however, that is proportional to the total flow. As shown schematically in Figure 25.14, a bypass flowmeter contains both a variable head element and a variable area element. The pressure drop across the fixed orifice or flow restrictor creates a proportionate flow through the parallel path containing the small rotameter. The scale on the rotameter generally reads directly in cubic feet per minute or liters per minute of total flow. In the versions used on portable high volume samplers, there is usually an adjustable bleed valve at the top of the rotameter that should be set initially, and periodically readjusted in laboratory calibrations so that the scale markings can indicate overall flow. If the rotameter tube accumulates dirt, or the bleed valve adjustment drifts, the scale readings may depart greatly from the true flows.

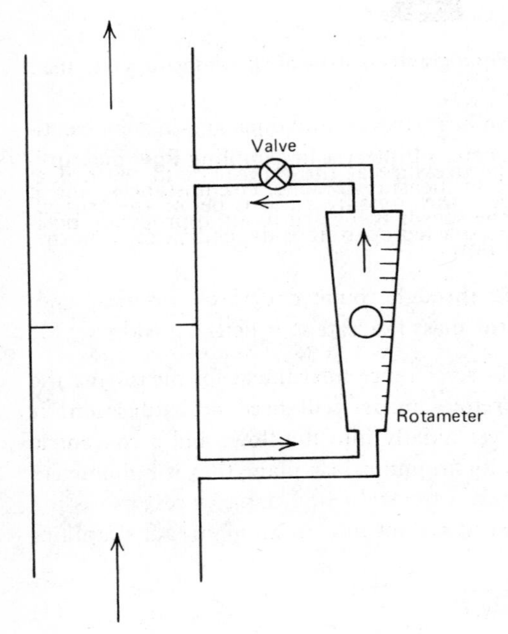

Figure 25.14 Bypass flow indicators.

4.3 Mass Flow and Tracer Techniques

4.3.1 Thermal Meters

A thermal meter measures mass air or gas flowrate with negligible pressure loss. It consists of a heating element in a duct section between two points at which the temperature of the air or gas stream is measured. The temperature difference between the two points is dependent on the mass rate of flow and the heat input.

4.3.2 Mixture Metering

The principle of mixture metering is similar to that of thermal metering. Instead of adding heat and measuring temperature difference, a contaminant is added and its increase in concentration is measured; or clean air is added and the reduction in concentration is measured. This method is useful for metering corrosive gas streams. The measuring device may react to some physical property such as thermal conductivity or vapor pressure.

4.3.3 Ion-Flow Meters

In the ion-flow meter illustrated in Figure 25.15, ions are generated from the central disk and flow radially toward the collector surface. Airflow through the cylinder causes an axial displacement of the ion stream in direct proportion to the mass flow. The instrument can measure mass flows from 0.1 to 150 scfm and velocities from 1 to 12,000 fpm.

4.4 Air Velocity Meters (Anemometers)

Air velocity is a parameter of direct interest in heat stress evaluations and in some ventilation evaluations. Though it is not the parameter of interest in sampling flow measurements, it may be the only feasible parameter to measure in some circumstances, and it usually can be related to flowrate, provided the sensor is located in an appropriate position and is suitably calibrated against overall flow.

4.4.1 Velocity Pressure Meters

Pitot Tube. The Pitot tube is often used as a reference instrument for measuring the velocity of air. A standard Pitot device, carefully made, will need no calibration. It consists of an impact tube whose opening faces axially into the flow, and a concentric static pressure tube with 8 holes spaced equally around it in a plane that is 8 diameters from the impact opening. The difference between the static and impact pressures is the velocity pressure. Bernoulli's theorem applied to a Pitot tube in an airstream simplifies to the dimensionless formula

$$V = \sqrt{2g_c P_v} \qquad (4)$$

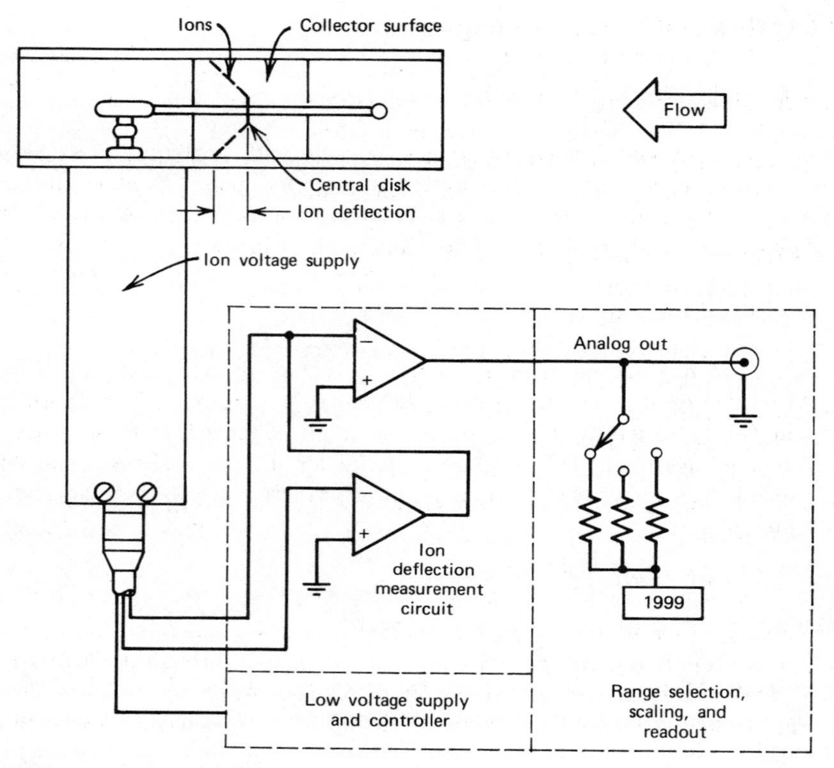

Figure 25.15 Ion-flow mass flowmeter. Schematic courtesy of Thermosystems, Inc., St. Paul, Minn., Leaflet No. TSI-54100671.

where V = linear velocity
g_c = gravitational constant
P_v = pressure head of flowing fluid or velocity pressure

Expressing V in linear feet per minute, P_v in inches of water, i.e., (h_v), and with:

$$g_c = 32.17 \frac{(\text{lb} - \text{mass})(\text{ft})}{(\text{lb} - \text{force})(\text{sec}^2)}$$

$$V = 1097 \left(\frac{h_v}{\rho}\right)^{1/2} \tag{5}$$

where ρ is the density of air or gas (lb/ft^3)

If the Pitot tube is to be used with air at standard conditions (70°F and 1 atm), Equation 5 reduces to

$$V = 4005\sqrt{h_v} \tag{6}$$

where V = velocity (fpm)
h_v = velocity pressure (in. H_2O)

There are several serious limitations to Pitot tube measurements in most sampling flow calibrations. It may be difficult to obtain or fabricate a small enough probe, and the velocity pressure may be too low to measure at the velocities encountered. Thus at 1000 fpm, $h_v = 0.063$ in. H_2O, a low value, even for an inclined manometer.

Heated Element Anemometers. Any instrument used to measure velocity can be referred to as an anemometer. In a heated element anemometer, the flowing air cools the sensor in proportion to the velocity of the air. Instruments are available with various kinds of heated element, including heated thermometers, thermocouples, films, and wires. They are all essentially nondirectional (i.e., with single element probes); they measure the air speed but not its direction. They all can accurately measure steady state air speed, and those with low mass sensors and appropriate circuits can also accurately measure velocity fluctuations with frequencies above 100,000 Hz. Since the signals produced by the basic sensors are dependent on ambient temperature as well as air velocity, the probes are usually equipped with a reference element that provides an output that can be used to compensate or correct errors due to temperature variations. Some heated element anemometers can measure velocities as low as 10 fpm and as high as 8000 fpm.

Other Velocity Meters. There are several means of utilizing the kinetic energy of a flowing fluid to measure velocity besides the Pitot tube. One way is to align a jeweled-bearing turbine wheel axially in the stream and count the number of rotations per unit time. Such devices are generally known as rotating vane anemometers. Some are very small and are used as velocity probes. Others are sized to fit the whole duct and become indicators of total flowrate; sometimes these are called turbine flowmeters.

The velometer, or swinging vane anemometer, is widely used for measuring ventilation airflows, but it has few applications in sample flow measurement or calibration. It consists of a spring-loaded vane whose displacement is indicative of velocity pressure.

4.5 Procedure for Calibrating Velocity; Flow and Volume Meters

In the limited space available, it is not possible to describe completely all the techniques available, or to go into great detail on those which are commonly used. This discussion is limited to selected procedures which should serve to illustrate recommended approaches to some commonly encountered calibration procedures.

4.5.1 Producing Known Velocity Fields

Known flow fields can be produced in wind tunnels of the type illustrated in Figure 25.16. The basic components needed have been described by Hama (6) as follows:

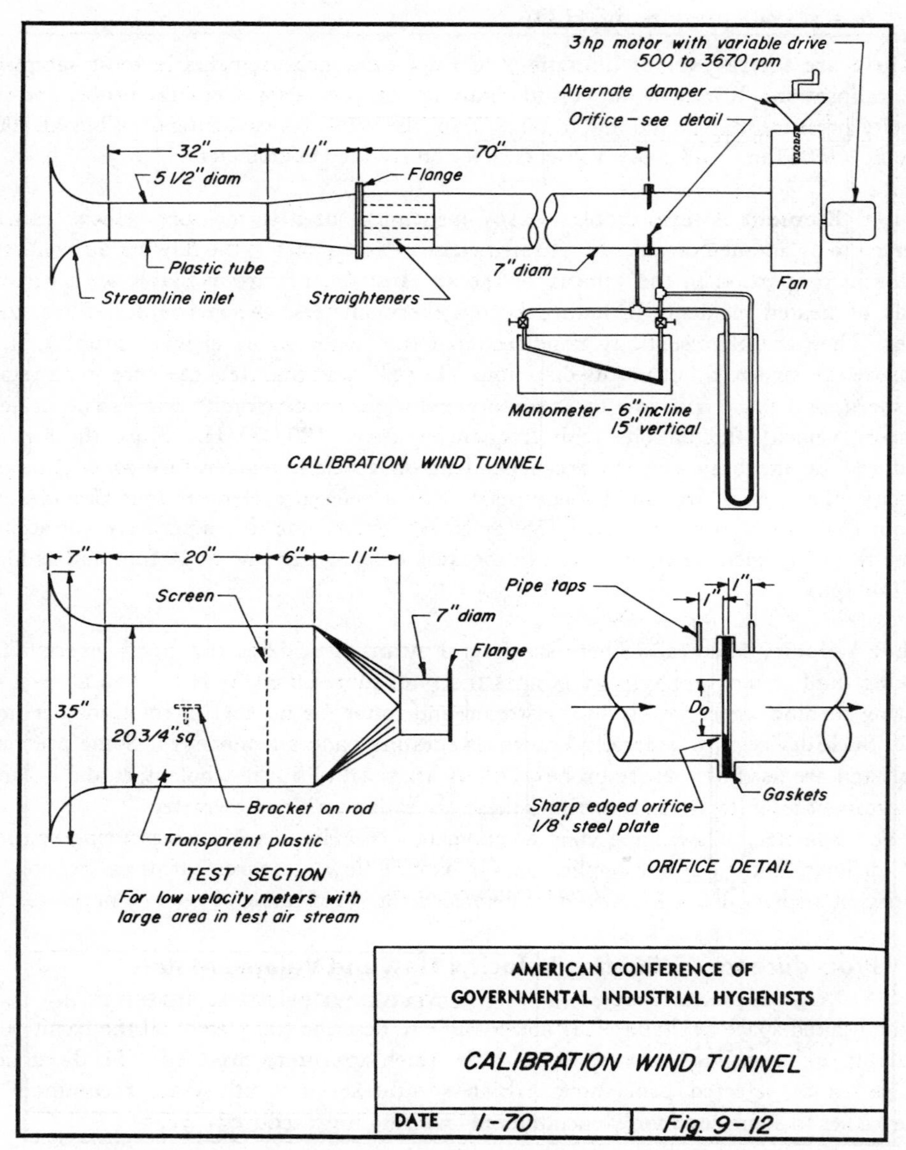

Figure 25.16 Wind tunnel and its use for calibration of anemometers. From *Industrial Ventilation.*

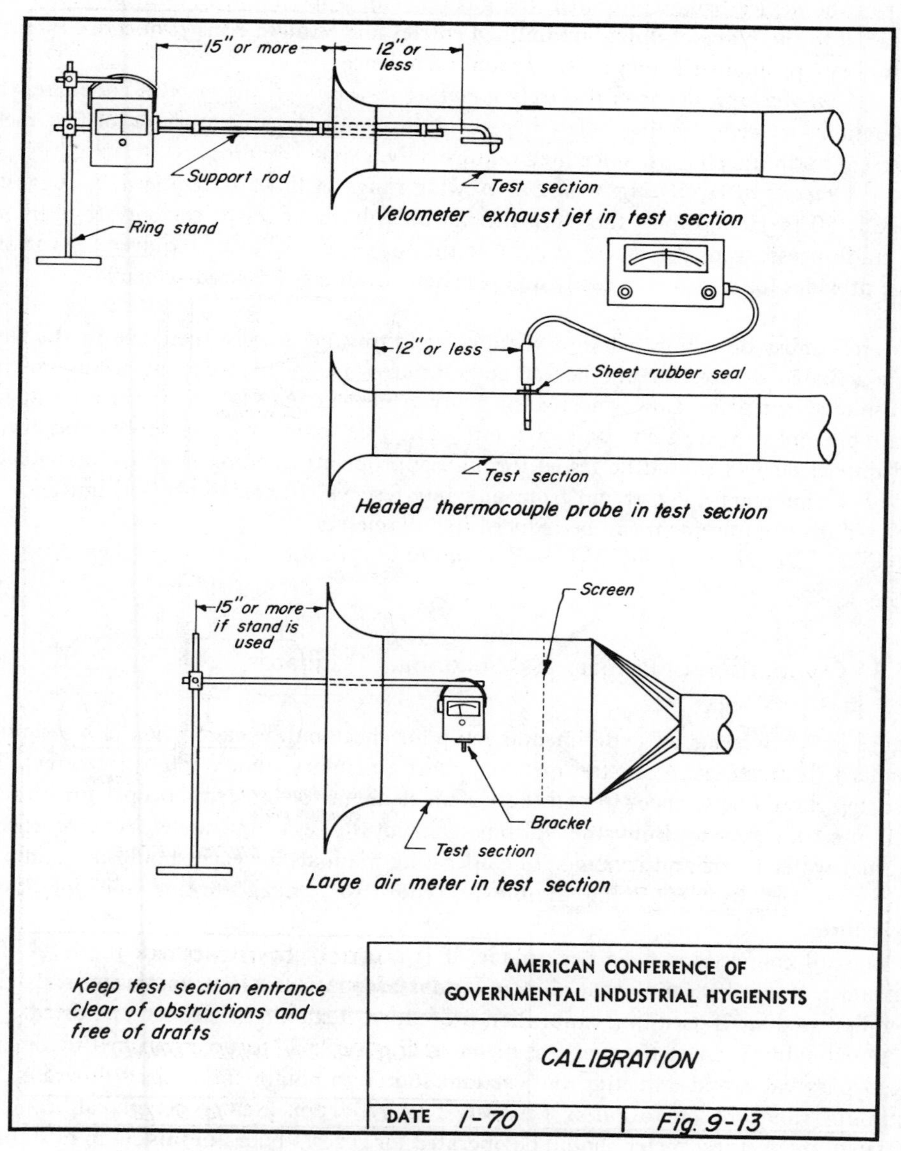
15" or more
12" or less
Support rod
Ring stand
Test section
Velometer exhaust jet in test section
12" or less
Sheet rubber seal
Test section
Heated thermocouple probe in test section
Screen
15" or more if stand is used
Bracket
Test section
Large air meter in test section
Keep test section entrance clear of obstructions and free of drafts
AMERICAN CONFERENCE OF GOVERNMENTAL INDUSTRIAL HYGIENISTS
CALIBRATION
DATE 1-70
Fig. 9-13

Figure 25.16 (Continued)

1. *A satisfactory test section.* Since this is the location of the probe or sensing element of the device being calibrated, the gas flows must be uniform, both perpendicular and axial to the plane of flow. Streamlined entries and straight runs of duct are essential to eliminate pronounced vena contracta and turbulence.
2. *A satisfactory means of precisely metering airflow.* A meter with adequate scale graduations to give readings of ±1 percent is required. A venturi or orifice meter represent optimum choices, since they require only a single reading.
3. *A means of regulating airflow.* A wide range of flows is required. A suggested range is 50 to 10,000 fpm; therefore the fan must have sufficient capacity to overcome the static pressure of the entire system at the maximum velocity required. A variable drive provides for a means of easily and precisely attaining a desired velocity.

Meters must be calibrated in a manner reflecting accurately their use in the field. Vane-actuated devices should be set on a bracket inside a large test section with a streamlined entrance. Low velocity, probe-type devices may be tested through appropriate openings in the same type of tunnel. High velocity ranges of probe-type devices and impact devices should be tested through appropriate openings in a circular duct at least 8.5 diameters downstream from any interference. If straighteners (Figure 25.16) are used, this requirement can be reduced to 7 diameters.

NOTE: *Devices must be calibrated at multiple velocities throughout their operating range.*

4.5.2 Comparison of Primary and Secondary Standards

Figure 25.17 presents an experimental setup for checking the calibration of a secondary standard (in this case, a wet test meter) against a primary standard (a spirometer). The first step should be to check out all the system elements for integrity, proper functioning, and interconnections. Both the spirometer and the wet test meter require specific internal water levels and leveling. The operating manuals for each should be examined, since they will usually outline simple procedure for leakage testing and operational procedures.

After all connections have been made, it is a good policy to recheck the level of all instruments and determine that all connections are clear and have minimum resistance. If compressed air is used in a calibration procedure, it should be cleaned and dried.

Actual calibration of the wet test meter in Figure 25.17 is accomplished by opening the bypass valve and adjusting the vacuum source to obtain the desired flowrate. The optimum range of operation is between 1 and 3 rpm. Before actual calibration is initiated, the wet test meter should be operated for several hours in this setup to stabilize the meter fluid with respect to temperature and absorbed gas, and to work in the bearings and mechanical linkage. After all elements of the system have been adjusted, zeroed, and stabilized, several trial runs should be made. During these runs, the cause of any indicated difference in pressure should be determined and corrected. The actual

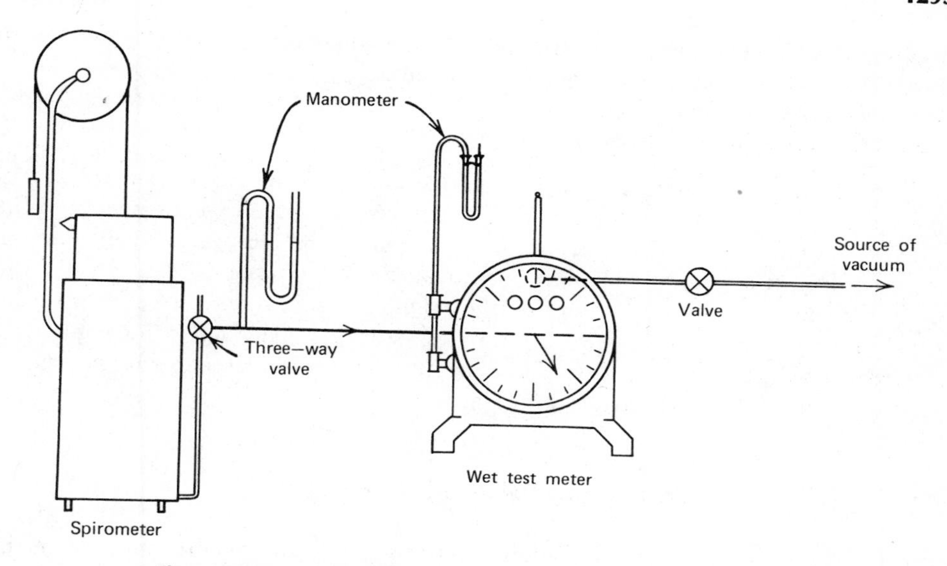

Figure 25.17 Calibration setup for calibrating a wet test meter.

procedure would be to instantly divert the air to the spirometer for a predetermined volume indicated by the wet test meter (minimum of one revolution), or to near capacity of the spirometer, then return to the bypass arrangement. Readings, both quantity and pressure of the wet test meter, must be taken and recorded while the device is in motion, unless a more elaborate system is set up. In the case of a rate meter, the interval of time that the air is entering the spirometer must be accurately timed. The bell should then be allowed to come to equilibrium before displacement readings are made. Enough different flowrates are taken to establish the shape or slope of the calibration curve, and the procedure being repeated three or more times for each point. For an even more accurate calibration, the setup should be reversed so that air is withdrawn from the spirometer. In this way any imbalance due to pressure differences would be canceled.

A permanent record should be made, consisting of a sketch of the setup and a list of data, conditions, equipment, results, and personnel associated with the calibration. All readings (volume, temperatures, pressures, displacements, etc.) should be legibly recorded, including trial runs or known faulty data, with appropriate comments. The identifications of equipment, connections, and conditions should be complete, enabling another person, solely by use of the records, to replicate the same setup, equipment, and connections.

After all the data have been recorded, the calculations (e.g., correction for variations in temperature, pressure, and water vapor) are made, using the ideal gas laws:

$$V_s = V_1 \times \frac{P_1}{760} \times \frac{273}{T_1} \tag{7}$$

where V_s = volume at standard conditions (760 mm and 0°C)
V_1 = volume measured at conditions P_1 and T_1
T_1 = temperature of V_1 (°K)
P_1 = pressure of V_1 (mm Hg)

In most cases the water vapor portion of the ambient pressure is disregarded. Also, the standard temperature of the gas is often referred to normal room temperature (i.e., 21°C rather than 0°C). The instruments, data reading and recording, calculaticns, and resulting factors or curves should be manipulated with extreme care. If a calibration disagrees with previous calibrations or the supplier's calibration, the entire procedure should be repeated and examined carefully to assure its validity. Upon completion of any calibration, the instrument should be tagged or marked in a semipermanent manner to indicate the calibration factor, where appropriate, the date, and the identity of the calibrater.

4.5.3 Reciprocal Calibration by Balanced Flow System

It is impractical to remove the flow-indicating device for calibration in many commercial instruments. This may be because of physical limitations, characteristics of the pump, unknown resistance in the system, or other limiting factors. In such situations it may be necessary to set up a reciprocal calibration procedure: that is, a controlled flow of air or gas is compared first with the instrument flow, then with a calibration source. Often a further complication is introduced by the static pressure characteristics of the air mover in the instrument. In such instances supplemental pressure or vacuum must be applied to the system to offset the resistance of the calibrating device. An example of such a system appears in Figure 25.18.

The instrument is connected to a calibrated rotameter and a source of compressed air. Between the rotameter and the instrument an open ended manometer is installed. The connections, as in any other calibration system, should be as short and resistance-free as possible.

In the calibration procedure the flow through the instrument and rotameter is adjusted by means of a valve or restriction at the pump until the manometer indicates

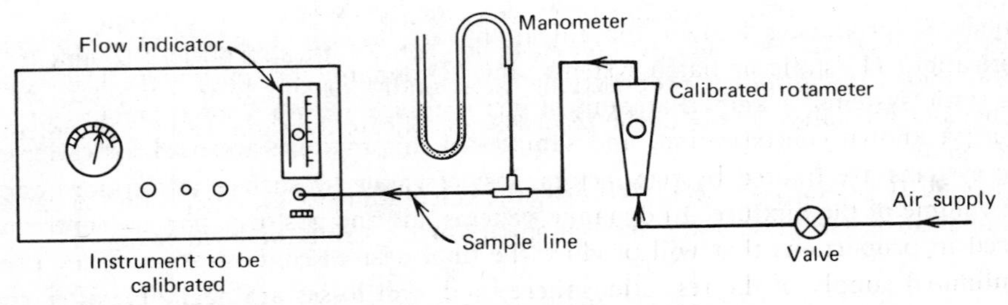

Figure 25.18 Setup for balanced flow calibration.

"0" pressure difference to the atmosphere. When this condition is achieved, both the instrument and the rotameter are operating at atmospheric pressure. The indicated and calibrated rates of flow are then recorded, and the procedure is repeated for other rates of flow.

4.5.4 Dilution Calibration

Normally gas dilution techniques are employed for instrument response calibrations; however several procedures (6–8) have been developed whereby sampling rates of flow can be determined. The principle is essentially the same except that different unknowns are involved. In airflow calibration a known concentration of the gas (e.g., carbon dioxide) is contained in a vessel. Uncontaminated air is introduced and mixed thoroughly in the chamber to replace that removed by the instrument to be calibrated. The resulting depletion of the agent in the vessel follows the theoretical dilution formula

$$C_t = C_0 e^{bt} \tag{8}$$

where C_t = concentration of agent in vessel at time t
C_0 = initial concentration at $t = 0$
e = base of natural logarithms
b = air changes in the vessel per unit time
t = elapsed time

The concentration of the gas in the vessel is determined periodically by an independent method. A linear plot should result from plotting concentration of agent against elapsed time on semilog paper. The slope of the line indicates the air changes per minute b, which can be converted to the rate Q of air withdrawn by the instrument from the following relationship: $Q = bV$, where V is the volume of the vessel.

This technique is advantageous in that virtually no resistance or obstruction is offered to the airflow through the instrument; however it is limited by the accuracy of determining the concentration of the agents in the air mixture.

4.6 Production of Known Vapor Concentrations

4.6.1 Introduction and Background

Methods of producing known concentrations are usually divided into two general classification: (1) static or batch systems, and (2) dynamic or continuous flow systems. With static systems, a known amount of gas is mixed with a known amount of air to produce a known concentration, and samples of this mixture are used for calibration. Static systems are limited by two factors, loss of vapor by surface adsorption, and the finite volume of the mixture. In dynamic systems, air and gas or vapor are continuously metered in proportions that will produce the final desired concentration. They provide an unlimited supply of the test atmosphere, and wall losses are negligible after equilibration has taken place.

In the field of industrial hygiene, gas or vapor concentrations are usually discussed in terms of parts per million (ppm). In this case, "parts per million" refers to a volume-to-volume relationship (i.e., so many liters of contaminant per liter of air). Thus by definition, both 1 μl of SO_2 per liter of air, and 1 ml of SO_2 per cubic meter of air are equal to 1 ppm SO_2. In the field of air pollution these concentrations may also be discussed as parts per 100 million or parts per billion, also on a volume-to-volume ratio.

Occasionally with direct reading instruments, and more frequently with chemical analysis of the atmosphere, confusion arises in converting milligrams per cubic meter to parts per million. Dimensional analysis is very useful in avoiding these errors. Thus if one has a concentration in milligrams per cubic meter of air, it must be converted to millimoles per cubic meter, and to milliliters per cubic meter or parts per million.

$$\left(\frac{\text{mg}_x}{\text{m}^3\text{ air}}\right)\left(\frac{\text{mmole}_x}{\text{mg}_x}\right)\left(\frac{22.4\text{ ml}_x}{\text{mmole}_x}\right)(F_t)\,(F_p) = \frac{\text{ml}_x}{\text{m}^3\text{ air}} = \text{ppm} \tag{9}$$

where F_t and F_p are the pressure and temperature conversion factors from the well-known gas laws, and the subscript x refers to a trace contaminant. Conversely

$$\text{ppm} = \left(\frac{\text{ml}_x}{\text{m}^3\text{ air}}\right)\left(\frac{\text{mmole}_x}{22.4\text{ ml}_x}\right)\left(\frac{\text{mg}_x}{\text{mmole}_x}\right)(F_t)\,(F_p) = \frac{\text{mg}_x}{\text{m}^3\text{ air}} \tag{10}$$

Chemical analysis of atmospheric samples is further complicated by procedures that call for some fixed volume of absorbing or reacting solution and sometimes for dilution. In this case it is convenient to determine the concentration of contaminant in solution and, by multiplying by the volume of solution, calculate the total amount of contaminant collected. This is then related to the volume of air sampled and converted to parts per million. For example, after bubbling 5 liters of air at 25°C and 755 mm Hg through 25 ml of an appropriate absorbing solution (100 percent collection efficiency) it was determined that the SO_2 (molecular weight = 64) concentration in solution was 5 μg/ml.

The total amount of SO_2 measured was

$$\frac{5\ \mu\text{g}}{\text{ml}} \times 25\text{ ml} = 125\ \mu\text{g} \tag{11}$$

The volume of 1 μmole of SO_2 at 25°C and 755 mm Hg is found as follows:

$$1\ \mu\text{mole} \times \frac{22.4\ \mu\text{l}}{\mu\text{mole}} \times \frac{298}{273} \times \frac{760}{755} = 24.6\ \mu\text{l} \tag{12}$$

The concentration in parts per million is

$$\frac{125\ \mu\text{g SO}_2}{5\text{ liters air}} \times \frac{\text{mole SO}_2}{64\ \mu\text{g SO}_2} \times \frac{24.6\ \mu\text{l SO}_2}{\mu\text{mole SO}_2} = \frac{9.6\ \mu\text{l SO}_2}{\text{liter of air}} = 9.6\text{ ppm} \tag{13}$$

When producing test atmospheres, many factors may interfere with the contaminant gas or the instrument or analytical procedure. These include (1) specificity of reagents or instrument being used to measure the particular contaminant and (2) loss of the contaminant by reaction with, or adsorption onto, other trace contaminants in the carrier gas or elements of the system. Thus before establishing test atmospheres, the dilution gas should be purified. In addition, the nature and chemistry of the material to be analyzed as well as the detection principle must be thoroughly understood.

4.6.2 Static Systems: Batch Mixtures

In static systems, a known amount of material is introduced into a container, either rigid or flexible, then diluted with an appropriate amount of clean air. If flexible systems are desired, bags of Mylar, aluminized Mylar, Teflon, or polyethyelene can be used. They provide the advantage that the entire volume of the bag is usable. Polyethylene is simple to use, but many pollutants either diffuse through it or are adsorbed onto the walls (7). Mylar and aluminized Mylar are less permeable (8), and since they are inelastic, offer the additional advantage of filling to a constant volume.

Rigid containers such as 5-gallon bottles (Figure 25.19) are commonly used for static systems. The bottles are usually equipped with a glass tube, a valved inlet, and a similar outlet. A third inlet or passthrough port for introduction of the contaminant may also be provided. In practice, after the mixture has come to equilibrium, samples are drawn from the outlet side while replacement air is allowed to enter through the inlet tube. Thus the mixture is being diluted while it is being sampled.

Under ideal conditions, the concentration remaining is a known function of the number of air changes in the bottle. If one assumes instantaneous and perfect mixing of the incoming air with the entire sample volume, the concentration change, as a small volume is withdrawn, is equal to the concentration times the percentage of the volume withdrawn:

$$dC = C\,\frac{dV}{V_0} \tag{14}$$

which integrates to

$$C = C_0 e^{-V/V_0} \qquad \text{or} \qquad 2.3\,\log_{(10)} \frac{C_0}{C} = \frac{V}{V_0} \tag{15}$$

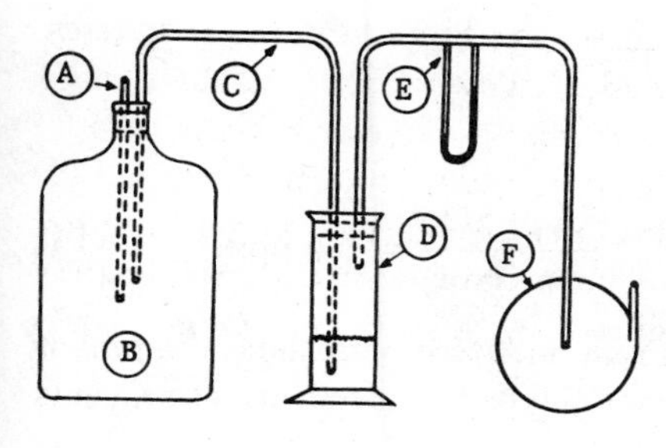

Figure 25.19 Five-gallon bottle for static calibration. *A*, intake tube; *B*, 5-gallon bottle; *C*, withdrawal tube; *D*, collecting device of direct reading instrument; *E*, flowmeter; *F*, suction pump. Schematic courtesy of Stead and Taylor.

where C = the concentration at any time
V = total volume of sample withdrawn
C_0 = original concentration
V_0 = the volume of the chamber

Thus if we remove one-tenth the volume, we can write

$$2.3 \log_{(10)} \frac{C_0}{C} = 0.1 \quad (16)$$

$$\log_{(10)} \frac{C_0}{C} = \frac{0.1}{2.3} = 0.0435 \quad (17)$$

$$\frac{C_0}{C} = 1.1053 \quad \text{or} \quad \frac{C}{C_0} = 0.9047 \quad (18)$$

The average concentration of the sample withdrawn is 0.9524. If instantaneous mixing does not occur, and the inlet and outlet port are separated, the average concentration may be even higher.

If one were interested in a maximum of 5 percent variation from the average concentration, only about 10 percent of the sample could be used. Setterlind (9) has shown that this limitation can be overcome by using two or more bottles of equal volume (V_0) in series, with the initial concentration in each bottle being the same. When the mixture is withdrawn from the last bottle, it is not displaced by air but by the mixture from the preceding bottle. If, as previously, a maximum of 5 percent variation in concentration can be tolerated, two bottles in series provides a usable sample of $0.6V_0$. With five bottles, the usable sample will increase to about $3V_0$. A table in Setterlind's paper gives both residual concentration and average concentration of the withdrawn sample as a function of the number of volumes withdrawn for each of five bottles in series.

A rigid system can also be modified to give greater usable volumes by attaching a balloon to the intake side inside the bottle. Air from the bottle can then be displaced without any dilution by merely blowing up the balloon.

Introduction of Material into Static Systems. Calibrated syringes provide a simple method for introduction of materials, either gaseous or liquid, into static systems. A wide variety of both gas and liquid syringes are available down to the microliter range (1). A second method is to produce glass ampoules containing a known amount of pure contaminant and break them within the fixed volume of the static system. Setterlind (9) has discussed the preparation of ampoules in detail. Other devices such as gas burettes, displacement manometers, and small pressurized bombs have been used successfully (10). Gaseous concentrations can also be produced by adding stoichiometrically determined amounts of reacting chemicals.

Finally, a standard cylinder can be evacuated, filled with a measured volume of gas or liquid, and repressurized with compressed air or other farrier gas to produce the concentrations required. This mixture can be used with further dilution if necessary. The techniques for filling cylinders have been discussed by Cotabish et al. (11). A number of various gases and vapors are available in different concentrations from several manufacturers (1). Analysis, usually gravimetric, is provided on request. These data should always be checked, since the trace gas may not be adequately mixed or may be partially lost because of wall adsorption.

4.6.3 Dynamic Systems: Continuous Flow

In dynamic systems the rate of airflow and the rate of addition of contaminant to the airstream are carefully controlled to produce a known dilution ratio. Dynamic systems offer a continuous supply of material, allow for rapid and predictable concentration changes, and minimize the effect of wall losses as the contaminant comes to equilibrium with the interior surfaces of the system. Both gases and liquids can be used with dynamic systems. With liquids, however, provision must be available for conversion to the vapor state.

Gas Dilution Systems. A simple schematic of a gas dilution system appears in Figure 25.20. Air and the contaminant gas are metered through restrictions and mixed. The output can be used as is or further diluted in a similar system. In theory this process can be repeated until the necessary dilution ratio is obtained. In practice, series dilution systems are subject to a variety of instabilities which make them difficult to control.

Saltzman (1) has described two flow dilution devices; Figure 25.12 shows an asbestos plug flowmeter, a device that assures that a restricting asbestos-plugged capillary receives a constant pressure of the contaminant gas. The contaminant gas flow, a function of the pressure, is controlled by the height of the column of water or oil. A second device (Figure 25.21) minimizes back pressure and includes a mixing chamber, since the airstream is split. The majority of the gas can be piped to waste through the larger tube. Immersing the end of this tube in water will provide a slightly positive pressure at the smaller sidearm delivery tube.

Cotabish et al. (11) have described a system originally patented by Mase for compensation of back pressure (Figure 25.22) in which both the air and contaminant gas flow are regulated by the height of a water column, which in turn is controlled by the back pressure of the calibration system. Thus an increase in back pressure causes an increase in the delivery pressure of both air and contaminant gas.

Calibrated Instruments, Inc., has two instruments available for calibration purposes: the ppm Maker (Figure 25.23) and the Stack Gas Calibrator (Figure 25.24). The ppm Maker consists of a four-output, positive displacement pump and two mechanized four-way stopcocks with single-bore plugs. The bore is normally aligned with the carrier gas

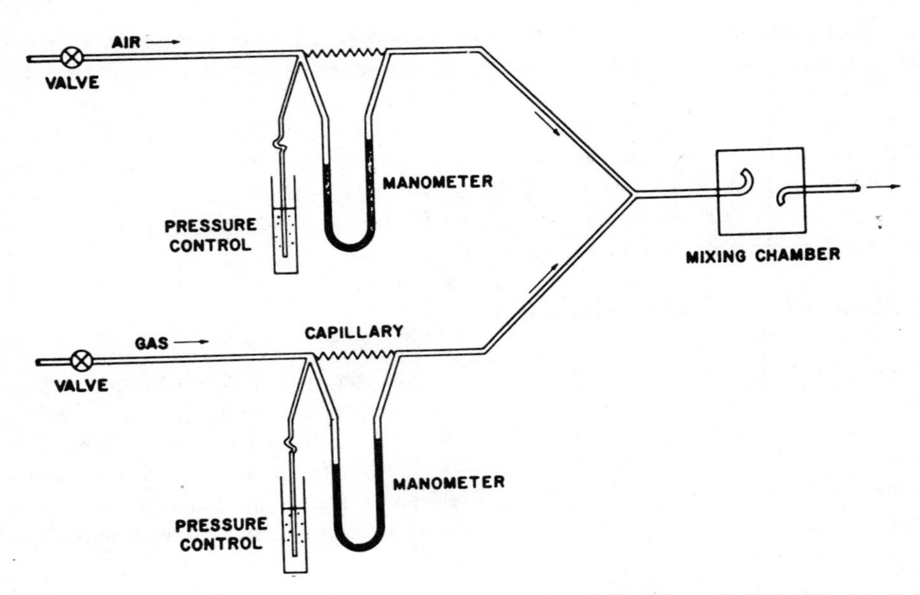

Figure 25.20 Continuous mixer for dynamic gas concentrations. From *Am. Ind. Hyg. Assoc. J.*, **22,** 393 (1961); courtesy of Williams & Wilkins Company, Baltimore, and Mine Safety Appliances Company, Pittsburgh.

flow. When activated, the stopcock is rotated 180°, momentarily aligning the bore with the contaminant gas airflow and delivering a precise volume to the carrier gas. A mixing chamber downstream mixes the carrier gas and the contaminant. The mixture is then pumped through a second identical system. By varying the flow-rates of the carrier gas, dilution ratios in the order of $1:10^9$ can be achieved. The stepwise increments of the

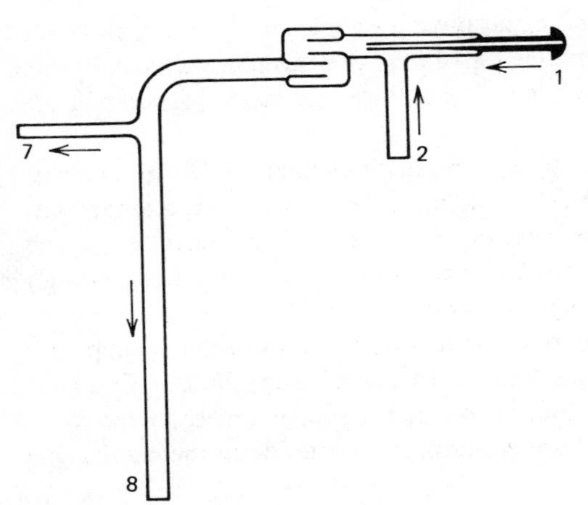

Figure 25.21 All-glass system. From *Anal. Chem.*, **33,** 1100 (1961).

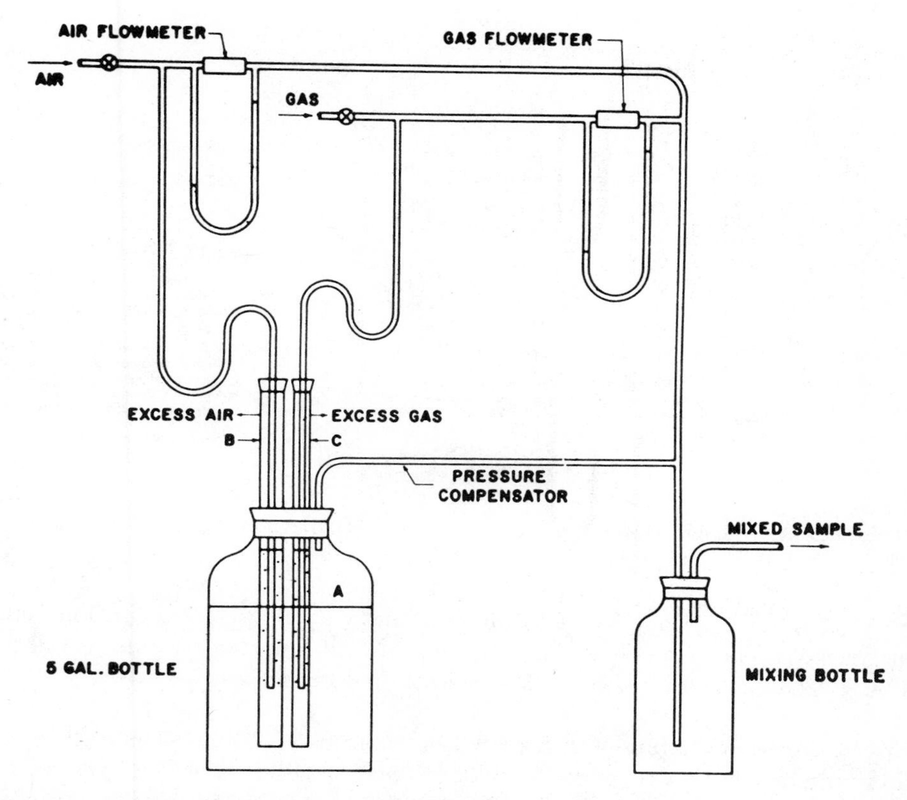

Figure 25.22. Modified Mase gas mixer for compensation of back pressure. From *Am. Ind. Hyg. Assoc. J.*, **22**, 392 (1961); courtesy of Williams & Wilkins, Baltimore, and Mine Safety Appliances Company, Pittsburgh.

pumps and the stopcocks provide more than 10,000 different concentration ratios. Lodge (7) has reported that European investigators have obtained excellent results with this type of system.

The Stack Gas Calibrator traps a fixed amount of gas by a series of valves and tubing and releases this volume into the carrier gas. The number of volumes released can be varied from 1 to 10 per minute. Depending on the size of the volume, three dilution ranges of ten steps are available, 200 to 2000 ppm, 660 to 6600 ppm, and 1320 to 13,200 ppm.

Another gas dilution system, the Dyna-Blender of Matheson Gas Products is represented schematically in Figure 25.25. Still another device for constant delivery of a pollutant gas has been described by Goetz and Kallai (12) (Figure 25.26). It consists of a large, gastight syringe with a centrifugal rotor attached to the piston so that the piston rotates around its axis. The rotation, caused by a jet of air directed tangentially toward

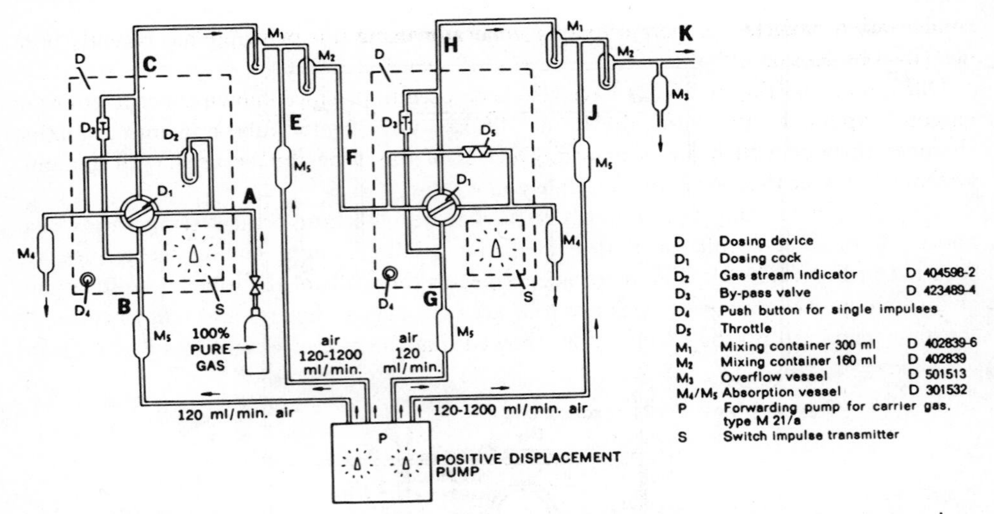

Figure 25.23 The ppm Maker. Schematic courtesy of Calibrated Instruments, Incorporated.

the rotor, is nearly friction-free and induces a constant pressure in the gas. The outlet of the syringe is connected on one side of a glass T-tube. Dilution air is piped into the base of the T, and the mixture exits the T-tube from the outer sidearm.

Liquid Dilution Systems. When the contaminant is a liquid at normal temperatures, a vaporization step must be included. One procedure is to use a motor-driven syringe (10, 11) and meter the liquid onto a wick or a heated plate in a calibrated airstream. Nelson and Griggs (13) have described a calibration apparatus that makes use of this principle (Figures 25.27 and 25.28). The system consists of an air cleaner, a solvent injection system, and a combination mixing and cooling chamber. A large range of solvent concentrations can be produced (2 to 2000 ppm). The device permits rapid changes in the concentrations and is accurate to about 1 percent. It can also be used to produce gas dilutions with an even wider range of available concentrations (0.05 to 2000 ppm). The Davis Products Division of Scott Aviation markets a "syringe calibrator assembly" with a motor-driven syringe.

A second vapor generation method is to saturate an airstream with vapor and dilute to the desired concentration with makeup air. The amount of vapor in the saturated airstream is dependent on both the temperature and the vapor pressure of the contaminant and can be precisely calculated. A simple vapor saturator is illustrated in Figure 25.29. The inert carrier gas passes through two gas-washing bottles in series, and these contain the liquid to be volatilized. The first bottle is kept at a higher temperature than the second one, which is immersed in a constant temperature bath. By using the two bottles in this fashion, saturation of the exit gas is assured. A filter is sometimes included to remove any droplets entrained in the airstream as well as any condensation particles. A mercury vapor generator using this principle has recently been described by Nelson (14).

Diffusion cells (Figure 25.30) have also been used to produce known concentrations of gaseous vapors. In this case, the liquid diffuses up a center tube and into a mixing chamber through which air is passed. Devices of this type can be used with dynamic systems, however they are limited to fairly low rates.

O'Keefe and Ortman (16) found that any material whose critical temperature was above 20 to 25°C could be sealed in Teflon tubing. Three sealing techniques are depicted in Figure 25.31. The material permeates the walls of the tube and diffuses out at a rate dependent on wall thickness and area (fixed parameters) and temperature. At constant temperature, the investigators showed that the rate of weight loss was constant

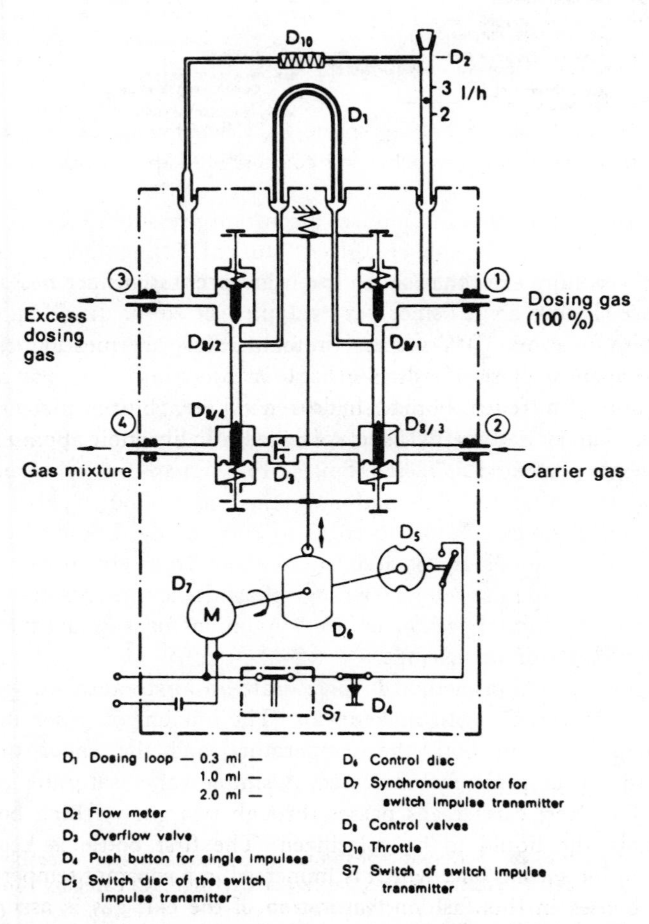

Figure 25.24 Stack gas calibrator. Schematic courtesy of Calibrated Instruments, Incorporated.

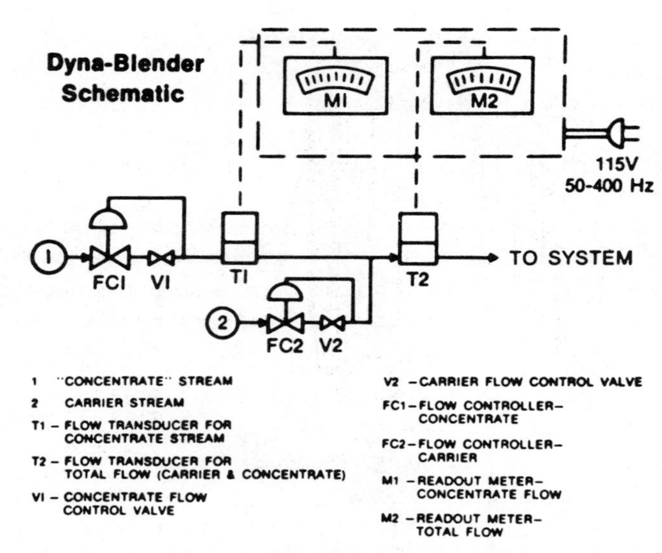

Figure 25.25 Matheson Dyna-Blender. Schematic courtesy of Matheson Gas Products, Division of Will Ross, Inc.

as long as there was liquid in the tube. In use, rigid precautions are necessary to assure fixed temperature (Figure 25.32) since, for example, the sulfur dioxide permeation rate more than doubles for every 10°C increase in temperature. Permeation tubes have been successfully used as primary standards (see Table 25.3).

Permeation tubes of nitrogen dioxide, hydrogen sulfide, chlorine, ammonia, propene, propane, butane, butene, and methyl mercaptan are also available. Permeation rates for these compounds are presented in Table 25.6.

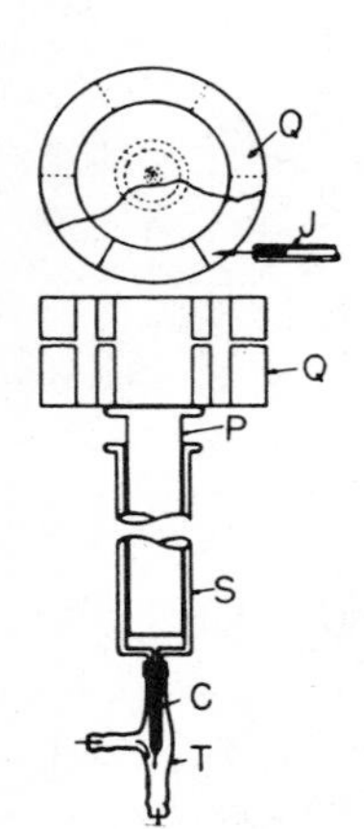

Figure 25.26 Schematic of spinning syringe calibrator assembly: *Q,* fan vanes; *J,* air jet; *P,* glass piston; *S,* large glass syringe; *C,* capillary tube; *T,* "T" tube. From *Journal of the Air Pollution Control Association,* **12,** 437 (1962), courtesy of the Air Pollution Control Association.

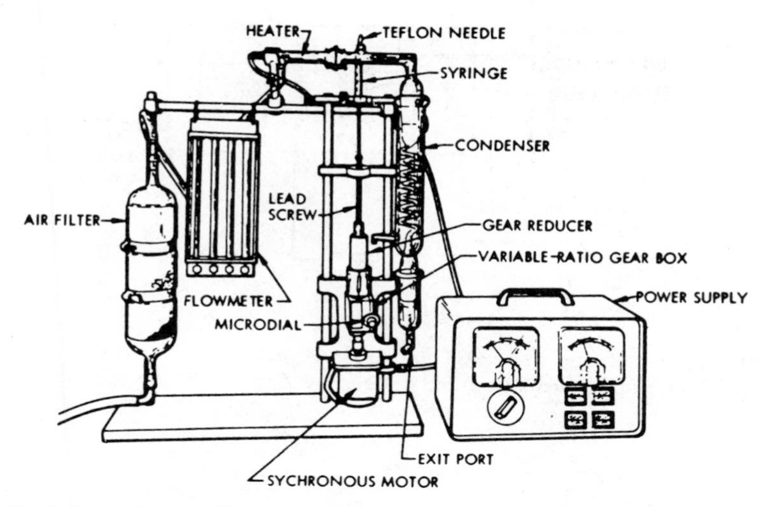

Figure 25.27 Syringe drive calibration assembly. Reprinted from U.C.R.L.-70394, courtesy of Lawrence Radiation Laboratory and the U.S. Atomic Energy Commission.

4.7 Production of Calibration Aerosols

4.7.1 Introduction

The generation of a test aerosol having the desired combination of physical and chemical properties is in most cases more difficult than the generation of gas of vapor test

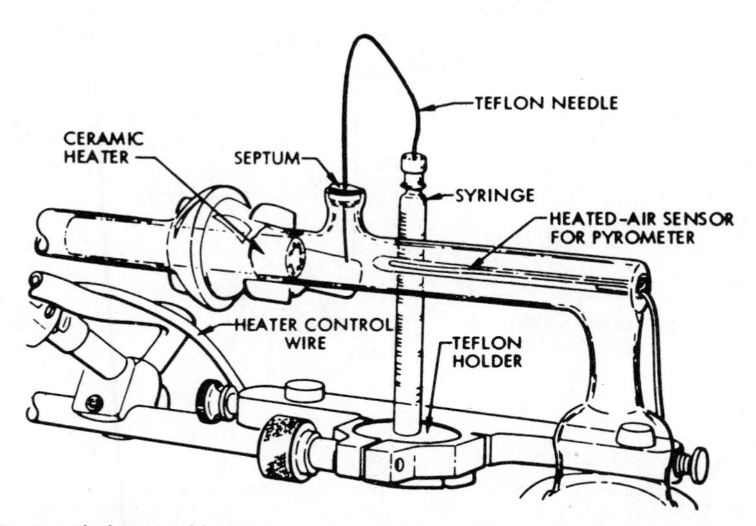

Figure 25.28 Detailed view of heating system and injection port. Reprinted from U.C.R.L.-70394, courtesy of Lawrence Radiation Laboratory and the U.S. Atomic Energy Commission.

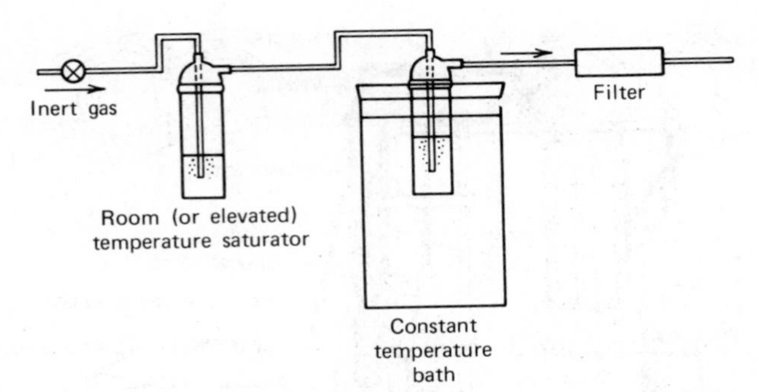

Figure 25.29 Vapor saturator. From *Am. Ind. Hyg. Assoc. J.*, **22,** 392 (1961), courtesy of the Williams & Wilkins Company, Baltimore, and Mine Safety Appliances Company, Pittsburgh.

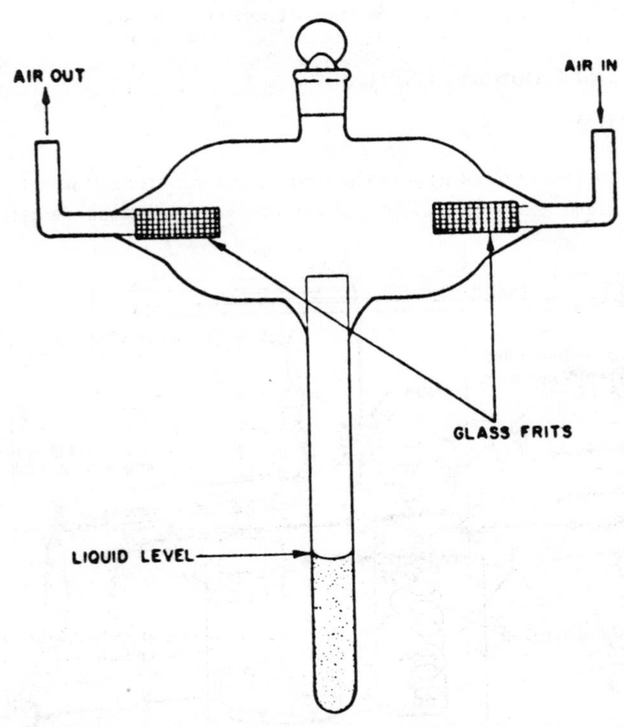

Figure 25.30 Diffusion cell. From *Anal. Chem.* **32,** 802 (1960); courtesy of the American Chemical Society.

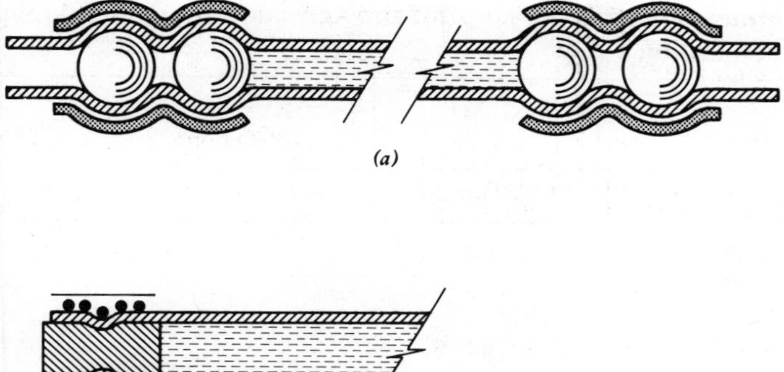

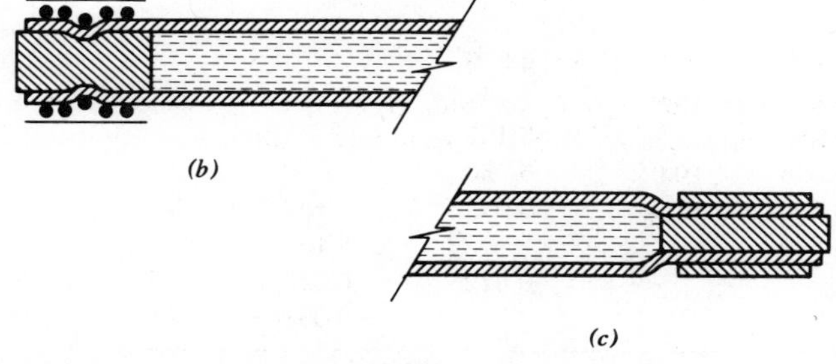

Figure 25.31 Three types of seal. (*a*) Steel on glass balls. (*b*) Teflon plug bound with wire. (*c*) Teflon plug held by a crimped metal band. From International Symposium on Identification and Measurement of Environmental Pollutants, National Research Council of Canada, Ottawa, Ont., 1971.

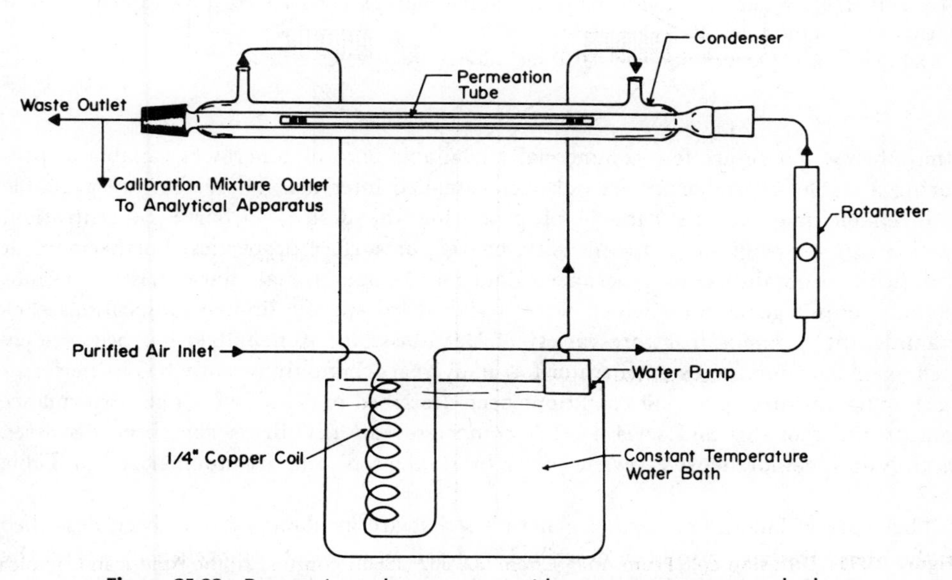

Figure 25.32 Permeation tube apparatus with constant temperature bath.

Table 25.6 Permeation Rates (μg/min-cm) and Activation Energies *E* (kcal/g-mole) for Some Teflon Permeation Tubes[a]

	Dynacel Tubes[b]		AID Tubes[c]	FEP Teflon[d]		TFE Teflon[d]	
Substance	Rate	*E*	Rate	Rate	*E*	Rate	*E*
SO_2	0.422	13.8	0.279				
NO_2	1.714	14.7	1.230	2.09	14.6		
HF	0.185						
H_2S	0.457	16.0	0.229				
Cl_2	2.418	14.0	1.430				
NH_3	0.280		0.165				
CH_3SH	0.036	10.0	0.030				
Propane	0.080	15.0	0.035	0.132	16.2	1.86	13.0
Propene	0.240			5.13	15.4		
n Butane			0.012	0.024	15.8	0.258	12.7
Butene-1				0.0316	14.8	0.368	12.4

[a] As reported by B. E. Saltzman, in: *The Industrial Environment—Its Evaluation and Control,* Government Printing Office, Washington, D.C., 1973, p. 133.
[b] Available from Metronics Associates, Inc., Palo Alto, Calif. 94304. Tubes are $\frac{3}{16}$ inch o.d., $\frac{1}{8}$ inch i.d. Rates are at 30°C.
[c] Available from Analytical Instrument Development, Inc., West Chester, Pa. 19308. Tubes are FEP Teflon, 0.250 inch o.d. and 0.062 inch wall thickness except for methyl mercaptan, which is 0.030 inch wall thickness. Rates are at 30°C.
[d] Wall thickness 0.250 inch o.d. and 0.030 inch. Rates are at 25°C.

atmospheres. There are few commercially available aerosol generators capable of producing a stable, reproducible aerosol over extended intervals. Those that are available and reliable may not be capable of producing the desired airborne concentration, particle size distribution, shape, density, charge, or surface properties. Furthermore, it is difficult to obtain useful generalized data on the operational characteristics of commercial aerosol generators, which were designed for specific limited applications. For example, the common drugstore variety of DeVilbiss No. 40 nebulizer has been widely used as a laboratory aerosol generator for many years, but only recently has its performance under varying operating conditions been described in detail (17). The performance characteristics of this and several other compressor air nebulizers have been discussed recently in a comprehensive review paper by Raabe (18) and are summarized in Table 25.7.

The types of laboratory aerosol generators that are available or have been described in the literature have also been described in detail in a recent comprehensive review by Kerker (19). Previous reviews were made by Fraser et al. (20), Silverman (10), and Lodge (7). A detailed review of techniques and equipment for producing monodisperse

Table 25.7 Representative Characteristics of Selected Compressed Air and Ultrasonic Nebulizers

Compressed Air Nebulizers[a]

Air Pressure (psig)	DeVilbiss, Selling No. 40 (17) (jet = 33 mil; vent closed) Output (evap.) (μl/liter)	Total Air (liters/min)	VMD (σ_g) (μm)	Lovelace[b] (jet = 9.2 mil) Output (evap.) (μl/liter)	Total Air (liters/min)	VMD (σ_g) (μm)	Dautrebande D-30 (17) (jet = 41 mil) Output (evap.) (μl/liter)	Total Air (liters/min)	VMD (σ_g) (μm)	Lauterbach (86) (jet = 13 mil) Output (evap.) (μl/liter)	Total Air (liters/min)	VMD (σ_g) (μm)	Collison (87) (3 jets = mil) Output (evap.) (μl/liter)	Total Air (liters/min)	VMD (σ_g) (μm)	Retec X-70/N (88) Output (evap.) (μl/liter)	Total Air (liters/min)	VMD (σ_g) (μm)
5	16.0	7.5	4.6 (1.8)	1.6	0.8		1.0 (9.7)	13.4		2.6	1.2							
10	16.0	10.8	4.2 (1.8)	15.3 (11)	1.2		1.6 (9.6)	17.9	1.7 (1.7)	3.9	1.7	3.8 (2.0)						
15	15.5 (8.6)	13.5	3.5 (1.8)	19.5	1.4					5.2	2.1							
20	14.0 (7.0)	15.8	3.2 (1.8)	30.0 (10)	1.7	5.8 (1.8)	2.3 (8.6)	25.4	1.4 (1.7)	5.7	2.4	2.4 (2.0)	7.7 (12.7)	7.1	2.0 (2.0)	53 (12)	5.4	5.7 (1.8)
30	12.1 (7.2)	20.5	2.8 (1.8)			4.7 (1.9)	2.4 (8.2)	32.7	1.3 (1.7)	5.9	3.2	2.4 (2.0)	5.9 (12.6)	9.4		54 (11)	7.4	3.6 (2.0)

Commercial Ultrasonic Nebulizers[a] (85)

Nebulizer	Output (Evap.) (μ/liter)	Total Air (liters/min)	VMD (σ_g) (μm)
DeVilbiss, setting No. 4	150 (33.1)	41.0	6.9 (1.6)
Mist-O_2-Gen, with reservoir	61.5 (22.2)	24.7	6.5 (1.4)

[a] Outputs are given in microliters of solution per liter of total aerosols (evaporation losses are in parentheses). Total volume of aerosol is indicated as total air in liters per minute. The droplet distribution of usable aerosol at initial formation is assumed to be log-normal with data given for the volume median diameters *VMD* and geometric standard deviations σ_g in parentheses. The sources of the values are indicated by reference numbers.

[b] Baffle setting has been optimized for operation at 20 psig. The data on the Lovelace Nebulizer are by Dr. Otto G. Raabe (88). G. J. Newton, and J. E. Bennick of the Lovelace Foundation.

aerosols has been prepared by Fuchs and Sutugin (21). From these and other sources, a condensed summary of techniques for generating monodisperse test aerosols has been constructed (Table 25.8). Sources of commercially available devices for producing polydisperse test aerosols are tabulated in Reference 1.

Aerosol generators can be divided into two types: those that produce condensation aerosols, and those that produce dispersion aerosols. In the former type, the material to be aerosolized is dispersed in the vapor phase and allowed to condense on airborne nuclei.

4.7.2 Generation of Monodisperse Condensation Aerosols

In an isothermal supersaturated environment, vapor molecules diffuse to and condense on airborne nuclei. Wilson and LaMer (22) demonstrated that the surface area of the resulting droplets increases linearly with time. Thus as the droplets become large in comparison to the nuclei, the size range becomes quite narrow, even when the nuclei on which the droplets grew may have varied widely in size.

The LaMer-Sinclair (23) aerosol generator, illustrated schematically in Figure 25.33,

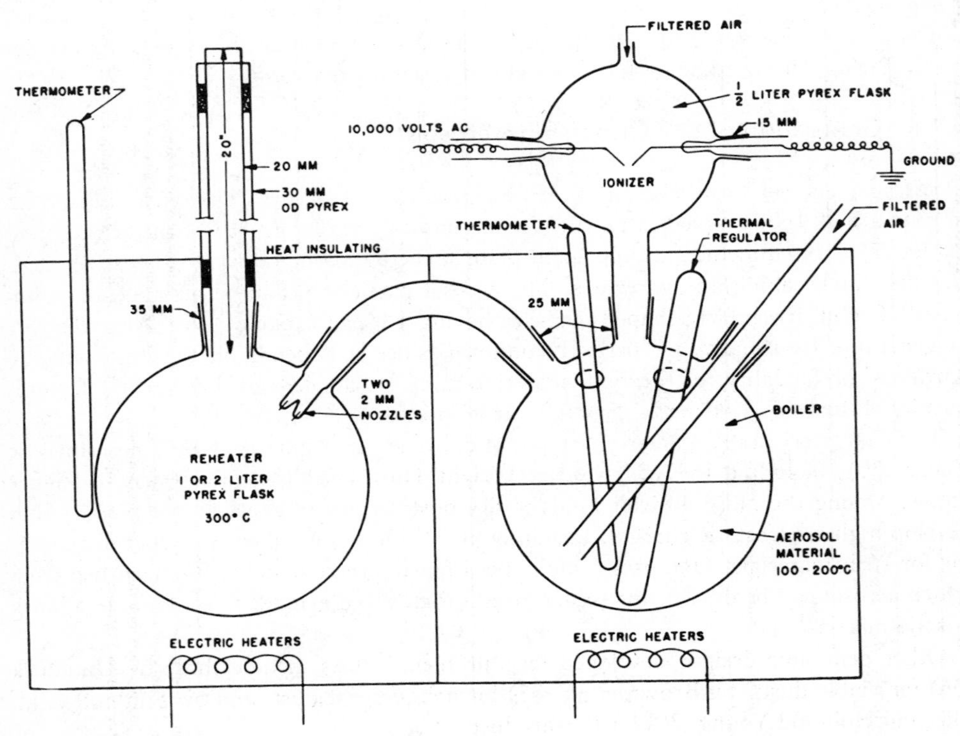

Figure 25.33 LaMer-Sinclair type of condensation aerosol generator. From U.S. Atomic Energy Commission. Handbook on Aerosols.

was based on these considerations. Improvements in the basic LaMer-Sinclair design have been described by Muir (24). Further refinements were made by Huang et al. (25). Rapaport and Weinstock (26) have described a condensation aerosol generator that is simpler and less expensive to produce, and requires less critical control of temperature and flowrate for the production of monodisperse aerosol. A more sophisticated version of this generator has been described by Liu and Lee (27). This type of generator is capable of producing high quality aerosols of high temperature boiling, low vapor pressure liquids (e.g., dioctyl phthalate, triphenylphosphate, and sulfuric acid in the size range of about 0.03 to 1.3 μm). Apparatus for producing monodisperse condensation aerosols of lead, zinc, cadmium, and antimony using a high frequency induction furnace has been described by Homma (28) and Movilliat (29). Matijevic et al. (30) and Kitani and Ouchi (31) have produced monodisperse condensation aerosols of sodium chloride.

The particles produced by condensation generators will be liquid and spherical unless the material vaporized has a melting point above ambient temperature. In this case, the particles will solidify and, if crystalline, may form nonspherical shapes. A summary of techniques for producing radioactively labeled monodisperse condensation aerosols with 18 organic compounds and 8 inorganic materials has been presented by Spurny and Lodge (32).

Kerker's review (19) provides the most complete summary of the state of knowledge on the factors affecting the performance of condensation generators.

4.7.3 Generation of Dry Dispersion Aerosols

Dispersion aerosol generators may be classified as wet or dry. Dry generators comminute a bulk solid or packed powder by mechanical means, usually with the aid of an air jet. They often include an impaction plate at the outlet for removal of oversize particles and for breaking up aggregates. The aerosol particles produced are typically composed of solid, irregularly shaped particles having a broad range of sizes. Also, the rate of generation is usually not perfectly uniform, since it depends on the uniformity of hardness, or friability, of the bulk material being subdivided, as well as on the uniformity of the feed-drive mechanism and air jet pressure.

The characteristics of a variety of types of dry dust generators have been described by Ebens (33), including the widely used Wright Dust Feed (34) illustrated in Figure 25.34. Among the more difficult kinds of dry dust aerosol to generate are plastics that develop high electrostatic charges. Laskin et al. (35) have described two types of generator for such materials. One uses a high speed fan to create a stable fluidized bed from which aerosol can be drawn; the second uses a high speed grinder to comminute a block of solid material.

Other generator designs developed for "problem" dusts include those by Dimmock (36) for viable dusts, by Brown et al. (37) for deliquescent dust, and by Timbrell et al. (38) and Holt and Young (39) for fibrous dust.

Useful aerosols of dry particles of metal and metal oxides are also produced with electrically heated (40, 41) or exploded wires (42, 43). These techniques have some

Table 25.8 Techniques for Generating Monodisperse Test Aerosols

Name or Type	Operational Mechanism of Generator	Types of Monodisperse Aerosol Produced	Typical Diameter Range, μm (σ_g)	Approximate Output (no./sec)	Approximate Flow (liters/min)	Utilities Required	Techniques for Tagging	Commercial Source or Reference
Uniform latex spheres	Nebulization	Latex spheres (18, 63) T3 *E. coli* phage (60) Type 3 Poliomyelitis Virus (65)	0.109–1.947 (1.02) 0.035 0.026	10^4	10	10 psig air	Emulsion polymerization (78–80)	Duke[d]
Atomizer-impactor	Nebulizer with impactor cutoff	Any liquid or solid residue	0.03^a–3 (1.4)	10^9	≥57	45 psig air	—[b]	Sierra[e]
Spinning Disk	Rotary atomizer	Any liquid or solid residue	1^a–30 (1.1)	10^7	≥283	60 Hz ac	—[b]	Sierra[e]
Spinning top	Rotary atomizer	Any liquid or solid residue	0.5^a–200 (1.1)	10^7	NA[c]	40 psig air		BGI[f]
Binek particle generator	Displacement of liquid from vibrating wetted whisker	Any liquid or solid residue	0.1–10 (1.07)	10^2	NA	50 Hz ac	—[b]	Sartorius[g]

Other vibrating reed or capillaries (transverse)	Displacement of liquid from reed or capillary in transverse vibration	Any liquid or solid residue	1[a]–200 (1.1)	10^2	NA	60 Hz ac	—[b]	(55, 57)
Vibrating orifice (axial)	Liquid filament disruption—mechanical instability	Any liquid or solid residue	1[a]–200 (<1.1)	10^5	NA	60 Hz ac	—[b]	TSI[h]
Electrostatic classifier	Mobility stripping	Any liquid or solid residue	0.01[a]–0.3 (<1.1)	10^6	3	60 Hz ac	—[b]	TSI[h]
Electrostatic nebulizer	Liquid filament disruption–electrical instability	Liquids with low electrical conductivity or their solid residue	<0.1–200 (NA)	10^8	NA	10 kV ac or dc	—[b]	(59, 60)
Condensation	Condensation on nuclei	DOP, TPP, other low vapor pressure, high boiling liquids, subliming solids	0.01–1 (1.2)	10^8	3	ac or dc Line	Use of radioactive nuclei	TSI[h]

[a] Lower size limit based on dried residue particles from dilute solutions or suspensions.
[b] Tags can be dissolved or suspended in feed liquid; see text for further discussion.
[c] Not available.
[d] Duke Standards Co., 445 Sherman Ave., Palo Alto, Calif. 94306
[e] Sierra Instruments, Inc., P.O. Box 909, Village Square, Carmel Valley, Calif. 93924
[f] BGI, Inc., 58 Guinan St., Waltham, Mass. 02154
[g] Sartorius Filters, Inc., 803 Grandview Drive, South San Francisco, Calif. 94080
[h] Thermo-Systems, Inc., 2500 Cleveland Ave. North, St. Paul, Minn. 55113

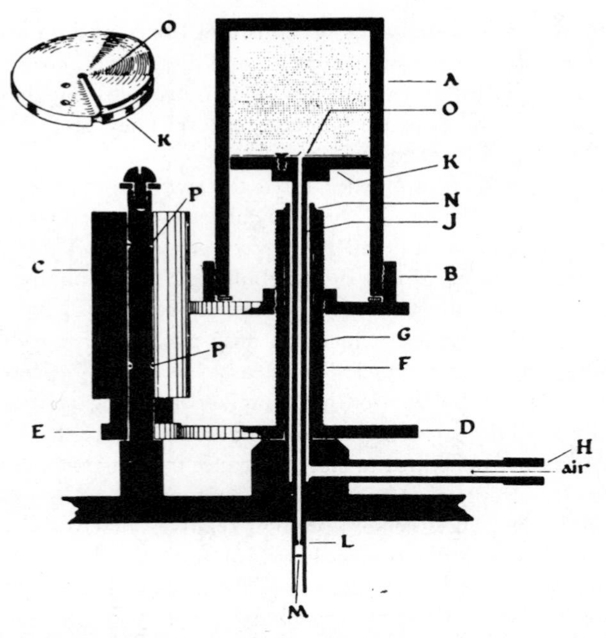

Figure 25.34 The Wright Dust Feed (34): *A*, dust cylinder; *B*, cap, with peripheral gear; *C*, pinion; *D*, wheel; *E*, pinion; *F*, threaded tube; *G*, tube, connected to *H*, compressed air line; *J*, small tube, carrying scraper head *K*, which communicates with jet *L*, which is above impaction plate *M* for breaking up aggregates; *O* is a spring disk with cutting edge.

disadvantages because of the very broad size distributions of the resulting particles and because of the tendency of particles to coalesce. There are applications, however, for this type of aerosol, and it is possible with a wire-heating method to produce spherical particles of many different metals or their oxides. Aerosols of very small particles have also been produced by arc vaporization (44).

4.7.4 Wet Dispersion Aerosol Generators

Wet dispersion generators break up bulk liquid into droplets. If the liquid is nonvolatile, the resulting aerosol is a mist or fog. If a volatile liquid is aerosolized, the resulting particles are composed of the nonvolatile residues in the feed liquid and are much smaller than the droplets dispersed from the generator. Solid particles can be produced by nebulizing salt or dye solutions or particle suspensions. Aqueous solutions, of course, produce water-soluble particles that may be hygroscopic. This may be an important factor, since the aerodynamic size for such aerosols will vary with ambient humidity.

A variety of techniques can be used to subdivide bulk liquid into airborne droplets. In most cases the liquid is accelerated by the application of mechanical, pneumatic, or centrifugal forces and drawn into filaments or films that break up into droplets because of surface tension. Centrifugal pressure nozzles and fan spray nozzles use hydraulic pressure to form a sheet of liquid that breaks up into droplets, but these generally have high liquid feed rates and produce very large droplets. They are seldom employed for producing aerosols for instrument calibration.

A commonly used type of aerosol generator is the two-fluid nozzle, which uses pneumatic energy to break up the liquid. Several laboratory scale compressed air-driven nebulizers have been described in detail by Mercer et al. (17). Table 25.7 summarizes the operational characteristics of six such nebulizers, including the Lauterbach and Lovelace designs which are illustrated in Figures 25.35 and 25.36 respectively. The DeVilbiss No. 40 is made of glass, which not only makes it fragile but also limits its precision of manufacture and reproducibility. Ready reproducibility led Whitby to select the British Collison (45) nebulizer for his atomizer-impactor aerosol generator (46). Other commercially available nebulizers, including those of Wright (47), (Figure 25.37) and Dautrebande (48), are machined to close tolerance from plastic materials.

Nebulizers produce droplets of many sizes, and resultant aerosol particles after evaporation are therefore polydisperse, although relatively narrow size dispersions can be obtained with Whitby's atomizer-impactor (46) and Dautrebande's D-30 (48). The droplet distributions described for nebulizers are the initial distributions at the instant of formation; droplet evaporation begins immediately, even at saturation humidity, since the vapor pressure on a curved surface is elevated (49). The rate of evaporation depends on many factors, including solute concentration, the hygroscopicity of the solute (50, 51), the presence of immiscible liquids or evaporation inhibitors (52), and the size of the droplets.

Evaporative losses cause an increase in the concentration of the solution or of the suspended particles, resulting in an increase in the size of the dry particles formed when the liquid evaporates. Evaporation occurs both from the surface of the liquid and from the droplets which evaporate slightly and hit the wall of the nebulizer to be returned to the reservoir, and is most important in nebulizers with small reservoirs but large volumetric airflows.

Rotary atomizers, such as the spinning disk, utilize centrifugal force to break up the liquid, which undergoes an acceleration as it spreads from the center to the edge of the disk. The liquid leaves the edge of the disk as individual droplets or as ligaments that disintegrate into droplets. Walton and Prewett (53) demonstrated that these atomizers can produce monodisperse aerosols when operated with low liquid feed rates and high peripheral speeds. A spinning disk generator designed specifically for the production of monodisperse test aerosols with radioactive tags has been described by Lippmann and Albert (54) and is illustrated in Figure 25.38.

Monodisperse test aerosols can also be produced by a variety of techniques that break up a laminar liquid jet into uniform droplets. Most of them vibrate a capillary at high speed with a variety of transducers and types of motion. Dimmock's (55) generator, for example, uses transverse vibrations, whereas Strom's (56) uses axial vibrations. Wolf (57) uses a vibrating reed, wetted to a constant length by passage through a liquid

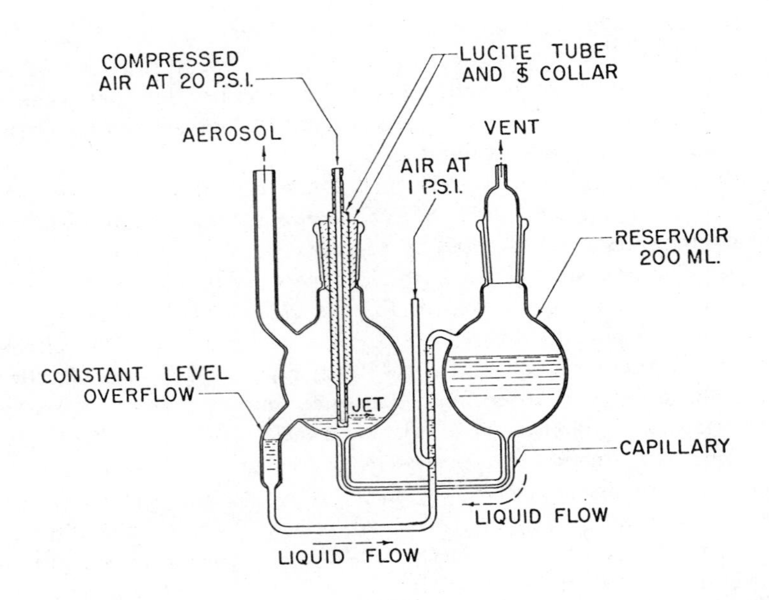

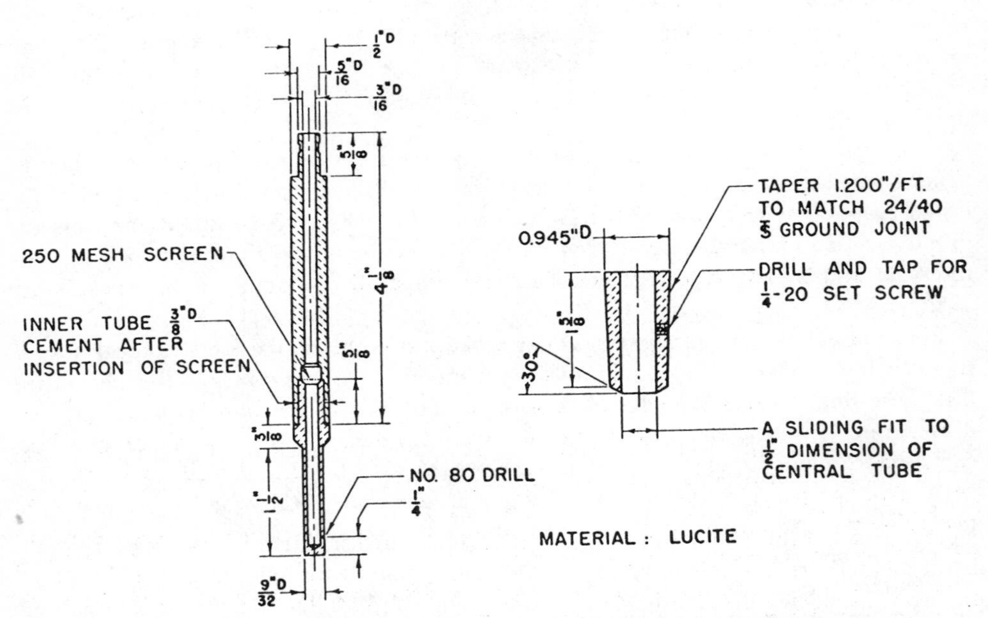

Figure 25.35 The Lauterbach (86) aerosol generator and its jet tube.

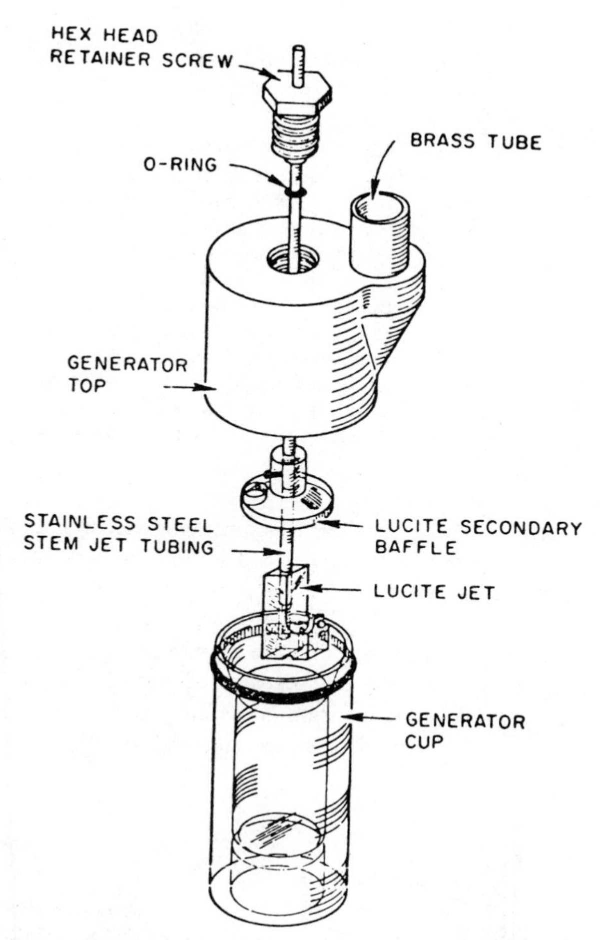

Figure 25.36 The Lovelace nebulizer, which operates with a liquid volume of ~ 4 ml and incorporates a jet baffle similar to that of Wright (47). Schematic courtesy of Dr. Otto G. Raabe.

reservoir, to create the droplet stream. The generator described by Raabe (18) has an air jet above the orifice and uses an ultrasonic transducer to convert a high frequency power signal into mechanical axial vibrations of the orifice. The vibrating orifice generator of Berglund and Liu (58), which is illustrated in Figure 25.39, uses a cylindrical piezoelectric ceramic to vibrate a thin orifice plate, with holes from 3 to 22 μm in diameter, producing droplet diameters from about 10 to 50 μm. The particles produced by these generators can be made to vary less than 1 percent in volume, less than the variation in size of aerosols generated by a spinning disk device.

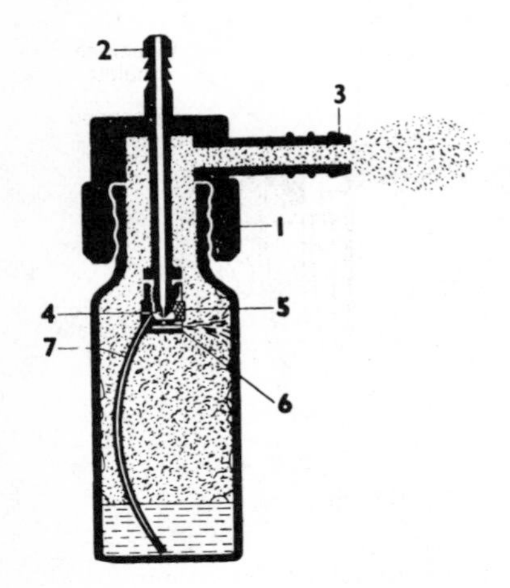

Figure 25.37 The Wright (47) nebulizer. It consists of a solid cap 1, into which can be screwed any suitable bottle; 2, inlet connection; 3, outlet connection. The inlet connection communicates with a fine jet 4 on to which is screwed a knurled nozzle 5. The nozzle carries a circular baffle plate 6 mounted on an eccentric pillar through which passes a flexible feed tube 7. As the air jet passes through the nozzle 5, a vacuum is created that draws liquid up the feed pipe 7. The resulting spray impacts against the baffle plate and the coarser droplets (more than about 8 μm diameter) are trapped, coalesce, and fall back into the liquid.

Electrostatic atomization can also produce monodisperse aerosols. Electric charges on a liquid surface act to decrease the surface tension. Liquid flowing through a capillary at high voltage is drawn into a narrow thread that breaks up into very small droplets (59–61).

Liu and Pui (62) developed a generator (Figure 25.40) for quite monodisperse submicrometer aerosol particles in which the polydisperse output of a compressed air nebulizer is classified electrostatically. A solution or colloid is aerosolized in a Collison atomizer, mixed with dry air to form a solid aerosol, and brought to a state of charge

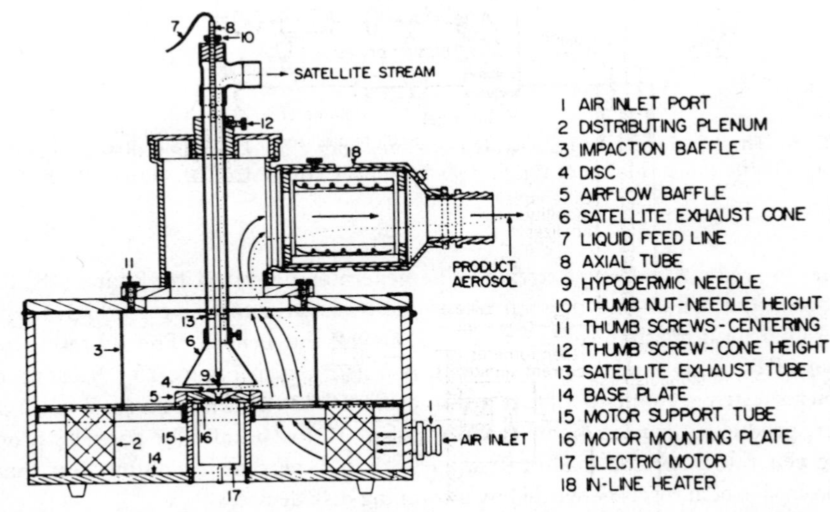

Figure 25.38 Electric motor driven spinning disk aerosol generator of Lippmann and Albert (54).

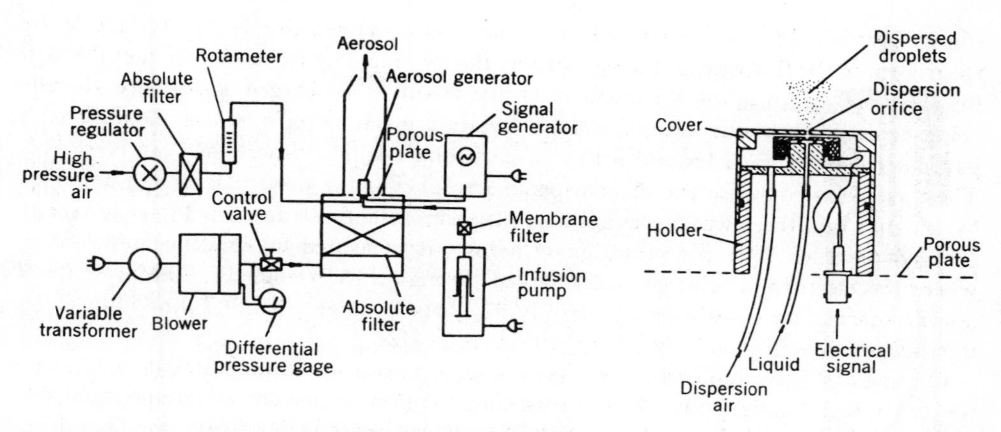

Figure 25.39 Vibrating orifice monodisperse aerosol generator. *Left:* schematic of system. *Right:* generator head. From *Air Pollut. Control Assoc. J.*, **24:** 12 (December 1974).

equilibrium with the aid of a ^{85}Kr source. This aerosol which is polydisperse (geometric standard deviation $\cong 2.7$, median particle diameter ranging from 0.009 to 0.65 μm) is introduced into a differential mobility analyzer, which functions as a particle size classifier, based on the electrical mobility of the different size categories. This apparatus consists of an inner cylindrical electrode along which flows a sheath of clean air sur-

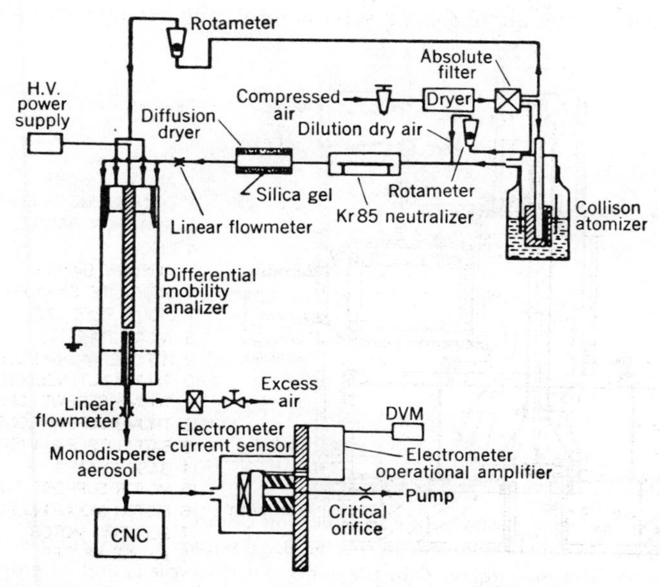

Figure 25.40 Apparatus for generating submicron aerosol standards.

rounded by an outer concentric sheath of the aerosol. Depending on the voltage of the electrode and the flowrate and the geometry, the more mobile particles will drift through the clean air sheath to the electrode, where they will be discharged and adhere, thereby being removed from the aerosol stream. Under a given set of operating conditions, a particular class of particles will drift to a particular position, where they can be vented. These particles comprise the monodisperse aerosol. For the particular design described by Liu and Pui, there was a coefficient of variance of .04 to .08 in particle size for singly charged particles. The concentration of aerosol is measured by collection on a filter, where the electrostatic charge is discharged through an electrometer. The presence of doubly charged particles increases greatly for particles larger than 0.3 μm. Thus above this size, the mobility no larger defines the particle size.

Commercially available ultrasonic aerosol generators can vibrate a liquid surface at a frequency high enough to result in the disintegration of the surface liquid into a polydisperse droplet aerosol. For mass median droplet diameters below 5 μm, the transducer must vibrate at a frequency greater than 1 MHz. The output characteristics of two commercial ultrasonic nebulizers are summarized in Table 25.7. An experimental ultrasonic generator (Figure 25.41) designed by G. J. Newton of the Lovelace Foundation has been described by Raabe (17).

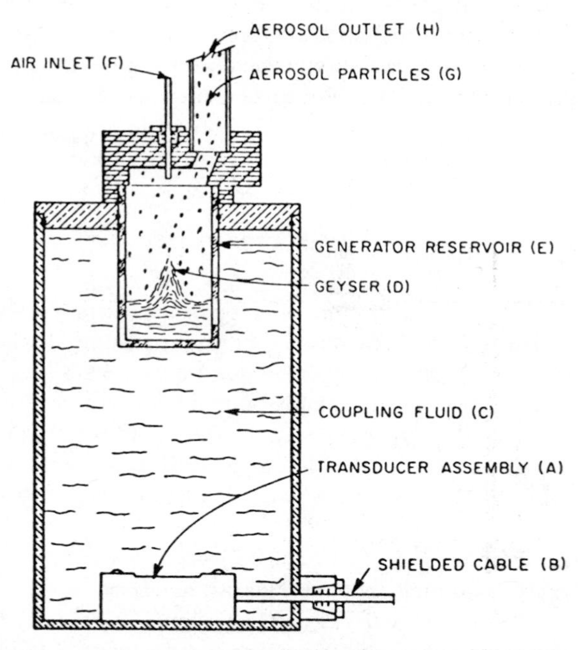

Figure 25.41 Sectional schematic view of an operating Ultrasonic Aerosol Generator: transducer assembly *A* receives power through shielded cable *B*, generates an acoustic field in the coupling fluid *C*, creating an ultrasonic geyser *D* in the generator reservoir *E*, and air entering at *F* carries away aerosol *G* through outlet *H*. Figure by G. J. Newton, reproduced from Raabe (18).

4.7.5 Generation of Solid Insoluble Aerosols with Wet Dispersion Generators

Solid insoluble aerosols can be produced by nebulizing particle suspensions. One technique for producing monodisperse test aerosols is to prepare a uniform suspension of the particles (latex, bacteria, etc.) in which the concentration is sufficiently dilute in the liquid phase that the probability of more than one particle being present in each droplet is acceptably small (63–65). This will result in a high vapor to particle ratio, thus limiting the mass concentration of aerosol produced. Another approach is to use a colloid as the feed liquid. In this case the diameter of the colloid particles can be orders of magnitude smaller than the particles in the resulting aerosol. Thus the volume of the droplet, and the size of the dried aggregate particles, is determined by the solids content of the sol.

An aerosol with chemical properties different from those of the feed material can be produced by utilizing suitable gas phase reactions such as polymerization or oxidation. Kanapilly et al. (66) describe the generation of spherical particles of insoluble oxides from aqueous solutions with heat treatment of the aerosols. This procedure involves (*a*) nebulizing a solution of metal ions in chelated form, (*b*) drying the droplets, (*c*) passing the aerosol through a high temperature heating column to produce the spherical oxide particles, and (*d*) cooling the aerosol with the addition of diluting air. Another example of aerosol alteration is the production of spherical aluminosilicate particles with entrapped radionuclides by heat fusion of clay aerosols (67). This method involves (*a*) ion exchange of the desired radionuclide cation into clay in aqueous suspension and washing away of the unexchanged fraction, (*b*) nebulization of the suspension yielding a clay aerosol, and (*c*) heat fusion of clay aerosol, removing water and forming an aerosol of smooth solid spheres.

4.7.6 Characterizing Aerosols

Size Dispersion. The size dispersion of a test aerosol produced by a laboratory generator is determined by the characteristics of the generator and feed materials. The data on size included in the preceding discussion on generator characteristics, and in Table 25.8, indicate the approximate range obtainable in normal operation. The actual size distribution in a given case should always be measured with appropriate techniques and instrumentation. Sampling for particle size analysis has been discussed by Peterson (1). Sampling and analytical techniques have also been reviewed by Raabe (18) and Giever (68).

The distribution of droplets produced by nebulizers and some dry dust generators can usually be described by assuming that the logarithms of size are normally distributed. This log normal distribution of sizes allows for simple mathematical transformations (69) and usually describes volume distributions satisfactorily (70). The characteristic parameters of a log normal distribution are the median (or geometric mean) and the geometric standard deviation σ_g. The median of a distribution of diameters is called the count median diameter CMD: the median diameter based on the surface area is called the surface median diameter *SMD*; the median of the mass or volume distribution of the droplets or particles is called either the mass median diameter *MMD* or the volume median diameter *VMD*. These are related as follows.

$$\ln(SMD) = \ln(CMD) + 2\ln^2\sigma_g$$
$$\ln(MMD) = \ln(CMD) + 3\ln^2\sigma_g$$

in which ln designates the natural logarithm. A representative log-normal distribution appears in Figure 25.42 for a *CMD* equal to 1 μm and a σ_g equal to 2.

When particles are classified on the basis of their airborne behavior, a parameter called aerodynamic diameter is often used. It refers to the size of a unit density sphere having the same terminal settling velocity as the particle in question. For radioactive particles, an ICRP task group (71) has used the parameter aerodynamic mass activity diameter (AMAD), which is the aerodynamic median size for airborne particulate activity.

Physical and Chemical Properties. An aerosol of a pure material having the desired physical and chemical characteristics can be prepared by dispersing that material into the air by any appropriate technique previously described. It is also possible to produce aerosols that differ in physical and/or chemical properties from the feed

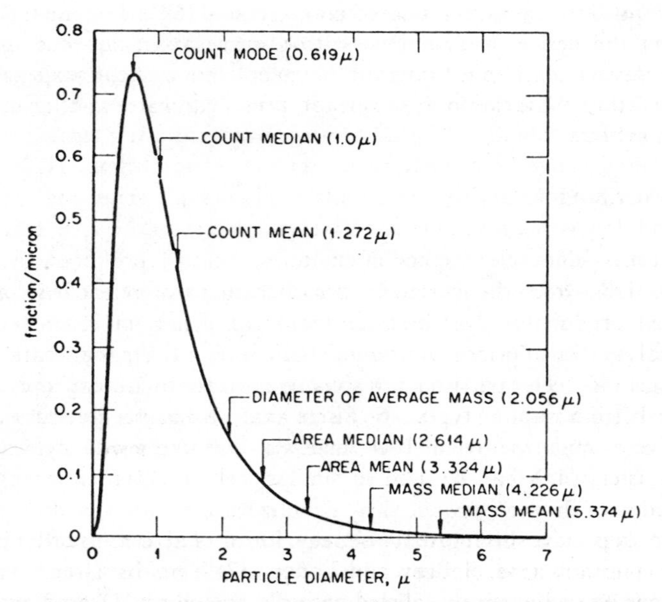

Figure 25.42 An example of the log-normal distribution function in normalized linear form for $CMD = 1.0$ μm and $\sigma_g = 2.0$ μm, showing the mode, median, and mean diameters, the mass distribution median and mean diameters, and the diameter of average mass. Graph courtesy Dr. Otto G. Raabe (18).

material. For example, particle size can be varied by dissolving or suspending the material in a suitable volatile solvent that evaporates in the air to leave residue particles smaller than the nebulized droplets.

Solid aerosols resulting from droplet evaporation are generally spherical, but not always. Too rapid solvent evaporation, low pH, and the presence of impurities may cause the dried particles to be wrinkled or to assume various shapes (72).

Aerosols produced from aqueous solutions (and some other methods) are charged by the random imbalance of ions in the droplets as they form (73). After evaporation, aerosol particles can be relatively highly charged; this may cause a small evaporating droplet to break up if the Rayleigh limit (74) is reached because of the repelling forces of the electrostatic charges overcoming the liquid surface tension (75). In some cases the net charge on a particle may be tens or even hundreds of electronic charge units, which will affect both the aerosol stability and behavior. Therefore a reduction in the net charges on aerosols produced by nebulization is desirable and in some experiments may be imperative. This can be accomplished either by mixing the aerosol with bipolar ions (76) or by passing it through a highly ionized volume near a radioactive source (77).

4.7.7 Detection of Aerosol Particles and Tagging Techniques

For many applications, such as efficiency testing of aerosol samplers or filters, it is often necessary to be able to measure concentrations that differ by several orders of magnitude. This type of testing can be done with untagged particles, such as polystyrene latex, using sensitive light scattering photometers for concentration measurements. When particles other than the test aerosol are present, however, as in many field test situations, this method should not be used. Also, light scattering techniques can be used over only a limited range of particle size, and the equipment is relatively expensive. Another approach to efficiency testing entails a microscopic count and/or size analyses of upstream and downstream samplers. However this procedure is so tedious and time-consuming that it is seldom the method of choice.

Particle detection is often facilitated by incorporating dye or radioisotope tags in the particles in their production. Test aerosols composed of or containing fluorimetric dyes that can be analyzed in solutions containing as little as 10^{-10} g/cm^3 have been used for such applications (46). The particles are soluble in water or alcohol and can be quantitatively leached from many types of filters and collection surfaces for analysis. Colorimetric dyes such as methylene blue, which is used in the British Standard Test for Respirator Canisters (45), can be used in similar fashion when extremes of sensitivity are not required.

Radioisotope tags have been used in many forms and can usually be detected at extremely low concentrations. Spurny and Lodge (32) have discussed a variety of techniques for preparing radioactively labeled aerosols, including (1) preparation by means of neutron activation of aerosols in a nuclear pile or other neutron source, (2) labeling by means of decay products of radon and thoron, (3) preparation by means of radioactively labeled elements and compounds (condensation aerosols, disperse aerosols, and plasma aerosols), and (4) preparation by means of radioactively labeled condensation nuclei.

Method 2 refers to a process in which the particle surface is tagged while the particle is airborne. Procedures for surface tagging of polystyrene latex particles with isotopes in liquid suspensions by emulsion-polymerization reactions have been described by Black and Walsh (78), Bogen (79), and Singer et al. (80). Flachsbart and Stöber (81) have described a technique for growing uniform silica particles in suspension and incorporating various radioactive tags.

Other insoluble test aerosols containing nonleaching radioisotope tags have been produced by several techniques. The technique of heat fusion of ion exchange clays (67) was discussed in the preceding section on insoluble aerosols. Techniques for producing insoluble spherical aggregate particles by nebulizing colloidal suspensions and plastics in solution have been described (54, 72). These aerosols made from colloids can be tagged with radioisotopes by mixing the nonradioactive colloid with a much lower mass concentration of an insoluble radioactive colloid before nebulization. The plastic particles can be tagged with radioisotopes in chelated form, dissolved in the plastic solution (72, 82–84).

4.8 Calibration of Sampler's Collection Efficiency

4.8.1 Use of Well-Characterized Test Atmospheres

To test the collection efficiency of a sampler for a given contaminant, it is necessary either (1) to conduct the test in the field using a proved reference instrument or technique as a reference standard, or (2) to reproduce the atmosphere in a laboratory chamber or flow system. Techniques and equipment for producing such atmospheres were discussed in Sections 4.6 and 4.7.

4.8.2 Analysis of Sampler's Collection and Downstream Total Collector

The best approach to use in the analysis of a sampler's collection is to operate the sampler under test in series with a downstream total collector, as illustrated in Figure 25.43. The sampler's efficiency is then determined by the ratio of the sampler's retention to the retention in the sampler and downstream collector combined. This approach is

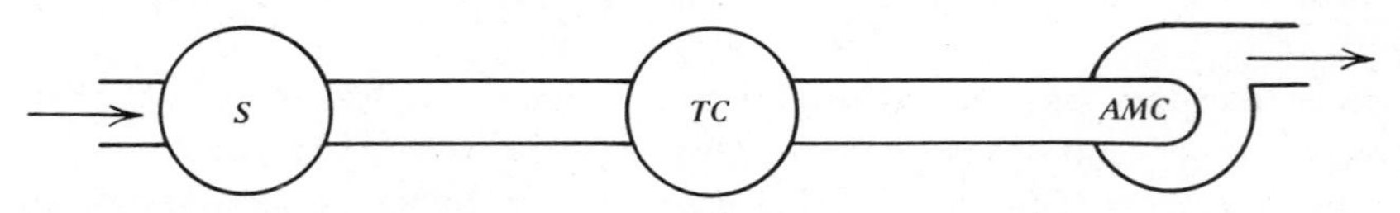

Figure 25.43 Sampler efficiency evaluation with downstream total collector: analysis of collections in sample under test *S* and total collector *TC*; *AMC*, air mover, flowmeter, and flow control.

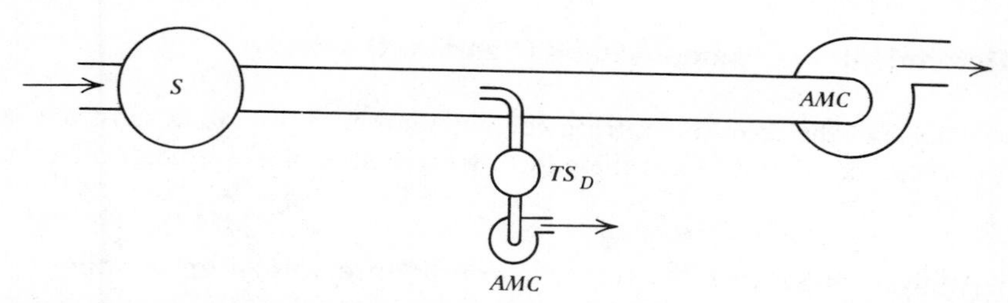

Figure 25.44 Sampler efficiency evaluation with downstream concentration sampler: analysis of collections in sample under test S and downstream sampler total collector TS_D.

not always feasible, however. When the penetration is estimated from downstream samples there may be additional errors if the samples are not representative.

4.8.3 Analysis of Sampler's Collection and Downstream Samples

It is not always possible or feasible to quantitatively collect all the test material that penetrates the sampler being evaluated. For example, a total collector might add too much flow resistance to the system, or be too bulky for efficient analysis. In this case, the degree of penetration can be estimated from an analysis of a sample of the downstream atmosphere, as illustrated in Figure 25.44. When this approach is used, it may be necessary to collect a series of samples across the flow profile rather than a single sample, to obtain a true average concentration of the penetrating atmosphere.

4.8.4 Analysis of Upstream and Downstream Samples

In some cases it is not possible to recover or otherwise measure the material trapped within elements of the sampling train such as sampling probes. The magnitude of such losses can be determined by comparing the concentrations upstream and downstream of the elements in question, as schematized in Figure 25.45.

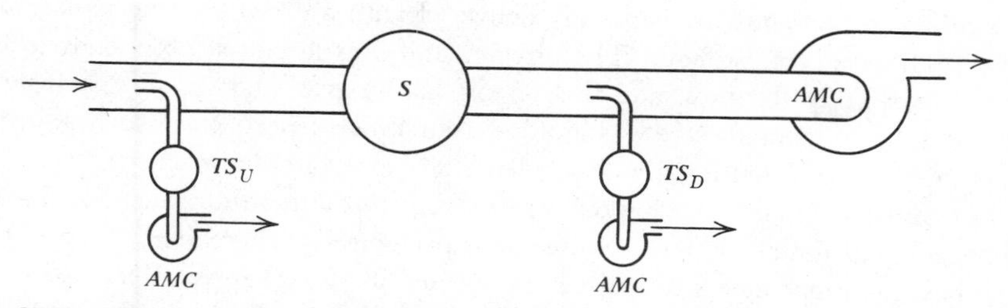

Figure 25.45 Sampler efficiency evaluation with upstream and downstream concentration samplers: analysis of collections in upstream and downstream samplers, total collection, TS_μ and TS_D.

4.9 Determination of Sample Stability and/or Recovery

The stability and the recovery of trace contaminants from sampling substrates are difficult to predict or control. Thus these factors are best explored by realistic calibration tests.

4.9.1 Analysis of Sample Aliquots at Periodic Intervals After Sample Collection

If the sample is divided into a number of aliquots that are analyzed individually at periodic intervals, it is possible to determine the long-term rate of sample degradation, or any tendency for reduced recovery efficiencies with time. These analyses would not, however, provide any information on losses that may have occurred during or immediately after collection because they had different rate constants. Such losses should be investigated using spiked samples.

Analysis of Spiked Samples. If known amounts of the contaminants of interest are intentionally added to the sample substrate, subsequent analysis of sample aliquots will permit calculation of sample recovery efficiency and rate of deterioration. These results will be valid only insofar as the added material is equivalent in all respects to the material in the ambient air. There are two basic approaches to spiked sample analyses: (1) the addition of known quantities to blank samples, and (2) the addition of radioactive isotopes to either blank or actual field collected samples.

When the material being analyzed is available in tagged form, the tag can be added to the sample in negligible or at least known low concentrations. If there are losses in sample processing or analysis, the fractional recovery of the tagged molecules will provide a basis for estimating the comparable loss that took place in the untagged molecules of the same species.

4.10 Calibration or Sensor Response

Direct reading instruments are generally delivered with a direct reading panel meter, a set of calibration curves, or both. The unwary and inexperienced user tends to believe the manufacturer's calibration, and this often leads to grief and error. Any instrument with calibration adjustment screws should of course be suspect, since such adjustments can easily be changed intentionally or accidentally—for example, in shipment.

All instruments should be checked against appropriate calibration standards and atmospheres immediately upon receipt and periodically thereafter. Procedures for establishing test atmospheres are discussed earlier in this chapter. Verification of the concentrations of such test atmospheres should be performed whenever possible using analytical techniques that are referee tested or otherwise known to be reliable.

With these techniques, calibration curves for direct reading instruments can be tested or generated. When environmental factors such as temperature, ambient pressure, and radiant energy may be expected to influence the results, these effects should be explored with appropriate tests whenever possible. Similarly, the effects of cocontaminants and water vapor on instrument response should also be explored.

5 ESTIMATION OF ERRORS

5.1 Sources of Sampling and Analytical Errors

The difference between the air concentration reported for an air contaminant on the basis of a meter reading or laboratory analysis, and true concentration at that time and place, represents the error of the measurement. The overall error is often due to a number of smaller component errors rather than to a single cause. To minimize the overall error, it is usually necessary to analyze each of its potential components, concentrating one's efforts on reducing the component error that is largest. It would not be productive to reduce the uncertainty in the analytical procedure from 10 to 1.0 percent when the error associated with the sample volume measurement is +15 percent.

Sampling problems are so varied in practice that it is possible only to generalize on the likely sources of error to be encountered in typical sampling situations. In analyzing a particular sampling problem, consideration should be given to each of the following:

1. Flowrate and sample volume.
2. Collection efficiency.
3. Sample stability under conditions anticipated for sampling, storage, and transport.
4. Efficiency of recovery from sampling substrate.
5. Analytical background and interferences introduced by sampling substrate.
6. Effect of atmospheric cocontaminants on samples during collection, storage, and analyses.

5.2 Cumulative Statistical Error

The most probable value of the cumulative error E_e can be calculated from the following equation:

$$E_e = [E_1^2 + E_2^2 + E_3^2 + \cdots + E_n^2]^{1/2}$$

For example, if accuracies of the flowrate measurement, sampling time, recovery, and analysis are ± 15, 2, 10, and 10 percent, respectively, and there are no other significant sources of error, the cumulative error would be

$$E_e = [15^2 + 2^2 + 10^2 + 10^2] = [429]^{1/2} = +20.7\%$$

It should be remembered that this provides an estimate of the deviation of the measured concentration from the true concentration at the time and place the sample was collected. As an estimate of the average concentration to which a workman was exposed in performing a given operation, it would have additional uncertainty, dependent on the variability of concentration with time and space at the work station.

6 SUMMARY AND CONCLUSIONS

Determinations of the concentrations of trace level contaminants in air and of heat stress, noise, and radiant energies are subject to numerous variables, many of them difficult to control. Thus it is prudent to perform frequent calibration checks on all industrial hygiene instruments. Such calibrations should be based on realistic simulations of the conditions encountered in the field.

The production of test atmospheres in the range of occupational threshold limits is often difficult. This chapter provides a review of available techniques for the production of test atmospheres of gases, vapors, and aerosols, with sketches of many of the more useful techniques.

Extreme care should be exercised in performing all calibration procedures. The following guidelines should be followed:

1. Use standard or reference atmospheres, instruments, and devices with care and attention to detail.
2. Check all standard materials and instruments and procedures periodically to determine their stability and/or operating condition.
3. Perform calibrations whenever a device has been changed, repaired, received from a manufacturer, subjected to use, mishandled, or damaged, and at any time when a question arises with respect to its accuracy.
4. Understand the operation of an instrument before attempting to calibrate it, and use a procedure or setup that will not change the characteristics of the instrument or standard within the operating range required.
5. When in doubt about procedures or data, make certain of their validity before proceeding to the next operation.
6. Keep all sampling and calibration train connections as short and free of constrictions and resistance as possible.
7. Exercise extreme care in reading scales, timing, adjusting, and leveling if needed, and in all other operations involved.
8. Allow sufficient time for equilibrium to be established, inertia to be overcome, and conditions to stabilize.
9. Obtain enough points or different rates of flow on a calibration curve to give confidence in the plot obtained. Each point should be made up of more than one reading whenever practical.
10. Maintain a complete permanent record of all procedures, data and results. This should include trial runs, known faulty data with appropriate comments, instrument identification, connection sizes, barometric pressure, and temperature.

11. When a calibration differs from previous records, determine the cause of change before accepting the new data or repeating the procedure.

12. Identify calibration curves and factors properly with respect to conditions of calibration, device calibrated and what it was calibrated against, units involved, range and precision of calibration, date, and name of the person who performed the actual procedure. Often it is convenient to indicate where the original data are filed and to attach a tag to the instrument indicating the items just listed.

REFERENCES

1. American Conference of Governmental Industrial Hygienists, *Air Sampling Instruments,* 5th Ed., ACGIH, Cincinnati, Ohio, 1978.
2. D. Craig, *Health Phys.,* **21,** 328–332 (1971).
3. J. H. Perry et al., Eds., *Chemical Engineering Handbook,* 4th ed., McGraw-Hill, New York, 1963.
4. American Society of Mechanical Engineers, "Flow Measurement by Means of Standardized Nozzles and Orifice Plates," ASME Power Test Code (PTC 19.5.4-1959), ASME, New York, 1959.
5. B. E. Saltzman, *Anal. Chem.,* **33,** 1100–1112 (1961).
6. G. Hama, *Air Eng.,* **9,** 18 (1967).
7. J. P. Lodge, in: *Air Pollution,* Vol. 2, 2nd ed., A. C. Stern, Ed., Academic Press, New York, 1968, Chapter 27.
8. W. D. Conner and J. S. Nader, *Am. Ind. Hyg. Assoc. J.,* **25,** 291–297 (1964).
9. A. N. Setterlind, *Am. Ind. Hyg. Assoc. Quart.,* **14,** 113–120 (1953).
10. L. Silverman, in: *Air Pollution Handbook,* P. L. Magill, F. R. Holden, and C. Ackley, Eds., McGraw-Hill, New York, 1956, pp. 12:1–12:48.
11. H. N. Cotabish, P. W. McConnaughey, and H. C. Messer, *Am. Ind. Hyg. Assoc. J.,* **22,** 392–402 (1961).
12. A. Goetz and T. Kallai, *J. Air Pollut. Control Assoc.,* **12,** 437–443 (1962).
13. G. O. Nelson and K. S. Griggs, *Rev. Sci. Instrum.,* **39,** 927–928 (1968).
14. G. O. Nelson, *Rev. Sci. Instrum.,* **41,** 776–777 (1960).
15. A. P. Altshuller and I. R. Cohen, *Anal. Chem.,* **32,** 802–810 (1960).
16. A. E. O'Keefe and G. O. Ortman, *Anal. Chem.,* **38,** 760–763 (1966).
17. T. T. Mercer, M. I. Tillery, and H. Y. Chow, *Am. Ind. Hyg. Assoc. J.,* **29,** 66–78 (1968).
18. O. G. Raabe, in: *Inhalation Carcinogenesis,* M. G. Hanna, P. Nettesheim and J. R. Gilbert, Eds., CONF-691001, Clearinghouse for Federal Scientific and Technical Information, NBS, U.S. Department of Commerce, Springfield, Va., April 1970.
19. M. Kerker, *Adv. Colloid Interface Sci.,* **5,** 105–172 (1975).
20. D. A. Fraser, R. E. Bales, M. Lippmann, and H. E. Stokinger, "Exposure Chambers for Research in Animal Inhalation," Public Health Monograph No. 57, Public Health Service Publication No. 662, Government Printing Office, Washington, D.C., 1959.
21. N. A. Fuchs and A. G. Sutugin, in: *Aerosol Science,* C. N. Davies, Ed., Academic Press, London, 1966, pp. 1–30.
22. B. Wilson and V. K. LaMer, *J. Ind. Hyg. Toxicol.,* **30,** 265–280 (1948).
23. V. K. LaMer and D. Sinclair, *An Improved Homogeneous Aerosol Generator,* OSRD Report No. 1668, Department of Commerce, Washington, D.C., 1943.

24. D. C. F. Muir, *Ann. Occup. Hyg.*, **8,** 233–238 (1965).
25. C. M. Huang, M. Kerker, E. Matijevic, and D. D. Cooke, *J. Colloid Interface Sci.*, **33,** 244 (1970).
26. E. Rapaport and S. G. Weinstock, *Experimentia,* **11**: 9, 363 (1955).
27. B. Y. H. Liu and K. W. Lee, *Am. Industr. Hyg. Assoc. J.*, **36,** 861–865 (1975).
28. K. Homma, *Ind. Health,* **4,** 129–137 (1966).
29. P. Movilliat, *Ann. Occup. Hyg.*, **4,** 275 (1962).
30. E. Matijevic, W. F. Espenscheid, and M. Kerker, *J. Colloid Interface Sci.*, **18,** 91–93 (1963).
31. S. Kitani and S. Ouchi, *J. Colloid Interface Sci.*, **23,** 200–202 (1967).
32. K. R. Spurny and J. P. Lodge, Jr., *Atmos. Environ.*, **2,** 429–440 (1968).
33. R. Ebens, *Staub,* **29,** 89–92 (1969).
34. B. M. Wright, *J. Sci. Instrum.*, **27,** 12–15 (1950).
35. S. Laskin, S. Posner, and R. Drew, paper presented at the annual meeting of the American Industrial Hygiene Association, St. Louis, May 1968.
36. R. L. Dimmock, *AMA Arch. Ind. Health,* **20,** 8–14 (July 1959).
37. J. R. Brown, J. Horwood, and E. Mastromatteo, *Ann. Occup. Hyg.*, **5,** 145–147 (1962).
38. V. Timbrell, A. W. Hyett, and J. W. Skidmore, *Ann. Occup. Hyg.*, **11,** 273–281 (1968).
39. P. F. Holt and D. K. Young, *Ann. Occup. Hyg.*, **2,** 249–256 (1960).
40. J. C. Couchman, *Metallic Microsphere Generation,* EG&G, Inc., Santa Barbara, Calif., 1966.
41. M. Polydorova, *Staub* (Engl. trans.), **29,** 38 (1969).
42. F. G. Karioris and B. R. Fish, *J. Colloid Sci.*, **17,** 155–161 (1962).
43. M. Tomaides and K. T. Whitby, *Proceedings of the Seventh International Conference on Condensation and Ice Nuclei,* Academia, Prague, 1969.
44. J. D. Holmgren, J. O. Gibson, and C. Sheer, *J. Electrochem. Soc.*, **3,** 362–369 (1964).
45. British Standards Institute, "Methylene Blue Particulate Test for Respiratory Canister," B.S. No. 2577, BSI, London, 1955.
46. K. T. Whitby, D. A. Lundgren, and C. M. Peterson, *Int. J. Air Water Pollut.*, **9,** 263–277 (1965).
47. B. M. Wright, *Lancet,* 24–25 (1958).
48. L. Dautrebande, *Microaerosols,* Academic Press, New York, 1962.
49. V. K. LaMer and R. Gruen, *Trans. Faraday Soc.*, **48,** 410–415 (1952).
50. C. Orr, F. K. Hurd, and W. J. Corbett, *J. Colloid Sci,* **13,** 472–482 (1952).
51. M. J. Pilat and R. J. Charlson, *J. Rech. Atmos.*, **2,** 165–170 (1966).
52. C. C. Snead and J. T. Zung, *J. Colloid Interface Sci.*, **27,** 25–31 (1968).
53. W. H. Walton and W. C. Prewett, *Proc. Phys. Soc.* (London), **62,** 341–350 (1949).
54. M. Lippmann and R. E. Albert, *Am. Ind. Hyg. Assoc. J.*, **28,** 501–506 (1967).
55. N. A. Dimmick, *Nature,* **166,** 686–687 (1950).
56. L. Strom, *Rev. Sci. Instrum.*, **40,** 778–782 (1969).
57. W. R. Wolf, *Rev. Sci. Instrum.*, **32,** 1124–1129 (1961).
58. R. N. Berglund and B. Y. H. Liu, *Environ. Sci. Technol.*, **7,** 147 (1973).
59. M. A. Nawab and S. G. Mason, *J. Colloid Sci.*, **12,** 179–187 (1958).
60. E. P. Yurkstas and C. J. Meisenzehl, "Solid Homogeneous Aerosol Production by Electrical Atomization," University of Rochester Atomic Energy Report No. UR-652, Rochester, N.Y., October 30, 1964.
61. V. A. Drozin, *J. Colloid Sci.*, **10,** 158 (1955).
62. B. Y. H. Liu and D. Y. H. Pui, *J. Colloid Interface Sci.*, **10,** 158 (1955).

63. P. C. Reist and W. A. Burgess, *J. Colloid Interface Sci.*, **24**, 271–273 (1967).
64. O. G. Raabe, *Am. Ind. Hyg. Assoc. J.*, **29**, 439–443 (1968).
65. S. C. Stern, J. S. Baumstark, A. I. Schekman, and R. K. Olson, *J. Appl. Phys.*, **30**, 952–953 (1959).
66. G. M. Kanapilly, O. G. Raabe, and G. J. Newton, *Am. Ind. Hyg. Assoc. J.*, **30**, 125 (1969) (abstract).
67. G. M. Kanipilly, O. G. Raabe, and G. J. Newton, *Aerosol Sci.*, **1**, 313 (1970).
68. P. M. Giever, in: *Air Pollution*, Vol. 2, 2nd ed., A. C. Stern, Ed., Academic Press, New York, 1968.
69. T. Hatch and S. P. Choate, *J. Franklin Inst.*, **207**, 369–387 (1929).
70. T. T. Mercer, R. F. Goddard, and R. L. Flores, *Ann. Allergy*, **23**, 314–326 (1965).
71. P. E. Morrow, *Health Phys.*, **12**, 173–208 (1966).
72. R. E. Albert, H. G. Petrow, A. S. Salam, and J. R. Spiegelman, *Health Phys.*, **10**, 933–940 (1964).
73. T. T. Mercer, *Health Phys.*, **10**, 873–887 (1964).
74. L. Rayleigh, *Phil. Mag.*, **14**, 184–186 (1882).
75. K. T. Whitby and B. Y. H. Liu, in: *Aerosol Science*, C. N. Davies, Ed., Academic Press, New York, 1966, pp. 59–86.
76. K. T. Whitby, *Rev. Sci. Instrum.*, **32**, 351–355 (1961).
77. S. L. Soong, M. S. thesis, University of Rochester, Rochester, N.Y., 1968.
78. A. Black and M. Walsh, *Ann. Occup. Hyg.*, **13**, 87–100 (1970).
79. D. C. Bogen, *Am. Industr. Hyg. Assoc. J.*, **31**, 349–352 (May–June 1970).
80. M. Singer, C. J. Van Oss, and W. Vanderhoff, *J. Reticuloendothel. Soc.*, **6**, 281–286 (1969).
81. H. Flachsbart and W. Stober, *J. Colloid Interface Sci.*, **30**, 568–573 (1969).
82. R. E. Albert, J. Spiegelman, M. Lippmann, and R. Bennett, *Arch. Environ. Health*, **17**, 50–58 (July 1968).
83. J. R. Spiegelman, G. D. Hanson, A. Lazarus, R. J. Bennett, M. Lippmann, and R. E. Albert, *Arch. Environ. Health*, **17**, 321–326 (1968).
84. D. V. Booker, A. C. Chamberlain, J. Rundo, D. C. F. Muir, and M. L. Thomson, *Nature*, **215**, 30–33 (1967).
85. T. T. Mercer, R. F. Goddard, and R. L. Flores, *Ann. Allergy*, **26**, 18–27 (1968).
86. K. E. Lauterbach, A. D. Hayes, and M. A. Coelho, *AMA Arch. Ind. Health*, **13**, 156–160 (1956).
87. K. R. May, *J. Aerosol Sci.*, **4**, 235–243 (1973).
88. O. G. Raabe, in: *Fine Particles*, B. Y. H. Liu, Ed., Academic Press, New York, 1976, pp. 60–110.

CHAPTER TWENTY-SIX

Industrial Sanitation

MAURICE A. SHAPIRO

In the second revised edition of this book, the chapter on industrial sanitation was introduced as follows:

> Industrial sanitation is essentially a specialized application of community environmental health services. Within the purview of industrial sanitation are the principles involved in controlling the spread of infection or other insults to the health of the employee that are not inherent in the manufacturing process per se.

This definition is still valid; and since the objective of industrial hygiene is to safeguard the health of working people and improve their work environments, industrial sanitation should be an intrinsic function of occupational safety and health. Nonoccupational exposures to pathogenic organisms and toxic substances can and do lead to illness among employees. For example, if an industrial establishment that provides its employees with a food service fails to supply its employees with sanitary food-handling facilities and practices affording maximum protection, it invites the disaster of widespread food-borne infection by harmful organisms and their toxins or by other poisonous materials. In the intervening years changes affecting the regulation of industrial sanitation have taken place. The Williams-Steiger Occupational Safety and Health Act of 1970 includes sections relating to "General Environmental Controls," "Sanitation," "Temporary Labor Camps," and "Non-Water Carriage Disposal Systems." Another significant regulatory change has been the passage of the Safe Drinking Water Act of 1974, and important regulations have been developed since its enactment.

The areas considered in this chapter are (*a*) the provision of a safe, potable, and adequate water supply; (*b*) the collection and disposal of liquid and solid wastes; (*c*) the assurance of a safe food supply; (*d*) the control and elimination of insects and rodents, especially those known to be vectors of disease; (*e*) the provision of adequate sanitary facilities and other personal services; and (*f*) the maintenance of general cleanliness of the industrial establishment.

1 PROVISION OF A SAFE, POTABLE, AND ADEQUATE WATER SUPPLY

1.1 Source and Regulatory Control: The Safe Drinking Water Act

In the case of large industries, the facilities for providing a safe and adequate water supply rival the size and complexity of many a community system. The advent of the Safe Drinking Water Act of 1974 (1) is causing fundamental changes in the regulation of drinking water quality. For example, the term "public water system" [Section 1401 (4)] as defined in the act now

> means a system for the provision to the public of piped water for human consumption, if such a system has at least fifteen service connections or regularly serves at least twenty-five individuals. Such term includes (A) any collection, treatment, storage, and distribution facilities under control of the operator of such system and used primarily in connection with such system, and (B) any collection or pretreatment storage facilities not under such control which are used primarily in connection with such system.

Privately owned as well as community systems are covered by the regulations. The U.S. Environmental Protection Agency (EPA) interprets service "to the public" to include "factories and private housing developments."

Thus the individual, private industrial or commercial supply previously designated as "private" or "semiprivate" and not specifically regulated by law now must meet the standards established for such systems under the Safe Drinking Water Act of 1974 and subject to the regulations promulgated by the EPA under the act. The regulations *do not* apply to or cover systems meeting *all* the following conditions:

1. System consists only of distribution and storage facilities.
2. System obtains all its water from, but is not owned or operated by, a public water system to which the regulations apply.
3. System does not sell water to any person.
4. System is not a carrier that conveys passengers in interstate commerce.

When a community water supply is not available to an industry at an economic cost, it must develop an adequate source of its own, and to assure potability and safety, it must treat the water in its private plant. The EPA, the designated agency administering the Safe Drinking Water Act of 1974, has promulgated "National Interim Primary Drinking Water Regulations" (2). The interim regulations, effective June 24, 1977, apply to all systems and set maximum contamination levels permitted for microorganisms, turbidity, and a number of organic and inorganic chemicals. In addition, monitoring and analytical requirements are defined for the several contaminants listed. A distinction between "community public water systems" and "noncommunity public water systems" acknowledges the fact that in most cases the consumption of water supplied by "noncommunity" systems was intermittent or transient. The original regulations would have applied all "maximum contaminant levels" (MCL) to noncommunity systems as well as to community systems. Although the proposed maximum contaminant levels for organic chemicals and most inorganic chemicals were based on the potential health effects of long-term exposure, this was not taken into account by the requirement.

Therefore the final regulations provide that maximum contaminant levels for organic chemicals, and for inorganic chemicals other than nitrates, are not applicable to "noncommunity" systems. To make clear which regulatory requirements apply to "community systems" and "noncommunity systems," the category covered is specifically indicated throughout the National Interim Primary Drinking Water Regulations.

The act provides that the states may be given primary enforcement responsibilities by the EPA administrator. If a system cannot reasonably meet the regulations, the state—or, in case a state has not assumed responsibility for regulation—the EPA, may grant variances and exemptions that do not impose unreasonable risk to the health of those served by the system. A schedule must be established for compliance, and a public hearing must be held before any public water system is granted such a schedule.

If a water supply system fails to comply with the primary regulations or fails to meet a compliance schedule established under a variance schedule granted by a state or the EPA, it must give notice to its customers or users of this failure. Furthermore, a supplier of water must report to the state or the EPA, within 48 hours, any such failure to comply with any drinking water regulations, including monitoring requirements. The regulations promulgated under the act are of two types: primary and secondary.

Primary regulations are those specifying contaminants that in the judgment of the EPA administrator "may have any adverse effect on the health of persons." The regulations apply to all "public water systems" as defined by the act, but they deal only with the basic legal requirements; a guidance manual for use by public water systems and the states is forthcoming.

Secondary regulations specify the maximum contaminant levels that in the judgment of the EPA administrator, will serve to protect the public welfare. Such contaminants are defined as those "(*a*) which may adversely affect the odor or appearance of such water and consequently may cause a substantial number of persons served by the public water system providing such water to discontinue its use, or (*b*) which may otherwise adversely affect the public welfare. Such regulations may vary according to geographic and other circumstances" (1).

A major provision of the Safe Drinking Water Act of 1974 of importance to industrial systems is "Part C, Protection of Underground Sources of Drinking Water," regulating the underground injection of fluids. The term "underground injection" refers to the subsurface emplacement of fluids by well injection. The act states that underground injection endangers drinking water sources (*a*) if it may result in the presence of contaminants in underground water supplies or (*b*) if it can reasonably be expected to do so, or (*c*) if the presence of such contaminant may result in a system failing to comply with any national primary drinking water regulation or may otherwise affect health of persons. The regulations being promulgated under the act are designed to allow states with different geological and other conditions to exercise judgment in their application to prevent underground injection practices from contaminating drinking water sources.

Section 1421(b)(2) of the Safe Drinking Water Act states that regulations may not prescribe a requirement that interferes with or impedes underground injection related to oil and natural gas production or secondary or tertiary recovery of oil or natural gas, unless such a requirement is essential to ensure that underground sources of drinking water will not be endangered by such injection.

1.2 Drinking Water Supply

The Safe Drinking Water Act of 1974 and the regulations promulgated at its initiation apply not only to community supplies but to any factory or plant possessing its own drinking water supply and employing 25 or more persons. The National Interim Drinking Water Standards (2), which became effective June 24, 1977, set maximum contaminant levels for inorganic chemicals, organic chemicals, turbidity, and microbiological organisms. A separate set of regulations, developed for radioactivity, became effective June 24, 1977. It appears that the regulations relating to radioactivity apply only to "community" supplies. However the hazard in industry should be of concern because of the danger of in-plant contamination with radioactive matter rather than from its source. Therefore a plant with its own drinking water supply should undertake an initial sampling to determine the level of radioactivity in its supply and, based on use, should order periodic reexaminations. The analytical methods utilized can be those detailed in the EPA bandbook, "Interim Radiochemical Methodology for Drinking Water" (3) or any subsequent revision.

In addition to assuring safety, industrial plants should provide water for drinking and cooking that is acceptable to the employee.

Drinking water should be supplied within a range of 40 to 80°F (the optimal range is 45 to 50°F). When cooling is needed, mechanical refrigeration or ice can be used: the ice must be produced from water meeting drinking water standards and maintained to prevent post production contamination. In non-food-purveying areas, ice that is used to cool drinking water should not be allowed to come in contact with the water. (In cafeterias and other food serving areas, the advent of ice-making machines makes it feasible to use ice to cool water and beverages.)

In addition to drinking water supply, industrial plants require water for cooling, processing, and cleaning, and different quality demands are associated with each. A pharmaceutical manufacturing plant, for example, may require deionized pyrogen-free water, whereas a steel mill uses very large amounts of cooling water with far different quality demands. Water supplied in drinking fountains and food preparation centers of the plant must be safe, clean, potable, and cool. Safety of the water may be threatened by contamination with disease-producing organisms, a wide variety of chemicals, and radioisotopes, which are in increasingly wide use.

Most industrial plants have requirements for water of varying quality. In the majority of circumstances, there is no need for these supplies to meet the high quality and bacteriological standards set for drinking and culinary water. The separate distribution and plumbing systems that convey nonpotable water are potential health hazards and must not be used as a source of drinking water. The outlets provided on the nonpotable distribution systems should be posted plainly with permanently attached, durable signs indicating that the water is unfit for drinking, culinary, or ablutionary purposes.

A sanitary drinking fountain of approved design is the most efficient method providing drinking water for employees. The American National Standard Minimum Requirements for Sanitation in Places of Employment (4) states that: "Sanitary drinking fountains shall be of a type and construction approved by the health authorities having jurisdiction" (or meet local plumbing code requirements). "New installations shall be constructed in accordance with the requirements of the health authorities having jurisdiction, or, if there are not such requirements, in accordance with American National Standards Institute Standard for Drinking Fountains, A112.11.1-1973, or the latest revision approved by the American National Standards Institute" (5). To keep refrigeration needs to a minimum, individual disposable drinking cups may be supplied. Whenever it is not feasible to have a drinking fountain connected to the supply, an approved drinking water container with an approved fountain or individual disposable cups should be furnished. In general, location of fountains may be determined by an overall standard of one drinking fountain for each 50 employees. However the distance the employee must travel to the nearest source may be a controlling factor in locating drinking water sources. (The ANSI Standard A211.11.1-1973 requires that this distance be no more than 200 feet.) Similarly, wherever the employees are subjected to above-normal heat stress, this fact should be the controlling criterion for the location of a drinking water source.

In a few industries, salt has been added to the drinking water supply to prevent heat exhaustion during the summer months. In one plant salt was added to the drinking water of one area beginning in 1941, to the remainder of the organization's drinking water in 1944. There has been a marked decrease in "heat sickness" since this program was adopted. Acceptance by employees has been good. There were numerous complaints when the program was inaugurated, and normally there are some calls in the late spring, when salinization is started. However the concentration of salt does not exceed 0.1 percent (1000 ppm). After the salination units are started up in the spring, the concentration is gradually increased, taking about 2 to 3 weeks to reach the maximum concentration. This practice, together with more general use of water coolers, has reduced complaints to a very low point. The equipment at the salinizing stations consists of a brine tank, a metering device, and a proportioner. Grade A rock salt is purchased in 100-pound paper bags. It is charged into the brine tank daily, and a saturated solution is maintained in the tank. The solution is drawn from the tank and injected into the drinking water feedline by means of the proportioner, and a fixed percentage of salt is maintained in the drinking waterline throughout its length (6).

The NIOSH Criteria Document on heat stress in the working environment (7) recommends a series of work practices to ensure that the employee body core temperature does not exceed 38.0°C (100.4°F). Among the seven work practices specified, the fifth relates to rest, water, and salt replacement:

V. Regular breaks, consisting of a minimum of one every hour, shall be prescribed for employees to get water and replacement salt. The employer shall provide a minimum of 8 quarts of cool potable 0.1% salted drinking water or a minimum of 8 quarts of cool potable water and salt tablets per shift. The water supply shall be located as near as possible to the position where the employee is regularly engaged in work, but never more than 200 feet (except where a variance had been granted) therefrom.

1.3 Water Uses in Industry

Manufacturing industries now withdraw about 50 billion gallons per day (bgd) from freshwater sources such as lakes, rivers, underground aquifers, and estuarine salt water areas. Kollar and Brewer (8), reporting on water resources planning for industry, assert that although manufacturing industries in the United States withdraw 50 bgd, the actual gross need is more than 125 bgd. This need is met through a combination of once-through use and recycling of treated industrial effluents. Ranking industrial water uses by category is as follows:

1. Cooling.
2. Steam generation.
3. Solvent.
4. Washing.
5. Conveying medium.
6. Air scrubbing.

Cooling is the largest volume use, and electrical power generation is the major user of cooling water. Since cooling water typically is separated from process water (i.e., water utilized in the manufacturing process), it usually does not come in contact with the product. Except for the addition of heat, the other quality characteristics of cooling water may not be significantly changed. Additives designed to reduce fouling due to bacterial and algal growths, when recycling is practiced, change its characteristics.

Water used for steam generation becomes contaminated with chemical additives or intermediate products which must be removed before reuse or discharge. In the petroleum refining industry, "sour water strippers" are used to remove sulfur and other polluting substances from condensed steam that has been employed to heat crude oil during the distillation process.

Water, the universal solvent, is used extensively in industry, to dissolve compounds. To effect reuse, the renovation of such water demands careful evaluation of its contaminants. Wash water picks up and entrains a wide variety of contaminants which, if the water is to be reused, must be removed or at least greatly reduced in concentration. In addition, washing and degreasing compounds and solvents may be introduced into the wash water. Water is used extensively as a conveying medium and in the process is contaminated with solids removed from the material conveyed (e.g., soil from agricultural products and any other biological, organic and inorganic matter that may be mechanically removed or dissolved).

With the advent of more stringent requirements to control the emission of air pollutants, air-scrubbing devices that use water to entrap and entrain pollutants are seeing increasing service. These devices use large quantities of water and are a good example of the "cross-media" problem engendered. The material scrubbed out of the airstream is essentially a concentrated pollutant that must be dewatered and disposed of or otherwise utilized as a solid waste.

To be able to establish the most efficient wastewater treatment and renovation processes, industrial establishments at all times must be able to determine the following:

1. Where wastewater streams originate.
2. The contaminants present, their concentrations, and the variability.
3. The diurnal, weekly, and so on, variability of the wastewater volume.

McLure (9) reports that in the case of steel mill cooling water containing a variety of suspended solids and iron, removal of the offending material is accomplished by precipitation and subsequent cyanide destruction. This followed by alkaline chlorination and phenol reduction by chlorination of the clarifier blowdown to enable utilization of a closed cycle cooling system. By so doing, only the blowdown from the clarifier or thickener has to be treated, thereby minimizing makeup water needs.

In general, the distribution by category of water use in industries that comprise manufacturing plants in the United States are as follows:

Use	Percentage
Process water	28.3
Boiler feed water; sanitary	4.8
Heat exchange	
Air conditioning	3.2
Cooling	
Steam, electric	12.1
Other, condensing	51.6

Some of this water is not returnable because of evaporation, incorporation into products, leaks, and other losses.

1.4 Conservation of Water in Industry

Another factor controlling the use of water in industry has been the growing necessity to conserve water resources, primarily but not exclusively in arid and semiarid regions. A new impetus of major importance is the advent of the Water Pollution Control Act

Amendments of 1972 (10). A most important element of the law and the program of water pollution control it has spawned, is the system of effluent limitations and required permits under the National Pollutant Discharge Elimination System (NPDES) applicable to discharges of industrial and other wastes into the navigable waters of the United States. The EPA has been empowered to issue wastewater discharge permits to individual industrial establishments, power generating plants, refineries, municipal wastewater treatment plants, and similar facilities, which are based on national effluent limitation guidelines. These guidelines designate the quantity and the chemical and physical and biological characteristics of effluent that industry may discharge into receiving bodies of water. Of crucial importance are the specific goals and objectives of Public Law 92-500 and the schedule of reaching water quality levels in the nation's waters, which provides for the protection and propagation of fish, shellfish, and wildlife, and for recreation in and on the water. Second, the Water Pollution Control Act Amendments of 1972 mandate the elimination of the discharge of pollutants into navigable waters by 1985.

A major review of the act and its goals was authorized by Congress, which established the National Commission on Water Quality and charged it to ". . . make a full and complete investigation and study of all the technological aspects of achieving, and all aspects of the total economic, social and environmental effects of achieving or not achieving, the effluent limitations and goals set forth for 1983 in section 301(b) (2) of this Act" (11).

Studies conducted by the National Commission on Water Quality have indicated that in-plant changes such as process modification and better internal control can be joined to play an important role in meeting abatement goals. Flow reduction measures by means of increased water recycling and reuse are common to all the postulated strategies, as are better housekeeping procedures. Table 26.1 illustrates the possibilities of some in-plant pollution abatement measures that the National Commission on Water Quality studies indicated may be feasible to retrofit or design into a significant number of plants by 1983.

An example of reuse technology reported by Renn (12) suggests that a relatively low technology level storage system for secondary treated wastewater is feasible, particularly when water is in short supply. In this case the recycled water was utilized for air conditioning heat exchange, cooling towers, and a variety of machine tool cutting operations. Successful operation required modification of the industrial operations to accept water containing varying concentrations of organic matter, color, suspended matter, and particulates generated during storage or in transmission.

> Storage and reuse of treated wastewaters during drought periods is a practical method of extending the available water supply for some industries and communities. It also is a device for achieving zero effluent discharge during critical, low stream flow periods. Ultimately, however, the storage system must be recharged (12).

Conservation of water in industry refers to reduction in "net-intake" water requirements. This should be differentiated from "gross water applied" or "gross water use," terms that are directly related to production. Significant reduction of "gross water use" is desirable and may be achieved by new technology such as substitution of other fluids for heat transfer or process water. Conservation entails both technical measures, such as leak detection and elimination, elimination of spills, and reduction of evaporation losses; it also demands awareness and alertness on the part of individuals and groups of employees. In general, the techniques utilized in water conservation practice may be classified as follows: (*a*) using less water by avoiding waste; (*b*) recycling (using water over again in the same process); (*c*) multiple or successive use (using water from one process for one or more additional processes in the same establishment, or using non-potable waters); and (*d*) nonevaporative cooling techniques (using special methods to reduce the amount of water required for cooling).

Use of reclaimed water from sewage and industrial wastes has been practiced for many years in many parts of the world. In industry, its use is not yet as widespread or as extensive as in agriculture. Reclaimed waters have in the past been most suited for cooling purposes, since in this instance the quantity and temperature of the water are of greater importance than its quality. Industry can employ reclaimed water for plant processing water, boiler feed waters, certain sanitary uses, fire protection, air conditioning, other miscellaneous cooling.

In addition to the general public health hazards involved in the utilization of a reclaimed effluent from a wastewater treatment plant, there are quality considerations that arise from the concentration effects of the treatment process. The concentration of total salts may be increased measurably, and the hardness of the water thus reclaimed can make its utilization impractical or costly.

The limitation that must be placed on the reuse of community wastewater and industrial wastes in plants manufacturing products for human food is severe, since the degree of treatment and the control required to keep the reclaimed water potable is high and makes such use costly.

However recent developments in wastewater treatment and so-called advanced wastewater treatment auger well for the possibility of use of renovated water in the growing of food crops, recharge of underground aquifers, and direct reuse in a wider variety of manufacturing plants such as poultry processing facilities. For example, studies performed on effluents from sewage treatment plants to determine the feasibility of recharge into the deep Cambrian-Ordovician aquifer in northeastern Illinois (13), as well as ongoing recharge in the Los Angeles Basin, are indicative of the extent of this reuse. A study (14) reports that in 1975 reuse of treated municipal wastewater in industry was continuous in 358 locations in the United States. Approximately 95 percent are located in the semiarid southwest. However only 133 billion gallons (0.503 billion m^3) is being reused in this manner. (It is difficult to tell the total use, after pumping from a recharged aquifer.)

Arnold et al. (15) in their report on the reclamation of water from sewage and industrial wastes in Los Angeles County in April 1949, drew the following conclusions:

(*a*) There are a number of important factors limiting the direct use of water reclaimed from wastewater in industry. In addition to public health hazards and total salt concentration,

Table 26.1 Summary of In-Plant Abatement Measures Possible by 1983 (existing plants)

Industry Category	Abatement Measures	Estimated Effect on Waste Load Needing Treatment	Assumed Implemented by	Expenditure for Installation as Percentage of Total Capital Expenditure for Abatement	Effect on Overall Plant Operating and Management Expenditures
Petroleum refining	Separation of waste water and storm water sewers; two-stage sour water stripping. Conversion of barometric to surface condensers. Spent caustic neutralization. Segregation of cooling and process water. Various water reuse and conservation measures.	Flow regulation: reduce ammonia and sulfides by 95%; reduce oil and grease emulsions; reduce sulfides, COD[a]; adjust pH; reduce flow being treated by 24–48%	1977	50–83%	Increase
Pulp and paper	Internal spill containment.	Eliminate shock loading	1977	17–26% for mills with pulping, 6–7% for mills without	Increase

	Clean cooling and service water segregation and reuse, other measures for water reuse. More efficient ("countercurrent") pulp washing techniques. Fiber recovery techniques.	Reduce flow being treated by 10–50%; reduce BOD[b] load by 5–30%, TSS[c] by 10–45%		12–24% (including spill containment)	Decrease (substantial savings in some cases)
Canned and preserved fruits and vegetables	Water conservation through use of "dry cleaning, washing, and peeling techniques; steam blanching; dry collection and transport of solid wastes; reuse of process waters; dry product transport. Miscellaneous water conservation and waste management measures (increased product yield results in some cases).	Reduce flow being treated by 5–7%; reduce BOD loading by 0–80%, TSS by 0–71%	1983	Not reported separately	Increase in operating and management expenses partially offset by savings in most cases; net decrease in the expenses realized in 6 of 24 subcategories

Source. National Commission on Water Quality, October 1975.
[a] Chemical oxygen demand.
[b] Biochemical oxygen demand.
[c] Total suspended solids.

considerations of the hardness of the water may be of material importance. (*b*) In general, the direct reuse in industry of acceptable water reclaimed from wastewater is tolerated and encouraged in manufacturing processes that do not involve contact between the reclaimed water and human beings or foods to be consumed by humans. (*c*) The most obvious direct reuse for reclaimed sewage water is for cooling and condensing operations.

With the advent of greatly increased quality requirements, recycling of industrial wastewater after treatment to meet future effluent guidelines should become more prevalent. Kollar and Brewer (8), reviewing the "twenty-best of file" plants with the current highest recycling rates, conclude that treated wastewater is being recycled at a high rate by many of the major water-using industries.

1.5 Cross-Connections

The great variety of water supplies and uses within a plant, as well as the growing need for reuse and recycling of industrial water, present actual and potential hazards to the cleanliness, safety, potability, and coolness of the drinking and culinary water supplies. Such danger occurs when there are any connections between a water supply of known potability (primary sourçe) and a supply of unknown or lesser quality, and when there are plumbing defects that may allow wastewater and toxic materials to enter the drinking water distribution system. There are countless possibilities for contaminating a potable supply by a nonpotable supply when unauthorized or unsafe connections are made. Pressure variations in distribution systems are not uncommon, and when reduced pressure conditions occur in a potable system connected with another of unknown potability, having an even momentary higher pressure, the flow will be from the system of unknown potability into the potable system. These and other potentially dangerous situations in water systems are termed cross-connections.

Spafford (16) reports that the 40 water-borne epidemics recorded from 1907 to 1953 in Illinois have resulted in 13,000 known cases of illness and 200 deaths. Eighteen of the outbreaks, almost half the total, were caused by some type of faulty piping or plumbing arrangement that permitted wastewater, or other contaminated water, to enter the safe water system, either at the source or in the distribution facilities. This experience is in agreement with the general observation that cross-connections between safe and raw or unsafe water systems have been the greatest single cause of water-borne epidemics.

Contamination of the potable water supply by means of cross-connections or defective plumbing fixtures may take place (*a*) by the occurrence of a vacuum in the supply lines, causing back siphonage of contaminated material; (*b*) by the development in the fixture, appliance, or piping system to which the water supply is connected of pressure that is greater than that in the supply system itself; and (*c*) through the activities or actions of vermin, birds, or small animals in parts of the supply system not under pressure, or by dust reaching water-holding devices.

Periods when reduced pressure or partial vacuum situations may occur within a potable water system result from the following circumstances or by a combination of a number of them:

1. The interruption of the supply for maintenance of the municipal or service supply main.
2. Excessive demand placed on the supply mains during fire or other emergencies.
3. Complete failure of the supply due to breaks in the mains, earthquake damage, interruption of pumping either by mechanical failure or power shutoff, or deficient water supply.
4. Freezing of mains or service lines in extremely cold weather.
5. Excessive friction losses due to inadequate size of mains or service lines, or due to reduction of the effective diameter of pipes caused by deposits and encrustations.
6. Occurrence of negative water hammer pressure waves.
7. Operation of long pump suction lines.
8. Condensation of steam within boilers, hot water systems, or units such as hospital sterilizers.

1.6 Elimination of Cross-Connections

The safest method of making two water supplies available to a building is by the use of an unobstructed vertical fall through the free atmosphere between the lowest opening from any pipe supplying water to a reservoir and the highest possible level the water may reach in the reservoir. The minimum air gap thus provided should be twice the diameter of the effective opening and in no case less than 4 inches. In addition, a float valve arrangement should be installed to cut off the supply when the water reaches the free overflow level. The overflow channel or pipe should allow free discharge to the atmosphere with no enclosed connection to the sewer (Figure 26.1). This is the recommended method of the Building Officials and Code Administrators International for safe water supply to tanks or cooling tower basins.

To prevent backflow from fixtures in which the outlet end may at times be submerged, such as a hose and spray, direct flushing valves, and other devices in which the surface of the water is exposed to atmospheric pressure, a vacuum breaker should be installed. There are two general types of vacuum breaker: moving part and nonmoving part. The type selected should meet test requirements of the American National Standard for Backflow Preventers in Plumbing Systems, ASA40.6-1943 and the American Water Works Association (AWWA) Manual M14, Recommended Practice for Backflow Prevention and Cross Connection Control (17), and the AWWA Standard for Backflow Prevention Devices (18).

All fixtures supplied by a faucet should have an air gap, which is measured vertically from the end of the faucet spout or supply pipe to the flood level rim of the fixture or vessel. The minimum air gap provided should be twice the diameter of the effective opening but no less than 1 inch. When affected by a near wall, the minimum air gap should be 1½ inches. Other conditions should meet the requirements of the American National Standard for Air Gaps in Plumbing Systems, ASA40.4-1942 (revised 1973).

Connections to condensers, cooling jackets, expansion tanks, overflow pans, and other devices that waste clear water only, should be discharged through a waste pipe connected to the drainage system with an air gap.

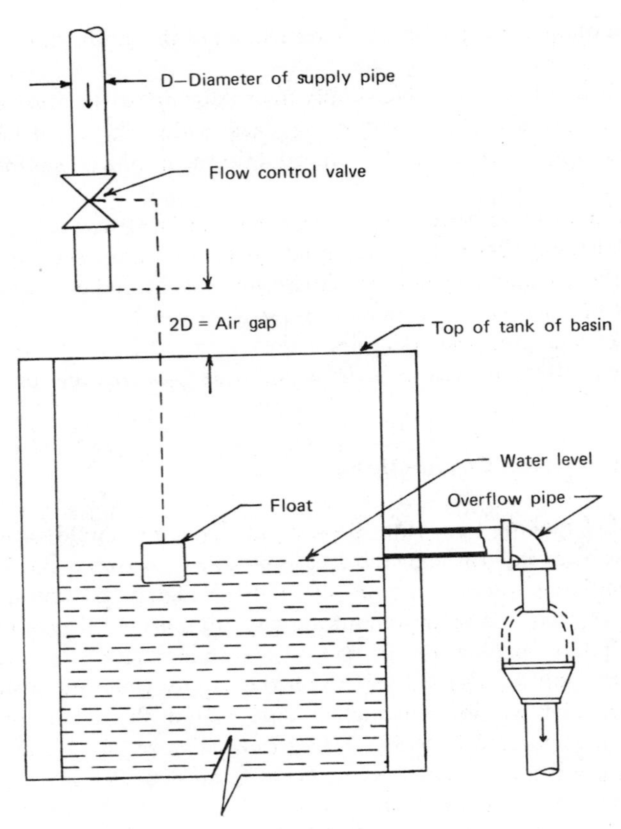

Figure 26.1

1.7 Cross-Connection Surveys

In an existing building or structure, it is advisable to conduct periodic comprehensive surveys of the entire plumbing system to determine the cross-connection hazards. The cross-connection survey should include an examination of all piping systems, plumbing fixtures, and their specifications, potable water supply connections to industrial equipment, pressure on the potable water distribution system, all backflow prevention devices, and adherence to a safety color code. The information thus obtained can then be profitably utilized whenever changes in the plumbing system have to be made.

A complete program of cross-connection control and backflow prevention includes instruction, inspection, repair, and improvement, which require a degree of expertise and training to develop inspectors. Such personnel should have qualifications equivalent to licensed plumbers with special training in cross-connection control.

The literature on cross-connection control has been evolving, and a number of manuals are available (17–21).

2 COLLECTION AND DISPOSAL OF LIQUID WASTES

There are end products to any manufacturing process, no matter how efficient or well designed its operation. These industrial wastes, as well as the wastes produced by plant personnel, must be rendered innocuous and disposed of, for they may harbor chemical toxins and disease-producing organisms that are hazardous to the health of the plant personnel, the population of the surrounding community, and the plant and animal life of the area. Accidental discharges or spills of hazardous materials are a constant source of concern. The Federal Water Pollution Control Act Amendments of 1972 deals with the discharge of hazardous materials into lakes and streams. The Marine Protection, Research and Sanctuaries Act of 1972 (Ocean Dumping) regulates ocean disposal operations. New problems of land disposal are the result of piecemeal regulation of hazardous substances, whereby a material is removed from one medium, say water or air, and diverted it to another. New toxic substances legislation now being considered by Congress is designed to permit a comprehensive national and state regulatory program. There is no current single definition of a hazardous material. Klee (22), for example, in an attempt to develop models for evaluating the "hazard" involved in a waste stream, suggests that the following attributes, among others, can be used: human, aquatic, plant and animal toxicities (mutagenicity and carcinogenicity should also be considered), flammability, persistence and mobility in the environment, and the quantity generated in a given time. To these one can add biological persistence and disease transmission potential and such specialized aspects as ground water toxicity. Outbreaks of typhoid fever and other enteric diseases traceable to contamination of a water supply with untreated wastes from human carriers are well known to environmental engineers. Of equal importance is the continual and more subtle chemical, physical, and biological degradation of the water resources of the country by indiscriminate release of untreated industrial waste.

In recent years a new field of industrial waste collection and disposal engineering has developed, with specialists devoting their full attention to the complex problem of reclamation, reduction, and proper disposal of wastes. These professionals cooperate with industrial and chemical engineers in the design of processes and equipment that reduce waste production to a minimum. This approach to the waste disposal problem in industry parallels the objectives of the industrial hygienist, who requires that hygienic factors receive the same basic consideration in the course of analysis and design as the requirements for satisfactory performance of the equipment.

In addition to the wastes of the more conventional industrial processes, the problem of proper disposal of wastes from the atomic energy industry is rapidly growing in magnitude and as a source of concern. Although the uniqueness of its hazards, operations, products, and wastes places the industry in a special category, and the attack on the problem requires specialized knowledge, it does come within the scope of the industrial waste engineer. A descriptive example of the scope and range of waste disposal problems is the steel industry. The liquid waste collection, treatment, and disposal problems in the steel industry are characterized by its use of very large quantities of water and variety of pollutants. The wastes from an integrated steel mill include suspended solids, oil and grease, ammonia, nitrate, phenols, cyanide, sulfide, total iron, ferrous iron, soluble iron, manganese, lead, zinc, tin, total chromium, hexavalent chromium, acids, alkalies, and sanitary sewage.

2.1 Industrial Wastewater

It was on October 18, 1972, that the Congress, over a presidential veto, passed the Federal Water Pollution Act Amendments of 1972 (10). This act is substantially different from previous water pollution legislation. Besides increasing the level of federal grants for the construction of publically owned wastewater treatment plants, it also supports planning for new systems and public participation in the planning process, and creates a new system of uniform effluent standards and a permit system (the National Pollutant Discharge Elimination System), for enforcement of effluent quality requirements on all point source dischargers.

The act has two primary goals:

- To reach, "wherever attainable," a water quality that "provides for the protection and propagation of fish, shellfish, and wildlife" and "for recreation in and on the water" by July 1, 1983.
- To eliminate the discharge of pollutants into navigable waters by 1985.

Eight policies were established.

1. To prohibit the discharge of toxic pollutants in toxic amounts.
2. To provide federal financial assistance for construction of publicly owned treatment works.
3. To develop and implement areawide waste treatment management planning.
4. To mount a major research and demonstration effort in wastewater treatment technology.
5. To recognize, preserve, and protect the primary responsibilities and roles of the states to prevent, reduce, and eliminate pollution.
6. To ensure, where possible, that foreign nations act to prevent, reduce, and eliminate pollution in international waters.
7. To provide for, encourage, and assist public participation in executing the act.
8. To pursue procedures that drastically diminish paperwork and interagency decision procedures and prevent needless duplication and unnecessary delays at all levels of government.

The act contains requirements and deadlines for achieving the goals and objectives in phases.

Phase I, an extension of the program embodied in many state laws and federal regulations based on prior legislation, required industry to install "best practicable control technology currently available" (BPT) by July 1, 1977" [Section 301(b) (1) (C)].

Phase II requirements are intended to be more rigorous and more innovative. Industries are to install "best available technology economically achievable (BAT) . . . which will result in reasonable further progress toward the national goal of eliminating the discharge of all pollutants; including reclaiming and recycling of water, and confined disposal of pollutants"—by July 1, 1983—as well as any water-quality-related effluent. The aim here is the achievement of the national goal of the elimination of the discharge of pollutants by 1985.

The act was intended to be more than a mandate for point source discharge control. It embodied an entirely new approach to the traditional way we have used—and abused—our water resources. Some of these mechanisms are found in Title I, the broad policy title; others can be perceived throughout the act in grants and planning, in standards and enforcement, and in permits.

The second section of the act calls for the development of comprehensive programs for preventing, reducing, and eliminating pollution, as well as research and development aimed at eliminating unnecessary water use. Section 208 directs the designation of areawide institutions (e.g., regional planning authorities) to plan, control, and maintain water quality and reduce pollution from all sources.

Construction grants for publicly owned treatment works are made available to encourage full waste treatment management, providing for the following:

1. The recycling of potential sewage pollutants through the production of agriculture, silviculture, and aquaculture products, or any combination thereof.
2. The confined and contained disposal of pollutants not recycled.
3. The reclamation of wastewater.
4. The ultimate disposal of sludge in a manner that will not result in environmental hazards.

The grantees are encouraged to combine with other facilities and utilize each other's processes and wastes. Facilities are to be designed and operated to produce revenues.

The sanitary sewage portion of the liquid wastes of an industry, if kept separate from the industrial wastes, offers no special problem to the engineer. The economics of any particular situation dictate whether the ultimate disposal facility will be the community's sewage treatment plant or a private plant. If the industry is located within reasonable distance of a municipal system, connection to it for ultimate disposal through the community's treatment plant is preferable to construction and operation of disposal facilities by industry.

The technology is generally available to control the quality of discharges to meet the 1977 and 1983 effluent limitations established by the Environmental Protection Agency. Whenever industry discharges liquid waste into a municipal system, the pretreatment standards promulgated under the act require the industry to reduce or eliminate "pollutants which are . . . not . . . susceptible to treatment by (publically owned) treatment works" or any "pollutant interferes with, passes through, or otherwise is incompatible with such works." Standards have not yet been promulgated for all industries, but the EPA's objective is to attempt to reach, as nearly as possible, the "best practicable control technology currently available" as the pretreatment standard for 1977 for major contributing industries with such "incompatible" pollutants.

The estimated and projected capital expenditures required of industry to meet BPT effluent guidelines have been variously estimated at range from 35 to 40 billion 1975 dollars.

The most prevalent pollutant characteristics of industrial wastewater are as follows:

Biochemical oxygen demand (BOD)
Chemical oxygen demand (COD)
Suspended solids (SS) and total suspended solids (TSS)
Heavy metals
Variation in pH

The EPA has established effluent guidelines and limitations for a number of industries. To illustrate, the parameters requiring control for 10 industries, which the EPA studied in depth, are given in Table 26.2.

Treatment of industrial wastewaters to remove pollutants before discharge into a receiving body of water is now commonly referred to as "end of the pipe abatement." This means that waste flows are combined at a wastewater treatment facility to be treated before discharge. Cost and treatability considerations induce reduction of pollutant production by various means, ranging from better housekeeping to reduce spills and leaks to education programs and, if necessary and possible, modification of the process itself. For example, an evaluation of the design requirements in the food processing industry as described by Eckenfelder (23) indicates that significant reductions in waste load, with corresponding cost reductions, can be achieved in the brewing industry by collecting the fermentation sediment, reducing BOD loads by modifying bottle and keg filling operations to minimize beer losses, collecting spilled grains, reusing rinse water, and establishing an educational program for the brewery workers.

The EPA has conducted plant-by-plant surveys of abatement measures. Reports of these surveys are available for the petroleum refining, nonferrous metals, iron and steel, and pulp and paper industries. Other surveys are progressing. This material serves as an excellent guide to current practice in these industrial segments.

2.2 Treatment

End of the pipe conventional treatment can be divided into two major categories: physical-chemical and biological. The two categories are employed individually or in combination. However physical processes such as straining, sedimentation, and flotation are common to all methods and categories of wastewater treatment. The particles in wastewater exhibit a large range of sizes. Therefore a system devised to remove them can consist of (1) coarse screening with vertical or inclined "bar screens" (openings 0.75 to 6 inches wide) to remove accumulated rags, sticks, and stones; (2) "grit chambers"

Table 26.2 Industrial Pollutant Parameters Controlled by Effluent Limitations Either Promulgated or Under Consideration by the EPA

Category	Parameters Controlled by 1977	Additional Parameters Controlled by 1983
Canned and preserved fruits and vegetables	BOD, TSS, Fecal col., pH, O & C	None
Inorganic chemicals	TSS, and pH for all subcategories; plus for specified products: ammonia, COD, BOD, heavy metals, fluoride, Iodate, sulfur, sulfite, sulfide, O & G, total organic carbon, cyanide	None[a]
Iron and steel	TSS, O & G, pH, cyanide, phenols, ammonia, heavy metals	Sulfides, fluoride, nitrates, manganese
Metal finishing	Heavy metals, TSS, pH, cyanide, fluoride, phosphorus	All other process water pollutants[b]
Organic chemicals	BOD, TSS, pH for all subcategories; plus for specified products: phenols, cyanide, heavy metals	COD
Petroleum refining	BOD, COD, TSS, O & G, pH, phenols, ammonia, sulfides, chromium	None
Plastics and synthetics	BOD, COD, TSS, pH for all subcategories; plus for specified products: O & G, heavy metals, phenols	None
Pulp and paper	BOD, TSS, pH	Color
Steam electric power	TSS, O & G, pH, polychlorinated biphenols, chlorine, heavy metals	Heat,[b] additional heavy metals
Textiles	BOD, COD, TSS, fecal col., pH, phenols, sulfides, chromium	Color

[a] The 1983 limitations correspond to complete elimination of discharge of the pollutants limited by 1977 for selected processes within the industry.
[b] The 1983 limitations for these parameters correspond to complete elimination of discharge of process water pollutants.
Source. National Commission on Water Quality, October 1975.

designed to remove more readily settled inorganic matter by reducing velocity of flow; (3) "flotation" in a sedimentation basin or "flotation unit" to remove organic materials that will float; (4) "sedimentation" to remove settleable material from incoming wastes or to remove materials rendered settleable by chemical or biological means ("sedimentation" or "settling" basins have two functions: (*a*) clarification–production of an effluent relatively free of settleable matter (*b*) thickening—production of an underflow that contains the solids removed in high concentration); and (5) "fine screening," which can also be utilized to remove settleable organics.

The manner in which particles settle out of a suspension allows them to be categorized. When the particles (e.g., sand, grit, coal, and fly ash) tend to remain discrete and do not coalesce when in contact, the settling or clarification is termed class I. When the particles to be removed have been flocculated or have a tendency to form flocs resulting in a change of size and settling rate, the clarification is termed class II. Design considerations are primarily (*a*) the rate at which clarification occurs and (*b*) the assurance that once a particle has settled, it will not be resuspended before being removed from the basin. The rate of sedimentation of particles in water is determined by their weight and the buoyant force and drag force opposing their subsidence.

The settling velocity of discrete particles, class I clarification, is influenced by fluid properties and the characteristics of the particle.

From a consideration of the forces acting on the particle (fluid density and viscosity and gravity), the relationship determining settling velocity of a discrete particle is

$$V_s = \sqrt{\frac{4}{3}\frac{g}{C_D}\frac{(\rho_s - \rho)}{\rho}d} \tag{1}$$

where g = gravitational constant
C_D = Newton's drag coefficient
ρ_s = density of the particles
ρ = density of the liquid
d = effective particle diameter

The coefficient of drag C_D at low (less than 1) Reynold's number, can be approximated from the equation

$$C_D = \frac{24}{N_{\mathrm{Re}}} \qquad \left(\text{Reynolds number} = N_{\mathrm{Re}} = \frac{Vd\rho}{\mu}\right) \tag{2}$$

and Equation 1 is transformed to become

$$V_s = \frac{g}{18}\frac{(\rho_s - \rho)}{\mu}d^2 \tag{3}$$

where

$$\mu = \text{dynamic viscosity}$$

Equation 3 is known as Stokes' law.

When particles are in high concentration in the liquid, streamlines about particles overlap so that their subsidence is retarded by the presence of neighboring particles. Under such conditions of hindered settling (class II settling), the subsidence rate of the suspension V_s becomes

$$V_s = V\rho\,\epsilon^{4.65} \tag{4}$$

where ϵ is the porosity (24).

When particles are agglomerated by flocculation, their settling characteristics are influenced by the nature of the floc. The larger the mass, the greater the settling rate. With high particle concentration, hindered settling is very pronounced. This causes the particles to settle in a fixed position relative to each other. The characteristic of this settling is the formation of a distinct liquid–solid interface, and it is termed zone settling.

If the concentration of a suspension is high enough that the particles are in contact, settling becomes dependent on deformation of particles or the destruction of the interparticle bonds. In this type of settling, compression takes place.

In biological treatment, the most commonly used processes employ bacteria as the primary organisms. The process may operate under aerobic conditions, (e.g., trickling filtration or activated sludge) or anaerobic conditions (e.g., anaerobic sludge digestion). The basic process is the same, consisting of microbial growth and energy utilization.

2.2.1 The Trickling Filter Process

Trickling filters consist of shallow, usually circular, tanks filled with crushed stone or, more recently, synthetic media. The liquid is applied by means of a distributor, continuously or intermittently, over the top surface. As the liquid percolates through the filter, it passes over the media and is collected at the bottom. The size of the voids permits the liquid to flow over the media and air to circulate. The designation of these units as "filters" is incorrect because the major removal is the result of an adsorption process that takes place on the surfaces of the biological growth encasing the filter media.

Since the organic substance is incorporated into the biological growth, it in turn releases to the liquid inorganic matter resulting from the oxidation of the adsorbed organics. Oxygen, supplied from the circulating air, allows aerobic oxidation in the surface layer of the biological growth.

The composition of the biological growths or film on the media is very similar to that of the flocculant material in activated sludge. Besides bacteria, it supports large numbers of higher organisms which feed on the organic matter. These are insect larvae, spiders, aquatic earthworms, and so on. In spring, when warm weather arrives, the film is sloughed off, and large numbers of organisms are dislodged. Trickling filters may be subject to infestation of a mothlike fly of the genus *Psychoda* which can become a distinct nuisance.

The biological mass on the media contains a wide variety of bacteria, the principal and in greatest number being heterotrophic bacteria. These, in turn, become the food for a number of saprobic organisms such as protozoa, rotifers, nematodes, and other invertebrates. Fungi also feed on chemical substances.

The trickling filter system includes a trickling filter and sedimentation sludge handling and treating and chlorination facilities.

2.2.2 Activated Sludge Processes

In activated sludge processes, in contrast with trickling filtration, the flocculated biological growths are in continuous circulation. The growths are always in contact with the wastewater being treated and with oxygen. Oxygen is supplied by a variety of mechanical means such as agitators and turbines and by injecting air into the sludge mixture, causing turbulent mixing with small diameter air bubbles. The aeration unit process is followed by liquid–solid separation, and a portion of the settled sludge is recirculated to the incoming wastewater. In the aeration process, the segment of the organics amenable to biological degradation is converted to inorganics, and the remainder is converted to additional activated sludge.

Activated sludge consists of microorganisms, generally similar to those found in the slime covering the media in trickling filters, non-living organic matter, and inorganic matter. The term "activated" is derived from the unique property of the sludge in which a series of steps occurs: (1) when in contact with activated sludge, suspended colloidal and soluble organic matter is rapidly adsorbed and flocculated; (2) there is a progressive oxidation and synthesis, proportional to the biologic oxidation, of organic matter that is continuously removed from the solution. When the complete storage capacity of the sludge has been taken up, the sludge floc is no longer active and can no longer adsorb organic matter. The "activity" of the sludge can be restored by aeration, during which the stored material is utilized in oxidation and synthesis. Lawrence and McCarty (25) have demonstrated that although the overall reaction in biological waste treatment differs in anaerobic processes such as sludge digestion, which contrast with aerobic processes such as activated sludge, the processes of microbial growth and energy utilization are similar. On this foundation it is possible to develop general relationships applicable in a wide variety of processes mediated by bacteria.

In anaerobic systems complex wastes are first hydrolyzed and fermented to organic acids. It has been demonstrated that the limiting step in anaerobic treatment of organic matter is the rate of fermentation of these acids to methane. McCarty (26) and O'Rourke (27) proposed that from a knowledge of the kinetics of fermentation of the key organic acids, it is possible to formulate the kinetics of the overall treatment process.

In practical terms, two clearly defined stages can be discerned as the mixture of incoming wastewater and return activated sludge flows through an aeration tank. The first is a clarification stage, in which the major portions of the colloidal and suspended organics are adsorbed to the surface of the floc. The second is a stabilization stage, which takes place in the major portion of the aeration basin or tank. It is during this period that the organics that are stored during the clarification stage are utilized in growth and oxidation.

Trickling filtration or activated sludge, with separate anaerobic digestion of sludge, are wastewater treatment processes widely utilized in industrial wastewater treatment. Alternative biological systems are extended aerobic digestion of sludge and ponds.

The pond treatment system has been selected to receive untreated municipal sewage, industrial wastes and primary treatment plant (screening, sedimentation, and flotation) effluents, secondary treatment plant (biological treatment) effluents, excess activated sludge, and other unsettled wastes containing settleable solids. In general, the treatment process in a pond system depends on the bacteria that pervade the system to degrade nonrefractive or putrescible organic matter and the algae that provide the major portion of the oxygen utilized by bacterial respiration in the aerobic portion of nonaerated ponds. There seems to be new evidence that the larger organisms inhabiting ponds also play a role in degradation and elimination of organic matter. The similarity among lakes and streams and ponds is seen from the manner in which ponds develop facultative bacterial systems of varying types, similar to those found in lakes and streams. Algae, which have the capacity for autotrophic and heterotrophic growth, usually perform as facultative chemical organotrophs, utilizing sugars and organic acids as a source of energy and reduced carbon.

During daylight, the photosynthetic algae production of oxygen and utilization of carbon dioxide may be as much as 20 times the reverse reaction, which takes place in the absence of light. The factors that are important in regulating the growth rate of algae in ponds and streams are light intensity and duration, temperature, and available nutrients. In industry, the canning, meat processing, and pulp and paper industries have made extensive use of ponds to treat their waste. The canning industry has long used ponds and pioneered in the use of sodium nitrate for odor control.

In the meat packing industry, ponds are employed as anaerobic systems to provide pretreatment, complete treatment preceded by the recovery of grease and settleable solids, and complete treatment in a series of anaerobic–aerobic ponds. They are also used to effect advanced treatment following trickling filtration or activated sludge treatment.

In the pulp and paper industry, the use of ponds to reduce the biochemical oxygen demand of the effluent is probably the form of secondary treatment most frequently employed. The pond treatment process has demonstrated a high degree of flexibility, permitting shifts and alteration in treatment parameters to achieve a high degree of pollutant removal at the most economic cost.

2.2.3 Physical-Chemical Processes

Physical-chemical processes, in addition to sedimentation, flotation, straining, filtration, coagulation, and flocculation, are increasingly employed in industrial wastewater treatment. Adsorption, the process in which a material is concentrated (i.e., becomes an adsorbate), takes place at the interface between two phases, (e.g., liquid–solid) and is employed to remove such pollutants as phenols from a liquid waste stream. In ion exchange, ions held by electrostatic forces or charged functional groups on the surface of a solid (a synthetic resin or natural solid) are exchanged for ions of similar charge in a solution in which the solid is immersed. Ion exchange is used extensively in water and wastewater treatment. In wastewater treatment it permits, apart from pollutant removal, recovery of reusable or otherwise valuable material. The problems to be overcome in the ion exchange treatment of industrial wastewater are clogging of the resins, destruction, and fouling. The metal plating industry employs ion exchange to recover such metals as hexavalent chromium (Cr^{6+}) from waste streams, and treated water is reused (27). The system makes use of a cation exchanger to remove other metals such as copper, zinc, iron, nickel, and trivalent chromium. The hexavalent chromium is passed through the effluent as CrO_4^{2-}, subsequently to be removed in an anion exchanger. This results in a demineralized effluent capable of reuse. For recovery of the Cr^{6+} the anion exchanger is regenerated with NaOH, and Na_2CrO_4 is released. The Na_2CrO_4 solution is moved through a cation exchanger in which sodium is exchanged for hydrogen releasing chromic acid (H_2CrO_4), which can then be recovered.

Separation processes such as reverse osmosis, electrodialysis, and ultrafiltration are three major membrane processes employed in wastewater treatment. As stated by Weber (27), reverse osmosis may provide a good technique for concentration, to reduce volume, to recover usable material, to be used in combination with other treatment processes, and to improve overall efficiency.

Chemical oxidation is employed in water and wastewater treatment to convert pollutant chemical species to species that are not harmful or otherwise undesirable. This implies that the process is not carried to completion. Therefore in water and wastewater treatment the process may be described as "a selective modification of objectionable and/or toxic substances" (27). In practice, only a few oxidizing agents meet treatability needs: economy, ease and safety of handling, and compatibility with the treatment train. The most widely used agents are air or oxygen, ozone, and chlorine. The finding that small amounts of haloforms are produced as a side reaction during chlorination of natural colored waters has led to investigations of other organic-containing waters. These studies have demonstrated the presence of an array of stable chlorinated organic compounds. The formation of halogenated organics in water supplies by chlorination was reported on by Morris (28) and Rook (29, 30), whose original finding was that upon chlorination of natural colored water, as a side reaction, a small amount of haloform (e.g., $CHCl_3$) was formed. Haloform formation upon chlorination is complicated because inorganic bromide ions are oxidized by chlorine to a valence that renders bromine suitable for bromination of organic matter. Rook (29) has found mixed bromochlorohaloforms, which he believes indicate that one type of reaction is involved—that is, the haloform reaction, which occurs when compounds containing the acetyl group bond to H or $R{-}CH_2$ or compounds such as ethanol that are oxidized to such substances when fluvic or yellow acids are combined with chlorine.

Basic and detailed information regarding the several processes employed and under investigation is provided in the general references listed at the end of the chapter. The annual Review of Literature published in the June issue of the *Journal of Water Pollution Control* covers a wide spectrum of the world's literature on matters pertaining to wastewater, water pollution, and treatment. Reports on studies and research conducted or contracted by government agencies such as the EPA are available through the National Technical Information Service.

Wastewater characteristics have a universal tendency to vary, often over wide ranges; it is therefore essential that studies be undertaken with representative samples to ascertain the costs associated with the available treatment possibilities. An example of such an undertaking was given by Zabban (31), reporting on defluoridization of wastewater. Significant quantities of fluorides are discharged in effluents from glass manufacture, electroplating operations, aluminum and steel production and processing, and fertilizer manufacture. Fluoride is also contained in the discharge from certain organic chemical manufacturing processes. For example, boron trifluoride (BF_3) is a catalyst of ore containing feldspar; it is present in the form of a complex ion in the effluents from coke-oven plants and in the effluents from the manufacture of semiconductor and electronic components.

The beneficial effects of optimal fluoride intake in prevention of dental caries are well known. On the other hand, detrimental dental fluorosis can result from excessive fluoride intake, usually in drinking water.

The "best practicable technology presently available" treatment process for defluoridization is the addition of excess of calcium salt, usually calcium hydroxide, with the attainment of a pH greater than 8 to precipitate the fluoride ion as insoluble calcium fluoride. Other processes utilize a mixture of calcium chloride and hydrated lime, or quicklime (CaO) and hydrated lime. These and other similar processes are capable of reducing the fluoride concentration to approximately 10 mg/liter. To reduce the concentration to approximately 1 mg/liter, other means have been employed, such as the use of apatite $n\mathrm{Ca_9(PO_4)_6 \cdot CaCO_3}$, tricalcium phosphate hydroxyapatite $\mathrm{Ca_5(PO_4)_3 \cdot OH}$, ion exchange resins, magnesia, aluminum sulfate, calcined (activated) alumina, and bone char (tricalcium phosphate and carbon). At least three of these systems, aluminum sulfate (alum), activated alumina, and the bone char columns process, have been proved to be practicable. From this wide array of processes, the selection of the most effective and efficient method cannot be accomplished theoretically; process trials with representative samples must be made before a process is recommended.

3 SOLID WASTES: COLLECTION AND DISPOSAL

Solid wastes are the discards of individuals, commerce, industry, and agriculture. Definitions are still evolving. However, a few have come into common use.

REFUSE: All putrescible and nonputrescible solid wastes except bodily wastes.

GARBAGE: The wasted or rejected food constituents.

RUBBISH: Nonputrescible solid waste constituents, including paper, metal, glass, and wool.

Thus, such discarded matter not capable of being transported in an air or liquid stream in spite of a high moisture content (e.g., as found in industrial sludges) is designated as a solid waste. The definitions are tenuous, especially when dealing with organic matter such as rejected food, or the peelings from a canning process. It is feasible to grind the organic matter into particles no larger than those in community wastewater, transport this material, and treat it in a wastewater treatment system.

Industrial solid wastes, apart from those resulting from usual human activity, are related to specific processes. It is estimated that more than 100 million tons per year of solid wastes is generated by industry. Mineral extraction and processing wastes have been estimated to amount to more than 10 times the weight of other industrial solid wastes. These wastes present special kinds of ecological, aesthetic, and hazard problems. The means of dealing with these wastes are evolving. State laws have been enacted, and Congress enacted the Resources Conservation and Recovery Act of 1976 (PL 94-580).

A special problem arises when an industry produces hazardous solid wastes (i.e., those that either individually or in combination with other wastes pose health, environmental, or ecological hazards). Such wastes may demand special handling because of their toxic, flammable, explosive, microbial, or radioactive properties. Pesticide residues, waste drugs, spent caustics and acids, and obsolete explosives are examples of such wastes. The most prevalent method of disposal is in landfills, but this creates possible problems of surface and groundwater pollution by the leachate. Water can infiltrate through the soil cover, groundwater may flow through laterally, or water may rise into the fill when the groundwater table rises seasonally.

The rating or ranking of hazardous wastes is dependent on the final disposition of the material. Thus the HCl in otherwise innocuous polyvinyl chloride becomes a problem when the plastic is incinerated. The Illinois Environmental Protection Agency has developed a categorization of landfill sites that relate to the management problem. Piskin (32) reports that Illinois is considering a new fivefold classification system based on site hydrology as well as type of waste. Only one of the site classifications will be allowed to receive hazardous wastes.

A number of toxicity ratings have been developed. Sax (33) proposed a quantitative rating system from 0 to 3, with 0 indicating no toxicity; 1, slight toxicity, 2, moderate toxicity; and 3, severe toxicity. To obtain a human toxicity rating, the rating compiled by Sax (33) was used as a guide, and each rating was multiplied by 13, resulting in a human toxicity range of 0 to 39. Synak et al. (34) described a statewide hazardous waste management system developed for Oregon.

Industry requires a system of hazardous waste identification, a continuing evaluation of the reuse and recycling possibilities, and treatment to reduce the relative hazard of the waste. The establishment of a waste "clearing-house" permits a continuing system of control and maximizes the possibility for reuse and recycling. The St. Louis Industrial Waste Exchange, recently established and now operated by the St. Louis Regional Commerce and Growth Association, assigns a code number to each listing received, and inquiries are referred to the appropriate number. The exchange acts strictly as a clearinghouse and refrains from involvement in subsequent transactions.

4 ASSURANCE OF A SAFE FOOD SUPPLY

The objective of food sanitation is the prevention of food-borne disease. Technological and economic considerations of purveying food play an important role in the application of effective control measures. However the individual food worker has a very significant and even crucial role in the safety of food that begins with its production, continues during processing and preparation, and ends when the food is served or otherwise made available to the employees. In addition, the aim of food sanitation programs is to prevent disease without impairing the nutritional value of food.

4.1 Food-Borne Disease

Disease caused by ingestion of contaminated food may be divided into four major classes: (*a*) infections caused by the ingestion of living pathogenic organisms (bacterial and viral) contained in the food, (*b*) intoxication resulting from ingestion of food in which preformed microbial toxins have developed, (*c*) animal parasitism, and (*d*) poisoning by chemically contaminated food. In addition, poisonous plants and animals are implicated.

Food-borne disease reporting in the United States began some 50 years ago, and in 1966 a system of food-borne disease surveillance was established by the Center for Disease Control (CDC). Reports of enteric disease outbreaks attributed to microbial or chemical contamination of food (and liquid vehicles) received by the CDC are incorporated into annual summaries. The Annual Summary of Food-Borne and Waterborne Disease Outbreaks (28) for 1974 lists 456 reported food-borne disease outbreaks involving 15,489 cases. Significantly, this is the largest number reported to the CDC Food-Borne Disease Surveillance Activity in a single year since publication of such outbreaks was initiated. An outbreak is defined as an incident in which (1) two or more persons experience a similar illness, usually gastrointestinal, after ingestion of a common food; and (2) epidemiologic analysis implicates the food as a source of the illness.

Estimates of the total annual number of food-borne disease cases in the United States are quite varied. However a reasonable one (29) cites the figure of 20 million per year.

In the 5-year period 1969–1974, the responsible pathogen was identified in 30 to 60 percent of food-borne disease outbreaks reported to CDC. As reported in the Annual Summary for 1974 (35), many pathogens are not identified because of late or incomplete laboratory investigation. In other cases the responsible pathogen may have escaped detection despite a thorough laboratory investigation because the pathogen is not yet appreciated as a cause of food-borne disease or because the pathogen cannot yet be identified by available laboratory techniques. There is a need for further clinical, epidemiological, and laboratory investigations to permit the identification of these pathogens or toxic agents and the institution of suitable measures to control diseases caused by them. The outbreak that resulted in 259 cases and 29 deaths among those attending an American Legion convention in Philadelphia during the summer of 1976 is a forceful case in point. Pathogens suspected of being, but not yet determined to be, etiologic agents in food-borne disease include group D *Streptococcus, Yersinia enterocoliticus Citrobacter, Enterobacter, Klebisiella, Pseudomonas,* and the presumably viral agents of acute infectious nonbacterial gastroenteritis.

4.1.1 Microbial Infections

The majority of food-borne disease outbreaks in the United States are caused by bacterial contamination. In 1974, for example, 17.4 percent of reported confirmed food-borne disease outbreaks were caused by *Salmonella* and 41.4 percent by *Staphylococcus*; and in the same reported outbreaks, 60.2 percent of the cases were caused by *Salmonella* and 17.1 percent by *Staphylococcus*. Other confirmed bacterial and viral causative agents were *Bacillus cereus, Clostridium botulinum, C. perfringens, Shigella,* group A *Streptococcus, Vibria cholerae,* and [suspect] group D *Streptococcus*. Viral disease has also been confirmed as being food borne, and in 1974 the CDC reported six such outbreaks with 282 cases. The relative incidence of food-borne disease of various etiologies is still unknown. Disorders characterized by short incubation periods, such as *Staphylococcus* infection, are more likely to be recognized as due to common source food-borne disease outbreaks than those with longer incubation periods. Viral disease, such as hepatitis A, with its typical incubation period of several weeks, is most likely to escape detection. Other detection problems are due to difficulties in the transportation and culturing of anaerobic specimens. Other problems relate to the fact that outbreaks caused by *Bacillus cereus* and *Escherichia coli* are probably less likely to be confirmed because these organisms are less often considered clinically, epidemiologically, and in the laboratory.

When contamination is heavy or when the particular bacteria are highly virulent, infection may follow ingestion of the food even though it has been stored under optimal conditions. More frequently, however, the bacteria originally contaminating the food multiply to dangerous numbers during prolonged storage at a temperature between 45 and 140°F (7 to 60°C), the so-called incubation or danger zone. Foods most commonly found to be vehicles of the bacterial infectious agents are milk and cream, icecream, seafoods, meats, poultry, eggs, salads, mayonnaise and salad dressings, custards, cream-filled pies, and eclairs and other filled pastries. Unfortunately, food contaminated by these disease agents is not necessarily decomposed or altered in taste, odor, or appearance.

Bacterial contamination of food may occur in different ways. Infected persons may transmit bacteria by droplets (sneeze or cough) from the respiratory tract, by discharges from skin infections of the hand, and by contamination of hands with feces, nasopharyngeal secretions, and discharges from open or draining sores elsewhere on the body (i.e., by failure to keep these adequately covered). In some instances, the infected person preparing food is without symptoms or signs of illness, therefore difficult to detect and control by, for example, assignment to other work. Another source of bacterial contamination of food is the multiplication of bacteria in improperly cleaned utensils, with transfer to foods subsequently prepared, stored, or served in them.

Mice, rats, and roaches may contaminate foods and utensils by bacteria carried mechanically on their feet and bodies from an infected to a noninfected area, or they may contaminate food with their urine and feces, which contain pathogenic organisms such as, in the case of rats, *Salmonella* and *Leptospira icterohemorrhagica* (cause of Weil's disease). The source of infection in meat and poultry may be infection before killing, or contamination may occur subsequent to killing, as with other foods.

The clinical symptoms of acute gastric enteritis due to bacterial infection usually have their onset 6 to 24 hours after ingestion of the infected food. Cramping, abdominal pain, diarrhea, nausea, and vomiting are the chief manifestations; often these symptoms are accompanied by headache, pyrexia, and general malaise (30).

4.1.2 Bacterial Food Intoxication

Bacterial food intoxication is caused by bacteria that although usually harmless to man, produce toxic substances when they grow in food. When ingested in sufficient amount, these toxins give rise to serious or fatal disease, even though after ingestion there is no further multiplication of the organisms in the body. Bacteria of the *Staphylococcus* group, often found in skin infections (pimples, boils, carbuncles) and respiratory tract infections, may form a virulent enterotoxin when allowed to grow in food. The symptoms of staphylococcal intoxication are nausea, vomiting, cramping, abdominal pain, and diarrhea. In severe cases, blood and mucus may be found in the stool and vomitus. The chief characteristic distinguishing food poisoning from toxins and enteric disease from that due to pathogenic bacteria carried in food is the time of onset of symptoms. In staphylococcal intoxication, this is often less than 3 hours after the ingestion of the contaminated food and rarely longer than 6 hours. On the other hand, in bacterial infections a delay of 6 to 24 hours between ingestion of food and onset of symptoms is more likely. The rapidity of onset and the uniform distribution of food poisoning symptoms in a group of people make it possible to determine the meal and particular foods in the meal responsible for poisoning.

4.1.3 Animal Parasitism

Some of the diseases caused by ingestion of food contaminated with animal parasites are amoebiasis, trichinosis, and tape worm infestation. In the instance of infection with *Entamoeba histolytica* (causative organism of amoebic dysentery), the source and transmission of infection is similar to that of other enteric pathogens such as *Shigella* and *Salmonella*. The problem is somewhat different for *Trichinella spiralis* and the tapeworms *Taenia saginata* (beef tapeworm) and *Taenia solium* (pork tapeworm), for in these cases the infectious organisms are present in cattle or swine before the time of killing. Prevention of disease caused by these parasites, and control of them, depends on inspection and selection of food at the time of purchase as well as proper preparation, since the heat of prolonged cooking will destroy the organisms.

4.1.4 Toxic Chemicals

Accidental contamination of foods by toxic chemicals is an additional hazard. Acute poisonings are characterized by sudden onset, from a few minutes to 2 hours after ingestion of the chemical. The more common poisons reported as borne by foods are cadmium (from metal-plated utensils), sodium fluoride (from roach powder), and arsenic (from insecticides). Other substances frequently mentioned are antimony, zinc, lead, and copper.

Compounds such as aliphatic and aromatic amines used as boiler feedwater conditioners may present an additional hazard in the preparation of foods. Wherever steam may come in direct contact with food, the toxicity, type, and quantity of boiler feedwater conditioners should be ascertained. Complete physical separation of the food and steam from the boiler is recommended. The OSHA standard in Section 1910.141 (Sanitation) states

(*g*) consumption of food and beverage on premises:
1. *Application.* This paragraph shall apply only where employees are permitted to consume food or beverages or both, on the premises.
2. *Eating and drinking areas.* No employee shall be allowed to consume food or beverages in a toilet room nor in any area exposed to a toxic material.
3. *Waste disposal containers.* Receptacles constructed of smooth, corrosion-resistant, easily cleanable or disposable materials, shall be provided and used for the disposal of waste food. The number, size, and location of such receptacles shall encourage their use and not result in overfilling. They shall be emptied not less frequently than once each working day, unless unused, and shall be maintained in a clean and sanitary condition. Receptacles shall be provided with a solid tight-fitting cover unless sanitary conditions can be maintained without use of a cover.
4. *Sanitary storage.* No food or beverages shall be stored in toilet rooms or in an area exposed to a toxic material.
h. *Food handling.* All employee food service facilities and operations shall be carried out in accordance with sound hygienic principles. In all places of employment where all or part of the food service is provided, the food dispensed shall be wholesome, free from spoilage, and shall be processed, prepared, handled, and stored in such a manner as to be protected against contamination.

The revised American National Standards Institute Minimum Requirements for Sanitation in Places of Employment (4) states in item 10.1.1 that:

In all places of employment where employees are permitted to lunch on the premises, an adequate space suitable for the purpose shall be provided for the maximum number of employees who may use such space at one time. Such space shall be separate from any location where there is exposure to toxic materials.

Furthermore, item 10.1.3 states that "No employee shall be permitted to store or eat any part of his lunch or eat other food at any time where there is present any toxic material or other substance that may be injurious to health." Item 10.1.4 is as follows:

In every establishment where there is exposure to injurious dusts or other toxic materials, a separate lunch room shall be maintained unless it is convenient for the employees to lunch away from the premises. The following number of square feet per person, based on the maximum number of persons using the room at one time, shall be required:

Persons	Area (ft^2)
25 and less	8
26–74	7
75–149	6
150–499	5
500 and more	4

4.2 Food Handling

Food handlers educated in the proper methods of handling foods, glassware, and utensils, and in good personal hygiene are the cornerstone of a safe food supply. In food sanitation, the health of the food handler is of primary importance, and every effort should be made to exclude from food handling work any sick or injured employee who has a discharging wound or lesion. This statement does not constitute recommendation of a system of frequent medical and laboratory examinations of food handlers. On the contrary, these procedures have not been found to be cost effective. Education of the food handler to promptly report any illness is a more effective means of reducing food-borne disease. In many localities health departments are prepared to provide preemployment or in-service food handler training programs, and a variety of training manuals and visual aid materials are available.

In the handling of foods, particularly those previously mentioned as being sources of infection or intoxication, care must be taken to keep the materials at a temperature out of the incubation zone; that is, they must be refrigerated at a temperature below 45°F or kept hot at 145°F or above. The exception to this rule is that during preparation, serving, or transportation, food may be kept at intermediate temperature for a period not exceeding 2 hours. Indicating thermometers of proved accuracy should be installed in all refrigerators. Common deficiencies found in refrigerated food storage are the overloading of the refrigerator, which prevents the free circulation of cold air, and the storage of foods in containers so deep that the total mass of the food cannot be cooled to 45°F or less within the prescribed 2 hours. Further provisions for continual vigilance against possible chemical contamination must be exercised by prohibiting the use of cadmium-plated utensils, prohibiting the use of galvanized utensils for cooking, and banning the use in food preparation and serving areas of roach powders containing sodium fluoride, as well as cyanide metal polish, and other hazardous substances (39).

In view of the widespread custom, especially in the United States, of providing cubed and crushed ice in drinking water and beverages, it is of interest to note two reports on the sanitary quality of such ice (40). Both studies report that such ice often fails to meet sanitary standards and is contaminated with *Escherichia coli,* clostridia, micrococci, and streptococci. A suggested control method is to provide a chlorine disinfecting solution (approximately 2 ppm) in which the ice is placed.

4.3 Kitchen and Kitchen Equipment Design

The ordinance and code prepared by the Public Health Service (41) and adopted by the American National Standards Institute (4) as a minimum standard whenever local regulations do not apply to the industrial food handling establishment, provides for certain materials of construction and gives minimum lighting and ventilation requirements. The design of kitchens represents a problem in materials handling, batch preparation, and small quantity distribution. The designer must provide for the greatest ease in performing a large variety of tasks, usually in quarters that are too crowded.

It behooves the kitchen designer to allow sufficient space in this important working area to permit the sanitary performance of the tasks of preparing food and washing of dishes and utensils. Separation of activities—food preparation, baking, and dish washing—is preferable to combining them in one room. The location of kitchen equipment in relation to ease of cleaning is of great importance. Too often stoves, steam kettles, peelers, and mixers are placed close to walls, allowing dirt to accumulate and providing excellent breeding and hiding places for roaches.

Tables and stands should have shelves no less than 6 inches off the floor. Fixed kitchen equipment should be sealed with impervious material when attached to the wall or, alternatively, kept far enough away from the wall to allow for easy cleaning. All work surfaces should be made of impervious, noncorrosive, and easily cleaned material. When wood is used, it should remain unpainted and should be scrubbed daily. Cutting, peeling, and other mechanical equipment should be designed to eliminate dirt-accumulating crevices and should be readily disassembled for cleaning. A voluntary organization, the National Sanitation Foundation, representing the public health profession, business, and industry, publishes a series of standards for soda fountains and luncheonette equipment, food service equipment, and spray-type dishwashing machines; others are in preparation (42). These standards are of value in identifying equipment that has been tested to meet exacting food sanitation requirements.

4.4 Food Vending Machines

Vending machines are available that serve hot beverages, cold carbonated and noncarbonated beverages, sandwiches, and pastries. A recommended ordinance and code has been published by the Department of Health, Education and Welfare (43). The following considerations may serve as a guide for the selection of automatic food vending machines.

All surfaces that come in contact with the food should be constructed of smooth, noncorrosive, nontoxic materials. All flexible piping should be nonabsorbant, nontoxic, and easily cleanable; treatment with the more common disinfecting agents should not harm it.

Vending machines purveying perishable food should be refrigerated. Their design should include an automatic temperature-controlled shutoff that will prevent the machine from operating whenever the temperature in the storage compartment rises above 45°F.

In general, the design of the machine should allow easy accessibility for cleaning and disinfecting all surfaces that come in direct contact with the beverage or food.

Whenever disposable cups are furnished they should be stored in protective devices, and it should be possible to reload them without touching the lips or interior surface of the cups.

The drip receptacles in beverage vending machines should be provided with a float switch that would prevent the machine from operating when the liquid in the receptacle reaches a certain level. The vending area of the machine should be protected from dust, dirt, vermin, and possible mishandling by patrons by means of self-closing gates or pans.

The cleaning of tanks, containers, and other demountable equipment should be done at a central location. Whenever cleaning is accomplished at the vending machine, a portable three-compartment sink should be provided, to permit proper washing, rinsing, and disinfecting of all demountable equipment. At such points water and sewer facilities should be provided.

The food and drink for the vending machines should be stored and handled as they would have to be to meet the requirements of food handling establishments. A good rule is to offer for sale only packaged foods that were prepared at a central commissary. All perishable food should be coded to show place of origin and date of preparation. Perishable food that remains unsold, even though it has been properly refrigerated, should be disposed of at frequent intervals.

No vending machines should be located in areas where there is exposure to toxic materials.

5 CONTROL AND ELIMINATION OF INSECTS AND RODENTS

5.1 Insect Control

Except for the food and other industries dealing with products that may serve as nourishment for insects, the problem of controlling these pests is dependent largely on good housekeeping. The more general problem of control of arthropods and arthropod-borne disease cannot be accomplished on a restricted scale as in a single industrial establishment. To be effective, it must be a community-wide or regional undertaking of considerable magnitude. It is axiomatic that an industry located in an infested area should cooperate with the agency responsible for effectuating control.

The number and prevalence of flies and roaches are good general indices of the "sanitation" practiced in the establishment. Of direct interest is the proof obtained of the relation between fly prevalence and the infections, diseases, and deaths caused by the bacillary dysentery organism (44). Field studies showed that bacillary dysentery was materially reduced in a community when effective fly control measures were undertaken.

The most direct method of dealing with an insect infestation problem is to eliminate breeding places. Wherever garbage, rubbish, or other organic matter is allowed to accumulate and become exposed to insects, it produces ideal conditions for their development. In food handling areas, nourishment is available, and cracks, crevices, voids, and other unused spaces harbor the pests.

The more common insect invaders are the roaches: German cockroach, *Blatella germanica*; American cockroach, *Periplaneta americana*; Australian cockroach, *Periplaneta australasiae*; Oriental cockroach, *Blatta orientalis*; brown-banded cockroach, *Supella supellectilium*; and flies, of which 90 to 95 percent are houseflies, *Musca domestica,* and blowflies, *Phaenicia* (44).

The use of insecticides, which should only be undertaken after the environmental conditions favoring breeding have been eliminated, poses health problems to the applicator and others. The problems and their prevention are extensively discussed in the literature. New insect control technology is being rapidly developed; for example, sex attractants and juvenile hormones are proposed as insect control materials. However they too may have adverse health effects on mammalian species, including man. On the other hand, biologic insect control may have beneficial public health and ecologic consequences because it could prevent the positive feedback aspects inherent in the use of chemical insecticides.

5.2 Rodent Control

Apart from the health hazards rodents produce as vectors of disease (bubonic plague, endemic or murine typhus fever, salmonellosis and Weil's disease), their infestation of an industrial plant can be a tremendous economic drain. As a general principle, in the community, the destruction of food, crops, merchandise, and property by rats is serious enough to justify suppressive measures, even if rats are not responsible for human disease. Because of their destructiveness and their pollution of food, rats have gained the unenvied fame of being the worst mammalian pest in civilization. Davis (45) estimated about one rat per 35 people in New York City and about one rat per 20 people in Baltimore. As pointed out by Davis, the regulatory factors in any problem of species management can be classed in three main groups: (*a*) environment, which includes availability of food and protective shelter; (*b*) predation, which includes animals that feed on the species, trapping, poisoning, and disease; and (*c*) competition, the fight for a limited supply of environmental necessities.

Control of the rodent population should be based primarily on a change of the environmental conditions so that the rodents are deprived of food and shelter. When this is accomplished, the rodent population will be reduced by intraspecies competition. Since the average reproductive rate of Norway rats (they reproduce during all seasons of the year, with a maximum in spring and fall occurring in many areas) is 20 rats per year, it is evident that killing procedures will be of little effect on reducing the population and maintaining it at low level if environmental conditions are favorable to rats (46).

5.3 Ratproofing (Building Out the Rat)

"Ratproofing" is a term applied to procedures for controlling the portals of entry of rats into a structure. Being extremely cautious, nocturnal mammals, rats prefer to enter a building through small openings. Only when they are subjected to peril will they venture away from normal pathways. The points of ingress and egress of rats that must be guarded against in particular are any openings around pipes, unused stacks and flues, ventilators, and hatches.

All new buildings should be designed and constructed to be ratproof. Foundations must be continuous, extending not less than 18 inches below the ground level, and they should always be flush with the under surface of the floor above. Floor joists should be embedded in the wall and the space between the joists filled in and completely closed to the floor level. All construction materials should be as ratproof as possible (46). In industry, the most frequent shelters for rats are areas surrounding the cafeteria, the lunch rooms, and other eating areas, although no place that affords attractive food and shelter is immune to rat infestation. The office in which employees are permitted to eat their lunch is often rat infested, the scraps left in the waste baskets providing the rats with a good food source. All other organic wastes, such as those found in sewers where they harbor, are a good additional source of food for rats.

5.3.1 Killing Rats

Trapping by means of snap and cage traps is a method of killing rats that requires great skill and ingenuity, since rats are rather wary mammals. Killing by use of rodenticides has advanced greatly in the past two decades. The most widely used and oldest poison is Red Squill, which has the advantage of having an emetic action that protects humans and all animals with the ability to vomit. Barium carbonate and phosphorus, utilized in the past, are not widely used now because of their unpredictable results. The newer rodenticides are Antu, a specific for Norway rats; sodium fluoroacetate (1080), an extremely powerful poison with all animals, which must be used with extreme caution and has been prohibited in many localities because of human deaths; and Warfarin (a derivative of dicumarol), an anticoagulant that causes internal hemorrhage in the rat. Although Warfarin affects other animals, it has a large factor of safety because of the low dosage (0.025 percent) used in the baits. Table 26.3 lists some characteristics of common rodenticides.

In certain enclosed structures, primarily aboard ships, fumigants such as hydrocyanic acid gas or carbon monoxide are used for quick and nearly complete elimination of rats.

Control of rats and other rodents requires (*a*) environmental control to eliminate sources of food and harborage, (*b*)effective ratproofing, and (*c*) efficient rat killing programs. Controlling rat populations, not individual rats, is the key to a successful rodent control program.

Table 26.3 Some Characteristics of Common Rodenticides

Poisons[a]	Lethal Dose (mg/kg)	Percent Used in Bait	Degree of Effectiveness	Acceptance	Reacceptance	Cumulative	Tolerance Developed	Odor	Taste	Chemical Deterioration in Baits
Anticoagulants Warfarin Fumarin Pival •••	1[c]	0.025	Good	Good	Good	Yes	No	None	Slight	None
Anticoagulant Diphacinone •••	0.5[c]	0.005	Good	Good	Good	Yes	No	None	Slight	None
Antu •	8[d]	1.5	Good	Good	Poor	No	Yes	Slight	Medium	Slight
Arsenic ••	100[e]	3.0	Fair	Fair	Fair	No	Yes	None	Medium[b]	None
Fluoroacetamide (1081) •••	15—Norway 51—Mice	2.0	Good	Good	Good	No	No	None	Slight	Slight
Phosphorus, yellow ••	1.7	0.05[f]	Good	Good	Fair	No	No	Strong	Strong	Fast
Red Squill •	500[g]	10.0	Fair	Fair	Poor	No	No	Medium	Strong[b]	Medium
Sodium fluoroacetate (1080) •••	5—Norway 2—Roof rat 10—Mice	½ oz/gal 1 Oz/28 lb	Good	Good	Good	No	No	None	Slight	Slight
Strychnine (alkaloid) ••••	6	0.6	Fair	Fair	Poor	No	Yes	None	Strong[b]	Slight
Strychnine (sulfate) ••••	8	0.8	Fair	Fair	Poor	No	Yes	None	Strong[b]	Slight
Zinc phosphide •••	40	1.0	Good	Good	Good	No	No	Strong	Strong	Fast

Source. "Control of Domestic Rats and Mice," Public Health Service Publication No. 583, Department of Health, Education and Welfare, Washington, D.C., 1968. Modified from Department of Interior, Fish and Wildlife Service Leaflet No. 337; revised December 1959.

[a] Effectiveness of poisons is keyed as follows:

- • Effective against Norway rats only
- • • Effective against Norway rats and roof rats
- • • • Effective against Norway rats, roof rats, and house mice
- • • • • Effective against mice only

[b] Action of poisons is keyed as follows:

- ■ Slow acting
- ■ ■ Fast acting
- ■ ■ ■ Very fast acting

[c] More or less; successive doses required for 5 to 10 days or more.

[d] Norway rats only, on first exposure.

[e] Particle sizes of U.S. Pharmacopoeia grades vary widely: some coarse powders test as high as 600 mg, micrometer size at 25 mg; latter are recommended, at 1.5 percent.

[f] Commercial preparations vary from 1 to 3 percent in paste form; use as label directs.

[g] Minimum acceptable level; more toxic squills give better results.

[h] Normally objectionable in rats.

[i] Can be taken through cuts or breaks in the skin; also danger of inhaling loose powder.

[j] Emetics used as first aid except as noted; speed is essential: 1 tablespoon of salt in a glass of warm water is usually effective. Call a physician immediately.

6 GENERAL ENVIRONMENTAL CONTROL AND PROVISION OF ADEQUATE SANITARY FACILITIES AND OTHER PERSONAL SERVICES

The Occupational Safety and Health Standards promulgated under the Williams-Steiger Act of 1970 (84 Stat. 1593) were published in the *Federal Register* of June 27, 1974 (47). Subpart J (General Environmental Controls) contains Section 191D.141 (Sanitation), Section 1910.142 (Temporary Labor Camps) and Section 1910.143 (Non-Water-Carriage Disposal Systems).

To be applicable in a wide variety of circumstances, the regulations in these sections are, for the most part, very general. For example, insect and rodent control, which is described as "Vermin Control," requires that "enclosed workplaces shall be constructed, equipped and maintained, so far as reasonably practicable, as to prevent the entrance and harborage of rodents, insects and other vermin. Furthermore, a continuing and effective extermination program shall be instituted where their presence is detected." Such a statement provides little that is helpful to the plant manager and less to OSHA personnel inspecting the premises. On the other hand, the requirements for toilet and washing facilities (e.g., lavatories and showers) are enumerated and relatively detailed.

The personal hygiene practices of employees carry many health implications without reference to the individuals' particular employment. For instance, the disease-transmitting potential from failure to clean the hands after defecating, or the skin diseases resulting from poor body cleansing, may not be related to the person's job. There are, however, health problems relating to personal hygiene practices that occur in industry exclusively. Such problems are related to ingestion of chemical toxins or disease-producing organisms and to local skin, conjunctival, and mucous membrane inflammation, due to sensitivity or direct irritative action of industrial chemicals.

Schwartz states that uncleanliness "is probably the most important predisposing cause of occupational dermatitis." Lack of cleanliness in the working environment exposes the worker to large doses of external irritants. Personal uncleanliness not only does the same, but also permits external irritants to remain in prolonged contact with the skin. Workers wearing or carrying to their homes their dirty work clothes may even cause dermatitis in other members of the family who come in contact with soiled clothes, or even among unsuspecting workers who clean them. Safeguards such as protective creams help, but personal hygiene is a necessity.

Where the employee is exposed to toxic materials, or if he is a food handler, the need for the optimum in clean, well-lighted, and well-ventilated washing and locker facilities becomes imperative. From the point of view of protecting the health of the individual employee and minimizing the possibility of his transmitting infections to others, toilet and washroom facilities should be easily cleanable, adequate in size and number, and accessible, or the personal hygiene habits of employees in industry will suffer. The American National Standards Institute Minimum Requirements for Sanitation (4) considers that "ready accessibility" has not been provided when an employee has to travel more than one floor-to-floor flight of stairs to or from a toilet facility. Some minimum standard on the basis of distance must be set. However the advent of complex automatic and remote controls and unit operations may bring about modification of some distance standards. In Section 1910.141 (Sanitation) of the OSHA June 27, 1974, Standards item (c) Toilet Facilities (ii) has taken this into account and exempts establishments in which mobile crews work in normally unattended work locations "so long as employees working in these locations have transportation immediately available to nearby toilet facilities which meet the usual requirements."

The OSHA standards state that whenever employees are required by a particular standard to wear protective clothing because of the possibility of contamination with toxic materials, change rooms equipped with storage facilities for the protective clothing shall be provided. Although recommended a number of years ago the following arrangement, which differs from the current OSHA requirement of locating work and street clothing lockers side by side, is good practice.

> A good arrangement is a room or building divided into two sections—a street clothes section and a work clothes section, with bathing and toilet facilities between. The street clothes section has an outside entrance and lockers for street clothes, toilets and wash basins or wash fountains, and changing facilities for supervisors. The work clothes section has rooms for work clothes, toilets and wash basins or wash fountains, showers and laundry. Three connections between the street clothes section and the work clothes section are (*a*) through the supervisors' change room, (*b*) through the main shower room, and (*c*) a hall with "one way traffic doors" from street to work clothes sections (48).

In the mineral, petroleum, and allied industries, the change house may be located at the entrance gate or near the parking lot to make it readily accessible. It may be a separate building connected to the working area by a covered walkway, and it should be constructed of fireproof or fire-resistant materials and be properly heated, lighted, and ventilated to meet the general requirements of Section 4 of the American Standard Minimum Requirements for Sanitation (4). The interior arrangement of the equipment and facilities in change houses and in locker and toilet rooms is not standardized but is a matter of proper design to meet the space, cost, and other requirements of the individual plant.

The minimum facilities required for places of employment are set forth in the various state regulations of departments of labor and divisions of industrial hygiene. In addition, various industrial associations have established, through industry-wide sanitation committees, practice codes related to the particular industry. An example is the Code of Recommended Practices for Industrial Housekeeping and Sanitation of the American Foundrymen's Association (49). The Association of Food Industry Sanitarians, in cooperation with the National Canners Association, has published a comprehensive manual, *Sanitation for the Food-Preservation Industries* (50). For all places of employment, minimum requirements for housekeeping, light and ventilation, water supply, toilet facilities, washing facilities, change rooms and retiring rooms for women, and food handling requirements are set forth in the American Standard Minimum Requirements for Sanitation (4).

The OSHA requirements (OSHA Standards June 27, 1974, Section 1910.141, Sanitation) are somewhat less stringent for establishments with fewer than 150 employees, nor does the OSHA regulation require the provision of retiring rooms. In general, minimum standards for toilet facilities can be delineated quite specifically with respect to their construction and provision for ease of cleaning. Critics point out that in some instances at least, the standard lavatories installed in washrooms fail to meet the need for ease of washing. The wash fountain with its foot pedal control supply, basin large enough to permit easy bathing of arms, face, and upper torso, free-flowing stream of thermostatically tempered water, and central dispenser of cleansing material offers a practical solution to the problem of providing good washing facilities. In addition, the wash fountain requires less floor space, initial installation is less complicated and costly, and it is maintained with greater ease. Experience with wash fountains in washrooms for female employees indicates that they are acceptable there.

7 MAINTENANCE OF GENERAL CLEANLINESS OF THE INDUSTRIAL ESTABLISHMENT

Housekeeping and plant cleanliness in their fullest sense imply not only the absence of clutter and debris in the working area and passageways, but also the maintenance of painted surfaces in a clean condition, the frequent removal of dust from lighting fixtures, and in general making the plant "a better place to work." The industrial plant that processes hazardous or potentially hazardous materials that cannot be prevented from escaping into the general atmosphere has a double responsibility to provide the utmost in housekeeping and plant cleanliness.

It is no longer sufficient to delegate responsibility for cleaniness of the plant and surroundings to a foreman or supervisor whose primary duties, interests, and training lie elsewhere. The complexity and cost of housekeeping machinery and supplies, the technical knowledge required for their proper and effective use, and the extensiveness of the work load make the establishment of a housekeeping department in large plants a virtual necessity. Working standards, schedules, and quality controls are also required in this routine plant operation (see Chapter 24). Additional evidence of the economic importance of this operation is the fact that fire insurance premiums are in part determined by the status of plant cleanliness.

Section 1910.22 (General Requirements) of the OSHA regulations sets very general standards for housekeeping. For a continuing process, too often relegated to a low level of managerial supervision, it is possible to establish quantitative measures of housekeeping effectiveness by such means as dust counts, bacteria counts, and color intensity and light transmission measurements.

The food industry presents special problems of frequent or even continuous cleaning operations as well as at the end of each shift. Thus this industry, with its combination of wet and dry cleaning requirements, represents a severe example for other industries. Perhaps the ultimate in "housekeeping" requirements are those for the "clean rooms" of the electronics and pharmaceutical industries, where continuous sanitation of the total environment is absolutely essential.

Housekeeping must be scheduled, and it must be the assigned responsibility of a trained group under proper supervision. A staggered work arrangement, for example, allows the use of a shift permitting the housekeeping crew to start work 1 to 2 hours after the production personnel. This permits some general cleaning during the morning before too much material has accumulated. It also permits scheduling cleaning during the regular lunch hour shutdown and completing the schedule after the day's operations have ended.

Housekeeping, a continuing indirect cost in industry, deserves uninterrupted scrutiny. The activity must be evaluated periodically with respect to effectiveness and cost.

8 THE ENVIRONMENTAL ENGINEER IN INDUSTRY

The environmental factors described in this chapter are inherent in varying degrees in every industrial establishment. The plants whose products are intended for human food may be placed in a special category, since they have a paramount responsibility to protect their product from contaminants that endanger the health of the consumer. Because of this responsibility, the roles of the environmental engineer and sanitarian in such industries should be obvious. There is great need to conserve the nation's water resources through effective control and treatment of industrial wastes, to provide production and management personnel with safe food and potable water at all times, and to maintain a sanitary working environment to protect the employee from the diseases that are not directly of occupational origin. Legislation such as the Water Pollution Control Act Amendments of 1972, the Clean Air Act of 1970, and the Safe Drinking Water Act of 1974 emphasizes the need for such specialized professionals in industry.

The standards and regulations promulgated to meet the goals and objectives are in a state of flux. Time tables are under administrative and legislative review, and the technology to achieve them is evolving. In addition, it is the intent of Congress to have increasing public participation in the standard setting process. Most generally, the environmental engineer is employed by industry to apply his knowledge to the solution of problems of water supply and their corollary, liquid waste disposal and air pollution control. However his engineering and public health skills are additionally applicable in such areas as solid wastes disposal, food sanitation, insect and rodent control, and plant maintenance. Although the environmental engineer may or may not be directly associated with the occupational health section of the industry, the roles filled by him and the industrial hygienist are directed toward the same end. The parallel may be further drawn by pointing out that just as the industrial hygienist aims to recognize hazardous working conditions in the design stage, to be able to eliminate them completely or to make the application of control measures more effective, convenient, and economical, the sanitary engineer in industry applies the same principle to the control of environmental factors other than those related to the occupation per se.

Many companies are too small to warrant the employment of a full-time environmental engineer or sanitarian. The possibility of a number of firms cooperating in the provision of environmental engineering services is worthy of consideration. In any event, state and local environmental department engineering personnel have a duty to cooperate with industry by supplying such services as review of plans and general consultation on environmental health problems within industry. This may be accomplished either through the industrial hygiene division or, if none exists, directly with the relevant environmental engineering division of the state health department, the EPA, or OSHA.

REFERENCES

1. S.433, Public Law 92-523, "Safe Drinking Water Act," 93rd Congress, Washington, D.C., December 16, 1974.
2. Protection of the Environment, Chapter 1, Subchapter D., Part 141, "National Interim Drinking Water Regulations," *Fed Reg.,* December 24, 1975.
3. "Interim Radiochemical Methodology for Drinking Water," *Environmental Monitoring and Support Laboratory,* EPA-600/4-75-008 U.S. Environmental Protection Agency, Cincinnati, Ohio.
4. American National Standard Minimum Requirement for Sanitation in Places of Employment [USASI] Z.41-1968, ANSI, New York.
5. American National Standards Institute, "Specifications for Drinking Fountains and Self-Contained, Mechanically Refrigerated Drinking–Water Coolers," A112.11.1-1973 ANSI, New York.
6. W. T. McCormick, personal communication.
7. "Criteria for a Recommended Standard: Occupational Exposure to Hot Environments," Catalog No. HSM 72-10269, U.S. Department of Health, Education and Welfare, National Institute of Occupational Safety and Health, Government Printing Office, Washington, D.C., 1972.
8. K. L. Kollar, and R. Brewer, *J. Am. Water Works Assoc.,* **67,** 686–690 (September 1975).
9. A. F. McClure, *J. Am. Water Works Assoc.,* **66,** 240–244 (April 1974).
10. Public Law 92-500, Water Pollution Control Act Amendments, 1972.
11. National Commission on Water Quality, Staff Draft Report, NCWQ, Washington, D.C., November 1975.
12. C. E. Renn, "Management of Recycled Waste–Process Water Ponds." Project WPD117, Environmental Technology Series, U.S. Environmental Protection Agency, R2-73-223, May 1973.
13. W. C. Ackerman, *J. Am. Water Works. Assoc.,* **12,** 691–693 (December 1975).
14. C. J. Schmidt et al., *J. Water Pollut. Control Fed.,* **47,** 2229 (September 1975).
15. C. E. Arnold, H. E. Hedger, and A. M. Rawn, "Report upon the Reclamation of Water from Sewage and Industrial Wastes in Los Angeles County, California," prepared for County of Los Angeles, 1949.
16. H. A. Spafford, *J. Am. Water Works Assoc.,* **46,** 993–998 (1954).
17. American Water Works Association *Recommended Practice for Backflow Prevention and Cross-Connection Control,* AWWA Manual M14, Denver, 1966.
18. American Water Works Association, AWWA Standard for Backflow Prevention Devices, "Reduced Pressure Principle and Double Check Valve Types (C506-69)," AWWA, Denver, 1969.
19. U.S. Environmental Protection Agency, *Cross-Connection Manual,* EPA-430/9-73-002, EPA Office of Water Programs-Water Supply Division 1973.

20. G. J. Angele, Sr., *Cross Connections and Backflow Prevention*. 2nd. ed., American Water Works Association, Denver, (1974).
21. American National Standard Institute, "Air Gaps in Plumbing Systems," A112.1.2-1942 (rev. 1973) ANSI, New York, (1973).
22. A. J. Klee, *Trans. Environ. Eng. Div., ASCE,* **102,** 111–125 (February 1976).
23. W. W. Eckenfelder, Jr., *Water Wastes Eng.,* **13,** 83–87 (September 1976).
24. J. F. Richardson, and W. N. Zaki, *Trans. Inst. Chem. Eng.,* **32,** 34–53 (January 1954).
25. A. W. Lawrence and P. L. McCarty, *Trans. Sanit. Eng. Div. ASCE,* **96:** SA3 (June 1970). Also, Proceeding Paper 7364, 757–778, June 1970.
26. P. L. McCarty, "Anaerobic Treatment of Soluble Wastes," in: *Advances in Water Quality Treatment,* E. F. Gloyna and W. W. Eckenfelder, Jr., Eds., University of Texas Press, Austin, 1968, pp. 336–352.
27. W. J. Weber, Jr., *Physicochemical Processes for Water Quality Control,* Wiley-Interscience, New York, 1972, 640 pp.
28. J. C. Morris, "Formation of Halogenated Organics by Chlorination of Water Supplies," U.S. Environmental Protection Agency Publication No. 600/1-75-002, March 1975.
29. J. J. Rook, *J. Water Treat. Exam.,* **23,** 234 (1974).
30. J. J. Rook, *J. Am. Water Works Assoc.,* **68,** 168 (March 1976).
31. W. Zabban and R. Helwick, "Defluoridization of Wastewater," *Proceedings, 30th Annual Purdue Industrial Waste Conference,* West Lafayette, Ind., 1975.
32. R. Piskin, *Waste Age,* **7,** 42 (July 1976).
33. N. I. Sax, *Dangerous Properties of Industrial Materials,* 4th ed., Van Nostrand, New York, 1975.
34. M. Synak, P. H. Wicks, and K. H. Spies, *Trans. Environ. Eng. Div. ASCE,* **101:** EE3 (June 1975).
35. Center for Disease Control, *Annual Summary of Food-Borne and Water-Borne Disease Outbreaks, 1974,* CDC, Atlanta, 1976.
36. *Preventive Medicine USA—Theory, Practice, and Application of Preventive Environmental Health Services,* Social Determination of Human Health, Prodist. New York, 1976.
37. K. F. Maxcy, "Contagious Disease Spread Largely Through Fecal Discharge," in: *Preventive Medicine and Hygiene,* 8th ed., K. F. Maxcy, Ed., Appleton-Century-Crofts, New York, 1956, Chapter 3.
38. Food and Drug Training Institute, *Current Concepts in Food Protection,* U.S. Department of Health, Education and Welfare, Cincinnati, Ohio, 1971.
39. Consumer Protection and Environmental Health Service, *Sanitary Food Service Instructor's Guide,* 1969 revision, U.S. Department of Health Education, and Welfare, Cincinnati, 45226.
40. E. W. Moore, E. W. Brown, and E. M. Hall, "Sanitation of Crushed Ice for Iced Drinks," *J. Am. Pub. Health Assoc.,* **43,** 1265–1269 (1953); V. D. Foltz, "Sanitary Quality of Crushed and Cubed Ice as Dispensed to the Consumer," *U.S. Pub. Health Rep.,* **68,** 949–954 (1953).
41. Public Health Service, *Ordinance and Code, "Regulating Eating and Drinking Establishments"* (recommended), PHS, Washington, D.C., 1962.
42. National Sanitation Foundation, "Soda Fountain and Luncheonette Equipment," Standard No. 1; "Food Service Equipment," Standard No. 2; "Spray-Type Dishwashing Machines," Standard No. 3, NSF, Ann Arbor, Mich.
43. "The Vending of Foods and Beverages," A Sanitation Ordinance and Code, Public Health Service Publication No. 546, Department of Health, Education and Welfare, 1965.
44. D. E. Lindsay, W. H. Stewart, and J. Watt, "Effect of Fly Control on Diarrheal Disease in an Area of Moderate Morbidity," *U.S. Pub. Health Rep.,* **68,** 361 (1953).
45. D. E. Davis, "Control of Rats and Other Rodents," in: *Preventive Medicine and Hygiene.* 8th ed., K. F. Maxcy, Ed., Appleton-Century-Crofts, New York, 1956, Chapter 8.

46. Communicable Disease Center, *Rat-Borne Disease; Prevention and Control,* CDC, Atlanta, 1949.
47. Occupational Safety and Health Standards, U.S. Department of Labor, Occupational Safety and Health Administration, *Fed. Reg.* **39**: 125 23502–23828 (June 24, 1974).
48. F. E. Cash, "Suggested Standards for Change Houses," paper presented at the annual meeting of the American Public Health Association, October 1951.
49. American Foundrymen's Association, *Code of Recommended Practices for Industrial Housekeeping and Sanitation,* AFA, Chicago, 1944.
50. Association of Food Industry Sanitarians and National Cancer Association, *Sanitation for the Food Preservation Industries,* AFIS–NCA, Berkeley, Calif.

GENERAL REFERENCES

American Water Works Association, *Water Quality and Treatment: A Handbook of Public Water Supplies,* 3rd ed., McGraw-Hill, New York, 1971.

G. M. Fair, J. C. Geyer, and D. A. Okun, *Water and Wastewater Engineering,* Vol. I, *Water Supply and Wastewater Removal*; Vol. 2, *Water Purification and Wastewater Treatment,* Wiley, New York, 1966 and 1968.

National Academy of Sciences–National Academy of Engineering, "Water Quality Criteria—1972," EPA-R3-73-033, Environmental Protection Agency, Washington, D.C., March 1973.

L. L. Ciaccio, Ed., *Water and Water Pollution Handbook,* Dekker, New York, 1971.

T. F. Yen, Ed., *Recycling and Disposal of Solid Wastes: Industrial, Agricultural, Domestic,* Ann Arbor Science Publishers, Ann Arbor, Mich., 1974.

J. W. Patterson, *Wastewater Treatments Technology,* Ann Arbor Science Publishers, Ann Arbor, Mich., 1975.

W. W. Eckenfelder, Jr., *Water Quality Engineering for Practicing Engineers,* Barnes & Noble, New York, 1970.

TRW Systems Group, *Recommended Methods of Reduction, Recovery or Disposal of Hazardous Wastes,* Vols 1–16, National Technical Information Service, Springfield, Va., 1973.

U.S. Environmental Protection Agency, *Technology Transfer Material: Technology Transfer,* EPA, Cincinnati, Ohio.

American Public Health Association, American Water Works Association, and Water Pollution Control Federation, *Standard Methods for the Examination of Water and Wastewater,* 14th ed., APHA, Washington, D.C., 1976.

Applied Foodservice Sanitation, National Institute for the Foodservice Industry, Chicago, 1974.

E. T. Chanlett, *Environmental Protection,* McGraw-Hill, New York, 1973.

CHAPTER TWENTY-SEVEN

Fire and Explosion Hazards of Combustible Gases, Vapors, and Dusts

JOSEPH GRUMER

1 GENERAL CONSIDERATIONS

Explosions and ensuing fires of combustible gases, vapors, and dusts are among the major hazards in homes, commerce, and industry. Understanding of these hazards and awareness of means of combating and preventing them are obvious necessities. The earlier literature is voluminous, and a total extract of it is outside the scope of this chapter. Here I want to update three prior reviews, which have summarized the literature into useful texts for those concerned with combating hazards due to the possible pressures of combustible gases, vapors, and dusts mixed with air; some consideration is also given to hazards due to these combustibles mixed with oxygen. The reader is referred to these reviews (1–3) for original references to the literature prior to 1962. To conserve space, this overview omits references to original sources of quoted data, provided the origins are cited in references 1, 2, or 3. Furthermore the present writing undertakes particularly to update Reference 2—namely, Chapter XVI, Section 1, by G. W. Jones, and Section 2 by I. Hartmann, of the second revised edition of Patty's *Industrial Hygiene and Toxicology*—and closely follows the organization of that earlier work. Other recommended references are those of the National Fire Protection Association, such as References 4 and 5.

Certain discussions of terminology are in order. Fires generally involve diffusion flames of a liquid or solid fuel bed in which there is no premixing of fuel and air before combustion (e.g., a wood fire or an oil tank fire). Explosion constitutes a sudden release of pressure, not necessarily as a consequence of combustion. For example, an explosion occurs when compressed air bursts its container suddenly. As a consequence, "explosive limits" is a poor term when applied to limits involving combustion; "flammable limits" or "limits of flammability" are preferable.

1.1 Limits of Flammability

Limits of flammability are the extremes in concentration between which homogeneous gas–air mixtures, vapor–air, or dust–air mixtures can be burned when subjected to an ignition source of adequate temperature and energy. For example, trace amounts of methane in air can be readily oxidized on a heated surface, but a flame will propagate from an ignition source at ambient temperatures and pressures only if the surrounding mixture contains at least 5 but less than 15 volume percent methane. The more dilute mixture is known as the lower limit, or the combustible-lean limit mixture; the more concentrated mixture is known as the upper limit, or combustible-rich limit mixture. In practice, the limits of flammability of a particular premixed system of gaseous combustibles and oxidant are affected by temperature, pressure, direction of flame propagation, and prefential diffusion due to gravity. The flammability limits of premixed combustible dusts and oxidants are affected by additional factors such as particle size.

It is impractical to differentiate between mixtures on the basis of the amount of violence produced, because more than mixture composition is involved. If confined in a long tube, open at one end and ignited at the open end, mixtures just within the limits of flammability will propagate flame quietly and slowly through the tube. (Propagation is usually at a uniform speed, and the speed for a given concentration of combustible in oxidant is faster for upward than for downward flame propagation.) If this mixture is confined in a large enough closed vessel and ignited, however, flame propagation will accelerate and propagate at speed several times that in the open tube, particularly if the mixture is initially in gentle or turbulent motion; pressures up to 30 psi or more may be attained. This is only to say that comfort should not be taken in expectations that at worst only a limit composition can be encountered in a specific operation. Any unwanted flammation is inherently disastrous.

Admittedly the rate of flame propagation through a flammable mixture and the maximum attainable pressure depend on a number of factors, including temperature, pressure, and mixture composition. For a given fuel–oxidant system, flame speed and peak pressure are relatively low at the limits of flammability and high near stoichiometric. However destruction can be fearful even from explosions due to relatively mild flames. Consider that at the lean limit of deflagration of 5 percent methane in air, peak pressures attainable in an adiabatic closed container are theoretically about 5 atmospheres or about 75 psi. This is enough to cause failure of structural elements, as evidenced by the critical peak overpressures given in Table 27.1.

The data in this survey are based on well-mixed, flammable, homogeneous mixtures. In homes, commerce, industry, and nature, heterogeneous single-phase gas and multi-

Table 27.1 Conditions of Failure of Peak Over-Pressure-Sensitive Elements (3)

Structural Element	Failure	Approximate Incident Blast Overpressure (psi)
Glass windows, large and small	Usually shattering, occasional frame failure	0.5–1.0
Corrugated asbestos siding	Shattering	1.0–2.0
Corrugated steel or aluminum paneling	Connection failure, followed by buckling	1.0–2.0
Wood siding panels, standard house construction	Usually failure occurs at main connections, allowing a whole panel to be blown in	1.0–2.0
Concrete or cinderblock wall panels, 8 or 12 in thick (not reinforced)	Shattering of the wall	2.0–3.0
Brick wall panel, 8 or 12 in. thick (not reinforced)	Shearing and flexure failures	7.0–8.0

phase (gas, liquid, and dust) incompletely mixed, locally flammable mixtures probably occur more frequently and are more important than are homogeneous mixtures. The occurrence and accounting of the hazardous characteristics of such heterogeneous mixtures is unfortunately still unpredictable in adequate detail. It is important to recognize, for example, that heterogeneous mixtures can ignite at overall concentrations (as if completely mixed) that would normally be nonflammable. For example, 1 liter of methane can form a flammable mixture with air near the top of a 100-liter container, although a 1.0 volume percent, nonflammable mixture would result if complete mixing occurred; only about 6 liters of the 100 liters of available air need mix with the methane in this hypothesized situation to form a flammable mixture. Once ignition occurs in a flammable zone in a given volume, unpredictable signatures of pressure, temperature, and composition rule, with possibly disastrous results. Not only lighter than air, roof-hugging flammable gases such as hydrogen and methane present layering hazards. Flammable gases that are heavier than air, such as propane, butane, and gasoline vapors, can produce disasters by way of ground-hugging layers. This is an important consideration, since layering can occur in both stationary and flowing mixtures. Consequently, good safety practice prohibits closely skirting flammability limits of homogeneous mixtures. Under the Mine Safety and Health Act of 1969 (6), precautionary measures such as shutting down machinery are mandated at methane gas concentrations in exhaust air of 20 percent of the lower flammability limits, and evacuation of the mine is required when gas concentration in exhaust air reaches 30 percent of this limit. Such margins avoiding hazardous concentrations (7) are well advised, though not necessarily sufficient in all instances.

Combustion may propagate through a mixture by deflagration or by detonation. If the propagation rate relative to the unburned gas is subsonic, the process is deflagration, and if the rate is supersonic, the process is detonation. If it is subsonic, the pressure caused by the deflagration process will equalize at the speed of sound throughout the enclosure in which combustion is taking place. The pressure drop across the flame front will be relatively small, generally in the order of millimeters of water pressure or less. If the flame propagation rate is supersonic, the pressure caused by the detonation process will equalize less rapidly than the flame propagation rate, and there will be an appreciable pressure drop across the flame front. Moreover, with most of air and combustibles at ordinary temperatures, the ratio of peak to initial pressure due to a deflagration within a closed vessel seldom exceeds 8 : 1 but that due to a detonation may be more than 40 : 1 (3). The pressure buildup is especially great when detonation follows a large pressure rise due to deflagration. The distance required for a deflagration to transit to a detonation varies with many factors, but the most important is the strength of the ignition; with a sufficiently powerful ignition, detonation may occur immediately on ignition, even in open air (3). Detonation limits are generally not as lean as the lean limit and not as rich as the rich limit for upward flame propagation. Available data suggest that the limits of detonability are within the limits of downward flame propagation (8).

The peak pressure that may be expected theoretically in an adiabatic situation in a closed vessel with central ignition is given by Equation 1:

$$P_b = P_1 \frac{\overline{M}_i T_b}{M_b T_i} \tag{1}$$

where P is pressure, M is molecular weight, and T is absolute temperature. The overbar indicates an average value, and the subscripts i and b indicate initial and burned gas parameters, respectively. The minimum elapsed time (in milliseconds) to reach peak pressure has been estimated (3) for parafin hydrocarbons and gasoline vapors to equal 75 times the cube root of the volume (in cubic feet) of the enclosure.

2 COMBUSTIBLE GASES AND VAPORS

2.1 Factors Affecting the Limits of Flammability

Some years ago the U.S. Bureau of Mines adopted a standard apparatus for limit-of-flammability determinations (1). This apparatus (Figure 27.1) was originally designed for determining limits of gases and vapors of liquids that are volatile enough at room temperatures to give flammable mixtures. In Figure 27.1, *a* is the 5 cm i.d. glass tube in which the mixture is tested. The lower end of the glass tube is closed by a lightly lubricated ground-glass plate *b* sealed with mercury *c*. After evacuation by pumping through tube *j*, gas is introduced from a gas holder through stopcock *r*, or if a vapor is to be tested, it is drawn from its liquid in container *p*. The amount of gas or vapor introduced is measured by the manometer *k*. Air or "other atmosphere" is then admitted through

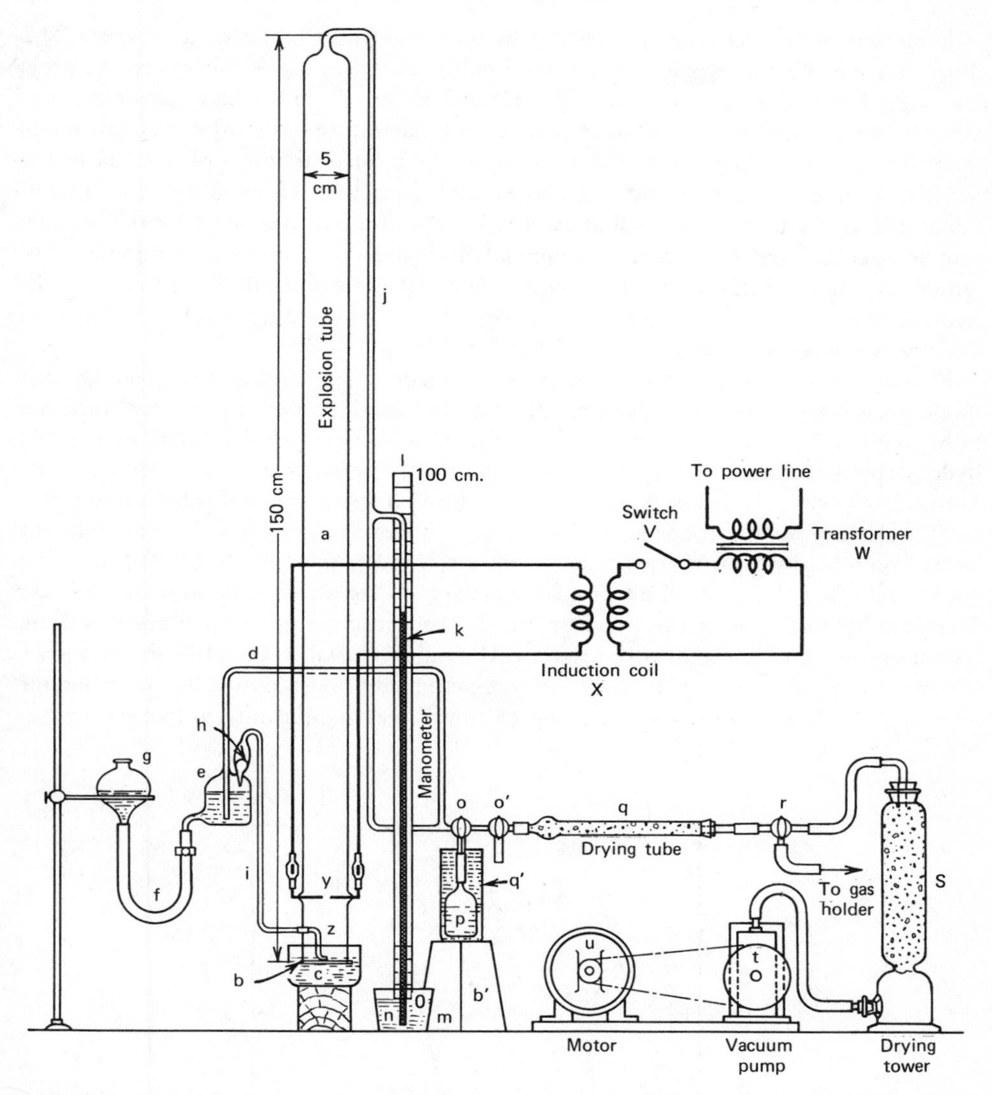

Figure 27.1 Apparatus for determining the limits of flammability of gases and vapors (1). Schematic courtesy of Bureau of Mines, U.S. Department of the Interior.

the drying tube *q* until atmospheric pressure is reached. Gaseous oxidant and fuel are thoroughly mixed by raising and lowering the mercury vessel *g* repeatedly for 10 to 30 minutes, depending on the density differences of the components in tube *a*. The mercury seal is then removed, glass plate *b* is slid off the tube, and the flammability of the mixture is tested by sparking at *y* or by passing a small flame across the open end of the tube.

If ignition is by sparking, the energy in the spark must be adequate. Figure 27.2 illustrates the effect of mixture composition on the electrical spark energy requirements for ignition of methane–air mixtures (3, 9). A 0.2 mJ spark cannot ignite any methane–air mixture at atmospheric pressure and room temperature. A 1 mJ spark can ignite mixtures containing between 6 and 11.5 volume percent methane; stronger sparks are needed to ignite leaner than 6 volume percent and richer than 11.5 volume percent flammable methane–air mixtures. Limit mixture compositions that depend on the ignition source strength are not limits of flammability; they are limits of ignitability, or ignitability limits. Limit mixtures, compositions that are essentially independent of the ignition source strength and support flames capable of propagating beyond the region of influence of the ignition source are limits of flammability.

Flammability limit determinations must be made in apparatus large enough that flame quenching by vessel walls is practically eliminated. A 5 cm i.d. vertical tube has been found to be large enough for determinations of flammability limits of paraffin hydrocarbons (methane, ethane, etc.) at atmospheric pressure and room temperature. However a 5 cm tube is not large enough for many halogenated and other compounds, nor for paraffin hydrocarbons at very low temperatures and pressures. For example, the lower flammability limit of trichloroethylene–air mixtures had to be measured in a 17.8 cm i.d. tube and the upper limit in a 20 cm sphere (3) to attain apparatus-independent flammability limits. Recently the Bureau of Mines apparatus for flammability limit measurements was changed to incorporate 10 and 30 cm i.d. tubes (10). Furthermore, the apparatus must be long enough to ensure continued propagation of flame beyond the zone wherein heat from the ignition source contributes significantly to flame propaga-

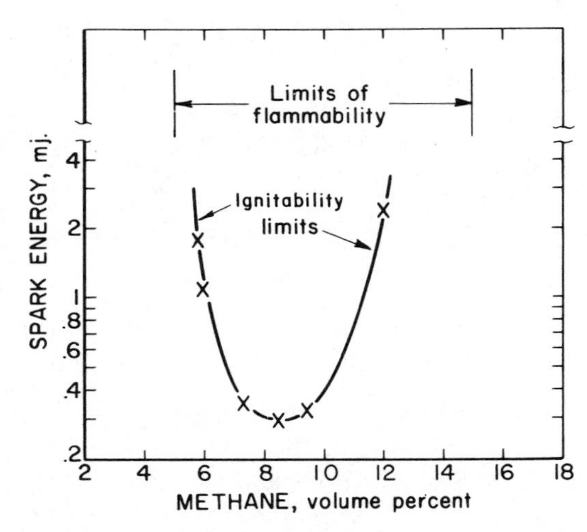

Figure 27.2 Ignitability curve and limits of flammability for methane–air mixtures at atmospheric pressure and 26°C (3). Graph courtesy of Bureau of Mines, U.S. Department of the Interior.

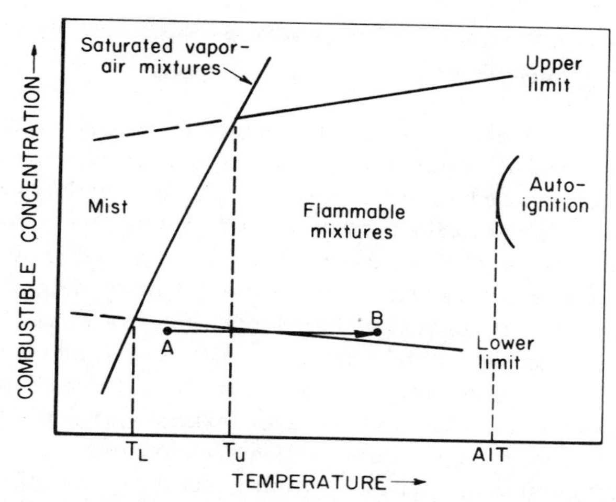

Figure 27.3 Effect of temperature on the limits of flammability of a combustible vapor in air at a constant initial pressure (3). Graph courtesy of Bureau of Mines, U.S. Department of the Interior.

tion. An apparatus about 1.25 meters long or more is generally long enough to allow observation of flammability limits free of influence of the ignition source. If limits are observed within very short distances close by the ignition source, the apparent limits can be forced leaner of the lower flammability limit and richer of the upper flammability limit. Therefore in some practical applications, the composite hazard due to ignition source and combustibles should be considered. Vessel shape (e.g., spherical vs. cylindrical) as well as size can affect observed limits, if the vessel's smallest dimension is less than needed to yield apparatus-free flammability limits; spheres about 10 cm in diameter should generally be large enough for attaining flammability limits.

Minor variations of temperature such as those in a laboratory do not significantly affect flammability limits, but major elevations of temperature will widen the composition range of fuel–oxidant between these limits. Consider for example, the shift of point *A* in Figure 27.3 from a nonflammable composition–temperature point to point *B*, which is in the flammable zone. Likewise, normal variations in barometric pressure have no appreciable effect, but limits—particularly those of paraffin hydrocarbon—at high pressure are generally wider than at atmospheric pressure.

The quantity of water vapor in air at room temperatures affects the lower limit of flammability very little because the mixture contains excess oxygen. Such is not the case at the upper limit of flammability, which is lowered because the mixture is oxygen short for stoichiometric combustion and the oxygen content of the mixture is reduced by the water vapor content.

Wider flammability limits are obtained for upward propagation of flame than for horizontal or downward propagation. From the safety point of view, flammability hazards should be evaluated on the basis of limits for upward propagation.

2.2 Limits of Flammability, Lower Temperature Limits, and Minimum Autoignition Temperatures

Most of the values of flammability limits of mixtures of air and combustibles given in Table 27.2 were obtained by using an apparatus such as described in Figure 27.1 and are believed to have been determined in adequately large vessels with an adequately strong ignition source. These limits are for upward propagation at room temperature and pressure, except as noted in footnotes. Table 27.2 also contains data on lower temperature limits and minimum temperatures for autoignition of combustible–air mixture. The three types of limit can be explained with reference to Figure 27.3.

Flammable mixtures considered in Figure 27.3 fall in one of three regions, defined by the parameters of initial temperatures and concentration of the combustible. To the left of the line labeled "saturated vapor–air mixtures" exists the region of flammable mists or flammable mixtures of droplets in air. A flammable mixture can form at temperatures below the flash point of a liquid combustible either if such a combustible is sprayed into air or any suitable oxidant, or if a mist or foam forms. Some details of the complex behavior of mists and sprays are given in Reference 3 and in the literature mentioned therein. The second region lies along the curve for saturated vapor–air mixtures. The intercept of this curve and the one for the lower limit of flammability defines the lowest temperature at which a homogeneous flammable vapor–air mixture can be formed from a liquid; this computed temperature is given in Table 27.2 as T_L. At a given pressure, the most common hazards are due to temperature-compositions in the region labeled "flammable mixtures." The autoignition curve defines limit temperature-composition point to the right of which mixtures can ignite spontaneously. The minimum temperature along this curve is the minimum autoignition temperature, (AIT) and is also listed in Table 27.2. It is the lowest temperature at which ignition can occur when a flammable mixture is heated to an elevated temperature. The time that elapses between the instant a mixture is raised to a given temperature and the formation of a flame is called the ignition lag or time delay before ignition. If the ignition lag is relatively short—certainly if it is less than a second, as may be the case in a flowing system—the ignition temperature increases above the minimum AIT as heating time decreases. From the standpoint of safety, the minimum temperature at which ignition can occur spontaneously is needed, and this temperature must be determined as a function of heating time lasting long enough (many minutes) that the temperatures for the appearance of flame are independent of increasing heating time. Again, more detail can be obtained in Reference 3 and also in Reference 2.

2.3. Prediction of Limits of Flammability

2.3.1 One Combustible

Although the limits of flammability of about 250 single-component combustibles are presented in Table 27.2, an estimate may be needed of the limits of a combustible not

Table 27.2 Summary of Limits of Flammability, Lower Temperature Limits (T_L), and Minimum Autoignition Temperatures (AIT) of Individual Gases and Vapors in Air at Atmospheric Pressure (3)

Combustible	Limits of Flammability (volume percent) L_{25}	U_{25}	T_L (°C)	AIT (°C)
Acetal	1.6	10	37	230
Acetaldehyde	4.0	60		175
Acetic acid	5.4[a]		40	465
Acetic anhydride	2.7[b]	10[a]	47	390
Acetanilide	1.0[d]			545
Acetone	2.6	13		465
Acetophenone	1.1[d]			570
Acetylacetone	1.7[d]			340
Acetyl chloride	5.0[d]			390
Acetylene	2.5	100		305
Acrolein	2.8	31		235
Acrylonitrile	3.0		−6	
Acetone cyanohydrin	2.2	12		
Adipic acid	1.6[d]			420
Aldol	2.0[d]			250
Allyl alcohol	2.5	18	22	
Allyl amine	2.2	22		375
Allyl bromide	2.7[d]			295
Allyl chloride	2.9		−32	485
o-Aminodiphenyl	.66	4.1		450
Ammonia	15	28		
n-Amyl acetate	1.0[a]	7.1[a]	25	360
n-Amyl alcohol	1.4[a]	10[a]	38	300
tert-Amyl alcohol	1.4[d]			435
n-Amyl chloride	1.6[e]	8.6[a]		260
tert-Amyl chloride	1.5[f]		−12	345
n-Amyl ether	0.7[d]			170
Amyl nitrite	1.0[d]			210
n-Amyl propionate	1.0[d]			380
Amylene	1.4	8.7		275
Aniline	1.2[g]	8.3[g]		615
Anthracene	0.65[d]			540
n-Amyl nitrate	1.1			195
Benzene	1.3[a]	7.9[a]		560
Benzyl benzoate	0.7[d]			480
Benzyl chloride	1.2[d]			585
Bicyclohexyl	0.65[a]	5.1[h]	74	245
Biphenyl	0.70[i]		110	540

Table 27.2 (Continued)

Combustible	Limits of Flammability (volume percent) L_{25}	U_{25}	T_L (°C)	AIT (°C)
2-Biphenylamine	0.8[d]			450
Bromobenzene	1.6[d]			565
Butadiene (1,3)	2.0	12		420
n-Butane	1.8	8.4	−72	405
1,3-Butandiol	1.9[d]			395
Butene-1	1.6	10		385
Butene-2	1.7	9.7		325
n-Butyl acetate	1.4[e]	8.0[a]		425
n-Butyl alcohol	1.7[a]	12[a]		
sec-Butyl alcohol	1.7[a]	9.8[a]	21	405
tert-Butyl alcohol	1.9[a]	9.0[a]	11	480
tert-Butyl amine	1.7[a]	8.9[a]		380
n-Butyl benzene	0.82[a]	5.8[a]		410
sec-Butyl benzene	0.77[a]	5.8[a]		420
tert-Butyl benzene	0.77[a]	5.8[a]		450
n-Butyl bromide	2.5[a]			265
Butyl cellosolve	1.1[h]	11[j]		245
n-Butyl chloride	1.8	10[a]		
n-Butyl formate	1.7	8.2		
n-Butyl stearate	0.3[d]			355
Butyric acid	2.1[d]			450
α-Butryolactone	2.0[h]			
Carbon disulfide	1.3	50		90
Carbon monoxide	12.5	74		
Chlorobenzene	1.4		21	640
m-Cresol	1.1[h]			
Crotonaldehyde	2.1	16[k]		
Cumene	0.88[a]	6.5[a]		425
Cyanogen	6.6			
Cycloheptane	1.1	6.7		
Cyclohexane	1.3	7.8		245
Cyclohexanol	1.2[d]			300
Cyclohexene	1.2[a]			
Cyclohexyl acetate	1.0[d]			335
Cyclopropane	2.4	10.4		500
Cymene	0.85[a]	6.5[a]		435
Decaborane	0.2			
Decalin	0.74[a]	4.9[a]	57	250
n-Decane	0.75[l]	5.6[m]	46	210
Deuterium	4.9	75		

Table 27.2 (Continued)

Combustible	Limits of Flammability (volume percent)		T_L (°C)	AIT (°C)
	L_{25}	U_{25}		
Diborane	0.8	88		
Diesel fuel (60 cetane)				225
Diethyl amine	1.8	10		
Diethyl analine	0.8[d]		80	630
1,4-Diethyl benzene	0.8[a]			430
Diethyl cyclohexane	0.75			240
Diethyl ether	1.9	36		160
3,3-Diethyl pentane	0.7[a]			290
Diethyl ketone	1.6			450
Diisobutyl carbinol	0.82[a]	6.1[j]		
Diisobutyl ketone	0.79[a]	6.2[a]		
2-4-Diisocyanate			120	
Diisopropyl ether	1.4	7.9		
Dimethyl amine	2.8			400
2,2-Dimethyl butane	1.2	7.0		
2,3-Dimethyl butane	1.2	7.0		
Dimethyl decalin	0.69[a]	5.3[i]		235
Dimethyl dichlorosilane	3.4			
Dimethyl ether	3.4	27		350
n,n-Dimethyl formamide	1.8[a]	14[a]	57	435
2,3-Dimethyl pentane	1.1	6.8		335
2,2-Dimethyl propane	1.4	7.5		450
Dimethyl sulfide	2.2	20		205
Dimethyl sulfoxide			84	
Dioxane	2.0	22		265
Dipentene	0.75[h]	6.1[h]	45	237
Diphenylamine	0.7[d]			635
Diphenyl ether	0.8[d]			620
Diphenyl methane	0.7[d]			485
Divinyl ether	1.7	27		
n-Dodecane	0.60[d]		74	205
Ethane	3.0	12.4	−130	515
Ethyl acetate	2.2	11		
Ethyl alcohol	3.3	19[k]		365
Ethyl amine	3.5			385
Ethyl benzene	1.0[a]	6.7[a]		430
Ethyl chloride	3.8			
Ethyl cyclobutane	1.2	7.7		210
Ethyl cyclohexane	2.0[n]	6.6[n]		260
Ethyl cyclopentane	1.1	6.7		260

Table 27.2 (Continued)

Combustible	Limits of Flammability (volume percent)		T_L (°C)	AIT (°C)
	L_{25}	U_{25}		
Ethyl formate	2.8	16		455
Ethyl lactate	1.5			400
Ethyl mercaptan	2.8	18		300
Ethyl nitrate	4.0			
Ethyl nitrite	3.0	50		
Ethyl propionate	1.8	11		440
Ethyl propyl ether	1.7	9		
Ethylene	2.7	36		490
Ethyleneimine	3.6	46		320
Ethylene glycol	3.5[d]			400
Ethylene oxide	3.6	100		
Furfural alcohol	1.8[o]	16[p]	72	390
Gasoline				
100/130	1.3	7.1		440
115/145	1.2	7.1		470
Glycerine				370
n-Heptane	1.05	6.7	−4	215
n-Hexadecane	0.43[d]		126	205
n-Hexane	1.2	7.4	−26	225
n-Hexyl alcohol	1.2[a]			185
n-Hexyl ether	0.6[d]			
Hydrazine	4.7	100		400
Hydrogen	4.0	75		
Hydrogen cyanide	5.6	40		
Hydrogen sulfide	4.0	44		360
Isoamyl acetate	1.1[a]	7.0[a]	25	350
Isoamyl alcohol	1.4[a]	9.0[a]		460
Isobutane	1.8	8.4	−81	
Isobutyl alcohol	1.7[a]	11[a]		430
Isobutyl benzene	0.82[a]	6.0[j]		
Isobutyl formate	2.0	8.9		465
Isobutylene	1.8	9.6		
Isopentane	1.4			460
Isophorone	0.84			
Isopropylacetate	1.7[d]			
Isopropyl alcohol	2.2			440
Isopropyl biphenyl	0.6[d]			
Jet fuel				240
JP-4	1.3	8		230
JP-6				210

Table 27.2 (Continued)

Combustible	Limits of Flammability (volume percent)		T_L (°C)	AIT (°C)
	L_{25}	U_{25}		
Kerosine				
Methane	5.0	15.0	−187	540
Methyl acetate	3.2	16		
Methyl acetylene	1.7			
Methyl alcohol	6.7	36[k]		385
Methyl amine	4.2[d]			430
Methyl bromide	10	15		
3-Methyl butene-1	1.5	9.1		
Methyl butyl ketone	1.2[e]	8.0[a]		
Methyl cellosolve	2.5	20[g]	46	380
Methyl cellosolve acetate	1.7[q]			
Methyl ethyl ether	2.2[d]			
Methyl chloride	7[d]			
Methyl cyclohexane	1.1	6.7		250
Methyl cyclopentadiene	1.3[a]	7.6[a]	49	445
Methyl ethyl ketone	1.9	10		
Methyl ethyl ketone peroxide			40	390
Methyl formate	5.0	23		465
Methyl cyclohexanol	1.0[d]			295
Methyl isobutyl carbinol	1.2[d]		40	
Methyl isopropenyl ketone	1.8[e]	9.0[e]		
Methyl lactate	2.2[a]			
α-Methyl naphthalene	0.8[d]			530
2-Methyl pentane	1.2[d]			
Methyl propionate	2.4	13		
Methyl propyl ketone	1.6	8.2		
Methyl styrene	1.0[d]		49	495
Methyl vinyl ether	2.6	39		
Methylene chloride				615
Monoisopropyl bicyclohexyl	0.52	4.1[r]	124	230
2-Monoisopropyl biphenyl	0.53[j]	3.2[r]	141	435
Monomethylhydrazine	4			
Naphthalene	0.88[s]	5.9[t]		526
Nicotine	0.75[a]			
Nitroethane	3.4			
Nitromethane	7.3			
1-Nitropropane	2.2			
2-Nitropropane	2.5		27	
n-Nonane	0.85[t]		31	205
n-Octane	0.95		13	220

Table 27.2 (Continued)

Combustible	Limits of Flammability (volume percent) L_{25}	U_{25}	T_L (°C)	AIT (°C)
Paraldehyde	1.3			
Pentaborane	0.42			
n-Pentane	1.4	7.8	−48	260
Pentamethylene glycol				335
Phthalic anhydride	1.2[g]	9.2[v]	140	570
3-Picoline	1.4[d]			500
Pinane	0.74[w]	7.2[w]		
Propadiene	2.16			
Propane	2.1	9.5	−102	450
1,2-Propandiol	2.5[d]			410
β-Propiolactone	2.9[c]			
Propionaldehyde	2.9	17		
n-Propyl acetate	1.8	8		
n-Propyl alcohol	2.2[l]	14[a]		440
Propyl amine	2.0			
Propyl chloride	2.4[d]			
n-Propyl nitrate	1.8[p]	100[p]	21	175
Propylene	2.4	11		460
Propylene dichloride	3.1[d]			
Propylene glycol	2.6[x]			
Propylene oxide	2.8	37		
Pyridine	1.8[k]	12[y]		
Propargyl alcohol	2.4[e]			
Quinoline	1.0[d]			
Styrene	1.1[z]			
Sulfur	2.0[aa]		247	
p-Terphenyl	0.96[d]			535
n-Tetradecane	0.5[d]			200
Tetrahydrofurane	2.0			
Tetralin	0.84[a]	5[h]	71	385
2,2,3,3-Tetramethyl pentane	0.8			430
Tetramethylene glycol				390
Toluene	1.2[a]	7.1[a]		480
Trichloroethane				500
Trichloroethylene	12[bb]	40[y]	30	420
Triethylene amine	1.2	8.0		
Triethylene glycol	0.9[h]	9.2[cc]		
2,2,3-Trimethyl butane	1.0			420
Trimethyl amine	2.0	12		
2,2,4-Trimethyl pentane	0.95			415

Table 27.2 (Continued)

Combustible	Limits of Flammability (volume percent)		T_L	AIT
	L_{25}	U_{25}	(°C)	(°C)
Trimethylene glycol	1.7[d]			400
Trioxane	3.2[d]			
Turpentine	0.7[a]			
Unsymmetrical dimethylhydrazine	2.0	95		
Vinyl acetate	2.6			
Vinyl chloride	3.6	33		
m-Xylene	1.1[a]	6.4[a]		530
o-Xylene	1.1[a]	6.4[a]		465
p-Xylene	1.1[a]	6.6[a]		530

[a] T = 100°C.
[b] T = 47°C.
[c] T = 75°C.
[d] Calculated.
[e] T = 50°C.
[f] T = 85°C.
[g] T = 140°C.
[h] T = 150°C.
[i] T = 110°C.
[j] T = 175°C.
[k] T = 60°C.
[l] T = 53°C.
[m] T = 86°C.
[n] T = 130°C.
[o] T = 72°C.
[p] T = 117°C.
[q] T = 125°C.
[r] T = 200°C.
[s] T = 78°C.
[t] T = 122°C.
[u] T = 43°C.
[v] T = 145°C.
[w] T = 160°C.
[x] T = 96°C.
[y] T = 70°C.
[z] T = 29°C.
[aa] T = 247°C.
[bb] T = 30°C.
[cc] T = 203°C.

yet tested. No hard and fast calculations of limits are possible, but past experience provides guidelines, which may be used to obtain rough estimates (1–3) of unmeasured lower limits in air. For paraffin hydrocarbons containing four or more carbons, the lower limit, in volume percent, is about (107 ÷ molecular weight) and about 48 mg per liter of air at 25°C. Another approximation for paraffinic hydrocarbons in air at 25°C is that the lower limit equals 0.55 times the volumetric percentage of the combustible for stoichiometric combustion. Table 27.3 contains a list by type of combustible of ratios of the volumetric percentage of the lower limit to the stoichiometric volumetric percentage and of the upper limit to the stoichiometric percentage.

Other rule-of-thumb techniques are available for estimating whether a mixture is flammable. The following three criteria (11, 12) pertain to carbon–hydrogen–oxygen–nitrogen systems with an average molecular weight of likely combustion products of about 29. An enthalpy release of 300 to 350 cal/g, an exothermicity of about 10 kcal per mole of product, or an adiabatic flame temperature of 1500 to 1600 K are marginal conditions for potential sustained combustion in most C–H–O–N systems. Among the exceptions are systems with unusually low combustion initiation temperatures, which manifest cool flames and smoldering combustion.

Table 27.3 Average Ratios of Volumetric Percentages of Lower and Upper Limits of Flammability to Stoichiometric Composition with Air at 25°C (1, 3)

	Ratios	
Combustible	Lower/Stoichiometric	Upper/Stoichiometric
Hydrocarbons		
Paraffinic	0.55	2.9
Olefinic	0.52	3.3
Acetylenic	0.33	—[a]
Aromatic	0.53[b]	3.4[b]
Cyclic	0.55	3.3
Other combustibles		
Alcohols	0.51[b]	3.0[b]
Aldehydes and ketones	0.54	3.2[b]
Ethers	0.54	5.7
Acids and esters	0.54	2.7
Hydrogen and deuterium	0.15	2.5
Organic oxides	0.51	5.2
Nitrogen compounds	0.56	3.6
Sulfur compounds	0.53	4.4
Halogen compounds	0.67	2.3
Selected exceptions		
Ethylene	0.42	4.4
Acetylene	0.32	10.4
Acetaldehyde	0.51	7.4
Diethylether	0.55	10.8
Carbon monoxide	0.42	2.5
Ethylene oxide	0.39	13.0
Hydrazine	0.27[b]	5.8[b,c]
Carbon disulfide	0.19	7.7

[a] Pure acetylene can propagate flame at atmospheric pressure in tubes with diameters over 12 cm.
[b] Initial temperature in the range of 50 to 100°C.
[c] Corresponds to 100 percent hydrazine.

2.3.2 Binary and Multiple-Component Combustibles

Limits of flammability of combustible mixtures can be calculated using the measured or estimated (Section 2.3.1) limits of each combustible and the percentage of each combustible in the mixture by means of the Chatelier mixture rule. By no means are these estimated values for mixtures a substitute for actual measurement, but they are better than no appraisal of hazard limits. This mixture rule is based on the premise that if separate limit combustible mixtures are mixed, the resultant mixture will also be a limit mixture. The Le Chatelier mixture rule is expressed as follows:

$$L = \frac{100}{C_1/N_1 + C_2/N_2 + \cdots} \tag{2}$$

where C is the volumetric percentage of the combustible gas in the air-free and inert gas-free mixture (so that $C_1 + C_2 + \ldots = 100$), N is the respective lower or upper limit of flammability of the combustible in air, and L is the lower or upper limit of flammability of the mixture in air. For example, a mixture containing 80 volume percent methane, 15 volume percent ethane, and 5 volume percent propane has a lower limit in air of

$$\frac{100}{80/5.0 + 15/3.0 + 5/2.1} = 4.3$$

Flammable liquid mixtures obeying Raoult's law (i.e., the partial pressure of each component equals the vapor pressure of that component when alone, multiplied by its mole fraction in the mixture) can also be treated by the Le Chatelier mixture rule. Exceptions to the rule include mixtures containing considerable excess nitrogen over that in air, mixtures of hydrogen–ethylene, hydrogen–acetylene, hydrogen sulfide–methane, methane–dichloroethylene, methyl-ethyl chlorides, and mixtures containing carbon disulfide (2). Generally, the Chatelier mixture rule does well with mixtures containing components having about the same (within 10 percent) ratio of the limit to the stoichiometric volume percentages (Table 27.3). More information about using the mixture rule, particularly with respect to complex mixtures containing a little air or air vitiated with nitrogen or carbon dioxide can be found in Reference 1. It is apparent from these exceptions that measured limits of flammability are preferred to those estimated by calculation.

2.4 Limits of Flammability and Minimum Autoignition Temperature of Gases and Vapors in Oxygen

The flammability of combustible gases and vapors in oxygen or in atmospheres of oxygen-enriched air is a vital consideration in many instances—for example, in space flight vehicles, in the compressed gas industry, and in hospitals. The limits of flammability and minimum autoignition temperatures in Table 27.4 are compiled from References 2, 3, and 5.

Oxygen-enriched atmospheres pose not only an explosion problem but also a fire problem, as in hospital and hyperbaric oxygen chambers (13–15). The partial pressure of oxygen, better than the percentage of oxygen, is directly related to the increased hazard with respect to fire spread rate, ignition temperature, and minimum ignition energies of sparks.

Table 27.4 Limits of Flammability and Minimum Autoignition Temperatures of Gases and Vapors in Oxygen (2, 3, 5)

	Limits of Flammability (volume percent)		Minimum Autoignition Temperatures (°C)
Combustible	Lower	Upper	
Acetaldehyde	4	93	159
Acetic acid	≤5.4	—	490
Acetone	≤2.6	60[a]	485
Acetylene	≤2.5	100	296
Ammonia	15.0	79	—
Aniline	≤1.2	—	—
n-Amyl acetate	≤1.0	—	234
Benzene	≤1.3	30	—
Bromochloromethane	10.0	85	368
Bromodifluoromethane	29.0	80	453
Butane	1.8	49	278
Isobutane	1.8	48	319
Butene-1	1.7	58	310
Butene-2	1.7	55	—
Butyl chloride	1.7	52[a]	235
Carbon disulfide	≤1.3	—	107
Carbon monoxide	12.5	94	588
2-Chloropropene	4.5	54	—
Isocrotyl bromide	6.4	50	—
Isocrotyl chloride	2.8	66	
Cyclopropane	2.5	60	454
Deuterium	4.9	94	—
Dichloroethylene	10.0	26	—
Diethyl ether	2.0	82	182
Dimethyl ether	3.9	61	252
Divinyl ether	1.8	85	166
Ethane	3.0	66	506
Ethyl acetate	≤2.2	—	—
Ethyl alcohol	≤3.3	—	329
Ethyl bromide	6.7	44	—
Ethyl chloride	4.0	67	468
Ethylene	2.9	80	485
Ethylene chloride	4.0	68	470
Ethylene oxide	≤3.6	100	—
Ethyl-*n*-propyl ether	2.0	78	—
Gasoline (100/130)	≤1.3	—	316
Glycol	≤3.5	—	—
n-Hexane	1.2	52[a]	218

Table 27.4 (Continued)

Combustible	Limits of Flammability (volume percent)		Minimum Autoignition Temperatures (°C)
	Lower	Upper	
Hydrogen	4.0	95	542
Hydrogen sulfide	≤4.0	—	220
Kerosene	0.7	—	216
Methane	5.1	61	556
Methyl alcohol	≤6.7	93	461
Methyl bromide	14.0	19	—
Methyl chloride	8.2	66	—
Methylene chloride	11.7[a]	68[a]	606
Naphtha (Stoddard)	≤1.0		216
n-Octane	≤0.8	—	208
Propane	2.2	52	468
Propylene	2.1	53	423
n-Propyl alcohol	≤2.2	—	328
Propylene oxide	≤2.8	—	—
Isopropyl ether	2.2	69	—
Toluene	1.2	—	—
Trichloroethane	5.5[a]	57[a]	418
Trichloroethylene	7.5[a]	91[a]	396
Vinyl chloride	4.0	70	396

[a] Determinations made at about 100°C.

2.5 Flash Points of Liquids and Temperature Range of Flammability

The concepts of minimum temperatures (T_L) for ignition by a source hot enough and energetic enough, and the minimum autoignition temperature (AIT) have been introduced and tabulated in Table 27.2 when air is the oxidant, and in Table 27.4 when oxygen is the oxidant: T_L is the intercept of the curve of the lean flammability limit of the vapor in air as a function of temperature, and the curve of the vapor pressure of the liquid as a function of temperature. The T_L involves ignition sources such as flames and sparks, with ignition occurring very rapidly, in much less than a second. Combustion can also be started by slow heating, resulting in autoignition at temperatures corresponding to the AIT values in Tables 27.2 and 27.4; the time scale is longer, of the order of seconds and many minutes. Flash points are related to T_L in that they are an attempt to approximate T_L values. Closed-cup and open-cup techniques are used to obtain these flash points. It must be realized that these two types of flash point in most instances are probably somewhat too high to be absolutely the minimum temperature at which a liquid fuel will ignite. In both the open-cup and closed-up methods, a given volume of liquid is heated in a prescribed container at a definite rate, and periodically a test flame or spark is passed across the surface of the liquid. The temperature at which a flame passes across the surface of the liquid is taken as the flash point. The container is open to the air in the open-cup method and closed to the air in the closed-cup method. A small opening is uncovered in the latter instance as the test flame is introduced. Lower values are obtained by means of the closed-cup method than by the open-cup method. Values listed in Table 27.5 were obtained by the closed-cup method. A far more extensive compendium, listing materials by trade names, has been published by the National Fire Protection Association (16), and by other organizations for lubricants (17) and jet aircraft fluids (18). The great value of flash point data lies in the ease of measurement, the standardization of open-cup and closed-cup methodology, and the comparative classification of the hazard of flammability of each listed liquid with respect to the others; the higher the flash point of a liquid, the safer it is with respect to fires and explosions. However flash points are not the absolute minimum temperature at which flammable mixtures may be produced by the vapors from a liquid; the flash point may be a few degrees to about 12°C above the minimum flammable temperature (2). As the temperature of some liquids is raised toward the boiling point, the liquid's vapors remain flammable until an upper temperature limit corresponding to the upper concentration limit of flammability is reached, at which point the percentage of vapor in saturated air is so high that the mixture is incapable of propagating flame. If this mixture becomes diluted with additional air, the resultant mixture will be flammable.

2.6 Methods of Minimizing Explosions and Fires

2.6.1 Reducing the Oxygen Content of the Atmosphere

Oxygen present in a flammable atmosphere or in air feeding a fire may be reduced by dilution with inert gases such as nitrogen, carbon dioxide, steam, or combinations of these inerts in the form of exhaust gases from flues, automobile engines, or jet engines. Carbon dioxide and compressed nitrogen fire extinguishers are frequently used to put out fires and to smother initial kernels of explosions. If the fuel bed is hot, it must be cooled generally by large quantities of water, if reignition is to be prevented on dissipation of the applied inert gases. Halons (Freons) may also be used to lower the oxygen content to levels not supporting combustion. In some applications, special reagents and catalysts can reduce the oxygen content of a potentially explosive atmosphere. More generally, the explosion has started and there is very little time to add inert material fast enough to quench the progressing explosion before structural damage occurs; fast-acting valves, operated often by safely contained explosive activators, are often used to release very rapidly the explosion-quenching agent. Halons are frequently employed as quenching agents because they can chemically inhibit combustion as well as lower the available oxygen concentration. It must be remembered that no toxic gas or vapor is suitable for use as a diluent in situations where it may be inhaled; likewise, atmospheres deficient in oxygen may not be respirable.

Table 27.5 Flash Point of Combustible Liquids and Vapors in Air (2)

Name	Flash point (°C)
Acetal	37
Acetaldehyde	−38
Acetanilide	174
Acetic acid	40
Acetic anhydride	53
Acetone	−18
Acetophenone	105
Acetyl acetone	79
Acetyl chloride	4
Acrolein	<−18
Acrylonitrile	−5
Adipic acid	196
Aldol	83
Allyl alcohol	22
Allyl chloride	−32
Amyl acetate	25
Isoamyl acetate	25
Amyl alcohol	38
tri-Isoamyl alcohol	42
tert-Amyl alcohol	19
Amylbenzene	66
Amyl chloride	13
tert-Amyl chloride	−12
Amylene (*n* and β)	−18
Amyl ether	57
Amyl propionate	41
Aniline	76
Anthracene	121
Benzaldehyde	64
Benzene	−11
Benzoic acid	121
Benzyl acetate	102
Benzyl alcohol	101
Benzyl benzoate	148
Benzyl chloride	60
Bromobenzene	65
Butyl acetate	22
Isobutyl acetate	18
Butyl alcohol	29
Isobutyl alcohol	28
sec-Butyl alcohol	21
tert-Butyl alcohol	11
Butylamine	8
Isobutylamine	−9
Butyl bromide	13
Butyl carbitol	78
Butyl cellosolve	61
Butyl chloride	−28
Butyl formate	18
Butyl propionate	32
Butyl stearate	160
Butyraldehyde	−18
Butyric acid	77
Carbon disulfide	−30
Cellosolve	40
Cellosolve acetate	51
Chlorobenzene	29
o-Cresol	81
m-Cresol	86
m-*p*-Cresol	86
Croton aldehyde	13
Cumene	39
Cyanamide	141
Cyclohexane	−17
Cyclohexanol	68
Cyclohexanone	64
p-Cymene	47
Decalin	58
Decane	46
Diisobutyl carbinol	—
Diisobutyl ketone	48
o-Dichlorobenzene	66
1,2-Dichloro-*n*-butane	52
sec-Dichloroethylene	14
Dichloroethyl ether	55
Diethanolamine	138
Diethylamine	<−18
Diethylene glycol	124
Dimethyl aniline	63
2,3-Dimethylbutane	<−29
Dimethyl ether	−41
Dimethyl formamide	57

Table 27.5 (Continued)

Name	Flash point (°C)
Dioxane	12
Dipentene	45
Dipenyl	113
Diphenylamine	153
Diphenylmethane	130
Dodecane	74
Ethyl acetate	−4
Ethyl alcohol	13
Ethylamine	<−18
Ethyl aniline	85
Ethylbenzene	18
Ethyl butyrate	26
Ethyl chloride	−50
Ethylene chlorohydrin	60
Ethylene dichloride	13
Ethylene glycol	111
Ethylene oxide	<−18
Ethyl ether	−45
Ethyl formate	−20
Ethyl lactate	46
Ethyl propionate	12
Formaldehyde	54
Furfural	75
Furfural alcohol	75
Gasoline, regular	−44
Glycerine	160
Heptane	−4
Hexane	−26
Isohexane	<−29
Hexyl alcohol	58
Hexyl ether	77
Hydrogen cyanide	−18
Isophorone	84
Isoprene	−54
Kerosene	38–74
Methyl acetate	−10
Methylal	−18
Methyl alcohol	12
Methyl amine	−18
Methylbutyl ketone	35
Methyl cellosolve	41
Methyl cyclohexane	−4
o-Methylcyclohexanol	68
Methyl ethyl ketone	−1
Methyl formate	−19
Methyl lactate	49
Methyl propionate	−2
Methyl propyl ketone	16
Methyl salicylate	101
Naphtha	−7–42
Naphthalene	80
Nitrobenzene	88
Nonane	31
Octane	13
Isooctane	−12
Oil	
Castor	229
Coconut	216
Corn	254
Cottonseed	252
Creosote	74
Fish	216
Gas	66
Lard	184
Linseed, raw	222
Linseed, boiled	206
Lubricating spindle	
Lubricating, turbine	204
Menhaden	224
Mineral seal	77.
Neat's-foot	243
Olive	225
Palm	162
Paraffin	229
Peanut	282
Pine	78
Pine tar	62
Rape	163
Rosin	130
Soybean	282
Sperm	220
Straw	157
Tallow	256
Transformer	146

Table 27.5 (Continued)

Name	Flash point (°C)
Tung	289
Turkey-red	247
Whale	230
Paraffin	199
Paraldehyde	17
Paraformaldehyde	70
Pentane	−40
Isopentane	<−51
Petroleum ether	−56
Phenol	79
Phthalic anhydride	152
Pinene	33
Propyl acetate	14
Isopropyl acetate	4
Propyl alcohol	15
Isopropyl alcohol	12
Isopropylamine	−26
Propylbenzene	30
Isopropylbenzene	39
Propyl chloride	<−18
Propylene diamine	22
Propylene dichloride	16
Propylene glycol	97
Propylene oxide	−37
Isopropyl ether	−28
Propyl formate	−3
Isopropyl formate	−6
Pyridine	23
Stoddard solvent	38–43
Stearic acid	196
Styrene	31
Tetradecane	100
Tetrahydrofurfural alcohol	75
2,2,3,3-Tetramethyl pentane	—
Toluene	4
o-Toluidine	85
p-Toluidine	87
Trichloroethylene	—
Triethylamine	−7
Triethylene glycol	177
Trimethylamine	—
Trimethylbenzene	—
2,2,4-Trimethylpentane	−12
Trioxane	45
Turpentine	35
Valeric acid	—
Valeric aldehyde	—
Vinyl acetate	−8
o-Xylene	17
m-Xylene	25
p-Xylene	25
o-Xylidine	97

Table 27.6 gives the minimum oxygen concentrations required to support flames of combustible gases and vapors, when the reduction of oxygen concentration is brought about by the addition of nitrogen or carbon dioxide. The critical oxygen concentrations for any combustible vary with the concentration of the combustible present; the oxygen concentrations in Table 27.6 are minimal, therefore covering any concentration of the combustible that may be present.

Figure 27.4 illustrates the relation between the concentrations of combustible and the minimum oxygen percentage for flammability when nitrogen or carbon dioxide is the diluent. The graph shows the flammable areas of possible relevant mixtures of butadiene, air, and added nitrogen or carbon dioxide. The line *a–d* represents mixtures of butadiene and pure air. Point *b* on this line is the lower limit of flammability (2.0 percent) and point *c* is the upper limit of flammability (11.5 percent); mixtures between

Table 27.6 Oxygen Percentages Below Which Flames of Combustible Gases and Vapors Are Extinguished (2)

Name	Oxygen Percentage Below Which No Mixture Is Flammable	
	Nitrogen as Diluent in Air	Carbon Dioxide as Diluent in Air
Methane	12.1	14.6
Ethane	11.0	13.4
Propane	11.4	14.3
Butane	12.1	14.5
Isobutane	12.0	14.8
Pentane	12.1	14.4
Isopentane	12.1	14.6
Hexane	11.9	14.5
2,2-Dimethylbutane	12.1	14.7
Heptane	11.6	14.2
Natural gas	12.0	14.4
Gasoline	11.6	14.4
Ethylene	10.0	11.7
Propylene	11.5	14.1
Butene-1	11.4	13.9
Butene-2	11.7	14.0
Isobutylene	12.1	14.8
3-Methylbutene-1	11.4	13.9
1,3-Butadiene	10.4	13.1
Benzene	11.2	13.9
Cyclopropane	11.7	13.9
Methyl alcohol	9.7	11.9
Ethyl alcohol	10.6	13.0
Dimethyl ether	10.3	13.1
Diethyl ether	10.3	13.2
Acetone	11.6	14.3
Methylisobutyl ketone	11.4	—
Methyl formate	10.1	12.5
Ethyl formate	10.4	12.8
Methyl acetate	10.9	13.6
Isobutyl formate	12.4	14.8
Methylamine	10.7	—
Allyl chloride	12.6	15.1
Hydrogen	5.0	5.9
Carbon disulfide	5.4	7.6
Carbon monoxide	5.6	5.9

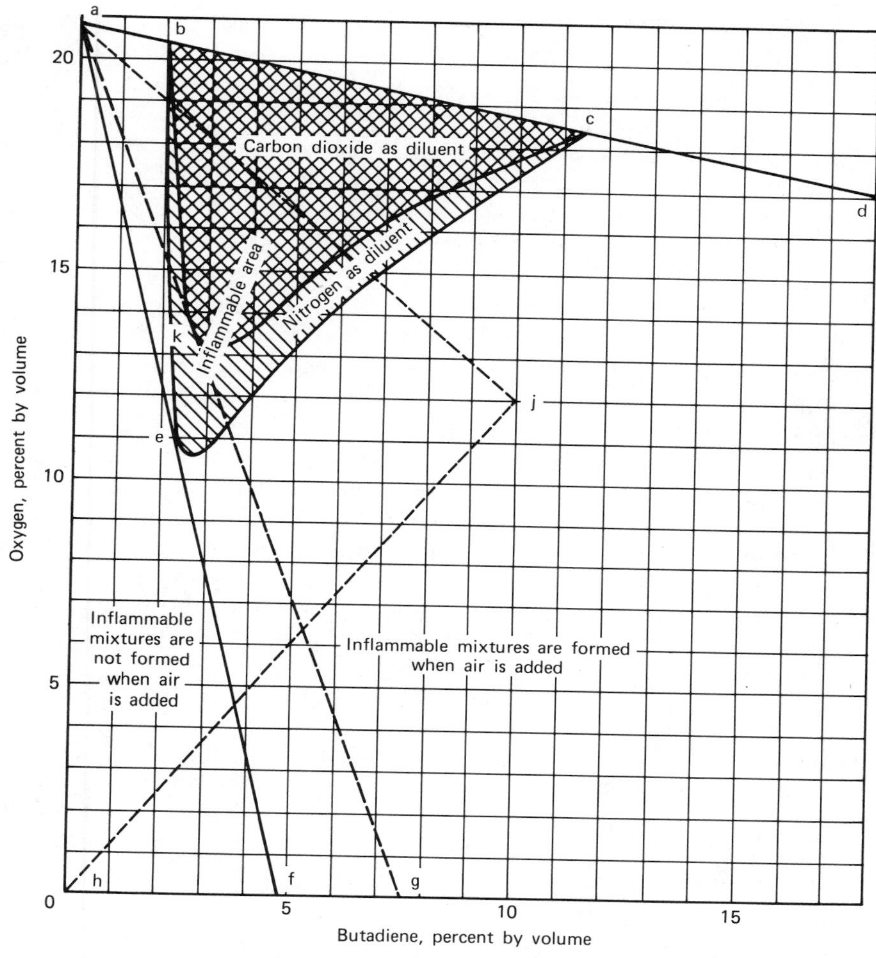

Figure 27.4 Flammability of butadiene, air, added nitrogen, and added carbon dioxide (2). Graph courtesy of Bureau of Mines, U.S. Department of the Interior.

b and *c* are flammable. As the oxygen concentration is lowered by the dilution with carbon dioxide (double-cross-hatched area) or with nitrogen (single-hatch area), the spread in composition between lower and upper limits narrows until the close or nose of the respective hatched area is reached, namely, at 10.4 percent oxygen when nitrogen is the diluent (point *e*) and at 13.1 percent oxygen when carbon dioxide is so employed (point *k*). The mixture that will propagate flame with a minimum of oxygen present (added nitrogen being the diluent) contains 2.6 percent butadiene. If the mixture contains 5.0 percent of butadiene, Figure 27.4 indicates that the oxygen concentration needs to be reduced with added nitrogen to 13.2 percent. A caution is in order; Figure 27.4 shows that if this mixture containing 5.0 percent butadiene and 13.2 percent oxygen is diluted with air, the nonflammable mixture becomes flammable. Only mixtures diluted with excess nitrogen falling along the line *a–e–f,* or diluted with excess carbon dioxide falling along the *a–k–g,* can be diluted with air and remain nonflammable.

A frequent problem is that of flushing out equipment containing a nonflammable gaseous mixture, without running through the flammable range. This operation can be accomplished successfully by using information such as that contained in Figure 27.4. Consider point *j,* which is a nonflammable mixture containing 12 percent oxygen, 10 percent butadiene, and 78 percent nitrogen. To stay out of the flammable area *b–c–e,* the oxygen content must be reduced to a value below 10.4 percent to pass safely from the composition of point *j* to compositions given by any point to the left of line *a–e–f.* Nitrogen can be added until the oxygen content is reduced to less than 4.6 percent, thereby shifting composition along the line *j–h.* When a mixture thus obtained is diluted with air, the composition will move along the line *a–e–f,* or to the left of it, and the mixture always will be nonflammable. At this state, air can be added to the container and the combustibles swept out without danger of explosion. Figure 27.4 applies only to butadiene. Similar diagrams for other flammable gases and vapors and dealing with the use of water vapor as the diluent appear in References 1 and 3. An extensive discussion in Reference 3 of the flushing of gasoline vapors with water vapor is particularly noteworthy; the logic is that just used to discuss the flushing of butadiene from a container.

2.6.2 Operating Outside the Range of Flammability

If possible, the concentration of combustibles should be kept below their lower limits of flammability, because any ingress of air into the mixture will not make the mixture flammable. If the concentrations of combustible must be above the lower limit of flammability, explosion hazards can be avoided if the combustibles are above the upper concentration limit of flammability; in this situation danger does arise when additional air finds its way into the mixture. With liquid combustibles, temperature control is another means of seeking safety. As Figure 27.3 indicates, temperature can be held low enough (T_L) that the corresponding low vapor pressure of the liquid does not produce a saturated vapor–air mixture with enough concentration of the combustible vapor to match the percentage at the lower limit of flammability; or more crudely, the temperature of the liquid's flash point is never reached for safety purposes. Nonflammability will be gained again when the temperature is raised to correspond to a vapor pressures equal to and above that corresponding to the concentration of the combustible at its upper limit of flammability (T_u, in Figure 27.3). It is noted again that combustion can occur at and above the upper temperature limit, if the air space is not saturated or if fresh air is admitted.

2.6.3 Use of Less Flammable Materials and Chemical Flame Inhibitors

Various halogenated hydrocarbons called Halons or Freons by their manufacturers can be used to advantage where nonflammable liquids are required. Not only are they nonflammable, but in many applications relatively small concentrations of these materials added to flammable mixtures can render the overall mixture nonflammable. Thus, for example, the National Fire Protection Association publication NFPA No. 12A (19) lists 2.0 volume percent of Halon 1301 (trifluorobromomethane) in the mixture as enough to inhibit the combustion of any methane–air mixture; other sources cite about 3.2 percent (20) or 4.7 percent (21). Whichever of the three values is accepted, the effectiveness of this agent exceeds that of carbon dioxide, since about 24 percent carbon dioxide is required to render all methane–air mixtures nonflammable (1).

2.6.4 Elimination of Ignition Sources

Combustible gases and vapors, mixed with an oxidant such as air or pure oxygen in proportions to give flammable mixtures and at temperatures below their AIT, will ignite if exposed to an ignition source. The atmospheres in the gasoline tanks of millions of automobiles and trucks on our road contain flammable mixtures, or mixtures that are not flammable because they are above the upper limit for gasoline but become flammable when air is admixed. Yet ignitions are rare because sources of ignition are rarely present in the tank or around the filler cap. Another way of describing ignition hazard is to say that ignitions of a fuel–gas or vapor–air mixture can occur only if the mixture composition is within the flammability range and a critical volume of that mixture is heated sufficiently to produce an exothermic chemical reaction capable of propagating flame beyond the initiation zone. Ignition sources are therefore of prime interest in safety, and representative ignition sources are as follows:

1. Electrical sparks and arcs, generated by static electricity, electrical shorts, or lightning.
2. Flames such as open pilot or burner flames, matches and cigarette lighters, flames in space heaters, cooking stoves, hot water heaters, and so on, burning materials, and incinerators.
3. Hot surfaces such as frictional sparks, incendiary particles, heated wires, rods or fragments, hot vessels or pipes, glowing metals, hot cinders, overheated bearings, and breaking electric light bulbs to expose hot filaments.
4. Hot gases brought about by shock compression, adiabatic compression, and hot gas jets.
5. Lasers.
6. Pyrophoricity or catalytic reaction.
7. Self-heating or spontaneous combustion.

Often temporal and spatial characteristics of heat sources are useful in evaluating ignition sources. In the case of an electrical spark, the ignition source is highly localized in space, and the duration of heating can be as little as a few microseconds; here temperatures are very high. However ignition is determined by the amount and rate of energy input into the flammable mixture. Lasers fall into this category when the medium readily absorbs the laser radiation. In contrast, a heated vessel represents a much more distributed ignition source spatially, and the duration of heating can extend to several minutes. Here temperature, rather than energy, is the critical factor and it is tied to the heating duration. Temperature for ignition is minimal when heating duration or ignition lag is so long that the temperature for ignition does not decrease with further increase in heating time; these are the AITs in Table 27.2. Both temperature and heating rate tend to be important with small hot surface sources as heated wires, hot metal fragments, and frictional sparks. Generally, ignition temperatures increase with a decrease in size or surface area of the heat source. Incendiary ignitions are a case of hot surface particle ignition augmented by the chemical reactivity of the heat sources. Hot gas ignitions also differ in their temporal and spatial characteristics, with shock waves being most localized in space, and heating time being less than milliseconds; adiabatic compression heating is more spatially distributed, extends over a few seconds, and can produce much lower temperature ignition than is possible by shock compression. Self-heating or spontaneous combustion involves slow oxidation or decomposition of the combustibles and requires hours or even days for ignition to occur. Pyrophoric or catalytic ignitions involve the combustible reacting with moisture or oxygen in the air; ignition lags are of the order of seconds.

Sources of ignitions just discussed can be controlled by proper safety practices, such as by using flameproof and vaporproof electrical equipment, and building industrial plants so that boilers, water heaters, incinerators, and other equipment with open flames and incandescent materials are isolated at safe distances from operations involving flammable materials. Static electricity is one of the most difficult ignition sources to control. Static electricity is a more serious hazard in dry atmospheres where the relative humidity is below 50 percent; where practical, it is advisable to maintain room humidity above 50 percent to reduce static sparks. Friction due to such situations as slipping belts or pulleys, revolving machinery, and passage of dust, pellets, liquids, or gases at high velocity through small openings, can generate static electricity, and so may processes of impact, cleavage, or induction. Proper grounding of machinery, pipes, and even personnel by means of semiconducting floors (9) is advisable, if charges may accumulate.

2.6.5 Segregation of Hazardous Operations

Operations involving large volumes of flammable gases or vapors should be separated from other operations. This requires installation of hazardous operations in building at a safe distance from others and suitably diked to avoid flow of hazardous liquids and vapors toward other structures in the event of spillage. Safe distance criteria are hard to define. Some guidance in this respect can be gotten from quantity–distance tables for liquid hydrogen developed by the Bureau of Mines (22), the Compressed Gas Association (23), and the Armed Services Explosives Safety Board (24). More definitive work is needed in this highly important area.

2.6.6 Adequate Ventilation

Adequate ventilation is necessary where flammable gases and vapors are used and transported—in the spaces around equipment in which the combustibles may be present for example, and in conduits, trenches, and tunnels carrying pipelines. Facile inspection for leaks should always be feasible. Whenever possible, operations should be carried out in the open air, as is done in certain processes in the petroleum industry, with closed structures being used to house instruments and equipment that must be protected from the weather.

2.6.7 Combustible Gas Indicators

Combustible gas indicators or recorders should be installed in all locations where hazardous concentrations of combustibles may occur, and if large locations are involved, sampling should be from different sites in the volume at risk. Flammables must not be lost to the sampling tubing walls because of improperly selected or maintained sampling lines, and temperature should be high enough to avoid condensation if heavy combustible vapors are being tested. Considerable progress has been made in combustible gas indicators, as, for example, those for hydrogen (25), methane (26), and several other gases (27).

2.6.8 Venting Explosions and Flame Arresting

In addition to procedures discussed under Sections 2.6.1 to 2.6.7, an installation in which an explosive mixture could conceivably be present should be protected by adequate release diaphragms for venting a possible explosion before destructive pressures develop. To properly protect an installation against such damage or destruction, determination must be made of the type and concentration range of combustibles that may be present, and the maximum pressure the enclosure can withstand versus the maximum pressure an explosion could develop. Then the engineer must determine the kind of material, area, thickness, and location of a release diaphragm or device that is capable of rupturing at a pressure safely below the maximum pressure the enclosure can withstand. For specific information on release or vent devices, the reader is referred to References 2 and 28 to 30. Such data are largely empirical and were obtained by means of experiments in relatively small vessels. Maisey (28) reviews simple theoretical relationships for the venting of closed vessel explosions of gas–air mixtures; these relationships are between flame speed, pressure and time, and the volume in which the explosion is occurring. Maisey (29) also discusses the similarities and differences between gas–air and dust–air explosions. He treats dust–air explosion venting by analogy to gas–air explosions and provides relationships predicting the characteristics of dust explosions and their venting. His starting points are measurements in equipment such as the Hartmann apparatus (Figure 27.5), the Godbert-Greenwald furnace, and related apparatus (31, 32); some 2500 materials, including dusts of plastics, chemicals, agricultural materials, carbonaceous materials, and metals, have been evaluated for their explosibility using such apparatus (33, 34).

The spread of explosion between physically connected equipment such as from flare stacks disposing of waste gases, ducting systems carrying solvent vapors to a recovery unit, pipes carrying gases to burners or furnaces, and exhaust pipes of internal combus-

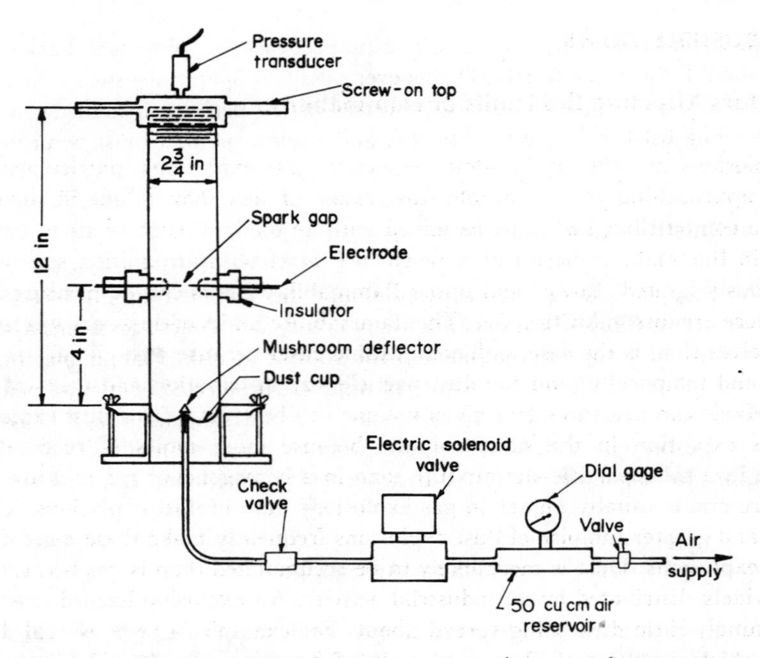

Figure 27.5 Hartmann tube, No. 2 (32). Schematic courtesy of Bureau of Mines, U.S. Department of the Interior.

tion engines operating in flammable atmospheres, can be prevented by means of flame arrestors or flame traps. These consist of an assembly of narrow passages or apertures through which gases or vapors can flow, yet small enough to quench any flame attempting to pass through. Such flame traps are generally made of wire gauze, crimped metal, sintered metal, perforated metal plates, or pebble beds. They are often used in conjunction with pressure release vents. The flame trap restricts the explosion from passing from one vessel to another but does not reduce the pressure developed by the explosion. The release vent can keep the pressure down but cannot prevent the explosion from propagating throughout a plant. A combination of both can therefore be used to limit the spread and effects of an explosion. Data on flame arrestors are given in References 34 and 35.

In spite of an extensive literature on explosion venting and flame arresting, the behavior of an explosion, once started, is not yet sufficiently predictable to make measures to combat an explosion entirely reliable; prevention is better than the cure (29).

3 COMBUSTIBLE DUSTS

3.1 Factors Affecting the Limits of Flammability

Dust explosions are similar in some respects to gas explosions, particularly for dust particles approaching the respirable dust range of less than 5 μm in diameter. To explode, a combustible dust must be mixed with an oxidant, such as air or oxygen, and must be in flammable concentration when in contact with an ignition source or when spontaneously ignited. Lower and upper flammability limits can be measured for most dusts. There are dissimilarities, too. The flammability limits of dusts are less meaningful by themselves than is the case with combustible gases because dust clouds are unstable, spatially and temporally, and the dust particle size distribution and dust volatility are factors. Maximum pressures in a given volume can be greater for a dust explosion than for a gas explosion in the same volume, because more moles of reactants can be contained in a solid particle–air mixture than in a homogeneous gas mixture. The rate of pressure rise is usually higher in gas explosions than in dust explosions. The longer duration and greater impulse of dust explosions frequently make these more destructive than gas explosions. Dust is more likely to be accumulated than is gas because dust can become widely distributed by an industrial activity. An explosion hazard can exist with even seemingly little dust being spread about. For example, a layer of coal dust about 0.002 cm thick, on the roof, floor, and walls of a passageway about 3 meters wide × 2 meters high is enough dust to form a stoichiometric mixture with air in the passageway. The existence of a dust cloud may be an inherent feature of a manufacturing process, such as within a grinding mill or a pneumatic transport system; flammable concentrations may occur, as in equipment handling flammable gases and air.

Any process handling combustible dust-producing stock or dusts directly should be considered as a possible source of a destructive explosion or fire, or both. Parameters strongly affecting the violence of a dust explosion are briefly discussed next.

3.1.1 Chemical Composition of Dust

The chemical composition of a combustible dust determines its oxygen requirement for complete or stoichiometric combustion and its ease of oxidations as reflected by its ignition temperature and burning rate. The chemical composition also determines the heat released on combustion, and the volumes of gaseous products of combustion have a direct bearing on the maximum pressure developed during an explosion. The gases remaining after the combustion of metal powders are chiefly nitrogenous, the oxygen of the air involved in the combustion being in the solid oxides formed; these remaining gases occupy less space at a given temperature than the initial air. For the combustion of metal powders resulting in solid oxides, therefore, the pressure developed is due to the expansion of nitrogen by the heat of combustion. On the other hand, when sugar or starch dust burns in air, the volume of gaseous combustion products, carbon dioxide, nitrogen, and water vapor exceeds that of the initial air. Accordingly, the maximum

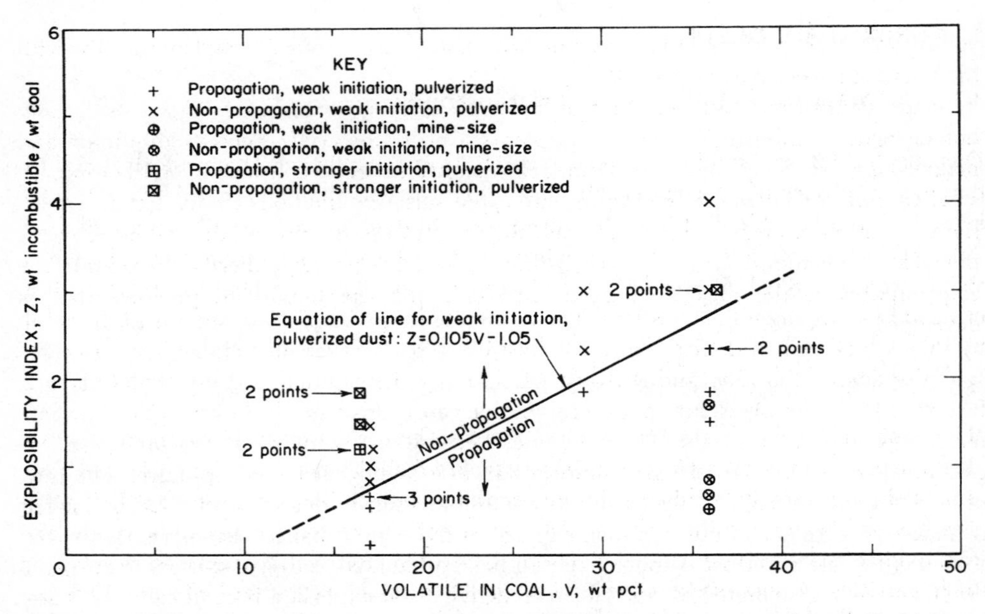

Figure 27.6 Explosibility index versus volatile in coal (36). Graph courtesy of Bureau of Mines, U.S. Department of the Interior.

pressure developed is due to a combination of increased volume of gases and temperature developed by the combustion.

Many dusts contain volatile matter, which can be evolved below the ignition temperature of the dust. This generally increases the combustion tendencies of the dust, as for example with coal dust (2). Figure 27.6 plots the explosibility of various coals, increasing with increasing volatile matter in the coal. (Regarding Figure 27.6, note that pulverized coal contains dust particles whose size is–100 mesh, and 75 to 80 percent is –200 mesh, and mine size coal is all –20 mesh, with 20 percent being –200 mesh.) The data in Figure 27.6 indicate that coal dust containing less than 10 percent volatiles will not propagate explosions. Indeed, anthracite coal with about 12 percent volatiles shows little if any explosibility (2, 36). Early investigators of coal dust explosions in mines were well aware that the explosibility of coal is directly related to the coal's content of volatile matter, and more recent studies (37, 38) continue to demonstrate that this is true. Speculation leads to the postulate that in mine scale coal dust explosions, volatiles are pyrolyzed out of the dust in the preheat zone of the explosions, and homogeneous gas phase combustion follows after the volatiles interdiffuse with air to form a more or less continuous unburned combustible gas–air zone: the residual coked coal particles burn at a later stage. This is the so-called predistillation theory of the ignition of pyrolytic dusts. However there exists evidence to support contrary speculation that in large-scale explosions of pyrolytic dusts, combustion proceeds heterogeneously, with the volatiles of each particle burning while still associated chemically and physically with the burning parent particle (39, 40).

Some combustible dusts vaporize on heating, and their explosions involve essentially homogeneous combustion. The boiling points of these materials, such as aluminum and magnesium, are lower than their flame temperatures, and their heats of vaporization are smaller than the energy needed to initiate rapid surface oxidation. Under certain conditions, aluminum, magnesium, iron, lead, thorium, titanium, zirconium, and other metal powders can combine with nitrogen in the air to form nitrides, thereby increasing the consumption of the dust during an explosion and the maximum pressure that is developed.

3.1.2 Fineness and Physical Structure of Dust

The explosibility of dusts generally increases with decreasing particle size because small dust particles can be thrown into suspension more readily; moreover, they disperse more uniformly and remain in suspension longer than coarse dust. Also, since smaller particles have a greater specific surface, they can absorb more oxygen per unit weight and can oxidize more rapidly than larger ones. Greater electrical charges can develop on finer particles because of greater electrical capacitance. The combined effect of these factors is a lower ignition temperature and a lower ignition energy (Figure 27.7). A lower

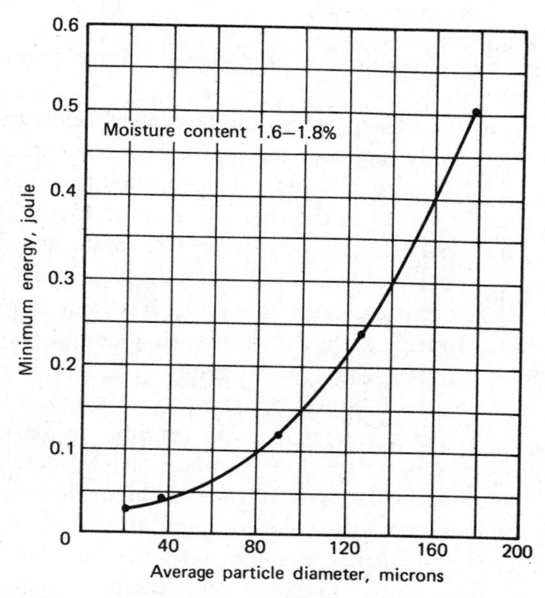

Figure 27.7 Effect of particle size on the minimum energy required to ignite dust clouds of cornstarch in air by electrical sparks (2). Graph courtesy of Bureau of Mines, U.S. Department of the Interior.

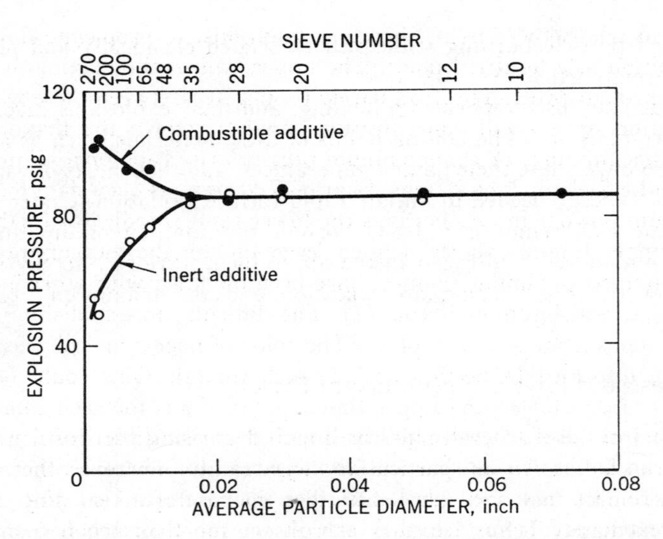

Figure 27.8 Effect of particle diameter of additive at 1.0 oz/ft^3 concentration on explosion pressure of cornstarch at 0.5 oz/ft^3 concentration (41). Graph courtesy of Bureau of Mines, U.S. Department of the Interior.

minimum explosive concentration, faster flame speed, and as shown in Figure 27.8, higher maximum pressures, are also observed. An extreme example of the effect of fineness on flammability is the pyrophoric combustion of very finely divided powders such as those of magnesium, lead, iron, and uranium. In addition to fineness, particle shape and surface characteristics resulting in differences in surface area, density, adsorbed gas, or protective surface layers can affect flammability. For example, flat, thin aluminum and magnesium particles produced by stamping are more flammable than nearly spherical particles of the same mesh produced by atomization or milling. These effects have been reported by Hartmann and co-workers (2), as well as by others.

Fineness and physical structure of dust enters into the aerodynamics of formation of dust clouds. These characteristics as they pertain to coal dust explosion in mines have been studied by Singer and co-workers (42, 43) and by Rae (44). Minimum velocities of about 5 to 30 m/sec were reported for coal and rock dust particles in the range of −74 to −14 μm dust. For very thin layers, the lowest required air velocity was observed around 30 μm; for smaller particles the minimum air velocities increased with decreasing particle size because of increasing immersion of the particles in the viscous boundary layer of the airflow (42).

3.1.3 Concentrations of Dust Clouds

There exist measured lower flammability limits and upper flammability limits of dusts (2). The lower limit is the minimum uniform dust concentration in an oxidant, generally air, in which there is just barely enough dust to propagate flame through a cloud after ignition at a localized point. The upper limit is the maximum uniform dust concentration through which flame can barely propagate. Between these two limits there is a concentration or a small concentration range at which the maximum pressure developed by that mixture, or the maximum pressure rise rate, or a related parameter, maximizes for the fuel–oxidant system. Figure 27.9 presents such data for the combustion of aluminum powder in air. Perhaps the lower limit is indicated by these data, but certainly the upper limit is poorly defined, even though the measurements extend to eightfold stoichiometric. Similar findings have been obtained with other metal powders, metallic hydrides, and bituminous coal (2). The difficulty in establishing flammability limits of dusts stems from several sources. The role of fineness and physical structure of dust have been discussed in Section 3.1.2; such complications could be handled by developing a family of lower and upper flammability limits for each combustible dust. But the uncertainty level of flammability limits of combustible dusts stems from the transient nature of dust clouds, particularly with respect to local concentrations in a cloud as a function of time and space. Consider the case of a coal dust explosion in a coal mine. Immediately before, dust is at rest on the floor, roof, and walls of the passageways. The initial ignition may be due to an electric or frictional spark or to an explosive shot that ignites a relatively small pocket of flammable methane–air mixture (firedamp), which may explode with relatively little violence. If this quantity of methane were the only fuel available, the damage caused by such an explosion would be relatively slight and localized. However with coal dust present, such a small explosion can generate an airflow capable of lifting coal dust into the air and can ignite the resulting cloud. Then the coupled processes of lifting dust ahead of the propagating explosion,

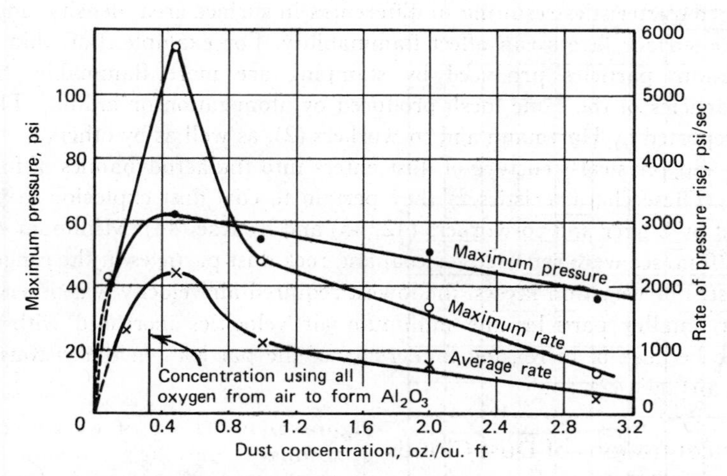

Figure 27.9 Maximum pressure and rates of pressure rise developed during explosions of aluminum powder at various concentrations (2). Graph courtesy of Bureau of Mines, U.S. Department of the Interior.

and propagation into the freshly formed cloud, can continue as long as coal dust is available. Dust concentration ahead of the combustion front builds until the dust supply is nearly consumed, and dust concentration is not uniform during a large-scale dust explosion. Even if the nominal or average concentration is below a lower flammability limit of the dust, the real concentration in part of the passageway can be above this limit, making flame propagation feasible. Likewise if the nominal or average concentration is above an upper flammability limit of the dust, the real concentration in part of the passageway can be low enough to support flame propagation. Such observations have been made in the Experimental Mine of the U.S. Bureau of Mines during explosions, with flame filling the passageway in strong explosions and not reaching the walls in weak explosions (40).

3.1.4 Composition of Atmosphere

The partial pressure of oxygen or the percentage of oxygen in the atmosphere carrying a dust cloud influences the flammability of the mixture; usually if the initial pressure is atmospheric, the percentage of oxygen is the parameter followed. Dust clouds ignite at lower temperatures, their minimum ignition energy is less, and the concentration at the lower flammability limit is less in oxygen than in air. Pressure rise rates and final pressures reached by explosions in closed vessels are generally higher in oxygen than in air. Vitiated air, with less oxygen than in normal air, generally leads to decreased ease of ignition as evidenced by the increase in minimum ignition energy of aluminum dust clouds (Figure 27.10). Such trends are also encountered in fires of solid fuels and are a

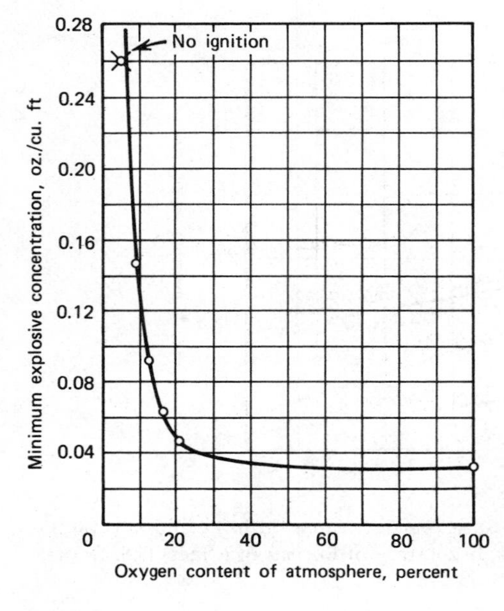

Figure 27.10 Effect of oxygen content of atmosphere on the minimum explosive concentration of atomized aluminum dust clouds (2). Graph courtesy of Bureau of Mines, U.S. Department of the Interior.

crucial factor in the safe operation of hyperbaric chambers (14, 15). Furthermore, with some dusts there exists a critical maximum oxygen concentration below which flame will not propagate. For example, the limiting oxygen concentration for flame propagation of bituminous coal dust clouds is between 16 and 17 percent oxygen (45). It must be remembered that elimination of oxygen does not prevent flame propagation of all dusts, since a number of metal powders including zirconium, magnesium, titanium, thorium, and some magnesium-aluminum alloys can be ignited by electric sparks and will propagate flame when dispersed in carbon dioxide. Diluents such as nitrogen, carbon dioxide, water vapor, argon, and helium vary in their effectiveness because of differences in specific heat, thermal conductivity, and thermal molecular dissociation. Atmospheric humidity introduces relatively minor concentrations of water vapor into a dust cloud, thus appearing to have little direct effect on the flammability of dust clouds. However high humidity favors agglomeration of particles and reduces their dispersibility. More important, high humidity also promotes leakage of static electricity and thereby reduces the hazard of ignition by static electrical sparks.

3.1.5 Effects of Ignition Source and Explosion Chamber

Dust clouds in contact with hot surfaces for long periods of time can often be ignited at lower temperatures than when exposed to nearly instantaneous contact. Large ignition sources are generally more effective than small ones. Electric sparks and arcs are less effective than open flames and hot surfaces; data in Tables 27.7 and 27.8 indicate that ignition in a furnace at 700°C is more readily achieved than by sparking and that layers of dust are ignited at lower temperatures than clouds of dust. However the temperatures in Table 27.8 cannot be regarded as fundamental properties of the fuel–oxidant mixture to the degree that minimum autoignition temperatures (AIT) and minimum ignition temperatures T_L in Table 27.2 may be regarded as fundamental properties of the mixture; the data in Table 27.8 are apparatus dependent and mixture dependent. Nevertheless, since the equipment and procedures were fairly constant throughout the many years of the investigations, the data are internally sound and constitute important guides on the relative explosibility of dusts examined. Ignition sources can strongly influence the possibility of ignition, the amount of inerting agent needed to prevent flame propagation, and the intensity of an explosion, particularly in large-scale chambers and passageways such as in coal mines (36, 44, 46). Turbulence in fast-flowing dust clouds leads to more violent explosions, as in a grinding mill, than in a slow-moving dust cloud of the same composition (2).

3.2 Laboratory Studies of Dust Explosibility

Potential explosion hazards of about 2500 dusts have been measured by means of laboratory tests and reported in a series of Bureau of Mines publications, including reports on the explosibility of agricultural (47), carbonaceous (48), chemical (49), metallic (50), plastic (51), and miscellaneous dusts (33). In addition, similar data have been published

Table 27.7 Description, Moisture Content, Particle Size, and Explosibility Index of Dusts (33)

Line No.	Moisture Content (percentage)	Particle size, (percentage finer than micrometer size)	Explosibility Index	Materials
		Agricultural Dusts		
1	—	80/74	—	Coconut shell charcoal (2 percent moisture, 18 percent volatile matter, 78 percent fixed carbon, 2 percent ash).
2	—	98/74	—	Coconut shell dust, raw
3	0	95/74	>10	Cork dust, drom factory dust collector
4	—	100/297	—	Cornstarch, 10 percent; stable metal oxide, 90 percent
5	—	100/297	—	Cornstarch, 20 percent; stable metal oxide, 80 percent
6	—	100/297	—	Cornstarch, 30 percent; stable metal oxide, 70 percent
7	—	100/297	—	Cornstarch, 50 percent; stable metal oxide, 50 percent
8	—	100/297	0.6	Cornstarch, 70 percent; stable metal oxide, 30 percent
9	0.6	100/44	6.3	Dextrin, U.S. Pharmacopoiea grade
10	4.9	42/74	—	Grain, mixed (oats, barley, wheat), with addition of linseed oil, cake meal, soybean oil, fish meal, bone meal, limestone, and salt
11	0	77/44	0.2	Grass seed
12	2.0	100/74	—	Horseradish
13	—	100/250	—	Rockweed, airdried
14	—	53/74	—	Wheat bran, with organic reducing material
		Cellulosics		
15	3.0	100/840	—	Cellucotton, coarse broke waste
16	4.5	100/74	—	Cellucotton, fluff, normal product
17	4.2	100/840	—	Cellucotton, from suction duct
18	4.0	100/840	—	Cellucotton, from beams and pipes in storeroom
19	—	70/74	—	Cellucotton, from dryer discharge
20	—	98/74	—	Cellucotton, contains paper, clay, starch, wood pulp, alum, casein, and color dyes, from balcony

Table 27.7 (Continued)

Line No.	Moisture Content (percentage)	Particle size, (percentage finer than micrometer size)	Explosibility Index	Materials
21	—	100/74	0.1	Cellucotton wadding, treated with zinc peroxide, from cyclone collector
22	—	100/74	2.0	Cellucotton wadding, treated with zinc peroxide, from air ducts of ventilation system
23	—	100/840	—	Celotex, fiberboard
24	2.3	100/74	6.0	Ethyl hydroxyethyl cellulose
25	2.2	100/74	6.9	Hydroxyethyl cellulose
26	2.3	100/74	>10	Hydroxypropyl cellulose
27	2.1	1/74	—	Hydroxypropyl methyl cellulose (29 percent methoxy, 6 percent hydroxy propoxy, 65 percent cellulose)
28	0	100/74	<0.1	Ligninsulfate paper waste
29	0	100/74	0.8	Ligninsulfite paper waste
30	0	100/74	—	Ligninsulfonate, desulfonated and demethylated
31	0	100/74	<0.1	Ligninsulfonate, sodium, Marasperse N, 76 percent, with 10 percent organic and 14 percent inorganic ingredients
32	0	100/74	<0.1	Ligninsulfonate, sodium, Maracell A, 44 percent, with 21 percent organic and 35 percent inorganic ingredients
33	0	100/74	<0.1	Ligninsulfonate, sodium, Maracell A, heat-treated
34	2.9	100/74	0.2	Ligninsulfonate mixture, RDA 49-116
35	0	100/74	0.2	Ligninsulfonic acid, calcium, spray-dried
36	0	100/74	0.1	Ligninsulfonic acid, calcium, standard
37	0	100/74	0.2	Ligninsulfonic acid, calcium, spray-dried
38	1.2	61/74	>10	Paper coating solids, from drying oven of paper coating operation
39	0	100/840	0.8	Paper dust, from dust collector
		Chemicals, Gums, and Resins		
40	0.1	100/74	—	Acetal copolymer, Celcon, molding material
41	0	100/74	0.3	Acrylamide and methyl methacrylate copolymer

Table 27.7 (Continued)

Line No.	Moisture Content (percentage)	Particle size, (percentage finer than micrometer size)	Explosibility Index	Materials
42	1.1	100/74	>10	Acrylonitrile-butadiene-styrene (22–18–60) copolymer
43	—	23/74	—	Batu gum regin
44	0	100/74	0.2	Bone glue, steer bones
45	—	—	—	Cream of tartar (potassium acid tartrate)
46	0	100/74	4.9	1,4-Cyclohexylene dimethylene isophthalate and 1,4-cyclohexylene dimethylene terephthalate copolymer
47	2.1	100/74	—	Dodecyl diphenyl oxide and disodium sulfonate
48	0	100/74	—	Formaldehyde–naphthalene sulfonic acid copolymer, sodium salt of, from cyclone dust collector
49	0.3	53/74	—	Latex mastic material (Neoprene, cement, marble chips, accelerator), for deck covering
50	0.5	100/44	—	Molybdenum disulfide, <5 μm
51	0.9	—	—	Polystyrene, rubber modified
52	0.7	100/44	>10	Rosin residue, extracted from pinewood
53	—	100/149	1.8	Santowax R, mixture of ortho-, meta-, and paraterphenyls
54	0	100/74	—	Tanning extract, vegetable, spray-dried

Detergents and Soaps

		Detergents		
55	4.6	100/149	—	Detergent (approximately 30 to 40 percent alkyl aromatic sulfonates combined with inorganic builders, including sodium sulfate, sodium tripolyphosphate, and sodium silicate)
56	3.9	100/149	—	Detergent (approximately 30 to 35 percent alkyl aromatic sulfonates combined with builders consisting chiefly of sodium sulfonate and small amounts of pyrophosphate)

Table 27.7 (Continued)

Line No.	Moisture Content (percentage)	Particle size, (percentage finer than micrometer size)	Explosibility Index	Materials
57	1.6	100/74	—	Detergent, NS-NR, S
58	0.6	96/74	—	Detergent, low active grade (38 percent sodium alkylaryl sulfonate, 62 percent sodium sulfate)
59	0	100/74	—	Detergent (70 percent sodium alkylaryl sulfonate, 27 percent sodium sulfate, 3 percent unsulfonated oil)
60	2.8	100/74	—	Detergent fines (85 percent alkyl sulfate plus inorganic chlorides and sulfates)
61	5.7	100/74	—	Detergent powder, S, derived from Alkane
62	3.8	72/74	—	Disinfectant rinse (33 percent monosodium phosphate, 15 percent sodium dodecylbenzene sulfonate, 13 percent trichloromelamine, 24 percent citric acid, 13 percent sodium bicarbonate, 2 percent polyethylene glycol)
			Soap Powder	
63	—	100/44	—	Soap powder (96 percent soap, 4 percent preservatives and additives)
64	—	100/44	—	Soap powder
65	2.8	100/74	—	Soap powder
66	3.1	100/74	—	Soap powder
67	1.2	100/74	—	Soap powder, fines (2 percent anhydrous soap)
68	1.0	100/74	0.7	Soap powder, fines (96 percent anhydrous soap)
69	3.6	100/74	0.2	Soap powder, from dust collectors
70	0.7	100/74	0.5	Soap powder, from dust collectors
71	1.1	100/74	1.0	Soap powder, from dust collectors
72	0.6	98/74	0.6	Soap powder, from filters
73	0	100/74	—	Soap powder aggregate (58 percent anhydrous soap)
74	1.7	100/74	0.4	Soap powder mixed with cadmium and barium silicates
75	0	100/149	<0.1	Soap product (65 percent anhydrous soap plus inorganic silicates and phosphates)

Table 27.7 (Continued)

Line No.	Moisture Content (percentage)	Particle size, (percentage finer than micrometer size)	Explosibility Index	Materials
				Explosives and Related Compounds
76	—	100/44	0.2	Aluminum, atomized, 35 percent (average 10 μm); barium nitrate, 65 percent (average 15 μm)
77	—	100/74	<0.1	Aluminum, grade B, 40 percent (average 16 μm); barium nitrate, 30 percent (average 50 μm); potassium perchlorate, 30 percent (average 18 μm)
78	—	100/74	—	Aluminum-magnesium alloy, 45.0 percent (60 percent finer than 74 μm); barium nitrate, 55.0 percent (95 percent finer than 74 μm)
79	—	100/840	—	Ammonium nitrate
80	—	100/840	—	Ammonium nitrate, 78 percent; potassium nitrate, 4 percent; charcoal, 16 percent; No. 2 fuel oil, 2 percent
81	0	100/30	—	Ammonium nitrate, 94 percent; carbon, 6 percent
82	4.0	100/840	—	Ammonium nitrate–fuel oil mixture (96 percent ammonium nitrate, 4 percent No. 2 fuel oil)
83	5.4	100/840	—	Ammonium nitrate–fuel oil mixture (94 percent ammonium nitrate, from 10 to 200 μm, 6 percent No. 2 fuel oil)
84	8.1	100/840	—	Ammonium nitrate–fuel oil mixture (90 percent ammonium nitrate, from 10 to 200 μm; 10 percent No. 2 fuel oil)
85	—	100/74	—	Ammonium nitrate No. 3, 73 percent; sodium nitrate, 10 percent; starch, 5 percent; wood pulp, 12 percent
86	0	100/74	—	Ammonium perchlorate, between 3 and 50 μm
87	—	100/74	—	Ammonium picrate, 50 percent; sodium nitrate, 45 percent; resin binder, 5 percent
88	—	—	—	Anthracene, 45 percent; potassium perchlorate, 55 percent

Table 27.7 (Continued)

Line No.	Moisture Content (percentage)	Particle size, (percentage finer than micrometer size)	Explosibility Index	Materials
89	—	100/74	—	Barium nitrate, average 15 μm
90	—	100/149	—	Black powder mixture (91 percent No. 6 grade A black powder, 9 percent Acrawax)
91	0.2	100/74	7.2	Dinitrobenzamide
92	0.1	100/74	4.0	Dinitrobenzoic acid
93	0.3	100/74	0.7	Dinitro-sym-diphenylurea (dinitrocarbanilide)
94	0.1	100/74	>10	Dinitrotoluamide (3,5-dinitro-ortho-toluamide)
95	—	100/149	—	Guanidine nitrate
96	0.8	100/74	—	Napalm (78 percent ferronapalm, 16 percent Pfister napalm, 6 percent caustic calcine magnesia)
97	—	100/840	—	Nitroguanidine
98	—	100/840	—	Nitroguanidine
99	—	100/840	—	Nitroguanidine
100	—	100/840	—	Nitroguanidine
101	—	100/840	—	Nitrostarch
102	—	100/840	—	Potassium chlorate, 50 percent; antimony sulfide, 30 percent; dextrin, 20 percent
103	—	100/44	—	Potassium perchlorate, <18 μm
104	—	100/840	—	Silicon, 20 percent; no binder; magnesium, grade B, 20 percent; lead chromate, 60 percent
105	—	100/840	—	Silicon, 20 percent; with binder; magnesium, grade B, 20 percent; lead chromate, 60 percent
106	—	100/840	<0.1	Silicon, 10 percent; lead oxide (red lead), 90 percent
107	—	100/74	—	Silicon, 6 percent; manganese, 40 percent; lead oxide (red lead), 54 percent
108	—	100/840	—	Silicon, 26 percent; aluminum, 13 percent; ferric oxide, 22 percent; potassium nitrate, 35 percent; carbon, 4 percent
109	—	100/149	—	Sodium nitrate
110	—	100/840	—	Thermite (28 percent aluminum, 72 percent iron oxide), 58 percent; plus igniter—barium 28 percent; nitrate

Table 27.7 (Continued)

Line No.	Moisture Content (percentage)	Particle size, (percentage finer than micrometer size)	Explosibility Index	Materials
				aluminum, fine grain, 12 percent; sulfur, 2 percent; castor oil, 0.2 percent
111	—	100/840	—	Trinitrotoluene (TNT)
				Feeds and Fertilizers
112	0	100/74	<0.1	Animal meal, defatted (bones, meat)
113	0	100/74	0.2	Animal meal, defatted (bones, meat), from machinery and ledges, processing plant
114	2.0	98/74	—	Bone meal, from machinery and ledges in the grinding room
115	2.0	94/74	—	Bone meal, from packaging department
116	2.3	95/74	1.0	Fermentation mash, spray-dried
117	—	100/74	—	Fertilizer (sodium nitrate, tankage, fish meal, potassium chloride, calcium triorthophosphate)
118	4.0	100/74	—	Guano, bat, from cave in Grand Canyon
119	—	100/74	—	Guano, Gaviota, containing free sulfur
120	—	100/74	—	Hoof and horn meal
121	2.6	100/74	—	Molasses fermentation residue (49 percent nitrogen free extract, 4 percent moisture, 28 percent crude protein, 4 percent crude fiber, 15 percent ash)
122	3.2	100/44	0.1	Organisms, 65 percent; dextrin, 20 percent; thiourea, 5 percent; ammonium chloride, 5 percent; ascorbic acid, 5 percent
123	—	100/44	0.8	*Serratia marcescens* (protein) cells, 90 percent; miscellaneous material, 8 percent; moisture, 2 percent
124	0	100/74	1.0	*Serratia marcescens* (protein) cells, 47 percent; sucrose, 34 percent; skim milk, 14 percent; thiourea, 5 percent
125	1.8	100/74	1.2	*Serratia marcescens* (protein) cells, 66 percent; dextrin, 19 percent; ammonium chloride, 5 percent; ascorbic acid, 5 percent; thiourea, 5 percent

Table 27.7 (Continued)

Line No.	Moisture Content (percentage)	Particle size, (percentage finer than micrometer size)	Explosibility Index	Materials
126	5.7	100/840	<0.1	Sewage, from ventilator openings in storage building of disposal plant
127	0	100/74	0.2	Sewage sludge, heat-dried, consisting of mixture of brown dust and varicolored chaff
128	1.3	100/74	—	Sewage sludge, black, dust deposit in breeching
129	0	100/74	—	Sewage sludge, dried, fertilizer product
130	0	100/74	—	Sheep manure
131	8.0	80/74	—	Water-soluble proteins and carbohydrates
132	7.6	87/74	—	Water-soluble proteins and carbohydrates
133	5.6	90/74	—	Water-soluble proteins and carbohydrates
134	5.2	43/74	—	Wood flour, 55 percent, plus wheat bran and organic reducing materials
135	4.6	23/74	—	Wood flour mixed with desiccated animal tissue
			Plant Dusts	
			Agricultural	
136	7.5	39/74	—	Corn dust, from cyclone collector in cleaning operation, prior to separation, cereal manufacturer
			Carbonaceous	
137	0.8	43/74	—	Boilerhouse dust, from vicinity of a porthole in a duct leading to a 135-foot stack, steel manufacturer
138	0	32/74	—	Boilerhouse dust (22 percent moisture), from bottom of stack, slurrylike, where dust had accumulated to a depth of 25 feet, steel manufacturer
139	—	—	—	Carbonaceous dust, particle size from 15 to over 100 μm, from pitch conveyor in paste plant, aluminum manufacturer

Table 27.7 (Continued)

Line No.	Moisture Content (percentage)	Particle size, (percentage finer than micrometer size)	Explosibility Index	Materials
140	—	—	—	Carbonaceous dust, particle size from 15 to over 100 μm, from I-beams of paste plant, third floor, carbon department, aluminum manufacturer
141	—	—	—	Carbonaceous dust, particle size from 15 to over 100 μm, from I-beams of paste plant, fourth floor, carbon department, aluminum manufacturer
142	—	—	—	Carbonaceous dust, particle size from less than 5 to about 70 μm, from screw-conveyor, paste plant, sixth floor, carbon department, aluminum manufacturer
143	0.8	79/74	—	Coal dust (1 percent moisture, 21 percent volatile matter, 51 percent fixed carbon, 27 percent ash (contains 1.6 percent sulfur)), from I-beam in electric power station, boiler room
144	1.0	40/74	—	Coal dust (54 percent ash), from fourth floor level above boilers, near head of coal dump, at forced draft fan level, steam generating plant, steel manufacturer
145	0.5	85/74	—	Coal dust–coke breeze mixture (52 percent ash), from above boiler, steam generating plant, steel manufacturer
146	0.8	52/74	—	Flue dust (55 percent ash), from blast furnace gas, collected from rafters on fourth floor level above boiler at forced draft fan level, steam generating plant, steel manufacturer
147	0.7	73/74	—	Flue dust (88 percent ash), from above is burned, steam generating plant, steel manufacturer
		Metal Dusts		
148	0.5	71/74	—	Aluminum (42 percent aluminum, 7.5 percent oil, plus lint and fibers), from

Table 27.7 (Continued)

Line No.	Moisture Content (percentage)	Particle size, (percentage finer than micrometer size)	Explosibility Index	Materials
				filters above cyclone collector used in connection with grinding aluminum castings, tool manufacturer
149	0	71/74	<0.1	Aluminum with oil removed
150	1.5	95/74	—	Aluminum, grained, plus dust from crane rail, metal alloy manufacturer
151	0.8	90/74	—	Aluminum, grained, plus dust on roof of small enclosure in main building, metal alloy manufacturer
152	0.4	91/74	—	Aluminum, from collector on grinding machines, metal alloy manufacturer
153	0.8	90/74	—	Aluminum, from bag in portable vacuum cleaner used around air conditioning machinery, metal alloy manufacturer
154	0.8	94/74	—	Iron (89 percent ash), from building trusses, malleable iron manufacturer
155	1.6	81/74	—	Iron (70 percent ash), from building trusses in coreroom, malleable iron manufacturer
156	0.9	99/74	—	Iron (83 percent ash), from building trusses in sand handling room, malleable iron manufacturer
157	0.5	95/74	—	Iron (43 percent ash), from carburizing department, malleable iron manufacturer
158	4.1	90/74	—	Iron (23 percent combustible), suspensions of dust settled on overhead surfaces above molding machine, iron foundry
159	0.1	77/74	—	Iron dust (21 percent iron plus lint and wool), tool manufacturer
160	—	100/840	—	Smelter dust, variety of metals, smelter and refining company
161	0.8	80/74	—	Titanium, contains wood, sand, paper, coal, concrete, slag, and flakes of paint, from around a collector, titanium plant
162	5.9	100/840	—	Titanium, contains wood, sand, paper, coal, concrete, slag, and flakes of paint, from outside bag collector, titanium plant

Table 27.7 (Continued)

Line No.	Moisture Content (percentage)	Particle size, (percentage finer than micrometer size)	Explosibility Index	Materials
163	0.5	100/74	—	Titanium, residual dust from abrasive grinding of titanium billets, titanium plant
164	0.2	100/74	—	Titanium, residue from grinding titanium sheets, titanium plant
165	1.4	100/74	—	Titanium, from area above crushers in sponge preparation room, titanium plant
166	0.8	100/74	—	Titanium, from area above blender in sponge preparation room, titanium plant
167	2.9	100/74	—	Titanium, from around feeders in melt shop, titanium plant
168	2.6	100/74	—	Titanium (24 percent titanium, 30 percent carbonaceous material, 13 percent silicon dioxide, 4 percent manganese, 13 percent iron, 16 percent metal oxides), from structural members in furnace building, titanium plant
169	0.6	100/74	<0.1	Titanium, mixed with iron, from grit blasting, titanium plant
170	0.8	91/74	0.2	Zinc (5 percent zinc, 4 percent silicon dioxide, 55 percent carbon, 36 percent ash), from beams and platform near hammermill coal crusher, zinc smelter plant
171	0.9	97/74	—	Zinc (32 percent zinc, 10 percent silicon dioxide, 26 percent carbon, 32 percent ash), from roof beams, zinc smelter plant
172	1.1	98/74	—	Zinc (22 percent zinc, 8 percent silicon dioxide, 38 percent carbon, 32 percent ash), from roof beams, zinc smelter plant
			Plastics	
173	1.3	29/74	—	Buffring machine dust, a mixture of Liquabrade 4787 (Lea Liquid buffing

Table 27.7 (Continued)

Line No.	Moisture Content (percentage)	Particle size, (percentage finer than micrometer size)	Explosibility Index	Materials
				compound), cotton buffing wheel lint, and cellulose acetate butyrate molding compound, electrical manufacturer
174	0.5	36/44	0.9	Residue, including polyvinyl chloride resins, compound fines, dioctylphthalate (plasticizer), calcium stearate, coloring pigments, clays, and titanium oxide, from structural framework in millroom, chemical company
175	0.7	7/44	—	Residue, including polyvinyl chloride resins, compound fines, dioctylphthalate (plasticizer), calcium stearate, coloring pigments, clays, and titanium oxide, from structural framework above baggers and cubers, chemical company
176	0.8	15/74		Residue, including polyvinyl chloride resins, compound fines, dioctylphthalate (plasticizer), calcium stearate, coloring pigments, clays, and titanium oxide, from structural framework above mills, chemical company
			Rubber	
177	—	100/840	—	Residue from rafters of tire recapping plant
178	—	100/840	—	Residue from under rafters of tire recapping plant
			Miscellaneous	
179	0.9	100/74	—	Blasting machine dust (quartz, hematite, magnetite, alpha iron) from collector used in cleaning airplane parts, aircraft maintenance depot
180	3.6	100/74	>10	Wood dust, from refinishing bowling alleys, first cut, with lacquer
181	5.6	100/74	>10	Wood dust, from refinishing bowling alleys, second or finishing cut

Table 27.8 Parameters Affecting Ignition (33)

Line No.[a]	Ignition Sensitivity	Minimum Ignition Temperature, (°C)		Minimum Ignition Energy (J)		Minimum Explosive Concentration (oz/ft³)	Relative Flammability (percentage inert)		Limiting Oxygen Concentration,[b] (percentage)	
		Cloud	Layer	Cloud	Layer		In Spark Apparatus	Furnace, 700°C	Spark	Furnace, 850°C
Agricultural Dusts										
1	—	730	—	—	—	—	—[c]	—	—	—
2	—	470	—	—	—	—	80	90+	—	—
3	3.6	460	210	0.035	—	0.035	85	90+	—	—
4	—	500	—	—	—	—	—[b]	—	—	—
5	—	495	—	—	—	—	Arc[d]	—	—	—
6	—	495	—	—	—	—	Arc[d]	—	—	—
7	<0.1	480	—	0.080	—	0.700	—[d]	—	—	—
8	0.7	485	—	0.060	—	0.105	—[d]	—	—	—
9	2.5	410	440	0.040	—	0.050	—[d]	—	N, 10; C, 14	—
10	—	460	—	—	—	0.120	75	90+	—	—
11	0.5	530	—	0.060	—	0.140	—[d]	—	—	—
12	—	—	—	—	—	<0.100	—[d]	—	—	—
13	—	520	—	—	—	—	—[c]	90+	—	—
14	—	550	—	—	—	—	60	90+	—	—
Cellulosics										
15	—	610	300	—	—	0.160	—[d]	—	—	—
16	—	650	—	—	—	<0.200	—[d]	—	—	—
17	—	440	—	—	—	0.075	—[d]	—	—	—

18	—	510	—	—	—	0.060	—[d]	—	—	—
19	—	590	—	—	—	—	—[c]	60	—	—
20	—	550	—	—	—	—	—[c]	75	—	—
21	0.2	480	280	0.320	—	0.055	80	90+	—	—
22	1.4	470	270	0.060	—	0.050	85	90+	—	—
23	—	—	—	—	—	—	85	—	—	—
24	8.6	390	—	0.030	—	0.020	—[d]	—	C, 16	—
25	4.9	410	—	0.040	—	0.025	—[d]	—	—	—
26	8.4	400	—	0.030	—	0.020	—[d]	—	—	—
27	—	430	—	—	—	0.800	5	85	—	—
28	<0.1	380	350	0.390	—	0.200	20	85	C, 18	C, 5
29	0.6	530	260	0.080	—	0.085	50	90+	—	—
30	—	390	—	—	—	0.230	—[d]	90+	—	C, 4
31	0.2	490	230	0.140	—	0.200	45	90+	C, 17	C, 4
32	<0.1	400	340	0.260	—	0.350	30	90+	C, 17	C, 4
33	0.2	390	350	0.140	—	0.250	45	90+	C, 17	C, 4
34	0.2	650	330	0.140	—	0.150	10	85	—	—
35	0.2	560	410	0.160	—	0.095	65	60	—	—
36	0.1	670	410	0.240	—	0.120	40	55	—	—
37	0.1	590	470	0.160	—	0.150	50	70	—	—
38	3.7	390	170	0.020	—	0.070	—[d]	—	—	—
39	1.4	440	270	0.060	—	0.055	—[d]	—	—	—

Chemicals, Gums, and Resins

40	—	470	—	—	—	0.060	—[d]	—	—	—
41	1.5	510	—	0.060	—	0.045	—[d]	—	—	—
42	7.1	470	—	0.030	—	0.020	90+	90+	—	—
43	—	420	—	—	—	—	75	90	—	—
44	0.3	550	—	0.140	—	0.030	70	90+	—	—
45	—	520	—	—	—	2.000		60	—	—
46	5.4	500	—	0.025	—	0.165	70	—	C, 13	—

Table 27.8 (Continued)

Line No.[a]	Ignition Sensitivity	Minimum Ignition Temperature, (°C)		Minimum Ignition Energy (J)		Minimum Explosive Concentration (oz/ft³)	Relative Flammability (percentage inert)		Limiting Oxygen Concentration,[b] (percentage)	
		Cloud	Layer	Cloud	Layer		In Spark Apparatus	Furnace, 700°C	Spark	Furnace, 850°C
47	<0.1	540	360	8.320	—	—[b]	hc[d]	—	—	—
48	—	620	420	—	—	—	20	80	—	—
49	—	640	440	—[b]	—	0.800	—[c]	45	—	—
50	—	570	290	—	—		—[c]	75	—	—
51	—	460	—	—[b]	—		50	90+	C, 17	C, 8
52	>10	470	—	0.015	—	0.016	90	90+	—	—
53	1.2	620	—	.080	—	.035	90+	90+	—	—
54	—	650	340	—	—	.200	—[d]	90	—	—
					Detergents and Soaps					
					Detergents					
55	—	510	260	—	—	—	gc[d]	90	—	—
56	—	520	250	—	—	—	gc[d]	90+	—	—
57	—	530	280	—	—	—	gc[d]	90+	—	—
58	—	540	300	—	—	—	—[d]	90+	—	—
59	—	300	570	—	—	0.130	30	90+	—	—
60	—	530	—	—	—	—	70	90	—	—
61	—	660	390	—	—	—	gc[d]	90	—	—
62	—		—	—	—	—	—[b]	90+	—	—

Soap Powder

63	—	635	—	—	—	—	—[c]	60	—	—
64	—	580	—	—	—	—	—[c]	80	—	—
65	—	—	—	—	—	0.045	—[d]	—	—	—
66	—	—	—	—	—	—	—[d]	—	—	—
67	—	560	480	—	—	—	gc[d]	90+	—	—
68	0.3	640	500	0.120	—	0.085	90	90+	—	—
69	0.4	600	380	0.120	—	0.075	55	85	—	—
70	0.6	430	600	0.100	—	0.085	70	90+	—	—
71	0.9	600	460	0.060	—	0.060	85	90+	—	—
72	0.8	540	450	0.100	—	0.045	80	90+	—	—
73	—	630	310	0.240	—	—	25	90+	—	—
74	0.4	380	260	0.120	—	0.125	80	90+	—	—
75	<0.1	650	430	0.960	—	0.070	70	90+	—	—

Explosives and Related Compounds

76	0.1	700	—	0.120	0.032	0.200	65	5	N, 5	C, 3
77	0.1	700	—	0.375	—	0.270	55	10	N, 11	C, 3
78	—	440	—	—	0.056	0.430	25	55	N, 11	C, 3
79	—	400	—	—	—	—	—[c]	30	—	—
80	—	490	190	—	—	—	—[c]	—	—	—
81	—	360	200	—	—	—	gc[d]	80	—	—
82	—	390	—	—	—	—	hc[d]	65	—	—
83	—	380	—	—	—	2.000	20	75	—	—
84	—	370	—	1.600	—	0.370	45	75	—	—
85	—	310	160	0.104	—	—	—[d]	65	—	—
86	—	—	260	—	—	—	—[c]	—	—	—
87	—	—	250	0.160	—	0.200	—[d]	—	—	—
88	—	530	700	—	—	0.160	65	90+	—	—

Table 27.8 (Continued)

Line No.[a]	Ignition Sensitivity	Minimum Ignition Temperature, (°C)		Minimum Ignition Energy (J)		Minimum Explosive Concentration (oz/ft^3)	Relative Flammability (percentage inert)		Limiting Oxygen Concentration,[b] (percentage)	
		Cloud	Layer	Cloud	Layer		In Spark Apparatus	Furnace, 700°C	Spark	Furnace, 850°C
89	—	—	—	—	—	—	—[c]	—	—	—
90	0.2	340	—	0.320	—	0.120	—[d]	—	—	—
91	2.2	500	—	0.045	—	0.040	85	90+	—	—
92	1.9	460	—	0.045	—	0.050	80	90+	—	—
93	0.6	550	—	0.060	—	0.095	85	90+	—	—
94	5.4	500	—	0.015	0.024	0.050	90	90+	C, 13	—
95	—	850	—	—	—	—	—[c]	—[c]	—	—
96	5.6	450	—	0.040	—	0.020	90	90+	N, 12	N, 8
97	—	400	—	—	—	—	—[d]	80	—	—
98	—	670	—	<7.200	—	—	—[d]	55	—	—
99	—	680	—	<7.200	—	—	—[d]	70	—	—
100	—	850	—	—	—	—	—[c]	(^{2})	—	—
101	3.8	190	165	0.040	—	0.070	—[d]	—	—	—
102	—	280	300	—[d]	—	0.070	25	80	—	—
103	—	—	—	—	—	—	—[c]	—	—	—
104	0.1	620	520	8.000	0.004	0.265	15	90	—	—
105	—	650	520	—[c]	0.320	<3.000	—[d]	90+	—	—
106	0.1	540	540	0.360	0.0016	<3.000	—[d]	25	—	—
107	0.1	540	450	0.350	—	0.650	10	48	—	—
108	—	<700	400	—[d]	—	—	10	15	—	—
109	—	—	—	—	—	0.760	—[c]	—	—	—
110	0.1	720	—	0.240	0.400	0.070	15	(^{2})	—	—
111	—	—	—	0.075	—		—[d]	—	—	—

112	0.3	530	350	0.180	—	0.065	75	90+	—	—
113	0.5	530	310	0.120	—	0.060	80	90+	—	—
114	—	490	230	—	—	—	gc[d]	85	—	—
115	—	560	250	—	—	—	gc[d]	75	—	—
116	0.7	500	310	0.080	—	0.070	80	90+	—	—
117	—	—[c]	380	—	—	—	—[c]	—	—	—
118	—	460	240	—	—	—	arc, gc[d]	90+	—	—
119	—	330	—	—	—	—	30	90+	—	—
120	—	660	—	—	—	—	45	85	—	—
121	—	660	240	—	—	—[d]	—[d]	90+	—	—
122	0.3	500	410	0.100	—	0.155	65	90+	—	—
123	0.6	490	200	0.080	—	0.080	—[d]	—	N, 14	—
124	0.4	470	180	0.080	—	0.130	—[d]	—	N, 14	—
125	0.6	490	200	0.080	—	0.080	—[d]	—	N, 14	—
126	0.3	390	190	0.120	—	0.165	—[d]	—	—	—
127	0.7	390	160	0.080	—	0.095	60	90+	—	—
128	—	530	180	0.960	—	—	—[d]	90	—	—
129	—	420	—	—	—	—	—[d]	90+	C, 16	C, 5
130	—	730	—	—	—	—	—[c]	—	—	—
131	—	550	—	—	—	—	45	90+	—	—
132	—	620	—	—	—	—	—[d]	90+	—	—
133	—	520	—	—	—	—	—[d]	90+	—	—
134	—	480	—	—	—	—	80	90+	—	—
135	—	470	—	—	—	—	80	90+	—	—
					Plant Dusts					
					Agricultural					
136	—	430	290	—	—	—	75	90+	—	—

Table 27.8 (Continued)

Line No.[a]	Ignition Sensitivity	Minimum Ignition Temperature, (°C)		Minimum Ignition Energy (J)		Minimum Explosive Concentration (oz/ft³)	Relative Flammability (percentage inert)		Limiting Oxygen Concentration,[b] (percentage)	
		Cloud	Layer	Cloud	Layer		In Spark Apparatus	Furnace, 700°C	Spark	Furnace, 850°C
					Carbonaceous					
137	—	—	510	—	—	—	—[c]	—	—	—
138	—	—	510	—	—	—	—[c]	—	—	—
139	—	690	—	—	—	—	—[c]	15	—	—
140	—	650	—	—	—	—	—[c]	45	—	—
141	—	630	—	—	—	—	60	80	—	—
142	—	660	—	—	—	—	—[c]	40	—	—
143	—	560	180	—	—	—	arc, gc[d]	85	—	—
144	—	800	520	—	—	—	—[c]	—	—	—
145	—	740	510	—	—	—	—[c]	—	—	—
146	—	800	550	—	—	—	—[c]	—	—	—
147	—	—	590	—	—	—	—[c]	—	—	—
					Metal Dusts					
148	—	470	580	—	—	0.200	15	85	—	—
149	<0.1	550	660	0.280	—	0.140	55	35	—	—
150	—	—	460	—	—	—	—[c]	—[c]	—	—
151	—	770	450	—	—	—	—[c]	—[c]	—	—
152	—	—	510	—	—	—	—[c]	—[c]	—	—
153	—	—	440	—	—	—	—[c]	—[c]	—	—
154	—	750	410	—	—	—	—[c]	—	—	—

155	—	470	220	—	—	—	gc[d]	85	—	—
156	—	560	320	—	—	—	—[c]	70	—	—
157	—	770	470	—	—	—	—[c]	—	—	—
158	—	780	430	—	—	—	arc[d]	—	—	—
159	—	—	520	—	—	—	gc[d]	—	—	—
160	—	700	—	—	—	—	gc[d]	10	—	—
161	—	650	460	—	—	—	—[c]	30	—	—
162	—	740	480	—	—	—	—[c]	—	—	—
163	—	950	380	—	—	—	arc, gc[d]	—	—	—
164	—	740	690	—	—	—	20	—	—	—
165	—	490	220	—	—	—	arc[d]	70	—	—
166	<0.1	490	260	6.400	—	0.800		70	—	—
167	—	500	240	—	—	—	30	60	—	—
168	—	640	250	—	—	—	50	50	—	—
169	<0.1	480	210	0.180	—	0.370	gc[d]	55	—	—
170	0.4	600	230	0.100	—	0.080	gc[d]	90	—	—
171	—	640	270	—	—	—		50	—	—
172	—	600	240	—	—	—		80	—	—
					Plastics					
173	<0.1	530	230	0.180	—	1.000	45	80	—	—
174	1.7	540	—	0.050	—	0.045	—[d]	—	—	—
175	—	580	—	—	—	—	—[d]	—	—	—
176	—	530	—	—	—	—	—[d]	—	—	—
					Rubber					
177	—	530	—	—	—	—	50	80	—	—
178	—	540	—	—	—	—	40	90	—	—

Table 27.8 (Continued)

Line No.[a]	Ignition Sensitivity	Minimum Ignition Temperature, (°C)		Minimum Ignition Energy (J)		Minimum Explosive Concentration (oz/ft³)	Relative Flammability (percentage inert)		Limiting Oxygen Concentration,[b] (percentage)	
		Cloud	Layer	Cloud	Layer		In Spark Apparatus	Furnace, 700°C	Spark	Furnace, 850°C
						Miscellaneous				
179	—	600	—	—	—	—	—[c]	—	—	—
180	4.6	360	180	0.035	—	0.035	90	90+	—	—
181	3.5	360	260	0.040	—	0.040	90	90+	—	—

[a] See Table 27.7 for identity of material.

[b] Prefix letter denotes diluent gas: C, carbon dioxide; N, nitrogen.

[c] No ignition.

[d] Dust ignited by spark except as noted: arc, carbon arc: gc, guncotton; hc, heated coil.

by other organizations such as the National Fire Protection Association (4). Some of these data are in Tables 27.7 to 27.9, based on Reference 33. Because the initial purpose of the Bureau of Mines investigators was to obtain laboratory results comparable to those obtained in tests in their Experimental Mine at Bruceton, Pennsylvania, it is likely that the equipment and procedures are tuned to simulate large-scale explosions, but one cannot be certain that such is the case for every measurement. The equipment used and procedures followed are given in Reference 32. The investigators (32, 33, 47–51), following the lead set largely by I. Hartmann at the Bureau of Mines, developed three empirical indices: the index of sensitivity of ignition, the index of explosion severity, and the overall explosibility index (47). Values of selected relative parameters are compared to those of Pittsburgh coal dust, which by definition has an explosibility index of unity; a dust with greater explosion hazard has an explosibility index greater than unity. The ignition sensitivity and explosion severity indices of a dust and the explosibility index are defined as follows:

$$\text{ignition sensitivity} = \frac{\left(\begin{array}{c}\text{ignition temperature} \times \\ \text{minimum ignition energy} \times \\ \text{minimum flammable concentration}\end{array}\right) \text{Pittsburgh coal dust}}{\text{(same parameters) other dust}}$$

$$\text{explosion severity} = \frac{\left(\begin{array}{c}\text{maximum explosion pressure} \times \\ \text{maximum rate of pressure rise}\end{array}\right) \text{Pittsburgh coal dust}}{\text{(same parameters) other dust}}$$

$$\text{explosibility index} = \text{(ignition sensitivity) (explosion severity)}$$

The characteristics of the Pittsburgh coal dust ($-74\ \mu m$) used as the reference dust are as follows:

Characteristic	Value
Explosibility index	1.0
Ignition sensitivity	1.0
Explosion severity	1.0
Minimum ignition temperature of dust cloud	610°C
Minimum ignition energy of dust cloud	0.06 J
Minimum explosive concentration	0.055 oz/ft^3
Maximum explosion pressure*	83 psig
Maximum rate of pressure rise*	2300 psi/sec

Explosion hazards were then classified in terms of these indices as follows. [Carbonaceous materials having a dust cloud ignition temperature of 730°C or more and not

* Dust concentration of 0.50 oz/ft^3.

Table 27.9 Explosion Severity, Pressures, and Rates of Pressure Rise of Dust Explosions (33)

Line No.[a]	Explosion Severity	Concentration 0.50 oz/ft^3		
		Maximum Pressure (psig)	Rate of Pressure Rise (psi/sec)	
			Average	Maximum
		Agricultural Dusts		
3	3.3	96	1700	6500
8	0.9	69	1800	2500
9	2.5	81	1800	6000
10	0.5	83	450	1200
11	0.3	52	300	900
12	0.4	78	600	900
		Cellulosics		
15	0.6	80	400	1500
16	<0.1	43	100	250
17	2.0	88	1300	4300
18	2.3	98	1600	4500
21	0.7	78	1200	1800
22	1.4	87	1500	3100
24	0.7	84	800	1500
25	1.4	106	1200	2600
26	1.3	84	1200	2900
28	0.1	42	300	500
29	1.4	105	1100	2500
30	<0.1	38	150	300
31	0.2	75	400	600
32	0.3	62	500	800
33	0.6	77	600	1400
34	1.2	99	1000	2300
35	1.0	79	1000	2300
36	1.0	84	1000	2400
37	1.8	94	1100	3700
38	3.7	80	900	2700
39	1.8	96	1300	3600

Table 27.9 (Continued)

		Concentration 0.50 oz/ft³		
			Rate of Pressure Rise (psi/sec)	
Line No.[a]	Explosion Severity	Maximum Pressure (psig)	Average	Maximum
		Chemicals, Gums, and Resins		
40	2.4	71	2100	6500
41	0.2	57	400	600
42	1.8	71	1700	4700
44	0.6	73	500	1500
46	0.9	79	800	2200
48	0.2	68	200	500
52	2.0	75	1600	5000
53	1.5	67	1200	4200
54	<0.1	15	<100	150
		Detergents and Soaps		
		Detergents		
55	*<0.1*	*5*	*<100*	*100*
57	*<0.1*	*5*	*<100*	*100*
59	*1.4*	*108*	*800*	*2400*
60	*1.1*	*90*	*700*	*2300*
61	*<0.1*	*10*	*150*	*350*
		Soap Powder		
65	0.9	97	900	1700
66	<0.1	30	100	200
67	<0.1	39	200	400
68	2.4	116	1200	4000
69	0.5	74	800	1200
70	0.9	67	1200	2800
71	1.1	84	1300	2600
72	0.7	78	900	1800

Table 27.9 (Continued)

		Concentration 0.50 oz/ft³		
			Rate of Pressure Rise (psi/sec)	
Line No.[a]	Explosion Severity	Maximum Pressure (psig)	Average	Maximum
73	0.2	61	250	500
74	0.9	69	900	2400
75	0.2	76	300	500
		Explosives and Related Compounds		
76	2.0	77	2400	5000
77	0.3	56	450	900
78	1.1	68	1200	3000
87	0.9	74	1400	2400
91	3.2	94	2600	6500
92	2.1	92	1800	4300
93	1.1	87	1000	2500
94	5.6	106	3200	10000+
101	6.1	116	6600	10000+
111	0.7	63	600	2100
		Feeds and Fertilizers		
112	0.2	58	300	700
113	0.4	61	400	1200
116	1.4	81	1200	3200
118	<0.1	6	100	200
121	<0.1	23	100	200
122	0.4	63	500	1200
123	1.3	76	1300	3200
124	2.5	87	1600	5500
125	2.0	88	1200	4400
126	<0.1	35	150	400
127	0.3	58	450	900
129	0.1	49	200	400
		Metal Dusts		
148	0.2	30	100	150
149	0.1	52	300	400
157	<0.1	12	100	150

Table 27.9 (Continued)

		Concentration 0.50 oz/ft^3		
			Rate of Pressure Rise (psi/sec)	
Line No.[a]	Explosion Severity	Maximum Pressure (psig)	Average	Maximum
169	<0.1	25	100	200
170	0.6	72	700	1700
171	<0.1	14	100	200
172	<0.1	19	100	150
		Plastics		
173	—	—	—	—
174	0.5	64	700	1500
175	<0.1	10	<100	100
176	0.1	45	400	600
		Rubber		
177	0.3	69	500	900
178	0.2	67	300	600
		Miscellaneous		
180	4.0	115	3100	6700
181	3.0	99	1600	5700

[a] See Table 27.7 for identity of material.

ignitable by an electric spark were rated primarily as fire hazards (33).] Data in Tables 27.7 to 27.9 pertain to some of the tested dusts that evidenced an explosion hazard. These

Explosion Hazard	Ignition Sensitivity	Explosion Severity	Explosibility Index
Weak	<0.2	<0.5	<0.1
Moderate	0.2–1.0	0.5–1.0	0.1–1.0
Strong	1.0–5.0	1.0–2.0	1.0–10
Severe	>5.0	>2.0	>10

dusts are grouped as agricultural dusts, cellulosics, chemicals, gums, and resins; detergents and soaps; explosions and related compounds; feeds and fertilizers; and industrial plant dusts. A line number identifying the dust, a description of the dust, explosibility index (when applicable), moisture content, and particle size are given in Table 27.7. Particle size is designated by two numbers; the number to the left of the diagonal stroke indicates the percentage passing, and the number to the right the sieve opening, in micrometers. The dusts were passed through a No. 20 U.S. Standard sieve (840 μm). Table 27.8 contains the calculated ignition sensitivity and the related parameters of minimum ignition temperature, minimum ignition energy, and minimum explosive concentration. Data are given also on layer ignition temperature, the relative flammability (which is the percent by weight of calcined fuller's earth required in admixture to prevent flame propagation), and the limiting atmospheric oxygen required to prevent ignition by electric spark and furnace at 850°C. Relative flammability data are given for both the spark and furnace (700°C) ignition sources. The column of relative flammability (spark) is used also to indicate those dusts not igniting by spark but by stronger sources. Table 27.9 contains the calculated explosion severities and the explosion pressures and rates of pressure rise at a dust concentration of 0.50 oz/ft^3 (500 mg/liter). For a few dusts that did not ignite by spark, data are included for ignition by flame from guncotton and heated coil.

Additional data are in References 33 and 47 to 51; Reference 33 contains information on materials that presented primarily a fire hazard and those which did not present a dust explosion hazard.

3.3 Fire Hazard

In addition to the laboratory data in Reference 33 on fire hazards due to dusts, it must be emphasized that the hazard of dust layers catching fire is greater than that of the same material in bulk. In general, the dusts ignite at lower temperatures and burn more rapidly, and the fires are more difficult to extinguish than is the case for larger pieces. Should the fire-fighting efforts generate a flammable dust cloud, an explosion may follow. The ignition temperature of undispersed layers and minimum spark energies for ignition of dust layers, such as those in Table 27.8, are indices of the potential fire hazard. Dust layers can smolder even in thin layers, presenting at times a hidden and prolonged hazard (2). Smoldering is possible at oxygen concentrations below that required for propagation of an explosion of that dust (45, 52); some indirect evidence of this hazard can be noted in Table 27.8, where the limiting oxygen concentrations are generally lower for furnace ignition of dust clouds than for their spark ignition. The minimum spark energies listed in Table 27.8 for ignition of dust layers are so low that fires obviously can be started in dusts by weak electrical sparks or arcs, including static sparks, as well as by hot surfaces, glowing particles, frictional sparks, open flames, or other ignition sources. Most dusts oxidize so slowly that no significant temperature rise occurs, but some dusts, including those of some agricultural products, activated carbons, and metal powders, can oxidize rapidly enough to combust spontaneously. In addition, some metal powders react with moisture, especially at elevated temperatures, and thereby facilitate their spontaneous ignition.

3.4 Prevention and Control of Dust Explosions and Fires

Dust explosions and fires constitute an ever-present hazard in industry whenever materials that engender combustible dusts are processed or stored. Too often a small dust explosion or a small dust fire starts a highly destructive chain of events. A small dust explosion disperses additional dust, which explodes more violently and in turn generates a still larger dust cloud, which explodes with greater force than before; the process continues till all the dust is exhausted or scattered fires in undispersed dust end the event. Alternatively, if a fire starts first in dust layers, enough dust may be stirred up to lead to an explosion, followed by more explosions. The problem of preventing dust explosions and fires has been investigated by several research organizations and has lead to codes such as that of the National Fire Protection Association (4) and related NFPA listings to be found therein. The reader is referred to these codes as a helpful guide in the design and construction of building and equipment and in regard to safe operating procedures.

Measures recommended to prevent fires and explosion of combustible dust include the following:

1. Ignition sources should be eliminated. Open flames, open lights, or smoking should be prohibited; electric or gas cutting or welding equipment are to be avoided, unless the vicinity is dustfree; all equipment that may produce electrical static sparks should be grounded; the National Electrical Code, especially as it pertains to hazardous locations, is to be followed in electrical installations and operations, and nonsparking fans, shafts, and belts are to be used wherever possible; magnetic separators should be used to prevent ignition of dusts by frictional sparks produced by tramp metal particles passing through equipment.
2. Buildings should be constructed to avoid collection of dust on beams, ledges, and other surfaces. Good housekeeping is necessary. In removal of dust, the use of compresed air is highly dangerous; vacuum or brushing is advisable.
3. Equipment in which dust clouds may be generated should be as dusttight as possible, but strong enough to contain explosion pressures; or it should be provided with venting adequate to prevent an explosion from being disastrous. Unrestricted openings, hinged windows, panels, light wall construction in rooms, and release diaphragms can be designed to release explosion pressures without damage (2). Recommendations on the proper area of vents and suitability of particular vent closures are in the National Fire Protection Guide for Explosion Venting (30) and in Reference 35. Recommended unrestricted venting areas range from 1 square foot for each 10 to 30 cubic feet of enclosure for small equipment and light construction, to 1 square foot per 80 cubic feet for large rooms and heavy reinforced concrete buildings (2).
4. Preferably dust collectors should be located outside buildings or in detached rooms with explosion vents. Ducts leading to and from collectors should be as short and straight as possible, and the blowers should be on the clean air side of collectors.
5. Grinding, conveying, and other equipment frequently can be protected by

maintaining an inert atmosphere in the equipment, sufficient to reduce the oxygen content below that at which the dust will explode. Some examples of the maximum oxygen percentage for flame propagation in dust clouds are given in Table 27.8, and more data are in References 48 to 52. Methods of producing and using inert gas are in the NFPA standard for fire and explosion prevention by inerting (53).

6. The explosion hazard can sometimes be combated by adding inert dusts to the combustible dust, as is done in coal mines to prevent the propagation of coal dust explosions (36), or by using flame inhibitors such as alkali bicarbonates or Halons (20, 45). Devices triggered by the early stages of an explosion to eject inert dust or water or flame inhibitors into the propagating explosion wave are presently being developed (39).

7. Equipment for fighting fires of flammable dusts should not be likely to produce a dust cloud, with a consequent explosion. Water pails, soda–acid extinguishers, hand-operated water pump tanks, hoses with spray-type or fog nozzles, foam nozzles, and automatic sprinklers are satisfactory for use with most dusts. Small fires in magnesium, aluminum, or other metal powders are best extinguished by sand, talc, limestone, soapstone, or other dry inert powders.

REFERENCES

1. H. F. Coward and G. W. Jones, "Limits of Flammability of Gases and Vapors," U.S. Bureau of Mines Bulletin No. 503, 1952, 155 pp.; NTIS [National Technical Information Service] AD701575.
2. F. A. Patty, Ed., *Industrial Hygiene and Toxicology,* 2nd rev. ed., Wiley-Interscience, New York, 1958, Chapter XVI, pp. 511–578.
3. M. G. Zabetakis, "Flammability Characteristics of Combustible Gases and Vapors," U.S. Bureau of Mines Bulletin No. 627, 1965, 121 pp.; NTIS AD701576.
4. National Fire Protection Association, "Fire Hazard Properties of Flammable Liquids, Gases and Volatile Solids," NFPA No. 325M, 1969, 139 pp.
5. National Fire Protection Association, "Fire Hazards in Oxygen-Enriched Atmospheres," NFPA No. 53M, 1974, 89 pp.
6. Federal Coal Mine Health and Safety Act of 1969, PL 91-173, December 30, 1969.
7. J. Grumer, A. Strasser, and R. A. Van Meter, "Safe Handling of Liquid Hydrogen," *Cryogen. Eng. News,* August 1967, pp. 60–63.
8. A. L. Furno, E. B. Cook, J. M. Kuchta, and D. S. Burgess, "Some Observations on Near-Limit Flames," Paper presented at the 13th Symposium (International) on Combustion, the Combustion Institute, Pittsburgh, 1971, pp. 593–599.
9. P. G. Guest, V. W. Sikora, and B. Lewis, "Static Electricity in Hospital Operating Suites: Direct and Related Hazards and Pertinent Remedies," U.S. Bureau of Mines, Report of Investigation No. 4833, 1952, 64 pp.
10. J. M. Kuchta, A. L. Furno, A. Bartkowiak, and G. H. Martindell, "Effect of Pressure and Temperature on Flammability Limits of Chlorinated Hydrocarbons in Oxygen–Nitrogen and Nitrogen Tetroxide–Nitrogen Atmospheres," *J. Chem. Eng. Data,* 13: 3, 421–428 (1968).
11. D. Burgess, "Thermochemical Criteria for Explosion Hazards," Paper No. 10b, presented at the 62nd Annual Meeting of the American Institute of Chemical Engineers, Washington, D.C., 1969, 44 pp.
12. D. Burgess and M. Hertzberg, "The Flammability Limits of Lean Full–Air Mixtures: Thermochemical and Kinetic Criteria for Explosion Hazards," *ISA Trans.,* **14:** 2, 129–136 (1975).

13. P. G. Guest, "Oily Fibers May Increase Oxygen Tent Fire Hazard," *Mod. Hosp.*, May 1965, pp. 180–182.
14. J. M. Kuchta, A. L. Furno, and G. H. Martindill, "Flammability of Fabrics and Other Materials in Oxygen-Enriched Atmospheres. Part I. Ignition Temperatures and Flame Spread Rates," *Fire Technol.*, **5**: 3, 203–215 (August 1969).
15. E. L. Litchfield and T. A. Kubala, Flammability of Fabrics and Other Materials in Oxygen-Enriched Atmospheres. Part II. Minimum Ignition Energies," *Fire Technol.*, **5**: 4, 341–345 (November 1969).
16. National Fire Protection Association, "Flash Point Index of Trade Name Liquids," NFPA No. 325A, 1972, 258 pp.
17. J. M. Kuchta and R. J. Cato, "Ignition and Flammability Properties of Lubricants, *Trans. SAE*, **77**, 1008–1020 (1968).
18. J. M. Kuchta, "Summary of Ignitions Properties of Jet Fuels and Other Aircraft Combustible Fluids," Report No. AFAPL-TR-75-70, 1975, 54 pp.
19. National Fire Protection Association, "Halogenated Fire Extinguishing Agent Systems—Halon 1301," NFPA No. 12A, 1973, 80 pp.
20. J. Grumer and A. E. Bruszak, "Inhibition of Coal Dust–Air Flames," U.S. Bureau of Mines, Report of Investigations No. 7552, 1971, 14 pp.
21. J. M. Kuchta, "Fire and Explosion Manual for Aircraft Accident Investigators," U.S. Bureau of Mines Report No. 4193, 1973, 117 pp.
22. M. G. Zabetakis and D. Burgess, Research on the Hazards Associated with the Production and Handling of Liquid Hydrogen," U.S. Bureau of Mines, Report of Investigations No. 5707, 1961, 50 pp.
23. Compressed Gas Association, "Standards for Liquefied Hydrogen Systems at Consumer Sites," Pamphlet No. G-5.2, CGA, 1966.
24. Armed Serviced Explosions Safety Board, "Quantity-Distance Criteria for Liquid Propellants," March 26, 1964.
25. A. Strasser, I. Liebman, and S. R. Harris, "Hydrogen Detectors," *Cryogen. Eng. News*, **2**: 12, 16–20 (1967).
26. M. C. Irani, A. Tall, B. M. Bench, and P. W. Heran, "A Continuous-Recording Methanometer for Exhaust Fan Monitoring," U.S. Bureau of Mines, Report of Investigations No. 7951, 1974, 18 pp.
27. H. B. Carroll, Jr., and F. E. Armstrong, "Accuracy and Precision of Several Portable Gas Indicators," U.S. Bureau of Mines, Report of Investigations No. 7811, 1973, 42 pp.
28. H. R. Maisey, "Gaseous and Dust Explosion Venting. Part I." *Chem. Process Eng.*, pp. 527–535, 563, October 1965.
29. H. R. Maisey, "Gaseous and Dust Explosion Venting. Part 2." *Chem. Process Eng.*, pp. 662–672, December 1965.
30. National Fire Protection Association, "Explosion Venting Guide," NFPA NO. 68, 1974, 84 pp.
31. J. Nagy and W. M. Portmann, "Explosibility of Coal Dust in an Atmosphere Containing a Low Percentage of Methane," U.S. Bureau of Mines, Report of Investigation No. 5815, 1961, 16 pp.
32. H. G. Dorsett, Jr., M. Jacobson, J. Nagy, and A. P. Williams, "Laboratory Equipment and Test Procedures for Evaluating Explosibility of Dusts," U.S. Bureau of Mines, Report of Investigation No. 5624, 1960, 21 pp.
33. J. Nagy, A. R. Cooper, and H. G. Dorsett, Jr., "Explosibility of Miscellaneous Dusts," U.S. Bureau of Mines, Report of Investigation No. 7208, 1968, 31 pp.
34. K. N. Palmer and Z. W. Rogowski, "The Use of Flame Arrestors for Protection of Enclosed Equipment in Propane–Air Atmospheres, I." Chemical Engineering Symposium Series No. 25, Institution of Chemical Engineers, London, 1968, pp. 76–85.
35. Ministry of Labour, "Guide to the Use of Flame Arresters and Explosion Reliefs," Safety, Health and Welfare Booklets New Series No. 34, Her Majesty's Stationery Office, London, 1965, 55 pp.

36. J. K. Richmond, I. Liebman, and L. F. Miller, "Effect of Rock Dust on Explosibility of Coal Dust," U.S. Bureau of Mines, Report of Investigation No. 8077, 1975, 34 pp.
37. J. K. Richmond and I. Liebman, "A Physical Description of Coal Mine Explosions," paper presented at the 15th Symposium (International) on Combustion, the Combustion Institute, Pittsburgh, 1975, pp. 115–126.
38. W. Cybulski, "Researches on the Relationship Between Coal Dust Explosibility and the Kind of Coal," Restricted Conference of Directors of Safety in Mines Research, Paper No. 1, 1961, 52 pp.
39. J. Grumer, "Recent Research Concerning Extinguishment of Coal Dust Explosions," paper presented at the 15th Symposium (International) on Combustion, the Combustion Institute, Pittsburgh, 1975, pp. 103–114.
40. I. Hartmann, "Studies on the Development and Control of Coal Dust Explosions in Mines," U.S. Bureau of Mines, Information Circular No. 7785, 1957, 27 pp.
41. J. Nagy, A. R. Cooper, and J. M. Stupar, "Pressure Development in Laboratory Explosions," U.S. Bureau of Mines, Report of Investigation No. 6561, 1964, 19 pp.
42. J. M. Singer, N. B. Greninger, and J. Grumer, "Some Aspects of the Aerodynamics of Formation of Float Coal Dust Clouds," U.S. Bureau of Mines, Report of Investigation No. 7252, 1969, 26 pp.
43. J. M. Singer, E. B. Cook, and J. Grumer, "Dispersal of Coal- and Rock-Dust Deposits," U.S. Bureau of Mines, Report of Investigations No. 7642, 1972, 32 pp.
44. D. Rae, "The Initiation of Weak Coal Dust Explosions in Long Galleries and the Importance of the Time Dependence of the Explosion Pressure," paper presented at the 14th Symposium (International) on Combustion, the Combustion Institute, Pittsburgh, 1973, pp. 1225–1234.
45. J. Grumer, L. F. Miller, A. E. Bruszak, and L. E. Dalverny, "Minimum Extinguishant and Maximum Oxygen Concentrations for Extinguishing Coal Dust–Air Explosions," U.S. Bureau of Mines, Report of Investigation No. 7782, 1973, 6 pp.
46. D. Rae, "Experimental Coal-Dust Explosions in the Buxton Full-Scale Surface Gallery. IV. The Influence of the Dust Deposit and Form of Initiation of Explosions in a Smooth Gallery," SMRE Research Report No. 277, 1972, 57 pp.
47. M. Jacobson, J. Nagy, A. R. Cooper, and F. J. Ball, "Explosibility of Agricultural Dusts," U.S. Bureau of Mines, Report of Investigation No. 5753, 1961, 23 pp.
48. J. Nagy, H. G. Dorsett, Jr., and A. R. Cooper, "Explosibility of Carbonaceous Dusts," U.S. Bureau of Mines, Report of Investigation No. 6597, 1965, 30 pp.
49. H. G. Dorsett, Jr., and J. Nagy, "Dust Explosibility of Chemicals, Drugs, Dyes and Pesticides," U.S. Bureau of Mines, Report of Investigation No. 7132, 1968, 23 pp.
50. M. Jacobson, A. R. Cooper, and J. Nagy, "Explosibility of Metal Powders," U.S. Bureau of Mines, Report of Investigation No. 6515, 1964, 25 pp.
51. M. Jacobson, J. Nagy, and A. R. Cooper, "Explosibility of Dusts Used in the Plastics Industry," U.S. Bureau of Mines, Report of Investigations, No. 5971, 1962, 30 pp.
52. D. Burgess and J. Murphy, "Some Experiments with the Application of Bromotrifluoromethane to Coal Fires," Paper No. 34, presented at the International Conference of Safety in Mines Research, Tokyo, 1969, 17 pp.
53. National Fire Protection Association, Explosion Prevention Systems, NFPA No. 69, 1973, 53 pp.

Index